SECOND EDITION

ENERGY EFFICIENCY AND RENEWABLE ENERGY HANDBOOK

MECHANICAL and AEROSPACE ENGINEERING

Frank Kreith & Darrell W. Pepper
Series Editors

RECENTLY PUBLISHED TITLES

Air Distribution in Buildings, *Essam E. Khalil*

Alternative Fuels for Transportation, *Edited by Arumugam S. Ramadhas*

Computer Techniques in Vibration, *Edited by Clarence W. de Silva*

Design and Control of Automotive Propulsion Systems, *Zongxuan Sun and Guoming (George) Zhu*

Distributed Generation: The Power Paradigm for the New Millennium, *Edited by Anne-Marie Borbely and Jan F. Kreider*

Elastic Waves in Composite Media and Structures: With Applications to Ultrasonic Nondestructive Evaluation, *Subhendu K. Datta and Arvind H. Shah*

Elastoplasticity Theory, *Vlado A. Lubarda*

Energy Audit of Building Systems: An Engineering Approach, *Moncef Krarti*

Energy Conversion, Second Edition, *Edited by D. Yogi Goswami and Frank Kreith*

Energy Efficiency and Renewable Energy Handbook, Second Edition, *Edited by D. Yogi Goswami and Frank Kreith*

Energy Efficiency in the Urban Environment, *Heba Allah Essam E. Khalil and Essam E. Khalil*

Energy Management and Conservation Handbook, Second Edition, *Edited by Frank Kreith and D. Yogi Goswami*

Essentials of Mechanical Stress Analysis, *Amir Javidinejad*

The Finite Element Method Using MATLAB®, Second Edition, *Young W. Kwon and Hyochoong Bang*

Fluid Power Circuits and Controls: Fundamentals and Applications, *John S. Cundiff*

Fuel Cells: Principles, Design, and Analysis, *Shripad Revankar and Pradip Majumdar*

Fundamentals of Environmental Discharge Modeling, *Lorin R. Davis*

Handbook of Hydrogen Energy, *Edited by S.A. Sherif, D. Yogi Goswami, Elias K. Stefanakos, and Aldo Steinfeld*

Heat Transfer in Single and Multiphase Systems, *Greg F. Naterer*

Heating and Cooling of Buildings: Design for Efficiency, Revised Second Edition, *Jan F. Kreider, Peter S. Curtiss, and Ari Rabl*

Intelligent Transportation Systems: Smart and Green Infrastructure Design, Second Edition, *Sumit Ghosh and Tony S. Lee*

Introduction to Biofuels, *David M. Mousdale*

Introduction to Precision Machine Design and Error Assessment, *Edited by Samir Mekid*

Introductory Finite Element Method, *Chandrakant S. Desai and Tribikram Kundu*

Large Energy Storage Systems Handbook, *Edited by Frank S. Barnes and Jonah G. Levine*

Machine Elements: Life and Design, *Boris M. Klebanov, David M. Barlam, and Frederic E. Nystrom*

Mathematical and Physical Modeling of Materials Processing Operations, *Olusegun Johnson Ilegbusi, Manabu Iguchi, and Walter E. Wahnsiedler*

Mechanics of Composite Materials, *Autar K. Kaw*

Mechanics of Fatigue, *Vladimir V. Bolotin*

Mechanism Design: Enumeration of Kinematic Structures According to Function, *Lung-Wen Tsai*

Mechatronic Systems: Devices, Design, Control, Operation and Monitoring, *Edited by Clarence W. de Silva*

The MEMS Handbook, Second Edition (3 volumes), *Edited by Mohamed Gad-el-Hak*

MEMS: Introduction and Fundamentals

MEMS: Applications

MEMS: Design and Fabrication

Multiphase Flow Handbook, *Edited by Clayton T. Crowe*

Nanotechnology: Understanding Small Systems, Third Edition, *Ben Rogers, Jesse Adams, Sumita Pennathur*

Nuclear Engineering Handbook, *Edited by Kenneth D. Kok*

Optomechatronics: Fusion of Optical and Mechatronic Engineering, *Hyungsuck Cho*

Practical Inverse Analysis in Engineering, *David M. Trujillo and Henry R. Busby*

Pressure Vessels: Design and Practice, *Somnath Chattopadhyay*

Principles of Solid Mechanics, *Rowland Richards, Jr.*

Principles of Sustainable Energy Systems, Second Edition, *Edited by Frank Kreith with Susan Krumdieck, Co-Editor*

Thermodynamics for Engineers, *Kau-Fui Vincent Wong*

Vibration and Shock Handbook, *Edited by Clarence W. de Silva*

Vibration Damping, Control, and Design, *Edited by Clarence W. de Silva*

Viscoelastic Solids, *Roderic S. Lakes*

Weatherization and Energy Efficiency Improvement for Existing Homes: An Engineering Approach, *Moncef Krarti*

SECOND EDITION

ENERGY EFFICIENCY AND RENEWABLE ENERGY HANDBOOK

EDITED BY

D. YOGI GOSWAMI

FRANK KREITH

CRC Press
Taylor & Francis Group
Boca Raton London New York

CRC Press is an imprint of the
Taylor & Francis Group, an informa business

CRC Press
Taylor & Francis Group
6000 Broken Sound Parkway NW, Suite 300
Boca Raton, FL 33487-2742

First issued in paperback 2017

ISBN-13: 978-1-4665-8508-9 (hbk)
ISBN-13: 978-1-138-74911-5 (pbk)

Library of Congress Cataloging-in-Publication Data

Handbook of energy efficiency and renewable energy
Energy efficiency and renewable energy handbook / [compiled by] D. Yogi Goswami and Frank Kreith. -- Second edition.
pages cm. -- (Mechanical and aerospace engineering series)
Revised edition of: Handbook of energy efficiency and renewable energy / edited by Frank Kreith, D. Yogi Goswami. 2007.
Includes bibliographical references and index.
ISBN 978-1-4665-8508-9
1. Energy conservation--Handbooks, manuals, etc. 2. Renewable energy sources--Handbooks, manuals, etc. I. Goswami, D. Yogi. II. Kreith, Frank. III. Title.

TJ163.3.H356 2015
333.79--dc23 2014045672

Visit the Taylor & Francis Web site at
http://www.taylorandfrancis.com

and the CRC Press Web site at
http://www.crcpress.com

Contents

Preface

Purpose

The goal of this handbook is to provide information necessary for engineers, energy professionals, and policy makers to plan a secure energy future. The time horizon of the handbook is limited to approximately 20 years because environmental conditions vary, new technologies emerge, and priorities of society continuously change. It is therefore not possible to make reliable projections beyond that period. Given this time horizon, the book deals only with technologies that are currently available or that are expected to be ready for implementation in the near future.

Energy is a mainstay of an industrial society. As the population of the world increases and people strive for a higher standard of living, the amount of energy necessary to sustain our society is ever increasing. At the same time, the availability of nonrenewable sources, particularly liquid fuels, is rapidly shrinking. Therefore, there is general agreement that to avoid an energy crisis, the amount of energy needed to sustain society will have to be contained and, to the extent possible, renewable sources will have to be used. As a consequence, conservation and renewable energy (RE) technologies are going to increase in importance and reliable, up-to-date information about their availability, efficiency, and cost is necessary for planning a secure energy future.

The timing of this handbook also coincides with a new impetus for the use of RE. This impetus comes from RE policies in Europe, Japan, China, India, and Brazil and the emergence of renewable portfolio standards (RPS) in many states of the United States. Germany introduced electricity feed-in laws that value electricity produced from RE resources much higher than that from conventional resources, which have created demand for photovoltaic and wind power. Following the success of Germany, other European countries introduced feed-in laws, which accelerated the deployment of RE in Europe. Other countries, such as China and India, have adopted modified versions of feed-in laws, whereby RE power companies bid discounts to the feed-in tariffs determined by the governments. RPS policies adopted by many states in United State require that a certain percentage of energy used be derived from renewable resources. RPSs and other incentives for RE are currently in place in 34 of the 50 states of the United States and the District of Columbia (DC) and Puerto Rico. The details of the RPS for RE and conservation instituted by state governments vary, but all of them essentially offer an opportunity for the industry to compete for the new markets. Thus, to be successful, renewable technologies will have to become more efficient, reliable, and cost-effective. RPSs have already demonstrated that they can reduce market barriers and stimulate the development of RE. The use of conservation and RE can help meet critical national goals for fuel diversity, price stability, economic development, environmental protection, and energy security and thereby play a vital role in national energy policy. The expected growth rate of RE from portfolio standards and other stimulants in the United States is impressive. As a result of various policy initiatives in the world, the global growth in solar photovoltaics (PV) production has averaged over 43% per year from 2000 to 2012 and 61% from 2007 to 2012, with Europe showing the maximum growth. The average annual growth in worldwide wind energy capacity from 2001 to 2012 was over 25%. The average annual growth in the United States over the same period was

37.7%. More recently, China has increased its capacity faster than any other country. China accounted for more than a quarter of the global wind capacity in 2012. With appropriate regulations and careful planning, the technical information in this handbook will ensure an orderly and peaceful transition to a sustainable energy future.

Organization and Layout

The book is essentially divided into three sections:

- General overviews, policy, and economics (Section I: Chapters 1 through 13)
- Energy efficiency, energy generation, infrastructure, and storage (Section II: Chapters 14 through 18; and Section III: Chapters 19 through 35)
- Renewable energy technologies (Section IV: Chapters 36 through 48; and Section V: Chapters 49 through 53).

The first chapter is a survey of current and future worldwide energy issues. A discussion of sound finance policies and stimulants for energy efficiency and RE is treated in Chapter 2. State and federal policies for RE in the United States are described in Chapter 3. Chapters 4 through 11 give an assessment of policies in Europe, China, India, Brazil, Israel, Australia, and Japan. Economic assessment methods for conservation and generation technologies are covered in Chapter 12, and the environmental costs of various energy generation technologies are discussed in Chapter 13. The use of renewables and conservation will initiate a paradigm shift toward distributed generation and demand-side management procedures, which are covered in Chapter 14 and 15. Although renewables, once in place, produce energy from natural resources and cause very little environmental damage, energy is required in their initial construction. One measure of the energy effectiveness of a renewable technology is the length of time required, after the system begins operation, to repay the energy used in its construction, called the energy payback period. Another measure is the energy return on energy investment ratio. The larger the amount of energy a renewable technology delivers during its lifetime compared to the amount of energy necessary for its construction, the more favorable its economic return on the investment will be and the less its adverse environmental impact. But during the transition to renewable sources, a robust energy production and transmission system from fossil and nuclear technologies is required to build the systems. Moreover, because there is a limit to how much of our total energy needs can be met economically in the near future, renewables will have to coexist with fossil and nuclear fuels for some time. Furthermore, the supply of all fossil and nuclear fuel sources is finite, and their efficient use in meeting our energy needs should be a part of an energy and CO_2 reduction strategy. Therefore, Chapters 16 and 17 give a perspective on the efficiencies, economics, and environmental costs of the key fossil and nuclear technologies. Finally, Chapter 18 provides projections for energy supply, demand, and prices of energy in the United States through the year 2040. Petroleum engineers predict that worldwide oil production will reach its peak within the next 10 years and then begin to decline. At the same time, demand for liquid fuel by an ever-increasing number of vehicles, particularly in China and India, is expected to increase significantly. As a result, gasoline prices will increase precipitously unless we reduce gasoline consumption by increasing the mileage of the vehicle fleet, reducing the number of vehicles on the road by using mass transport, and producing synthetic fuels from biomass and coal. The options to prevent an energy crisis in transportation include plug-in hybrid vehicles, biofuels, diesel engines, city planning, and mass-transport systems. These are treated in Chapter 19; biofuels and fuel cells are treated in Chapters 50 and 53, respectively. It is an unfortunate fact of life that the security of the energy supply and transmission system has recently been placed in jeopardy from various sources, including natural disasters and worldwide terrorism. Consequently, energy infrastructure security and risk analysis are an important aspect of planning future energy transmission and storage systems, and these topics are covered in Chapter 20. Energy efficiency is defined as the ratio of energy required to perform a specific task or service to the amount of energy used for the process. Improving energy efficiency increases the productivity of basic energy resources by providing the needs of society with less energy. Improving the efficiency across all sectors of the economy is therefore an important objective. The least expensive and most efficient means in this endeavor is energy conservation, rather than more energy production. Moreover, energy conservation is also the best way to protect the environment and reduce global warming.

Recognizing that energy conservation in its various forms is the cornerstone of successful national energy strategy, 11 chapters (22 through 32) are devoted to conservation. The topics covered include energy management strategies for industry and buildings, HVAC controls, co-generation, and advances in specific technologies, such as motors, lighting, appliances, and heat pumps. An important aspect of energy efficiency is efficient electric grid management, which includes energy storage, advanced concepts in transmission and distribution, and smart grid technology. These topics are covered in detail in Chapters 33 through 35.

The third section of the book deals with energy storage and energy generation from renewable sources. Chapters 36 through 39 present the availability of renewable sources: solar, wind, municipal waste, and biomass. The renewable generation technologies for solar thermal, wind power, PV, biomass, and geothermal are then covered in Chapters 40 through 53.

At this time, it is not clear whether hydrogen will play a major role in the national energy structure within the next 25 years, but there is an ongoing discussion about the feasibility and cost of what is called the hydrogen economy. Energy experts recognize that the generation and use of hydrogen has a critical inefficiency problem that is rooted in basic thermodynamics. There are also ground transportation options that are less expensive than using hydrogen vehicles powered by fuel cells. But there is substantial support for continuing research to eventually develop a viable place for hydrogen in a future energy structure. Therefore, the topics of hydrogen energy and fuel cells are included in Chapters 52 and 53, respectively. This information should be useful background for comparing competing options for energy generation, storage, and distribution.

We hope that this handbook will serve as a useful reference to all engineers in the energy field and pave the way for a paradigm shift from fossil fuels to a sustainable energy systems based on conservation and renewable technologies. But we also recognize the complexity of this task, and we invite readers to comment on the scope and the topics covered. A handbook such as this needs to be updated every 5–10 years, and we will respond to readers' comments and suggestions in the next edition.

D. Yogi Goswami
Frank Kreith
Editors-in-chief

Acknowledgments

At the time CRC suggested that the *Handbook of Energy Efficiency and Renewable Energy* deserved a second edition, I was approaching my 90th birthday. I was reluctant at this point in my life to undertake such a large and demanding assignment, but my long-time associate and co-editor, Yogi Goswami, and Bev Weiler, my long-term assistant, offered to help me in the endeavor. The new edition owes its existence to the invaluable contributions that these two have made in contacting previous authors and helping to find new qualified experts whenever necessary. I would also like to thank my wife Marion for having helped me in many tangible and intangible ways to continue an active professional life in my seventh decade as an engineering author.

Frank Kreith

Editors

Dr. D. Yogi Goswami is a university distinguished professor, the John and Naida Ramil Professor, and director of the Clean Energy Research Center at the University of South Florida, Tampa, Florida.

He conducts fundamental and applied research on solar thermal energy, thermodynamics, heat transfer, HVAC, photovoltaics, hydrogen, and fuel cells.

Dr. Goswami has served as an advisor and given testimonies on energy policy and the transition to renewable energy to the U.S. Congress and the Government of India, as well as provided consultant expertise to the U.S. Department of Energy, USAID, World Bank, and NIST, among others.

Professor Goswami is the editor-in-chief of the *Solar Energy* and *Progress in Solar Energy* journals. Within the field of RE, he has published as author/editor 16 books, 6 conference proceedings, and 393 refereed technical papers. He has delivered 52 keynote and plenary lectures at major international conferences. He holds 18 patents.

A recognized leader in professional scientific and technical societies, Professor Goswami has served as a governor of ASME-International (2003–2006), president of the International Solar Energy Society (2004–2005), senior vice president of ASME (2000–2003), vice president of ISES and president of the International Association for Solar Energy Education (IASEE, 2000–2002).

Dr. Goswami is a fellow of AAAS, ASME International, ASHRAE, the American Solar Energy Society, the National Academy of Inventors and a member of the Pan American Academy of Engineers. He is a recipient of the following awards:

- Technical Communities Globalization Medal, (ASME) 2013.
- Theodore and Venette Askounes-Ashford Distinguished Scholar Award, Univ. South Florida, 2011.
- Frank Kreith Energy Award, ASME, 2007
- Farrington Daniels Award, ISES, 2007 (highest award of ISES)
- Hoyt Clark Hottel Award, ASES, 2007
- Charles Greely Abbott Award for Outstanding Scientific, Technical and Human Contributions to the Development and Implementation of Solar Energy (highest award of the American Solar Energy Society), 1998.
- John Yellott Award for Outstanding Contributions to the Field of Solar Energy, ASME Solar Energy Division, 1995 (highest solar energy award from ASME).

He has also received more than 50 other awards and certificates from major engineering and scientific societies for his work in renewable energy.

Professor Frank Kreith is an internationally known energy consultant and professor emeritus of engineering at the University of Colorado, Boulder, Colorado. In 1945, after graduation from the University of California, Berkeley, he accepted a position at the Jet Propulsion Laboratory of the California Institute of Technology, where he developed a heat transfer laboratory and conducted research on building heat transfer. He received his MS in engineering in 1949 from the University of California, Los Angeles, and in 1950 was awarded a Guggenheim Fellowship to Princeton University, followed in 1951 by an appointment to the faculty of the University of California, Berkeley. From 1953 to 1959, he was associate professor of mechanical engineering at Lehigh University, where he did research on heat transfer in rotating systems and wrote the first edition of *Principles of Heat Transfer*, now in its seventh edition. In 1958, he received the Robinson Award from Lehigh University for excellence in teaching. In 1959, he joined the University of Colorado where he held appointments as professor of mechanical and chemical engineering. In 1962, he published a text on the design of solar power plants based on his consulting for the NASA space program. During his 20 years of tenure at the University of Colorado, he did research on heat transfer in biological systems and renewable energy. In 1964, he was awarded a doctorate in engineering from the University of Paris.

Dr. Kreith served as chief scientist and ASME legislative fellow at the National Conference of State Legislatures (NCSL) from 1988 to 2001, providing professional advice and assistance to all 50 state legislatures on energy and the environment. Prior to joining NCSL, he was senior research fellow at the Solar Energy Research Institute (now the National Renewable Energy Laboratory) where he participated in the Presidential Domestic Energy Review and served as energy advisor to the governor of Colorado. From 1974–1977, he was president of Environmental Consulting Services. Dr. Kreith has been an energy consultant to NATO, the U.S. Agency of International Development, and the United Nations. He has published more than a hundred peer-reviewed technical articles and more than 15 books, many translated into foreign languages and used extensively in engineering programs around the world. His books include *Principles of Heat Transfer* (now in its seventh edition), *Principles of Solar Engineering* (with J.F. Kreider), *Nuclear Impact* (with C.B. Wrenn), the *Handbook of Solid Waste Management*, the *CRC Handbook of Mechanical Engineering*, and *Principles of Sustainable Energy*, which is now in its second edition. He is a fellow of AAAS and was promoted to honorary member of ASME in 2004. Dr. Kreith's work has received worldwide recognition, including the Washington Award, Charles Greeley Abbot Award from ASES, the Max Jacob Award from ASME-AIChE, and the Ralph Coats Roe Medal from ASME for "significant contributions…through provision of information to legislators about energy and the environment." In 2004, ASME recognized Dr. Kreith's lifelong contributions to heat transfer and renewable energy by establishing the Frank Kreith Energy Award. He has recently completed an autobiography, *Sunrise Delayed: A Personal History of Solar Energy*, which is available on Amazon.com.

Contributors

Aníbal T. de Almeida
Department of Electrical and Computer Engineering
University of Coimbra
Coimbra, Portugal

Andrea L. Alstone
Energy Efficiency Standards Group
Lawrence Berkeley National Laboratory
Berkeley, California

Massoud Amin
Department of Electrical and Computer Engineering
University of Minnesota
Minneapolis, Minnesota

Anthony F. Armor
Electric Power Research Institute
Palo Alto, California

Barbara Atkinson
Energy Efficiency Standards Group
Lawrence Berkeley National Laboratory
Berkeley, California

Bilal M. Ayyub
Department of Civil and Environmental Engineering
University of Maryland
College Park, Maryland

Frano Barbir
Department of Electrical and Mechanical Engineering and Naval Architecture
University of Split
Split, Croatia

Dale E. Berg
Albuquerque, New Mexico

Peter Biermayer
Pacific Gas & Electric Co.
San Francisco, California

Robert C. Brown
Department of Mechanical Engineering
Iowa State University
Ames, Iowa

Stephan Buecheler
Laboratory for Thin Films and Photovoltaics
Swiss Federal Laboratories for Materials Science and Technology
Dübendorf, Switzerland

Barney L. Capehart
Department of Industrial and Systems Engineering
University of Florida
Gainesville, Florida

Jeffrey P. Chamberlain
Argonne National Laboratory
Lemont, Illinois

David E. Claridge
Department of Mechanical Engineering
Texas A&M University
College Station, Texas

John J. Conti
U.S. Energy Information Administration
U.S. Department of Energy
Washington, DC

Charles H. Culp
Energy Systems Laboratory
Texas A&M University
College Station, Texas

Michael Curley
Environmental Law Institute
Washington, DC

Mark J. Eshbaugh
U.S. Energy Information Administration
U.S. Department of Energy
Washington, DC

Ed Fouche
Global Energy Partners, LLC
Raleigh, North Carolina

Karina Garbesi
California State University, East Bay
Hayward, California

Clark W. Gellings
Electric Power Research Institute
Palo Alto, California

Brian F. Gerke
Energy Efficiency Standards Group
Lawrence Berkeley National Laboratory
Berkeley, California

Jose Gonzalez-Aguilar
IMDEA Energy
Madrid, Spain

D. Yogi Goswami
Clean Energy Research Center
University of South Florida
Tampa, Florida

Steve F. Greenberg
Lawrence Berkeley National Laboratory
University of California, Berkeley
Berkeley, California

Leonard M. Grillo
Grillo Engineering Co.
Hollis, New Hampshire

Gershon Grossman (Emeritus)
Department of Mechanical Engineering
Israel Institute of Technology
Haifa, Israel

Deepak Gupta
Ministry of New and Renewable Energy
Government of India
New Delhi, India

Roel Hammerschlag
Hammerschlag & Co., LLC
Olympia, Washington

Edwin Harvego
Idaho National Laboratory
Idaho Falls, Idaho

Aruna Ivaturi
School of Chemistry
University of Edinburgh
Edinburgh, United Kingdom

Masoud Kayhanian
Department of Civil and Environmental Engineering
University of California, Davis
Davis, California

Kevin Kitz
U.S. Geothermal, Inc.
Boise, Idaho

Eric Kleinert
FORTIS Colleges and Institutes
Lake Worth, Florida

Kenneth D. Kok
S&K Consulting
Richland, Washington

Moncef Krarti
Civil, Environmental and Architectural Engineering Department
University of Colorado
Boulder, Colorado

Jan F. Kreider
K&A, LLC
Boulder, Colorado

Frank Kreith (Emeritus)
Mechanical Engineering Department
University of Colorado
Boulder, Colorado

Andy S. Kydes
Z Inc.
Washington, DC

Alex Lekov
Lawrence Berkeley National Laboratory
University of California, Berkeley
Berkeley, California

Xianguo Li
Department of Mechanical Engineering
University of Waterloo
Waterloo, Ontario, Canada

James Lutz (Retired)
Lawrence Berkeley National Laboratory
University of California, Berkeley
Berkeley, California

Madhukar Mahishi
Cummins
Minneapolis, Minnesota

P.C. Maithani
National Informatics Centre
New Delhi, India

Thomas D. Marshall
KeyLogic Systems, Inc.
National Energy Technology Laboratory
Pittsburgh, Pennsylvania

Howard G. McIlvried
KeyLogic Systems, Inc.
National Energy Technology Laboratory
Pittsburgh, Pennsylvania

James E. McMahon
Better Climate Research and Policy Analysis
Moraga, California

Roger Messenger (Emeritus)
Department of Electrical Engineering
Florida Atlantic University
Boca Raton, Florida

Stephen Meyers
Lawrence Berkeley National Laboratory
University of California, Berkeley
Berkeley, California

Zhixin Miao
Department of Electrical Engineering
University of South Florida
Tampa, Florida

Jeffrey H. Morehouse (Professor Emeritus)
Mechanical Engineering Department
University of South Carolina
Columbia, South Carolina

Keibun Mori
Deloitte Tohmatsu Consulting, Co., Ltd.
Tokyo, Japan

Pedro Soares Moura
Department of Electrical and Computer Engineering
University of Coimbra
Coimbra, Portugal

Christopher Namovicz
U.S. Energy Information Administration
Washington, DC

Werner Niederle
Federal Environmental Agency
Fachgebiet Erneuerbare Energien
Dessau-Roßlau, Germany

Monica Oliphant
Monica Oliphant Research
Adelaide, Australia

Kelly E. Parmenter
Applied Energy Group, Inc.
Walnut Creek, California

Terry Penney (Retired)
National Renewable Energy Laboratory
Golden, Colorado

Sean I. Plasynski
National Energy Technology Laboratory
U.S. Department of Energy
Pittsburgh, Pennsylvania

Robert Pratt
Distribution and Demand Response Sector
Pacific Northwest National Laboratory
Richland, Washington

Ari Rabl (Emeritus)
Centre d'Energetique
Ecole des Mines
Paris, France

Prakash Rao
Lawrence Berkeley National Laboratory
University of California, Berkeley
Berkeley, California

Bryan P. Rasmussen
Department of Mechanical Engineering
Texas A&M University
College Station, Texas

T. Agami Reddy
School of Sustainable Engineering and the Built Environment
Arizona State University
Tempe, Arizona

Wesley M. Rohrer, Jr. (Deceased)

Manuel Romero
IMDEA Energy
Madrid, Spain

Greg Rosenquist
Lawrence Berkeley National Laboratory
University of California, Berkeley
Berkeley, California

Rosalie Ruegg (Retired)
TIA Consulting, Inc.
Emerald Isle, North Carolina

Ricardo Rüther
Universidade Federal de Santa Catarina
and
Técnico do Instituto IDEAL
Florianopolis, Brazil

Christopher P. Schaber (Retired)
Institute for Lifecycle Environmental Assessment
Seattle, Washington

Shelly H. Schneider
Franklin Associates, a Division of ERG
Prairie Village, Kansas

S.A. Sherif
Department of Mechanical and Aerospace Engineering
University of Florida
Gainesville, Florida

Walter Short (Retired)
National Renewable Energy Laboratory
Golden, Colorado

Craig B. Smith
Dockside Consultants, Inc.
Newport Beach, California

Joseph V. Spadaro
Basque Centre for Climate Change
Bilbao, Spain

Sesha S. Srinivasan
College of Innovation and Technology
Florida Polytechnic University
Lakeland, Florida

Rameshwar D. Srivastava
KeyLogic Systems, Inc.
National Energy Technology Laboratory
Pittsburgh, Pennsylvania

Herbert W. Stanford, III (Retired)
Stanford White, Inc.
Raleigh, North Carolina

Senthilarasu Sundaram
Environment and Sustainability Institute
University of Exeter
Penryn, United Kingdom

David M. Sweet
World Alliance for Decentralized Energy
Washington, DC

George Tchobanoglous (Retired)
University of California, Davis
Davis, California

Ayodhya N. Tiwari
Laboratory for Thin Films and Photovoltaics
Swiss Federal Laboratories for Materials Science and Technology
Zurich, Switzerland

Kirtan K. Trivedi
Exxon Mobil Research and Engineering Company
Fairfax, Virginia

W. Dan Turner
Building Energy Efficiency (Bee)
Austin, Texas

Hari M. Upadhyaya
Department of Mechanical, Aerospace, and Civil Engineering
Brunel University
London, United Kingdom

Alex J. Valenti
Energy Efficiency Standards Group
Lawrence Berkeley National Laboratory
Berkeley, California

Charles O. Velzy
White Haven, Pennsylvania

T. Nejat Veziroglu (Retired)
Clean Energy Research Institute
University of Miami
Coral Gables, Florida

Vagelis Vossos
Energy Analysis and Environmental Impacts Division
Lawrence Berkeley National Laboratory
Berkeley, California

Steve Widergren
Pacific Northwest National Laboratory
Richland, Washington

Mark M. Wright
Department of Mechanical Engineering
Iowa State University
Ames, Iowa

Yue (Ada) Wu
U.S.–China Energy Cooperation Program
Amcham, People's Republic of China

Ming Yi
Beijing, People's Republic of China

Eduardo Zarza
Plataforma Solar de Almería
Centro de Investigaciones Energéticas, Medioambientales y Tecnológicas
Madrid, Spain

Section I

Global Energy Systems, Policy, and Economics

1

Global Energy Systems

D. Yogi Goswami and Frank Kreith

CONTENTS

> A thing that will assume enormous importance quite soon is the exhaustion of our fuel resources. Coal and oil have been accumulating in the earth over five hundred million years, and at the present rates of demand for mechanical power, the estimates are that oil will be all gone in about a century, and coal probably in a good deal less than five hundred years. For the present purpose, it does not matter if these are under-estimates; they could be doubled or trebled and still not affect the argument. Mechanical power comes from our reserves of energy, and we are squandering our energy capital quite recklessly. It will very soon be all gone, and in the long run we shall have to live from year to year on our earnings.*

* Quote from *The Next Millenium*, 1953, by Charles Galton Darwin, the grandson of Charles Darwin, author of *On the Origin of Species*.

1.1 Global Energy Needs and Resources

Global energy consumption in the last half century has rapidly increased and is expected to continue to grow over the next 50 years, however, with significant differences. The past increase was stimulated by relatively "cheap" fossil fuels and increased rates of industrialization in North America, Europe, and Japan; yet while energy consumption in these countries continues to increase, additional factors make the picture for the next 50 years more complex. These additional factors include China's and India's rapid increase in energy use as they represent about a third of the world's population; the expected depletion of oil resources in the near future; and, the effect of human activities on global climate change. On the positive side, the renewable energy (RE) technologies of wind, biofuels, solar thermal, and photovoltaics (PV) are finally showing maturity and the ultimate promise of cost competitiveness.

Statistics from the International Energy Agency (IEA) World Energy Outlook 2004 and 2010 show that the total primary energy demand in the world increased from 5,536 MTOE in 1971 to 10,345 MTOE in 2002, representing an average annual increase of 2% (see Figure 1.1 and Table 1.1).*

By 2008, the world energy demand had increased to 12,271 MTOE representing an average annual increase of about 3%. The main reason for a 50% increase in the annual rate is the fast growing energy demand in Asia Pacific, more specifically China. Since the per capita energy used in the most populous countries, China and India is still very small,

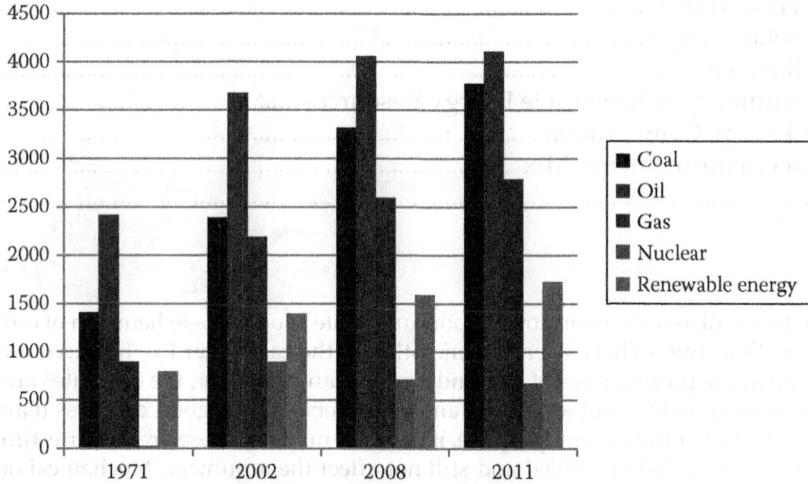

FIGURE 1.1
World primary energy demand (MTOE). (Data from IEA, *World Energy Outlook*, 2004; IEA, *World Energy Outlook 2010*, International Energy Agency, Paris, France, 2010; IEA, *World Energy Outlook 2013*, International Energy Agency, Paris, France, 2013.)

* The energy data for this chapter came from many sources, which use different units of energy, making it difficult to compare the numbers. The conversion factors are given here for a quick reference.
MTOE = Mega tons of oil equivalent; 1 MTOE = 4.1868×10^4 TJ (Terra Joules) = 3.968×10^{13} Btu.
GTOE = Giga tons of oil equivalent; 1 GTOW = 1000 MTOE.
Quadrillion Btu, also known as Quad: 10^{15} British Thermal Units or Btu; 1 Btu = 1055 J.
1 TWh = 10^9 kilowatt hours (kWh), 1 kWh = 3.6×10^6 J.

TABLE 1.1
World Total Energy Demand (MTOE)

Energy Source/Type	1971	2002	2008	2011	Annual Change 1971–2002 (%)	Annual Change 2002–2008 (%)	Annual Change 2008–2011 (%)
Coal	1,407	2,389	3,315	3,773	1.7	5.6	4.4
Oil	2,413	3,676	4,059	4,108	1.4	1.67	0.4
Gas	892	2,190	2,596	2,787	2.9	2.88	2.4
Nuclear	29	892	712	674	11.6	−3.7	−1.8
Hydro	104	224	276	300	2.5	3.6	2.8
Biomass and waste	687	1,119	1,225	1,300	1.6	1.6	2
Other renewables	4	55	89	127	8.8	8.46	12.6
Total	5,536	10,345	12,271	13,069	2.0	2.9	2.1

Sources: Data from IEA, *World Energy Outlook*, 2004; IEA, *World Energy Outlook 2010*, International Energy Agency, Paris, France, 2010; IEA, *World Energy Outlook 2013*, International Energy Agency, Paris, France, 2013.

their energy use may continue to increase at a high rate. From 2008 to 2011, the annual increase in energy use dropped back to 2.1% mainly because of a deep recession in United States and Europe where the energy use actually declined.

The last 10 years data for energy consumption from BP Corp. shows that during the most recent 10-year period even though the total primary energy use in North America and Europe has gone down, the global average increase has gone up to 2.8% (see Table 1.2). The rate of growth has risen mainly due to very rapid growth in Asia Pacific, which recorded an average annual increase of 6.1%. More specifically, China increased its primary energy consumption by approximately 10%/year from 2002 to 2012. Based on the current plans of China this trend will continue for at least another decade (IEA, 2013).

Even at a 2% increase per year, the primary energy demand of 12,271 MTOE in 2008 would double by 2043 and triple by 2063. Of course the global energy use cannot continue to increase at the same rate forever. IEA (2013) estimates that the global energy use will increase at an average annual rate of 1.2 up to 2035. Even at that optimistic slow growth rate of 1.2%, the global energy use will increase by 38% by 2035 reaching a value of 16,934 MTOE/year.

TABLE 1.2
Primary Energy Consumption (MTOE)[a]

Region	2002	2011	2012	2002–2012 Average Increase/Year (%)	2012 Change Over 2011 (%)
North America including United States	2741.1	2,774.3	2,725.4	−0.1	−2.0
United States	2295.5	2,265.2	2,208.8	−0.5	−2.8
South and Central America	474.9	649.5	665.3	3.5	2.2
Europe and Euro-Asia	2852	2,936.6	2,928.5	0.25	−0.5
Middle East	464.3	727.4	761.9	5.1	4.5
Africa	291.9	384.0	403.3	3.3	4.7
Asia Pacific	2773.7	4,753.2	4,992.2	6.1	4.7
China	1073.8	2,540.8	2,735.2	9.8	7.7
India	310.8	534.8	563.5	6.15	5.1
World	9487.9	12,225.0	12,477.0	2.8	1.8

Source: Data from BP Corp., London, U.K.

[a] This data does not include traditional biomass which was approximately 835 MTOE in 2011 according to IEA data.

Of the total world primary energy demand in 2002, fossil fuels accounted for about 80% with oil, coal, and natural gas being 36%, 23%, and 21%, respectively. Biomass accounted for 11% of all the primary energy in the world, with almost all of it being traditional biomass for cooking and heating in the developing countries, which is used very inefficiently. By 2011, fossil fuels contribution increased to approximately 82% of the global primary demand with oil, coal, and natural gas accounting for 31%, 29%, and 21%, respectively. Even though the oil use has continued to increase year after year, its overall share in the primary energy went down from 35% in 2002 to 31% in 2011. On the other hand, the share of coal in the primary energy increased from 23% in 2002 to 29% in 2011. The predominant reason for this shift is the rapid increase in power production in China where coal provides more than 75% of the electrical power (Table 1.3). The power capacity of China has been increasing at an annual rate of 12% since 2000 (Table 1.4) (Zhou, 2012) and has already overtaken the power capacity of United States.

With such high energy demand expected in the future, it is important to look at the available resources to fulfill the future demand 50 years from now, especially for electricity and transportation.

Although not a technical issue in the conventional sense, no matter what types of engineering scenarios are proposed to meet the rising demands of a growing world population, as long as that exponential growth continues, the attendant problems of energy and food consumption, as well as environmental degradation may have no long term solution (Bartlett, 2002). Under current demographic trends, the United Nations forecasts a rise in the global population to around 9 billion in the year 2050. This increase in 2.5 billion people will occur mostly in developing countries with aspirations for a higher standard of living. Thus, population growth should be considered as a part of the overall supply and demand picture to assure the success of future global energy and pollution strategy.

TABLE 1.3

Power Production in China by Energy Source

	1990	**%**	**2008**	**%**	**2011**	**%**
Coal	471	72.5	2759	79.0	3598	76.2
Oil	49	7.5	24	0.7	133.2	2.8
Gas	3	0.5	43	1.2	166.2	3.5
Nuclear	0	0.0	68	1.9	87.4	1.9
Hydro	127	19.5	585	16.7	662.6	14.0
Renewables	0	0.0	15	0.4	73.2	1.6
Total	650	100.0	3494	100.0	4720.6	100.0

TABLE 1.4

Power Capacity of China

Year	**GW**	**% Increase/Year**
1990	138	
2000	319	8.8
2008	793	12
2011	1056	11

1.2 Major Sectors of Primary Energy Use

The major sectors using primary energy sources include electrical power, transportation, heating and cooling, industrial and others, such as cooking. The IEA data shows that the electricity demand almost tripled from 1971 to 2002 and quadrupled by 2011. This is not unexpected as electricity is a very convenient form of energy to transport and use. Although primary energy use in all sectors has increased, their relative shares except for transportation and electricity have decreased (Figure 1.2). Figure 1.2 shows that the relative share of primary energy for electricity production in the world increased from about 20% in 1971 to about 40% in 2011. This is because electricity is becoming the preferred form of energy for all applications.

Figure 1.3 shows that coal is presently the largest source of electricity in the world. Consequently, the power sector accounted for almost 42% of all CO_2 emissions in 2011. Emissions could be reduced by increased use of RE sources. All RE sources combined accounted for about 20% share of electricity production in the world. Wind and solar power technologies have vastly improved in the last two decades and are becoming more cost effective. Therefore, their share of electricity production has been increasing at a very fast pace. Over the last decade wind power capacity has been increasing at an annual rate of close to 30% and solar photovoltaic power capacity has been increasing at an annual rate of close to 50%, which has resulted in wind and solar providing a combined 2% of all the electricity generation in the world in 2011, almost all of it coming online in less than two decades. Since solar and wind technologies are now mature, substituting fossil fuels with RE for electricity generation must be an important part of any strategy of reducing CO_2 emissions into the atmosphere and combating global climate change.

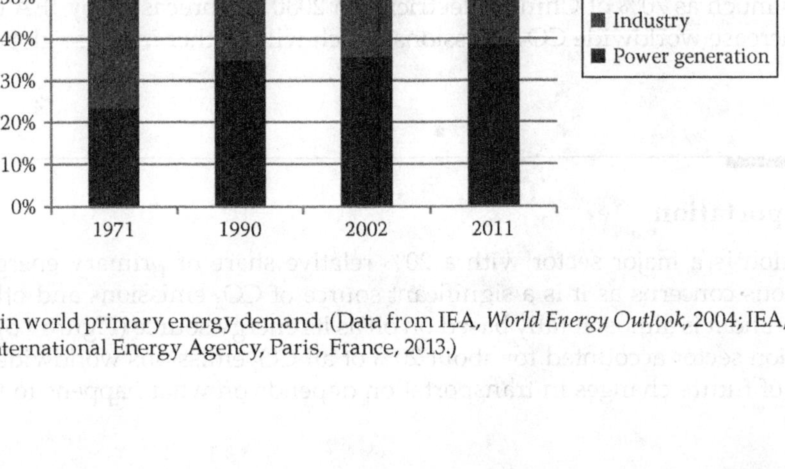

FIGURE 1.2
Sectoral shares in world primary energy demand. (Data from IEA, *World Energy Outlook*, 2004; IEA, *World Energy Outlook 2013*, International Energy Agency, Paris, France, 2013.)

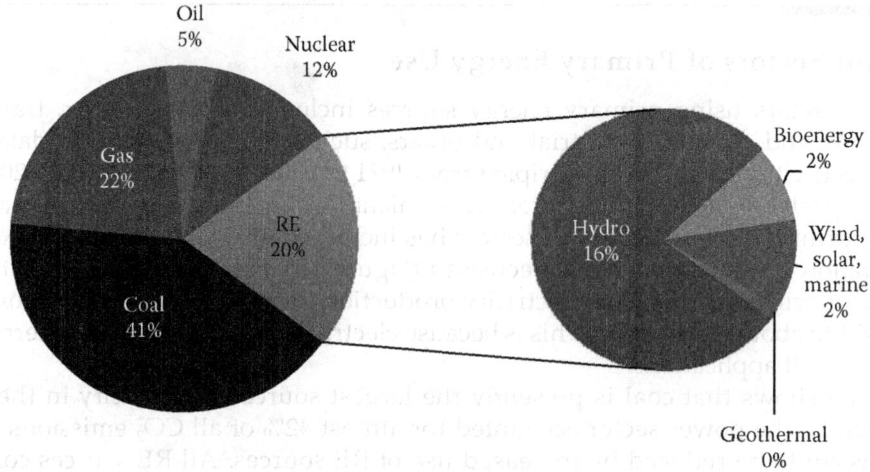

FIGURE 1.3
World electricity production by fuel in 2011. (Data from IEA, *World Energy Outlook 2013*, International Energy Agency, Paris, France, 2013.)

1.3 Electricity-Generating Capacity Additions to 2040

Figure 1.4 shows the global electricity-generating capacity in 2010 and additional electricity-generating capacity forecast by Energy Information Agency (EIA) of the U.S. Department of Energy for different regions in the world. The overall global annual increase of 1.6% in the electricity-generating capacity is in general agreement with the estimates from IEA (2013), which projects an average annual growth of 1.6% up to 2035. It is clear that of all countries, China will add the largest capacity with its projected electrical needs accounting for about 27.5% of the total world electricity-generating capacity. Non-OECD Asian countries (including China, India, Thailand, and Indonesia) combined will add about 60% of all the new capacity of the world. Therefore, what happens in these countries will have important consequences on the worldwide energy and environmental situation. If coal provides as much as 70% of China's electricity in 2030, as forecasted by IEA (2013), it will certainly increase worldwide CO_2 emissions which will further increase global warming.

1.4 Transportation

Transportation is a major sector with a 20% relative share of primary energy. This sector has serious concerns as it is a significant source of CO_2 emissions and other airborne pollutants—and it is almost totally based on oil as its energy source (Figure 1.5). In 2010, the transportation sector accounted for about 20% of all CO_2 emissions worldwide. An important aspect of future changes in transportation depends on what happens to the available

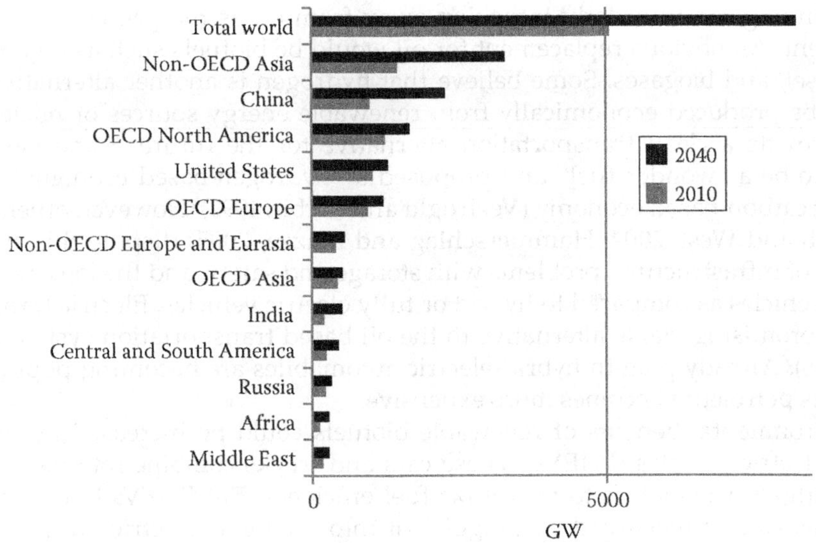

FIGURE 1.4
Electricity-generating capacity and projected additions to 2040 by region. (From EIA, *Annual Energy Outlook 2013*, U.S. Department of Energy, Washington, DC, 2013, www.eia.gov/ies.)

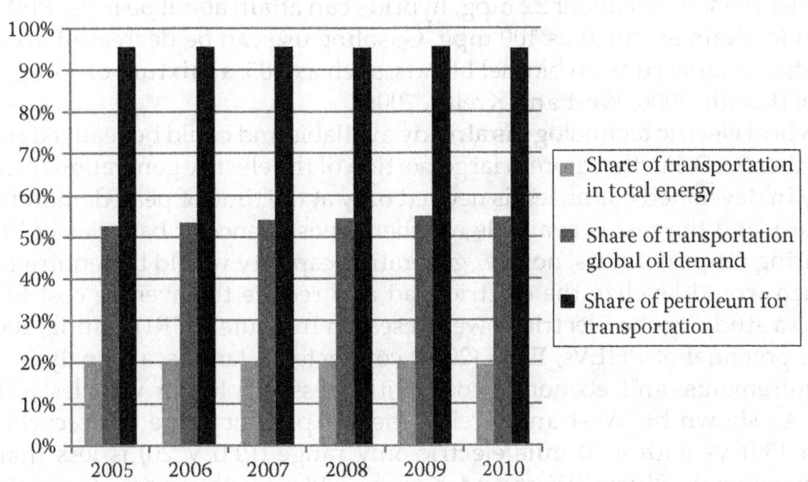

FIGURE 1.5
Share of transport in global oil demand and share of oil in transport energy demand. (Data and Forecast from EIA, *Annual Energy Outlook 2013*, U.S. Department of Energy, Washington, DC, 2013, www.eia.gov/aeo.)

oil resources, production, and prices. At present 95% of all energy for transportation comes from oil, and according to Figure 1.5, the EIA projects that petroleum will still provide 95% of all energy for transportation in 2040. However, with policy changes happening in the world due to serious concerns about global climate change and expected future technology developments, projections simply based on the past use will probably prove to be wrong.

As explained later in this chapter, irrespective of the actual amount of oil remaining in the ground, oil production will peak in the foreseeable future. Therefore, the need for careful planning for an orderly transition away from oil as the primary transportation fuel is urgent. An obvious replacement for oil would be biofuels such as ethanol, methanol, biodiesel, and biogases. Some believe that hydrogen is another alternative, because if it could be produced economically from renewable energy sources or nuclear energy, it could provide a clean transportation alternative for the future. Some have claimed hydrogen to be a "wonder fuel" and proposed a "hydrogen-based economy" to replace the present carbon-based economy (Veziroglu and Barbir, 1992). However, others (Shinnar, 2003; Kreith and West, 2004; Hammerschlag and Mazza, 2005) dispute this claim based on the lack of infrastructure, problems with storage and safety, and the lower efficiency of hydrogen vehicles as compared to hybrid or fully electric vehicles. Electric transportation presents a promising viable alternative to the oil-based transportation system (West and Kreith, 2006). Already plug-in hybrid-electric automobiles are becoming popular around the world as petroleum becomes more expensive.

The environmental benefits of renewable biofuels could be increased by using plug-in hybrid electric vehicles (PHEVs). These cars and trucks combine internal combustion engines with electric motors to maximize fuel efficiency. But PHEVs have more battery capacity that can be recharged by plugging it into a regular electric outlet. Then these vehicles can run on electricity alone for relatively short trips. The electric-only trip length is denoted by a number, for example, PHEV 20 can run on battery charge for 20 miles. When the battery charge is used up, the engine begins to power the vehicle. The hybrid combination reduces gasoline consumption appreciably. Whereas the conventional vehicle fleet has a fuel economy of about 22 mpg, hybrids can attain about 50 mpg. PHEV 20s have been shown to attain as much as 100 mpg. Gasoline use can be decreased even further if the combustion engine runs on biofuel blends, such as E85, a mixture of 15% gasoline and 85% ethanol (Kreith, 2006; West and Kreith, 2006).

Plug-in hybrid electric technology is already available and could be realized immediately without further R&D. Furthermore, a large portion of the electric generation infrastructure, particularly in developed countries, is needed only at the time of peak demand (60% in the United States), and the rest is available at other times. Hence, if batteries of PHEVs were charged during off-peak hours, no new generation capacity would be required. Moreover, this approach would levelize the electric load and reduce the average cost of electricity, according to a study by the Electric Power Research Institute (EPRI) (Sanna, 2005).

Given the potential of PHEVs, EPRI (2004) conducted a large-scale analysis of the cost, battery requirements, and economic competitiveness of plug-in vehicles today and in the future. As shown by West and Kreith, the net present value of lifecycle costs over 10 years for PHEVs with a 20 mile electric-only range (PHEV 20) is less than that of a similar conventional vehicle (West and Kreith, 2006). Furthermore, currently available nickel metal-hydride (NiMH) batteries are already able to meet required cost and performance specifications. More advanced batteries, such as lithium-ion (Li-ion) batteries, may improve the economics of PHEVs even further in the future.

1.5 World Energy Resources

With a view to meet the future demand of primary energy in 2050 and beyond, it is important to understand the available reserves of conventional energy resources including fossil fuels and uranium, and the limitations posed on them due to environmental considerations.

1.5.1 Conventional Oil

There is a considerable debate and disagreement on the estimates of "ultimate recoverable oil reserves," however, there seems to be a good agreement on the amount of "proven oil reserves" in the world. According to BP (2013), total identified or proven world oil reserves at the end of 2012 were 1668.9 billion barrels (bbl). This estimate is close to the reserves of 1700 billion bbl from other sources listed by IEA (2013). The differences among them are in the way they account for the unconventional oil sources. Considering the production rate of about 86.5 million bbl/day at the end of 2012, these reserves will last for about 53 years if there is no increase in production. Of course there may be additional reserves that may be discovered in the future. An analysis by the U.S. Energy Information Agency (2006) estimates the ultimately recoverable world oil reserves (including resources not yet discovered) at between 2.2×10^{12} and 3.9×10^{12} bbl. More recently, IEA has estimated that the ultimate remaining recoverable oil resources are as much as 2670 billion bbl of conventional oil (including Natural Gas Liquids), 345 billion bbl of light oil, 1880 billion of extra heavy oil and bitumen, and 1070 billion bbl kerogen oil. It is important to note that for this high estimate the IEA puts in a disclaimer, "However, resource estimates are inevitably subject to a considerable degree of uncertainty; this is particularly true for unconventional resources that are very large, but still relatively poorly known, both in terms of the extent of the resource in place and judgments about how much might be technically recoverable."

Ever since petroleum geologist M. King Hubbert correctly predicted in 1956 that U.S. oil production would reach a peak in 1973 and then decline (Hubbert, 1974), scientists and engineers have known that worldwide oil production would follow a similar trend. Today, the only question is when the world peak will occur. Bartlett (2002) has developed a predictive model based on a Gaussian curve similar in shape to the data used by Hubbert as shown in Figure 1.6. The predictive peak in world oil production depends on the assumed total amount of recoverable reserves.

If the BP estimated oil reserves are correct, we are close to the peak in the world oil production. If, however, estimates of the ultimate reserves (discovered and undiscovered)

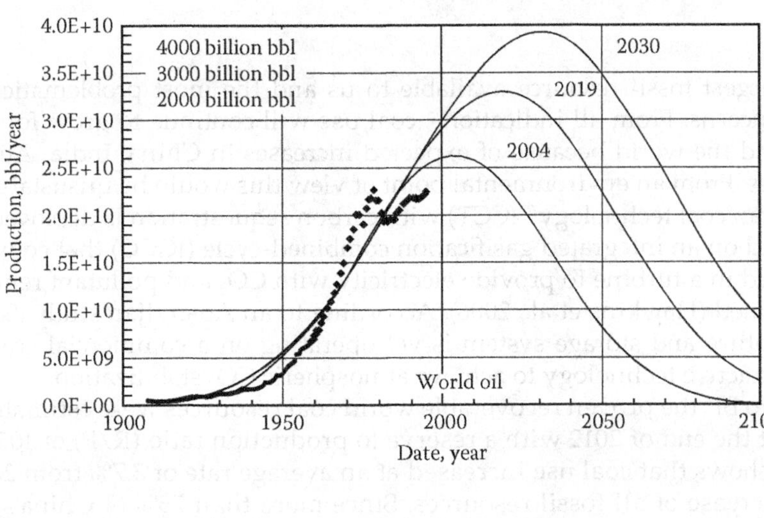

FIGURE 1.6
World oil production vs. time for various amounts of ultimate recoverable resource. (From Bartlett, A.A., *Math. Geol.*, 32, 1, 2002.)

are used, we may expect the oil production to increase a little longer before it peaks. But changing the total available reserves from 3×10^{12} to 4×10^{12} bbl increases the predicted time of peak production by merely 11 years, from 2019 to 2030. IEA World Energy Outlook 2013 estimates that under one policy scenario the oil production will peak at about 91 million bbl/day in 2020 while another policy scenario puts the peak at 101 million bbl/day in 2035. It is clear that no matter which scenario turns out to be true, the global oil production will peak sometime between 2019 and 2035. There is no question that once the world peak is reached and oil production begins to drop, either alternative fuel will have to make up the difference between demand and supply, or the cost of fuel will increase precipitously and create an unprecedented social and economic crisis for our entire transportation system.

The present trend of yearly increases in oil consumption, especially in China and India, shortens the window of opportunity for a managed transition to alternative fuels even further. Hence, irrespective of the actual amount of oil remaining in the ground, peak production will occur soon. Therefore, the need for starting to supplement oil as the primary transportation fuel is urgent because an orderly transition to develop petroleum substitutes will take time and careful planning.

1.5.2 Natural Gas

According to BP (2013) the total proven world natural gas reserves at the end of 2012 were 187.3 trillion m^3. Considering the production rate of gas in 2012, with no increase in production thereafter, these reserves would last for 55.7 years. However, production of natural gas has been rising at an average rate of 2.7% over the past 5 years. If production continues to rise because of additional use of CNG for transportation and increased power production from natural gas, the reserves would last for fewer years. Of course, there could be additional new discoveries. However, even with additional discoveries, it is reasonable to expect that all the available natural gas resources may last from about 50 to 80 years, with a peak in production occurring much earlier.

1.5.3 Coal

Coal is the largest fossil resource available to us and the most problematic from environmental concerns. From all indications, coal use will continue to grow for power production around the world because of expected increases in China, India, Australia, and other countries. From an environmental point of view this would be unsustainable unless advanced "clean coal technology" (CCT) with carbon sequestration is deployed.

CCT is based on an integrated gasification combined-cycle (IGCC) that converts coal to gas that is used in a turbine to provide electricity with CO_2 and pollutant removal before the fuel is burned (Hawkins et al., 2006). According to an Australian study (Sadler, 2004), no carbon capture and storage system is yet operating on a commercial scale, but may become an attractive technology to achieve atmospheric CO_2 stabilization.

According to BP, the proven recoverable world coal resources were estimated to be 861 billion tons at the end of 2012 with a reserve to production ratio (R/P) of 107 years. The BP data also shows that coal use increased at an average rate of 3.7% from 2007 to 2012, the largest increase of all fossil resources. Since more than 75% of China's electricity-generating capacity is based on coal and both China and India are continuing to build new coal power plants, it is reasonable to assume that coal use will continue to increase for at least some years in future. Therefore, the R/P ratio will decrease further from the present value of 107 years. The R/P ratio will decrease even more rapidly when clean coal technologies such as coal gasification and liquefaction are utilized instead of direct combustion.

1.5.4 Summary of Fossil Fuel Reserves

Even though there are widely differing views and estimates of the ultimately recoverable resources of fossil fuels, it is fair to say that they may last for around 50–100 years with a peak in production occurring much earlier. However, a big concern is the climatic threat of additional carbon that will be released into the atmosphere. According to the estimates from the IEA, if the present shares of fossil fuels are maintained up to 2040 without any carbon sequestration, a cumulative amount of approximately 1000 gigatons of carbon will be released into the atmosphere (based on Figure 1.7). This is especially troublesome in view of the fact that the present total cumulative emissions of about 500 gigatons of carbon have already raised serious concerns about global climate change.

1.5.5 Nuclear Resources

Increased use of nuclear power presents the possibility of additional carbon-free energy use and its consequent benefit for the environment. However, there are significant concerns about nuclear waste and other environmental impacts, the security of the fuel and the waste, and the possibility of their diversion for weapon production.

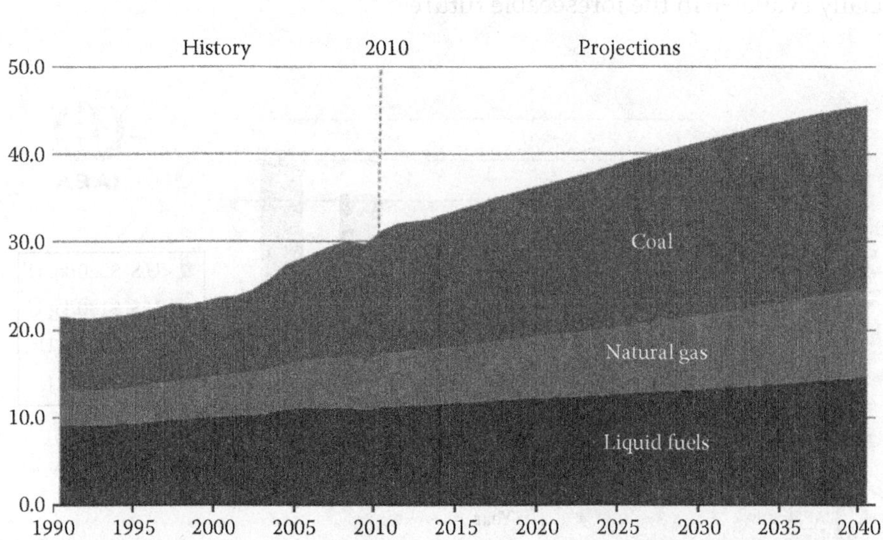

FIGURE 1.7
World energy-related CO_2 emissions by fuel (billion metric tons). (Data and forecast from IEA, *World Energy Outlook 2013*, International Energy Agency, Paris, France, 2013.)

Nuclear fission provided 14% of the electricity in the world in 2011 (IEA, 2013) and the worldwide nuclear capacity in 2011 was 375 GW (IAEA, 2011). Although a number of countries have decided to not build additional nuclear power plants after the Fukushima accident, nuclear power capacity is expected to continue to grow mainly because of the ongoing and planned construction in China and some other countries. IAEA estimates that the worldwide nuclear power capacity will increase at an average rate of 1.5%–2.7% until 2035 (IAEA, 2011). At present, uranium is used as the fissile material for nuclear power production. Thorium could also be used for nuclear fission; however, to date nobody has developed a commercial nuclear power plant based on thorium. Terrestrial deposits of both uranium and thorium are limited and concentrated in a few countries of the world. The International Atomic Energy Agency (IAEA) estimates the total identified recoverable uranium reserves in the world to be about 5 million tons which increase to about 7 million tons if the price of uranium goes up to $264/kg U (Figure 1.8). Additionally, there are nonconventional uranium resources, such as sea water which contains about 3 parts per billion uranium and some phosphate deposits (more than half of them in Morocco) which contain about 100 parts per million uranium. These resources are potentially huge; however, their cost effective recovery is not certain (Figure 1.9).

For generating 1 TWh of electricity from nuclear fission, approximately 22 tons of uranium are required (UNDP, 2004). Based on the 2011 world capacity of 375 GW, the identified reserves will last about 97 years if there is no change in the generation capacity. At an average annual growth rate of 2%, the uranium reserves of 7 million tons will last for about 60 years. This estimate does not consider regeneration of spent fuel. At present, nuclear fuel regeneration is not allowed in the United States. However, that law could be changed in future. Development of breeder reactors could increase the time period much further. The major impediment may be economic viability. Nuclear fusion could potentially provide a virtually inexhaustible energy supply; however, it is not expected to be commercially available in the foreseeable future.

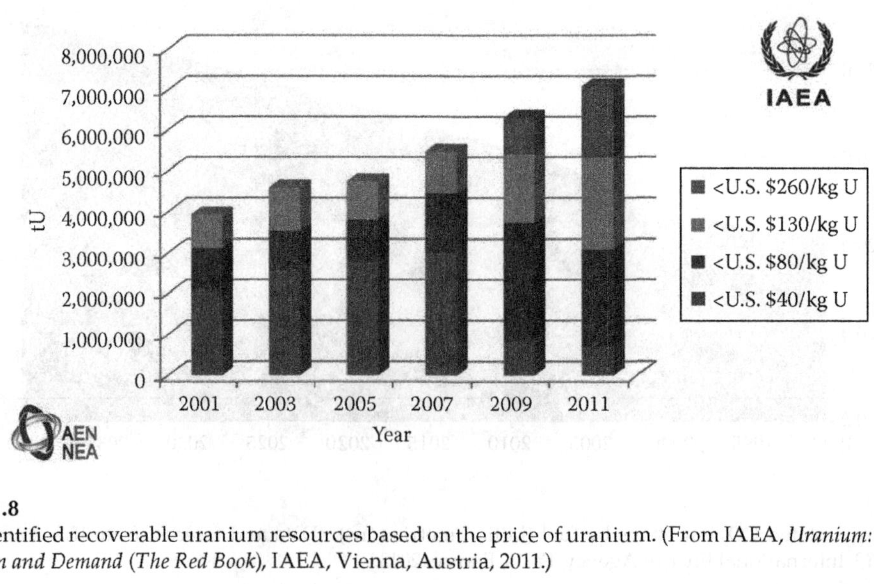

FIGURE 1.8
World identified recoverable uranium resources based on the price of uranium. (From IAEA, *Uranium: Resources, Production and Demand* (*The Red Book*), IAEA, Vienna, Austria, 2011.)

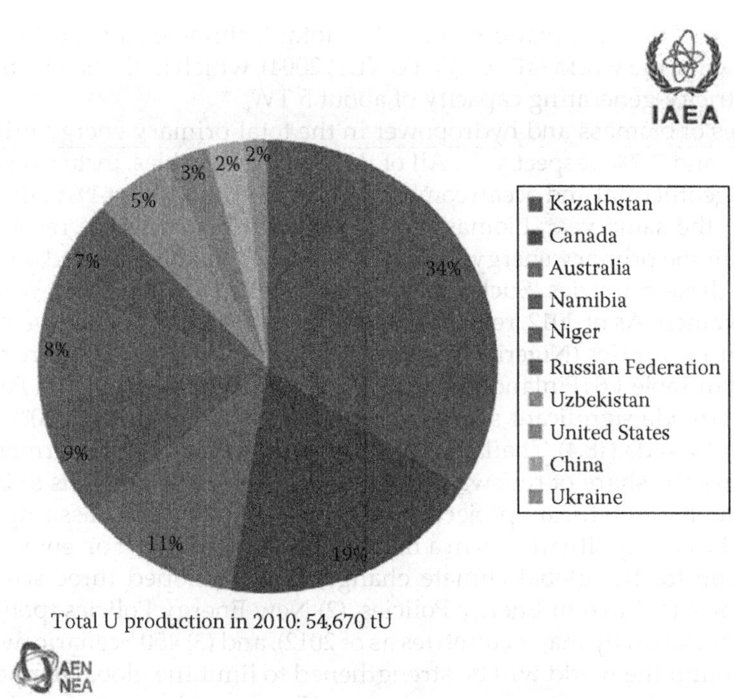

FIGURE 1.9
Top 10 uranium producing countries in 2010. (From IAEA, *Uranium: Resources, Production and Demand (The Red Book)*, IAEA, Vienna, Austria, 2011.)

1.6 Present Status and Potential of Renewable Energy

According to the data in Table 1.5, 13.2% of the world's total primary energy supply (TPES) came from RE in 2011. However, approximately 75% of the RE supply was from biomass, and in developing countries it is mostly converted by traditional open combustion, which is very inefficient. Because of its inefficient use, biomass resources presently supply only about 20% of what they could if converted by modern, more efficient, available technologies. As it stands, biomass provides only about 10% of the world total primary energy

TABLE 1.5
2011 Fuel Shares in World Total Primary Energy Supply

Source	Share (%)
Oil	31.4
Natural gas	21.3
Coal	28.9
Nuclear	5.2
Renewables	13.2

Source: IEA, *World Energy Outlook 2013*, International Energy Agency, Paris, France, 2013.

which is much less than it's real potential. The total technologically sustainable biomass energy potential for the world is 3–4 TW_e (UNDP, 2004), which is about 80% the entire present global electricity-generating capacity of about 5 TW_e.

In 2011, shares of biomass and hydropower in the total primary energy mix of the world were about 10% and 2.3%, respectively. All of the other renewables, including solar thermal, solar PV, wind, geothermal and ocean combined, provided only about 1% of the total primary energy. During the same year, biomass combined with hydroelectric resources provided almost 50% of all the primary energy in Africa. However, biomass is used very inefficiently for cooking in these countries. Such use has also resulted in significant health problems, especially for women. As of 2012, renewable energy contributes more than 40% of their total energy needs in 4 countries (Nigeria, Norway, Brazil, and Sweden) and more than 20% in 10 countries listed in Table 1.6 (Finland, Indonesia, India, Colombia, Chile, and Portugal). Other countries that provide significant shares of their energy from RE but <20% include, New Zealand (19.9%), Canada (18.4), Thailand (18.3%), Romania (15.2%), and Germany (14.2%).

Table 1.7 shows the share of renewable energy in 2011 and projections to 2020 and 2035. Keeping in mind that the future projections are only as good as the assumptions they are based on, and the energy situation is in a flux because of the impact on environment which is a major reason for the global climate change, IEA developed three scenarios for the future projections: (1) Current Energy Policies, (2) New Energy Policies (policies that have already been developed by major countries as of 2012), and (3) 450 Scenario, which assumes that policies around the world will be strengthened to limit the global temperature rise to 2°C or global atmospheric CO_2 concentrations to 450 ppm. Although there is considerable uncertainty about future policies, it is very likely that the future energy developments will lie somewhere in between the last two scenarios. According to these projections, the share of renewable energy will rise to as much as 18%–26% of the global primary energy and 31%–48% of the electricity-generating capacity by 2035. Based on the trends in the development and deployment of wind power and solar power in the last decade, there is reason to believe that values close to 450 scenario are achievable.

TABLE 1.6

Share of Renewable Energy in 2012 TPES for Top 10 Countries

Country	% Share of Renewables in TPES
Nigeria	80.5
Norway	47.2
Brazil	42.8
Sweden	40.0
Finland	30.6
Indonesia	26.2
India	24.3
Colombia	23.5
Chile	22.7
Portugal	22.5
New Zealand	19.9
Canada	18.4
Thailand	18.3
Romania	15.2
Germany	14.2
World	12.9

Source: Enerdata, *Enerdata Energy Statistical Yearbook 2013*, 2013.

TABLE 1.7

Share of Renewable Energy in 2011 and Projections for 2020 and 2035 Based on New Policies and 450 Scenario

		New Policies		450 Scenario	
	2011	2020	2035	2020	2035
Primary energy demand (MTOE)	**1,727**	**2,193**	**3,059**	**2,265**	**3,918**
United States	140	196	331	215	508
Europe	183	259	362	270	452
China	298	392	509	405	690
Brazil	116	148	207	150	225
Share of renewables in total primary energy (%)	*13*	*15*	*18*	*16*	*26*
Electricity generation (TWh)	**4,482**	**7,196**	**11,612**	**7,528**	**15,483**
Bioenergy	424	762	1,477	797	2,056
Hydro	3,490	4,555	5,827	4,667	6,394
Wind	434	1,326	2,774	1,441	4,337
Geothermal	69	128	299	142	436
Solar PV	61	379	951	422	1,389
CSP	2	43	245	56	806
Marine	1	3	39	3	64
Share of total generation (%)	*20*	*26*	*31*	*28*	*48*
Heat demand (MTOE)	**343**	**438**	**602**	**446**	**704**
Industry	209	253	316	248	328
Buildings and agriculture	135	184	286	198	376
Share of total final demand (%)	*8*	*10*	*12*	*10*	*16*
Biofuels (mboe/day)	**1.3**	**2.1**	**4.1**	**2.6**	**7.7**
Road transport	1.3	2.1	4.1	2.6	6.8
Aviation	0	0	0.1	0	0.9
Share of total transport (%)	*2*	*4*	*6*	*5*	*15*
Traditional biomass (MTOE)	**744**	**730**	**680**	**718**	**647**
Share of total bioenergy (%)	*57*	*49*	*37*	*47*	*29*
Share of renewable energy demand (%)	*43*	*33*	*22*	*32*	*17*

Source: IEA, *World Energy Outlook 2013*, International Energy Agency, Paris, France, 2013.

1.6.1 Wind Power

Wind energy technology has progressed significantly over the last two decades. The technology has been vastly improved and capital costs have come down to as low as $1000/kW. At this level of capital costs, wind power is already economical at locations with fairly good wind resources. Therefore, the average annual growth in worldwide wind energy capacity from 2001 to 2012 was over 25% (Figure 1.10). The average growth in the United States over the same period was 37.7%. The total worldwide installed wind power capacity which was 24 GW in 2001 (Figure 1.10), reached a level of 282 GW in 2012 (WWEA, 2013). The countries with the largest wind capacity in 2012 include China (75 GW), United States (60 GW), Germany (31 GW), Spain (23 GW), and India (18 GW) (Figure 1.11). The total theoretical potential for onshore wind power for the world is around 55 TW with a practical potential of at least 2 TW (UNDP, 2004), which is about 40% of the entire present worldwide generating capacity. The offshore wind energy potential is even larger.

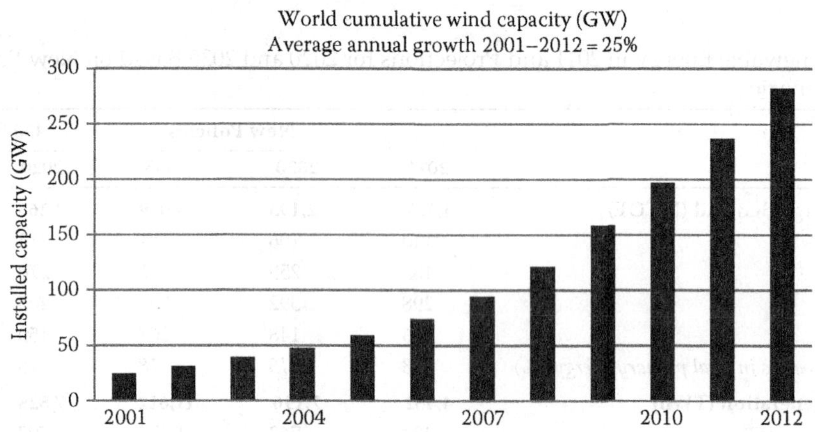

FIGURE 1.10
World wind energy installed capacity and growth rates. (Data from WWEA, World Wind Energy Association, 2012, http://www.wwindea.org/webimages/WorldWindEnergyReport2012_final.pdf.)

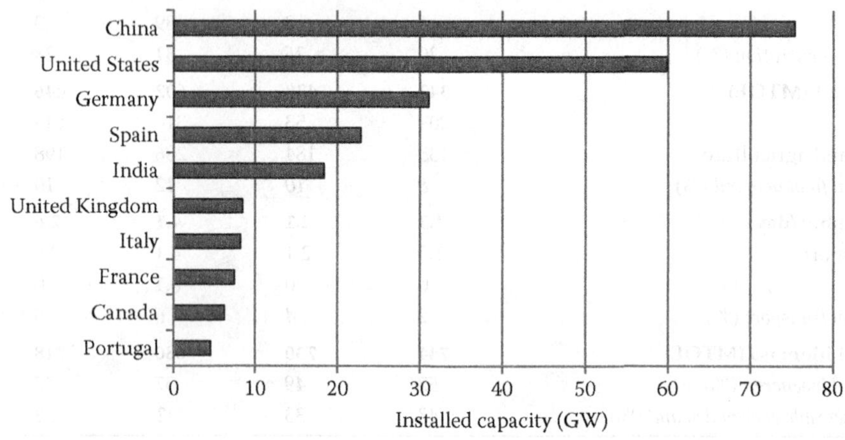

FIGURE 1.11
Top 10 countries with installed wind power capacity. (Data from WWEA, World Wind Energy Association, 2012, http://www.wwindea.org/webimages/WorldWindEnergyReport2012_final.pdf.)

1.6.2 Solar Energy

The amount of sunlight striking the earth's atmosphere continuously is 1.75×10^5 TW. Considering a 60% transmittance through the atmospheric cloud cover, 1.05×10^5 TW reaches the earth's surface continuously. If the irradiance on only 1% of the earth's surface could be converted into electric energy with a 10% efficiency, it would provide a resource base of 105 TW, while the total global energy needs for 2040 are projected to be about 8–9 TW. The present state of solar energy technologies is such that solar cell efficiencies have reached over 40% and solar thermal systems provide efficiencies of 40%–80%. With the present rate of technological development these solar technologies will continue to improve, thus bringing the costs down, especially with the economies of scale.

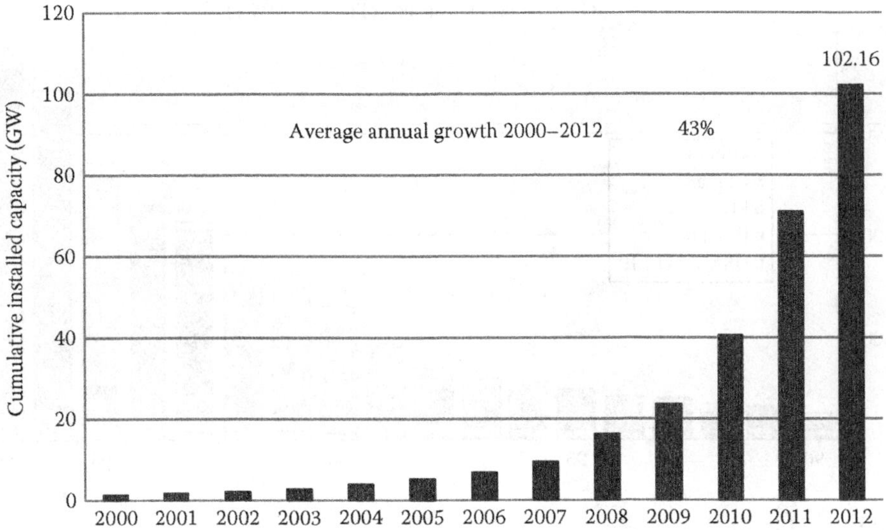

FIGURE 1.12
World solar PV production, 2000–2012 (GWp). (From EPIA, European Photovoltaic Industries Association, 2012, www.epia.org.)

Solar PV panels have come down in cost from about $30/W to about $0.50/W in the last three decades. At $0.50/W panel cost, the overall system cost is around $2/W, which is already lower than grid electricity in the Caribbean island communities. Of course, there are many off-grid applications where solar PV is already cost-effective. With net metering and governmental incentives, such as feed-in laws and other policies, grid-connected applications such as building integrated PV (BIPV) have become cost-effective even where grid electricity is cheaper. As a result, the worldwide growth in PV production has averaged over 43%/year from 2000 to 2012 and 61% from 2007 to 2012 (Figure 1.12) with Europe showing the maximum growth.

Solar thermal power using concentrating solar collectors was the first solar technology which demonstrated its grid power potential. A 354 MW_e concentrating solar thermal power (CSP) plant has been operating continuously in California since 1988. Progress in solar thermal power stalled after that time because of poor policy and lack of R&D. However, the last 10 years have seen a resurgence of interest in this area and a number of solar thermal power plants around the world are under construction. The largest CSP plant with a capacity of 400 MW came on line in Nevada in February 2014. The cost of power from these plants (which is so far in the range of 12–16 U.S. cents/kWh_e) has the potential to go down to 5 U.S. cents/kWh_e with scale-up and creation of a mass market. An advantage of solar thermal power is that thermal energy can be stored efficiently and fuels, such as, natural gas or biogas may be used as back up to ensure continuous operation. If this technology is combined with power plants operating on fossil fuels, it has the potential to extend the time frame of the existing fossil fuels.

Low temperature solar thermal systems and applications have been well developed for quite some time. They are being actively installed wherever the policies favor their deployment. Figure 1.13 gives an idea of the rate of growth of solar thermal systems in the world. In 2011, approximately 234 GW_{th} solar collectors were deployed around the world, a vast majority (65%) of those being in China (IEA, 2013) (Figure 1.14).

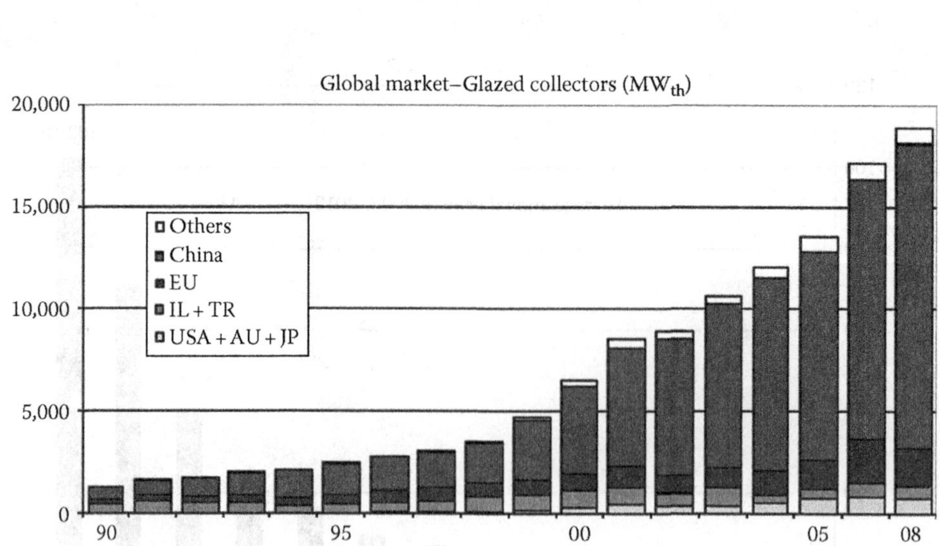

FIGURE 1.13
Deployment of solar heat (glazed) collectors, MW_{th}. (From ESIF, IEA SHC.)

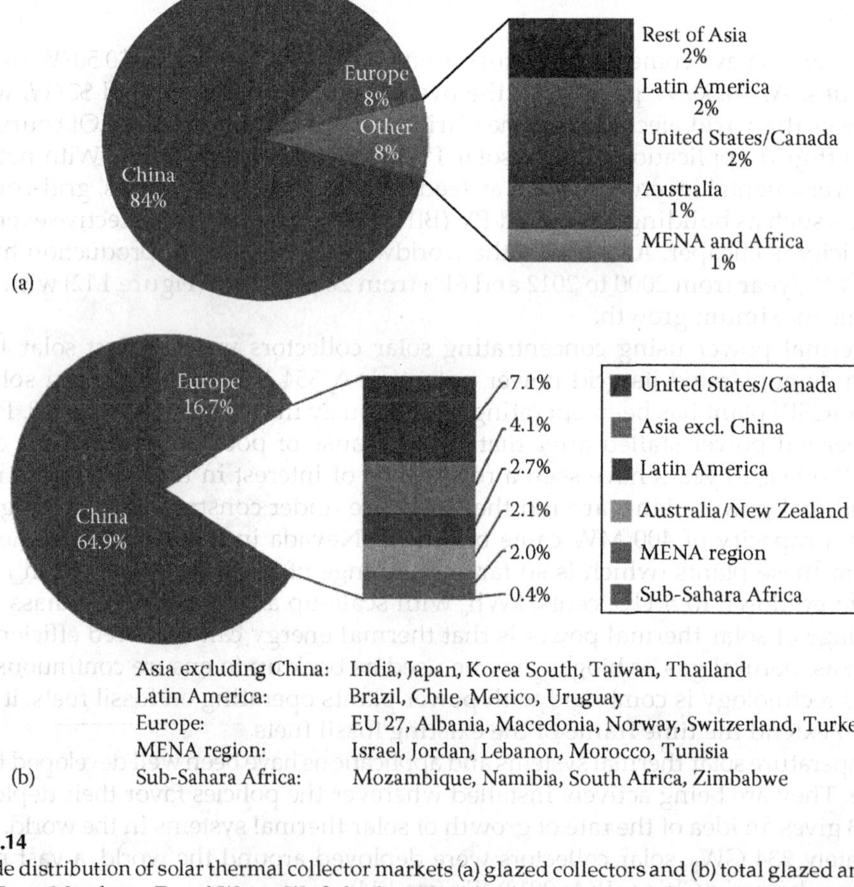

Asia excluding China:	India, Japan, Korea South, Taiwan, Thailand
Latin America:	Brazil, Chile, Mexico, Uruguay
Europe:	EU 27, Albania, Macedonia, Norway, Switzerland, Turkey
MENA region:	Israel, Jordan, Lebanon, Morocco, Tunisia
Sub-Sahara Africa:	Mozambique, Namibia, South Africa, Zimbabwe

(b)

FIGURE 1.14
Worldwide distribution of solar thermal collector markets (a) glazed collectors and (b) total glazed and unglazed in 2012. (From Mauthner, F. and Weiss, W., Solar heat worldwide—Markets and contribution to energy supply 2011, IEA Solar Heating and Cooling Program, Paris, France, May 2013.)

1.6.3 Biomass

Although theoretically harvestable biomass energy potential is of the order of 90 TW, the technical potential on a sustainable basis is of the order of 8–13 TW or 270–450 EJ/year (UNDP, 2005). This potential is 1.6–2.6 times the present electricity-generating capacity of the world. It is estimated that by 2025, even the municipal solid waste (MSW) could generate up to 6 EJ/year.

The biggest advantage of biomass as an energy resource is its relatively straightforward transformation into transportation fuels. Biofuels have the potential to replace as much as 75% of the petroleum fuels in use for transportation in the United States (Worldwatch, 2006). This is especially important in view of the declining oil supplies worldwide. Biofuels will not require additional infrastructure development. Therefore, development of biofuels is being viewed very favorably by governments around the world. Biofuels, along with other transportation options such as electric vehicles and hydrogen, will help diversify the fuel base for future transportation. Table 1.8 and Figure 1.15 show the global production of biofuels from 2001 to 2011. United States, Brazil, and Europe are the top producing countries and region of the world. Biofuel production grew more than five times in 10 years, although it started from a much

TABLE 1.8
Total Biofuels Production (1000 bbl/day)

	2001	2002	2003	2004	2005	2006	2007	2008	2009	2010	2011
United States	115.7	140.3	183.9	223.3	260.6	335.0	457.3	649.7	747.1	889.8	971.7
Brazil	197.6	216.9	249.4	251.7	276.4	307.3	395.7	486.3	477.5	527.1	438.1
Europe	21.2	29.3	39.3	48.9	76.8	123.9	153.8	198.1	233.2	255.2	250.5
Asia	3.1	8.3	17.2	21.1	28.2	44.9	49.2	75.6	93.8	99.8	118.2
Rest of the world	5.3	8.6	9.6	9.8	14.2	29.6	47.3	67.7	83.8	93.3	118.8
World	342.9	403.5	499.4	554.8	656.3	840.6	1,103.3	1,477.3	1,635.4	1,865.4	1,897.2

Sources: Enerdata, *Enerdata Energy Statistical Yearbook 2013*, 2013; IEA, *World Energy Outlook 2013*, International Energy Agency, Paris, France, 2013.

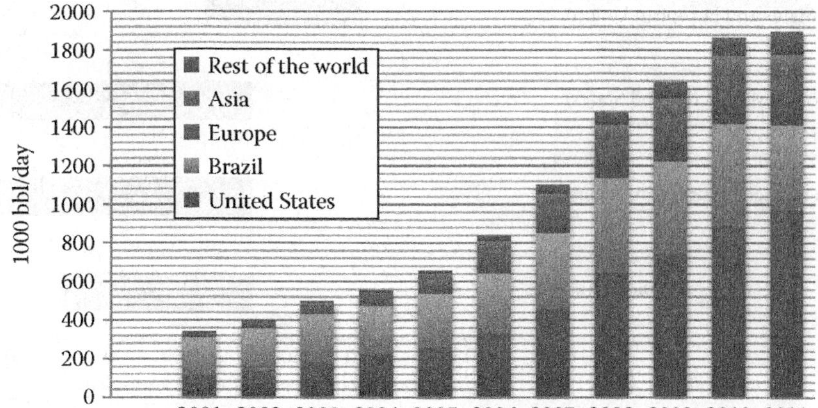

FIGURE 1.15
World biofuel production, 2001–2011. (From IEA, *World Energy Outlook 2013*, International Energy Agency, Paris, France, 2013.)

smaller base. In 2005, the world ethanol production had reached about 36 billion L/year while biodiesel production topped 3.5 billion L during the same year.

The present cost of ethanol production ranges from about €0.25 to about €1/gasoline equivalent L, as compared to the wholesale price of gasoline which is between €0.40 and €0.60/L (Figure 1.16). Biodiesel costs, on the other hand, range between €0.20 and €0.65/L of diesel equivalent (Figure 1.17). Figure 1.18 shows the feedstock used for these biofuels. An important consideration for biofuels is that the fuel not be produced at the expense of food while there are people going hungry in the world. This would not be of concern if biofuels were produced from MSW or nonfood forest resources.

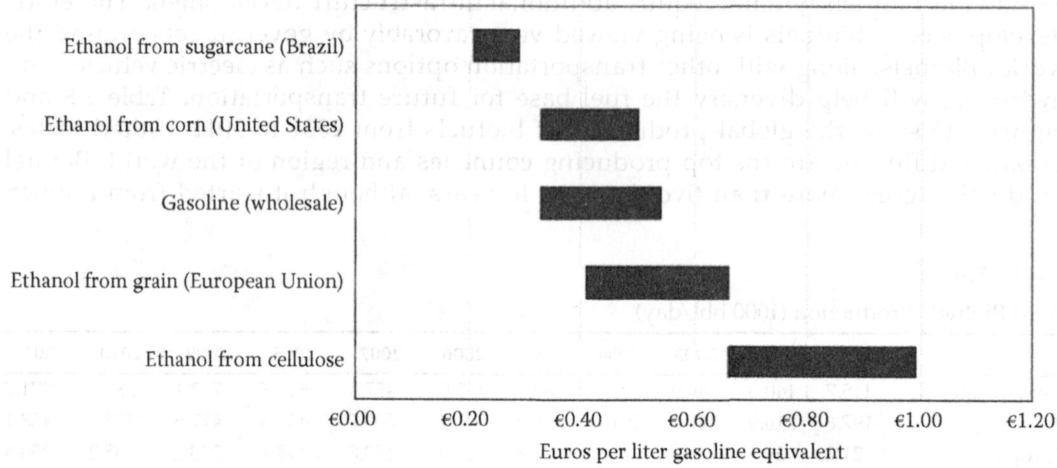

FIGURE 1.16
Cost ranges for ethanol and gasoline production, 2006. (From IEA, Reuters, DOE.)

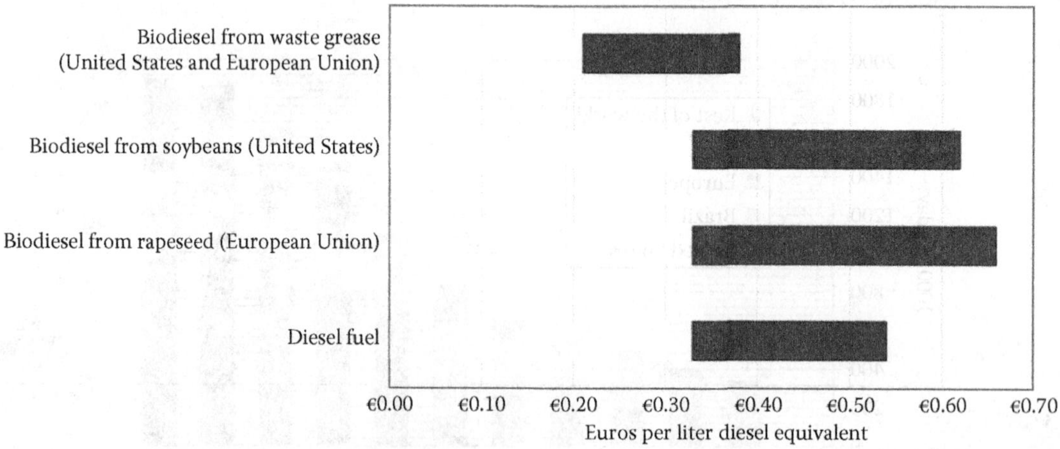

FIGURE 1.17
Cost ranges for biodiesel and diesel production, 2006. (From IEA, Reuters, DOE.)

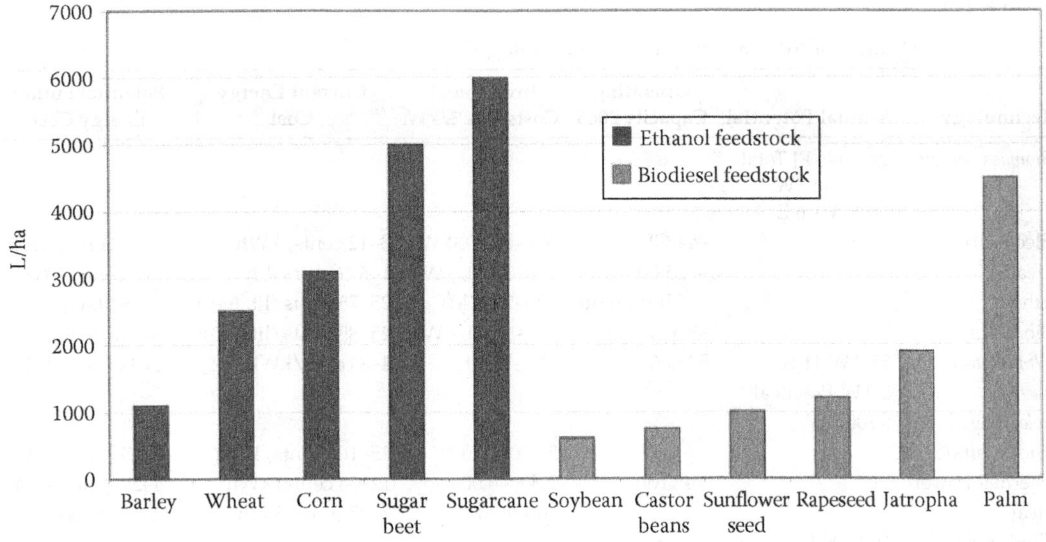

FIGURE 1.18
Biofuel yields of selected ethanol and biodiesel feedstocks. (From Hunt, S. and Forster, E. 2006. *Biofuels for Transportation: Global Potential and Implications for Sustainable Agriculture and Energy in the 21st Century*. Worldwatch Institute, Washington, DC.)

According to the Worldwatch report, a city of one million people produces about 1,800 tons of MSW and 1,300 tons of organic waste every day, which using the present-day technology could produce enough fuel to meet the needs of 58,000 persons in the United States, 360,000 in France, and nearly 2.6 million in China at current rates of per capita fuel use (Worldwatch, 2006).

1.6.4 Summary of Renewable Energy Resources

By definition, the term "reserves" does not apply to renewable resources. So we need to look at the annual potential of each resource. Table 1.9 summarizes the resource potential and the present costs and the potential future costs for each renewable resource.

As in the case of other new technologies, it is expected that cost competitiveness of the renewable energy technologies will be achieved with R&D, scale-up, commercial experience, and mass production. The experience curves in Figure 1.19 show industry-wide cost reductions in the range of 10%–20% for each cumulative doubling of production for wind power, photovoltaics, ethanol, and gas turbines (UNDP, 2004). Similar declines can be expected in solar thermal power and other renewable technologies. As seen from Figure 1.19, wind energy technologies have already achieved market maturity, and PV technologies are well on their way. Even though concentrating solar thermal power (CSP) is not shown in this figure, a GEF report estimates that CSP will achieve the cost target of about \$0.05/kWh by the time it has an installed capacity of about 40 GW (GEF, 2005). As a reference point, wind power achieved that capacity milestone in 2003.

TABLE 1.9

Potential and Status of Renewable Energy Technologies

Technology	Annual Potential	Operating Capacity 2005	Investment Costs U.S. $/kW	Current Energy Cost	Potential Future Energy Cost
Biomass energy	276–446 EJ Total or 8–13 TW MSW ~ 6 EJ				
Electricity		~44 GW	500–6,000/kW_e	3–12 cents/kWh	3–10 cents/kWh
Heat		~225 GW_{th}	170–1,000/kW_{th}	1–6 cents/kWh	1–5 cents/kWh
Ethanol		~36 billion lit.	170–350/kW_{th}	25–75 cents/lit. (ge)[a]	$6–$10/GJ
Biodiesel		~3.5 billion lit.	500–1,000/kW_{th}	25–85 cents/lit. (de)[b]	$10–$15/GJ
Wind power	55 TW Theo. 2 TW Practical	59 GW	850–1,700	4–8 cents/kWh	3–8 cents/kWh
Solar energy	>100 TW				
Photovoltaics		5.6 GW	5,000–10,000	25–160 cents/kWh	5–25 cents/kWh
Thermal Power		0.4 GW	2,500–6,000	12–34 cents/kWh	4–20 cents/kWh
Heat			300–1,700	2–25 cents/kWh	2–10 cents/kWh
Geothermal	600,000 EJ useful resource base				
Electricity	5,000 EJ economical in 40–50 years	9 GW	800–3,000	2–10 cents/kWh	1–8 cents/kWh
Heat		11 GW_{th}	200–2,000	0.5–5 cents/kWh	0.5–5 cents/kWh
Ocean energy					
Tidal	2.5 TW	0.3 GW	1,700–2,500	8–15 cents/kWh	8–15 cents/kWh
Wave	2.0 TW		2,000–5,000	10–30 cents/kWh	5–10 cents/kWh
OTEC	228 TW		8,000–20,000	15–40 cents/kWh	7–20 cents/kWh
Hydroelectric	1.63 TW Theo.				
Large	0.92 TW Econ.	690 GW	1,000–3,500	2–10 cents/kWh	2–10 cents/kWh
Small		25 GW	700–8,000	2–12 cents/kWh	2–10 cents/kWh

Sources: Data from UNDP, World Energy Assessment: Energy and the Challenge of Sustainability, 2004. Updated from other sources: Worldwatch, Biofuels for transportation—Global potential and implications for sustainable and energy in the 21st century, Report prepared for the German Federal Ministry for Food, Agriculture and Consumer Protection, Worldwatch Institute, Washington, DC, 2006; World Wind Energy Association Bulletin, 2006, www.wwindea.org; Photovoltoaic Barometer; EPIA, European Photovoltaic Industries Association, 2012, www.epia.org; World Geothermal Power Generation 2001–2005; GRC Bulletin; International Energy Annual; U.S. DOE-EIA.

Note: ge, gasoline equivalent liter; de, diesel equivalent liter; kWe, kilowatt electrical power; kWth, kilowatt thermal power.

1.7 Role of Energy Conservation

Energy conservation can and must play an important role in future energy strategy, because it can ameliorate adverse impacts on the environment rapidly and economically. Figures 1.20 and 1.21 give an idea of the potential of energy efficient improvements. Figure 1.20 shows that per capita energy consumption varies by as much as a factor of 3

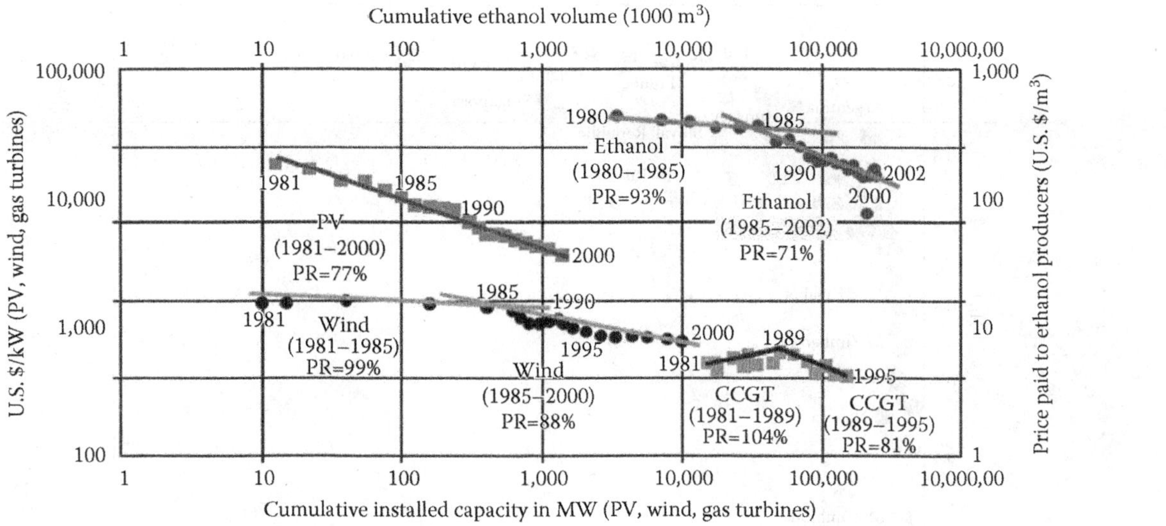

FIGURE 1.19
Experience curves for wind, PV, ethanol and gas turbines. (Adapted from UNDP, World Energy Assessment: Energy and the Challenge of Sustainability, 2004. For wind turbines: Neij, L. et al., Experience curves: A tool for Energy Policy Assessment, March 2003; For gas turbines: Claeson Colpier, U. and Cornland, D., *Energy Policy*, 30, 209, 2002; For photovoltaics: Parente, V. et al., *Prog. Photovolt. Res. Appl.*, 10(8), 571, 2002; For ethanol: Goldemberg, J. et al., *Biomass Energy*, in press.)

between the United States and some European countries with almost the same level of Human Development Index (HDI). Even taking just the OECD European countries combined, the per capita energy consumption in the United States is twice as much. It is fair to assume that the per capita energy of the United States could be reduced to the level of OECD Europe of 4.2 kW by a combination of energy efficiency improvements and changes in the transportation infrastructure. This is significant because the United States uses about 25% of the energy of the whole world. The present per capita energy consumption in the United States is 284 GJ, which is equivalent to about 9 kW/person, while the average for the whole world is 2 kW. Board of Swiss Federal Institutes of Technology has developed a vision of a 2 kW per capita society by the middle of the century (UNDP, 2004). The vision is technically feasible. However, to achieve this vision will require a combination of increased R&D on energy efficiency and policies that encourage conservation and use of high efficiency systems. It will also require some structural changes in the transportation systems. According to the 2004 World Energy Assessment by UNDP, a 25%–35% reduction in primary energy in the industrialized countries is achievable cost effectively in the next 20 years, without sacrificing the level of energy services. The report also concluded that similar reductions of up to 40% are cost effectively achievable in the transitional economies and more than 45% in developing economies. As a combined result of efficiency improvements and structural changes such as increased recycling, substitution of energy intensive materials, etc., energy intensity could decline at a rate of 2.5%/year over the next 20 years (UNDP, 2004).

McKinsey and Company conducted a comprehensive study of the energy conservation potential in United States in 2020. Figure 1.22 shows the potential in various sectors including the average cost of savings. According to this figure, the total U.S. economical potential of energy conservation to 2020 is 9500 trillion Btu or 25 GTOE.

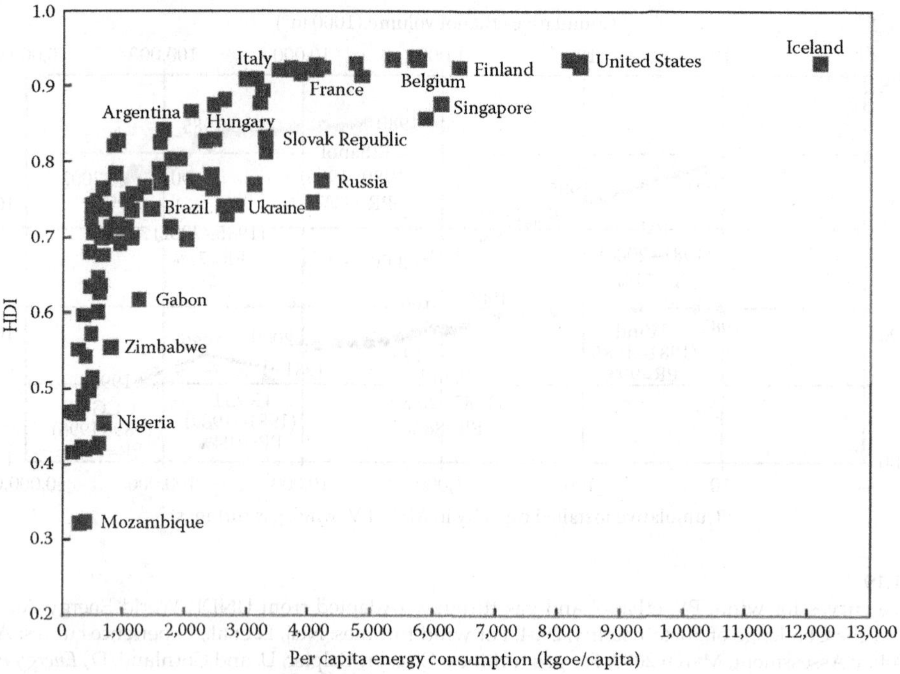

FIGURE 1.20
Relationship between Human Development Index (HDI) and per capita energy use, 1999–2000. (From UNDP, World Energy Assessment: Energy and the Challenge of Sustainability, 2004.)

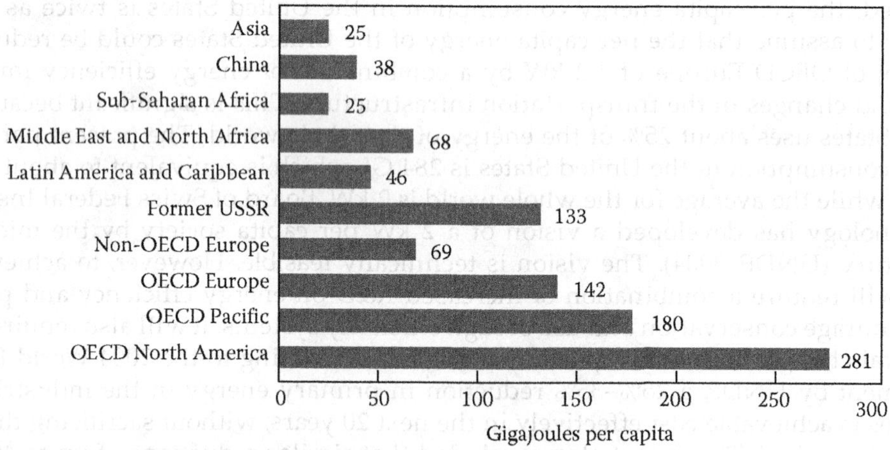

FIGURE 1.21
Per capita energy use by region (commercial and noncommercial) 2000. (From UNDP, World Energy Assessment: Energy and the Challenge of Sustainability, 2004.) *Note:* Asia excludes Middle East, China, and OECD countries; Middle East and North Africa comprises Algeria, Bahrain, Egypt, Iran, Iraq, Israel, Jordan, Kuwait, Lebanon, Libya, Morocco, Oman, Qatar, Saudi Arabia, Syria, Tunisia, United Arab Emirates, and Yemen; Latin America and Caribbean excludes Mexico; OECD Pacific comprises Australia, Japan, Korea, and New Zealand; Former USSR comprises Armenia, Azerbaijan, Belarus, Estonia, Georgia, Kazakhstan, Kyrgyztan, Latvia, Lithuania, Moldova, Russia, Tajikistan, Turkmenistan, Ukraine, and Uzbekistan; Non-OECD Europe comprises Albania, Bosnia and Herzegovina, Bulgaria, Croatia, Cyprus, Gibraltar, Macedonia, Malta, Romania, and Slovenia; OECD North America includes Mexico.

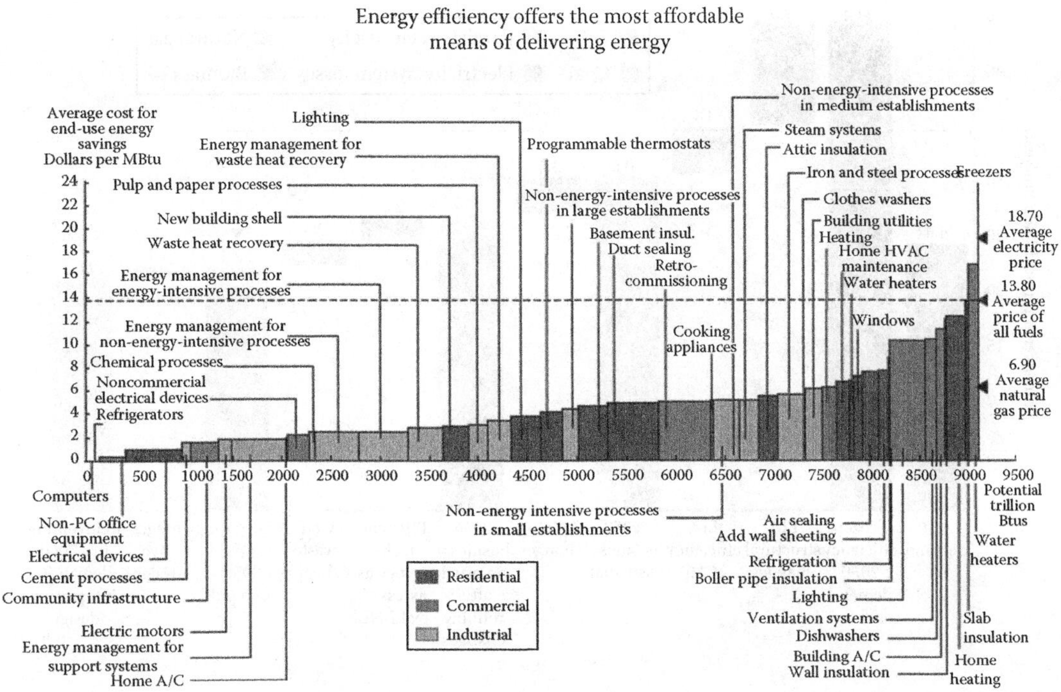

FIGURE 1.22
Energy saving potential of various sectors and cost of savings as compared to the price of electricity. (From Granade, H. et al., Unlocking energy efficiency in the U.S. economy, *2010 EIA Energy Conference*, Washington, DC, April 2010, www.eia.gov/conference/2010/session9/granade.pdf.)

Rocky Mountain Institute in Colorado, estimates that the total potential of energy savings due to efficiency improvements in the industry sector in United States by 2050 could be as much as 30% of energy use under the business as usual scenario (see Figure 1.23).

Improving energy efficiency across all sectors of the economy should become a worldwide objective (Energy Commission, 2004). It should be noted, however, that free market price signals may not always be sufficient to effect energy efficiency. Hence, legislation on the state and/or national level for energy efficiency standards for equipment in the residential and commercial sector may be necessary. There is considerable debate whether incentives or mandates are the preferred way to improve energy efficiency. Such measures may be necessary because surveys indicate that consumers consistently rank energy use and operating costs quite low on the lists of attributes they consider when purchasing an appliance or construct a building. Incentives may be the preferred option provided they induce decision makers to take appropriate action.

Figure 1.24 shows the projected energy savings from upgraded standards for products installed in the years 2010–2020. Outside the United States, over 30 countries have also adopted minimum energy performance standards. These measures have been shown to be economically attractive and can provide an appreciable reduction in adverse environmental impacts.

This handbook describes energy efficient improvements achievable with available technologies. The challenge is to adopt policies that accelerate the adoption of these technologies all over the world.

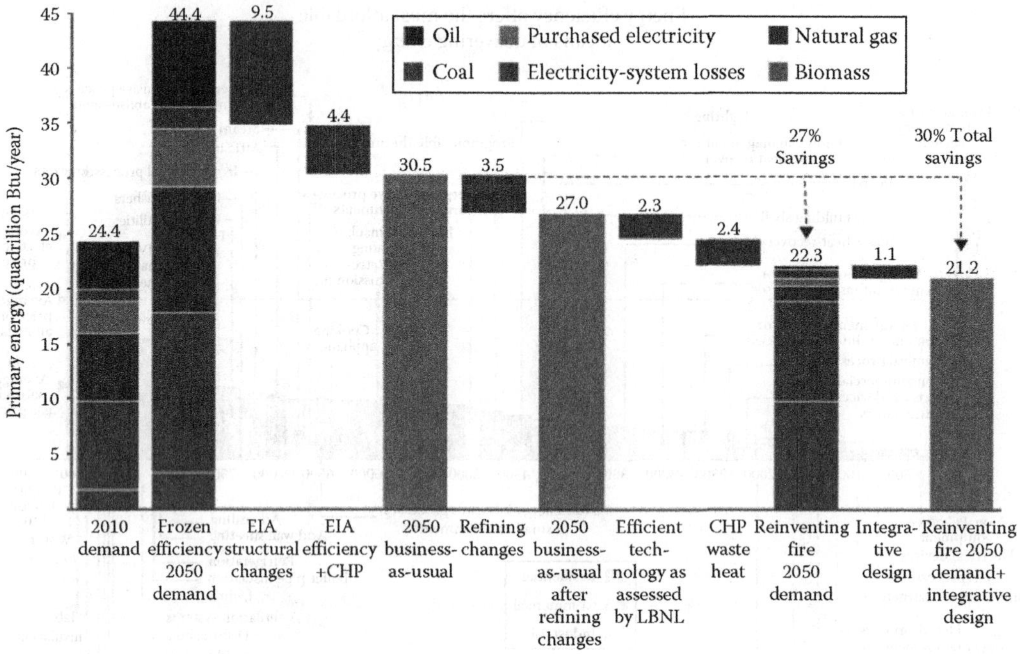

FIGURE 1.23
U.S. industry energy savings potential (2010–2050) as a percentage of the total industrial energy use in business as usual scenario. (From Rocky Mountain Institute, U.S. Industry Energy savings potential, 2010–2050, in: *Reinventing Fire: Bold Business Solutions for the New Energy Era*, RMI, Snowmass, CO, 2012, http://www.rmi.org/RFGraph-US_industry_energy_saving_potential; www.RMI.org./ReinventingFire.)

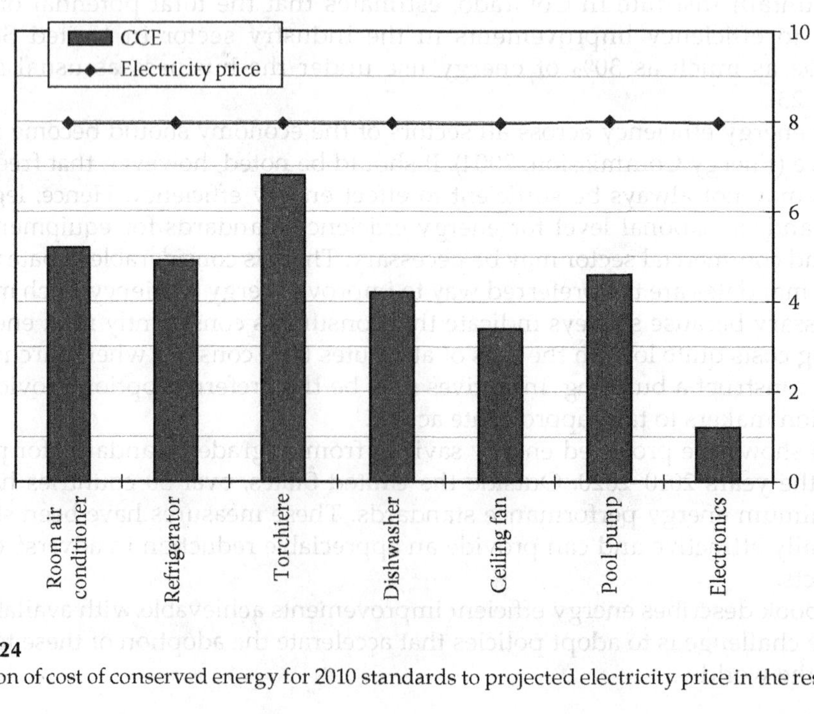

FIGURE 1.24
Comparison of cost of conserved energy for 2010 standards to projected electricity price in the residential sector.

1.8 Forecast of Future Energy Mix

As explained in this chapter, it is clear that oil production will peak in the near future and will start declining thereafter. Since oil comprises the largest share of world energy consumption, a reduction in availability of oil will cause a major disruption unless other resources can fill the gap. Natural gas and coal production may be increased to fill the gap, with the natural gas supply increasing more rapidly than coal. However, that will hasten the time when natural gas production also peaks. Additionally, any increase in coal consumption will worsen the global climate change situation. Although CO_2 sequestration is feasible, it is doubtful that there will be any large-scale application of this technology for existing plants. However, all possible measures should be taken to sequester CO_2 from new coal-fired power plants. Nuclear power does not produce CO_2, however, it is doubtful that nuclear power alone will be able to fill the gap. Forecasts from IAEA show that nuclear power around the world will grow at a rate of 1.2%–2.7% over the next 25 years (IAEA, 2013). This estimate is in the same range as that of IEA.

Based on this information it seems logical that the RE technologies of solar, wind, and biomass will not only be essential but will hopefully be able to fill the gap and provide a clean and sustainable energy future. Although wind and photovoltaic power have grown at rates of over 30%–35%/year over the last few years, this growth rate is based on very small existing capacities for these sources. There are many differing views on the future energy mix. The IEA gives forecasts based on different policy scenarios. Figure 1.25 shows the growth in primary energy demand and the corresponding CO_2 emissions for the

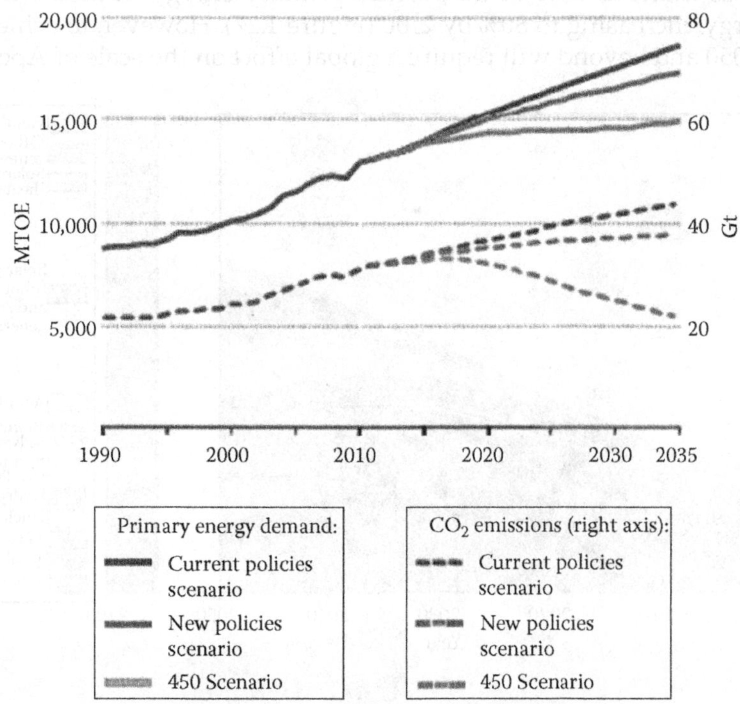

FIGURE 1.25
World primary energy demand by fuel types. (According to IEA, *World Energy Outlook 2013*, International Energy Agency, Paris, France, 2013.)

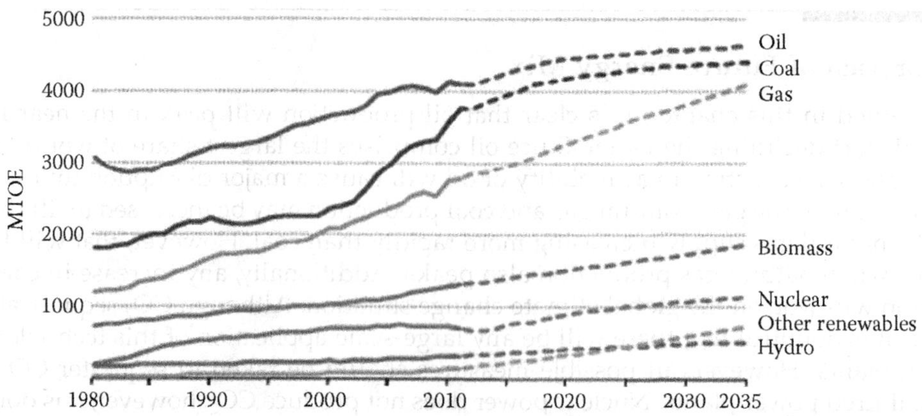

FIGURE 1.26
World primary energy demand by fuel in new policy scenario. (From IEA, *World Energy Outlook 2013*, International Energy Agency, Paris, France, 2013.)

three scenarios. Figure 1.26 shows the demand by fuel type in the "New Policy Scenario," in which renewable energy will provide 18% of the primary energy demand by 2035. However, in the "450 Scenario," renewable energy share goes up to 26% by 2035. This estimate is close to the estimate by the German Advisory Council on Global Change (WBGU), which performed a detailed analysis on combating global climate change with an orderly transition to increased energy efficiencies and increased use of renewable energy. WBGU estimates that as much as 50% of the world's primary energy in 2050 will come from renewable energy, increasing to 80% by 2100 (Figure 1.27). However, to achieve that level of RE use by 2050 and beyond will require a global effort on the scale of Apollo Project.

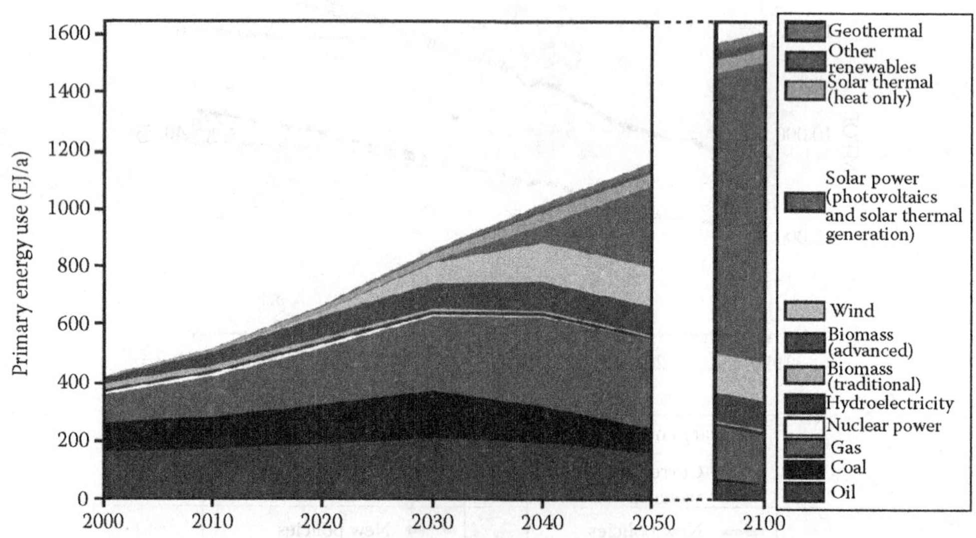

FIGURE 1.27
The global energy mix for year 2050 and 2100. (According to WBGU, World in transition—Towards sustainable energy systems, German Advisory Council on Global Change, Berlin, Germany, 2003, Report available at http://www.wbgu.de.)

References

Bartlett, A.A., 2002. An analysis of U.S. and world oil production patterns using Hubbert-style curves. *Mathematical Geology* 32, 1–17.

BP, 2013. *BP Statistical Review of World Energy 2013.* BP, London, U.K. The report is available on the web at www.bp.com/statisticalreview/.

EIA, 2013. *International Energy Outlook 2013.* Energy Information Agency, US Department of Energy, Washington, DC, http://www.eia.gov/forecasts/ieo/index.cfm.

Enerdata, 2013. *Enerdata Energy Statistical Yearbook 2013.* http://yearbook.enerdata.net.

Energy Commission, December 2004. *Ending the Energy Stalemate: A Bipartisan Strategy to Meet America's Energy Challenges.* The National Commission on Energy Policy, Washington, DC, www.energycommission.org.

EPIA, 2012. European Photovoltaic Industries Association, www.epia.org.

EPRI, 2004. Advanced batteries for electric drive vehicles: A technology and cost-effectiveness assessment for battery electric vehicles, power-assist hybrid electric vehicles, and plug-in hybrid electric vehicles. EPRI Technical Report No. 1009299. EPRI, Palo Alto, CA.

GEF, 2005. Assessment of the World Bank/GEF strategy for the market development of concentrating solar thermal power. Global Environmental Facility (GEF) Report, GEF/c.25/Inf.11, Washington, DC.

Granade, H. et al., April 2010. Unlocking energy efficiency in the U.S. economy. *2010 EIA Energy Conference*, Washington, DC, www.eia.gov/conference/2010/session9/granade.pdf.

Hammererschlag, R. and Mazza, P., 2005. Questioning hydrogen. *Energy Policy* 33, 2039–2043.

Hawkins, D.G., Lashof, D.A., and Williams, R.H., September 2006. What to do about coal—Carbon capture and storage. *Scientific American*, Special Issue, 68–75.

Hubbert, M.K., 1974. U.S. Energy Resources: A review as of 1972. A National Fuels and Energy Policy Study, Serial No. 93-40 (92-75). Part 1. U.S. Government Printing Office, Washington, DC.

Hunt, S. and Forster, E. 2006. *Biofuels for Transportation: Global Potential and Implications for Sustainable Agriculture and Energy in the 21st Century.* Worldwatch Institute, Washington, DC.

IAEA, 2011. *Uranium: Resources, Production and Demand (The Red Book).* IAEA, Vienna, Austria.

IEA, 2004. *World Energy Outlook 2004.* International Energy Agency, Paris, France.

IEA, 2010. *World Energy Outlook 2010.* International Energy Agency, Paris, France.

IEA, 2013. *World Energy Outlook 2013.* International Energy Agency, Paris, France.

Kreith, F., 2006. Plenary Lecture 2006. *ASES Conference*, Denver, CO, available from ASES, Boulder, CO.

Kreith, F. and West, R.E., 2004. Fallacies of a hydrogen economy. *Journal of Energy Resources Technology* 126, 249–257.

Mauthner, F. and Weiss, W., May 2013. Solar heat worldwide—Markets and contribution to energy supply 2011. IEA Solar Heating and Cooling Program, Paris, France.

Rocky Mountain Institute, 2012. U.S. Industry Energy savings potential, 2010–2050. In: *Reinventing Fire: Bold Business Solutions for the New Energy Era.* RMI, Snowmass, CO, http://www.rmi.org/RFGraph-US_industry_energy_saving_potential.

Sadler, H., Riedy, C., and Passey, R., September 2004. Geosequestration. Discussion Paper 72. The Australia Institute, Sydney, New South Wales, Australia. ISSN: 1322-5421.

Sanna, L., Fall 2005. Driving the solution—Plug-in hybrid vehicles. *EPRI Journal.*

Shinnar, R., 2003. The hydrogen economy, fuel cells, and electric cars. *Technology in Society* 25(4), 453–576.

UNDP, 2004. World Energy Assessment: Energy and the Challenge of Sustainability, UNDP, New York.

Veziroglu, T.N. and Barbir, F., 1992. Hydrogen: The wonder fuel. *International Journal of Hydrogen Energy* 17(6), 391–404.

WBGU, 2003. World in transition—Towards sustainable energy systems. German Advisory Council on Global Change, Berlin, Germany. Report available at http://www.wbgu.de.

West, R.E. and Kreith, F., 2006. A vision for a secure transportation system without hydrogen. *ASME Journal of Energy Resources Technologies* (to be published), JERT-06-1009-1. ASME, New York.
Worldwatch, 2006. Biofuels for transportation—Global potential and implications for sustainable and energy in the 21st century. Report prepared for the German Federal Ministry for Food, Agriculture and Consumer Protection, Worldwatch Institute, Washington, DC.
WWEA, 2012. World Wind Energy Association, http://www.wwindea.org/webimages/WorldWindEnergyReport2012_final.pdf.
WWEA, 2013. World Wind Energy Association, http://www.wwindea.org/wwea-publishes-half-year-report-2013-2/.
Zhou, X., September 2012. The present and future development of electric power in China. *Keynote Lecture at the International Conference for Energy for the Next Decades*. The Energy Institute, Hong Kong, China.

2

Sound Finance Policies for Energy Efficiency and Renewable Energy

Michael Curley

CONTENTS

This chapter presents examples that illustrate the economics of energy efficiency and renewable energy systems. Further sections of this chapter delve more deeply into United States and international financial energy policies.

Financial policy for energy efficiency and renewable energy technology is simply based on comparing the cost of the device with the monetary savings resulting from its deployment. In the present setting, the cost of power from the renewable energy system must be less than the cost of power from fossil fuels in order to be economically competitive.

The problem, of course, is that, at this stage in the development of civilization, these events seldom occur naturally. In most cases, the cost of energy efficiency devices exceeds the savings. Likewise with renewable energy: it usually costs more than fossil energy.

If you want people to buy energy efficient "on-demand" (tankless) water heaters, for example, which cost, let's say, $800 more than the conventional system, then the energy savings must be more than $800 over a reasonable period of time.

Same sort of thing is true about renewable energy. There is a gentleman who owns property on Lake Chautauqua in New York. He recently installed a geothermal heating system that cost him $60,000. The man wrote a check for $60,000! The savings on his electricity bill is $1000/month. He is very proud of the fact that—using straight accounting numbers—he will get his money back in 60 months, or 5 years.

Not many of us can write a check for $60,000. So, let us drop our example down an order of magnitude to $6000. Say we are offered a geothermal system for that sum that will save us $100 a month on our energy bill. Would you write a check for $6000? Would you be willing to wait 5 years to get your money back? Most people would not. Furthermore, if you put the $6000 into a CD at 3% for 5 years you would have $6955.64. So it's not really just $6000 that you're out; it's $6955.64. It will take you more than 5 years to break even. Closer to six.

Instead, you could get a $6000 second mortgage for 5 years at 7%? If so, you would be paying the bank $118.81 a month while you are saving only $100. In other words you'd be losing $18.81 a month for 5 years. So at the end of the 5 years, you'd still be out $1216.* You'd still have to wait another 13 months to break even.

The point here is that very, very few people will wait 60 or 73 months to get their renewable energy investment back. Likewise, people who buy energy efficiency devices don't want to wait several years to get their money back. So, if people want to get serious about energy efficiency and renewable energy they must—if a government or financial institution—create finance programs that deliver the goods at prices below the expected savings. If they are businesses or NGOs, they must seek out these programs and lobby their state, federal, and local legislators to create such programs.

When the wealthy gentleman in Chautauqua signed his $60,000 check and started saving his $1,000 a month, he was probably unaware that in several states other than New York there were government programs that would have financed his geothermal project for about $375 a month with $0 down. This means that in month one he would have spent $375, but saved $1,000.

In our lesser example, the county would have financed our $6,000 project and charged us $37.50 a month, meanwhile we would be saving $100 a month. Don't you think that many, many more people are likely to take on renewable energy projects when the savings kick-in on day one? So, now we just have to figure out how to design finance programs that do just that.

2.1 Some Sound Finance Principles for Creating Effective Finance Programs

Here are seven sound principles of finance for energy efficiency and renewable energy. First, we will list them and then discuss them individually.

1. *Leverage*: Leverage money. The most important word in improving environmental quality is "leverage."
2. *Guaranties*: Never make loans; guaranty them.
3. *Term*: Finance assets over the full term of their service lives.
4. *Subsidies*: Stop general subsidies, which waste billions of dollars. Target subsidies only to those who need them.
5. *Grants*: Never give grants unless absolutely necessary.

* Discounting at 3%.

6. *Cost/benefit analyses*: Make financial decisions based on strict cost/benefit analyses. Get politics and chance out of the decision making matrix.
7. *Full-cost pricing*: Insist on full-cost pricing of environmental services. Full cost pricing will drive technological innovation. In turn, new technologies will drive down costs.

2.1.1 Leverage

Many people think that leverage is some black art practiced by the charlatans on Wall Street. It's not. It's something we all use every day. We just don't call it that. For example, if you buy a $250,000 home with a $50,000 down payment and a $200,000 mortgage, you will have achieved a 5:1 leverage on your home purchase. If you can get away with putting only $25,000 down and get a $225,000 mortgage, then you'll have achieved a 10:1 leverage ratio. Sounds sinister to you? Of course not.

Same is true if you buy a $20,000 car with a $2,000 down payment and finance the balance. You will have achieved a 10:1 leverage again on this transaction. When car dealers get desperate, they often offer cars with no money down. In that case you can achieve infinite leverage. Black magic? No. Not even life in the fast lane. As a matter of fact, leverage is pretty common.

In these examples, the institution that provided the leverage was a bank. Think instead of the U.S. bond market, which was estimated to be $35.2 trillion in the second quarter of 2011 by the Securities Industry and Financial Markets Association, or the international bond market, which is close to $90 trillion. As far as energy efficiency and renewable energy projects are concerned, these international capital markets can provide almost a limitless source of funds.

Now, what's the impact on the projects—or in our examples, the homebuyer or the car buyer? Ask yourself this: if two car dealers offered you the same vehicle with financing for the same term at the same rate of interest but one wanted $4000, or 20%, down and the other wanted $0 down, which would you choose? Exactly! The one with the greatest leverage. And that is precisely why it is so important to design renewable energy and energy efficiency programs with the greatest possible leverage.

2.1.2 Guaranties

The best way to achieve truly extraordinary leverage is to use financial guaranties instead of making direct loans. Let us illustrate this by comparing four most common types of government finance programs: grants, subsidized loans, market-rate loans, and guaranties. To do this, we will set up a little game.

2.1.2.1 Rules of the Game

1. Government contributes $100,000,000 to the New Energy Fund (NEF)
2. Project size: $5,000,000
3. Term: 5 years
4. Payment terms: Level Principal Method
5. Interest rate on subsidized loans = 0%

6. Interest rate on market-rate loans = 10%
7. Interest rate on guaranty fund = 5%
8. Leverage ratio* paradigm†:

Number of Loans	Coverage (%)
0–20	100
20–30	90
30–40	80
40–50	70
50–60	60
60–70	50
70–80	40
80–90	30
90–100	20
100+	10

In our game, the legislature gives your finance agency $100 million and tells you to use it to finance as many energy efficiency and renewable energy projects as possible. So, you look at a grant program, a subsidized loan program, a market-rate loan program and a loan guaranty program.

The following table compares the effectiveness of these four financing techniques.

Year	0	1	2	3	4	5	6	7	8	9	10
Comparison of NEF financing schemes											
Total grants made	20	0	0	0	0	0	0	0	0	0	0
Total subsidized loans made	20	24	28	34	41	49	55	61	68	74	81
Total market rate loans made	20	26	33	42	53	66	79	93	109	126	145
Total guaranties issued	20	27	34	45	60	80	94	119	180	435	543

Let's turn these project numbers into project costs by multiplying the number of projects by $5 million:

Grants	20	$100,000,000
Subsidized loans	81	$405,000,000
Market-rate loans	145	$725,000,000
Loan guaranties	543	$2,715,000,000

Just look at that! $2,715,000,000 worth of energy projects! 543 projects financed! All for the same $100,000,000 that bought you 20 grant projects!

* In this example, the leverage is the ratio of guaranteed loans outstanding to money in the bank pledged to cover them.

† Estimating coverage between the minimum (>20) and the maximum (<100) loans is a totally inexact science; but what is presented certainly suffices for our illustrative purposes.

Loan guaranty programs are so efficient because they incorporate leverage. Leverage, as you already know, is the ability to increase the effect of the use of money. The same sum of money ($100,000,000) was used in each financing scheme. Under the subsidized loan program, there was modest leverage because the funds were paid and could be re-loaned for another project. Thus, the same amount of money could be used twice. This is leverage.

In market rate loan program, there was additional leverage because not only was the $100,000,000 of principal repaid, but 10% interest was also paid each year. So, this increased the leverage.

With loan guaranties, even greater leverage occurs because of the principle of insurance. A guaranty is the same as an insurance policy.

A loan guaranty program insures "prompt and full payment" of debt. It works just like an insurance policy. With $100,000,000, a fund such as the NEF should easily be able to guaranty over $1,000,000,000 of projects at any one time, because of the extreme unlikelihood that more than 10% of its projects would ever go into default at any one time. Thus, if the NEF were to guaranty commercial bank loans for $1,000,000,000 of projects, and 10% or $100,000,000 of them were to default, the NEF could still make good on its guaranties by paying the banks holding the defaulted $100,000,000 held in their reserve account.

Thus, from the point of view of a government wishing to finance energy efficiency or renewable energy projects, the creation of a loan guaranty program would be—by far—the most effective use of its funds.

2.1.3 Term

When most people talk about "low-cost" government finance programs they are almost always referring to a low interest rate. The world's most successful environmental finance program, the Clean Water State Revolving Fund (SRF) at EPA, routinely offers low-cost financing, which—to them and their borrowers—means loans with low interest rates. In SRF's case they offer loans at about 50% of market rates. So, if the market rate on a 20-year, AAA/Aaa tax-exempt municipal bond is 4%, an SRF will offer loans at 2%. If the market rate is 6%, SRFs will finance at 3%, and so on.

Now, to be able to offer sub-market rate loans requires the SRFs to subsidize such loans. This is an egregious waste of money. Here's why.

The reason for the low rates/subsidies is so that sewer-user fees can be minimized. Right? Now watch this.

> The annual payment on a *20-year* loan of $1 million at 2% is $61,157.
>
> The annual payment on a *30-year* loan of $1 million at *4%* is $57,830.

Even if the 30-year market rate is a bit higher, say 4.5%, the annual payment is $61,392, which is only $235 higher. That's $235 a year—not a month—on a $1,000,000 loan. Peanuts!

The point here is that you can have a much more effective program if you lengthen the term rather than subsidizing an interest rate. Furthermore, you save all of the money wasted on the needless subsidies.

Now we will compare annual payments on identical loans with different terms. We will do so with both types of loans: level principal payment loans and level payment loans.

The following table sets forth the first years' annual payments on a $100 loan at several different interest rates over several different terms. The first table deals with level principal payment loans. The second deals with level payment loans.

Interest Rates	0%	5%	10%	15%	20%
Term					
1 year	$100	$105	$110	$115	$120
2 years	$50	$55	$60	$65	$70
3 years	$33	$38	$43	$48	$53
4 years	$25	$30	$35	$40	$45
5 years	$20	$25	$30	$35	$40
10 years	$10	$15	$20	$25	$30
20 years	$5	$10	$15	$20	$25
30 years	$3	$8	$13	$18	$23
40 years	$2.50	$7.50	$12.50	$17.50	$22.50

Please note the hugely significant impact that term has on the annual debt service payments for these loans.

And, now we turn to the level payment method.

The impact of term on annual debt service payments is as significant when the level payment method is used as when the level principal payment method is used.

Interest Rates	0%	5%	10%	15%	20%
Term					
1 year	$100	$105	$110	$115	$120
2 years	$50	$54	$58	$62	$65
3 years	$33	$37	$40	$44	$47
4 years	$25	$28	$32	$35	$39
5 years	$20	$23	$26	$30	$33
10 years	$10	$13	$16	$20	$24
20 years	$5	$8	$12	$16	$21
30 years	$3	$7	$11	$15	$20
40 years	$2.50	$6	$10	$15	$20

A final word about lengthening term. How long is too long? The answer—which is an axiom of public finance—is that assets should be financed over terms commensurate with their service lives. Home insulation, for example, lasts as long as the home it insulates: so 30 years. I understand solar panels last 20 years. So, finance them over 20 years. Geothermal projects last forever, almost. So, again, 30 years should be no problem.*

2.1.4 Subsidies

When a normal person thinks of subsidies, one usually thinks of helping the poor pay for things they can't afford. Right? Well, in the energy infrastructure game, being poor has

* Although many assets have service lives longer than 30 years, the municipal bond market thins out considerably beyond that period. So, better to sell into a 30-year market with lots of buyers than a 30-plus-year market with relatively few.

nothing to do with subsidies…most of the time. Take the case of Loudoun County, Virginia, which has the highest Median Household Income (MHI) in the country. Their MHI was about $119,000 in 2011. That was the *median.* That means that 50% of the households earned *more than $119,000.* Let us now imagine the household with the absolute highest income in Loudoun County. Would it surprise you to learn that the families receive subsidies from the Commonwealth of Virginia? Well, they do.

A very wise mentor of mine breaks down subsidies into two main categories. The first is "general, supply-based subsidies." That's what you see in Loudoun County. The second category is "targeted, demand-based subsidies."

Think of a young family out west who lives in a double-wide, have household income of about $22,000 a year, has two young children, one with asthma, and lives in a valley that in terms of air quality is classified as a nonattainment area because of the wood smoke that hangs constantly over their valley like the sword of Damocles. The only heat the young couple has in their home is an old wood stove. When it gets cold, they have to light it. But when they light it, they are slowly killing their little girl with asthma. A few years ago, there was a $1500 tax credit program for—among other things—replacing old wood stoves. Now, our young couple can't use a tax credit. They don't pay taxes. Their income is too low. The tax credit is useless to them. On the other hand, their wealthy neighbors with weekend cabins on the tops of the mountains can use it. But not people, like our young couple, who really need help. So, instead of this wasteful general subsidy, how about a 100% cash rebate (subsidy) for people: (1) living in nonattainment areas, (2) below the poverty line, (3) with no other source of home heat, and (4) with a resident with a pulmonary ailment. Now, these exceedingly narrow criteria may be a bit over the top; but you see what we mean by a "targeted, demand-based" subsidy.

Now let us show you how truly wasteful general, supply-based subsidies are, by illustrating the point with the most infamous example of all: the tax-exempt municipal bond.

In 1895, the U.S. Supreme Court ruled that the Congress had no power under the Constitution to tax the interest on state or local debt.* In 1913, this position became statutory.

In the municipal bond market, the rule of thumb is that the best (AAA/Aaa) municipal bonds should have interest rates of about 75% of United States.† Treasury bonds of the same term. So let us envision an interest rate environment where, say, 20-year treasuries carried an 8% interest rate. Under these circumstances, you would expect AAA/Aaa tax-exempt municipal bonds to carry a rate of about 6%. Now, let us say that the 8% Treasury was bought by the richest householder in Loudoun County. We'll call her Ms Loudoun. If she bought $10,000 of these bonds, she would receive $800 a year in interest. Being in the 35% tax bracket, Ms Loudon would pay $280 in federal income taxes, and so have net earnings of $520.

Now, let us say, instead, that Ms Loudoun bought $10,000 of tax-exempt bonds at a 6% interest rate issued by the Commonwealth of Virginia. In this latter case, Ms Loudoun would receive $600/year in interest. Because this interest is exempt from federal income taxation, Ms Loudon gets to keep the whole $600. She doesn't pay a penny of that $600 in taxes.

So, we now know that if Ms Loudon bought the 6% tax-exempt bond, she would be $80 better off than if she had bought the 8% Treasury and paid taxes on it.‡

---

* This case, *Pollock vs. Farmers' Loan & Trust Co.*, was effectively overruled by the Supreme Court in 1988 in *South Carolina vs. Baker.* But the statutory prohibition remains intact.

† As of this writing, as the damage from the sub-prime mortgage crisis is still being felt in the financial markets, this 75% ratio isn't true. Today it's about 120%. But the 75% number is a decent historical ratio.

‡ Some Ms Loudouns might choose to forego the extra $80 because treasuries are a safer investment than munis. Or at least, that's how the story goes on Wall Street.

But what about the $280 that the U.S. Treasury *didn't get, that they missed out on*. Let's look at it this way. There are 535 members of Congress. Each one receives a salary of $174,000/year. That means there should be $93,090,000 in the Congressional payroll account at the beginning of the fiscal year. Let us say that this account is $280 short because Ms Loudon opted not to buy the treasury bond. If this were really the case, panic would sweep Capitol Hill! Congress would undoubtedly immediately pass a bill to raise some tax by $280 to make up the shortfall. Who would pay that tax? You and I would—as would the poorest tax-paying family in Buffalo County, South Dakota, the poorest county (by MHI)* in the country.

Targeted subsidies for our poor young couple with the sick little girl? Yes.

General subsidies for Loudon County and our wealthy cabin owners? No.†

2.1.5 Grants

Grants are subsidies. So they should only be used in relatively rare circumstances. And, as we learned earlier, they should be tightly targeted.

The four legitimate uses of grants are as follows:

1. *Paying for environmental services that are not affordable, either to individuals or communities*: Remember our young couple with the sick little girl. They desperately needed to buy a new wood stove to keep their home warm and their baby daughter healthy. What form of government assistance was available to them? A tax credit, which they were too poor to use. What did they need? A grant. Either a 100% grant, or something very close to 100%. That is an example of a good use of grants for individuals.

   A good example of a sound community grant is in the case of the USDA water and wastewater program. Here, when the USDA sees that a project costs a significant share of MHI and will make rates higher than in surrounding districts, then the USDA uses grant money to buy down the cost of the project to levels where it will be affordable to the average ratepayer. Targeted grants for communities. Good finance policy.
2. *Inducing people or businesses to make environmental improvements that they are not legally required to do*: When the American Reinvestment and Recovery Act (ARRA) was passed in 2009 to help us get out of the sub-prime mortgage disaster, renewables were just beginning to be a major buzzword. ARRA included substantial grant funds for those who would reduce their carbon footprint by installing solar panels. No one was required to install any kind of renewable or energy saving devices. So, the government paid them to do so. Good idea.

   Cover crops are another good example. After crops are harvested, a large amount of nitrogen remains in the soil. Over the Winter, with snows and rain on the now-bare soil, much of this nitrogen migrates to the nearest water body where it pollutes it. Cover crops are planted after the main crop is harvested. Their sole purpose is to sop up nitrogen left over in the soil to prevent it from polluting any streams or ponds. In the United States, we pay (i.e., give grants to) farmers who will plant cover crops. These grants come out of the "farm bills" that the Congress passes to maintain its elaborate scheme of subsidies for agriculture. In Germany,

* The MHI in Buffalo County, South Dakota is $12,692.

† Although the probability of ending tax-exempt bonds is close to zero.

they have a cover crop program too. It is also a grant program, but in Germany the grants come from a special tax that is collected each year. The point here is that planting cover crops is a good thing. But doing so is entirely voluntary. There is no law—anywhere—requiring farmers to plant cover crops. So, we should use grants to persuade them to do so.

3. *Creating or commercializing new environmental technologies.* Until most recently, it was not cost-effective to install solar panels. Even with generous government financing programs, the monthly cost of the panels exceeded the savings in electricity from the panels. In order to jump-start the solar industry, the U.S. Department of Energy started a grant (i.e., a subsidy program) for solar panel manufacturers. China did the same. However, the Chinese subsidies were so large that the Chinese manufacturers were able to sell their solar panels below the cost of production. The Chinese government created subsidies so that their manufacturers could export all over the world, creating thousands of good manufacturing jobs in China in the process. The United States, European Union, and China are now embroiled in a trade fight over this issue. Regardless of the trade issue, using grants to jump-start new, needed, environmental technologies is a very good idea.
4. *Environmental/Energy Education*: Giving out smallish grants is a good way to get community groups to take an avid interest in energy efficiency and the environmental issues such as climate change. This is especially so for poorer groups that can't readily raise funds for projects themselves. Maryland has a grant fund called the Chesapeake and Atlantic Coastal Bays Trust Fund, which most natives call the "green fund." This fund is relatively small but it gives out grants to community groups for such activities as streambed restoration, tree plantings and the like. Educating people about the value of the environment and the steps necessary to protect it is certainly a valid expenditure of public funds. And, in this case, it really needs to be grant funds.

2.1.6 Cost/Benefit Analyses

Cost/benefit analyses are extremely important for two allied reasons. First, to make sure money for projects or energy efficiency devices is wisely spent, and second, to convince the public of the same, so that they will support these types of programs.

In the field of financing water and wastewater facilities, these analyses can get complicated. They involve the collection of empirical data from the general public, which, in some less-developed countries, is not easy to come by. But for energy efficiency devices and renewable energy, they are quite simple—at least in theory.

In the case of energy efficiency devices, the benefit is the number of kilowatt hours saved. In the case of renewables, it is the number of kilowatt hours generated.

The cost, for energy efficiency devices, is the price of the device paid over the service life of the device at the lowest possible rate of interest.

This is easy for such devices as light bulbs. You know the four critical pieces of information to complete the cost/benefit analysis. You know the lumens the bulb will put out. You know the watts the bulb uses per hour. You know the service life of the bulb in hours (at least as estimated by the manufacturer). And you know the cost of the bulb. With these four pieces of information you can conduct a cost/benefit analysis on these bulbs and thus compare them to get the best value.

The cost for renewables is the installed cost of generating a kilowatt *hour* of electricity. This is a little different. Most people rate installed power by the kilowatt, not the kilowatt hour. But many renewables, such as wind or solar, don't generate power on a constant basis. They generate intermittently. Here is a very simplified example, omitting, among other things, the cost of maintenance.

Let us say two homeowners, one in Buffalo, New York, the other in Las Vegas, Nevada, each purchase a 4 kW solar array at an installed cost of $5/W, or $20,000/W. Let us say they each finance them for a 20-year term at an interest rate of 5% for an annual payment of $1605. Las Vegas has an average of 3825 h/year of sunlight. So, our homeowner in Las Vegas is paying $0.42/kW h.

On the other hand, Buffalo gets 2207 h of sunlight a year. So, our Buffalo homeowner pays $0.73/kW h.

So, you see why we need to use kilowatt HOURS, not just kilowatts to rate renewable energy projects.

2.1.7 Full Cost Pricing

As you already know, environmental utility costs are heavily subsidized. This does no one any good. If there are poor people who cannot afford their water or sewer bill, there are several strategies to effectively and compassionately deal with that.

Now the rest of the ratepayers who *can* afford the full cost of their service should definitely pay for it. No one likes to pay more. No board member or politician likes to raise rates. But rates can be raised gradually. (And when raising rates over, say, 5 years, the authorizing resolution should be passed today for all forthcoming increases. This will save the board and/or politicians the anguish of having to go back to the people every year for more money.)

Raising rates, whether to full-cost pricing levels, or not, will promote conservation. The higher the rate increase the more people will conserve. Think of raising our gasoline rates to the $7+ a gallon rates they charge in Europe. People would definitely find ways to drive less. Car manufacturers would smarten up too. Ditto when you raise power rates. People find ways to use less.

Finally, higher rates will also drive innovation, which will have the eventual effect of lowering costs. Take the example of installing a technology that costs $10 million at a system with full-cost pricing, that is, no subsidies. Engineers and scientists will know that if they can create a technology that does the same or better job at the same or lower cost, they can get into the game without being trumped by some hidden subsidy.

So, these seven principles should be used to guide the creation of any finance programs for renewable energy or energy efficiency.

Now that you know how to spend money most efficiently, please consider a few more principles—these to raise the money that you will spend most efficiently.

1. Raise money from many small charges, fees or taxes—not one big one. Many small sources of money are more stable than one large one. A small tax or surcharge on vehicle registration based on its fuel consumption. A carbon tax loaded into electricity bills.
2. Once collected, put all the environmental money in one basket. Do not fragment or piddle it away.

3. Change behavior while raising money. Do not tax all equally; tax the polluters more while rewarding energy efficiency and green practices. The vehicle/fuel consumption surcharge discussed earlier and the carbon tax will send effective messages to consumers, car manufacturers, and power companies.
4. Use "dedicated revenue streams" (such as annual taxes or fees) to finance capital, not operational, expenses. Such revenue streams that derive from charges like the vehicle/fuel consumption and carbon tax are highly regular and predictable. As such, they constitute an excellent, high quality source of repayment for bonds issued to finance energy efficiency and renewable energy projects. With proper structuring, they can be used to achieve the highest, AAA, ratings on bonds, which result in the lowest possible interest rates. Furthermore, these solid revenue streams don't actually even need to be used to repay bonds to achieve AAA ratings! They can just be pledged to repayment! In other words, let us take the example of a small business energy efficiency/renewable energy program. Small businesses are not known for stellar credit ratings. Moreover, since many small businesses rent their premises, there is no real property that can be used as collateral. In this case, a bond containing a portfolio of uncollateralized loans to small businesses would get either no rating or a triple zilch rating. In this case, the bonds could be structured with the loan receivables as the primary source of repayment and the revenue stream as a secondary source of repayment, that is, the revenue stream would be called upon if, and only if, a small business defaulted on its loan.
5. Make it as painless as possible. At the federal level, raising the rates on a general tax—like the income tax—will set up howls of protest across the country. It will mobilize armies of lobbyists in Washington who will roam the halls of Congress wheedling and bullying the members to oppose it. At the state level, the same phenomenon will occur if there is an income tax increase proposal or a real property tax increase proposal. Virtually, every newspaper in the state will editorialize against it. And the lobbyists will mob the state legislature. The more opposition, the less likely that an effective environmental finance program will be adopted.

The title of this chapter is "Sound Finance Policies for Energy Efficiency and Renewable Energy." These seven principles should be used not only to design programs that spend money on energy efficiency and renewable energy projects in a highly efficient manner, but also on how to raise money for such programs in a highly efficient manner as well.

3

State and Federal Policies for Renewable Energy

Christopher Namovicz

CONTENTS

3.1 Tax Incentives

Tax incentives have provided a key form of direct subsidy to renewable energy and energy efficiency in the United States at both the state and federal levels. These incentives can take several forms, including deductions from taxable income or a credit against tax liability. In addition, tax credits can be applied to the initial purchase or investment in a particular technology, or to the ongoing utilization of a technology or production of a covered commodity. In some cases, lower efficiency investments can be subject to additional taxes.

3.1.1 Investment

Renewable energy and energy efficient technologies are typically characterized by higher upfront costs resulting in significantly reduced fuel and/or operating costs (although not all technologies fit this characterization, e.g., biomass energy can involve substantial ongoing costs for fuel and operations). Many early policy incentives at both the Federal and state levels were intended to reduce the acquisition cost of these technologies, frequently through the use of tax credits proportional to capital investment costs. In some cases, such as some of the early deployment of wind generating technology in California during 1980s, it was believed that investment incentives provided insufficient incentive for high-quality technology or projects that would continue to operate once the initial incentive had been fully realized by the project owner. Such failures, however, may also be attributed to insufficient technology qualification measures, such as technology criteria or screening.[1]

Despite the apparent shortcomings of investment incentives in the early U.S. wind industry, these continue to see widespread use in both federal and state policies for other renewable energy and energy efficiency technologies.

Federal tax incentives proportional to the investment in renewable energy technologies played a significant role in the early adoption of these technologies during the 1980s. Originally adopted as part of the Energy Tax Act of 1978, permanently set at 10% for solar and geothermal facilities by the Energy Policy Act of 1992 (EPACT 92), and currently 30% for solar facilities through 2016, the investment tax credit directly offsets federal corporate income tax liability in proportion to the initial investment cost of the covered technology.[2] Current federal law also allows technologies eligible to receive the production tax credit (PTC) (see the following section) to instead receive a 30% investment tax credit. In addition, most renewable electricity generating technologies are also able to benefit from preferential federal tax depreciation allowance schedules. The Modified Accelerated Cost Recovery Schedule allows much faster depreciation of renewable generation investment costs than is allowed for other generation technologies, using a 5-year schedule rather than a 15- or 20-year schedule for combustion turbines or other thermal plants.[3]

Some states also have or have had tax incentives on the investment in renewable energy or energy efficiency. Additional investment tax credits in California during the 1980s, along with other policies such as the Public Utilities Regulatory Policy Act discussed in Section 3.2.2, helped spur the early adoption of wind and solar thermal generating capacity in that state.[4] Several states currently offer substantial investment tax credits to preferred renewable energy technologies, such as photovoltaic (PV) systems.[5] These credits, however, are not uniformly offered, vary significantly among states that do offer them, and may apply to electric generating technologies or to facilities that produce renewable fuels, such as ethanol. Rebates or exemptions from state-imposed sales taxes on both renewable technologies and energy-efficient appliances and equipment also offer a mechanism to reduce the first-cost of adopting these technologies by the end-user. Availability of such programs varies significantly among states, as do the sales-tax rates and the value and timing of a rebate or exemption where offered.[5] Sales-tax rebates may also, or instead, apply to a renewable fuel, such as biofuel. In this context, such a program may have an effect closer to that of a production incentive rather than an investment incentive. Certain vehicles with low gas mileage will incur a "gas guzzler" tax, which acts as a disincentive for low-efficiency technology investment.

3.1.2 Production/Utilization

Production-based tax incentives provide a tax credit proportional to the quantity of commodity, such as electric generation, produced or sold in a given year. Since production-based incentives reward project performance, they should tend to transfer project performance risk to the project owner, rather than the taxing authority, and without the need for extensive qualification criteria or screening of each project or technology. However, technologies that do not produce easily marketable (and hence taxable) output, such as most energy efficiency technologies, or where the output is generally consumed on-site (without a third-party transaction), such as on-site PV, may be not be amenable to a production-based incentive. In these cases, there may not be a sufficiently auditable record of production or the establishment of such an auditable record (such as internal metering of PV output) may add unwanted cost to a project.

The PTC for renewable electricity, Section 45 of the U.S. Internal Revenue Code, established by EPACT 92 and subsequently modified, provides an inflation-adjusted payment, 2.3¢/kW h in 2013, for the third-party electricity sales from the plant during the first 10-years of operation. The range of technologies eligible for the tax credit has been expanded since its inception, and now includes wind, several types of biomass resources, geothermal, and landfill-gas. However, some technologies do not receive the full credit amount or the same 10-year claim period.[6] The PTC has generally been credited with contributing to the significant growth in U.S. wind power since 1998. Having been allowed to expire and subsequently extended several times, the credit expired for projects starting development after December 31, 2013.[7] A number of states also offer tax credits on the production of preferred renewable energy sources.[5]

A tax credit of $1.00/gal for biodiesel and $0.50/gal for other qualified alternative fuels, including certain biomass-derived fuels, expired at the end of 2013.[8] A number of states also have tax credits for the production of ethanol or other renewable fuels. These credits may reduce income tax liability or, like the federal credit, be applied to a motor fuels tax (in effect, a sales-tax rebate). State programs vary by credit amount as well as by restrictions on local origin of the fuel.[5]

3.2 Regulatory

Regulatory mechanisms generally establish restrictions on market activity that are intended to result in increased adoption of policy-preferred technologies or limitation on policy-undesired technologies. Costs are typically borne directly by market participants, or by either energy producers, consumers, or both. Although regulatory policy may affect markets in many ways, this section will examine three major types of regulatory intervention: target-based standards, market facilitation or limitation policies, and technology specification standards.

3.2.1 Target-Based Standards

Target-based standards establish a target metric of renewable energy or energy efficiency achievement and require regulated industry to achieve the goal. The most important types of goal-based standards in U.S. energy policy are renewable electricity targets established by the various states, the Renewable Fuels Standard (RFS) for transportation fuels, and automotive fuel efficiency standards established by the federal government.

Renewable electricity targets can take the form of absolute levels of capacity (or generation) or of a specified fraction of some future level of total generation (or capacity). Generally called renewable portfolio standards (RPS), these targets can be targeted for a single future year or can be based on a gradually increasing compliance schedule. Renewable energy goals—found in a few states—can mimic RPS programs, but generally lack enforceability provisions, and thus cannot be considered as regulatory policy.[9] RPS policies can require absolute compliance by affected utilities, or, as frequently occurs, can allow the accumulation of "renewable energy credits" (RECs) that can facilitate either inter-temporal compliance "banking" (i.e., using RECs earned in 1 year to meet compliance targets in another year) and/or inter-utility or inter-state credit trading (whereby a utility that over-complies may sell RECs to a utility that cannot meet targets with native resources). Most states with RPS policies limit the geographic source of compliance to in-state resources, resources within the electric power pool(s) that service the state, or resources that can be "delivered" to the state or state power pool. The prevailing selling price of RECs may also be used to calibrate a penalty or alternative compliance payment, typically in the form of a price ceiling at which the state will provide RECs (without actual renewable capacity or generation) or otherwise waive actual compliance. Such "safety-valve" prices are generally intended to provide a clear maximum impact on general electricity prices. Other states may have a "safety valve" that explicitly limits compliance based on realized electricity rate impacts, and in some states compliance may also be waived or delayed for other, statutorily sanctioned reasons, such as protecting the financial solvency of affected utilities. Policies among states also show significant variation in resource eligibility, "grandfathering" of existing capacity, and mechanisms to show preferences among eligible technologies, such as awarding "bonus" credits or having differentiated targets for preferred technologies.[5,12]

The federal RFS was established by the Energy Policy Act of 2005 and the Energy Independence and Security Act of 2007. It establishes volume-based targets for ethanol and advanced biofuels, increasing each year through 2022. By this year, the RFS will require the use of 36 billion gal of renewable fuel. The law ensures the use of a variety of fuel types by limiting the amount of conventional ethanol (ethanol derived from corn) to be used for compliance and setting volumetric targets for various advanced biofuels. Advanced biofuels include fuels derived from "cellulosic" feedstocks and can include ethanol, biodiesel, "drop-in" fuels, and other qualifying formulations. Compliance is tracked through the use of Renewable Identification Numbers (RINs) assigned to each batch of qualifying fuel entering the market. RINs can be banked or traded to facilitate compliance.[10]

In 1975, the federal government established a target of doubling the fuel efficiency of the automobile fleet within 10 years. To implement this target, the aggregate sales of each manufacturer selling cars in the U.S. market had to achieve a set schedule for Corporate Average Fuel Economy (CAFE). In 2007, the law was updated, and, in 2012, regulations were issued to establish a target of over 40 miles/gal for passenger cars and light duty trucks by 2021 on a gasoline energy equivalent basis, potentially increasing to over 49 miles/gal by 2025.[11] With the current regulation, compliance for any given manufacturer is facilitated through credit banking and trading provisions. That is, excess credits earned in 1 year may be used to cover a shortfall in another year, or may be traded to another manufacturer to help cover their shortfall. Provisions to support the adoption of electric drive train vehicles may be adopted outside of the construct of the CAFÉ program.

3.2.2 Market Facilitation or Restriction

Regulatory policy can also be used to facilitate or hinder a preferred or undesirable renewable energy or energy efficiency technology from participating in the market. Facilitation can take many forms, including the target-based and technology-specification approaches discussed in Sections 3.2.1 and 3.2.3. Other types of market facilitation can require nondiscriminatory or even preferential market treatment of preferred technologies. Such policies operating at the federal or state level can include "feed-in tariff" (FIT) laws, net metering requirements, and interconnection standards.

In 1978, the Congress passed the Public Utilities Regulatory Policy Act (PURPA), which established the requirement that electric utilities must interconnect (i.e., accept generation feed from) small qualifying facilities that either co-generate process heat and electricity (combined heat and power or CHP) or utilize certain renewable resources.[12]

Furthermore, PURPA established a price floor for the power, known as "avoided cost," subsequently defined to mean the cost of electricity that the utility otherwise would have purchased. PURPA, in theory, established a non-discriminatory framework for adoption of efficient industrial CHP and renewable electricity, established by the federal government, but largely implemented by state regulatory authorities. Some of the non-discriminatory market features that PURPA specifically applied to renewable and CHP facilities were subsequently applied to the broad class of all power generation technologies as federal electricity policy moved toward deregulation of the wholesale power market.[13]

Many states have adopted regulations at the retail/distribution level to require the acceptance of some renewable electricity feeds at an established price floor.[5] Such net-metering laws typically require load serving utilities to facilitate end-user connection of renewable distributed generation technologies (especially solar, but sometimes wind or other renewable or non-renewable technologies) on the customer side of the meter. When instantaneous generation from the local resource exceeds instantaneous customer demand, the meter is allowed to "run backward," effectively causing the utility to purchase the excess generation at the prevailing retail rate. Most states limit the size of the distributed resource, sometimes by customer class, and may also provide limits on the total generation off-set allowed (e.g., the monthly or net annual bill may not be less than zero). Some states have also established limits on the number of customers or level of installed distributed capacity that may participate in net-metering.

More recently, a number of states, localities, and utilities have adopted FITs more similar to those found in Europe.[5] In the FIT model, the utility accepts the renewable feed, as with net metering, but also offers a premium payment over the consumer's retail value of the generation. In some cases, these FIT programs are established by a state or local government, but in other cases, the programs are voluntarily established by the utility itself, and thus may not be, strictly speaking, regulatory policy.

3.2.3 Technology Specification Standards

Another common form of regulatory intervention for renewable and energy efficient technologies is the establishment of minimum product specifications, either as voluntary targets or mandatory limits on product performance. Such standards are seen as an effective approach to improving energy efficiency among individual consumers. Commercial and industrial consumers presumably have significant incentive to optimize energy efficiency for their operations to maintain or improve profitability. However, individuals, while still sensitive to energy prices, may have less motivation to seek out products with higher upfront costs to achieve lower ongoing energy costs. In some cases, market structures may affect consumer decision-making with respect to energy efficiency.

The federal Energy Star program allows qualifying products—ranging from computer equipment to household appliances, to commercial building equipment—to display the "Energy Star" logo on product advertising and packaging.[14] This serves as a proxy for disclosure, in that the consumer is thus aware that the product is "best-in-class" for energy efficiency (although for products not displaying the logo, the consumer cannot tell if this is because the product did not meet the specification or because the manufacturer did not participate in the program). Through the Energy Policy and Conservation Act and its various amendments, the federal government also establishes mandatory energy efficiency specifications, such as minimum levels of energy efficiency, for a wide array of consumer appliances, such as furnaces, air conditioners, light bulbs and fixtures, and kitchen appliances.[15] At the state and local levels, energy efficiency standards may also be incorporated into building codes.

There are both federal and state regulations regarding transportation fuel composition that either directly or indirectly provide incentive for renewable fuels. In addition, the Clean Air Act Amendments of 1990 established a number of fuel specifications, including oxygenation, that vary by region and/or season.[16] Ethanol has emerged as a preferred oxygenate, especially in states with additional ethanol incentives or that have restricted the used of alternatives such as MTBE, but is also incentivized by the RFS. Restrictions on the sulfur content in diesel fuels may also encourage the use of "biodiesel" fuels derived from plant oils, if such fuels can be economically produced.

3.3 Research and Development

Government research and development (R&D) funding for renewable and energy efficiency technologies can support the adoption of these technologies by facilitating cost reductions, higher efficiency, and improved utilization. R&D funding may occur at all stages of the technology development cycle, including basic science, bench-scale technology development, proof-of-concept demonstration, and pilot applications.[17] Government funds may be directed toward government-owned research laboratories, academic institutions, or industry participants. For many projects, especially those developing technologies closer to commercialization, the government will leverage its contributions by requiring substantial cost-sharing (either financial or in-kind) with industry participants.

3.4 Financing

Government-assisted financing has also been used to support renewable energy and energy efficiency, both at the project level and at the manufacturing level. In particular, Section 1703 of the Energy Policy Act of 2005 and Section 1705 of the American Recovery and Reinvestment Act of 2009 established loan guarantee programs,[18] whereby the federal government would act as a third-party guarantor for qualified borrowers using the proceeds for allowed purposes, such as new project development or development of technology manufacturing capability. While no new loans may be authorized under Section 1705 authority, loans for advanced energy technologies (which may include some renewable technologies) may still be authorized under older Section 1703 authority.

The federal government has also provided financial assistance to publicly owned utilities and other governmental entities in the form of a tax-advantage bonding authority.[5] With such bonds, the government may borrow money, repaying only the principal to the bond holders. The bond holder receives interest payments in the form of income tax credits. Programs have been offered for Clean and Renewable Energy Bonds (CREBs) and Qualified Energy Conservation Bonds (QECBs). The CREB program had limited funding and is no longer accepting new project applications.

Several states also offer assistance with project or technology loans. A number of states and localities have also started financing distributed renewable energy projects (such as roof-top PV) using "Property Assessed Clean Energy" (PACE) financing. In PACE financing, the local government (typically) acts as the lender, allowing the project owner to repay the loan through an assessment attached to a property tax bill.

3.5 Other Direct Policy

Other common programs at the state and federal levels include direct payments (such as through grants or awards) and government purchase of these technologies. These mechanisms generally require continuing budgetary support, which may be provided from a dedicated revenue source, or may require periodic affirmation in appropriations process.

A number of states have established system benefit funds dedicated to supporting renewable energy and energy efficiency projects and technologies. Although varying greatly by state, these programs are typically structured to collect revenue based on an additional fee on retail generation or billing, commonly referred to as a systems benefit charge or public benefit fund.[5] As a result of EPACT and subsequent presidential orders, the various agencies of the federal government are required to obtain a share of their energy from renewable sources and reduce their consumption of energy per square foot of facility.[19] Finally, the federal-owned fleet of cars and other vehicles is required to meet requirements for both fuel economy and use of alternative fuels. Several local state governments have also established similar purchase or efficiency requirements for electricity or motor fuels.[5]

3.6 Indirect Policy

Numerous other policies at the state and federal levels, while not designed specifically to address renewable energy and energy efficiency markets, may have a significant or notable impact on these markets. Perhaps most significant among this broad category are efforts to regulate energy or other markets, manage government—or privately—owned lands, and protect the environment.

Efforts at the federal level to introduce competition in wholesale electricity generation markets, as well as in a number of states to introduce competitive retail electricity supply, have created the opportunity for electricity suppliers to sell "green" power—typically electricity produced from renewable, low-emission, or high efficiency technologies.[20] Such programs include competitive supply of clean or renewable power, special pricing for green power by regulated utilities, or the sale of the environmental attributes of renewable power apart from sale of electricity. In addition, the specific design of competitive wholesale markets for generation and transmission can impact the competitiveness of some renewable, especially intermittent resources such as wind.

Environmental regulation at the federal or state level, for air quality, water quality, solid waste disposal, land use, greenhouse gas emissions, and other pollution problems, can have substantial impact on both the cost and value of renewable energy and energy efficiency. The Clean Air Act Amendments of 1990 (CAA) provides the foundation for cap-and-trade regulation of sulfur dioxide, and for emission limits on other pollutants.[21] While these programs do not always directly address the use of renewables or efficiency as a pollution avoidance mechanism, they do not necessarily preclude their use to reduce overall emissions. Other CAA impacts on renewable energy and energy efficiency include reformulated gasoline requirements discussed earlier, which have interacted with state-level groundwater protection efforts to provide a preference (in some states) for ethanol as a preferred fuel additive for CAA compliance. More recently, EPA has begun to use authorities in the CAA to regulate greenhouse gas emissions, which may have a more pronounced, if still indirect, impact on renewable generation resources. As a result of the Resource Conservation and Recovery Act, some landfill operations have been required to install collection and flaring systems to prevent the dangerous build-up of methane-rich gas that results from the decomposition of organic matter in the landfills.[22] These systems have significantly reduced the cost of deploying small generators fueled by this off-gas. Impacts of land management policy at both the Federal and state levels can be significant factors in renewable energy policy, either to encourage or preclude its development on government owned land.

At the state level, a number of states, either working alone or in cooperation with other states, have established policies to control the emissions of greenhouse gases. For example, the California law known as AB32 (for Assembly Bill number 32, its ascension number in the legislative session) implements a number of policies to control or limit carbon emissions.[23] Some of these policies, such as a modification of the state's RPS or the low carbon fuel standard, directly address renewable energy or energy efficiency, while other policies, such as the cap on greenhouse gas emissions, may serve to encourage additional adoption of renewable energy resources and increased energy efficiency. The Regional Greenhouse Gas Initiative is a cooperative agreement among several states in the Northeast to limit greenhouse gas emissions, which may also incentivize renewable generation resources.

References

1. Geilecki, M. and J. Poling. *Policies to Promote Non-Hydro Renewable Energy in the United States and Selected Countries*. Energy Information Administration, Washington, DC, February 2005, p. 16. http://nrec.mn/data/uploads/Nom%20setguul%20xicheel/PV/nonhydrorenewablespaper_final.pdf (accessed January 23, 2014).
2. The American Recovery and Reinvestment Act of 2009 (Pub. L. 111-5), http://lobby.la.psu.edu/_107th/128_PURPA/Agency_Activities/EIA/Incentive_Mandates_and_Government.htm (accessed January 23, 2014).
3. Internal Revenue Service. How to depreciate property. Publication 946, 2004. http://www.irs.gov/publications/p946/index.html (accessed May 31, 2005).
4. Geilecki, M. and J. Poling. *Policies to Promote Non-Hydro Renewable Energy in the United States and Selected Countries*. Energy Information Administration, Washington, DC, February 2005, pp. 8–10. http://nrec.mn/data/uploads/Nom%20setguul%20xicheel/PV/nonhydrorenewablespaper_final.pdf (accessed January 23, 2014).
5. Interstate Renewable Energy Council. Database of state incentives for renewable energy. http://www.dsireusa.org/ (accessed January 23, 2014).

6. Energy Information Administration. Issues in focus: Production tax credit for renewable electricity generation. Annual Energy Outlook 2005. DOE/EIA-0383(2005), February 2005, pp. 58–62. http://www.eia.doe.gov/oiaf/aeo/pdf/issues.pdf (accessed May 31, 2005).
7. See Electricity produced from certain renewable resources, 26 U.S.C. §45 (2013), http://www.gpo.gov/fdsys/pkg/USCODE-2013-title26/pdf/USCODE-2013-title26-subtitleA-chap1-subchapA-partIV-subpartD-sec45.pdf (accessed March 9, 2015).
8. See Credit for alcohol fuel, biodiesel, and alternative fuel mixtures. 26 U.S.C. §6426 (2013). http://www.gpo.gov/fdsys/pkg/USCODE-2013-title26/html/USCODE-2013-title26-subtitleF-chap65-subchapB-sec6426.htm (accessed March 9, 2015).
9. Energy Information Administration. Annual Energy Outlook 2013. Legislation and regulations. http://www.eia.gov/forecasts/archive/aeo13/legs_regs_all.cfm#state (accessed January 23, 2014).
10. Energy Information Administration. U.S. ethanol production and the Renewable Fuel Standard RIN bank, June 5, 2013. http://www.eia.gov/todayinenergy/detail.cfm?id=11551 (accessed January 23, 2014).
11. National Highway Traffic Safety Administration (NHTSA). Corporate average fuel economy. http://www.nhtsa.gov/fuel-economy (accessed January 23, 2014).
12. Geilecki, M. and J. Poling. *Policies to Promote Non-Hydro Renewable Energy in the United States and Selected Countries*. Energy Information Administration, Washington, DC, February 2005, pp. 6–7. http://nrec.mn/data/uploads/Nom%20setguul%20xicheel/PV/nonhydrorenewablespaper_final.pdf (accessed January 23, 2014).
13. Federal Energy Regulatory Commission. Order No. 888, April 1996. See http://www.ferc.gov/legal/maj-ord-reg/land-docs/order888.asp (accessed January 23, 2014).
14. U.S. EPA and U.S. DOE. About Energy Star. http://www.energystar.gov/index.cfm?c=about.ab_index (accessed January 23, 2014).
15. U.S. DOE. Office of Energy Efficiency and Renewable Energy. See http://www1.eere.energy.gov/buildings/appliance_standards/index.html (accessed January 23, 2014).
16. U.S. EPA. Office of Transportation and Air Quality. See http://www.epa.gov/otaq/fuels/gasolinefuels/winterprograms/index.htm (accessed January 23, 2014).
17. U.S. DOE. Strategic Plan, May 2011. See http://energy.gov/sites/prod/files/2011_DOE_Strategic_Plan_.pdf (accessed January 23, 2014).
18. U.S. DOE. Office of Loan Guarantee Program. See http://lpo.energy.gov/programs/.
19. U.S. DOE. Office of Energy Efficiency and Renewable Energy. http://www.eere.energy.gov/femp/about/about.cfm (accessed January 23, 2014).
20. U.S. Department of Energy. Office of Energy Efficiency and Renewable Energy. See http://www.eere.energy.gov/greenpower/markets/index.shtml (accessed January 23, 2014).
21. U.S. EPA. Overview—The Clean Air Act Amendments of 1990. See http://epa.gov/oar/caa/caaa_overview.html (accessed January 23, 2014).
22. U.S. Environmental Protection Agency. Landfill Methane Outreach Program. http://www.epa.gov/lmop/faq/public.html (accessed January 23, 2014).
23. California Air Resources Board. Assembly Bill 32: Global Warming Solutions Act. See http://www.arb.ca.gov/cc/ab32/ab32.htm (accessed January 23, 2014).

4

Strategies and Instruments for Renewable Energy and Energy Efficiency: Internationally, in Europe, and in Germany

Werner Niederle*

CONTENTS

4.1 Framework for a Sustainable Development of Energy Supply Systems

4.1.1 Political Drivers toward the Transformation of Energy Supply

AGENDA 21, which was adopted at the United Nations Conference on Environment and Development in 1992 in Rio de Janeiro, sets out the requirements for sustainable development. Critical issues for sustainable development in energy supply are the continuous growth in human needs for energy in the face of limited fossil and nuclear resources, and their serious impacts on the environment and the climate.

Per capita consumption of energy varies widely between countries across the world. Cost-effective energy supplies are key to overcoming poverty. About 1.3 billion people have no

* Many thanks to Michael Bade, Marion Dreher, Harry Lehmann, Benjamin Lünenbürger, Jens Schuberth, Jan Seven, Stefan Rother, and Carla Vollmer for their contributions as well as to Gabriele Coan for translation work.

access to electricity, and 2.6 billion are without clean cooking facilities [1]. At the same time, the economic development in countries like China and India increases the global demand for energy, especially for fossil sources. The resources are unequally distributed; much of fossil and nuclear energy resources are in the hands of a few countries. This can present a considerable barrier to the economic development of resource-poor countries.

Costs are a key driver for the choice of energy source. The costs of fossil and nuclear energy supplies cannot be realistically evaluated unless the considerable environmental damage, climate change, and the risk posed by nuclear energy are taken into account. Insufficient attention continues to be paid to the external costs linked with fossil energy. Furthermore, the World Energy Outlook (WEO) reports energy production using fossil and nuclear sources as being supported by subsidies amounting to U.S. $544 billion in 2012 [2]. These must be systematically reduced.

On the other hand, nuclear energy can be seen to be subject to growing constraints to ensure operational safety, which leads to rising costs and therefore calls its economic viability into question. The costs for the use of renewable energy (RE) are falling, due to technological development. In addition to mitigating climate change, dynamic development of RE use offers access to comparably reliable, environmentally sound, and low-risk energy supplies as well as the opportunity for poor countries to develop socially and economically. Furthermore, significant cost-effective potential to save energy through efficient conversion and use exists worldwide and is only partly utilized today.

4.1.2 Successful Promotion of a Sustainable Energy Supply

The transformation of energy supply system requires a number of *framework conditions, instruments, and measures* which are capable of breaking down the decades-old structures of the existing energy supply system and promoting system changes which in part are fundamental. Important, if not an essential prerequisite for such a development in a country or an association of countries, is a basic commitment of government, administration, business, and the population. Generally, there must be a transparent strategy with legally binding development paths. The latter must refer to transparent and soundly calculated indicators with a clear reference to the short- and long-term goals. The development should be monitored continuously to make sure the trajectory is being complied with. The International Energy Agency has proposed general approaches to promoting efficient energy conversion and energy use [3]. In principle they can also be applied to the promotion of RE.

Generally, a set of instruments which specifically address the various market segments is needed. A strategy geared to a single goal, for example, climate protection, will not be sufficient to achieve the manifold sustainable-development goals. The instruments must be continuously adjusted as required for the long-term transformation of the energy system and must be coordinated in order to avoid contradictions between them. They should promote the development and application of the technical basis, encourage the integration of new products into the market, and activate cost reduction potential. One priority, alongside installations for the utilization of RE, should be the development and use of integrative elements such as energy grids, load management, and storage facilities in order to adapt fluctuating wind and solar energy to the given load profiles.

This is being discussed today mainly in the context of electricity supply, but in order to optimize the energy system as a whole, provision of electricity, heating and cooling as well as the energy needs for transport must be considered as a connected system. The implementation of cost-effective efficiency measures should be made mandatory through suitable legal norms, and longer amortization periods should be compensated through suitable support. The transformation of energy supply must be coordinated with the declining energy demand brought on by energy efficiency (EE) measures in order to ensure that the expansion of RE utilizition does not thwart subsequent efficiency measures in, for example, the buildings sector.

Making the energy supply system sustainable involves far-reaching changes. Policymakers, administrations, companies, and citizens must not only accept this transformation, they must also actively support it. International organizations and states committed to this task must therefore establish target-group-specific public relations activities in order to communicate the opportunities and challenges it involves. They must also devise new training and further-training contents to prepare the people and professions charged with this task for tackling it.

4.1.3 International Agreements on the Use of Renewable Energy

The debate about global sustainable development has also had an impact on organizations in the energy sector. Over the last 10 years, various global initiatives and organizations have been established in this context. They include, but are not limited to, those briefly described in the following section.

In 2011, United Nations Secretary-General Ban Ki-moon launched the *Sustainable Energy for All initiative* (SE4ALL). It has three objectives which are to be achieved by 2030: to ensure universal access to modern energy service, to double the global rate of improvement in EE, and to double the share of RE in the global energy mix. The initiative has developed a Global Action Agenda with 11 Action Areas: 7 "sectoral" areas (e.g., modern cooking appliances and fuels, large-scale renewable power, and buildings and appliances), and 4 "enabling" areas (energy planning and policies, business model and technology innovation, finance and risk management, capacity building and knowledge sharing). The initiative seeks to encourage governments and actors to initiate their own specific actions in these areas. Suitable metrics will be established to measure progress toward achievement of the objectives [4].

The proposal to set up an international organization for RE dates back to the United Nations Conference on New and Renewable Sources of Energy held in 1981 in Nairobi. It took three decades of manifold international efforts until on April 4, 2011, the International Renewable Energy Agency (IRENA) was established in Abu Dhabi, United Arab Emirates. By June 2013, 160 participants were registered, including 114 states and the EU as members and 45 states as signatories or in accession. Mandated by its Member States (MS), IRENA serves as a network hub of country, regional and global programs and activities, an advisory resource on planning, policy development and deployment, and as an authoritative, unified global voice for RE [5].

Numerous conferences paved the way for the activities and organizations addressed earlier. The World Summit for Sustainable Development in 2002 in Johannesburg stressed the importance of RE, and the foundation of IRENA was mentioned for the first time in the final declaration of the International Renewable Energy Conference in 2004 in Bonn. This was followed by a series of International Renewable Energy Conferences (IREC) in Beijing, Washington, Delhi, and Abu Dhabi. The 2013 IREC took place in conjunction with the third session of the IRENA Assembly and the Annual World Future Energy Summit during Sustainable Energy Week [6].

Last but not the least, the Renewable Energy Network (REN21) was established as a result of the Renewables 2004 Conference. It is a global network connecting actors from governments, international organizations, industry associations, science and civil society to support exchange of knowledge and data as well as global activities in the field of RE. REN21 annually publishes the Global Status Report, the Global Future Report as well as reports on regional activities and RE policies, and runs several websites [7]. The international activities outlined earlier cooperate closely and assist one another. UN Secretary-General Ban Ki-moon has welcomed IRENA as RE hub within the SE4ALL initiative [8], and the conferences served to discuss and prepare the global activities and were actively supported by REN21. Despite the growing intensity of this exchange, governments must take action themselves to promote RE within their own remit.

4.2 Strategies and Instruments in Europe

The global debate on climate change (e.g., reports by the Intergovernmental Panel on Climate Change [IPCC], Kyoto Protocol) and the dependency of the European Union (EU) on energy imports led to the adoption of the EU Climate and Energy Package in April 2009. It consists of several instruments designed to prevent or reduce greenhouse gas (GHG) emissions and sets three targets for the year 2020: a reduction of GHG emissions by 20% compared to 1990 levels (30% if other industrialized countries set own targets), an energy efficiency (EE) improvement of 20% compared to 2005, and an increase in the use of Renewable Energy (RE) to a share of 20% of gross final energy consumption. In addition, RE's share in the transport sector is to be increased to 10% [9].

4.2.1 European Energy Supply and Dependency on Energy Imports

In 2011, gross primary energy consumption in the 27 EU MS amounted to about 1,698 million tons of oil equivalent (toe), which corresponds to 71,092 terajoules (TJ) and is 6% less than in 2008. Gross final energy consumption saw a decrease of similar magnitude, to about 1103 toe. The energy dependency rate is about 54% of primary energy consumption, that is, more than half of energy consumption comes from imported sources. Consumption and dependency on imports vary greatly between MS. The largest energy consumers are—in line with economic performance—Germany, France, the United Kingdom, Spain, and Italy. As regards dependency on imports, small countries (many of them islands) like Malta, Zyprus, or Luxembourg head the list, but also Ireland, Italy, and Portugal have to import much of their energy requirement. On the other hand, Denmark is a net energy exporter [10].

4.2.2 Renewable Energy Directive (2009/28/EC) of April 23, 2009

Directive 2009/28/EC on the promotion of the use of energy from renewable sources (RE Directive, RED) is one of the instruments of the Climate and Energy Package and entered into force in June 2009 [11]. It establishes a framework for the further development of RE use in the electricity, heating, and transport sectors in the MS of the EU and provides recommendations for the design of the environment for RE promotion and expansion. The RED allocates the overall EU RE development target among the MS by setting MS-specific targets. Its main contents are

- A national action plan and progress reports serve for precise communication between MS, European Commission and European Parliament on progress made to achieve the Directive's goals. The MS submitted their national action plans by June 31, 2010. They contain the energy data structure needed to calculate GHG reduction effects (taking into account savings in consumption due to energy efficiency) and the expected expansion of RE until 2020 with 2-year intermediate values (trajectory). The report also describes the measures and instruments established to comply with the RED. From 2011 to 2019, the MS draw up a progress report every 2 years presenting an interim assessment of the development described in the action plan. The European Commission evaluates these reports and summarizes them in a progress report to the European Parliament and the Council. In the first progress report of March 27, 2013, the European Commission calls for improvements in MS, for example, better framework conditions for expansion of RE, and finds compliance with the trajectory unsatisfactory in some cases. It has even launched infringement proceedings against some MS. However, it also sees a need for further action in its own domain.

- The base data and methods needed to calculate GHG reduction effects.
- For cooperation between MS, the RED defines mechanisms for the statistical transfer of used amount of RE joint support schemes, and joint projects for production of electricity, heating, or cooling. Electricity production projects may also be carried out with non-EU countries. The produced amounts of energy are divided up between the countries involved for the purpose of counting them toward national targets.
- To reduce barriers to RE expansion, the RED addresses in detail the favorable framework conditions for project planning and implementation, access to and operation of grids as well as information and training, and recommends appropriate adjustment of relevant regulations.
- The system for guarantees of origin of renewable electricity defined by Directive 2001/77/EC was further developed and may now be extended to include heating and cooling. The system serves to ensure that the share of RE in electricity, heating or cooling product can be proven to final customers in a transparent and objective manner.
- The RED places great emphasis on sustainable provision of biofuels, addressing the global dimension of relevant markets. As a key requirement, biofuels are only accepted within the scope of the RED, for example, for fulfillment of the RE shares, if their use leads to a 35% GHG emission saving compared to the fossil-fuel reference. From 2017 the saving must be 50%. This aspect is discussed further in the following paragraphs.

The EU-wide *target for the share of RE in transport* applies equally to each MS. In addition to biofuels, fuels, or electricity from other RES may be counted toward this target. Since highly efficient technologies such as batteries and new conversion processes (e.g., power to gas) are still not available at feasible costs, this target has generated considerable pressure toward the production of biodiesel in particular.

The provision and use of biomass for energy production usually has adverse environmental impacts. In addition to other environmental impacts, intensified land use or the use of land previously used otherwise, referred to as land use change (LUC), for example, plowing up of meadows, can increase the eutrophication and acidification of soil and water bodies and lead to correspondingly higher emissions of highly potent GHG. GHG emissions from fossil-fuelled machinery or artificial fertilizers are hardly evitable. Competition with food and feed production is another relevant problem. In tropical countries, and especially in poor countries, the effects from this may be much more pronounced, for example, when food production is displaced to previously virgin rainforest areas (indirect land use change—ILUC). Competition with food production can have serious social consequences in these countries, for example, due to rising food prices [12].

These insights raise serious doubts as to the sustainability of biofuels (especially of biodiesel) or even their positive effects for the climate. Calculated GHG emissions do not yet include the effects of ILUC, as a commonly accepted calculation method does not yet exist. Therefore, the European Commission presented in autumn of 2012 a controversial initial proposal, on the basis of several experts' reports, on how ILUC could be accounted for in reporting under the RED on the promotion of RES [13]. The implementation of the proposal would lead to the situation that a reduction of GHG emissions could no longer be demonstrated for biofuels with a high oil-crop share [14]. This has led to an intense debate about the contribution of food-crop-based biofuels to the target for RE in the transport sector by 2020 under the RED, and for this option, a limitation to the current level is under discussion.

4.2.3 Instruments in EU Member States to Promote Renewable Energy for Electricity Generation

EU MS essentially use two instruments as main support instruments: feed-in payments with fixed tariffs or variable premiums, and quota models with tradable certificates in different variations. In addition, tendering, tax incentives, and mixed schemes are used. The support instruments are constantly changing, and their design and respective frameworks differ markedly between MS. Therefore, a comprehensive description would only have transitory relevance and would go beyond the scope of this contribution. Figure 4.1 shows the main support instruments applied by MS at the end of 2012 [15]. In the subsequent section, the basic schemes are briefly described and evaluated in terms of their economic efficiency.

With the *fixed feed-in tariff*, operators of RE installations receive fixed payments per kW h for the amount of electricity they produce. This makes electricity revenues readily calculable and provides high investment security. However, it eliminates the electricity price signal from being the decisive criterion for the RE installation operator; neither investment nor operating decision will be determined by the real-world shortage situation in the electricity market. With fixed feed-in payments, the speed of expansion can be controlled only indirectly and roughly via the funding rates. Adjusting the payments to the price development is difficult, especially when it is dynamic. Differentiating the payments according to technology may be appropriate, especially in the case of large cost differences, and may reduce or prevent the problem of windfall profits in RE promotion. When long term, the support can be adjusted to reflect the learning curves of RE technologies and windfall profits can be prevented.

With the *premium model*, the installation operator or a trader markets the renewable electricity via a power exchange. In addition to the market prices, RE installation operators are paid a premium per kW h for the amount of electricity produced. The premium is geared to the exchange price (e.g., the average monthly price) and therefore provides an incentive for market-adapted production. The behavior of a market premium model in terms of windfall profits and controllability of the speed of expansion is similar to that of the fixed feed-in payment. The uncertain electricity price reduces certainty for investors, but linkage with a fixed payment as in the case of the market premium provided for in Germany by the EEG limits this effect.

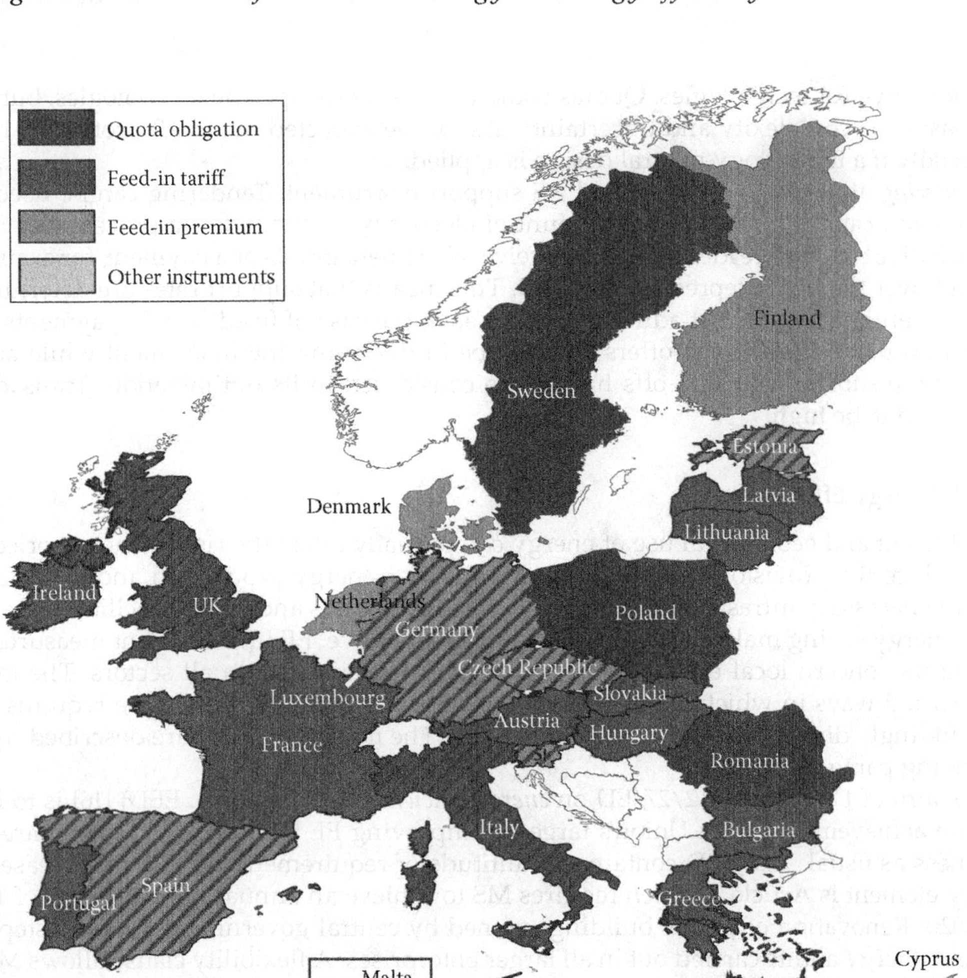

FIGURE 4.1
Overview of RE support instruments primarily applied in EU Member States. *Notes:* (1) The patterned colors represent a combination of instruments; (2) investment grants, tax exemptions, and fiscal incentives are not included in this picture unless they serve as the main support instrument. (From Ragwitz, M. et al., Review report on support schemes for renewable electricity and heating in Europe, report D8 compiled within the E European research project RE-Shaping [work package 3], January 2011, p. 21ff., www.reshaping-res-policy.eu, as of March 6, 2015.)

The *quota system with trading of certificates* requires market actors to provide a certain share of renewable electricity within their portfolio. Alternatively, they can buy certificates for renewable electricity generated by other market actors at variable market prices and might thus fulfill their quota in a more cost-effective way than in the case of own production. The quota system affords good controllability of the development path, provided that nonfulfillment of the required quota is sufficiently sanctioned. Since the revenues from renewable electricity in this scheme are directly dependent upon the electricity prices on the energy exchange, market integration is given. There is little certainty for investors, as not only the electricity prices on the energy exchange but also the prices for the certificates are volatile. RE investors must take high risk premiums into account, which reduces the efficiency of a quota system. On the other hand, the competition orientation has—in the short term—a beneficial effect on efficiency, since RE expansion is channeled to the most cost-effective RE technologies. Quotas could also be set for specific technologies, but this increases the complexity and uncertainty. As can be expected, windfall profits are high, especially if a technology-neutral design is applied.

Tendering alone is not a functional RE support instrument. Tendering can be used for production capacity or produced amount of electricity. In the bidding process, the most competitive bidders are identified and receive investment grants or a payment, for example, spread over the plant depreciation period. This means that support rates are determined by competition, and not set administratively as in the case of fixed feed-in payments and premium model. Tendering offers large scope in designing the instrument while at the same time substantial trade-offs have to be considered in its optimization. Transaction costs tend to be high.

4.2.4 Energy Efficiency

An efficient and economical use of energy can partially offset the rise in energy prices. It can reduce the provision of energy, installations for energy production and conversion, and the necessary infrastructure, for example, energy grids and storage facilities. In addition, energy saving makes the economy more competitive. EE improvement measures for end-users concern local energy provision, conversion, and use in all sectors. The many and varied ways in which energy losses can occur through inefficient use requires correspondingly diverse and small-scale measures. The most important are described in the following paragraphs.

The aim of Directive 2012/27/EU on *energy efficiency* (EE Directive, EED) [16] is to help ensure achievement of the Union's target of improving EE by 20% by 2020 compared to business as usual. The EED contains a multitude of requirements for MS to increase EE. A key element is Article 7, which requires MS to achieve an annual savings quota of 1.5% by 2020. Renovation of public buildings owned by central governments is to be stepped up and energy audits carried out in all larger enterprises. A flexibility clause allows MS to apply derogations, which may be counted toward the saving target in Article 7 at a rate of up to 25% (i.e., application of the derogations may not lead to a reduction of more than 25% of the energy savings resulting from the 1.5% target).

Directive 2009/125/EC, also known as *Energy-related Products* Directive, establishes a framework for the ecodesign of energy-related products [17]. It requires all manufacturers placing products on the market in EU MS to present a declaration of conformity indicating that the product's design complies with the provisions of the Directive and of regulations issued under it for homogenous groups of products. The regulations limit, for example, stand-by and off-mode power consumption of specific groups of devices like household and office equipment. Regulation 1275/2008/EC on electrical office and household equipment [18] alone will reduce unnecessary power losses in the EU by 35 billion kW h/year by 2020. This translates to a saving of 14 million tons of carbon dioxide emissions and nine 800 MW power plants [19]. Another regulation, 2013/801/EU, for data network equipment has entered into force recently. There are also regulations on the EE of televisions, lamps, electric motors, and other groups. Due to the factual ban on basic incandescent lamps, compact fluorescent lamps, also known as energy saving lamps, are being launched on the market in a large variety of forms and designs.

Many of appliances discussed earlier, as well as dishwashers, washing machines, refrigerators and freezers are labeled for their annual energy consumption under Directive 2010/30/EC [20]. Originally, starting in the 1990s, EE was divided into classes A to G for *labeling* purposes. This system had to be changed in response to the EE improvements which labeling brought about for these product groups. Appliances in classes C to G have largely disappeared from the market, and three further classes with one to three plus signs above class A (A+ to A+++) were introduced. The EE of labeled appliances improved 7% in the EU between 2005 and 2010 (ranging from 0% in the United Kingdom and 18% in Spain) [21].

Buildings account for a large proportion of final energy consumption. The European Commission has estimated a share of 40% and addresses existing saving potential in Directive 2010/31/EU on the *energy performance of buildings* [22]. The Directive requires EU MS *inter alia* to set minimum energy performance requirements for new buildings and existing buildings subject to major renovation with a view to achieving cost-optimal levels. By the end of 2020, all new buildings must be nearly zero-energy buildings. Buyers and tenants must be given information on the EE of the building; for example, it is now obligatory for EE parameters to be indicated in housing advertisements.

4.3 Strategies and Instruments in Germany

In Germany, the energy debate started in the 1970s, with the anti-nuclear movement as a part of the New Social Movement and received impetus from the oil-price crises in the 1970s and 1980s, forest dieback in Germany, and the Chernobyl nuclear disaster in 1986. After the UN-Conference in Rio 1992, increasing attention was given to the issues of sustainable development of the energy supply system as an overarching concept as well as climate protection. These discussions promoted various approaches to EE in the buildings sector and to the use of renewable energy (RE) in Germany. By the turn of the millennium, transforming the energy supply system had become an issue acknowledged by the German public as one of the most important tasks of the twenty-first century and which it now broadly supports.

4.3.1 Use of Renewable Energy in Germany

Primary energy consumption in Germany in 2011 amounted to 13,522 PJ. It has remained at similar levels since 1990, between 13,000 and 15,000 PJ. The share of fossil energy sources in primary energy consumption in 2011 was almost 79% (mineral oil 33.5%, natural gas 20.8%, hard coal 12.8%, and lignite 11.6%) and that of nuclear energy was <9%. The contribution of RES, in contrast, rose to a total of 11% [23]. In 2011, Germany had to import over 60% of primary energy consumption placing it in the upper mid-range of EU MS [10].

Provision of electricity, heat, and fuels from RES has almost quadrupled, from about 83 terawatt hours (TW h) in 1998 to around 318 TW h, or 12.3% of total final energy consumption, in 2012. Seventy seven terawatt hours of electricity from RE were provided in 2012. Their proportion in electricity production increased from 4.7% (1998) to 23.5%. Heat supply from RE in 2012 totaled approximately 140 TW h or 10.2% of total final energy consumption for heating (space heating, hot water, and industrial process heat). In the transport sector the contribution of RE was roughly 35 TW h in 2012, or about 5.7% of total fuel consumption by road transport [24] (Figure 4.2).

At the end of 2012, wind turbines with a capacity of about 31,300 megawatts (MW) were installed in Germany, which produced almost 50.7 TW h of electricity in that year.

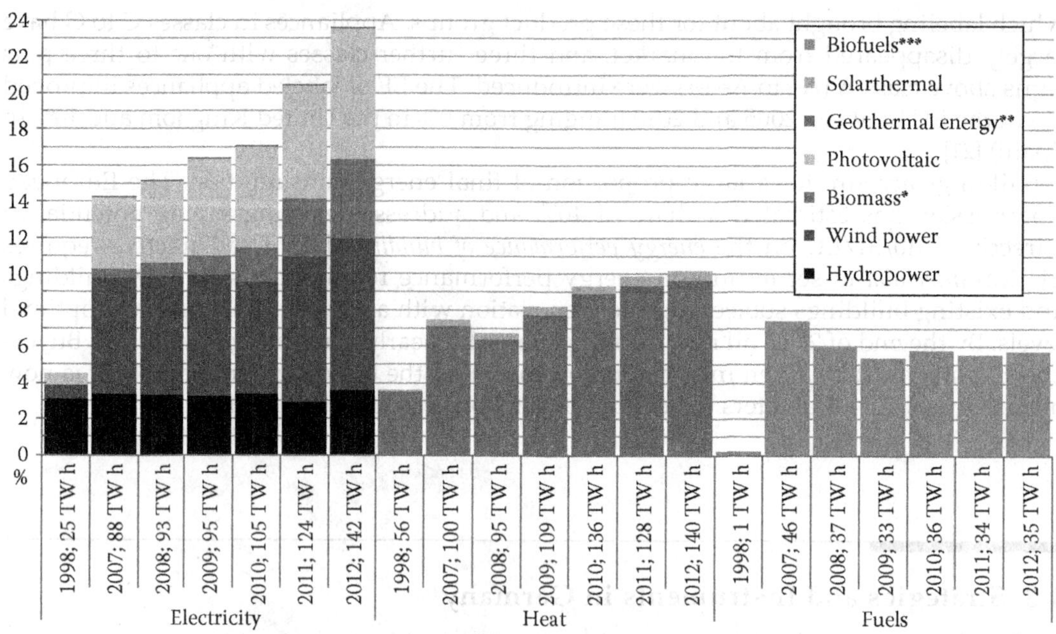

FIGURE 4.2
Contribution of RE to final energy consumption. *Biomass: solid and liquid biomass, biogas, landfill and sewage gas, biogenic share of waste. **Geothermal energy in the heat sector: deep geothermal energy, near-surface geothermal energy, ambient heat. ***Biofuels: biodiesel, bioethanol, vegetable oil, from 2008: biomethane; values rounded; as at August 2013. (From Federal Ministry for the Environment, Nature Conservation and Nuclear Safety (publisher), Renewable Energy Sources in Figures, July 2013; as at March 6, 2015 only the 2014 German update "Erneuerbare Energien in Zahlen", now published by Federal Ministry for Economic Affairs and Industry, is available on http://www.erneuerbare-energien.de/EE/Redaktion/DE/Downloads/erneuerbare-energien-in-zahlen.html; numbers might differ slightly due to statistical adjustments.)

In addition, photovoltaic systems with a capacity of over 32,600 MW were installed yielding almost 26.4 TW h. Bioenergy outdoes photovoltaics, with over 43.5 TW h from an installed capacity of almost 7600 MW, but is increasingly viewed critically due to harmful environmental and climate effects. Besides these sources, only hydropower plays an appreciable role, with 21.8 TW h from an installed capacity of 5600 MW. Old hydropower facilities in particular can have high impact on the environment; moreover hydropower's potential is largely exhausted in Germany. Electricity production from geothermal systems exhibits large technical potential, but the extent to which it is used is still small due to high costs and other hurdles.

In heat supply from RE, the use of bioenergy dominates, mainly from traditional log burning. This source is viewed critically from the viewpoint of sustainable resource use and because of its environmental impacts, for example, emissions of fine particulate matter. Heat provision from solar thermal and geothermal systems in 2012 only amounted to 6.7 and 7 TW h, respectively despite their large overall potential [24].

4.3.2 Renewable Energy Sources Act—Promotion in the Electricity Market

The Renewable Energy Sources Act (EEG) was enacted in 2000, the last overall amendment became effective in 2014. The purpose of the Act is to facilitate a sustainable development of energy supply by promoting the use of RES in the electricity sector. The objective is the

TABLE 4.1

Current Status and Targets under the *Energiewende* in Germany

	Basis	2011	2020	2030	2040	2050
Greenhouse gas emissions (reduction)						
Greenhouse gas emissions	1990	−26.4%	−40%	−55%	−70%	−80% to −95%
Efficiency (reduction/share of electricity generation)						
Primary energy consumption	2008	−6%	−20%		−50%	
Final energy consumption	2008	−2% per annum	−2.1% per annum (2008–2050)			
Gross electricity consumption	2008	−2.1%	−10%	—	—	−25%
Combined heat and power plants		15.4% (2010)	25%		—	—
Building stock (reduction/share in building stock)						
Heat requirement		—	−20%	—	—	—
Primary energy requirement		—	—	—	—	Appr. −80%
Building refurbishment		1% per annum		Increase to 2% per annum		
Transport (reduction/number of vehicles)						
Final energy consumption	2005	Appr. −0.5%	−10%	—	—	−40%
Electric vehicles		Appr. 6600	1 million	6 million	—	—
Renewable energies (share in consumption)						
Gross electricity consumption		20.3%	35% min	50% min	65% min	80% min
Final energy consumption		12.1%	18%	30%	45%	60%

Source: Adapted from Federal Ministry of Economics (BMWi), Federal Ministry for the Environment, Nature Conservation and Nuclear Safety (BMU), First monitoring report "Energy of the future", Summary (English), Berlin, Germany, December 2012, p. 4, See full report (in German) for detailed data, p. 21ff., p. 3, download from www.bundesnetzagentur.de/monitoringenergyofthefuture (as at March 6, 2015).

continuous expansion of RE to at least 80% of total electricity supply by 2050 at the latest, with corresponding intermediate targets (see Table 4.1). Due to the investment security for plant operators established by the EEG, the targets so far have always been exceeded. Many countries around the world are therefore adopting feed-in tariffs to promote RES in the electricity sector.

Basic elements in the EEG are the obligation to connect the facilities to the electricity grid and the obligation of grid operators to purchase and convey the electricity. The fixed feed-in tariffs per kilowatt hour paid by the grid operator for 20 years are based on the electricity production costs of the respective type of installation. They are continuously adjusted in line with cost developments.

Due to the growing market shares, *integration of electricity from RES in the energy market and system* is gaining in importance and gradually the EEG is being changed to this end. The amended EEG of 2012 introduces the possibility of direct marketing by the plant operator or a respective service provider on the spot market in conjunction with a market premium, which turned mandatory for plants over 500 kW installed capacity in the 2014 amendment. The plant operator receives a comparable total remuneration as in the fixed feed-in system, which now arises as the sum of revenues in the electricity market and the market premium. Yet facility operation considering the electricity price allows (slightly) higher profits and thus encourages the integration of RE. Corresponding potentials are primarily available in adjustable systems such as biogas facilities.

In addition, the 2014 amendment introduced controversially discussed auction procedures to determine the necessary funding of RE plants. There is great concern regarding the expectations of reduced costs of RE development. Private citizens and cooperatives, who have covered the main part of RE investments so far, might not be able to manage auctioning procedures and economic risks. RE development might be taken over by a few large companies and the current satisfaction with the Energiewende amongst German citizens could suffer.

The electricity that is not direct marketed is sold on the electricity exchange by transmission system operators. The remuneration paid to plant operators is reduced by the sales proceeds. The remaining sum and the paid premiums are allocated to the quantity of electricity sold in Germany and is thus paid by the consumers. This means, the financing of this instrument is independent of annual public budgets, which is a major reason for the stable development of the expansion of RE in the electricity sector [25].

The EEG must be linked with other climate protection instruments. For example, when setting emission caps the European emissions trading system takes into account the GHG emission reduction expected to arise from the EEG development targets and corresponding policies in other EU MS.

4.3.3 Renewable Energies Heat Act and Market Incentive Program

The Renewable Energies Heat Act (EEWärmeG) sets a target of 14% to increase the proportion of RE in final energy consumption for heating and cooling in Germany by 2020 [26]. This percentage is to be achieved by the expansion of RE for heat supply in buildings. For new buildings partial mandatory use applies: a portion of the building supply with heat and cooling must be covered by RES such as geothermal, solar thermal, or biomass. Alternatively, compensating measures for the efficient use of energy may be applied.

For existing buildings in general no mandatory use of RE applies. They are instead incentivized by financial support from the Market Incentive Program (MAP) [27]. Merely public buildings which are thoroughly renovated are for role model reasons obliged to use RE. The restriction of mandatory use to new buildings reduces the scope of the law considerably. Experts therefore advocate obligatory use of RE in the course of extensive refurbishment of all existing buildings. This is rejected by relevant stakeholders such as housing associations citing lack of economic viability. Regardless of this, however, use of RES in combination with EE measures in buildings must be strongly supported.

Additionally to the measures under the EEWärmeG, the MAP fosters the installation of RE facilities for heat supply in general by grants paid by the Federal Office of Economics and Export Control (BAFA). Moreover, the "Premium" Reconstruction Credit Institute (KfW) Renewable Energies Programme finances large commercial RE installations through inexpensive loans. In 2011, a total amount of €350 million were available. To achieve the Energiewende target of 14% by 2020, the MAP must be continuously developed.

4.3.4 Energy Efficiency Measures for Buildings

To achieve German and European targets, the entire building stock must undergo energy-saving renovation in the framework of regulatory elements and a long-term support strategy.

In Germany, the legal basis for this is the *Energy Saving Ordinance* (EnEV). It was formed in 2002 through merging of the Thermal Insulation Ordinance and the Heating Systems Ordinance to facilitate a coordination of technical and heat insulation requirements (i.e., a lower level of thermal insulation can be compensated for by better plant technology, and vice versa). Its central requirement for new buildings and extensively refurbished buildings is the limitation of the specific annual primary energy demand for space heating, ventilation, cooling, and water heating. This facilitates an adequate evaluation of the use of RE, among others. In addition, for new buildings the Ordinance requires compliance with a minimum level of thermal insulation by limiting transmission heat loss of the building shell. For major refurbishment of existing buildings (e.g., rendering refurbishment of external walls), it sets requirements for the heat transmission coefficient of the building components concerned. The energy quality of a building is described in an energy certificate, which must be issued and presented when a building is constructed, sold, or rented out.

The requirements of the EnEV were gradually further increased in 2007 and 2009. Another step will apply from 2016 to achieve the EU target for a nearly zero-energy standard for new buildings by the end of 2020. Refurbishments should be made using passive house components no later than 2018. For the funding of building refurbishment the KfW programme "Energy Efficient Refurbishment" was set up. It should continue to be equipped with a budget of at least €2 billion/year beyond 2020 in order to promote this important task.

4.3.5 Germany's *Energiewende*

In 2010 and 2011, the German government adopted a number of instruments, measures, and long-term goals for a permanent transformation of the energy supply system, subsuming them under the term *Energiewende*. The aim is for Germany to become one of the world's most energy-efficient and environmentally friendly economies which achieves a secure supply of energy with competitive prices and a high level of prosperity [28]. Instruments already in place prior to this, such as the EEG and its goals, were integrated into this concept. The *Energiewende* envisages a significant long-term reduction of energy consumption through much more efficient conversion and use of energy and the extensive replacement of fossil energy sources with RE (see Table 4.1). The complete phase-out of nuclear energy by 2022, which had already been adopted in 2004, was confirmed in the wake of the accident at the Fukushima nuclear power plant.

The implementation of the *Energiewende* is monitored through annual *monitoring reports* and a 3-year *progress report*, which is evaluated by independent experts. The public is closely involved in various developments in order to win the necessary acceptance. Barriers to the planning, funding, realization, and marketing of projects are addressed. On the technical side,

- Fossil-fuel power plants and RE installations are further developed with a view to improving their EE
- Grid expansion—regional, national and offshore—is systematically planned and promoted within the scope of annually revised grid concepts
- Concepts for the integration of heat and electricity use are developed
- Market integration is improved through development and promotion of load management and storage

German government supports, through research programs and other instruments for market development, investigations into cost reduction in the use of RE and EE, and also promotes marketing and the development of new market models to improve supply security.

The *economic effects* of the development that has been set in motion are, in principle, viewed positively. However, their assessment involves methodological problems. For instance, the current development as guided by Germany's energy policy must be compared with the unknown development that would have taken place without this guiding influence. The rise in energy prices brought on by the support must be contrasted with avoided energy imports, avoided external costs such as damage to the environment and climate, or industrial innovations from research and development. For example, in 2011 energy imports valued at €25 billion were avoided due to efficiency measures and RE expansion (energy imports in that year came to €89 billion). The investments triggered by the support, as well as employment and growth effects, also figure on the positive side of the balance sheet, but must be checked against the dampening impact of higher energy prices [30].

Criticism by independent experts and relevant institutions is based on various aspects of the development of individual key indicators. In the area of EE improvement in both the heat and transport sectors, a key prerequisite for GHG reduction, the limited progress is being criticized. In order for the *Energiewende* to succeed, it is essential that EE renovation of the building stock be speeded up. Development and system integration of RE does not happen automatically, not even in the electricity sector, but is progressing only slowly, especially in the heat and transport sectors. The implementation of integrated mobility concepts for passenger and freight transport, which help reduce dependency on fossil energy resources through more efficient traffic management, is essential to achieve this. Environmental protection, as a key requirement for sustainable development, continues to be relevant even when the use of RE (not just of biofuels) is substantially increased. Criticism is also leveled at the economic evaluation of energy supply, which is geared primarily to energy prices rather than macroeconomic indicators including external costs where appropriate. There are also reminders not to forget coordination with European climate policy and to ensure the functioning of the European emissions trading scheme. The drop in allowance prices has largely eliminated incentive for GHG reduction. The experts lament the lack of suitable indicators in several areas (e.g., EE, environmental protection, economic effects), which confounds proper assessment of the development [31].

Energy policy in Germany has embarked on a difficult, highly ambitious course, which is why it is followed closely worldwide. Despite all criticisms, the experts emphasize in their opinion that the process has only just begun and that this prohibits rash or overcritical judgment. It is important, however, that the advancement of the many and varied tasks remains sufficiently dynamic to prevent the possible loss of positive economic and technical development perspectives.

References

1. Global status of modern energy access, http://www.worldenergyoutlook.org/, Resources, energy access, energy access projections to 2030 (as at March 9, 2015).
2. Energy subsidies, http://www.worldenergyoutlook.org/, Resources, energy subsidies (as at December 18, 2013, now updated).
3. International Energy Agency. World Energy Outlook 2012—Executive summary, p. 3, http://www.worldenergyoutlook.org/, Publications, WEO-2012 (as at March 9, 2015).
4. United Nations Organisation's Secretary-General's High-Level Group on Sustainable Energy for All. Sustainable energy for all—A global action agenda, April 2012, http://www.se4all.org/, About us, Resources, Strategy Documents (as at March 9, 2015).
5. IRENA. 2013, www.irena.org, About IRENA, Creation of IRENA, Member states, and Vision and Mission (as at March 9, 2015).

6. The Renewables 2004 and the IRECs are documented on the REN21 website, http://www.ren21.net/, REN21 activities (as at March 9, 2015).
7. http://www.ren21.net/ (as at March 9, 2015).
8. Third session of the Assembly: Leaner, more effective IRENA will be global hub for energy transition, January 13, 2013, Abu Dhabi, United Arab Emirates, www.irena.org, Institutional Structure, Assembly, Third Session of the Assembly (as at March 9, 2015).
9. See homepage of European Commission's climate and energy package, http://ec.europa.eu/clima/policies/package/index_en.htm (as at March 6, 2015).
10. Eurostat news release 23/2013 of February 13, 2013, http://ec.europa.eu/eurostat, news, news releases (as at March 9, 2015).
11. Directive 2009/28/EC of the European Parliament and of the Council of April 23, 2009 on the promotion of the use of energy from renewable sources and amending and subsequently repealing Directives 2001/77/EC and 2003/30/EC, http://eur-lex.europa.eu/LexUriServ/LexUriServ.do?uri=OJ:L:2009:140:0016:0062:EN:PDF (as at March 9, 2015).
12. Jering, A. et al. Sustainable use of global land and biomass resources, Umweltbundesamt, June 2013, p. 53ff., http://www.umweltbundesamt.de/publikationen/sustainable-use-of-global-land-biomass-resources (as at March 9, 2015).
13. Proposal for a Directive of the European Parliament and of the Council amending Directive 98/70/EC relating to the quality of petrol and diesel fuels and amending Directive 2009/28/EC on the promotion of the use of energy from renewable sources, COM(2012) 595 final, Brussels, Belgium, October 17, 2012, http://ec.europa.eu/clima/policies/transport/fuel/docs/com_2012_595_en.pdf (as at March 9, 2015).
14. Laborde, B. Assessing the land use change consequences of European biofuel policies, study conducted by the International Food Policy Research Institute (IFPRI) commissioned by the European Commission, October 2011, p. 11ff., http://trade.ec.europa.eu/doclib/docs/2011/october/tradoc_148289.pdf (as at March 9, 2015). The cited paragraph summarizes several results on this issue documented in the report.
15. Ragwitz, M. et al. Review report on support schemes for renewable electricity and heating in Europe, report D8 compiled within the E European research project RE-Shaping (work package 3), January 2011, p. 21ff., www.reshaping-res-policy.eu (as at March 9, 2015).
16. Directive 2012/27/EU of the European Parliament and of the Council of October 25, 2012 on energy efficiency, amending Directives 2009/125/EC and 2010/30/EU and repealing Directives 2004/8/EC and 2006/32/EC, http://eur-lex.europa.eu/LexUriServ/LexUriServ.do?uri=OJ:L:2012:315:0001:0056:EN:PDF (as at March 9, 2015).
17. Directive 2009/125/EC of the European Parliament and of the Council of October 21, 2009 establishing a framework for the setting of ecodesign requirements for energy-related products, http://eur-lex.europa.eu/LexUriServ/LexUriServ.do?uri=OJ:L:2009:285:0010:0035:EN:PDF (as at March 9, 2015).
18. Commission Regulation (EC) No. 1275/2008 of December 17, 2008 implementing Directive 2005/32/EC of the European Parliament and of the Council with regard to ecodesign requirements for standby and off mode electric power consumption of electrical and electronic household and office equipment, http://eur-lex.europa.eu/LexUriServ/LexUriServ.do?uri=OJ:L:2008:339:0045:0052:EN:PDF (as at March 9, 2015).
19. Federal Environment Agency (publisher). Bye bye standby: EU Commission declares war on standby losses, press release no. 54/08, Dessau-Roßlau 2008, http://www.umweltbundesamt.de/en/press/pressinformation/bye-bye-standby-eu-commission-declares-war-on (as at March 9, 2015). The information originates from various sources used in the working process, which are not generally accessible.
20. Directive 2010/30/EU of the European Parliament and of the Council of May 19, 2010 on the indication by labelling and standard product information of the consumption of energy and other resources by energy-related products, http://eur-lex.europa.eu/LexUriServ/LexUriServ.do?uri=OJ:L:2010:153:0001:0012:EN:PDF.

21. GfK. Energieeffizienz ist in Europa angesagt (Energy efficiency is hot in Europe), press-release May 20, 2011, Nürnberg, Germany, 2011. The press release is no longer available at the GfK website, see http://www.lifepr.de/inaktiv/gfk/Energieeffizienz-ist-in-Europa-angesagt/boxid/232423 (as at March 9, 2015).
22. Directive 2010/31/EU of the European Parliament and of the Council of May 19, 2010 on the energy performance of buildings, http://eur-lex.europa.eu/LexUriServ/LexUriServ.do?uri=OJ:L:2010:153:0013:0035:EN:PDF (as at March 9, 2015).
23. Federal Ministry of Economics (BMWi), Federal Ministry for the Environment, Nature Conservation and Nuclear Safety (BMU). First monitoring report "Energy of the future", Summary (English), Berlin, Germany, December 2012, p. 4. See full report (in German) for detailed data, p. 21ff., www.bundesnetzagentur.de/monitoringenergyofthefuture (as at March 9, 2015).
24. Federal Ministry for the Environment, Nature Conservation and Nuclear Safety (publisher). Renewable Energy Sources in Figures, Berlin, July 2013; as at March 9, 2015 only updated charts are available in English, now published by Federal Ministry for Economic Affairs and Industry, http://www.bmwi.de/EN/root.html, Topics, Energy, Renewable Energy, Energy at a Glance; numbers might differ slightly due to statistical adjustments. See also Reference of Figure 4.2.
25. Act on the Development of Renewable Energy Sources (Renewable Energy Sources Act – EEG 2014), http://www.bmwi.de/EN/root.html, Topics, Energy, Renewable Energy, 2014 Renewable Energy Sources Act (as at March 9, 2015).
26. Consolidated version of the reasoning behind the Act on the Promotion of Renewable Energies in the Heat Sector (Erneuerbare-Energien-Wärmegesetz—EEWärmeG) of August 7, 2008, Federal Law Gazette (BGBl) 2008, Part I, No. 36 of August 18, 2008, p. 1658, http://www.erneuerbare-energien.de/en/, search for Acts and Ordinances, Heat Act (as at December 18, 2015).
27. In German: Marktanreizprogramm, http://www.erneuerbare-energien.de/EE/Navigation/DE/Home/home.html, Förderung, Beratung und Förderung, Marktanreizprogramm (as at March 9, 2015).
28. Federal Ministry of Economics (BMWi), Federal Ministry for the Environment, Nature Conservation and Nuclear Safety (BMU). First monitoring report "Energy of the future", Summary (English), Berlin, Germany, December 2012, p. 4. See full report (in German) for detailed data, p. 21ff., long version, p. 8, www.bundesnetzagentur.de/monitoringenergyofthefuture (as at March 9, 2015).
29. Federal Ministry of Economics (BMWi), Federal Ministry for the Environment, Nature Conservation and Nuclear Safety (BMU). First monitoring report "Energy of the future", Summary (English), Berlin, Germany, December 2012, p. 4. See full report (in German) for detailed data, p. 21ff., adapted p. 3, www.bundesnetzagentur.de/monitoringenergyofthefuture (as at March 9, 2015).
30. Federal Ministry of Economics (BMWi), Federal Ministry for the Environment, Nature Conservation and Nuclear Safety (BMU). First monitoring report "Energy of the future", Summary (English), Berlin, Germany, December 2012, p. 4. See full report (in German) for detailed data, p. 21ff., long version, p. 101ff, www.bundesnetzagentur.de/monitoringenergyofthefuture (as at March 9, 2015).
31. Löschel, A., Erdmann, G., Staiß, F., and Ziesing, H.-J. Expertenkommission zum Monitoring-Prozess "Energie der Zukunft". Stellungnahme zum ersten Monitoring-Bericht der Bundesregierung für das Berichtsjahr 2011, Berlin, Mannheim, Stuttgart, Germany, Dezember 2012, see http://www.ecologic.eu/8035, for General Information on the Expert Commission (as at March 9, 2015).

5

Energy Conservation and Renewable Energy Policies in China

Ming Yi and Yue (Ada) Wu

CONTENTS

5.1 Energy Conservation

As the world's second largest economy, China has remained at an average annual growth rate of about 10% for the last 30 years. With rapid growth of its economy, China's energy sector has also made remarkable progress. China is now the world's largest energy producer and consumer. It has built up a comprehensive energy supply system comprising coal, electricity, petroleum, natural gas, and new and renewable energy resources.[1] However, China has also paid a costly price and faced stiff challenges in the process of transformation from a backward developing country to one of the top economies in the world. During the 11th Five Year Plan period (FYP, 2006–2010), China consumed 40% of the coal, 50% of the cement, 60% of the iron and steel, and 9% of the oil produced in the world, but it only created 5% of the world GDP during the same period.[2] China's energy consumption per unit of GDP is five times the world average.[2] The rapid economic development also brought up serious consequences for the country's air, land, and water.

With limited fossil energy resources at home and huge emission control pressure from the international world, energy conservation and renewable energy have increasingly been a priority of the Chinese government. In 2009, China made a commitment that it would reduce the carbon emission per unit of GDP (carbon intensity) by 40%–45% by 2020, relative to 2005 intensity levels. In 2014, China announced that it will reach carbon emission peak in 2030. To achieve these goals and other related energy efficiency and renewable energy targets, the Chinese government has enacted various laws, rules, and regulations; implemented numerous energy efficiency and renewable energy programs; and also reconstructed its energy-governing agencies to make them function more efficiently.

China's energy sector has been mostly regulated and controlled by government agencies and state-owned enterprises (SOEs). China's National Development and Reform Commission (NDRC) is in charge of coordination of energy planning with the country's overall economic and social development. The National Energy Administration (NEA), which was established in 2008 and restructured in 2013, is under the NDRC and responsible for formulating and implementing energy development strategy, planning, and policies; advising energy system reform; and regulating the overall energy sector in China. Specifically, renewable energy development in China is mainly under the jurisdiction of NEA's New Energy and Renewable Energy Department. In addition to the NDRC, industry energy efficiency is also under the jurisdiction of Ministry of Industry and Information Technology (MIIT), whereas Ministry of Housing and Urban-Rural Development (MOHURD) also covers building energy efficiency. All these government agencies work with each other and other related government agencies such as the Ministry of Finance (MOF), the Ministry of Science and Technology (MOST), and the Standardization Administration in terms of fiscal, tax, and financial incentives; energy-related science and technology; and technical standards and codes for energy conservation and renewable energy development.

5.1.1 General Policy

In 2010, China reduced 19.1% of energy consumption per unit of GDP compared with the level of 2005, close to the 20% reduction target that the Chinese government set for the 11th FYP period that ended 2010. This is largely owing to the robust energy conservation policies and programs that the Chinese government implemented. In 1997, China introduced the country's first energy conservation law, which stipulated general regulations and guidelines for energy conservation in China. The law not only identified four focus areas of energy conservation—industry, building, transportation, and public institutions, but also defined main subjects of energy conservation—enterprises with annual energy consumption more than 10,000 tce (ton coal equivalence).

The law was amended in 2007 and one of the major revisions was the introduction of target responsibility system (TRS) concept. TRS uses a top-down approach to mandate the energy conservation target for the central, provincial, municipal, and county level governments. Local governments are accountable for energy conservation by signing an agreement with higher-level government. The outcomes of local energy conservation activities are directly linked to the performance evaluation of government officials.

The 12th FYP (2006–2010) for National Economy Development and Social Development, passed by the country's legislators in 2011, first proposed that China's energy intensity (energy consumption per unit of GDP) will be reduced by 16% and carbon intensity (carbon emissions per unit of GDP) will be reduced by 17% below 2010 levels by the end of 2015. The 16% reduction will bring the total reduction for the total 10-year period (2006–2015) to 32% below 2005 levels.[3]

Also in 2011, the State Council, China's cabinet, released the Comprehensive Work Plan on Energy Efficiency and Emissions Reduction for the 12th FYP (2011–2015), which details 50 specific measures to be carried out in support of the energy intensity target (as well as absolute reduction targets for criteria pollutants such as chemical oxygen demand, ammonia, sulfur dioxide, and nitric oxides).[4] In the following year, the State Council further issued the 12th FYP for Energy Conservation and Emission Control, which proposed major targets, and prioritized tasks and key projects of energy conservation. According to the plan, priorities will be given to restricting energy-intensive and high emission sectors, retiring outdated production capacity, upgrading traditional sectors, adjustment of energy consumption structure, and promotion of service and other newly emerging industries. Specifically, the plan listed prioritized tasks for energy conservation of industry, building, transportation, and public institutions.

5.1.2 Sector-Specific Energy Conservation Policies

5.1.2.1 Industry

With its energy consumption taking up about 70% of the national total, industry is the largest energy consumer in China. According to the 12th FYP Plan for Industry Energy Conservation issued by MIIT in 2012, China aims to reduce energy consumption per unit of industrial value-added output by 21% from 2011 to 2015 and achieve energy conservation of 670 million tce. The plan also sets specific energy consumption reduction targets for 9 energy-intensive sectors (including steel, nonferrous metals, petrochemical, chemical, building materials, mechanical, light industry, textile, and electronics) and 20 types of products.

To help achieve these targets, the same plan also identified key technologies and approaches to improve energy efficiency for each one of these 9 sectors and 10 types of prioritized energy efficiency projects including energy efficiency of industrial boilers and burners, internal combustion engines, generators, recovery and utilization of waste heat and pressure, combined heat and power, industrial by-product gas, enterprise energy management and control centers, and the combination of industry and information technology in energy conservation. In 2014, MIIT published the National Industry Energy Efficiency Guide (2014), which gives a comprehensive overview of the industry energy efficiency progress made since 2000. Interested readers can refer to the detailed effort China made in industry energy efficiency effort in the past decade.[5]

As a major initiative to help meet the energy conservation target of 670 million tce, the NDRC launched the Top 10,000 Energy-Consuming Enterprises Program, targeting enterprises that use more than 10,000 tce/year. The program, which is an expansion of the

Top 1000 Program that China implemented during the 11th FYP period, aims to achieve an absolute energy saving of 250 million tce. This is almost one-third of the country's total energy saving target in the 12th FYP.[6]

5.1.2.2 Building

Building accounts for nearly one-third of China's total primary energy consumption and carbon emissions. China has 40 billion m^2 of existing buildings, but only 1% is energy efficient. Between 2010 and 2020, China is expected to add 10–15 billion m^2 of residential buildings in urban areas. To improve building energy efficiency of the existing buildings and new buildings, the Chinese government has been actively engaged in the formulation and deployment of a series of legal and policy instruments.

The Renewable Energy Law, the Energy Conservation Law, and the Civil Building Energy Efficiency Code are three major laws and regulations covering building energy efficiency. In addition, more than a dozen of provinces and municipalities have also passed their own general energy efficiency codes and specific regulations. These laws and regulations, together with the 12th FYP on Building Energy Efficiency issued by MOHURD in May 2012 and the Green Building Work Plan issued by the general office of China's State Council released in January 2013, constitute the policy framework of building energy efficiency in China.

According to the 12th FYP on Building Energy Efficiency, China aims to reduce 116 million tce and plans to achieve the target through four prioritized areas: new buildings (45 million), heating supply reform and retrofitting in China's northern areas (27 million tce), government office and public buildings (14 million tce), and renewable energy adoption in buildings (30 million tce).[7]

As early as the 1990s, the Chinese government began to launch a series of policies to promote heat reform and retrofitting in existing buildings, especially in the northern areas where centralized heating is provided in most buildings. The aim of the heat reform is to reduce the amount of energy used through the reform of the heating pricing system and to establish a market mechanism to encourage heat suppliers' effort to improve the energy efficiency of their heat supply networks.[8] Given that the building in China's northern regions accounts for more than 40% of the country's total urban building energy consumption, the residential retrofitting in northern regions also plays a significant role in China's building energy efficiency efforts. From 2006 to 2010, China retrofitted 182 million m^2 of residential space in northern China.[8] It is predicted in the 12th FYP on Building Energy Efficiency that by 2015, China will complete heat supply measurement and retrofitting of 400 m^2 of existing buildings in the northern area.[9]

In addition, the Chinese government takes the initiative to implement various energy efficiency policies and measures for government office buildings, large-scale public buildings, and college and university buildings.[7] This effort is mainly focused on energy consumption monitoring and retrofit of public buildings. By 2015, China aims to retrofiting 120 million m^2 of government office buildings and public buildings.[9]

Application of renewable energy resources in buildings is also one of the government's priorities in building energy efficiency. During the 11th FYP, major initiatives included renewable energy building demo projects and demo cities, and renewable energy application in buildings of rural areas. At provincial and local levels, many supportive policies were introduced and implemented to promote the application of renewable energy technologies in building such as photovoltaic power generation, building integrated photovoltaic (BIPV), solar water heating, and geothermal heat pumps. According to the government's plan, by 2015, the newly added renewable energy building will amount to

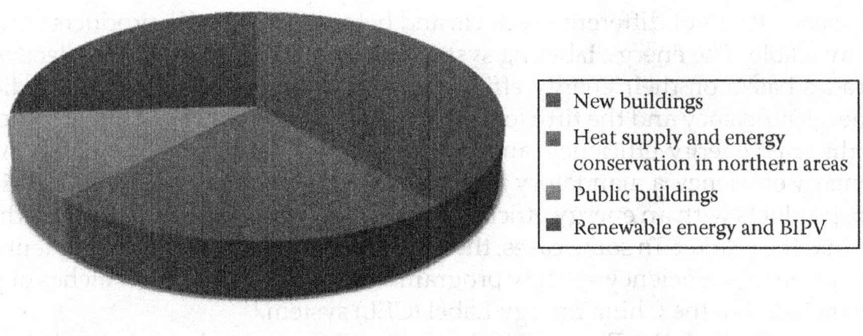

FIGURE 5.1
China's building energy efficiency target by 2015: 116 million tce. (From Ministry of Housing and Rural Urban Development, 12th FYP for Building Energy Efficiency, May 9, 2012.)

2.5 billion m², and renewable energy will account for 10% of the total energy consumption of buildings in renewable energy building demo cities (Figure 5.1).[9]

5.1.2.3 Transportation

The transportation sector accounts for one-fifth of China's total energy consumption, making it the nation's third largest greenhouse gas emissions industry. According to the 12th FYP for Transportation Energy Efficiency and Emission Control released by the Ministry of Transportation in 2011, the energy consumption of operational vehicles, ships, and ports will reduce 10%, 15%, and 8%, respectively, compared with the 2005 levels.[10]

According to the plan, energy efficiency in the transportation sector will focus on three aspects—transportation infrastructure, equipment, and network. Similar to the 12th FYPs for other sectors, the plan also identified major tasks and key energy efficiency projects in the transportation sector.

5.1.3 Standards and Labeling Programs

5.1.3.1 National Energy Efficiency Standards

With improvement of living conditions, the appliances and electronics of Chinese households have become a major drive of residential electricity use in China. This spurred the government to implement China's first mandatory equipment standards in 1990, which covered nine electronic products such as refrigerators, air conditioners, clothes washers, irons, rice cookers, televisions, radios, and fans. As of February 2013, China has complied and implemented 109 national energy efficiency standards. The standards have also expanded from those of household electronic appliances to those of energy-intensive industrial products, and energy measurement and management standards.[11] Very recently in 2015, the State Council issued Opinions on Strengthening Energy Conservation Standardization[12], which emphasizes a timely update of energy efficiency standards, the Top Runner program to promote the highest energy efficiency and mandatory Minimum Allowable Value of energy efficiency to outdate the backward 20% capacity and products.

5.1.3.2 China Energy Label (CEL) System

Since 2005, an energy label system has been introduced to illustrate the energy efficiency grade and typical energy consumption (TEC) values. The energy label, which usually appears on the surface or package of applicable products, allows customers to compare the energy efficiency levels of different products and helps them identify products with highest efficiency available. The energy labeling system categorizes appliances and electronics into several grades based on their energy efficiency performance. The first grade indicates the highest energy efficiency and the fifth (or third, depending on product categorization) grade indicates the least energy efficiency, and the least one is defined as Minimum Allowable Value of energy efficiency, a mandatory requirement for the product market access in China. Since 2005, products with an energy efficiency level lower than the fifth grade (or third) cannot be put into the market. In some cases, the first grade is required for government procurement or for an energy efficiency subsidy program. At the end of 2012, ten batches of products have been included in the China Energy Label (CEL) system.[13]

On December 31, 2014, the Energy Efficiency Top Runner Implementation Scheme[14] was jointly announced by seven ministries of Chinese government. This scheme was aimed to promote high energy efficiency product based on previous progress of CEL. The product has to meet the Grade 1 requirement of CEL, also be the highest energy efficiency in the same categorization. The criteria will be upgraded annually. The top runner logo will be added to the current CEL, and the certified year will also be specified. The government promised that further subsidy and promotion measures will be implemented to help R&D and product promotion.

5.1.3.3 Building Energy Efficiency Labeling and Evaluation and Green Building

China began to establish its building energy efficiency labeling and evaluation system in 2006. According to the building energy efficiency labeling regulation issued in 2008, the labeling system mainly covers new and existing government office buildings, large-scale public buildings, national and provincial building, energy efficiency demo projects, and green buildings. Building owners need to apply two types of labels—assets rating label and operational rating label. The former indicates the theoretical value of building energy efficiency evaluated during the acceptance stage, whereas the latter indicates the actual values of building energy efficiency evaluated during the operation of the building.[7] A five star rating system is also introduced in the regulation, with five stars representing the most energy efficiency building. This evaluation and labeling system was updated in 2014 and extended to residential buildings.

Since 2009, the MOHURD has promoted building energy efficiency labeling in newly built government office buildings and large-sized public buildings through pilot projects in selected provinces and cities. Building owners who apply for building energy efficiency labeling must comply with national mandatory standards, including building energy codes (design standards and the acceptance codes), before applying for building energy efficiency labels. As of 2010, 45 building projects had been approved and granted star ratings.[7]

While the U.S. LEED green building rating system is widely used in China, the country also developed its own three star green building rating system in 2004. This rating system is based on the Green Building Evaluation Standards—the first national standards for green buildings and technical guidelines for green building evaluation. Similar to energy efficiency building labeling, there are two types of green building labeling, with one covering building design and the other building operation. By the end of 2010, 113 projects were awarded three star green building label nationwide (Table 5.1).

5.1.3.4 Vehicle Fuel Efficiency Standard System

China's vehicle fuel efficiency standard system consists of fuel consumption test methods, fuel consumption limits, and labeling. China adopted its first nationwide fuel consumption

TABLE 5.1

Energy Efficiency Labeling Rating System Coverage

Energy Efficiency Labeling	Rating System	Coverage
China Energy label	1–5 stars	Ten batches of products ranging from air conditioners to personal computers
Energy Efficiency Building label	1–5 stars	New and existing government office buildings, large-scale public buildings, national and provincial building, energy efficiency demo projects, and green buildings
Green Building label	1–3 stars	Any building at the design, construction, or operation stage
Vehicle Consumption label	N/A	Passenger and commercial vehicles with a gross weight of 3.5 tons or less

limits for passenger vehicles in 2005. They are considered to be the world's third toughest, behind Japan's and Europe's.[7] In 2007 and 2012, fuel consumption limits for light-weight and heavy-weight commercial vehicles were also adopted. According to a new rule issued in March 2013, passenger cars' average fuel consumption is required to reduce to 6.9 L/100 km by 2015 and down further to 5 L by 2020.

Since 2010, vehicle fuel consumption labeling has been implemented for passenger and commercial vehicles with a gross weight of 3.5 ton or less. In addition to some generic information of a vehicle such as brand, make, and rated power, a vehicle fuel consumption label is required to contain information about fuel consumption per100 km and what test methods are used in determining the vehicle fuel consumption values.

5.1.4 Financial Support and Government Procurement

By the end of the 11th FYP period, China reduced 19.1% of its energy consumption per unit of GDP compared with 2005 level and successfully achieved the energy conservation targets set for the period. This is not only owing to the various energy efficiency policies and programs that the government issued and implemented, but also directly related to the massive capital investment that the government made in energy conservation during the period.

Over the 11th FYP period, about RMB 846.6 billion was invested to support energy conservation projects through different measures and channels. Among them, the central government invested RMB 101.7 billion, amounting to 12.1% of total energy efficiency investment in China, and provincial and local governments contributed RMB 48 billion, amounting to 5.7% of the total energy efficiency investment. And nongovernment sectors contributed RMB 696.9 billion, which comes from company investment, loan from banks, and funds raised by the stock market.

Among all the investments in energy efficiency, 95% was used to fund and subsidize energy efficiency–improving projects and the rest of 5% was used in other related fields such as policy and methodology research, institution and capacity building, promotion of energy efficiency products, and so on (Figure 5.2).

5.1.4.1 Financial Rewards for Energy-Saving Technical Retrofits

In 2007, the NDRC identified 10 types of key energy efficiency projects to receive the government rewards, such as coal-fired boiler retrofitting, waste heat and low temperature steam recovery, alternative oil, energy efficient motor systems, and energy system optimization. The program has been continued and expanded during the 12th FYP (2010–2015)

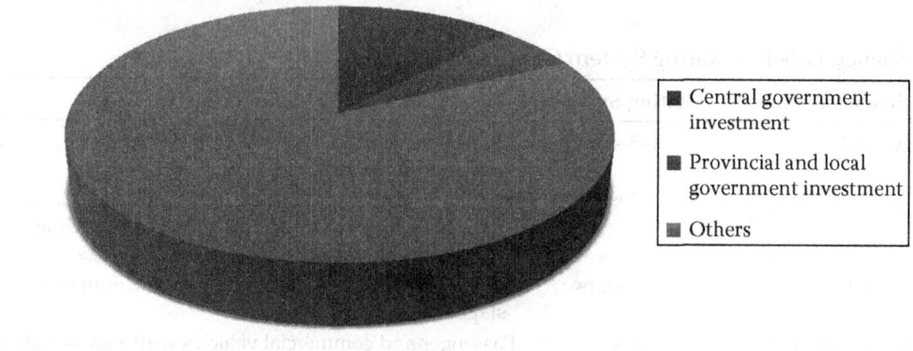

FIGURE 5.2
China's investment in energy efficiency during the 11th FYP (2006–2010). (From Dai, Y. et al., China Energy Efficiency Financing and Investment Report, 2010, China Science and Technology Press, Beijing, China, 2012.)

with raised levels of rewards and revised requirements for eligible projects and enterprises. Previously, eligible projects must have been one of those 10 projects, but the new program now covers more energy conservation technical retrofit projects. Energy-using companies used to be the only recipiens of the rewards, but in the new program, energy service companies (ESCOs) can also apply for the rewards.[15]

5.1.4.2 Energy Efficiency Product Subsidy Program

Promoting energy efficient products is the most direct and efficient way of energy conservation. During the 11th FYP period, the Chinese government provided RMB 16.5 billion to subsidize the public and businesses to purchase and provide energy efficiency products such as lights, air conditioners, and low-emission vehicles and engines.

The government initiated the Energy Efficient Product Subsidy Program in 2005 with household refrigerators and air conditioners among the first batch of products subsidized. By the end of 2012, 10 batches of products have been announced and the subsidized energy-efficient products range from rice cookers to automobiles, from household refrigerators to industry pumps and motors. Basically, the subsidies are either provided to purchasers or manufacturers of energy efficiency products. Also the actual amount of the subsidies ranges from RMB 150 to 3000. The program not only helped improve the energy efficiency level of energy-using products, especially electric appliances, but also raised the awareness of the public and business to use energy efficiency products. On May 12th, 2012, the State Council announced an allocation of RMB 36.3 billion for the subsidy program.

5.1.4.3 Fiscal Incentives for Retiring Outdated Capacity

Retiring the outdated capacity is considered as not only an important way to adjust the economic structure, but also a vital measure to achieve China's industry energy efficiency targets. During the 11th FYP period, the country invested 21.9 billion to wipe out the outdated capacity in 13 sectors. Some remarkable outcomes include the elimination of small thermal power plants with a generation capacity amounting to 80 MkW, and wiping out of outdated production capacities of 121.7 million tons of iron, 69.7 million tons of steel, 100 million tons of cement, and 10.3 tons of pulp.[2]

5.1.4.4 Financial Support for Building Energy Efficiency

During the 11th FYP period, to promote building energy conservation, the central government allocated RMB 15 billion in the following four major areas—new buildings, heat reform and retrofit projects in northern areas, government office buildings, large-scale public buildings, and renewable energy application in buildings.[3]

5.1.4.5 Energy Efficiency and New Energy Vehicles Subsidy

During the same period, RMB 2.22 billion was also allocated by the central government to promote energy efficiency and new energy vehicles in the public sector. Initially, only hybrid, electric, and fuel cell vehicles that are used in public transportation are eligible for government subsidy, but later, the subsidy scheme was expanded to include private cars. Also as an effort to speed up upgrading the auto industry, in 2009, the Chinese government launched a program to provide subsidies for replacing old cars and buses in the public domain with more fuel-efficient vehicles.

5.1.4.6 Energy Performance Contracting and Energy Service Companies

From 2010, the Chinese central government began to issue a series of supportive policies and incentives to foster the energy service industry in China. According to a reward policy for energy performance contact (EPC) projects released in 2010, qualified ESCO companies can receive RMB 240 for every ton of standard coal saved in their EPC projects from the central government and receive at least RMB 60 from the local government.

With government support and investment from financial institutions and private sectors, China's energy service sector has also seen tremendous growth in the past few years. At the end of 2012, there are 4175 companies in China engaging in energy service, among which 2339 are registered with the NDRC and the MOF. The total output of the energy service sector has also reached RMB 165.3 billion in 2012, 32.24% up from a year ago.[16] The Chinese government now considers the EPC model as one of the main market mechanisms for energy efficiency improvement and aims to build an advanced energy efficiency service system by 2015.[17]

5.1.4.7 Demand Side Management Pilot Cities

In July 2012, the MOF and the NDRC announced a new program to support adoption of energy efficiency power plants (EEPPs), demand response technologies and promotion of related scientific research, training and education, verifications, and evaluation work. In October 2012, the MOF and the NDRC announced the first four pilot cities to receive the incentive as the first step of the program.

According to the program, any pilot projects that use EEPP and load shifting technologies to achieve permanent load reductions and peak load shifting will be awarded RMB 400/kW reduced in eastern provinces or RMB 550/kW reduced in central and western provinces. For any pilot projects that lead to temporary reductions in peak load through demand response, there will be a reward of RMB 100/kW.[18]

5.1.4.8 Energy Efficiency Institution and Capacity-Building Subsidy Program

In addition to investment in specific sectors, the central government has also provided funding for institution and capacity building at provincial and local levels. The funding is mainly used to improve the energy efficiency monitoring and management capacity of provincial and municipal energy efficiency management institutions. These bodies, either called energy conservation supervision team or energy conservation center, not only help with the formulation of provincial and municipal energy efficiency planning, policies, and related research, but also facilitate the implementation of government's energy efficiency policies and regulations.

5.1.4.9 Government Procurement Program

Modeled after the U.S. Federal Energy Management Agency (FEMA), a government energy efficiency procurement policy was announced by the NDRC and the MOF in December 2004. Initially, the new policy specified that the products on the energy efficiency procurement list should be given *preferential* consideration in procurement.[19] In August 2007, the procurement policy was made mandatory at all levels of government. The energy efficiency procurement list has been updated several times and expanded from 9 types of products in 2004 to 24 by the end of 2012.

Similarly, the MOF and the State Environmental Protection Administration (now Ministry of Environmental Protection [MEP]) initiated a green purchase policy in December 2006. In July 2008, the MOF and MEP made the use of green purchase list mandatory at all levels of government.[19] There are some overlaps of the energy efficiency government procurement list and the green purchase list with the latter covering a wider range of environmentally friendly products. It is stipulated that the products listed on both lists will be given *preferential* consideration compared to those only listed on one list.

5.2 Renewable Energy

5.2.1 Market Overview

The combination of ambitious renewable energy targets, favorable government policies and entrepreneurial acumen has already made China a global leader in renewable energy. During the 13th FYP period (2015–2020), development of renewable energy and new energy sources will continue to be the government's priority given its importance in environmental protection, combating climate change and sustainable development. According to the 13th FYP for Renewable Energy which is being formulated, renewable energy will play an important role in optimizing China's energy structure and revolutionizing China's energy production and consumption.[20] The status of renewable energy will also shift from what is now called complementary energy sources to alternative energy, displacing a significant portion of fossil fuels.[21]

At the end of 2010, the installed capacities for wind and solar power both exceeded the targets that the Chinese government set in the 11th FYP (2006–2010). 2012 was another pivotal year for renewable energy development in China. The installed renewable energy reached 313 GW, up 11% from the previous year. This includes 248.9 GW of hydropower, 60.8 GW of grid-connected wind, and 3.3 GW of grid-connected solar power.[22] Most significantly, in 2012, wind power replaced nuclear power as the third largest energy source in China, after thermal and hydropower, and accounted for 2% of total energy power generation.[23] By the end of 2014, the installed capacity for renewable energy reached 430 GW, accounting for 32% of the country's total power capacity. The electricity generated by renewable energy reached 1.2 trillion kilowatt-hours, accounting for 22% of the total electricity generated during the same period.[24]

5.2.1.1 Wind Power

China ranks second in terms of wind energy resources, only next to the United States.[25] According to the findings of the wind energy resource survey and evaluation, China has 2380 and 200 GW land-based and offshore wind power potential at a height of 50 m.[26] The combined exploitable capacity is larger than that of hydropower.[1] While China is a latecomer to wind power development and deployment compared with many western countries, the country has witnessed tremendous growth in wind power in the past ten years.

At the end of 2010, China surpassed the US to become the world leader in installed wind power capacity, with 16 GW newly added wind power capacity and a total of 41.8 GW wind capacity.[27] According to the Global Wind Energy Council, the US installed about 5 GW of new wind power capacity in 2010 and its total installed capacity was 40.2 GW.[27] In terms of total grid-connected wind capacity, China also surpassed the US in 2012 with 50 GW on-grid wind capacity installed.[28] By the end of February 2014, the cumulative grid connected installed capacity for wind power reached 100.04 GW, which makes China the first country to top the 100GW wind power capacity milestone.[29] With the rapid and continuous increase of wind power installed capacity, China has already positioned itself as the largest wind power market in the world.

China's wind industry has mainly benefited from the favorable market conditions that government policies and regulations helped to foster. After several years of rapid growth, the wind power industry has been transitioning from frenzied initial stage with the emphasis placed on the maximum installed capacity to a new development phase focused more on quality, safety, reliability and efficiency. Manufacturing overcapacity, together with intense market competition and government policy to hold back funding and approval of wind projects have combined to cut the profit margins of many wind manufacturers in China. The growth rate of newly added wind power installed capacity started to slow down in 2011, from 18.92 GW in 2010 to 12.96 GW in 2012. However the downward trend didn't last very long. Data for 2013 show the wind power industry has regained momentum with 16.08 GW of new wind capacity installed during the year. In 2014, the newly added wind capacity reached an all time high with 19.81 GW new wind capacity installed.[30]

One of the challenges of China's wind power development is grid connection. Due to their weak capacity, many local grids are unable to integrate all the electricity generated by wind power. Large-scale wind power integration also creates serious problems in terms of power grid dispatch, reactive power regulation, grid safety and power quality. Furthermore, since the geographical distribution of wind energy resources does not match the country's power load profile, long distance transmission of electricity becomes necessary. Yet this is again restricted by the limited transmission capacity of power grids.

Largely due to insufficient power grid infrastructure, a great amount of electricity generated by wind power has to be discarded each year. National Energy Administration statistics show that more than 10 billion and 20 billion kWh of electricity generated from wind power was discarded in 2011 and 2012 respectively.[31] The grid connection problem not only needs a technical fix, but also calls for a fundamental change in China's overall power system. After the business of power generation was split from grid companies, which now only focused on power transmission and distribution, grid companies have little motivation to integrate renewable energy into their grids because the integration of renewables will cost them more money due to the extra expenses incurred in the grid connection and electricity purchases.

With the implementation of favorable wind power policies, improvement of grid transmission capacity and reduced amount of wind, the average rate of abandoned electricity fueled by wind power has been decreasing over the past two years from 11% in 2013 to 8% in 2014, which is at the lowest level in recent years.[32] However, the rate climbed again to 18.6% in the first quarter of 2015.[33] The fluctuation of the rate shows wind power grid integration and absorption have been and will continue to be a big challenge for the development of wind power in China.

5.2.1.2 Solar Power

Since 2007, China has also become the largest PV products manufacturer in the world. Among the world's top 10 solar PV module suppliers, six are based in China.[34] About 90% of PV panels manufactured in China were exported to countries with more favorable incentives, such as North American and European countries. From 2002–2008, there were only a few demonstration solar PV projects in China.[35] This is largely due to the high cost of PV systems and the barriers of grid connection, a problem also faced by China's wind power industry.

The cutback of government subsidies and declining PV demand in the European countries had a great impact on China's PV manufacturers, which rely heavily on the foreign market. What weighed further on China's PV industry are the anti-dumping and countervailing duties against Chinese PV products exported to the US, followed by similar trade investigations launched in the European Union, Canada and India. To offset their export losses and absorb manufacturing overcapacity, the Chinese PV manufacturers quickly turned to the domestic market which was largely untapped at the time and emerging markets which witnessed a surge in the growth of renewable energy in the past few years. Overcapacity has also triggered a new round of merging and restructuring among manufacturers in the PV sector. The Chinese government sees this as an opportunity to eliminate the outdated production capacity and upgrade the country's PV industry. It also released a variety of supportive policies to further encourage the expansion of the domestic PV market.

According to research released in the NPD Solarbuzz quarterly report, the demand for PV panels from the Chinese end-market has already risen to 33% of global demand during the final quarter of 2012.[36] In 2013, China has already become the world's largest market for solar power, outstripping Germany, Japan and the US with the installation of 12 GW of new PV capacity. This represents a 232% increase in generation capacity.[37] In 2014, China added 10.6 GW of newly installed PV capacity, which is one fourth of the world newly added installed capacity that year.[38]

China also made remarkable progress in concentrated solar power (CSP) generation in the past few years. According to the 12th Five Year Plan for Solar Power Generation Development, the installed capacity for CSP will reach 1 GW and 3 GW by 2015 and 2020.[39] In 2010, a 50 MW CSP commercial project in Inner Mongolia was launched through a public tending program. However, as of the time of writing, the project hasn't been constructed. It was not until July 2013 that the first CSP plant was connected to the power grid and began to generate electricity in Qinghai province. With a 50 MW installed capacity, the plant is expected to generate 112.5 million kWh electricity, equivalent to reduction of 394,000 tons of standard coal and about 103,000 tons of carbon dioxide.[40] The CSP plant symbolizes the transforming of CSP projects from a small-scale technology demonstration to a large-scale commercial project.

Compared with other forms of solar power utilization, CSP generation has lagged far behind. This is largely due to the lack of clear and robust government support for CSP. First, there's no fixed feed-in tariff (FIT) for CSP generation, and the FIT for a CSP project is decided on a case by case basis. In September 2014, the FIT was set at RMB 1.2/kWh for the 50 MW CSP plant in Qinghai. Secondly, the approval process of a CSP project can be very lengthy and sucks up a large amount of time and money from CSP project investors.[41]

Solar heat utilization has seen rapid development in China with favorable central government policies and incentives from provincial governments. Major forms of solar heat utilization include solar water heaters, centralized solar hot water supply, solar heating and cooling, industrial application of medium and high temperature solar energy, and solar cookers and solar houses, especially in remote and rural areas.[1] Among all, the solar water heater is the most widely used and commercialized form of solar heat utilization. After years of rapid development, China has already become the largest solar water heater manufacturer in the world with a complete value chain of solar water heater production. In 2012, China's solar water heater production has reached 63.9 million m^2 at a growth rate of approximately 10.7%, with total inventory of roughly 258 million m^2.[42] Due to overcapacity and shrinking demand, the solar water heater industry has also suffered a major decline since 2012. In 2014, solar heater production declined 17.6% from a year ago, which is the first negative growth in 17 years.[43]

5.2.1.3 Biomass

China has abundant biomass resources with an estimated annual amount of more than one billion tce. This is more than 1.3 times the country's annual energy consumption.[44] According to a report released by the Chinese Academy of Science on China's renewable energy development strategy, the capacity of biomass energy resources is twice as much as that of hydropower and 3.5 times that of wind power.[45] While biomass is a prioritized renewable energy together with wind and solar, the Chinese government failed to meet some of the targets for biomass energy set in the 11th FYP of Renewable Energy, such as the targets for methane utilization, nonfood fuel ethanol, biomass pellets and briquettes.[46]

There are many hurdles for large-scale utilization of biomass energy, one of which is the nature of biomass energy, such as low energy density and nonuniform consistency. The principal biomass feedstocks in China are wastes and residues from agriculture and forest industries; animal manure from medium- and large-scale livestock farms, and municipal solid waste.[44] So far, there's still no nationwide survey on the quantity, exploitable capacity and distribution of biomass energy resources. The feedstock of biomass energy is mainly collected manually with small-scale machinery. The collection, transportation and storage of feedstock of biomass energy are extremely inefficient. In addition, the technology of biomass utilization and equipment manufacturing remains a bottleneck for developing biomass energy such as the technologies of biomass gasification and second-generation fuel ethanol.

5.2.2 Renewable Energy Policies

5.2.2.1 General Policies

Renewable energy was first incorporated into the legislation list as early as 2003. One of the milestones of renewable energy is the Renewable Energy Law that came into force in 2006. The Law not only identifies the strategic role of renewable energy in energy security, environmental protection and sustainable development, but also serves as a framework for renewable energy related government work including resources investigation, target setting, planning, pricing and cost share, fiscal, financial and tax incentive mechanisms for renewable energy. After the Renewable Energy Law took effect, more than a dozen regulations were enacted to help enforce the law. Some of the provisions of the law were also amended in 2009. The majority of these renewable energy related policies are related to wind power and solar PV generation, whereas only a few are focused on bioenergy, geothermal and ocean energies.

The Renewable Energy Law stipulates that China's NRDC is in charge of overall planning of renewable energy development and energy pricing; the China Standardization

Administration manages technical standards and codes related to renewable energy; the MOF is responsible for fiscal and financial incentive mechanisms, like tax breaks and subsidized loans.[47] Established in 2008, the National Energy Administration is under the jurisdiction of NDRC and responsible for drafting and implementing energy development strategies, plans and policies, advising on energy and regulating sector.[48]

According to the Law and regulations, grid companies are mandated to sign grid integration agreements with eligible renewable energy power generation enterprises and to purchase the electricity that is generated. All the extra costs that grid companies paid to integrate and purchase the electricity generated by renewable energy compared to traditional power (also known as renewable energy surcharge or premium) need to be shared by end users of electricity nationwide. The renewable energy surcharge was first set at RMB 0.002/kWh under the Law and was later raised to RMB 0.004/kWh in 2009.

A renewable energy development fund was established under the Renewable Energy Law to promote renewable energy development and utilization. At first, the source of the fund is solely government allocations, but according to a 2009 amendment to the Law, part of the fund will come from the renewable surcharge paid by all electricity users around the country. With the rapid development of renewable energy installed capacity, the shortfall in funding the renewable energy projects has increasingly become a prominent problem. Statistics show the shortfall of the renewable energy development fund reached RMB 1.3 billion in 2009 and 2 billion in 2010.[49] By the end of 2014, the shortfall has expanded to RMB 14 billion. To make up for the shortfall, the renewable energy surcharge was raised twice to finance the fund from RBM 0.004/kWh to RMB 0.008 /kWh in 2010 and further to RMB 0.015 in 2013.[50]

In September 2007, more than eighteen months after China's Renewable Energy Law called for the establishment of overarching renewable energy targets, the NDRC released the Medium and Long-Term Plan for Renewable Energy Development. It was the first time the Chinese government set explicit quantified targets for renewable energy consumption. The plan established a national renewable energy target (including hydropower) of 15% of total primary energy consumption by 2020.[51] A renewable energy quota system was also introduced in the Plan. In the service range of large-scale power grid, nonhydro renewable energy power generation will surpass 3% by 2020. Power generators with self owned installed capacity over 5 GW are required to have nonhydro renewable energy installed capacity accounting for 8% of total self owned capacity by 2020.[52]

As of April 2015, a new version of the renewable energy quota system has been approved by NDRC and is now under the review of the State Council. According to this version, the renewable energy targets, which are categorized as basic targets and advanced targets, will be first broken down by province and then by city and county within a province.[53] Provincial governments and grid companies are responsible for meeting the targets of renewable energy, whereas local governments will exercise their management responsibility. Instead of setting specific and binding renewable energy targets, the new version allows provincial and local governments more flexibility to achieve their targets.[52] Furthermore, governments that are unable to meet their renewable energy targets will be punished, whereas those that meet the targets ahead of time will be rewarded. In addition to the renewable energy quota system at the national level, a few provinces, such as Hubei and Inner Mongolia, also released their own renewable energy quota systems.

In addition to large-scale renewable energy deployment, China has also been promoting distributed utilization of renewables based on the principle "self-generation, self consumption and feeding the surplus into the grid." This effort includes promoting distributed renewable energy technologies, standards and demonstrations, providing subsidies and tax incentives for distributed renewable energy projects.

To further drive comprehensive renewable energy use in urban and rural development, China plans to build 100 new energy demonstration cities and appoint 200 green energy counties according to the 12th FYP for Renewable Energy Development. One of the requirements for a new energy demonstration city is that by 2020, renewable energy will be no less than 6% of the candidate city's total primary energy consumption. According to the plan, China will also deploy 30 new energy microgrid demonstration projects as a way to explore the economic and technological feasibility of renewable energy-powered micrograms.

5.2.2.2 Wind Power Specific Policies

Wind power is the most commercially available nonhydro renewable energy in China. According to the 12th FYP for Renewable Energy Development released in 2012, China will have a total of 10 GW grid-connected wind installed capacity by 2015, and the annual wind power generation is expected to amount to 190 billion kWh by that time.[54] By the first quarter of 2015, the target for land-based wind installed capacity was achieved with a total installed capacity up to 10.1 GW.[55]

From 2003 to 2009, China mainly implemented a wind power concession program to promote large- scale deployment of wind projects. Under the program, wind power projects with installed capacity smaller than 50 MW is under the jurisdiction of provincial government, and the price of electricity generated from projects was determined through a competitive bidding process. Wind projects larger than 50 MW need to be approved by the central government, NDRC.

Under the Power Purchase Agreement (PPA) signed by a power grid company and a wind power project owner, the former is required to purchase all the electricity generated by the latter. A two-phase electricity rate is applied during the effective period of the PPA. Before the cumulative electricity production of wind power is equivalent to 30,000 full load hours, the electricity rate is the bid price specified in the PPA. Thereafter, the rate is the average price on the power market at the time.[56]

The wind power concession program not only plays an important role in wind power development from demonstration projects to scale up construction, but also helps the government to determine an appropriate FIT for land-based wind power. The price of nonconcession wind projects during the period was either based on the bidding results in the same area or required to be approved by the provincial government on a project by project basis.

In effect since August 2009, a wind power FIT policy for land-based projects was introduced in China. It divided China's territory into four regions with different FITs specified based on wind resources and construction conditions. The tariffs were first set per kilowatt hour at RMB 0.51, RMB 0.54, RMB 0.58 and RMB 0.61.[57] In January 2015, the FIT for the first three categories were lowered to RMB 0.49, 0.52 and 0.56.[58]

As of the end of 2014, China ranks fifth among the world's top offshore wind installation countries. After China's first offshore wind demonstration project was successfully completed and connected to grid in Shanghai, China launched the first concession program for an offshore wind project in China's southeast coastal province of Jiangsu in 2010. A few other provinces have also started working on offshore wind power development plans since then. By the end of 2014, China installed 229.3 MW new offshore wind capacity and the total installed capacity from offshore wind has amounted to 657.9 MW.[59] According to the 12th FYP for Wind Power Development, the installed capacity of offshore wind power will reach 5 GW by 2015 and 30 GW by 2020.[60] With less than one year left and 657.9 MW total offshore installed capacity so far, China is unlikely to meet its target of 5 GW offshore installed capacity by 2015 (Table 5.2).

TABLE 5.2
Targets of Wind Power Development by 2015 and 2020

Types of Targets	Main Targets	2015	2020
Installed Capacity	Land-based wind power	99 GW	170 GW
	Offshore wind power	5 GW	30 GW
	Total Installed Capacity	100.4 GW	200 GW
Generating Capacity	Electricity Generated by Wind Power	190 Billion kWh	390 billion kWh
	Percentage of Electricity Generated by Wind power	3%	5%

Source: Energy Research Observer Net. (2012) The 12th FYP for Wind Power Development. September 17. Retrieved April 30, 2015 from http://www.chinaero.com.cn/zcfg/xny/09/127069_4.shtml

In June 2014, NDRC also released the FIT scheme for offshore wind projects commissioned before 2017, which is RMB 0.75/kWh for intertidal and RMB 0.85/kWh for offshore wind projects.[61] Later that year, the NEA issued a national offshore wind power project construction plan to boost the development of offshore wind power. According to the plan, 44 offshore wind projects will be constructed from 2014 to 2016 with a total installed capacity of 10.53 GW.[62]

In addition to the FIT for land-based and offshore wind power, wind power generation enterprises can also enjoy a series of tax incentives such as 50% of value-added tax (VAT) return, exemption from corporate income tax during the first three years after the company earns profit from sales; and a 50% reduction of income tax during the 4–6 years after it earns a profit. Wind power equipment manufacturers can also enjoy some tax exemption or cut on import duty of certain key parts and raw materials of wind turbines. Complementing these favorable policies and regulations at the central government level, local governments also released a series of favorable policies on land use and taxation to encourage wind equipment manufacturers and wind project developers to invest in wind projects.

To address the grid integration and safety issues, the Chinese government has tightened the wind power project approval procedure from 2011, requiring all new wind power projects to be approved by the NDRC. Projects only approved by the provincial government will not be allowed to be connected to the power grid or enjoy renewable energy subsidies. In addition, projects in regions where over 20% of electricity generated by wind power is discarded due to limited integration and transmission capacity will no longer be approved. In May 2013, the NEA transferred authority over wind power project approval from NDRC to provincial or local governments.[63] However, a wind project still needs to be listed in the NEA project approval plan in order to receive government renewable energy subsidies.[64]

5.2.2.3 Solar Power Specific Policies

Solar power is the third most commercially viable renewable energy following hydropower and wind. According to the 12th FYP for Solar Power Development issued by the NEA under the NDRC, the target for installed solar power is 21 GW by 2015, among which 10 GW is distributed PV systems, 10 GW is grid connected PV and 1 GW is CSP system.[65] The plan also sets the target of installed solar power generation at 50 GW by 2020. The target of installed solar power by 2015 has been raised several times and the target was raised to 35 GW in 2013. In addition, the 12th FYP also set the total solar heat collection area at 400 m^2 by 2015 (Table 5.3).

Unlike wind power, early policies treated solar power more as one of the supplementary solutions to provide electricity to the remote and rural areas. This mindset was dominant among China's policy makers for quite a long time. It partly explained the reason why China prioritized wind over solar power up to 2008.

TABLE 5.3

China's Targets for Solar Power Installed Capacity by 2015 and 2020

Installed Capacity	2015	2020
Grid-connected PV	10 GW	20 GW
Distributed PV	10 GW	27 GW
Concentrated Solar Power (CSP)	1 GW	3 GW
Total	21 GW[a]	50 GW

Source: National Energy Administration. (2012) The 12th FYP for Solar Power Generation Development. July 7. Retrieved April 30, 2015 from http://zfxxgk.nea.gov.cn/auto87/201209/P020120912536329466033.pdf

[a] This target was raised to 35 GW in 2013.

To promote large-scale PV application and determine an appropriate FIT for solar energy, the Chinese government also adopted a concession program for large-scale grid-connected PV projects. In 2009, China issued its first tender for two 10 MW utility scale solar power plants in Dunhuang, Gansu province. In 2010, the Chinese government initiated a second round of concession bids for 13 large-scale PV projects for a total of 280 MW. The first project was granted RMB 1.09/kWh for the power fed into the grid. In the second round, the winning bids ranged from RBM 0.728/kWh to RMB 0.991/kWh. Largely due to the advocacy of PV manufacturers and local government, the FIT for solar power was established in July 2011. Under the scheme, projects approved for construction prior to July 1, 2011 is eligible for the FIT of RMB 1.15. Projects that finished construction and began the process of commencing generation prior to December 31, 2011 are also eligible for the FIT.[66] Later in 2011, the FIT was lowered to RMB 1/kWh. In addition to these concession projects, there are also other grid-connected PV projects implemented during the same period, and their power price is set by the government on a project by project basis.

In March 2012, a new solar FIT scheme for utility PV ground power plants was released for comments. The new solar power FITs were set at RMB 0.75/kWh, 0.85/kWh, 0.95/kWh and 1/kWh for each of four regions categorized based on different solar radiation level and construction conditions. In August 2013, NRDC announced a new FIT policy for solar power that was set at RMB 0.9/kWh, 0.95/kWh and 1/kWh for each of three types of regions.[67]

To address the overcapacity of solar power production while preventing further grid integration issues, the government has implemented robust subsidy programs for deployment of distributed PV. The Golden Sun program, initiated by the MOF, the Ministry of Science and Technology and the NEA in 2009, provides capital subsidies to electricity end users for solar PV installations. The Golden Sun program subsidies were granted before the construction of a project and there was very limited monitoring of an approved project's power generation. While the government tried hard to improve the implementation of the program, controversies around the program still exist. In March 2013, the government announced that the Golden Sun program would no longer accept applications. Later that year, the Golden Sun program was officially ended. Statistics show four batches of PV generation projects with a total 60 GW installed capacity were approved under the program. However, there is no public information on the status quo of the projects, the amount of electricity generated, and the total amount of subsidies received by these projects.

As a separate program, the Ministries of Finance and Housing and Urban–Rural Development (MOHURD) are providing subsidies to promote application of rooftop PV and building-integrated PV (BIPV) systems. Like Golden Sun program, this program is also one-time subsidy provided before the construction of the project. Differently, the Golden Roof program only focuses on BIPV whereas the Golden Sun program provides subsidies to both rooftop PV and BIPV. The two programs used to be one program jointly issued by the MOF, Ministry of Science and Technology, MOHURD and NEA, but was announced and implemented separately due to different views of MOHURD and NEA on distributed PV policies.[68]

For distributed PV projects, a subsidy of RMB 0.35/kWh was first introduced in 2012. In August 2013, NDRC raised the subsidy to RMB 0.42/kWh and made the eligibility period of the subsidy as long as 20 years.[65] In addition to the subsidy at the national level, provincial and local governments also provide additional subsidies to distributed PV projects in their jurisdictions.

During 2014, the Chinese government further issued a series of favorable policies to accelerate the growth of distributed PV. One of the most noteworthy policies is that rooftop distributed PV project owners can choose to sell either excess electricity or all electricity generated to the grid and that the electricity price will be the same as the local FIT for ground PV power plant. In the past, a distributed PV project owner could only sell excess electricity to a power grid after self consumption.[69] By guaranteeing all electricity purchased by the power grid and a fixed price paid to the project owner, the policy shift guarantees the stable revenue of a distributed PV project and thus is welcomed by the distributed PV project developers around the country.

At present, distributed PV projects mainly consist of PV power plants for residential use, rooftop PV projects of public facilities, and industrial and commercial buildings.[70] Due to reasons like difficulties in project financing, grid integration, and uncertainty in power load, distributed PV development has lagged far behind of that of large-scale ground PV power plants. Since China failed to meet its PV installed capacity targets in 2014 mainly because of distributed PV, NEA no longer sets a specific target for distributed PV for 2015.

Besides, a large scale distributed PV application demonstration area program was launched by the NEA in 2012, requiring provincial government to submit a plan on establishing pilot areas for distributed PV power generation. Following the NEA's efforts, the State Grid Corporation of China, the country's largest state-owned utility, released a series of documents to simplify the procedure of grid integration of residential distributed PV power generation no more than 6 MW. As of 2015 the NEA has announced 30 distributed PV application demonstration areas.[71]

5.2.2.4 Biomass Specific Policies

In China, biomass energy development is an integral part of the government's efforts to develop local economy, improve living conditions in rural areas, and protect the ecology. In the wake of the food crisis, the government issued strict rules to make sure that biofuel development (including fuel ethanol and biodiesel) does not compete with crops intended for human consumption and that the land for developing feedstock should not compete with land for crop production. The government also encourages experimentation with alternative crops such as sweet sorghum and cassava for new ethanol plants.[72]

According to the Chinese government's 12th FYP for Biomass Energy, installed capacity of biomass energy will reach 130 GW by 2015 with annual biomass power generation amounting to 78 billion kWh. The annual production of biogas will reach 22 billion m^3. In terms of various forms of biofuel, China's fuel ethanol utilization capacity will reach 4 million ton and the capacity of biodiesel and aviation biofuel will reach 1 million ton by 2015. In addition, the government aims at achieving full commercialization and large-scale utilization of biomass energy in its power sector, heat supply, and rural life. In the transportation sector, a larger portion of biomass energy is expected to replace fossil fuel by 2015.[73]

The FIT for biomass was first set at RMB 0.25/kWh premium plus the province-specific coal power generation price. The premium was then increased to RMB 0.35/kWh for biomass. The FIT for biomass energy was later raised to RMB 0.75/kWh in 2010 by NDRC. Like solar and wind, biomass energy can also enjoy a series of tax incentives. For instance, the electricity or heat generated from waste is eligible to receive tax rebates at the time the VAT is levied. The income tax for a biomass energy enterprise is calculated using 90% of the enterprise's total sales income as the base. Qualified biomass enterprises can enjoy a corporate income tax exemption during the first three years after it earns profit and enjoy a 50% cut of corporate income tax during the following three years.

References

1. Xinhua Net. (2012, October) Full text: China's energy policy 2012. Retrieved April 30, 2015 from http://news.xinhuanet.com/english/china/2012-10/24/c_131927649.html.
2. Dai, Y. et al. (2012) *China Energy Efficiency Financing and Investment Report, 2010*. China Science and Technology Press, Beijing, China.
3. Institute for Industrial Productivity. Energy intensity and carbon intensity targets during the 12th FYP. Industrial Energy Efficiency Policy Database. Retrieved May 10, 2015 from http://iepd.iipnetwork.org/policy/energy-and-carbon-intensity-targets-12th-five-year-plan.
4. PRC Central Government. (2011) Comprehensive Work Plan on 12th FYP Energy Efficiency and Emission Control. NDRC.Retrieved on May 10, 2015 from http://www.gov.cn/zwgk/2011-09/07/content_1941731.html.
5. PRC Central Government. (2014) National industry energy efficiency guide, Ministry of Industry and Information Technology. Retrieved April 15, 2015 from http://www.miit.gov.cn/n11293472/n11293832/n12843926/n13917012/16325136.html.
6. Institute for Industrial Productivity. (2013) Industrial Energy Efficiency Program Database: Top 10,000 Energy Consuming Enterprises Program. Retrieved on March 10, 2013 from http://iepd.iipnetwork.org/policy/top-10000-energy-consuming-enterprises-program.
7. PRC Central Government. (2012) 12th FYP for Building Energy Efficiency. Ministry of Housing and Rural Urban Development (May 9, 2012). Retrieved on May 10, 2015 from http://www.gov.cn/zwgk/2012-05/31/content_2149889.html.
8. Shui, B. and Li, J. (July 2012) Building energy efficiency policies in China: Status report. Retrievedon May 10, 2015 from http://wenku.baidu.com/view/701eef46852458fb770b56d7.html.
9. PRC Central Government. (2013) Green Building Work Plan (January 2013). General Office of State Council. Retrieved on May 10, 2015 from http://www.mohurd.gov.cn/zcfg/gwywj/201302/t20130204_212772.html.
10. PRC Central Government. (2011) 12th FYP for Transportation Energy Efficiency. Ministry of Transport, July. .Retrieved on March 10, 2013 from http://www.moc.gov.cn/zhuzhan/zhengcejiedu/guihuajiedu/shierwuguihuaJD/xiangguanzhengcefagui/201110/t20111010_1064457.html.
11. China News Agency. (2013) NDRC: 45 National standards released to promote energy efficiency standard initiative. February 21. Retrieved on March 10, 2013 from http://news.ifeng.com/mainland/detail_2013_02/21/22354433_0.shtml.
12. PRC Central Government. (2015) Conservation standardization. Retrieved on April 15, 2015 from http://www.gov.cn/zhengce/content/2015-04/04/content_9575.html.
13. Ask CI. (2012) The announcement of the 10th batch of products covered by energy efficiency label system. Nov. 11. Retrieved on March 10, 2013 from http://www.askci.com/news/201211/28/171149_21.shtml.

14. PRC Central Government. (2014) Energy efficiency top runner implementation scheme. National Development and Reform Commission. Retrieved on April 15, 2015 from http://www.ndrc.gov.cn/zcfb/zcfbtz/201501/t20150108_659703.html.
15. Gong, X. (July 11, 2011) Rewards for energy saving technical retrofits projects expanded. China Economic Herald. Retrieved on March 10, 2013 from http://news.emca.cn/n/20110711020518.html.
16. ESCO Committee of China Energy Conservation Association. (2012) *Journal of ESCO Summit 2012*.
17. Chen, K. and Zifeng, X. (2010) Energy contracting performance in China. King& Wood Mallesons. Retrieved on March 10, 2013 from http://www.kingandwood.com/article.aspx?id=Energy-Performance-Contracting-in-China&language=en.
18. Institute for Industrial Productivity (2013) Industrial Energy Efficiency Program database: Demand side management pilot cities. March 11. Retrieved on May 10, 2015 from http://iepd.iipnetwork.org/policy/demand-side-management-pilot-cities.
19. Zhou, N., Levin, M.D., and Price, L. (2010) Overview of current energy efficiency policies in China. *Energy Policy* 38(8).
20. PRC Central Government. (2015) The guiding opinions on formulating the 13th FYP for Renewable Energy Development. National Energy Administration, April 28. Retrieved April 30, 2015 from http://zfxxgk.nea.gov.cn/auto87/201504/t20150428_1911.html.
21. Liang Z., Renewable energy's replacement of fossil fuels has become main trend, (2015) Sina Finance, March 15. Retrieved April 30, 2015 from http://finance.sina.com.cn/hy/20150415/104221960784.shtml.
22. China Industry Information Net. (2013) China's renewable energy power generation continues to grow. June 13. Retrieved April 30, 2015 from. http://www.chyxx.com/industry/201306/208785.html.
23. Chen, Y. (2013) Wind power now no. 3 energy source. *People's Daily Online*, January 28. Retrieved April 30, 2015 from http://en.people.cn/90778/8109836.html.
24. National Energy Administration. (2015) Fifth IRENA Conference, Retrieved April 30, 2015 from http://www.nea.gov.cn/2015-01/21/c_133936365.html.
25. Peng, Y. et al. (2011) China wind sector shaped by globalization. An Accenture Report. March. Retrieved April 30, 2015 from http://wenku.baidu.com/link?url=Oo2g-LBHJwP2HQdZsE-HwZu2ybq8J8I43zGX97UsRubPgEVb3eMsbXM9JycbGgLbVI2o2FoxntazHbiTDc8-FAznr-lwbWH5NjBKXv_5U_3u.
26. Liu, T., ed. (2012) *Report on China's Energy Development for 2011*. Beijing: Economic Science Press, 2012.
27. People's Daily Online. China surpassed US on wind power capacity. (2012) January 14. Retrieved April 30, 2015 from http://en.people.cn/90001/90778/7260646.html.
28. Hexun News. (2012) China becomes the world leader in grid-connected wind power installation. December 28. Retrieved April 30, 2015 from http://news.hexun.com/2012-12-28/149603222.html.
29. China Power News Net. (2015) Cumulative on-grid wind installed capacity tops 100 GW. CPNN, April 28. Retrieved April 30, 2015 from http://www.cpnn.com.cn/zdyw/201504/t20150428_797857.html.
30. Chen, W. (2012) China's new wind power installation hits all-time high. Xinhua News Agency, February 12. Retrieved April 30, 2015 from http://www.gov.cn/xinwen/2015-02/12/content_2818566.html.
31. Sina Finance. (2013) 10 Billion yuan lost due to discarded wind power. March 22. Retrieved April 30, 2015 from http://finance.sina.com.cn/chanjing/cyxw/20130322/021514914817.shtml.
32. National Energy Administration. (2015) Overview of 2014 wind power industry. February 12. Retrieved April 30, 2015 from http://www.nea.gov.cn/2015-02/12/c_133989991.html.
33. National Energy Administration. (2015) Overview of wind power grid integration and operation in Q1 2015. April 28. Retrieved April 30, 2015 from http://www.nea.gov.cn/2015-04/28/c_134190553.html.
34. National Energy Administration. (2015) Overview of China's PV industry in 2014. February, 15. Retrieved April 30, 2015 from http://www.nea.gov.cn/2015-02/15/c_133997454.html.

35. Qi, Y., (Ed.). (2013). *Annual Review of Low-Carbon Development in China*. Beijing, China: Social Sciences Academics Press (China).
36. Renewable Energy Focus. (2013) China consumes 33% of global PV panel shipment in Q4 2012. January 22. Retrieved April 30, 2015 from http://www.renewableenergyfocus.com/view/30372/npd-solarbuzz-china-consumed-33-of-global-pv-panel-shipments-in-q4-2012/.
37. Clover, I. (2014) China surpassed Germany as world solar bright spot. *PV Magazine*, August 11. Retrieved April 30, 2015 from http://www.pv-magazine.com/news/details/beitrag/china-surpasses-germany-as-worlds-solar-bright-spot_100016051/#axzz3ZQQSRCuI.
38. Environmental Protection Online. (2015) Distributed PV welcomes new development opportunities. March 19. Retrieved April 30, 2015 from http://www.hbzhan.com/news/detail/96040.html.
39. PRC Central Government. (2012) Announcement on Release of the 12th FYP for Solar Power Generation Development. September 13. Retrieved April 30, 2015 from http://www.gov.cn/zwgk/2012-09/13/content_2223540.html.
40. National Energy Administration. (2013) The first CSP plant connected to power grid began to generate electricity in Qinghai. July 15. Retrieved April 30, 2015 from http://www.nea.gov.cn/2013-07/15/c_132542216.html.
41. Wang, H. (2014) Why China's CPS industry isn't hot enough? *China Energy News*, October 10. Retrieved April 30, 2015 from http://www.chinapower.com.cn/newsarticle/1221/new1221271.asp.
42. Environmental XPRT. (2013) Solar water heater industry report in China and around the world, 2013. January 25. Retrieved April 30, 2015 from http://www.environmental-expert.com/news/solar-water-heater-industry-report-in-china-and-around-the-world-2013-351467/view-comments.
43. ChinaNews.com. (2015) Solar power heat utilization industry saw negative growth in 17 years. March 17. Retrieved April 30, 2015 from http://finance.chinanews.com/cj/2015/03-17/7136581.shtml.
44. Zhu, X. et al. (2003) Biomass pyrolysis and its potential for China. International Conference on Bioenergy Utilization and Environment Protection September 24–26, Dalian, China. Retrieved April 30, 2015 from http://www.bioenergy-lamnet.org/publications/source/chi2/Session3-ZhuXifeng-LAMNET-WS-Dalian-0309.pdf.
45. State Grid News. (2011) Survey results of China's biomass energy released. April 30. Retrieved April 30, 2015 from http://www.indaa.com.cn/pl2011/nycj/201105/t20110504_635571.html.
46. Wang, E. (2011) China failed to meet biomass targets in the 11th FYP. China Chemical Industry News, August 9. Retrieved on April 30, 2015 from http://www.ccin.com.cn/ccin/news/2011/08/09/193079.shtml.
47. China Wind Power Center. (2005) Framework Policies. February. Retrieved April 30, 2015 from http://www.cwpc.cn/cwpp/en/information/wind-power-policy/wind-power-policy/.
48. Xinhua. (2013) China to restructure National Energy Administration. March 10. Retrieved April 30, 2015 from http://news.xinhuanet.com/english/china/2013-03/10/c_132221775.html.
49. Wang, L. (2013) Renewable energy surcharge to be doubling. Tecent Finance, June 26. Retrieved April 30, 2015 from http://finance.qq.com/a/20130626/000549.html.
50. Xinhua. (2011) Renewable Energy Surcharge Doubled. *People's Daily*, December 21. Retrieved April 30, 2015 from http://news.xinhuanet.com/fortune/2011-12/21/c_122457007.html.
51. PRC Central Government. (2007) Medium and long-term plan for renewable energy development. August. p. 18. Retrieved April 30, 2015 from http://www.ccchina.gov.cn/WebSite/CCChina/UpFile/2007/20079583745145.pdf.
52. PRC Central Government. (2007) Medium and long-term plan for renewable energy development. August. p. 30. Retrieved April 30, 2015 from http://www.ccchina.gov.cn/WebSite/CCChina/UpFile/2007/20079583745145.pdf.
53. Chen, S. (2015) National renewable energy quota system expected to be approved. *China Energy News*, April 22. Retrieved April 30, 2015 from http://guangfu.bjx.com.cn/news/20150422/610590.shtml.

54. National Energy Administration. (2012) The 12th FYP for Renewable Energy Development released. August 8. Retrieved April 30, 2015 from http://www.nea.gov.cn/2012-08/08/c_131767651.html.
55. National Energy Administration. (2015) Overview of wind power grid integration and operation. April 28. Retrieved April 30, 2015 from http://www.nea.gov.cn/2015-04/28/c_134190553.html.
56. Shi, P. (2006) Wind power concession projects and the issue of price. Greenpeace Research Team on Wind Power Price, October 20. Retrieved April 30, 2015 from http://www.docin.com/p-35027372.html.
57. China Wind Power Center. National Policy. Retrieved on March 10, 2013 from http://www.cwpc.cn/cwpc/en/node/6548#4o.
58. CNStock. (2015) NDRC lowers wind power FIT. January 7. Retrieved April 30, 2015 from http://www.cnstock.com/v_news/sns_bwkx/201501/3302627.html.
59. BJX Wind Power Generation Network. (2015) 2014 Wind power installed capacity statistics. March 20. Retrieved April 30, 2015 from http://news.bjx.com.cn/html/20150320/599956-3.shtml.
60. Energy Research Observer Net. (2012) Full Text: The 12th FYP for Wind Power Development. September 17. Retrieved April 30, 2015 from http://www.chinaero.com.cn/zcfg/xny/09/127069_4.shtml.
61. National Energy Administration. (2014) NDRC issued FIT for offshore wind power. June 20. Retrieved April 30, 2015 from http://www.sdpc.gov.cn/gzdt/201406/t20140619_615709.html.
62. Xinhua. (2015) 44 Offshore wind projects listed in the development plan. December 12. Retrieved April 30, 2013 from http://news.xinhuanet.com/fortune/2014-12/12/c_1113625576.html.
63. National Energy Administration. (2013) A batch of energy related administrative approval items removed or delegated. May 22. Retrieved April 30, 2015 from http://www.nea.gov.cn/2013-05/22/c_132399560.html.
64. OFweek. (2014) A wind project developer's recount—A "hard" road of project approval. July 28. Retrieved April 30, 2015 from http://windpower.ofweek.com/2014-07/ART-330002-8610-28858133_2.html.
65. PRC Central Government. (2012) Announcement of releasing the 12th FYP for Solar Power Generation Development. Retrieved April 30, 2015 from http://www.gov.cn/zwgk/2012-09/13/content_2223540.html.
66. PRC Central Government. (2011) NDRC's announcement on improving the FIT policy for solar power. July 24. Retrieved April 30, 2015 from http://www.gov.cn/zwgk/2011-08/01/content_1917358.html.
67. PRC Central Government. (2013) NDRC's announcement on promoting the healthy development of solar power. August 26. Retrieved on April 30, 2015 http://www.sdpc.gov.cn/zwfwzx/zfdj/jggg/dian/201308/t20130830_556127.html.
68. Ding, A. (2013) The golden roof vs. the golden sun. January 21. Retrieved April 30, 2015 from http://www.qlweekly.com/Finance/Expression/201301/217350.html.
69. National Energy Administration. (2014) Announcement on implementation of distributed pv related policies. September 2. Retrieved April 30, 2015 http://zfxxgk.nea.gov.cn/auto87/201409/t20140904_1837.html.
70. Environmental Protection Online. (2015) Distributed PV welcomes new development opportunities. March 19., Retrieved April 30, 2015 from http://www.hbzhan.com/news/detail/96040.html.
71. National Energy Administration. (2014) Announcement on promotion of distributed PV application demonstration areas. November 21. Retrieved April 30, 2015 from http://zfxxgk.nea.gov.cn/auto87/201412/t20141224_1874.html.
72. USDA. (2011) 2011 Biofuel annual report of China. July 21. Retrieved April 30, 2015 from. http://gain.fas.usda.gov/Recent%20GAIN%20Publications/Biofuels%20Annual_Beijing_China%20-%20Peoples%20Republic%20of_7-21-2011.pdf.
73. National Energy Administration. (2012) The 12th FYP for Biomass Energy Development. July 24. Retrieved April 30, 2015 from http://zfxxgk.nea.gov.cn/auto87/201212/t20121228_1568.html.

6

Renewable Energy and Energy Efficiency in India

Deepak Gupta and P.C. Maithani

CONTENTS

6.1 Background

India has traditionally been a low energy and electricity consumption country, notwithstanding its large and fast growing population. This was partly because of the so-called Hindu rate of growth during the several decades after independence. The economic policy changes undertaken since early 1990s with greater participation of private sector and deregulation of infrastructure and industrial sectors resulted in higher growth rates of the Indian economy, which grew in the last decade at an annual rate of about 8%* although it has declined progressively to about 5% in the last 2–3 years because of the

* From the Economic Surveys of the period, published by the Ministry of Finance, Government of India every year.

global recession. But the chances are that growth may continue to be between 7% and 9% in the long term. This process is being accompanied by large-scale urbanization, growth of the middle class, and changing life styles, leading to a huge growth in buildings and cars and sale of electrical appliances topped by air conditioners, fridges, geysers, and microwaves. Necessarily therefore, there has been, and will continue to be, significant increases in India's energy and electricity demand. In 2011, India was the fourth largest energy consumer in the world after the United States, China, and Russia.* India has been ranked by the IMF as the world's tenth largest economy and the third largest in terms of purchasing power parity. India will continue to climb up this ladder. An Indian Planning Commission study in 2006 estimated that based on 8%–9% growth, primary energy demand may go up 4–5 times by 2031–2032, while electricity generation requirements may go up by 6–7 times. It estimated that electricity-generating capacity may have to go up to 7–800 GW.†

Energy services have always remained in focus of successive Indian governments that has resulted in the expansion of the energy infrastructure within the country and steady expansion in total energy use. Commercial energy use increased 21 times and the power generation capacity went up by 100 times during the past 60 years. In 2012, the total commercial energy supply was 537 million tons of oil equivalent (mtoe) and involved coal, oil, gas, and electricity generated from nuclear, hydroelectric, and renewable sources.‡ These figures do not include the energy that is consumed from traditional sources by 56% of Indian households.§ Estimates of energy use from traditional sources tend to be approximate, but figures indicate that in 2012, 174 mtoe of energy came from such sources as fuel wood, dung, crop residue, biogas, and waste.¶ India's energy intensity has been declining over the years. From 1.09 kgoe per U.S. dollar, it has reduced to 0.62 kgoe per U.S. dollar in 2011.**

Nevertheless, the growth of energy sector could not and has not been able to match the growth in economy in spite of impressive progress in the last decade. The energy sector continues to be viewed as an important bottleneck to India's industrial growth that in turn is seen as critical to stimulating the country's economic and social development. While supporting around 17.8% of the world population, India's share in world energy use and electricity consumption is only 5.7% and 4.0%, respectively.†† The per-capita energy use at around 0.60 toe is far below that of industrialized countries, and, more importantly, is almost only a third of the world average. The situation in per-capita electricity consumption is even worse with a per-capita annual consumption of only 710 kWh, which is around a fourth of the world average.‡‡ In fact, in the event of achieving the capacities of 7–800 GW, the per-capita primary energy consumption will rise only to almost 1.25 toe, which would still be much lower than the current world average of about 1.88 toe/capita/annum.§§ Electricity deficit levels have remained consistently high

---

* US Energy Information Administration 2013.
† Planning Commission of India (2006). Integrated energy policy report; Available online: http://planningcommission.nic.in/reports/genrep/rep_intengy.pdf' last accessed on 22 March 2015.
‡ Planning Commission of India (2013): 12th Five Year Plan document (Vol 2) page 133. available online: http://planningcommission.gov.in/plans/planrel/12thplan/pdf/12fyp_vol2.pdf, last accessed on 22 March 2015.
§ Census of India 2011.
¶ Planning Commission of India (2013): 12th Five Year Plan document (Vol 2 p. 133).
** Planning Commission of India (2013): 12th Five Year Plan document (Vol 2 p. 130).
†† IEA Key World Energy Statistics 2013.
‡‡ IEA Key World Energy Statistics 2013.
§§ IEA Key World Energy Statistics 2013.

in recent years with supply trailing requirement by an estimated 8%–10%.* And this may not even be reflecting the total demand. Another major challenge continues to be providing access to modern energy sources to a large proportion of the country's population. Around 45% of rural households still rely on kerosene for meeting their lighting requirements.† Further, around 86% of rural households and more than 20% of the urban households still rely primarily on traditional fuels such as firewood, wood chips, or dung cakes to meet their cooking needs.

6.2 Growth of Electricity Capacity

The Indian power sector has grown significantly since 1947. The power-generating capacity has increased from 1.3 GW in 1947 to over 261 GW by January 2015.‡ However, despite significant growth in electricity generation over the years, shortage of power continues to exist primarily on account of growth in demand for power outstripping the growth in generation and capacity additions in power generation. The average energy deficit was 9.1% and the average peak power deficit was 12.8% between 2003 and 2012.§

The Electricity Act 1910 was the first act that was introduced to govern the Indian power sector. The Electricity (Supply) Act 1948 was introduced after independence, but it did not achieve the desired results. Following the liberalization and reform of the economy in 1991–1992, the electricity sector too witnessed major policy and regulatory initiatives. A regulatory framework was set up with independent regulators in the center and states recognizing that electricity and other infrastructure sectors required substantial investments in the face of resource constraints, investment by the private sector (including foreign capital) was allowed in electricity generation. Now 100% foreign direct investment (FDI) is allowed in generation, transmission, and distribution segments. The most important among all the policies announced by the government was the enactment of the Electricity Act 2003. It opened up the power generation sector and encouraged greater private participation. It unbundled the State Electricity Boards; separated generation, transmission, and distribution; and introduced open access. Over the past few years, the government of India undertook several further measures like the National Tariff Policy, National Electricity Plan, Competitive Bidding Guidelines, and Ultra Mega Power Projects. Incentives were also given to the sector through waiver of duties on capital equipment under the mega power policy. These all resulted in a surge in creation of capacity. A capacity of about 46 GW of thermal power was created between 2006 and 2012. However, this growth in capacity should be seen in the context of what has happened in China. The total capacity of China went up from 725 GW to about 1200 GW from 2007 to 2012, respectively.¶ The growth in power capacity over the years is given in Table 6.1.

Notwithstanding this recent growth, many problems are being faced, which may make it difficult to achieve the capacity that has been estimated to be required by 2032. The problems are caused by forest and environmental and logistical issues for coal. Problems related to the liability law and increasing people's opposition after the Fukushima incident

* Central Electricity Authority, All India Electricity Statistics, General Review 2012.

† Census of India 2011.

‡ Ministry of Power, Government of India.

§ Central Electricity Authority, All India Electricity Statistics 2012.

¶ U.S. Energy Information Administration.

TABLE 6.1
Growth in Electric Installed Capacity (in MW)

	Thermal	Large Hydro	Nuclear	Gas	Renewables	Total
1950	1,153	560	0	0	0	1,713
1990	41,421	18,307	1,565	2,343	0	63,636
2002	63,266	26,269	2,720	11,163	1,628	105,046
2007	72,323	34,654	3,900	13,692	7,760	132,329
2012	113,564	38,990	4,780	18,039	24,504	199,877
2017 (estimate)	183,364	49,887	10,080	20,579	54,503	318,414

Source: Central Electricity Authority, Bangalore, India (2002–2017 are 5-year plan periods).

in Japan suggest that nuclear may increase slowly perhaps reaching about 25–30 GW by 2032, instead of the targeted 60 MW. Gas reserves are not much, exploitation of new fields has been delayed, and production from existing fields has declined, leaving much of newly created gas power capacity stranded. Oil is facing similar problems. Large hydro was forecast to reach 150 GW, but there have been environmental and logistical problems to develop this potential in the Himalayas. Even reaching 100 GW, for which we must make our utmost efforts, may be a problem. With this scenario, meeting increasing demand will require other solutions apart from efforts to maximize use of all above sources.

The Power sector in India is also facing other problems and challenges apart from those of creating generating capacity in the future. First, irrespective of the capacity created, the supply availability of fuels is becoming a problem especially with respect to coal and gas. Second, these problems are necessitating higher imports of all fuels—gas, coal, and oil. Estimates including those made in the Integrated Energy Policy Report suggest that import dependence for energy in 2031–2032 could be as high as around 60%. There is great worry about the likely import dependency on oil of over 80%–90% and coal around 45%. Third, costs of all fuels are increasing. There is perhaps rise of 10%–15% in prices of coal because of imports, and the worry is that this trend may continue. Domestic prices for gas have also increased and imported prices will be more. The price per unit of electricity generated from gas is likely to become even higher than the current solar price. Rising capital costs of nuclear plants also suggest much higher generation costs in the future. Fourth, electricity tariffs have traditionally been lower than costs and have only been revised in the last couple of years in several states, though partially. Yet there is considerable political opposition to this and pressure to reduce. Rural tariffs have been lower still, and the rural irrigation power sector has been the lowest. This has resulted in most of the utilities being in very poor financial health. This restricts both investments in improving infrastructure as also in limiting purchases of power to meet normal demand. Fifth, the rising costs of imported fuels, caused by the increase in quantities has adversely impacted India's current account deficit. Huge subsidies for various oil products have caused serious problems for the revenue and fiscal deficits apart from creating other aberrations. The annual subsidy on kerosene alone is around U.S. $5 billion (though this is declining because of recent fall in oil prices). Domestic cooking gas has about 30%–40% subsidy.

It is clear from the aforementioned data that India's need for secure, affordable, and environmentally sustainable energy has become one of the principal economic and development challenges for the country. It is also clear that while energy conservation and demand management and energy efficiency will have important roles to play in the national energy strategy, renewable energy will become a key part of the solutions and is likely to play an increasingly important role for augmentation of grid power, providing energy access, reducing consumption of fossil fuels, and helping India pursue its low carbon developmental pathway.

6.3 New and Renewable Energy

After the oil shock of the seventies, there was a global recognition of the need to develop alternative energy technologies. Government of India created the Department of Non-Conventional Energy Sources (DNES) in 1982. The thrust areas identified included research, development, demonstration, and dissemination of renewable energy technologies for providing energy services in the rural areas and also for meeting energy needs of the country through renewable sources. Later in 1992, DNES was upgraded into an independent Ministry of Non-Conventional Energy Sources (MNES), which was further renamed as Ministry of New and Renewable Energy (MNRE) on October 20, 2006. Perhaps this is the only such ministry in the world.

In 1987, a separate financing institution called the Indian Renewable Energy Development Agency (IREDA) was set up as a public sector undertaking for providing institutional finance exclusively in the field of renewables and energy efficiency. IREDA entered the market when lending to the renewables sector was considered a high-risk and low-profit business. Over the years, IREDA has paved the way for broadening and deepening the market for renewables. In the initial years, it received bilateral and multilateral credits. Over the years, other financing institutions also started providing financial assistance for renewable energy projects and consequently, the market share of IREDA-financed projects has at present come to around 15%.

The programs started slowly with mostly concentration on small applications and rural areas—solar lanterns, solar cookers, solar water heating, water mills, etc. Two national programs were started on rural cooking energy viz National Programme on Biogas Development and National Programme on Improved Cook stoves and were initiated in 1981–1982 and 1984–1985, respectively. Alongside programs for deployment of solar energy devices were initiated. Small hydro (below 25 MW) was started in 1988. In the 1990s, path-breaking development started in wind power development, which gathered steam in the last decade and became the main driver of grid renewable energy in India. There were also developments in the cogeneration area with particular success in bagasse being utilized for power generation in sugar mills, particularly in those in the private sector. Biomass-based power plants also started being set up primarily based on rice husk.

The solar cooker program slowly withered away, not being able to establish a regular market. Solar dishes have since come in small sizes as well as large systems. They have also, however, not taken off in a big way, although there is a solar thermal system in the religious place of Shirdi where 20,000 meals are cooked every day. The biogas program continues at slow pace, installing about a lakh of plants annually, partly because of budgetary constraints and partly because of poor progress in northern states of UP, Bihar, and Haryana. The cook stove program was disbanded in the central sector and transferred to the states in 2002 after about 32 million improved stoves had been installed. After that it also quickly withered away. Improved cook stoves are continuing through efforts of some NGOs, etc., in small ways in different parts of the country. The solar water heating program has been expanded, but inroads into the industrial and commercial sector have been difficult and limited. There are some pilots of solar air conditioning, but this has to go a long way before it can be said to be mature or competitive. Solar lanterns have increased significantly, largely in the private market or locally funded initiatives though the central government and some states also distribute some. Urban waste to energy plants were set up in some towns, but they did not function properly because of difficulties of segregation of waste. This problem continues. There have been some medium and small plants using industrial waste and urban kitchen waste, but this has yet to really spread. One to two megawatt plants have been piloted based on fuel wood or other agri residues using the gasifier technology that has shown promise, though tariff continues to be a problem. There is a huge potential but not much progress. In the last few years, there was a large program to cover 10,000 remote villages where the grid was not likely to go. This was primarily covered by solar home lighting systems. Rural banks have also supported several thousand solar home lights through loans. The record, therefore, is mixed and patchy.

6.3.1 Solar Grid Developments

In 2008, India came out with a National Action Plan on Climate Change with 8 missions covering various areas of sustainable development.* As the Indian prime minister said: at the centerpiece was the proposed National Solar Mission, which would launch an ambitious plan for all-round development of solar energy. This mission was launched in January 2010, with an ambitious target of achieving 20 GW of grid power by 2022, 2 GW of off-grid solar power, covering 20 million households with solar lighting and installing 20 million m^2 of solar thermal collector area.† However, in the Budget 2015–2016, the Government of India has announced much more ambitious renewable energy targets of 175 GW by 2022. This includes scale-up Grid Connected (including roof top) Solar Power Projects from 20 GW to 100 GW. This program is well underway. From virtually zero, a capacity of 3000 MW grid power has already been installed. Both the center and the states have launched different policies and it is expected that the targets of the mission would be substantially met. The most significant development has been the reduction in costs globally for solar PV, and the reduction consequently in India achieved through competitive bidding rather than the prevalent feed-in-tariff (FIT) system. India has perhaps become the lowest cost solar power producer in the world, as acknowledged in a recent World Bank Report.[5] There will be challenges, particularly in relation to transmission infrastructure and land.

6.3.2 Capacity Creation

The following table provides details of the progress in renewable energy deployment up to January 31, 2015 (Table 6.2).

In between 2002 and 2013, share of renewable grid capacity has increased more than 6 times, from 2% to around 13% in only 13 years, and is contributing 6% to the electricity

* The National Action plan on Climate Change (NAPCC) was released on 30 June, 2008 to state India's contribution toward combating climate change. The plan outlines Eight National Missions. The NAPCC consists of several targets on climate change issues and addresses the urgent and critical concerns of the country through a directional shift in the development pathway. It outlines measures on climate change related adaptation and mitigation while simultaneously advancing development. The Missions form the core of the Plan, representing multi-pronged, long termed and integrated strategies for achieving goals in the context of climate change.

† The ultimate objective is to make solar energy competitive with fossil-based energy options. By increasing the share of solar energy in the total energy mix, it aims to empower people at the grass roots level. Another aspect of this Mission is to launch an R&D program facilitating international co-operation to enable the creation of affordable, more convenient solar energy systems and to promote innovations for sustained, long-term storage and use of solar power.

TABLE 6.2

Cumulative Progress of Renewable Energy Deployment as of December 31, 2013

S. No	Renewable Energy Programmes/ Systems	Cumulative Achievements
I. Power From Renewables		
Grid-Interactive Renewable Power (in MW)		
1.	Wind power	22,597.68
2.	Small hydro power	4,017.05
3.	Biomass power/cogeneration	4,183.55
4.	Waste to power	107.58
5.	Solar power	3,099.68
	Total	34,005.68
Off-Grid/Distributed Renewable Power (in $MW_{eq.}$)		
6.	Biomass (non-bagasse) cogeneration	569.75
7.	Biomass gasifiers	171.63
8.	Waste-to-energy	142.27
9.	Aero-generators/hybrid systems	2.43
10.	SPV systems	229.35
11.	Water mills/micro hydel	15.21
12.	Bio-gas based energy system	4.07
	Total	1,134.71
II. Remote Village Electrification		
	Villages/hamlets provided with electricity/lighting systems	11,308
III. Decentralized Energy Systems		
10.	Family type biogas plants (in lakh.)	47.95
11.	SPV street lighting system (in lakh.)	3.42
12.	SPV home lighting system (in lakh.)	11.94
13.	SPV lanterns (in lakh.)	9.85
14.	SPV pumps (nos.)	19,501
15.	Solar water heating (collection area in million m^2)	8.73

Sourse: Ministry of New and Renewable Energy, New Delhi, India.

generation mix. The grid renewable power has been dominated by wind, though now solar is making its presence felt. The important thing to notice in this progress is that the % age of renewables capacity has increased substantially at a time when there was maximum development of conventional power capacity. The high level of penetration of renewable power, after large hydro is added, in India compares favorably with that of the EU and far exceeds that of the United States.

6.3.3 Policy Support

The government has been promoting private investment in setting up of projects for power generation from renewable energy sources through an attractive mix of fiscal and financial incentives, in addition to the preferential tariffs being provided at the states' level. These include capital/interest subsidy, accelerated depreciation (AD) and nil/concessional excise, and customs duties. The level of capital subsidy being provided for off-grid depends on the renewable resource and region, and varies from about 10% to 90% of project cost, the higher level being given for projects in North-Eastern Region/special category states.

Electricity Act 2003 provided the necessary regulatory framework for growth of renewable power in India. The act required State Regulatory Commissions to specify renewable purchase obligations (RPOs) and also fix renewable resource–specific FIT. In addition solar-specific RPOs, starting with 0.25% in the first phase of the Solar Mission and leading to 3% by 2022, have also been introduced. The National Electricity Policy 2005 has further provided for progressive increase in these levels and purchases by distribution companies through competitive bidding process. The Tariff Policy 2006 requires fixation by State Electricity Regulatory Commissions (SERCs) of a minimum % age for purchase of energy from such sources taking into account availability of such resources in the region and its impact on retail tariffs and procurement by distribution companies at preferential tariffs determined by the SERCs. As of date, most of the SERCs have specified % ages for purchase of electricity from renewable sources of energy. Preferential tariff for grid interactive renewable power is being given in most potential states. Uniform guidelines by Central Electricity Regulatory Commission (CERC) for fixation of such preferential tariffs have been issued.

Other instruments include AD up to 80% to investors/developers in the first year. This was a tax benefit for profit-making companies and individuals, something akin to the production tax credit in the United States. This model led to substantial development of the wind sector. In 2010, generation-based incentive was also provided as an alternative for such power producers, which could not avail the benefit of AD in order to provide a level playing field. This resulted in more than 3 GW being installed in 2010–2011, the highest ever. However, the AD benefit for wind has since been discontinued, though the generation-based incentive (GBI) has recently been restored.

Renewable energy sources are not evenly spread across the country. On the one hand, there are states where the potential of renewable energy sources is not that significant; on the other, there are states where there is very high potential. Five southern states produce most of today's wind power and solar power's potential is similarly concentrated. This poses economic and operational challenges for these states. In 2010, tradable renewable energy certificates (RECs) have been introduced to address this gap and assist states in meeting RPOs. Renewable energy generators offer REC on the exchange, and the utilities, also of areas where with less renewable energy potential could buy and fulfill their RPO. However, in the absence for any focus on RPO compliance, the REC market is not taking off.

On December 19, 2014, the Government of India introduced the Electricity (Amendment) Bill, 2014 in the Lok Sabha (the National Parliament). It has many provisions for accelerating renewable power deployment in the country including provision for more robust RPO compliance and preparation, review, and notification of the National Renewable Energy Policy.

6.3.4 Solar Power Policy

Solar power was promoted with some additional measures. Under the Central initiative of the Solar Mission, a central agency was specified to buy the solar power produced by the developers. It then bundled it with cheaper thermal power available to its parent producer NTPC in the ratio 1:4 and sold to utilities with whom it entered into power purchase agreement (PPA). This bundling reduced the cost of power to the utilities tremendously bringing it near to the normal price of purchase. In addition, a separate solar purchase obligation was mandated for the utilities. At the start in 2010, solar PV tariff as calculated by the CERC was almost Rs. 18 per unit,* levelized over 25 years. But a bidding

* At present 1 U.S. $ is equivalent to around Rs. 60.

route was also adopted, asking developers to offer discounts to the CERC price. The two rounds of bidding brought down the lowest bids, first to about an average of Rs. 12 per unit (U.S. $20 cents) and later an average to about Rs. 7 perhaps more per unit (U.S. $12 cents). This was the policy for the first phase of 1.1 GW.

However, for the second phase of around 750 MW, the tariff has been fixed at Rs. 5.30 per unit. Developers were asked to bid for a minimum viability gap funding (VGF) to achieve this tariff. The bid results suggest that on an average VGF of Rs. 1 crore/MW installed capacity have been asked by the developers. In case of domestic solar cells, the VGF sought is in the order of Rs. 2.50 crore/MW. Power will be purchased and sold as earlier by a newly formed Solar Energy Corporation of India. Both phases had limited quantities and procedures for domestic content to support the hard-hit domestic industry because of low cost manufacture in China, which also damaged their export capability to Europe as also low cost financial support by the U.S. Exim bank for their local thin film manufacturers. This provision of domestic content was only for the National Solar Mission and not for state initiatives. However, the United States has gone out of its way to complain against Indian policies. It is felt that it has not only been overly aggressive and unfair but has also not realized the implications that would favor the Chinese industry rather than their own.

6.3.5 Future Projections and Trends

Renewable energy outlook for the country seeks to graduate from the present role of supplementing the fossil fuel based conventional energy sources, to becoming a mainstream source and eventually reaching a stage of substantially replacing them. There are different prognoses about the time frame when such sources could completely overtake conventional fossil fuel–based sources. The 12th Five Year Plan document of the Planning Commission, Government of India (for the period 2012–2017) has already depicted a 2030 outlook which suggests around 33% electricity installed capacity from renewable energy sources. Planning Commission, 12th Five Year Plan: online http://planningcommission.nic.in/plans/planrel/12thplan/pdf/12fyp_vol2.pdf

The Planning Commissions "Report of the Expert Group on Low Carbon Strategies for Inclusive Growth 2014" has estimated that under low carbon for inclusive growth scenario solar and wind power capacity by 2030 will be around 225 GW out of the total installed capacity of around 700 GW, and in energy terms these will contribute to around 14% in the electricity mix. Planning Commission "Report of the Expert Group on Low Carbon Strategies for Inclusive Growth 2014" online http://planningcommission.nic.in/reports/genrep/rep_carbon2005.pdf. Given the right mix of technology, finance, and necessary improvement in grid infrastructure these estimates could be achievable.

6.3.6 Renewable Energy in Off-Grid Mode

The problems with the grid in India, and the unique ability of solar to provide power from the smallest unit onward, offer almost unlimited potential for off-grid solar. There are four prime areas in India. First, simple roof top with the difference that this will be largely for self-consumption rather than being fed into the grid as is the usual practice so far in developed countries. This will both reduce demand for day-time electricity and partly meet the needs of small diesel generators, which are in use in plenty because of power shortages. Second, village electrification is a big problem in India, which we shall briefly discuss shortly. Kerosene supplied at highly subsidized rates meets the lighting through small ill-lanterns giving very poor light. Solar, whether through mini grids, or solar home lighting systems and lanterns, could provide a much better alternative. It will certainly save billions of liters of kerosene and huge amounts for the government in subsidies, possibly over Rs. 20,000 crores annually at present prices and subsidy rates if large-scale solarization happens. Third, there are a couple of hundred thousand telecom towers in rural areas. Because of unreliable supply of power in many areas, billions of liters of subsidized diesel is used. A lot of it can be saved by solar meeting power needs in the daytime when power cuts are the maximum. Fourth, India has more than 20 million irrigation pumps in the rural areas, with a large number running on diesel. Most of these pumps over time can be replaced by solar-powered pumps. And maybe there will be additional markets. There can be many more pumps, which are currently not installed, because either electricity is not there or diesel is too expensive. As we move toward eastern India from Delhi, the water table is high enough and electricity is scarce. This is a perfect combination and needs to be pursued as a separate mission. Altogether these can create a market of 50 GW, with no land and transmission requirements, no involvement of the grid, and no losses as generation will be at consumption points. But solar companies need to become oriented to provide such small and dispersed services. In fact, this will allow many professional small companies and generate employment and incomes in rural areas. It will simultaneously help India's energy security by saving very large amounts of nonessential use of kerosene and diesel. Besides, huge amounts of undeserved subsidy would also be saved. And the utilities will save a lot of their losses arising out of rural supplies supplied at low tariffs, both for households, and particularly for irrigation. Therefore, in the Indian situation, which may find similarities in many other developing countries, solar off-grid is a completely win–win situation. The best part is that all the beneficiaries will actually be spending less on their fuels than currently, leave alone doing so in the future.

6.3.7 Energy Access

Energy access includes both electricity and cooking energy. The position in India in respect of both is quite dismal. This is also primarily a rural problem. Though it may be difficult to put exact numbers, it is generally expected that 300 million people in India do not have electricity access and 700 million people use biomass in traditional stoves for cooking. India has made rapid progress in rural electrification through the Rajiv Gandhi Grameen Vidyutikaran Yojna in the last several years and we claim that over 90% of the villages have been electrified. However, over 40% of the above poverty line (APL) rural households do not have electricity connections (a large number of below PL households were given connections under the official program). The problem has been the definition, which allows a village to be deemed electrified if 10% of the households have connections. There are additional problems of the number of hours of supply, which sometimes are not many and mostly not in the evening hours when needed most. There are worries that these problems may continue with power supply being short and the utility finances continuing to be affected adversely by low tariffs. It is, therefore, believed by many that solar minigrids or solar home lighting systems may be the answer to India's electricity access problems.[6] Many pilots have been conducted, and there is subsidy support. However, a regular business model is yet to be found as also who could be the possible entrepreneurs. This is going to be an important area for policy development in the next few years. This is also a good area for international funding.

6.3.8 Cooking Energy

As far as cooking is concerned this is a serious problem although there has not been enough recognition of it by the policy makers, and indeed even in public discussion. In 2014, the Global Disease burden study showed that indoor air pollution from combustion of biomass in inefficient cook stoves was the biggest health hazard for India. In general, there has been a belief that the real solution is the supply of cooking gas. However, studies have noted the difficulties of shifting to cleaner fuels, even if subsidized, in a situation where biomass is largely free. Besides, the country simply cannot afford the subsidy either of cooking gas or kerosene for this solution to work. Therefore, biomass will remain the principal fuel. And it is imperative that we move to ultraclean stoves, which will burn all types of biomass and also last for at least 5 years. Some efforts in this direction are ongoing. Efforts are also on to finalize proper standards. Cost of such stoves could be an issue, and they may require subsidies. Had the carbon market been working properly, distribution of improved cook stoves could probably be fully financed. In its absence, the government would have to provide some subsidy as we develop slowly a proper market. The market size could be 100 million, and if the life is about 4 years, there will be a huge replacement market. Unfortunately, the international system has done precious little to deal with this problem, except express concern. More, and decisive, action is needed.

6.3.9 Carbon Emissions—Issue of Energy Intensity

It may be useful in the context of this scenario to briefly examine the issue of India's contribution to carbon emissions. In the climate change negotiations, often India has been painted as a villain. It is true that India has argued for the developed nations to accept their historical responsibility for filling up the carbon space and therefore reduce their emissions substantially even as there needs to be space for her to grow to address the problems of development and removal of poverty. The developed countries initially accepted this argument and promised differentiated responsibilities. However, they have hardly delivered on any of their promises. And, they have now started arguing that the developing countries will become the larger emitters in the future and, therefore, they must restrict or also reduce their emissions. Nothing could be more insidious and ethically unacceptable. But the real problem is that the fingers are pointed at China and India and both of them are put in the same boat. But consider the differences. China emits about 26% of the global emissions now and India about 5%–6%, roughly same as Japan and the Middle East, which is also growing rapidly. The United States is about 16% and the EU about 12%. Moreover, in per-capita emission terms, India is well below the per-capita global average of 4.5 tons, while the United States is well above. Therefore, India is not the problem. The problems of power supply mentioned earlier also suggest that we will not be in the future. Ideally, it would only be equitable if the first ceiling for emissions for countries should be the current global per-capita average multiplied by the current population. This would give India enough space to grow in a reasonably carbon-constrained manner, which would require strong reliance on renewable energy, energy efficiency, and a new transport model. If we did this, we should also be able to meet our commitment of 20%–25% reduction in emission intensity of GDP by 2020. Needless to say, it will also be a much more sustainable model designed to meet more economic and social equity principles also.

6.4 Energy Efficiency

National interest in energy conservation in India was triggered by the oil crisis of 1973 but was largely concerned initially with petroleum products. The Petroleum Conservation Research Institute (PCRA) was set up in 1978. Much later, in 1989, the Energy Management Centre (EMC) was set up to promote conservation in power. After the opening up of the economy in 1991, and as electricity became important, the need was felt for proper legislative and institutional backing to promote electricity efficiency. In 2001, the Energy Conservation Act was passed. In 2002, the EMC was reconstituted as the Bureau of Energy Efficiency (BEE), which became a statutory body. The Electricity Act 2003 mandated efficiency in generation, transmission, and distribution. The National Electricity Policy 2004 gave demand side management (DSM) a high priority. Periodic energy audits were made compulsory for power intensive industries; emphasis was placed on labeling of appliances and high efficiency pumps in agriculture. The Integrated Energy Policy Report 2006 emphasized that investments in energy efficiency were as important as investments in generation as energy saving was more advantageous in many ways than energy generated. One could say that a view emerged that energy efficiency should be India's primary fuel. Under the National Action Plan for Climate Change in 2008, a new National Mission on Enhanced Energy Efficiency (NMEEE) was proposed that was approved in 2010 and is under implementation.

Industrial and domestic sectors are the two largest consumers of utility generated power, with the latter growing rapidly because of economic growth, rapid urbanization, and changing lifestyles, which is leading to huge growth in personalized transport, residential buildings, and electric appliances. It is estimated that while India's energy consumption may go up by 6 times, its per-capita emissions may only have space for a little over doubling. Therefore, it has become imperative to identify areas where huge reductions are possible, as also where power supply from conventional sources can be replaced by renewable energy sources. Some of the important areas are briefly discussed in the following text.

6.4.1 Buildings

Almost 40% of the total energy is utilized in the building sector. As urbanization is growing rapidly, this sector will create great demand for energy in the future. In 2009, the buildings sector consumed one-third of the total electricity, with the residential sector being 25% (BEE). Eighty percent of the electricity is for cooling and lighting, with ACs accounting for 40%–50% having more than doubled in the last few years. This will only increase as it is estimated that 60%–70% of the building stock is yet to be built by 2030. Forty percent savings could accrue if we have energy-efficient buildings.

There are many policy drivers to ensure this. There is a National Building Code. There is the Energy Conservation Building Code that specifies some norms for certain types and size of buildings. This needs to become more stringent and mandatory for more sizes of buildings. Larger complexes require prior environmental clearances, but this is somewhat loosely administered in terms of ensuring final compliances. There are two voluntary rating systems for green buildings. The U.S. LEED-inspired IGBC-LEED system, which is more western oriented and the TERI—MNRE GRIHA system, which is based on traditional Indian architectural and solar passive principles. The latter is more suited to Indian conditions and seeks to reduce also AC and lighting needs. The former focuses on efficient systems and encourages glass facades. The Government of India took a decision that all their new buildings should be at least GRIHA 3*. This norm has also been accepted by the

Central Public Works Department, which leads the issues regarding specifications and building codes. None of these instruments, however, have created the type of huge impact that is needed, and builders and developers are clearly ignoring these needs; there is neither sufficient knowledge nor interest for consumers to demand such buildings as people do not follow life cycle cost principles. There is an urgent need for municipal regulations and to move tax concession on house building loans toward energy-efficient housing. A few municipalities have taken some steps.

6.4.1.1 *Electrical Appliances*

This is going to be a high growth area. Mckinsey (2009) has said that household energy demand will spike up to 1300 TWh by 2030. There has been a voluntary labeling program in India for some years. Currently, it is incrementally increasing the norms for each level of star rating. However, even though moderately successful, star label penetration is still low. In 2008, star AC's penetration was 1.5% and fridges 7.5%. This must have gone up since then. Since this will be a huge growth area, and 30%–40% savings are possible, and the use of appliances will spread to washing machines, microwaves, dish washers, etc., it is imperative that norms become stricter and are mandated. Other products should simply not be allowed in the market.

6.4.2 Lighting

Lighting is a huge electricity load, particularly at evening peak time. The CFL market in India has been growing at about 20% annually. LED is just beginning, including their assembly/manufacturing in India. A lighting/LED revolution must happen in India—both for domestic or institutional lighting and for street lights. There was a very unique CDM project for the period 2009–2011 targeting replacement of 400 million light points converting bulbs to CFLs with a potential saving of 20,000 MW. Twenty million lights were changed, but then the carbon market collapsed. We believe that a program such as this, probably now shifting to LED, and a project for 100 million cook stoves would have been the best possible carbon projects, and should be funded from the voluntary market. Such lighting would make even solar more attractive as lesser modules would be required to provide the same amount of light. A huge program for change in municipal street lighting to LEDs is also required funded by a special tax if need be. Twenty-five to thirty percent of the lighting load of municipalities is from street lights. Steps in this direction are underway.

6.4.3 Industry

As per Central Electricity Authority, 23% of total electricity is used by industry in the country, of which >60% is by few large and heavy industrial consumers. India's energy intensity in manufacturing sector has fallen consistently over the last 20 years, but there is still a saving potential of up to 25% even as the total energy consumption naturally increases. Under the NMEEE, a Perform, Achieve and Trade (PAT) scheme has been launched that covers the 9 most energy-consuming industries—thermal power plants, iron and steel, cement, fertilizer, aluminum, textiles, pulp and paper, and chlor alkali. Energy consumption reduction of 5%–10% is targeted between 2011 and 2014. All this would lead to 5623 MW of avoided capacity. Clearly, we have to do more and include more industries while also doing everything possible in the medium and small sector also. Motors, pumps, and boilers are the key.

The activities under the PAT scheme provide opportunities for new markets as it devises cost-effective energy efficient strategies for end-use demand-side management leading to ecological sustainability.

6.4.4 Agriculture

This sector uses almost one-third of the electricity while contributing to only about 8%–10% of the revenue. This is predicted to grow at 2%–5% annually till 2020. Seventy percent of irrigated area is from ground water. Since the tariff is extremely low and it is really not metered, the marginal cost of pumping water is zero and farmers have no incentive to restrict or reduce their usage of electricity (and water) or invest in energy efficient pump sets. Rationalization of tariff and mandating production of energy efficient pumps is required. Even in municipalities, energy efficient pumping is important, because 15%–30% of their electricity load comes from this area.

We have suggested that solar irrigation pumps could be a good alternative in the long run.

6.4.5 Future Directions

There are several barriers that are preventing large-scale adoption of energy efficient measures. Some of the most important are as follows:

1. Price distortions including subsidized prices of petroleum products and electricity. Proper energy pricing and reductions of subsidies, though politically difficult, will be very important.
2. High costs of upfront investments. This may be applicable for government agencies or organizations. Absence of development of the RESCO model, where roof top solar or energy efficiency measures can be investments of others paid back in rental mode. Also lack of successful examples and knowledge of interventions.
3. Policies are not fully in place. Mandating is not easy. The mindset, whether of individuals or organizations, is still not sensitized to both the urgent need for saving energy and its conservation, as well as the benefits, both financial and otherwise, which would accrue. This also underlines the need for both behavioral and attitudinal change at all levels.
4. And, as part of the chicken or egg coming story, there is a lack of successful examples and knowledge of interventions that create confidence as well as mitigate an environment of risk.

As far as financing is concerned, there are a number of options to accelerate pace of energy efficiency program. Establishment of state level Clean Energy Funds using the Public Benefit Charge concept could be an area of action. Such funds may be established as special purpose funds by national or state governments and regulators for financing clean energy projects. Internationally also the most common, reliable, and sustainable source of funding is a tariff surcharge, cess, or levy established by the regulator and collected by the utility via the customer's electricity bill. Such a surcharge or levy is known as a Public Benefit Charge. The funds could be utilized for leveraging commercial financing, interest *buy-down* on commercial loans, loan guarantees, grants for public sector projects, and rebate. Other options could be development of regulatory schemes to acquire energy efficiency resources using a Standard Offer Programme, promoting utility financing of energy efficiency projects by establishing energy efficiency obligations, encouraging Indian banks and financial institutions to mainstream energy efficiency in corporate loans, creation of a facility to provide energy savings insurance, establishment of a *clean Energy Financing Facility* for debt financing of energy efficiency projects, and designation of energy efficiency financing as Priority Sector Lending.

Simultaneously, there is a need to have a much more comprehensive program of awareness building and education.

6.5 Conclusion

Historically, India has initiated systematic programs for renewables including for research and development and also for energy efficiency. The challenge for India is gigantic and exciting. Considerable progress has been made but actions are required on many fronts. Policy issues related to all sectors of conventional electricity generation are urgently needed. Renewable energy has already caught the imagination and with proper policy framework and planning, India could be in a position to meet significant portion of its energy needs through these sources. There must also be a large-scale off-grid program. The successful implementation of the National Mission on Enhanced Energy Efficiency could also yield similar results, but simultaneously, its coverage needs to be substantially increased. Curtailment of consumption of oil products by various methods in the transport sector and some suggested here in the electricity sector have to be prioritized. All these would also lead to substantial reduction of, or less increase in, carbon dioxide emission. While budgetary support for renewable energy and energy efficiency have progressively increased over the years, particularly for large-scale grid connected power, these need to be further enhanced for both grid and off-grid to achieve the large upscaling that is required. Subsidy reform would automatically provide substantial funds. But some critical policy and structural gaps remain, particularly for decentralized distribution in the areas of access to capital, technology development and adaptation, innovation induction, and strategies to upscale deployment. However, new opportunities and compulsions will lead to think beyond the government-centric programs and to create new instruments, strategies, and pathways.

References

1. Planning Commission of India (2013). 12th five year plan document, Vol. 2, p. 130.
2. Planning Commission of India (2006). Integrated energy policy report.
3. Planning Commission of India (2013). 12th five year plan document, Vol. 2.
4. Planning Commission of India (2013). 12th five year plan document, Vol. 2, p. 133.
5. World Bank/ESMAP report (2013). Paving the way for a transformational future lessons from Jawaharlal Nehru National Solar Mission Phase I.
6. Maithani, P.C. and D. Gupta (2015). *Achieving Universal Energy Access in India: Challenges and the Way Forward*. London, U.K.: SAGE.

7

Renewable Energy Policies in Brazil: Bioenergy, Photovoltaic Generation, and Transportation

Ricardo Rüther

CONTENTS

Bioenergy is regarded as one of the key options to mitigate greenhouse gas (GHG) emissions and to substitute fossil fuels [1,2]. While there are many renewable energy technologies available for large-scale electricity production with low or no direct GHG emissions, the transport sector, which is responsible for 23% of GHG emissions [3–5], is almost entirely based on fossil fuels, and has fewer alternatives.

Transportation represents some 27% of the world's secondary energy consumption (21% of primary), grows 1% per year on average, and is almost exclusively fuelled by petroleum [4,5]. Biofuels can play an important role in addressing both the GHG emissions of transport and the dependency on petroleum. However, currently marketed, first-generation biofuels represent only around 3% of global road transport fuels, and can be regarded as a transition technology. This limited penetration of first-generation biofuels is due in part to the potential competition with food crops and market prices (sugar × ethanol prices in international markets). Concerns that these technologies are falling short of expectations have led to the development of second-generation (cellulosic ethanol, biomass-to-liquids, pyrolysis oil, dimethyl ether) and third-generation (algal-biodiesel, biofuels from third-generation processes) biofuel technologies [4], which still need to be scaled up in order to become cost-competitive.

With the exception of sugarcane ethanol, the traditional biofuels have a number of severe disadvantages that are related to the feedstock. The current costs of rapeseed biodiesel and ethanol from cereals or beets are much higher than the costs of gasoline and diesel, and substantial subsidies are needed to make them competitive. These high costs are a result of the low net energy yield of most annual crops (100–200 GJ/ha year in the long term [6]), the high-quality agricultural land required, and the intensive management. The lower productivity per hectare and high fertilizer requirement also limit the well-to-wheel reduction of fossil energy use, which limit the environmental benefits [7,8]. The net energy of perennial crops (220–550 GJ/ha year), grasses (220–260 GJ/ha year), and sugar cane (500–650 GJ/ha year) is considerably higher, and Brazil has been a world leader in promoting biofuels for over 30 years under its *Proálcool* program.

7.1 *Proálcool* Program in Brazil

Since 1975, Brazil has mandated that ethanol be blended with all gasoline sold in the country. Although the required blend level is adjusted frequently, it has been in the range of 20%–25%, and all filling stations are required to sell gasohol (E25) and pure ethanol (E100). Tax incentives have been given to vehicles that run on pure ethanol in the early days of the *Proálcool* program, and more recently, the introduction of the so-called *flex fuel* vehicles by most of the local automakers allows for any proportion between ethanol and gasoline (E25–E100) to be used any time. This has led to the growing share that *flex fuel* cars hold in the Brazilian market [9,10], which should reach over 96% by the end of 2014 [11,12]. In 10 years, U.S. $16 billion were invested in genetic research for sugar cane improvement, alcohol subsidies, and agrobusiness, and policies have included blending mandates, retail distribution requirements, production subsidies, and other measures [13].

After a rocky period in the 1980s and 1990s, when high demand resulted in a seller's market, which led to high alcohol prices and delivery shortages [13], the escalating oil prices and the event of the *flex fuel* car have brought attention back to ethanol. Advances in electronic fuel injection technology solved many of the technological problems associated with ethanol engines (in cool mornings, drivers had to wait precious minutes for their car engines to warm up before they could drive). In 1985, with falling oil prices, the Brazilian government was not able to keep up with the required subsidies, and in 1989, ethanol shortages and high prices resulted in very disappointed drivers and the near collapse of this promising technology. In 1990, the sector was deregulated, and the *Alcohol and Sugar Institute* (IAA—Instituto do Açúcar e do Álcool), which regulated export quotas and subsidies, was closed down. In a free, deregulated market, ethanol producers chose to turn back to sugar when the international price of that commodity recovered, and the automakers reduced alcohol-driven car production to negligible levels. More recently, with the boom in *flex fuel* car sales, ethanol prices soared once again. This time however, with *flex fuel* cars came the choice for consumers to avoid abusive price increases from alcohol producers, and whenever ethanol prices go over the mark of 70% of gasoline price, *flex fuel* car drivers fill up with the traditional fossil fuel. While at present 9 out of 10 cars produced in Brazil run on both fuels, ethanol has not been the choice of most drivers. This is mostly because the price difference between ethanol and gasoline has been below 20% in the last few years, due to the more profitable sugar market. Furthermore, because *flex fuel* cars running on pure ethanol have a ~30% lower mileage than when running on gasohol, there is no real advantage in filling up with the so-called green fuel.

7.2 Bio-Oil

Brazil has more recently begun to target increased use of biodiesel fuels, derived primarily from domestically produced soybean oil, with recent legislation mandating blends of 5% biodiesel in diesel fuels (B5). Brazil has the potential to be a world leader in biodiesel production, as it will be able to produce bio oil from many different sources. The issue fuel × food, however, will always persist, and might be even more pronounced in the case of biodiesel than it has been so far in the case of bioethanol.

7.3 Photovoltaic Generation and the Net Metering Law in Brazil

In early 2012, Brazil has enacted new legislation allowing for solar photovoltaic (PV) generators of up to 1 MWp to be connected to the public distribution grid and to feed excess electricity to the grid under a net metering scheme [14]. The excellent solar radiation resource availability in the country, ranging from over 1,500 to nearly 2,200 kWh/m^2/year [15], and the extensive rooftop availability in the residential, industrial, commercial, and public sectors, complemented by considerably high electricity tariffs in most parts of the country, should make building-integrated, grid-connected, and net metered PV generation an attractive option for a large number of consumers in the near future.

7.4 Prospects

In addition to Brazil, mandates for blending biofuels into vehicle fuels have been appearing in several other countries in recent years, mostly driven by GHG reduction targets. In the long term, however, it can be argued that biofuels will play a transitional role toward electric vehicles and solar PV generation. In terms of energy conversion efficiency, biofuels for transportation can be largely regarded as a transition technology. A direct comparison between sugar cane ethanol + *flex fuel* cars, and sunlight-to-electricity (PV) conversion + electric cars is illustrated in the following figure: a 1 ha (100 m × 100 m) sugar cane plantation will yield from 6,800 to 8,000 L of ethanol per year, allowing a medium-size *flex fuel* car to drive up to 53,900 km/year in a best-case scenario. If the same area is covered with commercially available solar PV modules at 15% conversion efficiency and a 1,500 kWh/kWp/year energy yield, the electricity produced over 1 year on the same 1 ha will allow the electric version of the same medium-size car to drive over 14,000,000 km/year (over 260 times more!). The reasons for this overwhelming difference are related to both the orders of magnitude larger efficiency of PV conversion over photosynthesis, and the considerably larger efficiency of electric motors over internal combustion engines. While fully electric, plug-in vehicles are still in the early stages of large-scale production, battery technologies are continuously being improved, and volume production should lead to the necessary cost reductions for this benign technology to be adopted worldwide. In a sunny and large country like Brazil, building-integrated PV in the urban environment can not only cover the additional electricity needs represented by a new fleet of electric vehicles, but can also supply on site a considerable fraction of the urban electricity needs, without any further area requirements and without competing with arable land.

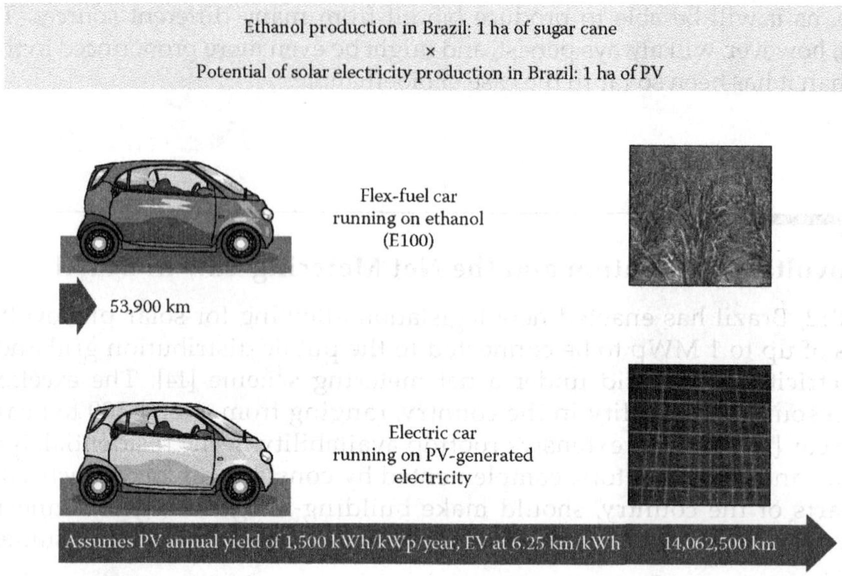

References

1. Hall, D.O., Rosillo-Calle, F., Williams, R.H., and Woods, J. 1993. Biomass for energy: Supply prospects. In: Johansson, T.B., Kelly, H., Amulya, K.N.R., Williams, R.H. (Eds.), *Renewable Energy, Sources for Fuels and Electricity*. Island Press, Washington, DC, pp. 593–653.
2. Goldemberg, J. 2000. *World Energy Assessment*, Preface. United Nations Development Programme, New York.
3. IEA—International Energy Agency. 2012. *CO_2 Emissions from Fuel Combustion 2012—Highlights*. OECD/IEA, Paris, France. Available at http://www.iea.org/publications/freepublications/publication/name,4010,en.html (accessed on October 1, 2013).
4. Fiorese, G., Catenacci, M., Verdolini, E., and Bosettim, V. 2013. Advanced biofuels: Future perspectives from an expert elicitation survey. *Energy Policy* 56, 293–311.
5. EIA. 2013. *International Energy Outlook 2013*. Energy Information Administration, Office of Integrated Analysis and Forecasting, Washington, DC. Available at http://www.eia.gov/forecasts/ieo/transportation.cfm.
6. Hamelinck, C.N. and Faaij, A.P.C. 2006. Outlook for advanced biofuels. *Energy Policy* 34, 3268–3283.
7. Ranney, J.W. and Mann, L.K. 1994. Environmental considerations in energy crop production. *Biomass and Bioenergy* 6, 221–228.
8. Berndes, G., Hoogwijk, M., and Van den Broek, R. 2003. The contribution of biomass in the future global energy supply: A review of 17 studies. *Biomass and Bioenergy* 25, 1–28.
9. Martinot, E., Sawin, J.L. et al. 2013. *REN21 Renewable Energy Policy Network—Renewables 2013 Global Status Report*. Worldwatch Institute, Washington, DC, 177pp.
10. Guandalini, G. and Silva, C. 2006. *A Dupla Conquista*, Veja, 01/02/2006, pp. 90–94.
11. Nascimento, G. 2006. *A Febre dos Carros Flex*, Isto É, 01/03/2006, pp. 72–73.
12. Santos, C.T., Favaro, F., and Parente, V. 2010. Previsão de Fabricação de Carros Bicombustíveis e de Demanda de Etanol no Brasil em 2014. *Future Studies Research Journal* 2, 85–98.

13. Rosillo-Calle, F. and Cortez, L.A.B. 1998. Towards Proalcool II—A review of the Brazilian bioethanol programme. *Biomass and Bioenergy* 14, 115–124.
14. ANEEL—Agência Nacional de Energia Elétrica, Resolução Normativa 482/2012. Available at http://www.aneel.gov.br/cedoc/ren2012482.pdf.
15. Martins, F.R., Rüther, R., Pereira, E.B., and Abreu, S.L. 2008. Solar energy scenarios in Brazil. Part two: Photovoltaics applications. *Energy Policy* 36, 2865–2877.

8

Energy in Israel: A Case for Renewables

Gershon Grossman

CONTENTS

8.1 Introduction

Israel is a country of 8.0 million inhabitants in an area about 20,000 km^2 located on the eastern Mediterranean coast. Israel's electric grid and overall energy economy are isolated from those of its neighboring countries. The precarious political situation has led in the past to the country being subjected to energy and other economic boycotts. Until recently, Israel's economy has depended almost entirely on imported coal and petroleum. Despite the political difficulties, Israel's economy is the fastest growing in the Middle East, leading to an ever-increasing demand for energy. Israel's total energy consumption has almost doubled in the last 20 years, and electric power production and consumption have increased by a factor of 2.75 in that period of time [1].

Israel has learned to use the one abundant and inexhaustible energy resource it has—solar energy. Israel has been a pioneer in developing solar technology, leading the world (together with Greece and Cyprus), and until recently is number one in the world in per capita utilization of solar energy. According to the Ministry of Energy and Water Resources (MEW) Division of Energy Conservation [2], about 85% of households (i.e., about 2,100,000 households) use domestic solar water heating, accounting for 3% of the country's primary energy consumption. Government encouragement and regulations, along with over 50 years of proven experience, have made this application widespread. Other forms of solar energy utilization are considered, with the potential of supplying a much larger portion of the country's energy demand.

Israel faces a number of choices in securing reliable, clean, and efficient sources of energy over the next 25 years. Recent discoveries of offshore natural gas have generated interest in developing this resource for domestic consumption to reduce dependence on imported coal and petroleum, which involves both economic and environmental risks. According to a MEW-commissioned forecasting study on Israel's energy demand [3], the overall merits of developing these gas reserves entail a number of trade-offs involving energy security, the environment, and cost. Development of renewable sources is free of these concerns and holds a great promise to supply the country with long-term clean and secure energy.

8.2 Israel's Energy Economy

The Israeli economy is largely fueled by imported coal and petroleum. Prior to 1982, only a small amount of coal was consumed and the country's primary energy source was oil and its derivatives. At present, coal accounts for about 35% of Israel's primary energy. It is almost entirely used in electric power generation. Recent discoveries of offshore natural gas fields have added this important resource, which at present is used mainly for electric power production.

Central Bureau of Statistics (CBS) data [1] show that the total primary energy supply to the country for the year 2009 was about 17.4 million tons oil equivalent (MTOE). Total end use was about 13 MTOE, consisting roughly of 8 MTOE of petroleum products and 5 MTOE of electric power. Among petroleum products, gas oil and diesel oil comprise the largest category. Electric consumption is about 1/3 for residential and 2/3 for industrial use (goods and services), the latter comprising a significant portion for agriculture and water pumping. Fuel consumption other than for electric power generation is split at 60%, 36%, and 4% among the transportation, industrial, and residential sectors, respectively. The same consumption pattern has been in effect for the last 10 years. According to forecast [3], it is likely to change in the coming years, with the introduction of natural gas, which will replace heavy oil in the electric power and industrial sectors.

Another component of the energy economy—relatively small in quantity but important in significance—is solar energy, utilized mainly for domestic water heating. As mentioned earlier, it accounts for about 3% of the country's energy demand and saves about 5% of the electric power. Israel is currently one of the three largest per capita users of solar energy in the world, challenging the notion that solar energy is not yet economical.

The government MEW's stated goals and objectives with regard to energy are the following [2]:

- Expansion of the supply and diversification of energy sources for the national economy and consumers
- Advancement of the well-being of the national energy and mining economy, while improving product quality and prices and the service provided
- Minimization of environmental effects created by energy facilities in the national economy
- Aiming to satisfy demands reliably and stably while conducting activity to reduce demand through economizing in the use of energy resources
- Analysis of opportunities in natural resource inventories in Israel and securing their availability while conducting licensing processes and encouraging exploration and excavation activity
- Activity for securing Israel's social, economic, and physical strength, which will allow for coping with the inevitable occurrence of strong earthquakes, while minimizing possible loss of life, reduction of the extent of damage to property and infrastructures, and rapid transition to an ordinary life routine
- Promotion and support of studies for expanding knowledge in soil and sea scientific fields

A number of measures have been initiated to promote renewable energies in electric power generation. A government resolution dated January 2009 set an objective to produce at least

10% of the total electric power from renewables by the year 2020. An intermediate goal is to produce 5% of the total electric power from renewables by the year 2014. Accordingly, the Public Utilities Authority has instituted a set of premiums for licensed electricity producers employing renewables, proportional to the achieved reduction in pollution. Another government resolution dated August 2008 endeavors to develop technologies for electric power generation from renewables, with dedicated funding under a 5-year plan. A further government resolution dated September 2008 promotes energy conservation and reduction in national electricity consumption. Energy conservation by the public is encouraged by market transformation—increasing public awareness of energy efficiency in appliances and the like.

A number of laws, regulations, directives, and legal instruments have been introduced to implement energy policy. Most of them regulate the use of petroleum and petroleum products, natural gas, and electric power and deal with licensing and safety issues. Some deal with energy efficiency, minimum standards, and labeling. Recent regulations encourage cogeneration and distributed generation. Noteworthy is Article 9 of the Law for Planning and Building (1970), mandating the installation of solar water heating systems in all new constructions [4]. This article has contributed a great deal to the advancement of solar energy utilization and will be discussed further in the next section.

8.3 Solar Energy Utilization in Israel—The (Incomplete) Success

A visitor to Israel will unavoidably notice the urban landscape bursting with solar collectors and hot water storage tanks covering the roofs of buildings. Almost all residences in Israel are equipped with solar water heaters. The most common is the thermosyphonic system, a completely passive, stand-alone unit consisting of one or two flat plate solar collectors and an insulated storage tank. Large multistory apartment buildings often use a central system with a collector array on the roof and a storage tank in the basement, employing a pump controlled by a differential thermostat. Other arrangements are also available. In most of the country, the solar system will supply the full demand for hot water during 9–10 months per year, with an electric resistance backup employed the rest of the time. Freeze protection is never required, except in some isolated locations. The economics: the installed cost of a typical single-family system comprising a 150 L storage tank and 2–3 m^2 of flat plate collectors is about $700; an equivalent electric-powered system costs about $300. The difference of $400 is recovered by the owner in about 4 years (on a simple-payback basis); these systems carry a manufacturer's warranty for 6–8 years, and if properly maintained can last over 12 years. Several decades of nationwide experience have generated consumer confidence and acceptance to the point that a domestic solar water heater is perceived as a common, reliable household appliance.

There is no single legislation concerning solar energy utilization in Israel. The aforementioned Article 9 of the Law for Planning and Building (1970) [4] is probably the most important solar legislation, and has been the government's predominant contribution to Israel's success in the solar area. The law requires the builder (not the homeowner), since 1980, to install a solar water heating system in every new building up to nine stories tall. The MEW is currently leading a move designed to extend the requirement to multistory buildings. This is driven by the clear tendency in recent years to build high-rise buildings. Other laws and regulations describe in detail the size of the installation required for the various types of buildings, set minimum standards for the quality of the solar equipment and installation, and provide the regulations for retrofit installation of solar water heaters in existing multiapartment buildings. Based on government data [5], an average single-family domestic solar water heater saves 1250 kWh of electric power per year; the total contribution to the country is about 1.6 billion kWh/year, 21% of the electricity for the domestic sector or 5.2% of the national electricity consumption, providing for 3% savings on primary energy consumption. This amounts to about 270 kWh/year/capita—one of the highest in the world.

Israel's example in domestic solar water heating provides an impressive demonstration of what can be achieved (in countries with similar or even more favorable climates), if the government makes a commitment to clean and environment-friendly technologies [5]. However, while solar for residential use has become an everyday reality in Israel, the much larger industrial/commercial sector uses very little solar energy, despite the fact that the industrial user is much better suited to do so than a homeowner. Some key considerations are: industry works mostly during the day, requiring little storage relative to a residence; the economy of scale provides a significant capital advantage to large industrial installations; industry generally has plenty of roof area in one-story buildings located in areas where architectural considerations do not hinder the installation of solar collectors; the industrial user is equipped to perform small maintenance jobs, thus eliminating the need for a full-proof system and reducing first cost. While some industries require high-temperature process heat, there are many who need the same temperature range as the domestic user; these include textile, food, pharmaceutical, chemical, and many more. The same applies, of course, to the commercial sector. It is estimated that widespread solar energy utilization in industry for process heat and the like, and in commercial applications, could increase the country's utilization by a factor of five, if not more.

Unfortunately, current tax considerations create a negative incentive for businesses to use solar energy. An industry burning polluting fuel can write off the cost as a business expense, thus reducing its tax liability, whereas an investment in a solar heating system can only be amortized over 8–12 years, making it considerably less attractive economically. Moreover, the law currently exempts industrial plants, shops, hospitals, and high-rise buildings (height over 27 m) from the requirement to install a solar water heating system in new buildings [5]. The government could play an important role in changing this situation, by introducing appropriate measures, closing tax loopholes, and creating positive incentives for renewable energy. This can be achieved within a short time—there is no need for long-term investments and development of new technologies. Solar energy is a reality here and now, as already demonstrated by the country's residential sector.

References

1. Israel Central Bureau of Statistics, Energy balance of Israel, http://www.cbs.gov.il/energy/new_energy/new_enr_nach_eng_new_huz.html (last accessed March 9, 2015).
2. Israel Ministry of National Infrastructures, Renewable energy, http://energy.gov.il/English/Subjects/RenewableEnergy/Pages/GxmsMniRenewableEnergyLobby.aspx (last accessed March 9, 2015).
3. T. J. Considine, An energy demand forecasting model for Israel and the market potential for natural gas. Draft Equation Chapter 1 Section 1 Final Report. Pennsylvania State University, State College, PA, May 19, 2003.
4. Planning and Building Law, Application for permit, conditions thereof, fees and regulations, 5730/1970. System Article 9: Installation of a solar water heating system, 1970.
5. J. Nowarski, Solar Israel—A practical and legislative model. *Renewable Energy World*, March–April 2000, 92–99.

9

Renewable Energy in Australia

Monica Oliphant

CONTENTS

9.1 Energy Sector in Australia

Australia is resource rich in both minerals and energy. It is a net energy exporter with 68% of total energy production being exported and with a value of AUD$71.5B in 2012/2013. It is the eighth largest energy producer in the world but is a net importer of crude oil and refined petroleum products. Coal is Australia's largest energy export earner, with a value of around AUD$40B in 2012–2013, down AUD$8B from 2011 to 2012. China is the largest importer of Australia's coal. Australia is also one of the world's largest exporters of uranium, though it does not have any installed nuclear power plants of its own.

Over the past 10 years, Australia's real export earnings have increased, on average 10% a year. However, they fell 12% in 2012–2013 largely as a result of lower coal prices but then rose by 6% in 2013–2014 supported by higher LNG prices [1].

In 2012–2013, fossil fuels, coal, oil, and natural gas made up about 94.4% of energy consumption. However, there has been a gradual shift away from coal to natural gas. The penetration of renewable energy is still small on a primary energy level at 5.6%—bioenergy and hydro making up 4.7% of this. See Table 9.1 and Figure 9.1 [1].

The high fossil fuel penetration in the energy mix has led to Australia to become one of the world's highest greenhouse gas emitters per capita. However, the contribution from

TABLE 9.1
Australian Energy Consumption by Fuel Type

	2012–2013		Average Annual Growth	
	PJ	**Share (%)**	**2012–2013 (%)**	**10 Years (%)**
Coal	1946	33.1	−5.9	1.4
Oil	2221	37.7	1.3	2.4
Gas	1387	23.6	2.2	3.3
Renewables	330	5.6	11.5	1.9
Total	**5884**	**100**	**−0.5**	**1.1**

Source: BREE 2014, Australian Energy Statistics, Table C.

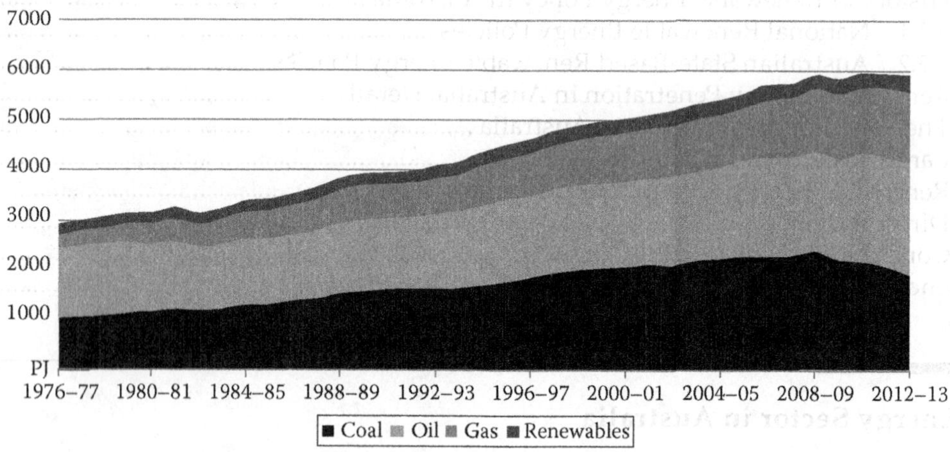

FIGURE 9.1
Australian energy consumption by fuel type. (From BREE 2014, Australian Energy Statistics, Table C.)

wind and solar is starting to increase particularly in the electricity sector (see the following). It is well endowed with a range of possible renewable resources—solar, wind, geothermal, wave, and tidal power. Currently, hydro is the largest source of renewable electricity, but its growth is limited and confined mainly to the State of Tasmania and some parts of the eastern seaboard, and the biomass at present is mainly bagasse from sugarcane in Queensland and landfill gas from around the country.

The electricity sector is also dominated by fossil fuels (86.9% in 2012–2013)—see Table 9.2 and Figure 9.2 that was derived from Table 9.2 [1]—but renewable energy is on the increase, in particular large-scale wind and rooftop solar. A number of successful policy initiatives were put in place between 2000 and 2013 that enabled the growth of renewables in Australia. However, between 2013 and 2015, growth and investment in renewables has stalled due to government policy uncertainties—see later.

The electricity network in Australia is divided into two main sections—the national grid in the East, extending over 5,000 km with 40,000 km of transmission line and cable—and the western Australian (WA) grid. The two grids are not connected because population between the two grids is extremely small (Figure 9.3).

TABLE 9.2

Australia's Electricity Generation, by Energy Source, 2012–2013

	2012–2013		Average Annual Growth	
	GWh	**Share (%)**	**2012–2013 (%)**	**10 Years (%)**
Fossil Fuels	**216509**	**86.91**	**−3.4**	**0.3**
Black coal	111491	44.8	−4.4	−0.9
Brown coal	47555	19.1	−13.6	−1.7
Gas	51053	20.5	5.1	5.7
Oil	4464	1.9	65.2	13.9
Other[a]	1945	0.8	78.8	0.8
Renewable Energy	**32566**	**13.1**	**26.2**	**6.2**
Bioenergy	3151	1.3	26.2	6.2
Wind	7328	2.9	19.9	29.7
Hydro	18270	7.3	29.7	1.3
Solar PV	3817	1.5	49.2	56.4
Geothermal	1.0	0.0	0.0	0.0
Total	**249075**	**100**	**−0.3**	**0.9**

Source: BREE 2014, Australian Energy Statistics, Table O.

[a] Includes multifuel fired power plants.

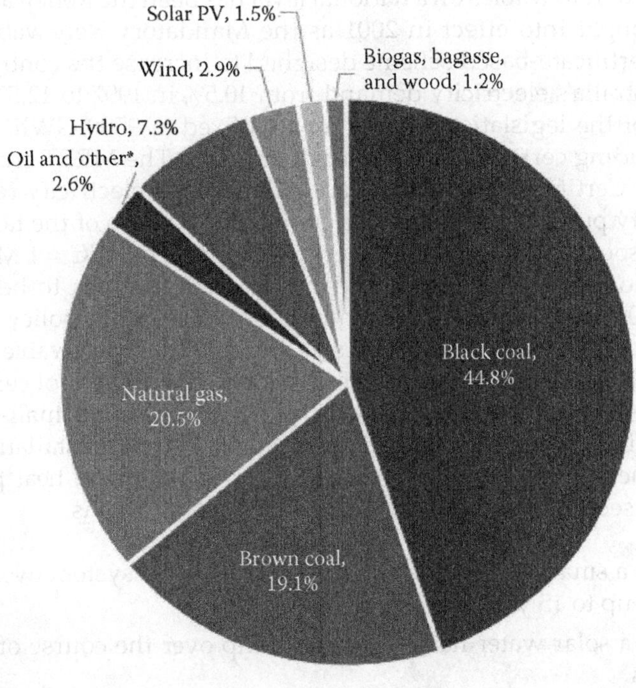

FIGURE 9.2
Australia's electricity generation, by energy source, 2012–2013.

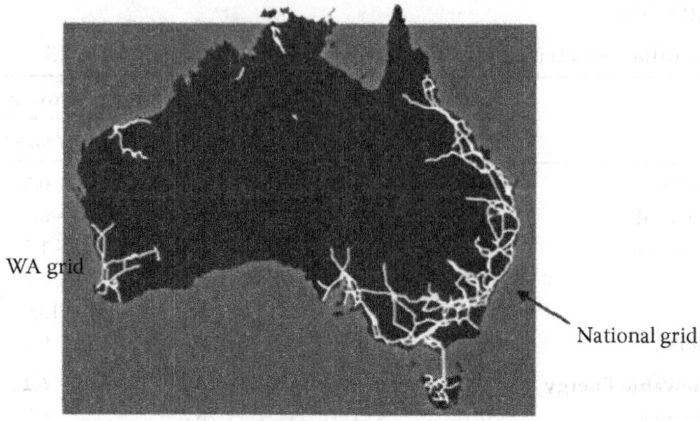

FIGURE 9.3
Australian electricity transmission network. (From Briefing Note: Parsons Brinkerhoff to ENA, March 2009, Energy network infrastructure and the climate change challenge, http://www.aph.gov.au/~/media/wopapub/senate/committee/climate_ctte/submissions/su b307a_pdf.ashx, accessed April 1–16, 2015.)

9.2 History of Renewable Energy Policy in Australia

9.2.1 National Renewable Energy Policies

The main support for renewables on a national level has been the Renewable Energy Target (RET) scheme. Brought into effect in 2001 as the Mandatory Renewable Energy Target (MRET), it was a certificate-based scheme designed to increase the contribution of renewable energy to Australia's electricity demand from 10.5% in 1997 to 12.5% in 2010. During the development of the legislation, the target was fixed at 9500 GWh. Fixing the target was aimed at providing certainty to the energy industry. The MRET created a market for Renewable Energy Certificates by placing an obligation on electricity retailers and other wholesale electricity purchasers to source a proportional share of the target from eligible renewable energy sources (1 Renewable Energy Certificate or REC = 1 MWh).

In 2009, the national RET was expanded by nearly five times to become a target of 45,000 GWh in 2020 to support the Commonwealth Government's policy commitment that "at least 20 per cent of Australia's electricity should come from renewable sources by 2020." At this time, a Solar Credit "multiplier" was applied to the number of certificates received from certain small-scale generation technologies. The creation of Small-scale Technology Certificates (STCs) is a tradable commodity attached to eligible installations of renewable energy systems (including solar panels, solar water heaters, and heat pumps), with the number of STCs based on the amount of electricity in MWh that is

- Generated by a small-scale solar panel, wind, or hydro system over the course of its lifetime of up to 15 years or
- Displaced by a solar water heater or heat pump over the course of its lifetime of up to 10 years

The number of certificates that can be claimed may vary, depending on geographic location.

At the beginning of the program, the PV solar multiplier was 4× and delivered quite a significant capital cost reduction together with the 15-year deeming. In July 2013, the multiplier was reduced to 1 STC per MWh.

The STC multiplier program resulted in a large number of "virtual" certificates being generated, especially when the added state and territories feed in tariffs (FiTs) were introduced around 2009 thereby escalating growth in residential PV. The overall effect was to depress REC prices and discourage investment in large-scale projects, wind and solar farms, etc. So in June 2010, two separate markets were established—the Large-scale Renewable Energy Target (LRET) and the Small-scale Renewable Energy Scheme (SRES). These new markets began operating on January 1, 2011 with the LRET being capped at 41,000 MWh.

The SRES is uncapped, but with an implicit target of 4000 MWh. All PV systems up to 100 kWp are, currently, able to claim STCs up-front for up to 15 years of deemed generation, based on location. This means that the STCs for such systems act as an initial capital cost reduction.

From 2011, the two schemes no longer competed with one another.

On July 1, 2012, a price on carbon was introduced in Australia as part of a broad energy reform package called the Clean Energy Plan, which aimed to reduce greenhouse gas emissions by 5% below 2000 levels by 2020 and 80% below 2000 levels by 2050 plus provide funding to achieve the 20% of electricity from renewables by 2020. The price on carbon aimed to accomplish these targets by encouraging Australia's largest emitters to increase energy efficiency and invest in sustainable energy. The price started at $23.50 in 2012/2013 and was to reach $24.40 in 2014/2015 and then transition to an emissions trading scheme. However, in 2013 there was a change in Federal Government, from Labor to Liberal. One of the first things the Liberals did on achieving power was to fulfil their election promise of "axing the carbon tax," and a "Direct Action" program was introduced instead—see later. They also introduced in 2014 a scheduled RET review headed by a well-known climate skeptic who recommended that the RET be considerably cut back. One rationale for this was that the growth in the electricity sector had slowed down, and it was anticipated that the target of 41,000 GWh in 2020 would be greater than 20% of sales. At the time of writing, April 2015, there were intense political negotiations as to what the new target should be. The new government does not want to go above 32,000 MWh, which would have quite a significant impact on the RE industry. Opponents want a higher target but have suggested a compromise of 33,500 GWh—which has not been accepted.

Historically, the Large-Scale Generation Certificates (LGCs) commanded prices between $40 and $50/MWh but dropped to almost $20/MWh with the uncertainty of the RET, but the price has been steadily increasing in the hope of an imminent resolution of the target. The STCs have remained around $40/MWh since mid-2014 when the number of certificates reduced due to the phasing out of the multiplier and FiT in most states.

The RET uncertainty has had a significant impact on RE investment in Australia, and though globally renewable energy sales increased 16% in 2014 [3] in Australia, Kane Thornton, CEO of the Clean Energy Council, announced in January 2015 that "Investment in large-scale renewable energy projects such as wind and solar farms last year was down 88 per cent to just $240 million (it was $2.1 billion in 2013)—the worst levels we have seen for more than a decade. Australia's renewable energy investment is now lagging behind countries such as Panama, Honduras and Myanmar" [4].

9.2.2 Australian State-Based Renewable Energy Policies

Each state and territory in Australia has its own renewable energy policy, incentives, and energy efficiency programs.

By far, the most effective program has been the FiT—mainly confined to small-scale PV programs. The FiT was first legislated in South Australia in 2008 and came into being in 2009. Most other states then followed. There was no federal FiT program and no consistency in programs around the country. As what happened in many places globally, high incentives were initially introduced, but then as expected targets were rapidly exceeded, the FiT was either reduced or phased out—having a damaging effect on PV installation businesses.

A review of FiT incentives for mid-2013 can be found at http://www.ecosmartelectricians.com.au/index.php?page=what-is-the-feed-in-tariff.

The FiT has been criticized as being the source of increased electricity tariffs and being a mechanism whereby the low-income population subsidize the rich. However, in reality, the FiT contribution to the average retail electricity price in Australia has been estimated to be about 2.3% and the combined RET scheme and FiT about 5%, compared to a network cost contribution of 45.7% [5]. In addition, the majority of the purchasers of PV systems are the middle- to low-income demographic rather than those with high-income demographic; nevertheless, programs are underway to try and reduce cost impacts on disadvantaged sectors of the community.

Though Australia has, as a whole, a target of 20% of its electricity from renewable sources by 2020, two regions of the country have higher targets. The Australian Capital Territory—90% by 2020 and South Australia—50% by 2025 [8] and http://www.environment.act.gov.au/__data/assets/pdf_file/0004/581701/Renewable-energy-brochure_ACCESS.pdf.

9.3 Renewable Energy Penetration in Australia: Detail

The penetration of renewable energy in the primary energy mix of Australia has been reasonably constant at around 4%–6% for about two decades (see Table 9.1). However, the composition has changed from predominantly hydro and biomass around 2000 to currently hydro and wind electricity generation from large-wind farms. The latter having been installed as a result of RET program support—wind being more cost effective than solar technologies to date. So far, no large-scale solar thermal electricity generation projects have been installed in Australia but many projects have been proposed. Though most of the growth in renewable electricity generation has been from wind, its penetration as a percent of total electricity generation in Australia is small, ~3% of total. However, in the state of South Australia (SA), wind averaged 31% of electricity generation during 2013/2014 [6] and 33% by the end of 2014—http://www.renewablessa.sa.gov.au/—up from zero 10 years ago. Additionally, the contribution of rooftop solar to electricity generation has been 6% from 25% of SA households over 2013/2014 giving a total of 39% of intermittent generation on the SA grid. Per capita, this is one of the highest penetrations of intermittent generation globally and has come about not only due to the exploitation of a good wind resource but because of good State Government policy support and legislation in the early years.

The growth of wind in Australia is shown in Figure 9.4 and now is approaching 3500 GW, with about 40% of installed capacity being in South Australia [7].

(For those interested, daily performance data of wind generators on the National Grid can be seen at http://windfarmperformance.info/.)

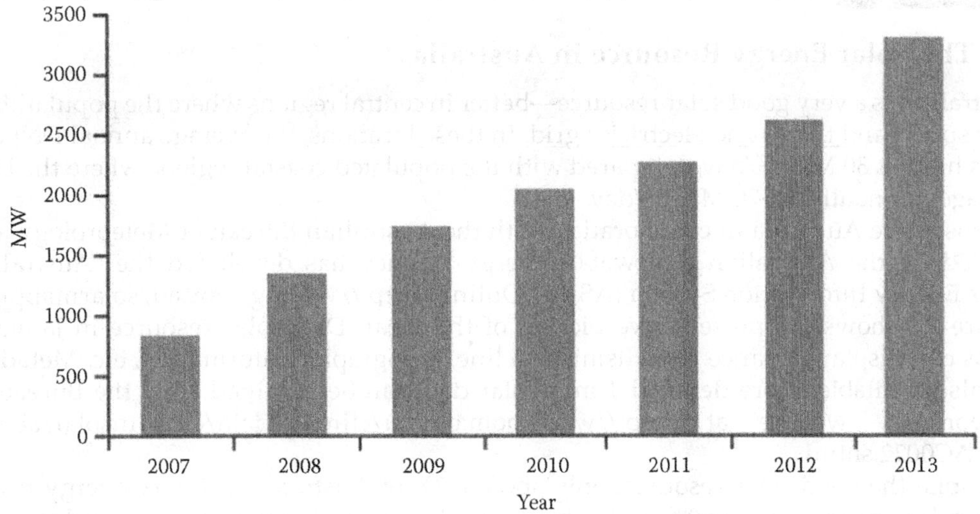

FIGURE 9.4
Cumulative installed wind capacity in Australia.

The installed renewable electricity generating capacity from hydro, wind, and PV (mainly rooftop solar) on the National Energy Market (NEM) grid is shown in Figure 9.5 [7] and totals, excluding large hydro, 12,875 and 28,476 MW, including hydro (August 2014). Adding in Western Australia and the Northern Territory to get an Australia wide total, increases installed renewable capacity by about another 1000 MW. See [8].

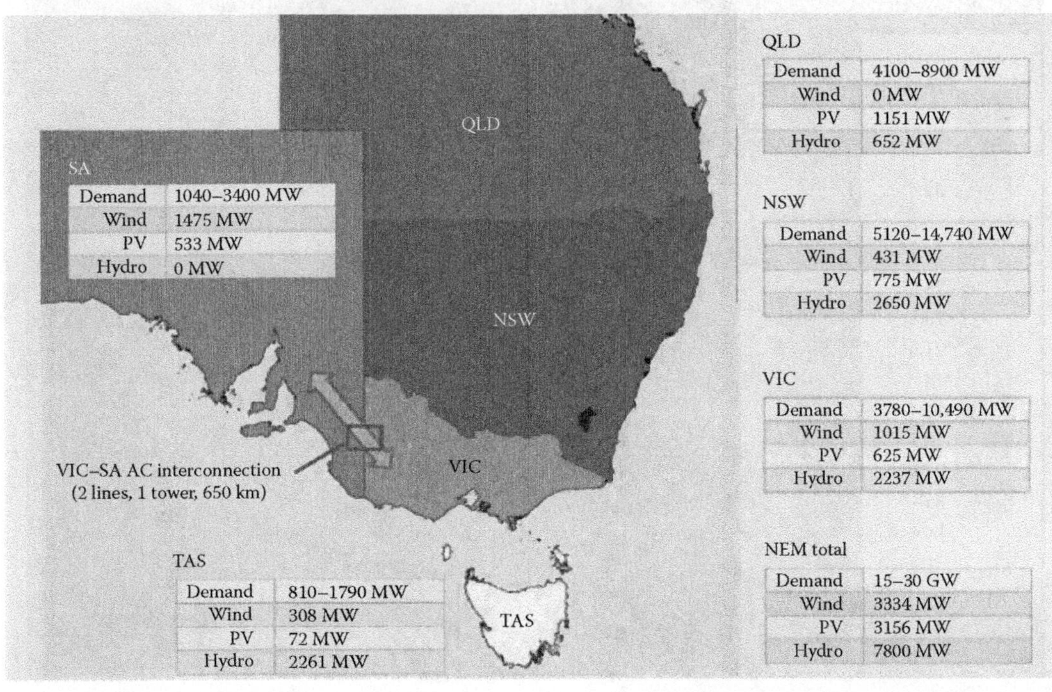

FIGURE 9.5
Renewable generation on the NEM (National Energy Market grid).

9.4 The Solar Energy Resource in Australia

Australia has a very good solar resource—better in central regions where the population is very sparse and there is no electricity grid. In these locations, the average annual DNI can be as high as 30 $MJ/m^2/day$, compared with the populated coastal regions where the DNI averages, annually, 18–20 $MJ/m^2/day$.

Geoscience Australia in collaboration with the Australian Bureau of Meteorology and funded by the Australian Renewable Energy Agency has developed the "Australian Solar Energy Information System (ASEIS) Online" http://www.ga.gov.au/solarmapping. Figure 9.6 shows a representative picture of the mean DNI solar resource in January. Maps can display distance to transmission line, topographic information, etc. Metadata are also available. More detailed 1 min solar data can be obtained from the Bureau of Meteorology website at http://www.bom.gov.au/climate/data/oneminsolar/about-IDCJAC0022.shtml.

Despite the good solar resource, only about 0.2% of Australia's primary energy use is from solar energy—about 90% of this being used in the residential sector for solar water heating and rooftop PV. Over the 10-year period to 2012–2013, electricity generation from solar PV grew at an average annual rate of 56.4%, from a very low base to 1.5 TWh in 2012–2013—see Table 9.2 [1].

Figure 9.7 shows very clearly how the introduction of the FiT in Australia in 2009 increased the number of residential grid-connected PV installations. In 2013, typical PV module prices were \$0.50–\$0.75/W AUD and installed costs \$1.8–\$2.5/W AUD [9].

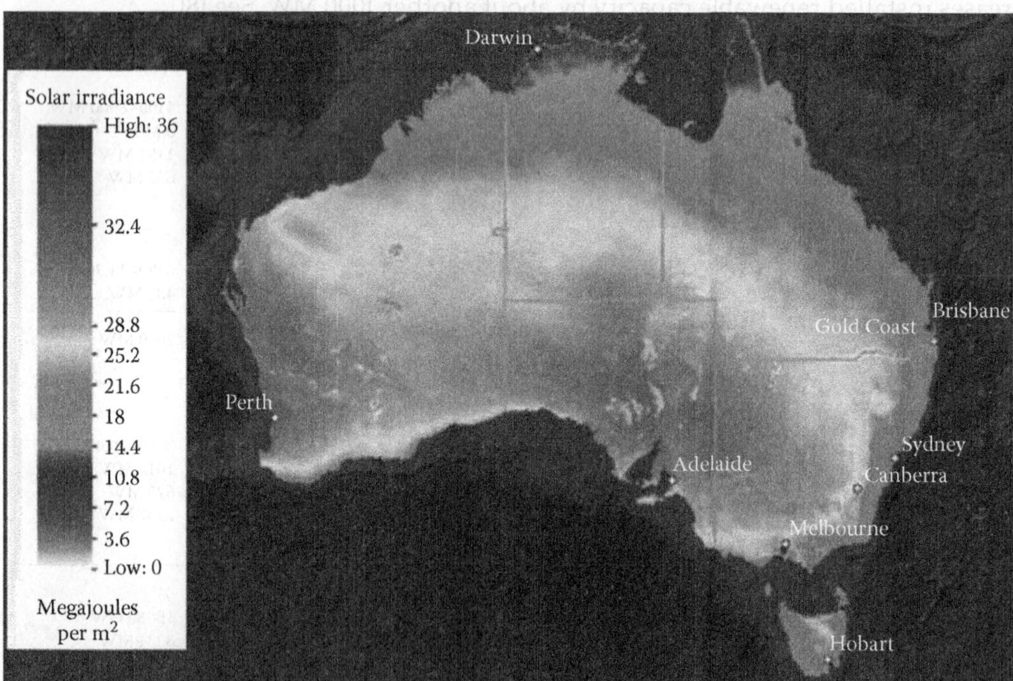

FIGURE 9.6
January mean DNI—Direct Normal Irradiance on a horizontal plane.

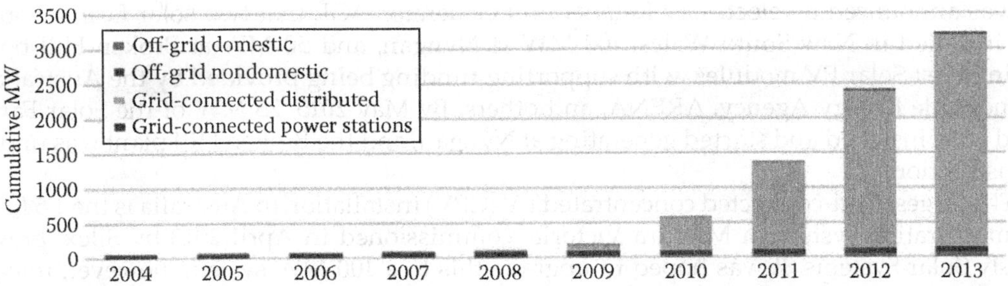

FIGURE 9.7
Cumulative installation of PV in Australia 2004–2013. (From Muriel Watt et al., June 2014, APVI Report: National Survey Report of PV Power Applications in Australia, 2013, http://apvi.org.au/wp-content/uploads/2014/07/PV-in-Australia-Report-2013.pdf, accessed April 1–16, 2015.)

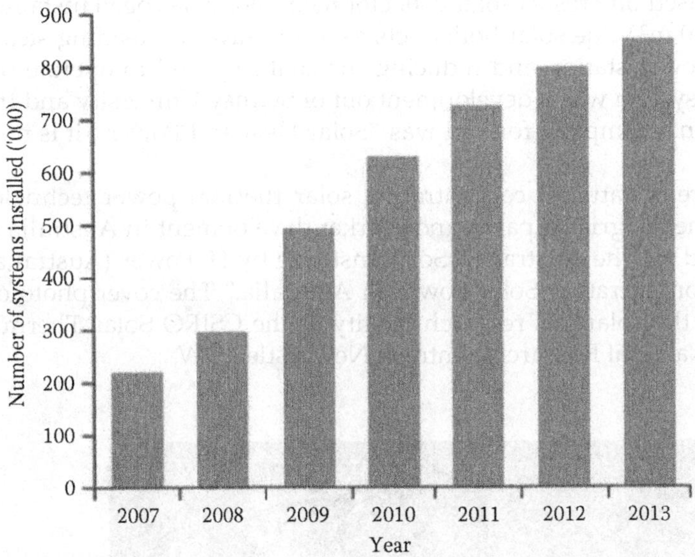

Year	Cumulative number of installations
2007	212,422
2008	297,806
2009	492,497
2010	619,589
2011	724,639
2012	794,105
2013	847,700

FIGURE 9.8
Cumulative installed solar water heaters in Australia.

Australia has no local commercial manufacturer of PV cells and one module assembler—Tindo Solar (South Australia).

There has also been a rapid growth in recent years of solar water heaters—see Figure 9.8. This growth has been helped by the phasing out of high emission storage electric and low-efficiency gas water heaters [10].

9.5 Large-Scale Solar Installations

In October 2012, the 10 MW Greenough River PV Solar Farm near Geraldton, Western Australia was completed using First Solar modules. At that time, it was the largest PV installation and the only one with greater capacity than 1 MW in Australia. In July 2013, it was announced by electricity Retailer and Generator AGL that two solar farms would be installed in New South Wales, 100 MW at Nyngan, and 50 MW at Broken Hill both using First Solar PV modules, with supporting funding being provided by the Australian Renewable Energy Agency, ARENA, and others. By May 2015, 25 MW of the Solar Farm had been installed and started generating at Nyngan, and the Broken Hill plant was under construction.

The largest grid-connected concentrated PV (CPV) installation in Australia is the 1.5 MW demonstration system in Mildura Victoria, commissioned in April 2013 by Silex, previously Solar Systems. It was hoped to upgrade this to a 100 MW system; however, uncertainty in the Federal Government's RET has meant that this upgrade has been abandoned. Project used ultra-high-efficiency (43%) "multi-junction" PV cells, technology originally developed by Boeing to power satellites.

At present, there are no Solar Thermal Electric Power Stations in Australia. However, at the 2000 MW Liddell coal-fired power station 260 km north of Sydney, a 9.3 MW thermal solar-concentrating system based on Fresnel solar collector technology has been installed. (Total mirror surface of 18,500 m^2.) The solar boiler acts as a fuel saver by feeding steam into the existing coal-fired power station and reducing the coal required to operate the facility. The Fresnel collector system was a development out of Sydney University and the University of NSW. The original company formed was "Solar Heat and Power"; it is now owned by Novatech.

A good review of the current status of concentrating solar thermal power technologies, their potential costs, benefits, grid impacts, and market development in Australia is the May 2012 report prepared for the Australian Solar Institute by IT Power (Australia), "Realising the potential of Concentrating Solar Power in Australia." The cover photo on the report, Figure 9.9, shows the solar gas research facility at the CSIRO Solar Thermal Research Hub of the CSIRO National Research Centre at Newcastle NSW.

FIGURE 9.9
Solar gas research facility at the CSIRO solar thermal research hub of the CSIRO National Research Centre at Newcastle NSW.

9.6 Renewable Energy Research in Australia

Australia has a long history of renewable energy research particularly in Solar Energy with groundbreaking work having been conducted on the following:

- High-efficiency PV research—University of New South Wales
- Evacuated tubular collectors—Sydney University
- Solar Thermal "Big Dish"—Australian National University
- Solar Gas Research —CSIRO
- Cooperative Research Centre for Low Carbon Living

to name but a few.

In July 2012, the Australian Renewable Energy Agency (ARENA) was established with a budget of AUD$3 billion extending to 2022 for

- Research, development, demonstration, deployment, and commercialization of renewable energy and related technologies
- Storage and sharing of knowledge and information about renewable energy technologies

The website http://www.arena.gov.au outlines the projects, initiatives, and reports funded by ARENA.

In mid-April 2015, the Australian Federal Government released its Energy White Paper [11] in which less than one page of the 74-page document was devoted to Renewable Energy. An extract from the page is given in the following:

> The Australian Government is providing over $1 billion toward the research, development and demonstration of renewable energy projects. The Australian Renewable Energy Agency (ARENA) was established as a statutory entity to make renewable energy solutions more affordable and increase the amount of renewable energy used in Australia. It has currently committed over $1 billion to more than 200 projects. The Clean Energy Finance Corporation invests in projects that use a commercial approach to overcoming market barriers and mobilising investment in renewable energy, energy efficiency and low emissions technologies. The Australian Government has announced that it will abolish these agencies, but maintain a commitment to existing projects.

The Energy White Paper does not address climate change; this is left to the "Direct Action" program in the Federal Department of the Environment.

9.7 Direct Action

The Australian Government intends to reach its emissions reduction target through its Direct Action Plan "to efficiently and effectively source low cost emissions reductions and improve Australia's environment. This will be done primarily through the Emissions Reduction Fund." See http://www.environment.gov.au/clean-air.

The Emissions Reduction Fund (ERF) is the centerpiece of the Government's climate action policy. It will work with other incentives under the Direct Action Plan to help meet Australia's target of reducing emissions by 5% below 2000 levels by 2020. Design options for the Safeguard Mechanism for the ERF fund are currently in the consultation phase.

A large part of the ERF will be soil carbon and carbon offsets through revegetation programs.

9.8 Conclusion

Australia's energy sector is dominated by fossil fuels, in particular coal. Although the contribution by renewables to primary energy in Australia is small, about 5.6% in 2013, they represented 13.1% of electricity generation. There is a very large renewable energy potential that could be exploited in Australia, with significant natural resources for solar, wind, geothermal, and marine energy and with more limited resources in hydro and bioenergy.

Electricity consumption in Australia has been declining over the past few years due to the increased implementation of energy efficiency programs, installation of rooftop solar and solar water heater systems, and a downturn in the industry sector. In addition, this has led to the shutdown or reduced usage of some coal-fired power stations around the country. It is also behind the reasoning of the Federal Government desire to reduce the RET.

Up to 2013, the RET had already attracted about $18.5 billion of investment in RE into Australia and created about 21,000 jobs. However, after over a year with no target having been decided on and with uncertainty in Government support for renewable energy programs and R & D, investors and investment in large-scale renewable energy programs have significantly declined and there have been many job losses. There has, however, been an increasing interest in community-driven programs, and it is likely that change will be driven by the people. Additionally, it has been found that the electricity grid can support high penetrations of intermittent renewables—up to almost 40% in South Australia already exists. Though problems have been created they can be solved and a paradigm shift in how electricity distributors and retailers will operate in future is likely to come sooner rather than later.

Many are following with interest how renewable energy in Australia will develop in the next few years. There is potential that some areas of Australia could be a model of how to integrate successfully and reliably a range of renewable energy sources into the energy mix if old business models are not followed.

References

1. Australian Government Bureau of Resources and Energy Economics (BREE), July 2014, 2014 Australian energy statistics update, http://www.industry.gov.au/industry/Office-of-the-Chief-Economist/Publications/Documents/energy-in-aust/bree-energyinaustralia-2014.pdf (accessed April 1–16, 2015).
2. Briefing Note: Parsons Brinkerhoff to ENA, March 2009, Energy network infrastructure and the climate change challenge, http://www.aph.gov.au/~/media/wopapub/senate/committee/climate_ctte/submissions/su b307a_pdf.ashx (accessed April 1–16, 2015).

3. AEMC (Australian Energy Market Commission) Review, March 22, 2013, p. 107, Electricity Price Trends Final Report—Possible future retail electricity price movements: July 1, 2012–June 30, 2015, http://www.aemc.gov.au/media/docs/ELECTRICITY-PRICE-TRENDS-FINAL-REPORT-609e9250-31cb-4a22-8a79-60da9348d809-0.PDF (accessed April 1–16, 2015).
4. Bloomberg New Energy Finance, January 9, 2015, http://about.bnef.com/press-releases/rebound-clean-energy-investment-2014-beats-expectations/ (accessed April 1–16, 2015).
5. Kane Thornton, CEC, January 13, 2015, Federal government review collapses renewable energy investment: New analysis, https://www.cleanenergycouncil.org.au/media-centre/media-releases/january-2015/150113-bloomberg.html (accessed April 1–16, 2015).
6. Climate Council, The Australian Renewable Energy Race: Which stats are winning or losing?, 2014, http://www.climatecouncil.org.au/uploads/ee2523dc632c9b01df11ecc6e3dd2184.pdf (accessed April 1–16, 2015).
7. AEMO (Australian Energy Market Operator), August 8, 2014, 2014 South Australian Electricity Report.
8. Joint AEMO and ElectraNet Report, October 2014, Renewable Energy Integration in South Australia, http://www.aemo.com.au/Electricity/Planning/Integrating-Renewable-Energy (accessed April 1–16, 2015).
9. Muriel Watt et al., June 2014, APVI Report: National Survey Report of PV Power Applications in Australia, 2013, http://apvi.org.au/wp-content/uploads/2014/07/PV-in-Australia-Report-2013.pdf (accessed April 1–16, 2015).
10. Clean Energy Council, Clean Energy Australia Report, 2013, https://www.cleanenergycouncil.org.au/policy-advocacy/reports/clean-energy-australia-report.html (accessed April 1–16, 2015).
11. Australian Government Department of Industry and Science, April 2015, 2015 Energy white paper, http://ewp.industry.gov.au/ (accessed April 1–16, 2015).

10

Japan's Post-Fukushima Energy Policy

Keibun Mori

CONTENTS

10.1 Introduction

Japan's approach to greenhouse gas (GHG) mitigation long advocated not only for energy efficiency and renewable energy, but also for nuclear energy. This three-pronged approach, along with the contributions from Kyoto mechanisms and economic stagnation, had been effective in meeting the Kyoto Protocol obligation. However, in response to the massive earthquake and associated nuclear accident in March 2011, the Japanese government issued a moratorium on the restart of nuclear reactors and new construction, and the share of nuclear power in electricity generation declined sharply from 29% in FY 2010 to 11% in FY 2011, and to only 1.0% in FY 2013.[1] A nationwide electricity-saving campaign slashed demand by 5%, and Japan was able to avert an electricity shortage due largely to successful nationwide energy saving campaign. Still, the sustained moratorium on nuclear energy caused the increased mobilization of thermal plants fueled by natural gas and petroleum, resulting in adverse impact on electricity costs and GHG emissions[2] (Figure 10.1).

Under these circumstances, the Noda administration of the Democratic Party of Japan (DPJ) in 2012 drafted a new energy plan called *the Innovative Strategy for Energy and the Environment*, which aspires to phase out all nuclear power plants during the 2030s through accelerated renewable energy deployment (17.7% in total electricity generation by 2020 (Table 10.1)) and rigorous energy conservation measures (total energy consumption 19% below 2010 levels by 2030).[3] Nonetheless, its forecast however shows that without nuclear power, the emissions in 2020 will be below 1990 levels only by 5% at most, far short of the reduction goal of 25% below 1990 levels by 2020, set in 2009. Furthermore, electric utilities and energy intensive industries raised the concerns over the grid stability and economic costs of such a drastic change in the electricity generation fuel mix.

For these reasons, the subsequent Abe administration of the Liberal Democratic Party (LDP) shelved this ambitious phase-out plan and instead revised *the Basic Energy Plan*,

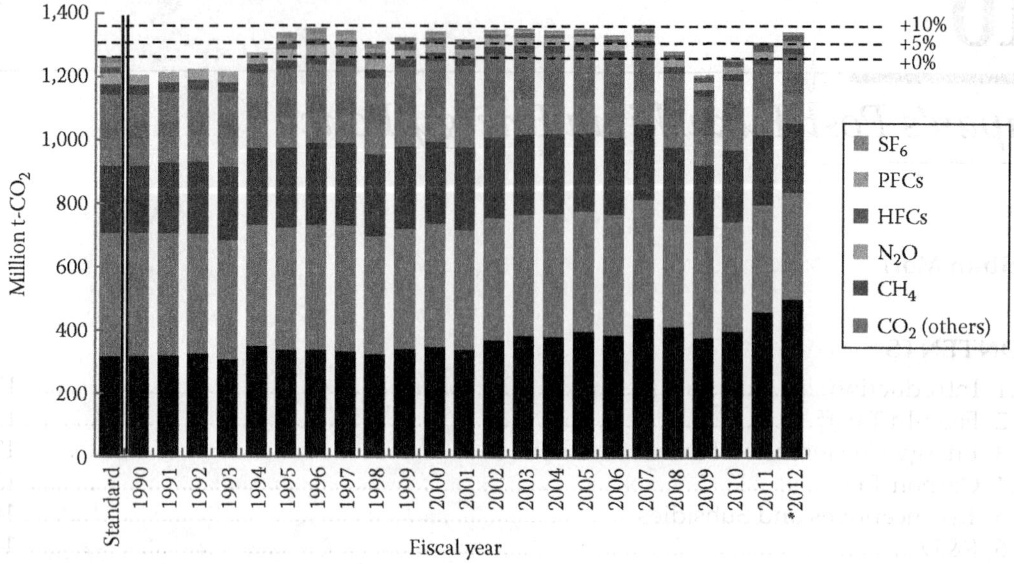

FIGURE 10.1
GHG emission contributions and trend. * The figure for 2012 is. (From Greenhouse Gas Inventory Office of Japan, National greenhouse gas inventory report of Japan, 2013, http://www-gio.nies.go.jp/aboutghg/nir/nir-j.html.)

TABLE 10.1
Renewable Energy Target in the Innovative Strategy for Energy and the Environment

	2010		2020		2030	
Source	**TWh**	**%**	**TWh**	**%**	**TWh**	**%**
Renewable	114.5	10.5	192.4	17.7	308.0	28.7
Hydropower	89.4	8.2	109.1	10.0	117.5	11.0
Photovoltaic	3.8	0.3	35.1	3.2	66.6	6.2
Wind	4.3	0.4	16.9	1.6	66.3	6.2
Geothermal	2.6	0.2	7.5	0.7	21.9	2.0
Biomass	14.4	1.3	23.6	2.2	32.8	3.1
Marine	0.0	0.0	0.0	0.0	3.0	0.3
Others	976.3	89.5	897.5	82.3	764.3	71.3
Total	1,090.8	—	1,089.9	—	1,072.3	—

Source: Cabinet Office of Japan, Innovative strategy for energy and the environment, 2012, http://www.cas.go.jp/jp/seisaku/npu/policy09/sentakushi/database/index.html.
Note: No nuclear in 2020 and 2030, robust growth, and before additional measures.

a long-term vision previously drafted in 2010. The revised *Basic Energy Plan* repositions nuclear power as *an important baseload electricity source* and describes the prospect of renewable energy to be *further above* the past forecasts, footnoting 13.5% by 2020 and 20% by 2030 as a reference, in lieu of the explicit numerical target set by the Noda administration. In 2015 the Abe administration plans to establish new targets of electricity generation fuel mix as well as GHG emissions for 2030, and the new mix is likely to have a significant share of nuclear energy while showing less ambitious role of renewable energy.

In practice, in spite of the first approval of the restart of a reactor under the new safety regulation by the politically-independent Nuclear Regulation Authority in 2014, a more risk-averse public makes it unrealistic to count on the restart of high-risk reactors or new construction for the foreseeable future. Moreover, the authority established the so-called 40-year rule which requires *special safety checks* on the restart of reactors after 40 years of operation that in principle makes the restart impossible without costly retrofits; in fact, several utilities are leaning toward decommissioning outdated reactors after experiencing the complex and lengthy approval process for relatively new reactors. When combined with the continuing needs to reduce GHG emissions, these tight limits on Japan's nuclear future prompted the country to roll out a number of policy actions accelerating renewable energy development and energy efficiency investment.

10.2 Feed-In Tariff

A feed-in tariff (FIT) program for renewable energy, introduced in July 2012, is perhaps the most drastic policy change after the accident, replacing a facile renewable portfolio standard (RPS). The FIT aims at stimulating renewable energy development through government-guaranteed 20-year power purchase contracts* with a predetermined purchase price (tariff). An independent expert panel determines the tariff rates for each technology type and facility scale to expedite investment in a diverse array of renewable energy projects, and it revises the rates every year to reflect various changes such as a fall in solar panel costs (Figure 10.2).

Due to the urgent need for new power generation amid the moratorium on the restart of nuclear power plants, the expert panel granted premium rates in the first year to enable investors to secure financing promptly. The rates were then trimmed each year for PV, after the premium rates and the legally-binding purchase system resulted in extraordinary growth of utility-scale PV, whereas new categories were created in FY 2014 and FY 2015 for several costly but promising technologies such as offshore wind and small-scale biomass.

The outcome of the FIT has been mixed thus far. According the Agency for Natural Resources and Energy (ANRE), the cumulative installed capacity of renewable energy jumped from 20.3 GW at the program start to 35.2 GW in November 2014, plus 58.6 GW worth of authorized projects in the pipeline. PV was responsible for most of this growth, showing that other energy sources have yet to gain traction (Figure 10.3). The likely causes of the apparent favoritism for PV are (1) favorable tariff structure, (2) an environmental assessment waiver, and (3) siting advantages. The basis of the tariff rates was the national study of life cycle costs of electricity generation in 2011,[4] but the costs of solar panels have fallen sharply since then, enabling the investors to exploit the price differentials between the actual and assumed costs. The second factor favors utility-scale PV by waiving the lengthy environmental assessment process, which normally takes at least 3 years for most renewable energy development, for PV projects covering an area of 50 ha or less and having no adverse impact on land use. Utility-scale PV also has siting advantage as it can be built on most vacant property, whereas wind, geothermal, and biomass power plants typically have to be located in remote areas with little transmission infrastructure to major energy consumption centers. These factors gave PV competitive advantage over other renewable sources, causing the concentration of investment in a single generation technology, at least for now.

The heavy growth in PV is causing several complications. First of all, the capacity factor, a ratio of its actual output over a period of time (e.g., kWh) to its potential output (e.g., kW), is

* For small-scale photovoltaic and geothermal, the contract duration is 10 and 15 years, respectively.

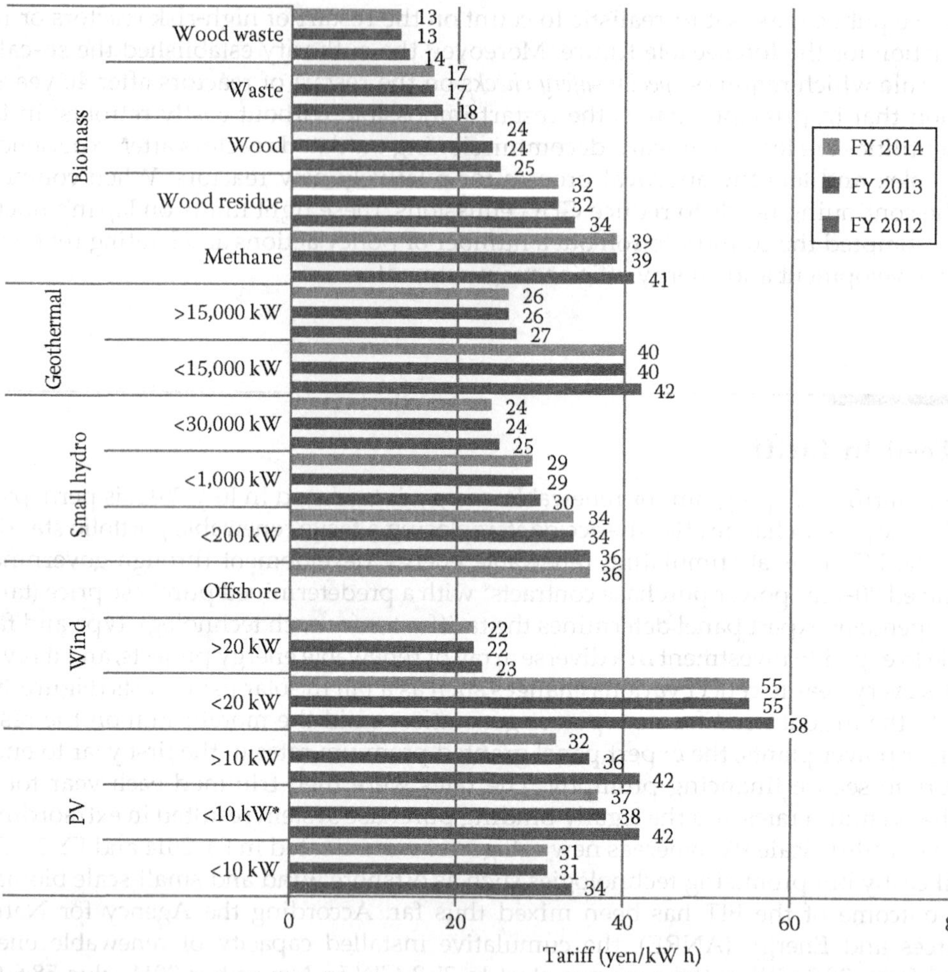

FIGURE 10.2
Tariff rate of the FIT program. (From Agency for Natural Resources and Energy, Avoidable cost calculation method and facility certification system reform, 2014, http://www.meti.go.jp/committee/sougouenergy/shoene_shinene/shin_ene/kaitoriseido_wg/pdf/003_s01_00.pdf.)

so low for PV that the development figure in capacity basis tends to mislead the public about the impact on national fuel mix. In fact, despite the massive introduction of renewable energy in capacity basis, the share of renewable energy (excluding large-scale hydropower) in electricity generation increased only by 0.8% between FY 2011 and FY 2013, from 1.4% to 2.2%.[1]

Lastly and most importantly, the impact on grid stability is also a concern for electric utilities, as the current intraregional grid system is designed to balance the load and generation within a single region and thus can integrate only a limited amount of variable energy sources. In fact, in late 2014, several major electric utilities declared the moratorium for granting grid-access for new renewable energy developments despite the legal obligation, citing the limitation of the grid ability to handle variable energy sources. In response, ANRE established new rules in January 2015, allowing electric utilities to curtail grid-access for PV and wind energy producers for up to 360 hours per year without compensation, along with possible unlimited curtailment without compensation for future project developers if the

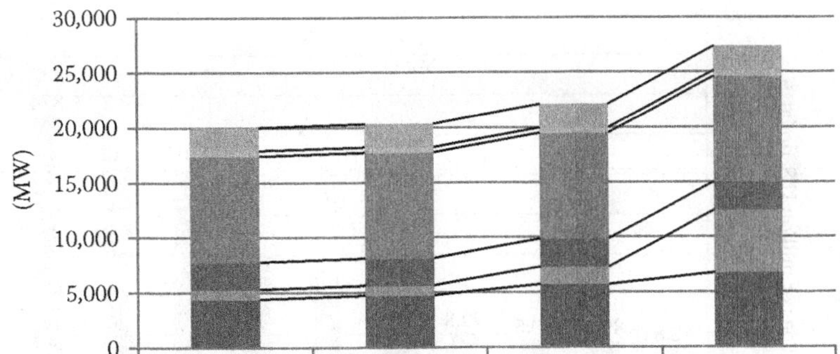

	March 2012	June 2012 (FIT start)	March 2013	December 2013
Biomass	2,100	2,106	2,136	2,225
Geothermal	500	500	501	501
Small hydro	9,600	9,602	9,605	9,608
Wind	2,500	2,500	2,563	2,574
PV (Utility-scale)	900	902	1,606	5,731
PV (rooftop)	4,400	4,700	5,669	6,716

FIGURE 10.3
Cumulative renewable energy capacity development. (From Agency for Natural Resources and Energy, Avoidable cost calculation method and facility certification system reform, 2014, http://www.meti.go.jp/committee/sougouenergy/shoene_shinene/shin_ene/kaitoriseido_wg/pdf/003_s01_00.pdf.)

utility determines that its grid is saturated with existing generation capacity. While major curtailment is unlikely to occur in near future due to the slow approval process for restart of nuclear reactors, this new rule crowds the long-term financial certainty for variable energy developers and thus is expected to stagnate new developments, especially utility-scale PV.

These circumstances suggest that there needs to a fine-tuning to provide more incentives for other renewable energy sources and new mechanisms to promote load and generation shaping measures that enable grid integration of more variable resources, such as expanded use of demand management strategies, flexible backup energy sources, interregional power interchange, and energy storage facilities. Secondly, the high costs of PV could possibly compromise Japan's global economic competitiveness through increased electricity prices, particularly in energy-intensive sectors such as the steel industry. ANRE reports the national costs of the FIT program amount to 130 billion yen (approx. U.S. $1.3 billion) for FY 2012, which is anticipated to grow to 313 billion yen (approx. U.S. $3.1 billion) in FY 2013,[5] and the increasing economic burden could adversely impact low-income households and energy-intensive industries in the coming years.

10.3 Energy Conservation Law

Since the 1970s oil crisis, Japan has made massive investments in energy conservation technologies and practices, leading Japan to achieve one of the lowest energy intensities (per GDP) in the world[6] despite the presence of energy-intensive manufacturing industries

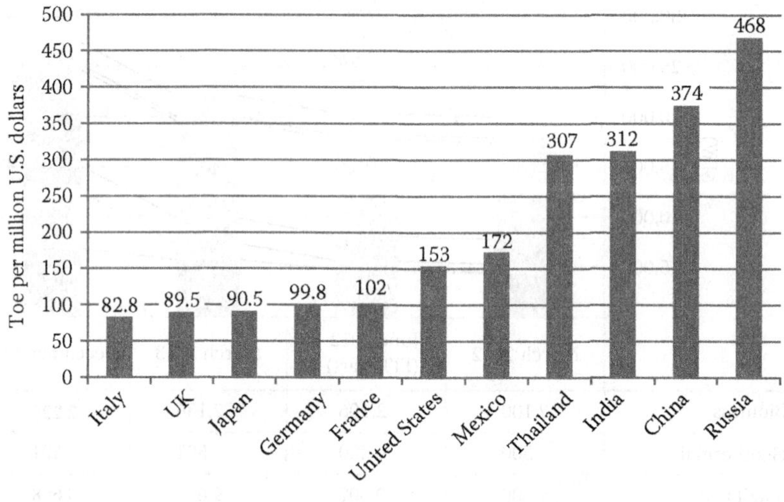

FIGURE 10.4
Energy use per nominal GDP in 2010. (From Energy Data and Modeling Center, *Handbook of Energy and Economic Statistics in Japan*, 2013.)

(Figure 10.4). The energy conservation law has been the primary policy instrument propelling energy efficiency investment; it was first enacted in 1979 and amended in 1998, 2002, 2005, 2008, and 2013 to expand the target and enhance the enforcement mechanism. The law not only targets specific emitting equipment such as trucks and refrigerators, but also requires businesses to draft and implement energy conservation plans for their factories, buildings, and fleets. For example, businesses reporting total energy use of 1,500 tons of oil equivalent (toe) per year must draft and implement a long-term plan to improve efficiency by 1% or more annually, and the latest amendment in 2013 encourages a *peak cut*—curbing electricity demand during peak hours through measures such as demand shift, cogeneration, and storage battery installation.

For equipment, the law establishes *the Top Runner Program*, which sets energy efficiency standards based on the performance of the best available technologies (*top runners*) for each item.[7] The program began in 1999 with standards for 11 items, and expanded the coverage over time to 27 items (Table 10.2). The focus of the program has been on automobiles and office and home appliances, but the latest amendment in 2013 added building materials as an enabling item to save energy used in other equipment. The impact of the program has been outstanding: for instance, the energy use of a standard air conditioner decreased by 43.3% between 1995 and 2012,[6] and the average fuel efficiency of a passenger car rose by 58.5% between 1993 and 2012.[8]

The regulatory approaches under the energy conservation law have been successful in Japan due largely to the unique cooperative relationship between businesses and regulators, and the law gained popularity among the general public as a plain and transparent regulation.[9] Nonetheless, some businesses claim the uniform reduction mandate on commercial energy use is unfair, as the difficulty in achieving the target depends on the baseline energy use, where those who acted early to reduce their energy use now have to work harder to meet the requirement. The *Top Runner Program* now applies to roughly half of the energy use in the residential and commercial sectors, but misses some energy-intensive equipment such as washers, dryers, commercial refrigerators, and commercial freezers. The program sets separate standards for size-based subcategories,

TABLE 10.2

Target Items in Top Runner Program

Start Year	Item		
FY 1999	Passenger vehicle	Freight truck	Air conditioner
	Television	Fluorescent lamp	Copying machine
	Magnetic disk unit	Video cassette recorder	Computer
	Freezer	Refrigerator	
FY 2002	Space heater	Gas stove	Electric toilet seat
	Petroleum water heater	Gas water heater	Vending machine
	Transformer		
FY 2006	Rice cooker	Microwave oven	DVD recorder
FY 2008	Router	Switching unit	
FY 2012	Printer	Multifunction printer	Electric water heater
FY 2013	Thermal insulator		

Source: Agency for Natural Resources and Energy, Comprehensive energy and environmental strategy research report, 2013, http://www.meti.go.jp/meti_lib/report/2013fy/E003456.pdf.

and some analysts have argued that it may be discouraging downsizing of passenger cars, televisions, or other items. These criticisms suggest that, though the 36-year-old law has made important contributions to energy conservation in Japan, it still has room for improvement.

10.4 Carbon Tax

In response to the oil crisis, Japan imposed heavy taxes on petroleum, but not on other fossil fuel sources. In October 2012, however, Japan introduced a carbon tax (*tax for global warming countermeasures*), whose rate is based on carbon content, and thus giving largest impact on coal prices due to its high carbon intensity. The tax rate began at 95 yen (approx. U.S. $1) per t-$CO_2$ and is slated to reach a cap of 289 yen (approx. U.S. $3) in 2015. While it is still relatively small compared to CAN $30 per t-$CO_2$ in British Columbia, Canada, Japan's carbon tax covers all fuel uses, including aviation and industrial use, which are often exempted or taxed differently in other areas, and is in addition to a variety of other high energy taxes (Figures 10.5 and 10.6). It is expected to raise 262 billion yen (approx. U.S. $2.62 billion) in FY 2015 and thereafter, and its revenues are dedicated for energy efficiency and renewable energy programs, unlike British Columbia's revenue-neutral approach where the revenues are offset by reductions in other taxes.

The major intent of the tax is to incentivize energy efficiency and renewable energy investment by sending price signals to favor energy-saving technologies and behaviors. In addition, the elevated costs of fossil fuels are expected to impact electricity generation fuel mix in favor of other sources, and the revenues allocated to various R&D and rebate programs are expected to accelerate the development and deployment of new, low-carbon technologies. The Ministry of the Environment (MOE) estimates that the tax will lower 2020 emissions by 0.5%–2.2% relative to a no-action baseline,[10] but this estimate does not account for impacts on fuel mix and behavioral change, so the impact could be substantially larger if accounting for these factors.[11]

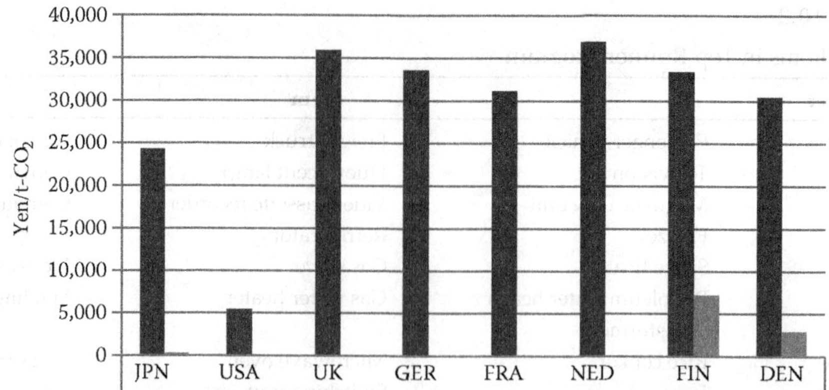

	JPN	USA	UK	GER	FRA	NED	FIN	DEN
Total tax	24,241	5,424	35,892	33,614	31,167	36,887	33,403	30,452
Carbon tax	289	0	0	0	0	0	7,189	2,950

FIGURE 10.5
Energy taxes on gasoline (2014). (From Ministry of the Environment, Interim report from the environmental tax commission, 2013, https://www.env.go.jp/policy/tax/conf01.html.)

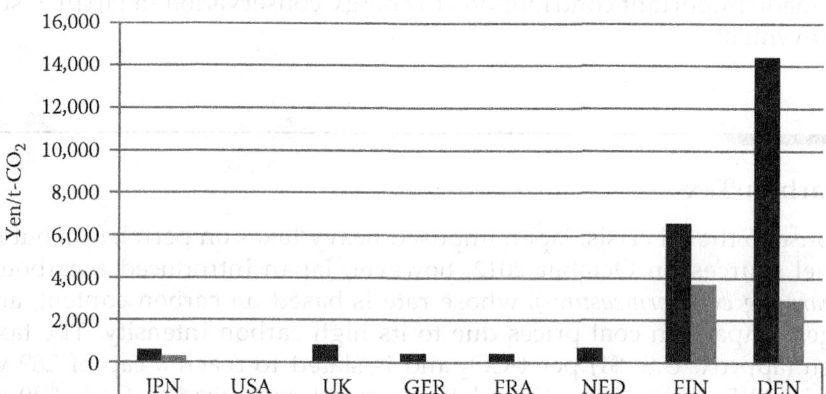

	JPN	USA	UK	GER	FRA	NED	FIN	DEN
Total tax	590	0	816	434	434	692	6561	14,384
Carbon tax	289	0	0	0	0	0	3708	2,947

FIGURE 10.6
Energy taxes on coal (2014). (From Ministry of the Environment, Interim report from the environmental tax commission, 2013, https://www.env.go.jp/policy/tax/conf01.html.)

A carbon tax is an economy-wide policy tool to curb GHG emissions, and the resulting price signal can incentivize low-carbon technologies and behaviors without the significant implementation cost and economic inefficiency of regulatory approaches. While MOE's expert panel recommends more extended use of a carbon tax,[12] energy-intensive industries argue that a higher tax could lead to a leakage problem in which economic activities, and their GHG emissions, are shipped overseas. For this reason, worldwide implementation or some form of a border tax may be needed for more extended use of a carbon tax.[13]

TABLE 10.3

Target Items in Green Investment Tax Break

Area	Item	
Renewable energy	Photovoltaic	Wind, hydrothermal
	Snowpack storage	Biomass
Transportation	Hybrid electric truck	Plug-in hybrid vehicle, electric vehicle
	EV charging system	Hybrid electric construction machinery
Commercial Building	Insulated window	Efficient ventilation system, solid state lighting (LED)
	Efficient air conditioner	Building energy management system (BEMS)
Industrial and power sector	Cogeneration system	Combined-cycle gas turbine, efficient industrial furnace
	Efficient processing tool	High-voltage transmission system

10.5 Tax Incentives and Subsidies

The government has long adopted various tax incentive and subsidy programs for individual items such as the Eco-car Tax Break and the Residential PV Installation Support Subsidy. The Green Investment Tax Break* is perhaps the most comprehensive tax incentive program for corporations, targeting a variety of items in renewable energy, transportation, building, and industrial efficiency (Table 10.3). The government launched the program in 2011 to reduce the tax burden through bonus depreciation—30% of purchase and installation costs—or 7% tax credit for corporate income tax. This tax scheme is similar to the Business Energy Investment Tax Credit in the United States but covers a broader set of low-carbon technologies.

The total tax break from the program amounted to 97.6 billion yen (approx. U.S. $976 million) in the first year, and a survey estimates it induced 7%– 45% of additional investment, depending on the item.[14] The same study forecasts the annual GHG reduction to total 3.1 million t-CO_2 in 2015, equivalent to 0.2% of the national emissions, from the FY 2011 tax break alone. Although the impact of the program will become larger as it continues to induce more investment, the major concern is that the beneficiaries are limited to profit-making firms paying corporate income tax. The worldwide recession and the 2011 earthquake collaborated to put over 70% of Japanese enterprises reporting losses or accumulated losses in FY 2011,[15] so the vast majority of businesses were simply not eligible for the program. The economic recovery is broadening the coverage, but a fundamental problem remains that many small and startup businesses tend to report losses. For this reason, a tax break on other taxes such as property tax may be a better policy instrument to attract investment from all businesses.

10.6 R&D

Japan has been one of the R&D leaders in low-carbon technologies, successfully commercializing various low-carbon technologies such as hybrid and electric vehicles, heat pump water heaters, PV cells, carbon fiber, power electronics, hydrogen fuel cells, and

* Translated from the original name, *Green Toshi Genzei*.

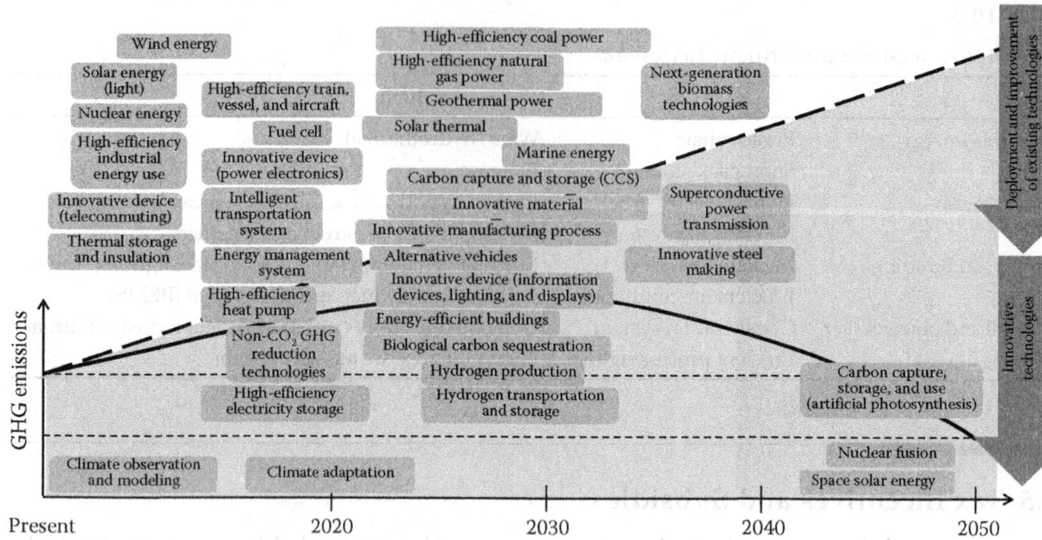

FIGURE 10.7
Focus areas in *the Environmental Energy Technology Innovation Plan*. (From Council for Science and Technology Policy, The Environmental Energy Technology Innovation Plan, 2013, http://www8.cao.go.jp/cstp/gaiyo/kankyo/20130809.html.)

LED light bulbs. To build on these successes, in 2013, the Abe administration revised a comprehensive R&D roadmap for low-carbon technologies called the *Environmental Energy Technology Innovation Plan*.* The plan is a basis for government R&D funding allocation, and lists 37 technology areas to cut today's world GHG emissions in half by 2050 (Figure 10.7),[16] of which five are new for 2013: (1) innovative materials such as carbon-fiber-reinforced plastic (CFRP) and high tensile strength steel (HTSS), (2) artificial photosynthesis to produce various chemicals such as hydrogen and olefin, (3) geothermal energy including enhanced geothermal system (EGS) and advanced exploration and drilling techniques, (4) solar thermal energy such as concentrated solar power (CSP) and direct use of thermal energy in heating and cooling, and (5) ocean energy such as tidal and current power.

Shifting attention to ongoing R&D activities, the government is boosting funding for the following prospective technologies: (1) offshore wind, particularly massive floating wind farms tested off the shore of Fukushima and Nagasaki, (2) hydrogen and fuel cells for residential application branded as Ene-Farm and for automobile use, rolled-out for the mass in 2014, (3) advanced thermal power such as advanced ultrasupercritical (A-USC) and integrated coal gasification combined cycle (IGCC) demonstration, (4) advanced electronics such as normally off computing, organic light-emitting diodes (OLED), and ultralow voltage devices, and (5) cost reduction and capacity enhancement of various types of battery such as Li-ion, sodium-sulfur, and redox flow batteries. The government envisions these technologies to play a pivotal role in curbing the emissions not only in Japan but also elsewhere in the world through export and technology transfer.

* Translated from the original name, *Kankyo Energy Gizyutsu Kakushin Keikaku*.

10.7 Conclusion

The earthquake and associated nuclear incident in 2011 triggered a variety of policy changes to propel energy efficiency and renewable energy investment. The policy instruments covered in this section will be crucial in curbing GHG emissions in Japan, but it is difficult to assess if the nation as a whole will be able to achieve significant reduction amid the uncertain future of nuclear energy. To maximize Japan's contribution to the world's efforts to reduce the emissions, Japan needs to actively share its technologies through export, licensing, and mostly notably a technology transfer instrument called Joint Crediting Mechanism (JCM). Moreover, the lessons learned from policy implementation in Japan, including the benefits, drawbacks, and areas for improvement, could be valuable for other nations to craft optimal strategies to encourage energy efficiency and renewable energy investment.

References

1. Federation of Electric Power Companies of Japan (2013). Electricity generation fuel mix. http://www.fepc.or.jp/about_us/pr/sonota/__icsFiles/afieldfile/2013/05/17/kouseihi_2012.pdf.
2. Greenhouse Gas Inventory Office of Japan (2013). National greenhouse gas inventory report of Japan. http://www-gio.nies.go.jp/aboutghg/nir/nir-j.html.
3. Cabinet Office of Japan (2012). Innovative strategy for energy and the environment. http://www.cas.go.jp/jp/seisaku/npu/policy09/sentakushi/database/index.html.
4. Cabinet Office of Japan (2011). Cost examination commission report. http://www.cas.go.jp/jp/seisaku/npu/policy09/archive02_hokoku.html.
5. Agency for Natural Resources and Energy (2014). Avoidable cost calculation method and facility certification system reform. http://www.meti.go.jp/committee/sougouenergy/shoene_shinene/shin_ene/kaitoriseido_wg/pdf/003_s01_00.pdf.
6. Energy Data and Modeling Center (2013). *Handbook of Energy and Economic Statistics in Japan.*
7. Agency for Natural Resources and Energy (2010). Top runner program. http://www.enecho.meti.go.jp/policy/saveenergy/toprunner2011.03en-1103.pdf.
8. Ministry of Land, Infrastructure, Transport and Tourism (2013). Automobile fuel efficiency. http://www.mlit.go.jp/jidosha/jidosha_fr10_000010.html.
9. Nordqvist, J. (2006). Evaluation of Japan's top runner programme. https://lup.lub.lu.se/search/publication/1026063.
10. Ministry of the Environment (2012). Introduction of a tax for global warming countermeasures. http://www.env.go.jp/policy/tax/about.html#sec06.
11. Mori, K. (2012). Modeling the impact of a carbon tax in Washington State. *Energy Policy* 48, 627–639.
12. Ministry of the Environment (2013). Interim report from the environmental tax commission. https://www.env.go.jp/policy/tax/conf01.html.
13. Morgenstern, R.D., J.E. Aldy, E.M. Herrnstadt, M. Ho, and W.A. Pizer (2007). Competitiveness impacts of carbon dioxide pricing policies on manufacturing, Assessing US Climate Policy Options, Washington, DC. http://rff.org/RFF/Documents/CPF_COMPLETE_REPORT.pdf.
14. Agency for Natural Resources and Energy (2013). Comprehensive energy and environmental strategy research report. http://www.meti.go.jp/meti_lib/report/2013fy/E003456.pdf.
15. National Tax Agency (2013). Corporate sample survey. http://www.nta.go.jp/kohyo/tokei/kokuzeicho/kaishahyohon2011/pdf/kekka.pdf.
16. Council for Science and Technology Policy (2013). The environmental energy technology innovation plan. http://www8.cao.go.jp/cstp/gaiyo/kankyo/20130809.html.

11

Policies for Distributed Energy Generation

David M. Sweet

CONTENTS

11.1 Introduction

Decentralized energy* (DE) is showing ever-increasing promise as a cost-effective method of electricity supply, resulting in significant environmental benefits. Though every sector of the DE industry is showing impressive growth, the total share of DE in overall global electricity capacity remains at only about 7.2%.[1] Lack of clear policy in the energy sector is one of the main barriers preventing greater investment in DE. It is clear that adopting policies that ensure an independent regulator combined with open

* DE technologies consist of the following forms of power generation systems that produce electricity at or close to the point of consumption:

- High efficiency cogeneration/CHP
- Onsite renewable energy systems
- Energy recycling systems, including the use of waste gases, waste heat, and pressure drops to generate electricity onsite

WADE classifies such systems as DE regardless of project size, fuel or technology, or whether the system is on-grid or off-grid.

markets that reward investors for noneconomic benefits of DE will be key in driving future growth in the sector.[2]

In June 2004, the largest renewables energy event to date was held in Bonn, Germany. Over 3000 attendees from 154 nations were present. Though the conference dealt only with renewables, to a large extent, the recommendations would also hold true for all forms of DE, including cogeneration. Arising from that event were some guidelines for policy makers as to what must be put in place in order for DE policy to be effective. For DE to become an integral part of the energy mix, policymakers must*

1. Integrate DE into overall energy policy
2. Establish clear goals
3. Establish transparent market conditions
4. Reduce or eliminate subsidies for conventional centralized energy
5. Address cost issues
6. Create incentives that are eventually phased out
7. Ensure that energy issues form part of the basis of decisions in *nonenergy* sectors such as urban planning and infrastructure, etc.
8. Educate the public on the benefits and costs of DE
9. Promote DE jobs and training
10. Develop public organizations to promote DE
11. Strengthen regional and international cooperation on DE matters
12. Secure grid access for DE
13. Do not omit thermal energy in decision making
14. Support research and development in DE
15. Involve a range of actors in promoting DE including local governments, financiers, international bodies, and multilateral and bilateral banks and institutions
16. Harness the power of public procurement

However, in general, it is important that a policy be stable enough in the long term to foster confidence to everyone that the rules will not change after they have committed to one plan of action. Involving locals in the establishment of DE policies is the other key to both garnering public support and establishing a basis of shared experience on which future policy revisions can be made.

The aforementioned list illustrates some of the goals of policies to promote DE. There are a multitude of approaches that can be used to achieve these goals and the following section provides many examples from all over the world. For each of the nations covered in this chapter, there will be a short summary paragraph followed by a table with three major headings: Technical, Financial, and Others.

The Technical heading will discuss policies that address technical issues such as mandated technologies, interconnection procedures, manufacturing and interconnection

* http://www.renewables2004.de/pdf/policy_recommendations_final.pdf.

standards, and safety rules. Technical policies if designed carefully can be important drivers for DE investment just as poorly designed policies or lack of policies can stifle investment in DE.

The Financial heading will cover policies that exist in the area that are based on economic incentives such as rebates for investing in certain technologies, tax breaks for eligible investments, and pricing arrangements for power produced by on-site generators. Although project economics is affected by a multitude of factors (including fuel prices, capital expenditures, opportunity costs, etc.) financial policies can often make the difference between a project that is feasible and one that is not.

The Others section will discuss all remaining issues relating to DE policy including more general policy guidance and legislation mandating DE use or energy efficiency/renewables.

The general organization of each section will take the form of the following table.

In some cases, no examples for the specific category of policy/technology were found for a given jurisdiction in which case the given cell was left blank.

General Organizational Structure for Country Policy Summaries

Technical	DE in general
	Large-scale cogeneration
	Domestic cogeneration
	PV
	Wind
	Hydro
Financial	DE in general
	Large-scale cogeneration
	Domestic cogeneration
	On-site PV
	On-site Wind
	Hydro
Others	DE in general
	Large-scale cogeneration
	Domestic cogeneration
	PV
	Wind
	Hydro

It is very difficult to separate policies that aim to promote renewables and efficiency in general from those aimed at promoting DE. For example, many policies aim to promote renewables in general including large-scale wind farms, which should not necessarily be considered as DE. Likewise some policies are aimed at only cogeneration and do not affect renewables. In other cases, for example, in the case of rooftop PV, renewable policy and DE policy is one in the same. In most cases, however, policies that affect DE will be framed in the context of policies to promote renewables or general efficiency (in the case of cogeneration). Currently, there remains a shortage of concrete policy examples that target on-site power or DE. In order for DE to truly reach its potential, laws and policies will have to be increasingly aimed at DE specifically.

11.2 Canada

With a low population density and wide climatic variations, Canada has one of the world's highest energy use per capita. The Canadian government has taken some steps to promote DE using financial mechanisms, but coordinated policy cooperation between the various provincial governments and the federal government may prove most effective for promoting DE.

Canadian Distributed Generation Policy by Technology and Type

Technical	DE in general	The federal government in Canada has control over interprovincial gas pipelines and transmission lines crossing international borders. Because the Canadian constitution enshrines energy as a provincial jurisdiction in all other cases, each utility tends to have its own interconnection policy. Interconnection policies tend not to be technology specific. The government funds the *micropower* connect initiative, which aims to harmonize interconnection rules in the provinces.
	Large-scale cogeneration	Interconnection rules vary by province.
	Domestic cogeneration	Interconnection rules vary by province
	PV	Interconnection rules vary by province.
	Wind	Interconnection rules vary by province.
	Hydro	Interconnection rules vary by province.
Financial	DE in general	There is no federal law, which guarantees retail electricity prices. In most cases, on-site generators feeding excess power to the grid obtain wholesale electricity price or less, and in some cases, retail price is earned. Various funding schemes exist to promote energy innovation including: the technology Early Action Measure program, the Climate Change Technology and Innovation Program, the Federation of Canadian Municipalities Green Fund, and a program to promote on-site generation at government facilities. In many cases, the utility companies of a province have much influence over provincial electricity policy so that policy is often determined by the private sector.
	Large-scale cogeneration	Under Canadian renewable and conservation expenses (CRCE), regime development of a cogeneration equipment may have tax benefits. CRCE generally includes intangible expenses incurred by a *principal-business corporation* (as defined in the act) and payable to an arm's length party in connection with the development of an energy project wherein at least 50% of the capital cost of the depreciable property in the renewal energy project is property described in Class 43.1 (a *Class 43.1 Asset*) or Class 43.2 (a *Class 43.2 Asset*), under the Canadian taxation system for capital cost allowance (*CCA*) under Schedule II to the Regulations. Recent amendments have expanded Class 43.2 with respect to waste-fuelled thermal energy equipment and equipment of a district energy system that uses thermal energy provided primarily by eligible waste-fuelled thermal energy equipment. These amendments have also expanded Class 43.2 to include equipment that uses the residue of plants, generally produced by the agricultural sector, to generate electricity and heat (biowaste). These measures should encourage investment in technologies that can contribute to a reduction in emissions of greenhouse gasses and air pollutants in support of Canada's targets as set forth in the Federal Sustainable Development Strategy. These measures may also contribute to the diversification of Canada's energy supply.

(*Continued*)

Canadian Distributed Generation Policy by Technology and Type

		The 2013 Canadian Federal Budget proposes to expand Class 43.2 by making biogas production equipment that uses more types of organic waste eligible for inclusion in Class 43.2, including pulp and paper waste and wastewater, beverage industry waste and wastewater and separated organics from municipal waste, and the range of cleaning and upgrading equipment used to treat eligible gases from waste that is eligible for inclusion in Class 43.2, to all such cleaning and upgrading equipment.
		These proposed measures will apply to property acquired on or after March 21, 2013, that was not used or acquired for use before March 21, 2013.[a]
	Domestic cogeneration	Cogeneration equipment is listed as Class 43.1 assets: they are eligible for accelerated depreciation rates for capital costs (30%).
	On-site PV	Various feed-in tariffs (FiTs) exist depending on the type (rooftop, free field) and the capacity of the panels.
	On-site wind	Financial incentives vary province by province, some offer net metering (which allows small wind turbine owners to offset their electricity consumption, or earn credit from their local utility if they produce excess power through their turbine), others have special loans, or FiTs. The Canadian federal government also supports the research, development, marketing, and deployment of small wind turbines in Canada through R&D initiatives.
	Hydro	In 2009, Ontario enacted FiT for hydropower. Under the new FiT rules, hydropower producers are paid 13.1 cents/kW h for up to 10 MW over a 40-year term. The rate is 12.2 cents for 10–50 MW over the same period. According to the order of the Minister of Energy in fall 2013, changes to the FiT program are expected.
Others	DE in general	The Government of Canada has set a goal of generating 90% of Canada's electricity from zero-emitting sources by 2020, while various provinces have their own goals to improve their zero emission sources. (e.g., Ontario plans to end coal generation by 2014, and is expecting to reach 2650 MW of solar PV by 2015.)
		In 2011, Canada announced its withdrawal from the Kyoto Accords, but participated in Durban talks, which were leading to a new binding treaty with targets for all countries to take effect in 2020.
	Large-scale cogeneration	
	Domestic cogeneration	
	PV	Between 2014 and 2018, installed solar capacity in Canada is expected to reach 3.48 GW, which presumes an annual growth of 450 MW.
	Wind	Current installed wind capacity in Canada is 6500 MW, which is only 1.3% of the total energy demand.
	Hydro	Canada is the world's largest producer of hydroelectricity in the world (before China and Brazil), and one of a few countries to generate the majority of its electricity from hydroelectricity. Some provinces and territories, such as British Columbia, Manitoba, Newfoundland and Labrador, Quebec, and Yukon, produce over 90% of their electricity in this manner.

[a] http://www.millerthomson.com/en/publications/newsletters/tax-notes/april-2013/canadian-renewable-conservation-expense.

11.3 Mexico

Mexico is resource rich for many DE sources including natural gas, sunshine, wind, and biomass, but little of that potential has been realized. To date there has been limited success in developing a strong policy foundation for DE. The recently announced intention of the Mexican government on energy reform, to open the market for foreign investors, may result in fundamental changes on power generation as well.

Mexican Distributed Generation Policy by Technology and Type

Technical	DE in general	In 1992, a law was established which permitted nonutility generation and on-site generation for the first time. The *Ley del Servicio Público de Energía Eléctrica* (LSPEE) forbids bilateral trading of electricity between individuals but allows companies to generate for their own demand and cogeneration with a permit from the Comisión Reguladora de Energía (CRE) provided they meet the conditions as defined by Art. 36 of LSPEE. However, in 2012, certain amendments were made to this law, and in 2008, the *Ley para el aprovechamiento de energías renovables y el financiamiento de la transicion energetica* has been approved, there are still serious barriers that exist for DE.
	Large-scale cogeneration	The Mexican government defined cogeneration in the 1992 law permitting private sector involvement in the electricity sector[a] but stated that the power must be used on-site. Interconnection, wheeling, and backup services are allowed but require permits from the CFE and the energy secretariat (SENER). Interconnection standards for cogeneration plants are based on the U.S. IEEE 1547.
	Domestic cogeneration	
	PV	
	Wind	
	Hydro	
Financial	DE in general	Mexico's income tax law (ITL) provides a 100% deduction incentive for taxpayers who carry out investments in renewable energy equipment. Qualifying sources like sun, wind, water, and geothermal energies, as well as biomass fuel equipment, are eligible for this incentive. To finance sustainability projects, the Fund for Energy Transition and Sustainable Exploit of Energy was created in 2009. Companies or individuals may request incentives from the fund by submitting proposals for projects that involve renewable energies and energy transition. Every fiscal year, the Ministry of Energy (SE) and the National Council of Science and Technology (CONACYT) establish a special fund for energy sustainability projects in which universities and research centers are the potential participants and beneficiaries. The resources for the fund are provided by the Mexican oil company PEMEX.[3]
	Large-scale cogeneration	The Mexican government incentivizes cogeneration by treating it as a renewable energy source. Pollution control equipment or equipment for research and technological development in renewable energy are exempt from general import and export taxes. Mexican policy outlines accelerated depreciation for investments in renewable energy, allowing 100% depreciation on investment for machinery and equipment for generating energy from renewable sources.

(Continued)

Mexican Distributed Generation Policy by Technology and Type

		Incentives and funding programs for cogeneration projects can also be found through the Mexican Secretariat of the Environment and Natural Resources, SEMARNAT.[b]
	Domestic cogeneration	
	On-site PV	The country offers tax incentives for solar projects and a so-called net metering system, as well as the possibility of long-term power purchase deals with CFE.
	On-site wind	
	Small hydro	
Others	DE in general	Investors have to secure a wide array of permits, which can be very costly. In some cases, more than 100 permits are required (an existing plan is to reduce to 31) including local/municipal planning permission, environmental permitting, gas supply and connection permits, power generation authorization and network interconnection permits, etc. According to a recently published analysis by Frost and Sullivan, insufficient centralized electricity, which has compelled several consumers to generate their own power, along with incomplete grid facilities in large isolated areas, offers huge scope for the distributed power generation (DPG) market in Mexico; however, DE providers' reliance on electricity subsidies coupled with grid integration issues in isolated areas makes market development uncertain. Energy subsidies have added to the challenge, as artificially low electricity prices for most consumers in the country reduce opportunities for DE technologies, which have become a costly alternative to grids. In fact, although DE technologies reduce transmission costs and provide the benefit of local energy management and economies of scale, their start-up costs are higher than that of centralized electricity production, further affecting adoption.[c] A 2012 law requires 35% of electricity from renewable resources by 2024. At the Solar Power Mexico conference in 2012, it was said that PV electricity and solar will comprise up to 5% of Mexico's energy by 2030 and up to 10% by 2050. By the announcement of the Mexican government at the beginning of August 2013, Mexico is ready to change its constitution to let foreign private companies to find and produce oil and gas in the country. The initiative also seeks to liberalize Mexico's electricity sector by allowing private firms to produce and sell electricity to consumers. These changes might open up more opportunities for cogeneration to flourish in the country.
	Large-scale cogeneration	In March 2013, SENER, Spain, published an *Initiative for the development of the cogeneration potential in Mexico*, which accepted an action plan with main actions such as: energy surplus remuneration methodology, interconnection process standardization, private sector participation in PEMEX projects, and development of transportation network and natural gas distribution, and secondary actions such as measures for financial support for combined Heat and Power (CHP) projects in the short term and communication plan and pilot projects to eliminate current cultural and information limitations.[d]
	Domestic cogeneration	
	PV	By the end of 2012, installed PV capacity was 38 MW, which is expected to increase up to 60 MW or even more in 2013.
	Wind	By the end of 2012, installed wind capacity was 1370 MW.

(*Continued*)

Mexican Distributed Generation Policy by Technology and Type	
Hydro	Recently Mexico's congress has changed its renewable energy and energy transition law to include larger hydroelectric projects in the country's definition of *renewable*, provided they satisfy a certain criteria. According to the modifications, hydropower plants with capacities about 30 MW are to be classified as *renewable*, but only if they meet a generation density standard.
	Mexico's current total installed hydropower capacity accounts for about 11.5 GW, which is expected to increase by as much as 4 GW with regard to its small hydroelectric capacity and 40 GW with regard to its large capacity.

[a] WADE Market Analysis 2005.

[b] http://www.semarnat.gob.mx/informacionambiental/publicaciones/Publicaciones/Guia%20de%20Programas%20de%20Fomento%20de%20Energ%C3%ADas%20Renovables.pdf.

[c] http://www.frost.com/prod/servlet/press-release.pag?docid=278940767.

[d] http://www.sener.gob.mx/webSener/res/0/Cogeneration_01.pdf.

11.4 United States

The United States remains highly influential in international energy policy development. As the single largest energy-using nation, policy to promote DE is clearly needed to develop capacity to meet the United States' rising demand. Many innovative policy developments have been introduced at the state level that have effectively buoyed DE markets, and the United States remains one of the world leaders in renewable energy development. Recent results of shale gas and oil research, impacts of climate change, and express government intention to increase the ratio of unconventional energy may elevate DE to a more important position. However, steps to harmonize state level policy with policy at the federal level still will be the main factor for successfully advancing DE.

USA Distributed Generation Policy by Technology and Type		
Technical	DE in general	Section 1254 of the Energy Policy Act of 2005 requires that electric utilities offer grid interconnection based on a nationwide standard. Currently to connect DE units must be IEEE P1547 compliant, which requires *automatic and rapid disconnection in the event of a fault*.
	Large-scale cogeneration	
	Domestic cogeneration	
	PV	
	Wind	
	Hydro	
Financial	DE in general	At a state level, the United States has many promising policies in support of renewables and distributed generation. Grant programs, financing, R&D funds, net metering, FiTs, and tax breaks are some of the financial measures that have proven successful for promoting DE technologies at a state level. The Energy Act of 2005 grants tax credits for developers who build energy efficient buildings that could have ramifications for DE especially microcogeneration.

(Continued)

USA Distributed Generation Policy by Technology and Type	
	In details, the following support schemes are available: **PTC** is applicable for wind, geothermal, landfill gas, trash combustion, open-loop biomass, closed-loop biomass, hydropower, and wave tide. The PTC provides a tax credit for the production of electricity from renewable sources and the sale of that electricity to an unrelated party. Credit amount is 2.2 cents/kW h for wind, closed-loop biomass, and geothermal, and is 1.1 cents/kW h for other renewable energy resources. PTC is available for facilities placed in service before January 1, 2014 (2013 for wind), and available for a 10-year period beginning with the year the facility is placed in service. **Investment tax credit (ITC)** is applicable for solar, geothermal, qualified fuel cell or microturbine property, CHP systems, small wind, geothermal heat pumps, and PTC-eligible facilities placed in service after 2008 and before 2014 (2013 for wind). The ITC provides a credit for qualifying energy property. For any taxable year, the ITC is the energy percentage of the basis of each energy property placed in service during the taxable year. Credit amount is 30% of eligible costs for fuel cell, solar, and small wind property, 10% of eligible costs for CHP, microturbine property, and geothermal heat pumps. The ITC is generally available for eligible property placed in service on or before December 13, 2016. **Grant in lieu of PTC and ITC** is applicable for tangible personal property or other property that is an integral part of a qualified facility (as defined by the PTC and ITC rules). The American Recovery and Reinvestment Act (ARRA) enacted a new grant program, which provides a cash grant in lieu of the PTC or ITC. ARRA permits PTC or ITC projects to elect a grant of up to 30% of costs of construction of PTC or ITC energy property in lieu of tax credits. Projects must begin construction before 2012 and submit a grant application no later than October 1, 2012. Projects must be placed in service before their PTC or ITC credit expires, PTC before 2014 (2013 for wind), ITC before 2017. **Renewable portfolio standards (RPSs)** generally place an obligation on electric supply companies to produce a specified fraction of their electricity from renewable energy sources and enumerates mechanisms that are permitted to achieve compliance, such as renewable energy credits (RECs). Currently, no federal RPS legislation has been enacted. A total of 29 states and the District of Columbia have an RPS.[3]
Large-scale cogeneration	Microturbines and CHPs are eligible to corporate tax incentive with the following conditions: Credit for microturbines equals 10% of expenditures, with no maximum credit limit stated (explicitly). The credit for microturbines is capped at $200/kW of capacity. Eligible property includes microturbines up to 2 MW in capacity that have an electricity-only generation efficiency of 26% or higher.

(*Continued*)

USA Distributed Generation Policy by Technology and Type

		Credit for CHP equals 10% of expenditures, with no maximum limit stated. Eligible CHP property generally includes systems up to 50 MW in capacity that exceeds 60% energy efficiency, subject to certain limitations and reductions for large systems. The efficiency requirement does not apply to CHP systems that use biomass for at least 90% of the system's energy source, but the credit may be reduced for less efficient systems. This credit applies to eligible property placed in service after October 3, 2008.
	Domestic cogeneration	
	On-site PV	No federal FiT exists; however, certain states or voluntary companies offer FiT both for domestic or commercial PV generators.
		Solar generators are eligible to corporate tax incentive. The credit is equal to 30% of expenditures, with no maximum credit. Eligible solar energy property includes equipment that uses solar energy to generate electricity, to heat or cool (or provide hot water for use in) a structure, or to provide solar process heat. Hybrid solar lighting systems, which use solar energy to illuminate the inside of a structure using fiber-optic distributed sunlight, are eligible. Passive solar systems and solar pool-heating systems are not eligible.
	On-site wind	No federal FiT exists; however, certain states offer FiT for wind power producers.
		Wind power generators may apply for corporate tax incentives. The credit is equal to 30% of expenditures, with no maximum credit for small wind turbines placed in service after December 31, 2008. Eligible small wind property includes wind turbines up to 100 kW in capacity.
		On January 1, 2013, the production tax credit (PTC) was extended for another year.
	Small hydro	Various incentives are available for hydropower plants. These can be PTCs, stimulus funding, and clean renewable energy bonds.
Others	DE in general	PURPA of 1978 provided the foundation for competition in the electricity sector allowing DE investors to participate in the market for the first time. The Energy Act of 2005 made several amendments to PURPA including the retraction of the mandate that utilities must buy power from *qualifying facilities* (which included many cogeneration facilities). Section 1253 states that as long as the utility can prove that the facility has fair access to the grid, there is no longer any obligation to buy power (nondiscriminatory access to the grid is guaranteed in section 1231). The bill also requires all utilities to offer net metering to small on-site generators (section 1251) and offer time-differentiated rate schedules upon request so that customers can take advantage of technologies such as *smart metering*, which also must be supplied by the utility (section 1252). And, without proving any clear guidance the utilities are disallowed from relying on a single fuel, that is, there is a federal mandate to diversify energy portfolios.

(*Continued*)

USA Distributed Generation Policy by Technology and Type	
	A total of 29 states and the District of Columbia have enacted RPSs according to section 203 of the Energy Act of 2005, the federal government must ensure that not less than 3% of its total energy consumption is derived from renewables in the fiscal years of 2007–2009, not less than 5% in the fiscal years of 2010–2012, and not less than 7.5% in the fiscal year of 2013 and each fiscal year thereafter.
	There is a federal mandate to study the potential from renewable energy (Section 201). There is a goal to develop at least 10,000 MW of nonhydro renewables on public lands by 2015.
Large-scale cogeneration	According to Environmental Protection Agency in 2012, the United States had an installed capacity of 82 GW of CHP, with 87% of those in manufacturing plants. This number may tremendously increase in the future due to the executive order of President Barack Obama in 2012, that would increase the number of cogeneration plants in the United States by 50% by 2020.
	The economic success of shale gas in the United States since 2000 might strongly motivate CHP developments, presenting cheap combustible gas for a long time. It has been postulated that there may be a 100-year supply of natural gas in the United States (however, only 11 years of gas supply is in the form of proved reserves).
Domestic cogeneration	
PV	According to North America PV Market Report, in 2013, the United States passed the 10 GW installed solar PV milestone, and now only Germany, Italy, and China have more installed PV capacity.
	Solar PV has become one of America's fastest-growing energy sources in recent years. The solar PV market has expanded at a compound annual growth rate of 50% since 2007, and 83% of the 10 GW capacity was installed within the past 14 quarters. Only in the first half of 2013, more than 1.8 GW of new solar PV capacity was installed. In the future, further significant growth of an additional 80% by the end of 2014 is expected, putting the United States on track to pass 17 GW installed solar PV capacity.
Wind	Construction of new wind power generation capacity in the fourth quarter of 2012 totaled 8,380 MW bringing the cumulative installed capacity to 60,007 MW by the end of 2012. This capacity is exceeded only by China.
	The U.S. Department of Energy's report *20% Wind Energy by 2030* envisioned that wind power could supply 20% of all U.S. electricity, which included a contribution of 4% from offshore wind power.
Hydro	Hydropower is the largest renewable energy generation resource in the United States, providing nearly 8% of the electricity generated. Including pumped storage facilities, there are approximately 100,000 MW of current installed hydropower capacity in the United States.

11.5 Argentina

The energy sector in Argentina, though subject to wholesale competition, remains largely dominated by a few large players. In recent years, the additional investment required to keep the power on has been shortcoming as result of market uncertainties, but plentiful natural gas as well as some new policy developments promise to buoy DE investments in the near future.

Argentinean Distributed Generation Policy by Technology and Type[4]

Technical	DE in general	Ley 24065 guarantees access to the grid for generators including investors in DE. All generators must meet the same technical standards to participate in the power pool including reactive power, frequency control, etc., and must bear the cost of real-time metering equipment.[5]
	Large-scale cogeneration	Rules have been established to allow larger DE investors to take advantage of seasonal, weekly, and daily electricity markets (including spot markets) with the argentine power pool, the Mercado Electrico Mayorista MEM, required to take and provide fair compensation for power fed into the grid.
	Domestic cogeneration	
	PV	
	Wind	
	Hydro	
Financial	DE in general	In Argentina, various supports are available for renewable energy sources including biofuels, solar, wind, hydro, and geothermal, among others. At the local tax level, there are anticipated VAT refunds for the new depreciable property included in the project and accelerated income tax depreciation. The property used for the project will not be part of the minimum presumed income tax taxable base. In addition, biofuel producers will not be subject to the hydric infrastructure tax, the tax on liquid fuels, and the gasoil tax for the amount of fuel that is marketed in the national territory. At the provincial level, there are real estate tax exemption, stamp tax exemption turnover tax exemption/deferral, and tax stability. The type of benefit depends on the geographic area in which the renewable energy plant operates, so the plant's specific location must be supplied for a proper tax classification. Subsidies at the national level are for wind: 0.015 Argentine peso (ARS)/kW h, for solar: 0.9 ARS/kW h, and for hydro for less than 30 MW installed capacity: 0.015 ARS/kW h, while for others: 0.015 ARS/kW h. Several provinces have different incentive FiTs according to the kind of energy they want to promote. There is also a quota obligation with the aim to reach a contribution of sources of renewable energy equal to 8% of the total national consumption of electric energy within a term of 10 years, starting in 2006, the effective date of the regime.[3]
	Large-scale cogeneration	
	Domestic cogeneration	

(*Continued*)

Argentinean Distributed Generation Policy by Technology and Type[4]

	On-site PV	The PERMER project, cofinanced by World Bank, GEF, and local governments (national and provinces), promoted investment in solar panels for rural households and small public buildings. The project was closed on December 31, 2012, and is considered to be very successful.
	On-site wind	Law 190/2006, passed in December 2006, introduced FiTs for wind. The FiTs, set in 2009, provide generation developers with \$5–\$10/MW h above the spot price on the wholesale energy market. As this FiT scheme failed, in 2009, the government established a state utility (ENARSA) to purchase at least 1 GW of renewable energy capacity. Renewable energy developers bid for power purchase agreements (PPAs) at fixed rates, guaranteed for 15 years. (GENREN program). The first PPAs were signed under GENREN in 2010. Seventeen wind farms with a combined capacity of 754 MW were awarded PPAs for prices between \$120 and \$134/MW h. The second stage of GENREN was established under Law 108/2011. It enables CAMMESA to sign sPPAs via ENARSA with renewable energy providers. Under this program, wind project owners will be paid around U.S. \$125/MW h for their power, guaranteed for 15 years.
	Small hydro	FiT introduced by Law 160/2006 was also applicable for small-scale hydropower, and those may also be awarded under GENREN program.
Others	DE in general	As a response to the economic situation of the country Law 26.190, Promotion of Renewable Sources of Energy for Electricity Production, has been passed on December 6, 2006. The new law includes financial incentives in terms of deferred tax payments and defined FiT premiums, but for a larger range of renewable technologies and with an entitlement period of 15 years. Thereafter, the Argentine government launched the GENREN program (Generación por Energías Renovables), another initiative to promote public–private investment in renewable energy. The state electricity company ENARSA (Argentine Energy, PLC, of the National Secretary of Energy) launched GENREN in 2009 in order to tackle the shortfall in energy supply, deal with organic waste (e.g., bagasse in the provinces of Tucumán and sawdust in Corrientes), reduce GHG emissions, and increase electricity access to rural areas (Río Negro, 2009). The initiative attracted unexpected interest from national and international investors, and before the end of 2009, more than half of the projects (49 in total) had been sold (Renou, 2009). It is predicted that, due to the low generation power of each individual project, the GENREN scheme will have an impact on both remote rural populations and people with low-energy demand.
	Large-scale cogeneration	
	Domestic cogeneration	
	PV	
	Wind	
	Hydro	

11.6 Brazil

Brazil has for a long time been a world leader in alcohol fuels, and some recent policy developments suggest that it may be well on its way to showing similar leadership in development of cogeneration. The plentiful biomass resources along with newly accessible supplies of onshore and offshore natural gas, combined with feed-in laws for renewables and gas-fired cogeneration, are likely to drive significant investment in DE in the near future.

Brazilian Distributed Generation Policy by Technology and Type

Technical	DE in general	All DE plants including gas and biomass are guaranteed a market for surplus power based on law 10848.
	Large-scale cogeneration	
	Domestic cogeneration	On April 17, 2012, the Brazilian Federal Energy Regulatory Agency (ANEEL) enacted new rules aimed at reducing barriers for the incorporation of DPG (DG Regulation) into utility procurement and into Brazil's distribution planning processes. To achieve the goal of enabling production of renewable energy on a broad scale but without reliance on long-distance transmission lines, the DG Regulation establishes the following measures: A net metering program (Sistema de Compensação de Energia) allowing small-scale power production generators of 1 MW or less to offset their electricity bills with credits from the energy they provide to the grid. Ease the interconnection regulatory burden and improve grid integration for small-scale renewable energy production generators
	PV	Although the DG Regulation covers other *incentivized* renewable energy sources such as wind and biomass, it was essentially designed to stimulate solar power generation. Brazil has made it clear that it wants to add solar energy generation to its energy mix at a fast pace. Since net metering might significantly spur consumer investment in solar without the need to construct new long distance transmission lines, it is not surprising that ANEEL released another resolution in conjunction with the DG Regulation that provides specific tax benefits for solar power deployment.
	Wind	
	Hydro	
Financial	DE in general	Law 10848 guarantees DE systems access to the grid and are guaranteed the average price of the larger market.

(*Continued*)

Brazilian Distributed Generation Policy by Technology and Type

		The PROINFA (Alternative Sources of Energy Incentive program) program implemented in 2002 and still ongoing, sought to establish 3300 MW from renewable energy (wind, biomass cogeneration, and microhydropower 1100 MW each) by 2007. Purchase contracts for qualifying facilities would have their contracts guaranteed by Electrobras for 20 years.[6] Qualifying projects are also eligible for grants from the National Economic and Social Development Bank (BNDES). The *Luz para Todos* program aimed to subsidize the electrification of some 2.5 million households in Brazil by 2008, and there is a target to employ renewable energy for at least 200,000 of those households. Until today the program brought electricity to almost 12 million people, is still ongoing, and is now serving communities living in isolated areas.[a]
	Large-scale cogeneration	Biomass CHP plants are expected to benefit from the PROINFA program.
	Domestic cogeneration	
	On-site PV	ANEEL will increase the existing discounts in transmission and distribution system usage charges (TUST and TUSD, respectively) from 50% to 80% for qualifying solar systems up to 30 MW and those applicable up to the first 10 years of operation.
	On-site wind	
	Small hydro	
Others	DE in general	In 2004, the Brazilian government passed law 10848 with regard to electricity sector reform. The law in effect created two separate market structures for electricity exchange: the bilateral contract environment (ACL), where individual generators and consumers can negotiate PPAs, and the regulated contract environment (ACR) where distribution companies must purchase the power they need to meet their contract from public auctions. The Decree 5163 dealt specifically with distributed generation, and states that in the ACR, scenario distribution companies must also buy power from *alternative sources* at prices set by the government and are permitted to buy up to 10% of their required supply. The clause was designed to create incentive for CHP and other designated alternative sources. The decree states specifically that cogeneration plants with an efficiency of over 75%, as well as hydro plants with a capacity below 30,000 kW and other renewable sources all qualify as *alternative energy* for the purposes of the law. The additional costs incurred by the grid owners are, by law, built into the formula used to calculate end-user tariffs.[6]

(Continued)

Brazilian Distributed Generation Policy by Technology and Type

	The new law requires each distribution company to disclose 5 year plans for power supply and up to 10% of the supply needed to meet total loads can come from direct bilateral agreements with cogeneration plants instead of the typical power auctions.[b]
	In addition, the revised auction structure is likely to favor cogeneration plants, because winning tenders for supply must have supply on line within 3 years—too short a lead time for most central plants.
	Brazil has set a target of 3.3 GW of capacity from wind, biomass, and small hydro as a national target for 2007.[7] By early 2005, the first phase of PROINFA was completed and 3300 MW were completed (1266 MW Solar, 655 MW Biomass, 1379 Wind). Second phase of PROINFA is to reach 90% nationalization rate and a 10% Brazilian electrical energy annual consumption rate is to be supplied by renewable sources. Time frame for the second phase is 2009–2030.
Large-scale cogeneration	
Domestic cogeneration	
PV	Currently there is no FiT policy for PV.
Wind	Currently there is no FiT policy for wind.
Hydro	Currently there is no FiT policy for hydropower.

[a] http://www.brasil.gov.br/news/history/2012/03/28/with-technological-innovation-luz-para-todos-takes-electricity-to-isolated-areas/newsitem_view?set_language=en.

[b] http://gastopowerjournal.com/regulationapolicy/item/1257-germany-frees-up-%E2%82%AC750-m-budget-to-raise-chp-market-share-to-25-by-2020#axzz2aSREnklp.

11.7 Chile

Chile was the first nation in the world to break up vertically integrated monopolies into separate markets for generation and wires. Strong policy is still lacking to directly promote DE, which partly explains the lack of investment in the area. Gas shortages in the early 2000s in Chile have been creating somewhat of an energy crisis that should encourage gas conservation (and thus efficient gas cogeneration) but will likely reduce interest in gas-fired cogeneration in the near future. Other DE sources such as biomass and on-site renewables will likely gain added attention and perhaps policy support as a result. In 2013, certain investments were initiated but those are mostly to supply self-consumption, thus may not have significant effect on the energy sector.

Chilean Distributed Generation Policy by Technology and Type		
Technical	DE in general	The *Superintendencia de Electricidad y Combustibles* (Secretariat of Electricity and Fuels) sets and enforces the technical standards of the system. The Chilean Electricity Law guarantees open access to the transmission system, but must contribute to investment in the *system*. It is unclear if DE is included in the system.
	Large-scale cogeneration	
	Domestic cogeneration	
	PV	
	Wind	
	Hydro	
Financial	DE in general	In 2008, Chile enacted the Non-Conventional Renewable Energy (NCRE) Law number 20,257, which made it mandatory for electricity companies selling energy to final customers to assure that a minimum of 5% of the energy they sell comes from NCRE sources, directly or indirectly. This percentage will increase by 0.5% per year as from 2015 to reach 10% in 2024. The Chilean Economic Development Agency (CORFO) offers various incentives for investments in NCRE projects: Grants to fund preliminary preinvestment feasibility studies or specialized consultancy required before the materialization of NCRE projects that involve investments of more than U.S. $400,000. The amount of the subsidy is up to 50% of the total cost of the study or consultancy with a cap of U.S. $60,000 per project submission, and provided that it does not exceed 2% of the total estimated investment. The wind prospecting studies have a maximum subsidy of U.S. $20,000 for monitoring in a point, or U.S. $30,000 for two points. Grants to fund advanced preinvestment feasibility studies. The grant covers up to 50% of the total cost of the study or consultancy with a maximum of 5% of the estimated investment, provided that it does not exceed U.S. $160,000 per project submission. This grant applies for projects that have been submitted to the preliminary processes and does not apply to studies measuring the availability of resources or for prefeasibility studies. During the second half of 2011, the Ministry of Energy implemented a fund to promote NCRE projects; specifically, it will finance instruments that will directly support initiatives of this type.
	Large-scale cogeneration	There are certain regulatory barriers of the commercialization of the energy surpluses. At present, there is no legal regulation for the sale of electricity surpluses, but a net metering law is under discussion in the Chilean congress in July 2013.
	Domestic cogeneration	
	On-site PV	No FiT exists, but PV investments are supported by CORFO, financing feasibility studies for projects. According to published forecasts, by 2016, installed solar capacity is expected to reach 1520 MW.

(Continued)

Chilean Distributed Generation Policy by Technology and Type

	On-site wind	No FiT policy exists, but instead a long-term mechanism granting renewables priority access to the grid, together with a solid structure for long-term PPAs, serves as a financial incentive for wind investments.
	Small hydro	
Others	DE in general	The 1982 Energy Act law DFL N° 1 provided the legal foundation for separate generation, transmission, and distribution, including wholesale competition in the generation side. Smaller generators are not allowed representation on the Centro de Despacho Económico de Carga (Centre for Economic Load Dispatch) committees, which has raised some questions of fairness.[a] The Chilean National Energy Commission (CNE) is responsible for implementing regulations outlined in Chile's General Electricity Services Law. It proposes regulated tariffs twice a year in April and October, and prepares the overall plan for new generation capacity. The SEC is responsible for ensuring all generators, distributors, and transmission operators are in compliance with the law.[b] Chile is a participant in TECH4CDM project, developed over 2008 and 2009, and financed by the European Union under the Sixth Framework program of R&D, with a primary goal, to promote renewable and efficient energy technologies, paying special attention to overcoming technological barriers, as well as the analysis of the Clean Development Mechanisms (CDM) of the Kyoto Protocol that may assist in projects based on wind energy, cogeneration, solar thermal, and rural electrification through renewable energies. Both European and Latin American institutions participate in the project, which is coordinated by the Spanish Institute for Energy Diversification and Saving (IDAE).[c] Law 20.257 was published on April 1, 2008, and modified the LGSE (Ley General de Servicios Electricos/General Law on Electricity Services) with respect to power generation based on ENERC sources. The law states that companies trading electric energy on the grid, and have a installed capacity of more than 200 MWs, are obliged to verify annually: that a certain percentage of the total of their trade on the grid, originates from renewable and non conventional sources, either it is their own generation or from any contracted partner. This percentage is 5% between 2010 and 2014, and from 2015 that shall be increased gradually, 0.5% every year, until 10% is reached in 2024.
	Large-scale cogeneration	The *Ley Corta* law also sets new efficiency standards, which may benefit cogeneration.[d] According to the applicable laws, it can be stated that cogeneration is not differentiated from other conventional power generation. Furthermore, cogeneration plants can get access to the grid and sell the generated power if the installed capacity is higher than 9 MW. This hinders the access to the grid of the majority of the possible cogen plants, as most of them would be installed in the 0.1–5 MW range. This kind of regulations could not serve as a catalyst for improvement of this technology. On the other hand, the Chilean electricity policy made great efforts to promote the improvement of energy efficiency, where cogeneration could be an ideal alternative of the conventional power generation.

(Continued)

Chilean Distributed Generation Policy by Technology and Type

	Domestic cogeneration	
	PV	Ley 20.365 established the *tax-free tributary* with respect to PV systems. The objective of the law is to create the necessary conditions for the improvement of the market of PV systems, for warm water production in newly constructed buildings, through the instrument of a financial subsidy.
	Wind	Installed wind capacity as of January 2013 reached 200 MW.
	Hydro	

[a] According to the EEG effective as of mid-2013.
[b] According to the EEG effective as of mid-2013.
[c] http://www.tech4cdm.com/uploads/documentos/documentos_Solar_Thermal_in_Chile_b873ac46.pdf.
[d] According to the EEG effective as of mid-2013.

11.8 Czech Republic

The Czech Republic's DE heritage remains largely as a leftover from the Soviet era. There is much existing cogeneration, especially in district heating applications, but much of it will be in need of refurbishment in the coming years. In the early 1990s, Czech Republic saw a boom in the development of smaller cogeneration plants and policy is now playing an important part in once again fostering cogeneration investment in the country. As an EU member, Czech Republic is bound to the EU legislation, which pushes on DE improvement.

Czech Distributed Generation Policy by Technology and Type[8]

Technical	DE in general	
	Large-scale cogeneration	All DE systems are guaranteed *nondiscriminatory* access to the grid.
	Domestic cogeneration	A European standard for microcogeneration is currently being drafted under the auspices of Cenelec technical committee TC8X WG2. The standard has been worked out, implementation process is going on.
	PV	
	Wind	
	Hydro	
Financial	DE in general	All DE is guaranteed to be purchased by the utility. The rates depend on technology. In addition, distributed technologies are awarded an extra amount depending on what voltage level they are interconnected to.
	Large-scale cogeneration	Since January 1, 2006, a new support scheme has been introduced for CHP units, based on a feed-in premium on top of the market price of electricity for cogenerated electricity paid by network operators (distribution or transmission). The premiums are divided into three categories according to the installed electric capacity: up to 1 MWe, 1–5 MWe, and above 5 MWe. The premium is higher if producers sell electricity only in peak time. Producers can sell electricity to the market or use it themselves. The system of price regulation is controlled by the Energy Regulatory Office (not as state aid).

(Continued)

Czech Distributed Generation Policy by Technology and Type[8]		
	Domestic cogeneration	
	On-site PV	Currently a FiT of 21.91–58.44 cent (EUR)/MW h shall apply for PV power in the Czech Republic.[a]
	On-site wind	Currently, a FiT of 8.71–10.94 cent (EUR)/MW h shall apply for wind power in the Czech Republic.[b]
	Small hydro	Currently, a FiT of 7.46–11.95 cent (EUR)/MW h shall apply for small hydro power in the Czech Republic.[c]
Others	DE in general	Currently, the utility CEZ holds about two-thirds of the generation market in the republic and also owns a majority stake in six of the eight existing distribution companies. The government-owned CEPS operates the transmission system.
	Renewables in general	March/April 2005 law passed to promote renewables. Sets out an 8% target of total gross consumption by 2010 and 13% by 2020. Guarantees 15 years of FiT for renewables, a premium for green electricity will also be paid. Prices are not set in law but are to be determined by the Energy Regulatory office.[d]
		The EU and the Czech government provide special incentives for biogas plant projects in the Czech Republic. One of the main reasons is that the carbon dioxide emissions per capita are rather high compared to other countries. Czech farmers receive financial support for the establishment of biogas plants from an EU environmental fund and an EU rural development fund. By 2015, biomass is to become the Czech Republic's primary source of renewable energy.
		Since 2005, the feed-in law for decentralized eco-power has resulted in an increase in the energy production from regenerative sources. In 2010, a share of about 10% of the energy was already produced from alternative sources, compared to only 4% in 2008.[e]
	Large-scale cogeneration	The Czech Republic is bound by the EU cogeneration Directive 2004/8/EC, which states that member states are obliged to address key market barriers such as ensuring grid access for cogeneration. The directive also states that in order for cogeneration to be considered high efficiency, it must provide an energy saving of at least 10% compared to separate production of heat and power. Czech Republic is also bound to EU Directive 2012/27/EU, which set as a target to strengthen the regulations of Directive 2004/8/EC, and ordered member states to complete until December 31, 2015, the evaluation of the feasibility potential of high efficiency cogeneration and efficient district heating/cooling.
	Domestic cogeneration	
	PV	
	Wind	
	Hydro	

[a] www.map.ren21.net.
[b] www.map.ren21.net.
[c] www.map.ren21.net.
[d] http://www.edie.net/news/news_story.asp?id=9710&channel=6.
[e] www.worldofcogeneration.com.

11.9 Denmark

Denmark is one of the world's leading renewable energy champions and has had strong policy in support of renewables longer than most countries. Furthermore, Denmark is one of the few countries with policies established to specifically encourage wind cooperatives. Denmark is also one of the world's leaders in cogeneration. Despite its strong track record in DE, there is still much room for improvement and the evolution of Denmark's policy on DE reflects this. Economic competitiveness and energy security could prove to be the most important drivers for DE along with environmental concerns.

Danish Distributed Generation Policy by Technology and Type

Technical	DE in general	
	Large-scale cogeneration	
	Domestic cogeneration	A European standard for microcogeneration is currently being drafted under the auspices of Cenelec technical committee TC8X WG2. The standard has been worked out, implementation process is going on.
	PV	
	Wind	
	Hydro	
Financial	DE in general	Denmark established the first feed-in law requiring utilities to buy electricity from onsite renewable generators at a set price in 1999. Currently, Denmark does not have a coherent FiT scheme. Biogas finds one solution, onshore wind a second that may be quite different from offshore. For PV, the net metering principle is the rule. The tariff for wind energy in Denmark depends on several variables: • On which year the turbine went into operation • How many full-load hours they already delivered • Whether they are offshore or onshore The tariff comprises a market power price element, power balance compensation, and a government subsidy. There is no governmental support for fossil fuel CHP.
	Large-scale cogeneration	A state subsidy for small cogeneration was introduced in 1992 for all plants fuelled by waste incineration, natural gas, and renewables. At the start of the program, the tariff was 10 øre/kW h but was subsequently reduced to 7 øre/kW h, for all but plants smaller than 3 MW.[3] In 2005, Denmark updated the way it provides incentive for CHP, whereas before distribution companies were obligated to purchase all power from CHP facilities via a FiT, the FiT has been phased out with the 2005 market reforms to better reflect free market principles. Now CHP plants are guaranteed the right to sell power to the grid but only if they can find interested buyers. According to a recent decision, in late 2018, state subsidies will be removed.
	Domestic cogeneration	
	On-site PV	

(*Continued*)

Danish Distributed Generation Policy by Technology and Type		
	On-site wind	New wind turbines as well onshore as offshore receive a price premium of 3.3 cent (EUR)/kW h for 22,000 full-load hours. Additional 0.3 cent (EUR)/kW h in the entire lifetime of the turbine to compensate for the cost of balancing, etc. Household wind turbines below 25 kW receive a fixed FiT of 8 cent (EUR)/kW h. For special wind parks at sea, the support is settled by a tendering procedure. In previous tenders, the Horns Rev II wind park of 200 MW ended at a fixed FiT of 6.9 cent (EUR)/kW h in 50,000 full-load hours, while the Rødsand II wind park in the Baltic Sea of 200 MW ended at a fixed tariff of €8.4/kW h for 50,000 full-load hours. The replacement scheme for wind turbines on land is very complicated. A scrapping certificate can be earned by replacing old and inappropriately situated wind turbines with new and more efficient turbines. This grants the right to an extra price supplement. The energy policy agreement of March 2004 on wind energy and decentralized CHP is the background for the scrapping scheme. The scrapping scheme provides an extra price supplement to new onshore wind turbines, provided the owner has a scrapping certificate for a wind turbine with an installed capacity of 450 kW or less, which was decommissioned in the period December 15, 2004 to December 15, 2010. A price supplement of 1.6 cent (EUR)/kW h is granted for electricity production corresponding to 12,000 full-load hours for double the installed capacity of the decommissioned wind turbine. However, the total amount of payment should not exceed 6.4 cent (EUR)/kW h; then the price supplement will be reduced.[a]
	Small hydro	
	DE in general	Denmark is the most energy-efficient country in the world, and also considered the leader in the co-gen sector, with about **55%** of its total energy use coming from cogeneration. Energy liberalization began in 1996 in Denmark and as of 2003, all electricity customers are free to choose their supplier. Denmark's overall goal is to increase its share of renewables from 8% in 1996 to 30% in 2025.[b] Municipal waste and geothermal energy receive special mention in the plan. Denmark has introduced a demand-side RPS: all customers must source at least 20% of their electricity from certified renewables sources.[c]
Others	Large-scale cogeneration	The Danish Energy Authority and two main utilities have developed a research and development strategy for a range of renewable energies including fuel cells, biomass, wind energy, and photovoltaics. Strategies for biofuels, wave energy, hydrogen, and system integration are being developed. The 1988 Heat Supply Act encourages renewable fuelled district heating using by prohibiting electric heating in specified residential areas. Denmark is bound by the EU cogeneration Directive 2004/8/EC, which states that member states are obliged to address key market barriers such as ensuring grid access for cogeneration. The directive also states that in order for cogeneration to be considered high efficiency, it must provide an energy saving of at least 10% compared to separate production of heat and power.

(Continued)

Danish Distributed Generation Policy by Technology and Type		
		Denmark is also bound to EU Directive 2012/27/EU, which set as a target to strengthen the regulations of Directive 2004/8/EC, and ordered member states to complete until December 31, 2015, the evaluation of the feasibility potential of high efficiency cogeneration and efficient district heating/cooling.
	Domestic cogeneration	
	PV	A program has been established, funded by a carbon tax on electricity, which pays up to 40% of the total cost of a PV system including materials cost and installation.[d]
	Wind	The Danish *Agreement on wind turbines* agreed in 1996 set a goal of 200 MW of nonutility owned wind capacity and an additional 1500 MW owned by private and public utilities. The goal was met 5 years ahead of schedule, and the wind market remains strong. By the end of 2012, installed wind capacity was 4162 MW.
	Hydro	

[a] http://www.wcre.de/en/images/stories/pdf/WCRE_Maegaard_Danish%20RE%20Policy.pdf.
[b] Shinichi Nakane, Japan Cogeneration Center.
[c] Shinichi Nakane, Japan Cogeneration Center.
[d] Shinichi Nakane, Japan Cogeneration Center.

11.10 Germany

Germany is one of the world's leaders in renewables with especially strong growth over the last decade. There is evidence that the German policy makers continue to work at creating the right policy environment for DE to thrive. Nevertheless, there is still much unrealized potential especially for small on-site renewable applications and cogeneration, whereas most policies to date have focused on larger-scale, often remote, renewables. Climate change negotiations and security of supply concerns will continue to be the key drivers for DE in the coming years.

Germany already gets 25% of its electricity from renewable sources and is headed for 80% by 2050.

German Distributed Generation Policy by Technology and Type		
Technical	DE in general	The federal Ministry of Education and Research heads up a program aimed at integrating DE successfully into networks. Federal states also undertake their own research funding.
	Large-scale cogeneration	
	Domestic cogeneration	A European standard for microcogeneration is currently being drafted under the auspices of Cenelec technical committee TC8X WG2. The standard has been worked out, and implementation process is going on. Germany is also developing their own regulations for the installation of microcogeneration.
	PV	
	Wind	
	Hydro	
	Biomass	

(*Continued*)

German Distributed Generation Policy by Technology and Type		
Financial	DE in general	By 2013, the German government has allocated a €750 million budget to underpin its objective of raising the market share of CHP installations to 25% by 2020.[a]
	Large-scale cogeneration	To obtain tax relief, operators need to prove that the efficiency level of the CHP plant exceeds 70%. According to current rules (valid up to 2020), operators of bigger plants are entitled for proportional compensation, whereby they get 5.41 cent for the first 50 kW; 4 cent for the next 200 kW, 2.41 cent for the next 1750 kW; and 1.8 cent for the exceeding power capacity. If the plant is subject to the EU emission trading rules, operators can claim for 2.1 cent/kW h. MiniCHP (≤50 kW) operators can opt for 30,000 full operation hours support. A new class of small-scale of CHP plant between 50 and 250 kW has also been introduced.[b]
	Domestic cogeneration	
	Renewables in general	German FiTs for the generation of electricity from renewable sources, under the Act on Granting Priority to Renewable Energy Sources (*the EEG*)—accepted in 2012, are entering their third phase of existence. In Phase One (2000–2009), Germany focused on scaling up domestic renewable electricity generation. During Phase Two (2009–2011), rapid declines in the cost of solar PV modules prompted Germany to more actively adjust its PVFIT in order to manage the volume of annual PV installations under its FiT programs (e.g., linking FiT degressions for PV to the volume of PV installations in previous periods and reviewing the PV policy more frequently). In Phase Three (2012–),continued cost declines are making solar PV, wind, and biomass increasingly competitive with traditional sources of electricity; in response, the key elements of Germany's 2012 EEG—including reduced FiT payments, a market premium option, a 90% cap on FiT-eligible PV electricity, and addition of a 52 GW PV capacity threshold—all mark an evolution of German FiTs toward a *grid parity* future where policy is more flexible and may offer less TLC (Transparency, Longevity, Certainty) to investors.[c] However, FiTs are dropped in 2013 compared to 2012, no retroactive effect on existing facilities.
	On-site PV	FiT rates for PV electricity vary depending on the size and locations of the systems. Since 2009, there are additional tariffs for electricity immediately consumed rather than supplied to the grid with increasing returns if more than 30% of overall production is consumed on-site. This is to incentivize a demand side management and help develop solutions to the intermittency of solar power. Duration of tariff is usually 20 calendar years plus the year of installation. Systems receive the tariff in effect at the time of installation for the entire duration.

(*Continued*)

German Distributed Generation Policy by Technology and Type

	On-site wind	**Onshore FiTs** According to the EEG, the initial tariff of 8.8 cents (EUR)/kW h is paid for at least 5 years. This period is extended according to location and reference yield. Thereafter, the final tariff of 4.8 cents (EUR)/kW h is paid for the time remaining. The law also provides bonuses of 0.47 cents (EUR)/kW h for improved network integration and 0.49 cents (EUR)/kW h for wind energy facility repowering. For installations commissioned in the subsequent calendar years, the tariffs and bonuses will be reduced degressively each year by 1.5%. **Offshore FiTs** The initial 20-year long guaranteed EEG FiT is equivalent to 15 cents (EUR) for the first 12 years or 19 cents (EUR) for the first 8 years and extended subject to location. The larger the water depth of the wind turbine and the further from shore it is located, the longer the higher initial tariff compensation paid level. Thereafter, the sum payable amounts to 3.5 cents (EUR)/kW h. The annual percentage degression for tariffs and bonuses for electricity generated from offshore wind installations shall be 7% from the year 2018 onward (commissioning date).[d]
	Small hydro	Rates are depending on the share of capacity of the plant and differing for existing and newly built facilities. A yearly degression of 1.0% shall apply.[e]
Others	DE in general	The German government has played a lead role in the renewable energy policy network an international policy network arising from the Bonn 2004 renewable energy conference.[9] Disclosure law states that all generators must state generation portfolio including cogeneration and renewables. In September 2010, the federal government adopted the Energy Concept, a comprehensive new strategy for a long-term integrated energy pathway to 2050. Following Fukushima accident, Germany plans to close all of its nuclear power plants by 2022. This decision resulted in the adoption of a suite of new policy measures and determined renewable energy as the cornerstone of future energy supply, a set of policy instruments commonly known as the Energiewende. Presently 25% of electricity is from renewable sources and Germany set up a plan to reach 80% by 2050. In 2008, new CHP Law has been published in Germany. The changes entered into effect from August 21, 2009. The changes mainly concerned the clarification that belong to the eligible consumers leaving the heat network law to accelerate the development of high voltage electricity network

(*Continued*)

German Distributed Generation Policy by Technology and Type

	The main support for CHP in Germany is a stepped FiT in which the first kW h generated in a given year are rewarded at a higher level than subsequent generation. The result of such a mechanism is that fixed costs can be accounted for during the initial run time, and therefore, the investment presents less of a risk for developers. The model used here is unable to reflect the value of this FiT design as it smoothes value over total annual generation.
Large-scale cogeneration	During the process of electricity market restructuring in Germany, the main goal with respect to CHP was to maintain the existing capacity and renovate existing plant. Germany is bound by the EU cogeneration Directive 2004/8/EC, which states that member states are obliged to address key market barriers such as ensuring grid access for cogeneration. The directive also states that in order for cogeneration to be considered high efficiency, it must provide an energy saving of at least 10% compared to separate production of heat and power. Germany is also bound to EU Directive 2012/27/EU, which set as a target to strengthen the regulations of Directive 2004/8/EC, and ordered member states to complete by December 31, 2015, the evaluation of the feasibility potential of high efficiency cogeneration and efficient district heating/cooling.
Domestic cogeneration	
PV	As a result of the earlier improvements, on June 16, 2013, 60% of the total electricity supply was produced by PV and wind plants.
Wind	As a result of the earlier improvements, on June 16, 2013, 60% of the total electricity supply was produced by PV and wind plants.
Hydro	At The 2004 Renewable Energy Conference in Bonn, Germany announced its Mini-Hydro Programme.[10]

[a] http://gastopowerjournal.com/regulationapolicy/item/1257-germany-frees-up-%E2%82%AC750-m-budget-to-raise-chp-market-share-to-25-by-2020#axzz2aSREnklp.

[b] http://gastopowerjournal.com/regulationapolicy/item/1257-germany-frees-up-%E2%82%AC750-m-budget-to-raise-chp-market-share-to-25-by-2020#axzz2aSREnklp.

[c] http://www.dbresearch.com/PROD/DBR_INTERNET_EN-PROD/PROD0000000000294376/The+German+Feed-in+Tariff%3A+Recent+Policy+Changes.pdf.

[d] According to the EEG effective as of mid 2013.

[e] According to the EEG effective as of mid 2013.

11.11 Hungary

The majority of Hungary's existing DE capacity is in the form of district heating applications, though many have already reached the end of their life. However, in the past 6 years, some new installments have been connected to the grid. Due to recent regulatory decisions, the proportion of cogeneration in the total national generation has fallen down from 20.9% in 2010 to 10% in the first half of 2013, whereas, pursuant to the latest EU regulations, member states should elaborate long-term national plans on the possible ways to improve cogeneration and on the conceptions of the subsidies thereto.

Hungarian Distributed Generation Policy by Technology and Type

Technical	DE in general	However, technical requirements of grid connections are clearly set out; in certain cases, DSOs may hinder the connection by their noncooperative behavior.
	Large-scale cogeneration	
	Domestic cogeneration	A European standard for microcogeneration is currently being drafted under the auspices of Cenelec technical committee TC8X WG2. The standard has been worked out; implementation process is going on.
	PV	
	Wind	
	Hydro	
Financial	DE in general	Pursuant to the Electricity Act (86 of 2007) and the Governments Decree nr. 389/2007 (XII.23), the Hungarian Energy and Public Utility Regulatory Authority (HEPURA) has the right and is obliged to publish until the 7th workday of December every year, the applicable FiTs for the subsequent year. Tariffs shall be adjusted by the average of the yearly consumer price index.
	Large-scale cogeneration	HEPURA publishes tariffs for cogeneration in eight categories. Rates depend on the combination of installed capacity, usage of the generated heat, and the type of the permission issued to the power plant.
	Domestic cogeneration	Domestic generators have free access to the grid; DSOs shall grant free and nondiscriminatory access. Very few programs are existing to support domestic generation; thus low FiTs, long return periods, do not promote such kind of investments.
	On-site PV	
	On-site wind	
	Small hydro	
Others	DE in general	In 2012, the government published its new National Energy Strategy until 2030. Aim of the strategy is to increase the proportion of renewable energy within primary energy use from 7% (in 2012) to 20% until 2030—with an interim proportion of 14.56% until 2020. Among renewable, biogas and biomass CHPs and technologies using thermal energy shall have priority. The strategy also states that in order to define directions of district heating and combined generation development, a district heating action plan has to be elaborated; furthermore, improvement of the level of supply shall not be delayed any more. The strategy is partly in line with the effective European goals, as the EU, however, earlier established goals until 2010 could not be fulfilled by the members states, is still committed to keep the 20% ratio of renewable energy use within the entire consumption until 2020.

(Continued)

Hungarian Distributed Generation Policy by Technology and Type	
Large-scale cogeneration	Hungary is bound by the EU cogeneration Directive 2004/8/EC, which states that member states are obliged to address key market barriers such as ensuring grid access for cogeneration. The directive also states that in order for cogeneration to be considered high efficiency, it must provide an energy saving of at least 10% compared to separate production of heat and power. Hungary is also bound to EU Directive 2012/27/EU, which set as a target to strengthen the regulations of Directive 2004/8/EC, and ordered member states to complete until December 31, 2015, the evaluation of the feasibility potential of high efficiency cogeneration and efficient district heating/cooling.
Domestic cogeneration	
PV	
Wind	
Hydro	

11.12 Portugal

After joining the EU in 1986, Portugal's economy grew steadily. However, since the financial crisis, Portugal is continuously listed between the critical financial status countries. Nevertheless in the first quarter of 2013, Portugal had recorded green power production.

Portuguese Distributed Generation Policy by Technology and Type		
Technical	DE in general	However, free access to the grid is secured by the law to any suppliers, TSOs and DSOs have hindering practice of refusing access to the grid with the reason of security of the transmission and distribution system.
	Large-scale cogeneration	
	Domestic cogeneration	A European standard for microcogeneration is currently being drafted under the auspices of Cenelec Technical Committee TC8X WG2. The standard has been worked out; implementation process is going on.
	PV	
	Wind	
	Hydro	
Financial	DE in general	In Portugal, the generation of electricity from renewable energy sources is mainly promoted through a guaranteed FiT. Operators of renewable energy plants are contractually entitled against the grid operator to payment for electricity exported to the grid. (The grid operator is obliged to enter into a contract on the purchase of electricity at a price set by law). There is also a regime for micro- and a miniproduction units, which is also under review and had a few changes recently introduced by DL 25/2013. The guaranteed FiT, which is calculated by a formula, is the only means of promotion. The calculation is based on various factors like plants' output and capacity. The formulas and payment rates for some technologies have been revised in 2007.

(Continued)

Portuguese Distributed Generation Policy by Technology and Type

		There are individual tariffs for electricity generated from renewable sources by so-called microproduction units and for electricity combined with heating systems. Microproduction units are installations that use a single production technology and have a single-phase or three-phase load operating at a low voltage, and a capacity of no more than 5.75 kW solar energy installations, wind power plants, hydroelectric power plants, or biomass-fuelled CHP plants whose capacity is <3.68 kW are eligible for a special tariff. The operators of microproduction units receive the special tariff for 15 years.
	Large-scale cogeneration	Cogenerated electricity exported to the grid benefits from FiTs. These FiTs are applied for a period of 10 years and it is indexed to the price of oil, to reduce fuel-price risks. As the price of oil rises, so will the FiT for cogenerated electricity.
	Domestic cogeneration	Micro-CHP in Portugal benefits from a higher level of support than larger cogeneration.
	On-site PV	FiTs for mini (3.68 kW) PV generators in 2013: 0.196 EUR/kW h for the first 8 years and 0.165 EUR/kW h for the following 7 years (FIT granted for total of 15 years); FiTs for micro (3–68 kW—20 kW) PV generators in 2013: 0.151 EUR/kW h for a period of 15 years. In comparison, these tariffs are 30% lower than the previous year's tariffs.
	On-site wind	In case of micro-and miniproduction units, the producer receives a tariff based on the reference tariff in place at the time of the issuance of the certification of exploitation. Microproduction units: in general, the reference tariff in 2013 is €272/MW h for the first period of 8 years and €150/MW h for the second period of 7 years. Miniproduction units: in general, the reference tariff in 2013 is €185/MW h: Microproduction unit: 80% of the reference tariff Miniproduction unit: 80% of the reference tariff
	Small hydro	
Others	DE in general	Decree-law 23/2010 transposes into Portuguese Law Directive 2004/8/CE on the promotion of cogeneration. Decree-Law 34/2011, called the Mini Production Law, regulates the production of electricity from renewable energy sources in small units of less or equal to 250 kW, utilizing only one type of technology and excluding the microproduction, cogeneration units and innovation and proof of concept systems. ESCOs are allowed to apply to this law. The regulation simplifies the licensing regime through the new SRMini electronic platform managed by DGEG. Any entity that has a contract for purchasing electricity with a relevant consumption of electricity can apply to this law, provided that injected power does not exceed 50% of the contracted power. The law included two types of FiT systems for the remuneration of renewable electricity. First, the general regime whereby the tariff depends on the conditions of the market and no special tariff is applied. Second, the *bonafide tariff*, where a special reference tariff of EUR 250/MW h is provided for systems below 20 kW. As for systems larger than 20 kW, a tender process is used to select the systems that offer better discount against the reference tariff. An annual ceiling of 50 MW is used for the Bonified Regime. The Bonified Regime is also dependent on the realization of energy efficient audits in the place of the installation and implementation of the suggested energy efficiency measures. Annually 1% of the registered installations will be fiscalized.

(Continued)

Portuguese Distributed Generation Policy by Technology and Type	
Large-scale cogeneration	Portugal is bound by the EU cogeneration Directive 2004/8/EC, which states that member states are obliged to address key market barriers such as ensuring grid access for cogeneration. The directive also states that in order for cogeneration to be considered high efficiency, it must provide an energy saving of at least 10% compared to separate production of heat and power. Portugal is also bound to EU Directive 2012/27/EU, which set as a target to strengthen the regulations of Directive 2004/8/EC, and ordered member states to complete by December 31, 2015, the evaluation of the feasibility potential of high efficiency cogeneration and efficient district heating/cooling.
Domestic generation	An annual quota of 25 MW was implemented in the microgeneration segment (up to 3.68 kW inverter capacity), which represents approximately the requested capacity. In 2012, the capacity was lowered to 10 MW.
PV	Limitation in microgeneration segment created serious problems for the PV sector, as it is below the demand for licenses. The main barrier is the application of technical rules, which creates high costs for the change of the grid connection. Nevertheless each consumer can install a system with a capacity of 50% of the capacity contracted for consumption. The production capacity is limited to 25% of the medium voltage transformer capacity.
Wind	
Hydro	

11.13 United Kingdom

The United Kingdom was the first in Europe to experiment with competitive electricity markets, and some DE investment arose as a direct result of that policy. Like other European nations, key DE drivers in the near future for the U.K. remain climate change and energy security. The United Kingdom has pioneered some interesting policy in defining high-quality cogeneration, but there is much more room for policy reform to spur investment in DE to meet environmental and national security objectives.

British Distributed Generation Policy by Technology and Type		
Technical	DE in general	There are two key documents that regulate connection to the United Kingdom grid—Engineering Recommendations G83 and G59. Published by the Electricity Networks Association, these documents set out what is expected by the distribution network operator (DNO). ER G83 covers systems up to 16A per phase; ER G59 covers larger systems.
	Large-scale cogeneration	
	Small scale	Amendment P81 of the electricity trading rules allows small generators (16 A per phase on the low voltage 230 V single phase or multiphase 400 V supply) to use existing meters rather than having to use otherwise mandated half-hour interval meters.

(Continued)

British Distributed Generation Policy by Technology and Type		
	Domestic cogeneration	A European standard for microcogeneration is currently being drafted under the auspices of Cenelec Technical Committee TC8X WG2. The standard has been worked out; implementation process is going on. The United Kingdom has also developed changes in legislation, and published Engineering Recommendations in this area.[a]
	PV	Grid interconnection of PV systems is governed by G.59/2. Latest version is 3.4, issued in January 2013.
	Wind	Grid interconnection of wind systems is governed by G.59/3. Latest version is 3.4, issued in January 2013.
	Hydro	Grid interconnection of hydropower systems is governed by G.59/2. Latest version is 3.4, issued in January 2013.
Financial	DE in general	The Climate Change Levy is a tax on energy introduced in 2001. All businesses have to pay 0.456 p/kW h unless the producer has a Levy Exemption Certificate. A Carbon Price Floor, from April 1, 2013, has been introduced and is applied as a levy for electricity generators based on the carbon content of each fuel type. Such supplies are charged at the relevant carbon price support rate depending on the type of fossil fuel used, which is determined by the average carbon content of each fossil fuel equivalent to GBP4.94/tCO_2 for 2013–2014. Proposed rates for 2014–2015 and 2015–2016 are GBP7.28/tCO_2 and GBP9.86/tCO_2.[b] The U.K. government announced details of reforms to the U.K. market in late June 2011. Key features of the paper included • Introduction of a two-way FiT with Contracts for Difference for each low-carbon generation technology, likely to replace the Renewable Obligation Scheme by 2017 • Disincentives on fossil fuel generators such as the Carbon Price Floor (a proposal that the effective price of carbon should be GBP70/tCO_2 in 2020) and an Emissions Performance Standard (EPS) (set at 450 g CO_2/kW h).
	Large-scale cogeneration	EU Emissions trading scheme will have an uncertain effect on cogeneration investment. Cogeneration schemes that have been certified as *good quality* (see the following) are exempt from the climate change levy. From April 2009 under the U.K. renewables Obligation (RO), electricity generated by Good-Quality CHP fuelled by biomass is eligible for additional renewables Obligation Certificates (ROCs) for each MW h of renewable electricity generated. As a rule, revisions to the RO have aimed to award CHP generators with a premium of 0.5 ROCs per MW h over power-only plants, except in the case of so-called Advanced Conversion Technologies, where there is no marginal incentive. In addition, grants for capital purchases of biomass-fuelled heat-only or CHP equipment have been made available through the Bioenergy capital grants scheme run by the U.K. Department for Environment, Food and Rural Affairs (Defra).[c]

(*Continued*)

British Distributed Generation Policy by Technology and Type

	Domestic cogeneration	FiTs are available for small-scale, low-carbon electricity generated by private/business users (maximum capacity 5 MW) providing payment of up to 41.3 p/kW h generated (depending on the type and size of the system used to generate renewable energy) plus a guaranteed 3 p/kW h sold on to the U.K. electricity grid. Typically, the tariffs last for 20 years (the exception is the Solar PV tariff, which currently lasts for 25 years).[d]
	On-site PV	Exempt from climate change levy. Each MW h generated via renewable energy also creates a tradable *Renewable Obligation Credit*, which can be sold on the market to generators who have not reached their legal obligation to generate every-time prescribed volume of their output from renewable. VAT on PV installations systems has been set at the reduced rate of 5% since April 2000. In February 2012, the U.K. government reduced the FiT from 41.3 to 21 p/kW h and stipulated an efficiency requirement for Solar PV schemes registered after March 2012.
	On-site wind	Exempt from climate change levy. Each MW h generated via renewable energy also creates a tradable *Renewable Obligation Credit*, which can be sold on the market to generators who have not reached their legal obligation to generate every-time prescribed volume of their output from renewable.
	Small hydro	Exempt from climate change levy. Each MW h generated via renewable energy also creates a tradable *Renewable Obligation Credit*, which can be sold on the market to generators who have not reached their legal obligation to generate every-time prescribed volume of their output from renewable.
Others	DE in general	Generating electricity from renewable and energy-efficient sources is a key part of the British Government's strategy to tackle climate change. The 2009 Renewable Energy Directive sets a target for the United Kingdom to achieve 15% of its energy consumption from renewable sources by 2020, implying substantial growth in distributed generation and investment in the network infrastructure. Many stakeholders have expressed concern that the government's targets may not be met because of major difficulties relating to unit pricing, delays in obtaining planning consent and the fact that the current regulatory framework does not incentivize DNOs to connect distributed generation.
	Large-scale cogeneration	The United Kingdom is bound by the EU cogeneration Directive 2004/8/EC, which states that member states are obliged to address key market barriers such as ensuring grid access for cogeneration. The directive also states that in order for cogeneration to be considered high efficiency, it must provide an energy saving of at least 10% compared to separate production of heat and power. The United Kingdom is also bound to EU Directive 2012/27/EU, which set as a target to strengthen the regulations of Directive 2004/8/EC, and ordered member states to complete by December 31, 2015, the evaluation of the feasibility potential of high efficiency cogeneration and efficient district heating/cooling.

(*Continued*)

British Distributed Generation Policy by Technology and Type

		Nevertheless, the U.K.'s own model for defining cogeneration deserves mention. The United Kingdom has established a cogeneration quality assurance scheme,[e] a voluntary methodology for determining which arrangements or projects can be defined as cogeneration and which are therefore eligible for cogeneration incentives and financial support. The scheme assesses cogeneration projects using two thresholds. The power efficiency threshold states that when a project's power efficiency is greater than 20%, then all fuel used in the project is *good quality*. The second threshold, the quality index threshold, considers both power and heat efficiency, which encourages good environmental practice. Though complex, the cogeneration QA scheme is robust in its consideration of technologies and fuels.
	Domestic cogeneration	
	PV	
	Wind	
	Hydro	

[a] Micro-cogeneration needs specific treatment in the European Directive on Cogeneration, Cogen Europe.
[b] http://www.kpmg.com/Global/en/IssuesAndInsights/ArticlesPublications/Documents/taxes-incentives-renewable-energy-2012.pdf.
[c] http://www.iea.org/media/files/chp/profiles/UK.pdf.
[d] http://www.kpmg.com/Global/en/IssuesAndInsights/ArticlesPublications/Documents/taxes-incentives-renewable-energy-2012.pdf.
[e] http://www.cogenerationqa.com.

11.14 China

Market potential for DE in China is enormous, but clear policies will be required in order for it to be realized. Demand for electricity has been growing at rates and scales that dwarf those of Europe and North America. China has issued a series of policies to promote CHP/DHC. As a result, China has become the second-largest country in terms of installed CHP capacity, and now installed capacity exceeds 100 GW and getting very close to 200 GW. Increasingly, high electricity prices, rotating blackouts, coal shortages, emerging gas availability, and the growing appeal of renewables for their environmental benefits are all factors that will drive Chinese investment in DE. Policies to promote renewables have recently had a high profile, but other forms of DE still require attention. International climate change negotiations may also rise in influence in Chinese energy policy.

Chinese Distributed Generation Policy by Technology and Type

Technical	DE in general	
	Large-scale cogeneration	
	Domestic cogeneration	
	PV	
	Wind	
	Hydro	

(Continued)

Chinese Distributed Generation Policy by Technology and Type

Financial	DE in general	With the Renewable Energy Law as revised in April 2010, the State Bureau of Energy and other departments of the State Council will promulgate guidelines on the full purchase of electricity generated by new energies. According to the revised law, the price of on-grid electricity generated by renewable energies shall be determined by the competent price department of the State Council. The council will consider the difference in areas and the electricity generated by different types of renewable energy companies. In China, some financial funds and financial subsidies are available for renewable energy projects and for energy conservation technologies improvement. A reduced corporate income tax rate of 15% is given for qualified advanced and new technology enterprises. Applicable fields include solar energy, wind energy, biomaterial energy, and geothermal energy.
	Large-scale cogeneration	Throughout the 1990s, the government provides capital grants and tax benefits for cogeneration, but the program has since been withdrawn.
	Domestic cogeneration	
	On-site PV	In March 2009, the China government introduced the *Solar Roofs Plan* for promoting the application of solar PV building. The Ministry of Finance in July of the same year re-introduced the *Golden Sun Project* with more specific details of the related policy. The policy provides that for the grid-connected photovoltaic power generation project and its supporting transmission and distribution system, the state will provide subsidies of 50% of the total investment. The subsidy will rise to 70% for solar power systems in remote areas that are not currently connected to the grid. Projects with a minimum capacity of 500 MW would be eligible for the related incentive. The corporate income tax reduction shall also be applied.
	On-site wind	Wind projects currently receive a 50% reduction in sales tax as well as corporate and income tax.
	Small hydro	
Others	DE in general	From 2008 to January 2012, China held the top spot in clean energy investment. The Renewable Energy Law passed in 2005 explicitly states in its first chapter that the development and the usage of renewable energy is a prioritized area in energy development. Detailed incentive policies and programs include Golden Sun program providing financial subsidies, technology support, and market incentives to facilitate the development of solar power industry; the suggestions on promoting wind electricity industry in 2006 offering preferential policies for wind power development; and many other policies. Besides promoting policies, China has enacted a number of other policies to standardize renewable energy products, to prevent environmental damage, and to regulate price of green energy. These policies include, but are not limited to, Renewable Energy Law, Safety Regulations of Hydropower Dams, and National Standard of Solar Water Heater. The Twelfth Five-Year Plan, the current plan, also gives great emphasis on green energy. According to the plan, China will build up microgrid demonstration zones in areas redundant of solar and wind power. Meanwhile, China will push forward the construction of 100 new energy demonstration cities. National Energy Administration proposed that up to 2020, China will promote the use of distributed generation systems all over the country with the installed capacity to hit 50 GW. Besides, China plans to build 10 distributed energy demonstration zones of various characteristics.

(Continued)

Chinese Distributed Generation Policy by Technology and Type	
Large-scale cogeneration	Besides the aggressive growth of energy need, China is committed to reduce its emission, but renewable sources will not be able to keep up with the needs, large-scale cogeneration shall have an important role in securing China's need for energy and in avoiding energy shortages in the next years. Coal-fired plants still will have a significant role; the aim is to reduce participation of such kind of plants in power generation. However, earlier gas-fired power plants were traditionally used in offshore oil drilling platforms, or as peak load power plants; in recent years, a strong trend toward gas-fired power plants as a main electricity and heat resource has been recorded. In 2011, gas-fired power plant capacity reached 33 GW: a growth of 24% compared with 2010. By 2015, the installed capacity of gas-fired power plants will reach at least 60 GW, implying a compound annual growth rate of around 20%. District heating is also expected to drive considerable growth in gas-fired equipment demand. As of 2010, district heating with CHP supplies over a third of total construction areas in northern cities and towns, the rest being heat-only boiler stations fuelled by coal or natural gas. To reduce energy waste and pollution caused by coal-fired localized boilers and improve energy efficiency, cities including Beijing, Tianjin, and Taiyuan have planned to restrict new heat plants to gas-fired only. As shown in currently released plans, new heating plants in these areas have all adopted gas-fired cogeneration.
Domestic cogeneration	
PV	According to plans unveiled by the National Development and Reform Commission in 2007, the country's installed solar capacity was to grow to 1800 MW by 2020. Due to recent developments in May 2011, the National People's Congress (NPC) set 5 GW as an official minimum PV target for 2015, with a longer-term target of 20–30 GW by 2020. Topping the plan only in 2012, China added 5.0 GW of panels, bringing installed capacity to 8300 MW and, in 2013, may add 6.8 GW. According to the European Photovoltaic Industry Association, the total installed capacity could grow from 47 to 66 GW by 2017!
Wind	In 2012, Chinese wind power plants generated 100.8 billion kW h of wind power, compared with 71.5 billion kW h in 2011, according to the data of the National Energy Administration. Total grid-connected installed capacity increased 31% to 62.7 GW from 47.9 GW in 2011. The goal is to install 100 GW of grid-connected wind farms by 2015 and to generate 190 billion kW h of power. Wind energy has become the Asian nation's third-biggest energy resource, following coal and hydropower, and accounts for about 2% of total electricity, according to data from China's State Electricity Regulatory Commission.
Hydro	By 2015, China's hydropower installations are targeted to reach around 325 GW, and to reach 430 GW (up from 380 GW) by 2020. As part of this project in July 2013, the second largest hydropower plant started the operation with one turbine of a capacity of 770,000 kW. Total capacity of the plant will be 13.86 GW when all of its 18 units go into operation in 2014.

11.15 India

India is sure to be a major market for DE in the coming years given its large population and the urgent need for investment in the power sector in order to meet growing demand for power. The blackout in July 2012 leaving 700 million people without power demonstrated the inefficiency of Indian power generation and the vulnerability of centralized supply compared to decentralized generation. Recently, upcoming possibility of LNG trade with the United States might also have a catalytic effect on future DE improvements.

Indian Distributed Generation Policy by Technology and Type

Technical	DE in general	According to Section 7 of Electricity Act: "Any generating company may establish, operate and maintain a generating station without obtaining a licence under this Act if it complies with the technical standards relating to connectivity with the grid ..."
	Large-scale cogeneration	
	Domestic cogeneration	
	PV	
	Wind	
	Hydro	To set up a hydrogenerating station, a scheme with a prescribed content shall be prepared and filed into the controlling authority for concurrence.
Financial	DE in general	In India, various supports and subsidies exist to help improving DE.
		Foreign direct investment (*FDI*) is permitted up to 100% in the sector under the automatic route in Renewable Energy Generation and Distribution projects subject to the provisions of the Electricity Act.
		Undertakings engaged in generation or generation and distribution of power were offered a 10-year tax holiday for renewable energy plants if it begun to generate power before March 31, 2013. (However, they have to pay a minimum alternative tax at the rate of approximately 20%, which can be offset in future years.)
		According to the new regulations effective as of April 1, 2013, alternative incentive mechanism provides for expenditure-based incentive to business of generation, transmission, or distribution of power. As regards this incentive, all revenue and capital expenditure (except few) will be allowed as tax deduction upfront instead of claiming amortization/depreciation on the capital expenditure and there would be no tax holiday available.
		Under Ministry for Non-Conventional Energy Sources, The Indian Renewable Energy Development Agency has been set up and is a specialized financing agency to promote and finance renewable energy projects.
		In order to attract foreign investors, the Indian Government has taken several initiatives such as introducing generation-based incentives (GBI) scheme to promote projects under Independent Power Producers (*IPP*) mode (wind power INR 0.50/MW, solar power INR 12.41/kW).
		Under the domestic income-tax law, power companies have been provided with an option to claim depreciation under straight-line method. However, a company can claim either accelerated depreciation or GBI, but not both.

(Continued)

Indian Distributed Generation Policy by Technology and Type

		India aims to derive 15% of its energy requirements from renewable energy sources by the year 2020. Renewable Purchase Obligation (RPO) is one of the tools of implementing this ambitious goal. Under these rules, distribution companies, open access consumers, and captive consumers are obligated to buy a certain percentage of their power from renewable sources of energy.
		The Indian Government has given various incentives on setting up the renewable energy power project, which includes exemption from customs and excise duties on specific goods required for setting up the renewable energy projects.[3]
	Large-scale cogeneration	Tax and duty structure for CHP capital equipment is not as attractive as for other renewable energy technologies.
	Domestic cogeneration	
	On-site PV	Under the domestic income-tax law, renewable companies (solar as well as wind power) are provided with accelerated depreciation at 80%.
		In January 2010, Jawaharlal Nehru National Solar Mission (JNNSM) was launched with a mission target of 20,000 MW of solar-generating capacity by the end of the Thirteenth Five-Year Plan (2022).
	On-site wind	In 2012, the Indian Government has restricted the accelerated depreciation of 80% to windmills installed on or before March 31, 2012. Windmills installed after March 31, 2012, will be eligible for depreciation of 15% instead of 80% on written-down value method.
	Small hydro	
Others	DE in general	Electricity Act of 2003 includes favorable provisions for DE[11] and guarantees access to all generators regardless of size. The act has moved India toward an electricity market with separate generation, transmission, and distribution of power, with increasing potential for competition. A milestone in this progress is the introduction of the *Availability Based Tariff (ABT)*, which is an intermediate step in the effort to develop a true spot market for electricity. An ABT has been introduced at all five electrical regions of the country at the interstate level.[a]
		Section 30 of the 2003 Electricity Act is designed to facilitate on-site generation. The Bill Section 38, 2(d) of the 2003 Electricity Act makes it obligatory to provide *nondiscriminatory open access*. There is a similar obligation on the State Transmission Utility, Section 39-2(d), to provide nondiscriminatory open access.
		According to the current (12th: 2012–2017) 5-year plan of India, it is planned to add a grid interactive renewable capacity addition of about 30,000 MW, comprising of 15,000 MW wind, 10,000 MW solar, 2,100 small hydro, while the balance is planned primarily from biomass. So the share of renewables in electricity generated is expected to rise from around 6% in 2012 to 9% in 2017 and 16% in 2030.
	Large-scale cogeneration	CHP could be one of the beneficiaries of the deep shock that followed India's July 2012 blackout. India already leads in biomass CHP in its sugar industry but has plenty of unrealized potential for tapping waste heat for CHP.
		The IEA estimates that India could grow from its current base of less than 10 GW to almost 28 GW of CHP in 2015 and 85 GW in 2030. CHP and district heating and cooling could be implemented in smaller industrial parks, special economic zones, and other areas with a concentration of large commercial and software establishments needing secure, low-cost heat, cooling, and power, according to the agency.

(Continued)

Indian Distributed Generation Policy by Technology and Type		
	Domestic cogeneration	
	PV	Grid-interactive solar power generation capacity stands at 1761 MW with about 557 MW installed by Q2 in 2013. With most of the CSP projects that were due to be commissioned in May 2013 delayed as the forecast for installations in 2013 looks flat compared to 2012. Only about 60% of the targeted installation goal has been achieved by Q2 in 2013 despite the commissioning deadline of Phase 1 of JNNSM ending in May.
	Wind	India has the fifth largest installed wind power capacity in the world. In 2009–2010, India's growth rate was highest among the other top four countries. As of January 31, 2013, the installed capacity of wind power in India was 19,564.95 MW. It is estimated that 6000 MW of additional wind power capacity will be installed in India by 2012.Wind power accounts for 8.5% of India's total installed power capacity, and it generates 1.6% of the country's power.
	Hydro	Hydropower makes up about 22% of the total installed capacity in India. However, the country has a total hydropower capacity of 68% yet to be developed. The utilization of hydropower potential is only 39,449 MW out of the total capacity of 145,320 MW

[a] Distributed Generation—a Strategy for Optimal Future Power Generation in India Ajit Kapadia and K.N. Naik Centre for Fuel Studies and Research, Pune, India.

11.16 Japan

Japan has a long history of DE leadership and boasts the world's lowest energy consumed per GDP ratios. Still Japan is highly dependent on energy imports, and additional investment in DE may go a long way in addressing this. Japan's PV manufacturing infrastructure is the world's most developed, and there appears to be ever-increasing policy support for DE. There remains much room for improvement, and additional policy support will be key in realizing the potential of DE. Major drivers for DE growth will continue to be climate change and energy security. Following Fukushima catastrophe, DE became far more important than before, as a result, Japanese Government is focusing even much stronger on supporting DE.

Japanese Distributed Generation Policy by Technology and Type		
Technical	DE in general	Technical guidelines for grid interconnection of small DE applications have been established.[a]
	Large-scale cogeneration	
	Domestic cogeneration	
	PV	
	Wind	
	Hydro	
Financial	DE in general	In Japan, investment subsidies and tax benefits are used as the main tools, rather than a FiT approach. Subsidies are regularly reviewed in the light of technological and economic developments.

(Continued)

Japanese Distributed Generation Policy by Technology and Type

		The following subsidies and tax incentives are available: **Subsidies for High-Efficiency Natural Gas CHP (10–3000 kW):** The Support Programme for New Energy Users provides subsidies for businesses that introduce qualifying new energy systems such as natural gas CHP systems and fuel cells. **The Programme for the Promotion of New Energy in Local Areas:** This program provides subsidies for local public entities that plan to introduce qualifying new energy systems, which are close to being commercial but still have high system costs. Eligible technologies include energy efficient applications such as clean energy vehicles, natural gas–fuelled CHP systems and fuel cells, as well as renewables. **Accelerated tax depreciation of CHP investment:** The Taxation System for the Promotion of Investment in Energy Supply–Demand offers a 7% tax exemption for small- and medium-sized businesses or an accelerated tax depreciation of 30% of the standard acquisition value of the equipment. **R&D on high-efficiency natural gas CHP and fuel cells:** The Japanese government also actively supports R&D, demonstration, and commercialization of gas-engine and fuel cell CHP systems for residential use.[3]
	Large-scale cogeneration	
	Domestic cogeneration	
	On-site PV	The 1992 net-metering policy required utilities to purchase excess power from PV systems at the retail rate the same year ambitious targets for PV were set. In 1994 the *70,000 Roofs* program was launched. Incentives included low interest loans, a comprehensive education and outreach program, and declining rebates for interconnected residential systems. Rebates were initially 50% of the installed cost for end-users but declined annually and were phased out completely in 2002. In 1997, the program was altered to also extend rebates to landlords and housing developers. The government promoted PV with print and television advertising campaigns. By the end of the program in 2002, it had exceeded its goals.[12] The FiT originally introduced in 1997. On June 18, 2012, a new FiT was approved, of 42 yen/kW h, about 0.406 Euro/kW h or U.S. $0.534/kW h. The tariff covers the first 10 years of excess generation for systems less than 10 kW, and generation for 20 years for systems over 10 kW. It became effective on July 1, 2012. In April 2013, the FiT was reduced to 37.8 yen/kW h.
	On-site wind	FiT in 2013 is 23.1 yen/kW h—which is 2.5 times more than the relevant German tariff—while for small wind at less than 20 kW capacity gets even higher tariffs—they are set at 57.75 yen. However, there was zero new capacity deployed under these tariffs until now
	Small hydro	
Others	DE in general	In 2012, Japan has published its latest Strategic Energy Plan (SEP), originally formulated in 2003 and reviewed every 3 years by conference bodies as well as various opinion polls. The latest revision includes a road map of energy-based economic growth and reformation of the energy infrastructure. The strategy is to ensure growth and encourage multifaceted international trade by taking into account the social cost and price differences between nuclear, thermal, and renewable energy generation.

(Continued)

Japanese Distributed Generation Policy by Technology and Type		
	Large-scale cogeneration	SEP set up a goal of increasing the use of cogeneration from 30 to 150 billion kW h until 2030.
	Domestic cogeneration	
	PV	By the end of 2012, Japan had installed 7000 MW of PV. Due to the new FiT, Japan is expected to install 5300 MW in 2013.
	Wind	Wind power in Japan generates a small but increasing proportion of the country's electricity, as the installed capacity has been growing in recent years. Current installed capacity is 2304 MW.
	Hydro	Main renewable energy source with a potential of 34.7 GW.

[a] Shinichi Nakane, Japan Cogeneration Center.

11.17 South Korea

Like many OECD countries, South Korea is almost totally reliant on energy imports to meet domestic demand. Energy independence, reliability, and economic competitiveness will continue to be an important driver for DE for some time to come. It is clear that policy will play an important role in realizing South Korea's ambitious goals in developing DE.

Korean Distributed Generation Policy by Technology and Type		
Technical	DE in general	
	Large-scale cogeneration	
	Domestic cogeneration	
	PV	
	Wind	
	Hydro	
Financial	DE in general	South Korea is ranked at fifth on KPMG's Green Tax Index, which provides an indication of which countries are most active in using green tax incentives and penalties to drive sustainable corporate behavior and achieve green policy objectives. South Korea in common with the United States has a green tax system weighted toward incentives rather than penalties. South Korea leads the ranking for green innovation, which suggests that South Korea is especially active in using its tax code to encourage green research and development. At the end of 2011, the FiT was abrogated due to introduction of a RPS in 2012, where power plants over 500 MW have an obligation to buy renewable energy. (The government maintains a FiT only for existing recipients.) There is a research and development tax credit program, which is applied for renewable energy technologies. Import duties are reduced by 50% for all components and/equipment used in renewable energy power plants. The government also provides subsidies up to 60% to local governments for the installation of renewable energy facilities, and it offers low interest loans (5.5%–7.5%) to renewable energy projects, including a 5-year grace period followed by a 10-year payment period.[3]

(Continued)

Korean Distributed Generation Policy by Technology and Type

	Large-scale cogeneration	Through the government's Integrated Energy Supply Policy (IESP), the most significant support for CHP uses urban planning policy to designate new developments as Integrated Energy Supply Areas (IESAs), thereby creating a captive market for DHC CHP. Tax incentives and low-interest loans are available for businesses installing CHP equipment. CHP plants over 100 MWe can buy natural gas directly from the Korea Gas Corporation (KOGAS) at the wholesale price.[a]
	Domestic cogeneration	
	On-site PV	Due to the lack of a domestic track record for solar energy during the enactment of the new law on RPS, the standard price for solar energy was fixed at 120% of the German price.
	On-site wind	Among the 11 energy sources selected by the act on RPS wind energy, water energy has its standard price calculated by the government.
	Small hydro	Among the 11 energy sources selected by the act on RPS, water energy has its standard price calculated by the government.
Others	DE in general	Under RPS, which came into force on January 1, 2013, the 13 power companies that operate plants over 500 MW will have an obligation to buy renewable energy. Percentage in 2013 is 4%, which shall be increased by 0.5% point yearly until 2016 and by 1% point after that to reach 10% by 2022.
	Large-scale cogeneration	
	Domestic cogeneration	
	PV	Under RPS, a solar installation of 220 MW has been set and gradually increasing to 1.2 GW in 2015.
	Wind	Offshore wind energy is set to grow tremendously, with a 2.5 GW offshore farm being planned on the southwest coast of the country by 2019. The project comes at a cost of U.S. $9 billion, and will be implemented in three stages, with the first 100 MW being brought online by 2014.
	Hydro	In 2012, Korea Western Power (WP) announced their intention to build 20 tidal power plants from now to 2014, with a total capacity of 520 MW.

[a] http://www.iea.org/media/files/chp/profiles/Korea.pdf.

References

1. WADE, *World Survey of Decentralized Energy*, Edinburgh, U.K., 2005.
2. WADE, *Seven Guiding Principles for Effective Electricity Market Regulation*, Scotland, U.K., 2003.
3. KPMG International, Taxes and incentives for renewable energy, http://www.kpmg.com/Global/en/IssuesAndInsights/ArticlesPublications/Documents/taxes-incentives-renewable-energy-2012.pdf (accessed October 1, 2013).
4. M. Vignolo, The new electricity supply industry in Argentina and Chile, Facultad de Ingeniería—IIE, Montevideo, Uruguay.

5. CAMMESA, Annex 12 Autogeneradores y cogeneradores, version 6, Oct. 1999.
6. A. Maia, Brazil weighs up new potential, *Cogeneration and Onsite Power Production Magazine*, July–August 2005.
7. E. Martinot, Renewables 2005 global status report, The Worldwatch Institute, Washington, DC, 2005, p. 25.
8. Position of DE and cogeneration in the Czech market: Former, present and future. *Josef Jelecek Cogen Europe Conference*, 2005.
9. Anniversary of renewables 2004, Status of the Implementation of the Conference Results, BMU.
10. *Renewables 2004—International Conference for Renewable Energies*, Bonn, Germany, Conference Report, Outcomes and Documentation—Political Declaration/International Action Programme/Policy Recommendations for Renewable Energies, June 1–4, 2004.
11. *World Survey for Decentralized Energy*, WADE, Edinburgh, U.K., 2005, p. 19.
12. IEA PVPS annual report 2004.

12

*Economics Methods**

Walter Short and Rosalie Ruegg

CONTENTS

* Modified by Walter Short, retired from the National Renewable Energy Laboratory (NREL), from the original text prepared by Rosalie Ruegg for the 2007 *Handbook of Energy Efficiency and Renewable Energy.*

12.1 Introduction

Economic-evaluation methods facilitate comparisons among energy technology investments. Generally, the same methods can be used to compare investments in energy supply or energy efficiency. All sectors of the energy community need guidelines for making economically efficient energy-related decisions.

This chapter provides an introduction to some basic methods that are helpful in designing and sizing cost-effective systems, and in determining whether it is economically efficient to invest in specific energy efficiency or renewable energy projects. The targeted audience includes analysts, architects, engineers, designers, builders, codes and standards writers, and government policy makers—collectively referred to as the "design community."

The focus is on microeconomic methods for measuring cost-effectiveness of individual projects or groups of projects, with explicit treatment of uncertainty. The chapter does not treat macroeconomic methods and national market-penetration models for measuring economic impacts of energy efficiency and renewable energy investments on the national economy. It provides sufficient guidance for computing the measures of economic performance for relatively simple investment choices, and it provides the fundamentals for dealing with complex investment decisions.

12.2 Making Economically Efficient Choices*

Economic-evaluation methods can be used in a number of ways to increase the *economic efficiency* of energy-related decisions. There are methods that can be used to obtain the largest possible savings in energy costs for a given energy budget; there are methods that can be used to achieve a targeted reduction in energy costs for the lowest possible efficiency/renewable energy investment; and there are methods that can be used to determine how much it pays to spend on energy efficiency and renewable energy to lower total lifetime costs, including both *investment costs* and energy costs.

The first two ways of using economic-evaluation methods (i.e., to obtain the largest savings for a fixed budget and to obtain a targeted savings for the lowest budget) have more limited applications than the third, which aims at minimizing total costs or maximizing *net benefits* (*NB*) (net savings (NS)) from expenditure on energy efficiency and renewables. As an example of the first, a plant owner may budget a specific sum of money for the purpose of retrofitting the plant for energy efficiency. As an example of the second, designers may be required by state or federal building standards and/or codes to reduce the design energy loads of new buildings below some specified level. As an example of the third, engineers may be required by their clients to include, in a production plant, those energy efficiency and renewable energy features that will pay off in terms of lower overall production costs over the long run.

Note that economic efficiency is not necessarily the same as engineering thermal efficiency. For example, one furnace may be more "efficient" than another in the engineering technical sense, if it delivers more units of heat for a given quantity of fuel than another. Yet, it may not be economically efficient if the first cost of the higher output

* This section is based on a treatment of these concepts provided by Marshall and Ruegg (1980a).

furnace outweighs its fuel savings. The focus in this chapter is on economic efficiency, not engineering efficiency.

Economic efficiency is conceptually illustrated in Figures 12.1 through 12.3 with an investment in energy efficiency. Figure 12.1 shows the level of energy conservation, Q_c, that maximizes NB from energy conservation—that is, the level that is most profitable over the long run. Note that it corresponds to the level of energy conservation at which the curves are most distant from one another.

Figure 12.2 shows how "marginal analysis" can be used to find the same level of conservation, Q_c, that will yield the largest NB. It depicts changes in the total benefits and cost curves (i.e., the derivatives of the curves in Figure 12.1) as the level of energy conservation is increased. The point of intersection of the marginal curves coincides with the most profitable level of energy conservation indicated in Figure 12.1. This is the point at which the cost of adding one more unit of conservation is just equal to the corresponding benefits in terms of energy savings (i.e., the point at which "marginal costs" and "marginal benefits" are equal). To the left of the point of intersection, the additional

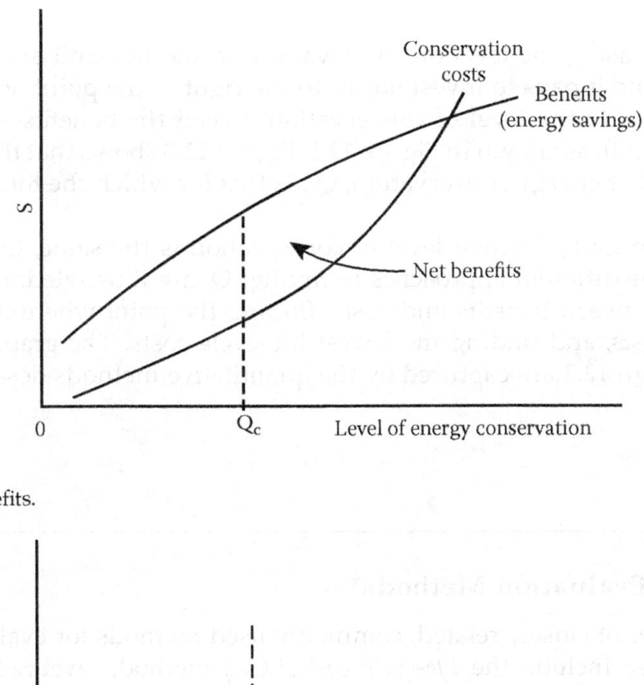

FIGURE 12.1
Maximizing net benefits.

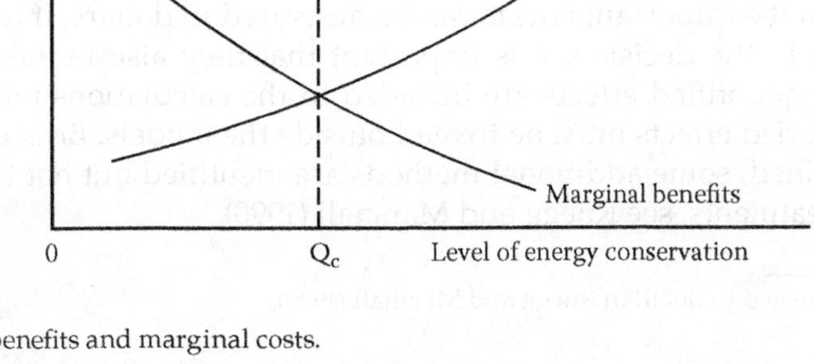

FIGURE 12.2
Equating marginal benefits and marginal costs.

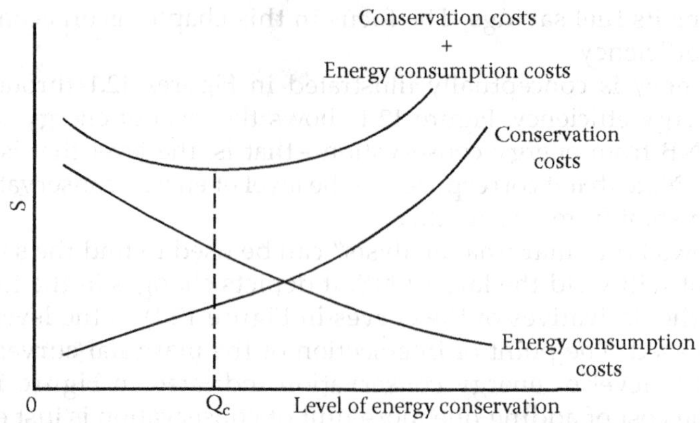

FIGURE 12.3
Minimizing LCC.

benefits from increasing the level of conservation by another unit are greater than the additional costs, and it pays to invest more. To the right of the point of intersection, the costs of an addition to the level of conservation exceed the benefits—and the level of total NB begins to fall, as shown in Figure 12.1. Figure 12.3 shows that the most economically efficient level of energy conservation, Q_c, is that for which the total cost curve is at a minimum.

The most economically efficient level of conservation is the same, Q_c, in Figures 12.1 through 12.3. Three different approaches to finding Q_c are illustrated: finding the maximum difference between benefits and costs; finding the point where marginal benefits equal marginal costs; and finding the lowest life-cycle costs. The graphical methods of Figures 12.1 through 12.3 are captured by the quantitative methods described in the section that follows.

12.3 Economic Evaluation Methods*

There are a number of closely related, commonly used methods for evaluating economic performance. These include the *life-cycle cost* (LCC) method, levelized cost of energy (LCOE) method, net present value (NPV) or NB (net present worth) method, benefit/cost (or savings-to-investment) ratio (SIR) method, internal rate-of-return (IRR) method, overall rate-of-return (ORR) method, and discounted payback (DPB) method. All of these methods are used when the important effects can be measured in dollars. If incommensurable effects are critical to the decision, it is important that they also be taken into account. But, because only quantified effects are included in the calculations for these economic methods, unquantified effects must be treated outside the models. Brief treatments of the methods are provided; some additional methods are identified but not treated. For more comprehensive treatments, see Ruegg and Marshall (1990).

* These methods are treated in detail in Ruegg and Marshall (1990).

12.3.1 Life-Cycle Cost (LCC) Method

The life-cycle costing method sums, for each investment alternative, the costs of acquisition, maintenance, repair, replacement, energy, and any other monetary costs (less than any income amounts, such as salvage value) that are affected by the investment decision. The *time value of money* must be taken into account for all amounts, and the amounts must be considered over the relevant period. All amounts are usually measured either in *present value* or annual value dollars. This is discussed later in Sections 12.5.2 and 12.5.3. At a minimum, for comparison, the investment alternatives should include a "base-case" alternative of not making the energy efficiency or renewable investment, and at least one case of an investment in a specific efficiency or renewable system. Numerous alternatives may be compared. The alternative with the lowest LCC that meets the investor's objective and constraints is the preferred investment. This least-cost solution is analogous to the least cost presented in Figure 12.3.

The following is a formula for finding the LCCs of each alternative:

$$LCC_{A1} = I_{A1} + E_{A1} + M_{A1} + R_{A1} - S_{A1} \tag{12.1}$$

where

LCC_{A1} = life-cycle cost of alternative A1

I_{A1} = present-value investment costs of alternative A1

E_{A1} = present-value energy costs associated with alternative A1

M_{A1} = present-value nonfuel operating and maintenance cost of A1

R_{A1} = present-value repair and replacement costs of A1

S_{A1} = present-value resale (or salvage) value less disposal cost associated with alternative A1

The LCC method is particularly useful for decisions that are made primarily on the basis of cost-effectiveness, such as whether a given energy efficiency or renewable energy investment will lower total cost (e.g., the sum of investment and operating costs). It can be used to compare alternative designs or sizes of systems, as long as the systems provide the same service. The method, if used correctly, can be used to find the overall cost-minimizing combination of energy efficiency investments and energy supply investments within a given facility. However, in general, it cannot be used to find the best investment, because totally different investments do not provide the same service.

12.3.2 Levelized Cost of Energy (LCOE) Method

The LCOE is similar to the LCC method, in that it considers all the costs associated with an investment alternative and takes into account the time value of money for the analysis period. However, it is generally used to compare two alternative energy supply technologies or systems, for example, two electricity production technologies that may or may not provide exactly the same service ,that is, the same level of energy production. It differs from the LCC in that it usually considers taxes, but like LCC, frequently ignores financing costs.

The LCOE is the value that must be received for each unit of energy produced to ensure that all costs and a reasonable profit are made. Profit is ensured by discounting future revenues at a discount rate that equals the rate of return that might be gained on other investments of comparable risk, that is, the opportunity cost of capital. This can be represented in the following equation:

$$\sum_{t=1}^{t=N} \frac{LCOE * Q_t}{(1+d')^t} = \sum_{t=0}^{t=N} \frac{C_t}{(1+d)^t} \tag{12.2}$$

where

- N = the analysis period
- Q_t = the amount of energy production in period t
- C_t = the cost incurred in period t
- d' = the discount rate or opportunity cost of capital; if d' is a real discount rate (excludes inflation) then the LCOE will be in real (constant) dollar terms, while if d' is a normal discount rate, the LCOE will be in nominal (current) dollar terms
- d = the discount rate used to bring future costs back to their present value. If those costs are expressed in real dollars, then the discount rate d should be a real discount rate; while, if they are in nominal dollars, the discount rate should be a nominal discount rate

12.3.3 Net Present Value (NPV) or Net Benefits (NB) Method

The NPV method finds the excess of benefits over costs, where all amounts are discounted for their time value. (If costs exceed benefits, net losses result.)

The NPV method is also often called the "net present worth" or "NS" method. When this method is used for evaluating a cost-reducing investment, the cost savings are the benefits, and it is often called the "NS" method.

Following is a formula for finding the NPV from an investment, such as an investment in energy efficiency or renewable energy systems:

$$NPV_{A1:A2} = \sum_{t=0}^{N} \frac{B_t - C_t}{(1+d)^t} \tag{12.3}$$

where

- $NPV_{A1:A2}$ = NB, that is, present value benefits (savings) net of present value costs for alternative A1 as compared with alternative A2
- B_t = benefits in year t, which may be defined to include energy savings associated with using alternative A1 instead of alternative A2
- C_t = costs in year t associated with alternative A1 as compared with a mutually exclusive alternative A2
- d = discount rate

The NPV (NB) method is useful for deciding whether to make a given investment and for designing and sizing systems. It is not appropriate for comparing investments that provide different services.

12.3.4 Benefit-to-Cost Ratio (BCR) or Savings-to-Investment Ratio (SIR) Method

This method divides benefits by costs or, equivalently, savings by investment. When used to evaluate energy efficiency and renewable energy systems, benefits are in terms of energy cost savings. The numerator of the SIR is usually constructed as energy savings, and net of maintenance and repair costs; and the denominator as the sum of investment costs and the present value of replacement costs less salvage value (capital cost items). However, depending on the objective, sometimes only initial investment costs are placed in the denominator and the other costs are subtracted in the numerator—or sometimes only the investor's equity capital is placed in the denominator. Like the three preceding methods, this method is based on discounted cash flows.

Unlike the three preceding methods that provided a performance measure in dollars, this method gives the measure as a dimensionless number. The higher the ratio, the more the dollar savings realized per dollar of investment. In particular, a value greater than 1 is generally required for an investment to be considered economically efficient.

Following is a commonly used formula for computing the ratio of savings-to-investment costs:

$$SIR_{A1:A2} = \frac{\sum_{t=0}^{N}\left(CS_t(1+d)^{-t}\right)}{\sum_{t=0}^{N}\left(I_t(1+d)^{-t}\right)} \tag{12.4}$$

where

$SIR_{A1:A2}$ = savings-to-investment ratio for alternative A1 relative to mutually exclusive alternative A2

CS_t = cost savings (excluding those investment costs in the denominator) plus any positive benefits of alternative A1 as compared with mutually exclusive alternative A2

I_t = additional investment costs for alternative A1 relative to A2

Note that the particular formulation of the ratio with respect to the placement of items in the numerator or denominator can affect the outcome. One should use a formulation appropriate to the decision maker's objectives.

The ratio method can be used to determine whether or not to accept or reject a given investment on economic grounds. It also can be used for design and size decisions and other choices among mutually exclusive alternatives, if applied incrementally (i.e., the investment and savings are the difference between the two mutually exclusive alternatives). A primary application of the ratio method is to set funding priorities among projects competing for a limited budget. When it is used in this way—and when project costs are "lumpy" (making it impossible to fully allocate the budget by taking projects in order according to the size of their ratios)—SIR should be supplemented with the evaluation of alternative sets of projects using the NPV or NB method.

12.3.5 Internal Rate-of-Return (IRR) Method

The IRR method solves for the discount rate for which dollar savings are just equal to dollar costs over the analysis period; that is, the rate for which the NPV is zero. This discount rate is the rate of return on the investment. It is compared to the investor's minimum acceptable rate of return to determine whether the investment is desirable. Unlike the preceding three techniques, the IRR does not call for the inclusion of a prespecified discount rate in the computation, but, rather, solves for a discount rate.

The rate of return is typically calculated by a process of trial and error, by which various compound rates of interest are used to discount cash flows until a rate is found for which the NPV of the investment is zero. The approach is the following: compute NPV using Equation 12.3, except substitute a trial interest rate for the discount rate, d, in the equation. A positive NPV means that the IRR is greater than the trial rate; a negative NPV means that the IRR is less than the trial rate. Based on the information, try another rate. By a series of iterations, find the rate at which NPV equals zero.

Computer algorithms, graphical techniques, and—for simple cases—discount-factor tabular approaches are often used to facilitate IRR solutions (Ruegg and Marshall, 1990, pp. 71–72). Expressing economic performance as a rate of return can be desirable for ease in comparing the returns on a variety of investment opportunities, because returns are often expressed in terms of annual rates of return. The IRR method is useful for accepting or rejecting individual investments or for allocating a budget. For designing or sizing projects, the IRR method, like the SIR, must be applied incrementally. It is not recommended for selecting between mutually exclusive investments with significantly different lifetimes (e.g., a project with a high annual return of 35% for 20 years is a much better investment than a project with the same 35% annual return for only 2 years).

IRR is a widely used method, but it is often misused, largely due to shortcomings that include the possibility of

- No solution (the sum of all nondiscounted returns within the analysis period are less than the investment costs)
- Multiple solution values (some costs occur later than some of the returns)
- Failure to give a measure of overall return associated with the project over the analysis period (returns occurring before the end of the analysis are implicitly assumed to be reinvested at the same rate of return as the calculated IRR. This may or may not be possible).

12.3.6 Overall Rate-of-Return (ORR) Method

The ORR method corrects for the last two shortcomings expressed earlier for the IRR. Like the IRR, the ORR expresses economic performance in terms of an annual rate of return over the analysis period. But unlike the IRR, the ORR requires, as input, an explicit reinvestment rate on interim receipts and produces a unique solution value.* The explicit reinvestment rate makes it possible to express net cash flows (excluding investment costs) in terms of their future value at the end of the analysis period. The ORR is then easily computed with a closed-form solution as shown in Equation 12.5.

* As shown in Equation 12.5, the reinvestment rate is also used to bring all investments back to their present value. Alternatively, investments after time zero can be discounted by the overall growth rate. In this case, a unique solution is not guaranteed, and the ORR must be found iteratively (Stermole and Stermole, 2000).

$$ORR_{A1:A2} = \left[\frac{\left[\sum_{t=0}^{N}(B_t - C_t)(1+r)^{N-t}\right]}{\sum_{t=0}^{N}\left[\frac{I_t}{(1+r)^t}\right]}\right]^{1/N} - 1 \tag{12.5}$$

where

$ORR_{A1:A2}$ = overall rate of return on a given investment alternative A1 relative to a mutually exclusive alternative A2 over a designated study period

B_t = benefits from a given alternative relative to a mutually exclusive alternative A2 over time period t

C_t = costs (excluding that part of investment costs on which the return is to be maximized) associated with a given alternative relative to a mutually exclusive alternative A2 over time t

r = the reinvestment rate at which net returns can be reinvested, usually set equal to the discount rate

N = the length of the study period

I_t = investment costs in time t on which the return is to be maximized

The ORR is recommended as a substitute for the IRR, because it avoids some of the limitations and problems of the IRR. It can be used for deciding whether or not to fund a given project, for designing or sizing projects (if it is used incrementally), and for budget-allocation decisions.

12.3.7 Discounted Payback (DPB) Method

This evaluation method measures the elapsed time between the time of an initial investment and the point in time at which accumulated discounted savings or benefits—net of other accumulated discounted costs—are sufficient to offset the initial investment, taking into account the time value of money. (If costs and savings are not discounted, the technique is called "simple payback.") For the investor who requires a rapid return of investment funds, the shorter the length of time until the investment pays off, the more desirable is the investment.

To determine the *DPB* period, find the minimum value of Y (year in which payback occurs) such that the following equality is satisfied.

$$\sum_{t=1}^{Y}\frac{B_t - C'_t}{(1+d)^t} = I_0 \tag{12.6}$$

where

B_t = benefits associated in period t with one alternative as compared with a mutually exclusive alternative

C'_t = costs in period t (not including initial investment costs) associated with an alternative as compared with a mutually exclusive alternative in period t

I_0 = initial investment costs of an alternative as compared with a mutually exclusive alternative, where the initial investment cost comprises total investment costs

DPB is often—correctly—used as a supplementary measure when project life is uncertain. It is used to identify feasible projects when the investor's time horizon is constrained. It is used as a supplementary measure in the face of uncertainty to indicate how long capital is at risk. It is a rough guide for accept/reject decisions. It is also overused and misused. Because it indicates the time at which the investment just breaks even, it is not a reliable guide for choosing the most profitable investment alternative, as savings or benefits after the payback time could be significant.

12.3.8 Other Economic-Evaluation Methods

A variety of other methods have been used to evaluate the economic performance of energy systems, but these tend to be hybrids of those presented here. One of these is the required revenue method, which computes a measure of the before-tax revenue in present or annual value dollars required to cover the costs on an after-tax basis of an energy system (Ruegg and Short, 1988, pp. 22–23). Mathematical programming methods have also been used to evaluate the optimal size or design of projects, as well as other mathematical and statistical techniques.

12.4 Risk Assessment

Many of the inputs to the evaluation methods mentioned earlier will be highly uncertain at the time an investment decision must be made. To make the most informed decision possible, an investor should employ these methods within a framework that explicitly accounts for risk and uncertainty.

Risk assessment provides decision makers with information about the "risk exposure" inherent in a given decision—that is, the probability that the outcome will be different from the "best-guess" estimate. Risk assessment is also concerned with the "risk attitude" of the decision maker, which describes his/her willingness to take a chance on an investment of uncertain outcome. Risk assessment techniques are typically used in conjunction with the evaluation methods outlined earlier; and not as stand-alone evaluation techniques.

The risk assessment techniques range from simple and partial to complex and comprehensive. Though none takes the risk out of making decisions, the techniques—if used correctly—can help the decision maker make more informed choices in the face of uncertainty.

This chapter provides an overview of the following probability-based risk assessment techniques:

- Expected value (EV) analysis
- Mean-variance criterion (MVC) and coefficient of variation (CV)
- Risk-adjusted discount rate (RADR) technique
- Certainty equivalent (CE) technique
- Monte Carlo simulation
- Decision analysis
- Real options analysis (ROA)
- Sensitivity analysis

There are other techniques that are used to assess the risks and uncertainty (e.g., CAP_M and break-even analysis), but those are not treated here.

12.4.1 Expected Value (EV) Analysis

EV analysis provides a simple way of taking into account uncertainty about input values, but it does not provide an explicit measure of risk in the outcome. It is helpful in explaining and illustrating risk attitudes.

How to calculate EV: An "expected value" is the sum of the products of the dollar value of alternative outcomes, a_i (i = 1,..., n), and their probabilities of occurrence, p_i. The EV of the decision is calculated as follows:

$$EV = a_1 p_1 + a_2 p_2 + \cdots + a_n p_n \tag{12.7}$$

Example of EV analysis: The following simplified example illustrates the combining of EV analysis and NPV analysis to support a purchase decision.

Assume that a not-for-profit organization must decide whether to buy a given piece of energy-saving equipment. Assume that the unit purchase price of the equipment is $100,000, the yearly operating cost is $5,000 (obtained by a fixed-price contract), and both costs are known with certainty. The annual energy cost savings, on the other hand, are uncertain, but can be estimated in probabilistic terms as shown in Table 12.1 in the columns headed a_1, p_1, a_2, and p_2. The present-value calculations are also given in Table 12.1.

If the equipment decision was based only on NPV, calculated with the "best-guess" energy savings (column a_1), the equipment purchase would be found to be uneconomic with a NPV of $ − 4812. But if the possibility of greater energy savings is taken into account by using the EV of savings rather than the best guess, the conclusion is that, over repeated applications, the equipment is expected to be *cost-effective*. The expected NPV of the energy-saving equipment is $25,000 per unit.

Advantages and disadvantages of the EV technique: An advantage of the technique is that it predicts a value that tends to be closer to the actual value than a simple "best-guess" estimate over repeated instances of the same event, provided, of course, that the input probabilities can be estimated with some accuracy.

A disadvantage of the EV technique is that it expresses the outcome as a single-value measure, such that there is no explicit measure of risk. Another is that the estimated outcome

TABLE 12.1

Expected Value (EV) Example

Year	Equipment Purchase $1000	Operating Costs $1000	Energy Savings a_1 $1000	p_1	a_2 $1000	p_2	PV[a] Factor	PV $1000
0	−100	—	—	—	—	—	1	−100
1		−5	25	0.8	50	0.2[b]	0.926	23.1
2		−5	30	0.8	60	0.2	0.857	26.6
3		−5	30	0.7	60	0.3	0.794	27.0
4		−5	30	0.6	60	0.4	0.735	27.2
5		−5	30	0.8	60	0.2	0.681	21.1
								25.0

Note: Expected NPV.

[a] Present-value calculations are based on a discount rate of 8%.

[b] Probabilities sum to 1.0 in a given year.

is predicated on many replications of the event, with the EV, in effect, a weighted average of the outcome over many like events. But the EV is unlikely to occur for a single instance of an event. This is analogous to a single coin toss: the outcome will be either heads or tails, not the probabilistic-based weighted average of both.

EV and risk attitude: EVs are useful in explaining risk attitude. Risk attitude may be thought of as a decision maker's preference between taking a chance on an uncertain money payout of known probability versus accepting a sure money amount. Suppose, for example, a person were given a choice between accepting the outcome of a fair coin toss where heads means winning $10,000 and tails means losing $5000 and accepting a certain cash amount of $2000. EV analysis can be used to evaluate and compare the choices. In this case, the EV of the coin toss is $2500, which is $500 more than the certain money amount. The "risk-neutral" decision maker will prefer the coin toss because of its higher EV. The decision maker who prefers the $2000 certain amount is demonstrating a "risk-averse" attitude. On the other hand, if the certain amount were raised to $3000 and the first decision maker still preferred the coin toss, he or she would be demonstrating a "risk-taking" attitude. Such trade-offs can be used to derive a "utility function" that represents a decision maker's risk attitude.

The risk attitude of a given decision maker is typically a function of the amount at risk. Many people who are risk averse when faced with the possibility of significant loss, become risk neutral—or even risk taking, when potential losses are small. Because decision makers vary substantially in their risk attitudes, there is a need to assess not only risk exposure (i.e., the degree of risk inherent in the decision) but also the risk attitude of the decision maker.

12.4.2 Mean-Variance Criterion (MVC) and Coefficient of Variation (CV)

These techniques can be useful in choosing among risky alternatives, if the mean outcomes and standard deviations (variation from the mean) can be calculated.

Consider a choice between two projects—one with higher mean NB and a lower standard deviation than the other. This situation is illustrated in Figure 12.4. In this case, the project whose probability distribution is labeled B can be said to have stochastic dominance over the project labeled A. Project B is preferable to Project A, both on grounds that its output is likely to be higher and that it entails less risk of loss. But what if Project A, the alternative with higher risk, has the higher mean NB, as illustrated in Figure 12.5? If this were the case, the MVC would provide inconclusive results.

When there is no stochastic dominance of one project over the other(s), it is helpful to compute the CV to determine the relative risk of the alternative projects. The CV indicates

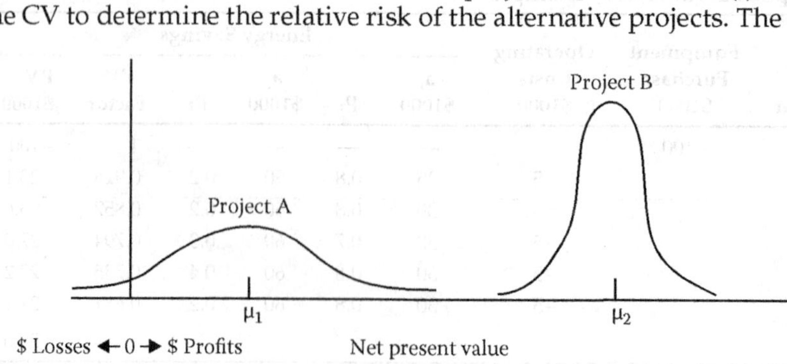

FIGURE 12.4
Stochastic dominance as demonstrated by mean-variance criterion.

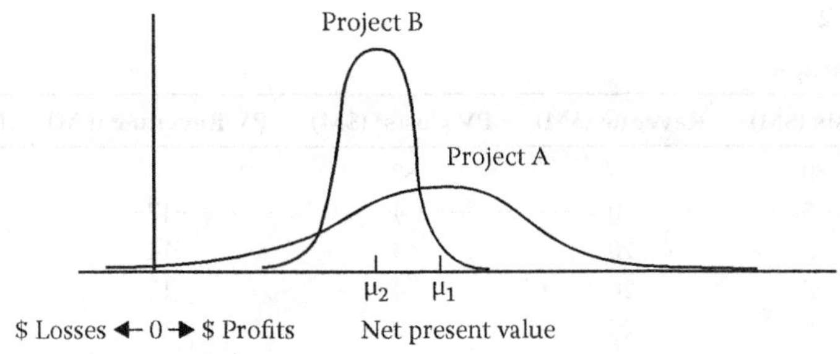

FIGURE 12.5
Inconclusive results from mean-variance criterion.

which alternative has the lower risk per unit of project output. Risk-averse decision makers will prefer the alternative with the lower CV, other things being equal. The CV is calculated as follows:

$$CV = \frac{\sigma}{\mu} \tag{12.8}$$

where
CV = coefficient of variation
σ = standard deviation
μ = mean

The principal advantage of these techniques is that they provide quick, easy-to-calculate indications of the returns and risk exposure of one project relative to another. The principal disadvantage is that the MVC does not provide a clear indication of preference when the alternative with the higher mean output has the higher risk, or vice versa.

12.4.3 Risk-Adjusted Discount Rate (RADR) Technique

The RADR technique takes account of risk through the discount rate. If a project's benefit stream is riskier than that of the average project in the decision maker's portfolio, a higher-than-normal discount rate is used; if the benefit stream is less risky, a lower-than-normal discount rate is used. If costs are the source of the higher-than-average uncertainty, a lower-than-normal discount rate is used and vice versa. The greater the variability in benefits or costs, the greater the adjustment in the discount rate.

The RADR is calculated as follows:

$$RADR = RFR + NRA + XRA \tag{12.9}$$

where
RADR = risk-adjusted discount rate
RFR = risk-free discount rate, generally set equal to the treasury bill rate
NRA = "normal" risk adjustment to account for the average level of risk encountered in the decision maker's operations
XRA = extra risk adjustment to account for risk greater or less than normal risk

TABLE 12.2
RADR Example

Year	Costs ($M)	Revenue ($M)	PV Costs[a] ($M)	PV Revenue[a] ($M)	NPV ($M)
0	80	—	80	—	−80
1	5	20	4	17	13
2	5	20	4	14	10
3	5	20	4	12	8
4	5	20	3	10	7
5	5	20	3	9	6
6	5	20	3	7	4
7	5	20	2	6	4
Total NPV					−28

[a] Costs are discounted with a discount rate of 12%; revenue with a discount rate of 18%.

An example of using the RADR technique is the following: A company is considering an investment in a new type of alternative energy system with high payoff potential and high risk on the benefits side. The projected cost and revenue streams and the discounted present values are shown in Table 12.2. The treasury bill rate, taken as the risk-free rate, is 8%. The company uses a normal risk adjustment of 5% to account for the average level of risk encountered in its operations. This investment is judged to be twice as risky as the company's average investment, so an additional risk adjustment of 5% is added to the risk-adjusted discount rate. Hence, the RADR is 18%. With this RADR, the NPV of the investment is estimated to be a loss of $28 million. On the basis of this uncertainty analysis, the company would be advised not to accept the project.

Advantages of the RADR technique are that it provides a way to account for both *risk exposure* and *risk attitude.* Moreover, RADR does not require any additional steps for calculating NPV once a value of the RADR is established. The disadvantage is that it provides only an approximate adjustment. The value of the RADR is typically a rough estimate based on sorting investments into risk categories and adding a "fudge factor" to account for the decision maker's risk attitude. It generally is not a fine-tuned measure of the inherent risk associated with variation in cash flows. Further, it typically is biased toward investments with short payoffs because it applies a constant RADR over the entire analysis period, even though risk may vary over time.

12.4.4 Certainty Equivalent (CE) Technique

The CE technique adjusts investment cash flows by a factor that will convert the measure of economic worth to a "CE" amount—the amount a decision maker will find equally acceptable to a given investment with an uncertain outcome. Central to the technique is the derivation of the certainty equivalent factor (CEF), which is used to adjust net cash flows for uncertainty.

Risk exposure can be built into the CEF by establishing categories of risky investments for the decision maker's organization and linking the CEF to the CV of the returns—greater variation translating into smaller CEF values. The procedure is as follows:

1. Divide the organization's portfolio of projects into risk categories. Examples of investment risk categories for a private utility company might be the following: low-risk investments—expansion of existing energy systems and equipment

replacement; moderate-risk investments—adoption of new, conventional energy systems; and high-risk investments—investment in new alternative energy systems.

2. Estimate the CVs (see Section 12.4.2) for each investment-risk category (e.g., on the basis of historical risk-return data).
3. Assign CEFs by year, according to the coefficients of variation, with the highest-risk projects being given the lowest CEFs. If the objectives are to reflect only risk exposure, set the CEFs such that a risk-neutral decision maker will be indifferent between receiving the estimated certain amount and the uncertain investment. If the objective is to reflect risk attitude as well as risk exposure, set the CEFs such that the decision maker with his or her own risk preference will be indifferent.

To apply the technique, proceed with the following steps:

4. Select the measure of economic performance to be used—such as the measure of NPV (i.e., NB).
5. Estimate the net cash flows and decide in which investment-risk category the project in question fits.
6. Multiply the yearly net cash flow amounts by the appropriate CEFs.
7. Discount the adjusted yearly net cash flow amounts with an RFR (an RFR is used because the risk adjustment is accomplished by the CEFs).
8. Proceed with the remainder of the analysis in the conventional way.

In summary, the CE NPV is calculated as follows:

$$NPV_{CE} = \sum_{t=0}^{N} \left[\frac{CEF_t(B_t - C_t)}{(1+RFD)^t} \right], \tag{12.10}$$

where
NPV_{CE} = NPV adjusted for uncertainty by the CE technique
B_t = estimated benefits in time period t
C_t = estimated costs in time period t
RFD = risk-free discount rate

Table 12.3 illustrates the use of this technique for adjusting NPV calculations for an investment in a new, high-risk alternative energy system. The CEF is set at 0.76 and is assumed to be constant with respect to time.

A principal advantage of the CE Technique is that it can be used to account for both risk exposure and risk attitude. Another is that it separates the adjustment of risk from discounting and makes it possible to make more precise risk adjustments over time. A major disadvantage is that the estimation of CEF is only approximate.

12.4.5 Monte Carlo Simulation

Monte Carlo simulation entails the iterative calculation of the measure of economic worth from probability functions of the input variables. The results are expressed as a probability

TABLE 12.3

CE Example (Investment-Risk Category; High-Risk—New-Alternative Energy System)

Yearly Net Cash Flow ($M)		CV	CEF	RFD Discount Factors[a]	NPV ($M)
1	−100	0.22	0.76	0.94	−71
2	−100	0.22	0.76	0.89	−68
3	20	0.22	0.76	0.84	13
4	30	0.22	0.76	0.79	18
5	45	0.22	0.76	0.75	26
6	65	0.22	0.76	0.7	35
7	65	0.22	0.76	0.67	33
8	65	0.22	0.76	0.63	31
9	50	0.22	0.76	0.59	22
10	50	0.22	0.76	0.56	21
Total NPV					60

[a] The RFD is assumed equal to 6%.

density function and as a cumulative distribution function. The technique, thereby, enables explicit measures of risk exposure to be calculated. One of the economic-evaluation methods treated earlier is used to calculate economic worth; a computer is employed to sample repeatedly—hundreds of times—from the probability distributions and make the calculations. Monte Carlo simulation can be performed by the following steps:

1. Express variable inputs as probability functions. Where there are interdependencies among input values, multiple probability density functions, tied to one another, may be needed.
2. For each input for which there is a probability function, draw randomly an input value; for each input for which there is only a single value, take that value for calculations.
3. Use the input values to calculate the economic measure of worth and record the results.
4. If inputs are interdependent, such that input X is a function of input Y, first draw the value of Y, then draw randomly from the X values that correspond to the value of Y.
5. Repeat the process many times until the number of results is sufficient to construct a probability density function and a cumulative distribution function.
6. Construct the probability density function and cumulative distribution function for the economic measure of worth, and perform statistical analysis of the variability.

The strong advantage of the technique is that it expresses the results in probabilistic terms, thereby providing explicit assessment of risk exposure. A disadvantage is that it does not explicitly treat risk attitude; however, by providing a clear measure of risk exposure, it facilitates the implicit incorporation of risk attitude in the decision. The necessity of expressing inputs in probabilistic terms and the extensive calculations are also often considered disadvantages.

12.4.6 Decision Analysis

Decision analysis is a versatile technique that enables both risk exposure and risk attitude to be taken into account in the economic assessment. It diagrams possible choices, costs, benefits, and probabilities for a given decision problem in "decision trees," which are useful in understanding the possible choices and outcomes.

Although it is not possible to capture the richness of this technique in a brief overview, a simple decision tree, shown in Figure 12.6, is discussed to give a sense of how the technique is used. The decision problem is whether to lease or build a facility. The decision must be made now, based on uncertain data. The decision tree helps to structure and analyze the problem. The tree is constructed left to right and analyzed right to left. The tree starts with a box representing a decision juncture or node—in this case, whether to lease or build a facility. The line segments branching from the box represent the two alternative paths: the upper one the lease decision and the lower one the build decision. Each has a cost associated with it that is based on the expected cost to be incurred along the path. In this example, the minimum expected cost of $6.26 million is associated with the option to build a facility.

An advantage of this technique is that it helps to understand the problem and to compare alternative solutions. Another advantage is that, in addition to treating risk exposure, it can also accommodate risk attitude by converting benefits and costs to utility values (not addressed here). A disadvantage is that the technique, as typically applied, does not provide an explicit measure of the variability of the outcome.

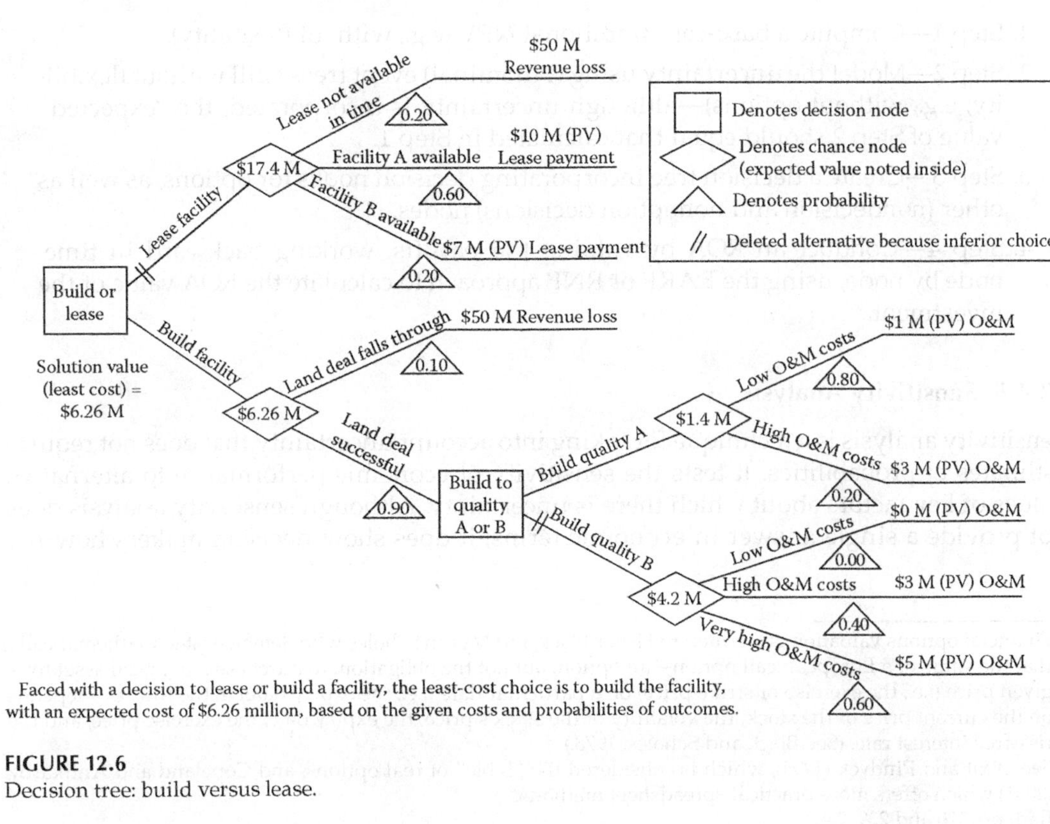

FIGURE 12.6
Decision tree: build versus lease.

12.4.7 Real Options Analysis (ROA)

ROA is an adaptation of financial options valuation techniques* to real asset investment decisions. ROA is a method used to analyze decisions in which the decision maker has one or more options regarding the timing or sequencing of an investment. It explicitly assumes that the investment is partially or completely *irreversible*, that there exists leeway or *flexibility* about the timing of the investment, and that it is subject to *uncertainty* over future payoffs. Real options can involve options (and combinations) to: defer, sequence, contract, shut down temporarily, switch uses, abandon, or expand the investment. This is in contrast to the NPV method that implies the decision is a "now or never" choice.

The value of an investment with an option is said to equal the value of the investment using the traditional NPV method (that implicitly assumes no flexibility or option) plus the value of the option. The analysis begins by construction of a decision tree with the option decision embedded in it. There are two basic methods to solve for the option value: the risk-adjusted replicating portfolio (RARP) approach and the risk-neutral probability (RNP) approach. The RARP discounts the expected project cash flows at a RADR, while the RNP approach discounts CE cash flows at a risk-free rate. In other words, the RARP approach takes the cash flows essentially as is, and adjusts the discount rate per time period to reflect that fact that the risk changes as one moves through the decision tree (e.g., risk declines with time as more information becomes available). In the RNP approach, the cash flows themselves are essentially adjusted for risk and discounted at a risk-free rate.

Copeland and Antikarov provide an overall four-step approach for ROA†:

1. Step 1—Compute a base-case traditional NPV (e.g., without flexibility).
2. Step 2—Model the uncertainty using (binominal) event trees (still without flexibility; e.g., without options)—although uncertainty is incorporated, the "expected" value of Step 2 should equal that calculated in Step 1.
3. Step 3—Create a decision tree incorporating decision nodes for options, as well as other (nondecision and nonoption decisions) nodes.
4. Step 4—Conduct an ROA by valuing the payoffs, working backward in time, node by node, using the RARP or RNP approach to calculate the ROA value of the investment.

12.4.8 Sensitivity Analysis

Sensitivity analysis is a technique for taking into account uncertainty that does not require estimates of probabilities. It tests the sensitivity of economic performance to alternative values of key factors about which there is uncertainty. Although sensitivity analysis does not provide a single answer in economic terms, it does show decision makers how the

* Financial options valuation is credited to Fisher Black and Myron Scholes who demonstrated mathematically that the value of a European call option—an option, but not the obligation, to purchase a financial asset for a given price (i.e., the exercise or strike price) on a particular date (i.e., the expiry date) in the future—depends on the current price of the stock, the volatility of the stock's price, the expiry date, the exercise price, and the risk-free interest rate. (See Black and Scholes, 1973.)

† See Dixit and Pindyck (1994), which is considered the "bible" of real options, and Copeland and Antikarov (2001) which offers more practical spreadsheet methods.
Ibid, pp. 220 and 239–240.

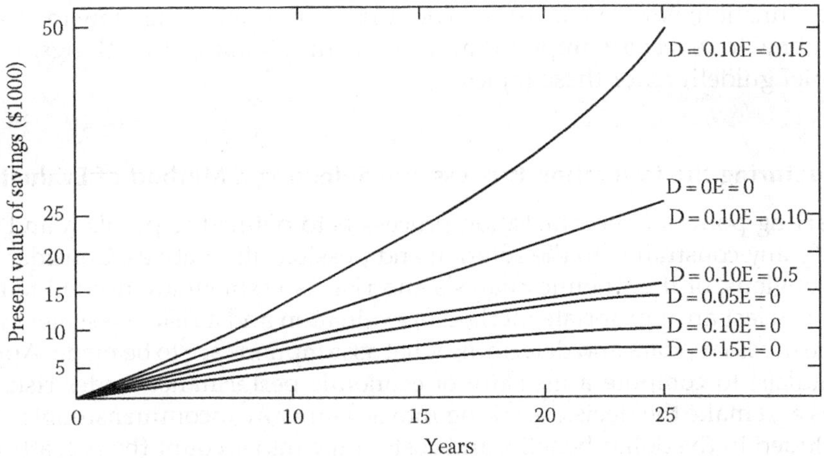

FIGURE 12.7
Sensitivity of present-value energy savings to time horizons, discount rates, and energy price escalation rates.

economic viability of a renewable energy or efficiency project changes as fuel prices, discount rates, time horizons, and other critical factors vary.

Figure 12.7 illustrates the sensitivity of fuel savings realized by a solar energy heating system to three critical factors: time horizons (0–25 years), discount rates (D equals 0%, 5%, 10%, and 15%), and energy escalation rates (E equals 0%, 5%, 10%, and 15%). The present value of savings is based on yearly fuel savings valued initially at $1000.

Note that, other things being equal, the present value of savings increase with time—but less with higher discount rates and more with higher escalation rates. The huge impact of fuel price escalation is most apparent when comparing the top line of the graph (D = 0.10, E = 0.15) with the line next to the bottom (D = 0.10, E = 0). The present value of savings at the end of 25 years is approximately $50,000 with a fuel escalation rate of 15%, and only about $8000 with no escalation, other things being equal. Whereas the quantity of energy saved is the same, the dollar value varies widely, depending on the escalation rate.

This example graphically illustrates a situation frequently encountered in the economic justification of energy efficiency and renewable energy projects: The major savings in energy costs, and thus the bulk of the benefits, accrue in the later years of the project and are highly sensitive to both the assumed rate of fuel-cost escalation and the discount rate. If the two rates are set equal, they will be offsetting as shown by the straight line labeled D = 0 E = 0 and D = 0.10 E = 0.10.

12.5 Building Blocks of Evaluation

Beyond the formula for the basic evaluation methods and risk assessment techniques, the practitioner needs to know some of the "nuts-and-bolts" of carrying out an economic analysis. He or she needs to know how to structure the evaluation process; how to choose a method of evaluation; how to estimate dollar costs and benefits; how to perform discounting operations; how to select an analysis period; how to choose a discount rate; how to adjust for inflation; how to take into account taxes and financing; how to treat residual values; and how to reflect assumptions and constraints, among other things. This section provides brief guidelines for these topics.

12.5.1 Structuring the Evaluation Process and Selecting a Method of Evaluation

A good starting point for the evaluation process is to define the problem and the objective. Identify any constraints to the solution and possible alternatives. Consider if the best solution is obvious, or if economic analysis and risk assessment are needed to help make the decision. Select an appropriate method of evaluation and a risk assessment technique. Compile the necessary data and determine what assumptions are to be made. Apply appropriate formula(s) to compute a measure of economic performance under risk. Compare alternatives and make the decision, taking into account any incommensurable effects that are not included in the dollar benefits and costs. Take into account the risk attitude of the decision maker, if it is relevant.

Although the six evaluation methods given earlier are similar, they are also sufficiently different, in that they are not always equally suitable for evaluating all types of energy investment decisions. For some types of decisions, the choice of method is more critical than for others. Figure 12.8 categorizes different investment types and the most suitable evaluation methods for each. If only a single investment is being considered, the "accept/reject" decision can often be made by any one of several techniques, provided the correct criterion is used.

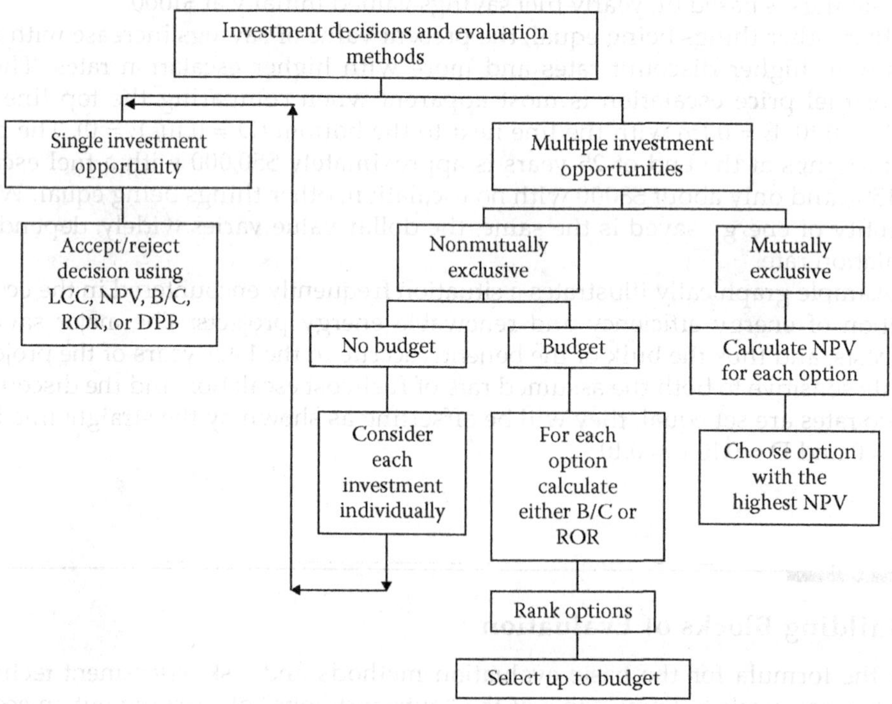

FIGURE 12.8
Investment decisions and evaluation methods.

Accept/reject criteria:
LCC technique—LCC must be lower as a result of the energy efficiency or renewable energy investment than without it.

- NPV (NB) technique—NPV must be positive as a result of the investment.
- B/C (SIR) technique—B/C (SIR) must be greater than 1.
- IRR technique—the IRR must be greater than the investor's minimum acceptable rate of return.
- DPB technique—the number of years to achieve DPB must be less than the project life or the investor's time horizon, and there are no cash flows after payback is achieved that would reverse payback.

If multiple investment opportunities are available, but only one investment can be made (i.e., they are mutually exclusive), any of the methods (except DPB) will usually work, provided they are used correctly. However, the NPV method is usually recommended for this purpose, because it is less likely to be misapplied. The NPV of each investment is calculated and the investment with the highest present value is the most economic. This is true even if the investments require significantly different initial investments, have significantly different times at which the returns occur, or have different useful lifetimes. Examples of mutually exclusive investments include different system sizes (e.g., three different photovoltaic array sizes are being considered for a single rooftop), different system configurations (e.g., different turbines are being considered for the same wind farm), and so forth.

If the investments are not mutually exclusive, then (as shown in Figure 12.8) one must consider whether there is an overall budget limitation that would restrict the number of economic investments that might be undertaken. If there is no budget (i.e., no limitation on the investment funds available), than there is really no comparison to be performed and the investor simply makes an "accept/reject" decision for each investment individually as described earlier.

If funds are not available to undertake all of the investments (i.e., there is a budget), then the easiest approach is to rank the alternatives, with the best having the highest benefit-to-cost ratio or rate of return. (The investment with the highest NPV will not necessarily be the one with the highest rank, because present value does not show return per unit investment.) Once ranked, those investments at the top of the priority list are selected until the budget is exhausted.

In the case where a fast turnaround on investment funds is required, DPB is recommended. The other methods, although more comprehensive and accurate for measuring an investment's lifetime profitability, do not indicate the time required for recouping the investment funds.

12.5.2 Discounting

Some or all investment costs in energy efficiency or renewable energy systems are incurred near the beginning of the project and are treated as "first costs." The benefits, on the other hand, typically accrue over the life span of the project in the form of yearly energy saved or produced. To compare benefits and costs that accrue at different points in time, it is necessary to put all cash flows on a time-equivalent basis. The method for converting cash flows to a time-equivalent basis is often called "discounting."

The value of money is time-dependent for two reasons: First, inflation or deflation can change the buying power of the dollar; and second, money can be invested over time to yield a return over and above inflation. For these two reasons, a given dollar amount today will be worth more than that same dollar amount a year later. For example, suppose a person were able to earn a maximum of 10% interest per annum risk-free. He or she would require $1.10 a year from now to be willing to forego having $1 today. If the person were indifferent between $1 today and $1.10 a year from now, then the 10% rate of interest would indicate that person's time preference for money. The higher the time preference, the higher the rate of interest required to make future cash flows equal to a given value today. The rate of interest for which an investor feels adequately compensated for trading money now for money in the future is the appropriate rate to use for converting present sums to future equivalent sums and future sums to present equivalent sums (i.e., the rate for discounting cash flows for that particular investor). This rate is often called the "discount rate."

To evaluate correctly the economic efficiency of an energy efficiency or renewable energy investment, it is necessary to convert the various expenditures and savings that accrue over time to a lump-sum, time-equivalent value in some base year (usually the present), or to annual values. The remainder of this section illustrates how to discount various types of cash flows.

Discounting is illustrated by Figure 12.9 in a problem of installing, maintaining, and operating a heat pump, as compared to an alternative heating/cooling system. The life-cycle cost calculations are shown for two reference times. The first is the present, and it is therefore called a present value. The second is based on a yearly time scale and is called an annual value. These two reference points are the most common in economic evaluations of investments. When the evaluation methods are derived properly, each time basis will give the same relative ranking of investment priorities.

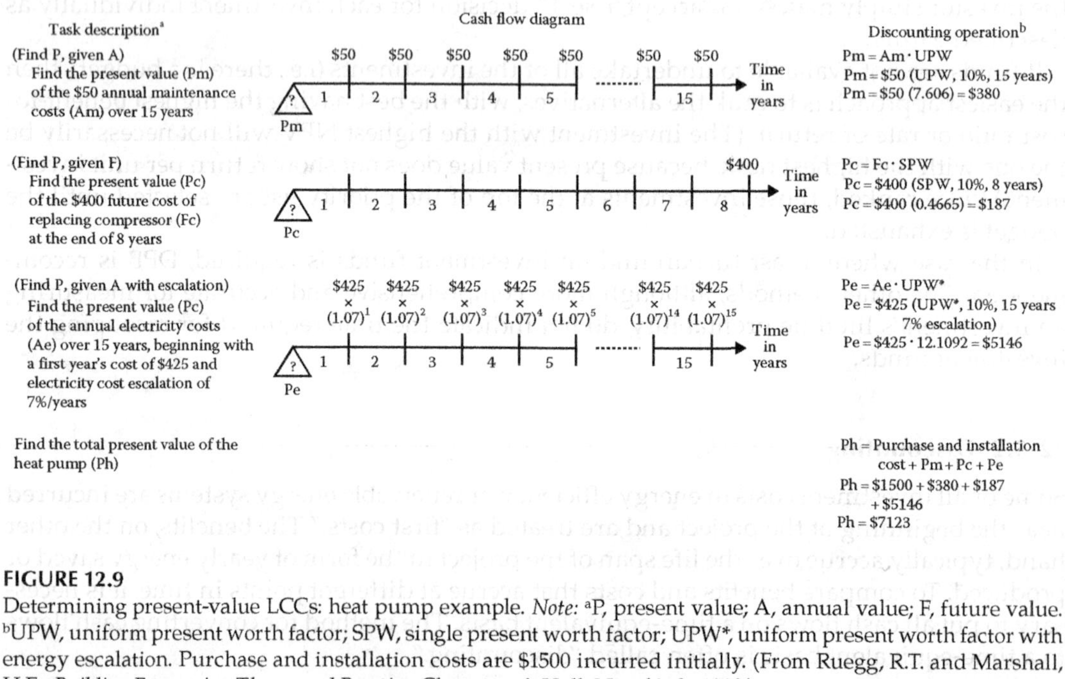

FIGURE 12.9
Determining present-value LCCs: heat pump example. *Note:* [a]P, present value; A, annual value; F, future value. [b]UPW, uniform present worth factor; SPW, single present worth factor; UPW*, uniform present worth factor with energy escalation. Purchase and installation costs are $1500 incurred initially. (From Ruegg, R.T. and Marshall, H.E., *Building Economics: Theory and Practice*, Chapman & Hall, New York, 1990.)

The assumptions for the heat pump problem—which are given only for the sake of illustration and not to suggest actual prices—are as follows:

1. The residential heat pump (not including the ducting) costs $1500 to purchase and install.
2. The heat pump has a useful life of 15 years.
3. The system has annual maintenance costs of $50 every year during its useful life, fixed by contractual agreement.
4. A compressor replacement is required in the eighth year at a cost of $400.
5. The yearly electricity cost for heating and cooling is $425, evaluated at the outset, and increased at a rate of 7% per annum due to rising electricity prices.
6. The discount rate (a nominal rate that includes an inflation adjustment) is 10%.
7. No salvage value is expected at the end of 15 years.

The LCCs in the sample problem are derived only for the heat pump and not for alternative heating/cooling systems. Hence, no attempt is made to compare alternative systems in this discounting example. To do so would require similar calculations of life-cycle costs for other types of heating/cooling systems. Total costs of a heat pump system include costs of purchase and installation, maintenance, replacements, and electricity for operation. Using the present as the base-time reference point, we need to convert each of these costs to the present before summing them. If we assume that the purchase and installation costs occur at the base reference point (the present), the $1500 is already in present value terms.

Figure 12.9 illustrates how to convert the other cash flows to present values. The first task is to convert the stream of annual maintenance costs to present value. The maintenance costs, as shown in the cash flow diagram of Figure 12.9, are $50 per year, measured in current dollars (i.e., dollars of the years in which they occur). The triangle indicates the value to be found. Here we follow the practice of compounding interest at the end of each year. The present refers to the beginning of year one.

The discounting operation for calculating the present value of maintenance costs (last column of Figure 12.9) is to multiply the annual maintenance costs times the uniform present worth (UPW) factor. The UPW is a multiplicative factor computed from the formula given in Table 12.4, or taken from a look-up table of factors that have been published in many economics textbooks. UPW factors make it easy to calculate the present values of a uniform series of annual values. For a discount rate of 10% and a time period of 15 years, the UPW factor is 7.606. Multiplying this factor by $50 gives a present value maintenance cost equal to $380. Note that the $380 present value of $50 per year incurred in each of 15 years is much less than simply adding $50 for 15 years (i.e., $750). Discounting is required to achieve correct statements of costs and benefits over time.

The second step is to convert the one-time future cost of compressor replacement, $400, to its present value. The operation for calculating the present value of compressor replacement is to multiply the *future value* of the compressor replacement times the single-payment present worth (SPW) factor, which can be calculated from the formula in Table 12.4, or taken from a discount factor look-up table. For a discount rate of 10% and a time period of 15 years, the SPW factor is 0.4665. Multiplying this factor by $400 gives a present-value cost of the compressor replacement of $187, as shown in the last column of Figure 12.9.

TABLE 12.4
Discount Formulas

Standard Nomenclature	Use When	Standard Notation	Algebraic Form
Single compound amount	Given P; to find F	(SCA, d%, N)	$F = P(1+d)^N$
Single present worth	Given F; to find P	(SPW, d%, N)	$P = F\frac{1}{(1+d)^N}$
Uniform compound amount	Given A; to find F	(UCA, d%, N)	$F = A\frac{(1+d)^N - 1}{d}$
Uniform sinking fund	Given F; to find A	(USF, d%, N)	$A = F\frac{d}{(1+d)^N - 1}$
Uniform capital recovery	Given P; to find A	(UCR, d%, N)	$A = P\frac{d(1+d)^N}{(1+d)^N - 1}$
Uniform present worth	Given A; to find P	(UPW, d%, N)	$P = A\frac{(1+d)^N - 1}{d(1+d)^N}$
Uniform present worth modified	Given A escalating at a rate e; to find P	(UPW*, d%, e, N)	$P = A\frac{(1+e)\left[1-\left(\frac{1+e}{1+d}\right)^N\right]}{d-e}$

Note: P, a present sum of money; F, a future sum of money, equivalent to P at the end of N periods of time at a discount rate of d; N, number of interest periods; A, an end-of-period payment (or receipt) in a uniform series of payments (or receipts) over N periods at *discount rate d*, usually annually; e, a rate of escalation in A in each of N periods.

Again, note that discounting makes a significant difference in the measure of costs. Failing to discount the $400 would result in an overestimate of cost, in this case of $213.

The third step is to convert the annual electricity costs for heating and cooling to present value. A year's electricity costs, evaluated at the time of installation of the heat pump, are assumed to be $425. Electricity prices, for purposes of illustration, are assumed to increase at a rate of 7% per annum. This is reflected in Table 12.4 by multiplying $425 times $(1.07)^t$ where $t = 1, 2, \ldots, 15$. The electricity cost at the end of the fourth year, for example, is $\$425(1.07)^4 = \$55$.

The discounting operation for finding the present value of all electricity costs (shown in Figure 12.9) is to multiply the initial, yearly electricity costs times the appropriate UPW* factor. (An asterisk following UPW denotes that a term for price escalation is included.) The UPW or UPW* discount formulas in Table 12.4 can also be used to obtain present values from annual costs or multiplicative discount factors from look-up tables can be used. For a period of 15 years, a discount rate of 10%, and an escalation rate of 7%, the UPW* factor is 12.1092. Multiplying the factor by $425 gives a present value of electricity costs of $5146. Note, once again, that failing to discount (i.e., simply adding annual electricity expenses in current prices) would overestimate costs by $1229 ($6376–$5146). Discounting with a UPW factor that does not incorporate energy price escalation would underestimate costs by $1913 ($5146–$3233).

The final operation described in Figure 12.9 is to sum purchase and installation cost and the present values of maintenance, compressor replacement, and electricity costs. Total LCCs of the heat pump in present value terms are $7213. This is one of the amounts that a designer would need for comparing the cost-effectiveness of heat pumps to alternative heating/cooling systems.

Only one discounting operation is required for converting the present value costs of the heat pump to annual value terms. The total present value amount is converted to the total annual value simply by multiplying it by the uniform capital recovery (UCR) factor—in this case the UCR for 10% and 15 years. The UCR factor, calculated with the UCR formula given in Table 12.4, is 0.13147. Multiplying this factor by the total present value of $7213 gives the cost of the heat pump as $948 in annual value terms. The two figures—$7213 and $948 per year—are time-equivalent values, made consistent through the discounting.

Figure 12.9 provides a model for the designer who must calculate present values from all kinds of benefit or cost streams. Most distributions of values occurring in future years can be handled with the SPW, the UPW, or the UPW* factors.

12.5.3 Discount Rate

Of the various factors affecting the NB of energy efficiency and renewable energy investments, the discount rate is one of the most dramatic. A project that appears economic at one discount rate will often appear uneconomic at another rate. For example, a project that yields NS at a 6% discount rate might yield net losses if evaluated with a 7% rate.

As the discount rate is increased, the present value of any future stream of costs or benefits is going to become smaller. High discount rates tend to favor projects with quick payoffs over projects with benefits deferred further in the future.

The discount rate should be set equal to the rate of return available on the next-best investment opportunity of similar risk to the project in question—that is, it should indicate the opportunity cost of the investor.

The discount rate may be formulated as a "real rate" exclusive of general price inflation or as a "nominal rate" inclusive of inflation. The former should be used to discount cash flows that are stated in *constant dollars*. The latter should be used to discount cash flows stated in *current dollars*.

12.5.4 Inflation

Inflation is a rise in the general price level. Because future price changes are unknown, it is frequently assumed that prices will increase at the rate of inflation. Under this assumption, it is generally easier to conduct all economic evaluations in constant dollars and to discount those values using "real" discount rates. For example, converting the constant dollar annual maintenance costs in Figure 12.9 to a present value can be easily done by multiplying by a UPW factor (calculated using a real discount rate) because the maintenance costs do not change over time. However, some cash flows are more easily expressed in current dollars—for example, equal loan payments, tax depreciation, etc. These can be converted to present values using a nominal discount rate.

12.5.5 Analysis Period

The analysis period is the length of time over which costs and benefits are considered in an economic evaluation. The analysis period need not be the same as either the "useful life" or the "economic life," two common concepts of investment life. The useful life is the period over which the investment has some value; that is, the investment continues to conserve or provide energy during this period. Economic life is the period during which the investment in question is the least-cost way of meeting the requirement. Often, economic life is shorter than useful life.

The selection of an analysis period will depend on the objectives and perspective of the decision maker. A speculative investor who plans to develop a project for immediate sale, for example, may view the relevant time horizon as that short period of ownership from planning and acquisition of property to the first sale of the project. Although the useful life of a solar domestic hot water heating system, for example, might be 20 years, a speculative home builder might operate on the basis of a 2-year time horizon, if the property is expected to change hands within that period. Only if the speculator expects to gain the benefit of those energy savings through a higher selling price for the building, will the higher first cost of the solar energy investment likely be economic.

If an analyst is performing an economic analysis for a particular client, that client's time horizon should serve as the analysis period. If an analyst is performing an analysis in support of public investment or a policy decision, the life of the system or building is typically the appropriate analysis period.

When considering multiple investment options, it is best with some evaluation methods (such as LCC, IRR, and ORR) to use the same analysis period. With others like NPV and BCR, different analysis periods can be used. If an investment's useful life is shorter than the analysis period, it may be necessary to consider reinvesting in that option at the end of its useful life. If an investment's useful life is longer than the analysis period, a salvage value may need to be estimated.

12.5.6 Taxes and Subsidies

Taxes and subsidies should be taken into account in economic evaluations, because they may affect the economic viability of an investment, the return to the investor, and the optimal size of the investment. Taxes, which may have positive and negative effects, include—but are not limited to—income taxes, sales taxes, property taxes, excise taxes, capital gain taxes, depreciation recapture taxes, tax deductions, and tax credits.

Subsidies are inducements for a particular type of behavior or action. They include grants—cash subsidies of specified amounts; government cost sharing; loan-interest reductions; and tax-related subsidies. Income tax credits for efficiency or renewable energy expenditures provide a subsidy by allowing specific deductions from the investor's tax liability. Property tax exemptions eliminate the property taxes that would otherwise add to annual costs. Income tax deductions for energy efficiency or renewable energy expenses reduce annual tax costs. The imposition of higher taxes on nonrenewable energy sources raises their prices and encourages efficiency and renewable energy investments.

It is important to distinguish between a before-tax cash flow and an after-tax cash flow. For example, fuel costs are a before-tax cash flow (they can be expensed), while a production tax credit for electricity from wind is an after-tax cash flow.

12.5.7 Financing

Financing of an energy investment can alter the economic viability of that investment. This is especially true for energy efficiency and renewable energy investments that generally have large initial investment costs with returns spread out over time. Ignoring financing costs when comparing these investments against conventional sources of energy can bias the evaluation against the energy efficiency and renewable energy investments.

Financing is generally described in terms of the amount financed, the loan period, and the interest rate. Unless specified otherwise, a uniform payment schedule is usually assumed. Generally, financing improves the economic effectiveness of an investment if the after-tax nominal interest rate is less than the investor's nominal discount rate.

Financing essentially reduces the initial outlay in favor of additional future outlays over time—usually equal payments for a fixed number of years. These cash flows can be treated like any other: The equity portion of the capital cost occurs at the start of the first year, and the loan payments occur monthly or annually. The only other major consideration is the tax deductibility of the interest portion of the loan payments.

12.5.8 Residual Values

Residual values may arise from salvage (net of disposal costs) at the end of the life of systems and components, from reuse values when the purpose is changed, and from remaining value when assets are sold prior to the end of their lives. The present value of residuals can generally be expected to decrease, other things equal, as (1) the discount rate rises, (2) the equipment or building deteriorates, and (3) the time horizon lengthens.

To estimate the residual value of energy efficiency or renewable energy systems and components, it is helpful to consider the amount that can be added to the selling price of a project or building because of those systems. It might be assumed that a building buyer will be willing to pay an additional amount equal to the capitalized value of energy savings over the remaining life of the efficiency or renewable investment. If the analysis period is the same as the useful life, there will be no residual value.

12.6 Economic Analysis Software for Renewable Energy Investments

Over the last three to four decades, a large number of models for the economic analysis of renewable energy systems have been developed. In the early years, the emphasis was on simpler models for the analysis of solar hot water and space heating. In the last decade, the emphasis has shifted to models of renewable electric technologies as those technologies have become more and more cost competitive. At the same time due to the complexities of the electric system, the models have become more and more sophisticated with respect to both system performance and system economics and financing. The need for reliable power and the variability of wind and solar are dealt with in many models by performance projections down to the hourly level. Similarly, the complex ownership and regulation of power plants and electric utilities has led to increasingly sophisticated financing and ownership structures in today's models. We will examine these intricacies as we briefly review a handful of the more prominent models used in the United States with an emphasis on the economic measures used, the technologies treated, and the different areas of emphasis of the different models.

System advisor model (available at https://sam.nrel.gov/) SAM is probably the most sophisticated of the tools available today for the analysis of renewable energy technologies in the electric sector. Developed by the National Renewable Energy Laboratory (NREL), "SAM makes performance predictions and cost of energy estimates for grid-connected power projects based on installation and operating costs and system design parameters that you specify as inputs to the model." (Gilman and Dobos 2012). It calculates the performance for each of the 8760 hours of the year which can be viewed at the hourly level or through more aggregated measures like capacity factors and seasonal or annual output. The technologies it can evaluate include

- Photovoltaic systems (flat plate and concentrating)
- Parabolic trough concentrating solar power systems
- Power tower concentrating solar power systems
- Linear Fresnel concentrating solar power systems
- Dish-stirling concentrating solar power systems
- Conventional fossil-fuel thermal systems
- Solar water heating for residential or commercial buildings*
- Large and small wind power projects
- Geothermal power and coproduction
- Biomass power

SAM can evaluate the economics of these systems from the perspective of different owners and developer perspectives to include

- Residential rooftop
- Commercial rooftop
- Utility scale (power purchase agreement)
  - Single owner
  - Leveraged partnership flip
  - All equity partnership flip
  - Sale leaseback

SAM calculates the following measures based on the cash flows for the ownership, financing, and other descriptors input by the user:

- Payback period (buildings only)
- Revenue with and without renewable energy system
- LCOE
- NPV
- Power purchase agreement price (electricity sales price)
- IRR

In making these financial calculations, SAM can account for a wide range of incentives including

- Investment-based incentives
- Capacity-based incentives
- Production-based incentives

* This is the only non-electric-production technology that can be evaluated with SAM.

- Investment tax credits
- Production tax credits
- Depreciation (MACRS, straight-line, custom)

Cost of Renewable Energy Spreadsheet Tool (CREST is available at https://financere.nrel.gov/finance/content/crest-cost-energy-models) CREST was developed for NREL by Sustainable Energy Advantage. It is distinguished from SAM primarily in that it is easier to use, is spreadsheet based, and has fewer technical performance, financing, and economic-metric details and options. It calculates the first year cost of energy or the LCOE for photovoltaics, concentrating solar power, wind, and geothermal electricity technologies, as well as anaerobic digestion technologies. It can account for different cost-based incentives and several ownership structures.

HOMER (available at: http://homerenergy.com/index.html) HOMER can be used to design and analyze hybrid power systems, that can include storage, conventional generators and combined heat and power systems along with photovoltaics, wind, hydropower, and biomass. HOMER was also originally developed at NREL and is now supported by HOMER Energy. It is distinguished from NREL's SAM and CREST models described earlier primarily in that it can be used for grid and off-grid applications of hybrid and distributed energy systems. To analyze these systems, HOMER's performance calculations are made on an hourly basis for every hour of a year and show the changing mix over time of contributing generators from the hybrid system.

RETScreen (available from the Canadian government at: http://www.retscreen.net/ang/home.php): RETScreen is actually two separate programs—RETScreen 4, an Excel spreadsheet program (available at: http://www.retscreen.net/ang/version4.php), and RETScreen Plus, a Windows-based performance measurement and verification program. We focus here on the economic analysis part, RETScreen 4. It differs from the NREL suite of models described earlier primarily in that it addresses both electric and nonelectric renewable energy technologies as well as efficiency and cogeneration projects. It includes a large database of international resource and weather data for analyzing systems in almost all global locations, although it makes annual performance approximations, not hourly estimates. It calculates the standard set of economic metrics, for example, IRR, NPV, payback, through cash flow analysis that considers financing and tax provisions.

12.7 Summary

There are multiple methods for evaluating economic performance and multiple techniques of risk analysis that can be selected and combined to improve decisions in energy efficiency and renewable energy investments. Economic performance can be stated in a variety of ways, depending on the problem and preferences of the decision maker: as NPV, as LCCs, as the cost of energy, as a rate of return, as years to payback, or as a ratio. To reflect the reality that most decisions are made under conditions of uncertainty, risk assessment techniques can be used to reflect the risk exposure of the project and the risk attitude of the decision maker. Rather than expressing results in single, deterministic terms, they can be expressed in probabilistic terms, thereby revealing the likelihood that the outcome will differ from the best-guess answer. These methods and techniques can be used to decide whether or not to invest in a given energy efficiency or renewable energy system; to determine which system design or size is economically efficient; to find the combination of components and systems that are expected to be cost-effective; to estimate how long before a project will break even; and to decide which energy-related investments are likely to provide the highest rate of return to the investor. The methods support the goal of achieving economic efficiency—which may differ from engineering technical efficiency. There are many models available today that can assist in evaluating the economic and financial viability of an investment in renewable energy.

12.8 Defining Terms

Analysis period—Length of time over which costs and benefits are considered in an economic evaluation.

Benefit/cost (B/C) or saving-to-investment (SIR) ratio—A method of measuring the economic performance of alternatives by dividing present-value benefits (savings) by present-value costs.

Constant dollars—Values expressed in terms of the general purchasing power of the dollar in a base year. Constant dollars do not reflect price inflation or deflation.

Cost-effective investment—The least-cost alternative for achieving a given level of performance.

Current dollars—Values expressed in terms of actual prices of each year (i.e., current dollars reflect price inflation or deflation).

Discount rate—Based on the opportunity cost of capital, this minimum acceptable rate of return is used to convert benefits and costs occurring at different times to their equivalent values at a common time.

Discounted payback period—The time required for the discounted annual net benefits derived from an investment to pay back the initial investment.

Discounting—A technique for converting cash flows that occur over time to equivalent amounts at a common point in time using the opportunity cost for capital.

Economic efficiency optimization—Maximizing net benefits or minimizing costs for a given level of benefits (i.e., "getting the most for your money").

Economic life—That period of time over which an investment is considered to be the least-cost alternative for meeting a particular objective.

Future value (worth)—The value of a dollar amount at some point in the future, taking into account the opportunity cost of capital.

Internal rate of return—The discount rate that equates total discounted benefits with total discounted costs.

Investment costs—The sum of the planning, design, and construction costs necessary to obtain or develop an asset.

Levelized cost of energy—The before-tax revenue required per unit of energy to cover all costs plus a profit/return on investment equal to the discount rate used to levelize the costs.

Life-cycle cost—The total of all relevant costs associated with an asset or project over the analysis period.

Net benefits—Benefits minus costs.

Present value (worth)—Past, present, or future cash flows all expressed as a lump sum amount as of the present time, taking into account the time value of money.

Real options analysis—Method used to analyze investment decisions in which the decision maker has one or more options regarding the timing or sequencing of investment.

Risk assessment—As applied to economic decisions, the body of theory and practice that helps decision makers assess their risk exposures and risk attitudes in order to increase the probability that they will make economic choices that are best for them.

Risk attitude—The willingness of decision makers to take chances on investments with uncertain outcomes. Risk attitudes may be classified as risk averse, risk neutral, and risk taking.

Risk exposure—The probability that a project's economic outcome will be less favorable than what is considered economically desirable.

Sensitivity analysis—A non-probability-based technique for reflecting uncertainty that entails testing the outcome of an investment by altering one or more system parameters from the initially assumed values.

Time value of money—The amount that people are willing to pay for having money today rather than some time in the future.

Uncertainty—As used in the context of this chapter, a lack of knowledge about the values of inputs required for an economic analysis.

References

Black, F. and Scholes, M. 1973. The pricing of options and corporate liabilities. *Journal of Political Economy* 81: 637–659.

Copeland, T. and Antikarov, V. 2001. *Real Options: A Practitioner's Guide*. Texere, New York.

Dixit, A.K. and Pindyck, R.S. 1994. *Investment under Uncertainty*. Princeton University Press, Princeton, NJ.

Gilman, P. and Dobos, A. 2012. System advisor model, SAM 2011.12.2: General description, NREL/TP-6A20-53437. National Renewable Energy Laboratory, Golden, CO, February 2012.

Marshall, H.E. and Ruegg, RT. 1980a. Principles of economics applied to investments in energy conservation and solar energy systems. In *Economics of Solar Energy and Conservation Systems*, eds. F. Kreith and R. West, pp. 123–173. CRC Press, Boca Raton, FL.

Ruegg, R.T. and Marshall, H.E. 1990. *Building Economics: Theory and Practice*. Chapman & Hall, New York.

Ruegg, R.T. and Short, W. 1988. Economic methods. In *Economic Analysis of Solar Thermal Energy Systems*, eds. R. West and F. Kreith, pp. 18–83. The MIT Press, Cambridge, MA.

Stermole, F.J. and Stermole, J.M. 2000. *Economic Evaluation and Investment Decision Methods*. Investment Evaluations Corporation, Lakewood, CO.

13

Environmental Impacts and Costs of Energy

Ari Rabl and Joseph V. Spadaro

CONTENTS

13.1 Introduction

Air pollution causes considerable damage to human health, flora and fauna, and materials. The damage costs are external costs, to the extent that they are not included in the prices of goods, that is, they are external to the market. The external costs can be internalized via taxes, tradable permits, or other environmental regulations. Here we use the term *damage costs*, to avoid possible ambiguities about the extent to which the damage costs are already internalized.

Since 1990, there has been much progress in the analysis of damage costs, thanks to several major projects to evaluate the external costs of energy in Europe (ExternE 1995, 1998, 2005, 2008) and also in the United States (ORNL/RFF 1994, Rowe et al. 1995, Abt 2004, NRC 2010). This chapter, by key participants of the ExternE project series, presents an overview of the methodology and the results.

The quantification of damage costs has many important applications:

- Guidance for environmental regulations (e.g., determining the optimal level of the limit for the emission of a pollutant)
- Finding the socially optimal level of a pollution tax
- Identifying technologies with the lowest social cost (e.g., coal, nuclear, or wind for the production of electricity?)
- Evaluating the benefits of improving the pollution abatement of an existing installation such as a waste incinerator
- Optimizing the dispatching of power plants
- "Green accounting," that is, including corrections for environmental damage in the traditional accounts of GDP

13.2 Methodology

13.2.1 Impact Pathway Analysis (IPA)

To calculate the damage costs, one needs to carry out an impact pathway analysis (IPA), tracing the passage of a pollutant from where it is emitted to the affected receptors (population, crops, forests, buildings, etc.). The principal steps of an IPA can be grouped as follows, as shown in Figure 13.1,

1. *Emission*: Specification of the relevant technologies and pollutants, for example, kg of NO_x per GWh_e emitted by a power plant
2. *Dispersion*: Calculation of increased pollutant concentrations in all affected regions, for example, incremental concentration of ozone, using models of atmospheric dispersion and chemistry for ozone formation due to NO_x (this step is also called environmental fate analysis, especially when it involves more complex pathways that pass through the food chain)

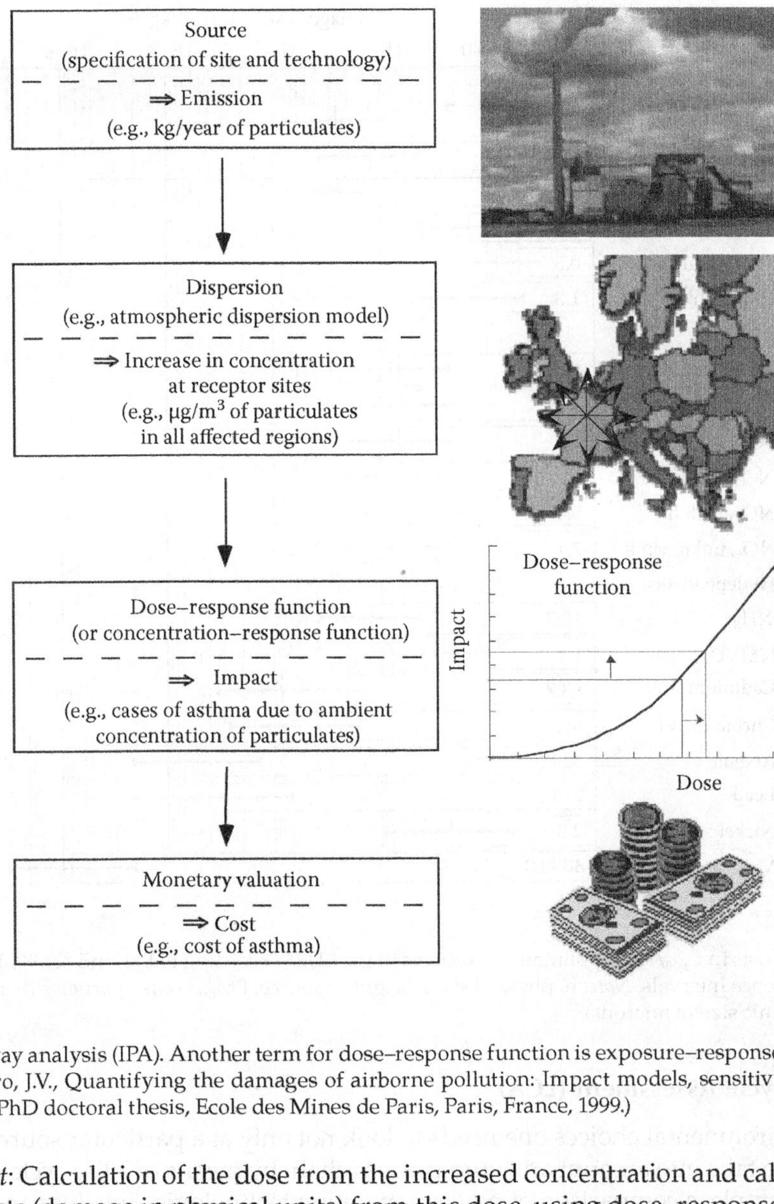

FIGURE 13.1
Impact pathway analysis (IPA). Another term for dose–response function is exposure–response function (ERF). (From Spadaro, J.V., Quantifying the damages of airborne pollution: Impact models, sensitivity analyses and applications, PhD doctoral thesis, Ecole des Mines de Paris, Paris, France, 1999.)

3. *Impact*: Calculation of the dose from the increased concentration and calculation of impacts (damage in physical units) from this dose, using dose–response functions (also known as exposure–response functions [ERFs]), for example, cases of asthma due to this increase in ozone
4. *Cost*: Monetary valuation of these impacts, for example, multiplication by the cost of a case of asthma

The impacts and costs are summed over all receptors of concern. The work involves a multidisciplinary system analysis, with inputs from engineers, dispersion modelers, epidemiologists, ecologists, and economists. The result of an IPA is the damage cost per kg of emitted pollutant, as shown in Figure 13.2.

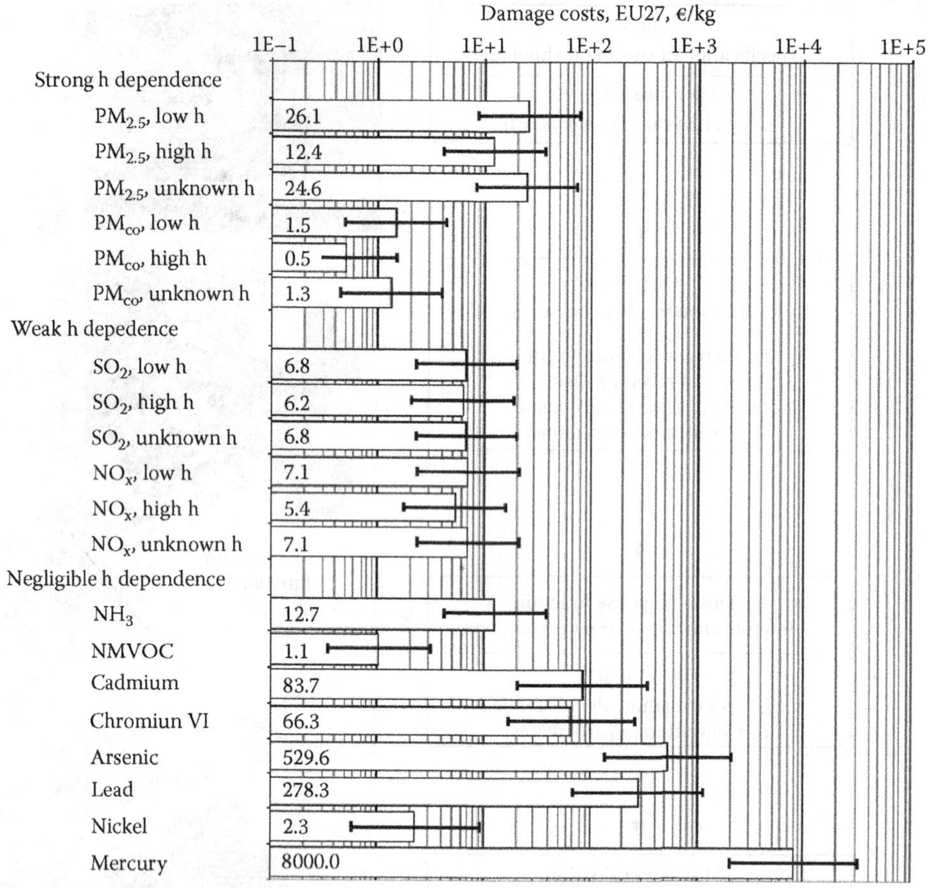

FIGURE 13.2
Typical damage costs in €$_{2000}$/kg of pollutant for sources in the EU27, according to ExternE (2008). The error bars show 68% confidence intervals. *Note*: h, physical stack height of source; PM_{co}, coarse particles (between 2.5 and 10 μm; aerodynamic size in microns).

13.2.2 Life Cycle Assessment (LCA)

For many environmental choices one needs to look not only at a particular source of pollutants, but has to take into account an entire process chain by means of a life cycle assessment (LCA). For example, a comparison of power generation technologies involves an analysis of the entire fuel chain, as sketched in Figure 13.3. Whether an IPA of a single source or an LCA of an entire process or product chain is required, depends on the policy choice in question. For finding the optimal limit for the emission of NO_x from an incinerator, an IPA is sufficient, but the choice between coal, nuclear, and wind for electric power requires an LCA.

In principle, the damages and costs for each pollution source in the life cycle should be evaluated by a site-specific IPA. But in practice, most LCA studies have taken the shortcut of first summing the emissions over all stages and then multiplying the result by site-independent indices of *potential impact*. Also, most practitioners of LCA reject the concept of monetary valuation. By contrast, ExternE and analogous work in the United States are based on a realistic IPA assessment of the impacts and damage costs.

IPA: → **Real impacts** for each stage (site-specific) **Goal:** Evaluate the entire matric

Steps of IPA → / **Stage of process chain** ↓	Emission	Dispersion	Exposure–response function	Monetary valuation
Production of the materials				
Assembly of the materials				
Fuel feedstock				
Fuel supply				
Utilization of the car				
Disposal of the car				

Most conventional **LCA**: First sum over emissions ↓ $\sum$ then → × multiplication by **potential impact** indices

FIGURE 13.3
Relation between impact pathway analysis (IPA) and current practice of most LCA, illustrated for the example of electricity production with fossil fuel. (From Spadaro, J.V., Quantifying the damages of airborne pollution: Impact models, sensitivity analyses and applications, PhD doctoral thesis, Ecole des Mines de Paris, Paris, France, 1999; Spadaro, J.V. and Rabl, A., *Int. J. Life Cycle Assess.*, 4(4), 229, 1999.)

13.2.3 Atmospheric Dispersion and Chemistry

The principal greenhouse gases, CO_2, CH_4, and N_2O, stay in the atmosphere long enough to mix uniformly over the entire globe. For most other air pollutants, in particular PM_{10},* NO_x, and SO_2, atmospheric dispersion is significant over hundreds to thousands of km, so both local and regional effects are important. One needs, therefore, a combination of local and regional dispersion models to account for all significant damages. For the local range (<50 km from the source), Gaussian plume models can be used.

At the regional scale, one needs to take into account the chemical reactions that lead to the transformation of primary pollutants (i.e., the pollutants as they are emitted) to secondary pollutants, for example, the creation of sulfates from SO_2.

Whereas, only the inhalation dose matters for the classical air pollutants (PM_{10}, NO_x, SO_2, and O_3), certain micropollutants (toxic metals and persistent organic pollutants) affect us also through food and drink. A relatively simple model for the passage of such pollutants through the food chain has been developed by Spadaro and Rabl (2004). A general finding of such calculations is that the dose from ingestion can be about two orders of magnitude larger than that from inhalation. However, for electric power technologies, the emitted quantities of these micropollutants are so small that they make at most a small contribution to the total damage cost (even for the dirtiest fuels, coal, and heavy oil).

* Usually PM (particulate matter) has a subscript, indicating the largest size in μm, of the particles that are included in the measured PM concentration. The most common measures are PM_{10} and $PM_{2.5}$.

13.2.4 Exposure–Response Functions (ERFs)

The dose–response function, also known as exposure–response function (ERF), relates the pollution exposure that affects a receptor (e.g., population) to the physical impact on this receptor (e.g., incremental number of hospitalizations). A damage can be quantified only if the corresponding ERF is known. Such functions are known for the most important impacts on human health, building materials, and crops, caused by the most important air pollutants such as primary and secondary (i.e., nitrates, sulfates) particles, ozone, CO, SO_2, NO_x, benzene, dioxins, As, Cd, Cr, Hg, Ni, and Pb.

13.2.5 Monetary Valuation

The goal of the monetary valuation of damages is to account for all costs, market and nonmarket. For example, the valuation of an asthma attack has to include not only the cost of the medical treatment but also the willingness-to-pay (WTP) to avoid the remaining suffering. If the WTP for a nonmarket good has been determined correctly, it is like a price, consistent with prices paid for market goods. Economists have developed several tools for determining nonmarket costs; of these tools, contingent valuation (CV) has enjoyed increasing popularity (Mitchell and Carson 1989). The basic idea of a CV is to ask people how much they would be willing to pay for a certain good if they could buy it. Countless CV studies have been carried out and can be used for the assessment of damage costs.

As it turns out, damage costs of air pollution are dominated by nonmarket goods, especially the valuation of mortality. The single most important parameter is the so-called *value of statistical life* (VSL). This term often evokes hostile reactions from people who think that economists try to measure the value of life. The value of life is limitless (to save an individual in danger no means are spared). Really VSL is the *willingness to pay for avoiding the risk of an anonymous premature death.* A better term is *value of prevented fatality* (VPF).

13.2.6 Health Impacts

In monetary terms, the health impacts contribute the largest part of the damage costs. The most important cost comes from chronic mortality due to particulate matter (PM), calculated on the basis of data obtained by Pope et al. (2002) (this term, chosen by analogy with acute and chronic morbidity impacts, indicates that the total or long-term effects of pollution on mortality have been included, by contrast to acute mortality impacts that are observed within a few days of exposure to pollution). Table 13.1 shows the most important pollutants and impacts on health. Primary pollutants are the pollutants as they are emitted, secondary pollutants are created in the environment by chemical transformation of certain primary pollutants. For instance, SO_2 is transformed to sulfate aerosols, and the assessments by ExternE, Abt (2004), and NRC (2010) assume that their health impacts are like those of PM.

13.2.7 Global Warming

Estimating the cost of global warming is difficult because of the large number of different impacts in all countries of the world that should be taken into account. Furthermore, as these impacts will occur in future decades and centuries one needs to estimate how these impacts and costs will evolve into the distant future (IPCC 2007). On top of the scientific uncertainties, there are controversial ethical issues related to the valuation of mortality in developing countries (where most of the impacts will occur) and the choice of the discount rate for intergenerational costs.

TABLE 13.1

Air Pollutants and Their Effects on Health

Primary Pollutants	Secondary Pollutants	Impacts
Particulate matter (PM_{10}, $PM_{2.5}$)		Mortality, cardiopulmonary morbidity, (cerebrovascular hospital admissions, congestive heart failure, chronic bronchitis, chronic cough in children, lower respiratory symptoms, cough in asthmatics)
SO_2		Mortality, cardiopulmonary morbidity (hospitalization, consultation of doctor, asthma, sick leave, restricted activity)
SO_2	Sulfates	Assumed to be like PM
NO_x		Morbidity
NO_x	Nitrates	Assumed to be like PM
NO_x + VOC	Ozone	Mortality, morbidity (respiratory hospital admissions, restricted activity days, asthma attacks, symptom days)
CO		Mortality (congestive heart failure), morbidity (cardiovascular)
PAH (diesel soot, benzene, 1,3,-butadiene, dioxins)		Cancers
As, Cd, Cr^{VI}, Ni		Cancers, other morbidity
Hg, Pb		Morbidity (neurotoxic), mortality

Many studies have been published with estimates of the cost per tonne of CO_2. Tol (2005) carried out a comprehensive review of studies of climate change costs. Combining the over 100 estimates for the marginal damage cost of CO_2 to form an overall probability density function, he finds that the distribution is strongly right-skewed, with a median of $3.8/$t_{CO_2}$, a mean of $25.4/$t_{CO_2}$, and upper bound of the 95% confidence interval of $95/$t_{CO_2}$. According to Tol, under standard assumptions of time discounting, equity weighting, and risk aversion, the marginal damage cost is unlikely to exceed $14/$t_{CO_2}$ and is probably even smaller. This value is significantly lower than the $85/$t_{CO_2}$* reported by the widely publicized Stern Review (Stern et al. 2006). The discrepancy can to a large extent be explained by the very low value of the intergenerational discount rate employed by Stern et al., in contrast to the conventional social discount rate used by most other studies.

The conventional social discount rate has two components: the growth rate of the economy and the so-called pure rate of time preference (basically the impatience of people to consume now rather than in the future). We have shown that inclusion of the pure rate of time preference in the intergenerational discount rate leads to an inconsistency in any cost-benefit analysis of abatement options for climate change: the moneys assumed for future cost-benefit trade-offs will not be there (Rabl 1996, Rabl et al. 2014). Since the discount rate of the Stern Review is very close to taking the pure rate of time preference as zero, we believe that its damage cost estimate is closer to the truth than estimates obtained with the conventional discount rate, in particular the value of 21 €/$t_{CO_{2eq}}$ used by the current phase of ExternE (2008).†

* 65 €/t_{CO_2} at the exchange rate of $1.3/€ during 2012.

† The subscript "eq" indicates that the result can also be used for other greenhouse gases if their masses are multiplied by their global warming potential (GWP).

13.3 Assumptions and Models of ExternE

13.3.1 Assumptions

The key assumptions for the calculations of ExternE are summarized in Table 13.2. They have been implemented in the EcoSense software that has been used to calculate the damage costs of ExternE. The atmospheric modeling is based on source–receptor (SR) matrices that have been provided specifically for ExternE by EMEP, the official modeling organization for transboundary pollution in Europe (http://www.emep.int/). The grid cell size is approximately 50 km × 50 km for the model of the European region, that is, the continent plus adjacent areas.

For PM, the emissions are provided in terms of $PM_{2.5}$ and PM_{co} (= PM fraction between 2.5 and 10 μm) rather than PM_{10}, and so are the SR matrices and the damage costs. Damage costs for PM_{10} can be calculated if one knows the respective fractions are $f_{2.5}$ and f_{co} of $PM_{2.5}$ and PM_{co} in the PM_{10}.

$$€/kg_{PM10} = f_{2.5}\ €/kg_{PM2.5} + f_{co}\ €/kg_{PMco}. \tag{13.1}$$

TABLE 13.2

Key Assumptions for the Calculations of ExternE

Atmospheric dispersion models	
Local range	Gaussian plume models ISC (for elevated sources) and ROADPOL (for ground level sources).
Regional range (Europe)	Source–receptor (SR) matrices calculated by EMEP were used for ExternE (2008), without any local modeling. SR matrices were calculated for primary and secondary pollutants, including O_3.
Impacts on health	
Form of ERFs	Straight line for all health impacts, without threshold for PM (primary and secondary) and with threshold for O_3.
Chronic mortality	The ERF for life expectancy loss is derived from increase in age-specific mortality due to $PM_{2.5}$ (Pope et al. 2002), by integrating over age distribution.
Acute mortality	For ozone, assuming 0.75 YOLL per death.
Nitrate and sulfate aerosols	The ERFs for $PM_{2.5}$ and PM_{10} are applied to nitrates and sulfates on the basis of mass and size range, without distinction of composition or other characteristics.
	Sulfates are assumed to be entirely $PM_{2.5}$ whereas for nitrates the $PM_{2.5}$ and PM_{co} fractions are calculated by EMEP.
Radionuclides	Linear ERFs without threshold: 0.05 Fatal cancers/person-Sv; 0.12 Nonfatal cancers/person-Sv; 0.01 Severe effects hereditary/person-Sv.
Micropollutants	Cancers due to As, Cd, Cr, Ni, dioxins, benzene, butadiene, and formaldehyde. Neurotoxic impacts of Hg and Pb.
Impacts on plants	ERFs for crop loss due to SO_2 and ozone.
Impacts on buildings and materials	Corrosion and erosion due to SO_2 and soiling due to particles.
Monetary valuation	
Accidents	Value of prevented fatality (VPF) = 1.5 M€
Loss of life expectancy	Proportional to loss of life expectancy, with value of life year (VOLY) = €40,000 for chronic, and €60,000 for acute mortality.
Cancers	0.48 M€ per nonfatal cancer, 1.12 M€ per fatal cancer.
Discount rate	3% until 2030, 2% thereafter.

For example, what is the damage cost of PM_{10} emissions from low stacks in the EU27 if they have a composition 60% $PM_{2.5}$ and 40% PM_{co}? Inserting the respective numbers from Figure 13.2 into Equation 13.1 we obtain 0.6×26.1 €/$kg_{PM2.5}$ + 0.4*1.5 €/kg_{PMco} = 16.3 €/kg_{PM10}.

13.3.2 ERFs and Unit Costs

The ERF slopes and unit costs of ExternE (2008) are listed in Table 13.3 (Preiss et al. 2008). The entries in the ERF column include the fraction of the total population that is affected by the respective pollutant and endpoint, thus they are applied directly to the general population.

As EMEP provides SR matrices for $PM_{2.5}$ and PM_{co} but not for PM_{10}, one has to decide how to apply the ERFs for PM_{10}. For that one would need to know how much of the impact of the

TABLE 13.3

Exposure–Response Function Slopes and Unit Costs, Assumed by ExternE (2008)

Pollutant	Endpoint	ERF Slope (cases/year) per (person·µg/m³)	Unit Cost €$_{2000}$/case	Exposure Cost (€/year) per (person·µg/m³)	*% of Total*
$PM_{2.5}$	Chronic mortality, YOLL	6.51E-04	40,000[a]	26.04	*79.4%*
	Net restricted activity days (netRAD)	9.59E-03	130	1.25	*3.8%*
	Work loss days (WLD)	1.39E-02	295	4.10	*12.5%*
	Minor restricted activity days (MRAD)	3.69E-02	38	1.40	*4.3%*
	Total			**32.79**	***100%***
PM_{10}	Infant mortality, deaths	6.84E-08	3,000,000	0.205	*3.4%*
	New cases of chronic bronchitis	1.86E-05	200,000	3.710	*62.2%*
	Respiratory hospital admissions	7.03E-06	2,000	0.014	*0.2%*
	Cardiac hospital admissions	4.34E-06	2,000	0.009	*0.1%*
	Medication/bronchodilator use, children	4.03E-04	1	0.000	*0.0%*
	Medication/bronchodilator use, adults	3.27E-03	1	0.003	*0.1%*
	LRS, children, days	2.08E-02	38	0.792	*13.3%*
	LRS, adults, days	3.24E-02	38	1.230	*20.6%*
	Total			**5.963**	***100%***
O_3, SOMO35	Acute mortality, YOLL	2.23E-06	60,000[b]	0.134	*15.1%*
	Respiratory hospital admissions	1.98E-06	2,000	0.004	*0.4%*
	Minor restricted activity days (MRAD)	7.36E-03	38	0.280	*31.7%*
	Medication use/ bronchodilator use	2.62E-03	1	0.003	*0.3%*
	LRS excluding cough, days	1.79E-03	38	0.068	*7.7%*
	Cough days	1.04E-02	38	0.396	*44.8%*
	Total			**0.885**	***100%***

Note: YOLL, years of life lost; LRS, lower respiratory symptoms; SOMO35, sum of ozone means over 35 ppb (Equation 13.3).

[a] Value of life year lost due to chronic exposure (discounted because of delay between exposure and death).

[b] Value of life year lost due to acute exposure (not discounted).

PM_{10} ERFs is due to $PM_{2.5}$ and how much due to PM_{co}. Such information is not available, and so the following somewhat literal interpretation has been used by ExternE (2008) for applying the ERFs for PM_{10}: both $PM_{2.5}$ and PM_{co} being by their definition also PM_{10}, the PM_{10} exposure costs are applied to both $PM_{2.5}$ and PM_{co}. To calculate the damage cost of $PM_{2.5}$, the concentration of $PM_{2.5}$ is multiplied by each of the exposure costs for $PM_{2.5}$ and for PM_{10}, and the resulting terms are summed; thus the $PM_{2.5}$ concentrations are multiplied by (32.79 + 5.963) (€/year) per (person μg/m³). For the damage cost of PM_{co}, the concentration of PM_{co} is multiplied by the exposure costs for PM_{10}; thus the PM_{co} concentrations are multiplied by 5.963 (€/year) per (person μg/m³). This procedure, documented in Preiss et al. (2008), can be summarized as follows:

$$\text{Cost per } PM_{2.5} \text{ exposure} = (32.79 + 5.963)\ (€/\text{year})/(\text{person}\ \mu g/m^3)$$

$$= 38.753\ (€/\text{year})/(\text{person}\ \mu g/m^3) \quad (13.2a)$$

and

$$\text{Cost per } PM_{co} \text{ exposure} = 5.963\ (€/\text{year})/(\text{person}\ \mu g/m^3). \quad (13.2b)$$

For O_3, ExternE (2008) assumes a hockey stick ERF with zero impact below a threshold of 35 ppb and linear increase above. This is implemented by stating the ERF slope in terms of SOMO35, the sum of ozone means over 35 ppb. More explicitly SOMO35 is defined as the sum of the differences between maximum daily 8 h moving average concentrations greater than 35 ppb (= 70 μg/m³) and 35 ppb:

$$\text{SOMO35} = \sum_{i=1}^{365} \text{Max}\left[0, c_i - 35\text{ppb}\right] \quad (13.3)$$

where c_i is the maximum daily 8 h running mean ozone concentration in ppb and the summation is over all days per calendar year. SOMO35 has a dimension of ppb·days (or (μg/m³)·days).

Multiplying the ERF slope by the unit cost, one obtains the exposure cost, that is, the incremental cost per 1/(μg/m³) per person per year, to be multiplied by the population in the grid cells of EcoSense and by the concentration increase per kg of emitted pollutant to calculate the cost per kg of pollutant. In the uniform world model (UWM), the incremental cost per 1/(μg/m³) per person per year is multiplied by the regional population density and divided by the depletion velocity v_{dep}, see the example in Section 13.3.4.

13.3.3 Damage Costs of ExternE (2008)

Figure 13.2 shows the damage costs calculated by ExternE (2008) for typical emissions in the EU27. The error bars indicate the uncertainties, as 68% confidence intervals (corresponding to one geometric standard deviation), as estimated by Spadaro and Rabl (2008).

Not shown in Figure 13.2 are the greenhouse gases. The cost per tonne of CO_2 depends on when it is emitted. The value used by ExternE (2008), 21 €/$t_{CO_{2eq}}$ for emissions in 2010.

13.3.4 UWM: A Simple Model for Damage Cost Estimation

A simple and convenient tool for the development of typical values is the UWM, first presented by Curtiss and Rabl (1996) and further developed, with detailed validation studies, by Spadaro (1999), Spadaro and Rabl (2002), and Rabl et al. (2014). The UWM is a product of a few factors; it is simple and transparent, showing at a glance the role of the most important parameters of the IPA. It is exact for tall stacks in the limit where the distribution of either the sources or the receptors is uniform and the key atmospheric parameters do not vary with location. In practice, the agreement with detailed models is usually within a factor of two for stack heights above 50 m. For policy applications, one needs typical values and the UWM is more relevant than a detailed analysis for a specific site.

The UWM for the damage cost rate $\dot{D}_{uni}$ in €/year of a particular impact due to the emission rate $\dot{m}$ of a primary pollutant is

$$\dot{D}_{uni} = p\, S_{ERF}\, \dot{m}\, \rho / v_{dep} \tag{13.4}$$

where

p = unit cost per case (*price*) (€/case)
S_{ERF} = ERF slope ([cases/year]/[person·(μg/m³)])
$\dot{m}$ = emission rate of pollutant (kg/year)
ρ = average (land and water) population density (person/km²) within 1,000 km of source
v_{dep} = depletion velocity of pollutant (m/s)

The latter quantity accounts for the removal rate of the pollutant from the atmosphere, due to dry and wet deposition. For secondary pollutants, the equation has the same form, but with an effective depletion velocity that includes the transformation rate of the primary into the secondary pollutant. With this model it is easy to transfer the results from one region to another (if ERF and depletion velocity are the same): simply rescale the result in proportion to receptor density and unit cost.

Example 13.1

Use the UWM to estimate the contribution of chronic mortality to the cost of $PM_{2.5}$ if the regional population density ρ = 100 persons/km² and the depletion velocity v_{dep} = 0.005 m/s.

Solution

Inserting the entries for chronic mortality in Table 13.3 and an emission rate of $\dot{m}$ = 1 kg/year into Equation 13.4, we find an impact rate of

$$\dot{D}_{uni} = \frac{40{,}000\text{ €}}{\text{YOLL}} \times \frac{6.51\text{E–}04\text{ YOLL/year}}{\text{person}\cdot(\mu\text{g/m}^3)} \times \frac{10^{-4}\text{ persons}}{\text{m}^2} \times \frac{1\text{ kg}}{365.25\times 24\times 3600\text{ s}} \times \frac{10^9\ \mu\text{g/kg}}{0.005\text{ m/s}}$$

$$= 16.50\text{ €/year.}$$

Since both emission rate and impact rate are stated per year, the cost per kg is 16.50 €/kg.

13.4 Results for Electricity Production

13.4.1 General Remarks

Since the 1990s, several assessments of fuel chains have been carried out, both in the United States (Levy et al. 1999, 2009, Abt 2004, NRC 2010) and the EU (ExternE 1995, 1998, 2008, Rabl et al. 2004). They all use an IPA, generally based on the same literature, although with somewhat different interpretations. The focus in all these studies is on greenhouse gases and the classical air pollutants. Here we summarize key results for the United States and EU27.

To put the damage costs in perspective, we show in Table 13.4 the average selling price of electricity in the United States and in France, for industrial and for residential consumers.

TABLE 13.4

Average Selling Price of Electricity in the United States and in France around 2010 for Industrial and for Residential Consumers

	Residential	Industrial
United States, ¢/kWh	11.0	6.7
France, €cent/kWh	11.0	6.0

13.4.2 Damage Costs of Power in the EU27

The emissions inventory of ExternE (2008) for coal and natural gas plants is summarized in Table 13.5. It is based on the ecoinvent database (Frischknecht et al. 2007) and distinguishes four LCA stages: construction of the power plant, operation of the power plant (for fossil plants combustion of the fuel), dismantling of the power plant, and fuel (for fossil plants production and transport of the fuel).

In Figures 13.4 and 13.5 we show selected results, with two entries for nuclear and wind, respectively. For wind, the first entry is for a system to supply base load power, consisting of wind turbine plus natural gas combined cycle (NGCC). The second entry is for wind turbines alone, assuming that sufficient storage is available; this entry does not include any costs for storage. The first entry for nuclear is taken directly from the data CD of ExternE (2008), for the second entry we have added the estimate of Rabl and Rabl (2013) for the cost of a catastrophic accident and waste storage. The choice between the two estimates for nuclear power involves problematic subjective judgments: is the future accident risk for reactors under consideration as bad as the worldwide average in the past, and is the cost of waste management fully internalized by the payments to a reserve fund that are currently paid by the nuclear industry?

The arrows in Figure 13.4 indicate the total cost where it exceeds the scale of the graph. For photovoltaics (PV), there has been a dramatic decrease of the private cost since 2011 and we do not try to estimate the current cost per kWh. The private cost of nuclear was estimated by EdF (Electricité de France), the world's largest nuclear utility company which in the past has been able to build nuclear power plants for impressively low cost, thanks to

TABLE 13.5

Typical Emissions by Coal and Gas Power Plants in the EU-27, in g/kWh$_e$

	Hard Coal, Condensing Power Plant			Natural Gas Combined Cycle		
	Upstream	Plant	Total	Upstream	Plant	Total
CO_2	30.9	730.2	761.1	46.9	355.5	402.4
CH_4	2.2	0.0	2.2	1.0	0.0	1.0
CO_{2eq}[a]	84.8	730.6	815.4	71.6	355.6	427.3
$PM_{2.5}$	0.01	0.00	0.01	0.00	0.00	0.01
PM_{co}	0.02	0.03	0.04	0.00	0.00	0.00
NO_x	0.33	0.55	0.88	0.15	0.27	0.42
SO_2	0.24	0.55	0.79	0.14	0.01	0.15

Source: ExternE, 2008, With this reference we cite the methodology and results of the latest phases of ExternE: NEEDS (2004–2008) and CASES (2006–2008). The results can be found in the data CD distributed with the CASES book edited by Markandya, A., Bigano, A., and Porchia, R. in 2010: *The Social Cost of Electricity: Scenarios and Policy Implications*, Edward Elgar Publishing Ltd., Cheltenham, U.K.

[a] Assuming GWP of 25 for CH_4.

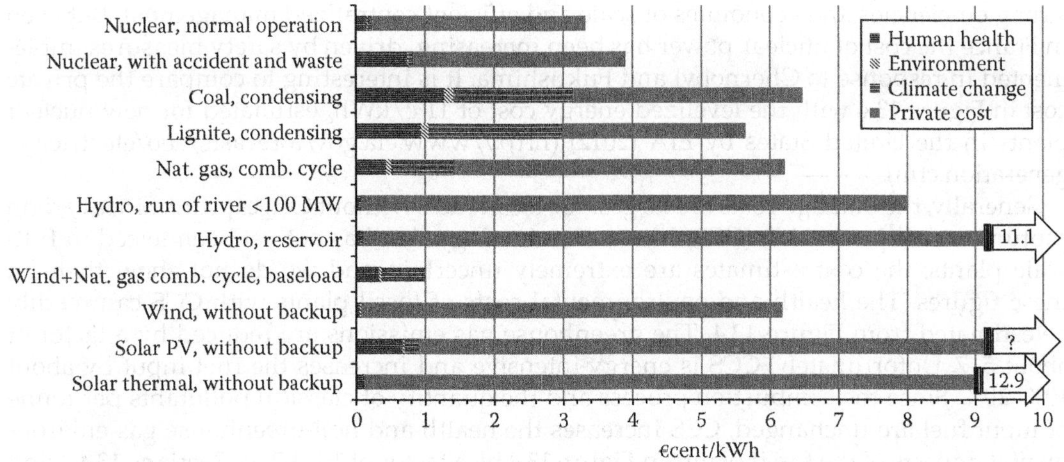

FIGURE 13.4
Selected results for the costs, both external and private, of current power technologies in the EU27. The arrows indicate the estimated total cost. For the second version of nuclear, the estimate of Rabl and Rabl (2013) for accident and waste has been added. For wind, the first version is for a combination wind + NGCC for base load (without private costs), the second version is for wind with storage (but without costs for the latter). (Data of ExternE, 2008, with this reference we cite the methodology and results of the latest phases of ExternE: NEEDS (2004–2008) and CASES (2006–2008). The results can be found in the data CD distributed with the CASES book edited by Markandya, A., Bigano, A., and Porchia, R., in 2010: *The Social Cost of Electricity: Scenarios and Policy Implications*, Edward Elgar Publishing Ltd., Cheltenham, U.K.)

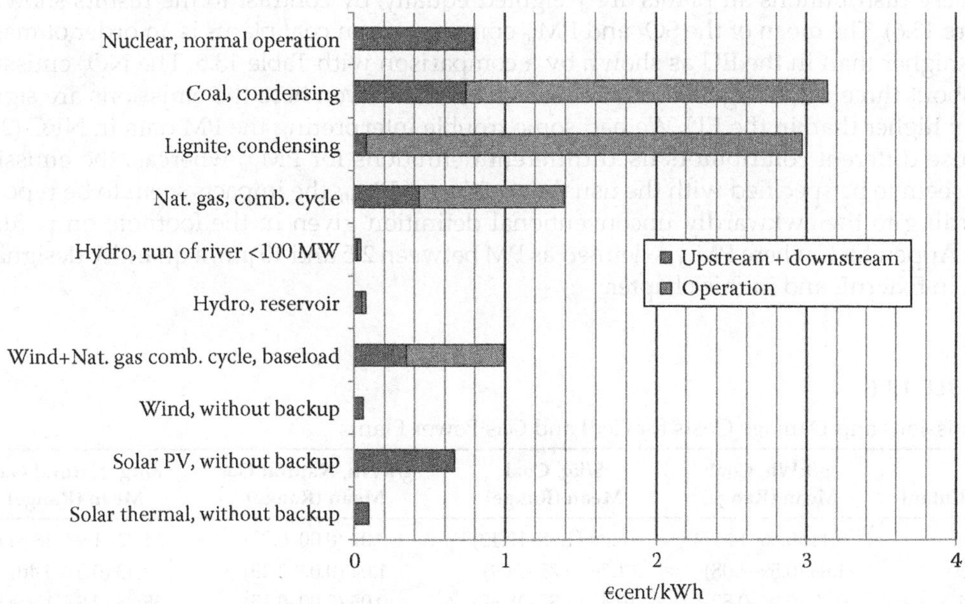

FIGURE 13.5
Breakdown of damage costs of Figure 13.4 by stages of fuel chain. (Data of ExternE, 2008, with this reference we cite the methodology and results of the latest phases of ExternE: NEEDS (2004–2008) and CASES (2006–2008). The results can be found in the data CD distributed with the CASES book edited by Markandya, A., Bigano, A., and Porchia, R., in 2010: *The Social Cost of Electricity: Scenarios and Policy Implications*, Edward Elgar Publishing Ltd., Cheltenham, U.K.)

strong efficiencies and economies of scale and efficient centralized management. But even in France the cost of nuclear power has been increasing, driven by safety measures implemented in response to Chernobyl and Fukushima. It is interesting to compare the private cost in Figure 13.4 with the levelized energy cost of 11 ¢/kWh_e estimated for new nuclear plants in the United States by EIA (2012) (http://www.eia.gov/forecasts/aeo/electricity_generation.cfm).

Generally, the damage costs are largest for fossil fuels without CO_2 capture. Since carbon capture and sequestration (CCS) is a new technology that has not yet been tested in full-scale plants, the cost estimates are extremely uncertain and we do not show them in these figures. The health and environmental costs of fossil plants with CCS can readily be estimated from Figure 13.4. The greenhouse gas emissions are reduced by a factor of about 5–7. Unfortunately, CCS is energy-intensive and increases the fuel input by about 10%–20%. Since the combustion process and the quantity of classical pollutants per tonne of input fuel are unchanged, CCS increases the health and non-greenhouse gas environmental impacts of the fossil plants in Figure 13.4 by a factor of 1.1–1.2. In Sections 13.4.4 and 13.4.5, we provide further detail on the nuclear and renewable technologies.

13.4.3 Damage Costs of Power in the United States

The externality study by NRC (2010) uses a methodology very similar to ExternE. In Table 13.6, we show the emissions and damage costs of coal and gas-fired power plants; they do not include upstream emissions. Both emission per kWh and damage cost per kg vary widely, and the indicated range includes all values from the 5th percentile to the 95th percentile (in these distributions all plants are weighted equally, by contrast to the results shown in Figure 13.6). The mean of the SO_2 and PM_{10} emissions from coal plants is an order of magnitude higher than in the EU as shown by a comparison with Table 13.5. The NO_x emissions are about three times higher. Gas is much cleaner, but even here the emissions are significantly higher than in the EU. We had some trouble interpreting the PM data in NRC (2010) because different contributors used different definitions for PM_{10}; whereas, the emissions data seem to be specified with the usual definition of PM_{10}, the impacts seem to be reported according to the awkwardly unconventional definition given in the footnote on p. 313 of their Appendix C where PM_{10} is defined as PM between 2.5 and 10 μm, a quantity designated PM_{co} in ExternE and in this chapter.

TABLE 13.6

Emissions and Damage Costs for Coal and Gas Power Plants

Pollutant	g/kWh, Coal Mean (Range)	$/kg, Coal Mean (Range)	g/kWh, Natural Gas Mean (Range)	$/kg, Natural Gas Mean (Range)
SO_2	5.44 (0.68–14.97)	6.39 (1.98–12.13)	0.02 (0.00–0.07)	14.33 (1.98–48.51)
NO_x	1.86 (0.59–4.08)	1.76 (0.75–3.09)	1.04 (0.02–2.49)	2.43 (0.51–5.40)
$PM_{2.5}$	0.27 (0.04–0.82)	10.47 (2.87–28.67)	0.05 (0.00–0.13)	35.28 (2.87–176.41)
PM_{co}	0.06 (0.05–0.95)	0.51 (0.15–1.43)	0.00 (0.00–0.15)	1.87 (0.19–8.60)

Source: Data of NRC, Hidden costs of energy: Unpriced consequences of energy production and use, National Research Council of the National Academies, Washington, DC, 2010, Available from National Academies Press, http://www.nap.edu/catalog.php?record_id=12794.

Note: The range covers the distribution from 5% to 95%; it indicates geographical variability, not uncertainty. These data and Figure 13.6 include only the power plants, without upstream emissions.

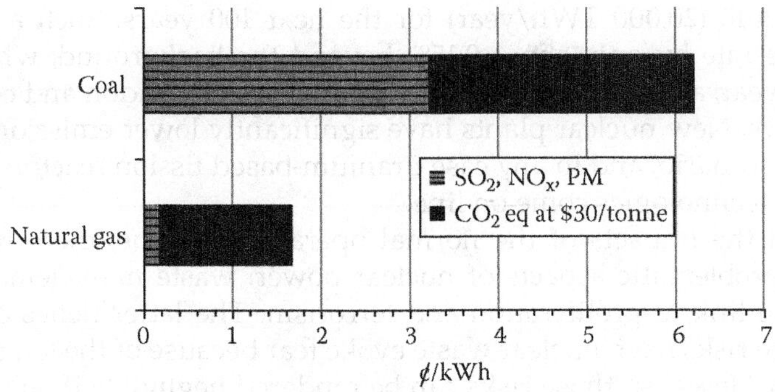

FIGURE 13.6
Results of NRC (2010) for damage costs of electric power in the United States, in ¢/kWh. For illustrative purposes the study indicates climate damage with a hypothetical \$30/t_{CO_2eq}.

Finally, Figure 13.6 shows the average damage cost of coal and gas power plants in the United States, weighted according to their production. For the cost of greenhouse gases, NRC (2010) does not want to make a definitive choice but uses \$30/t_{CO_2eq} as a hypothetical value for the purpose of illustration (25 €/t_{CO_2eq} at an exchange rate of \$1.3/€).* The damage costs in Table 13.6 and Figure 13.6 include only the power plant emissions, not upstream or downstream impacts. As for nuclear power, NCR only reviews the studies by ORNL/RFF (1994) and ExternE (1995) without trying a new assessment. No damage cost estimates are provided for renewable energy sources. Like in Europe the damage costs for fossil fuel power plants are very significant compared to the market price (see Table 13.4).

NRC also has some interesting data on combustion products of coal power plants. The total production of coal wastes is about 120 million tonnes/year, of which 43% is utilized, mostly for construction materials. The quantity of combustion products is 0.06 kg/kWh of electricity from coal.

13.4.4 Nuclear Fuel Chain

13.4.4.1 Normal Operation

Several assessments of the nuclear fuel cycle have been published since 1990, including ExternE (1995), ORNL/RFF (1994), Ontario Hydro (1993), and ExternE (2008). All of these are based on IPA and LCA, and they all agree that the damage cost due to radionuclides from normal operation is very small. The only significant radiological contribution to the damage cost comes from human health, in particular cancers and hereditary effects. Some studies find significant non-radiological contributions if isotope separation is done in countries where much of the electricity comes from fossil fuels, especially coal. But even those studies find that the total external costs from the normal operation of the nuclear fuel chain are small compared to fossil fuels.

To illustrate how small the radiological doses from normal operation are, Rabl et al. (2014) look at the dose per person that would result if the nuclear technology of France of the 1990s were used to provide electricity at a rate equal to the entire electricity production

* The NRC report uses the same "tons" for both short tons (= 907 kg) and metric tonnes. Since the climate change literature is based on tonnes, we believe that NRC means tonnes for CO_2, by contrast to results for other pollutants which are presumably in short tons because they are calculated with the APEEP model which works with short tons (Nick Muller, personal communication, May 21, 2013).

of the world in 2010 (20,000 TWh/year) for the next 100 years. Such a scenario would increase the dose rate by only about 0.15% beyond the background, which is generally around 2.5 mSv/year, almost entirely due to natural sources (radon and cosmic radiation) and medical x-rays. New nuclear plants have significantly lower emissions than what we assumed for this scenario, and in any case uranium-based fission reactors are only a stop-gap until cleaner technologies come on line.

However small the impacts of the normal operation may be, there remains concern about the three problematic aspects of nuclear power: waste management, catastrophic accidents, and the link to proliferation and terrorism. The latter defies quantification in terms of costs. The risks from nuclear waste evoke fear because of the long lives of some of the radionuclides. However, those risks can be rendered negligible if one chooses permanently safeguarded retrievable storage, the option currently planned for France. Even with sealed *leave alone* storage such had been planned at Yucca Mountain in the United States, the impacts of a leak would be limited to the local zone because any groundwater would not move far, and the resulting damage costs would be small, by contrast to emissions from a major nuclear accident which can disperse widely through air or ocean.

13.4.4.2 Nuclear Accidents

Usually the cost of a large nuclear accident has been estimated by combining probabilistic safety assessments (PSAs) of nuclear reactors, to determine the probability of an accident, with programs to evaluate the health and economic consequences of an accident. Such calculations have been carried out by ORNL/RFF (1994), ExternE (1995), and Dreicer et al. (1995), and the costs per kWh were found to be quite small because of the low probability estimated by PSAs based on engineering fault tree analysis.

But is the probability of severe accidents really so low? Since they are rare occurrences, historical data do not provide much guidance. There have been only two severe accidents with significant release of core material: Chernobyl and Fukushima. They are rated seven, the highest number on the scale of accident severity which is logarithmic. The accident of Three Mile Island rates only five on this scale. Each time one learns and improves the safety. No one builds Chernobyl-type reactors anymore, and the protection against earthquakes and floods will be improved after Fukushima. But what unexpected risks remain lurking in the background? Fear of another catastrophic accident where operators and designers have underestimated risk is why many people are calling for an end to nuclear power and even the shutdown of all existing nuclear plants, arguing that energy efficiency and renewables are clean and cost-effective alternatives.

It would not be wise to retire nuclear plants precipitously, if the alternatives entail total (private + damage) costs that are even higher. Rabl and Rabl (2013) have compared the damage costs of nuclear with those of the alternatives. Unlike the earlier assessments with complex and opaque computer calculations, they present a very simple and transparent calculation based on the actual track record of nuclear power plants. It is summarized in Table 13.7 for nuclear and Table 13.8 for the cleanest alternative; for detailed justification of the assumptions and parameters we refer to their paper. For their central estimate they assume impacts comparable to Chernobyl and Fukushima. The release of radioactive material and the ensuing health impacts of Chernobyl were much worse than for Fukushima. The total worldwide cancer deaths due to Chernobyl have been estimated as 24,000 by Garwin (2005) and as 16,000 (95% uncertainty interval 6,700–38,000) by Cardis et al. (2006). For Fukushima, Ten Hoeve and Jacobson (2012) put the total expected death toll at 130 (uncertainty range of 15–1100), based on detailed modeling of dispersion and doses.

TABLE 13.7

Assumptions and Results for the External Costs of Nuclear Power

Cost Elements	Parameter[a]	Units	Central	Low	High
Fatal cancers	10,000				
Cost per cancer	5	M€/cancer			
Discount factor for cancers[b]	0.38				
Cost of cancers		G€	*18.8*	10	50
Lost reactors, 1 GW each	6				
Cost per reactor	5	G€/reactor			
Cost of reactors		G€	*30*	20	40
Cost of cleanup		G€	*30*	20	200
Displaced persons	500,000				
Cost per displaced person	0.5	M€/person			
Cost of displaced persons		G€	*250*	100	1000
Area lost for agricultural production	1,000	km^2			
Yield, cereals, 5 tonnes/ha per year	500	tonnes/km^2/year			
Price, cereals 150 €/tonne	150	€/tonne			
Loss €/km^2/year	75,000	€/km^2/year			
Loss duration	100	Years			
Cost of lost agriculture		G€	*7.5*	5	50
Lost power production	90	TWh			
Value per kWh	0.2	€/kWh			
Cost of lost power		G€	*18*	10	50
Total cost of accident, if now		G€	*354*	165	1390
Years without accident			25	40	15
Discount rate	0.05				
Discount factor[c]			0.56	0.43	0.69
Nuclear production, 2008	2100	TWh/year			
Damage cost of accident, per kWh[d]		€cent/kWh	*0.38*	0.08	2.29
Damage cost of normal operation[e]		€cent/kWh	*0.21*	0.07	0.63
External cost of waste management		€cent/kWh	*0.20*	0.10	0.30
Total external cost of nuclear		€cent/kWh	*0.79*	0.25	3.22

Source: From Rabl, A. and Rabl, V.A., *Energy Policy*, 57, 575, 2013.
Note: All but the last 3 lines is for a catastrophic accident. T, tera; G, giga; M, mega; k, kilo.
[a] The parameter values are used only for the central estimate.
[b] To account for delay between accident and occurrence of cancer.
[c] To account for an accident occurring between now and 25 (or 15 or 40) years from now.
[d] Cost if now × discount factor/(years without accident × TWh/year).
[e] Damage cost of nuclear estimated by ExternE (2008).

Rabl and Rabl also include an item for the cost of waste management. To account for the general tendency to underestimate future costs, Rabl and Rabl take 0.2 €cent/kWh as a damage cost of providing a permanent nuclear waste disposal site even though much of that is already internalized because plant operators are required to put sufficient money into a waste management fund. Since the 0.2 €cent/kWh for the cost of waste management in Table 13.7 is not a damage cost but an estimate of insufficient payment by utilities for future waste management, we label it external cost. By contrast, the cost of an accident and the 0.21 €cent/kWh for pollution from normal operation are damage costs.

Of course, any assessment of the external costs of nuclear is controversial, in particular with regard to accidents, proliferation, terrorism, and waste management. Subjective

TABLE 13.8

Calculations of Rabl and Rabl (2013) for the Damage Costs of Producing Base Load Power by Wind Turbines with NGCC Plants as Cleanest Fossil Backup

Cost Elements	Units	Central	Low	High
Cost of GHG[a]	€/$t_{CO_{2eq}}$	21	7	63
Fraction of base load power provided by wind		0.34	0.38	0.27
Damage cost of wind turbines	€cent/kWh	*0*	*0*	*0*
GHG emissions from NGCC[a]	kg/kWh	0.43	0.43	0.43
Cost of GHG emissions from NGCC[a]	€cent/kWh	0.90	0.3	2.7
Cost of classical air pollutants from NGCC[a]	€cent/kWh	0.50	0.17	1.49
Fraction of base load power provided by NGCC		0.66	0.62	0.73
NGCC contribution due to GHG costs[a]	€cent/kWh	*0.59*	*0.19*	*1.97*
NGCC contribution due to classical air pollutants[a]	€cent/kWh	*0.33*	*0.10*	*1.09*
Total damage cost[a]	€cent/kWh	*0.92*	*0.29*	*3.05*

Note: GHG, greenhouse gases.
[a] These numbers have been modified slightly to be consistent with ExternE (2008).

choices are inevitable and any specific assumption can and will be criticized. Rabl and Rabl offer their assessment as basis for discussion because it is better to base decisions on an explicit analysis rather than vague impressions.

Table 13.7 shows what such assumptions imply for the central estimate of the accident cost. The cost per kWh is obtained by dividing the cost of an accident by the number of years between accidents and the annual electricity production of the countries considered here (EU, United States, Canada, Japan, South Korea, and Taiwan); in addition it is multiplied by a discount factor to account for an accident occurring between now and 25 years from now. The values listed in the parameter columns are used only for the central results. The last two columns indicate lower and upper bounds; they are rough estimates of plausible ranges about the central values. For comparison, note that the accident database of Hirschberg et al. (2004) shows an estimate of G\$ 339_{1996} (approximately 360 G€$_{2010}$) for the cost of Chernobyl; it is within the range shown in Table 13.7.

The historical accident probability assumed for Table 13.7 is about an order of magnitude higher than what has been estimated by the standard PSAs with detailed fault tree modeling for the majority of reactors that have been built. That is the main reason why the accident cost in Table 13.7 is so much higher than the estimates by ORNL/RFF (1994), ExternE (1995), and Dreicer et al. (1995). Thus any estimate of the external costs of nuclear power involves problematic subjective judgments: is the future risk for reactors under consideration as bad as the worldwide average in the past, and is the cost of waste management fully internalized by the payments to a reserve fund that are currently paid by the nuclear industry? If one believes that waste management is fully internalized and that the future accident risk is as estimated by PSAs, the bottom line in Table 13.7 would be three times smaller, about 0.25 €cent/kWh.

13.4.5 Renewable Energy Technologies

Among renewable electricity sources there is a wide variety of technologies such as hydro, wind, biomass, and various forms of direct solar energy utilization, in particular PV and solar thermal power plants. Hydro, wind, and solar have special appeal, being not only inexhaustible but generally with little environmental impact or health risk. Of course, detailed studies are needed to check whether the impacts of these technologies are really benign.

For electricity from biomass, the main technologies that have been considered are combustion with steam turbine or gasification with gas turbine. There are significant health impacts from the air pollution emitted by the power plant and by the machinery needed for the production and transport of the fuel. The net greenhouse gas emissions from the biomass itself are zero, but there are emissions from the associated machines and vehicles, as well as chemical inputs (fertilizers, pesticides). Combustion with steam turbine is not the most efficient way of using biomass because temperature and pressure of the produced steam, and thus the power plant efficiency, are relatively low. In addition such plants are usually too small to maximize economies and efficiencies of scale. A better alternative is to add wood chips to coal in conventional coal-fired power plants (up to 5% of the total fuel can be added in this form), as already practiced in some power plants.

For hydro, PV, and wind there are of course no emissions from power generation. There are upstream emissions from the production of the materials, although the corresponding damage costs are small. However, one must not overlook the variability of wind or insolation and the corresponding variability in the amount of electricity they supply. To achieve a reliable power supply, backup capacity must be available, especially if solar and wind provide a high fraction of the total electricity production. Of course, energy storage would be an attractive solution, but for most applications storage of the required magnitude and duration is still too expensive or the potential sites (for the most cost-effective option, pumped hydro) are too limited. Without sufficient storage the backup capacity requirement of wind and solar implies that part of the replaced electricity will come from fossil fuels, with the attendant costs for health and environment.

Rabl and Rabl (2013) calculate the damage costs of producing base load power by wind turbines with NGCC plants as cleanest fossil backup. Their results are shown in Table 13.8, although slightly modified to make them consistent with the numbers of ExternE (2008). The fraction of base load power provided by wind is around 34%, a number for very good wind sites. Rabl and Rabl make this calculation in the context of proposals for premature retirement of existing nuclear plants and note that the damage cost of a wind plus NGCC base load plant, 0.92 €cent/kWh in Table 13.8, is higher than the damage cost of nuclear with the costs of accident and waste management, 0.79 €cent/kWh for the central estimate shown in Table 13.7; however, the uncertainty ranges are large.

Amenity impacts of renewables are highly dependent on local conditions, in particular the population near the site. The impacts of hydro are so variable that specific results cannot be taken as general guidelines. The impacts can range from beneficial (for instance due to flood control or recreational facilities) to extremely harmful if large populations are displaced without compensation or if a dam breaks. The decomposition of submerged vegetation in new reservoirs can cause significant emissions of greenhouse gases. Such emissions are highly variable with site; Gagnon and van de Vate (1997) indicate 15 $g_{CO_{2eq}}$/kWh as typical value for cold climates, and possibly much higher values in tropical zones.

13.5 Waste Treatment and Energy Recovery

13.5.1 Assumptions

Despite all the efforts to increase the recycling of waste, there remains much waste that is not suitable for recycling or biotreatment, and it has to be either sent to landfill or incinerated (pyrolysis could be a cleaner alternative to incineration which can produce fuels, but

TABLE 13.9

Assumptions for the Emissions from Incineration of MSW

Pollutant	mg/Nm³	g/t_{waste}	€/$kg_{pollutant}$	€/t_{waste}
PM_{10}	10	51.5	7.6	0.39
SO_2	50	258	6.2	1.60
NO_2	200	1,030	5.4	5.52
CO_2		807,000	0.021	16.95
As (2.8% of 0.5 mg/Nm³)	0.014	0.072	529.6	0.04
Cd (81.2% of 0.05 mg/Nm³)	0.0406	0.21	83.7	0.02
Cr^{VI} (6.5% of 0.2 × 0.5 mg/Nm³)[a]	0.0065	0.033	66.3	0.00
Hg (0.05 mg/Nm³)	0.05	0.26	8,000	2.06
Ni (33.8% of 0.5 mg/Nm³)	0.169	0.87	2.3	0.00
Pb (22% of 0.5 mg/Nm³)	0.11	0.57	278.3	0.16
Dioxins	1.00E−07	5.15E−07	37,000,000	0.02

Note: They are taken as the limit values of the flue gas concentrations, in Directive EC (2000), assuming 5150 Nm^3/t_{waste}. For metals other than Hg the directive specifies only certain sums: 0.5 mg/Nm³ for the sum of As + Co + Cr + Cu + Mn + Ni + Pb + Sn + Sb + V, and 0.05 mg/Nm³ for the sum of Cd + Tl; % within these sums based on ETSU (1996). The last two columns show the damage costs per kg of pollutant assumed for this chapter and the resulting cost per t_{waste} of the direct incinerator emissions.

[a] Assuming that 20% of Cr from incinerators is Cr^{VI}.

so far it is not widely used). Waste can be an important source of energy, and so we present results of an assessment of the damage costs of waste treatment and the benefits of energy recovery, for the principal technologies for municipal solid waste, namely incineration and landfill. These results are based on the paper by Rabl et al. (2008), recently updated in the book by Rabl et al. (2014); like the rest of this chapter they are based on ExternE (2008).

For incinerators we assume the emission limits of Directive EC (2000), as summarized in Table 13.9. This table also shows our assumptions for the damage cost per kg of pollutant and the implications for the cost per tonne of waste due to emissions from the incinerator.

The principal emissions to air from landfill are CH_4 and CO_2. Figure 13.7 shows the total greenhouse gas emissions of a municipal solid waste landfill versus time. CH_4 is expressed as equivalent CO_2, using a global warming potential (GWP) of 25. Note that a modern landfill is divided into a large number of individual compartments; they are filled one after another and sealed when full. The data of ADEME (2003) are plotted in Figure 13.7, where the time is measured from the date that a compartment is sealed. In practice, it is impossible to capture all of the CH_4, and capture rates around 70% are commonly assumed (although measured data seem to be difficult to find). Here we assume a capture rate of 70% for the first 40 years, on average, after closure of a compartment; after 40 years we assume that all the remaining CH_4 escapes to the atmosphere. This last assumption reflects uncertainty in the way that sites will be managed around the time of closure is approached. One possibility is that regulators will simply sign off a site as *inert* after a certain period of time. Another is that regulators would require operators to provide evidence, such as data on emission rates, to show that a site is inert. The latter would provide a higher level of environmental protection.

13.5.2 Recovery of Energy and Materials

The damage costs and the comparison between landfill and incineration turn out to be extremely sensitive to assumptions about energy recovery. For that reason we consider a fairly large number of options, for typical installations in France, according to ADEME (2000). In Section 13.5.3, we indicate the options with labels where the letters E and H refer

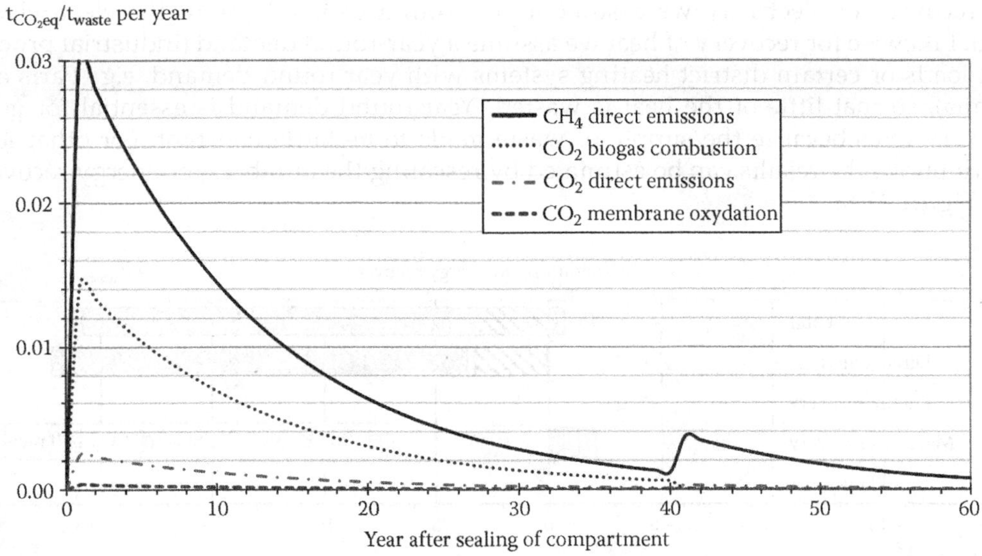

FIGURE 13.7
Greenhouse gas emissions from a municipal solid waste landfill versus time, t_{CO_2eq}/t_{waste} if 70% of the CH_4 is captured. (Based on ADEME, Outil de calcul des émissions dans l'air de CH_4, CO_2, SO_x, NOx issues des centres de stockage de déchets ménagers et assimilés (Tool for calculation of emissions of CH_4, CO_2, SO_x, NO_x from municipal and similar solid waste.), Agence Française de l'Environnement et de la Maîtrise de l'Energie, Paris, France, 2003.)

to heat and electricity and the letters c, g, n, and o to the fuel (coal, gas, nuclear, and oil, respectively) displaced by energy recovery. For example, (E = c&o, H = c&o) designates a system where heat and electricity are produced, each displacing a fuel mixture of coal and oil, 50% of each. The following options are considered for incineration:

- Recovery of heat and electricity (E = …, H = …)
- Recovery of electricity only (E = …)
- Recovery of heat only (H = …)

The following options are considered for landfill:

- No energy recovery
- Recovery of electricity, by motor (reciprocating engine) (E = …)
- Recovery of electricity, by turbine (E = …)
- Recovery of heat (H = …)

For each of these options we consider several suboptions:

- The recovered electricity displaces coal and oil-fired power plants, 50% each (E = c&o)
- The recovered electricity displaces nuclear power plants (E = n)
- The recovered heat displaces gas and oil-fired heating systems, 50% each (H = g&o)
- The recovered heat displaces only oil-fired heating systems (H = o)

For recovery of electricity we assume a year-round demand, so all the electricity is used. Likewise for recovery of heat we assume a year-round demand (industrial process heat loads or certain district heating systems with year-round demand, e.g., Paris and Vienna), so that little of the heat is wasted. Year-round demand is essential for good recovery rates because the supply of waste tends to be fairly constant. For other load distributions the results can be estimated by rescaling the numbers for energy recovery (see Figure 13.8).

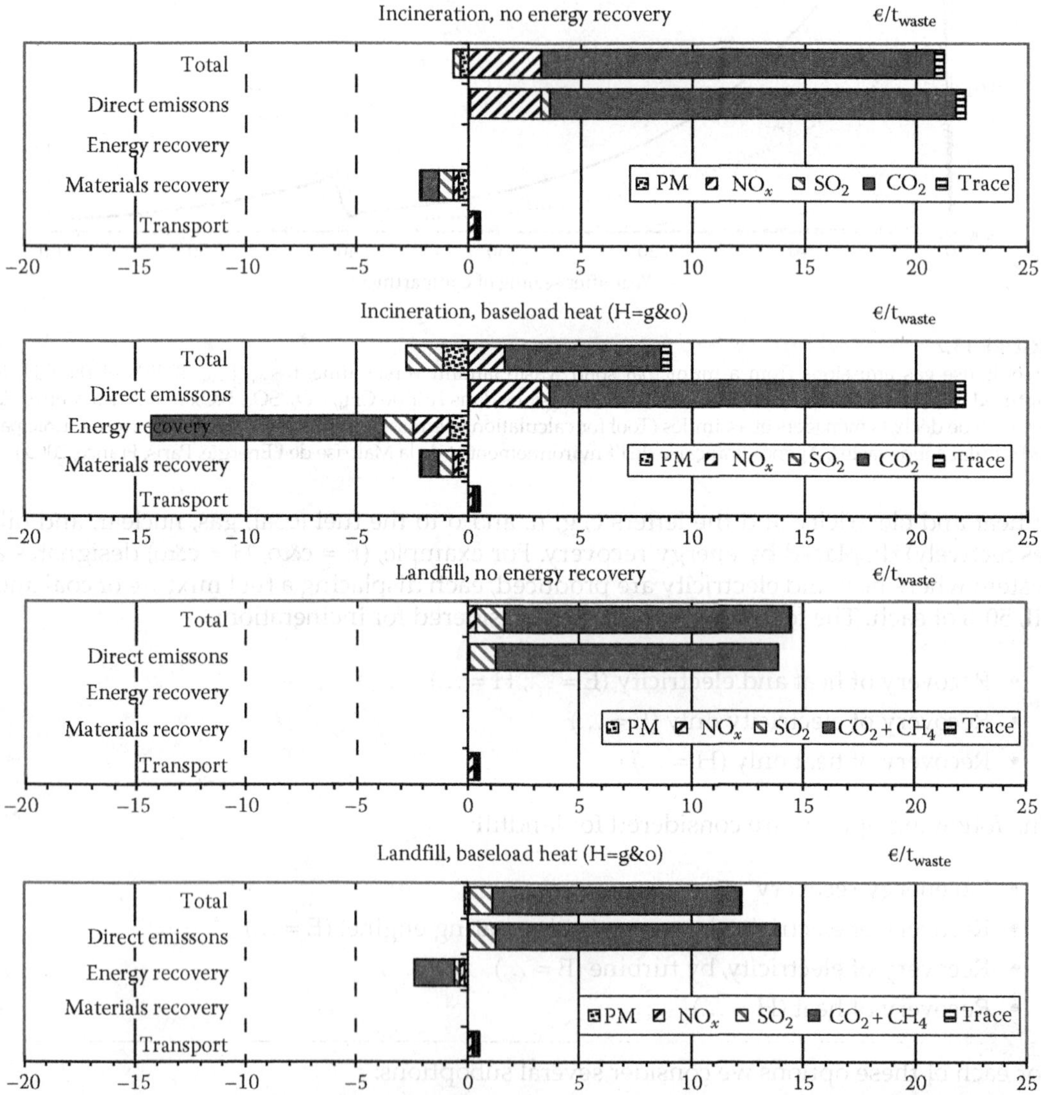

FIGURE 13.8
Some detailed results, by stage and pollutant. Trace = dioxins and toxic metals. Benefits from energy recovery can be quite different for different countries. The direct incinerator emissions are based on the actual data for France in 2011.

TABLE 13.10

Assumptions for Energy Recovery

kWh/t_{waste}	Electricity	Heat
Part load heat and electricity	202	607
Base load electricity only	270	0
Base load heat	0	1,850

Source: Based on Rabl, A. et al., *Waste Manag. Res.*, 26, 147, 2008.

TABLE 13.11

Assumptions for Recovery Rates and Avoided Damage Costs for Materials Recovery from Incinerators

	kg/t_{waste}	€/kg	$€/t_{waste}$
Slag	230	0	0
Iron	20.2	0.107	2.16
Aluminum	1.5	0.539	0.80

Source: Based on Rabl, A. et al., *Waste Manag. Res.*, 26, 147, 2008.

Note that for the purpose of this analysis the benefit of recovered electricity is essentially zero if it displaces nuclear because the damage costs of nuclear are very small compared to those of oil or coal; thus this option is almost equivalent to no electricity production at all as far as damage costs are concerned. Our assumptions for energy recovery rates, to be taken through to the next section, are shown in Table 13.10. The avoided emissions are based on the Large Combustion Plant Directive (EC 2001).

Recovery of materials is practical only for incinerators, not for landfills. Our assumptions for recovery rates and avoided damage costs are listed in Table 13.11; the avoided damage costs are based on the LCA inventory of Delucchi (2003).

13.5.3 Results for Damage Cost per Tonne of Waste

The real emissions are likely to be lower than the limit values of EC (2000). In Table 13.12, we compare the incinerator emissions and damage costs for limit values and with those for the actual emissions in France, based on the data kindly provided to us by Olivier Guichardaz of Dechets-Infos. For most of the pollutants the emissions have been greatly reduced. CO_2, by contrast, has increased slightly since the composition of waste has been changing with recycling.

A summary of the total damage cost for all the options, assuming the actual emissions in France in 2011, is shown in Figure 13.9. More detailed results for some of the options can be found in Figure 13.8, showing the contribution of each stage and of the major pollutants (dioxins and toxic metals are shown as *trace*). The damage costs of waste transport are negligible, as illustrated in Figure 13.8 with an arbitrary choice of 100 km roundtrip by a 16 tonne truck for which they amount to 0.45 $€/t_{waste}$. By far, the most important contributions come from direct emissions (of the landfill or incinerator) and energy recovery.

TABLE 13.12

Comparison of Incinerator Emissions and Damage Costs for Limit Values and Actual Emissions in France

	Limit Values EC (2000)		Actual Emissions, France 2011		
Pollutant	**g/t_{waste}**	**$€/t_{waste}$**	**g/t_{waste}**	**$€/t_{waste}$**	**Ratio Actual/Limit**
PM_{10}	51.5	0.39	7.3	0.06	0.14
SO_2	257.5	1.60	54.3	0.34	0.21
NO_2	1,030	5.52	600	3.22	0.58
CO_2	807,000	16.95	863,900	18.14	1.07
As	0.072	0.04	0.013	0.01	0.18
Cd	0.209	0.02	0.007	0.00	0.03
Cr^{VI}	0.033	0.00	0.016	0.00	0.48
Hg	0.258	2.06	0.053	0.43	0.21
Ni	0.870	0.00	0.000	0.00	0.00
Pb	0.567	0.16	0.093	0.03	0.16
Dioxins	5.15E–07	0.02	8.7E–08	0.00	0.17
Total		26.76		22.22	0.83

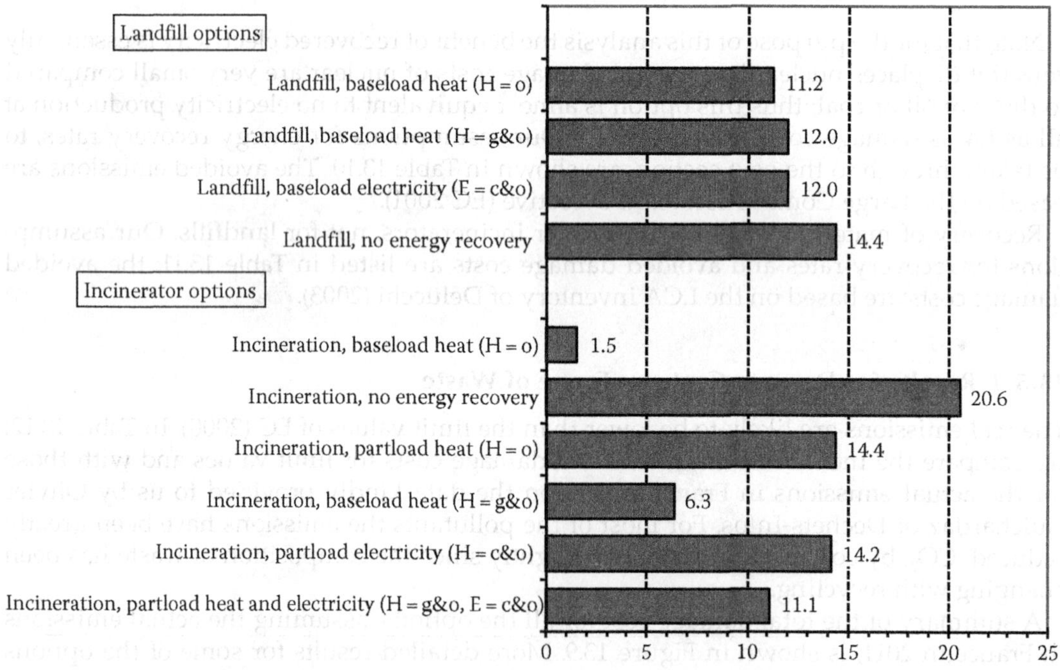

FIGURE 13.9
Results for total damage cost for waste treatment options. If electricity displaces nuclear, damage costs are essentially the same as for the case without energy recovery. Benefits from energy recovery can be quite different for different countries. Amenity costs are not included; they are very site-specific and could make a contribution on the order of 1 $€/t_{waste}$. The direct incinerator emissions are based on the actual data for France in 2011. For comparison, the private costs of landfill are around 50 $€/t_{waste}$, those of incineration around 100 $€/t_{waste}$.

Materials recovery is relevant for incineration only and brings a relatively small benefit. For incineration energy recovery can be crucial and can reduce the total damage cost to 1.5 €/t_{waste}.

For the direct emissions of the incinerator, the total is 22.22 €/t_{waste}, most of that being due to NO_x and CO_2. Toxic metals and dioxins, shown in Figure 13.8 as *trace*, contribute about 2% of the total (0.46 €/t_{waste}), mostly because of Hg and Pb. The contribution of dioxins is negligible, only 0.003 €/t_{waste}, thanks to the low emission limit of the Waste Incineration Directive (EC 2000).

For landfill, the cost is dominated by CH_4 emissions because only about 70% of the CH_4 can be captured. Energy recovery from a landfill is not very significant (and because of NO_x from the electricity production, this option increases the damage cost if the electricity displaces nuclear). By contrast, energy recovery is crucial for the damage cost of incineration. Under favorable conditions (all heat produced by incinerator displaces coal and oil), the total damage cost could even be negative in some situations, that is, a net benefit. By contrast to most other countries, in France recovery of electricity does not bring significant benefits, because it is base load power and all the base load power is produced by nuclear. In any case, electricity production brings far lower benefits than heat because of the poor conversion efficiency of incinerator heat (due to its low temperature) compared to central station power plants.

13.6 Conclusions

The damage costs are worst for coal, oil, and lignite. If internalized by a pollution tax, they would significantly increase the market price of electricity. Natural gas is much cleaner, with damage costs intermediate between coal and the renewables. Nuclear has low damage costs. Even if the cost of catastrophic accidents is included, with a pessimistic assessment based on the historical record, its damage cost is lower for baseload than for wind with NGCC as backup. For the fossil fuels, the largest contribution to the damage cost comes from greenhouse gases, here valued at 21 €/t_{CO_2eq}. However, as explained in Section 13.2.6, we believe that this is an underestimate, the true cost being closer to the 65 €/t_{CO_2eq} of the Stern Review (Stern et al. 2006). That implies a much larger damage cost for the fossil fuel chains. For waste treatment, energy recovery can bring important benefits, especially for waste incineration where it can greatly reduce the damage costs.

The uncertainties of damage costs are large, about a factor of 3 in either direction, according to Spadaro and Rabl (2008). The reader may wonder whether it is meaningful to use them as basis for decisions. The first reply is that even a threefold uncertainty is better than infinite uncertainty without analysis. Second, in many cases the benefits are either so much larger or so much smaller than the costs that the implication for a decision is clear even in the face of uncertainty. Third, if in the other cases the decisions are made without a significant bias in favor of either costs or benefits, some of the resulting decisions will err on the side of costs, others on the side of benefits. Rabl et al. (2005) have examined the consequences of such unbiased errors and found a very reassuring result: the extra social cost incurred because of uncertain damage costs is remarkably small, less than 10%–20% in most cases even if the damage costs are in error by a factor of 3. But without any knowledge of the damage costs, the extra social cost (compared to the minimal social cost that one would incur with perfect knowledge) could be very large.

Glossary

CBA	Cost-benefit analysis
CCS	carbon capture and sequestration
CO_{2eq}	Quantity of a greenhouse gas expressed as equivalent quantity of CO_2, using the GWP of the gas
CV	Contingent valuation
Discount rate	Rate r that allows comparison of monetary values incurred at different times, defined such that an amount P_n in year n has the same utility as an amount $P_0 = P_n (1 + r)^{-n}$ in year 0.
ERF	Exposure–response function
EU27	European Union with 27 member states
External costs	Costs that arise when the social or economic activities of one group of people have an impact on another for which the first group does not fully account, for example, when a polluter does not compensate others for the damage imposed on them.
GDP	gross domestic product
GWP	Global warming potential
IPA	Impact pathway analysis
LCA	Life cycle assessment
N	Nitrogen
NGCC	natural gas combined cycle
NO_x	Unspecified mixture of NO and NO_2
O_3	Ozone
PAH	Polycyclic aromatic hydrocarbons
PM_d	Particulate matter with aerodynamic diameter smaller than d μm
PV	photovoltaics
PSA	probabilistic safety assessment
S	Sulfur
S_{ERF}	Slope of exposure–response function [cases/(person·year·μg/m³)]
UWM	Uniform world model (for simplified estimation of damage costs)
v_{dep}	Depletion velocity (m/s)
VOC	Volatile organic compounds
VOLY	Value of a life year
VPF	value of prevented fatality = VSL
VSL	Value of statistical life
WTP	Willingness-to-pay
YOLL	Years of life lost (reduction of life expectancy)

References

Abt. 2004. Power plant emissions: Particulate matter-related health damages and the benefits of alternative emission reduction scenarios. Prepared for EPA by Abt Associates, Inc., Bethesda, MD.

ADEME. 2000. Analyse environnementale de systèmes de gestion de déchets ménagers, Phase 1: analyse des parameters determinants pour les impacts environnementaux des différents modules (Environmental system analysis of municipal waste treatment, Phase 1: analysis of the key parameters for the environmental impacts). Study by BIO Intelligence and Ecobilan for ADEME and Eco-Emballages. Agence Française de l'Environnement et de la Maîtrise de l'Energie, Paris, France.

ADEME. 2003. Outil de calcul des émissions dans l'air de CH_4, CO_2, SO_x, NO_x issues des centres de stockage de déchets ménagers et assimilés. (Tool for calculation of emissions of CH_4, CO_2, SO_x, NO_x from municipal and similar solid waste). Agence Française de l'Environnement et de la Maîtrise de l'Energie, Paris, France.

Cardis E, Krewski D, Boniol M et al. 2006. Briefing document: The cancer burden from Chernobyl in Europe. World Health Organization, International Agency for Research on Cancer, Lyon, France, April 2006. http://www.iarc.fr/en/media-centre/pr/2006/IARCBriefingChernobyl.pdf.

Curtiss PS and Rabl A. 1996. Impacts of air pollution: General relationships and site dependence. *Atmospheric Environment* 30, 3331–3347.

Delucchi MA. 2003. A lifecycle emissions model (LEM): Lifecycle emissions from transportation fuels, motor vehicles, transportation modes, electricity use, heating and cooking fuels, and materials. UCD-ITS-RR-03-17, Institute of Transportation Studies, University of California, Davis, CA.

Dreicer M, Tort V, and Margerie H. 1995. Nuclear fuel cycle: Implementation in France. Final report for ExternE Program, contract EC DG12 JOU2-CT92-0236. CEPN, Fontenay-aux-Roses, France.

EC. 2000. Directive 2000/76/EC of the European Parliament and of the Council of 4 December 2000 on the incineration of waste. European Commission, Brussels.

EC. 2001. Directive 2001/80/EC of the European Parliament and of the Council of 23 October 2001 on the limitation of emissions of certain pollutants into the air from large combustion plants. European Commission, Brussels.

EIA. 2012. Levelized cost of new generation resources in the Annual Energy Outlook 2012. US Energy Information Administration. Washington, DC. http://www.eia.gov/forecasts/aeo/electricity_generation.cfm.

ETSU. 1996. Economic evaluation of the draft incineration directive. Report for the European Commission DG11. ETSU, Harwell Laboratory, Oxfordshire, U.K.

ExternE. 1995. *ExternE: Externalities of Energy*. Vol. 1: *Summary* (EUR 16520); Vol. 2: *Methodology* (EUR 16521); Vol. 3: *Coal and Lignite* (EUR 16522); Vol. 4: *Oil and Gas* (EUR 16523); Vol. 5: *Nuclear* (EUR 16524); Vol. 6: *Wind and Hydro Fuel Cycles* (EUR 16525). European Commission, Directorate-General XII, Science Research and Development, Office for Official Publications of the European Communities, Luxembourg.

ExternE. 1998. *ExternE: Externalities of Energy*. Vol. 7: *Methodology* 1998 Update (EUR 19083); Vol. 8: *Global Warming* (EUR 18836); Vol. 9: *Fuel Cycles for Emerging and End-Use Technologies, Transport and Waste* (EUR 18887); Vol. 10: *National Implementation* (EUR 18528). European Commission, Directorate-General XII, Science Research and Development, Office for Official Publications of the European Communities, Luxembourg. Results are also available at http://ExternE.jrc.es/publica.html.

ExternE. 2005. *ExternE: Externalities of Energy, Methodology 2005 Update*. Edited by P. Bickel and R. Friedrich. European Commission, Directorate-General for Research, Sustainable Energy Systems, Office for Official Publications of the European Communities, Luxembourg.

ExternE. 2008. With this reference we cite the methodology and results of the latest phases of ExternE: NEEDS (2004–2008) and CASES (2006–2008). The results can be found in the data CD distributed with the CASES book edited by Markandya A, Bigano A, and Porchia R in 2010: *The Social Cost of Electricity: Scenarios and Policy Implications*, Edward Elgar Publishing Ltd., Cheltenham, U.K. They can also be downloaded from the corresponding website of the CASES project http://www.feem-project.net/cases/project.php. Here we use the numbers of the data CD because they are a fixed reference unlike the website where some numbers have been changed in the meantime.

Frischknecht R, Jungbluth N, Althaus H-J, Doka G, Dones R, Hischier R, Hellweg S, Nemecek T, Rebitzer G, and Spielmann M. 2007. Overview and methodology. Final report ecoinvent data v2.0, no. 1. Swiss Centre for Life Cycle Inventories, Dübendorf, CH. www.ecoinvent.ch

Gagnon L and van de Vate JF. 1997. Greenhouse gas emissions from hydropower: The state of research in 1996. *Energy Policy* 25, 7–13.
Garwin RL. 2005. Europe features: Chernobyl's real toll. United Press International. http://www.fas.org/rlg/051109-chernobyl.pdf.
Hirschberg S, Burgherr P, Spiekerman G, and Dones R. 2004. Severe accidents in the energy sector: Comparative perspective. *Journal of Hazardous Materials* 111, 57–65.
IPCC. 2007. Climate change 2007: Synthesis report. Intergovernmental Panel on Climate Change. Downloaded October 1, 2010 from http://www.ipcc.ch/publications_and_data/ar4/syr/en/contents.html.
Levy JI, Baxter, LK, and Schwartz, J. 2009. Uncertainty and variability in health-related damages from coal-fired power plants in the United States. *Risk Analysis* 29(7), 1000–1014.
Levy JI, Hammitt, JK, Yanagisawa Y, and Spengler JD. 1999. Development of a new damage function model for power plants: Methodology and applications. *Environmental Science and Technology* 33(24), 4364–4372.
Mitchell RC and Carson RT. 1989. *Using Surveys to Value Public Goods: The Contingent Valuation Method.* Resources for the Future, Washington, DC.
NRC. 2010. Hidden costs of energy: Unpriced consequences of energy production and use. National Research Council of the National Academies, Washington, DC. Available from National Academies Press, http://www.nap.edu/catalog.php?record_id=12794.
Ontario Hydro. 1993. *Full Cost Accounting for Decision Making*. Ontario Hydro, Toronto, Ontario, Canada.
ORNL/RFF. 1994. *External Costs and Benefits of Fuel Cycles*. Prepared by Oak Ridge National Laboratory and Resources for the Future. Edited by Russell Lee, Oak Ridge National Laboratory, Oak Ridge, TN.
Pope CA, Burnett RT, Thun MJ, Calle EE, Krewski D, Ito K, and Thurston GD. 2002. Lung cancer, cardiopulmonary mortality, and long term exposure to fine particulate air pollution. *Journal of the American Medical Association* 287(9), 1132–1141.
Preiss P, Friedrich R, and Klotz V. 2008. Report on the procedure and data to generate averaged/aggregated data. NEEDS project, FP6, Rs3a_D1.1—Project no: 502687. Institut für Energiewirtschaft und Rationelle Energieanwendung (IER), Universität Stuttgart, Stuttgart, Germany.
Rabl A. 1996. Discounting of long term costs: What would future generations prefer us to do? *Ecological Economics* 17, 137–145.
Rabl A and Rabl VA. 2013. External costs of nuclear: Greater or less than the alternatives? *Energy Policy* 57, 575–584.
Rabl A, Spadaro J, Bickel P et al. 2004. ExternE-Pol. Externalities of energy: Extension of accounting framework and policy applications. Final report contract No. ENG1-CT2002-00609. EC DG Research. Available at: www.arirabl.org.
Rabl A, Spadaro JV, and Holland M. 2014. *How Much Is Clean Air Worth: Calculating the Benefits of Pollution Control.* Cambridge University Press, July 2014.
Rabl A, Spadaro JV, and van der Zwaan B. 2005. Uncertainty of pollution damage cost estimates: To what extent does it matter? *Environmental Science and Technology* 39(2), 399–408.
Rabl A, Spadaro JV, and Zoughaib A. 2008. Environmental impacts and costs of municipal solid waste: A comparison of landfill and incineration. *Waste Management and Research* 26, 147–162.
Rowe RD, Lang CM, Chestnut LG, Latimer D, Rae D, Bernow SM, and White D. 1995. *The New York Electricity Externality Study.* Oceana Publications, Dobbs Ferry, New York.
Spadaro JV. 1999. Quantifying the damages of airborne pollution: Impact models, sensitivity analyses and applications. PhD doctoral thesis, Ecole des Mines de Paris, Paris, France.
Spadaro JV and Rabl A. 1999. Estimates of real damage from air pollution: Site dependence and simple impact indices for LCA. *International Journal of Life Cycle Assessment* 4(4), 229–243.
Spadaro JV and Rabl A. 2002. Air pollution damage estimates: The cost per kilogram of pollutant. *International Journal of Risk Assessment and Management* 3(1), 75–98.

Spadaro JV and Rabl A. 2004. Pathway analysis for population—Total health impacts of toxic metal emissions. *Risk Analysis*, 24(5), 1121–1141.

Spadaro JV and Rabl A. 2008. Estimating the uncertainty of damage costs of pollution: A simple transparent method and typical results. *Environmental Impact Assessment Review* 28(2), 166–183.

Stern N et al. 2006. The economics of climate change: The Stern Review. Cambridge University Press, Cambridge, U.K. Available at: http://www.hm-treasury.gov.uk/stern_review_report.htm.

Ten Hoeve JE and Jacobson MZ. 2012. Worldwide health effects of the Fukushima Daiichi nuclear accident. *Energy and Environmental Science* 5, 8743–8756.

Tol RSJ. 2005. The marginal damage costs of carbon dioxide emissions: An assessment of the uncertainties. *Energy Policy* 33, 2064–2074.

Section II

Energy Generation through 2025

14

Distributed Generation Technologies through 2025

Aníbal T. de Almeida and Pedro Soares Moura

CONTENTS

14.1 Distributed Generation Technologies

14.1.1 Introduction

Distributed generation (DG) can be defined as a source of electric power connected to a distribution network or a customer site, representing an innovative and efficient way to both generate and deliver electricity, since it generates electricity right where it is going to be used. Nowadays, technological improvements allow power generation systems to be built in smaller sizes with high efficiency, low cost, and minimal environmental impact.

DG can serve as a supplement to electricity generated by huge power plants and delivered through the electric grid. However, located at a customer's site, DG can also be used to manage energy service needs or help meet increasingly rigorous requirements for power quality (PQ) and reliability. DG has the potential to provide site-specific reliability improvement, as well as transmission and distribution (T&D) benefits including: shorter and less extensive outages, lower reserve margin requirements, improved PQ, reduced lines losses, reactive power control, mitigation of T&D congestion, and increased system capacity with reduced T&D investment.

DG also provides economic benefits because its technologies are modular and provide location flexibility and redundancy as well as short lead times. Economic benefits can also be gained by using DG technologies for peak-shaving purposes, for combined heat and power (CHP) (cogeneration), and for standby power applications. In addition, many DG technologies provide environmental benefits including reduced land requirements, lower or no greenhouse gas emissions, and lower environmental compliance costs.

DG technologies can be divided in two different kinds, according to their availability: firm and intermittent power. The firm power technologies are those that enable the power control of DG units that can be managed as a function of the load requirements. Firm DG plants can be utilized as backup, working only in situations of grid unavailability, in periods of high consumption (when the electricity is more expensive), working continuously or dispatched to meet the variable load in an optimal manner. The intermittent power technologies do not allow the management of the produced energy by themselves having a random generation character. Examples of this kind of technologies are wind power or solar power that only produces energy when the wind or the sun is available. These technologies can be installed aggregated with energy storage, which by filtering the energy generation fluctuation enables the management of the delivered energy by the combined system.

In DG applications, traditional technologies can be used, such as internal combustion engines (ICEs), gas turbines, and in large installations steam turbines and combined cycle turbines. Other kinds of technologies such as microturbines, Stirling engines, fuel cells or renewable energies, including solar power, geothermal power, or wind power, can also be utilized (Figure 14.1).

14.1.2 Internal Combustion Engines

ICEs were one of the first technologies that used fossil fuels for electricity generation. Developed more than a century ago, ICEs are the most common of all DG technologies. They are available from sizes of a few kW for residential backup generation to generators on the order of 10 MW.

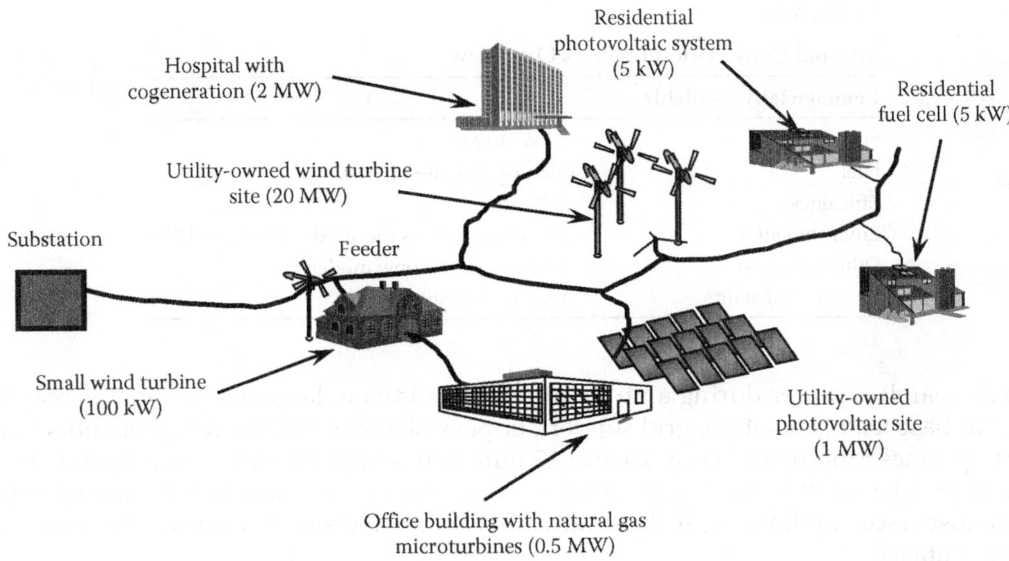

FIGURE 14.1
Power system with multiple energy sources.

An ICE uses the thermal energy of fuel combustion to move a piston inside a cylinder, converting the linear motion of the piston to rotary motion of a crankshaft and uses that rotation to turn an AC electric generator. ICEs are also called reciprocating engines because of the reciprocating linear motion of the pistons. ICEs can be fuelled with gasoline, natural gas, diesel fuel, heavy oil, biodiesel, or biogas.

The two main types of ICEs used for DG applications are

1. Four cycle-spark-ignited engines (Otto cycle), which use an electrical spark introduced into the cylinder. This explosion engine, or ignition by spark, that uses the Otto cycle, was invented by Nikolaus August Otto, in 1867. Fast-burning fuels, such as gasoline and natural gas, are commonly used in these engines. Biofuels such as alcohols and biogas may also be used.
2. Compression-ignited (diesel cycle) engines, in which the combustion is initiated by compression. This engine was developed by Rudolph Diesel, in 1892, in which the compression of the fuel–air mixture inside the piston cylinder raises it to a temperature where it spontaneously ignites work best with slow-burning fuels, such as diesel. Biofuels, such as biodiesel and vegetable oils, may also be used.

DG engines haves efficiencies that range from 25% to 45% (Table 14.1). In general, diesel engines are more efficient than Otto engines because they operate at higher compression ratios. Engine manufacturers are targeting lower fuel consumption and shaft efficiencies up to 50%–55% in large engines (>1 MW). Efficiencies of Otto engines using natural gas are expected to improve and approach those of diesel engines.

ICE generators for distributed power applications, commonly called gensets, are found universally in sizes from less than 5 kW to over 10 MW. Gensets are frequently used as a backup power supply in residential, commercial, and industrial applications. When used in combination with a 1–5 min UPS (uninterruptible power supply), the system is able to

TABLE 14.1

Internal Combustion Engines Overview

Commercially Available	Yes
Size range	5 kW–10 MW
Fuel	Natural gas, diesel, heavy fuel, biogas
Efficiency	25%–45%
Environmental	Emission controls required for NOx and CO
Other features	Cogeneration (some models)
Commercial status	Widely available

supply seamless power during a utility outage. In addition, large ICE generators may be used as base load generation, grid support, or peak-shaving devices. ICE generators have start-up times ranging between 0.5 and 15 min, and a high tolerance for frequent starts and stops. The smaller engines, available in sizes as small as a few kW, are intended for much dispersed applications, such as individual homes and small businesses to cope with power outages.

14.1.3 Gas Turbines

Gas turbines are an electricity production technology used very often in the last decades. The first studies of gas utilization to actuate turbines started at the end of the nineteenth century. However, the first efficient gas turbines started to operate in 1930. A gas turbine consists of a compressor, a combustion chamber, and a turbine coupled to the generator. The turbines with only one axis have all the pieces associated with a continuous axis, all rotating at the same speed. This kind of architecture is used when variations in the turbine speed are not foreseeable. The rotor that drives the generator can be mechanically separated from the rotor driven by the combustion of gases, with more flexibility in the operation speed.

Unlike ICEs, the gas turbines work in continuous process and not in a repetition of a sequence of different operations. However, the operation can be viewed as a set of four stages similar to the four strokes of the ICEs (Figure 14.2).

1. A compressor drives a rotor that directs the work fluid (air) to the combustion chamber, where the air is compressed, increasing the pressure up to 10 bar and the temperature to 300°C.
2. The compressed air is mixed with the burning fuel, achieving temperatures of 1250°C. This combustion occurs with controlled conditions to maximize the fuel efficiency and minimize the emissions.
3. The air at high pressure is passed through the turbine, which converts the air energy into mechanical energy. Part of this energy is transmitted to the compressor and the remainder is used for electricity generation through a generator.
4. The exhaust gases are released to the atmosphere, or may be used for generation of process heat, or to increase the electricity generated as described in the following text.

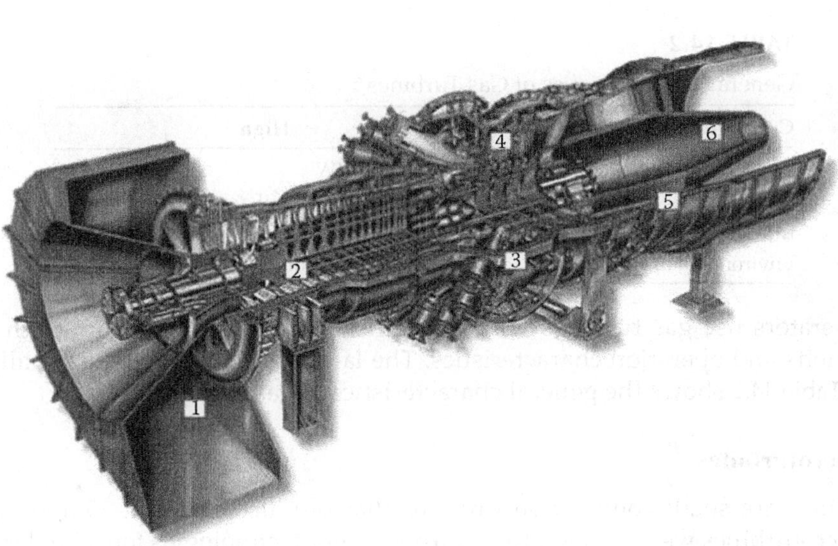

FIGURE 14.2
Combustion turbine. (1) Air intake section; (2) compression section; (3) combustion section; (4) turbine section; (5) exhaust section; and (6) exhaust diffuser. (From Siemens, Combustion turbine, 2015, http://www.energy.siemens.com/hq/en/fossil-power-generation/, accessed October 12, 2015. With permission.)

Since gas turbines produce a large volume of exhaust gases at high temperatures, the energy of these gases can be utilized for steam production for industrial processes (cogeneration) or for electricity production through combined cycle. In a combined cycle, the gas turbine is used as the first cycle, where the exhaust gases of the gas turbine are used to produce steam in a heat recovery steam generator. This steam is then used to drive a steam turbine, increasing the electrical global efficiency of the system to values up to 60%.

In the Cheng cycle, the steam is injected in the expansion chamber of the gas turbine (superheated steam injection). In the expansion chamber, the steam is mixed with the gases of the combustion, which expands and produces additional work increasing the electrical efficiency.

The conversion of mechanical energy into electricity is made almost always through synchronous generators. In DG applications, the rotation speed of the turbines can be higher than the generator synchronous speed which requires a gearbox, reducing the conversion efficiency by about 3%. The generator also works like an auxiliary engine to start the turbine.

Natural gas is the fuel that enables the best efficiency to be reached in gas turbines. However, gas turbines can work with other fuels, such as fuel oil, diesel, propane, J-5 (used in aeronautics), kerosene, methane, and biogas. The heavy oil utilization decreases the efficiency and the power of the turbine by 5%–8%. Since heavy oil is less expensive, it can decrease the electricity production costs, but the emissions are higher than with other fuels.

The gas turbine generators are available in a wide power range, corresponding at three types of generators:

1. Microturbines (20–500 kW)
2. Medium turbines (500–10,000 kW)
3. Large turbines (more than 10 MW)

TABLE 14.2
General Characteristics of Gas Turbines

Commercial Availability	High
Size range	500 kW–250 MW
Fuel	Natural gas, biogas, oil derivatives
Efficiency	25%–45%
Environmental emissions	Very low when controls are used, high noise

These generators use gas turbines with the same working principle, but with different configurations and operation characteristics. The large turbines are not normally considered DG. Table 14.2 shows the general characteristics of gas turbines.

14.1.4 Microturbines

Microturbines are small combustion turbines that produce between 25 and 500 kW of power. Microturbines were derived from turbocharger technologies found in large trucks or in the turbines found in aircraft auxiliary power units. Only the largest class of gas turbine generators, those made for central station utility applications, are designed specifically for electric power production. Microturbines have undergone significant innovations, enabling the energy production with high quality and reliability, with low greenhouse gas emissions, and with moderate costs, thus becoming a competitive technology.

Microturbines have the same working principle as the gas turbines, with various modifications in the system configuration. In these turbines the rotor rotates at a very high speed (up to 100,000 rpm) and one of the innovations consists in the adoption of a unique shaft, on which the compressor, the turbine, and the generator are assembled (Figure 14.3).

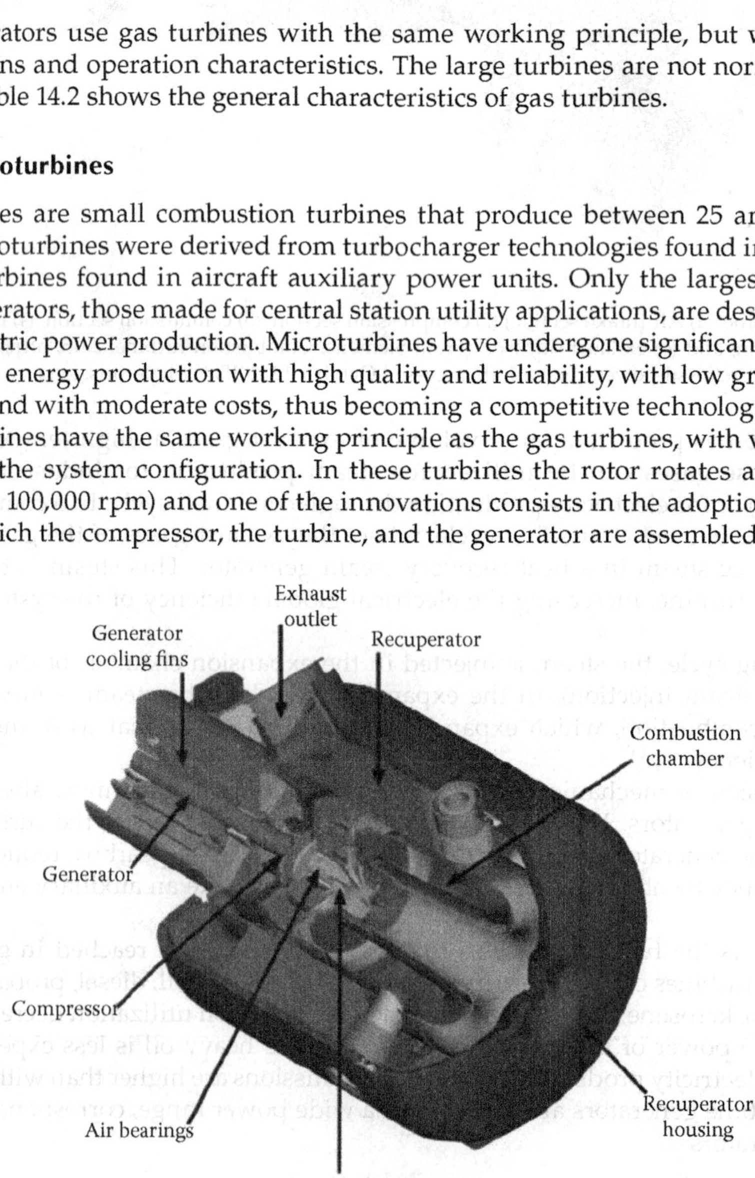

FIGURE 14.3
Microturbine. (From Capstone, Microturbine, 2005, http://www.capstoneturbine.com/, accessed October 12, 2005. With permission.)

These systems eliminate the gearbox, reducing the cost and increasing the reliability, but with a reduced overall efficiency. Another innovation is the utilization of air bearings, avoiding the need of a fluid for refrigeration and lubrication, because the unique used element is the air. The air can be continuously renovated and will never be contaminated by the materials wastage and by the combustion products.

One of the key characteristics of the microturbines is the heat recovery, that utilizes the thermal energy of the exhaust gases (at high temperatures), for preheating the air supply to the compressor. The mechanical energy is converted to electrical energy by a permanent magnet AC generator, which includes a low inertia rotor rotating at the turbine speed.

Because of the high rotor speed, the AC output has a frequency of approximately 2 kHz. To connect the microturbine generator with a 50 or 60 Hz grid, the microturbine voltage output must be connected to an AC–DC–AC converter. In this converter, the microturbine voltage output is rectified, filtered, and converted to an AC voltage through an inverter system synchronized with the 50 or 60 Hz supply.

Microturbines can also operate with a wide variety of fuels, such as natural gas (at high or low pressure), propane, diesel, gasoline, biogas (methane), or kerosene. The electrical efficiency of microturbines is between 20% and 35%, with heat recovery utilization. If the heat recovery does not exist, this value can decrease to 15%. In cogeneration systems, the global efficiency can reach 85% (Table 14.3).

Microturbine generators can be divided into two general classes:

- Heat recovery microturbines recover heat from the exhaust gas to boost the temperature of the air stream supplied to the combustion and increase the efficiency. Further exhaust heat recovery can be used in a cogeneration configuration.
- Microturbines without heat recovery (or simple cycle) have lower efficiencies, but also have lower capital costs.

Other applications of microturbine technology include the following:

- Core power conversion element of vehicles, such as buses, trucks, helicopters, and so on. Automotive companies are interested in microturbines to provide a lightweight and efficient fuel-based energy source for hybrid electric vehicles.
- Standby power, PQ, peak shaving, and cogeneration applications. Some types of microturbines are well-suited for small commercial building establishments, such as restaurants, hotels, small offices, retail stores, and many others.
- Utilization of by-products of processes in oil-processing, gas-transferring, petroleum production, industrial waste utilization for the purpose of optimizing the use of natural gas, associated gas, biogas, landfill gas, and others.

TABLE 14.3

Microturbines Overview

Commercial Availability	**Yes (Only a Few Manufacturers)**
Size range	30–1000 kW
Fuel	Natural gas, hydrogen, propane, diesel
Efficiency	20%–35% (recuperated)
Environmental emissions	Low (<9–50 ppm) NOx
Other features	Cogeneration (50°C–80°C water)
Commercial status	Medium volume production

The development of microturbines is ongoing in a variety of areas:

- Heat recovery/cogeneration
- Use of waste heat for absorption cooling
- Increase of the efficiency
- Fuel flexibility
- Vehicles
- Hybrid systems (e.g., fuel cell/microturbine, flywheel/microturbine)

14.1.5 Stirling Engines

Stirling engines (Figure 14.4), invented in 1816 by Robert Stirling, are a class of reciprocating piston engines and are classed as external combustion engines. They constitute an efficient thermodynamic machine, for the direct conversion of heat into mechanical work, with a theoretical efficiency of 40%. Stirling engines were commonly used prior to beginning of the twentieth century. As steam engines improved and the competing compact Otto cycle engine was invented, Stirling engines lost favor. The developments in DG and concentrated solar thermal power have revived interest in Stirling engines. The principles of Stirling engines are described in Chapter 42 and will not be covered here.

The operation is reversible, that is, by supplying thermal energy, mechanical energy is produced, and by supplying mechanical energy, thermal energy is produced. Stirling

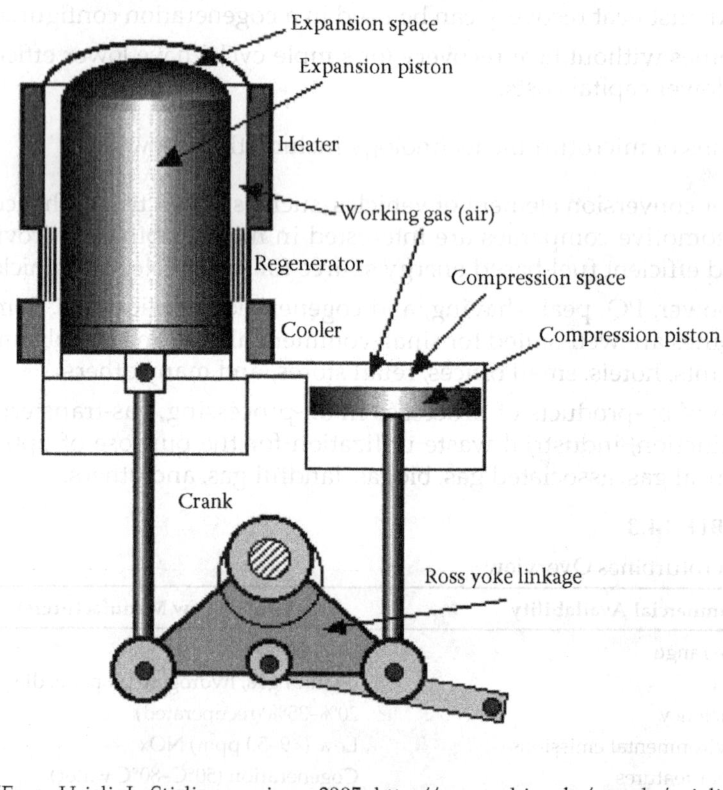

FIGURE 14.4
Stirling engine. (From Urieli, I., Stirling engines, 2005, http://www.ohio.edu/people/urieli/stirling/engines/engines.html, accessed October 12, 2005.)

TABLE 14.4
Stirling Engines Overview

Commercially Available	**On a Limited Scale**
Size range	<1 to 25 kW
Fuel	Fuel flexibility—fossil or renewable heat is possible
Efficiency	12%–30%
Environmental	Potential for low emissions
Other features	Cogeneration (some models)
Commercial status	Availability for specialized applications

engines can be fuelled by any source of heat (fossil fuel or renewable) and some models have possibilities to perform cogeneration. The continuous combustion process in a Stirling engine, which is easier to optimize and control, results in lower emissions compared to the intermittent explosions of fuel–air mixtures in internal combustion Otto and diesel engines.

Usually, Stirling engines are found in sizes from 1 to 25 kW (Table 14.4) and are produced in small quantities for specialized applications. Large 25 kW Stirling engines have an electrical efficiency of approximately 30%. The primary challenges faced by Stirling engines over the last two decades have been their long-term durability/reliability and their relatively high cost.

Stirling engines are ideally suited for solar thermal power. Using a gas as an operating fluid, there is no practical limit placed on the solar unit's upper temperature due to its operating fluid. Maximum temperature would be limited only by the materials used in its construction.

Stirling engine developments have been directed at a wide range of applications, including

- Small scale—Residential or portable power generation.
- Solar dish applications—Heat reflected from concentrating dish reflectors is used to drive the Stirling engine.
- Vehicles—Auto manufacturers have investigated utilizing Stirling engines in vehicles to improve the fuel economy.
- Refrigeration—Stirling engines are being developed to provide cooling for applications such as microprocessors and superconductors.
- Aircrafts—Stirling engines could provide a quieter-operating engine for small aircraft.
- Space—Power generation units aboard space ships and vehicles.

14.1.6 Fuel Cells

Fuel cells are a technology for power generation that is quiet and highly efficient with no moving parts. Fuel cells generate electricity through an electrochemical process in which the energy stored in a fuel is converted directly into DC electricity and thermal energy. The chemical energy typically comes from hydrogen contained in various types of fuels (hydrocarbon fuels, such as natural gas, methanol, ethanol, biogas, etc.), including pure hydrogen.

Several hundred phosphoric acid fuel cells (PAFCs) demonstration and test plants have been built in the mid-1990s to early twenty-first century, mostly with 200 kW capacity, appropriate for DG applications in many commercial buildings to provide premium PQ for demanding loads. The operating temperature is about 200°C, which is suitable for cogeneration applications in buildings and in small industrial plants. They do not offer opportunity of self-reforming and they require precious platinum for their catalyst. PAFCs' efficiency and peak output capability deteriorate by about 2% per year.

One of the most promising developments is the design of hybrid systems. Solid oxide fuel cells (SOFCs) have high efficiency, particularly in a combined cycle operation mode, in which they can surpass conventional combined cycle gas turbine plants. High efficiencies under part load operation also result in high overall efficiency.

For applications in industrial heat-power cogeneration and public electricity supply, high-temperature fuel cells, such as SOFCs and molten-carbonate fuel cells (MCFCs) are most suitable. Both systems permit the use of a wide range of fuels. Such fuel cell systems compete in the lower rating range with gas turbines and motor cogeneration plants, and in the upper rating range with combined gas and steam turbine power plants. Conventional plants have a clear advantage in terms of practical experience and in terms of comparatively low capital costs compared with fuel cell plants. High-temperature fuel cells are likely to gain market penetration due to a decrease in specific need for primary fuels and also due to a sharp decrease of specific pollutant emissions, when compared with conventional generation. This last factor is a key advantage in DG applications in urban areas.

A detailed description of the fuel cells technology is given in Chapter 53.

14.1.7 Solar Photovoltaic

Photovoltaic (PV) cells, or solar cells, convert sunlight directly into electricity. PV technology has several applications, including

- Off grid/remote
- Grid attached residential and commercial buildings
- Remote communication systems
- Central power plants (above 1 MW)

Traditionally, PV has been used in remote areas where the grid is not available; such systems store electricity in batteries for use when the sun is not shining and are called stand-alone power systems. PV-generated power is less expensive than conventional power where the load is small and the area is too difficult to serve by electric utilities. However, solar power is now appearing more in urban areas due to the decreasing costs of PV panels and due to policy mechanisms to promote PV generation. Here, the surplus or all the generated electricity is injected into the grid. These are called grid-connected solar systems.

Distributed PV systems that provide electricity at the point of use have reached widespread commercialization. Chief among these distributed applications are PV power systems for individual buildings. Interest in the building integration of PVs, where

FIGURE 14.5
Building-integrated photovoltaics in a façade and in a roof.

the PV elements actually become an integral part of the building, often serving as the façade or exterior weather skin, is growing worldwide. PV specialists and innovative designers in Europe, Japan, and the United States are now exploring creative ways of incorporating solar electricity into their work. A building-integrated photovoltaics (BIPV) system consists of integrating PVs modules into the building envelope, such as the roof or the façade (Figure 14.5). By simultaneously serving as building envelope material and power generator, BIPV systems can provide savings in materials and in electricity costs.

PVs can be integrated into many different assemblies within a building envelope:

- Incorporated into the façade of a building, complementing or replacing traditional glass
- Incorporated in the external layers of the wall of a building façade
- Use in roofing systems providing a direct replacement for different types of roofing material
- Incorporated in skylight systems in which part of the solar light is transmitted to the inside of the building and the other part is converted into electricity.

PV is the most modular and operationally simple of the clean distributed power technologies, with benefits that include the ability to provide power during summer peak periods, distribution congestion benefits, environmental benefits, and reduced fuel price risk.

A detailed description of the PV technology is given in Chapter 44.

14.1.8 Wind Power

Wind energy became a significant research area in the 1970s during the energy crisis and the resulting search for potential renewable energy sources. Modern wind turbine technology has made significant advances over the last 25 years. Today, wind power technology is available as a mature, environmentally sound, and convenient alternative. Generally, individual wind turbines are grouped into wind farms containing several turbines. Many wind farms are MW scale, ranging from a MW to tens of MW. Wind turbines

FIGURE 14.6
Building-integrated wind power.

may be connected directly to utility distribution systems and the larger wind farms are often connected to transmission lines.

The small-scale wind farms and individual units are typically defined as distributed generation. Residential systems (1–15 kW) are available (Figure 14.6). However, they have low efficiency and are generally not suitable for urban or small-lot suburban homes due to space requirements and reduced wind velocity in urban environments. Single small turbines, below 50 kW, are also available and used in telecommunications stations, or for water pumping.

A detailed description of the wind power technology is given in Chapter 37.

14.1.9 Cogeneration

CHP, also known as cogeneration, is an efficient, clean, and reliable approach to producing both electricity and usable thermal energy (heating and/or cooling) at high efficiency and near the point of use, from a single fuel source. Since CHP is highly efficient, it reduces traditional air pollutants and carbon dioxide emissions, the leading greenhouse gas associated with climate change, as well.

CHP can use a variety of technologies to meet an energy user's needs. The range of technologies available allows the design of cogeneration facilities to meet specific onsite heat and electrical requirements. CHP systems consist of a number of individual components—prime mover, generator, heat recovery, and electrical interconnection—configured into an integrated system. The type of equipment that drives the overall system (i.e., the prime mover) typically identifies the CHP system.

Typical CHP prime movers include

- Combustion turbines
- Reciprocating engines
- Boilers with steam turbines
- Microturbines
- Fuel cells

CHP may be used in a variety of applications ranging from small 10 kW systems to very large utility-scale applications approaching 1000 MW. The first step in assessing which CHP application is right for a particular facility is to identify whether there is

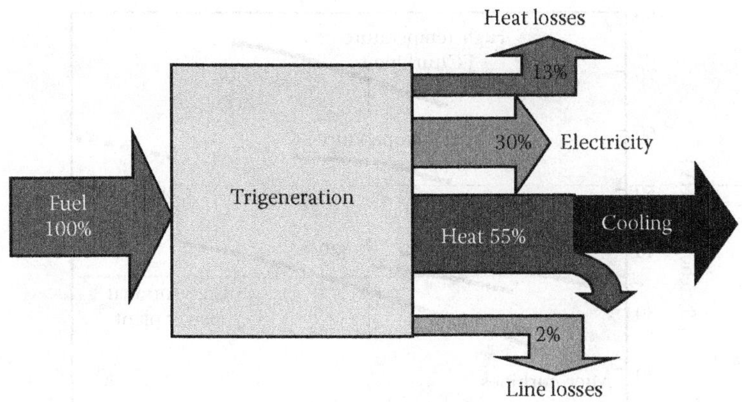

FIGURE 14.7
Trigeneration technology.

coincident demand of electrical and thermal energy at the host site. The CHP project will be most economically viable when the system provides the maximum amount of energy that can be used. Therefore, CHP project development begins with an analysis of site electrical and thermal load profiles. Based on these profiles, the type of CHP technology which most closely matches the facility's power and demand will be chosen.

In developed countries, about 10% of all electricity is generated in CHP plants, leading to huge primary energy savings and reduction of emissions. CHP in some industries, such as the pulp and paper, uses biomass byproduct as fuel input.

Trigeneration can provide even greater efficiency than cogeneration. Trigeneration is the conversion of a single fuel source into three useful energy products: electricity, steam or hot water, and chilled water (Figure 14.7). Trigeneration converts and distributes up to 90% of the energy contained in the fuel burned in a turbine or engine into usable energy. Introducing an absorption chiller into a cogeneration system means that the site is able to increase the operational hours of the plant with an increased utilization of heat, particularly in summer periods. Trigeneration has been applied with very positive results in buildings, such as hotels and hospitals, which feature a large space conditioning load during most of the year.

A detailed description of cogeneration technologies is given in Chapter 32.

14.1.10 Conclusions

The utilization of DG technologies enables the creation of a power system with multiple energy sources, allowing the integration of conventional central power plants with dispersed DG fossil fuel–based generation, as well as with dispersed DG renewable generation. It is anticipated that DG growth and its large-scale application may lead to an improvement in reliability, to improved security of supply, to the decrease of power costs, and to the minimization of environmental impacts. In Figure 14.8 and Table 14.5, a characterization of the most important DG technologies is presented showing typical parameters associated with each technology.

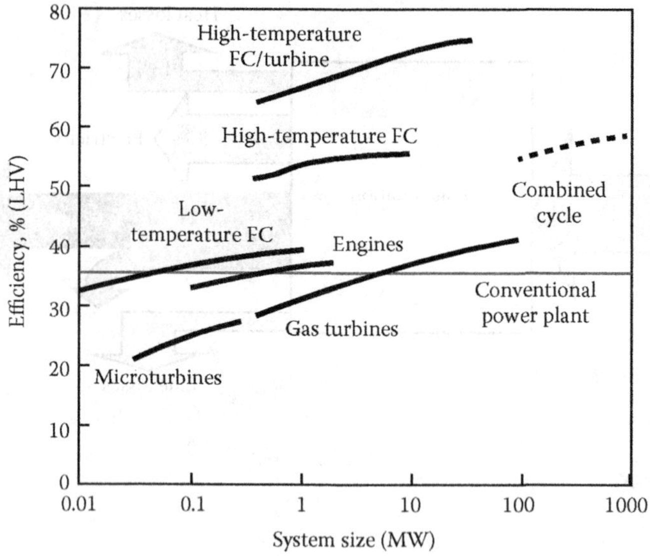

FIGURE 14.8
DG technologies comparative efficiency range.

14.2 Integration of Distributed Generation and Smart Grids

14.2.1 Introduction

Connecting a distributed power system to the electrical grid has potential impacts on the safety and reliability of the grid, which is one of the most significant barriers to the installation of DG technologies. Electric utilities have understandably always placed a high priority on the safety and reliability of their electrical systems. Faced with the interconnection of potentially large number of distributed generators, utilities have perceived DG as a threat. This has led some utilities to place overly conservative restrictions on interconnected systems, causing added costs that may make an installation economically unfeasible. Several techniques may reduce adverse network impacts allowing DG connection, but those techniques can be project specific and may be expensive, and adversely affect project economics.

Connection of DG fundamentally affects the operation of distribution networks with changes and impacts such as

- Voltage fluctuations
- Increased fault levels
- Degraded protection
- Bidirectional power flow
- Altered transient stability

To reduce the impact in the power grid, several requirements are needed. Typical requirements include equipment that prevents power from being fed to the grid when the grid

TABLE 14.5

DG Technologies Characterization

	Internal Combustion Engines	Gas Turbines	Microturbines	Stirling Engine	Fuel Cell	Photovoltaic	Wind Power
Power range (kW)	5–10,000	500–250,000	30–1000	1–25	1–10,000	0.07–1000	0.3–5000
Electric efficiency (%) (LHV)	25–45	25–45	20–35	12–30	30–70	5–15	25–40
Efficiency with partial load	Reasonable until 35%–40% of the rated load	Reasonable until 40% of the rated load	Bad below 40% of the rated load	(a)	Good/Reasonable until 35%–40% of the rated load	(a)	(a)
Load following capacity	Very good	Good	Reasonable/low (b)	(a)	Very good	(a)	(a)
Start time	10 s–15 min	2 min–1 h	60 s	(a)	(a)	NA	(a)
Availability (%)	90–98	90–98	90–98	90–98	90–95	5–25 (c)	10–40 (c)
Interval between the maintenance stops (×1000 h)	0.5–2	30	5–8	(a)	10–40	NA	4
Useful lifetime (years)	15–20	20–25	10	15–20	20	20–30	20
Fuel flexibility	Good	Good	Good	Excellent	Good	NA	NA
Noise	High	Moderate to high	Moderate	Low	Low	NA	Moderate
Commercial availability	High	High	Moderate	Very low	Low	High	High
Acquisition costs (€/kW)	300–900	300–1,000	900–1,250	2,000–50,000	3,000–4,000	2,250–3,500	1,350–7,000

Notes: NA, not applicable.
(a) Dependent upon conditions; (b) despite present a high potential to load following, the actual models do not present a good fulfillment at this level; (c) depending on the climate condition at the location.

is de-energized, manual disconnects and PQ requirements such as limits on the interconnected system's effects on "flicker," harmonic distortion, and other types of waveform disturbance. Systems may also be required to automatically shut down in the event of electrical failures, to provide an isolation transformer for the system, as well as to provide liability insurance.

Interconnection requirements for large DG installations (~10 MW) are well understood because they are very similar to the interconnections required for central power stations. Interconnection requirements for smaller installations are more difficult because the utility must balance the desire for a safe interconnection with the plant owner's desire to have a "quick and easy" interconnection design to get the DG up and running. Interconnection complexity generally increases with project size and is technology dependent.

The grid interconnection is important for three main reasons:

1. The number of small generators seeking interconnection to the grid has been increasing.
2. DG advocates contend that the current interconnection requirements and processes are effectively increasing costs unfairly and pricing DG out of the market.
3. Distribution companies are concerned that DG will negatively impact the safety and reliability of the grid and unfairly increase the distribution companies' costs.

14.2.2 Power Distribution

Electric grid is broadly divided into two systems: the transmission system that transfers bulk power at high voltages, from power plants to utility-owned substations and a few very large customers, and the distribution system that delivers power at medium and low voltages from the substation to the majority of customers.

The utility distribution systems can be categorized as either radial or networked. Power system design is a trade-off between complexity and cost in order to maximize economy and reliability. As a result, the general structure of the power delivery system has a networked nature at the transmission level and a more radial nature at the distribution level.

Radial distribution refers to a system where the power lines extend from a common substation to the customer loads coming off at single nodes along the line (Figure 14.9). In these distribution systems, power can only flow in one direction: from the substation

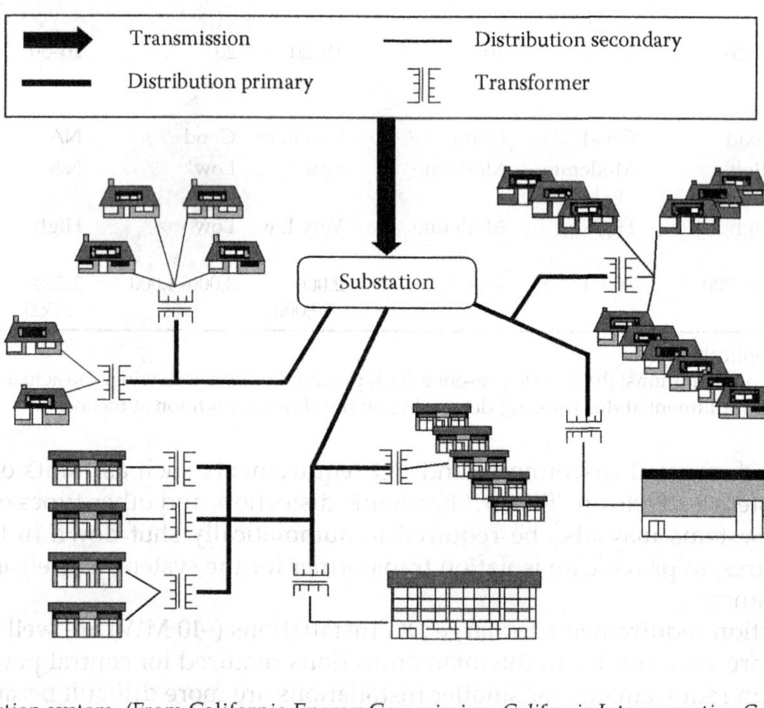

FIGURE 14.9
Radial distribution system. (From California Energy Commission, *California Interconnection Guidebook: A Guide to Interconnecting Customer-Owned Electric Generation Equipment to the Electric Utility Distribution System Using California's Electric Rule 21*, California Energy Commission, Sacramento, CA, 2003, http://www.energy.ca.gov/reports/2003-11-13_500-03-083F.PDF. With permission.)

to a load. While there can be many radial distribution lines emanating from a substation, each load is typically served by only one line. A disruption to that feed or substation will typically affect all customer loads on that line. A radial system generally offers a less reliable power source than a networked system because it lacks redundancy. However, the radial system and its protection equipment are less complex and less expensive than the networked system. The introduction of an energy source such as DG within the radial distribution system will affect the load distribution in the system, and may even cause reverse power flow if it is large relative to the load. Introduction of a sufficiently large power source within the radial distribution normally requires some modification to the protection system.

Networked distribution refers to a system where numerous separate lines form a grid so that customer loads can tap off of multiple independent feeds, which are then tied to a common bus on the secondary side of the transformers (Figure 14.10). These can be separate lines from a common substation or they can be from independent substations.

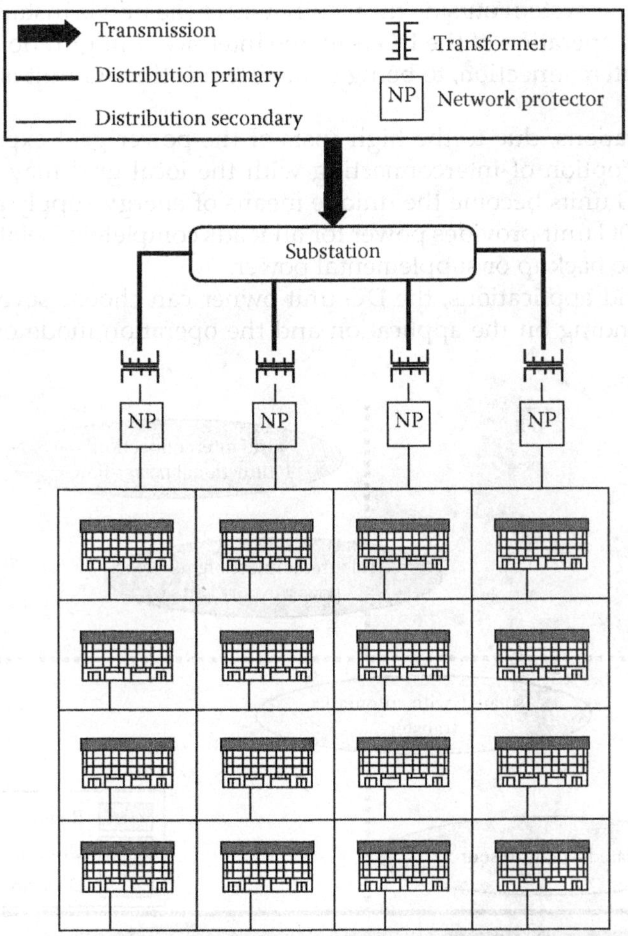

FIGURE 14.10
Networked distribution system. (From California Energy Commission, *California Interconnection Guidebook: A Guide to Interconnecting Customer-Owned Electric Generation Equipment to the Electric Utility Distribution System Using California's Electric Rule 21*, California Energy Commission, Sacramento, CA, 2003, http://www.energy.ca.gov/reports/2003-11-13_500-03-083F.PDF. With permission.)

The networked system offers reliability advantages over the radial system because it provides multiple power sources for loads. This multipath design is sometimes referred to as a "looped" system. These systems use network protectors, which quickly isolate faults to protect the grid and shift customer loads onto the remaining feeds. Understandably, utilities are reluctant to allow the interconnection of anything that they feel will endanger the integrity or safety of this system. System protection in a networked distribution system is more complex and expensive than in the radial distribution system, due to the extra intelligence needed for reliable, effective protection. Since the networked system is specifically designed to deliver energy from multiple transformers to loads, it is capable of dealing with reverse power flow.

14.2.3 Types of Grid Connections

The electric power system (EPS) interface is the means by which the DG unit electrically connects to the power system outside the facility where the unit is installed. Depending on the application and operation of the DG unit, the interface configuration can range from a complex parallel interconnection, to being nonexistent if the DG unit is operated in isolation (Figure 14.11).

In remote applications, due to the high costs of the power grid expansion to the consumption site, the option of interconnecting with the local grid may be impractical. In these cases, the DG units become the unique means of energy supply at low cost. In this configuration the DG unit provides power for all loads completely isolated from grid, providing the utility no backup or supplemental power.

In near to the grid applications, the DG unit owner can choose several types of interconnections. Depending on the application and the operation mode of the DG unit, the

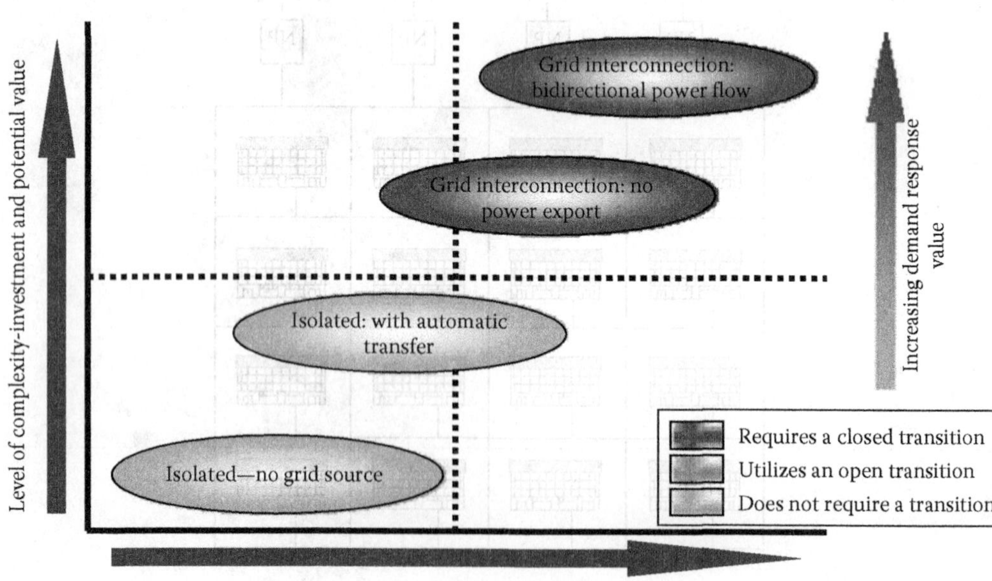

FIGURE 14.11
Interface complexity vs. interaction. (From Little, A.D., Distributed generation: System interfaces, ADL Publishing, Boston, MA, 1999, http://www.encorp.comdwnld/pdf/whitepaper/ADLittleWhitePaperDGSystemInterfaces.pdf. With permission.)

connection system with the grid can represent a complex parallel interconnection or can be nonexistent if the DG is operating isolated from the grid. The complexity of the interconnection systems increases with the required interaction level between the DG unit and the distribution grid.

For most customers, DG systems are most cost-effective and efficient when they are interconnected with the utility grid. In simplest terms, interconnected with the grid means that both the DG system and the grid supply power to the facility at the same time. Paralleled systems offer added reliability, because when the DG system is down for maintenance, the grid meets the full electrical load, and vice-versa.

DG systems can be designed to keep a facility up and running without an interruption if the grid experiences an outage. Also, grid-interconnected systems can be sized smaller to meet the customer's base load as opposed to its peak load. Not only is the smaller base load system cheaper, it also runs closer to its rated capacity and, therefore is more fuel efficient and cost effective.

Two different types of grid interconnection are possible: parallel or roll-over. With the parallel operation, the DG system and the grid are interconnected and both are connected to the load. In the roll-over operation, the two sources are interconnected, but only one is connected with the load.

A typical interconnection system includes three kinds of equipment:

1. Control equipment for regulating the output of the DG
2. A switch and circuit breaker (including a "visible open") to isolate the DG unit
3. Protective relaying mechanisms to monitor system conditions and to prevent dangerous operating conditions

14.2.3.1 Isolated Operation

In remote applications, the DG units become the unique means of energy supply at low cost. In this configuration, the DG unit provides power for all loads completely isolated from grid, providing the utility no backup or supplemental power (Figure 14.12). Isolated operation is also possible in sites which are normally connected to the grid, but in which continuous supply is required in the event of an outage. Some generating facilities, such as a hospital emergency generator, power the customer's partial or entire load isolated from the utility.

Isolated operation involves no interaction with the utility's distribution system, since the generator does not operate in parallel with the utility. In some isolated systems, the

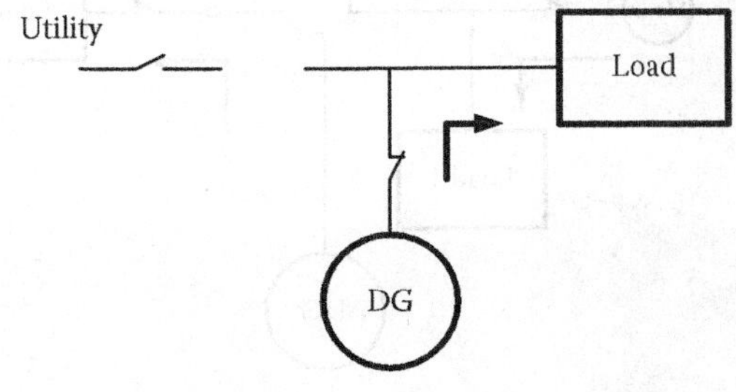

FIGURE 14.12
Isolated operation.

generator is sized for a specific load that is always powered from the generator and never from the utility. There are two ways of transferring load to isolated operation:

1. A break-before-make transfer switch (also known as open transition switching) disconnects the load from the utility prior to making the new connection with the on-site electric generating facility
2. A momentary-parallel (or closed transition) switch, a control system starts the customer's generator and parallels it with the utility's distribution system, quickly ramps the generator output power to meet the customer's load demand and then disconnects the load from the utility.

14.2.3.2 Roll-Over Operation

When a roll-over connection exists (Figure 14.13), the load can be connected only to one of the two sources (grid or DG system) at any given moment. Both the sources are connected to a load control center with a load transfer switch. When the source that is feeding the load fails, this device makes the commutation, ensuring that the other source feeds the load. This commutation can be automatic or manual, to allow the interchange between the two sources due to technical or economic reasons, even when a failure does not exist. In this configuration, the DG unit provides power to load 2 for peaking, baseload, or backup power and the utility provides power to load 1 and occasionally to load 2.

The automatic commutation devices achieve the fast transition between sources, but there is always a time period in which the load is not fed. This kind of connection does not allow the decrease of the frequency of interruption, but reduces considerably their duration. To eliminate the feeding interruption the installation of energy storage devices, between the switch and the load, would be necessary. In this configuration, the transfer time of the load switch is typically 0.1–0.15 s.

The roll-over operation is cheaper and simpler, because it does not need a high number of control, protection, and coordination equipment. This type of operation can easily ensure the impossibility of the DG system injecting energy into the grid when the grid is out of service. This phenomenon called backfeed or islanding can be harmful to the grid operation, and can put people and goods at risk. For example, if a line is disconnected due to technical reasons a worker can be electrocuted, thinking that the line does not have voltage.

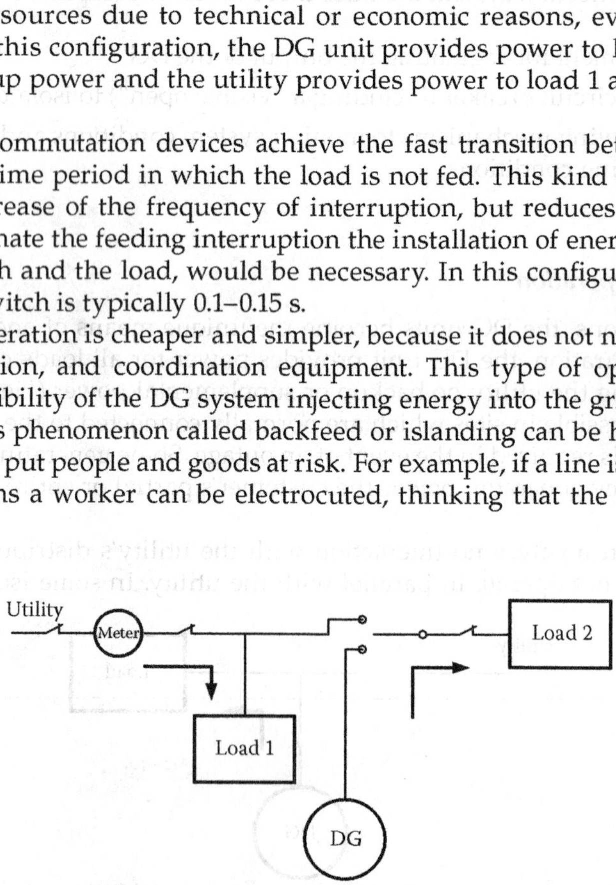

FIGURE 14.13
Roll-over operation.

14.2.3.3 Parallel Operation

When a parallel operation exists, the sources are interconnected and both are connected with the load. If one of the sources fails, the load passes to be instantaneously fed exclusively by the other, without any interruption in the load supply. The fact of the two sources operating in parallel implies that the DG unit will be in operation and in synchronism with the grid, aggregating the necessary conditions to feed the load at the moment of the grid failure. This kind of operation is more expensive because besides the necessary additional protection and control equipment, there are additional fuel and equipment wear out costs, which occur in generating equipment even without electricity generation.

The parallel operation requires a large quantity of monitoring, control, synchronization, and protection devices. Both the sources must be protected against the failures of the other, including the backfeed phenomenon. This kind of connection is necessary in the cases in which the DG unit owner wants to sell energy to the grid.

Several types of parallel connections are available (Figure 14.14), depending on the DG unit localization and the possibility of selling energy to the grid. In the first configuration, the DG unit operates in parallel with the grid supplying energy to all the loads or

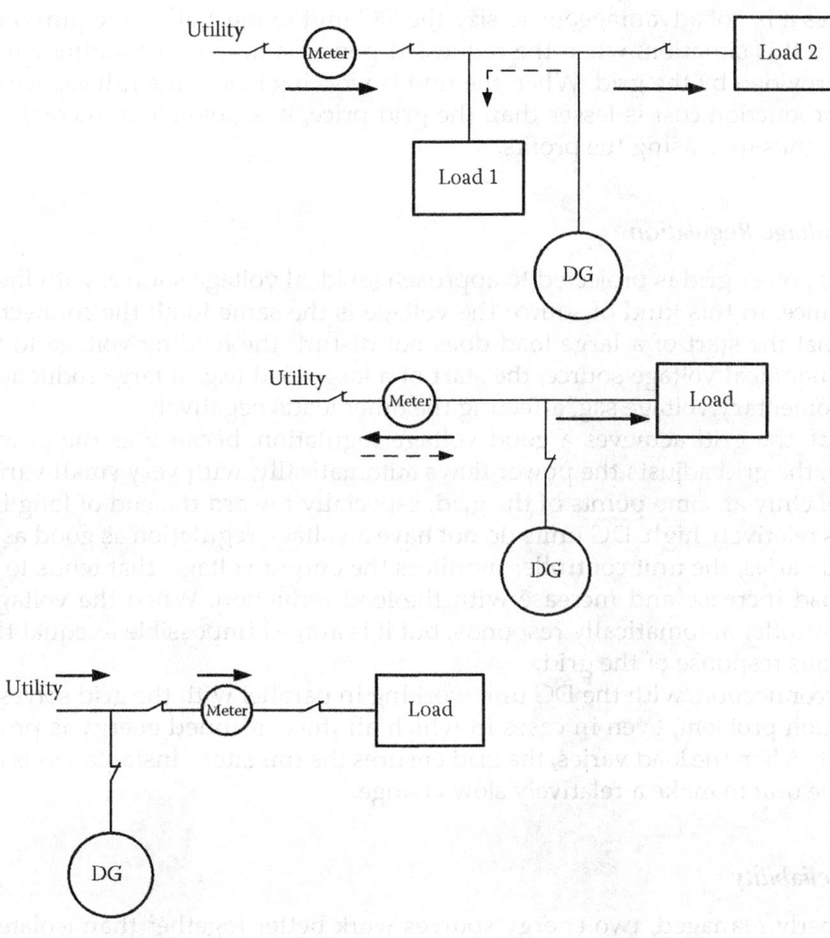

FIGURE 14.14
Parallel operation.

to some loads, particularly providing the utility supplemental or backup power. In this configuration, it is impossible to supply energy to the grid. In the second configuration, the DG unit operates in parallel with the grid supplying energy to all the loads. With this configuration, it is possible to supply energy to the grid. The DG unit provides peak or baseload power to load and exports power to the grid, providing the utility supplemental and backup power. In the third configuration the DG unit operates in parallel with the grid supplying energy to the grid and to the consumer. In this configuration, the DG unit usually does not belong to the consumer.

14.2.4 Advantages of the Grid Interconnection

14.2.4.1 Economic Advantages

The cost of the electricity provided by the grid can be smaller than the cost of the local production in some time periods. Thus, during the periods in which the marginal production cost in the local unit is superior to the grid electricity cost, it makes sense to use the grid energy. During a peak period, the cost of the grid electricity is higher and therefore the local production can become advantageous.

Sometimes it is not advantageous to size the DG unit to meet all the required power by the loads. In this situation, when the requested power is higher, the additional required energy is provided by the grid. When the unit is working below the full capacity and the marginal production cost is lesser than the grid price, it is possible to increase the local production, thus increasing the profits.

14.2.4.2 Voltage Regulation

The electric power grid is projected to approach an ideal voltage source, with lower internal impedance. In this kind of source the voltage is the same to all the connected loads, ensuring that the start of a large load does not disturb the feeding voltage to the other loads. In a nonideal voltage source, the start of a large load (e.g., a large induction motor) causes a momentary voltage sag, affecting the other loads negatively.

In general, the grid achieves a good voltage regulation, because as the power of the load varies, the grid adjusts the power flows automatically, with very small variations in the voltage. Only at some points of the grid, especially toward the end of long lines, this variation is relatively high. DG units do not have a voltage regulation as good as the grid. As the load varies, the unit controller monitors the output voltage, that tends to decrease with the load increase and increase with the load reduction. When the voltage varies, the unit controller automatically responds, but it is almost impossible to equal the nearly instantaneous response of the grid.

The interconnection, with the DG unit working in parallel with the grid solves the voltage regulation problem, even in cases in which all the consumed energy is provided by the DG unit. When the load varies, the grid ensures the transitory instantaneous response, allowing the unit to make a relatively slow change.

14.2.4.3 Reliability

When properly managed, two energy sources work better together than isolated. Either with DG reserve units or with normal operating units, the DG system owner can view the grid like a reserve energy source. In fact, in almost all the sites, the grid reliability is higher than the reliability of any isolated DG unit. In the DG projects, it is common to use values of 92%–93% for the units' availability. Considering an availability of 99%, which is extremely difficult to obtain, the corresponding unavailability will be higher than 80 h by year, which is unacceptable to most of the appliances.

Even in areas with poor performance of the electrical grid, the grid availability is normally higher than the DG unit, being many times higher than the availability of a DG system, even with several units. The grid utilization as reserve source makes sense in most of the cases, if the charged costs by the grid operator are reasonable.

14.2.5 Disadvantages of the Grid Interconnection

14.2.5.1 Costs with the Grid Operator

In any kind of market, the grid operator will charge a considerable value for interconnection with the grid. The DG unit owner, making the interconnection, will normally pay, not only the energy supplied by the grid, but also a charge dependent upon the maximum power delivered by the grid.

14.2.5.2 Additional Equipment and Maintenance

DG system operation with interconnection with the grid is more complex than an isolated operation. Besides all the necessary equipment for the system operation, it is necessary to install additional control, metering, and protection devices to isolate the DG system from the grid. Usually the additional equipment, besides increasing the cost, increases the system complexity, thus increasing the maintenance needs.

14.2.5.3 Increasing Maintenance Needs to Ensure High Reliability Levels

The potential reliability improvement achieved by the interconnection does not appear automatically. It is necessary to have a constrained management of the DG/grid combination to obtain the expected reliability levels. There are three essential points to pay attention to.

In the first place, the DG system operation in parallel with the grid needs an exhaustive monitoring of the grid operation conditions of the DG system, the load, and the interconnection. The existence of two sources and a load cause a complex control problem that is easily solved with the modern electronic devices, but it is a potential weakness. Problems can occur either due to an incorrectly programmed system or due to a failure in a key element of the system that can deactivate the entire system. Usually, in the critical DG units interconnected with the grid, redundant control equipment is installed with auto-monitoring.

Second, problems at grid points distant from the DG installation can cause perturbations in the DG unit operation. The most common problems are the atmospheric discharges. Reaching a line, a lightning discharge causes a current impulse that, without the appropriated protection, can reach the DG system. To mitigate this problem, since it is impossible to avoid it completely, additional protection equipment is needed, increasing the total cost of the system.

Third, undesirable events in the grid can disturb the DG unit operation, especially if the control and protection equipment is very sensitive. Momentary failure of a line can cause an abrupt fall in the voltage (voltage sag), that can be interpreted by the control system as a risk situation to the DG unit, removing it from service, or it can also be interpreted as an abrupt increase in the load. In the last situation, the control device responds by increasing the power of the DG unit. After the automatic reclosure of the line, the voltage level increases very fast and the unit is not able to react in time, while it continues to try follow what seems a load increase. In this situation, an overvoltage occurs that is detected by the control system and which can remove the DG unit from service.

14.2.6 Ancillary Services

Ancillary Services is the designation given to a number of functions, which are necessary to support the reliable and efficient operations of the power system network. Besides energy (kW h) and power (kW), DG can provide other additional benefits including spinning reserve capacity, peaking, load following, reactive power and voltage support, and other ancillary services.

Generally, a customer can use DG together with the traditional utility service or as a separate service. There are two ways of DG utilization: to supply power and energy during peak periods or during the entire demand period. DG equipment can also be used as backup or standby power. Some ancillary services provided by the conventional generators can also be provided by DG, thus minimizing the cost of supplying ancillary services.

In addition to generating energy, DG operation can provide the following benefits:

- Eliminate the need to upgrade the size of feeders
- Improve voltage levels at the feeder ends
- Eliminate the need for capacitor banks
- Provide reactive power compensation
- Eliminate the need for voltage regulators
- Reduce feeder loading and delay replacement
- Reduce line losses and transmission system load

The main types of ancillary services, which can be provided by generators, include

- Regulation service and frequency response
- Spinning reserves
- Supplementary (nonspinning) reserves
- Replacement or operating reserves
- Voltage support
- Reactive power support
- Black-start generation capability

To the regulation service and frequency response it is provided capacity that is available and running, and that can be used to maintain real-time balance in the transmission system. As system loads fluctuate minute-by-minute, generators must be available to match instantly the fluctuations due to the increase or decrease of the loads. An automated generation control (AGC) reacts to perceived system fluctuations by adjusting its output to oppose or dampen the fluctuation, whether it is caused by load changes or changes in the output of other bulk system generators as they ramp up or down. Generation units equipped with AGC can follow load variations on the timescale of seconds. The load balance is a critical service to the stability of day-to-day grid operation.

The load following capability service is associated with the "regulation service and spinning reserves." Load following is the use of available generation capacity to meet the variations in system load. Spinning reserves refers to supplemental generation capacity that is ready to quickly ramp up at short notice. Spinning reserves is the incremental generating supply that an active unit can ramp up to within 10 min and then sustain, typically for 30–120 min. An amount of generating capacity must be kept fully warmed up and ready to take over within seconds in the event of a generator or transmission line failure. The term spinning refers to the fact that the generator is on, spinning at rated speed (in the case of turbine generators), and synchronized to the grid. It only needs to adjust its power output to the prescribed level.

Large DG and aggregated small DG alike can provide spinning reserve service. Implicit in the definition, however, is the availability of the capacity to be called upon at any time. Therefore, for example, a DG unit cannot use its full capacity for peak shaving a local load, and at the same time qualify that capacity for spinning reserve. This limitation is true for nonspinning and replacement reserve services as well. A generator designed to run at 80% of its normal capacity for local purposes can qualify the remaining 20% capacity for spinning reserve, as long as it is synchronized to the grid for the defined reserve period. Some quick-release hydro units allow a change from 0 to full power in 1 min. Alternatively, quick-response loads (using demand response controls) can also contribute to achieve a fast balance between supply and demand.

Supplementary (nonspinning) reserves refer to generation that is available but not running. Generation is kept on standby so that it can be started rapidly in the event that generators or lines suddenly fail, but not as rapidly as spinning reserves above. The incremental generation that can be achieved by units with slower responses, and those requiring start-up, is considered nonspinning reserve. Jurisdictions differ as to the ramping time allowed, varying from 10 to 30 min. Generators that can start, synchronize, and ramp to full power in short time periods can therefore participate in the quick-response reserve market without running at all times. Fast-start combustion turbines can serve this function. In addition, customers in the form of medium fast-response load may provide nonspinning reserve services.

Nonspinning reserve in most cases will be a more appropriate choice over spinning reserve for unused DG capacity. Most distribution level DG technologies do not require 10 min to start up, and therefore would not gain from remaining synchronized to the grid when not needed. Nonspinning reserves further provide ample opportunity for generators installed as emergency backup systems to participate in the reserve market, where they would not under spinning reserve. These generators are designed to remain off under normal circumstances and serve the customer's load only if the utility experiences an outage, so their capacity during normal utility operation is always available.

Replacement or operating reserve is the incremental generation that can be obtained in the next hour to replace spinning and nonspinning reserves used in the current hour. Replacement reserve is very similar to nonspinning reserve with the exception that the generator has 60 min to start and ramp up instead of only 10. Each resource providing replacement reserve must be capable of supplying any level of output up to and including its full reserved capacity within 60 min after issue of dispatch instructions by the Independent System Operator (ISO). Each resource providing replacement reserve must be capable of sustaining the required output for at least 2 h.

Replacement reserve may be supplied from resources already providing another ancillary service, such as spinning reserve. However, the sum of the ancillary services capacity plus the replacement reserve cannot exceed the capacity of such resource. Replacement reserve can be provided by large and aggregated small DG that requires more than 10 min (and less than 1 h) to start and ramp to full power. This would be appropriate in cases where the generator technology itself has ramping limitations, or where the generator starting functions are not automated in response to a signal from the ISO, and therefore require delayed manual intervention.

Voltage support services are required to maintain transmission voltage level margins within the criteria in force. Dispatchers at the control center alter the settings on transformers, transmission lines, and other downstream grid-connected equipment, as well as provide sufficient reactive power in areas where needed.

Reactive power support is the injection or absorption of reactive power from generators to maintain transmission-system voltage within required ranges. Generators and loads may be dispatched and operated within a prescribed power factor range to boost the voltage during heavy load periods, or reduce the voltage during light load periods. The service can be provided by generators, loads, and utility distribution companies alike, as long as they have the proper power factor adjustment capabilities.

Black-start generation capability is the ability of a generating unit to go from a shutdown condition to an operating condition without assistance from the electrical grid and to then energize the grid to help other units start after a blackout occurs. Generators are started in a sequence so that each subsequent generator has an energized bus with which to synchronize. Strategically located black-start generators are a key factor for ensuring timely restoration after a major outage. Each black-start generating unit must be able to start up with a dead primary and station service bus within 10 min of issue of a dispatch instruction by the ISO requiring a black start. Each unit must also provide sufficient reactive capability to keep the energized transmission bus voltages within emergency voltage limits over the range of no-load to full load and must be capable of sustaining its output for a minimum period of 12 h from the time when it first starts delivering energy.

The other characteristics that may influence the adoption of DG technologies for ancillary service applications will vary according to the service performed and the ultimate shape of the ancillary service market. Start-up time for all electrical generators is an extremely important parameter to determine if the particular unit can be used as the reserve or can operate in the load following mode. The part-load capabilities of DG technologies and the start-up time periods of each are presented in Table 14.6.

14.2.7 Power Quality Applications

PQ-related issues are of most concern nowadays. The widespread use of electronic equipment, such as information technology equipment, power electronics such as adjustable speed drives (ASD), programmable logic controllers (PLC), energy-efficient lighting, led to a complete change of electric loads nature. These loads are simultaneously the major causers and the major victims of PQ problems. Due to their nonlinearity, all these loads cause disturbances in the voltage waveform.

Along with technology advance, the organization of the worldwide economy has evolved toward globalization and the profit margins of many activities tend to decrease. The increased sensitivity of the vast majority of processes (industrial, services, and even residential) to PQ problems turns the availability of electric power with quality as a crucial factor for competitiveness in every activity sector. The most critical areas are the continuous

TABLE 14.6

Summary Table of Some Performance Characteristics by Distributed Generation Technology Type

Technology	Part-Load	Start-Up Time
PV	Fair to good (dependent on resource availability)	Less than 10 s
Wind	Fair to good (dependent on resource availability)	30 s
Steam turbine	Satisfactory	1 h–1 day
Diesel engine	Good	10 s
Nat. gas engine	Satisfactory	10 s
Gas turbine	Poor	10 min–1 h
Microturbine	Satisfactory	60 s
PAFC	Satisfactory	1–4 h
MCFC	Poor	More than 10 h
SOFC tubular	Poor	5–10 h
SOFC planar	Poor	Not available
PEMFC	Satisfactory	<0.1 h

Source: U.S. Environmental Protection Agency, 2005, Introduction to CHP technologies, http://www.epa.gov/chp/, accessed October 12, 2005.

process industry and the information technology services. When a disturbance occurs, huge financial losses may happen, with the consequent loss of productivity and competitiveness.

Although many efforts have been taken by utilities, some consumers require a level of PQ higher than the level provided by modern electric networks. This implies that some measures must be taken in order to achieve higher levels of PQ. The most common types of PQ problems are presented in Table 14.7.

Even the most advanced T&D systems are not able to provide electrical energy with the desired level of reliability for the proper functioning of the loads in the modern society. Modern T&D systems are projected for 99.9%–99.99% availability. This value is highly dependent of redundancy level of the network, which is different according to the geographical location and the voltage level (availability is higher at the HV network). In some remote sites, availability of T&D systems may be as low as 99%. Even with a 99.99% level, there is an equivalent interruption time of 52 min/year. The most demanding processes in the modern digital economy need electrical energy with 99.9999999% availability (9-nines reliability) to function properly.

The mitigation of PQ problems may take place at different levels: transmission, distribution, and end-use equipment. As seen in Figure 14.15, several measures can be taken at these levels. Many PQ problems have origin in the transmission or distribution grid. Thus, a proper T&D grid, with adequate planning and maintenance, is essential to minimize the occurrence of PQ problems.

Interest in the use of distributed energy resources has increased substantially over the last few years because of their potential to provide increased reliability. These resources include DG and energy storage systems. Energy storage systems, also known as restoring technologies, are used to provide the electric loads with ride-through capability in poor PQ environment. Recent technological advances in power electronics and storage technologies are turning the restoring technologies as one of the premium solutions to mitigate PQ problems (Figure 14.16).

DG units can be used to provide clean power to critical loads, isolating them from disturbances with origin in the grid. DG units can also be used as backup generators to ensure

TABLE 14.7

Most Common Power Quality Problems

1. Voltage sag (or dip) 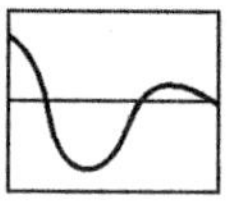	*Description*: A decrease of the normal voltage level between 10% and 90% of the nominal RMS voltage at the power frequency, for durations of 0.5 cycle to 1 min. *Causes*: Faults in the transmission or distribution network (most of the times on parallel feeders). Faults in consumer's installation. Connection of heavy loads and start-up of large motors. *Consequences*: Malfunction of information technology equipment, namely microprocessor-based control systems (PCs, PLCs, ASDs, etc.) that may lead to a process stoppage. Tripping of contactors and electromechanical relays. Disconnection and loss of efficiency in electric rotating machines.
2. Very short interruptions 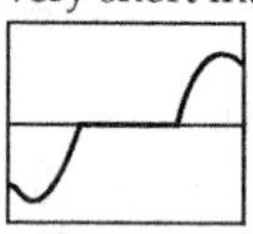	*Description*: Total interruption of electrical supply for duration from few milliseconds to 1 or 2 s. *Causes*: Mainly due to the opening and automatic reclosure of protection devices to decommission a faulty section of the network. The main fault causes are insulation failure, lightning, and insulator flashover. *Consequences*: Tripping of protection devices, loss of information and malfunction of data processing equipment. Stoppage of sensitive equipment, such as ASDs, PCs, PLCs, if they are not prepared to deal with this situation.
3. Long interruptions 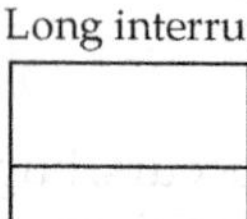	*Description*: Total interruption of electrical supply for duration greater than 1–2 s *Causes*: Equipment failure in the power system network, storms and objects (trees, cars, etc.) striking lines or poles, fire, human error, bad coordination, or failure of protection devices. *Consequences*: Stoppage of all equipment.
4. Voltage spike 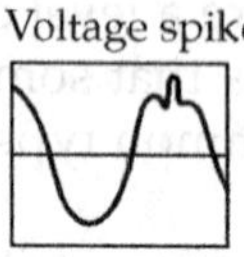	*Description*: Very fast variation of the voltage value for durations from several microseconds to few milliseconds. These variations may reach thousands of volts, even in low voltage. *Causes*: Lightning, switching of lines or power factor correction capacitors, disconnection of heavy loads. *Consequences*: Destruction of components (particularly electronic components) and of insulation materials, data processing errors or data loss and electromagnetic interference.
5. Voltage swell 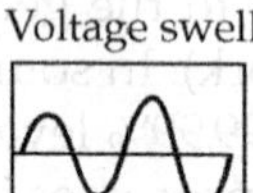	*Description*: Momentary increase of the voltage, at the power frequency, outside the normal tolerances, with duration of more than one cycle and typically less than a few seconds. *Causes*: Start/stop of heavy loads, badly dimensioned power sources, badly regulated transformers (mainly during off-peak hours). *Consequences*: Data loss, flickering of lighting and screens, stoppage or damage of sensitive equipment, if the voltage values are too high.
6. Harmonic distortion 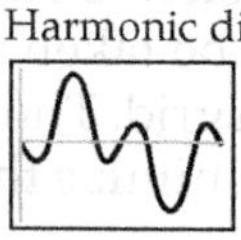	*Description*: Voltage or current waveforms assume nonsinusoidal shape. The waveform corresponds to the sum of different sine-waves with different magnitudes and phases, having frequencies that are multiples of power-system frequency. *Causes*: Classic sources: electric machines working above the knee of the magnetization curve (magnetic saturation), arc furnaces, welding machines, rectifiers, and DC brush motors. Modern sources: all nonlinear loads, such as power electronics equipment including ASDs, switched mode power supplies, data processing equipment, high efficiency lighting. *Consequences*: Increased probability of occurrence of resonance, neutral overload in three-phase systems, overheating of all cables and equipment, loss of efficiency in electric machines, electromagnetic interference with communication systems, errors in measures when using average reading meters, nuisance tripping of thermal protections.

(*Continued*)

TABLE 14.7 (*Continued*)

Most Common Power Quality Problems

7. Voltage fluctuation 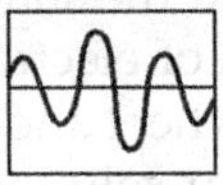	*Description*: Oscillation of voltage value, amplitude modulated by a signal with frequency of 0–30 Hz. *Causes*: Arc furnaces, frequent start/stop of electric motors (for instance, elevators), oscillating loads. *Consequences*: Most consequences are common to undervoltages. The most perceptible consequence is the flickering of lighting and screens, giving the impression of unsteadiness of visual perception.
8. Noise 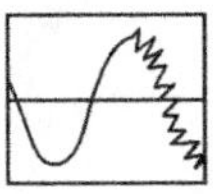	*Description*: Superimposition of high frequency signals on the waveform of the power-system frequency. *Causes*: Electromagnetic interferences caused by Hertzian waves, such as microwaves, television diffusion, and radiation due to welding machines, arc furnaces, and electronic equipment. Improper grounding may also be a cause. *Consequences*: Disturbances on sensitive electronic equipment, usually not destructive. May cause data loss and data processing errors.
9. Voltage unbalance 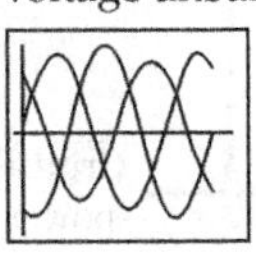	*Description*: A voltage variation in a three-phase system in which the three voltage magnitudes or the phase-angle differences between them are not equal. *Causes*: Large single-phase loads (induction furnaces, traction loads), incorrect distribution of all single-phase loads by the three phases of the system (this may be also due to a fault). *Consequences*: Unbalanced systems imply the existence of a negative sequence that is harmful to all three-phase loads. The most affected loads are three-phase induction machines.

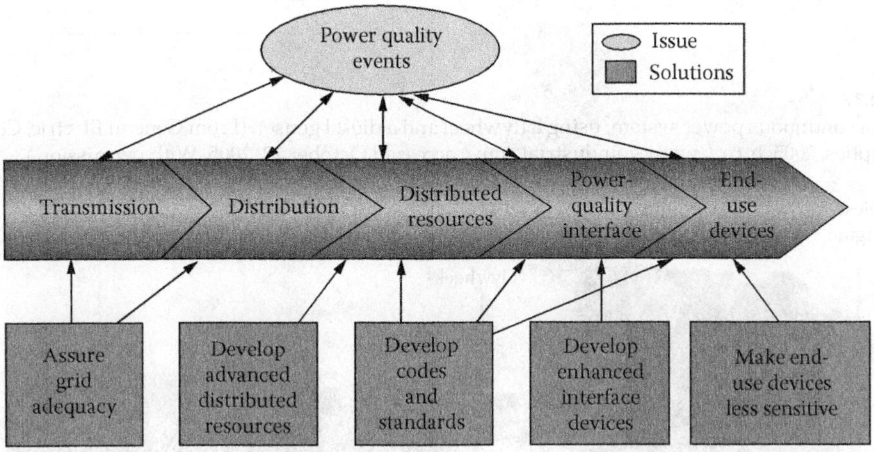

FIGURE 14.15
Solutions for power quality.

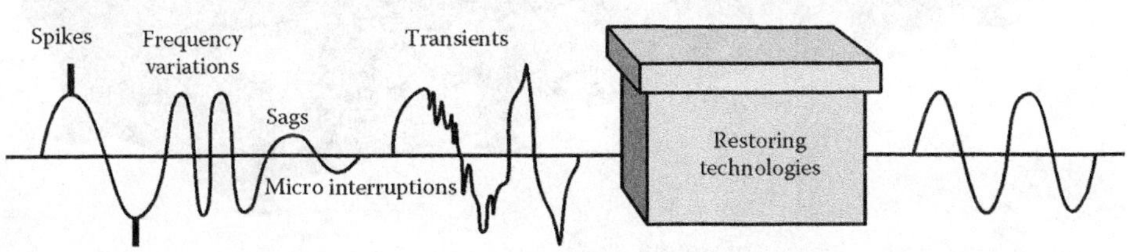

FIGURE 14.16
Restoring technologies principle.

energy supply to critical loads during sustained outages. Additionally, DG units can be used for load management purposes to decrease the peak demand.

If DG units are to be used as backup generation, a storage unit must be used to provide energy to the loads during the period between the origin of the disturbance and the start-up of the emergency generator. The most common solution is the combination of electrochemical batteries' UPS and a diesel genset. At present, the integration of a flywheel and a diesel genset in a single unit (Figures 14.17 and 14.18) is also becoming a popular solution, offered by many manufacturers.

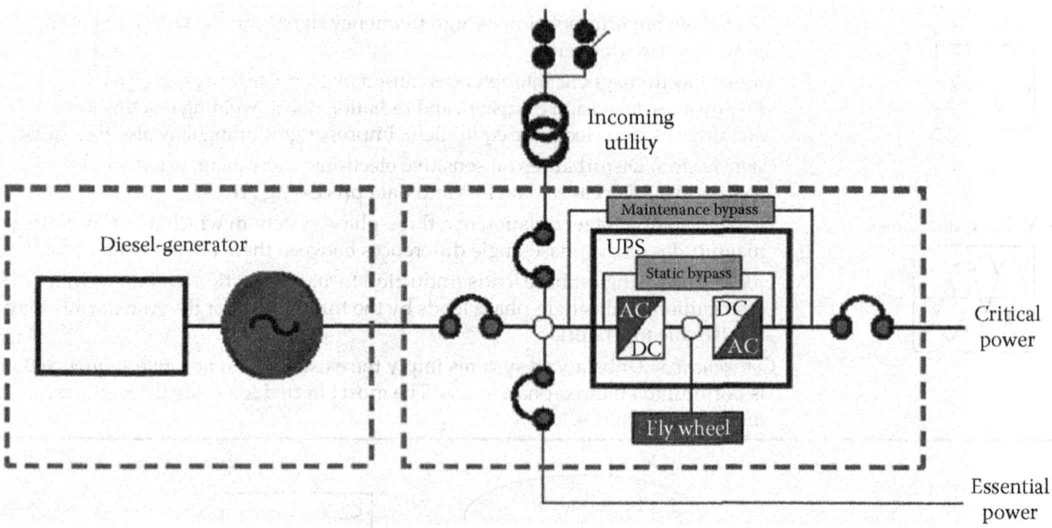

FIGURE 14.17
Scheme of a continuous power system, using a flywheel and a diesel genset. (From General Electric, Continuous power supplies, 2005, http://www.geindustrial.com/, accessed October 12, 2005. With permission.)

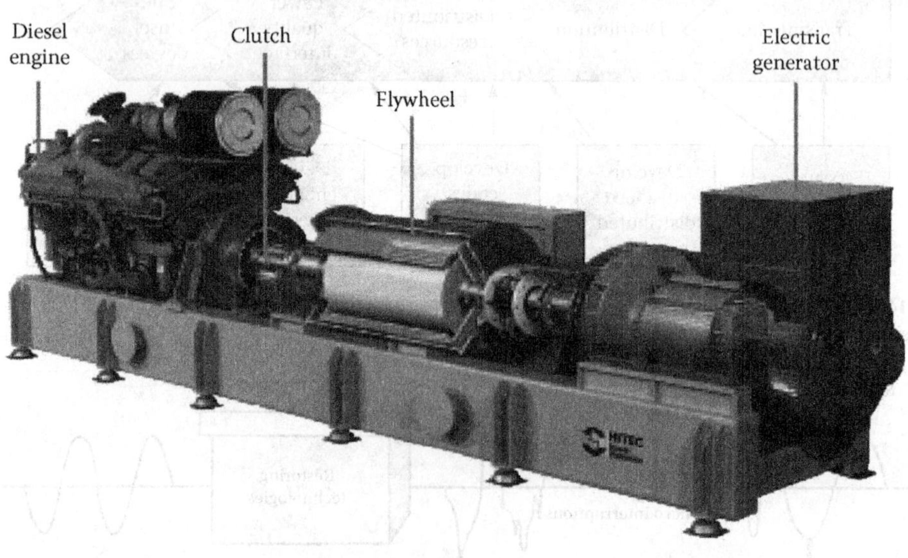

FIGURE 14.18
Dynamic UPS. (From HITEC Power Protection, Dynamic UPS, 2005, http://www.hitecups.com/, accessed October 12, 2005. With permission.)

14.2.8 Standards for Interconnecting Distributed Resources

Many of the technical concerns that the companies of distributed electricity usually raise for the distributed resources (DRs) interconnection with the grid are related to reliability, security, and quality of service. *The IEEE 1547 Standard for Interconnecting Distributed Resources with Electric Power Systems* defines interconnection technical specifications and requirements that are universally needed for interconnection of DR. This standard was approved by the IEEE Standards Board in June 2003 and approved as an American National Standard in October 2003, being an important step to overcome the barriers and increase the development of DG installations.

This standard provides requirements relevant to the performance, operation, testing, safety considerations, and maintenance of the interconnection. General requirements are related to voltage regulation, integration with area EPS grounding, synchronization, DR on secondary grid and spot networks, inadvertent energization and reconnection to area EPS, monitoring and isolation devices. The standard applies to all DR technologies, with aggregate capacity of 10 MVA or smaller at the point of common coupling, and to all EPS at typical primary and/or secondary distribution voltages.

Abnormal conditions can arise on the area EPS that require a response from the connected DR. This response contributes to the safety of utility maintenance personnel and the general public, as well as the avoidance of damage to the connected equipment, including the DR. The abnormal conditions of concern are voltage and frequency excursions above or below the values stated in the IEEE 1547 (Table 14.8), and the isolation of a portion of the area EPS with some DR, presenting the potential for an unintended island.

PQ issues are also addressed in the IEEE 1547, namely, limitation of DC injection, voltage flicker induced by the DR, harmonic current injection (Table 14.9), immunity protection and surge capabilities, as well as the islanding considerations.

The standard also provides test requirements for an interconnection system to demonstrate that it meets all the requirements. The following tests are required for all interconnection systems:

- Interconnection test
- Production tests
- Interconnection installation evaluation
- Commissioning tests
- Periodic interconnection tests

TABLE 14.8
Interconnection System Response to Abnormal Voltages

Voltage Range (% of Base Voltage)	Clearing Time (s)[a]
V < 50	0.16
50 ≤ V < 88	2
110 < V < 120	1
V ≥ 120	0.16

Source: From Institute of Electric and Electronics Engineers (IEEE), *IEEE 1547 Standard for Interconnecting Distributed Resources with Electric Power Systems,* IEEE Standards Coordinating Committee 21 (IEEE SCC21) on Fuel Cells, Photovoltaics, Dispersed Generation, and Energy Storage of the IEEE Standards Association, New York, 2003.

[a] Clearing time: Time between the start of the abnormal condition and the DR ceasing to energize the area EPS.

TABLE 14.9

Maximum Harmonic Current Distortion in Percent of Current (I)[a,b]

Individual Harmonic Order (Odd Harmonics)	<11	11 ≤ h < 17	17 ≤ h < 23	23 ≤ h < 35	35 ≤ h	TDD
Percent (%)	4.0	2.0	1.5	0.6	0.3	5.0

Source: Institute of Electric and Electronics Engineers (IEEE), *IEEE 1547 Standard for Interconnecting Distributed Resources with Electric Power Systems*, IEEE Standards Coordinating Committee 21 (IEEE SCC21) on Fuel Cells, Photovoltaics, Dispersed Generation, and Energy Storage of the IEEE Standards Association, New York, 2003.

Note: TDD, total demand distortion.

[a] I = the greater of the local EPS maximum load current integrated demand (15–30 min) without the DR unit, or the DR unit rated current capacity (transformed to the PCC when a transformer exists between the DR unit and the PCC.

[b] Even harmonics are limited to 25% of the odd harmonic limits above.

IEEE P1547 is part of a family of IEEE interconnection standards for DR. The other standards are as follows (IEEE, 2013):

- *IEEE P1547.1 Standard for Conformance Tests Procedures for Equipment Interconnecting Distributed Resources with Electric Power Systems* (published in 2005)—This standard specifies the type, production, and commissioning tests that shall be performed to demonstrate that the interconnection functions and equipment of a DR conform to IEEE Standard 1547.
- *IEEE 1547.2 Application Guide for IEEE 1547 Standard for Interconnecting Distributed Resources with Electric Power Systems* (published in 2008)—This guide provides technical background and application details to support the understanding of *IEEE 1547 Standard for Interconnecting Distributed Resources with Electric Power Systems.*
- *IEEE 1547.3 Guide for Monitoring, Information Exchange, and Control of Distributed Resources Interconnected with Electric Power Systems* (published in 2007)—This document provides guidelines for monitoring, information exchange, and control for DRs interconnected with EPSs.
- *IEEE 1547.4 Guide for Design, Operation, and Integration of Distributed Resource Island Systems with Electric Power Systems* (published in 2011)—This document provides alternative approaches and good practices for the design, operation, and integration of DR island systems with EPSs.
- *IEEE P1547.5 Draft Technical Guidelines for Interconnection of Electric Power Sources Greater than 10 MVA to the Power Transmission Grid* (withdrawn in 2011)—This document provides guidelines regarding the technical requirements, including design, construction, commissioning acceptance testing, and maintenance/performance requirements, for interconnecting dispatchable electric power sources with a capacity of more than 10 MVA to a bulk power transmission grid.
- *IEEE 1547.6 Recommended Practice for Interconnecting Distributed Resources with Electric Power Systems Distribution Secondary Networks* (published in 2011)—This standard establishes recommended criteria, requirements, and tests, and provides guidance for interconnection of distribution secondary network system types of area EPSs with DRs providing electric power generation in local EPSs.

- *IEEE P1547.7 Draft Guide to Conducting Distribution Impact Studies for Distributed Resource Interconnection*—This guide describes criteria, scope, and extent for engineering studies of the impact on area EPSs of a DR or an aggregate DR interconnected to an area electric power distribution system.
- *IEEE P1547.8 Recommended Practice for Establishing Methods and Procedures that Provide Supplemental Support for Implementation Strategies for Expanded Use of IEEE Standard 1547*—This recommended practice applies to the requirements set forth in IEEE 1547 and provides recommended methods that may expand the usefulness and uniqueness of IEEE 1547 through the identification of innovative designs, processes, and operational procedures.

Nowadays, there are a Working Group (IEEE P1547a) to establishing updates to voltage regulation, response to area EPSs abnormal conditions of voltage and frequency, and considering if other changes to IEEE Standard 1547. At the same time, there are other Working Groups (IEEE P1547.1a) to establishing updates to IEEE Standard 1547.1 in relation to the details of the IEEE P1547a revisions.

14.2.9 Smart Grids

The efforts to reduce greenhouse gas emissions related to electricity generation have been leading to a fast increase in the deployment of renewable generation, in particular PV and wind power. Such renewable energy sources are being deployed not only as bulk generation facilities but also as distributed local generation facilities connected to the electricity distribution grids or even to private consumption infrastructures. DG has the potential to provide site-specific reliability improvement, as well as T&D benefits. However, the renewable generation resources have characteristics that differ from conventional energy sources. The output of wind and solar power is determined by random meteorological processes, outside the control of the generators or the system operators. Therefore, unlike conventional capacity, wind and solar generated electricity cannot be reliably dispatched or perfectly forecasted, and exhibits significant temporal variability.

EPSs have been designed for more than a century ago in a top-down perspective, based on highly predictable bulk generation power plants following load variations, controlled by centralized systems. The basic topology of the existing electrical grid has remained unchanged, being a strictly hierarchical system with clear demarcations between its generation, transmission, and distribution subsystems. It is unidirectional in nature, with an energy flow from the power plant to the customer without any real-time share of information between the consumption and generation points.

However, the defined scenarios for decarbonizing the energy sector will lead to important impacts on the electricity grid, namely high penetration of intermittent renewable technologies. In addition, such new generation resources are deployed mostly as DG, creating a more decentralized system, where the figure of the "prosumer" emerges. With the increasing connection of renewable generation sources at all voltage levels due to the mismatching between local generation (mainly PV) and consumption (Figure 14.19), in some distribution grids, the power flow can invert the way, flowing from the distribution network to the transmission network. As a result, the scalability of the grid

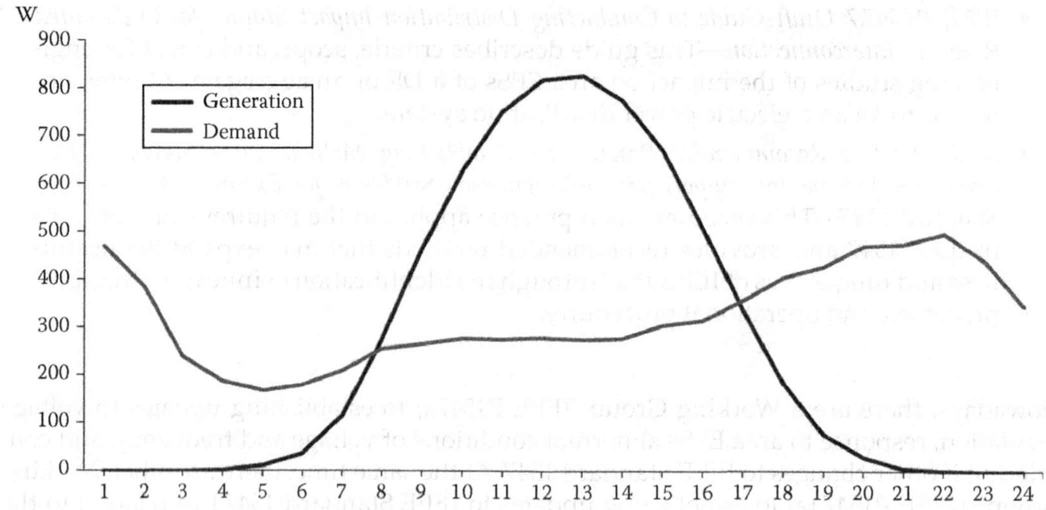

FIGURE 14.19
Typical photovoltaic generation and demand profiles in a residential building.

is improved and the energy flows, and so the losses, are reduced, but the complexity of managing such an infrastructure increases dramatically.

As penetration rates of variable generation increase over levels of 15%–20%, and depending on the electricity system in question, it can become increasingly difficult to ensure the reliable and stable management of electricity systems relying solely on conventional grid architectures and limited flexibility. Therefore, new monitoring and control tools are needed to increase the system flexibility and maintain stability and balance at the distribution grid level.

The increasing energy consumption, in most countries, must be also properly addressed by the electrical grid. To achieve it, a change on the paradigm of the electric grid management is needed, moving from an approach where the generation must always follow the load to a new approach where the load is also considered a manageable resource, which can be controlled to follow the available generation, using demand response programs. Therefore, the electrical grid must have new tools to optimize the resources and to promote energy efficiency, both in the supply and in the demand sides, through monitoring and control infrastructures based on sensors and actuators networks and demand response programs.

Part of the new electricity consumption will be caused by deployment of electric vehicles. However, they can also be an important tool to the grid management with the possible control of the charging cycles and the use of the battery as a local energy storage device. At the same time, the improvements achieved on other energy storage technologies bring new solutions to the grid management. Thus, the control of consumption together with energy storage will ensure the needed flexibility to ensure the generation/consumption matching.

To deal with such challenges and take advantage of the new resources, new technology has been developed and deployed. ICT and M2M (machine-to-machine) communications will definitely play a key role to make the smart grid a reality, since they will enable the required bidirectional communication between the high volume of devices to be monitored and controlled, and the end users and the information systems where the intelligence resides.

The metering side of the distribution system has been the focus of the smart grid investments so far, with the initial introduction of one-way AMR (automated meter reading) systems to read meter data. However, due to its one-way communication system, AMR systems do not allow the transition to the smart grid, due to the absence of control capabilities. In a second stage, the AMI (advanced metering infrastructure) has been used to provide a two-way communication system to the meter, as well as the ability to ensure load management and revenue protection.

The next step is the leverage of the AMI to implement distributed command and control strategies with pervasive control and intelligence across all geographies, components, and functions of the system. To ensure it several different smart grid technological areas have to be developed, such as wide-area monitoring and control, information and communications technology integration, renewable and DG integration, transmission enhancement applications, distributed grid management, AMI, electric vehicles charging infrastructure, and customer-side systems (Figure 14.20).

Smart grids are expected to address the major shortcomings of the existing grid and explore the new resources. The smart grid still depends on the support of large central-station generation, but it includes energy storage and renewable energy generation facilities in the different grid levels. In addition, the smart grid needs to provide the utility companies with enhanced sensory and control capability to ensure the full visibility and pervasive control over their assets and services.

The aim of smart grids is to maximize the system reliability, resilience and stability, and minimize costs and environmental impacts by coordinating the needs and resources of end-users and generation, grid and market operators. Due to the available technology,

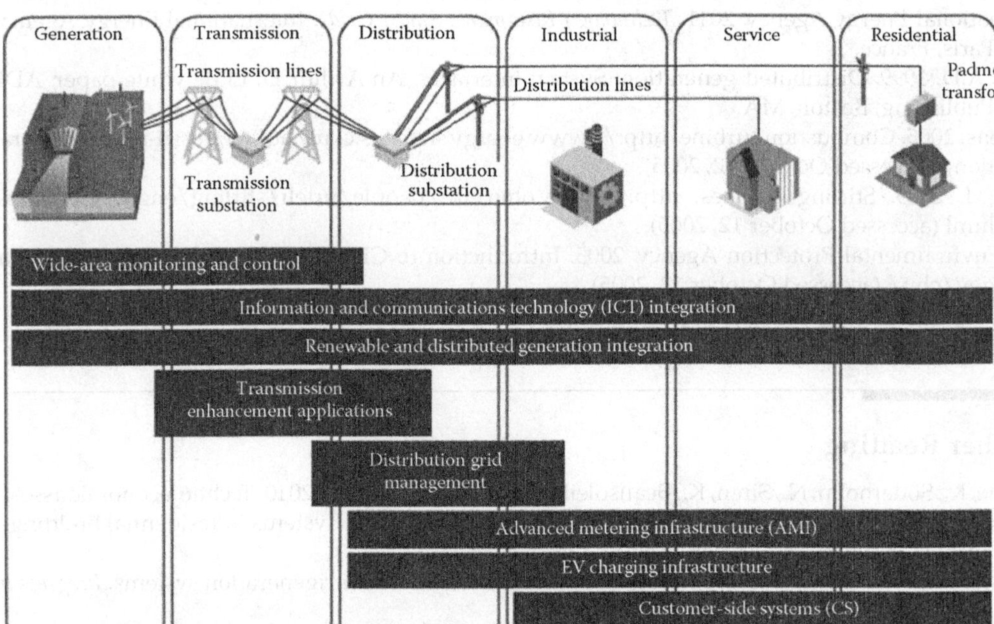

FIGURE 14.20
Smart grid technology areas. (From International Energy Agency, *Technology Roadmaps: Smart Grids*, International Energy Agency, Paris, France, 2011.)

the smart grids are no longer just a future vision, but already a reality. A high number of implemented projects and pilots have been proving the impacts of smart grids which can ensure multiple benefits, not only to the utilities but also to the users and society. smart grids will facilitate the integration of diverse supply-side resources, support the integration of distributed and on-site generation on the customer side, promote more active engagement of demand-side resources and participation of customers' loads, and allow the widespread permeation of dynamic pricing to beyond-the-meter applications.

References

California Energy Commission. 2003. *California Interconnection Guidebook: A Guide to Interconnecting Customer-Owned Electric Generation Equipment to the Electric Utility Distribution System Using California's Electric Rule 21*. California Energy Commission, Sacramento, CA.

Capstone. 2005. Microturbine. http://www.capstoneturbine.com/ (accessed October 12, 2005).

General Electric. 2005. Continuous power supplies. http://www.geindustrial.com/ (accessed October 12, 2005).

HITEC Power Protection. 2005. Dynamic UPS. http://www.hitecups.com/ (accessed October 12, 2005).

Institute of Electric and Electronics Engineers (IEEE). 2003. *IEEE 1547 Standard for Interconnecting Distributed Resources with Electric Power Systems*. IEEE Standards Coordinating Committee 21 (IEEE SCC21) on Fuel Cells, Photovoltaics, Dispersed Generation, and Energy Storage of the IEEE Standards Association, New York.

Institute of Electric and Electronics Engineers (IEEE). 2013. IEEE 1547 standard for interconnecting distributed resources with electric power systems, http://grouper.ieee.org/groups/scc21/1547/1547_index.html (accessed December 13, 2011).

International Energy Agency. 2011. *Technology Roadmaps: Smart Grids*. International Energy Agency, Paris, France.

Little, A.D. 1999. Distributed generation: System interfaces. An Arthur D. Little white paper, ADL Publishing, Boston, MA.

Siemens. 2005. Combustion turbine. http://www.energy.siemens.com/hq/en/fossil-power-generation/ (accessed October 12, 2005).

Urieli, I. 2005. Stirling engines. http://www.ohio.edu/people/urieli/stirling/engines/engines.html (accessed October 12, 2005).

U.S. Environmental Protection Agency. 2005. Introduction to CHP technologies. http://www.epa.gov/chp/ (accessed October 12, 2005).

Further Reading

Alanne, K., Söderholm, N., Sirén, K., Beausoleil-Morrison, I. December 2010. Techno-economic assessment and optimization of Stirling engine micro-cogeneration systems in residential buildings. *Energy Conversion and Management* 51(12), 2635–2646.

Angrisani, G., Roselli, C., Sasso, M. August 2012. Distributed microtrigeneration systems. *Progress in Energy and Combustion Science* 38(4), 502–521.

Bathie, W.W. 1995. *Fundamentals of Gas Turbines*, 2nd edn. Wiley, New York.

Borbely, A.-M. Kreider, J.F. 2001. *Distributed Generation: The Power Paradigm of the New Millennium*. CRC Press, Boca Raton, FL.

Eltawil, M.A., Zhao, Z. January 2010. Grid-connected photovoltaic power systems: Technical and potential problems—A review. *Renewable and Sustainable Energy Reviews* 14(1), 112–129.

Ilic, M., Black, J.W., Prica, M. July 2007. Distributed electric power systems of the future: Institutional and technological drivers for near-optimal performance. *Electric Power Systems Research* 77(9), 1160–1177.

Ismail, M.S., Moghavvemi, M., Mahlia, T.M.I. May 2013. Current utilization of microturbines as a part of a hybrid system in distributed generation technology. *Renewable and Sustainable Energy Reviews* 21, 142–152.

Jagaduri, R.T., Radman, G. January 2007. Modeling and control of distributed generation systems including PEM fuel cell and gas turbine. *Electric Power Systems Research* 77(1), 83–92.

Kolanowski, B.F. 2004. *Guide to Microturbines*. Marcel Dekker, New York.

Moreno-Munoz, A., de-la-Rosa, J.J.G., Lopez-Rodriguez, M.A., Flores-Arias, J.M., Bellido-Outerino, F.J., Ruiz-de-Adana, M. December 2010. Improvement of power quality using distributed generation. *International Journal of Electrical Power and Energy Systems* 32(10), 1069–1076.

Passey, R., Spooner, T., MacGill, I., Watt, M., Syngellakis, K. October 2011. The potential impacts of grid-connected distributed generation and how to address them: A review of technical and non-technical factors. *Energy Policy* 39(10), 6280–6290.

Pepermans, G., Driesen, J., Haeseldonckx, D., Belmans, R., D'haeseleer, W. April 2005. Distributed generation: Definition, benefits and issues. *Energy Policy* 33(6), 787–798.

Siva Reddy, V., Kaushik, S.C., Ranjan, K.R., Tyagi, S.K. November 2013. State-of-the-art of solar thermal power plants—A review. *Renewable and Sustainable Energy Reviews* 27, 258–273.

Thornton, A., Monroy, C.R. December 2011. Distributed power generation in the United States. *Renewable and Sustainable Energy Reviews* 15(9), 4809–4817.

Toledo, O.M., Filho, D.O., Diniz, A.S.A.C. January 2010. Distributed photovoltaic generation and energy storage systems: A review. *Renewable and Sustainable Energy Reviews* 14(1), 506–511.

Viral, R., Khatod, D.K. September 2012. Optimal planning of distributed generation systems in distribution system: A review. *Renewable and Sustainable Energy Reviews* 16(7), 5146–5165.

Wang, Y., Chen, K.S., Mishler, J., Cho, S.C., Adroher, X.C. April 2011. A review of polymer electrolyte membrane fuel cells: Technology, applications, and needs on fundamental research. *Applied Energy* 88(4), 981–1007.

Wee, J.-H. December 2011. Molten carbonate fuel cell and gas turbine hybrid systems as distributed energy resources. *Applied Energy* 88(12), 4252–4263.

Willis, H.L., Scott, W.G. 2000. *Distributed Power Generation, Planning and Evaluation*. Marcel Dekker, New York.

Wolsink, M. January 2012. The research agenda on social acceptance of distributed generation in smart grids: Renewable as common pool resources. *Renewable and Sustainable Energy Reviews* 16(1), 822–835.

15

Demand-Side Management

Clark W. Gellings and Kelly E. Parmenter

CONTENTS

15.1 Introduction

Since the mid-1980s, demand-side management has been an important element of the electric utility planning approach referred to as *integrated resource planning*. At that time, annual demand-side management expenditures in the United States were measured in billions of dollars, energy savings were measured in billions of kWh, and peak load reductions were stated in thousands of MW. While activities nationally slowed during the couple of decades that followed the 1980s, activity has again been on the rise since the turn of the millennium. Expenditures for utility and third-party administered electric efficiency programs were nearly \$6 billion in 2012, up from a low of \$0.9 billion in 1998; in addition, estimated electric savings nationwide exceeded 22 billion kWh in 2011 (Hayes et al. 2013). Therefore, demand-side management practices have continued to persevere and have recently regained a considerable and growing influence on the demand for energy resources. This chapter defines demand-side management, describes the role demand-side management plays in integrated resource planning, discusses the main elements of demand-side management programs, and summarizes the key best practices for program design and delivery. It then presents case studies of four successful demand-side management programs.

15.2 What Is Demand-Side Management?

The term *demand-side management* is the result of a logical evolution of planning processes used by utilities in the late 1980s. One of the first terms, *demand-side load management* was introduced by the author, Clark W. Gellings, in an article for *IEEE Spectrum* in 1981. Shortly after the publication of this article, at a meeting of The Edison Electric Institute (EEI) Customer Service and Marketing Executives in 1982, Mr. Gellings altered the term to *demand-side planning*. This change was made to reflect the broader objectives of the planning process. Mr. Gellings coined the term *demand-side management* and continued to popularize the term throughout a series of more than 100 articles since that time, including the five volume set *Demand-Side Management* which is widely recognized as a definitive and practical source of information on the demand-side management process.

Perhaps the most widely accepted definition of demand-side management is the following:

> Demand-side management is the planning, implementation, and monitoring of those utility activities designed to influence customer use of electricity in ways that will produce desired changes in the utility's load shape, i.e., changes in the time pattern and magnitude of a utility's load. Utility programs falling under the umbrella of demand-side management include: load management, new uses, strategic conservation, electrification, customer generation, and adjustments in market share (Gellings 1984–1988).

However, demand-side management is even more encompassing than this definition implies because it includes the management of all forms of energy at the demand-side, not just electricity. In addition, groups other than just electric utilities (including natural gas suppliers, government organizations, nonprofit groups, and private parties) implement demand-side management programs.

In general, demand-side management embraces the following critical components of energy planning:

1. Demand-side management *will influence customer use.* Any program intended to influence the customer's use of energy is considered demand-side management.
2. Demand-side management *must achieve selected objectives.* To constitute a desired load-shape change, the program must further the achievement of selected objectives; that is, it must result in reductions in average rates, improvements in customer satisfaction, achievement of reliability targets, etc.
3. Demand-side management *will be evaluated against non-demand-side management alternatives.* The concept also requires that selected demand-side management programs further these objectives to at least as great an extent as non-demand-side management alternatives, such as generating units, purchased power, or supply-side storage devices. In other words, it requires that demand-side management alternatives be compared to supply-side alternatives. It is at this stage of evaluation that demand-side management becomes part of the integrated resource planning process.
4. Demand-side management *identifies how customers will respond.* Demand-side management is pragmatically oriented. Normative programs (we ought to do this) do not bring about the desired; positive efforts (if we do this, that will happen) are required. Thus, demand-side management encompasses a process that identifies how customers will respond not how they should respond.

5. Demand-side management *value is influenced by load shape*. Finally, this definition of demand-side management focuses upon the load shape. This implies an evaluation process that examines the value of programs according to how they influence costs and benefits throughout the day, week, month, and year.

Subsets of these activities have been referred to in the past as *load management*, *strategic conservation*, and *marketing*.

15.3 Demand-Side Management and Integrated Resource Planning

A very important part of the demand-side management process involves the consistent evaluation of demand-side to supply-side alternatives and vice versa. This approach is referred to as *integrated resource planning*. Figure 15.1 illustrates how demand-side management fits into the integrated resource planning process. For demand-side management to be a viable resource option, it has to compete with traditional supply-side options.

15.4 Demand-Side Management Programs

A variety of programs have been implemented since the introduction of demand-side management in the early 1980s. Mr. Gellings and the EPRI (Electric Power Research Institute) have been instrumental in defining a framework for utilities and other implementers to follow when planning demand-side management programs (Gellings 1984–1988; Evans et al. 1993; Gellings and Chamberlin 1993; Gellings 2002; Siddiqui et al. 2008). This section describes the main elements of the demand-side management planning framework. It then discusses the types of end-use sectors, buildings, and end-use technologies targeted during program development. It also lists the various entities typically responsible for implementing programs, along with several program implementation methods. Lastly, this section summarizes specific characteristics of successful program design and delivery.

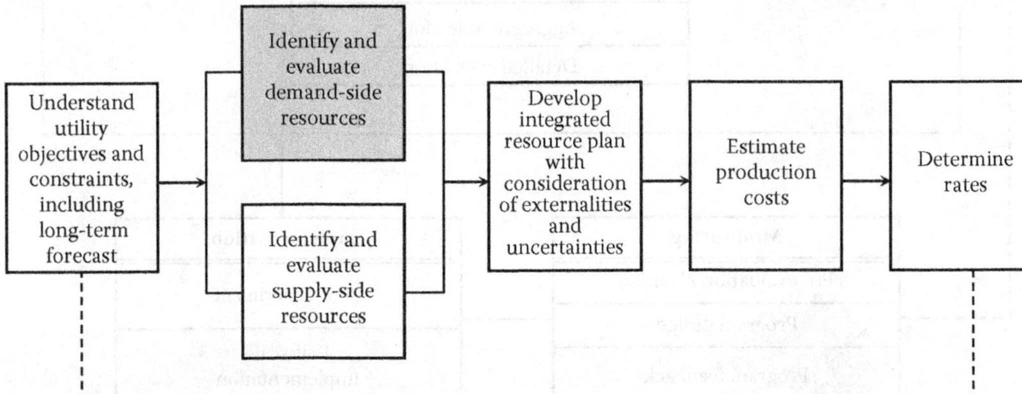

FIGURE 15.1
How demand-side management fits into integrated resource planning.

15.4.1 Elements of the Demand-Side Management Planning Framework

Figure 15.2 illustrates the five main elements of the demand-side management planning framework. These five elements are summarized as follows:

1. *Set objectives*: The first step in demand-side management planning is to establish overall organizational objectives. These strategic objectives are quite broad and generally include examples such as conserving energy resources, reducing peak demand (thereby deferring need to build new power plants), decreasing

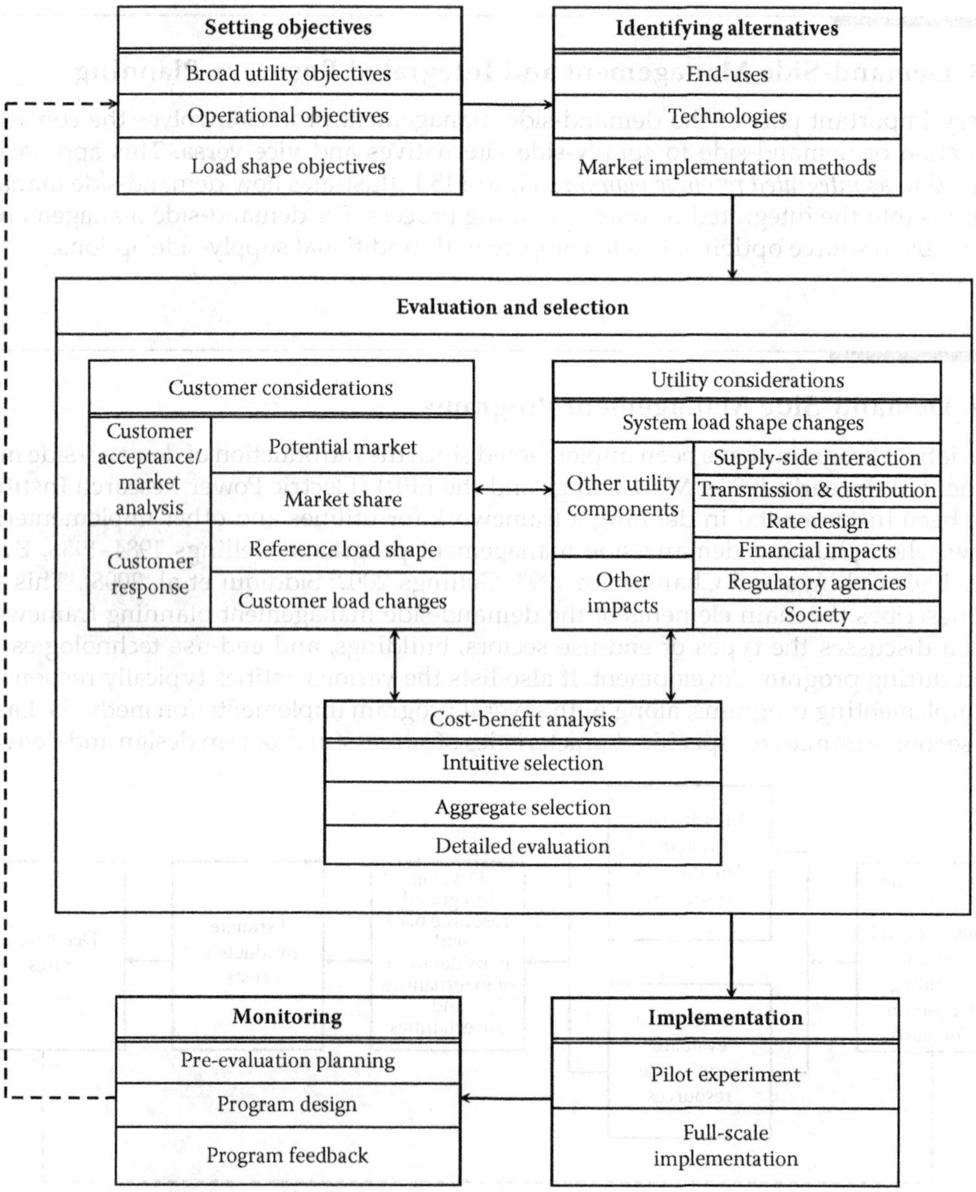

FIGURE 15.2
Elements of the demand-side management planning framework.

greenhouse gas emissions, reducing dependence on foreign imports, improving cash flow, increasing earnings, and improving customer and employee relations. In this level of the formal planning process, the planner needs to operationalize broad objectives to guide policymakers to specific actions. It is at this operational level or tactical level that demand-side management alternatives should be examined and evaluated. For example, an examination of capital investment requirements may show periods of high investment needs. Postponing the need for new construction through a demand-side management program may reduce investment needs and stabilize the financial future of an energy company, or a utility and its state or country. Specific operational objectives are established on the basis of the conditions of the existing energy system—its system configuration, cash reserves, operating environment, and competition. Once designated, operational objectives are translated into desired demand-pattern changes or load-shape changes that can be used to characterize the potential impact of alternative demand-side management programs. Although there is an infinite combination of load-shape-changing possibilities, six have been illustrated in Figure 15.3 to show the range of possibilities, namely peak clipping, valley filling, load shifting, strategic conservation, strategic load growth, and flexible load shape. These six are not mutually exclusive, and may frequently be employed in combinations.

2. *Identify alternatives*: The second step is to identify alternatives. The first dimension of this step involves identifying the appropriate end-uses whose peak load and energy consumption characteristics generally match the requirements of the load-shape objectives established in the previous step. In general, each

Peak clipping—or the reduction of the system peak loads, embodies one of the classic forms of load management and is now commonly referred to as ***demand response***. Peak clipping is generally considered as the reduction of peak load by using time-based rate options or incentive-based strategies, with or without enabling technologies. While many utilities consider this as a means to reduce peaking capacity or capacity purchases and consider strategies only during the most probable days and times of system peak, these strategies can be used to reduce operating cost and dependence on critical fuels by economic dispatch.

Valley filling—is the second classic form of load management and applies to both gas and electric systems. Valley filling encompasses building off-peak loads. This may be particularly desirable where the long-run incremental cost is less than the average price of energy. Adding properly priced off-peak load under those circumstances decreases the average price. Valley filling can be accomplished in several ways, one of the most popular of which displaces loads served by fossil fuels with electric loads that are operated during off-peak periods (e.g., water heating and/or space heating).

Load shifting—is the last classic form of load management and also applies to both gas and electric systems. This involves shifting load from on-peak to off-peak periods. Popular applications include use of storage water heating, storage space heating, coolness storage (the most common type of thermal energy storage), and customer load shifts. The load shift from storage devices involves displacing what would have been conventional appliances.

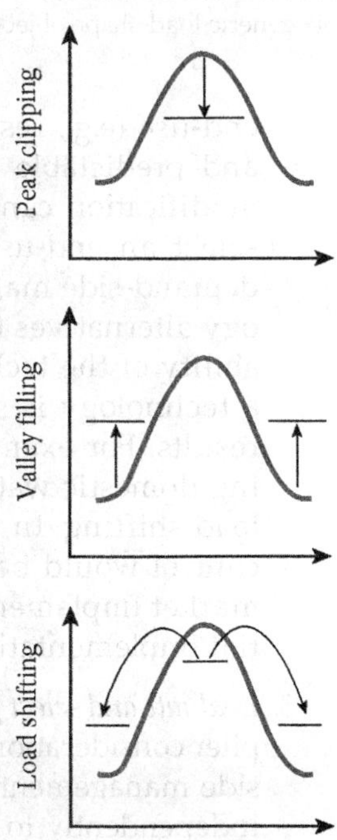

FIGURE 15.3
Six generic load-shape objectives that can be considered during demand-side management planning. (*Continued*)

Strategic conservation—is the load-shape change that results from programs directed at end-use consumption. Not normally considered load management, the change reflects a modification of the load shape involving a reduction in consumption as well as a change in the pattern of use. In employing energy conservation, the planner must consider what conservation actions would occur naturally and then evaluate the cost-effectiveness of possible intended programs to accelerate or stimulate those actions. Examples include weatherization and appliance efficiency improvement.

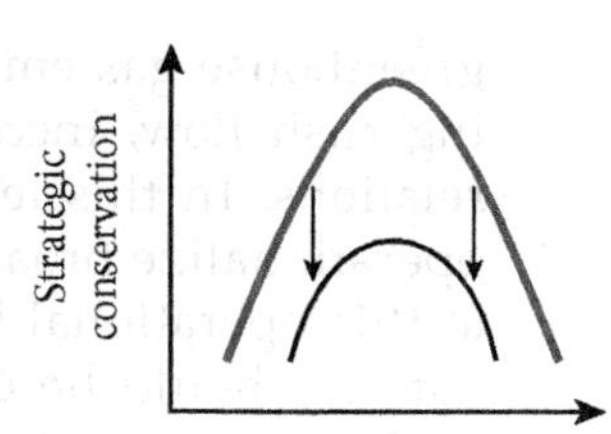

Strategic load growth—is the load-shape change that refers to a general increase in sales beyond the valley filling described previously. Load growth may involve increased market share of loads that are or can be, served by competing fuels, as well as economic development. Load growth may include electrification. Electrification is the term being employed to describe the new emerging electric technologies surrounding electric vehicles, industrial process heating, and automation. These have a potential for increasing the electric energy intensity of the industrial sector. This rise in intensity may be motivated by reduction in the use of fossil fuels and raw materials resulting in improved overall productivity.

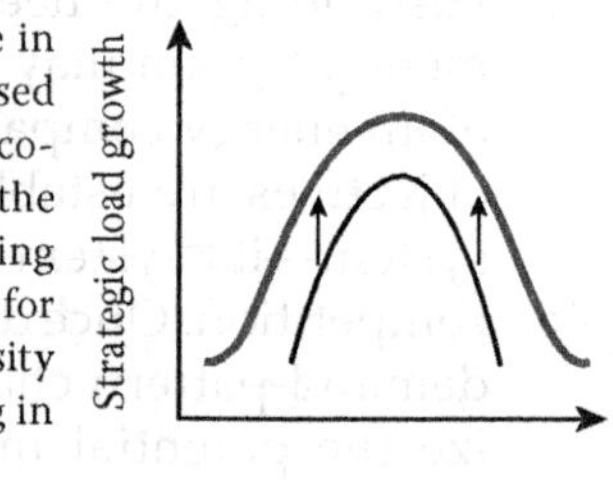

Flexible load shape—is a concept related to electric system reliability, a planning constraint. Once the anticipated load shape, including demand-side activities, is forecast over some horizon, the power supply planner studies the final optimum supply-side options. Among the many criteria he or she uses is reliability. Load shape can be flexible—if customers are presented with options as to the variations in quality of service that they are willing to allow in exchange for various incentives. The program involved can be variations of interruptible or curtailable load; concepts of pooled, integrated energy management systems; or individual customer load control devices offering service constraints.

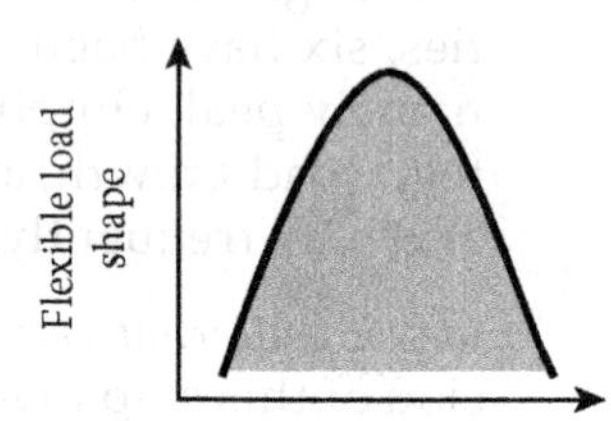

FIGURE 15.3 *(Continued)*
Six generic load-shape objectives that can be considered during demand-side management planning.

end-use (e.g., residential space heating, commercial lighting) exhibits typical and predictable demand or load patterns. The extent to which load pattern modification can be accommodated by a given end-use is one factor used to select an end-use for demand-side management. The second dimension of demand-side management alternatives involves choosing appropriate technology alternatives for each target end-use. This process should consider the suitability of the technology for satisfying the load-shape objective. Even though a technology is suitable for a given end-use, it may not produce the desired results. For example, although water-heater wraps are appropriate for reducing domestic water-heating energy consumption, they are not appropriate for load shifting. In this case, an option such as electric water-heating direct load control would be a better choice. The third dimension involves investigating market implementation methods (see Section 15.4.5 for a description of potential implementation methods).

3. *Evaluate and select program(s)*: The third step balances customer considerations, supplier considerations, and cost/benefit analyses to identify the most viable demand-side management alternative(s) to pursue. Although customers and suppliers act independently to alter the pattern of demand, the concept of demand-side management implies a supplier/customer relationship that produces mutually beneficial results. To achieve that mutual benefit, suppliers must carefully consider such

factors as the manner in which the activity will affect the patterns and amount of demand (load shape), the methods available for obtaining customer participation, and the likely magnitudes of costs and benefits to both supplier and customer prior to attempting implementation.

4. *Implement program(s)*: The fourth step is to implement the program(s), which takes place in several stages. As a first step, a high level, demand-side management project team should be created with representation from the various departments and organizations, and with the overall control and responsibility for the implementation process. It is important for implementers to establish clear directives for the project team, including a written scope of responsibility, project team goals and time frame. When limited information is available on prior demand-side management program experiences, a pilot experiment may precede the program. Pilot experiments can be a useful interim step toward making a decision to undertake a major program. Pilot experiments may be limited either to a subregion or to a sample of consumers throughout an area. If the pilot experiment proves cost-effective, then the implementers may consider initiating the full-scale program.
5. *Monitor program(s)*: The fifth step is to monitor the program(s). The ultimate goal of the monitoring process is to identify deviations from expected performance and to improve both existing and planned demand-side management programs. Monitoring and evaluation processes can also serve as a primary source of information on customer behavior and system impacts, foster advanced planning and organization within a demand-side management program, and provide management with the means of examining demand-side management programs as they develop.

15.4.2 Targeted End-Use Sectors/Building Types

The three broad categories of end-use sectors targeted for demand-side management programs are residential, commercial, and industrial. Each of these broad categories includes several subsectors. In some cases, the program will be designed for one or more broad sectors; in other cases, it may be designed for a specific subsector. For example, the residential sector can be divided into several subsectors including single family homes, multifamily homes, mobile homes, low-income homes, etc. In addition, the commercial sector can be split into subsets, such as offices, restaurants, health care facilities, educational facilities, retail stores, grocery stores, hotels/motels, etc. There are also numerous specific industrial end-users that may be potentially targeted for a demand-side management program. Moreover, the program designer may want to target a specific type or size of building within the chosen sector. The program could focus on new construction, old construction, renovations and retrofits, large customers, small customers, or a combination. Crosscutting programs target multiple end-use sectors and/or multiple building types. Figure 15.4 illustrates the broad types of end-use sectors and building types and how they relate to other aspects of demand-side management program planning.

15.4.3 Targeted End-Use Technologies/Program Types

There are several end-use technologies or program types targeted in demand-side management programs (see Figure 15.4). Some programs are comprehensive, and crossover between end-use technologies. Other programs target specific end-use equipment, such as lighting,

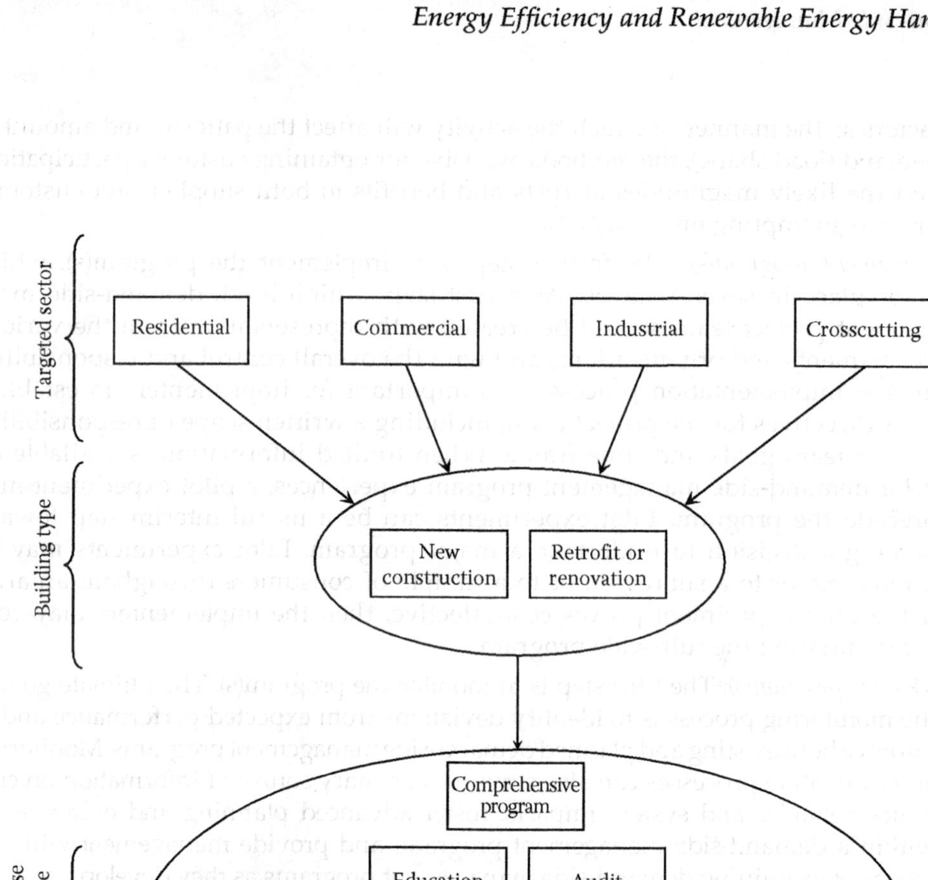

FIGURE 15.4
Relationship between end-use sectors, building types, end-use programs, and program implementers.

air conditioners, dishwashers, etc. Still others target load control measures, such as demand response programs whereby customers temporarily curtail loads in response to peak demand events or programs that permanently shift loads to off-peak hours (e.g., via thermal energy storage). Figure 15.4 shows representative end-use technologies or program types and how they relate to other aspects of demand-side management program planning.

15.4.4 Program Implementers

Implementers of demand-side management programs are often utilities. However, other possible implementers include government organizations, nonprofit groups, private parties, or a collaboration of several entities (see Figure 15.4). Utilities and governments, in particular, have a special interest in influencing customers' demand—treating it not as fate but as choice—in order to provide better service at lower cost while increasing their own profits and reducing their business risks. Energy planners can choose from a wide range of market push and pull methods designed to influence consumer adoption and reduce barriers, as discussed in the next section.

15.4.5 Implementation Methods

Among the most important dimension in the characterization of demand-side alternatives is the selection of the appropriate market implementation methods. Planners and policy makers can select from a variety of methods for influencing customer adoption and acceptance of demand-side management programs. The methods can be broadly classified into six categories. Table 15.1 lists examples for each category of market implementation method. The categories include the following:

1. *Customer education*: Many energy suppliers and governments have relied on some form of customer education to promote general customer awareness of programs. Websites, brochures, bill inserts, information packets, clearinghouses, educational curricula, and direct mailings are widely used. Customer education is the most basic of the market implementation methods available and should be used in conjunction with one or more other market implementation method for maximum effectiveness.
2. *Direct customer contact*: Direct customer contact techniques refer to face-to-face communication between the customer and an energy supplier or government representative to encourage greater customer acceptance of programs. Energy suppliers have for some time employed marketing and customer service representatives to provide advice on appliance choice and operation, sizing of heating/cooling systems, lighting design, and even home economics. Direct customer contact can be accomplished through energy audits, specific program services (e.g., equipment servicing), store fronts where information and devices are displayed, workshops, exhibits, on-site inspection, etc. A major advantage of these methods is that they allow the implementer to obtain feedback from the consumer, thus providing an opportunity to identify and respond to major customer concerns. They also enable more personalized marketing, and can be useful in communicating interest in and concern for controlling energy costs.

TABLE 15.1

Examples of Market Implementation Methods

Market Implementation Method	Illustrative Objective	Examples
Customer education	• Increase customer awareness of programs • Increase perceived value of energy services	• Websites • Bill inserts • Brochures • Information packets • Displays • Clearinghouses • Direct mailings
Direct customer contact	• Through face-to-face communication, encourage greater customer acceptance and response to programs	• Energy audits • Direct installation • Store fronts • Workshops/energy clinics • Exhibits/displays • Inspection services
Trade ally cooperation (i.e., architects, engineers, appliance dealers, heating/cooling contractors)	• Increase capability in marketing and implementing programs • Obtain support and technical advice on customer adoption of demand-side technologies	• Cooperative advertising and marketing • Training • Certification • Selected product sales/service
Advertising and promotion	• Increase public awareness of new programs • Influence customer response	• Mass media (Internet, radio, TV, and newspaper) • Point-of-purchase advertising
Alternative pricing	• Provide customers with pricing signals that reflect real economic costs and encourage the desired market response	• Demand rates • Time-of-use rates • Real-time pricing • Critical peak pricing • Off-peak rates • Seasonal rates • Inverted rates • Variable levels of service • Promotional rates • Conservation rates
Direct incentives	• Reduce up-front purchase price and risk of demand-side technologies to the customer • Increase short-term market penetration • Provide incentives to employees to promote demand-side management programs	• Low- or no-interest loan • Cash grants • Subsidized installation/modification • Rebates • Buyback programs • Rewards to employees for successful marketing of demand-side management programs

3. *Trade ally cooperation*: Trade ally cooperation and support can contribute significantly to the success of many demand-side management programs. A trade ally is defined as any organization that can influence the transactions between the supplier and its customers or between implementers and consumers. Key trade ally groups include home builders and contractors, local chapters of professional societies, technology/product trade groups, trade associations, and associations representing wholesalers and retailers of appliances and energy consuming devices. Depending on the type of trade ally organization, a wide range of services are performed, including

development of standards and procedures, technology transfer, training, certification, marketing/sales, installation, maintenance, and repair. Generally, if trade ally groups believe that demand-side management programs will help them (or at least not hinder their business), they will likely support the program.

4. *Advertising and promotion*: Energy suppliers and government energy entities have used a variety of advertising and promotional techniques. Advertising uses various media to communicate a message to customers in order to inform or persuade them. Advertising media applicable to demand-side management programs include the Internet, radio, television, magazines, newspapers, outdoor advertising and point-of-purchase advertising. Promotion usually includes activities to support advertising, such as press releases, personal selling, displays, demonstrations, coupons, and contest/awards.
5. *Alternative pricing*: Pricing as a market-influencing factor generally performs three functions: (a) transfers to producers and consumers information regarding the cost or value of products and services being provided, (b) provides incentives to use the most efficient production and consumption methods, and (c) determines who can afford how much of a product. These three functions are closely interrelated. Alternative pricing, through innovative schemes can be an important implementation technique for utilities promoting demand-side options. For example, rate incentives for encouraging specific patterns of utilization of electricity can often be combined with other strategies (e.g., direct incentives) to achieve electric utility demand-side management goals. Pricing structures include time-of-use rates, real-time pricing, critical peak pricing, inverted rates, seasonal rates, variable service levels, promotional rates, off-peak rates, etc. A major advantage of alternative pricing programs over some other types of implementation techniques is that the supplier has little or no cash outlay. The customer receives a financial incentive, but over a period of years, so that the implementer can provide the incentives as it receives the benefits.
6. *Direct incentives*: Direct incentives are used to increase short-term market penetration of a cost control/customer option by reducing the net cash outlay required for equipment purchase or by reducing the payback period (i.e., increasing the rate of return) to make the investment more attractive. Incentives also reduce customer resistance to options without proven performance histories or options that involve extensive modifications to the building or the customer's lifestyle. Direct incentives include cash grants, rebates, buyback programs, billing credits, and low-interest or no-interest loans. One additional type of direct incentive is the offer of free, or very heavily, subsidized, equipment installation or maintenance in exchange for participation. Such arrangements may cost the supplier more than the direct benefits from the energy or demand impact, but can expedite customer recruitment, and allow the collection of valuable empirical performance data.

Energy suppliers, utilities, and government entities have successfully used many of these marketing strategies. Typically, multiple marketing methods are used to promote demand-side management programs. The selection of the individual market implementation method or mix of methods depends on a number of factors, including the following:

- Prior experience with similar programs
- Existing market penetration
- The receptivity of policy makers and regulatory authorities

- The estimated program benefits and costs to suppliers and customers
- Stage of buyer readiness
- Barriers to implementation

Some of the most innovative demand-side marketing programs started as pilot programs to gauge consumer acceptance and evaluate program design before large-scale implementation.

The objective of the market implementation methods is to influence the marketplace and to change the customer behavior. The key question for planners and policy makers is the selection of the market implementation method(s) to obtain the desired customer acceptance and response. Customer acceptance refers to customer willingness to participate in a market implementation program, customer decisions to adopt the desired fuel/appliance choice and efficiency, and behavior change as encouraged by the supplier, or state. Customer response is the actual load-shape change that results from customer action, combined with the characteristics of the devices and systems being used.

Customer acceptance and responses are influenced by the demographic characteristics of the customer, income, knowledge, and awareness of the technologies and programs available, and decision criteria, such as cash flow and perceived benefits and costs, as well as attitudes and motivations. Customer acceptance and response are also influenced by other external factors, such as economic conditions, energy prices, technology characteristics, regulation, and tax credits.

15.4.6 Characteristics of Successful Programs

Numerous demand-side management programs are implemented in the United States yearly by various organizations including utilities, state agencies, and third-party implementers. In addition, some businesses implement their own internal or corporate-wide energy efficiency programs. Over the years, groups such as the EPRI, American Council for an Energy-Efficient Economy (ACEEE), Energy Trust of Oregon, and California Best Practices Project Advisory Committee with contractor, Quantum Consulting Inc., have reviewed and compared demand-side management programs to identify exemplary programs and best practices. The results of those studies are included in an assortment of reports (Peters 2002; Quantum Consulting 2004; Wikler et al. 2008; Nowak et al. 2013; Young and Mackres 2013).

EPRI conducted an assessment of best practices in demand-side management programs (Wikler et al. 2008). The study included interviews with utility program staff, an online survey of utility representatives, and a comprehensive review of recent best practices studies in the literature. The study revealed a number of elements associated with successful program design and delivery. Some of the elements vary by customer segment and program type, while others are common across multiple program types. Table 15.2 lists the challenges faced by selected types of Residential and Commercial & Industrial (C&I) programs and identifies the key factors that have contributed to programs' success. The table also includes examples of programs that have applied these best practices. The following subsections distill some of the more generally applicable best practices for program design and delivery.

TABLE 15.2

Synthesis of Key Challenges and Success Factors for Selected Energy Efficiency and Load Management Programs

Program Type	Key Challenges	Key Success Factors	Program Example
Residential conservation	• Technology developments • Easy access to pay stations	• Enhancing customer knowledge and control over their consumption	Salt River Project—M Power
Residential lighting	• Low customer motivation, especially if electricity prices are low • Utility resource allocation toward marketing and outreach efforts • Overcome customer experience or perception of poor quality products • CFL disposal issues	• Aggressive customer awareness and outreach efforts • Strong publicity campaigns • Leverage ENERGY STAR® brand • Build and maintain relationships with manufacturers and retailers	Georgia Power—ENERGY STAR Change a Light (CAL) Campaign Snohomish County PUD—Residential CFL Program
Residential load management	• Obtain customer trust to install equipment • Address customer complaints unrelated to the program • Low level of customer motivation if electricity prices are low	• Simple program design • Widespread customer awareness and outreach efforts • Establish system reliability • Maintain continuity in customer participation • Build strong customer relationships • Provide customer choice and control on electricity usage	Florida Power and Light—Residential On Call Gulf Power—Good Cents Select
Residential new construction	• Generate interest in the program by all parties • Maintain consistency with builders and energy raters	• ENERGY STAR name recognition • Build and maintain relationships with builders and raters	Center Point Energy—ENERGY STAR New Home Oncor Electric Delivery—ENERGY STAR Homes Program
Low income	• Build customer trust and confidence • Rapid turnaround in contractor workforce responsible for delivery	• Deliver nonenergy benefits, such as comfort and safety • Engender participant trust through program stability and continuity • Successful education and outreach efforts • Partner with community-based organizations for effective delivery	Pacific Gas and Electric—Energy Partners
C&I energy efficiency	• Incorporate flexible incentives to fit different customer requirements, while keeping program design simple	• Simplicity in program design • Strong financial incentives • Keep customer's financial bottom-line in mind • Strong customer relationships/maintaining close contact with customers directly and through contractors	Alliant Energy—Shared Savings Program

(Continued)

TABLE 15.2 (*Continued*)

Synthesis of Key Challenges and Success Factors for Selected Energy Efficiency and Load Management Programs

Program Type	Key Challenges	Key Success Factors	Program Example
C&I new construction	• Generate interest from designers, builders, and owners • Shortage of qualified staff	• Build broad awareness • Start with a pilot program and expand • Staff development and training • Flexible approach • Full energy simulation capability	We Energies—C/I New Construction Program
C&I retrofit	• Need to conform to new/increased codes and standards • Interactions with regulators	• Flexibility in accommodating different measures • Stakeholder collaborative process in designing programs • Highly skilled utility staff • Contractor capacity building • Contractor–customer relationships	Connecticut Light and Power - Energy Opportunities
C&I niche	• Accommodate industry requirements • Obtain customer interest for participation • Differences in using the new technology	• Target niche, high-growth industries • Dynamic program design to fit market requirements • Form partnerships and collaborations with related groups • Integrate energy efficiency into customers' business strategies • Achieve more than just the energy benefits	Pacific Gas and Electric—High Tech Energy Efficiency Program Salt River Project—Pre Rinse Spray Valves
Small business	• Develop web-based delivery infrastructure • Bring vendors up to speed	• Make program participation as simple as possible • Build a strong network of vendors	Southern California Edison—Express Efficiency Rebate Program

Source: Wikler, G. et al., Best practices in energy efficiency and load management programs, 1016383, EPRI, Palo Alto, CA, 2008.

15.4.6.1 Key Elements of Program Design

Successful program design effectively translates the program design theories into a practical program structure with actionable policies and procedures aimed at meeting program goals and objectives. Key elements of success include the following:

- Maintain simplicity in program design
- Design incentive structure to fit customer requirements
- Maintain flexibility in program design to accommodate measures
- Maintain dynamism in program design to fit market requirements in specific industries

- Develop sound performance tracking mechanisms
- Incorporate customer choice and control features in program design
- Ensure resource allocation is commensurate with program tasks
- Obtain stakeholder support right at the design stage
- Maintain high quality of products
- Establish program branding
- Undertake program improvements over time

15.4.6.2 Key Elements of Program Delivery

Program delivery consists of marketplace actions to promote demand-side management practices and to increase adoption of technologies and measures. Specific delivery activities encompass the implementation methods discussed in Section 15.5. There are several key elements of successful program delivery:

- Foster trade ally relationships and partnerships
- Undertake contractor capacity building efforts
- Build networks and alliances with other relevant parties and groups
- Undertake program publicity campaigns
- Establish strong customer education and outreach efforts
- Foster utility–customer relationships
- Coordinate with other utilities and program administrators
- Build customer–contractor relationships
- Maintain strong in-house capabilities
- Integrate energy efficiency into customer's business strategy
- Deliver nonenergy benefits
- Maintain consistency over time
- Enhance program delivery through collaborative efforts

15.5 Case Studies

This section examines case studies of four successful demand-side management programs and a smart grid demonstration project. Each of the four programs targets a different sector and set of end-uses. In addition, each program represents a different U.S. geographical region and has a different implementation structure. The smart grid demonstration project is a widespread effort involving a collaboration of over 20 utilities in the United States and abroad with the common goal of advancing the integration of distributed energy resources into the electric power grid.

Table 15.3 presents the first case study, which is a comprehensive industrial energy efficiency program administered by Bonneville Power Administration (BPA). The second

TABLE 15.3

Case Study #1—2012 Bonneville Power Administration (BPA) Energy Smart Industrial (ESI) Program

Description:
BPA's ESI program was designed and rolled out in 2009 with partner Cascade Energy to help BPA meet aggressive industrial energy savings goals. The program targets the industrial sector in BPA's service territory in the Pacific Northwest and incorporates a wide range of retrofit and operation and maintenance measures as well as continuous improvement options aimed at behavioral changes. BPA's participating utility customers can choose from this assortment of measures and several different market delivery options to serve their industrial customers. Customers receive incentives for projects and many are eligible for subsidized technical consulting services. A unique program aspect is the use of a qualified ESI partner (ESIP) assigned by the ESI program to act as a single point of contact for the program to the utilities and industrial customers. In addition, a pilot feature of the program incorporates three innovative Energy Management options to improve operation and maintenance and management aspects of energy use so that savings persist: (1) co-funding for an Energy Project Manager to serve as a staff resource to promote energy efficiency projects at qualifying facilities, (2) support for no-cost/low-cost operation and maintenance improvements including tools for data collection and tracking and incentive funding (referred to as Track and Tune), and (3) use of behavior-based and continuous improvement approaches to train industrial facilities to incorporate energy management into core business practices (referred to as High Performance Energy Management). The program's success is reflected in its high performance over a short period of time, cost-effectiveness, utility and industrial customer satisfaction with offerings, and simplified communications approach.

Targeted sector/building type:
Industrial customers of BPA's multiple utility customers in the Pacific Northwest.
Any industrial customer of a participating utility is eligible.
Targeted industries include pulp and paper, wood products, food processing, and water and wastewater.

Targeted end-use technology/program type:
Targets multiple end-users through various measure types: custom and prescriptive retrofit projects; no-cost/low-cost operation and maintenance improvements; and behavior-based measures

Program administrator/implementer:
Administered by Bonneville Power Administration
Implemented by Cascade Energy

Program expenditures for 2012 (estimated):
$15.2 million

Program results for 2012 (estimated):
Net energy savings: 91,980 MWh
Net demand savings: 10.50 aMW
Participation: 105 enrolled utilities; 86 engaged utilities; 478 end-users

Source: Nowak, S. et al., Leaders of the pack: ACEEE's third national review of exemplary energy efficiency programs, Report No. U132, ACEEE, Washington, DC, 2013.
Personal communication with Jennifer Eskil, Agriculture/Industrial Sector Lead, Bonneville Power Administration, August 20, 2013.

case study (Table 15.4) is a retrocommissioning (RCx) program offered to the commercial sector by Commonwealth Edison (ComEd) with a focus on controls optimization and other operational energy efficiency improvements. The third case study (Table 15.5) is a Southern California Edison (SCE) pilot program that was offered to water utilities and targeted both water and energy savings. These first three programs received the distinction of *exemplary* by ACEEE in 2013 for their excellence in program design and delivery. Table 15.6 presents the fourth case study, a smart grid-enabled demand response program offered to residential and small commercial customers by Oklahoma Gas and

TABLE 15.4

Case Study #2—2011/2012 Commonwealth Edison (ComEd) Smart Ideas for Your Business Retrocommissioning (RCx) and Monitoring-Based Commissioning (MBCx) Program

Description:
In ComEd's Smart Ideas for Your Business RCx and MBCx programs, approved engineering firms (Service Providers) provide onsite assessments and analysis of building energy systems that focus on controls optimization and other operational improvements. ComEd pays the Service Providers to do the study, at no cost to the customer. The customer in exchange agrees to invest at least a minimum amount on operational improvements with paybacks of 18 months or less. A newer feature of the program is the inclusion of natural gas measures in the analysis, which is enabled by partnerships with local gas companies. Investigating electric and gas measures simultaneously optimizes the Service Providers' time at the site and provides additional value to the customers and utilities. Another enhancement to the program is the addition of an MBCx option in which the customer receives an incentive to help offset the cost of installing advanced building automation software to monitor energy use and identify opportunities for operational improvements over a period of 18 months or more. Then, the customer is paid an additional incentive for verified energy savings resulting from these improvements. Some of the reasons for the program's success include the use of preapproved Service Providers as a sales channel, the partnerships with natural gas utilities and ability to identify electric and gas measures, and the program's continued efforts to enhance processes and adapt to reach more customers and savings opportunities.

Targeted sector/building type:
Commercial sector in Northern Illinois: retail and office buildings; commercial real estate; hospitals; education; hospitality; and other buildings with > 150,000 sq. ft. of air-conditioned floor space

Targeted end-use technology/program type:
RCx and MBCx programs
Targets operation and maintenance improvements such as economizer and ventilation control, equipment scheduling, and fan optimization and air distribution modifications
Include electric and gas saving measures

Program administrator/implementer:
Administrated by ComEd in partnership with Nicor Gas, North Shore Gas, and Peoples Gas
Implemented by Nexant

Program expenditures for program year 4 (2011/2012):
$4.84 million

Program results for program year 4 (2011/2012):
Net electricity savings: 25,021 MWh
Participation: 50 customers

Source: Nowak, S. et al., Leaders of the pack: ACEEE's third national review of exemplary energy efficiency programs, Report No. U132, ACEEE, Washington, DC, 2013.
Personal communication with Rick Tonielli, Sr. Energy Efficiency Program Manager, ComEd, August 14, 2013.

Electric (OG&E). OG&E was recognized as Utility of the Year in 2011 by *Electric Light & Power*, in part due to OG&E's successful demand management programs and their high customer satisfaction rating during a period of very high Smart Meter deployment. The smart grid demonstration project (Table 15.7) is a large-scale EPRI initiative focused on demonstrating grid integration of distributed energy resources, such as demand response technologies, electric vehicles, thermal energy storage, electric storage, solar photovoltaics, wind generation, conservation voltage reduction, and distributed generation. The project just completed its fifth year of a 7-year effort. As Tables 15.3 through 15.7 show, all of the programs and initiatives presented in these case studies have yielded impressive results. Even more importantly, their innovative approaches have contributed to the advancement of demand-side management practices.

TABLE 15.5

Case Study #3—Southern California Edison (SCE) Leak Detection Pilot Program

Description:

The SCE Leak Detection Pilot Program was one of nine *Embedded Energy in Water Pilots* established by the California Public Utility Commission (CPUC) to study alternatives for reducing water-related energy use. The pilot program involved conducting audits and repairing leaks in water distribution systems. Saving water through leak repair also reduces the embedded energy requirements associated with the supply, conveyance, treatment, and distribution of water to end-users. The pilot consisted of three partners: SCE, participating water agencies, and the audit implementation contractor, Water Systems Optimization Inc. The program demonstrated the largest energy savings potential for relatively low cost compared to the CPUC pilots and was successful as measured by the satisfaction of program partners. Success was due to a combination good planning, dedicated involvement by the SCE program manager, good communication between partners, and professional and high quality work. The potential for larger scale deployment of the program hinges on the CPUC and California Energy Commission. Of particular relevance is whether or not the CPUC will publish a rule allowing electric utilities to claim embedded energy savings arising from water saving measures.

Targeted sector/building type:

Municipal water utilities in Southern California

Targeted end-use technology/program type:

Targets leaks in water distribution systems

Pilot program that includes audits of distribution systems and repairs of leaks

Program administrator/implementer:

SCE

Program expenditures for an 18-month period (July 08–December 09):

$300,000

Program results:

Annual energy savings: 498 MWh

Annual water savings: 83 million gallons

Participation: three water agencies (Las Virgenes Municipal Water District, Apple Valley Ranchos Water Company, and Lake Arrowhead Community Services District)

Source: Young, R. and Mackres, E., Tackling the nexus: Exemplary programs that save both energy and water, Report No. E131, ACEEE, Washington, DC, 2013.

Personal communication with Gene Rodrigues, Director of DSM Strategy, Portfolio Oversight and Technical Support, Southern California Edison Company, August 13, 2013.

TABLE 15.6

Case Study #4—2012 OG&E SmartHours Program

Description:

OG&E's SmartHours Program is a multilevel voluntary dynamic pricing program designed to encourage participants to reduce on-peak energy use by offering them lower rates during off-peak periods. The on-peak hours are defined as 2:00–7:00 p.m. on weekdays during the summer from June 1 through September 30, excluding holidays. The rest of the time is considered off-peak. During the on-peak, there are several pricing levels based on system demand and weather conditions: low, standard, high, and critical. Pricing events are called via e-mail, text message, and/or voice mail the day ahead for the first three levels. For the critical level, pricing events can be called a day ahead or there is also a provision for a critical price overcall event (CPE) in which participants receive a minimum of 2-h notice and then the price rises to the critical level. A CPE lasts 2–8 h and the maximum number of CPE hours per year is 80.

OG&E has deployed Smart Meters throughout its service territory. All SmartHours participants have Smart Meters and access to their energy information through a website. Program participants are also offered a free programmable communicating thermostat (PCT) to help them automate their response to pricing events (62% of 2012 participants accepted the PCT). OG&E does not directly control the thermostat; instead, customers set thresholds based on pricing events and their personal preferences.

Aspects of the program that have contributed to its success include free enabling technology and services, customer empowerment of thermostat control and overall energy use, pricing that reflects the true cost of electricity, and the no-lose proposition provided to customers for the first year to guarantee they will not pay more on the SmartHours rate than they would have on the standard flat rate.

Targeted sector/building type:

Residential and small commercial customers

Targeted end-use technology/program type:

Program targets reduction in on-peak loads through time-based pricing and smart grid products and services

Program implementer:

OG&E

Program expenditures for 2012 (approximate):

$20 million

Program results for 2012:

Overall maximum demand reduction: 51.4 MW on day of the system peak at hour-ending 4:00 p.m.

System peak demand reduction: 44.1 MW on day of system peak at system peak hour (hour-ending 5:00 p.m.)

Participation: 35,144 participants as of September 2012

Participants without PCTs responded significantly less than those with PCTs

Source: Marrin, K. and Williamson, C., *2012 Evaluation of OG&E SmartHours Program*, Prepared for Oklahoma Gas & Electric, EnerNOC, Walnut Creek, CA, 2012.
Personal communication with Mike Farrell, Director of Customer Programs, Oklahoma Gas and Electric, October 1, 2013.

TABLE 15.7

Case Study #5—EPRI Smart Grid Demonstration Initiative

Overview of initiative:

EPRI's Smart Grid Demonstration Initiative is a 7-year collaborative research effort involving two dozen utilities from Australia, Canada, France, Ireland, Japan, and the United States. The focus of the initiative is to design, deploy, and evaluate ways to integrate distributed energy resources into the electric grid and into market operations. The goal of the collaborative effort is to leverage electric utility Smart Grid investments and share research and demonstration results with smart grid technologies and applications to create a *smarter* grid. Each project that is undertaken by a member of the initiative contributes to the collective knowledge. With the initiative recently completing its fifth year, many projects addressing a wide range of research questions have been completed or are underway. EPRI and utility members of the initiative have created hundreds of reports and case studies describing the results (see www.smartgrid.epri.com).

Collaborators:

The collaborators consist of host sites and non-host sites. Host site collaborators are conducting the majority of the field projects.

Host sites: American Electric Power, Con Edison, Duke Energy, Electricité de France, Ergon, ESB Networks, Exelon (ComEd/PECO), First Energy, Hawaiian Electric Company, Hydro Québec, Kansas City Light and Power, PNM Resources, Sacramento Municipal Utility District, Southern California Edison, Southern Company

Non-host sites: Ameren, Central Hudson Gas and Electric, CenterPoint Energy, Entergy, Salt River Project, Tennessee Valley Authority, Tokyo Electric Power Company, Wisconsin Public Service

Distributed energy resources under investigation:

The initiative is studying grid integration of eight different types of distributed energy resource technologies: demand response technologies, electric vehicles, thermal energy storage, electric storage, solar photovoltaics, wind generation, conservation voltage reduction, and distributed generation

Communications and standards under investigation:

Customer Domain (SEP, WiFi, etc.), Distribution (DNP3, IEC 61850, etc.), Cyber Security, Advanced Metering Infrastructure or Automated Meter Reading (AMI or AMR), Radio Frequency (RF) Mesh or Tower, Public or Private Internet, Cellular 3G (GPRS, CDMA, etc.), and Cellular 4G (WiMAX, LTE, etc.)

Grid management under investigation:

Volt/VAR optimization (VVO) and conservation voltage reduction (CVR), Distribution Automation, and Grid Management System (DMS, DERMS, DRMS)

Demand-side management programs under investigation:

Price-based (time of use, critical peak pricing, real-time pricing, etc.)

Incentive-based (direct load control, interruptible, etc.)

Operations and planning under investigation:

System Operations Integration, System Planning Integration, and Modeling and/or Simulation Tools

Source: Wakefield, M. and Horst, G., EPRI smart grid demonstration initiative: 5 year update, 3002000778, EPRI, Palo Alto, CA, 2013.

15.6 Conclusions

Since the early 1970s, economic, political, social, technological, and resource supply factors have combined to change the energy industry's operating environment and its outlook for the future. Many utilities are faced with staggering capital requirements for new plants, significant fluctuations in demand and energy growth rates, declining financial performance, and political or regulatory and consumer concern about rising prices and the environment. While demand-side management is not a cure-all for these difficulties, it does provide for a great many additional alternatives that have myriad nonenergy benefits as well as the more obvious energy-related benefits. These demand-side alternatives are equally appropriate for consideration by utilities, energy suppliers, energy-service suppliers, and government entities. Implementation of demand-side measures not only benefits the implementing organization by influencing load characteristics, delaying the need for new energy resources, and in general improving resource value, but also provides benefits to customers such as reduced energy bills and/or improved performance from new technological options. In addition, society as a whole receives economic, environmental, and national security benefits. For example, since demand-side management programs can postpone the need for new power plants, the costs and emissions associated with fossil-fueled electricity generation are avoided. Demand-side management programs also tend to generate more jobs and expenditures within the regions where the programs are implemented, boosting local economies. Moreover, demand-side management programs can help reduce a country's dependence on foreign oil imports, improving national security. Demand-side management alternatives will continue to hold an important role in resource planning in the United States and abroad, and will be a critical element in the pursuit of a sustainable energy future.

References

Evans, M., P. Meagher, A. Faruqui, and J. Chamberlin. 1993. Principles and practice of demand-side management. TR-102556. Palo Alto, CA: EPRI.

Gellings, C. W. 1984–1988. *Demand-Side Management: Volumes 1–5*. Palo Alto, CA: EPRI.

Gellings, C. W. 2002. Using demand-side management to select energy efficient technologies and programs. In *Efficient Use and Conservation of Energy*, C. W. Gellings (ed.). Encyclopedia of Life Support Systems (EOLSS). Oxford, U.K.: EOLSS Publishers. http://www.eolss.net (accessed March 3, 2015).

Gellings, C. W. and J. H. Chamberlin. 1993. *Demand-Side Management: Concepts and Methods*, 2nd edn. Lilburn, GA: The Fairmont Press, Inc.

Hayes, S., N. Baum, and G. Herndon. 2013. Energy efficiency: Is the United States improving? White Paper. Washington, DC: ACEEE.

Marrin, K. and C. Williamson. 2012. *2012 Evaluation of OG&E SmartHours Program*. Prepared for Oklahoma Gas & Electric. Walnut Creek, CA: EnerNOC.

Nowak, S., M. Kushler, P. Witte, and D. York. 2013. Leaders of the pack: ACEEE's third national review of exemplary energy efficiency programs. Report No. U132. Washington, DC: ACEEE.

Peters, J. 2002. *Best Practices from Energy Efficiency Organizations and Programs*. Portland, OR: Energy Trust of Oregon.

Quantum Consulting. 2004. *National Energy Efficiency Best Practices Study.* Multiple Volumes. Submitted to California Best Practices Project Advisory Committee. Berkeley, CA: Quantum Consulting Inc. http://www.eebestpractices.com (accessed March 3, 2015).

Siddiqui, O., P. Hurtado, K. Parmenter et al. 2008. Energy efficiency planning guidebook: Energy efficiency initiative. 1016273. Palo Alto, CA: EPRI.

Wakefield, M. and G. Horst. 2013. EPRI smart grid demonstration initiative: 5 year update. 3002000778. Palo Alto, CA: EPRI.

Wikler, G., D. Ghosh, K. Smith, and O. Siddiqui. 2008. Best practices in energy efficiency and load management programs. 1016383. Palo Alto, CA: EPRI.

Young, R. and E. Mackres. 2013. Tackling the nexus: Exemplary programs that save both energy and water. Report No. E131. Washington, DC: ACEEE.

16

Fossil Fuels

Anthony F. Armor, Rameshwar D. Srivastava, Howard G. McIlvried, Thomas D. Marshall, and Sean I. Plasynski

CONTENTS

16.1 Fossil Fuels*

This section discusses the generation of electric power based on combustion of fossil fuels including coal, natural gas, and fuel oil derived from petroleum (Perry 2008; Riegel 2007). Though the scope of this review is worldwide, the emphasis is on the United States, where

* Though this chapter was largely written in 2013, Table 16.6 was updated in review to reflect 2015 data.

a wealth of available information and the nation's leading economic position make it a frequent example. Similarly, the worldwide abundance of coal makes it a frequent fuel of choice, though there are environmental-regulatory issues that must be addressed in order for its use in electric power generation to continue.

16.1.1 Introduction

The generation of electric power from fossil fuels has increasingly been called upon to be clean and safe while maintaining affordability. The drive to reduce emissions and increase efficiency, have led to significant technical advances over the past several decades in order to keep electric power generation from fossil fuels economically competitive in the marketplace. Globally, fossil fuel–fired electric generation is about two-thirds of all power produced. More large power plants worldwide burn coal than any other fuel, although many generating companies are adding natural gas–fired plants, particularly where the cost of gas-fired generation and the long-term supply of gas appear favorable.

It is likely, particularly in the developed world, that gas turbine–based plants will continue to be added to the new-generation fleet in the immediate future. The most advanced combustion turbines now achieve more than 40% lower heating value (LHV) efficiency in simple cycle mode and greater than 50% efficiency in combined cycle mode. (See Section 16.1.5.6 for a discussion of combined cycles.) In addition, natural gas combined cycle (NGCC) plants offer siting flexibility, rapid construction schedules, and capital costs between \$728/kW and \$1511/kW (June 2011 basis), less than half that of other plant types (NETL 2012b). These advantages, coupled with adequate natural gas supplies (especially in light of the rapid development of shale gas in the United States) and the potential of coal gasification backup, have made NGCC technology an important choice for green-field and repowered plants in the United States and Europe.

However, new capacity in the developing nations of Asia, such as India and China, as well as existing fleets in the Americas and Europe, remains dominated by pulverized coal (PC) plants. For developing economies, coal-fired power plants become a primary choice for power generation: Coal is plentiful and inexpensive, and sulfur dioxide (SO_2) scrubbers have proven to be reliable and effective. In fact, up to 99% SO_X removal efficiency is now possible. Removal of nitrogen oxides (NO_X) is also well advanced, with removal levels of over 90% possible using selective catalytic reduction (SCR). Technologies for mercury removal are available, and the effort to capture carbon dioxide (CO_2) from fossil plants and send it to geologic storage is receiving attention as a means to reduce greenhouse gas (GHG) emissions.

Coal is converted to electric power by three basic technologies: PC-fired boilers, combustion in a fluidized bed of coal and other solids, and integrated gasification combined cycle (IGCC) plants. The PC plant, the most common coal-fired power plant, has the potential for much improved efficiency even with full flue gas desulfurization (FGD). Ferritic materials technology has now advanced to the point where higher steam pressures and temperatures are possible. Advanced supercritical and ultrasupercritical PC plants are moving ahead commercially, particularly in Japan and Europe. (For a discussion of steam cycles, including subcritical, supercritical, and ultrasupercritical, see Section 16.1.5.)

Worldwide, the application of atmospheric fluidized-bed combustion (AFBC) plants has increased, and such plants offer reductions in both SO_2 and NO_X while permitting the efficient combustion of low rank fuels such as lignites, of which there are vast deposits. Since the early 1990s, AFBC boiler technology has become established worldwide as a mature, reliable technology for the generation of steam and electric power with its advantage of in-bed SO_2 capture with limestone. In fact, a major driving force in the widespread deployment, since the mid-1980s, of this relatively new boiler technology has been its capability for in situ SO_2 capture, which would eliminate or reduce the need for FGD. Another advantage is the ability of fluidized-bed combustion (FBC) to handle a range of coal qualities.

IGCC power plants are operating at the 250–600 MW level. Much of the impetus for their development came from the U.S. DOE's clean coal program, which spawned two integrated gasification projects that achieved successful commercial operation. Large gasification plants for power are in operation in Europe and Asia. Gasification with combined cycle operation not only leads to minimum SO_2, NO_X, and particulate emissions, but also provides an opportunity to take advantage of gas turbine advances.

Emergent technologies also include ultrasupercritical steam cycles; improved AFBC, including pressurized FBC; oxycombustion; chemical looping combustion; and fuel cell power plants, potentially including fuel cell combined cycle.

16.1.2 Fuels for Electric Power

The U.S. Energy Information Administration provides an outlook for existing fossil-fueled electricity-generating capacity from various energy sources for both the United States and the world (see Figures 16.1 and 16.2).

The U.S. electric power industry burns about 930 million short tons of coal each year, accounting for 10%–20% of the power costs of coal-fired plants. Options in the fuel mix for power plants include mixtures of eastern high-sulfur coal and low-sulfur, low-cost coals, often from the Powder River Basin (PRB) in Montana and Wyoming. Compared to eastern bituminous coals, PRB subbituminous coals have reduced LHV and sulfur and higher ash and moisture content. Other mixes include 10%–20% gas with coal in a boiler designed for coal firing, and mixtures of coal and petroleum coke—a by-product of refining with a (currently) low cost but high sulfur content. Table 16.1 shows world fossil fuel reserves.

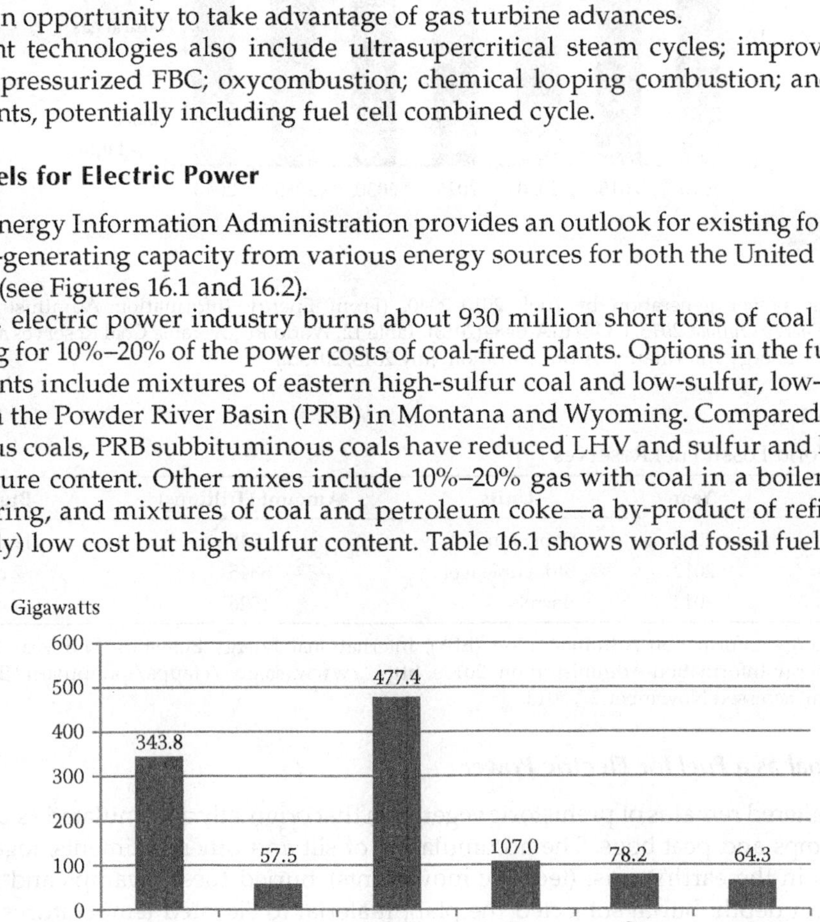

FIGURE 16.1
U.S. electric generating capacity by fuel. (From Energy Information Administration (EIA), *Electric Power Annual 2011*, Table 4.3, Existing Capacity by Energy Source, 2011 (Megawatts), p. 70., U.S. Energy Information Administration, January 2013, 2013a.)

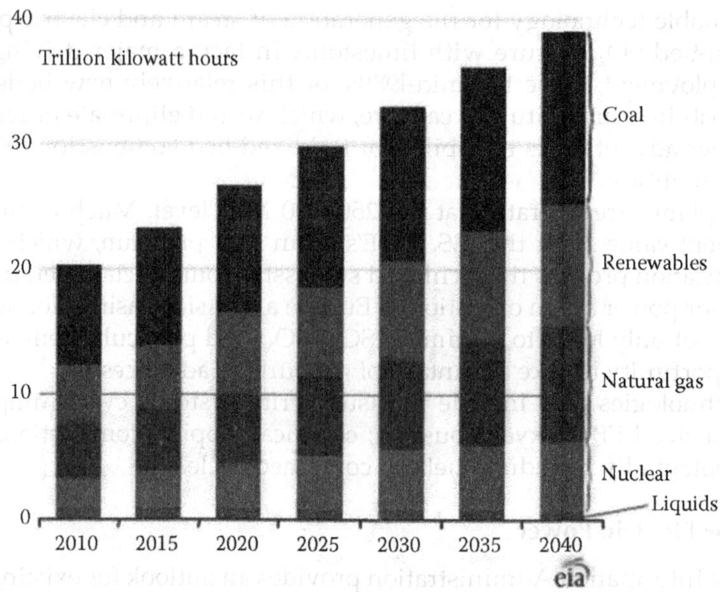

FIGURE 16.2
World electric power generation by fuel, 2010–2040. (From Energy Information Administration (EIA), *International Energy Outlook 2013*, DOE/EIA-0484(2013), Table 12, World Recoverable Coal Reserves as of January, 2009, p. 85, U.S. Energy Information Administration, July 2013, 2013d.)

TABLE 16.1
Proved World Fossil Fuel Reserves

Fuel	Year	Units	Amount (Trillions)	Btu (approx.)
Coal	2008	Short tons	948	19 quintillion
Natural gas	2012	Std. cubic feet	6845	7 quintillion
Crude oil	2012	Barrels	1526	9 quintillion

Source: Energy Information Administration (EIA), International Energy Statistics, End year 2012, U.S. Energy Information Administration, 2013e, http://www.eia.gov/cfapps/ipdbproject/IEDIndex3.cfm, accessed November 27, 2013.

16.1.2.1 Coal as a Fuel for Electric Power

Coal is the altered remains of prehistoric vegetation that originally accumulated as plant material in swamps and peat bogs. The accumulation of silt and other sediments, together with movements in the earth's crust (tectonic movements), buried these swamps and peat bogs, often to great depth. Burial subjected the plant material to elevated temperatures and pressures, which caused physical and chemical changes in the vegetation, transforming it into coal by a process known as coalification. Initially, peat, a precursor of coal, was converted into lignite or brown coal—coal types with low organic maturity. Over time, the continuing effects of temperature and pressure produced additional changes, converting the lignite progressively into subbituminous coal, bituminous coal, and anthracite. As this process continues, the coal becomes increasingly harder, its carbon content increases, and its oxygen content decreases.

The degree of metamorphism or coalification undergone by a coal, as it matures from peat to anthracite, is referred to as the *rank* of the coal; rank has an important bearing on the physical and chemical properties. Low-rank coals, such as lignite and subbituminous, are typically soft, friable materials with a dull, earthy appearance and are characterized by high moisture level and low carbon content, and hence low energy content. Higher rank coals are harder and stronger and often have a black vitreous luster. Increasing rank is accompanied by a rise in the carbon and energy contents and a decrease in the moisture content. Anthracite ranks at the top of the scale and has higher carbon and energy content and low moisture.

Formation of large coal deposits commenced following the evolution of land plants in the Devonian period, some 400 million years ago. Significant accumulations of coal occurred in the Northern Hemisphere during the Carboniferous period (350–280 million years ago), in the Southern Hemisphere during the Carboniferous/Permian periods (350–225 million years ago) and, more recently, in the late Cretaceous to early Paleogene periods (approximately 100–60 million years ago) in areas as diverse as the United States, South America, Indonesia, and New Zealand.

Worldwide, coal is the most plentiful fossil fuel. It is geographically dispersed, being spread over 100 countries and all continents. Coal reserves have been identified that confirm over 200 years of resource availability. U.S. reserves categorized by type are depicted in Figure 16.3. Over half (54%) is lignite and subbituminous coal used primarily for power generation. Table 16.2 shows world coal reserves by rank.

As it comes from the mine, coal typically contains extraneous mineral matter, often with a high sulfur content. Removal of as much of this material as possible is an important step in preparing coal for shipment to a power plant. Cleaning not only increases the coal's heating value (typically by about 10% but occasionally by 30% or higher) and lowers sulfur (up to 70% SO_2 emissions reduction is possible), but also reduces shipping costs. Coal cleaning is well established in the United States with more than 400 operating cleaning plants, mostly located at mines. Sulfur in coal occurs in two forms: mineral and organic.

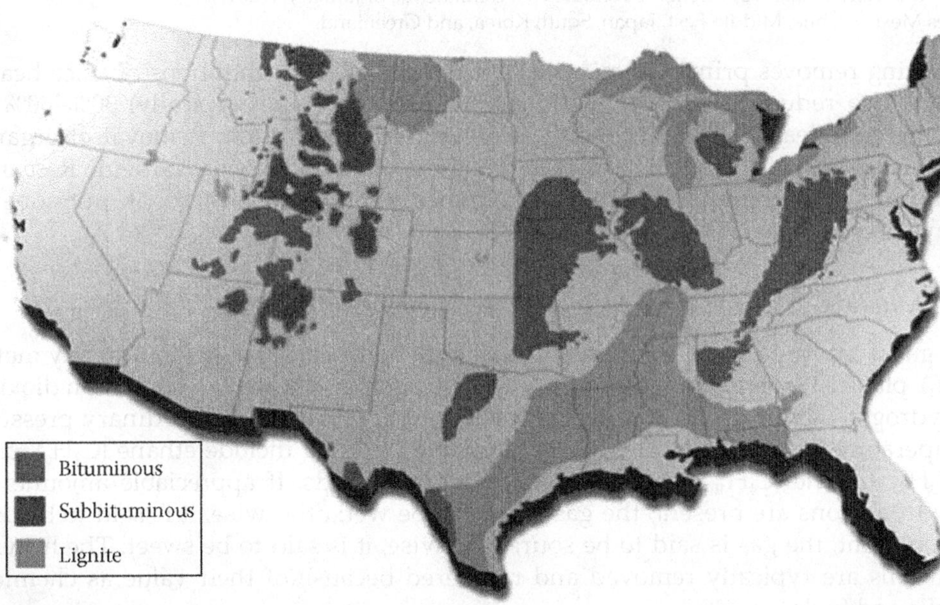

FIGURE 16.3
U.S. coal basins.

TABLE 16.2

World Recoverable Coal Reserves as of January 1, 2009

Region/Country	Recoverable Reserves by Coal Rank (Billion Short Tons) Bituminous and Anthracite	Subbituminous	Lignite	Totals	2010 Production	Reserves-to-Production Ratio (Years)
World total	**445.0**	**285.9**	**215.2**	**946.1**	**7.954**	**119**
United States[a]	118.4	107.2	33.1	**258.6**	1.084	238
Russia	54.1	107.4	11.5	**173.1**	0.359	482
China	68.6	37.1	20.5	**126.2**	3.506	36
Other non-OECD Europe and Eurasia	42.2	18.9	39.9	**100.9**	0.325	311
Australia and New Zealand	40.9	2.5	41.4	**84.8**	0.473	179
India	61.8	0.0	5.0	**66.8**	0.612	109
OECD Europe	6.2	0.9	54.5	**61.6**	0.620	99
Africa	34.7	0.2	0.0	**34.9**	0.286	122
Other non-OECD Asia	3.9	3.9	6.8	**14.7**	0.508	29
Other Central and South America	7.6	1.0	0.0	**8.6**	0.085	101
Canada	3.8	1.0	2.5	**7.3**	0.075	97
Brazil	0.0	5.0	0.0	**5.0**	0.006	842
Others[b]	2.6	0.8	0.1	**3.6**	0.015	233

Source: Energy Information Administration (EIA), *International Energy Outlook 2013*, DOE/EIA-0484(2013), Table 12, World Recoverable Coal Reserves as of January 1, 2009, p. 85, U.S. Energy Information Administration, July 2013, 2013d.

[a] Data for the United States represent recoverable coal estimates as of January 1, 2012.

[b] Includes Mexico, Chile, Middle East, Japan, South Korea, and Greenland.

Coal cleaning removes primarily mineral (pyritic) sulfur. Concentrations of trace heavy metals are also reduced, with conventional cleaning removing (typically) 30%–80% of arsenic, mercury, lead, nickel, antimony, selenium, and chromium. Removal of organic sulfur, which is bound chemically and is part of the coal matrix, is more difficult. Research has been pursued using microorganisms, enzymes, and chemical methods, but none of these have been commercially successful.

16.1.2.2 Natural Gas as a Fuel for Electric Power Generation

Natural gas is a naturally occurring mixture of light hydrocarbons, predominantly methane (CH_4), plus a few other compounds, such as nitrogen (N_2), water (H_2O), carbon dioxide (CO_2), hydrogen sulfide (H_2S), and heavier hydrocarbons. It is a gas at ordinary pressure and temperature. Heavier hydrocarbons that may be present include ethane (C_2H_6), propane (C_3H_8), butane (C_4H_{10}), and possibly other compounds. If appreciable amounts of other hydrocarbons are present, the gas is said to be wet; otherwise, it is said to be dry. If H_2S is present, the gas is said to be sour; otherwise, it is said to be sweet. The heavier hydrocarbons are typically removed and recovered because of their value as chemical feedstocks and fuels.

Natural gas, like coal and petroleum, is the result of the decomposition of buried plant material. In most cases, it has migrated from its point of formation to a geologic trap, where it remains until being discovered and produced. Natural gas is produced both as a single

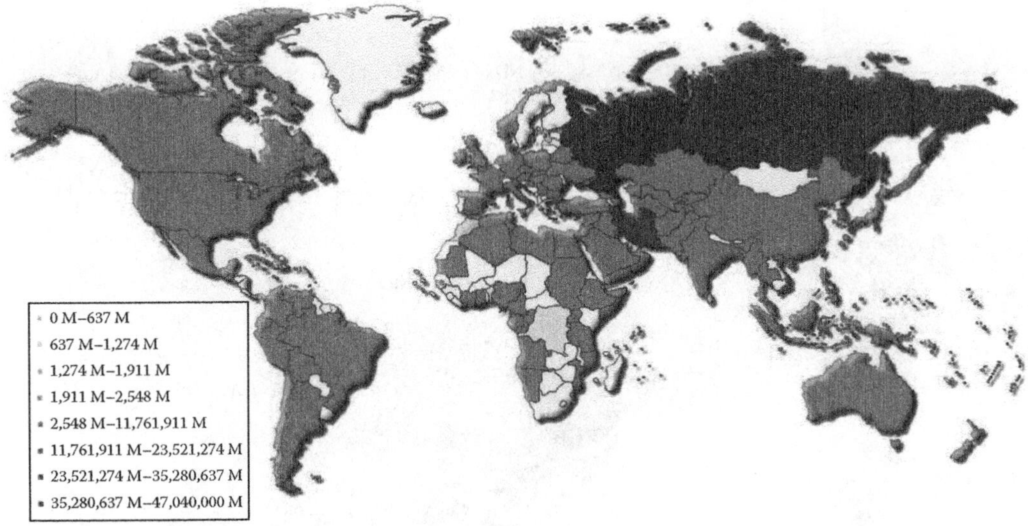

FIGURE 16.4
World proved natural gas reserves, 2012. (From Central Intelligence Agency (CIA), *The World Factbook*, 2013, https://www.cia.gov/library/publications/the-world-factbook/, Accessed November 27, 2013.)

product and as a by-product of crude oil production. Virtually, all natural gas is processed to meet the following pipeline specifications:

- Higher heating value must be within a specified range; a typical specification is 1035 ± 50 Btu/standard cubic foot.
- Gas must meet dew point temperature at delivery point to avoid condensation.
- It must contain no more than trace levels of H_2S, CO_2, N_2, H_2O, and O_2.
- It must be free of solid particulates and liquid water that could cause equipment damage.

Figure 16.4 shows world proved reserves of natural gas by country. Figure 16.5 shows the changes in U.S. wet natural gas proved reserves from 2008 to 2009, with the single largest difference arising from shale gas exploration.

16.1.2.3 Petroleum-Based Fuels for Electric Power Generation

Petroleum is a naturally occurring mineral consisting of a wide boiling range of mainly hydrocarbons, but also containing varying amounts of sulfur, nitrogen, and oxygen compounds and trace elements. Petroleum (also known as crude oil or simply oil) is the fossilized remains of plants and animals buried millions of years ago and trapped in geologic formations with impervious cap rocks. Petroleum reserves are widely distributed around the world (see Figure 16.6) and vary widely in quality from light crude with essentially no asphalt to tar sands, which are practically 100% asphalt. Crude oil is also classified as sweet (low sulfur) and sour (high sulfur).

Petroleum is the major source of transportation fuels (gasoline, diesel fuel, jet fuel, and kerosene). The fraction of each of these components varies widely from crude to crude, and many processes have been developed to increase the yield of desired products.

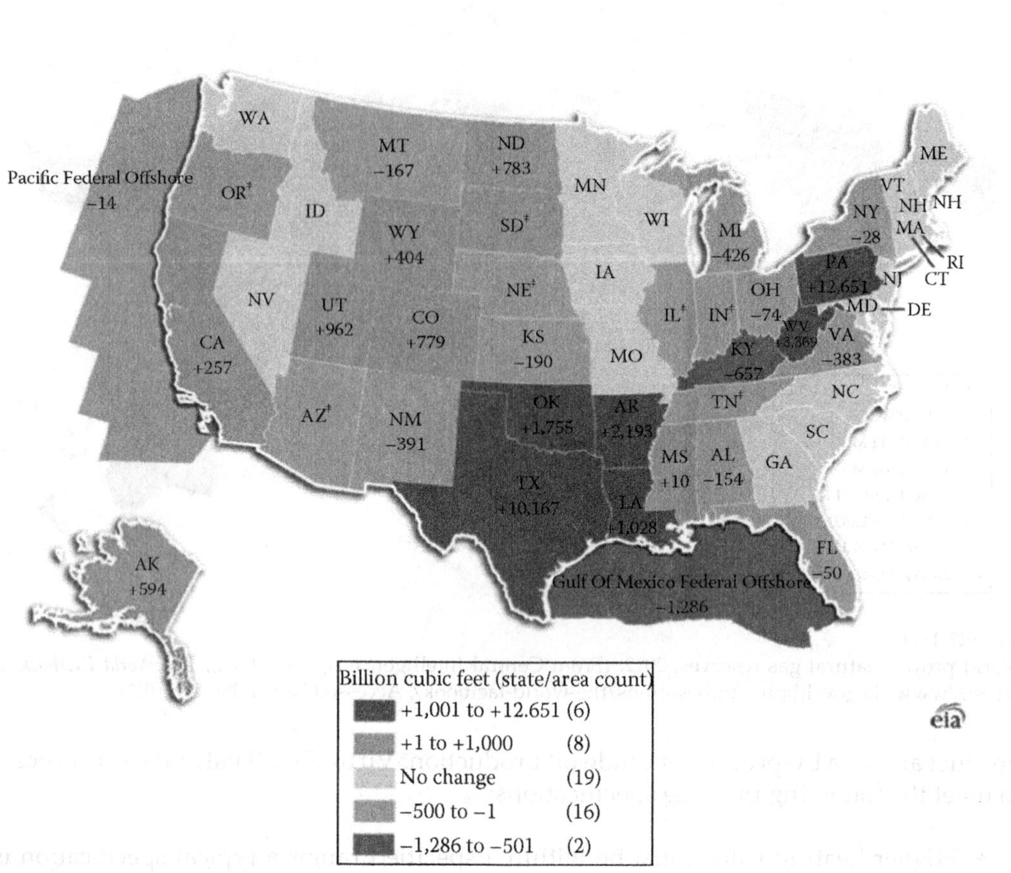

FIGURE 16.5
U.S. wet natural gas proved reserves with changes from 2010 to 2011. Total U.S. increase: 31.2 trillion cubic feet. (From Energy Information Administration (EIA), Natural Gas, "U.S. Crude Oil and Natural Gas Proved Reserves, 2011", Figure 18, Changes in wet natural gas proved reserves by state/area, 2010 to 2011, p. 22, U.S. Energy Information Administration, August 2013, 2013f.) *Note*: ‡ indicates that data is withheld to avoid disclosure of individual company data.

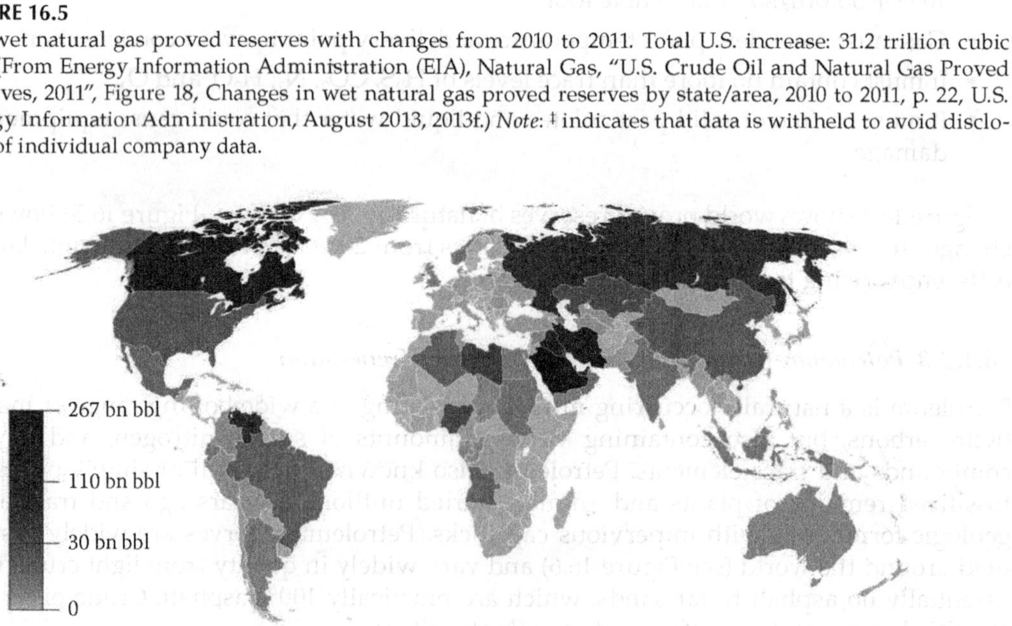

FIGURE 16.6
World petroleum reserves. (From Central Intelligence Agency (CIA), *The World Factbook*, 2013, https://www.cia.gov/library/publications/the-world-factbook/, accessed November 27, 2013.)

The first step in the refining of petroleum is distillation, which separates the crude into a number of fractions by boiling range. These fractions are then further treated using processes such as catalytic cracking, hydrocracking, reforming, alkylation, and coking. The heaviest fraction from fractionation is referred to as resid. It is the material that is too high boiling to distill. A small amount is used as a fuel for power production, but this usage has declined in recent years, and only about 20,000 barrels a day is now burned for electric power production.

In some refineries, the resid is subjected to a process called coking, which involves heating the resid to a high temperature, which results in decomposition, with part of the resid being converted to gas and liquid products and the rest remaining as petroleum coke or petcoke. In a few instances, this petcoke may be mixed with coal and burned in a PC boiler, or gasified.

16.1.3 Current Status of Power Production from Fossil Fuels

Coal is the most widely used fossil fuel for the production of electric power, with natural gas second. Only small quantities of fuel oil are currently used for electric power production.

The U.S. Energy Information Administration listed 874 GW of fossil fuel-fired generating nameplate capacity in the United States in 2012, with approximately 680 GW capacity operated by electric utilities. Coal-fired units predominated, with 1290 units capable of generating over 252 GW, followed by 240 GW gas-fired capacity. Fossil fuel-fired plants represented more than 75% of all electric energy–generating capacity in the United States, and more than 77% of electric utility-generating capacity, with coal-fired plants making up 28% and 37%, respectively. Within the electric utility industry, fossil-fired steam generating plants provided 312 GW or roughly 46% of the generating capacity (EIA 2013c). Although these units are aging—approximately half first went operational before 1966—they will remain a foundation of the power industry for the immediate future.

16.1.3.1 Pulverized Coal Power Plants

In 2012, as shown in Table 16.3, more than one-fourth of total U.S. electricity generation (including industrial and commercial generation, combined heat and power, and auxiliary power supplies as well as generation by electric utilities) was supplied by 1046 generating units burning PC. Of electric utility generation alone, approximately one-third was supplied by 688 generating units burning PC. Roughly half of these units have a capacity

TABLE 16.3

U.S. PC-Generating Capacity

Sector	Number of Generators Firing PC	PC Capacity (GW)	PC Percentage of Sector Electricity Generation[a]
Electric utilities	688	230	33.8
All[b]	1046	302	25.8

Source: Energy Information Administration (EIA), Electricity data browser: Net generation by energy source: All sectors, U.S. Energy Information Administration, 2013c, http://www.eia.gov/electricity/data/browser/, accessed November 27, 2013.

[a] PC generation as a proportion of electric power generation from all sources within the specified sector. This includes nuclear, hydropower, and renewables as well as all types of fossil fuel generation.

[b] Includes industrial combined heat and power (CHP) and non-CHP; commercial CHP and non-CHP; independent power producers CHP and non-CHP; and electric utilities.

of 250 MW or greater. Like the encompassing fossil-fired steam plant category, the plants utilizing these PC-fired generators are aging, with half built before 1969, and they compete with other regional coal-fired plants at the cost margin. Fuel cost is frequently the determining factor in assuring cost-competitiveness. For this reason—and also to reduce the price of emissions control—many eastern coal-fired units now burn a mixture of eastern coal and coal from the low-sulfur deposits of the Powder River Basin of Wyoming and Montana.

Raising the temperature and pressure of a fossil-fired steam unit into the supercritical (above 373°C, 22 MPa) and ultrasupercritical (up to 760°C, 35 MPa) steam conditions is a way to increase plant efficiency. The selection of a supercritical versus a subcritical cycle for a fossil-fired steam unit is dependent on many site-specific factors including fuel cost, emissions regulations, capital cost, load factor, duty, local labor rates, and estimated reliability and availability. In fact, the use of subcritical cycles for the limited number of fossil-fired steam plants that have been built in the United States since 1995 has been mainly due to relatively low fuel costs, which eliminated the cost justification for the somewhat higher capital costs of the higher efficiency cycles. However, in some international markets where fuel cost is a higher fraction of the total cost, the higher efficiency cycles offer a favorable cost-of-electricity comparison and provide lower emissions compared to a subcritical plant. This is true in both Europe and Asia.

Since 2007, supercritical cycles have gradually increased their share of new steam-based generation capacity even in the United States. The reduction of CO_2 emissions with the supercritical cycle could be a deciding factor as ways are sought to reduce GHG emissions. Ultrasupercritical cycles (discussed in Section 16.1.5.1) have also been introduced in recent years, first in Asia and then in Europe. One such plant is in operation in the United States.

16.1.3.2 Fluidized Bed Power Plants

A variation on PC combustion is FBC. A fluidized bed is an apparatus that mixes solid fuel (typically coal) and air with a sorbent, such as limestone. Combustion occurs in the bed, and the mixing action in the bed results in high heat and mass transfer rates. Often, a component is included to help catalyze combustion. The bed operates at relatively low temperatures, which creates less NO_X and allows for greater SO_2 capture by the limestone. More than 95% of the sulfur pollutants in coal can be captured in the bed by the sorbent. Additionally, this technology has generated interest because fluid bed combustors are well suited to cofiring of biomass. A 460 MW supercritical fluidized bed plant went into operation in 2009 in Poland. The largest FBC unit constructed in the United States to date is rated at 700 MW.

Following commercial introduction in the early 1980s, the AFBC boiler (Figure 16.7) has found growing application for power generation. From the first FBC boiler in 1967, generating 5000 lb/h of steam, the technology has matured to the 350 MW units available today. Atmospheric fluidized beds are available in two forms: bubbling fluidized beds and circulating fluidized beds (CFBs). Pressurized fluidized beds, discussed in Section 16.1.5, have also entered service. In 2012, 139 generating units in the United States utilized fluidized beds, including 26 generator units at 14 electric utility plants (EIA 2013c).

In the bubbling bed version of the AFBC, fuel and inert solids, together with limestone or dolomite for SO_2 capture, are suspended through the action of fluidizing air, which flows at a velocity of 3–8 ft/s in essentially a one-pass system. CFBs differ from bubbling beds in that much of the bed material passes through a cyclone separator

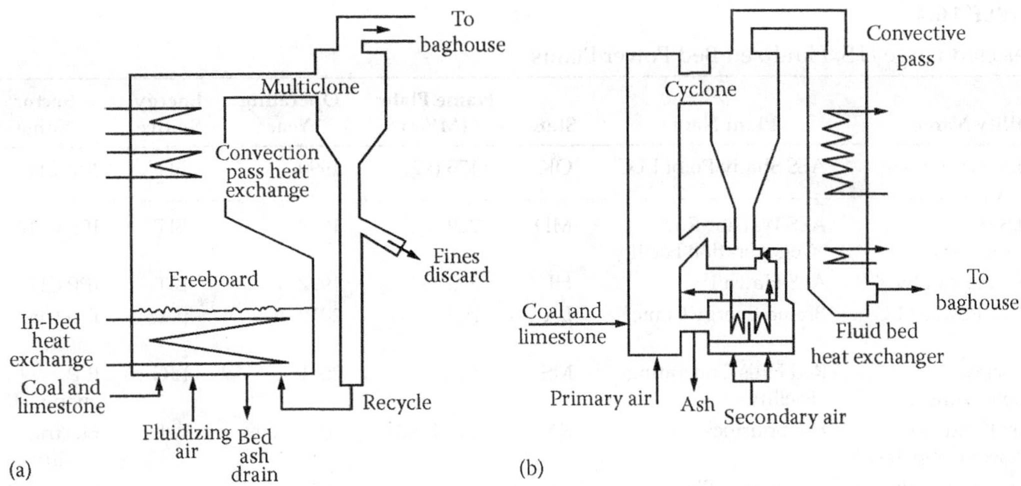

FIGURE 16.7
The primary types of atmospheric fluidized beds are (a) bubbling bed and (b) circulating bed.

before being circulated back to the boiler. In-bed tubes are generally not used with CFB units, permitting a much higher fluidizing velocity of 16–26 ft/s. Experience has shown that bubbling fluidized bed boilers are best suited for wet fuels (greater than 60% moisture) and CFB boilers for high calorific fuels and large-scale power generation (Rainio 2010).

Since the early AFBC designs, attention has been directed toward increasing unit efficiency, and reheat designs are now usual in large units. When SO_2 capture is important, a key parameter is the ratio of calcium in the limestone to sulfur in coal. Typical calcium to sulfur ratios for 90% SO_2 reduction are in the range of 3.0–3.5 for bubbling beds and 2.0–2.5 for circulating beds. Nitrogen oxide levels in AFBCs are inherently low due to the low bed temperature, and nominally less than 0.2 lb/million Btu. It is important to note that for CFBs, boiler efficiencies can be as high as those of a PC unit. In fact, plants in Poland and Russia combine CFBs with supercritical steam conditions, and there are prospects for cycles up to 30 MPa (4500 psia), 593°C (1100°F) with double reheat.

In the United States, 139 FBC units generate in excess of 11,000 MW. Burning coal in a suspended bed with limestone or dolomite permits effective capture of sulfur. Fuel flexibility allows a broad range of opportunity fuels, including coal wastes (culm from anthracite, gob from bituminous coal), peat, petroleum coke, and a wide range of coals from bituminous to lignite. A low (816°C; 1500°F) combustion temperature leads to low NO_X formation.

Examples of large generating FBC plants in the United States are shown in Table 16.4.

16.1.3.3 Power Plants Burning Oil and Gas Fuels

In the United States in 2012, approximately 4% (27.5 GW) of the electricity generated by electric utilities was produced by firing fuel oil. Several technologies are used to generate electricity from fuel oil. Two such technologies are to burn fuel oil in an internal combustion engine to run a generator, or in a boiler to produce steam, which is sent to a steam turbine to generate electricity. A more common method is to burn the oil in a combustion

TABLE 16.4

Selected Large U.S. Fluidized Bed Power Plants

Utility Name	Plant Name	State	Name Plate[a] (MWe)	Operating Year	Energy Source	Sector Name
AES Shady Point LLC	AES Shady Point LLC	OK	175 (×2)	1990	BIT	IPP CHP
AES WR Ltd Partnership	AES Warrior Run Cogeneration Facility	MD	229	1999	BIT	IPP CHP
AES Hawaii Inc.	AES Hawaii	HI	203	1992	BIT	IPP CHP
Cleco Power LLC	Brame Energy Center	LA	703.8	2010	petc	Electric utility
Choctaw Generating LP	Red Hills Generating Facility	MS	513.7	2001	LIG	IPP non-CHP
East Kentucky Power Coop, Inc.	H L Spurlock	KY	329.4 (×2)	2005; 2009	BIT	Electric utility
Formosa Plastics Corp	CFB Power Plant	TX	155 (×2)	2012	petc	Industrial non-CHP
JEA	Northside Generating Station	FL	297.5 (×2)	1966; 1972	petc	Electric utility
Tennessee Valley Authority	Shawnee	KY	175	1956	BIT	Electric utility
Virginia Electric & Power Co	Virginia City Hybrid Energy Center	VA	668	2012	BIT	Electric utility
NRG Wholesale Generation LP	Seward	PA	585	2004	WC	IPP non-CHP
Optim Energy LLC	Twin Oaks Power One	TX	174.6 (×2)	1990; 1991	LIG	IPP non-CHP
Cedar Bay Operating Services LLC	Cedar Bay Generating Company LP	FL	291.6	1994	BIT	IPP CHP
Sandow Power Co LLC	Sandow No 5	TX	661.5	2010	LIG	IPP non-CHP

Source: Energy Information Administration (EIA), Electricity data browser: Net generation by energy source: All sectors, U.S. Energy Information Administration, 2013b, http://www.eia.gov/electricity/data/browser/, accessed November 27, 2013.

Note: BIT, bituminous coal; petc, petroleum coke; LIG, lignite; WC, waste/other coal (culm, gob, coke, and breeze); IPP CHP, independent power provider/combined heat and power; IPP non-CHP, independent power provider/non-combined heat and power; industrial non-CHP, industrial user/noncombined heat and power.

[a] Rated generator capacity in megawatts.

turbine, which is similar to a jet engine. An advance on this technology is to send the hot exhaust gas from the combustion turbine to a heat recovery steam generator (HRSG) to make steam to drive a steam turbine. This technology is called a *combined cycle* and is more efficient, because it recovers more of the energy in the fuel oil. (For more discussion of combined cycle operations, see Section 16.1.6.)

As with fuel oil, natural gas can be burned in an internal combustion engine or in a boiler to raise steam. Alternatively, the natural gas can be burned in a combustion turbine. In most recent plants, the hot exhaust gas from the combustion turbine is sent to an HRSG to raise steam to drive a second turbine, employing the more efficient combined cycle technology.

16.1.4 Environmental Controls for Fossil-Steam Plants

Because of the many advances in technology over the past several decades, meeting even stricter emissions regulations is much easier now. In the United States by 2012, SO_2 scrubbers were operational on 695 units. These are valuable additions that will permit plants to operate in compliance for many more years. Typically, a 450 MW coal-fired plant will emit 75 tons of SO_2 per day without a scrubber and perhaps 8 tons per day with a 90% FGD system in place.

For the case of NO_X, where the same plant might emit 10–35 tons/day, NO_X reductions of up to 90% can reduce emissions to the 1–3.5 tons/day range. Control options range from burner optimization to the use of SCR. Regarding CO_2, the plant referenced earlier emits about 9000 tons/day at a plant efficiency of 38%, which translates to 2452 tons of carbon. Such emissions are of increasing concern, and potential future carbon taxes must be considered. A combined cycle gas plant, for comparison, emits about half the amount of CO_2 on a per-MWh basis due to the higher efficiency of NGCC plants and the lower carbon content of natural gas. The removal of mercury from the products of fossil fuel combustion will be an increasingly important task for fossil plant operators. Current plant testing for mercury control is exploring several promising options. An overall perspective of emissions control technologies applicable to modern PC-fired plants is shown in Figure 16.8.

16.1.4.1 Clean Coal Technology Development

At an increasing rate in the last few years, innovations have been developed and tested, aimed at reducing emissions through improved combustion and environmental controls in the near term, and in the longer term by fundamental changes in the way coal is

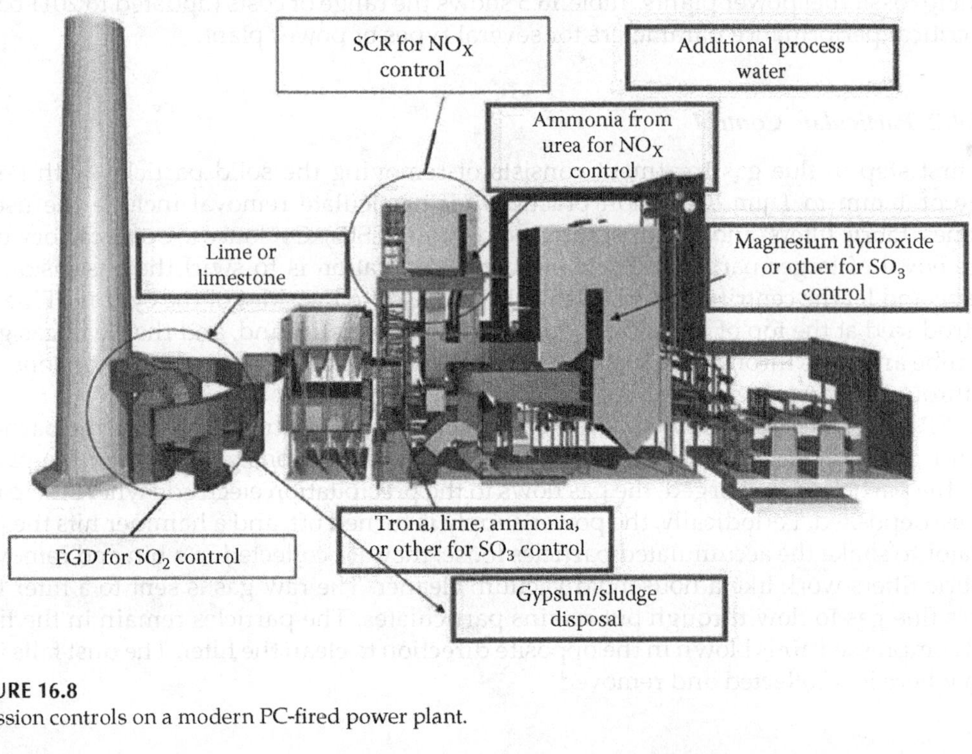

FIGURE 16.8
Emission controls on a modern PC-fired power plant.

preprocessed before converting its chemical energy to electricity. Such approaches are referred to as clean coal technologies (CCTs); many of these projects were sponsored by the Department of Energy's National Energy Technology Laboratory (NETL). They consist of a family of precombustion, combustion, postcombustion, and conversion technologies designed to provide the coal user with added technical capabilities and flexibility, and the world with an opportunity to exploit the most abundant fossil resource while minimizing environmental impacts. They can be categorized as follows:

- *Precombustion*: This refers to technologies that remove CO_2 and/or sulfur, ash, and other impurities from a fuel before it is burned. This applies chiefly to cleaning raw syngas from a gasifier before the gas is burned.
- *Combustion*: This refers to techniques applied in the boiler, such as low-NO_X burners, to reduce the production of pollutants resulting from combustion of the fuel.
- *Postcombustion*: This refers to technologies such as solvent, sorbent, and membrane-based processes that treat flue gas to reduce pollutant emissions.
- *Conversion*: This refers to technologies such as gasification or coal-to-liquids processes that change coal into a form that can be cleaned and used as a fuel or chemical feedstock.

A variety of products of combustion are produced when PC is burned in a boiler. The most important include fly ash from the mineral matter in the coal; SO_X (mainly SO_2, but also some SO_3) from sulfur; NO_X (mainly NO) from nitrogen and oxygen in air; mercury; and CO_2 from the carbon in the coal. To meet environmental regulations, most of these materials must be removed. Most of the CCT projects were conducted to develop improved processes to reduce pollutants. The following sections discuss environmental controls pertinent to fossil fuel power plants. Table 16.5 shows the range of costs (updated to 2011 basis) and critical performance parameters for several types of power plant.

16.1.4.2 Particulate Control

The first step in flue gas treatment consists of removing the solid particles with a size range of 1 mm to 1 μm. Common practices for particulate removal include the use of cyclones, fabric filters, and electrostatic precipitators (ESPs). Cyclones are satisfactory only for removal of larger particles. Their method of operation is to swirl the gas inside the cyclone and let the centrifugal force drive the particles to the sides of the cyclone. Flue gas is introduced at the top of the cone-shaped cyclone, swirls around, and the clean gas goes up a tube and exits through the top. The particulates drop to the bottom of the cyclone and exit through an opening.

In ESPs, the gas passes between high voltage plates that electrically charge the particles contained in the flue gas due to adsorption of ions and electrons generated by the plates. After the particles are charged, the gas flows to the precipitation electrode where the particles are deposited. Periodically, the power supply is turned off, and a hammer hits the precipitator to shake the accumulated particles loose; they are collected in a bin and removed.

Fabric filters work like a household vacuum cleaner. The raw gas is sent to a filter that allows flue gas to flow through but retains particulates. The particles remain in the filter until compressed air is blown in the opposite direction to clean the filter. The dust falls into a bin where it is collected and removed.

TABLE 16.5

Performance Comparisons of Four CCTs

Technology	Units	PC-Sub	PC-SC	AFBC	IGCC
Potential capacity	MW/Unit	250–700	350–800	30–300	300–600
Fuel range/diversity		All grades; lignite to anthracite	All grades; lignite to anthracite	All grades; biomass and wastes	All grades; bituminous preferred
Build time	Years	3	3	3	3
Manning levels					
Operators	Number/MW	0.16	0.16	0.18	0.18
Total staff	Number/MW	0.31	0.31	0.46	0.33
Expected availability					
Planned outage	%	11.1	11.1	5.7	4.7
Forced outage	%	3.7	3.9	4.1	10.1
Equiv. availability	%	85.7	85.4	90.4	85.7
Expected efficiencies	% (HHV)	34.4–35.7	36.4–37.7	34.6–35.6	39.3–41.1
Expected heat rate	kJ/kW h (HHV)	9570–9930	9050–9380	9600–9870	8310–8680
Emission ranges					
SO_2	kg/MW h	0.66–0.68	0.62–0.65	0.66–0.68	0.04–0.22
NO_X	kg/MW h	0.66–0.68	0.62–0.65	0.66–0.68	0.23–0.24
CO_2	kg/MW h	831–862	786–815	834–857	723–754
Particulates	kg/MW h	0.10	0.10	0.10	0.01
Solid waste (total)	kg/MW h	62–113	59–107	55–141	42–44
Ash	kg/MW h	45–48	45–48	47–44	42–44
Spent sorbent	kg/MW h	12–66	11–62	11–94	0
Cooling water requirements	Cu M/h/MW	244	236	249	185
Flexibility load range	%	30–100	30–100	30–100 per unit	50–100 per unit 25–100 for 2 trains
Total plant cost[a]	$/kW	1953–3596	1981–3563	n/a	2392–3299
Cost of electricity[a]	$/MW h	82.27–142.04	80.95–137.28	n/a	101.16–133.07

[a] Updated to 2011 costs (NETL 2012b).

ESPs and fabric filters have a similar operational area requirement, with filters showing slightly better performance for particles smaller than 1 μm. Cyclones alone cannot achieve the emission levels required for modern power plants. Cyclones, however, can be utilized as a complement with other flue gas treatment stages.

16.1.4.3 Sulfur Dioxide Control

The need for removal of sulfur dioxide from flue gases has led to the installation of FGD units on much of the coal-fired capacity in the United States (Figure 16.9).

More than 200 coal-fired units in the United States use wet or dry scrubbing to remove sulfur. Of these, the majority use wet lime or limestone scrubbers with perhaps between 20 and 30 using dry scrubbing when the sulfur content of the coal is generally lower. Lime may be used alone or in combination with magnesia, carbides, or with alkaline fly ash if the boiler burns subbituminous coal or lignite.

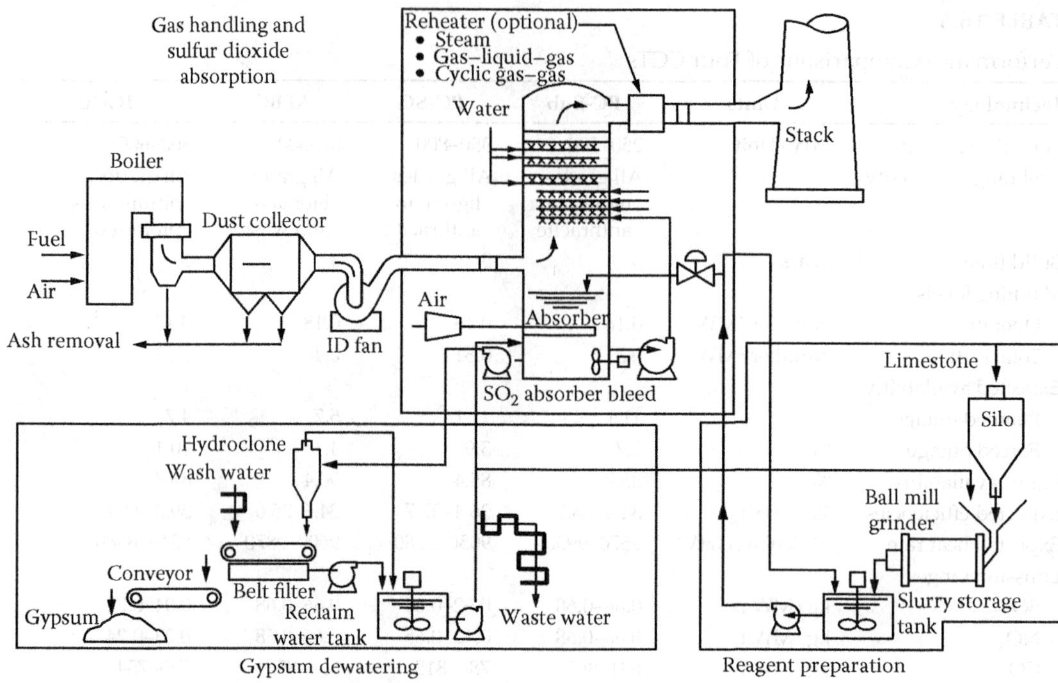

FIGURE 16.9
Wet limestone flue gas desulfurization system.

In the most common wet systems, flue gas contacts lime or limestone slurry in a spray tower. Sulfur dioxide in the flue gas reacts with the lime or limestone in the slurry and is absorbed according to the following reactions:

$$SO_2 + Ca(OH)_2 \rightarrow CaSO_3 + H_2O$$

$$SO_2 + CaCO_3 \rightarrow CaSO_3 + CO_2$$

$$2CaSO_3 + O_2 \rightarrow 2CaSO_4$$

$$SO_3 + CaCO_3 \rightarrow CaSO_4 + CO_2$$

The sulfite/sulfate is collected in a reaction tank, where it precipitates. A portion of the slurry is then pumped to a thickener, where the crystals settle out before going to a filter for final dewatering. Some of the calcium sulfate (gypsum) is used to make wallboard, cement, or agricultural products, and the rest is typically disposed of in a landfill.

Dry FGD systems employ calcium or sodium reagents that are either injected as dry powders or in spray drying systems as slurries that dry on contact with flue gas. Dry injection systems are most economical for space-constrained sites or applications that require only moderate emissions reductions. Spray drying systems, which can achieve higher SO_2 removal efficiencies, have mainly been applied at units burning low to medium-sulfur coals.

16.1.4.4 NO_X Control

Nitrogen oxides are produced during fossil fuel combustion by two main mechanisms: (1) oxidation of nitrogen in the fuel and (2) reaction between oxygen and nitrogen in the combustion air. The options for NO_X control may be grouped into two broad categories: (1) combustion modifications and (2) postcombustion methods (Figure 16.10).

Combustion modifications include low-NO_X burners, which are designed to control the air to fuel ratio, which in turn limits the temperature and reduces thermal NO_X production. Another option is use of overfire air, which involves introducing part of the air downstream in the furnace, again to reduce temperature and NO_X. Low-NO_X burners, especially when combined with staging of overfired air, are currently deployed most often. Combustion optimization techniques appear to offer a low cost alternative to hardware options, particularly where modest reductions (up to 30%) are needed, and many units are now operating with some form of optimization of air and fuel flows, perhaps utilizing advanced flame diagnostics or software based on neural networks to do the optimizing.

Several processes have been developed to reduce NO_X in flue gas. One option is SCR, which is used widely in Europe (especially Germany, where it is installed on more than 30,000 MW of coal-fired boilers), Japan, and increasingly in the United States. In an SCR, ammonia (NH_3) is injected into the boiler exhaust gases ahead of the catalyst bank (at about 288°C–400°C; 550°F–750°F). Nitrogen oxides and NH_3 then react to produce nitrogen and water according to the following chemical reactions:

$$4NO + 4NH_3 + O_2 \rightarrow 4N_2 + 6H_2O$$

$$6NO_2 + 8NH_3 \rightarrow 7N_2 + 12H_2O$$

$$2NO_2 + 4NH_3 + O_2 \rightarrow 3N_2 + 6H_2O$$

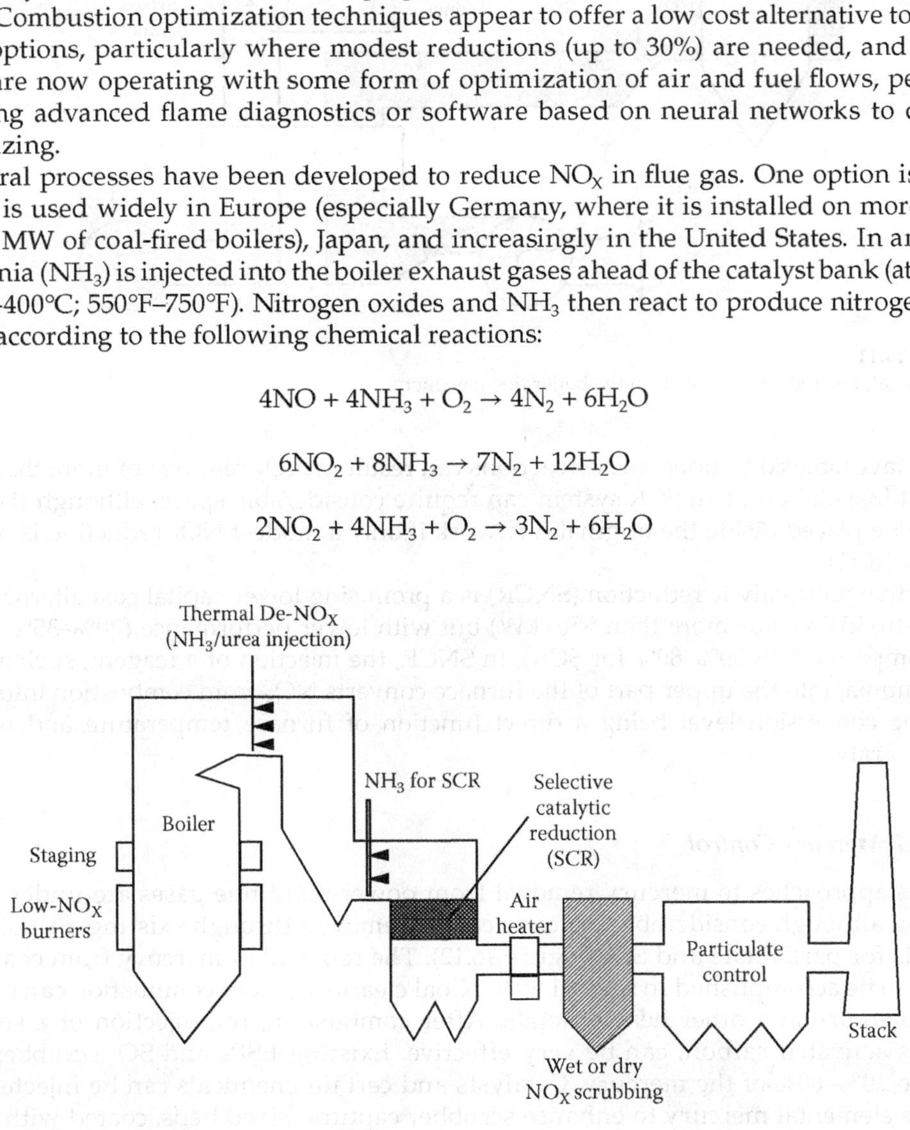

FIGURE 16.10
NO_X control options for fossil fuel–fired boilers.

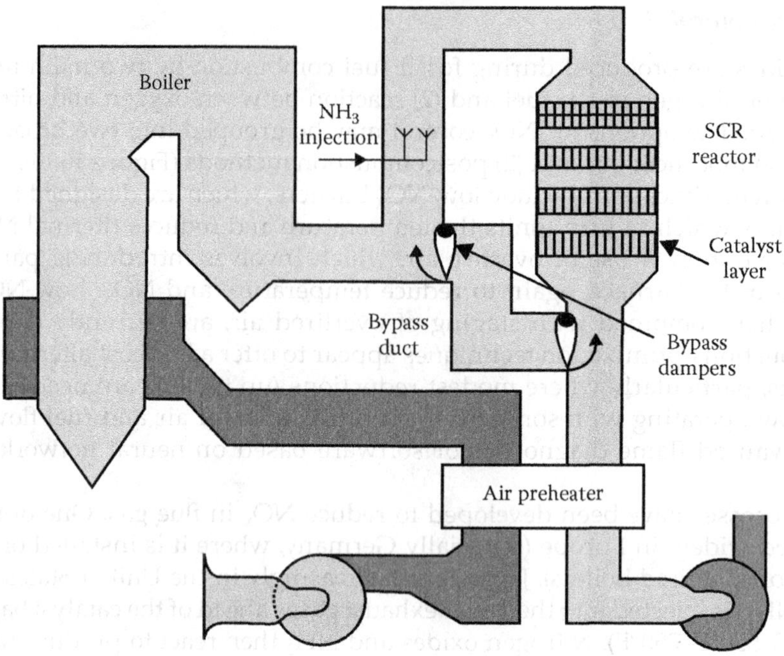

FIGURE 16.11
Selective catalyst reduction system in the boiler flue gas duct.

Under favorable conditions, these reactions can result in NO_X removal of more than 90%. Retrofit installation of an SCR system can require considerable space, although the reactor can be placed inside the original ductwork if only a modest NO_X reduction is needed (Figure 16.11).

Selective noncatalytic reduction (SNCR) is a promising lower capital cost alternative to SCR ($10/kW versus more than $50/kW) but with lower performance (20%–35% reduction compared with 50%–80% for SCR). In SNCR, the injection of a reagent, such as urea or ammonia, into the upper part of the furnace converts NO_X from combustion into nitrogen, the conversion level being a direct function of furnace temperature and reagent injection rate.

16.1.4.5 Mercury Control

Various approaches to mercury removal from power plant flue gases are under development, although considerable mercury can be removed through existing air pollution controls for particulate and SO_2 (Figure 16.12). The removal of mercury from coal-fired units can be accomplished in several ways. Coal cleaning before combustion can remove some mercury and other heavy metals. After combustion, the injection of a sorbent, such as activated carbon, can be very effective. Existing ESPs and SO_2 scrubbers can capture 20%–60% of the mercury. Catalysts and certain chemicals can be injected that oxidize elemental mercury to enhance scrubber capture. Fixed beds, coated with materials such as gold, can form amalgams with mercury, but may prove too expensive at a utility scale.

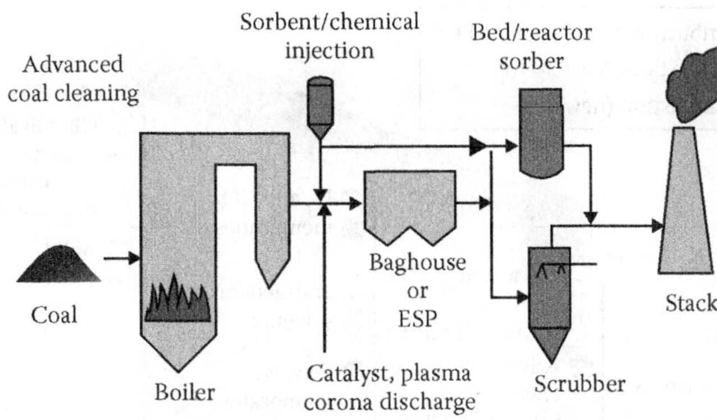

FIGURE 16.12
Options for the removal of mercury.

16.1.4.6 Carbon Dioxide Control and Storage

Carbon dioxide production is a result of the combustion of any fuel containing carbon. The amount released per unit of energy varies from fuel to fuel. Of the common fossil fuels, coal produces the most CO_2 and natural gas the least. Most climate scientists believe that the buildup of CO_2 in the atmosphere is contributing to global climate change. To reduce these emissions, considerable effort is being expended to develop technologies for the capture, transport, and geologic storage of CO_2, often referred to as carbon capture and storage (CCS). Plant emissions of CO_2 shown in Table 16.5 (Section 16.1.4.1) are a function of the plant heat rate (efficiency) and the type of fuel. For plants that use limestone for sulfur capture, there are some minor additional contributions of CO_2 associated with the decomposition of the limestone.

Carbon dioxide capture from coal-fueled power generation can be achieved by various approaches, including postcombustion capture, precombustion capture, and oxycombustion. All three pathways are under investigation, some at an early stage of development. A summary of the inherent advantages and disadvantages of each of these approaches is discussed in detail in Figueroa et al. (2008). Although there have been advances in CO_2 capture, there is still an impact on overall plant efficiency and the cost of electricity (COE). Thus, an important question when considering CO_2 control is how the additional control equipment will affect the power output and efficiency of the power plant. This is particularly significant when considering the retrofit of control equipment to an existing unit. There are approximately 1400 large coal-fired plants in the United States that potentially could be retrofitted. Estimates of the impact of 90% CO_2 control have been developed for several current and advanced plant types.

Studies show that, should CO_2 control be required, removal from the product gas from a coal gasifier prior to combustion is much less expensive than from the flue gas of a PC power plant where the CO_2 is at a concentration of only about 15% at atmospheric pressure. Generally, when CO_2 control is included, the COE increases. This is particularly true for PC plants, where postcombustion capture has to deal with a high volume of flue gas at atmospheric pressure that contains a relatively low concentration (about 15%) of CO_2. This results in a low driving force for CO_2 removal compared to removal from a much smaller volume of raw syngas from a gasifier, where the CO_2 is at about 60% concentration under a pressure of several hundred psi.

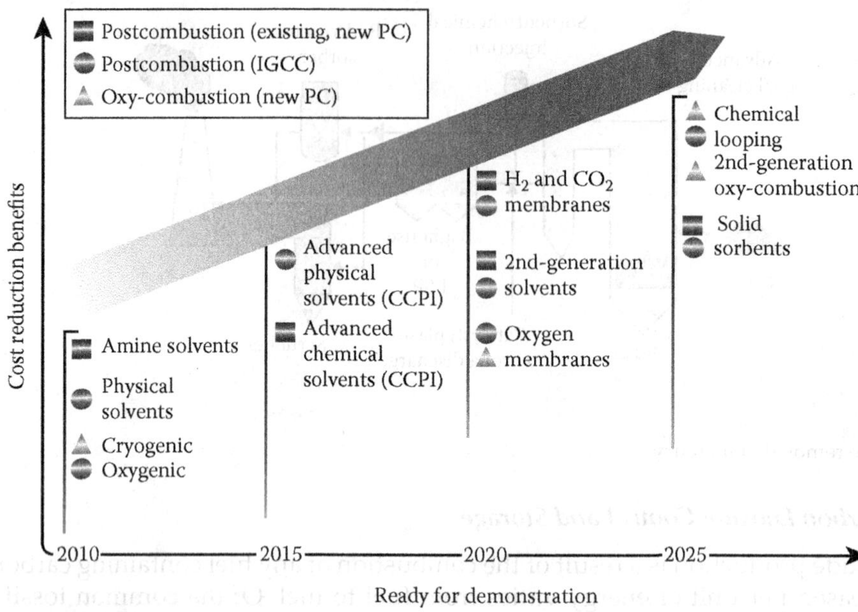

FIGURE 16.13
Perspectives on CO_2 control technology development to 2030. (Modified from Figueroa, J.D. et al., *Int. J. Greenhouse Gas Control*, 2(1), 9, 2008.)

Figure 16.13 puts into perspective expected advances in technology and the approximate time frame for their implementation. As advances in innovative CO_2 recovery technologies are made and commercialized, significant cost reductions should be realized. Current state-of-the-art technology for capturing CO_2 in industrial situations involves absorption using solvents—either reactive, such as amines, or physical, such as methanol—to remove CO_2 from various process streams. The figure also indicates that cryogenic oxygen is available for the IGCC option. Considerable effort is being expended on advanced solvents, both physical and chemical, which should be available in 3–5 years. Another area receiving intensive effort is the development of membranes. Membranes can be used in several ways to enhance CO_2 capture. One approach is to use dense ceramic membranes to remove CO_2 from syngas produced by a gasifier; alternatively, one can use a hydrogen-permeable membrane to remove the H_2. Another use for membranes is to separate O_2 from air. This technology is expected to be available by about 2020. Finally, alternative flow schemes, which inherently produce a concentrated CO_2 stream, such as oxycombustion and chemical looping, are expected to be viable options in the 2020–2025 time frame.

However, as of the end of 2013, there was no U.S. regulation requiring CO_2 removal from flue gas. There are few installations doing this, and those that do are recovering the CO_2 for industrial use, not for geologic storage.

16.1.4.7 CO_2 Storage Technologies Overview

One way to reduce emissions of CO_2 is to increase generating plant efficiency so that less fuel is burned. Another approach is to switch fuels. Burning natural gas produces less CO_2 than burning coal. However, these approaches alone will not reduce GHG emissions sufficiently. Therefore, other solutions are needed. One approach that shows promise is CCS, in which CO_2 is captured from a large stationary source, compressed, shipped by pipeline to a suitable storage site, and injected into a geologic formation, such as a saline reservoir, with an impervious cap rock, which will contain the CO_2 for thousands of years. An early implementation of this approach is enhanced oil recovery (EOR), in which a fluid such as CO_2 or water is injected into depleted oil fields to yield additional crude oil while leaving the CO_2 trapped in the formation. Carbon dioxide injection reduces oil viscosity, establishing an advantage as a fluid for this application.

Geologic formations have been identified that have the potential to store hundreds of years of GHG production at current production rates (NETL 2012a).

16.1.5 Advanced Power Production Technologies

Improvements to power production processes and equipment are under constant development. The following sections discuss some of these developments.

16.1.5.1 Advanced Ultrasupercritical (A-USC) Power Plants

In any PC combustion plant, steam raised from combustion of coal is used to generate electricity. More specifically, electricity is produced when high-pressure steam from the boiler is expanded through a steam turbine. After expansion through the high-pressure turbine, steam is typically sent back to the boiler to be reheated before expanding through the intermediate- and low-pressure turbines. Sometimes there is a second reheat between the intermediate- and low-pressure turbines. Reheating, single or double, increases cycle efficiency by raising the higher average temperature at which heat is converted to work.

The main difference among PC technologies is the operating conditions. Moving from subcritical to supercritical to ultrasupercritical conditions results in an increasingly severe operating environment, with both temperature and pressure increasing. Typical operating conditions and efficiencies are shown in Table 16.6 and Figure 16.14.

Generation efficiency can be further increased by designing new coal-burning units to operate at even higher steam temperature and pressure. Although a number of supercritical units were built in the United States through the 1970s and early 1980s, most of the existing U.S. fleet is in the subcritical category. Since 2007, most new PC plants in the

TABLE 16.6
Impact of Steam Conditions on Pulverized Coal Plant Efficiencies

Steam Conditions	Temperature	Pressure	Net Plant Efficiency (% HHV)[a]
Subcritical	~565°C (1050°F)	~16 MPa (2400 psig)	~39%
Supercritical (SC)	~593°C (1100°F)	~24 MPa (3500 psig)	~40-41%
Ultrasupercritical (USC)[b]	>600°C (>1112°F)	>24 MPa (>3500 psig)	≥41%
Advanced USC (DOE Program Goals)	~760°C (1400°F)	~35 MPa (5000 psig)	~44%

Source: NETL 2015, *Cost and Performance Baseline for Fossil Energy Plants, Volume 1: Bituminous Coal and Natural Gas to Electricity*. Revision 3, 2015 draft. National Energy Technology Laboratory, Albany, OR; Anchorage, AK; Morgantown, WV; Pittsburgh, PA; and Sugar Land, TX, March 2015.

[a] Net plant efficiencies are based on a reference plant burning bituminous coal, at ISO conditions, with 50°F reheat, wet flue gas desulfurization, and wet cooling towers. Other design parameters and site conditions impact plant efficiency.

[b] No precise definition exists of what constitutes USC conditions, and criteria are not consistent (especially internationally); rather, USC represents a broad range of steam conditions. Commercially available USC technology results in efficiencies similar to or slightly above the state-of-the-art supercritical pulverized coal plant referenced here.

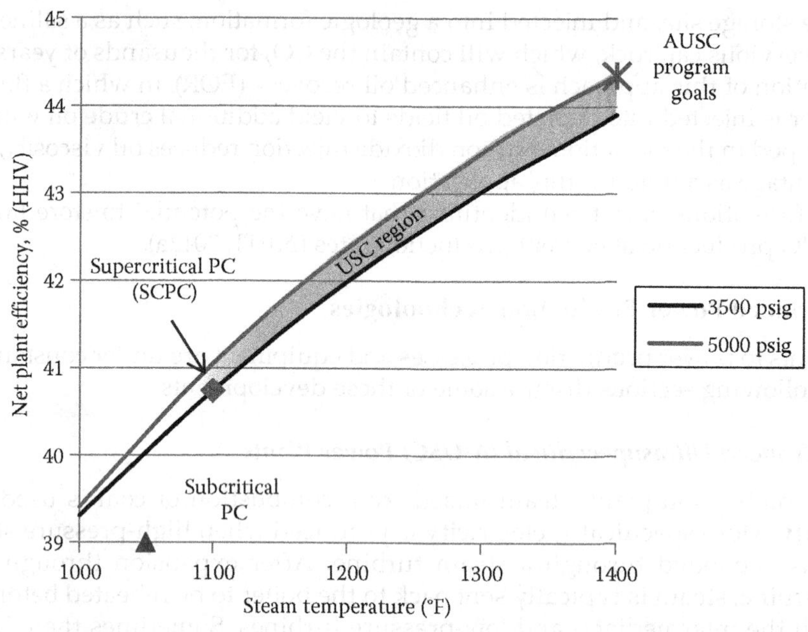

FIGURE 16.14
Impact of steam conditions on pulverized coal plant efficiencies. (From NETL 2015, *Cost and Performance Baseline for Fossil Energy Plants, Volume 1: Bituminous Coal and Natural Gas to Electricity*, Revision 3, 2015 draft, National Energy Technology Laboratory, Albany, OR; Anchorage, AK; Morgantown, WV; Pittsburgh, PA; and Sugar Land, TX, March 2015.)

United States are higher efficiency supercritical designs. Improvements in metallurgy have led to emergence of ultrasupercritical designs.

16.1.5.2 Advanced Fluidized-Bed Technologies

Fluidized bed combustion (FBC) is a variation on PC combustion that is receiving considerable attention because of its flexibility in the properties of the fuel it can handle. FBC is discussed in Section 16.1.3.2. An interesting advance in fluidized-bed technology is the use of a pressurized (PFBC) system. As discussed earlier, a fluidized bed operates at a considerably lower temperature than a PC boiler. To make up for this and increase the gas turbine firing temperature, natural gas is introduced in addition to vitiated air from the PFB combustor. This mixture is burned in a topping combustor to provide higher inlet temperatures for greater combined cycle efficiency. One drawback is that the method uses natural gas, which historically has cost more than coal.

A further advance in fluidized-bed combustion can be realized by incorporating a pressurized carbonizer to heat the feed coal and convert it into fuel gas and char, resulting in an advanced circulating pressurized fluidized-bed combustion (APFBC) combined cycle system. The char is burned in the PFBC to raise steam and heat combustion air for the gas turbine. The carbonizer fuel gas is burned in a topping combustor that heats fuel gas to the combustion turbine's rated firing temperature. Heat recovered from the gas turbine exhaust raises steam in a heat recovery steam generator (HRSG). The steam is used to drive a conventional steam turbine, resulting in improved overall efficiency for the combined cycle. These systems are entirely coal fueled. Another advance, the Gasification fluidized-bed combustion combined cycle systems (GFBCC), has a pressurized circulating fluidized-bed (PCFB) partial gasifier which feeds syngas to the gas turbine topping combustor. The gas turbine exhaust supplies combustion air for the atmospheric circulating fluidized-bed combustor that burns the char from the PCFB partial gasifier.

16.1.5.3 Oxycombustion Power Plants

Oxycombustion involves the use of oxygen, with a purity of 95%–99%, instead of air, to combust coal. Since the nitrogen present in air is removed and not heated in the boiler, higher flame temperatures are potentially possible with accompanying improved efficiency and reduced fuel consumption.

Oxycombustion cannot simply be substituted for air combustion in existing fossil-fueled power plants due to differences in combustion characteristics. In particular, combustion temperature would be too high for conventional materials of construction. To use oxycombustion in an existing plant, a thermal diluent is required to replace nitrogen and moderate combustion temperature. The diluent is typically recycled CO_2 that is added to oxygen produced by air separation to approximate the combustion characteristics of air. Oxycombustion produces approximately 75% less flue gas than air fueled combustion and produces exhaust consisting primarily of CO_2 and H_2O.

A major attraction of oxycombustion is that a highly concentrated CO_2 stream is produced, reducing the cost of separation from the flue gas. Water vapor is readily separated from the CO_2 by condensation through cooling and compression. Further treatment of the flue gas may be needed to remove pollutants such as SO_X and NO_X before the CO_2 is sent to storage.

A low-cost supply of oxygen is required for oxycombustion to be a cost-effective power generation option. In the most frequently proposed version of this concept, a cryogenic air separation unit is used to supply high-purity oxygen to the boiler. This commercially available technology is both capital and energy intensive and could reduce both the affordability of electricity and plant efficiency. However, novel technologies currently under development, such as ion transport membranes (ITMs), have the potential to significantly reduce the cost of oxygen production.

In an ITM, oxygen is adsorbed on one surface of the membrane and splits into two oxygen atoms. These atoms pick up two electrons each and diffuse through the membrane. At the other surface, the oxygen ions shed their electrons and recombine to form oxygen gas. The driving force is the partial pressure difference across the membrane. Only oxygen passes through the membrane, so selectivity is very high. If successful, ITM could be an economical replacement for cryogenic oxygen.

The justification for using oxycombustion is to produce a CO_2-rich flue gas ready for storage. Oxyfuel combustion has significant advantages over traditional air-fired plants. Among these are

- The mass and volume of flue gas are reduced by approximately 75%, which results in less heat lost in the flue gas and also smaller processing equipment.
- The flue gas is primarily CO_2 suitable for geologic storage.
- The concentration of pollutants in the flue gas is higher, making separation easier.
- Because nitrogen is largely excluded, nitrogen oxide production is greatly reduced.

16.1.5.4 Chemical Looping Power Plants

A major difficulty associated with oxygen-blown IGCC is the need for an oxygen plant, which is a costly item. A concept for producing electric power from fossil fuels while producing a concentrated CO_2 stream without the need for an oxygen plant is chemical looping combustion. Chemical looping splits combustion into separate oxidation and reduction stages. A metal (e.g., iron, nickel, copper, or manganese) oxide or other compound is used as an oxygen carrier. The plant typically consists of two vessels, an oxidizer and a combustor. The carrier is sent to the combustor where it reacts with the fuel to release heat and produce CO_2. The carrier is then recycled to the oxidation chamber, where the metal oxide is regenerated by contact with air. The advantage of using two chambers for the combustion process is that the flue gas is a highly concentrated CO_2 stream, once water is removed, and not diluted with nitrogen as is the case in a typical PC plant. An alternative flow stream for chemical looping combustion is to maintain the solids in fixed beds while periodically switching flows so that the oxygen carrier is alternatively oxidized and reduced. The benefit of chemical looping is that no air separation plant or external CO_2 separation equipment is required.

16.1.5.5 Fuel Cell Power Plants

A fuel cell is a device for converting chemical energy directly into electric power. Its operation is similar to that of a battery. Since a fuel cell is not a heat engine, its operation is not bound by the restrictions inherent in the second law of thermodynamics that limit the efficiency of a steam turbine. Thus, quite high efficiencies are theoretically possible. In a fuel cell, fuel (typically hydrogen or syngas) is introduced at the anode and oxygen (air) is introduced at the cathode. Oxygen ions flow from the cathode, through the electrolyte that separates the electrodes to the anode, where they react with the fuel to produce water vapor and CO_2. Electrons flow through an external circuit back to the cathode, thus producing an electric current.

Various types of fuel cells are under development and use for a variety of applications. The types that see the most use for stationary electric power production—on scales ranging from auxiliary residential and commercial power to electric utilities—are the polymer electrolyte membrane fuel cell (PEMFC), molten carbonate fuel cell (MCFC), and solid oxide fuel cell (SOFC). Interest in fuel cells is prompted by their potential high efficiency and clean operation. Some fuel cells can feed methane; others must have hydrogen or syngas. One option for a power station would be to produce syngas by coal gasification as fuel for a fuel cell. In 2012, installed fuel cell capacity for all stationary applications amounted to about 125 MW, nearly evenly divided among PEMFC, MCFC, and SOFC types (Fuel Cell Today 2013). Although the number of fuel cell power plants and their proportion of installed capacity remain low, capacity is rapidly growing, and includes some installations by electric utilities.

16.1.5.6 Combined Cycle Power Plants

Natural gas or syngas produced by coal gasification is typically burned in a combustion turbine to drive a generator to produce electric power. However, the gas leaving the turbine is still relatively hot and contains considerable energy. In a combined cycle, this waste heat is captured in a HRSG to produce steam that can be expanded in a steam turbine, thus increasing the efficiency of the power plant by generating additional electricity from the same fuel. Such an arrangement is called a combined cycle, because it combines both a Rankine cycle and a Brayton cycle to maximize plant efficiency. Virtually, all modern natural gas–fueled power plants employ a combined cycle to attain greater efficiency.

16.1.5.6.1 *Integrated Gasification Combined Cycle Power Plants*

Coal gasification has been employed for some time, and there have been substantial advancements in gasification-based power systems during the last few decades.

Development programs have been in place in the United States and across the world. Two areas of research that are likely to produce significant improvements in coal gasification technology by improving efficiency and reducing costs are advances in air separation for oxygen production and in CO_2 control.

Gasification of coal produces synthesis gas, or syngas, a mixture of H_2 and CO, which is used primarily for two purposes: the production of chemicals and the generation of electricity by combustion in a gas turbine. Since, as noted earlier, a single-cycle gas turbine will not achieve the efficiency attainable when combined with an HRSG and steam turbine, power plants involving coal gasification invariably employ a combined cycle. In addition to coal, petroleum coke is also frequently used as gasifier feed. Other petroleum waste products, asphalt, biomass, and municipal waste are also used.

At present about 14 such plants are in operation worldwide (see Table 16.7). Six gasify coal alone or in a mix with three located in the United States, the largest of which (Duke Energy's Edwardsport Integrated Gasification Combined Cycle plant in Indiana,

TABLE 16.7

Operating IGCC Plants 100 MW or Greater, Worldwide

Plant	Fuel	Gasification Technology	Plant Size (MW)	Startup Date (Full IGCC Mode)
Wabash River, Indiana, USA	Petroleum coke	CB&I E-Gas (formerly owned by Destec, then ConocoPhillips)	262	1995
Tampa Electric, Florida, USA	Coal/petroleum coke	Texaco (now GE Energy)	250	1996
SUAS Vresova, Czech Republic	Coal (lignite)	Lurgi	400	1996
ELCOGAS, Puertollano, Spain	Coal/natural gas	Krupp-Uhde Prenflo	310	1998
ISAB Energy IGCC Project, Italy	Asphalt	GE Energy	1203	1999
SARLUX IGCC Plant	Petroleum visbreaker residue	GE Energy	1271	2000
api Energia S.p.A. IGCC Plant	Petroleum visbreaker residue	GE Energy	526	2001
Chawan IGCC Plant, Singapore	Residual oil	GE Energy	364	2001
Negishi Refinery, Japan	Residual oil	Texaco (now GE Energy)	342	2003
Agip IGCC	Petroleum visbreaker residue	Shell	457	2006
Nakoso, Japan	Coal	Mitsubishi Heavy Industries	250	2007
Thermoselece Vresova	Petroleum tars, oils, phenol concentrates	Siemens	140	2008
Greengen, Tianjin City, China	Coal	ECUST	250	2012
Duke Energy Edwardsport, Indiana, USA	Coal	GE Energy	618	2013

Source: Operator websites.
Note: Several other IGCC plants worldwide are in development. (See NETL Gasification Portal.)

rated at 618 MW) began commercial operation in June 2013. A larger IGCC plant has been built in the Netherlands (NUON's Magnum; 1311 MW), but an arrangement to burn only natural gas until 2020, rather than gasifying coal, was made in part to allay concerns about CO_2 emissions from gasification. The two smaller U.S. combined cycle projects were supported under the U.S. Department of Energy's (DOE) CCT demonstration program.

The Tampa Electric Integrated Gasification Combined-Cycle Plant, shown in Figure 16.15, is typical of an IGCC plant. PC is slurried with water and pumped into the Texaco (now GE Energy) gasifier where it reacts with oxygen piped in from a cryogenic oxygen plant. The gasifier operates at a high temperature and modest pressure to produce syngas with an LHV of approximately 245 Btu/standard cubic foot. Molten ash flows out of the bottom of the gasifier into a water-filled sump where it forms a solid slag. The syngas flows from the gasifier to a radiant syngas cooler and a convective syngas cooler (CSC), which cool the syngas while generating high-pressure steam. The cooled gases flow to a water-wash scrubber for particulate removal. Next, a hydrolysis reactor converts carbonyl sulfide (COS) in the raw syngas to H_2S that is more easily removed. The raw syngas is further cooled before entering a conventional amine sulfur removal system. The recovered H_2S is converted to sulfuric acid in a sulfuric acid plant (SAP). The cleaned gases are reheated and routed to a combined-cycle system for power generation. A GE MS 7001FA gas turbine generates 192 MWe. Thermal NO_X is controlled to 0.7 lb/MWh by injecting nitrogen from the oxygen plant. A steam turbine uses steam produced by cooling the syngas and superheated by the gas turbine exhaust gases in the HRSG to produce an additional 123 MWe. The air separation unit consumes 55 MW and auxiliaries require 10 MW, resulting in 250 MWe net power to the grid. The plant heat rate is 9650 Btu/kWh (HHV).

FIGURE 16.15
The 250 MW coal gasification unit at the Polk Plant of Tampa Electric.

16.1.6 Clean Coal Options to 2040

An overall comparison of the performance and cost of CCTs discussed earlier—PC supercritical, fluidized bed, oxycombustion, chemical looping combustion, and IGCC—to current subcritical technology is shown in Table 16.5. It should be recognized that comparisons of this type do not capture various site-specific factors, such as variations in availability and cost of fuel, site conditions, and operational and business requirements that ultimately determine the choice of technology for a particular project.

The exploding global demand for electricity, particularly in the developing world, implies practical electric-generating options and significant increases in efficiency throughout the entire energy chain. A strong portfolio of advanced power generation options would include fossil, renewable and nuclear, all essential to meet these growth requirements for power, both domestically and globally.

Through 2040, the developed world will meet the need for additional capacity largely with coal- and gas-fired plants, while the developing world will continue to rely on indigenous resources, particularly coal in the cases of China and India. In the United States, the balance between coal- and gas-fired new generation is ongoing. While gas-fired additions currently predominate, the relative fuel prices (coal to gas) and the timing of additional emissions requirements will ultimately determine the preferred choice between these two fossil fuels. In the longer term, noncarbon sources will begin to supplant both.

Coal and gas fuels are expected to continue to dominate both world and U.S. central stations in the next decade, with gas-fired combined cycles supplanting several older fossil steam stations in the United States. Timing is still a question, though. In 2012, six new coal-fired plants were being planned in as many U.S. states. Based on the expected mix of coal, oil, gas, nuclear, and renewables through the year 2025, U.S. central generation options for fossil fuels are expected to continue to depend on Rankine cycles.

> Existing coal, oil, and gas-fired plants of conventional design, mostly Rankine cycles, are expected to continue to dominate the electric power generation market for the next ten years. These represent the majority of plants currently in operation. On average they are 45 years old and many (more than 100,000 MW) are equipped with SO_2 scrubbers, NO_X control additions, and other environmental upgrades. They provide the bulk of our electricity needs, are extremely reliable, and are increasingly in demand as evidenced by an average capacity factor above 70% in the early years of this century.

There will be a lot of action in repowered plants based on gas firing and combined cycle operation. Many of the coal- and gas-fired steam plants, especially in the United States, that have changed hands, are targeted for repowering—that is, combustion turbines will replace boilers, generating electricity and providing exhaust heat for HRSGs to power the existing steam turbines. This combination of gas and steam turbine cycles adds megawatts, reduces emissions, and improves efficiencies by 5% or more.

Advancing steam temperatures and pressures in PC steam plants greatly improve overall efficiency. Such ultrasupercritical cycles are already in operation both within and outside the United States. Increasing steam temperatures above 700°C (1292°F) from current levels of about 590°C (1100°F) enhances efficiency and reduces emissions. When used in coal combined cycles with temperatures increased to 750°C (1382°F) or higher, efficiencies beyond 55% can theoretically be attained. Significant challenges still exist in materials technology. The fuel cell is an exciting advance that will change the energy picture in the long term. In the shorter term, incremental advances are being made in both mobile and stationary applications.

References

Central Intelligence Agency (CIA) 2013. *The World Factbook*. https://www.cia.gov/library/publications/the-world-factbook/. Accessed November 27, 2013.

Energy Information Administration (EIA) 2013a. *Electric Power Annual 2011*. Table 4.3. Existing Capacity by Energy Source, 2011 (Megawatts). Page 70. U.S. Energy Information Administration. January 2013.

Energy Information Administration (EIA) 2013b. Electricity data browser: Net generation by energy source: All sectors. 2013.

Energy Information Administration (EIA) 2013c. Electricity Form EIA860 detailed data, 2012. U.S. Energy Information Administration. http://www.eia.gov/electricity/data/eia860/index.html. Accessed December 19, 2013.

Energy Information Administration (EIA) 2013d. *International Energy Outlook 2013*. DOE/EIA-0484(2013). Table 12, World Recoverable Coal Reserves as of January 1, 2009, p. 85. U.S. Energy Information Administration. July 2013.

Energy Information Administration (EIA) 2013e. International Energy Statistics. End year 2012. U.S. Energy Information Administration. http://www.eia.gov/cfapps/ipdbproject/IEDIndex3.cfm. Accessed November 27, 2013.

Energy Information Administration (EIA) 2013f. Natural Gas. "U.S. Crude Oil and Natural Gas Proved Reserves, 2011." Figure 18, Changes in wet natural gas proved reserves by state/area, 2010 to 2011, p. 22. U.S. Energy Information Administration. August 2013.

Figueroa, J.D., T. Fout, S. Plasynski, H. McIlvried, and R.D. Srivastava. 2008. Advances in CO_2 capture technology—The U.S. department of energy's carbon sequestration program. *International Journal of Greenhouse Gas Control*, 2(1), 9–20.

Fuel Cell Today. 2013. *The Fuel Cell Industry Review 2013*. Royston, UK: Johnson Matthey Plc, p. 33.

Goswami, D.Y. and F. Kreith (eds.). 2007. *Handbook of Energy Efficiency and Renewable Energy*. Boca Raton, FL: CRC Press.

NETL 2015. Cost and performance baseline for fossil energy plants, Volume 1: Bituminous coal and natural gas to electricity. Revision 3, 2015 draft. National Energy Technology Laboratory, Albany, OR; Anchorage, AK; Morgantown, WV; Pittsburgh, PA; and Sugar Land, TX, March 2015.

NETL 2012a. *The United States 2012 Carbon Utilization and Storage Atlas*, 4th edn. (Atlas IV), pp. 15–29. U.S. Department of Energy, National Energy Technology Laboratory, Albany, OR; Anchorage, AK; Morgantown, WV; Pittsburgh, PA; and Sugar Land, TX, December 2012.

NETL 2012b. Updated costs (June 2011 basis) for selected bituminous baseline cases, DOE/NETL-341-082312. U.S. Department of Energy, National Energy Technology Laboratory, Albany, OR; Anchorage, AK; Morgantown, WV; Pittsburgh, PA; and Sugar Land, TX, August 2012.

NETL Gasification Portal. Undated. http://www.netl.doe.gov/gasification-portal.html. Accessed November 27, 2013.

Perry, R.H. 2008. *Perry's Chemical Engineers' Handbook*, 8th edn. D.W. Green and R.H. Perry (eds.). The McGraw-Hill Companies, Inc. 2008, pp. 24-21–24-45

Rainio, A. 2010. ASME, Fluidized bed technologies for biomass combustion. *Energy-Tech Magazine*, November 2013, http://www.energy-tech.com/article.cfm?id=27790. Accessed November 21, 2013.

Riegel, R. 2007. *Kent and Riegel's Handbook of Industrial Chemistry and Biotechnology*, 11th edn. New York: Springer Science+Business Media, LLC.

17

Nuclear Power Technologies through Year 2035

Kenneth D. Kok and Edwin Harvego

CONTENTS

17.1 Introduction

Nuclear power is derived from the fission of heavy element nuclei or the fusion of light element nuclei. This chapter will discuss nuclear power derived from the fission process, since fusion as a practical power source will not reach the stage of commercial development in the next 20–25 years. In a nuclear reactor, the energy available from the fission process is captured as heat which is transferred to working fluids that are used to generate electricity. Uranium-235 is the primary fissile fuel currently used in nuclear power plants. It is an isotope of uranium that occurs naturally at about 0.72% of all natural uranium deposits. When uranium-235 is "burned" (fissioned) in a reactor, it provides about one megawatt day of energy for each gram of uranium-235 fissioned (3.71E+10 Btu/lb).

Nuclear power technology includes not only the nuclear power plants which produce electric power but also the entire nuclear fuel cycle. Nuclear power begins with the mining of uranium. The ore is processed and converted to a form that can be enriched in the U235 isotope so that it can be used efficiently in today's light-water moderated reactors. The reactor fuel is then fabricated into appropriate fuel forms for use in nuclear power plants. Spent fuel can then be either reprocessed or stored for future disposition. Radioactive waste materials are generated in all of these operations and must be disposed of. The transportation of these materials is also a critical part of the nuclear fuel cycle.

In this chapter, the development, current use, and future of nuclear power will be discussed. The first section is a brief review of the development of nuclear energy as a source for the production of electric power. The second section looks at nuclear power as it is deployed today both in the United States and worldwide. The third section examines the next generation of nuclear power plants that will be built. The fourth section reviews concepts being proposed for a new generation of nuclear power plants. The fifth section provides a brief introduction to small modular reactors. The sixth section describes the nuclear fuel cycle beginning with the availability of fuel materials and ending with a discussion of fuel reprocessing technologies. The seventh section discusses nuclear waste and the options for managing this waste. The eighth section addresses nuclear power economics. Conclusions are presented in Section 17.9.

17.2 Development of Current Power Reactor Technologies

The development of nuclear reactors for power production began following World War II, when engineers and scientists involved in the development of the atomic bomb recognized that controlled nuclear chain reactions could provide an excellent source of heat for the production of electricity. Early research on a variety of reactor concepts culminated in President Eisenhower's 1953 address to the United Nations in which he gave his famous "Atoms for Peace" speech, where he pledges the United States "to find the way by which the miraculous inventiveness of man shall not be dedicated to his death, but consecrated to his life." In 1954, President Eisenhower signed the 1954 Atomic Energy Act, which fostered the cooperative development of nuclear energy by the Atomic Energy Commission (AEC) and private industry. This marked the beginning of the commercial nuclear power program in the United States.

The world's first large-scale nuclear power plant was the Shipping port Atomic Power Station in Pennsylvania, which began its operation in 1957. This reactor was a pressurized water reactor (PWR) nuclear power plant designed and built by the Westinghouse Electric Company and operated by the Duquesne Light Company. The plant produced 68 MWe and 231 MWt.

The first commercial size boiling water reactor (BWR) was the Dresden Nuclear Power Plant which began its operation in 1960. This 200-MWe plant was owned by the Commonwealth Edison Company and was built by the General Electric Company at Dresden, Illinois, about 50 miles southwest of Chicago.

While other reactor concepts, including heavy-water moderated, gas-cooled, and liquid-metal-cooled reactors, have been successfully operated, the PWR and BWR designs have dominated the commercial nuclear power market, particularly in the United States.

These commercial power plants rapidly increased in size from tens of MWe generating capacity to over 1000 MWe. Today nuclear power plants are operating in 33 countries. The following section presents the current status of nuclear power plants operating or under construction around the world.

17.2.1 Current Nuclear Power Plants Worldwide

At the end of 2012, there were 433 individual nuclear power reactors operating throughout the world. More than half of these nuclear reactors are PWRs. The distribution of current reactors by type is listed in Table 17.1. As shown in Table 17.1, there are six types of reactors currently used for electricity generation throughout the world. The following sections provide a more detailed description of these different reactor types.

17.2.1.1 Pressurized Water Reactors

PWRs represent the largest number of reactors used to generate electricity throughout the world. They range in size from about 400 to 1500 MWe. The PWR, shown in Figure 17.1, consists of a reactor core contained within a pressure vessel and is cooled by water under high pressure. The nuclear fuel in the core consists of uranium dioxide fuel pellets enclosed in zircaloy rods that are held together in fuel assemblies. There are 200–300 rods in an assembly and 100–200 fuel assemblies in the reactor core. The rods are arranged vertically and contain 80–100 tons of enriched uranium.

TABLE 17.1

Nuclear Power Units by Reactor Type, Worldwide

Reactor Type	Main Countries	# Units Operational	GWe	Fuel
Pressurized light-water reactors (PWR)	United States, France, Japan, Russia	271	251	Enriched UO_2
Boiling light-water reactors (BWR) and Advanced boiling light-water reactors (AWBR)	United States, Japan, Sweden	83	78	Enriched UO_2
Pressurized heavy-water reactors—CANDU (PHWR)	Canada	48	24	Natural UO_2
Gas-cooled reactors (Magnox and AGR)	United Kingdom	15	8	Natural U (metal), enriched UO_2
Graphite-moderated light-water reactors (RMBK)	Russia	15	10	Enriched UO_2
Liquid-metal-cooled fast-breeder reactors (LMFBR)	Japan, France, Russia	1	1	PuO_2 and UO_2
		433	**371**	

Source: Information taken from *Nuclear News*, American Nuclear Society, 15th Annual Reference Issue, March 2013.

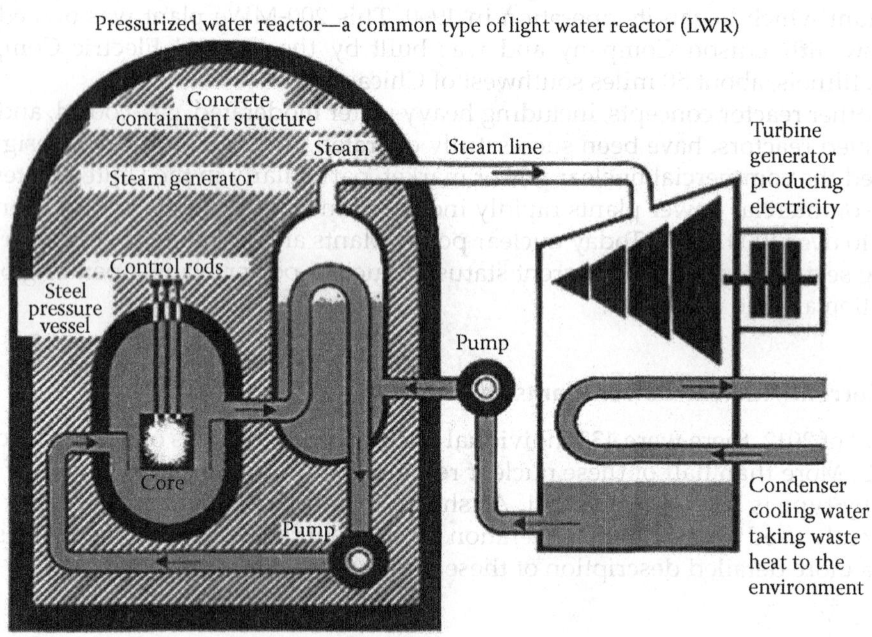

FIGURE 17.1
Sketch of a typical PWR power plant. (Courtesy of World Nuclear Association.)

The pressurized water at 315°C is circulated to the steam generators. The steam generator is a tube and shell type of heat exchanger with the heated high pressure water circulating through the tubes. The steam generator isolates the radioactive reactor cooling water from the steam which turns the turbine generator. Water enters the steam generator shell side and is boiled to produce steam which is used to turn the turbine generator producing electricity. The pressure vessel containing the reactor core and the steam generators are

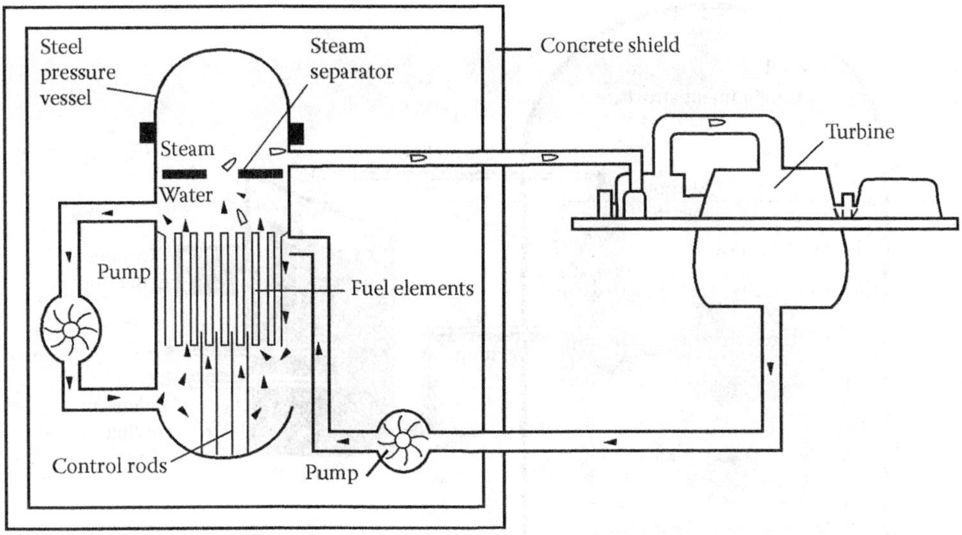

FIGURE 17.2
Sketch of a typical BWR power plant. (Courtesy of World Nuclear Association.)

located in the reactor containment structure. The steam leaving the turbine is condensed in a condenser and returned to the steam generator. The condenser cooling water is circulated to cooling towers where it is cooled by evaporation. The cooling towers are often pictured as an identifying feature of a nuclear power plant.

17.2.1.2 Boiling Water Reactors

The BWR power plants represent the second largest number of reactors used for generating electricity. BWRs range in size from 400 MWe to the largest being about 1200 MWe. The BWR, shown in Figure 17.2, consists of a reactor core located in a reactor vessel that is cooled by circulating water. The cooling water is heated to 285°C in the reactor vessel and the resulting steam is sent directly to the turbine generators. There is no intermediate loop as in PWR. The reactor vessel is contained in the reactor building. The steam leaving the turbine is condensed in a condenser and returned to the reactor vessel. The condenser cooling water is circulated to the cooling towers, where it is cooled by evaporation.

17.2.1.3 Pressurized Heavy-Water Reactor

The so-called "CANDU reactor" was developed in Canada, beginning in the 1950s. It consists of a large tank called calandria containing the heavy-water moderator. The tank is penetrated horizontally by pressure tubes which contain the reactor fuel assemblies. Pressurized heavy water is passed over the fuel and heated to 290°C. As in the PWR, this pressurized water is circulated to a steam generator where light water is boiled to form the steam, used to drive the turbine generators.

The pressure tube design allows the CANDU reactor to be refueled while it is in operation. A single pressure tube can be isolated and the fuel can be removed and replaced while the reactor continues to operate. The heavy water in the calandria is also circulated and the heat is recovered from it. The CANDU reactor is shown in Figure 17.3.

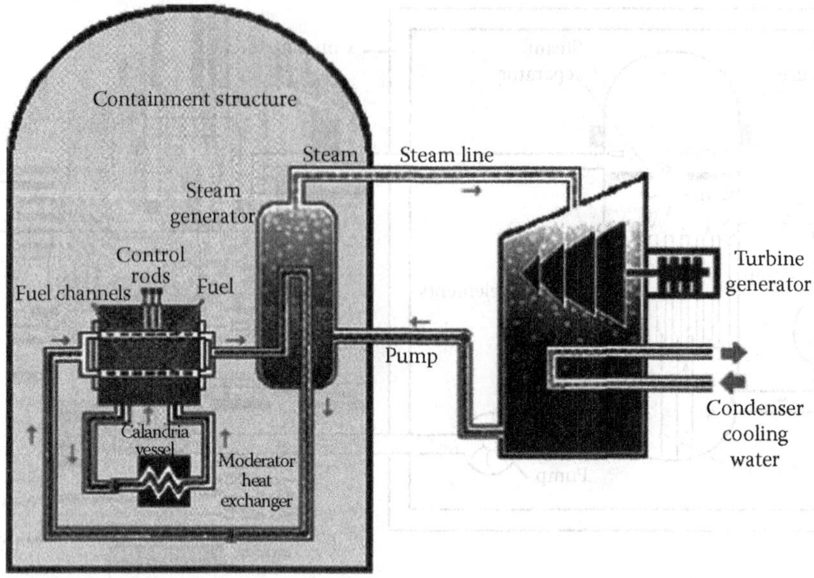

FIGURE 17.3
Sketch of a typical CANDU reactor power station. (Courtesy of World Nuclear Association.)

17.2.1.4 Gas-Cooled Reactors

Gas-cooled reactors were developed and implemented in the United Kingdom. The first generation of these reactors was called "Magnox" and they were followed by the advanced gas-cooled reactor (AGR). These reactors are graphite moderated and cooled by CO_2. Magnox reactors are fueled with uranium metal fuel, while the AGRs use enriched UO_2 as the fuel material. The CO_2 coolant is first circulated through the reactor core and then to a steam generator. The reactor and the steam generators are located in a concrete pressure vessel. As with the other reactor designs, the steam is used to turn the turbine generator to produce electricity. Figure 17.4 shows the configuration for a typical gas-cooled reactor design.

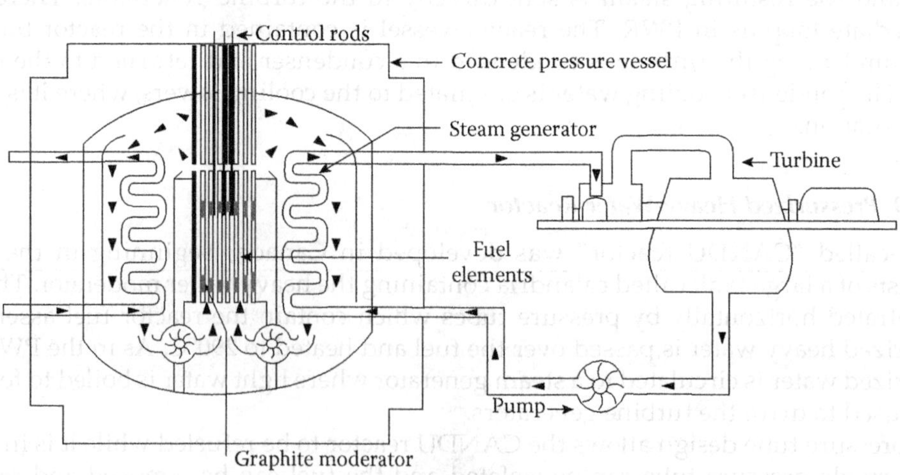

FIGURE 17.4
Sketch of a typical gas-cooled reactor power station. (Courtesy of World Nuclear Association.)

17.2.1.5 Other Power Reactors

The remaining reactors listed in Table 17.1 are the light-water graphite-moderated reactors used in Russia, and the liquid-metal-cooled fast-breeder reactors (LMFBRs) in Japan, France, and Russia. In the light-water graphite-moderated reactors, the fuel is contained in vertical pressure tubes where the cooling water is allowed to boil at 290°C, and the resulting steam is circulated to the turbine generator system as in BWR. In the case of the LMFBR, sodium is used as the coolant and a secondary sodium cooling loop is used to provide heat to the steam generator.

17.2.2 Growth of Nuclear Power

The growth of nuclear power generation is being influenced by three primary factors. These factors are (1) current plants are being modified to increase their generating capacity, (2) the life of old plants is being lengthened by life extension practices that include relicensing, and (3) new construction is adding to the number of plants operating worldwide. According to the IAEA, following the Fukushima accident, the World Energy Outlook 2011 New Policies scenario has a 60% increase in nuclear capacity to 2035, compared with about 90% the year before. At the end of 2012, there were 433 nuclear power plants in operation with a total net installed capacity of 371 GWe. They now anticipate the installed capacity to reach 630 GWe by 2035. This implies that over 250 new nuclear plants will be built over the next 20 years and this does not account for replacement of current plants that reach the end of life and it assumes that the overall average capacity will remain the same.

17.2.2.1 Increased Capacity

Operating nuclear plants are being modified to increase their generating capacity. Reactors in the United States, Belgium, Sweden, Switzerland, Spain, and Finland are being uprated. In the United States, the Nuclear Regulatory Commission has approved 140 uprates totaling over 6500 MWe since 1977 with some of them having capacity increased up to 20%. The number of operating reactors in the United States as of June 2013 is 102, which have a generating capacity exceeding 100 GWe. The generating capacity increase was due to both power uprating and improvements in operation and maintenance practices to produce higher plant availability. Switzerland increased the capacity of its plants by over 13%, while in Spain uprating has added 11% to that country's nuclear capacity. The uprating process has proven to be a very cost-effective way to increase overall power production capacity, while avoiding the high capital cost of new construction.

17.2.2.2 Plant Life Extension

Life extension is the process by which the life of operating reactors is increased beyond the original planned and licensed life. Most reactors were originally designed and licensed for an operational life of 40 years. Without life extension, many of the reactors that were built in the 1970s and 1980s would reach the end of their operational lives during the years 2010–2030. If they are not replaced with new plant construction, there would be a significant decrease in nuclear-based electricity generation as these plants reached the end of their useful life.

Engineering assessments of current nuclear plants have shown that they are able to operate for longer than their original planned and licensed lifetime. Over 70 plants in the

United States have been granted 20-year extensions to their operating licenses by the U.S. Nuclear Regulatory Commission (NRC). The operators of most of the remaining plants are also expected to apply for license extensions. This will give the plants an operating life of 60 years. In Japan, operating lifetimes of 70 years are envisaged.

The oldest nuclear power stations in the world were operated in Great Britain. Chalder Hall and Chaplecross were built in the 1950s and were expected to operate for 20–25 years. They were authorized to operate for 50 years but were shut down in 2003 and 2004 for economic reasons. In 2000, the Russian government extended the lives of their 12 oldest reactors by 15 years for a total of 45 years.

While life extension has become the norm throughout the world, many reactors have been shut down due to economic, regulatory, and political reasons. Many of these reactors were built early in the development of nuclear power. They tended to be smaller in size and were originally built for demonstration purposes. However the political and regulatory process in some countries has led to the termination of nuclear power programs and the shutdown of viable reactor plants. Germany is probably the best example having made a political decision to phase out nuclear power. Plants such as San Onofre Units 2 and 3, in southern California, and Kewaunee, in Wisconsin, are being shut down for economic reasons.

17.2.2.3 New Nuclear Plant Construction

New nuclear power plants are currently being constructed in several countries. The majority of these constructions are in Asia. Plants currently under construction are listed in Table 17.2.

17.3 Next-Generation Technologies

The next generation, Generation III, nuclear power reactors are being developed to meet the power production needs throughout the world. These reactors incorporate the lessons that have been learned by operation of nuclear power systems since the 1950s. The reactors are designed to be safer, more economic, and more fuel-efficient. The first of these reactors were built in Japan and began operation in 1996.

The biggest change in the Generation III reactors is the addition of passive safety systems. Earlier reactors relied heavily on operator actions to deal with a variety of operational upset conditions or abnormal events. The advanced reactors incorporate passive or inherent safety systems that do not require operator intervention in the case of a malfunction. These systems rely on such things as gravity, natural convection, and resistance to high temperatures.

Generation III reactors also have the following advantages:

- Standardized designs with many modules of the reactor being factory constructed and delivered to the construction site leading to expedited licensing, reduction of capital cost, and reduced construction time.
- Simpler designs with fewer components that are more rugged, easier to operate, and less vulnerable to operational upsets.

TABLE 17.2

Power Reactors under Construction

Start Operation	Country, Organization	Reactor	Type	MWe (Net)
2013	Iran, AEOI	Bushehr 1[a]	PWR	950
2013	India, NPCIL	Kudankulam 1	PWR	950
2013	India, NPCIL	Kudankulam 2	PWR	950
2013	China, CGNPC	Hongyanhe 1[a]	PWR	1080
2013	China, CGNPC	Ningde 1[a]	PWR	1080
2013	Korea, KHNP	Shin Wolsong 2	PWR	1000
2013	Korea, KHNP	Shin-Kori 3	PWR	1350
2013	Russia, Rosenergoatom	Leningrad II-1	PWR	1070
2013	Argentina, CNEA	Atucha 2	PHWR	692
2013	China, CGNPC	Ningde 2	PWR	1080
2013	China, CGNPC	Yangjiang 1	PWR	1080
2013	China, CGNPC	Taishan 1	PWR	1700
2013	China, CNNC	Fangjiashan 1	PWR	1080
2013	China, CNNC	Fuqing 1	PWR	1080
2013	China, CGNPC	Hongyanhe 2	PWR	1080
2014	Russia, Rosenergoatom	Novovoronezh II-1	PWR	1070
2015	Russia, Rosenergoatom	Rostov 3	PWR	1070
2014	Slovakia, SE	Mochovce 3	PWR	440
2014	Slovakia, SE	Mochovce 4	PWR	440
2014	Taiwan Power	Lungmen 1	ABWR	1300
2014	China, CNNC	Sanmen 1	PWR	1250
2014	China, CPI	Haiyang 1	PWR	1250
2014	China, CGNPC	Ningde 3	PWR	1080
2014	China, CGNPC	Hongyanhe 3	PWR	1080
2014	China, CGNPC	Yangjiang 2	PWR	1080
2014	China, CGNPC	Taishan 2	PWR	1700
2014	China, CNNC	Fangjiashan 2	PWR	1080
2014	China, CNNC	Fuqing 2	PWR	1080
2014	Korea, KHNP	Shin-Kori 4	PWR	1350
2014?	Japan, Chugoku	Shimane 3	ABWR	1375
2014	India, Bhavini	Kalpakkam	FBR	470
2014	Russia, Rosenergoatom	Beloyarsk 4	FNR	750
2015	United States, TVA	Watts Bar 2	PWR	1180
2015	Taiwan, Power	Lungmen 2	ABWR	1300
2015	China, CNNC	Sanmen 2	PWR	1250
2015	China, CGNPC	Hongyanhe 4	PWR	1080
2015	China, CGNPC	Yangjiang 3	PWR	1080
2015	China, CGNPC	Ningde 4	PWR	1080
2015	China, CGNPC	Fangchenggang 1	PWR	1080
2015	China, CNNC	Changjiang 1	PWR	650
2015	China, CNNC	Changjiang 2	PWR	650
2015	China, CNNC	Fuqing 3	PWR	1080
2015	India, NPCIL	Kakrapar 3	PHWR	640
2015?	Japan, EPDC/J Power	Ohma 1	ABWR	1350

(Continued)

TABLE 17.2 (*Continued*)

Power Reactors under Construction

Start Operation	Country, Organization	Reactor	Type	MWe (Net)
2016	Finland, TVO	Olkilouto 3	PWR	1600
2016	France, EdF	Flamanville 3	PWR	1600
2016	Russia, Rosenergoatom	Novovoronezh II-2	PWR	1070
2016	Russia, Rosenergoatom	Leningrad II-2	PWR	1200
2016	Russia, Rosenergoatom	Vilyuchinsk	PWR x 2	70
2016	India, NPCIL	Kakrapar 4	PHWR	640
2016	India, NPCIL	Rajasthan 7	PHWR	640
2016	Pakistan, PAEC	Chashma 3	PWR	300
2016	China, China Huaneng	Shidaowan	HTR	200
2016	China, CPI	Haiyang 2	PWR	1250
2016	China, CGNPC	Yangjiang 4	PWR	1080
2016	China, CGNPC	Hongyanhe 5	PWR	1080
2015	China, CNNC	Hongshiding 1	PWR	1080
2015	China, CGNPC	Fangchenggang 2	PWR	1080
2016	China	Several others	PWR	
2017	United States, Southern	Vogtle 3	PWR	1200
2017	Russia, Rosenergoatom	Baltic 1	PWR	1200
2017	Russia, Rosenergoatom	Rostov 4	PWR	1200
2017	Russia, Rosenergoatom	Leningrad II-3	PWR	1200
2017	Ukraine, Energoatom	Khmelnitsky 3	PWR	1000
2017	Korea, KHNP	Shin-Ulchin 1	PWR	1350
2017	India, NPCIL	Rajasthan 8	PHWR	640
2017	Romania, SNN	Cernavoda 3	PHWR	655
2017?	Japan, JAPC	Tsuruga 3	APWR	1538
2017	Pakistan, PAEC	Chashma 4	PWR	300
2017	United States, SCEG	Summer 2	PWR	1200
2017	China	Several		
2018	Korea, KHNP	Shin-Ulchin 2	PWR	1350

Source: Information taken from Plans for new reactors worldwide, World Nuclear Association Information Paper, www.world-nuclear.org, March, 2013.

[a] Is the latest announced date as of March 2013; ? for the reactors in Japan is because of the uncertainty following the Fukushima event.

- Longer operating lives of 60 years and designed for higher availability.
- Reduced probability of accidents leading to core damage.
- Safety systems that allow 72 h after an accident before active intervention is required.
- Resistance to serious damage that would allow release of radioactivity following an aircraft impact.
- Higher fuel burnup reducing refueling outages and increasing fuel utilization with less waste produced.

The following sections describe the different types of Generation III reactors being developed worldwide. Many of them are larger than their predecessors and involve consortiums of international companies. Certification of the design concepts continues on a country-by-country basis.

In Europe, there is an effort to harmonize requirements for nuclear plant licensing. Plants certified to comply with the European Utilities Requirements (EUR) include the Westinghouse AP1000, Gidropress' AES-92, Areva's EPR, GE-s ABWR, Areva's Kerena, and Westinghouse BWR 90. In the United Kingdom, the Office of Nuclear Regulation is processing the Areva EPR and the Westinghouse AP1000.

In the United States, the commercial nuclear industry in conjunction with the U.S. Department of Energy (DOE) has developed four advanced light-water reactor designs. Two of these are based on experience obtained from operating reactors in the United States, Japan, and Western Europe. These reactors will operate in the 1300 MW range. One of the designs is the advanced boiling water reactor (ABWR). It was designed in the United States and is already being constructed and operated in Asia. The NRC gave final design certification to the ABWR in 1997. It was noted that the design exceeded NRC "safety goals by several orders of magnitude." The other type, designated System 80+, is an advanced PWR. This reactor system was ready for commercialization but the sale of this design is not being pursued.

Advanced thermal reactors currently being marketed are summarized in Table 17.3, and the various types of reactors and specific identified designs will be discussed in the following sections.

17.3.1 Light-Water Reactors

Generation III advanced light-water reactors are being developed in several countries.

EPR: Areva in France has developed a large 1630-MWe PWR, which is designated as the European pressurized water reactor (EPR). It is designed with double containment with four separate redundant active safety systems. The first EPR is being built in Olkiluoto, Finland, the second two will be built in France. Areva has also designed a US-EPR and has been submitted to the USNRC for design certification. It is known as the Evolutionary PWR (EPR).

AP1000, CAP1400: The Westinghouse AP1000 (scaled up from the AP-600) received final design approval from the NRC and has received full design certification 2011. The passive safety systems in this reactor design lead to a large reduction in components including 50% fewer valves, 35% fewer pumps, 80% less pipe, 45% less seismic building volume, and 70% less cable. The AP1000 is designated the CAP1000 in China. Another aspect of the AP1000 is the construction process. Once the plant is ordered, the plant will be constructed in a modular fashion with modules being fabricated in a factory setting and then transported to the reactor site. The anticipated design construction time for the plant is 36 months. The construction cost of an AP1000 is expected to be greater than $5000 per kW in the United States and less than $2000 per kW in China. The plant is designed to have a 60 year operating life. AP1000 plants are under construction in China and the United States.

ABWR: Advanced boiling water reactor is offered in slightly different versions by GE Hitachi, Hitachi-GE, and Toshiba. This design produces 1350–1600 MWe. Two of the Hitachi versions and two of the Toshiba versions have been built in Japan and are in commercial operation, Hitachi is promoting the UK-ABWR while Toshiba has the EU-ABWR. The design can operate with MOX fuel and has a design life of 60 years. The emergency cooling system uses pump mounted at the bottom of the reactor vessel and are part of an active safety system.

TABLE 17.3
Advanced Thermal Reactors Being Marketed

Country and Developer	Reactor	Size (MWe)	Design Progress	Main Features
United States–Japan (GE-Hitachi-Toshiba)	ABWR	1380	Commercial operation in Japan since 1996–1997. In the United States: NRC certified 1997, first-of-a-kind engineering (FOAKE).	Evolutionary design More efficient, less waste Simplified construction (48 months) and operation
South Korea (derived from Westinghouse)	APR-1400 (PWR)	1450	Under construction—Shin Kori 3 and 4. Sold to UAE.	Evolutionary design Increased reliability Simplified construction and operation
United States (Westinghouse)	AP-600 AP1000 (PWR)	600 1200	AP-600: NRC certified 1999, FOAKE AP1000 NRC design approval 2011.	Passive safety features Simplified construction and operation 3 years to build 60-year plant life
Japan (Utilities, Westinghouse, Mitsubishi)	APWR US-APWR EU-APWR	1530 1700 1700	Basic design in progress, planned for Tsuruga U.S.-design certification application.	Hybrid safety features Simplified construction and operation
Europe (Areva NP)	EPR US-EPR (PWR)	1750	Future French standard. French design approval. Being built in Finland, France, and China. Undergoing certification in the United States.	Evolutionary design Improved safety features High fuel efficiency Low cost electricity
United States (GE)	ESBWR	1600	Developed from the ABWR, pre-certification in the United States.	Evolutionary design Short construction time Enhanced safety features
	Atmea1 (PWR)	1150	French design approval February 2012, ready for deployment.	Innovative design High fuel efficiency
Russia (Gidropress)	VVER-1200 (PWR)	1200	Under construction at Leningrad, Novovoronezh and Baltic plants.	Evolutionary design 50-year plant life High fuel efficiency
Canada (CANDU Energy)	Enhanced CANDU-6—EC6	750	Improved model Design review in progress.	Evolutionary design Flexible fuel requirements
	HTR-PM	2 × 105 (module)	Demonstration plant being built at Shidaowan.	Modular plant, low cost High temperature High fuel efficiency

Source: World Nuclear Association, Advanced nuclear power reactors. Information paper, www.world-nuclear.org, September 2013.

ESBWR: GE Hitachi is working on this design which is known as the Economic and Simplified BWR. It is in the final stages of certification by the USNRC. This design eliminates the need for pumps for emergency cooling and uses passive systems and stored energy instead. It has a design life of 60 years and produces 1535 MWe.

APWR: APWR is an advanced PWR marketed by Mitsubishi. The first of these reactors was planned for construction at Tsurga in Japan. The U.S. version of this reactor will have a 1620 MWe output. Another version of this reactor was planned for construction in Texas but the request for certification has been delayed because the utility is looking at other electrical generating systems. It combines active and passive emergency cooling systems in a double containment and features high burnup fuel.

APR1400: The APR1400 is an advanced design, earlier known as Korean Next-Generation Reactor. The design was certified by Korean Institute of Nuclear Safety in 2003 and the first two units have been constructed in Korea. The reactor is being marketed in the Middle East as a heat source for large desalinization systems. It has a generating capacity of 1350–1400 MWe.

Atmea1: The Atmea1 was developed by the Atmea joint venture by Areva and Mitsubishi to produce an evolutionary PWR which would use the same steam generators as the EPR. The reactor was designed to have load-following capability and frequency control capacity. The design has been approved by the French regulator ASN and is designed for sale to countries that do not have previous nuclear experience. It has three redundant active and passive safety systems and a backup cooling system similar to the EPR.

Areva-EdF-CGNPC: This is a proposed effort between the French companies and China Guangdong Nuclear Power Group to develop a mid-sized PWR with third-generation safety features. No reactor design has yet come from this effort.

Kerena: This reactor design was developed by Areva in conjunction with two German utilities. Kerena is based on the design of a German plant designed and built by Siemens and completed in 1999. This reactor can operate with MOX fuel and also has load-following capability.

AES-92, V392: These are late model VVER-1000 with enhanced safety systems and are designed by Gidropress. This reactor model is being built in China and India. AES-92 is certified as meeting EUR.

AES-2006, MIR-1299, VVER-TOI: These are third-generation standardized VVER-1200 reactors that will operate at 1070 MWe. Models of this system are being built in Russia and are expected to begin operation in 2014. The systems are designed for 60 years of operation at a 90% capacity factor. They have enhanced safety systems including those related to earthquakes and aircraft impact with some passive safety features. The MIR-1200 (Modernized International Reactor) is intended for sale and operation outside of Russia.

ACPR1000 and ACP1000: These are essentially Chinese designed versions of French designs with modern features. The ACP1000 is an 1100-MWe plant with load-following capability. The APCR1000 is also an 1100-MWe plant with a double containment and a core catcher.

IRIS: An international project being led by Westinghouse is designing a modular 335 MWe reactor known as the "International Reactor Innovative & Secure" (IRIS). This PWR is designed with integral steam generators and a primary cooling system that are all contained in the reactor pressure vessel. The goal of this system is to reach an 8-year refueling cycle using 10% enriched fuel with an 80,000 MWd/t burnup.

VBER-300: The VBER is a small PWR based on navel reactors and was designed to be a barge-mounted floating nuclear power plant (295–325 MWE unit) with an expected life of 90 years with a 90% capacity factor. The first land-based reactor of this design will be built in Kazakhstan.

RMWR, RBWR: The reduced moderation water reactor (RMWR) is a light-water reactor with the fuel arry packed more tightly then the typical PWR and BWR of today. These Hitachi-designed reactors are expected to produce more plutonium, and thus make better use of the uranium resource. The breeding ratio is near 1 instead of 0.6 of today's reactors. Reprocessing through an advanced fluoride volatility process leaves a uranium–plutonium mixture that can be made directly into MOX fuels of standard light-water reactors. Thor Energy in Norway is exploring the use of an RBWR with a thorium oxide (Th-MOX) fuel because high conversion factors for production of U233 can be achieved with this design. The core of the reactor is flatter with tightly packed fuel rods to ensure sufficient fast neutron leakage and a negative void reactivity coefficient.

17.3.2 Heavy-Water Reactors

Heavy-water reactors are primarily designed and used in Canada and India. Both countries are actively taking steps in developing and making advances in this technology. In 2011 AECL sold the reactor division which now operates as Candu Energy, Inc.

EC6: The EC6 is an advancement of the CANDU-9, which was designed in the late 1990s and never built. This design is proposed for building in China after the successful construction of CANDU-6 units and is known as the "Enhanced CANDU-6." It is a versatile reactor which will be able to use natural uranium fuel as well as reprocessed PWR uranium and various mixtures of plutonium, uranium, and thorium. The reactor operates at 690 MWe.

ACR: The advanced CANDU reactor was designed as a more innovative concept, but the design and regulatory approval efforts have now been terminated.

AHWR: Advance heavy-water reactor is being developed in India. The purpose of this reactor is to use a thorium-based fuel cycle. The thorium-based fuel will be seeded with both U233 and Pu239. It is a 284-MWe reactor moderated with heavy water and cooled with boiling light water. It is designed for a 100-year plant life. The AHWR-LEU is an export version of this design. It will use low-enriched uranium and thorium as fuel.

17.3.3 High-Temperature Gas-Cooled Reactors

These reactors use high-temperature helium as a coolant at up to 950°C which is used to make steam via a steam generator or is used directly to drive a gas turbine. The fuel is in the form of small particles coated with carbon and silicon carbide which will provide containment of the fission products to temperatures up to 1600°C or more.

HTR-PM: The first commercial version of the HTGR will be China's HTR-PM. This reactor is being built in China and will consist of two 105-MWe modules. The reactor will have an outlet temperature of 750°C. This will be a demonstration system for a complete 18-module power station.

PMRB: The pebble bed modular reactor is being developed by a consortium in South Africa led by Eskom along with Mitsubishi Heavy Industries. Production units were planned to be 165 MWe with a direct cycle gas turbine generator. Development of this system has ceased due to lack of funds and customers.

GT-MHR: The gas turbine—modular helium reactor is a U.S. design with modules of 285 MWe each, generated by a directly driven gas turbine at a 48% efficiency. The reactor was being developed by General Atomics in partnership with Russia's OKBM Afrikantov and supported by Fuji in Japan. The preliminary design stage was completed in 2001, but the program has stalled then.

17.3.4 Fast Neutron Reactors

Several countries are working on the development of fast breeder reactors (FBR), which are fast neutron reactor (FNR) configured with a breeding ratio greater than 1.0. About 20 of these reactors have been built and operated since the 1950s. The FBR can use uranium much more efficiently than the light-water reactor but they are more expensive to construct; so long as the price of uranium remains low, they are not economic.

India has operated a 40 mMWt fast breeder test reactor since 1985. Construction of 500-MWe prototype fast reactor was started in 2004 with expected operation in 2012. The reactor is fueled with Ur–Pu fuel with a thorium blanket to breed U233. This will provide support for India's efforts to make use of the abundant supply of thorium available there.

The Russian BN-600 reactor has been supplying electricity to the grid since 1981 and has the best operating and production record of any reactor in the country. The BN-350 reactor is operated in Kazakhstan for 27 years and about half of the output was used for water desalinization.

BN-800: The BN-800 is a new and more powerful FBR designed by OKBM. It has a wide fuel flexibility, enhanced safety, and improved economy. It will be used to burn plutonium from dismantled weapons and to test the recycle of minor actinides. Two units of this reactor design have been sold to China and are under construction.

BN-1200: The BN1200 is designed to operate with MOX fuel and also U–Pu nitride in a closed cycle. It can be sodium or lead cooled and will produce 1220 MWe. The BN-1200 is being considered to be a Gen-IV design.

BREST: Russia has used lead–bismuth cooling for 40 years in submarine reactors. BREST is a new fast neuron reactor designed to produce 300 MWe with lead coolant. Construction of the first reactor of this type is expected to begin in 2015 with operation in 2020. It is designed to have an equilibrium core that will allow it to use its own recycled fuel but will not produce excess material for additional reactors.

ELSY: The European Lead-Cooled System is a 600-MWe fast reactor being designed in Italy. ELSY is designed to be fueled by either depleted uranium or thorium and also burn the actinides from light-water reactor fuel. It is cooled by either lead or lead–bismuth eutectic.

PRISM: PRISM is a GE-Hitachi design fast reactor of the pool type with passive cooling for decay heat removal. PRISM is designed to produce 311 MWe and will operate in blocks of two reactors using a metal fuel of depleted uranium and plutonium. It is designed to be used with the electrometallurgical reprocessing so that all the transuranics are recycled with the fuel.

KALIMER: The Korea Advanced Liquid Metal Reactor is the Korean-designed fast reactor system. This is a 600-MWe pool type system that is expected to be developed as a Gen-IV reactor.

17.3.5 Summary of Generation III Reactors

As can be seen from the discussion earlier, there are many reactor systems of many types under development. The key feature of all of these reactors is the enhancement of safety systems. Some of these reactors have already been built and are in operation while others are under construction. This activity indicates that there will be a growth of nuclear reactor generated electricity during the next 20 years. Table 17.3, taken from World Nuclear Association information on advanced nuclear power reactors shows the advanced thermal reactors that are being marketed around the world.

17.4 Small Modular Reactors

In the early days of nuclear power reactors, most of the reactors had small power outputs and many had unique first of a kind designs. Most of the nuclear power plants had electric power outputs below 100 MW and were considered demonstration plants which used a variety of technologies. Table 17.9 is a partial list of these small reactors. They include sodium-cooled fast reactors, heavy-water moderated reactors, BWRs, PWRs, and gas-cooled reactors.

Today the concept of the small modular reactor (SMR) is being developed using modern technologies with advanced safety systems. The International Atomic Energy Agency (IAEA) defines "small" as under 300 MWe; it also defines a medium-sized reactor as the one up to 700 MWe. The driving force behind the ideas for reactors of this size is based on several factors including

- Smaller reactors can be produced in a factory and moved as a single unit to the reactor site to be installed. This will shorten the time for construction of the power station and should be less costly.
- If a large power station is needed in the future, it could be made up of multiple small units each being added as the need arises. This is not unlike the manner in which fossil-fired plants were originally built with multiple smaller units.
- Smaller reactors can be used to serve remote area, where large amounts of power are not required and where service via transmission from larger plants is difficult.
- Smaller reactors can more easily be designed to operate with passive safety systems since the residual heat loads are smaller.
- They can also be built underground which would protect them from attack and also make siting possible in more urban areas.

The development of SMRs is in process across the world by nuclear system supplier. This development is shown in Table 17.10. The whole industry in this area is in a state of flux and details on specific reactor are changing rapidly. Specific current information of designs being considered is best found in the World Nuclear Association Information Library.[8]

17.5 Generation IV Technologies

As discussed earlier, the development of nuclear power occurred in three general phases. The initial development of prototype reactor designs occurred in the 1950s and 1960s, development and deployment of large commercial plants occurred in the 1970s and 1980s, and development of advanced light-water reactors occurred in the 1990s.

While the earlier generations of reactors have effectively demonstrated the viability of nuclear power, the nuclear industry still faces a number of challenges that need to be overcome in order for nuclear power to achieve its full potential. Among these challenges are (1) public concern about the safety of nuclear power in the wake of the Three Mile Island accident in 1979 and the Chernobyl accident in 1986, (2) high capital costs and licensing uncertainties associated with the construction of new nuclear power plants, (3) public concern over potential vulnerabilities of nuclear power plants to terrorist attacks, and (4) issues associated with the accumulation of nuclear waste and the potential for nuclear material proliferation in an environment of expanding nuclear power production.

To address these concerns and to fully realize the potential contributions of nuclear power to future energy needs in the United States and worldwide, the development of a new generation of reactors, termed Generation IV, was initiated in 2001. The intent or objective of this effort is to develop multiple Generation IV nuclear power systems that would be available for international deployment before 2030. The development of the Generation IV reactor systems is an international effort, initiated by the U.S. Department of Energy (DOE) with participation from 10 countries. These countries established a formal organization referred to as the Generation IV International Forum (GIF). The GIF countries included Argentina, Brazil, Canada, France, Japan, the Republic of Korea, the Republic of South Africa, Switzerland, the United Kingdom, and the United States. The intent of the GIF is "... to develop future-generation nuclear energy systems that can be licensed, constructed, and operated in a manner that will provide competitively priced and reliable energy products while satisfactorily addressing nuclear safety, waste, proliferation, and public perception concerns."

The process used by the GIF to identify the most promising reactor concepts for development (referred to as the Generation IV Technology Roadmap) consisted of three steps. These steps were (1) to develop a set of goals for new reactor systems, (2) solicit proposals from the worldwide nuclear community for new reactor systems to meet these goals, and (3) using experts from around the world, evaluate the different concepts to select the most promising candidates for further development.

The eight goals developed by the GIF for Generation IV nuclear systems [Ref. 7] were

- *Sustainability-1.* Generation IV nuclear energy systems will provide sustainable energy generation that meets clean air objective and promotes long-term availability of systems and effective fuel utilization for worldwide energy production.
- *Sustainability-2.* Generation IV nuclear energy systems will minimize and manage their nuclear waste and notably reduce the long-term stewardship burden in the future, thereby improving protection for the public health and the environment.
- *Economics-1.* Generation IV nuclear energy systems will have a clear life-cycle cost advantage over other energy sources.
- *Economics-2.* Generation IV nuclear energy systems will have a level of financial risk comparable to other energy projects.

- *Safety and Reliability-1.* Generation IV nuclear energy systems operations will excel in safety and reliability.
- *Safety and Reliability-2.* Generation IV nuclear energy systems will have a very low likelihood and degree of reactor core damage.
- *Safety and Reliability-3.* Generation IV nuclear energy systems will eliminate the need for offsite emergency response.
- *Proliferation Resistance and Physical Protection-1.* Generation IV nuclear energy systems will increase the assurance that they are very unattractive and the least desirable route for diversion or theft for weapons-usable materials, and provide increased physical protection against acts of terrorism.

Over 100 Generation IV candidates were evaluated by experts from the GIF countries and six reactor systems were selected for further evaluation and potential development. These six reactor systems selected are described as follows.

17.5.1 Gas-Cooled Fast Reactor (GFR) System

Gas-cooled fast reactor (GFR) System is a fast neutron spectrum reactor that uses helium as the primary coolant. It is designed to operate at relatively high helium outlet temperatures, making it a good candidate for the high-efficiency production of electricity or hydrogen. As shown in Figure 17.5, a direct Brayton cycle is used for the production of electricity with the helium gas delivered from the reactor outlet to a high-temperature gas turbine connected to a generator that produces electricity. In alternative designs, the high-temperature helium can also be used to produce hydrogen using either a thermochemical process or high-temperature electrolysis, or for other high-temperature process heat applications.

The reference plant is designed to produce 288 MWe using the direct Brayton cycle with a reactor outlet temperature of 850°C. The fuel forms being considered for high-temperature operation include composite ceramic fuel, advanced fuel particles, or ceramic clad elements of actinide compounds. Alternative core configurations include prismatic blocks, pin- or plate-based assemblies. GFR's fast neutron spectrum also makes it possible to efficiently use available fissile and fertile materials in a once-through fuel cycle.

17.5.2 Very-High-Temperature Reactor

Very-high-temperature reactor (VHTR) is a helium-cooled reactor designed to provide heat at very high temperatures, in the range of 1000°C for high-temperature process heat applications. In particular, the 1000°C reactor outlet temperature makes it a good candidate for the production of hydrogen using either thermochemical or high-temperature electrolysis processes. As shown in Figure 17.6, heat for the production of hydrogen is delivered through an intermediate heat exchanger that serves to isolate the reactor system from the hydrogen production process.

The reference design for VHTR is a 600 MW(t) reactor with an outlet temperature of 1000°C. The reactor core uses graphite as a moderator to produce thermal neutrons for the fission process. The core configuration can be either graphite blocks or pebbles about the size of billiard balls in which fuel particles are dispersed. For electricity production, either a direct Brayton cycle gas turbine using the primary helium coolant as the working

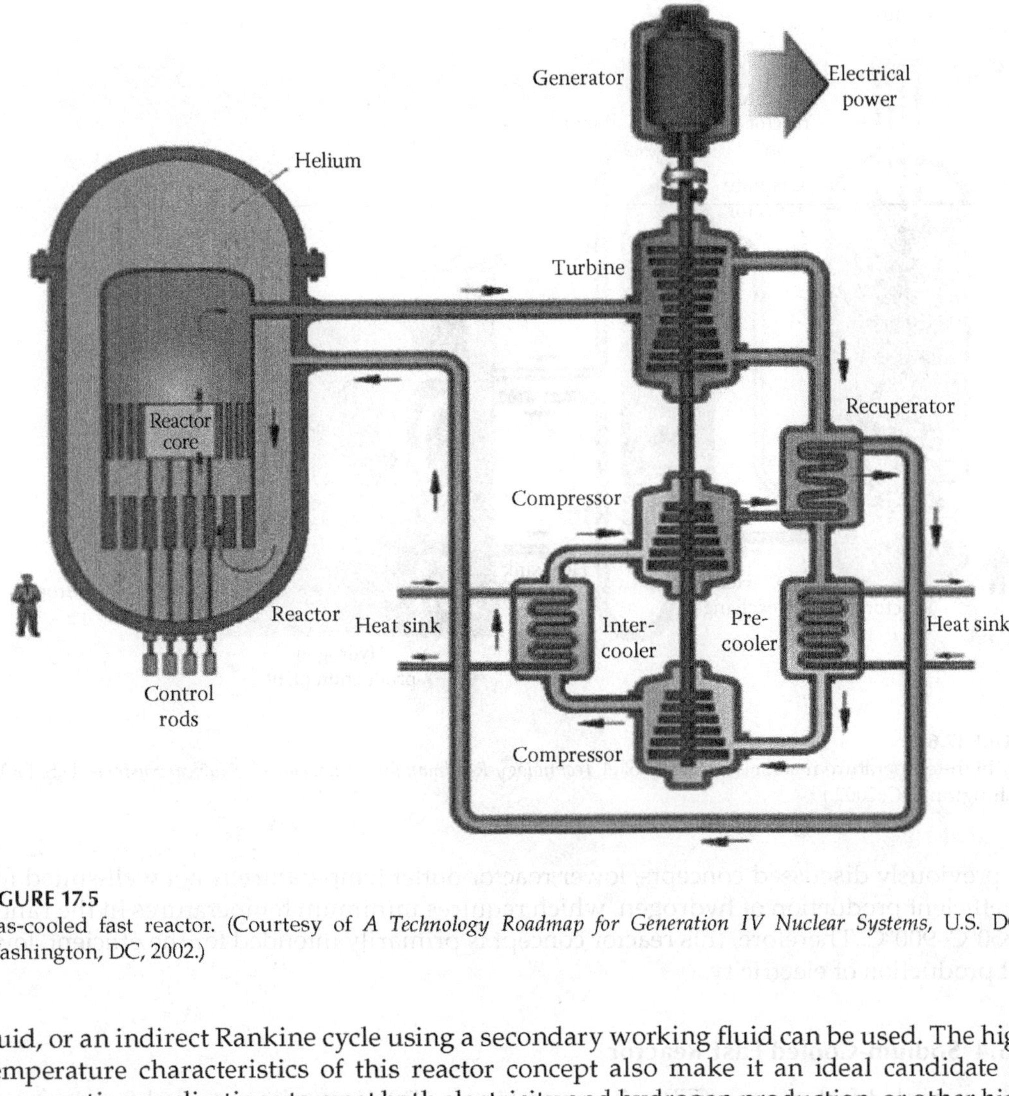

FIGURE 17.5
Gas-cooled fast reactor. (Courtesy of *A Technology Roadmap for Generation IV Nuclear Systems,* U.S. DOE, Washington, DC, 2002.)

fluid, or an indirect Rankine cycle using a secondary working fluid can be used. The high-temperature characteristics of this reactor concept also make it an ideal candidate for cogeneration applications to meet both electricity and hydrogen production, or other high-temperature process heat needs.

17.5.3 Supercritical-Water-Cooled Reactor (SWR)

Supercritical-water-cooled reactor (SWR) is a relatively high-temperature, high-pressure reactor designed to operate above the thermodynamic critical point of water, which is 374°C and 22.1 MPa. Since there is no phase change in the supercritical coolant water, the balance of plant design, shown in Figure 17.7, utilizes a relatively simple, direct cycle power conversion system. The reference design for this concept is a 1700 MWe reactor operating at a pressure of 25 MPa with a reactor outlet temperature ranging between 510°C and 550°C. This reactor can be designed as either a fast neutron spectrum or thermal neutron spectrum reactor. The relatively simple design also allows for the incorporation of passive safety features similar to those of the simplified BWR (discussed earlier). However, unlike

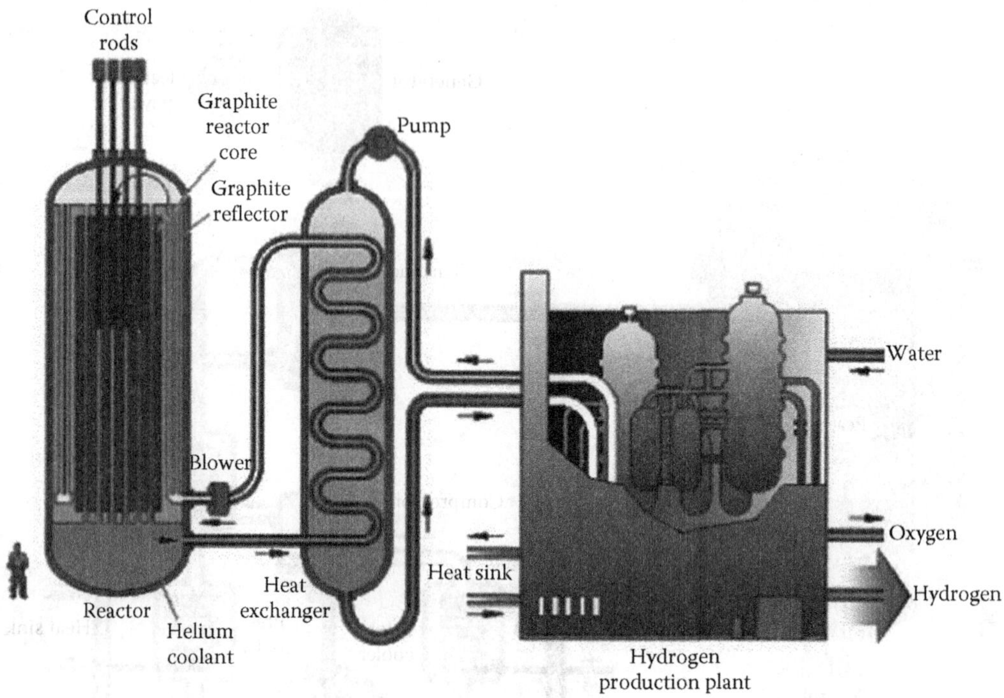

FIGURE 17.6
Very-high-temperature reactor. (Courtesy of *A Technology Roadmap for Generation IV Nuclear Systems*, U.S. DOE, Washington, DC, 2002.)

the previously discussed concepts, lower reactor outlet temperature is not well-suited for the efficient production of hydrogen, which requires minimum temperatures in the range of 850°C–900°C. Therefore, this reactor concept is primarily intended for the efficient, low-cost production of electricity.

17.5.4 Sodium-Cooled Fast Reactor

Sodium-cooled fast reactor (SFR), shown in Figure 17.8, is a sodium-cooled fast-neutron-spectrum reactor designed primarily for the efficient management of actinides and conversion of fertile uranium in a closed fuel cycle. Two reference designs to support different fuel reprocessing options have been defined for this concept. First is a medium-sized, sodium-cooled reactor with a power level between 150 and 500 MW(e) that utilizes uranium-plutonium-minor-actinide-zirconium metal alloy fuel. This reactor concept is supported by a fuel cycle based on pyrometallurgical processing in which the processing facilities are an integral part of the reactor plant design.

The second reactor reference design is a large sodium-cooled reactor with a power output capability between 500 and 1500 MWe that utilizes uranium–plutonium oxide fuel. This reactor design is supported by a fuel cycle based on an advanced aqueous process that would include a centrally located processing facility supporting a number of reactors.

Both versions of this reactor concept would operate at coolant outlet temperatures in the range of 550°C, and are intended primarily for the management of high-level waste

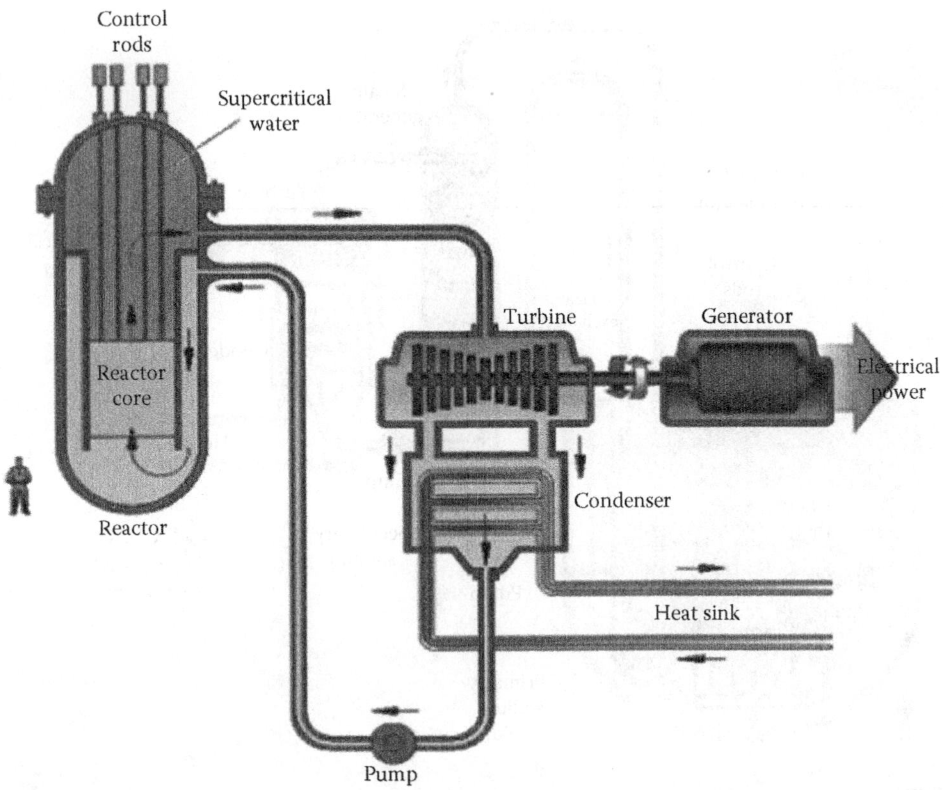

FIGURE 17.7
Supercritical water-cooled reactor. (Courtesy of *A Technology Roadmap for Generation IV Nuclear Systems*, U.S. DOE, Washington, DC, 2002.)

(HLW) and the production of electricity. In addition to design innovations to reduce capital costs, these reactors incorporate a number of enhanced safety features that include long thermal response time, a large margin to coolant boiling, a primary system that operates near atmospheric pressure, and an intermediate sodium system between the radioactive sodium in the primary system and the water and steam in the power plant.

17.5.5 Lead-Cooled Fast Reactor

Lead-cooled fast reactor (LFR) is a fast-neutron spectrum reactor cooled by either molten liquid or a lead-bismuth eutectic liquid metal. It is designed for the efficient conversion of fertile uranium and management of actinides in a closed fuel cycle. The reactor core for this design, shown in Figure 17.9, utilizes a metal- or nitride-based fuel containing fertile uranium and transuranics. As shown in Figure 17.9, LFR relies on natural convection to cool the reactor core. Outlet temperature for the current reactor concept is about 550°C, but with advanced materials, reactor outlet temperatures of 800°C may be possible. An indirect gas Brayton cycle is used to produce electrical power.

There are currently three versions of reference designs for this concept. The smallest design, rated at 50–150 MW(e), is intended for distributed power applications or electricity production on small grids. This reactor design, referred to as a "battery," features modular

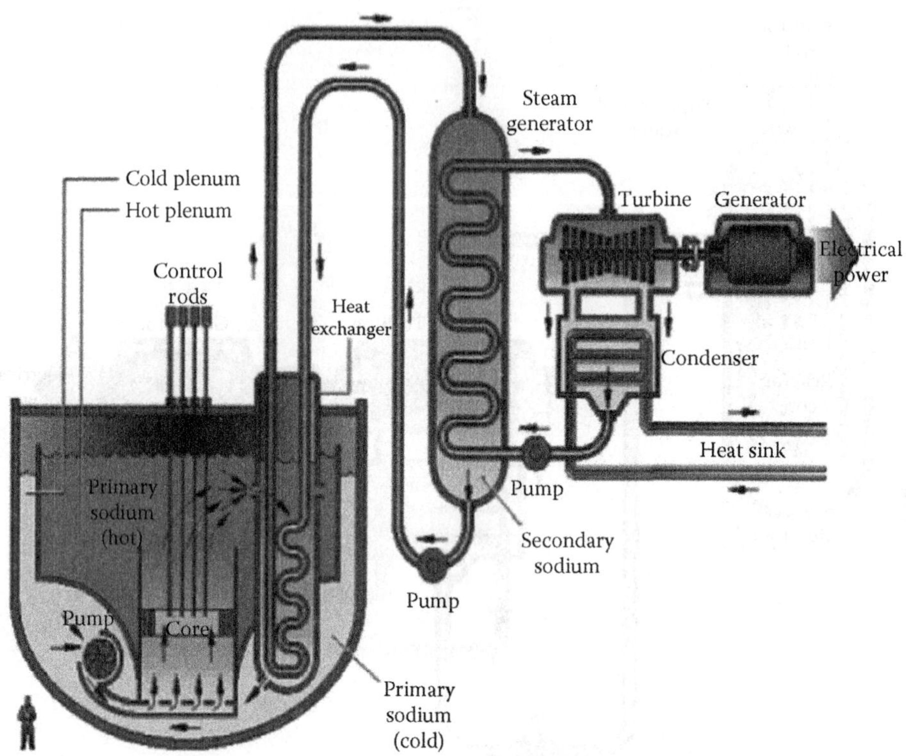

FIGURE 17.8
Sodium-cooled fast reactor. (Courtesy of *A Technology Roadmap for Generation IV Nuclear Systems*, U.S. DOE, Washington, DC, 2002.)

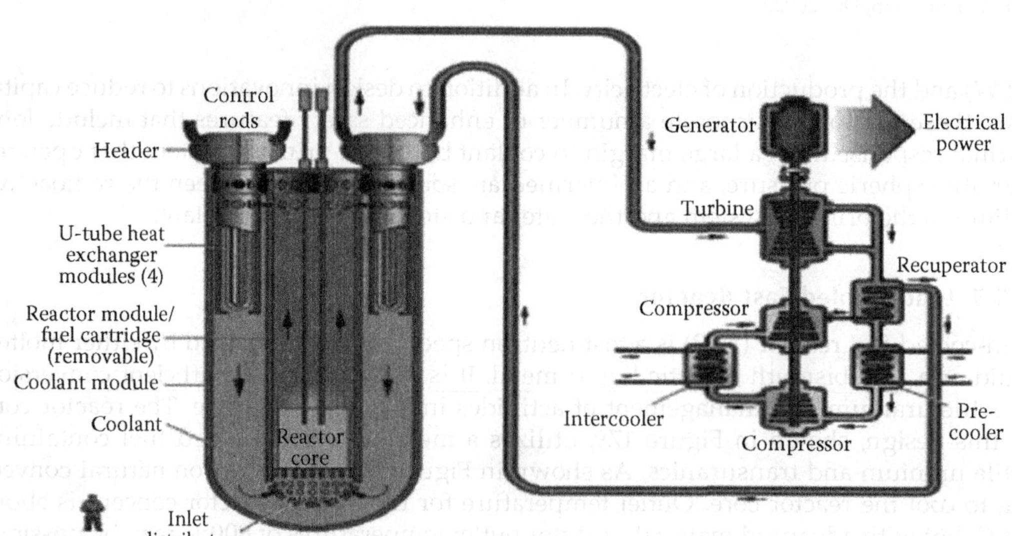

FIGURE 17.9
Lead-cooled fast reactor. (Courtesy of *A Technology Roadmap for Generation IV Nuclear Systems*, U.S. DOE, Washington, DC, 2002.)

design with a factory fabrication "cassette" core. The reactor is designed for very long refueling interval (15–20 years), with refueling accomplished by replacement of the cassette core or reactor module.

The other two versions of this design are a modular system rated at 300–400 MWe and a large plant rated at 1200 MWe. The different power options for this design are intended to fulfill different needs or opportunities in the power market, and be economically competitive with comparable alternative power sources.

17.5.6 Molten Salt Reactor

The molten salt reactor (MSR), shown in Figure 17.10, produces power by circulating a molten salt and fuel mixture through graphite core flow channels. The slowing down of neutrons by the graphite moderator in the core region provides the epithermal neutrons necessary to produce the fission power for sustained operation of the reactor. The heat from the reactor core is then transferred to a secondary system through an intermediate heat exchanger and then through a tertiary heat exchanger to the power conversion system that produces the electric power. The circulating coolant flow for this design is a mixture of sodium, uranium, and zirconium fluorides. In a closed fuel cycle, actinides such as plutonium can be efficiently burned by adding these constituents to the liquid fuel without the need for special fuel fabrication. The reference design for this concept is a 1000 MWe power plant with a coolant outlet of 700°C. To achieve higher thermal efficiencies for this concept, coolant outlet temperatures as high as 800°C may also be possible.

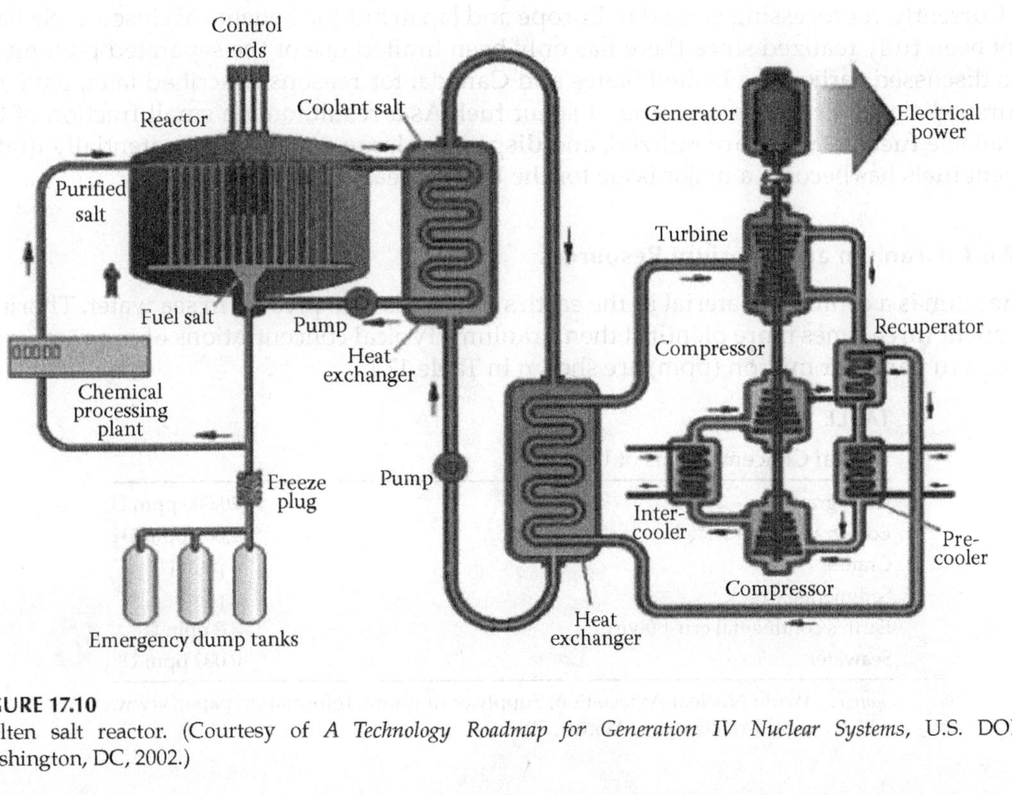

FIGURE 17.10
Molten salt reactor. (Courtesy of *A Technology Roadmap for Generation IV Nuclear Systems*, U.S. DOE, Washington, DC, 2002.)

17.6 Fuel Cycle

The process of following the fuel material from uranium or thorium mine through processing and reactor operation until it becomes waste is called the fuel cycle for nuclear systems. After a discussion of fuel cycle in general, it will be examined by looking at uranium and thorium resources, mining and milling, enrichment, reactor fuel use, spent fuel storage, nuclear materials transportation, and reprocessing. Nuclear waste will be addressed in a separate section.

General discussion of the fuel cycle often includes the terms "open" or "closed." The open fuel cycle is also called the once-through cycle. In the once-through fuel cycle, the uranium fuel is fabricated and run through the reactor once and then disposed-off as waste. There is no reprocessing of the fuel. In the closed cycle, the fuel is reprocessed after leaving the reactor so that it can be reused to improve overall fuel utilization.

In the open cycle, the fuel is introduced into the reactor for 1–2 years. It is then removed and put into long-term storage for eventual disposal. The impact of this cycle is wastage of about 99% of energy contained in the fuel. The United States adopted the open cycle in 1977 when President Jimmy Carter issued an executive order to stop reprocessing as a part of the fuel cycle. Canada has also adopted the open cycle.

The closed cycle was envisioned when the development of nuclear power began. Uranium and plutonium removed from reactors would be reprocessed and returned to reactors as fuel. Breeder reactors would be used to breed additional plutonium for use in thermal reactors. Thorium could also be used as a breeding material to generate U233 as a reactor fuel. The intent of the closed fuel cycle was to maximize the use of available reactor fuel resources while minimizing the waste generated by operating reactors.

Currently, reprocessing is used in Europe and Japan but the benefits of closed cycle have not been fully realized since there has only been limited use of the separated plutonium. As discussed earlier, the United States and Canada, for reasons described later, have not pursued closed cycle reprocessing of spent fuel. As a result, only a small fraction of the available fuel resources are utilized, and disposal of large quantities of potentially usable spent fuels has become a major issue for the U.S. nuclear industry.

17.6.1 Uranium and Thorium Resources

Uranium is a common material in the earth's crust. It is also present in sea water. Thorium is about three times more plentiful then uranium. Typical concentrations of uranium measured in parts per million (ppm) are shown in Table 17.4.

TABLE 17.4

Typical Concentrations of Uranium

High-grade ore—2% U	20,000 ppm U
Low-grade ore—0.1% U	1,000 ppm U
Granite	4 ppm U
Sedimentary rock	2 ppm
Earth's continental crust (avg)	2.8 ppm U
Seawater	0.003 ppm U

Source: World Nuclear Association, Supply of uranium. Information paper, www.world-nuclear.org, August, 2012.

The amount of recoverable uranium is dependent on the price. As the price increases, more material is economically recoverable. Also, more exploration will occur and it is likely that additional orebodies will be discovered. An orebody is defined as an occurrence of mineralization from which the metal, in this case uranium, can be recovered economically. Because of the uncertainties of price and its impact on exploration, any statement of recoverable amounts of uranium is simply a picture at an instant in time and is likely to change many times in the future. There is also a store of highly enriched uranium that is being recovered as nuclear weapons are dismantled. In addition, there are millions of tons of U238 which are the results of previous enrichment activities around the world. U238 can be blended with highly enriched uranium or plutonium to make fuel for nuclear power plants. It can also be used to breed plutonium in FBR fuel cycles.

Table 17.5 presents a list of recoverable resources of uranium. The table is taken from information gathered by the World Nuclear Association from other sources and was generated in 2004.

The 3.5 Mt is enough to fuel the world's current reactors for 50 years assuming the same fuel cycles currently in use. IAEA estimates the world supply at over 14 Mt which provides a supply exceeding 200 years at the current rate of use. This estimate does not include the uranium in phosphate deposits estimated at 22 Mt or the uranium available in sea water estimated at 1400 Mt. In addition, the ability of nuclear reactors to achieve higher burnups (utilize more of the uranium in the fuel) has also increased. This increases the efficiency of uranium use. Since thorium is not included in these fuel supply numbers, and as noted earlier is about three times as plentiful as uranium, there does not appear to be a fuel supply limitation for nuclear power in the foreseeable future.

TABLE 17.5

Known Recoverable Resources of Uranium

	Uranium (Tons)	Total (%)
Australia	1,661,000	31
Kazakhstan	629,000	12
Russia	487,200	9
Canada	468,700	9
Niger	421,000	8
South Africa	279,100	5
Brazil	276,700	5
Namibia	261,000	5
United States	207,400	4
China	166,100	3
Ukraine	119,600	2
Uzbekistan	96,200	2
Mongolia	55,700	1
Jordan	33,800	1
Other	164,000	3
World total	5,327,200	

Source: World Nuclear Association, Supply of uranium. Information paper, www.world-nuclear.org, August 2012.

17.6.2 Mining and Milling

Uranium is being mined using traditional underground and open-pit excavation technologies, and also using in situ leaching or solution mining techniques.

Underground mining is used when the orebody is deep underground, usually greater than 120 m deep. In underground mines, only the orebody material is extracted. Underground mining is hazardous and made more so by high concentrations of radon from the radioactive decay of uranium. Once mined, the extracted ore is sent to a mill where uranium in the ore is concentrated.

Open-pit technology is used when the orebody is near the surface. This leads to the excavation of large amounts of material that does not contain the ore itself. The ore that is recovered is also sent to a mill for further processing.

Solution mining involves the introduction of an aqueous solution into the ore body. The solution, oxygenated ground water, is pumped into the porous orebody and the uranium is dissolved. The uranium-rich solution is then extracted and sent to the mill for further processing.

The milling process for the solid ore material involves crushing the ore and then subjecting it to a highly acidic or alkaline solution to dissolve the uranium. Mills are normally located close to the mining activity and a single mill will often support several mines. The solution containing uranium goes through a precipitation process which yields a material called yellow-cake. The yellow-cake contains about 80% uranium oxide. The yellow-cake is packaged and sent to a conversion and enrichment facility for further processing.

17.6.3 Conversion and Enrichment

Prior to entering the enrichment process, the impure U_3O_8 is converted through a series of chemical processing steps to UF_6. During these processes, the uranium is purified. Conversion facilities are operating commercially in the United States, Canada, France, the United Kingdom, and Russia. UF_6 is solid at room temperature but converts to its gaseous form at moderate temperature levels making the compound suitable for use in the enrichment process. UF_6 is very corrosive and reacts readily with water. It is transported in large cylinders in its solid state.

Conversion of U_3O_8 to UO_2 is also done at conversion facilities. Natural UO_2 is used in reactors such as CANDU which do not require enriched uranium as fuel.

The first enrichment facilities were operated during the 1940s. Electromagnetic isotope separation process was used to separate the U235 used in the first atomic bomb. The process used a magnetic field to separate U235 from U238 as the ions were accelerated and turned, and they moved differently because of the mass difference. Multiple stages were required and the process was very difficult to run efficiently so it was soon abandoned.

Today, only two processes, gaseous diffusion and gas centrifuge are used commercially. The capacity of enrichment plants is measured in separative work units (SWU). SWU is a complex term that is dependent on the amount of uranium processed, and the concentration of U235 in the product and tails. It is a measure of the amount of energy used in the process.

The first commercial enrichment was carried out in large gaseous diffusions plants in the United States. It has also been used in Russia, the United Kingdom, France, China, and Argentina. Today operating plants exist in the United States, France, and China with a total nominal capacity of 30 million SWU.

In the gaseous diffusion process UF_6 is pumped through a series of porous walls or membranes that allow more of the light U235 to pass through. Because the lighter U235 particles travel faster than heavier U238 particles, more of them penetrate the membrane. This process continues through a series of membranes with the concentration of U235 increasing each time. For commercial reactor fuel, the process continues until U235 concentration is 3%–5%. The slower U238 particles are left behind and collect as a product referred to as "tails." The tails have a reduced concentration of U235 and are commonly referred to as depleted uranium. This process uses a very large amount of energy and thus is very expensive to operate.

In the centrifuge enrichment process, the gaseous UF_6 is placed in a high speed centrifuge. The spinning action forces out the heavier U238 particles while the lighter U235 particles remain closer to the center. To obtain the enrichment required for power reactor fuel many stages of separation are required. The arrangement is known as a cascade. Again, the process is continued until U235 concentration is 3%–5%. The centrifuge process uses only about 2% of the energy required by gaseous diffusion.

Table 17.6 shows the location and size of enrichment facilities around the world.

17.6.4 Fuel Fabrication and Use

Following enrichment, UF_6 is shipped to a fuel fabrication facility where it is converted to UO_2 and pressed into cylindrical ceramic pellets. The pellets are sintered, heated to high temperature, and inserted in the fuel-cladding tubes. The tubular material is zircaloy, an alloy of zirconium. The tubes are sealed forming fuel rods which are assembled into fuel assemblies and shipped to a reactor for use. All the dimensions of pellets and fuel rods are very carefully controlled to assure uniformity throughout the fuel assemblies.

The primary hazard in the fabrication facility is the potential for an accidental criticality because they are working with enriched uranium. Therefore, all of the processing quantities and the dimensions of the processing vessels must be controlled. This must be done even with the low-enriched uranium.

A typical 1000 MWe reactor will use about 27 tons of UO_2 each year. Typical burnup in current reactors is 33 GWd/t of uranium fed to the reactor. The energy available from the fission of uranium is 1 MW/g of uranium or 1000 GW/t. Using these numbers the actual amount of uranium burned is only 3%–5%. This means that the unused energy available from the spent fuel, if it could be completely burned, is over 95%. During the operation of the reactor some of the U238 is converted to plutonium and that also contributes to the thermal energy of the reactor.

Advanced fuel use in reactors is estimated to be up to 200 GWd/t. In this case, about 80% of the energy available from the uranium remains in the spent fuel. These facts are the driving force behind the questions regarding reprocessing. In the once-through fuel cycle the spent fuel will be disposed of as waste. In the closed cycle, the spent fuel is reprocessed and the remaining uranium and plutonium are recovered.

17.6.5 Reprocessing

In 1940s reactors were operated solely for the production of plutonium to be used in weapons. The fuels from the production reactors were reprocessed in order to recover plutonium. Chemical processes were developed to separate the fission products and uranium from plutonium. The most common process was the PUREX process. This is the process that is used today by countries that reprocess power reactor fuels.

TABLE 17.6
Location, Size, and Type of Enrichment Facilities around the World

Country	Owner/Controller	Plant Name/Location	Capacity (1,000 SWU)
Gaseous diffusion plants			
Argentina	CNEA ⇨	Pilcaniyeu [ISSUES]	20
China	CNNC ⇨	Lanzhou	900?
France	EURODIF	Tricastin (closed since June 7, 2012)	(10,800)
United States	U.S.-Enrichment Corp. ⇨	Paducah, Kentucky ⇨ (closed since May 31, 2013)	(11,300)
		Portsmouth, Ohio ⇨ (closed since May 11, 2001)	(7,400)
Subtotal			920
Centrifuge plants			
Brazil	INB ⇨	Resende	?
China	CNNC	Hanzhong	500
		Lanzhou	500
France	Eurodif	Georges Besse II, Tricastin (under constr.)	2,800
Germany	UrencoDeutschlandGmbH	Gronau	4,200
India	DAE Nuclear Fuel Complex ⇨	Ratnahalli, Karnataka	4.5
Iran	AEOI ⇨	Natanz [ISSUES]	?
		Qom [ISSUES]	?
Japan	JNC	Ningyo Toge	200
	Japan Nuclear Fuel Limited (JNFL)	Rokkasho-mura	1,050
Korea, DPR		Yongbyon [ISSUES]	8
		Tongchang [ISSUES]	?
Netherlands	Urenco Netherland BV	Almelo	5,550
Pakistan	Pakistan Atomic Energy Commission (PAEC)	Kahuta	5
Russia	Rosatom ⇨	Urals Electrochemical Integrated Enterprise (UEIE), ⇨ Novouralsk (formerly Sverdlovsk-44, near Ekaterinburg) [ISSUES]	7,000
		Siberian Chemical Combine (SKhK), ⇨ Seversk (formerly Tomsk-7) [ISSUES]	4,000
		Electrochemical Plant (ECP), ⇨ Zelenogorsk (formerly Krasnoyarsk-45) [ISSUES]	3,000
		Angarsk Electrolytic Chemical Combine (AEKhK), ⇨ Angarsk [ISSUES]	2,600
United Kingdom	Urenco UK Ltd.	Capenhurst	5,000
United States	Urenco USA	Lea County, NM (under constr.) [ISSUES]	2,200
Subtotal			38,567.5
Total			39,487.5

Source: WISE Uranium Project, World nuclear fuel facilities, www.wise-uranium.org, November 2013.

The purpose of reprocessing is to recover the uranium and plutonium in the spent fuel. As discussed earlier these materials contain a large amount of potential energy if they are reused as reactor fuel. Plutonium separated in the PUREX process can be mixed with uranium to form a mixed oxide (MOX) fuel. Plutonium from the dismantlement of weapons can be used in the same way.

The potential availability of separated plutonium is seen by some as a potential mechanism for the proliferation of nuclear weapons. This was the basis of the U.S. decision to halt reprocessing. In the 1970s, research on ways to modify the chemical process began, so that plutonium and uranium would remain together at the end of the process. In this method, called co-processing, the short-lived fission products would be separated and the remaining uranium, plutonium, and other actinide elements would remain together. This remaining mixture would be highly radioactive but could be remotely processed into new reactor fuel. A blend of fast neutron and thermal reactors could be used to maximize the use of this material.

The current world wide reprocessing capability is shown in Table 17.7. All of these facilities use the PUREX technology. More than 80,000 tons of commercial fuel have been reprocessed in these facilities.

Three processes are considered to be the mature options for reprocessing fuel. They are PUREX, UREX+, and Pyroprocessing. Each of them has certain advantages and disadvantages.

17.6.5.1 PUREX

The PUREX process is the oldest and most common reprocessing option. It uses liquid–liquid extraction to process light-water reactor spent fuel. The spent fuel is dissolved in nitric acid, and then the acid solution is mixed with an organic solvent consisting of tributyl phosphate in kerosene. Uranium and plutonium are extracted in the organic phase and the fission products remain in the aqueous phase. Further processing allows the separation of uranium and plutonium. The advantage of this process is the long-term experience with the process. The disadvantage is that it cannot separate fission products such as technetium, cesium, and strontium nor can it separate actinides such as neptunium, americium, and curium.

TABLE 17.7
World Commercial Reprocessing Capacity

	Location	Tons/Year
LWR fuel	France, La Hague	1700
	United Kingdom, Sellafield (THORP)	900
	Russia, Ozersk (Mayak)	400
	Japan (Rokkasho)	800
	Subtotal	3800
Other nuclear fuels	United Kingdom, Sellafield	1500
	India	330
	Subtotal	1830
Civilian capacity	Total	5630

Source: Uranium Information Centre, Processing of used nuclear fuel, UIC Nuclear Issues Briefing Paper 72, www.uic.com.au, March 2005.

17.6.5.2 UREX+

The UREX+ process is a liquid–liquid extraction process like PUREX. It can be used for light-water reactor fuels and it includes additional extraction steps that allow separation of neptunium/plutonium, technetium, uranium, cesium/strontium, americium, and curium. The advantage of this process is that it meets the requirements for continuous recycle in light-water reactors and it builds on current technology. The disadvantage is that it cannot be used to process short-cooled fuels and it cannot be used for some specialty fuels being developed for advanced reactors.

17.6.5.3 Pyroprocessing

This process was developed and tested at Experimental Breeder Reactor-2 (EBR-2) by Argonne National Laboratory in the United States. It is an electrochemical process rather than a liquid–liquid extraction process. Oxide fuels are first converted to metals in order to be processed. The metallic fuel is then treated to separate uranium and the transuranic elements from the fission products. The advantage of this process is the ability to process short-cooled and specialty fuels designed for advanced reactors. The disadvantage is that it does not meet the requirements for continuous recycle from thermal reactors but it is ideal for fuel for fast neutron reactors.

17.6.6 Spent Fuel Storage

Spent fuel is routinely discharged from operating reactors. As it is discharged it is moved to the spent fuel storage pool that is an integral part of the reactor facility. Reactors are built with storage pools that will hold fuel from many years of operation. The pools are actively cooled by circulating cooling water. The fuel stored at many older reactors is reaching the capacity of the on-site storage pools. At this point, the fuel is being transferred to dry storage. Dry storage takes place in large metal or concrete storage facilities. These dry facilities are passively cooled by the air circulating around them.

17.6.7 Spent Fuel Transportation

Spent fuel is transported in large engineered containers designated as Type B containers (casks). Casks provide shielding for the highly radioactive fuel so that they can be safely handled. They are made of cast iron or steel. Many of them use lead as the shielding material. They are also designed to protect the environment by maintaining their integrity in case of accident. They are designed to withstand severe accidents including fires, impacts, immersion, pressure, heat, and cold, and are tested as part of the design certification process.

Casks have been used to transport radioactive materials for over 50 years. The IAEA has published advisory regulations for safe transportation of radioactive materials since 1961. Casks are built to standards, designed to meet the IAEA advisory regulations, specified by licensing authorities such as NRC in the United States.

Spent fuel is shipped from reactor sites by road, rail, or water. Larger casks can weigh up to 110 tons and hold about 6 tons of spent fuel. Since 1971, about 7000 shipments of spent fuel (over 35,000 tons) have been transported over 30 million km with no property damage or personal injury, no breach of containment, and a very low-dose rate to the personnel involved.

17.7 Nuclear Waste

Radioactive wastes are produced throughout the reactor fuel cycle. The costs of managing these wastes are included in the costs of the nuclear fuel cycle and thus are part of the electricity cost.

Since these materials are radioactive, they decay with time. Each radioactive isotope has a half-life, which is the time it takes for half of the material to decay away. Eventually, these materials decay to a stable nonradioactive form.

The process of managing radioactive waste involves the protection of people from the effects of radiation. The longer-lived materials tend to emit alpha and beta particles. It is relatively easy to shield people from this radiation but if these materials are ingested the alpha and beta radiation can be harmful. The shorter-lived materials usually emit gamma rays. These materials require greater amounts of shielding.

17.7.1 Types of Radioactive Wastes

The strict definitions of types of radioactive waste may vary from country to country. In the following discussion, more generally accepted terminology will be used.

17.7.1.1 Mine Tailings

Mining and milling of uranium produces a sandy type of waste which contains naturally occurring radioactive elements that are present in uranium ore. The decay of these materials produces radon gas which must be contained. This is often done by covering the tailings piles with clay to contain the radon gas. Technically, tailings are not classified as radioactive waste.

17.7.1.2 Low-Level Wastes

Low-level wastes (LLW) is generated from medical and industrial uses of radioactive materials as well as from the nuclear fuel cycle. In general, these wastes include materials such as paper, clothing, rags, tools filters, soils, etc., which contain small amounts of radioactivity. This radioactivity tends to be short-lived. These materials generally do not have to be shielded during transport and they are suitable for shallow land burial. The volume of these materials may be reduced by compacting or incineration prior to disposal. They make up about 90% of the volume of radioactive waste but contain only about 1% of the radioactivity of all the radioactive waste.

17.7.1.3 Intermediate-Level Wastes

Intermediate-level wastes (ILW) are generated during the operation of nuclear reactors, in the reprocessing of spent fuel, and from the decommissioning of nuclear facilities. These materials contain higher amounts of radioactivity and generally require some shielding during storage and transportation. ILW is generally made up of resins, chemical sludge, fuel cladding, and contaminated materials from decommissioning nuclear facilities. Some of these materials are processed before disposal by solidifying them in concrete or bitumen. They make up about 7% of the volume and have about 4% of the radioactivity of all the radioactive waste.

17.7.1.4 High-Level Wastes

High level waste is generated during the operation of a nuclear reactor. This waste consists of fission products and transuranic elements generated during the fission process. This material is highly radioactive and is also thermally hot so that it must be both shielded and cooled. It accounts for 95% of the radioactivity produced by nuclear power reactors.

17.7.2 Managing HLW from Spent Fuel

The form of HLW from spent fuel is either the spent fuel itself or the waste products from reprocessing. The level of radioactivity from spent fuel falls to about one thousandth of the level it was when removed from the reactor in 40–50 years. This means the heat generated is also greatly reduced.

Currently, there are about 270,000 tons of spent fuel in storage at reactor sites around the world. An additional 12,000 tons are generated each year and about 3,000 tons of this are sent for reprocessing.

When spent fuel reprocessing is used, uranium and plutonium are first removed during reprocessing, and then the much smaller volume of remaining HLW is solidified using a vitrification process. In this process, the fission products are mixed in a glass material, vitrified in stainless steel canisters, and stored in shielded facilities for later disposal.

High-level waste will eventually be disposed-off in deep geologic facilities. Several countries have selected sites for these facilities and they are expected to be commissioned for use after 2015.

17.7.3 Managing Other Radioactive Wastes

Generally, ILW and LLW are disposed by burial. ILW generated from fuel reprocessing will be disposed in deep geological facilities. Some low-level liquid wastes from reprocessing plants are discharged to the sea. These liquids include some distinctive materials such as technetium-99 that can be discerned hundreds of kilometers away. Such discharges are tightly controlled and regulated so that the maximum dose any individual receives is a small fraction of natural background radiation.

Nuclear power stations and reprocessing facilities release small quantities of radioactive gases to the atmosphere. Gases such as krypton-85 and xenon-133 are chemically inert and gases such as iodine-131 have short half-lives. The net effect of these gases is too small to warrant further consideration.

Table 17.8 provides a summary of waste management adopted by countries throughout the world.

17.8 Nuclear Power Economics

Any discussion of the economics of nuclear power involves a comparison with other competitive electric generation technologies. The competing technologies are usually coal and natural gas.

Nuclear power costs include capital costs, fuel cycle costs, waste management costs and the cost of decommissioning after operation. The costs vary widely depending on the

TABLE 17.8

Waste Management for Used Fuel and HLW from Nuclear Power Reactors

Country	Policy	Facilities and Progress Toward Final Repositories
Belgium	Reprocessing	Central waste storage at Dessel
		Underground laboratory established in 1984 at Mol
		Construction of repository to begin about 2035
Canada	Direct disposal	Nuclear Waste Management Organisation set up in 2002
		Deep geological repository confirmed as policy, retrievable
		Repository site search from 2009, planned for use in 2025
China	Reprocessing	Central used fuel storage at LanZhou
		Repository site selection to be completed by 2020
		Underground research laboratory from 2020, disposal from 2050
Finland	Direct disposal	Program started in 1983, two used fuel storages in operation
		Posiva Oy set up in 1995 to implement deep geological disposal
		Underground research laboratory Onkalo under construction
		Repository planned from this, near Olkiluoto, to open in 2020
France	Reprocessing	Underground rock laboratories in clay and granite
		Parliamentary confirmation in 2006 of deep geological disposal, containers to be retrievable and policy "reversible"
		Bure clay deposit is likely repository site to be licensed in 2015, operating from 2025
Germany	Reprocessing but moving to direct disposal	Repository planning started in 1973 Used fuel storage at Ahaus and Gorleben salt dome Geological repository may be operational at Gorleben after 2025
India	Reprocessing	Research on deep geological disposal for HLW
Japan	Reprocessing	Underground laboratory at Mizunami in granite since 1996
		Used fuel and HLW storage facility at Rokkasho since 1995
		Used fuel storage under construction at Mutsu, start-up in 2013
		NUMO set up in 2000, site selection for deep geological repository underway to 2025, operation from 2035, retrievable
Russia	Reprocessing	Underground laboratory in granite or gneiss in Krasnoyarsk region from 2015, may evolve into repository
		Sites for final repository under investigation on Kola peninsula
		Pool storage for used VVER-1000 fuel at Zheleznogorsk since 1985
		Dry storage for used RBMK and other fuel at Zheleznogorsk from 2012
		Various interim storage facilities in operation
South Korea	Direct disposal, maybe change	Waste program confirmed in 1998, KRWM set up in 2009 Central interim storage planned from 2016
Spain	Direct disposal	ENRESA established in 1984, its plan was accepted in 1999
		Central interim storage at Villar de Canas from 2016 (volunteered location)
		Research on deep geological disposal, decision after 2010
Sweden	Direct disposal	Central used fuel storage facility—CLAB—in operation since 1985
		Underground research laboratory at Aspo for HLW repository
		Osthammar site selected for repository (volunteered location)
Switzerland	Reprocessing	Central interim storage for HLW and used fuel at ZZL Wurenlingen since 2001
		Smaller used fuel storage at Beznau
		Underground research laboratory for high-level waste repository at Grimsel since 1983
		Deep repository by 2020, containers to be retrievable

(Continued)

TABLE 17.8 (*Continued*)

Waste Management for Used Fuel and HLW from Nuclear Power Reactors

Country	Policy	Facilities and Progress Toward Final Repositories
United Kingdom	Reprocessing	Low-level waste repository in operation since 1959 HLW from reprocessing is vitrified and stored at Sellafield Repository location to be on basis of community agreement New NDA subsidiary to progress geological disposal
United States	Direct disposal	DoE responsible for used fuel from 1998, accumulated $32 billion waste fund Considerable research and development on repository in welded tuffs at Yucca Mountain, Nevada The 2002 Congress decision that geological repository be at Yucca Mountain was countered politically in 2009 Central interim storage for used fuel now likely

Source: World Nuclear Association, Waste management in the nuclear fuel cycle. Information paper, www.world-nuclear.org, December 2012.

location of the generating plant. In countries such as China, Australia, and the United States, coal remains economically attractive because of large accessible coal resources. This advantage could be changed if a charge is made on carbon emissions. In other areas nuclear energy is competitive with fossil fuels even though nuclear costs include the cost of all waste disposal and decommissioning.

17.8.1 Comparison of Generation Technologies

As previously stated, nuclear power costs include spent fuel management, plant decommissioning, and final waste disposal. These costs are not generally included in the costs of other power generation technologies.

Decommissioning costs are estimated to be 9%–15% of the initial cost of a nuclear plant. Since these costs are discounted over the life of the plant, they contribute only a few percent to the investment cost of the plant and have an even lower impact on the electricity generation cost. This impact in the United States is about 0.1–0.2 cent/kWh or about 5% of the cost of electricity produced.

Spent fuel interim storage and ultimate disposal in a waste repository contribute another 10% to the cost of electricity produced. This cost is reduced if the spent fuel is disposed-off directly. This does not account for the energy that could be extracted from the fuel if it was reprocessed.

Costs for nuclear-based electricity generation have been dropping over the last decade. This reduction in the cost of nuclear generated electricity is a result of reductions in nuclear plant fuel, operating, and maintenance costs. However, the capital construction costs for nuclear plants are significantly higher than coal and gas-fired plants. Because the capital cost of nuclear plants contribute more to the cost of electricity than for coal- or gas-fired generation the impact of changes in fuel, operation, and maintenance costs on the cost of electricity generation is less than for coal- or gas-fired generation.

One of the primary contributors to the capital cost of nuclear plants has been the cost of money used to finance nuclear plant construction. The financing costs increase when the time required to license and construct a plant increases. Two factors are leading to the reduction in this portion of the cost. First, especially in the United States, the licensing process is changing so that a plant receives both the construction permit and the operating license prior to the start of construction. Under this process, there is no large investment in plant hardware prior to completion of a significant portion of the licensing process, leading to a reduction in time required for the plant to begin producing revenue. Second, the new generation of nuclear plants will be highly standardized and modularized. This will allow a significant reduction in the time required to construct a new plant. It is estimated that the time from the start of construction to the start of operation will be reduced from nearly 10 years to 4–5 years. This will have a significant impact on capital costs.

The reduced capital costs associated with the licensing and construction of new nuclear power plants, and the fact that nuclear power is inherently less susceptible to large fluctuations in fuel costs, have made nuclear power an attractive energy option for many countries seeking to diversify their energy mix in the face of rising fossil fuel costs.

17.9 Conclusions

The development of nuclear power began after World War II and continues today. The first power-generating plants were constructed in the late 1950s. During the 1960s and 1970s, there was a large commitment to nuclear power until the accidents occurred at Three Mile Island in 1979 and then at Chernobyl in 1986. The new safety requirements and delays caused by these accidents drove up the costs and at the same time caused a loss of public acceptance. In the United States, many plant orders were canceled and in other countries the entire nuclear programs were canceled.

The ability of nuclear reactors to produce electricity economically and safely without the generation of greenhouse gases has revitalized the interest in nuclear power as an alternative energy source. Many lessons have been learned from the operation of current power plants that have allowed the safety of newly designed plants to be improved. This coupled with the desire of many nations to develop secure energy sources and a diversity of energy options has resulted in the continuing development of a whole new generation of nuclear plants to meet future energy needs.

Nuclear power is also not as susceptible to fluctuation in fuel costs as petroleum and natural gas. As discussed, the supply of uranium is very large, and if it is supplemented with thorium, the fuel supply will seemingly be unlimited. This drives many other aspects of the fuel cycle, such as the choice between closed and open fuel cycles discussed earlier. For example, because of the large uranium resource and the fears of nuclear proliferation, the once-through (open) fuel cycle is favored by many. This will require large, deep, geologic waste repositories for the disposal of large quantities of spent fuel. However, when reprocessing is included in the closed fuel cycle, the amount of needed repository space is greatly reduced but the expense of operation is increased. Finally, it may be possible to essentially eliminate the need for repositories by utilizing advanced fuel cycles that utilize almost all of the energy available in uranium and the other transuranic products of reactor operation.

The need for energy and the use of electricity as the primary energy source for the end-user will drive the increase in electricity generation around the world. The drive to reduce the production of greenhouse gases will contribute to a wider use of nuclear power for electricity generation. The recognition that nuclear power can safely provide large base-load generating capacity at a reasonable cost using known technologies will also be a major factor in its future development.

References

1. World Nuclear Association. Plans for new reactors worldwide. Information paper, www.world-nuclear.org, March 2013.
2. World Nuclear Association. Advanced nuclear power reactors. Information paper, www.world-nuclear.org, September 2013.
3. WISE Uranium Project. World nuclear fuel facilities. www.wise-uranium.org, November 2013.
4. World Nuclear Association. Waste management in the nuclear fuel cycle. Information paper, www.world-nuclear.org, December 2012.
5. UIC Nuclear Issues Briefing Paper #72. Processing of used nuclear fuel. www.uic.com.au, March 2005.
6. World Nuclear Association. Supply of uranium. Information paper, www.world-nuclear.org, August 2012.
7. U.S. DOE Nuclear Energy Research Advisory Committee and the Generation IV International Forum. A technology roadmap for the Generation IV nuclear energy systems, GIF-002–00, December 2002.
8. World Nuclear Association. Small nuclear power reactors. Information paper, www.world-nuclear.org, February 2014.

18

Outlook for U.S. Energy Consumption and Prices, 2011–2040

Andy S. Kydes, Mark J. Eshbaugh, and John J. Conti

CONTENTS

18.1 Executive Summary

The *Annual Energy Outlook 2013* (*AEO2013*) projection [2] to 2040 summarizes the following:

- The projected growth in U.S. energy production outpaces the growth in energy consumption, particularly for natural gas, and leads to a decline in the net import share of total U.S. energy consumption from 19% in 2011 to 9% in 2040. For reference, in 2005, the import share was 30%. However, the United States does not achieve energy independence by 2040.
- The continued improvement of advanced crude oil and natural gas production technologies—particularly for shale and other tight formations—continues to lift projected U.S. supply, with domestic crude oil production peaking at about 7.5 million barrels per day (bbl/day) in 2019. Domestic natural gas production outpaces domestic consumption and becomes the dominant source of domestic energy production by 2040, reaching 33 trillion cubic feet (Tcf) and accounting for 34% of total U.S. primary energy production.
- The decline in energy imports reflects increased domestic petroleum and natural gas production, an increased use of biofuels (much of which are produced domestically), and decreased demand resulting from rising energy prices and the adoption of new efficiency standards for vehicles;
- Relatively low natural gas prices, facilitated by growing shale gas production, spur increased natural gas use in the industrial and electric power sectors. The share of natural gas in electricity generation rises from 25% in 2011 to 30% in 2040. The share of electricity generated by coal declines from 42% in 2011 to 35% in 2040. Natural gas also makes gains in other areas, such as exports, as a fuel for heavy-duty freight transportation, and as a feedstock for producing diesel and other fuels.
- Electricity generation from solar, and to a lesser degree wind energy sources, grows as recent cost declines coupled with state-level renewable portfolio standards (RPS) and federal tax incentives make renewables a preferred option to other feedstocks such as coal and natural gas.
- EIA projects domestic coal production to grow modestly between 2011 and 2040, but its market share of total energy production falls from about 28% in 2011 to 24% in 2040. Natural gas displaces most of the expansion opportunities for coal in power generation, thereby freeing more coal for potential export. U.S. coal exports grow from 107 million short tons in 2011 to 159 million short tons in 2040.
- The *AEO2013* Reference case projects improvements in vehicle efficiency, with total energy demand in the transportation sector falling to its 2011-level by 2040, after increasing early in the projection period.
- Efficiency gains in the residential and commercial sectors offset household and floor-space growth in each sector that would otherwise increase overall energy consumption.

- U.S. energy-related CO_2 emissions remain more than 5% below their 2005 level through 2040, reflecting increased efficiency and the shift to a less carbon-intensive fuel mix in both transportation and power generation [3,5]. Market factors, such as technological improvements and competition with lower-cost renewables, and policy factors, such as the Renewable Fuel Standard [6] and the new CAFÉ [4] standard for light-duty vehicles (LDV), which raised the new car efficiency to 47.5 miles per gallon by 2025, are the principal drivers for the decline in emissions. See Section 18.6.1 for more details on CAFÉ.

18.2 Introduction: Outlook for U.S. Energy, 2011–2040*

18.2.1 Background

This chapter provides a summary of the U.S. Energy Information Administration's (EIA's) *Annual Energy Outlook 2013* (*AEO2013*) Reference case (the central scenario with respect to crude oil prices, macroeconomic growth, and technological progress). The *AEO2013* provides projections of a number of potential energy futures and explores the impact of different assumptions through the comparison of a series of side cases analyses to the Reference case. Key side cases include different crude oil import prices, differences in domestic resource availabilities (oil and natural gas), different economic growth assumptions, assumed policy changes, and technological changes.

EIA's *AEO2013* Reference case is different from many private projections in that the EIA does not attempt to include a "forecast" of policy changes. EIA understands that while changes will occur, the exact nature of those changes is far from certain. EIA uses the term "projection," *not* "prediction" or "forecast," because all the midterm outlooks—including the *AEO2013* Reference case—are highly conditioned on the specific assumptions made for those particular cases.

Central to the Reference case is the assumption that current technological and demographic trends, and current laws and regulations, will continue as usual into the future. EIA does not propose, advocate, or speculate on future legislative or regulatory changes. Thus, the *AEO2013* projections provide a policy-neutral basis for analyzing policy initiatives and alternative assumptions regarding economic growth and technological progress, to name a few. For more detail and a more useful treatment of uncertainties related to key assumptions of the Reference case projection, the reader can visit the EIA website at http://www.eia.gov/forecasts/aeo/er/index.cfm and observe some of the other cases included in the *AEO2013*.

* The outlook of U.S. energy markets through 2040, as presented in Sections 18.1 through 18.6, uses the EIA's National Energy Model (NEM)[1] (http://www.eia.doe.gov/oiaf/aeo/overview/index.html). This chapter describes the reference case projection for *AEO2013*, which is available on the EIA website (http://www.eia.doe.gov/oiaf/aeo/index.html) along with another 30 side cases. See http://www.eia.doe.gov/oiaf/aeo/assumption/pdf/0554(2013).pdf for a detailed description of the assumptions of the *AEO2013* Reference case. The outlook and commentary described herein represent solely the views of the authors and do not necessarily reflect the views of EIA, the U.S. Department of Energy, the Administration, or any agency of the U.S. government.

18.2.2 The AEO2013 Reference Case

The prevailing assumption of the *AEO2013* Reference case is that current laws and regulations continue as promulgated. One notable exception is that the *AEO2013* Reference case adds a premium to the cost of financing carbon dioxide (CO_2)-intensive technologies to reflect current perceived risks regarding possible future policies that may be used to mitigate greenhouse gas (GHG) emissions.

Change is the norm in the outlooks for U.S. energy markets, as such the Reference case projection in *AEO2013* is quite different from the Reference case projections from past *AEOs*. This is largely due to changes in the assumptions underlying each of the projections (e.g., there are differences in the outlook for crude oil prices, economic growth, technological improvements, etc.) and changes to U.S. supply expectations as a result of improved technology and processes (such as fracking). Additionally, crude oil supply and demand have changed substantially since 2005, thereby shifting the expected crude oil price path (Figure 18.1).

Although EIA does not expect the United States to achieve energy independence by 2040 in the Reference case, technological and process advances in oil and gas supply have altered the outlook for U.S. energy imports and exports. The growth in domestically produced crude oil and natural gas from tight formations proves to be an important driver of change across all sectors. For example, natural gas production increases over the projection period, driven primarily by the growth in shale gas development. That development keeps natural gas prices relatively low throughout most of the projection period, compared to previous projections, increasing natural gas use in the industrial sector in the short term and helping natural gas stay competitive with coal in the electricity sector.

New sources of oil and natural gas, coupled with lower economic growth, new end-use efficiency regulations, new regulations on fuel use in the transportation sector, and

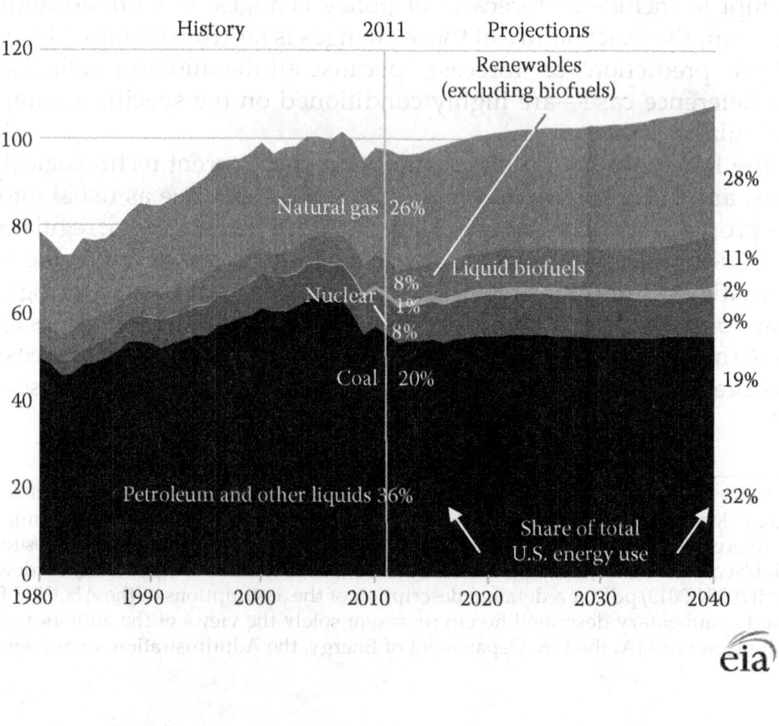

FIGURE 18.1
Primary energy use by fuel, 1980–2040 (quadrillion Btu).

higher CAFE standards through 2025 result in lower energy demand from LDVs, particularly for gasoline, in the longer term. All of these changes result in much lower CO_2 emission expectations than in previous *AEOs*. The *AEO2013* Reference case projects that CO_2 emissions never exceed the emissions-levels of 2005 throughout the projection's horizon (2040).

18.3 AEO2013 Key Assumptions and Uncertainties

All projections of energy-economic systems, particularly those that incorporate consumer and producer behavior, are inherently uncertain. However, consumer and producer behavior are not the only sources of uncertainty. Some of the other critical uncertainties on which projections depend include the following:

- The rate of technological progress for end-use and supply technologies
- Changes in energy market regulations and efficiency standards
- The quantity and location of energy resources in the ground (e.g., coal, crude oil, natural gas, nuclear material)
- The costs to explore, locate, and ultimately produce energy resources
- The willingness of financial markets to make energy investments

Consequently, the projections described in this chapter are not a statement of what will happen, but what might happen given the assumptions and methodologies used by EIA.

18.3.1 Economy

In the *AEO2013* Reference case, annual real gross domestic product (GDP) growth averages 2.5% per year between 2011 and 2040, down by 0.2 percentage points when compared with the 30-year period prior to the projection period, largely reflecting both lower labor productivity and lower levels of consumption. U.S. real consumption growth averages 2.2% per year in EIA's projections. In EIA's projections, U.S. economic output depends on the size of the labor force, the amount of capital in the economy, and the productivity of both labor and capital. Over the past two decades, the labor force has grown faster than the population. Moving forward, not only does EIA project that both the labor force and population will grow more slowly, but the labor force will grow more slowly than population as larger segments of the population reach retirement age. The slower growth in the labor force is offset somewhat by projected faster growth in spending on research and development and business fixed investment; EIA projects the share of GDP devoted to business fixed investment to range from 10% to 17% of GDP through 2040 compared to an average rate of approximately 10% over the past 30 years.

In the *AEO2013* Reference case, EIA expects the economic recovery to continue, and increased energy production from domestic resources improves U.S. competitiveness. Growth in U.S. manufacturing output accelerates through 2020, with energy-intensive industries growing at a rate of 1.8% per year from 2011 to 2020. From 2020 to 2040, energy-intensive industries will grow by just 0.6% per year, as manufacturing output declines in the face of stronger foreign competition and higher energy prices (which impact energy-intensive industries more significantly).

Total U.S. energy expenditures decline as a percentage of GDP in the *AEO2013* Reference case. After peaking in 2011, the projected ratio of energy expenditures to GDP falls through 2040, averaging 6.8% from 2011 to 2040, which is below the historical average of 8.8% from 1970 to 2010. In the Reference case, oil's value as a share of U.S. GDP falls from 5% in 2011 to 4% in 2040.

Issues such as financial market reform, fiscal policies, and the solvency of Eurozone member countries will affect both short-run and long-run growth, adding uncertainty to these projections.

18.3.2 Oil Prices

The definition of the Reference case benchmark crude oil price changed in the *AEO2013* from the West Texas Intermediate (WTI) spot oil price to the Brent crude oil price. This change was in response to the significant divergence of WTI prices from similar internationally traded crudes, and the observed impact that Brent crude oil prices have had on the price that many U.S. consumers pay for motor gasoline.

The future path of world oil prices is a major uncertainty facing world oil markets. World crude oil prices reached an average high of about $105 per barrel (in 2011 dollars) in 2008 before declining to an average of $81.31 per barrel in 2010. Brent crude oil prices increase in the Reference case by an average of 1.3% per year from 2011 to 2040, from $111.26 to $162.68 per barrel.

EIA's crude oil price paths represent equilibrium market conditions between supply and demand. Because there is significant uncertainty in future crude oil prices, EIA has developed side cases to help examine how different oil price paths could impact U.S. energy markets. The following graph (Figure 18.2) shows a range of uncertainty surrounding

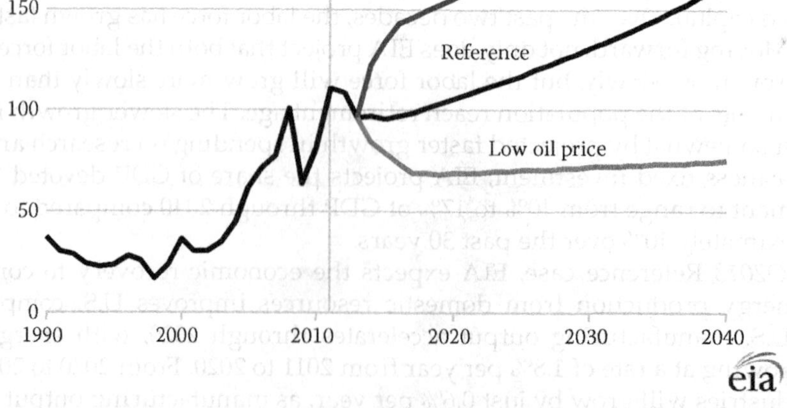

FIGURE 18.2
Brent crude oil spot prices in three cases, 1990–2040 (2011 dollars per barrel).

crude oil prices (Brent light), reflected in the price differential between the *AEO2013*'s Reference case, High Price side case, and Low Price side case. These are annual average prices and do not show the volatility that occurs in oil prices in the short term. Over the last two decades, volatility within a single year averaged about 30%. There are no probabilities associated with any of the crude oil price cases.

18.4 Petroleum and Other Liquids

18.4.1 Global Petroleum Market

The *AEO2013* Reference case projects total global liquids consumption to grow from 88 million bbl/day in 2011 to 113 million bbl/day in 2040, a 28% increase by 2040. Crude oil represents 85% of the total liquids supply in 2011, but EIA projects a decline to 81% in 2040 as other fuels increase their market share.

Of the major producers, the share of global supply by the Organization of the Petroleum Exporting Countries (OPEC) increases slightly, from 40% in 2011 to 43% in 2040. The Organisation for Economic Cooperation and Development (OECD) share of global supply remains fairly steady at about 25% through 2040, as the United States and Canada increase their liquids production and European North Sea and Mexican production decline. Meanwhile, other non-OECD producers such as Brazil, China, the former Soviet Union countries, and small African producers see their share of total supply fall by an amount roughly equal to the OPEC increase between 2011 and 2040.

18.4.2 Global Liquid Prices

Petroleum prices are set in a worldwide marketplace and reflect the interaction of supply and demand forces. Strong growth in global oil demand, particularly in China and other developing countries, is one of the key factors behind the sharp price increases seen since 2002. Another factor contributing to the upward trend in world oil prices is a tight supply situation that—until recently—has had only limited production response by countries outside of OPEC. Finally, concerns about economic and political situations in the Middle East, Venezuela, Nigeria, China, and the former Soviet Union could manifest themselves through higher prices.

Significant uncertainties reside in the long-term trend for both the supplies and prices of crude oil. The sources of uncertainty include the following:

- Size and location of potential resources
- Access to potential resources which vary on geopolitical factors
- Availability and cost of technologies to explore, discover, and develop the resources
- Rate of technological change for resource development technologies
- Willingness to invest in specific resource development projects and the rate of investment
- Rate of growth and location of growth of the demand for oil products
- Cartel behavior
- Other aboveground factors (e.g., geopolitical factors)

In the *AEO2013* Reference case, the projection of crude oil import prices of Brent crude in 2011 real dollars continues to rise, as global demand grows and both belowground (geology, resource availability, and technology) and aboveground (economic, geopolitical, and access) factors constrain supply. In addition, tighter U.S. fuel economy standards, increased use of biofuels, and rising energy prices act together to temper the domestic demand growth, and higher prices and mandates spur increased domestic production of crude oil and biofuels.

As discussed in Section 18.2.2, global crude oil prices were updated in the *AEO2013* Reference case to represent the projected cost of Brent light crude oil. In the Reference case, EIA projects Brent crude oil prices to increase by 1.3% per year from 2011 to 2040, to $162.68 per barrel (see Table 18.1) as a result of belowground and aboveground factors. Liquids prices to all domestic end users are shown in Table 18.2.

18.4.3 Domestic Liquids

One of the central story lines in the *AEO2013* Reference case is the growth in domestic liquids production, particularly from tight formations. In the *AEO2013* Reference case, total production of petroleum and other liquids in the United States grows from 10.4 million bbl/day in 2011 to 13.1 million bbl/day at its peak in 2019. Most of the growth is the result of higher production of crude oil and natural gas liquids (NGL) from tight oil formations, such as the Bakken and Eagle Ford plays.

EIA projects domestic crude oil production, particularly from tight oil plays, to rise sharply over the next decade. Total domestic crude oil production reaches a peak of 7.5 million bbl/day in 2019 before eventually falling to 6.1 million bbl/day in 2040 (still higher than the levels of 2011). The growth in domestic oil production is due to the application of new processes and technologies (such as fracking) and the relatively attractive world Brent crude oil price (Figure 18.3).

Total U.S. tight oil production grew to over 1.7 million bbl/day in July 2012, an increase of 690 thousand bbl/day over July 2011. EIA projects tight oil production to increase to 2.8 million bbl/day in 2020. Relatively high initial flow rates from tight formations allow for significant growth in total tight oil production in the short term as drilling activity increases. However, growth in production will slow as producers move to less productive and more expensive areas in known plays unless producers can discover and exploit new productive tight formations—another uncertainty. Following the initial surge, EIA expects tight oil production to decline to 2.2 million bbl/day in 2030 and ends up at 2.0 million bbl/day in 2040. Nevertheless, tight oil represents a large proportion of the growth in U.S. oil production in the Reference case.

Total U.S. liquid fuels consumption sits at approximately 19.0 million bbl/day in both 2011 and 2040, but peaks in 2019 at 19.8 million bbl/day (Table 18.3). U.S. consumption of liquid fuels increases by 6.4% from 2011 to 2020 and then gradually decreases to the 2011 level (~19.0 million bbl/day) by 2040 primarily because of declining energy use in LDVs. The petroleum content of liquid fuels declines at an average of 0.6% per year over the period, with the difference largely made up by natural gas plant liquids and biofuels (with the latter increasing from 0.9 million bbl/day in 2011 to 1.3 million bbl/day in 2040). As a result, U.S. reliance on net petroleum imports—which was at approximately 60% in 2005 and roughly 45% in 2011—falls to 37% in 2040. Nonetheless, imports continue to meet a large part of total U.S. liquids demand throughout the projection period.

Consumption of domestically produced biofuels increases significantly between 2011 and 2040, rising from 1.3 quadrillion British thermal unit (Btu) in 2011 to 2.1

TABLE 18.1

Total Energy Supply, Disposition, and Price Summary

Supply, Disposition, and Prices	2010	2011	2015	2020	2025	2030	2035	2040	Annual Growth 2011–2040 (%)
Production									
Crude oil and lease condensate	11.59	12.16	15.58	15.95	14.50	13.47	13.40	13.12	0.3
Natural gas plant liquids	2.78	2.88	3.74	4.14	4.20	3.85	3.87	3.89	1.0
Dry natural gas	21.82	23.51	24.56	27.19	29.22	30.44	32.04	33.87	1.3
Coal[a]	22.04	22.21	21.12	21.74	22.54	23.25	23.60	23.54	0.2
Nuclear/uranium[b]	8.43	8.26	8.57	9.25	9.54	9.49	9.14	9.44	0.5
Hydropower	2.54	3.17	2.76	2.83	2.86	2.87	2.90	2.92	−0.3
Biomass[c]	4.05	4.05	4.25	5.00	5.27	5.42	5.83	6.96	1.9
Other renewable energy[d]	1.31	1.58	2.12	2.22	2.32	2.50	2.91	3.84	3.1
Other[e]	0.76	1.20	0.76	0.83	0.85	0.88	0.90	0.89	−1.0
Total	**75.31**	**79.02**	**83.46**	**89.16**	**91.29**	**92.18**	**94.59**	**98.46**	**0.8**
Net Imports									
Crude oil	20.14	19.46	16.08	15.02	15.57	16.33	16.43	16.89	−0.5
Liquid fuels and other petroleum[f]	0.40	−0.84	−0.41	0.18	0.33	0.08	−0.42	−0.89	n/a
Natural gas[g]	2.68	2.02	1.49	−0.09	−1.56	−2.08	−2.54	−3.55	n/a
Other[h]	0.52	0.43	0.21	0.11	0.17	0.13	0.48	0.84	2.4
Coal	−2.10	−2.75	−3.00	−3.13	−3.18	−3.51	−3.80	−3.79	−1.1
Total	**21.64**	**18.32**	**14.37**	**12.09**	**11.33**	**10.95**	**10.15**	**9.50**	**−2.2**
Consumption									
Liquid fuels and other petroleum[l]	37.76	37.02	37.04	37.54	36.87	36.08	35.82	36.07	−0.1
Natural gas	24.32	24.91	25.86	26.77	27.28	27.95	29.06	29.83	0.6
Coal[m]	20.81	19.66	18.18	18.59	19.35	19.70	20.09	20.35	0.1
Nuclear/uranium[b]	8.43	8.26	8.57	9.25	9.54	9.49	9.14	9.44	0.5
Hydropower	2.54	3.17	2.76	2.83	2.86	2.87	2.90	2.92	−0.3
Biomass[n]	2.87	2.74	2.88	3.53	3.82	3.94	4.23	4.91	2.0

(Continued)

TABLE 18.1 (*Continued*)

Total Energy Supply, Disposition, and Price Summary

Supply, Disposition, and Prices	2010	2011	2015	2020	2025	2030	2035	2040	Annual Growth 2011–2040 (%)
Other renewable energy[d]	1.31	1.58	2.12	2.22	2.32	2.50	2.91	3.84	3.1
Other[o]	0.31	0.35	0.32	0.31	0.30	0.28	0.26	0.29	−0.6
Total	**98.35**	**97.70**	**97.72**	**101.04**	**102.34**	**102.81**	**104.41**	**107.64**	**0.3**
Prices (2011 dollars per unit)									
Brent spot price (dollars per barrel)[p]	81.31	111.26	95.91	105.57	117.36	130.47	145.41	162.68	1.3
West Texas Intermediate spot price (dollars per barrel)	81.08	94.86	88.21	103.57	115.36	128.47	143.41	160.68	1.8
Natural gas at Henry Hub (dollars per MMBtu)	4.46	3.98	3.12	4.13	4.87	5.40	6.32	7.83	2.4
Coal, Minemouth (dollars per ton)[q]	36.37	41.16	45.68	49.26	52.02	55.64	58.57	61.28	1.4
Coal, Minemouth (dollars per MMBtu)[q]	1.80	2.04	2.27	2.45	2.60	2.79	2.94	3.08	1.4
Coal, delivered (dollars per MMBtu)[r]	2.42	2.57	2.64	2.77	2.94	3.10	3.25	3.42	1.0
Electricity (cents per kWh)	10.0	9.9	9.2	9.4	9.5	9.7	10.1	10.8	0.3
Energy use and related statistics									
Delivered energy use	71.49	71.01	71.79	74.01	74.40	74.38	75.41	77.63	0.3
Total energy use	98.35	97.70	97.72	101.04	102.34	102.81	104.41	107.64	0.3
Population (millions)	310.06	312.38	324.59	340.45	356.46	372.41	388.35	404.39	0.9
U.S. GDP (billion 2005 dollars)	13,062	13,299	14,679	16,859	18,984	21,355	24,094	27,276	2.5
Carbon dioxide emissions (million metric tons carbon dioxide equivalent)	5633.6	5470.7	5380.9	5454.6	5501.4	5522.8	5606.7	5691.1	0.1
Energy intensity (thousand Btu per 2005 dollars of GDP)	7.53	7.35	6.66	5.99	5.39	4.81	4.33	3.95	−2.1
Carbon intensity (mtCo_2e per million 2005 dollars of GDP)	431.27	411.36	366.56	323.53	289.78	258.62	232.69	208.64	−2.3
Nonenergy GHG emissions	1234.41	1248.69	1321.28	1435.29	1430.69	1434.30	1419.90	1408.60	0.4
Greenhouse gas intensity	525.76	505.25	456.57	408.67	365.14	325.78	291.62	260.28	−2.3
% Decline from 2002	12.29	15.71	23.84	31.83	39.09	45.65	51.35	56.58	4.5

(*Continued*)

TABLE 18.1 (*Continued*)

Total Energy Supply, Disposition, and Price Summary

Sources: 2010 natural gas supply values: U.S. Energy Information Administration (EIA), Natural Gas Annual 2010, DOE/EIA-0131(2010), Washington, DC, December 2011. 2011 natural gas supply values and natural gas wellhead price: EIA, Natural Gas Monthly, DOE/EIA-0130(2012/07), Washington, DC, July 2012; 2010 natural gas spot price at Henry Hub based on daily data from Natural Gas Intelligence; 2010 and 2011 coal minemouth and delivered coal prices: EIA, Annual Coal Report 2011, DOE/EIA-0584(2011), Washington, DC, November 2012;. 2011 petroleum supply values and 2010 crude oil and lease condensate production: EIA, Petroleum Supply Annual 2011, DOE/EIA-0340(2011)/1, Washington, DC, August 2012; Other 2010 petroleum supply values: EIA, Petroleum Supply Annual 2010, DOE/EIA-0340(2010)/1, Washington, DC, July 2011; 2010 and 2011 crude oil spot prices: Thomson Reuters; Other 2010 and 2011 coal values: EIA, Quarterly Coal Report, October–December 2011, DOE/EIA-0121(2011/4Q), Washington, DC, March 2012; Other 2010 and 2011 values, EIA, Annual Energy Review 2011, DOE/EIA-0384(2011), Washington, DC, September 2012; Projections: EIA, AEO2013 National Energy Modeling System run ref2013.d102312a.

Notes: Totals may not equal sum of components due to independent rounding. Data for 2010 and 2011 are model results and may differ slightly from official EIA data reports. Quadrillion Btu, unless otherwise noted.

Btu, British thermal unit; MMBtu, million Btu; n/a, not applicable.

[a] Includes waste coal.

[b] These values represent the energy obtained from uranium when it is used in light water reactors. The total energy content of uranium is much larger, but alternative processes are required to take advantage of it.

[c] Includes grid-connected electricity from wood and wood waste; biomass, such as corn, used for liquid fuels production; and nonelectric energy demand from wood. Refer to Table A17 (http://www.eia.gov/forecasts/aeo/pdf/0383%282013%29.pdf, pp. 153--154) for details.

[d] Includes grid-connected electricity from landfill gas; biogenic municipal waste; wind; photovoltaic and solar thermal sources; and nonelectric energy from renewable sources, such as active and passive solar systems. Excludes electricity imports using renewable sources and nonmarketed renewable energy. See Table 17 for selected nonmarketed residential and commercial renewable energy data.

[e] Includes non-biogenic municipal waste, liquid hydrogen, methanol, and some domestic inputs to refineries.

[f] Includes imports of finished petroleum products, unfinished oils, alcohols, ethers, blending components, and renewable fuels such as ethanol.

[g] Includes imports of liquefied natural gas that are later reexported.

[h] Includes coal, coal coke (net), and electricity (net). Excludes imports of fuel used in nuclear power plants.

[i] Includes crude oil, petroleum products, ethanol, and biodiesel.

[j] Includes reexported liquefied natural gas.

[k] Balancing item. Includes unaccounted for supply, losses, gains, and net storage withdrawals.

[l] Includes petroleum-derived fuels and nonpetroleum-derived fuels, such as ethanol and biodiesel, and coal-based synthetic liquids. Petroleum coke, which is a solid, is included. Also included are natural gas plant liquids and crude oil consumed as a fuel. Refer to Table 17 for detailed renewable liquid fuels consumption.

[m] Excludes coal converted to coal-based synthetic liquids and natural gas.

[n] Includes grid-connected electricity from wood and wood waste, nonelectric energy from wood, and biofuels heat and coproducts used in the production of liquid fuels, but excludes the energy content of the liquid fuels.

[o] Includes non-biogenic municipal waste, liquid hydrogen, and net electricity imports.

[p] Annual average price for Brent crude oil on the Intercontinental exchange.

[q] Includes reported prices for both open market and captive mines. Prices weighted by production, which differs from average minemouth prices published in EIA data reports where it is weighted by reported sales.

[r] Prices weighted by consumption; weighted average excludes residential and commercial prices, and export free-alongside-ship (f.a.s.) prices.

TABLE 18.2

Energy Prices by Sector and Source

Sector and Source	2010	2011	2015	2020	2025	2030	2035	2040	Annual Growth 2011–2040 (%)
Residential									
Propane	27.61	25.06	21.35	23.41	24.77	25.73	26.70	27.99	0.4
Distillate fuel oil	21.77	26.38	24.95	26.91	29.08	31.26	33.71	36.54	1.1
Natural gas	11.36	10.80	10.16	11.78	12.67	13.37	14.60	16.36	1.4
Electricity	34.52	34.34	33.06	33.62	33.96	34.56	35.42	37.10	0.3
Commercial									
Propane	24.10	22.10	17.54	20.04	21.74	22.97	24.23	25.94	0.6
Distillate fuel oil	21.35	25.87	22.20	24.26	26.51	28.51	30.91	33.74	0.9
Residual fuel	11.39	19.17	13.27	14.82	16.60	18.77	20.89	23.41	0.7
Natural gas	9.40	8.84	8.10	9.47	10.19	10.70	11.68	13.21	1.4
Electricity	30.49	29.98	27.73	28.57	28.49	28.65	29.66	31.75	0.2
Industrial[a]									
Propane	23.73	22.54	17.81	20.51	22.33	23.64	24.97	26.78	0.6
Distillate fuel oil	21.87	26.50	22.50	24.67	27.02	28.91	31.31	34.16	0.9
Residual fuel oil	11.30	18.86	15.63	17.19	18.96	21.09	23.25	25.78	1.1
Natural gas[b]	5.48	4.89	4.42	5.53	6.15	6.56	7.45	8.88	2.1
Metallurgical coal	5.96	7.01	8.00	8.75	9.36	10.09	10.69	11.11	1.6
Other industrial coal	2.77	3.43	3.33	3.44	3.56	3.71	3.88	4.06	0.6
Coal to liquids	0.00	0.00	0.00	0.00	2.30	2.55	2.76	2.95	n/a
Electricity	20.26	19.98	17.88	18.72	19.18	19.73	20.80	22.74	0.4
Transportation									
Propane	27.52	26.06	22.39	24.48	25.83	26.80	27.77	29.07	0.4
E85[c]	25.56	25.30	24.94	29.64	27.27	26.94	29.19	30.58	0.7

(Continued)

TABLE 18.2 (*Continued*)

Energy Prices by Sector and Source

Sector and Source	2010	2011	2015	2020	2025	2030	2035	2040	Annual Growth 2011–2040 (%)
Motor gasoline[d]	23.18	28.70	25.99	27.84	29.26	30.73	32.99	36.18	0.8
Jet fuel[e]	16.57	22.49	19.52	21.50	23.73	26.03	28.52	31.07	1.1
Diesel fuel (distillate fuel oil)[f]	22.38	26.15	24.39	26.61	28.98	30.81	33.19	36.05	1.1
Residual fuel oil	10.62	17.83	13.42	14.91	16.58	18.34	20.25	22.45	0.8
Natural gas[g]	16.51	16.14	15.74	16.87	17.97	18.90	19.86	21.20	0.9
Electricity	33.91	32.77	29.84	29.60	30.40	31.53	32.84	35.07	0.2
Electric power[h]									
Distillate fuel oil	19.22	23.30	20.45	22.45	24.61	26.80	29.23	32.03	1.1
Residual fuel oil	12.11	15.97	21.74	24.94	27.29	29.36	31.85	34.54	2.7
Natural gas	5.26	4.77	3.84	4.90	5.58	6.05	6.98	8.38	2.0
Steam coal	2.30	2.38	2.40	2.52	2.69	2.87	3.03	3.20	1.0
Average price to all users[i]									
Propane	16.23	17.13	10.94	13.69	16.07	18.14	20.43	23.79	1.1
E85[c]	25.56	25.30	24.94	29.64	27.27	26.94	29.19	30.58	0.7
Motor gasoline[d]	23.06	28.47	25.99	27.84	29.26	30.72	32.99	36.17	0.8
Jet fuel[e]	16.57	22.49	19.52	21.50	23.73	26.03	28.52	31.07	1.1
Distillate fuel oil	22.17	26.18	24.05	26.25	28.62	30.48	32.88	35.73	1.1
Residual fuel oil	11.06	17.65	14.51	15.97	17.72	19.59	21.61	23.95	1.1
Natural gas	7.27	6.68	5.93	7.07	7.76	8.27	9.31	10.94	1.7
Metallurgical coal	5.96	7.01	8.00	8.75	9.36	10.09	10.69	11.11	1.6
Other coal	2.33	2.45	2.46	2.57	2.74	2.92	3.08	3.25	1.0
Coal to liquids	0.00	0.00	0.00	0.00	2.30	2.55	2.76	2.95	n/a
Electricity	29.40	29.03	26.90	27.50	27.79	28.41	29.55	31.58	0.3

(*Continued*)

TABLE 18.2 (*Continued*)

Energy Prices by Sector and Source

Sources: 2010 and 2011 prices for motor gasoline, distillate fuel oil, and jet fuel are based on prices in the U.S. Energy Information Administration (EIA), Petroleum Marketing Monthly, DOE/EIA-0380(2012/08), Washington, DC, August 2012; 2010 residential, commercial, and industrial natural gas delivered prices: EIA, Natural Gas Annual 2010, DOE/EIA-0131(2010), Washington, DC, December 2011; 2011 residential, commercial, and industrial natural gas–delivered prices: EIA, Natural Gas Monthly, DOE/EIA-0130(2012/07), Washington, DC, July 2012; 2010 transportation sector natural gas–delivered prices are based on EIA, Natural Gas Annual 2010, DOE/EIA-0131(2010), Washington, DC, December 2011 and estimated state taxes, federal taxes, and dispensing costs or charges; 2011 transportation sector natural gas–delivered prices are model results. 2010 and 2011 electric power sector distillate and residual fuel oil prices: EIA, Monthly Energy Review, DOE/EIA-0035(2011/09), Washington, DC, September 2010; 2010 and 2011 electric power sector natural gas prices: EIA, Electric Power Monthly, April 2011 and April 2012, Table 4.2, and EIA, State Energy Data System 2010, DOE/EIA-0214(2010), Washington, DC, June 2012; 2010 and 2011 coal prices based on: EIA, Quarterly Coal Report, October–December 2011, DOE/EIA-0121(2011/4Q), Washington, DC, March 2012 and EIA, AEO2013 National Energy Modeling System run ref2013.d102312a. 2010 and 2011 electricity prices: EIA, Annual Energy Review 2011, DOE/EIA-0384 (2011), Washington, DC, September 2012; 2010 and 2011 E85 prices derived from monthly prices in the Clean Cities Alternative Fuel Price Report; Projections: EIA, AEO2013 National Energy Modeling System run ref2013.d102312a.

Notes: Data for 2010 and 2011 are model results and may differ slightly from official EIA data reports. 2011 dollars per million Btu, unless otherwise noted.
Btu, British thermal unit; n/a, not applicable.

[a] Includes combined heat and power plants that have a non-regulatory status, and small on-site generating systems.

[b] Excludes use for lease and plant fuel.

[c] E85 refers to a blend of 85% ethanol (renewable) and 15% motor gasoline (nonrenewable). To address cold starting issues, the percentage of ethanol varies seasonally. The annual average ethanol content of 74% is used for forecast.

[d] Sales weighted-average price for all grades. Includes federal, state, and local taxes.

[e] Kerosene-type jet fuel. Includes federal and state taxes while excluding county and local taxes.

[f] Diesel fuel for on-road use. Includes federal and state taxes while excluding county and local taxes.

[g] Natural gas used as a vehicle fuel. Includes estimated motor vehicle fuel taxes and estimated dispensing costs or charges.

[h] Includes electricity-only and combined heat and power plants that have a regulatory status.

[i] Weighted averages of end-use fuel prices are derived from the prices shown in each sector and the corresponding sectoral consumption.

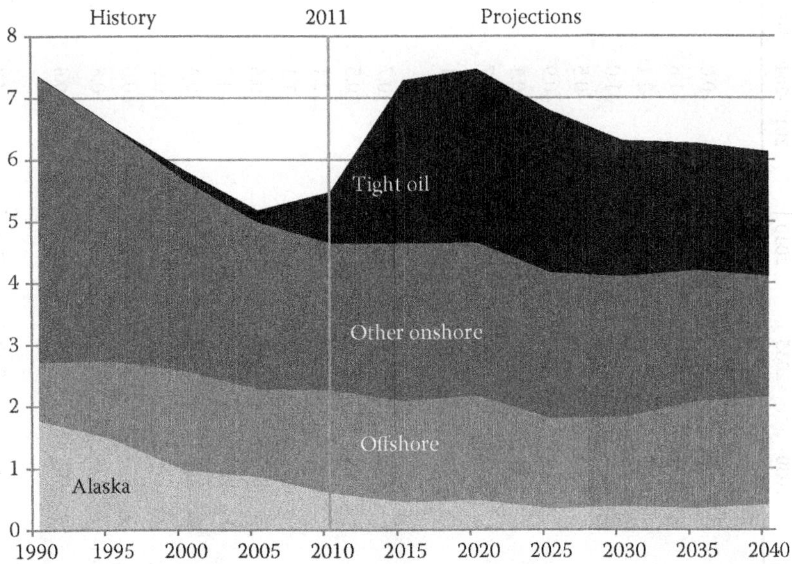

FIGURE 18.3
U.S. oil production by source, 1990–2040 (million bbl/day).

quadrillion Btu in 2040, largely as a result of the incentives provided by the Renewable Fuel Standard and the Low Carbon Fuel Standard of California.* The share of total U.S. liquid fuels consumption from domestically produced biofuels grows from 3.5% in 2011 to almost 6% in 2040.

18.4.4 Liquids Refining and Processing

18.4.4.1 Light versus Heavy Crudes and the Changing Product Slate

Refineries generally process a mix of crude oils with a range of API gravities† in order to optimize refinery operations. Over the past 15 years, the API gravity of crude oil processed in U.S. refineries averaged between 30° and 31°. As U.S. refiners process more domestic light crude produced from tight formations, they need less imported light oil crude to maintain an optimal API gravity. With increasing U.S. production of light crude oil in the Reference case, EIA expects the average API gravity of crude oil imports to decline overall, as domestic light crudes displace some of the lighter imported crudes. In addition, refinery operations might need to be modified and/or investments made to produce the market desired product mix.

In the *AEO2013* Reference case, the trend toward increasing imports of heavier crude oils continues through 2035 before stabilizing. The increase in demand for diesel fuel in the projection, from 3.5 to 4.3 million bbl/day, leads to an increase in distillate hydrocracking capacity (which increases diesel production capability) from 1.6 to 3.0 million bbl/day from 2011 to 2040. In the *AEO2013* Reference case, the ratio of gasoline-to-diesel production at

* For a summary of the Renewable Fuel Standard, see http://www.fas.org/sgp/crs/misc/R40155.pdf. For a summary of Low Carbon Fuel Standard of California, see http://www.energy.ca.gov/low_carbon_fuel_standard/.

† API gravity is a measure of the specific gravity, or relative density, of a liquid, as defined by the American Petroleum Institute (API). It is expressed in "degrees," where a higher number indicates lower density.

TABLE 18.3

Energy Consumption by Sector and Source

Sector and Source	2010	2011	2015	2020	2025	2030	2035	2040	Annual Growth 2011–2040 (%)
Residential									
Propane	0.53	0.53	0.53	0.52	0.52	0.52	0.52	0.52	0.0
Kerosene	0.03	0.02	0.02	0.01	0.01	0.01	0.01	0.01	−1.8
Distillate fuel oil	0.58	0.59	0.58	0.51	0.45	0.40	0.36	0.32	−2.1
Liquid fuels subtotal	1.14	1.14	1.12	1.05	0.98	0.93	0.89	0.86	−1.0
Natural gas	4.89	4.83	4.76	4.62	4.54	4.46	4.34	4.23	−0.5
Coal	0.01	0.01	0.01	0.01	0.01	0.01	0.01	0.00	−0.9
Renewable energy[a]	0.44	0.45	0.43	0.44	0.44	0.45	0.45	0.45	0.1
Electricity	4.93	4.86	4.66	4.84	5.08	5.36	5.67	6.03	0.7
Delivered energy	**11.41**	**11.28**	**10.99**	**10.95**	**11.04**	**11.20**	**11.35**	**11.57**	**0.1**
Commercial									
Propane	0.14	0.14	0.16	0.16	0.16	0.16	0.17	0.17	0.7
Motor gasoline[b]	0.06	0.05	0.05	0.05	0.06	0.06	0.06	0.06	0.5
Kerosene	0.00	0.00	0.00	0.00	0.00	0.00	0.00	0.01	2.1
Distillate fuel oil	0.41	0.42	0.35	0.34	0.33	0.32	0.31	0.30	−1.1
Residual fuel oil	0.08	0.07	0.10	0.09	0.09	0.09	0.09	0.09	0.6
Liquid fuels subtotal	0.69	0.69	0.67	0.65	0.64	0.64	0.63	0.63	−0.3
Natural gas	3.17	3.23	3.40	3.40	3.43	3.50	3.59	3.68	0.4
Coal	0.06	0.05	0.05	0.05	0.05	0.05	0.05	0.05	0.0
Renewable energy[c]	0.11	0.13	0.13	0.13	0.13	0.13	0.13	0.13	0.0
Electricity	4.54	4.50	4.52	4.72	4.97	5.22	5.47	5.72	0.8
Delivered energy	**8.57**	**8.60**	**8.76**	**8.95**	**9.22**	**9.54**	**9.86**	**10.21**	**0.6**
Industrial[d]									
Liquefied petroleum gases	2.12	2.10	2.20	2.46	2.54	2.47	2.40	2.30	0.3
Propylene	0.41	0.40	0.46	0.56	0.56	0.52	0.49	0.46	0.6
Motor gasoline[b]	0.28	0.27	0.30	0.32	0.32	0.32	0.32	0.32	0.6

(Continued)

TABLE 18.3 (*Continued*)

Energy Consumption by Sector and Source

Sector and Source	2010	2011	2015	2020	2025	2030	2035	2040	Annual Growth 2011–2040 (%)
Distillate fuel oil	1.19	1.21	1.23	1.22	1.19	1.18	1.19	1.22	0.0
Residual fuel oil	0.12	0.11	0.10	0.11	0.11	0.11	0.11	0.11	–0.1
Petrochemical feedstocks	0.94	0.88	0.90	1.03	1.08	1.08	1.08	1.09	0.7
Other petroleum[e]	3.70	3.61	3.50	3.54	3.48	3.46	3.53	3.65	0.0
Liquid fuels subtotal	8.76	8.57	8.68	9.25	9.28	9.14	9.11	9.16	0.2
Natural gas	6.67	6.92	7.26	7.86	8.00	7.97	8.02	8.08	0.5
Natural-gas-to-liquids heat and power	0.00	0.00	0.00	0.13	0.16	0.21	0.27	0.33	n/a
Lease and plant fuel[f]	1.31	1.42	1.44	1.57	1.68	1.73	1.84	1.97	1.1
Natural gas subtotal	7.98	8.34	8.70	9.56	9.84	9.91	10.13	10.38	0.8
Metallurgical coal	0.55	0.56	0.61	0.60	0.58	0.52	0.48	0.46	–0.7
Other industrial coal	1.06	1.04	0.99	1.00	1.00	1.00	1.02	1.05	0.0
Coal-to-liquids heat and power	0.00	0.00	0.00	0.00	0.07	0.09	0.12	0.15	20.5
Net coal coke imports	–0.01	0.01	0.00	–0.01	–0.03	–0.04	–0.06	–0.05	n/a
Coal subtotal	1.60	1.62	1.61	1.58	1.63	1.57	1.56	1.61	0.0
Biofuels heat and coproducts	0.85	0.67	0.75	0.82	0.82	0.85	0.97	1.37	2.5
Renewable energy[g]	1.47	1.51	1.56	1.72	1.85	1.97	2.11	2.28	1.4
Electricity	3.31	3.33	3.61	3.95	4.05	3.96	3.90	3.91	0.6
Delivered energy	**23.98**	**24.04**	**24.91**	**26.87**	**27.46**	**27.40**	**27.77**	**28.71**	**0.6**
Transportation									
Propane	0.04	0.06	0.06	0.06	0.06	0.07	0.08	0.08	1.3
E85[h]	0.01	0.05	0.12	0.08	0.14	0.16	0.15	0.17	4.3
Motor gasoline[b]	16.79	16.31	15.39	14.88	13.86	13.06	12.69	12.64	–0.9
Jet fuel[i]	3.07	3.01	3.03	3.11	3.20	3.28	3.35	3.42	0.4
Distillate fuel oil[j]	5.82	5.91	6.78	7.28	7.52	7.61	7.73	7.90	1.0
Residual fuel oil	0.88	0.82	0.83	0.84	0.85	0.86	0.86	0.87	0.2
Other petroleum[k]	0.17	0.17	0.15	0.15	0.15	0.16	0.16	0.16	–0.1

(Continued)

TABLE 18.3 (*Continued*)

Energy Consumption by Sector and Source

Sector and Source	2010	2011	2015	2020	2025	2030	2035	2040	Annual Growth 2011–2040 (%)
Liquid fuels subtotal	26.78	26.32	26.36	26.42	25.79	25.20	25.01	25.24	−0.1
Pipeline fuel natural gas	0.68	0.70	0.69	0.71	0.73	0.74	0.76	0.78	0.4
Compressed/liquefied natural gas	0.04	0.04	0.06	0.08	0.12	0.26	0.60	1.05	11.9
Liquid hydrogen	0.00	0.00	0.00	0.00	0.00	0.00	0.00	0.00	n/a
Electricity	0.02	0.02	0.02	0.03	0.04	0.04	0.06	0.07	3.9
Delivered energy	**27.52**	**27.09**	**27.13**	**27.24**	**26.68**	**26.25**	**26.43**	**27.14**	**0.0**
Electric power[n]									
Distillate fuel oil	0.08	0.06	0.08	0.08	0.08	0.08	0.08	0.08	0.9
Residual fuel oil	0.31	0.23	0.13	0.10	0.10	0.10	0.10	0.11	−2.6
Liquid fuels subtotal	0.39	0.30	0.21	0.18	0.18	0.18	0.18	0.19	−1.6
Natural gas	7.55	7.76	8.25	8.40	8.63	9.08	9.64	9.70	0.8
Steam coal	19.13	17.99	16.52	16.95	17.66	18.07	18.48	18.68	0.1
Nuclear/uranium[o]	8.43	8.26	8.57	9.25	9.54	9.49	9.14	9.44	0.5
Renewable energy[p]	3.85	4.74	4.89	5.49	5.77	5.93	6.38	7.44	1.6
Electricity imports	0.09	0.13	0.09	0.08	0.07	0.05	0.03	0.06	−2.4
Total[q]	**39.67**	**39.40**	**38.75**	**40.57**	**42.07**	**43.02**	**44.08**	**45.73**	**0.5**

Sources: 2010 and 2011 consumption based on U.S. Energy Information Administration (EIA), Annual Energy Review 2011, DOE/EIA-0384(2011), Washington, DC, September 2012; 2010 and 2011 population and gross domestic product: IHS Global Insight Industry and Employment models, August 2012; 2010 carbon dioxide emissions: EIA, Monthly Energy Review, DOE/EIA-0035(2011/10), Washington, DC, October 2011; 2011 carbon dioxide emissions and emission factors: EIA, Monthly Energy Review, DOE/EIA-0035(2012/08), Washington, DC, August 2012; Projections: EIA, AEO2013 National Energy Modeling System run ref2013.d102312a.

Notes: Totals may not equal sum of components due to independent rounding. Data for 2010 and 2011 are model results and may differ slightly from official EIA data reports. Quadrillion Btu, unless otherwise noted.

Btu, British thermal unit; n/a, not applicable.

[a] Includes wood used for residential heating. See Table A4 and/or Table A17 (http://www.eia.gov/forecasts/aeo/pdf/0383%282013%29.pdf, pp. 129--130 and/or 153--154, respectively) for estimates of nonmarketed renewable energy consumption for geothermal heat pumps, solar thermal water heating, and electricity generation from wind and solar photovoltaic sources.

[b] Includes ethanol (blends of 15% or less) and ethers blended into gasoline.

(*Continued*)

TABLE 18.3 (*Continued*)

Energy Consumption by Sector and Source

[c] Excludes ethanol. Includes commercial sector consumption of wood and wood waste, landfill gas, municipal waste, and other biomass for combined heat and power. See Table A5 and/or Table A17 (http://www.eia.gov/forecasts/aeo/pdf/0383%282013%29.pdf, pp. 131--132 and/or 153--154, respectively) for estimates of nonmarketed renewable energy consumption for solar thermal water heating and electricity generation from wind and solar photovoltaic sources.

[d] Includes energy for combined heat and power plants that have a nonregulatory status, and small on-site generating systems.

[e] Includes petroleum coke, asphalt, road oil, lubricants, still gas, and miscellaneous petroleum products.

[f] Represents natural gas used in well, field, and lease operations, in natural gas processing plant machinery, and for liquefaction in export facilities.

[g] Includes consumption of energy produced from hydroelectric, wood and wood waste, municipal waste, and other biomass sources. Excludes ethanol blends (15% or less) in motor gasoline.

[h] E85 refers to a blend of 85% ethanol (renewable) and 15% motor gasoline (nonrenewable). To address cold starting issues, the percentage of ethanol varies seasonally. The annual average ethanol content of 74% is used for this forecast.

[i] Includes only kerosene type.

[j] Diesel fuel for on- and off- road use.

[k] Includes aviation gasoline and lubricants.

[l] Includes unfinished oils, natural gasoline, motor gasoline blending components, aviation gasoline, lubricants, still gas, asphalt, road oil, petroleum coke, and miscellaneous petroleum products.

[m] Includes electricity generated for sale to the grid and for own use from renewable sources, and nonelectric energy from renewable sources. Excludes ethanol and nonmarketed renewable energy consumption for geothermal heat pumps, buildings photovoltaic systems, and solar thermal water heaters.

[n] Includes consumption of energy by electricity-only and combined heat and power plants that have a regulatory status.

[o] These values represent the energy obtained from uranium when it is used in light water reactors. The total energy content of uranium is much larger, but alternative processes are required to take advantage of it.

[p] Includes conventional hydroelectric, geothermal, wood and wood waste, biogenic municipal waste, other biomass, wind, photovoltaic, and solar thermal sources. Excludes net electricity imports.

[q] Includes non-biogenic municipal waste not included above.

[r] Includes conventional hydroelectric, geothermal, wood and wood waste, biogenic municipal waste, other biomass, wind, photovoltaic, and solar thermal sources. Excludes ethanol, net electricity imports, and nonmarketed renewable energy consumption for geothermal heat pumps, buildings photovoltaic systems, and solar thermal water heaters.

petroleum refineries declines from 2.3 in 2012 to 1.6 after 2035. The transition to lower gasoline and higher diesel production has a significant effect on petroleum refinery operations, investments, and infrastructure.

The large increase in domestic production of light crude oil and the increase in imports of heavier crude oils have prompted significant investments in crude midstream infrastructure, including pipelines that will bring higher quantities of light sweet crudes to petroleum refineries along the U.S. Gulf Coast. In addition, there have been significant investments to move crude oil to refineries by rail. The Reference case assumes that sufficient infrastructure investments will be made through 2040 to move both light and heavy crude oils.

18.4.4.2 Biofuels and Ethanol

The *AEO2013* Reference case, relative to the *Annual Energy Outlook 2012* (*AEO2012*), projects lower production levels for cellulosic biofuels, resulting from changes to a number of important assumptions, including the following: lower near-term crude oil price assumption, fewer flex-fuel vehicle sales, lower E85 availability, as well as a downward adjustment in the projection of "drop-in" biofuels.

Biofuels consumption grows in the *AEO2013* Reference case but falls well short of the Renewable Fuels Standard (RFS) target of 36 billion gallon (gal) ethanol equivalent in 2022, largely because of a decline in gasoline consumption, which results from a newly enacted CAFE standard and lowered expectations for sales of vehicles capable of using E85 (flex-fueled vehicles). Ethanol consumption falls from 16.4 billion gal in 2022 to 14.9 billion gal in 2040 in the *AEO2013* Reference case, as gasoline demand continues to drop and E85 consumption levels off. However, domestic consumption of drop-in cellulosic biofuels grows from 0.3 billion gal to 9.0 billion gal ethanol equivalent per year from 2011 to 2040, as rising oil prices lead to price increases for diesel fuel, heating oil, and jet fuel, while production costs for advanced biofuels technologies fall. From 2011 to 2022, demand for motor gasoline ethanol blends (E10 and E15) falls from 8.7 to 8.1 million bbl/day which limits the amount of ethanol that can be splash blended with gasoline—an illustration of the "blend wall" effect.

Domestic ethanol production remains relatively flat throughout the projection, as consumption of motor gasoline decreases and the penetration of ethanol in the gasoline pool is slowed by the limitations of the E10 and E15 blend market and the limited availability of flex-fuel vehicles, few E15- and E85-capable filling stations, and a reluctance of consumers to search out and use higher ethanol blends. Total biofuels production increases by 0.4 million bbl/day in the projection, as drop-in fuels (e.g., biodiesel) from biomass enter the market. Other emerging technologies capable of producing liquids—such as xTL, which includes coal-to-liquids (CTL) and gas-to-liquids technologies—also become more economic as more plants are built through the manufacturing "learning effect" or "technological progress." In 2040, liquids production from xTL plants total 0.3 million bbl/day. High capital costs and the risk that xTL production will not remain price-competitive with crude oil slow investment in xTL technologies.

18.4.5 Domestic Liquids Prices

Domestic liquids prices (Table 18.2) reflect the trends in global petroleum pricing as well as shifts in domestic liquids demand. EIA expects that the rate of technological progress for extracting crude oil from tight formations and natural gas from shale, the implementation of new appliance efficiency standards, and a new LDV efficiency standard in the United States that begins in 2017 and continues through 2025 will all temper U.S. crude oil demand for imports and impact crude oil prices.

In addition, potential new international emission control protocols and significant global concerns about the security of the petroleum supply represent additional factors which may also reduce the upward pressure on crude oil prices. Finally, technological progress, which may lead to cost reductions in xTL technologies and the development of new methods to economically develop domestic oil shale resources, and the discovery of crude oil in other relatively unexplored regions of the world, could also influence oil prices. For a full listing of delivered liquids prices, see Table 18.2.

18.5 Natural Gas

In the *AEO2013* Reference case, natural gas becomes the dominant source of U.S. energy production, growing from a 30% market share in 2011 to a 34% share by 2040, primarily as a result of increased production from tight plays. Natural gas consumption rises by more than 20% by 2040, primarily due to its increased use in electric power generation, the industrial sector, and as a transportation fuel in heavy-duty trucks. The growth in natural gas production (and consumption) is another of the major story lines of the *AEO2013* Reference case, as the increase in domestic production has a number of positive follow-on effects in the U.S. energy-economic system (Figure 18.4).

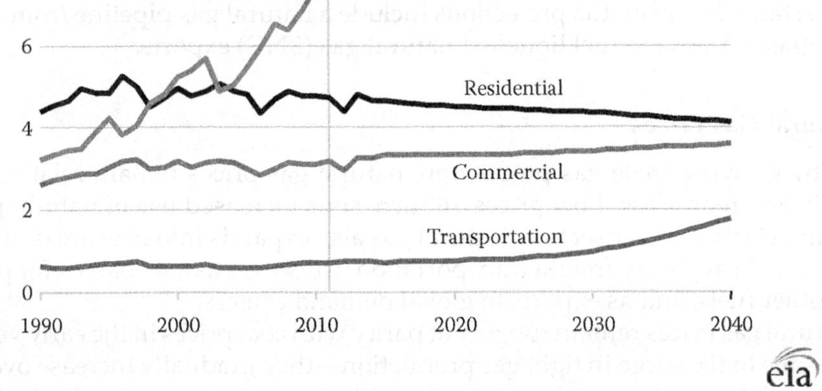

FIGURE 18.4
Natural gas consumption by sector, 1990–2040 (Tcf).

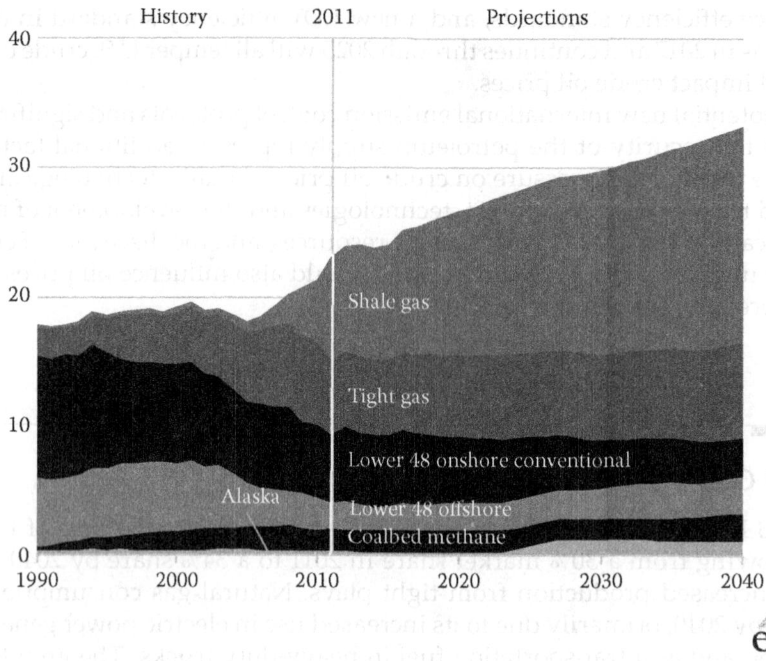

FIGURE 18.5
Natural gas production by source, 1990–2040 (Tcf).

18.5.1 Natural Gas Supply

Shale gas production is the major contributor to increased natural gas production, increasing from 7.8 Tcf in 2011 (34% of total U.S. dry gas production) to 16.7 Tcf in 2040 (50% of total production) as shown in Figure 18.5. Much of the growth in natural gas production is the result of drilling in shale plays with high concentrations of NGL and crude oil. High oil prices make drilling attractive to producers and make unconventional plays economical.

U.S. natural gas production exceeds consumption by 2020, and the resulting surplus spurs net exports of natural gas. Unlike the case in previous projections, the *AEO2013* Reference case does not include the construction of a natural gas pipeline from Alaska to the lower 48 states; however, the projections include a natural gas pipeline from the North Slope to southern Alaska to fuel liquefied natural gas (LNG) exports.

18.5.2 Natural Gas Prices

Facilitated by growing shale gas production, natural gas prices remain relatively low in the *AEO2013* Reference case. Low prices, in turn, spur increased use of natural gas in the industrial and electric power sectors. Natural gas also expands into new areas of the economy as a fuel for heavy-duty freight transportation (trucking), as a feedstock for producing diesel and other fuels, and as exports to global demand centers.

While natural gas prices remain roughly at parity with coal prices in the early years of the projection—due to the surge in tight gas production—they gradually increase over time as drilling moves into more expensive resources with lower recovery rates. In the Reference case, natural gas prices at the Henry Hub rise from $3.98 per million Btu (MMBtu) in 2011 to $7.83 per MMBtu in 2040 (see Table 18.1).

18.6 Energy Consumption, by Sector

In the *AEO2013* Reference case, total U.S. energy consumption grows from approximately 98 quadrillion Btu in 2011 to 108 quadrillion Btu in 2040, an average annual growth rate of just 0.3%. Across sectors, the biggest shifts in demand occur in the industrial and commercial sectors, each of which grows at an average of 0.5% per year to 2040. Please refer to Table 18.3 and Figure 18.6 for a summary of energy demand by sector.

18.6.1 Transportation Sector

In the *AEO2013* Reference case, the transportation sector consumes 27.1 quadrillion Btu of energy in 2040, the same level of energy demand as 2011, largely as a result of declining demand from LDVs. This projection differs markedly from the historical trend, which saw 1.1% average annual growth from 1975 to 2011. Energy demand for LDVs, which includes cars and light trucks, declines from 16.1 quadrillion Btu in 2011 to 13.0 quadrillion Btu in 2040, in contrast to 0.9% average annual growth from 1975 to 2011.

EIA's projection includes the National Highway Traffic Safety Administration (NHTSA) and the Environmental Protection Agency (EPA) standards for LDVs, which serve to raise the fuel economy of new LDVs from 32.5 miles per gallon (mpg) in 2012 to 47.3 mpg in 2025. The GHG emissions and CAFE standards are held constant after 2025, but LDV fuel economy continues to rise, to 49.0 mpg in 2040. Passenger car fuel economy averages 56.1 mpg and light-duty truck fuel economy averages 40.5 mpg. LDVs that use diesel, other alternative fuels, hybrid-electric, or all-electric systems play a significant role in meeting more stringent GHG emissions and CAFE standards over the projection period.

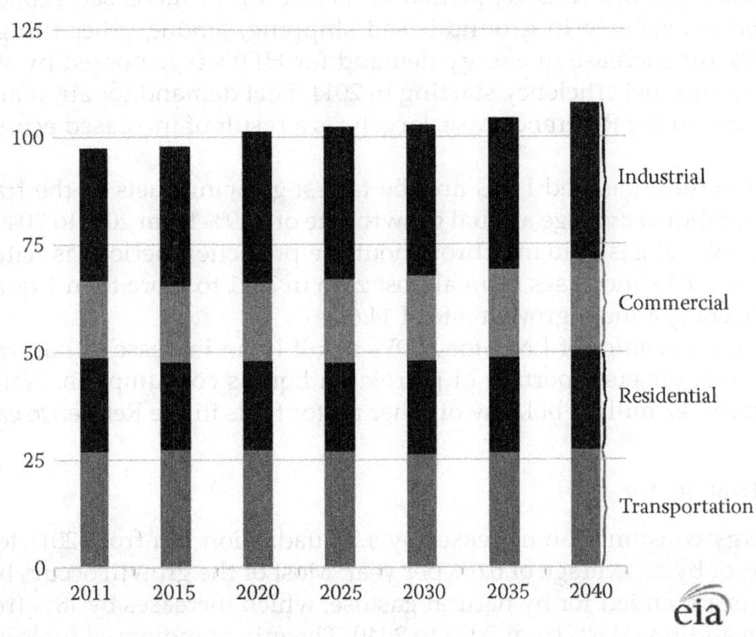

FIGURE 18.6
Primary energy use by end-use sector, 2011–2040 (quadrillion Btu).

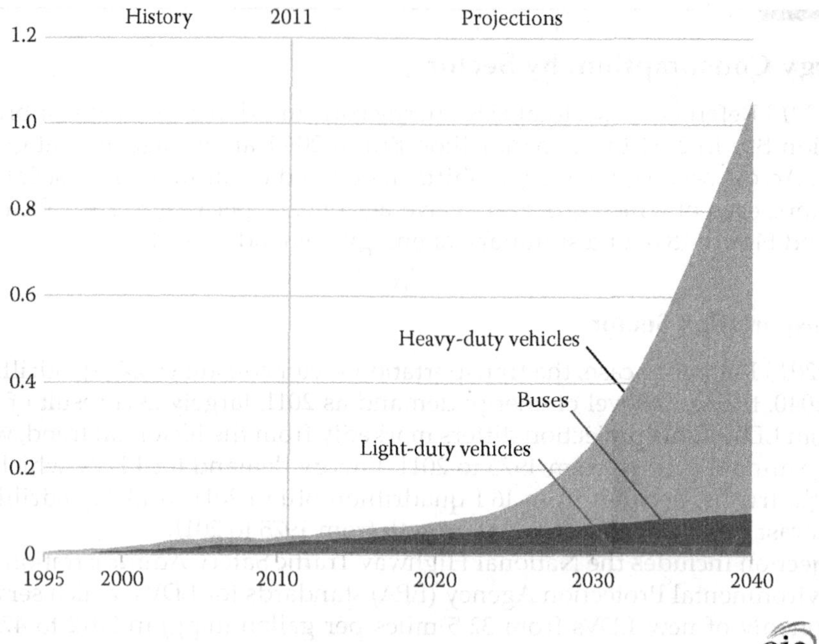

FIGURE 18.7
Natural gas consumption in the transportation sector, 1995–2040 (quadrillion Btu).

Energy demand for heavy duty vehicles (HDVs) increases from 5.2 quadrillion Btu in 2011 to 7.6 quadrillion Btu in 2040, primarily as a result of increased economic output. HDVs are used extensively in ground-based shipping, among other things. As is the case with LDVs, the increase in energy demand for HDVs is tempered by standards for GHG emissions and fuel efficiency starting in 2014. Fuel demand for air, marine, and rail traffic all increase in the Reference case, largely as a result of increased economic output (Figure 18.7).

Compressed natural gas and LNG are the fastest-growing fuels in the transportation sector, with a combined average annual growth rate of 11.9% from 2011 to 2040. HDVs lead the growth in natural gas demand throughout the projection period, as natural gas fuel consumption by HDVs increases from almost zero in 2011 to more than 1 quadrillion Btu in 2040, at an average annual growth rate of 14.6%.

The improved economics of LNG for HDVs result in an increase in natural gas use in transportation that offsets a portion of petroleum liquids consumption. Natural gas use in HDVs displaces 0.7 million bbl/day of other motor fuels in the Reference case in 2040.

18.6.2 Industrial Sector

Industrial energy consumption increases by 4.7 quadrillion Btu from 2011 to 2040 in the Reference case, or by an average of 0.6% per year. Most of the growth occurs between 2011 and 2025, and is accounted for by natural gas use, which increases by 18% from 2011 and 2025 and by an additional 6% from 2025 to 2040. The mix of industrial fuels changes relatively slowly, however, reflecting limited capability for fuel switching in most industries. Despite a 76% increase in industrial shipments, industrial-delivered energy consumption increases by only 19% from 2011 to 2040 in the *AEO2013* Reference case, reflecting a continuing transition to less energy-intensive industries.

Industrial consumption of NGL, consumed predominantly as feedstocks in the bulk chemicals industry, increases by 21% from 2011 to 2025, followed by a 9% decline from 2025 to 2040. NGL usage declines starting in 2025 as shipments of bulk chemicals fall in the face of increased international competition. Consumption of renewable fuels in the industrial sector grows by 22% from 2011 to 2025 and by 37% from 2025 to 2040. The paper industry remains the predominant consumer of renewable energy (mostly biomass) in the industrial sector.

In the projection from 2011 to 2040, the energy-intensive industries' (including chemicals, petroleum refineries, paper products, iron and steel, and food products) share of total industrial output continues declining at a rate of 1.0% per year, a decline that continues the shift of the U.S. economy to less energy-intensive industries. The decline in energy intensity for the less energy-intensive industries is almost twice that for the more energy-intensive industries. While energy-intensive manufacturing shows stable energy intensity, the nonenergy-intensive manufacturing industries—such as plastics, computers, and transportation equipment—show a 33% decline in energy intensity from 2011 to 2040, or an average decline of about 1.4% per year.

Refining is the only industry subsector (energy intensive) that shows an increase in energy intensity. Shipments from refineries are essentially flat in the early years and then decline slightly after 2019, with a 4% decline in shipments overall from 2011 to 2040. In contrast, energy use for refining increases by 13% over the same period, as CTL production and the use of heavy crude feedstock, both of which require more energy to process than typical crude oil, increase after 2022.

18.6.3 Residential Sector

The Reference case projection reflects a general population geographical shift to warmer climates. As a result, the average electricity demand per household declines by 6% in the Reference case, from 12.3 MWh in 2011 to 11.5 MWh in 2040. Despite the per-household reduction in electricity demand, the increasing number of households over the projection period results in an increase of 24% of total delivered electricity consumption in the residential sector. Over the same period, residential use of natural gas falls by 12%, and use of petroleum and other liquids falls by 25%. In 2040, space cooling and "other uses" consume 42% and 52% more electricity, respectively, than in 2011, and they remain the largest residential uses of electricity. These shifts reflect not only population geographical shifts to warmer climates but also efficiency changes. Total energy demand for most electric end uses increases, even as it declines on a per-household basis.

EIA projects combined photovoltaic (PV) and wind capacity in the residential sector to grow from 1.5 GW in 2011 to 9.6 GW in 2016. After 2016, expiration of the investment tax credit (ITC) results in slower growth, with only 4.1 GW added from 2017 through 2040. Similarly, ground-source heat pumps and solar water heaters grow from a combined 1.3 million units in 2011 to 2.4 million units in 2016; but after 2016, with expiration of the ITCs, only 1.4 million additional units are added through 2040.

In the *AEO2013* Reference case, the energy intensity of residential demand (defined as annual energy use per household) declines from 97.2 MMBtu in 2011 to 75.5 MMBtu in 2040. The projected 22% decrease in intensity occurs along with a 32% increase in the number of homes (about 1% per year), and total residential square footage increasing by

41% from 2011 to 2040. Without efficiency improvements, energy demand for uses such as heating, cooling, and lighting would increase at rates similar to the increases of households and square footage.

18.6.4 Commercial Sector

While total delivered commercial energy grows over the projection period and commercial floor space grows at about 1% per year in EIA's projections, average delivered energy consumption per square foot of commercial floorspace declines at an annual rate of 0.4% from 2011 to 2040 in the *AEO2013* Reference case. Natural gas consumption increases at about one-half the growth rate of delivered electricity consumption, which grows by 0.8% per year. These changes are largely due to higher efficiency standards and higher efficiency uptake, particularly in electric appliances.

Renewable technologies, including solar PV and wind, account for 61% of commercial distributed generation capacity in 2040 in the *AEO2013* Reference case. Exponential growth of PV capacity occurs as a result of utility incentives, new financing options, and the 30% federal ITC that reverts to 10% in 2017. In the Reference case, commercial PV capacity increases by 5.4% annually from 2011 to 2040. Small-scale wind capacity increases more than eightfold from 2011 to 2040 because it is not hampered by rising fuel prices that offset the effects of the 10% ITC on nonrenewable technologies for distributed generation after 2017. Microturbines using natural gas and fuel cells account for nearly all of the remaining distributed generation in the commercial sector.

The average energy intensity of commercial buildings declines from 105.2 thousand Btu/sq. ft. in 2012 to 93.8 thousand Btu/sq. ft. in 2040, driven largely by federal efficiency standards. However, miscellaneous electric loads—which include medical equipment and video displays—are not all subject to federal standards, and increase the energy intensity from "other" end uses by 34%.

Increasing energy use for miscellaneous electric loads, many of which currently are not subject to federal efficiency standards, leads to a 34% increase in energy intensity from 2011 to 2040 for "other" end uses. Miscellaneous electric loads in the commercial sector include medical equipment and video displays, among many other devices.

With ongoing improvements in equipment efficiency and building shells, the average energy intensity of commercial buildings declines over time by 10.8%, from 105.2 thousand Btu/ft^2 in 2012 to 93.8 thousand Btu/ft^2 in 2040. Federal efficiency standards in end uses such as space heating and cooling, water heating, refrigeration, and lighting, and act to limit growth in energy consumption to less than the growth in commercial floorspace.

18.6.5 Electricity Sector

The growth of U.S. electricity demand (including retail sales and direct use) slowed in each decade since the 1950s, from a 9.8% annual rate of growth from 1949 to 1959 to only 0.7% per year in the first decade of the twenty-first century. Electricity demand growth in the Reference case is slow, as increasing demand for electricity services is offset by efficiency gains from new appliance standards and investments in energy-efficient equipment. Total electricity demand grows by 28% in the projection (0.9% per year), from 3839 billion kilowatt-hours (kWh) in 2011 to 4930 billion kWh in 2040. The share of electricity generated by coal declines over the projection period by an annual average of 0.5%, primarily due to growth in generation from natural gas and renewables.

The decline in coal-fired electricity generation and the rise of natural gas-fired generation in recent years was driven primarily by two factors: (1) the huge fall in natural gas prices that occurred as U.S. shale gas resources were successfully developed and (2) the recession that began in late 2007, which contributed to a 2-year decline in electricity generation (the first time that has occurred since EIA began tracking the data). Coal absorbed most of the negative impacts of the reduction, as natural gas prices fell sharply in the period to keep volumes relatively stable.

Between 1990 and 2008, coal-fired power plants accounted for 50% or more of U.S. electricity generation each year. However, since 2008, coal's share of generation declined every year, falling to 42% in 2011. The story for natural gas is almost the complete opposite. After falling to less than 10% of total generation in 1988, natural gas's share of generation increased to nearly 25% in 2011. EIA expects the natural gas share of electricity generation to be much higher in 2012 when final data are available. While natural gas prices remain low in the early years of the projection due to the surge in tight gas production, they gradually increase over time as drilling moves into more expensive resources with lower recovery rates.

The *AEO2013* Reference case expects the share of natural gas in generation to continue growing as prices slowly increase. EIA expects goal-fired generation to recover as natural gas prices rise, but not enough to increase its market share. Partly as a result of the crowding out effect natural gas has on coal, U.S. coal exports in the *AEO2013* Reference case grow from 107 million short tons in 2011 to 159 million short tons in 2040.

Natural gas and renewables account for the largest share of the growth in domestic energy production between 2011 and 2040 (76%) and their share of total production grows over the projection, from 42% in 2011 to 49% in 2040. Non-hydro renewable generation shows rapid growth, as state and federal programs spur growth in the near term, and they become increasingly competitive in the long term.

18.6.5.1 Coal

In the *AEO2013* Reference case, coal-fired power plants continue to be the largest source of electricity generation throughout the projection period, but their market share declines from 42% in 2011 to 35% in 2040. The United States retires approximately 15% of the coal-fired capacity active in 2011 by 2040, while only 3% of total new generating capacity added by 2040 is coal-fired. Existing coal-fired units that have undergone environmental equipment retrofits continue to operate throughout the projection.

Coal production grows modestly between 2011 and 2040 even though coal consumption for power generation falls through 2016 with the retirement of over 30 GW of coal capacity. After 2016, coal consumption grows as the remaining coal capacity is used more intensely and growth in coal use in CTL production accelerates.

18.6.5.2 Natural Gas

As a result of the new outlook for natural gas production, generation from natural gas increases by an average of 1.6% per year from 2011 to 2040, and its share of total generation grows from 24% in 2011 to 30% in 2040. The relatively low cost of natural gas makes the dispatching of existing natural gas plants more competitive with coal plants and, in combination with relatively low capital costs of new natural gas units, makes plants fueled by natural gas an attractive option for new generation capacity.

18.6.5.3 Nuclear

Nuclear generation grows by 14% in the Reference case, due to 11 GW of additional nuclear capacity and 8 GW of uprates of existing capacity offsetting over 7 GW of retirements. Nuclear's share of total U.S. electricity generation declines from 19% in 2011 to 17% in 2040, as the growth in natural gas-fired generation outpaces electricity generation from nuclear power plants (NPPs). Generation from U.S. NPPs increases by 0.5% per year on average from 2011 to 2040, with most of the growth between 2011 and 2025.

18.6.5.4 Renewables

In the *AEO2013* Reference case, renewable generating capacity accounts for nearly one-fifth of total generating capacity (but not total generation) in 2040. Nearly all renewable capacity additions over the period consist of non-hydropower capacity, which grows by more than 150% from 2011 to 2040. Renewable capacity additions are supported by state renewable portfolio standards, the federal RFS, and federal tax credits.

Generation from renewable sources grows by 1.7% per year on average, and its share of total generation rises from 13% in 2011 to 16% in 2040 (Figure 18.8). Renewable generation increases from 524 billion kWh in 2011 to 858 billion kWh in 2040. The non-hydropower share of total renewable generation increases from 38% in 2011 to 65% in 2040. The increase in wind-powered generation from 2011 to 2040 of 134 billion kWh, or 2.6% per year, represents the largest absolute increase in renewable generation. However, generation from solar energy grows by 92 billion kWh over the same period, representing the highest annual average growth at 9.0% per year. Biomass generation increases by 95 billion kWh over the projection period, for an average annual increase of 4.5%.

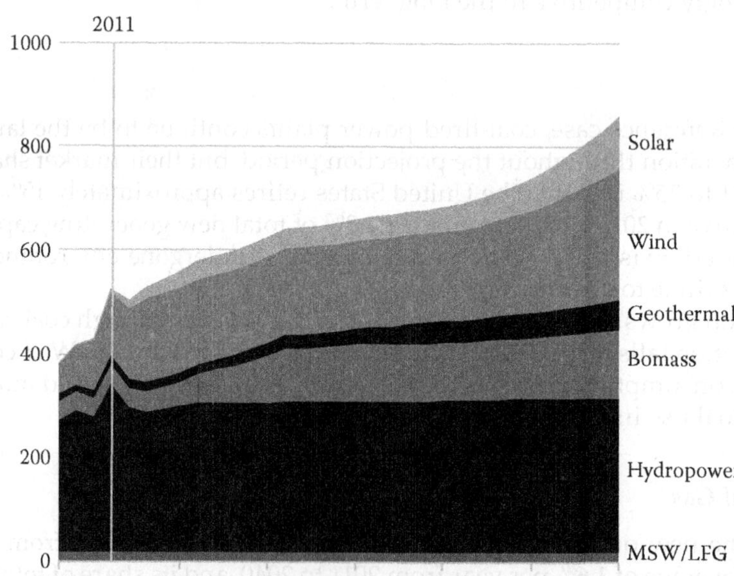

FIGURE 18.8
Renewable electricity generation by type, including end-use generation, 2008–2040 (billion kWh).

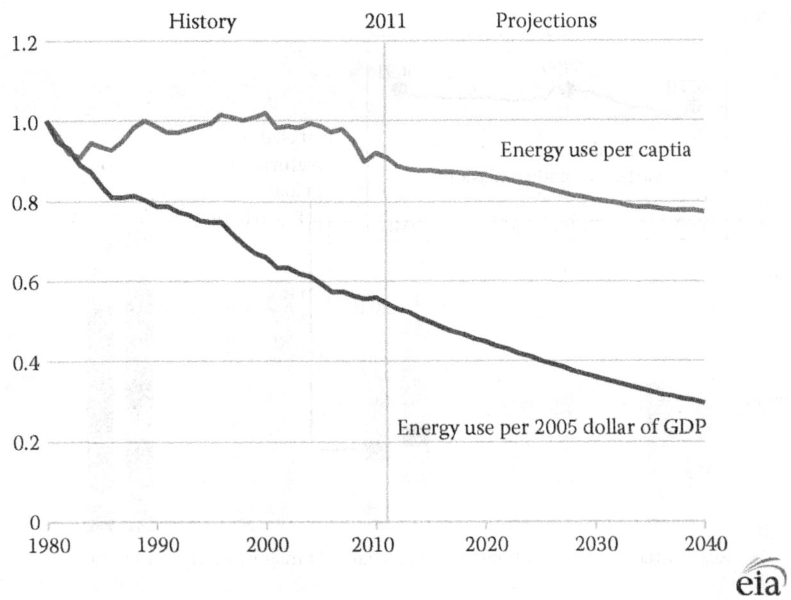

FIGURE 18.9
Energy use per capita and per dollar of gross domestic product, 1980–2040 (index, 1980 = 1).

18.7 Energy Intensity and CO_2 Emissions

The potential for more energy conservation continues to receive attention as energy prices rise. The Reference case projection does not assume policy-induced conservation measures beyond those in existing legislation and regulation, nor does it assume behavioral changes beyond those experienced in the past.

Energy intensity (thousand Btu per 2005 dollars of GDP) in the United States declined by an average of 1.9% per year between 1992 and 2003, as the role of energy-intensive industries in the U.S. economy fell sharply (the share of industrial output from the energy-intensive industries declined on average by 1.3% per year from 1992 to 2003) (Figure 18.9). On a national basis, energy intensity falls by 2.1% per year from 2011 to 2040 in the Reference case.

Historically, energy use per person varies over time with the level of economic growth, weather conditions, and energy prices, among other factors. During the late 1970s and early 1980s, energy consumption per capita fell in response to high energy prices and weak economic growth. Starting in the late 1980s and lasting through the mid-1990s, energy consumption per capita increased with declining energy prices and strong economic growth. Per capita energy use declines in the Reference case projection by nearly 0.6% per year between 2011 and 2040, with growth in the demand for energy services offset by efficiency gains (Figure 18.10).

Efficiency gains in the transportation sector help to moderate the growth of CO_2 emissions from energy use, increasing from 5471 million metric tons in 2011 to 5501.1 million metric tons in 2025, and to 5691.1 million metric tons in 2040. The CO_2 emissions-intensity of the U.S. economy falls from 411 thousand metric tons per million dollars of GDP in

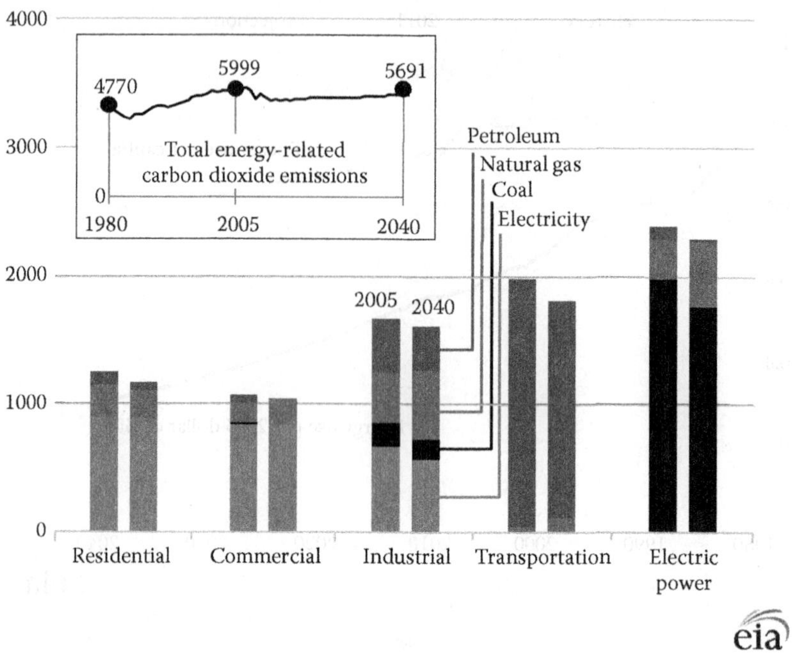

FIGURE 18.10
U.S. energy-related carbon dioxide emissions by sector and fuel, 2005 and 2040 (million metric tons).

2005 dollars in 2011 to 290 thousand metric tons per million dollars GDP in 2025. Further, they continue to drop to 209 thousand metric tons per million dollars (2005 USD) of GDP in 2040.

By sector—and including the emissions associated with the electricity consumed—CO_2 emissions for 2011 to 2040 remain largely unchanged in the residential sector, increase by 0.3% per year for the commercial sector, increase by 0.3% per year for the industrial sector, and decline slightly for transportation. Power sector CO_2 emissions grow at 0.2% per year but these are already included in the sectoral emissions listed earlier.

Acknowledgments

We are indebted to the staff of the Office of Energy Analysis at the U.S. Energy Information Administration (EIA) for their integral part in helping to develop this projection. Credits for this effort are shared with the staff and management who produced the *Annual Energy Outlook 2013* (*AEO2013*). Like all projections, this projection represents what might occur under laws, regulations, policies, and other key assumptions of the scenarios at the time of their development. It is not a forecast of what will occur. All errors remain the sole responsibility of the authors.

References

1. U.S. Energy Information Administration, *Annual Energy Outlook 2012*, Washington, DC, June 2012.
2. U.S. Energy Information Administration, *Annual Energy Outlook 2013*, Washington, DC, April 2013.
3. California Legislative Information, Assembly Bill No. 32: California Global Warming Solutions Act of 2006 (Sacramento, CA: September 27, 2006), http://www.leginfo.ca.gov/pub/05–06/bill/asm/ab_0001–0050/ab_32_bill_20060927_chaptered.pdf.
4. U.S. Environmental Protection Agency and Department of Transportation, National Highway Traffic Safety Administration, 2017 and Later Model Year Light-Duty Vehicle Greenhouse Gas Emissions and Corporate Average Fuel Economy Standards; Final Rule, *Federal Register*, Vol. 77, No. 199 (Washington, DC: October 15, 2012), https://www.federalregister.gov/articles/2012/10/15/2012–21972/2017-and-later-model-year-light-duty-vehicle-greenhouse-gas-emissions-and-corporateaverage-fuel.
5. California Environmental Protection Agency, Air Resources Board, Subchapter 10. Climate Change, Article 4. Regulations to Achieve Greenhouse Gas Emission Reductions, Subarticle 7. Low Carbon Fuel Standard. http://www.arb.ca.gov/fuels/lcfs/CleanFinalRegOrder112612.pdf.
6. U.S. Environmental Protection Agency, Final Rule to Identify Additional Fuel Pathways under the Renewable Fuel Standard Program (RFS), *Federal Register*, Vol. 78, No. 43, Tuesday, March 5, 2013, Rules and Regulations. http://www.gpo.gov/fdsys/pkg/FR-2013-03-05/pdf/2013-04929.pdf.

Section III

Energy Infrastructure and Storage

19

Transportation

Terry Penney and Frank Kreith

CONTENTS

19.1 Introduction

A viable transportation system is a crucial part of a sustainable energy future. Transportation is a complex interdisciplinary topic, which really deserves a book unto itself. However, some of the main issues related to transportation have to do with fuels and energy storage, which are the topics covered in this book. This section does not purport to be exhaustive, but examines some of the key issues related to a viable transportation future.

Gasoline and diesel not only are very convenient fuels for ground transportation, but also have a high energy density that permits storage in a relatively small volume—an important asset for automobiles. For example, these liquid fuels have a volumetric specific energy content of about 10,000 kWh/m^3, compared to hydrogen compressed to 100 bar at about 300 kWh/m^3. But known petroleum resources worldwide are being consumed rapidly, and future availability of these resources is bound to decline. At present, more than 97% of the fuel used for ground transportation in the United States is petroleum based. Importation of fossil fuels in the United States has recently been decreasing, primarily because of domestic exploration and production of oil and natural gas. The increase in the cost of gas and oil has become of growing concern to average citizens, and the emission from current transportation systems is a major component of CO_2 pollution that produces global warming.

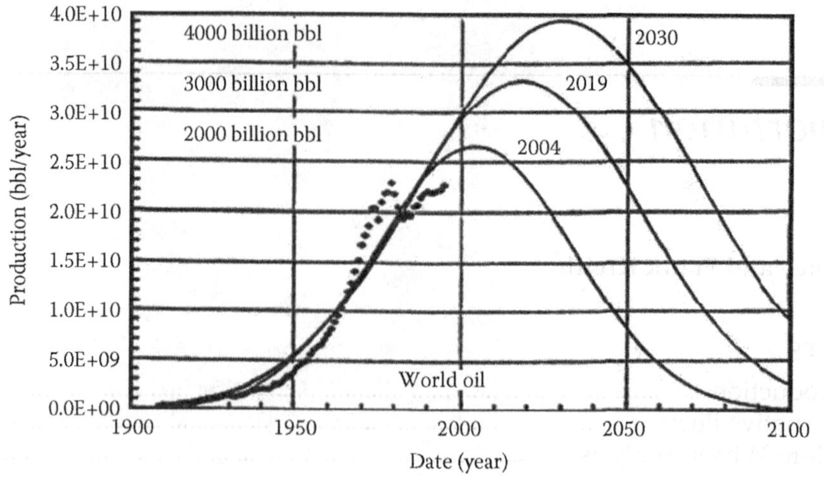

FIGURE 19.1
Oil production vs. time for various recoverable amounts of petroleum. (According to Bartlett, A.A., *Math. Geol.*, 32(1), 2000. With permission.)

There is worldwide agreement among oil experts that global oil production will reach a peak sometime between 2020 and 2030. The predictions for the date of peak world oil production according to various estimates [1] are demonstrated in Figure 19.1. The oil production is shown as a function of time for three total amounts of recoverable oil that span the entire range of assumptions by experts. Although new oil fields may be discovered in the future, it is not expected that they will substantially increase the total recoverable amounts. Hence, it is believed that the total amounts lie somewhere between 3000 and 4000 billion barrels of oil (bbl). The obvious conclusion to be drawn from these predictions is that the production peak is imminent and the price of oil will continue to escalate as supplies decline. Thus, planning for a sustainable transportation system that does not depend entirely on petroleum resources is an imperative segment of a sustainable energy future.

19.2 Alternative Fuels

Alternative fuels available to supplement oil as well as their feedstock are shown in Table 19.1 [2]. Inspection of the table shows that biodiesel, electricity, ethanol, and hydrogen (via electricity) are the fuels potentially independent of a petroleum resource such as oil or natural gas. The potential of producing liquid fuel from biomass has been treated elsewhere in this handbook. There is every reason to believe that biomass will provide an increasing percentage of the future transportation fuel, especially if ethanol can be produced from cellulosic materials such as switchgrass or bio-waste, or diesel from algae. The reason for this is that ethanol produced by traditional methods from corn kernels has only an energy return on energy investment on the order of 1.25, whereas the EROI for cellulosic ethanol is claimed to be about 6. Diesel from algae may be even better. As shown in Figure 19.2, even cellulosic ethanol would not be able to replace oil, because growing it would require too large a percentage of all the arable lands in the United States, and the

TABLE 19.1
Feedstocks for Alternative Fuels

Fuel	Feedstock
Propane (LPG)	Natural gas (NG), petroleum
CNG	NG
Hydrogen	NG or (water + electricity)
FT diesel	NG, coal, or algae
Methanol (M85)	NG or coal
Ethanol (E85)	Corn, sugarcane, or cellulosic biomass
Electricity	NG, coal, uranium, or renewables

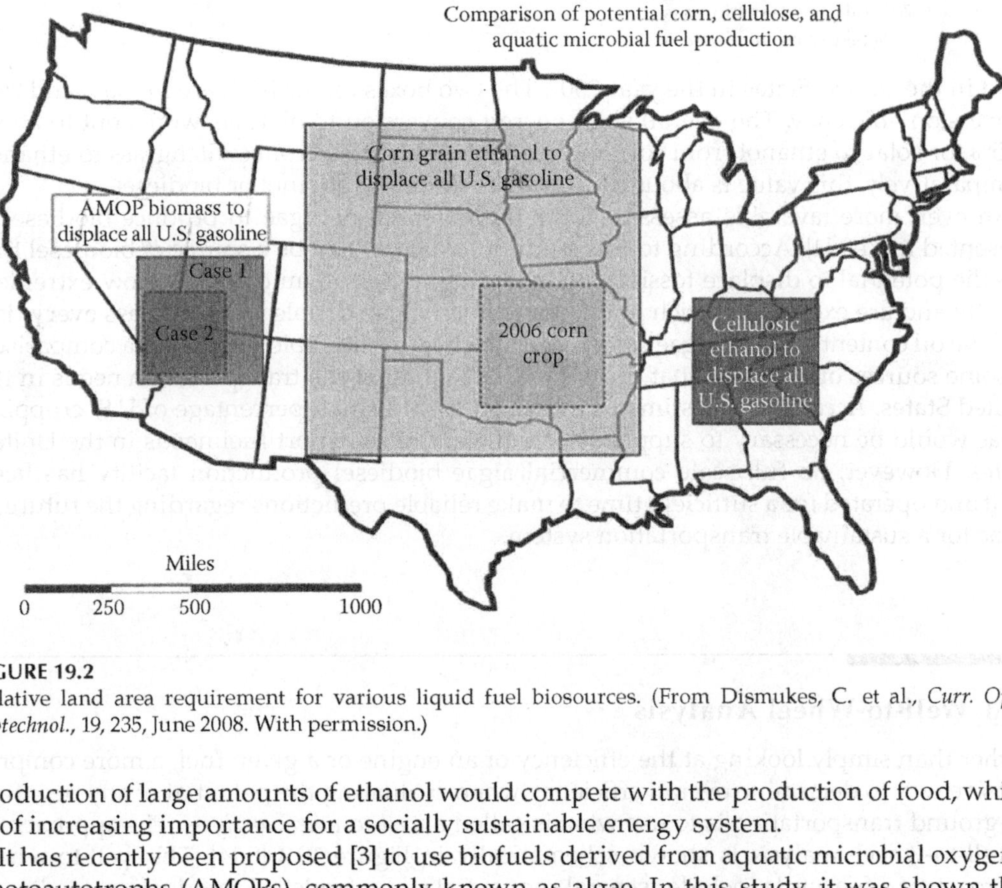

FIGURE 19.2
Relative land area requirement for various liquid fuel biosources. (From Dismukes, C. et al., *Curr. Opin. Biotechnol.*, 19, 235, June 2008. With permission.)

production of large amounts of ethanol would compete with the production of food, which is of increasing importance for a socially sustainable energy system.

It has recently been proposed [3] to use biofuels derived from aquatic microbial oxygenic photoautotrophs (AMOPs), commonly known as algae. In this study, it was shown that AMOPs are inherently more efficient solar collectors, use less or no land, can be converted to liquid fuels using simpler technologies than cellulose, and offer secondary uses that fossil fuels do not provide. AMOPs have a 6- to 12-fold energy advantage over terrestrial plants because of their inherently higher solar energy conversion efficiency, which is claimed to be between 3% and 9%. Figure 19.2 compares the area needed for three different biomass sources. The data are for corn grain, switchgrass, mixed prairie grasses, and AMOPs. Each box superimposed on the map of the United States represents the area needed to produce a sufficient amount of biomass to generate enough liquid fuel to displace all the gasoline

TABLE 19.2

Comparison of Some Sources of Biodiesel

Crop	Oil Yield (L/ha)	Land Area Needed (Mha)[a]	Percentage of Existing U.S. Cropping Area[a]
Corn	172	1540	846
Soybean	446	594	326
Canola	1,190	223	122
Oil palm	5,950	45	24
Microalgae[b]	136,900	2	1.1
Microalgae[c]	58,700	4.5	2.5

Source: Abstracted from Christi, Y., *Biotechnol. Adv.*, 25, 294, 2007.

[a] For meeting 50% of all transport fuel needs of the United States.

[b] 70% oil (by wt) in biomass.

[c] 30% oil (by wt) in biomass.

used in the United States in the year 2007. The two boxes for AMOPs are for 30% and 70% conversion efficiency. The overall solar energy conversion to biofuels works out to about 0.05% for solar to ethanol from corn grain and roughly 0.5% from switchgrass to ethanol. Comparatively, this value is about 0.5%–1% for AMOPs to ethanol or biodiesel.

An even more favorable assessment for the potential of algae to produce biodiesel is presented in Ref. [4]. According to this study, microalgae may be a source of biodiesel that has the potential to displace fossil fuel. According to Ref. [4], microalgae grow extremely rapidly and are exceedingly rich in oil. Some microalgae double their biomass every day, and the oil content of microalgae can exceed 80% by weight. Table 19.2 shows a comparison of some sources of biodiesel that could meet 50% of all of the transportation needs in the United States. According to estimates in Ref. [4], only a small percentage of U.S. cropping areas would be necessary to supply 50% of the entire transport fuel needs in the United States. However, no full-scale commercial algae biodiesel production facility has been built and operated for a sufficient time to make reliable predictions regarding the future of algae for a sustainable transportation system.

19.3 Well-to-Wheel Analysis

Rather than simply looking at the efficiency of an engine or a given fuel, a more comprehensive way to determine overall efficiency when evaluating the potential of any new fuel for ground transportation is to use what is called a *well-to-wheel analysis*. The approach to a well-to-wheel analysis is shown schematically in Figure 19.3 [5,14]. The well-to-wheel approach of a fuel cycle includes several sequential steps: feedstock production; feedstock transportation and storage; fuel production; transportation, storage, and distribution (T&S&D) of fuel; and finally vehicle operation. This approach is essential for a fair comparison of different options because each step entails losses. For example, whereas a fuel cell has a much higher efficiency than an internal combustion (IC) engine, for its operation, it depends on a supply of hydrogen, which must be produced by several steps from other sources. Moreover, there is no infrastructure for transporting and storing hydrogen, and this step in the overall well-to-wheel analysis has large losses and contributes to much larger energy requirements compared to a gasoline or electrically driven vehicle.

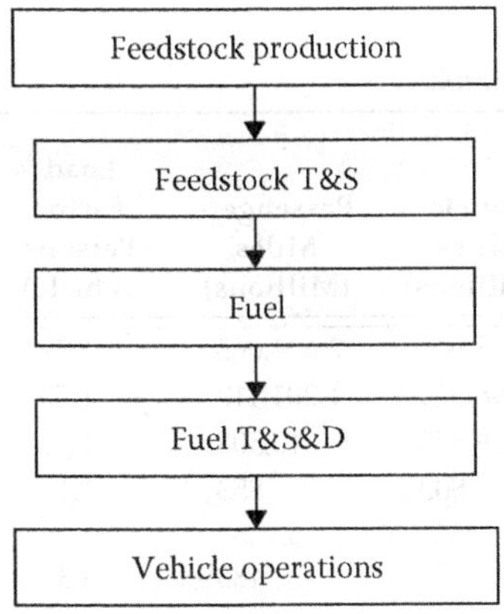

FIGURE 19.3
Steps in a well-to-wheel analysis for ground transportation vehicles.

19.4 Mass Transportation

An opinion held widely among state governments and environmentalists is that the mass transportation would greatly reduce the total energy consumption for the transportation sector. However, as shown in Table 19.3, based upon data collected by the U.S. Transportation Department in this country, the energy intensity of intercity rail and transit buses, that is, the energy spent per passenger-mile traveled, is virtually the same as the energy intensity of automobiles with current use. This is due to the urban sprawl in major cities that makes it difficult to reach outlying areas by a mass transport network. In other words, unless there are incentives for mass transport or disincentives to use the automobile, thereby achieving higher mass transport load factors—that is, more passengers per mile—on transit buses and light rail, the availability of mass transport systems will not materially change the overall energy use by the transportation sector. Moreover, installing new light rail is very capital-intensive and may not always be worth the energy and/or the money invested in its construction. Although passenger mass transit does not appear to offer large untapped opportunities, using ship or rail instead of air for shipment of freight and commercial goods is a source of enormous fuel reduction potential.

19.5 Hybrid Electric Vehicles

Another obvious approach to ameliorating the expected increase in price and lack of availability of petroleum fuel is to increase the mileage of the vehicles. This can be achieved by improving the efficiency of the IC engine, for instance, by using advanced diesel engines that have a higher compression ratio than spark ignition (SI) engines, or by using hybrid electric vehicles (HEVs). Increasing the efficiencies of IC or diesel engines is a highly specialized topic and is not discussed here. However, HEVs offer a near-term option for utilizing improved battery technology.

An HEV is powered by the combination of a battery pack and electric motor—like that of an electric vehicle—and a power generation unit (PGU), which is normally an IC or

TABLE 19.3

Passenger Travel and Energy Use, 2002

	No. of Vehicles (Thousands)	Vehicle-Miles (Millions)	Passenger-Miles (Millions)	Load Factor (Persons/ Vehicle)	Energy Intensities (Btu/ Vehicle-Mile)	(Btu/ Passenger-Mile)	Energy Use (Trillion Btu)
Automobiles	135,920.7	1,658,640	2,604,065	1.57	5,623	3,581	9,325.9
Personal trucks	65,268.2	698,324	1,201,117	1.72	6,978	4,057	4,872.7
Motorcycles	5,004.2	9,553	10,508	1.22	2,502	2,274	23.9
Demand response	34.7	803	853	1.1	14,449	13,642	11.6
Vanpool	6.0	77	483	6.3	8,568	1,362	0.7
Buses	[a]	[a]	[a]	[a]	[a]	[a]	191.6
Transit	76.8	2,425	22,029	9.1	37,492	4,127	90.0
Intercity[b]	[a]	[a]	[a]	[a]	[a]	[a]	29.2
School[b]	617.1	[a]	[a]	[a]	[a]	[a]	71.5
Air	[a]	[a]	[a]	[a]	[a]	[a]	2,28.9
Certified route[c]	[a]	5,841	559,374	95.8	354,631	3,703	2071.4
General aviation	211.2	[a]		[a]	[a]	[a]	141.5
Recreation boats	12,409.7	[a]	[a]	[a]	[a]	[a]	187.2
Rail	18.2	1,345	29,913	22.2	74,944	3,370	100.8
Intercity[d]	0.4	379	5,314	14.0	67,810	4,830	25.7
Transit[e]	8.5	682	15,095	22.1	72,287	3,268	49.3
Commuter	5.3	284	9,504	33.5	90,845	2,714	25.8

Source: Davis, S. and Diegel, S., *Transportation Energy Data Book,* Oak Ridge National Laboratory/U.S. Department of Energy, Oak Ridge, TN, 2004.

[a] Data are not available.

[b] Energy use is estimated.

[c] Includes domestic scheduled service and half of international scheduled service. These energy intensities may be inflated because all energy use is attributed to passengers; cargo use not taken into account.

[d] Amtrak only.

[e] Light and heavy rail.

diesel engine. Unlike electric vehicles, however, HEV batteries can be recharged by an onboard PGU, which can be fueled by existing fuel infrastructure.

HEVs can be configured in a parallel or a series design. The parallel design enables the HEV to be powered by both the PGU and the motor, either simultaneously or separately. The series design uses the PGU to generate electricity, which recharges the HEV battery pack and produces power via an electric motor. The key element of either design is that the battery pack, as well as the PGU, can be much smaller than those of a typical electric vehicle or a vehicle powered by an IC engine because the IC engine can be operated at its maximum efficiency nearly all the time.

Currently available HEVs, such as the Toyota Prius, use a parallel configuration, as shown in Figure 19.4. A parallel HEV has two propulsion paths: one from the PGU and one from the motor, while computer chips control the output of each. A parallel-configuration HEV has a direct mechanical connection between the PGU and the wheels, as in a conventional vehicle (CV), but also has an electric motor that can drive the wheels. For example, a parallel vehicle could use the electric motor for highway cruising and the power from

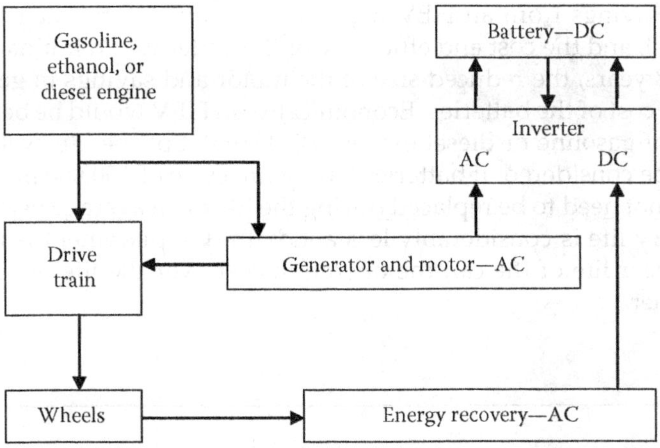

FIGURE 19.4
Schematic of a parallel-configuration hybrid electric vehicle.

(a)

(b)

FIGURE 19.5
Fuel-efficient cars (a) 2010 Toyota Prius: Toyota's third-generation Prius is a *parallel* hybrid; the 80 hp (60 kW) electric motor and gas engine work together to produce an average 50 mpg. (b) 2010 Ford Fusion: The Fusion can travel up to 47 mph (75 kph) on electric power alone. The motor is powered by a nickel metal hydride battery pack, and the car's engine is a 2.5 L Atkinson Cycle I-4 gasoline engine. The Fusion gets 41 mpg in the city and 36 mpg highway. (Courtesy of *Solar Today*, November/December 2009.)

IC engine for accelerating. The power produced by the PGU also drives the generator, which in turn can charge the battery as needed. The system to transfer electricity from the generator to the battery pack is exactly like that of an electric vehicle, with alternating current converted to DC by the inverter. HEV parallel designs also use a regenerative braking feature that converts energy stored in the inertia of the moving vehicle into electric power during deceleration (see Figure 19.5).

Some benefits of a parallel configuration versus a series configuration include the following:

- The vehicle has more power because both the engine and the motor provide power simultaneously.
- Parallel HEVs do not need a separate generator.
- Power is directly coupled to the road, thus operating the vehicle more efficiently.

Energy and cost savings from an HEV depend on many factors, such as the overall car design, cost of fuel, and the cost and efficiency of the batteries. Preliminary estimates indicate that over 5–8 years, the reduced size of the motor and savings in gasoline could pay for the additional cost of the batteries. Economically, an HEV would be beneficial when the price per gallon of gasoline or diesel exceeds $3.00/gal, but the life cycle of the batteries will also have to be considered. If batteries have a life cycle of 150,000 miles, as claimed by Toyota, they will not need to be replaced during the life of an average vehicle. On the other hand, if the battery life is considerably less and battery replacement is necessary during the expected 10-year life of the car, the operation cost over the life of an HEV would be considerably higher.

19.6 Plug-In Hybrid Electric Vehicles

Given the current state of technology, probably the most promising near-term solution to the ground transportation crisis is the use of plug-in HEVs (PHEVs). PHEVs have the potential of making the leap to the mainstream consumer market because they require neither a new technology nor a new distribution infrastructure. Like hybrids, which are already widely available, PHEVs have a battery and an IC engine for power, but the difference is that a PHEV has a larger battery capacity and a plug-in charger with which the battery can be recharged whenever the car is parked near a 110 or 220 V outlet. A so-called series PHEV40 can travel the first 40 miles on grid-supplied electric power when fully charged. When that charge is depleted, the gas or diesel motor kicks in, and the vehicle operates like a conventional hybrid. Because most commuter trips are less than 40 miles, it is estimated that a PHEV could reduce gasoline usage by 50% or more for many U.S. drivers. Moreover, using electric energy is cheaper and cleaner than using gasoline in automobile-type ground transportation. Figure 19.6 shows a Prius with

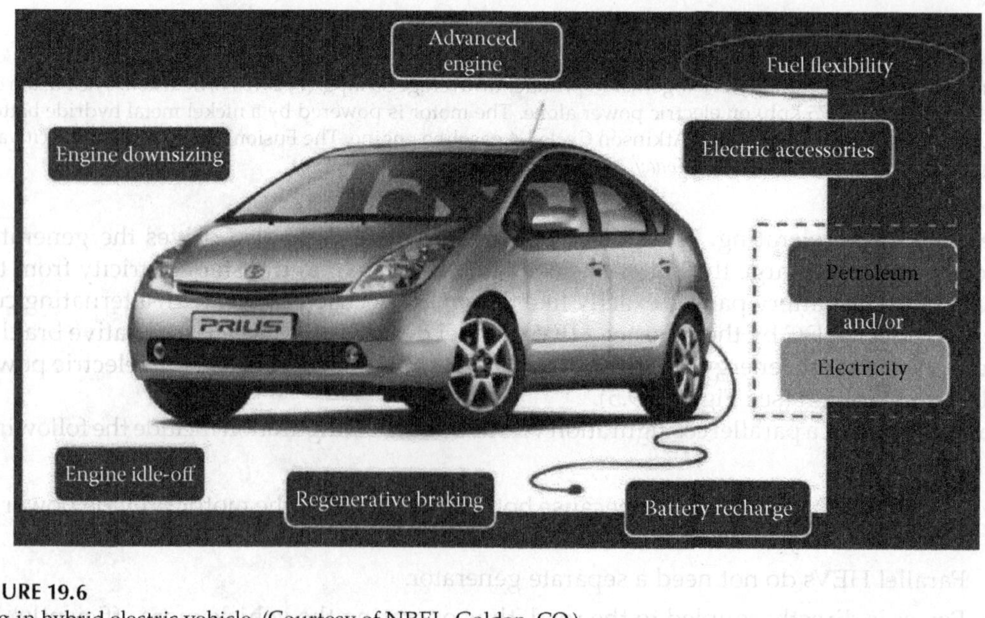

FIGURE 19.6
Plug-in hybrid electric vehicle. (Courtesy of NREL, Golden, CO.)

plug-in potential added. These early plug-in additions are available as aftermarket conversions, but because they are not in mass production, their cost is high, and Toyota does not honor their warranty when plug-in features are added aftermarket. More recently, all major original equipment manufacturers (OEMs) have introduced plug-in vehicles with various battery storage capacities.

An important feature of PHEVs is that their batteries can be charged at night when utilities have excess power available. Utilities have taken notice of the potential energy charging and storage capabilities of PHEVs because off-peak charging would help utilities to use low-cost baseload generation more fully. Furthermore, more advanced vehicle-to-grid concepts would allow utilities to buy back energy from the batteries of vehicle owners during peak demand periods and thus make a fleet of PHEVs into a large distributed storage-generation network. The arrangement would also enable renewable energy storage by charging PHEVs using solar- or wind-generated excess capacity. As mentioned in the previous section on batteries, lithium-ion and lithium-polymer batteries have the potential to store large charges in a lightweight package, which would make the HEVs even more attractive than the current technology, which uses nickel metal hydride (NiMH) batteries.

The efficiency of a PHEV depends on the number of miles the vehicle travels on liquid fuel and electricity, respectively, as well as on the efficiency of the prime movers according to the following equation:

$$\eta_{PHEV} = \frac{\text{Energy to wheels}}{\text{Energy from primary source}} = f_1\eta_1\eta_2 + f_2\eta_3\eta_4 \tag{19.1}$$

where

η_1 is the efficiency from the primary energy source to electricity
η_2 is the efficiency of transmitting energy to the wheels
f_1 is the fraction of energy supplied by electricity
f_2 is the fraction of energy supplied by fuel = $(1 - f_1)$
η_3 is the well-to-wheel efficiency
η_4 is the tank-to-wheel efficiency

PHEVs can be designed with different all-electric ranges. The distance, in miles, that a PHEV can travel on batteries alone is denoted by a number after PHEV. Thus, a PHEV20 can travel 20 miles on fully charged batteries without using the gasoline engine. According to a study by the Electric Power Research Institute (EPRI) [7], on average, one-third of the annual mileage of a PHEV20 is supplied by electricity and two-thirds by gasoline. The percentage depends, of course, on the vehicle design and the capacity of the batteries on the vehicle. A PHEV60 can travel 60 miles on batteries alone, and the percentage of electric miles will be greater as will the battery capacity and weight. The tank-to-wheel (more appropriately battery-to-wheel) efficiency for a battery all-electric vehicle according to EPRI [7] is 0.82.

Given the potentials for plug-in hybrid vehicles, the EPRI [7] conducted a large-scale analysis of the cost, the battery requirements, and the economic competitiveness of plug-in vehicles today and within the near-term future. Table 19.4 presents the net present value of life cycle costs over 10 years for a midsized IC engine vehicle such as the Ford Focus [IC], HEVs such as the Prius [HEV], and a future PHEV20 plug-in electric vehicle. The battery module cost in dollars per kWh is the cost at which the total life cycle costs of all three vehicles would be the same. According to projections for the production of NiMH battery

TABLE 19.4

Net Value of Life Cycle Costs over 117,000 Miles/10 Years for Conventional Gasoline (IC), HEV and PHEV20 Midsize Vehicles with Gasoline Costs at $1.75/gal

Vehicle Type	IC	HEV	PHEV 20
Battery unit cost ($/kWh)		385[a]	316[a]
Incremental vehicle cost ($)		547	224
Battery pack cost ($)	60	3,047	3,893
Fuel costs ($)	5,401	3,725	2,787
Maintenance costs ($)	5,445	4,733	4,044
Battery salvage costs ($)		54	43
Total life cycle costs ($)	10,906	10,904	10,905

Source: Extracted from EPRI, Advanced batteries for electric drive vehicles: A technology and cost-effectiveness assessment for battery electric vehicles, power assist hybrid electric vehicles and plug-in hybrid electric vehicles, EPRI Tech. Report 1009299, EPRI, Palo Alto, CA, 2004.

[a] Battery module price at which life cycle parity with CV occurs.

modules, a production volume of about 10,000 units per year would achieve the necessary cost reduction to make both an HEV and a PHEV20 economically competitive. The EPRI analysis was conducted in 2004 and is, therefore, extremely conservative because it assumed a gasoline cost of $1.75/gal. A reevaluation of the analysis based upon a gasoline cost of $2.50/gal showed that the permitted battery price at which the net present value of conventional IC vehicles and battery vehicles are equal for battery module costs $1135 for an HEV and $1648 for a PHEV20, respectively.

Table 19.5 shows the electric and plug-in hybrid vehicle battery requirements (module basis for the cost estimates in Table 19.4), and it is apparent that, even with the currently available NiMH batteries, the cost of owning and operating HEVs and PHEVs is competitive with that of an average IC engine vehicle.

TABLE 19.5

NiMH Battery Cost Assumptions for Table 19.4

	ARB 2000 Report for BEVs		
Assumption	2003	Volume	EPRI Assumptions
Module cost ($/kWh)[a]	300	235	Varied[b]
Added cost for pack ($/kWh)	40	20	680 + 13
Multiplier for manufacturer and dealer markup	1.15	1.15	Varies[c]
Battery life assumptions (years)	6	10	10

[a] Equivalent module costs for an HEV 0 battery is $480 for 2003 and $384 for volume. Equivalent module costs for a PHEV 20 battery is $376 for 2003 and $301 for volume. HEV 0 and PHEV 20 batteries have a higher power-to-energy ratio and are more costly. These figures are based on data shown in the biomass chapter of this handbook.

[b] Battery module costs were varied in this analysis to determine the effect of battery module cost on life cycle cost.

[c] Manufacturer and dealer mark-up for HEV 0 battery modules estimated at $800, PHEV 20 battery modules $850, pack hardware mark-up assumed to be 1.5. Method documented in 2001 EPRI HEV report.

A cautionary note in the expectation of future ground transportation systems is the reduction in the rate of petroleum consumption that can be expected as HEVs and PHEVs is introduced into the fleet [8]. In these estimates, a rate of new vehicle sales of 7% of the fleet per year, a retirement rate of 5% per year resulting in a net increase in total vehicles of 2% per year was assumed. This increase is in accordance with previous increase rates between 1966 and 2003. Based upon the existing mileage for IC engine, HEVs, and PHEVs, it was estimated that *even if all new cars were HEVs or PHEVs*, after 10 years, the annual gasoline savings as a percentage of the gasoline usage by an all-gasoline fleet in the same year would only be about 30% for the HEVs and 38% for the PHEVs. These relatively small reductions in the gasoline use are due to the fact that despite introduction of more efficient vehicles, it takes time to replace the existing fleet of cars, and the positive effects will not be realized for some years.

19.7 Advanced Ground Transportation with Biomass Fuel

The previous section analyzed the potential of using batteries combined with traditional engines to reduce the petroleum consumption in the transportation system. However, the scenario used for this analysis can also be extended to determine the combination of PHEVs with biofuels, particularly ethanol made from corn or cellulosic biomass. No similar analysis for using diesel from algae is available at this time.

It is important to note that in October 2010, the U.S. Environmental Protection Agency granted a partial waiver for the use of E15 (15% ethanol and 85% gasoline) for use in light-duty motor vehicles in cars newer than model year 2007. In January 2011, a second waiver was granted that allowed for use of E15 in light-duty vehicles manufactured in 2001–2006. The decision to grant the waivers was the result of testing performed by the Department of Energy (DOE) and information regarding the potential effect of 15 on vehicle emissions [9]. According to the Renewable Fuels Association, the E15 waivers pertain to over 62% of vehicles currently on the roads in the United States. If all passenger cars and pickup trucks were to switch to E15, this would represent 17.5 gal of ethanol use annually [10]. There remain some practical infrastructure barriers, as the current fuel pumps at gasoline retail stations do not support this higher blend. However, as consumers begin to demand more options in biofuels, it is likely that E15 will be made more available in the foreseeable future.

A scenario for a sustainable transportation system based on fuel from biomass has been presented in Ref. [11]. In this analysis, the following four vehicle types combined with various fuel options have been calculated. The preferred mixture in an economy based largely on ethanol (E85) would be 85% ethanol and 15% gasoline that could be used in Flex Fuel automobiles. Currently, the United States is using E10, a mixture of 90% gasoline and 10% ethanol, with ethanol produced from corn. The fuel types used in this analysis are gasoline only, E10 with ethanol made from either corn or cellulosic materials, and E85 with ethanol from either corn or cellulosic materials. The four vehicle combinations are a convention SI engine, an HEV similar to the Toyota Prius, a PHEV20, and a PHEV30. The analysis was based upon an average 2009 performance on the U.S. light vehicle fleet at approximately 20 miles per gallon (mpg), for an HEV at 45 mpg, and for a PHEV20 at 65 mpg, according to Ref. [7]. For ethanol-/gasoline-blended fuels, it was assumed that the gasoline and ethanol are utilized with the same efficiency. That is, the mileage per unit of fuel energy is the same for gasoline and ethanol.

Based on the earlier assumption, the following parameters were calculated:

1. The miles per gallon of fuel, including the gasoline used to make ethanols (mpg).
2. The petroleum required to drive a particular distance for a case vehicle as a percentage of the petroleum required to drive the same distance by a gasoline-fueled SI vehicle.
3. The carbon dioxide emission rate for case vehicles as a percentage of that for SI gasoline only.

Using the earlier assumptions, one can calculate the mpg of fuel as

$$MF_{ij} = \text{miles/gal gas)} \times \left[\text{(gal gas/gal fuel)} + (1 - \text{gal gas/gal fuel)} \times \frac{\text{(energy, LHV/gal ethanol)}}{\text{(LHV/gal gasoline)}} \right]$$

$$= MGO_i \times [FG_j + (1 - FG_j) \times (\text{LHV ratio})] \tag{19.2}$$

where

MF_{ij} is the mpg of fuel for vehicles type i and fuel type j
MGO_i is the mpg gasoline-only for vehicle i (see Table 19.4)
FG_j is the volume fraction of gasoline in fuel type
$j(1 - FG_j)$ is the volume fraction of ethanol in fuel type j
LHV ratio is the ratio of lower heating values; LHV ratio = (LHV/gal ethanol)/(LHV/gal gasoline) = 0.6625

The index i indicates the vehicle type as shown in Table 19.4, and the index j denotes the volume fraction of gasoline in an ethanol–gasoline blend as follows: for gasoline only, $FG_1 = 1$; for E10 (i.e., 10 vol% ethanol and 90 vol% gasoline), $FG_2 = 0.90$; and for E85 (85 vol% ethanol, 15 vol% gasoline), $FG_3 = 0.15$. Other ethanol concentrations could be used (Figure 19.7).

The mpg of gasoline in the fuel, including the petroleum-based fuels used in the making of the ethanol in the fuel (by counting the energy of all petroleum-based fuels as gasoline), is given by

$$MG_{ijk} = \frac{MF_{ij}}{[FG_j + (1 - FG_j) \times (0.6625) \times (\text{MJ gasoline used in making 1 gal ethanol})/\text{MJ/gal ethanol}]}$$

$$= \frac{MF_{ij}}{FG_j + (1 - FG_j) \times (0.6625) \times Rk} \tag{19.3}$$

where Rk denotes the gallons of gasoline used to make 1 gal of ethanol. For corn-based ethanol, $k = 1$, $R1 = 0.06$, while for cellulosic-based ethanol, $k = 2$, $R2 = 0.08$, according to Ref. [12].

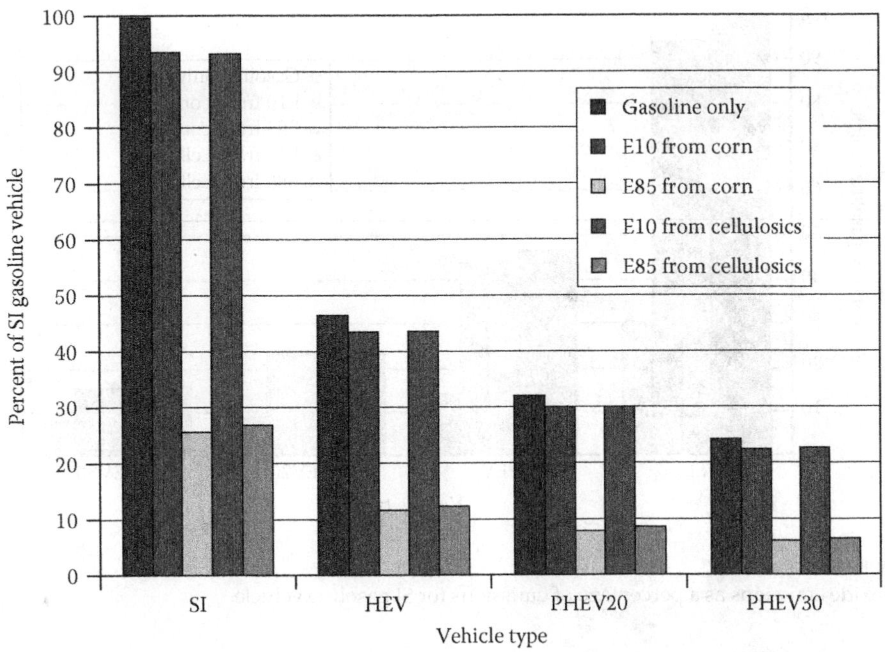

FIGURE 19.7
Petroleum requirement as a percentage of that for SI gasoline vehicle. (Calculated by Kreith, F. and West, R.E., *ASME J. Energy Res. Technol.*, 128(9), 236, September 2006.)

19.7.1 Petroleum Requirement

The petroleum required to produce an ethanol–gasoline blend, including the petroleum used to make the ethanol, is expressed as a percentage of the petroleum required for the same miles traveled by the same vehicle type using gasoline only (Figure 19.7). For a gasoline-only-fueled vehicle of any type, this percentage is 100%.

$$\text{Petroleum requirement (\%)} = 100 \times \frac{(MGO_i)}{(MG_{ijk})} \tag{19.4}$$

19.7.2 Carbon Dioxide Emissions

The CO_2 emissions, including those from making the ethanol and generating the electricity used from the grid by the vehicle, are expressed as a percentage of the emissions produced by the same type of vehicle fueled by gasoline only traveling the same number of miles (Figure 19.8).

$$\begin{aligned} &= 100 \times [FE_m \times (\text{kWh/miles}) \times (\text{g-carbon/kWh})] + (1/\text{MGO}_i) \times (1 - FE_m) \\ &\quad \times [FG_j \times (\text{g-carbon/MJ gasoline}) \times (\text{MJ/gal gasoline})] + (1 - FG_j) \\ &\quad \times \frac{[(\text{MJ/gal ethanol}) \times (\text{g-carbon/MJ ethanol})]}{[(\text{g-carbon/MJ gasoline used}) \times (\text{MJ/gal gasoline})/(\text{MGO}_1)]} \end{aligned} \tag{19.5}$$

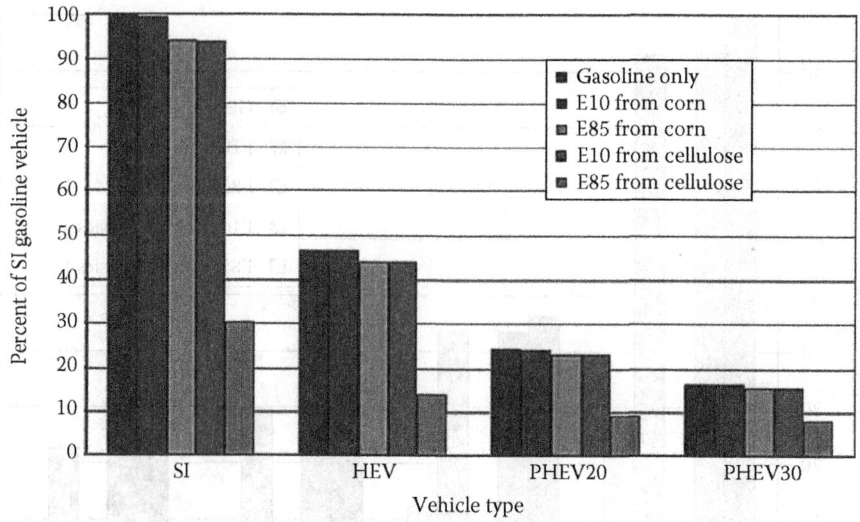

FIGURE 19.8
Carbon dioxide emissions as a percentage of emissions for SI gasoline vehicle.

where FE_m is the fraction of the miles driven by electricity from the grid for a plug-in hybrid vehicle. For $m = 0$, any non-PHEV, $FE_0 = 0$; for a PHEV20, $m = 1$, and $FE_1 = 0.327$ [7]; and for a PHEV30, $m = 2$, and $FE_2 = 0.50$ estimated by analogy with PHEV20. According to Ref. [7], the kWh/mile from the grid = 0.2853, the g-carbon emitted/kWh = 157 (146 average for all electricity generation [13] divided by 0.93, the average transmission efficiency), the g-carbon emitted/MJ gasoline = 94 [7], the MJ/gal gasoline = 121, the MJ/gal ethanol = 80.2, and CE_k the g-carbon emitted/MJ ethanol. For $k = 1$ (corn), $CE_1 = 87$, and for $k = 2$, cellulosics, $CE_2 = 11$.

So

CO_2 production as % of that for a gasoline-only vehicle

$$= \frac{100 \times \{FE_m \times (0.2853) \times (157) + (1/MF_{ij}) \times (1 - FE_m) \times [FG_j \times (94) \times 121 + (1 - FG_j) \times CE_k \times (80.2)]\}}{(94) \times (121)/21} \quad (19.6)$$

The final question to be asked in the utilization of PHEVs is whether or not the existing electricity system of the United States could handle the charging of the batteries in a PHEV-based ground transportation system. This question has recently been answered by an analysis in Ref. [14]. This analysis clearly showed that with a normal commuting scenario, which was based upon statistical information from a major city, if charging occurred during off-peak hours, no additional generational capacity or transmission requirements would be needed to charge a significant portion of the automotive fleet with PHEVs in the system, as illustrated in Figure 19.9. The analysis also showed that the off-peak charging scenario would also add to the profit of the electric utilities.

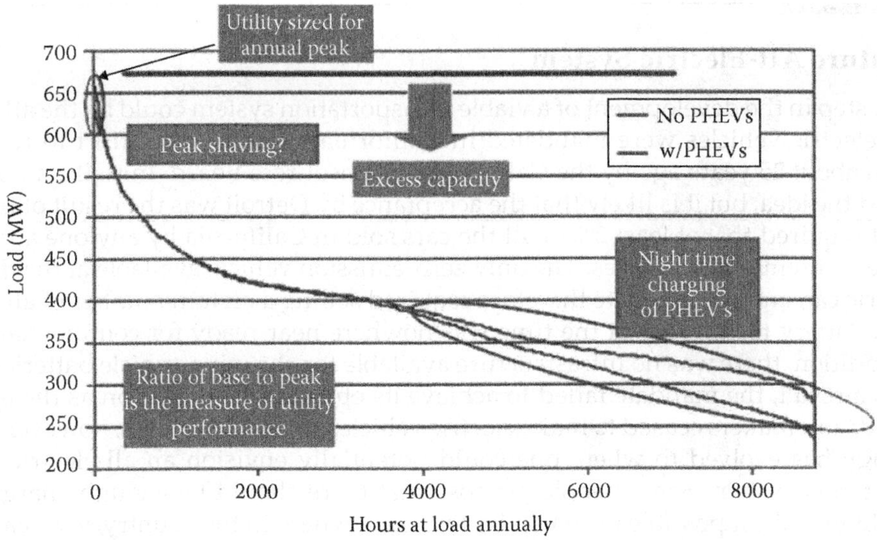

FIGURE 19.9
PHEV impact on utilities' load profile. (Courtesy of NREL, Golden, CO.)

An unexpected result of the analysis was that, although the amount of CO_2 emitted by the electric utility will increase from PHEV charging, if this is compared to the corresponding reduction in tailpipe emission to assess the overall environmental impact, the generation/PHEV transportation system would substantially decrease CO_2 emission even if the current mix of coal, nuclear, and renewable generation facilities were unchanged. This result can be explained, however, because the average efficiency of the electric power system is on the order of 43%, whereas the average efficiency of IC engines is only on the order of 22%. Consequently, the net emission of CO_2 would be reduced by a hybrid system consisting of PHEVs and electric charging during off-peak hours. Moreover, the utility generation profile would be evened out, and this would also contribute to reducing CO_2 emissions, as well as the cost of producing electricity.

In summary, in a PHEV transportation system, if charging of batteries is limited to off-peak hours, this hybrid arrangement can

- Reduce the amount of petroleum consumed by the transportation sector
- Reduce the cost of driving
- Reduce CO_2 emission
- Improve the load profile of electric utilities

Although no quantitative analyses are as yet available, it is believed that the availability of electric storage in the batteries of a fleet of PHEVs could also be used to reduce the peak demand on electric utilities by utilizing the vehicle's batteries as a distributed storage system. Details of such an arrangement would have to be worked out by differential charges, incentives, and taxation arranged between utilities and owners of PHEVs.

19.8 Future All-Electric System

The next step in the development of a viable transportation system could be the all-electric car. All-electric vehicles were mandated in California as part of an effort to reduce air pollution about 25 years ago by the California Air Resources Board. Initially, automakers embraced the idea, but it is likely that the acceptance by Detroit was the result of the mandate that required that at least 2% of all the cars sold in California by any one automaker had to be zero-emission vehicles. The only zero-emission vehicle available at the time was the electric car, and the mandate therefore required selling a certain number of all-electric vehicles. Battery technology at the time was nowhere near ready for commercialization, and in addition, there was no infrastructure available for charging vehicle batteries on the road. As a result, the mandate failed to achieve its objective, and as soon as the mandate was lifted, automakers ceased to make electric vehicles. In the meantime, however, battery technology has evolved to where one could potentially envision an all-electric ground transportation system. Some people propose that there should be enough charging stations built to make it possible to charge batteries anywhere in the country, whereas others propose that there should be, instead of gas stations, battery exchange stations that would simply replace batteries as they reach the end of their charge. At present, batteries take too long to be charged during a trip, and it may be necessary to combine the two ideas to evolve an all-battery transportation system sometime in the near future.

City pollution could be significantly reduced with the use of electric cars due to their zero tailpipe emissions. It is postulated that carbon dioxide emissions could be reduced by up to 40%, depending upon the country's current energy mix. The information from Electric Drive Transportation Association (EDTA) as of January 2015 (Figure 19.10) shows how the market for PHEVs and EVs are growing over time [15].

At this time, it is not possible to assess which electrified powertrain will achieve the bigger market domination over time as the price of fuel, batteries, and other infrastructure all play into that equation. The earlier summary is a snapshot at the time of preparing this text and is a continuous state of flux. The reader is encouraged to follow developments in the current literature included in the bibliography. The EDTA is an excellent source for timely developments regarding this technology.

19.9 Hydrogen for Transportation

Hydrogen was touted as a potential transportation fuel after former President George W. Bush said in his 2001 inaugural address that "a child born today will be driving a pollution-free vehicle... powered by hydrogen." This appeared to be welcome news. However, to analyze whether or not hydrogen is a sustainable technology for transportation, one must take into account all the steps necessary to make the hydrogen from a primary fuel source, get it into the fuel tank, and then power the wheels via a prime mover and a drive train. In other words, one must perform a well-to-wheel analysis, as shown in Figure 19.3. There is a loss in each step, and to obtain the overall efficiency, one must multiple the efficiencies of all the steps. Using natural gas as the primary energy source, a well-to-wheel analysis [16] showed that a hybrid SI car would have a wheel-to-energy

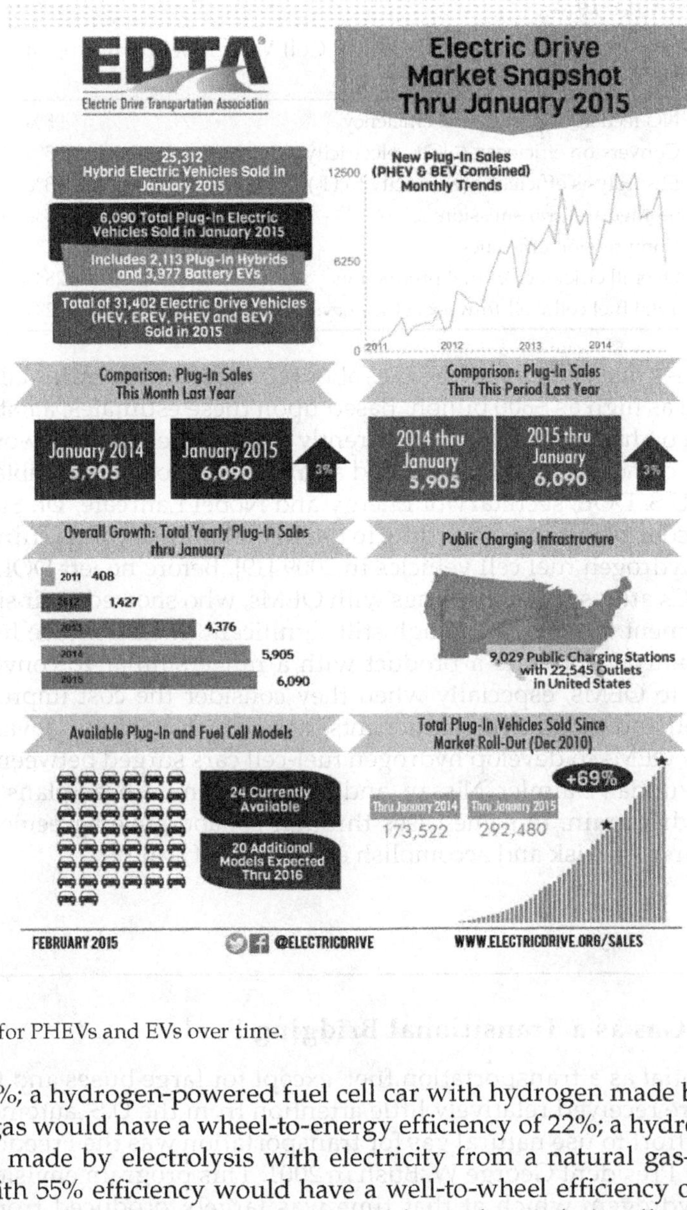

FIGURE 19.10
Growth of market for PHEVs and EVs over time.

efficiency of 32%; a hydrogen-powered fuel cell car with hydrogen made by steam reforming of natural gas would have a wheel-to-energy efficiency of 22%; a hydrogen fuel cell car with hydrogen made by electrolysis with electricity from a natural gas–combined cycle power plant with 55% efficiency would have a well-to-wheel efficiency of 12%, as shown in Table 19.6; and if the hydrogen were produced from photovoltaic cells, the well-to-wheel efficiency of the automobiles would be less than 5%. The estimates in Table 19.6 assume a fuel cell stack efficiency at a peak load of 44.5%, a part load efficiency factor of 1.1, and a transmission efficiency of 90%. For details, see [17].

In addition to the low overall efficiency of a hydrogen transportation system, as well as the high cost compared to other options, it should also be noted that before a hydrogen transportation system could be put into practice, an infrastructure for the distribution and storage of the hydrogen would have to be constructed. An extensive study of the comparative costs of fuel distribution systems conducted at Argonne National Laboratory in 2001 estimated that for a market penetration by the year 2030, 40% of hydrogen vehicles with a

TABLE 19.6

Well-to-Wheel Efficiency of Fuel Cell Vehicle with Hydrogen Produced by Electrolysis

NG feedstock production efficiency	95%
Conversion efficiency (NG to electricity)	55%
Electrolysis efficiency (electricity to H_2)	63%
Storage and transmission	97%
Compression efficiency	87%
Overall efficiency of fuel production	28%
Total fuel cell well-to-wheel efficiency: (0.28 × 445 × 1.1 × 0.9)	12%

mileage of 2.5 times that of average CVs (i.e., about 55 mpg) would minimally cost $320 billion, but could be as high as $600 billion. Based upon these estimates, a national transportation plan based on hydrogen with any currently available technology would be wasteful and inefficient and should not be considered as a pathway to a sustainable energy future [16]. The former U.S. DOE, secretary of Energy and Nobel Laureate, Dr. Steven Chu, concurred with this conclusion and, according to the *New York Times*, cutoff funds for development related to hydrogen fuel cell vehicles in 2009 [19]. Before he left DOE, Chu reversed his support for FCs after several meetings with OEMs, who showed their significant progress and commitment to FCVs. Although still significant infrastructure hurdles exist for hydrogen, the ability to provide a product with a range similar to conventional ICE is very compelling to OEMs, especially when they consider the cost improvement potential of the fuel cell and associated components including new and innovative packaging. Commitments by OEMs to develop hydrogen fuel-cell cars surged between 2012 and 2014. BMW, Toyota, Hyundai, Daimler, Nissan, and Honda all announced plans to commercialize the fuel cell drivetrain, in some cases through collaborative agreements in order to spread early technology risk and accomplish economies of scale.

19.10 Natural Gas as a Transitional Bridging Fuel

Despite its potential as a transportation fuel, except for large buses and trucks, natural gas has heretofore received relatively little attention from the U.S. automobile industry. The only major effort to use natural gas for transportation was the Freedom CAR initiative proposed by President George W. Bush in 2001. This program envisioned replacing gasoline with hydrogen, which at that time was largely produced from natural gas. More than a billion dollars was provided for R&D as well as generous tax incentives. But as shown in Table 19.6, the efficiency of a natural gas/hydrogen vehicle based on a well-to-wheel analysis is less than that of hybrid SI/natural gas vehicle (NGV), and the construction of a hydrogen distribution system would be extremely expensive. Thus, the hydrogen via natural gas effort for transportation was a failure and was terminated in 2010. However, since 2009, new supplies of natural gas have become available in the United States, primarily as a result of fracking technology for natural gas extraction from oil shale deposits. This development has heightened awareness of the potential of natural gas as a bridging fuel for transportation for an eventual zero carbon future. Within this context, MIT has conducted an interdisciplinary investigation that addresses the question, "What is the role of natural gas in a carbon-constrained economy?" [20].

Natural gas is likely to find increased use in the transportation sector with compressed natural gas (CNG) playing an important role, particularly for high-mileage fleets. But the advantage of liquid fuel in transportation indicates that the chemical conversion of the gas into some form of liquid fuel may be a more desirable pathway for the future. It should be noted that a basic infrastructure for distributing natural gas exists, and if CNG were to be used to fuel automobiles, the only major addition at gas stations that have a natural gas outlet would be a compressor to increase the gas pressure above that in the natural gas automobile tank. However, the vast majority of natural gas supplies are delivered to markets by pipeline, and delivery costs typically represent a relatively large fraction of the total cost in the supply chain.

Natural gas vehicles have been in use for trucks and other large vehicles in many parts of the world for some time as shown in Table 19.7. With a starting price of $26,305, a 2013 natural gas Civic costs $8,100 more than the base gasoline model. Big trucks that burn 20,000–40,000 gal a year can easily make up that difference, but it takes far longer for regular consumers, who may use only 500 gal/year. Home fueling stations add $4000–$6000 to the cost.

Range is also a concern. The United States has 1100 natural gas fueling stations, but only about half are open to the public. A natural gas Civic can go around 200 miles on a tank. That's better than an electric car, which might go 100 miles on a charge. But it's less than the 300–350 miles a driver can go on a tank of gasoline in a regular Civic.

All those things could change, but GE is trying to develop a $500 home fueling station, and the federal government could encourage sales with tax credits, as it has done with its $7500 electric vehicle credit. Some states are already giving tax credits to CNG vehicle buyers, including West Virginia—which gives up to $7,500 for smaller vehicles and $20,000 for trucks—and Colorado, which gives up to $6,000.

As availability of natural gas wanes, a number of renewable sources for natural gas or biomethane are available. Biomethane can be produced from any organic matter. Nature produces it naturally in landfills, but it can also be produced, as discussed in Chapter 13, in anaerobic digesters or through pyrolysis from sewage, industrial, animal, or crop wastes, or from specific energy crops. The biomethane from landfills used in NGVs reduces

TABLE 19.7
Top 11 NGV Countries

Country	No. of NGVs
1. Pakistan	2,850,500
2. Iran	2,070,930
3. Argentina	1,927,007
4. Brazil	1,667,038
5. India	1,100,000
6. Italy	754,659
7. China	550,000
8. Colombia	340,000
9. Thailand	238,583
10. Ukraine	200,019
11. United States	112,000

Source: *Gas Vehicle Rep.*, 10(4), June 2011.

greenhouse gases by 90% according to the California Air Resources Board. After biomethane is produced, it can be injected into natural gas pipeline systems and sold to NGV station operators. A 5% or 10% blend of biomethane with natural gas would add to NGV's greenhouse gas potential. While many other alternative fuels are still in the R&D stage, NGVs are not in that category and are ready to go now.

References

1. Kreith, F. (1999) *Ground Transportation for the 21st Century*. National Conference of State Legislatures, Denver, CO; ASME Press, New York.
2. Kreith, F., West, R.E., and Isler, B. (2002) Legislative and technical perspectives for advanced ground transportation system. *Transportation Quarterly* 96(1), 51–73, Winter 2002.
3. Dismukes, C. et al. (June 2008) Aquatic phototrophs: Efficient alternatives to land-based crops for biofuels. *Current Opinions in Biotechnology* 19, 235–240.
4. Christi, Y. (2007) Biodiesel from microalgae. *Biotechnology Advances* 25, 294–306.
5. Kreith, F. and West, R.E. (2003) Gauging efficiency: Well-to-wheel. *Mechanical Engineering Power*, 20–23.
6. Davis, S. and Diegel, S. (2004) *Transportation Energy Data Book*. Oak Ridge National Laboratory/U.S. Department of Energy, Oak Ridge, TN.
7. EPRI (2004) Advanced batteries for electric drive vehicles: A technology and cost-effectiveness assessment for battery electric vehicles, power assist hybrid electric vehicles and plug-in hybrid electric vehicles. EPRI Tech Report 1009299, EPRI, Palo Alto, CA.
8. Kreith, F. and West, R.E. (September 2006) A vision for a secure transportation system without hydrogen or oil. *ASME Journal of Energy Resources Technology* 128(9), 236–243.
9. U.S. EPA (2013). E15 (a blend of gasoline and ethanol). www.epa.gov/otaq/regs/fuels/additive/e15. Accessed April 8, 2014.
10. Renewable Fuels Association. The new fuel: E15. www.ethanolrfa.org/pages/E15. Accessed April 8, 2014.
11. Kreith, F. and West, R.E. (May 2008) A scenario for a secure transportation system based on fuel from biomass. *Journal of Solar Energy Engineering* 130, 1–6.
12. Farrell, A.E., Plevin, R.J., Turner, B.T., Jones, A.D., O'Hare, M., and Kammen, D.M. (2006) Ethanol can contribute to energy and environmental goals. *Science* 311, 506–508.
13. EIA (2007) www.cia.doe.gov/fuelelectric/electricityinfocard2005.
14. Himelich, J.B. and Kreith, F. (2008) Potential benefits of plug-in hybrid electric vehicles for consumers and electric power utilities. In: *Proceedings of ASME 2008 IMEC*, Boston, MA, October 31–November 6, 2008.
15. Electric Drive Transportation Association (EDTA). (2015). www.electricdrive.org. Accessed April 1, 2015.
16. Kreith, F. and West, R.E. (2004) Fallacies of a hydrogen economy: A critical analysis of hydrogen production and utilization. *Journal of Energy Resources Technology* 126, 249–257.
17. Kreith, F., West, R.E., and Isler, B.E. (2002) Efficiency of advanced ground transportation technologies. *Journal of Energy Resources Technology* 24, 173–179.
18. Mince, M. (2001) Infrastructure requirement of advanced technology vehicles. In: *NCSL/TRB Transportation Technology and Policy Symposium*, Argonne National Laboratory, Argonne, IL.
19. Wald, M.L. (2009) *New York Times* News Service, April 8, 2009.
20. Moniz, E. et al. (2011). The future of natural gas: An interdisciplinary MIT study. *MIT Energy Initiative*. http://web.mit.edu/mitei/research/studies/natural-gas-2011.shtml.
21. Bartlett, A.A. (2000) An analysis of U.S. and world oil production patterns using Hubbard-style curves. *Mathematical Geology* 32(1).

Online Resources

http://www.nissanusa.com/leaf-electric-car/index.
http://www.teslamotors.com/models.
http://www.nrel.gov/sustainable_nrel/transportation.html.
http://www.nrel.gov/learning/re_biofuels.html.
http://www.nrel.gov/vehiclesandfuels/energystorage/batteries.html.
http://cta.ornl.gov/vtmarketreport/index.shtml.
http://cta.ornl.gov/data/download31.shtml (a large source of data that can be downloaded for free).

20

Infrastructure Risk Analysis and Security

Bilal M. Ayyub

CONTENTS

20.1 Introduction

Risk is associated with all projects and business ventures taken by individuals and organizations regardless of their sizes, their natures, and their time and place of execution and utilization. Risk is present in various forms and levels even in small domestic projects such as adding a deck to a residential house and in large multibillion-dollar projects such as developing and producing a space shuttle. These risks could result in significant budget overruns, delivery delays, failures, financial losses, environmental damages, and even injuries and loss of life. Risks are taken even though they could lead to devastating consequences because of potential benefits, rewards, survival, and future return on investment. This chapter defines and discusses risk and its dimensions, risk analysis, risk management and control, and risk communication.

20.2 Risk Terminology

Definitions that are needed for presenting risk-based technology methods and analytical tools are presented in this section.

20.2.1 Hazards

A hazard is an act or phenomenon posing potential harm to some person(s) or thing(s), which is a source of harm, and its potential consequences. For example, uncontrolled fire, water, and strong wind are hazards. For the hazard to cause harm, it needs to interact with person(s) or thing(s) in a harmful manner. The magnitude of the hazard is the amount of harm that might result, including the seriousness and the exposure levels of people and the environment. Hazards need to be identified and considered in projects' life cycle analyses because they could pose threats and could lead to project failures.

The interaction between a person (or a system) and a hazard can be voluntary or involuntary. For example, exposing a marine vessel to a sea environment might lead to its interaction with extreme waves in an uncontrollable manner, that is, an involuntary manner. Although the decision of a navigator of the vessel to go through a storm system that is developing can be viewed as a voluntary act in nature, and might be needed to meet schedule constraints or other constraints, the potential rewards of delivery of shipment or avoidance of delay charges offer an incentive that warrants such an interaction. Other examples can be constructed where individuals interact with hazards for potential financial rewards, fame, and self-fulfillment and satisfaction ranging from investment undertaking to climbing cliffs.

20.2.2 Reliability

Reliability can be defined for a system or a component as its ability to fulfill its design functions under designated operating or environmental conditions for a specified time period. This ability is commonly measured using probabilities. Reliability is, therefore, the occurrence probability of the complementary event to failure as provided in the following expression:

$$\text{Reliability} = 1 - \text{Failure probability} \tag{20.1}$$

20.2.3 Event Consequences

For an event of failure, *consequences* can be defined as the degree of damage or loss from some failure. Each failure of a system has some consequence(s). A failure could cause economic damage, environmental damage, injury, loss of human life, or other possible events.

Consequences need to be quantified in terms of failure consequence severities using relative or absolute measures for various consequence types to facilitate risk analysis.

For an event of success, consequences can be defined as the degree of reward or return or benefits from success. Such an event could cause economic outcomes, environmental effects, or other possible events. Consequences need to be quantified using relative or absolute measures for various consequence types to facilitate risk analysis.

20.2.4 Risk

The concept of risk can be linked to uncertainties associated with events. Within the context of projects, risk is commonly associated with an uncertain event or condition that, if it occurs, has a positive or a negative effect on the objectives of a project. Risk originates from the Latin term *risicum,* which means the challenge presented by a barrier reef to a sailor. The *Oxford Dictionary* defines risk as the chance of hazard, bad consequence, loss, etc., or risk can be defined as the chance of a negative outcome.

Risk is commonly associated with a system and commonly defined as the potential loss resulting from an uncertain exposure to a hazard or resulting from an uncertain event that exploits the system's vulnerability. Risk should be based on identified risk events or event scenarios.

In 2009, the International Organization of Standardizations (ISO) provided a broadly based definition of risk in its standard (ISO 31000:2009) as the "effect of uncertainty on objectives" in order to cover the following considerations as noted in the standard:

- An effect is a deviation from the expected that can be positive and/or negative.
- Objectives can have different aspects, such as financial, health and safety, and environmental goals, and can apply at different levels, such as strategic, organization-wide, project, product, and process.
- Risk is often characterized by reference to potential events and consequences, or a combination of these as provided in the commonly used definition.
- Risk is often expressed in terms of a combination of the consequences of an event, including changes in circumstances, and the associated likelihood of occurrence as provided in the commonly used definition.

Providing two definitions of risk should not cause any confusion since most of the coverage in this book focuses on the adverse domain of effects, that is, using the former definition; however, readers must become familiar and comfortable with the latter broadly based definition.

Risk context is the external and internal parameters or considerations to be taken into account when managing risk and setting the scope and risk criteria for the risk management policy as follows (ISO 31000:2009):

- *External*—the cultural, social, political, legal, regulatory, financial, technological, economic, natural, and competitive environment, whether international, national, regional, or local; key drivers and trends having impact on the objectives of the organization; and relationships with, and perceptions and values of, external stakeholders.
- *Internal*—governance, organizational structure, roles, and accountabilities; policies, objectives, and the strategies that are in place to achieve them; the capabilities,

understood in terms of resources and knowledge (e.g., capital, time, people, processes, systems, and technologies); information systems, information flows, and decision-making processes (both formal and informal); relationships with, and perceptions and values of, internal stakeholders; the organization's culture; standards, guidelines, and models adopted by the organization; and the form and extent of its contractual relationships.

In the context of the former risk definition, risk can be viewed to be a multidimensional quantity that includes event occurrence probability, event occurrence consequences, consequence significance, and the things at risk including populations, properties, and environmental concerns; however, it is commonly measured as a pair of the probability of occurrence of an event and the outcomes or consequences associated with the event's occurrence. Another common representation of risk is in the form of an exceedance probability function of consequences. Sometimes, risk is shown schematically as the intersection of a hazard (or a threat) defined by scenarios with a system of interest that exploit its vulnerabilities and could impact things at risk including populations, properties, and environmental concerns. Such a representation shows how each aspect of risk can be controlled or managed by countermeasures, system hardening, and mitigations as discussed in detail in later sections of this chapter.

Risk results from an event or sequence of events referred to as a *scenario*. The event or scenario can be viewed as a cause and, if it occurs, results in consequences with various severities. Sometimes, these events or scenarios are called risk factors. For example, an event or cause may be a shortage of personnel necessary to perform a task required to produce a project. The event, in this case, of a personnel shortage for the task will have consequences in regard to the project cost, schedule, and/or quality. The events can reside in the project environment, which may contribute to project success or failure through project management practices, or in external partners or subcontractors.

Risk has defining characteristics that should be recognized in a risk assessment process. Risk is a characteristic of an uncertain future and is a characteristic of neither the present nor the past. Once uncertainties are resolved and/or the future is attained, the risk becomes nonexistent; therefore, we cannot describe risks for historical events or risks for events that are currently being realized. Moreover, risks cannot be directly associated with a success. Although risk management through risk mitigation of selected events could result in project success, leading to rewards and benefits, these rewards and benefits cannot be considered as outcomes of only the nonoccurrence of events associated with the risks. The occurrence of particular events leads to adverse consequences that are clearly associated with their occurrence; however, their nonoccurrences are partial contributors to the project success that leads to rewards and benefits. The credit in the form of rewards and benefits cannot be given solely to the nonoccurrence of these events. Some risk assessment literature defines risk to include both potential losses and rewards, which should be treated separately as (1) risks leading to adverse consequences and (2) risks, if appropriately and successfully managed, contributing to benefits or rewards. Such a treatment utilizes the latter risk definition where a threat (or opportunity or a factor) could affect adversely (or favorably) the achievement of the objectives of a project and associated outcomes.

Developing an economic, analytical framework for a decision situation involving risks requires examining the economic and financial environments of a project. These environments can have significant impacts on the occurrence probabilities of events associated with risks. This added complexity might be necessary for particular projects in order to obtain justifiable and realistic results. The role of such environments in risk analysis is discussed in subsequent sections and chapters.

Risk, as the potential of losses for a system resulting from an uncertain exposure to a hazard or resulting from an uncertain event, can be viewed to be a multidimensional quantity that includes event occurrence probability, event occurrence consequences, consequence significance, and commonly, the population at risk; however, common practices minimally define it in terms of event probability and event outcomes or consequences. This pairing can be represented by the following equation:

$$\text{Risk} \equiv [(p_1,c_1),(p_2,c_2),\ldots,(p_i,c_i),\ldots,(p_n,c_n)] \tag{20.2}$$

where

p_i is the occurrence probability of an outcome or event i
c_i is the occurrence consequences or outcomes of the event

A generalized definition of risk is sometime expressed as

$$\text{Risk} \equiv [(l_1,o_1,u_1,cs_1,po_1),(l_2,o_2,u_2,cs_2,po_2),\ldots,(l_n,o_n,u_n,cs_n,po_n)] \tag{20.3}$$

where

l is the likelihood
o is the outcome
u is the utility (or significance)
cs is a causal scenario
po is the population affected by the outcome
n is the number of outcomes

The definition provided by Equation 20.3 covers key attributes measured in risk assessment that are described in this chapter and offers a practical description of risk, starting with the causing event to the affected population and consequences. The population-size effect should be considered in risk studies as society responds differently for risks associated with a large population in comparison to a small population. For example, a fatality rate of 1 in 100,000 per event for an affected population of 10 results in an expected fatality of 10^{-4} per event, whereas the same fatality rate per event for an affected population of 10,000,000 results in an expected fatality of 100 per event. Although the impact of the two scenarios might be the same on society (same expected risk value), the total number of fatalities per event or accident is a factor in risk acceptance. Plane travel may be safer than, for example, recreational boating, but 200–300 injuries per accident in the case of air travel are less acceptable to society. Therefore, the size of the population at risk and the number of fatalities per event should be considered as factors in setting acceptable risk.

The dimension of likelihood can be illusive in nature due to two of its aspects: (1) the means of quantification and (2) the effect of time. The most common means of quantification are as follows:

- *Frequency* defined as the count of an outcome of interest from a number of repeated observations of identical experiments or systems. If expressed as a fraction or percent, it is called *relative frequency*.

- *Rate* commonly defined as the count of an outcome of interest for a system occurring within a time period. The rate itself can be time dependent due to changes in the system's state, for example, due to aging. The term "frequency" is sometimes incorrectly used to mean the rate.
- *Probability* defined as a measure of chance or likelihood.

The effects of time on these three quantification means are discussed in the following, respectively:

- As for the frequency, by increasing the observation time, an estimate of the frequency tends toward a value, and for cases involving unbiased consistent estimators, the estimate tends to the true value.
- As for the rate, by increasing the observation time, an estimate of the rate tends toward a value, and for cases involving unbiased consistent estimators, the estimate tends to the true value.
- As for the probability, we are interested in the probability of an event in a time period. By increasing the length of this time period, this probability tends to 1. As long as the event is possible, it has a sure eventual occurrence; otherwise, it goes against the premise of being possible.

Risk matrices, also called heat maps, are basically tools for representing and displaying risks by defining ranges for consequence and likelihood as a 2D presentation of likelihood and consequences. According to this method, risk is characterized by categorizing probabilities and consequences on the two axes of a matrix. Risk matrices have been used extensively for the screening of various risks. They may be used alone or as a first step in a quantitative analysis. Regardless of the approach used, risk analysis should be a dynamic process—that is, a living process where risk assessments are reexamined and adjusted. Actions or inactions in one area might affect risk in another; therefore, continuous updating is necessary.

To quantify risk, we must accordingly assess its defining components and measure the chance, its negativity, and potential rewards or benefits. Estimation of risk is commonly approximated by a point estimate as the expected value resulting from the multiplication of the conditional probability of the event occurring by the consequence of the event given that it has occurred as follows with a loss in information in terms of associated dispersion or variability:

$$\text{Risk} = \text{Likelihood} \times \text{Impact} \tag{20.4}$$

In Equation 20.4, the measurement scales, as the bases for quantifying likelihood, impact, and risk, are likelihood measured on an event rate scale in units of count of events per time period of interest, for example, events per year; impact measured on a loss scale, such as monetary units or fatalities or any other units suitable for analysis or multiple units per event, for example, dollars per event; and risk is (event per unit time) × (loss units per event) producing loss units per unit time. The likelihood in Equation 20.4 can also be expressed as a probability. Equation 20.4 presents risk as an expected value of loss per unit time or an average loss.

The product in Equation 20.4 is sometimes interpreted as the Cartesian product for scoping the space defined by the two dimensions of likelihood and impact for all underlying events and scenarios, which is a preferred interpretation. This interpretation preserves the complete nature of risk. Ideally, the entire probability distribution of consequences should be estimated.

A plot of occurrence probabilities and consequences is a *risk profile* or a *Farmers curve*. An example Farmers curve is given in Figure 20.1a based on a nuclear case study provided herein for illustration purposes (Kumamoto and Henley 1996). It should be noted that the abscissa provides the number of fatalities, and the ordinate provides the annual

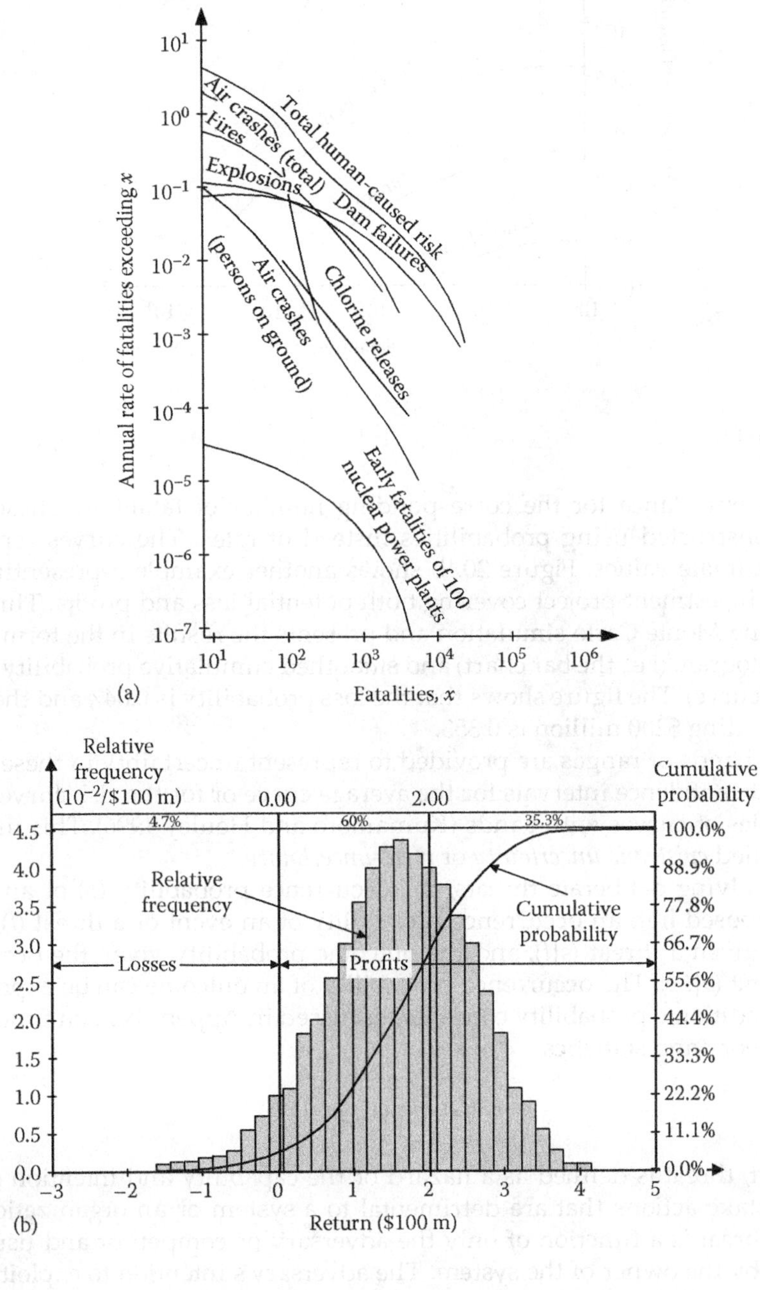

FIGURE 20.1
(a) Example risk profile. (b) Example project risk profile.

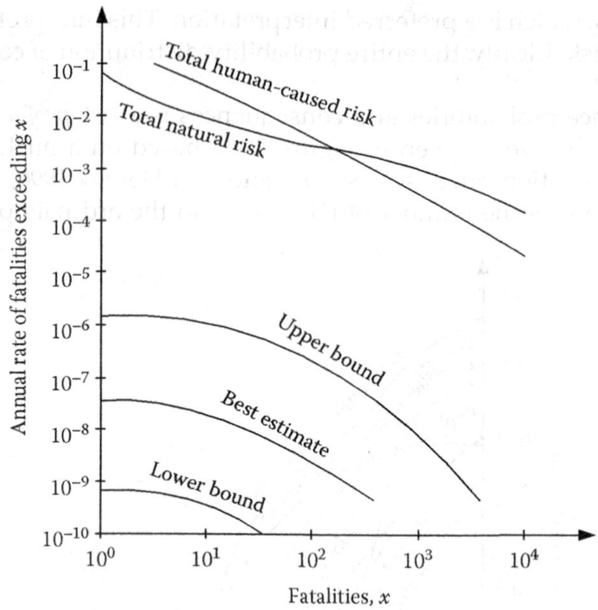

FIGURE 20.2
Uncertain risk profile.

frequency of exceedance for the corresponding number of fatalities. These curves are sometimes constructed using probabilities instead of rates. The curves represent average or best estimate values. Figure 20.1b shows another example representing the gross margin of an investment project covering both potential loss and profits. This figure was generated using Monte Carlo simulation and presents the results in the form of a relative frequency histogram (i.e., the bar chart) and smoothed cumulative probability distribution (i.e., the solid curve). The figure shows that the loss probability is 0.047, and the probability of profits exceeding $200 million is 0.353.

Sometimes, bands or ranges are provided to represent uncertainty in these curves, and they represent confidence intervals for the average curve or for the risk curve. Figure 20.2 shows examples of curves with bands (Kumamoto and Henley 1996). This uncertainty is sometimes called *epistemic uncertainty* or *meta-uncertainty*.

In cases involving deliberate threats, the occurrence probability (p) of an outcome (o) can be decomposed into an occurrence probability of an event or a threat (t), probability of success (s) given a threat ($s|t$), and the outcome probability given the occurrence of a successful event ($o|t,s$). The occurrence probability of an outcome can be expressed as follows using conditional probability concepts discussed in Appendix A on fundamentals of probability theory and statistics:

$$p(o) = p(t)p(s|t)p(o|t,s) \tag{20.5}$$

In this context, threat is defined as a hazard or the capability and intention of an adversary to undertake actions that are detrimental to a system or an organization's interest. In this case, threat is a function of only the adversary or competitor and usually cannot be controlled by the owner of the system. The adversary's intention to exploit his capability may, however, be encouraged by the vulnerability of the system or discouraged by an owner's countermeasures. The probability $p(o|t)$ can be decomposed further into two components of success probability of the adversary and a conditional probability on this success of consequences. This probability $p(o|t)$ can then be computed as the success probability of the adversary times the conditional probability of consequences given this success.

Risk register—is a record of information about identified risks, sometimes called *risk log*.

Risk profile—is a description of any set of risks that may relate to the whole organization, part of the organization, or a group of stakeholders.

Risk aggregation—is a combination of a number of risks into one risk to develop a more complete understanding of an overall risk, whereas *risk segregation* is the decomposition of an overall risk profile into a number of underlying risk profiles.

20.2.5 Performance

The performance of a system or component can be defined as its ability to meet functional requirements. The performance of an item can be described by various elements, including items such as speed, power, reliability, capability, efficiency, and maintainability. The design and operation of the product or system influence performance.

20.2.6 Risk-Based Technology

Risk-based technologies (RBT) are methods or tools and processes used to assess and manage the risks of a component or system. RBT methods can be classified into risk management that includes risk assessment/risk analysis and risk control using failure prevention and consequence mitigation, and risk communication as shown in Figure 20.3.

Risk assessment consists of hazard identification, event probability assessment, and consequence assessment. Risk control requires the definition of acceptable risk and comparative evaluation of options and/or alternatives through monitoring and decision analysis. Risk control also includes failure prevention and consequence mitigation. Risk communication involves perceptions of risk, which depends on the audience targeted, hence, classified into risk communication to the media, the public, and to the engineering community.

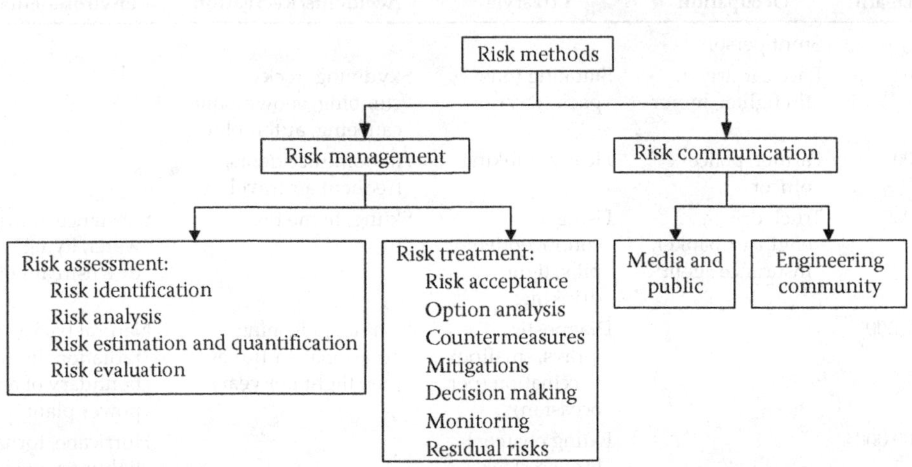

FIGURE 20.3
Risk-based technology methods.

20.2.7 Safety

Safety can be defined as the judgment of risk acceptability for the system. Safety is a relative term since the decision of risk acceptance may vary depending on the individual making the judgment. Different people are willing to accept different risks as demonstrated by different factors such as location, method or system type, occupation, and lifestyle. The selection of these different activities demonstrates an individual's safety preference despite a wide range of risk values. Table 20.1 identifies varying annual risks for different activities based on typical exposure times for these activities. Also Figure 20.4 from the Imperial Chemical Industries, Ltd. shows the variation of risk exposure during a typical day that starts by waking up in the morning from sleep and getting ready to go to work, then commuting and working during morning hours, followed by a lunch break, then additional work hours followed by commuting back to having dinner, and round-trip on a motorcycle to a local pub. The ordinate in this figure is the fatal accident frequency rate (FAFR) with a FAFR of 1.0 corresponding to one fatality in 11,415 years, or 87.6 fatalities per one million years. The figure is based on an average number of deaths in 10^8 h of exposure to a particular activity.

Risk perceptions of safety may not reflect the actual level of risk in some activity. Table 20.2 shows the differences in risk perception by three groups of the league of women voters, college students, and experts of 29 risk items. Only the top items are listed in the table. Risk associated with nuclear power was ranked as the highest type by women voters and college students, whereas it was placed as the 20th by experts. Experts place motor vehicles as the first risk. Public perception of risk and safety varies by age, gender, education, attitudes, and culture among other factors. Individuals sometimes do not recognize uncertainties associated with risk event or activity causing an unwarranted confidence in an individual's perception of risk or safety. Rare causes of death are often overestimated and common causes of death are often underestimated. Perceived risk is often biased by the familiarity of the hazard. The significance or the

TABLE 20.1
Relative Risk of Different Activities

Risk of Death	Occupation	Lifestyle	Accidents/Recreation	Environmental Risk
1 in 100	Stunt person			
1 in 1,000	Race car driver, firefighter, miner	Smoking (one pack/day)	Skydiving, rock climbing, snowmobile, canoeing, automobile	
1 in 10,000	Farmer, police officer	Heavy drinking	All home accidents, frequent air travel	
1 in 100,000	Truck driver, engineer, banker, insurance agent	Using contraceptive pills, light drinking	Skiing, home fire	Substance in drinking water, living downstream of a dam
1 in 1,000,000		Diagnostic x-rays, smallpox vaccination (per occasion)	Fishing, poisoning, occasional air travel (one flight per year)	Natural background radiation, living at the boundary of a nuclear power plant
1 in 10,000,000		Eating charcoal-broiled steak (once a week)		Hurricane, tornado, lightning, animal bite or insect sting

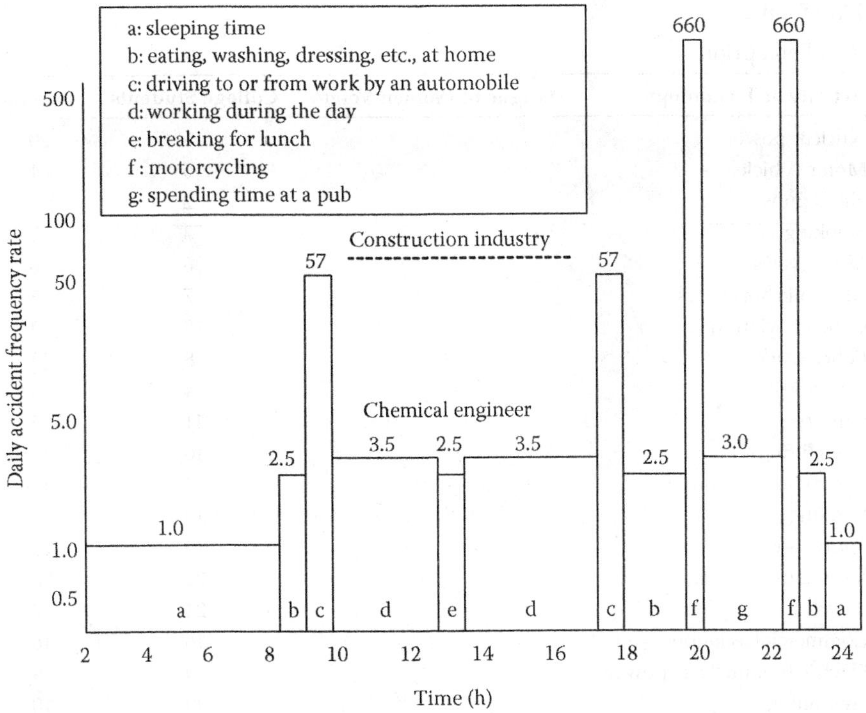

FIGURE 20.4
Daily death risk exposure for a working healthy adult.

impact of safety perceptions stems from those decisions that are often made on subjective judgments. If the judgments hold misconceptions about reality, this bias affects the decision. For example, the choice of a transportation mode—train, automobile, motorcycle, bus, bicycle, etc.—results in a decision based on many criteria including items such as cost, speed, convenience, and safety. The weight and evaluation of the decision criteria in selecting a mode of transportation rely on the individual's perception of safety that may deviate sometimes significantly from the actual values of risks. Understanding these differences in risk and safety perceptions is vital to performing risk management decisions and risk communications as provided in subsequent sections on risk management and control.

20.3 Risk Assessment

20.3.1 Risk Assessment Methodologies

Risk studies require the use of analytical methods at the system level that considers subsystems and components in assessing their failure probabilities and consequences. Systematic, quantitative, qualitative, or semiquantitative approaches for assessing the failure probabilities and consequences of engineering systems are used for this purpose.

TABLE 20.2

Risk Perception

Activity or Technology	League of Women Voters	College Students	Experts
Nuclear power		1	20
Motor vehicles		5	1
Hand guns		2	4
Smoking		3	2
Motorcycles		6	6
Alcoholic beverages		7	3
General aviation		15	12
Police work		8	17
Pesticides		4	8
Surgery		11	5
Firefighting		10	18
Large construction		14	13
Hunting		18	23
Spray cans		13	25
Mountain climbing		22	28
Bicycles		24	15
Commercial aviation		16	16
Electric (nonnuclear) power		19	9
Swimming		29	10
Contraceptives		9	11
Skiing		25	29
X-rays		17	7
High school or college sports		26	26
Railroads		23	19
Food preservatives		12	14
Food coloring		20	21
Power mowers		28	27
Prescription antibiotics		21	24
Home applications		27	22

A systematic approach allows an analyst to evaluate expediently and easily complex systems for safety and risk under different operational and extreme conditions. The ability to quantitatively evaluate these systems helps cut the cost of unnecessary and often expensive redesign, repair, strengthening, or replacement of components, subsystems, and systems. The results of risk analysis can also be utilized in decision analysis methods that are based on cost–benefit trade-offs.

Risk assessment is a technical and scientific process by which the risks of a given situation for a system are modeled and quantified. Risk assessment can require and/or provide both qualitative and quantitative data to decision makers for use in risk management.

Risk assessment or risk analysis provides the process for identifying hazards, event-probability assessment, and consequence assessment. The risk assessment process answers three basic questions: (1) What can go wrong? (2) What is the likelihood that it will go wrong? (3) What are the consequences if it does go wrong? Answering these questions requires the utilization of various risk methods as discussed in this chapter.

A risk assessment process should utilize experiences gathered from project personnel including managers, other similar projects and data sources, previous risk assessment models, experiences from other industries and experts, in conjunction with analysis and damage evaluation/prediction tools. A risk assessment process is commonly a part of a risk-based or risk-informed methodology that should be constructed as a synergistic combination of decision models, advanced probabilistic reliability analysis algorithms, failure consequence assessment methods, and conventional performance assessment methodologies that have been employed in related industry for performance evaluation and management. The methodology should realistically account for the various sources and types of uncertainty involved in the decision-making process (Ayyub and McCuen 2003; Ayyub and Klir 2006).

In this section, a typical overall methodology is provided in the form of a workflow or block diagram. The various components of the methodology are described in subsequent sections. Figure 20.5 provides an overall description of a methodology for risk-based management of structural systems for the purpose of demonstration. The methodology consists of the following primary steps:

1. Definition of analysis objectives and systems
2. Hazard analysis, definition of failure scenarios, and hazardous sources and their terms
3. Collection of data in a life cycle framework
4. Qualitative risk assessment
5. Quantitative risk assessment
6. Management of system integrity through failure prevention and consequence mitigation using risk-based decision making.

These steps are briefly described in the following with additional background materials provided in subsequent sections.

The first step of the methodology is to define the system. This definition should be based on a goal that is broken down into a set of analysis objectives. A system can be defined as an assemblage or combination of elements of various levels and/or details that act together for a specific purpose. Defining the system provides the risk-based methodology with the information it needs to achieve the analysis objectives. The system definition phase of the proposed methodology has four main activities. The activities are to

1. Define the goal and objectives of the analysis
2. Define the system boundaries
3. Define the success criteria in terms of measurable performances
4. Collect information for assessing failure likelihood
5. Collect information for assessing failure consequences

For example, structural systems require a structural integrity goal that can include objectives stated in terms of strength, performance, serviceability, reliability, cost-effectiveness, and environmental soundness. The objectives can be broken down further to include other structural integrity attributes, such as alignment and water tightness in case of marine vessels. A system can be defined based on a stated set of objectives. The same system can be defined differently depending on these stated objectives. A marine

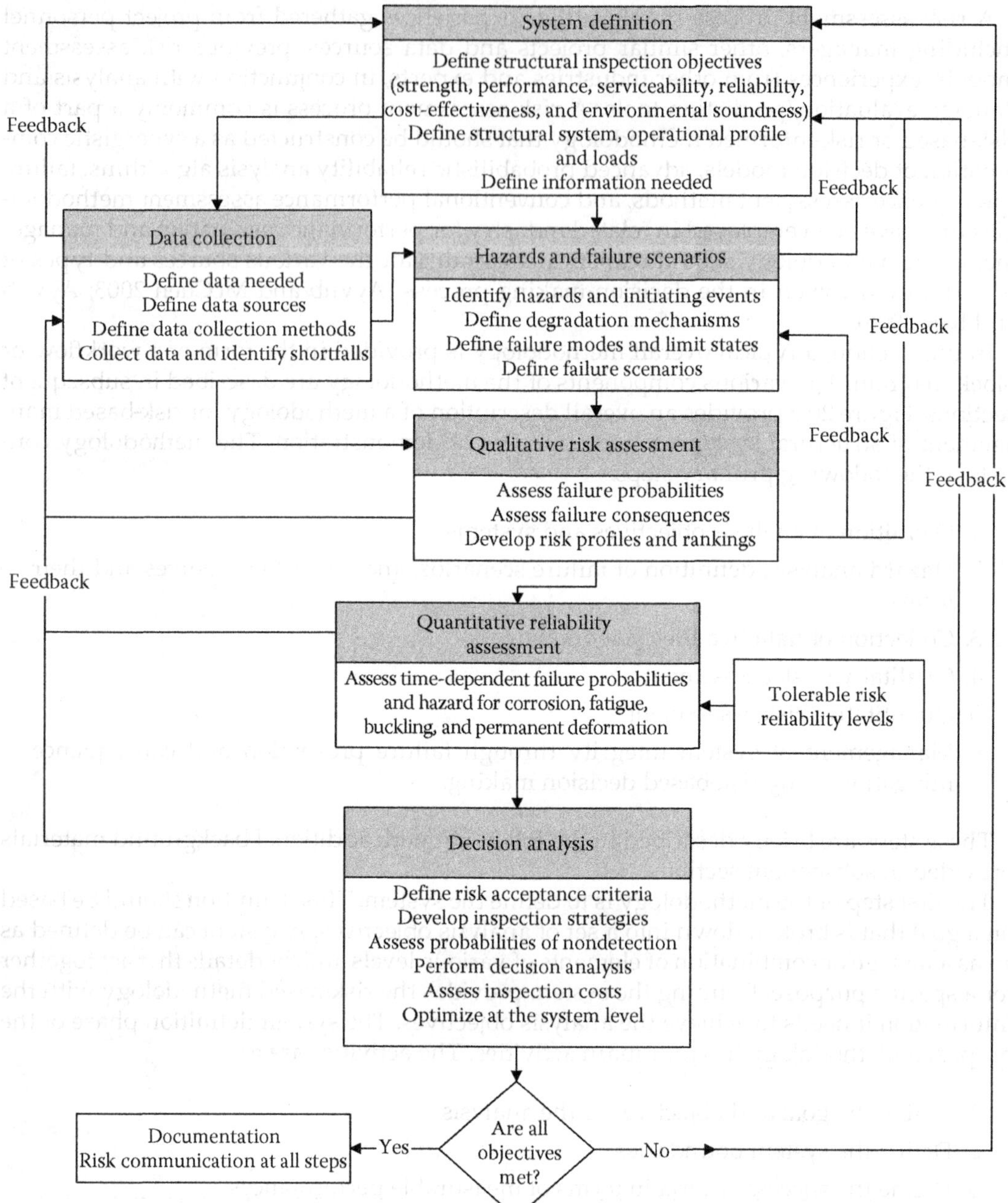

FIGURE 20.5
Methodology for risk-based life cycle management of structural systems.

vessel structural system can be considered to contain individual structural elements such as plates, stiffened panels, stiffeners, longitudinals, etc. These elements could be further separated into individual components and/or details. Identifying all of the elements, components, and details allows an analysis team to collect the necessary operational, maintenance, and repair information throughout the life cycle of each item so that failure rates, repair frequencies, and failure consequences can be estimated. The system definition might need to include nonstructural subsystems and components that would be affected in case of failure. The subsystems and components are needed to assess the consequences.

To understand failure and the consequences of failure, the states of success need to be defined. For the system to be successful, it must be able to perform its designed functions by meeting measurable performance requirements. But the system may be capable of various levels of performance, all of which might not be considered a successful performance. While a marine vessel maybe able to get from point A to point B only at a reduced speed due to a fatigue failure that results in excessive vibration at the engine room, its performance would probably not be considered successful. The same concept can be applied to individual elements, components, and details. It is clear from this example that the vessel's success and failure impacts should be based on the overall vessel performance that can easily extend beyond the structural systems.

With the development of the definition of success, one can begin to assess the likelihood of occurrence and causes of failures. Most of the information required to develop an estimate of the likelihood of failure might exist in maintenance and operating histories available on the systems and equipment and based on judgment and expert opinion. This information might not be readily accessible, and its extraction from its current source might be difficult. Also, assembling it in a manner that is suitable for the risk-based methodology might be a challenge.

Operation, maintenance, engineering, and corporate information on failure history needs to be collected and analyzed for the purpose of assessing the consequences of failures. The consequence information might not be available from the same sources as the information on the failure itself. Typically, there are documentations of repair costs, reinspection or recertification costs, lost person-hours of labor, and possibly even lost opportunity costs due to system failure. Much more difficult to find and assess are costs associated with the effects on other systems, the cost of shifting resources to cover lost production, and things like environmental, safety loss, or public relations costs. These may be attained through carefully organized discussions and interviews with cognizant personnel including the use of expert-opinion elicitation.

20.3.2 Risk Events and Scenarios

To adequately assess all risks associated with a project, the process of identification of risk events and scenarios is an important stage in risk assessment. Risk events and scenarios can be categorized as follows:

- Technical, technological, quality, or performance risks, such as unproven or complex technology, unrealistic performance goals, and changes to the technology used or to the industry standards during the project.
- Project management risks, such as poor allocation of time and resources, inadequate quality of the project plan, and poor use of project management disciplines.
- Organizational risks, such as cost, time, and scope objectives that are internally inconsistent, lack of prioritization of projects, inadequacy or interruption of funding, resource conflicts with other projects in the organization, errors by individuals or by an organization, and inadequate expertise and experience by project personnel.

- External risks, such as shifting legal or regulatory environment, labor issues, changing owner priorities, country risk, and weather.
- Natural hazards, such as earthquakes, floods, strong wind, and waves generally require disaster recovery actions in addition to risk management. Within these categories, several risk types can be identified.

20.3.3 Identification of Risk Events and Scenarios

The risk assessment process starts with the question "What can go wrong?" The identification of what can go wrong entails defining hazards, risk events, and risk scenarios. The previous section provided categories of risk events and scenarios. Risk identification involves determining which risks might affect the project and documenting their characteristics. The risk identification generally requires the participation from a project team, risk management team, subject matter experts from other parts of the company, customers, end users, other project managers, stakeholders, and outside experts on as needed basis. Risk identification can be an iterative process. The first iteration may be performed by selected members of the project team or by the risk management team. The entire project team and primary stakeholders may take a second iteration. To achieve an unbiased analysis, persons who are not involved in the project may perform the final iteration. Risk identification can be a difficult task, because it is often highly subjective, and there are no unerring procedures that may be used to identify risk events and scenarios other than relying heavily on the experience and insight of key project personnel.

The development of the scenarios for risk evaluation can be created deductively (e.g., fault tree) or inductively (e.g., failure mode and effects analysis [FMEA]) as provided in Table 20.3. The table shows methods of multiple uses including likelihood or frequency estimation expressed either deterministically or probabilistically. Also, they can be used to assess varying consequence categories including items such as economic loss, loss of life, or injuries.

The risk identification process and risk assessment require the utilization of these formal methods as shown in Table 20.3. These different methods contain similar approaches to answer the basic risk assessment questions; however, some techniques may be more appropriate than others for risk analysis depending on the situation.

20.3.4 Risk Breakdown Structure

Risk sources for a project can be organized and structured to provide a standard presentation that would facilitate understanding, communication, and management. The previously presented methods can be viewed as simple linear lists of potential sources of risk, providing a set of headings under which risks can be arranged. These lists are sometimes called risk taxonomy. A simple list of risk sources might not provide the richness needed for some decision situations since it only presents a single level of organization. Some applications might require a full hierarchical approach to define the risk sources, with as many levels as are required to provide the necessary understanding of risk exposure. Defining risk sources in such a hierarchical structure is called a risk breakdown structure (RBS). The RBS is defined as a source-oriented grouping of project risks organized to define the total risk exposure of a project of interest.

TABLE 20.3

Risk Assessment Methods

Method	Scope
Safety/review audit	Identifies equipment conditions or operating procedures that could lead to a casualty or result in property damage or environmental impacts.
Checklist What-If	Ensures that organizations are complying with standard practices.
Hazard and operability study (HAZOP)	Identifies hazards, hazardous situations, or specific accident events that could result in undesirable consequences.
Preliminary hazard analysis (PrHA)	Identifies system deviations and their causes that can lead to undesirable consequences and determine recommended actions to reduce the frequency and/or consequences of the deviations.
Probabilistic risk analysis (PRA)	Identifies and prioritizes hazards leading to undesirable consequences early in the life of a system. It determines recommended actions to reduce the frequency and/or consequences of the prioritized hazards. This is an inductive modeling approach.
Failure modes and effects analysis (FMEA)	Methodology for quantitative risk assessment developed by the nuclear engineering community for risk assessment. This comprehensive process may use a combination of risk assessment methods. Identifies the components' (equipment) failure modes and the impacts on the surrounding components and the system. This is an inductive modeling approach.
Fault tree analysis (FTA)	Identifies combinations of equipment failures and human errors that can result in an accident. This is a deductive modeling approach.
Event tree analysis (ETA)	Identifies various sequences of events, both failures and successes that can lead to an accident. This is an inductive modeling approach.
The Delphi technique	Assists to reach consensus of experts on a subject such as project risk while maintaining anonymity by soliciting ideas about the important project risks that are collected and circulated to the experts for further comment. Consensus on the main project risks may be reached in a few rounds of this process.
Interviewing	Identifies risk events by interviews of experienced project managers or subject matter experts. The interviewees identify risk events based on experience and project information.
Experience based identification	Identifies risk events based on experience including implicit assumptions.
Brainstorming	Identifies risk events using facilitated sessions with stakeholders, project team members, and infrastructure support staff.

Each descending level represents an increasingly detailed definition of risk sources for the project. The value of the RBS can be in aiding an analyst to understand the risks faced by the project.

An example RBS is provided in Table 20.4. In this example, four risk levels are defined as shown in the table. The project's risks are viewed as level 0. Three types of level 1 risks are provided in the table for the purpose of demonstration. The number of risk sources in each level varies and depends on the application at hand. The subsequent level 2 risks are provided in groups that are detailed further in level 3. The RBS provides a means to systematically and completely identify all relevant risk sources for a project.

The RBS should not be treated as a list of independent risk sources since commonly they have interrelations and common risk drivers. Identifying causes behind the risk sources is a key step toward an effective risk management plan including mitigation actions. A process of risk interrelation assessment and root-cause identification can be utilized to potentially lead to identifying credible scenarios that could lead to snowball effects for risk management purposes.

TABLE 20.4
Risk Breakdown Structure for a Project

Level 0	Level 1	Level 2	Level 3
		Corporate	History, experiences, culture, personnel
			Organization structure, stability, communication
			Finances conditions
			Other projects
			M
	Management	Customers and Stakeholders	History, experiences, culture, personnel
			Contracts and agreements
			Requirement definition
			Finances and credit
			M
		Natural environment	Physical environment
			Facilities, site, equipment, materials
			Local services
			M
Project Risks	External	Cultural	Political
			Legal, regulatory
			Interest groups
			Society and communities
			M
		Economic	Labor market, conditions, competition
			Financial markets
			M
	Technology	Requirements	Scope and objectives
			Conditions of use, users
			Complexity
			M
		Performance	Technology maturity
			Technology limitations
			New technologies
			New hazards or threats
			M
		Application	Organizational experience
			Personnel skill sets and experience
			Physical resources
			M

20.3.5 System Definition for Risk Assessment

Defining the system is an important first step in performing a risk assessment. A system can be defined as a deterministic entity comprising an interacting collection of discrete elements and commonly defined using deterministic models.

The word "deterministic" implies that the system is identifiable and not uncertain in its architecture. The definition of the system is based on analyzing its functional and/or performance requirements. A description of a system may be a combination of functional and physical elements. Usually, functional descriptions are used to identify high information levels on a system. A system may be divided into subsystems that interact. Additional detail leads to a description of the physical elements, components, and various aspects of the system.

The examination of a system needs to be made in a well-organized and repeatable fashion so that risk analysis can be consistently performed, therefore insuring that important elements of a system are defined and extraneous information omitted. The formation of system boundaries is based upon the objectives of the risk analysis.

The establishment of system boundaries can assist in developing the system definition. The decision on what the system boundary is partially based on what aspects of the system's performance are of concern.

The selection of items to include within the external boundary region is also reliant on the goal of the analysis. Beyond the established system boundary is the external environment of the system.

Boundaries beyond the physical/functional system can also be established. For example, time may also be a boundary since an overall system model may change, as a product is further along in its life cycle. The life cycle of a system is important because some potential hazards can change throughout the life cycle. For example, material failure due to corrosion or fatigue may not be a problem early in the life of a system; however, this may be an important concern later in the life cycle of the system.

Along with identifying the boundaries, it is also important to establish a resolution limit for the system. The selected resolution is important because it limits the detail of the analysis. Providing too little detail might not provide enough information for the problem. Too much information may make the analysis more difficult and costly due to the added complexity. The depth of the system model needs to be sufficient for the specific problem. Resolution is also limited by the feasibility of determining the required information for the specific problem. For failure analysis, the resolution should be to the components level where failure data are available. Further resolution is not necessary and would only complicate the analysis.

The system breakdown structure is the top-down division of a system into subsystems and components. This architecture provides internal boundaries for the system. Often the systems/subsystems are identified as functional requirements that eventually lead to the component level of detail. The functional level of a system identifies the function(s) that must be performed for the operation of the system. Further decomposition of the system into "discrete elements" leads to the physical level of a system definition identifying the hardware within the system. By organizing a system hierarchy using a top-down approach rather than fragmentation of specific systems, a rational, repeatable, and systematic approach to risk analysis can be achieved.

Further system analysis detail is addressed from modeling the system using some of the risk assessment methods described in Table 20.3. These techniques develop processes that can assist in decision making about the system. The logic of modeling based on the interaction of a system's components can be divided into induction and deduction. This difference in the technique of modeling and decision making is significant. Induction logic provides the reasoning of a general conclusion from individual cases. This logic is used when analyzing the effect of a fault or condition on a systems operation. Inductive analysis answers the question, "What are the system states due to some event?" In reliability and risk studies, this "event" is some fault in the system. Several approaches using the inductive approach include preliminary hazard analysis (PrHA), FMEA, and event tree analysis (ETA). Deductive approaches provide reasoning for a specific conclusion from general conditions. For system analysis, this technique attempts to identify what modes of a system/subsystem/component failure can be used to contribute to the failure of the system. This technique answers the question,

"How a system state can occur?" Inductive reasoning provides the techniques for fault tree analysis (FTA) or its complement success tree analysis (STA).

20.3.6 Selected Risk Assessment Methods

Qualitative versus quantitative risk assessment: The risk assessment methods can be categorized according to how the risk is determined: by quantitative or qualitative analysis. Qualitative risk analysis uses judgment and sometimes "expert" opinion to evaluate the probability and consequence values. This subjective approach maybe sufficient to assess the risk of a system, depending on the available resources.

Quantitative analysis relies on probabilistic and statistical methods and databases that identify numerical probability values and consequence values for risk assessment. This objective approach examines the system in greater detail to assess risks.

The selection of a quantitative or qualitative method depends upon the availability of data for evaluating the hazard and the level of analysis needed to make a confident decision. Qualitative methods offer analyses without detailed information, but the intuitive and subjective processes may result in differences in outcomes by those who use them. Quantitative analysis generally provides a more uniform understanding among different individuals, but requires quality data for accurate results. A combination of both qualitative and quantitative analyses can be used depending on the situation.

Risk assessment requires estimates of the failure likelihood at some identified levels of decision making. The failure likelihood can be estimated in the form of lifetime failure likelihood, annual failure likelihood, mean time between failures, or failure rate. The estimates can be in numeric or nonnumeric form. An example numeric form for an annual failure probability is 0.00015 and for a mean time between failures is 10 years. An example nonnumeric form for "an annual failure likelihood" is large and for a "mean time between failures" is medium. In the latter nonnumeric form, guidance needs to be provided regarding the meaning of terms such as large, medium, small, very large very small, etc. The selection of the form should be based on the availability of information, the ability of the personnel providing the needed information to express it in one form or another, and the importance of having numeric versus nonnumeric information in formulating the final decisions.

The types of failure consequences that should be considered in a study need to be selected. They can include production loss, property damage, environmental damage, and safety loss in the form of human injury and death. Approximate estimates of failure consequences at the identified levels of decision making need to be determined. The estimates can be in numeric or nonnumeric form. An example numeric form for production loss is 1000 units. An example nonnumeric form for production loss is large. In the latter nonnumeric form, guidance needs to be provided regarding the meaning of terms such as large, medium, small, very large, very small, etc. The selection of the form should be based on the availability of information, the ability of the personnel providing the needed information to express it in one form or another, and the importance of having numeric versus nonnumeric information in formulating the final decisions.

Risk estimates can be determined as a pair of the likelihood and consequences and computed as the arithmetic multiplication of the respective failure likelihood and consequences for the equipment, components, and details. Alternatively, for all cases, plots of failure likelihood versus consequences can be developed. Then, approximate ranking of them as groups according to risk estimates, failure likelihood, and/or failure consequences can be developed.

Preliminary hazard analysis: PrHA is a common risk-based technology tool with many applications. The general process is shown in Figure 20.6. This technique requires experts to

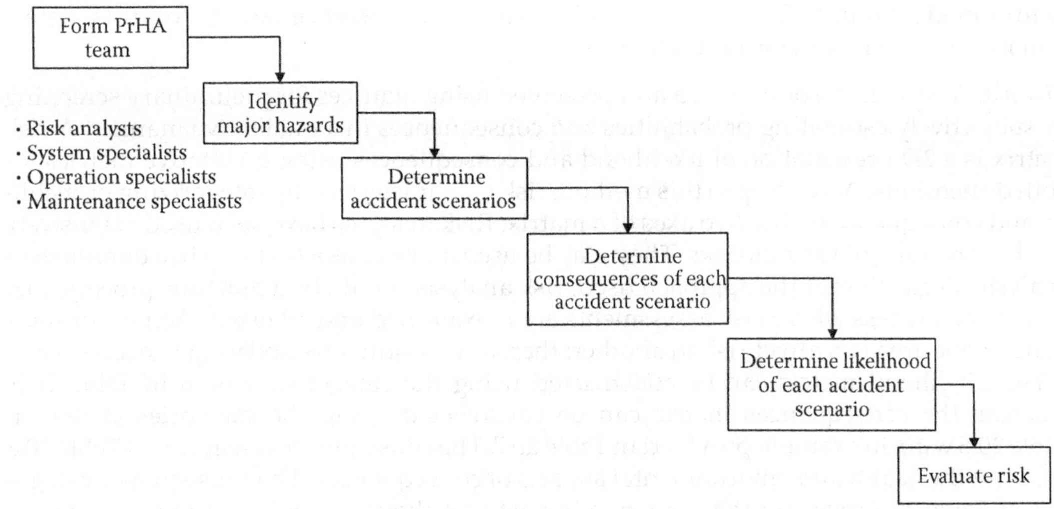

FIGURE 20.6
Preliminary hazard analysis (PrHA) process.

identify and rank the possible accident scenarios that may occur. It is frequently used as a preliminary method to identify and reduce the risks associated with major hazards of a system.

Failure mode and effects analysis: FMEA is another popular risk-based technology tool as shown in Figure 20.7. This technique has been introduced in both the national and international regulations for the aerospace (US MIL-STD-1629A), processing plant, and marine industries. The Society of Automotive Engineers, in its recommended practice, introduces two types of FMEA: design and process FMEA. This analysis tool assumes that a failure mode occurs in a system/component through some failure mechanism; the effect of this failure on other systems is then evaluated. A risk ranking can be developed for each failure mode for the effect on the overall performance of the system.

The various terms used in FMEA with examples based on the manufacturing of personal flotation devices (PFDs) are provided under subsequent headings to include

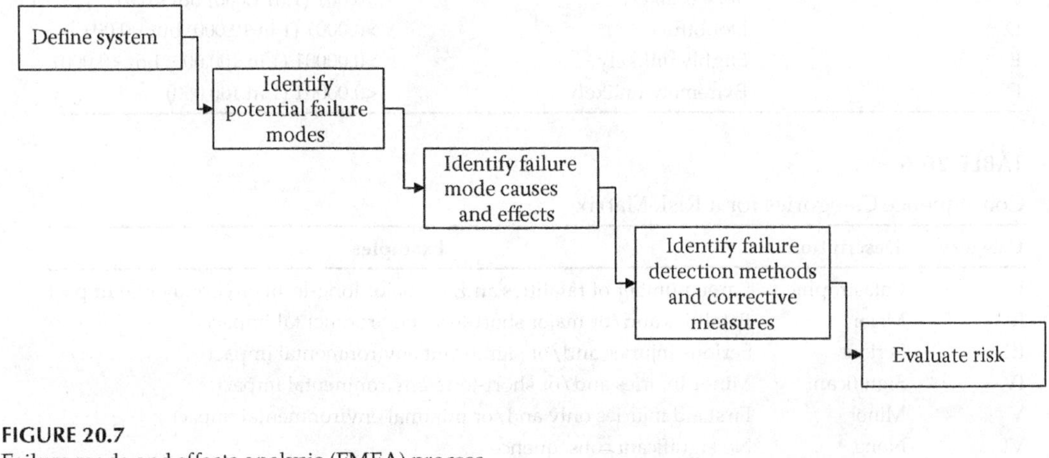

FIGURE 20.7
Failure mode and effects analysis (FMEA) process.

failure mode, failure effect, severity rating, causes, occurrence rating, controls, detection rating, and risk priority number.

Risk matrices: Risk can be assessed and presented using matrices for preliminary screening by subjectively estimating probabilities and consequences in a qualitative manner. A risk matrix is a 2D presentation of likelihood and consequences using qualitative metrics for both dimensions. According to this method, risk is characterized by categorizing probability and consequence on the two axes of a matrix. Risk matrices have been used extensively for the screening of various risks. They may be used alone or as a first step in a quantitative analysis. Regardless of the approach used, risk analysis should be a dynamic process, that is, a living process where risk assessments are reexamined and adjusted. Actions or inactions in one area can affect risk in another; therefore, continuous updating is necessary.

The likelihood metric can be constructed using the categories shown in Table 20.5, whereas the consequences metric can be constructed using the categories shown in Table 20.6 with an example provided in Table 20.7. The consequence categories of Table 20.6 focus on the health and environmental aspects of consequences. The consequence categories of Table 20.7 focus on the economic impact and should be adjusted to meet specific needs of industry and/or applications. An example risk matrix is shown in Figure 20.8. In the figure, each boxed area is shaded depending on a subjectively assessed risk level. Three risk levels are used herein for illustration purposes: low (L), medium (M), and high (H). Other risk levels may be added using a scale of five levels instead of three levels if needed. These risk levels are also called *severity factors*. The high (H) level can be considered as unacceptable risk level, the medium (M) level can be treated as either undesirable or acceptable with review, and the low (L) level can be treated as acceptable without review.

Event modeling: event, success trees, and fault trees: Event modeling is a systematic—and often the most complete—way to identify accident scenarios and quantify risk for

TABLE 20.5

Likelihood Categories for a Risk Matrix

Category	Description	Annual Probability Range
A	Likely	>0.1 (1 in 10)
B	Unlikely	>0.01 (1 in 100) but <0.1
C	Very unlikely	>0.001 (1 in 1,000) but <0.01
D	Doubtful	>0.0001 (1 in 10,000) but <0.001
E	Highly unlikely	>0.00001 (1 in 100,000) but <0.0001
F	Extremely unlikely	<0.00001 (1 in 100,000)

TABLE 20.6

Consequence Categories for a Risk Matrix

Category	Description	Examples
I	Catastrophic	Large number of fatalities and/or major long-term environmental impact
II	Major	Fatalities and/or major short-term environmental impact
III	Serious	Serious injuries and/or significant environmental impact
IV	Significant	Minor injuries and/or short-term environmental impact
V	Minor	First aid injuries only and/or minimal environmental impact
VI	None	No significant consequence

TABLE 20.7

Example Consequence Categories for a Risk Matrix in 2003 Monetary Amounts (US$)

Category	Description	Cost
I	Catastrophic loss	>$10,000,000,000
II	Major loss	>$1,000,000,000 but <$10,000,000,000
III	Serious loss	>$100,000,000 but <$1,000,000,000
IV	Significant loss	>$10,000,000 but <$100,000,000
V	Minor loss	>$1,000,000 but <$10,000,000
VI	Insignificant loss	<$1,000,000

Probability category							
	A	L	M	M	H	H	H
	B	L	L	M	M	H	H
	C	L	L	L	M	M	H
	D	L	L	L	L	M	M
	E	L	L	L	L	L	M
	F	L	L	L	L	L	L
		VI	V	IV	III	II	I
		Consequence category					

FIGURE 20.8
Example risk matrix.

risk assessment. This risk-based technology tool provides a framework for identifying scenarios to evaluate the performance of a system or component through system modeling. The combination of ETA, STA, and FTA can provide a structured analysis to system safety.

ETA is often used if the successful operation of a component/system depends on a discrete (chronological) set of events. The initiating event is first followed by other events leading to an overall result (consequence). The ability to address a complete set of scenarios is developed because all combinations of both the success and failure of the main events are included in the analysis. The probability of occurrence of the main events of the event tree can be determined using a fault tree or its complement, the success tree. The scope of the analysis for event trees and fault trees depends on the objective of the analysis.

ETA is appropriate if the operation of some system/component depends on a successive group of events. Event trees identify the various combinations of event successes and failures as a result of an initiating event to determine all possible scenarios. The event tree starts with an initiating event followed by some reactionary event. This reaction can be either a success or a failure. If the event succeeds, the most commonly used indication is the upward movement of the path branch. A downward branch of the event tree marks the failure of an event. The remaining events are evaluated to determine the different possible scenarios. The scope of the events can be functions/systems that can provide some reduction to the possible hazards from the initiating event. The final outcome of a sequence of events identifies the overall state resulting from the scenario of events. Each path represents a failure scenario with varying levels of probability and risk. Different event trees can be created for different event initiators. Figure 20.9 shows an example event tree for the basic elements of a sprinkler system that might be critical for maintaining the integrity of a marine vessel.

Based on the occurrence of an initiating event, ETA examines possible system outcomes or consequences. This analysis tool is particularly effective in showing interdependence

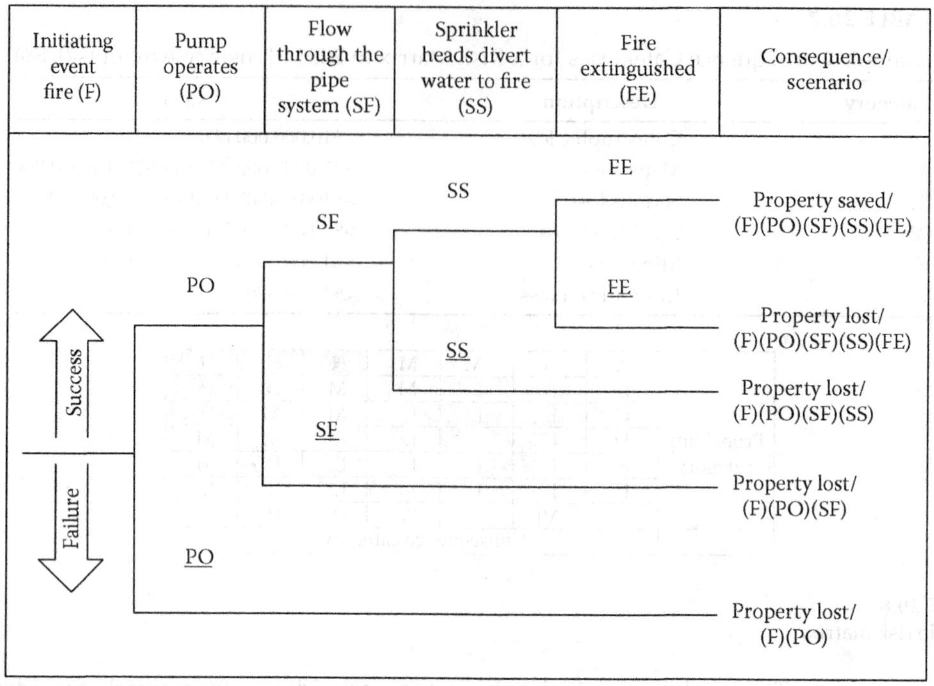

FIGURE 20.9
Event tree example for sprinkler system.

of system components, which is important in identifying events, that at first might appear insignificant, but due to severity factors. The high (H) level can be considered as unacceptable risk level, the medium (M) level can be treated as either undesirable or acceptable with review, and the low (L) level can be treated as acceptable without review. The interdependence leads to devastating results. ETA is similar to FTA because both methods use probabilistic reliability data of the individual components and events along each path to compute the likelihood of each outcome.

A quantitative evaluation of event tree probability values can be used for each event to evaluate the probability of the overall system state. Probability values for the success or failure of the events can be used to identify the probability for a specific event tree sequence. The probabilities of the events in a sequence can be provided as an input to the model or evaluated using fault trees. These probabilities for various events in a sequence can be viewed as conditional probabilities and therefore can be multiplied to obtain the occurrence probability of the sequence. The probabilities of various sequences can be summed up to determine the overall probability of a certain outcome. The addition of consequence evaluation of a scenario allows for generation of a risk value. For example, the occurrence probability of the top branch, that is, scenario, in Figure 20.9 is computed as the product of the probabilities of the composing events to this scenario, that is, F fi POH SFfi SSfi FE or (F)(PO)(SF)(SS)(FE) for short.

Complex systems are often difficult to visualize and the effect of individual components on the system as a whole is difficult to evaluate without an analytical tool. Two methods of modeling that have greatly improved the ease of assessing system reliability/risk are fault trees (FT) and success trees (ST). A fault tree is a graphical model created by deductive reasoning leading to various combinations of events that lead to the occurrence of some top event failure. A success tree shows the combinations of successful events leading to the success of the top event. A success tree can be produced as the complement (opposite) of the fault tree as illustrated in this section. FT and ST are used to further analyze the event tree headings (the main events in an event tree) to provide further detail to understand system complexities. In constructing the FT/ST, only those failure/success events that are considered significant are modeled. This determination is assisted by defining system boundaries. For example, the event "pump operates (PO)" in Figure 20.9 can be analyzed by developing a top-down logical breakdown of failure or success using fault tress or event trees, respectively.

FTA starts by defining a top event that is commonly selected as an adverse event. An engineering system can have more than one top event. For example, a ship might have the following top events for the purpose of reliability assessment: power failure, stability failure, mobility failure, or structural failure. Then, each top event needs to be examined using the following logic: in order for the top event to occur, other events must occur. As a result, a set of lower-level events is defined. Also, the form in which these lower-level events are logically connected (i.e., in parallel or in series) needs to be defined. The connectivity of these events is expressed using "AND" or "OR" gates. Lower-level events are classified into the following types:

1. Basic events: These events cannot be decomposed further into lower-level events. They are the lowest events that can be obtained. For these events, failure probabilities need to be obtained.
2. Events that can be decomposed further: These events can be decomposed further to lower levels. Therefore, they should be decomposed until the basic events are obtained.
3. Undeveloped events. These events are not basic and can be decomposed further. However, because they are not important, they are not developed further. Usually, the probabilities of these events are very small or the effect of their occurrence on the system is negligible, or can be controlled or mediated.
4. Switch (or house) events. These events are not random and can be turned on or off with full control. The symbols shown in Figure 20.10 are used for these events. Also, a continuation symbol is shown that is used to break up a fault tree into several parts for the purpose of fitting it in several pages.

FTA requires the development of a tree-looking diagram for the system that shows failure paths and scenarios that can result in the occurrence of a top event. The construction of the tree should be based on the building blocks and the Boolean logic gates.

The outcome of interest from the FTA is the occurrence probability of the top event. Because the top event was decomposed into basic events, its occurrence can be stated in the form of "AND" and "OR" of the basic events. The resulting statement can be restated by replacing the "AND" with the intersection of the corresponding basic events, and the "OR" with the union of the corresponding basic events. Then, the occurrence probability of the top event can be computed by evaluating the probabilities of the unions and intersections of the basic events. The dependence between these events also affects the resulting probability of the system.

For large fault trees, the computation of the occurrence probability of the top event can be difficult because of their size. In this case, a more efficient approach is needed for assessing the reliability of a system, such as the minimal cut set approach. According to this

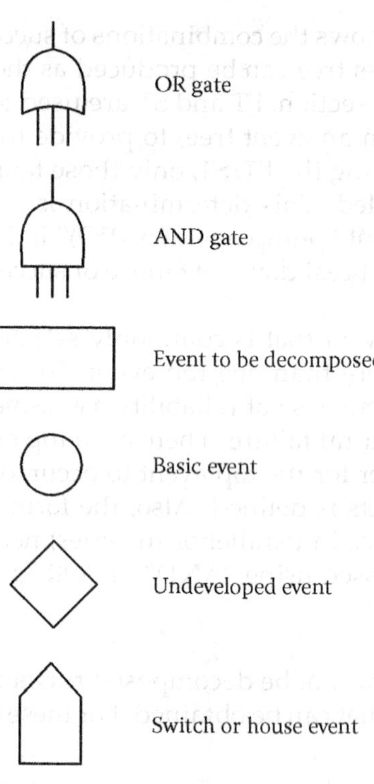

FIGURE 20.10
Symbols used in fault tree analysis.

approach, each cut set is defined as a set of basic events where the joint occurrence of these basic events results in the occurrence of the top event. A minimal cut set is a cut set with the condition that the nonoccurrence of any one basic event from this set results in the nonoccurrence of the top event. Therefore, a minimal cut set can be viewed as a subsystem in parallel. In general, systems have more than one minimal cut set. The occurrence of the top event of the system can, therefore, be due to any one of these minimal cut sets. As a result, the system can be viewed as the union of all the minimal cut sets for the system. If probability values are assigned to the cut sets, a probability for the top event can be determined.

A simple example of this type of modeling is shown in Figure 20.11 for a pipe system using a reliability block diagram. If the goal of the system is to maintain water flow from one end of the system to the other, then the individual pipes can be related with a Boolean logic. Both pipe (a) and pipe (d) and pipe (b) or pipe (c) must function for the system to meet its goal as shown in the success tree (Figure 20.12a). The complement of the success tree

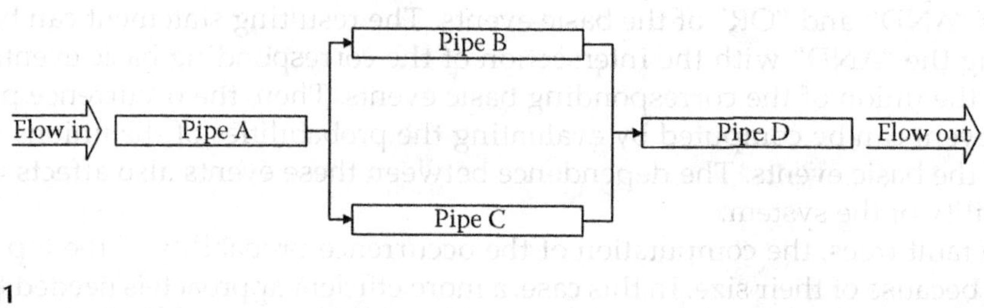

FIGURE 20.11
A reliability block diagram for a piping system.

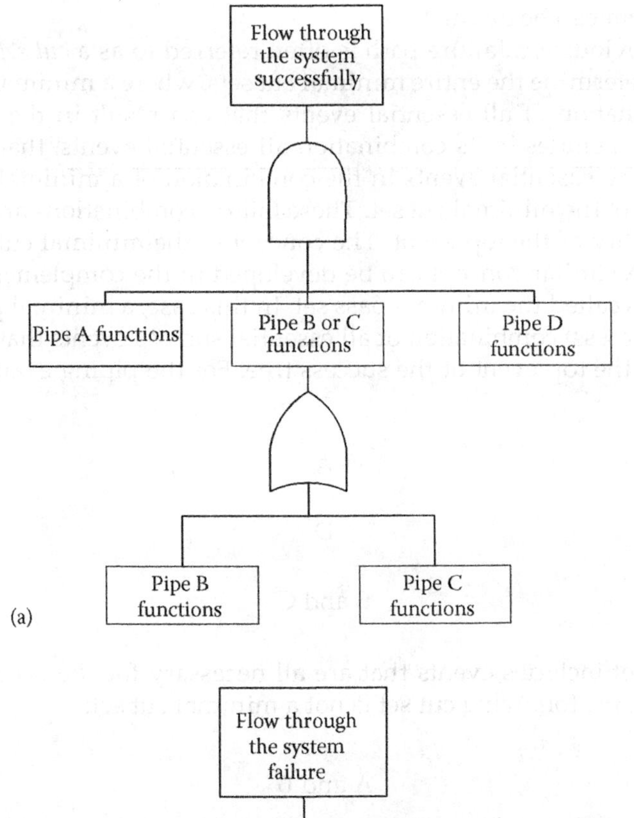

FIGURE 20.12
(a) Success tree for the pipe system example. (b) Fault tree for the pipe system example.

is the fault tree. The goal of the fault tree model is to construct the logic for system failure as shown in Figure 20.12b. After these tree elements have been defined, possible failure scenarios of a system can be defined.

As described previously, a failure path is often referred to as a *cut set*. One objective of the analysis is to determine the entire minimal cut sets, where a minimal cut set is defined as a failure combination of all essential events that can result in the failure top event. A minimal cut set includes in its combination all essential events, that is, the nonoccurrence of any of these essential events in the combination of a minimal cut set results in the nonoccurrence of the minimal cut set. These failure combinations are used to compute the failure probability of the top event. The concept of the minimal cut sets applies only to the fault trees. A similar concept can be developed in the complementary space of the success trees and is called the minimal pass set. In this case, a minimal pass set is defined as a survival (or success) combination of all essential success events that can result in success as defined by the top event of the success tree. For the piping example, the minimal cut sets are

$$\text{A} \tag{20.6a}$$

$$\text{D} \tag{20.6b}$$

$$\text{B and C} \tag{20.6c}$$

A minimal cut set includes events that are all necessary for the occurrence of the top event. For example, the following cut set is not a minimal cut set:

$$\text{A and B} \tag{20.7}$$

Example 20.1 Trends in Fault Tree Models and Cut Sets

This example demonstrates how the cut sets can be identified and constructed for different arrangements of OR and AND gates logically defining a top event occurrence. Generally, the number of cut sets increases by increasing the number of OR gates in the tree. For example, Figure 20.13 shows this trend by comparing cases a, b, and d. On the other hand, increasing the number of AND gates results in increasing the number of events included in the cut sets as shown in case c of Figure 20.13.

Common cause scenarios are events or conditions that result in the failure of seemingly separate systems or components. Common cause failures complicate the process of conducting risk analysis because a seemingly redundant system can be rendered ineffective by a common cause failure. For example, an emergency diesel generator fed by the same fuel supply as the main diesel engine will fail with the main diesel generator, if the fuel supply is the root source of the failure. The redundant emergency diesel generator is not truly redundant due to a common cause failure. Another example of common cause events is the failure of two separate but similar pieces of machinery due to a common maintenance problem, two identical pieces of equipment failing due to a common manufacturing defect, or two pieces of equipment failing due to a common environmental condition such as the flooding of a compartment or a fire in the vicinity of both pieces of machinery. A method for calculating the reliability of a system while taking into account common cause effects is the beta-factor model. Other methods include multiple Greek letter model, alpha-factor model, and beta-binomial failure-rate model.

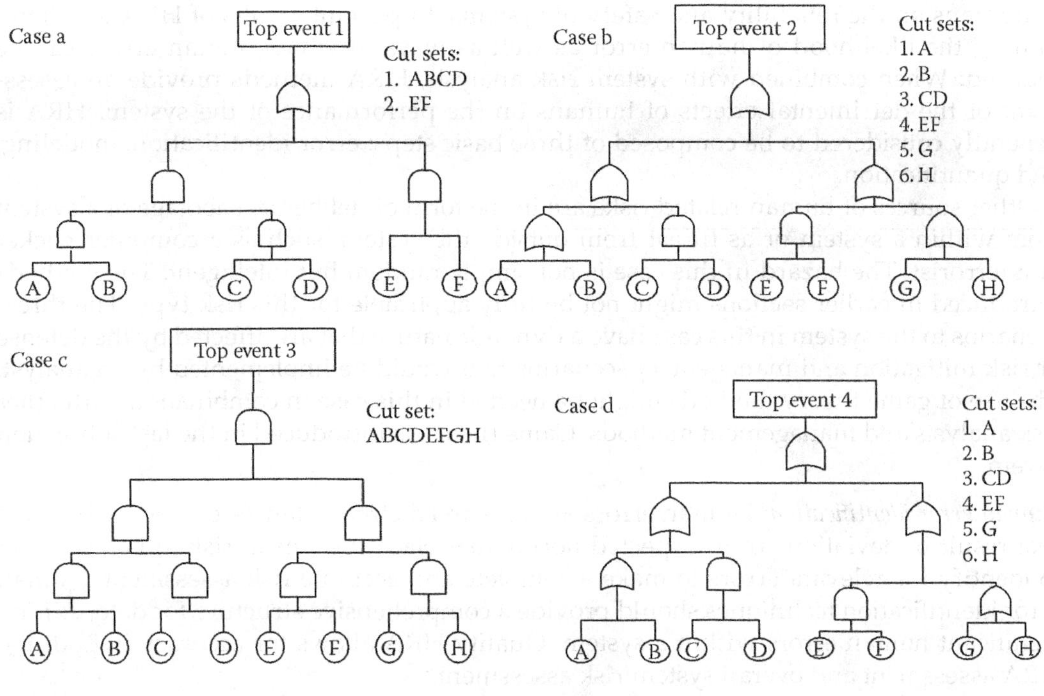

FIGURE 20.13
Trends in fault tree models and cut sets. (From Maryland Emergency Management Agency (MEMA), *State of Maryland Guide for the Protection of Critical Infrastructure and Key Resources for Homeland Security, Volume 1: Critical Asset & Portfolio Risk Assessment (CAPRA) Methodology*, Office of Homeland Security, Annapolis, MD, 2006.)

Part of risk-based decision analysis is pinpointing the system components that result in high-risk scenarios. Commercial system reliability software provides this type of analysis in the form of system reliability sensitivity factors to changes in the underlying component reliability values. In performing risk analysis, it is desirable to assess the importance of events in the model, or the sensitivity of final results to changes in the input failure probabilities for the events. Several sensitivity or importance factors are available and can be used. The most commonly used factors include (1) Fussell–Vesely factor and (2) Birnbaum factor. Also, a weighted combination of these factors can be used as an overall measure.

20.3.7 Human-Related Risks

Risk assessment requires the performance analysis of an entire system composed of a diverse group of components. The system definition readily includes the physical components of the system; however, humans are also part of most systems and provide significant contributions to risk. It has been estimated that nearly 90% of the accidents at sea are contributed to human error. The human contribution to risk can be estimated from an understanding of behavioral sciences. Both the "hardware failure" and human error should be addressed in the risk assessment since they both contribute to risks associated with the system. After the human error probabilities are determined, human error/failures are treated in the same fashion as hardware failures in performing risk assessment quantification.

The determination of the human error contribution to risk is determined by human reliability analysis (HRA) tools. HRA is the discipline that enables the analysis and impact of humans on the reliability and safety of systems. Important results of HRA are determining the likelihood of human error as well as ways in which human errors can be reduced. When combined with system risk analysis, HRA methods provide an assessment of the detrimental effects of humans on the performance of the system. HRA is generally considered to be composed of three basic steps: error identification, modeling, and quantification.

Other sources of human-related risks are in the form of deliberate sabotage of a system from within a system or as threat from outside the system, such as a computer hacker or a terrorist. The hazard in this case is not simply random but intelligent. The methods introduced in earlier sections might not be fully applicable for this risk type. The threat scenarios to the system in this case have a dynamic nature that are affected by the defense or risk mitigation and management scenarios that would be implemented by an analyst. The use of game theory methods might be needed in this case in combination with other risk analysis and management methods. Game theory is introduced in the last subsection herein.

Human error identification: Human errors are unwanted circumstances caused by humans that result in deviations from expected norms that place systems at risk. It is important to identify the relevant errors to make a complete and accurate risk assessment. Human error identification techniques should provide a comprehensive structure for determining significant human errors within a system. Quality HRA allows for accuracy in both the HRA assessment and overall system risk assessment.

Identification of human errors requires knowledge about the interactions of humans with other humans or machines (the physical world). It is the study of these interfaces that allow for the understanding of human errors. Potential sources of information for identifying human error may be determined from task analysis, expert judgment, laboratory studies, simulation, and reports. Human errors may be considered active or latent depending on the time delay between when the error occurs and when the system fails.

It is important to note the distinction between human errors and human factors. Human errors are generally considered separately from human factors that provide information about human behavior, abilities, limitations, and other characteristics to the design of tools, machines, systems tasks, jobs, and environments for productive, safe, comfortable, and effective human use. Human factors are determined from performing descriptive studies for characterizing populations and experimental research. However, human factors analysis may contribute to the HRA.

Human error modeling: After human errors have been identified, they must be represented in a logical and quantifiable framework along with other components that contribute to the risk of the system. This framework can be determined from the development of a risk model. Currently, there is no consensus on how to model human reliably. Many of these models utilize human event trees and fault trees to predict human reliability values. The identifications of human failure events can also be identified using failure mode and effects analysis. The human error rate estimates are often based on simulation tests, models, and expert estimation.

Human error quantification: Quantification of human error reliability promotes the inclusion of the human element in risk analysis. This is still a developing science requiring understanding of human performance, cognitive processing, and human perceptions. Because an exact model for human cognition has not been developed, much of the current human reliability data rely on accident databases, simulation, and other empirical approaches.

Many of the existing data sources were developed for from specific industry data such as nuclear and aviation industries. The application of these data sources for a specific problem should be thoroughly examined prior to application for a specific model. The result of the quantification of human reliability in terms of probability of occurrence is typically called a *human error probability (HEP)*. There are many techniques that have been developed to help predict the HEP values. The technique for human error rate prediction (THERP) is one of the most widely used methods for HEP. This technique is based on data gathered from the nuclear and chemical processing industries. THERP relies on HRA event tree modeling to identify the events of concern. Quantification is performed from data tables of basic HEP for specific tasks that may be modified based on the circumstances affecting performance.

The degree of human reliability is influenced by many factors often called *performance shaping factors (PSFs)*. PSFs are those factors that affect the ability of people to carry out required tasks. For example, the knowledge people have on how to don/activate a PFD will affect the performance of this task. Training (another PSF) in donning PFDs can also assist in the ability to perform this task. Another example is the training that is given to passengers on airplanes before takeoff on using seatbelts, emergency breathing devices, and flotation devices. Often, the quantitative estimates of reliability are generated from a base error rate that is then altered based on the PSFs of the particular circumstances. Internal performance shaping factors are an individual's own attributes (experience, training, skills, abilities, attitudes) that affect the ability of the person to perform certain tasks. External PSFs are the dynamic aspects of situation, tasks, and system that affect the ability to perform certain tasks. Typical external factors include environmental stress factors (such as heat, cold, noise, situational stress, time of day), management, procedures, time limitations, and quality of person–machine interface. With these PSFs, it is easy to see the dynamic nature of HEP evaluation based on the circumstances of the analysis.

Reducing human errors: Error reduction is concerned with lowering the likelihood for error in an attempt to reduce risk. The reduction of human errors may be achieved by human factors interventions or by engineering means. Human factors interventions include improving training or improving the human–machine interface (such as alarms, codes, etc.) based on an understanding of the causes of error. Engineering means of error reduction may include automated safety systems or interlocks. The selection of the corrective actions to take can be done through decision analysis considering cost–benefit criteria.

Game theory for intelligent threats: Game theory can be used to model human behavior as a threat to a system. Generally, game theory utilizes mathematics, economics, and the other social and behavioral sciences to model human behavior.

An example of intelligent threats is terrorism and sabotage as an ongoing battle between coordinated opponents representing a two-party game, where each opponent seeks to achieve their own objectives within a system. In the case of terrorism, it is a game of a well-established political system as a government versus an emerging organization that uses terrorism to achieve partial or complete dominance. Each player in this game seeks a utility, that is, benefit, that is a function of the desired state of the system. In this case, maintaining system survival is the desired state for the government, whereas the opponent seeks a utility based on the failure state of the system. The government, as an opponent, is engaged in risk mitigation whose actions seek to reduce the threat, reduce the system vulnerability, and/or mitigate the consequences of any successful attacks. The terrorists, as an opponent, can be viewed as the aggressor who strives to alter or damage their opponent's desired system state. This game involves an intelligent threat and is dynamic. The game is ongoing until the probability of a successful disruptive attempt of the aggressor reaches an acceptable level of risk—a stage where risk is considered under control—and the game is brought to an end. Classical game theory can be used in conjunction with probabilistic risk analysis to determine optimal mitigation actions that maximize benefits.

A classical example used to introduce game theory is called the *prisoners' dilemma* and is based on two suspects that are captured near the scene of a crime and are questioned separately by authority such as the police. Each has to choose whether or not to confess and implicate the other. If neither person confesses, then both will serve, for example, 1 year on a charge of carrying a concealed weapon. If each confesses and implicates the other, both will go to prison for, say, 10 years. However, if one person confesses and implicates the other, and the other person does not confess, the one who has collaborated with the police will go free, while the other person will go to prison for, say, 20 years on the maximum penalty. The strategies in this case are as follows: confess or do not confess. The payoffs, herein *penalties,* are the sentences served. The problem can be expressed compactly in a payoff table of a kind that has become standard in game theory as provided in Table 20.8. The entries of this table mean that each prisoner chooses one of the two strategies, that is, the first suspect chooses a row and the second suspect chooses a column. The two numbers in each cell of the table provide the outcomes for the two suspects for the corresponding pair of strategies chosen by the suspects as an ordered pair. The number to the left of the comma is the payoff to the person who chooses the rows, that is, the first suspect, whereas the number to the right of the comma is the payoff to the person who chooses the columns, that is, the second suspect. Thus, reading down the first column, if they both confess, each gets 10 years, but if the second suspect confesses and the first suspect does not, the first suspect gets 20 years and second suspect goes free. This example is not a zero-sum game because the payoffs are all losses. However, many problems can be cast with losses (negative numbers) and gains (positive numbers) with a total for each cell in the payoff table. A problem with a payoff table such that the payoffs in each cell add up to zero is called a *zero-sum game.*

The solution to this problem needs to be based on identifying rational strategies that can be based on both persons wanting to minimize the time they spend in jail. One suspect might reason that "two things can happen: either the other suspect confesses or keeps quiet. Suppose the second suspect confesses, I will get 20 years if I don't confess, 10 years if I do; therefore, in this case, it's best to confess. On the other hand, if the other suspect doesn't confess, and I don't either, I get a year; but in that case, if I confess I can go free. Either way, it's best if I confess. Therefore, I'll confess." But the other suspect can and presumably will reason in the same way. In this case, they both confess and go to prison for 10 years each, although if they had acted irrationally, and kept quiet, they each could have gotten off with 1 year each. The rational strategies of the two suspects have fallen

TABLE 20.8

Payoff Table in Years for the Prisoners' Dilemma Game

		Second Suspect	
		Confess	**Don't Confess**
First suspect	Confess	(10, 10)	(0, 20)
	Don't confess	(20, 0)	(1, 1)

into something called *dominant strategy equilibrium.* The meaning of the term "dominant strategy equilibrium" requires defining the term "dominant strategy" that results from an individual player (the suspect, in this case) in a game evaluating separately each of the strategy combinations being faced, and, for each combination, choosing from these strategies the one that gives the greatest payoff. If the same strategy is chosen for each of the different combinations of strategies the player might face, that strategy is called a dominant strategy for that player in that game. The dominant strategy equilibrium occurs if, in a game, each player has a dominant strategy, and each player plays the dominant strategy, then that combination of (dominant) strategies and the corresponding payoffs are said to constitute the dominant strategy equilibrium for that game. In the prisoners' dilemma game, to confess is a dominant strategy, and when both suspects confess, dominant strategy equilibrium is established. The dominant strategy equilibrium is also called *Nash equilibrium.* The definition of Nash equilibrium is a set of strategies with the property that no player can benefit by changing his/her strategy while the other players keep their strategies unchanged, then that set of strategies and the corresponding payoffs constitute the Nash equilibrium.

The prisoners' dilemma game is based on two strategies per suspect that can be viewed as deterministic in nature, that is, nonrandom. In general, many games, especially those permitting repeatability in choosing strategies by players, can be constructed with strategies that have associated probabilities. For example, strategies can be constructed based on probabilities of 0.4 and 0.6 that sum to one. Such strategies with probabilities are called *mixed strategies* as opposed to *pure strategies* that do not involve probabilities of the prisoners' dilemma game. A mixed strategy occurs in a game if a player chooses among two or more strategies at random according to specific probabilities.

In general, gaming could involve more than two players. In the prisoners' dilemma game, a third player that could be identified is the authority and its strategies. The solution might change as a result of adding the strategies of this third player. The use of these concepts in risk analysis and mitigation needs further development and exploration.

Risk methods for protecting infrastructure and key resources: The protection of critical infrastructure and key resources (CI/KR) for homeland security requires choosing among a large set of protective, response, and recovery actions for reducing risk to an acceptable level. The selection of investment alternatives for improving asset security and increasing infrastructure resilience depends on two factors: their cost to implement and relative cost-effectiveness. To accomplish this task, the Department of Homeland Security (DHS) has identified risk methods as the primary underlying framework for system evaluations, operational assessments, technology assessments, resource and support analyses, and field operations analyses. According to the draft DHS National Infrastructure Protection Plan, cost–benefit analysis is the hallmark of critical infrastructure protection decision making. Figure 20.14 shows a methodology for informing decisions relating to CI/KR protection (MEMA 2006). It is a quantitative, asset-driven framework for assessing the risk of consequential CI/KR exploitation and disruption due to a variety of plausible threat scenarios and hazard events. This process is divided into five phases as shown in Figure 20.14. The results from each phase support decision making of some type, whether it be asset screening and selection based on consequences and their criticality, or countermeasure selection and evaluation based on the results from the security vulnerability assessment phase.

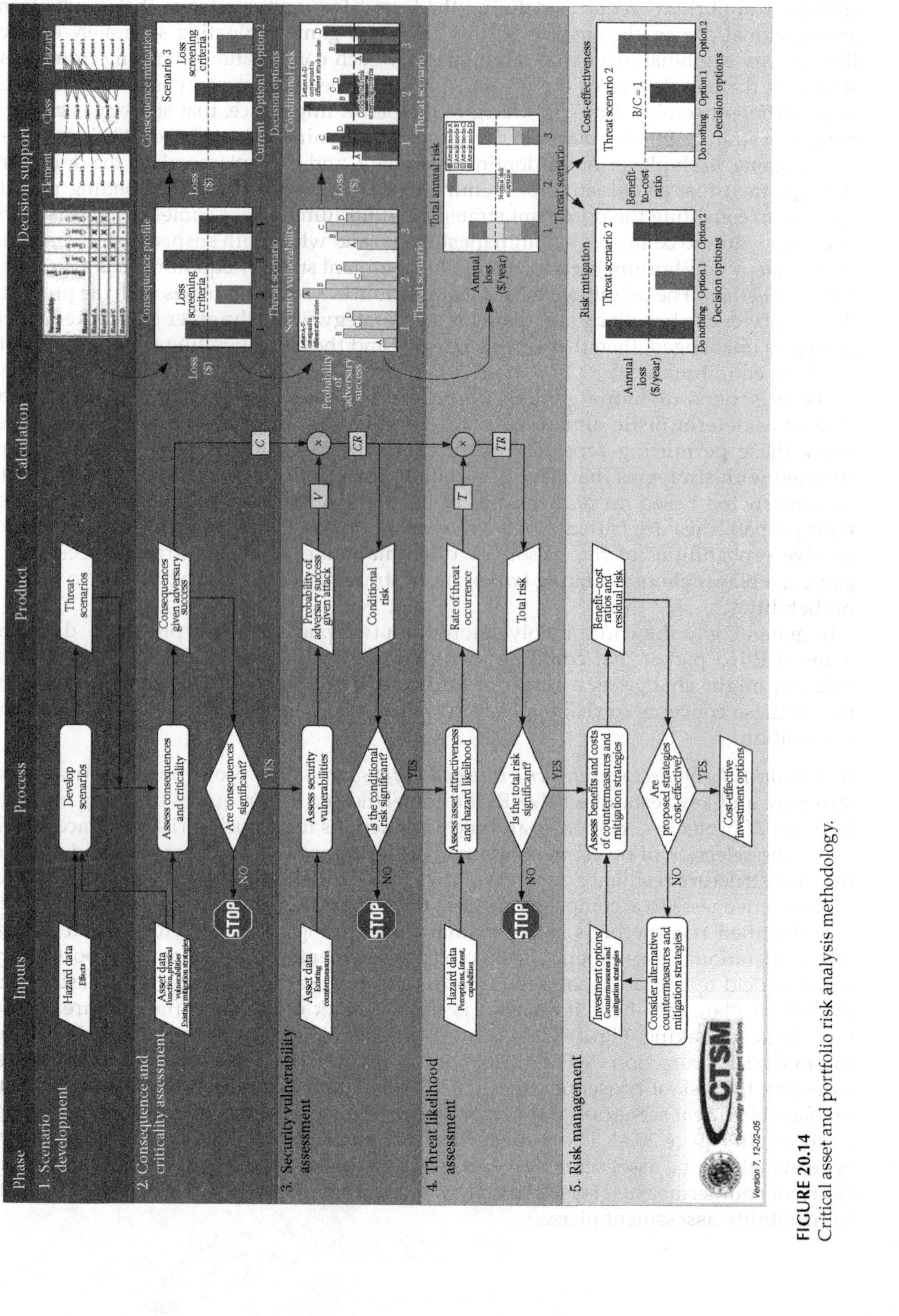

FIGURE 20.14
Critical asset and portfolio risk analysis methodology.

20.3.8 Economic and Financial Risks

Economic and financial risks can be grouped into categories that include market risks, credit risks, operation risks, and reputation risks. These four categories are described in subsequent sections.

Market risks: Governments and corporations operate in economic and financial environments with some levels of uncertainty and instability. A primary contributor to defining this environment is interest rates. Interest rates can have significant impact on the costs of financing a project and corporate cash flows and asset values. For example, interest rates in the United States shot up in 1979 and peaked in 1981, followed by gradual decline with some fluctuations until 2002.

For projects that target global markets, exchange rate instability can be a major risk source. Exchange rates have been volatile ever since the breakdown of the Bretton Woods system to fixed exchange rates in the early 1970s. An example of a bust-up in exchange rates is the fall of the value of the British sterling and Italian lira as a result of the failure of the exchange rate mechanism in September 1992.

Many projects are dependent on the availability of venture capital and the stock performance of corporation thereby introducing another risk source related to stock market volatility. Stock prices rose significantly in the inflationary booms of the early 1970s, then fell considerably a little later. They recovered afterward and fell again in early 1981. The market rose to a peak until it crashed in 1987, followed by an increase with some swings till it reached reaching a new peak fueled by Internet technologies and finally collapsed in 2001.

Other contributing factors to economic and finance instability are commodity prices in general and energy prices in particular, primarily crude oil. The hikes in oil prices in the 1973–1974 affected commodity prices greatly and posed a series of challenges to countries and corporations.

Another contributing source to volatility is derivatives for commodities, foreign currency exchange rates, and stock prices and indices, among others. Derivatives are defined as contracts whose values or payoffs depend on those of other assets, such as the options to buy commodities in the future or options to sell commodities in the future. They offer not only opportunities for hedging positions and managing risks that can be stabilizing, but also speculative opportunities to others that can be destabilizing and a contributor to volatility.

Credit risks: Credit risks are associated with potential defaults on notes or bonds, as examples, by corporations, including subcontractors. Also, credit risks can be associated with market sentiments that determine a company likelihood of default that could affect its bond rating and ability to purchase money and maintain projects and operations.

Operational risks: Operational risks are associated with several sources that include out-of-control operations risk that could occur when a corporate branch undertakes significant risk exposure that is not accounted for by a corporate headquarters leading potentially to its collapse, an example being the British Barings Bank that collapsed primarily as a result of its failure to control the market exposure being created within a small overseas branch of the bank.

Another risk source in this category is liquidity in which a corporation needing funding more than it can arrange. Also, it could include money transfer risks and agreement breach.

Operational risks include model risks. Model risks are associated with the models and underlying assumptions used to incorrectly value financial instruments and cash flows.

Reputation risks: The loss of business attributable to decrease in a corporation's reputation can pose another risk source. This risk source can affect its credit rating, ability to maintain clients, workforce, etc. This risk source usually occurs at a slow attrition rate. It can be an outcome of poor management decisions and business practices.

20.3.9 Data Needs for Risk Assessment

In risk assessment, the methods of probability theory are used to represent engineering uncertainties. In this context, it refers to event occurrence likelihoods that occur with periodic frequency, such as weather, yet also to conditions that are existent but unknown, such as probability of an extreme wave. It applies to the magnitude of an engineering parameter, yet also to the structure of a model. By contrast, probability is a precise concept. It is a mathematical concept with an explicit definition. The mathematics of probability theory are used to represent uncertainties, despite that those uncertainties are of many forms.

The term *probability* has a precise mathematical definition, but its meaning when applied to the representation of uncertainties is subject to differing interpretations. The frequentist view holds that probability is the propensity of a physical system in a theoretically infinite number of repetitions; that is, the frequency of occurrence of an outcome in a long series of similar trials (e.g., the frequency of a coin landing heads-up in an infinite number of flips is the probability of that event). In contrast, the Bayesian view holds that probability is the rational degree of belief that one holds in the occurrence of an event or the truth of a proposition; probability is manifest in the willingness of an observer to take action upon this belief. This latter view of probability, which has gained wide acceptance in many engineering applications, permits the use of quantified professional judgment in the form of subjective probabilities. Mathematically, such subjective probabilities can be combined or operated on as any other probability.

Data are needed to perform quantitative risk assessment or provide information to support qualitative risk assessment. Information maybe available if data have been maintained on a system and components of interest. The relevant information for risk assessment included the possible failures, failure probabilities, failure rates, failure modes, possible causes, and failure consequences. In the case of a new system, data maybe used from similar systems if this information is available. Surveys are a common tool used to provide some means of data. Statistical analysis can be used to assess confidence intervals and uncertainties in estimated parameters of interest. Expert judgment may also be used as another source of data (Ayyub 2002). The uncertainty with the quality of the data should be identified to assist in the decision-making process.

Data can be classified to include generic- and project- or plant-specific types. Generic data are information from similar systems and components. This information may be the only information available in the initial stages of system design. Therefore, potential differences due to design or uncertainty may result from using generic data on a specific system. Plant-specific data are specific to the system being analyzed. This information is often developed after the operation of a system. Relevant data need to be identified and collected as data collection can be costly. The data collected can then be used to update the risk assessment. Bayesian techniques can be used to combine objective and subjective data.

Data can be classified as failure probability data and failure consequence data. The failure probability data can include failure rates, hazard functions, times between failures, results from reliability studies, and any influencing factors and their effects. Failure consequence data include loss reports, damages, litigation outcomes, repair costs, injuries, and human losses. Also included are influencing factors and effects of failure prevention and consequence mitigation plans. Areas of deficiency in terms of data availability should be identified, and sometimes failure databases need to be constructed. Data deficiency can be used as a basis for data collection and expert-opinion elicitation.

20.4 Risk Management and Control

Adding risk control to risk assessment produces risk management. Risk management is the process by which system operators, managers, and owners make safety decisions, regulatory changes, and choose different system configurations based on the data generated in the risk assessment. Risk management involves using information from the previously described risk assessment stage to make educated decisions about system safety. Risk control includes failure prevention and consequence mitigation.

Risk management requires the optimal allocation of available resources in support of group goals. Therefore, it requires the definition of acceptable risk, and comparative evaluation of options and/or alternatives for decision making. The goals of risk management are to reduce risk to an acceptable level and/or prioritize resources based on comparative analysis. Risk reduction is accomplished by preventing an unfavorable scenario, reducing the frequency, and/or reducing the consequence. A graph showing the risk relationship is shown in Figure 20.15 as linear contours of constant risk, although due to risk aversion these lines are commonly estimated as nonlinear curves and should be treated as nonlinear curves. Moreover, the vertical axis is termed as probability, whereas it is commonly expressed as an annual exceedence probability or frequency as shown in Figure 20.1. In cases involving qualitative assessment, a matrix presentation can be used as shown

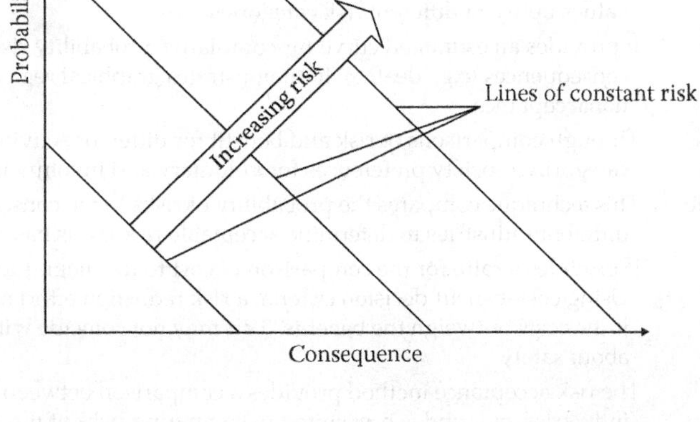

FIGURE 20.15
Risk plot.

in Figure 20.8. The figure shows probability categories, severity categories, and risk ratings. A project's base value is commonly assumed as zero. Each risk rating value requires a different mitigation plan.

20.4.1 Risk Acceptance

Risk acceptance constitutes a definition of safety as discussed in previous sections. Therefore, risk acceptance is considered a complex and controversial subject that is often subject to debate. The determination of acceptable levels of risk is important to determine the risk performance a system needs to achieve to be considered safe. If a system has a risk value above the risk acceptance level, actions should be taken to address safety concerns and improve the system through risk reduction measures. One difficulty with this process is defining acceptable safety levels for activities, industries, structures, etc. Because the acceptance of risk depends upon society's perceptions, the acceptance criteria do not depend on the risk value alone. This section describes several methods that have been developed to assist in determining acceptable risk values as summarized in Table 20.9.

Risk managers make decisions based on risk assessment and other considerations including economical, political, environmental, legal, reliability, producibility, safety, and other factors. The answer to the question "How safe is safe enough?" is difficult and constantly changing due to different perceptions and understandings of risk. To determine "acceptable risk," managers need to analyze alternatives for the best choice. In some industries, an acceptable risk has been defined by consensus. For example, the U.S. Nuclear Regulatory Commission requires that reactors be designed such that the probability of a large radioactive release to the environment from a reactor incident shall be less than 1×10^{-6} per year. Risk levels for certain carcinogens and pollutants have also been given acceptable concentration levels based on some assessment of acceptable risk. However, risk acceptance for many other activities is not stated.

TABLE 20.9

Methods for Determining Risk Acceptance

Risk Acceptance Method	Summary
Risk conversion factors	This method addresses the attitudes of the public about risk through comparisons of risk categories. It also provides an estimate for converting risk acceptance values between different risk categories
Farmers curve	It provides an estimated curve for cumulative probability risk profile for certain consequences (e.g., deaths). It demonstrates graphical regions of risk acceptance/nonacceptance
Revealed preferences	Through comparisons of risk and benefit for different activities, this method categorizes society preferences for voluntary and involuntary exposure to risk.
Evaluation of magnitude of consequences	This technique compares the probability of risks to the consequence magnitude for different industries to determine acceptable risk levels based on consequence
Risk effectiveness	It provides a ratio for the comparison of cost to the magnitude of risk reduction. Using cost–benefit decision criteria, a risk reduction effort should not be pursued if the costs outweigh the benefits. This may not coincide with society values about safety
Risk comparison	The risk acceptance method provides a comparison between various activities, industries, etc., and is best suited to comparing risks of the same type

For example, qualitative implications for risk acceptance are identified in the several existing maritime regulations. The International Maritime Organization High-Speed Craft Code and the U.S. Coast Guard Navigation and Vessel Inspection Circular (NVIC) 5–93 for passenger submersible guidance both state that if the end effect is hazardous or catastrophic, a backup system and a corrective operating procedure are required. These references also state that a single failure must not result in a catastrophic event, unless the likelihood is extremely remote.

Often the level of risk acceptance with various activities is implied. Society has reacted to risks through the developed level of balance between risk and potential benefits. Measuring this balance of accepted safety levels for various risks provides a means for assessing society values. These threshold values of acceptable risk depend on a variety of issues including the activity type, industry, and users, and the society as a whole.

Target risk or reliability levels are required for developing procedures and rules for ship structures. For example, the selected reliability levels determine the probability of failure of structural components. The following three methods were used to select target reliability values:

1. Agreeing upon a reasonable value in cases of novel structures without prior history
2. Calibrating reliability levels implied in currently, successfully used design codes
3. Choosing target reliability level that minimizes total expected costs over the service life of the structure for dealing with design for which failure results in only economic losses and consequences

The first approach can be based on expert-opinion elicitation. The second approach, called *code calibration,* is the most commonly used approach as it provides the means to build on previous experiences. For example, rules provided by classification and industry societies can be used to determine the implied reliability and risk levels in respective rules and codes, then target risk levels can be set in a consistent manner, and new rules and codes can be developed to produce future designs and vessels that are of similar levels that offer reliability and/or risk consistency. The third approach can be based on economic and trade-off analysis. In subsequent sections, the methods of Table 20.9 for determining risk acceptance are discussed.

Risk conversion factors: Analysis of risks shows that there are different taxonomies that demonstrate the different risk categories, often called *risk factors.* These categories can be used to analyze risks on a dichotomous scale comparing risks that invoke the same perceptions in society. For example, the severity category may be used to describe both ordinary and catastrophic events. Grouping events that could be classified as ordinary and comparing the distribution of risk to a similar grouping of catastrophic categories yields a ratio describing the degree of risk acceptance of ordinary events as compared to catastrophic events. The comparison of various categories determined the risk conversion values as provided in Table 20.10. These factors are useful in comparing the risk acceptance for different activities, industries, etc. By computing the acceptable risk in one activity, an estimate of acceptable risk in other activities can be calculated based on the risk conversion factors. A comparison of several common risks based on origin and volition is shown in Table 20.11.

Farmers curve: The Farmers curve is a graph of the cumulative probability versus consequence for some activity, industry, or design, as shown in Figures 20.1 and 20.2. This curve

TABLE 20.10

Risk Conversion Values for Different Risk Factors

Risk Factors	Risk Conversion (RF) Factor	Computed RF Value
Origin	Natural/human made	20
Severity	Ordinary/catastrophic	30
Volition	Voluntary/involuntary	100
Effect	Delayed/immediate	30
Controllability	Controlled/uncontrolled	5–10
Familiarity	Old/new	10
Necessity	Necessary/luxury	1
Costs	Monetary/nonmonetary	NA
Origin	Industrial/regulatory	NA
Media	Low profile/high profile	NA

Note: NA, not available.

TABLE 20.11

Classification of Common Risks

		Voluntary		Involuntary	
Source	**Size**	**Immediate**	**Delayed**	**Immediate**	**Delayed**
Human	Catastrophic	Aviation		Dam failure, building fire, nuclear accident	Pollution, building fire
Made	Ordinary	Sports, boating, automobiles	Smoking, occupation, carcinogens	Homicide	
Natural	Catastrophic			Earthquakes, hurricanes, tornadoes, epidemics	
	Ordinary			Lighting animal bites	Disease

introduces a probabilistic approach in determining acceptable safety limits. Probability (or frequency) and consequence values are calculated for each level of risk generating a curve that is unique to the hazard of concern. The area to the right (outside) of the curve is generally considered unacceptable because the probability and consequence values are higher than the average value delineated by the curve. The area to the left (inside) of the curve is considered acceptable because probability and consequence values are less than the estimated valve of the curve.

Method of revealed preferences: The method of revealed preferences provides a comparison of risk versus benefit and categorization for different risk types. The basis for this relationship is that risks are not taken unless there is some form of benefit. Benefit may be monetary or some other item of worth such as pleasure. The different risk types are for the risk category of voluntary versus involuntary actions as shown in Figure 20.16.

Magnitudes of risk consequence: Another factor affecting the acceptance of risk is the magnitude of consequence of the event that can result from some failure. In general, the larger the consequence, the less the likelihood that this event may occur. This technique has been used in several industries to demonstrate the location of the industry within societies' risk acceptance levels based on consequence magnitude as shown in Figure 20.17. Further evaluation has resulted in several estimates for the relationship between the accepted

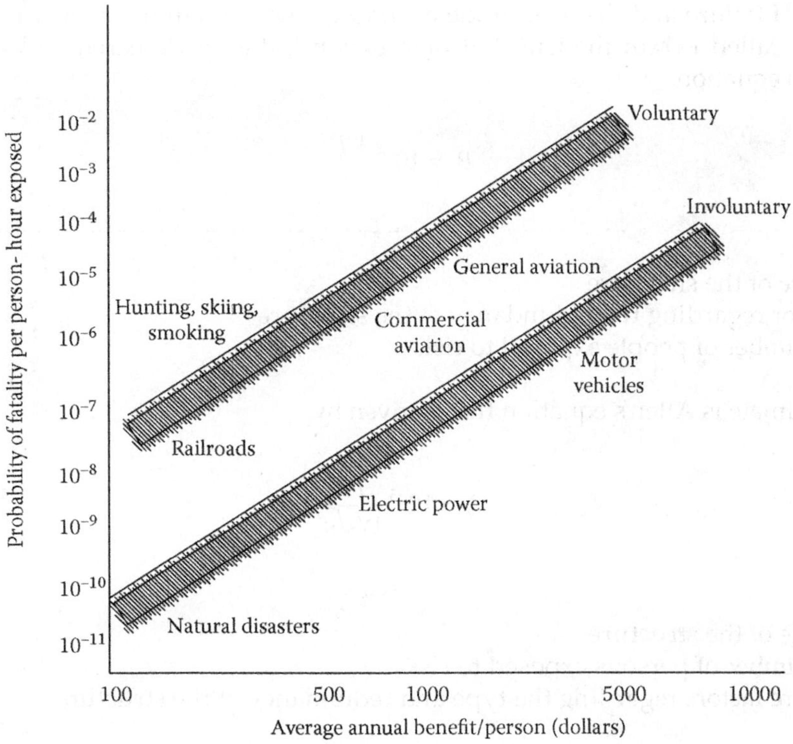

FIGURE 20.16
Accepted risk of voluntary and involuntary activities.

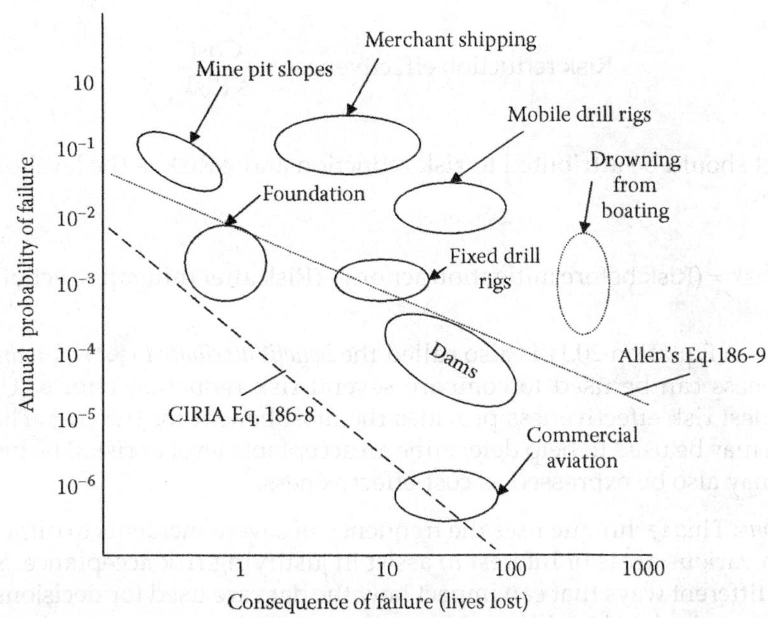

FIGURE 20.17
Comparison of risk and control costs.

probability of failure and the magnitude of consequence for failure as provided by Allen in 1981 and called herein the CIRIA (Construction Industry Research and Information Association) equation:

$$P_f = 10^{-4}\frac{KT}{n} \tag{20.8}$$

where
- T is the life of the structure
- K is a factor regarding the redundancy of the structure
- n is the number of people exposed to risk

Another estimate is Allen's equation that is given by

$$P_f = 10^{-5}\frac{TA}{W\sqrt{n}} \tag{20.9}$$

where
- T is the life of the structure
- n is the number of persons exposed to risk
- A and W are factors regarding the type and redundancy of the structure

Equation 20.8 offers a lower bound, whereas Equation 20.9 offers a middle line.

Risk reduction cost-effectiveness ratio: Another measuring tool to assess risk acceptance is the determination of risk reduction effectiveness:

$$\text{Risk reduction effectiveness} = \frac{\text{Cost}}{\Delta\,\text{Risk}} \tag{20.10}$$

where the cost should be attributed to risk reduction and ΔRisk is the level of risk reduction as follows:

$$\Delta\,\text{Risk} = (\text{Risk before mitigation action}) - (\text{Risk after mitigation action}) \tag{20.11}$$

The difference in Equation 20.11 is also called the *benefit attributed to a risk reduction action*. Risk effectiveness can be used to compare several risk reduction efforts. The initiative with the smallest risk effectiveness provides the most benefit for the cost. Therefore, this measurement may be used to help determine an acceptable level of risk. The inverse of this relationship may also be expressed as cost-effectiveness.

Risk comparisons: This technique uses the frequency of severe incidents to directly compare risks between various areas of interest to assist in justifying risk acceptance. Risks can be presented in different ways that can impact how the data are used for decisions. Often values of risk are manipulated in different forms for comparison reasons as demonstrated in Table 20.12. Comparison of risk values should be taken in the context of the values' origin and uncertainties involved.

TABLE 20.12

Ways to Identify Risk of Death

Ways to Identify Risk of Death	Summary
Number of fatalities	This measure shows the impact in terms of the number of fatalities on society. Comparison of these values is cautioned since the number of persons exposed to the particular risk may vary. Also, the time spent performing the activity may vary. Different risk category types should also be considered to compare fatality rates.
Annual mortality rate/individual	This measure shows the mortality risk normalized by the exposed population. This measure adds additional information about the number of exposed persons; however, the measure does not include the time spent on the activity.
Annual mortality	This measure provides the most complete risk value since the risk is normalized by the exposed population and the duration of the exposure.
Loss of life exposure (LLE)	This measure converts a risk into a reduction in the expected life of an individual. It provides a good means of communicating risks beyond probability values.
Odds	This measure is a layman format for communicating probability, for example, 1 in 4.

This technique is most effective for comparing risks that invoke the same human perceptions and consequence categories. Comparing risks of different categories is cautioned because the differences between risk and perceived safety may not provide an objective analysis of risk acceptance. The use of risk conversion factors may assist in transforming different risk categories. Conservative guidelines for determining risk acceptance criteria can be established for voluntary risks to the public from the involuntary risk of natural causes.

20.4.1.1 Rankings Based on Risk Results

Another tool for risk management is the development of risk ranking. The elements of a system within the objective of analysis can be analyzed for risk and consequently ranked. This relative ranking may be based on the failure probabilities, failure consequences, risks, or other alternatives with concern toward risk. Generally, risk items ranked highly should be given high levels of priority; however, risk management decisions may consider other factors such as costs, benefits, and effectiveness of risk reduction measures. The risk ranking results may be presented graphically as needed.

20.4.1.2 Decision Analysis

Decision analysis provides a means for systematically dealing with complex problems to arrive at a decision. Information is gathered in a structured manner to provide the best answer to the problem. A decision generally deals with three elements: alternatives, consequences, and preferences. The alternatives are the possible choices for consideration. The consequences are the potential outcomes of a decision. Decision analysis provides methods for quantifying preference trade-offs for performance along multiple decision attributes while taking into account risk objectives. Decision attributes are the performance scales that measure the degree to which objectives are satisfied. For example, one possible attribute is reducing the lives lost for the objective of increasing safety. Additional examples of objectives may include minimizing the cost, maximizing utility, maximizing reliability, and maximizing profit. The decision outcomes may be affected by uncertainty; however, the goal is to choose the best alternative with the proper consideration of uncertainty. The analytical depth and rigor for decision analysis depends on the desired detail in making the decision. Cost–benefit analysis, decision trees, influence diagrams, and the analytic hierarchy process are some of the tools to assist in decision analysis. Also, decision analysis should consider constraints, such as availability of system for inspection, availability of inspectors, preference of certain inspectors, and availability of inspection equipment.

20.4.1.3 Cost–Benefit Analysis

Risk managers commonly weigh various factors including cost and risk. The analysis of three different alternatives is shown graphically in Figure 20.18 as an example. The graph shows that alternative (C) is the best choice because the level of risk and cost is less than alternatives (A) and (B). However, if the only alternatives were A and B, the decision would be more difficult. Alternative (A) has higher cost and lower risk than alternative (B); alternative (B) has higher risk but lower cost than alternative (A). A risk manager needs to weigh the importance of risk and cost in making this decision and availability of resources and make use of risk-based decision analysis.

Risk–benefit analysis can also be used for risk management. Economic efficiency is important to determine the most effective means of expending resources. At some point, the costs for risk reduction do not provide adequate benefit. This process compares the costs and risk to determine where the optimal risk value is on a cost basis. This optimal value occurs, as shown in Figure 20.19, when costs to control risk are equal to the risk cost due to the consequence (loss). Investing resources to reduce low risks below this equilibrium point is not providing a financial benefit. This technique may be used when cost values can be attributed to risks. This analysis might be difficult to perform for certain risks such as risk to human health and environmental risks because the monetary values are difficult to estimate for human life and the environment.

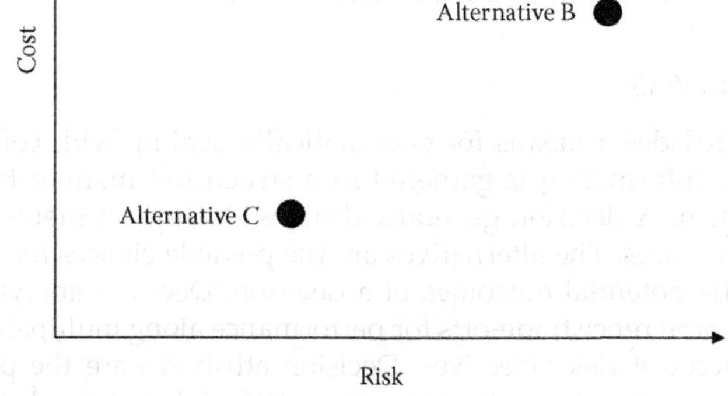

FIGURE 20.18
Risk benefit for three alternatives.

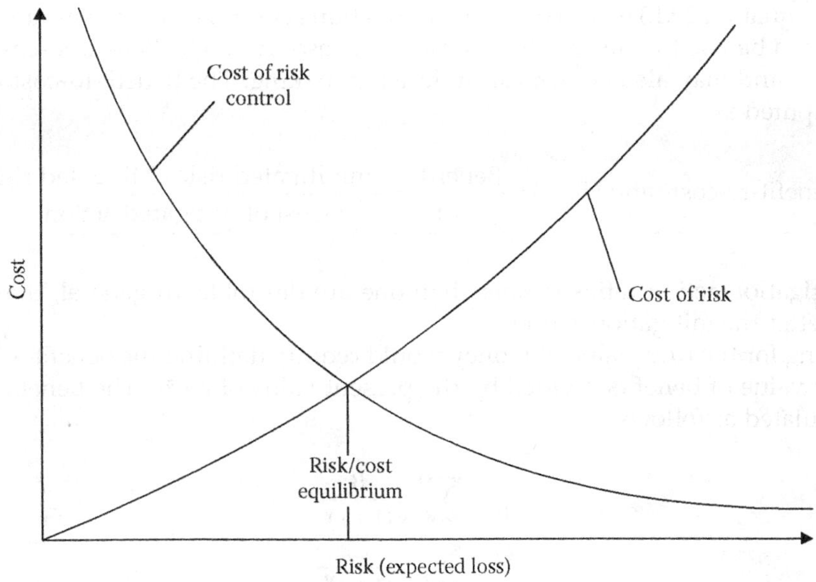

FIGURE 20.19
Comparison of risk and control costs.

The present value of incremental costs and benefits can be assessed and compared among alternatives that are available for risk mitigation or system design. Several methods are available to determine which, if any, option is most worth pursuing. In some cases, no alternative will generate a net benefit relative to the base case. Such a finding would be used to argue for pursuit of the base case scenario. The following are the most widely used present value comparison methods: (1) NPV, (2) benefit–cost ratio, (3) internal rate of return, and (4) payback period. NPV method requires that each alternative need to meet the following criteria to warrant investment of funds: (1) having a positive NPV and (2) having the highest NPV of all alternatives considered. The first condition insures that the alternative is worth undertaking relative to the base case, for example, it contributes more in incremental benefits than it absorbs in incremental costs. The second condition insures that maximum benefits are obtained in a situation of unrestricted access to capital funds. The NPV can be calculated as follows:

$$\text{NPV} = \sum_{t=0}^{k} \frac{(B-C)_t}{(1+r)^t} = \sum_{t=0}^{k} \frac{B_t}{(1+r)^t} - \sum_{t=0}^{k} \frac{C_t}{(1+r)^t} \tag{20.12}$$

where
- B is the future annual benefit in constant dollars
- C is the future annual cost in constant dollars
- r is the annual real discount rate
- k is the number of years from the base year over which the project will be evaluated
- t is an index running from 0 to k representing the year under consideration

The benefit of a risk mitigation action can be assessed as follows:

$$\text{Benefit} = \text{Unmitigated risk} - \text{Mitigated risk}. \tag{20.13}$$

The cost in Equation 20.13 is the cost of the mitigation action. The benefit minus the cost of mitigation can be used to justify the allocation of resources. The benefit-to-cost ratio can be computed and may also be helpful in decision making. The benefit-to-cost ratio (B/C) can be computed as

$$\text{Benefit-to-cost ratio } (B/C) = \frac{\text{Benefit}}{\text{Cost}} = \frac{\text{Unmitigated risk} - \text{Mitigated risk}}{\text{Cost of mitigated action}} \tag{20.14}$$

Cost of mitigation action ratios greater than one are desirable. In general, the larger the ratio, the better the mitigation action.

Accounting for the time value of money would require defining the benefit–cost ratio as the present value of benefits divided by the present value of costs. The benefit–cost ratio can be calculated as follows:

$$\frac{B}{C} = \frac{\sum_{t=0}^{k} \frac{B_t}{(1+r)^t}}{\sum_{t=0}^{k} \frac{C_t}{(1+r)^t}} \tag{20.15}$$

where

B_t is the future annual benefit in constant dollars
C_t is the future annual cost in constant dollars
r is the annual real discount rate
t is an index running from 0 to k representing the year under consideration

A proposed activity with a B/C ratio of discounted benefits to costs of 1 or more is expected to return at least as much in benefits as it costs to undertake, indicating that the activity is worth undertaking.

The internal rate of return (*IRR*) is defined as the discount rate that makes the present value of the stream of expected benefits in excess of expected costs zero. In other words, it is the highest discount rate at which the project will not have a negative *NPV*. To apply the IRR criterion, it is necessary to compute the IRR and then compare it with a base rate of, say, a 7% discount rate. If the real IRR is less than 7%, the project would be worth undertaking relative to the base case. The IRR method is effective in deciding whether or not a project is superior to the base case; however, it is difficult to utilize it for ranking projects and deciding among mutually exclusive alternatives. Project rankings established by the IRR method might be inconsistent with those of the NPV criterion. Moreover, a project might have more than one IRR value, particularly when a project entails major final costs, such as cleanup costs. Solutions to these limitations exist in capital budgeting procedures and practices that are often complicated or difficult to employ in practice and present opportunities for error.

The payback period measures the number of years required for net undiscounted benefits to recover the initial investment in a project. This evaluation method favors projects with near-term and more certain benefits and fails to consider benefits beyond the payback period. The method does not provide information on whether an investment is worth undertaking in the first place.

The previous models for cost–benefit analysis presented in this section do not account for the full probabilistic characteristics of B and C in their treatment. Concepts from reliability assessment 4 can be used for this purpose. Assuming *B* and *C* to be normally distributed, a benefit–cost index ($\beta_{B/C}$) can be defined as follows:

$$\beta_{B/C} = \frac{\mu_B - \mu_C}{\sqrt{\sigma_B^2 + \sigma_C^2}} \tag{20.16}$$

where μ/J and σ are the mean and standard deviation. The failure probability can be computed as

$$P_{f,B/C} = P(C > B) = 1 - \Phi(\beta) \tag{20.17}$$

In the case of lognormally distributed *B* and *C*, the benefit–cost index ($\beta_{B/C}$) can be computed as

$$\beta_{B/C} = \frac{\ln\left(\frac{\mu_B}{\mu_C}\sqrt{\frac{\delta_C^2 + 1}{\delta_B^2 + 1}}\right)}{\sqrt{\ln\left[(\delta_B^2 + 1)(\delta_C^2 + 1)\right]}} \tag{20.18}$$

where δ is the coefficient of variation. Equation 20.18 also holds for the case of lognormally distributed *B* and *C*. In the case of mixed distributions or cases involving basic random variables of *B* and *C*, the advanced second moment method or simulation method can be used. In cases where benefit is computed as revenue minus cost, benefit might be correlated with cost requiring the use of other methods.

Example 20.2 Protection of Critical Infrastructure

This example is used to illustrate the cost of cost–benefit analysis using a simplified decision situation. As an illustration, assume that there is a 0.01 probability of an attack on a facility containing hazardous material during the next year. If the attack occurs, the probability of a serious release to the public is 0.01 with a total consequence of \$100B. The total consequence of an unsuccessful attack is negligible. The unmitigated risk can therefore be computed as

$$\text{Unmitigated risk} = 0.01(\$0.01)(100\text{B}) = \$10\text{M}.$$

If armed guards are deployed at each facility, the probability of attack can be reduced to 0.001 and the probability of serious release if an attack occurs can be reduced to 0.001. The cost of the guards for all plants is assumed to be \$100M per year. The mitigated risk can therefore be computed as

$$\text{Mitigated risk} = 0.001(0.001)(\$100\text{B}) = \$0.10\text{M}.$$

The benefit in this case is

$$\text{Benefit} = \$10\text{M} - \$0.1\text{M} \quad \text{or} \quad \sim \$10\text{M}.$$

The benefit-to-cost ratio is about 0.1. Therefore, the \$100M cost might be difficult to justify.

20.4.2 Risk Mitigation

A risk mitigation strategy can be presented from a financial point of view. Risk mitigation in this context can be defined as an action to either reduce the probability of an adverse event occurring or reduce the adverse consequences if it does occur. This definition captures the essence of an effective management process of risk. If implemented correctly a successful risk mitigation strategy should reduce any adverse (or downside) variations in the financial returns from a project, which are usually measured by either (1) the NPV defined as the difference between the present value of the cash flows generated by a project and its capital cost and calculated as part of the process of assessing and appraising investments or (2) the IRR defined as the return that can be earned on the capital invested in the project, that is, the discount rate that gives an NPV of zero, in the form of the rate that is equivalent to the yield on the investment.

Risk mitigation involves direct costs like increased capital expenditure or the payment of insurance premiums, hence it might reduce the average overall financial returns from a project. This reduction is often a perfectly acceptable outcome, given the risk aversion of many investors and lenders. A risk mitigation strategy is the replacement of an uncertain and volatile future with one where there is less exposure to adverse risks and so less variability in the return, although the expected NPV or IRR may be reduced. These two aspects are not necessarily mutually exclusive. Increasing risk efficiency by simultaneously improving the expected NPV or IRR and simultaneously reducing the adverse volatility is sometimes possible and should be sought. Risk mitigation should cover all phases of a project from inception to closedown or disposal.

Four primary ways are available to deal with risk within the context of a risk management strategy as follows:

- Risk reduction or elimination
- Risk transfer, for example, to a contractor or an insurance company
- Risk avoidance
- Risk absorbance or pooling

These four methods are described in subsequent sections.

Risk reduction or elimination: Risk reduction or elimination is often the most fruitful form for exploration. For example, could a design of a system be amended so as to reduce or eliminate either the probability of occurrence of a particular risk event or the adverse consequences if it occurs? Alternatively, could the risks be reduced or eliminated by retaining the same design but using different materials or a different method of assembly? Other possible risk mitigation options in this category include as examples: a better labor relations policy to minimize the risk of stoppages, training of staff to avoid hazards, better site security to avoid theft and vandalism, a preliminary investigation of possible site pollution, advance ordering of key components, noise abetment measures, good signposting, and liaisons with the local community.

Risk transfer: A general principle of an effective risk management strategy is that commercial risks in projects and other business ventures should be borne wherever possible by the party that is best able to manage them, and thus mitigate the risks. Contracts and financial agreements are the principal forms to transfer risks. Companies specializing in risk transfer can be consulted that could appropriately meet the needs of a project. Risks can be transferred alternately to an insurance company which, in return for a payment (i.e., premium) linked to the probability of occurrence and severity associated with the risk, is obliged by the contract to offer compensation to the party affected by the risk. Insurance coverage can range from straight insurance for expensive risks with a low probability, such as fire, through performance bonds, which ensure that the project will be completed if the contractor defaults, to sophisticated financial derivatives such as hedge contracts to avoid risks such as unanticipated losses in foreign exchange markets.

Risk avoidance: A most intuitive way of avoiding a risk is to avoid undertaking the project in a way that involves that risk. For example, if the objective is to generate electricity but a nuclear power source, although cost efficient, is considered to have a high risk due to potentially catastrophic consequences, even after taking all reasonable precautions, the practical solution is to turn to other forms of fuel to avoid that risk. Another example would be the risk that a particularly small contractor would go bankrupt. In this case, the risk could be avoided by using a well-established contractor for that particular job.

Risk absorbance and pooling: Cases where risks cannot, or cannot economically, be eliminated, transferred, or avoided, they must be absorbed if the project is to proceed. Normally, a sufficient margin in the project's finances needs to be created to cover the risk event should it occur. However, it is not always essential for one party alone to bear all these absorbed risks. Risks can be reduced through pooling possibly through participation in a consortium of contractors, when two or more parties are able to exercise partial control over the incidence and impact of risk. Joint ventures and partnerships are other examples of organizational forms for pooling risks.

Uncertainty characterization: Risk can be mitigated through proper uncertainty characterization. The presence of improperly characterized uncertainty could lead to higher adverse event occurrence likelihood and consequences. Also, it could result in increasing estimated cost margins as a means of compensation. Therefore, risk can be reduced by a proper characterization of uncertainty. The uncertainty characterization can be achieved through data collection and knowledge construction.

20.5 Risk Communication

Risk communication can be defined as an interactive process of exchange of information and opinion among stakeholders such as individuals, groups, and institutions. It often involves multiple messages about the nature of risk or expressing concerns, opinions, or reactions to risk managers or to legal and institutional arrangements for risk management. Risk communication greatly affects risk acceptance and defines the acceptance criteria for safety.

Risk communication provides the vital link between the risk assessors, risk managers, and the public to understand risk. However, this does not necessarily mean that risk communication will always lead to agreement among different parties. An accurate perception of risk provides for rational decision making. The Titanic was deemed an unsinkable ship, yet was lost on its maiden voyage. Space shuttle flights were perceived to be safe enough for civilian travel until the Space Shuttle Challenger disaster. These disasters obviously had risks that were not perceived as significant until after the disaster. Risk communication is a dynamic process that must be considered prior to management decisions.

The communication process deals with technical information about controversial issues. Therefore, it needs to be skillfully performed by risk managers and communicators who might be viewed as adversaries to the public. Risk communication between risk assessors and risk managers is necessary to effectively apply risk assessments in decision making. Risk managers must participate in determining the criteria for determining what risk is acceptable and unacceptable. This communication between the risk managers and risk assessors is necessary for a better understanding of risk analysis in making decisions.

Risk communication also provides the means for risk managers to gain acceptance and understanding by the public. Risk managers need to go beyond the risk assessment results and consider other factors in making decisions. One of these concerns is politics, which is largely influenced by the public. Risk managers often fail to convince the public that risks can be kept to acceptable levels. Problems with this are shown by the public's perception of toxic waste disposal and nuclear power plant operation safety. As a result of the public's perceived fear, risk managers may make decisions that are conservative to appease the public.

The value of risk calculated from risk assessment is not the only consideration for risk managers. All risks are not created equal and society has established risk preferences based on public preferences. Decision makers should take these preferences into consideration when making decisions concerning risk.

To establish a means of comparing risks based on the society preferences, risk conversion factor (RCF) may be used. The RCF expresses the relative importance of different attributes concerning risk. An example of possible risk conversion factors is shown in Table 20.10. These values were determined by inferences of public preferences from statistical data with the consequence of death considered.

For example, the voluntary and involuntary classification depends on whether the events leading to the risk are under the control of the persons at risk or not, respectively. Society, in general, accepts a higher level of voluntary risk than involuntary risk by an estimated factor of 100. Therefore, an individual will accept a voluntary risk that is 100 times greater than an involuntary risk.

The process of risk communication can be enhanced and improved in three aspects: (1) the process, (2) the message, and (3) the consumers. The risk assessment and management process needs to have clear goals with openness, balance, and competence. The contents of the message should account for audience orientation and uncertainty, provide risk comparison, and be complete. There is a need for consumer's guides that introduce risks associated with a specific technology, the process of risk assessment and management, acceptable risk, decision making, uncertainty, costs and benefits, and feedback mechanisms. Improving risk literacy of consumers is an essential component of the risk communication process.

The USACE has a 1992 Engineering Pamphlet (EP) on risk communication (EP 1110-2-8). The following are the guiding considerations in communicating risk:

- Risk communication must be free of jargon
- Consensus of expert needs to be established
- Materials must be cited and their sources must be credible
- Materials must be tailored to audience
- The information must be personalized to the extent possible
- Motivation discussion should stress a positive approach and the likelihood of success
- Risk data must be presented in a meaningful manner

References

Ayyub, B. M. 2002. *Elicitation of Expert Opinions for Uncertainty and Risks*, CRC Press, Boca Raton, FL.

Ayyub, B. M. and Klir, G. J. 2006. *Uncertainty Modeling and Analysis for Engineers and Scientists*, Chapman & Hall/CRC Press, Boca Raton, FL.

Ayyub, B. M. and McCuen, R. 2003. *Probability, Statistics and Reliability for Engineers and Scientists*, 2nd edn., Chapman & Hall/CRC Press, Boca Raton, FL.

Kumamoto, H. and Henley, E. J. 1996. *Probabilistic Risk Assessment and Management for Engineers and Scientists*, 2nd edn., IEEE Press, New York.

Maryland Emergency Management Agency (MEMA). 2006. *State of Maryland Guide for the Protection of Critical Infrastructure and Key Resources for Homeland Security, Volume 1: Critical Asset & Portfolio Risk Assessment (CAPRA) Methodology*, Office of Homeland Security, Annapolis, MD.

21

Electricity Infrastructure Resilience and Security

Massoud Amin

CONTENTS

21.1 Introduction

The North American power network may be considered the largest and most complex machine in the world—its transmission lines connect all the electric generation and distribution on the continent. In that respect, it exemplifies many of the complexities of electric power infrastructure and how technological innovation combined with efficient markets and policies can address them. This network represents an enormous investment, including over 15,000 generators in 10,000 power plants, and hundreds of thousands of miles of transmission lines and distribution networks, whose estimated worth is over U.S. $800 billion. In 2000, transmission and distribution cables alone were valued at U.S. $358 billion [9–16]. The National Academy of Engineering classified North American electric power grid as the greatest engineering achievement of the twentieth century. Today, the North American electric power grid is a network of approximately 10,000 power plants, 450,000 miles of high-voltage (>100 kV) transmission lines, over 6 million miles of lower voltage distribution lines in the United States, and more than 15,000 substations. The transmission system is an interstate grid whose primary purpose is to connect generating plants with electrical load centers like cities and with high demand commercial and industrial facilities. In turn, the local distribution system provides for service to residential, commercial, and small business customers.

Through the North American electricity infrastructure every user, producer, distributor, and broker of electricity buys and sells, competes, and cooperates in an *electric enterprise*. Every industry, every business, every store, and every home is a participant, active or passive, in this continent-scale conglomerate. Over the last decade and during the next few years, the electric enterprise will undergo dramatic transformation as its key participants—the traditional electric utilities—respond to deregulation, competition, renewable energy portfolio standards, tightening environmental/land-use restrictions, and other global trends.

Most of the systems currently in place were built by and for the regulated monopoly utility industry and are not fully prepared to handle the increasingly larger and faster changes in markets and technologies on both the supply and consumer side. It is essential to assume that these changes will continue and the infrastructure must become more flexible and adaptable to deal with them. Our long-term vision is to tie generation, transmission, distribution, and end use into a coherent network with ubiquitous secure data, information, and knowledge to enable best decisions for each stakeholder.

In addition, this network has evolved without formal analysis of the system-wide implications of this evolution, including its diminished transmission and generation shock-absorber capacity under the forces of deregulation, the digital economy, and interaction with other infrastructures. Only recently, with the advent of deregulation, unbundling, and competition in the electric power industry, has the possibility of power delivery beyond neighboring areas become a key design and engineering consideration. Yet we still expect the existing grid to handle a growing volume and variety of long-distance, bulk power transfers. To meet the needs of a pervasively digital world that relies on microprocessor-based devices in vehicles, homes, offices, and industrial facilities, grid congestion and atypical power flows are increasing, as are customer reliability expectations.

In the months after Hurricane Sandy and over a decade after the massive power outages in 2003 in the United States, Canada, the United Kingdom, and Italy, the electric infrastructure's vulnerabilities were increasingly apparent [1–11]. Since then, progress, however slow, is being made to mitigate risks.

Much of the current federal effort is focused on transmission. Yet, it is essential that market design and grid expansion programs for both the transmission and distribution systems work together to maintain adequate levels of grid reliability and to provide customers with the services they demand. At the minimum, the system must support the addition of both conventional and renewable generators along with demand response and enable implementation of technologies like electric vehicles and solar on the distribution system. Much of the renewable energy and natural gas potential in the United States is located in areas that are remote from population centers, lack high demand for energy, and are not well connected to our national infrastructure for transmission of bulk electric power. The recent expansion of natural gas production in the United States has also affected the development of the grid. To achieve public policy objectives, sufficient transmission capacity must link new natural gas-generating plants, onshore or offshore wind farms, solar plants, and other renewables to customers in order to serve the energy needs of homes and businesses and have the potential to replace significant portions of the oil used today in vehicle transportation.

New transmission will play a critical role in the transformation of the electric grid to enable public policy objectives, accommodate the retirement of older generation resources, increase transfer capability to obtain greater market efficiency for the benefit of consumers, and continue to meet evolving national, regional, and local reliability standards. To optimize the use of these natural gas sites and renewable energy resources, the necessary electric infrastructure must be installed, requiring both significant financial investments and cooperation at all levels on politically challenging items such as the sitting of facilities and the routing of new transmission lines.

21.1.1 What Is the Problem?

It is fair to ask: why should we make significant, ongoing investments in upgrading the electric grid?

Hurricane Sandy's widespread damage and the resulting extended outages certainly put a fine point on the answer, but the context is much broader than investing to withstand the occasional horrific superstorm. Simply put, if the United States is to remain economically competitive on a global scale, it must modernize its grid just as other nations are doing.

Currently, outages from all sources cost the U.S. economy somewhere between $80 and $188 billion annually. A 2011 competitiveness report by the World Economic Forum ranked

U.S. infrastructure below 20th among the world's nations in most of nine categories and below 30th for the quality of our air transport and electric power sectors. Clearly, the United States needs to invest in grid modernization simply to catch up with its global rivals.

21.1.2 What Is Needed to Speed Up Restoration?

In the months since Hurricane Sandy struck the East Coast with unprecedented fury, much discussion has focused on questions about power restoration in the Northeast: *Did* the smart grid help? Or *would* a smart grid have helped?

The questions are valid, and the short answer is unsatisfying: it depends. A longer answer is more helpful, because it allows us to consider the drivers of grid modernization, the concepts governing a *self-healing* grid, and what we need to do to maximize the benefits of future investments.

Detailed, postevent analysis will be needed to ascertain whether smart grid technology did in fact soften Sandy's impact or speed up power restoration. Meanwhile, let's place the storm and its impacts in context.

First, it needs to be understood that a massive, physical assault on the scale of the October 2012 superstorm is bound to overwhelm the power infrastructure, at least temporarily. No amount of money or technology can guarantee uninterrupted electric service under such circumstances.

Second, the power industry in the United States is just beginning to adapt to a wider spectrum of risk. It is noteworthy that both the number and frequency of annual, weather-caused, major outages have increased since the 1950s. Between the 1950s and 1980s, those outages increased from 2 to 5 each year. In the period 2008–2012, those outages increased to between 70 and 130 per year. In that 5-year period, weather-related outages accounted for 66% of power disruptions, which affected up to 178 million customers (meters).

This adaptation process continues as we implement strategies, technologies, and practices that will harden the grid and improve restoration performance after a physical disturbance. The investments so far in advanced metering infrastructure and the coming wave of investment in distribution automation (DA) are but the beginning of a multidecade, multibillion-dollar effort to achieve an end-to-end, intelligent, secure, resilient, and self-healing system.

Third, cost-effective investments to harden the grid and support resilience will vary by region, by utility, by the legacy equipment involved, and even by the function and location of equipment within a utility's service territory.

In Sandy's case, coastal areas were subject to storm surges and flooding, while inland, high winds and lashing rain produced the most damage. Improved hardening and resilience for distribution systems in those different environments would take different forms. Underground substations along the coasts may have to be rebuilt on the surface, while it might be cost-effective to perform *selective undergrounding* for some overhead lines further inland.

The one generalization we can make, however, is that the pursuit of an intelligent, self-healing grid has some common characteristics that will make the grid highly reliable in most circumstances—certainly in cases where disruptions are less catastrophic than Hurricane Sandy. Additional, location-specific steps based on rational risk assessment also can be taken by utilities and customers.

The economic benefits of a modernized grid will accrue as investments are made. Indeed, in my view, our twenty-first-century digital economy depends on us making these investments, regardless of the prognosis for more extreme weather to come as our climate changes.

In this chapter, the security, agility, and robustness/survivability of large-scale power delivery infrastructure facing new threats and unanticipated conditions is presented. In addition, we focus in part on challenges associated with development of a smart self-healing electric power grid for enhanced system security, reliability, efficiency, and quality.

21.2 Electricity Enterprise: Today and Tomorrow

21.2.1 Where Are We?

The existing power delivery system is vulnerable to natural disasters and intentional attack. Regarding the latter, a successful terrorist attempt to disrupt the power delivery system could have adverse effects on national security, the economy, and the lives of every citizen.

Both the importance and difficulty of protecting power systems have long been recognized. In 1990, the Office of Technology Assessment (OTA) of the U.S. Congress issued a detailed report, *Physical Vulnerability of the Electric System to Natural Disasters and Sabotage*, concluding: "Terrorists could emulate acts of sabotage in several other countries and destroy critical [power system] components, incapacitating large segments of a transmission network for months. Some of these components are vulnerable to saboteurs with explosives or just high-powered rifles." The report also documented the potential cost of widespread outages, estimating them to be in the range of \$1–\$5 per kW h of disrupted service, depending on the length of outage, the types of customers affected, and a variety of other factors. In the New York City outage of 1977, for example, damage from looting and arson alone totaled about \$155 million—roughly half of the total cost.

The reality of a coordinated attack has raised the issue of security to be considered along with power systems' reliability, which posits more random and independent failures. The system's vulnerability to natural disasters and physical attacks has long been recognized, but this vulnerability has significantly increased in recent years, in part because the system is operating closer to its capacity and in part because terrorist attacks are no longer hypothetical.

The situation has become even more complex because accounting for all critical assets includes thousands of transformer, line reactors, series capacitors, and transmission lines. Protection of *all* the widely diverse and dispersed assets is impractical because there are so many involved:

- Over 450,000 miles of HV lines (100 kV and above)
- Over 6644 transformers in the Eastern Interconnection
- Over 6000 HV transformers in the North American Interconnection (of which <300 are critical assets)
- Control centers
- Interdependence with gas pipelines
- Compressor stations
- Dams
- Rail lines
- Telecommunication equipment (monitoring and control of the system)

As an example, trends show that worldwide cyber attacks are on the rise; the number of documented attacks and intrusions has been rising very rapidly in recent years. Due to the increasingly sophisticated nature and speed of some malicious code, intrusions, and denial-of-service attacks, human response may be inadequate. Relatively little infrastructure (especially transmission lines) has been added to handle increased demand with an adequate safety cushion.* Some trends and documented nonsensitive modes of attack are shown in Figures 21.1 and 21.2.

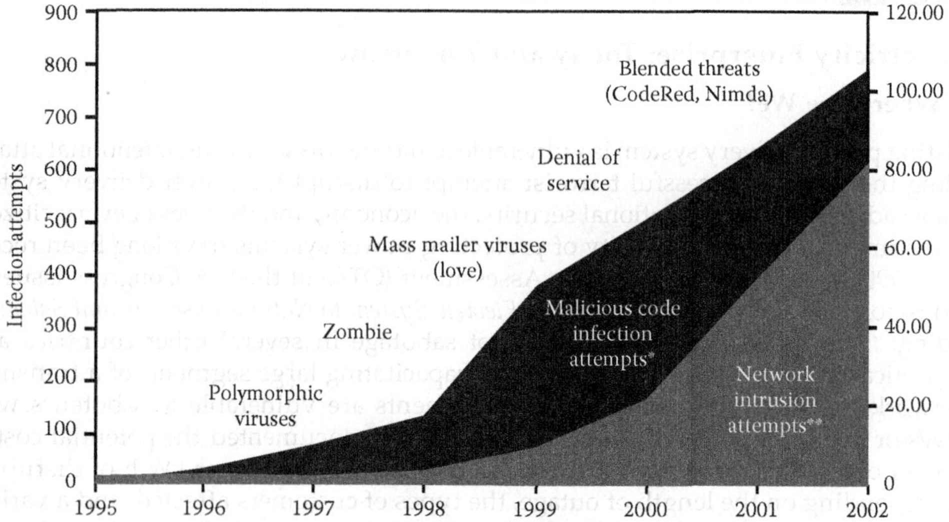

FIGURE 21.1
Attack trends: Documented network intrusion attempts (**, from CERT) and infection attempts (*, analysis by Symantec Security Response using data from Symantec, IDC, and ICSA).

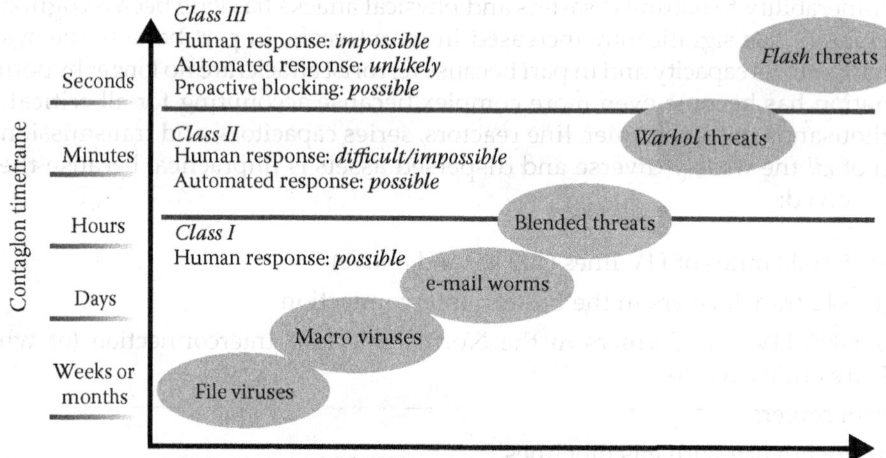

FIGURE 21.2
Malicious codes' threat evolution: increased sophistication and much faster than a few years ago.

* NERC (2001), Reliability Assessment 2001–2010.

Secure and reliable operation of these systems is fundamental to national and international economy, security, and quality of life. Their very interconnectedness makes them more vulnerable to global disruption, initiated locally by material failure, natural calamities, intentional attack, or human error.

In addition to the security problems associated with electric transmission systems, additional concerns arise with nuclear power systems. Nuclear power serves the United States as well as many countries around the world as a fairly reliable source of electricity that will not contribute to global warming producing carbon dioxide. In that sense, it is a *clean* source of electricity. But there are ways in which nuclear power can be a hazardous technology.

There are two pressing safety concerns with nuclear power in this age of terrorism. The first has to do with the spent fuel storage areas located outside and near the reactors. These spent fuel *swimming pools* are laden with a far higher inventory of radioactive material than is present in the reactor itself and are far less protected from attack than the reactor. If a *9/11* plane had hit the Indian Point nuclear plant spent fuel storage area (about 35 miles from New York City), it is possible that millions of people would have had to be evacuated to escape radiation and that parts of New York would have had to be abandoned. There might of course be solutions to avoid this catastrophic scenario, but they would add to the cost of nuclear power.

The other long-term safety concern is the eventual transportation of thousands of high-level nuclear waste canisters across open roads of the United States (through 43 states) on their way to a final resting place, for example, Yucca Mountain. These shipments would be enticing targets for terrorists. Technology may be able to ameliorate this type of accident, but it is currently not in place.

North American power network's transmission lines connect all generation and distribution on the continent to form a vertically integrated hierarchical network. The question is raised as to whether there is a unifying paradigm for the simulation, analysis, and optimization of time-critical operations (both financial transactions and actual physical control) in these multiscale, multicomponent, and distributed systems. In addition, mathematical models of interactive networks are typically vague (or may not even exist); moreover, existing and classical methods of solution are either unavailable, or are not sufficiently powerful. For the most part, no present methodologies are suitable for understanding their behavior.

Another important dimension is the effect of deregulation and economic factors on a particular infrastructure. While other and more populous countries, such as China and India, will have greater potential electricity markets and demands, the United States has the largest national market for electric power. Its electric utilities have been mostly privately owned, vertically integrated, and locally regulated. National regulations in areas of safety, pollution, and network reliability also constrain their operations to a degree, but local regulatory bodies, mostly at the state level, have set their prices and their return on investment (ROI) and have controlled their investment decisions while protecting them from outside competition. That situation is now rapidly changing, state regulators are moving toward permitting and encouraging a competitive market in electric power.

The electric power grid was historically operated by separate utilities, each independent in its own control area and regulated by local bodies, to deliver bulk power from generation to load areas reliably and economically—as a noncompetitive, regulated monopoly,

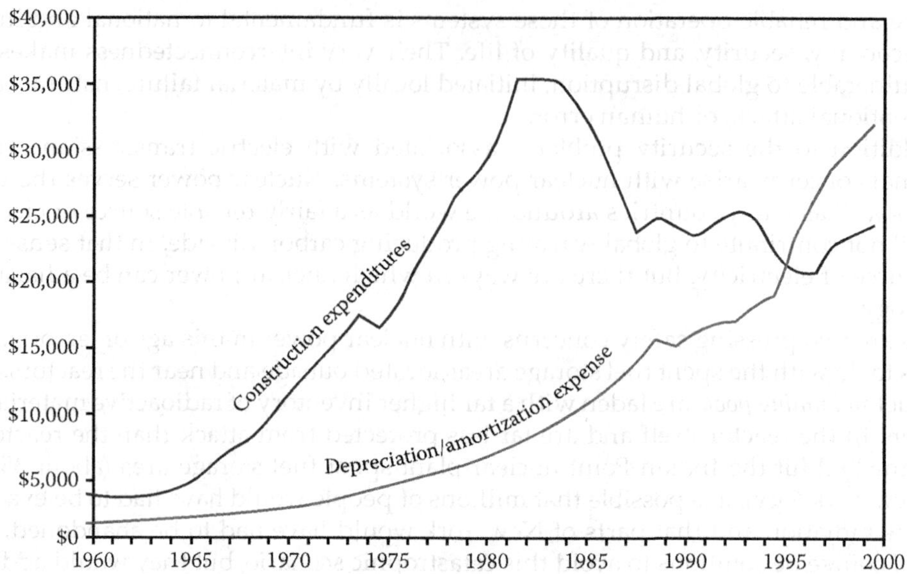

FIGURE 21.3
Since the *crossover* point in about 1995, utility construction expenditures have lagged behind asset depreciation. This has resulted in a mode of operation of the system analogous to *harvesting the farm far more rapidly than planting new seeds. Utility construction expenditures and depreciation/amortization expense*: In recent years, the investor-owned utility industry's annual depreciation expenses have exceeded construction expenditures. The industry is now generally in a *harvest the assets* mode rather than an *invest in the future of the business* mode. (Data provided by EEI, *Historical Statistics of the Electric Utility Industry* and *EEI Statistical Yearbook*, © 2003, EPRI Inc., Palo Alto, CA. All rights reserved; Graph courtesy of EPRI, Palo Alto, CA.)

emphasis was on reliability (and security) at the expense of economy. Competition and deregulation have created multiple energy producers that must share the same regulated energy delivery network. Traditionally, new delivery capacity would be added to handle load increases, but because of the current difficulty in obtaining permits and the uncertainty about achieving an adequate rate of ROI, total circuit miles added annually are declining while total demand for delivery resources continues to grow. In recent years, the *shock absorbers* have been shrinking; for example, during the 1990s, actual demand in the United States increased some 35%, while capacity has increased only 18%, the most visible parts of a larger and growing U.S. energy crisis that is the result of years of inadequate investments in the infrastructure. According to Electric Power Research Institute (EPRI) analyses, since 1995 to the present, the amortization/depreciation rate exceeds utility construction expenditures (Figure 21.3).

As a result of these *diminished shock absorbers*, the network is becoming increasingly stressed, and whether the carrying capacity or safety margin will exist to support anticipated demand is in question. The complex systems used to relieve bottlenecks and clear disturbances during periods of peak demand are at great risk to serious disruption, creating a critical need for technological improvements.

NORTH AMERICAN ELECTRICITY INFRASTRUCTURE VULNERABILITIES AND COST OF CASCADING FAILURES

Attention to the grid has gradually increased after several cascading failures. The August 10, 1996, blackout cost was over $1.5 billion and included all aspects of interconnected infrastructures and even the environment. Most recently, the August 13, 2003, outage is estimated to have a cost in the range of $6–$10 billions. Past disturbances in both the power grid give you some idea of how cascading failures work:

- November 1965—A cascaded system collapse blackout in 10 states in the Northeast United States affected about 30 million people.
- 1967—The Pennsylvania–New Jersey–Maryland (PJM) blackout occurred.
- May 1977—15,000 square miles and one million customers in Miami lost electricity.
- July 1977—In New York's suburbs, lightning caused overvoltages and faulty protection devices, which caused 10 million people to lose power for over 24 h, resulting in widespread looting, over 4000 arrests, and ultimately, the ouster of New York City's mayor.
- December 1978—Blackout in part of France due to voltage collapse.
- January 1981—1.5 million customers in Idaho, Utah, and Wyoming were without power for 7 h.
- March 1982—Over 900,000 lost power for 1.5 h due to high-voltage line failure in Oregon.
- December 1994—two million customers from Arizona to Washington state lost power.
- July 1996—A high-voltage line touched a tree branch in Idaho and fell. The resulting short circuit caused blackouts for two million customers in 14 states for approximately 6 h.
- August 1996—Following the July 2 blackout, two high-voltage lines fell in Oregon and caused cascading outages affecting over seven million customers in 11 Western states and two Canadian provinces.
- January 1998—Ice storms caused over three million people to lose power in Canada, New York, and New England.
- December 1998—San Francisco, California Bay Area blackout.
- July 1999—New York City blackout caused 300,000 people to be without power for 19 h.
- 1998–2001—Summer price spikes affect customers (infrastructure's inadequacy affecting markets).
- Industry-wide Y2K readiness program identified telecommunication failure as the biggest source of risk of the lights going out on rollover to 2000.
- Western states' suffered power crises in summer 2001 and its aftermath.
- Eastern United States and Canada face cascading outages on August 14, 2003.

21.3 Reliability Issues

Several cascading failures during the past 40 years spotlighted our need to understand the complex phenomena associated with power network systems and the development of emergency controls and restoration. Widespread outages and huge price spikes during the past few years raised public concern about grid reliability at the national level [6–10,16]. According to data from the North American Electric Reliability Corporation (NERC) and analyses from the EPRI, average outages from 1984 to the present have affected nearly 700,000 customers per event annually. Smaller outages occur much more frequently and affect tens to hundreds of thousands of customers every few weeks or months, while larger outages occur every 2–9 years and affect millions. Much larger outages affect seven million or more customers per event each decade. These analyses are based on data collected for the U.S. Department of Energy (DOE), which requires electric utilities to report system emergencies that include electric service interruptions, voltage reductions, acts of sabotage, unusual occurrences that can affect the reliability of bulk power delivery systems, and fuel problems [1,3–5,9,10,16,22].

Coupling these analyses with diminished infrastructure investments, and noting that the crossover point for the utility construction investment versus depreciation occurred in 1995 (Figure 21.3), we analyzed the number and frequency of major outages along with the number of customers affected during the decade 1991–2000, splitting it into the two time periods 1991–1995 and 1996–2000 (Figure 21.4a and b). Based on EPRI's analyses [1,14] of data in NERC's Disturbance Analysis Working Group (DAWG) database [1,9,10], 41% more outages affected 50,000 or more consumers in the second half of the 1990s than in the first half (58 outages in 1996–2000 versus 41 outages in 1991–1995). The average outage affected 15% more consumers from 1996 to 2000 than from 1991 to 1995 (average size per event was 409,854 customers affected in the second half of the decade versus 355,204 in the first half of the decade). In addition, there were 76 outages of size 100 megawatts (MW) or more in the second half of the decade, compared to 66 such occurrences in the first half. During the same period, the average lost load caused by an outage increased by 34%, from 798 MW during 1991–1995 to 1067 MW during 1996–2000 (Figure 21.4) [1,9,10,14].

Furthermore, there were 41% more outages affecting 50,000 or more consumers in the second half of the 1990s than in the first half (58 outages in 1996–2000 versus 41 outages in 1991–1995). In addition, between 1996 and 2000, outages affected 15% more consumers than they did between 1991 and 1995 (the average size per event was 409,854 customers affected in the second half of the decade versus 355,204 in the first half of the decade). Similar results were determined for a multitude of additional statistics such as the kilowatt magnitude of the outage and average load lost. These trends have persisted in this decade. NERC data show that during 2001–2005, we had 140 occurrences of over 100 MW dropped, and 92 occurrences of over 50,000 or more consumers affected.

The U.S. electric grid has been plagued by ever more and ever worse blackouts over the past 15 years. In an average year, outages total 92 min/year in the Midwest and 214 min in the Northeast. Japan, by contrast, averages only 4 min of interrupted service each year. The outage data exclude interruptions caused by extraordinary events such as fires or extreme weather.

We analyzed two sets of data, one from the U.S. DOE's EIA and the other from the NERC. Generally, the EIA database contains more events, and the NERC database gives more information about the events, including the date and time of an outage, the utility

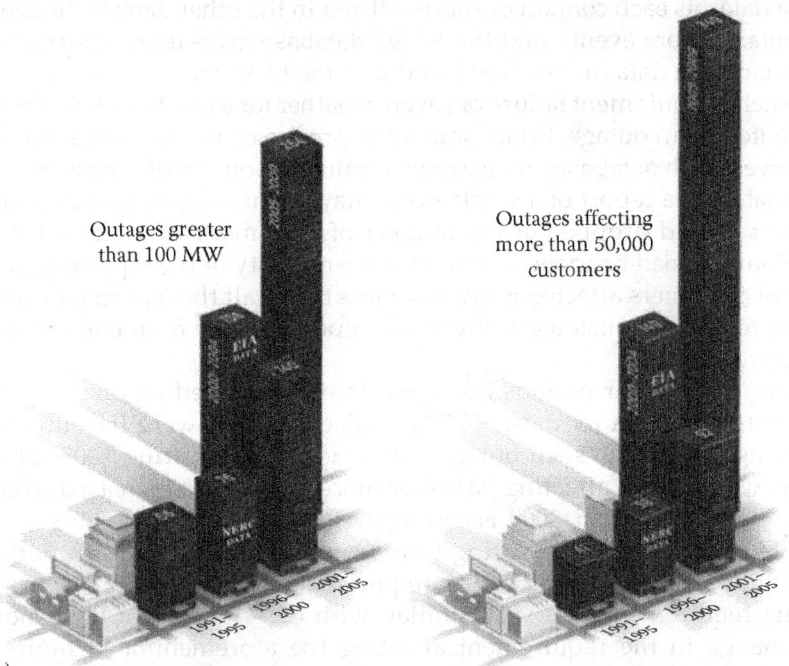

(a)

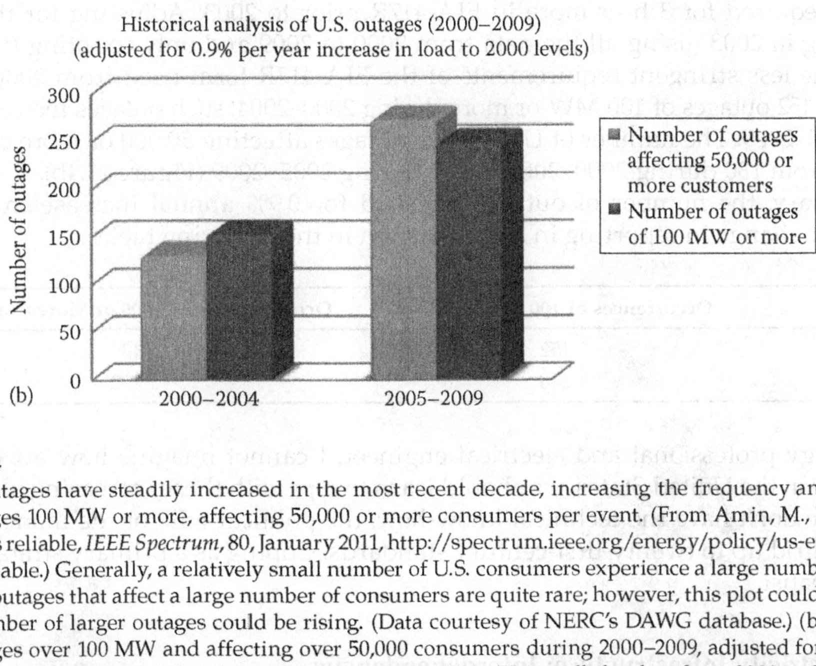

(b)

FIGURE 21.4
(a) Power outages have steadily increased in the most recent decade, increasing the frequency and size of U.S. power outages 100 MW or more, affecting 50,000 or more consumers per event. (From Amin, M., U.S. electrical grid gets less reliable, *IEEE Spectrum*, 80, January 2011, http://spectrum.ieee.org/energy/policy/us-electrical-grid-gets-less-reliable.) Generally, a relatively small number of U.S. consumers experience a large number of outages; conversely, outages that affect a large number of consumers are quite rare; however, this plot could also indicate that the number of larger outages could be rising. (Data courtesy of NERC's DAWG database.) (b) U.S. electric power outages over 100 MW and affecting over 50,000 consumers during 2000–2009, adjusted for 0.9% annual increase in load and adjusted for change in reporting in 2003. (Using all the data from 2000 to 2009 and only counting the outages that met the less stringent requirements of the EIA-417R form used during 2000–2002.)

involved, the region affected, the quantity of load dropped, the number of customers affected, the duration of the outage, and some information about the nature of the event.

These two datasets each contain events not listed in the other dataset. In general, the EIA database contains more events, and the NERC database gives more information about the events. The narrative data in the NERC (and also the EIA) databases are sufficient to identify factors such as equipment failure or severe weather (or a combination of both!) that may have contributed to an outage. Establishment of precise cause is beyond the scope of most of the narratives. Both databases are extremely valuable sources of information and insight.

In both databases, a report of a single event may be missing certain data elements such as the amount of load dropped or the number of customers affected. In the NERC database, the amount of load dropped is given for the majority of the reported events, whereas the number of customers affected is given for less than half the reported events. In the EIA database, the number of customers affected is reported more frequently than the amount of load dropped.

In both sets, each 5-year period was worse than the preceding one: according to data assembled by the U.S. EIA for most of the past decade, there were 156 outages of 100 MW or more during 2000–2004; such outages increased to 264 during 2005–2009. The number of U.S. power outages affecting 50,000 or more consumers increased from 149 during 2000–2004 to 349 during 2005–2009, according to EIA (Figure 21.2).

In 2003, EIA changed their reporting form from EIA-417R to OE-417. Both forms were attached with descriptions of reporting requirements (pp. 3 and 6, respectively). In all, the reporting requirements are very similar, with OE-417 being a little more stringent. The main change in the requirement affecting the aforementioned figures is that all outages greater than 50,000 customers for 1 h or more be reported in OE-417, where it was only required for 3 h or more in EIA-417R prior to 2003. Adjusting for the change in reporting in 2003 (using all the data from 2000 to 2009 and only counting the outages that met the less stringent requirements of the EIA-417R form used from 2000 to 2002): there were 152 outages of 100 MW or more during 2000–2004; such outages increased to 248 during 2005–2009. The number of U.S. power outages affecting 50,000 or more consumers increased from 130 during 2000–2004 to 272 during 2005–2009 (Figure 21.4b).

In summary, the number of outages adjusted for 0.9% annual increase in load and adjusted for change in reporting in 2003 is shown in the following table:

	Occurrences of 100 MW or More	Occurrences of 50,000 or More Consumers
2000–2004	152	130
2005–2009	248	272

As an energy professional and electrical engineer, I cannot imagine how anyone could believe that in the United States, we should learn to *cope* with these increasing blackouts—and that we don't have the technical know-how, the political will, or the money to bring our power grid up to twenty-first-century standards. Coping as a primary strategy is ultimately defeatist.

21.3.1 Electricity Infrastructure: Interdependencies with Cyber and Digital Infrastructures

As the world grows more interconnected, we are becoming surrounded by complex networked systems. These systems consist of numerous components interlinked in complicated webs. As a result of the number of components and their intricate interconnections, complex networked systems are extremely difficult to design, analyze, control, and protect. Despite these challenges, understanding complex networked systems is becoming critical. Many of our nation's critical infrastructures are complex interdependent/networked dynamical system of systems, including

- Electric power grids with overlays of sensor/communications/control systems and markets
- Fuel supply networks together with energy/oil/gas pipelines
- Telecommunications and satellite systems
- The Internet, computer networks, and the *cyber infrastructure*
- Transportation systems
- Energy markets, banking, and finance systems
- Environmental, consumer, societal, and policy macrosystems
- State and local water supply, emergency, and other services

Secure and reliable operation of complex increasingly interdependent infrastructure systems such as these is fundamental to our economy, security, and quality of life. Of particular importance is the uninterrupted availability of inexpensive, high-quality electric power and reliable, high-performance communication networks.

Electric power utilities typically own and operate at least parts of their own telecommunications systems that often consist of backbone fiber-optic or microwave connecting major substations, with spurs to smaller sites. Increased use of electronic automation raises significant issues regarding the adequacy of operational security. As is true of other critical infrastructures, increased use of automated technologies raises significant security issues, however:

- Reduced personnel at remote sites makes the sites more vulnerable to hostile threats.
- Interconnecting automation and control systems with public data networks makes them accessible to individuals and organizations, from any worldwide location using an inexpensive computer and a modem.
- Use of networked electronic systems for metering, scheduling, trading, or e-commerce imposes numerous financial risks associated with network failures.

In what follows we shall provide a brief overview of some key areas and present selected security aspects of operational systems, without discussing potentially sensitive material; these aspects include the following:

- Operational systems rely very heavily on the exchange of information among disparate systems.
- Utilities rely on very extensive private and leased telecommunication systems.
- Networking of these systems is expanding rapidly.
- This networking is expanding beyond utility doors, to encompass other utilities, corporations, and customers.

- Standard communication protocols and integration techniques are a *must*, despite the increased security risks.
- Increased security concerns in the aftermath of tragic events of September 11, 2001.
- Deregulation is increasing the incentives for unauthorized access to information.

As the power grids become heavily loaded with long-distance transfers, the already complex system dynamics becomes even more important. The potential for rare events but high-impact cascading phenomena represents just a few of many new science and technology concepts that are under development. Analysis and modeling of interdependent infrastructures (e.g., the electric power grid, together with protection systems, telecommunications, oil/gas pipelines, and energy markets) is especially pertinent. Regarding the latter, during the past four decades, much effort has been committed to better understand the dynamics of large-scale power systems in order to enhance security, quality, reliability, and availability (SQRA) of the overall system.

Specific attributes of SQRA are needed for electricity to meet the needs of the evolving digital society. Reliability and uninterruptability are practical necessities in digital enterprises. The digital society is expected (and designed) to be continuously operational, without interruption or denial of service. The interface between digital systems, processes, and enterprises and electric power delivery must support this reliability, with innovations spanning from the generation sources to the microchip. Similarly, availability of power is also a necessity. While higher reliability is nearly always a key objective for electric power suppliers, availability is the parameter with which users of sensitive digital equipment and processes are most concerned. Other concepts considered in our ongoing work include the following:

- *Self-healing and redundant*: Digital communications use a network of high-capacity nodes to route data efficiently and expeditiously. The interface between digital systems, processes, and enterprises and electric power delivery needs the same capability. A parallel or self-healing system may actually be less reliable (i.e., suffering more individual component failures) but still provide higher availability of power to critical loads.
- *Massive customization*: The digital revolution has revealed an inexhaustible appetite for instant and user-specific customization of service and support. The interface between digital systems, processes, and enterprises and electric power delivery must have this capability to support this revolution.
- *Rapid reconfiguration*: The digital world has demonstrated its ability to rapidly adjust to changing requirements. For example, the Internet evolved from a simple document-exchange system just a few years ago to the now familiar capabilities of e-mail, online transactions, and data presentation. Even newer technologies are now being implemented, such as video conferencing, voice-over IP, and civic systems such as web-based voting. The interface between digital systems, processes, and enterprises and electric power delivery must be able to emulate this rapid adaptive behavior to support the ongoing evolution of the digital society.

Today's accepted industry practice for power system performance measurements is the starting point for introducing new and more integrated performance metrics. The challenge is to bridge between independent mathematical models and performance metrics that are used at different levels of the power system. In some cases, the models and these independent metrics may conflict with each other to the extent that improving performance in one area may detract from others. This independence stems from several distinct performance areas such as distribution customer availability or outage reporting, grid operating contingencies, monitoring of power quality variations, and measurements of transmission reliability. The way that each of these performance elements had been applied in the past depended on the area of interest in the power system. In addition, utility industry performance measurement practices and standards evolved via groups that represented their interests in either one of end use, distribution, transmission, or generation.

21.3.2 Electric Power Grids

Electric infrastructure, increasingly linked with other critical and tightly coupled interdependencies noted earlier, is becoming increasingly interconnected and complex, thus posing new challenges for its secure, reliable, and efficient operation. The infrastructure is a complex dynamic network, geographically dispersed, nonlinear, and interacting both among itself with communication systems, fuel supplies, and markets and with its human owners, operators, and users. No single entity has complete control over its operation, nor does any such entity have the ability to evaluate, monitor, and manage it in real time. In fact, the conventional mathematical methodologies that underpin today's modeling, simulation, and control paradigms are unable to handle the complexity in its dynamics, and its increasing interconnectedness.

At its most fundamental level, the network's transmission lines form a vertically integrated hierarchical network consisting of the generation layer and then three network levels [17]:

- The first level is the transmission network, which is a meshed network combining extrahigh voltage (above 300 kV) and high voltage (100–300 kV) connected to large generation units, very large customers, and neighboring transmission networks and to the subtransmission level via tie-lines.
- The second level is the subtransmission, which consists of a radial or weakly coupled network including some high voltage (100–300 kV) but typically 5–15 kV that is connected to large customers and medium-sized generators.
- The third level is distribution, which is typically a radial network including low voltage (110–115 or 220–240 V) and medium voltage (1–100 kV) connected to small generators, medium-sized customers, and local low-voltage networks for small customers.

Several pertinent theories on power system operating conditions have been provided in the literature; these contributions not only provide mathematical foundations but also include some guidance on how to measure and adapt to disturbances.

The varying nature and types of extreme disturbances coupled with their anticipated infrequent occurrences makes it desirable to take automated actions to stabilize the system, which may include system separation in a controlled and coordinated manner. Inadequacy of a well-coordinated overall defense plan makes it more difficult to prevent spreading of the disturbance.

As a subset, disturbance classification lends itself to quick reaction or even the ability to predict events. At the very least, a *snapshot* of the event will have been taken. This will mean that no event will go unnoticed. In the past, events have gone unnoticed.

Furthermore, the ability to predict and react would indicate that problems could be detected and mitigated much sooner. A system operator could be trained accordingly, while taking into account both communication delays and computer server status.

Besides these many operational, spatial, and energy levels, power systems are also multiscaled in the time domain, from nanoseconds to decades. The relative time of action for different types of events, from normal to extreme, varies depending on the nature and speed of the disturbance, and the need for coordination. The aforementioned system is multiscale in time, operational space, and its dynamics. The timescale of actions and operations within the power grid (often continental in scale) ranges from microseconds to milliseconds for wave effects and fast dynamics (such as lightning); milliseconds for switching overvoltages; 100 ms or a few cycles for fault protection; 1–10 s for tie-line load frequency control; 10 s to 1 h for economic load dispatch; 1 h to a day or longer for load management, load forecasting, and generation scheduling; and several years to a decade for new transmission or generation planning and integration.

21.4 Infrastructures under Threat

The recent natural disasters and the terrorist attacks of September 11 have exposed critical vulnerabilities in America's essential infrastructures: never again can the security of these fundamental systems be taken for granted. Electric power systems constitute *the* fundamental infrastructure of modern society. A successful terrorist attempt to disrupt electricity supplies could have devastating effects on the national security, the economy, and the lives of every citizen. Yet power systems have widely dispersed assets that can never be absolutely defended against a determined attack.

Because critical infrastructures touch all of us, the growing potential for infrastructure problems stems from multiple sources. These sources include system complexity, deregulation, economic effects, power-market impacts, terrorism, and human error. The existing power system is also vulnerable to natural disasters and intentional attacks. Ongoing efforts at NERC, DOE, and EPRI—including EPRI's Enterprise Information Security Program and Infrastructure Security Initiative—have highlighted utility-specific threats and the technologies to counteract them. Regarding the latter, a November 2001 EPRI assessment developed in response to the September 11, 2001, attacks highlights three different kinds of potential threats to the U.S. electricity infrastructure [1–3,12]:

- *Attacks upon the power system*: In this case, the electricity infrastructure itself is the primary target—with ripple effects, in terms of outages, extending into the customer base. The point of attack could be a single component, such as a critical substation, or a transmission tower. However, there could also be a simultaneous, multipronged attack intended to bring down the entire grid in a region of the United States. Similarly, the attack could target electricity markets, which because of their transitional status is highly vulnerable.
- *Attacks by the power system*: In this case, the ultimate target is the population, using parts of the electricity infrastructure as a weapon. Power plant cooling towers, for example, could be used to disperse chemical or biological agents.

- *Attacks through the power system*: In this case, the target is the civil infrastructure. Utility networks include multiple conduits for attack, including lines, pipes, underground cables, tunnels, and sewers. An electromagnetic pulse, for example, could be coupled through the grid to with the intention of damaging computer and/or telecommunications infrastructure.

Protection against all three modes of attack thus represents a potentially critical *showstopper* for realizing a digital economy and enabling customer-managed service networks. Indeed, if infrastructure security is not ensured, even maintaining current levels of productivity and service will be jeopardized. Conversely, deploying some of the advanced technologies needed to enhance security will have a positive effect on efforts to improve grid reliability and coordinate power system operations with those of other infrastructures.

Therefore, the imperative for enhancing security in the electric power system has reached a new level that demands industry attention. In order to address this imperative in a logical and deliberate way, we need to understand what is involved and how to measure the current and future levels of secure performance.

The technologies that support the operational control of electric networks range from energy management systems (EMSs) to remote field devices. Critical systems include the following.

21.4.1 Energy Management System

The objective of the EMS is to manage the production, purchasing, transmission, distribution, and sale of electric energy in the power system at a minimal cost with respect to safety and reliability. Management of the real-time operation of an electric power system is a complex task requiring interaction of human operators, computer systems, communications networks, and real-time data-gathering devices in power plants and substations. An EMS consists of computers, display devices, software, communication channels, and remote terminal units (RTUs) that are connected to other RTUs, control actuators, and transducers in power plants and substations. The main tasks that an EMS performs have to do with generator control and scheduling, network analysis, and operator training. Control of generation requires that the EMS maintain system frequency and tie-line flows while economically dispatching each generating unit. The management of the transmission network requires that the EMS monitor up to thousands of telemetered values, estimate the electrical state of the network, and inform the operator of the best strategy to handle potential outages that could result in an overload or voltage-limit violation. EMSs can have real-time two-way communication links between substations, power plants, independent system operators, and other utility EMSs.

21.4.2 Supervisory Control and Data Acquisition System

A supervisory control and data acquisition (SCADA) system supports operator control of remote (or local) equipment, such as opening or closing a breaker. A SCADA system provides three critical functions in the operation of an electric power system: data acquisition, supervisory control, and alarm display. It consists of one or more computers with appropriate applications software connected by a communications system to a number of RTUs placed at various locations to collect data, perform intelligent control of electrical system devices, and report results back to an EMS. SCADAs can also be used for similar applications in natural gas pipeline transmission and distribution applications. A SCADA can have real-time communication links with one or more EMSs and hundreds of substations.

21.4.3 Remote Terminal Unit

RTUs are special-purpose microprocessor-based computers that contain analog-to-digital converters (ADCs) and digital-to-analog converters (DACs), digital inputs for status, and digital output for control. There are transmission substation RTUs and DA RTUs. Transmission substation RTUs are deployed at substation and generation facilities, where a large number of status and control points are required. DA RTUs are used to

- Control air switches and Var-compensation capacitor banks on utility poles
- Control pad-mounted switches
- Monitor and automate feeders
- Monitor and control underground networks
- Monitor, control, and automate smaller distribution substations

RTUs can also be used as indicated earlier in natural gas transmission and distribution. RTUs can be configured and interrogated using telecommunication technologies. They can have hundreds of real-time communication links with other substations, EMSs, and power plants.

21.4.4 Programmable Logic Controller

Programmable logic controllers (PLCs) have been used extensively in manufacturing and process industries for many years and are now being used to implement relay and control systems in substations. PLCs have extended I/O systems similar to transmission substation RTUs. The control outputs can be controlled by software residing in the PLC and via remote commands from a SCADA system. The PLC user can make changes in the software stored in EEPROM without making any major hardware or software changes. In some applications, PLCs with RTU-reporting capability may have advantages over conventional RTUs. PLCs are also used in many power plant and refinery applications. They were originally designed for use in discrete applications like coal handling. They are now being used in continuous control applications such as feedwater control. PLCs can have many real-time communication links inside and outside substations or plants.

21.4.5 Protective Relays

Protective relays are designed to respond to system faults and short circuits. When faults occur, the relays must signal the appropriate circuit breakers to trip and isolate the faulted equipment. Distribution system relaying must be coordinated with fuses and reclosures for faults while ignoring cold-load pickup, capacitor-bank switching, and transformer energization. Transmission-line relaying must locate and isolate a fault with sufficient speed to preserve stability, reduce fault damage, and minimize the impact on the power system. Certain types of *smart* protective relays can be configured and interrogated using telecommunication technologies.

21.4.6 Automated Metering

Automated metering is designed to upload residential and/or commercial gas and/or electric meter data. These data can then be automatically downloaded to a PC or other device and transmitted to a central collection point. With this technology, real-time communication links exist outside the utility infrastructure.

21.4.7 Plant-Distributed Control Systems

Plant-distributed control systems (DCSs) are plant-wide control systems that can be used for control and/or data acquisition. The I/O count can be higher than 20,000 data points. Often, the DCS is used as the plant data highway for communication to and from intelligent field devices, other control systems (such as PLCs), RTUs, and even the corporate data network for enterprise resource planning (ERP) applications. The DCS traditionally has used a proprietary operating system. Newer versions are moving toward open systems such as Windows NT and Sun Solaris. DCS technology has been developed with operating efficiency and user configurability as drivers, rather than system security. Additionally, technologies have been developed that allow remote access, usually via PC, to view and potentially reconfigure the operating parameters.

21.4.8 Field Devices

Examples of field devices are process instrumentation such as pressure and temperature sensors and chemical analyzers. Other standard types of field devices include electric actuators. Intelligent field devices include electronics to enable field configuration and upload calibration data. These devices can be configured off-line. They also can have real-time communication links between plant control systems, maintenance management systems, stand-alone PCs, and other devices inside and outside the facility.

Security of these cyber and communication networks is fundamental to the reliable operation of the grid. As power systems rely more heavily on computerized communications and control, system security has become increasingly dependent on protecting the integrity of the associated information systems. Part of the problem is that existing control systems, which were originally designed for use with proprietary, stand-alone communication networks, were later connected to the Internet (because of its productivity advantages and lower costs), but without adding the technology needed to make them secure. Communication of critical business information and controlled sharing of that information are essential parts of all business operations and processes.

As the deregulation of the energy industry unfolds, information security will become more important. For the energy-related industries, the need to balance the apparently mutually exclusive goals of operating system flexibility with the need for security will need to be addressed from a business perspective. Key electric energy operational systems depend on real-time communication links (both internal and external to the enterprise). The functional diversity of these organizations has resulted in a need for these key systems to be designed with a focus on open systems that are user configurable to enable integration with other systems (both internal and external to the enterprise). In many cases, these systems can be reconfigured using telecommunication technologies. In nearly all cases, the systems dynamically exchange data in real time. This results in a need for highly reliable, secure control and information-management systems.

Power plant DCSs produce information necessary for dispatch and control. This requires real-time information flow between the power plant and the utility's control center, system dispatch center, regulatory authorities, and so on. A power plant operating as part of a large wholesale power network may have links to an independent system operator, a power pool, and so on. As the generation business moves furthermore into market-driven competitive operation, both data integrity and confidentiality will become major concerns for the operating organizations.

Any telecommunication link that is even partially outside the control of the organization that owns and operates power plants, SCADA systems, or EMSs represents a potentially insecure pathway into the business operations of the company as well as a threat to the grid itself. The interdependency analyses done by most companies during Y2K preparations have identified these links and the system's vulnerability to their failures. Thus, they provide an excellent reference point for a cyber-vulnerability analysis.

In particular, monitoring and control of the overall grid system is a major challenge. Existing communication and information system architectures lack coordination among various operational components, which usually is the cause for the unchecked development of problems and delayed system restoration. Like any complex dynamic infrastructure system, the electricity grid has many layers and is vulnerable to many different types of disturbances. While strong centralized control is essential to reliable operations, this requires multiple, high-data-rate, two-way communication links, a powerful central computing facility, and an elaborate operation-control center, all of which are especially vulnerable when they are needed most—during serious system stresses or power disruptions. For deeper protection, intelligent-distributed control is also required, which would enable parts of the network to remain operational and even automatically reconfigure in the event of local failures or threats of failure.

21.5 The Dilemma: Security and Quality Needs

The specter of terrorism raises a profound dilemma for the electric power industry: How to make the electricity infrastructure more secure without compromising the productivity advantages inherent in today's complex, highly interconnected electric networks? Resolving this dilemma will require both short-term and long-term technology development and deployment, affecting some of the fundamental characteristics of today's power systems:

- *Centralization/decentralization of control*: For several years, there has been a trend toward centralizing control of electric power systems. Emergence of regional transmission organizations (RTOs) as agents of wide-area control, for example, offers the promise of greatly increased efficiency and improved customer service. But if terrorists can exploit the weaknesses of centralized control, security would seem to demand that smaller, local systems become the system configuration of choice. In fact, strength and resilience in the face of attack will increasingly rely upon the ability to bridge simultaneous top-down and bottom-up decision making in real time.
- *Increasing complexity*: The North American electric power system has been called the *most complex machine ever built*. System integration helps move power more efficiently over long distances and provides redundancy to ensure reliable service, and it also makes the system more complex and harder to operate. In response,

new mathematical approaches are needed to simplify the operation of complex power systems and to make them more robust in the face of natural or man-made interruptions.

- *Dependence on Internet communications*: Today's power systems could not operate without tightly knit communications capability—ranging from high-speed data transfer among control centers to interpretation of intermittent signals from remote sensors. Because of the vulnerability of Internet communications, however, protection of the electricity supply system requires new technology to enhance the security of power system command, control, and communications, including both hardware and software.
- *Accessibility and vulnerability*: Because power systems are so widely dispersed and relatively accessible, they are particularly vulnerable to attack. Although *hardening* of some key components, such as power plants and critical substations, is certainly desirable, it is simply not feasible or economic to provide comprehensive physical protection to all components. Probabilistic assessments can offer strategic guidance on where and how to deploy security resources to greatest advantage.

A survey of electric utilities revealed real concerns about grid and communications security. Figure 21.5 ranks the perceived threats to utility control centers. The most likely threats were bypassing controls, integrity violations, and authorization violations. Concern about the potential threats generally increased as the size of the utility (peak load) increased.

The system's equipment and facilities are dispersed throughout the North American continent, which complicates protection of the system from a determined terrorist attack. In addition, we must consider another complexity—the power delivery systems' physical

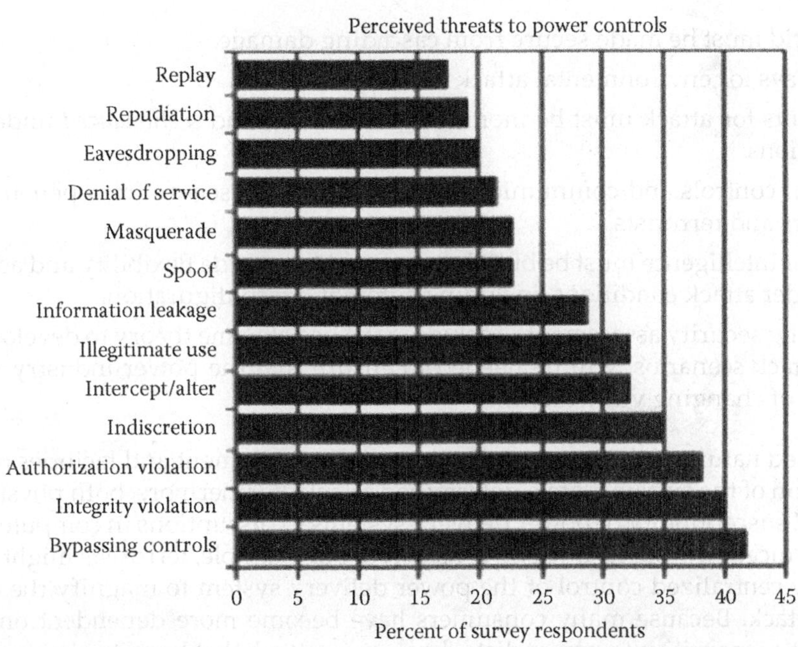

FIGURE 21.5
Threats to power supply control. (From EPRI 2000, *Communication Security Assessment for the United States Electric Utility Infrastructure,* EPRI Report 1001174, pp. 4–11, December 2000.)

vulnerabilities and susceptibility to disruptions in computer networks and communication systems. For example, terrorists might exploit the increasingly centralized control of the power delivery system to magnify the effects of a localized attack. Because many consumers have become more dependent on electronic systems that are sensitive to power disturbances, an attack that leads to even a momentary interruption of power can be costly. A 20 min outage at an integrated circuit fabrication plant, for example, could cost U.S. $30 million.

Despite increasing concerns in these areas as well as the continuing erosion of reserve margins, the infrastructures of today do generally function well. The electricity is on, the phones work, and traffic flows nearly all of the time. But more and more, the traditional level of performance is no longer good enough; more robust infrastructures are needed for the *digital society* envisioned for tomorrow. For example, in the electric power area, there is a need for an increase in reliability from today's average of about 99.9% (approximately 8 h of outage/year) to 99.9999% (about 32 s outage/year) or even 99.9999999% (one outage lasting less than a single AC cycle/year). Such near-perfect power is needed today for error-free operation of the microprocessor chips finding their way into just about everything, including billions of embedded applications.

Fortunately, the core technologies needed to strategically enhance system security are the same as those needed to resolve other areas of system vulnerability, as identified in the *Electricity Technology Roadmap*. These result from open access, exponential growth in power transactions, and the reliability needed to serve a digital society.

The North American electric power system needs a comprehensive strategy to prepare for the diverse threats posed by terrorism. Such a strategy should both increase protection of vital industry assets and ensure the public that they are well protected. A number of actions will need to be considered in formulating an overall security strategy:

- The grid must be made secure from cascading damage.
- Pathways for environmental attack must be sealed off.
- Conduits for attack must be monitored, sealed off, and *sectionalized* under attack conditions.
- Critical controls and communications must be made secure from penetration by hackers and terrorists.
- Greater intelligence must be built into the grid to provide flexibility and adaptability under attack conditions, including automatic reconfiguration.
- Ongoing security assessments, including the use of game theory to develop potential attack scenarios, will be needed to ensure that the power industry can stay ahead of changing vulnerabilities.

The dispersed nature of the power delivery system's equipment and facilities complicates the protection of the system from a determined attack. Furthermore, both physical vulnerabilities and susceptibility of power delivery systems to disruptions in computer networks and communication systems must be considered. For example, terrorists might exploit the increasingly centralized control of the power delivery system to magnify the effects of a localized attack. Because many consumers have become more dependent on electronic systems that are sensitive to power disturbances, an attack that leads to even a momentary interruption of power can be costly.

In summary, the existing end-to-end energy and power delivery system is vulnerable to natural disasters and intentional cyber attacks. Regarding the latter, a successful attempt to disrupt the power delivery system could have adverse effects on national security, the economy, and the lives of every citizen. For decades, our industry has been dedicated to the secure and reliable operation of the electric system that is fundamental to the national and international economy, security, and quality of life. Challenges to the security of the electric infrastructure include the following:

- *Physical security*: The size and complexity of the North American electric power grid makes it impossible both financially and logistically to physically protect the entire end-to-end and interdependent infrastructure. There currently exist over 450,000 miles of 100 kV or higher transmission lines and many more thousands of miles of lower voltage lines. As an increasing amount of electricity is generated from distributed renewable sources, the problem will be only be exacerbated.
- *Cybersecurity*: Threats from cyberspace to our electric grid are rapidly increasing and evolving. While there have been no publicly known major power disruptions due to cyber attacks, public disclosures of vulnerabilities are making these systems more attractive as targets.

Due to the increasingly sophisticated nature and speed of some malicious code, intrusions, and denial-of-service attacks, human response may be inadequate. Furthermore, currently more than 90% of successful intrusions and cyber attacks take advantage of known vulnerabilities and misconfigured operating systems, servers, and network devices. Technological advances targeting system awareness, cryptography, trust management, and access controls are underway, and continued attention is needed on these key technological solutions.

Security of cyber and communication networks is fundamental to the reliable operations of the grid. As power systems rely more heavily on computerized communication and control, system security has become increasingly dependent on protecting the integrity of the associate information systems. As an example, any utility system connected directly or indirectly to the public Internet is disputed territory and is vulnerable to cyber intrusions and impersonating field devices to spoof bad data into control centers.

Cyber has *weakest link* issues. Awareness, education, and pragmatic tool developments in this vital area also continue to remain challenges. Educating stakeholders and colleagues in the cyber-physical interdependencies has often been difficult, as those who are distinguished members of the community and understand power systems well, but not the cyber threats, routinely minimize emphasis on these persisting novel threats.

Cyber threats are dynamic, evolving quickly and often combined with lack of training and awareness: cyber connectivity has increased the complexity of the control system and facilities it is intended to safely and reliably control. Thus, in order to defend electric infrastructure against the impacts of cyber-physical attacks, significant challenges must be overcome before extensive deployment and implementation of smart grid technologies can begin. Cybersecurity and interoperability are two of the key challenges of the smart grid transformation. As for security, it must be built in as part of its design and *not* glued on as afterthought.

One important constraint on regulatory oversight of security protection is the split jurisdiction over the grid. The bulk electric system is under federal regulation, but the distribution grid, metering, electric vehicles, and other aspects of the grid are regulated by individual states. As a result, the oversight of cybersecurity is split along with other regulatory functions.

21.6 Human Performance

Since humans interact with these infrastructures as managers, operators, and users, human performance plays an important role in their efficiency and security. In many complex networks, the human participants themselves are both the most susceptible to failure and the most adaptable in the management of recovery. Modeling and simulating these networks, especially their economic and financial aspects, will require modeling the bounded rationality of actual human thinking, unlike that of a hypothetical *expert* human as in most applications of artificial intelligence (AI). Even more directly, most of these networks require some human intervention for their routine control and especially when they are exhibiting anomalous behavior that may suggest actual or incipient failure.

Operators and maintenance personnel are obviously *inside* these networks and can have direct, real-time effects on them. But the users of a telecommunication, transportation, electric power, or pipeline system also affect the behavior of those systems, often without conscious intent. The amounts, and often the nature, of the demands put on the network can be the immediate cause of conflict, diminished performance, and even collapse. Reflected harmonics from one user's machinery degrade power quality for all. Long transmissions from a few users create Internet congestion. Simultaneous lawn watering drops the water pressure for everyone. In a very real sense, no one is *outside* the infrastructure.

Given that there is some automatic way to detect actual or immanent local failures, the obvious next step is to warn the operators. Unfortunately, the operators are usually busy with other tasks, sometimes even responding to previous warnings. In the worst case, the detected failure sets off a multitude of almost simultaneous alarms as it begins to cascade through the system, and, before the operators can determine the real source of the problem, the whole network has shut itself down automatically.

Unfortunately, humans have cognitive limitations that can cause them to make serious mistakes when they are interrupted. In recent years, a number of systems have been designed that allow users to delegate tasks to intelligent software assistants (*softbots*) that operate in the background, handling routine tasks and informing the operators in accordance with some protocol that establishes the level of their delegated authority to act independently. In this arrangement, the operator becomes a supervisor, who must either cede almost all authority to subordinates or be subject to interruption by them. At present, we have very limited understanding of how to design user interfaces to accommodate interruption.

21.7 Broader Technical Issues

In response to the aforementioned challenges, several enabling technologies and advances are/will be available that can provide necessary capabilities when combined in an overall system design. Among them are the following:

- Flexible AC transmission system (FACTS) devices, which are high-voltage thyristor-based electronic controllers that increase the power capacity of transmission lines and have already been deployed in several high-value applications. At peak demand, up to 50% more power can be controlled through existing lines.

- Fault current limiters (FCLs), which absorb the shock of short circuits for a few cycles to provide adequate time for a breaker to trip. It is noteworthy that preliminary results of post–August 14 outage show that FCLs could have served as large electrical *shock absorbers* to limit the size of blackouts.
- Wide-area measurement systems (WAMSs), which integrate advanced sensors with satellite communication and time stamping using global positioning systems (GPSs) to detect and report angle swings and other transmission system changes.
- Innovations in materials science and processing, including high-temperature superconducting (HTS) cables, oxide-powder-in-tube technology for HTS wire, and advanced silicon devices and wide-bandgap semiconductors for power electronics.
- Distributed resources such as small combustion turbines, solid oxide and other fuel cells, photovoltaics, superconducting magnetic energy storage (SMES), and transportable battery energy storage systems (TBESSs).
- Information systems and online data processing tools such as the Open Access Same-Time Information System (OASIS) and Transfer Capability Evaluation (TRACE) software, which determines the total transfer capability for each transmission path posted on the OASIS network while taking into account the thermal, voltage, and interface limits.
- *Monitoring and use of IT*: WAMSs, OASIS, SCADA systems, and EMSs.
- *Analysis tools*: Several software systems for dynamic security assessment of large/wide-area networks augmented with market/risk assessment.
- *Control*: FACTS; FCL.
- Intelligent electronic devices (IEDs) with security provisions built in combining sensors, computers, telecommunication units, and actuators; integrated sensor; two-way communication; *intelligent agent* functions (assessment, decision, learning); and actuation, enabled by advances in several areas including semiconductors and resource-constrained encryption.

However, if most of the aforementioned technologies are developed, still the overall systems' control will remain a major challenge. This is a rich area for research and development (R&D) of such tools, as well as to address systems and infrastructure integration issues of their deployment in the overall network—especially now because of increased competition, the demand for advanced technology to gain an advantage, and the challenge of providing the reliability and quality consumers demand.

21.8 Western States Power Crises: A Brief Overview of Lessons Learned

An example of *urgent* opportunities is within the now seemingly calm California energy markets; the undercurrents that led to huge price spikes and considerable customer pain in recent years are yet to be fully addressed and alleviated. Such *perfect storms* may appear once again during another cycle of California economic recovery and growth. The California power crisis in 2000 was only the most visible parts of a larger and growing U.S. energy crisis that is the result of years of inadequate investments in the infrastructure.

For example, at the root of the California crisis was declining investment in infrastructure components that led to a fundamental imbalance between growing demand for power and an almost stagnant supply. The imbalance had been brewing for many years and is prevalent throughout the nation (see EPRI's Western States Power Crises white paper; www.epri.com/WesternStatesPowerCrisisSynthesis.pdf).

California is a good downside example of a societal test bed for the ways that seemingly *good* theories can fail in the real world. For example, inefficient markets provide inadequate incentives for infrastructure investment:

- Boom-bust cycle may be taking shape in generation investment
- Transmission investment running at one-half of 1975 level
- Congestion in transmission network is rising, as indicated by an increased number of transmission loading reliefs (TLRs) during the last 3 years

Cost of market failure can be also very high, as indicated by the exercise of market power in California during the summer of 2000 that cost consumers $4 billion initially, while the ongoing intermediate loss to businesses may well be considerably higher.

More specifically regarding the electricity under investments and persisting undercurrents, very specific *investments* by the state were made, on the order of $10 billion, paid to subsidize (hold down) electricity prices and to bail out bankrupt companies through long-term noncompetitive contracts that did not address the undercurrents and shortcomings of the earlier policies.

To address these issues, there are both tactical and strategic needs: for example, the so-called low-hanging fruits to improve transmission networks include the following:

- Deploy existing technologies to improve use of already in place transmission assets (e.g., FACTS, dynamic thermal circuit rating, and energy storage–peak shaving technologies): for example, through the integration of load management technologies saving nearly 5000 MW, which amounts to about 10% of total demand, combined with a more precise control enabled by the use of FACTS devices, which enable nearly 50% more transfer capability over existing transmission lines.
- Develop and deploy new technologies to improve transmission reliability and throughput (e.g., low-sag composite conductors, HTS cables, extrahigh-voltage AC and DC transmission systems, hierarchical control systems).
- Improve real-time control of network via monitoring and data analysis of dynamic transmission conditions.
- Develop and deploy self-healing grid tools to adaptively respond to overload and emergency conditions.
- Digital control of the power delivery network (reliability, security, and power quality).
- Integrated electricity and communications for the user.
- Transformation of the meter into a two-way energy/information portal.
- Integration of distributed energy resource (DER) into the network.
- The complex grid can operate successfully *if* technology is deployed and operated in an integrated manner (there is no *silver bullet*!).

In addition, longer term strategic considerations must be addressed:

- Greater fuel diversity including renewable energy technologies—regional and national priorities.
- Risk assessment of long-term U.S. reliance—analysis of the value of risk management through fuel diversity.
- Introduce time-varying prices and competitive market dynamics for all customers.
- Create a planning process and in silico testing of designs, devices, and power markets.
- Model market efficiencies, environmental constraints, and renewables.
- Develop advanced EM threat detection, shielding, and surge-suppression capabilities.
- Develop the tools/procedures to ensure a robust and secure marketplace for electricity.
- Develop the portfolio of advanced power generation technologies to assure energy security.
- Transmission network expansion and RTOs: for example, would an RTO complement a competitive wholesale power market and result in a sustainable and robust system? How large should they be?
- Comprehensive architecture for power supply and delivery infrastructure that anticipates rapidly escalating demands of digital society.
- Enable self-healing power delivery infrastructure.
- Significant investment in R&D, transmission, generation, and conservation resources is needed.
- Incentives for technology innovation and accountability for R&D.
- Revitalize the national public/private electricity infrastructure partnership needed to fund the *self-healing grid* deployment.
- The *law of unintended consequences* should be considered in crafting any solution.

Having discussed the aforementioned technology-intensive *push*, we must also consider the fact that adoption of new technologies often creates equally new markets. For example, wireless communication creates the market of spectrum, and broadband technologies create the market of bandwidth. Reduced regulation of major industries has required new markets wherever the infrastructure is congested: airlines compete for landing rights, power generators for transmission rights, and oil and gas producers for pipeline capacity.

From a national perspective, a key grand challenge is how to redesign, retrofit, and upgrade the nearly 200,000 miles of electromechanically controlled system into a smart self-healing grid that is driven by a well-designed market approach.

In addressing this challenge, as technology progresses, and the economy becomes increasingly dependent on markets, infrastructures such as electric power, oil/gas/water pipelines, telecommunications, and financial and transportation networks become increasingly critical and complex. In particular, since it began in 1882, electric power has grown to become a major industry essential to a modern economy.

Over the past two decades, governments around the globe have introduced increasing amounts of competition into network industries. With the advent of restructuring in the electric power industry, we are witnessing the onset of a historical transformation of the energy infrastructure in the context of global trends:

- Increasing electricity demand as a consequence of economic and population growth
- Technological innovations in power generation, delivery, control, and communications
- Increasing public acceptance of market mechanisms
- Growing public concerns about environmental quality and depletion of exhaustible resources

Services previously supplied by vertically integrated, regulated monopolies are now provided by multiple firms. The transition to competition has fundamentally altered important aspects of the engineering and economics of production. This presents unique opportunities and challenges. Clearly, this change will have far-reaching implications for the future development of the electricity industry. More fundamentally, as we look beyond the horizon, this change will further power the information revolution and increasing global interdependence. The long-term socioeconomic impacts of such a transformation will be huge, and the tasks are just as daunting, going well beyond the boundary of existing knowledge.

To meet such a challenge, collaborative research between engineers and economists is critical to providing a holistic and robust basis that will support the design and management of complex technological and economic systems in the long-term. The electric power industry offers an immediate opportunity for launching such research, as new ways are being sought to improve the efficiency of electricity markets while maintaining the reliability of the network. Complexity of the electric power grid combined with even more intricate interactions with markets offers a plethora of new and exciting research opportunities. In what follows we provide our vision and approach to enabling a smart self-healing electric power system that can respond to a broad array of destabilizers.

21.8.1 Complex System Failure

Beyond the human dimension, there is a strategic need to understand the societal consequences of infrastructure failure risks along with benefits of various tiers of increased reliability. From an infrastructure interdependency perspective, power, telecommunications, banking and finance, transportation and distribution, and other infrastructures are becoming more and more congested, and are increasingly vulnerable to failures cascading through and between them. A key concern is the avoidance of widespread network failure due to cascading and interactive effects. Moreover, interdependence is only one of several characteristics that challenge the control and reliable operation of these networks. Other factors that place increased stress on the power grid include dependencies on adjacent power grids (increasing because of deregulation), telecommunications, markets, and computer networks. Furthermore, reliable electric service is critically dependent on the whole grid's ability to respond to changed conditions instantaneously.

More specifically, secure and reliable operation of critical infrastructures poses significant theoretical and practical challenges in analysis, modeling, simulation, prediction, control, and optimization. To address these challenges, a research initiative—the EPRI/DOD Complex Interactive Networks/Systems Initiative (CIN/SI)—was undertaken during

1998–2001 to enable critical infrastructures to adapt to a broad array of potential disturbances, including terrorist attacks, natural disasters, and equipment failures.

The CIN/SI overcame the long-standing problems of complexity, analysis, and management for large interconnected systems—and systems of systems—by opening up new concepts and techniques. Dynamical systems, statistical physics, information and communication science, and computational complexity were extended to provide practical tools for measuring and modeling the power grid, cell phone networks, Internet, and other complex systems. For the first time, global dynamics for such systems can be understood fundamentally (Figure 21.6).

As an example, related to numerous major outages, narrowly programmed protection devices have contributed to worsening the severity and impact of the outage—typically performing a simple on/off logic that locally acts as preprogrammed while destabilizing a larger regional interconnection.

From a broader perspective, any critical national infrastructure typically has many layers and decision-making units and is vulnerable to various types of disturbances. Effective, intelligent, distributed control is required that would enable parts of the constituent networks to remain operational and even automatically reconfigure in the event of local failures or threats of failure. In any situation subject to rapid changes, completely centralized control requires multiple, high-data-rate, two-way communication links, a powerful central computing facility, and an elaborate operations control center. But all of these are liable to disruption at the very time when they are most needed (i.e., when the system is stressed by natural disasters, purposeful attack, or unusually high demand).

When failures occur at various locations in such a network, the whole system breaks into isolated *islands*, each of which must then fend for itself. With the intelligence distributed,

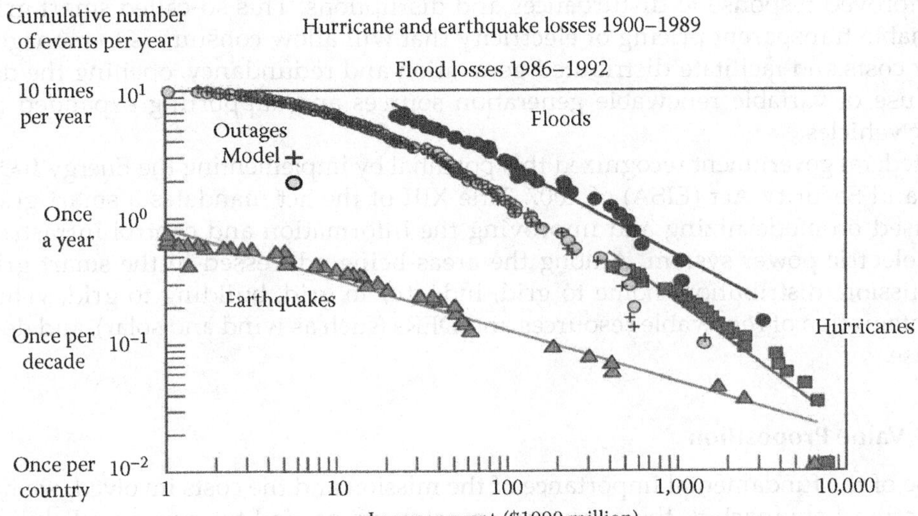

FIGURE 21.6
Understanding complex systems and global dynamics. Economic losses from disasters were found to follow a power law distribution—for hurricanes, floods, earthquakes, and even electrical outages. Fundamental power law distributions also were found for forest fires, Internet congestion, and other systems. CIN/SI results such as these translate in new approaches for optimizing complex systems in terms of productivity and robustness to disaster. (From the EPRI/DOD CIN/SI. Our goal is to move the power outage curve down toward the origin, that is, to make outages less frequent and with smaller impact on customers.)

and the components acting as independent agents, those in each island have the ability to reorganize themselves and make efficient use of whatever local resources remain to them in ways consonant with the established global goals to minimize adverse impact on the overall network. Local controllers will guide the isolated areas to operate independently while preparing them to rejoin the network, without creating unacceptable local conditions either during or after the transition. A network of local controllers can act as a parallel, distributed computer, communicating via microwaves, optical cables, or the power lines themselves, and intelligently limiting their messages to only that information necessary to achieve global optimization and facilitate recovery after failure.

21.8.2 How to Make an Electric Power System Smart

To add intelligence to an electric power transmission system, independent processors in each component and each substation and power plant are needed. These processors must have a robust operating system and be able to act as independent agents that can communicate with and cooperate with other systems, forming a large distributed computing platform. Each agent must be connected to sensors associated with its own component or its own substation so that it can assess its own operating conditions and report them to its neighboring agents via the communications paths. Thus, for example, a processor associated with a circuit breaker would have the ability to communicate with sensors built into the breaker and communicate those sensor values using high-bandwidth fiber communications connected to other such processor agents.

Adding and utilizing existing intelligence—sensors, communications, monitors, optimal controls, and computers—to our electric grid can substantially improve its efficiency and reliability through increased situational awareness, reduced outage propagation, and improved response to disturbances and disruptions. This so-called smart grid can also enable transparent pricing of electricity that will allow consumers to manage their energy costs and facilitate distributed generation and redundancy, opening the door to wider use of variable renewable generation sources and supporting expanded use of electric vehicles.

The federal government recognized this potential by implementing the Energy Independence and Security Act (EISA) of 2007. Title XIII of the act mandates a smart grid that is focused on modernizing and improving the information and control infrastructure of the electric power system. Among the areas being addressed in the smart grid are transmission, distribution, home to grid, industry to grid, building to grid, vehicle to grid, integration of renewable resources and DERs (such as wind and solar), and demand response.

21.8.3 Value Proposition

Because of the fundamental importance of the mission and the costs involved, we have to take a critical approach to the enormous investments needed to improve reliability and resiliency and enable economic competitiveness. One metric is *payback*, or ROI.

Having studied this in depth each $1 invested garners a return of $2.80–$6 to the broader economy. The ROI begins immediately with job creation and economic stimulus. To reach these numbers, we used a very narrow definition of *smart grid*. If the definition is broadened, the benefits increase.

The cost of a smarter grid would depend on how much instrumentation you actually put in, such as the communications backbone, enhanced security, and increased resilience.

The total price tag ranges around $340–$480 billion, which, over a 20-year period, would be something like $20–$25 billion/year. But right off the bat, the benefits are $70 billion/year in reduced costs from outages, and on a year where there are lots of hurricanes, lots of ice storms, and other disturbances, that benefit even goes further. Currently, outages from all sources cost the U.S. economy somewhere between $80 and $188 billion annually. Costs of outages reduced by about $49 billion/year, and reduced CO_2 emissions by 12%–18% by 2030. In addition, it would increase system efficiency by over 4%—that's another $20.4 billion a year.

The costs cover a wide variety of enhancements to bring the power delivery system to the performance levels required for a smart grid. They include the infrastructure to integrate DERs and achieve full customer connectivity but exclude the cost of generation, the cost of transmission expansion to add renewables and to meet load growth, and a category of customer costs for smart-grid-ready appliances and devices. Despite the costs of implementation, investing in the grid would pay for itself, to a great extent.

But this is also about (1) increased cyber/IT security, and overall energy security, with security built in the design as part of a layered defense system architecture and (2) job creation and an economic benefit. With the actual investment, for every dollar, the return is about $2.80–$6 to the broader economy. And this figure is very conservative.

21.8.4 What Are the Economic Benefits of Upgrading the Grid?

The economic benefits of a modernized grid will accrue as investments are made. Indeed, in my view, our twenty-first-century digital economy depends on us making these investments, regardless of the prognosis for more extreme weather to come as our climate changes.

Yet the economic argument is clear: the payback of smart grid technologies in the United States is three to six times greater than the money invested, and grows with each sequence of grid improvement. For example, the 2009 government stimulus plan funding and matching support from utilities and the private sector in the Smart Grid Investment Grant (SGIG) and Smart Grid Demonstration Project (SGDP) programs generated a significant impact on the U.S. economy:

- As of March 2012, the total invested value of $2.96 billion to support smart grid projects generated at least $6.8 billion in total economic output.
- Smart grid deployment positively impacted employment and labor income throughout the economy. Overall, about 47,000 full-time equivalent jobs were supported by investments. Among smart grid vendors—an ecosystem of manufacturers, information technology, and technical services providers—about 12,000 direct jobs were supported, with the remaining jobs being in those companies' respective supply chains and induced by the money spent throughout the broader economy.
- Investment in core smart grid industries supports high-paying jobs. Industrial sectors that benefit directly include computer systems' design, technical and scientific services and consulting, and electrical/wireless equipment and component manufacturing. Industrial sectors that experience indirect and induced benefits include real estate, wholesale trade, financial services, restaurants, and health care. The U.S. DOE smart grid American Recovery and Reinvestment Act (ARRA) investments also supported employment in personal service sectors such as health

care, financial services, real estate, and food/restaurants through indirect and induced impacts.

- The smart grid gross domestic product (GDP) multiplier is higher than many forms of government investment. For every $1 million of direct spending, which includes both government ARRA funds and private sector matching, the GDP increased by $2.5–$2.6 million, depending on the scenario being evaluated, which compares favorably against other potential government investments in general spending or other types of infrastructure.

21.8.5 What Role Does Renewable Energy Play in the Smart Grid and What Needs to Be Done to Access Sustainable Energy Resources?

Much of the renewable energy and natural gas potential in the United States is located in areas that are remote from population centers, lack high demand for energy, and are not well connected to our national infrastructure for the transmission of bulk electric power. The recent expansion of natural gas production in the United States has also affected the development of the grid. To achieve public policy objectives, sufficient transmission capacity must link new natural gas generating plants, onshore or offshore wind farms, solar plants, and other renewables to customers if those resources are to serve the energy needs of homes and businesses and have the potential to replace significant portions of the oil used today in vehicle transportation.

New transmission will play a critical role in the transformation of the electric grid to enable public policy objectives, accommodate the retirement of older generation resources, increase transfer capability to obtain greater market efficiency for the benefit of consumers, and continue to meet evolving national, regional, and local reliability standards. With a stronger and smart grid, 40% of our electricity in the United States can come from wind by 2030.

Much of the renewable energy and natural gas potential in the United States is located in areas that are remote from population centers, lack high demand for energy, and are not well connected to our national infrastructure for transmission of bulk electric power. The recent expansion of natural gas production in the United States has also affected development of the grid. To achieve public policy objectives, sufficient transmission capacity must link new natural gas generating plants, onshore or offshore wind farms, solar plants, and other renewables to customers if those resources are to serve the energy needs of homes and businesses, and have the potential to replace significant portions of the oil used today in vehicle transportation.

New transmission will play a critical role in the transformation of the electric grid to enable public policy objectives; accommodate the retirement of older generation resources, increase transfer capability to obtain greater market efficiency for the benefit of consumers, and continue to meet evolving national, regional, and local reliability standards.

New types of generation and new uses of electricity are creating both challenges and opportunities. Examples of new generation types include wind and solar, both of which vary in output and predictability. While most wind installations are grid-scale generation sources, many solar, and a few wind, sources are also in place in distribution grids, presenting additional challenges, which are discussed in the following. An example of a technology that serves both as an electricity use (load) and as an electricity source is energy storage, affording the opportunity to compensate for varying grid conditions by providing or absorbing energy to help correct system voltage or frequency. Placing an energy storage device in the distribution grid to serve both as a load and as a DER also offers new integration challenges and opportunities for increased reliability: the electric vehicle, for example, presents challenges in minimizing the grid impact of its charging and also in the opportunity for its use as a DER.

Another technology that has received renewed interest is direct current (DC), especially in localized grids called *microgrids*. For example, solar photovoltaic produces DC, batteries store DC, and loads such as computer equipment and variable speed motors operate on DC. The grid operates mainly on alternating current (AC), and conversions need to take place between AC and DC to interconnect DC generation or loads to the AC grid. Efficiency considerations suggest minimizing the number of such individual conversions, leading to exploration of new concepts for managing electricity at locations involving these generation sources, storage methods, and loads. Local electricity generation and storage and distribution systems should be improved next, to increase the self-sufficiency of end users. Longer term, flow-directing technologies would be added to address fluctuations and differences between energy supply and demand. Electricity might be redirected at times of peak load, for instance.

Cost-effective solutions will vary by region, utility, equipment, and threats involved. In Hurricane Sandy's case, coastal areas that were subject to storm surges and flooding might need underground substations rebuilding on the surface. Inland, where high winds and lashing rain produced the most damage, overhead lines could be buried.

Customer demand combined with sound policies supportive of this modernization along with innovation-based new business opportunities will drive the market for these generation, storage, and distribution technologies. Consumers, as surveys show, are increasingly driven by an interest in energy efficiency, digital demand, and awareness of the cost of energy disruptions. Utilities will have to fix their networks once people question why power cuts are preventing them from working on their computers, for instance.

Manufacturers in turn must integrate customer feedback into their R&D roadmaps and improve coordination of standards, R&D, and funding to drive down costs and broaden the market. Related, enabling technologies will be needed, including EMS and communication technologies.

21.8.6 Self-Healing Power Grid

Smart grid is not a very precise term. I prefer smart *self-healing* grid, because it more precisely describes the desired outcome of the investments I advocate for grid modernization.

A self-healing grid uses digital components and real-time communications technologies installed throughout a grid to monitor the grid's electrical characteristics at all times and constantly tune itself so that it operates at an optimum state. It has the intelligence to constantly look for potential problems caused by storms, catastrophes, human error, or even sabotage. It will react to real or potential abnormalities within a fraction of a second, just as a military fighter jet reconfigures itself to stay aloft after it is damaged. The self-healing grid isolates problems immediately as they occur, before they snowball into major blackouts, and reorganizes the grid and reroutes energy transmissions so that services continue for all customers while the problem is physically repaired by line crews.

A self-healing smart grid can provide a number of benefits that lend to a more stable and efficient system. Three of its primary functions include the following:

- *Real-time monitoring and reaction*, which allows the system to constantly tune itself to an optimal state. Real-time monitoring can alert grid operators to the precursors or signatures of impending faults, based on probabilistic analysis.

Real-time monitoring has been enabled by a leap from traditional SCADA systems to phasor measurement units (PMUs), aka synchrophasors. This technology improves the resolution of data polled from field devices from 2×–4× per s with SCADA to 20×–50× per s with PMUs. PMUs also provide precise, GPS-based time stamping so that events on the system can be analyzed accurately and chronologically in a WAMS. This allows operators to see *how the dominoes fell in the dark*.

- *Anticipation*, which enables the system to automatically look for problems that could trigger larger disturbances. Real-time monitoring enables operators to react swiftly to restore balance to the system or to program field devices to respond automatically. This allows constant tuning of the grid's many components to achieve an optimal, highly efficient state.
- *Rapid isolation*, which allows the system to isolate parts of the network that experience failure from the rest of the system to avoid the spread of disruption and enables a more rapid restoration.

As a result of these functions, a self-healing smart grid system is able to reduce power outages and minimize their length when they do occur. The smart grid is able to detect abnormal signals, make adaptive reconfigurations, and isolate disturbances, eliminating or minimizing electrical disturbances during storms or other catastrophes. And, because the system is self-healing, it has an end-to-end resilience that detects and overrides human errors that result in some of the power outages, such as when a worker error left millions of California residents without electricity in September 2011.

21.8.7 End-to-End Technologies

An end-to-end system that anticipates problems, supports operator decisions, or reacts automatically has a few elements worthy of emphasis.

At the customer end are the interval meters that provide usage data, serve as end-of-line sensors for voltage conservation, and emit *last gasps* as they (and the customer) lose power. Upstream of the meters, but downstream of the operators, we will see a proliferation of advanced sensors (IEDs) that facilitate real-time monitoring and control of critical assets. Advanced protective relays, for instance, provide improved isolation of faults. All of the technologies discussed in this chapter are supported by two-way communication networks that bring the real-time monitoring data back to operators and allow the latter to send commands back to assets in the field. Finally, visualization tools such as dashboards convert data into color-coded graphics and automated alerts that provide decision support.

21.8.8 Three-Tiered Intelligence

The self-healing grid can be thought of as having three tiers of intelligence.

The bottom layer, closest to devices in the field, is distributed intelligence. It is akin to the reptilian brain, with simple responses to environmental stimuli. At a substation, for instance, an intelligent device monitors the health of the asset and communicates that to the middle layer, where the validation of incoming data and coordination of various functions takes place in milliseconds. This is somewhat akin to a mammalian neocortex that can strategize, act, and be upgraded through experience to higher functionality. The top layer contains the centralized command-and-control functions directed by human operators.

21.8.9 Why Do We Need a Smart Self-Healing Grid?

Grid modernization is a global phenomenon, but in any day in the United States, about a half million people are without power for 2 or more hours. The number of weather-caused, major outages in the United States has risen since the 1950s, from between 2 and 5 each year by the 1980s to 70–130 between 2008 and 2012. Two-thirds of weather-related power disruptions have occurred in the past 5 years, affecting up to 178 million customers (meters), as changing weather patterns impact aging infrastructure.

The U.S. electric power system still relies on 1960s and 1970s technology. The sector is second from the bottom of major industries in terms of research and development (R&D) spending as a fraction of revenue; only pulp and paper is worse. Electricity R&D received just 0.17% of net sales from between 2001 and 2006, and has not risen since.

Meanwhile, electricity needs are changing and growing fast. Tweeting, and the devices and infrastructure needed to operate the underpinning communication network, data centers, and storage alone, adds more than 2500 megawatt hours (MW h) of demand globally per year that did not exist 5 years ago. Kilowatt hour (kW h) is commonly used by power companies for billing, since the monthly energy consumption of a typical residential customer ranges from a few hundred to a few thousand kilowatt hours. One MW h is equal to 1000 kW of electricity used continuously for 1 h. One MW h is the amount of electricity used by approximately 330 homes in 1 h. On average, 2500 MW h is equivalent to the electricity used by about 825,000 homes. Factor in Internet TV, video streaming, online gaming and the digitization of medical records, and the world's electricity supply will need to triple by 2050 to keep up.

21.8.10 How Does a Smart Self-Healing Grid Benefit Consumers?

Beyond managing power disturbances, a smart grid system has the ability to measure how and when consumers use the most power. This information allows utility providers to charge consumers variable rates for energy based upon supply and demand. Ultimately, this variable rate will incentivize consumers to shift their heavy use of electricity to times of the day when demand is low and will contribute to a healthier environment by helping consumers better manage and more efficiently use energy.

21.8.11 What Is Involved in Creating a Smart Self-Healing Grid?

To transform our current infrastructure into a self-healing smart grid, several technologies must be deployed and integrated.

The ideal smart grid system consists of microgrids, which are small, mostly self-sufficient power systems, and a stronger, smarter high-voltage power grid, which serves as the backbone to the overall system.

Upgrading the grid infrastructure for self-healing capabilities requires replacing traditional analog technologies with digital components, software processors, and power electronics technologies. These must be installed throughout a system so that it can be digitally controlled, which is the key ingredient to a grid that is self-monitoring and self-healing.

Much of the technology and systems thinking behind self-healing power grids comes from the military aviation sector, where I worked for 14 years on damage-adaptive flight systems for F-15 aircraft, optimizing logistics and studying the survival of squadrons and mission effectiveness. In January 1998, when I joined the EPRI, I helped bring these concepts to electricity power systems and other critical infrastructure networks, including energy, water, telecommunications, and finance (http://massoud-amin.umn.edu/publications/Amin_Computer_Aug_00.pdf). Following the September 11, 2001, terrorist attacks, resilience and security has become even more important.

Today in the United States, 16 programs on smart grids are ongoing, amounting to several billions per year, at EPRI, NSF, DHS, DOE, and DOD. Over 100 public and private projects, many on smart meters, address the electricity system in piecemeal fashion. There is no coordinated national decision-making body.

Jurisdiction over the grid is split: the bulk electric system is under federal regulation; the distribution grid is under the purview of state public utility commissions. Local regulations essentially kill the motivation for any utility to lead a regional or nationwide effort. Government policies shift with election cycles, variously championing energy independence, clean energy, environmental protection, jobs, and so on.

21.8.12 Is the Smart Grid Secure, Private, and Safe?

It is noteworthy that smart grid is modernization of the entire end-to-end system, from fuel source to smart homes. At the end use, a smart meter, which is a sensor node, is a digital upgrade to the decades-old mechanical meters used in homes and businesses. Some smart meters use wireless technologies that transmit radio frequencies (RFs) to provide two-way secure communication of the aggregate data on the electricity usage to the electric company through remote communication technologies.

This means electric companies will no longer need to send meter readers to read our meters on a monthly basis, and instead as consumers, we receive the information as we choose, on an hourly, daily, weekly, or monthly basis, rather than receiving the bill at the end of the month with no ability to adjust our usage.

Some consumers have expressed concerns about security/confidentially of their information (important to note that as customers we own our own data/information):

- Cybersecurity threats can cause disruptions in the flow of power and other problems if cyber intruders send computer signals to the electronic controls used in some electric generation and transmission infrastructure. The electric power industry takes cybersecurity threats very seriously. In fact, electric companies must meet mandatory cybersecurity standards that require companies to implement training, physical security, and asset recovery plans to protect against the threat of cyber attacks.
- As the smart grid is built, electric companies are incorporating cybersecurity protections into both the grid architecture and the new smart grid technologies. The electric power industry is working closely with vendors, manufacturers, and government agencies to ensure that the smart grid is secure. These measures also help to ensure that customer data remain protected from cybersecurity threats.

21.8.13 Are There Potential Health Impacts of the Radiofrequency Signal Emitted by Wireless Smart Meters?

The RF exposure levels from smart meters are far below the levels permitted by the U.S. Federal Communications Commission (FCC), which sets health standards for RF exposure, based on extensive reviews of the biological and health literature. All scientific studies, including a request by the California State Assembly for a study conducted by the *California Council on Science and Technology* (CCST) on potential health impacts from smart meters.

No causations or correlations and no health impacts were found, based both on the lack of scientific evidence of harmful effects from RF signal and on the observation that the RF exposure of people in their homes to smart meters is likely to be minuscule compared to RF exposure to items such as cell phones, microwave ovens, baby monitors, wireless routers, televisions, and laptop computers, which most of us already have in our homes [28].

According to the EPRI, the *relatively weak* strength of the RF signals generated by smart meters means that any impact of RF exposure would be minimal—similar to the levels of the exposure from televisions and radios [29].

- RF exposure from a smart meter is far below—and more infrequent—than other common electric devices. In fact, smart meters typically broadcast their signal for less than a minute at a time and usually less than a total of 15 min each day. The communication is usually from outside the customer's home, so exposure to radio waves is minimal. In addition, the electric panel and wall behind the meter actually block much of the radio signal from entering the home.
- RF is measured in units of microwatts per square centimeter. A microwatt is very small—it's one-millionth of a watt.
- Held at your ear, a cell phone's RF signal would be 1000–5000 $\mu W/cm^2$.
- Standing 2 ft from a microwave oven, the RF signal would be 50–200 $\mu W/cm^2$.
- Standing 10 ft from a smart meter, the RF signal would be 4 $\mu W/cm^2$.

In summary, the RF signal emitted by a smart meter is one order of magnitude less than proximity to the microwave oven, and 1000-fold less than cell phones held to our ears.

21.8.14 What Will It Take to Address Concerns That Communications Linked to Energy Services Will Invade Consumers' Privacy?

Customer concerns are of vital importance. When it comes right down to it, what would the power supply or power grid be without consumers? If there is any compromise of the privacy or security of the service, it will undermine everything. An incident would not only create a breach of confidentiality for the information that has been compromised, but it might also compromise the potential future markets the technology might have been able to create if the service had been secure.

The bottom line is that security cannot be added to a system as an afterthought. We need to start from scratch, at the very beginning of any microgrid project, and consider privacy and security in all design criteria. Strategic consideration of these issues will make a huge difference in the confidence and protection that the overall system provides. This is necessary whether the design effort is focusing on silicon chips, network components, end user devices, the architecture, or the system as whole.

In our work, we have proposed and tested several different layers of technologies that monitor and support the privacy of customer information. Security technologies are employed for traffic analyzers, signal analyzers, and agents that monitor voltages, frequency, current (along with their rates of changes), and user behavior. Each component is secured independently and locally so the security precautions cannot be reverse engineered.

This is not a hierarchical system that can be destroyed or taken down. If one or two layers fail, the system does not fail. It's essentially a self-reconfiguring, self-healing architecture. If anybody attacks it or tries to compromise one part of it, the system reconfigures to not only protect itself but to localize and fend off such attacks.

21.8.15 Risk Assessment

The initial step, before implementing these strategies and technologies, is risk assessment. Risk is dynamic, local, and specific. National policies will help, but achieving hardening and resiliency on the ground will be specific to a utility's customers' needs, its legacy systems, location, and technology roadmap.

A dynamic risk landscape requires annual updating to ensure protection of the right assets. How has the risk portfolio or the spectrum of risk changed? With climate change, the variability of weather events has increased. We are going to see more and more extreme events that have never happened before and we'll see them with greater frequency. Hurricane Sandy appears to be an example of this challenge.

So, a clear sense of dynamic risk should guide our investments in hardening and resilience, based on evidence and data. We need a new set of tools and a fresh set of approaches to system upgrades in order to be more dynamic and more adaptive to achieve resiliency and security.

21.8.16 Conclusions: Toward a Secure, Resilient, and Efficient Infrastructure

How to control a heterogeneous, widely dispersed, yet globally interconnected system is a serious technological problem in any case. It is even more complex and difficult to control it for optimal efficiency and maximum benefit to the ultimate consumers while still allowing all its business components to compete fairly and freely. A similar need exists for other infrastructures, where future advanced systems are predicated on the near-perfect functioning of today's electricity, communications, transportation, and financial services.

From a strategic R&D viewpoint, agility and robustness/survivability of large-scale dynamic networks that face new and unanticipated operating conditions will be presented. A major challenge is posed by the lack of a unified mathematical framework with robust tools for modeling, simulation, control, and optimization of time-critical operations in complex multicomponent and multiscaled networks.

Our immediate and critical goal is to avoid widespread network failure, but the longer-term vision is to enable adaptive and robust infrastructure. Achieving this vision and sustaining infrastructure reliability, robustness, and efficiency are critical long-term issues that require strategic investments in research and development.

However, three factors hinder improvements to the U.S. system. First, investment is too low. Since 2010, President Obama's stimulus plan has channeled $3.4 billion toward a U.S. smart grid; industry has added another $4.3 billion. The full cost will be around $400 billion, or $25–$30 billion a year for 20 years. Second, technological challenges for matching supply and demand remain. And third, fragmentation of the U.S. electricity system and projects across states and funding agencies requires a national strategy.

Smart grid systems must interact across centralized and decentralized electric networks and support advanced services (such as net metering, load aggregation, and real-time energy monitoring). They must meet the evolving needs of a dynamic digital future. To enable this transition, a systemic, sound, and sustainable policy framework for the regulatory oversight of grid modernization and security will be needed to provide incentives for collaboration between state utilities and federal agencies. I am often asked: "Government can't do it alone and the industry, which owns over 90% of the infrastructure, is unable to do alone; what are we to do?" A key enabler is creation of National Infrastructure Banks, which are focused on addressing both the much needed repairs today (to modernize existing aging infrastructure) *and* to bridge to more advanced, smarter, more secure, and sustainable lifeline infrastructures envisioned for the next 10–20 years. Created as public/private partnership enterprises, these banks would lend money on a sustainable basis with clear cost/benefit, performance metrics and include fees for quality of services provided by the modernized infrastructures.

In summary, a balanced, cost-effective approach to investments and use of technology can make a sizable difference in mitigating the risk. Electricity shall prevail at the quality, efficiency, and reliability that customers demand and are willing to pay for. On the one hand, the question is, *Who provides it?*; on the other hand, it is important to note that achieving the grid performance, security, and reliability is a profitable national investment, not a cost burden on the taxpayer. The economic payback is three to six times greater than the money invested. Further, the payback starts with the completion of each sequence of grid improvement. The issue is not merely who invests money, because that is ultimately the public, but whether it's invested through taxes or kWh rates. Considering the impact of regulatory agencies, they should be capable of inducing the electricity producers to plan and fund the process; this may be the most efficient way to get it in operation. The current absence of a coordinated national decision-making body is a major obstacle. State's rights and state PUC regulators have removed the individual state's utility motivation for a national plan.

One important constraint on regulatory oversight of the reliability-centered upgrades and security protection is the split jurisdiction over the grid. The bulk electric system is under federal regulation, but the distribution grid, metering, and other aspects of the grid are regulated by individual states. And those local regulations essentially kill the motivation for any utility or utility group to lead a regional or nationwide effort. We therefore need a policy framework to provide incentives for collaboration in grid modernization—making the system smarter and stronger—and for R&D. In addition, as a result the oversight of cybersecurity is split along with other regulatory functions. Going forward, we must bridge the jurisdictional gap between federal/NERC and the state commissions on much needed grid modernization and the associated cybersecurity.

Because of the hefty budget and long timescales involved, the main investment for this infrastructure will have to come from the public purse. The regulatory agencies would need to decide whether it is best invested through taxes or power usage rates and induce the electricity producers to plan and cofund the process.

When the United States has made such strategic commitments in the past, the payoffs have been huge. Think of the interstate highway system, the lunar landing project, the Internet. Meeting each of those challenges has produced world-leading economic growth by enabling commerce and technology development and producing a highly skilled workforce in the process. The electricity smart grid could do the same. In my view, the U.S.'s twenty-first-century digital economy depends on these investments. Given the economic, societal, and quality-of-life challenges and the ever-increasing interdependencies among infrastructures we have today, we must decide whether to build electric power and energy infrastructures that support a twenty-first century's digital society, or be left behind as a twentieth-century industrial relic.

Acknowledgments

I developed some of the foundational parts of the material and findings presented here while I was at the EPRI in Palo Alto, CA. I am most grateful for EPRI's support and feedback from numerous colleagues at EPRI, universities, industry, and government agencies.

References

1. M. Amin, North America's electricity infrastructure: Are we ready for more perfect storms? *IEEE Security and Privacy Magazine*, 1(5), 19–25, September/October 2003.
2. M. Amin, Security challenges for the electricity infrastructure, Special Issue of the *IEEE Computer Magazine on Security and Privacy*, April 2002.
3. M. Amin, Toward self-healing energy infrastructure systems, Cover Feature in the *IEEE Computer Applications in Power*, 14(1), 20–28, January 2001.
4. M. Amin, Toward self-healing infrastructure systems, *IEEE Computer Magazine*, 33(8), 44–53, August 2000.
5. M. Amin, Special Issues of *IEEE Control Systems Magazine on Control of Complex Networks*, 21(6), December 2001; 22(1), February 2002.
6. Committee Hearing of the House Committee on Energy and Commerce, *Blackout 2003: How Did It Happen and Why?* September 3–4, 2003, http://energycommerce.house.gov.
7. *Critical Foundations: Protecting America's Infrastructures*, The report of the President's Commission on Critical Infrastructure Protection, Washington, DC, October 1997, www.ciao.ncr.gov.
8. DOE, *National Transmission Grid Study*, U.S. Department of Energy, Washington, DC, May 2002, http://tis.eh.doe.gov/ntgs/gridstudy/main_screen.pdf.
9. Energy Information Administration, *Annual Energy Outlook 2003*, DOE, Washington, DC, http://www.eia.doe.gov/oiaf/aeo/figure_3.html.
10. North American Electric Reliability Council (NERC), Disturbance Analysis Working Group (DAWG) Database.
11. EPRI 2003, Complex Interactive Networks/Systems Initiative: Final Summary Report—Overview and Summary Final Report for Joint EPRI and U.S. Department of Defense University Research Initiative, EPRI, Palo Alto, CA, December 2003, 155pp.
12. EPRI 2001, *Electricity Infrastructure Security Assessment*, Vols. I and II, EPRI, Palo Alto, CA, November and December 2001.
13. EPRI 2000, *Communication Security Assessment for the United States Electric Utility Infrastructure*, EPRI Report 1001174, EPRI, Palo Alto, CA, December 2000, pp. 4–11.
14. EPRI 2003, *Electricity Technology Roadmap: Synthesis Module on Power Delivery System and Electricity Markets of the Future*, EPRI, Palo Alto, CA, July 2003.
15. EPRI 1999, *Electricity Technology Roadmap: 1999 Summary and Synthesis*, Technical Report, CI-112677-V1, EPRI, Palo Alto, CA, July 1999, 160pp., http://www.epri.com/corporate/discover_epri/roadmap/index.html.
16. F.F. Hauer and J.E. Dagle, *Review of Recent Reliability Issues and System Events*, Consortium for Electric Reliability Technology Solutions, Transmission Reliability Program, Office of Power Technologies, U.S. DOE, Washington, DC, August 30, 1999.
17. P. Kundur, *Power System Stability and Control*, EPRI Power System Engineering Series, McGraw-Hill, Inc., 1994.
18. T.E. Dy Liacco, The adaptive reliability control system, *IEEE on PAS*, 86(5), 517–561, May 1967.
19. L.H. Fink and K. Carlsen, Operating under stress and strain, *IEEE Spectrum*, 15(3), 48–53, March 1978.
20. National Science Foundation, Division of Science Resources Statistics, *Research and Development in Industry: 2000*, Arlington, VA (NSF 03-318), June 2003, http://www.nsf.gov/sbe/srs/nsf03318/pdf/taba19.pdf.
21. S. Silberman, The energy web, *Wired*, 9(7), July 2001.
22. M. Amin, Special Issue of the *Proceedings of the IEEE on Energy Infrastructure Defense Systems*, 2005, forthcoming.
23. M. Samotyj, C. Gellings, and M. Amin, Power system infrastructure for a digital society: Creating the new frontiers, *Proceedings and Keynote Address at the GIGRE/IEEE-PES Symposium on Quality and Security of Electric Power Delivery*, Montreal, Quebec, Canada, October 7–10, 2003, 10pp.

24. C.W. Gellings and K.E. Yeager, Transforming the electric infrastructure, *Physics Today*, 45–51, December 2004.
25. M. Amin, V. Gerhart, and E.Y. Rodin, System identification via artificial neural networks: Application to on-line aircraft parameter estimation, *Proceedings of AIAA/SAE 1997 World Aviation Congress*, Anaheim, CA, Paper #975612, October 13–16, 1997, 22pp.
26. J. Marburger, Testimony before the House Committee on Science, June 14, 2002.
27. C. Starr and M. Amin, Global transition dynamics unfolding the full social implications of national decision pathways, Submitted to the President of the U.S. National Academy of Engineering, September 2003, 11pp., available at: http://cdtlnet.cdtl.umn.edu/Amin/GlobalTransition.pdf.
28. D.J. Hess and J. Coley, Wireless smart meters and public acceptance: The environment, limited choices, and precautionary politics, *Public Understanding of Science*, 2013.
29. *An Investigation of Radiofrequency Fields Associated with the Itron Smart Meter*, Electric Power Research Institute, Palo Alto, CA, December 2010.

Further Readings

Frequency of Outages

Amin, M. U.S. electrical grid gets less reliable, *IEEE Spectrum*, January 2011, 80, http://spectrum.ieee.org/energy/policy/us-electrical-grid-gets-less-reliable.

Costs of Outages

Amin, S.M., Securing the electricity grid, *The Bridge*, quarterly publication of the U.S. National Academy of Engineering, 40(1), 13–20, Spring 2010, http://massoud-amin.umn.edu/publications/Securing-the-Electricity-Grid.pdf.

Amin, S.M. and A.M. Giacomoni, Smart grid-safe, secure, self-healing: Challenges and opportunities in power system security, resiliency, and privacy, *IEEE Power & Energy Magazine*, January/February 2012, 33–40, http://massoud-amin.umn.edu/publications/IEEE_P&E_1-2012-amin.pdf.

EPRI, EPRI/Primen Study: The Cost of Power Disturbances to Industrial & Digital Economy Companies, 2002.

EPRI, Gellings, EPRI ups estimates of smart grid's investment benefits, *IEEE Smart Grid Newsletter*, May 2011, at http://smartgrid.ieee.org/may-2011/102-epri-ups-estimates-of-smart-grids-investment-benefits.

Lawrence Berkeley National Laboratory, ETO, Temporal trends in U.S. electricity reliability, *IEEE Smart Grid Newsletter*, October 2012, at http://smartgrid.ieee.org/october-2012/687-temporal-trends-in-u-s-electricity-reliability.

Pacific Northwest National Laboratory, Regarding the increased efficiency and reduced emissions by 12–18 percent per year, please see PNNL-19112 report, January 2010, at http://www.pnl.gov/main/publications/external/technical_reports/PNNL-19112.pdf.

U.S. DOE, Economic impact of recovery act investments in the smart grid, Smart Grid Investment Grant Program, U.S. Department of Energy, Washington, DC, Electricity Delivery and Energy Reliability, April 2013.

22

Electrical Energy Management in Buildings

Craig B. Smith and Kelly E. Parmenter

CONTENTS

22.1 Principal Electricity Uses in Buildings

22.1.1 Introduction: The Importance of Energy Efficiency in Buildings

A typical building is designed for a 40-year economic life. This implies that the existing inventory of buildings—with all their good and bad features—is turned over very slowly. Today we know it is cost-effective to design a high degree of energy efficiency into new buildings, because the savings on operating and maintenance costs will repay the initial investment many times over. Many technological advances have occurred in the last few decades, resulting in striking reductions in the energy usage required to operate buildings safely and comfortably. Added benefits of these developments are reductions in air pollution and greenhouse gas emissions, which have occurred as a result of generating less electricity.

There are hundreds of building types, and buildings can be categorized in many ways—by use, type of construction, size, or thermal characteristics, to name a few. For simplicity, two designations will be used here: residential and commercial. Industrial facilities are not included here, but are discussed in Chapter 29.

The residential category includes features common to single family dwellings, condominiums and townhouses, and multifamily apartments. In 2012, there were approximately 115 million occupied housing units in the United States (U.S. Census Bureau 2012). The commercial category includes a major emphasis on office buildings, as well as a less detailed discussion of features common to retail stores, hospitals, restaurants, and laundries. The extension to other types is either obvious or can be pursued by referring to the literature. There are roughly five million commercial buildings in the United States, estimated to total 83 billion square feet in 2012 (EIA 2013a). Three-quarters of these buildings are 25 or more years old, and the average building age is about 50 years (SMR Research Corporation 2011). Most of this space is contained in buildings larger than 10,000 ft^2.

Total energy consumption in the two sectors has evolved over time and since the previous three editions of this book (see Table 22.1):

There has been a remarkable shift in the energy sources used by the residential and commercial sectors in the decades since 1975. Natural gas, which increased rapidly in these sectors prior to 1975, flattened out and has remained more or less constant. The use of petroleum has decreased. The big change has been the dramatic increase in electricity sales to the residential and commercial sectors, more than doubling from 1980 to 2011 and increasing by 2.6 times since 1975.

TABLE 22.1

Total Energy Consumption Trends, Quadrillion BTU

	Year						
	1955	1965	1975	1985	1995	2005	2011
Residential	7.3	10.6	14.8	16.0	18.5	21.6	21.6
Commercial	3.9	5.8	9.5	11.5	14.7	17.9	18.0
Total	11.2	16.5	24.3	27.5	33.2	39.5	39.6

Source: EIA, Annual energy review 2011. DOE/EIA-0384(2011), U.S. Energy Information Administration, Washington, DC, 2012, http://www.eia.gov/totalenergy/data/annual/pdf/aer.pdf.

The approach taken in this chapter is to list two categories of specific strategies that are cost-effective methods for using electricity efficiently. The first category includes those measures that can be implemented at low capital cost using existing facilities and equipment in an essentially unmodified state. The second category includes technologies that require retrofitting, modification of existing equipment, or new equipment or processes. Generally, moderate to substantial capital investments are also required.

22.1.2 Electricity Use in Residential and Commercial Buildings

Figures 22.1 and 22.2 illustrate estimated breakdowns of purchased electricity by major end use for the residential and commercial sectors, respectively. The data are from the Energy Information Administration's (EIA's) most recent estimates (EIA 2013a,b).

The single most significant residential end use of electricity is space cooling (19%), followed by lighting (13%), water heating (9%), refrigeration (8%), televisions and related equipment (6%), and space heating (6%). The combination of other uses such as clothes washers and dryers, personal computers, dishwashers, fans, pumps, etc. is also substantial, accounting for nearly 40% of residential electricity use.

In the commercial sector, space conditioning—that is, the combination of heating, ventilating, and air conditioning (HVAC)—is the top user of electricity, accounting for 28%. When HVAC equipment are considered as separated end uses, lighting rises to the top as the largest single end use of electricity in the commercial sector at 21%. Refrigeration (8%), non–personal computer (non-PC) office equipment (5%), and PC office equipment (5%) are also significant end uses of electricity. In addition, other end uses such as water heating, transformers, medical equipment, elevators and escalators, fume hoods, etc. account for the remaining third of electricity use in commercial buildings.

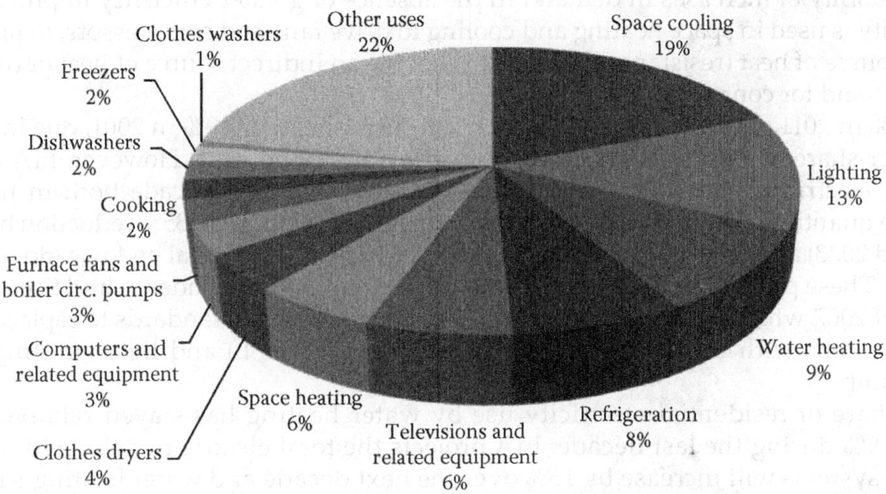

FIGURE 22.1
Breakdown of residential electricity end use, 2011, 1424 billion kWh. (From EIA, Annual energy outlook 2013 with projections to 2040. Residential sector key indicators and consumption. DOE/EIA-0383(2013), U.S. Energy Information Administration, Washington, DC, 2013b, http://www.eia.gov/oiaf/aeo/tablebrowser/.)

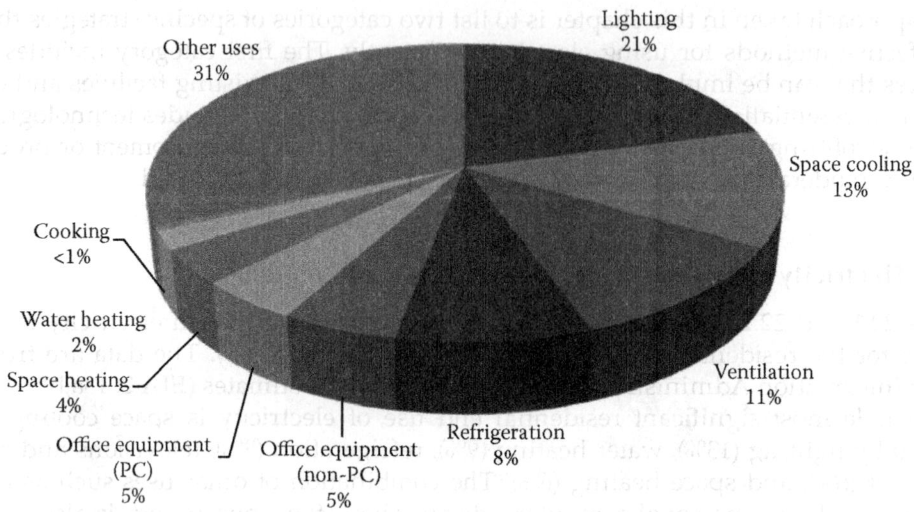

FIGURE 22.2
Breakdown of commercial electricity end use, 2011, 1319 billion kWh. (From EIA, Annual energy outlook 2013 with projections to 2040. Commercial sector key indicators and consumption. DOE/EIA-0383(2013), U.S. Energy Information Administration, Washington, DC, 2013a, http://www.eia.gov/oiaf/aeo/tablebrowser/.)

22.1.2.1 Residential Electricity Use

Space conditioning (including both space cooling and heating and associated fans and pumps) is the most significant end use of electricity in residential buildings, accounting for more than one-quarter of total electricity purchases today. EIA projects space conditioning's share of total residential electricity use will continue to be substantial with the possibility of increases in demand in the absence of greater efficiency improvements. Electricity is used in space heating and cooling to drive fans and compressors, to provide a direct source of heat (resistance heating), to provide an indirect source of heat or *cool* (heat pumps),* and for controls.

At 13% in 2011, residential lighting electricity use is up from 9% in 2001, due in part to a smaller share of electricity being consumed by other end uses. However, EIA expects lighting electricity use to decline considerably over the next decade both in terms of absolute quantity of electricity delivered to homes to power lights (35% reduction between 2011 and 2023) and in terms of lighting's relative share of residential end use (down to 8% in 2023). These projected reductions are due to the Energy Independence and Security Act (EISA) of 2007, which as of January 1, 2013, has been phasing in standards to replace incandescent lamps with more efficient compact fluorescent (CFL) and light-emitting diode (LED) lamps.

The share of residential electricity use by water heating has stayed relatively constant at 9% during the last decade. EIA projects the total electricity delivered to water heating systems will increase by 15% over the next decade and water heating's relative share of electricity use will rise to over 10% by 2023. Electricity use for this purpose currently occurs in electric storage tank water heaters, tankless electric water heaters,

* Heat pumps are discussed in detail in Chapter 27 of this handbook.

and heat pump water heaters. Solar water heating (discussed in Chapters 40–42) is another alternative that is used on a limited basis.

Refrigerators are another important energy end use in the residential sector, accounting for 8% of residential electricity consumption in 2011, down from 14% in 2001. EIA projects refrigeration's share of residential electricity use as well as the absolute quantity of electricity delivered to refrigerators will maintain roughly constant between now and 2023. For the last half century, virtually every home in the United States has had a refrigerator. Therefore, refrigerators have fully penetrated the residential sector for some time. However, significant changes related to energy use have occurred during this period as new standards have been implemented. For one, the average size of refrigerators has more than doubled from less than 10 ft^3 in 1947 to over 20 ft^3 today. Meanwhile, the efficiency of refrigerators has increased dramatically. The net result is that current refrigerators use about one-quarter as much energy (~450 kWh/year in 2011) as smaller units did 40 years ago (~1800 kWh/year in 1972), according to data from the Association of Home Appliance Manufacturers. Furthermore, when new efficiency standards take effect in Fall 2014, new refrigerators will use about one-fifth of the energy use of those in the early 1970s.

TVs, computers, and related equipment collectively account for over 9% of residential electricity use. These types of electronic systems have proliferated in homes during the last few decades. Clothes washers and dryers, cooking, dishwashers, and freezers account for another 11%, while *other* uses (including small electric devices, heating elements, and motors not included earlier) make up the balance (22%) of electricity use in the residential sector.

22.1.2.2 Commercial Electricity Use

For the commercial sector as a whole, HVAC dominates electricity use, with space cooling, space heating, and ventilation systems accounting for over one-quarter of electricity use. This trend is also true for most types of commercial buildings where space conditioning is used. There are exceptions of course—in energy-intensive facilities such as laundries, the process energy will be most important. Electricity is used in space conditioning to run fans, pumps, chillers, cooling towers, and heat pumps. Other uses include electric resistance heating (e.g., in terminal reheat systems) or electric boilers.

Commercial lighting is generally next in importance to HVAC for total use of electricity, except in those facilities with energy-intensive processes. Interior lighting is predominantly fluorescent, with an increasing portion of metal halide, a small fraction of incandescent, and a growing use of LED lamps. High-efficiency fluorescent lamps, electronic ballasts, CFL lamps, and improved lighting controls are now the norm. Incandescent lamps still see use in retail for display lighting as well as in older buildings or for decorative or esthetic applications. Lighting—estimated by the EIA to represent 21% of commercial sector electricity use in 2011—shows a reduction over the last two decades, dropping from 27.7% in 1992. The EIA predicts lighting's share of commercial building electricity use will continue to drop, reaching an estimated 15% by 2040 (EIA 2013a).

Refrigeration is an important use of energy in supermarkets and several other types of commercial facilities. As in residential applications, commercial refrigeration's share of electricity use has decreased in the past few decades due to efficiency gains. For example, in 2011, refrigeration accounted for 8% of commercial electricity use, down from 10% in 1992; EIA predictions estimate its share will drop to 7% over the next decade (EIA 2013a).

Commercial electricity use by office equipment has changed substantially over the last 20 years; its share grew from 7% in 1992 to 18% in 1999 and has now decreased to 10% in 2011. Much of the increase at the turn of the millennium was due to a greater use of computers. Computers have now greatly penetrated the commercial sector, and efficiency is constantly improving. As a result, EIA estimates the absolute quantity of energy used by computers will stay relatively constant over the next decade.

Water heating is another energy use in commercial buildings, but here circulating systems (using a heater, storage tank, and pump) are more common. Many possibilities exist for using heat recovery as a source of hot water. The share of electricity use for water heating in commercial buildings has grown from about 1% in 1999 to 2% in 2011. The EIA predicts the absolute quantity of electricity use for water heating will stay relatively constant over the next couple of decades (EIA 2013a).

In commercial buildings, the balance of the electricity use is for elevators, escalators, and miscellaneous items.

22.2 Setting Up an Energy Management Program

The general procedure for establishing an energy management program in buildings involves five steps:

- Review historical energy use.
- Perform energy audits.
- Identify energy management opportunities (EMOs).
- Implement changes to save energy.
- Monitor the energy management program, set goals, and review progress.

Each step will be described briefly. These steps have been designed for homeowners, apartment owners, and commercial building owners or operators to carry out. Many electric utilities also provide technical and financial assistance for various types of energy efficiency studies and improvements, so building owners and operators are encouraged to seek support from their local utility.

22.2.1 Review of Historical Energy Use

Utility records can be compiled to establish electricity use for a recent 12-month period. These should be graphed (see Figure 22.3) so that annual variations and trends can be evaluated. By placing several years (e.g., last year, this year, and next year projected) on the graph, past trends can be reviewed and future electricity use can be compared with goals. Alternatively, several energy graphs can be compared, or energy use versus production determined (meals served, for a restaurant, kilograms of laundry washed, for a laundry, etc.).

Previously, the norm was to review monthly data to look for energy use trends; however, daily, hourly, 15 min, and even 1 min energy data is now much more available due to the increasing deployment of interval meters. Studying trends in interval data allows building owners to have a greater understanding of exactly when electricity is used in addition to how much is used in a given billing cycle. This additional knowledge can

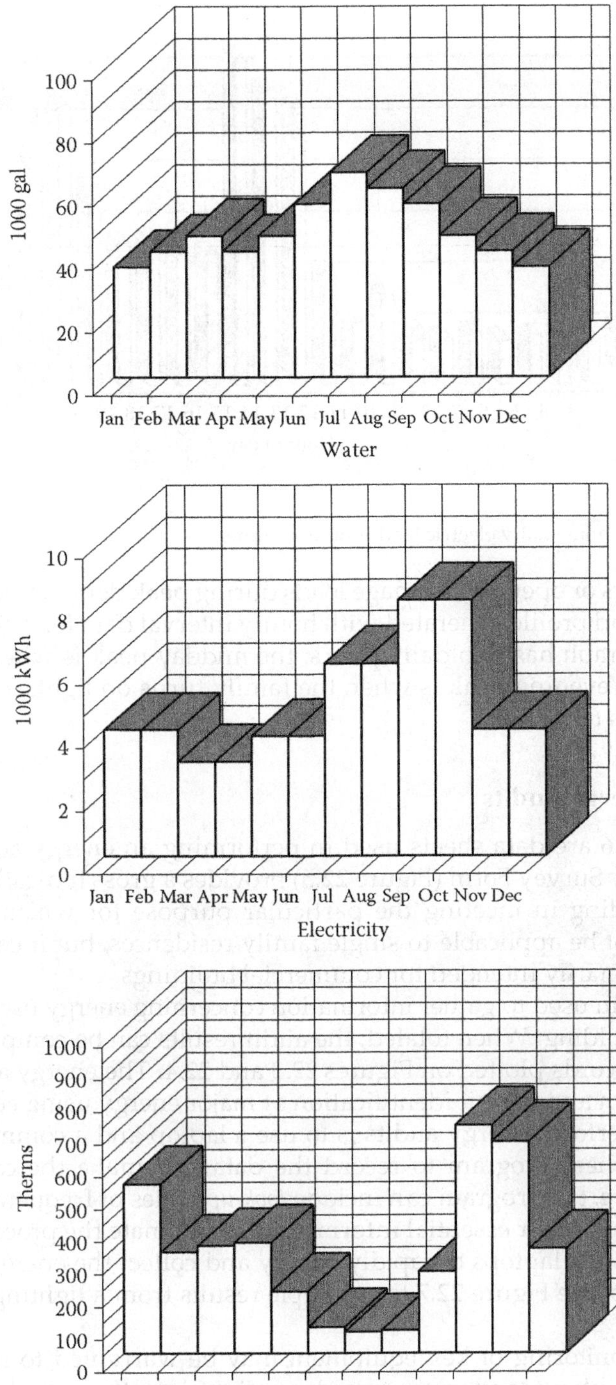

FIGURE 22.3
Sample graph: Historical energy use in an office building.

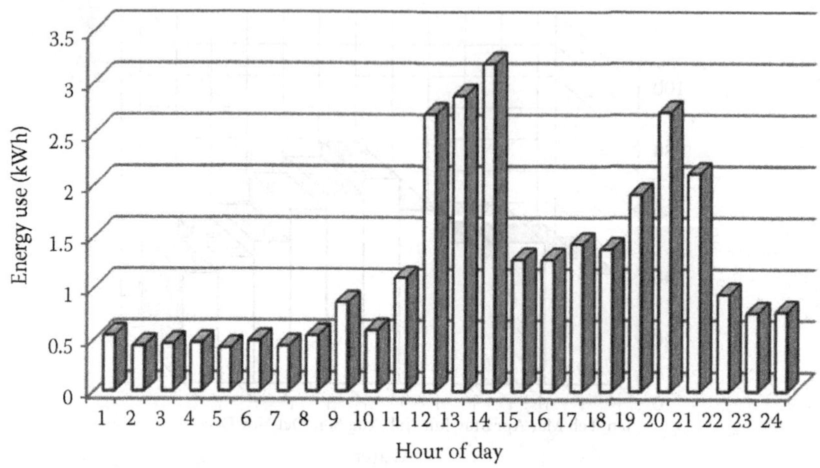

FIGURE 22.4
Example of single family home daily electric load profile, summer.

help building owners or operators manage loads during peak demand periods. Figure 22.4 illustrates a daily load profile generated with hourly interval data for a single family home in California. The graph has two daily peaks: the midday peak is when the home's pool pump is on and the evening peak is when the family turns on lights, cooks dinner, uses the spa, and watches television.

22.2.2 Perform Energy Audits

Figures 22.5 and 22.6 are data sheets used in performing an energy audit of a building. The Building Energy Survey Form (Figure 22.5) provides a gross indication of how energy is used in the building in meeting the particular purpose for which it was designed. This form would not be applicable to single family residences, but it could be used with apartments. It is primarily intended for commercial buildings.

Figure 22.6 is a form used to gather information concerning energy used by each piece of equipment in the building. When totaled, the audit results can be compared with the historical energy use records plotted on Figures 22.3 and 22.4. The energy audit results show a detailed breakdown and permit identification of major energy using equipment items.

Another way to perform energy audits is to use a laptop and a commercially available database or spreadsheet program to record the data and make the calculations. If the workload is extensive, the program can include *look-up* tables of frequently used electrical loads, utility rates, and other essential information to automate the process. We have used teams of engineers with laptops to rapidly survey and collect the energy data from large commercial facilities. See Figure 22.7 for example results from a lighting audit conducted several years ago.

In some cases, monitoring of key equipment may be warranted to improve the accuracy of the audit results. Monitoring provides insight into the actual time-based loads of equipment. It can also be used to simply indicate when loads are turned on or off. It is particularly valuable for highly variable or weather-dependent loads, including HVAC equipment. Monitoring can be accomplished with temporary data loggers that are placed on equipment for a representative period of time (e.g., 2–4 weeks). Permanent monitoring devices integrated with building management system can also be used to characterize

Building Description

- Name____________________Age______years Construction material____________________
- Location____
- No. of floors____________Gross floor area____________m² (ft²) Net floor area__________m² (ft²)
- Percentage of surface area that is glazed____________________% Double or single pane (circle)
- Type of air conditioning system (describe)____________________________________
- Type of heating system (describe)____
- Cooling degree days____________________Heating degree days____________________
- Percentage breakdown of lighting equipment: Fluorescent__________% HID__________%
- Incandescent__________% Other__________% Lighting controls?____________________
- Other unique attributes (describe)____

Building Mission

- What is facility used for____
- Full time occupancy (employees)____________________persons
- Transient occupancy (visitors or public)____________________persons
- Hours of operations per year____________________
- Unit of production per year____________Unit is____________

Installed Capacity

- Installed capacity for lighting____________________kW
- Total installed capacity of electric drives greater than 7.5 kW (10 hp) (motors, pumps, fans, elevators, chillers, etc.)__________hp × 0.746 =____________________kW
- Total steam requirements__________lbs/day or__________kg/day
- Total gas requirements__________ft³/day or BTU/h or__________m³/day
- Total other fuel requirements____

Annual Energy End Use

Energy Form × Conversion			kBTU/year	Metric Units		Conversion	MJ/year
• Electricity	____ kWh/year	× 3.41	=____	____ kWh/year	×	3.6	=____
• Steam	____ lb/year	× 1.00	=____	____ kg/year	×	2.32	=____
• Natural gas	____ cf/year	× 1.03	=____	____ m3/year	×	38.4	=____
• Oil	____ gls/year	× {#2 139; #6 150}	=____	____ l/year	×	{# 2 38.9; # 6 41.8}	=____
• Coal	____ tons/year	× 24,000	=____	____ kg/year	×	28.0	=____
• Other	____	×	=____		×		=____
	Totals		____				____

Energy Use Performance Factors (EUPFs) for Building

- EUPF 1 = MJ/year(kBTU/year) ÷ Net floor area = ____________MJ/m²year(kBTU/ft² year)
- EUPF 2 = MJ/year(kBTU/year) ÷ Average annual occupancy = __________MJ/person · year (kBTU/person · year)
- EUPF 3 = MJ/year(kBTU/year) ÷ Annual units of production = __________MJ/unit · year (kBTU/unit · year)

FIGURE 22.5
Building energy survey form.

energy use patterns on a continuous basis. Major advantages of continuous monitoring include the ability to spot energy use anomalies early and to measure performance of energy efficiency improvements against savings targets.

Energy audits can vary from simple 1-day walkthroughs to comprehensive multiday studies. Often significant opportunities are captured in a simple audit, but more detailed

Conversion Factors

Facility Name__________	Date______ By______ Sheet__ Of_____	Multiply	by	to get
Location__________	Period of survey: 1 day 1 week 1 month 1 year	kWh	3.6	MJ
Symbols: K = 10^3 M = 10^6	Notes____________________	Btu/h	0.000293	kW
		hp	0.746	kW

Item No.	Equipment Description	Power			Est. Hrs Use Per Period	kWh	Conv. Factor	Total Energy Use Per Period (MJ)
		Name Plate Rating (Btu/h, kW, hp, etc.)	Conv. Factor to kW	Est.% Load (100%, 50%, etc.)				

FIGURE 22.6
Energy audit data sheet.

studies may be necessary to justify large capital projects. (See Chapter 31 for further information on energy audits.)

22.2.3 Identify Energy Management Opportunities

An overall estimate should be made of how effectively the facility uses its energy resources. This is difficult to do in many cases, because so many operations are unique. An idea can be obtained, however, by comparing similar buildings located in similar climates. Several online benchmarking tools exist, including EPA's ENERGY STAR Portfolio Manager. Table 22.2 shows representative values of energy use intensity (EUI), which is a measure of energy use per square foot of building area. The table includes median values for several types of commercial buildings and illustrates the range in performance factors that exists across these different building types.

Next, areas or equipment that use the greatest amounts of electricity should be examined. Each item should be reviewed and these questions asked:

- Is this actually needed?
- How can the same equipment be used more efficiently?
- How can the same purpose be accomplished with less energy?
- Can the equipment be modified to use less energy?
- Would new, more efficient equipment be cost-effective?

Lighting Energy Savings Summary				Prepare by:	
Existing annual kWh:	181,828 kWh	Existing kW draw:	36.37 kW	Annual Energy $$ saved:	$12,094
Proposed annual kWh:	95,234 kWh	Proposed kW draw:	18.52 kW	Estimated PRE-REBATE cost:	$13,592
Annual kWh savings:	86,594 kWh	kW savings:	17.85 kW		
% kWh savings:	47.6%	% kW savings:	49.1%		

Lighting Inventory, Recommendations, and Savings

		Existing				Recommended			Savings			Estimated	
Item #	Location	Weekly Hours	Qty	Fixtures	W/Fix	Qty	Fixture	W/Fix	kW	Annual kWh	Annual Energy ($)	Unit Cost ($)	Total Cost ($)
1	Presidents office	60	8	75 Watt INC Spotlight	75	8	18 Watt CFL/SI/Ref.	18	0.5	1368	219	22	176
2	Presidents office	60	6	2-F40T12(40W)/STD	96	6	2-F32T8(32W)/ELEC	61	0.2	630	101	48	288
3	V.P. office	60	8	75 WATT INC Spotlight	75	8	18 Watt CFL/SI/Ref.	18	0.5	1368	219	22	176
4	V.P. office	60	4	2-F40T12(40W)/STD	96	4	2-F32T8(32W)/ELEC	61	0.1	420	67	48	192
5	Night lighting	168	4	2-F40T12(40W)/STD	96	4	2-F32T8(32W)/ELEC	61	0.1	1176	145	48	192
6	Women's restroom mirror	60	12	25 Watt INC	25		None						
7	Women's restroom	60	6	2-F40T12(34W)/U/STD	94	6	2-F40T12(34W)/U/ELEC	60	0.2	612	98	48	288
8	Men's restroom	60	6	2-F40T12(34W)/U/STD	94	6	2-F40T12(34W)/U/ELEC	60	0.2	612	98	48	288
9	Main office area	60	56	3-F40T12(40W)/2-Class 1 1	136	56	3-F32T8(32W)/1-ELEC	90	2.6	7728	1240	48	2688
10	Storage room	25	1	100 Watt INC	100	1	28 Watt PL CFL/SI	30	0.1	88	19	32	32
11	Parking garage	168	26	500 Watt Quartz	350	26	175 Watt MH	205	3.8	31,668	3906	200	5200
12	Parking garage	168	8	2-F96T12(75W)/STD	173	8	20F96T8(50W)/ELEC	104	0.6	4637	572	60	480
13	Physical plant	80	28	2-F96T12(215W) WHO/STD	450	28	2-F96T12(95W)/HO/ELEC	166	8.0	31,808	4741	110	3080
14	Physical plant	80	162	100 Watt INC	100	16	28 Watt PL CFL/SI	30	1.0	4480	668	32	512
	Total								17.9	86,594	$12,094	N/A	$13,592

FIGURE 22.7
Sample energy audit results XYZ corporation.

TABLE 22.2

Median EUI Values for Representative Building Types

Building Type	Source EUI[a] (kBTU/ft²)	Site EUI[b] (kBTU/ft²)
Fast food	1170	418
Food sales	570	193
Restaurant/cafeteria	434	207
Strip and enclosed mall	247	94
College/university	244	104
Lodging	163	72
Outpatient and health care	163	62
Retail (nonmall)	139	53
Public assembly	89	42

Source: EPA, CBECS national median source energy use and performance comparisons by building type, 2011, http://www.energystar.gov/ia/business/tools_resources/new_bldg_design/2003_CBECSPerformanceTargetsTable.pdf. (Accessed August 13, 2013).

[a] Source EUI represents the quantity of raw fuel used to operate a building; it captures energy losses associated with the generation, transmission, and distribution of electricity.

[b] Site EUI represents the quantity of energy used directly on site.

22.2.4 Implement Changes

Once certain actions to save energy have been identified, an economic analysis will be necessary to establish the economic benefits and to determine if the cost of the action is justified. (Refer to Chapter 12 for guidance.) Those changes that satisfy the economic criteria of the building owner (or occupant) will then be implemented. Economic criteria might include a minimum return on investment (e.g., 25%), a minimum payback period (e.g., 2 years), or a minimum benefit–cost ratio (e.g., 2.0).

22.2.5 Monitor the Program, Establish Goals

This is the final—and perhaps most important—step in the program. A continuing monitoring program is necessary to ensure that energy savings do not gradually disappear as personnel return to their old ways of operation, equipment gets out of calibration, needed maintenance is neglected, etc. Also, setting goals (they should be realistic) provides energy management personnel with targets against which they can gauge their performance and the success of their programs.

22.2.6 Summary of Energy Management Programs

The foregoing has been outlined in two tables to provide a step-by-step procedure for electrical energy management in buildings. Table 22.3 is directed at the homeowner or apartment manager, while Table 22.4 has been prepared for the commercial building owner or operator. Industrial facilities are treated separately; refer to Chapter 29. As mentioned to previously, prior to undertaking an independent energy management program, ask your electric utility if you qualify for technical support or financial assistance.

One problem in performing the energy audit is determining the energy used by each item of equipment. In many cases, published data are available—as in Table 22.5 for residential appliances. In other cases, engineering judgments must be made, the manufacturer consulted, or instrumentation provided to actually measure energy use.

TABLE 22.3

Energy Management Plan for the Homeowner or Apartment Manager

Review Historical Data

1. Collect utility bills for a recent 12-month period.
2. Add up the bills and calculate total kWh, total $, average kWh (divide total by 12), average $, and note the months with the lowest and highest kWh. If you have access to interval data, investigate daily load profiles and note the hours of peak usage.
3. Calculate a seasonal variation factor (svf) by dividing the kWh for the greatest month by the kWh for the lowest month.

Perform Energy Audits

4. Identify all electrical loads greater than 1 kW (1000 W). Refer to Table 22.5 for assistance. Most electrical appliances have labels indicating the wattage. If not, use the relation W = V × A.
5. Estimate the number of hours per month each appliance is used.
6. Estimate the percentage of full load (pfl) by each device under normal use. For a lamp, it is 100%; for water heaters and refrigerators, which cycle on and off, about 30%, for an electric range, about 25% (only rarely are *all* burners *and* the oven used), etc.
7. For each device, calculate kWh by multiplying: kW × hours/month × pfl = kWh/month.
8. Add up all kWh calculated by this method. The total should be smaller than the average monthly kWh calculated in Step 2.
9. Note: if the svf is greater than 1.5, the load shows strong seasonal variation, for example, summer air conditioning, winter heating, etc. If this is the case, make two sets of calculation, one for the lowest month (when the fewest loads are operating) and one for the highest month.
10. Make a table listing the wattage of each lamp and the estimated numbers of hours of use per month for each lamp. Multiply Watts times hours for each, sum, and divide by 1000. This gives kWh for the lighting loads. Add this to the total shown.
11. Add the refrigerator, television, and all other appliances or tools that use 5 kWh/month or more.
12. By this process, you should now have identified 80%–90% of electricity using loads. Other small appliances that are used infrequently can be ignored. The test is to now compare with the average month (high or low month if svf is greater than 1.5). If your total is too high, you have overestimated the pfl or the hours of use.
13. Now rank each appliance in descending order of kWh used per month. Your list should read approximately like this:

    First: Air conditioning (hot climates)
    Second: Lighting
    Third: Water heating
    Fourth: Refrigeration
    Fifth: Televisions and related equipment
    Sixth: Space heating (electric)
    Seventh to last: All others

Identify Opportunities and Apply Energy Management Principles

14. Attack the highest priority loads first. There are three general things that can be done to save energy and/or lower energy bills: (1) reduce kW (more efficient lamps and appliances); (2) reduce hours of use (turn lights off, etc.); (3) shift operation of loads to off-peak periods (if your electricity rates are higher during peak hours). Refer to the text for detailed suggestions.
15. *Educate* and get support from family members or other occupants of the space.

Monitor Program, Calculate Savings

16. After the energy management program has been initiated, examine subsequent utility bills to determine if you are succeeding.
17. Calculate energy and cost savings by comparing utility data.
18. Continue to set goals and try to meet them.

TABLE 22.4

Energy Management Plan for Commercial Building Operator

Review Historical Data

1. Collect utility bills for a recent 12-month period.
2. Add up the bills and calculate total kWh, total $, average kWh (divide total by 12), average $, and note the months with the lowest and highest kWh. If you have access to interval data, investigate daily load profiles and note hours of peak usage.
3. Calculate an svf by dividing the kWh for the greatest month by the kWh for the lowest month.
4. Prepare a graph of historical energy use (see Figures 22.3 and 22.4).

Perform Energy Audits

5. Evaluate major loads. In commercial buildings, loads can be divided into four categories:
   a. HVAC (fans, pumps, chillers, heaters, cooling towers)
   b. Lighting
   c. Office equipment and appliances (elevators, computers, cash registers, copy machines, hot water heaters, etc.)
   d. Process equipment (as in laundries, restaurants, bakeries, shops, etc.)

   Items a–c are common to all commercial operations and will be discussed here. Item d overlaps with industry and the reader should also refer to Chapter 29. Generally items a, b, and d account for the greatest use of electricity and should be examined in that order.
6. In carrying out the energy audit, focus on major loads. Items that together comprise less than 1% of the total connected load in kW can often be ignored with little sacrifice in accuracy.
7. Use the methodology described earlier and in Chapter 31 for conducting the audit.
8. Compare audit results with historical energy use. If 80%–90% of the total (according to the historical records) has been identified, this is generally adequate.

Formulate the Energy Management Plan

9. Secure management commitment. The need for this varies with the size and complexity of the operation. However, any formal program will cost something, in terms of salary for the energy coordinator as well as (possibly) an investment in building modifications and new equipment. At this stage, it is very important to project current energy usage and costs ahead for the next 3–5 years, make a preliminary estimate of potential savings (typically 10%–50% per year), and establish the potential payback or return on investment in the program.
10. Develop a list of EMOs and estimate the cost of each EMO and its payback. Methods for economic analysis are given in Chapter 12. For ideas and approaches useful for identifying EMOs, refer to the text.
11. Communicate the plan to employees, department heads, equipment operators, etc. Spell out who will do what, why there is a need, what are the potential benefits and savings. Make the point (if appropriate) that *the energy you save may save your job*. If employees are informed, understand the purpose, and realize that the plan applies to everyone, including the president, cooperation is increased.
12. Set goals for department managers, building engineers, equipment operators, etc., and provide monthly reports so that they can measure their performance.
13. Enlist the assistance of all personnel in (1) better *housekeeping and operations* (e.g., turning off lights, keeping doors closed) and (2) locating obvious wastes of electricity (e.g., equipment operating needlessly, better methods of doing jobs).

Implement Plan

14. Implementation should be done in two parts. First, carry out operational and housekeeping improvements with a goal of, say, 10% reduction in electricity use at essentially no cost and no reduction in quality of service or quantity of production. Second, carry out those modifications (retrofitting of buildings, new equipment, process changes) that have been shown to be economically attractive.
15. As changes are made, it is important to continue to monitor electricity usage to determine if goals are being realized. Additional energy audits may be justified.

Evaluate Progress, Management Report

16. Compare actual performance to the goals established in Item 12. Make corrections for weather variations, increases or decreases in production or number of employees, addition of new buildings, etc.
17. Provide a summary report of energy quantities and dollars saved.
18. Prepare new plans for the future.

TABLE 22.5

Typical Residential Energy Usage for Common Appliances

Electric Appliances	Power (W)
Home entertainment	
Radio	10–20
Stereo	50–400
Speakers	50
Television	
Tube, 25″–27″	90–120
LCD, >40″	150–220
DLP, >40″	200–240
Plasma, >40″	400–480
CD player	10–20
DVD player	20–25
Gaming stations	20–210
Personal computer	
Laptop	20–50
Desktop CPU	30–120
Desktop monitor	30–150
Computer printer	100
Aquarium	50–1210
Food preparation	
Blender	300
Coffee maker	400–1,200
Dishwasher	1,200–2,400
Frying pan	1,200
Hot plate	1,200
Mixer	127
Microwave oven	750–1,100
Range	
Oven bake unit	2,300–3,200
Broil unit	3,600
Self-cleaning feature	4,000
Toaster	750–1,400
Trash compactor	400
Waffle iron	1200
Waste dispenser	450
Refrigerator/freezer	
Top freezer, <10 years old	440–600 kWh/year
Side by side, <10 years old	600–1200 kWh/year
Laundry	
Electric clothes dryer	1,800–5,000
Clothes washer	350–500
Iron (hand)	1,000–1,800
Water heater	2,500–5,500

(Continued)

TABLE 22.5 (*Continued*)

Typical Residential Energy Usage for Common Appliances

Electric Appliances	Power (W)
Housewares	
Clock	2
Floor polisher	305
Sewing machine	75–100
Vacuum cleaner	630–1,400
Comfort conditioning	
Air conditioner (room)	600–2,000
Central air conditioner	2,000–5,000
Electric blanket	60–180
Heating pad	65
Dehumidifier	257–785
Fan	
Attic	370
Ceiling	50–175
Furnace	300–1,000
Portable	55–250
Heater (portable, baseboard)	750–1,500
Electric furnace with fan	10,500
Heat pump	2,900–10,000
Humidifier	177
Health and beauty	
Hair dryer	1,000–1,875
Curling iron	50
Heat lamp (infrared)	250
Shaver	15
Tooth brush	1
Swimming pool and spa	
Sweep pump (3/4 hp)	560
Filter pump (1–1/2 hp)	1,120
Filter pump (2 hp)	1,500
Spa electric heater	1,500–5,500

22.3 Electricity-Saving Techniques by Category of End Use

This section discusses strategies for saving energy in the major end uses found in residential and commercial buildings. Projects that can be implemented in a short time at zero or low capital cost are presented along with retrofit and new design strategies. The ordering of topics corresponds approximately to their importance in terms of building energy use. Smaller end uses and processes specific to select building types are excluded.

22.3.1 Residential HVAC

Residential HVAC units using electricity are generally air conditioning systems, electric resistance heaters, heat pumps, and ventilation fans. Cooling systems range from window air conditioning units to central air conditioning systems. Evaporative coolers are also used in some climates. Heater types range from electric furnaces, small radiant heaters, duct heaters, and strip or baseboard heaters, to embedded floor or ceiling heating systems. Heat pump systems can be used for both heating and cooling and are a highly efficient alternative.

Principal operational and maintenance strategies for existing equipment include the following:

- System maintenance and cleanup
- Thermostat calibration and set-back
- Night cool down
- Improved controls and operating procedures
- Heated or cooled volume reduction
- Reduction of infiltration and exfiltration losses

System maintenance is an obvious but often neglected energy-saving tool. Dirty heat transfer surfaces decrease in efficiency. Clogged filters increase pressure drops and pumping power. Inoperable or malfunctioning dampers can waste energy and prevent proper operation of the system.

In residential systems, heating and cooling are generally controlled by a central or local thermostat. Thermostats should be set to about 26°C (78°F) or higher for cooling and 20°C (68°F) or lower for heating when occupants are home and awake. During periods when occupants are away or asleep, the temperature set points should be adjusted to reduce heating and cooling energy use. As a first step, check the calibration of the thermostat, since these low cost devices can be inaccurate by as much as ±5°C. Programmable thermostats are now widely available. These can be programmed by the user to set-back or set-forward temperature automatically depending on the time of day and day of week. By eliminating the need for manual control, they ensure that the settings will indeed be changed, whereas manual resetting of thermostats depends on occupant diligence. Some utilities have set up demand response programs in which they can communicate with smart thermostats, also referred to as programmable communicating thermostats (PCTs). Participating customers can program their PCTs to control the temperature in response to pricing signals or other demand response event notifications from the utility, thereby reducing peak demand and lowering energy costs. In some programs, the utility uses the PCT to control the customer's equipment directly during high peak periods. A general rule of thumb is that for every 1°C of thermostat set-back (heating) or set-forward (cooling) during an 8 h period, there is a 1% savings in annual heating or cooling energy costs. (The energy savings are generally lower in more severe climates.)

Sometimes simple changes in controls or operating procedures will save energy. In cooling, use night air for summer cool-down. When the outside air temperature is cool, turn off the air conditioner and circulate straight outside air or simply open windows. If fan units have more than one speed, use the lowest speed that provides satisfactory operation. Also, check the balance of the system and the operation of dampers and vents to insure that heating and cooling is provided in the correct quantities where needed.

Reducing the volume of the heated or cooled space can yield energy savings. This can be accomplished by closing vents, doors, or other appropriate means. Usually, it is not necessary to heat or cool an entire residence; the spare bedroom is rarely used, halls can be closed off, etc.

A major cause of energy wastage is air entering or leaving a home. Unintentional air transfer toward the inside is referred to as *infiltration*, and unintentional air transfer toward the outside is referred to as *exfiltration*. However, *infiltration* is often used to imply air leakage both into and out of a home, and this is the terminology used in this chapter. In a poorly *sealed* residence, infiltration of cold or hot air will increase heating or cooling energy use. A typical home loses 25%–40% of its HVAC energy through infiltration. Infiltration also affects concentrations of indoor pollutants and can cause uncomfortable drafts. Air can infiltrate through numerous cracks and spaces created during building construction, such as those associated with electrical outlets, pipes, ducts, windows, doors, and gaps between ceilings, walls, floors, and so on. Infiltration results from temperature and pressure differences between the inside and outside of a home caused by wind, natural convection, and other forces. Major sources of air leakage are attic bypasses (paths within walls that connect conditioned spaces with the attic), fireplaces without dampers, leaky ductwork, window and door frames, and holes drilled in framing members for plumbing, electrical, and HVAC equipment. According to the U.S. Department of Energy, the most significant source for infiltration is the combination of walls, ceilings, and floors, which comprises 31% of the total infiltration in a typical home. Ducts (15%), fireplaces (14%), plumbing penetrations (13%), doors (11%), and windows (10%) are also substantial contributors to infiltration. Of lesser consequence are fans and vents (4%) and electrical outlets (2%).

To combat infiltration, builders of energy-efficient homes use house wraps, caulking, foam insulation, tapes, and other seals. Sealing ducts in the home is also important to prevent the escape of heated or cooled air. Homeowners should also check for open doors and windows, open fireplace dampers, inadequate weather stripping around windows and doors, and any other openings that can be sealed. Caution must be exercised to provide adequate ventilation, however. Standards vary, depending on the type of occupancy. Ventilation rates specified in the builder guidelines for the American Lung Association's Health House program state that for healthy homes "continuous general ventilation should be at least 1.0 cfm (cubic feet per minute) per 100 ft^2 of floor area plus at least 15 cfm for the first bedroom and 7.5 cfm for each additional bedroom." In addition, intermittent ventilation for the kitchen should be at least 100 cfm. For the bathrooms, rates should be 50 cfm intermittent or 20 cfm continuous. The Health House ventilation rates comply with ASHRAE standard 62.2.

In retrofit or new design projects, the following techniques will save energy:

- Site selection and building orientation
- Building envelope design
- Selection of efficient heating/cooling equipment

Site selection and building orientation are not always under the control of the owner/occupant. Where possible, though, select a site sheltered from temperature extremes and wind. Orient the building (in cold climates) with a maximum southerly exposure to take advantage of direct solar heating in winter. Use earth berms to reduce heat losses on northerly exposed parts of the building. Deciduous trees provide summer shading but permit winter solar heating.

Building envelope design can improve heat absorption and retention in winter and summer coolness. The first requirement is to design a well-insulated, thermally tight structure. Insulation reduces heating and cooling loads by resisting the transfer of heat through ceilings, walls, floors, and ducts. Reductions are usually proportionately higher for heating than for cooling because of generally larger indoor-to-outdoor temperature differences in winter than in summer. Insulation is available in a variety of forms including batts, rolls, boards, blocks, loose-fill, or sprayed foam. The appropriate insulation material is selected on the basis of climate, building type, and recommended R-value. Higher R-values indicate better insulating properties. It is typically cost-effective to use greater than recommended R-values to improve energy efficiency above and beyond standard building practice.

Windows are an important source of heat gain and loss. The heat loss for single pane glazing is around 5–7 W/m²°C. For double glazing, the comparable value is in the range of 3–4 W/m²°C. Window technology is constantly improving. Newer windows often have low emissivity (low-E) or spectrally selective coatings to prevent heat gain and/or loss. Low-E windows with an argon gas fill have a heat loss rate of about 2 W/m²°C. They have a higher visible transmittance, and are available with a low solar heat gain coefficient to reduce cooling loads in the summer. Low-E windows are available with an internal plastic film, which essentially makes them triple glazed. The heat loss rate for these windows is on the order of 1 W/m²°C.

Windows equipped with vinyl, wood, or fiberglass frames, or aluminum frames with a thermal barrier, provide the best insulation. It is also important to seal windows to prevent infiltration, as well as to use window coverings to minimize heat loss by radiation to the exterior during the evening. The appropriate placement of windows can also save energy by providing daylighting.

In general, the most efficient electric heating and cooling system is the heat pump. Common types are air-to-air heat pumps, either a single package unit (similar to a window air conditioner) or a split system where the air handling equipment is inside the building and the compressor and related equipment are outdoors. Commercially available equipment demonstrates a wide range of efficiency. Heating performance is measured in terms of a heating seasonal performance factor (HSPF), in BTUs of heat added per Watt-hour of electricity input. Typical values are 6.8–10.0 for the most efficient heat pumps. Cooling performance of residential heat pumps, air conditioners, and packaged systems is measured in terms of a seasonal energy efficiency ratio (SEER), which describes the ratio of cooling capacity to electrical power input. Typical values are 10.0–14.5 and even higher for the most efficient systems. The Federal standards set in 2006 for air conditioners, heat pumps, and residential packaged units require a minimum SEER of 13 and a minimum HSPF of 7.7. Many existing older units have SEERs of 6–7, or roughly half the new minimum requirement. Therefore, substantial efficiency improvements are possible by replacing older equipment. In purchasing new equipment, consider selecting systems with the highest HSPF and SEER. The operating savings almost always justify the higher initial cost of these units. In addition, many utilities offer rebates for installing the more efficient units.

Sizing of equipment is important, since the most efficient operation generally occurs at or near full load. Selection of oversized equipment is thus initially more expensive, and will also lead to greater operating costs.

The efficiency of heat pumps declines as the temperature difference between the heat source and heat sink decreases. Since outside air is generally the heat source, heat is most difficult to get when it is most needed. For this reason, heat pumps often have electrical backup heaters for extremely cold weather.

An alternate approach is to design the system using a heat source other than outside air. Examples include heated air (such as is exhausted from a building), a deep well (providing water at a constant year-round temperature), the ground, or a solar heat source. There are a great many variations on solar heating and heat pump combinations.

22.3.2 Nonresidential HVAC

HVAC systems in commercial buildings and other nonresidential installations may involve package rooftop or ground mounted units, or a central plant. Although the basic principles are similar to those discussed earlier in connection with residential systems, the equipment is larger and control more complex.

Efficiency of many existing HVAC systems can be improved. Modifications can reduce energy use by 10%–15%, often with building occupants unaware that changes have been made. Instituting an energy management program that includes evaluating and fine-tuning the operation of HVAC systems to ensure they are operating as originally designed (referred to as retrocommissioning in existing buildings) can uncover many inefficiencies and often yields significant energy savings and improved comfort.

The basic function of HVAC systems is to heat, cool, dehumidify, humidify, and provide air mixing and ventilating. The energy required to carry out these functions depends on the building design, its duty cycle (e.g., 24 h/day use as in a hospital vs. 10 h/day in an office), the type of occupancy, the occupants' use patterns and training in using the HVAC system, the type of HVAC equipment installed, and finally, daily and seasonal temperature and weather conditions to which the building is exposed.

A complete discussion of psychometrics, HVAC system design, and commercially available equipment types is beyond the scope of this chapter.
Energy management strategies will be described in three parts:

- Equipment modifications (control, retrofit, and new designs)
  - Fans
  - Pumps
  - Packaged air conditioning units
  - Chillers
  - Ducts and dampers
  - Systems
- Economizer systems and enthalpy controllers
- Heat recovery techniques

22.3.2.1 Equipment Modifications (Control, Retrofit, and New Designs)

22.3.2.1.1 Fans

All HVAC systems involve some movement of air. The energy needed for this motion can make up a large portion of the total system energy used. This is especially true in moderate weather when the heating or cooling load drops off, but the distribution systems often operate at the same level.

22.3.2.1.1.1 Control Simple control changes can save electrical energy in the operation of fans. Examples include turning off large fan systems when relatively few people are in the building or stopping ventilation a half hour before the building closes. The types of changes that can be made will depend upon the specific facility. Some changes involve more sophisticated controls, which may already be available in the HVAC system.

22.3.2.1.1.2 Retrofit The capacity of the building ventilation system is usually determined by the maximum cooling or heating load in the building. This load may change over time due to reduced outside air requirements, more efficient lighting, and fluctuations in building occupancy. As a result, it may be feasible to decrease air flow in existing commercial buildings as long as adequate indoor air quality is maintained.

The volume rate of air flow through the fan, Q, varies directly with the speed of the impeller's rotation. This is expressed as follows for a fan whose speed is changed from N_1 to N_2.

$$Q_2 = \left(\frac{N_2}{N_1}\right) \times Q_1 \tag{22.1}$$

The pressure developed by the fan, P, (either static or total) varies as the square of the impeller speed.

$$P_2 = \left(\frac{N_2}{N_1}\right)^2 \times P_1 \tag{22.2}$$

The power needed to drive the fan, H, varies as the cube of the impeller speed.

$$H_2 = \left(\frac{N_2}{N_1}\right)^3 \times H_1 \tag{22.3}$$

The result of these laws is that for a given air distribution system (specified ducts, dampers, etc.), if the air flow is to be doubled, eight (2^3) times the power is needed. Conversely, if the air flow is to be cut in half, one-eighth ($½^3$) of the power is required. This is useful in HVAC systems, because even a small reduction in air flow (say 10%) can result in significant energy savings (27%).

The manner in which the air flow is reduced is critical in realizing these savings. Sizing the motor exactly to the requirements helps to yield maximum savings. Simply changing pulleys to provide the desired speed will also result in energy reductions according to the cubic law. Note the efficiency of existing fan motors tends to drop off below the half load range.

If variable volume air delivery is required, it may be achieved through inlet vane control, outlet dampers, variable speed drives (VSDs), controlled pitch fans, or cycling. Energy efficiency in a retrofit design is best obtainable with VSDs on motors. This can be seen by calculating the power reduction that would accompany reduced flow using different methods of control, as noted in Table 22.6. Numbers in the table are the percent of full-flow input power:

22.3.2.1.1.3 New Design The parameters for new design are similar to those for fan retrofit. It is desirable, when possible, to use a varying ventilation rate that will decrease as the load decreases. A system such as variable air volume incorporates this in the interior zones of a building. In some cases, there will be a trade-off between power saved by running the fan slower and the additional power needed to generate colder air. The choices should be determined on a case-by-case basis.

22.3.2.1.2 Pumps

Pumps are found in a variety of HVAC applications such as chilled water, heating hot water, and condenser water loops. They are another piece of peripheral equipment that can use a large portion of HVAC energy, especially at low system loads.

TABLE 22.6

Comparison of Flow Control Alternatives for Fans and Pumps

	Percent of Full-Flow Input Power				
	Fans			Pumps	
Flow (%)	Inlet Vanes	Dampers	VSDs	Throttle Valve	VSDs
100	102	103	102	101	103
90	86	98	76	96	77
80	73	94	58	89	58
70	64	88	43	83	41
60	56	81	31	77	30
50	50	74	22	71	19
40	46	67	15	65	13
30	41	59	9	59	8

22.3.2.1.2.1 Control The control of pumps is often neglected in medium and large HVAC systems where it could significantly reduce the demand. A typical system would be a three chiller installation where only one chiller is needed much of the year. Two chilled water pumps in parallel are designed to handle the maximum load through all three chillers. Even when only one chiller is on, both pumps are used. When not needed to meet demand, two chillers could be bypassed and one pump turned off. All systems should be reviewed in this manner to ensure that only the necessary pumps operate under normal load conditions.

22.3.2.1.2.2 Retrofit Pumps follow laws similar to fan laws, the key being the cubic relationship of power to the volume pumped through a given system. Small decreases in flow rate can save significant portions of energy.

In systems in which cooling or heating requirements have been permanently decreased, flow rates may be reduced also. A simple way to do this is by trimming the pump impeller. The pump curve must be checked first, however, because pump efficiency is a function of the impeller diameter, flow rate, and pressure rise. After trimming, one should ensure that the pump will still be operating in an efficient region. This is roughly the equivalent of changing fan pulleys in that the savings follow the cubic law of power reduction.

Another method for decreasing flow rates is to use a *throttle* (pressure reducing) valve. The result is equivalent to that of the discharge damper in the air-side systems. The valve creates an artificial use of energy, which can be responsible for much of the work performed by the pumps. VSDs are more efficient for varying flow, as shown in Table 22.6.

22.3.2.1.2.3 New Design In a variable load situation, common to most HVAC systems, more efficient systems with new designs are available, rather than the standard constant volume pump. (These may also apply to some retrofit situations.)

One option is the use of several pumps of different capacity so that a smaller pump can be used when it can handle the load and a larger pump used the rest of the time. This can be a retrofit modification as well when a backup pump provides redundancy. Its impeller would be trimmed to provide the lower flow rate.

Another option is to use VSD pumps. While their initial cost is greater, they offer a significant improvement in efficiency over the standard pumps. The economic desirability of this or any similar change can be determined by estimating the number of hours the system will operate under various loads. Many utilities also offer rebates for installing variable speed pumps.

22.3.2.1.3 *Package Air Conditioning Units*

The most common space conditioning systems for commercial buildings are unitary equipment, either single package systems or split systems. These are used for cooling approximately two-thirds of the air-conditioned commercial buildings in the United States. For very large buildings or building complexes, absorption chillers or central chiller plants are used. (The following section describes Chillers.)

Air conditioner efficiency is typically rated by one or more of three parameters: the energy efficiency ratio (EER), the SEER as described previously, and the integrated energy efficiency ratio (IEER) that replaces a former metric used called the integrated part-load value (IPLV). The EER is easy to understand; it is the ratio of cooling capacity, expressed in BTU/hour (kilojoules/hour) to the power input required in Watts. The SEER is a calculated ratio of the total annual cooling produced per annual electrical energy input in Watt-hours for units rated at less than 65,000 BTU/h. The IEER is a new part-load metric introduced in ASHRAE Standard 90.1-2010 to replace the IPLV for rating commercial unitary loads of more than 65,000 BTU/h.

Great improvements in packaged air conditioner efficiency have been made in the last few decades, and new standards continue to increase efficiency even further. This is illustrated by the 30% increase in minimum SEER requirement that took effect in June 2008 for commercial three-phase central air conditioners and heat pumps under 65,000 BTU/h as the result of the EISA (2007). The current standard is a minimum SEER of 13.0. These commercial units are similar to the residential central air conditioners and heat pumps discussed earlier in the chapter in the residential HVAC section. Standards for larger units (>65,000 BTU/h) also continue to increase. All current standards are published in ASHRAE Standard 90.1-2010. Manufacturers sell systems with a broad range of efficiencies. Units with high EERs are typically more expensive, as the greater efficiency is achieved with larger heat exchange surface, more efficient motors, and so on.

To evaluate the economic benefit of the more efficient units, it is necessary to determine an annual operating profile, which depends in part on the nature of the load and on the weather and temperature conditions at the site where the equipment will be installed. Or, an approximate method can be used. ASHRAE publishes tables that show typical *equivalent full-load operating hours* for different climate zones. These can be used to estimate the savings in electrical energy use over a year and thereby determine if the added cost of a more efficient unit is justified. (It almost always is.)

Since the more efficient unit is almost always more cost-effective (except in light or intermittent load conditions), one might wonder why the less efficient units are sold. The reason is that many commercial buildings are constructed and sold by developers whose principal concern is keeping the initial cost of the building as low as practicable. They do not have to bear the annual operating expense of the building once it is sold, and therefore have less incentive to minimize operating expenses. However, the demand for energy-efficient and LEED-certified buildings is growing, causing many developers to rethink design approaches.*

22.3.2.1.4 *Chillers*

Chillers are often the largest single energy user in the HVAC system. The chiller cools the water used to extract heat from the building and outside air. Optimizing chiller operation improves the performance of the whole system.

* LEED stands for Leadership in Energy & Environmental Design, http://www.usgbc.org/leed.

Two basic types of chillers are found in commercial and industrial applications: absorption and mechanical chillers. Absorption units boil water, the refrigerant, at a low pressure through absorption into a high concentration lithium bromide solution. Mechanical chillers cool through evaporation of a refrigerant at a low pressure after it has been compressed, cooled, and passed through an expansion valve. Hydrochlorofluorocarbon (HCFC) products such as R22 are among the most common refrigerants used; however, they are currently being phased out, because they contribute to ozone depletion. The Environmental Protection Agency (EPA) lists acceptable refrigerant substitutes for various types of chillers on their website. Ammonia is a type of refrigerant that has been increasing in favor due to its efficiency advantages and the fact that it will continue to be available.

There are three common types of mechanical chillers. They have similar thermodynamic properties, but use different types of compressors. Reciprocating and screw-type compressors are both positive displacement units. The centrifugal chiller uses a rapidly rotating impeller to pressurize the refrigerant.

All of these chillers must reject heat to a heat sink outside the building. Some use air-cooled condensers, but most large units operate with evaporative cooling towers. Cooling towers have the advantage of rejecting heat to a lower-temperature heat sink, because the water approaches the ambient wet-bulb temperature, while air-cooled units are limited to the dry-bulb temperature. As a result, air-cooled chillers have a higher condensing temperature, which lowers the efficiency of the chiller. In full-load applications, air-cooled chillers require about 1–1.3 kW or more per ton of cooling, while water-cooled chillers usually require between 0.4 and 0.9 kW/ton. Air-cooled condensers are sometimes used, because they require much less maintenance than cooling towers and have lower installation costs. They can also be desirable in areas of the country where water is scarce and/or water and water treatment costs are high, since they do not depend on water for cooling.

Mechanical cooling can also be performed by direct expansion (DX) units. These are very similar to chillers except that they cool the air directly instead of using the refrigerant as a heat transfer medium. They eliminate the need for chilled water pumps and also reduce efficiency losses associated with the transfer of the heat to and from the water. DX units must be located close (~30 m) to the ducts they are cooling, so they are typically limited in size to the cooling required for a single air handler. A single large chiller can serve a number of distributed air handlers. Where the air handlers are located close together, it can be more efficient to use a DX unit.

22.3.2.1.4.1 Controls Mechanical chillers operate on a principle similar to the heat pump. The objective is to remove heat from a low-temperature building and deposit it in a higher temperature atmosphere. The lower the temperature rise that the chiller has to face, the more efficiently it will operate. It is useful, therefore, to maintain as warm a chilled water loop and as cold a condenser water loop as possible.

Using lower-temperature water from the cooling tower to reject the heat can save energy. However, as the condenser temperature drops, the pressure differential across the expansion valve drops, starving the evaporator of refrigerant. Many units with expansion valves, therefore, operate at a constant condensing temperature, usually 41°C (105°F), even when more cooling is available from the cooling tower. Field experience has shown that in many systems, if the chiller is not fully loaded, it can be operated with a lower cooling tower temperature.

22.3.2.1.4.2 Retrofit Where a heat load exists and the wet-bulb temperature is low, cooling can be done directly with the cooling tower. If proper filtering is available, the cooling tower water can be piped directly into the chilled water loop. Often a direct heat exchanger between the two loops is preferred to protect the coils from fouling. Another technique is to turn off the chiller but use its refrigerant to transfer heat between the two loops. This *thermocycle* uses the same principles as a heat pipe, and only works on chillers with the proper configuration.

A low wet-bulb temperature during the night can also be utilized. It requires a chiller that handles low condensing temperatures and a cold storage tank. This thermal energy storage (TES) technique is particularly desirable for consumers with access to time-based electricity rates that reward peak-shaving or load-shifting.

22.3.2.1.4.3 New Design In the purchase of a new chiller, an important consideration should be the load control feature. Since the chiller will be operating at partial load most of the time, it is important that it can do so efficiently. Variable speed chillers are an efficient alternative. In addition to control of single units, it is sometimes desirable to use multiple compressor reciprocating chillers. This allows some units to be shut down at partial load. The remaining compressors operate near full load, usually more efficiently. Good opportunities to install a high-efficiency chiller are when an old unit needs to be replaced, or when it is necessary to retire equipment that uses environmentally unacceptable refrigerants.

22.3.2.1.5 Ducting-Dampers

22.3.2.1.5.1 Controls In HVAC systems using dual ducts, static pressure dampers are often placed near the start of the hot or cold plenum run. They control the pressure throughout the entire distribution system and can be indicators of system operation. In an overdesigned system, the static pressure dampers may never open more than 25%. Reducing the fan speed and opening the dampers fully can eliminate the previous pressure drop. The same volume of air is delivered with a significant drop in fan power.

22.3.2.1.5.2 Retrofit Other HVAC systems use constant volume mixing boxes for balancing that create their own pressure drops as the static pressure increases. An entire system of these boxes could be overpressurized by several inches of water without affecting the air flow, but the required fan power would increase. (One inch of water pressure is about 250 N/m^2 or 250 Pa.) These systems should be monitored to ensure that static pressure is controlled at the lowest required value. It may also be desirable to replace the constant volume mixing boxes with boxes without volume control to eliminate their minimum pressure drop of approximately 1 in. of water. In this case, static pressure dampers will be necessary in the ducting.

Leakage in any dampers can cause a loss of hot or cold air. Adding or replacing seals on the blades can slow leakage considerably. If a damper leaks more than 10%, it can be less costly to replace the entire damper assembly with effective positive-closing damper blades rather than to tolerate the loss of energy.

22.3.2.1.5.3 New Design In the past, small ducts were installed because of their low initial cost despite the fact that the additional fan power required offset the initial cost on a life cycle basis. ASHRAE 90.1 guidelines now set a maximum limit on the fan power that can be used for a given cooling capacity. As a result, the air system pressure drop must be low enough to permit the desired air flow. In small buildings, this pressure drop is often largest across filters, coils, and registers. In large buildings, the duct runs may be responsible for a significant fraction of the total static pressure drop, particularly in high velocity systems. New designs should incorporate ducting that optimizes energy efficiency.

22.3.2.1.6 Systems

The use of efficient equipment is only the first step in the optimum operation of a building. Equal emphasis should be placed upon the combination of elements in a system and the control of those elements. This section discusses some opportunities for equipment modifications. Chapter 23 describes HVAC control systems in greater detail.

22.3.2.1.6.1 Control Some systems use a combination of hot and cold to achieve moderate temperatures. Included are dual duct, multizone, and terminal reheat systems, and some induction, variable air volume, and fan coil units. Whenever combined heating and cooling occurs, the temperatures of the hot and cold ducts or water loops should be brought as close together as possible, while still maintaining building comfort.

This can be accomplished in a number of ways. Hot and cold duct temperatures are often reset on the basis of the temperature of the outside air or the return air. Another approach is to monitor the demand for heating and cooling in each zone. For example, in a multizone building, the demand of each zone is communicated back to the supply unit. At the supply end, hot air and cold air are mixed in proportion to this demand. The cold air temperature should be just low enough to cool the zone calling for the most cooling. If the cold air were any colder, it would be mixed with hot air to achieve the right temperature. This creates an overlap in heating and cooling not only for the zone but for all the zones, because they would all be mixing in the colder air.

If no zone calls for total cooling, then the cold air temperature can be increased gradually until the first zone requires cooling. At this point, the minimum cooling necessary for that multizone configuration is performed. The same operation can be performed with the hot air temperature until the first zone is calling for heating only.

Note that simultaneous heating and cooling is still occurring in the rest of the zones. This is not an ideal system, but it is a first step in improving operating efficiency for these types of systems.

The technique for resetting hot and cold duct temperatures can be extended to the systems that have been mentioned. Ideally, it would be performed automatically with a control system, but it could also be done manually. In some buildings, it will require the installation of more monitoring equipment (usually only in the zones of greatest demand), but the expense should be relatively small and the payback period short.

Nighttime temperature set-back is another control option that can save energy without significantly affecting the comfort level. Energy is saved by shutting off or cycling fans. Building heat loss may also be reduced, because the building is cooler and no longer pressurized.

In moderate climates, complete night shutdown can be used with a morning warm-up period. In colder areas where the overall night temperature is below 4°C (40°F), it is usually necessary to provide some heat during the night. Building set-back temperature is partially dictated by the capacity of the heating system to warm the building in the morning. In some cases, it may be the mean radiant temperature of the building rather than air temperature that determines occupant comfort.

Some warm-up designs use *free heating* from people and lights to help attain the last few degrees of heat. This also provides a transition period for the occupants to adjust from the colder outdoor temperatures.

In some locations during the summer, it is desirable to use night air for a cool-down period. This *free cooling* can decrease the temperature of the building mass that has accumulated heat during the day. In certain types of massive buildings (such as libraries or buildings with thick walls), a long period of night cooling may decrease the building mass temperature by a degree or two. This represents a large amount of cooling that the chiller will not have to perform the following day.

22.3.2.1.6.2 Retrofit Retrofitting HVAC systems may be an easy or difficult task depending upon the possibility of using existing equipment in a more efficient manner. Often retrofitting involves control or ducting changes that appear relatively minor but will greatly increase the efficiency of the system. Some of these common changes, such as decreasing air flow, are discussed elsewhere in this chapter. This section will describe a few changes appropriate to particular systems.

Both dual duct and multizone systems mix hot and cold air to achieve the proper degree of heating or cooling. In most large buildings, the need for heating interior areas is essentially nonexistent, due to internal heat generation. A modification that adjusts for this is simply shutting off air to the hot duct. The mixing box then acts as a variable air volume box, modulating cold air according to room demand as relayed by the existing thermostat. (It should be confirmed that the low volume from a particular box meets minimum air requirements.)

Savings from this modification come mostly from the elimination of simultaneous heating and cooling, depending on fan control strategies. That is, if fans in these systems are controlled by static pressure dampers in the duct after the fan, they do not unload very efficiently and would represent only a small portion of the savings.

22.3.2.2 Economizer Systems and Enthalpy Controllers

The economizer cycle is a technique for introducing varying amounts of outside air to the mixed air duct. Basically, it permits mixing warm return air at 24°C (75°F) with cold outside air to maintain a preset temperature in the mixed air plenum (typically 10°C–15°C, 50°F–60°F). When the outside temperature is slightly above this set point, 100% outside air is used to provide as much of the cooling as possible. During very hot outside weather, minimum outside air will be added to the system.

A major downfall of economizer systems is poor maintenance. The failure of the motor or dampers may not cause a noticeable comfort change in the building, because the system is often capable of handling the additional load. Since the problem is not readily apparent, corrective maintenance may be put off indefinitely. In the meantime, the HVAC system will be working harder than necessary for any economizer installation.

Typically, economizers are controlled by the dry-bulb temperature of the outside air rather than its enthalpy (actual heat content). This is adequate most of the time, but can lead to unnecessary cooling of air. When enthalpy controls are used to measure wet-bulb temperatures, this cooling can be reduced. However, enthalpy controllers are more expensive and less reliable.

The rules that govern the more complex enthalpy controls for cooling-only applications are as follows:

- When outside air enthalpy is greater than that of the return air or when outside air dry-bulb temperature is greater than that of the return air, use minimum outside air.
- When the outside air enthalpy is below the return air enthalpy and the outside dry-bulb temperature is below the return air dry-bulb temperature but above the cooling coil control point, use 100% outside air.

- When outside air enthalpy is below the return air enthalpy and the outside air dry-bulb temperature is below the return air dry-bulb temperature and below the cooling coil controller setting, the return and outside air are mixed by modulating dampers according to the cooling set point.

These points are valid for the majority of cases. When mixed air is to be used for heating and cooling, a more intricate optimization plan will be necessary, based on the value of the fuels used for heating and cooling.

22.3.2.3 Heat Recovery

Heat recovery is often practiced in industrial processes that involve high temperatures. It can also be employed in HVAC systems.

Systems are available that operate with direct heat transfer from the exhaust air to the inlet air. These are most reasonable when there is a large volume of exhaust air, for example, in once-through systems, and when weather conditions are not moderate.

Common heat recovery systems are broken down into two types: regenerative and recuperative. Regenerative units use alternating air flow from the hot and cold stream over the same heat storage/transfer medium. This flow may be reversed by dampers or the whole heat exchanger may rotate between streams. Recuperative units involve continuous flow; the emphasis is upon heat transfer through a medium with little storage.

The rotary regenerative unit, or heat wheel, is one common heat recovery device. It contains a corrugated or woven heat storage material that gains heat in the hot stream. This material is then rotated into the cold stream where the heat is given off again. The wheels can be impregnated with a desiccant to transfer latent as well as sensible heat. Purge sections for HVAC applications can reduce carryover from the exhaust stream to acceptable limits for most installations.

The heat transfer efficiency of heat wheels generally ranges from 60% to 85% depending upon the installation, type of media, and air velocity. For easiest installation, the intake and exhaust ducts should be located near each other.

Another system that can be employed with convenient duct location is a plate type air-to-air heat exchanger. This system is usually lighter though more voluminous than heat wheels. Heat transfer efficiency is typically in the 60%–75% range. Individual units range from 1,000 to 11,000 SCFM and can be grouped together for greater capacity. Almost all designs employ counterflow heat transfer for maximum efficiency.

Another option to consider for nearly contiguous ducts is the heat pipe. This is a unit that uses a boiling refrigerant within a closed pipe to transfer heat. Since the heat of vaporization is utilized, a great deal of heat transfer can take place in a small space.

Heat pipes are often used in double wide coils, which look very much like two steam coils fastened together. The amount of heat transferred can be varied by tilting the tubes to increase or decrease the flow of liquid through capillary action. Heat pipes cannot be *turned off*, so bypass ducting is often desirable. The efficiency of heat transfer ranges from 55% to 75%, depending upon the number of pipes, fins per inch, air face velocity, etc.

Runaround systems are also used for HVAC applications, particularly when the supply and exhaust plenums are not physically close. Runaround systems involve two coils (air-to-water heat exchangers) connected by a piping loop of water or glycol solution and a small pump. The glycol solution is necessary if the air temperatures in the inlet coils are below freezing. Standard air conditioning coils can be used for the runaround system. Precaution should be used when the exhaust air temperature drops below 0°C (32°F), which would cause freezing of the condensed water on its fins. A three-way bypass valve will maintain the temperature of the solution entering the coil at just above 0°C (32°F). The heat transfer efficiency of this system ranges from 60% to 75% depending upon the installation.

Another system similar to the runaround in layout is the desiccant spray system. Instead of using coils in the air plenums, it uses spray towers. The heat transfer fluid is a desiccant (lithium chloride) that transfers both latent and sensible heat—desirable in many applications. Tower capacities range from 7,700 to 92,000 SCFM; multiple units can be used in large installations. The enthalpy recovery efficiency is in the range of 60%–65%.

22.3.2.4 Thermal Energy Storage

TES systems are used to reduce the on-peak electricity demand caused by large cooling loads. TES systems utilize several different storage media, with chilled water, ice, or eutectic salts being most common. Chilled water requires the most space, with the water typically being stored in underground tanks. Ice storage systems can be aboveground insulated tanks with heat exchanger coils that cause the water to freeze, or can be one of several types of ice-making machines.

In a typical system, a chiller operates during off-peak hours to make ice (usually at night). Since the chiller can operate for a longer period of time than during the daily peak, it can have a smaller capacity. Efficiency is greater at night, when the condensing temperature is lower than it is during the day. During daytime operation, chilled water pumps circulate water through the ice storage system and extract heat. Systems can be designed to meet the entire load, or to meet a partial load, with an auxiliary chiller as a backup.

This system reduces peak demand and can also reduce energy use. With ice storage, it is possible to deliver water at a lower temperature than is normally done. This means that the chilled water piping can be smaller and the pumping power reduced. A low-temperature air distribution system will allow smaller ducts and lower capacity fans to deliver a given amount of cooling. Careful attention must be paid to the system design to ensure occupant comfort in conditioned spaces. Some government agencies and electric utilities offer incentives to customers installing TES systems. The utility incentives could be in the form of a set rebate per ton of capacity or per kW of deferred peak demand, or a time-of-use pricing structure that favors TES.

22.3.3 Water Heating

22.3.3.1 Residential Systems

Residential storage water heaters typically range in size from 76 L (20 gal) to 303 L (80 gal). Electric units generally have one or two immersion heaters, each rated at 2–6 kW depending on tank size. Energy input for water heating is a function of the temperature at which water is delivered, the supply water temperature, and standby losses from the water heater, storage tanks, and piping.

Tankless water heaters are an energy-efficient alternative that has achieved greater market penetration over the last few years. These systems can be electric or gas-fired and provide hot water on demand, eliminating energy losses associated with a storage tank.

The efficiency of water heaters is referred to as the energy factor (EF). Higher EF values equate to more efficient water heaters. Typical EF values range from about 0.9–0.95 for electric resistance heaters, 0.6–0.86 for natural gas units, 0.5–0.85 for oil units, and 1.5–2.2

for heat pump water heaters. The higher efficiency values for each fuel type represent the more advanced systems available, while the lower efficiency values are for the more conventional systems.

In single tank residential systems, major savings can be obtained by

- Thermostat temperature set back to 60°C (140°F)
- Automated control
- Supplementary tank insulation
- Hot water piping insulation

The major source of heat loss from electric water theaters is standby losses through the tank walls and from piping, since there are no flame or stack losses in electric units. The heat loss is proportional to the temperature difference between the tank and its surroundings. Thus, lowering the temperature to 60°C will result in two savings: (1) a reduction in the energy needed to heat water and (2) a reduction in the amount of heat lost. Residential hot water uses do not require temperatures in excess of 60°C; for any special use which does, it would be advantageous to provide a booster heater to meet this requirement when needed, rather than maintain 100–200 L of water continuously at this temperature with associated losses. The temperature should be set back even lower when occupants are away during long periods of times.

When the tank is charged with cold water, both heating elements operate until the temperature reaches a set point. After this initial rise, one heating element thermostatically cycles on and off to maintain the temperature, replacing heat that is removed by withdrawing hot water or that is lost by conduction and convection during standby operation.

Experiments indicate that the heating elements may be energized only 10%–20% of the time, depending on the ambient temperature, demand for hot water, water supply temperature, etc. By carefully scheduling hot water usage, this time can be greatly reduced. In one case, a residential water heater was operated for 1 h in the morning and 1 h in the evening. The morning cycle provided sufficient hot water for clothes washing, dishes, and other needs. Throughout the day, the water in the tank, although gradually cooling, still was sufficiently hot for incidental needs. The evening heating cycle provided sufficient water for cooking, washing dishes, and bathing. Standby losses were eliminated during the night and much of the day. Electricity use was cut to a fraction of the normal amount. This method requires the installation of a time clock or other type of control to regulate the water heater. A manual override can be provided to meet special needs.

Supplementary tank insulation can be installed at a low cost to reduce standby losses. The economic benefit depends on the price of electricity and the type of insulation installed. However, paybacks of a few months up to a year are typical on older water heaters. Newer units have better insulation and reduced losses. Hot water piping should also be insulated, particularly when hot water tanks are located outside or when there are long piping runs. If copper pipe is used, it is particularly important to insulate the pipe for the first 3–5 m where it joins the tank, since it can provide an efficient heat conduction path.

Since the energy input depends on the water flow rate and the temperature difference between the supply water temperature and the hot water discharge temperature, reducing either of these two quantities reduces energy use. Hot water demand can be reduced by cold water clothes washing, and by providing hot water at or near the use temperature, to avoid the need for dilution with cold water. Supply water should be provided at the warmest temperature possible. Since reservoirs and underground piping systems are generally warmer than the air temperature on a winter day in a cold climate, supply piping should be buried, insulated, or otherwise kept above the ambient temperature.

Solar systems are available today for heating hot water. Simple inexpensive systems can preheat the water, reducing the amount of electricity needed to reach the final temperature. Alternatively, solar heaters (some with electric backup heaters) are also available, although initial costs may be prohibitively high, depending on the particular installation.

Heat pump water heaters may save as much as 25%–30% of the electricity used by a conventional electric water heater. Some utilities have offered rebates of several thousand dollars to encourage customers to install heat pump water heaters.

The microwave water heater is an interesting technology that is just beginning to emerge in both residential and commercial applications. Microwave water heaters are tankless systems that produce hot water only when needed, thereby avoiding the energy losses incurred by conventional water heaters during the storage of hot water. These heaters consist of a closed stainless steel chamber with a silica-based flexible coil and a magnetron. When there is a demand for hot water, either because a user has opened a tap or because of a heater timing device, water flows into the coil and the magnetron bombards it with microwave energy at a frequency of 2450 MHz. The microwave energy excites the water molecules, heating the water to the required temperature.

Heat recovery is another technique for preheating or heating water, although opportunities in residences are limited. This is discussed in more detail under commercial water heating.

Apartments and larger buildings use a combined water heater/storage tank, a circulation loop, and a circulating pump. Cold water is supplied to the tank, which thermostatically maintains a preset temperature, typically 71°C (160°F). The circulating pump maintains a flow of water through the circulating loop, so hot water is always available instantaneously upon demand to any user. This method is also used in hotels, office buildings, etc.

Adequate piping and tank insulation is even more important here, since the systems are larger and operate at higher temperature. The circulating hot water line should be insulated, since it will dissipate heat continuously otherwise.

22.3.3.2 Heat Recovery in Nonresidential Systems

Commercial/industrial hot water systems offer many opportunities for employing heat recovery. Examples of possible sources of heat include air compressors, chillers, heat pumps, refrigeration systems, and water-cooled equipment. Heat recovery permits a double energy savings in many cases. First, recovery of heat for hot water or space heating reduces the direct energy input needed for heating. The secondary benefit comes from reducing the energy used to dissipate waste heat to a heat sink (usually the atmosphere). This includes pumping energy and energy expended to operate cooling towers and heat exchangers. Solar hot water systems are also finding increasing use. Interestingly enough, the prerequisites for solar hot water systems also permit heat recovery. Once the hot water storage capacity and backup heating capability has been provided for the solar hot water system, it is economical to tie in other sources of waste heat, for example, water jackets on air compressors.

22.3.4 Lighting

There are seven basic techniques for improving the efficiency of lighting systems:

- Delamping
- Relamping
- Improved controls
- More efficient lamps and devices
- Task-oriented lighting
- Increased use of daylight
- Room color changes, lamp maintenance

The first two techniques and possibly the third are low cost and may be considered operational changes. The last four items generally involve retrofit or new designs. (Chapter 25 contains further details on energy-efficient lighting technologies.)

The first step in reviewing lighting electricity use is to perform a lighting survey. An inexpensive handheld light meter can be used as a first approximation; however, distinction must be made between raw intensities (lux or foot-candles) recorded in this way and illumination quality.

Many variables can affect the *correct* lighting values for a particular task: task complexity, age of employee, glare, etc. For reliable results, consult a lighting specialist or refer to the literature and publications of the Illuminating Engineering Society.

The lighting survey indicates those areas of the building where lighting is potentially inadequate or excessive. Deviations from illumination levels that are adequate can occur for several reasons: over design; building changes; change of occupancy; modified layout of equipment or personnel, more efficient lamps, improper use of equipment, dirt buildup, etc.

Once the building manager has identified areas with potentially excessive illumination levels, he or she can apply one or more of the seven techniques listed previously. Each of these will be described briefly.

Delamping refers to the removal of lamps to reduce illumination to acceptable levels. With incandescent lamps, remove unnecessary bulbs. With fluorescent or high intensity discharge (HID) lamps, remove lamps and disconnect ballasts, since ballasts account for 10%–20% of total energy use. Note that delamping is not recommended if it adversely affects the distribution of the lighting; instead, consider relamping.

Relamping refers to the replacement of existing lamps by lamps of lower wattage and lower lumen output in areas with excessive light levels. More efficient lower wattage fluorescent tubes are available that require 15%–20% less wattage while producing 10%–15% less light. In some types of HID systems, it is possible to substitute a more efficient lamp with lower lumen output directly. However, in most cases, ballasts must also be changed.

Improved controls permit lamps to be used only when and where needed. For example, certain office buildings have all lights for one floor or large area on a single contactor. These lamps will be switched on at 6 a.m. before work begins, and are not turned off until 10 p.m. when maintenance personnel finish their cleanup duties. In such cases, energy usage can be cut by as much as 50% by better control and operation strategies: installing individual switches for each office or work area; installing timers, occupancy sensors, daylighting controls, and/or dimmers; and instructing custodial crews to turn lights on as needed and turn them off when work is complete.

Sophisticated building-wide lighting control systems are also available, and today are being implemented at an increasing rate, particularly in commercial buildings.

There is a great variation in the efficacy (a measure of light output per electricity input) of various lamps. Incandescent lamps have the lowest efficacy, typically 5–20 lm/W. Wherever possible, substitute incandescent lamps with fluorescent lamps, or with other efficient alternatives. This not only saves energy, but offers substantial economic savings as well, since fluorescent lamps last 10–50 times longer. Conventional fluorescent lamps have efficacies in the range of 30–70 lm/W; high-efficiency fluorescent systems yield about 85–100 lm/W.

CFLs and LED lamps are also good substitutes for a wide range of incandescent lamps. They are available in a variety of wattages and will replace incandescent lamps with a fraction of the energy consumption. Typical efficacies are 55–70 lm/W for CFLs and 60–100 lm/W for LEDs. In addition to the energy savings, they have longer rated lifetimes and thus do not need to be replaced as often as incandescent lamps. As mentioned previously, EISA (2007) put standards in place to phase out common incandescent lamps (40–100 W) with more efficient CFL and LED options. Replacement lighting must use at least 27% less energy. The first phase began in January 2012 and affects 100 W bulbs; the second phase took effect in January 2013, affecting 75 W bulbs.

Still greater improvements are possible with HID lighting, particularly metal halide lamps. While they are generally not suited to residential use (high light output and high capital cost), they are increasingly being used in commercial and industrial buildings for their high efficiency and long life. It should be noted that HID lamps are not suited for any area that requires lamps to be switched on-and-off, as they still take several minutes to restart after being turned off.

Improving ballasts is another way of saving lighting energy. One example of a significant increase in efficacy in the commercial sector is illustrated in the transition from T12 (1.5 in. diameter) fluorescent lamps with magnetic ballasts to T8 (1 in. diameter) fluorescent lamps with electronic ballasts. This transition began to occur in the late 1970s and early 1980s. Now T8 electronic ballast systems are the standard for new construction and retrofits. The efficacy improvement, depending on the fixture, is roughly 20%–40% or even more. For example, a two-lamp F34T12 fixture (a fixture with two 1.5 in. diameter, 34 W lamps) with energy-saving magnetic ballast requires 76 W, whereas a two-lamp F32T8 fixture (a fixture with two 1 in. diameter, 32 W lamps) with electronic ballast requires only 59 W, which is an electricity savings of 22%. The savings is attributable to the lower wattage lamps as well as the considerably more efficient ballast.

In recent years, T5 (5/8 in. diameter) fluorescent lamps have been gaining a foothold in the fluorescent market. They are smaller and have a higher optimum operating temperature than T8s, which makes them advantageous in certain applications. One common use is for high bay lighting, where they are displacing HID alternatives because of their better coloring rendering, longer life, shorter warm-up time, and greater lumen maintenance properties.

Task-oriented lighting is another important lighting concept. In this approach, lighting is provided for work areas in proportion to the needs of the task. Hallways, storage areas, and other nonwork areas receive less illumination.

Task lighting can replace the so-called uniform illumination method sometimes used in offices and other types of commercial buildings. The rationale for uniform illumination was based on the fact that the designer could never know the exact layout of desks and equipment in advance, so designs provided for uniform illumination and the flexibility it offers. With today's higher electricity costs, a more customized task lighting approach is often a cost-effective alternative.

Daylighting was an important element of building design for centuries before the discovery of electricity. In certain types of buildings and operations today, daylighting can be utilized to at least reduce (if not replace) electric lighting. Techniques include windows, an atrium, skylights, etc. There are obvious limitations such as those imposed by the need for privacy, 24 h operation, and building core locations with no access to natural light.

The final step is to review building and room color schemes and decor. The use of light colors can substantially enhance illumination without modifying existing lamps.

An effective lamp maintenance program also has important benefits. Light output gradually decreases over lamp lifetime. This should be considered in the initial design and when deciding on lamp replacement. Dirt can substantially reduce light output; simply cleaning lamps and luminaries more frequently can gain up to 5%–10% greater illumination, permitting some lamps to be turned off.

In addition to the lighting energy savings, efficient lighting also reduces energy requirements for cooling since efficient systems produce less heat. This is a yearlong benefit in many commercial buildings, since space conditioning equipment often operates year-round. However, it is an energy penalty when buildings operate heating systems, since the heating systems have to work a little harder.

22.3.5 Refrigeration

The refrigerator, at roughly 40–140 kWh/month depending on the size and age of the model, is among the top four residential users of electricity. In the last 60 years, the design of refrigerator/freezers has changed considerably, with sizes increasing from 5–10 ft^3 to 20–25 ft^3 today. At the same time, the energy input per unit increased up until the oil embargo, after which efforts were made that led to a steady decline in energy use per unit between the mid-1970s through today, despite increases in average refrigerator size. As noted earlier, current refrigerators use about one-quarter the energy (average of ~450 kWh/year in 2011) smaller units did 40 years ago (average of ~1800 kWh/year in 1972).

Energy losses in refrigerators arise from a variety of sources. The largest losses are due to heat gains through the walls and frequent door openings. Since much of the energy used by a refrigerator depends on its design, care should be used in selection. Choose a refrigerator that is the correct size for the application and look for ENERGY STAR models to maximize efficiency. Refrigerators with freezers on the top or bottom are more efficient than side-by-side models. In addition, refrigerators without in-door ice and water dispensers use less energy, as do models with automatic moisture control and manual defrost.

Purchase of a new, more efficient unit is not a viable option for many individuals who have a serviceable unit and do not wish to replace it. In this case, the energy management challenge is to obtain the most effective operation of the existing equipment. Techniques include the following:

- Reducing operation of automatic defrost and antisweat heaters
- Providing a cool location for the refrigerator
- Reducing the number of door openings
- Maintaining temperature settings recommended levels for food safety (not lower)
- Precooling foods before refrigerating

Commercial refrigeration systems are found in supermarkets, liquor stores, restaurants, hospitals, hotels, schools, and other institutions—about one-fifth of all commercial facilities. Systems include walk-in dairy cases, open refrigerated cases, and freezer cases. In a typical supermarket, lighting, HVAC, and miscellaneous uses account for half the electricity use, while refrigerated display cases, compressors, and condenser fans account for the other half. Thus, commercial refrigeration can be an important element of electric energy efficiency.

It is a common practice in some types of units to have the compressor and heat exchange equipment located remotely from the refrigerator compartment. In such systems, try to locate the compressor in a cool location rather than placing it next to other equipment that gives off heat. Modern commercial refrigerators often come equipped with heat recovery systems, which recover compressor heat for space conditioning or water heating.

Walk-in freezers and refrigerators lose energy though door openings; refrigerated display cases have direct transfer of heat. Covers, strip curtains, air curtains, glass doors, or other thermal barriers can help mitigate these problems. In addition, it is important to use the most efficient light sources in large refrigerators and freezers; elimination of 1 W of electricity to produce light also eliminates two additional Watts required to extract the heat. Other potential energy saving improvements include high-efficiency motors, VSDs, more efficient compressors, and improved refrigeration cycles and controls, such as the use of floating head pressure controls.

22.3.6 Cooking

Consumer behavior toward cooking has changed dramatically since the first edition of this handbook. Consumers today are cooking less in the home and dining out or picking up prepared food more often. Consumers are also purchasing more foods that are easier to prepare—convenience is key to the modern family. In response, the food processing industry offers a wide variety of pre-prepared, ready-to-eat products. Recent end use electricity data illustrate this change in behavior. Cooking accounted for about 7% of residential electricity use in 2001; in 2011, it accounted for only 2%.

Though habits are changing and more cooking is occurring outside the home (in restaurants and food processing facilities), reductions in energy use are still important. In general, improvements in energy use efficiency for cooking can be divided into three categories:

- More efficient use of existing appliances
- Use of most efficient existing appliances
- More efficient new appliances

The most efficient use of existing appliances can lead to substantial reductions in energy use. While slanted toward electric ranges and appliances, the following observations also apply to cooking devices using other sources of heat.

First, select the right size equipment for the job. Do not heat excessive masses or large surface areas that will needlessly radiate heat. Second, optimize heat transfer by ensuring that pots and pans provide good thermal coupling to the heat sources. Flat-bottomed pans should be used on electric ranges. Third, be sure that pans are covered to prevent heat loss and to shorten cooking times. Fourth, when using the oven, plan meals so that several dishes are cooked at once. Use small appliances (electric fry pans, *slow* cookers, toaster ovens, etc.) whenever they can be substituted efficiently for the larger appliances such as the oven.

Different appliances perform similar cooking tasks with widely varying efficiencies. For example, the electricity used and cooking time required for common foods items can vary as much as ten to one in energy use and five to one in cooking times, depending on the method. As an example, four baked potatoes require 2.3 kWh and 60 min in an oven (5.2 kW) of an electric range, 0.5 kWh and 75 min in a toaster oven (1.0 kW), and 0.3 kWh and 16 min in a microwave oven (1.3 kW). Small appliances are generally more efficient when used as intended. Measurements in a home indicated that a pop-up toaster cooks two slices of bread using only 0.025 kWh. The toaster would be more efficient than using the broiler in the electric range oven, unless a large number of slices of bread (more than 17 in this case) were to be toasted at once.

If new appliances are being purchased, select the most efficient ones available. Heat losses from some ovens approach 1 kW, with insulation accounting for about 50%; losses around the oven door edge and through the window are next in importance. These losses are reduced in certain models. Self-cleaning ovens are normally manufactured with more insulation. Careful design of heating elements can also contribute to better heat transfer. Typically, household electric ranges require around 2300–3200 W for oven use, 3600 W for broiler use, and 4000 W for use of the self-cleaning feature.

Microwave cooking is highly efficient for many types of foods, since the microwave energy is deposited directly in the food. Energy input is minimized, because there is no need to heat the cooking utensil. Although many common foods can be prepared effectively using a microwave oven, different methods must be used as certain foods are not suitable for microwave cooking. A typical microwave oven requires 750–1100 W. Convection ovens and induction cook tops are two additional electric alternatives that can reduce cooking energy use.

Commercial cooking operations range from small restaurants and cafes where methods similar to those described previously for residences are practiced, to large institutional kitchens in hotels and hospitals. Many of the same techniques apply. Careful scheduling of equipment use, and provision of several small units rather than a single large one, will save energy. For example, in restaurants, grills, soup kettles, bread warmers, etc., often operate continuously. Generally, it is unnecessary to have full capacity during off-peak hours; one small grill might handle midmorning and midafternoon needs, permitting the second and third units to be shut down. The same strategy can be applied to coffee warming stations, hot plates, etc.

22.3.7 Residential Appliances

A complete discussion of EMOs associated with all the appliances found in homes is beyond the scope of this chapter. However, the following subsections discuss several of the major ones and general suggestions applicable to the others are provided.

22.3.7.1 Clothes Drying

Clothes dryers typically use about 2.5 kWh or more per load, depending on the unit and size of the load. A parameter called the EF, which is a measure of the lbs of clothing dried per kWh of electricity consumed, can be used to quantify the efficiency of clothes drying. The current minimum EF for a standard capacity electric dryer is 3.01. New standards based on a new metric that incorporates standby energy use (combined energy factor [CEF]) are scheduled to take effect in January 2014. ENERGY STAR certified models are not yet available. In the United States, new dryers are not required to display energy use information, so it is difficult to compare models. In fact, most electric dryers in the market are pretty comparable in their construction and the basic heating technology. However, the actual energy consumption of the dryer varies with the types of controls it has, and how the operator uses those controls. Models with moisture-sensing capability can result in the most energy savings—savings on the order of 15% compared to conventional operation are common.

In addition, electric clothes dryers operate most efficiently when fully loaded. Operating with one-third to one-half load costs roughly 10%–15% in energy efficiency.

Locating clothes dryers in heated spaces could save 10%–20% of the energy used by reducing energy needed for heating up. Another approach is to save up loads and do several loads sequentially, so that the dryer does not cool down between loads.

The heavier the clothes, the greater the amount of water they hold. Mechanical water removal (pressing, spinning, and wringing) generally requires less energy than electric heat. Therefore, be certain the washing machine goes through a complete spin cycle (0.1 kWh) before putting clothes in the dryer.

Solar drying, which requires a clothesline (rope) and two poles or trees, has been practiced for millennia and is very sparing of electricity. The chief limitation is, of course, inclement weather. New technologies such as microwave or heat pump clothes dryers may help reduce clothes drying energy consumption in the future.

22.3.7.2 Clothes Washing

The modified energy factor (MEF) can be used to compare different models of clothes washers. It is a measure of the machine energy required during washing, the water heating energy, and the dryer energy needed to remove the remaining moisture. A higher MEF value indicates a more efficient clothes washer. According to Federal standards, all new clothes washers manufactured or imported after January 2007 are required to have an MEF of at least 1.26. In addition, as of February 2013, to be qualified as an ENERGY STAR unit, residential clothes washers must have an MEF of at least 2.0.

Electric clothes washers are designed for typical loads of 3–7 kg. Surprisingly, most of the energy used in clothes washing is for hot water; the washer itself only requires a few percent of the total energy input. Therefore, the major opportunity for energy management in clothes washing is the use of cold or warm water for washing. Under normal household conditions, it is not necessary to use hot water. Clothes are just as clean (in terms of bacteria count) after a 20°C wash as after a 50°C wash. If there is concern for sanitation (e.g., a sick person in the house), authorities recommend use of chlorine bleach. If special cleaning is required, such as removing oil or grease stains, hot water (50°C) and detergent will emulsify oil and fat. There is no benefit in a hot rinse.

A secondary savings can come from using full loads. Surveys indicate that machines are frequently operated with partial loads, even though a full load of hot water is used.

22.3.7.3 Dishwashers

The two major energy uses in electric dishwashers are the hot water and the dry cycle. Depending on the efficiency of the model and operation, dishwashers use between about 2 and 5 kWh/load.

The water heating often accounts for 80% of the total energy requirement of a dishwasher. The volume of hot water used ranges from about 5 gal for the more efficient units to more than double that for less efficient models. The water volume can be varied on some machines depending on the load.

Since 1990, new models have been required to allow for a no-heat drying option. If you are using a very old unit in which that option is not available, stop the cycle prior to the drying step and let the dishes air-dry. Operating the dishwasher with a full load and using a cold water prerinse are additional ways to minimize energy use.

New standards that took effect in mid-2013 require standard-size residential dishwashers to use a maximum of 307 kWh/year and 5.0 gal/cycle and compact dishwashers to use a maximum of 222 kWh/year and 3.5 gal/cycle. ENERGY STAR models are also available.

22.3.7.4 General Suggestions for Residential Appliances and Electrical Equipment

Many electrical appliances (pool pumps, televisions, stereos, DVD and CD players, electronic gaming systems, aquariums, blenders, floor polishers, hand tools, mixers, etc.) perform unique functions that are difficult to duplicate. This is their chief value. Attention should be focused on those appliances that use more than a few percent of annual electricity use. General techniques for energy management include

- Reduce use of equipment where feasible (e.g., turn off entertainment systems when not in use).
- Perform maintenance to improve efficiency (e.g., clean pool filters to reduce pumping power).
- Schedule use for off-peak hours (evenings).

The last point requires further comment and applies to large electric appliances such as washers, dryers, and dishwashers as well. Some utilities now offer time-based pricing that includes a premium charge for usage occurring on-peak (when the greatest demand for electricity takes place) and lower energy costs for off-peak electricity use. By scheduling energy-intensive activities for off-peak hours (e.g., clothes washing and drying in the evening), the user helps the utility reduce its peaking power requirement, thereby reducing generating costs. The utility then returns the favor by providing lower rates for off-peak use.

22.4 Closing Remarks

This chapter has discussed the management of electrical energy in buildings. Beginning with a discussion of energy use in buildings, we next outlined the major energy using systems and equipment, along with a brief description of their features that influence energy use and efficiency. A systematic methodology for implementing an energy management program was then described. The procedure has been implemented in a wide variety of situations including individual homes, commercial buildings, institutions, multinational conglomerates, and cities, and has worked well in each case. Following the discussion of how to set up an energy management program, a series of techniques for saving electricity in each major end use were presented. The emphasis has been on currently available, cost-effective technology. There are other techniques available, some of which are provided in other chapters of this handbook, but we have concentrated on those of we know will work in today's economy, for the typical energy consumer.

The first edition of this book was published in 1980. Much of the data in the first edition dated from the 1975 to 1980 time frame, when the initial response to the oil embargo of 1973 was gathering momentum and maturing. It is remarkable to return to those data and look at the progress that has been made. In 1975, total U.S. energy use was 72 quads (1 quad = 10^{15} BTU = 1.055 EJ). Most projections at that time predicted U.S. energy use in excess of 100 quads by 1992; instead, we saw that it only reached 85 quads by 1992. In fact, by 2004, total U.S. energy use had just reached the 100 quad mark. Estimates for 2012 show that usage has actually declined to a value of 95 quads (EIA 2013c). In addition, between 1975 and 2012, energy consumption per real dollar of the U.S. gross domestic product (GDP) decreased by half, from 14.76 to 7.00 thousand BTU per chained 2005 dollar (EIA 2013c). Some of this is due to a decrease in domestic energy-intensive industry and the recent economic recession, but much of it represents a remarkable improvement in overall energy efficiency.

As noted earlier in this chapter (Table 22.1), a significant growth in total energy consumption in the residential and commercial sectors has occurred in the intervening decades, but efficiency improvements have helped to moderate this growth rate considerably. The improvement in energy efficiency in lighting, refrigerators, air conditioning, and other devices has been truly remarkable. Today the local hardware or home improvement store has supplies of energy-efficient devices that were beyond imagination in 1973.

This is a remarkable accomplishment, technically and politically, given the diversity of the residential/commercial market. Besides the huge economic savings this has meant to millions of homeowners, apartment dwellers, and businesses, think of the environmental benefits associated with avoiding the massive additional amounts of fuel, mining, and combustion, which otherwise would have been necessary.

References

EIA. 2012. Annual energy review 2011. DOE/EIA-0384(2011). Washington, DC: U.S. Energy Information Administration. http://www.eia.gov/totalenergy/data/annual/pdf/aer.pdf (accessed March 6, 2015).

EIA. 2013a. Annual energy outlook 2013 with projections to 2040. Commercial sector key indicators and consumption. DOE/EIA-0383(2013). Washington, DC: U.S. Energy Information Administration. http://www.eia.gov/oiaf/aeo/tablebrowser/ (accessed March 6, 2015).

EIA. 2013b. Annual energy outlook 2013 with projections to 2040. Residential sector key indicators and consumption. DOE/EIA-0383(2013). Washington, DC: U.S. Energy Information Administration. http://www.eia.gov/oiaf/aeo/tablebrowser/ (accessed March 6, 2015).

EIA. 2013c. Monthly energy review: July 2013. DOE/EIA-0035(2013/07). Washington, DC: U.S. Energy Information Administration. http://www.eia.gov/totalenergy/data/monthly/pdf/mer.pdf (accessed March 6, 2015).

EPA. 2011. CBECS national median source energy use and performance comparisons by building type. http://www.energystar.gov/ia/business/tools_resources/new_bldg_design/2003_CBECSPerformanceTargetsTable.pdf (accessed August 13, 2013).

SMR Research Corporation. 2011. *The Commercial Building Inventory. Age of Commercial Buildings Spreadsheet*. Hackettstown, NJ: SMR Research Corporation.

U.S. Census Bureau. 2012. Housing vacancies and home ownership. Annual statistics: 2012. (Table 22.11. Estimates of the Total Housing Inventory for the United States: 2011 and 2012.) http://www.census.gov/housing/hvs/data/ann12ind.html (accessed March 6, 2015).

23

Heating, Ventilating, and Air Conditioning Control Systems

Bryan P. Rasmussen, Jan F. Kreider, David E. Claridge, and Charles H. Culp

CONTENTS

23.1 Introduction: The Need for Control

This chapter describes the essentials of control systems for heating, ventilating, and air conditioning (HVAC) of buildings designed for energy conserving operation. Of course, there are other renewable and energy conserving systems that require control. The principles described herein for buildings also apply with appropriate and obvious modification to these other systems. For further reference, the reader is referred to several standard references in the list at the end of this chapter.

HVAC system controls are the information link between varying energy demands on a building's primary and secondary systems and the (usually) approximately uniform demands for indoor environmental conditions. Without a properly functioning control system, the most expensive, most thoroughly designed HVAC system will be a failure. It simply will not control indoor conditions to provide comfort.

The HVAC control system must

- Sustain a comfortable building interior environment
- Maintain acceptable indoor air quality
- Be as simple and inexpensive as possible and yet meet HVAC system operation criteria reliably for the system lifetime
- Result in efficient HVAC system operation under all conditions
- Be commissioned, including the building, equipment, and control systems
- Be fully documented, so that the building staff successfully operates and maintains the HVAC system

A considerable challenge is presented to the HVAC system designer to design a control system that is energy efficient and reliable. Inadequate control system design, inadequate commissioning, and inadequate documentation and training for the building staff often create problems and poor operational control of HVAC systems. This chapter develops the basics of HVAC control and follows through with the operational needs for successfully maintained operation. The reader is encouraged to review the following references on the subject: ASHRAE (2002, 2003, 2004, 2005), Haines (1987), Honeywell (1988), Levine (1996), Sauer et al. (2001), Stein and Reynolds (2000), and Tao and Janis (2005).

To achieve proper control based on the control system design, the HVAC system must be designed correctly and then constructed, calibrated, and commissioned according to the mechanical and electrical systems drawings. These must include properly sized primary and secondary systems. In addition, air stratification must be avoided, proper provision for control sensors is required, freeze protection is necessary in cold climates, and proper attention must be paid to minimizing energy consumption subject to reliable operation and occupant comfort.

The principal and final controlled variable in buildings is zone temperature (and to a lesser extent humidity and/or air quality in some buildings). This chapter will therefore focus on methods to control temperature. Supporting the zone temperature control, numerous other control loops exist in buildings within the primary and secondary HVAC systems, including boiler and chiller control, pump and fan control, liquid and airflow control, humidity control, and auxiliary system control (e.g., thermal energy storage control). This chapter discusses only *automatic* control of these subsystems. Honeywell (1988)

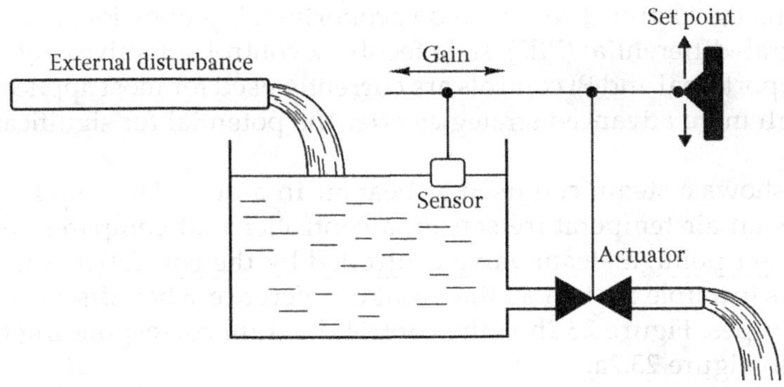

FIGURE 23.1
Simple water level controller. The set point is the full water level; the error is the difference between the full level and the actual level.

defines an automatic control system as "a system that reacts to a change or imbalance in the variable it controls by adjusting other variables to restore the system to the desired balance."

Figure 23.1 defines a familiar control problem with feedback. The water level in the tank must be maintained under varying outflow conditions. The float operates a valve that admits water to the tank as the tank is drained. This simple system includes all the elements of a control system:

Sensor—float; reads the controlled variable, the water level.

Controller—linkage connecting float to valve stem; senses difference between full tank level and operating level and determines needed position of valve stem.

Actuator (controlled device)—internal valve mechanism; sets valve (the final control element) flow in response to level difference sensed by controller.

Controlled system characteristic—water level; this is often termed the *controlled variable.*

This system is called a *closed-loop* or *feedback* system because the sensor (float) is directly affected by the action of the controlled device (valve). In an open-loop system, the sensor operating the controller does not directly sense the action of the controller or actuator. An example would be a method of controlling the valve based on an external parameter such as the time of day, which may have an indirect relation to water consumption from the tank.

There are four common methods of control of which Figure 23.1 shows, but one. In the next section, we will describe each with relation to an HVAC system example.

23.2 Modes of Feedback Control

Feedback control systems adjust an output control signal based on feedback. The feedback is used to generate an error signal, which then drives a control element. Figure 23.1 illustrates a basic control system with feedback. Both off-on (i.e., two-position) control and analog (i.e., variable) control can be used. Numerous methodologies have been developed to implement analog control. These include proportional, proportional–integral (PI), proportional–integral–differential (PID), state feedback control, adaptive control, and predictive control. Proportional and PI controls are currently used for most applications in HVAC control, although more advanced strategies offer the potential for significantly improved performance.

Figure 23.2a shows a steam coil used to heat air in a duct. The simple control system shown includes an air temperature sensor, a controller that compares the sensed temperature to the set point, a steam valve controlled by the controller, and the coil itself. We will use this example system as the point of reference when discussing the various control system types. Figure 23.2b is the control diagram corresponding to the physical system shown in Figure 23.2a.

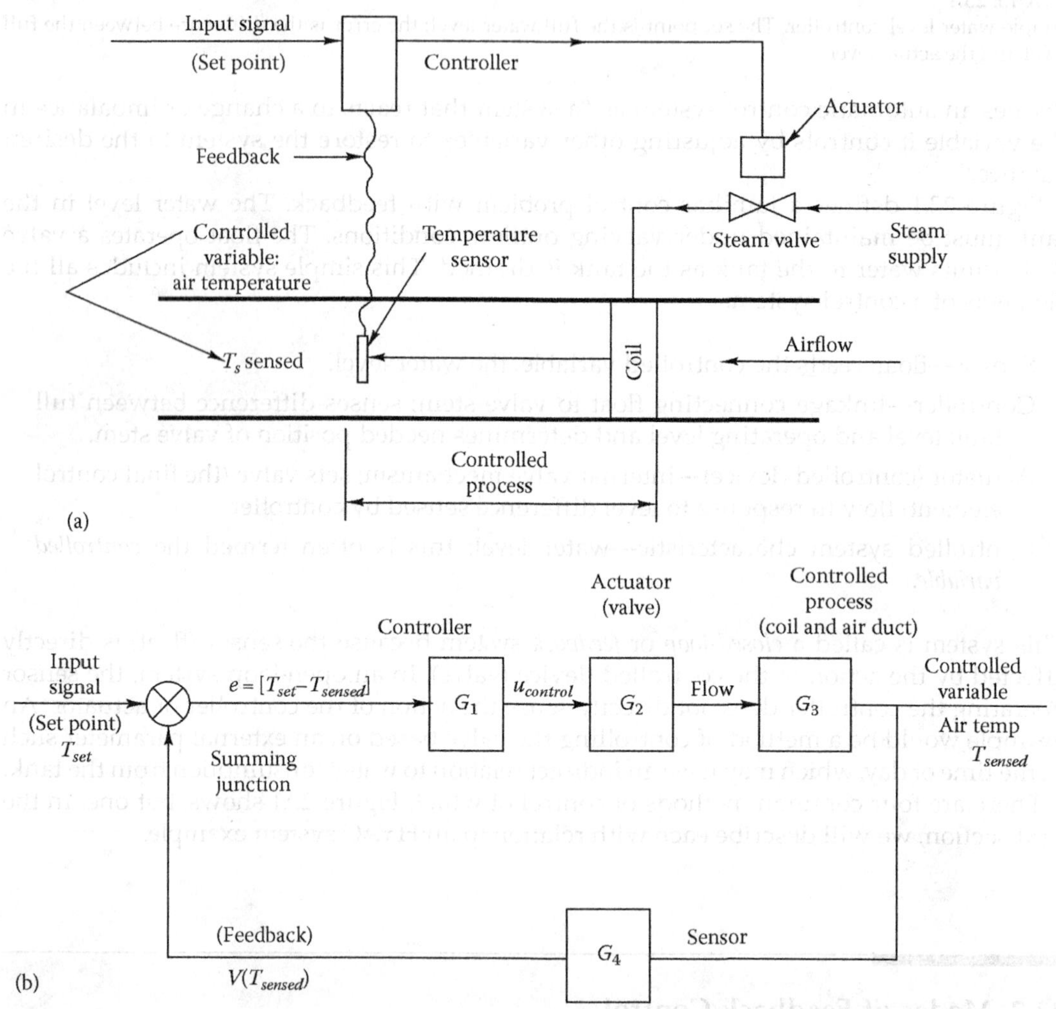

FIGURE 23.2
(a) Simple heating coil control system showing the process (coil and short duct length), controller, controlled device (valve and its actuator), and sensor. The set point entered externally is the desired coil outlet temperature. (b) Equivalent control diagram for heating coil. The Gs represent functions relating the input to the output of each module.

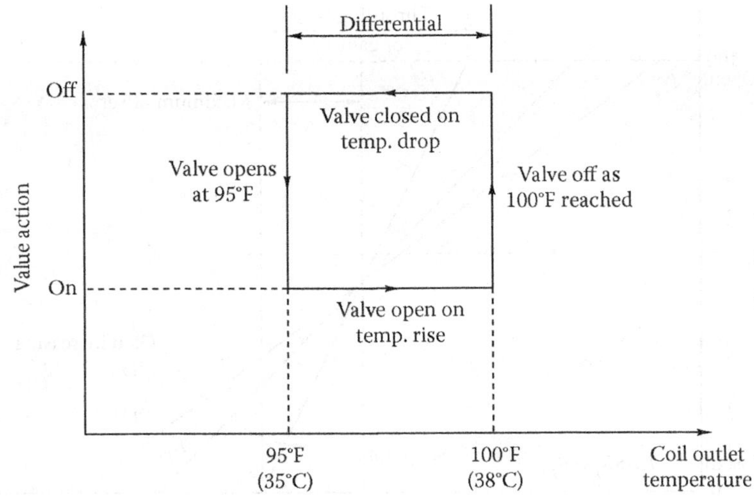

FIGURE 23.3
Two-position (on–off) control characteristic.

Two-position control applies to an actuator that is either fully open or fully closed. This is also known as *on-off, bang-bang,* or *hysteretic* control. In Figure 23.2a, the valve is a two-position valve if two-position control is used. The position of the steam valve is determined by the value of the coil outlet temperature. Figure 23.3 depicts two-position control of the valve. If the air temperature drops below 95°F, the valve opens and remains open until the air temperature reaches 100°F. The differential is usually adjustable, as is the temperature setting itself. Two-position control is the least expensive method of automatic control and is suitable for control of HVAC systems with large time constants and where an oscillatory response is acceptable. Examples include residential space and water heating systems. Systems that are fast reacting should not be controlled using this approach, since overshoot and undershoot may be excessive.

Proportional control adjusts the controlled variable in proportion to the difference between the controlled variable and the set point. For example, a proportional controller would increase the coil heat rate in Figure 23.2 by 10% if the coil outlet air temperature dropped by an amount equal to 10% of the temperature range specified for the heating to go from off to fully on. The following equation defines the behavior of a proportional control loop:

$$u = u_0 + K_p e \tag{23.1}$$

where
u is the controller output
u_0 is the constant value of controller output when no error exists
K_p is the proportional *control gain*; it determines the rate or proportion at which the control signal changes in response to the error
e is the *error*; in the case of the steam coil, it is the difference between the air temperature set point and the sensed supply air temperature:

$$e = T_{set} - T_{sensed} \tag{23.2}$$

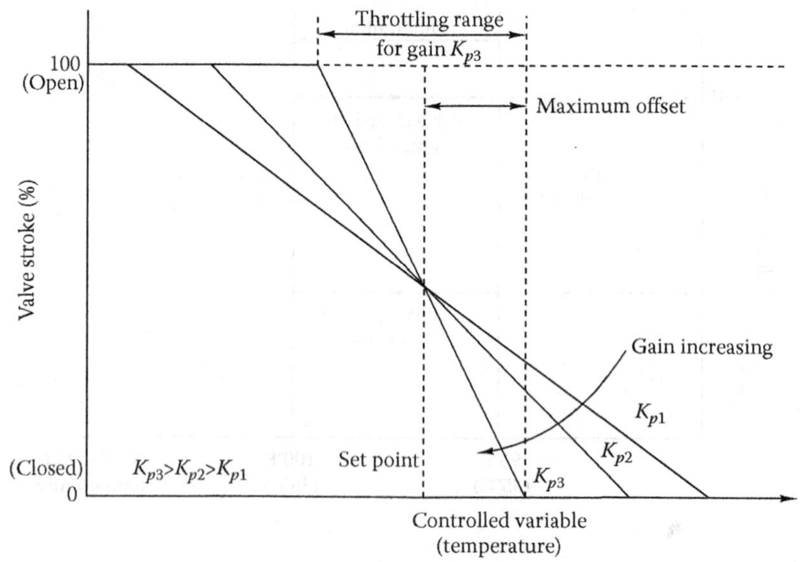

FIGURE 23.4
Proportional control characteristic showing various throttling ranges and the corresponding proportional gains K_p. This characteristic is typical of a heating coil temperature controller.

As coil air outlet temperature drops farther below the set temperature, error *e* increases leading to increased control action—an increased steam flow rate. Note that the temperatures in Equations 23.1 and 23.2 are often replaced by voltages or other variables, particularly in electronic controllers.

The *throttling range* is the total change in the controlled variable that is required to cause the actuator or controlled device to move between its limits. For example, if the nominal temperature of a zone is 72°F and the heating controller throttling range is 6°F, then the heating control undergoes its full travel between a zone temperature of 69°F and 75°F. This control, whose characteristic is shown in Figure 23.4, is *reverse acting*; that is, as the temperature (controlled variable) increases, the heating valve position decreases.

The throttling range is inversely proportional to the gain as shown in Figure 23.4. Beyond the throttling range, the system is out of control. In actual hardware, one can set the set point and either the gain or the throttling range (most common), but not both of the latter. Proportional control by itself is not capable of reducing the error to zero, since an error is needed to produce the capacity required for meeting a load as we will see in the following example. This unavoidable value of the error in proportional systems is called the offset. From Figure 23.4, it is easy to see that the offset is larger for systems with smaller gains. There is a limit to which one can increase the gain to reduce offset, because high gains can produce control instability.

Example 23.1 Proportional Gain Calculation

If the steam heating coil in Figure 23.2a has a heat output that varies from 0 to 20 kW as the outlet air temperature varies from 35°C to 45°C in an industrial process, what is the coil gain and what is the throttling range? Find an equation relating the heat rate at any sensed air temperature to the maximum rate in terms of the gain and set point.

Given

$$\dot{Q}_{max} = 20 \text{ kW}$$

$$\dot{Q}_{min} = 0 \text{ kW}$$

$$T_{max} = 45°\text{C}$$

$$T_{min} = 35°\text{C}$$

Figure: See Figure 23.2a.
Assumptions: Steady-state operation
Find: K_p, ΔT_{max}
Solution: The throttling range is the range of the controlled variable (air temperature) over which the controlled system (heating coil) exhibits its full capacity range. The temperature varies from 35°C to 45°C. Assuming that a temperature of 35°C corresponds to heating rate of 0 kW, then the throttling range is

$$\Delta T_{max} = 45°\text{C} - 35°\text{C} = 10°\text{C} \tag{23.3}$$

The proportional gain is the ratio of the controlled system (coil) output to the throttling range. For this example, the Q controller output is $\dot{Q}$ and the gain is

$$K_p = \frac{\dot{Q}_{max} - \dot{Q}_{min}}{\Delta T_{max}} = \frac{(20-0) \text{ kW}}{10 \text{ K}} = 2.0 \text{ kW/K} \tag{23.4}$$

The controller characteristic can be found by inspecting Figure 23.4. It is assumed that the average air temperature (40°C) occurs at the average heat rate (10 kW). The equation of the straight line shown is

$$\dot{Q} = K_p(T_{set} - T_{sensed}) + \frac{\dot{Q}_{max} - \dot{Q}_{min}}{2} = K_p e + \frac{\dot{Q}_{max} - \dot{Q}_{min}}{2} \tag{23.5}$$

Note that the quantity ($T_{set} - T_{sensed}$) is the **error** e and a nonzero value indicates that the set temperature is not met. However, the proportional control system used here requires the presence of an error signal to fully open or fully close the valve.

Inserting the numerical values, we have

$$\dot{Q} = 2.0 \text{ kW/K}(40 - T_{sensed}) + 10 \text{ kW} \tag{23.6}$$

Comments: In an actual steam coil control system, it is the steam valve that is controlled directly to indirectly control the heat rate of the coil. This is typical of many HVAC system controls in that the desired control action is achieved indirectly by controlling another variable that in turn accomplishes the desired result. That is why the controller and actuator are often shown separately as in Figure 23.2b.

This example illustrates with a simple system how proportional control uses an error signal to generate an offset and how the offset controls an output quantity. Using a bias value, the error can be set to be zero at one value in the control range. Proportional control generally results in a nonzero error over the remainder of the control range, as shown later.

A common approach for modeling dynamic HVAC systems where only one input–output is considered is a process control model (Equation 23.7). The basic form of this model includes a gain, k_0, a first-order dynamic characterized by the time constant, τ, and a pure time delay, T_d. This basic model is easily augmented with additional time constants as needed:

$$G(s) = \frac{k_0 e^{-T_d s}}{\tau s + 1} \tag{23.7}$$

Assuming the basic feedback control loop (Figure 23.2b) where the controller is selected to be a proportional control and the system model is given as in Equation 23.7, then the relationship between the desired reference signal and the system error is given as follows:

$$\frac{E(s)}{R(s)} = \frac{1}{1 + C(s)G(s)} = \frac{1}{\tau s + 1 + K_p k_0 e^{-T_d s}} \tag{23.8}$$

Assuming a unit step change in the reference signal, then the steady-state value of the system output is given by

$$e_{ss} = \lim_{s \to 0}\left[s\left(\frac{E(s)}{R(s)}\right) R(s)\right] = \frac{1}{1 + C(0)G(0)} = \frac{1}{1 + K_p k_0} \tag{23.9}$$

Thus, using proportional control, the system error (i.e., the difference between the reference signal and system output) becomes smaller as the proportional gain is increased, but the error will never be zero. For more details regarding this type of analysis, see Franklin et al. (2006) for a detailed discussion of transfer functions, final value theorem, and steady-state error analysis.

Real systems also have a dynamic response. This means that proportional control is best suited to slow-response systems, where the throttling range can be set so that the system achieves stability. Typically, slow-responding mechanical systems include pneumatic thermostats for zone control and air handler unit damper control.

Integral control is often added to proportional control to eliminate the offset inherent in proportional-only control. The result, proportional plus integral control, is identified by the acronym PI. Initially, the corrective action produced by a PI controller is the same as for a proportional-only controller. After the initial period, a further adjustment due to the integral term reduces the offset to zero. The rate at which this occurs depends on the timescale of the integration. In equation form, the PI controller is modeled by

$$V = V_0 + K_p e + K_i \int e\,dt \tag{23.10}$$

in which K_i is the integral gain constant. It has units of reciprocal time and is the number of times that the integral term is calculated per unit time. This is also known as the reset rate; reset control is an older term used by some to identify integral control.

Today, most PI control implementations use electronic sensors, analog-to-digital converters (A/Ds), and digital logic to implement the PI control. Integral windup must be taken into account when using PI control, which occurs when actuators reach hardware or software limitations, and the tracking error is nonzero. The integral term will continue to grow, or *wind up,* and creates a large offset. This can prevent the control loop from performing as desired for long periods until the integral term recovers. Various methods exist to minimize or eliminate the windup problem.

The integral term in Equation 23.10 has the effect of adding a correction to the output signal V as long as the error term exists. The continuous offset produced by the proportional-only controller can thereby be reduced to zero because of the integral term. For HVAC systems, the timescale (K_P/K_i) of the integral term is often in the range of 10+ s to 10+ min. Using large integral gains will allow the control system to converge quickly to the desired set point value. However, large gains also tend to increase the oscillations in the response, and thus a balance must be found.

PI control is used for fast-acting systems for which accurate control is needed. Examples include mixed-air controls, duct static pressure controls, and coil controls. Because the offset is eventually eliminated with PI control, the throttling range can be set rather wide to ensure stability under a wider range of conditions than good control would permit with proportional-only control. Hence, PI control is also used on almost all electronic thermostats.

Derivative control is used to speed up the action of PI control. When derivative control is added to PI control, the result is called PID control. The derivative term added to Equation 23.10 generates a correction signal proportional to the time rate of change of error. This term has little effect on a steady proportional system with uniform offset (time derivative is zero) but initially, after a system disturbance, produces a larger correction more rapidly. The derivative control action is anticipatory in nature and can be effective in dampening oscillations caused by an aggressive PI controller. Equation 23.11 includes the derivative term in the mathematical model of the PID controller

$$V = V_0 + K_p e + K_i \int e dt + K_d \frac{de}{dt} \tag{23.11}$$

in which K_d is the derivative gain constant. The timescale (K_d/K_p) of the derivative term is typically in the range of 0.2–15 min. Since HVAC systems do not often require rapid control response, the use of PID control is less common than use of PI control. Since a derivative is involved, any noise in the error (i.e., sensor) signal must be avoided, or this noise will be amplified in the control action, creating undesirable fluctuations. One application in buildings where PID control has been effective is in duct static pressure control, a fast-acting subsystem that has a tendency to be unstable otherwise.

Derivative control has limited application in HVAC systems because it requires correct tuning for each of the three gain constants (*K*s) over the performance range that the control loop will need to operate. Another serious limitation centers on the fact that most facility operators lack training and skills in tuning PID control loops.

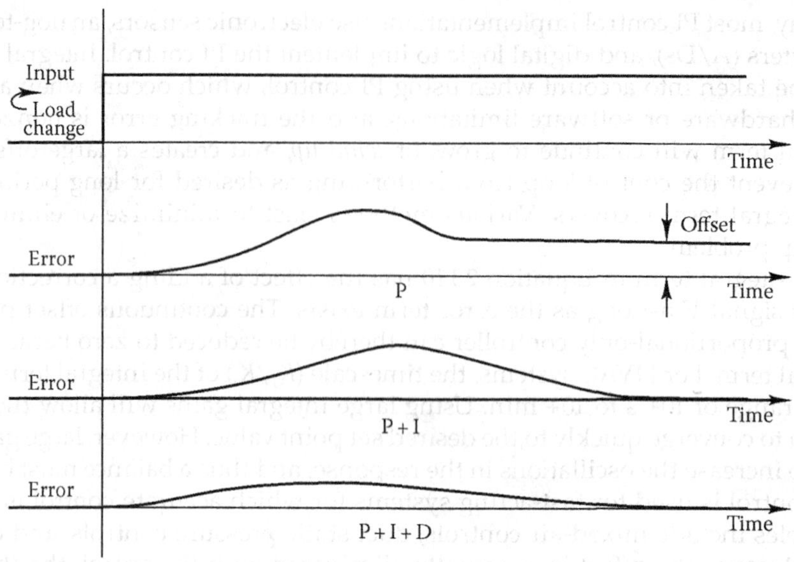

FIGURE 23.5
Performance comparison of P, PI, and PID controllers when subjected to a uniform, input step change.

As a result, many PID controllers are *detuned* and use artificially low control gains. This has the effect of sacrificing good performance, so as to ensure a stable, albeit slow, response.

Figure 23.5 illustrates the loop response for three correctly configured systems when a step function change, or disturbance, occurs. Note that the PI loop achieves the same final control as the PID, and only the PI error signal is larger. An improperly configured PID loop can oscillate from the high value to the low value in a continuous oscillatory manner.

23.3 Basic Control Hardware

In this section, the various physical components needed to achieve the actions required by the control strategies of the previous section are described. Since there are two fundamentally different control approaches—pneumatic and electronic—the following material is so divided. Sensors, controllers, and actuators for principal HVAC applications are described.

23.3.1 Pneumatic Systems

The first widely adopted automatic control systems used compressed air as the signal transmission medium. Compressed air had the advantage that it could be *metered* through various sensors and could power large actuators. The fact that the response of a normal pneumatic loop could take several minutes often worked as an advantage. Pneumatic controls use compressed air (approximately 20 psig in the United States) for the operation of sensors

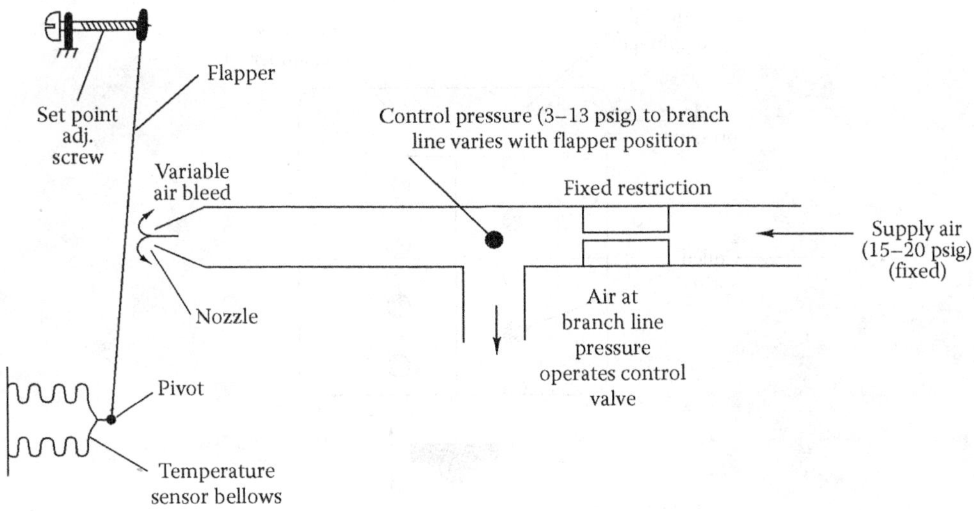

FIGURE 23.6
Drawing of pneumatic thermostat showing adjustment screw used to change temperature setting.

and actuators. Though most new buildings use electronic controls, many existing buildings use pneumatic controls. This section provides an overview of how these devices operate.

Temperature control and damper control comprise the bulk of pneumatic loop controls. Figure 23.6 shows a method of sensing temperature and producing a control signal. Main supply air, supplied by a compressor, enters a branch line through a restriction. The zone thermostat bleeds a variable amount of air out, depending on the position of the flapper, controlled by the temperature sensor bellows. As more air bleeds out, the branch line pressure (control pressure) drops. This reduction in the total pressure to the control element changes the output of the control element. This control can be forward acting or reverse acting. The restrictions typically have hole diameters on the order of a few thousandths of an inch and consume very little air. Typical pressures in the branch lines range between 3 and 13 psig (20–90 kPa). In simple systems, this pressure from a thermostat could operate an actuator such as a control valve for a room heating unit. In this case, the thermostat is both the sensor and the controller—a rather common configuration.

Many other temperature sensor approaches can be used. For example, the bellows shown in Figure 23.6 can be eliminated and the flapper can be made of a bimetallic strip. As temperature changes, the bimetal strip changes curvature, opening or closing the flapper/nozzle gap. Another approach uses a remote bulb filled with either liquid or vapor that pushes a rod (or a bellows) against the flapper to control the pressure signal. This device is useful if the sensing element must be located where direct measurement of temperature by a metal strip or bellows is not possible, such as in a water stream or high-velocity ductwork. The bulb and connecting capillary size may vary considerably by application.

Pressure sensors may use either bellows or diaphragms to control branch line pressure. For example, the motion of a diaphragm may replace that of the flapper in Figure 23.6 to control the bleed rate. A bellows similar to that shown in the same figure may be internally pressurized to produce a displacement that can control air bleed rate. A bellows produces significantly greater displacements than a single diaphragm.

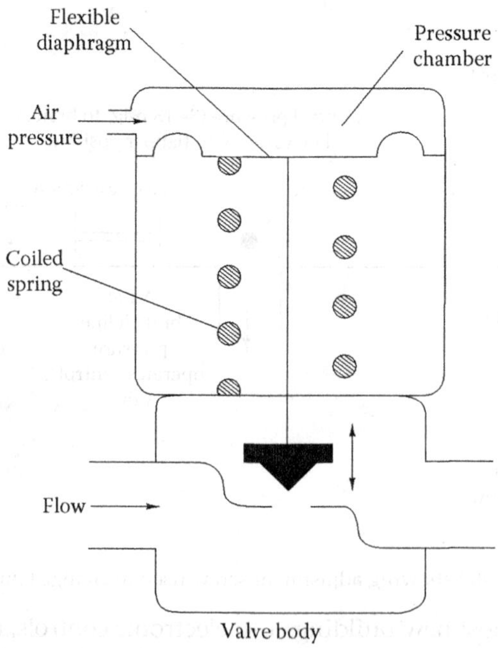

FIGURE 23.7
Pneumatic control valve showing counterforce spring and valve body. Increasing pressure closes the valve.

Humidity sensors in pneumatic systems are made from materials that change size with moisture content. Nylon or other synthetic hygroscopic fibers that change size significantly (i.e., 1%–2%) with humidity are commonly used. Since the dimensional change is relatively small on an absolute basis, mechanical amplification of the displacement is used. The materials that exhibit the desired property include nylon, hair, and cotton fibers. Human hair exhibits a much more linear response with humidity than nylon; however, because the properties of hair vary with age, nylon has much wider use (Letherman, 1981). Humidity sensors for electronic systems are quite different and are discussed in the next section.

An *actuator* converts pneumatic energy to motion—either linear or rotary. It creates a change in the controlled variable by operating control devices such as dampers or valves. Figure 23.7 shows a pneumatically operated control valve. The valve opening is controlled by the pressure in the diaphragm acting against the spring. The spring is essentially a linear device. Therefore, the motion of the valve stem is essentially linear with air pressure. However, this does not necessarily produce a linear effect on flow as discussed later. Figure 23.8 shows a pneumatic damper actuator. Linear actuator motion is converted into rotary damper motion by the simple mechanism shown.

Pneumatic *controllers* produce a branch line (see Figure 23.6) pressure that is appropriate to produce the needed control action for reaching the set point. Such controls are manufactured by a number of control firms for specific purposes. Classifications of controllers include the sign of the output (direct or reverse acting) produced by an error, by the control action (proportional, PI, or two-position), or by number of inputs or outputs. Figure 23.9 shows the essential elements of a dual-input, single-output controller. The two inputs could be the heating system supply temperature and the outdoor temperature sensors, used to control the output water temperature setting of a boiler in a building heating system. This is essentially a boiler *temperature reset* system that reduces heating water temperature with increasing ambient temperature for better system control and reduced energy use.

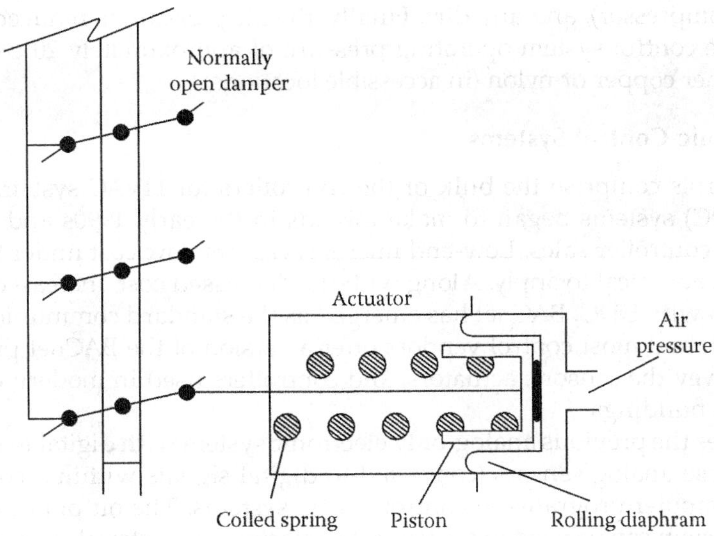

FIGURE 23.8
Pneumatic damper actuator. Increasing pressure closes the parallel-blade damper.

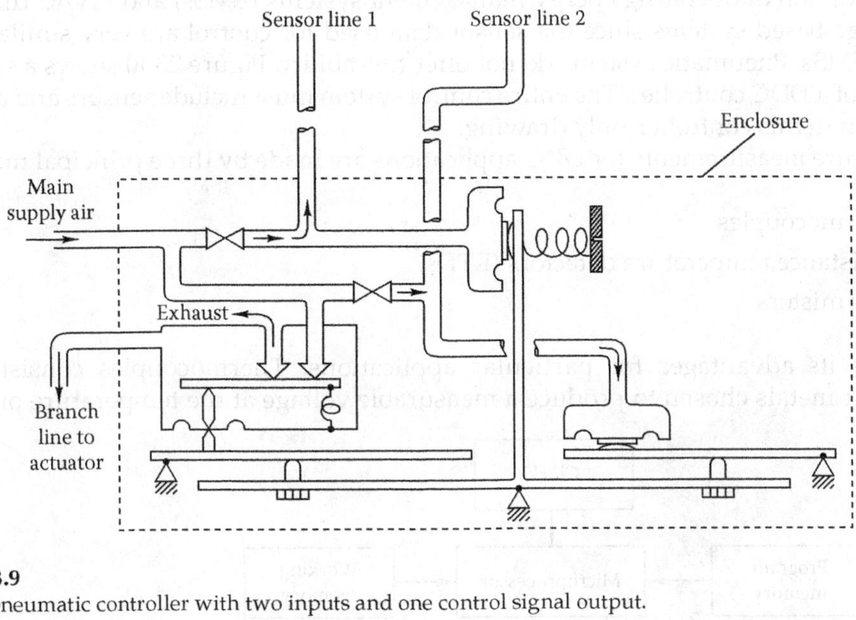

FIGURE 23.9
Example pneumatic controller with two inputs and one control signal output.

The air supply for pneumatic systems must produce very clean, oil-free, dry air. A compressor producing 80–100 psig is typical. Compressed air is stored in a tank for use as needed, avoiding continuous operation of the compressor. The air system should be oversized by 50%–100% of estimated, nominal consumption. The air is then dried to avoid moisture freezing in cold control lines in air-handling units (AHUs) and elsewhere. Dried air should have a dew point of −30°F or less in severe heating climates. In deep cooling climates, the lowest temperature to which the compressed air lines are exposed may be the building cold air supply. Next, the air is filtered to remove water droplets, oil (from the compressor), and any dirt. Finally, the air pressure is reduced in a pressure regulator to the control system operating pressure of approximately 20 psig. Control air piping uses either copper or nylon (in accessible locations).

23.3.2 Electronic Control Systems

Electronic controls comprise the bulk of the controllers for HVAC systems. Direct digital control (DDC) systems began to make inroads in the early 1990s and now make up over 80% of all controller sales. Low-end microprocessors now cost under $0.50 each and are thus very economical to apply. Along with the decreased cost, increased functionality can be obtained with DDC. BACnet has emerged as the standard communication protocol (ASHRAE, 2001) and most control vendors offer a version of the BACnet protocol. In this section, we survey the sensors, actuators, and controllers used in modern electronic control systems for buildings.

DDC enhances the previous analog-only electronic system with digital features. Modern DDC systems use analog sensors (converted to digital signals within a computer) along with digital computer programs to control HVAC systems. The output of this microprocessor-based system can be used to control either electronic, electrical, or pneumatic actuators or a combination. DDC systems have the advantage of reliability and flexibility that others do not. For example, accurately setting control constants in computer software is easier than making adjustments at a controller panel with a screwdriver. DDC systems offer the option of operating energy management systems (EMSs) and HVAC diagnostic, knowledge-based systems since the sensor data used for control are very similar to that used in EMSs. Pneumatic systems do not offer this ability. Figure 23.10 shows a schematic diagram of a DDC controller. The entire control system must include sensors and actuators not shown in this controller-only drawing.

Temperature measurements for DDC applications are made by three principal methods:

1. Thermocouples
2. Resistance temperature detectors (RTDs)
3. Thermistors

Each has its advantages for particular applications. Thermocouples consist of two dissimilar metals chosen to produce a measurable voltage at the temperature of interest

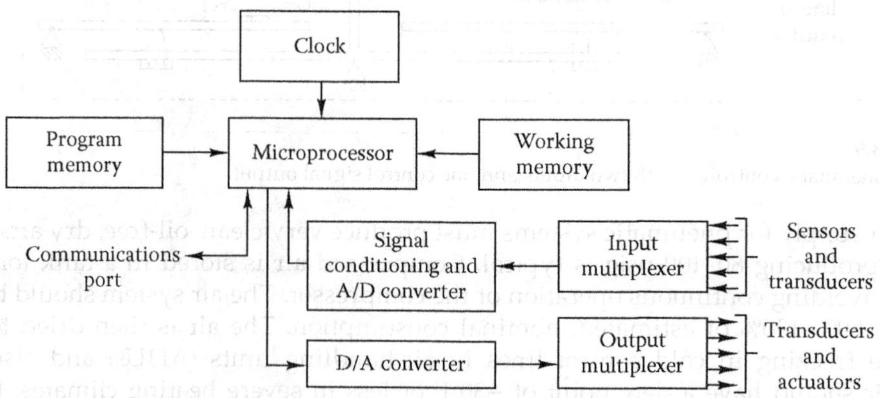

FIGURE 23.10
Block diagram of a DDC controller.

(i.e., Seebeck effect). The voltage output is low (millivolts) but is a well-established function of the junction temperature. By themselves, thermocouples generally produce voltages too small to be useful in most HVAC applications (e.g., a type J thermocouple produces only 5.3 mV at 100°C). However, modern signal conditioning equipment can easily amplify these signals, as well as provide calibration (also known as cold junction compensation).

RTDs use small, responsive sensing sections constructed from metals whose resistance–temperature characteristic is well established and reproducible. To first order,

$$R = R_0(1 + kT) \tag{23.12}$$

where
R is the resistance, ohms
R_0 is the resistance at the reference temperature (0°C), ohms
k is the temperature coefficient of resistance, °C^{-1}
T is the RTD temperature, °C

This equation is easy to invert to find the temperature as a function of resistance. Although complex higher order expressions exist, their use is not needed for HVAC applications.

Two common materials for RTDs are platinum and Balco (a 40% nickel, 60% iron alloy). The nominal values of k, respectively, are 3.85×10^{-3}°C^{-1} and 4.1×10^{-3}°C^{-1}.

Modern electronics measure current and voltage and then determine the resistance using Ohm's law. The measurement causes power dissipation in the RTD element, raising the temperature and creating an error in the measurement. This Joule self-heating can be minimized by minimizing the power dissipated in the RTD. Raising the resistance of the RTD helps, but the most effective approach requires pulsing the current and making the measurement in a few milliseconds. Since one measurement per second will generally satisfy the most demanding HVAC control loop, the power dissipation can be reduced by a factor of 100 or more. Modern digital controls can easily handle the calculations necessary to implement piecewise linearization and other curve-fitting methods to improve the accuracy of the RTD measurements. In addition, lead wire resistance can cause lack of accuracy for the class of platinum RTDs whose nominal resistance is only 100 ohms because the lead resistance of 1–2 ohms is not negligible by comparison to that of the sensor itself.

Thermistors are semiconductors that exhibit a standard exponential dependence for resistance versus temperature given by

$$R = Ae^{(B/T)} \tag{23.13}$$

A is related to the nominal value of resistance at the reference temperature (77°F) and is of the order of several thousands of ohms. The exponential coefficient B (a weak function of temperature) is of the order of 5400–7200 R (3000–4000 K). The nonlinearity inherent in thermistors can be reduced by connecting a properly selected fixed resistor in parallel with it. The resulting linearity is desirable from a control system design viewpoint. Thermistors can have a problem with long-term drift and aging; the designer and control manufacturer should consult on the most stable thermistor design for HVAC applications. Some manufacturers provide linearized thermistors that combine both positive and negative resistive dependence on temperature to yield a more linear response function.

Humidity measurements are needed for control of enthalpy economizers or may also be needed to control special environments such as clean rooms, hospitals, and areas housing computers. Relative humidity, dew point, and humidity ratio are all indicators of the moisture content of air. An electrical, capacitance-based approach using a polymer with interdigitated electrodes has become the most common sensor type. The polymer material absorbs moisture and changes the dielectric constant of the material, changing the capacitance of the sensor. The capacitance of the sensor forms part of a resonant circuit so when the capacitance changes, the resonant frequency changes. This frequency then can be correlated to the relative humidity and provide reproducible readings if not saturated by excessive exposure to high humidity levels (Huang, 1991). The response times of tens of seconds easily satisfy most HVAC application requirements. These humidity sensors need frequent calibration, generally yearly. If a sensor becomes saturated or has condensation on the surface, they become uncalibrated and exhibit an offset from their calibration curve. Older technologies used ionic salts on gold grids. These expensive sensors frequently failed.

Pressure measurements are made by electronic devices that depend on a change of resistance or capacitance with imposed pressure. Figure 23.11 shows a cross-sectional drawing

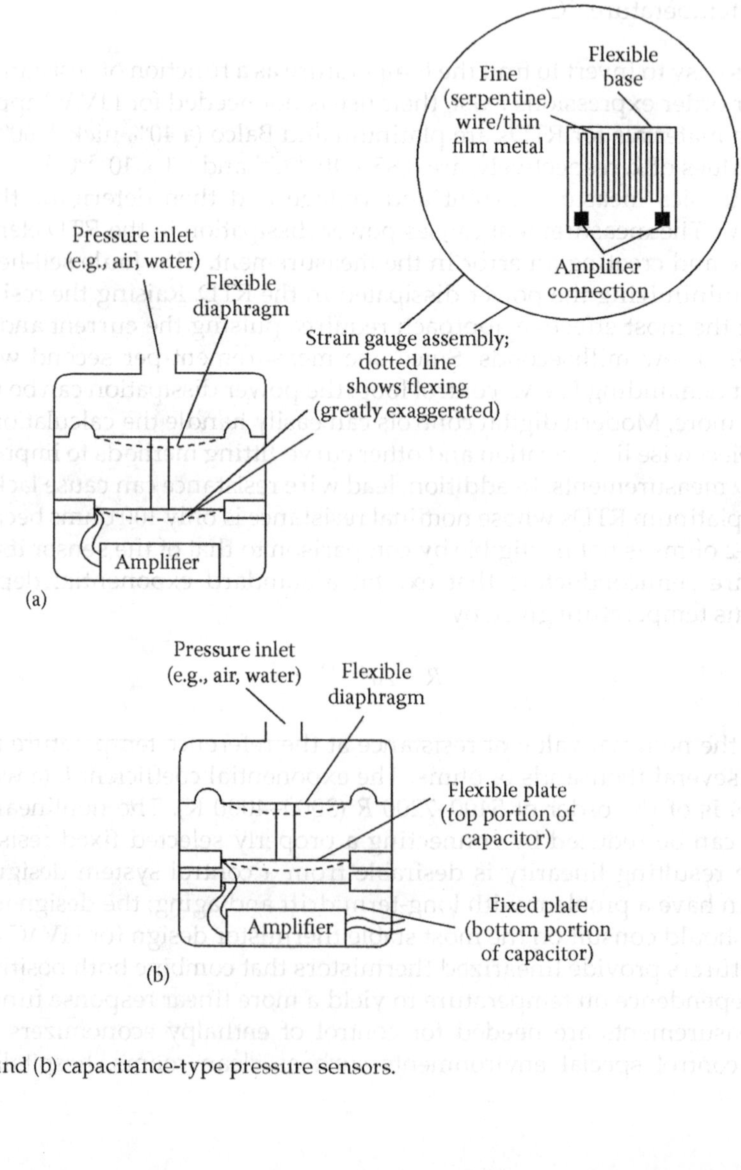

FIGURE 23.11
(a) Resistance- and (b) capacitance-type pressure sensors.

of each. In the resistance type, stretching of the membrane lengthens the resistive element thereby increasing resistance. This resistor is an element in a Wheatstone bridge; the resulting bridge voltage imbalance is linearly related to the imposed pressure. The capacitive-type unit has a capacitance between a fixed and a flexible metal that decreases with pressure. The capacitance change is amplified by a local amplifier that produces an output signal proportional to pressure. Pressure sensors can burst from overpressure or a water hammer effect. Installation needs to carefully follow the manufacturer's requirements.

DDC systems require *flow* measurements to determine the energy flow for air and water delivery systems. Pitot tubes (or arrays of tubes) and other flow measurement devices can be used to measure either air or liquid flow in HVAC systems. Airflow measurements allow for proper flow in variable air volume (VAV) system control, building pressurization control, and outside air control. Water flow measurements enable chiller and boiler control and monitoring and various water loops used in the HVAC system. Some controls only require the knowledge of flow being present. Open–closed sensors fill this need and typically have a paddle that makes a switch connection under the presence of flow. These types of switches can also be used to detect *end of range*, that is, fully open or closed for dampers and other mechanical control elements.

Temperature, humidity, and pressure *transmitters* are often used in HVAC systems. They amplify signals produced by the basic devices described in the preceding paragraphs and produce an electrical signal over a standard range thereby permitting standardization of this aspect of DDC systems. The standard ranges are

Current: 4–20 mA (dc)

Voltage: 0–10 V or 0–5 V (dc)

Although the majority of transmitters produce such signals, the noted values are not universally used.

Figure 23.10 shows the elements of a DDC controller. The heart of the controller is the microprocessor that can be programmed in either a standard or system-specific language. Control algorithms (linear or not), sensor calibrations, output signal shaping, and historical data archiving can all be programmed as the user requires. A number of firms have constructed controllers on standard personal computer platforms. Describing the details of programming HVAC controllers is beyond the scope of this chapter, since each manufacturer uses a different approach. The essence of any DDC system, however, is the same as shown in the figure. Honeywell (1988) discusses DDC systems and their programming in more detail.

Actuators for electronic control systems include

Motors—operate valves, dampers
Variable speed controls—pump, fan, chiller drives
Relays and motor starters—operate other mechanical or electrical equipment (pumps, fans, chillers, compressors), electrical heating equipment

Additional components provide necessary functionality, such as transducers that convert signal types (e.g., electrical to pneumatic) and visual displays that inform system operators of control and HVAC system functions.

Pneumatic and DDC systems have their own advantages and disadvantages. Pneumatics possess increasing disadvantages of cost, hard-to-find replacements, requiring an air compressor with clean oil-free air, sensor drift, imprecise control, and a lack of automated monitoring. The retained advantages include explosion-proof operation and a fail-soft degradation of performance. DDC systems have emerged and have taken the lead for HVAC systems over pneumatics because of the ability to integrate the control system into a large energy management and control system (EMCS), the accuracy of the control, and the ability to diagnose problems remotely. Systems based on either technology require maintenance and skilled operators.

23.4 Basic Control System Design Considerations

This section discusses selected topics in control system design including control system zoning, valve and damper selection, and control logic diagrams. The following section shows several HVAC system control design concepts. Bauman (1998) may be consulted for additional information.

The ultimate purpose of an HVAC control system is to control zone temperature (and secondarily air motion and humidity) to conditions that assure maximum comfort and productivity of the occupants. From a controls viewpoint, the HVAC system is assumed to be able to provide comfort conditions if controlled properly. Basically, a zone is a portion of a building that has loads that differ in magnitude and timing sufficiently from other areas so that separate portions of the secondary HVAC system and control system are needed to maintain comfort.

Having specified the zones, the designer must select the thermostat (and other sensors, if used) location. Thermostat signals are either passed to the central controller or used locally to control the amount and temperature of conditioned air or coil water introduced into a zone. The air is conditioned either locally (e.g., by a unit ventilator or baseboard heater) or centrally (e.g., by the heating and cooling coils in the central air handler). In either case, a flow control actuator is controlled by the thermostat signal. In addition, airflow itself may be controlled in response to zone information in VAV systems. Except for variable speed drives used in variable-volume air or liquid systems, flow is controlled by valves or dampers. The design selection of valves and dampers is discussed next.

23.4.1 Steam and Liquid Flow Control

The flow through valves such as that shown in Figure 23.7 is controlled by valve stem position, which determines the flow area. The variable flow resistance offered by valves depends on their design. The flow characteristic may be linear with position or not. Figure 23.12 shows flow characteristics of the three most common types. Note that the plotted characteristics apply only for *constant valve pressure* drop. The characteristics shown are idealizations of actual valves. Commercially available valves will resemble but not necessarily exactly match the curves shown.

The *linear* valve has a proportional relation between volumetric flow V and valve stem position z:

$$\dot{V} = kz \tag{23.14}$$

The flow in *equal percentage* valves increases by the same fractional amount for each increment of opening. In other words, if the valve is opened from 20% to 30% of full travel, the

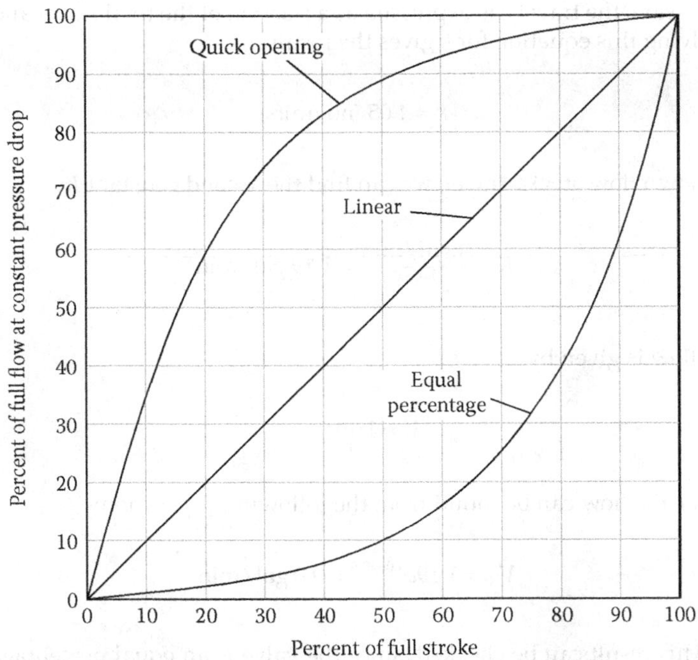

FIGURE 23.12
Quick-opening, linear, and equal percentage valve characteristics.

flow will increase by the same percentage as if the travel had increased from 80% to 90% of its full travel. However, the absolute volumetric flow increase for the latter case is much greater than for the former. The equal percentage valve flow characteristic is given by

$$\dot{V} = Ke^{(kz)} \tag{23.15}$$

where k and K are proportionality constants for a specific valve. Quick-opening valves do not provide good flow control but are used when rapid action is required with little stem movement for on/off control.

Example 23.2: Equal Percentage Valve

A valve at 30% of travel has a flow of 4 gal/min. If the valve opens another 10% and the flow increases by 50% to 6 gal/min, what are the constants in Equation 23.15? What will be the flow at 50% of full travel?

Figure: See Figure 23.15.
Assumptions: Pressure drop across the valve remains constant.

Find: k, K, $\dot{V}_{50}$

Solution: Equation 23.15 can be evaluated at the two flow conditions. If the results are divided by each other, we have

$$\frac{\dot{V}_2}{\dot{V}_1} = \frac{6}{4} = e^{k(z_2 - z_1)} = e^{k(0.4-0.3)} \tag{23.16}$$

In this expression, the travel z is expressed as a fraction of the total travel and is dimensionless. Solving this equation for k gives the result

$$k = 4.05 \text{ (no units)}$$

From the known flow at 30% travel, we can find the second constant K:

$$K = \frac{4 \text{ gal/min}}{e^{4.05 \times 0.3}} = 1.19 \text{ gal/min} \tag{23.17}$$

Finally, the flow is given by

$$\dot{V} = 1.19e^{4.05z} \tag{23.18}$$

At 50% travel, the flow can be found from the following expression:

$$\dot{V}_{50} = 1.19e^{4.05 \times 0.5} = 9.0 \text{ gal/min} \tag{23.19}$$

Comments: This result can be checked, since the valve is an equal percentage valve. At 50% travel, the valve has moved 10% beyond its 40% setting at which the flow was 6 gal/min. Another 10% stem movement will result in another 50% flow increase from 6 to 9 gal/min, confirming the solution.

The plotted characteristics of all three valve types assume constant pressure drop across the valve. In an actual system, the pressure drop across a valve will not remain constant, but if the valve is to maintain its control characteristics, the pressure drop across it must be the majority of the entire loop pressure drop. If the valve is designed to have a full-open pressure drop equal to that of the balance of the loop, good flow control will exist. This introduces the concept of valve *authority* defined as the valve pressure drop as a fraction of total system pressure drop:

$$A \equiv \frac{\Delta p_{v,open}}{(\Delta p_{v,open} + \Delta p_{system})} \tag{23.20}$$

For proper control, the full-open valve authority should be at least 0.50. If the authority is 0.5 or more, control valves will have installed characteristics not much different from those shown in Figure 23.12. If not, the valve characteristic will be distorted upward, since the majority of the system pressure drop will be dissipated across the system at high flows.

Valves are further classified by the number of connections or ports. Figure 23.13 shows sections of typical *two-way* and *three-way* valves. Two-port valves control flow through coils or other HVAC equipment by varying valve flow resistance as a result of flow area changes. As shown, the flow must oppose the closing of the valve. If not, near closure, the valve would slam shut or oscillate, both of which cause excessive wear and noise. The three-way valve shown in the figure is configured in the *diverting* mode. That is, one stream is split into two depending on the valve opening.

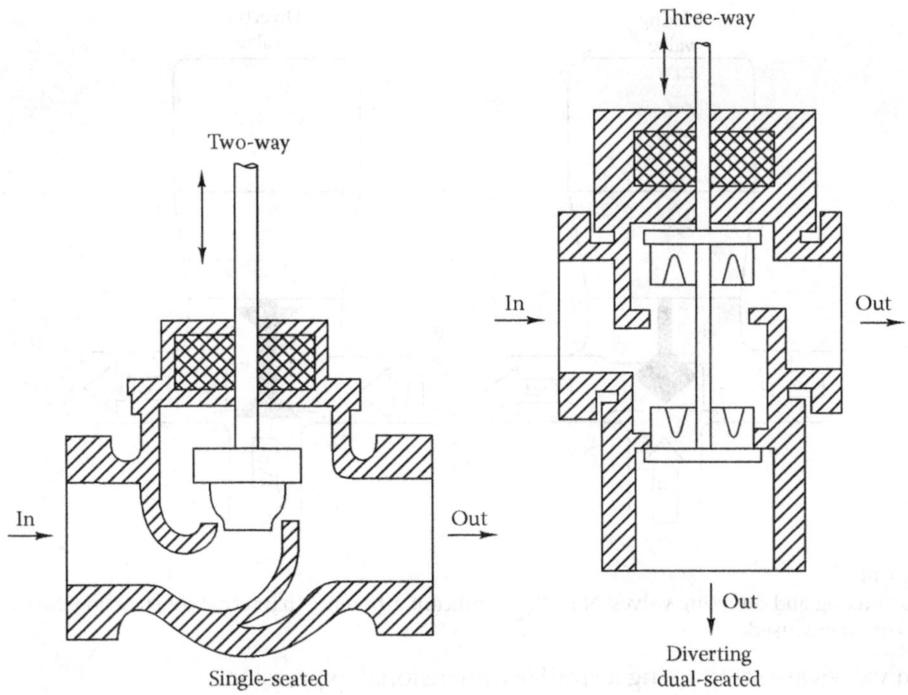

FIGURE 23.13
Cross-sectional drawings of direct-acting, single-seated two-way valve and dual-seated, three-way, diverting valve.

The three-way valve shown is double seated (single-seated three-way valves are also available); it is therefore easier to close than a single-seated valve, but tight shutoff is not possible.

Three-way valves can also be used as *mixing* valves. In this application, two streams enter the valve and one leaves. Mixing and diverting valves *cannot be used interchangeably,* since their internal design is different to ensure that they can each seat properly. Particular attention is needed by the installer to be sure that connections are made properly; arrows cast in the valve body show the proper flow direction. Figure 23.14 shows an example of three-way valves for both mixing and diverting applications.

Valve flow capacity is denoted in the industry by the dimensional flow coefficient C_v defined by

$$\dot{V}\,(\text{gal/min}) = C_v\left[\Delta p\,(\text{psi})\right]^{0.5} \tag{23.21}$$

Δp is the pressure drop across the fully open valve, so C_v is specified as the flow rate of 60°F water that will pass through the fully open valve if a pressure difference of 1.0 psi is imposed across the valve. If SI units (m^3/s and Pa) are used, the numerical value of C_v is 17% larger than in USCS units. Once the designer has determined a value of C_v, manufacturer's tables can be consulted to select a valve for the known pipe size. If a fluid other than water is to be controlled, the C_v, found from Equation 23.21, should be multiplied by the square root of the fluid's specific gravity.

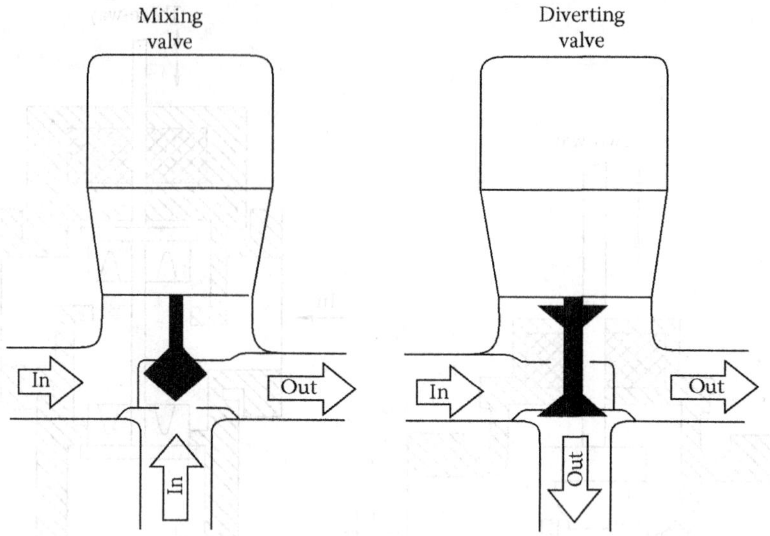

FIGURE 23.14
Three-way mixing and diverting valves. Note the significant difference in internal construction. Mixing valves are more commonly used.

Steam valves are sized using a similar dimensional expression:

$$\dot{m}\ (\text{lb/h}) = 63.5C_v \left[\frac{\Delta p\ (\text{psi})}{v\ (\text{ft}^3/\text{lb})}\right]^{0.5} \tag{23.22}$$

in which v is the steam-specific volume. If the steam is highly superheated, multiply C_v found from Equation 23.22 by 1.07 for every 100°F of superheat. For wet steam, multiply C_v by the square root of the steam quality. Honeywell (1988) recommends that the pressure drop across the valve to be used in the equation be 80% of the difference between steam supply and return pressures (subject to the sonic flow limitation discussed later). Table 23.1 can be used for preliminary selection of control valves for either steam or water.

The type of valve (linear or not) for a specific application must be selected so the controlled system is as nearly linear as possible. Control valves are very commonly used to control the heat transfer rate in coils. For a linear system, the combined characteristic of the actuator, valve, and coil should be linear. This will require quite different valves for hot water and steam control, for example, as we shall see.

Figure 23.15 shows the part load performance of a hot water coil used for air heating; at 10% of full flow, the heat rate is 50% of its peak value. The heat rate in a cross-flow heat exchanger increases roughly in exponential fashion with flow rate, a highly nonlinear characteristic. This heating coil nonlinearity follows from the longer water residence time in a coil at reduced flow and the relatively large temperature difference between air being heated and the water heating it.

However, if one were to control the flow through this heating coil by an equal percentage valve (positive exponential increase in flow with valve position), the combined valve plus the coil characteristic would be roughly linear. Referring to Figure 23.15, we see that 50% of stem travel corresponds to 10% flow. The third graph in the figure is the combined characteristic. This approximately linear subsystem is much easier to control than if a linear valve were used with the highly nonlinear coil. Hence the general rule: use equal percentage valves for heating coil control.

TABLE 23.1

Quick Sizing Chart for Control Valves

	Steam Capacity (lb/h)						Water Capacity (gal/min)						
	Vacuum Return Systems[a]			Atmospheric Return Systems			Differential Pressure (psig)						
	2 psi Supply Press.	5 psi Supply Press.	10 psi Supply Press.	2 psi Supply Press.	5 psi Supply Press.	10 psi Supply Press.							
C_v	3.2 psi Press. Drop[b]	5.6 psi Press. Drop[b]	9.6 psi Press. Drop[b]	1.6 psi Press. Drop[b]	4.0 psi Press. Drop[b]	8.0 psi Press. Drop[b]	2	4	6	8	10	15	20
0.33	7.7	11.0	16.0	5.4	9.3	14.6	0.41	0.66	0.81	0.93	1.04	1.27	1.47
0.63	14.6	20.9	30.5	10.4	17.7	27.8	0.89	1.26	1.54	1.78	1.99	2.4	2.81
0.73	17.0	24.3	35.4	12	20.5	32.2	1.0	1.46	1.78	2.06	2.3	2.8	3.25
1.0	23.0	33.2	48.5	16.4	28	44	1.4	2.0	2.44	2.82	3.16	3.9	4.46
1.6	37.09	53.1	77.6	26.8	45	70.6	2.25	3.2	3.9	4.51	5.06	6.2	7.13
2.5	58.25	82.9	121.2	41.9	70.25	110.25	3.53	5.0	6.1	7.05	7.9	9.68	11.15
3.0	69.9	99.5	145.5	50.2	84.3	132.3	4.23	6.0	7.32	8.46	9.48	11.61	13.38
4.0	93.2	132.2	194.0	67	112.4	177.4	5.6	8.0	9.76	11.28	12.6	15.5	17.87
5.0	116.2	165.2	242.5	82.7	140.5	220.5	7.1	10.0	12.2	14.1	15.8	19.4	22.3
6.0	139	200	291.0	99	168	265	8.5	12.0	14.6	16.92	18.9	23.2	27.0
6.3	146	209	311.5	104	177	278	8.9	12.6	15.4	17.78	19.9	24.4	28.1
7.0	162	233	339.5	115	196	309	9.9	14.0	17.1	19.74	22.1	27.1	31
8.0	186.5	264.4	388.0	131.2	224.8	352.8	11.3	16.0	19.5	22.56	25.3	31.6	35.7
10.0	232	332	485.0	164	281	441	14.1	20	24.4	28.2	31.6	38.7	44.6
11.0	256	366	533.5	181	309	486	15.5	22	27	31.02	34.4	42.5	49
13.0	303	434	630.5	213.7	365.3	373.3	18.3	27	31.7	36.7	41.1	50.3	58
14.0	326	465	679.0	232	393	617	19.7	28	34	39	44	54	62
15.0	349.3	497.6	727.5	246	421.5	661.5	21.1	30	36.6	42.3	47.4	58	66.9
16.0	370.9	531	776.0	268	450	706	22.5	32	39	45.1	50.6	62	71.3
18.0	419	597	873.0	301	505	794	25	36	44	51	57	70	80
20.0	466	664	970.0	335	562	882	28	40	49	56	63	77	89
23.0	541	763	1,115	385	646	1,014	32	46	56	65	73	89	103
25.0	582.5	829	1,212	419	702.5	1,102.5	35.3	50	61	70.5	79	96.8	111.5

(Continued)

TABLE 23.1 (*Continued*)
Quick Sizing Chart for Control Valves

	Steam Capacity (lb/h)						Water Capacity (gal/min)						
	Vacuum Return Systems[a]			Atmospheric Return Systems			Differential Pressure (psig)						
	2 psi Supply Press.	5 psi Supply Press.	10 psi Supply Press.	2 psi Supply Press.	5 psi Supply Press.	10 psi Supply Press.							
C_v	3.2 psi Press. Drop[b]	5.6 psi Press. Drop[b]	9.6 psi Press. Drop[b]	1.6 psi Press. Drop[b]	4.0 psi Press. Drop[b]	8.0 psi Press. Drop[b]	2	4	6	8	10	15	20
38.0	885	1,257	1,833	636	1,069	1,676	53	76	93	107	120	147	169
40.0	932	1,322	1,940	670	1,124	1,764	56	80	97.6	112.8	126	155	178.7
50.0	1,162	1,652	2,425	827	1,405	2,205	71	100	122	141	158	194	223
56.0	1,305	1,851	2,716	938	1,574	2,469	79	112	137	158	177	217	250
63.0	1,460	2,090	3,056	1,043	1,770	2,778	89	126	154	178	199	244	281
75.0	1,748	2,481	3,637	1,230	2,107	3,307	106	150	183	212	237	290	335
80.0	1,865	2,644	3,880	1,312	2,248	3,528	113	160	195	225.6	253	316	357
90.0	2,096	2,980	4,365	1,476	2,529	3,969	127	180	220	254	284	348	401
97.0	2,229	3,204	4,703	1,590	2,725	4,277	137	196	231	274	307	375	432
100.0	2,330	3,319	4,850	1,640	2,816	4,410	141	200	244	282	316	387	446
105.0	2,442	3,481	5,092	1,722	2,950	4,630	148	210	256	296	332	406	468
130.0	3,030	4,340	6,305	2,137	3,653	5,733	183	270	317	367	411	503	580
150.0	3,493	4,976	7,275	2,460	4,215	6,615	211	300	366	423	474	280	699
160.0	3,709	5,310	7,760	2,680	4,500	7,060	225	320	390	451	560	620	713
170.0	3,960	5,642	8,245	2,788	4,777	7,497	240	340	415	479	537	658	758
190.0	4,450	6,310	9,215	3,116	5,339	8,379	268	360	464	536	600	735	847
244.0	5,670	7,930	11,834	4,001	6,856	10,760	344	488	595	688	771	944	1,088
250.0	5,825	8,290	12,125	4,190	7,025	11,025	353	500	610	705	790	968	1,115
270.0	6,282	8,960	13,095	4,525	7,587	11,907	381	540	659	761	853	1,045	1,204
300.0	6,990	9,950	14,550	5,025	8,430	13,230	423	600	732	846	948	1,161	1,338
350.0	8,160	11,590	16,975	5,860	9,835	15,435	494	700	854	987	1,106	1,355	1,561

(*Continued*)

TABLE 23.1 (*Continued*)
Quick Sizing Chart for Control Valves

	Steam Capacity (lb/h)						Water Capacity (gal/min)						
	Vacuum Return Systems[a]			Atmospheric Return Systems			Differential Pressure (psig)						
	2 psi Supply Press.	5 psi Supply Press.	10 psi Supply Press.	2 psi Supply Press.	5 psi Supply Press.	10 psi Supply Press.							
C_v	3.2 psi Press. Drop[b]	5.6 psi Press. Drop[b]	9.6 psi Press. Drop[b]	1.6 psi Press. Drop[b]	4.0 psi Press. Drop[b]	8.0 psi Press. Drop[b]	2	4	6	8	10	15	20
480.0	11,180	15,860	23,280	8,045	13,408	21,168	677	960	1,171	1,353	1,517	1,858	2,141
640.0	14,910	21,180	31,040	10,496	17,984	28,224	902	1,280	1,561	1,805	2,022	2,477	2,854
760.0	17,700	25,120	36,860	12,464	21,356	33,516	1,071	1,520	1,854	2,143	2,401	2,941	3,390
1,000.0	23,300	33,190	48,500	16,400	28,160	44,100	1,410	2,000	2,440	2,820	3,160	3,870	4,460
1,200.0	27,150	39,790	58,200	19,680	33,720	52,920	1,692	2,400	2,928	3,384	2,792	4,644	5,352
1,440.0	33,290	47,160	69,840	23,616	40,464	63,504	2,030	2,880	3,514	4,061	4,550	5,573	6,422

Source: From Honeywell, Inc., *Engineering Manual of Automatic Control*, Honeywell, Inc., Minneapolis, MN, 1988.
[a] Assuming a 4–8 in. vacuum.
[b] Pressure drop across fully open valve taking 80% of the pressure difference between supply and return main pressures.

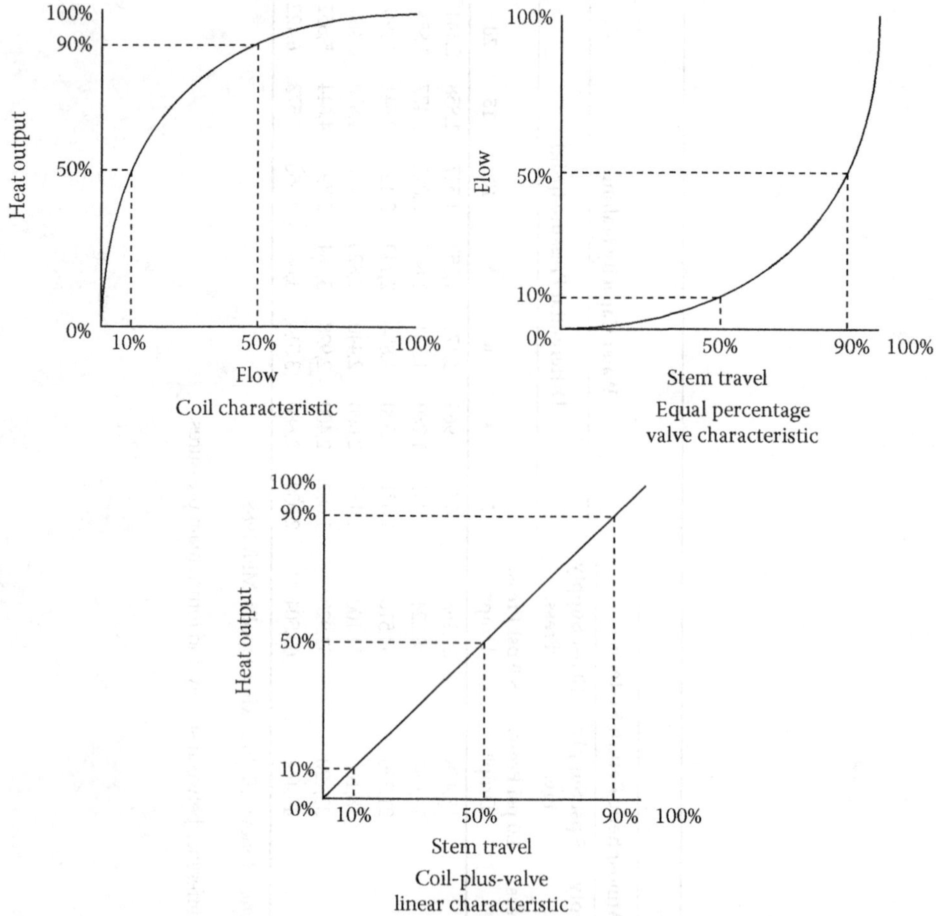

FIGURE 23.15
Heating coil, equal percentage valve, and combined coil-plus-valve linear characteristic.

Linear, two-port valves are to be used for steam flow control to coils, since the transfer of heat by steam condensation is a linear, constant temperature process—the more steam supplied, the greater the heat rate in exact proportion. Note that this is a completely different coil flow characteristic than for hot water coils. However, steam is a compressible fluid, and the sonic velocity sets the flow limit for a given valve opening when the pressure drop across the valve is more than 60% of the steam supply line absolute pressure. As a result, the pressure drop to be used in Equation 23.22 is the *smaller* of (1) 50% of the absolute stream pressure upstream of the valve and (2) 80% of the difference between the steam supply and return line pressures. The 80% rule gives good valve modulation in the subsonic flow regime (Honeywell, 1988).

Chilled water control valves should also be linear, since the performance of chilled water coils (smaller air–water temperature difference than in hot water coils) is more similar to steam coils than to hot water coils.

Either two- or three-way valves can be used to control flow at part load through heating and cooling coils as shown in Figure 23.16. The control valve can either be controlled from coil outlet water or air temperature. Two- or three-way valves achieve the same local result at the coil when used for part load control. However, the designer must consider the effects on the balance of the secondary system when selecting the valve type.

In essence, the two-way valve flow control method results in variable flow (tracking variable loads) with constant coil water temperature change, whereas the three-way valve approach results in roughly constant secondary loop flow rate but smaller coil water temperature change (beyond the local coil loop itself). In large systems, a primary/secondary design with two-way valves is preferred, unless the primary equipment can handle the range of flow variation that will result without a secondary loop. Since chillers and boilers require that flow remain within a restricted range, the energy and cost savings that could accrue due to the two-way valve, variable-volume system are difficult to achieve in small

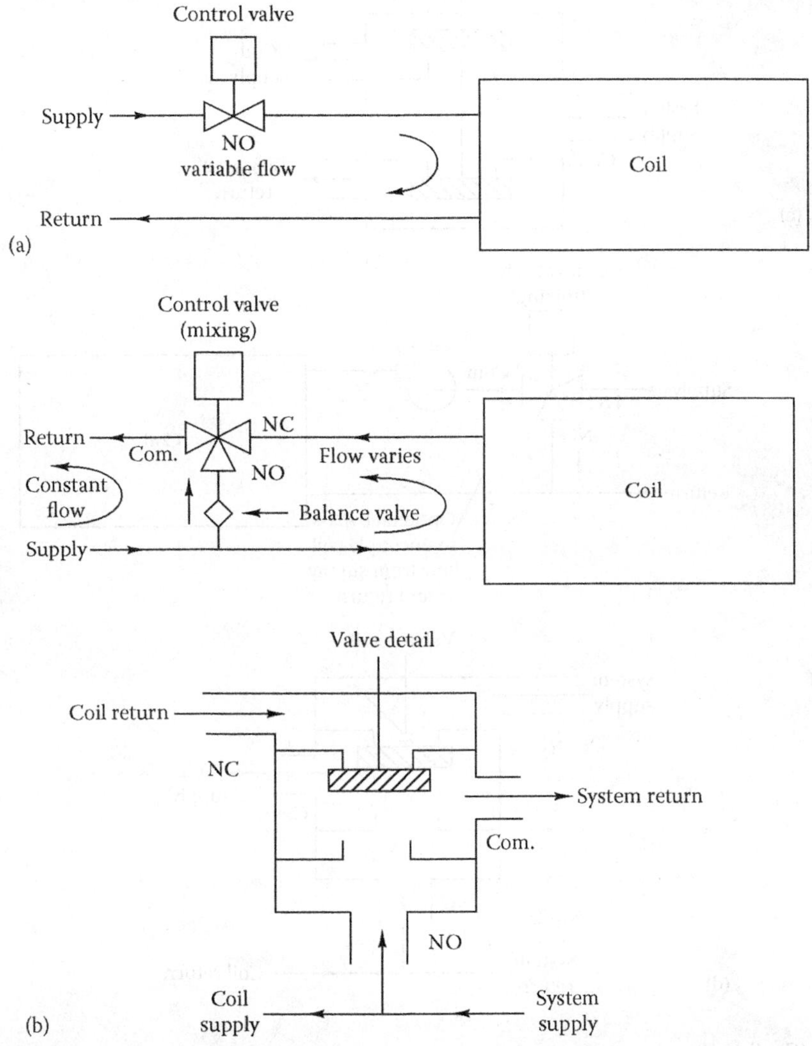

FIGURE 23.16
Various control valve piping arrangements: (a) two-way valve, (b) three-way mixing valve. *(Continued)*

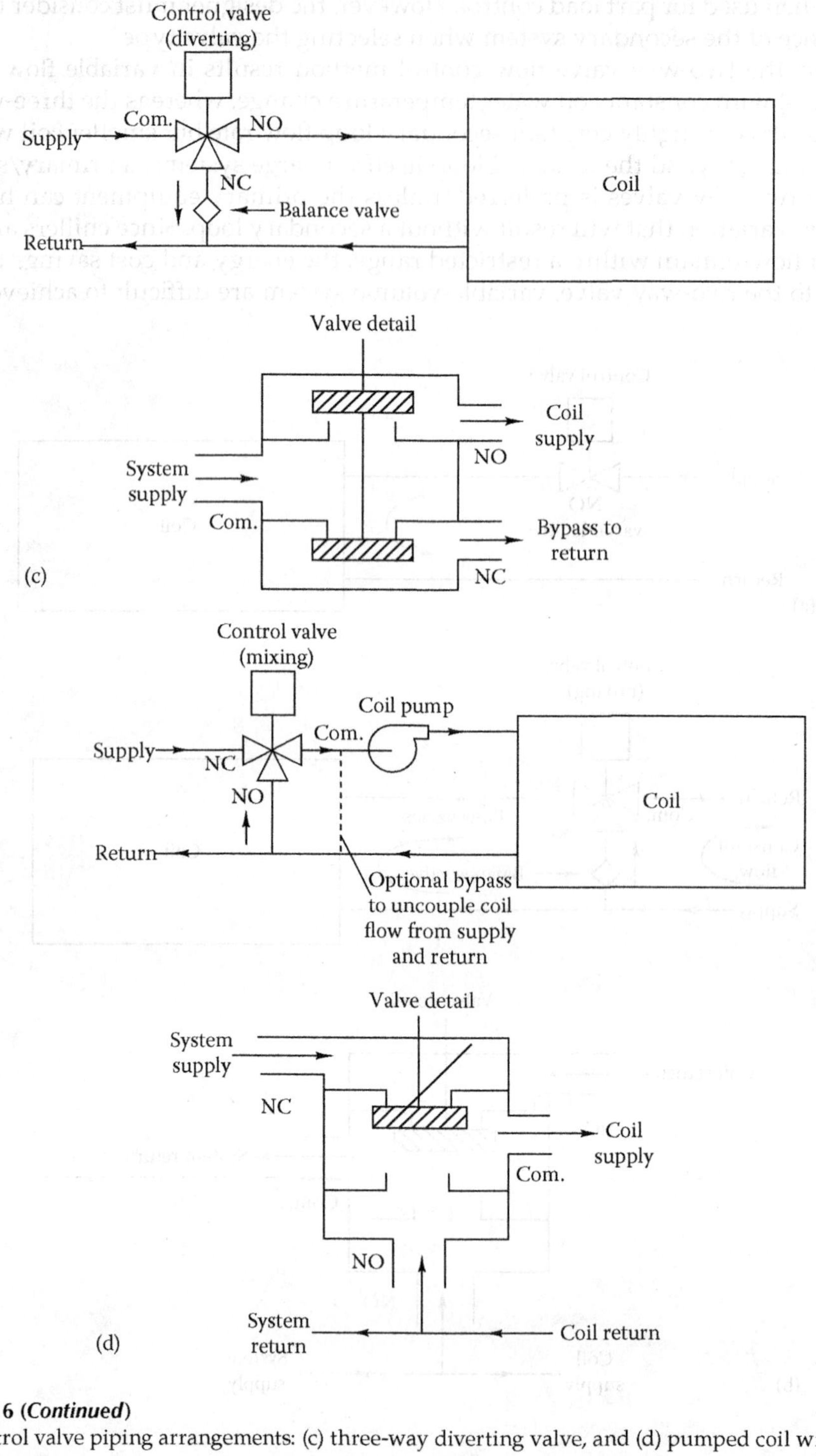

FIGURE 23.16 ***(Continued)***
Various control valve piping arrangements: (c) three-way diverting valve, and (d) pumped coil with three-way mixing valve.

systems unless a two-pump, primary/secondary loop approach is employed. If this dual-loop approach is not used, the three-way valve method is required to maintain required boiler or chiller flow.

The location of the three-way valve at a coil must also be considered by the designer. Figure 23.16b shows the valve used downstream of the coil in a mixing, bypass mode. If a balancing valve is installed in the bypass line and set to have the same pressure drop as the coil, the local coil loop will have the same pressure drop for both full and zero coil flows. However, at the valve mid-flow position, overall flow resistance is less, since two parallel paths are involved, and the total loop flow increases to 25% more than that at either extreme.

Alternatively, the three-way valve can also be used in a diverting mode as shown in Figure 23.16c. In this arrangement, essentially the same considerations apply as for the mixing arrangement discussed earlier.* However, if a circulator (small pump) is inserted as shown in Figure 23.16d, the direction of flow in the branch line changes and a mixing valve is used. The reason that pumped coils are used is that control is improved. With constant coil flow, the highly nonlinear coil characteristic shown in Figure 23.15 is reduced, since the residence time of hot water in the coil is constant independent of load. However, this arrangement appears to the external secondary loop the same as a two-way valve. As load is decreased, flow into the local coil loop also decreases. Therefore, the uniform secondary loop flow normally associated with three-way valves is not present unless the optional bypass is used.

For HVAC systems requiring precise control, high-quality control valves are required. The best controllers and valves are of *industrial quality*; the additional cost for these valves compared to conventional building hardware results in more accurate control and longer lifetime.

23.4.2 Airflow Control

Dampers are used to control airflow in secondary HVAC air systems in buildings. In this section, we discuss the characteristics of dampers used for flow control in systems where constant-speed fans are involved. Figure 23.17 shows cross sections of the two common types of dampers used in commercial buildings. Parallel-blade dampers use blades that rotate in the same direction. They are most often applied to two position locations—open or closed. Use for flow control is not recommended. The blade rotation changes airflow direction, a characteristic that can be useful when airstreams at different temperatures are to be effectively blended.

Opposed-blade dampers have adjacent counterrotating blades. Airflow direction is not changed with this design, but pressure drops are higher than for parallel blading. Opposed-blade dampers are preferred for flow control. Figure 23.18 shows the flow characteristics of these dampers to be closer to the desired linear behavior. The parameter α on the curves is the ratio of system pressure drop to fully open damper pressure drop.

* A little-known disadvantage of three-way valve control has to do with the *conduction* of heat from a closed valve to a coil. For example, the constant flow of hot water through two ports of a *closed* three-way heating coil control valve keeps the valve body hot. Conduction from the closed, hot valve mounted close to a coil can cause sufficient air heating to actually decrease the expected cooling rate of a downstream cooling coil during the cooling season. Three-way valves have a second practical problem; installers often connect three-way valves incorrectly given the choice of three pipe connections and three pipes to be connected. Both of these problems are avoided by using two-way valves.

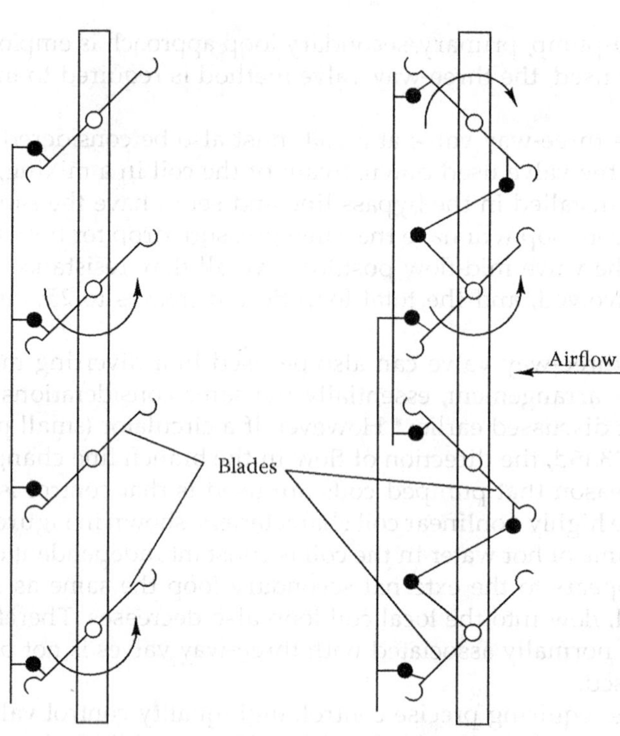

FIGURE 23.17
Diagram of parallel and opposed-blade dampers.

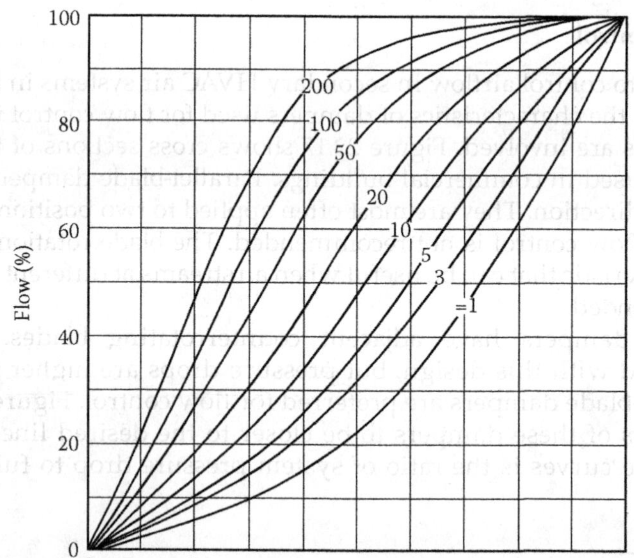

FIGURE 23.18
Flow characteristics of opposed-blade dampers. The parameter α is the ratio of system resistance (not including the damper) to damper resistance. An approximately linear damper characteristic is achieved if this ratio is about 10 for opposed-blade dampers.

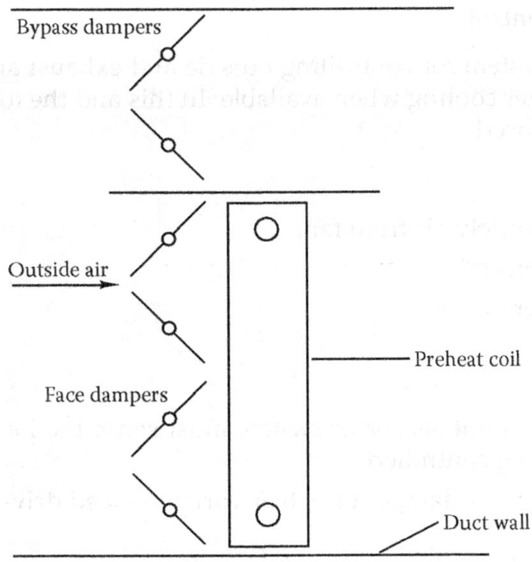

FIGURE 23.19
Face and bypass dampers used for preheating coil control.

A common application of dampers controlling the flow of outside air uses two sets in a *face and bypass* configuration as shown in Figure 23.19. For full heating, all air is passed through the coil and the bypass dampers are closed. If no heating is needed in mild weather, the coil is bypassed (for minimum flow resistance and fan power cost, flow through fully open face and bypass dampers can be used if the preheat coil water flow is shut off). Between these extremes, flow is split between the two paths. The face and bypass dampers are sized so that the pressure drop in full bypass mode (damper pressure drop only) and full heating mode (coil plus damper pressure drop) is the same.

23.5 Example HVAC System Control Systems

Several widely used control configurations for specific tasks are described in this section. These have been selected from the hundreds of control system configurations that have been used for buildings. The goal of this section is to illustrate how control components described previously are assembled into systems and what design considerations are involved. For a complete overview of HVAC control system configurations, see ASHRAE (2002, 2003, 2004), Grimm and Rosaler (1990), Tao and Janis (2005), Sauer et al. (2001), and Honeywell (1988). The illustrative systems in this section are drawn in part from the latter reference.

In this section, we will discuss seven control systems in common use. Each system will be described using a schematic diagram, and its operation and key features will be discussed in the accompanying text.

23.5.1 Outside Air Control

Figure 23.20 shows a system for controlling outside and exhaust air from a central AHU equipped for economizer cooling when available. In this and the following diagrams, the following symbols are used:

C—cooling coil
DA—discharge air (supply air from fan)
DX—direct-expansion coil
E—damper controller
EA—exhaust air
H—heating coil
LT—low-temperature limit sensor or switch; must sense the lowest temperature in the air volume being controlled
M—motor or actuator (for damper or valve), variable speed drive
MA—mixed air
NC—normally closed
NO—normally open
OA—outside air
PI—proportional plus integral controller
R—relay
RA—return air

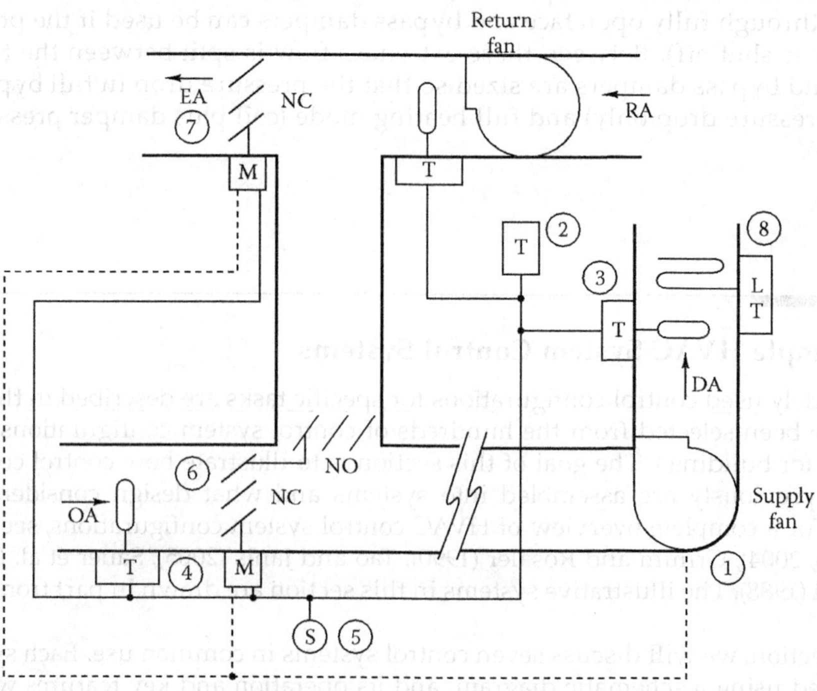

FIGURE 23.20
Outside air control system with economizer capability.

S—switch

SP—static pressure sensor used in VAV systems

T—temperature sensor; must be located to read the average temperature representative of the air volume being controlled

This system is able to provide the minimum outside air during occupied periods, to use outdoor air for cooling when appropriate by means of a temperature-based economizer cycle, and to operate fans and dampers under all conditions. The numbering system used in the figure indicates the sequence of events as the air-handling system begins operation after an off period:

1. The fan control system turns on when the fan is turned on. This may be by a clock signal or a low- or high-temperature space condition.
2. The space temperature signal determines if the space is above or below the set point. If above, the economizer feature will be activated if the OA temperature is below the upper limit for economizer operation and control the outdoor and mixed-air dampers. If below, the outside air damper is set to its minimum position.
3. The discharge air PI controller controls both sets of dampers (OA/RA and EA) to provide the desired mixed-air temperature.
4. When the outdoor temperature rises above the upper limit for economizer operation, the outdoor air damper is returned to its minimum setting.
5. Switch S is used to set the minimum setting on outside and exhaust air dampers manually. This is ordinarily done only once during building commissioning and flow testing.
6. When the supply fan is off, the outdoor air damper returns to its NC position and the return air damper returns to its NO position.
7. When the supply fan is off, the exhaust damper also returns to its NC position.
8. Low temperature sensed in the duct will initiate a freeze-protect cycle. This may be as simple as turning on the supply fan to circulate warmer room air. Of course, the OA and EA dampers remain tightly closed during this operation.

23.5.2 Heating Control

If the minimum air setting is large in the preceding system, the amount of outdoor air admitted in cold climates may require preheating. Figure 23.21 shows a preheating system using face and bypass dampers. (A similar arrangement is used for DX cooling coils.) The equipment shown is installed upstream of the fan in Figure 23.20. This system operates as follows:

1. The preheat subsystem control is activated when the supply fan is turned on.
2. The preheat PI controller senses temperature leaving the preheat section. It operates the face and bypass dampers to control the exit air temperature between 45°F and 50°F.
3. The outdoor air sensor and associated controller controls the water valve at the preheat coil. The valve may be either a modulating valve (better control) or an on-off valve (less costly).
4. The low-temperature sensors (LTs) activate coil freeze protection measures including closing dampers and turning off the supply fan.

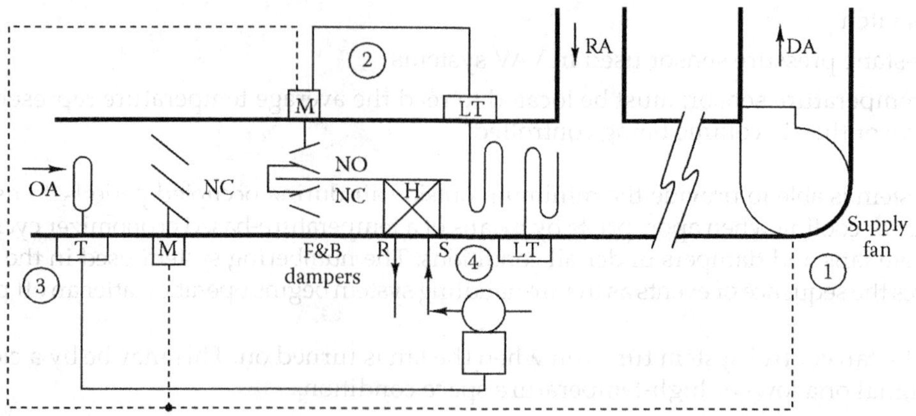

FIGURE 23.21
Preheat control system. Counterflow of air and hot water in the preheat coil results in the highest heat transfer rate.

Note that the preheat coil (as well as all coils in this section) is connected so that the hot water (or steam) flows counter to the direction of airflow. Counterflow provides a higher heating rate for a given coil than does parallel flow. Mixing of heated and cold bypass air must occur upstream of the control sensors. Stratification can be reduced by using sheet metal *air blenders* or by propeller fans in the ducting. The preheat coil should be located in the bottom of the duct. Steam preheat coils must have adequately sized traps and vacuum breakers to avoid condensate buildup that could lead to coil freezing at light loads.

The face and bypass damper approach enables air to be heated to the required system supply temperature without endangering the heating coil. (If a coil were to be as large as the duct—no bypass area—it could freeze when the hot water control valve cycles are opened and closed to maintain the discharge temperature.) The designer should consider pumping the preheat coil as shown in Figure 23.19d to maintain water velocity above the 3 ft/s needed to avoid freezing. If glycol is used in the system, the pump is not necessary but heat transfer will be reduced.

During winter, in heating climates, heat must be added to the mixed airstream to heat the outside air portion of mixed air to an acceptable discharge temperature. Figure 23.22 shows a common heating subsystem controller used with central air handlers. (It is assumed that the mixed-air temperature is kept above freezing by the action of the preheat coil, if needed.) This system has the added feature that coil discharge temperature is adjusted for ambient temperature, since the amount of heat needed decreases with increasing outside temperature. This feature, called coil discharge reset, provides better control and can reduce energy consumption. The system operates as follows:

1. During operation, the discharge air sensor and PI controller control the hot water valve.
2. The outside air sensor and controller reset the *set point* of the discharge air PI controller up as ambient temperature drops.
3. Under sensed low-temperature conditions, freeze protection measures are initiated as discussed earlier.

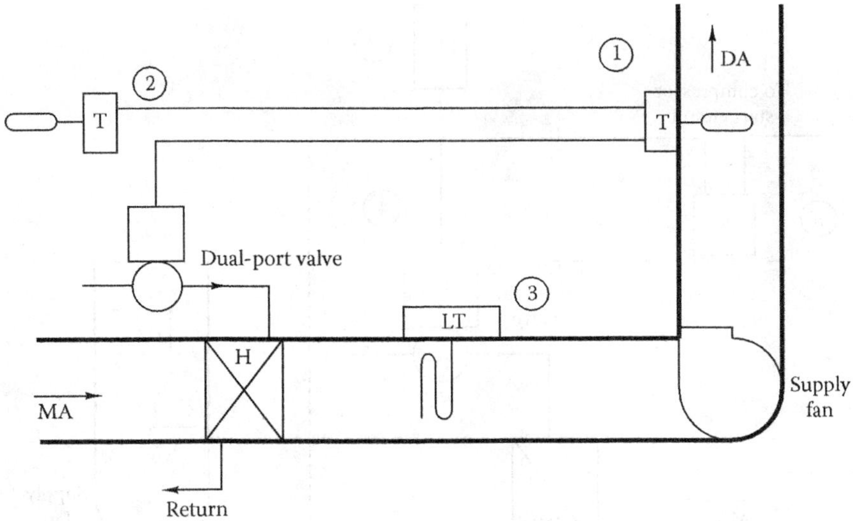

FIGURE 23.22
Heating coil control subsystem using two-way valve and optional reset sensor.

Reheating at zones in VAV or other systems uses a system similar to that just discussed. However, boiler water temperature is reset and no freeze protection is normally included. The air temperature sensor is the zone thermostat for VAV reheat, not a duct temperature sensor.

23.5.3 Cooling Control

Figure 23.23 shows the components in a cooling coil control system for a single-zone system. Control is similar to that for the heating coil discussed earlier except that the zone thermostat (not a duct temperature sensor) controls the coil. If the system were a central system serving several zones, a duct sensor would be used. Chilled water supplied to the coil

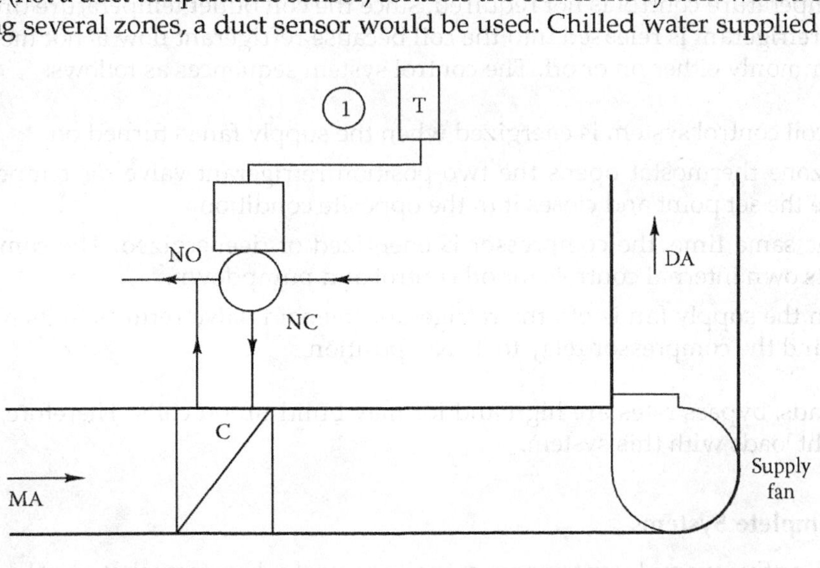

FIGURE 23.23
Cooling coil control subsystem using three-way diverting valve.

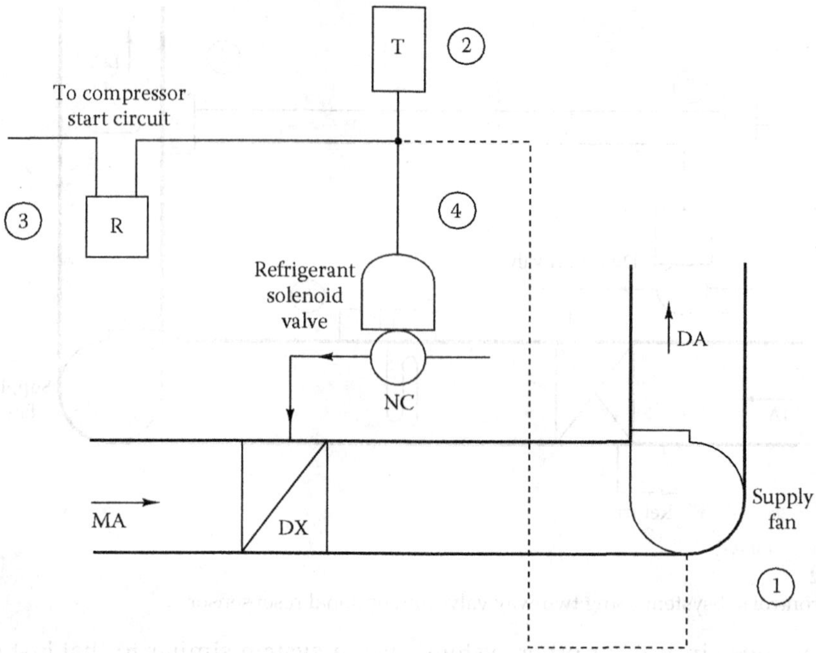

FIGURE 23.24
DX cooling coil control subsystem (on-off control).

partially bypasses and partially flows through the coil, depending on the coil load. The use of three- and two-way valves for coil control has been discussed in detail earlier. The valve NC connection is used as shown so that valve failure will not block secondary loop flow.

Figure 23.24 shows another common cooling coil control system. In this case, the coil is a DX refrigerant coil and the controlled medium is refrigerant flow. DX coils are used when precise temperature control is not required, since the coil outlet temperature drop is large whenever refrigerant is released into the coil because refrigerant flow is not modulated; it is most commonly either on or off. The control system sequences as follows:

1. The coil control system is energized when the supply fan is turned on.
2. The zone thermostat opens the two-position refrigerant valve for temperatures above the set point and closes it in the opposite condition.
3. At the same time, the compressor is energized or deenergized. The compressor has its own internal controls for oil control and pumpdown.
4. When the supply fan is off, the refrigerant solenoid valve returns to its NC position and the compressor relay to its NO position.

At light loads, bypass rates are high and ice may build up on coils. Therefore, control is poor at light loads with this system.

23.5.4 Complete Systems

The preceding five example systems are actually control subsystems that must be integrated into a single control system for the HVAC system's primary and secondary systems. In the

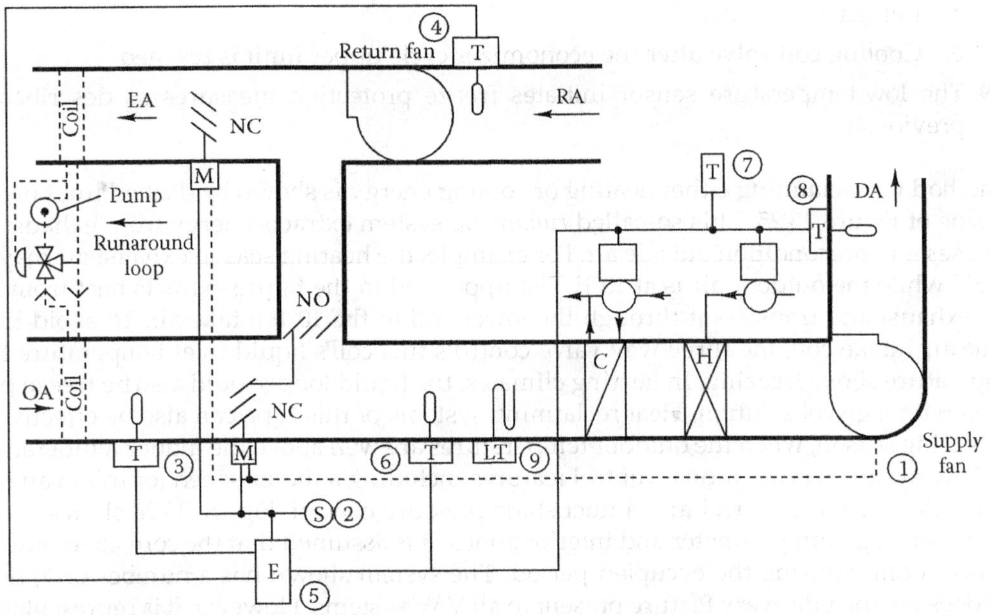

FIGURE 23.25
Control for a complete, constant-volume HVAC system. Optional runaround heat recovery system is shown to left by dashed lines.

remainder of this section, we will describe briefly two complete HVAC control systems widely used in commercial buildings. The first is a constant-volume system, whereas the second is a VAV system.

Figure 23.25 shows a constant-volume central system air-handling system equipped with supply and return fans, heating and cooling coils, and economizer for a single-zone application. If the system were to be used for multiple zones, the zone thermostat shown would be replaced by a discharge air temperature sensor. This constant-volume system operates as follows:

1. When the fan is energized, the control system is activated.
2. The minimum outside air setting is set (usually only once during commissioning as described earlier).
3. The OA temperature sensor supplies a signal to the damper controller.
4. The RA temperature sensor supplies a signal to the damper controller.
5. The damper controller positions the dampers to use outdoor or return air depending on which is cooler.
6. The mixed-air low-temperature controller controls the outside air dampers to avoid excessively low-temperature air from entering the coils. If a preheating system were included, this sensor would control it.
7. The space temperature sensor resets the coil discharge air PI controller.
8. The discharge air controller controls the
    a. Heating coil valve
    b. Outdoor air damper
    c. Exhaust air damper

d. Return air damper
e. Cooling coil valve after the economizer cycle upper limit is reached

9. The low-temperature sensor initiates freeze protection measures as described previously.

A method for reclaiming either heating or cooling energy is shown by dashed lines on the left side of Figure 23.25. This so-called *runaround* system extracts energy from exhaust air and uses it to precondition outside air. For example, the heating season exhaust air may be at 75°F while the outdoor air is at 10°F. The upper coil in the figure extracts heat from the 75°F exhaust and transfers it through the lower coil to the 10°F intake air. To avoid icing of the air intake coil, the three-way valve controls this coil's liquid inlet temperature to a temperature above freezing. In heating climates, the liquid loop should also be freeze protected with a glycol solution. Heat reclaiming systems of this type can also be effective in the cooling season, when the outdoor temperatures are well above the indoor temperature.

A VAV system has additional control features including a motor speed (or inlet vanes in some older systems) control and a duct static pressure control. Figure 23.26 shows a VAV system serving both perimeter and interior zones. It is assumed that the core zones always require cooling during the occupied period. The system shown has a number of options and does not include every feature present in all VAV systems. However, it is representative of VAV design practice. The sequence of operation during the heating season is as follows:

1. When the fan is energized, the control system is activated. Prior to activation during unoccupied periods, the perimeter zone baseboard heating is under control of room thermostats.
2. Return and supply fan interlocks are used to prevent pressure imbalances in the supply air ductwork.

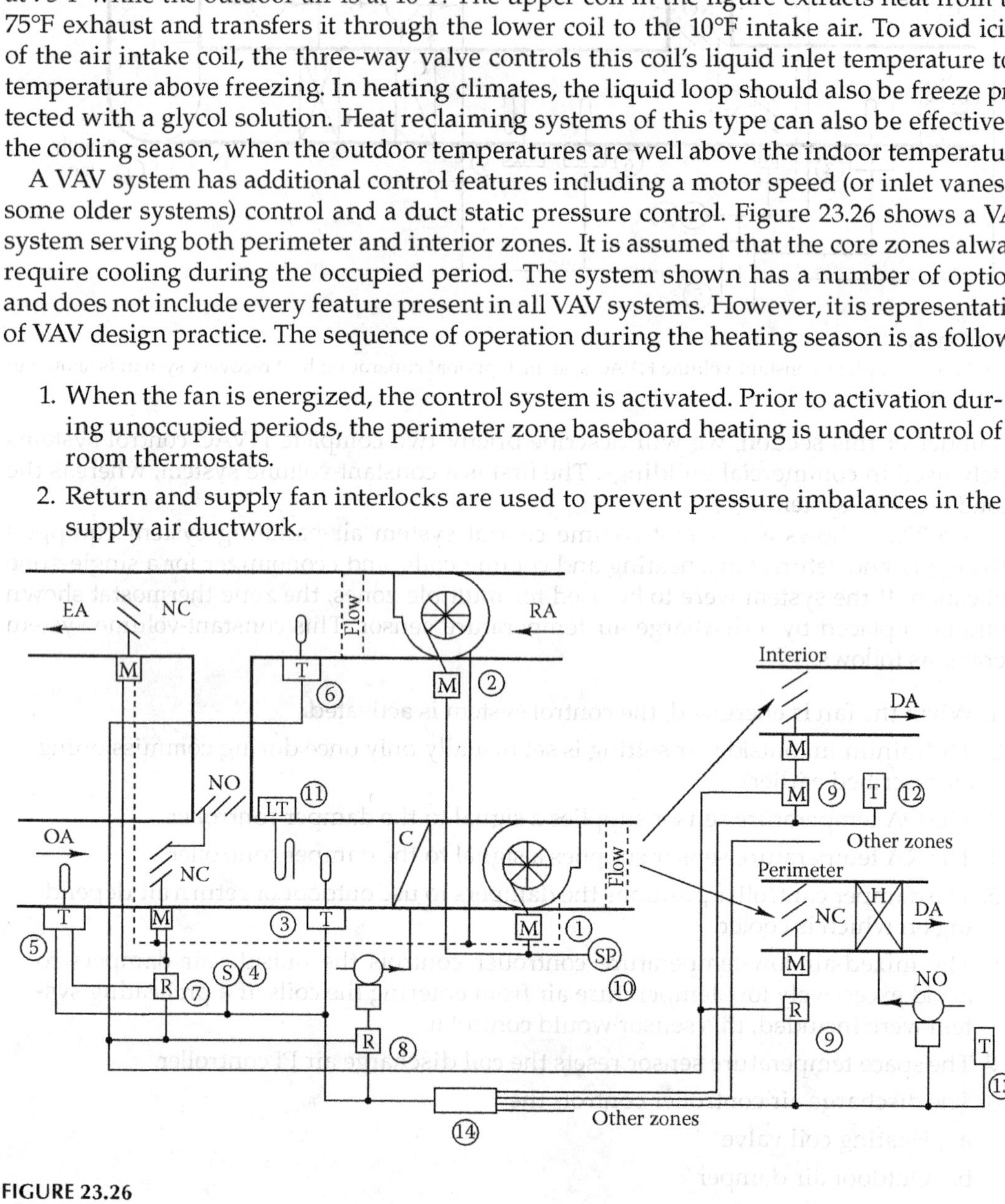

FIGURE 23.26
Control for complete, VAV system. Optional supply and return flow stations shown by dashed lines.

3. The mixed-air sensor controls the outdoor air dampers (and/or preheat coil not shown) to provide proper coil air inlet temperature. The dampers will typically be at their minimum position at about 40°F.
4. The damper minimum position controls the minimum outdoor airflow.
5. As the upper limit for economizer operation is reached, the OA dampers are returned to their minimum position.
6. The return air temperature is used to control the morning warm-up cycle after night setback (option present only if night setback is used).
7. The outdoor air damper is not permitted to open during morning warm-up by the action of the relay shown.
8. Likewise, the cooling coil valve is deenergized (NC) during morning warm-up.
9. All VAV box dampers are moved full open during morning warm-up by the action of the relay override. This minimizes the warm-up time. Perimeter zone coils and baseboard units are under control of the local thermostat.
10. During operating periods, the PI static pressure controller controls both the supply and return fan speeds (or inlet vane positions) to maintain approximately 1.0 in. WG of static pressure at the pressure sensor location (or optionally to maintain building pressure). An additional pressure sensor (not shown) at the supply fan outlet will shut down the fan if fire dampers or other dampers should close completely and block airflow. This sensor overrides the duct static pressure sensor shown.
11. The low-temperature sensor initiates freeze protection measures.
12. At each zone, room thermostats control VAV boxes (and fans if present); as zone temperature rises, the boxes open more.
13. At each perimeter zone room, thermostats close VAV dampers to their minimum settings and activate zone heat (coil and/or perimeter baseboard) as zone temperature falls.
14. The controller, using temperature information for all zones (or at least for enough zones to represent the characteristics of all zones), modulates outdoor air dampers (during economizer operation) and the cooling control valve (above the economizer cycle cutoff) to provide air sufficiently cooled to maintain acceptable zone humidity and meet the load of the warmest zone.

The duct static pressure controller is critical to the proper operation of VAV systems. The static pressure controller must be of PI design, since a proportional-only controller would permit duct pressure to drift upward as cooling loads drop due to the unavoidable offset in P-type controllers. In addition, the control system should position inlet vanes (if present) closed during fan shutdown to avoid overloading on restart.

Return fan control is best achieved in VAV systems by an actual flow measurement in supply and return ducts as shown by dashed lines in the figure. The return airflow rate is the supply rate less local exhausts (fume hoods, toilets, etc.) and exfiltration needed to pressurize the building.

VAV boxes are controlled locally, assuming that adequate duct static pressure exists in the supply duct and that supply air is at an adequate temperature to meet the load (this is the function of the controller described in item 14). Figure 23.27 shows a *local* control system used with a series-type, fan-powered VAV box. This particular system delivers a

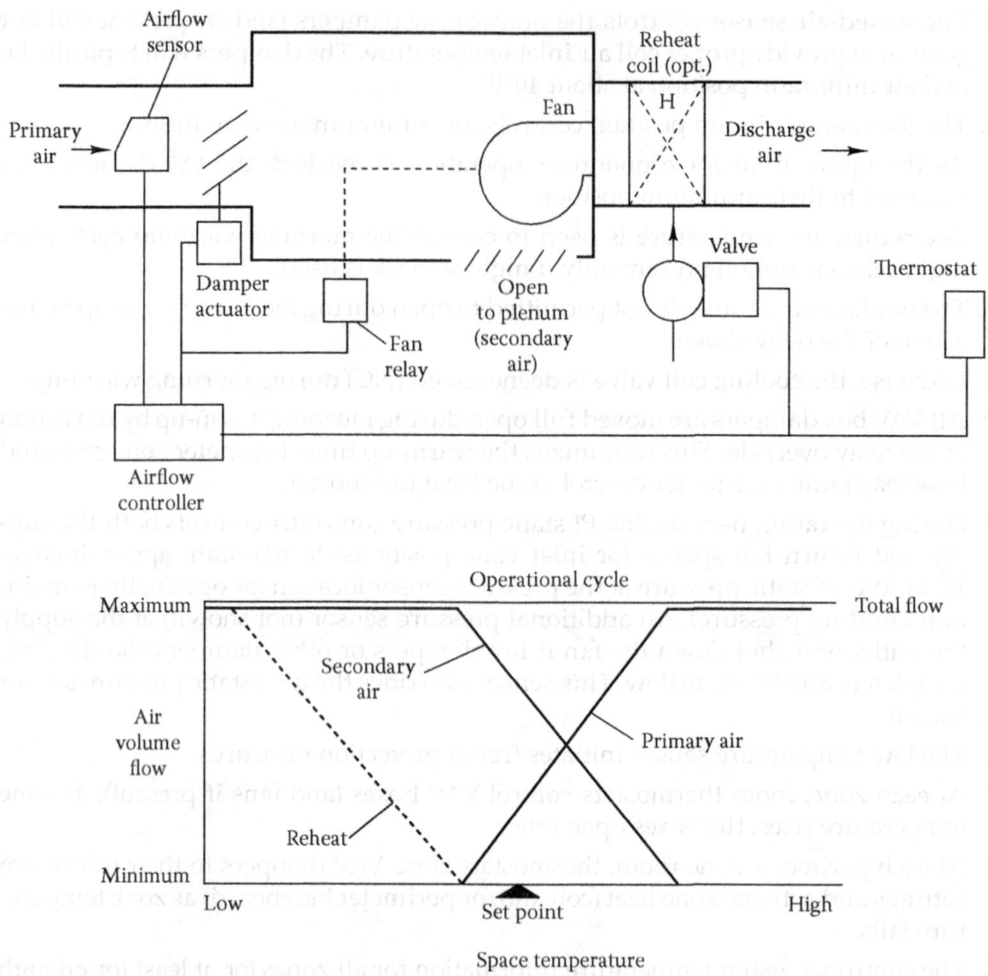

FIGURE 23.27
Series-type, fan-powered VAV box control subsystem and primary flow characteristic. The total box flow is constant at the level identified as *maximum* in the figure. The difference between primary and total airflow is secondary air recirculated through the return air grille. Optional reheat coil requires airflow shown by dashed line.

constant flow rate to the zone, to assure proper zone air distribution, by the action of the airflow controller. Primary air varies with cooling load as shown in the lower part of the figure. Optional reheating is provided by the coil shown.

23.5.5 Other Systems

This section has not covered the control of central plant equipment such as chillers and boilers. Most primary system equipment controls are furnished with the equipment and as such do not offer much flexibility to the designer. However, Braun et al. (1989) have shown that considerable energy savings can be made by properly sequencing cooling tower stages on chiller plants and by properly sequencing chillers themselves in multiple chiller plants.

Fire and smoke control are important for life safety in large buildings. The design of smoke control systems is controlled by national codes. The principal concept is to eliminate smoke from the zones where it is present while keeping adjacent zones pressurized to avoid smoke infiltration. Some components of space conditioning systems (e.g., fans) can be used for smoke control, but HVAC systems are generally not smoke control systems by design.

Electrical systems are primarily the responsibility of the electrical engineer on a design team. However, HVAC engineers must make sure that the electrical design accommodates the HVAC control system. Interfaces between the two occur where the HVAC controls activate motors on fans or chiller compressors, pumps, electrical boilers, or other electrical equipment.

In addition to electrical specifications, the HVAC engineer often conveys electrical control logic using a *ladder diagram*. An example is shown in Figure 23.28 for the control of the supply and return fans in a central system. The electrical control system is shown at the bottom and operates on low voltage (24 or 48 VAC) from the control transformer shown. The supply fan is manually started by closing the *start* switch. This activates the motor starter coil labeled 1M, thereby closing the three contacts labeled 1M in the supply fan circuit. The fourth 1M contact (in parallel with the start switch) holds the starter closed after the start button is released.

The hand–off–auto switch is typical and allows both automatic and manual operations of the return fan. When switched to the *hand* position, the fan starts. In the *auto* position, the fans will operate only when the adjacent contacts 3M are closed. Either of these actions activates the relay coil 2M, which in turn closes the three 2M contacts in the return fan motor starter. When either fan produces actual airflow, a flow switch is closed in the ducting, thereby completing the circuit to the pilot lamps L. The fan motors are protected by fuses and thermal overload heaters. If motor current draw is excessive, the heaters shown in the figure produce sufficient heat to open the normally closed thermal overload contacts.

This example ladder diagram is primarily illustrative and is not typical of an actual design. In a fully automatic system, both fans would be controlled by 3M contacts actuated by the HVAC control system. In a fully manual system, the return fan would be activated by a fifth 1M contact, not by the 3M automatic control system.

23.6 Commissioning and Operation of Control Systems

This chapter emphasizes the importance of making sound decisions in the design of HVAC control systems. Ensuring that the control system be commissioned and used properly is extremely important. The design process requires many assumptions about the building and its use. The designer must be sure that the systems will provide comfort under extreme conditions, and the sequence of design decisions and construction decisions often leads to systems that are substantially oversized. Operation at loads far below design conditions is generally much less efficient than at larger loads. Normal control practice can be a major contributor to this inefficiency. For example, it is quite common to see variable-volume air handler systems that operate at minimum flow as constant-volume systems almost all the time due to design flows that sometimes are twice as large as the maximum flow used in the building.

Hence it is very important that following construction, the control system and the rest of the HVAC system be *commissioned*. This process (ASHRAE, 2005) normally seeks to ensure that the control system operates according to design intent. This is really a minimum requirement to be sure that the system functions as designed. However, after construction, the control system setup can be modified to meet the loads actually present in

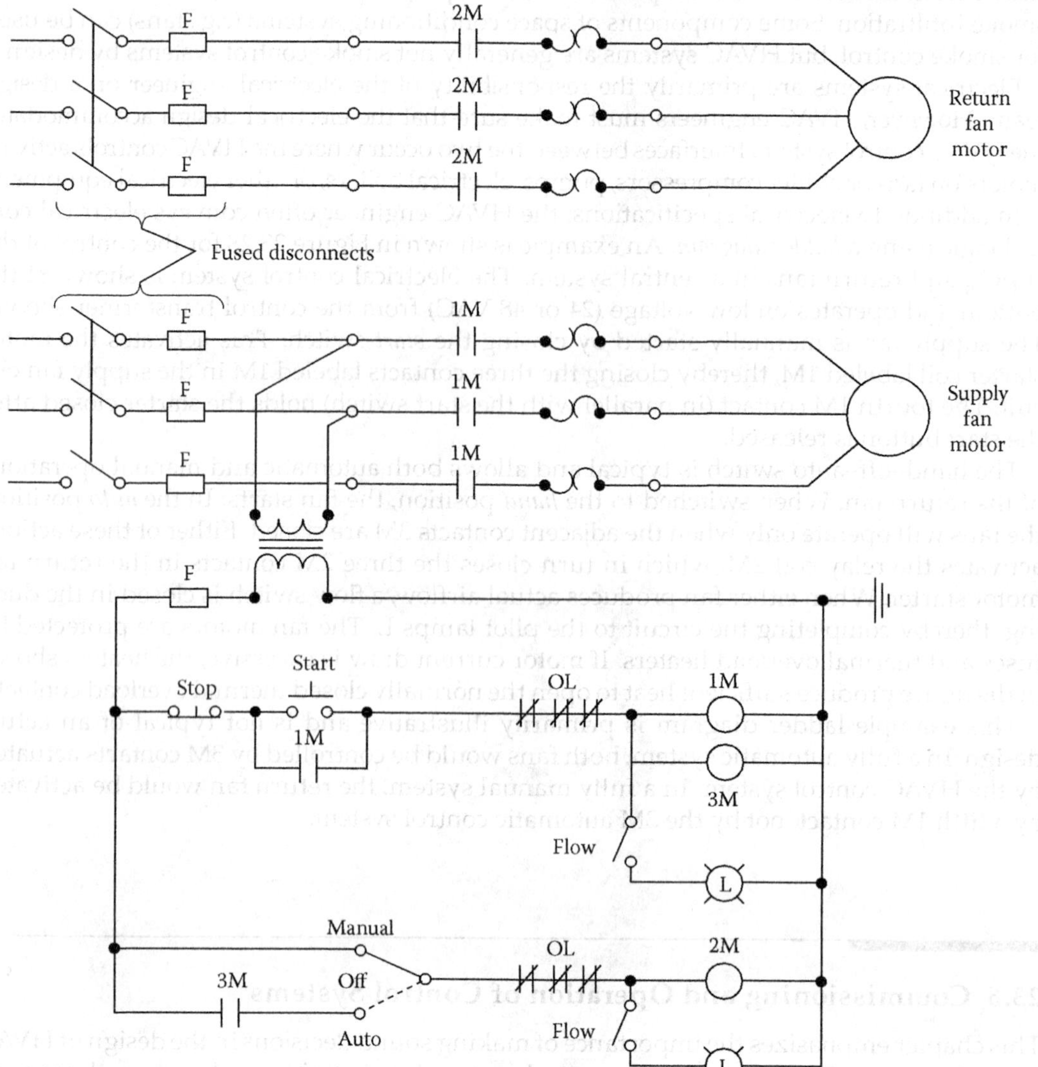

FIGURE 23.28
Ladder diagram for supply and return fan control. Hand–off–auto switch permits manual or automatic control of the return fan.

the building and fit the way the building is actually being used rather than basing these decisions on the design assumptions. If the VAV system is designed for more flow than is required, minimum flow settings of the terminal boxes can be reduced below the design value so that the system will operate in the VAV mode most of the time. Numerous other adjustments may be made as well. Such adjustments, as commonly made during the version of commissioning known as *Continuous Commissioning*®* (CC®), can frequently reduce the overall building energy use by 10% or more (Liu et al., 2002). If the process is applied to an older building where control practices have drifted away from design intent and undetected component failures have further eroded system efficiency, energy savings often exceed 20% (Claridge et al., 2004).

23.6.1 Control Commissioning Case Study

A case study in which this process was applied to a major army hospital facility located in San Antonio, TX (Zhu et al., 2000a–c), is provided. The Brooke Army Medical Center (BAMC) was a relatively new facility when the CC process was begun. The facility was operated for the army by a third-party company, and it was operated in accordance with the original design intent (Figure 23.29).

BAMC is a large, multifunctional medical facility with a total floor area of 1,349,707 ft^2. The complex includes all the usual in-patient facilities as well as out-patient and research areas. The complex is equipped with a central energy plant, which has four 1200-ton water-cooled electric chillers. Four primary pumps (75 hp each) are used to pump water through the chillers. Two secondary pumps (200 hp each), equipped with VFDs, supply chilled water from the plant to the building entrance. Fourteen chilled water risers equipped with 28 pumps totaling 557 hp are used to pump chilled water to all of the

FIGURE 23.29
The BAMC in San Antonio, TX.

* Continuous Commissioning and CC are registered trademarks of the Texas A&M Engineering Experiment Station.

AHUs and small fan coil units. All of the chilled water riser pumps are equipped with VFDs. There are four natural gas-fired steam boilers in this plant. The maximum output of each boiler is 20 MMBtu/h. Steam is supplied to each building at 125 psi (prior to CC) where heating water is generated.

There are 90 major AHUs serving the whole complex with a total fan power of 2570 hp. VFDs are installed on 65 AHUs, while the others are constant-volume systems. There are 2700 terminal boxes in the complex of which 27% are dual-duct variable-volume boxes, 71% are dual-duct constant-volume boxes, and 2% are single-duct variable-volume boxes.

The HVAC systems (chillers, boilers, AHUs, pumps, terminal boxes, and room conditions) are controlled by a DDC system. Individual controller-field panels are used for the AHUs and water loops located in the mechanical rooms. The control program and parameters can be changed either by the central computers or by the field panels.

23.6.1.1 Design Conditions

The design control program was being fully utilized by the EMCS. It included the following features:

1. Hot deck reset control for AHUs
2. Cold deck reset during unoccupied periods for some units
3. Static pressure reset between high and low limits for VAV units
4. Hot water supply temperature control with reset schedule
5. VFD control of chilled water pumps with ΔP set point (no reset schedule)
6. Terminal box level control and monitoring

It was also determined that the facility was being well maintained by the facility operator in accordance with the original design intent. The building is considered energy efficient for a large hospital complex.

The commissioning activities were performed at the terminal box level, AHU level, loop level, and central plant level. Several different types of improved operation measures and energy solutions were implemented in different HVAC systems due to the actual function and usage of the areas and rooms. Each measure will be discussed briefly, starting with the AHUs.

23.6.1.2 Optimization of AHU Operation

EMCS trending complemented by site measurements and use of short-term data loggers found that many supply fans operated above 90% of full speed most of the time. Static pressures were much higher than needed. Wide room temperature swings due to AHU shutoff lead to hot and cold complaints in some areas. Through field measurements and analysis, the following opportunities to improve the operation of the two AHUs were identified:

1. Zone air balancing and determination of new static pressure set points for VFDs
2. Optimize the cold deck temperature set points with reset schedules
3. Optimize the hot deck temperature reset schedules
4. Control of outside air intake and relief dampers during unoccupied periods to reduce ventilation during these periods

5. Optimized time schedule for fans to improve room conditions
6. Improve the preheat temperature set point to avoid unnecessary preheating

Implementation of these measures improved comfort and reduced heating, cooling, and electric use.

23.6.1.3 Optimization at the Terminal Box Level

Field measurements showed that many VAV boxes had minimum flow settings that were higher than necessary, and some boxes were unable to supply adequate hot air due to specific control sequences. A new control logic was developed, which increased hot air capacity by 30% on average, in the full heating mode, and reduced simultaneous heating and cooling. During unoccupied periods, minimum flow settings on VAV boxes were reduced to zero and flow settings were reduced in constant-volume boxes.

During commissioning, it was found that some terminal boxes could not provide the required airflow either before or after the control program modification. Specific problems were identified in about 200 boxes, with most being high flow resistance due to kinked flex ducts.

23.6.1.4 Water Loop Optimization

There are 14 chilled water risers equipped with 28 pumps that provide chilled water to the entire complex. During the commissioning assessment phase, the following were observed:

1. All the riser pumps were equipped with VFDs and they were running from 70% to 100% of full speed.
2. All the manual balancing valves on the risers were only 30%–60% open.
3. The ΔP sensor for each riser was located 10–20 ft from the far-end coil of the AHU on the top floor.
4. Differential pressure set points for each riser ranged from 13 to 26 psi.
5. There is no control valve on the return loop.
6. Although most of the cold deck temperatures were holding well, there were 13 AHUs whose cooling coils were 100% open but could not maintain cold deck temperature set points.

Since the risers are equipped with VFDs, traditional manual balancing techniques are not appropriate. All the risers were rebalanced by initially opening all of the manual balancing valves. The actual pressure requirements were measured for each riser, and it was determined that the ΔP for each riser could be reduced significantly. Pumping power requirements were reduced by more than 40%.

23.6.1.5 Central Plant Measures

Boiler system: Steam pressure was reduced from 125 to 110 psi and one boiler was operated instead of two during summer and swing seasons.

Chilled water loop: Before the commissioning, the blending valve separating the primary and secondary loops at the plant was 100% open. The primary and secondary pumps were both running. The manual valves were partially open for the secondary loop although the secondary loop pumps are equipped with VFDs. After the commissioning assessment and investigations, the following were implemented:

1. Open the manual valves for the secondary loop.
2. Close the blending stations.
3. Shut down the secondary loop pumps.

As a result, the primary loop pumps provide required chilled water flow and pressure to the building entrance for most of the year, and the secondary pumps stay offline for most of the time. The operator drops the online chiller numbers according to the load conditions and the minimum chilled water flow can be maintained to the chillers. At the same time, the chiller efficiency is also increased.

23.6.1.6 *Results*

For the 14-month period following initial CC implementation, measured savings were nearly $410,000, or approximately $30,000/month, for a reduction in both electricity and gas use of about 10%. The contracted cost to meter, monitor, commission, and provide a year's follow-up services was less than $350,000. This cost does not include any time for the facilities operating staff who repaired kinked flex ducts, replaced failed sensors, implemented some of the controls and subroutines, and participated in the commissioning process.

23.6.2 Commissioning Existing Buildings

The savings achieved from commissioning HVAC systems in older buildings are even larger. In addition to the opportunities for improving efficiency similar to those in new buildings, opportunities come from the following:

1. Control changes that have been made to *solve* problems, often resulting in lower operating efficiency
2. Component failures that compromise efficiency without compromising comfort
3. Deferred maintenance that lowers efficiency

Mills et al. (2004, 2005) surveyed 150 existing buildings that had been commissioned and found median energy cost savings of 15%, with savings in one-fourth of the buildings exceeding 29%. Over 60% of the problems corrected were control changes, and another 20% were related to faulty components that prevented proper control.

This suggests that relatively few control systems are operated to achieve the efficiency they are capable of providing.

23.7 Advanced Control System Design Topics

Modern control techniques can offer significant benefits compared to basic control algorithms such as PID. This section provides a short introduction to two approaches that are being used with increasing frequency to improve the stability and performance of HVAC control systems. These include nonlinear compensation and model predictive control (MPC).

23.7.1 Nonlinear Compensation

Most HVAC and refrigeration systems exhibit nonlinear dynamics. The physics of these systems are described by nonlinear thermo-fluid relationships that result in dynamic behavior that changes based on operating condition, external conditions, and controllable inputs. Despite the many factors that contribute to the system nonlinearity, the dominant source tends to be the system actuators that manipulate the flow of mass; compressors, valves, pumps, and fans all are intended to vary the flow of primary/secondary fluids. In contrast, the control objectives are generally posed in terms of regulating the flow or state of energy, such as regulating chiller outlet water temperature or specifying a desired amount of cooling. As mass flow devices, HVAC actuators are typically designed to provide a linear relationship between control input and flow rate. However, the remaining portion of the system generally exhibits a nonlinear relationship between flow rate and system outputs (temperatures, pressures, etc.).

There are many established methods or ways to account for these dynamic nonlinearities. Although these can be approximated with linear models, they are typically only valid for a small operating range. Common approaches to extend model validity include Hammerstein/Wiener models (i.e., identified linear models with a nonlinear input or output function, Figure 23.30) and linear parameter-varying models where parameter values are determined at multiple conditions and then scheduled based on a measured signal:

$$G(s,\theta) = \frac{k(\theta)}{\tau(\theta)s+1} \tag{23.23}$$

Fixed controllers (such as PID) are typically tuned for a single operating condition and thus perform poorly at off-design conditions because the fundamental behavior of the system is different. Oscillations, hunting behavior, and sluggish responses are all symptomatic of this problem. Common nonlinear control techniques include the following:

- *Gain-scheduled control*: An approach where multiple linear controllers are designed to cover the range in operating conditions and then interpolated appropriately (Khalil, 1996). Examples of this approach applied to HVAC systems include Outtagarts et al. (1997) and Finn and Doyle (2000).

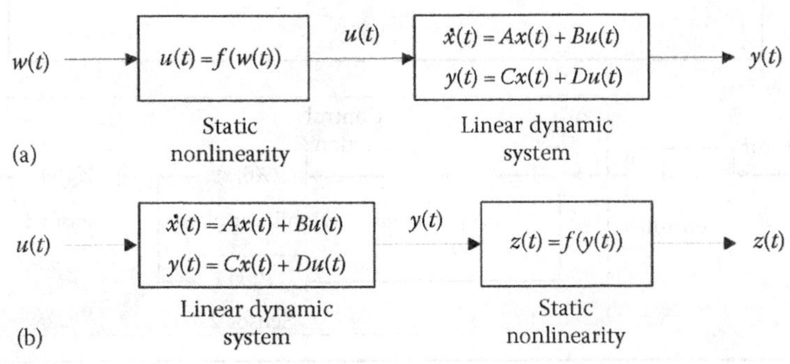

FIGURE 23.30
Nonlinear process models: (a) Hammerstein model and (b) Wiener model.

- *Adaptive control*: A technique for updating the dynamic model using real-time measurements and then using a model-based control strategy that is likewise updated (e.g., Astrom and Wittenmark, 1995).
- *Static nonlinearity compensation*: For systems whose dominant nonlinearity is a change in system gain, for example, $k(\theta)$. Various methods have been proposed, including inverse modeling (Franklin et al., 2006) and nonlinear mapping (Singhal and Salsbury, 2007). The interested reader can find a survey of linearization through feedback in Guardabassi and Savaresi (2001).
- *Cascaded feedback loops*: Using multiple feedback loops has immediate practical benefits in terms of inherent robustness (Skogestad and Postlethwaite, 1996) and can partially compensate for nonlinear system gains (Elliott and Rasmussen, 2010).

23.7.1.1 Case Study: Nonlinear Compensation for Air Conditioning Expansion Valves

In this section, we present an example of a particularly simple and effective technique for compensating for the static nonlinearities present in many HVAC systems. The system being considered is a vapor compression cycle, which is used extensively for air conditioning and refrigeration systems. During operation, an expansion valve modulates the refrigerant flow through the evaporator and regulates the temperature at the evaporator outlet (i.e., superheat). In these systems, superheat must not only be kept low to ensure efficient operation, but also remain high enough to prevent liquid refrigerant from entering the compressor.

As with many HVAC actuators, the relationship between valve input and system outputs is highly nonlinear, with the system gain, $k(\theta)$, varying several hundred percent between low-flow and high-flow conditions. This can lead to poor performance in practice. If fixed-gain controllers (e.g., PID) are designed for the high-flow conditions, then the system will oscillate at low- and medium-flow conditions (i.e., valve hunting). However, if the controller gains are selected based on the low-flow conditions, the performance will be extremely sluggish at medium- and high-flow conditions.

One simple method for compensating for the nonlinear system gain is to utilize cascaded control loops (Figure 23.31). An inner loop controller utilizes a high proportional

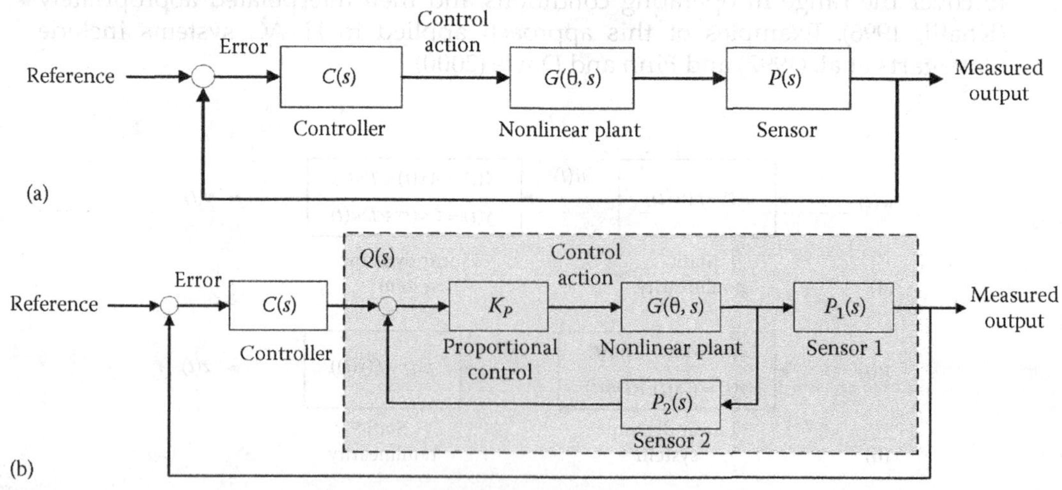

FIGURE 23.31
Feedback control architectures: (a) standard feedback control and (b) cascaded feedback control.

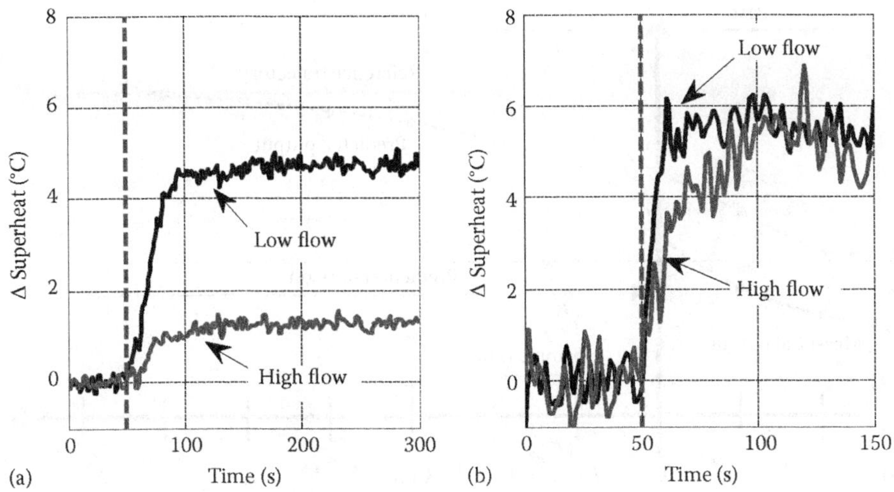

FIGURE 23.32
Superheat response to step change in valve position for high and low flows: (a) standard feedback control and (b) cascaded feedback control. Note the difference in speed of response, as well as the nonlinear compensation.

gain, which significantly lessens the sensitivity of the resulting closed-loop system, $Q(s)$, to changes in the system gain. A standard fixed controller, such as PID, is then designed and implemented in cascade. The resulting system generally exhibits improved performance over a wide range of operating conditions. This approach was advocated by Elliott et al. (2009) and Elliott and Rasmussen (2010) who analyzed the effects for nonlinear electronic expansion valve (EEV) controlled evaporator using temperature and pressure sensors. The resulting step responses under different operating conditions are shown in Figure 23.32 and clearly show that the use of cascaded control loops improves the response time, as well as virtually eliminates the dynamic differences due to system nonlinearities.

23.7.2 Model Predictive Control

MPC is an overarching term for a suite of control strategies first developed for the process industries during the 1970s and 1980s. With the increased computational capabilities of embedded controllers, MPC is increasingly being used in a wide range of control applications. MPCs use an explicit dynamic model to optimize a user-defined cost function. System constraints, such as actuator limitations, can be explicitly accounted for in the optimization, which permits better operation than mere saturation of actuator signals. Additionally, output constraints can be imposed upon the controller, which will keep the system operating in a safe range. Since for many industrial systems, the most efficient operating point is at or near a set of operating constraints, MPC has been successfully implemented as a cost and energy minimization technology (Garcia et al., 1989; Qin and Badgwell, 2003).

MPC is a discrete-time control approach. At each time step, the controller uses a system model to predict outputs over a finite-time prediction horizon (Figure 23.33a) and then determine the optimal choice of the control inputs for a defined control horizon (Figure 23.33b). The controller can also account for the effect of external disturbances, input constraints, and, to some extent, output constraints. The controller then implements the first

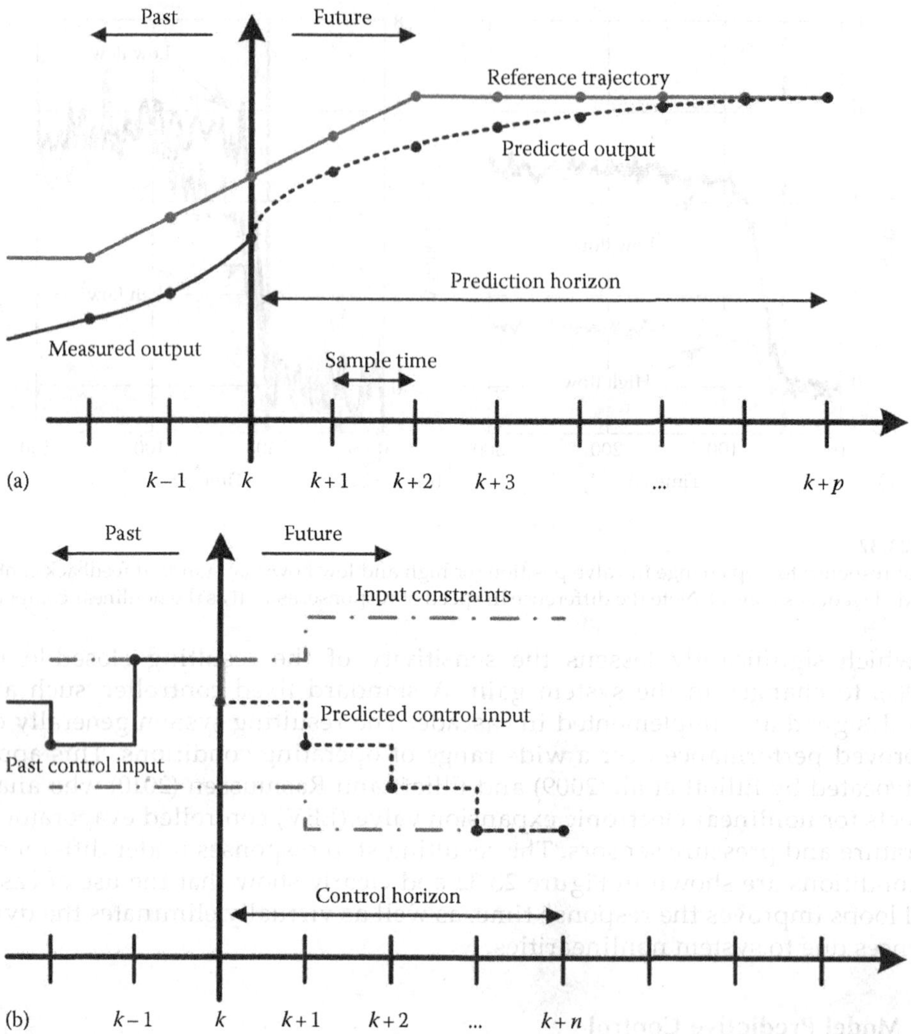

FIGURE 23.33
MPC: (a) prediction horizon and (b) control horizon.

control action in this optimal sequence. At the next time step, the process is repeated, accounting for any new information. The ability to anticipate future outcomes and optimize the response while conforming to constraints is unique to MPC and is not found in basic PID controllers, or even fixed optimal controllers such as the linear quadratic regulator.

The majority of MPC techniques assume a linear model of the system dynamics, usually a discrete-time model based on empirical data. A representative model is defined in Equation 23.24, where x are the dynamic states of the system, u are the control inputs, w and v are the external disturbances, y is the measured output, r is the desired reference, e is the tracking error, and k is the sampling index. The controller uses this model to determine the sequence of control actions that minimizes the cost function $J(x,u)$ (Equation 23.25), without violating the constraints (Equation 23.26). In this formulation, Q and R are matrices that define the relative penalty of tracking errors and actuator effort, p and n are the prediction and control horizons, and T and P are the matrices that define the hard actuator constraints and soft state constraints, respectively:

$$x(k+1) = Ax(k) + B_u u(k) + B_w w(k)$$

$$y(k) = Cx(k) + Dv(k)$$

$$e(k) = r(k) - y(k) \tag{23.24}$$

$$J(x,u) = \sum_0^p e(k)^T Qe(k) + \sum_0^n u(k)^T Ru(k) \tag{23.25}$$

$$Tu \leq b$$

$$Px \leq c \tag{23.26}$$

MPC is attractive for many applications because of its capability to optimize performance while explicitly handling constraints. MPC is a mature technology with extensive research literature addressing issues of stability, constrained feasibility, robustness, and nonlinearities. Since optimal operating conditions often lie at the intersection of constraints, MPC offers a safe way to drive the system to the optimum while not violating performance or actuation limitations. The interested reader is referred to Garcia et al. (1989), Clarke (1994), Kothare et al. (1996), Scokaert and Rawlings (1999), Mayne et al. (2000), Qin and Badgwell (2003), Rossiter (2003), Camacho and Bordons (2004), Allgower (2005), and Rawlings and Mayne (2009), and the references therein for additional details on each of these topics.

Given these advantages, it is not surprising that MPC quickly found success beyond its origins in the chemical industries and is applied with increasing frequency to HVAC systems, including the following:

- Direct application of MPC to basic HVAC components, such as AHUs, VAVs, or heating systems (e.g., MacArthur and Woessner, 1993; He et al., 2005; Yuan and Perez, 2006; Xi et al., 2007; Freire et al., 2008; Huang et al., 2010; Xu et al., 2010; May-Ostendorp et al., 2011; Privara et al., 2011)
- Indirect application as a method of tuning standard HVAC control algorithms (e.g., Dexter and Haves, 1989; MacArthur and Woessner, 1993; Sousa et al., 1997; Xu et al., 2005)
- In the form of the generalized predictive control algorithm, which is perhaps most widely used in HVAC applications (Clarke, 1987)
- Vapor compression systems (e.g., Leducq et al., 2006; Elliott and Rasmussen, 2012, 2013)
- As a supervisory control for HVAC applications (Wang and Ma, 2008)

23.7.2.1 Case Study: MPC for Air Conditioning Expansion Valves

The ability to explicitly account for constraints when determining appropriate control actions is a unique capability of MPC. This capability has numerous potential applications for HVAC equipment. This section presents an experimental example of MPC's constraint handling.

Again we examine the case of expansion valve control for vapor compression systems. Although most valves are designed to regulate superheat to a fixed level, the exact superheat set point is not of particular importance for system efficiency; as long as superheat is kept in a reasonable band, the coefficient of performance (COP) does not vary significantly. By using an MPC-based controller, the superheat can be kept in a band around an optimal point without exerting a large amount of actuator effort, achieving efficient operation while minimizing actuator wear. The following experimental results demonstrate that an MPC controller can be used with an EEV to keep superheat in a band, exerting little effort until superheat begins to leave the defined range.

Figures 23.33 and 23.34 present the results of experiments conducted on a residential air conditioning system. For these tests, an MPC controller is used to keep the evaporator superheat between 5°C and 13°C, with a set point of 9°C. As long as the superheat does not leave the designated band, the controller will not react. While similar tracking results can be had with a very nonaggressive PID controller, the construction of the MPC controller allows the actuator to react very strongly to large disturbances while responding slowly to small disturbances, so long as the user-defined constraints are not violated. This capability is not

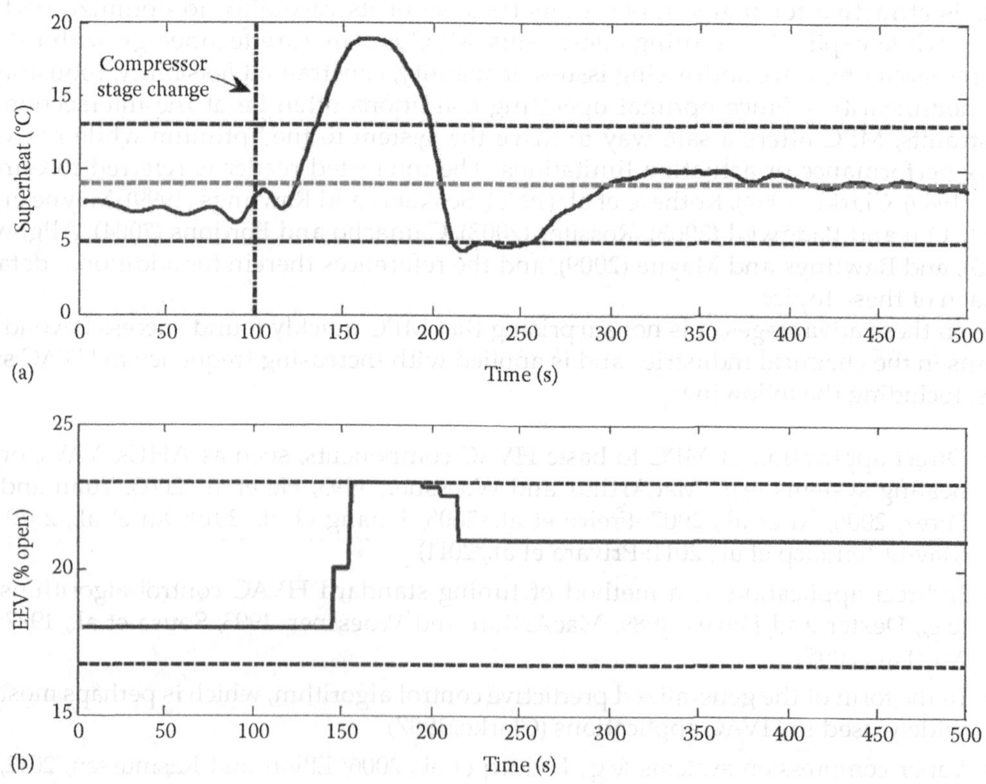

FIGURE 23.34
Superheat response using MPC of the EEV with compressor stage change: (a) superheat response and soft output constraints, (b) EEV actuation with hard input constraints.

available under a PID control paradigm. An illustration of this is provided in the test displayed in Figure 23.33a. At 100 s, the compressor changes from stage 1 (low cooling) to stage 2 (high cooling). As the superheat leaves the prescribed band, the EEV reacts strongly, opening up to the EEV maximum constraint, which is set to 23% open. As the superheat drops down toward its minimum constraint, the valve reacts again by closing, bringing superheat back within the desired band. As Figure 23.33b shows, very little actuator effort is expended, even for such a large disturbance as a compressor step change.

Figure 23.34 displays the system efficiency under two different control paradigms. The left-hand side shows the superheat, actuator effort, evaporator fan speed, and system COP under the same MPC controller as that used in Figure 23.33. The right-hand side shows the same data for a standard PID controller. The evaporator fan speed oscillated as shown, with a period of 20 min. In the MPC-controlled case, the superheat oscillates within the prescribed limits, with the valve only moving as necessary to respect the constraints. In the PID-controlled case, the superheat is kept within a much tighter band, as is expected. The bottom figures show that the mean COP that each control paradigm achieves is virtually the same (4.9 for MPC, 4.8 for PID). Additionally, the MPC controller requires much less motion from the EEV—a total of 14 steps as opposed to 254 steps over 2500 s of test time. Despite the relatively poor set point tracking of the MPC controller, neither system efficiency nor compressor safety is compromised, and the actuator is used much less heavily, implying a longer service life for the valve (Figure 23.35).

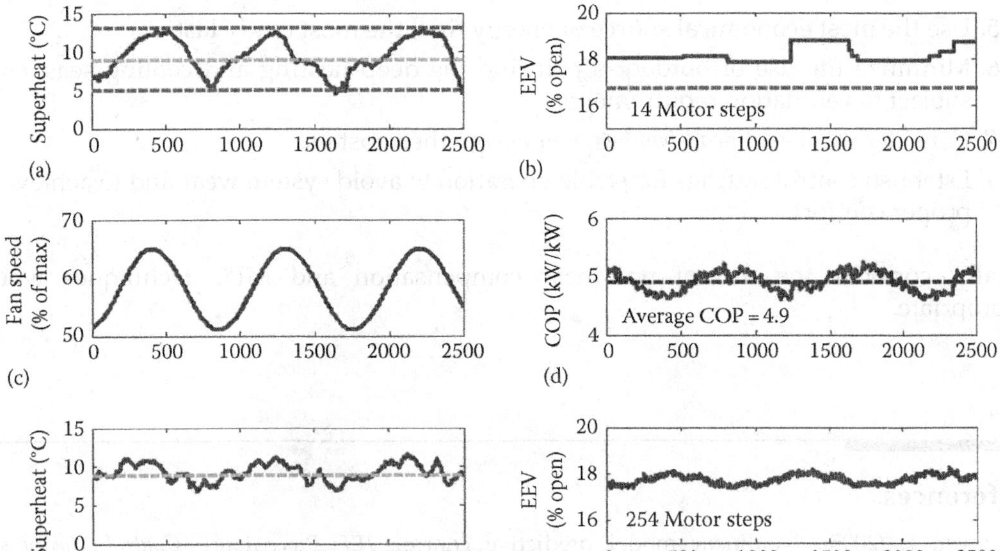

FIGURE 23.35
Comparison of superheat response for MPC- and PID-controlled EEV : (a–d) evaporator superheat, EEV actuation, fan speed, and COP using MPC control; (e–h) evaporator superheat, EEV actuation, fan speed, and COP using standard PID control

23.8 Summary

This chapter has introduced the important features of properly designed control systems for HVAC applications. Sensors, actuators, and control methods have been described. The method for determining control system characteristics either analytically or empirically has been discussed.

The following rules (ASHRAE, 2003) should be followed to ensure that the control system is as energy efficient as possible:

1. Operate HVAC equipment only when the building is occupied or when heat is needed to prevent freezing.
2. Consider the efficacy of night setback vis-à-vis building mass. Massive buildings may not benefit from night setback due to overcapacity needed for the morning pickup load.
3. Do not supply heating and cooling simultaneously. Do not supply humidification and dehumidification at the same time.
4. Reset heating and cooling air or water temperature to provide only the heating or cooling needed.
5. Use the most economical source of energy first, the most costly last.
6. Minimize the use of outdoor air during the deep heating and cooling seasons subject to ventilation requirements.
7. Consider the use of *dead-band* or *zero-energy* thermostats.
8. Establish control settings for stable operation to avoid system wear and to achieve proper comfort.

Finally, consider the use of nonlinear compensation and MPC techniques where appropriate.

References

Allgower, F. (2005). Nonlinear model predictive control. *IEE Proceedings—Control Theory and Applications* **152**(3): 257–258.

ASHRAE (2001). ANSI/ASHRAE standard 135-2001, BACnet: A data communication protocol for building automation and control networks. ASHRAE, Atlanta, GA.

ASHRAE (2002). *HVAC Applications*. ASHRAE, Atlanta, GA.

ASHRAE (2003). *HVAC Systems and Equipment*. ASHRAE, Atlanta, GA.

ASHRAE (2004). *Handbook of Fundamentals*. Chapter 15, Fundamentals of control. ASHRAE, Atlanta, GA.

ASHRAE (2005). Guideline 0, commissioning. ASHRAE, Atlanta, GA.

Astrom, K. J. and B. Wittenmark (1995). *Adaptive Control*. Addison-Wesley, Reading, MA.

Bauman, H. D. (2004). *Control Valve Primer*, 4th edn. Instrument Society of America, North Carolina.

Braun, J. E., S. A. Klein et al. (1989). Applications of optimal control to chilled water systems without storage. *ASHRAE Transactions* **95**(Pt. 1): 663–675.

Camacho, E. F. and C. Bordons (2004). *Model Predictive Control*. Springer, New York.

Claridge, D. E., W. D. Turner et al. (2004). Is commissioning once enough? *Energy Engineering* **101**(4): 7–19.

Clarke, D. (1987). Generalized predictive control—Part I. The basic algorithm. *Automatica* **23**(2): 137–148.

Clarke, D. (1994). *Advances in Model-Based Predictive Control*. Oxford University Press, New York.

Dexter, A. and P. Haves (1989). A robust self-tuning predictive controller for HVAC applications. *ASHRAE Transactions* **95**(2): 431–437.

Elliott, M., B. Bolding et al. (2009). Superheat control: A hybrid approach. *HVAC&R Research* **15**(6): 1021–1044.

Elliott, M. S. and B. P. Rasmussen (2010). On reducing evaporator superheat nonlinearity with control architecture. *International Journal of Refrigeration* **33**(3): 607–614.

Elliott, M. S. and B. P. Rasmussen (2012). Neighbor-communication model predictive control and HVAC systems. *American Control Conference (ACC)*, Montreal, Quebec, Canada, June 27–29, 2012.

Elliott, M. S. and B. P. Rasmussen (2013). Decentralized model predictive control of a multi-evaporator air conditioning system. *Control Engineering Practice* **21**(12): 1665–1677.

Finn, D. P. and C. J. Doyle (2000). Control and optimization issues associated with algorithm-controlled refrigerant throttling devices. *ASHRAE Transactions* **106**(1): 524–533.

Franklin, G. F., J. D. Powell et al. (2006). *Feedback Control of Dynamic Systems*. Pearson Prentice Hall, Upper Saddle River, NJ.

Freire, R., G. Oliveira et al. (2008). Predictive controllers for thermal comfort optimization and energy savings. *Energy and Buildings* **40**: 1353–1365.

Garcia, C. E., D. M. Prett et al. (1989). Model predictive control: Theory and practice—A survey. *Automatica* **25**(3): 335–348.

Grimm, N. R. and R. C. Rosaler (1990). *Handbook of HVAC Design*. McGraw-Hill, New York.

Guardabassi, G. O. and S. M. Savaresi (2001). Approximate linearization via feedback—An overview. *Automatica* **37**: 1–15.

Haines, R. W. (1987). *Control Systems for Heating, Ventilating and Air Conditioning*, 4th edn. Van Nostrand Reinhold, New York.

He, M., W. Cai et al. (2005). Multiple fuzzy model-based temperature predictive control for HVAC systems. *Information Sciences* **169**: 155–174.

Honeywell, Inc. (1988). *Engineering Manual of Automatic Control*. Honeywell, Inc., Minneapolis, MN.

Huang, G., S. Wang et al. (2010). Robust model predictive control of VAV air-handling units concerning uncertainties and constraints. *HVAC&R Research* **16**(1): 15–33.

Huang, P. H. (1991). Humidity measurements and calibration standards. *ASHRAE Transactions* **97**(Pt. 2): 298–304.

Khalil, H. K. (1996). *Nonlinear Systems*. Prentice-Hall Inc., Upper Saddle River, NJ.

Kothare, M., V. Balakrishnan et al. (1996). Robust constrained model predictive control using linear matrix inequalities. *Automatica* **32**(10): 1361–1379.

Leducq, D., J. Guilpart et al. (2006). Non-linear predictive control of a vapour compression cycle. *International Journal of Refrigeration* **29**: 761–772.

Letherman, K. M. (1981). *Automatic Controls for Heating and Air Conditioning*. Pergamon Press, New York.

Levine, W. S. (1996). *The Control Handbook*. CRC Press Inc., Boca Raton, FL.

Liu, M., D. E. Claridge et al. (2002). *Continuous Commissioning SM Guidebook: Maximizing Building Energy Efficiency and Comfort*. Federal Energy Management Program, U.S. Department of Energy, 144pp.

MacArthur, J. and M. Woessner (1993). Receding horizon control: A model-based policy for HVAC applications. *ASHRAE Transactions* **99**(1): 139–148.

May-Ostendorp, P., G. Henze et al. (2011). Model-predictive control of mixed-mode buildings with rule extraction. *Building and Environment* **46**(2): 428–437.

Mayne, D., J. Rawlings et al. (2000). Constrained model predictive control: Stability and optimality. *Automatica* **36**(6): 789–814.

Mills, E., N. Bourassa et al. (2005). The cost-effectiveness of commissioning new and existing commercial buildings: Lessons from 224 buildings. *Proceedings of the 13th National Conference on Building Commissioning*, New York, May 4–6, CD.

Mills, E., H. Friedman et al. (2004). The cost-effectiveness of commercial-buildings commissioning: A meta-analysis of energy and non-energy impacts in existing buildings and new construction in the United States. Lawrence Berkeley National Laboratory Report No. 56637. Available at http://eetd.lbl.gov/emills/PUBS/Cx-Costs-Benefits.html (accessed April 6, 2015).

Outtagarts, A., P. Haberschill et al. (1997). Transient response of an evaporator fed through an electronic expansion valve. *International Journal of Energy Research* **21**(9): 793–807.

Privara, S., J. Siroky et al. (2011). Model predictive control of a building heating system: The first experience. *Energy and Buildings* **43**: 564–572.

Qin, S. and T. Badgwell (2003). A survey of model predictive control technology. *Control Engineering Practice* **11**(7): 733–764.

Rawlings, J. and D. Mayne (2009). *Model Predictive Control: Theory and Design*. Nob Hill, Madison, WI.

Rossiter, J. A. (2003). *Model-Based Predictive Control: A Practical Approach*. CRC Press, Boca Raton, FL.

Sauer, H. J., R. H. Howell et al. (2001). *Principles of Heating, Ventilating and Air Conditioning*. American Society of Heating, Refrigerating and Air-Conditioning Engineers, Inc., Atlanta, GA.

Scokaert, P. and J. Rawlings (1999). Feasibility issues in linear model predictive control. *AIChE Journal* **45**(8): 1649–1659.

Singhal, A. and T. Salsbury (2007). Characterization and cancellation of static nonlinearity in HVAC systems. *ASHRAE Transactions* **113**(1): 391–399.

Skogestad, S. and I. Postlethwaite (1996). *Multivariable Feedback Control: Analysis and Design*. John Wiley & Sons, New York.

Sousa, J., R. Babuska et al. (1997). Fuzzy predictive control applied to an air-conditioning system. *Control Engineering Practice* **5**(10): 1395–1406.

Stein, B. and J. S. Reynolds (2000). *Mechanical and Electrical Equipment for Buildings*. John Wiley & Sons, Inc., New York.

Tao, W. K. Y. and R. R. Janis (2005). *Mechanical and Electrical Systems in Buildings*. Pearson Prentice Hall, Upper Saddle River, NJ.

Wang, S. and Z. Ma (2008). Supervisory and optimal control of building HVAC systems: A review. *HVAC&R Research* **14**(1): 3–32.

Xi, X.-C., A.-N. Poo et al. (2007). Support vector regression model predictive control on a HVAC plant. *Control Engineering Practice* **15**: 897–908.

Xu, M., S. Li et al. (2005). Practical receding-horizon optimization control of the air handling unit in HVAC systems. *Industrial & Chemical Engineering Research* **44**(8): 2848–2855.

Xu, X., S. Wang et al. (2010). Robust MPC for temperature control of air-conditioning systems concerning on constraints and multitype uncertainties. *Building Services Engineering Research & Technology* **31**(1): 39–55.

Yuan, S. and R. Perez (2006). Model predictive control of supply air temperature and outside air intake rate of a VAV air-handling unit. *ASHRAE Transactions* **112**(2): 145–161.

Zhu, Y., M. Liu et al. (2000a). Integrated commissioning for a large medical facility. *Proceedings of the 12th Symposium on Improving Building Systems in Hot and Humid Climates*, San Antonio, TX, May 15–16, 2000.

Zhu, Y., M. Liu et al. (2000b). A simple and quick chilled water loop balancing for variable flow systems. *Proceedings of the 12th Symposium on Improving Building Systems in Hot and Humid Climates*, San Antonio, TX, May 15–16, 2000.

Zhu, Y., M. Liu et al. (2000c). Optimization control strategies for HVAC terminal boxes. *Proceedings of the 12th Symposium on Improving Building Systems in Hot and Humid Climates*, San Antonio, TX, May 15–16, 2000.

24

Stirling Engines

Frank Kreith

CONTENTS

The efficiency of the ideal Sterling cycle is the same as that of a Carnot cycle operating between the same temperatures. The Sterling cycle was originally proposed by a Scottish minister, Reverend Robert Sterling, as an alternative to a steam engine. Sterling engines have recently received increased attention because they can utilize concentrated solar energy, which can be produced by parabolic concentrators. Since a Sterling engine is an external combustion engine, it can use any fuel or concentrated energy. Sterling engines can operate at high temperatures, typically between 600°C and 800°C, resulting in conversion efficiency of 30%–40%. Sterling engines have also recently been developed for cryogenic applications [1], and the advances in that field can be applied to solar-driven Sterling systems.

24.1 Thermodynamics of a Sterling Cycle

Figure 24.1 shows the thermodynamic diagram of an ideal Sterling cycle with a perfect gas as the working fluid. The gas is compressed isothermally (at constant temperature) from state 1 to 2 by means of heat rejection at the low temperature of the cycle, T_L. The gas is then heated at constant volume from state 2 to 3, followed by an isothermal expansion from state 3 to 4. During this expansion process, heat is added at the high temperature in the cycle, T_H. Finally, the gas is cooled at constant volume from T_H to T_L during the process from state 4 to 1. Figure 24.1 shows the thermodynamic diagrams of an ideal Sterling engine on pressure–volume and temperature–entropy diagrams. The cross-hatched areas in the temperature–entropy diagram during the constant-volume process between states 2 and 3 represent the heat addition to the working gas while raising its temperature from T_L to T_H. Similarly, the cross-hatched area between states 4 and 1 represents the heat rejection as the gas is cooled from T_H to T_L. Note that the heat addition from state 2 to 3 is equal

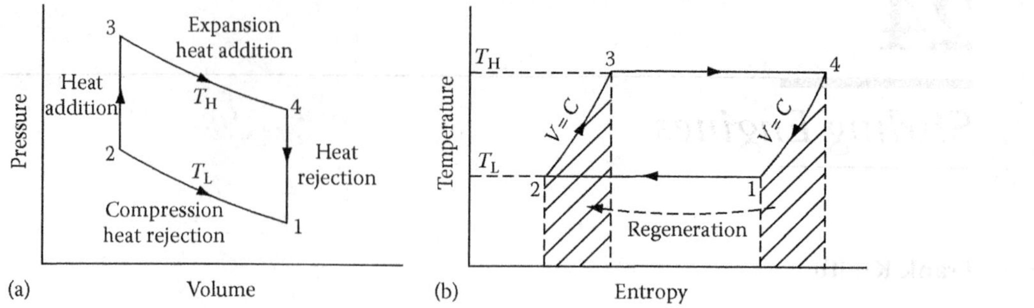

FIGURE 24.1
Thermodynamic diagrams of an ideal Stirling engine cycle. (a) Pressure–volume and (b) temperature–entropy. (From Goswami, Y. et al., *Principles of Solar Engineering*, 2nd edn., Taylor & Francis, Philadelphia, PA, 2000. With permission.)

to the heat rejection from state 4 to 1 and that the processes occur between the same temperature limits. In the ideal cycle, the heat rejected between 4 and 1 is stored and transferred by perfect regeneration to the gas in processes 2–3. Hence, the only external heat addition in the cycle occurs in the process between states 3 and 4, and is given by

$$Q_{3-4} = -W_{3-4} = \int_3^4 p dV = mRT_H \ln \frac{V_4}{V_3}. \tag{24.1}$$

The work input during compression from state 1 to 2 is

$$W_{1-2} = -\int_1^2 p dV = mRT_L \ln \frac{V_1}{V_2}$$

$$= -mRT_L \ln \frac{V_2}{V_1}. \tag{24.2}$$

Noting that the ratio $V_2/V_1 = V_3/V_4$, and combining the earlier two equations, the net work output is

$$mR \ln \frac{V_3}{V_4}(T_H - T_L). \tag{24.3}$$

Therefore, the cycle efficiency

$$\eta = \frac{\text{Net work out}}{\text{Heat input}} = \frac{T_H - T_L}{T_L} \tag{24.4}$$

This efficiency, which is equal to the Carnot cycle efficiency, is based on the assumption that regeneration is perfect, which is not possible in practice. Therefore, the cycle efficiency would be lower than that indicated by the earlier equation. For a regeneration effectiveness *e* as defined later, the efficiency is given by

$$\eta = \frac{T_H - T_L}{T_H + [(1-e)/(k-1)][T_H - T_L/\ln(V_1/V_2)]}, \quad (24.5)$$

where

$e = (T_R - T_L)/(T_H - T_L)$ and T_R = regenerator temperature
$k = C_p/C_v$ for the gas

For perfect regeneration (e = 1), this expression reduces to the Carnot efficiency. It is also seen from the earlier equation that regeneration is not necessary for the cycle to work because even for e = 0, the cycle efficiency is not zero.

Example 24.1

A Stirling engine with air as the working fluid operates at a source temperature of 400°C and a sink temperature of 80°C. The compression ratio is 5.

Assuming perfect regeneration, determine the following:

1. Expansion work
2. Heat input
3. Compression work
4. Efficiency of the machine

If the regenerator temperature is 230°C, determine

5. The regenerator effectiveness
6. Efficiency of the machine
7. If the regeneration effectiveness is zero, what is the efficiency of the machine?

Solution

1. Expansion work per unit mass of the working fluid
Assuming air as an ideal gas,

$$w_{34} = -\int_3^4 p dv = RT_H \ln \frac{v_4}{v_3} = (0.287)(400 - 273)\ln 5 = -310.9 \text{ kJ/kg}.$$

Minus sign shows work output.
2. Heat input per unit mass of the working fluid

$$q_{34} = w_{34} = 310.9 \text{ kJ/kg}.$$

3. Compression work per unit mass of the working fluid

$$w_{12} = -\int_1^2 P dv = RT_L \ln \frac{v_2}{v_1} = -(0.287)(80 + 273)\ln\left(\frac{1}{5}\right) = 163.1 \text{ kJ/kg}.$$

4. Efficiency of the machine

$$\eta = \frac{T_H - T_L}{T_H} = \frac{400-80}{(400-273)} = 0.475 = 47.5\%.$$

5. The regenerator effectiveness

$$e = \frac{T_R - T_L}{T_H - T_L} = \frac{230-80}{400-80} = 0.469.$$

6. Efficiency of the machine

$$\eta = \frac{T_H - T_L}{T_H + \dfrac{(1-e)(T_H - T_L)}{(k-1)\ln(\upsilon_1/\upsilon_2)}} = \frac{400-800}{(400+273) + \dfrac{(1-0)}{(1.4-1)}\dfrac{(400-80)}{\ln(5)}} = 0.341,$$

$$\eta = 34.1\%.$$

7. If the regeneration effectiveness is zero, the efficiency of the machine is

$$\eta = \frac{T_H - T_L}{T_H + \dfrac{(1-e)(T_H - T_L)}{(k-1)\ln(\upsilon_1/\upsilon_2)}} = \frac{400-80}{(400+273) + \dfrac{(1-0)}{(1.4-1)}\dfrac{(400-80)}{\ln(5)}} = 0.273,$$

$$\eta = 27.3\%.$$

24.2 Examples of Solar Stirling Power Systems

In order to understand how the Stirling cycle shown in Figure 24.1 may operate, in practice, the simple arrangement and sequence of processes shown in Figure 24.2 are helpful. In the proposed arrangement, two cylinders with pistons are connected via a porous media, which allows gas to pass through from one cylinder to the other. As the gas passes through the porous media, it exchanges heat with the media. The porous media, therefore, serves as the regenerator. In practice, this arrangement can be realized in three ways, as shown in Figure 24.3, by alpha, beta, and gamma types.

The choice of a working fluid for Stirling engine depends mainly on the thermal conductivity of the gas in order to achieve high heat transfer rates. Air has traditionally been used as the working fluid. Helium has a higher ratio of specific heats (k), which lessens the impact of imperfect regeneration.

In the alpha configuration, there are two cylinders and pistons on either side of a regenerator. Heat is supplied to one cylinder, and cooling is provided to the other. The pistons move at the same speed to provide constant-volume processes. When all the gas has moved to one cylinder, the piston of that cylinder moves with the other remaining fixed to provide expansion or compression. Compression is done in the cold cylinder and

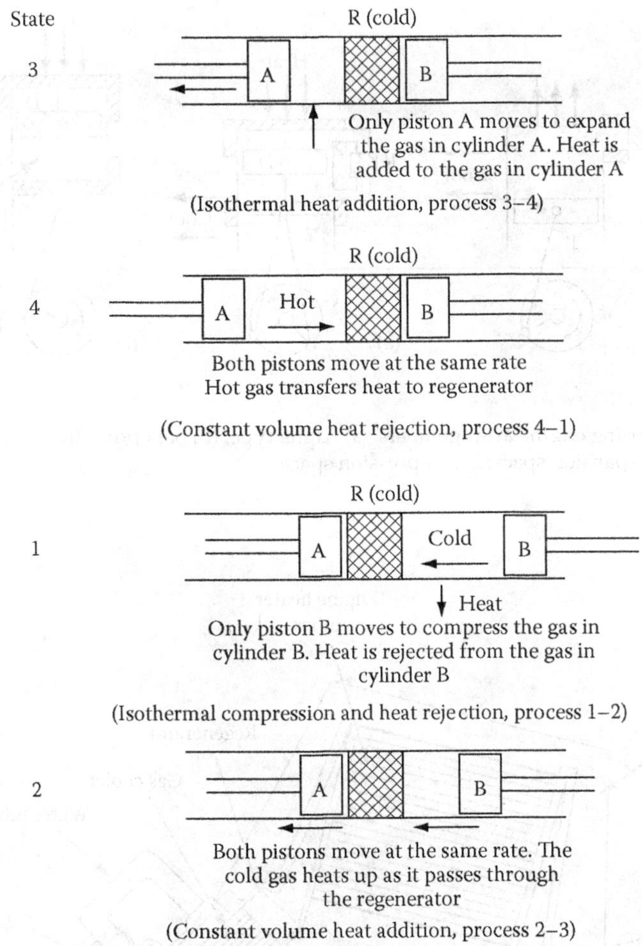

FIGURE 24.2
Stirling cycle states and processes with reference to Figure 24.1.

expansion in the hot cylinder. The Stirling Power Systems V-160 engine (Figure 24.4) is based on an alpha configuration.

The beta configuration has a power piston and a displacer piston, which divides the cylinder into hot and cold sections. The power piston is located on the cold side and compresses the gas when the gas is in the cold side and expands it when it is in the hot side. The original patent of Robert Stirling was based on beta configuration, as are free-piston engines.

The gamma configuration also uses a displacer and a power piston. In this case, the displacer is also the regenerator, which moves gas between the hot and cold ends. In this configuration, the power piston is in a separate cylinder.

In a piston/displacer drive, the power and displacer pistons are designed to move according to a simple harmonic motion to approximate the Stirling cycle. This is done by a crankshaft or a bouncing spring/mass second-order mechanical system [2].

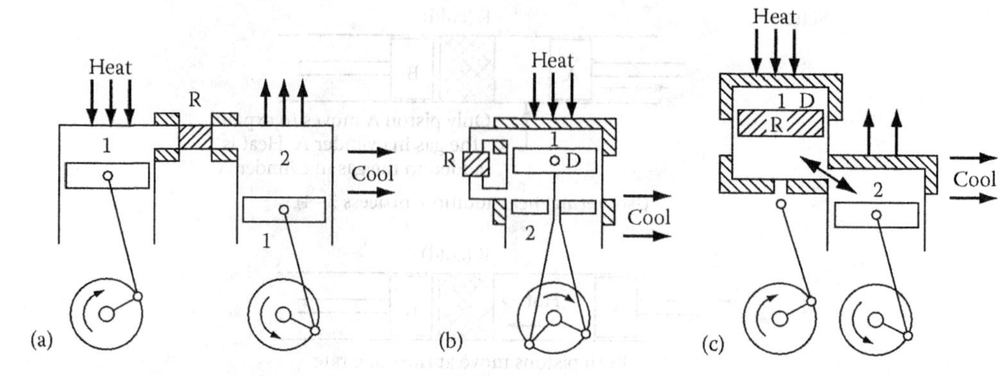

FIGURE 24.3
Three basic types of Stirling engine arrangements. (a) Alpha type, (b) beta type, and (c) gamma type. R, regenerator; D, displacer; 1, expansion space; 2, compression space.

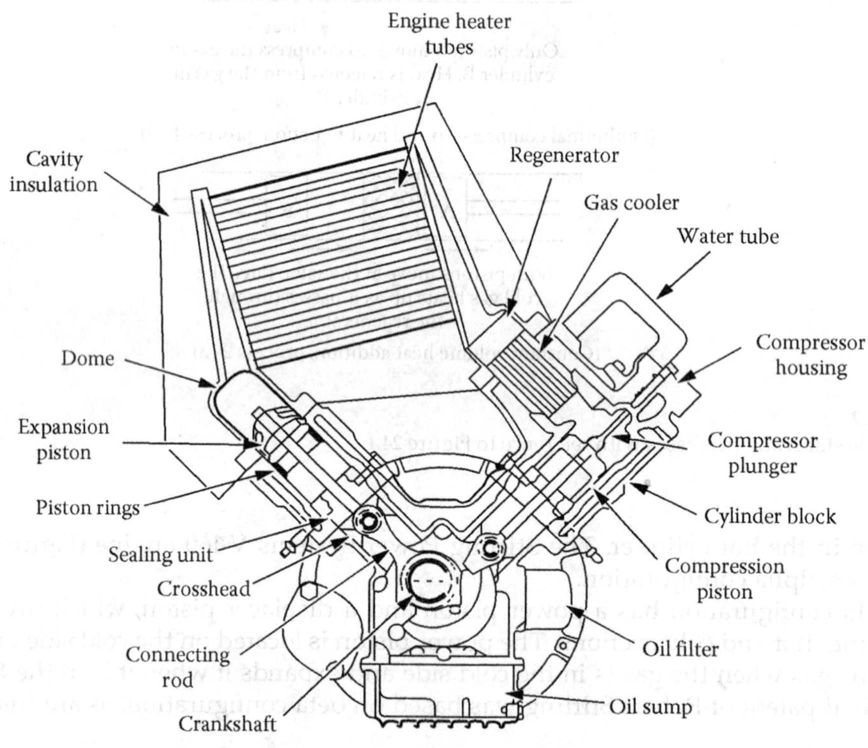

FIGURE 24.4
Stirling power systems/Solo Kleinmotoren V-160 alpha-configuration Stirling engine. (Adapted from Kistler, B.L., A user's manual for DELSOL3: A computer code for calculating the optical performance and optimal system design for solar thermal central receiver plants, Sandia National Labs, Albuquerque, NM, 1986, SAND86-80112.1.)

FIGURE 24.5
The McDonnell Douglas/United Stirling dish–Stirling 25 kW_e module. (From Goswami, Y. et al., *Principles of Solar Engineering*, 2nd edn., Taylor & Francis, Philadelphia, PA, 2000. With permission.)

In a kinematic engine, the power piston is connected to the output shaft by a connecting rod crankshaft arrangement. Free-piston arrangement is an innovative way to realize the Stirling cycle. In this arrangement, the power piston is not connected physically to an output shaft. The piston bounces between the working gas space and a spring (usually a gas spring). The displacer is also usually free to bounce. This configuration is called the Beale free-piston Stirling engine after its inventor, William Beale [2]. Since a free-piston Stirling engine has only two moving parts, it offers the potential of simplicity, low cost, and reliability. Moreover, if the power piston is made magnetic, it can generate current in the stationary conducting coil around the engine as it moves. This is the principle of the free piston/linear alternator in which the output from the engine is electricity.

Stirling engines can provide very high efficiencies with high-concentration solar collectors. Since practical considerations limit the Stirling engines to relatively small sizes, a Stirling engine fixed at the focal point of a dual tracking parabolic dish provides an optimum match, as shown in Figure 24.5. Therefore, all of the commercial solar developments to date have been in parabolic dish–Stirling engine combination. The differences in the commercial systems have been in the construction of the dish and the type of Stirling engine. A thorough description of past and current dish–Stirling technologies is presented in Ref. [3].

References

1. www.stirlingcryogenics.com (accessed March 31, 2015).
2. Stine, W.B. (1998) Stirling engines, in: *The CRC Handbook of Mechanical Engineering*, F. Kreith. (Ed.), CRC Press, Boca Raton, FL.
3. Mancini, T. et al. (May 2003) Dish-stirling systems: An overview of development and status, *JSEE* 125, 135–151.
4. Kistler, B.L. (1986) A user's manual for DELSOL3: A computer code for calculating the optical performance and optimal system design for solar thermal central receiver plants, SAND86-80112.1. Sandia National Labs, Albuquerque, NM.
5. Goswami, Y. et al. (2000) *Principles of Solar Engineering*, 2nd edn., Taylor & Francis, Philadelphia, PA.

25

Energy-Efficient Lighting Technologies and Their Applications in the Residential and Commercial Sectors

Karina Garbesi, Brian F. Gerke, Andrea L. Alstone, Barbara Atkinson, Alex J. Valenti, and Vagelis Vossos

CONTENTS

25.1 Introduction

Electricity for lighting accounts for approximately 15% of global power consumption and 5% of worldwide greenhouse gas (GHG) emissions [1]. According to the United Nations Environment Programme's en.lighten initiative,* replacing all inefficient grid-connected lighting globally would reduce global electricity consumption by approximately 5% and CO_2 emissions by 490 million tons per year [2]. Grid-connected *space* lighting accounts for 99% of total lighting energy use, with vehicle lighting (0.9%) and off-grid fuel-based lighting (0.1%) constituting the remainder [3]. Together, residential and commercial lighting constitute about three quarters of global grid-connected lighting (Figure 25.1).

Per-capita consumption of artificial light averages about 20 megalumen-hours per year globally but varies widely. North Americans consume about 100 megalumen-hours per year on average. By comparison, Chinese consume about 10, and Indians consume only 3, megalumen-hours per year [3]. Given global demographic trends, lighting efficiency improvements not only offer significant energy savings in developed countries, but more importantly, the opportunity to contain large growth in lighting energy use in less developed countries.

Figure 25.2 shows the historic efficacies of different light source technologies (solid lines). Incandescent lamps, including halogen incandescent lamps, are the lowest efficiency light source type. (Lighting efficiency, referred to as efficacy, is quantified in units of lm/W.)

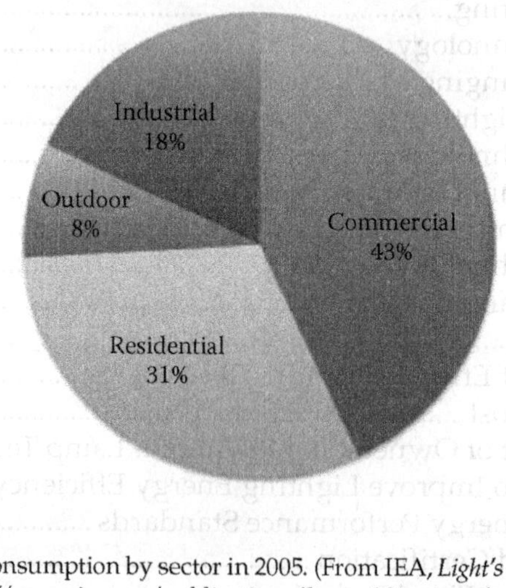

FIGURE 25.1
Global grid-connected lighting consumption by sector in 2005. (From IEA, *Light's Labour's Lost*, International Energy Agency, Paris, France, 2006, http://www.iea.org/publications/freepublications/publication/light2006.pdf.)

* The en.lighten initiative was created in 2009 as a partnership between the United Nations Environment Programme, OSRAM AG, Philips Lighting, and the National Lighting Test Centre of China, with the support of the Global Environment Facility.

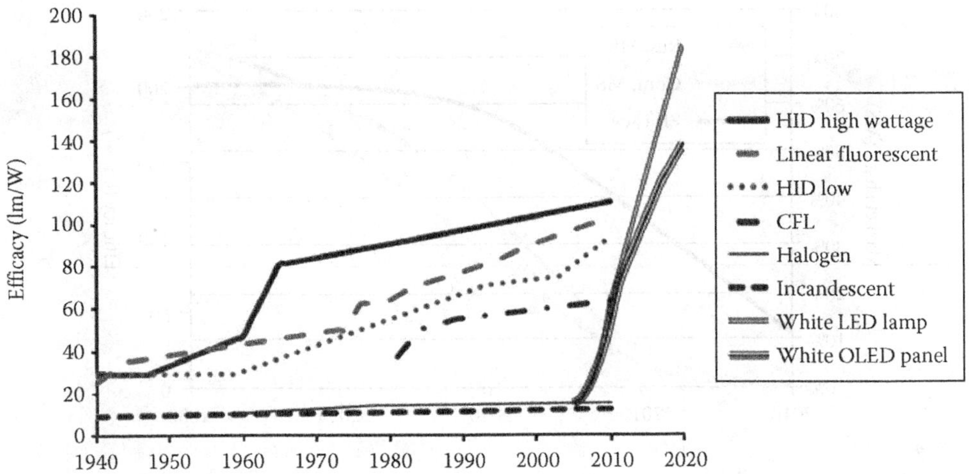

FIGURE 25.2
Historic best-on-market lamp efficacy trends by light source and lamp type and projected trends for solid-state lighting. (From DOE, Solid-state lighting research and development: Multi-year Program Plan, U.S. Department of Energy, Energy Efficiency & Renewable Energy Building Technologies Program, Washington, DC, April 2012.)

Compact fluorescent lamps (CFLs) are the next highest with efficacies approximately four times those of incandescent lamps. Linear fluorescents have trended with an efficacy intermediate that of low and high wattage high-intensity discharge (HID) lamps. Light-emitting diodes (LEDs) are now starting to exceed the efficacy of fluorescent lamps, and expected to reach about 200 lm/W by 2020—approximately twice that of today's most efficacious conventional lighting technologies.

Up to the present, different lighting technologies have dominated different sectors. For example, in 2010 in the United States, an estimated 86% of outdoor lighting was provided by HID lamps, 72% of commercial lighting was provided by linear fluorescent lamps, and 79% of residential lighting was provided by incandescent lamps [5]. Given the large differences in efficiency of the different light sources (Figure 25.2), this has resulted in large sectoral differences in efficiency. Solid-state lighting is poised to change that equation with LED rapidly entering all sectors of the lighting market.

LEDs are expected to replace essentially all other lighting technologies in the coming decades. Navigant Research anticipates LED's global market share of replacement lamps to grow from 5% in 2013 to 63% by 2021 [6]. The U.S. Department of Energy expects LEDs to represent 70% of the U.S. residential and commercial lighting markets by 2030 (Figure 25.3). In combination with an expected doubling in LED lamp efficacy anticipated by 2020 (also shown in Figure 25.3), the total energy savings are expected to be very large. This is particularly true in the residential sector, where the dominant incumbent technology (incandescent lamps) has an efficacy of only about 15 lm/W. In addition to the trend toward LED lighting, increased use of lighting controls is expected to further reduce energy use.

Because LED lights are significantly longer lived than conventional residential and commercial lamps, global lamp sales are expected to peak and decline as a result of the LED takeover of global lighting markets. Table 25.1 compares the typical lifetimes of various lamp technologies currently on the market. Figure 25.4 shows projected revenues from technology-specific lamps sales through 2021. This implies that, not only will consumers

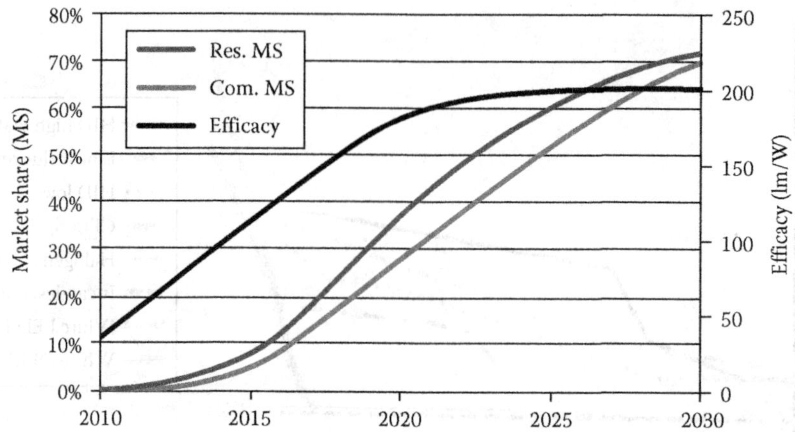

FIGURE 25.3
The projected market share of LED lamps in the U.S. residential and commercial markets and projected LED lamp efficacy. (From Navigant Consulting, Inc., Energy savings potential of solid-state lighting in general illumination applications, U.S. Department of Energy, Solid State Lighting Program, Navigant Consulting, Inc., January 2012, http://apps1.eere.energy.gov/buildings/publications/pdfs/ssl/ssl_energy-savings-report_10-30.pdf.)

TABLE 25.1
Typical Lifetimes and Approximate Best-on-Market Efficacies in 2012, by Lamp Type

Lamp Type	Lifetime (h)	Efficacy (lm/W)
Incandescent	1,000	15
Halogen incandescent	5,000	22
Compact fluorescent	12,000	65
Linear fluorescent	25,000	120
HID (high wattage)	30,000	120
LED (commercial)	50,000	130
LED (residential)	25,000	100

Sources: DOE, Solid-state lighting research and development: Multi-year program plan, U.S. Department of Energy, Energy Efficiency & Renewable Energy Building Technologies Program, Washington, DC, April 2012; The U.S. Department of Energy's LED Lighting Facts Database, http://www.lightingfacts.com/Products.

save money as a result of efficiency-induced electricity savings resulting from increasing use of LEDs, they will also save as a result of reduced lamps purchases.

This paper examines these lighting efficiency trends, focusing on opportunities in the residential and commercial sectors, which includes indoor and outdoor lighting associated with residential and commercial buildings. Because there is some sectoral overlap in lighting technologies, much of the technology discussion also applies to stationary source applications for industrial and roadway lighting. Section 25.2 introduces lighting concepts used throughout the remainder of the chapter. Section 25.3 discusses the different major lamp technologies by light source type, focusing on factors that affect their energy efficiency and related market trends. Section 25.4 discusses the cost-effectiveness of the different light source technologies from both the societal and consumer perspectives. Section 25.5 describes policies to stimulate energy lighting efficiency improvements. Section 25.6 presents conclusions.

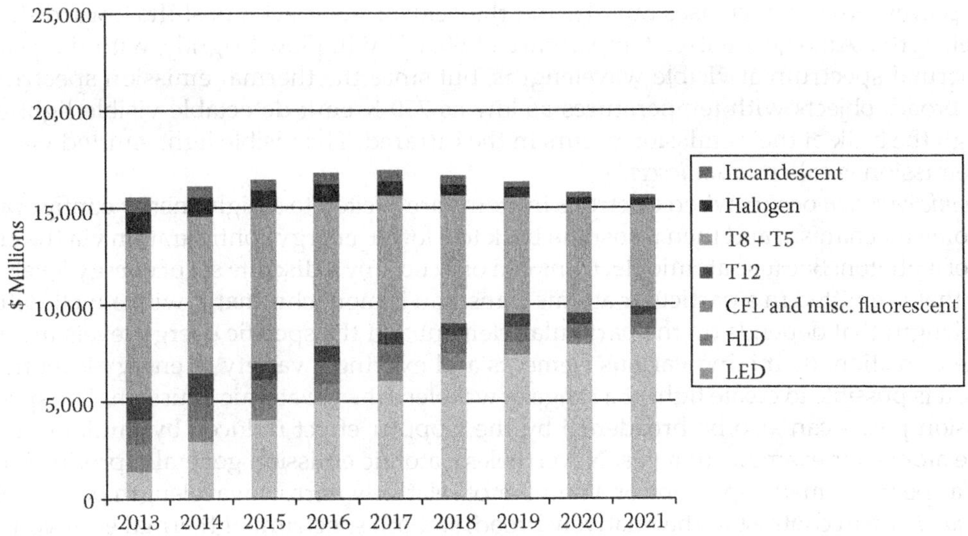

FIGURE 25.4
Projected revenues in world lamps market by technology type. The bar segments in each bar are ordered from top to bottom at shown in the legend. (From Navigant Research, Inc., Energy efficient lighting for commercial markets—LED lighting adoption and global outlook for LED, fluorescent, halogen and HID lamps and luminaires in commercial buildings: Market analysis and forecasts (Executive Summary), Navigant Consulting, 2013, http://www.navigantresearch.com/research/energy-efficient-lighting-for-commercial-markets. With permission.)

25.2 Lighting Concepts

25.2.1 Principles of Light Production

Visible light is made up of electromagnetic waves whose wavelengths fall within a narrow band of the electromagnetic spectrum, between 390 and 700 nm. The perceived color of visible light corresponds to its wavelength, with violet and blue light having shorter wavelengths, while orange and red light have longer wavelengths. Light with wavelengths longer than visible light, from 700 nm up to 1 mm, is called *infrared radiation* and is generally perceived as heat. Light with wavelengths shorter than visible light, from 390 nm down to 10 nm is called *ultraviolet radiation*. On an atomic level, light emission occurs in discrete units called *photons*, which are wave packets that carry energy that is inversely proportional to their wavelength.

Common electric light sources use one of two basic physical mechanisms to produce visible light: thermal and atomic emission. The mechanism that has been used longest is *thermal emission*, which is electromagnetic radiation emitted by all objects with temperatures above absolute zero. The spectrum of thermal emission for most objects is very well approximated by the spectrum from a *blackbody*, which is a theoretical object that perfectly absorbs all incident light and emits only thermal radiation. *Blackbody radiation* has a broad spectrum, with an intensity that scales as the fourth power of the object's temperature and a peak wavelength that varies inversely as the temperature. Objects at room temperature emit most of their thermal radiation in the infrared part of the spectrum, and the visible light emitted is too faint to be readily detected. As temperature is increased, the total power radiated increases rapidly, and the peak of the spectrum shifts toward shorter wavelengths. An object with a temperature of 5000 K will glow brightly, with the peak of its thermal spectrum at visible wavelengths, but since the thermal emission spectrum is very broad, objects with temperatures as low as 750 K emit detectable visible light even though the bulk of their emission occurs in the infrared. The visible light emitted via thermal emission is called *incandescence.*

Atomic emission occurs when electrons in atoms are excited to a high-energy configuration via some mechanism and then transition back to a lower-energy configuration via the emission of a photon. Because atomic electrons can only occupy a discrete set of energy levels, the light that is emitted in a particular atomic transition is monochromatic, with a well-defined wavelength that depends on the particular element and the specific energy levels involved in the transition. By mixing various elements and exciting a variety of energy-level transitions, it is possible to create light at a range of wavelengths via atomic emission. The spectral emission peaks can also be broadened by the Doppler effect induced by random motion of the atoms, for example, in a gas. Nevertheless, atomic emission generally produces light with a spectrum made up of one or a number of relatively narrow wavelength peaks (called *emission lines*) in contrast to the relatively broad, smooth spectrum of thermal emission.

Figure 25.5 shows example spectra for blackbody radiation, atomic emission, and, for reference, daylight. Solid lines show the theoretical spectrum of a blackbody at various different temperatures (from bottom to top, 3750 K, 4500 K, 5250 K, and 6000 K). As the temperature rises, the source intensity increases, and the spectral peak shifts to shorter wavelengths. The filled light gray region shows the spectrum of sunlight, whose broad, smooth shape is similar to a blackbody spectrum. The filled dark gray region shows atomic emission from a mercury vapor (MV) lamp, characterized by narrow peaks (emission lines) at characteristic wavelengths. The curves plotted are not on the same scale, except for the relative intensities of the different blackbody temperatures. (Spectral data for sunlight and MV lamp taken from McNamara [8].)

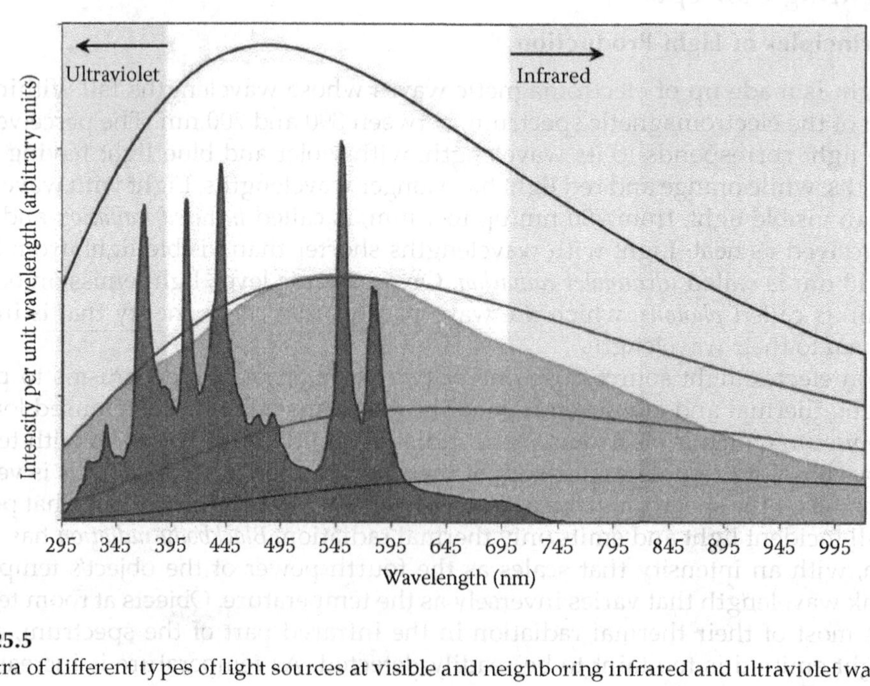

FIGURE 25.5
The spectra of different types of light sources at visible and neighboring infrared and ultraviolet wavelengths.

Atomic emission can take many forms, depending on the mechanism by which the electrons are excited. *Fluorescence* occurs when an atom absorbs light at a particular wavelength, which excites an electron, and then emits light at the same or a different wavelength as that electron relaxes to some lower energy level. If the electron relaxes to a higher energy than it had initially, the emitted photon will have a longer wavelength than the absorbed photon; this mechanism is commonly used to convert ultraviolet to visible light, for example, in fluorescent lamps. *Plasma discharge emission* occurs when an electric current passes through plasma (i.e., an ionized gas) in a phenomenon known as an *electric arc*. The free electrons collide with atoms in the gas and excite their electrons, which then transition to lower energy levels and emit photons. This is the primary light-generation method for fluorescent and HID lamps. *Diode emission* from a LED occurs when an electrical current is applied to a semiconductor made of two adjoining materials with different electrical properties. Excited electrons, which are free to move through the semiconducting material, are added to one material and removed from the other by the current. Those electrons that are conducted across the junction between the two materials then relax to the empty energy levels in the electron-depleted atoms, emitting photons in the process.

25.2.2 Design of Energy-Efficient Lighting Systems

Residential and commercial lighting systems typically are designed to provide both functional lighting and aesthetic value at the same time. The aim of energy-efficient lighting design is to provide the desired quality of light and other aesthetic properties of the lighting system with the least energy use. Design choices made for aesthetic reasons may in some cases reduce efficiency. For example, adding a diffuser to a luminaire to reduce glare will reduce luminaire efficacy. In other cases, these choices may improve efficiency. For examples, lighting controls used to avoid the delivery of undesirable excess light are aesthetically preferable and save energy.

The goal of energy-efficient lighting design is to provide only as much artificial light as is desirable at the locations where it is needed, when it is needed. The optimal system incorporates daylighting, provides proper illumination for the design application, and addresses changes in occupancy. In the future, systems may even modulate the spectral distribution throughout the day to improve human performance and well-being (see Section 25.2.3).

The key elements of efficient lighting design are (1) making maximum use of natural light, (2) using the most efficient artificial lighting technologies, (3) incorporating automated lighting controls, and (4) controlling high-illumination task lighting separately from lower-illumination ambient lighting. While energy-efficient lighting technologies and controls are addressed at length in the following, this section provides additional background on the benefits and pitfalls of daylighting. Proper daylighting design requires climate-sensitive building design to avoid potential adverse consequences. Specifically, excessive direct penetration of sunlight can result in discomfort from glare and heat loading and increase cooling energy demand, which could offset or outweigh lighting energy savings. On the other hand, because energy-efficient lights produce less heat, they can reduce cooling energy demand, which is an advantage in regions where peak electric loads are dominated by cooling. Achieving an optimal balance requires building energy modeling.

The potential to offset the need for artificial light with daylight is greatly affected by the overall architectural design of a building and the design of the interior space. Passive solar building design, which optimizes building orientation and window placement and minimizes extreme thermal excursions using appropriately located shading devices, thermal mass, and phase change materials, can facilitate daylighting while avoiding excessive heat loading. Skylights, reflective surfaces inside of and outside of the building, and other measures can be used to bounce natural light farther into a building. Light pipes can be used to provide natural light deeper to the interior. Light pipes may even incorporate a hybrid design that includes artificial lights when natural light is insufficient. Advanced window technologies with dynamic solar control can limit heat losses in winter and avoid excessive heat loading in summer while allowing natural light to enter buildings. The newest technologies incorporate zonal controls that can block glare to building occupants while allowing unaltered light to pass where it will not cause discomfort to occupants. Smart blinds are being developed that automatically adjust to address light level, glare, and thermal loading. Interior design and furnishings that place high light-need activities near windows and that allows the maximum penetration of light from windows into the building interior, can also reduce the need for artificial light during the day.

Another environmentally promising design option that has recently emerged in the market is solar-assisted lighting—products that have integrated photovoltaic (PV) power supply into lighting systems. NexTek Power Systems pioneered this approach with efficient commercial lighting that runs directly off of the direct current (DC) generated by the PV system when PV power is available, and off of rectified alternating current (AC) power when it is not. The approach avoids DC to AC to DC power conversion losses as well as allowing for flexible lighting design. The lights run off of a 24 V bus integrated into the metal grid that supports ceiling tiles in standard commercial drop ceiling (Armstrong Ceiling's DC Flexzone system). The design allows lights to be removed and installed without an electrician and to be placed anywhere on the grid, reducing the cost and greatly enhancing the flexibility of commercial lighting design. At the opposite end of the spectrum, small solar-assisted lighting systems that include a small PV module, a battery, and LED lights (e.g., Fenix International's ReadySet Solar Kit) are now in the market. The systems were designed for developing country markets. While particularly appropriate for off-grid areas, they may also become popular in the many regions with highly unreliable grid power.

25.2.3 Lighting, Human Health, and the Environment

Exposure to light, both natural and artificial, has both beneficial and adverse effects on human health that depend strongly on the wavelength of the light, the timing of exposure, and the organs exposed [9]. Exposure to the natural diurnal cycle of darkness and light promotes mental and physical health by maintaining the body's circadian rhythms [9]. Exposure to blue light in the morning triggers the production of an array of hormones that control stress and impulsive behavior, hunger, and metabolism, and promote reproductive health and alertness. Lack of circadian blue light in the morning can cause sadness, food cravings, decreased energy and libido, anxiety, and depression. The risks of cancer and infectious disease are also increased due to insufficient blue light exposure in the morning and the absence of darkness at night because of disruption of immune response pathways that destroy viruses and cancer.

In addition to being damaging to vision [10], exposure to blue light at night is particularly disruptive to well-being [11–13]. Exposure to darkness stimulates the body's production of hormones that promote sleep, reduce blood pressure, and cellular repair. Eliminating the exposure of geriatric and Alzheimer patients to blue light at night was demonstrated to improve sleep, reduce illness and blood pressure, reduce aggression, and improve lucidity and social participation in both populations [14].

While to date, lighting design has tended to ignore the health impacts of light color, avoidance of light pollution has become a major issue. The concept is that light should hit its intended target, and it should not shine elsewhere. The BUG System, which quantifies *backlight, uplight,* and *glare,* is a system that quantifies undesirable stray light from outdoor lighting [15]. The system was adopted into the Illuminating Engineering Society's Model Lighting Ordinance [16].

While high-end lighting designers have begun in recent years to incorporate spectral dynamics in their lighting systems to alter the mood of spaces throughout the day, off-the-shelf technologies are now becoming available, enabling residential customers to control the hue of lighting. For example, the new Philips Hue lighting system is designed to be controlled by the Apple iPhone. The technologies are likely to become more popular with increased awareness of the impact of lighting on human health and productivity.

25.3 Lighting Technologies

The following sections discuss the technologies and changing markets of the four major lamp types: incandescent lamps, including halogen (Section 25.3.1); fluorescent lamps (Section 25.3.2), plasma lamps (Section 25.3.3), and LEDs and organic light-emitting diodes (OLEDs) (Section 25.3.4). Following that, two additional technologies are described in the context of their impact on energy efficiency: lighting controls (Section 25.3.5) and luminaires (Section 25.3.6).

25.3.1 Incandescent

An *incandescent lamp* produces light by passing an electric current through a conductive filament, which is heated by electrical resistance to the point of incandescence. The filament is generally housed in a glass vessel (bulb) that has been evacuated and typically filled with an inert gas. The bulb typically has a metal base that can be attached to an electrical fixture, most often via a screw-in connection.

25.3.1.1 Technology

Incandescent lamps are the oldest electric lighting technology, having been invented independently by Thomas Edison in the United States and Joseph Swann in England in the late 1800s. In modern lamps, the filament is made of tungsten, although filaments of carbon, tantalum, and osmium were used in early incandescent lamps produced around the turn of the twentieth century. (Upon their introduction, tungsten filaments represented a substantial improvement in efficacy over these earlier filament materials.) Incandescent lamp filaments produce light whose spectrum is well approximated by a blackbody spectrum. Typical incandescent lamps have operating temperatures at which the overwhelming majority of the radiated power (typically 90% or more) falls in the infrared part of the spectrum, where it is sensed as radiant heat rather than visible light. This makes incandescent lamps significantly less efficacious than other lamp types. Nevertheless, incandescent lamps have remained very popular, especially in residential applications.

The continuing popularity of incandescent lamps owes partly to their familiarity but also to a wide range of desirable features. Incandescent lamps have warm color temperatures and excellent color rendering index. They are easily dimmed with no adverse effect on the lamp. Compared with other lighting technologies, incandescent lamps are inexpensive, small, lightweight, and can be used with inexpensive fixtures with no need for ballasts or other controllers. They work equally well on AC or DC power, and they have no need for high-quality power input. In a properly designed fixture, they permit excellent optical control. In addition, incandescent lamps are easy to install and maintain, produce no audible noise, create no electromagnetic interference, and contain few toxic chemicals, allowing their disposal in the general waste stream.

The two primary types of incandescent lamps are general-service and reflector lamps. General-service lamps (also known as A-lamps) are pear-shaped, common household lamps. Reflector lamps are typically conical in shape with a reflective coating applied to part of the bulb surface that has been specially contoured to control the light direction and distribution. Common types of reflector lamps include flood and spot lights, which are often used to illuminate outdoor areas, highlight indoor retail displays and artwork, and improve the optical efficiency of track lights or downlights.

Halogen lamps operate on the same principle as standard incandescent lamps, by heating a filament to the point of incandescence, but they are made more efficacious by the utilization of the tungsten-halogen cycle. During operation of a standard incandescent lamp, tungsten is evaporated from the filament and deposited on the interior surface of the glass envelope. This process blackens the bulb over time and causes the filament to grow thinner until it eventually fails. Operating the lamp at a higher temperature would accelerate this process. In a halogen lamp, the filament is surrounded by quartz capsule filled with a small amount of halogen gas, such as iodine or bromine. At moderate temperatures, the halogen gas binds to the tungsten that has evaporated from the filament, preventing deposition onto the quartz capsule. At the higher temperatures found in the vicinity of the filament, the tungsten-halogen bond is broken, and the tungsten is redeposited onto the filament. This tungsten-halogen cycle thus allows halogen lamps to be operated at higher temperatures without adversely affecting their lifetimes (indeed, many halogen lamps have longer lifetimes than standard incandescent lamps). The higher-temperature blackbody spectrum thus emitted by the filament has a larger overlap with the response curve of the human eye, so that more lumens are produced at a fixed power draw, resulting in a higher luminous efficacy.

Even more efficacious than the standard tungsten-halogen lamp is the *halogen infrared reflecting* (*HIR*) *lamp*. In such lamps, the halogen capsule, or the lamp reflector, is coated with an optically transparent but infrared-reflective coating, which reflects some portion of the infrared radiation back onto the filament. This increases the operating temperature of the filament without the need for additional watts of electrical power. Because the increased temperature provides more lumens at a fixed wattage, HIR lamps have a higher luminous efficacy than non-HIR halogens. HIR lamps have historically had small market share due to their high initial cost, even though they generally have lifetimes that are two to three times longer than standard halogens.

Because they are operating at higher temperature, halogen lamps produce bright white light with color temperatures slightly higher than those of standard incandescent lamps, with similarly high CRI values. In addition, they tend to have longer rated lifetimes, can be much more compact, are slightly more efficacious, and have better lumen maintenance than standard incandescent lamps. Halogen technology has historically been used heavily in reflector lamps; minimum efficiency standards have led them to capture an increasing share of this market in the United States and elsewhere. General-service halogen lamps are also available in the market. Historically, these have seen limited use, but recent improvements in their efficacy coupled with new regulatory standards are expected to increase their adoption in the near term.

25.3.1.2 Changing Market

General-service incandescent lamps have long been the workhorse of residential lighting applications, so much so that they are commonly referred to as household light bulbs. Incandescent and halogen reflector lamps have also been widely used in the residential sector, especially for outdoor lighting and directional indoor lighting. They have also seen significant use in the commercial sector in applications that require directional lighting, such as retail display.

There are now a variety of significantly more efficacious substitutes for incandescent and halogen lamp technology. The past two decades have seen a substantial portion of the market for general-service incandescent lamps shift to CFL technology. However, CFLs' unfamiliar appearance, higher price, inferior CRI, and limited capacity for dimming have limited the further growth of their market share. In recent years, LED-based lamps are entering the market intended to be direct replacements for many incandescent lamps. The directional nature of LED sources lends itself to reflector-lamp applications, and such lamps have gained a small foothold in the commercial sector. Recent advances in omnidirectional LED lamps have produced products with similar efficacy and color rendering to CFLs; these can be used as replacements for general-service incandescent lamps. Their very high prices have significantly limited their adoption to date, but as discussed in Section 25.4, LED prices have been falling rapidly. As discussed in Section 25.3.4, the color rendering and efficacy of LED lamps have simultaneously been improving, and LED lamps are expected to capture much of the current market for incandescent lamps over the coming decades [7]. As mentioned earlier, recent efficiency regulations in many markets are phasing out traditional general-service incandescent lamps; so a faster shift toward halogen, CFL, and LED technologies is expected in the general-service market.

25.3.2 Fluorescent

A typical fluorescent lamp system consists of a lamp and an electrical regulating ballast. The lamp is comprised of a glass tube, two electrodes, and a lamp base. The glass tube contains low pressure MV and an inert gas, which is usually argon, and a phosphor coating on the inside of the tube, which determines the light spectrum of the fluorescent lamp and thus its CCT and CRI. The electrodes are made of tungsten wire and are coated with a mix of alkaline oxides that, when heated, emit the electrons that excite and ionize the mercury atoms inside the tube. The base holds the lamp in place and provides the electrical connection to the lamp.

25.3.2.1 Technology

25.3.2.1.1 Ballasts

A fluorescent ballast provides the necessary voltage to start and maintain the arc in the lamp tube, because the electrical resistance of a plasma goes down as its temperature increases and a ballast is required to regulate the electric current through the lamp. Some ballast types are also designed to provide a specified amount of energy in the form of heat to the lamp electrodes. This minimizes stress on the electrode coating materials and thus extends lamp lifetime. The current delivered by the ballast to the lamp determines its light output. For a particular lamp-ballast system, light output is characterized by the ballast factor, which is the lumen output of the lamp (or lamps) operated with the ballast, relative to the lumen output of the same lamp (or lamps) operated with a reference ballast. Ballasts can be integral to the lamp or external. For example, medium screw base CFLs have integral ballasts, whereas in most linear fluorescent lamps, the ballast is external to the lamp.

There are two main types of fluorescent ballasts: magnetic and electronic. Magnetic ballasts operate at the electric power system frequency, while electronic ballasts operate at higher frequency and have many advantages over magnetic ballasts. These include higher efficiency, the ability to drive multiple lamps in series or parallel, lower weight, reduced lamp flicker, and quieter operation. These characteristics, as well as regulatory measures, have led to the phase-out of most types of magnetic fluorescent ballasts.

Ballasts are usually categorized by their starting method. Typical starting methods for electronic ballasts are instant start, rapid start, and programmed start. Instant start ballasts use a high voltage to strike the arc of the lamp, without preheating of the electrodes, having a negative impact on lamp life. As a result, instant start ballasts are very energy efficient but are more appropriate for applications with less frequent switching cycles (on–off). Rapid start ballasts heat the electrodes and apply starting voltage at the same time. They use more energy compared to instant start ballasts (~2 W/lamp), but they allow more lamp switching cycles (on–off) before lamp failure. Programmed start ballasts delay the starting voltage until the electrodes are heated to an optimum temperature to minimize the impact of starting on lamp lifetime. They are most suitable for frequent starting applications, such as areas with occupancy sensors. Table 25.2 summarizes characteristics of electronic ballast starting types.

Fluorescent lamps can be dimmed with the use of dimmable ballasts. Dimmable ballasts operate by reducing the electric current through the lamp. While doing so, the ballast must maintain the electrode and starting voltages and regulate electrode heating to maintain rated lamp lifetime. The power required to maintain these functions under dimming conditions can lead to higher ballast energy use at dimmed levels compared to full output levels. However, dimming still typically reduces the overall energy consumption of the lamp-ballast system because of the reduced power consumption of the lamp.

Because fluorescent lamps cannot operate without ballasts, their energy use must account for energy losses in the ballast. Ballast energy use is characterized by two metrics: ballast efficacy factor (BEF) and ballast luminous efficiency (BLE). The BEF, also sometimes referred to as the ballast efficiency factor, is the ratio of a lamp's ballast factor to the ballast input power. BEF is a lamp-ballast performance metric, which accounts for the efficiency of the lamp-ballast system compared to other systems with the same type and number of lamps.

TABLE 25.2

Electronic Ballast Starting Type Characteristics

Ballast Starting Type	Start Time (s)	Lamp Switch Cycles
Instant start	<0.1	10,000–15,000
Rapid start	0.5–1.0	15,000–20,000
Programmed start	1.0–1.5	50,000

Source: Philips, The ABCs of electronic fluorescent ballasts, A guide to fluorescent ballasts, Philips Advance, 2011, http://www.siongboon.com/projects/2010-08-22 electronic ballast/ABC of electronic fluorescent ballast.pdf.

BEF requires the measurement of several (lamp-related) parameters, which can lead to inaccuracies in its measurement. To reduce measurement variation and testing burden, BLE was introduced in 2012. BLE is immaterial to the lamp and only considers ballast characteristics. It is the total lamp arc power divided by the ballast input power and multiplied by a frequency adjustment factor, which depends on the ballast operating frequency.

25.3.2.1.2 Lamps

The two main categories of fluorescent lamps are linear fluorescent lamps and CFLs. Linear fluorescent lamps, or fluorescent tubes, are typically categorized based on their tube diameter. The most common types of these lamps are T12, T8, and T5 lamps (with tube diameters of 12/8, 8/8, and 5/8 of 1 in., respectively). They are usually available in lengths of 4 ft (1.2 m) lamps and 8 ft (2.4 m) lamps, although 2, 3, 5, and 6 ft lengths can also be found. CFLs have small diameter tubes (typically ¼ of 1 in.) that are bent into two to six sections or into a spiral shape. The tube(s) are sometimes covered with a diffuser that makes the assembly look more like a general-service incandescent lamp. They have a compact size and various base types, including common household-type screw bases. This allows them to substitute for incandescent lamps in many fixtures.

The physical characteristics (length, diameter, and shape) of fluorescent lamps can influence their efficacy. Several other lamp characteristics, such as the emissive quality of the electrode coatings and the ability of the inert gas fill to improve the mobility of the mercury atoms, can also impact fluorescent lamp efficacy. Another important factor that can significantly affect efficacy, as well as color quality and lumen maintenance, is the phosphor coating. High efficacy fluorescent lamps include *triband* phosphors, which contain oxides of certain rare earth elements—lanthanum, cerium, europium, terbium, and yttrium. Rare earth oxides account for a significant portion of the manufacturing cost of fluorescent lamps. This percentage is higher in high efficacy fluorescent lamps (e.g., 800-series T8 and T5 linear fluorescent), which use 100% triband phosphor coatings, whereas lower efficacy linear fluorescents (700-series) contain only about 30% triband phosphor. Between 2010 and 2011, the price of rare earth oxides increased sharply, causing concerns about the resulting price impacts on high efficacy fluorescent lamps, but has since been reduced considerably. The future trajectory of rare earth oxide prices is highly uncertain and will depend on global supply and demand of rare earth elements.

25.3.2.2 Changing Market

Linear fluorescent lighting is used predominantly in the commercial (and industrial) building sectors. In the residential sector, linear fluorescent lighting is primarily found in specific areas, such as kitchens, bathrooms, garages, and workshops. T12 lamp/ballast systems are less efficacious and more costly to operate compared to T8 and T5 lamp/ballast systems, and thus are currently being phased out of the market. CFLs have seen increased adoption in recent years for residential applications as replacements of incandescent and halogen technologies because of their decreasing first costs, higher efficacies, longer life times, and improved color quality.

Continuous performance improvements and reducing costs of LED-based technologies, in particular in LED troffer luminaires as replacements for linear fluorescents [18], and in reflector and omnidirectional LED lamps as replacements for CFLs, are causing a market shift in the fluorescent lamp market to LEDs. LEDs are projected to replace more than 60% of the linear fluorescent market and to reach 70% market penetration in residential applications by 2030 [7].

25.3.3 Plasma Lighting

Plasma lamps emit light via plasma discharge emission from electrically excited gases contained within an arc tube. A typical lamp design consists of a ballast, igniter, electrodes, an alumina glass arc tube filled with pressurized gases, and a lamp base. The operation of a plasma lamp is unique, as it involves a warm-up period of 2–10 min and, when power is interrupted, a restrike time of up to 10 min depending on the technology.

25.3.3.1 Technology

25.3.3.1.1 Ballasts

Plasma lighting requires a ballast to initiate the lamp and to regulate the current during operation, similar to fluorescent ballasts. In addition to a ballast, some plasma lamps require an igniter to initiate the necessary high voltage during start-up. As with fluorescent lighting, there are both magnetic and electronic ballast technologies and similarly, electronic ballasts offer improved lamp lifetime, lumen maintenance, ballast efficiency, higher lamp efficacy, and lower ballast losses, offering better light quality and energy savings. For more description on basic ballast function, see fluorescent Section 25.3.2.

Electronic ballast technology offers higher efficiencies for HID technologies over magnetic ballasts, nearly a 15% higher ballast efficiency for some lamp and ballast types [19]. Industry experts agree that there are limited efficacy gains remaining in HID ballast technology, that only a few percent gains are technologically feasible but may be practically infeasible to attain.

In lieu of further efficiency gains, the use of electronic ballasts allows for the implementation of lighting controls, which reduce hours of use and in turn energy consumption. Due to the nature of HID lamp technology, these control systems utilize a stepwise dimming function where discreet dimming levels are designated, instead of continuous systems, as in incandescent and fluorescent lamp systems. Some systems offering dimming capabilities near 40% of total light output, which in certain applications translates to nearly a 30% reduction in energy consumption.

25.3.3.1.2 Lamps

The most common plasma lamps are mercury vapor (MV), high-pressure sodium (HPS), and metal halide (MH); these are classified as HID. Low-pressure sodium (LPS) lamps are also plasma lamps and are often grouped together with HID, since they are high lumen output technologies used in similar applications (exterior and large interior commercial and industrial spaces). LPS lamp technology is not discussed in depth, as it is an older technology that constitutes a very small fraction of commercial and residential markets.*

MV lamps, first developed in 1901, use ionized mercury as the primary discharge element, producing a bluish light with better CRI than sodium lamps, but with relatively low efficacy. Phosphors can be used to improve the color and CRI of MV lamps' relatively blue light, but they yield relatively poor color rendering compared to other HID technologies. These lamps are now used primarily in legacy lighting designs or special applications, and are being replaced by MH lamps that have twice the efficacy and far better CRI.

HPS lamps and LPS lamps both utilize mixtures of ionized sodium and mercury gases, and this difference in pressure leads to unique light quality characteristics. LPS lamps

* For example, in the United States, LPS constitutes less than half of 1% of the total lumen output of HID lamps [5].

TABLE 25.3

Typical Characteristics of Plasma Lamps

Lighting Type	Typical Efficacies (lm/W)	Lifetime (h)	CRI	CCT (K)	Application
LPS	60–150	12,000–18,000	−44 (very poor)	1,800 (warm)	Exterior
HPS	50–140	16,000–24,000	25 (poor)	2,100 (warm)	Exterior
MV	25–60	16,000–24,000	50 (poor to fair)	3,200–7,000 (warm to cold)	Exterior
MH	70–115	5,000–20,000	70 (fair)	3,700 (cool)	Interior/exterior
Light-emitting plasma	60–130	25,000–100,000	70–95 (good)	2,000–10,000 (warm to cold)	Interior/exterior

Sources: luxim.com; Topanga.com; http://www.eere.energy.gov/basics/buildings/low_pressure_sodium.html; http://www.eere.energy.gov/basics/buildings/high_intensity_discharge.html.

offer the highest efficacy, but the monochromatic condition of the light spectra reduces the visual acuity in mesopic (relatively low-light) conditions, thus reducing the effectiveness of the light in outdoor applications. HPS produces a broader light spectrum than LPS by subjecting the gas to higher pressures. This broadening of spectral emission creates a better quality light than LPS but with reduced efficacy. The characteristics of available lamps are highlighted in Table 25.3.

MH lamps are similar to MV lamps, but MH gases are added to the mercury gas in the discharge tube, yielding higher lumen output and efficacy with higher CRI. Improvements in the starting technology of MH lamps have had considerable effect on lamp quality. Innovation from probe start, which involves a third electrode to initiate the lamp, to pulse-start, which utilizes an exterior igniter, allows for higher pressures in the arc tube, which leads to a shorter start-up and restrike time, longer lamp life, lumen maintenance, better cold weather operation, and up to a 20% increase in efficacy compared to traditional MH lamps. Ceramic metal halide (CMH) lamps, utilizing a ceramic glass arc tube and the pulse starting method, have some of the highest efficacies for MH lamps.

A new HID technology, known as light-emitting plasma, uses induction starting to excite the gas vapors instead of an electrical discharge. This electrodeless induction lamp utilizes fill gases that are similar to those in MH lamps but uses a solid-state radio frequency driver to excite the gases inside the lamp. Manufacturers report long rated lifetimes of up to 100,000 h, which far exceed those of other HID technologies. Another advantage of this technology is the relatively small size of the lamp capsule, which allows for greater directionality of the light from the luminaire, leading to significantly higher luminaire efficacies than traditional HID luminaires. The induction starting technique allows for a quicker start-up time (45 s) and restrike time (2 min) than other HID lamps, which allows for easier integration of occupancy controls. Light-emitting plasma (LEP) also has dimming functionality that could allow for significant energy savings compared to other HID technologies.

25.3.3.2 Changing Market

While HID lamps are primarily used for exterior lighting, they are also widely used for large indoor spaces in the commercial and industrial sectors and in rare cases in the residential sector. In the United States in 2010, 24% of the installed stock of HID lamps served the commercial sector, while only 1% served the residential sector [5]. Commercial lighting applications include exterior security lighting, warehouse high bay and low bay lighting, walkway lighting, and, more recently, retail lighting due to an influx of smaller sized CMH lamps with high CRI. Residential HID lighting consists mainly of exterior security lighting with rare specialty interior lighting utilizing CMH lamps.

Advancements in HID efficacy have been leveling off in recent years, and with the advancement of LED replacement technology (see Figure 25.2), HID market share is projected to decrease. LED replacement efficacies are projected to double from 100 lm/W to over 200 lm/W in the next 20 years (Figure 25.2), offering significant energy savings [4]. LED replacements also have the potential for spectral control, are amenable to occupancy-based controls, and are more conducive to dimming. But in applications where an intense point-source light is needed, LEDs may never replace plasma lamps. With the persistence of these high-output applications and the limitations of current LED technology, continued improvements are expected for MH lamp technology, both traditional and induction, whereas there is no expectation of the same for MV, LPS, and HPS lamps [20].

25.3.4 Solid-State Lighting (LED and OLED)

LEDs and OLEDs turn electricity into narrow-band light via diode emission and can produce varying wavelengths ranging from infrared through the visible spectrum to ultraviolet. LEDs consist of a semiconducting die, typically a form of silicon or germanium, a circuit board, and a lens or diffuser. In some cases, phosphors and a heat sink are also utilized depending on the application and desired light output. OLEDs consist of a wide variety of organic light-emitting compounds, electrodes (one usually being transparent), and typically a plastic or glass envelope that holds all of the components in place and acts as a lens or diffuser for the unit.

25.3.4.1 Technology

There are currently three general methods for creating the white light necessary for general-service lighting from LEDs: Phosphor conversion (PC) where phosphors are used in conjunction with blue or violet LEDs to produce white light via fluorescence, color mixing (CM) where LEDs with discrete colors (red, green, blue, and sometimes amber) are mixed in an array to create white light, and hybrid where PC and CM are combined to create white light.

LED lamps utilizing PC are the most common on the market today. Typically, phosphor (or a mixture of phosphors) is deposited directly onto a blue or near-ultraviolet LED die and encapsulated into the LED package. Locating phosphors away from the LED package, known as remote phosphor conversion, allows for better light dispersion and can create larger uniform light-emitting surfaces that better accommodate current form factors of existing luminaires. Phosphor-converted LEDs can have a range of color temperatures ranging from cool to warm white light. Color rendering of PC-LEDs ranges from moderate to excellent, with CRIs ranging from 70 to greater than 90, depending on the phosphors used. Higher CRIs are achieved at the cost of efficacy. Efficacies have been demonstrated as high as 140 lm/W for cool white emitters and 110 lm/W for warm white LEDs [4]. In 2013, lamps were available in the market with efficacies ranging from 75 to 92 lm/W.

Currently, cool-colored LEDs are about 20% more efficacious than warm-colored LEDs, for both CM and PC LEDs, but the gap is decreasing and it is expected to become negligible over the next decade [18]. White light achieved by CM has the potential for higher efficacies owing to fewer conversion losses compared to PC-LED. However, the green and sometimes amber colored LEDs used for CM currently have a lower efficiency than blue LEDs, which means less light output at a given input power compared to most other LED colors. Improvements made to the efficiency of green and amber LEDs could lead to large improvements in the efficacy of CM white LED lamps. Current mixed-color lamps on the market have efficacies around 90 lm/W. Color rendering of mixed-color lamps is good to excellent: mixed-color lamps currently exist in the market, with CRI greater than 90.

Another unique feature of CM lamps is the ability to tune the CCT of the output light. This ability can allow for color correcting as lamps age, since LEDs of different colors have different rates of lumen depreciation. It can also allow users to change the color of lamps via appropriately integrated controls.

The third approach, the hybrid method, uses PC with the addition of amber LED to fill in gaps in the PC-LED spectrum. This enables high-CRI lamps with greater efficacies. The light mixing can take place at the LED die level or in the luminaire, depending on the application.

The advantages of LEDs compared to incumbent lighting technologies include the following:

- High efficacies, resulting in lower operating costs; long lifetimes, with reported lifetime values of 30,000 h and projections of up to 50,000 h, which lowers annualized life-cycle costs
- Improved operation in low-temperature applications, since LEDs are more efficient at lower temperatures
- Dimmability, with compatible dimmers designed to work with LED drivers
- Ability to tune colors with CM technologies

Disadvantages of LEDs compared to incumbent technologies include the following:

- High first cost, although LED prices have been dropping very quickly
- Changes in color temperature over time
- Heat management issues: LED lifetimes and efficacies are greatly reduced when operating at high temperature, limiting their ability to be used in existing enclosed fixtures
- Incompatibility with incumbent dimmers: Dimmers that operate by varying the incoming line current to an LED lamp or fixture often do not work well with LED lamps
- Flicker problems with some LED lamps, especially when dimming
- High glare, due to the compact nature of LED packages, though this can be remedied by use of a diffuser

Despite the limitations of current LED technology, none of the current disadvantages appear to be technologically insurmountable. Industry is working, often with the support of governments, to address them.

OLEDs use a thin film of organic materials to emit diffuse light over a planar surface. They have been used mainly for displays in handheld electronic devices such as smart phones, but hold potential for general and niche lighting applications, in particular for diffuse lighting. The main difference between OLEDs and LEDs is form factor, the way light emission is distributed. OLEDs produce relatively diffuse light over a large area compared to the high-intensity compact source of light given off by an LED. OLEDs can be deposited on a number of different substrates including glass, plastic, and metal. This flexibility allows for the creation of luminaires with a wide variety of shapes and sizes. However, there are few OLED luminaires currently on the market and those that are available are not cost-competitive with other lighting technologies.

Currently, OLEDs lag behind LEDs in many performance metrics; in 2012, the best reported efficacies for OLEDs were approximately 60 lm/W. Color rendering by OLEDs is good to excellent, with CRIs up to 95. However, the lifetime of an OLED is only 10,000 h and they are very sensitive to moisture and heat. Care must be taken in the manufacturing process to properly seal OLEDs or their lifetime is shortened dramatically [4,21].

25.3.4.2 Changing Market

Single colored LEDs, developed in the 1960s, were the original technology used in nongeneral lighting applications such as traffic signals, exit signs, and outdoor displays. In 1993, innovations in blue LEDs made it possible to create products that could serve general purpose lighting applications that require white light as it became possible to create a white light by CM with the newly invented blue LED. As indicated in the introduction, LEDs are poised to become the dominant lighting technology for all general lighting applications as technology improves and prices continue to drop. Manufacturers' commitments to improving incumbent technologies are waning with most of their research and development efforts shifting toward LED lighting and lighting controls. It is currently unclear the role that OLEDs will play in that market.

25.3.5 Lighting Controls

Lighting controls are used to reduce energy consumption, to enhance security, and for aesthetic design. Lighting controls systems can include motion sensors, infrared sensors, photo sensors, timers, transmitters and receivers, and computer control systems. They can range in complexity from simple on/off switches that control a single luminaire to sophisticated integrated lighting control systems that control the lighting of entire buildings or campuses of buildings.

25.3.5.1 Technology

There are three main strategies for lighting control: manual controls, sensor-based controls, and scheduled controls. Manual controls are the most simple; they allow users to control light levels either with on/off switches or dimmers to desired levels. Sensor-based controls rely on sensors to initiate changes in light levels. Scheduled controls adjust light levels according to a predetermined schedule.

Although manual controls, in the form of simple on/off switches, have been used since the inception of electric lighting, they are sometimes incorporated in more advanced lighting strategies (e.g., providing the capacity for individuals in large commercial work places to control the light in their own work space). Multiple strategies may also be included in a single luminaire: for example, stair-well lighting with multi-light-level switching controlled by occupancy sensors.

Sensor technologies fall into two main categories: occupancy sensors and photosensors. Occupancy sensors use several different strategies to determine whether a space is occupied.

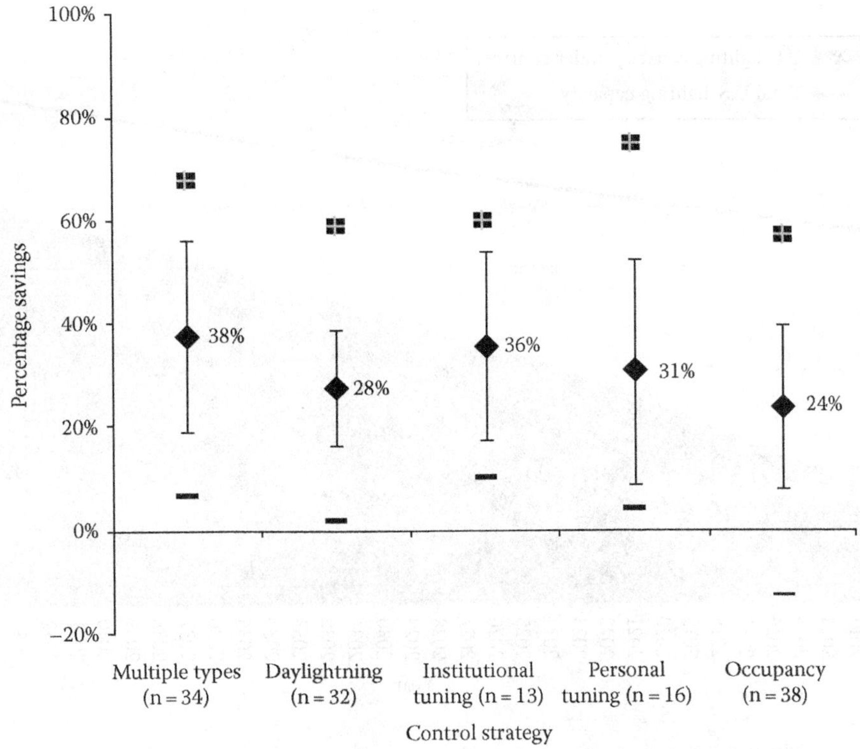

FIGURE 25.6
Energy savings for lighting controls and lighting energy in actual installations. The symbols indicate the maximum, average, and minimum values found in the study. The vertical bars indicated one standard deviation. (From Williams, A. et al., *LEUKOS* 8(3), 161, 2012.)

Passive infrared (PIR) sensors react to moving emitters of infrared radiation at wavelengths near the peak thermal emission given off by people. Ultrasonic and microwave sensors use the Doppler effect to detect movement, by emitting ultrasonic waves and microwaves, respectively, and measuring frequency shifts in the reflected signal to detect movement. Dual technology sensors use both PIR and ultrasonic technology to detect occupancy, which reduces the false triggering events that can occur with either technology used singly.

A recent metastudy of energy savings from lighting controls [22] found that providing the capacity for manual dimming can save as much energy as more advanced lighting strategies. As shown in Figure 25.6, based on field studies of actual energy savings, personal tuning (using dimmers) saved about the same amount of lighting energy (31%, on average) compared to automated controls, including occupancy sensors (saved 24%, on average) and daylight sensors (28%, on average). While applying multiple strategies simultaneously saved somewhat more energy (38%, on average), the biggest savings are obtained by the first control installed.

25.3.5.2 Changing Market

The market for lighting controls is expected to grow rapidly in the coming decades. Building codes have recently begun to incorporate requirements for lighting controls as a means to reduce lighting energy use beyond the more traditional lighting power density limits.

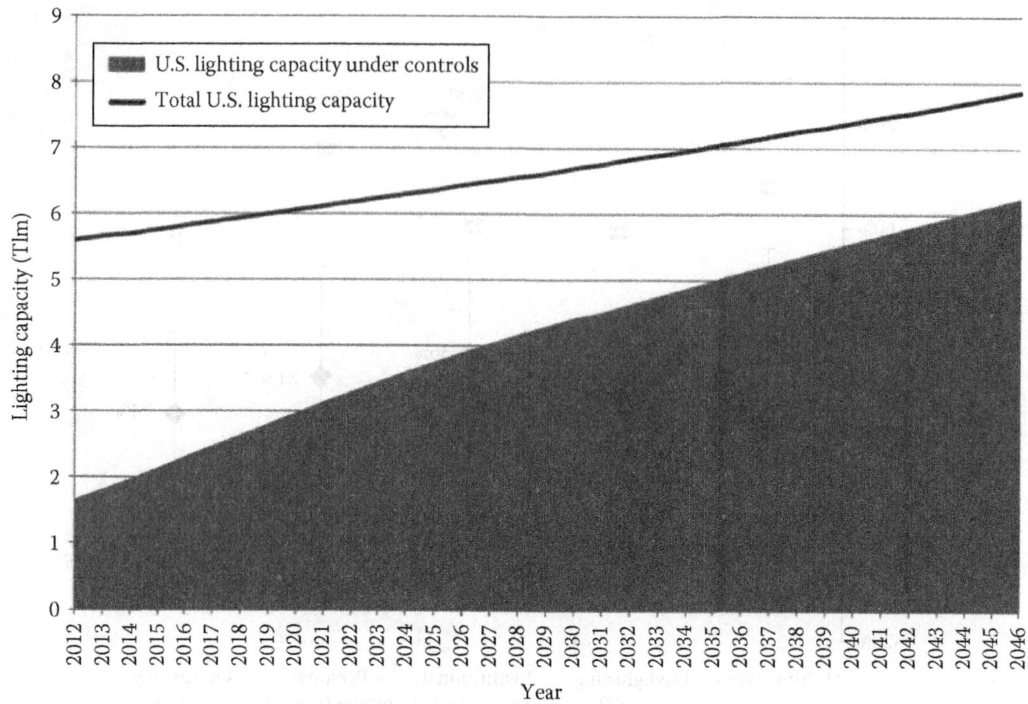

FIGURE 25.7
U.S. Department of Energy projections of the total annual commercial lighting capacity and the capacity projected to be under lighting controls. (From DOE, GSFL-IRL preliminary analysis technical support document, Appendix 9A, Lighting controls, U.S. Department of Energy, 2013, http://www.regulations.gov/#!documentDetail;D=EERE-2011-BT-STD-0006-0022, accessed February 28, 2013.)

For example, based primarily on building code requirements now in place, the United States Department of Energy estimates that lighting controls will cover approximately 75% of U.S. commercial floorspace by 2050, up from approximately 30% in 2012, see Figure 25.7 [23].

25.3.6 Luminaires

A number of factors can affect lighting efficiency at the luminaire level: the choice of light source type (as described earlier), the design of luminaire optical elements (reflectors, lenses, shades, and louvers), and the use of luminaire-based controls. In addition to energy efficiency, light quality must be considered in luminaire design (e.g., the need for glare control in certain applications). While there can be trade-offs between efficiency and light quality—for example, diffusers, which control glare, also reduce efficiency—advanced lighting technologies may eliminate such trade-offs (e.g., by using an inherently diffuse and efficiency light sources, like OLEDs). Similarly, the incorporation of increasingly advanced controls may allow the customization of light levels that reduce energy use.

Various metrics have been developed to quantify luminaire efficiency, and they have become increasingly sophisticated. The simplest metric is luminaire efficiency, the ratio of light leaving the fixture to the light emitted by its light source(s), which accounts for light absorbed in the luminaire. The luminaire efficacy rating (LER) system, developed by the National Electrical Manufacturers Association (NEMA), also accounts for light absorbed in the luminaire, but the metric is quantified as the light output from the luminaire (lumens) divided by the power input to the luminaire.* In 2008, NEMA introduced its target efficacy rating (TER) system, which was designed to supersede LER [24]. TER, also quantified in lumens per watt, credits only light that hits its intended target surface. TER covers 22 widely used types of commercial, residential, and industrial luminaires.

Because luminaire efficacy depends on lamp efficacy, luminaire efficacy must be measured with lamps in place. If those lamps are replaced by consumers with lamps that have a different efficacy, the luminaire will no longer perform at the rated level. Luminaire design can protect against this possibility to a significant degree—a factor that is worth considering in the design of lighting efficiency policies and programs. Luminaires with integral LED lighting (also called inseparable solid-state lighting) do not have replaceable lamps, so their efficacies are known with a high degree of certainty over their lifetimes. The socket type used in the luminaire may also ensure that lamp replacement cannot result in large efficiency losses. For example, ANSI Standard ANSI_ANSLG C81.62-2009, GU-24 bases are not allowed to be used for incandescent lamp technologies; therefore, the efficacies of replacement lamps will be at least at the fluorescent or LED levels.

25.4 Cost-Effectiveness of Efficient Lighting Technology

25.4.1 Life-Cycle Cost

While lighting provides obvious benefits to consumers, it also imposes costs to society, only some of which are borne by the consumer. While the costs of production are typically included in the price paid by the consumer, environmental and social costs are typically not included in the price. Therefore, lighting policies are increasingly incorporating, to the degree possible, total life-cycle costs.

The environmental costs of lighting depend on lamp technology. The U.S. Department of Energy has produced a three-part study of the life-cycle environmental and resource costs in the manufacturing, transport, use, and disposal of LED lighting products in relation to comparable traditional lighting technologies.† Figure 25.8, from part 3 of that series [25], compares the life-cycle impacts of different CFLs and LED lamps manufactured in 2012, and those expected to be in the market by 2017, relative to those of general-service incandescent lamps. The life-cycle environmental impacts of the current generation of LEDs are approximately comparable to the impacts of CFLs, but the impacts of the next generation of LED technology are expected to be significantly smaller.

25.4.2 Current Cost of Ownership of Different Lamp Technologies

In contrast to policymakers, typical lighting consumers generally consider only those costs associated with the ownership of lighting: the purchase price and possibly, if relevant information is available, the costs associated with operation, maintenance, and disposal. This section compares the current cost of ownership of different general-service lighting technologies.

High-efficacy lighting technologies, (e.g., fluorescent or LED) lamps generally have longer lifetimes (Table 25.1) and higher prices than lower-efficacy (incandescent) options.

* The LER system was developed by the National Electrical Manufacturers Association. The method for measuring LER was published as a series of three standards for fluorescent lighting (LE5), for commercial and residential downlights (LE5A), and for HID industrial luminaires (LE5B).

† These reports are available online at http://www1.eere.energy.gov/buildings/ssl/tech_reports.html.

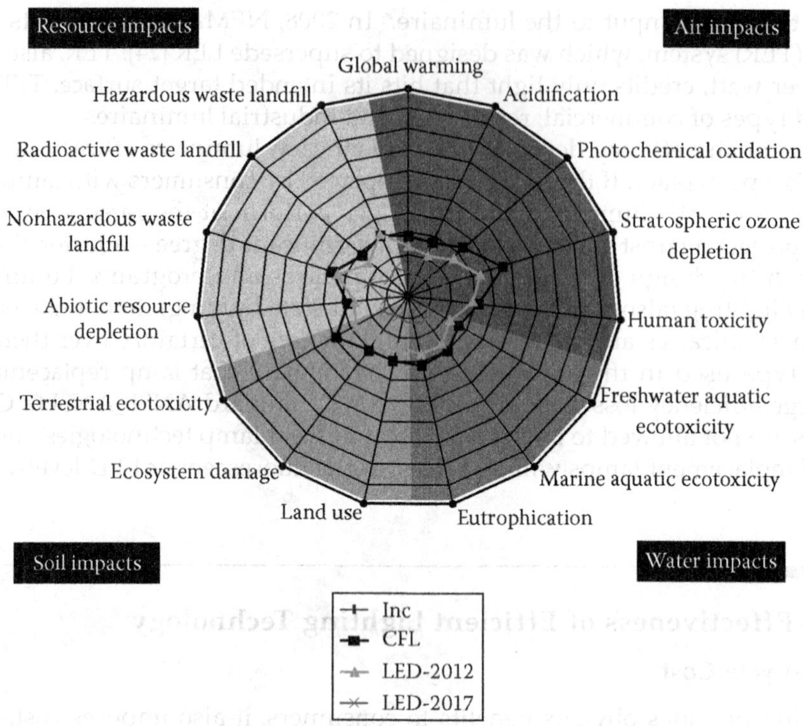

FIGURE 25.8
Life-cycle impacts of different general-service lamp technologies normalized to the impact of an incandescent lamp. (From Tuenge, J. et al., Life-cycle assessment of energy and environmental impacts of LED lighting products, Part 3: LED environmental testing, U.S. Department of Energy, Solid State Lighting Program, 2013, http://apps1.eere.energy.gov/buildings/publications/pdfs/ssl/lca_webcast_03-28-2013.pdf, accessed March 23, 2013.)

Determining the cost-effectiveness of a given technology, by balancing the higher initial price of a lamp against its energy cost savings and longer service life, can be a complex task for the consumer. Different technologies also often have different typical lumen outputs, further complicating the comparison. As a simplifying approach to determine the cost-effectiveness of a particular lighting technology, one can compare the total cost *per year* and *per unit light output* of owning and operating a particular lamp. Additional technical considerations, such as the need for a particular light distribution, CRI, or color temperature, must also be taken into account when choosing a lighting technology, but this approach can be used to compare the cost-effectiveness of a set of lighting options that have been prescreened to meet the technical needs of a particular application.

To determine the total cost of owning and operating a lamp, one first estimates the number of hours for which the lamp will be used in the course of a year. Since lamp lifetimes are typically measured in hours, this roughly determines the lamp's lifetime in years (although the calculation can be more complicated for technologies, like fluorescent lamps, whose lifetimes are shortened by short-term and intermittent use). The annual energy cost is then calculated by multiplying the lamp wattage by the annual hours of use and by the price per watt-hour of electricity. The lifetime operating cost is determined by summing the annual energy cost over the lifetime of the lamp, often multiplied by a discount factor that accounts for the reduced value of future savings compared to up-front costs (e.g., owing to financing costs). The total cost of ownership is then the sum of the price, the

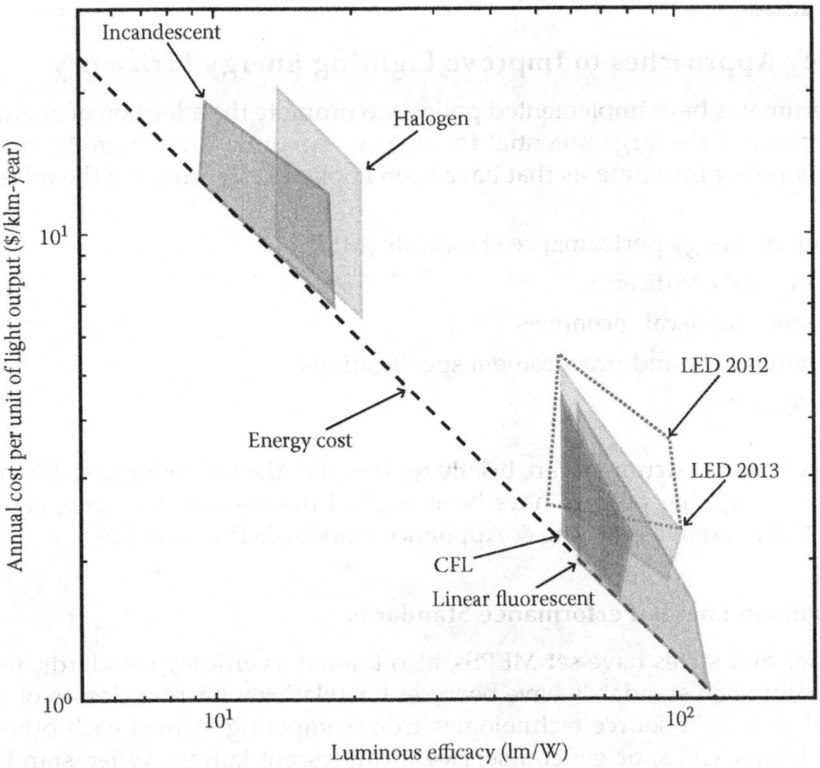

FIGURE 25.9
Relative cost-effectiveness of different general-service lighting technologies.

lifetime operating cost, and any end-of-life disposal costs. This value can then be divided by the lamp's lifetime in years and the lamp's total light output in kilolumens to produce a value that can be used to compare the cost of different lighting technologies on an equal basis. Cost per kilolumen-year is a particularly convenient set of units, because it typically yields values in the order of $1–$10 to be used for comparison (Figure 25.9).

For the purposes of the figure, we have assumed 1095 annual hours of use, corresponding to 3 h/day, which is an industry-standard value for computing lamp lifetimes and is fairly typical in residential applications. As shown in the figure, linear fluorescent lamps are the most cost-effective and efficacious general-service lighting technology in 2013, yielding a significant savings in both energy and cost over incandescent technologies and a small savings compared to CFL and LED lamps. LED lamps are evolving rapidly, however, and have seen extremely rapid improvements in both cost and efficacy in recent years, as illustrated by the dotted region in the figure. They are expected over the coming decade to surpass all competing technologies in both efficacy and cost. Zero discounting of future energy savings has been assumed in this figure, since the particular choice of discount rate is highly dependent on the individual consumer. Zero disposal cost is also assumed, which is typical in most applications. The impact of discounting would be to slightly reduce the cost-effectiveness of longer-lifetime technologies relative to shorter-lifetime technologies, so that, for example, LEDs would become slightly less cost-effective relative to CFLs. For typical discount rates, however, the overall ranking of the technologies by cost-effectiveness shown in the figure would not change.

25.5 Policy Approaches to Improve Lighting Energy Efficiency

Many governments have implemented policies to promote the adoption of energy-efficient lighting because of the large potential for energy, economic, and environmental savings. The primary policy instruments that have been applied to lighting are the following:

- Minimum energy performance standards (MEPSs)
- Labeling and certification
- Economic and fiscal incentives
- Bulk purchasing and procurement specifications
- Building codes

The primary policy instruments are briefly reviewed in the following text. For more information on how lighting policies have been applied in different countries, see en.lighten [26,27] and Collaborative Labeling & Appliance Standards Program [28].

25.5.1 Minimum Energy Performance Standards

Many nations and states have set MEPSs, also known as efficacy standards, for lighting. Conventionally, such standards have been set for relatively narrow classes of lamps that prevent different light source technologies from competing against each other (e.g., 4 ft fluorescent lamps, CFLs, or general-service incandescent lamps). When standards operate on broader categories of lighting equipment, like luminaires or general-service lamps, larger gains are possible.

Recently, a significant number of nations have adopted such standards on general-service lamps that have the effect of eliminating conventional incandescent lamps from the market. The International Energy Agency's 4E program examined the early impact on the lighting market of such these standards in nine countries and the European Union [29]. IEA concluded that the standards were proving successful in Australia, Korea, and the United Kingdom, with the average efficacy of lamp sales having risen by up to 50% in 3 years, with no significant effect observable at the time in the other jurisdictions, although it is important to note that the requirements had not yet gone into force in some jurisdictions. Minimum efficacy standards for luminaires could have a similar effect, necessitating widely used luminaires to use high efficacy lamps, incorporate lighting controls, or both.

Also relatively new are standards for LED lamps, which in some cases also include quality standards [26]. For example, the International Electrotechnical Commission (IEC)'s IEC/PAS 62612 standard contains performance requirements for *self-ballasted* LED lamps (LEDs lamps with integrated drivers) for general-service lighting. Quality standards can be important to avoid the LED market being spoiled by flooding with low-quality products that undermine consumer confidence.

25.5.2 Labeling and Certification

Labeling and certification programs educate the public about the efficiency of lighting products and provide assurance that products perform as advertised. Information programs include mandatory and voluntary product labeling and certification.

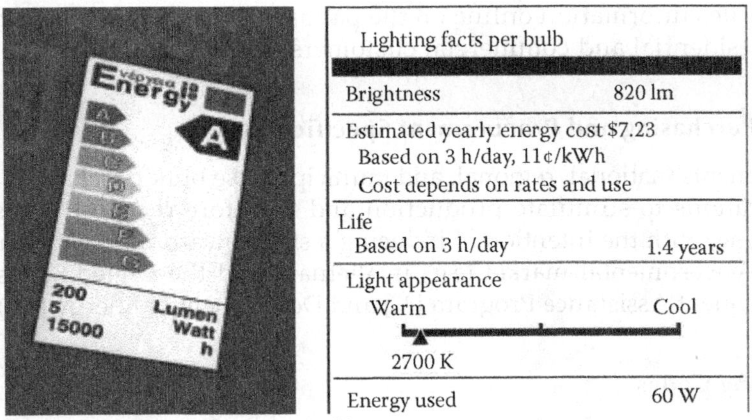

(a) (b)

FIGURE 25.10
(a) EU energy label for a high-efficiency lamp. (b) Lighting Facts label developed by the U.S. FTC, required for use on general-service lamps sold in the United States.

Certification programs like U.S. Energy Star and Brazil's Selo Procel brand efficient products with an easily recognized symbol ensuring consumer that the product is efficient [30,31]. Energy-efficiency labels, on the other hand, provide information on the efficiency of the actual efficiency of the product or how that efficiency compares with other products.

There has been much discussion about appropriate labeling protocols for lighting, because residential customers in particular are unfamiliar with the lumen metric, and have generally used wattage as an indicator of brightness. With the increased using of high efficiency, low wattage lamps, consumers need clearly conveyed information on brightness and efficiency. Comparative labels, like the European Union's EU Energy Label (Figure 25.10a), clearly indicate the performance of the labeled product relative to other products in the market. The label assigns a letter grade, with high grades corresponding to high efficacy. The consumer does not need to understand efficacy units to understand the performance of the lamps relative to other products in the market. While the Federal Trade Commission's (FTC) Light Facts label (Figure 25.10b) provides information on light output (lumens) and energy consumption (watts), to determine the efficacy, the consumer would need to know to divide the two and then would not know how this product performs relative to other products in the market. The FTC label does provide easily understood information on color temperature, however.

25.5.3 Economic and Fiscal Incentives

Many jurisdictions offer fiscal incentives for the adoption of high-efficiency lighting. Rebates are the most common incentive for efficient lighting, which are often implemented as part of utility demand-side management programs. In recent years, rebates have begun to focus on lighting controls as well as on efficient lamps. Low interest loans for commercial and residential lighting systems upgrades are also widely available. Tax incentives are also widely offered at the state and federal levels for lighting system upgrades. The Database of State Incentives for Renewables and Efficiency

(DSIRE) compiles information online on the panoply of economic and fiscal incentives available to residential and commercial customers in U.S. states.

25.5.4 Bulk Purchasing and Procurement Specifications

Many governments (national, regional, and municipal) use bulk purchasing and procurement specifications to stimulate production and therefore reduce the cost of energy-efficient lighting, with the intention of inducing a shift toward more efficient products in the larger on-governmental market (e.g., in Vietnam and the United States, see Energy Sector Management Assistance Program [32] and Department of Energy [33]).

25.5.5 Building Codes

Energy codes are a subset of building codes, which govern the design and construction of residential and commercial structures. Energy codes generally include lighting provisions. Code provisions are generally applicable only to new construction and to renovations that affect some threshold fraction of a building's floor area. Therefore, incorporation of code provisions into the building stock depends on construction and renovation rates, which can result in relatively slow penetration rates, depending on economic conditions. On the other hand, building codes have the advantage of being able to address lighting efficiency at the application level, accommodating different lighting needs in different spaces.

Lighting provisions in building codes may be prescriptive or performance based. The prescriptive approach stipulates minimum efficacy requirements for lighting products used in different spaces. The more flexible, performance-based approach specifies the maximum allowable lighting power density in watts/square meter (square foot) for different spaces. The lighting designer can then choose any equipment that together meets the overall requirement for the space.

Lighting controls have been increasingly incorporated in building code requirements. This includes increasing the building space required to incorporate lighting controls, incorporation of increasingly sophisticated controls, and the layering of multiple types of controls. See, for example, updates to ASHRAE/IESNA 90.1 [34,35]. In some cases, lighting controls have been used as an alternative compliance path, allowing the use of less efficient lamps. For example, the State of California's Title 24 provisions require that all permanently installed outdoor luminaires using lamps rated over 100 W either have a lamp efficacy of at least 60 lm/W or that they be controlled by a motion sensor [36].

Because all of these policy instruments described earlier rely on accurate ratings of product efficiency, a sound underlying national infrastructure of product testing, monitoring, and enforcement is needed for lighting policies to be effective.

Governments may also support product development to spur efficiency innovations. The L-Prize is creative and highly successful example, in which the U.S. Department of Energy offered a $10 M cash prize for the development of a 90 lm/W solid-state lamp to replace the standard A19 60-W incandescent. This represented an enormous advancement in LED efficiency when the prize was announced in May 2008. The L-Prize resulted in a fierce competition, which Philips won in August 2011. But the benefits of the competition continued after the prize was won. Today there are eight products in the market that are certified by DOE's Lighting Facts program to meet or exceed the 90 lm/W requirements, with the best omnidirectional A-lamps achieving almost 100 lm/W.

25.6 Conclusions

With the advent of LED lighting, efficiency improvements can now be achieved without the degradation of light quality. Indeed, new LED lighting technologies can improve light quality by offering color-tuning capacity as well as all of the conventional standard and automated controls. Not only aesthetically pleasing, this feature could offer a potent tool to provide light that is health enhancing, by better mimicking the color dynamics of the diurnal cycle of natural light, essential for well-being.

While at this point falling prices and improved light quality alone may be enough to induce the shift to an LED lighting future, well-crafted lighting policies can continue to drive energy and environmental savings. Energy efficiency standards, coupled with quality standards, can ensure that low-efficiency and low quality LEDs are removed from the market. Bulk purchasing requirements, building codes, and luminaire standards can be used to drive the adoption of lighting controls and earlier adoption of high-efficiency lamps.

In conclusion, global lighting is poised for great increases in energy efficiency and reductions in total life-cycle costs, as a result of solid-state lighting products entering the market and policies aimed at removing low-efficiency incandescent lamps from the market. This is especially true in the residential market where new LED lamps are more than six times as efficient as the incumbent incandescent lamps. Increased use of lighting controls is also likely to significantly reduce energy use. Further opportunities to reduce environmental impacts are available by integrating PV power into lighting systems. Taken together, these opportunities suggest a bright future for lighting.

Glossary

Color rendering index (CRI) is a measure of the degree of color shift objects undergo when illuminated by the light source as compared with the color of those same objects when illuminated by a reference source of comparable color temperature. [RP-16-10] CRI is used to quantify how well a light source renders color across the color spectrum. For color temperatures below 5000 K (which applies to most indoor lighting), the reference source is a blackbody radiator, which is most nearly approximated by an incandescent lamp. For color temperatures above 5000 K, the reference source is a standard daylight condition of the same color temperature. Therefore, effectively, this metric quantifies how closely other light source types mimic an incandescent light source or a daylight source.

Correlated color temperature (CCT) is the absolute temperature of a blackbody radiator, expressed in degrees Kelvin, whose chromaticity (color appearance) most nearly resembles that of the light source [adapted from RP-16-10]. In lighting design, this metric is used to describe how warm or cool a light source appears. Light with a CCT at or above 4000 K, with a greater proportion of its flux at shorter (blue–violet) wavelengths, is considered cool. Light with a CCT at or below 3000 K, with more flux at longer (red–yellow) wavelengths, is considered warm.

Efficacy is the ratio of the lumens to the power (watts) drawn by a lighting system. The concept may be applied to the lamp alone (lamp efficacy), to the lamp-and-ballast system (system efficacy) for fluorescent or HID systems, to the lamp-and-driver system for LED lamps (system efficacy), or to the luminaire (luminaire efficacy). The distinction between these is which luminous flux is included and where the power input is measured.

Lumen (lm) is the SI unit for measuring luminous flux.

Lumen maintenance is the lumen output at a given time in the life of the lamp and expressed as a percentage of the initial lumen output of the light source.

Luminaire is a complete lighting unit consisting of the light fixture, in combination with its light source(s), any necessary ballasts or drivers, and all parts connecting the power source to the light source, including any integrated controls.

Luminous Flux is the rate of flow of electromagnetic radiation emitted by a light source in all directions, weighted by the spectral sensitivity of the human visual system [37].

References

1. en.lighten. 2013. Lighten initiative, United Nations Environment Program. Accessed June 5, 2013. Available online: http://www.enlighten-initiative.org/portal/Home/tabid/56373/Default.aspx.
2. en.lighten. 2013. en.lighten initiative. Country lighting assessments. Accessed April 2013. Available online: http://www.enlighten-initiative.org/portal/CountrySupport/CountryLightingAssessments/tabid/104272/Default.aspx.
3. IEA. 2006. *Light's Labour's Lost.* International Energy Agency, Paris, France. Accessed September 13, 2013. Available online: http://www.iea.org/publications/freepublications/publication/light2006.pdf.
4. DOE. April 2012. Solid-state lighting research and development: Multi-year Program Plan. U.S. Department of Energy, Energy Efficiency & Renewable Energy Building Technologies Program, Washington, DC.
5. DOE. January 2010. 2010 U.S. lighting market characterization. U.S. Department of Energy, Energy Efficiency & Renewable Energy Building Technologies Program, Washington, DC. Accessed September 13, 2013. Available online: http://apps1.eere.energy.gov/buildings/publications/pdfs/ssl/2010-lmc-final-jan-2012.pdf.
6. Navigant Research, Inc. 2013. Energy efficient lighting for commercial markets—LED lighting adoption and global outlook for LED, fluorescent, halogen and HID lamps and luminaires in commercial buildings: Market analysis and forecasts (Executive Summary). Navigant Consulting. Accessed September 13, 2013. Available online: http://www.navigantresearch.com/research/energy-efficient-lighting-for-commercial-markets.
7. Navigant Consulting, Inc. January 2012. Energy savings potential of solid-state lighting in general illumination applications. U.S. Department of Energy, Solid State Lighting Program. Navigant Consulting, Inc. Washington, DC. Accessed September 13, 2013. Available online: http://apps1.eere.energy.gov/buildings/publications/pdfs/ssl/ssl_energy-savings-report_10-30.pdf.
8. McNamara, G. 2012. PubSpectra—Open data access fluorescence spectra. Selected works. Accessed September 13, 2013. Available online: http://works.bepress.com/gmcnamara/9.
9. Roberts, J.E. December 2012. Light and dark and human health. *J R Astron Soc Can* 11–14 (Special report). Available online: http://www.rasc.ca/sites/default/files/LPA_Special_Issue_LR.pdf
10. Wielgus, A.R. and Roberts, J.E. 2012. Invited review: Retinal photodamage by endogenous and xenobiotic agents. *Photochem Photobiol* 88: 1320–1345.
11. Eastman, C.I. and Martin, S.K. 1999. How to use light and dark to produce circadian adaptation to night shift work. *Ann Med* 31: 87–98.
12. Roberts, J.E. 2000. Light and immunomodulation. *N Y Acad Sci* 917: 435–445.

13. Wehr, T.A. et al. 2001. Evidence for a biological dawn and dusk in the human circadian timing system. *J Physiol* 535(Pt 3): 937–951.
14. Mercier, K. October 2012. Maximizing health and sleep in the elderly. *LD+A*, Illuminating Engineering Society, pp. 42–47.
15. International Dark Sky Association. 2009. The BUG system—A new way to control stray light from outdoor luminaires. Specifier Bulletin for Dark Sky Applications, International Dark-Sky Association, 2(1), Tucson, AZ. Accessed September 21, 2014. Available online: http://www.aal.net/content/resources/files/BUG_rating.pdf.
16. IES. 2011. Model lighting ordinance (MLO) with user's guide. Illuminating Engineering Society. Accessed June 15, 2011. Available online: http://www.ies.org/PDF/MLO/MLO_FINAL_June2011.pdf.
17. Philips. 2011. The ABCs of electronic fluorescent ballasts. A guide to fluorescent ballasts. Philips Advance. Philips Lighting Electronics, N.A., Rosemont, IL. Accessed September 13, 2013. Available online: http://www.siongboon.com/projects/2010-08-22 electronic ballast/ABC of electronic fluorescent ballast.pdf.
18. DOE. 2013. Energy efficiency of LEDs, building technologies program solid-state lighting technology fact sheet. U.S. Department of Energy. National Academies Press, Washington, DC. Accessed September 13, 2013. Available online: http://www1.eere.energy.gov/buildings/ssl/factsheets.html.
19. DOE. January 2006. 2005 High intensity discharge lighting technology workshop report. U.S. Department of Energy, Energy Efficiency & Renewable Energy Building Technologies Program R&D, Washington, DC.
20. Scholand, M. 2012. Max Tech and beyond: High-intensity discharge lamps. Lawrence Berkeley National Laboratory, Berkeley, CA.
21. NAS. 2013. Assessment of advanced solid state lighting. National Academy of Sciences, Committee on Assessment of Solid State Lighting; Board on Energy and Environmental Systems; Division on Engineering and Physical Sciences; National Research Council. Accessed September 13, 2013. Available online: http://www.nap.edu/openbook.php?record_id=18279.
22. Williams, A. et al. 2012. A meta-analysis of energy savings from lighting controls in commercial buildings. *LEUKOS* 8(3): 161–180. Accessed September 13, 2013. Available online: http://www.ies.org/leukos/samples/1_Jan12.pdf.
23. DOE. 2013. GSFL-IRL preliminary analysis technical support document, Appendix 9A. Lighting controls, U.S. Department of Energy. Accessed February 28, 2013. Available online: http://www.regulations.gov/#!documentDetail;D=EERE-2011-BT-STD-0006-0022.
24. NEMA. 2008. Procedure for determining target efficacy ratings for commercial, industrial, and residential luminaires. National Electrical Manufacturers Association, NEMA Standards Publication LE 6, Rosslyn, VA.
25. Tuenge, J. et al. 2013. Life-cycle assessment of energy and environmental impacts of LED lighting products, Part 3: LED environmental testing. U.S. Department of Energy, Solid State Lighting Program. Accessed March 23, 2013. Available online: http://apps1.eere.energy.gov/buildings/publications/pdfs/ssl/lca_webcast_03-28-2013.pdf.
26. en.lighten. 2013. en.lighten initiative. Achieving the global transition to energy efficient lighting toolkit. Available online: http://www.thegef.org/gef/sites/thegef.org/files/publication/Complete EnlightenToolkit_1.pdf.
27. en.lighten. 2013. en.lighten initiative. Global policy map. Accessed September 13, 2013. Available online: http://www.enlighten-initiative.org/portal/CountrySupport/GlobalPolicyMap/tabid/104292/Default.aspx.
28. CLASP. 2013. Collaborative labeling and appliance standards project. Accessed September 13, 2013. Available online: http://www.clasponline.org/.
29. IEA-4E. July 2011. Impact of 'phase-out' regulations on lighting markets, Benchmarking report. International Energy Agency 4E, Efficient Electric End-Use Equipment. Accessed September 13, 2013. Available online: http://mappingandbenchmarking.iea-4e.org/shared_files/231/download.

30. EPA and DOE. 2013. Find ENERGY STAR products. Energy Star Program, U.S. Environmental Protection Agency and U.S. Department of Energy. Accessed September 13, 2013. Available online: http://www.energystar.gov/index.cfm?c=products.pr_find_es_products.
31. PROCEL. 2013. Selo PROCEL: Apresentação. Programa Nacional de Conservação de Energia Elétrica, Eletrobras. Accessed September 13, 2013. Available online: http://www.eletrobras.com/elb/procel/main.asp?TeamID0{95F19022-F8BB-4991-862A-1C116F13AB71}.
32. ESMAP. 2012. Case study Vietnam compact fluorescent lamp Program. Energy Sector Management Assistance Program. Accessed September 13, 2013. Available online: http://www.org//sites/.org/files/18.%20Vietnam_CFL_Case_Study.pdf.
33. DOE. 2013. Energy-efficient product procurement. U.S. Department of Energy, Federal Energy Management Program. Accessed September 13, 2013. Available online: http://www1.eere.energy.gov/femp/technologies/procuring_eeproducts.html.
34. Lighting Controls Association. 2013. ASHRAE Releases 90.1-2010—Part 2: Lighting controls. Accessed September 13, 2013. Available online: http://lightingcontrolsassociation.org/ashrae-releases-90-1-2010-part-2-lighting-controls/.
35. DiLaura, D. et al. 2011. *The Lighting Handbook: Reference and Application*, 10th edn. Illuminating Engineering Society of North America, New York.
36. CEC. December 2008. 2008 Building energy efficiency standards for residential and non-residential buildings. California Energy Commission. CEC-400-2008-001-CMF. Accessed September 13, 2013. Available online: http://www.energy.ca.gov/2008publications/CEC-400-2008-001/CEC-400-2008-001-CMF.PDF.
37. ALG. 2013. Advanced lighting guidelines. Sources & auxiliaries. Accessed April 2013. Available online: http://algonline.com/index.php?sources-auxiliares.

26

Energy-Efficient Technologies: Major Appliances and Space Conditioning Equipment

Eric Kleinert, James E. McMahon, Greg Rosenquist, James Lutz, Alex Lekov, Peter Biermayer, and Stephen Meyers

CONTENTS

26.1 Introduction

The annual electricity consumption in the U.S. residential sector in 2012 was 3.76 billion kW h/day. By the end of 2013, the total was 3.77 billion kW h/day. In 2014, the U.S. residential sector used approximately 3.73 billion kW h/day. The average annual residential electricity usage per customer in 2012 was 10,834 kW h. By the end of 2013, it was approximately 10,786 kW h and in 2014, it was 10,612 kW h (EIA short-term outlook, October 2013). That is roughly an annual cost of $1200.00 in 2012, $1300.00 in 2013, and $1320.00 in 2014.

The U.S. Energy Information Administration (EIA) uses the British thermal unit (Btu) as a common energy unit to compare or add up energy consumption across the different energy sources that produce electricity. The United States uses several kinds of energy (petroleum, coal, natural gas, propane, nuclear, hydroelectric, wind, wood, biomass waste, geothermal, and solar). It is much easier to convert the various types of energy into Btus and come up with a total for electricity consumption for the various sectors and products.*

In April, 2013, the EIA came out with a scenario for the next 30 years of electrical consumption in the residential sector indicating that the United States will be consuming more electricity. This does not mean that our products are consuming more electricity; it means that there are going to be more homes, people,† and electrical products over the next 30 years (see Table 26.1).

Reducing the energy consumption of residential appliances and space conditioning equipment depends on replacing older equipment with the much more efficient models available now and on continuing to design even more energy-efficient appliances. National energy efficiency standards for appliances have driven efficiency improvements over the last 20 years, and appliances have become significantly more efficient as a result. Further improvement of appliance efficiency represents a significant untapped technological opportunity.

* Electricity 1 kilowatt hour (kW h) = 3412 Btu (but on average, it takes about three times the Btu of primary energy to generate the electricity [EIA]).

† U.S. population will kept increasing, see http://www.census.gov/popclock/.

TABLE 26.1

Residential Sector Key Indicators and Consumption (Quadrillion, Btu, Unless Otherwise Noted)

ref2013.d102312a	2010	2011	2012	2013	2014	2015	2016	2017	2018	2019	2020	2021	2022	2023	2024	2025	2026	2027	2028	2029	2030	2031	2032	2033	2034	2035	2036	2037	2038	2039	2040
	Report	Annual Energy Outlook 2013																													
	Scenario	ref2013	Reference Case																												
	Datekey	d102312a																													
	Release Date	April 2013																													
4. Residential Sector Key Indicators and Consumption (quadrillion Btu, unless otherwise noted)																															
Key Indicators and Consumption	2010	2011	2012	2013	2014	2015	2016	2017	2018	2019	2020	2021	2022	2023	2024	2025	2026	2027	2028	2029	2030	2031	2032	2033	2034	2035	2036	2037	2038	2039	2040
Key indicators																															
Households (millions)																															
Single-family	82.85	83.56	84.53	85.57	86.08	86.86	87.73	88.63	89.52	90.39	91.25	92.07	92.87	93.68	94.52	95.37	96.22	97.03	97.81	98.57	99.34	100.09	100.82	101.54	102.28	103.03	103.80	104.55	105.30	106.03	106.77
Multifamily	25.78	26.07	26.43	26.82	27.19	27.57	27.98	28.44	28.90	29.36	29.82	30.27	30.72	31.16	31.60	32.05	32.52	33.01	33.52	34.03	34.54	35.05	35.55	36.06	36.55	37.05	37.55	38.04	38.54	39.04	39.53
Mobile homes	6.60	6.54	6.50	6.46	6.40	6.36	6.35	6.36	6.39	6.42	6.45	6.48	6.51	6.54	6.57	6.60	6.63	6.66	6.69	6.72	6.75	6.77	6.80	6.83	6.85	6.88	6.91	6.93	6.96	6.99	7.02
Total	**115.23**	**116.17**	**117.46**	**118.86**	**119.67**	**120.78**	**122.05**	**123.42**	**124.80**	**126.18**	**127.52**	**128.83**	**130.10**	**131.38**	**132.69**	**134.02**	**135.37**	**136.71**	**138.02**	**139.32**	**140.63**	**141.91**	**143.17**	**144.42**	**145.69**	**146.96**	**148.25**	**149.53**	**150.80**	**152.06**	**153.32**
Average house square footage	**1653**	**1659**	**1665**	**1671**	**1676**	**1682**	**1687**	**1691**	**1696**	**1700**	**1704**	**1708**	**1712**	**1716**	**1720**	**1724**	**1728**	**1731**	**1734**	**1737**	**1740**	**1743**	**1746**	**1748**	**1751**	**1754**	**1757**	**1759**	**1762**	**1764**	**1767**
Energy intensity (million Btu per household)																															
Delivered energy consumption	99.2	97.2	91.6	94.7	92.2	91.1	89.9	88.8	87.8	87.0	86.0	85.1	84.4	83.7	83.0	82.5	81.9	81.3	80.7	80.2	79.7	79.2	78.7	78.2	77.8	77.3	76.9	76.5	76.1	75.8	75.5
Total energy consumption	189.0	185.0	174.1	175.4	171.9	169.2	166.6	164.9	163.9	163.3	161.7	160.5	159.6	158.8	158.0	157.4	156.5	155.9	155.2	154.6	154.0	153.4	152.8	152.3	151.9	151.4	151.0	150.7	150.5	150.5	150.6
(thousand Btu per square foot)																															
Delivered energy consumption	60.0	58.6	55.0	56.6	55.0	54.1	53.3	52.5	51.8	51.2	50.4	49.8	49.3	48.8	48.3	47.8	47.4	47.0	46.6	46.2	45.8	45.4	45.1	44.7	44.4	44.1	43.7	43.5	43.2	43.0	42.7
Total energy consumption	114.3	111.5	104.5	104.9	102.6	100.6	98.8	97.5	96.6	96.1	94.9	93.9	93.2	92.5	91.9	91.3	90.6	90.1	89.5	89.0	88.5	88.0	87.6	87.1	86.7	86.3	86.0	85.7	85.4	85.3	85.2
Delivered energy consumption by fuel																															
Purchased electricity																															
Space heating	0.30	0.27	0.25	0.29	0.28	0.28	0.28	0.29	0.29	0.29	0.29	0.30	0.30	0.30	0.30	0.30	0.31	0.31	0.31	0.31	0.31	0.32	0.32	0.32	0.32	0.32	0.32	0.32	0.32	0.32	0.32
Space cooling	0.92	0.93	0.93	0.80	0.89	0.89	0.90	0.91	0.92	0.94	0.95	0.97	0.98	1.00	1.02	1.04	1.06	1.08	1.10	1.12	1.14	1.15	1.17	1.19	1.21	1.23	1.25	1.27	1.28	1.30	1.32
Water heating	0.45	0.45	0.46	0.47	0.48	0.48	0.49	0.49	0.49	0.50	0.50	0.51	0.51	0.52	0.52	0.52	0.53	0.53	0.53	0.53	0.53	0.53	0.53	0.53	0.53	0.54	0.54	0.54	0.54	0.55	0.55
Refrigeration	0.38	0.38	0.38	0.38	0.38	0.37	0.37	0.38	0.38	0.38	0.38	0.38	0.38	0.39	0.39	0.39	0.39	0.40	0.40	0.41	0.41	0.41	0.42	0.42	0.43	0.43	0.43	0.44	0.44	0.45	0.45

(Continued)

TABLE 26.1 (*Continued*)

Residential Sector Key Indicators and Consumption (Quadrillion, Btu, Unless Otherwise Noted)

ref2013.d102312a	2010	2011	2012	2013	2014	2015	2016	2017	2018	2019	2020	2021	2022	2023	2024	2025	2026	2027	2028	2029	2030	2031	2032	2033	2034	2035	2036	2037	2038	2039	2040
Cooking	0.11	0.11	0.11	0.11	0.11	0.12	0.12	0.12	0.12	0.12	0.12	0.13	0.13	0.13	0.13	0.13	0.13	0.14	0.14	0.14	0.14	0.14	0.15	0.15	0.15	0.15	0.15	0.15	0.16	0.16	0.16
Clothes dryers	0.20	0.20	0.20	0.20	0.21	0.21	0.21	0.21	0.21	0.21	0.22	0.22	0.22	0.22	0.22	0.23	0.23	0.23	0.23	0.24	0.24	0.24	0.24	0.25	0.25	0.25	0.25	0.25	0.26	0.26	0.26
Freezers	0.08	0.08	0.08	0.08	0.08	0.08	0.08	0.08	0.08	0.08	0.08	0.08	0.08	0.08	0.08	0.08	0.08	0.08	0.08	0.08	0.08	0.08	0.08	0.08	0.08	0.08	0.08	0.08	0.08	0.08	0.09
Lighting	0.65	0.63	0.63	0.55	0.52	0.50	0.50	0.49	0.49	0.49	0.45	0.43	0.42	0.41	0.40	0.40	0.39	0.39	0.39	0.38	0.38	0.38	0.37	0.37	0.37	0.37	0.37	0.37	0.38	0.38	0.38
Clothes washers[a]	0.03	0.03	0.03	0.03	0.03	0.03	0.03	0.03	0.03	0.03	0.03	0.03	0.03	0.03	0.03	0.02	0.02	0.02	0.02	0.02	0.02	0.02	0.02	0.02	0.02	0.02	0.02	0.02	0.03	0.03	0.03
Dishwashers[a]	0.10	0.10	0.10	0.10	0.10	0.10	0.10	0.10	0.10	0.10	0.10	0.10	0.10	0.10	0.10	0.10	0.11	0.11	0.11	0.11	0.11	0.11	0.11	0.12	0.12	0.12	0.12	0.12	0.13	0.13	0.13
Televisions and related equipment[b]	0.32	0.32	0.32	0.33	0.33	0.33	0.34	0.34	0.34	0.35	0.35	0.36	0.36	0.37	0.37	0.37	0.38	0.38	0.39	0.39	0.40	0.41	0.41	0.42	0.42	0.43	0.43	0.44	0.44	0.45	0.45
Computers and related equipment[c]	0.16	0.16	0.15	0.14	0.14	0.13	0.13	0.13	0.13	0.13	0.13	0.13	0.13	0.13	0.13	0.12	0.12	0.12	0.12	0.12	0.12	0.12	0.12	0.12	0.12	0.12	0.12	0.12	0.12	0.13	0.13
Furnace fans and boiler circulation pumps	0.13	0.13	0.13	0.14	0.14	0.14	0.14	0.14	0.14	0.14	0.14	0.14	0.14	0.14	0.14	0.14	0.14	0.14	0.14	0.14	0.14	0.14	0.14	0.14	0.14	0.14	0.14	0.14	0.14	0.14	0.14
Other uses[d]	1.11	1.07	0.95	0.99	0.97	0.98	1.00	1.01	1.03	1.06	1.08	1.11	1.13	1.16	1.18	1.21	1.23	1.26	1.28	1.30	1.33	1.36	1.38	1.41	1.43	1.46	1.48	1.52	1.55	1.59	1.62
Delivered energy	**4.93**	**4.86**	**4.72**	**4.64**	**4.65**	**4.66**	**4.68**	**4.72**	**4.77**	**4.83**	**4.84**	**4.88**	**4.92**	**4.97**	**5.02**	**5.08**	**5.13**	**5.19**	**5.25**	**5.30**	**5.36**	**5.42**	**5.48**	**5.54**	**5.61**	**5.67**	**5.73**	**5.80**	**5.88**	**5.95**	**6.03**
Natural gas																															
Space heating	3.32	3.25	2.89	3.32	3.16	3.15	3.13	3.10	3.06	3.04	3.02	3.00	2.98	2.95	2.94	2.92	2.91	2.89	2.88	2.87	2.85	2.84	2.82	2.81	2.79	2.77	2.75	2.73	2.70	2.68	2.67
Space cooling	0.00	0.00	0.00	0.00	0.00	0.00	0.00	0.00	0.00	0.00	0.00	0.00	0.00	0.00	0.00	0.00	0.00	0.00	0.00	0.00	0.00	0.00	0.00	0.00	0.00	0.00	0.00	0.00	0.00	0.00	0.00
Water heating	1.30	1.30	1.33	1.33	1.34	1.33	1.32	1.32	1.32	1.32	1.33	1.33	1.33	1.33	1.33	1.33	1.33	1.33	1.32	1.32	1.31	1.30	1.29	1.28	1.28	1.27	1.27	1.26	1.26	1.26	1.26
Cooking	0.22	0.22	0.22	0.22	0.22	0.22	0.22	0.22	0.22	0.22	0.22	0.22	0.22	0.22	0.22	0.22	0.22	0.22	0.23	0.23	0.23	0.23	0.23	0.23	0.23	0.23	0.23	0.23	0.23	0.23	0.24
Clothes dryers	0.06	0.06	0.06	0.06	0.06	0.06	0.06	0.06	0.06	0.06	0.06	0.06	0.06	0.06	0.06	0.06	0.07	0.07	0.07	0.07	0.07	0.07	0.07	0.07	0.07	0.07	0.07	0.07	0.07	0.07	0.07
Delivered energy	**4.89**	**4.83**	**4.49**	**4.93**	**4.78**	**4.76**	**4.73**	**4.70**	**4.66**	**4.64**	**4.62**	**4.61**	**4.59**	**4.57**	**4.55**	**4.54**	**4.52**	**4.51**	**4.49**	**4.47**	**4.46**	**4.44**	**4.41**	**4.39**	**4.37**	**4.34**	**4.31**	**4.29**	**4.27**	**4.25**	**4.23**
Distillate fuel oil																															
Space heating	0.49	0.50	0.49	0.55	0.51	0.50	0.49	0.48	0.47	0.46	0.45	0.44	0.43	0.42	0.41	0.40	0.39	0.38	0.37	0.36	0.36	0.35	0.34	0.33	0.33	0.32	0.31	0.31	0.30	0.29	0.29
Water heating	0.10	0.09	0.08	0.08	0.08	0.07	0.07	0.06	0.06	0.06	0.06	0.06	0.05	0.05	0.05	0.05	0.05	0.05	0.04	0.04	0.04	0.04	0.04	0.04	0.04	0.04	0.04	0.04	0.03	0.03	0.03
Delivered energy	0.58	0.59	0.57	0.63	0.59	0.58	0.56	0.55	0.53	0.52	0.51	0.49	0.48	0.47	0.46	0.45	0.44	0.43	0.42	0.41	0.40	0.39	0.38	0.37	0.36	0.36	0.35	0.34	0.33	0.33	0.32
Propane																															
Space heating	0.28	0.27	0.25	0.28	0.27	0.27	0.26	0.26	0.26	0.25	0.25	0.25	0.25	0.24	0.24	0.24	0.24	0.23	0.23	0.23	0.23	0.23	0.22	0.22	0.22	0.22	0.22	0.22	0.21	0.21	0.21
Water heating	0.07	0.07	0.07	0.06	0.06	0.06	0.06	0.05	0.05	0.05	0.05	0.05	0.05	0.05	0.05	0.05	0.05	0.05	0.05	0.05	0.05	0.04	0.04	0.04	0.04	0.04	0.04	0.04	0.04	0.04	0.04
Cooking	0.03	0.03	0.03	0.03	0.03	0.03	0.03	0.03	0.03	0.03	0.03	0.03	0.03	0.03	0.03	0.03	0.03	0.03	0.03	0.03	0.03	0.03	0.03	0.03	0.03	0.03	0.03	0.03	0.03	0.03	0.02
Other uses[e]	0.15	0.16	0.17	0.17	0.18	0.18	0.18	0.19	0.19	0.19	0.19	0.20	0.20	0.20	0.20	0.21	0.21	0.21	0.21	0.22	0.22	0.22	0.23	0.23	0.23	0.23	0.24	0.24	0.24	0.25	0.25
Delivered energy	**0.53**	**0.53**	**0.51**	**0.54**	**0.53**	**0.53**	**0.53**	**0.53**	**0.53**	**0.53**	**0.52**	**0.52**	**0.52**	**0.52**	**0.52**	**0.52**	**0.52**	**0.52**	**0.52**	**0.52**	**0.52**	**0.52**	**0.52**	**0.52**	**0.52**	**0.52**	**0.52**	**0.52**	**0.52**	**0.52**	**0.52**
Marketed renewables (wood)[f]	0.44	0.45	0.42	0.48	0.44	0.43	0.43	0.43	0.43	0.43	0.44	0.44	0.44	0.44	0.44	0.44	0.44	0.44	0.44	0.45	0.45	0.45	0.45	0.45	0.45	0.45	0.45	0.45	0.45	0.45	0.45
Other fuels[g]	0.04	0.02	0.02	0.03	0.02	0.02	0.02	0.02	0.02	0.02	0.02	0.02	0.02	0.02	0.02	0.02	0.02	0.02	0.02	0.02	0.02	0.02	0.02	0.02	0.02	0.02	0.02	0.02	0.02	0.02	0.02
Delivered energy consumption by end use																															
Space heating	4.86	4.76	4.32	4.94	4.69	4.66	4.62	4.58	4.54	4.50	4.47	4.44	4.41	4.38	4.35	4.32	4.30	4.28	4.26	4.24	4.22	4.19	4.17	4.15	4.12	4.09	4.07	4.04	4.01	3.98	3.96

(*Continued*)

TABLE 26.1 (*Continued*)

Residential Sector Key Indicators and Consumption (Quadrillion, Btu, Unless Otherwise Noted)

ref2013.d102312a	2010	2011	2012	2013	2014	2015	2016	2017	2018	2019	2020	2021	2022	2023	2024	2025	2026	2027	2028	2029	2030	2031	2032	2033	2034	2035	2036	2037	2038	2039	2040
Space cooling	0.92	0.93	0.93	0.80	0.89	0.89	0.90	0.91	0.92	0.94	0.95	0.97	0.98	1.00	1.02	1.04	1.06	1.08	1.10	1.12	1.14	1.15	1.17	1.19	1.21	1.23	1.25	1.27	1.28	1.30	1.32
Water heating	1.91	1.91	1.94	1.95	1.96	1.94	1.93	1.93	1.93	1.93	1.94	1.94	1.95	1.95	1.95	1.95	1.95	1.95	1.94	1.94	1.93	1.92	1.91	1.90	1.89	1.89	1.88	1.88	1.88	1.88	1.89
Refrigeration	0.38	0.38	0.38	0.38	0.38	0.37	0.37	0.38	0.38	0.38	0.38	0.38	0.38	0.39	0.39	0.39	0.39	0.40	0.40	0.41	0.41	0.41	0.42	0.42	0.43	0.43	0.43	0.44	0.44	0.45	0.45
Cooking	0.36	0.36	0.36	0.36	0.36	0.36	0.36	0.37	0.37	0.37	0.37	0.37	0.38	0.38	0.38	0.38	0.38	0.39	0.39	0.39	0.40	0.40	0.40	0.40	0.40	0.41	0.41	0.41	0.41	0.42	0.42
Clothes dryers	0.25	0.25	0.26	0.26	0.27	0.27	0.27	0.27	0.27	0.28	0.28	0.28	0.28	0.29	0.29	0.29	0.29	0.30	0.30	0.30	0.30	0.31	0.31	0.31	0.32	0.32	0.32	0.32	0.33	0.33	0.33
Freezers	0.08	0.08	0.08	0.08	0.08	0.08	0.08	0.08	0.08	0.08	0.08	0.08	0.08	0.08	0.08	0.08	0.08	0.08	0.08	0.08	0.08	0.08	0.08	0.08	0.08	0.08	0.08	0.08	0.08	0.08	0.09
Lighting	0.65	0.63	0.63	0.55	0.52	0.50	0.50	0.49	0.49	0.49	0.45	0.43	0.42	0.41	0.40	0.40	0.39	0.39	0.39	0.38	0.38	0.38	0.37	0.37	0.37	0.37	0.37	0.37	0.38	0.38	0.38
Clothes washers[a]	0.03	0.03	0.03	0.03	0.03	0.03	0.03	0.03	0.03	0.03	0.03	0.03	0.03	0.03	0.03	0.02	0.02	0.02	0.02	0.02	0.02	0.02	0.02	0.02	0.02	0.02	0.02	0.02	0.03	0.03	0.03
Dishwashers[a]	0.10	0.10	0.10	0.10	0.10	0.10	0.10	0.10	0.10	0.10	0.10	0.10	0.10	0.10	0.10	0.10	0.11	0.11	0.11	0.11	0.11	0.11	0.11	0.12	0.12	0.12	0.12	0.12	0.13	0.13	0.13
Televisions and related equipment[b]	0.32	0.32	0.32	0.33	0.33	0.33	0.34	0.34	0.34	0.35	0.35	0.36	0.36	0.37	0.37	0.37	0.38	0.38	0.39	0.39	0.40	0.41	0.41	0.42	0.42	0.43	0.43	0.44	0.44	0.45	0.45
Computers and related equipment[c]	0.16	0.16	0.15	0.14	0.14	0.13	0.13	0.13	0.13	0.13	0.13	0.13	0.13	0.13	0.13	0.12	0.12	0.12	0.12	0.12	0.12	0.12	0.12	0.12	0.12	0.12	0.12	0.12	0.12	0.13	0.13
Furnace fans and boiler circulation pumps	0.13	0.13	0.13	0.14	0.14	0.14	0.14	0.14	0.14	0.14	0.14	0.14	0.14	0.14	0.14	0.14	0.14	0.14	0.14	0.14	0.14	0.14	0.14	0.14	0.14	0.14	0.14	0.14	0.14	0.14	0.14
Other uses[h]	1.26	1.23	1.12	1.17	1.14	1.16	1.18	1.20	1.22	1.25	1.28	1.30	1.33	1.36	1.38	1.41	1.44	1.47	1.50	1.52	1.55	1.58	1.61	1.64	1.66	1.69	1.72	1.76	1.79	1.83	1.87
Delivered energy	**11.41**	**11.28**	**10.75**	**11.24**	**11.02**	**10.99**	**10.96**	**10.95**	**10.95**	**10.97**	**10.95**	**10.96**	**10.97**	**10.99**	**11.01**	**11.04**	**11.07**	**11.11**	**11.14**	**11.17**	**11.20**	**11.23**	**11.26**	**11.29**	**11.32**	**11.35**	**11.39**	**11.43**	**11.47**	**11.52**	**11.57**
Electricity related losses	**10.35**	**10.20**	**9.69**	**9.59**	**9.54**	**9.44**	**9.36**	**9.40**	**9.50**	**9.63**	**9.66**	**9.71**	**9.79**	**9.87**	**9.95**	**10.04**	**10.11**	**10.20**	**10.28**	**10.36**	**10.45**	**10.53**	**10.61**	**10.70**	**10.80**	**10.90**	**11.00**	**11.10**	**11.23**	**11.36**	**11.50**
Total energy consumption by end use																															
Space heating	5.49	5.33	4.83	5.55	5.26	5.23	5.19	5.15	5.11	5.08	5.05	5.03	5.00	4.97	4.95	4.93	4.90	4.88	4.86	4.84	4.83	4.81	4.78	4.76	4.74	4.71	4.68	4.65	4.62	4.60	4.57
Space cooling	2.84	2.88	2.83	2.46	2.71	2.70	2.69	2.72	2.76	2.81	2.86	2.90	2.94	2.99	3.04	3.10	3.15	3.20	3.24	3.29	3.35	3.40	3.45	3.50	3.55	3.60	3.65	3.69	3.74	3.79	3.84
Water heating	2.85	2.85	2.89	2.93	2.94	2.92	2.90	2.90	2.91	2.93	2.95	2.96	2.97	2.98	2.98	2.99	2.99	2.99	2.98	2.97	2.97	2.95	2.94	2.93	2.92	2.92	2.92	2.92	2.92	2.93	2.94
Refrigeration	1.16	1.16	1.15	1.16	1.15	1.13	1.12	1.12	1.13	1.13	1.14	1.14	1.15	1.15	1.16	1.16	1.17	1.18	1.19	1.20	1.21	1.22	1.23	1.23	1.24	1.25	1.26	1.27	1.28	1.30	1.31
Cooking	0.58	0.59	0.59	0.59	0.59	0.60	0.60	0.60	0.61	0.61	0.62	0.62	0.63	0.63	0.64	0.65	0.65	0.66	0.66	0.67	0.67	0.68	0.68	0.69	0.69	0.70	0.70	0.71	0.71	0.72	0.72
Clothes dryers	0.66	0.66	0.67	0.68	0.69	0.69	0.68	0.69	0.69	0.70	0.71	0.71	0.72	0.73	0.73	0.74	0.74	0.75	0.76	0.76	0.77	0.77	0.78	0.79	0.79	0.80	0.80	0.81	0.82	0.82	0.83
Freezers	0.25	0.26	0.25	0.26	0.26	0.25	0.25	0.25	0.25	0.25	0.25	0.25	0.25	0.25	0.25	0.25	0.25	0.25	0.25	0.25	0.25	0.24	0.24	0.24	0.24	0.24	0.24	0.25	0.25	0.25	0.25
Lighting	2.02	1.97	1.93	1.69	1.58	1.52	1.49	1.47	1.47	1.48	1.35	1.29	1.25	1.23	1.21	1.19	1.17	1.16	1.14	1.13	1.11	1.10	1.10	1.09	1.09	1.09	1.09	1.09	1.09	1.10	1.10
Clothes washers[a]	0.10	0.10	0.10	0.10	0.10	0.10	0.09	0.09	0.09	0.08	0.08	0.08	0.08	0.08	0.08	0.07	0.07	0.07	0.07	0.07	0.07	0.07	0.07	0.07	0.07	0.07	0.07	0.07	0.07	0.07	0.07
Dishwashers[a]	0.32	0.32	0.32	0.32	0.32	0.31	0.31	0.31	0.31	0.31	0.31	0.31	0.31	0.31	0.31	0.31	0.31	0.32	0.32	0.32	0.33	0.33	0.34	0.34	0.35	0.35	0.35	0.36	0.36	0.37	0.37
Televisions and related equipment[b]	0.98	0.98	0.99	1.01	1.01	1.01	1.01	1.01	1.02	1.04	1.05	1.06	1.08	1.09	1.10	1.12	1.13	1.14	1.15	1.17	1.18	1.19	1.21	1.22	1.23	1.25	1.26	1.27	1.29	1.30	1.32
Computers and related equipment[c]	0.49	0.49	0.45	0.43	0.41	0.40	0.40	0.39	0.39	0.39	0.39	0.39	0.38	0.38	0.37	0.37	0.37	0.37	0.36	0.36	0.36	0.36	0.36	0.36	0.36	0.36	0.36	0.36	0.36	0.36	0.36
Furnace fans and boiler circulation pumps	0.42	0.42	0.39	0.43	0.42	0.42	0.42	0.41	0.42	0.42	0.42	0.42	0.42	0.42	0.42	0.42	0.42	0.42	0.42	0.42	0.42	0.42	0.42	0.42	0.42	0.42	0.42	0.41	0.41	0.41	0.41
Other uses[h]	3.60	3.48	3.06	3.22	3.13	3.15	3.17	3.22	3.28	3.36	3.44	3.51	3.59	3.66	3.73	3.80	3.86	3.93	4.00	4.07	4.14	4.22	4.28	4.35	4.42	4.49	4.57	4.66	4.76	4.86	4.97
Total	**21.76**	**21.48**	**20.43**	**20.83**	**20.56**	**20.42**	**20.32**	**20.35**	**20.44**	**20.59**	**20.62**	**20.66**	**20.76**	**20.85**	**20.96**	**21.08**	**21.18**	**21.31**	**21.42**	**21.53**	**21.65**	**21.76**	**21.88**	**21.99**	**22.12**	**22.25**	**22.38**	**22.53**	**22.69**	**22.88**	**23.08**

(Continued)

TABLE 26.1 (*Continued*)

Residential Sector Key Indicators and Consumption (Quadrillion, Btu, Unless Otherwise Noted)

ref2013.d102312a	2010	2011	2012	2013	2014	2015	2016	2017	2018	2019	2020	2021	2022	2023	2024	2025	2026	2027	2028	2029	2030	2031	2032	2033	2034	2035	2036	2037	2038	2039	2040
Nonmarketed renewables[i]																															
Geothermal heat pumps	0.01	0.01	0.01	0.01	0.01	0.02	0.02	0.02	0.02	0.02	0.02	0.02	0.02	0.02	0.02	0.02	0.02	0.02	0.02	0.02	0.02	0.02	0.02	0.02	0.03	0.03	0.03	0.03	0.03	0.03	0.03
Solar hot water heating	0.01	0.01	0.01	0.01	0.02	0.02	0.02	0.02	0.02	0.02	0.02	0.02	0.02	0.02	0.02	0.02	0.02	0.02	0.02	0.02	0.02	0.02	0.02	0.02	0.02	0.02	0.02	0.02	0.02	0.02	0.02
Solar photovoltaic	0.01	0.02	0.04	0.07	0.09	0.11	0.14	0.14	0.14	0.14	0.14	0.15	0.15	0.15	0.15	0.15	0.16	0.16	0.16	0.16	0.17	0.17	0.17	0.18	0.18	0.18	0.19	0.19	0.20	0.21	0.21
Wind	0.00	0.00	0.00	0.00	0.01	0.01	0.01	0.01	0.01	0.01	0.01	0.01	0.01	0.01	0.01	0.01	0.01	0.01	0.01	0.01	0.01	0.01	0.01	0.01	0.01	0.01	0.01	0.01	0.01	0.01	0.01
Total	**0.03**	**0.04**	**0.07**	**0.10**	**0.12**	**0.16**	**0.19**	**0.19**	**0.19**	**0.19**	**0.20**	**0.20**	**0.20**	**0.20**	**0.20**	**0.21**	**0.21**	**0.21**	**0.21**	**0.22**	**0.22**	**0.22**	**0.23**	**0.23**	**0.24**	**0.24**	**0.25**	**0.25**	**0.26**	**0.26**	**0.27**
Heating degree Days																															
New England	5944	6138	5796	6490	6215	6200	6186	6172	6158	6144	6131	6117	6103	6089	6075	6062	6048	6034	6020	6006	5992	5978	5964	5950	5936	5922	5907	5893	5879	5865	5850
Middle Atlantic	5453	5413	5038	5779	5459	5443	5427	5411	5394	5378	5362	5346	5330	5314	5298	5281	5265	5249	5233	5217	5201	5185	5169	5153	5137	5121	5105	5089	5073	5058	5042
East North Central	6209	6187	5462	6358	6137	6127	6116	6105	6094	6084	6073	6062	6051	6040	6030	6019	6008	5997	5986	5976	5965	5954	5943	5932	5922	5911	5900	5889	5878	5867	5856
West North Central	6585	6646	5633	6619	6372	6360	6348	6336	6323	6310	6297	6284	6271	6257	6244	6230	6216	6202	6189	6175	6161	6147	6133	6119	6105	6091	6077	6063	6049	6035	6020
South Atlantic	3183	2555	2337	2800	2699	2693	2687	2680	2674	2667	2660	2654	2647	2640	2634	2627	2621	2615	2608	2602	2596	2590	2584	2578	2572	2566	2560	2554	2549	2543	2538
East South Central	4003	3397	2934	3599	3438	3435	3431	3428	3424	3421	3417	3414	3410	3407	3403	3400	3396	3393	3389	3386	3382	3378	3375	3371	3367	3364	3360	3356	3352	3349	3345
West South Central	2503	2203	1829	2280	2084	2076	2068	2060	2052	2044	2036	2028	2020	2012	2004	1996	1988	1980	1972	1964	1956	1948	1940	1932	1924	1916	1908	1900	1892	1884	1876
Mountain	4882	5054	4624	5021	4666	4647	4628	4608	4587	4566	4545	4522	4500	4476	4453	4430	4406	4383	4359	4336	4312	4288	4264	4240	4216	4192	4168	4144	4119	4095	4071
Pacific	3202	3411	3170	3239	3112	3109	3106	3104	3101	3097	3094	3091	3087	3083	3079	3076	3072	3068	3064	3061	3057	3053	3050	3046	3043	3039	3036	3033	3029	3026	3022
United States	**4388**	**4240**	**3811**	**4355**	**4146**	**4131**	**4116**	**4101**	**4085**	**4070**	**4054**	**4039**	**4024**	**4008**	**3993**	**3978**	**3963**	**3947**	**3932**	**3918**	**3903**	**3888**	**3873**	**3858**	**3844**	**3829**	**3814**	**3799**	**3785**	**3770**	**3756**
Cooling degree Days																															
New England	655	607	604	465	559	564	569	573	578	583	588	592	597	602	607	611	616	621	625	630	635	640	644	649	654	659	664	668	673	678	683
Middle Atlantic	997	887	889	688	834	841	848	854	861	868	875	882	889	896	902	909	916	923	930	937	944	950	957	964	971	978	984	991	998	1005	1011
East North Central	978	898	1011	746	793	795	797	799	801	803	805	807	809	811	813	815	817	819	821	822	824	826	828	830	832	834	836	838	840	842	844
West North Central	1123	1116	1241	957	987	988	990	991	992	994	995	997	998	1000	1002	1003	1005	1007	1008	1010	1012	1014	1015	1017	1019	1021	1022	1024	1026	1028	1030
South Atlantic	2289	2357	2240	2041	2177	2185	2193	2202	2210	2219	2228	2236	2245	2254	2262	2271	2279	2288	2296	2305	2313	2322	2330	2339	2347	2356	2364	2372	2380	2389	2397
East South Central	1999	1811	1817	1597	1740	1746	1753	1760	1766	1773	1779	1786	1792	1799	1805	1812	1818	1825	1831	1838	1845	1851	1858	1864	1871	1877	1884	1890	1897	1903	1910
West South Central	2755	3194	2881	2512	2772	2784	2797	2810	2822	2835	2847	2860	2873	2885	2898	2911	2923	2936	2948	2961	2974	2986	2999	3011	3024	3037	3049	3062	3074	3087	3099
Mountain	1490	1396	1522	1420	1628	1639	1650	1662	1674	1686	1698	1711	1725	1739	1752	1766	1780	1794	1808	1822	1837	1851	1866	1880	1895	1910	1925	1940	1955	1970	1985
Pacific	746	809	888	844	898	901	903	906	908	910	913	915	918	920	923	925	928	930	933	935	938	940	942	945	947	950	952	954	957	959	961
United States	**1498**	**1528**	**1518**	**1319**	**1444**	**1453**	**1462**	**1471**	**1480**	**1489**	**1499**	**1508**	**1517**	**1526**	**1535**	**1545**	**1554**	**1563**	**1572**	**1582**	**1591**	**1600**	**1610**	**1619**	**1628**	**1638**	**1647**	**1657**	**1666**	**1676**	**1685**

Source: 2010 and 2011 consumption based on: U.S. Energy Information Administration (EIA), Annual Energy Review 2011, DOE/EIA-0384(2011) (Washington, DC, September 2012). 2010 and 2011 degree days based on state-level data from the National Oceanic and Atmospheric Administration's Climatic Data Center and Climate Prediction Center.
Projections: EIA, AEO2013 National Energy Modeling System run ref2013.d102312a.

Note: Totals may not equal sum of components due to independent rounding. Data for 2010 and 2011 are model results and may differ slightly from official EIA data reports. Btu = British thermal unit. - - - = Not applicable.

[a] Does not include water heating portion of load.
[b] Includes televisions, set-top boxes, and video game consoles.
[c] Includes desktop and laptop computers, monitors, printers, speakers, networking equipment, and uninterruptible power supplies.
[d] Includes small electric devices, heating elements, and motors not listed above. Electric vehicles are included in the transportation sector.
[e] Includes such appliances as outdoor grills and mosquito traps.
[f] Includes wood used for primary and secondary heating in wood stoves or fireplaces as reported in the Residential Energy Consumption Survey 2005.
[g] Includes kerosene and coal.
[h] Includes all others not listed above.
[i] Consumption determined by using the fossil fuel equivalent of 9756 Btu/kW h.

26.2 Description of Major Appliances and Space Conditioning Equipment

26.2.1 Refrigerator-Freezers and Freezers

Refrigerators, refrigerator-freezers, and freezers keep food cold by transferring heat from the air in the appliance cabinet to the outside of the cabinet. A refrigerator is a well-insulated cabinet used to store food at 34°F (1.1°C) or above, a refrigerator-freezer is a refrigerator with an attached freezer compartment that stores food below 0°F (−17.7°C), and a standalone freezer is a refrigerated cabinet to store and freeze foods at −12°F (−24.4°C) or below. Almost all refrigerators are fueled by electricity. The refrigeration system includes an evaporator, a condenser, a metering device, and a compressor. The system uses a vapor compression cycle, in which the refrigerant changes phase (from high pressure vapor to high pressure liquid and back to low pressure vapor) while circulating in a closed loop system. The refrigerant absorbs or discharges heat as it changes phase. Although most refrigerants and insulating materials once contained chlorofluorocarbons (CFCs), all U.S. models sold after January 1, 1996, are CFC-free. Under the Montreal Protocol, in 1996, refrigerator-freezer manufacturers are using R134a (1,1,1,2-tetrafluoroethane) as the replacement for R-12 (dichlorodifluoromethane) refrigerant. R134a is also sold under other names such as: Dymel 134a, Forane 134a, Genetron 134a, HFA-134a, HFC-134a, R-134a, Suva 134a, and Norflurane.

There are over 170 million refrigerators and refrigerator-freezers currently in use today in U.S. homes with over 60 million refrigerators over 10 years old and are costing consumers billions in excessive energy costs.

26.2.2 Water Heaters

A water heater is an appliance that is used to heat potable water for use outside the heater upon demand. Water heaters supply water to sinks, bathtubs and showers, dishwashers, and clothes washing machines. Most water heaters in the United States are storage water heaters, which continuously maintain a tank of water at a thermostatically controlled temperature. The most common storage water heaters consist of a cylindrical steel tank that is lined with glass in order to prevent corrosion. Most hot water tanks manufactured today are insulated with polyurethane foam and wrapped in a steel jacket. Although some use oil, almost all storage water heaters are fueled by natural gas (or LPG) or electricity.

Rather than storing water at a controlled temperature, instantaneous water heaters (tankless) heat water as it is being drawn through the water heater. Both gas-fired and electric instantaneous water heaters (tankless) are available. Instantaneous water heaters are quite popular in Europe and Asia. Although they are not commonly used in the United States, their presence does seem to be increasing.

Like refrigerators, water heaters are present in almost all U.S. households. Approximately 54% of households have gas-fired water heaters, and approximately 38% have electric water heaters. Hot water use varies significantly from household to household, mostly due to differences in household size and occupant behavior.

The Department of Energy (DOE) final rule effective on April 16, 2015, mandates will require higher energy factor (EF) ratings on virtually all residential gas, electric, oil, and tankless gas water heaters. The EF is the ratio of useful energy output from the water heater to the total amount of energy delivered to the water heater. The higher the EF, the more efficient is the water heater (see Table 26.2).

TABLE 26.2

2015 Energy Conservation Standards for Residential Water Heaters

Product Classes Affected by Change	Rated Storage Volume/Inputs Affected by Change	New EF Requirements
Electric	≥20 and ≤55 gal, ≤12 kW input	0.960 – (0.0003 × V)
	>55 and ≤120 gal, ≤12 kW input	2.057 – (0.00113 × V)
Gas fired	≥20 and ≤55 gal, ≤75,000 Btu/h	0.675 – (0.0015 × V)
	>55 and ≤100 gal, ≤75,000 Btu/h	0.8012 – (0.00078 × V)
Oil fired	≤50 gal, ≤105,000 Btu/h	0.68 – (0.0019 × V)
Instantaneous electric[a]	≤2 gal, ≤12 kW input	0.93 – (0.00132 × V)
Instantaneous gas fired	≤2 gal, ≤200,000 Btu/h	0.82 – (0.0019 × V)

[a] No change.

26.2.3 Furnaces and Boilers

Furnaces and boilers are major household appliances used to provide central space heating. Both fuel-burning and electric furnaces and boilers are available. A typical gas furnace installation is composed of the following basic components: (1) a cabinet or casing; (2) heat exchangers; (3) a system for obtaining air for combustion; (4) a combustion system including burners and controls; (5) a venting system for exhausting combustion products; (6) a circulating air blower and motor; and (7) an air filter and other accessories. (Furnaces that burn oil and liquid petroleum gas (LPG) are also available, though not as common.) In an electric furnace, the casing, air filter, and blower are very similar to those used in a gas furnace. Rather than receiving heat from fuel-fired heat exchangers, however, the air in an electric furnace receives heat from electric heating elements. Controls include electric overload protection, contactor, limit switches, and a fan switch. Furnaces provide heated air through a system of ducts leading to spaces where heat is desired. In a better system, hot water or steam is piped to terminal heating units placed throughout the household. The boiler itself is typically a pressurized heat exchanger of cast iron, steel, or copper in which water is heated.

According to the U.S. Census Bureau in its report *Annual 2012 Characteristics of New Housing* report, 59% of new family homes completed in 2012 used natural or LPG gas for heating, followed by 39% that use electricity for heating, 1% that use oil, and 2% that use other forms.

26.2.4 Central Air Conditioning, Room Air Conditioners, and Ductless Minisplit Air Conditioners

A central air conditioning (AC) system is an appliance designed to provide cool air to an enclosed space. Typically, central AC systems consist of an indoor unit and an outdoor unit (split system). Central air conditioning (AC) units are also available in a package unit. A packaged unit air conditioning system has all of the components contained in one unit. This type of air conditioning unit can be installed on a rooftop with duct work added. A package unit air conditioner operates the same as a central unit. The outdoor condenser unit contains a compressor, condenser coil (outdoor heat exchanger coil), condenser fan, and condenser fan motor; the indoor unit consists of an evaporator coil (indoor heat exchanger coil) and a refrigerant flow control device (a capillary tube, thermostatic expansion valve, or orifice) residing either in a forced-air furnace or an air handler. Refrigerant tubing connects the two units. A central AC system provides conditioned air by drawing warm air from the living area space and blowing it through the evaporator coil; as it is passing through the evaporator coil, the air gives up its heat content to the refrigerant. The conditioned air is then delivered back to the living area space (via a ducted system) by the blower residing in the furnace or air handler. The compressor takes the vaporized refrigerant aiming out of the evaporator and raises it to a temperature exceeding that of the outside air. The refrigerant then passes on to the condenser unit (outside coil), where the condenser coil rejects the heat from the refrigerant to the cooler outside air, and condenses. The liquid refrigerant passes through the flow control device, and its pressure and temperature are reduced. The refrigerant reenters the evaporator coil, where the refrigeration cycle is repeated.

Unlike the two-unit, central AC system, a room air conditioner is contained within one cabinet and is mounted in a window or a wall so that part of the unit is outside the building and part is within the occupied space. The two sides of the cabinet are typically separated by an insulated divider wall in order to reduce heat transfer. The components in the outdoor portion of the cabinet are the compressor, condenser coil, condenser fan, fan motor, and capillary tube. The components in the indoor portion of the cabinet are the evaporator coil and evaporator fan. The fan motor drives both the condenser and evaporator fans. A room AC provides conditioned air in the same manner described for a central AC system but without air ducts.

Like conventional central air conditioners, minisplit air conditioners use an outside compressor/condenser unit and an indoor evaporator coil/ductless air handler unit. The difference is that each room or zone to be cooled has its own ductless air handler. Each indoor ductless air handler unit is connected to the outdoor condensing unit via a conduit carrying the electrical power, refrigerant lines, and condensate lines. The primary advantage is that by providing dedicated units to each occupied space or zones, it is easier to meet the varying comfort needs of different rooms of the residence. By avoiding the use of ductwork, minisplit AC also avoid energy losses associated with central AC. Some minisplit air conditioner condensers are designed to handle up to five ductless air handlers at one time.

Approximately 89% of U.S. households had an AC system in 2009; room ACs is used in 6% of households, and 5% without AC. According to the U.S. Census Bureau in its report *Annual 2012 Characteristics of New Housing* report, 89% of new family homes completed in 2012, followed by 11% completed without AC installed.

AC manufacturers are beginning to offer HFC-410A refrigerant in AC systems as an alternative to HCFC-22 (R-22) refrigerant units. The EPA has established the phase-out of the HCFC-22 refrigerant with no production or importing beginning in 2020. However, manufacturers of AC equipment must phase out the use of HCFC-22 refrigerant in new AC equipment by January 1, 2010. In general, existing R-22 systems will probably be converted to R-407C, an alternative refrigerant; however, new AC equipment will be designed to operate on R-410A.

26.2.5 Heat Pumps

Unlike air conditioners, which provide only space cooling, heat pumps use the same equipment to provide both space heating and cooling. A heat pump draws heat from the outside air into a building during the heating season and removes heat from a building to the outside during the cooling season. An air source heat pump contains the same components and operates in the same way as a central AC system but is able to operate by reversing the refrigerant directional flow as well, in order to provide space heating. In providing space heat, the indoor coil acts as the condenser while the outdoor coil acts as the evaporator. When the outside air temperature drops below 2°C (35.6°F) during the heating season, the available heat content of the outside air significantly decreases; in this case, a heat pump will utilize supplementary electric-resistance backup heat. A ground-source heat pump operates on the same principle as air-source equipment except that heat is rejected or extracted from the ground instead of the air. Since ground temperatures do not vary over the course of a day or a year as much as the ambient air temperature, more stable operating temperatures are achieved. The ground loop for a ground-source heat pump is a closed system that uses a pressurized, sealed piping system filled with a water/antifreeze mixture. The indoor mechanical equipment of a ground-source system includes a fan coil unit with an indoor coil, a compressor, and a circulation pump for the ground loop. Almost all heat pumps are powered by electricity.

Heat pumps were used in approximately 11% of U.S. households in 2001. The energy consumption of heat pumps varies according to the same user characteristics discussed earlier for AC systems. According to the U.S. Census Bureau in its report *Annual 2012 Characteristics of New Housing* report, 38% of new family homes completed in 2012 have heat pump units installed.

The EPA has established the phase-out of the HCFC-22 refrigerant for heat pumps with no production or importing beginning in 2020. However, manufacturers of heat pump equipment must phase out the use of HCFC-22 refrigerant in new heat pump equipment by January 1, 2010. In general, existing R-22 systems will probably be converted to R-407C, an alternative refrigerant; however, new heat pump equipment will be designed to operate on R-410A.

26.2.6 Clothes Washers

A clothes washer is an appliance that is designed to clean fabrics by using water, detergent, and mechanical agitation. The clothes are washed, rinsed, and spun within the perforated basket that is contained within a water-retaining tub. Top-loading washers move clothes up and down, and back and forth, typically about a vertical axis. Front-loading machines move clothes around a horizontal axis. Electricity is used to power an electric motor that agitates and spins the clothes, as well as a pump that is used to circulate and drain the water in the washer tub. Some washer models use a separate water heater element to heat the water used in the washer.

Approximately 84% of households had clothes washers in 2009. Most of the clothes washers sold in the United States are top-loading, vertical-axis machines. The majority of energy used for clothes washing (85%–90%) is used to heat the water. User behavior significantly affects the energy consumption of clothes washers. The user can adjust the amount of water used by the machine to the size of the load, and thereby save water and energy. Choosing to wash with cold water rather than hot water reduces energy consumption by the water heater. Similarly, rinsing with cold water rather than warm can reduce energy consumption. Energy consumption depends on how frequently the washer is used. The DOE's test procedure assumes clothes washers are used 300 times a year on average. According to the U.S. Energy Star program, if every washer purchased in the United States in 2013 earned the Energy Star rating, we would save about $250 million in electricity, water usage, and gas every year thereafter.

26.2.7 Clothes Dryers

A clothes dryer is an appliance that is designed to dry fabrics by tumbling them in a cabinet like drum with forced-air circulation. The source of heated air may be powered either by electricity or natural gas. The motors that rotate the drum and drive the fan are powered by electricity. Approximately 79% of U.S. households have automatic clothes dryers and 21% do not use a clothes dryer in 2009. An automatic dryer in U.S. homes consists of 80% electric dryers, 19% natural gas dryers, and 1% propane gas (LPG) dryers.

26.2.8 Dishwashers

A dishwasher is an appliance that is designed to wash and dry kitchenware by using water and detergent. Typically, in North America, hot water is supplied to the dishwasher by an external water heater. In addition, an internal electric heater further raises the water temperature within the dishwasher. Electric motors pump water through spray arms impinging on the kitchenware in a series of wash and rinse cycles. An optional drying function is also enabled by electric heaters and sometimes a fan. In recent years, some dishwashers incorporate soil sensors that determine when the dishes are clean and the washing cycle can be stopped. Approximately 69% of U.S. households had dishwashers in 2011.

26.2.9 Cooktops and Ovens

A cooktop is a horizontal surface on which food is cooked or heated from below; a conventional oven is an insulated, cabinet like appliance in which food is surrounded by heated air. When a cooktop and an oven are combined in a single unit, the appliance is referred to as a range. Both gas and electric ranges are available. Cooktops and ovens are present in almost all households. Almost 60% of households use electric cooktops and ovens, and the remaining 40% of households use gas cooktops and ovens.

In a microwave oven, nonionizing microwaves directed into the oven cabinet cavity cause the water molecules inside the food to vibrate. Movement of the water molecules heats the food from the inside out. The fraction of households with microwave ovens has increased dramatically in recent years. Appliance manufacturers have sold over 5 million microwave ovens in 2013.

26.3 Current Production

Table 26.3 shows the number of various appliances that was shipped by manufacturers in 2012 and 2013. Shipments have been increasing for most of the major appliances and air conditioners.

26.4 Efficient Designs

State and federal standards requiring increased efficiency for residential appliances, utility programs, and labels (such as Energy Star) have improved appliance efficiency dramatically since the late 1970s (Meyers et al., 2004). For example, the annual energy consumption (according to the DOE test procedure) of a new refrigerator in 2003 was less than half the consumption in 1980. Because of the slow turnover rate of appliances, however, the older, less efficient equipment remains in use for a long time. Promising design options for further improving the efficiency of residential appliances are discussed next.

The DOE has set standards and test procedures for appliances and air conditioners and has required manufacturers to comply with their rulings. See website for further details on the appliances and air conditioners listed in this chapter at: http://www1.eere.energy.gov/buildings/appliance_standards/standards_test_procedures.html.

TABLE 26.3
Shipments of Major Appliances and Space Conditioning Equipment in the United States

	YTD-2012	YTD-2013	%Chg
Major home appliances and space conditioning equipment (thousands of units)			
All major appliances	36,981.10	37,971.10	2.70
Cooking—total	9,014.60	9,166.70	1.70
Electric cooking—total	2,386.60	2,555.60	7.10
Electric ranges	1,902.70	2,009.00	5.60
Electric ovens	316.5	365.9	15.60
Surface cooking units	167.4	180.7	8.00
Gas cooking—total	1,412.80	1,521.50	7.70
Gas ranges	1,228.30	1,304.20	6.20
Gas ovens	17.1	17.7	3.70
Surface cooking units	167.5	199.5	19.20
Microwave ovens	5,215.20	5,089.70	−2.40
Home laundry—total	7,320.90	8,258.00	12.80
Automatic washers	4,104.40	4,591.90	11.90
Dryers—total	5,215.20	3,666.00	14.00
Electric	2,615.80	2,963.50	13.30
Gas	600.7	702.5	17.00
Kitchen cleanup—total	6,621.40	7,253.00	9.50
Disposers	3,419.80	3,725.00	8.90
Dishwashers—total	3,181.30	3,505.00	10.20
Built in	3,153.70	3,480.20	10.40
Portable	27.6	24.8	−10.10
Compactors	20.3	23	12.90
Food preservation—total	5,976.30	6,090.00	1.90
Refrigerators 6.5 and over	4,897.90	5,134.10	4.80
Freezers—total	1,078.50	955.9	−11.40
Chest	714.1	581.2	−18.60
Upright	364.4	374.6	2.80
Gas water heaters—total	3,186.50	3,535.50	11.00
Electric water heaters—total	3,077.30	3,330.70	8.20
Gas furnaces—total	1,787.30	2,093.90	17.20
Oil furnaces—total	25,567.00	24,593.00	−10.80
Air conditioners and heat pumps—total	5,019.00	5,540.40	10.40
Air conditioners only—total	3,526.60	3,807.50	8.00
Heat pumps only—total	1,492.30	1,732.90	16.10
Home comfort—total	8,047.90	7,206.50	−10.50
Room air conditioners	7,224.70	6,436.10	−10.90
Dehumidifiers	823.2	770.4	−6.40

Source: Association of Home Appliance Manufacturers (AHAM), Washington, DC.

Notes: Figures (in units) include shipments for the U.S. market, whether imported or domestically produced. Export shipments are not included; Industry figures are estimates derived from the best available figures supplied by a sample of AC manufacturers and are subject to revision.

26.4.1 Refrigerators and Freezers

Relative to the 2001 U.S. federal efficiency standard, achieving a 15% energy use reduction for refrigerator-freezers is possible with the use of a high-efficiency compressor, high-efficiency motors for the evaporator and condenser fans, and adaptive defrost control. Models at this level of efficiency account for a modest market share in the United States. Achieving a 25% energy use reduction generally would require a reduction in load transmitted through the unit's walls and doors, which might require the use of vacuum panels.

26.4.2 Improved Fan Motors

The evaporator and condenser fans of large refrigerators are powered by motors. The most common motor used for this purpose is a shaded-pole motor. Large efficiency gains are possible in refrigerators and freezers by switching to electronically commutated motors (ECMs), also known as brushless permanent-magnet motors, which typically demand less than half as much power as shaded-pole motors.

26.4.3 Vacuum Insulation Panels

The use of vacuum insulation panels (VIPs) can significantly reduce heat gain in a refrigerated cabinet and thereby decrease the amount of energy necessary to maintain a refrigerator or freezer at a low temperature. When using VIP, a partial vacuum is created within the walls of the insulation panels. Because air is conductive, the amount of heat transfer from the outside air to the refrigerated cabinet is reduced as the amount of air within the panels is reduced. Evacuated panels are filled with low-conductivity powder, fiber, or aerogel in order to prevent collapse. Energy savings associated with the use of vacuum panel insulation range from 10% to 20%. Vacuum panel technology still faces issues regarding cost and reliability before it can come into widespread use in refrigeration applications (Malone and Weir, 2001).

26.4.4 Water Heaters

26.4.4.1 Gas-Fired Storage Water Heater

The current models of gas-fired storage water heaters have a central flue that remains open when the water heater is not firing. This leads to large off-cycle standby losses. It should be possible to dramatically reduce off-cycle losses with relatively inexpensive technical modifications to the water heater. Energy savings derived from models with these modifications are expected to be about 25% compared to the 2004 U.S. standards.

The amount of heat extracted from the fuel used to fire a gas appliance can be increased by condensing the water vapor in the flue gases. In a condensing storage water heater, the flue is lengthened by coiling it around inside the tank. The flue exit is located near the bottom of the tank where the water is coolest. Because the flue gases are relatively cool, a plastic venting system may be used. A drain must be installed in condensing systems. Energy savings associated with the use of a gas-fired condensing water heater are approximately 40% compared to the 2004 efficiency standard. At this time, the high cost of the water heater results in a payback time that exceeds the typical lifetime of a water heater, but it is reasonable to assume that the cost could be reduced to the point that the water heater would be cost effective. In applications with heavy hot water use, such as laundromats or hotels, they may already be cost effective. Currently, a few companies produce condensing storage water heaters for commercial markets and they are sometimes sold as combined water heater/space heating systems for residential use.

26.4.4.2 Heat Pump Water Heaters

Heat pumps used with water heaters capture heat from the surrounding air or recycle waste heat from AC systems and then transfer the heat to the water in the storage tank. In this way, less energy is used to bring the water to the desired temperature. The heat pump can be a separate unit that can be attached to a standard electric water heater. Water is circulated out of the water heater storage tank, through the heat pump, and back to the storage tank. The pump is small enough to sit on the top of a water heater but could be anywhere nearby. Alternatively, the heat pump can be directly integrated into the water heater. Research indicates that this technology uses 60%–70% less energy than conventional electric resistance water heaters. Field and lab tests have been completed for several prototypes. A few models are currently available for sale.

26.4.4.3 Solar Water Heaters

Technological improvements in the last decade have improved the quality and performance of both passive and active solar water heaters. Research indicates that, in general, solar water heaters use 60% less energy than conventional electric resistance water heaters. There are several types of solar water heaters commercially available today.

26.4.5 Furnaces

26.4.5.1 Condensing Furnaces

The efficiency of a conventional gas furnace can be increased by using an additional heat exchanger to capture the heat of the flue gases before they are expelled to the outside. The secondary heat exchanger is typically located at the outlet of the circulating air blower, upstream of the primary heat exchanger. A floor drain is required for the condensate. A condensing furnace has efficiency up to 96% AFUE, well above the 80% AFUE rating of a standard noncondensing gas furnace. Condensing furnaces have been on the market since the 1980s and now constitute one-third of all gas furnace sales. They are particularly popular in colder areas of the United States, where the cost of heating is high. An early technical problem, corrosion of the secondary heat exchanger, has been resolved by the industry.

26.4.5.2 Integrated Water Heaters and Furnaces

Traditionally, water heating and space heating have required two separate appliances—a hot water heater and a furnace. Combining a water heater and a furnace into a single system can potentially provide both space heating and hot water at a lower overall cost. Integrated water and space heating is most cost effective when installed in new buildings, because gas connections are necessary for only one appliance rather than two.

Combination of space- and water-heating appliances fall into two major classes: (1) boiler/tankless-coil combination units, and (2) water-heater/fancoil combination units. A great majority of boiler/tankless-coil combination units are fired with oil, whereas most water-heater/fancoil combination units are fired with natural gas. In the latter units, the primary design function is domestic water heating. Domestic hot water is circulated through a heating coil of an air-handling system for space heating. Usually the water heater is a tank-type gas-fired water heater, but instantaneous gas-fired water heaters can be used as well.

The efficiencies of these integrated systems are determined largely by the hot water heating component of the system. Compared to a system using a standard water heater or boiler, an integrated system using a condensing water heater or condensing boiler can reduce energy consumption by as much as 25%.

26.4.6 Central Air Conditioning, Room Air Conditioners, and Ductless Minisplit Air Conditioners

26.4.6.1 Electric Variable-Speed Air Conditioning

Variable-speed central air conditioners use ECMs, which are more efficient than the induction motors used in a single-speed system. In addition, the speed of the ECM can be varied to match system capacity more precisely to a building load. Cycling losses, which are associated with a system that is continually turned off and on in order to meet building load conditions, are thus reduced. Unlike induction motors, ECMs retain their efficiency at low speeds; consequently, energy use is also reduced at low-load conditions. A variable-speed AC system uses approximately 40% less energy than a standard single-speed AC system. Although these AC systems are now available from major manufacturers, they account for a small fraction of sales.

26.4.6.2 Electric Two-Speed Air Conditioning

Two-speed induction motors are not as efficient as variable-speed ECMs, but they are less expensive. Like variable-speed air conditioners, two-speed air conditioners reduce cycling losses. When two-speed induction motors are used to drive compressor and fans, the system can operate at two distinct capacities. Cycling losses are reduced, because the air conditioner can operate at a low speed to meet low building loads. In some models, two-speed compressors are coupled with variable-speed indoor blowers to improve system efficiency further. A two-speed AC system reduces energy consumption by approximately one-third. Although these AC systems are available from several major manufacturers, they account for a small fraction of sales.

26.4.6.3 Room Air Conditioners

The most efficient room air conditioners have relatively large evaporator and condenser heat-exchanger coils, high-efficiency rotary compressors, and permanent split-capacitor fan motors. Compared to standard room ACs, highly efficient room ACs reduce energy consumption by approximately 25%. Such room ACs are available from several manufacturers. Units with Energy Star designation, which must exceed federal minimum efficiency standards by 10%, accounted for 35% of sales in 2004.

26.4.6.4 Ductless Minisplit Air Conditioners

Ductless minisplit air conditioners are available in numerous mix-and-match capacities and configurations. This type of AC system does not require bulky ductwork or complicated installation or expensive modifications to be installed in a home or office. Multiple indoor units (zones) with varying capacities can be connected to one condenser unit. These units are available in heat and/or cool. These high-efficiency air conditioners can save up to 20% energy usage.

26.4.7 Heat Pumps

26.4.7.1 Variable-Speed and Two-Speed Heat Pumps

Like central air conditioners, heat pumps can be made more efficient by the use of two-speed and variable-speed motors (see the earlier discussion of efficient central air conditioners). Both two-speed and variable speed air-source heat pumps are available; variable-speed ground-source heat pumps are not commercially available at this time. Compared to standard models, two-speed air-source heat pumps reduce energy consumption by approximately 27%, variable-speed air-source heat pumps reduce energy consumption by 35%, and two-speed ground-source heat pumps reduce energy consumption by 46%. Variable-speed and two-speed air-source heat pumps are made by the same companies that make variable speed and two-speed AC systems. Several manufacturers produce efficient air-source and two-speed ground-source heat pumps, but they account for a small fraction of all heat pump sales.

26.4.7.2 Gas-Fired Heat Pumps

Currently, all residential heat pumps are electric, but researchers have been developing gas heat pumps. The Gas Research Institute (GRI) and a private corporation jointly developed a natural gas, engine-driven, variable-speed heat pump in which the compressor is driven by an internal combustion spark-ignition engine and heat is recovered in the space-heating mode. This engine-driven heat pump was put on the market in 1994, but was withdrawn in the late 1990s due to lack of a maintenance infrastructure. In addition, the DOE has been funding the development of a gas-fired ammonia-water absorption-cycle heat pump. Gas-driven heat pumps have the potential to reduce heat pump energy consumption by approximately 35%–45%.

26.4.8 Distribution Systems

When assessing the efficiency of a space conditioning system, it is important to consider the efficiency of the distribution system as well as the appliance. It is not uncommon for air ducts to have distribution losses of 20%–40% due to conduction as well as leakage. Better insulation as well as more careful duct sealing can reduce these losses. In general, the most effective strategy for reducing distribution losses is to include the distribution system in the conditioned space so that any losses due to conduction or air leakage go directly into the space to be conditioned. This requires careful attention by the architect in the design of new buildings and is typically expensive as a retrofit measure.

26.4.9 Clothes Washers

26.4.9.1 Horizontal-Axis Washers

Although horizontal-axis clothes washers dominate the European market, the vast majority of clothes washers sold in the United States are top-loading, vertical-axis machines. However, this is expected to change by 2007 when more stringent minimum efficiency regulations on clothes washers will take effect. Horizontal-axis washers, in which the tub spins around a horizontal axis, use much less water than their vertical-axis counterparts, and less hot water is therefore required from water heaters. As mentioned earlier, the majority of energy used for clothes washing is used for heating water, so a significant amount of energy can be saved by using horizontal-axis washers. Research has indicated that horizontal-axis washers are more than twice as efficient as vertical-axis washers of comparable size (U.S. Department of Energy, 2000).

26.4.9.2 High-Spin-Speed Washers

Clothes washers can be designed so that less energy is required to dry clothes after they have been washed. Extracting water from clothes mechanically in a clothes washer uses approximately 70 times less energy than extracting the water with thermal energy in electric clothes dryer. Thus, by increasing the speed of a washer's spin cycle, one can reduce the energy required to dry clothes.

Because gas clothes dryers require so much less energy than electric dryers, based on energy consumption including that consumed at the electric power station, energy savings are much more significant when a high-spin-speed washer is used with an electric dryer. In a vertical-axis clothes washer, an increase in spin speed from 550 to 850 rpm reduces moisture retention from 65% to 41%. In an electric dryer, this reduces the energy consumption by more than 40%. In a horizontal-axis washer, an increase in spin speed from 550 to 750 rpm reduces moisture retention from 65% to 47%; in an electric dryer, energy consumption is reduced by more than 30%. High-spin-speed washers have been common on European horizontal axis machines for some time, and are becoming more common in the United States.

26.4.10 Clothes Dryers

26.4.10.1 Microwave Dryers

In conventional clothes dryers, hot air passes over wet clothes and vaporizes the surface water. During the later stages of drying, the surface dries out and heat from the hot air must be transferred to the interior, where the remaining moisture resides. In contrast, in microwave drying, water molecules in the interior of a fabric absorb electromagnetic energy at microwave wavelengths, thereby heating the water and allowing it to vaporize. Several U.S. appliance manufacturers have experimented with microwave clothes dryers, and a few small companies have built demonstration machines. Issues of arcing on metal objects and possible fire within the dryer would need to be resolved before microwave clothes dryers are likely to be sold.

26.4.10.2 Heat Pump Dryers

A heat pump dryer is essentially clothes dryer and an air conditioner packaged as one appliance. In a heat pump dryer, exhaust heat energy is recovered by recirculating all the exhaust air back to the dryer; the moisture in the recycled air is removed by a refrigeration–dehumidification system. A drain is required to remove the condensate; because washers and dryers are usually located side by side, a drain is generally easily accessible. Heat pump dryers can be 50%–60% more efficient than conventional electric dryers. Introduced in Europe in 1999, heat pump dryers are available both in the United States and Europe.

26.5 Cost Effectiveness of Energy-Efficient Designs

The cost effectiveness of energy-efficient designs for major appliances and space conditioning equipment depends on consumer energy prices and also on per-unit costs when new technologies are manufactured on a large scale. Therefore, the cost effectiveness of such designs varies.

A U.S. study, conducted in 2004 for the National Commission on Energy Policy (NCEP), reviewed the literature on energy-efficient designs for major appliances and space conditioning equipment and estimated their costs and energy savings relative to baseline designs (Rosenquist et al., 2004). For a number of products, it determined the efficiency level that provides users with the lowest life-cycle cost over the appliance lifetime. Table 26.4 lists the selected efficiency levels (expressed in terms of annual energy consumption in some cases), along with their associated cost of conserved energy (CCE) for consumers. The CCE in terms of $/kW h or MBtu (gas) is an expression of the extra first cost incurred to save a unit of energy. Calculation of CCE requires application of a present worth factor (PWF) to spread the initial incremental cost over the lifetime of the equipment. The PWF uses a discount rate to effectively amortize costs over time. The CCEs for electric appliances in Table 26.4 are all below the average U.S. residential electricity price.

26.6 Conclusion

Residential appliances consume significant amounts of electricity and natural gas. This chapter described the basic engineering principles of the major appliances and space conditioning equipment as well as promising energy-efficient designs.

Significant potential energy savings are possible beyond the models typically sold in the marketplace today. Among the various end-uses, energy savings range from 10% to 50%. Many of these efficient appliances appear to be cost effective at currently projected

TABLE 26.4

Cost Effectiveness of Efficient Designs for Selected Appliances and Space Conditioning Equipment

Product	Baseline Technology	Efficient Technology	CCE[a]
Refrigerator	484 kW h/year	426 kW h/year	4.9 cents/kW h
Room air conditioner	9.85 EER	10.11 EER	5.2 cents/kW h
Electric water heater	92 EF	Heat pump	3.9 cents/kW h
Clothes washer[b]	3.23 kW h/cycle	Horizontal axis and high spin speed (1.87 kW h/cycle)	5.0 cents/kW h
Gas furnace[c]	80% AFUE	Condensing furnace (90% AFUE)	$9.30/MBtu

[a] The CCE may be compared to expected residential energy prices. In the United States, long-term forecasts (as of 2005) place average prices in the range of 8.5–9.0 cents/kW h for electricity and $7.7–$8.4/MBtu for natural gas.

[b] Includes the energy for water heating and drying associated with clothes washing. Refers to electric water heater and dryer. Baseline refers to the 1994 DOE standard level. Additional consumer benefits derive from reduction in water usage.

[c] The CCE value is a U.S. national average. In cold climates, the amount of saved energy is higher, so the CCE is lower by a third or more.

manufacturing costs, with simple payback times that are shorter than the typical appliance lifetimes of 10–20 years. If costs of efficiency improvements decrease, or if future energy prices increase, more of the potential energy savings that are already technically possible will become economically attractive. In addition, future research is likely to identify additional technological opportunities to save energy.

Acknowledgments

This work was supported by the Assistant Secretary for Energy Efficiency and Renewable Energy, Office of Buildings Technology, Office of Code and Standards, of the U.S. DOE, under Contract No. DE-AC03-76SF00098. The opinions expressed in this chapter are solely those of the authors and do not necessarily represent those of the Ernest Orlando Lawrence Berkeley National Laboratory, or the U.S. DOE.

Author Note

Eric Kleinert an experienced professional in the major appliance, air-conditioning, refrigeration, and service industry for 45 years. He is accomplished on all sides of the industry: owning and operating successful sales, service and parts companies; teaching vocational students preventive, diagnostic services and techniques; and, the author of the leading major appliance and HVAC repair textbooks on the market, Troubleshooting and Repairing Major Appliances, McGraw-Hill and HVAC and Refrigeration Preventive Maintenance, McGraw-Hill.

References

Energy Information Administration (EIA), 2005a. *Household Energy Consumption and Expenditures 2001*. U.S. Department of Energy, Washington, DC.

Energy Information Administration (EIA), 2005b. *Annual Energy Outlook 2005*. U.S. Department of Energy, Washington, DC.

Energy Information Administration (EIA), 2013. *Annual Energy Outlook 2013*. U.S. Department of energy, Washington, DC.

Malone, B. and Weir, K., 2001. State of the art for VIP usage in refrigeration applications. *International Appliance Manufacturing*.

27

Heat Pumps

Herbert W. Stanford, III

CONTENTS

27.1 Heat Pump Concept

Heat pumps are *reverse cycle* building heating/cooling units or systems that can extract heat *from a building* and reject that heat to the environment, providing cooling for a building, and can switch from providing cooling to providing heating by extracting heat from the environment and rejecting that heat *into a building.* This heat transfer cycle can be accomplished using a vapor-compression refrigeration cycle or an absorption refrigeration cycle; though by far, vapor-compression systems are more widely used (see Section 27.4 for more discussion of absorption cycle systems).

All refrigeration cycles hinge on one common physical characteristic: if a chemical compound (which we can call a *refrigerant*) changes phase from a liquid to a gas, a process called *evaporation*, the compound must absorb heat to do so. Likewise, if refrigerant changes phase back from a gas to a liquid, the process of *condensation*, the absorbed heat must be rejected. Thus, all refrigeration cycles depend on circulating a refrigerant between a heat *source* (from which heat is removed, thus producing cooling) and heat *sink* (somewhere to which the collected heat can be rejected).

The *vapor-compression refrigeration cycle*, wherein a chemical substance alternately changes from liquid to gas and from gas to liquid, actually consists of four distinct steps:

1. *Compression*: Low-pressure, low-temperature refrigerant gas is compressed, thus raising its pressure by expending mechanical energy. There is a corresponding increase in temperature along with the increased pressure.
2. *Condensation*: The high-pressure, high-temperature gas is cooled by air or water that serves as a *heat sink* and condenses into liquid form.
3. *Expansion*: The high-pressure, high-temperature liquid flows through an orifice in the expansion valve, thus reducing both liquid pressure and temperature.
4. *Evaporation*: The low-pressure, low-temperature liquid absorbs heat from an air or water that serves as a *heat source* and evaporates as a gas or vapor. The low-pressure, low-temperature vapor flows to the compressor and the process repeats.

As shown in Figure 27.1, the vapor-compression refrigeration system consists of four components that perform these four steps. The *compressor* raises the pressure of the initially low-pressure refrigerant gas. The *condenser* is a heat exchanger that cools the high-pressure gas so that it changes phase to liquid. The *expansion valve* (or other pressure-reducing device) controls the pressure ratio, and thus flow rate, between the high- and low-pressure regions of the system. The *evaporator* is a heat exchanger that heats the low-pressure liquid, causing it to change phase from liquid to vapor.

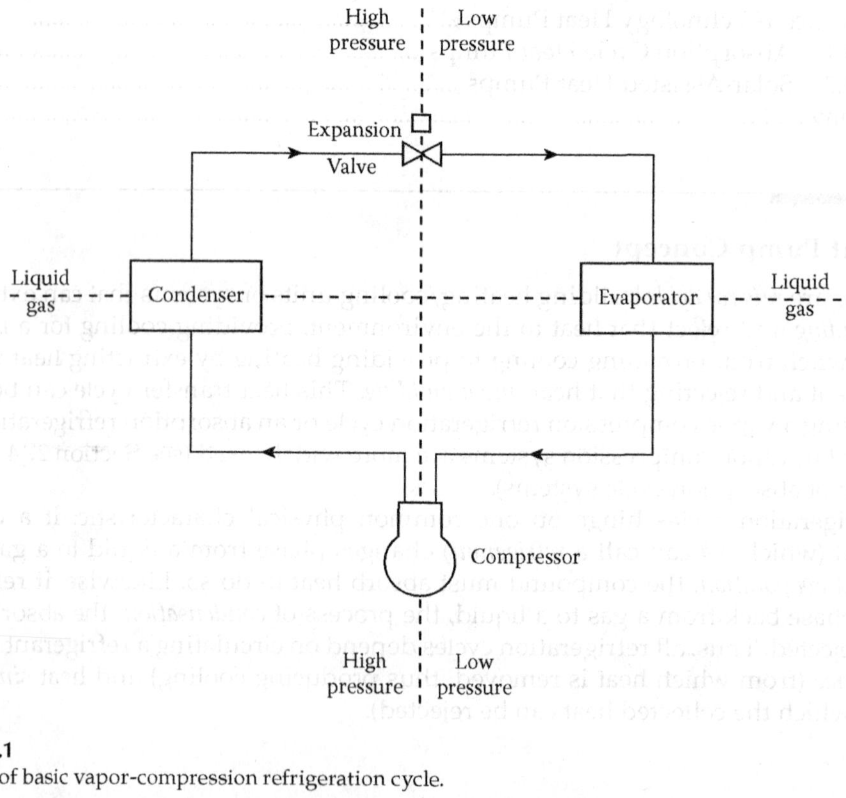

FIGURE 27.1
Schematic of basic vapor-compression refrigeration cycle.

For any heat pump utilizing the vapor-compression refrigeration cycle, there is a fifth element required: a *reversing valve*. This allows the condenser and evaporator heat exchangers to switch roles as the system switches from providing cooling to providing heating and vice versa by reversing the direction of refrigerant flow between these two components.

The amount of energy consumed by the compressor is dictated by the *lift* or pressure increase that the compressor must provide to raise the pressure/temperature of the low-pressure, low-temperature vapor that enters to the high-pressure, high-temperature vapor that leaves. On the low-pressure, low-temperature side of the system, refrigerant conditions depend on the temperature needed to provide required sensible cooling and dehumidification (typically about 45°F [7°C]), while the high-pressure/hot side refrigerant must be at conditions that allow it to condense at ambient temperatures of 105°F–125°F (51°C–65°C). If condensing temperatures are lowered to the heat sink's temperature condition, then the required lift is decreased and required energy input to the compressor also decreases.

The *type* of heat pump is typically defined by the environmental component that is used as a heat *source*, from which heat is extracted when providing heating, and heat *sink*, to which heat is rejected when providing cooling. The heat source/sink may be atmospheric air, directly or indirectly; the ground; or large water sources, such as lakes, rivers, or oceans. Subtypes of heat pumps are also sometimes defined on the basis of heat transfer media utilized: air, refrigerant, or water.

Vapor-compression cycle heat pumps have been available since the 1950s and factory-built heat pump units are widely used in residential and light commercial applications in mild climate regions, mostly in the South. However, since the early 2000s, the energy efficiency of these units, both air source and water source, has been improved dramatically:

Variable air volume supply air: Variable indoor airflow provided by the use of *electronically commutated motors* (ECMs). ECMs are DC motors that function using a built-in inverter and a magnet rotor and as a result are able to achieve greater efficiency in airflow systems than conventional *permanent split capacitor* (PSC) motors used in the past. Initially, silicon rectifier controllers were added to PSC motors to provide variable speed control. But, as PSC motor speed was reduced, efficiency suffered, falling from 65%–70% to as low as 12%. ECMs, on the other hand, maintain a high level (65%–75%) of efficiency over almost its full speed range. Additionally, unlike PSC motors, ECMs are not prone to overheating and do not require additional measures to offset the generation of heat.

By varying the indoor airflow rate in response to the imposed cooling and heating loads, indoor fan energy consumption is reduced significantly since no heat pump system is required to operate at peak capacity more than a few hours each year. Additionally, when providing cooling, variable airflow results in much better humidity control than provided by older, single-speed systems, a real boon for installations in the South.

Variable speed compressor(s): Variable compressor speed/load control provides a better match between the imposed load and the compressor capacity. Older, single-speed compressors cycled on and off as the imposed cooling/heating load was reduced, resulting in thermal losses at the beginning and end of each run cycle. The next level of improvement was to provide two-speed compressors and, for the last 10 years or so, fully variable speed compressors. Some very high-efficiency systems even use multiple variable speed compressor configurations in which one compressor is used during normal *part load* operation and the second is used only during *peak load* conditions.

Variable heat sink/source medium flow: For both air-source and water-source heat pumps, variable sink/source flow is being applied to reduce energy consumption required for air or water transport.

Demand defrost control: For air-source heat pumps, demand defrost control is incorporated to minimize energy losses associated with the buildup of frost or ice on the outdoor heat exchanger when the unit is providing heating.

And vapor-compression cycle heat pumps now typically utilize R-410A or R-134a refrigerant in lieu of the older R-22 refrigerant that is no longer available, significantly reducing the potential for atmospheric ozone depletion and warming in the event of a refrigerant leak. The refrigerant change has resulted in the heat transfer area of both indoor and outdoor heat exchanges being increased, improving both cooling/heating performance and energy consumption even further.

27.2 Air-Source Heat Pumps

Air-source heat pumps utilize atmospheric (outdoor) air as their ultimate heat source/sink, either extracting heat from outdoor air to provide heating or rejecting excess heat into it when cooling. Thus, the performance and seasonal efficiency of these heat pumps is significantly impacted by outdoor air temperature.

27.2.1 Premium Efficiency Air-Source Heat Pumps

Standard 90.1-2010, promulgated by the American Society of Heating, Refrigerating, and Air-Conditioning Engineers, Inc. (ASHRAE), forms the basis of most current energy conservation building codes in the United States. For air-source heat pumps, this standard defines efficiency on the basis of specific required minimum performance measurements:

For cooling by units rated with a cooling output of less than 65,000 Btu/h (19 kW), *seasonal energy efficiency ratio* (SEER) is defined as the total cooling output of the heat pump during its normal cooling usage period (in Btu) divided by the total electric energy input during that same period (in watt hours).

For cooling by units rated with a cooling output of 65,000 Btu/h (19 kW) or greater, *energy efficiency ratio* (EER) is defined as the ratio of net cooling capacity (in Btu/h) to the total rate of electrical input (in watts) under designated operating conditions.

For heating by units rated with a cooling output of less than 65,000 Btu/h (19 kW), *heating seasonal performance factor* (HSPF) is defined as the total heating output of the heat pump during its normal heating usage period (in Btu) divided by the total electric energy input during that same period (in watt hours).

For heating by units rated with a cooling output of 65,000 Btu/h (19 kW) or greater, heating efficiency is defined in terms of *coefficient of performance* (COP). COP is a dimensionless value defined as the net heating capacity of the heat pump (in Btu/h [kW]) divided by the energy input (in Btu/h [kW]) under designated operating conditions.

ASHRAE has established current minimum performance requirements for packaged (all components in one outdoor enclosure) and split system (separate indoor and outdoor components) air-source heat pumps at SEER = 13.0 and HSPF = 7.7 for heat pumps rated with a cooling output of less than 65,000 Btu/h (19 kW). All air-source heat pumps with a cooling output of less than 65,000 Btu/h (19 kW) available from larger equipment manufacturers will meet or exceed the minimum ASHRAE performance standards. Additionally, most manufacturers offer *high*-efficiency units that have SEER = 15–16 and HSPE = 8–9 and some offer *premium* efficiency units that have SEER = 19–20 and HSPE = 9–10, which comply generally with the EPA's Energy Star requirements.

For larger-capacity air-source heat pump units, the following table summarizes ASHRAE's current minimum requirements:

System Cooling Capacity (Btu/h)	Minimum Cooling Performance (EER)	Minimum Heating Performance (COP)
65,000–134,999 (19–39 kW)	11.0	3.3 at 47°F (9°C)/2.25 at 17°F (−8°C)
135,000–239,999 (40–70 kW)	10.6	3.2 at 47°F (9°C)/2.05 at 17°F (−8°C)
≥240,000 (≥70 kW)	9.5	

Again, though, manufacturers typically offer larger-capacity premium efficiency equipment with EER = 12–12.5 and COP (at 47°F [9°C]) = 3.5–4.5.

An additional important part of achieving high-efficiency air-source heat pump energy performance is to select each heat pump capacity as close as possible to the anticipated peak load. Since most heat pumps are installed in mild climate areas where cooling is more of a consideration than heating, units are typically selected on the basis of the peak imposed cooling load in order to minimize oversizing and loss of cooling efficiency. For example, a residence or commercial area may require 40,800 Btu/h (12 kW) or 3.4 tons of cooling. Because heat pump units are manufactured in specific capacity ranges, typically 2–6 tons at 1 ton increments, 7–1/2 tons, and 10 tons, contractors would normally recommend installation of a 4.0 tons heat pump for this example.

But the more efficient and less costly approach would be to install a 3.0 tons heat pump. Doing so would result in better match between imposed load and available capacity during part load operating periods, reducing energy consumption. The shortfall in capacity at peak load may result in indoor temperatures rising 1°F–2°F (1°C) above cooling set point temperature, but since the peak cooling period occurs for only a few hours each year, the negative comfort impact is minimal.

For applications located where heating requirements are more significant, sizing the heat pump unit on the basis of heating need rather than cooling need may result in improved heat pump performance and improved energy efficiency. While this selection may result in some oversizing relative to cooling and a small loss in cooling efficiency, the larger system will be able to provide compressor-based heating for longer periods, reducing the amount of low-efficiency supplement heating required, saving energy, and reducing operating costs.

Careful review of imposed heating and cooling loads is required to select an air-source heat pump unit and maximize energy efficiency. It is important that consumers and designers evaluate alternative unit selections and make the final decision on the basis of the lowest life-cycle cost, considering both first cost and anticipated energy costs of the anticipated life of the unit, typically 15 years.

27.2.2 Cold Climate Air-Source Heat Pumps

For all air-source heat pumps, and based on system thermodynamics, as the outdoor air temperature falls, the heat pump is able to extract less and less heat from the environment, while requiring more and more energy input to the compressor. Ultimately, the heat pump will be unable to extract the amount of heat needed to offset building losses and additional *supplemental* heating will be required. Typically, the supplemental heating capacity is provided by electric resistance heating coils that, by themselves, have a COP = 0.98. When the outdoor air is too cold for the refrigerant to extract any heat from it, there is no need to operate the compressor at all and all of the heating is done by the supplemental heater, at low efficiency and high expense.

However, since the 1990s, research to develop an air-source heat pump that would work in colder climates has been underway. One manufacturer now offers a cold-climate heat pump that features a two-speed, two-cylinder compressor for efficient operation, a backup booster compressor that allows the system to operate efficiently down to 15°F (−9°C), and a plate heat exchanger called an *economizer* that further extends the performance of the heat pump to well below 0°F (−18°C).

27.2.3 Dual Fuel Air-Source Heat Pumps

An alternative to the use of electric resistance supplemental heating, the dual fuel heat pump, has been in use since about 1980. This type of heat pump utilizes, today, a high-efficiency condensing furnace firing natural gas or LPG (propane) to provide heating when fossil fuel firing is more cost-effective than operating the unit compressor(s) and consuming electricity.

While the COP of a condensing furnace is essentially the same as for electric resistance heating, the much lower cost of natural gas or LPG relative to cost of electricity can result in much more economical heating. For example, with a typical residential electricity cost of $0.12/kWh, the cost of electric resistance heating will be $34 per 1,000,000 Btu ($0.116/kW). In December 2014, residential natural gas sold for $9.98 (U.S. average) per 1,000,000 Btu ($0.034/kW). Therefore, the heat pump system must maintain a minimum heating COP of 3.41, well above the ASHRAE minimum, to provide heating at a cost that is equal to or less than the cost of fossil fuel firing. Typically, the *tipping* point for switching from the heat pump's compressor-based heating cycle to fossil fuel–fired heating is when the outdoor temperature falls below about 30°F (−1°C).

27.3 Water-Source Heat Pumps

Air-source heat pumps units cannot easily be applied to larger, particularly multistoried, buildings that require a large number of independently controlled thermal zones. Larger apartment buildings, hotels, schools, office buildings, etc. simply will not have the outdoor space available for a large number of independent air-source heat pumps.

The alternative, then, is to go to water-source heat pumps. Typically, a water-source heat pump system consists of one or more individual single zone, water-cooled packaged units with a supply fan, indoor air coil, compressor, and water-to-refrigerant heat exchanger, connected to a common water supply system that acts as the heat source/sink for each individual heat pump unit.

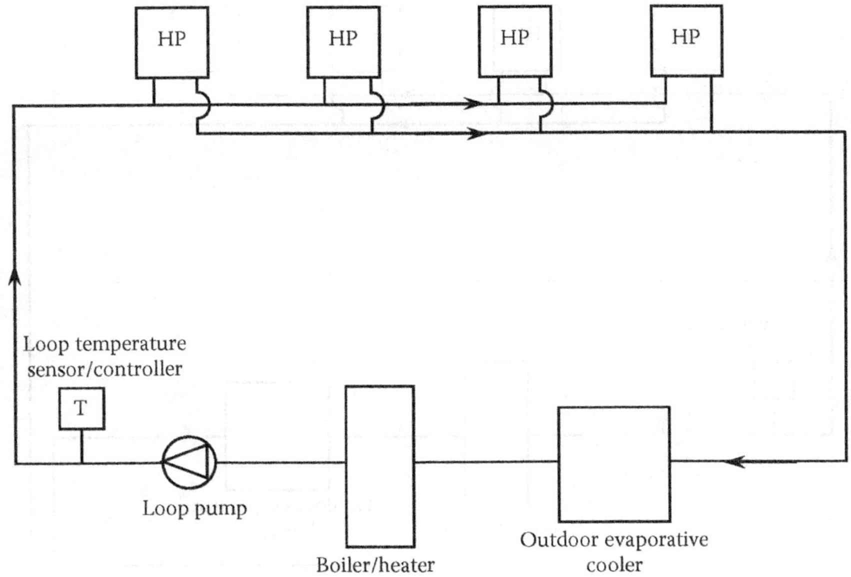

FIGURE 27.2
Schematic of basic closed-circuit water-source heat pump system.

27.3.1 Closed-Circuit Water-Source Heat Pump Systems

The most common, and very efficient, approach to using water-source heat pumps is to utilize a closed-circuit water heat exchanger as illustrated by Figure 27.2.

In this configuration, each packaged water-source heat pump refrigerant system reverses the direction of refrigerant flow from summer to winter with the following results: during summer, heat is removed from the space via the air coil (evaporator) and rejected to the central circulating water heat exchanger via the water-to-refrigerant heat exchanger (condenser). Excess heat rejected to this heat exchanger is, in turn, rejected to the outdoors through a closed-circuit evaporative cooler in order to maintain the heat exchanger temperature at or below 85°F (29°C).

During winter, heat is removed from the circulating water heat exchanger via the water-to-refrigerant heat exchanger (evaporator) and rejected to the space via the air coil (condenser). If enough heat is removed from the heat exchanger to lower its temperature to below 65°F (18°C), an auxiliary heater (typically a high-efficiency condensing hot water boiler) adds supplemental heat to the heat exchanger to maintain the required temperature.

ASHRAE stipulates minimum energy performance requirements only for closed heat exchanger water-source heat pump units of less than 135,000 Btu/h (40 kW) cooling capacity. The minimum requirements are a cooling EER = 12.0 and a heating COP = 4.2. Again, manufacturers offer *premium* performance units that can produce a cooling EER of 16 or higher and a heating COP of almost 5.0.

27.3.2 Closed-Circuit Geothermal Heat Pump Systems

Closed circuit geothermal heat pumps are a variation to the basic closed-circuit heat pump system, as discussed in the previous subsection, that are designed to take advantage of the earth (ground) or a large body of water as heat source/sink. The supplemental heater(s)

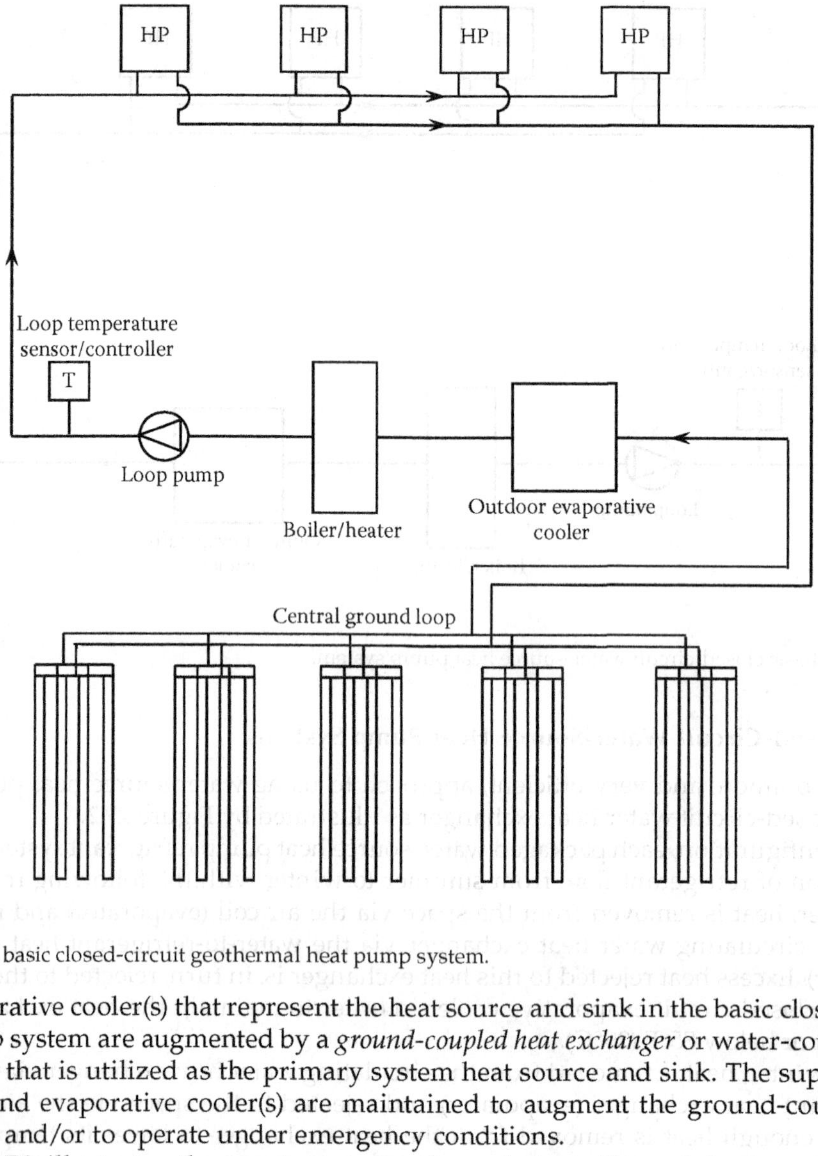

FIGURE 27.3
Schematic of basic closed-circuit geothermal heat pump system.

and evaporative cooler(s) that represent the heat source and sink in the basic closed-circuit heat pump system are augmented by a *ground-coupled heat exchanger* or water-coupled heat exchanger that is utilized as the primary system heat source and sink. The supplemental heater(s) and evaporative cooler(s) are maintained to augment the ground-coupled heat exchanger and/or to operate under emergency conditions.

Figure 27.3 illustrates the common ground-coupled geothermal heat pump system configuration.

The ground-coupled heat exchanger may be installed either horizontally directly in the ground or vertically through one or more wells, typically referred to as *boreholes*. The fluid could be either water or a refrigerant.

Geothermal heat pumps take advantage of the natural constant temperature of the ground 5–6 ft [1.5–1.8 m] below grade where the ground mean annual temperature remains between 45°F and 65°F [7°C–18°C], depending on location and water table. The ground temperature is warmer than the air above it in the winter and cooler than the air in the summer, resulting in the need for less compressor lift and more efficient heating and cooling.

The geothermal heat pump system has three major parts: the ground-coupled heat exchanger, the heat pump unit, and the air delivery system (ductwork and air outlets/inlets). The heat exchanger is a system of pipes that is buried in the shallow ground near the

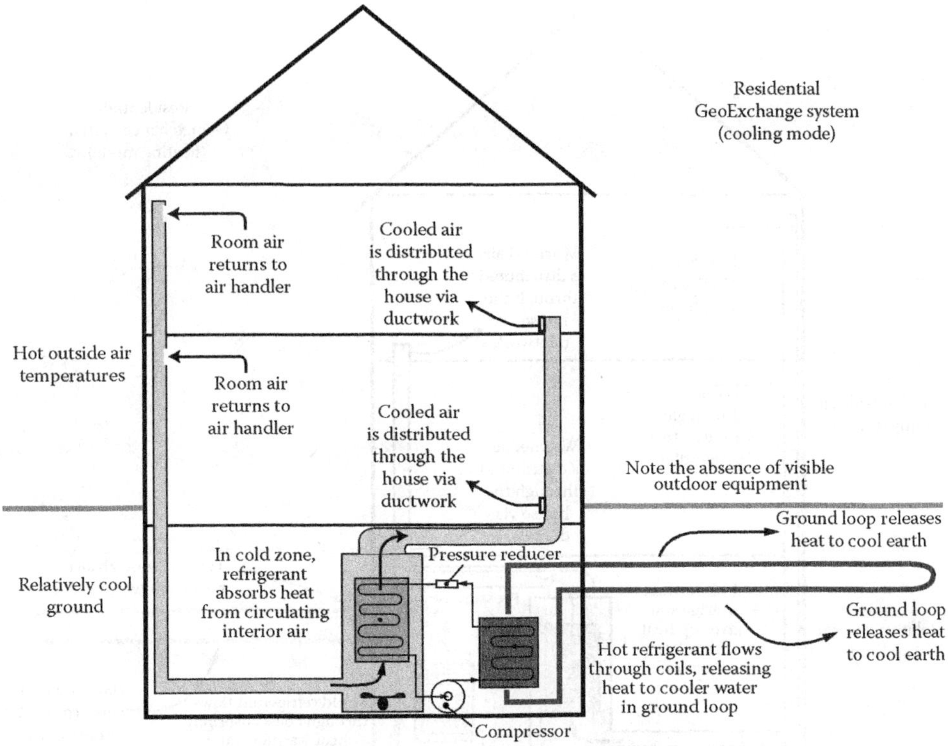

FIGURE 27.4
Example of a residential geothermal heat pump configuration in the summer months. (Courtesy of the Geothermal Heat Pump Consortium.)

building. A fluid (usually water or a mixture of water and antifreeze) circulates through the pipes to absorb or deposit heat within the ground (Figure 27.4).

In the winter, the heat pump removes heat from the heat exchanger and pumps it into the indoor air delivery system. In the summer, the process is reversed, and the heat pump moves heat from the indoor air into the heat exchanger. The heat removed from the indoor air during the summer can also be used to heat water (Figure 27.5).

The ground-coupled heat exchanger provides the means of transferring heat to the earth in summer and extracting heat from the earth in winter. Physically, the heat exchanger consists of several lengths of plastic pipe typically installed either in horizontal trenches or vertical boreholes that are backfilled to provide close contact with the earth.

Fluid inside the heat exchanger is pumped through a refrigerant heat exchanger in the geothermal heat pump. In the summer, it absorbs heat from the refrigerant and carries it to the ground through the ground-coupled heat exchanger piping to be rejected. In winter, it absorbs heat from the earth and transfers that heat to be extracted by the refrigerant.

Once installed, the heat exchanger remains out of sight beneath the surface. The ground heat exchanger consists of high-density polyethylene piping with an anticipated service life of at least 50 years.

Horizontal closed heat exchangers are the most cost-effective configuration when (1) adequate land is available, (2) soil conditions are such that trenches are easy to dig (e.g., no rock), and (3) trench depths are not defined by frost lines exceeding 36 in. [1 m] below grade. Trenching machines or backhoes are utilized to dig the trenches normally 5–6 ft

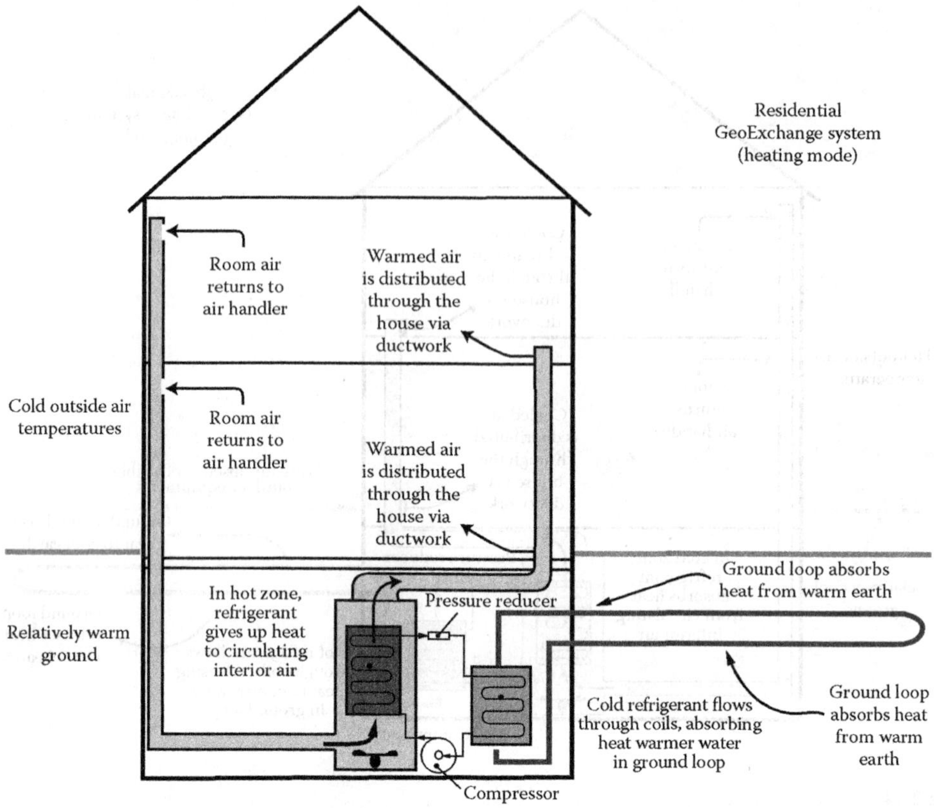

FIGURE 27.5
Example of a residential geothermal heat pump configuration in the winter months. (Courtesy of the Geothermal Heat Pump Consortium.)

below finished grade and then lay a series of parallel plastic pipes. Backfilling the trench requires that great care is taken to eliminate sharp rocks or debris that may damage the heat exchanger piping. Typically, compacted sand or a mix of sand and bentonite is backfilled for the first 6–12 in. [15–30 cm] to create good contract between the piping and earth. A typical horizontal heat exchanger will be about 400–600 ft [120–180 m] long per ton of required cooling capacity (Figure 27.6).

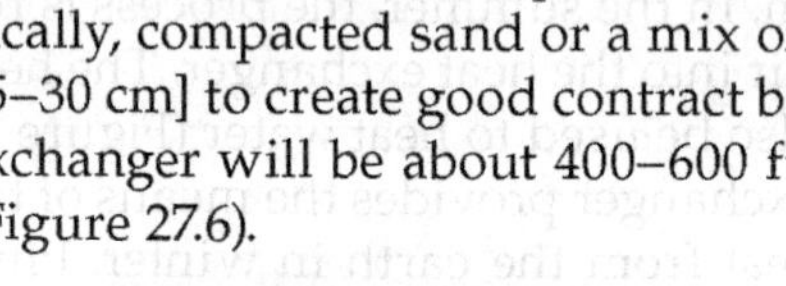

FIGURE 27.6
Example of a horizontal heat exchanger configuration in a residential installation. (Courtesy of the Geothermal Heat Pump Consortium.)

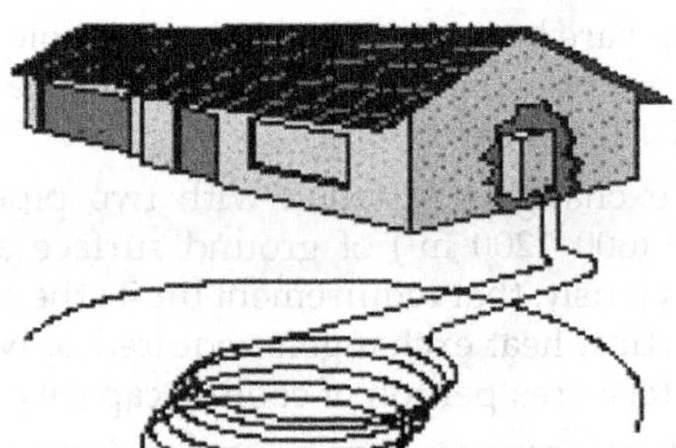

FIGURE 27.7
Example of a *slinky* heat exchanger configuration in a residential installation. (Courtesy of the Geothermal Heat Pump Consortium.)

The pipe may be curled into a slinky shape in order to fit more of it into shorter trenches. While this reduces the amount of land space needed, it may require more pipes (Figure 27.7).

For vertical heat exchangers, boreholes are drilled into the ground from 150 to 450 ft [45–140 m] deep. Each hole contains a single heat exchanger of pipe with a U-bend at the bottom. After the pipe is inserted, the hole is backfilled or grouted with a slurry of mixture of sand and bentonite. Each vertical pipe is then connected to a horizontal pipe, which is also concealed underground. The horizontal pipe then carries fluid in a closed system to and from each geothermal heat pump unit (Figure 27.8).

ASHRAE stipulates minimum performance requirements only for ground-coupled heat pump systems of less than 135,000 Btu/h [40 kW] cooling capacity, requiring a cooling EER = 13.4 and a heating COP = 3.6.

Despite their excellent energy efficiency, the application of ground-coupled heat pump systems has been relatively limited due to a number of *challenges* that significantly reduce their cost effectiveness, especially as conventional air-source and water-source heat pumps continue to improve their energy performance. These challenges include:

Land type and space for a ground-coupled heat exchanger: The soil in which a horizontal ground-coupled heat exchanger is installed must be reasonably finely textured, no rock, and soil that contains little clay. The soil in which boreholes and vertical heat

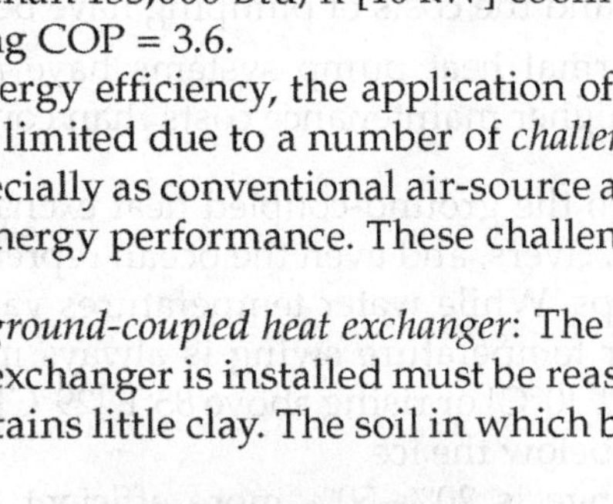

FIGURE 27.8
Example of a vertical heat exchanger configuration in a residential installation. (Courtesy of the Geothermal Heat Pump Consortium.)

exchangers are installed are less critical since the borehole is backfilled with a mix of sand and bentonite after the piping is installed to insure good heat transfer with the surrounding undisturbed earth.

Horizontal trench heat exchangers installed with two pipes per trench typically require 2000–4000 ft² (600–1200 m²) of ground surface area per ton of cooling capacity required. Obviously, that requirement limits the applicability of this type of heat exchanger. Vertical heat exchangers required only about 200–400 ft² (60–120 m²) of ground surface area per ton of cooling capacity required.

Costs of ground-coupled heat exchanger: Horizontal trenching, assuming no rock is present near the ground surface, is relatively inexpensive since it is accomplished by conventional backhoes or trenching machines fairly quickly. Drilling boreholes for vertical heat exchangers is more difficult and expensive than trenching. And, even with vertical heat exchangers, some horizontal trenching is required to connect each borehole piping heat exchanger to a common piping system.

Another cost element associated with vertical heat exchangers is the need to drill a test bore prior to designing the system in order to determine actual subsurface conditions. The soil thermal conductivity can vary as much as 400% depending on the type of soil and rock encountered, ground water hydrology, etc. This means that the number and depth of geothermal boreholes may vary significantly from site to site for the same system capacity.

Circuit pumping energy consumption: Geothermal heat pumps require a flow ranging from 2.5 to 3.5 gal/min (9.4–13.2 L/min) for each ton of cooling load. In older systems, the pumps to provide this flow operated at all times at a constant speed and, thus, significant pumping energy was consumed. Today's systems utilized variable flow pumping and the costs of pumping have been reduced.

System complexity: Geothermal heat pump systems have more components, more complex controls, and higher maintenance costs than conventional systems.

A more efficient alternative to the ground-coupled heat exchanger is a *water-coupled heat exchanger*. Large ponds, lakes, rivers, and even the ocean represent excellent heat sources/sinks for use with heat pumps. While water temperatures vary in these bodies of water summer to winter, the water temperature swing is always more moderate than for air, rarely falling below about 32°F [0°C] or rising above 85°F [29°C]. Even water that freezes on the surface is above freezing below the ice.

Water-to-water heat exchange is 20%–50% more efficient than water-to-ground heat exchange, significantly improving the efficiency of heat exchange of water-coupled systems over ground-coupled systems, and these systems typically have a cooling EER of 18–20 and a heating COP of 3–4.

Water-coupled heat exchangers may be configured with horizontal coiled piping submerged in the body of water. The alternative is to install a high-efficiency plate-and-frame heat exchanger with closed-circuit flow to the heat pump unit(s) on one side and flow from the body of water on the other. This second configuration, despite having somewhat higher pumping energy consumption, is preferred to avoid the problems with silting, drift, and mechanical damage that can occur to submerged piping loops.

While the applicability of water-coupled heat pump systems is limited simply because availability of usable water sources/sinks is more limited, the systems themselves are both less expensive and more energy efficient than ground-coupled systems and represent a very cost-effective approach when applicable.

27.3.3 Open-Circuit Geothermal Heat Pump Systems

Open-circuit geothermal heat pump systems, often called *groundwater heat pump systems,* use two or more groundwater wells for their heat source/sink, one or more wells from which groundwater is withdrawn and one or more wells into which the water is *injected* back into the ground after passing through the heat pump units. These systems are as efficient, sometimes even more so, than water-coupled heat pump systems.

However, these systems have a significant negative environmental impact. Since the water flow rate required for heat pump operation is relatively high, the amount of water withdrawn from an available aquifer amounts for a significant demand on the aquifer. Then, since the water reinjection point is typically far outside the *recharge zone* for the aquifer, simply pumping the water back into the ground really does not benefit the aquifer from which it was removed.

Since most aquifer levels in the United States have already fallen significantly over the past 50 years and continue to fall, the use of this heat pump system configuration and further stress on aquifers is not recommended.

27.3.4 Gas-Fired Engine-Driven Heat Pumps

To many, the use of electricity as a primary energy source is not desirable. The electrical power industry routinely releases huge quantities of CO_2 into the atmosphere, along with large quantities of SO_2 and NO_x. On a comparative basis, these poor greenhouse gas emissions make grid electricity a poor sustainable choice for any building heating and cooling system.

In the early 1990s, a U.S. manufacturer introduced a small (5 tons) natural-gas-fired engine-driven packaged air-source heat pump for the residential and light commercial failure. While the engine itself is no more efficient than grid electricity power production, its use of clean-burning natural gas or LPG (propane) resulted in significantly reduced greenhouse gas emissions and a lower operating cost due to the low cost of gas fuels compared to electricity.

But that heat pump was a dismal failure! Of over 4000 units sold, 100% of the units failed within the first 3 years of their operation and the manufacturer was forced to replace almost all of these systems with conventional units. The problem with this concept was not with the heat pump, but with the engine. In this case, the manufacturer attempted to use a modified engine designed for lawn tractors and small engine–generators designed for 150–500 h annual use in an application that required the engine to operate 4000–5000 h/year.

Theoretically, an engine-driven heat pump, coupled with recovered heat from the engine jacket and exhaust, can achieve a heating COP of 1.4–2.0. And, today, manufacturers have teamed with automotive engine manufacturers to offer gas-engine-driven packaged air-source heat pumps in the 8–20 tons cooling capacity range, primarily for commercial applications. These units have an advertised SEER of 18+ and heating COP of 1.5.

The development of smaller gas-engine-driven systems is reportedly underway in both Japan and at Oak Ridge National Laboratory, and domestic-scale units could be brought to market within a few years.

27.3.5 Heat Recovery Chiller/Heat Pump System

Large buildings are often cooled by chilled water systems that circulated water at 40°F–45°F [4°C–7°C] to cooling coils located in air-handling units located throughout the building. While small residential or light commercial buildings require cooling *or* heating, large buildings may require simultaneous cooling *and* heating. And, there is almost always a need to generate service hot water in large buildings.

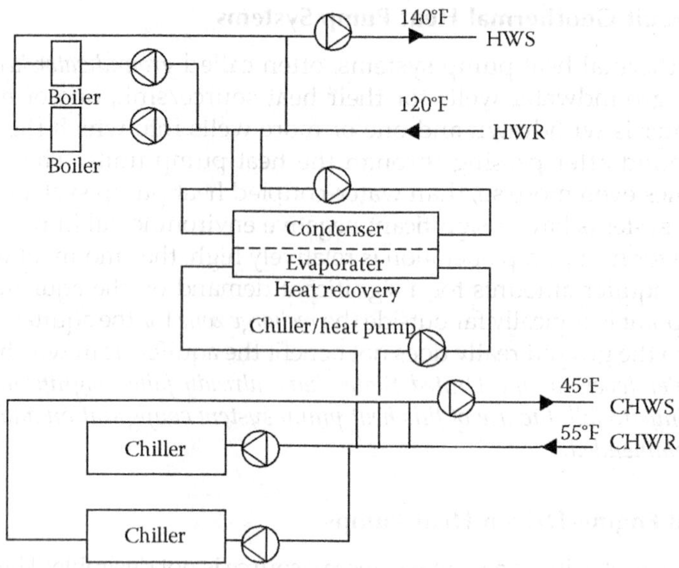

FIGURE 27.9
Schematic of HRC/heat pump system.

When providing cooling, the chilled water system rejects the excess collected heat to the outdoor air via cooling towers via condenser water at 85°F–95°F [29°C–35°C] temperature. These temperatures are typically too low to allow condenser water to be used for heating, so the *dedicated heat recovery chiller* (*HRC*) *heat pump* system is sometimes applied as illustrated by Figure 27.9.

Though not a truly *reverse cycle system*, this system is a heat pump configuration. A *lead* HRC is configured with a primary/secondary piping loop on the chilled water return piping upstream of the remaining cooling-only chillers. The HRC capacity is selected based on the need to provide heating during the summer. Thus, the HRC removes heat from the return chilled water, lowering its temperature, and rejects that heat to the hot water system at 130°F–140°F [54°C–60°C]. The effect is to reduce the imposed load on the cooling-only chillers by *precooling* the return chilled water while simultaneously producing hot water at sufficiently high temperature to be useful.

For this concept to be cost effective, there must be a need for heating during the summer cooling period. This heat may be needed for reheat associated with space temperature or humidity control, service hot water heating, pool heating, etc. To maximize the energy efficiency of this concept, though, there must be a need for chilled water cooling for as long as feasible. This means that the use of airside economizer cycles, required by most energy conservation building codes, needs to be carefully evaluated to determine when it is more efficient to provide mechanical cooling and capture the rejected heat versus using the outdoor air to provide *free* cooling and operating boilers to provide heat.

27.3.6 Variable Refrigerant Flow Heat Pump System

Since about 2000, the use of *variable refrigerant flow* (VRF) systems that achieve a better level of thermal zoning in buildings while using a common outdoor air unit has increased significantly. With VRF systems, up to 16 indoor evaporator units of varying capacity can be coupled with a single outdoor condensing unit. Currently widely applied in large buildings such as offices and hospitals outside the United States, especially in Japan, Europe, and Canada, these systems are just *catching on* in the United States.

These systems use multiple compressors, including inverter-driven variable speed units, and deliver excellent part-load performance and zoned temperature control, resulting in excellent occupant comfort. The basic difference between these systems and conventional HVAC systems is that they circulate refrigerant directly to multiple indoor units, rather than using water (as in a chiller) or air (as in a ducted DX system) to achieve heat transfer to the space.

VRF systems are extremely flexible, enabling a single condensing unit to be connected to a large number of indoor units of varying capacity and configuration. The exact number of indoor units varies according to the manufacturer, but one typical manufacturer allows connection of up to 16 indoor units to one condensing unit or up to 30 indoor units on a single refrigerant circuit supplied by 3 outdoor units.

Typically, each condensing unit uses two or three compressors, one of which is an inverter-driven variable speed compressor. Systems are commonly designed by combining multiple condensing units to achieve system capacities of up to several hundred tons.

Energy savings are due to several factors:

High part-load efficiency: Because VRF systems consist of multiple compressors, some of which are variable speed, the system's part load efficiency is excellent. A typical dual-compressor system can operate at 21 capacity steps. Since most HVAC systems spend most of their operating hours between 30% and 70% of their maximum capacity, a load range in which the COP of the system is very high and, thus, the seasonal energy efficiency of these systems is excellent.

Effective zone control: Indoor units can easily be turned off in locations needing no cooling, while the system retains highly efficient operation.

Heat pump operation: In buildings where simultaneous heating and cooling are needed, such as many office buildings, the system is configured as a three-pipe heat recovery system. In this configuration, refrigerant flow control is used to circulate refrigerant from the discharge of the evaporators in space being cooled to the evaporators of zones needing heat and vice versa. By using refrigerant to move heat between zones, a very high heating COP can be realized (perhaps as high as 4.0).

Because the energy savings of VRF systems are so application dependent, it is difficult to make definitive, general statements about their energy efficiency. Initial estimates of energy savings are in the range of 5%–15%, with higher savings in hot, humid climates, and lower savings in cold climates.

27.4 Advanced-Technology Heat Pumps

27.4.1 Absorption Cycle Heat Pumps

Absorption cycle heat pumps are essentially air-source heat pumps driven not by electricity but by a heat source such as natural gas, propane, solar-heated water, or geothermal-heated water. Because natural gas is the most common heat source for absorption heat pumps, they are also referred to as *gas-fired heat pumps*.

Residential absorption heat pumps use an ammonia–water absorption cycle to provide heating and cooling. As in a vapor-compression cycle heat pump, the refrigerant (in this case, ammonia) is condensed in one heat exchanger to release its heat; its pressure is then reduced and the refrigerant is evaporated to absorb heat.

The absorption cycle heat pump has no compressor. Evaporated ammonia is absorbed into water, and a relatively low-power pump then pumps the ammonia–water solution to a heat exchanger at a slightly higher pressure where heat is added to essentially boil the ammonia out of the water.

The *generator absorber heat exchanger* boosts the efficiency of the unit by recovering the heat that is released when the ammonia is absorbed into the water. Other innovations include high-efficiency vapor separation, variable ammonia flow rates, and low-emissions, variable-capacity combustion of the natural gas.

Absorption cycle cooling only is now commercially available in the 5+ ton capacity range and absorption cycle heat pumps are under development.

27.4.2 Solar-Assisted Heat Pumps

Solar-assisted (sometimes called *solar-augmented*) heat pumps utilize rooftop solar heat collectors configured to augment air-source or ground-source geothermal heat exchangers. These systems are generally series configured with solar panels that are arranged to act as an additional heat source for the heat pump.

For air-source or water-source heat pumps, solar panels can provide higher temperature heat during the winter, reducing the need for supplemental heating by electric coils or boilers. This configuration can also reduce surface land requirements for ground-coupled heat pumps, expanding their range of applicability.

Bibliography

American Society of Heating, Refrigerating, and Air-Conditioning Engineers. *Commercial/Institutional Ground-Source Heat Pump Engineering Manual*, American Society of Heating, Refrigerating, and Air-Conditioning Engineers, Atlanta, GA, 1995.

American Society of Heating, Refrigerating, and Air-Conditioning Engineers. *ASHRAE Handbook: HVAC Applications*, American Society of Heating, Refrigerating, and Air-Conditioning Engineers, Atlanta, GA, 2011 (Chapter 34).

American Society of Heating, Refrigerating, and Air-Conditioning Engineers. *ASHRAE Handbook: HVAC Systems and Equipment*, American Society of Heating, Refrigerating, and Air-Conditioning Engineers, Atlanta, GA, 2012 (Chapters 7 and 49).

ANSI/ASHRAE/IES Standard 90.1-2010. *Energy Standard for Buildings Except Low-Rise Residential Buildings*, American Society of Heating, Refrigerating, and Air-Conditioning Engineers, Atlanta, GA, 2012.

Staffell I., Brett D., Brandon N., and Hawkes, A. A review of domestic heat pumps, *Energy Environmental Science*, 2012, 5: 9291–9306.

Stanford, H.W., III. *Analysis and Design of Heating, Ventilating, and Air-Conditioning Systems*, Prentice-Hall, Englewood Cliffs, NJ, 1988.

Stanford, H.W., III. *HVAC Water Chillers and Cooling Towers: Fundaments, Application, and Operation*, 2nd edn. CRC Press/Taylor & Francis Group, Boca Raton, FL, 2010.

28

Electric Motor Systems Efficiency

Aníbal T. de Almeida, Steve F. Greenberg, and Prakash Rao

CONTENTS

28.1 Introduction

Motor systems are by far the most important type of electrical load, ranging from small fractional horsepower motors incorporated in home appliances to multi-megawatt motors driving pumps and fans in power plants. Motors consume over half of total U.S. electricity, and in industry they are responsible for about two-thirds of the electricity consumption. In the commercial and residential sectors, motors consume slightly less than half of the electricity. The cost of powering motors is immense, roughly $100 billion a year in the United States alone. There is a vast potential for saving energy and money by increasing the efficiency of motors and motor systems.

28.1.1 Motor Types

Motors produce useful work by causing the shaft to rotate. Motors have a rotating part, the rotor, and a stationary part, the stator. Both parts produce magnetic fields, either through windings excited by electric currents or through the use of permanent magnets (PMs). It is the interaction between these two magnetic fields that is responsible for the torque generation, as shown in Figure 28.1.

There are a wide variety of electric motors, based on the type of power supply (AC or DC) that feeds the windings, as well as on different methods and technologies to generate the magnetic fields in the rotor and in the stator. Figure 28.2 presents the most important types of motors.

Because of their low cost, high reliability, and fairly high efficiency, most of the motors used in large home appliances, industry, and commercial buildings are induction motors. Figure 28.3 shows the operating principle of a three-phase induction motor.

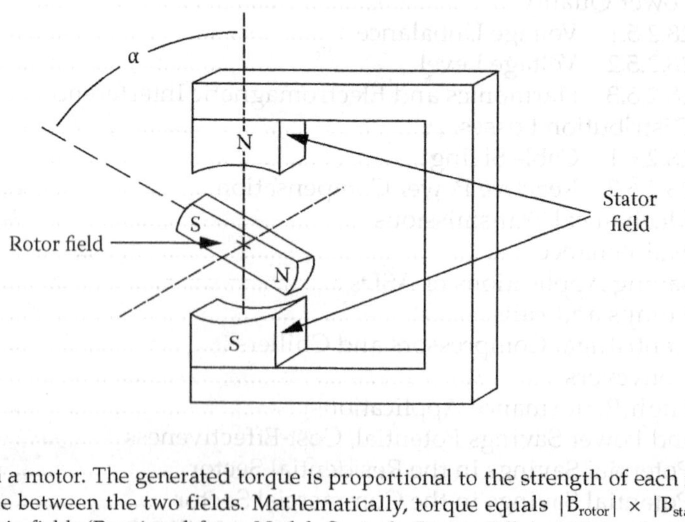

FIGURE 28.1
Torque generation in a motor. The generated torque is proportional to the strength of each magnetic field and depends on the angle between the two fields. Mathematically, torque equals $|B_{rotor}| \times |B_{stator}| \times \sin \alpha$, where B refers to a magnetic field. (Reprinted from Nadel, S. et al., *Energy-Efficient Motor Systems: A Handbook on Technologies, Programs, and Policy Opportunities*, 1st edn., American Council for an Energy-Efficient Economy, Washington, DC, 1992. With permission.)

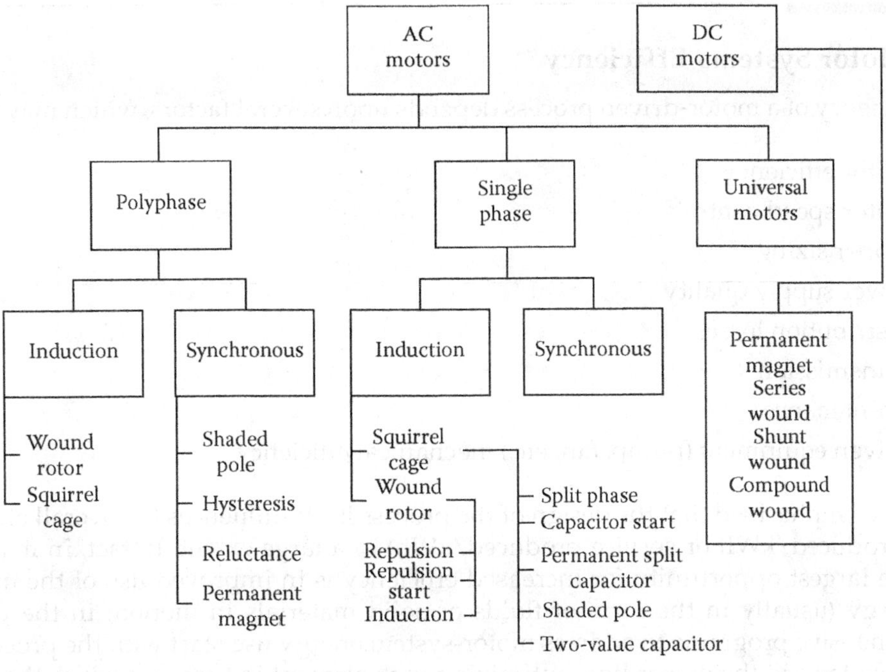

FIGURE 28.2
Motor types. (From EPRI, Electric motors: Markets, trends and application, EPRI Report TR-100423, Electric Power Research Institute, Palo Alto, CA, 1992.)

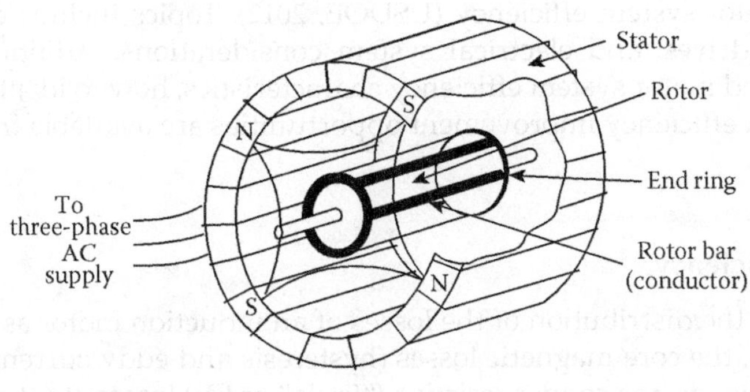

FIGURE 28.3
Operation of a four-pole squirrel-cage induction motor. Rotating magnetic field is created in the stator by AC currents carried in the stator winding. Three-phase voltage source results in the creation of north and south magnetic poles that revolve or *move around* the stator. The changing magnetic field from the stator induces current in the rotor conductors, in turn creating the rotor magnetic field. Magnetic forces in the rotor tend to follow the stator magnetic fields, producing rotary motor action.

Synchronous motors, where the rotor spins at the same speed as the stator magnetic field, are used in applications requiring constant speed, high operating efficiency, and controllable power factor. Efficiency and power factor are particularly important above 1000 hp. Although DC motors are easy to control, both in terms of speed and torque, they are expensive to produce and have modest reliability. DC motors are used for some industrial and electric traction applications, but their importance is dwindling.

28.2 Motor Systems Efficiency

The efficiency of a motor-driven process depends upon several factors, which may include

- Motor efficiency
- Motor speed controls
- Proper sizing
- Power supply quality
- Distribution losses
- Transmission
- Maintenance
- Driven equipment (pump, fan, etc.) mechanical efficiency

It must be emphasized that the design of the process itself influences the overall efficiency (units produced/kWh or service produced/kWh) to a large extent. In fact, in many systems the largest opportunity for increased efficiency is in improved use of the mechanical energy (usually in the form of fluids or solid materials in motion) in the process. Comprehensive programs to address motor-system energy use start with the process and work back toward the power line, optimizing each element in turn, as well as the overall system. Outlining such a program is beyond the scope of this discussion; see, for example, Baldwin (1989) for the benefits that propagate all the way back to the power plant. Additionally, the U.S. Department of Energy (U.S. DOE) has published several *Energy Tips* sheets on motor system efficiency (USDOE, 2012). Topics include estimating motor efficiency, motor drives, and electrical system considerations. Additional publications detailing motor and motor system efficiency characteristics, how to identify motor system losses, and system efficiency improvement opportunities are available from the U.S. DOE (USDOE, 2014a–c).

28.2.1 Motor Efficiency

Figure 28.4 shows the distribution of the losses of an induction motor as a function of the load. At low loads, the core magnetic losses (hysteresis and eddy currents) are dominant, whereas at higher loads, the copper resistive ("Joule" or I^2R) losses are the most important. Mechanical losses are also present in the form of friction in the bearings and windage.

28.2.1.1 Motor Efficiency: Energy Efficient, Premium Efficient, and Beyond

After World War II and until the early 1970s, there was a trend to design inefficient motors, which minimized the use of raw materials (copper, aluminum, and silicon steel). These induction motors had lower initial costs and were more compact than the previous generations of motors, but their running costs were higher. When electricity prices started escalating rapidly in the mid-1970s, most of the large motor manufacturers added a line of higher-efficiency motors to their selection. Such motors feature optimized design, more generous electrical and magnetic circuits, and higher quality materials (Baldwin, 1989). Incremental efficiency improvements are still possible with the use of superior materials (e.g., amorphous silicon steel) and optimized computer-aided design techniques.

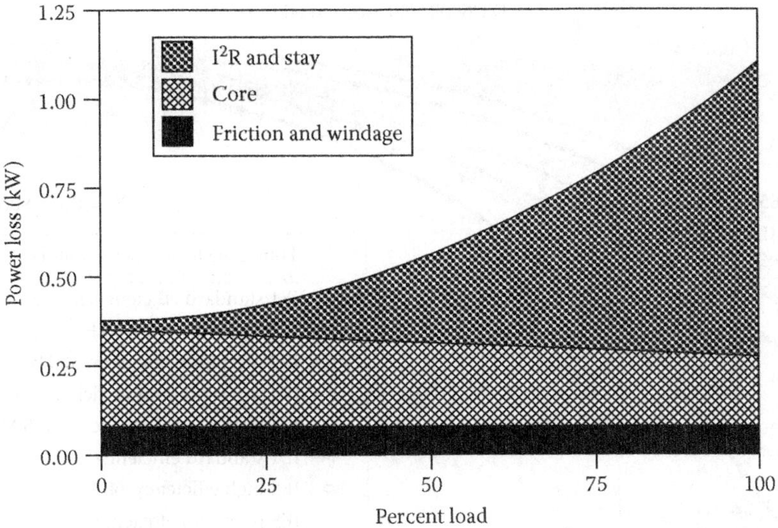

FIGURE 28.4
Variation of losses with load for a 10 hp motor. (Reprinted from Nadel, S. et al., *Energy-Efficient Motor Systems: A Handbook on Technologies, Programs, and Policy Opportunities*, 1st edn., American Council for an Energy-Efficient Economy, Washington, DC, 1992. With permission.)

In 1997, the U.S. Energy Policy Act (EPAct) put in place mandatory efficiency standards for many general-purpose motors. Also in the 1990s, the Consortium for Energy Efficiency (CEE) developed a voluntary premium-efficiency standard, which evolved into the NEMA Premium designation (Nadel et al., 2002; NEMA, 2000). Premium efficiency motors offer an efficiency improvement over EPAct, which typically ranges from 4% for a 1 hp motor to 2% for a 150 hp motor (Figure 28.5). Due to long motor lives, many motors in use are less efficient than EPAct, and thus there is an even larger difference between them and premium efficiency motors. Premium efficiency motors normally cost around 15%–25% more than standard motors, which translates into a price premium of $8–$40/hp. In new applications, and for motors with a large number of operating hours, the paybacks are normally under 4 years for Premium vs. EPAct motors, and under 2 years for Premium vs. older, standard-efficiency motors.

In 2007, the U.S. Energy Independence and Security Act (EISA) raised the minimum efficiency standards set by EPAct to meet efficiency requirements equivalent to NEMA Premium effective December 2010. Additionally, EISA also expanded the number of motors covered by EPAct. In 2014, U.S. DOE passed a final rule expanding the number of motors required to meet the energy efficiency guidelines outlined in NEMA Table 12-12, which is the table used to designate a premium efficiency motor. For example, where EISA requires 1–200 hp general purpose Design B motors to meet NEMA Premium efficiency levels, the 2014 rule expands the covered size range to 500 hp (USDOE 2014c).

An important measure for the wide market acceptance of high efficiency motors is the availability of harmonized standards, dealing with motor performance testing, efficiency classification, and display of ratings (Brunner et al., 2013, De Almeida 2013). This also applies to adjustable-speed drives (ASDs; see later section). In PR China and in the EU, high-efficiency/IE2 motors are mandatory since 2011, and Premium/IE3 motors will be mandatory in 2015 in the EU (De Almeida et al., 2014). The IEC60034-30 efficiency

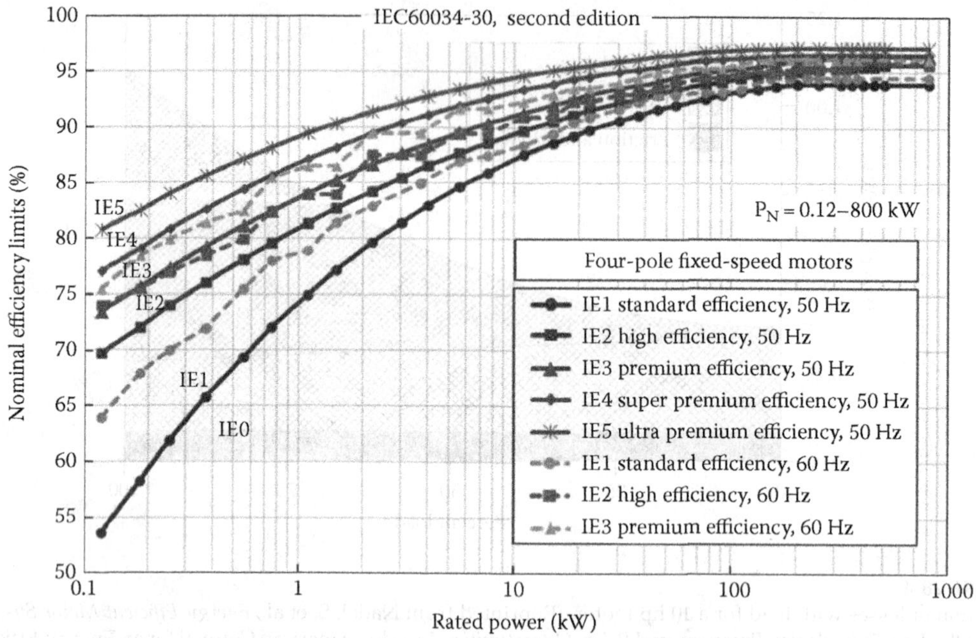

FIGURE 28.5
Planned revision (second edition) of IEC60034-30 nominal efficiency classes limits, for four-pole motors (0.12–800 kW power range). (From De Almeida, A. et al., *IEEE Indust. Appl. Mag.*, 17(1), 12, January/February 2011; De Almeida, A. et al., *IEEE Trans. Indust. Appl.*, 50(2), March/April, 1274, 2014; IEC60034-2-1, 2nd edn., 2/1687/CDV: Rotating electrical machines—Part 2-1: Standard method for determining losses and efficiency from tests [excluding machines for traction vehicles], 2013.)

classification standard (IEC60034-30, 2008) is being revised, and while the IE4* Super-Premium efficiency class is now well established, a new IE5 Ultra-Premium efficiency class is being considered in the Second Edition† (IEC/TS60034-31, 2010). However, there is a question with regard to its achievability (De Almeida et al., 2013).

In Figure 28.5, the IE1, IE2, IE3, IE4, and IE5 classes of the revised IEC60034-30 standard are shown for four-pole, 50/60 Hz motors (IEC60034-30, 2008). The motor nominal efficiency should be determined according to IEC60034-2-1 (IEC60034-2-1, 2007; IEC6003-2-1, 2013).

28.2.1.2 Efficiency of Rewound Motors

When a motor fails, the user has the options of having the motor rebuilt, buying a new standard motor, or buying a NEMA Premium motor. Except for large motors with low annual operating hours, it is typically very cost-effective to replace the failed motor with a Premium motor. Although motor rebuilding is a low-cost alternative, the efficiency of a rebuilt motor can be substantially decreased by the use of improper methods for stripping the old winding. On average, the efficiency of a motor decreases by about 1% each time the motor is rewound.

* The designation of the energy-efficiency class in the IEC60034-30 standard consists of the letters "IE" (International Energy-Efficiency), directly followed by a numeral representing the classification.

† The Second Edition will be denoted as IEC60034-30-1, for line-start motors and IEC60034-30-2 for ASD-fed motors.

The use of high temperatures (above 350°C) during the rewinding process can damage the interlaminar insulation and distort the magnetic circuit with particular impact on the air gap shape, leading to substantially higher core and stray losses. Before any motor is rewound, it should be checked for mechanical damage and the condition of the magnetic circuit should be tested with an electronic iron core loss meter. There are techniques available to remove the old windings, even the ones coated with epoxy varnish, which do not exceed 350°C (Dreisilker, 1987).

28.2.2 Recent Motor Developments

In the low horsepower range, the induction motor is being challenged by new developments in motor technology such as PM and reluctance motors, which are as durable as induction motors and are more efficient. These advanced motors do not have losses in the rotor and feature higher torque and power/weight ratio. In fractional-horsepower motors, such as the ones used in home appliances, the efficiency improvements can reach 10%–15%, compared with single-phase induction motors. Compared to the shaded-pole motors commonly used in small fans, improved motor types can more than double motor efficiency.

28.2.2.1 Permanent-Magnet Motors

Over the last few decades, there has been substantial progress in the area of PM materials. Figure 28.6 shows the relative performance of several families of magnetic materials. High-performance PM materials such as neodymium–iron–boron alloys, with a large energy

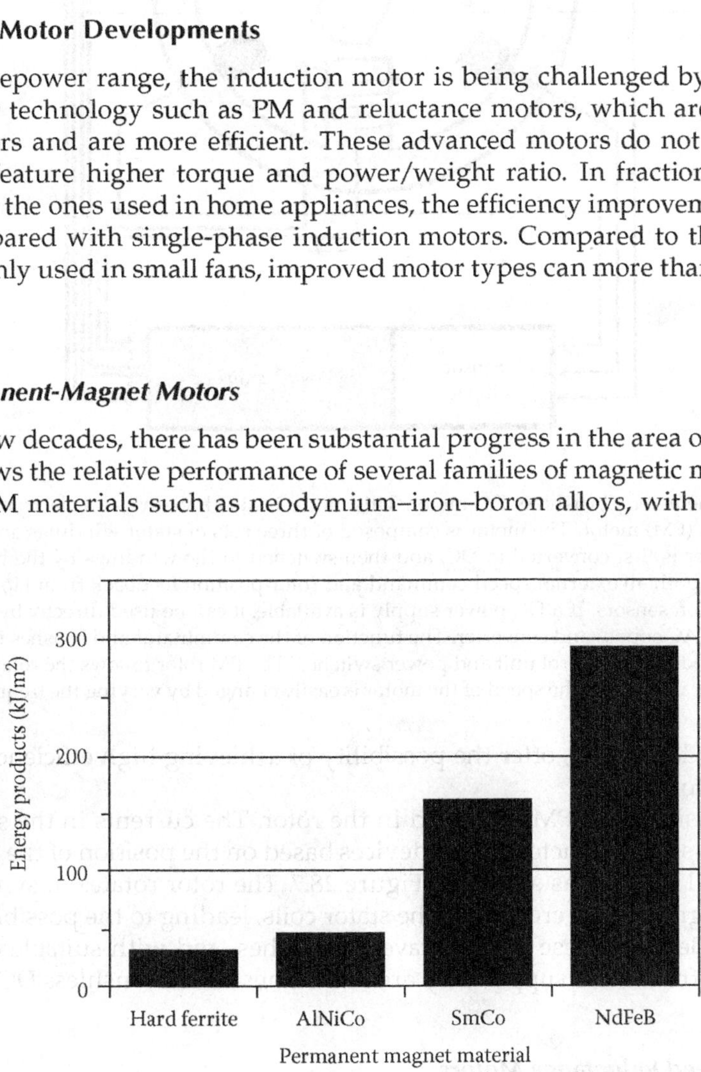

FIGURE 28.6
The evolution of permanent-magnet materials, showing the increasing magnetic energy density (*energy product*). Ferrites were developed in the 1940s and AlNiCos (aluminum, nickel, and cobalt) in the 1930s. The rare-earth magnets were developed beginning in the 1960s (samarium–cobalt) and in the 1980s (neodymium–iron–boron). The higher the energy density, the more compact the motor design can be for a given power rating. (Reprinted from Nadel, S. et al., *Energy-Efficient Motor Systems: A Handbook on Technologies, Programs, and Policy Opportunities*, 1st edn., American Council for an Energy-Efficient Economy, Washington, DC, 1992. With permission.)

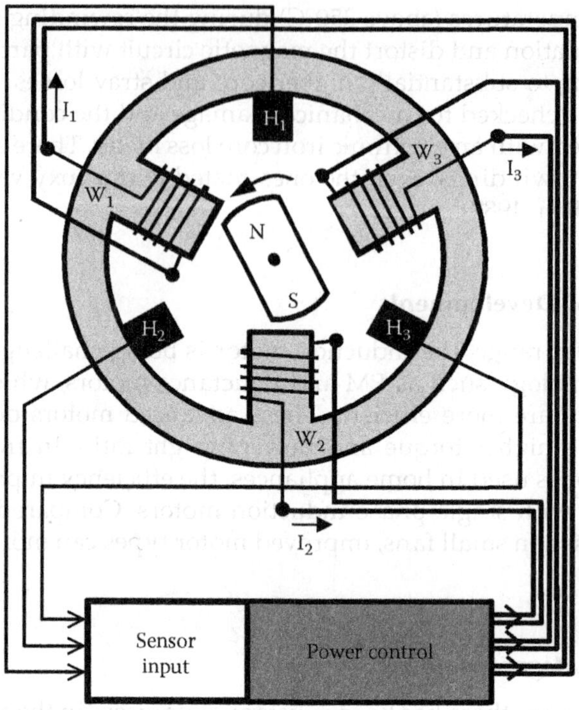

FIGURE 28.7
Control system scheme of a brushless DC motor. This motor is also known as an electronically commutated permanent-magnet (PM) motor. The motor is composed of three sets of stator windings arranged around the PM rotor. AC power is first converted to DC, and then switched to the windings by the power control unit, which responds to both an external speed command and rotor position feedback from H_1, H_2, and H_3, which are magnetic position sensors. If a DC power supply is available, it can be used directly by the power control unit in place of the AC supply and converter. The function of the commutator and brushes in the conventional DC motor is replaced by the control unit and power switches. The PM rotor follows the rotating magnetic field created by the stator windings. The speed of the motor is easily changed by varying the frequency of switching.

density and moderate cost, offer the possibility of achieving high efficiency and compact lightweight motors.

In modern designs, the PMs are used in the rotor. The currents in the stator windings are switched by semiconductor power devices based on the position of the rotor, normally detected by Hall sensors, as shown in Figure 28.7. The rotor rotates in synchronism with the rotating magnetic field created by the stator coils, leading to the possibility of accurate speed control. Because these motors have no brushes, and with suitable control circuits they can be fed from a DC supply, they are sometimes called brushless DC motors.

28.2.2.2 Switched Reluctance Motors

Switched reluctance motors are also synchronous motors whose stator windings are commutated by semiconductor power switches to create a rotating field. The rotor has no windings, being made of iron with salient poles. The rotor poles are magnetized by the influence of the stator rotating field. The attraction between the magnetized poles and the rotating field creates a torque that keeps the rotor moving at synchronous speed. Figure 28.8 shows the structure of a switched reluctance motor.

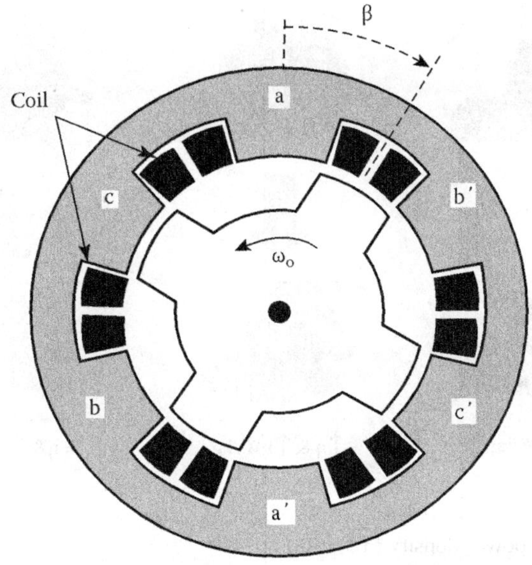

FIGURE 28.8
Schematic view of a switched reluctance motor. The configuration shown is a 6/4 pole. A rotating magnetic field is produced by switching power on and off to the stator coils in sequence, thus magnetizing poles a–a′, b–b′, and c–c′ in sequence. Switching times are controlled by microprocessors with custom programming.

Switched reluctance motors are more efficient than induction motors, are simple to build, and are robust, and if mass produced, their price can compete with induction motors. Switched reluctance motors can also be used in high-speed applications (above the 3600 rpm possible with induction or synchronous motors operating on a 60 Hz AC supply) without the need for gears.

28.2.2.3 Synchronous Reluctance Motors

In synchronous reluctance motors (SynRMs), the stator has windings (similar to those used in the squirrel-cage induction motors [SCIMs]) and the rotor is cylindrical (as distinguished from Figure 28.8) but with anisotropic magnetic structure (magnetic reluctance varies with the flux direction) (ABB, 2012, 2013; Boglietti et al., 2005). In these motors, when fed by an ASD, speed and torque control is possible and large energy savings may be realized in variable-flow pump and fan applications.

Similar to permanent-magnet synchronous motors (PMSMs), it is possible to design SynRMs for (as illustrated in Figure 28.9 from left to right) the same frame size and better efficiency, an in-between solution with slight higher efficiency and slightly smaller frame size, or smaller frame size and equal efficiency.

Figure 28.10 shows the expected loss reduction when comparing the same and smaller frame sizes to a reference SCIM frame size (De Almeida et al., 2013). The new rotor has neither magnets nor windings, and thus suffers virtually no power losses—which makes it uniquely cool. For the same frame size, there is typically a 25% reduction in the overall losses with respect to IE3-Class/NEMA Premium SCIMs. Therefore, there is the possibility of a significant efficiency gain if the standard frame size is used.

It should be mentioned that the commercially available models of SynRM and drive packages are optimized for variable-speed pump and fan applications to achieve the

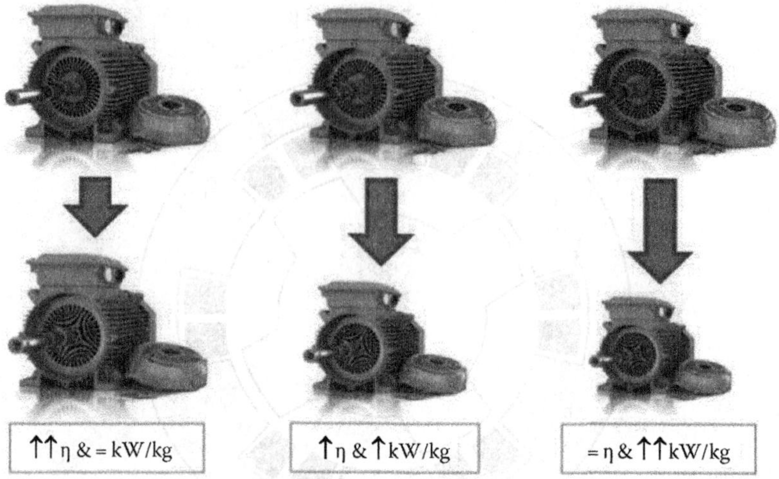

FIGURE 28.9
Frame size vs. efficiency and power density in SynRMs.

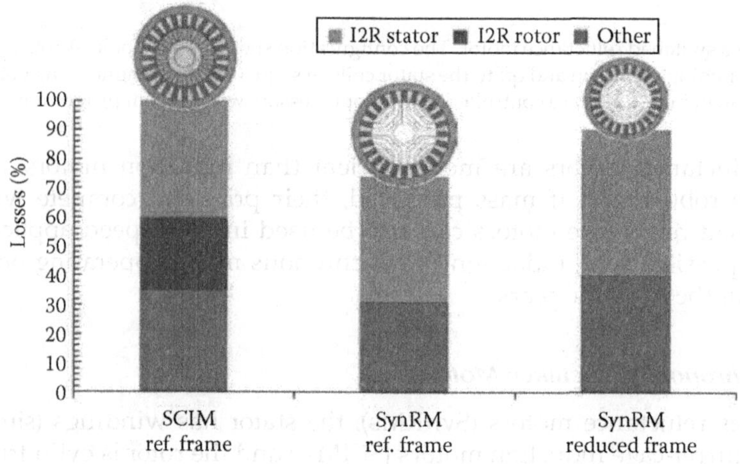

FIGURE 28.10
Same stator size, different rotor type: loss reduction (note: stator copper losses slightly increase and rotor losses are reduced to zero).

highest efficiency. Since SynRM have no rotor cage losses, their overall losses are lower and, consequently, the efficiency is higher. For the same efficiency, SynRM can be up to two frame sizes smaller than conventional SCIM. The output and efficiency performance of SynRM are comparable to a PMSM drive, but with construction techniques similar to those used in SCIMs, bringing the best of both worlds to users.

28.2.2.4 Promising Industrial Motor Technologies

The Super-Premium efficiency class (NEMA Premium or IE3 mentioned earlier) motor technologies can be divided into fixed- and variable-speed applications. For fixed-speed applications, the existing SCIM and line-start (motors that can start and run directly from grid power without an ASD) permanent magnet (LSPM) technologies are already within the IE4 class. Although not commercially available, SynRM technology is likely to evolve to a line-start solution by means of embedding an auxiliary cage in the rotor (e.g., embedded in the air spaces of the rotor), as was done for the PM technology. This type of motor would be useful in constant speed applications and would not require an ASD.

Presently for variable-speed applications, the existing ASD-fed PMSMs (with rare-earth PMs and silicon steel core or with ferrite PMs and amorphous metal core) and SynRMs are the best options. In such applications, it is important to note that the efficiency at partial speed and torque is very important since the operation cycle typically includes long periods with speeds and torques lower than nominal. Therefore (as with any variable load motor system), when analyzing the energy performance of the drive system, the efficiency maps are much more relevant than a single efficiency point for the rated speed and torque.

28.2.3 Motor Speed Controls

AC induction and synchronous motors are essentially constant-speed motors. Most motor applications would benefit if the speed could be adjusted to the process requirements. This is especially true for new applications where the processes can be designed to take advantage of the variable speed. The potential benefits of speed variation include increased productivity and product quality, less wear in the mechanical components, and substantial energy savings.

In many pump, fan, and compressor applications, the mechanical power grows roughly with the cube of the fluid flow; to move 80% of the nominal flow only half of the power is required. Fluid-flow applications are therefore excellent candidates for motor speed control.

Conventional methods of flow control have used inefficient throttling devices such as valves, dampers, and vanes. These devices have a low initial cost but introduce high running costs due to their inefficiency. Figure 28.11 shows the relative performance of different techniques to control flow produced by a fan. In Figure 28.11, it should be noted that in VAV systems static pressure typically acts as a feedback mechanism for controlling fan speed. Consequently, there will be some power consumption at 0 CFM with the result that overall savings is less than the theoretical condition.

Motor system operation can be improved through the use of several speed-control technologies, such as those covered in the following three sections.

28.2.3.1 Mechanical and Eddy-Current Drives

Mechanical speed control technologies include hydraulic transmissions, adjustable sheaves, and gearboxes. Eddy-current drives work as induction clutches with controlled slip (Magnusson, 1984).

Both mechanical drives and eddy drives have relatively low importance. They suffer from low efficiency, bulkiness, limited flexibility, or limited reliability when compared with other alternatives; in the case of mechanical drives, they may require regular maintenance.

Mechanical and eddy drives are not normally used as a retrofit due to their space requirements. Their use is more and more restricted to the low horsepower range where their use may be acceptable due to the possible higher cost of ASDs.

28.2.3.2 Multispeed Motors

In applications where only a few operating speeds are required, multispeed motors may provide the most cost-effective solution. These motors are available with a variety of

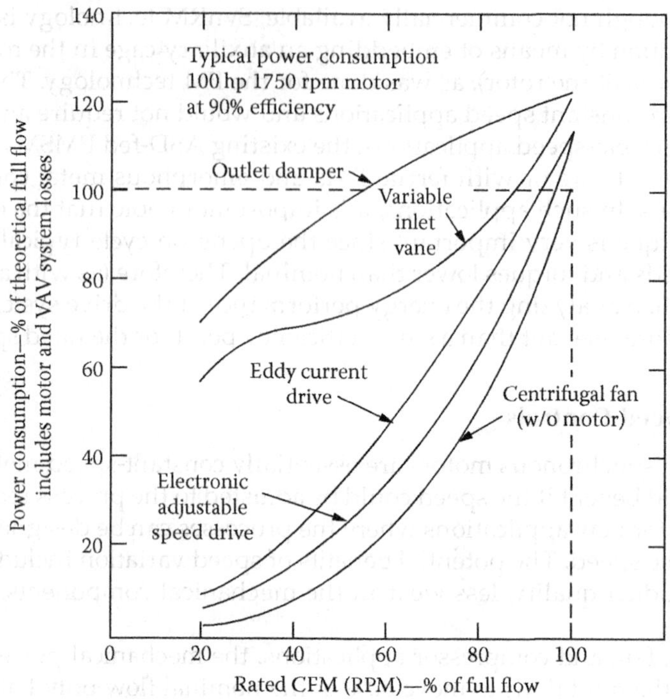

FIGURE 28.11
Comparison of several techniques for varying air flow in a variable air volume (VAV) ventilation system. The curve on the lower right represents the power required by the fan itself, not including motor losses. Electronic adjustable-speed drives are the most efficient VAV control option, offering large savings compared to outlet dampers or inlet vanes, except at high fractions of the rated fan speed. (From Greenberg, S. et al., Technology assessment: Adjustable speed motors and motor drives, Lawrence Berkeley Laboratory Report LBL-25080, Lawrence Berkeley Laboratory, University of California, Berkeley, CA, 1988.)

torque-speed characteristics (variable torque, constant torque, and constant horsepower) (Andreas, 1992), to match different types of loads. Double-winding motors can provide up to four speeds but they are normally bulkier (one frame size larger) than single-speed motors for the same horsepower rating. Pole-amplitude modulated (PAM) motors are single-winding, two-speed SCIMs that provide a wide range of speed ratios (Pastor, 1986). Because they use a single winding they have the same frame size of single-speed motors for the same horsepower and are thus easy to install as a retrofit. PAM motors are available with a broad choice of speed combinations (even ratios close to unity), being especially suited and cost-effective for those fan and pump applications that can be met by a two-speed duty cycle.

28.2.3.3 Electronic Adjustable-Speed Drives

Induction motors operate with a torque-speed relation as shown in Figure 28.12. The speed of the motor is very nearly proportional to the frequency of the AC power supplied to it; thus, the speed can be varied by applying a variable-frequency input to the motor. Electronic ASDs (Bose, 1986) achieve this motor input by converting the fixed frequency power supply (50 or 60 Hz), normally first to a DC supply and then to a continuously variable frequency/variable voltage (Figure 28.13). ASDs are thus able to continuously change the speed of AC motors. Electronic ASDs have no moving parts (sometimes with the

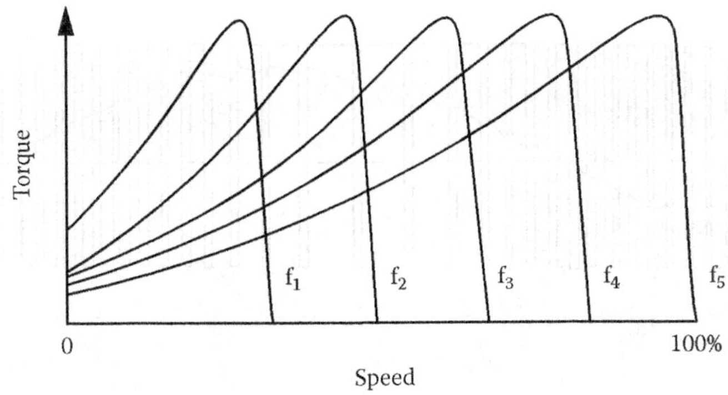

FIGURE 28.12
Speed–torque curves for an induction motor ($f_1 < f_2 < f_3 < f_4 < f_5$ and f_5 = normal line frequency). Normal operation of the motor is in the nearly vertical part of the curves to the right of the *knee* (known as the *breakdown* or *pullout* torque).

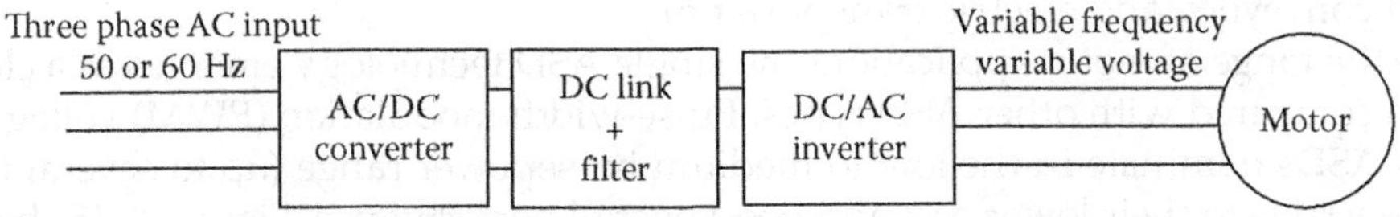

FIGURE 28.13
General inverter-based adjustable-speed drive power circuit with motor load.

exception of a cooling fan), presenting high reliability and efficiency and low maintenance requirements. Because ASDs are not bulky and have flexible positioning requirements, they are generally easy to retrofit.

Electronic ASDs are the dominant motor speed control technology at the present and for the foreseeable future. Developments in the past two decades in the areas of microelectronics and power electronics make possible the design of efficient, compact, and increasingly cost-competitive electronic ASDs. As ASDs control the currents/voltages fed to the motor through power semiconductor switches, it is possible to incorporate motor protection features, soft-start and remote control, at a modest cost. By adding additional power switches and controlling circuitry (Figure 28.14), ASDs can provide regenerative braking,

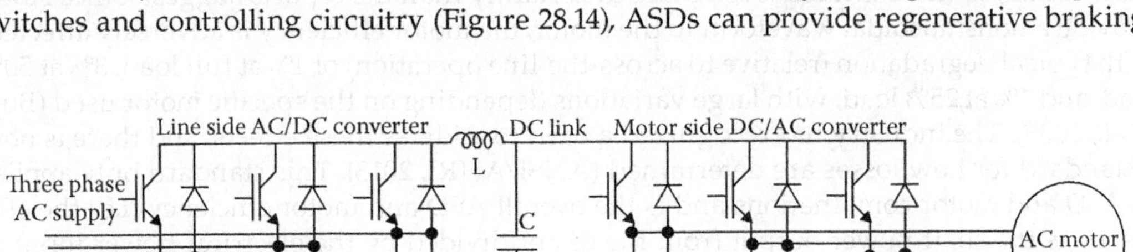

FIGURE 28.14
Power circuitry of a pulse-width modulation variable speed drive with regenerative capacity and power factor control. Whereas conventional adjustable-speed drives use diode rectifiers in the input stage, regenerative units use insulated gate bipolar transistors (IGBTs) at both the input and output stages to enable bidirectional power flow.

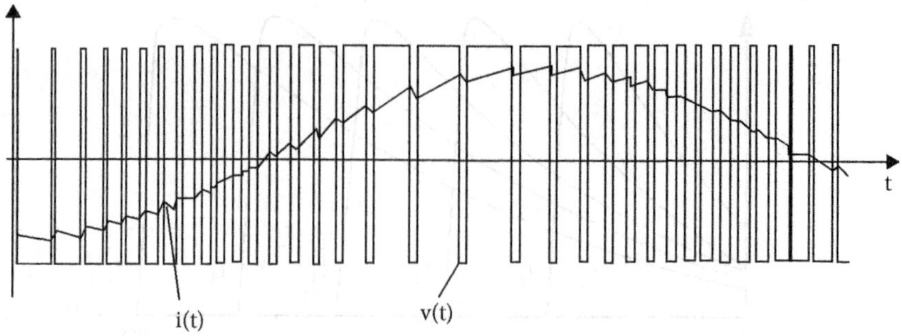

FIGURE 28.15
Pulse-width modulation for synthesizing a sinusoidal output. Output voltage, the average of which resembles the current waveform i(t), is varied by changing the width of the voltage pulses v(t). Output frequency is varied by changing the length of the cycle.

slowing the driven load and feeding power back into the AC supply. Such regenerative braking capability can increase the energy efficiency of applications such as elevators, downhill conveyors, and electric transportation.

Across the range of motor applications, no single ASD technology emerges as a clear winner when compared with other ASD types. Pulse-width modulation (PWM) voltage-source inverters ASDs dominate in the low to medium horsepower range (up to several hundred horsepower) due to their lower cost and good overall performance. Figure 28.15 shows how the variable-frequency/variable-voltage waveform is synthesized by a PWM ASD.

In the range above several hundred horsepower, the choice of ASD technology depends on several factors including the type of motor, horsepower, speed range, and control requirements (Greenberg et al., 1988). Table 28.1 presents a general classification of the most widely used adjustable-speed motor drive technologies.

It should be noted that electronic ASDs have internal power losses, and they also adversely affect the efficiency of the motors they control. The losses within the drive are due to the AC to DC conversion losses, internal resistance in the DC bus filter, and switching losses in the DC to AC inverter section. As a function of rated output power, they are typically 94%–97% efficient at full load, 91%–95% at 50% load, and 86%–94% at 25% load, over the power range of 3–100 hp, respectively, though there is significant variation in reported losses (Krukowski and Wray, 2013). There is no standard for how the losses are determined, so there is likely even more uncertainty than the reports suggest. Since ASDs provide a nonsinusoidal waveform to the motor, the motor efficiency is adversely affected, with typical degradation (relative to across-the-line operation) of 1% at full load, 3% at 50% load, and 1% at 25% load, with large variations depending on the specific motor used (Burt et al., 2008). The industry has recognized a need to address these effects, and there is now a standard for how losses are determined (ANSI/AHRI, 2013). This standard only applies to ASD and motor combinations and is the overall ASD and motor efficiency (i.e., the efficiency is the shaft power output from the motor divided by the electrical power input to the ASD); the standard does not attempt to segregate the losses within the system.

28.2.4 Motor Oversizing

Motors are often oversized as a result of the compounding of successive safety factors in the design of a system (Smeaton, 1988). The magnetic losses, friction, and windage losses are practically constant as a function of the load. Therefore, motors that are oversized

TABLE 28.1

Adjustable-Speed Motor Drive Technologies

Technology	Applicability (R = Retrofit; N = New)	Cost[b]	Comments
Motors			
Multispeed (incl. PAM[a]) motors	Fractional—500 hp PAM; fractional—2,000 + hp R, N	1.5–2 times the price of single-speed motors	Larger and less efficient than one-speed motors. PAM is more promising than multiwinding. Limited number of available speeds.
Direct-current motors	Fractional—10,000 hp N Shaft-applied drives (on motor output)	Higher than AC induction motors	Easy speed control. More maintenance required.
Mechanical			
Hydraulic drive	5–10,000 hp N	Large variation	5:1 speed range. Low efficiency below 50% speed.
Eddy-current drive	Fractional ~2,000 + hp N Wiring-applied drives (on motor input)	\$900–\$60/hp (for 1–150 hp)	Reliable in clean areas, Relatively long life. Low efficiency below 50% speed.
Electronic ASDs			
Voltage-source inverter	Fractional—1500 hp R, N	\$300–\$80/hp (for 1–300 hp)	Multimotor capability. Can generally use existing motor. PWM[c] appears most promising.
Current-source inverter	100–100,000 hp R, N	\$120–\$50/hp (for 100–20,000 hp)	Larger and heavier than VSI Industrial applications, including large synchronous motors.
Others	Fractional—100,000 hp R, N	Large variation	Includes cycloconverters, wound rotor, and variable voltage. Generally for special industrial applications.

[a] PAM means pole amplitude modulated.

[b] The prices are listed from high to low to correspond with the power rating, which is listed from low to high. Thus, the lower the power rating, the higher the cost per horsepower.

[c] PWM means pulse width modulation.

(working all the time below 50% of capacity) present not only lower efficiency (as shown in Figure 28.16) but also poor power factor (NEMA, 1999). The efficiency drops significantly when a motor operates lightly loaded (below 40% for a standard motor). The power factor drops continuously from full load. The decrease in performance is especially noticeable in small motors and standard-efficiency motors.

It is therefore essential to size new motors correctly and to identify motors that run grossly underloaded all the time. In the last case, the economics of replacement by a correctly sized motor should be considered. Considerations include the load at which the motor is most efficient, difference in efficiency between current and correctly sized motor, and practicality of mounting smaller sized motor. In medium or large industrial plants, where a stock of motors is normally available, oversized motors may be exchanged for the correct size versions.

28.2.5 Power Quality

Electric motors, and in particular induction motors, are designed to operate with optimal performance when fed by symmetrical three-phase sinusoidal waveforms with the

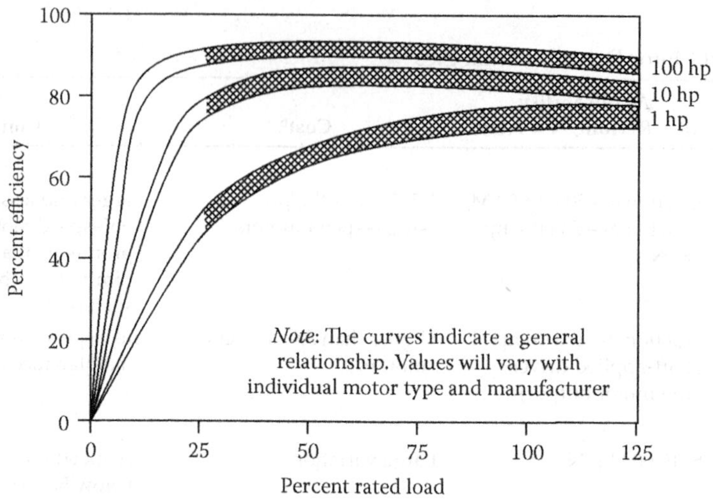

FIGURE 28.16
Typical efficiency vs. load curves for 1800 rpm, three-phase 60 Hz Design B squirrel-cage induction motors. (Reprinted from Nadel, S. et al., *Energy-Efficient Motor Systems: A Handbook on Technologies, Programs, and Policy Opportunities*, 2nd edn., American Council for an Energy-Efficient Economy, Washington, DC, 2002. With permission.)

nominal voltage value. Deviations from these ideal conditions may cause significant deterioration of the motor efficiency and lifetime. Possible power quality problems include voltage unbalance, undervoltage or overvoltage, and harmonics and interference. Harmonics and interference can be caused by, as well as affect, motor systems.

28.2.5.1 Voltage Unbalance

Induction motors are designed to operate at their best with three-phase balanced sinusoidal voltages. When the three-phase voltages are not equal, the losses increase substantially. Phase unbalance is normally caused by an unequal distribution of the single-phase loads (such as lighting) on the three phases or by faulty conditions. An unbalanced supply can be mathematically represented by two balanced systems rotating in opposite directions. The system rotating in the opposite direction to the motor induces currents in the rotor that heat the motor and decrease the torque. Even a modest phase unbalance of 2% can increase the losses by 25% (Cummings et al., 1985).

When a phase unbalance is present, the motor must be derated according to Figure 28.17.

28.2.5.2 Voltage Level

When an induction motor is operated above or below its rated voltage, its efficiency and power factor change. If the motor is underloaded, a voltage reduction may be beneficial, but for a properly sized motor, the best overall performance is achieved at the rated voltage. The voltage fluctuations are normally associated with ohmic (IR) voltage drops or with reactive power (poor power factor) flow in the distribution network (see Section 28.2.6).

28.2.5.3 Harmonics and Electromagnetic Interference

When harmonics are present in the motor supply, they heat the motor and do not produce useful torque. This in turn affects the motor lifetime and causes a derating of the motor

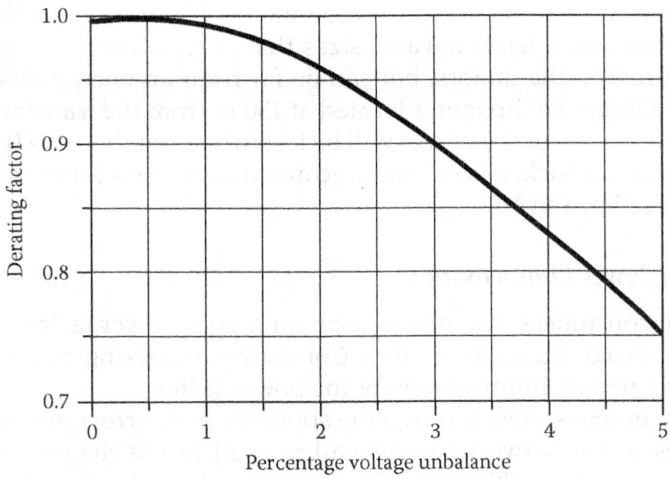

FIGURE 28.17
Derating factor due to unbalanced voltage for integral-horsepower motors. (Reprinted from Nadel, S. et al., *Energy-Efficient Motor Systems: A Handbook on Technologies, Programs, and Policy Opportunities*, 1st edn., American Council for an Energy-Efficient Economy, Washington, DC, 1992. With permission.)

capacity. This is also true when motors are supplied by ASDs that generate the harmonics themselves. The use of premium-efficiency motors can alleviate these problems due to their higher efficiency and thermal capacity; there are also motors specially designed for use with ASDs known as inverter-duty motors.

Reduction of harmonics is also important for the benefit of other consumer and utility equipment. Harmonics, caused by nonlinear loads such as the semiconductor switches in ASDs, should be reduced to an acceptable level as close as possible to the source. The most common technique uses inductive/capacitive filters at the ASD input circuit to provide a shunt path for the harmonics and to perform power factor compensation.

IEEE Standard No. 519 (IEEE, 1992) contains guidelines for harmonic control and reactive power compensation of power converters. The cost of the harmonic filter to meet this standard is typically around 5% of the cost of the ASD.

ASD power semiconductor switches operate with fast switching speeds to decrease energy losses. The fast transitions in the waveforms contain high-frequency harmonics, including those in the radio-frequency range. These high-frequency components can produce electromagnetic interference (EMI) through both conduction and radiation. The best way to deal with EMI is to suppress it at the source. Radiated EMI is suppressed through shielding and grounding of the ASD enclosure. Proper ASD design, the use of a dedicated feeder, and the use of a low-pass input filter (an inductor; often called a *line reactor*), will normally suppress conducted EMI.

28.2.6 Distribution Losses

28.2.6.1 Cable Sizing

The currents supplied to the motors in any given installation will produce Joule (I^2R) losses in the distribution cables and transformers of the consumer. Correct sizing of the cables will not only allow a cost-effective minimization of those losses but also help decrease the voltage drop between the transformer and the motor. The use of the National Electrical Code for sizing conductors leads to cable sizes that prevent overheating and allow adequate starting current to the motors, but can be far from an energy-efficient design. For example, when feeding a 100 hp motor located at 150 m from the transformer with a cable sized using NEC, about 4% of the power will be lost in heating the cable (Howe et al., 1999). Considering a 2-year payback, it is normally economical to use a cable one wire size larger than the one required by the NEC.

28.2.6.2 Reactive Power Compensation

In most industrial consumers, the main reason for a poor power factor is the widespread application of oversized induction motors. Correcting oversizing can thus contribute in many cases to a significant improvement of the power factor.

Reactive power compensation, through the application of correction capacitors, not only reduces the losses in the network but also allows full use of the power capacity of the power system components (cables, transformers, circuit breakers, etc.). In addition, voltage fluctuations are reduced, thus helping the motor to operate closer to its design voltage.

28.2.7 Mechanical Transmissions

The transmission subsystem transfers the mechanical power from the motor to the motor-driven equipment. To achieve overall high efficiency, it is necessary to use simple, properly maintained transmissions with low losses. The choice of transmission is dependent upon many factors including speed ratio desired, horsepower, layout of the shafts, and type of mechanical load.

Transmission types available include direct shaft couplings, gearboxes, chains, and belts. Belt transmissions offer significant potential for savings. About one-third of motor transmissions use belts (Howe et al., 1999). Several types of belts can be used such as V-belts, cogged V-belts, and synchronous belts.

V-belts have efficiencies in the 90%–96% range. V-belt losses are associated with flexing, slippage, and a small percentage due to windage. With wear, the V-belt stretches and needs retensioning, otherwise the slippage increases and the efficiency drops. Cogged V-belts have lower flexing losses and have better gripping on the pulleys, leading to 2%–3% efficiency improvement when compared with standard V-belts.

Synchronous belts can be 98%–99% efficient as they have no slippage and have low flexing losses; they typically last over twice as long as V-belts, leading to savings in avoided replacements that more than offset their extra cost. Figure 28.18 shows the relative performance of V-belts and synchronous belts. The efficiency gains increase with light loads.

28.2.8 Maintenance

Regular maintenance (such as inspection, adjustment, cleaning, filter replacement, lubrication, and tool sharpening) is essential to maintain peak performance of the mechanical parts and to extend their operating lifetime. Both under- and overlubrication can cause higher friction losses in the bearings and shorten the bearing lifetime. Additionally, overgreasing can cause the accumulation of grease and dirt on the motor windings, leading to overheating and premature failure.

The mechanical efficiency of the driven equipment (pump, fan, cutter, etc.) directly affects the overall system efficiency. Monitoring wear and erosion in this equipment is

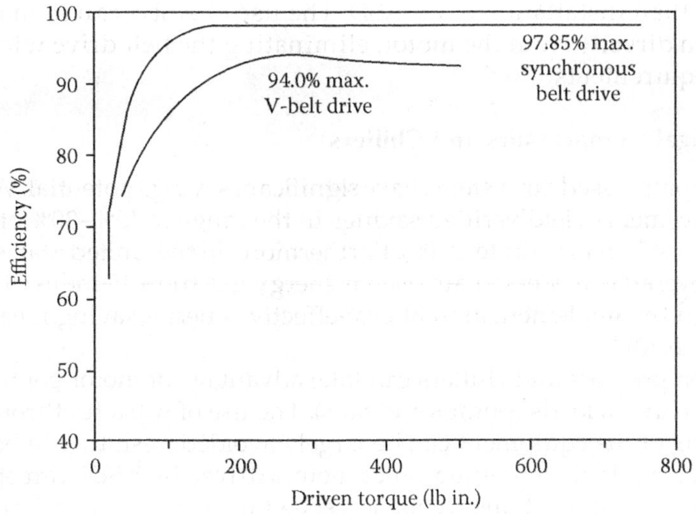

FIGURE 28.18
Efficiency vs. torque for V-belts and synchronous belts in a typical application. (Reprinted from Nadel, S. et al., *Energy-Efficient Motor Systems: A Handbook on Technologies, Programs, and Policy Opportunities*, 1st edn., American Council for an Energy-Efficient Economy, Washington, DC, 1992. With permission.)

especially important as its efficiency can be dramatically affected. For example, in chemical process industries, the erosion of the pump impeller will cause the pump efficiency to drop sharply; a dull cutter will do the same to a machine tool.

Cleaning the motor casing is also relevant because its operating temperature increases as dust and dirt accumulates on the case. The same can be said about providing a cool environment for the motor. A temperature increase leads to an increase of the windings' resistivity and therefore to larger losses. An increase of 25°C (45°F) in the motor temperature increases the Joule losses by 10%.

28.3 Energy-Saving Applications of ASDs

Typical loads that may benefit from the use of ASDs include those covered in the following four sections.

28.3.1 Pumps and Fans

In many pumps and fans where there are significant variable-flow requirements (i.e., where a significant portion of the system operating hours occurs at various partial loads), substantial savings can be achieved, as the power is roughly proportional to the cube of the flow (and thus speed of the motor). The use of ASDs instead of throttling valves with pumps shows similar behavior to that for fans in Figure 28.11. In applications where the majority of the system operation occurs at a single partial load, other techniques, such as pump impeller trimming and changing fan speed, may lead to greater savings than installation of an ASD. The use of ASDs can in many cases allow fans to be driven directly from the motor, eliminating the belt drive with its losses and maintenance requirements.

28.3.2 Centrifugal Compressors and Chillers

Most industrial compressed air systems have significant savings potential. Well-engineered efficiency improvements yield verified savings in the range of 15%–30% of system energy consumption (Fraunhofer Institute, 2000). Furthermore, in the United States, it is estimated that 21% of the overall compressed air system energy use from 15 industrial sectors could be saved through the implementation of cost-effective energy-saving measures (McKane and Hasanbeigi, 2010).

Centrifugal compressors and chillers can take advantage of motor controls in the same way as other centrifugal loads (pumps and fans). The use of wasteful throttling devices or the on–off cycling of the equipment can be largely avoided, resulting in both energy savings and extended equipment lifetime. Since motors driven by ASDS can operate at higher speeds than those operating at line frequency, speed-up gearboxes can be eliminated, saving energy, first cost, and maintenance cost.

Savings from compressed air measures are mostly coincident with electric system peak periods. Plant air systems normally have a large duty factor, typically operating 5000–8000 h/year. Thus, energy and demand reductions are very likely to occur at system peaks and contribute to system reliability. Compressed air system efficiency improvements are highly cost-effective and additionally lead to reduced plant downtime. Many projects have been identified with significant energy and demand reduction potential and paybacks less than 2 years (XENERGY, 2001).

28.3.3 Conveyors

The use of speed controls, in both horizontal and inclined conveyors, allows the matching of speed to material flow. As the conveyor friction torque is constant, energy savings are obtained when the conveyor is operated at reduced speed. In long conveyors, such as those found in power plants and in the mining industry, the benefits of soft-start without the need for complex auxiliary equipment are also significant (De Almeida et al., 2005).

For horizontal conveyors (Figure 28.19), the torque is approximately independent of the transported load (it is only friction dependent). Typically, the materials handling output of a conveyor is controlled through regulating the material input, and the torque and speed are roughly constant. But if the materials input to the conveyor is changed, it is possible to

FIGURE 28.19
Power required by a conveyor.

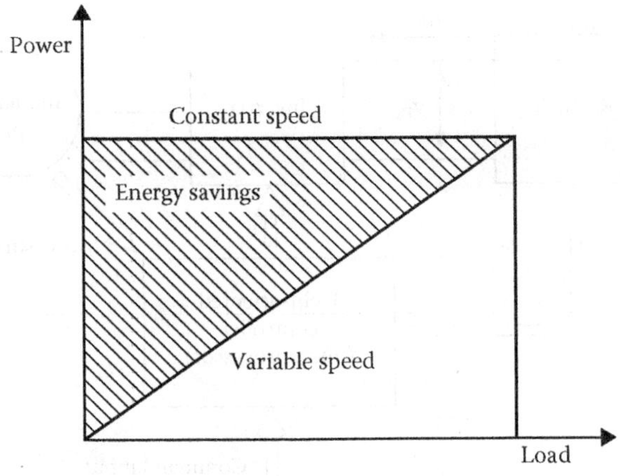

FIGURE 28.20
Energy savings in a conveyor using speed control, in relation to the typical constant speed.

reduce the speed (the torque is the same), and, as can be seen in Figure 28.20, significant energy savings can be achieved, proportional to the speed reduction.

28.3.4 High-Performance Applications

AC motors have received much attention in recent years as a proposed replacement for DC motors in high-performance speed control applications, where torque and speed must be independently controlled. Induction motors are much more reliable, more compact, more efficient, and less expensive to buy and maintain than DC motors. As induction motors have no carbon brush commutation, they are especially suitable for corrosive and explosive environments. In the past, induction motors have been difficult to control as they behave as complex nonlinear systems. However, the appearance on the market of powerful and inexpensive microprocessors has made it possible to implement in real time the complex algorithms required for induction motor control.

Field-oriented control, also called vector control, allows accurate control of the speed and torque of induction motors, in a way similar to DC motor control (Leonhard, 1984). The motor current and voltage waveforms, together with motor position feedback, are processed in real time, allowing the motor current to be decomposed into speed producing component and into a torque producing component. Vector control operation principle is represented in Figure 28.21 and is being applied to a wide variety of high-performance applications described next.

Rolling mills were one of the strongholds of DC motors, due to the accurate speed and torque requirements. With present ASD technology, AC drives can outperform DC drives in all technical aspects (reliability, torque/speed performance, maximum power, efficiency) and are capable of accurate control down to zero speed.

The availability of large diameter, high torque, and low speed AC drives makes them suitable for use in applications like ball mills and rotary kilns without the need for gearboxes. This area was also a stronghold of DC drives. Again, AC drives have the capability to offer superior performance in terms of reliability, power density, overload capability, efficiency, and dynamic characteristics.

AC traction drives can also feature regenerative braking. AC traction drives are already being used in trains, rapid transit systems, and ship propulsion and are the proper choice for the electric automobile.

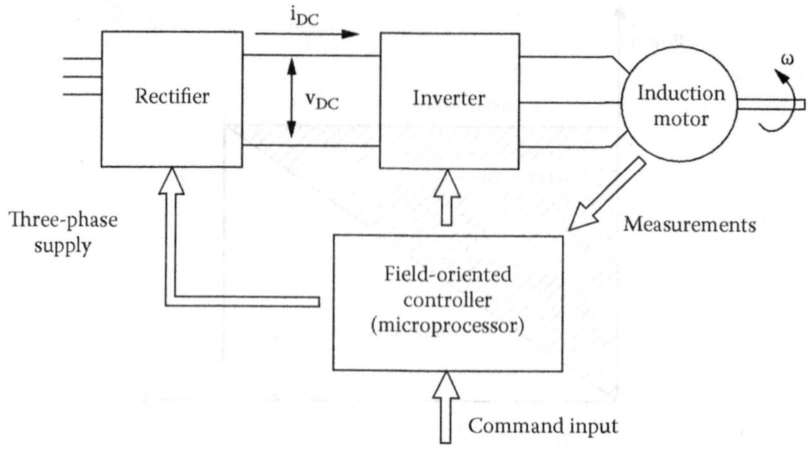

FIGURE 28.21
Schematic of a vector-control drive (also known as a field-oriented control). (Reprinted from Nadel, S. et al., *Energy-Efficient Motor Systems: A Handbook on Technologies, Programs, and Policy Opportunities*, 1st edn., American Council for an Energy-Efficient Economy, Washington, DC, 1992. With permission.)

DC drives have traditionally been used with winders in the paper and steel industry in order to achieve the constant tension requirements as the winding is performed. Sometimes the constant tension is obtained by imposing friction, which wastes energy. AC drives can also be used to replace DC drives, saving energy, and achieving better overall performance.

The use of field-oriented AC drives in machine tools and robotics allows the stringent requirements regarding dynamic performance to be met. Positioning drives can produce a peak torque up to 10 times the rated torque and make it possible to adjust the speed to very low values.

In machine tools, AC drives can provide accurate higher spindle speeds than DC drives without the need for gearboxes and their associated losses. The ASDs can also adjust the voltage level when the spindle drive is lightly loaded, providing further savings. In robotics, the higher power density and superior dynamics of AC drives are important advantages.

28.4 Energy and Power Savings Potential; Cost-Effectiveness

The energy and peak power savings potential for any motor-related technology depends on a number of factors, including the characteristics of the motor, the motor drive, the driven equipment, and the load (Nadel et al., 2002). Since all of this information is seldom available, it is difficult to determine the effect of even a single application; it is far more difficult to determine the savings potential for diverse applications, across all sectors, for an entire nation.

A 2010 UNIDO study made an initial effort to categorize the energy saving potential—both technical and cost effective—of measures frequently undertaken to improve the energy efficiency of industrial compressed air, pumping, and fan systems in several countries/regions (the United States, Canada, the European Union, Thailand, Vietnam, and Brazil) through combining available data and expert opinion. For example, the study finds that 25% of 2008 motor system energy used in the United States could be saved through implementing cost-effective measures. However, the study highlighted the lack of current installed motor system efficiency data as a substantial barrier to the analysis. The full report is available online (McKane and Hasanbeigi, 2010).

A 2013 study by the U.S. DOE estimated that the technical annual energy savings potential from upgrading motors and implementing ASDs in the residential sector is approximately 0.54 quads of the 4.73 quads of motor driven energy consumption (includes electricity generation, transmission, and distribution losses) in the sector. Similarly, the report estimated that the technical annual energy savings from upgrading motors in the commercial sector is 0.46 quads of the total 4.87 quads (includes electricity generation, transmission, and distribution losses) of motor driven energy consumption in the sector and 0.53 quads could be saved through implementing ASDs. The report states that total savings potential in the commercial sector from upgrading motors and implementing ASDs would be less than the sum of the individual saving estimates.

This section estimates the energy and power savings that could be realized nationwide in the United States through the application of certain efficiency technologies in the residential, commercial, industrial, and utility sectors. Table 28.2 lists the major motor-driven end-uses, estimates of their energy use as a percentage of total national electricity use (based on Howe et al., 1999; Nadel et al., 2002), and the potential energy savings from ASDs expressed as a fraction of existing use.

TABLE 28.2
Electric Motor Usage and Potential Adjustable-Speed Drive Savings by Sector and End Use

Sector	End Use	Usage (% of Total U.S. Usage)	Potential ASD Savings (% of Usage for Each End Use)
Residential	Refrigeration	3.0	10
	Space heating	2.1	20
	Air-conditioning	3.4	15
	HVAC dist. fan	1.8	25
	Others	1.6	5
	Residential total	11.9	15
Commercial	Refrigeration	2.2	10
	HVAC compressors	3.3	15
	HVAC distribution	4.2	25
	Others	0.7	20
	Commercial total	10.4	18
Industrial	Refrigeration	1.5	10
	Pumps	5.7	20
	Fans	3.2	20
	Compressed air	3.6	15
	Material handling	2.8	15
	Material process	5.2	15
	Others	1.0	0
	Industrial total	23.1	16
Utilities	Pumps and fans	4.9	15
	Material handling and processing	2.2	15
	Utilities total	7.1	19

28.4.1 Potential Savings in the Residential Sector

About 38% of residential electricity is used in motor systems. The primary motor technology for realizing energy and power savings in the residential sector is the electronic ASD. Heat pumps, air conditioners, forced-air furnaces, and washing machines with ASDs have already been introduced into the market. Other appliances, such as refrigerators, freezers, heat-pump water heaters, and evaporative coolers, are also potential candidates for adjustable-speed controls. Most of the energy-saving potential of ASDs in the home is associated with the use of refrigerant compressors for cooling or heating (as in heat pumps, air conditioners, refrigerators, and freezers). In all of these applications, ASDs can reduce energy consumption by matching the speed of the compressor to the instantaneous thermal load. Given the assumed savings potential for each end use, the overall savings is about 15% of the sector's motor electricity.

Several improvements in home appliance technology that are likely to become common over the next few years will complement the use of ASDs:

- *High-efficiency compressor and fan motors*: The use of PM and reluctance motors can increase the efficiency of the motor by 5%–15%, when compared with conventional squirrel-cage single-phase induction motors; as noted earlier, even larger savings are possible with many small fan motors. PM AC motors are used in the latest ASD-equipped furnace and heat pump.
- *Rotary and scroll compressors*: The use of rotary (in small applications) or scroll compressors in place of reciprocating compressors can take full advantage of the speed variation potential of ASDs.
- *Larger heat exchangers with improved design*: Improved heat exchangers increase the efficiency by decreasing the temperature difference of the compressor thermal cycle.

28.4.2 Potential Savings in the Commercial Sector

An estimated 37% of commercial electricity use is for motor-driven end uses (Howe et al., 1999); the percentage for peak power is higher. The savings potential of ASDs in air-conditioning and ventilation applications was estimated by running DOE-2 computer simulations on two representative building types in five U.S. cities (Eto and de Almeida, 1988). Comparisons were made between the cooling and ventilation systems with and without ASDs. The results indicate ventilation savings of approximately 25% for energy and 6% for peak power, and cooling savings of about 15% and 0%, respectively. In Table 28.2, it is assumed that these energy results can be applied nationwide.

The estimated 10% energy savings for refrigeration are shown in Table 28.2; an estimated 5% savings in peak power should also be attainable. Other motor efficiency measures (discussed in Section 28.2) combined can capture approximately 10% more energy and demand savings, with an overall potential savings of about 18% of the sector's motor system electricity.

28.4.3 Potential Savings in the Industrial and Utility Sectors

About 70% of industrial and 89% of the utility sector electricity use is for motor systems. In Table 28.2, the fluid moving end-use savings are estimated at 20%, except for utilities, where the system requirements of municipal water works limit the savings. Compressed air and materials applications are assumed to have 15% potential.

As most industries are nonseasonal, with flat load profiles during operating hours, the peak savings are similar to energy savings. When other motor efficiency measures (see Section 28.2) are combined, approximately 10% more energy and demand savings can be obtained, resulting in a total of 18% combined savings for the motor systems in these sectors.

28.4.4 Cost-Effectiveness of ASDs

The price of ASD equipment, in terms of cost/horsepower, is a function of the horsepower range, the type of AC motor used, and the additional control and protection facilities offered by the electronic ASD. ASD installation costs vary tremendously depending on whether the application is new or retrofit, available space, weather protection considerations, labor rates, etc. Thus, there is a huge range of installed costs possible for any given ASD size, and the costs listed in Table 28.3 necessarily have large uncertainties. These numbers are presented in their original units of Euros and kW.

To determine whether an ASD is cost-effective for any given application, the following need to be taken into account:

- First cost (acquisition and installation)
- System operating load profile (number of hours per year at each level of load)
- Motor and drive performance curves as a function of load
- Cost of electricity
- Maintenance requirements
- Reliability
- Secondary benefits (less wear on equipment, less operating noise, regeneration capability, improved control, soft-start, and automatic protection features)
- Secondary problems (power factor, harmonics, and interference)

A careful analysis should weigh the value of the benefits offered by each option against the secondary problems, such as power quality, that may impose extra costs for filters and power factor correction capacitors.

Comparing the cost of conserved energy to the cost of electricity is a crude way to assess the cost-effectiveness of energy efficiency measures. More accurate calculations would account for the time at which conservation measures save energy relative to the utility system peak demand, and relate these *load shape characteristics* to baseload, intermediate, and peaking supply resources. See Koomey et al. (1990a,b) for more details.

TABLE 28.3

Typical ASD Unit and Installation Costs for Various Sizes

Power (kW)	Unit Price (€)	Installation Costs (€)
0.12–0.75	200	50%–300%
0.75–2.1	280	50%–250%
7.5–45	1,130	50%–200%
76–110	5,320	50%–175%
375–1000	41,790	50%–150%

Source: CEMEP, Personal communication with Bernhard Sattler, Secretary CEMEP Working Group Low Voltage AC Motors, 2013.

References

ABB Review, 2012. Motoring ahead, *The Corporate Journal*, 1, 11, 56–61.

ABB, 2013. High output synchronous reluctance motor and drive package—Optimized cost of ownership for pump and fan applications.

Andreas, J., 1992. *Energy-Efficient Electric Motors: Selection and Application*. New York: Marcel Dekker.

ANSI/AHRI, October 2013. ANSI/AHRI standard 1210 (I-P) with addendum 1: 2011 standard for performance rating of variable frequency drives. Arlington, VA: Air-Conditioning, Heating & Refrigeration Institute.

Baldwin, S. F. 1989. Energy-efficient electric motor drive systems. In: Johansson, T. B. et al., eds., *Electricity: Efficient End-Use and New Generation Technologies, and Their Planning Implications*. Lund, Sweden: Lund University Press.

Boglietti, A., Cavagnino, A., Pastorelli, M., and Vagati, A., 2005. Experimental comparison of induction and synchronous reluctance motors performance. In: *Industry Applications Conference, 2005. Fortieth IAS Annual Meeting*, Hong Kong, Vol. 1, pp. 474–479.

Bose, B., 1986. *Power Electronics and AC Drives*. Englewood Cliffs, NJ: Prentice Hall.

Brunner, C., Evans, C., and Werle, R., 2013. Standard format for IEC standards—Learning from motor standards for the other electric equipment. In: *Eighth Energy Efficiency in Motor Driven Systems (EEMODS'13)*, Rio de Janeiro, Brazil.

Brunner, C., Waide, P., and Jakob, M., 2011. Harmonized standards for motors and systems—Global progress report and outlook. In: *Seventh Energy Efficiency in Motor Driven Systems (EEMODS'11)*, Alexandria, VA.

Burt, C. M., Piao, X., Gaudi, F., Busch, B., and Taufik, N. F. N., 2008. Electric motor efficiency under variable frequencies and loads. *Journal of Irrigation and Drainage Engineering*, 134(2), 129–136.

CEMEP, 2013. Personal communication with Bernhard Sattler, Secretary CEMEP Working Group Low Voltage AC Motors.

Cummings, P., Dunki-Jacobs, J., and Kerr, R., 1985. Protection of induction motors against unbalanced voltage operation. *IEEE Transactions on Industry Applications*, IA-21, 4.

De Almeida, A., Ferreira, F., and Baoming, G., 2013. Beyond induction motors—Technology trends to move up efficiency. In: *IEEE Industrial & Commercial Power Systems Conference*, Stone Mountain, GA, April 30–May 3, 2013 (*IEEE Transactions on Industry Applications*, 2014, accepted for publication).

De Almeida, A., Ferreira, F., and Both, D., January–February 2005. Technical and economical considerations in the application of variable speed drives with electric motor systems. *IEEE Industrial Applications Transactions*, 41(1), 188–199.

De Almeida, A., Ferreira, F., and Duarte, A., 2014. Technical and economical considerations on super high-efficiency three-phase motors. *IEEE Transactions on Industry Applications*, 50(2), March/April, 1274–1285

De Almeida, A., Ferreira, F. J. T. E., and Fong, J., January/February 2011. Standards for efficiency of electric motors. *IEEE Industry Applications Magazine*, 17(1), 12–19.

Dreisilker, H., August 1987. Modern rewind methods assure better rebuilt motors. *Electrical Construction and Maintenance*, 84, 31, 36.

EPRI, 1992. Electric motors: Markets, trends and application. EPRI Report TR-100423. Palo Alto, CA: Electric Power Research Institute.

Eto, J. and de Almeida, A. 1988. Saving electricity in commercial buildings with adjustable speed drives. *IEEE Transactions on Industrial Applications*, 24(3), 439–443.

Fraunhofer Institute, 2000. Compressed air systems market transformation study XVII/4.1031/Z/98-266. Prepared for the European Commission, Brussels, Belgium.

Greenberg, S., Harris, J. H., Akbari, H., and de Almeida, A. 1988. Technology assessment: Adjustable speed motors and motor drives. Lawrence Berkeley Laboratory Report LBL-25080. Berkeley, CA: Lawrence Berkeley Laboratory, University of California.

Howe, B., Lovins, A., Houghton, D., Shepard, M., and Stickney, B. 1999. *Drivepower Technology Atlas*. Boulder, CO: E-Source.

IEC60034-2-1, 2007. 1st edn.: Rotating electrical machines—Part 2-1: Standard method for determining losses and efficiency from tests (excluding machines for traction vehicles).

IEC60034-2-1, 2013. 2nd edn., 2/1687/CDV: Rotating electrical machines—Part 2-1: Standard method for determining losses and efficiency from tests (excluding machines for traction vehicles).

IEC60034-30, 2008. 1st edn.: Rotating electrical machines—Part 30: Efficiency classes of single-speed, three-phase, cage-induction motors (IE-code).

IEC60034-30, November 2011. 2nd edn., Draft, WG31/2CD: Rotating electrical machines—Part 30: Efficiency classes of single-speed, three-phase, cage-induction motors (IE-code).

IEC/TS60034-31, 2010. 1st edn.: Rotating electrical machines—Part 31: Selection of energy-efficient motors including variable speed applications—Application guide.

IEEE, 1992. IEEE guide for harmonic control and reactive compensation of static power converters. IEEE Standard 519. New York: Institute of Electrical and Electronics Engineers.

Koomey, J., Rosenfeld, A., and Gadgil, A., 1990a. Conservation screening curves to compare efficiency investments to power plants: Applications to commercial sector conservation programs. In: *Proceedings of the 1990 ACEEE Summer Study on Energy Efficiency in Buildings*. Asilomar, CA: American Council for an Energy Efficient Economy.

Koomey, J., Rosenfeld, A. H., and Gadgil, A. K., 1990b. Conservation screening curves to compare efficiency investments to power plants. *Energy Policy*, 18(8), 774.

Kurkowski, A. and Wray, C., June 2013. Standardizing data for VFD efficiency. *ASHRAE Journal*, 55(6), 16.

Leonhard, W., 1984. *Control of Electrical Drives*. New York: Springer-Verlag.

Magnusson, D., 1984. Energy economics for equipment replacement. *IEEE Transactions on Industry Applications*, IA-20, 2.

McKane, A. and Hasanbeigi, A., 2010. *Motor Systems Efficiency Supply Curves*. Vienna, Austria: UNIDO.

Nadel, S., Elliot, R.N., Shepard, M., Greenberg, S., Katz, G., and de Almeida, A., 1992. *Energy-Efficient Motor Systems: A Handbook on Technologies, Programs, and Policy Opportunities*, 1st edn. Washington, DC: American Council for an Energy-Efficient Economy.

Nadel, S., Elliot, R.N., Shepard, M., Greenberg, S., Katz, G., and de Almeida, A., 2002. *Energy-Efficient Motor Systems: A Handbook on Technologies, Programs, and Policy Opportunities*, 2nd edn. Washington, DC: American Council for an Energy-Efficient Economy.

NEMA, 1999. NEMA Standards Publication MG10-1994 (R1999), Energy management guide for selection and use of polyphase motors. Washington, DC: National Electrical Manufacturers' Association.

NEMA, 2000. NEMA Standards Publication MG1-1998, Revision 1. Motors and generators. Washington, DC: National Electrical Manufacturers' Association.

Pastor, C. E., April 30–May 2, 1986. *Motor Application Considerations: Single Speed, Multi-Speed, Variable Speed*. Oakland, CA: Pacific Gas and Electric Company Conference.

Smeaton, R., 1988. *Motor Application and Maintenance Handbook*. New York: McGraw-Hill.

USDOE, 2012. Motor systems tip sheets. http://www.energy.gov/eere/amo/motor-systems (accessed on March 6, 2015).

USDOE, 2013. *Energy Savings Potential and Opportunities for High Efficiency Electric Motors in Residential and Commercial Equipment*. Washington, DC.: United States Department of Energy.

USDOE, 2014a. *Improving Motor and Drive System Performance—A Sourcebook for Industry*. Washington, D.C.: United States Department of Energy.

USDOE, 2014b. *Continuous Energy Improvement in Motor Driven Systems—A Guidebook for Industry*. Washington, D.C.: United States Department of Energy.

USDOE, 2014c. *Premium Efficiency Motor Selection and Application Guide—A Handbook for Industry*. Washington, D.C.: United States Department of Energy.

XENERGY Inc., 2001. National market assessment: Compressed air system efficiency services. Washington, DC: U.S. Department of Energy.

Bibliography

Arthur D. Little, Inc., 1999. Opportunities for energy savings in the residential and commercial sectors with high-efficiency electric motors, final report. Prepared for the U.S. Department of Energy, Washington, D.C.

BPA/EPRI, 1993. Electric motor systems sourcebook. Olympia, WA: Bonneville Power Administration, Information Clearinghouse.

CEE, 2013. *Motor Efficiency, Selection, and Management: A Guidebook for Industrial Efficiency Programs.* Boston, MA: Consortium for Energy Efficiency.

De Almeida, A., Ferreira, F., Fong, J., and Conrad, B., 2008. Electric motor ecodesign and global market transformation. In: *Proceedings of the IEEE Industrial & Commercial Power Systems Conference*, Clearwater Beach, FL, May 4–8, 2008.

De Almeida, A., Ferreira, F., Fonseca, P., Falkner, H., Reichert, J., West, M., Nielsen, S., and Both, D., 2001. *VSDs for Electric Motor Systems.* Institute of Systems and Robotics, University of Coimbra, Coimbra, Portugal (European Commission, Directorate-General for Transport and Energy, SAVE II Programme).

EN50347, August 2001. General purpose three-phase induction motors having standard dimensions and outputs—Frame numbers 56 to 315 and flange numbers 65 to 740. EN Standards.

EPRI, 1992. *Adjustable Speed Drive Directory*, 3rd edn. Palo Alto, CA: Electric Power Research Institute.

EPRI, 1993. *Applications of AC Adjustable Speed Drives.* Palo Alto, CA: Electric Power Research Institute.

IEC60072-1, 1991. 6th edn.: Dimensions and output series for rotating electrical machines—Part 1: Frame numbers 56 to 400 and flange numbers 55 to 1080.

Jarc, D. and Schieman, R., 1985. Power line considerations for variable frequency drives. *IEEE Transactions on Industry Applications*, IA-21, 5.

Lovins, A. and Howe, B., 1992. *Switched Reluctance Motor Systems Poised for Rapid Growth.* Boulder, CO: E-Source.

Moore, T., March 1988. The advanced heat pump: All the comforts of home… and then some. *EPRI Journal.* Palo Alto, CA: Electric Power Research Institute, Vol. 13, pp. 4–13.

Waide, P. and Brunner, C., 2011. Energy efficiency policy opportunities for electric motor-driven systems. Paris, France: International Energy Agency.

XENERGY Inc., 1998. United States industrial electric motor market opportunities assessment. Washington, DC: U.S. Department of Energy.

29

Industrial Energy Efficiency and Energy Management

Craig B. Smith, Barney L. Capehart, and Wesley M. Rohrer, Jr.*

CONTENTS

* Deceased.

29.1 Introduction

The industrial sector in the United States is highly diverse—consisting of manufacturing, mining, agriculture, and construction activities—and consumes one-third of the nation's primary energy use, at an annual cost of around $200 billion.[1] The industrial sector encompasses more than three million establishments engaged in manufacturing, agriculture, forestry, construction, and mining. These industries require energy to light, heat, cool, and ventilate facilities (end uses characterized as energy needed for comfort). They also use energy to harvest crops, process livestock, drill and extract minerals, power various manufacturing processes, move equipment and material, raise steam, and generate electricity. Some industries require additional energy fuels for use as raw materials—or feedstocks—in their production processes. Many industries use by-product fuels to satisfy part or most of their energy requirements. In the more energy-intensive manufacturing and nonmanufacturing industries, energy used by processes dwarfs the energy demand for comfort.

The U.S. sector energy use for 2014 is shown in Figure 29.1, industrial energy use is shown in Figure 29.2, and industrial electricity use is shown in Figure 29.3.

Manufacturing companies, which use mechanical or chemical processes to transform materials or substances to new products, account for about 69% of the total industrial sector use. The "big three" in energy use are petroleum, chemicals, and paper; these manufacturers together consume almost one-half of all industrial energy. The "big six," which adds the primary metals group, the food and kindred products group, as well as the stone, clay, and glass group, together accounts for 88% of manufacturing energy use, and over 60% of all industrial sector energy consumption.[1]

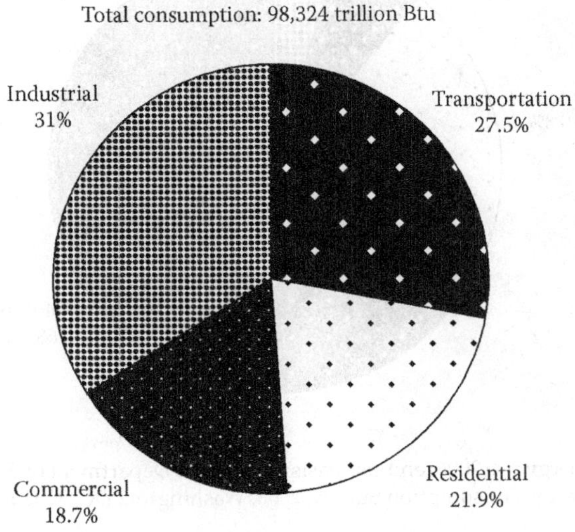

FIGURE 29.1
The U.S. energy use by sector, 2014. (From U.S. DOE, Energy Information Agency, Monthly Energy Review, Washington, DC, March 27, 2014; U.S. Department of Energy, Office of Industrial Technologies, Industrial Technology Program, Washington, DC, 2005.)

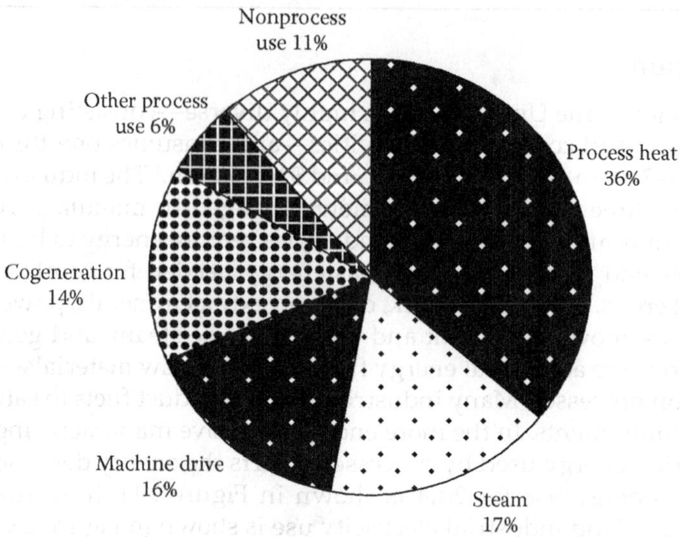

FIGURE 29.2
Manufacturing energy use 2006 (end-use basis). (From U.S. Department of Energy, Energy Information Agency, Manufacturing Energy Consumption Survey, Washington, DC, September 2012.)

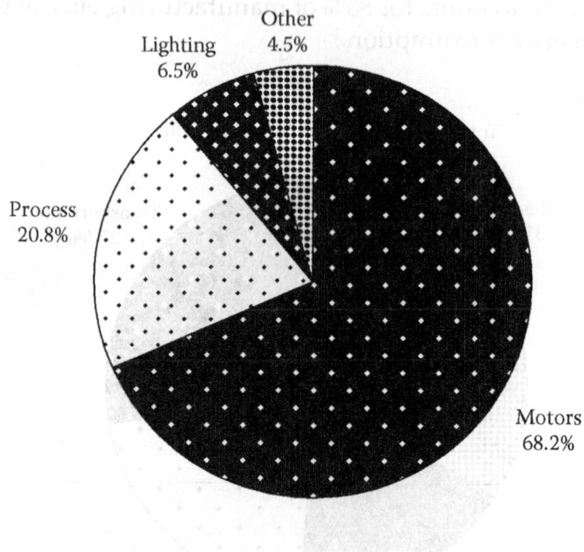

FIGURE 29.3
Manufacturing electrical energy use 2006 (end-use basis) (From U.S. Department of Energy, Energy Information Agency, Manufacturing Energy Consumption Survey, 2006, Washington, DC, 2002.)

According to the U.S. Energy Information Administration, energy efficiency in the manufacturing sector improved by 25% over the period 1980–1985.[2] During that time, manufacturing energy use declined 19%, and output increased 8%. These changes resulted in an overall improvement in energy efficiency of 25%. However, the "big five" did not match this overall improvement; although their energy use declined 2%, their output decreased by 5%, resulting in only a 17% improvement in energy efficiency during 1980–1985.

This 5-year record of improvement in energy efficiency of the manufacturing sector came to an end, with total energy use in the sector growing by 10% from 1986 to 1988, and overall manufacturing energy intensity stagnated during 1985–1994 due to falling and low energy prices, economic recession during part of this period, and a recovery in some of the more energy-intensive groups such as steel and aluminum production. However, industrial energy intensity again declined significantly during the late 1990s due to high capital investment, rapid industrial modernization, and explosive growth of "high-tech" industries that are not energy-intensive.

Since 1980, the overall value of industrial output increased through 2003, while the total energy consumed by the industrial sector has fallen overall throughout 2003.[3] This relationship is shown in Figure 29.4, where the consumption index for both primary and site energy is greater than the output index before 1980, and less afterward, with the gap consistently widening in the late 1980s. New energy-efficient technology and the changing production mix from the manufacture of energy-intensive products to less intensive products are responsible for this change.

Continuing this overall record of energy efficiency improvements in industry will require emphasis on energy management activities, as well as making capital investments in new plant processes and facilities improvements. Reducing the energy costs per unit of manufactured product is one way that the United States can become more competitive in the global industrial market. The U.S. Department of Energy has formally recognized these multiple benefits to the country by including the following statement in its 2004 Industrial Technology Program Report: "By developing and adopting more energy efficiency technologies, U.S. industry can boost its productivity and competitiveness while strengthening national energy security, improving the environment, and reducing emissions linked to global climate change."[4] Additionally, it is interesting to note that

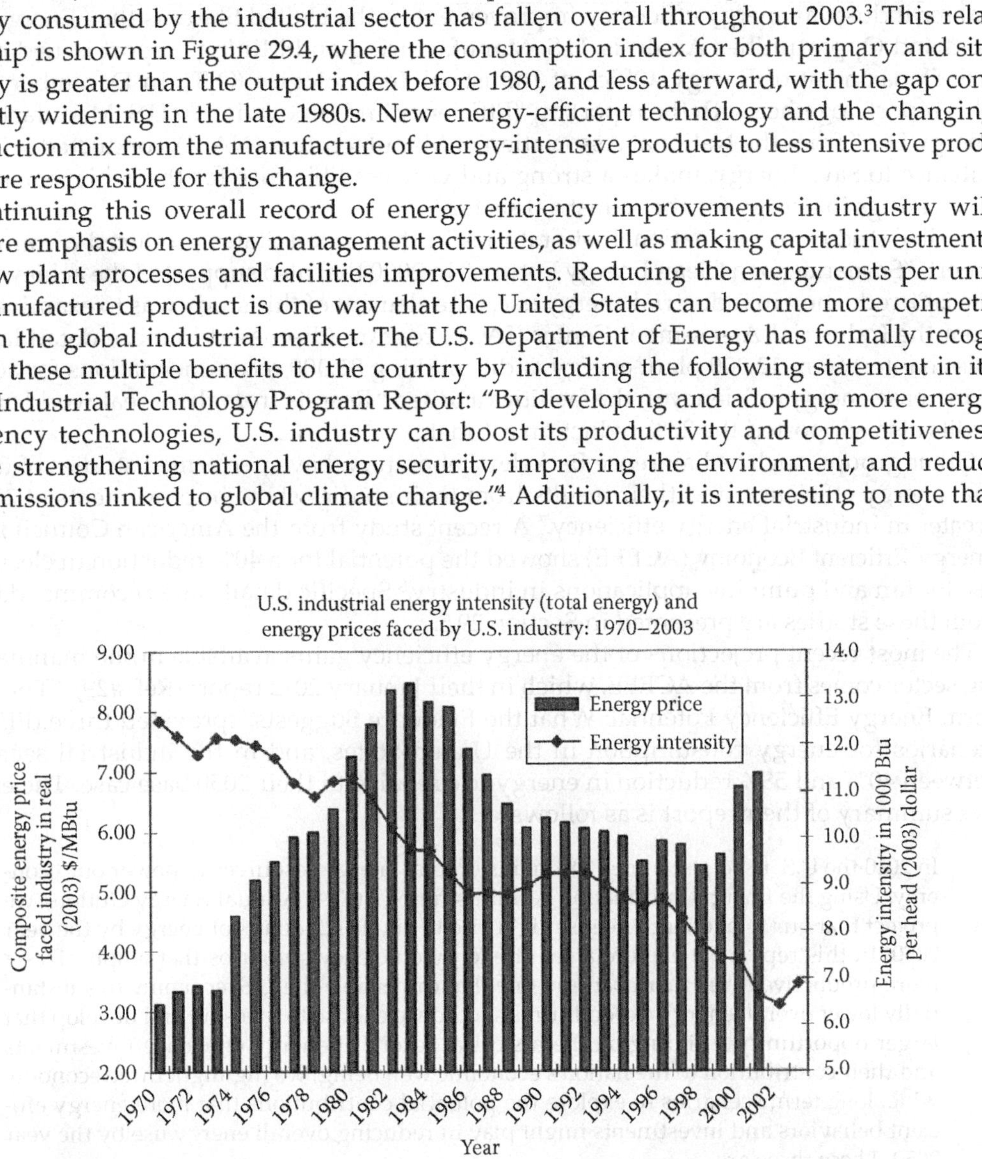

FIGURE 29.4
Industrial energy intensity 1980–2003. (From NAM, *Efficiency and Innovation in U.S. Manufacturing Energy Use*, National Association of Manufacturers, Washington, DC, 2005. With permission.)

Japan—one of the United States' major industrial competitors—has a law that says every industrial plant must have a full-time energy manager.[5]

Manufacturing energy intensity continued its gradual drop, as documented by the National Renewable Energy Laboratory report in 2009 showing the industrial use of Btu per dollar of gross domestic product fell by 20% from 1998 to 2007 (Ref. #7).

Several studies of industrial energy efficiency have been performed in recent years, and the results from the studies show that there is a readily achievable, cost-effective, 20% reduction in industrial consumption using good energy management practices and energy-efficient equipment. One highly credible study that has been done of the potential for industrial energy efficiency improvement in the United States is the "Energy Use, Loss and Opportunities Analysis: U.S. Manufacturing and Mining" report issued by the U.S. Department of Energy, Office of Industrial Technologies (OIT), in December 2004.[1] This report, together with the report on "Efficiency and Innovation in U.S. Manufacturing Energy Use" from the National Association of Manufacturers (NAM), conducted with the Alliance to Save Energy, makes a strong and very credible case for the achievement of a 20% savings in energy for the industrial sector.[3]

Earlier studies by the national laboratories in the United States produced a composite report, "Scenarios for a Clean Energy Future," in 2000 that also supported the achievement of a 20% reduction in industrial energy use. One element of this study was to examine data from the Industrial Assessment Center (IAC) Program (discussed in more detail later in this chapter) from 12,000 plant energy audits making 82,000 recommendations for actions to increase energy efficiency in the facilities audited.[6] Results from these real-world energy audits also supported the 20% reduction estimate.

Other groups, such as Lawrence Berkeley Laboratory, have performed studies of industrial energy efficiency, and their results from their studies show improvements of 20% or greater in industrial energy efficiency.[7] A recent study from the American Council for an Energy-Efficient Economy (ACEEE) showed the potential for a 40% reduction in electricity use for fan and pumping applications in industry.[8] Specific details and recommendations from these studies are presented in Section 29.5.

The most recent projections of the energy efficiency gains available in the manufacturing sector comes from the ACEEE, which in their January 2012 report (Ref. #29), "The Long Term Energy Efficiency Potential: What the Evidence Suggests," provided three different scenarios for energy consumption in the United States, and in the industrial sector, of between 40% and 58% reduction in energy use relative to their 2050 base case. The executive summary of their report is as follows:

> In 2010 the U.S. used just under 100 quads of total energy resources to power our economy. Using the Energy Information Administration's (EIA) Annual Energy Outlook we project that our total energy needs might rise to about 122 quads of energy by the year 2050. In this report we explore a set of energy efficiency scenarios that emphasizes a more productive investment pattern, one that can enable the U.S. economy to substantially lower overall energy expenditures should we choose to invest in and develop that larger opportunity. Building on the historical record of energy efficiency investments and their contribution to the nation's economic well being, we highlight three economy wide, long term scenarios to explore the potential contributions that more energy efficient behaviors and investments might play in reducing overall energy use by the year 2050. These three are
>
> 1. *Reference case*: A continuation of trends projected by EIA for the 2030–2035 period
> 2. *Advanced scenario*: It includes penetration of known advanced technologies

3. *Phoenix scenario*: In addition to advanced technologies also includes greater infrastructure improvements and some displacement of existing stock to make way for newer and more productive energy efficiency technologies, as well as configurations of the built environment that reduce energy requirements for mobility

For the industrial sector, energy intensity is projected to improve in the Reference Case by about 1%/year through 2050, but leading companies have been achieving continued improvements at more than double this rate. In our Advanced Scenario we project a 2%/year improvement rate for overall industrial energy intensity, which increases to 2.75% in our Phoenix Scenario. Future energy efficiency opportunities will come less from seeking out individual sources of waste and more from optimization of complex systems enabled by advances in information, communication and computational infrastructure. Most of the energy use in industry is in processes, not individual equipment, so improved processes represent the largest opportunity for energy intensity improvements. Current focus has been on process optimization, but we anticipate that even greater opportunities exist in the optimization of entire supply chains that may span many companies and supply chain integration that allows for efficient use of feedstocks and elimination of wasted production.

The really good news in the ACEEE study is that there is a tremendous potential for saving energy in the industrial sector, which is mainly the manufacturing sector. For their Advanced Scenario, which projects a 2%/year reduction in industrial energy intensity, this means that energy use in the industrial sector would fall to half its value in 2010 by 2045, a 35-year period.

29.2 Industrial Energy Management and Efficiency Improvement

29.2.1 Setting Up an Energy Management Program

The effectiveness of energy utilization varies with specific industrial operations because of the diversity of the products and the processes required to manufacture them. The organization of personnel and operations involved also varies. Consequently, an effective energy management program should be tailored for each company and its plant operations. There are some generalized guidelines, however, for initiating and implementing an energy management program. Many large companies have already instituted energy management programs and have realized substantial savings in fuel and electric costs. Smaller industries and plants, however, often lack the technical personnel and equipment to institute and carry out effective programs. In these situations, reliance on external consultants may be appropriate to initiate the program. Internal participation, however, is essential for success. A well planned, organized, and executed energy management program requires a strong commitment by top management.

Assistance also can be obtained from local utilities. Utility participation would include help in getting the customer started on an energy management program, technical guidance, or making information available. Most electric and gas utilities today have active programs that include training of customer personnel or provision of technical assistance. Table 29.1 summarizes the elements of an effective energy management program. These are discussed further in more detail.

TABLE 29.1

Elements of an Energy Management Program

Phase 1: Management Commitment	
1.1	Commitment by management to an energy management program
1.2	Assignment of an energy management coordinator
1.3	Creation of an energy management committee of major plant and department representatives
Phase 2: Audit and Analysis	
2.1	Review of historical patterns of fuel and energy use
2.2	Facility walk-through survey
2.3	Preliminary analyses, review of drawings, data sheets, equipment specifications
2.4	Development of energy audit plans
2.5	Conduct facility energy audit, covering a. Processes b. Facilities and equipment
2.6	Calculation of annual energy use based on audit results
2.7	Comparison with historical records
2.8	Analysis and simulation step (engineering calculations, heat and mass balances, theoretical efficiency calculations, computer analysis, and simulation) to evaluate energy management options
2.9	Economic analysis of selected energy management options (life cycle costs, rate of return, benefit–cost ratio)
Phase 3: Implementation and Submeters	
3.1	Establish energy effectiveness goals for the organization and individual plants
3.2	Determine capital investment requirements and priorities
3.3	Establish measurement and reporting procedures, install monitoring and recording instruments as required
3.4	Institute routine reporting procedures ("energy tracking" charts) for managers and publicize results
3.5	Promote continuing awareness and involvement of personnel
3.6	Provide for periodic review and evaluation of overall energy management program

29.2.1.1 Phase I: Management Commitment

A commitment by the directors of a company to initiate and support a program is essential. An energy coordinator is designated, and an energy management committee is formed. The committee should include personnel representing major company activities utilizing energy. A plan is formulated to set up the programs with a commitment of funds and personnel. Realistic overall goals and guidelines in energy savings should be established based on overall information in the company records, projected activities, and future fuel costs and supply. A formal organization as described earlier is not an absolute requirement for the program; smaller companies will simply give the energy management coordination task to a staff member.

29.2.1.1.1 Organizing for Energy Conservation Programs

The most important organizational step that will affect the success of an energy management (EM) program is the appointment of one person who has full responsibility for its operation. Preferably that person should report directly to the top management position and be given substantial authority in directing technical and financial resources within the bounds set by the level of management commitment. It is difficult to stress enough the importance of making the position of plant energy manager a full-time job. Any diversion of interest and attention to other aspects of the business is bound to badly affect the EM program. One reason is that the greatest opportunity for energy cost control and energy efficiency gains is in improved operational and maintenance practices. Implementing and sustaining good operational and maintenance procedures is an exceedingly demanding job and requires a constant attention and a dedication to detail that is rarely found in corporate business life. The energy manager should be energetic, enthusiastic, dedicated, and political.

The second step is the appointment of the plant EM committee. This should consist of one group of persons who are able to and have some motivation for cutting fuel and electric costs and a second group who have the technical knowledge or access to data needed for the program department managers or their assistants. Thus, the EM committees should include labor representatives, the maintenance department head, a manager of finance or data storage, some engineers, and a public relations person. The energy manager should keep up to date on the energy situation daily, convene the committee weekly, and present a definitive report to top management at least monthly and at other times when required by circumstance. It is suggested also that several subcommittees be broken out of the main committee to consider such important aspects as capital investments, employee education, operator-training programs, external public relations, and so on. The committee will define strategy, provide criticism, publish newsletters and press releases, carry out employee programs, argue for the acceptance of feasible measures before management, represent the program in the larger community, and be as supportive as possible to the energy coordinator. This group has the most to risk and the most to gain. They must defend their own individual interests against the group, but at the same time must cooperate in making the program successful and thus be eligible for rewards from top management for their good work and corporate success.

As the EM program progresses to the energy audit and beyond, it will be necessary to keep all employees informed as to its purposes, its goals, and how its operation will impact plant operations and employee routine, comfort, and job performance. The education should proceed through written and oral channels as best benefits the organizational structure. Newsletters, posters, and employee meetings have been used successfully.

In addition to general education about energy conservation, it may prove worthwhile to offer specialized courses for boiler, mechanical, and electrical equipment operators and other workers whose jobs can affect energy utilization in the plant. The syllabuses should be based on thermodynamic principles applied to the systems involved and given on an academic level consistent with the workers' backgrounds. Long-range attempts to upgrade job qualifications through such training can have very beneficial effects on performance. The courses can be given by community colleges, by private enterprises professional societies, or by in-house technical staff, if available.

The material presented here on organization is based on the presumption that a considerable management organization already exists and that sufficient technical and financial resources exist for support of the energy management program as outlined. Obviously, very small businesses cannot operate on this scale; however, we have found many small companies that have carried out effective energy management efforts.

29.2.1.1.2 *Setting Energy Conservation Goals*

It is entirely appropriate and perhaps even necessary to select an energy reduction goal for the first year of the program very early in the program. The purpose is to gain the advantage of the competitive spirit of those employees that can be aroused by a target goal. Unfortunately, the true potential for conservation and the investment costs required to achieve it are not known until the plant energy audit is completed and a detailed study made of the data. Furthermore, a wide variety of energy-use patterns exists even with a single industry.

However, looking at the experience of other industries that have set goals and met them can provide some useful guidance.

An excellent example of a long-term successful energy management program in a large industrial corporation is that of the 3M Company, headquartered in St. Paul, Minnesota.[9] 3M is a large, diversified manufacturing company with more than 50 major product lines; it makes some 50,000 products at over 50 different factory locations around the country. The corporate energy management objective is to use energy as efficiently as possible in all operations; the management believes that all companies have an obligation to conserve energy and all other natural resources.

Energy productivity at 3M improved 63% from 1973 to 2004. They saved over $70 million in 2004 because of their energy management programs and saved a total of over $1.5 billion in energy expenses from 1973 to 2004. From 1998 to 2004, they reduced their overall energy use by 27% in their worldwide operations.[10] Their program is staffed by three to six people who educate and motivate all levels of personnel on the benefits of energy management. The categories of programs implemented by 3M include conservation, maintenance procedures, utility operation optimization, efficient new designs, retrofits through energy surveys, and process changes.

Energy efficiency goals at 3M are set, and then the results are measured against a set standard to determine the success of the programs. The technologies that have resulted in the most dramatic improvement in energy efficiency include heat recovery systems, high-efficiency motors, variable-speed drives (VSDs), computerized facility management systems, steam-trap maintenance, combustion improvements, variable-air-volume systems, thermal insulation, cogeneration, waste-steam utilization, and process improvements. Integrated manufacturing techniques, better equipment utilization, and shifting to non-hazardous solvents have also resulted in major process improvements.

The energy management program at 3M has worked very well, but the company's management is not yet satisfied. They have historically set a goal of further improving energy efficiency at a rate of 2%–3%/year for the next 5 years. This goal has produced a 10%–15% reduction in energy use per pound of product or per square foot of building space. They expect to substantially reduce their emissions of waste gases and liquids, to increase the energy recovered from wastes, and to constantly increase the profitability of their operations. 3M continues to stress the extreme importance that efficient use of energy can have on their industrial productivity.

In 2010, 3M met their 20% goal of reducing energy use from 2005. Their new goal is a 15% reduction in energy use per pound of product by 2015.

29.2.1.2 Phase 2: Audit and Analysis

29.2.1.2.1 Energy Audit of Equipment and Facilities

Historical data for the facility should be collected, reviewed, and analyzed. The review should identify gross energy uses by fuel types, cyclic trends, fiscal year effects, dependence on sales or workload, and minimum energy-use ratios. Historical data are graphed in a form similar to the one shown in Chapter 22. Historical data assist in planning a detailed energy audit and alert the auditors as to the type of fuel and general equipment to expect. A brief facility walk-through is recommended to establish the plant layout, major energy uses, and primary processes or functions of the facility.

The energy audit is best performed by an experienced or at least trained team, since visual observation is the principal means of information gathering and operational assessment. A team would have from three to five members, each with a specific assignment for the audit. For example, one auditor would check the lighting, another the HAVC system, another the equipment and processes, another the building structure (floor space, volume, insulation, age, etc.), and another the occupancy use schedule, administration procedures, and employees' general awareness of energy management.

The objectives of the audit are to determine how, where, when, and how much energy is used in the facility. In addition, the audit helps to identify opportunities to improve the energy-use efficiency and its operations. Some of the problems encountered during energy audits are determining the rated power of equipment, determining the effective hours of use per year, and determining the effect of seasonal, climatic, or other variable conditions on energy use. Equipment ratings are often obscured by dust or grease (unreadable name-plates). Complex machinery many not have a single nameplate listing the total capacity, but several giving ratings for component equipment. The effect of load is also important because energy use in a machine operating at less than full load may be reduced along with a possible loss in operating efficiency.

The quantitative assessment of fuel and energy use is best determined by actual measurements under typical operational conditions using portable or installed meters and sensing devices. Such devices include light meters, ammeters, thermometers, air flowmeters, and recorders. In some situations, sophisticated techniques such as infrared scanning or thermography are useful. The degree of measurement and recording sophistication naturally depends on available funds and the potential savings anticipated. For most situations, however, nameplate and catalog information are sufficient to estimate power demand. Useful information can be obtained from operating personnel and their supervisors particularly as it relates to usage patterns throughout the day.

The first two columns of the form are self-explanatory. The third column is used for the rated capacity of the device (e.g., 5 kW). The sixth column is used if the device is operated at partial load. Usage hours (column 7) are based on all work shifts and are corrected to account for the actual operating time of the equipment. The last three columns are used to convert energy units to a common basis (e.g., MJ or Btu).

Data recorded in the field are reduced easily by the use of specialized software or spread sheets that provide uniform results and summaries in a form suitable for review or for further analysis. Computer analysis also provides easy modification of the results to reflect specific management reporting requirements or to present desired comparisons for different energy use, types of equipment, and so on.

29.2.1.2.2 In-Plant Metering

Submetering reduces the work and time required for an energy audit; indeed, it does much more than that. Because meters are tools for assessing production control and for measuring equipment efficiency, they can contribute directly to energy conservation and cost containment. Furthermore, submetering offers the most effective way of evaluating the worth of an energy efficiency measure. Too many managers accept a vendor's estimate of fuel savings after buying a recuperator. They may scan the fuel bills for a month or two after the purchase to get an indication of savings—usually in vain—and then relax and accept the promised benefit without ever having any real indication that it exists. It may well be that, in fact, it does not yet exist. The equipment may not be adjusted correctly or it may be operated incorrectly, and there is no way of knowing without directly metering the fuel input. It is estimated that at least 2.5% waste is recoverable by in-plant metering.

Oil meters are just as effective as gas meters used in the same way and are even less expensive on an energy-flow basis. Electric meters are particularly helpful in monitoring the continued use of machines or lighting during shutdown periods and for evaluating the efficacy of lubricants and the machinability of feedstock. The use of in-plant metering can have its dark side too. The depressing part is the requirement for making periodic readings.

It does not stop even there. Someone must analyze the readings so that something can be done about them. If full use is to be made of the information contained in meter readings, it must be incorporated into the energy information portion of the management information system. At the very least, each subreading must be examined chronologically to detect malfunctions or losses of efficiency. Better still, a derived quantity such as average energy per unit of production should be examined.

29.2.1.2.3 A Special Case: Energy Audit of a Process

In some manufacturing and process industries, it is of interest to determine the energy content of a product. This can be done by a variation of the energy audit techniques described earlier. Since this approach resembles classical financial accounting, it is sometimes called energy accounting. In this procedure, the energy content of the raw materials is determined in a consistent set of energy units. Then, the energy required for conversion to a product is accounted for in the same units. The same is done for energy in the waste streams and the by-products. Finally, the net energy content per unit produced is used as a basis for establishing efficiency goals.

In this approach, all materials used in the product or used to produce it are determined. Input raw materials used in any specific period are normally available from plant records. Approximations of specific energy content for some materials can be found in the literature or can be obtained from the U.S. Department of Commerce or other sources. The energy content of a material includes that due to extraction and refinement as well as an inherent heating value it would have as a fuel prior to processing. Consequently, nonfuel-type ores in the ground are assigned zero energy, and petroleum products are assigned their alternate value as a fuel prior to processing in addition to the refinement energy. The energy of an input metal stock would include the energy due to extraction, ore refinement to metal, and any milling operations.

Conversion energy is an important aspect of the energy audit, since it is under direct control of plant management. All utilities and fuels coming into the plant are accounted for. They are converted to consistent energy units (joules or Btu) using the actual data available on the fuels or using approximate conversions.

Electrical energy is assigned the actual fuel energy required to produce the electricity. This accounts for power conversion efficiencies. A suggested approach is to assume (unless actual values are available from your utility) that 10.8 MJ (10,200 Btu) is used to produce 3.6 MJ (1 kW h), giving a fuel conversion efficiency of 3.6 + 10.8 = 0.33 or 33%.

The energy content of process steam includes the total fuel and electrical energy required to operate the boiler as well as line and other losses. Some complexities are introduced when a plant produces both power and steam, since it is necessary to allocate the fuel used to the steam and power produced. One suggested way to make this allocation is to assume that there is a large efficient boiler feeding steam to a totally condensing vacuum turbine. Then, one must determine the amount of extra boiler fuel that would be required to permit the extraction of steam at whatever pressure while maintaining the constant load on the generator. The extra fuel is considered the energy content of the steam being extracted.

Waste disposal energy is that energy required to dispose of or treat the waste products. This includes all the energy required to bring the waste to a satisfactory disposal state. In a case where waste is handled by a contractor or some other utility service, it would include the cost of transportation and treatment energy.

If the plant has by-products or coproducts, then energy credit is allocated to them. A number of criteria can be used. If the by-product must be treated to be utilized or recycled (such as scrap), then the credit would be based on the raw material less the energy expended to treat the by-product for recycle. If the by-product is to be sold, the relative value ratio of the by-product to the primary product can be used to allocate the energy.

29.2.1.2.4 *Analysis of Audit Results and Identification of Energy Management Opportunities*

Often, the energy audit will identify immediate energy management opportunities, such as unoccupied areas that have been inadvertently illuminated 24 h/day, equipment operating needlessly, and so on. Corrective housekeeping and maintenance action can be instituted to achieve short-term savings with little or no capital investment.

An analysis of the audit data is required for a more critical investigation of fuel waste and identification of the potential for conservation. This includes a detailed energy balance of each process, activity, or facility. Process modification and alternatives in equipment design should be formulated, based on technical feasibility and economic and environmental impact. Economic studies to determine payback, return on investment, and net savings are essential before making capital investments.

29.2.1.3 *Phase 3: Implementation and Submeters*

At this point, goals for saving energy can be established more firmly, and priorities set on the modification and alterations to equipment and the process: effective measurement and monitoring procedures are essential in evaluating progress in the energy management program. Routine reporting procedures between management and operations should be established to accumulate information on plant performance and to inform plant supervisors of the effectiveness of their operation. Time-tracking charts of energy use and costs can be helpful. Involving employees and recognizing their contributions facilitates the achievement of objectives. Finally, the program must be continually reviewed and analyzed with regard to established goals and procedures.

29.2.2 Energy Audit Report

Energy audits do not save money and energy for companies unless the recommendations are implemented. Audit reports should be designed to encourage implementation. The goal in writing an audit report should not be the report itself; rather, it should be to achieve implementation of the report recommendations and thus achieve increased energy efficiency and energy cost savings for the customer. In this section, the authors discuss their experience with writing industrial energy audit reports and suggest some ways to make the reports more successful in terms of achieving a high rate of the recommendations.[11]

- *Present information visually*: The authors present their client's energy-use data visually with graphs showing the annual energy and demand usage by month. These graphs give a picture of use patterns. Any discrepancies in use show up clearly.
- *Make calculation sections helpful*: The methodology and calculations used to develop specific energy management opportunity recommendations are useful in an audit report. Including the methodology and calculations gives technical personnel the ability to check the accuracy of one's assumptions and one's work. However, not every reader wants to wade through pages describing the methodology and showing the calculations. Therefore, the authors provide this information in a technical supplement to the audit report. Because this section is clearly labeled as the technical supplement, other readers are put on notice as to the purpose of this section.

- *Use commonly understood units*: When preparing the report, be sure to use units that the client will understand. Discussing energy savings in terms of Btu (British thermal units) may be meaningful only to the engineers and more technical readers. For management and operating personnel, kilowatt-hours (for electricity) or therms (for natural gas) are better units, because most energy bills use these units.
- *Explain your assumptions*: A major problem with many reports is the failure to explain the assumptions underlying the calculations. For example, when the authors use operating hours in a calculation, it is always carefully shown how the number was figured, for example, "Your facility operates from 7:30 a.m. to 8:00 p.m., 5 days a week, 51 weeks/year. Therefore, we will use 3188 h in our calculations."

When basic assumptions and calculations are shown, the reader can make adjustments if those facts change. In the example given, if the facility decided to operate 24 h/day, the reader would know where and how to make changes in operating hours, because that calculation had been clearly labeled.

The authors use one section of their report to list the standard assumptions and calculations. Thus, explanations for each of the recommendations do not have to be repeated. Some of the standard assumptions/calculations included in this section are operating hours, average cost of electricity, demand rate, off-peak cost of electricity, and the calculation of the fraction of air-conditioning load attributable to lighting.

- *Be accurate and consistent*: The integrity of a report is grounded in its accuracy. This does not just mean correctness of calculations. Clearly, inaccurate calculations will destroy a report's credibility, but other problems can also undermine the value of a report. Use the same terminology so that the reader is not confused. Make sure that the same values are used throughout the report. Do not use two different load factors for the same piece of equipment in different recommendations. This, for example, could happen if one calculated the loss of energy due to leaks from a compressor in one recommendation and the energy savings due to replacing the compressor motor with a high-efficiency motor in another recommendation.
- *Proofread the report carefully*: Typographical and spelling errors devalue an otherwise good product. With computer spell checkers, there is very little excuse for misspelled words. Nontechnical readers are likely to notice this type of error, and they will wonder if the technical calculations are similarly flawed.

29.2.2.1 Report Sections

The authors have found that the following report format meets their clients' needs and fits the authors' definition of a user-friendly report.

Executive summary: The audit report should start with an executive summary that basically lists the recommended energy conservation measures and shows the implementation cost and dollar savings amount. This section is intended for the readers who want to see only the bottom line. Although the executive summary can be as simple as a short table, the authors add some brief text to explain the recommendations and sometimes include other special information needed to implement the recommendations. They also copy the executive summary on colored paper so that it stands out from the rest of the report.

Energy management plan: Following the executive summary, some information is provided to the decision makers on how to set up an energy management program in their facility. The authors view this section as one that encourages implementation of the report, so every attempt is made to try to make it as helpful as possible.

Energy action plan: In this subsection, the authors describe the steps that a company should consider in order to start implementing the report's recommendations.

Energy financing options: The authors also include a short discussion of the ways that a company can pay for the recommendations. This section covers the traditional use of company capital, loans for small businesses, utility incentive programs, and the shared savings approach of the energy service companies.

Maintenance recommendations: The authors do not usually make formal maintenance recommendations in the technical supplement, because the savings are not often easy to quantify. However, in this section of the report, energy savings maintenance checklists are provided for lighting, heating/ventilation/air conditioning, and boilers.

The technical supplement: The technical supplement is the part of the report that contains the specific information about the facility and the audit recommendations. The authors' technical supplement has two main sections: one includes the report's assumptions and general calculations and the other describes the recommendations in detail, including the calculations and methodology. The authors sometimes include a third section that describes measures that were analyzed and determined not to be cost effective, or that have payback times beyond the client's planning horizon.

Standard calculations and assumptions: This section was briefly described earlier when the importance of explaining assumptions was discussed. Here, the reader is provided with the basis for understanding many of the authors' calculations and assumptions. Included is a short description of the facility: square footage (both air-conditioned and unconditioned areas); materials of construction; type and level of insulation; etc. If the authors are breaking the facility down into subareas, those areas are described, and each area is assigned a number that is then used throughout the recommendation section.

Standard values calculated in this section include operating hours, average cost of electricity, demand rate, off-peak cost of electricity, and the calculation of the fraction of air-conditioning load attributable to lighting. When a value is calculated in this section, the variable is labeled with an identifier that remains consistent throughout the rest of the report.

Audit recommendations: This section contains a discussion of each of the energy management opportunities the authors have determined to be cost effective. Each energy management recommendation (or EMR) that was capsulized in the executive summary is described in depth in this section.

Again, the authors try to make the EMRs user-friendly. To do this, the narrative discussion is placed at the beginning of a recommendation, and the technical calculations are left for the very end. In this manner, the authors allow the readers to decide for themselves whether they want to wade through the calculations.

Each EMR starts with a table that summarizes the energy, demand and cost savings, implementation cost, and simple payback period. Then follows a short narrative section that provides some brief background information about the recommended measure and explains how it should be implemented at the facility in question. If the authors are recommending installation of more than one item (lights, motors, air-conditioning units, etc.), a table is often used to break down the savings by unit or by area.

The final section of each EMR is the calculation section. Here the authors explain the methodology that was used to arrive at the report's savings estimates. The equations are provided, and it is shown how the calculations are performed so that the clients can see what has been done. If they want to change the report's assumptions, they can. If some of the data the authors have used are incorrect, they can replace it with the correct data and recalculate the results. However, by placing the calculations away from the rest of the discussion rather than intermingling it, the authors do not scare off the readers who need to know the other information.

Appendix: The authors use an appendix for lengthy data tables. For example, there is a motor efficiency table that is used in several of the authors' EMRs. Instead of repeating it in each EMR, it is printed in the appendix. The authors also include a table showing the facility's monthly energy-use history and a table listing the major energy-using equipment. Similar to the calculation section of the EMRs, the appendix allows the authors to provide backup information without cluttering up the main body of the report.

29.3 Improving Industrial Energy Audits

29.3.1 Preventing Overestimation of Energy Savings in Audits

A frequent criticism of energy audits is that they overestimate the savings potential available to the facility. This section addresses several problem areas that can result in overly optimistic savings projections and suggests ways to prevent mistakes.[12] This possibility of overestimation concerns many of the people and organizations that are involved in some part of this energy audit process. It concerns utilities that do not want to pay incentives for demand-side management programs if the facilities will not realize the expected results in energy or demand savings. Overestimates also make clients unhappy when their energy bills do not decrease as much as promised. The problem multiplies when a shared savings program is undertaken by the facility and an energy service company. Here, the difference between the audit projections and the actual metered and measured savings may be so significantly different that either there are no savings for the facility or the energy service company makes no profit.

More problems are likely with the accuracy of the energy audits for industrial and manufacturing facilities than for smaller commercial facilities or even large buildings because the equipment and operation of industrial facilities are more complex. However, many of the same problems discussed here in terms of industrial and manufacturing facilities can occur in audits of large commercial facilities and office buildings. Based on the authors' auditing experience for industrial and manufacturing facilities over the last 5 years, it is possible to identify a number of areas where problems are likely to occur, and a number of these are presented and discussed. In addition, the authors have developed a few methods and approaches to dealing with these potential problems, and a few ways have been found to initiate energy audit analyses that lead the authors to improved results. One of these approaches is to collect data on the energy-using equipment in an industrial or manufacturing facility and then to perform both an energy and a demand balance to help ensure that reasonable estimates of energy uses—and therefore, energy savings—are available for this equipment.

In addition, unfortunately, some analysts use the average cost of electricity to calculate energy savings. This can give a false picture of the actual savings and may result in overly optimistic savings predictions. This section also discusses how to calculate the correct values from the electricity bills and when to use these values. Finally, this section discusses several common energy savings measures that are frequently recommended by energy auditors. Some of these may not actually save as much energy or demand as expected, except in limited circumstances. Others have good energy-saving potential but must be implemented carefully to avoid increasing energy use rather than decreasing it.

29.3.2 Calculating Energy and Demand Balances

The energy and demand balances for a facility are an accounting of the energy flows and power used in the facility. These balances allow the energy analyst to track the energy and power inputs and outputs (uses) and see whether they match. A careful energy analyst should perform an energy-and-demand balance on a facility before developing and analyzing any EMRs.[13] In this way, the analyst can determine what the largest energy users are in a facility, can find out whether all—or almost all—energy uses have been identified, and can see whether more savings have been identified than are actually achievable. Making energy-use recommendations without utilizing the energy and demand balances is similar to making budget-cutting recommendations without knowing exactly where the money is currently being spent.

When the authors perform an energy survey (audit), all of the major energy-using equipment in the facility is inventoried. Then, the authors list the equipment and estimate its energy consumption and demand using the data gathered at the facility, such as nameplate ratings of the equipment and operating hours. The energy balance is developed by major equipment category, such as lighting, motors, HVAC, and air compressors. There is also a category called *miscellaneous* to account for loads that were not individually surveyed, such as copiers, electric typewriters, computers, and other plug loads. The authors typically allocate 10% of the actual energy use and demand to the *miscellaneous* category in the demand and energy balances. (For an office building instead of a manufacturing facility, this miscellaneous load might be 15%–20%.) Then, the energy and demand for each of the other categories is calculated.

29.3.2.1 Lighting

The first major category analyzed is lighting, because this is usually the category in which the authors have the most confidence for knowing the actual demand and hours of use. Thus, they believe that the energy and demand estimates for the lighting system are the most accurate and can then be subtracted from the total actual use to let the authors continue to build up the energy and demand balance for the facility. The authors record the types of lamps and number of lamps used in each area of the facility and ask the maintenance person to show them the replacement lamps and ballasts used. With this lamp and ballast wattage data, together with a good estimate of the hours that the lights are on in the various areas, they can construct what they believe to be a fairly accurate description of the energy and demand for the lighting system.

29.3.2.2 Air Conditioning

There is generally no other "easy" or "accurate" category to work on, so the authors proceed to either air conditioning or motors. In most facilities, there will be some air conditioning, even if it is just for the offices that are usually part of the industrial or manufacturing facility. Many facilities—particularly in the hot and humid southeast—are fully air conditioned. Electronics, printing, medical plastics and devices, and many assembly plants are common ones seen to be fully air conditioned. Boats, metal products, wood products, and plastic pipe-manufacturing facilities are most often not air conditioned. Air-conditioning system nameplate data are usually available and readable on many units, and efficiency ratings can be found from published ARI data[14] or from the manufacturers of the equipment. The biggest problem with air conditioning is to get runtime data that will allow the author(s) of the report to determine the number of full-load equivalent operating hours for the air-conditioning compressors or chillers. From the authors' experience in north and north-central Florida, about 2200–2400 h is used per year of compressor runtime for facilities that have air conditioning that responds to outdoor temperature. Process cooling requirements are much different and would typically have much larger numbers of full-load equivalent operating hours. With the equipment size, the efficiency data, and the full-load equivalent operating hours, it is possible to construct a description of the energy and demand for the air-conditioning system.

29.3.2.3 Motors

Turning next to motors, the authors begin looking at one of the most difficult categories to deal with in the absence of fully metered and measured load factors on each motor in the facility. In a 1-day plant visit, it is usually impossible to get actual data on the load factors for more than a few motors. Even then, that data are only good for the 1 day that it was taken. Very few energy-auditing organizations can afford the time and effort to make long-term measurements of the load factor on each motor in an industrial or manufacturing facility. Thus, estimating motor load factors becomes a critical part of the energy and demand balance, and also a critical part of the accuracy of the actual energy audit analysis. Motor nameplate data show the horsepower rating, the manufacturer, and sometimes the efficiency. If not, the efficiency can usually be obtained from the manufacturer, or from standard references such as the *Energy-Efficient Motor Systems Handbook*,[15] or from software databases such as MotorMaster produced by the Washington State Energy Office.[16] The authors inventory all motors over 1 hp and sometimes try to look at the smaller ones if there is enough time.

Motor runtime is another parameter that is very difficult to obtain. When the motor is used in an application where it is constantly on is an easy case. Ventilating fans, circulating pumps, and some process-drive motors are often in this class because they run for a known constant period of time each year. In other cases, facility operating personnel must help provide estimates of motor runtimes. With data on the horsepower, efficiency, load factor, and runtimes of motors, it is possible to construct a detailed table of motor energy and demands to use in the report's balances. Motor load factors will be discussed further in a later section of this chapter.

29.3.2.4 Air Compressors

Air compressors are a special case of motor use with most of the same problems. Some help is available in this category because some air compressors have instruments showing the load factor and some have runtime indicators for hours of use. Most industrial and manufacturing facilities will have several air compressors, and this may lead to some questions as to which air compressors are actually used and how many hours they are used. If the air compressors at a facility are priority-scheduled, it may turn out that one or more of the compressors are operated continuously, and one or two smaller compressors are cycled or unloaded to modulate the need for compressed air. In this case, the load factors on the larger compressors may be unity. Using these data on the horsepower, efficiency, load factor, and runtimes of the compressors, the authors develop a detailed table of compressor energy use and demand for the report's energy and demand balances.

29.3.2.5 Other Process Equipment

Specialized process equipment must be analyzed on an individual basis because it will vary tremendously depending on the type of industry or manufacturing facility involved. Much of this equipment will utilize electric motors and will be covered in the motor category. Other electrically powered equipment, such as drying ovens, cooking ovens, welders, and laser and plasma cutters, are nonmotor electric uses and must be treated separately. Equipment nameplate ratings and hours of use are necessary to compute the energy and demand for these items. Process chillers are another special class that are somewhat different from the comfort air-conditioning equipment, because the operating hours and loads are driven by the process requirements and not the weather patterns and temperatures.

29.3.2.6 Checking the Results

After the complete energy and demand balances are constructed for the facility, the authors check to see if the cumulative energy/demand for these categories plus the miscellaneous category is substantially larger or smaller than the actual energy usage and demand over the year. If it is, and it is certain that all of the major energy uses have been identified, the authors know that a mistake was made somewhere in their assumptions. As mentioned earlier, one area that has typically been difficult to accurately survey is the energy use by motors. Measuring the actual load factors is difficult on a 1-day walk-through audit visit, so the authors use the energy balance data to help estimate the likely load factors for the motors. This is done by adjusting the load factor estimates on a number of the motors to arrive at a satisfactory level of the energy and demand from the electric motors. Unless this is done, it is likely that the energy used by the motors will be overestimated and thus overestimate the energy savings from replacing standard motors with high-efficiency motors.

As an example, the authors performed an energy audit for one large manufacturing facility with a lot of motors. It was first assumed that the load factors for the motors were approximately 80%, based on what the facility personnel explained. Using this load factor gave a total energy use for the motors of over 16 million kW h/year and a demand of over 2800 kW. Because the annual energy use for the entire facility was just over 11 million kW h/year and the demand never exceeded 2250 kW, this load factor was clearly wrong. The authors adjusted the average motor load factor to 40% for most of the motors, which reduced the energy-use figure to 9 million kW h and the demand to just under 1600 kW. These values are much more reasonable with motors making up a large part of the electrical load of this facility.

After the energy/demand balances have been satisfactorily compiled, the authors use a graphics program to draw a pie chart showing the distribution of energy/demand between the various categories. This allows visual representation of which categories are responsible for the majority of the energy use. It also makes it possible to focus the energy savings analyses on the areas of largest energy use.

29.3.3 Problems with Energy Analysis Calculations

Over the course of performing 120 industrial energy audits, the authors have identified a number of problem areas. One lies with the method of calculating energy cost savings: whether to use the average cost of electricity or break the cost down into energy and demand cost components. Other problems include instances where the energy and demand savings associated with specific energy efficiency measures may not be fully realized or where more research should go into determining the actual savings potential.

29.3.3.1 On-Peak and Off-Peak Uses: Overestimating Savings by Using the Average Cost of Electricity

One criticism of energy auditors is that they sometimes overestimate the dollar savings available from various energy efficiency measures. One way overestimation can result is when the analyst uses only the average cost of electricity to compute the savings. Because the average cost of electricity includes a demand component, using this average cost to compute the savings for companies who operate on more than one shift can overstate the dollar savings. This is because the energy cost during the off-peak hours does not include a demand charge. A fairly obvious example of this type of problem occurs when the average cost of electricity is used to calculate savings from installing high-efficiency security lighting. In this instance, there is no on-peak electricity use, but the savings will be calculated as if all the electricity was used on-peak.

The same problem arises when an energy efficiency measure does not result in an expected—or implicitly expected—demand reduction. Using a cost of electricity that includes demand in this instance will again overstate the dollar savings. Examples of energy efficiency measures that fall into this category are occupancy sensors, photosensors, and adjustable-speed drives (ASDs). Although all of these measures can reduce the total amount of energy used by the equipment, there is no guarantee that the energy use will only occur during off-peak hours. While an occupancy sensor will save lighting kW h, it will not save any kW if the lights come on during the peak load period. Similarly, an ASD can save energy use for a motor, but if the motor needs its full-load capability—as an air-conditioning fan motor or chilled-water pump motor might—during the peak load period, the demand savings may not be there. The reduced use of the device or piece of equipment on peak load times may introduce a diversity factor that produces some demand savings. However, even this savings will be overestimated by using the average cost of electricity in most instances.

On the other hand, some measures can be expected to provide their full demand savings at the time of the facility's peak load. Replacing 40 W T12 fluorescent lamps with 32 W T8 lamps will provide a verifiable demand savings because the wattage reduction will be constant at all times and will specifically show up during the period of peak demand. Shifting loads to off-peak times should also produce verifiable demand savings. For example, putting a timer or energy management system control on a constant-load electric drying oven to ensure that it does not come on until the off-peak time will result in the full demand savings. Using high-efficiency motors also seems like it would also produce verifiable savings because of its reduced kW load, but in some instances, there are other factors that tend to negate these benefits. This topic is discussed later.

To help solve the problem of overestimating savings from using the average cost of electricity, the authors divide their energy savings calculations into a demand savings and an energy savings. In most instances, the energy savings for a particular piece of equipment is calculated by first determining the demand savings for that equipment and then multiplying by the total operating hours of the equipment. To calculate the annual cost savings (*CS*), the following formula is used:

$$CS = [\text{Demand savings} \times \text{Average monthly demand rate} \times 12 \text{ months/year}]$$
$$+ [\text{Energy savings} \times \text{Average cost of electricity without demand}]$$

If a recommended measure has no demand savings, then the energy cost savings is simply the energy savings times the average cost of electricity without demand (or off-peak cost of electricity). This procedure forces us to think carefully about which equipment is used on-peak and which is used off-peak.

To demonstrate the difference in savings estimates, consider replacing a standard 30 hp motor with a high-efficiency motor. The efficiency of a standard 30 hp motor is 0.901 and a high-efficiency motor is 0.931. Assume the motor has a load factor of 40% and operates 8760 h/year (three shifts). Assume also that the average cost of electricity is \$0.068/kW h (including demand), the average demand cost is \$3.79/kW/month, and the average cost of electricity without demand is \$0.053/kW h. The equation for calculating the demand of a motor is

$$D = HP \times LF \times 0.746 \times \frac{1}{Eff}$$

The savings on demand (or demand reduction) from installing a high-efficiency motor is

$$DR = HP \times LF \times 0.746 \times \left(\frac{1}{Eff_S} - \frac{1}{Eff_H} \right)$$

$$= 30 \text{ hp} \times 0.40 \times 0.746 \text{ kW/hp} \times \left(\frac{1}{0.901} - \frac{1}{0.931} \right) = 0.32 \text{ kW}$$

The annual energy savings (*ES*) is

$$ES = DR \times H = 0.32 \text{ kW} \times 8760 \text{ h/year} = 2803.2 \text{ kW h/year}$$

Using the earlier average cost of electricity, the cost savings (CS_1) are calculated as

$$CS_1 = ES \times (\text{Average cost of electricity}) = 2803.2 \text{ kW h/year} \times 0.068/\text{kW h} = 190.62/\text{year}$$

Using the earlier recommended formula,

$$CS = [\text{Demand savings} \times \text{Average monthly demand rate} \times 12 \text{ months/year}]$$

$$\times [\text{Energy savings} \times \text{Average cost of electricity without demand}]$$

$$= (0.32 \text{ kW} \times 3.79/\text{month} \times 12 \text{ months/year}) + (2803.2 \text{ kW h/year} \times 0.053/\text{kW h})$$

$$= (14.55 + 148.57)/\text{year} = 163.12/\text{year}$$

In this example, using the average cost to calculate the energy cost savings overestimates the cost savings by $27.50/year, or 17%. Although the actual amount is small for one motor, if this error is repeated for all the motors for the entire facility as well as all other measures that reduce the demand component only during the on-peak hours, then the cumulative error in cost savings predictions can be substantial.

29.3.3.2 Motor Load Factors

Many in the energy-auditing business started off assuming that motors ran at full load or near full load, and based their energy consumption analysis and energy savings analysis on that premise. Most books and publications that give a formula for finding the electrical load of a motor do not even include a term for the motor load factor. However, since experience soon showed the authors that few motors actually run at full load or near full load, they were left in a quandary about what load factor to actually use in calculations, because good measurements on the actual motor load factor are rarely to be had. A classic paper by Hoshide shed some light on the distribution of motor load factors from his experience.[17] In this paper, Hoshide noted that only about one-fourth of all three-phase motors run with a load factor greater than 60%, with 50% of all motors running at load factors between 30% and 60%, and one-fourth running with load factors less than 30%. Thus, those auditors who had been assuming that a typical motor load factor was around 70% or 80% had been greatly overestimating the savings from high-efficiency motors, adjustable-speed drives, high-efficiency belts, and other motor-related improvements.

The energy and demand balances discussed earlier also confirm that overall motor loads in most facilities cannot be anywhere near 70%–80%. The authors' experience in manufacturing facilities has been that motor load factors are more correctly identified as being in the 30%–40% range. With these load factors, one obtains very different savings estimates and economic results than when one assumes that a motor is operating at a 70% or greater load factor, as shown in the example earlier.

One place where the motor load factor is critical—but often overlooked—is in the savings calculations for ASDs. Many motor and ASD manufacturers provide easy-to-use software that will determine savings with an ASD if you supply the load profile data. Usually a sample profile is included that shows calculations for a motor operating at full load for some period of time and at a fairly high overall load factor—for example, around 70%. If the motor has a load factor of only 50% or less to begin with, the savings estimates from a quick use of one of these programs may be greatly exaggerated. If the actual motor use profile with the load factor of 50% is use, one may find that the ASD will still save some energy and money, but often not as much as it looks like when the motor is assumed to run at the higher load factor. For example, a 20 hp motor may have been selected for use on a 15 hp load to ensure that there is a "safety factor." Thus, the maximum load factor for the motor would be only 75%. A typical fan or pump in an air-conditioning system that is responding to outside weather conditions may operate at its maximum load only about 10% of the time. Because that maximum load here is only 15 hp, the average load factor for the motor might be more like 40% and will not be even close to 75%.

29.3.3.3 High-Efficiency Motors

Another interesting problem area is associated with the use of high-efficiency motors. In Hoshide's paper mentioned earlier, he notes that, in general, high-efficiency motors run at a faster full-load speed than standard-efficiency motors. This means that when a standard motor is replaced by a high-efficiency motor, the new motor will run somewhat faster than the old motor in almost every instance. This is a problem for motors that drive centrifugal fans and pumps, because the higher operating speed means greater power use by the motor. Hoshide provides an example where he shows that a high-efficiency motor that should be saving about 5% energy and demand actually uses the same energy and demand as the old motor. This occurs because the increase in speed of the high-efficiency motor offsets the power savings by almost exactly the same 5% due to the cube law for centrifugal fans and pumps.

Few energy auditors ever monitor fans or pumps after replacing a standard motor with a high-efficiency motor; therefore, they have not realized that this effect has cancelled the expected energy and demand savings. Since Hoshide noted this feature of high-efficiency motors, the authors have been careful to make sure that their recommendations for replacing motors with centrifugal loads carry the notice that it will probably be necessary to adjust the drive pulleys or drive system so that the load is operated at the same speed to achieve the expected savings.

29.3.3.4 Motor Belts and Drives

The authors have developed some significant questions about the use of cogged and synchronous belts, and the associated estimates of energy savings. It seems fairly well accepted that cogged and synchronous belts do transmit more power from a motor to a load than if standard smooth V-belts are used. In some instances, this should certainly result in some energy savings. A constant-torque application like a conveyor drive may indeed save energy with a more efficient drive belt because the motor will be able to supply that torque with less effort. Consider also a feedback-controlled application, such as a thermostatically controlled ventilating fan or a level-controlled pump. In this case, the greater energy transmitted to the fan or pump should result in the task being accomplished faster than if less drive power were supplied, and some energy savings should exist. However, if a fan or a pump operates in a nonfeedback application—as is common for many motors—then there will not be any energy savings. For example, a large ventilating fan that operates at full load continuously without any temperature or other feedback may not use less energy with an efficient drive belt, because the fan may run faster as a result of the drive belt having less slip. Similarly, a pump that operates continuously to circulate water may not use less energy with an efficient drive belt. This is an area that needs some monitoring and metering studies to check the actual results.

Whether efficient drive belts result in any demand savings is another question. Because, in many cases, the motor is assumed to be supplying the same shaft horsepower with or without high-efficiency drive belts, a demand savings does not seem likely in these cases. It is possible that using an efficient belt on a motor with a constant-torque application that is controlled by an ASD might result in some demand savings. However, for the most common applications, the motor is still supplying the same load and thus would have the same power demand. For feedback-controlled applications, there might be a diversity factor involved so that the reduced operation times could result in some demand savings—but not the full value otherwise expected. Thus, using average cost electricity to quantify the savings expected from high-efficiency drive belts could well overestimate the value of the savings. Verification of the cases where demand savings are to be expected is another area where more study and data are needed.

29.3.3.5 Adjustable-Speed Drives

The authors would like to close this discussion with a return to ASDs because these are devices that offer a great potential for savings, but have far greater complexities than are often understood or appreciated. Fans and pumps form the largest class of applications where great energy savings is possible from the use of ASDs. This is a result again of the cube law for centrifugal fans and pumps where the power required to drive a fan or pump is specified by the cube of the ratio of the flow rates involved. According to the cube law, a reduction in flow to one-half the original value could now be supplied by a motor using only one-eighth of the original horsepower. Thus, whenever an air flow or liquid flow can be reduced, such as in a variable-air-volume system or with a chilled-water pump, there is a dramatic savings possible with an ASD. In practice, there are two major problems with determining and achieving the expected savings.

The first problem is the one briefly mentioned earlier, and that is determining the actual profile of the load involved. Simply using the standard profile in a piece of vendor's software is not likely to produce very realistic results. There are so many different conditions involved in fan and pump applications that taking actual measurements is the only way to get a good idea of the savings that will occur with an ASD. Recent papers have discussed the problems with estimating the loads on fans and pumps and have shown how the cube law itself does not always give a reasonable value.[18–20] The Industrial Energy Center at Virginia Polytechnic Institute and Virginia Power Company have developed an approach wherein they classify potential ASD applications into eight different groups and then estimate the potential savings from the analysis of each system and from measurements of that system's operation.[21] Using both an analytical approach and a few measurements allows them to get a reasonable estimate of the motor load profile and thus a reasonable estimate of the energy and demand savings possible.

The second problem is achieving the savings predicted for a particular fan or pump application. It is not sufficient to just identify the savings potential and then install an ASD on the fan or pump motor. In most applications, there is some kind of throttling or bypass action that results in almost the full horsepower still being required to drive the fan or pump most of the time. In these applications, the ASD will not save much, unless the system is altered to remove the throttling or bypass device and a feedback sensor is installed to tell the ASD what fraction of its speed to deliver. This means that in many air flow systems, the dampers or vanes must be removed so that the quantity of air can be controlled by the ASD changing the speed of the fan motor. In addition, some kind of feedback sensor must be installed to measure the temperature or pressure in the system to send a signal to the ASD or a PLC controller to alter the speed of the motor to meet the desired condition. The additional cost of the alterations to the system and the cost of the control system needed greatly change the economics of an ASD application compared to the case where only the purchase cost and installation cost of the actual ASD unit are considered.

For example, a dust collector system might originally be operated with a large 150 hp fan motor running continuously to pick up the dust from eight saws. However, because production follows existing orders for the product, sometimes only two, three, or four saws are in operation at a particular time. Thus, the load on the dust collector is much lower at these times than if all eight saws are in use. An ASD is a common recommendation in this case, but estimating the savings is not easy to begin with, and after the costs of altering the collection duct system and of adding a sophisticated control system to the ASD are considered, the bottom-line result is much different than the cost of the basic ASD with installation. Manual or automatic dampers must be added to each duct at a saw so that it can be shut off when the saw is not running. In addition, a PLC for the ASD must be added to the new system, together with sensors added to each damper so that the PLC will know how many saws are in operation and therefore what speed to tell the ASD for the fan to run to meet the dust collection load of that number of saws. Without these system changes and control additions, the ASD itself will not save any great amount of energy or money. Adding them in might well double the cost of the basic ASD and double the payback time that may have originally been envisioned.

Similarly, for a water or other liquid flow application, the system piping or valving must be altered to remove any throttling or bypass valves, and a feedback sensor must be installed to allow the ASD to know what speed to operate the pump motor. If several sensors are involved in the application, then a PLC may also be needed to control the ASD. For example, putting an ASD on a chilled-water pump for a facility is much more involved, and much more costly, than simply cutting the electric supply lines to the pump motor and inserting an ASD for the motor. Without the system alterations and without the feedback control system, the ASD cannot provide the savings expected.

29.3.4 General Rules

New energy auditors often do not have the experience to have engineering judgment about the accuracy of their analyses. That is, they cannot look at the result and immediately know that it is not within the correct range of likely answers. Because the authors' IAC program has a fairly steady turnover of students, they find the same type of errors cropping up over and over as draft audit reports are reviewed. To help new team members develop the engineering judgment that they will eventually gain through experience, the authors are developing "rules of thumb" for energy analyses. The rules of thumb are intended to provide a ballpark estimate of the expected results. For example, if the rule of thumb for the percentage for installing high-efficiency motors says that the savings range is 3%–5% of the energy use by the motors, then a student who comes up with a savings of 25% will immediately know that the calculations are wrong and will know to check the assumptions and data entry to see where the error lies. Without these rules of thumb, the burden for checking these results is shifted to the team leaders and program directors. Although this does not obviate the need for report review, it minimizes the likelihood that errors will occur. The authors suggest that other organizations who frequently utilize and train new energy auditors consider developing such rules of thumb for the major types of facilities or geographic areas that they audit.

Energy auditing is not an exact science, but a number of opportunities are available for improving the accuracy of the recommendations. Techniques that may be appropriate for small-scale energy audits can introduce significant errors into the analyses for large complex facilities. This chapter began by discussing how to perform an energy-and-demand balance for a company. This balance is an important step in doing an energy-use analysis, because it provides a check on the accuracy of some of the assumptions necessary to calculate savings potential. It also addressed several problem areas that can result in overly optimistic savings projections and suggested ways to prevent mistakes. Finally, several areas where additional research, analysis, and data collection are needed were identified. After this additional information is obtained, everyone can produce better and more accurate energy audit results.

29.3.4.1 Decision Tools for Improving Industrial Energy Audits: OIT Software Tools

The OIT—a program operated by the U.S. Department of Energy, Division of Energy Efficiency and Renewable Energy (EERE)—provides a series of computer software tools that can be obtained free from their website or by ordering a CD at no cost from them.

With the right know-how, these powerful tools can be used to help identify and analyze energy system savings opportunities in industrial and manufacturing plants. Although the tools are accessible at the U.S. DOE website for download, they also encourage users to attend a training workshop to enhance their knowledge and take full advantage of opportunities identified in the software programs. For some tools, advanced training is also available to help further increase expertise in their use.

29.3.4.2 Decision Tools for Industry: Order the Portfolio of Tools on CD

The Decision Tools for Industry CD contains the MotorMaster+ (MM+), Pump System Assessment Tool, Steam System Tool Suite, 3E Plus, and the new AIRMaster+ software packages described here. In addition, it includes MM+ training. The training walks the user through both the fundamentals and the advanced features of MM+ and provides examples for using the software to make motor purchase decisions. The CD can be ordered via e-mail from the EERE Information Center or by calling the EERE Information Center at 1-877-EERE-INF (877-337-3463).

DOE industry tools

- AIRMaster+
- Chilled Water System Analysis Tool (CWSAT)
- Combined Heat and Power Application Tool (CHP)
- Fan System Assessment Tool (FSAT)
- MotorMaster+ 4.0
- MotorMaster+ International
- NO_x and Energy Assessment Tool (NxEAT)
- Plant Energy Profiler for the Chemical Industry (ChemPEP Tool)
- Process Heating Assessment and Survey Tool (PHAST)
- Pumping System Assessment Tool 2004 (PSAT)
- Steam System Tool Suite

Other industry tools

- ASDMaster: Adjustable-Speed Drive Evaluation Methodology and Application

AIRMaster+: AIRMaster+ provides comprehensive information on assessing compressed-air systems, including modeling existing and future system upgrades, and evaluating savings and effectiveness of energy efficiency measures.

Chilled Water System Analysis Tool Version 2.0: Use the CWSAT to determine energy requirements of chilled-water distribution systems and to evaluate opportunities for energy and cost savings by applying improvement measures. Provide basic information about an existing configuration to calculate current energy consumption and then select proposed equipment or operational changes for comparison. The results of this analysis will help the user quantify the potential benefits of chilled-water system improvements.

Combined Heat and Power Application Tool: The CHP Application Tool helps industrial users evaluate the feasibility of CHP for heating systems such as fuel-fired furnaces, boilers, ovens, heaters, and heat exchangers. It allows the analysis of three typical system types: fluid heating, exhaust-gas heat recovery, and duct burner systems. Use the tool to estimate system costs and payback period, and to perform "what-if" analyses for various utility costs. The tool includes performance data and preliminary cost information for many commercially available gas turbines and default values that can be adapted to meet specific application requirements.

Fan System Assessment Tool: Use the FSAT to help quantify the potential benefits of optimizing fan system configurations that serve industrial processes. FSAT is simple and quick, and requires only basic information about the fans being surveyed and the motors that drive them. With FSAT, one can calculate the amount of energy used by one's fan system, determine system efficiency, and quantify the savings potential of an upgraded system.

MotorMaster+ 4.0: An energy-efficient motor selection and management tool, MotorMaster+ 4.0 software, includes a catalog of over 20,000 AC motors. This tool features motor inventory management tools, maintenance log tracking, efficiency analysis, savings evaluation, energy accounting, and environmental reporting capabilities.

MotorMaster+ International: MotorMaster+ International includes many of the capabilities and features of MotorMaster+; however, now it can help evaluate repair/replacement options on a broader range of motors, including those tested under the Institute of Electrical and Electronic Engineers standard and those tested using International Electrical Commission methodology. With this tool, analyses can be conducted in different currencies, and it will calculate efficiency benefits for utility rate schedules with demand charges, edit and modify motor rewind efficiency loss defaults, and determine "best available" motors. The tool can be modified to operate in English, Spanish, and French.

NO_x and Energy Assessment Tool: The NxEAT helps plants in the petroleum refining and chemical industries to assess and analyze NO_x emissions and application of energy efficiency improvements. Use the tool to inventory emissions from equipment that generates NO_x and then compare how various technology applications and efficiency measure affect overall costs and reduction of NO_x. Perform "what-if" analyses to optimize and select the most cost-effective methods for reducing NO_x from systems such as fired heaters, boilers, gas turbines, and reciprocating engines.

Plant Energy Profiler for the Chemical Industry: The ChemPEP Tool provides chemical plant managers with the information they need to identify savings and efficiency opportunities. The ChemPEP Tool enables energy managers to see overall plant energy use, identify major energy-using equipment and operations, summarize energy cost distributions, and pinpoint areas for more detailed analysis. The ChemPEP Tool provides plant energy information in an easy-to-understand graphical manner that can be very useful to managers.

Process Heating Assessment and Survey Tool: The PHAST provides an introduction to process heating methods and tools to improve thermal efficiency of heating equipment. Use the tool to survey process-heating equipment that uses fuel, steam, or electricity, and identify the most energy-intensive equipment. It can also help perform an energy (heat) balance on selected equipment (furnaces) to identify and reduce nonproductive energy use. Compare performance of the furnace under various operating conditions and test "what-if" scenarios.

Pumping System Assessment Tool 2004: The PSAT helps industrial users assess the efficiency of pumping system operations. The PSAT uses achievable pump performance data from Hydraulic Institute standards and motor performance data from the MotorMaster+ database to calculate potential energy and associated cost savings.

Steam System Tool Suite: In many industrial facilities, steam system improvements can save 10%–20% in fuel costs. To help tap into potential savings in typical industrial facilities, DOE offers a suite of tools for evaluating and identifying steam system improvements.

- *Steam System Assessment Tool (SSAT) Version 2.0.0*: The SSAT allows steam analysts to develop approximate models of real steam systems. Using these models, SSAT can be applied to quantify the magnitude—energy, cost, and emissions savings—of key potential steam improvement opportunities. SSAT contains the key features of typical steam systems. The enhanced and improved version includes features such as a steam demand savings project, a user-defined fuel model, a boiler stack loss worksheet for the SSAT fuels, a boiler flash steam recovery model, and improved steam trap models.
- *3E Plus, Version 3.2*: The program calculates the most economical thickness of industrial insulation for user input operating conditions. Calculations can be made using the built-in thermal performance relationships of generic insulation materials or supply conductivity data for other materials.
- *Steam Tool Specialist Qualification Training*: Industry professionals can earn recognition as Qualified Specialists in the use of the BestPractices Steam Tools. DOE offers an in-depth (2–1/2)-day training session for steam system specialists, including 2 days of classroom instruction and a written exam. Participants who complete the workshop and pass the written exam are recognized by DOE as Qualified Steam Tool Specialists. Specialists can assist industrial customers in using the BestPractices Steam Tools to evaluate their steam systems.

ASDMaster: Adjustable-Speed Drive Evaluation Methodology and Application. This Windows® software program helps plant or operations professionals determine the economic feasibility of an ASD application, predict how much electrical energy may be saved by using an ASD, and search a database of standard drives. The package includes two 3.5 in. diskettes, a user's manual, and a user's guide. Please order from EPRI. For more information, see the ASDMaster website.

29.3.4.3 Energy-Auditing Help from Industrial Assessment Centers

The IACs, sponsored by the U.S. Department of Energy, EERE Division Industrial Technologies Program (ITP), provide eligible small- and medium-sized manufacturers with no-cost energy assessments. Additionally, the IACs serve as a training ground for the next generation of energy-savvy engineers.

Teams composed mainly of engineering faculty and students from the centers, located at 24 universities around the country, conduct energy audits or industrial assessments and provide recommendations to manufacturers to help them identify opportunities to improve productivity, reduce waste, and save energy. Recommendations from industrial assessments have averaged about $55,000 in potential annual savings for each manufacturer.

As a result of performing these assessments, upper-class and graduate engineering students receive unique hands-on assessment training and gain knowledge of industrial process systems, plant systems, and energy systems, making them highly attractive to employers.

FIGURE 29.5
Industrial Assessment Center (IAC) locations, http://www1.eere.energy.gov/manufacturing/tech_assistance/iacs_locations, and service areas. (From U.S. Department of Energy, Office of Industrial Technologies, Washington, DC, 2005, http://www.energy.gov/eere/amo/industrial-assessment-centers-iacs.)

To be eligible for an IAC assessment, a manufacturing plant must meet the following criteria:

- Within Standard Industrial Codes 20-39
- Generally be located within 150 miles of a host campus
- Gross annual sales below $100 million
- Fewer than 500 employees at the plant site
- Annual energy bills more than $100,000 and less than $2.5 million
- No professional in-house staff to perform the assessment

Presently (2014), there are 26 schools across the country participating in the IAC Program. For additional information or to apply for an assessment, go to http://www.energy.gov/eere/amo/industrial-assessment-centers-iacs and click on one of the school names for contact information. A map of the IAC centers and their service areas is shown in Figure 29.5.

29.4 Industrial Electricity End Uses and Electrical Energy Management

29.4.1 Importance of Electricity in Industry

Electricity use in industry is primarily for electric drives, electrochemical processes, space heating, lighting, and refrigeration. Table 29.2 lists the relative importance of the use of electricity in the industrial sector. Timely data on industrial and manufacturing energy use are difficult to obtain, since its frequency of collection is only every 3 years, and then it takes EIA

TABLE 29.2

Manufacturing Electricity by End Use, 2006

	GW h	Percent Use
Machine drives (motors)	422,408	53.8
Electrochemical processes	60,323	10.1
Process heat	101,516	10.4
Facility HVAC (motors)	77,768	8.3
Facility lighting	58,013	6.5
Process cooling and refrigeration	60,389	6.0
On-site transportation (motors)	2,197	0.1
Other process	13,816	0.4
Other	45,000	
Total	8,40,787	100

Source: U.S. Department of Energy, Energy Information Agency, Manufacturing Energy Consumption Survey, 2006, Washington, DC, 2012.

another 3 years to process the results. In addition, it is only a sample survey, so the accuracy of the data is less than what would be optimal. However, it is better data than is available for most countries in the world. The most recent, detailed data on manufacturing energy end use (as of early 2013) are still the 2006 Manufacturing Energy Consumption Survey (MECS) data. Table 29.2 shows the electrical energy consumed for different end uses in 2006.

Because several of the categories mentioned earlier involve motor use, the total manufacturing energy use for motors is actually the sum of several categories:

Machine drives	50.5%
Facility HVAC	7.2%
Process cooling and refrigeration	9.3%
On-site transportation	0.1%
Total motors	67%

This gives the same results shown as the pie chart shown in Figure 29.3.

29.4.2 Electric Drives

Electric drives of one type or another use 68% of industrial electricity. Examples include electric motors, machine tools, compressors, refrigeration systems, fans, and pumps. Improvements in these applications would have a significant effect on reducing industrial electrical energy.

Motor efficiency can be improved in some cases by retrofit (modifications, better lubrication, improved cooling, heat recovery), but generally requires purchasing of more efficient units. For motors in the sizes 1–200 hp, manufacturers today supply a range of efficiency. Greater efficiency in a motor requires improved design, more costly materials, and generally greater first cost. Losses in electric drive systems may be divided into four categories:

	Typical Efficiency (%)
Prime mover (motor)	10–95
Coupling (clutches)	80–99
Transmission	70–95
Mechanical load	1–90

TABLE 29.3

Typical Electric Motor Data (1800 rpm Motors)

	Full Load Efficiency (%)	
Size (hp)	**1975**	**1993 (EPACT)**
1	76	82.5–85.5
5	84	87.5–89.5
10	85	90.2–91.7
20	85	91.7–93.0
40	85	93.0–94.1
100	91	94.5–95.4
200	90	95.4–96.2

Each category must be evaluated to determine energy management possibilities. In many applications, the prime mover will be the most efficient element of the system. Table 29.3 shows typical induction motor data, illustrating the improvement in efficiency due to the 1992 Energy Policy Act. Note that both efficiency and power factor decrease with partial load operation, which means that motors should be sized to operate at or near full load ratings.

Manufacturers have introduced new high-efficiency electric motors in recent years. Many utilities offer rebates of $5–$20/hp for customers installing these motors.

29.4.3 Electrochemical Processes

Industrial uses of electrochemical processes include electrowinning, electroplating, electrochemicals, electrochemical machining, fuel cells, welding, and batteries.

A major use of electrolytic energy is in electrowinning—the electrolytic smelting of primary metals such as aluminum and magnesium. Current methods require on the order of 13–15 kW h/kg; efforts are under way to improve electrode performance and reduce this to 10 kW h/kg. Recycling now accounts for one-third of aluminum production; this requires only about 5% of the energy required to produce aluminum from ore.

Electrowinning is also an important low-cost method of primary copper production. Another major use of electrochemical processes is in the production of chlorine and sodium hydroxide from salt brine. Electroplating and anodizing are two additional uses of electricity of great importance. Electroplating is basically the electrodeposition of an adherent coating upon a base metal. It is used with copper, brass, zinc, and other metals. Anodizing is roughly the reverse of electroplating, with the workpiece (aluminum) serving as the anode. The reaction progresses inward from the surface to form a protective film of aluminum oxide on the surface.

Fuel cells are devices for converting chemical energy to electrical energy directly through electrolytic action. Currently, they represent a small use of energy, but research is directed at developing large systems suitable for use by electric utilities for small dispersed generation plants. Batteries are another major use of electrolytic energy, ranging in size from small units with energy storage in the joule or fractional joule capacity up to units proposed for electric utility use that will store 18×10^9 J (5 MW h). Electroforming, etching, and welding are forms of electrochemical used in manufacturing and material shaping. The range of applications for these techniques stretches from microcircuits to aircraft carriers. In some applications, energy for machining is reduced, and reduction of scrap also saves energy. Welding has benefits in the repair and salvage of materials and equipment, reducing the need for energy to manufacture replacements.

29.4.4 Electric Process Heat

Electricity is widely used as a source of process heat due to ease of control, cleanliness, wide range in unit capacities (watts to megawatts), safety, and low initial cost. Typical heating applications include resistance heaters (metal sheath heaters, ovens, furnaces), electric salt bath furnaces, infrared heaters, induction and high-frequency resistance heating, dielectric heating, and direct arc electric furnaces.

Electric arc furnaces in the primary metals industry are a major use of electricity. Typical energy use in a direct arc steel furnace is about 2.0 kW h/kg. Electric arc furnaces are used primarily to refine recycled scrap steel. This method uses about 40% of the energy required to produce steel from iron ore using basic oxygen furnaces. Energy savings can be achieved by using waste heat to preheat scrap iron being charged to the furnace.

Glass making is another process that uses electric heat. An electric current flows between electrodes placed in the charge, causing it to melt. Electric motors constitute a small part of total glass production. Major opportunities for improved efficiency with electric process heat applications in general include improved heat transfer surfaces, better insulation, heat recovery, and improved controls.

29.4.5 HVAC

Heating, ventilating, and air conditioning (HVAC) is an important use of energy in the industrial sector. The environmental needs in an industrial operation can be quite different from residential or commercial operations. In some cases, strict environmental standards must be met for a specific function or process. More often, the environmental requirements for the process itself are not limiting, but space conditioning is a prerequisite for the comfort of production personnel. Chapters 22 and 23 have a more complete discussion of energy management opportunities in HVAC systems.

29.4.6 Lighting

Industrial lighting needs range from low-level requirements for assembly and welding of large structures (such as shipyards) to the high levels needed for the manufacture of precision mechanical and electronic components such as integrated circuits. Lighting uses about 20% of the U.S. electrical energy and 7% of all energy. Of all lighting energy, about 20% is industrial, with the balance being sizes of systems; energy management opportunities in industrial lighting systems are similar to those in residential/commercial systems (see Chapter 22).

29.4.7 Electric Load Analysis

The energy audit methodology is a general tool that can be used to analyze energy use in several forms and over a short or long period of time. Another useful technique, particularly for obtaining a short-term view of industrial electricity use, is an analysis based on the evaluation of the daily load curve. Normally, this analysis uses metering equipment installed by the utility and therefore available at the plant. However, special metering equipment can be installed if necessary to monitor specific process or building.

For small installations, both power and energy use can be determined from the kilowatt hour meter installed by the utility. Energy in kW h is determined by

$$E = (0.001)(K_h P_t C_t N) \text{ (kW h)} \tag{29.1}$$

where

E is the electric energy used, kW h
K_h is the meter constant, W h/revolution
P_t is the potential transformer ratio
C_t is the current transformer ratio
N is the number of revolutions of the meter disk

(The value of K_h is usually marked on the meter. P_t and C_t are usually 1.0 for small installations.) To determine energy use, the meter would be observed during an operation and the number of revolutions of the disk counted. Then, the equation can be used to determine E.

To determine the average load over some period p (h), determine E as earlier for time p and then use the relation that

$$L = \frac{E}{p} \text{ (kW)} \tag{29.2}$$

where

E is in kW h
p is in h
L is the load in kW

Larger installations will have meters with digital outputs or strip charts. Often, these will provide a direct indication of kW h and kW as a function of time. Some also indicate the reactive load (kVARs) or the power factor.

The first step is to construct the daily load curve. This is done by obtaining kW h readings each hour using the meter. The readings are then plotted on a graph to show the variation of the load over a 24 h period. Table 29.4 shows a set of readings obtained over a 24 h period in the XYZ manufacturing plant located in Sacramento, California, and operating one shift per day. These readings have been plotted in Figure 29.6.

Several interesting conclusions can be immediately drawn from this figure:

- The greatest demand for electricity occurs at 11:00.
- Through the lunch break, the third highest demand occurs.
- The ratio of the greatest demand to the least demand is approximately 3:1.
- Only approximately 50% of the energy used actually goes into making a product (54% on-shift use, 46% off-shift use).

When presented to management, these facts were of sufficient interest that a further study of electricity use was requested. Additional insight into the operation of a plant (and into the cost of purchase of electricity) can be obtained from the load analysis. Following a brief discussion of electrical load parameters, a load analysis for the XYZ company will be described.

TABLE 29.4

Kilowatt Hour Meter Readings for XYZ Manufacturing Company

Time Meter Read	Elapsed (kW h)	Notes Concerning Usage	Percentage of Total Usage (%)
1:00 (a.m.)	640		
2:00	610		
3:00	570		
4:00	570	7 h preshift use	
5:00	640		
6:00	770		
7:00	1,120		
Subtotal	4,920		17
8:00	1,470		
9:00	1,700		
10:00	1,790		
11:00	1,850		
	(Peak)	9 h on-shift use	
12:00 (noon)	1,830		
13:00 (1 p.m.)	1,790		
14:00 (2 p.m.)	1,790		
15:00 (3 p.m.)	1,760		
16:00 (4 p.m.)	1,690		
Subtotal	15,670		54
17:00 (5 p.m.)	1,470		
18:00 (6 p.m.)	1,310		
19:00 (7 p.m.)	1,210		
20:00 (8 p.m.)	1,090	8 h postshift use	
21:00 (9 p.m.)	960		
22:00 (10 p.m.)	800		
23:00 (11 p.m.)	730		
24:00 (12 a.m.)	640		
Subtotal	8,210		29
Grand totals	28,800		100

Any industrial electrical load consists of lighting, motors, chillers, compressors, and other types of equipment. The sum of the capacities of this equipment, in kW, is the *connected* load. The actual load at any point in time is normally less than the connected load since every motor is not turned on at the same time; only part of the lights may be on at any one time, and so on. Thus, the load is said to be diversified, and a measure of this can be found by calculating a *diversity factor*:

$$DV = \frac{(D_{m1} + D_{m2} + D_{m3} + \cdots)}{(D_{max})} \tag{29.3}$$

where

D_{m1}, D_{m2}, and so on are the sum of maximum demand of individual loads in kW

D_{max} is the maximum demand of plant in kW

If the individual loads do not occur simultaneously (usually they do not), the diversity factor will be greater than unity. Typical values for industrial plants are 1.3–2.5.

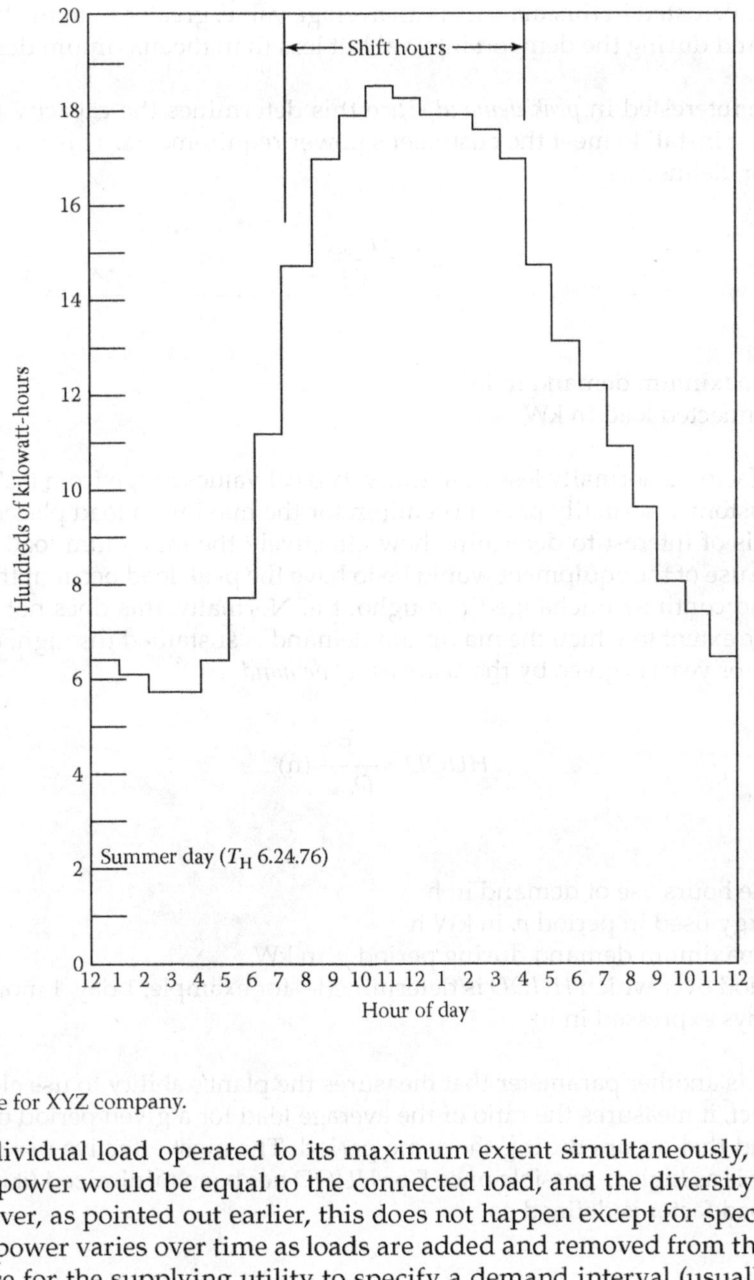

FIGURE 29.6
Daily load curve for XYZ company.

If each individual load operated to its maximum extent simultaneously, the *maximum* demand for power would be equal to the connected load, and the diversity factor would be 1.0. However, as pointed out earlier, this does not happen except for special cases. The demand for power varies over time as loads are added and removed from the system. It is usual practice for the supplying utility to specify a demand interval (usually 0.25, 0.5, or 1.0 h) over which it will calculate

$$D = \frac{E}{p} \text{ (kW)} \tag{29.4}$$

where
- D is the demand in kW
- E is the kW h used during p
- p is the demand interval in h

The demand calculated in this manner is an average value, greater than the lowest instantaneous demand during the demand interval but less than the maximum demand during the interval.

Utilities are interested in *peak demand* since this determines the capacity of the equipment they must install to meet the customer's power requirements. This is measured by a demand factor, defined as

$$DF = \frac{D_{max}}{CL} \tag{29.5}$$

where

D_{max} is the maximum demand in kW
CL is the connected load in kW

The demand factor is normally less than unity; typical values range from 0.25 to 0.90.

Since the customer normally pays a premium for the maximum load placed on the utility system, it is of interest to determine how effectively the maximum load is used. The most effective use of the equipment would be to have the peak load occur at the start of the use period and continue unchanged throughout it. Normally, this does not occur, and a measure of the extent to which the maximum demand is sustained throughout the period (a day, month, or year) is given by the *hours use of demand*:

$$HUOD = \frac{E}{D_{max}} \text{ (h)} \tag{29.6}$$

where

$HUOD$ is the hours use of demand in h
E is the energy used in period p, in kW h
D_{max} is the maximum demand during period p, in kW
p is the period over which $HUOD$ is determined—for example, 1 day, 1 month, or 1 year (p is always expressed in h)

The *load factor* is another parameter that measures the plant's ability to use electricity efficiently. In effect, it measures the ratio of the average load for a given period of time to the maximum load that occurs during the same period. The most effective use results when the load factor is as high as possible once E or $HUOD$ has been minimized (it is always less than 1). The load factor is defined as

$$LF = \frac{E}{(D_{max})(p)} \tag{29.7}$$

where

LF is the load factor (dimensionless)
E is the energy used in period p in kW h
D_{max} is the maximum demand during period p in kW
p is the period over which load factor is determined (e.g., 1 day, 1 month, or 1 year) in h

Another way to determine *LF* is from the relation

$$LF = \frac{HUOD}{p} \tag{29.8}$$

Still another method is to determine the average load, *L*—kW h/*p* during *p* divided by *p*—and then use the relation

$$LF = \frac{L}{D_{max}} \tag{29.9}$$

These relations are summarized for convenience in Table 29.5.

Returning to the XYZ plant, the various load parameters can now be calculated. Table 29.6 summarizes the needed data and the results of the calculations. The most striking

TABLE 29.5

Summary of Load Analysis Parameters

Formulas	Definitions
$E = \frac{K_h P_t C_t N}{1000}$	E = Electric energy used in period p (kW h)
$L = \frac{E}{p}$	E_{max} = Maximum energy used during period p (kW h)
$DV = \frac{(D_{m1} + D_{m2} + D_{m3})}{D_{max}}$	K_h = Meter constant (W h/revolution)
$D = \frac{E}{p}, D_{max} = \frac{E_{max}}{p}$	P_t = Potential transformer ratio
$DF = \frac{D_{max}}{CL}$	C_t = Current transformer ratio
$HUOD = \frac{E}{D_{max}}$	N = Number of revolutions of the meter disk
$LF = \frac{E}{(D_{max})(p)} = \frac{HUOD}{p} = \frac{L}{D_{max}}$	L= Average load (kW)
	p = Period of time used to determine load, demand, electricity use, etc., normally 1 h, day, month, or year; measured in hours
	DV = Diversity factor (dimensionless)
	D_{max} = Maximum demand in period p (kW)
	$D_{m1}, D_{m2}, \ldots$ = Maximum demand of individual load (kW)
	D = Demand during period p (kW)
	DF = Demand factor for period p (dimensionless)
	CL = Connected load (kW)
	$HUOD$ = Hours use of demand during period p (h)
	LF = Load factor during period p (dimensionless)

TABLE 29.6

Data for Load Analysis of XYZ Plant

p = 24 h
E = 28,800 kW h/day
E_{max} = 1850 kW h
CL = 2792 kW
$D_{m1}, D_{m2}, \ldots$ = 53, 62, 144, 80, 700, 1420 kW

thing shown by the calculations is the hours use of demand, equal to 15.6. This is a surprise, since the plant is only operating one shift. The other significant point brought out by the calculations is the low load factor.

An energy audit of the facility was conducted, and the major loads were the evaluated number of energy management opportunities whereby both loads (kW) and energy use (kW h) could be reduced. The audit indicated that inefficient lighting (on about 12 h/day) could be replaced in the parking lot. General office lighting was found to be uniformly at 100 fc; by selective reduction and task lighting, the average level could be reduced to 75 fc or less. The air-conditioning load would also be reduced. Improved controls could be installed to automatically shut down lighting during off-shift and weekend hours (the practice had been to leave the lights on). Some walls and ceilings were selected for repainting to improve reflectance and reduced lighting energy. It was found that the air-conditioning chillers operated during weekends and off-hours; improved controls would prevent this. Also, the ventilation rates were found to be excessive and could be reduced. In the plant, compressed air system leaks, heat losses from plating tanks, and on-peak operation of the heat treat furnace represented energy and load management opportunities.

The major energy management opportunities were evaluated to have the following potential savings, with a total payback of 5.3 months, as shown in Table 29.7. The average daily savings of electricity amounted to approximately 4400 kW h/day. This led to savings of $80,000/year, with the cost of the modification being $36,000.

This can be compared to the original situation (Table 29.4). See also Figure 29.6, which shows the daily load curve after the changes have been made. The percentage of use on shift is now higher. Note that D_{max} has been improved significantly (reduced by 13%); the *HUOD* has improved slightly (about 3% lower now); and the *LF* is slightly lower. Furthermore, improvements are undoubtedly still possible in this facility; they should be directed first at reducing nonessential uses, thereby reducing *HUOD*.

So far, the discussion has dealt entirely with power and has neglected the reactive component of the load. In the most general case, the apparent power in kV A that must be supplied to the load is the sum of the active power in kW and the *reactive power* in kVAR (the reader who is unfamiliar with these terms should refer to a basic electrical engineering text):

$$|S| = \sqrt{P^2 + Q^2} \tag{29.10}$$

where

S is the apparent power in kV A
P is the active power in kW
Q is the reactive power in kVAR

In this notation, the apparent power is a vector of magnitude S and angle θ where θ is commonly referred to as the phase angle and given as

$$\theta = \tan^{-1}(\text{kVAR/kW}) \tag{29.11}$$

Another useful parameter is the *power factor*, given by

$$pf = \cos\theta \tag{29.12}$$

TABLE 29.7

Sample Calculations for XYZ Plant

1. $D_{max} = \frac{E_{max}}{p} = \frac{1850\text{ kW h}}{1\text{ h}} = 1850\text{ kW}$

2. $DV = \frac{D_{m1} + D_{m2} + D_{m3} + \cdots}{D_{max}} = \frac{2459}{1850} = 1.33$

3. $DF = \frac{D_{max}}{CL} = \frac{1850}{2792} = 0.66$

4. $HOUD_{(daily)} = \frac{E}{D_{max}} = \frac{28{,}800\text{ kW h/day}}{1{,}850\text{ kW}} = 15.6\text{ h/day}$

5. $LF_{(daily)} = \frac{HUOD}{p} = \frac{15.6}{24.0} = 0.65$

The load parameters after the changes were made can be found:

$$D_{max} = \frac{1850 - 235}{1\text{ h}} = 1615\text{ kW}$$

$$HOUD = \frac{24{,}400\text{ kW h/day}}{1{,}615\text{ kW}} = 15.1\text{ h/day}$$

$$LF = \frac{15.1}{24} = 0.63$$

	Savings	
Calculated Savings	**kW**	**kW h/year**
More efficient parking lot lighting	16	67,000
Reduce office lighting	111	495,000
Office lighting controls to reduce off-shift use	—	425,000
Air-conditioning controls and smaller fan motor	71	425,000
Compressed air system repairs and reduction of heat losses from plating tanks	—	200,000
Shift heat treat oven off-peak	37	—
Totals	235	1,612,000
The revised electricity use was found to be	kW h	%
Preshift	4,400	18
On shift	14,400	59
Postshift	5,600	23
Totals	24,400	100

The power factor is also given by

$$pf = \frac{|P|}{|S|} \tag{29.13}$$

The power factor is always less than or equal to unity. A high value is desirable because it implies a small reactive component to the load (Figure 29.7). A low value means the reactive component is large.

The importance of the power factor is related to the reactive component of the load. Even though the reactive component does not dissipate power (it is stored in magnetic or electric fields), the switch gear and distribution system must be sized to handle the current required by the apparent power, or the vector sum of the active and reactive components. This results in a greater capital and operating expense. The operating expense is increased due to the standby losses that occur in supplying the reactive component of the load.

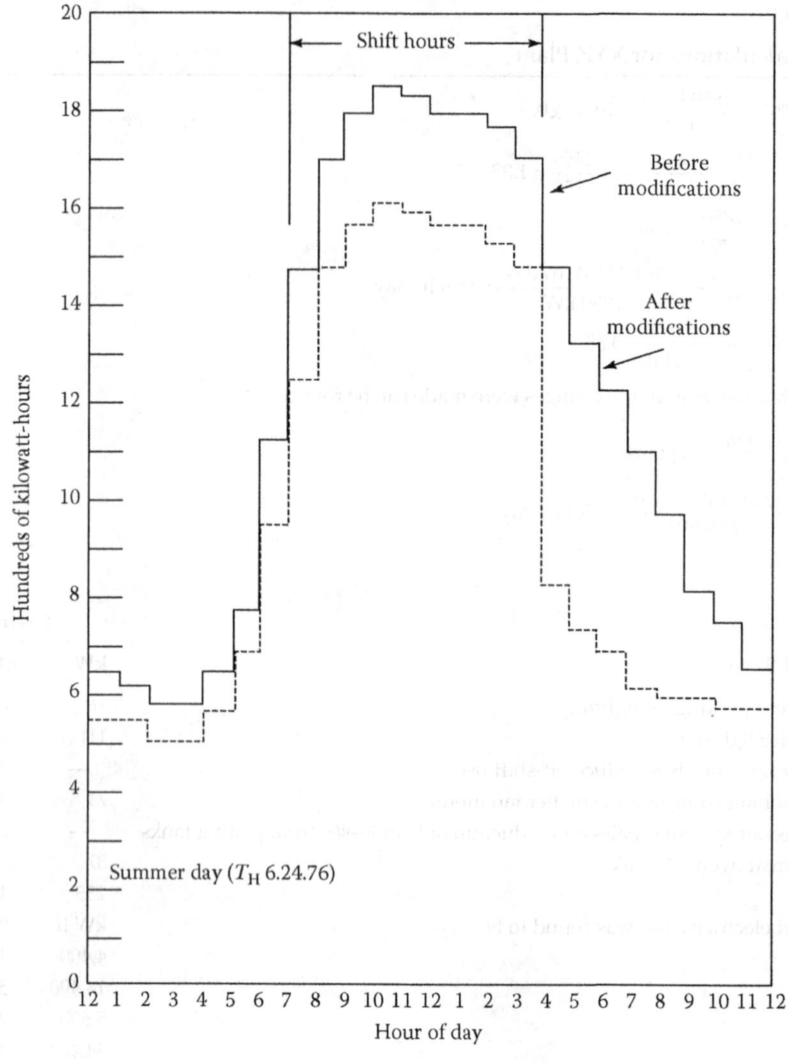

FIGURE 29.7
Daily load curve for XYZ company after modifications.

The power factor can be improved by adding capacitors to the load to compensate for part of the inductive reactance. The benefit of this approach depends on the economics of each specific case and generally requires a careful review or analysis.

These points can be clarified with an example. Consider the distribution system shown in Figure 29.8. Four loads are supplied by a 600 A bus. Load A is a distant load that has a large reactive component and a low power factor (pf = 0.6). To supply the active power requirement of 75 kW, an apparent power of 125 kV A must be provided, and a current of 150 A is required.

The size of the wire to supply the load is dictated by the current to be carried and voltage drop considerations. In this case, #3/0 wire that weighs 508 lb/1000 ft and has a resistance

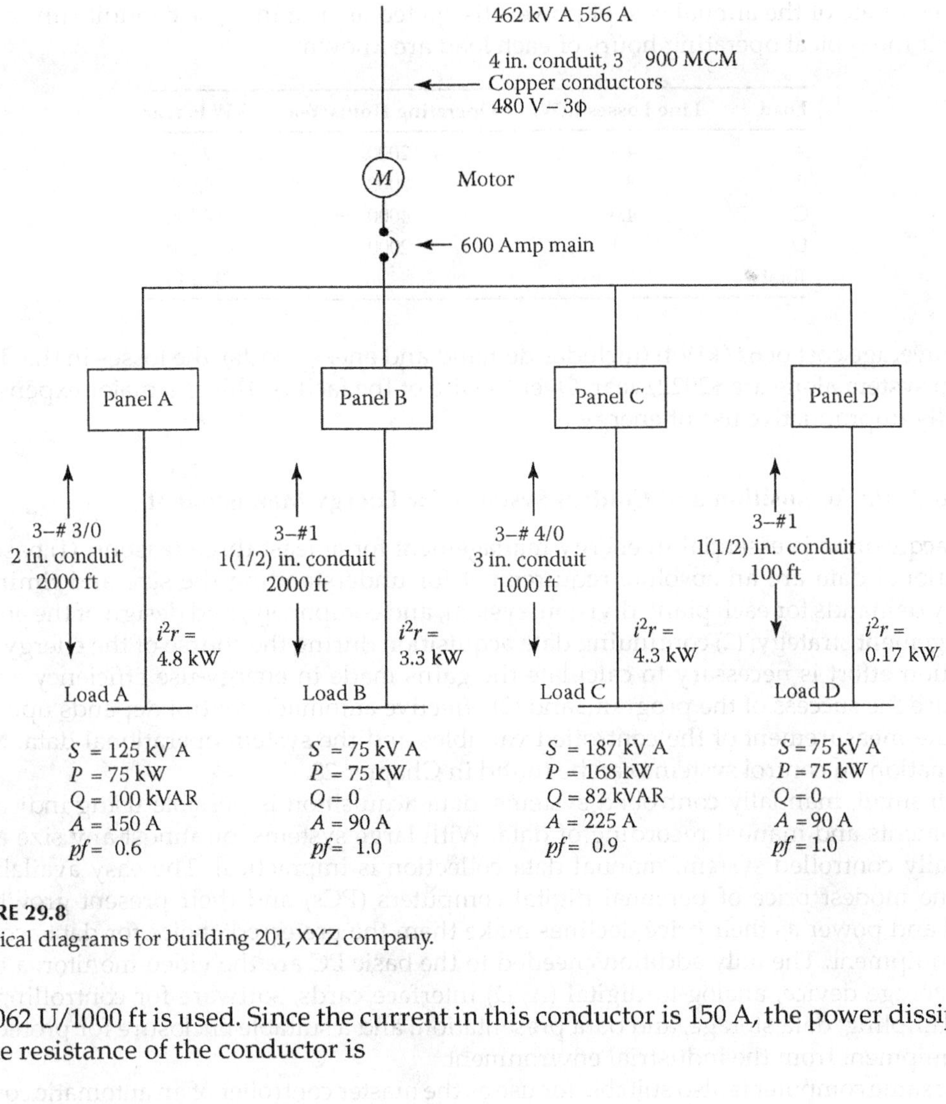

FIGURE 29.8
Electrical diagrams for building 201, XYZ company.

of 0.062 Ω/1000 ft is used. Since the current in this conductor is 150 A, the power dissipated in the resistance of the conductor is

$$P = \sqrt{3} i^2 r = (1.732)(150^2)(0.124)\ \text{W}$$

$$P = 4.8\ \text{kW}$$

Similar calculations can be made for load B, which uses #1 wire, at 253 lb/1000 ft and 0.12 Ω/1000 ft.

Now the effect of the power factor is visible. Although the active power is the same for both load A and load B, load A requires 150 A vs. 90 A for load B. The i^2r standby losses are also higher for load A as opposed to load B. The installation cost to service load B is roughly half that of load A, due to the long conduit run of load B compared to load D. For large loads that are served over long distances and operate continuously, consideration should be given to using larger wire sizes to reduce standby losses.

An estimate of the annual cost of power dissipated as heat in these conduit runs can be made if the typical operating hours of each load are known:

Load	Line Losses (kW)	Operating Hours/Year	kW h/Year
A	4.8	2000	9,600
B	3.3	2000	6,600
C	4.3	4000	17,200
D	0.17	2000	340
Total			33,740

At an average cost of 6¢/kW h (includes demand and energy costs), the losses in the distribution system alone are $2022/year. Over the life of the facility, this is a major expense for a totally unproductive use of energy.

29.4.8 Data Acquisition and Control Systems for Energy Management

Data acquisition is essential in energy management for at least three reasons: (1) baseline operational data are an absolute requirement for understanding the size and timing of energy demands for each plant, division, system, and component, and design of the energy management strategy; (2) continuing data acquisition during the course of the energy conservation effort is necessary to calculate the gains made in energy-use efficiency and to measure the success of the program; and (3) effective automatic control depends upon the accurate measurement of the controlled variables and the system operational data. More information on control systems can be found in Chapter 23.

With small, manually controlled systems, data acquisition is possible using indicating instruments and manual recording of data. With large systems, or almost any size automatically controlled system, manual data collection is impractical. The easy availability and the modest price of personal digital computers (PCs) and their present growth in speed and power as their price declines make them the preferred choice for data acquisition equipment. The only additions needed to the basic PC are the video monitor; a mass data storage device, analog-to-digital (A/D) interface cards; software for controlling the data sampling, data storage, and data presentation; and a suitable enclosure for protecting the equipment from the industrial environment.

That same computer is also suitable for use as the master controller of an automatic control system. Both functions can be carried on simultaneously, with the same equipment, with a few additions necessary for the control part of the system. These additional electronic components include multiplexers to increase the capacity of the A/D cards; direct memory access (an addition to the A/D boards), which speeds operation by allowing the data transfer to bypass the CPU and programmable logic controllers (PLCs); and digital input/output cards (I/Os), both used for controlling the equipment controllers. The most critical parts of either the data acquisition or the control system are the software. The hardware can be ordered off the shelf, but the software must be either written from scratch or purchased and modified for each particular system in order to achieve the reliable, high-performance operation that is desired. Although many proprietary program languages exist for the PCs and PLCs used for control purposes, BASIC is the most widely used. A schematic diagram of a typical data acquisition and control system is shown in Figure 29.9.

A major application has been in the control of mechanical and electrical systems in commercial and industrial buildings. These have been used to control lighting, electric demand,

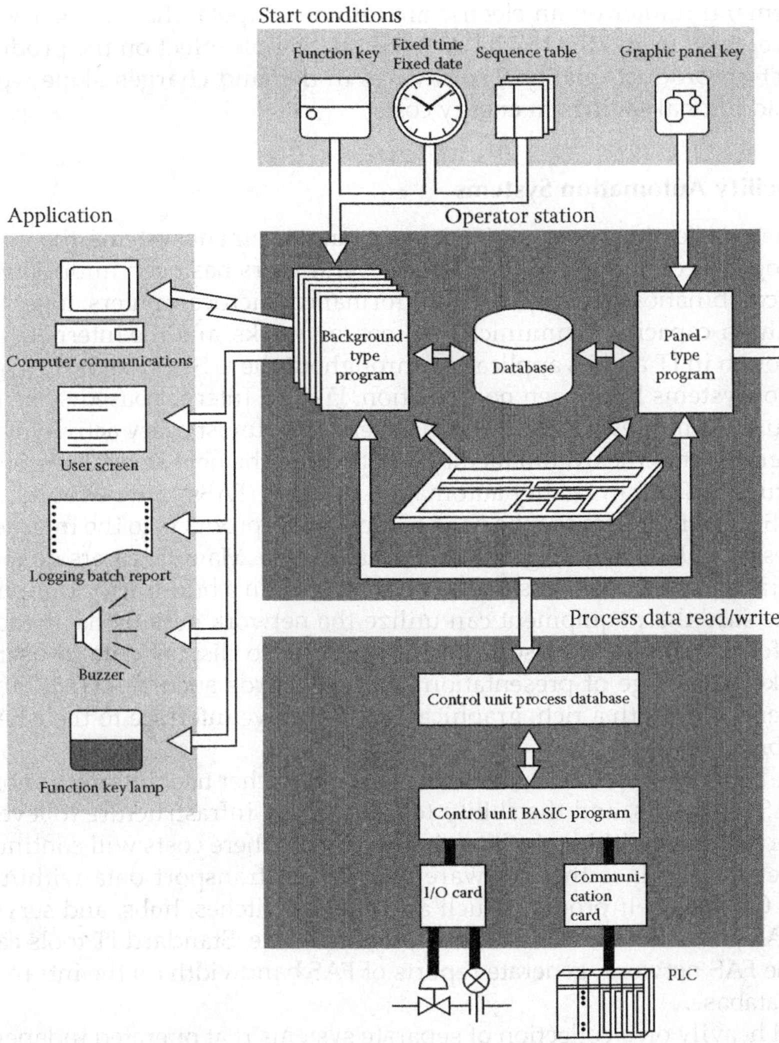

FIGURE 29.9
Typical data acquisition and control system.

ventilating fans, thermostat setbacks, air-conditioning systems, and the like. These same computer systems can also be used in industrial buildings with or without modification for process control. Several manufactures of the computer systems will not only engineer and install the system but will maintain and operate it from a remote location. Such operations are ordinarily regional. For large systems in large plants, one may be able to have the same service provided in-plant. However, there are also many small stand-alone analog control systems that can be used advantageously for the control of simple processes. Examples of these are a temperature controller using a thermocouple output to control the temperature of a liquid storage tank; a level controller, which keeps the liquid level in a tank constant by controlling a solenoid valve; and an oxygen trim system for a small boiler, which translates the measurement of the oxygen concentration in the exhaust stream into the jack shaft position, which regulates the combustion air flow to the burner.

After installing a demand limiter on an electric arc foundry cupola, the manager was able to reduce the power level from 7100 to 4900 kW with negligible effect on the production time and no effect on product quality. The savings in demand charges alone were $4400/month with an additional savings in energy costs.

29.4.9 Web-Based Facility Automation Systems

Of all recent developments affecting computerized energy management systems, the most powerful new technology to come into use in the last several years has been information technology, or IT. The combination of cheap, high-performance microcomputers, together with the emergence of high-capacity communication lines, networks, and the Internet, has produced explosive growth in IT and its application throughout the U.S. economy. Energy information and control systems have been no exception. IT- and Internet-based systems are the wave of the future. Almost every piece of equipment and almost every activity will be connected and integrated into the overall facility operation in the next several years.[22]

In particular, the future of DDC in facility automation systems (FASs) can be found on the web. Almost all FAS manufacturers see the need to move their products to the Internet. Tremendous economies of scale and synergies can be found there. Manufacturers no longer have to create the transport mechanisms for data to flow within a building or campus. They just need to make sure their equipment can utilize the network data paths already installed or designed for a facility. Likewise, with the software to display data to users, manufacturers that take advantage of presentation layer standards such as HTML and JAVA can provide the end user with a rich, graphical, and intuitive interface to their FAS using a standard web browser.

Owners will reap the benefits of Internet standards through a richer user interface, more competition among FAS providers, and the ability to use their IT infrastructure to leverage the cost of transporting data within a facility. Another area where costs will continue to fall in using Internet standards is the hardware required to transport data within a building or a campus. Off-the-shelf products such as routers, switches, hubs, and server computers make the FAS just another node of the IT infrastructure. Standard IT tools can be used to diagnose the FAS network, generate reports of FAS bandwidth on the intranet, and back up the FAS database.

The FAS of old relied heavily on a collection of separate systems that operated independently, and often with proprietary communication protocols that made expansion, modification, updating, and integration with other building or plant information and control systems very cumbersome, if not impossible. Today, the FAS is not only expected to handle all of the energy- and equipment-related tasks, but also to provide operating information and control interfaces to other facility systems, including the total facility or enterprise management system.

Measuring, monitoring, and maximizing energy savings is a fundamental task of all FAS and is the primary justification for many FAS installations. Improving facility operations in all areas, through enterprise information and control functions, is fast becoming an equally important function of the overall FAS or facility management system. The web provides the means to share information easier, quicker, and cheaper than ever before. There is no doubt that the web is having a huge impact on the FAS industry. The FAS of tomorrow will rely heavily on the web, TCP/IP, high-speed data networks, and enterprise-level connectivity. If it has not already been done, it is a good time for energy managers to get to know their IT counterparts at their facility, along with those in the accounting and maintenance departments. The future FAS will be here sooner than you think.[23]

29.4.9.1 Energy Management Strategies for Industry

Energy management strategies for industry can be grouped into three categories:

1. Operational and maintenance strategies
2. Retrofit or modification strategies
3. New design strategies

The order in which these are listed corresponds approximately to increased capital investment and increased implementation times. Immediate savings at little or no capital cost can generally be achieved by improved operations and better maintenance. Once these "easy" savings have been realized, additional improvements in efficiency will require capital investments.

29.4.9.2 Electric Drives and Electrically Driven Machinery

About 68% of industrial electricity use is for electrical motor–driven equipment. Integral horsepower (<0.75 kW) motors are more numerous. Major industrial motor loads are, in order of importance, pumps, compressors, blowers, fans, and miscellaneous integral motor applications including conveyors, DC drives, machine tools, and fractional horsepower applications.

Numerous examples of pumping in industry can be observed. These include process pumping in chemical plants, fluid movement in oil refineries, and cooling water circulation. An example of compressors is in the production of nitrogen and oxygen, two common chemicals. Large amounts of electricity are used to drive the compressors, which supply air to the process.

The typical industrial motor is a polyphase motor rated at 11.2 kW (15 hp) and having a life of about 40,000 h. The efficiencies of electric motors have increased recently as a result of higher energy prices, conservation efforts, and new government standards (the Energy Policy Act). High-efficiency motors cost roughly 20%–30% more than standard motors, but this expense is quickly repaid for motors that see continuous use.

Most efficient use of motors requires that attention be given to the following:

Optimum power: Motors operate most efficiently at rated voltage. Three-phase power supplies should be balanced; an unbalance of 3% can increase losses 25%.[1]

Good motor maintenance: Provide adequate cooling, keep heat transfer surfaces and vents clean, and provide adequate lubrication. Improved lubrication alone can increase efficiency a few percentage points.

Equipment scheduling: Turn equipment off when not in use; schedule large motor operation to minimize demand peaks.

Size equipment properly: Match the motor to the load and to the duty cycle. Motors operate most efficiently at rated load.

Evaluate continuous vs. batch processes: Sometimes a smaller motor operating continuously will be more economical.

Power factor: Correct if economics dictate savings. Motors have the best power factor at rated load.

Retrofit or new designs permit use of more efficient motors. For motors up to about 10–15 kW (15–20 hp), there are variations in efficiency. Select the most efficient motor for the job. Check to verify that the additional cost (if any) will be repaid by the savings that will accrue over the life of the installation.

In addition to reviewing the electric drive system, consider the power train and the load. Friction results in energy dissipation in the form of heat. Bearings, gears, and belt drives all have certain losses, as do clutches. Proper operation and maintenance can reduce energy wastage in these systems and improve overall efficiency.

Material shaping and forming, such as is accomplished with machine tools, requires that electrical energy be transformed into various forms of mechanical energy. The energy expenditure related to the material and to the depth and speed of the cut. By experimenting with a specific process, it is possible to establish cutting rates that are optimum for the levels of production required and are most efficient in terms of energy use. Motors are not the only part of the electric drive system that sustains losses. Other losses occur in the electric power systems that supply the motor. Electric power systems include substations, transformers, switching gear, distribution systems, feeders, power and lighting panels, and related equipment. Possibilities for energy management include the following:

Use highest voltages that are practical: For a given application, doubling the voltage cuts the required current in half and reduces the i^2r losses by a factor of 4.

Eliminate unnecessary transformers: They waste energy. Proper selection of equipment and facility voltages can reduce the number of transformers required and cut transformer losses. Remember, the customer pays for losses when the transformers are on his side of the meter. For example, it is generally better to order equipment with motors of the correct voltage, even if this costs more, than to install special transformers.

Energy losses are an inherent part of electric power distribution systems: This is primarily due to i^2r losses and transformers. The end-use conversion systems for electrical energy used in the process also contribute to energy waste. Proper design and operation of an electrical system can minimize energy losses and contribute to the reduction of electricity bills. Where long feeder runs are operated at near-maximum capacities, check to see if larger wire sizes would permit savings and be economically justifiable.

The overall power factor of electrical systems should be checked for low power factor: This could increase energy losses and the cost of electrical service, in addition to excessive voltage drops and increased penalty charges by the utility. Electrical systems studies should be made, and consideration should be given to power factor correction capacitors. In certain applications, as much as 10%–15% savings can be achieved in a poorly operating plant.

Check load factors: This is another parameter that measures the plant's ability to use electrical power efficiently. It is defined as the ratio of the actual kW h used to the maximum demand in kW times the total hours in the time period. A reduction in demand to bring this ratio closer to unity without decreasing plant output means more economical operation. For example, if the maximum demand for a given month (200 h) is 30,000 kWe and the actual kW h is 3.6×10^6 kW h, the load factor is 60%. Proper management of operations during high-demand periods, which may extend only 15–20 min, can reduce the demand during that time without curtailing production. For example, if the 30,000 kWe could be reduced to 20,000 kWe, this would increase the load factor to about 90%. Such a reduction could amount to a $20,000–$50,000 reduction in the electricity bill.

Reduce peak loads wherever possible: Many nonessential loads can be shed during the demand peak without interrupting production. These loads would include such items as air compressors, heaters, coolers, and air conditioners. Manual monitoring and control is possible but is often impractical because of the short periods of time that are normally involved and the lack of centralized control systems. Automatic power demand control systems are available.

Provide improved monitoring or metering capability, submeters, or demand recorders: While it is true that meters alone will not save energy, plant managers need feedback to determine if their energy management programs are taking effect. Often, the installation of meters on individual processes or buildings leads to immediate savings of 5%–10% by virtue of the ability to see how much energy is being used and to test the effectiveness of corrective measures.

29.4.9.3 Fans, Blowers, and Pumps

Simple control changes are the first thing to consider with these types of equipment. Switches, time clocks, or other devices can ensure that they do not operate except when needed by the process. Heat removal or process mass flow requirements will determine the size of fans and pumps. Often, there is excess capacity, either as a result of design conservatism or because of process changes subsequent to the installation of equipment. The required capacity should be checked, since excess capacity leads to unnecessary demand charges and decreased efficiency.

For fans, the volume rate of air flow Q varies in proportion to the speed of the impeller:

$$Q = c_f N \text{ (m}^3\text{/s)} \tag{29.14}$$

where

Q is the air flow, m^3/s
c_f is a constant, m^3/r
N is the fan speed, r/s

The pressure developed by the fan varies as the square of the impeller speed. The important rule, however, is that the power needed to drive the fan varies as the cube of the speed:

$$P = P_c N^3 \text{ (W)} \tag{29.15}$$

where

P is the input power, W
P_c is a constant, W s^3/r^3

The cubic law of pumping power indicates that if the air flow is to be doubled, eight (2^3) times as much power must be supplied. Conversely, if the air flow is to be cut in half, only one-eighth ($1/2^3$) as much power must be supplied. Air flow (and hence power) can be reduced by changing pulleys or installing smaller motors.

Pumps follow laws similar to fans, the key being the cubic relationship of power to the volume pumped through a given system. Small decreases in flow rate, such as might be obtained with a smaller pump or gotten by trimming the impeller, can save significant amounts of energy.

VSDs are another technique for reducing process energy use. VSDs permit fans, blowers, and pumps to vary speed depending on process requirements. This can lead to significant savings on noncontinuous processes. Recent improvements in solid-state electronics have caused the price of VSDs to drop substantially. This is another technology that is supported by utility rebates in many areas.

29.4.9.4 Air Compressors

Compressed air is a major energy use in many manufacturing operations. Electricity used to compress air is converted into heat and potential energy in the compressed air stream. Efficient operation of compressed air systems therefore requires the recovery of excess heat where possible, as well as the maximum recovery of the stored potential energy.

Efficient operation is achieved in these ways:

Select the appropriate type and size of equipment for the duty cycle required: Process requirements vary, depending on flow rates, pressure, and demand of the system. Energy savings can be achieved by selecting the most appropriate equipment for the job. The rotary compressor is more popular for industrial operations in the range of 20–200 kW, even though it is somewhat less efficient than the reciprocal compressor. This has been due to lower initial cost and reduced maintenance. When operated at partial load, reciprocating units can be as much as 25% more efficient than rotary units. However, newer rotary units incorporate a valve that alters displacement under partial load conditions and improves efficiency. Selection of an air-cooled vs. a water-cooled unit would be influenced by whether water or air was the preferred medium for heat recovery.

Proper operation of compressed air systems can also lead to improved energy utilization: Obviously, air leaks in lines and valves should be eliminated. The pressure of the compressed air should be reduced to a minimum. The percentage saving in power required to drive the compressor at a reduced pressure can be estimated from the fan laws described previously. For example, suppose the pressure were reduced to one half the initial value. Since pressure varies as the square of the speed, this implies the speed would be 70.7% of the initial value. Since power varies as the cube of the speed, the power would now be $0.707^3 = 35\%$ of the initial value. Of course, this is the theoretical limit; actual compressors would not do as well, and the reduction would depend on the type of compressor. Measurements indicate that actual savings would be about half the theoretical limit; reducing pressure 50% would reduce brake horsepower about 30%. To illustrate this point further, for a compressor operating at 6.89×10^5 N/m^2 (100 psi) and a reduction of the discharge pressure to 6.20×10^5 N/m^2 (90 psi), a 5% decrease in brake horsepower would result. For a 373 kW (500 hp) motor operating for 1 year, the 150,000 kW h savings per year would result in about $9000/year in electric power costs.

The intake line for the air compressor should be at the lowest temperature available: This normally means outside air. The reduced temperature of air intake results in a smaller volume of air to be compressed. The percentage horsepower saving relative to a 21°C (70°F) intake air temperature is about 2% for each 10°F drop in temperature. Conversely, input power increases by about 2% for each 10°F increase in intake air temperature.

Leakage is the greatest efficiency offender in compressed air systems: The amount of leakage should be determined and measures should be taken to reduce it. If air leakage in a plant is more than 10% of the plant demand, a poor condition exists. The amount of leakage can be determined by a simple test during off-production hours (when air-using equipment is shut down) by noting the time that the compressor operates under load compared with the total cycle. This indicates the percentage of the compressor's capacity that is used to supply the plant air leakage. Thus, if the load cycle compared with the total cycle were 60 s compared with 180 s, the efficiency would be 33%, or 33% of the compressor capacity is the amount of air leaking in m^3/min (ft^3/min).

Recover heat where feasible: There are sometimes situations where water-cooled or air-cooled compressors are a convenient source of heat for hot water, space heating, or process applications. As a rough rule of thumb, about 300 J/m^3 min of air compressed (~10 Btu/ft^3 min) can be recovered from an air-cooled rotary compressor.

Substitute electric motors for air motor (pneumatic) drives: Electric motors are far more efficient. Typical vaned air motors range in size from 0.15 to 6.0 kW (0.2–8 hp), cost $300–$1500, and produce 1.4–27 N m (1–20 ft lb) of torque at 620 kN/m^2 (90 psi) air pressure. These are used in manufacturing operations where electric motors would be hazardous, or where light weight and high power are essential. Inefficiency results from air system leaks and the need (compared to electric motors) to generate compressed air as an intermediate step in converting electric to mechanical energy.

Review air usage in paint spray booths: In paint spray booths and exhaust hoods, air is circulated through the hoods to control dangerous vapors. Makeup air is constantly required for dilution purposes. This represents a point of energy rejection through the exhaust air.

Examination should be made of the volumes of air required in an attempt in reduce flow and unnecessary operation. Possible mechanisms for heat recovery from the exhaust gases should be explored using recovery systems.

29.4.9.5 Electrochemical Operations

Electrochemical processes are an industrial use of electricity, particularly in the primary metals industry, where it is used in the extraction process for several important metals. Energy management opportunities include the following:

Improve design and materials for electrodes: Evaluate loss mechanisms for the purpose of improving efficiency.

Examine electrolysis and plating operations for savings: Review rectifier performance, heat loss from tanks, and the condition of conductors and connections.

Welding is another electrochemical process. Alternating current welders are generally preferable when they can be used, since they have a better power factor, better demand characteristics, and more economical operation.

Welding operations can also be made more efficient by the use of automated systems, which require 50% less energy than manual welding. Manual welders deposit a bead only 15%–30% of the time the machine is running. Automated processes, however, reduce the no-load time to 40% or less. Different welding processes should be compared in order to determine the most efficient process. Electroslag welding is suited only for metals over 1 cm (0.5 in.) thick but is more efficient than other processes.

Two other significant applications of electrolysis of concern to industry are batteries and corrosion. Batteries are used for standby power, transportation, and other applications. Proper battery maintenance and improved battery design contribute to efficient energy use.

Corrosion is responsible for a large loss of energy-intensive metals every year and thus indirectly contributes to energy wastage. Corrosion can be prevented and important economies realized, by use of protective films, cathodic protection, and electroplating or anodizing.

29.4.9.6 Electric Process Heat and Steam Systems

In as much as approximately 40% of the energy utilized in industry goes toward the production of process steam, it presents a large potential for energy misuse and fuel waste from improper maintenance and operation. Even though electrically generated steam and hot water is a small percentage of total industrial steam and hot water, the electrical fraction is likely to increase as other fuels increase in price. This makes increased efficiency even more important. Examples are as follows:

Steam leaks from lines and faulty valves result in considerable losses: These losses depend on the size of the opening and the pressure of the steam, but can be very costly. A hole 0.1 ft in diameter with steam at 200 psig can bleed \$1000–\$2000 worth of steam (500 GJ) in a year.

Steam traps are major contributors to energy losses when not functioning properly: A large process industry might have thousands of steam traps, which could result in large costs if they are not operating correctly. Steam traps are intended to remove condensate and noncondensable gases while trapping or preventing the loss of steam. If they stick open, orifices as large as 6 mm (0.25 in.) can allow steam to escape. Such a trap would allow 1894 GJ/year (2000 MBtu/year) of heat to be rejected to the atmosphere on a 6.89×10^5 N/m (100 psi) pressure steam line. Many steam traps are improperly sized, contributing to an inefficient operation. Routine inspection, testing, and a correction program for steam valves and traps are essential in any energy program and can contribute to cost savings.

Poor practice and design of steam distribution systems can be the source of heat waste up to 10% or more: It is not uncommon to find an efficient boiler or process plant joined to an inadequate steam distribution system. Modernization of plants results from modified steam requirements. The old distribution systems are still intact, however, and can be the source of major heat losses. Large steam lines intended to supply units no longer present in the plant are sometimes used for minor needs, such as space heating and cleaning operations, that would be better accomplished with other heat sources.

Steam distribution systems operating on an intermittent basis require a start-up warming time to bring the distribution system into proper operation. This can extend up to 2 or 3 h, which puts a demand on fuel needs. Not allowing for proper ventilating of air can also extend the start-up time. In addition, condensate return can be facilitated if it is allowed to drain by gravity into a tank or receiver and is then pumped into the boiler feed tank.

Proper management of condensate return: Proper management can lead to great savings. Lost feedwater must be made up and heated. For example, every 0.45 kg (1 lb) of steam that must be generated from 15°C feedwater instead of 70°C feedwater requires an additional 1.056×10^5 J (100 Btu) more than 1.12 MJ (1063 Btu) required or a 10% increase in fuel. A rule of thumb is that a 1% fuel saving results for every 5°C increase in feedwater temperature. Maximizing condensate recovery is an important fuel laying procedure.

Poorly insulated lines and valves due either to poor initial design or a deteriorated condition: Heat losses from a poorly insulated pipe can be costly. A poorly insulated line carrying steam at 400 psig can lose ~1000 GJ/year (10^9 Btu/year) or more per 30 m (100 ft) of pipe. At steam costs of $2.00/GJ, this translates to a $2000 expense/year.

Improper operation and maintenance of tracing systems: Steam tracing is used to protect piping and equipment from cold weather freezing. The proper operation and maintenance of tracing systems will not only ensure the protection of traced piping but also saves fuel. Occasionally, these systems are operating when not required. Steam is often used in tracing systems, and many of the deficiencies mentioned earlier apply (e.g., poorly operating valves, insulation, leaks).

Reduce losses in process hot water systems: Electrically heated hot water systems are used in many industrial processes for cleaning, pickling, coating, or etching components. Hot or cold water systems can dissipate energy. Leaks and poor insulation should be repaired.

29.4.9.7 Electrical Process Heat

Industrial process heat applications can be divided into four categories: direct-fired, indirect-fired, fuel, or electric. Here we shall consider electric direct-fired installations (ovens, furnaces) and indirect-fired (electric water heaters and boilers) applications. Electrical installations use metal sheath resistance heaters, resistance ovens or furnaces, electric salt bath furnaces, infrared heaters, induction and high-frequency resistance heaters, dielectric heaters, and direct arc furnaces. From the housekeeping and maintenance point of view, typical opportunities would include the following:

Repair or improve insulation: Operational and standby losses can be considerable, especially in larger units. Remember that insulation may degrade with time or may have been optimized to different economic criteria.

Provide finer controls: Excessive temperatures in process equipment waste energy. Run tests to determine the minimum temperatures that are acceptable, then test instrumentation to verify that it can provide accurate process control and regulation.

Practice heat recovery: This is an important method, applicable to many industrial processes as well as HVAC systems and so forth. It is described in more detail in the next section.

29.4.9.8 Heat Recovery

Exhaust gases from electric ovens and furnaces provide excellent opportunities for heat recovery. Depending on the exhaust gas temperature, exhaust heat can be used to raise steam or to preheat air or feedstocks. Another potential source of waste-heat recovery is the exhaust air that must be rejected from industrial operations in order to maintain health and ventilation safety standards. If the reject air has been subjected to heating and cooling processes, it represents an energy loss inasmuch as the makeup air must be modified to meet the interior conditions. One way to reduce this waste is through the use of heat wheels or similar heat exchange systems.

Energy in the form of heat is available at a variety of sources in industrial operations, many of which are not normally derived from primary heat sources. Such sources include electric motors, crushing and grinding operations, air compressors, and drying processes. These units require cooling in order to maintain proper operation. The heat from these systems can be collected and transferred to some appropriate use such as space heating or water heating.

The heat pipe is gaining wider acceptance for specialized and demanding heat transfer applications. The transfer of energy between incoming and outgoing air can be accomplished by banks of these devices. A refrigerant and a capillary wick are permanently sealed inside a metal tube, setting up a liquid-to-vapor circulation path. Thermal energy applied to either end of the pipe causes the refrigerant to vaporize. The refrigerant vapor then travels to the other end of the pipe, where thermal energy is removed. This causes the vapor to condense into liquid again, and the condensed liquid then flows back to the opposite end through the capillary wick.

Industrial operations involving fluid flow systems that transport heat such as in chemical and refinery operations offer many opportunities for heat recovery. With proper design and sequencing of heat exchangers, the incoming product can be heated with various process steams. For example, proper heat exchanger sequence in preheating the feedstock to a distillation column can reduce the energy utilized in the process.

Many process and air-conditioning systems reject heat to the atmosphere by means of wet cooling towers. Poor operation can contribute to increased power requirements.

Water flow and airflow should be examined to see that they are not excessive: The cooling tower outlet temperature is fixed by atmospheric conditions if operating at design capacity. Increasing the water flow rate or the air flow will not lower the outlet temperature.

The possibility of utilizing heat that is rejected to the cooling tower for other purposes should be investigated: This includes preheating feedwater, heating hot water systems, space heating, and other low-temperature applications. If there is a source of building exhaust air with a lower wet bulb temperature, it may be efficient to supply this to a cooling tower.

29.4.9.9 Power Recovery

Power recovery concepts are an extension of the heat recovery concept described earlier. Many industrial processes have pressurized liquid and gaseous streams at 150°C–375°C (300°F–700°F) that present excellent opportunities for power recovery. In many cases, high-pressure process stream energy is lost by throttling across a control valve.

The extraction of work from high-pressure liquid streams can be accomplished by means of hydraulic turbines (essentially diffuser-type or volute-type pumps running backward). These pumps can be either single or multistage. Power recovery ranges from 170 to 1340 kW (230–1800 hp). The lower limit of power recovery approaches the minimum economically justified for capital expenditures at present power costs.

29.4.9.10 Heating, Ventilating, and Air-Conditioning Operation

The environmental needs in an industrial operation can be quite different from those in a residential or commercial structure. In some cases, strict environmental standards must be met for a specific function or process. More often, the environmental requirements for the process itself are not severe; however, conditioning of the space is necessary for the comfort of operating personnel, and thus large volumes of air must be processed. Quite often, opportunities exist in the industrial operation where surplus energy can be utilized in environmental conditioning. A few suggestions follow:

Review HVAC controls: Building heating and cooling controls should be examined and preset.

Ventilation, air, and building exhaust requirements should be examined: A reduction of air flow will result in a savings of electrical energy delivered to motor drives and additionally reduce the energy requirements for space heating and cooling. Because pumping power varies as the cube of the air flow rate, substantial savings can be achieved by reducing air flows where possible.

Do not condition spaces needlessly: Review air-conditioning and heating operations, seal off sections of plant operations that do not require environmental conditioning, and use air-conditioning equipment only when needed. During nonworking hours, the environmental control equipment should be shut down or reduced. Automatic timers can be effective.

Provide proper equipment maintenance: Ensure that all equipment is operating efficiently. (Filters, fan belts, and bearings should be in good condition.)

Use only equipment capacity needed: When multiple units are available, examine the operating efficiency of each unit and put operations in sequence in order to maximize overall efficiency.

Recirculate conditioned (heated or cooled) air where feasible: If this cannot be done, perhaps exhaust air can be used as supply air to certain processes (e.g., a paint spray booth) to reduce the volume of air that must be conditioned.

For additional energy management opportunities in HVAC systems, see Chapter 23.

29.4.9.11 Lighting

Industrial lighting needs range from low-level requirements for assembly and welding of large structures (such as shipyards) to the high levels needed for the manufacture of precision mechanical and electronic components (e.g., integrated circuits). There are four basic housekeeping checks that should be made:

Is a more efficient lighting application possible? Remove excessive or unnecessary lamps.

Is relamping possible? Install lower-wattage lamps during routine maintenance.

Will cleaning improve light output? Fixtures, lamps, and lenses should be cleansed periodically.

Can better controls be devised? Eliminate turning on more lamps than necessary. For modification, retrofit, or new design, consideration should be given to the spectrum of high-efficiency lamps and luminaries that are available. For example, high-pressure sodium lamps are finding increasing acceptance for industrial use, with savings of nearly a factor of five compared to incandescent lamps. See Chapter 22 for additional details.

29.4.9.12 New Electrotechnologies

Electricity has certain characteristics that make it uniquely suitable for industrial processes. These characteristics include electricity's suitability for timely and precise control, its ability to interact with materials at the molecular level, the ability to apply it selectively and specifically, and the ability to vary its frequency and wavelength so as to enhance or inhibit its interaction with materials. These aspects may be said to relate to the *quality* of electricity as an energy form. It is important to recognize that different forms of energy have different qualities in the sense of their ability to perform useful work. Thus, although the Btu content of two energy forms may be the same, their ability to transform materials may be quite different.

New electrotechnologies based on the properties of electricity are now finding their way into modern manufacturing. In many cases, the introduction of electricity reduces manufacturing costs, improves quality reduces pollution, or has other beneficial results. Some examples include the following:

Microwave heating	Ion nitriding
Induction heating	Infrared drying
Plasma processing	UV drying and curing
Magnetic forming	Advanced finishes
RF drying and heating	Electron beam heating

Microwave heating is a familiar technology that exhibits the unique characteristics of electricity described earlier. First, it is useful to review how conventional heating is preformed to dry paint, anneal a part, or remove water. A source of heat is required, along with a container (oven, furnace, pot, etc.) to which the heat is applied. Heat is transferred from the container to the workpiece by conduction, radiation, convection, or a combination of these. There are certain irreversible losses associated with heat transfer in this process. Moreover, since the container must be heated, more energy is expended than is really required. Microwave heating avoids these losses due to the unique characteristics of electricity.

Timely control: There is no loss associated with the warm-up or cooldown of ovens. The heat is applied directly when needed.

Molecular interaction: By interacting at the molecular level, heat is deposited directly in the material to be heated, without having to preheat an oven, saving the extra energy required for this purpose and avoiding the losses that result from heat leakage from the oven.

Selective application: By selectively applying heat only to the material to be heated, parasitic losses are avoided. In fact, the specificity of heat applied this way can improve quality by not heating other materials.

Selective wavelength and frequency: A microwave frequency is selected that permits the microwave energy to interact with the material to be heated, and not with other materials. Typically, the frequency is greater than 2000 MHz.

Microwave heating was selected for this discussion, but similar comments could be made about infrared, ultraviolet (UV), dielectric, induction, or electron beam heating. In each case, the frequency or other characteristics of the energy form are selected to provide the unique performance required.

UV curing (now used for adhesive and finishes) is another example. The parts to be joined or coated can be prepared and the excessive adhesive removed without fear of prehardening. Then, the UV energy is applied, causing the adhesive to harden.

Induction heating is another example. It is similar to microwave heating except that the energy is applied at a lower frequency. Induction heating operates on the principle of inducing electric currents to flow in materials, heating them by the power dissipated in the material. The method has several other advantages. In a conventional furnace, the workpiece has to be in the furnace for a sufficient time to reach temperature. Because of this, some of the material is oxidized and lost as scale. In a typical high-temperature gas furnace, this can be 2% of the throughput. Additional product is scrapped as a result of surface defects caused by uneven heating and cooling. This can amount to another 1% of throughput. Induction heating can reduce these losses by a factor of four.

The fact that electricity can be readily controlled and carries with it a high information content through digitization or frequency modulation also offers the potential for quantum improvements in efficiency. A slightly different example is the printing industry.

Today, the old linotype technology has been replaced by electronic processes. The lead melting pots that used to operate continuously in every newspaper plant have been removed, eliminating a major energy use and an environmental hazard. Books, magazines, and newspapers can be written composed and printed entirely by electronic means. Text is processed by computer techniques. Camera-ready art is prepared by computers directly or prepared photographically and then optically scanned to create digital images. The resulting electronic files can be used in web offset printing by an electronic photochemical process. The same information can be transmitted electronically, via satellite, to a receiving station at a remote location where a high-resolution fax machine reconstitutes the image. This method is being used to simultaneously and instantaneously distribute advertising copy to multiple newspapers, using a single original. Previously, to insert an advertisement in 25 newspapers, 25 sets of photographic originals would have to be prepared and delivered, by messenger, air express, or mail, to each newspaper.

Some of the other applications of the new electrotechnologies include RF drying of plywood veneers, textiles, and other materials; electric infrared drying for automobile paint and other finishes; electric resistance melting for high-purity metals and scrap recovery; and laser cutting of wood, cloth, and other materials.

29.4.9.13 General Industrial Processes

The variety of industrial processes is so great that detailed specific recommendations are outside the scope of this chapter. Useful sources of information are found in trade journals, vendor technical bulletins, and manufacturers' association journals. These suggestions are intended to be representative, but by no means do they cover all possibilities.

In machining operations, eliminate unnecessary operations and reduce scrap: This is so fundamental from a purely economic point of view that it will not be possible to find significant improvements in many situations. The point is that each additional operation and each increment of scrap also represent a needless use of energy. Machining itself is not particularly energy intensive. Even so, there are alternate technologies that can not only save energy but reduce material wastage as well. For example, powder metallurgy generates less scrap and is efficient if done in induction-type furnaces.

Use stretch forming: Forming operations are more efficient if stretch forming is used. In this process, sheet metal or extrusions are stretched 2%–3% prior to forming, which makes the material more ductile so that less energy is required to form the product.

Use alternate heat treating methods: Conventional heat treating methods such as carburizing are energy intensive. Alternate approaches are possible. For example, a hard surface can be produced by induction heating, which is a more efficient energy process. Plating, metallizing, flame spraying, or cladding can substitute for carburizing, although they do not duplicate the fatigue strengthening compressive skin of carburization or induction hardening.

Use alternative painting methods: Conventional techniques using solvent-based paints require drying and curing at elevated temperature. Powder coating is a substitute process in which no solvents are used. Powder particles are electrostatically charged and attracted to the part being painted so that only a small amount of paint leaving the spray gun misses the part and the overspray is recoverable. The parts can be cured rapidly in infrared ovens, which require less energy than standard hot air systems. Water-based paints and high-solids coatings are also being used and are less costly than solvent-based paints. They use essentially the same equipment as the conventional solvent paint spray systems so that the conversion can be made at minimum costs. New water-based emulsion paints contain only 5% organic solvent and require no afterburning. High-solids coatings are already in use commercially for shelves, household fixtures, furniture, and beverage cans, and require no afterburning. They can be as durable as conventional finishes and are cured by either conventional baking or UV exposure.

Substitute for energy-intensive processes such as hot forging: Hot forging may require a part to go through several heat treatments. Cold forging with easily wrought alloys may offer a replacement. Lowering the preheat temperatures may also be an opportunity for savings. Squeeze forging is a relatively new process in which molten metal is poured into the forging dye. The process is nearly scrap free, requires less press power, and promises to contribute to more efficient energy utilization.

Movement of materials through the plant creates opportunities for saving energy. Material transport energy can be reduced by the following methods:

Combining processes or relocate machinery to reduce transport energy: Sometimes merely relocating equipment can reduce the need to haul materials.

Turning off conveyors and other transport equipment when not needed: Look for opportunities where controls can be modified to permit shutting down of equipment not in use.

Using gravity feeds wherever possible: Avoid unnecessary lifting and lowering of products.

29.4.9.14 Demand Management

The cost of electrical energy for medium-to-large industrial and commercial customers generally consists of two components. The first component is the *energy charge*, which is based on the cost of fuel to the utility, the average cost of amortizing the utility generating plant, and on the operating and maintenance costs experienced by the utility.

Energy costs for industrial users in the United States are typically in the range of \$0.05–\$0.10/kW h.

The second component is the *demand charge,* which reflects the investment cost the utility must make to serve the customer. Besides the installed generating capacity needed, the utility also provides distribution lines, transformers, substations, and switches whose cost depends on the size of the load being served. This cost is recovered in a demand charge, which typically is \$2–\$10/kW month.

Demand charges typically account for 10%–50% of the bill, although wide variations are possible depending on the type of installation. Arc welders, for example, have relatively high demand charges, since the installed capacity is great (10–30 kW for a typical industrial machine) and the energy use is low.

From the utility's point of view, it is advantageous to have its installed generating capacity operating at full load as much of the time as possible. To follow load variations conveniently, the utility operates its largest and most economical generating units continuously to meet its base load and then brings smaller (and generally more expensive) generating units online to meet peak load needs.

Today, consideration is being given to time-of-day or *peak load pricing* as a means of assigning the cost of operating peak generating capacity to those loads that require it. See Chapter 18 for a discussion of utility pricing strategies. From the viewpoint of the utility, *load management* implies maintaining a high capacity factor and minimizing peak load demands. From the customer's viewpoint, *demand management* means minimizing electrical demands (both on- and off-peak) so as to minimize overall electricity costs.

Utilities are experimenting with several techniques for load management. Besides rate schedules that encourage the most effective use of power, some utilities have installed remotely operated switches that permit the utility to disconnect nonessential parts of the customer's load when demand is excessive. These switches are actuated by a radio signal, through the telephone lines, of over the power grid itself through a harmonic signal (ripple frequency) that is introduced into the grid.

Customers can control the demand of their loads by any of several methods:

- Manually switching off-loads ("load shedding")
- Use of timers and interlocks to prevent several large loads from operating simultaneously
- Use of controllers and computers to control loads and minimize peak demand by scheduling equipment operation
- Energy storage (e.g., producing hot or chilled water during off-peak hours and storing it for use on-peak)

Demand can be monitored manually (by reading a meter) or automatically using utility-installed equipment or customer-owned equipment installed in parallel with the utility meter. For automatic monitoring, the basic approach involves pulse counting.

The demand meter produces electronic pulses, the number of which is proportional to demand in kW. Demand is usually averaged over some interval (e.g., 15 min) for calculating cost. By monitoring the pulse rate electronically, a computer can project what the demand will be during the next demand measurement interval and can then follow a preestablished plan for shedding loads if the demand set point is likely to be exceeded.

Computer control can assist in the dispatching of power supply to the fluctuating demands of plant facilities. Large, electrically based facilities are capable of forcing large power demands during peak times that exceed the limits contracted with the utility or cause penalties in increased costs. Computer control can even out the load by shaving peaks and filling in the valleys, thus minimizing power costs. In times of emergency or fuel curtailment, operation of the plant can be programmed to provide optimum production and operating performance under prevailing conditions. Furthermore, computer monitoring and control provide accurate and continuous records of plant performance.

It should be stressed here that many of these same functions can be carried out by manual controls, time clocks, microprocessors, or other inexpensive devices. Selection of a computer system must be justified economically on the basis of the number of parameters to be controlled and the level of sophistication required. Many of the benefits described here can be obtained in some types of operations without the expense of a computer.

29.5 Thermal Energy Management in Industry

29.5.1 Importance of Fuel Use and Heat in Industry

The U.S. manufacturing sector depends heavily on fuels for the conversion of raw materials into usable products. Industry uses a wide range of fuels, including natural gas, petroleum, coal, and renewables. The petroleum forms used include distillate fuel oil, residual fuel oil, gasoline, LPG, and others. How efficiently energy is used, its cost, and its availability consequently have a substantial impact on the competitiveness and economic health of the U.S. manufacturers. More efficient use of fuels lowers production costs, conserves limited energy resources, and increases productivity. Efficient use of energy also has positive impacts on the environment—reductions in fuel use translate directly into decreased emissions of pollutants such as sulfur oxides, nitrogen oxides, particulates, and greenhouse gases (e.g., carbon dioxide).

From Figure 29.10, it can be seen that fuel use in manufacturing is just over 70% of the total energy used in manufacturing on an end-use basis. Figure 29.11 shows the percentage of each fuel used for boiler fuel and process heat combined. Of this total, 53% is used for boiler fuel, and 47% for direct process heating. Thus, there is a huge potential for energy management and energy efficiency improvement related to the use fuels and thermal energy in industry. Section 29.5 discusses the use of improved and new equipment and technology to accomplish some of these reductions in energy use and cost.

Energy efficiency can be defined as the effectiveness with which energy resources are converted into usable work. Thermal efficiency is commonly used to measure the efficiency of energy conversion systems such as process heaters, steam systems, engines, and power generators. Thermal efficiency is essentially the measure of the efficiency and completeness of fuel combustion, or, in more technical terms, the ratio of the net work supplied to the heat supplied by the combusted fuel. In a gas-fired heater, for example, thermal efficiency is equal to the total heat absorbed divided by the total heat supplied; in an automotive engine, thermal efficiency is the work done by the gases in the cylinder divided by the heat energy of the fuel supplied.[1]

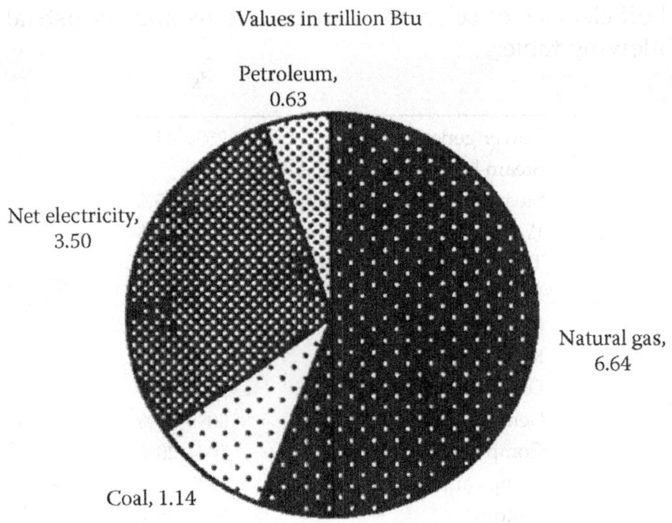

FIGURE 29.10
Manufacturing energy consumption by fuel (end-use data). (From U.S. Department of Energy, Energy Information Agency, Manufacturing Energy Consumption Survey, 2006, Washington, DC, 2012.)

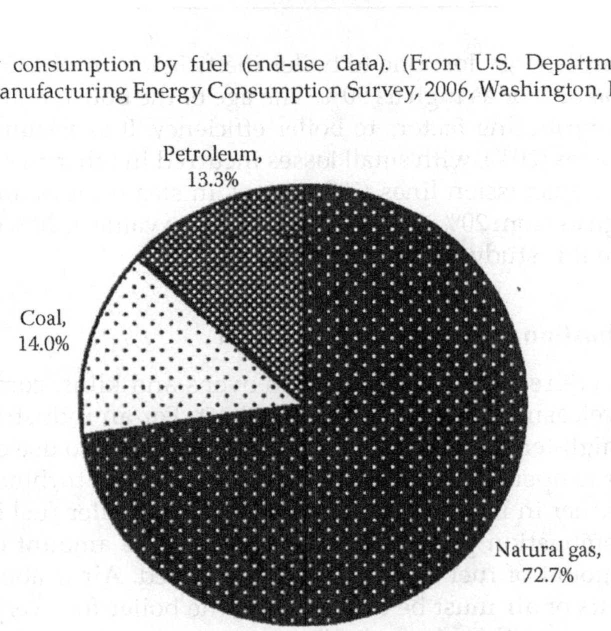

FIGURE 29.11
Manufacturing energy use for boiler fuel and process heating end-use basis. (From U.S. Department of Energy, Energy Information Agency, Manufacturing Energy Consumption Survey, 1998, Washington, DC, 2002.)

Energy efficiency varies dramatically across industries and manufacturing processes, and even between plants manufacturing the same products. Efficiency can be limited by mechanical, chemical, or other physical parameters, or by the age and design of equipment. In some cases, operating and maintenance practices contribute to lower-than-optimum efficiency. Regardless of the reason, less-than-optimum energy efficiency implies that not all of the energy input is being converted to useful work—some is released as lost energy. In the manufacturing sector, these energy losses amount to several quadrillion Btus (quadrillion British thermal units, or quads) and billions of dollars in lost revenues every year.

Typical thermal efficiencies of selected energy systems and industrial equipment[23] are provided in the following table:

Power generation	25%–44%
Steam boilers (natural gas)	80%
Steam boilers (coal and oil)	84%–85%
Waste-heat boilers	60%–70%
Thermal cracking (refineries)	58%–61%
EAF steelmaking	56%
Paper drying	48%
Kraft pulping	60%–69%
Distillation column	25%–40%
Cement calciner	30%–70%
Compressors	10%–20%
Pumps and fans	55%–65%
Motors	90%–95%

Boiler losses represent energy lost due to boiler inefficiency. In practice, boiler efficiency can be as low as 55%–60% or as high as 90%. The age of the boiler, maintenance practices, and fuel type are contributing factors to boiler efficiency. It is assumed that the greater losses are in steam pipes (20%), with small losses incurred in other fuel transmission lines (3%) and electricity transmission lines (3%). Losses in steam pipes and traps have been reported to be as high as from 20% to 40%. A conservative value of 20% was used for steam distribution losses in this study.[1]

29.5.2 Boiler Combustion Efficiency Improvement

Boilers and other fuel-fired equipment, such as ovens and kilns, combust fuel with air for the purpose of releasing chemical energy as heat. For an industrial boiler, the purpose is to generate high-temperature and high-pressure steam to use directly in a manufacturing process or to operate other equipment such as steam turbines to produce shaft power. As shown earlier in Figure 29.11, the predominant boiler fuel is natural gas. The efficiency of any combustion process is dependent on the amount of air that is used in relation to the amount of fuel and how they are mixed. Air is about 20% oxygen, so approximately 5 units of air must be brought into the boiler for every 1 unit of oxygen that is needed. Controlling this air–fuel mixture, and minimizing the amount of excess air while still obtaining safe mixing of the air and fuel, is key to ensuring a high combustion efficiency in the boiler.[24]

29.5.2.1 Combustion Control

The stoichiometric equation for the combustion of methane, the principal constituent of natural gas, with air is

$$CH_4 + 2O_2 + 7.52N_2 \rightarrow CO_2 + 2H_2O + 7.52N_2 \quad (29.16)$$

The stoichiometric equation is the one representing the exact amount of air necessary to oxidize the carbon and hydrogen in the fuel to carbon dioxide and water vapor. However, it is necessary to provide more than the stoichiometric amount of air since the mixing of fuel and air is imperfect in the real combustion chamber. Thus, the combustion equation for hydrocarbon fuels becomes

$$C_xH_y + \phi\left(\frac{x+y}{4}\right)O_2 + 3.76\phi\left(\frac{x+y}{4}\right)N_2 \rightarrow xCO_2 + y/2H_2O$$

$$+ (\phi - 1)\left(\frac{x+y}{4}\right)O_2 + 3.76\phi\left(\frac{x+y}{4}\right)N_2 \quad (29.17)$$

Note that for a given fuel, nothing in the equation changes except the parameter *f*, the equivalence ratio, as the fuel–air ratio changes.

As ϕ is increased beyond the optimal value for good combustion, the stack losses increase, and the heat available for the process decreases. As the equivalence ratio increases for a given flue temperature and a given fuel, more fuel must be consumed to supply a given amount of heat to the process.

The control problem for the furnace or boiler is to provide the minimum amount of air for good combustion over a wide range of firing conditions and a wide range of ambient temperatures. The most common combustion controller uses the ratio of the pressure drops across orifices, nozzles, or venturis in the air and fuel lines. Since these meters measure volume flow, a change in the temperature of combustion air with respect to fuel, or vice versa, will affect the equivalence ratio of the burner. Furthermore, since the pressure drops across the flowmeters are exponentially related to the volume flow rates, control dampers must have very complicated actuator motions. All the problems of ratio controllers are eliminated if the air is controlled from an oxygen meter. These are now coming into more general use as reasonably priced, high-temperature oxygen sensors become available. It is possible to control to any set value of percentage oxygen in the products; that is,

$$\%O_2 = \frac{\phi - 1}{((x + y/2)/(x + y/4)) + 4.76\phi - 1} \quad (29.18)$$

Figure 29.12 is a nomograph from the Bailey Meter Company[32] that gives estimates of the annual dollar savings resulting from the reduction of excess air to 15% for gas-, oil-, or coal-fired boilers with stack temperatures from 300°F to 700°F. The fuel savings are predicted on the basis that as excess air is reduced, the resulting reduction in mass flow of combustion gases results in reduced gas velocity and thus a longer gas residence time in the boiler. The increased residence time increases the heat transfer from the gases to the water. The combined effect of lower exhaust gas flows and increased heat exchange effectiveness is estimated to be 1.5 times greater than that due to the reduced mass flow alone.

As an example, assume the following data pertaining to an oil-fired boiler. Entering the graph at the top abscissa with 6.2% O_2, we drop to the oil fuel line and then horizontally to the 327°C (620°F) flue gas temperature line. Continuing to the left ordinate, we can see that 6.2% O_2 corresponds to 37.5% excess air. Dropping vertically from the intersection of

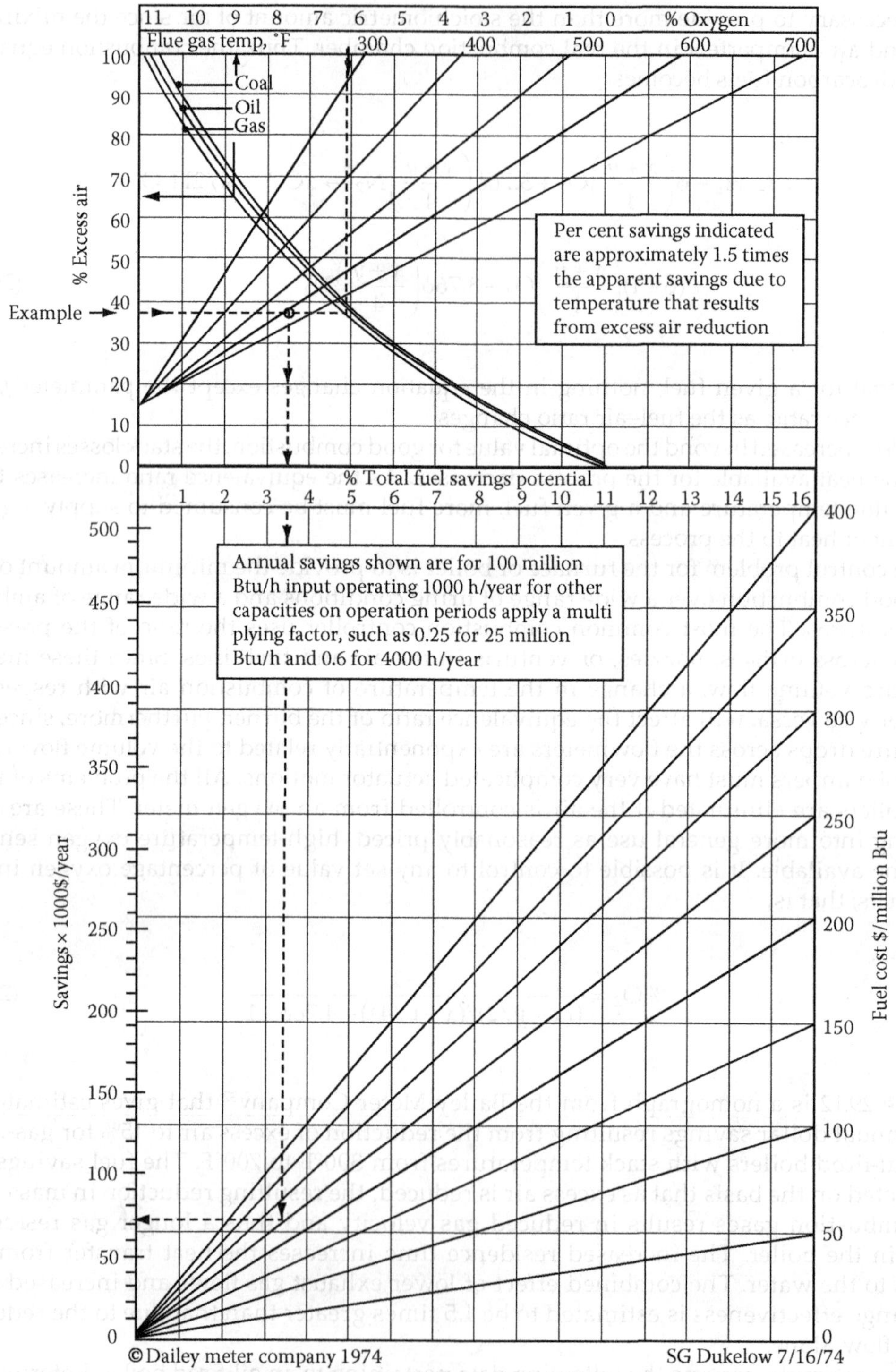

FIGURE 29.12
Nomograph for estimating savings from the adjustment of burners. (From Dukelow, S.G., Bailey Meter Company, Wickliffe, OH, 1974. With permission.)

the flue gas temperature line and the excess air line, we note a 3.4% total fuel savings. Fuel costs are as follows:

Burner capacity	63 GJ/h (60 × 10^6 Btu/h)
Annual operating hours	6,200
Fuel cost	\$0.38/L (\$1.44/gal)
Heating value fuel	42.36 MJ/L (152,000 Btu/gal)
Percent O_2 in exhaust gases	6.2%
Stack temperature	327°C (620°F)

$$10^9 \text{ J/GJ } (10^6 \text{ Btu/million Btu}) \times \frac{\$0.38/\text{L}}{74{,}236{,}500 \text{ J/L}} \left(\frac{1.44/\text{gal}}{152{,}000 \text{ Btu/gal}} \right)$$

$$= \$8.96/\text{GJ } (9.48/\text{million Btu})$$

Continuing the vertical line to intersect the \$2.50/million Btu and then moving to left ordinate show a savings of \$140,000/year for 8000 h of operation, 100 × 10^6 Btu/h input, and \$5.00/million Btu fuel cost. Adjusting that result for the assumed operating data,

$$\text{Annual savings} = \frac{6{,}200}{8{,}000} \times \frac{60 \times 10^6}{100 \times 10^6} \times \frac{9.48}{5.00} \times \$140{,}000 = \$123{,}430/\text{year}$$

This savings could be obtained by installing a modern oxygen controller, an investment with approximately a 1-year payoff, or from heightened operator attention with frequent flue gas testing and manual adjustments. Valuable sources of information concerning fuel conservation in boilers and furnaces are given by the DOE/OIT website.

29.5.2.2 Waste-Heat Management

Waste heat as generally understood in industry is the energy rejected from any process to the atmospheric air or to a body of water. It may be transmitted by radiation, conduction, or convection, but often it is contained in gaseous or liquid streams emitted to the environment. Almost 50% of all fuel energy used in the United States is transferred as waste heat to the environment, causing thermal pollution as well as chemical pollution of one sort or another. It has been estimated that half of that total may be economically recoverable for useful heating functions.

What must be known about waste-heat streams in order to decide whether they can become useful? Here is a list along with a parallel list of characteristics of the heat load that should be matched by the waste-heat supply.

Waste-heat supply
Quantity
Quality
Temporal availability of supply
Characteristics of fluid
Heat load
Quantity required
Quality required
Temporal availability of load
Special fluid requirements

Let us examine the particular case of a plant producing ice-cream cones. All energy quantities are given in terms of 15.5°C reference temperature. Sources of waste heat include the following:

- Products of combustion from 120 natural gas jets used to heat the molds in the carousel-type baking machines. The stack gases are collected under an insulated hood and released to the atmosphere through a short stack. Each of six machines emits 236.2 m^3/h of stack gas at 160°C. Total source rate is 161,400 kJ/h or 3874 MJ/day for a three-shift day.
- Cooling water from the jackets, intercoolers, and aftercoolers of two air compressors used to supply air to the pneumatic actuators of the cone machines; 11.36 L/min of water at 48.9°C is available. This represents a source rate of 96 MJ/h. The compressors run an average of 21 h/production day. Thus, this source rate is 2015 MJ/day.
- The water chillers used to refrigerate the cone batter make available—at 130°F—264 MJ/h of water heat. This source is available to heat water to 48.9°C using desuperheaters following the water chiller compressors. The source rate is 6330 MJ/day.
- At 21.2°C, 226 m^3/min of ventilating air is discharged to the atmosphere. This is a source of rate of less than 22.2 MJ/h or 525 MJ/day.

Uses for waste heat include the following:

- For cleanup operations during 3 h of every shift or during 9 of every 24 h, 681 L/h of hot water at 82.2°C is needed. Total daily heat load is 4518 MJ.
- Heating degree-days total in excess of 3333 annually. Thus, any heat available at temperatures above 21.1°C can be utilized with the aid of runaround systems during the 5(1/2)-month heating season. Estimated heating load per year is 4010 GJ.
- Total daily waste heat available—12.74 GJ/day.
- Total annual waste heat available—3.19 TJ/year.
- Total annual worth of waste heat (at $5.00/GJ for gas)—$15,893.
- Total daily heat load—this varies from a maximum of 59.45 GJ/day at the height of the heating season to the hot water load of 4.52 GJ/day in the summer months.

Although the amount of waste heat from the water chillers is 40% greater than the load needed for hot water heating, the quality is insufficient to allow its full use, since the hot water must be heated to 82°C and the compressor discharge is at a temperature of 54°C.

However, the chiller waste heat can be used to preheat the hot water. Assuming 13°C supply water and a 10°C heat exchange temperature approach, the load that can be supplied by the chiller is

$$\frac{49-13}{82-13} \times 4.52 = 2.36 \text{ GJ/day}$$

Since the cone machines have an exhaust gas discharge of 3.87 GJ/day at 160°C, the remainder of the hot water heating load of 2.17 GJ/day is available. Thus, a total saving of 1129 GJ/year in fuel is possible with a cost saving of $5645 annually based on $5.00/GJ gas. The investment costs will involve the construction of a common exhaust heater for the cone machines, a desuperheater for each of the three water chiller compressors, a gas-to-liquid

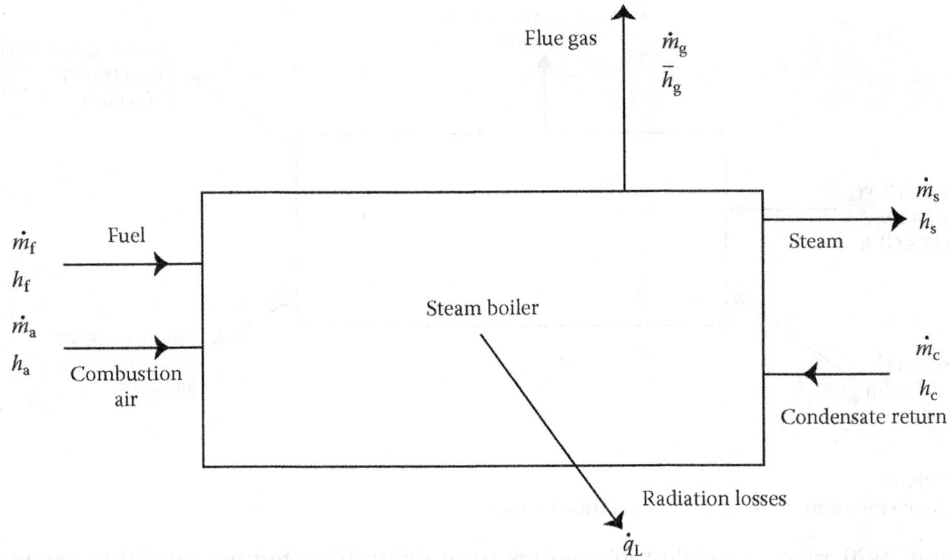

FIGURE 29.13
Heat balance on steam boiler.

heat exchanger following the cone-machine exhaust heater, and possibly an induced draft fan following the heat exchanger, since the drop in exhaust gas temperature will decrease the natural draft due to hot-gas buoyancy.

It is necessary to almost match four of four characteristics. Not exactly, of course, but the closer means thermodynamic availability of the waste heat. Unless the energy of the waste stream is sufficiently hot, it will be impossible to even transfer it to the heat load, since spontaneous heat transfer occurs only from higher to lower temperature.

The quantity and quality of energy available from a waste-heat source or for a heat load are studied with the aid of a heat balance. Figure 29.13 shows the heat balance for a steam boiler. The rates of enthalpy entering or leaving the system fluid streams must balance with the radiation loss rate from the boiler's external surfaces. Writing the first law equation for a steady-flow-steady-state process

$$\dot{q}_L = \dot{m}_f h_f + \dot{m}_a h_a + \dot{m}_c h_c - \dot{m}_s h_s - \dot{m}_g h_g \tag{29.19}$$

and referring to the heat-balance diagram, one sees that the enthalpy flux $\dot{m}_g h_g$, leaving the boiler in the exhaust gas stream, is a possible source of waste heat. A fraction of that energy can be transferred in a heat exchanger to the combustion air, thus increasing the enthalpy flux $\dot{m}_g h_a$ and reducing the amount of fuel required. The fraction of fuel that can be saved is given in the equation

$$\frac{\dot{m}_f \dot{m}_{f'}}{\dot{m}_f} = 1 - \left[\frac{K_1 - (1+\phi)\bar{C}_p T_g}{K_1 - (1+\phi)\bar{C}_p' T_g'}\right] \tag{29.20}$$

where the primed values are those obtained with waste-heat recovery. K_1 represents the specific enthalpy of the fuel–air mixture, $h_f + \phi h_a$, which is presumed to be the same with or without waste-heat recovery, ϕ is the molar ratio of air to fuel, and $\bar{C}_p$ is the specific heat averaged over the exhaust gas components. Figure 29.14, which is derived from

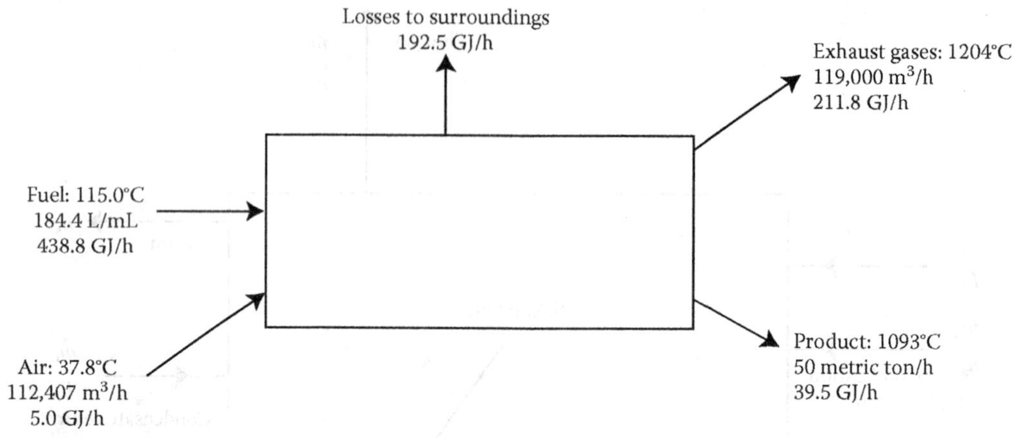

FIGURE 29.14
Heat balance for a simple continuous steel tube furnace.

Equation 29.20, gives possible fuel savings from using high-temperature flue gas to heat the combustion air in industrial furnaces.

It should be pointed out that the use of recovered waste heat to preheat combustion air, boiler feedwater, and product to be heat treated confers special benefits not necessarily accruing when the heat is recovered to be used in another system. The preheating operation results in less fuel being consumed in the furnace, and the corresponding smaller air consumption means even smaller waste heat being ejected from the stacks.

Table 29.8 shows heat balances for a boiler with no flue gas–heat recovery, with a feedwater preheater (economizer) installed and with an air preheater, respectively.

It is seen that air preheater alone saves 6% of the fuel and the economizer saves 9.2%. Since the economizer is cheaper to install than the air preheater, the choice is easy to make for an industrial boiler. For a utility boiler, both units are invariably used in series in the exit gas stream.

Table 29.9 is an economic study using 2005 prices for fuel, labor, and equipment. At that time, it was estimated that a radiation recuperator fitted to a fiberglass furnace would cost $820,200 and effect a savings of $90,435/year, making for a payoff period of approximately 2.8 years. This assumes of course that the original burners, combustion control system, and so on could be used without modification.

At this point, we can take some time to relate waste-heat recovery to the combustion process itself. We can first state categorically that the use of preheated combustion air improves combustion conditions and efficiency at all loads and in a newly designed installation permits a reduction in size of the boiler or furnace. It is true that the increased mixture temperature that accompanies air preheat results in some narrowing of the mixture stability limits, but in all practical furnaces, this is of small importance.

In many cases, low-temperature burners may be used for preheated air, particularly if the air preheat temperature is not excessive and if the fuel is gaseous. However, the higher volume flow of preheated air in the air piping may cause large enough increases in pressure drop to require a larger combustion air fan. Of course, larger diameter air piping can prevent the increased pressure drop, since air preheating results in reduced quantities of fuel and combustion air. For high preheat temperatures, alloy piping, high-temperature insulation, and water-cooled burners may be required. Since many automatic combustion-control systems sense volume flows of air and/or fuel, the correct control settings will

TABLE 29.8

Heat Balances for a Steam Generator

	Input Streams				Output Streams			
Case	**Name**	**Temperature (°C)**	**Flow Rate**	**Energy**	**Name**	**Temperature (°C)**	**Flow Rate**	**Energy**
Without economizer or air preheater	Natural gas	26.7	3,611 m^3/h	134.78 GJ/h	Steam	185.6	45,349 kg/h	126.23
	Air	26.7	35,574 m^3/h	560.32 MJ/h	Flue gas	372.2	41,185 m^3/h	19.49
	Makeup water	10.0	7,076 kg/h	297.25 MJ/h	Surface losses	—	—	1.21
	Condensate return	82.2	41,277 kg/h	14.21 GJ/h	Blowdown	185.6	2,994 kg/h	2.36
With air preheater	Natural gas	26.7	3,395 m^3/h	126.72 GJ/h	Steam	185.6	45,359 kg/h	126.23
	Air	232.2	32,114 m^3/h	478.90 GJ/h	Flue gas	260.0	35,509 m^3/h	11.42
	Makeup water	10	70,767 kg/h	297.25 MJ/h	Surface losses	—	—	1.21
	Condensate return	82.2	41,277 kg/h	14.21 GJ/h	Blowdown	185.6	2,994 kg/h	2.36
With economizer	Natural gas	26.7	3278 m^3/h	122.37 GJ/h	Steam	185.6	45,359 kg/h	126.23
	Air	26.7	31,013 m^3/h	462.48 MJ/h	Flue gas	176.7	34,281 m^3/h	7.08
	Makeup water	101	7,076 kg/h	297.25 GJ/h	Surface loss	—	—	1.21
	Condensate water	82.2	41,277 kg/h	12.21 GJ/h	Blowdown	185.6	2,994 kg/h	2.36

TABLE 29.9

Cost–Fuel Savings Analysis of a Fiberglass Furnace Recuperator

Operation	Continuous
Fuel input	19.42 GJ/h
Fuel	No. 3 fuel oil
Furnace temperature	1482°C
Flue gas temperature entering recuperator	1204°C
Air preheat (at burner)	552°C
Fuel savings = 37.4%	
Q = 7.26 GJ/h or 173.6 L/h of oil	
Fuel cost savings estimation	
Per GJ = $5.00	
Per hour = $36.30	
Per year (8,000 h) = $290,435	
Cost of recuperator	$421,400
Cost of installation, related to recuperator	$398,800
Total cost of recuperator installation	$820,200
Approximate payback time	2.82 years

change when preheated air is used. Furthermore, if air preheat temperature varies with furnace load, then the control system must compensate for this variation with an auxiliary temperature-sensing control. On the other hand, if control is based on the oxygen content of the flue gases, the control complications arising from gas volume variation with temperature is obviated. This is the preferred control system for all furnaces, and only cost prevents its wide use in small installations. Burner operation and maintenance for gas burners is not affected by preheating, but oil burners may experience accelerated fuel coking and resulting plugging from the additional heat being introduced into the liquid fuel from the preheated air. Careful burner design, which may call for water cooling or for shielding the fuel tip from furnace radiation, will always solve the problems. Coal-fired furnaces may use preheated air up to temperatures that endanger the grates or burners. Again, any problems can be solved by special designs involving water cooling if higher air temperatures can be obtained and/or desired.

The economics of waste-heat recovery today range from poor to excellent depending upon the technical aspects of the application as detailed earlier, but the general statement can be made that, at least for most small industrial boilers and furnaces, standard designs and/or off-the-shelf heat exchangers prove to be the most economic. For large systems, one can often afford to pay for special designs, construction, and installations. Furthermore, the applications are often technically constrained by material properties and space limitations and, as shall be seen later, always by economic considerations. For further information on heat exchangers, see Chapter 29.

29.5.2.3 Heating, Ventilating, and Air Conditioning

HVAC, while not usually important in the energy-intensive industries, may be responsible for the major share of energy consumption in the light manufacturing field, particularly in high-technology companies and those engaged primarily in assembly.

Because of air pollution from industrial processes, many HVAC systems require 100% outside ventilating air. Furthermore, ventilating air requirements are often much in excess of those in residential and commercial practice. An approximate method for calculating the total heat required for ventilating air in kJ per heating season is given by

$$E_v \text{ (kJ)} = 60 \times 24 \text{ (min/day)} \times (1.2 \times 0.519) \text{ (kJ/m}^3\text{K)} \times \text{SCMM} \times \text{DD}$$
$$= 896.8 \times \text{SCMM} \times \text{DD} \quad (29.21)$$

where

SCMM is the standard cubic meter per minute of total air entering plant including unwanted infiltration

DD is the heating degree-days (°C)

This underestimates the energy requirement, because degree-days are based on 18.33°C reference temperature, and indoor temperatures are ordinarily held 1.6°–3.9° higher. For a location with 3333 degree-days, each year the heating energy given by Equation 29.13 is about 17% low.

Savings can be effected by reducing the ventilating air rate to the actual rate necessary for health and safety and by ducting outside air into direct-fired heating equipment such as furnaces, boilers, ovens, and dryers. Air infiltration should be prevented through a program of building maintenance to replace broken windows, doors, roofs, and siding, and by campaigns to prevent unnecessary opening of windows and doors.

Additional roof insulation is often economic, particularly because thermal stratification makes roof temperatures much higher than average wall temperatures. Properly installed vertical air circulators can prevent the vertical stratification and save heat loss through the roof. Windows can be double glazed, storm windows can be installed, or windows can be covered with insulation. Although the benefits of natural lighting are eliminated by this measure, it can be very effective in reducing infiltration and heat transfer losses.

Waste heat from ventilating air itself, from boiler and furnace exhaust stacks, and from air-conditioning refrigeration compressors can be recovered and used to preheat make-up air. Consideration should also be given to providing spot ventilation in hazardous locations instead of increasing general ventilation air requirements.

As an example of the savings possible in ventilation air control, a plant requiring 424.5 CMM outside air flow is selected. A gas-fired boiler with an energy input of 0.0165 GJ is used for heating and is supplied with room air.

$$\text{Combustion air} = \frac{16.5 \times 10^6}{37{,}281} \times \frac{12}{60} = 88.52 \text{ CMM}$$

for a fuel with 37,281 kJ/m^3 heating value and an air fuel ratio of 12 m^3 air/m^3 fuel. The number of annual degree-days was 3175.

A study showed that the actual air supplied through infiltration and air handlers was 809 m^3/min. An outside air duct was installed to supply combustion air for the boiler, and the actual ventilating air supply was reduced to the required 424 m^3/min. The fuel saving that resulted using Equation 29.21 was

$$896.8(809 - 424)3175 = 1096.2 \text{ GJ}$$

worth $5482 in fuel at $5.00/GJ for natural gas.

29.5.2.4 Modifications of Unit Processes

The particular process used for the production of any item affects not only the cost of production but also the quality of the product. Since the quality of the product is critical in customer acceptance and therefore in sales, the unit process itself cannot be considered a prime target for the energy conservation program. That does not say that one should ignore the serendipitous discovery of a better and cheaper way of producing something. Indeed, one should take instant advantage of such a situation, but that clearly is the kind of decision that management could make without considering energy conservation at all.

29.5.2.5 Optimizing Process Scheduling

Industrial thermal processing equipment tends to be quite massive compared to the product treated. Therefore, the heat required to bring the equipment to steady-state production conditions may be large enough to make start-ups fuel intensive. This calls for scheduling this equipment so that it is in use for as long periods as can be practically scheduled. It also may call for idling the equipment (feeding fuel to keep the temperature close to production temperature) when it is temporarily out of use. The fuel rate for idling may be between 10% and 40% of the full production rate for direct-fired equipment. Furthermore, the stack losses tend to increase as a percentage of fuel energy released. It is clear that overfrequent start-ups and long idling times are wasteful of energy and add to production costs. The hazards of eliminating some of that waste through rescheduling must not be taken lightly. For instance, a holdup in an intermediate heating process can slow up all subsequent operations and make for inefficiency down the line. The breakdown of a unit that has a very large production backlog is much more serious than that of one having a smaller backlog. Scheduling processes in a complex product line is a very difficult exercise and perhaps better suited to a computerized PERT program than to an energy conversation engineer. That does not mean that the possibilities for saving through better process scheduling should be ignored. It is only a warning to move slowly and take pains to find the difficulties that can arise thereby.

A manufacturer of precision instruments changed the specifications for the finishes of over half of his products, thereby eliminating the baking required for the enamel that had been used. He also rescheduled the baking process for the remaining products so that the oven was lighted only twice a week instead of every production day. A study is now proceeding to determine if electric infrared baking will not be more economic than using the gas-fired oven.

29.5.2.6 Cogeneration of Process Steam and Electricity

In-plant (or on-site) electrical energy cogeneration is nothing new. It has been used in industries with large process steam loads for many years, both in the United States and in Europe. It consists of producing steam at a higher pressure than required for process use, expanding the high-pressure steam through a backpressure turbine to generate electrical energy and then using the exhaust steam as process steam. Alternatively, the power system may be a diesel engine that drives an electrical generator. The diesel engine exhaust is then stripped of its heat content as it flows through a waste-heat boiler where steam is generated for plant processes. A third possibility is operation of a gas turbine generator to supply electric power and hot exhaust gases, which produce process steam in a waste-heat boiler. As will be seen later, the ratio of electric power to steam-heat rate varies markedly from one of these types of systems to the next. In medium-to-large industrial plants, the cogeneration of electric power and process steam is economically feasible provided certain plant energy characteristics are present. In small plants or in larger plants with small process steam loads, cogeneration is not economic because of the large capital expenditure involved. Under few circumstances is the in-plant generation of electric power economic without a large process steam requirement. A small industrial electric plant cannot compete with an electric utility unless the generation required in-plant exceeds the capacity of the utility. In remote areas where no electric utility exists, or where its reliability is inferior to that of the on-site plant, the exception can be made.

Cogeneration if applied correctly is not only cost effective, but also fuel conserving. That is, the fuel for the on-site plant is less than that used jointly by the utility to supply the plant's electric energy and that used on-site to supply process steam. Figures 29.15 and 29.16 illustrate the reasons for and the magnitude of the savings possible. However, several conditions must be met in order that an effective application is possible. First, the ratio of process steam heat rate to electric power must fall close to these given in the following table:

Heat Engine Type	E_{stream}/E_{elect}
Steam turbine	2.3
Gas turbine	4.0
Diesel engine	1.5

The table is based upon overall electric plant efficiencies to 30%, 20%, and 40% respectively, for steam turbine, gas turbine, and diesel engine. Second, it is required that the availability

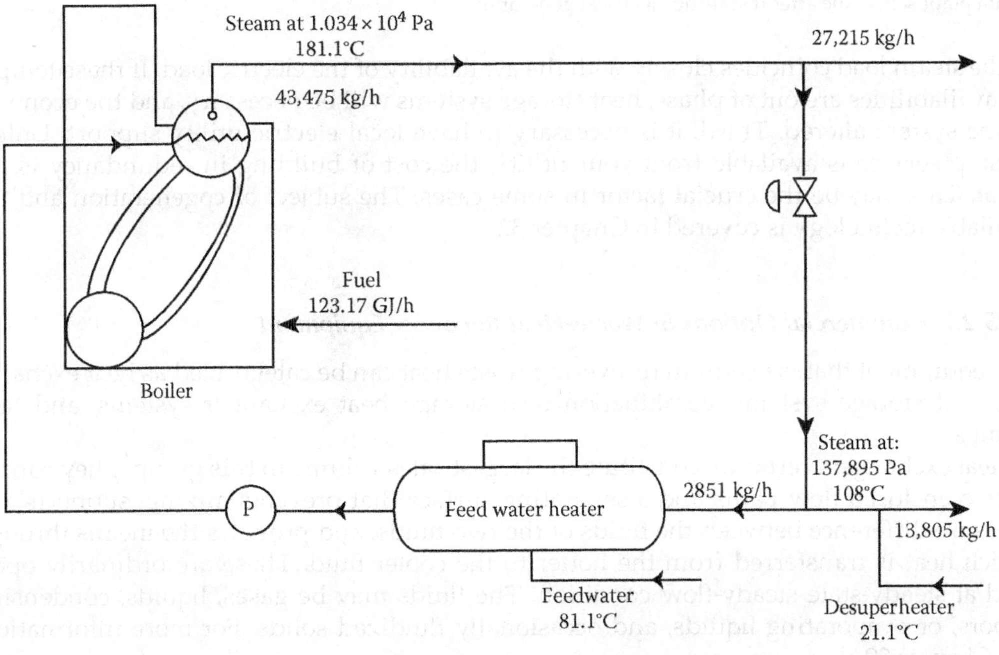

FIGURE 29.15
Steam plant schematic before adding electrical generation.

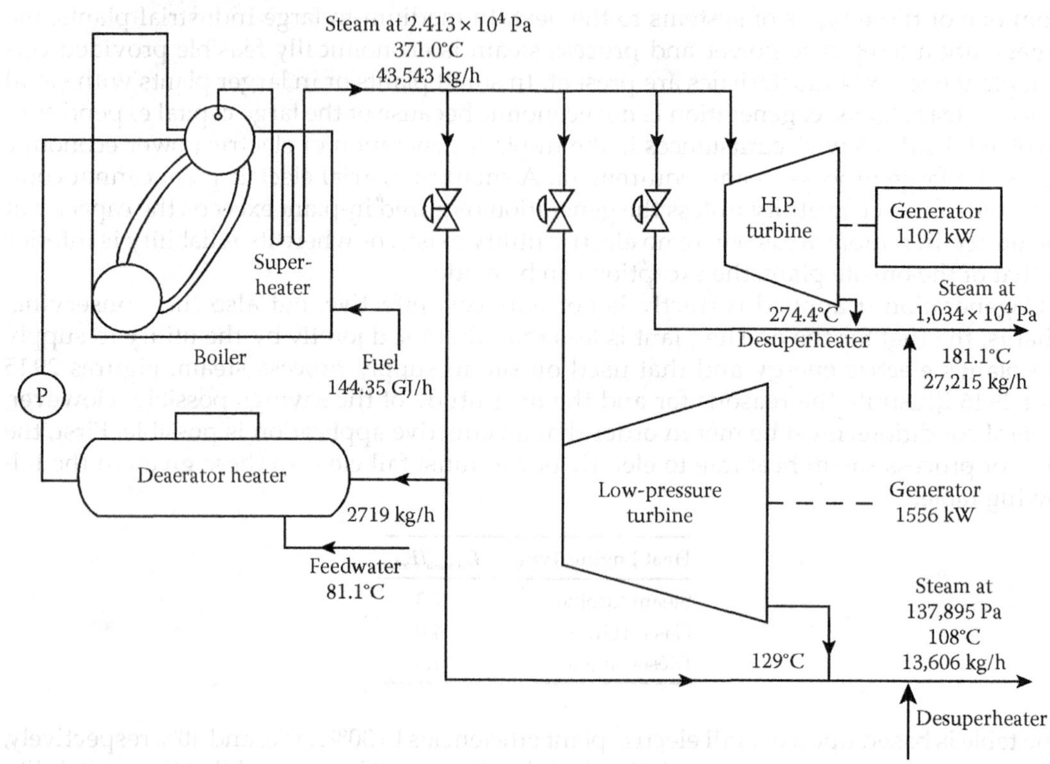

FIGURE 29.16
Steam plant schematic after installing electrical generation.

of the steam load coincides closely with the availability of the electric load. If these temporal availabilities are out of phase, heat storage systems will be necessary, and the economy of the system altered. Third, it is necessary to have local electric utility support. Unless backup service is available from your utility, the cost of building in redundancy is too great. This may be the crucial factor in some cases. The subject of cogeneration and the available technology is covered in Chapter 32.

29.5.2.7 Commercial Options in Waste-Heat Recovery Equipment

The equipment that is useful in recovering waste heat can be categorized as heat exchangers, heat storage systems, combination heat storage–heat exchanger systems, and heat pumps.

Heat exchangers certainly constitute the largest sales volume in this group. They consist of two enclosed flow paths and a separating surface that prevents mixing, supports any pressure difference between the fluids of the two fluids, and provides the means through which heat is transferred from the hotter to the cooler fluid. These are ordinarily operated at steady-state steady-flow condition. The fluids may be gases, liquids, condensing vapors, or evaporating liquids, and occasionally fluidized solids. For more information, see Chapter 29.

Radiation recuperators are high-temperature combustion-air preheaters used for transferring heat from furnace exhaust gases to combustion air. As seen in Figure 29.17, they

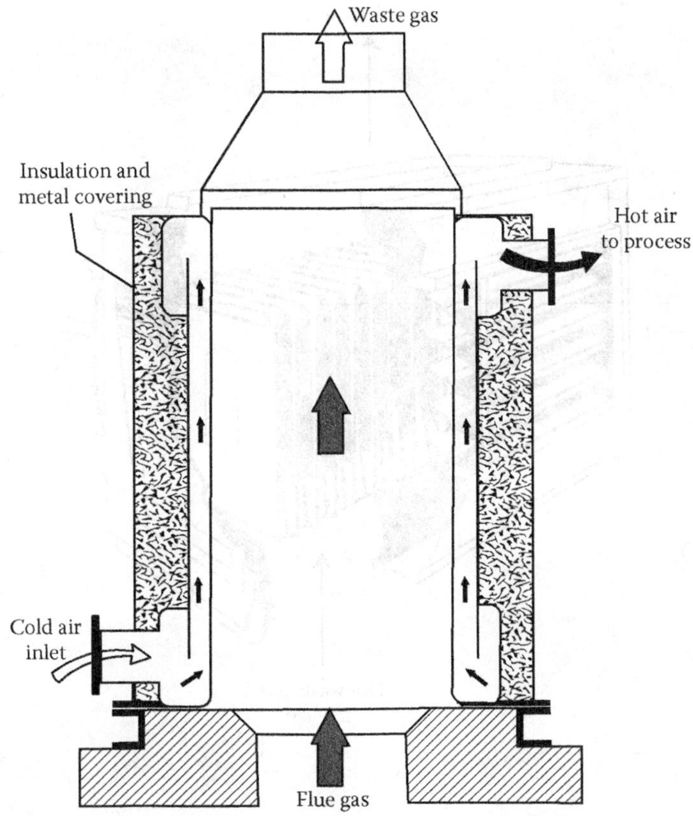

FIGURE 29.17
Metallic radiation recuperator.

consist of two concentric cylinders, the inner one as a stack for the furnace and the concentric space between the inner and outer cylinders as the path for the combustion air, which ordinarily moves upward and therefore parallel to the flow of the exhaust gases. With special construction materials, these can handle 1355°C furnace gases and save as much as 30% of the fuel otherwise required. The main problem in their use is damage due to overheating for reduced air flow or temperature excursions in the exhaust gas flow.

Convective air preheaters are corrugated metal or tubular devices that are used to preheat combustion air in the moderate temperature range (121°C–649°C) for ovens, furnaces, boilers, and gas turbines, or to preheat ventilating air from sources as low in temperature as 21°C. Figures 29.18 and 29.19 illustrate typical construction. These are often available in modular design so that almost any capacity and any degree of effectiveness can be obtained by multiple arrangements. The biggest problem is keeping them clean.

Economizer is the name traditionally used to describe the gas-to-liquid heat exchanger used to preheat the feedwater in boilers from waste heat in the exhaust gas stream. These often take the form of loops, spiral, or parallel arrays of finned tubing through which the feedwater flows and over which the exhaust gases pass. They are available in modular form to be introduced into the exhaust stack or into the breeching. They can also be used in reverse to heat air or other gases with waste heat from liquid streams.

A more recent development is the use of condensing economizers that are placed in the exhaust stream following high-temperature economizers. They are capable of extracting

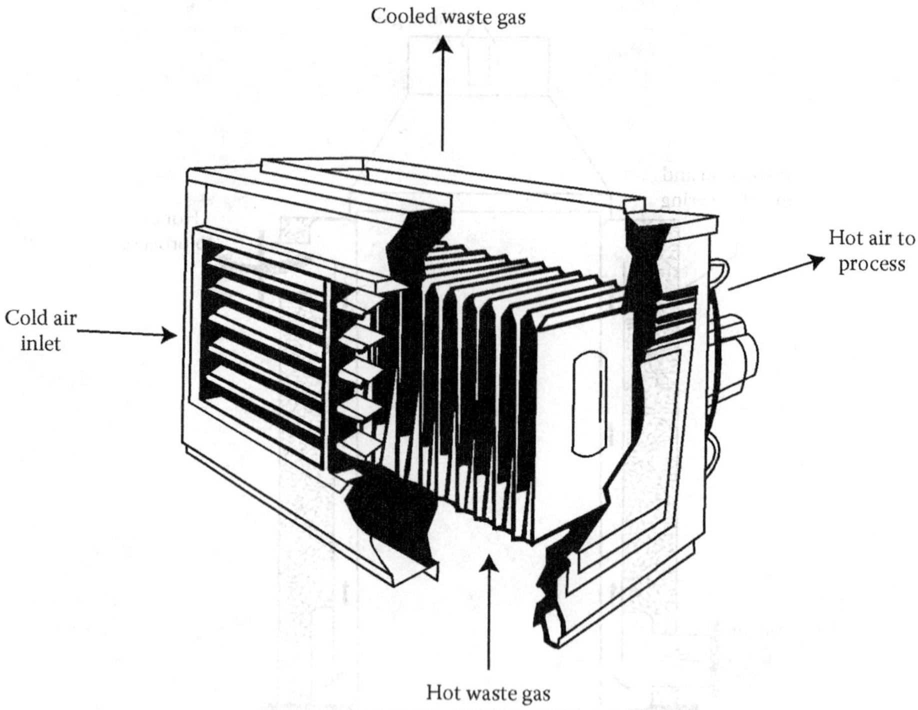

FIGURE 29.18
Air preheater.

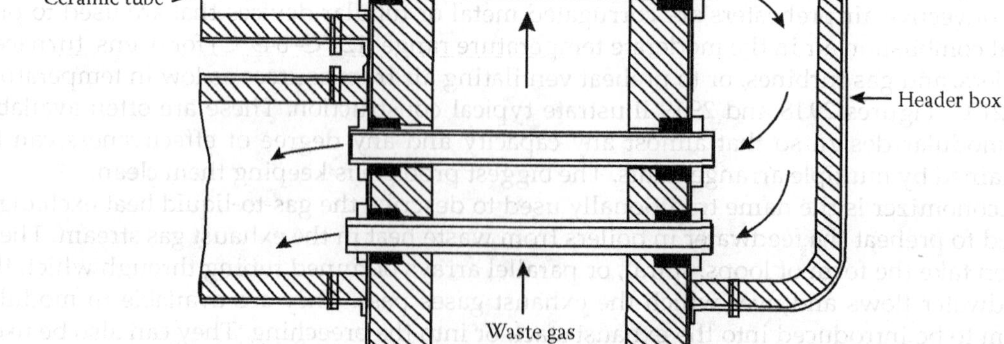

FIGURE 29.19
Ceramic tube recuperator.

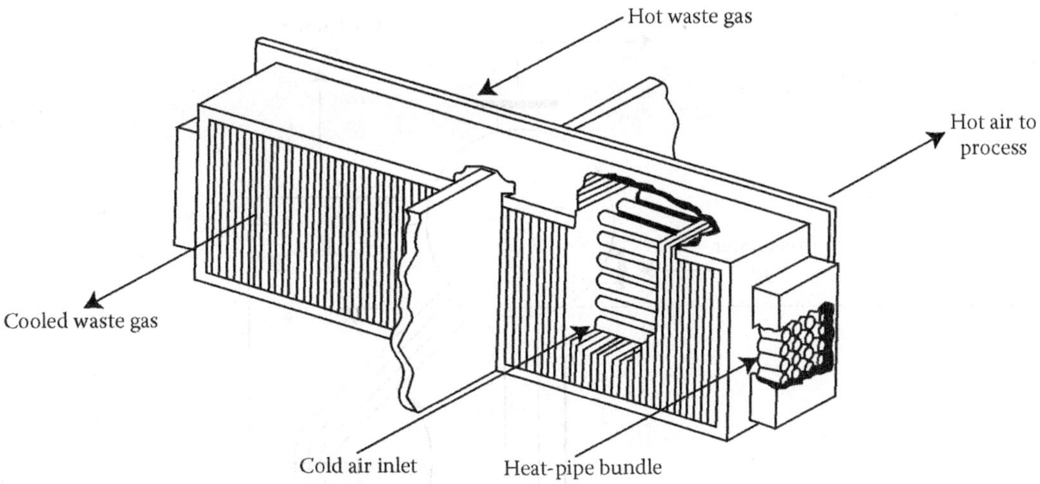

FIGURE 29.20
Heat pipe recuperator.

an additional 6%–8% of the fuel input energy from the boiler exhaust gases. However, they are used only under certain restricted conditions. Obviously, the cooling fluid must be at a temperature below the dew point of the exhaust stream. This condition is often satisfied when boilers are operated with 100% make-up water. A second, less restrictive condition is that the flue gases be free of sulfur oxides. This is normally the case for natural gas-fired boilers. Otherwise, the economizer tubes will be attacked by sulfurous and/or sulfuric acid. Acid corrosion can be slowed down markedly by the use of all-stainless steel construction, but the cost of the equipment is increased significantly.

Heat-pipe arrays are often used for air-to-air heat exchangers because of their compact size. Heat-transfer rates per unit area are quite high. A disadvantage is that a given heat pipe (i.e., a given internal working substance) has a limited temperature range for efficient operation. The heat pipe transfers heat from the hot end by evaporative heating and at the cold end by condensing the vapor. Figure 29.20 is a sketch of an air preheater using an array of heat pipes.

Waste-heat boilers are water-tube boilers, usually prefabricated in all but the largest sizes, used to produce saturated steam from high-temperature waste heat in gas streams. The boiler tubes are often finned to keep the dimensions of the boiler smaller. They are often used to strip waste heat from diesel engine exhausts, gas turbine exhausts, and pollution control incinerators or afterburners. Figure 29.21 is a diagram of the internals of a typical waste-heat boiler. Figure 29.22 is a schematic diagram showing a waste-heat boiler for which the evaporator is in the form of a finned-tube economizer. Forced water circulation is used giving some flexibility in placing the steam drum and allowing the use of smaller tubes. It also allows the orientation of the evaporator to be either vertical or horizontal. Other advantages of this design are the attainment of high boiler efficiencies, a more compact boiler, less cost to repair or retube, the ability to make superheated steam using the first one or more rows of downstream tubes as the superheater, and the elimination of thermal shock, since the evaporator is not directly connected to the steam drum.

Heat storage systems, or regenerators, once very popular for high-temperature applications, have been largely replaced by radiation recuperators because of the relative simplicity of the latter. Regenerators consist of twin flues filled with open ceramic checkerwork.

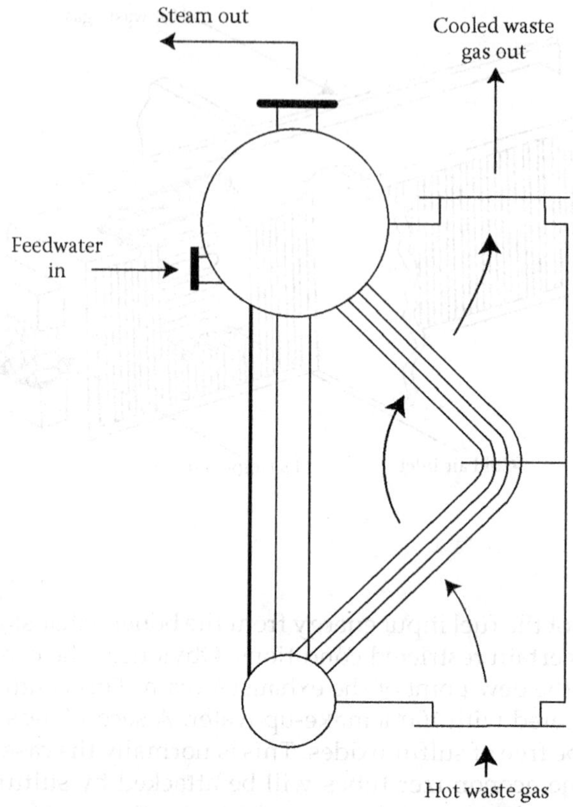

FIGURE 29.21
Waste-heat boiler.

The high-temperature exhaust of a furnace flowing through one leg of a swing valve to one of the flues heated the checkerwork, while the combustion air for the furnace flowed through the second flue in order to preheat it. When the temperatures of the two masses of checkerwork were at proper levels, the swing valve was thrown and the procedure was continued, but with reversed flow in both flues. Regenerators are still used in some glass- and metal-melt furnaces, where they are able to operate in the temperature range 1093°C–1649°C. It should be noted that the original application of the regenerators was to achieve the high-melt temperatures required with low-heating-value fuel.

A number of ceramic materials in a range of sizes and geometric forms are available for incorporation into heat storage units. These can be used to store waste heat in order to remedy time discrepancies between source and load. A good example is the storage of solar energy in a rock pile so that it becomes available for use at night and on cloudy days. Heat storage, other than for regenerators in high-temperature applications, has not yet been used a great deal for waste-heat recovery but will probably become more popular as more experience with it accumulates.

Combination heat storage unit–heat exchangers called heat wheels are available for waste-heat recovery in the temperature range 0°C–982°C. The heat wheel is a porous flat cylinder that rotates within a pair of parallel ducts, as can be observed in Figure 29.23. As the hot gases flow through the matrix of the wheel, they heat one side of it, which then gives up that heat to the cold gases as it passes through the second duct. Heat-recovery

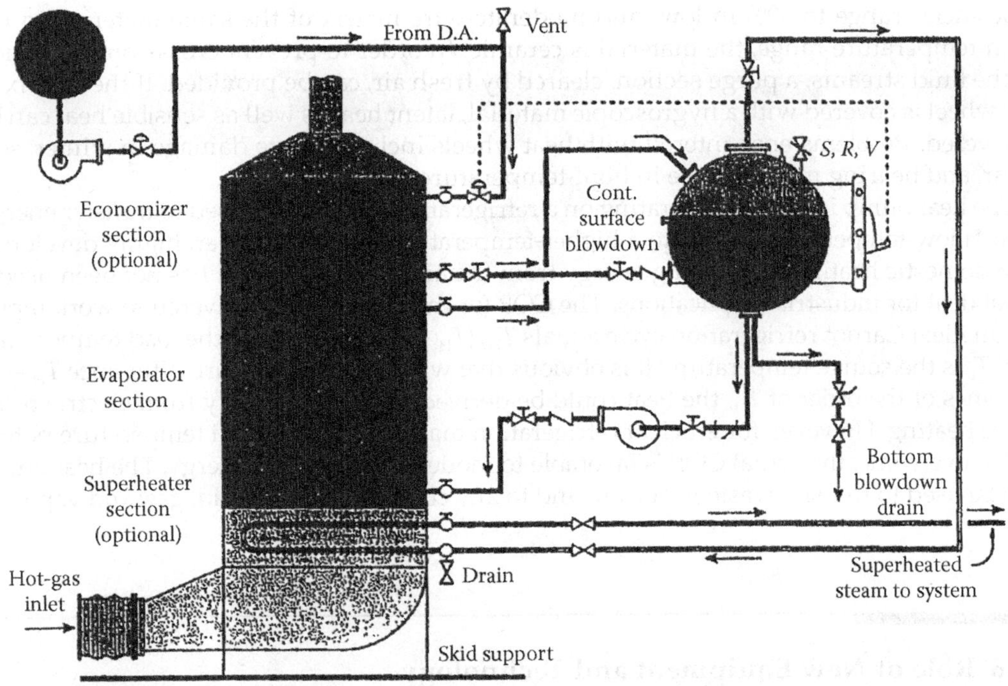

FIGURE 29.22
Schematic diagram of a finned tube waste-heat boiler. (Courtesy of Canon Technology.)

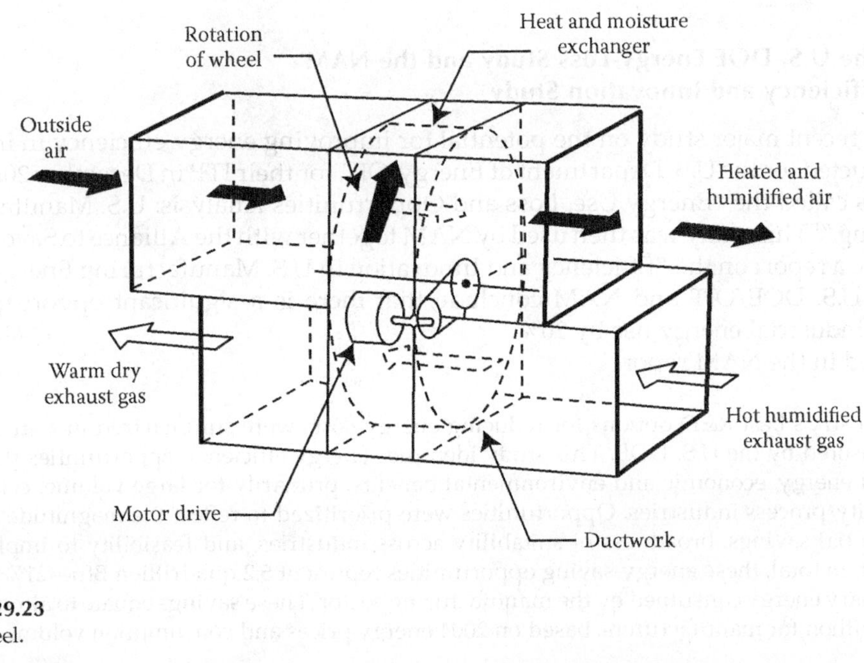

FIGURE 29.23
Heat wheel.

efficiencies range to 80% in low- and moderate-wire matrix of the same material. In the high-temperature range, the material is ceramic. In order to prevent cross-contamination of the fluid streams, a purge section, cleared by fresh air, can be provided. If the matrix of the wheel is covered with a hygroscopic material, latent heat as well as sensible heat can be recovered. Problems encountered with heat wheels include freeze damage in winter, seal wear, and bearing maintenance in high-temperature applications.

The heat pump is a device operating on a refrigeration cycle that is used to transfer energy from a low-temperature source to a higher-temperature load. It has been highly developed as a domestic heating plant using energy from the air or from well but has not been used a great deal for industrial applications. The COP (or the ratio of heat delivered to work input) for an ideal Carnot refrigeration cycle equals $T_H/(T_H - T_L)$, where T_H is the load temperature and T_L is the source temperature. It is obvious that when the temperature difference $T_H - T_L$ becomes of the order of T_H, the heat could be derived almost as cheaply from electric resistance heating. However, for efficient refrigeration machines and a small temperature potential to overcome, the actual COP is favorable to moderate-cost waste energy. The heat pump can be used to transfer waste heat from and to any combination of liquid, gas, and vapor.

29.6 Role of New Equipment and Technology in Industrial Energy Efficiency

29.6.1 Industrial Energy Savings Potential

The beginning of this chapter mentioned several studies of industrial energy efficiency that showed that a 20% reduction could be accomplished in a relatively easy and cost-effective manner. The purpose of this section is to provide the equipment, technology, and operational changes that could lead to industrial energy savings on the order of 20% or more.

29.6.2 The U.S. DOE Energy-Loss Study and the NAM Efficiency and Innovation Study

The most recent major study on the potential for improving energy efficiency in industry was conducted for the U.S. Department of Energy, OIT, for their ITP in December 2004. This study was called the "Energy Use, Loss and Opportunities Analysis: U.S. Manufacturing and Mining."[1] This study was then used by NAM together with the Alliance to Save Energy to produce a report on the "Efficiency and Innovation in U.S. Manufacturing Energy Use."[3] Both the U.S. DOE/OIT and NAM conclude that there is a significant opportunity for reducing industrial energy use by 20%.

As stated in the NAM report

> [i]ndustry's best R&D options for reducing energy costs were summarized in a study sponsored by the U.S. DOE. This study identifies energy efficiency opportunities that yield energy, economic and environmental benefits, primarily for large volume, commodity/process industries. Opportunities were prioritized to reflect the magnitude of potential savings, broadness of suitability across industries, and feasibility to implement. In total, these energy-saving opportunities represent 5.2 quadrillion Btu—21% of primary energy consumed by the manufacturing sector. These savings equate to almost $19 billion for manufacturers, based on 2004 energy prices and consumption volumes.

TABLE 29.10

Best Opportunities for Future Industrial Energy Savings

Type of Opportunity	Total Energy Savings (trillion Btu)	% of Total	Total Cost Savings ($mill.)	% of Total
Top R&D opportunities for energy savings in commodity/process manufacturing; initiatives that provide the largest energy and dollar savings				
Waste-heat and energy recovery	1831	35	6,408	34
Improvements to boilers, fired systems, process heaters, and cooling opportunities	907	17	3,077	16
Energy system integration and best practices opportunities	1438	28	5,655	30
Energy source flexibility and combined heat and power	828	16	3,100	16
Improved sensors, controls, automation and robotics for energy systems	191	4	630	3
Totals	5195		18,870	

Source: U.S. DOE, *Annual Report: Technology, Delivery, Industry of the Future*, U.S. Department of Energy, Office of Industrial Technologies, Industrial Technology Program, Washington, DC, 2004, see references section (DOE-TP); NAM, *Efficiency and Innovation in U.S. Manufacturing Energy Use*, National Association of Manufacturers, Washington, DC, 2005.

Table 29.10 summarizes these leading opportunities. An expanded version of this information appears in Table 29.11.

29.6.3 The ACEEE Fan and Pump Study

In April 2003, the ACEEE released a report on their study "Realizing Energy Efficiency Opportunities in Industrial Fan and Pump Systems." They concluded that fans and pumps account for more than a quarter of industrial electricity consumption and that optimization of the operation of these fans and pumps could achieve electricity savings ranging from 20% to well over 50% of this category of use.[10] This report says that most optimization projects involve greater engineering costs than equipment costs, but the average payback for a good optimization project is about 1.2 years, with the cost of saved energy on the order of $0.012/kW h. In addition, these estimates do not account for productivity gains known to exist at many of the plant sites, which are sometimes as much as two to five times the energy savings.

This ACEEE report contains some excellent data on motor systems end use that is very difficult to find in general. Their data show that 40% of industrial motor use is for fans and pumps. Because the MECS data from EIA show that about 68% of electric use in industry is motors, this leads to the fraction of industrial electric use from fans and pumps to be (0.68) × (0.4) = 0.32, or 32%. The end-use data from ACEEE are reproduced in Table 29.12.

To see the final impact of this estimate of industrial energy savings, now apply the ACEEE estimates of 20%–50% savings on fan and pump energy to the 32% fraction of fan and pump contribution to the total industrial electricity use. Thus, the overall savings are in the range (0.2) × (0.32) = 6.4% to (0.5) × (0.32) = 16%. Just this opportunity alone could result in achieving over half of the 20% savings contained in the U.S. DOE/OIT study.

Table 29.13 presents the ACEEE estimates of the relative magnitude of electricity consumed by fans and pump systems for the important industries on a national basis.

TABLE 29.11

Top R&D Opportunities for Industrial Energy Savings

Top R&D Opportunities for Energy Initiatives That Provide the Largest Energy	Savings in Commodity/Process Manufacturing and Dollar Savings	Total Energy Savings		Total Cost Savings	
Type of Opportunity	Leading Industry Recipients	(trillion Btu)	(% of Total)	($mil.)	(% of Total)[a]
Waste-heat and energy recovery		1,831	35	6,408	34
...From gases and liquids, including hot gas cleanup and dehydration of liquid waste streams	Chemicals, petroleum, forest products	851	16	2,271	12
...From drying processes	Chemicals, forest products, food processing	377	7	1,240	7
...From gases in metals and nonmetallic minerals manufacture (excluding calcining), including hot gas cleanup	Iron and steel, cement	235	5	1,133	6
...From by-product gases	Petroleum, iron, and steel	132	3	750	4
...Using energy export and colocation (fuels from pulp mills, forest bio-refineries, colocation of energy sources/sinks)	Forest products	105	2	580	3
...From calcining (not flue gases)	Cement, forest products	74	1	159	1
...From metal quenching/cooling processes	Iron and steel, cement	57	1	275	1
Improvements to boilers, fired systems, process heaters, and cooling opportunities		907	17	3,077	16
Advanced industrial boilers	Chemicals, forest products, petroleum, steel, food processing	400	8	1,090	6
Improved heating/heat transfer systems (heat exchangers, new materials, improved heat transport)	Petroleum, chemicals	260	5	860	5
Improved heating/heat transfer for metals, melting, heating, annealing (cascade heating, batch to continuous process, improved heat channeling, modular systems)	Iron and steel, metal casting, aluminum	190	4	915	5
Advanced process cooling and refrigeration	Food processing, chemicals, petroleum, and forest products	57[b]	1	212	1
Energy system integration and best practices opportunities		1,438	28	5,655	30
* Steam best practices (improved generation, distribution, and recovery), not including advanced boilers	All manufacturing	310	6	850	5
* Pump system optimization	All manufacturing	302[b]	6	1,370	7

(Continued)

TABLE 29.11 ***(Continued)***

Top R&D Opportunities for Industrial Energy Savings

Top R&D Opportunities for Energy Initiatives That Provide the Largest Energy	Savings in Commodity/Process Manufacturing and Dollar Savings	Total Energy Savings		Total Cost Savings	
Type of Opportunity	**Leading Industry Recipients**	**(trillion Btu)**	**(% of Total)**	**($mil.)**	**(% of Total)[a]**
* Energy system integration	Chemicals, petroleum, forest products, iron and steel, food, aluminum	260	5	860	5
* Energy-efficient motors and rewind practices	All manufacturing	258[b]	5	1,175	6
* Compressed air system optimization	All manufacturing	163[b]	3	740	4
* Optimized materials processing	All manufacturing	145[b]	3	660	3
Energy source flexibility and combined heat and power		828	16	3,100	16
* Combined heat and power on-site in manufacturers' central plants, producing both thermal and electricity needs	Forest products, chemicals, food processing, metals, machinery	634	12	2,000	11
Energy source flexibility (heat-activated power generation, waste steam for mechanical drives, indirect vs. direct heat vs. steam)	Chemicals, petroleum, forest products, iron and steel	194	4	1,100	6
Improved sensors, controls, automation and robotics for energy systems	Chemicals, petroleum, forest products, iron and steel, food, cement, aluminum	191	4	630	3
Totals		5,195		18,870	

Source: U.S. DOE, *Annual Report: Technology, Delivery, Industry of the Future,* U.S. Department of Energy, Office of Industrial Technologies, Industrial Technology Program, Washington, DC, 2004.

Note: All are R&D opportunities except for items denoted by an asterisk (*), which are near-term best practices, applicable to current assets.

[a] Totals may not add up due to rounding.

[b] Energy savings figures include the corresponding recapture of losses inherent in electricity generation, transmission, and distribution.

TABLE 29.12

National Industrial Motor Systems Energy End Use

Pumps	25%
Materials processing	22%
Compressed air	16%
Fans	14%
Material handling	12%
Refrigeration	7%
Other	4%

Source: Elliott, R.N. and Nadel, S., *Realizing Energy Efficiency Opportunities in Industrial Fan and Pump Systems,* Report A034, American Council for an Energy-Efficient Economy, Washington, DC, 2003.

TABLE 29.13

Characterization of Industrial Fan and Pump Load in the United States

NAICS	Industry	Electricity Demand 1997	Pumps (%)	Fans and Blowers (%)	Total Motors (%)	Motor Electricity	Fans/Pumps Share of Electricity (%)	Fans/Pumps Electricity Use
11	Agriculture	16,325	25	20	75	12,244	45	7,346
22	Mining	85,394	7	21	90	76,854	29	24,363
311	Food mfg.	66,166	11	5	81	53,756	16	10,809
314	Textile product mills	5,135	14	15	82	4,221	30	1,523
321	Wood product mfg.	21,884	4	10	80	17,464	14	3,064
322	Paper mfg.	119,627	28	16	84	101,078	44	52,636
324	Petroleum and coal products mfg.	69,601	51	13	85	59,369	63	44,061
325	Chemical mfg.	212,709	18	8	73	154,693	26	54,797
326	Plastics and rubber mfg.	52,556	9	4	66	34,847	13	6,729
327	Nonmetallic minerals product mfg.	37,416	4	4	65	24,328	8	3,037
331	Primary metal mfg.	172,518	2	4	26	44,855	6	10,351
332	Fabricated metal product mfg.	49,590	7	5	65	32,462	12	6,149
333	Machinery mfg.	27,295	8	4	67	18,391	12	3,330
334	Computer and electronic product mfg.	40,099	2	3	54	21,783	4	1,801
336	Transportation equipment mfg.	54,282	4	6	64	34,629	11	5,753
	Total	1,030,598				690,974		235,750
			Fraction of total elec.			67%		23%

29.6.4 The LBL/ACEEE Study of Emerging Energy-Efficient Industrial Technologies

In October 2000, ACEEE released a report of a study it did in conjunction with staff from Lawrence Berkeley Laboratories (LBL), where they identified 175 emerging energy-efficient technologies, and honed this list down to 32 technologies that had a high likelihood of success and a high energy savings.[25] An interesting aspect of this study is that it shows that the United States is not running out of technologies to improve energy efficiency and economic and environmental performance, and will not run out in the future. The study shows that many of the technologies have important nonenergy benefits, ranging from reduced environmental impact to improved productivity. Several technologies have reduced capital costs compared to the current technology used by those industries. Nonenergy benefits such as these are frequently a motivating factor in bringing this kind of technology to market.

The LBL/ACEEE list of 32 most beneficial technologies is shown in Table 29.14.

TABLE 29.14

Technologies with High Energy Savings and a High Likelihood of Success

Technology	Code	Total Energy Savings	Likelihood of Success	Recommended Next Steps
Efficient cell retrofit designs	Alum-2	High	High	Demo
Advanced lighting technologies	Lighting-1	High	High	Dissemination, demo
Advance ASD designs	Motorsys-1	High	High	R&D
Membrane technology wastewater	Other-3	High	High	Dissemination, R&D
Sensors and controls	Other-5	High	High	R&D, demo, dissemination
Black liquor gasification	Paper-1	High	High	Demo
Near-net-shape casting/strip casting	Steel-2	High	High	R&D
New EAF furnace processes	Steel-3	High	High	Field test
Oxy-fuel combustion in reheat furnace	Steel-4	High	High	Field test
Advanced CHP turbine systems	Utilities-1	High	High	Policies
Autothermal reforming—ammonia	Chem-7	High	Medium	Dissemination
Membrane technology—food	Food-3	High	Medium	Dissemination, R&D
Advanced lighting design	Lighting-2	High	Medium	Dissemination, demo
Compressed air system management	Motorsys-3	High	Medium	Dissemination
Motor system optimization	Motorsys-5	High	Medium	Dissemination, training
Pump efficiency improvement	Motorsys-6	High	Medium	Dissemination, training
High-efficiency/low-NO_x burners	Other-2	High	Medium	Dissemination, demo
Process integration (pinch analysis)	Other-4	High	Medium	Dissemination
Heat recovery—paper	Paper-5	High	Medium	Demo
Impulse drying	Paper-7	High	Medium	Demo
Smelting reduction processes	Steel-5	High	Medium	Demo
Advanced reciprocating engines	Utilities-2	High	Medium	R&D, demo
Fuel cells	Utilities-3	High	Medium	Demo
Microturbines	Utilities-4	High	Medium	R&D, demo
Inert anodes/wetted cathodes	Alum-4	High	Medium	R&D
Advanced forming	Alum-1	Medium	High	R&D
Plastics recovery	Chem-8	Medium	High	Demo
Continuous melt silicon crystal growth	Electron-1	Medium	High	R&D
100% Recycled glass cullet	Glass-1	Medium	High	Demo
Anaerobic waste water treatment	Other-1	Medium	High	Dissemination, demo
Dry sheet forming	Paper-4	Medium	High	R&D, demo
Biodesulfurization	Refin-1	Medium	High	R&D, demo

Note: Technologies in this table are listed in alphabetical order based on industry sector.

29.7 Conclusion

Energy is the lifeblood of industry; it is used to convert fuels to thermal, electric, or motive energy to manufacture all the products of daily life. Using this energy efficiently is a necessity to keep industries competitive, clean, and at their peak of productivity. Energy management programs that improve the operational efficiency and the technological efficiency of industry are critical to the long-term success of industry and manufacturing in the United States. One important result in this area has been a recognition that the United States is not running out of technologies to improve industrial energy efficiency, productivity, and environmental performance, and it is not going to run out in the foreseeable future. A substantial opportunity to the country's industrial energy use by over 20% is currently available using better operational procedures and using improved equipment in industrial plants. These savings to industry are worth almost $19 billion at 2004 energy prices. With crude oil prices edging toward $70 in late summer 2005, this dollar savings amount should be substantially higher. It is time to capture the benefits of this opportunity.

References

1. U.S. Department of Energy (U.S. DOE), September 2012. EIA annual energy review for 2011, Washington, DC.
2. U.S. DOE Energy Information Administration, 1995. Manufacturing energy consumption survey: Changes in energy efficiency 1985–1991. U.S. Department of Energy, Office of Industrial Technologies, Industrial Technology Program, Washington, DC.
3. NAM, 2005. Efficiency and innovation in U.S. manufacturing energy use. National Association of Manufacturers, Washington, DC.
4. U.S. DOE, 2004. Annual report: Technology, delivery, industry of the future. U.S. Department of Energy, Office of Industrial Technologies, Industrial Technology Program, Washington, DC.
5. U.S. DOE, 1990. The National energy strategy, Chapter on Industrial energy use. U.S. Department of Energy, Washington, DC.
6. Oak Ridge National Laboratory, 2000. Scenarios for a clean energy future, Interlaboratory Working Group. Oak Ridge National Laboratory, Oak Ridge, TN.
7. Price, L. and Worrell, E. 2004. Improving industrial energy efficiency in the U.S.: Technologies and policies for 2010–2050. *Proceedings of the 10–50 Solution: Technologies and Policies for a Low Carbon Future*. Lawrence Berkeley National Laboratory, Berkeley, CA.
8. Elliott, R. N. and Nadel, S. 2003. Realizing energy efficiency opportunities in industrial fan and pump systems, Report A034. American Council for an Energy-Efficient Economy, Washington, DC.
9. Aspenson, R. L. 1989. Testimony to the U.S. Department of Energy on the National Energy Strategy, hearings on energy and productivity, Providence, RI.
10. 3M Company. April 2013. Improving energy efficiency. http://www.MMM.com.
11. Capehart, L. C. and Capehart, B. L. 1994. Writing user-friendly energy audit reports. *Strategic Planning for Energy and the Environment*, 14(2), 17–26 (Published by the Association of Energy Engineers, Atlanta, GA).
12. Capehart, B. L. and Capehart, L. C. 1994. Improving industrial energy audit analyses. *Proceedings of the ACEEE Summer Study of Industrial Energy Use*. ACEEE, Washington, DC.
13. Pawlik, K.-D., Capehart, L. C., and Capehart, B. L. 2001. Analyzing facility energy use: A balancing act. *Strategic Planning for Energy and the Environment*, 21(2), 8–23.
14. Air-Conditioning & Refrigeration Institute. June 2005. ARI Unitary Directory. http://www.ari.org.
15. Nadel, S., Elliott, R. N., Shepard, M., Greenberg, S., and Katz, G. 2002. *Energy-Efficient Motor Systems: A Handbook on Technology, Program and Policy Opportunities*, 2nd edn. American Council for an Energy-Efficient Economy, Washington, DC.
16. Washington State Energy Office. June 2005. MotorMaster electric motor selection software and database. http://www.oit.doe.gov/bestpractices/software_tools.shtml.
17. Hoshide, R. K. 1994. Electric motor do's and don't's. *Energy Engineering*, 91(1), 6–24 (Published by the Association of Energy Engineers, Atlanta, GA).

18. Stebbins, W. L. 1994. Are you certain you understand the economics for applying ASD systems to centrifugal loads? *Energy Engineering*, 91(1), 25–44 (Published by the Association of Energy Engineers, Atlanta, GA).
19. Vaillencourt, R. R. 1994. Simple solutions to VSD pumping measures. *Energy Engineering*, 91(1), 45–59 (Published by the Association of Energy Engineers, Atlanta, GA).
20. Kempers, G. 1995. DSM pitfalls for centrifugal pumps and fans. *Energy Engineering*, 92(2), 15–23 (Published by the Association of Energy Engineers, Atlanta, GA).
21. Webb, M. (Senior Engineer). January 1995. Personal communication. Virginia Power Company, Roanoke, VA.
22. Capehart, B. L., ed. 2004. *Information Technology for Energy Managers: Understanding Web Based Energy Information and Control Systems*. Fairmont Press, Atlanta, GA.
23. Capehart, B. L. and Capehart, L. C., eds. 2005. *Web Based Energy Information and Control Systems: Case Studies and Applications*. Fairmont Press, Atlanta, GA.
24. Turner, W. C., ed. 2012. *Energy Management Handbook*. Fairmont Press, Atlanta, GA.
25. Martin, N. and Elliott, R. N. 2000. *Emerging Energy Efficient Technologies*. American Council for an Energy-Efficient Economy, Washington, DC.
26. U.S. Department of Energy, Energy Information Agency. 2012. Manufacturing energy consumption survey, 2006, Washington, DC.
27. U.S. DOE, Energy Information Agency. September 2014. Monthly energy review, Washington, DC.
28. U.S. Department of Energy. 2005. Office of Industrial Technologies, Industrial Technology Program, Washington, DC.
29. U.S. Department of Energy, Energy Information Agency. 2002. Manufacturing Energy Consumption Survey, 2006, Washington, DC.
30. U.S. Department of Energy. 2005. Office of Industrial Technologies, Washington, DC, http://www.oit.doe.gov/iac/schools.shtml.
31. U.S. Department of Energy, Energy Information Agency. 2002. Manufacturing Energy Consumption Survey, 1998, Washington, DC.
32. Dukelow, S.G. 1974. Bailey Meter Company, Wickliffe, OH.

30

Process Energy Efficiency: Pinch Technology

Kirtan K. Trivedi, Ed Fouche, and Kelly E. Parmenter

CONTENTS

30.1 Pinch Technology in Theory

Many industrial processes require that heat be added to and removed from streams within the process. This is notably the case in the chemical process industry. Raw materials are heated or cooled to the appropriate reaction temperature. Heat may be added or removed to carry out the reaction at the specified condition of temperature and pressure. The product separation is achieved by heat addition (to a reboiler in a distillation column) and removal (from a condenser of the distillation column). The product may be heated or cooled to the correct temperature for storage and transportation. Thus, at all times heat is either added or removed to a variety of process streams by utilities or by heat exchange between process streams. In an integrated plant this is normally achieved with a *heat exchanger network* (HEN). A large fraction of the capital cost of many chemical process plants is attributed to heat recovery networks.

The aim of a designer is to synthesize a near optimal configuration for a process. This indicates that a proper trade-off between the capital invested and the operating cost of the plant should be achieved. The capital cost depends on the type, number, and size of units utilized to satisfy the design objectives. A substantial part of the operating cost usually depends upon the utilities consumed. To reduce these costs, the process designer should aim for an economic combination having near the theoretical minimum number of heat exchanger units and aim to recover the maximum possible heat with them. An obvious way to recover heat is by exchanging it between hot process streams that need to be cooled and cold process streams that need to be heated, in addition to the heating and cooling utilities. Furthermore, the designer should also investigate the operability of the final design.

These objectives can be achieved by synthesizing a good heat exchanger network. However, a very large number of alternatives exist. Because of this, it will be highly rewarding to synthesize quickly and systematically the best possible alternatives. It is now possible to do this with the aid of "pinch technology."

Pinch technology refers to a large and growing set of methods for analyzing process energy requirements in order to find economically optimal and controllable designs. Considerable development has taken place in pinch technology during the last three decades, mainly due to the efforts of Linnhoff and co-workers. Pinch technology has proved to be effective and is successfully applied to process integration[1] that encompasses overall plant integration and includes heat exchanger networks, heat and power integration or cogeneration, and thermal integration of distillation columns. To date, there are more than 65 concepts used by this technology. However, due to the limited scope of this chapter, only the most important concepts are discussed here.

Industrial applications of this technology include capital cost reduction, energy cost reduction, emissions reduction, operability improvement, and yield improvement for both new process design and revamp process design. Imperial Chemical Industries (ICI) where this technology was first developed reported an averaged energy saving of about MM$11/year, (about 30%) in processes previously thought optimized.[2] The payback time was typically in the order of 12 months. Union Carbide showed even better results.[3] Studies conducted by Union Carbide in nine projects showed an average savings of 50% with an average payback period of 6 months. BASF reports energy savings of 25% obtained by application of pinch technology to their Ludwigshafen site in Germany.[4] Over a period of 3 years about 150 projects were undertaken by BASF. Fraser and Gillespie[5] reported energy savings of 10% with a payback period of 2 years by applying pinch technology to the Caltex Refinery situated in Milnerton, South Africa. The energy consumption before the study was 100 MW for the whole refinery. A newsfront article in *Chemical Engineering*[6] gives more details about the experience of various companies in using pinch technology and the benefits obtained.

30.1.1 Fundamental Principles and Basic Concepts

Pinch technology is based on thermodynamic principles. Hence, in this section we will review the important thermodynamic principles and some basic concepts.

30.1.1.1 Temperature-Enthalpy Diagram

Whenever there is a temperature change occurring in a system, the enthalpy of the system will change. If a stream is heated or cooled, then the amount of heat absorbed or liberated can be measured by the amount of the enthalpy change. Thus,

$$Q = \Delta H \tag{30.1}$$

For sensible heating or cooling at constant pressure where $CP = mc_p$,

$$\Delta H = CP\Delta T \tag{30.2}$$

For latent heating or cooling

$$\Delta H = m\lambda \tag{30.3}$$

If we assume that the temperature change for latent heating or cooling is 1°C, then

$$CP = m\lambda \tag{30.4}$$

Equation 30.2 enables us to represent a heating or cooling process on a temperature-enthalpy diagram. The abscissa is the enthalpy and the ordinate is the temperature. The slope of the line is (1/CP). Figure 30.1a shows a cold stream being heated from 20°C to 80°C with CP = 2.0 kW/°C.

As enthalpy is a relative function, the stream can be drawn anywhere on the enthalpy scale as long as it is between its starting and target temperatures and has the same enthalpy change. Figure 30.1b shows such a case. Thus, one of the important advantages of representing a stream on the temperature-enthalpy plot is that the stream can be moved horizontally between the same temperature intervals. Figure 30.1c shows two different streams in the same temperature interval.

Two streams can be easily added on the temperature-enthalpy plot to represent a composite of the two streams. Figure 30.2 shows how to obtain the composite of two streams on this plot. This feature will be used in later sections to predict the minimum utility requirement for multistream problems.

A heat exchanger is represented by two heat curves on the temperature-enthalpy diagram as shown in Figure 30.3. This figure also shows how we will represent heat exchangers in a grid diagram form and the conventional form. In the grid diagram form, the hot stream goes from left to right and the cold stream goes from right to left. If the flow

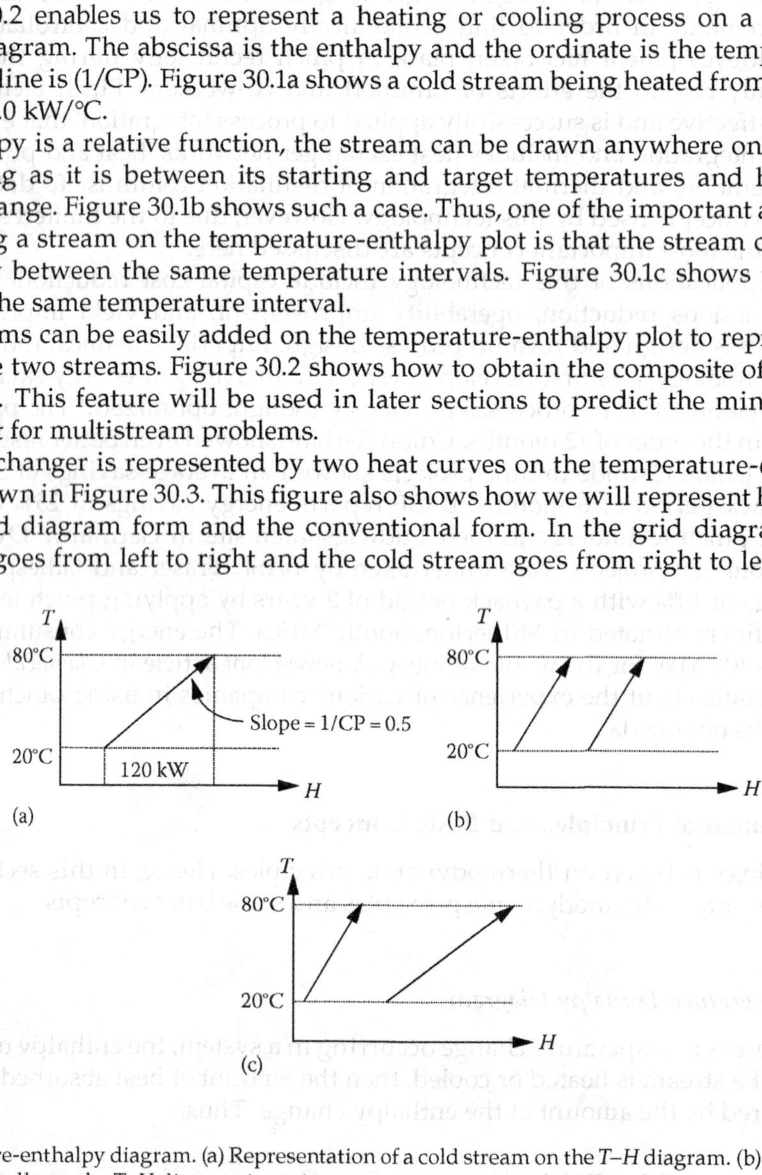

FIGURE 30.1
The temperature-enthalpy diagram. (a) Representation of a cold stream on the *T–H* diagram. (b) A stream can be moved horizontally on the *T–H* diagram in a given temperature interval. (c) Two different streams in the same temperature interval.

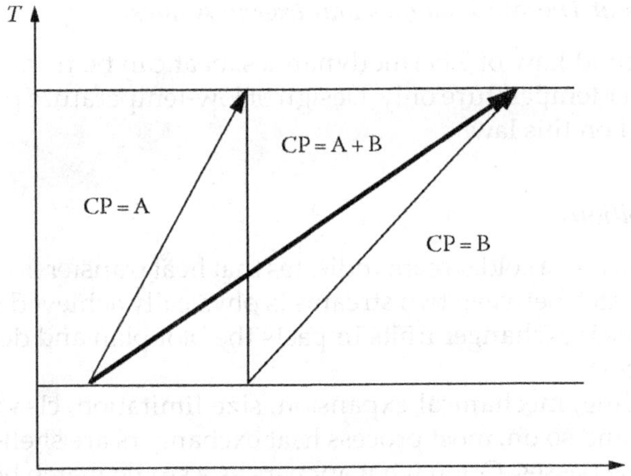

FIGURE 30.2
Composite stream obtained from two different streams in the same temperature interval.

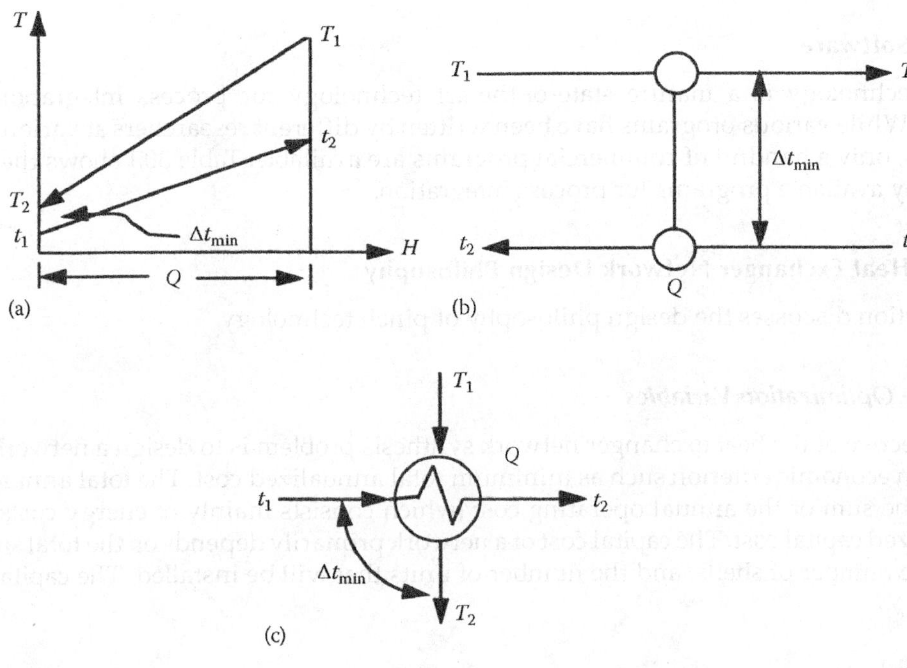

FIGURE 30.3
Heat exchanger representation. (a) Representation of an exchanger on the *T–H* diagram. (b) Grid diagram representation of an exchanger. (c) Conventional representation of an exchanger.

is counter-current in the exchanger, then the temperature will decrease from left to right. The exchanger is represented by two circles connected by a vertical line. The advantages of the grid diagram will become apparent when we discuss the design of heat exchanger networks. The exchanger has a temperature difference at the hot end and another at the cold end of the heat exchanger. The smaller of these two temperature differences is called Δt_{min}.

30.1.1.2 Second Law of Thermodynamics and Exergy Analysis

According to the Second Law of Thermodynamics, heat can be transferred from a higher temperature to a lower temperature only. Design of low-temperature processes uses exergy analysis that is based on this law.

30.1.1.3 Some Definitions

A *match* between a hot and a cold stream indicates that heat transfer is taking place between the two streams. A match between two streams is physically achieved via a heat exchanger *unit*. The number of heat exchanger units impacts the plot plan and determines the piping and the foundation cost.

For reasons of fouling, mechanical expansion, size limitation, cleaning, improved heat transfer coefficients, and so on, most process heat exchangers are shell-and-tube type with 1 shell pass and 2 tube passes. Often what appears as a single match between two streams in a heat exchanger network representation is actually installed as several 1–2 exchangers in series or parallel. The term *shell* will be used to represent a single 1–2 shell-and-tube heat exchanger. See Chapter 41 for more discussion on 1–2 shell-and-tube heat exchangers.

30.1.2 Software

Pinch technology is a mature state-of-the-art technology for process integration and design. While various programs have been written by different researchers at various universities, only a handful of commercial programs are available. Table 30.1 shows the commercially available programs for process integration.

30.1.3 Heat Exchanger Network Design Philosophy

This section discusses the design philosophy of pinch technology.

30.1.3.1 Optimization Variables

The objective of the heat exchanger network synthesis problem is to design a network that meets an economic criterion such as minimum total annualized cost. The total annualized cost is the sum of the annual operating cost (which consists mainly of energy costs) and annualized capital cost. The capital cost of a network primarily depends on the total surface area, the number of shells, and the number of units that will be installed. The capital cost

TABLE 30.1

Commercially Available Computer Programs for Pinch Technology

Software Name	Marketed By
Supertarget	KBC Advanced Technologies plc, Head Office and Registered Office, 42-50 Hersham Road, Walton on Thames, Surrey KT12 1RZ, UK. Tel.: +44 (0) 1932 242 424; Fax: +44 (0) 1932 224 214; Web: www.kbcat.com
Advent	Aspen Technology, Inc., 20 Crosby Drive, Bedford, Massachusetts 01730, USA. Tel.: +1-781-221-6400; Fax: +1-781-221-6410; Web: www.aspentech.com
i-Heat	Process Integration Limited, Station House, Stamford New Road, Altrincham, Cheshire, WA14 1EP, United Kingdom. Tel.: +44 (0) 161 974 0090; Web: www.processint.com

will also depend on the individual type of heat exchangers and their design temperature, pressure, and material of construction.

If we include the pressure drop incurred in a heat exchanger, then the capital and operating cost of the pumps will also have to be taken into account. We will limit the scope of this chapter by not including stream pressure drop constraints.

Figure 30.4 shows the various variables that affect the optimization of a heat exchanger network. To make the network operable or flexible, extra cost may be incurred. This cost may be in the form of added equipment, use of extra utilities, or use of additional area on some of the exchangers in the network.

From this discussion, it is clear that the synthesis of heat exchanger networks is a multi-variable optimization problem. The design can never violate the laws of thermodynamics. Hence, the philosophy adopted in pinch technology is to establish targets for the various optimization variables based on thermodynamic principles. These targets set the boundaries and constraints for the design problem. Further, targets help us identify the various trade-offs between the optimization parameters. They help us in obtaining a bird's-eye view of the solution space and identify the optimal values of the optimization parameters.

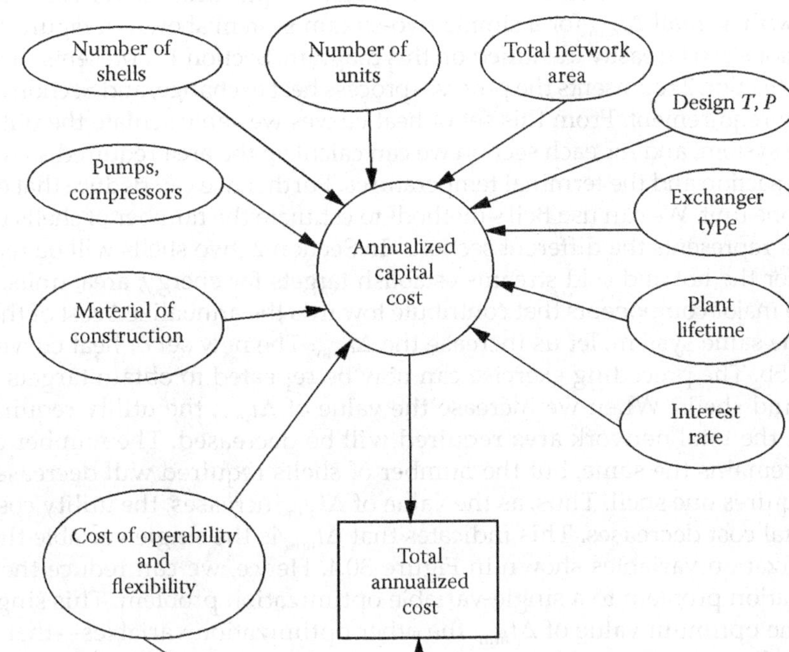

FIGURE 30.4
Optimization variables in the design of heat exchanger networks.

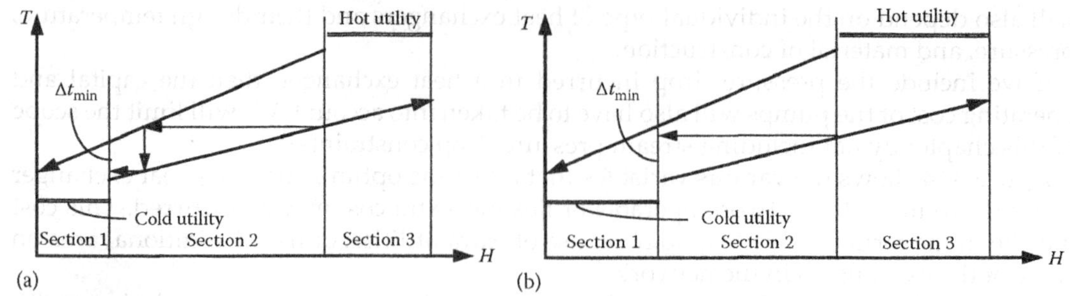

FIGURE 30.5
Effects of Δt_{min} on the energy, area, units, and shells required for a heat exchange system. (a) Small ΔT_{min}; (b) large ΔT_{min}.

Once the optimal values are identified, design the network at these values. This approach will always lead to an optimal design.

To understand the interaction between the different optimization variables consider the heat curves with a small Δt_{min} for a simple two-stream system shown in Figure 30.5a. Three different sections can be easily identified on this diagram. Section 1 represents the cold utility requirement, Section 2 represents the process–process heat exchange, and Section 3 represents the hot utility requirement. From this set of heat curves we can calculate the utility requirements for the system, and for each section we can calculate the area required as we know the duty for each section and the terminal temperatures. Further, we can deduce that each section will require one unit. We can use Bell's method[7] to estimate the number of shells required for each unit that represents the different sections. In Section 2, two shells will be required. The heat curves for the hot and cold streams establish targets for energy, area, units, and shells. These are the major components that contribute towards the annualized cost of the network.

Now for the same system, let us increase the Δt_{min}. The new set of heat curves is shown in Figure 30.5b. The preceding exercise can now be repeated to obtain targets for energy, area, units, and shells. When we increase the value of Δt_{min}, the utility requirement will increase and the total network area required will be decreased. The number of units for this system remains the same, but the number of shells required will decrease. Section 2 now only requires one shell. Thus, as the value of Δt_{min} increases, the utility cost increases and the capital cost decreases. This indicates that Δt_{min} is the single variable that fixes the major optimization variables shown in Figure 30.4. Hence, we can reduce the multivariable optimization problem to a single-variable optimization problem. This single variable is Δt_{min}. At the optimum value of Δt_{min}, the other optimization variables—that is, number of shells, number of units, total network area requirement, and the hot and cold utility requirements—will also be optimal. For a multistream problem the same conclusion can be easily derived. Hence, for the multistream problem optimization discussed in the next sections, we will develop methods to optimize the value of Δt_{min}.

30.1.4 Design Problem

Consider an illustrative plant flowsheet shown in Figure 30.6. The feed is heated to the reaction temperature. The reactor effluent is further heated, and the products are separated in a distillation column. The reboiler and condenser use external utilities for control purposes. The overhead and bottoms products are cooled and sent for further processing. We shall use this problem to illustrate the concepts of pinch technology.

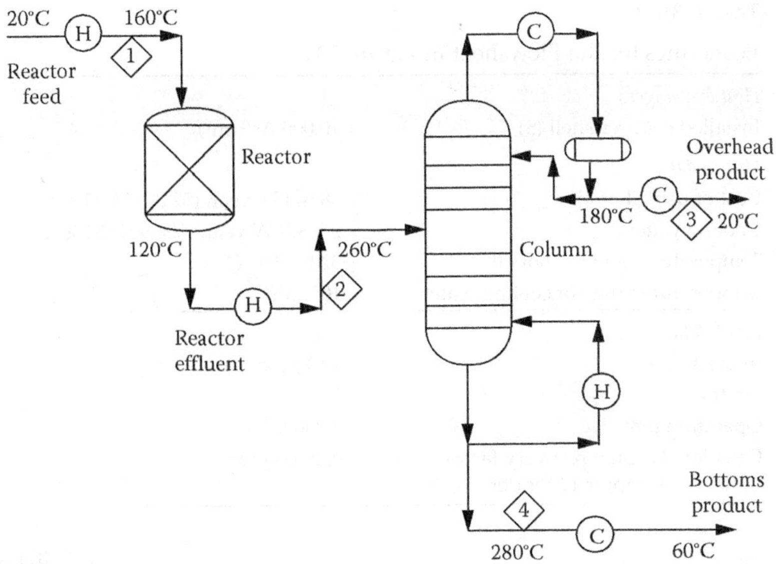

FIGURE 30.6
A typical plant flowsheet.

TABLE 30.2
Stream Data for the Flowsheet in Figure 30.6

Stream Number	Stream Name	Stream No.	T_s (°C)	T_t (°C)	CP (kW/°C)
1	Reactor feed	1	20	160	40
2	Reactor effluent	2	120	260	60
3	Overhead product	3	180	20	45
4	Bottoms product	4	280	60	30

Table 30.2 lists the starting and target temperatures of all the streams involved in the flowsheet. It also shows the CP values. While pressure drop constraints on the individual streams determine the film heat transfer coefficients, we shall not take into account this constraint. Instead, we use (unrealistically) the same heat transfer coefficient value of 1 kW/m^2 K for all streams. Our objective is to design an optimum heat exchanger network for this process using the economics outlined in Table 30.3. Further, let Δt_{min} denote the minimum temperature difference between any hot stream and any cold stream in any exchanger in the network. It is desired that the final network should be able to control the target temperature of all the streams at the design temperature.

30.1.5 Targets for Optimization Parameters

30.1.5.1 Energy Targets

30.1.5.1.1 Composite Curves

Let us plot the heat curves for all the hot streams on the *T–H* diagram (see Figure 30.7). We can divide the diagram into a number of temperature intervals, defined by the starting and target temperatures for all the streams. Between two adjacent temperatures we

TABLE 30.3
Economics for the Flowsheet in Figure 30.6

Heat exchangers	
Installed cost per shell ($)	= 10,000 $A^{0.6}$ (m^2)
Utility data	
Cost of using hot oil	= 68 $/kW year, ($2.36/MMBtu)
Cooling water	= 2.5 $/kW year, ($0.09/MMBtu)
Temperature range of hot oil	= 320°–310°C
Temperature range of cooling water	= 10°–20°C
Plant data	
Interest rate	= 10%/year
Lifetime	= 5 years
Operation time	= 8,000 h/year
Calculated capital recovery factor, CRF (see Chapter 12 for details)	= 0.2638/year

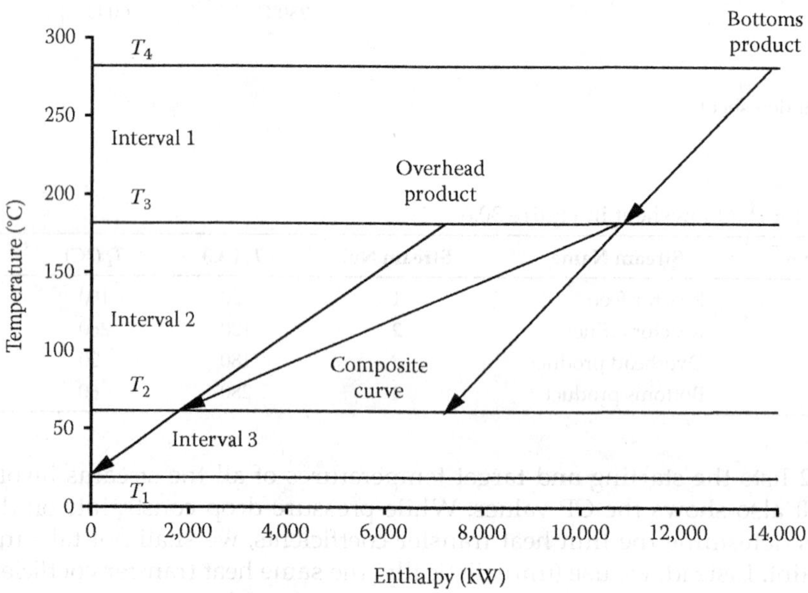

FIGURE 30.7
Construction of the hot composite curve.

can calculate the total heat content of all the streams that are present in this temperature interval. For example, between 180°C and 60°C, the sum total of the heat available is calculated as

$$\Delta H = (T_3 - T_2)\Sigma_j \mathrm{CP}_j$$
$$\Delta H_2 = (180 - 60)(40 + 30) = 8400 \text{ kW} \tag{30.5}$$

A composite curve that represent the heat content of all the hot streams is obtained by summing the heat available in each of these temperature intervals as shown in Figure 30.7.

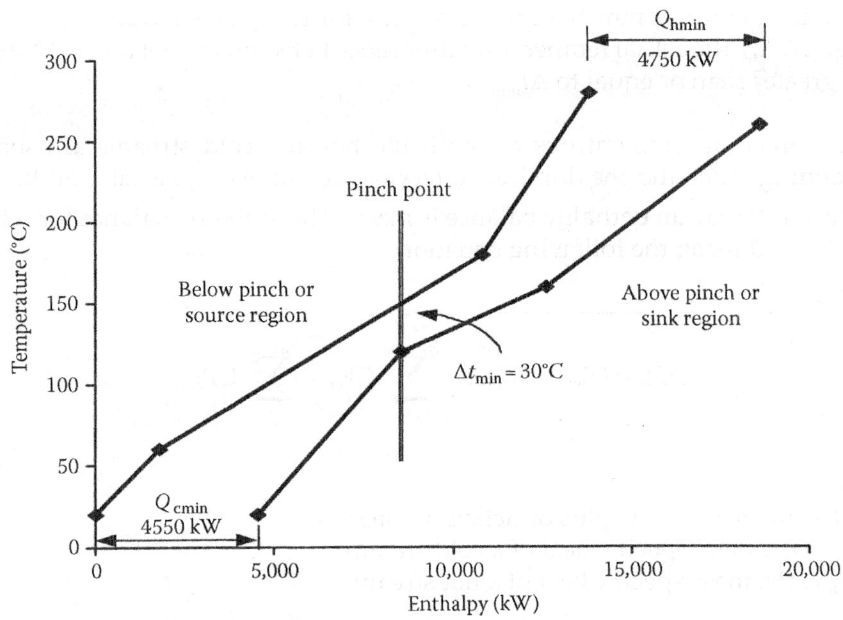

FIGURE 30.8
Composite curves for the flowsheet in Figure 30.9.

Similarly, a composite curve for all the cold streams can be obtained. These two composite curves can be used as the heat curves for the whole process.

To obtain the energy target, fix the hot composite curve and move the cold composite curve horizontally till the shortest vertical distance between the two curves is equal to the value of Δt_{min}. The overshoot of the cold composite curve is the minimum hot utility requirement and the overshoot of the hot composite curve is the minimum cold utility requirement.[8] For $\Delta t_{min} = 30°C$, the minimum hot utility requirement is 4750 kW and the cold utility requirement is 4550 kW (see Figure 30.8).

30.1.5.1.2 The Problem Table

The preceding procedure for obtaining the energy targets using composite curves is time-consuming and clumsy. An alternative method based on thermodynamic principles is developed. Hohmann[8] called it the feasibility table. Linnhoff and Flower[9] independently developed the problem table. The problem table algorithm is easy and involves no trial and error.

The algorithm consist of the following steps:

- Select a value of Δt_{min}. Since we have already established the targets using composite curves for $\Delta t_{min} = 30°C$, we shall use that value here.
- Convert the actual terminal temperatures into interval temperatures as follows:

  For the hot streams: $T_{int} = T_{act} - \Delta t_{min}/2$

  For the cold streams: $T_{int} = T_{act} + \Delta t_{min}/2$

where

T_{int} is the interval temperature
T_{act} is the actual stream temperature

The interval temperatures now have the allowance for Δt_{min}. This modification guarantees that for a given T_{int}, the actual temperature difference between the hot and cold stream will always be greater than or equal to Δt_{min}.

- All the interval temperatures for both the hot and cold streams are sorted in descending order and the duplicate intervals are removed (see Table 30.4).
- For each interval, an enthalpy balance is made. The enthalpy balance for interval i is calculated using the following equation:

$$\Delta H_i = (T_{int,i} - T_{int,i+1})\left(\sum_j^{N_{streams}} CP_{cj} - \sum_j^{N_{streams}} CP_{hj}\right) \tag{30.6}$$

where

ΔH_i is the net heat surplus or deficit in interval i
CP_c is the mass specific heat of a cold stream
CP_h is the mass specific heat of a hot stream

For the example problem this calculation is shown in Table 30.5. With each interval the enthalpy balance will indicate that there is either heat deficit or surplus or the interval is in heat balance. A heat surplus is negative and a heat deficit is positive.

The Second Law of Thermodynamics only allows us to cascade heat from a higher temperature to a lower temperature. Thus, a heat deficit from any interval can be satisfied in two possible ways—by using an external utility or by cascading the surplus heat from a higher temperature interval. If no external utility is used, the heat cascade column of Table 30.5 can be constructed. All intervals in this heat cascade have negative heat flows. This is thermodynamically impossible in any interval (it would mean heat flowing from lower to higher temperature). To correct this situation, we take the largest negative flow in the cascade and supply that amount of external hot utility at the highest temperature interval. This modification will make the heat cascade feasible (i.e., none of the intervals will have negative heat flows). The amount of external heat supplied to the first interval is minimum amount of hot utility required. The surplus heat in the last interval is the minimum amount of cold utility required.

TABLE 30.4
Generation of Temperature Intervals for Δt_{min} = 30°C

	Actual Temperature (°C)		Interval Temperature (°C)			
Stream No.	T_s	T_t	T_s	T_t	Interval Number	Ordered Interval Temperatures (°C)
1	20	160	35	175	1	275
2	120	260	135	275	2	265
3	180	20	165	5	3	175
4	280	60	265	45	4	165
					5	135
					6	45
					7	35
					8	5

TABLE 30.5

The Problem Table

Stream No.		**1**	**2**	**3**	**4**					
CP		**40**	**60**	**45**	**30**					
Temperature Interval	**Interval Number**					ΔT_{int}	$\Sigma CP_{cj} - \Sigma CP_{hj}$	ΔH_{int}	**Heat Cascade**	**Corrected Heat Cascade**
Hot utility–275	0								0	4750
275–265	1		■			10	60	600	−600	4150
265–175	2		■		■	90	30	2700	−3300	1450
175–165	3	■	■		■	10	70	700	−4000	750
165–135	4	■	■	■	■	30	25	750	−4750	0
135–45	5	■	■	■	■	90	−35	−3150	−1600	3150
45–35	6	■		■	■	10	−5	−50	−1550	3200
35–5	7	■		■		30	−45	−1350	−200	4550
5–Cold utility	8			■						4550

For the example under consideration, the minimum hot utility required is 4750 kW and the minimum cold utility required is 4550 kW. These are the same targets obtained from the composite curves. It is clear, however, that the problem table algorithm is an easier, quicker, and more exact method for setting the energy targets. One can easily set up a spreadsheet to implement the problem table algorithm. Further, once the energy targets are obtained, the composite curves can be easily drawn to visualize the heat flows in the system. It should be noted that the absolute minimum utility targets for a fixed flowsheet are determined by the case of $\Delta t_{min} = 0$.

30.1.5.2 Capital Cost Targets

To find the optimal value of Δt_{min} ahead of design, we need to set the targets for the capital and energy costs. As seen before, the capital cost target for a given Δt_{min} depends on the total area of the network, the number of shells required in the network, and the number of units required in the network. We will now establish these targets.

30.1.5.2.1 Target for Minimum Total Area for the Network

The composite curves can be divided into different enthalpy intervals at the discontinuities in the hot and cold composite curves. Between any two adjacent enthalpy intervals, if the hot streams in that interval transfer heat only to the cold streams in that interval and vice versa, then we say that vertical heat transfer takes place along the composite curves.[10] This mode of heat transfer models the pure countercurrent heat exchange.

The equation for establishing the minimum area target is based on a complex network called the "spaghetti" network. This network models the vertical heat transfer on the composite curves. In this network, for any enthalpy interval defined by the discontinuities in the composite curves, each hot stream is split into the number of cold streams in that enthalpy interval. Each cold stream in the enthalpy interval is also split into the number of hot streams in that interval. A match is made between each

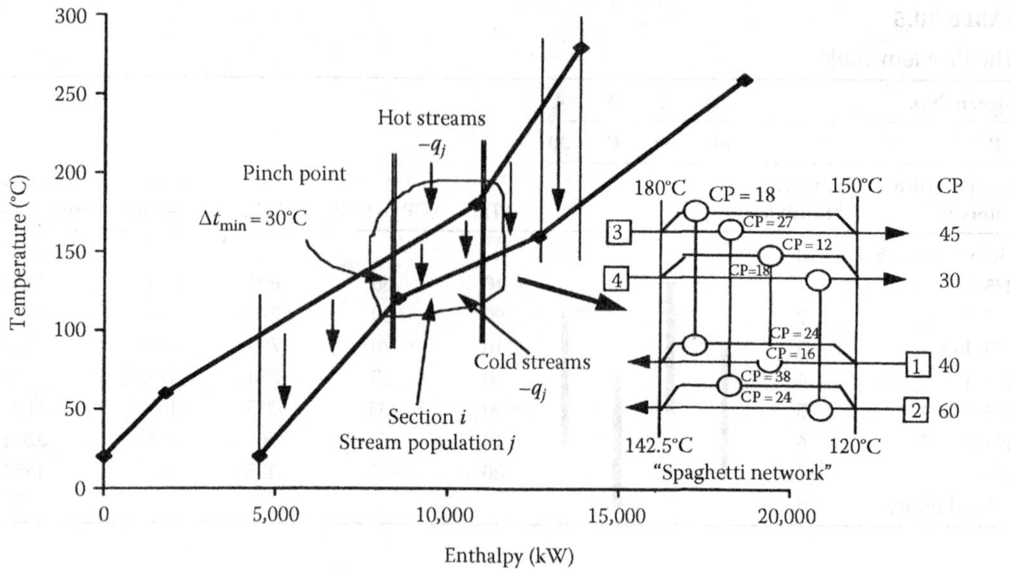

FIGURE 30.9
Enthalpy intervals for area targeting and the spaghetti network that mimics vertical heat transfer.

hot stream and each cold stream. Figure 30.9 shows this network. The area target is obtained by the following equation:

$$A_{min} = \sum_{i}^{N_{intervals}} \frac{1}{\Delta T_{LMTDi}} \left(\sum_{j}^{N_{streams}} \frac{q_j}{h_j} \right)_i \tag{30.7}$$

where

A_{min} is the total minimum area for the network
ΔT_{LMTDi} is the ΔT_{LMTD} for enthalpy interval i
q_j is the heat content of stream j in enthalpy interval i
h_f is the film plus fouling heat transfer coefficient of stream j in enthalpy interval i

The equation gives a minimum total surface area for any system where the process streams have uniform heat transfer coefficients. Townsend and Linnhoff[10] claim that for non uniform heat transfer coefficients the equation gives an useful approximation of the minimum area, with errors being typically within 10%.

30.1.5.2.2 Minimum Number of Units Target

The minimum number of units is given by[11]:

$$u = N + L - s \tag{30.8}$$

where

u is the number of units including heaters and coolers
N is the number of streams including utilities
L is the number of loops
s is the number of separate networks

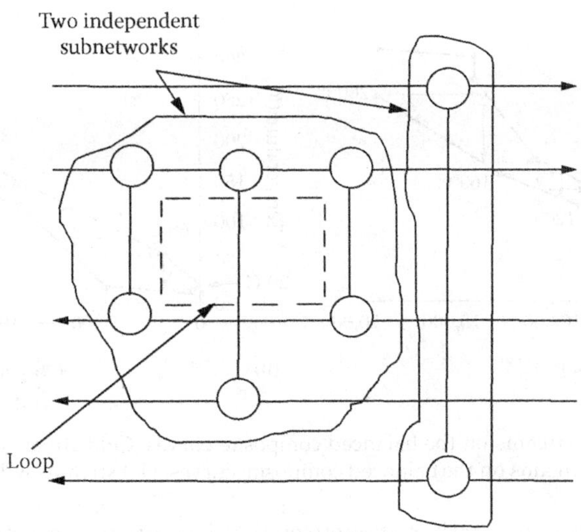

FIGURE 30.10
Loops and independent subnetworks in a heat exchanger network.

A loop is a closed path through the network, and its effect is to increase the number of units. For a network to have the minimum number of units, the number of loops should be zero. Figure 30.10 shows a loop on the grid diagram. Normally for a given stream system, only a single network exists and if the number of loops is assumed to be zero, then the equation can be reduced to[8]

$$u = N - 1 \tag{30.9}$$

Figure 30.10 illustrates the occurrence of independent networks. The total network consists of two independent subnetworks. Thus, for the system shown in this figure, $N = 5$, $L = 1$, $s = 2$, and $u = 4$.

30.1.5.2.3 Target for Minimum Number of Shells

Trivedi et al.[12] have developed a method for estimating the total number of shells required for the heat exchanger network. The method starts by setting the energy targets for the selected value of Δt_{min}. The hot and cold utilities are added to the process stream system and the composite curves for the process, and the utility streams are constructed. No external utility requirement will be needed for these set of composite curves. Hence, these are called the balanced composite curves.[13] The method estimates the number of shells based on the balanced composite curves. The method involves two steps:

Step 1: Estimate the total number of shells required by the cold process streams and the cold utility streams. See Figure 30.11a.

- Commencing with a cold stream target temperature, a horizontal line is drawn until it intercepts the hot composite curve. From that point a vertical line is dropped to the cold composite curve. This section, defined by the horizontal line, represents a single exchanger shell in which the cold stream under consideration gets heated without the possibility of a temperature cross. In this section, the cold stream will have at least one match with a hot stream. Thus, this section implies that the cold stream will require at least one match

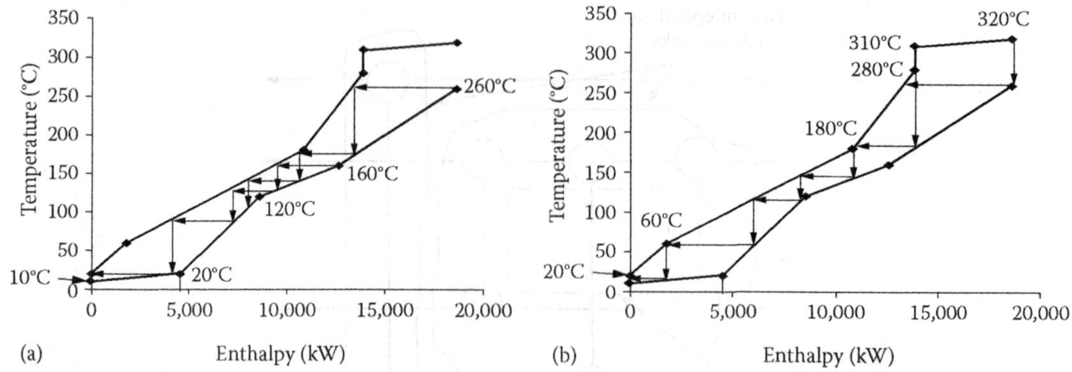

FIGURE 30.11
(a) Shell targeting for cold streams on the balanced composite curves. Cold streams will require eight shells. (b) Shell targeting for hot streams on the balanced composite curves. Hot streams will require nine shells.

with a hot stream. Further, it ensures that log mean temperature (LMTD) correction factor, $F_t \geq 0.8$.

- Repeat the procedure until a vertical line intercepts the cold composite curve at or below the starting temperature of that particular stream.
- The number of horizontal lines will be the number of shells the cold stream is likely to require to reach its target temperature.
- Repeat the procedure for all the cold streams including the cold utility streams.
- The sum of the number of shells for all the cold streams is the total number of shells required by the cold streams to reach their respective target temperatures.

Step 2: Estimate the total number of shells required by the hot process streams and the hot utility streams. See Figure 30.11b.

- Starting from the hot stream initial temperature, drop a vertical line on the balanced composite curve until it intercepts the cold composite curve. From this point construct the horizontal and vertical lines until a horizontal line intercepts the hot composite curve at or below the hot stream target temperature.
- The number of horizontal lines will be the number of shells required by the hot streams for heat exchange in the network.
- Repeat the procedure for all the hot streams including the hot utility streams.
- The sum of the number of shells required by the hot streams will be the total number of shells required by the hot streams to reach their respective target temperatures.

The quasi-minimum number of shells required in the network would be the larger of either the total number of shells required by the hot streams or the total number of shells required by the cold streams.

Ahmad and Smith[14] have pointed out that this procedure works for the problems whose composite curves are wide. They suggest that the procedure should be applied independently to the above and below pinch subnetworks. (Above and below pinch subnetworks are discussed in Section 30.1.7.2) They have also proposed another method for finding the number of shells in a network. Their method is more sophisticated and is not within the scope of this chapter.

30.1.5.3 Optimum Δt_{min} Value

The procedures presented so far establish, for a selected Δt_{min} value, targets for minimum energy, minimum area, minimum number of units, and minimum number of shells. These targets along with the cost data can be translated into capital and energy cost for the network. The targets can be evaluated at different values of Δt_{min} to obtain a bird's-eye view of the solution space and the optimal value of Δt_{min} ahead of design. This philosophy was first proposed by Hohmann[8] and later developed by Linnhoff and Ahmad[15] as "supertargeting."

For the process flowsheet illustrated in Figure 30.6, the different targeting curves are shown in Figure 30.12a through f. It is seen from the total annualized cost curve,

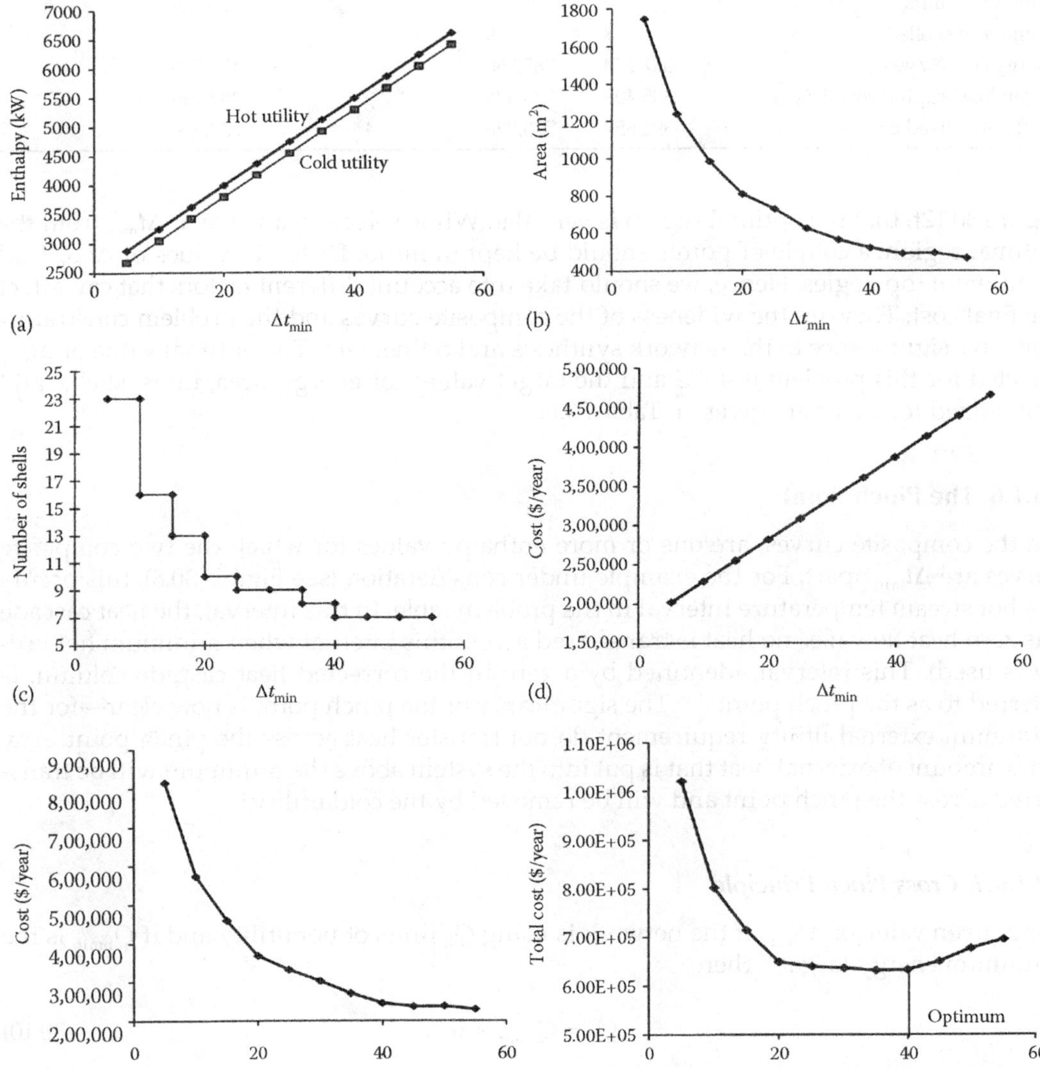

FIGURE 30.12
(a) Minimum energy targeting plot. (b) Minimum network area targeting plot. (c) Targeting plot for minimum number of shells. (d) Minimum energy cost targeting plot. (e) Minimum annualized capital cost targeting plot. (f) Targeting plot for total annualized cost.

TABLE 30.6

Comparison of Target and Design Values for Different Network Optimization Variables at $\Delta t_{min} = 40°C$

	Target	Figure 30.24 MER Design		Figure 30.25 Design	
			% Deviation from Target		% Deviation from Target
Total network area (m²)	687	779	13	708	3
Total hot utility consumption (kW)	5,500	5,500		6,400	16
Total cold utility consumption (kW)	5,300	5,300		6,200	17
Number of units	7	7		6	
Number of shells	8	9		7	
Energy cost ($/year)	387,250	387,250		450,700	16
Annualized capital cost ($/year)	305,407	333,576	9	282,996	−7
Total annualized cost ($/year)	692,657	720,826	4	733,696	6

Figure 30.12f, that the optimal region is very flat. While selecting a value of Δt_{min} from the optimal region, a couple of points should be kept in mind. Different values of Δt_{min} lead to different topologies. Hence, we should take into account different factors that can affect the final cost. They are the wideness of the composite curves and the problem constraints that have significance in the network synthesis and refinement. The optimal value of Δt_{min} selected for this problem is 40°C and the target values for energy, area, units, shells, and annualized total cost are given in Table 30.6.

30.1.6 The Pinch Point

On the composite curves, are one or more enthalpy values for which the two composite curves are Δt_{min} apart. For the example under consideration (see Figure 30.8), this occurs at a hot stream temperature interval in the problem table. In this interval, the heat cascade has zero heat flow (i.e., no heat is transferred across this interval when minimum hot utility is used). This interval, identified by a zero in the corrected heat cascade column, is referred to as the pinch point.[11,16] The significance of the pinch point is now clear—for the minimum external utility requirement do not transfer heat across the pinch point. Any extra amount of external heat that is put into the system above the minimum will be transferred across the pinch point and will be removed by the cold utility.[17]

30.1.6.1 Cross Pinch Principle

For a given value of Δt_{min}, if the network is using Q_h units of hot utility and if Q_{hmin} is the minimum energy target,[17] then

$$Q_h = Q_{hmin} + \alpha \tag{30.10}$$

If the network uses Q_c units of cold utility and if Q_{cmin} is the minimum energy target, then

$$Q_c = Q_{cmin} + \alpha \tag{30.11}$$

where α is the amount of cross-pinch heat transfer.

30.1.6.2 The Pinch Point and Its Significance

The pinch point divides the stream system into two independent subsystems. The subsystem above the pinch point is a net heat sink, and the subsystem below the pinch point is a net heat source. Thus, for a system of hot and cold streams, to design a network that meets the minimum utility targets the following rules set by the pinch principle should be follows[17]:

- Do not transfer heat across the pinch point.
- Do not use hot utility below the pinch point.
- Do not use cold utility above the pinch point.

We shall use these principles to design the network for the optimal value of Δt_{min}.

30.1.7 Network Design

30.1.7.1 Network Representation on the Grid Diagram

If we attempt to design the network using the conventional flowsheet format, any changes in the design may lead to redrawing the flowsheet. Hence, we shall use the grid diagram to represent and design the network. Figure 30.13 shows the grid diagram for the flowsheet shown in Figure 30.6. On this grid diagram we can place a heat exchanger match between two streams without redrawing the whole stream system. The grid representation reflects the countercurrent nature of heat transfer that makes it easier to check temperature feasibility of the match that is placed. Furthermore, we can easily represent the pinch point on the grid diagram as shown in Figure 30.13.

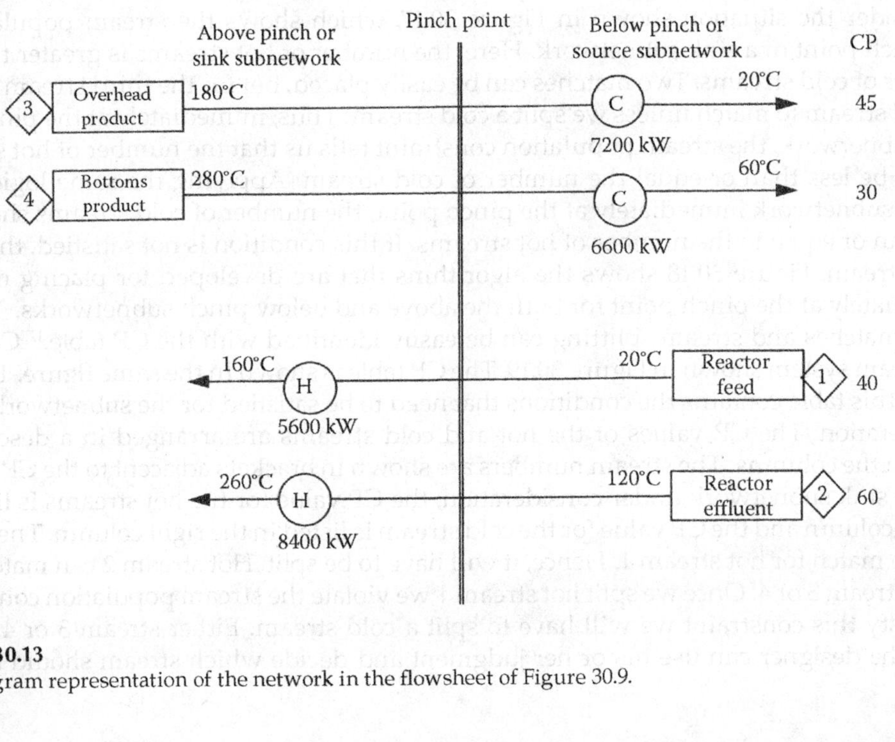

FIGURE 30.13
Grid diagram representation of the network in the flowsheet of Figure 30.9.

30.1.7.2 Pinch Point and Network Design

We now continue with the design of the example problem. Figure 30.8 shows the composite curves with $\Delta t_{min} = 30°C$ for the process flowsheet. It also shows the pinch point. The pinch point is the most constrained region of the composite curves. From optimization principles, we know that for a constrained problem, the optimal solution is located at the point formed by the intersection of multiple constraints. Hence, if we can satisfy the constraints at that point then we are guaranteed an optimal design. Thus, we should start the design where the problem is most constrained, that is, the pinch point.[17] The pinch point divides the problem into two independent subnetworks—an above pinch subnetwork that requires only hot utility and a below pinch subnetwork that requires only cold utility.

30.1.7.2.1 Pinch Design Rules and Maximum Energy Recovery

Let us make some observations at the pinch point. Immediately above the pinch point,[17]

$$\Sigma CP_h \leq \Sigma CP_c \tag{30.12}$$

and immediately below the pinch point,

$$\Sigma CP_h \geq \Sigma CP_c \tag{30.13}$$

This condition defines the pinch point, that is, the constraint which the designer has to satisfy. Thus, each and every match in the sink subnetwork at the pinch point should be placed such that $CP_h \leq CP_c$. Similarly, each and every match in the source subnetwork at the pinch point should be placed such that $CP_c \leq CP_h$. Figure 30.14 shows these conditions on the *T–H* diagram.

Two situations arise where the above condition may not be satisfied. The first condition is shown in Figure 30.15. The solution is to split the hot stream. Another situation is shown in Figure 30.16. Here we will have to split the cold stream.

Consider the situation shown in Figure 30.17, which shows the stream population at the pinch point in a sink subnetwork. Here, the number of hot streams is greater than the number of cold streams. Two matches can be easily placed, but for the third stream there is no cold stream to match unless we split a cold stream. Thus, immediately at the pinch for a sink subnetwork, the stream population constraint tells us that the number of hot streams should be less than or equal the number of cold stream. Applying the same logic to the source subnetwork immediately at the pinch point, the number of cold streams should be less than or equal to the number of hot streams. If this condition is not satisfied, then split a hot stream. Figure 30.18 shows the algorithms that are developed for placing matches immediately at the pinch point for both the above and below pinch subnetworks.

The matches and stream splitting can be easily identified with the CP table.[17] Consider the stream system shown in Figure 30.19. The CP table is shown in the same figure. The first row in this table contains the conditions that need to be satisfied for the subnetwork under consideration. The CP values of the hot and cold streams are arranged in a descending order in the columns. The stream numbers are shown in brackets adjacent to the CP values. For the sink subnetwork under consideration, the CP value for the hot streams is listed in the left column and the CP value for the cold stream is listed in the right column. There is no feasible match for hot stream 1. Hence, it will have to be split. Hot stream 2 can match with either stream 3 or 4. Once we split hot stream 1 we violate the stream population constraint. To satisfy this constraint we will have to split a cold stream. Either stream 3 or 4 can be split. The designer can use his or her judgment and decide which stream should be split

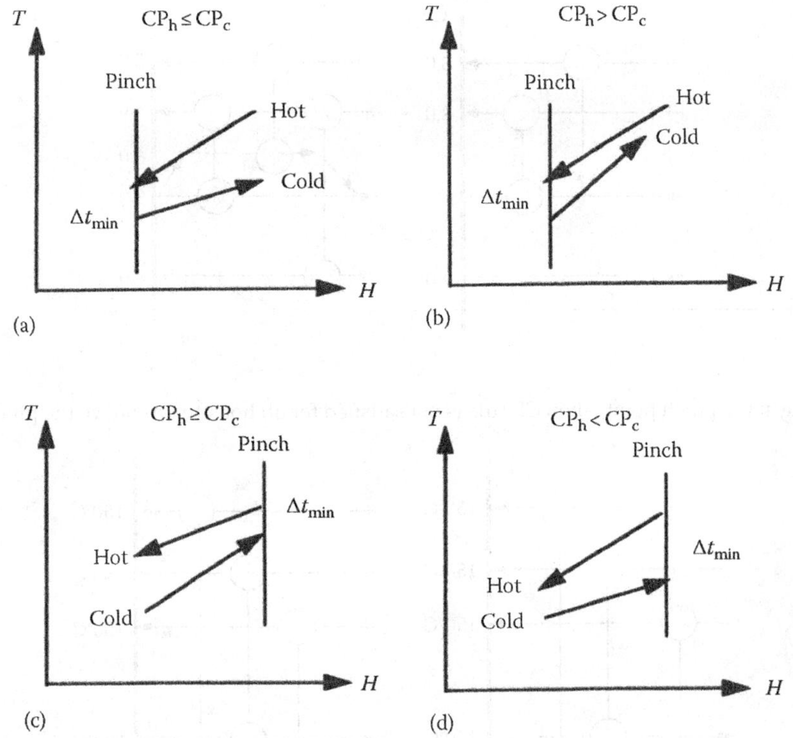

FIGURE 30.14
CP matching rules at the pinch point. (a) Match ok. (b) Match not feasible above pinch. (c) Match ok. (d) Match not feasible below pinch.

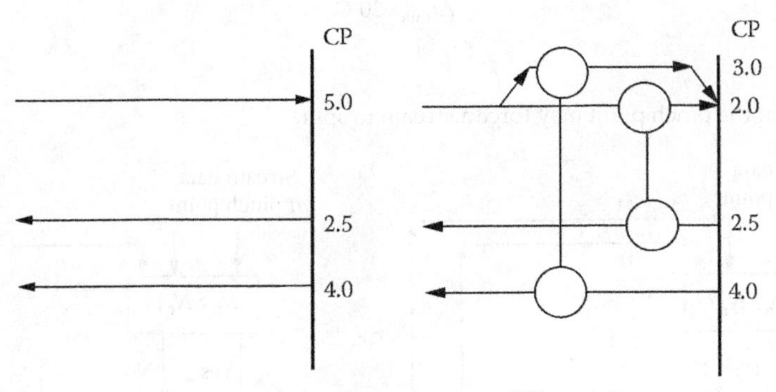

FIGURE 30.15
Stream splitting when CP rule is not satisfied.

depending on controllability and other physical constraints. For example, it is not advisable to split a stream that may have a two-phase flow.

Two different designs will be obtained depending on which stream is split. The hot stream is split into two streams having CP values of X and $7 - X$. In the first alternative, we split stream 4 into streams having CP values of Y and $6 - Y$. In the second option we split stream 3 into streams having CP values of Z and $5 - Z$. To find the initial values of X, Y, and Z it is recommended that all matches except for one are set for CP equality. Thus, for

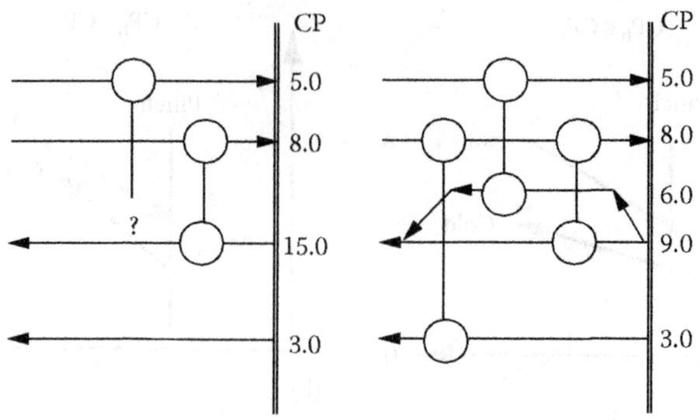

FIGURE 30.16
Stream splitting at the pinch point when CP rule is not satisfied for all hot streams above the pinch point.

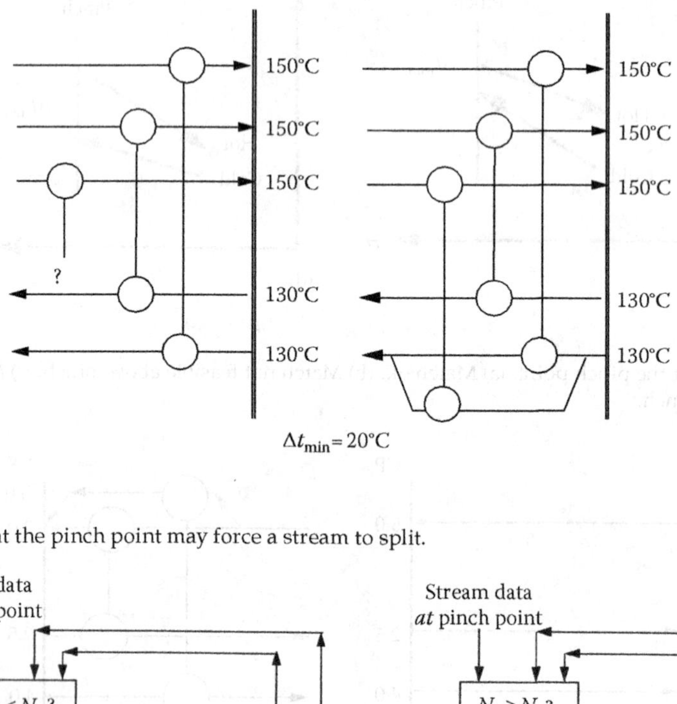

FIGURE 30.17
Stream population at the pinch point may force a stream to split.

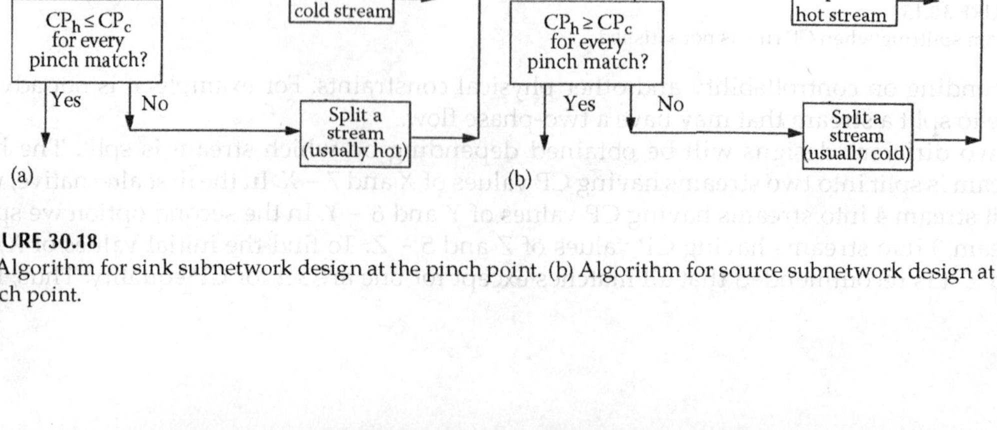

FIGURE 30.18
(a) Algorithm for sink subnetwork design at the pinch point. (b) Algorithm for source subnetwork design at the pinch point.

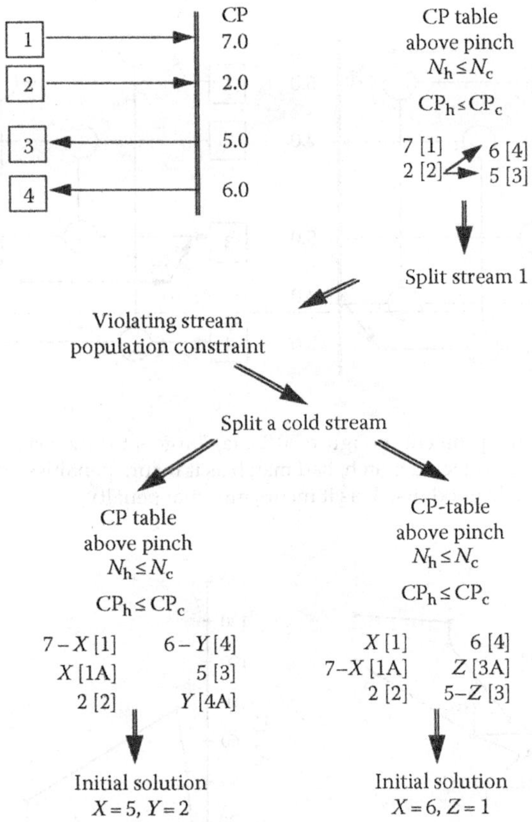

FIGURE 30.19
Identifying matches and stream splitting using the CP table.

the first option set $X = 5$ and $Y = 2$. For the second option set $X = 6$ and $Z = 1$. This is an initial solution. The values of X, Y, and Z can be adjusted to obtain an optimal design. The two different topologies obtained for this example are shown in Figure 30.20.

Once the matches are identified at the pinch, the heat loads for these matches are fixed to maximize the heat exchange to the limit of heat load of either the hot stream or the cold stream. This will eliminate that stream from the analysis. Fixing the heat loads for match with this heuristic will also help us to minimize the number of units and thus the installation cost.[17]

Matches away from the pinch point do not have to satisfy the condition just outlined, as the system is not constrained in this region. Matches are easily identified away from the pinch point. The network obtained will satisfy the energy targets set. Such a network is called maximum energy recovery (MER) network.[17]

30.1.7.3 $\Delta T - T_c$ Plot

To appropriately place matches in the network that utilize the correct driving forces, the driving force plot is very useful design aid.[18] A plot of the temperature difference ΔT between the hot composite curve and the cold composite curve vs. the cold composite temperature is derived from the composite curves as shown in Figure 30.21. Different versions of this plot can also be used such as the $\Delta T - T_h$ plot or $T_h - T_c$ plot. To meet the area targets established

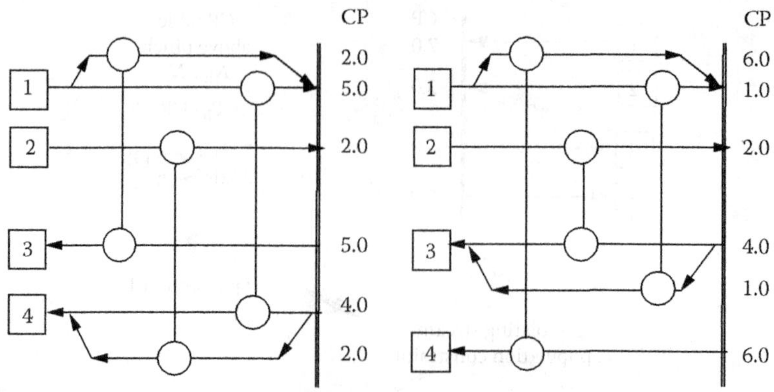

FIGURE 30.20
Two different topologies for the problem in Figure 30.23. (a) Targets for the original problem. (b) Targets for the remaining problem with the selected match; bad match as it incurs penalties. (c) Targets for the remaining problem with the selected match; good match as it incurs minimal penalty.

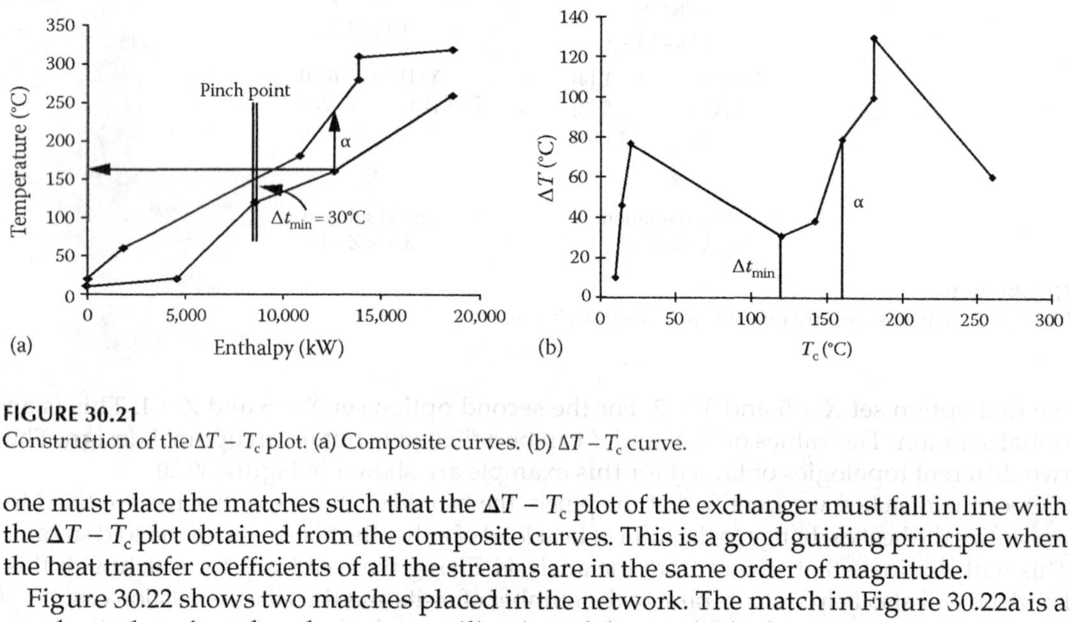

FIGURE 30.21
Construction of the $\Delta T - T_c$ plot. (a) Composite curves. (b) $\Delta T - T_c$ curve.

one must place the matches such that the $\Delta T - T_c$ plot of the exchanger must fall in line with the $\Delta T - T_c$ plot obtained from the composite curves. This is a good guiding principle when the heat transfer coefficients of all the streams are in the same order of magnitude.

Figure 30.22 shows two matches placed in the network. The match in Figure 30.22a is a good match, as it makes the correct utilization of the available driving forces. The match in Figure 30.22b is a bad match, as it uses more ΔT than available. If a match uses more driving force than is available, it will force some other match in the network to use less driving force than available. This results in an overall increase in the total surface area of the network. Badly placed matches do not waste energy but waste surface area.[3]

30.1.7.4 Remaining Problem Analysis

As we have noted, the rules that are established for design guidelines can identify multiple options. The option that satisfies the targets established ahead of design will be the optimal design. To help screen different options, Remaining Problem Analysis (RPA) technique[18] should be used. RPA evaluates the impact of the design decision made on the

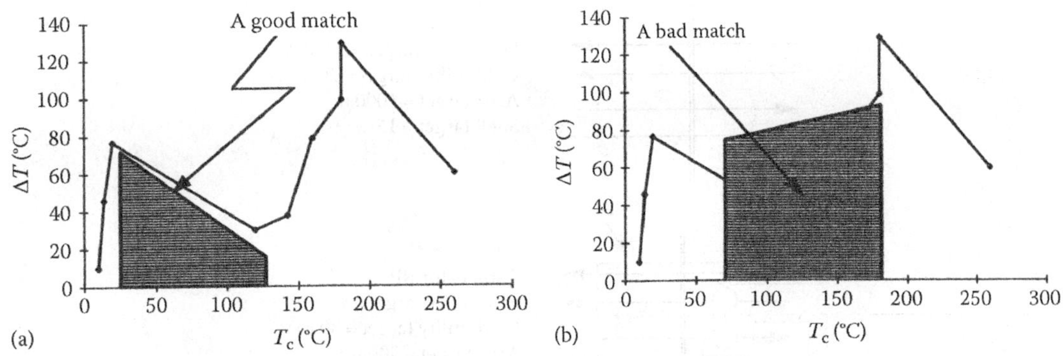

FIGURE 30.22
Identifying good and bad matches using the $\Delta T - T_c$ plot. (a) Good match. (b) Bad match.

remaining problem. Once a matching decision is made, the remaining stream data excluding the match is retargeted. The targets for the remaining problem should be in line with the original targets. In case they are not, look for other options such as decreasing the heat duty on the match or finding a new match.

Figure 30.23 shows how RPA is undertaken. For the original problem the targets are

$$Q_{\text{hmin}} = 30 \text{ MW}$$

$$Q_{\text{cmin}} = 20 \text{ MW}$$

$$\text{Minimum area} = 1000 \text{ m}^2$$

$$\text{Minimum number of shells} = 15$$

A match is placed that has a duty of 40 MW, has a 200 m^2 area, and needs three shells (see Figure 30.23b). The targets for the remaining stream data that excluded the match are

$$Q_{\text{hmin}} = 40 \text{ MW}$$

$$Q_{\text{cmin}} = 30 \text{ MW}$$

$$\text{Minimum area} = 900 \text{ m}^2$$

$$\text{Minimum number of shells} = 14$$

Thus, by placing the selected match, 10 MW of extra energy will be required, 100 m^2 more area will be required, and two extra shells will be needed. If we place a new match as shown in Figure 30.23c the remaining problem targets are in line with the original targets (i.e., only a small penalty is associated with this match). Hence this match should be selected.

The procedure is repeated after placing each and every match. This way the designer gets continuous feedback on the impact of the matching decisions and the heat load assigned to the match. The final design will meet all the targets set before we start the design.

30.1.7.5 Local Optimization—Energy Relaxation

The optimal region is generally very flat, and the optimal value of Δt_{min} is not a single point but a region. The targeting exercise helps identify this region. Once a value of Δt_{min} is selected, and the network is designed using the methodology outlined earlier, there is still some scope for further optimization. This is achieved by a process called energy relaxation.[17]

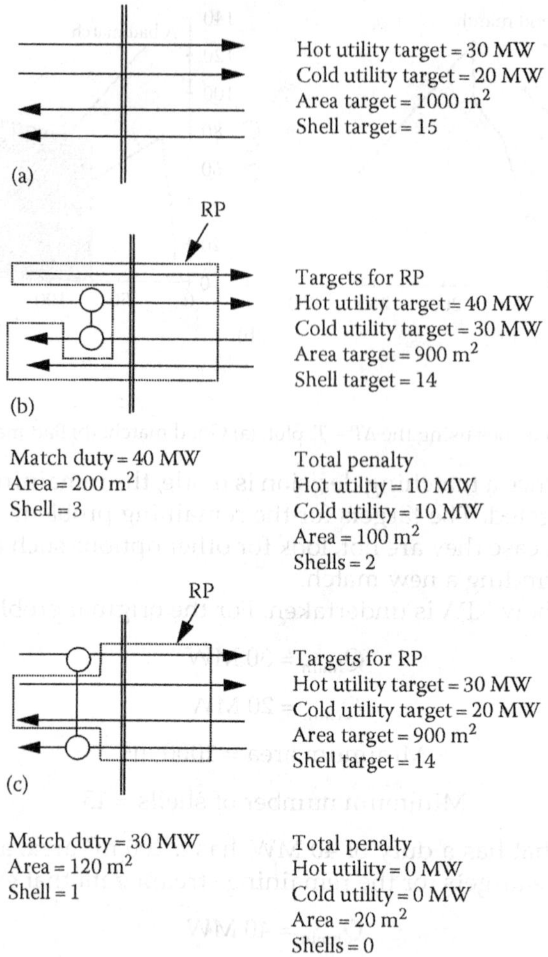

FIGURE 30.23
Remaining problem analysis.

The number of units in the design obtained using this procedure will generally be greater than the target value. This is due to the presence of loops. If we can eliminate the loops in a network, then the heat transfer area is concentrated on fewer matches and hence will decrease the piping and foundation requirements. This will tend to decrease the capital cost of the network. However, the same energy penalty may be incurred.

Identifying loops in a large network cannot be done visually as we had done in the example problem. Various algorithms exist to identify loops in a given network.[19,20] Trivedi et al.[21] have proposed LAPIT (Loop and Path Identification Tree) for identifying loops and paths present in the network. A path is the connection between a heater and a cooler via process–process matches.

Loops can be broken by removal of unit in a loop and redistributing the load of the unit among the remaining units of the loop. Some exchangers may result having a very small Δt_{min} when a loop is broken. The Δt_{min} across such exchangers can be increased by increasing the utility consumption along a path that consists of a heater, the unit having a small value of Δt_{min}, and a cooler.

Loop breaking is a very complex optimization process. Trivedi et al.[21,22] have proposed a detail method that systematically breaks loop and identifies options available at each step of the process. The method is based on loop network interaction and load transfer analysis (LONITA)[22] and on a best-first-search procedure.[21]

30.1.7.6 Summary of the Design Procedure

The pinch design procedure can be summarized now. Establish the value of the optimum Δt_{min} using the targeting techniques. This identifies the region in which the design should be initialized. Using the pinch design procedure, design the network. Reduce the number of units using the loop breaking and energy relaxation techniques outlined earlier.

30.1.7.7 Example

The principles and procedures just discussed are used for the design of the network for the flowsheet of Figure 30.6. The value of the optimum Δt_{min} is 40°C. The final network MER design is shown in Figure 30.24. Note that the annualized capital cost of this network is only about 4% higher than the target value of $692,700. This design is in the neighborhood of the predicted optimum.

The design after energy relaxation is shown in Figure 30.25. This design has one less unit than the MER design in Figure 30.24. Hence, this will decrease the piping, installation, and maintenance cost. The cost of the new design is about 2% higher than the MER design. Since we are using simplified models for costing, the cost of the design in Figure 30.25 is within the errors of cost estimating. However, this design will be cheaper when detail cost is evaluated, as it has less number of units.

Table 30.6 compares the target values of the various optimization parameters with actual design values for Figures 30.24 and 30.25.

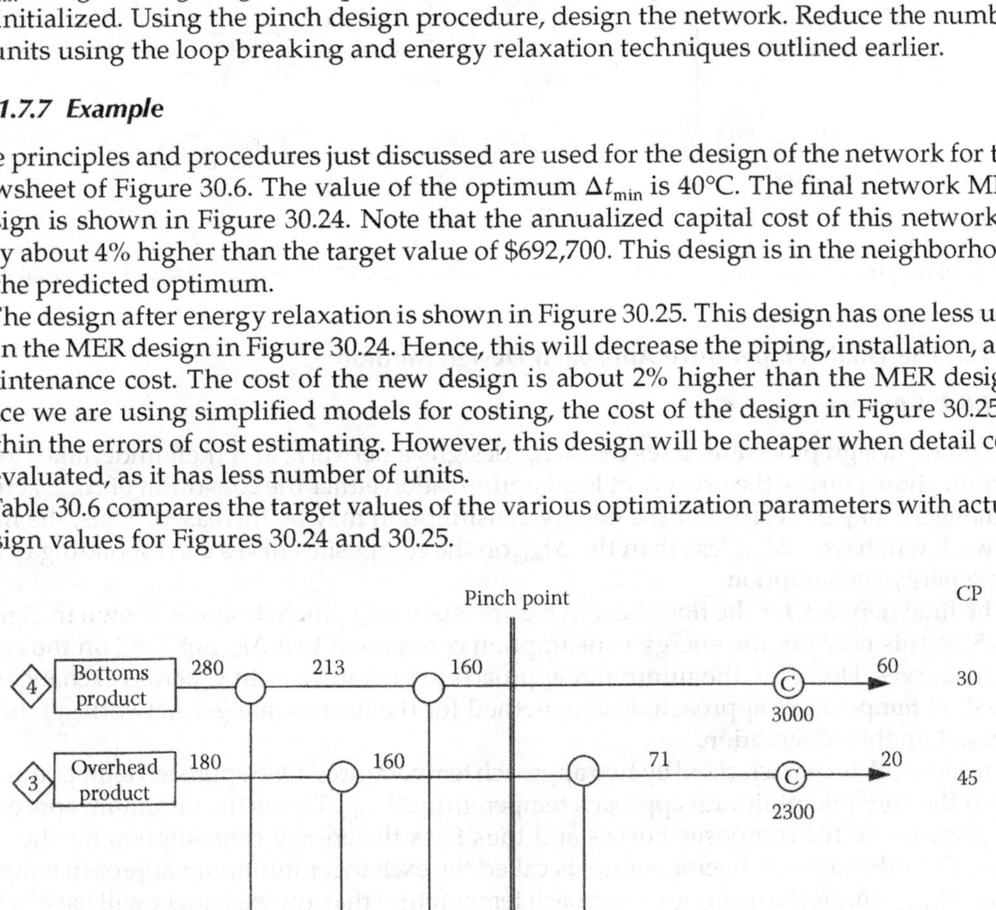

FIGURE 30.24
Pinch design for the flowsheet in Figure 30.9. Total annualized cost is $720,800.

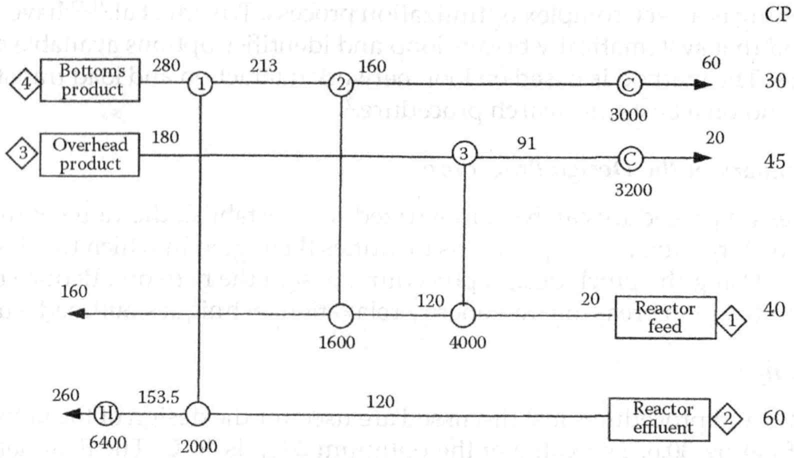

FIGURE 30.25
Energy of the pinch design in Figure 30.30. Total annualized cost is \$773,700. Δt_{Nmin} = 52 K and Δt_{Emin} = 40 K.

30.1.8 The Dual Temperature Approach Design Method

30.1.8.1 Concept

The pinch design procedure fixes the Δt_{min}, designs a network, and then undertakes local optimization. During the process of local optimization either the constraint of Δt_{min} across exchangers may be relaxed or the energy consumption may be increased. Thus, the final network will have a Δt_{min} less than the Δt_{min} on the composite curves corresponding to the final energy consumption.

The final network for the flowsheet in Figure 30.6 using pinch design is shown in Figure 30.25. In this network the energy consumption correspond to a Δt_{min} of 52°C on the composite curves. However, the minimum approach temperature is 40°C across exchanger 2. The dual temperature approach design method for the heat exchanger network synthesis is based on this observation.

The network is characterized by two approach temperatures. One approach temperature is called the network minimum approach temperature Δt_{Nmin}. This is the minimum approach temperature on the composite curves and thus fixes the energy consumption for the network. The other approach temperature is called the exchanger minimum approach temperature Δt_{Emin}. This is the minimum approach temperature that any exchanger will have in the network. In the network of Figure 30.25, Δt_{Nmin} = 52° C and Δt_{Emin} = 40° C. Generally, $\Delta t_{Emin} \leq \Delta t_{Nmin}$. When $\Delta t_{Emin} = \Delta t_{Nmin}$ the network has a global minimum approach temperature just like the Δt_{min} parameter of the pinch design procedure. This is in contrast to the pinch design method, which utilizes a global value of Δt_{min} (i.e., $\Delta t_{Nmin} = \Delta t_{Emin}$) and then relaxes the energy consumption of the network (with the obvious implication of increase in Δt_{min}).

The concept of using dual approach temperatures for the design of the network is very powerful as the designer retains complete control over the energy consumption of the network. This is in contrast with the pinch design method where energy consumption increases following loop breaking and is determined by the new topology and not the designer. Another advantage in using the dual temperature method is the reduction in the number of shells and units. As well, unnecessary stream splits may be avoided, making a more practical network with better operability characteristics. This may result in substantial decrease (~20%) in the capital cost for a fixed operating cost, representing millions of dollars savings in industrial applications.[23]

The design procedure proposed by Trivedi et al.[24] incorporates the best features of both the pinch design method and the dual temperature approach design procedure. Their design procedure is based on the concept of a pseudo-pinch point.

30.1.8.2 Location of the Pseudo-Pinch Point

Initially two sets of composite curves are generated using Δt_{Nmin} and Δt_{Emin}, with Δt_{Nmin} selected such that $\Delta t_{Nmin} \geq \Delta t_{Emin}$. Both sets will have different energy consumption denoted as EC. As $\Delta t_{Emin} \leq \Delta t_{Nmin}$, $EC(\Delta t_{Nmin}) \geq EC(\Delta t_{Emin})$.

Let us define the energy difference a (see Figure 30.26a,b) as

$$\alpha = EC(\Delta t_{Nmin}) - EC(\Delta t_{Emin}) \quad (30.14)$$

As the network is designed employing two approach temperatures, a new interpretation of the problem could be this: As Δt_{Emin} is less than Δt_{Nmin}, an amount of energy α traverses the Δt_{Emin} conventional pinch point. This can be compensated by energy carried upward across the pinch because heat exchange is allowed at temperature differences as low as Δt_{Emin} between streams. The maximum amount of this upward carriage that can be achieved is the value of α defined earlier. This represents additional flexibility in stream matching by providing for upward and downward carriage, which is permitted by the reduced Δt_{Emin} between streams, while the total energy requirement remains fixed at $EC(\Delta t_{Nmin})$, which in turn is defined by the network approach temperature Δt_{Nmin}.

A pseudo-pinch point (actually a set of stream temperatures) is defined so that the stream temperatures at this point allow the problem to be partitioned as shown in Figure 30.27. Thus, the two parts of the network are in enthalpy balance with the utility consumption determined by Δt_{Nmin}.

The pseudo-pinch point is defined based on the observation that the changes in slope of the composite curve that occur at the conventional pinch point require (at least) one hot or cold stream entering at a real pinch point temperature.[25] This stream's starting temperature is chosen to determine the pseudo-pinch point. Let the subscripts pp denote the pseudo-pinch temperatures. To determine the pseudo-pinch point temperatures with α units of

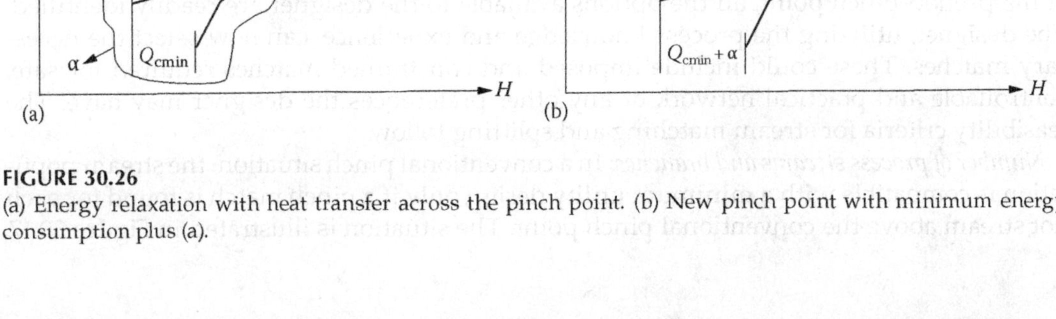

FIGURE 30.26
(a) Energy relaxation with heat transfer across the pinch point. (b) New pinch point with minimum energy consumption plus (a).

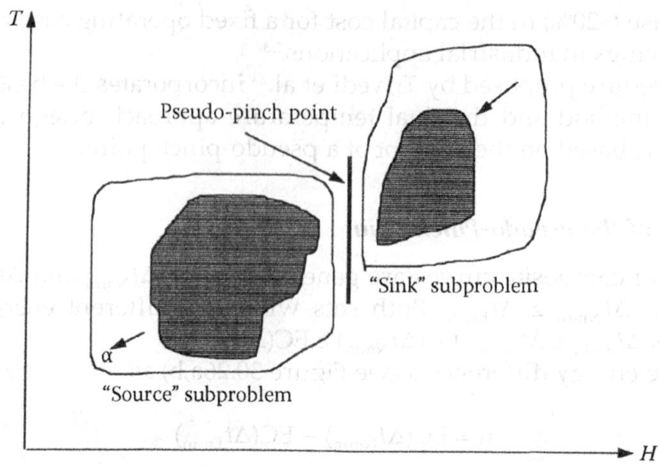

FIGURE 30.27
Pseudo-pinch division of the system.

heat transferred across the conventional pinch point at Δt_{Emin}, various strategies for allocating the α units are proposed by Trivedi et al.[24] Generally the following heuristics are used:

1. Pseudo-pinch point temperatures having $\Delta T = \Delta t_{Emin}$: If a hot stream with a starting temperature T_{hs} determines the conventional pinch point for Δt_{Nmin} or Δt_{Emin}, then this stream is also assumed to be at the pseudo-pinch point with all the cold stream temperatures given by $T_{hs} - \Delta t_{Emin}$. Likewise, if a cold stream entrance determines the conventional pinch point, then the pseudo-pinch temperatures of all the hot streams are given by $T_{cs} + \Delta t_{Emin}$. It is possible that different streams may determine the conventional pinch point for both Δt_{Emin} and Δt_{Nmin}. In such a situation different topologies will result and all possible configurations should be investigated.
2. Pseudo-pinch point temperatures with $\Delta T > \Delta t_{Emin}$: The α units of heat carried across the conventional pinch point will increase the pseudo-pinch point ΔT's for some stream combinations. Assuming that a hot stream determines the conventional pinch point, various strategies are used. Generally, if there are N other hot streams with $T_{hsj} = T_{hspp}$, then $\alpha_j = \alpha/N$ is allocated to each of these streams so that their pseudo-pinch point temperatures are increased by α/NCP_j.

30.1.8.3 Design of the Sink Subproblem

30.1.8.3.1 Feasibility Criteria at the Pseudo-Pinch

As the problem constraints are relaxed at the conventional pinch point by passing heat across it, a wide variety of network topologies can be generated. The method commences the design at the pseudo-pinch point. As there are only a few essential matches to be made at the pseudo-pinch point, all the options available to the designer are readily identified. The designer, utilizing the process knowledge and experience, can now select the necessary matches. These could include imposed and constrained matches required for safe, controllable and practical network or any other preferences the designer may have. The feasibility criteria for stream matching and splitting follow.

Number of process streams and branches: In a conventional pinch situation, the stream population is compatible with a minimum utility design only if a pinch match is found for each hot stream above the conventional pinch point. The situation is illustrated in Figure 30.17,

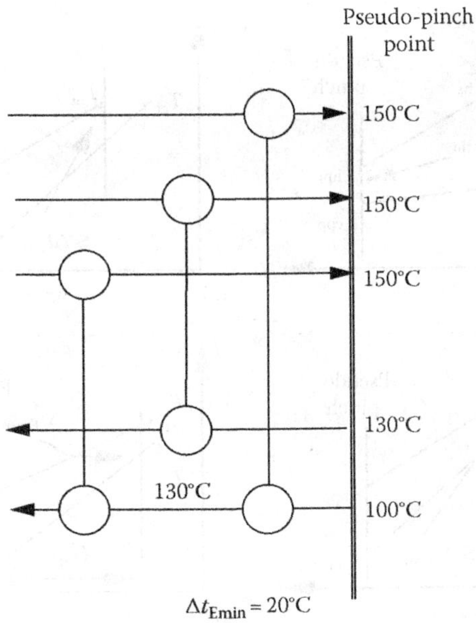

FIGURE 30.28
Stream splitting may not be necessary in pseudo-pinch problems.

where stream splitting is unavoidable. However, for the pseudo-pinch situation, splitting may not be necessary as demonstrated in Figure 30.28. This results from the relaxation of the conventional pinch temperatures resulting in an increase in the available driving forces for the pseudo-pinch matches.

It is exceedingly difficult to determine in advance if stream splitting is necessary. Hence, two approaches are suggested. The first approach is identical to the pinch design method. Stream splits may be removed once the initial design is generated. This will be discussed later. In the second approach, we place the pseudo-pinch matches and see if the unmatched streams at the pseudo-pinch can be satisfied. If the streams are matching violating Δt_{Emin}, then a stream is split and the pseudo-pinch matching is repeated.

Temperature feasibility criteria: Four temperature profiles are possible for a match at the pseudo-pinch point as shown in Figure 30.29. These profiles are for the sink subproblem. The match illustrated in Figure 30.29a possesses the same characteristics as that of an above conventional pinch match. However, for a pseudo-pinch, other matches of the type illustrated in Figure 30.29b through d are also possible. Trivedi et al.[24] have outlined the various criteria to be used at the pseudo-pinch that determine the maximum heat load Q_{max} for a match between a hot and a cold stream.

30.1.8.4 Design of the Source Subproblem

For the source subproblem analogous feasibility criteria can readily be developed. These depend both on the stream populations and temperature feasibility shown in Figure 30.30. The complete design procedure is summarized in Figure 30.31a and b.

30.1.8.5 Design of the Remaining Network

Once the pseudo-pinch matches are placed, the remaining problem can be designed employing the rules mentioned earlier depending on the subproblem. It should be noted

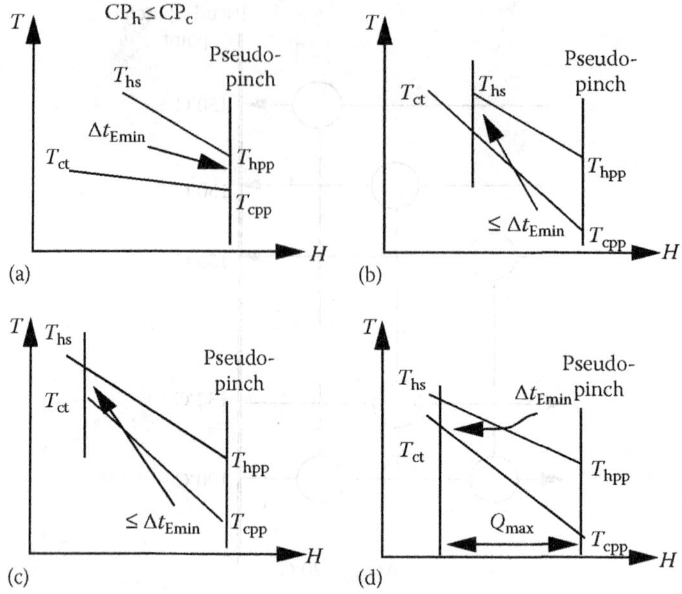

FIGURE 30.29
Temperature feasibility for stream splitting immediately adjacent to the pseudo-pinch point for the sink subproblem. (a) Temperature feasibility at pseudo-pinch point. (b) Temperature feasibility away from pseudo-pinch point; Hot stream starting temperature controls the duty of the match. (c) Temperature feasibility away from pseudo-pinch point; Cold stream target temperature controls the duty of the match. (d) Temperature feasibility away from pseudo-pinch point; Maximum feasible duty of the match.

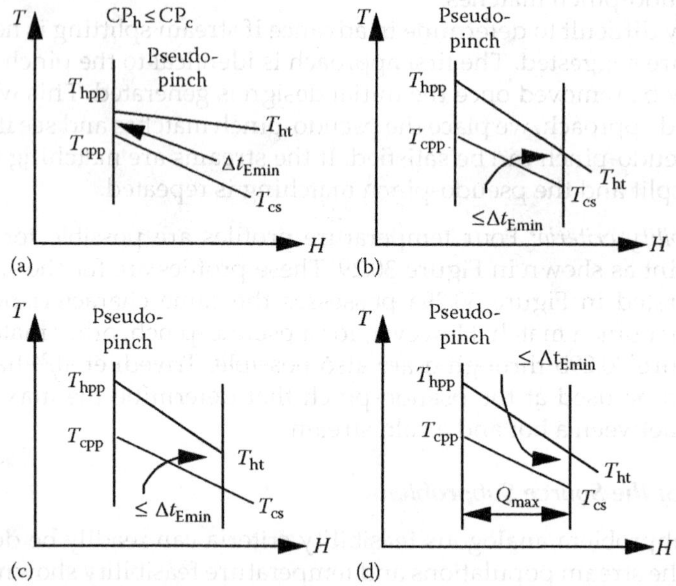

FIGURE 30.30
Temperature feasibility for stream splitting immediately adjacent to the pseudo-pinch point for the source subproblem. (a) Temperature feasibility at pseudo-pinch point. (b) Temperature feasibility away from pseudo-pinch point; Cold stream starting temperature controls the duty of the match. (c) Temperature feasibility away from pseudo-pinch point; Hot stream target temperature controls the duty of the match. (d) Temperature feasibility away from pseudo-pinch point; Maximum feasible duty of the match.

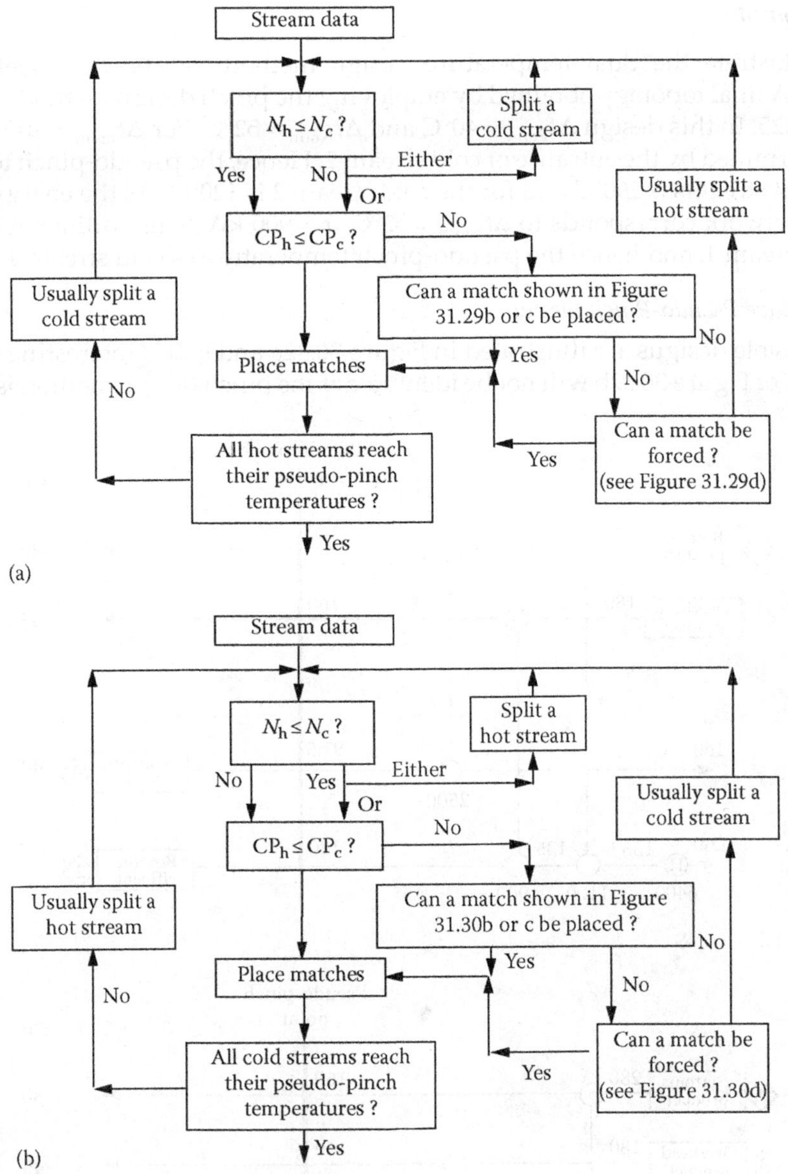

FIGURE 30.31
(a) Above pseudo-pinch design algorithm. (b) Below pseudo-pinch design algorithm.

that the remaining problem is still in enthalpy balance and only requires either a hot or cold utility, depending on the subproblem. The utilities can be placed on either end or at an appropriate temperature level. (The rules outlined by Trivedi et al.[24] can also be employed in selecting the nonpinch matches when the pinch design method is employed.)

Once an initial topology is synthesized, its capital cost can be further reduced by decreasing the number of units and removing stream splits keeping the energy consumption constant. A linear program can be used to remove stream splits.[24] To reduce the number of units, LONITA can be used.

30.1.8.6 Example

We shall illustrate the dual temperature design method for the flowsheet shown in Figure 30.6. A final topology obtained by employing the pinch design method is illustrated in Figure 30.25. In this design $\Delta t_{Emin} = 40°C$ and $\Delta t_{Nmin} = 52°C$. For $\Delta t_{Emin} = 40°C$, the pinch point is determined by the entrance of cold stream 2. Hence, the pseudo-pinch temperature for all the hot streams is 160°C and for the cold stream 2 is 120°C. As the energy consumption of the network corresponds to $\Delta t_{Nmin} = 52°C$, $\alpha = 900$ kW. This additional heat load is carried by stream 1, and hence the pseudo-pinch temperature of cold stream 1 is 97.5°C.

30.1.8.6.1 Above Pseudo-Pinch Design

The two possible designs are illustrated in Figure 30.32a and b. It is interesting to note that the topology of Figure 30.32b will not be identified if the pinch design method is employed.

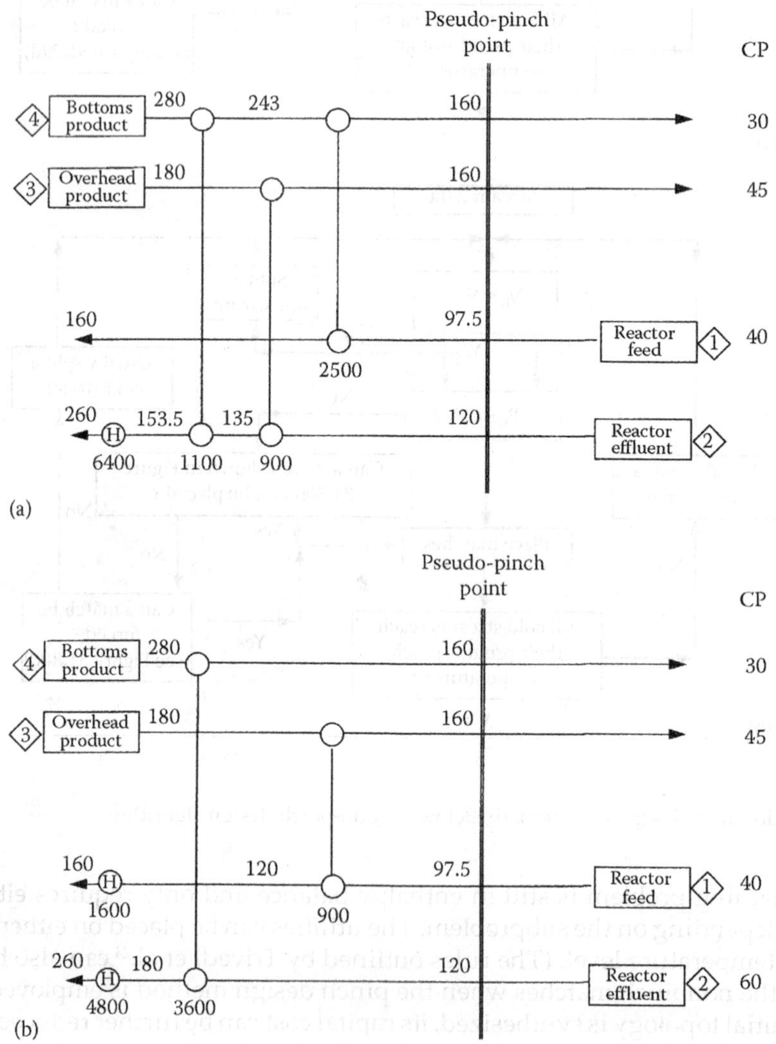

FIGURE 30.32
(a) Above pseudo-pinch design—1. (b) Above pseudo-pinch design—2. *(Continued)*

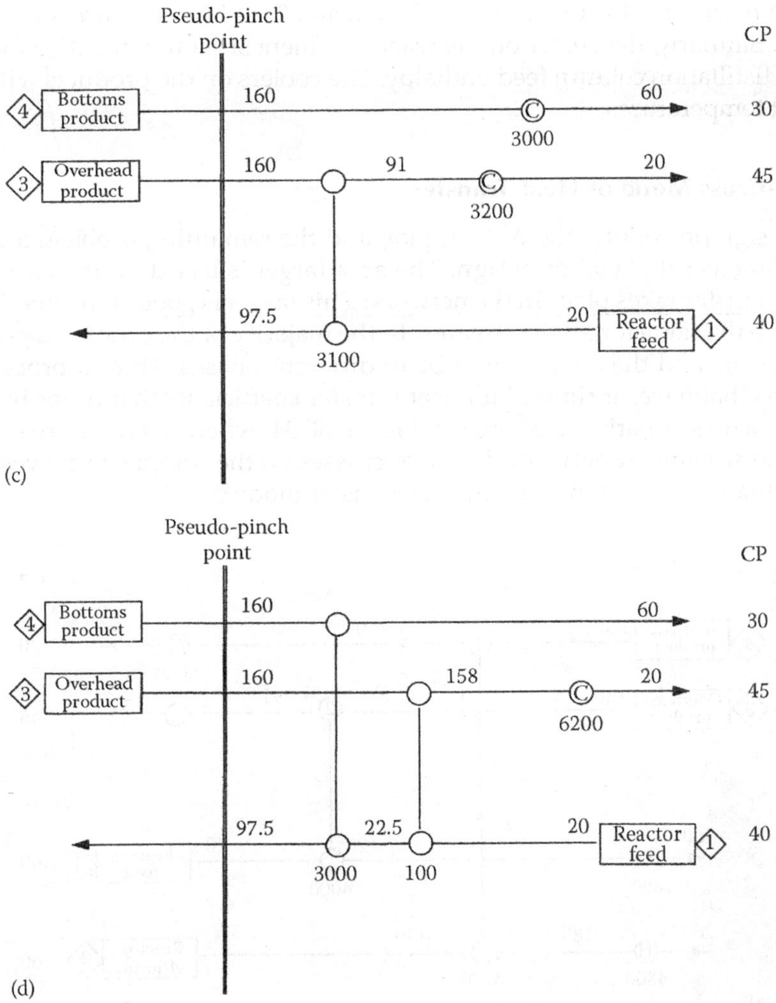

FIGURE 30.32 ***(Continued)***
(c) Below pseudo-pinch design—1. (d) Below pseudo-pinch design—2.

30.1.8.6.2 Below Pseudo-Pinch Design

Again two designs are possible as illustrated in Figure 30.32c and d. The matches suggested in Figure 30.32c will not be identified by the use of pinch design rules.

The above and below pseudo-pinch designs can be combined to create four different fixed-energy designs. Combining Figure 30.32a and c gives the same design as shown in Figure 30.25. The other designs are illustrated in Figure 30.33a through c. The capital cost of the designs obtained so far are within 3% of the one shown in Figure 30.24. As the cost of all these networks is very close, we will choose the final design based on other factors such as operability, controllability, and flexibility. A detailed analysis may have to be under taken on all these designs.

The design shown in Figure 30.33a has utility matches for all the streams. This design will probably be the most controllable design. Hence it is selected. The final process flow diagram for the flowsheet in Figure 30.6 is shown in Figure 30.33d. Observe that there is a heater on the reactor feed before it enters the reactor. This heater controls the reactor feed temperature. Similarly, the heater on the reactor effluent stream to the distillation column controls the distillation column feed enthalpy. The coolers on the products will control the final product temperatures.

30.1.9 Criss-Cross Mode of Heat Transfer

The pinch design procedure, the $\Delta T - T_c$ plot, and the remaining problem analysis try to achieve the targets set ahead of design. The area target is based on the assumption that vertical heat transfer takes place in the network. This target is good if the film heat transfer coefficients are the same for all the streams. In the majority of the processes phase changes will be occurring, and the streams will be in different phases. Thus, a process may consist of streams that have, at times, film heat transfer coefficients that differ by an order of magnitude. Such as situation is shown in Figure 30.34, where there are two gas streams and two liquid streams. A network that criss-crosses on the composite curves may have a smaller area than predicted by vertical heat transfer mode.[26]

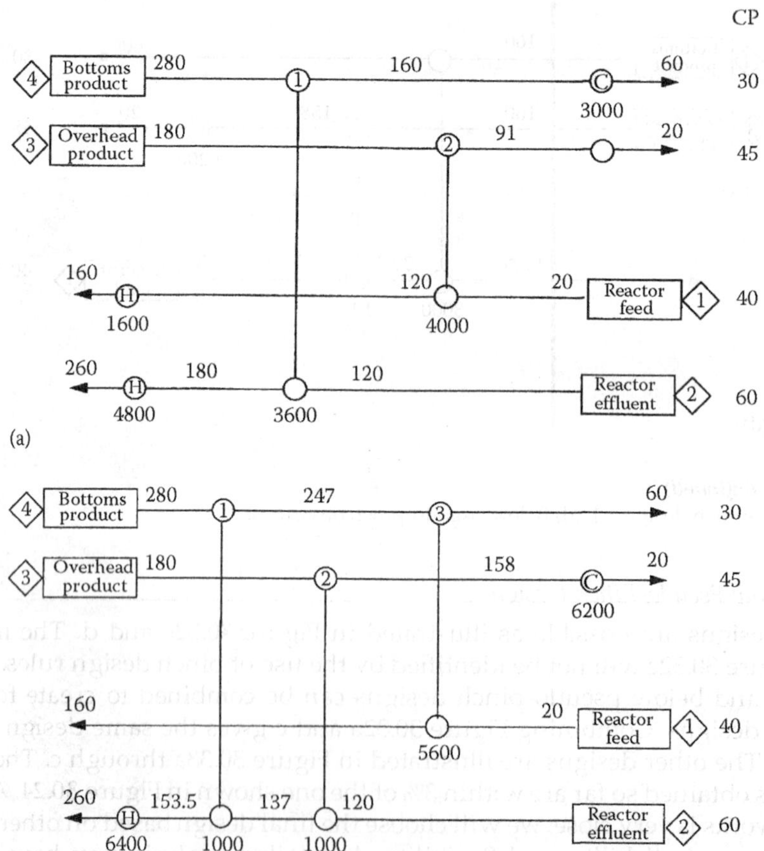

FIGURE 30.33
(a) Pseudo-pinch design—2. Total annualized cost is $743,400. (b) Pseudo-pinch design—3. Total annualized cost is $727,600.
(Continued)

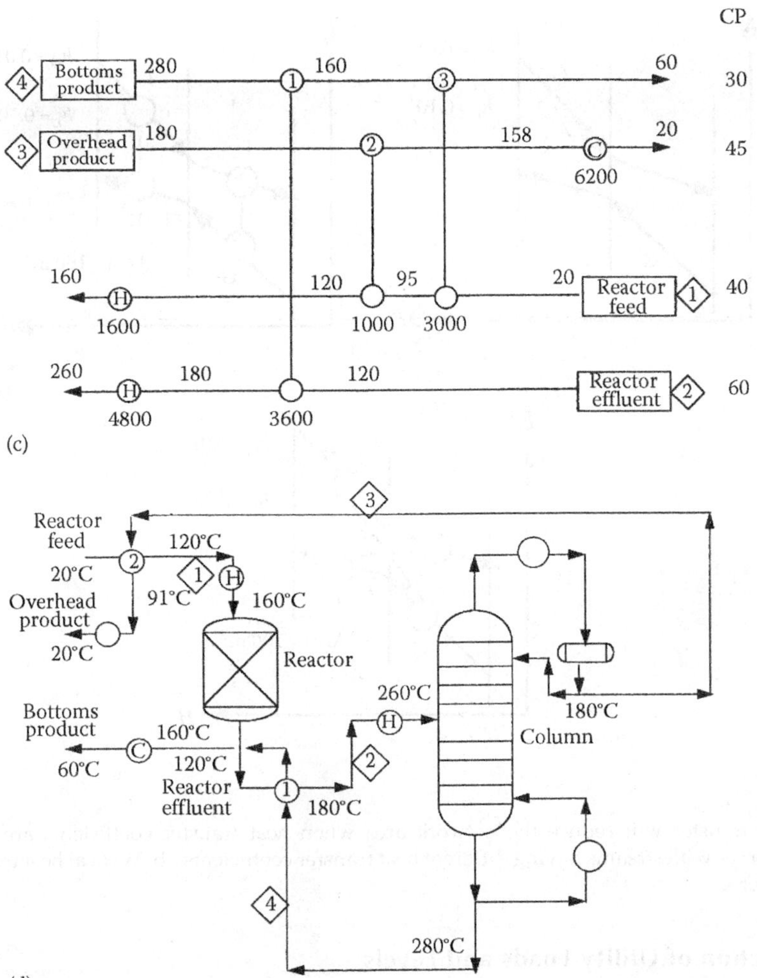

FIGURE 30.33 ***(Continued)***
(c) Pseudo-pinch design—4. Total annualized cost is $740,100. (d) Final flow diagram for the plant in Figure 30.9.

A network will generally have different types of exchangers. Further, each exchanger will have different material of construction and design temperature and pressure. These factors contribute to the capital cost of the network. Thus each stream match will have different coefficients in the cost equation. To optimize the network one will have to undertake criss-crossing on the composite curves.

Stream matching constraints may also require criss-crossing on the composite curve. For example, a match between two streams may be prohibited to maintain process safety or to avoid incurring excessive piping costs associated with a particular match. When constraints are imposed, extra utilities may be required. As a result, one may have to use cold utilities above the pinch point, use hot utilities below the pinch point, or transfer heat across the pinch point. This is totally in contrast to the pinch design rules.

All these scenarios force the designer to criss-cross on the composite curves. Various design methods are proposed in the literature to obtain optimal designs using the criss-cross mode of heat transfer in the network and is not with the scope of this work.

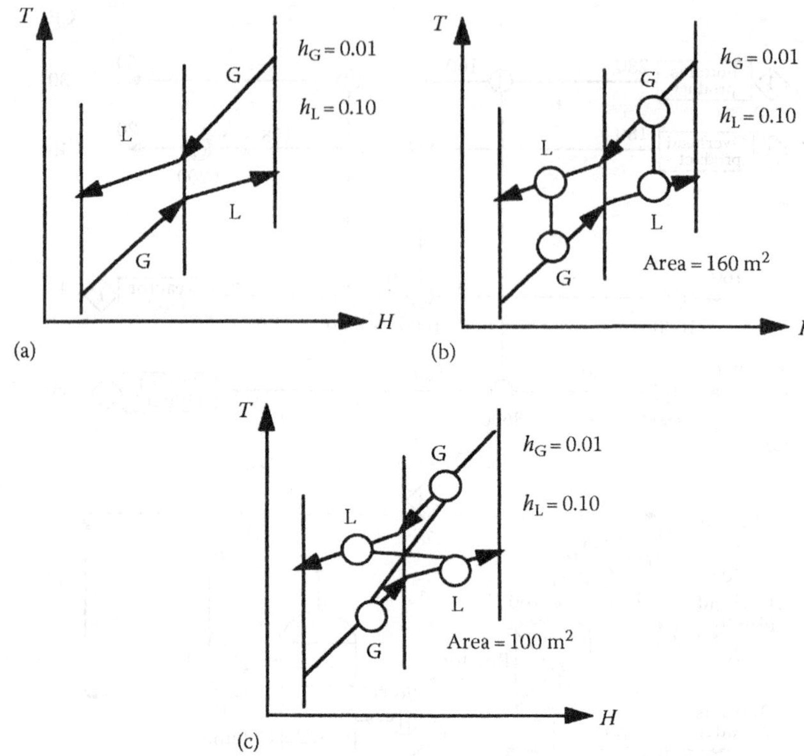

FIGURE 30.34
Criss-cross heat transfer will reduce the network area when heat transfer coefficients are very different. (a) Composite curves with streams having different heat transfer coefficients. (b) Vertical heat transfer. (c) Criss-cross heat transfer.

30.1.10 Selection of Utility Loads and Levels

The annual operating cost depends on the amount and type of utilities used. In a complex process such as an ethylene plant, there would be about four or five steam levels and about seven or eight refrigeration levels. High-pressure steam generated within the process is let down to other steam levels via steam turbines that generate power. Steam from these levels is used for heating the process. The power generated from the turbines is used by the compressors and pumps and to generate various levels of refrigeration needed. The question faced is what pressure levels of steam to use and what the load is on each level. Also, what are the best temperature levels of refrigeration and what will be their respective loads?

Pinch technology helps us answer these questions in a very simple manner using the Grand Composite Curves (GCC).[27] The GCC is the curve that shows the heat demand and supply within each temperature interval. This curve is derived from the problem table (refer to Table 30.5). In the problem table, we had modified the stream starting and target temperatures depending on the value of Δt_{min}. From now on we shall refer these modified temperatures as shifted temperature. The heat flows between two adjacent shifted temperature intervals can be plotted on the shifted temperature-enthalpy plot. Figure 30.35 shows the heat cascade and how the grand composite curve is developed for the flowsheet shown in Figure 30.6. The grand composite curve gives us a graphical representation of heat flows taking place in the system. At the pinch point the heat flow is zero.

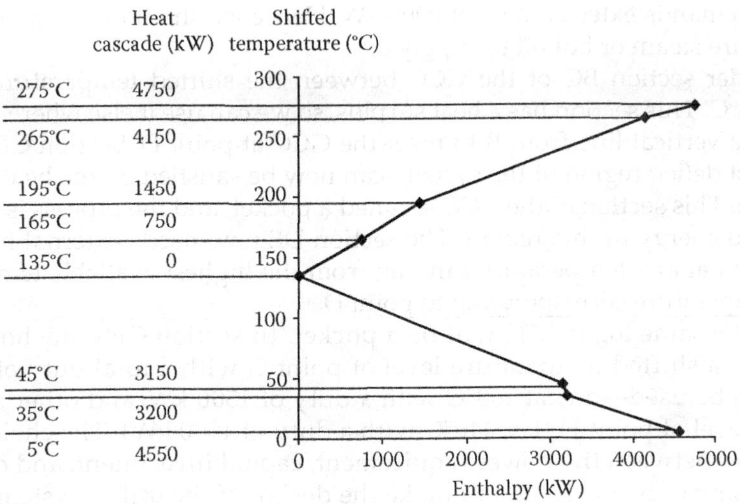

FIGURE 30.35
Grand composite curve for the flowsheet of Figure 30.9.

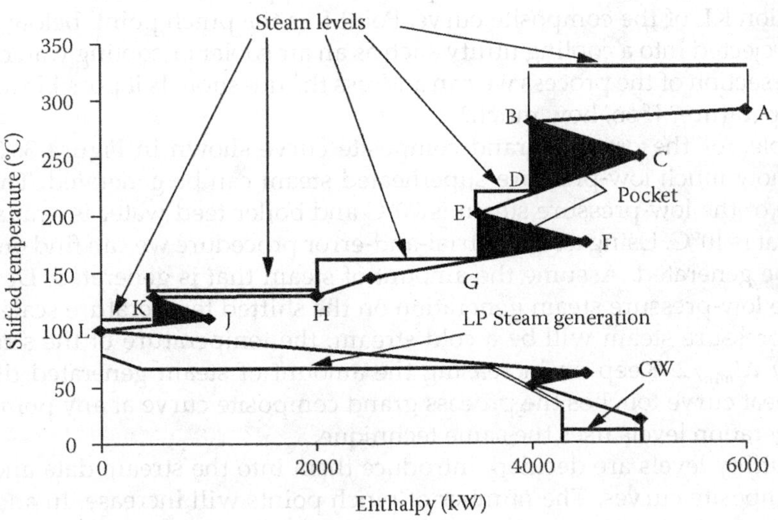

FIGURE 30.36
Selection of utility loads and levels.

The grand composite curve is piecewise linear. The slope of this curve also changes from interval to interval. A line with a positive slope indicates that the system in that region needs external heat. A line with a negative slope indicate that there is surplus heat available within that temperature interval that can be cascade down within the system and used at a lower temperature interval. It is very clear from the grand composite curve that the above pinch region is a heat sink and the below pinch region is a heat source.

To further explain the importance of the grand composite curve and the kind of information that can be extracted from it, consider the curve shown in Figure 30.36. Consider the section AB that is between the shifted temperature interval of 295°C and 280°C.

This section demands external heat of 2000 kW. Hence we should place a hot utility such as high-pressure steam or hot oil to supply it.

Next, consider section BC of the GCC between the shifted temperature intervals of 280°C and 250°C. This section has a heat surplus, so we can use it else where in the system. We can drop a vertical line from B to meet the GCC at point D. Section CD of the GCC, which is a heat deficit region of the system, can now be satisfied by the heat surplus from the BC section. This section of the GCC is called a pocket, and the process is self-sufficient with respect to energy in this region. The section DE now needs external heat. This heat can be supplied at any temperature ranging from the highest available temperature to a minimum temperature corresponding to point D.

Following the same logic, EFG will be a pocket. In section GHI one hot utility level can be used at a shifted temperature level of point G with a total duty of 3000 kW, or two levels can be used—one at 160°C with a duty of 1500 kW, and other at the shifted temperature level of point H (i.e., 140°C with a duty of 1500 kW). The choice is dictated by the trade-off between the power requirement, capital investment, and complexity of the design. Using only one level will make the design of the utility system simpler and the capital cost of the heat exchanger smaller due to higher-temperature approaches. On the other hand, if there is demand for power, then using two levels will produce more power.

A similar economic trade-off will be required for supplying the external heating requirement in section KL of the composite curve. Point L is the pinch point. Below point L heat needs to be rejected into a cooling utility such as an air cooler or cooling water. Also, in the below pinch section of the process we can address the question, Is it possible to raise steam at some temperature? If so, how much?

For example, for the process grand composite curve shown in Figure 30.36, we want to find out how much low-pressure superheated steam can be generated. The saturation temperature of the low-pressure steam is 70°C and boiler feed water is available at 30°C. The superheat is 10°C. Using a simple trial-and-error procedure we can find out how much steam will be generated. Assume the amount of steam that is generated. Develop a heat curve for the low-pressure steam generation on the shifted temperature scale. As generation of low-pressure steam will be a cold stream, the temperature of the stream will be increased by $\Delta t_{min}/2$. Keep on increasing the amount of steam generated till the steam generation heat curve touches the process grand composite curve at any point. The selection of refrigeration levels uses the same technique.

Once the utility levels are decided, introduce them into the stream data and obtain the balanced composite curves. The number of pinch points will increase. In addition to the original process pinch point, each utility level will introduce at least one pinch point. A balanced grid diagram that includes all the utility streams and all the pinch points identified on the balanced composite curve can now be used along with the network design algorithms to develop a network that achieves the target set. Advances have made the task of selecting and optimizing multiple utilities easier.[28,29]

30.1.11 Process Integration

Until now we have discussed the design and optimization of heat exchanger networks. But heat exchanger networks are a part of a whole process. Processes consist of reactors, distillation columns, utilities, and so on. For the process to be optimally designed, all the unit operations should be properly integrated. Generally each unit operation is individually optimized. It is a misconception that if each individual unit is individually optimized then the resulting

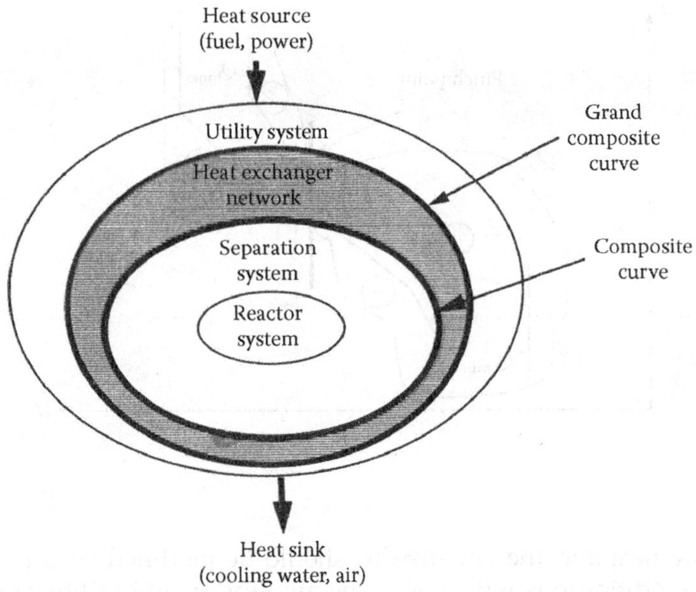

FIGURE 30.37
Onion diagram.

process is also optimized. Each unit operation interacts with the other in the process. Hence, for the process to be optimized, each unit operation should be properly integrated.

Figure 30.37 shows the onion diagram proposed by Linnhoff et al.[30] that represents the hierarchy of process design. Generally, the reactors are designed first that fix the heat and material balance for the whole process. The products, by-products, and reactants are then separated using the most common unit operation—distillation. A number of columns may be used if a large number of components need to be separated. To achieve the separation and to carry out the reaction in the reactor, the heat exchanger network, using heat exchangers, heats and cools the process streams. External heating and cooling is supplied by the utility system. The utility system may consist of a simple boiler system or a cogeneration unit that produces the necessary steam for the process and the power requirement. All these layers interact. An optimal process design must take into account these interactions and must be properly integrated.

Process integration uses principles that we have already established. Process integration starts by optimizing stand-alone distillation columns with respect to reflux ratio, feed conditioning, side-reboilers, and side-condensers.[31] To properly integrate the distillation column with the background process, it should be placed above or below but not across the background process pinch point.[32] The same principle applies for the appropriate integration of heat engines. Heat pumps should be placed across the pinch point.[27] Linnhoff and de Leur[33] have developed procedures for integrating furnaces with the rest of the process.

30.1.12 Process Modification

In the above pinch region, hot streams should be modified such that they can transfer more heat to the cold streams. The cold streams should be modified such that they require less heat. Both of these modifications will decrease the amount of external hot utility requirement. Similarly, in the below pinch region, the cold streams should be modified such that

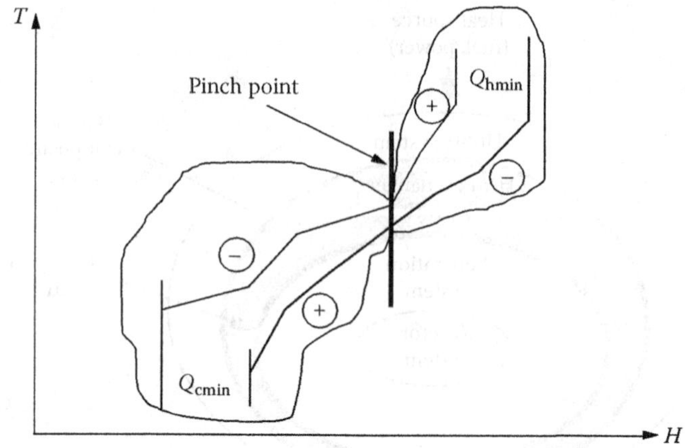

FIGURE 30.38
The plus/minus principle.

they require more heat and the hot streams should be modified such that they transfer less heat. These modifications will reduce the amount of cold utility requirement. This principle is called the plus/minus principle[3] and is illustrated in Figure 30.38.

30.1.13 Shaftwork Targets

Recently, Linnhoff and Dhole[34] have proposed shaftwork targets based on exergy analysis of the stream data for a plant. The shaftwork targeting procedure calculates the change in the total shaftwork requirement of the system due to any changes in the base case. They use exergy composite and grand composite curves that are obtained from the composite and grand composite curves by changing the temperature axis to the carnot factor. Shaftwork targets are very important in the design of low-temperature processes.

30.1.14 Sitewide Integration

Generally, individual process are located in a site. The site consist of multiple units. For example a refinery may consist of a crude unit, vacuum unit, naphtha hydrotreater, diesel hydrotreater, fluidized catalytic cracking (FCC) unit, visbreaker unit coker unit, and so on. A petrochemical complex may consist of an ethylene unit, polyethylene unit, and so on. All these units will have one utility system that provides hot and cold utilities as well as power to the whole site. Each unit will have steam demands at different temperatures. At the same time, each unit may be producing steam at different levels. It is possible to directly integrate different units, but that may cause political problems or piping problems. Hence, different units are generally integrated indirectly through the utility system. The problem that sitewide integration addresses is what is the correct level of steam for a site and what is the trade-off between heat and power. Should power be imported or cogenerated?[35]

The most important benefit of the sitewide analysis is that correct pricing for different levels of steam are no longer required. The impact of any modification in terms of fuel or power can be easily evaluated. There is no need for "cost of steam," as energy pricing is only done with respect to either fuel or power at the battery limit.

Sitewide analysis has increased the understanding of global emissions associated with any processing industry. To minimize the emissions associated with a process, sharper separations are required. Since distillation is the workhorse of the chemical industry, to obtain sharp separation, the reboiler and condenser duties will increase or additional processes may be added that require additional heating and cooling. External energy is obtained by burning fuel, and when fuel is burned emissions are generated. If extra power is used, then the emissions in the utility company will increase. Thus, to decrease the emissions in the process we might end up increasing the emissions associated with burning the fuel. The net effect might be that we increase the global emissions.[1] Recently, Smith and Delaby[36] have established techniques to target for CO_2 emissions associated with energy. Combining these CO_2 targeting techniques with sitewide analysis will help in trading off different emissions.[37]

30.1.15 Data Extraction

Process integration studies start from a base-case flowsheet. This flowsheet may be existing or may be developed from designer's experience. To conduct pinch analysis properly, it is important to extract the flow rate, temperature, and heat duty data correctly.

Stream target and starting temperatures should be chosen so that we do not generate the original flowsheet.[30] To illustrate this, consider the flowsheet shown in Figure 30.39. If we extract the data as two streams, then we might end up with the original flowsheet. If the drum temperature is not important then we can consider it to be one stream and we stand a chance for finding new matches. The drum and the pump can then be kept as a natural break point in the system.

While extracting data, extra care should be taken when streams are mixing nonisothermally.[30] Consider the system shown in Figure 30.40. Stream A is being cooled to the mixed temperature, and stream B is being heated to the mixed temperature. This happens due to mixing the streams. The mixed stream is then heated to a higher temperature. If we extract the data as shown in Figure 30.40a and if the pinch temperature is 70°C then we are inherently transferring heat across the pinch point by mixing the two streams. In the process of mixing we are cooling stream A and then subsequently heating it up again. The correct way to extract the data is shown in Figure 30.40b.

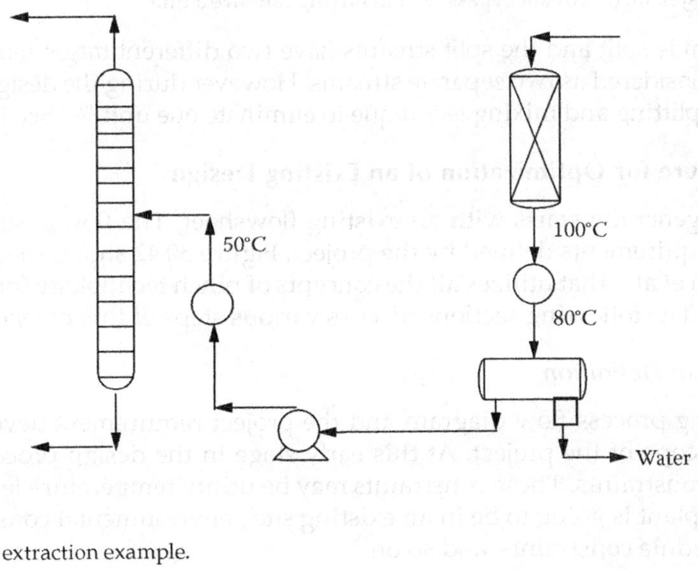

FIGURE 30.39
Flowsheet for data extraction example.

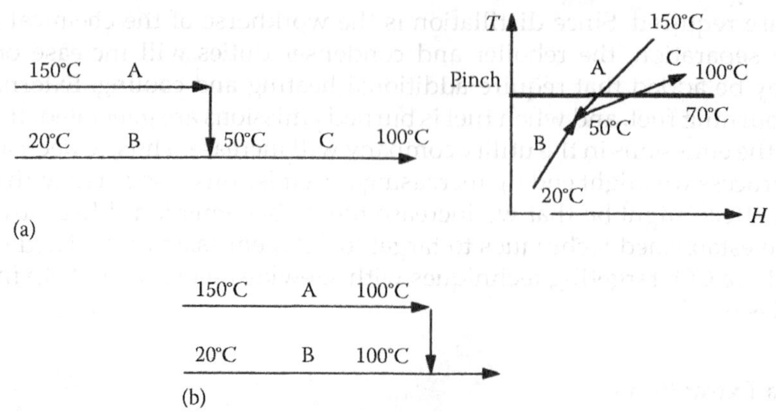

FIGURE 30.40
Data extraction for streams that are mixing. (a) A hot stream mixed with a cold stream reaching a combined target temperature. (b) Hot stream and cold stream reaching the same target temperature.

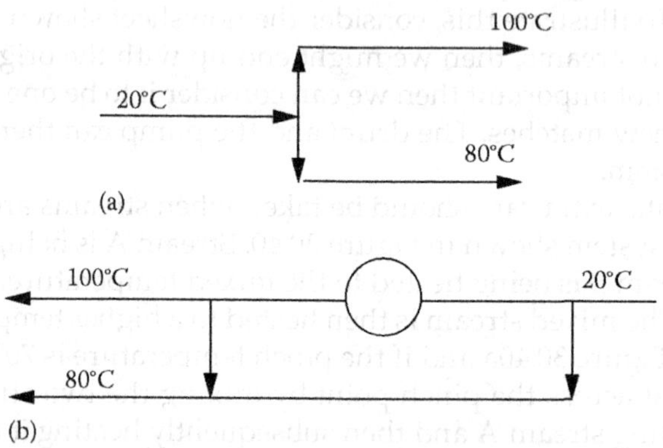

FIGURE 30.41
(a) Original stream system. (b) Stream bypassing and mixing can save a unit.

When a stream is split and the split streams have two different target temperatures then each stream is considered as two separate streams. However during the design phase we can use the stream splitting and mixing technique to eliminate one unit.[30,38] See Figure 30.41.

30.1.16 Procedure for Optimization of an Existing Design

Process design generally starts with an existing flowsheet. The flowsheet is modified to meet the new requirements defined by the project. Figure 30.42 shows the procedure proposed by Trivedi et al.[39] that utilizes all the concepts of pinch technology for optimizing an existing design. The following sections discuss various steps of this procedure.

30.1.16.1 Problem Definition

From the existing process flow diagram and the project requirement develop a problem definition and scope of the project. At this early stage in the design process, identify all the key project constraints. These constraints may be utility temperature levels (if the construction of the plant is going to be in an existing site), environmental constraints, budget constraints, schedule constraints, and so on.

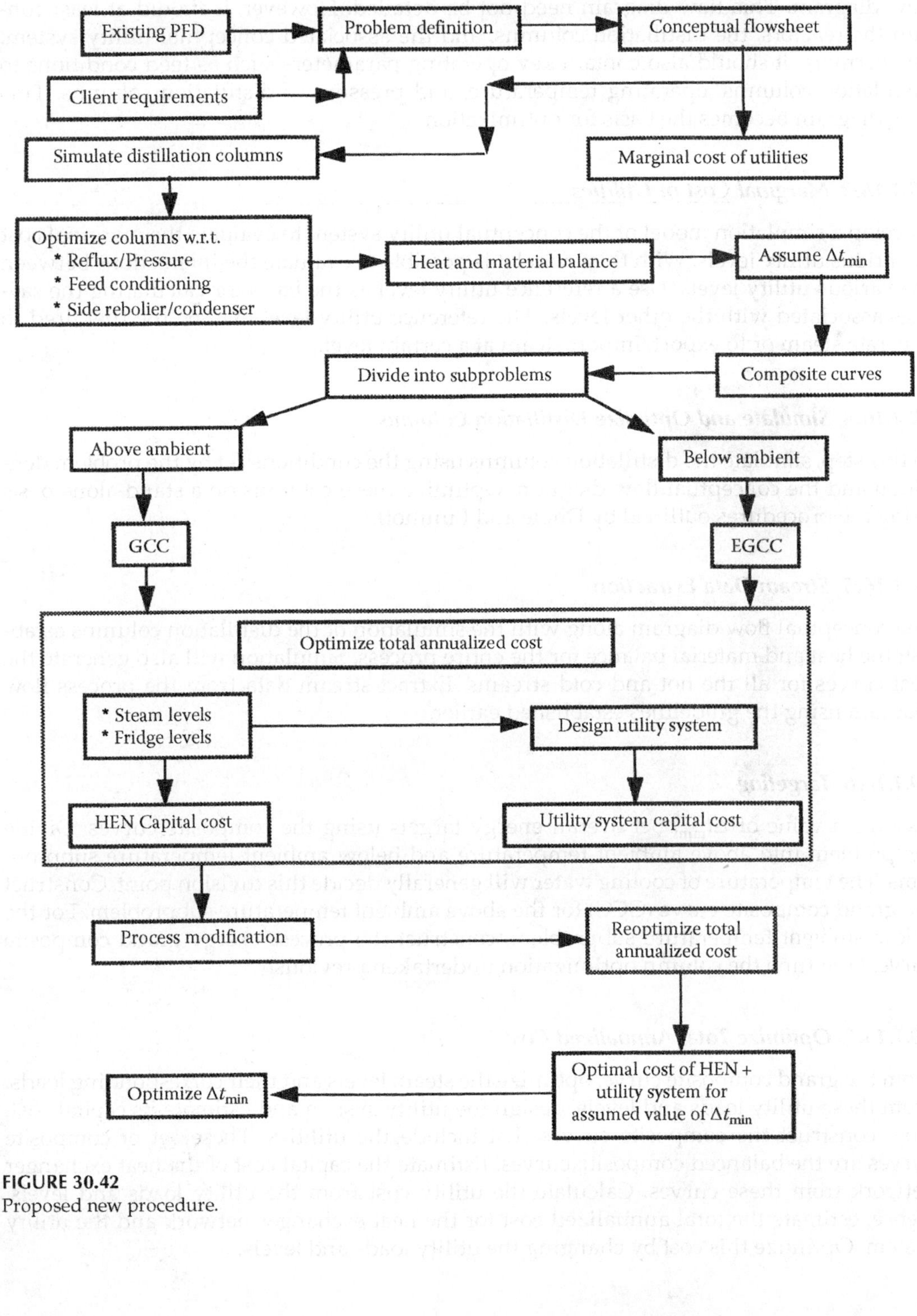

FIGURE 30.42
Proposed new procedure.

30.1.16.2 Conceptual Flow Design

On the basis of the problem definition and project constraints, develop a new conceptual flow diagram. This flow diagram need not be detailed. However, it should at least contain the reactors, the distillation columns, and the associated conceptual utility system. Furthermore, it should also contain key operating parameters such as feed conditions to distillation columns, operating temperature, and pressure of distillation columns. This flow diagram becomes the basis for optimization.

30.1.16.3 Marginal Cost of Utilities

Develop a simulation model of the conceptual utility system to evaluate the marginal cost of various utility levels. With this model it is possible to evaluate the interactions between the various utility levels. Use a reference utility level as the basis for calculating the savings associated with the other levels. The reference utility level may be the fuel fired to generate steam or to export/import steam at a certain level.

30.1.16.4 Simulate and Optimize Distillation Columns

In this step, simulate the distillation columns using the conditions set by the problem definition and the conceptual flow diagram. Optimize these columns on a stand-alone basis using the procedures outlined by Dhole and Linnhoff.[31]

30.1.16.5 Stream Data Extraction

The conceptual flow diagram along with the simulation of the distillation columns establish the heat and material balance for the entire process. Simulation will also generate the heat curves for all the hot and cold streams. Extract stream data from the process flow diagram using the guidelines established earlier.

30.1.16.6 Targeting

Assume a value of Δt_{min}. Set overall energy targets using the composite curves. Divide the problem into above ambient temperature and below ambient temperature subproblems. The temperature of cooling water will generally decide this division point. Construct the grand composite curve (GCC) for the above ambient temperature subproblem. For the below ambient temperature subproblem, construct the process exergy grand composite curve. Fine-tune the column optimization undertaken previously.

30.1.16.7 Optimize Total Annualized Cost

From the grand composite curve, optimize the steam levels and their corresponding loads. From these utility loads and levels, design the utility system and estimate its capital cost. Now construct the composite curves that include the utilities. These set of composite curves are the balanced composite curves. Estimate the capital cost of the heat exchanger network from these curves. Calculate the utility cost from the utility loads and levels. Hence, estimate the total annualized cost for the heat exchanger network and the utility system. Optimize this cost by changing the utility loads and levels.

30.1.16.8 Process Modification

Use the balanced composite curves to identify process modifications that can potentially reduce the total annualized cost. Changing the operating pressure of a distillation column or modifying the flow rate of a stream are some of the potential process modifications that can be made. The total annualized cost needs to be reoptimized if the utility system design changes significantly.

30.1.16.9 Optimize the Value of Δt_{min}

At this stage we have the optimal total annualized cost targets for the heat exchanger network and the utility system for the assumed value of Δt_{min}. Assuming another value of Δt_{min} will give another optimized total annualized capital cost. Repeat the exercise over a range of Δt_{min} values to find the optimal value. Design the heat exchanger network at the optimal value of Δt_{min}.

Trivedi et al.[39] have reported a case study that discusses the application of the above outlined approach to optimize Brown and Root's state-of-the-art Ethylene process. They achieved 12% savings in energy consumption. The capital cost increased by 2% with a simple payback period of 2 years. (Typical specific energy consumption of an ethylene plant for cracking ethane through naphtha ranges from 3000 to 5000 kcal/kg. The total world production capacity of ethylene in 1989 was 61 million MTY[40]).

30.1.17 Recent Developments

Recent developments have been made to solve different process design problems using pinch technology principles and concepts. We shall discuss these developments in brief.

Normally plants have many variables that change during the course of time. For example, depending on market conditions the through put may change or the feed stock can change. Process conditions can change from summer to winter operations. Catalytic reactors lose their effectiveness and heat exchangers get fouled over time. The final process design should be able to handle such changes. Sensitivity analysis and multiple base-case design procedures are developed that address the issue of flexibility.[8,41]

The pressure drop in a heat exchanger network is also important. Polley et al.[42] have developed methods that account for stream pressure drops during the targeting and detail design phase. Stream pressure drop constraints are very important during revamp projects. Pinch technology concepts are also applied to retrofit projects, heat integrate batch processes, minimize waste water, integrate evaporator systems, and design power cycles.[1]

30.2 Pinch Technology in Practice

The approach typically taken in a pinch study begins with the formation of a team consisting of pinch experts, key plant personnel, and possibly utility representatives. The team then visits the site to gather data pertaining to the processes, as well as economic and utility information. With data in hand, the pinch experts characterize the heating and cooling requirements of the plant's processing units using pinch technology. They next quantify the scope for potential improvement in each unit and the overall site. Then, the team identifies specific projects based on process change, process heat recovery, etc. that can be implemented to achieve target energy savings. The target energy savings is the summation of savings from practical, cost-effective projects identified to move the energy use at the site as close as possible to the minimum total process heating and cooling energy defined by the pinch analysis.

The Electric Power Research Institute (EPRI) and its utility members in the United States and overseas have sponsored a variety of energy-related projects in the process industries through EPRI's Industrial Technology Application Service. The Industrial Technology Application Service is now owned and operated by Global Energy Partners. Pinch studies have historically been a key element of EPRI's process industries programs, particularly in the Pulp, Paper, and Forest Products program (see Section 30.2.2) and the Chemicals, Pharmaceuticals, Petroleum Refining, and Natural Gas program (see Section 30.2.3). Electric Power Research Institute has also carried out numerous studies for the food and beverage industry and several studies for the textile and fiber industry. In all, EPRI has conducted more than 50 pinch studies in the last 15 years that, on average, have identified energy cost savings of more than 20% with 1–3 year payback periods.

30.2.1 Pinch Technology in Pulp and Paper Industries

Pulping and papermaking are energy-intensive. Rising energy costs have triggered a new awareness of energy usage and have spawned a number of energy efficiency studies across the industry. The U.S. Department of Energy cofunds several programs for improving energy efficiency. Its short-term initiatives include the popular plant-wide energy assessments where energy audits identify cost-reduction opportunities. Longer term, its Agenda 2020 program focuses on developing major technology jumps meant to reduce the energy costs of papermaking and to provide environmental benefits. Energy and environmental issues are intertwined; lower energy use reduces emissions from both the mill and from the power plant that generates the energy.

EPRI and associates have conducted over 30 pinch studies for the pulp, paper, and forest products industry alone. Collectively, the studies have pinpointed potential energy savings of over $80 million/year, with many projects having a simple payback period of about 1 year. Table 30.7 summarizes results from some of these studies. The table shows the target, actual, and scope energy values, as well as practical energy and cost savings by mill type. Target values represent the minimum energy required to run a process; actual values represent the current, actual energy use; scope values represent the maximum possible heat recovery; and practical values represent what is realistically feasible considering constraints. As the table shows, practical energy savings per mill range from about 50 to 350 million Btu/h (~15–100 MW), and practical cost savings range from about $1 to $6 million/year.

Table 30.8 lists the average cost savings per ton of product for several types of mills. Savings range from $1.3 per bone dry short ton (BDST) ($1.4/metric ton) for nonintegrated papermaking to $5.5/BDST ($6.1/metric ton) for bleached market pulp.

Two specific examples of pinch technology being applied to the pulp, paper, and forest products industry are described in the following case studies.

TABLE 30.7

Summary of Results from Pinch Studies in the Pulp, Paper, and Forest Products Industry

Mill Type	Target 10^6 Btu/h (MW)	Actual 10^6 Btu/h (MW)	Scope 10^6 Btu/h (MW)	Practical Savings 10^6 Btu/h (MW)	Practical Savings $\$10^6$/year
Semisulfite/OCC linerboard	297 (87)	417 (122)	120 (35)	45–80 (13–23)	1.4–2.8
Kraft/NSSC/OCC linerboard	939 (275)	1250 (366)	311 (91)	177 (52)	1.37
Continuous bleached kraft	NA	NA	177 (52)	127 (37)	3.2
Bleached kraft	428 (125)	553 (162)	125 (37)	85 (25)	3.2
Bleached kraft	732 (215)	866 (254)	134 (39)	80 (23)	1
Newsprint kraft	516 (151)	921 (270)	405 (119)	160 (47)	1.6
Bleached kraft	292 (86)	442 (130)	150 (44)	100 (29)	3
Continuous bleached kraft HWD	NA	NA	140 (41)	118 (35)	2.5
Bleached kraft/TMP/groundwood	792 (232)	1154 (338)	362 (106)	300 (88)	4.0
Bleached kraft	826 (242)	1096 (321)	270 (79)	160 (47)	2.0
Bleached kraft	1265 (371)	1904 (558)	639 (187)	235 (69)	2.8
Bleached semisulfite	235 (69)	314 (92)	79 (23)	37 (11)	1.4
Bleached kraft/TMP	625 (183)	965 (283)	340 (100)	294 (86)	2.1
Bleached kraft and NSSC	499 (146)	688 (202)	189 (55)	100 (29)	2.1
TMP/recycle mill greenfield design	Capital cost avoided ~$5 million				
Bleached kraft	531 (156)	693 (203)	162 (47)	79 (23)	2
TMP	NA	NA	137 (40)	70 (21)	1.5
Bleached kraft	1212 (355)	1387 (407)	175 (51)	100 (29)	1.8
TMP/Deink Mill	19 (6)	156 (46)	137 (40)	67 (20)	1.3
Bleached kraft/TMP/groundwood	Multiple case scenario			350 (103)	5.8
Bleached kraft	542 (159)	864 (253)	322 (94)	175 (51)	2.3
Bleached kraft	NA	NA		80 (23)	3.5
Continuous bleached kraft northern	NA	NA	190 (56)	110 (32)	3.1
Groundwood/coaters	255 (75)	360 (106)	105 (31)	53 (16)	0.7
Kraft/NSSC/OCC linerboard	933 (273)	1178 (345)	245 (72)	225 (66)	4.8

TABLE 30.8

Practical Steam Cost Savings Identified by Pinch Technology in the Pulp, Paper, and Forest Products Industry

Mill Type	\$/BDST (\$/metric ton)
Bleached kraft/NSSC	2.9 (3.2)
Bleached kraft/TMP and other	5.0 (5.5)
Bleached market pulp	5.5 (6.1)
Nonintegrated papermaking	1.3 (1.4)
Kraft/NSSC/OCC	2.9 (3.2)
Sulfite or semisulfite	4.4 (4.9)

30.2.1.1 Case Study 1: Minimizing Process Energy Use in a Large Thermomechanical Pulp (TMP) Mill*

EPRI and American Process, Inc. conducted an energy targeting scoping study using pinch analysis at a large TMP mill in Canada. A local utility company cofunded the project. The mill produces bleached, unbleached, and semibleached kraft market pulp and standard and offset newsprint. Fiber is prepared in kraft continuous and batch digesters, groundwood, and TMP departments. The kraft digesters separate wood chips into fiber by use of chemicals that act on the wood's binding agent (lignin). In the groundwood process, wood chips are mechanically pressed against a grindstone and, with the addition of water, are separated into fibers. The TMP process uses both heat and mechanical forces to convert wood chips into fibers.

Spent chemicals and lignin, which are referred to as *black liquor,* are recovered from the kraft process in a recovery cycle that includes three evaporators and three recovery boilers, a recausticizing line, and two kilns. First, the black liquor is removed from the pulp in washers; next, it is thickened in the evaporators; then, it is burned in the recovery boilers. During the burning process, the chemicals and lignin are recovered and the steam raised in the boilers is used to support operations. Two lime kilns provide calcium oxide for causticizing. Chlorine dioxide is produced onsite and used in a six-stage bleach plant.

The mill also has power boilers that use oil and hog fuels. The steam from the power boilers is used in steam turbines to generate a portion of the mill's power requirement, as well as to support various operations. The balance of the mill's power is purchased.

The intent of the energy targeting scoping study using pinch analysis was to identify ways for the mill to reduce energy costs as part of a long-term improvement in competitiveness. In particular, operators of the mill wanted to reduce process steam consumption so that they could decrease steam production from the power boilers.

Pinch analysis yielded the hot and cold composite curves shown in Figure 30.43 and the grand composite curve shown in Figure 30.44. The figures show that the pinch interval temperature is 246°F (119°C) and the thermal energy target (minimum thermal energy required to run the extracted process steam) is 1023 million Btu/h (300 MW). Because the existing process thermal consumption is 1,336 million Btu/h (392 MW), the maximum potential process steam savings (scope) is about 313 million Btu/h (~310,000 lb/h, 92 MW, 140,000 kg/h).

To accomplish the objective of reducing steam consumption, two options were developed for the recovery of TMP dirty steam. Option 1 uses TMP steam in the mill's pulp *and* paper companies, and option 2 restricts the use of TMP steam to the paper company. For option 1, implementation of TMP heat recovery in conjunction with several other recommended projects was found to potentially save 292,400 lb/h (132,600 kg/h) of process steam; for option 2, the potential savings were found to be 232,800 lb/h (105,600 kg/h). These steam savings equate to cost savings of about Can$9.5 million/year (Canadian dollars per year) for option 1 and Can$8.7 million/year for option 2 as a result of decreased steam production in the power boilers. The simple payback periods were estimated to be 2.5 years for option 1 and 2.1 years for option 2.

Additionally, a target cogeneration analysis showed that with the existing steam consumption, there was an opportunity to increase onsite power generation by about 21 MW (which equates to a savings of about Can$6.3 million/year) by correcting

* Data from EPRI, Minimizing process energy use for a large TMP mill with pinch technology, EPRI, Palo Alto, CA, 2000.

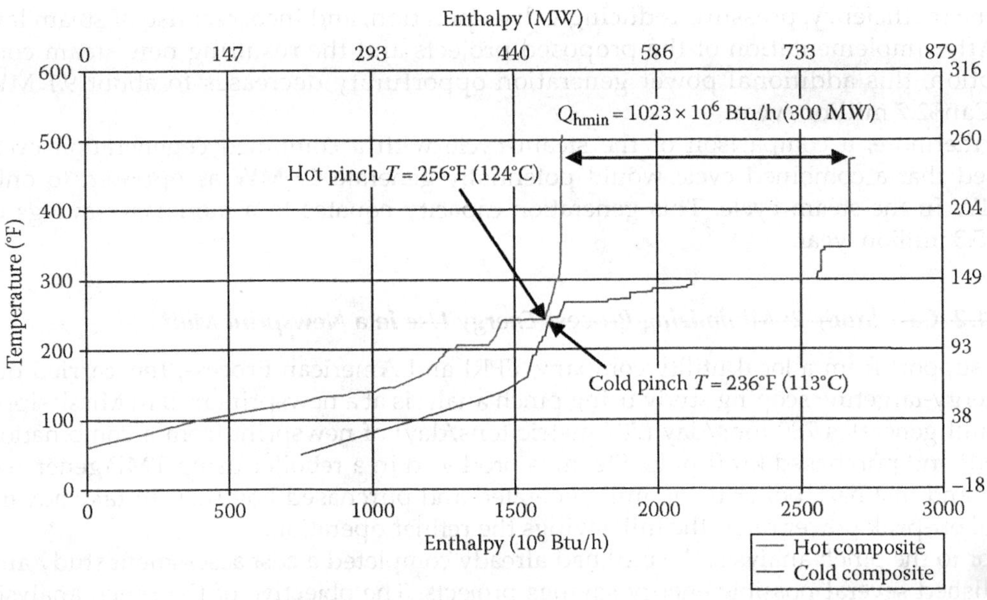

FIGURE 30.43
Hot and cold composite curves for thermomechanical pulping (TMP) mill.

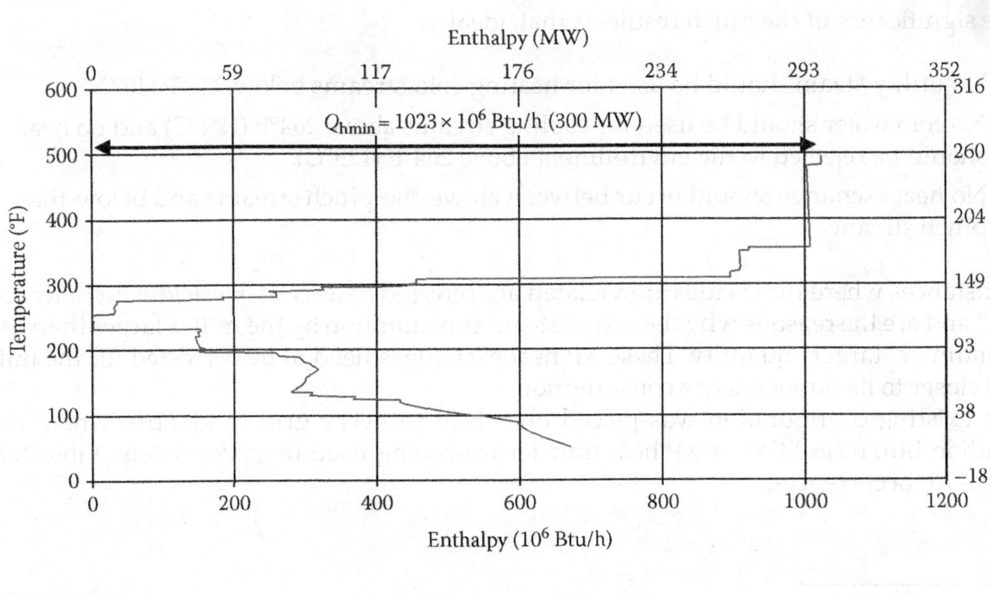

FIGURE 30.44
Grand composite curve for thermomechanical pulping (TMP) mill.

turbine inefficiency, pressure-reducing valve operation, and incorrect use of steam levels. After implementation of the proposed projects and the resulting new steam consumption, this additional power generation opportunity decreases to about 9.1 MW and Can$2.7 million/year.

Furthermore, a comparison of the steam cycle with a combined cogeneration cycle showed that a combined cycle would potentially generate 49 MW as opposed to only 8.6 MW in the steam cycle. This generation capacity equates to a monetary savings of Can$5.3 million/year.

*30.2.1.2 Case Study 2: Minimizing Process Energy Use in a Newsprint Mill**

With support from a local utility company, EPRI and American Process, Inc. carried out an energy-targeting scoping study using pinch analysis at a newsprint mill in Mississippi. The mill generates 780 tons/day (707 metric tons/day) of newsprint from a combination of TMP and purchased kraft pulp. Steam is produced in a reboiler using TMP-generated steam and in a bark boiler using mill-generated and purchased hog fuel. To take advantage of off-peak power rates, the mill swings the refiner operation.

Prior to the pinch analysis, the mill had already completed a cost assessment study and established several possible energy savings projects. The objective of the pinch analysis was to augment previous efforts and identify additional possible projects, as well as to prioritize all projects and ensure that the savings were additive.

For the steam duties evaluated, the hot and cold composite curves and the grand composite curve yielded a pinch interval temperature of 244°F (118°C) for the minimum temperature driving force of the Δt_{min} = 20°F (11°C), and a thermal energy target (minimum thermal energy required to run the extracted process streams) of 42 million Btu/h (12 MW). Because the existing process thermal consumption of the streams is 96 million Btu/h (28 MW), the scope (or maximum potential to reduce process heat demand) is 54 million Btu/h (16 MW).

The significance of the pinch results is that, ideally:

- No utility steam should be used for heating cold streams below 224°F (107°C).
- No cold water should be used for cooling streams above 264°F (129°C) and no heat should be rejected to the environment above 264°F (129°C).
- No heat exchange should occur between above-the-pinch streams and below-the-pinch streams.

The instances where these rules are violated are referred to as *cross-pinch heat exchanges* or "XPs," and are the reasons why the actual steam consumption by the mill is larger than the minimum or "target" quantity. These XP heat exchanges need to be corrected for the mill to get closer to its target energy consumption.

The existing configuration was placed on a heat recovery grid to identify where the 54 million Btu/h (16 MW) of XP heat transfer was being used inappropriately. Table 30.9 lists the XP occurrences.

* Data from EPRI, Energy targeting scoping study using pinch analysis: Bowater newsprint, Grenada, MS, EPRI, Palo Alto, CA, 2004.

TABLE 30.9

Existing Cross-Pinch Heat Transfer Occurrences in a Newsprint Mill

Stream	Process-Process XP 10^6 Btu/h (MW)	Stream XP 10^6 Btu/h (MW)	Waste XP 10^6 Btu/h (MW)	Total XP 10^6 Btu/h (MW)
Bark boiler air heating		3.78 (1.11)		
Demin water heating		7.88 (2.31)		
Condensate return heating		2.90 (0.85)		
White water heating		3.70 (1.08)		
Off machine silo heating		5.64 (1.65)		
Warm water heating		1.30 (0.38)		
Paper machine PV air heating		14.12 (4.14)		
Paper machine building air heating		0.86 (0.25)		
Primary and secondary refiners steam			12.20 (3.58)	
Primary and secondary refiners steam condensate	1.36 (0.40)		0.63 (0.18)	
Subtotals	1.36 (0.40)	40.18 (11.78)	12.83 (3.76)	54.37 (15.93)

The team developed a variety of projects for the mill to consider to achieve energy savings. For the energy savings to correspond to realistic monetary savings, the energy savings by steam reduction must lead to a decrease in purchased hog fuel. Some of the projects initially developed would have had good payback periods if the steam were produced with fossil fuel; however, they yielded poor returns based on the cost of marginal steam from hog fuel. Six additional low-cost projects were proposed that would result in steam savings of 9,000 lb/h (4,080 kg/h) in the summer and 29,000 lb/h (13,150 kg/h) in the winter with payback periods of only 5–20 months. An additional project was recommended to temper incoming mill water with effluent during the winter months to reduce the seasonal variation in steam demand, yielding a potential savings of 15,000 lb/h (6,800 kg/h) during the winter. Table 30.10 summarizes the energy costs savings, capital costs, and payback periods

TABLE 30.10

Summary of Costs and Benefits of Projects Developed for a Newsprint Mill

No.	Description	Savings ($/year)	Capital Cost ($)	Payback (years)
1	Scrubber—condensate storage tank vent	30,000	25,000	0.83
2	Sewer pressate purge after pressate coolers	25,000	10,300	0.41
3	Turpentine condenser—heated water generation	30,000	50,100	1.67
4	Optimize reject refinery vent heat recovery	39,900	55,500	1.39
5	Ejector for vent steam to chip bin	58,400	103,200	1.77
6	Series water flow through pressate coolers	15,000	10,700	0.71
7	Tempering mill water with process effluent	74,800	207,300	2.77
1–7	Totals	273,100	462,100	1.69

TABLE 30.11

Representative Results from Pinch Studies in the Chemical and Petroleum Industries

Type of Facility	Study Highlights
Petroleum refinery	$840,000/year savings at 1.5-year payback, plus capital savings of $390,000 for planned retrofit
Petroleum refinery	Scope for 32% reduction in net fuel consumption
Petroleum refinery	Revised heat exchanger network in crude distillation unit to reduce costs by $1 million/year at less than 1.5-year payback
Petroleum refinery	Revised heat exchanger network in hydrocracker to save $2.7 million/year in steam/fuel costs at 1.6-year payback
Petroleum refinery	Steam cost savings of $5 million/year with 1–4-year payback
Pharmaceutical company	Increased heat recovery to reduce fuel costs by 13% at less than 1-year payback
Pharmaceutical company	Identified measures to reduce thermal energy requirements by 30%; annual savings of $70,000/year at 1.3–2.5-year payback
Specialty chemical plant: kelp harvesting and extraction	Steam savings of $2.8 million/year at 1–2-year payback
Wax extraction plant	Cost savings of $930,000/year at 2.3-year payback

for all of the recommended projects. If implemented, the potential savings would be about $273,100/year with a capital cost of $462,100 and a simple payback period of 1.69 years.

30.2.2 Pinch Technology in Chemical and Petroleum Industries

The chemical and petroleum industries offer tremendous opportunities for energy savings from pinch analysis. EPRI and associates have carried out several dozen pinch studies in these industries since 1988. The calculated payback periods for most of the projects have been below 2 years and energy cost savings have typically been over $1 million. Table 30.11 summarizes a few of the studies conducted. Two more detailed examples are provided in the following case studies.

*30.2.2.1 Case Study 1: Minimizing Process Energy Use in a Petroleum Refinery**

With cofunding from local utility company, EPRI and Aspen Technology, Inc. conducted a sitewide pinch study at an oil refinery in Illinois. The objective of the study was to recommend ways to improve the refinery processes and utility system to achieve plant-wide energy savings. The project was conducted in two phases. The first phase consisted of a targeting and scoping study of the overall site. The second phase involved revising the basic design of selected process units. The following processing units were included in the study:

- Atmospheric crude and vacuum distillation unit (VDU)
- Two delayed coker units
- Reformer and naphtha hydrotreater
- CCR reformer and naphtha hydrotreater
- Isom and LSR hydrotreater

* Data from EPRI, Pinch technology/process optimization: Volume 9: Case study, EPRI, Palo Alto, CA, 1997.

TABLE 30.12

Sitewide Energy Targets for a Petroleum Refinery

Unit	Rating	Total Hot Existing	Total Hot Target	Hot Delta	Total Cold Existing	Total Cold Target	Cold Delta	Sum Delta
Crude/Vacuum	1	199.9	181.5	18.4	106.9	88.0	18.9	−0.5
FCC	1	20.8	0.0	20.8	177.9	157.1	20.8	0.0
Isom	1	15.0	11.0	4.0	17.8	13.8	4.0	0.0
CCR reformer	1	196.4	167.5	28.9	115.1	86.2	28.9	0.0
Coker A	1	34.1	31.6	2.5	52.2	49.7	2.5	0.0
Coker B	1	39.1	38.1	1.0	53.7	52.7	1.0	0.0
Reformer	1	159.9	139.1	20.8	87.0	66.2	20.8	0.0
Hydrocracker	1	65.4	37.1	28.3	140.0	111.7	28.3	0.0
HF alkylation	2	62.3	59.8	2.5	81.2	78.7	2.5	0.0
DHT	2	31.9	21.0	10.9	71.7	60.8	10.9	0.0
MTBE	2	30.6	29.3	1.3	31.6	30.3	1.3	0.0
Amine	3	22.1	22.1	0.0	21.3	21.3	0.0	0.0
Naphtha splitter	3	12.2	12.2	0.0	12.0	12.0	0.0	0.0
BT/PFS towers	3	26.2	26.2	0.0	22.1	22.1	0.0	0.0
Sat gas 23	3	22.1	21.9	0.2	23.0	22.7	0.3	−0.1
Sat gas 8	3	31.0	31.1	−0.1	32.0	32.1	−0.1	0.0
SWS	3	24.2	24.2	0.0	23.1	23.1	0.0	0.0
Sulfur plants	3	6.9	2.2	4.7	51.4	46.7	4.7	0.0
Total		1000.0	856.0	144.0	1119.9	975.3	144.6	−0.6
Total site		0.0	853.6		0.0	968.2		

- HF alkylation
- FCCU and light ends recovery
- Diesel hydrotreater
- MTBE
- Two saturated gas plants
- Two sulfur recovery trains and tail gas units
- Hydrocracker
- Utility system

During the first phase, the team performed an overall site analysis that included energy targeting, utility placement, and total site targeting. Table 30.12 summarizes the results of the study for each of the individual processes and for the total site. The table compares the total existing hot and cold utility usage for each of the processes to the total hot and cold utility targets. The final rows repeat this comparison for the total site.

The estimated existing process hot utility usage for the refinery is 1000 energy units and the energy target for the complete refinery is 856 energy units. Therefore, the scope for energy reduction is 144 energy units, or 14.4%.

The table indicates that of the 18 units included in the study, there are eight that are key to the overall refinery efficiency improvement, three others have an impact but are less important, and the remaining seven need not be considered further. The key units are the crude/vacuum unit, FCC, Isom, CCR reformer, two cokers, reformer, and hydrocracker. The eight key units account for 73% of the existing hot utility usages, or 730 energy units, and 71% of the energy target, or 606 energy units. Therefore, concentrating on the eight key units reduces the scope for energy reduction to 124 energy units, or 12% of the site.

Phase 2 of the project involved using pinch analysis to study the key units in detail. Six of the eight key units were chosen:

1. Coker A
2. Coker B
3. Crude/vacuum unit
4. CCR reformer
5. Hydrocracker
6. Reformer

For reasons mainly associated with future investment plans and plant turnaround schedules, the FCC and Isom were not studied further.

Phase 2 analysis resulted in the identification of several projects with energy cost savings ranging from $200,000/year to $1,410,000/year and simple payback periods of as low as 0.5 years up through 16 years. The projects with the biggest energy savings were associated with the longest payback periods; however, these projects also offer significant, cost-effective debottlenecking potential for the units. Implementation of all mutually exclusive projects would result in energy savings of about $4.6 million/year.

A plant-side steam balance was also conducted. The results from the steam balance showed an overall energy savings potential in the steam system of about $5 million/year with a payback of 1–4 years. However, it is important to note that projects identified within the units and within the steam system sometimes compete. Therefore, energy cost savings for the projects and steam system are not directly additive.

*30.2.2.2 Case Study 2: Minimizing Process Energy Use in a Wax Extraction Plant**

Together with support from a local utility company, EPRI and Linnhoff March, Inc. conducted a comprehensive energy study using pinch analysis at a wax extraction plant in Pennsylvania. The plant produces high-grade waxes for use in a variety of commercial and consumer markets. The objective of the study was to identify ways for the plant to reduce processing costs per barrel of throughput. The plant's first priority was to implement quick payback projects such as insulation of storage tanks. The plant was also interested in increasing manufacturing flexibility and achieving compliance with a nitrogen oxide (NO_x) abatement order at a minimum cost.

The waxes produced at the plant are petroleum based. Waxy crudes and distillates are first fractionated in a VDU to separate the oil into several distinct cuts. The VDU bottoms stream is treated with propane in the propane deasphalting (PDA) unit to separate out the asphaltenes (or heavy resin fraction), and the main oil fraction is recovered for further processing. The side cuts from VDU and recovered oil from PDA are then subjected to solvent extraction using methyl ethyl ketone (MEK) to separate wax from oil. The waxes are purified by bauxite filtration. The residual oils are sold as by-products to lube oil processors and heating oil dealers, and asphalt is used onsite as fuel.

The project team was charged with improving operating economics, debottlenecking processing capacity, evaluating the feasibility of hydrotreating vs. bauxite filtration for the wax

* Data from EPRI, Pinch technology/process optimization: Volume 7: Case study—A wax extraction plant, EPRI, Palo Alto, CA, 1995.

TABLE 30.13

Economics of Developed Projects for a Wax Extraction Plant

Project	Savings ($/year)	Capital Cost ($)	Payback (years)
VDU retrofit	300,000	300,000	1.0
MEK retrofit	350,000	1,500,000	4.2
Tank insulation	280,000	300,000	1.0
Total	930,000	2,100,000	2.3

finishing operation, and determining the optimum combined heat and power strategy for the site, paying particular heed to compliance with rather stringent NO_x abatement requirements.

The results of the study yielded a comprehensive process improvement strategy that improved wax yield by 8%, debottlenecked refrigeration capacity by 10%, and reduced total steam demand by 22%. The steam savings would allow shut down of two of the plant's four boilers, automatically reducing NO_x emissions, enabling compliance with the NO_x abatement order without installing expensive end-of-pipe controls. Implementation of suggested retrofits in the VDU and MEK process areas along with tank insulation would potentially yield operating savings of $930,000 million/year with a cost of $2.1 million and a simple payback period of 2.3 years. Evaluations related to cogeneration and to replacing the existing bauxite process with a hydrotreater showed that the economics did not favor either measure. Table 30.13 summarizes the economics of the proposed retrofits. Note that although the stand-alone economics of the MEK revamp are not particularly attractive (i.e., the payback period is 4.2 years), this project is essential to enable shutdown of two of the boilers, thereby achieving a relatively low-cost NO_x reduction.

References

1. Linnhoff, B. 1994. Use pinch analysis to knock down capital costs and emissions, *Chem. Eng. Prog.*, 90, 32–57.
2. Linnhoff, B. and Turner, J. A. 1981. Heat recovery networks: New insights yield big savings, *Chem. Eng.*, November 2, 56.
3. Linnhoff, B. and Vredeveld, D. R. 1984. Pinch technology has come of age, *Chem. Eng. Prog.*, 80(7), 33.
4. Korner, H. 1988. Optimal use of energy in the chemical industry, *Chem. Eng. Tech.*, 60(7), 511.
5. Fraser, D. M. and Gillespie, N. E. 1992. The application of pinch technology to retrofit energy integration of an entire oil refinery, *Trans. IChemE*, 70(Part A), 395.
6. Samdani, G. and Moore, S. 1993. Pinch technology: Doing more with less, *Chem. Eng.*, 43.
7. Bell, K. J. 1978. Estimate S & T exchanger design fast, *Oil Gas J.*, 59, 4.
8. Hohmann, E. C. 1971. Optimum networks for heat exchange, PhD thesis, University of Southern California, Los Angeles, CA.
9. Linnhoff, B. and Flower, J. R. 1978. Synthesis of heat exchanger networks, *AIChE J.*, 24, 633.
10. Townsend, D. W. and Linnhoff, B. April 1984. Surface area targets for heat exchanger networks, *IChemE 11th Annual Research Meeting*, Bath University, Bath, U.K.
11. Linnhoff, B., Mason, D. R., and Wardle, I. 1979. Understanding heat exchanger networks, *Comput. Chem. Eng.*, 3, 295.
12. Trivedi, K. K., Roach, J. R., and O'Neill, B. K. 1987. Shell targeting in heat exchanger networks, *AIChE. J.*, 33(12), 2087.

13. Linnhoff, B. March 1986. The process/utility interface, Paper presented at *the Second International Meeting on "National Use of Energy,"* Liege, Belgium.
14. Ahmad, S. and Smith, R. 1989. Targets and design for minimum number of shells in heat exchanger networks, *Chem. Eng. Res. Des. Dev.*, 67, 481.
15. Linnhoff, B. and Ahmad, S. 1989. Supertargeting: Optimum synthesis of energy management systems, *J. Energy Resour. Technol.*, 111, 121.
16. Umeda, T., Itoh, J., and Shiroko, K. 1978. Heat exchange system synthesis, *Chem. Eng. Prog.*, 74, 70.
17. Linnhoff, B. and Hindmarsh, E. 1983. The pinch design method for heat exchanger networks, *Chem. Eng. Sci.*, 38(5), 745.
18. Linnhoff, B. and Ahmad, S. 1990. Cost optimum heat exchanger networks—Part I: Minimum energy and capital using simple models for capital cost, *Comput. Chem. Eng.*, 14(7), 729.
19. Su, L. J. 1979. A loop breaking evolutionary method for the synthesis of heat exchanger networks, MS thesis, Sever Institute of Washington University, St. Louis, MO.
20. Forder, G. J. and Hutchison, H. P. 1969. The analysis of chemical flowsheets, *Chem. Eng. Sci.*, 24, 771.
21. Trivedi, K. K., O'Neill, B. K., Roach, J. R., and Wood, R. M. 1990. A best-first search method for energy relaxation, *Eng. Optim.*, 16, 291.
22. Trivedi, K. K., O'Neill, B. K., Roach, J. R., and Wood, R. M. 1990. Systematic energy relaxation in MER heat exchanger networks, *Comput. Chem. Eng.*, 14(6), 601.
23. O'Reilly, M. 1985. Personal view, *Chem. Eng.*, 410, 46.
24. Trivedi, K. K., O'Neill, B. K., Roach, J. R., and Wood, R. M. 1989. A new dual-temperature design method for the synthesis of heat exchanger networks, *Comput. Chem. Eng.*, 13(6), 667.
25. Grimes, L. E., Rychener, M. D., and Westerberg, A. W. 1982. The synthesis and evolution of networks of heat exchanges that feature the minimum number of units, *Chem. Eng. Commun.*, 14, 339.
26. Townsend, D. W. Surface area and capital cost targets for process energy systems, PhD thesis, UMIST, Manchester, U.K.
27. Townsend, D. W. and Linnhoff, B. 1983. Heat and power networks in process design: Part I: Criteria for placement of heat engines and heat pumps in process networks; part II: Design procedure for equipment selection and process matching. *AIChE J.*, 29(5), 742.
28. Hall, S. G. 1989. Targeting for multiple utilities in pinch technology, PhD thesis, UMIST, Manchester, U.K.
29. Parker, S. J. 1989. Supertargeting for multiple utilities, PhD thesis, UMIST, Manchester, U.K.
30. Linnhoff, B., Townsend, D. W., Boland, D., Hewitt, G. F., Thomas, B. E. A., Guy, A. R., and Marshland, R. H. 1982. Users guide on process integration for the efficient use of energy, *IChemE*, Rugby, U.K.
31. Dhole, V. R. and Linnhoff, B. 1993. Distillation column targets, *Comput. Chem. Eng.*, 17(5/6), 549.
32. Linnhoff, B., Dunford, H., and Smith, R. 1983. Heat integration of distillation columns into overall processes, *Chem. Eng. Sci.*, 38(8), 1175.
33. Linnhoff, B. and de Leur, J. 1988. Appropriate placement of furnaces in the integrated process, Paper presented at *the IChemE Symposium—Understanding Process Integration—II*, UMIST, Manchester, U.K.
34. Linnhoff, B. and Dhole, V. R. 1992. Shaftwork targets for low temperature process design, *Chem. Eng. Sci.*, 47(8), 2081.
35. Raissi, K. 1994. Total site integration, PhD thesis, UMIST, Manchester, U.K.
36. Smith, R. and Delaby, O. 1992. Targeting flue gas emissions, *Trans. IChemE*, 69(Part A), 492.
37. Linnhoff, B. and Dhole, V. R. 1993. Targeting for CO_2 emissions for total sites, *Chem. Eng. Technol.*, 16, 256.
38. Wood, R. M., Wilcox, R. J., and Grossmann, I. E. 1985. A note of the minimum number of units for heat exchanger network synthesis, *Chem. Eng. Commun.*, 39, 371.

39. Trivedi, K. K., Pang, K. H., Klavers, H. R., O'Young, D. L., and Linnhoff, B. April 1994. Integrated ethylene process design using pinch technology, Presented at *the AIChE Meeting*, Atlanta, GA.
40. Ma, J. L. James. October 1991. Ethylene supplement E, Report 29E, SRI International, Menlo Park, CA.
41. Kotjabasakis, E. and Linnhoff, B. 1986. Sensitivity tables for the design of flexible processes, part I: How much contigency in heat exchange networks is cost-effective? *Chem. Eng. Res. Des.*, 64, 197.
42. Polley, G. T., Panjeh Shahi, M. H., and Jegede, F. O. 1990. Pressure drop consideration in the retrofit of heat exchanger networks, *Trans. IChemE*, 68(Part A), 211.

31

Analysis Methods for Building Energy Auditing

Moncef Krarti

CONTENTS

31.1 Introduction

Since the oil embargo of 1973, significant improvements have been made in the energy efficiency of new buildings. However, the vast majority of the existing building stock is more than 20 years old and does not meet current energy efficiency construction standards (EIA, 2008). Therefore, energy retrofits of existing buildings will be required for decades to come if the overall energy efficiency of the building stock is to meet the standards.

Investing to improve the energy efficiency of buildings provides an immediate and relatively predictable positive cash flow resulting from lower energy bills. In addition to the conventional financing options available to owners and building operators (such as loans and leases), other methods are available to finance energy retrofit projects for buildings. One of these methods is performance contracting in which payment for a retrofit project is contingent upon its successful outcome. Typically, an energy services company (ESCO) assumes all the risks for a retrofit project by performing the engineering analysis and obtaining the initial capital to purchase and install equipment needed for energy efficiency improvements. Energy auditing is an important step used by energy service companies to insure the success of their performance contracting projects.

Moreover, several large industrial and commercial buildings have established internal energy management programs based on energy audits to reduce waste in energy use or to comply with the specifications of some regulations and standards. Other building owners and operators take advantage of available financial incentives typically offered by utilities or state agencies to perform energy audits and implement energy conservation measures (ECMs).

In the 1970s, building energy retrofits consisted of simple measures such as shutting off lights, turning down heating temperatures, turning up air-conditioning temperatures, and reducing the hot water temperatures. Today, building energy management includes a comprehensive evaluation of almost all the energy systems within a facility. Therefore, the energy auditor should be aware of key energy issues such as the subtleties of electrical utility rate structures and of the latest building energy efficiency technologies and their applications.

This chapter describes a general but systematic procedure for energy auditing suitable for both commercial buildings and industrial facilities. Some of the commonly recommended ECMs are briefly discussed. A case study for an office building is presented to illustrate the various tasks involved in an energy audit. Finally, an overview is provided to outline the existing methods for measurement and verification of energy savings incurred by the implementation of ECMs.

31.2 Types of Energy Audits

The term *energy audit* is widely used and may have different meanings depending on the energy service companies. Energy auditing of buildings can range from a short walk-through of the facility to a detailed analysis with hourly computer simulation. Generally, four types of energy audits can be distinguished as briefly described in the following (Krarti, 2010).

31.2.1 Walk-Through Audit

This audit consists of a short on-site visit of the facility to identify areas where simple and inexpensive actions can provide immediate energy use and/or operating cost savings. Some engineers refer to these types of actions as operating and maintenance (O&M) measures. Examples of O&M measures include setting back heating set point temperatures, replacing broken windows, insulating exposed hot water or steam pipes, and adjusting boiler fuel–air ratio.

31.2.2 Utility Cost Analysis

The main purpose of this type of audit is to carefully analyze the operating costs of the facility. Typically, the utility data over several years are evaluated to identify the patterns of energy use, peak demand, weather effects, and potential for energy savings. To perform this analysis, it is recommended that the energy auditor conducts a walk-through survey to get acquainted with the facility and its energy systems.

It is important that the energy auditor understands clearly the utility rate structure that applies to the facility for several reasons:

- To check the utility charges and insure that no mistakes were made in calculating the monthly bills. Indeed, the utility rate structures for commercial and industrial facilities can be quite complex with ratchet charges and power factor penalties.
- To determine the most dominant charges in the utility bills. For instance, peak demand charges can be significant portion of the utility bill especially when ratchet rates are applied. Peak shaving measures can be then recommended to reduce these demand charges.
- To identify whether or not the facility can benefit from using other utility rate structures to purchase cheaper fuel and reduce its operating costs. This analysis can provide a significant reduction in the utility bills especially with implementation of the electrical deregulation and the advent of real-time pricing rate structures.

Moreover, the energy auditor can determine whether or not the facility is prime for energy retrofit projects by analyzing the utility data. Indeed, the energy use of the facility can be normalized and compared to indices (for instance, the energy use per unit of floor area—for commercial buildings—or per unit of a product, for industrial facilities).

31.2.3 Standard Energy Audit

The standard audit provides a comprehensive energy analysis for the energy systems of the facility. In addition to the activities described for the walk-through audit and for the utility cost analysis described earlier, the standard energy audit includes the development of a baseline for the energy use of the facility and the evaluation of the energy savings and the cost-effectiveness of appropriately selected ECMs. The step-by-step approach of the standard energy audit is similar to that of the detailed energy audit that is described later on in the following section.

Typically, simplified tools are used in the standard energy audit to develop baseline energy models and to predict the energy savings of ECMs. Among these tools are the degree-day methods and linear regression models (Fels, 1986). In addition, a simple payback analysis is generally performed to determine the cost-effectiveness of ECMs.

31.2.4 Detailed Energy Audit

This audit is the most comprehensive but also time-consuming energy audit type. Specifically, the detailed energy audit includes the use of instruments to measure energy use for the whole building and/or for some energy systems within the building (for instance, by end uses, lighting systems, office equipment, fans, and chillers). In addition, sophisticated computer simulation programs are typically considered for detailed energy audits to evaluate and recommend energy retrofits for the facility.

The techniques available to perform measurements for an energy audit are diverse. During on-site visit, handheld and clamp-on instruments can be used to determine the variation of some building parameters such as the indoor air temperature, the luminance level, and the electrical energy use. When long-term measurements are needed, sensors are typically used and connected to a data-acquisition system so measured data can be stored and be remotely accessible. Recently, nonintrusive load monitoring (NILM) techniques have been proposed (Shaw et al., 1998). The NILM technique can determine the real-time energy use of the significant electrical loads in a facility using only a single set of sensors at the facility service entrance. The minimal effort associated with using the NILM technique when compared to the traditional submetering approach (which requires separate set of sensors to monitor energy consumption for each end use) makes the NILM a very attractive and inexpensive load monitoring technique for energy service companies and facility owners.

The computer simulation programs used in the detailed energy audit can provide typically the energy use distribution by load type (i.e., energy use for lighting, fans, chillers, and boilers). They are often based on dynamic thermal performance of the building energy systems and require typically high level of engineering expertise and training. These simulation programs range from those based on the bin method (Knebel, 1983) to those that provide hourly building thermal and electrical loads such as DOE-2 (LBL, 1980). The reader is referred to Krarti (2010) for more detailed discussion of the energy analysis tools that can be used to estimate energy and cost savings attributed to ECMs.

In the detailed energy audit, more rigorous economical evaluation of the ECMs is generally performed. Specifically, the cost-effectiveness of energy retrofits may be determined based on the life-cycle cost (LCC) analysis rather than the simple payback period analysis. LCC analysis takes into account a number of economic parameters such as interest, inflation, and tax rates. Krarti (2010) describes some of the basic analysis tools that are often used to evaluate energy efficiency projects.

31.3 General Procedure for a Detailed Energy Audit

To perform an energy audit, several tasks are typically carried out depending on the type of the audit and the size and function of the audited building. Some of the tasks may have to be repeated, reduced in scope, or even eliminated based on the findings of other tasks. Therefore, the execution of an energy audit is often not a linear process and is rather iterative. However, a general procedure can be outlined for most buildings.

31.3.1 Step 1: Building and Utility Data Analysis

The main purpose of this step is to evaluate the characteristics of the energy systems and the patterns of energy use for the building. The building characteristics can be collected from the architectural/mechanical/electrical drawings and/or from discussions with building operators. The energy use patterns can be obtained from a compilation of utility bills over several years. Analysis of the historical variation of the utility bills allows the energy auditor to determine if there are any seasonal and weather effects on the building energy use. Some of the tasks that can be performed in this step are presented here with the key results expected from each task and are noted:

- Collect at least 3 years of utility data (*to identify a historical energy use pattern*).
- Identify the fuel types used (electricity, natural gas, oil, etc.) (*to determine the fuel type that accounts for the largest energy use*).

- Determine the patterns of fuel use by fuel type (*to identify the peak demand for energy use by fuel type*).
- Understand utility rate structure (energy and demand rates) (*to evaluate if the building is penalized for peak demand and if cheaper fuel can be purchased*).
- Analyze the effect of weather on fuel consumption (*to pinpoint any variations of energy use related to extreme weather conditions*).
- Perform utility energy use analysis by building type and size (building signature can be determined including energy use per unit area) (*to compare against typical indices*).

31.3.2 Step 2: Walk-Through Survey

From this step, potential energy savings measures should be identified. The results of this step are important since they determine if the building warrants any further energy auditing work. The following are some of the tasks involved in this step:

- Identify the customer concerns and needs.
- Check the current O&M procedures.
- Determine the existing operating conditions of major energy use equipment (lighting, HVAC systems, motors, etc.).
- Estimate the occupancy, equipment, and lighting (energy use density and hours of operation).

31.3.3 Step 3: Baseline for Building Energy Use

The main purpose of this step is to develop a base-case model that represents the existing energy use and operating conditions for the building. This model is to be used as a reference to estimate the energy savings incurred from appropriately selected ECMs. There are major tasks to be performed during this step:

- Obtain and review architectural, mechanical, electrical, and control drawings.
- Inspect, test, and evaluate building equipment for efficiency, performance, and reliability.
- Obtain all occupancy and operating schedules for equipment (including lighting and HVAC systems).
- Develop a baseline model for building energy use.
- Calibrate the baseline model using the utility data and/or metered data.

31.3.4 Step 4: Evaluation of Energy Savings Measures

In this step, a list of cost-effective ECMs is determined using both energy savings and economical analysis. To achieve this goal, the following tasks are recommended:

- Prepare a comprehensive list of ECMs (using the information collected in the walk-through survey).

- Determine the energy savings due to the various ECMs pertinent to the building using the baseline energy use simulation model developed in phase 3.
- Estimate the initial costs required to implement the ECMs.
- Evaluate the cost-effectiveness of each ECM using an economical analysis method (simple payback or LCC analysis).

Tables 31.1 and 31.2 provide summaries of the energy audit procedure recommended, respectively, for commercial buildings and for industrial facilities. Energy audits for thermal and electrical systems are separated since they are typically subject to different utility rates.

TABLE 31.1

Energy Audit Summary for Residential and Commercial Buildings

Phase	Thermal Systems	Electrical Systems
Utility analysis	• Thermal energy use profile (building signature). • Thermal energy use per unit area (or per student for schools or per bed for hospitals). • Thermal energy use distribution (heating, DHW, process, etc.). • Fuel types used. • Weather effect on thermal energy use. • Utility rate structure.	• Electrical energy use profile (building signature). • Electrical energy use per unit area (or per student for schools or per bed for hospitals). • Electrical energy use distribution (cooling, lighting, equipment, fans, etc.). • Weather effect on electrical energy use. • Utility rate structure (energy charges, demand charges, power factor penalty, etc.).
On-site survey	• Construction materials (thermal resistance type and thickness). • HVAC system type. • DHW system. • Hot water/steam use for heating. • Hot water/steam for cooling. • Hot water/steam for DHW. • Hot water/steam for specific applications (hospitals, swimming pools, etc.).	• HVAC system type. • Lighting type and density. • Equipment type and density. • Energy use for heating. • Energy use for cooling. • Energy use for lighting. • Energy use for equipment. • Energy use for air handling. • Energy use for water distribution.
Energy use baseline	• Review architectural, mechanical, and control drawings. • Develop a base-case model (using any baselining method ranging from very simple to more detailed tools). • Calibrate the base-case model (using utility data or metered data).	• Review architectural, mechanical, electrical, and control drawings. • Develop a base-case model (using any baselining method ranging from very simple to more detailed tools). • Calibrate the base-case model (using utility data or metered data).
ECMs	• Heat recovery system (heat exchangers). • Efficient heating system (boilers). • Temperature setback. • EMCS. • HVAC system retrofit. • DHW use reduction. • Cogeneration.	• Energy-efficient lighting. • Energy-efficient equipment (computers). • Energy-efficient motors. • HVAC system retrofit. • EMCS. • Temperature setup. • Energy-efficient cooling system (chiller). • Peak demand shaving. • TES system. • Cogeneration. • Power factor improvement. • Reduction of harmonics.

TABLE 31.2

Energy Audit Summary for Industrial Facilities

Phase	Thermal Systems	Electrical Systems
Utility analysis	• Thermal energy use profile (building signature). • Thermal energy use per unit of a product. • Thermal energy use distribution (heating, process, etc.). • Fuel types used. • Analysis of the thermal energy input for specific processes used in the production line (such as drying). • Utility rate structure.	• Electrical energy use profile (building signature). • Electrical energy use per unit of a product. • Electrical energy use distribution (cooling, lighting, equipment, process, etc.). • Analysis of the electrical energy input for specific processes used in the production line (such as drying). • Utility rate structure (energy charges, demand charges, power factor penalty, etc.).
On-site survey	• List of equipment that use thermal energy. • Perform heat balance of the thermal energy. • Monitor the thermal energy use of all or part of the equipment. • Determine the by-products of thermal energy use (such as emissions and solid waste).	• List of equipment that use electrical energy. • Perform heat balance of the electrical energy. • Monitor the electrical energy use of all or part of the equipment. • Determine the by-products of electrical energy use (such as pollutants).
Energy use baseline	• Review mechanical drawings and production flow charts. • Develop a base-case model (using any baselining method). • Calibrate the base-case model (using utility data or metered data).	• Review electrical drawings and production flow charts. • Develop a base-case model (using any baselining method). • Calibrate the base-case model (using utility data or metered data).
ECMs	• Heat recovery system. • Efficient heating and drying system. • EMCS. • HVAC system retrofit. • Hot water and steam use reduction. • Cogeneration (possibly with solid waste from the production line).	• Energy-efficient motors. • Variable speed drives. • Air compressors. • Energy-efficient lighting. • HVAC system retrofit. • EMCS. • Cogeneration (possibly with solid waste from the production line). • Peak demand shaving. • Power factor improvement. • Reduction of harmonics.

31.4 Common Energy Conservation Measures

In this section, some ECMs commonly recommended for commercial and industrial facilities are briefly discussed. It should be noted that the list of ECMs presented in this section does not pretend to be exhaustive nor comprehensive. It is provided merely to indicate some of the options that the energy auditor can consider when performing an energy analysis of a commercial or an industrial facility. More discussion of energy efficiency measures for various building energy systems is provided in later chapters of this book. However, it is strongly advised that the energy auditor keeps abreast of any new technologies that can improve the building energy efficiency. Moreover, the energy auditor should recommend the ECMs only based on a sound economical analysis for each ECM.

31.4.1 Building Envelope

For some buildings, the envelope (i.e., walls, roofs, floors, windows, and doors) can have an important impact on the energy used to condition the facility. The energy auditor should determine the actual characteristics of the building envelope. During the survey, a descriptive sheet for the building envelope should be established to include information such as materials of construction (for instance, the level of insulation in walls, floors, and roofs) and the area and the number of building envelope assemblies (for instance, the type and the number of panes for the windows). In addition, comments on the repair needs and recent replacement should be noted during the survey.

The following are some of the commonly recommended ECMs to improve the thermal performance of building envelope:

(a) *Addition of thermal insulation*: For building surfaces without any thermal insulation, this measure can be cost-effective.
(b) *Replacement of windows*: When windows represent a significant portion of the exposed building surfaces, using more energy-efficient windows (high *R*-value, low-emissivity glazing, air tight, etc.) can be beneficial in both reducing the energy use and improving the indoor comfort level.
(c) *Reduction of air leakage*: When infiltration load is significant, leakage area of the building envelope can be reduced by simple and inexpensive weather-stripping techniques.

The energy audit of the envelope is especially important for residential buildings. Indeed, the energy use from residential buildings is dominated by weather since heat gain and/or loss from direct conduction of heat or from air infiltration/exfiltration through building surfaces accounts for a major portion (50%–80%) of the energy consumption. For commercial buildings, improvements to building envelope are often not cost-effective due to the fact that modifications to the building envelope (replacing windows, adding thermal insulation in walls) are typically considerably expensive. However, it is recommended to systematically audit the envelope components not only to determine the potential for energy savings but also to insure the integrity of its overall condition. For instance, thermal bridges—if present—can lead to heat transfer increase and to moisture condensation. The moisture condensation is often more damaging and costly than the increase in heat transfer since it can affect the structural integrity of the building envelope.

31.4.2 Electrical Systems

For most commercial buildings and a large number of industrial facilities, the electrical energy cost constitutes the dominant part of the utility bill. Lighting, office equipment, and motors are the electrical systems that consume the major part of energy in commercial and industrial buildings.

(a) *Lighting*: Lighting for a typical office building represents on average 40% of the total electrical energy use. There are a variety of simple and inexpensive measures to improve the efficiency of lighting systems. These measures include the use of energy-efficient lighting lamps and ballasts, the addition of reflective devices, delamping (when the luminance levels are above the recommended levels by

TABLE 31.3
Typical Efficiencies of Motors

Motor Size (HP)	Standard Efficiency (%)	Premium Efficiency (%)
1	73.0	85.5
2	75.0	86.5
3	77.0	86.5
5	80.0	89.5
7.5	82.0	89.5
10	85.0	91.7
15	86.0	92.4
20	87.5	93.0
30	88.0	93.6
40	88.5	93.6
50	89.5	94.1

the standards), and the use of daylighting controls. Most lighting measures are especially cost-effective for office buildings for which payback periods are less than 1 year.

(b) *Office equipment*: Office equipment constitutes the fastest-growing part of the electrical loads especially in commercial buildings. Office equipment includes computers, fax machines, printers, and copiers. Today, there are several manufacturers that provide energy-efficient office equipment (such those that comply with the U.S. EPA Energy Star specifications). For instance, energy-efficient computers automatically switch to a low-power *sleep* mode or off mode when not in use.

(c) *Motors*: The energy cost to operate electric motors can be a significant part of the operating budget of any commercial and industrial building. Measures to reduce the energy cost of using motors include reducing operating time (turning off unnecessary equipment), optimizing motor systems, using controls to match motor output with demand, using variable speed drives for air and water distribution, and installing energy-efficient motors. Table 31.3 provides typical efficiencies for several motor sizes.

In addition to the reduction in the total facility electrical energy use, retrofits of the electrical systems decrease space cooling loads and therefore further reduce the electrical energy use in the building. These cooling energy reductions as well as possible increases in thermal energy use (for space heating) should be accounted for when evaluating the cost-effectiveness of improvements in lighting and office equipment.

31.4.3 Daylighting Controls

Several studies indicated that daylighting can offer a cost-effective alternative to electrical lighting for commercial and institutional buildings. Through sensors and controllers, daylighting can reduce and even eliminate the use of electrical lighting required to provide sufficient illuminance levels inside office spaces. Recently, a simplified calculation method has been developed by Krarti et al. (2005) to estimate the reduction in the total lighting energy use due to daylighting with dimming controls for office buildings. The method has been shown to apply for office buildings in the United States as well as in Egypt (El Mohimem et al., 2005). The simplified calculation method is easy to use and can be used as a predesign tool to assess the potential of daylighting in saving electricity use associated with artificial lighting for office buildings.

To determine the percent savings, f_d, in annual use of artificial lighting due to implementing daylighting using daylighting controls in office buildings, Krarti et al. (2005) found that the following equation can be used:

$$f_d = b[1 - \exp(-a\tau_w A_w / A_p)]\frac{A_p}{A_f} \tag{31.1}$$

where

A_w/A_p is the window to perimeter floor area. This parameter provides a good indicator of the window size relative to the daylit floor area

A_p/A_f is the perimeter to total floor area. This parameter indicates the extent of the daylit area relative to the total building floor area. Thus, when $A_p/A_f = 1$, the whole building can benefit from daylighting

a and b are the coefficients that depend only on the building location and are given by Table 31.4 for various sites throughout the world

τ_w is the visible transmittance of the glazing

TABLE 31.4

Coefficients a and b of Equation 31.1 for Various Locations throughout the World

Location	a	b	Location	a	b
Atlanta	19.63	74.34	Casper	19.24	72.66
Chicago	18.39	71.66	Portland	17.79	70.93
Denver	19.36	72.86	Montreal	18.79	69.83
Phoenix	22.31	74.75	Quebec	19.07	70.61
New York City	18.73	66.96	Vancouver	16.93	68.69
Washington, DC	18.69	70.75	Regina	20.00	70.54
Boston	18.69	67.14	Toronto	19.30	70.48
Miami	25.13	74.82	Winnipeg	19.56	70.85
San Francisco	20.58	73.95	Shanghai	19.40	67.29
Seattle	16.60	69.23	Kuala Lumpur	20.15	72.37
Los Angeles	21.96	74.15	Singapore	23.27	73.68
Madison	18.79	70.03	Cairo	26.98	74.23
Houston	21.64	74.68	Alexandria	36.88	74.74
Fort Worth	19.70	72.91	Tunis	25.17	74.08
Bangor	17.86	70.73	Sao Paulo	29.36	71.19
Dodge City	18.77	72.62	Mexico 91	28.62	73.63
Nashville	20.02	70.35	Melbourne	19.96	67.72
Oklahoma City	20.20	74.43	Roma	16.03	72.44
Columbus	18.60	72.28	Frankfurt	15.22	69.69
Bismarck	17.91	71.50	Kuwait	21.98	65.31
Minneapolis	18.16	71.98	Riyadh	21.17	72.69
Omaha	18.94	72.30			

31.4.4 HVAC Systems

The energy use due to HVAC systems can represent 40% of the total energy consumed by a typical commercial building. The energy auditor should obtain the characteristics of major HVAC equipment to determine the condition of the equipment, their operating schedule, their quality of maintenance, and their control procedures. A large number of measures can be considered to improve the energy performance of both primary and secondary HVAC systems. Some of these measures are listed:

(a) *Setting up/back thermostat temperatures*: When appropriate, setting back heating temperatures can be recommended during unoccupied periods. Similarly, setting up cooling temperatures can be considered.
(b) *Retrofit of constant air volume systems*: For commercial buildings, variable air volume (VAV) systems should be considered when the existing HVAC systems rely on constant volume fans to condition part or the entire building.
(c) *Installation of heat recovery systems*: Heat can be recovered from some HVAC equipment. For instance, heat exchangers can be installed to recover heat from air handling unit (AHU) exhaust air streams and from boiler stacks.
(d) *Retrofit of central heating plants*: The efficiency of a boiler can be drastically improved by adjusting the fuel–air ratio for proper combustion. In addition, installation of new energy-efficient boilers can be economically justified when old boilers are to be replaced.
(e) *Retrofit of central cooling plants*: Currently, there are several chillers that are energy efficient and easy to control and operate and are suitable for retrofit projects.

It should be noted that there is a strong interaction between various components of heating and cooling system. Therefore, a whole-system analysis approach should be followed when retrofitting a building HVAC system. Optimizing the energy use of a central cooling plant (which may include chillers, pumps, and cooling towers) is one example of using a whole-system approach to reduce the energy use for heating and cooling buildings.

31.4.5 Compressed Air Systems

Compressed air has become an indispensable tool for most manufacturing facilities. Its uses range from air-powered hand tools and actuators to sophisticated pneumatic robotics. Unfortunately, staggering amounts of compressed air are currently wasted in a large number of facilities. It is estimated that only a fraction of 20%–25% of input electrical energy is delivered as useful compressed air energy. Leaks are reported to account for 10%–50% of the waste, while misapplication accounts for 5%–40% of loss in compressed air (Howe and Scales, 1998).

To improve the efficiency of compressed air systems, the auditor can consider several issues including whether or not compressed air is the right tool for the job (for instance, electric motors are more energy efficient than air-driven rotary devices), how compressed air is applied (for instance, lower pressures can be used to supply pneumatic tools), how it is delivered and controlled (for instance, the compressed air needs to be turned off when the process is not running), and how compressed air system is managed (for each machine or process, the cost of compressed air needs to be known to identify energy and cost savings opportunities).

31.4.6 Energy Management Controls

With the constant decrease in the cost of computer technology, automated control of a wide range of energy systems within commercial and industrial buildings is becoming increasingly popular and cost-effective. An energy management and control system (EMCS) can be designed to control and reduce the building energy consumption within a facility by continuously monitoring the energy use of various equipment and making appropriate adjustments. For instance, an EMCS can automatically monitor and adjust indoor ambient temperatures, set fan speeds, open and close AHU dampers, and control lighting systems.

If an EMCS is already installed in the building, it is important to recommend a system tune-up to insure that the controls are properly operating. For instance, the sensors should be calibrated regularly in accordance with manufacturers' specifications. Poorly calibrated sensors may cause increase in heating and cooling loads and may reduce occupant comfort.

31.4.7 Indoor Water Management

Water and energy savings can be achieved in buildings by using water-savings fixtures instead of the conventional fixtures for toilets, faucets, showerheads, dishwashers, and clothes washers. Savings can also be achieved by eliminating leaks in pipes and fixtures.

Table 31.5 provides typical water use of conventional and water-efficient fixtures for various end uses. In addition, Table 31.5 indicates the hot water use by each fixture as a fraction of the total water. With water-efficient fixtures, savings of 50% of water use can be achieved for toilets, showers, and faucets.

31.4.8 Advanced Technologies

The energy auditor may consider the potential of implementing and integrating new technologies within the facility. It is therefore important that the energy auditor understands these new technologies and knows how to apply them. The following are among the new technologies that can be considered for commercial and industrial buildings:

(a) *Building envelope technologies*: Recently, several materials and systems have been proposed to improve the energy efficiency of building envelope and especially windows including

- Spectrally selective glasses that can optimize solar gains and shading effects

TABLE 31.5

Usage Characteristics of Water-Using Fixtures

End Use	Conventional Fixtures	Water-Efficient Fixtures	Usage Pattern	Hot Water (%)
Toilets	3.5 gal/flush	1.6 gal/flush	4 flushes/day	0
Showers	5.0 gal/min	2.5 gal/min	5 min/shower	60
Faucets	4.0 gal/min	2.0 gal/min	2.5 min/day	50
Dishwashers	14.0 gal/load	8.5 gal/load	0.17 loads/day	100
Clothes washers	55.0 gal/load	42.0 gal/load	0.3 loads/day	25
Leaks	10% of total use	2% of total use	N/A	50

- Chromogenic glazings that change its properties automatically depending on temperature and/or light level conditions (similar to sunglasses that become dark in sunlight)
- Building integrated photovoltaic (PV) panels that can generate electricity while absorbing solar radiation and reducing heat gain through building envelope (typically roofs)

(b) *Light pipe technologies*: While the use of daylighting is straightforward for perimeter zones that are near windows, it is not usually feasible for interior spaces, particularly those without any skylights. Recent but still emerging technologies allow to *pipe* light from roof or wall-mounted collectors to interior spaces that are not close to windows or skylights.

(c) *HVAC systems and controls*: Several strategies can be considered for energy retrofits:

- Heat recovery technologies such as rotary heat wheels and heat pipes can recover 50%–80% of the energy used to heat or cool ventilation air supplied to the building.
- Desiccant-based cooling systems are now available and can be used in buildings with large dehumidification loads during long periods (such as hospitals, swimming pools, and supermarket fresh produce areas).
- Geothermal heat pumps can provide an opportunity to take advantage of the heat stored underground to condition building spaces.
- Thermal energy storage (TES) systems offer a mean of using less expensive off-peak power to produce cooling or heating to condition the building during on-peak periods. Several optimal control strategies have been developed in recent years to maximize the cost savings of using TES systems.

(d) *Cogeneration*: This is not really a new technology. However, recent improvements in its combined thermal and electrical efficiency made cogeneration cost-effective in several applications including institutional buildings such as hospitals and universities.

31.5 Net-Zero Energy Retrofits

While the concept of the zero net energy (ZNE) buildings has been mostly applied to new construction, it can be considered for retrofit projects including residential buildings. Typically, ZNE is defined in terms of either site energy or resource energy. Site energy consists of energy produced and consumed at the building site. Source or primary energy includes site energy as well as the energy used to generate, transmit, and distribute this site energy. Therefore, source energy provides a better indicator of the energy use of buildings and their impact on the environment and the society and thus is better suited for ZNE building analysis. An analysis based on source energy effectively allows different fuel types, such as electricity and natural gas commonly used in buildings, to be encompassed together.

It is generally understood that ZNE buildings produce as much energy as they consume on-site annually. These buildings typically include aggressive energy efficiency measures and active solar water heating systems. Moreover, ZNE buildings employ grid-tied, net-metered renewable energy generation technologies, generally PV systems, to produce electricity. Effectively, ZNE buildings use the grid as *battery storage* to reduce required generation system capacity.

Unlike the case for new construction, analysis methods and case studies of net zero energy (NZE) retrofits of existing buildings are limited. A more detailed description of NZE retrofitting approaches, analysis techniques, and some case studies is provided by Krarti (2012). In this section, a case study for NZE retrofitting of existing homes in Mexico is provided. Specifically and in a study by Griego et al. (2012), NZE retrofits of Mexican residential building have been carried out. In the study, various combinations of energy efficiency are applied to arrive at an optimum set of recommendations for existing residential and new construction residential buildings. The optimum point is the minimum annualized energy-related costs and the corresponding annual source energy savings. Two separate optimizations are performed for an existing-unconditioned home and an existing-conditioned home.

The baseline annualized energy-related costs for the conditioned and unconditioned homes are US$797 and US$557, respectively. These costs are obtained before any of the energy efficiency and thermal comfort measures are applied. Therefore, it is determined that the cost of improved thermal comfort for a typical home in Salamanca is roughly US$240/year. The optimum point for the unconditioned case occurs at 17.3% annual energy savings and a corresponding minimum cost of US$433 as shown in Figure 31.1. The conditioned case on the other hand has a greater opportunity for energy savings and achieves a minimum cost of US$542 at 35.0% energy savings as indicated in Figure 31.2. The optimum point for the conditioned case includes implementing methods to reduce miscellaneous plug loads, *R*-1.4 m²- K/W roof assembly, low-flow showerheads and sinks, an electric stove, 100% compact fluorescent lamps, and *R*-0.35 m²- K/W trunk–branch domestic hot water (DHW) pipe distribution. The optimum point for the unconditioned case includes all of the same measures as the conditioned model with the exception of the added roof insulation.

The comparison between the unconditioned and the conditioned optimization results outlined in Figure 31.3 reveals that the optimum point for the conditioned case (US$542) is roughly the cost neutral point for the unconditioned case (US$557). It is also useful to compare energy end uses to gauge measures with highest potential of the energy savings. Figures 31.4 and 31.5 include a summary of the total annual source energy consumption by

FIGURE 31.1
Retrofit optimization path for an unconditioned home in Salamanca, Mexico.

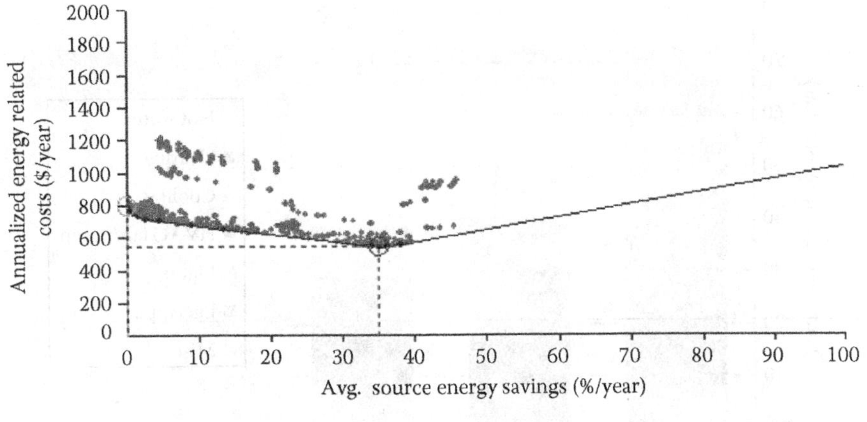

FIGURE 31.2
Retrofit optimization path for an air-conditioned home in Salamanca, Mexico.

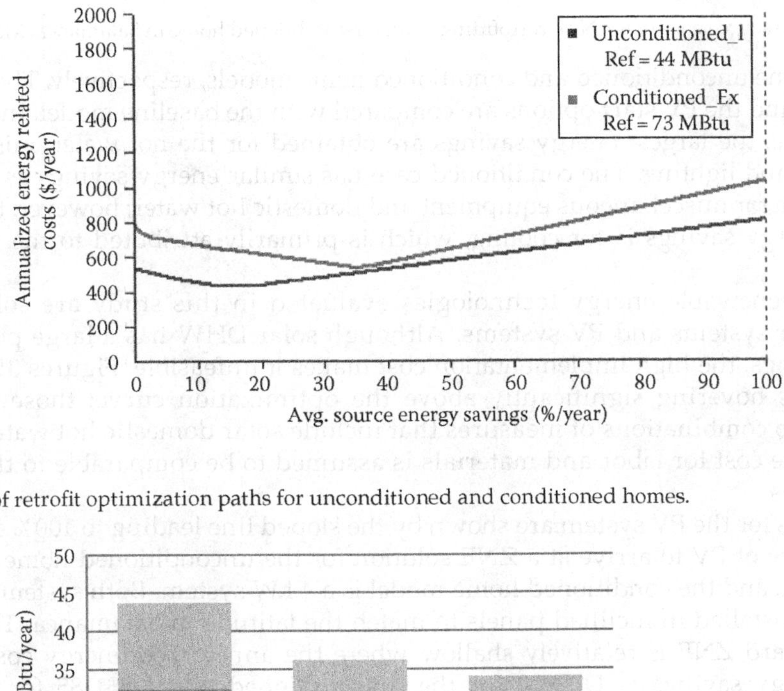

FIGURE 31.3
Comparison of retrofit optimization paths for unconditioned and conditioned homes.

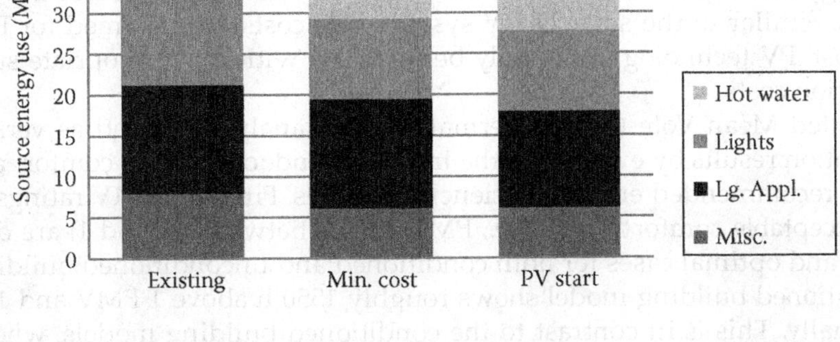

FIGURE 31.4
Annual end-use source energy for NZE retrofitting of an unconditioned home in Salamanca, Mexico.

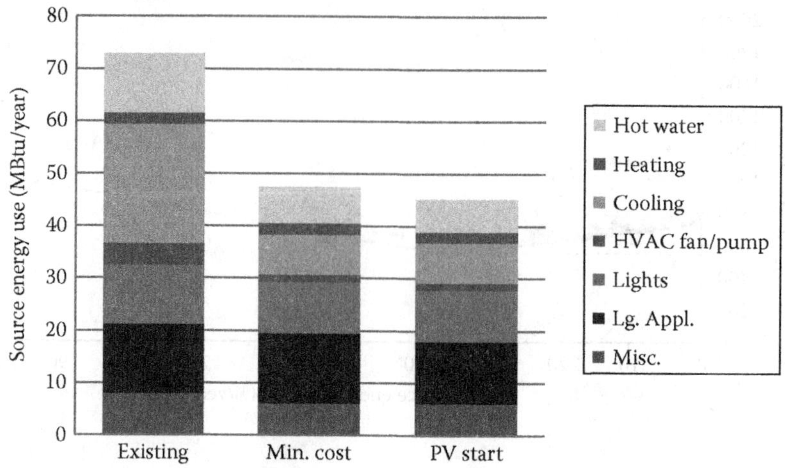

FIGURE 31.5
Annual end-use source energy for NZE retrofitting of an air-conditioned home in Salamanca, Mexico.

end use for the unconditioned and conditioned home models, respectively. The minimum cost option and the PV start options are compared with the baseline model. In the unconditioned case, the largest energy savings are obtained for the hot water, miscellaneous equipment, and lighting. The conditioned case has similar energy savings as the unconditioned case for miscellaneous equipment and domestic hot water; however, the greatest area for energy savings is for cooling, which is primarily attributed to the use of roof insulation.

The two renewable energy technologies evaluated in this study are solar domestic hot water systems and PV systems. Although solar DHW has a large potential for energy savings, the high implementation cost makes it unfeasible. Figures 31.1 and 31.2 show points hovering significantly above the optimization curve; those points are associated to combinations of measures that include solar domestic hot water systems. Note that the cost for labor and materials is assumed to be comparable to those in the United States.

The results for the PV system are shown by the sloped line leading to 100% energy savings. The size of PV to arrive at a ZNE solution for the unconditioned home model is a 3 kW system, and the conditioned home model is a 4 kW system. Both systems are south facing and installed in inclined panels to match the latitude in Salamanca. The slope of the line toward ZNE is relatively shallow where the annualized energy cost for 100% annual energy savings is US$920 for the unconditioned and US$1185 for the conditioned cases. Similar to the solar DHW system, U.S. costs are assumed for PV material and labor cost. PV technology may only be desirable with the appropriate subsidies for implementation costs.

The Predicted Mean Vote (PMV) thermal comfort analysis is used as verification for the optimization results by evaluating the improved indoor thermal comfort after implementing the recommended energy efficiency measures. First, the PMV ratings above and below the acceptable comfort range (i.e., PMV values between –1 and 1) are determined, the baseline and optimal cases for both conditioned and unconditioned building models. The unconditioned building model shows roughly 1550 h above 1 PMV and 150 h below 1 PMV annually. This is in contrast to the conditioned building models, where thermal comfort is maintained throughout the year, as expected.

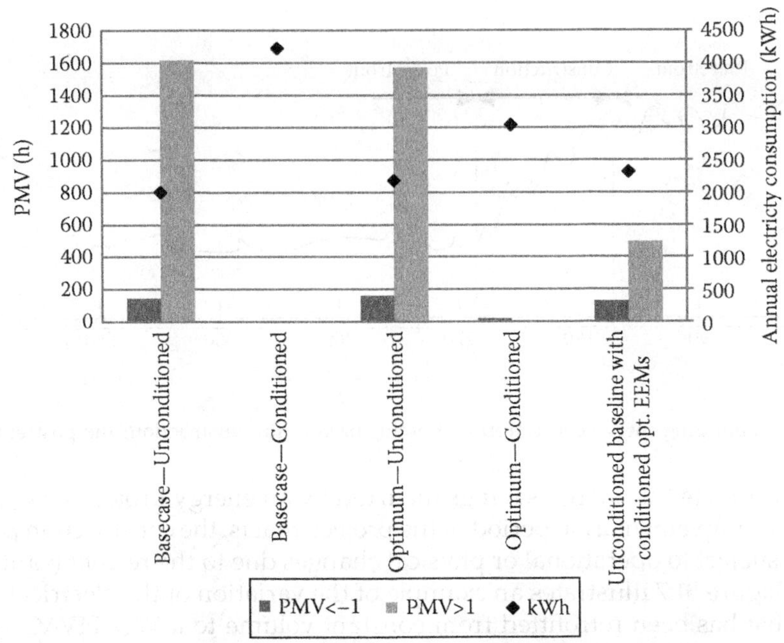

FIGURE 31.6
Comparison of various existing home configurations with and without NZE retrofits including thermal comfort analysis.

The thermal comfort analysis is also applied at the optimum building models. The annual energy consumption in the conditioned new construction and existing building models decreases relative to the baseline when thermal insulation is added to the building and, as predicted, the annual PMV ratings remain relatively constant. However, when the optimum set of ECMs from the conditioned case are applied to the unconditioned baseline home model, the number of hours outside of the thermal comfort zone decreases significantly as indicated in Figure 31.6.

In the conditioned baseline retrofit case, the cost of installing an electric heat pump is estimated at US$4394 for a 3.5 tons unit. However, when roof insulation is added to the unconditioned retrofit building, the number of hours outside of the thermal comfort zone decreases by almost 60% for a much lower initial cost of roughly US$426.

31.6 Verification Methods of Energy Savings

Energy conservation retrofits are deemed cost-effective based on predictions of energy and cost savings. However, several studies have found that large discrepancies exist between actual and predicted energy savings. Due to the significant increase in the activities of ESCOs, the need became evident for standardized methods for measurement and verification of energy savings. This interest has led to the development of the *North American Energy Measurement and Verification Protocol* published in 1996 and later expanded and revised under the *International Performance Measurement and Verification Protocol* (IPMVP, 1997, 2002, 2007).

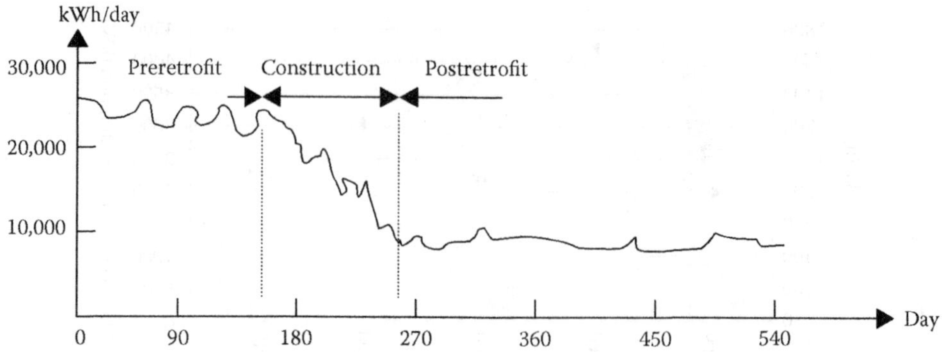

FIGURE 31.7
Daily variation of a building energy consumption showing preretrofit, construction, and postretrofit periods.

In order to estimate the energy savings incurred by an energy project, it is important to first identify the implementation period of the project, that is, the construction phase where the facility is subject to operational or physical changes due to the retrofit (Guiterman and Krarti, 2011). Figure 31.7 illustrates an example of the variation of the electrical energy use in a facility that has been retrofitted from constant volume to a VAV HVAC system. The time-series plot of the facility energy use clearly indicated the duration of the construction period, the end of the preretrofit period, and the start of the postretrofit period. The duration of the construction period depends on the nature of the retrofit project and can range from few hours to several months.

In principle, the measurement of the retrofit energy savings can be obtained by simply comparing the energy use during pre- and postretrofit periods. Unfortunately, the change in energy use between the pre- and postretrofit periods is not only due to the retrofit itself but also to other factors such as changes in weather conditions, levels of occupancy, and HVAC operating procedures. It is important to account for all these changes to determine accurately the retrofit energy savings.

Several methods have been proposed to measure and verify energy savings of implemented ECMs in commercial and industrial buildings. Some of these techniques are briefly described:

Regression models: The early regression models used to measure savings adapted the variable-base degree-day method. Among these early regression models, the Princeton Scorekeeping Method is the method that uses measured monthly energy consumption data and daily average temperatures to calibrate a linear regression model and determine the best values for non-weather-dependent consumption, the temperature at which the energy consumption began to increase due to heating or cooling (the change-point or base temperature), and the rate at which the energy consumption increased. Several studies have indicated that the simple linear regression model is suitable for estimating energy savings for residential buildings. However, subsequent work has shown that the PRISM model does not provide accurate estimates for energy savings for most commercial buildings (Ruch and Claridge, 1992). Single-variable (temperature) regression models require the use of at least four-parameter segmented-linear or change-point regressions to be suitable for commercial buildings. Figure 31.8 illustrates the basic functional forms commonly used for ambient-temperature linear

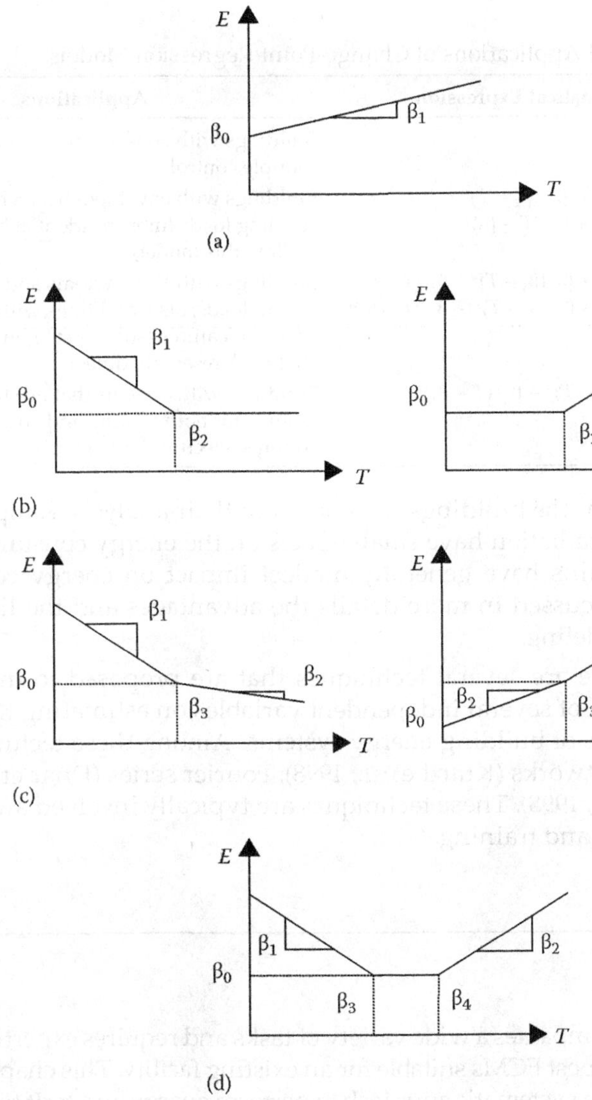

FIGURE 31.8
Basic forms of single-variable regression models using (a) 2 parameters (2P), (b) 3 parameters (3P), (c) 4 parameters (4P), and (d) 5 parameters (5P).

regression models. There are regression models that are also called change-point or segmented-linear models. Table 31.6 summarizes the mathematical expressions of four change-point models and their applications. In general, the change-point regression models are more suitable for predicting heating rather than cooling energy use. Indeed, these regression models assume steady-state conditions and are insensitive to the building dynamic effects, solar effects, and nonlinear HVAC system controls such as on–off schedules.

Katipamula et al. (1994) proposed multiple linear regression models to include as independent variables internal gain, solar radiation, wind, and humidity ratio, in addition to

TABLE 31.6

Mathematical Expressions and Applications of Change-Point Regression Models

Model Type	Mathematical Expression	Applications
Two-parameter (2-P) model	$E = \beta_0 + \beta_1 \cdot T$	Buildings with constant air volume systems and simple controls
Three-parameter (3-P) model	Heating $E = \beta_0 + \beta_1 \cdot (\beta_2 - T)^+$ Cooling $E = \beta_0 + \beta_1 \cdot (T - \beta_2)^+$	Buildings with envelope-driven heating or cooling loads (most residential buildings follow this model)
Four-parameter (4-P) model	Heating $E = \beta_0 + \beta_1 \cdot (\beta_3 - T)^+ - \beta_2 \cdot (T - \beta_3)^+$ Cooling $E = \beta_0 - \beta_1 \cdot (\beta_3 - T)^+ + \beta_2 \cdot (T - \beta_3)^+$	Buildings with VAV systems and/or with high latent loads; also, buildings with nonlinear control features (such as economizer cycles and hot deck reset schedules)
Five-parameter (5-P) model	$E = \beta_0 + \beta_1 \cdot (\beta_3 - T)^+ + \beta_2 \cdot (T - \beta_4)^+$	Buildings with systems that use the same energy source for both heating and cooling (i.e., heat pumps, electric heating and cooling systems)

the outdoor temperature. For the buildings considered in their analysis, Katipamula et al. found that wind and solar radiation have small effects on the energy consumption. They also found that internal gains have generally modest impact on energy consumption. Katipamula et al. (1998) discussed in more details the advantages and the limitations of multivariate regression modeling.

Time-variant models: There are several techniques that are proposed to include the effect of time variation of several independent variables on estimating the energy savings due to retrofits of building energy systems. Among these techniques are the artificial neural networks (Krarti et al., 1998), Fourier series (Dhar et al., 1998), and NILM (Shaw et al., 1998). These techniques are typically involved and require high level of expertise and training.

31.7 Summary

Retrofitting of buildings encompasses a wide variety of tasks and requires expertise in a number of areas to determine the best ECMs suitable for an existing facility. This chapter provided a description of a general but a systematic approach to perform energy audits. If followed carefully, the approach helps facilitate the process of analyzing a seemingly endless array of alternatives and complex interrelationships between building and energy system components. In particular, the chapter discussed net-zero energy retrofitting of buildings using detailed simulation analysis and optimization techniques to ensure that over a period of 1 year, the retrofitted building would effectively consume as much source energy as it produces.

References

Dhar, A., Reddy, T.A., and Claridge, D.E. 1998. Modeling hourly energy use in commercial buildings with fourier series functional forms. *ASME Solar Energy Engineering Journal*, 120(3), 217.

EIA. 2008. *2005 Residential Energy Consumption Survey*. Washington, DC: U.S. Energy Information Administration.

El Mohimem, M.A., Hanna, G., and Krarti, M. 2005. Analysis of daylighting benefits for office building in Egypt. *ASME Journal of Solar Energy Engineering*, 127(3), 366–370.

Fels, M. 1986. Special issue devoted to measuring energy savings: The scorekeeping approach. *Energy and Buildings*, 9(2), 127–136.

Griego, D., Krarti, M., and Hernandez-Guerrero, A. 2012. Optimization of energy efficiency and thermal comfort for residential buildings in Salamanca, Mexico. *Energy and Buildings*, 50, 550–549.

Guiterman, T. and Krarti, M. 2011. Analysis of measurement and verification methods for energy retrofits applied to residential buildings. *ASHRAE Transactions*, 117(2), 382–394.

Howe, B. and Scales, B. 1998. Beyond leaks: Demand-side strategies for improving compressed air efficiency. *Energy Engineering*, 95, 31.

IPMVP. 1997. *International Performance Monitoring and Verification Protocol*. U.S. Department of Energy DOE/EE-0157. Washington, DC: U.S. Government Printing Office.

IPMVP. 2002. *International Performance Monitoring and Verification Protocol, Concepts and Options for Determining Energy and Water Savings*, Volume 1. U.S. Department of Energy DOE/GO-102002-1554. Washington, DC: U.S. Government Printing Office.

IPMVP. 2007. *International Performance Monitoring and Verification Protocol, Concepts and Options for Determining Energy and Water Savings*. Washington, DC: U.S. Government Printing Office. http://www.evo-world.org (accessed February 5, 2014).

Katipamula, S., Reddy, T.A., and Claridge, D.E. 1994. Development and application of regression models to predict cooling energy use in large commercial buildings. *Proceedings of the ASME/JSES/JSES International Solar Energy Conference*, San Francisco, CA, p. 307.

Katipamula, S., Reddy, T.A., and Claridge, D.E. 1998. Multivariate regression modeling. *ASME Solar Energy Engineering Journal*, 120(3), 177.

Knebel, D.E. 1983. *Simplified Energy Analysis Using the Modified Bin Method*. Atlanta, GA: American Society of Heating, Refrigeration, and Air-Conditioning Engineers.

Krarti, M. 2010. *Energy Audit of Building Systems: An Engineering Approach*, 2nd Ed. Boca Raton, FL: CRC Press.

Krarti, M. 2012. *Weatherization and Energy Efficiency Improvement for Existing Homes: An Engineering Approach*, 1st Ed. Boca Raton, FL: CRC Press.

Krarti, M., Erickson, P., and Hillman, T. 2005. A simplified method to estimate energy savings of artificial lighting use from daylighting. *Building an Environment*, 40, 747–754.

Krarti, M., Kreider, J.F., Cohen, D., and Curtiss, P. 1998. Estimation of energy savings for building retrofits using neural networks. *ASME Journal of Solar Energy Engineering*, 120(3), 211.

LBL. 1980. *DOE-2 User Guide*, Version 2.1, LBL report No. LBL-8689 Rev. 2. Berkeley, CA: Lawrence Berkeley Laboratory.

Ruch, D. and Claridge, D.E. 1992. A four-parameter change-point model for predicting energy consumption in commercial buildings. *ASME Journal of Solar Energy Engineering*, 104, 177.

Shaw, S.R., Abler, C.B., Lepard, R.F., Luo, D., Leeb, S.B., and Norford L.K. 1998. Instrumentation for high performance non-intrusive electrical load monitoring. *ASME Journal of Solar Energy Engineering*, 120(3), 224.

32

Cogeneration

W. Dan Turner

CONTENTS

32.1 Introduction

32.1.1 Background

The term "cogeneration" as used in this chapter is defined as the combined production of electrical power* and useful thermal energy by the sequential use of a fuel or fuels. Each term in this definition is important. Combined means that the production processes of the electric power and thermal energy are linked, and often are accomplished in a series or parallel fashion. Electrical power is the electricity produced by an electrical generator, which is most often powered by a prime mover such as a steam turbine, gas turbine, or reciprocating engine. Thermal energy is that product of the process which provides heating or cooling. Forms of this thermal energy include hot exhaust gases, hot water, steam, and chilled water. Useful means that the energy is directed at fulfilling an existing need for heating or cooling. Simply exhausting hot exhaust gases does not meet the definition of useful thermal energy. In other words, cogeneration is the production of electrical power and the capture of coexisting thermal energy for useful purposes.

The importance of cogeneration is monetary and energy savings. Any facility that uses electrical power and needs thermal energy is a candidate for cogeneration. Although many considerations are involved in determining if cogeneration is feasible for a particular

* Although the power output of a cogeneration system could be mechanical power as well as electrical power, the majority of systems produce electrical power. This chapter will consider only cogeneration systems that produce electrical power.

facility, the basic consideration is if the saving on thermal energy costs is sufficient to justify the capital expenditures for a cogeneration system. Facilities that may be considered for cogeneration include those in the industrial, commercial, and institutional sectors.

The technology for cogeneration is for the most part available and exists over a range of sizes: from less than 100 kW to over 100 MW. The major equipment requirements include a prime mover, electrical generator, electrical controls, heat recovery systems, and other typical power plant equipment. These components are well developed, and the procedures to integrate these components into cogeneration systems are well established.

In addition to the economic and technical considerations, the application of cogeneration systems involves an understanding of the governmental regulations and legislation on electrical power production and on environmental impacts. With respect to electrical power production, certain governmental regulations were passed during the late 1970s that remove barriers and provide incentives to encourage cogeneration development. Finally, no cogeneration assessment would be complete without an understanding of the financial arrangements that are possible.

32.1.2 Overall Scope of the Chapter

The objective of this chapter is to provide a complete overview of the important aspects necessary to understand cogeneration. Specifically, this chapter includes information on the technical equipment and components, technical design issues, regulatory considerations, economic evaluations, financial aspects, computer models and simulations, and future technologies. As briefly described earlier, a thorough discussion of cogeneration includes many aspects of engineering, economics, law, finance, and other topics. The emphasis of this chapter is on the technical and engineering considerations. The descriptions of the economic, legal, and governmental aspects are provided for completeness, but should not be considered a substitute for consulting with appropriate specialists such as attorneys, accountants, and bankers.

This chapter is divided into sections on basic cogeneration systems and terminology, technical components, design issues, regulatory considerations, economic evaluations, and financial aspects. Also, two case studies are presented that illustrate the application of cogeneration to an industrial and an institutional facility. Finally, the chapter concludes with some comments on the future technologies of cogeneration. This introductory section ends with a brief review of the history of cogeneration.

32.1.3 History of Cogeneration

At the beginning of the twentieth century, electrical power generation was in its infancy. Most industrial facilities generated all their own electrical power and often supplied power to nearby communities. They used the thermal energy that was available during the electrical power production to provide or supplement process or building heat. These industrial facilities, therefore, were the first "cogenerators." The dominant prime mover at this time was the reciprocating steam engine, and the low-pressure exhaust steam was used for heating applications.

Between the early 1920s and through the 1970s, the public electric utility industry grew rapidly because of increasing electrical power demands. Coincident with this rapid growth was a general reduction in the costs to produce electrical power, mainly due to the economies of scale, more efficient technologies, and decreasing fuel costs. During this period, industry often abandoned their own electrical power generation because of

(1) the decreasing electrical rates charged by public utilities, (2) income tax regulations that favored expenses instead of capital investments, (3) increasing costs of labor, and (4) the desire of industry to focus on their product rather than the side issue of electrical power generation. Estimates are available that suggest that industrial cogenerated electrical power decreased from about 25% to 9% of the total electrical power generated in the country between the years of 1954 and 1976. Since about the mid-1980s, this percentage has been fairly constant at about 5%. For example, at the end of 1992, 5.1% of the total U.S. electrical capacity was due to cogeneration systems.

During the 1960s and 1970s, the natural gas industry promoted a "total energy" concept of cogeneration. This effort was not very successful due to the relatively poor economics (e.g., relatively inexpensive electricity and expensive fuels), and the lack of governmental regulations to ease the interface with public utilities.

In late 1973 and again in 1979, America experienced major "energy crises" that were largely a result of reduced petroleum imports. Between 1973 and 1983, the prices of fuels and electrical power increased by a factor of approximately 5. Any facility purchasing electrical power began to consider (or reconsider) the economic savings associated with cogeneration. These considerations were facilitated by Federal regulations that were enacted to ease or remove barriers to cogeneration.

In 1978, the government passed the National Energy Act (NEA) that included several important pieces of legislation. The NEA included the Fuel Use Act, the Natural Gas Policy Act, and the Public Utility Regulatory Policies Act (PURPA). Each of these acts had a direct impact on cogeneration, but PURPA was the most significant. In particular, PURPA defined cogeneration systems to include those power plants that supplied a specified fraction of their input energy as useful thermal output in addition to a mechanical or electrical output. This legislation is discussed in more detail in later sections of this chapter.

In addition, other regulatory legislation passed beginning in the late 1950s and continuing through the early 1990s impacts the installation of cogeneration systems. In particular, Federal legislation directed at managing air and water quality significantly affects the installations of cogeneration systems. For managing air pollution, the original legislation is the Air Quality Act of 1967 that was amended with the Clean Air Act Amendments in 1970, 1977, and 1990. The main legislative basis for managing water pollution is the Federal Water Pollution Control Act of 1956, as amended by the Water Quality Act of 1965, the Federal Water Pollution Control Act amendments of 1972, and Clean Water Act of 1977. These acts and other legislation, including the Energy Policy Act of 2005, and their impacts on the development of cogeneration projects, are discussed in a subsequent section of this chapter.

The term *cogeneration* lost favor in the U.S. in the 1980s and 1990s, particularly with the electric utilities. Instead, a new term, "combined heat and power," or CHP, became popular. A CHP Association was formed, and the U.S. Department of Energy began funding demonstration projects. A typical system might include a small combustion turbine producing electrical power used in conjunction with an absorption chiller that is direct-fired with the gas turbine exhaust gases. These smaller CHP systems are also being touted for use in buildings, called *building CHP* or *BCHP*. There has been an increased interest in distributed generation since parts of the U.S. and Canada suffered a massive electrical blackout in August 2003 that left eight states in the northeast and parts of Canada without power for several days.

The number of smaller CHP systems increased during the early part of the twenty-first century; however, relatively cheap electricity prices and moderate natural gas prices still hindered the widespread applications of CHP. Two natural disasters in the summer of 2005,

Hurricanes Katrina and Rita, caused widespread damage to offshore natural gas wells in the Gulf of Mexico. The price of natural gas virtually doubled over a 2-month period. This could have a profound impact on new cogeneration systems because, despite the increased system efficiency, the economics of most CHP systems depend on moderate gas prices (relative to electricity). The increased demand for natural gas, both in the U.S. and worldwide, and the price volatility make the near-term future of CHP highly uncertain. However, new technologies, transmission-grid problems forcing a greater reliance on distributed generation, and favorable electricity prices will make CHP systems more prominent in the twenty-first century. The overall CHP system efficiency will eventually drive the market to more widespread acceptance.

32.2 Basic Cogeneration Systems

32.2.1 Advantages of Cogeneration

To illustrate the fuel and monetary savings of a cogeneration system, the following simple comparison is presented. Figure 32.1 shows the details of this comparison. The power and thermal needs to be satisfied are 30 energy units* of electrical power and 34 energy units of heat. These power and thermal needs may be satisfied by either a cogeneration system or by conventional systems. Conventional systems for this example would be an electric utility and a standard boiler.

At the left of Figure 32.1 is the cogeneration system, and at the right of Figure 32.1 is the conventional system. All the components of the cogeneration system are represented by the box. Assuming an electrical generation efficiency of 30% for the cogeneration system, then 30 energy units are produced with 100 energy units of fuel energy. This cogeneration system then supplies 34 energy units of heat. These are conservative numbers for the electrical and

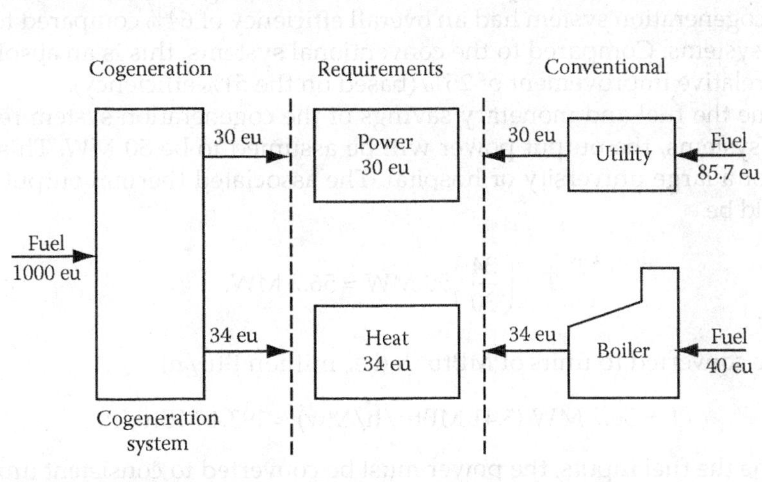

FIGURE 32.1
A schematic illustration of the advantages of cogeneration relative to conventional systems.

* For this example, an energy unit is defined as any consistent unit of energy such as kW, MW, Btu, MBtu, or Btu/h. The use of arbitrary units avoids confusion and the need for unit conversions.

thermal outputs of the cogeneration system, and many systems may have higher efficiencies. The overall thermodynamic efficiency* of the cogeneration system is defined as

$$\eta_0 = \frac{(P+T)}{F} \tag{32.1}$$

where
P represents the power
T represents the thermal or heat energy rate
F represents the fuel input rate (all in consistent units)

This overall efficiency also may be found as the sum of the individual efficiencies of the electrical and thermal production. For this example, the cogeneration system has an overall efficiency of

$$\eta_0 = \frac{(30+34)}{100} = 64\% \tag{32.2}$$

The conventional systems, diagrammed at the right of Figure 32.1, satisfy the power and thermal needs by the use of an electric utility and a boiler. In this case, the utility is assumed to be able to deliver the 30 energy units of electrical power with a plant efficiency of 35% (which is on the high side for most utilities). This results in a fuel input of 85.7 energy units. The boiler supplies the 34 energy units of heat with an 85% efficiency, which requires a fuel input of 40 energy units. The overall efficiency is, therefore,

$$\eta_0 = \frac{(P+T)}{F} = \frac{(30+34)}{(85.7+40)} = 51\% \tag{32.3}$$

This simple example demonstrates the thermodynamic advantage of a cogeneration system relative to conventional systems for accomplishing the same objectives. In this example, the cogeneration system had an overall efficiency of 64% compared to 51% for the conventional systems. Compared to the conventional systems, this is an absolute increase of 13% and a relative improvement of 25% (based on the 51% efficiency).

To determine the fuel and monetary savings of the cogeneration system relative to the conventional systems, the output power will be assumed to be 50 MW. This is a typical power level for a large university or hospital. The associated thermal output rate for this example would be

$$T = \left(\frac{34}{30}\right) 50 \text{ MW} = 56.7 \text{ MW}. \tag{32.4}$$

This may be converted to units of MBtu/h (i.e., million Btu/h)

$$T = 56.7 \text{ MW } (3.41 \text{ MBtu/h/Mw}) = 193.4 \text{ MBtu/h}. \tag{32.5}$$

To determine the fuel inputs, the power must be converted to consistent units

$$P = 50 \text{ MW } (3.41 \text{ MBtu/h/MW}) = 170.7 \text{ MBtu/h}. \tag{32.6}$$

* This overall thermodynamic efficiency should not be confused with the "PURPA" efficiency, which is a legislated definition and is described in a later subsection.

Now, the fuel inputs are

$$(\text{Fuel})_{\text{cogen}} = \frac{(P+T)}{(\eta_0)_{\text{cogen}}} = \frac{(170.7+173.4)}{0.64} = 569 \text{ MBtu/h.} \tag{32.7}$$

$$(\text{Fuel})_{\text{conv}} = \frac{(P+T)}{(\eta_0)_{\text{conv}}} = \frac{(170.7+193.4)}{0.51} = 714 \text{ MBtu/h.} \tag{32.8}$$

The fuel savings is the difference between these two numbers:

$$(\text{Fuel savings}) = (\text{Fuel})_{\text{conv}} - (\text{Fuel})_{\text{cogen}} = 714 - 569 = 145 \text{ MBtu/h.} \tag{32.9}$$

If the plant operates 6,000 h/year (this is about 68% of the time on average—a conservative estimate), then the fuel energy savings per year is 870,000 MBtu. If the fuel price is \$8.00/MBtu, then the monetary saving is \$6.96 × 10^6 per year. Although this is a simple economic analysis, the general trends are relevant. A later section of this chapter will describe more comprehensive economic analyses used in actual assessments of cogeneration applications.

In summary, the use of cogeneration is an effective way to more efficiently use fuels. The energy released from the fuels is used to both produce electrical power as well as provide useful thermal energy. The savings in energy and money can be substantial, and such systems have been shown to be technically and economically feasible in a wide range of applications.

32.2.2 Topping and Bottoming Cycles

A cogeneration system may be classified as either a topping-cycle system or a bottoming-cycle system. Figure 32.2 is a schematic illustration of a topping-cycle system. As shown, a prime mover uses fuel to power an electrical generator to produce electricity. This electricity may be used completely on-site or may be tied into an electrical distribution network for sale to the local utility or other customers. The hot exhaust gases are directed to a heat

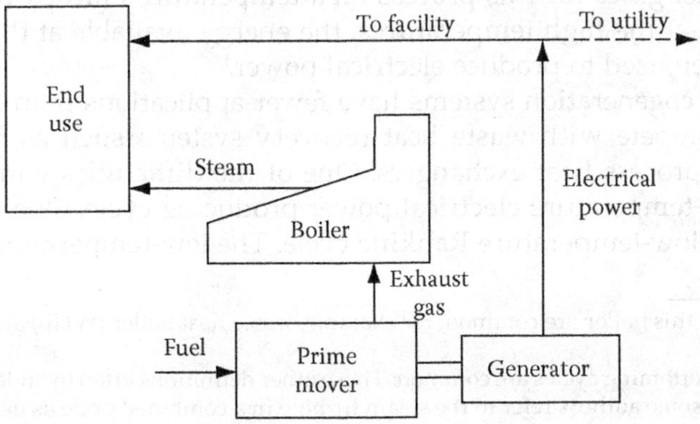

FIGURE 32.2
A schematic illustration of a cogeneration topping-cycle system.

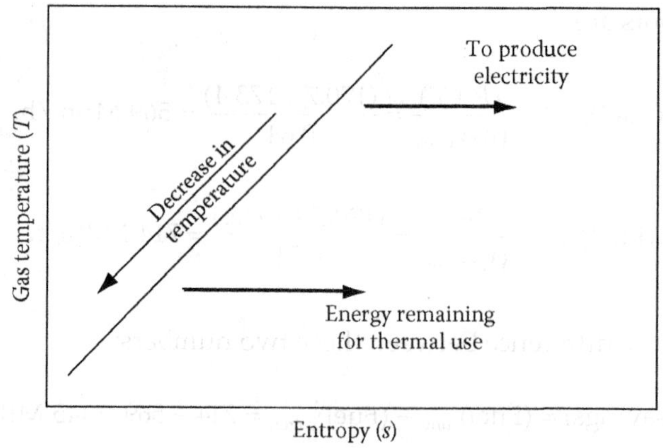

FIGURE 32.3
The gas temperature as a function of entropy for a topping-cycle system.

recovery boiler (HRB)* to produce steam or hot water. This steam or hot water is used on-site for process or building heat.

This cogeneration system is classified as a topping cycle because the electrical power is first generated at the higher (top) temperatures associated with the fuel combustion process and then the rejected or exhausted energy is used to produce useful thermal energy (such as the steam or hot water in this example). Figure 32.3 shows the thermodynamic states of the exhaust gases for this process on a temperature-entropy diagram. For this process, as energy is removed from the combustion gases, the temperature and entropy decrease (since the process involves energy removal). As shown, the top or high-temperature gases are used first to produce the electrical power and then the lower-temperature gases (exhaust) are used to produce useful thermal energy. The majority of cogeneration systems are based on topping cycles.

The other classification of cogeneration systems is bottoming cycle systems. Figure 32.4 is a schematic illustration of a bottoming cycle system. As shown, the high-temperature combustion gases are used first in a high-temperature thermal process (such as high-temperature metal treatment) and then the lower-temperature gases are used in a special low-temperature cycle to produce electrical power. Figure 32.5 shows the thermodynamic states of the exhaust gases for this process on a temperature-entropy diagram. After the energy is removed at the high temperatures, the energy available at the bottom or lower temperatures is then used to produce electrical power.†

Bottoming-cycle cogeneration systems have fewer applications than topping-cycle systems and must compete with waste heat recovery systems such as feedwater heaters, recuperators, and process heat exchangers. One of the difficulties with bottoming cycle systems is the low-temperature electrical power producing cycle. One example, depicted in Figure 32.4, is a low-temperature Rankine cycle. The low-temperature Rankine cycle is

* Many other terms for this boiler are common: for example, waste heat boiler (WHB) and heat recovery steam generator (HRSG).

† Other definitions of bottoming cycles are common. These other definitions often include any second use of the energy. For example, some authors refer to the steam turbine in a combined cycle as using a bottoming cycle. The more precise thermodynamic definition employed here is preferred, although fewer applications meet this definition.

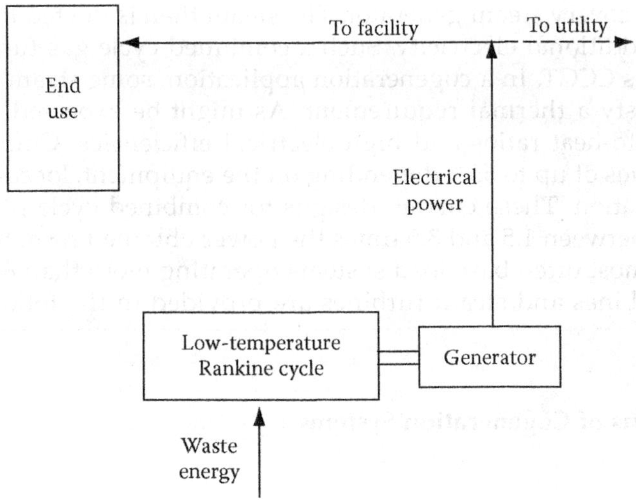

FIGURE 32.4
A schematic illustration of a cogeneration bottoming-cycle system.

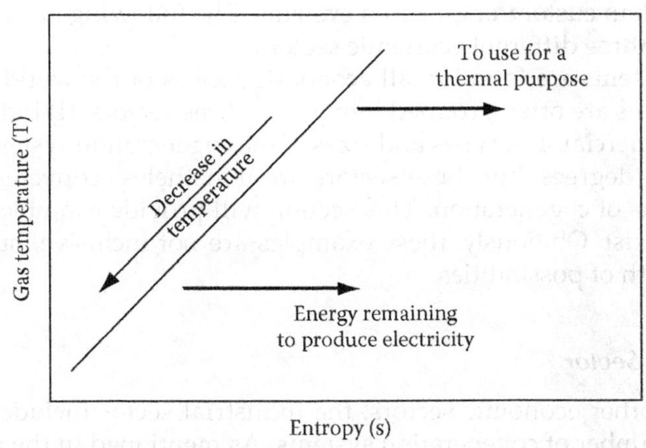

FIGURE 32.5
The gas temperature as a function of entropy for a bottoming-cycle system.

a power cycle similar to the conventional steam Rankine cycle, but a special fluid such as an organic substance (like a refrigerant) is used in place of water. This fluid vaporizes at a lower temperature than water, so this cycle is able to utilize the low-temperature energy. These cycles are generally much less efficient than conventional power cycles, often involve special equipment, and use more expensive working fluids. The majority of this chapter on cogeneration is based on topping cycle systems.

32.2.3 Combined Cycles

One power plant configuration that is based on a form of a topping cycle and is widely used in industry and by electrical utilities is known as a combined cycle. Typically in this configuration, a gas turbine is used to generate electricity and the exhaust gas is ducted to a heat recovery steam generator. The steam then is ducted to a steam turbine, which produces additional electricity. Such a combined cycle gas turbine power plant is often denoted as CCGT. In a cogeneration application, some steam would then need to be used to satisfy a thermal requirement. As might be expected, combined cycles have high power-to-heat ratios and high electrical efficiencies. Current designs have electrical efficiencies of up to 55% depending on the equipment, location, and details of the specific application. These current designs for combined cycle plants result in gas turbine power of between 1.5 and 3.5 times the power obtained from the steam turbine. These plants are most often base load systems operating more than 6000 h/year. More details on gas turbines and steam turbines are provided in the following sections on the prime movers.

32.2.4 Applications of Cogeneration Systems

32.2.4.1 General

Cogeneration systems may involve different types of equipment and may be designed to satisfy specific needs at individual sites. On the other hand, many sites have similar needs and packaged (pre-engineered) cogeneration systems may satisfy these needs and are more economical than custom-engineered systems. The following are examples of cogeneration systems in three different economic sectors.

Cogeneration systems are found in all economic sectors of the world. For convenience, cogeneration systems are often grouped into one of three sectors: (1) industrial, (2) institutional, and (3) commercial. The types and sizes of the cogeneration systems in these sectors overlap to varying degrees, but these sectors are nonetheless convenient for describing various applications of cogeneration. This section will provide examples of the variety of applications that exist. Obviously, these examples are not inclusive, but are intended to illustrate the breadth of possibilities.

32.2.4.2 Industrial Sector

Compared to the other economic sectors, the industrial sector includes the oldest, largest, and greatest number of cogeneration systems. As mentioned in the preceding history, cogeneration was first used by industry in the early 1900s to supply both electrical and thermal needs in an efficient manner. Many industries have a rich and continuous history of cogeneration applications. The industrial sector is dominant in cogeneration for several reasons. Industrial facilities often operate continuously, have simultaneous electrical and thermal requirements, and already have a power plant and operating staff. Those industries that are particularly energy intensive, such as the petrochemical and paper and pulp industries, are significant cogenerators. Many of these industries are cogenerating hundreds of megawatts of electrical power at one site. In addition, of course, mid-sized and smaller industries also use cogeneration.

32.2.4.3 Institutional Sector

The institutional sector includes a wide range of largely not-for-profit enterprises including universities, colleges, other schools, government building complexes, hospitals, military bases, and not-for-profit institutes. Many of these enterprises operate for a majority of the day, if not continuously. Some, such as hospitals, may already have provisions for

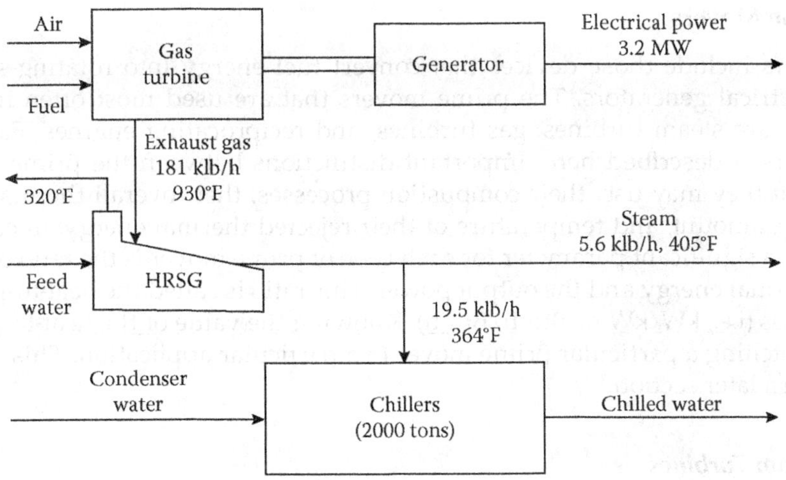

FIGURE 32.6
A schematic diagram of the topping-cycle cogeneration system installed in 1986 at Rice University in Houston, Texas.

emergency backup power and have the staff to operate a cogeneration system. Although not as large as the largest industrial cogenerators, a large hospital complex or university may require 50 or more megawatts of cogenerated electrical power.

Figure 32.6 is a schematic of a cogeneration system installed in 1986 at Rice University in Houston, Texas. As shown, this is a topping-cycle cogeneration system that uses a 3.2 MW gas turbine, a heat recovery steam generator, and two 1000 ton absorption chillers. Because this system was so successful, a second cogeneration system using a combined cycle was installed in 1989. It included a 3.7 MW gas turbine, a 400 kW steam turbine, and a 1500 ton absorption chiller. These two cogeneration systems supply the university with 90% of its electrical power, heating, and cooling requirements.

32.2.4.4 Commercial Sector

The commercial sector includes a wide range of for-profit enterprises including businesses, hotels, motels, apartment and housing complexes, restaurants, shopping centers, laundries, and laboratories. Generally, this sector includes the smallest cogenerators, and unless the electrical rates are unusually high, the economics are often less favorable than for the other sectors. There has been a greater interest in building cogeneration systems with the push for distributed generation and increased efficiency of electrical generation. Fuel cells are also being used for cogeneration applications in buildings.

32.3 Equipment and Components

Cogeneration systems consist of several major pieces of equipment and many smaller components. This section will describe and provide guidance on the selection of this equipment and components. The following discussion is grouped into four subsections: (1) prime movers, (2) electrical equipment, (3) heat recovery devices, and (4) absorption chillers.

32.3.1 Prime Movers

Prime movers include those devices that convert fuel energy into rotating-shaft power to drive electrical generators. The prime movers that are used most often in cogeneration systems are steam turbines, gas turbines, and reciprocating engines. Each of these prime movers is described here. Important distinctions between the prime movers are the fuels that they may use, their combustion processes, their overall thermal efficiency, and the type, amount, and temperature of their rejected thermal energy. In cogeneration applications, a significant parameter for each type of prime mover is the ratio of the rate of supplied thermal energy and the output power. This ratio is called the heat-to-power ratio, and is unitless (i.e., kW/kW or Btu h/Btu h). Knowing the value of the heat-to-power ratio assists in matching a particular prime mover to a particular application. This matching is discussed in a later section.

32.3.1.1 Steam Turbines

Steam turbines are widely used in power plants throughout industry and electric utilities. Steam turbines use high-pressure, high-temperature steam from a boiler. The steam flows through the turbine, forcing the turbine to rotate. The steam exits the turbine at a lower pressure and temperature. A major difference of the steam turbine relative to the reciprocating engines and gas turbines is that the combustion occurs externally in a separate device (boiler). This allows a wide range of fuels to be used including solid fuels such as coal or solid waste materials. The exit steam, of course, can be used for thermal heating or to supply the energy to an absorption chiller.

Steam turbines are available in a multitude of configurations and sizes. This description will highlight only the major possibilities, but more complete information is available elsewhere (Wood 1982). A major distinction is whether the machine is a condensing or noncondensing (back-pressure) steam turbine. Condensing steam turbines are steam turbines designed so that the steam exits at a low pressure (less than atmospheric) such that the steam may be condensed in a condenser at near ambient temperatures. Condensing steam turbines provide the maximum electrical output, and hence, are most often used by central plants and electric utilities. Since the exiting steam possesses little available energy, the application of condensing steam turbines for cogeneration is negligible.

Noncondensing steam turbines are steam turbines designed such that the exiting steam is at a pressure above atmospheric. These steam turbines are also referred to as back-pressure steam turbines. The exiting steam possesses sufficient energy to provide process or building heat. Either type of steam turbine may be equipped with one or more extraction ports so that a portion of the steam may be extracted from the steam turbine at pressures between the inlet and exit pressures. This extracted steam may be used for higher temperature heating or process requirements.

Noncondensing steam turbines are available in a wide range of outputs beginning at about 50 kW and increasing to over 100 MW. Inlet steam pressures range from 150 to 2000 psig, and inlet temperatures range from 500°F to 1050°F. Depending on the specific design and application, the heat-to-power ratio for steam turbines could range from 4 to over 10. The thermal efficiency increases with size (or power level) from typically 8% to 20%. Although the major source of thermal energy is the exit or extracted steam, the boiler exhaust may be a possible secondary source of thermal energy in some cases.

32.3.1.2 Gas Turbines

As with steam turbines, stationary gas turbines are major components in many power plants. Stationary gas turbines share many of the same components with the familiar aircraft gas turbines. In fact, both stationary (or industrial) and aircraft gas turbines are used in cogeneration systems. This brief description will highlight the important characteristics of gas turbines as applied to cogeneration.

Many configurations, designs, and sizes of gas turbines are available. The simple-cycle gas turbine uses no external techniques such as regeneration to improve its efficiency. The thermal efficiency of simple-cycle gas turbines may therefore be increased by the use of several external techniques, but the designs and configurations become more complex. Many of these modifications to the simple-cycle gas turbine are directed at using the energy in the exhaust gases to increase the electrical output and efficiency. Of course, such modifications will decrease the available energy in the exhaust. For cogeneration applications, therefore, the most efficient gas turbine may not always be the appropriate choice.

Gas turbines are available in a wide range of outputs beginning at about 100 kW and increasing to over 100 MW. Depending on the specific design, the heat-to-power ratio for gas turbines could range from about 1 to 3. The design point thermal efficiency increases with size (or power level) and complexity from typically 20% to 45%. The higher thermal efficiencies, compared to steam turbines, is the reason the heat-to-power ratio is lower than for steam turbines. These high efficiencies are for full-load (design point) operation. At part load, a gas turbine's efficiency decreases rapidly. As mentioned earlier, the use of regeneration, intercooling, reheating, and other modifications are used to improve the overall performance of the simple-cycle gas turbine.

Due to the large amount of excess air (the total air mass may be on the order of 100 times the fuel mass) used in the combustion process of gas turbines, the exiting exhaust gas contains a relatively high concentration of nitrogen and oxygen. Hence, the gas turbine exhaust may be characterized as mostly heated air and is nearly ideal for process or heating purposes. Gas turbines may use liquid fuels such as jet fuel or kerosene, or they may use gaseous fuels such as natural gas or propane. The highest performance is possible with liquid fuels, but the lowest emissions have been reported for natural gas operation.

Power ratings for gas turbines are provided for continuous and intermittent duty cycles. The continuous ratings are slightly lower than the intermittent ratings to provide long life and good durability. In both cases, the power ratings are provided for a set of standard operating conditions known as ISO conditions. ISO conditions include 1 atm, 59°F (15°C), sea level, and no inlet or exhaust pressure losses. For a specific application, the local conditions must be used to adjust the rated power to reflect the actual operating conditions. Most manufacturers will provide potential users with their recommended adjustments, which depend on the difference between the local and standard conditions. In particular, the local ambient inlet air temperature affects the gas turbine output in a significant manner. As the ambient air temperature increases, the performance of a gas turbine decreases due to the lower air density.

Gas turbines require more frequent and specialized maintenance than steam turbines. Major overhauls are required at 20,000–75,000 h of operation, depending on the service duty and the manufacturer. Aircraft gas turbines are designed to have their "hot section" returned to the manufacturer and replaced with a conditioned unit to minimize downtime, although the overhaul is more expensive. Industrial gas turbines are designed to be overhauled in the field, and this is generally a less expensive procedure.

32.3.1.3 Reciprocating Engines

A third category of prime movers for cogeneration systems is internal combustion (IC), reciprocating engines. Although rotary engines could be used in cogeneration systems, no significant applications are known. The remaining discussion will focus on reciprocating engines. These engines are available in several forms. Probably the most common form of the reciprocating engine is the typical spark-ignited, gasoline engine used in automobiles. For cogeneration applications, the spark-ignited gasoline engine has been converted to operate in a stationary, continuous mode with fuels such as natural gas. The majority of reciprocating engines for mid- to large-sized cogeneration systems are stationary diesel engines operating either with diesel fuel or in a dual-fuel mode with natural gas. These engines share some common characteristics for cogeneration applications and have some distinctive features as well.

Power ratings for reciprocating engines are similar to those for gas turbines in that both continuous and intermittent duty cycle ratings are provided. As with the gas turbines, these power ratings are provided for a set of standard conditions for ambient temperature, pressure, and elevation. The standard power ratings need to be adjusted for the local conditions at the site of the installation. For cogeneration applications, reciprocating engines are available in many power levels and designs. These power levels range from less than 50 kW to over 200 MW. Some manufacturers even offer "mini" cogeneration systems with outputs as low as 6 kW.

For those reciprocating, internal combustion engines that are liquid cooled (the majority of the engines considered here), the cooling liquid is a secondary source of thermal energy. Although not at the high temperatures of exhaust gas, this energy can be used to produce hot water or low-pressure steam. Several designs are available for recovering the energy in the cooling liquid. These designs use one or more direct or indirect heat exchangers to generate the hot water or low-pressure steam. Liquid-to-liquid heat exchangers can have high efficiencies, and most of this energy is recoverable (but at relatively low temperatures). Other energy sources from reciprocating engines include oil-coolers and turbocharger after-coolers. This energy is usually at temperatures below 160°F, and would only be practical to recover for low-temperature requirements. Another benefit of the reciprocating engine is that the maintenance and repair is less specialized than for gas turbines. On the other hand, the maintenance may be more frequent and more costly.

32.3.1.4 Heat-to-Power Characteristics

As mentioned earlier, an important feature in the selection of a prime mover for a cogeneration system is the heat-to-power ratio. This ratio may be a fairly constant characteristic of a family of a particular type of prime mover. Figures 32.7 and 32.8 show the electrical power output as a function of the potential available thermal energy rate output for a specific family of gas turbines and a specific family of reciprocating engines, respectively. The symbols represent the power and thermal output for individual gas turbines and engines in the two "families." The solid lines are the linear best fits to this data and represent the average heat-to-power (H/P) ratio for each family. As shown, the data does scatter some about the linear line, but the characteristic for each prime mover family is a good average. Also as shown, the average heat-to-power ratio is 2.0 for the gas turbine and 0.5 for the reciprocating engine. These are representative values for these two types of prime movers. These characteristic heat-to-power ratios will be used here in matching the prime mover to a specific application.

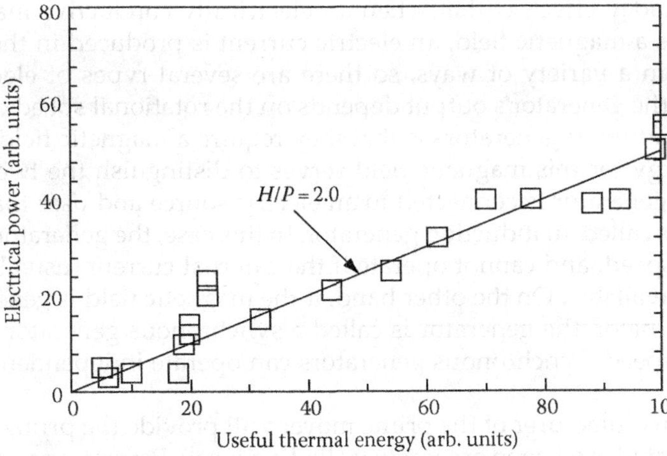

FIGURE 32.7
The electrical power as a function of the useful thermal energy (in arbitrary units) for a number of simple-cycle gas turbines. The average heat-to-power ratio for this family of gas turbines is 2.0, and is shown as the solid line.

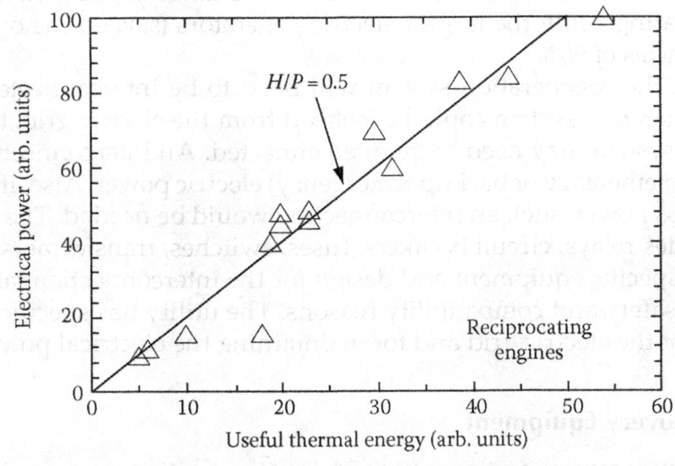

FIGURE 32.8
The electrical power as a function of the useful thermal energy (in arbitrary units) for similar diesel engines. The average heat-to-power ratio for this family of reciprocating engines is 0.5, and is shown as the solid line.

32.3.2 Electrical Equipment

The electrical equipment for cogeneration systems includes electrical generators, transformers, switching components, circuit breakers, relays, electric meters, controls, transmission lines, and related equipment. In addition to the equipment that supports electrical production, cogeneration systems may need equipment to interconnect with an electric utility to operate in parallel for the use of backup (emergency) power or for electrical sales to the utility. This section will briefly highlight some of the aspects regarding electrical generators and interconnections, but will not be able to review the other important electrical considerations.

The electric generator is a device for converting the rotating mechanical energy of a prime mover to electrical energy (i.e., electricity). The basic principle for this process, known as the Faraday effect, is that when an electrically conductive material such as a wire moves across a magnetic field, an electric current is produced in the wire. This can be accomplished in a variety of ways, so there are several types of electric generators. The frequency of the generator's output depends on the rotational speed of the assembly.

An important feature of generators is that they require a magnetic field to operate. The source of the energy for this magnetic field serves to distinguish the two major types of generators. If the generator is connected to an electric source and uses that source for the magnetic field, it is called an induction generator. In this case, the generator operates above the synchronous speed, and cannot operate if the external current (usually from the electric utility) is not available. On the other hand, if the magnetic field is generated internally using a small alternator, the generator is called a synchronous generator and operates at the synchronous speed. Synchronous generators can operate independent of the external electric grid.

Most often the manufacturer of the prime mover will provide the prime mover and generator as an integrated, packaged assembly (called a genset). Performance characteristics of generators include power rating, efficiency, voltage, power factor, and current ratings. Each of these performance characteristics must be considered when selecting the proper generator for a given application. Electric generators may have conversion efficiencies of between about 50% to over 98%. The efficiency increases in this range as the generator increases in size and power rating. Only the largest electric generators (say, on the order of 100 MW) can attain efficiencies of 98%.

In many cases, the cogeneration system will need to be interconnected to the utility. Although a cogeneration system could be isolated from the electric grid, there are several reasons why the system may need to be interconnected. An interconnection is necessary for receiving supplementary or backup (emergency) electric power. Also, if the cogenerator elects to sell excess power, such an interconnection would be needed. The interconnection equipment includes relays, circuit breakers, fuses, switches, transformers, synchronizers, and meters. The specific equipment and design for the interconnection largely is dictated by the utility for safety and compatibility reasons. The utility has specific responsibilities for the integrity of the electric grid and for maintaining the electrical power quality.

32.3.3 Heat Recovery Equipment

The primary heat recovery equipment used in cogeneration systems includes several types of steam and hot-water production facilities. In addition, absorption chillers could be considered in this section, but for organizational reasons, chillers will be discussed in the following section. Several configurations of heat recovery devices are available. As mentioned earlier, these devices may be referred to as "heat recovery steam generators" or HRSGs. HRSGs are often divided into the following categories: (1) unfired, (2) partially fired, and (3) fully fired. An unfired HRSG is essentially a convective heat exchanger. A partially fired HRSG may include a "duct burner," which often uses a natural gas burner upstream of the HRSG to increase the exhaust gas temperature. A fully fired HRSG is basically a boiler that simply uses the exhaust gas as preheated air.

For most of these heat recovery devices, exhaust gas flows up through the device and exits at the top. Energy from the exhaust gas is used to heat and vaporize the water, and to superheat the steam. Figure 32.9 shows this process on a temperature diagram. The top line shows the exhaust gas temperature decreasing from left to right as energy is removed from the gas to heat the water. The lower line represents the water heating up from right to left in the diagram. The lower-temperature exhaust is used to preheat the water to

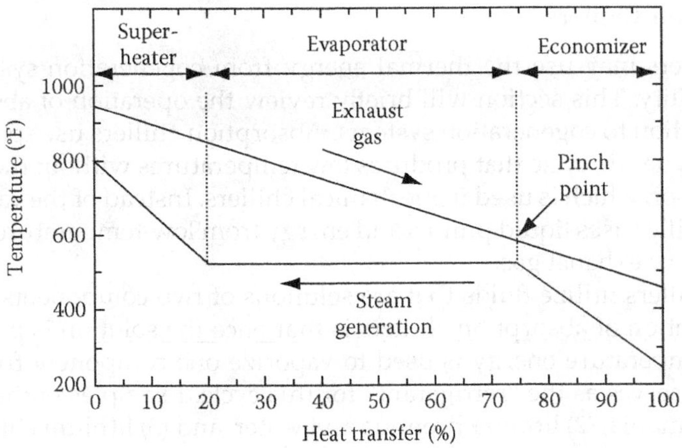

FIGURE 32.9
The temperatures of the exhaust gas and water/steam as a function of the heat transfer coordinate for the steam generation process of an HRSG.

saturation conditions in the economizer. The intermediate-temperature exhaust is used to vaporize (or boil) the water to form saturated steam. Finally, the highest-temperature exhaust is used to superheat the steam.

The temperature difference between the exhaust gas and the water where the water first starts to vaporize is referred to as the pinch point temperature difference. This is the smallest temperature difference in the HRSG and may limit the overall performance of the heat recovery device. Since the rate of heat transfer is proportional to the temperature difference, the greater this difference the greater the heat transfer rate. On the other hand, as this temperature difference increases the steam flow rate must decrease and less of the exhaust gas energy will be utilized. To use smaller temperature differences and maintain higher heat transfer rates, larger heat exchanger surfaces are required. Larger heat transfer surface areas result in higher capital costs. These, then, are the types of trade-offs that must be decided when incorporating a heat recovery device into a cogeneration system design.

The proper selection of HRSG equipment depends on the prime mover, the required steam conditions, and other interdependent factors. Some of these considerations are described next. Most of these considerations involve a trade-off between increased performance and higher initial capital cost.

The "back-pressure" of the HRSG unit affects the overall system performance. As the back-pressure decreases, the HRSG efficiency increases, but the cost of the HRSG unit increases. Successful units often have 10–15 in. water pressure of back-pressure. As mentioned earlier, the selection of the "pinch point" temperature difference affects the performance and cost. Typical pinch point temperature differences are between 30°F and 80°F. High-efficiency, high-cost units may use a temperature difference as low as 25°F.

The final outlet stack gas temperature must be selected so as to avoid significant acid formation. This requires the outlet stack temperature to be above the water condensation temperature. This is typically above 300°F. The amount of sulfur in the fuel affects this decision, since the sulfur forms sulfuric acid. As the amount of sulfur in the fuel increases, the recommended stack gas outlet temperature increases.

The final consideration is the steam temperature and pressure. This is a complex decision that depends on many factors such as the application for the steam, the source of the exhaust gas, the exhaust gas temperature and flow rate, and inlet water condition and temperature.

32.3.4 Absorption Chillers

Absorption chillers may use the thermal energy from cogeneration systems to provide cooling for a facility. This section will briefly review the operation of absorption chillers and their application to cogeneration systems. Absorption chillers use special fluids and a unique thermodynamic cycle that produces low temperatures without the requirement of a vapor compressor, which is used in mechanical chillers. Instead of the vapor compressor, an absorption chiller uses liquid pumps and energy from low-temperature sources such as hot water, steam, or exhaust gas.

Absorption chillers utilize fluids that are solutions of two components. The basic principle of the operation of absorption chillers is that once the solution is pumped to a high pressure, low-temperature energy is used to vaporize one component from the solution. This component serves as the "refrigerant" for this cycle. Examples of these solutions are (1) water and ammonia, (2) lithium bromide and water, and (3) lithium chloride and water. In the first case the ammonia serves as the refrigerant, and in the latter two cases the water serves as the refrigerant.

For cogeneration applications, the important feature of absorption chillers is that they use relatively low-temperature energy available directly or indirectly from the prime mover and produce chilled water for cooling. The use of absorption chillers is particularly advantageous for locations where space and water heating loads are minimal during a good part of the year. For these situations, the thermal output of a cogeneration system can be used for heating during the colder part of the year and, using an absorption chiller, for cooling during the warmer part of the year. Furthermore, by not using electric chillers, the electric loads are more constant throughout the year. In warm climates, absorption chillers are often an important, if not an essential, aspect of technically and economically successful cogeneration systems.

Some machines are designed as indirect-fired units using hot water or steam. As examples of typical numbers, a single-stage unit could use steam at 250°F to produce a ton of cooling for every 18 lb of steam flow per hour. A dual-stage unit would need 365°F steam to produce a ton of cooling for every 10 lb of steam flow per hour. If hot water is available, a ton of cooling could be produced for every 220 lb of 190°F hot water per hour.

Other machines use the exhaust gas directly and are called direct-fired units. In these cases, the exhaust gas temperature needs to be 550°F–1000°F. The higher the exhaust temperature, the less energy (or exhaust gas flow) is needed per ton of cooling. For example, for 1000°F exhaust gas, a ton of cooling requires 77 lb/h of flow whereas for 550°F exhaust gas a ton of cooling requires 313 lb/h of flow.

32.4 Technical Design Issues

32.4.1 Selecting and Sizing the Prime Mover

The selection of a prime mover for a cogeneration system involves the consideration of a variety of technical and nontechnical issues. Technical issues, which often dominate the selection process, include the operating mode or modes of the facility, the required heat-to-power ratio of the facility, the overall power level, and any special site considerations (e.g., low noise). Other issues, which may play a role in the selection process, include matching existing equipment and utilizing the skills of existing plant personnel. Of course, the final decision is often dominated by the economics.

Steam turbines and boilers usually are selected for a cogeneration system if the fuel of choice is coal or another solid fuel. Occasionally, for very large (>50 MW) systems operating at base load, a steam turbine system may be selected even for a liquid or gaseous fuel. Also, steam turbines and boilers would be selected if a high heat-to-power ratio is needed. Steam turbines also may be selected for a cogeneration system in certain specialized cases. For example, a large pressure reduction valve in an existing steam system could be replaced with a steam turbine and thereby provide electrical power and thermal energy. In many applications, however, steam turbines are selected to be used in conjunction with a gas turbine in a combined cycle power plant to increase the power output. Combined cycle gas turbine power plants were described in an earlier section.

Gas turbines are selected for many cogeneration systems where the required heat-to-power ratio and the electrical power need are high. Also, gas turbines are the prime mover of choice where minimal vibration or a low weight-to-power ratio (such as for a roof installation) is required. Reciprocating engines are selected where the heat-to-power ratio is modest, the temperature level of the thermal energy is low, and the highest electrical efficiency is necessary for the economics. Additionally, reciprocating engines may be selected if the plant personnel are more suited to the operation and maintenance of these engines.

Selecting the appropriate size prime mover involves identifying the most economic cogeneration operating mode. This is accomplished by first obtaining the electrical and thermal energy requirements of the facility. Next, various operating modes are considered to satisfy these loads. By conducting a comprehensive economic analysis, the most economic operating mode and prime mover size can be identified. The process of matching the prime mover and the loads is described next.

32.4.2 Matching Electrical and Thermal Loads

To properly select the size and operating mode of the prime mover, the electric and thermal loads of the facility need to be obtained. For the most thorough "matching," these loads are needed on an hourly, daily, monthly, and yearly basis. Figure 32.10 is an example of the hourly electrical and thermal loads of a hypothetical facility for a typical work day in the winter. As shown, the heating and power demands begin to increase at 6:00 a.m.

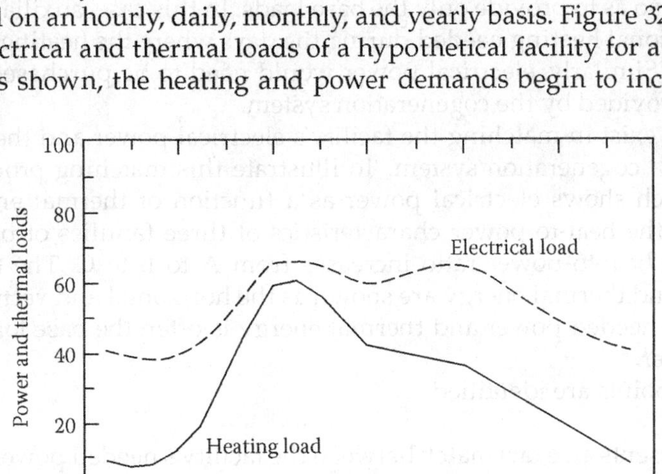

FIGURE 32.10
The power and thermal loads (in arbitrary units) as a function of the hour of the day for a hypothetical facility for one work day in the winter.

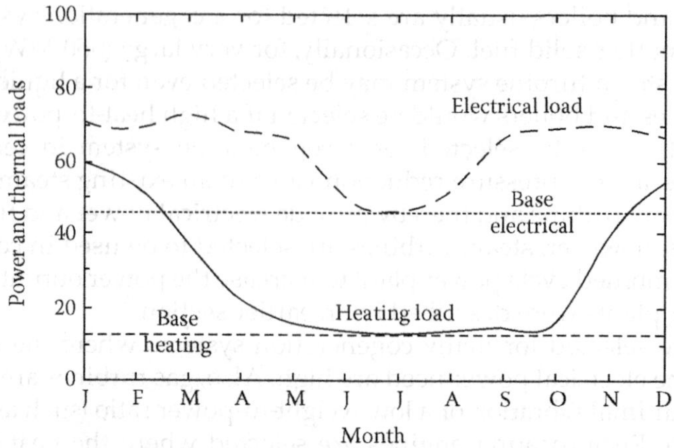

FIGURE 32.11
The total power and thermal loads (in arbitrary units) as a function of the month of the year for a hypothetical facility.

as the day's activities begin. The heating load is shown to peak shortly after 8:00 a.m. and then decrease for the remainder of the day. The electrical load remains nearly constant during the working part of the day, and is somewhat lower during the evening and nighttime hours.

Figure 32.11 shows the month totals for the electrical and thermal loads for the same hypothetical facility. For the summer months, the heating and electrical loads are minimum. In addition, this figure shows dotted and dashed lines, which represent the "base loads" for the electrical and heating loads, respectively. The base loads are the minimum loads during the year and form a floor or base for the total loads. Often a cogeneration system may be sized so as to provide only the base loads. In this case, auxiliary boilers would provide the additional heating needed during the days where the heating needs exceeded the base amount. Similarly, electrical power would need to be purchased to supplement the base power provided by the cogeneration system.

Several options exist in matching the facility's electrical power and thermal needs to a prime mover for a cogeneration system. To illustrate this matching procedure, consider Figure 32.12, which shows electrical power as a function of thermal energy. The three dashed lines are the heat-to-power characteristics of three families of prime movers. In this example, the heat-to-power ratio increases from A to B to C. The facility's needed electrical power and thermal energy are shown as the horizontal and vertical dotted lines, respectively. This needed power and thermal energy is often the base load of the facility as explained earlier.

Five operating points are identified:

1. Point 1 represents an exact match between the facility's needed power and thermal energy, and the power and thermal energy available from a prime mover with the heat-to-power ratio characteristic B. Such an exact match is not likely or desired, so the other operating points must be considered.
2. Point 2 represents a case where a prime mover is selected with a characteristic C such that the thermal needs are matched but the electrical power supplied is too low. In this case, the facility would need to purchase supplementary electrical

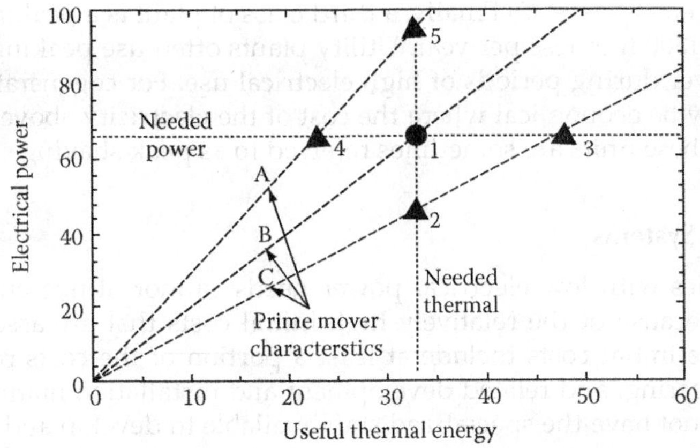

FIGURE 32.12
The electrical power as a function of the useful thermal energy (in arbitrary units).

power usually from the local electrical utility. This is a common case that is often the best economic and operational choice.

3. Point 3 represents a case where a prime mover is selected with a characteristic C such that the electrical power needs are matched, but excess thermal energy is available. Generally, this is not an economical choice because the overall system efficiency is low. For certain situations, however, this case might be selected—for example, if future thermal loads are expected to increase or if a customer can be identified to purchase the excess thermal energy.
4. Point 4 represents a case where a prime mover is selected with a characteristic A such that the electrical power needs are matched but the available thermal energy does not satisfy the needs of the facility. In this case, supplementary boilers are often used to supply the additional thermal energy. This may be the best economic choice, particularly if existing boilers are available and the desire is to size the cogeneration system such that all the thermal energy is used.
5. Point 5 represents a case where a prime mover is selected with a characteristic A such that the thermal needs are matched but the electrical power supplied is too high. This would only be economical if an external customer was available to purchase the excess electrical power. This is often the situation with third-party cogenerators, who own and operate a cogeneration system and sell the thermal energy and electrical power to one or more customers.

These cases represent distinct operating mode selections. In practice, other technical and nontechnical aspects must be considered in the final selection. These cases are presented to illustrate the global decision process.

The possible overall operating modes for a cogeneration power plant are often categorized into one of three classes. (1) The plant may operate as a base load system with little or no variation in power output. Base load plants operate in excess of 6000 h/year. Power needs above the base load are typically provided by interconnections to a local utility or by an auxiliary power plant. (2) The plant may operate as an intermediate system for 3000–4000 h/year. These systems are less likely than base load systems, but if the economics are positive, may have application for facilities that are not continuously operated such as some commercial enterprises. (3) Finally, a third class of plant is a peaking system, which operates only for 1000 h or less per year. Utility plants often use peaking systems to provide peaking power during periods of high electrical use. For cogeneration applications, peaking units may be economical where the cost of the electricity above a certain level is unusually high. These units are sometimes referred to as peak shaving systems.

32.4.3 Packaged Systems

In general, facilities with low electrical power needs cannot utilize customized cogeneration systems because of the relatively high initial costs that are associated with any size system. These initial costs include at least a portion of the costs related to the initial design, engineering, and related development and installation matters. Also, smaller facilities often do not have the specialized staff available to develop and operate complex power plants.

To solve some of these above problems, pre-engineered, factory-assembled, "packaged" cogeneration systems have been developed. Packaged cogeneration systems range from less than 50 kW to over 1 MW. Some larger units (say, 20 MW or larger) using gas turbines or reciprocating engines are sometimes described as packaged cogeneration systems, and are, at least to some degree, "packaged." For example, a gas turbine and generator may be factory assembled, tested, and skid mounted for shipping. These units, although packaged, do not completely eliminate the problems of project development, installation, and operation. The remainder of this section will consider only the smaller packaged cogeneration systems.

The major advantage of packaged cogeneration systems is that the initial engineering, design, and development costs can be spread over many units, which reduces the capital cost (per kW) for these systems. Other advantages of packaged cogeneration systems include factory assembly and testing of the complete system. If there are any problems, they can be fixed while the system is still at the manufacturer's plant. The standard design and reduced installation time result in short overall implementation times. In some cases, a packaged cogeneration system could be operational within a few months after the order is received. This short implementation time reduces the project's uncertainty, which eases making decisions and securing financing.

Another advantage of packaged cogeneration systems is that the customer only interacts with one manufacturer. In some cases, the packaged cogeneration system manufacturer will serve as project engineer and take the project from initial design to installation to operation. The customer often may decide to purchase a Turnkey system. This provides the customer with little uncertainty and places the burden of successful project completion on the manufacturer. Also, the manufacturer of a packaged cogeneration system will have experience interacting with regulating boards, financing concerns, and utilities, and may assist the customer in these interactions.

The major disadvantage of packaged cogeneration systems is that the system is not customized for a specific facility. This may mean some compromise and lack of complete optimization. Specialized configurations may not be available. Also, beyond a certain size, packaged cogeneration systems are simply not offered and a customized unit is the only alternative.

Although the initial capital costs of packaged cogeneration systems are low (on a per kW basis), the other site-specific costs could be relatively high and result in unsatisfactory economics. These other costs depend on the selected fuel, the available space at the site, and the compatibility of existing electrical and mechanical facilities. In particular, the interconnection costs associated with the electric utility and the mechanical systems can be prohibitively high. The electric interconnection costs depend on the electric utility's requirements. The fuel system costs depend on whether the selected fuel is already available on site at the appropriate conditions (such as gas pressure for gaseous fuels). Other miscellaneous costs include those for services and assistance with engineering, permitting, zoning, financing, legal review, and construction oversight.

In summary, packaged cogeneration systems offer a cost-effective solution for facilities with low power and thermal requirements such as small to medium hospitals, schools, hotels, and restaurants. The majority of current packaged cogeneration systems use reciprocating engines and standard components. They are factory assembled and tested, and are often sold as Turnkey installations. Although the packaged cogeneration systems' costs can be attractive, each facility must be assessed for total costs to determine the economic feasibility. Future developments of packaged cogeneration systems will include even more efficient and cost-effective engines, more advanced microprocessor control systems, and more effective integration of the various components.

32.5 Regulatory Considerations

This section includes brief overviews of the relevant Federal regulations on electric power generation and the related environmental constraints. It should be pointed out that some sections of PURPA were repealed by the Energy Policy Act of 2005; however, its importance to cogeneration is so significant that it deserves the recognition and discussion as noted here.

32.5.1 Federal Regulations Related to Cogeneration—Early History

As mentioned earlier, the passage of PURPA, the Public Utility Regulatory Policies Act, in 1978, helped make cogeneration a much more attractive option for electric power generation for a variety of facilities. PURPA is one of the most controversial bills relating to power generation that has been passed by Congress. PURPA was one of five separate acts comprising the National Energy Act passed by Congress in 1978:

1. National Energy Conservation Policy Act (NECPA)
2. Natural Gas Policy Act (NGPA)
3. Powerplant and Industrial Fuel Use Act (FUA)
4. Energy Tax Act (ETA)
5. Public Utility Regulatory Policies Act (PURPA)

The Federal government, and for that matter the entire nation, was concerned about the oil crisis in 1973, the rapidly increasing price of energy, the increased dependence on foreign oil, and our lack of concern for energy efficiency and the use of renewable energy. The National Energy Act addressed most of those concerns, and PURPA, in particular, dealt with independent power generation, energy efficiency, and the use of renewables for power generation. Although all five acts had an impact on our nation's energy consumption and energy use patterns, PURPA had the greatest overall impact and led to a resurgence of interest in cogeneration. Prior to the passage of PURPA, the three most common barriers to cogeneration were (Laderoute 1980) (1) no general requirement by electric utilities to purchase electric power from cogenerators, (2) discriminatory backup power for cogenerators, and (3) fear on the part of the cogenerator that they might become subject to the same state and Federal regulations as electric utilities. The passage of PURPA helped to remove these obstacles to development of cogeneration facilities (Spiewak 1980, 1987).

Although PURPA was passed in 1978, it was not until 1980 that the Federal Energy Regulatory Commission (FERC) issued its final rulemakings and orders on PURPA. In the National Energy Act, FERC was designated as the regulatory agency for implementation of PURPA. The regulations dealing with PURPA are contained in Part 292 of the FERC regulations, and Sections 201 and 210 are the two primary sections relevant to small power production and cogeneration (FERC 1978).

Section 201 contains definitions of cogeneration and sets annual efficiency standards for new topping cycle* cogeneration facilities that use oil or natural gas.† For a cogenerating facility to qualify for the privileges and exclusions specified in PURPA, the facility must meet these legislated standards. These standards define a legislated or artificial "efficiency," which facilities must equal or exceed to be considered a "qualified facility" (QF). A qualified facility is eligible to use the provisions outlined in PURPA regarding nonutility electric power generation. This legislated "efficiency" is defined as

$$\eta_{PURPA} = \frac{\left(P + \frac{1}{2}T\right)}{F} \tag{32.10}$$

where
- P is the electrical energy output
- T is the useful thermal energy
- F is the fuel energy used

Table 32.1 lists the standards for the PURPA efficiencies. These standards state that a facility must produce at least 5% of the site energy in the form of useful thermal energy, and must meet the efficiency standards in Table 32.1. Values for the thermal fraction and the PURPA efficiency are based upon projected or estimated annual operations.

The purpose of introducing the "artificial" standards was to insure that useful thermal energy was produced on-site in sufficient quantities to make the cogenerator more

TABLE 32.1

Required Efficiency Standards for Qualified Facilities (QFs)

Useful Thermal Energy Fraction	Required η_{PURPA}
≥5.0%	≥45.0%
≥15.0%	≥42.5%

* Since bottoming cycle cogeneration facilities do not use fuel for the primary production of electrical power, these facilities are only regulated when they use oil or natural gas for supplemental firing. The standards states that during any calendar year the useful power output of the bottoming cycle cogeneration facility must equal or exceed 45% of the energy input if natural gas or oil is used in the supplementary firing. The fuels are used first in the thermal process prior to the bottoming cycle cogeneration facility are not taken into account for satisfying PURPA requirements.

† For topping cycle cogeneration facilities using energy sources other than oil or natural gas (or facilities installed before March 13, 1980), no minimum has been set for efficiency.

efficient than the electric utility. Any facility that meets or exceeds the required efficiencies will be more efficient than any combination of techniques producing electrical power and thermal energy separately. Section 201 also put limitations on cogenerator ownership; that is, electric utilities could not own a majority share of a cogeneration facility, nor could any utility holding company, nor a combination thereof. This section also defined the procedures for obtaining QF status.

Section 210 of the PURPA regulations specifically addressed major obstacles to developing cogeneration facilities. In fact, the regulations in Section 210 not only leveled the playing field, they tilted it in favor of the cogenerator. The principal issues in Section 210 include the following legal obligations of the electric utility toward the cogenerator:

- Obligation to purchase cogenerated energy and capacity from QFs.
- Obligation to sell energy and capacity to QFs.
- Obligation to interconnect.
- Obligation to provide access to transmission grid to "wheel" to another electric utility.
- Obligation to operate in parallel with QFs.
- Obligation to provide supplementary power, Backup power, maintenance power, and interruptible power.

Section 210 also exempted QFs from utility status and established a cost basis for purchase of the power from QFs. FERC specified that the price paid to the QF must be determined both on the basis of the utility's avoided cost for producing that energy and, if applicable, on the capacity deferred as a result of the QF power (e.g., cost savings from not having to build a new power plant). Other factors, such as QF power dispatchability, reliability, and cooperation in scheduling planned outages, could also be figured into the price paid to the QFs by the electric utilities. The state public utility commissions were responsible for determining the value of these avoided cost rates.

PURPA also gave special consideration to power produced from small (less than 80 MW) power production facilities (SPPFs). These must be fueled by biomass, geothermal, wastes, renewable resources, or any combination thereof. H.R. 4808, passed by the 101st Congress, lifted the 80 MW size restriction on wind, solar, geothermal cogeneration, and some waste facilities.* These SPPFs could become QFs and have the same status as a large gas-fired industrial cogeneration facility. Facilities between 1 kW and 30 MW may receive exemptions from the Federal Power Act (FPA) and the Public Utility Holding Company Act (PUHCA), and SPPFs using biomass are exempt from PUHCA and certain state restrictions, but not from the FPA.

32.6 Regulatory Developments of the 1990s and Early Twenty-First Century

New and restored terms in the power generation area were introduced in the early 1990s: nonutility generators (NUGs), independent power producers (IPPs), and exempt wholesale generators (EWGs). An EWG is an entity that wishes to build a

* Serial 101–160, Hearing of the Committee on Energy and Commerce, U.S. House of Representatives, June 14, 1990.

new nonrate-based power plant under PUHCA and is only in the business of selling wholesale power. A revised PUHCA and a 1992 FERC ruling gave more open access to transmission lines and limited what a utility could charge third-party transmission users for that access.

The Energy Policy Act of 1992 enacted into law many of FERC's rules on EWGs and defined their legal status. In addition to opening up transmission-line access, the Energy Act also opened the possibility of retail wheeling, an area strongly favored by industry and heavily opposed by the electric utilities. An example of retail wheeling might be an industrial cogenerator in the Texas Gulf Coast area selling power to an industry in the Dallas area. Two or more electric utilities would be affected on their transmission lines, and the Dallas-area utility would lose the revenue from the customer in their service territory. From industry's viewpoint, they would like the ability to buy power wherever they could purchase it cheaper. The Energy Policy Act of 1992 was, in part, responsible for the deregulation of the electric utility industry.

The California Public Utility Commission held hearings in August 1994 on a proposal to have individual consumers choose their own electric company. California led the way "for the now-inevitable downfall of regulated electric utility monopolies…."* The California experience with deregulation was not pleasant. Electric rates remained high for consumers, brownouts occurred, and price fixing was prevalent among some of the state's power generators. The California experiences in 2000 and 2001 discouraged many states from deregulating their electric utilities. However, in 2005 approximately half of the states in the lower 48 have some form of deregulation already in place or a time schedule for deregulation. States in the northeastern United States, anchored by Pennsylvania, and the state of Texas accomplished deregulation without experiencing many of the problems that occurred in California.

The Energy Policy Act of 2005 modified a number of provisions in PURPA. Because so many states have already deregulated their electrical utilities, some of the PURPA provisions discussed above were no longer needed. Section 1251 of the EPACT of 2005 amended PURPA by adding a "Net Metering" requirement for cogenerating facilities; Section 1252 added a requirement for "Smart Meters"; Section 1253 terminated the mandatory purchase of cogenerated power and power produced by small power cogenerators. Section 1254 of the 2005 act requires electric utilities to interconnect with customers who are also cogenerators. Section 1263 of the EPACT 2005 repeals PUHCA in its entirety and shifts all PUHCA responsibilities to FERC.

The Energy Policy Act of 2005 is a very broad energy act, placing more emphasis on energy efficiency, distributed generation, improved transmission systems, and fuel cells. Overall, it is favorable to cogeneration.

32.7 Environmental Considerations, Permitting, Water Quality

As described in the introduction, air and water regulations have been legislated since the 1950s. The Clean Air Act Amendments of 1977 established a new source of performance standards and placed environmental enforcement in the hands of the Environmental

* *USA Today*, Electricity consumers may get power of choice, August 5, 1994, p. 10A.

Protection Agency (EPA). The National Energy Policy Act of 1992 and the revised Clean Air Act Amendments (1990) placed even more stringent regulations on environmental pollutants. Concerns over acid rain, global warming, and the depletion of the ozone layer have prompted these tougher regulations.

The Clean Air Act Amendment of 1990 also produced new challenges and opportunities for power generation—trading and selling of emissions. Certain areas in the U.S. were designated as nonattainment zones where no new emissions sources could be located. Industries could, however, trade or even sell emissions (i.e., so much per ton of CO_2 or SO_2). This could become very important if a cogenerator wanted to locate a plant in a nonattainment zone and could buy emissions from a company that reduced production or somehow cut emissions from their plant.

A number of certifications are required to get a cogeneration plant approved. These include not only a FERC certificate but also various state permits. Since each state will set its own permitting requirements, it is not possible to generalize what is required on each application, but the following information will likely be required:

1. Nature of pollutant source.
2. Type of pollutant.
3. Level of emissions.
4. Description of process technology.
5. Process flow diagram.
6. Certification that the "Best Available Control Technology" (BACT) is being used.

Different requirements must be met if the proposed site is in an attainment or nonattainment zone. An attainment area will require sufficient modeling of ambient air conditions to insure that no significant deterioration of existing air quality occurs. This is PSD modeling or "prevention of significant deterioration." For a nonattainment area, no new emissions can be added unless they are offset by the removal of existing emissions. This has led to the selling or trading of emissions by industrial facilities and utilities.

32.7.1 Water Quality and Solid Waste Disposal

Water quality is usually not a problem with natural gas-fired cogeneration plants but will probably be monitored and will require both state and Federal permits if the wastewater is discharged into a public waterway. In extreme conditions it may be necessary to control both wastewater temperature and pH.

Solid waste disposal is not a problem with either natural gas or oil-fired cogeneration plants but could be a major problem for a coal-fired, coal gasification, or waste-to-energy cogeneration system. In some states the bottom ash from waste-to-energy plants was considered hazardous waste and the ash disposal cost per ton was more expensive than the refuse disposal cost. In 1994, bottom ash was declared by the EPA to be nonhazardous and therefore was allowed for land filling in standard landfills. Again, all states have different standards for both the quality of the water discharged and the requirements for solid waste disposal, and all project planners should check with the appropriate regulatory agencies in the state where the project is planned to be sure what air, water, and solid waste (if applicable) permits are required.

32.8 Economic Evaluations

This section will present a brief summary of the important aspects of economic evaluations of cogeneration systems. This section is not intended to be a complete presentation of engineering economics, which would be beyond the scope of this chapter. Rather, those aspects of economic evaluations that are specific or especially important to cogeneration systems will be described. These aspects include the baseline (or "business as usual") financial considerations, cogeneration economics including capital and operating costs, and the available economic and engineering computer programs. The two case studies presented at the end of this chapter will further illustrate some of the economic principles discussed in this section.

32.8.1 Baseline Considerations

Because a cogeneration project is capital intensive, it is extremely important to have accurate data to determine current and estimated (projected) future energy consumption and costs. In a large industrial plant with known process steam needs and 15 min electric data available, it is usually fairly straightforward to establish the baseline case (i.e., the no-cogeneration case). For universities, hospitals, small manufacturing plants, and so on, the data are not always available. Typically whole campus/facility electrical data are available on a monthly basis only (from utility bills), and hourly electrical profiles may have to be "constructed" from monthly energy and demand data. Hourly thermal data also may not exist. Boiler operators may have daily logs, but these are typically not in electronic form. Constructing accurate hourly thermal energy profiles can involve hours of tedious work pouring over graphs and boiler operator logs. In the worst-case scenario, only monthly gas bills may be available, and it may be necessary to construct hourly thermal profiles from monthly bills, boiler efficiencies, and Btu content of the fuel. Since the monthly dollars have to be matched, there is often a great deal of trial and error involved before there is a match between assumed hourly energy profiles and monthly energy consumption and costs from the utility bills.

While a first-cut energy and cost analysis may be done on a monthly basis for an industrial facility that operates 7 days a week, 24 h a day, that is not the case for a university, a commercial building complex, or a one- or two-shift manufacturing facility. Energy loads are highly variable, and hourly analyses are required. The possible exception may be hospitals, where 7-day-a-week operation is normally the case. However, there will often be a significant difference between weekday and weekend loads, which would require some sort of "day typing" in the analysis.

Assuming a good energy profile (hourly preferred) can be constructed, the "business as usual" scenario costs are determined. This is the annual cost of doing business without the cogeneration system, including annual purchased electrical energy sales, electrical demand charges, boiler fuel costs, and boiler maintenance. That total number is the baseline to which the cogeneration economics are compared.

32.8.2 Cogeneration Economics

If a simple payback approach is used, the cogeneration system costs divided by the savings over the baseline costs, would be the simple payback:

$$\text{Annual savings} = \text{Baseline costs} - \text{Cost with cogeneration system}$$

$$\text{Simple payback} = \frac{\text{Cogeneration system capital costs}}{\text{Annual savings}} \tag{32.11}$$

For many analyses, a simple payback approach is often enough. Once an acceptable payback period is defined, the cogeneration decision can be made. If the simple payback period is fairly short (2–3 years), then small variations in energy prices will have little effect on the decision. As the payback period lengthens (e.g., 4–8 years), then other factors should be examined. The time value of money has to be considered, as well as projected energy rates, and projected changes in energy needs for the facility. See Chapter 12 for more information.

A typical scenario for a project financed over 15–20 years might require an economic analysis that would include the following steps:

1. Determine current energy costs and current thermal and electrical energy requirements.
2. Project energy costs into the 15- to 20-year future using forecasts from the local utility or from the state public utility commission.
3. Project thermal and electrical needs into the future, considering possible growth (or declines) in energy requirements.

   Note: Any projections of future thermal and electrical loads should carefully consider the impact of energy conservation on the total energy needs. Energy conservation is extremely important in the cost analysis because it represents dollars that do not have to be spent. The term "negawatt" has been applied to demand-side management programs as a means of expressing a decrease in electrical demand at a facility.
4. Project increasing maintenance costs associated with an expected increase in thermal loads (if appropriate).

Summing these costs will provide a more realistic "business as usual" case over the expected financing period of the cogeneration system.

Once the base case is determined, various sizing and operating scenarios for the cogeneration system should then be evaluated. These scenarios include (1) sizing a base-load cogeneration system and, if necessary, purchasing additional required electrical energy and producing supplemental required thermal energy, (2) following the thermal load profile and purchasing electrical energy needs, (3) following the electrical load profile and producing supplemental thermal energy, and (4) following the thermal profile with excess electricity produced. Note that with any potential scenario considered, all the thermal energy produced by the cogeneration system is used. It is the use of all the thermal energy that generally makes a cogeneration system feasible. If electricity production were the primary output, the cogenerator could not compete with the local electric utility. It is the simultaneous production of and need for the thermal energy that makes cogeneration attractive.

The next step is to determine the costs associated with a cogeneration system. This will depend on the type and size of system selected—diesel engine/generator, dual fuel diesel engine/generator, natural gas-fired engine/generator, or gas turbine/generator. As described in a previous section, a variety of possibilities exist in sizing the cogeneration system for a particular application (see Section 32.4.2). Ideally, the cogeneration system

TABLE 32.2

Possible Scenarios to Consider for Cogeneration System Sizing

Scenario	Result	Application
Baseline cogeneration system that meets a portion of the facility's electrical and thermal requirements	Requires production of supplemental thermal energy and purchase of additional electrical energy	Universities/commercial buildings/thermal energy plants that have variable loads and do not want to get into the power sales business
Cogeneration system where the thermal load exceeds the equivalent electrical output of the generator-thermal load sizing	Electrical energy will have to be purchased from the local utility	Industrial facility or manufacturing plant with relatively constant thermal load
Cogeneration system designed to match electrical loads	Thermal energy to be produced by a boiler	Industrial facility with fairly high and constant electrical loads and lower, but variable, thermal loads
Cogeneration system sized to meet high thermal loads, but with lower equivalent electrical requirements	Excess electrical power sales to the local utility	Industrial or university facility that has lower or variable electrical needs, which require electricity sales to make the project economical

could be sized to match the electrical and thermal loads exactly; however, seldom is there an exact match. Table 32.2 summarizes four possible outcomes.

32.8.3 Operating and Capital Costs

In any scenario, the savings and income from the cogeneration system must be determined as well as the additional costs such as the additional fuel, and the maintenance. Steps for determining the savings and income of the cogeneration system include

1. Savings in electricity generated (avoided purchase) = kWh × 4/kWh.
2. Savings in demand charges (decreased demand) = kW × $/kW.
3. Savings in HRSG-produced steam.
4. Income from excess electrical power sales (if applicable).
5. Income from excess thermal energy sales (if applicable).

There will be additional costs for the cogeneration system. These include:

- Turbine/generator annual maintenance—use manufacturer's recommended fees, typically expressed in terms of ¢4/kWh of electricity produced. The range may be from 1.0 ¢/kWh for gas turbines to as much as 3.0 ¢/kWh for small engines.
- Fuel cost for the prime mover (fuel oil and/or natural gas) at cost per MBtu
- Maintenance cost for HRSG—use manufacturer's recommended fee schedule, typically around $1/klb of steam.
- Any electrical utility charges for standby power, maintenance energy or power, etc., expressed in $/kW or ¢/kWh

As an example of the savings and costs of a cogeneration system, see Table 32.4 in the following case study of the Austin State Hospital. This table provides detailed cost and savings data for that application.

Finally, the system or capital costs must be determined. A general "rule of thumb" for estimating these system costs is as follows:

Major equipment packages (combustion turbine and HRSG)	40%
Balance of plant equipment	25%
Engineering and construction	15%
Other (interest during construction, permitting, contingency, etc.)	20%
	100%

Any cogeneration analysis is very complicated, and the results will vary greatly depending upon interest rates, cost of fuel, permitting requirements, cost of electricity sales, and other factors.

32.8.4 Final Comments on Economic Evaluations

Economic evaluations for assessing the feasibility of cogeneration projects is an important aspect of any development. Although many of the considerations are similar to other engineering economic evaluations, several aspects are unique to cogeneration projects. Obtaining detailed electric and thermal energy data is a prerequisite to completing accurate economic assessments. After the data are available, the savings and income of a cogeneration project must be compared to the additional costs of operating the system. If the net savings will "pay back" the capital expense of the cogeneration system in a reasonable time, then the project may make economic sense. A variety of sizes and operating modes of the cogeneration system should be examined to determine the most economical scenario. Computer programs are available to help in this process. Other detailed considerations that are often necessary in a thorough economic evaluation include tax and financing arrangements, but these considerations are beyond the scope of this chapter.

32.9 Financial Aspects

32.9.1 Overall Considerations

Cogeneration facilities can be financed by a variety of options. The traditional approach to financing is owner financing; however, cogeneration facilities are expensive to construct, and a company may choose not to use its own capital. When deciding on the most favorable financial arrangement, companies often consider the following factors:

- Shortage of capital.
- Effect on credit rating (if borrowing the money to finance the plant).
- Impact on balance sheet.
- Desire to fix savings (minimize risk).
- Inability to utilize available tax benefits.
- Lack of interest in operating or owning the plant (not main line of business).

Financing is critical to the success of a cogeneration facility, and it is best to determine, early on, the financing method to be used. Various financial structures are described below.

There may be other, more creative ways, to finance projects, including combinations of these, but most financing will fall under one of the following:

1. Conventional ownership and operation.
2. Joint venture partnership.
3. Lease.
4. Third-party ownership.
5. Guaranteed savings.

32.9.2 Conventional Ownership and Operation (100% Ownership)

The heat load owner has two basic financing options in a conventional owner/operate structure: (1) to fund the project internally from profits in other areas of the business, or (2) to fund part of the project from internal sources and borrow the remainder from a conventional lending institution. Figure 32.13 shows one such financial structure.

Most businesses have a minimum internal rate of return on equity that they require for any investment. They may not be willing to fund any project that does not meet the internal hurdle rate using 100% equity (internal) financing. With 100% internal financing the company avoids the problems of arranging external financing (perhaps having to add partners). If there is a marginal return on equity, however, the company will not finance the project, especially if the money could be used to expand a product line or create a new product that could provide a greater return on equity.

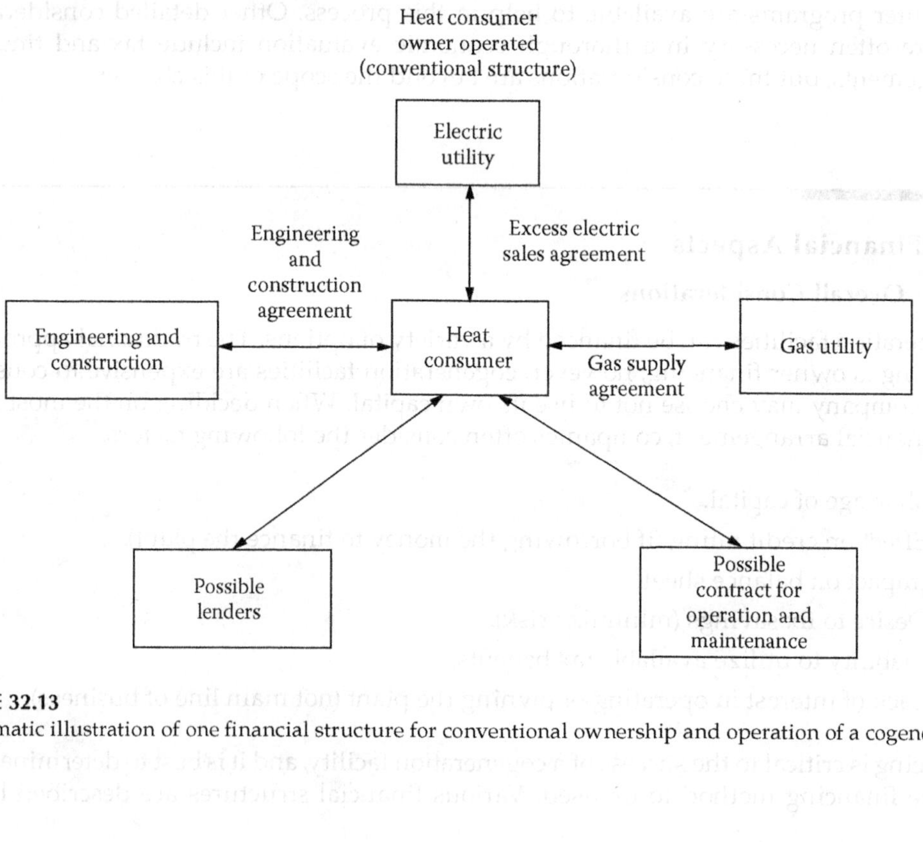

FIGURE 32.13
A schematic illustration of one financial structure for conventional ownership and operation of a cogeneration facility.

Because the cost of borrowed money is typically lower than a business's own return on equity requirements, the combination of partial funding internally and conventional borrowing is often used. By borrowing most of the funds, the internal funds can be leveraged for other projects, thus magnifying the overall return on equity.

External contract issues are simpler in conventional ownership. Contracts will be required for the gas supply, for excess power sales to the local utility, possible O&M agreements, plus any agreements with possible lenders, but no contracts may be needed for the steam and electricity if all is used internally.

32.9.3 Joint Venture

One alternative to 100% ownership is to share ownership. Figure 32.14 shows the outline of one possible arrangement of a joint venture project. PURPA regulations allow electric utilities to own up to a 50% share in a cogeneration facility. Profits shared with a utility in a partial ownership role are unregulated. Other partners might include a gas utility or a major equipment vendor, such as a major gas turbine producer.

The major advantages of a joint venture are the sharing of risks and credit. The disadvantages are that profits are also shared and that contract complexity increases.

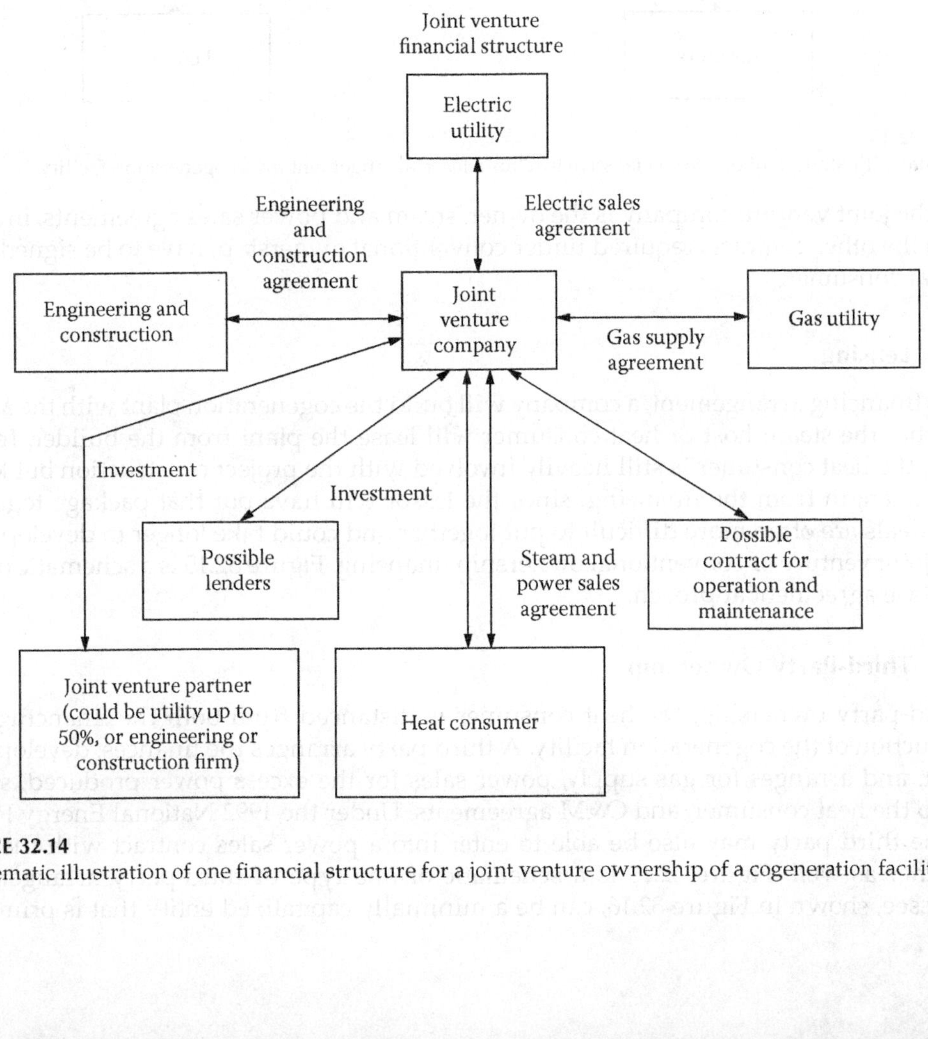

FIGURE 32.14
A schematic illustration of one financial structure for a joint venture ownership of a cogeneration facility.

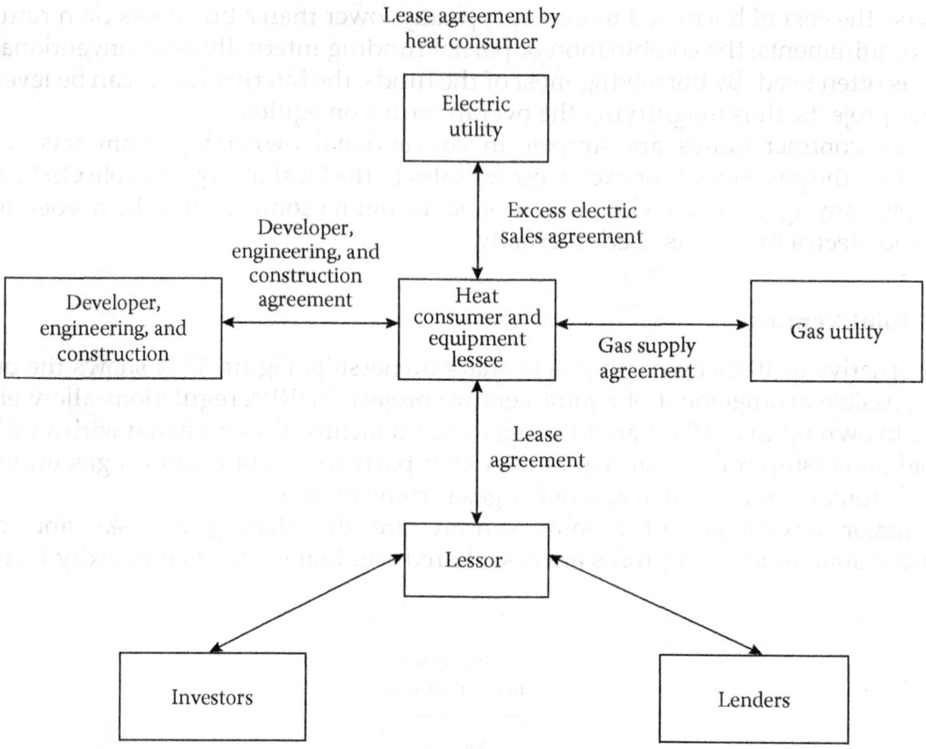

FIGURE 32.15
A schematic illustration of one financial structure for a lease arrangement for a cogeneration facility.

Since the joint venture company is the owner, steam and power sales agreements, in addition to the other contracts required under conventional ownership, have to be signed with the heat consumer.

32.9.4 Leasing

In this financing arrangement, a company will build the cogeneration plant with the agreement that the steam host or heat consumer will lease the plant from the builder. In this model, the heat consumer is still heavily involved with the project construction but keeps at arm's length from the financing, since the lessor will have put that package together. These deals are often more difficult to put together and could take longer to develop than either joint venture or conventional ownership financing. Figure 32.15 is a schematic of one such lease agreement approach.

32.9.5 Third-Party Ownership

In third-party ownership, the heat consumer is distanced from both the financing and construction of the cogeneration facility. A third party arranges the finances, develops the project, and arranges for gas supply, power sales for the excess power produced, steam sales to the heat consumer, and O&M agreements. Under the 1992 National Energy Policy Act, the third party may also be able to enter into a power sales contract with the heat consumer as well. Figure 32.16 is a schematic of one type of third-party arrangement. The lessee, shown in Figure 32.16, can be a minimally capitalized entity that is primarily

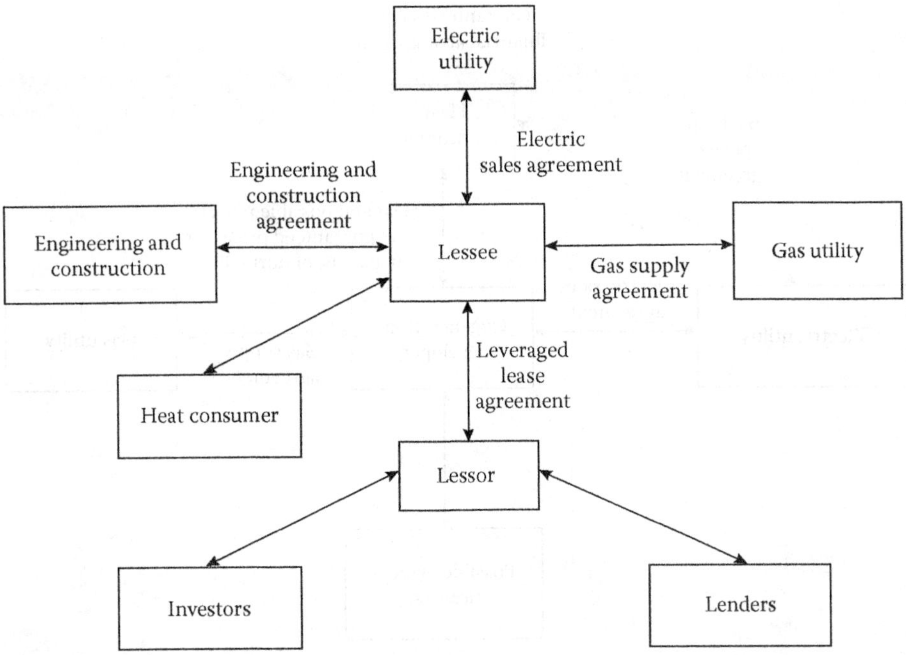

FIGURE 32.16
A schematic illustration of one financial structure for third-party ownership and operation of a cogeneration facility.

in the business of plant operation and handling energy sales once the plant is built. The contracts for firm energy sales are key to this type of financing operation. This type of financing option can be complex, requiring more development time, so development costs could be much higher.

32.9.6 Guaranteed Savings Contracts

These types of contracts are more common in smaller sized cogeneration systems, and typically may use packaged facilities. Figure 32.17 is a schematic of one possible guaranteed savings arrangement. A developer may pay all costs for construction of a cogeneration facility and then operate and maintain the system. Further, a guarantee of savings would be made to the owner over a specified length of time. Typically, a guaranteed savings contract will run from 5 to 10 years, with an average of 8 years, and will guarantee a fixed savings per year. The "split" between the guaranteed savings contractor usually ranges from a 75/25 division to a 95/5 division, with a "typical" split of 85/15; that is, 85% of the savings to the guaranteed savings contractor and 15% to the owner. Any additional savings above the guarantee are often split between the guaranteed savings contractor and the owner.

Guaranteed savings contracts have a distinct advantage to the owner in that all the capital and most of the risks are taken by the guaranteed savings contractor. The owner is guaranteed a percentage savings each year. These types of contracts should be considered when (1) the owner does not have capital to invest in a cogeneration system, or (2) when the owner does not have the necessary experience in-house to operate/manage such a project. If the capital is available, and if the in-house expertise exists, the company would be ahead financially to do the project itself, as the net return would be much higher.

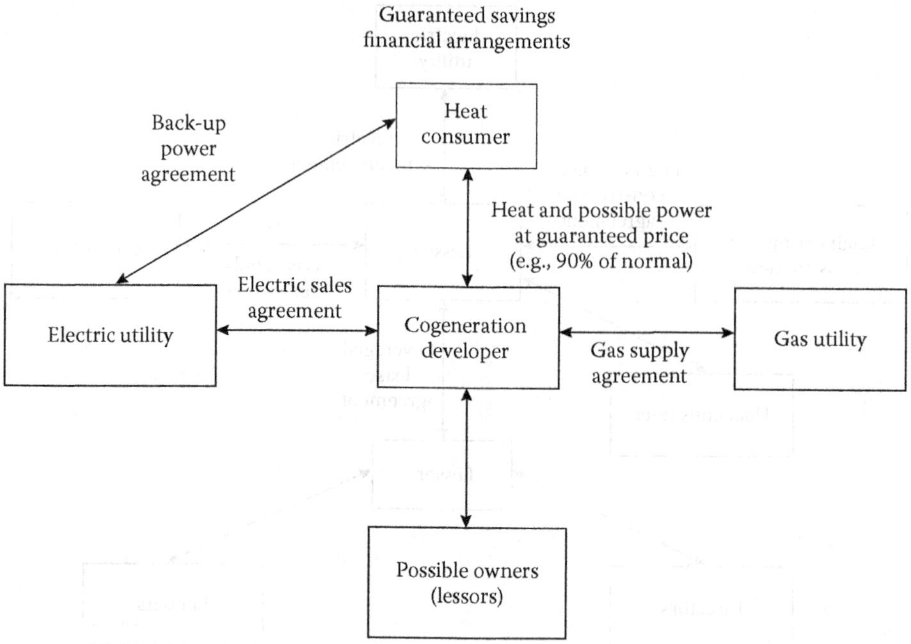

FIGURE 32.17
A schematic illustration of one financial structure for a guaranteed savings arrangement of a cogeneration facility.

32.9.7 Final Comments on Financial Aspects

Financing arrangements are a crucial aspect of most cogeneration developments. These arrangements may range from simple to highly complex. They are affected by internal factors such as ownership arrangements, credit ratings, and risk tolerance. In addition, these financial aspects are affected by external factors such as the financial and credit markets, tax laws, and cogeneration regulations. A variety of initial financial arrangements have been outlined in this section to illustrate the nature of these arrangements. Much more detailed arrangements are possible and often necessary, but these are beyond the scope of this chapter.

32.10 Case Studies

Case studies will be presented. One is a state hospital where a small gas turbine was installed, and the other is a utility application that generates power and sells steam to a local paper mill.

32.10.1 The Austin State Hospital Case Study

The Austin State Hospital was a facility operated by the Texas Mental Health and Mental Retardation (MHMR) state agency. It was one of 15 sites selected for a potential cogeneration site and was analyzed by Texas A&M's Energy Systems Laboratory (ESL) in 1985. A large laundry and cafeteria provided most of the thermal loads needed for the project, but there were also significant heat loads during the winter. In 1985, the ESL study recommended that gas turbines totaling 1.5 MW be installed, at an estimated cost of $1.5 million. An engineering study was then initiated in 1986–1987, and steam data was taken to determine the steam loads more accurately. The consulting engineering study recommended a 1 MW system at an estimated cost of $1.3 million. A $1.5 million grant application was submitted to and approved by Department of Energy (DOE). Specifications were prepared, the project was bid, and all bids came in too high. The project was then stalled for nearly 2 years until additional funding was secured and a long-term gas contract could be signed with the Texas General Land Office (GLO). The project was rebid in 1990, and a contractor was selected (Caton et al. 1991).

The 1 MW gas turbine, heat-recovery steam generator, and controls were completed in May 1992, at a cost of over $1.8 million, or slightly more than $1800/kW. The system supplies steam to the Austin State Hospital complex and provides base-load electricity for the facility. The cogeneration system was projected to save $225,000 per year, for a simple payback of 8 years. A long-term gas contract with the GLO assured the long-term economics of the system. The GLO established a maximum gas cost escalation rate of 5% per year for the duration of the payback period. Because DOE was providing a $1.5 million grant for the system, the benefit to the state of Texas was large. The state contribution of $300,000 would be paid back in less than 2 years. Table 32.3 summarizes 2 years of operation of the cogeneration system. The system was monitored by the ESL at Texas A&M University.

Shortly after the installation of the cogeneration system was completed (in May 1992), lightning struck the transformer and power line into the Austin State Hospital and burned

TABLE 32.3
Energy Usage for the Austin State Hospital Cogeneration System

Date	Purchased Electricity (kWh/month)	Cogenerated Electricity (kWh/month)	Total Electricity (kWh/month)	HRSG Heating (MMBtu/month)
August 1992	994,909	629,714	1,624,623	3851
September 1992	1,159,110	395,537	1,554,647	1182
October 1992	694,243	655,740	1,349,983	4030
November 1992	475,120	550,046	1,025,166	3449
December 1992	303,951	721,222	1,025,173	5168
January 1993	251,741	704,480	956,221	4862
February 1993	268,584	663,623	932,207	4755
March 1993	346,492	698,975	1,045,467	4910
April 1993	541,877	557,241	1,099,118	3947
May 1993	601,993	649,361	1,251,354	5066
June 1993	795,057	601,731	1,396,788	4923
July 1993	880,342	629,613	1,509,955	4913
August 1993	887,767	641,948	1,529,715	5221
September 1993	756,969	610,222	1,367,191	5016
October 1993	605,129	588,025	1,193,154	4812
November 1993	284,797	684,472	969,269	5077
December 1993	305,742	703,319	1,009,061	5535
January 1994	320,072	676,443	996,515	4708
February 1994	246,905	656,497	903,402	4643
March 1994	326,711	661,360	988,071	4841
April 1994	409,269	641,899	1,051,168	5009
May 1994	515,273	655,417	1,170,690	5437
June 1994	866,959	582,881	1,449,840	5220

TABLE 32.4

Summary of Savings Calculations for the Austin State Hospital (1992 Dollars)

Electrical energy cost		\$0.02609/kWh
Average demand cost		\$11.69/kW
Natural gas fuel cost (August 1992–June 1994)		\$3.19/MCF
	Energy usage (per year)	**Savings**
a. Electricity	14,599,766 kWh	\$198,209
b. Demand (avoided)	23,190 kW	\$141,183
c. Heating (HRSG)	106,477 Mbtu	\$229,565
d. Maintenance of conventional boiler at \$1.10/klb of steam (not required)		\$57,496
	Total savings	\$626,452
Energy costs for the cogeneration system	**Expenses (per year)**	
Natural gas for the gas turbine	\$332,394	
Standby charge	30,600	
Maintenance of the cogeneration system (\$0.004/kWh)	30,365	
Maintenance of HRSG (\$1.00/klb)	52,261	
Total costs	\$445,620	
Net savings	\$181,000/year	

out many of the controls of the cogeneration system. This caused several weeks of downtime during that summer. Additional start-up problems were attributed to the waste heat system. Overall, the cogeneration system performed exceptionally well, with a turbine/generator availability of 96.8% for a 2-year period (July 1992 through June 1994). This 96.8% availability includes downtime for scheduled maintenance. Excluding scheduled maintenance downtime, the turbine/generator availability was nearly 100%.

Problems with the waste-heat boiler reduced its availability to approximately 92% over the same period. This reduced the expected steam recovery below the assumed availability. This is not the reason, however, for the lower-than-projected savings. The city of Austin changed the type of electrical service and increased the demand charges in 1992. An extra standby charge added approximately \$30,000 annually to the bill. Without this charge, the cogeneration system would be saving nearly \$210,000 annually, almost exactly the annual amount predicted by the savings analysis. Table 32.4 is a summary based on actual metered data for the ASH cogeneration system for 23 months of operation (August 1992 through June 1994). The annual dollar savings are approximately \$181,000 per year, which gives a simple payback of approximately 10 years. Because the bulk of the money was provided by a Federal grant, the state portion has already been repaid. Therefore, the cogeneration project saved the state over \$180,000 annually based on 1992 electricity and gas prices. The Austin State Hospital cogeneration system was shut down after 10 years of operation. The gas turbine remained reliable throughout most of the period of operation. The waste-heat recovery system had many problems during its operating period. However, Tables 32.3 and 32.4 give actual data and measured performance of a small-scale system.

32.10.2 Klamath Cogeneration Project—An Industrial/Utility Case Study

The Klamath cogeneration project (KCP) is a 484 MW natural-gas fired, combined cycle plant that was completed in 2001. This plant is an example of community and state involvement that is prevalent in twenty-first century plants. State involvement came from the state of Oregon's Energy Facility Siting Council (EFSC), which approved the project based on reduced CO_2 emissions from the more efficient cogeneration system. Community involvement was also important because a nearby industry (wood processing facility) was to be a steam host for the cogeneration plant. The facility's boilers were to be shut down, thus reducing CO_2 emissions. Although there are other aspects of the overall project, the bottom line is a significant offset of CO_2 emissions, largely from the cogeneration project.

32.10.3 Project Overview

The original intent of the project was to generate 484 MW of electricity and supply 200,000 lb/h of steam to the wood processing plant. The Oregon EFSC approved the permitting based on shutting down the plant's boilers, which would guarantee an offset of 4,464,395 short tons of CO_2 over 30 years. Unfortunately, after the permitting was approved, the wood processing plant shut down one of their process lines, which reduced their steam needs to less than half on the original design.

Table 32.5 gives the steam supplied to the wood processing plant since 2001. Although the steam usage is increasing, the steam supplied to the plant is far short of the 200,000 lb/h required in the permitting. Thus, the Klamath cogeneration project will have to find another large steam host or purchase additional CO_2 offsets in order to meet the state of Oregon environmental requirements.

The city of Klamath Falls also had to guarantee the project's heat rate. If it fell short, then additional carbon offsets had to be provided to cover any shortfalls.

This case study is being provided to show the links between cogeneration projects and the environment. Even in a very strict environmental state such as Oregon, a highly efficient cogeneration project can be permitted and licensed to operate. It also points to a problem area discussed earlier, i.e., the need to have a thermal load for the excess steam. When the project was first initiated, there was a large steam host capable of accepting huge quantities of steam. The needs of the wood processing plant changed, however, requiring only one-third of the original steam, and this has created a problem for the cogeneration system. The economics could change dramatically if the success of the project hinged on selling 200,000 lb/h of steam, and the environmental impact by reducing the CO_2 offset could place severe environmental constraints on the city. At the end of 2004, the city was over three million tons of CO_2 short of achieving its offset goal.

TABLE 32.5

KCP Facility's Provision of Steam to Collins

Year	Months of Operation	Steam Sent to Collins (lb)	Hours of Operation (Approximately)	Flow Rate (lb/h)	CO_2 Offsets (Short Tons)
2001	July–December	181,288,984	3884	46,675	15,301
2002	January–December	533,655,766	8551	62,408	50,077
2003	January–December	623,281,664	8758	71,159	53,324
2004	January–December	623,759,744	8713	71,214	53,154

Source: TRC Global Management Solutions, Klamath Cogeneration Project, 2005, Annual report to the Energy Facility Siting Council, Revision 1.

32.11 Small-Scale Cogeneration Applications in Buildings

Large-scale cogeneration systems have been shown to be very attractive in increasing thermal efficiency and reducing fuel costs and emissions. This technology also offers opportunities for cogeneration systems to be applied to commercial and residential buildings. Recent development of equipment such as microturbines, small-sized reciprocating engines, Stirling engines, and fuel cells makes cogeneration suitable for building use. Building-oriented, small-scale CHP is called *BCHP*. BCHP uses the waste heat for heating, cooling, and humidity control of buildings. The overall thermal efficiency of small-scale CHP can be as much as 85%–90%.

32.11.1 Technology Status of Small-Scale CHP

Microscale CHP started in western Europe and Japan, and is gaining popularity in these countries, Driven by the potential market and advanced technology, the U.S. DOE and the U.S. EPA initiated the "CHP Challenge" in December 1998 to double the installed capacity of CHP systems in the U.S. by 2010. In 1998, the installed CHP capacity was 46 GW, and the goal set for 2010 is 92 GW. As of the end of December 2003, the installed capacity of CHP in the U.S. was approximately 71 GW. About 0.04% of this installed capacity is coming from systems with less than 100 kW of power output, and much less coming from systems with less than 10 kW of power output (Bernstein and Knoke 2004). However, with technology advances, the small-scale cogeneration systems are expected to experience greater growth in the near future.

Table 32.6 compares some of the important performance characteristics of the various types of small-scale systems.

Natural gas is the preferred fuel for the spark-ignition engines, but they can also run on propane or gasoline. Compression-ignition engines can operate on diesel fuel or heavy oil, or they can be set up in a dual-fuel configuration that burns primary natural gas with a small amount of diesel fuel. The engines used for micro-CHP units are normally designed as packaged units. The size of micro-CHP units, which use internal combustion engine as a prime mover, varies from 10 to 200 kW. The unit can be setup easily on-site. However, some drawbacks exist in the reciprocating-engine package units because they require regular maintenance, are noisy without proper noise abatement, and have potentially high emissions. They are not as attractive for residential applications.

Microturbines are a newly developed small-sized gas turbine with power generation from 25 to 250 kW. Like the lager gas turbines, the microturbine generator consists of a compressor, a combustion chamber, a one-stage turbine, and a generator. The rotating speed of the generator can be up to 10,000 rpm. The high-frequency electricity output is first rectified and then converted to 60 Hz. A rectifier and a transformer are invaluable devices in the generation set. Several companies are developing packaged microturbine CHP systems. The 30 and 60 kW units are already in the marketplace. The electricity efficiency of microturbines is about 25%–30%. The advantages of microturbine CHP are low noise and relatively low NO_x emission. The disadvantages are low electricity efficiency and high cost. Microturbine CHP are suitable for taking a base-load of electricity, heating, and cooling because of their inflexibility in handling load changes.

The Stirling engines are small-scale engines ranging from 2 to 50 kW that are targeting the future residential CHP needs. The Stirling engine uses the Stirling cycle and is a reciprocating engine. However, unlike the internal-combustion engine, the Stirling engine

TABLE 32.6

Technical Features of Small-Scale CHP Devices

	Reciprocating Engines	Microturbines	Stirling Engines	PEM Fuel Cells
Electrical power (kW)	10–200	25–250	2–50	2–200
Electrical efficiency, full load (%)	24–45	25–30	15–35	40
Electrical efficiency, half load (%)	23–40	20–25	35	40
Total efficiency (%)	75–85	75–85	75–85	75–85
Heat/electrical power ratio	0.9–2	1.6–2	1.4–3.3	0.9–1.1
Output temperature level (°C)	85–100	85–100	60–80	60–80
Fuel	Natural or biogas, diesel fuel oil	Natural or biogas, diesel, gasoline, alcohols	Natural or biogas, LPG, several liquid or solid fuels	Hydrogen, gases, including hydrogen, methanol
Interval between maintenance (h)	5,000–20,000	20,000–30,000	5000	N/A
Investment cost ($/kW)	800–1,500	900–1,500	1300–2000	2500–3500
Maintenance costs (¢/kW)	1.2–2.0	0.5–1.5	1.5–2.5	1.0–3.0

Source: Alanne, K. and Saari, A., *Renewable Sustain. Energy Rev.*, 8, 401, 2004.

is an external-combustion engine. The working gases inside the engine cylinder are typically helium or hydrogen and never leave the engine. The combustion takes place outside the cylinder. The piston is driven by compression or expansion of working gases due to the alternating heating and cooling of the cylinder by external heat sources. The engine converts the temperature difference into electricity. The Stirling engine can run on various fuels, both gas fuels and solid fuels, due to its external combustion. The advantages of Stirling engines are its quiet operation, little maintenance, and low NO_x emission. The disadvantages are relative lower electricity efficiency, typically about 25%–30%. When the Stirling engine is used as prime mover in micro-CHP systems, the total thermal efficiency is about the same level of other micro-CHP applications. Stirling-engine CHP packaged units are being developed. Several demonstration projects have shown the system efficiency can be over 90%.

A fuel cell is an electrochemical device that produces electricity by an oxidation-reduction reaction between a fuel, typically hydrogen, and atmospheric oxygen. A fuel cell consists of an anode, a cathode, and an electrolyte. The fuel (hydrogen) is fed to the anode and releases electrons on the anode that are conducted to the cathode via an external circuit. The electron current can be utilized to generate electricity. The hydrogen protons diffuse through the electrolyte material to the cathode. Oxygen or air passes through the cathode in the fuel cell, where oxygen is reacted with the hydrogen proton to form water and release heat. A typical fuel cell system consists of three parts: a fuel reformer section, power generation section, and power conditioning section. The heat rejected from the fuel reformer and power generation section can be recovered and used to drive absorption chillers and heat exchanges to supply hot water and chilled water for building air conditioning systems. The fuel cell technologies can be classified as polymer electrolyte membrane (PEM), phosphoric acid fuel cell (PAFC), solid oxide fuel cell (SOFC), and molten carbonate fuel cell (MCFC), according to their types of electrolyte. The PEM and PAFC are lower-temperature fuel cells, with operating temperatures of 200°F and 400°F, respectively. The SOFC and MCFC are high-temperature fuel cells with operating temperatures of 1750°F and 1200°F, respectively. The advantages of fuel cell technology are its high electrical efficiency, which can be as high as 45%–55%, and low emissions. Fuel cells are applicable for various purposes, based on the operating temperature levels of the cells. The major disadvantage of a fuel cell is the high price. In 2005, the 200 kW PAFC became available in the market.

32.11.2 Building Combined Heat and Power Systems

Commercial and residential buildings are responsible for about 37% of total U.S. primary energy consumption. Ninety-five percent of the commercial buildings are powered by electricity. Ninety-seven percent of these buildings use electricity for air conditioning. Approximately 73.9% of the electricity is generated from combustion of coal, petroleum, and natural gas in the U.S. Because many commercial buildings have a simultaneous need for electricity and thermal energy, there is a potential application for BCHP. The energy consumed by the BCHP system will be utilized at a higher efficiency than conventional means.

Modular CHP units or packaged systems have been developed in recent years to cater to the needs of buildings. Modular cogeneration systems are compact, and can be manufactured economically. These systems, ranging in size from 20 to 650 kW, produce electricity and hot water from engine waste heat. Figure 32.18 shows an IC-engine-driven BCHP system. The engine exhaust gases pass through the heat recovery steam generator, which generates superheated steam to drive the steam-driven absorption chiller and hot-water heat exchanger. The condensed water first passes through the engine to take away the engine heat, then passes through HRSG. This system makes use of engine exhaust heat

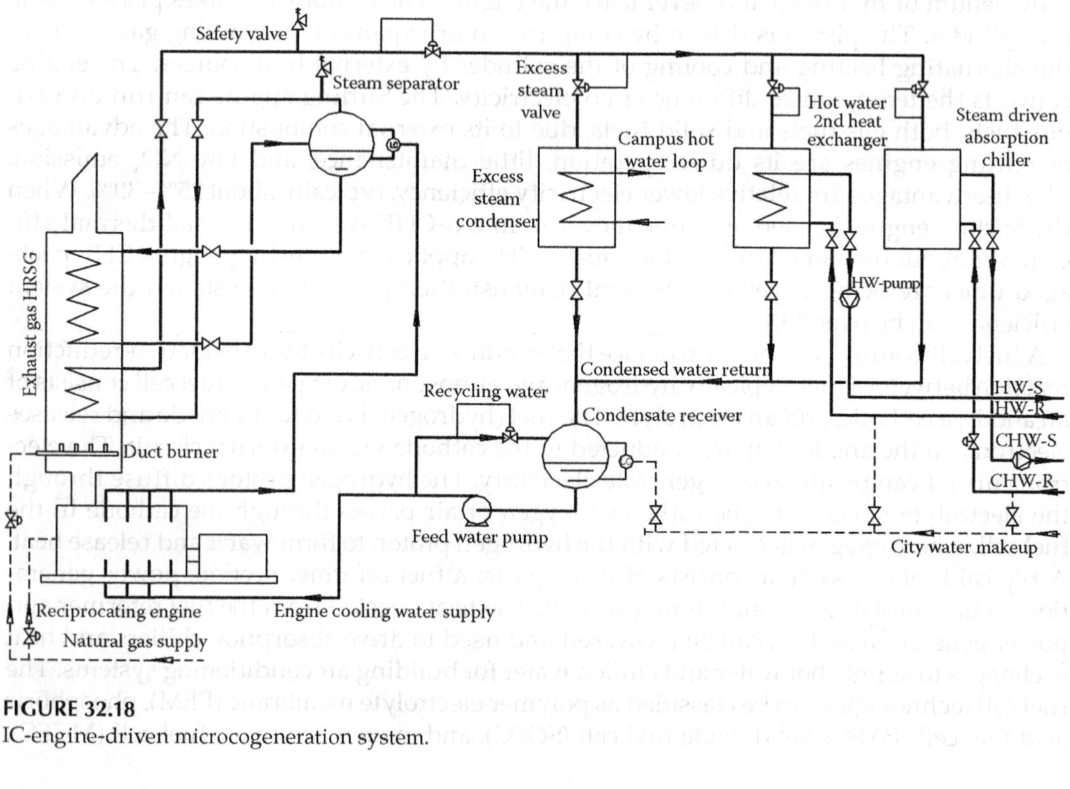

FIGURE 32.18
IC-engine-driven microcogeneration system.

and engine cooling heat to supply hot water and chilled water for the building air conditioning system. It is usually better to size the systems to meet the basic electricity load and use an auxiliary boiler or duct burner to adjust CHP thermal output. Therefore, this type of BCHP can run continuously. The best CHP applications for buildings are hospitals or restaurants that have a year-round need for hot water or steam.

Microscale CHP systems using different prime movers are also being developed by several companies to satisfy the electricity and thermal need of residential buildings. These micro-CHP package have a capacity of up to 16 kW, and are capable of providing most of the heating and electrical needs for a home. Currently, there are several micro-CHP demonstration units running, in which the prime movers vary as IC engine, microturbine, and fuel cell. The major barriers in commercializing these systems in the U.S. market are cost, reliability, and regulations.

32.11.3 Market Barriers and Drivers for Building CHP

Electricity deregulation has opened up the traditional power generation market for competition. On the scale of BCHP, deregulation takes the form of net metering in many states, which requires utilities to pay the customer retail price for the electricity exported to the grid. Deregulation, environmental concerns, and incentive programs of the government are the major market drivers for the installation of BCHP. However, several factors also affect the growth of installation of BCHP. These factors include the initial cost of buying and installing a micro-CHP system, maintenance costs, and environmental control requirements. Some electric utilities charge stand-by fees for the installation of micro-CHP so that the utilities may recover the costs of the stranded assets when customers turn to on-site generation. Customers using micro-CHP typically require a backup source of power to meet load requirements during outages or scheduled maintenance. The stand-by charge sometimes can kill a micro-CHP project.

Currently, the high cost is a major barrier preventing the installation of BCHP. High initial cost relative to the central station electric power generation can be a deterrent to the investment in micro-CHP. Although micro-CHP products normally have a higher efficiency compared to the traditional power generation, the cost of the electricity and heat generated by micro-CHP products highly depends on the cost of the micro-CHP products and the cost of fuel. Technology development will certainly reduce the price of the micro-CHP products in the future.

32.12 Future of Cogeneration

The future of cogeneration will include refined and better versions of the technology already described. In addition, several advanced cogeneration cycles are currently in various stages of development. These include (1) the Cheng cycle, (2) the Kalina cycle, and (3) coal gasification combined cycles.

32.12.1 Cheng Cycle

The Cheng cycle for optimizing the performance of gas-turbine cogeneration systems was first identified by Dr. Dan Yu Cheng in 1974. It uses the steam produced in the HRSG for reinjection into the combustion chamber to provide additional power from the gas turbine.

Because the gas turbine is a mass-flow device, combining the steam with the combustion gases will increase the total electrical power produced. This cycle has an advantage over the traditional combined cycle (Rankine cycle) because of its simplicity. There is no need for steam turbine condensers and pumps. Part of the steam produced in the HRSG can also be used for process steam. Following electrical loads or following process steam loads is possible using steam injection in the Cheng cycle. Substantial increases in both power output and generating efficiency are possible.

An added benefit from the steam injection is reduced NO_x emissions from the system. Strasser (1991) summarizes the results of several Cheng cycle installations in the U.S. and abroad, the oldest of which was an installation at San Jose State University in San Jose, California. That particular site had accumulated over 46,000 operating hours through November 1991.

32.12.2 Kalina Cycle

The Kalina cycle has been proposed by Dr. Alex Kalina as a means of improving the overall efficiency of combined-cycle plants. Its application to cogeneration systems would be to replace the conventional Rankine cycle power generation with a more efficient Kalina cycle. Dr. Kalina proposed the use of an ammonia-water mixture as the working fluid instead of water. The advantages include a 10%–20% increased thermal efficiency over a Rankine cycle with the same boundary conditions.

Ibrahim and Kovach (1993) describe a power generation application using the Kalina cycle, and Zervos et al. (1992) describe a 3 MW demonstration plant that uses the Kalina cycle. The complexity of the plant increases with the use of the NH_3–H_2O mixture. The conventional boiler is replaced by a vapor generator, the steam turbine by a vapor turbine, and the condenser by a distillation/condensation system. In the DOE demonstration project (Zervos et al. 1992), the liquid ammonia concentration is 70% by weight as it leaves the distillation/condensation system and enters the simulated waste-heat boiler. The superheated vapor then enters a specially designed vapor turbine, which drives a generator and produces electrical power. The Kalina cycle demonstration project operates between 2000 and 20 psia with ammonia. Mixture concentrations vary from 30% to 70%. The vapor turbine inlet conditions include a pressure of 1600 psia, a temperature of 960°F, an enthalpy of 1221 Btu/lb, an ammonia concentration of 70%, and a flow rate of 31,000 lb/h. The distillation/condensation system consists of a low-pressure condenser, vaporizer, preheater, and pump; an intermediate pressure shell-and-tube heat exchanger, falling film evaporator, flash drum, and pump; and a high-pressure condenser, preheater, pump, and solids removal system. Obviously, the management of the distillation/condensation system is crucial to the operation of a Kalina cycle system.

The vapor turbine has to withstand the high-temperature corrosion of the ammonia-water mixture and is made of type-316 stainless steel. Pressurized steam is injected into the labyrinth seals to contain the ammonia-water mixture and prevent its leakage during the vapor expansion phase in the turbine. This demonstration plant using the Kalina cycle began operation in December 1991.

32.12.3 Coal Gasification Combined Cycle Applications

The United States Clean Coal Technology (CCT) program was established in 1984 with the set-aside of $750 million for research and development of commercial processes to accelerate the use of clean coal technologies for power production. The Clean Air Act amendments of 1990 prescribed reductions in acid rain effects and emissions, aimed primarily at the electric utility industry. Coal gasification projects, as well as fluidized-bed combustion, have been the focus of the DOE CCT program (Hemenway et al. 1991).

The purpose of the coal gasification project is to produce a syngas that can be burned in a combustion turbine, hopefully at a price that is competitive with natural gas. The obstacles are coal cleanup, including desulfurization and particulate removal, development of components capable of operating at higher temperatures, disposal of solids and particulates from the process, high-temperature cleaning of the gas stream, and optimization of combustor design for the lower-temperature gas.

Sulfur removal is a must for these units to meet the standards of the Clean Air Act for electric power production. There are various approaches used to remove sulfur, including the addition of limestone or dolomite to the gasifier. Particulate cleanup is usually accomplished by a series of cyclone separators and particle filters. The tricky problem with the gas cleanup is that it should be accomplished in the heated gas stream that may at a temperature of 1000°F or more. Waste heat from the combustion turbine exhaust is routed through the HRSG for steam production. The steam may then be piped to a steam turbine to produce additional electrical power or used partly for process heating. If some stage in the gasification process requires a fixed temperature to be maintained, it is possible that additional steam could be produced in the gasifier that could also be used for process heating or for additional power production.

Ruth and Bedick (1992) summarize five major integrated gasification combined cycle (IGCC) applications sponsored by the DOE. All of the projects are scheduled for start-up/demonstration in the 1995–1996 calendar years. The five activities are (1) Combustion Engineering IGCC Repowering, (2) Tampa Electric IGCC, (3) Pinion Pine IGCC, (4) Toms Creek IGCC Demonstration, and (5) Wabash River Coal Gasification Repowering Project.

Of the five planned projects discussed by Ruth and Bedick, three have been built and are in operation. The Wabash (Indiana) plant is a repowering plant, another application of the IGCC technology—to replace aging coal-fired plants with more efficient and more environmentally friendly IGCC plants. The first "greenfield" (or new plant) is Tampa's 313 MW facility, dedicated in 1997, received many awards for its gas cleaning technology (DOE Fossil Energy website 2006). In February 1997 there were nine IGCC plants operating worldwide, including three in the United States (The third U.S. plant is outside Reno, Nevada, the Pinion Pine IGCC project.). By 2000 nearly 4 Gigawatts (GW) were in use worldwide (Coal 21 website). Most of the commercial IGCC plants are in the 250–300 MW capacity, currently restricted by the size of the pressurized gasifiers. Since the gasifiers are pressure vessels, they cannot be fabricated on site, and transport restrictions currently limit size (IEA Clean Coal Centre).

Other authors have examined the economics of obtaining and marketing the by-products produced by the gasification process, including methanol, acetic acid, nitric acid, and formaldehyde, as a means of enhancing revenues (Bahmann et al. 1992). Several scenarios were analyzed for cost effectiveness depending on the type of coproduct produced. The IGCC research at the U.S. DOE and throughout the world is focused on high temperature gas cleanup and improved turbines, both of which will improve IGCC efficiency. There is still a huge emphasis on improved methods to remove harmful pollutants that are emitted from the IGCC process. A good overview of the current IGCC technology is a summary by Maurstad (2005).

32.13 Summary and Conclusions

The importance of cogeneration is monetary, environmental, and energy savings. Any facility that uses electrical power and has thermal energy needs is a candidate for cogeneration. Facilities that may be considered for cogeneration include those in the industrial, commercial, and institutional sectors. The technology for conventional cogeneration systems is for the most part available and exists over a range of sizes: from less than 100 kW to over 100 MW. The major equipment requirements include a prime mover, electrical generator, electrical controls, heat recovery systems, and other typical power plant equipment. These components are well developed, and the procedures to integrate these components into cogeneration systems are well established.

In addition to the economic and technical considerations, the application of cogeneration systems involves an understanding of the governmental regulations and legislation on electrical power production and on environmental impacts. With respect to electrical power production, certain governmental regulations were passed during the late 1970s, again in the 1990s, and in 2005 that remove barriers and provide incentives to encourage cogeneration development. Finally, no cogeneration assessment would be complete without an understanding of the financial arrangements that are possible.

This chapter has provided a brief survey of the important aspects necessary to understand cogeneration. Specifically, this chapter has included information on the technical equipment and components, technical design issues, regulatory considerations, economic evaluations, financial aspects, and future cogeneration technologies. The chapter included two case studies to illustrate the range of technologies available for cogeneration systems, the types of considerations involved in selecting a cogeneration system, and the overall economic and environmental evaluations that may be completed.

Cogeneration will continue to experience progressive growth because of its inherent efficiency over conventional power generation and thermal energy production. As distributed generation becomes more widespread, we will see advances in smaller systems, sized for single buildings, including residential applications.

Acknowledgments

The author wishes to acknowledge the assistance of Gary Gong, PhD candidate at Texas A&M University, for his assistance with this chapter. Also, Dr. Jerald Caton of the Mechanical Engineering Department at Texas A&M University was a contributing author to the earlier chapter.

References

Alanne, K. and Saari, A. 2004. Sustainable small-scale CHP technologies for building: The basis for multiperspective decision-making. *Renewable and Sustainable Energy Reviews*, 8(5), 401–431.

Bahmann, P. D., Epstein, M., and Kern, E. E. 1992. Coal gasification-based integrated coproduction energy facilities. In: *1992 ASME Cogen-Turbo Conference, IGTI*, Vol. 7, pp. 69–74.

Bernstein, S. and Knoke, S. 2004. Micro-CHP: U.S. market potential and complex challenges. *Refocus*, 5(2), 36–39.

Caton, J. A., Muraya, N., and Turner, W. D. 1991. Engineering and economic evaluations of a cogeneration system for the Austin State Hospital. In: *Proceedings of the 26th Intersociety Energy Conversion Engineering Conference*, Boston, MA, Vol. 5, pp. 438–443.

Coal 21 website. http://www.coal21.com.au/IGCC.php.

DOE Fossil Energy website. 2006. http://www.fe.doe.gov/programs/powersystems/gasification.

FERC Regulations. 1978. Part 292—Regulations under sections 201 and 210 of the Public Utility Regulatory Policies Act of 1978 with regard to small power production and gogeneration.

Hemenway, A., Williams, W. A., and Huber, D. A. 1991. Effects of the Clean Air Act Amendments of 1990 on the commercialization of fluidized bed technology. In: *ASME Fluidized Bed Combustion*, pp. 219–224.

Ibrahim, M. and Kovach, R. M. 1993. A Kalina cycle application for power generation. *Energy*, 18(9), 961–969.

IEA Clean Coal Centre. http://www.iea-coal.org.uk.

Laderoute, C. D. 1980. The FERC's PURPA cogeneration rules: Economics, rate design, and policy aspects. In: *Annual UMM DNR Seventh Conference Energy Proceedings*, University of Missouri, Rolla, MO, pp. 114–122.

Maurstad, O. 2005. An overview of coal based integrated gasification combined cycle (IGCC) technology, MIT LFEE 2005-002 WP. Massachusetts Institute of Technology, Cambridge, MA, September 2005. http://lfee.mit.edu, click on publications, then click on working papers.

Paffenbarger, J. A. 1991. A GCC power plant with methanol storage for intermediate-load duty. *Journal of Engineering for Gas Turbines and Power*, 113, 151–157.

Ruth, L. K. and Bedick, R. C. 1992. Research and development efforts at the Department of Energy (DOE) supporting integrated gasification combined cycle (IGCC) demonstrations. In: *1992 ASME Cogen-Turbo Conference, IGTI*, Vol. 7, pp. 87–94.

Shipley, A. M. and Elliot, R. N. 2003. Phantom power: The status of fuel cell technology markets. In: *Industrial Energy Technology Conference Proceedings*, Houston, TX, May 13–16, pp. 119–127.

Spiewak, S. 1980. Regulation of cogeneration. In: *McGraw-Hill Conference on Industrial Power—Electrical Rates, Reliability, and Energy Management*, October 7–8.

Spiewak, S. A. 1987. *Cogeneration and Small Power Production Manual*. Fairmont Press, Atlanta, GA.

Strasser, A. 1991. The Cheng cycle cogeneration system: Technology and typical applications. In: *ASME Cogen-Turbo Conference, JGTI*, Vol. 6, pp. 419–428.

Swenson, A. 1998. *A Look at Commercial Buildings in 1995: Characteristics, Energy Consumption, and Energy Expenditures*. U.S. Department of Energy, Washington, DC.

The Energy Information Administration (EIA) at the U.S. Department of Energy (DOE). 2005. Building energy data. http://www.eia.doe.gov.

TRC Global Management Solutions. 2005. Klamath Cogeneration Project. 2005 Annual report to the energy facility siting Council. Revision 1.

Turbine-based cogeneration systems. In: *Planning Cogeneration Systems*, D. R. Limaye, ed., pp. 119–143. Fairmont Press, Atlanta, GA.

United States Combined Heat and Power Association. 2001. National CHP roadmap.

Wood, B. D. 1982. *Applications of Thermodynamics*, 2nd edn. Addison-Wesley, Reading, MA.

Zervos, N. G., Leibowitz, H. M., and Robinson, K. S. 1992. Startup and demonstration experience of the Kalina cycle demonstration plant. In: *ASME Cogen-Turbo Conference, IGTI*, Vol. 7, pp. 187–191.

33

*Energy Storage Technologies**

Jeffrey P. Chamberlain, Roel Hammerschlag, and Christopher P. Schaber

CONTENTS

* Updated by Jeffrey P. Chamberlain from Chapter 18.1 of *Handbook of Energy Efficiency and Renewable Energy* (F. Kreith and D.Y. Goswami, eds.), Boca Raton, FL: CRC Press, 2007.

33.1 Overview of Storage Technologies

The availability of affordable and reliable electrical energy storage technologies is crucial to the worldwide effort to transform our electricity and transportation systems and break society's century-long dependence on fossil fuels.

Concerns about energy security and climate change are driving demand for renewable energy generation and storage systems as an alternative to current technologies. While grid energy generation using wind and solar technologies is becoming more common, the lack of cost-effective, high-capacity storage systems has severely limited the use of these technologies. Similarly, shortcomings in storage systems have limited the shift in transportation fuel from petroleum to electricity.[1] Because the current state of knowledge does not allow us to overcome these limitations, revolutionary advances in science and engineering are needed.[2]

The electric grid is undergoing a transformation that requires electricity storage in order to realize its objectives of low-carbon, reliable operation. Tomorrow's grid uses a highly diverse generation mix with inflexible (i.e., unpredictable) renewable energy sources in both centralized and distributed deployments. While battery-based storage would avoid reliability issues, the cost of batteries (~$500/kWh for a Na–S battery) is five times that of other storage technologies.[3]

In parallel, widespread market penetration of electric vehicles (EVs) will require a five-fold cost reduction, from $500–$600/kWh to $100–$125/kWh. To achieve a 350-mile driving range, EV batteries also must operate at pack-level energy densities of 400 Wh/kg or more, rather than today's 80–100 Wh/kg.

Energy storage will play a more and more critical role in an efficient and renewable energy future, much more so than it does in today's fossil-based energy economy. There are two principal reasons that energy storage will grow in importance with increased development of renewable energy:

1. Many important renewable energy sources are intermittent, and generate when weather dictates, rather than when energy demand dictates.
2. Many transportation systems require energy to be carried with the vehicle.*

* This is almost always true for private transportation systems, and usually untrue for public transportation systems, which can rely on rails or overhead wires to transmit electric energy. However, some public transportation systems such as buses do not have fixed routes and also require portable energy storage.

Energy can be stored in many forms: as mechanical energy in rotating, compressed, or elevated substances, as thermal or electrical energy waiting to be released from chemical bonds, or as electrical charge ready to travel from positive to negative poles on demand.

Storage media that can take and release energy in the form of electricity have the most universal value, because electricity can efficiently be converted either to mechanical or heat energy, while other energy conversion processes are less efficient. Electricity is also the output of three of the most promising renewable energy technologies: wind turbines, solar thermal, and photovoltaics. Storing this electricity in a medium that naturally accepts electricity is favored, because converting the energy to another type usually has a substantial efficiency penalty.

Still, some applications can benefit from mechanical or thermal technologies. Examples are when the application already includes mechanical devices or heat engines that can take advantage of the compatible energy form, lower environmental impacts that are associated with mechanical and thermal technologies, or low cost resulting from simpler technologies or efficiencies of scale.

In this chapter, we group the technologies into five categories: direct electric, electrochemical, mechanical, direct thermal, and thermochemical. Table 33.1 is a summary of all of the technologies covered. Each is listed with indicators of appropriate applications, which are further explained in Section 33.3.

33.2 Principal Forms of Stored Energy

The storage media discussed in this chapter can accept and deliver energy in three fundamental forms: electrical, mechanical, and thermal. Electrical and mechanical energies are both considered high-quality energy, because they can be converted to either of the other two forms with fairly little energy loss (e.g., electricity can drive a motor with only about 5% energy loss, or a resistive heater with no energy loss).

The quality of thermal energy storage depends on its temperature. Usually, thermal energy is considered low quality, because it cannot be easily converted to the other two forms. The theoretical maximum quantity of useful work W_{max} (mechanical energy) extractable from a given quantity of heat Q is

$$W_{max} = \frac{(T_1 - T_2)}{T_1 \times Q},$$

where

T_1 is the absolute temperature of the heat

T_2 is the surrounding, ambient absolute temperature

Any energy storage facility must be carefully chosen to accept and produce a form of energy consistent with either the energy source or the final application. Storage technologies that accept and/or produce heat should as a rule only be used with heat energy sources or with heat applications. Mechanical and electric technologies are more versatile, but in most cases, electric technologies are favored over mechanical, because electricity is more easily transmitted, because there is a larger array of useful applications, and because the construction cost is typically lower.

TABLE 33.1
Overview of Energy Storage Technologies and Their Applications

	Utility Shaping	Power Quality	Distributed Grid	Automotive
Direct electric				
Ultracapacitors		✓		✓
SMES		✓		
Electrochemical				
Batteries				
Lead–acid	✓	✓	✓	
Lithium-ion	✓	✓	✓	✓
Nickel–cadmium	✓	✓		
Nickel metal hydride				✓
Zebra				✓
Sodium–sulfur	✓	✓		
Flow batteries				
Vanadium redox	✓			
Polysulfide bromide	✓			
Zinc bromide	✓			
Electrolytic hydrogen				✓
Mechanical				
Pumped hydro	✓			
Compressed air	✓			
Flywheels		✓		✓
Direct thermal				
Sensible heat				
Liquids			✓	
Solids			✓	
Latent heat				
Phase change	✓		✓	
Hydration–dehydration	✓			
Chemical reaction	✓		✓	
Thermochemical				
Biomass solids	✓		✓	
Ethanol	✓			✓
Biodiesel				✓
Syngas	✓			✓

33.3 Applications of Energy Storage

In Table 33.1, each technology is classified by its relevance in one to four different, principal applications.

Utility shaping is the use of very large capacity storage devices in order to answer electric demand, when a renewable resource is not producing sufficient generation. An example would be nighttime delivery of energy generated by a solar thermal plant during the prior day.

Power quality is the use of very responsive storage devices (capable of large changes in output over very short timescales) to smooth power delivery during switching events, short outages, or plant run-up. Power quality applications can be implemented at central generators, at switchgear locations, and at commercial and industrial customers' facilities. Uninterruptible power supplies (UPS) are an example of this category.

Distributed grid technologies enable energy generation and storage at customer locations, rather than at a central (utility) facility. The distributed grid is an important, enabling concept for photovoltaic technologies, which are effective at a small scale and can be installed on private homes and commercial buildings. When considered in the context of photovoltaics, the energy storage for the distributed grid is similar to the utility shaping application insofar that both are solutions to an intermittent, renewable resource, but distributed photovoltaic generation requires small capacities in the neighborhood of a few tens of MJ, while utility shaping requires capacities in the TJ range.* Renewable thermal resources (solar and geothermal) can also be implemented on a distributed scale, and require household-scale thermal storage tanks. For the purposes of this chapter, district heating systems are also considered a distributed technology.

Automotive applications include battery-electric vehicles (EVs), hybrid gasoline-electric vehicles (HEVs), plug-in hybrid electric vehicles (PHEVs), and other applications that require mobile batteries larger than those used in today's internal combustion engine cars. A deep penetration of automotive batteries also could become important in a distributed grid. Large fleets of EVs or PHEVs that are grid-connected when parked would help enable renewable technologies, fulfilling utility shaping and distributed grid functions as well as their basic automotive function.

Additional energy storage applications exist, most notably portable electronics and industrial applications. However, the four applications described here make up the principal components that will interact in a significant way with the global energy grid.

33.4 Specifying Energy Storage Devices

Every energy storage technology, regardless of category, can be roughly characterized by a fairly small number of parameters. Self-discharge time, unit size, and efficiency serve to differentiate the various categories. Within a category, finer selections of storage technology can be made by paying attention to cycle life, specific energy, specific power, energy density, and power density.

Self-discharge time is the time required for a fully charged, noninterconnected storage device to reach a certain depth of discharge (DOD). DOD is typically described as a percentage of the storage device's useful capacity, so that, for instance, 90% DOD means 10% of the device's energy capacity remains. The relationship between self-discharge time and DOD is rarely linear, so self-discharge times must be measured and compared at a uniform DOD. Acceptable self-discharge times vary greatly: from a few minutes for some power quality applications to years for devices designed to shape annual power production.

Unit size describes the intrinsic scale of the technology, and is the least well defined of the parameters listed here. If the unit size is small compared to the total required capacity

* Storage capacities in this chapter are given in units of MJ, GJ and TJ. 1 MJ = 0.28 kWh, 1 GJ = 280 kWh, and 1 TJ = 280 MWh.

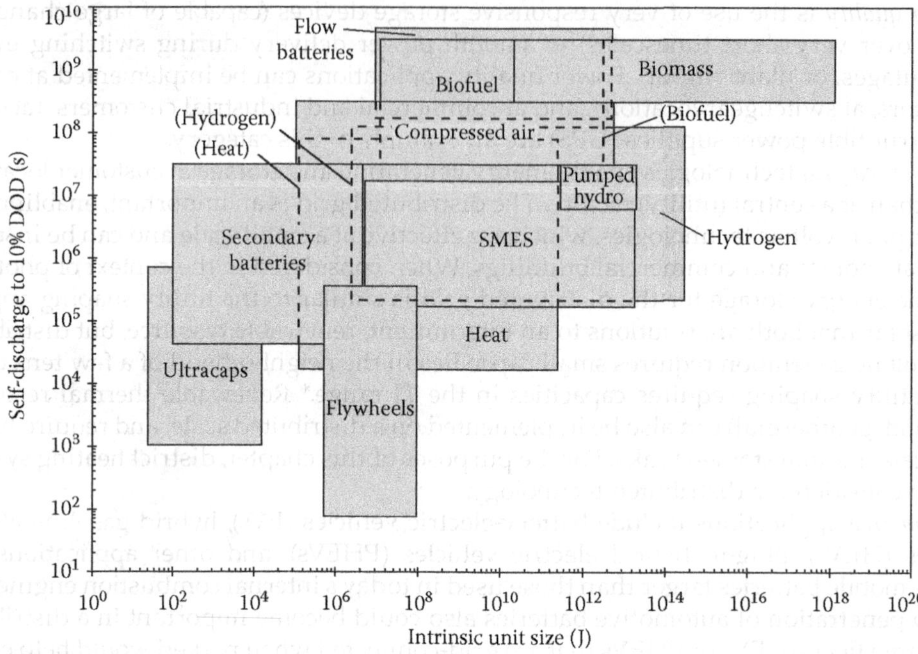

FIGURE 33.1
All storage technologies, mapped by self-discharge time and unit size. Not all hidden lines are shown. Larger self-discharge times are always more desirable, but more or less important depending on the application. Intrinsic unit size does not have a desirability proportional to its value, but rather must be matched to the application.

of a project, complexity and supply shortages can increase the cost relative to technologies with a larger unit size. Some technologies have a fairly large unit size that prohibits small-scale energy storage.

Figure 33.1 maps all of the technologies discussed in this chapter, according to their unit size and 10% self-discharge time. The gamut of technologies available covers many orders of magnitude on each axis, illustrating the broad choice available. Utility shaping applications require a moderate self-discharge time and a large unit size; power quality applications are much less sensitive to self-discharge time but require a moderate unit size. Distributed grid and automotive applications both require a moderate self-discharge time and a moderate unit size.

Efficiency is the ratio of energy output from the device to the energy input. Like energy density and specific energy, the system boundary must be carefully considered when measuring efficiency. It is particularly important to pay attention to the form of energy required at the input and output interconnections, and to include the entire system necessary to attach to those interconnections. For instance, if the system is to be used for shaping a constant-velocity, utility wind farm, then presumably, both the input and output will be AC electricity. When comparing a battery with a fuel cell in this scenario, it is necessary to include the efficiencies of an AC-to-DC rectifier for the battery, an AC-powered hydrogen generation system for the fuel cell system, and DC-to-AC converters associated with both systems.

Efficiency is related to self-discharge time. Technologies with a short self-discharge time will require constant charging in order to maintain a full charge; if discharge occurs much later than charge in a certain application, the apparent efficiency will be lower, because a significant amount of energy is lost in maintaining the initial, full charge.

Cycle life is the number of consecutive charge–discharge cycles a storage installation can undergo while maintaining the installation's other specifications within certain, limited ranges. Cycle life specifications are made against a chosen DOD depending on the application of the storage device. In some cases, for example, pressurized hydrogen storage in automobiles, each cycle will significantly discharge the hydrogen canister and the appropriate DOD reference might be 80% or 90%. In other cases, for example, a battery used in a HEV, most discharge cycles may consume only 10% or 20% of the energy stored in the battery. For most storage technologies, cycle life is significantly larger for shallow discharges than deep discharges, and it is critical that cycle life data be compared across a uniform DOD assumption.

Specific energy is a measure of how heavy the technology is. It is measured in units of energy per mass, and in this chapter, we will always report this quantity in MJ/kg. The higher the specific energy, the lighter the device. Automotive applications require high specific energies; for utility applications, specific energy is relatively unimportant, except where it impacts construction costs.

Energy density is a measure of how much space the technology occupies. It is measured in units of energy per volume, and in this chapter, we will always report this quantity in megajoule per liter (MJ/L). The higher the energy density, the smaller is the device. Again, this is most important for automotive applications, and rarely important in utility applications. Typical values for energy density associated with a few automotive-scale energy technologies are listed in Table 33.2, together with cycle life and efficiency data.

Energy density and specific energy estimates are dependent on the system definition. For example, it might be tempting to calculate the specific energy of a flow battery technology by dividing its capacity by the mass of the two electrolytes. But it is important to also include the mass of the electrolyte storage containers, and of the battery cell for a fair and comparable estimate of its specific energy. Thus, the energy density and specific energy are dependent on the size of the specific device; large devices benefit from efficiency of scale with a higher energy density and specific energy.

Specific power and *power density* are the power correlates to specific energy and energy density.

TABLE 33.2

Nominal Energy Density, Cycle Life, and Efficiency of Automotive Storage Technologies

	Energy Density (MJ/L)	Cycle Life at 80% DOD[a]	Electric Efficiency (%)
Ultracapacitors	0.2	50,000	95
Li-ion batteries	1.8	2,000	85
NiMH batteries	0.6	1,000	80
H_2 at 350 bar	3.0	n/a[b]	47
H_2 at 700 bar	5.0	n/a	45
Air at 300 bar	<0.1	n/a	37
Flywheels	<0.1	20,000	80
Ethanol	23.4	n/a	n/a

Note: Electric efficiencies are calculated for electric-to-electric conversion and momentary storage.

[a] Depth of discharge.

[b] Not applicable.

33.5 Specifying Fuels

A *fuel* is any (relatively) homogenous substance that can be combusted to produce heat. Though the energy contained in a fuel can always be extracted through combustion, other processes may be used to extract the energy (e.g., reaction in a fuel cell). A fuel may be gaseous, liquid, or solid. All energy storage technologies in the thermochemical category store energy in a fuel. In the electrochemical category, electrolytic hydrogen is a fuel.

A fuel's *lower heating value* (*LHV*) is the total quantity of sensible heat released during combustion of a designated quantify of fuel. For example, in the simplest combustion process, that of hydrogen,

$$2H_2 + O_2 \rightarrow 2H_2O(\text{vapor}) + \text{LHV}$$

or for the slightly more complex combustion of methane,

$$CH_4 + 2O_2 \rightarrow CO_2 + 2H_2O(\text{vapor}) + \text{LHV}.$$

In this chapter, the quantity of fuel is always expressed as a mass, so that LHV is a special case of specific energy. Like specific energy, LHV is expressed in units of MJ/kg in this chapter.

Higher heating value (*HHV*) is the LHV, plus the latent heat contained in the water vapor resulting from combustion.* For the examples of hydrogen and methane, this means

$$2H_2 + O_2 \rightarrow 2H_2O(\text{liquid}) + \text{HHV}$$

and

$$CH_4 + 2O_2 \rightarrow CO_2 + 2H_2O(\text{liquid}) + \text{HHV}$$

The latent heat in the water vapor can be substantial, especially for the hydrogen-rich fuels typical in renewable energy applications. Table 33.3 lists LHVs and HHVs of fuels

TABLE 33.3

Properties of Fuels

	Chemical Formula	Density (g/L)	LHV (MJ/kg)	HHV (MJ/kg)
Methanol	CH_3OH	794	19.9	22.7
Ethanol	C_2H_5OH	792	26.7	29.7
Methane	CH_4	0.68	49.5	54.8
Hydrogen	H_2	0.085	120	142
Dry syngas, airless process[a]	$40H_2 + 21CO + 10CH_4 + 29CO_2$	0.89	11.2	12.6
Dry syngas, air process[a]	$25H_2 + 16CO + 5CH_4 + 15CO_2 + 39N_2$	0.99	6.23	7.01

Source: All except syngas from U.S. Department of Energy, Properties of fuels, Alternative Fuels Data Center 2004.

[a] Chemical formulae and associated properties of syngas are representative; actual composition of syngas will vary widely according to manufacturing process.

* The concepts of sensible and latent heat are explained further in Section 33.9.

discussed in this chapter; in the most extreme case of molecular hydrogen, the HHV is some 18% higher than the LHV. Recovery of the latent heat requires controlled condensation of the water vapor; technologies for doing so are described in Chapter 13.

In this chapter, all heating values are reported as HHV rather than LHV. HHV is favored for two reasons. One, its values allow easier checking of energy calculations with the principle of energy conservation. Two, when examining technologies for future implementation, it is wise to keep an intention of developing methods for extracting as much of each energy source's value as possible.

33.6 Electrochemical Energy Storage

33.6.1 Secondary Batteries

A secondary battery allows electrical energy to be converted into chemical energy, stored, and converted back to electrical energy. Batteries are made up of three basic parts: a negative electrode, positive electrode, and an electrolyte (Figure 33.2). The negative electrode gives up electrons to an external load, and the positive electrode accepts electrons from the load. The electrolyte provides the pathway for charge to transfer between the two electrodes. Chemical reactions between each electrode and the electrolyte remove electrons from the positive electrode and deposit them on the negative electrode. This can be written as an overall chemical reaction that represents the states of charging and discharging of a battery. The speed at which this chemical reaction takes place is related to the *internal resistance* that dictates the maximum power at which the batteries can be charged and discharged.

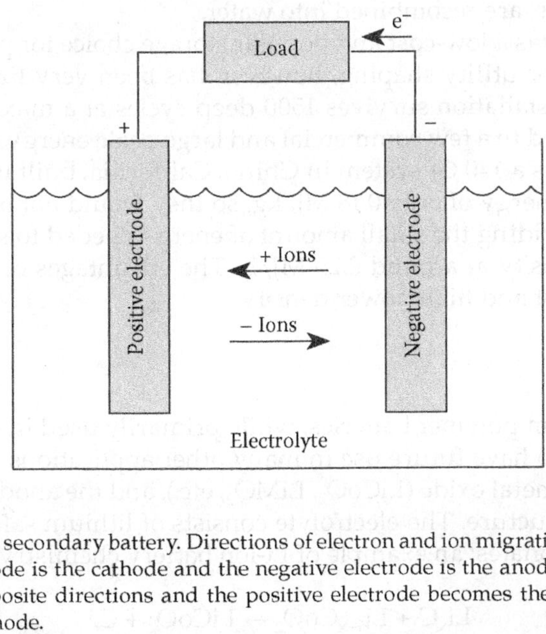

FIGURE 33.2
Schematic of a generalized secondary battery. Directions of electron and ion migration shown are for discharge, so that the positive electrode is the cathode and the negative electrode is the anode. During charge, electrons and ions move in the opposite directions and the positive electrode becomes the anode, while the negative electrode becomes the cathode.

Some batteries suffer from the *memory effect* in which a battery exhibits a lower discharge voltage under a given load than is expected. This gives the appearance of lowered capacity but is actually a voltage depression. Such a voltage depression occurs when a battery is repeatedly discharged to a partial depth and recharged again. This builds an increased internal resistance at this partial DOD, and the battery appears as a result to only be dischargeable to the partial depth. The problem, if and when it occurs, can be remedied by deep discharging the cell a few times. Most batteries considered for modern renewable applications are free from this effect, however.

33.6.1.1 Lead-Acid

Lead-acid is one of the oldest and most mature battery technologies. In its basic form, the lead-acid battery consists of a lead (Pb) negative electrode, a lead dioxide (PbO_2) positive electrode, and a separator to electrically isolate them. The electrolyte is dilute sulfuric acid (H_2SO_4), which provides the sulfate ions for the discharge reactions. The chemistry is represented by

$$PbO_2 + Pb + 2H_2SO_4 \leftrightarrow 2PbSO_4 + 2H_2O.$$

(In all battery chemistries listed in this chapter, left-to-right indicates battery discharge and right-to-left indicates charging.)

There are three main types of lead-acid batteries: the flooded cell, the sealed gel cell, and the sealed absorbed glass mat (AGM) lead-acid battery. The wet cell has a liquid electrolyte, which must be replaced occasionally to replenish the hydrogen and oxygen that escape during the charge cycle. The sealed gel cell has a silica component added to the electrolyte to stiffen it. The AGM design uses a fiberglass-like separator to hold electrolyte in close proximity to the electrodes, thereby increasing efficiency. For both the gel and AGM configurations, there is a greatly reduced risk of hydrogen explosion and corrosion from disuse. These two types do require a lower charging rate, however. Both the gel cells and the AGM batteries are sealed and pressurized so that oxygen and hydrogen, produced during the charge cycle, are recombined into water.

The lead-acid battery is a low-cost and popular storage choice for power quality applications. Its application for utility shaping, however, has been very limited due to its short cycle life. A typical installation survives 1500 deep cycles at a maximum.[4] Yet, lead-acid batteries have been used in a few commercial and large-scale energy management applications. The largest one is a 140 GJ system in Chino, California, built in 1988. Lead-acid batteries have a specific energy of only 0.18 MJ/kg, so they would not be a viable automobile option apart from providing the small amount of energy needed to start an engine. It also has a poor energy density at around 0.25 MJ/L. The advantages of the lead-acid battery technology are low cost and high power density.

33.6.1.2 Lithium Ion

Lithium ion and lithium polymer batteries, while primarily used in the portable electronics market, are likely to have future use in many other applications. The cathode in these batteries is a lithiated metal oxide ($LiCoO_2$, $LiMO_2$, etc.), and the anode is made of graphitic carbon with a layer structure. The electrolyte consists of lithium salts (such as $LiPF_6$) dissolved in organic carbonates; an example of Li-ion battery chemistry is

$$Li_xC + Li_{1-x}CoO_2 \leftrightarrow LiCoO_2 + C.$$

When the battery is charged, lithium atoms in the cathode become ions and migrate through the electrolyte toward the carbon anode where they combine with external electrons and are deposited between carbon layers as lithium atoms. This process is reversed during discharge.

The lithium polymer variation replaces the electrolyte with a plastic film that does not conduct electricity but allows ions to pass through it. The 60°C operating temperature requires a heater, reducing overall efficiency slightly.

Lithium ion batteries have a high energy density of about 0.72 MJ/L and have low internal resistance; so, they will achieve efficiencies in the 90% range and above. They have an energy density of around 0.72 MJ/kg. Their high energy efficiency and energy density make lithium-ion batteries excellent candidates for storage in all four applications we consider here: utility shaping, power quality, distributed generation, and automotive.

33.6.1.3 Nickel–Cadmium

Nickel–cadmium (NiCd) batteries operate according to the chemistry

$$2NiOOH + 2H_2O + Cd \leftrightarrow 2Ni(OH)_2 + Cd(OH)_2.$$

NiCd batteries are not common for large stationary applications. They have a specific energy of about 0.27 MJ/kg, an energy density of 0.41 MJ/L, and an efficiency of about 75%. Alaska's Golden Valley Electric Association commissioned a 40 MW/290 GJ nickel–cadmium battery in 2003 to improve reliability and to supply power for essentials during outages.[5] Resistance to cold and relatively low cost were among the deciding factors for choosing the NiCd chemistry.

Cadmium is a toxic heavy metal and there are concerns relating to the possible environmental hazards associated with the disposal of NiCd batteries. In November 2003, the European Commission adopted a proposal for a new battery directive that includes recycling targets of 75% for NiCd batteries. However, the possibility of a ban on rechargeable batteries made from nickel–cadmium still remains, and hence, the long-term viability and availability of NiCd batteries continues to be uncertain. NiCd batteries can also suffer from *memory effect*, where the batteries will only take full charge after a series of full discharges. Proper battery management procedures can help to mitigate this effect.

33.6.1.4 Nickel Metal Hydride

The nickel metal hydride (NiMH) battery operates according to the chemistry

$$MH + NiOOH \leftrightarrow M + Ni(OH)_2,$$

where M represents one of a large variety of metal alloys that serve to take up and release hydrogen. NiMH batteries were introduced as a higher energy density and more environmentally friendly version of the nickel–cadmium cell. Modern NiMH batteries offer up to 40% higher energy density than nickel–cadmium. There is potential for yet higher energy density, but other battery technologies (lithium ion in particular) may fill the same market sooner.

NiMH is less durable than nickel–cadmium. Cycling under heavy load and storage at high temperature reduces the service life. NiMH suffers from a higher self-discharge rate than the nickel–cadmium chemistry. NiMH batteries have a specific energy of 0.29 MJ/kg, an energy density of about 0.54 MJ/L, and an energy efficiency about 70%. These batteries have been an important bridging technology in the portable electronics and hybrid automobile markets. Their future is uncertain, because other battery chemistries promise higher energy storage potential and cycle life.

33.6.1.5 Sodium–Sulfur

A sodium–sulfur (NaS) battery consists of a liquid (molten) sulfur positive electrode and liquid (molten) sodium negative electrode, separated by a solid beta-alumina ceramic electrolyte (Figure 33.3). The chemistry is as follows:

$$2Na + xS \leftrightarrow Na_2S_x.$$

When discharging, positive sodium ions pass through the electrolyte and combine with the sulfur to form sodium polysulfides. x in the equation is 5 during early discharging, but once free sulfur has been exhausted, a more sodium-rich mixture of polysulfides with lower average values of x develops. This process is reversible as charging causes sodium polysulfides in the positive electrode to release sodium ions that migrate back through the electrolyte and recombine as elemental sodium. The battery operates at about 300°C. NaS batteries have a high energy density of around 0.65 MJ/L and a specific energy of up to 0.86 MJ/kg. These numbers would indicate an application in the automotive sector, but warm-up time and heat-related accident risk make its use there unlikely. The efficiency of this battery chemistry can be as high as 90% and would be suitable for bulk storage applications while simultaneously allowing effective power smoothing operations.[6]

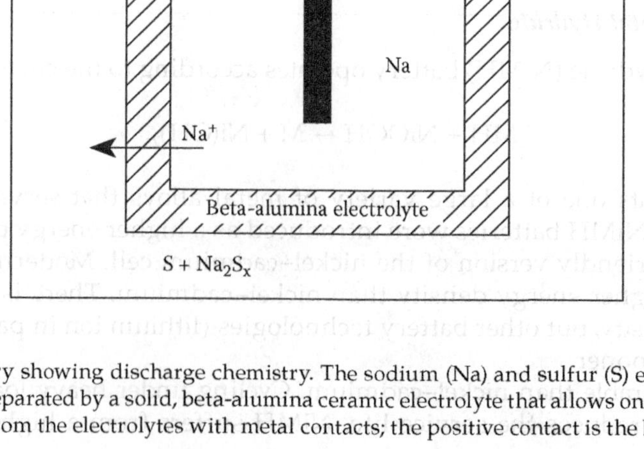

FIGURE 33.3
Sodium–sulfur battery showing discharge chemistry. The sodium (Na) and sulfur (S) electrodes are both in a liquid state and are separated by a solid, beta-alumina ceramic electrolyte that allows only sodium ions to pass. Charge is extracted from the electrolytes with metal contacts; the positive contact is the battery wall.

33.6.1.6 Zebra

Zebra is the popular name for the sodium nickel chloride battery chemistry:

$$NiCl_2 + 2Na \leftrightarrow Ni + 2NaCl.$$

Zebra batteries are configured similarly to sodium–sulfur batteries (see Figure 33.3), and also operate at about 300°C. Zebra batteries boast a greater than 90% energy efficiency, a specific energy of up to 0.32 MJ/kg, and an energy density of 0.49 MJ/L.[7] Its tolerance for a wide range of operating temperature and high efficiency, coupled with a good energy density and specific energy, make its most probable application the automobile sector, and as of 2003, Switzerland's MES-DEA is pursuing this application aggressively.[8] Its high energy efficiency also makes it a good candidate for the utility sector.

33.6.2 Flow Batteries

Most secondary batteries use electrodes both as an interface for gathering or depositing electrons, and as a storage site for the products or reactants associated with the battery's chemistry. Consequently, both energy and power density are tied to the size and shape of the electrodes. Flow batteries store and release electrical energy by means of reversible electrochemical reactions in two liquid electrolytes. An electrochemical cell has two compartments, one for each electrolyte, physically separated by an ion-exchange membrane. Electrolytes flow into and out of the cell through separate manifolds and undergo chemical reaction inside the cell, with ion or proton exchange through the membrane and electron exchange through the external electric circuit. The chemical energy in the electrolytes is turned into electrical energy and vice versa for charging. They all work in the same general way but vary in chemistry of electrolytes.[9]

There are some advantages to using the flow battery over a conventional secondary battery. The capacity of the system is scaleable by simply increasing the amount of solution. This leads to cheaper installation costs, as the systems get larger. The battery can be fully discharged with no ill effects and has little loss of electrolyte over time. Because the electrolytes are stored separately and in large containers (with a low surface area to volume ratio), flow batteries show promise to have some of the lowest self-discharge rates of any energy storage technology available.

Poor energy densities and specific energies remand these battery types to utility-scale power shaping and smoothing, though they might be adaptable for distributed generation use. There are three types of flow batteries that are closing in on commercialization: vanadium redox, polysulfide bromide, and zinc bromide. There is a fourth type of flow battery in early stages of R&D that may enable a significant increase in energy density over aqueous systems by enabling operation in a wider voltage window than allowed by water: nonaqueous flow batteries.

33.6.2.1 Vanadium Redox

The vanadium redox flow battery (VRB) was pioneered at the University of New South Wales, Australia, and has shown potentials for long cycle life and energy efficiencies of over 80% in large installations.[10] The VRB uses compounds of the element vanadium in both electrolyte tanks. The reaction chemistry at the positive electrode is

$$V^{5+} + e^- \leftrightarrow V^{4+}$$

and at the negative electrode is

$$V^{2+} \leftrightarrow V^{3+} + e^{-}.$$

Using vanadium compounds on both sides of the ion-exchange membrane eliminates the possible problem of cross-contamination of the electrolytes and makes recycling easier.[11] As of 2005, two small, utility-scale VRB installations are operating: one 2.9 GJ unit on King Island, Australia, and one 7.2 GJ unit in Castle Valley, Utah.

33.6.2.2 Polysulfide Bromide

The polysulfide bromide battery (PSB) utilizes two salt solution electrolytes, sodium bromide (NaBr) and sodium polysulfide (Na_2S_x). PSB electrolytes are separated in the battery cell by a polymer membrane that only passes positive sodium ions. Chemistry at the positive electrode is

$$NaBr_3 + 2Na^{+} + 2e \leftrightarrow 3NaBr$$

and at the negative electrode is

$$2Na_2S_2 \leftrightarrow Na_2S_4 + 2Na^{+} + 2e.$$

The PSB battery is being developed by Canada's VRB Power Systems, Inc.[12] This technology is expected to attain energy efficiencies of approximately 75%.[13] Though the salt solutions themselves are only mildly toxic, a catastrophic failure by one of the tanks could release highly toxic bromine gas. Nevertheless, the Tennessee Valley Authority released a finding of no significant impact for a proposed, 430 GJ facility and deemed it safe.[14]

33.6.2.3 Zinc Bromide

In each cell of a zinc bromide (ZnBr) battery, two different electrolytes flow past carbon-plastic composite electrodes in two compartments separated by a microporous membrane. Chemistry at the positive electrode follows the equation:

$$Br_2(aq) + 2e^{-} \leftrightarrow 2Br$$

and at the negative electrode:

$$Zn \leftrightarrow Zn^{2+} + 2e^{-}.$$

During discharge, Zn and Br combine into zinc bromide. During charge, metallic zinc is deposited as a thin film on the negative electrode. Meanwhile, bromine evolves as a dilute solution on the other side of the membrane, reacting with other agents to make thick bromine oil that sinks down to the bottom of the electrolytic tank. During discharge, a pump mixes the bromine oil with the rest of the electrolyte. The zinc bromide battery has an energy efficiency of nearly 80%.[15]

Exxon developed the ZnBr battery in the early 1970s. Over the years, many GJ-scale ZnBr batteries have been built and tested. Meidisha demonstrated a 1 MW/14 GJ ZnBr battery in 1991 at Kyushu Electric Power Company. Some GJ-scale units are now available preassembled, complete with plumbing and power electronics.

33.6.2.4 Nonaqueous Redox Flow

Current flow battery technologies depend on aqueous electrodes, a choice that limits energy densities to <40 Wh/L as a consequence of low solubilities of redox species and operating voltages that are bounded by water electrolysis.[16] Employing nonaqueous solvents offers a wider window of electrochemical stability that enables cell operation at dramatically higher potentials. Higher cell voltage (>3 V) leads to higher energy density, and typically to higher roundtrip efficiency as well, which together reduce the total cost of energy. Compared to aqueous flow systems, fewer stack layers, lower flow velocities, smaller tanks, and fewer ancillaries are required, which significantly reduces hardware costs and enhances system reliability. In order to capitalize on the inherent benefits of nonaqueous redox flow, several fundamental science challenges must be overcome.

Nonaqueous flow batteries fall into two broad classes of early investigation: solution phase systems, in which electroactive materials are dissolved in the electrolyte,[17,18] and suspensions of active intercalant hosts, which are essentially fluidized versions of standard solid-state electrodes found in current lithium ion batteries.[19,20]

33.6.3 Electrolytic Hydrogen

Diatomic, gaseous hydrogen (H_2) can be manufactured with the process of electrolysis; an electric current applied to water separates it into components O_2 and H_2. The oxygen has no inherent energy value, but the HHV of the resulting hydrogen can contain up to 90% of the applied electric energy, depending on the technology.[21] This hydrogen can then be stored, and later combusted to provide heat or work, or power a fuel cell (see Chapter 27).

The gaseous hydrogen is low density and must be compressed to provide useful storage. Compression to a storage pressure of 350 bar, the value usually assumed for automotive technologies, consumes up to 12% of the hydrogen's HHV if performed adiabatically, though the loss approaches a lower limit of 5% as the compression approaches an isothermal ideal.[22] Alternatively, the hydrogen can be stored in liquid form, a process that costs about 40% of HHV using current technology, and that at best would consume about 25%. Liquid storage is not possible for automotive applications, because mandatory boil-off from the storage container cannot be safely released in closed spaces (i.e., garages).

Hydrogen can also be bonded into metal hydrides using an absorption process. The energy penalty of storage may be lower for this process, which requires pressurization to only 30 bar. However, the density of the metal hydride can be between 20 and 100 times the density of the hydrogen stored. Carbon nanotubes have also received attention as a potential hydrogen storage medium.[23] Hydrogen storage technologies are covered in more detail in Chapter 53.

33.7 Direct Electric Storage

33.7.1 Ultracapacitors

A capacitor stores energy in the electric field between two oppositely charged conductors. Typically, thin conducting plates are rolled or stacked into a compact configuration with a dielectric between them. The dielectric prevents arcing between the plates and allows the plates to hold more charge, increasing the maximum energy storage.

The ultracapacitor—also known as supercapacitor, electrochemical capacitor, or electric double layer capacitor (EDLC)—differs from a traditional capacitor in that it employs a thin electrolyte, in the order of only a few angstroms, instead of a dielectric. This increases the energy density of the device. The electrolyte can be made of either an organic or an aqueous material. The aqueous design operates over a larger temperature range, but has a smaller energy density than the organic design. The electrodes are made of a porous carbon that increases the surface area of the electrodes and further increases energy density over a traditional capacitor.

Ultracapacitors' ability to effectively equalize voltage variations with quick discharges makes them useful for power quality management and for regulating voltage in automotive systems during regular driving conditions. Ultracapacitors can also work in tandem with batteries and fuel cells to relieve peak power needs (e.g., hard acceleration) for which batteries and fuel cells are not ideal. This could help extend the overall life and reduce lifetime cost of the batteries and fuel cells used in HEV and EV. This storage technology also has the advantage of very high cycle life of greater than 500,000 cycles and a 10–12 year lifespan.[24] The limitations lie in the inability of ultracapacitors to maintain charge voltage over any significant time, losing up to 10% of their charge per day.

33.7.2 SMES

A superconducting magnetic energy storage (SMES) system is well suited in storing and discharging energy at high rates (high power.) It stores energy in the magnetic field created by direct current in a coil of cryogenically cooled, superconducting material. If the coil were wound using a conventional wire such as copper, the magnetic energy would be dissipated as heat due to the wire's resistance to the flow of current. The advantage of a cryogenically cooled, superconducting material is that it reduces electrical resistance to almost zero. The SMES recharges quickly and can repeat the charge–discharge sequence thousands of times without any degradation of the magnet. A SMES system can achieve full power within 100 ms.[25] Theoretically, a coil of around 150–500 m radius would be able to support a load of 18,000 GJ at 1000 MW, depending on the peak field and ratio of the coil's height and diameter.[26] Recharge time can be accelerated to meet specific requirements, depending on system capacity.

Because no conversion of energy to other forms is involved (e.g., mechanical or chemical), the energy is stored directly and round-trip efficiency can be very high.[5] SMES systems can store energy with a loss of only 0.1%; this loss is due principally to energy required by the cooling system.[6] Mature, commercialized SMES is likely to operate at 97%–98% round-trip efficiency and is an excellent technology for providing reactive power on demand.

33.8 Mechanical Energy Storage

33.8.1 Pumped Hydro

Pumped hydro is the oldest and largest of all of the commercially available energy storage technologies, with existing facilities up to 1000 MW in size. Conventional pumped hydro uses two water reservoirs, separated vertically. Energy is stored by moving water from the lower to the higher reservoir, and extracted by allowing the water to flow back to the lower reservoir. Energy is stored according to the fundamental physical principle of potential energy. To calculate the stored energy in joules, use the formula:

$$E_s = Vdgh,$$

where
- V is the volume of water raised (m^3)
- d is the density of water (1000 kg/m^3)
- g is the acceleration of gravity (9.8 m/s^2)
- h is the elevation difference between the reservoirs (m) and is often referred to as the *head*

Though pumped hydro is by nature a mechanical energy storage technology, it is most commonly used for electric utility shaping. During off-peak hours, electric pumps move water from the lower reservoir to the upper reservoir. When required, the water flow is reversed to generate electricity. Some high dam hydro plants have a storage capability and can be dispatched as pumped hydro storage. Underground pumped storage, using flooded mine shafts or other cavities, is also technically possible but probably prohibitively expensive. The open sea can also be used as the lower reservoir if a suitable upper reservoir can be built at close proximity. A 30 MW seawater pumped hydro plant was first built in Yanbaru, Japan, in 1999.

Pumped hydro is most practical at a large scale with discharge times ranging from several hours to a few days. There is over 90 GW of pumped storage in operation worldwide, which is about 3% of global electric generation capacity.[27] Pumped storage plants are characterized by long construction times and high capital expenditure. Its main application is for utility shaping. Pumped hydro storage has the limitation of needing to be a very large capacity to be cost effective, but can also be used as storage for a number of different generation sites.

Efficiency of these plants has greatly increased in the last 40 years. Pumped storage in the 1960s had efficiencies of 60% compared with 80% for new facilities. Innovations in variable speed motors have helped these plants to operate at partial capacity, and greatly reduced equipment vibrations, increasing plant life.

33.8.2 Compressed Air

A relatively new energy storage concept that is implemented with otherwise mature technologies is compressed air energy storage (CAES). CAES facilities must be coupled with a combustion turbine, so they are actually a hybrid storage/generation technology.

A conventional gas turbine consists of three basic components: a compressor, combustion chamber, and an expander. Power is generated when compressed air and fuel burned in the combustion chamber drive turbine blades in the expander. Approximately 60% of the mechanical power generated by the expander is consumed by the compressor supplying air to the combustion chamber.

A CAES facility performs the work of the compressor separately, stores the compressed air, and, at a later time, injects it into a simplified combustion turbine. The simplified turbine includes only the combustion chamber and the expansion turbine. Such a simplified turbine produces far more energy than a conventional turbine from the same fuel, because there is potential energy stored in the compressed air. The fraction of output energy beyond what would have been produced in a conventional turbine is attributable to the energy stored in compression.

The net efficiency of storage for a CAES plant is limited by the heat energy loss occurring at compression. The overall efficiency of energy storage is about 75%.[28]

CAES compressors operate on grid electricity during off-peak times, and use the expansion turbine to supply peak electricity when needed. CAES facilities cannot operate without combustion, because the exhaust air would exit at extremely low temperatures, causing trouble with brittle materials and icing. If 100% renewable energy generation is sought, biofuel could be used to fuel the gas turbines. There might still be other emissions issues, but the system could be fully carbon neutral.

The compressed air is stored in appropriate underground mines, caverns created inside salt rocks or possibly in aquifers. The first commercial CAES facility was a 290 MW unit built in Hundorf, Germany, in 1978. The second commercial installation was a 110 MW unit built in McIntosh, Alabama, in 1991. The third commercial CAES is a 2700 MW plant under construction in Norton, Ohio. This nine-unit plant will compress air to about 100 bar in an existing limestone mine 2200 ft (766 m) underground.[29] The natural synergy with geological caverns and turbine prime movers dictates that these be on the utility scale.

33.8.3 Flywheels

Most modern flywheel energy storage systems consist of a massive rotating cylinder (comprised of a rim attached to a shaft) that is supported on a stator by magnetically levitated bearings that eliminate bearing wear and increase system life. To maintain efficiency, the flywheel system is operated in a low vacuum environment to reduce drag. The flywheel is connected to a motor/generator mounted onto the stator that, through some power electronics, interacts with the utility grid.

The energy stored in a rotating flywheel, in joules, is given by

$$E = \tfrac{1}{2} I\omega^2,$$

where

I is the flywheel's moment of inertia (kg-m^2)

ω is its angular velocity (1/s^2)

I is proportional to the flywheel's mass, so energy is proportional to mass and the square of speed. In order to maximize energy capacity, flywheel designers gravitate toward increasing the flywheel's maximum speed rather than increasing its moment of inertia. This approach also produces flywheels with the higher specific energy.

Some of the key features of flywheels are low maintenance, a cycle life of better than 10,000 cycles, a 20-year lifetime, and environmentally friendly materials. Low speed, high mass flywheels (relying on I for energy storage) are typically made from steel, aluminum, or titanium; high speed, low mass flywheels (relying on ω for energy storage) are constructed from composites such as carbon fiber.

Flywheels can serve as a short-term ride-through before long-term storage comes online. Their low energy density and specific energy limit them to voltage regulation and UPS capabilities. Flywheels can have energy efficiencies in the upper 90% range, depending on frictional losses.

33.9 Direct Thermal Storage

Direct thermal technologies, though they are storing a lower grade of energy (heat, rather than electrical or mechanical energy), can be useful for storing energy from systems that provide heat as a native output (e.g., solar thermal, geothermal), or for applications where the energy's commodity value is heat (e.g., space heating, drying).

While thermal storage technologies can be characterized by specific energy and energy density like any other storage technology, they can also be characterized by an important, additional parameter, the delivery temperature range. Different end uses have more or less allowance for wide swings of the delivery temperature. Also, some applications require a high operating temperature that only some thermal storage media are capable of storing.

Thermal storage can be classified into two fundamental categories: sensible heat storage and latent heat storage. Applications that have less tolerance for temperature swings should utilize a latent heat technology.

Input to and output from heat energy storage is accomplished with heat exchangers. The following discussion focuses on the choice of heat storage materials; the methods of heat exchange will vary widely depending on properties of the storage material, especially its thermal conductivity. Materials with higher thermal conductivity will require a smaller surface area for heat exchange. For liquids, convection or pumping can reduce the need for a large heat exchanger. In some applications, the heat exchanger is simply the physical interface of the storage material with the application space (e.g., phase change drywall, see the following).

33.9.1 Sensible Heat

Sensible heat is the heat that is customarily and intuitively associated with a change in temperature of a massive substance. The heat energy E_s stored in such a substance is given by

$$E_s = (T_2 - T_1)cM,$$

where

c is the specific heat of the substance (J/kg-°C)
M is the mass of the substance (kg)
T_1 and T_2 are the initial and final temperatures, respectively (°C)

The specific heat c is a physical parameter measured in units of heat per temperature per mass: substances with the ability to absorb heat energy with a relatively small increase in temperature (e.g., water) have a high specific heat, while those that get hot with only a little heat input (e.g., lead) have a low specific heat. Sensible heat storage is best accomplished with materials having a high specific heat.

33.9.1.1 Liquids

Sensible heat storage in a liquid is with very few exceptions accomplished with water. Water is unique among chemicals in having an abnormally high specific heat of 4186 J/kg-K, and furthermore has a reasonably high density. Water is also cheap and safe. It is the preferred choice for most nonconcentrating solar thermal collectors.

Liquids other than water may need to be chosen if the delivery temperature must be higher than 100°C, or if the system temperature can fall below 0°C. Water can be raised to temperatures higher than 100°C, but the costs of storage systems capable of containing the associated high pressures are usually prohibitive. Water can be mixed with ethylene glycol or propylene glycol to increase the useful temperature range and prevent freezing.

When a larger temperature range than that afforded by water is required, mineral, synthetic, or silicone oils can be used instead. The trade-offs for the increased temperature range are higher cost, lower specific heat, higher viscosity (making pumping more difficult), flammability, and, in some cases, toxicity.

For very high temperature ranges, salts are usually preferred, which balance a low specific heat with a high density and relatively low cost. Sodium nitrate has received the most prominent testing for this purpose, in the U.S. Department of Energy's *Solar Two* project located in Barstow, California.

Liquid sensible heat storage systems are strongly characterized not just by the choice of heat transfer fluid, but also by the system architecture. Two-tank systems store the cold and hot liquids in separate tanks (Figure 33.4). Thermocline systems use a single tank with cold fluid entering or leaving the bottom of the tank and hot fluid entering or leaving the top (Figure 33.5). Thermocline systems can be particularly low cost, because they minimize the required tank volume, but require careful design to prevent mixing of the hot and cold fluid.

One particularly interesting application of the thermocline concept is nonconvecting, salinity-gradient solar ponds, which employ the concept in reverse. Solar ponds are both an energy collection and energy storage technology. Salts are dissolved in the water to introduce a density gradient, with the densest (saltiest) water on the bottom and lightest (freshest) on top. Solar radiation striking the dark bottom of the pond heats the densest water, but convection of the heated water to the top cannot occur, because the density gradient prevents it. Salinity-gradient ponds can generate and store hot water at temperatures approaching 95°C.[30]

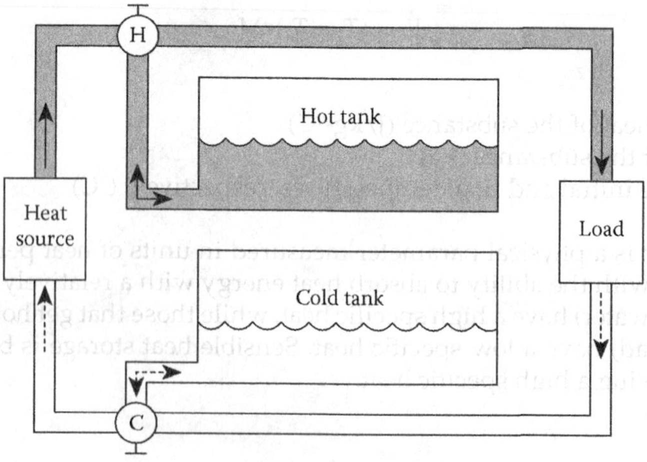

FIGURE 33.4
Two-tank thermal storage system; hot water is shown in gray and cold water is shown in white. When the heat source is producing more output than required for the load, valve H is turned to deposit hot liquid in the tank. When it is producing less than required for the load, the valve is turned to provide supplemental heat from the storage tank. Note that each tank must be large enough to hold the entire fluid capacity of the system.

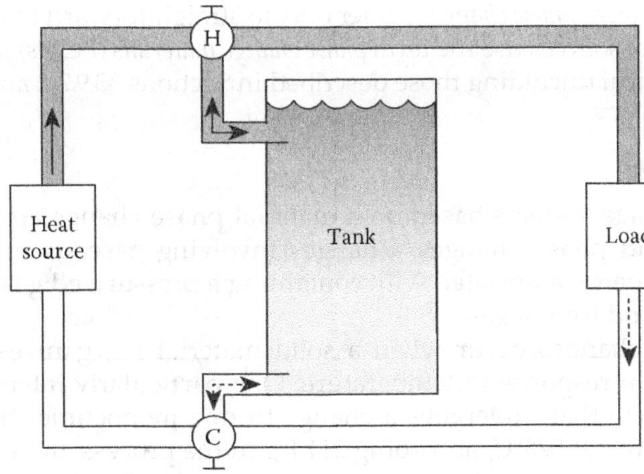

FIGURE 33.5
Thermocline storage tank. Thermocline storage tanks are tall and narrow to encourage the gravity-assisted separation of hot and cold fluid, and include design features (especially at the input/output connectors) to prevent mixing in the stored fluid.

33.9.1.2 Solids

Storage of sensible heat in solids is usually most effective when the solid is in the form of a bed of small units, rather than a single mass. The reason is that the surface-to-volume ratio increases with the number of units, so that heat transfer to and from the storage device is faster for a greater number of units. Energy can be stored or extracted from a thermal storage bed by passing a gas (such as air) through the bed. Thermal storage beds can be used to extract and store the latent heat of vaporization from water contained in flue gases.

Though less effective for heat transfer, monolithic solid storage has been successfully used in architectural applications and solar cookers.

33.9.2 Latent Heat

Latent heat is absorbed or liberated by a phase change or a chemical reaction, and occurs at a constant temperature. A phase change means the conversion of a homogenous substance among its various solid, liquid, or gaseous phases. One very common example is boiling water on the stovetop: though a substantial amount of heat is absorbed by the water in the pot, the boiling water maintains a constant temperature of 100°C. The latent heat E_s stored through a phase change is

$$E_s = lM,$$

where

M is the mass of material undergoing a phase change (kg)

l is the latent heat of vaporization (for liquid–gas phase changes) or the latent heat of fusion (for solid–liquid phase changes), in units of energy per mass (J/kg)

Conservation of energy dictates that the amount of heat absorbed in a given phase change is equal to the amount of heat liberated in the reverse phase change.

Though we use the term *phase change* to refer only to straightforward freezing and melting (Section 33.9.2.1), many sources use the term *phase change materials* (*PCMs*) to refer to any substance storing latent heat (including those described in Sections 33.9.2.2 and 33.9.2.3 as well).

33.9.2.1 Phase Change

Practical energy storage systems based on a material phase change are limited to solid–solid and solid–liquid phase changes. Changes involving gaseous phases are of little interest due to the expense associated with containing a pressurized gas, and difficulty of transferring heat to and from a gas.

Solid–solid phase changes occur when a solid material reorganizes into a different molecular structure in response to temperature. One particularly interesting example is lithium sulfate (Li_2SO_4) that undergoes a change from a monoclinic structure to a face-centered cubic structure at 578°C, absorbing 214 J/g in the process, more than most solid–liquid phase changes.[31]

Some common chemicals, their melting points, and heats of fusion are listed in Table 33.4. Fatty acids and paraffins received particular attention in the 1990s as candidate materials for the heat storage component of phase change drywall, a building material designed to absorb and release heat energy near room temperature for the purpose of indoor temperature stabilization.[32] In this application, solids in the drywall maintain the material's structural integrity even though the PCMs are transitioning between solid and liquid states.

TABLE 33.4
Melting Points and Heats of Fusion for Solid–Liquid Phase Changes

	Melting Point (°C)	Heat of Fusion (J/g)
Aluminum bromide	97	42
Aluminum iodide	191	81
Ammonium bisulfate	144	125
Ammonium nitrate	169	77
Ammonium thiocyanate	146	260
Anthracine	96	105
Arsenic tribromide	32	37
Beeswax	62	177
Boron hydride	99	267
Metaphosphoric acid	43	107
Naphthalene	80	149
Naphthol	95	163
Paraffin	74	230
Phosphoric acid	70	156
Potassium	63	63
Potassium thiocyanate	179	98
Sodium	98	114
Sodium hydroxide	318	167
Sulfur	110	56
Tallow	76	198
Water	0	335

Source: From Kreith, F. and Kreider J.F., *Principles of Solar Engineering*, Taylor & Francis, 1978. With permission.

33.9.2.2 Hydration–Dehydration Reactions

In this process, a salt or similar compound forms a crystalline lattice with water below a *melting point* temperature, and at the melting point, the crystal dissolves in its own water of hydration. Sodium sulfate (Na_2SO_4) is a good example, forming a lattice with ten molecules of water per molecule of sulfate ($Na_2SO_4 \cdot 10H_2O$) and absorbing 241 J/g at 32°C.[33]

Hydration–dehydration reactions have not found significant application in renewable energy systems, though they too have been a candidate for phase change drywall.

33.9.2.3 Chemical Reaction

A wide variety of reversible chemical reactions are available that release and absorb heat (e.g., Ref. [34]). The principal feature of this category of latent heat storage technologies is the ability to operate at extremely high temperatures, in some cases over 900°C. Extremely high temperature applications have focused primarily on fossil and advanced nuclear applications; to date, none of these chemical methods of heat storage have been deployed in commercial renewable energy applications.

33.10 Thermochemical Energy Storage

This section provides an overview of biomass storage technologies from an energetic perspective only. Additional details on biomass fuels are presented in Chapters 39, 49, and 50.

33.10.1 Biomass Solids

Plant matter is a storage medium for solar energy. The input mechanism is photosynthesis conversion of solar radiation into biomass. The output mechanism is combustion of the biomass to generate heat energy.

Biologists measure the efficiency of photosynthetic energy capture with the metric *net primary productivity* (*NPP* , which is usually reported as a yield in units similar to dry Mg/ha-yr (dry metric tons per hectare per year.) However, to enable comparisons of biomass with other solar energy storage technologies, it is instructive to estimate a *solar efficiency* by multiplying the NPP by the biomass heating value (e.g., MJ/dry Mg) and then dividing the result by the average insolation at the crop's location (e.g., MJ/ha-yr). The solar efficiency is a unitless value describing the fraction of incident solar energy ultimately available as biomass heating value. Most energy crops capture between 0.2% and 2% of the incident solar energy in heating value of the biomass; Table 33.5 shows examples of solar efficiencies estimated for a number of test crops.

The principal method for extracting useful work or electricity from biomass solids is combustion. Hence, the solar efficiencies listed in Table 33.5 need to be multiplied by the efficiency of any associated combustion process to yield a net solar efficiency. For example, if a boiler-based electric generator extracts 35% of the feedstock energy as electricity, and the generator is sited at a switchgrass plantation achieving 0.30% solar capture efficiency on a mass basis, the electric plant has a net solar efficiency of 0.30% × 35% = 0.11%. Because biomass is a very low efficiency collector of solar energy, it is very land-intensive compared to photovoltaic or solar thermal collectors, which deliver energy at solar efficiencies

TABLE 33.5

Primary Productivity and Solar Efficiency of Biomass Crops

Location	Crop	Yield (Dry Mg/ha-Year)	Average Insolation (W/m^2)	Solar Efficiency (%)
Alabama	Johnsongrass	5.9	186	0.19
Alabama	Switchgrass	8.2	186	0.26
Minnesota	Willow and hybrid poplar	8–11	159	0.30–0.41
Denmark	Phytoplankton	8.6	133	0.36
Sweden	Enthropic lake angiosperm	7.2	106	0.38
Texas	Switchgrass	8–20	212	0.22–0.56
California	*Euphorbia lathyris*	16.3–19.3	212	0.45–0.54
Mississippi	Water hyacinth	11.0–33.0	194	0.31–0.94
Texas	Sweet sorghum	22.2–40.0	239	0.55–0.99
Minnesota	Maize	24.0	169	0.79
West Indies	Tropical marine angiosperm	30.3	212	0.79
Israel	Maize	34.1	239	0.79
Georgia	Subtropical saltmarsh	32.1	194	0.92
Congo	Tree plantation	36.1	212	0.95
New Zealand	Temperate grassland	29.1	159	1.02
Marshall Islands	Green algae	39.0	212	1.02
New South Wales	Rice	35.0	186	1.04
Puerto Rico	*Panicum maximum*	48.9	212	1.28
Nova Scotia	Sublittoral seaweed	32.1	133	1.34
Colombia	Pangola grass	50.2	186	1.50
West Indies	Tropical forest, mixed ages	59.0	212	1.55
California	Algae, sewage pond	49.3–74.2	218	1.26–1.89
England	Coniferous forest, 0–21 years	34.1	106	1.79
Germany	Temperate reedswamp	46.0	133	1.92
Holland	Maize, rye, two harvests	37.0	106	1.94
Puerto Rico	*Pennisetum purpurcum*	84.5	212	2.21
Hawaii	Sugarcane	74.9	186	2.24
Java	Sugarcane	86.8	186	2.59
Puerto Rico	Napier grass	106	212	2.78
Thailand	Green algae	164	186	4.90

Source: From Klass, D.L., *Biomass for Renewable Energy, Fuels, and Chemicals*, Academic Press, San Diego, CA, 1998. With permission.

over 20% (see Chapters 43 and 47 for a full discussion). However, the capacity of land to store standing biomass over time is extremely high, with densities up to several hundred Mg/ha (and therefore several thousand GJ/ha), depending on the forest type. Standing biomass can serve as very long-term storage, though multiple stores need to be used in order to accommodate fire risk. For short-term storage, woody biomass may be dried, and is frequently chipped or otherwise mechanically treated to create a fine and homogenous fuel suitable for burning in a wider variety of combustors.

33.10.2 Ethanol

Biomass is a more practical solar energy storage medium if it can be converted to liquid form. Liquids allow for more convenient transportation and combustion, and enable extraction on demand (through reciprocating engines) rather than through a less dispatchable, boiler- or turbine-based process. This latter property also enables its use in automobiles.

Biomass grown in crops or collected as residue from agricultural processes consists principally of cellulose, hemicellulose, and lignin. The sugary or starchy by-products of some crops such as sugarcane, sugar beet, sorghum, molasses, corn, and potatoes can be converted to ethanol through fermentation processes, and these processes are the principal source of ethanol today. Starch-based ethanol production is low efficiency, but does succeed in transferring about 16% of the biomass heating value to the ethanol fuel.[35]

When viewed as a developing energy storage technology, ethanol derived from cellulose shows much more promise than the currently prevalent starch-based ethanol.[36] Cellulosic ethanol can be manufactured with two fundamentally different methods: either the biomass is broken down to sugars using a hydrolysis process, and then the sugars are subjected to fermentation, or the biomass is gasified (see the following), and the ethanol is subsequently synthesized from this gas with a thermochemical process. Both processes show promise to be far cheaper than traditional ethanol manufacture via fermentation of starch crops, and will also improve energy balances. For example, it is estimated that dry sawdust can yield up to 224 L/Mg of ethanol, thus recovering about 26% of the HHV of the sawdust.[37] Since the ethanol will still need to be combusted in a heat engine, the gross, biomass-to-useful-work efficiency will be well below this. In comparison, direct combustion of the biomass to generate electricity (per the discussion in Section 33.10.1) makes much more effective use of the biomass as an energy storage medium. Hence, the value of ethanol as an energy storage medium lies mostly in the convenience of its liquid (rather than solid) state.

33.10.3 Biodiesel

As starch-based ethanol is made from starchy by-products, most biodiesel is generated from oily by-products. Some of the most common sources are rapeseed oil, sunflower oil, and soybean oil. Biodiesel yields from crops like these range from about 300 to 1000 kg/ha-year, but the crop as a whole produces about 20 Mg/ha-year, meaning that the gross solar capture efficiency for biodiesel from crops ranges between 1/20 and 1/60 the solar capture efficiency of the crop itself. Because of this low solar capture efficiency, biomass cannot be the principal energy storage medium for transportation needs.[38]

Biodiesel can also be manufactured from waste vegetable or animal oils; however, in this case, the biodiesel is not functioning *per se* as a solar energy storage medium, so it is not further treated in this work.

33.10.4 Syngas

Biomass can be converted to a gaseous state for storage, transportation, and combustion (or other chemical conversion).[39] Gasification processes are grouped into three different classes: *pyrolysis* is the application of heat in anoxic conditions; *partial oxidation* is combustion occurring in an oxygen-starved environment; *reforming* is the application of heat in the presence of a catalyst. All three processes form *syngas*, a combination of methane, carbon monoxide, carbon dioxide, and hydrogen. The relative abundances of the gaseous products can be controlled by adjusting heat, pressure, and feed rates. The HHV of the resulting gas can contain up to 78% of the original HHV of the feedstock, if the feedstock is dry.[40] Compositions and heating values of two example syngases are listed in Table 33.3.

The equivalent of up to 10% of the gas HHV will be lost when the gas is pressurized for transportation and storage. Even with this loss, gasification is a considerably more efficient method than ethanol manufacture for transferring stored solar energy to a nonsolid medium.

References

1. Srinivasan, V., Batteries for vehicular applications, *Physics of Sustainable Energy, AIP Conference Proceedings*, 1044, 283–296, 2008.
2. Goodenough, J.B., Abruna, H.D., Buchanan, M.V., *Basic Research Needs in Electrical Energy Storage*, Bethesda, MD, 2007. Basic energy sciences workshop report. web.anl.gov/energy-storage-science/publications/ees.rpt.pdf.
3. Grid-scale rampable intermittent dispatchable storage, U.S. Department of Energy—Advanced Research Projects Agency—Energy (ARPA-E), Funding Opportunity Number: DE-FOA-0000290, 2010.
4. EA Technology, Review of electrical energy storage technologies and systems and of their potential for the UK, United Kingdom Department of Trade and Industry, London, U.K., 2004 (URN 04/1876).
5. DeVries, T., World's biggest battery helps stabilize Alaska, *Modern Power Systems*, 22, 40, 2002.
6. Nourai, A., NaS battery demonstration in the USA, *Electricity Storage Association Spring Meeting*, Electricity Storage Association, Morgan Hill, CA, 2003.
7. Sudworth, J.L., Sodium/nickel chloride (Zebra) battery, *Journal of Power Sources*, 100, 149, 2001.
8. The Zebra, *Fleets & Fuels*, February 17, 2003, p. 8.
9. Price, A., Technologies for energy storage—Present and future: Flow batteries, *2000 Power Engineering Society Meeting*, Seattle, WA, July 16–20, 2000.
10. Skyllas-Kazacos, M., *Recent Progress with the Vanadium Redox Battery*, University of New South Wales, Sydney, Australia, 2000.
11. Menictas, C. et al., Status of the vanadium battery development program, *Proceedings of the Electrical Engineering Congress*, Sydney, Australia, 1994.
12. VRB Power Systems, Inc., http://www.vrbpower.com.
13. Wilks, N., Solving current problems, *Professional Engineering*, 13, 27, 2000.
14. Scheffler, P., *Environmental Assessment for the Regenesys Energy Storage System*, Tennessee Valley Authority, 2001.
15. Lex, P. and Jonshagen, B., The zinc/bromide battery system for utility and remote applications, *Power Engineering Journal*, 13, 142, 1999.
16. Weber, A.Z., Mench, M.M., Meyers, J.P., Ross, P.N., Gostick, J.T., and Liu, Q., Redox flow batteries: A review, *Journal of Applied Electrochemistry*, 41(10), 1137–1164, 2011.
17. Brushett, F.R., Jansen, A.N., Vaughey, J.T., and Zhang, Z., Exploratory research of non-aqueous flow batteries for renewable energy storage, *220th Electrochemical Society Meeting*, Boston, MA, October 9–14, 2011.
18. Huskinson, B., Marshak, M.P., Suh, C., Er, S., Gerhardt, M.R., Galvin, C.J., Chen, X., Aspuru-Guzik, A., Gordon, R.G., and Aziz, M.J., A metal-free organic–inorganic aqueous flow battery, *Nature*, 505, 195–198, 2014.
19. Duduta, M., Ho, B.Y., Wood, V.C., Limthongkul, P., Brunini, V.E., Carter, W.C., and Chiang, Y.M., Semi-solid lithium rechargeable flow battery, *Advanced Energy Materials*, 1, 511–516, 2011.
20. Ho, B.Y., An experimental study on the structure-property relationship of composite fluid electrodes for use in high energy density semi-solid flow cells, PhD thesis, MIT, Cambridge, MA, February 2012.

21. Kruger, P., Electric power requirement for large-scale production of hydrogen fuel for the world vehicle fleet, *International Journal of Hydrogen Energy*, 26, 1137, 2001.
22. Bossel, U., Eliasson, B., and Taylor, G., The future of the hydrogen economy: Bright or bleak?, *European Fuel Cell Forum*, Oberrohrdorf, Switzerland, 2003.
23. Dillon, A. et al., Storage of hydrogen in single-walled carbon nanotubes, *Nature*, 386, 377, 1997.
24. Linden, D. and Reddy, T.B., *Handbook of Batteries*, 3rd edn., McGraw-Hill, New York, 2002.
25. Luango, C.A., Superconducting storage systems: An overview, *IEEE Transactions on Magnetics*, 32, 1996.
26. Cheung, K.Y.C. et al., *Large Scale Energy Storage Systems*, Imperial College London, London, U.K., 2003 (ISE2 2002/2003).
27. Donalek, P., Advances in pumped storage, *Electricity Association Spring Meeting*, Chicago, IL, 2003.
28. Kondoh, J. et al., Electrical energy storage systems for energy networks, *Energy Conservation and Management*, 41, 1863, 2000.
29. van der Linden, S., The case for compressed air energy system, *Modern Power Systems*, 22, 19, 2002.
30. Hull, J. et al., *Salinity Gradient Solar Ponds*, CRC Press, 1988.
31. Sørensen, B., *Renewable Energy: Its Physics, Engineering, Environmental Impacts, Economics & Planning*, 3rd edn., Elsevier Academic Press, Burlington, MA, 2004.
32. Neeper, D.A., Thermal dynamics of wallboard with latent heat storage, *Solar Energy*, 68, 393, 2000.
33. Goswami, D.Y., Kreith, F., and Kreider, J.F., *Principles of Solar Engineering*, 2nd edn., Taylor & Francis, Philadelphia, PA, 2000.
34. Hanneman, R., Vakil, H., and Wentorf Jr., R., Closed loop chemical systems for energy transmission, conversion and storage, *Proceedings of the Ninth Intersociety Energy Conversion Engineering Conference*, American Society of Mechanical Engineers, New York, 1974.
35. Shapouri, H., Duffield, J.A., and Wang, M., The energy balance of corn ethanol: An update, USDA, Office of Energy Policy and New Uses, Agricultural Economics, 2002 (Rept. No. 813).
36. Hammerschlag, R., Ethanol's energy return on investment: A survey of the literature 1990–present, *Environmental Science & Technology* (submitted).
37. Klass, D.L., *Biomass for Renewable Energy, Fuels, and Chemicals*, Academic Press, San Diego, CA, 1998.
38. Bockey, D. and Körbitz, W., *Situation and Development Potential for the Production of Biodiesel—An International Study*, Union zur Forderung von Oel- und Proteinpflanzen, e.V., Berlin, Germany, 2003.
39. Bridgwater, A.V., The technical and economic feasibility of biomass gasification for power generation, *Fuel*, 74, 631, 1995.
40. Klass, *Biomass for Renewable Energy, Fuels, and Chemicals*, 302.
41. U.S. Department of Energy, Properties of fuels, Alternative Fuels Data Center, 2004.
42. Kreith, F. and Kreider J.F., *Principles of Solar Engineering*, Taylor & Francis, 1978.

34

Advanced Concepts in Transmission and Distribution

Robert Pratt, Christopher P. Schaber, and Steve Widergren

CONTENTS

34.1 Introduction

Global prosperity depends upon efficient and affordable energy. Without a major shift in the way energy systems are planned, built, and operated, the world will invest trillions of dollars in conventional electric infrastructure over the next 20 years to meet expected growth.[32,33]

The future electric system will incorporate information technology in revolutionary ways, much as information technology has transformed other aspects of business because, fundamentally, "bits are cheaper than iron." The impact of the information age on the electric system will allow nations to realize the benefits already achieved by leading-edge industries that use real-time information, distributed e-business systems and market efficiencies to minimize the need for inventory and infrastructure and to maximize productivity, efficiency, and reliability.

With the help of information technologies, the creation of a distributed, yet integrated, system will empower consumers to participate in energy markets—the key to stabilizing prices. Market participants from utilities to new third parties to consumers will create value by developing and deploying solutions that cross enterprise and regulatory boundaries. At the same time, this transformation of the energy system addresses the urgent need to enhance national security. A distributed, network-based electric system will reduce single-point vulnerabilities and allow the grid to become "self-healing" by incorporating autonomic system reconfiguration in response to human-caused or natural disruptions.

Deploying information technology in a highly connected world will maximize the use of existing assets and minimize the need for new assets. To do this, the future electric system will

- Provide the incentive for customer and third-party assets to collaborate with existing grid assets to control costs and improve reliability by revealing the true time- and location-dependent value of electricity
- Provide the basis for collaboration by allowing the revealed values to be shared in real time by leveraging broadband communications that are rapidly becoming ubiquitous
- Provide the means to take advantage of the opportunities for collaboration and capture value in return through rapid advances in distributed controls and e-business applications

Much of the challenge and opportunity presented by these solutions is that their value proposition often involves crossing enterprise and regulatory boundaries. This means that nontechnical challenges must be addressed to support policy changes and business models that allow new technology and operational practices to flourish.

Information technology will profoundly transform the planning and operation of the power grid, just as it has changed business, education, and entertainment. It will form the "nervous system" that integrates new distributed technologies—demand response, distributed generation, and storage—with traditional grid generation, transmission, and distribution assets to share responsibility for managing the grid as a collaborative "society" of devices.[35–38]

34.2 Technology Advancements for Electricity Delivery

The following sections describe new technologies, or concepts that are becoming more economically feasible, for application to improved operation of the electric system. These are summarized in three general categories: distributed resources, transmission and distribution resources, information and communications technology, and control, applications, and operational paradigms.

34.2.1 Distributed Resources

Distributed resources are physical, capital assets that generate, store, or consume electric power at the fringes of the electric delivery system. Traditionally, these resources have not been broadly considered or deployed as transmission and distribution assets; however, significant efforts are underway to take advantage of such existing resources, and to provide motivation for future installations that can further benefit the system.[34] This section describes several of the resources that fall into this category.

End use control: Manual or automated control of end use systems to reduce or defer electric power use is an important element in active measures for enhancing delivery operations. Automatic load shedding (under-frequency, under-voltage), operator-initiated interruptible load, demand-side management programs, voltage reduction, and other load curtailment strategies have long been an integral part of coping with unforeseen contingencies as a last resort, and/or as a means of assisting the system during high-stress, overloaded conditions. Advances in load control technology will allow end use systems to play a more active role in the day-to-day operations of the electric system as well as create a more graceful response to emergency conditions.

Intelligent building systems: Intelligent building systems are emerging in factories, commercial buildings, and now residential facilities. These systems save energy by increasing efficient operations. Coordinated utilization of cooling, heating, and electricity in these establishments can significantly reduce energy consumption. Operated in a system that supports price-responsive demand programs, intelligent building systems can locally optimize operations to the benefit of larger system concerns.

System-sensitive appliances: System-sensitive appliances are emerging with the ability to measure and respond to frequency or voltage conditions or to react to simple signals, such as energy prices or emergency conditions. Washers, dryers, HVAC units, and other devices now come equipped with onboard intelligent systems that have the potential to respond to energy signals (see Section 34.3).

Distributed generation (DG): Fuel cells, microturbines, reciprocating engines, and small renewable generation such as wind and solar power, are increasingly being deployed across the system. DG can supply local load or sell into the system and offers owners self-determination. As these distributed resources increase in number, they can become a significant resource for reliable system operations. Their vast numbers teamed with end use systems adds a further dimension to the advantages of controllable load discussed above.

Combined-heating power (CHP): Combined-heating-power and buildings-cooling-heating-power (BCHP) are systems that combine distributed fuel-fired generation with equipment that recovers waste heat for heating or cooling at the for a building, facility, or district, maximizing the utilization of fuel.

Renewable generation: Renewable generation includes small wind-power, photovoltaic solar, and other renewable generation systems connected at the distribution level. Note that large wind farms are similar to central generation in that they do not displace the need for transmission and distribution assets. Unlike distributed fuel-fired generation, renewable generation is difficult to dispatch coincident with the needs of the grid, unless a storage system is involved.

Local energy storage: Systems that store energy locally for the benefit of end use systems can also be aggregated to the benefit of larger system operation concerns.

Electric storage: Systems that store electricity for later discharge include traditional batteries, flow batteries, supercapacitors, flywheels, and extend to uninterruptible power supplies modified to support local system operations.

Thermal storage: These systems store energy used for heat or cooling, usually in the form of hot or chilled water or ice, for later use in a building or an industrial process. The ability to "charge" to or "draw" from these resources provides flexibility to the timing of electricity use.

Direct DC end use: Returning to the days of Edison, equipment within customer premises are emerging that distribute and use DC power directly (instead of converting it from AC with attendant cost and inefficiencies) such as computers and electronic equipment, fans, pumps and other equipment with variable speed drives, battery chargers, and uninterruptible power supplies. DC distribution is especially useful if onsite generation inherently involves DC power directly or at an intermediate stage, such as from photovoltaics, microturbines, and fuel cells.

Local voltage regulation: Transformers, capacitors, or power electronics equipment that maintains voltage at the customer meter at a specified nominal level, can reduce wasted consumption from over-voltage while allowing higher-than-nominal distribution voltages to increase system throughput during times of peak demand.

Advanced energy efficiency: Investments in improved equipment, building structure, or industrial processes can reduce the quantity of electricity consumed yet provide equivalent service. Efficiency benefits occur with sensitivity to the end use being served, but are indifferent to the needs of the grid.

34.2.2 Transmission and Distribution Resources

The transmission and distribution (T&D) system of the future must not only have increased capacity to support the market demand for energy transactions, it must also be flexible to adapt to alterations in energy delivery patterns. These patterns change at various timescales: hourly, daily, weekly, and seasonally. The delivery system must also adapt to operation patterns dictated by the evolving geographical distribution of load and generation. As generation siting and dispatch decision making become less predictable and demand-side resources play a larger role in system operations, new technologies that afford transmission and distribution planners a wider range of alternatives for deployment of power become more attractive. This section discusses some of the newer hardware technologies that are being researched and deployed to reinforce grid operations. The list presented is not exhaustive.[39–41]

Conductors: The siting of the new transmission will continue to be a major challenge. Getting the most out of existing rights of way minimizes the need for new lines and rights of way and can minimize the societal concerns associated with visual pollution and high-energy EMFs.

Advanced composite conductors: Usually, transmission lines contain steel-core cables that support strands of aluminum wires that are the primary conductors of electricity. New cores developed from composite materials are proposed to replace the steel core. A new core consisting of composite glass-fiber materials shows promise as stronger than steel-core aluminum conductors while being 50% lighter in weight with up to 2.5 times less sag. The reduced weight and higher strength equate to greater current-carrying capability. This technology can be integrated in the field by most existing reconductoring equipment.

High-temperature superconducting (HTSC) technology: The conductors in HTSC devices operate at extremely low resistances. They require refrigeration (generally liquid nitrogen) to supercool ceramic superconducting materials. The technology is applicable to transmission lines, transformers, reactors, capacitors, and current limiters. HTSC cable occupies less space (AC transmission lines bundle three phase together; transformers and other equipment occupy smaller footprint for same level of capacity). Exposure to EMFs can be reduced with buried cables and this also counteracts visual pollution issues. Transformers can reduce or eliminate cooling oils that, if spilled, can damage the environment. For now, the maintenance costs remain high.

Below-surface cables: The underground cable advancements are employing fluid-filled polypropylene paper laminate (PPL) and extruded dielectric polyethylene (XLPE) cable technologies. Other approaches, such as gas-insulated transmission lines (GIL), are being researched and hold promise for future applications. Although there have been significant improvements in the technology, manufacturing costs remain high.

T&D configurations: Advances are being made in the configuration of transmission lines. New design processes coupled with powerful computer programs can optimize the height, strength, and positioning of transmission towers, insulators, and associated equipment in order to meet engineering standards appropriate for the conductor (e.g., distance from ground and tension for a given set of weather parameters).

Tower and pole design tools: A set of tools is emerging to analyze upgrades to existing transmission facilities or the installation of new facilities to increase their power-transfer capacity and reduce maintenance. Unused potential can be discovered in existing facilities to enhance upgrades, whereas new facilities can be engineered to closer design tolerances. Visualization techniques are adding the regulatory and public review cycles.

Modular equipment: One way to gain flexibility for changing market and operational situations is to develop standards for the manufacture and integration of modular equipment. This can reduce overall the time and expense for transmission systems to adapt to the changing economic and reliability landscape.

T&F system devices: Implemented throughout the system, these devices include capacitors, phase shifters, static VAR compensators (SVCs), thyristor-controlled series capacitors (TCSC), thyristor-controlled dynamic brakes, and other similar devices. Used to adjust system impedance, these devices can increase the delivery system's transfer capacity, support bus voltages by providing reactive power, or enhance dynamic or transient stability.

HVDC: With active control of real and reactive power transfer, HVDC can be modulated to damp oscillations or provide power-flow dispatch independent of voltage magnitudes or angles (unlike conventional AC transmission). HVDC runs independent of system frequency and can control the amount of power sent through the line. This latter benefit is the same as for FACTS devices discussed below. The high cost of converter equipment and its maintenance limits the application; however, as these costs continue to drop, the number of implementations will rise. Proposals have been made for DC application at the distribution level. The military continues to advance the technology in shipboard systems and mobile land-based facilities.

Flexible AC transmission system (FACTS): FACTS devices use power electronics to adjust the apparent impedance of the system. Capacitor banks are applied at loads and substations to provide capacitive reactive power to offset the inductive reactive power typical of most power system loads and transmission lines. With long intertie transmission lines, series capacitors are used to reduce the effective impedance of the line. By adding thyristors to both of these types of capacitors, actively controlled reactive power is available using static VAR compensators and thyristor-controlled series compensators, which are shunt-and series-controlled capacitors, respectively. The thyristors are used to adjust the total impedance of the device by switching individual modules. Unified power-flow controllers (UPFCs) also fall into this category. Phase shifters are transformers configured to change the phase angle between buses; they are particularly useful for controlling the power flow on the transmission network. Adding thyristor control to the various tap settings of the phase-shifting transformer, permits continuous control of the effective phase angle (and thus control of power flow). As with HVDC, the power electronics used in these devices are expensive in price and maintenance, but costs are dropping.

T&D energy storage: The traditional function of an energy storage device is to save production costs by holding cheaply generated off-peak energy that can be dispatched during peak consumption periods. By virtue of its attributes, energy storage can also provide effective power system control with modest incremental investment. Different dispatch modes can be superimposed on the daily cycle of energy storage, with additional capacity reserved for the express purpose of providing these control functions. Storage at the bulk system level also benefits the integration of intermittent renewable resources such as wind and solar power. The loss of efficiency between converting electricity into and out of storage is an important consideration in deploying these resources.

Batteries: Large battery systems use converters to transform the DC in the storage device to the AC of the power grid. Converters also operate in the opposite direction to recharge the batteries. Battery converters use power electronics that, by the virtue of their ability to change the power exchange rapidly, can be utilized for a variety of real-time control applications ranging from enhancing transient stability to preconditioning the area control error for automatic generator control enhancement. The expense of manufacturing and maintaining batteries has limited their impact in the industry.

Superconducting magnetic energy storage (SMES): SMES uses cryogenic technology to store energy by circulating current in a superconducting coil. SMES devices are efficient and compact because of their superconductive properties. As with the superconducting equipment mentioned above, SMES entails costs for the cooling system, special protection, and the specialized skills required to maintain the device.

Pumped hydro and compressed air storage: Pumped hydro consists of large ponds with turbines that can be run in either pump or generation modes. During periods of light load (e.g., night) excess, inexpensive capacity drives the pumps to fill the upper pond. During heavy-load periods, the water generates electricity into the grid. Compressed air storage uses the same principle except that large, natural underground vaults are used to store air under pressure during light-load periods. Pumped hydro, like any hydro generation project, requires significant space and has corresponding ecological impact. Compressed air storage systems require special geological formations.

Flywheels: Flywheels spin at high velocity to store energy. As with pumped hydro or compressed air storage, the flywheel is connected to a motor that either accelerates the flywheel to store energy or draws energy to generate electricity. Superconductivity technology can also be deployed to increase efficiency.

34.2.3 Information and Communications Technology

Automation and the ability to coordinate intelligent systems across wide areas are transforming all areas of our economy. The electric system is leveraging advances in generally available information and communication resources to enhance its operational effectiveness and incorporate new participants at the fringes of the system in operations.

Communication media: Wireless, power line carrier, cable, fiber, and other forms of broadband communication to delivery system components and end use premises is beginning to carry the information needed to implement new applications such as, consumption data, sensor data on grid conditions, and e-commerce information including service offerings, prices, contract terms, and incentives.

Information security: Technology that provides secure and reliable communications is essential to prevent intrusion by unauthorized parties or those without a need or right to know specific information.

Privacy and authentication: Grid operators, consumers and businesses all demand technology that ensures the privacy of information is maintained and the identity of participants is unambiguous.

E-commerce transactions: Technology that manages a multitude of small financial transactions in an auditable and traceable way will provide the financial incentive for day-to-day, moment-by-moment collaboration of end users with the grid (e.g., micropayments).

Real-time monitoring: The capability of the electricity grid is restricted through a combination of the limitations on individual devices and the composite capacity of the system. Improving monitoring to determine these limits in real time and to measure the system state directly can increase grid capability.

Power-system device sensors: Advancement in sensor technology is enabling dynamic ratings on the use of T&D resources. This includes the measurement of conductor sag, transformer coil temperature, and underground cable monitoring and diagnostics.

System-state sensors: Technology advancements are improving the operational view of large regions of the power network. Power-system monitors collect essential signals (key power flows, bus voltages, alarms, etc.) from local monitors and make them available to site operation functions. This provides regional surveillance over important parts of the control system to verify system performance in real time. Sample rates and data quality is increasing as costs decrease. For example, phasor measurement units (PMUs) are synchronized digital transducers that stream data, in real time, to phasor data concentrators (PDC). The general functions and topology for this network resemble those for dynamic monitor networks. Data quality for phasor technology is trustworthy, and secondary processing of the acquired phasor information can provide a broad range of signal types.

Advanced meters: Electric usage meters at customer premises are being replaced with technology that not only measures the energy usage, but also provides flexible interval energy monitoring and real-time power measurement to address needs such as peak usage, and power quality.

34.3 Control, Applications, and Operational Paradigms

The technologies summarized in the previous sections can be integrated into the electric system in a cooperative manner to address many applications: existing, proposed, and yet to be invented. There are many potential applications; some emerging ones are listed below and organized by end use, distribution, transmission, electricity service provider, and market operations.

34.3.1 End Use Operations and Planning

The management of end use resources in factories, commercial buildings, and residential facilities offer arguably the greatest area for advancement and contribution to system operations. Some important applications include the following:

- *Customer information gateways*: Information portals that support two-way communications, including real-time price signals and long-term contracts, across the customer enterprise boundary to suppliers, grid operators, and other third parties.
- *Demand response and energy management systems*: Controls that optimize the scheduling of energy use by appliances, equipment, and processes to minimize overall costs for electricity and respond to incentives from the service providers to curtail loads at times of peak demand or grid distress.
- *System-sensitive appliances, equipment, and processes*: Controls that autonomously sense disturbances in grid frequency and voltage, directing end use devices to immediately curtail their demand for periods of up to a few minutes (or, conversely, turning things like heating elements on to soak up momentary fluctuations of excess power) to prevent or arrest cascading blackouts. They also delay device restart after a voltage collapse or a blackout, easing service restoration by preventing the surge in demand from devices that have been without power for a while.
- *Autonomous agents for power purchases*: Software serving as the trading agent on behalf of customers, searching market opportunities and arranging for the lowest cost suppliers, including subsequently auditing power bills to ensure fair play.

34.3.2 Distribution Operations and Planning

- *Automated meter reading*: A smart meter allows the customer and their suppliers to access the electric meter to enable automated meter reading for billing and other purposes as well as access to consumption data by end use energy management systems.
- *Distribution automation*: Advanced distribution grid management systems that optimize supply voltage, manage peak demands by reconfiguring feeders to switch customers from one supply point to another, minimize the number of customers

affected by outages, and accelerate outage restoration. This includes advanced protection schemes that adapt protection device settings based on operating conditions.

- *Fault location and isolation*: Use advanced sensing and communications to better identify the location of faults, then invoke distribution automation capabilities to reconfigure feeders, safely reenergize as much of the affected region as possible, and inform work crews to repair the problem. Diagnostics and fault location tools for underground cables represent an important aspect of this area.
- *Distribution capacity marketplaces*: Software that projects the need for new distribution capacity, posting the cost of required upgrades and allowing customers and third parties to offer distributed resource projects (such as distributed generation, efficiency, and demand response) to defer or avoid construction; then operating those resources through local market signals or incentives to manage net demand at the capacity constraint.
- *Distributed generation for reactive power support and other ancillary services*: Software that manages customer-owned distributed generators to support the power grid by supplying reactive power in addition to real power and other ancillary services like spinning and nonspinning reserves, via direct dispatch based on prearranged contracts or indirect dispatch through markets or incentives.

34.3.3 Transmission Operations and Planning

- *Transactive control for transmission grids*: Systems for trading real-time transmission rights, analogous to automated stock market trading systems, that allow market signals to cause suppliers and load serving entities to shift or curtail wheeling of power from one region to another, effectively reconfiguring the power grid in real time in response to a potential or pending crisis.
- *Substation automation*: Device-level intelligence combines with high-speed, reliable communications to improve operations effectiveness. For example, transformer tap changes will coordinate with capacitor banks to perform voltage or reactive power control functions. More and better quality information is used locally for adaptive relay protection schemes that change their settings based on present system configuration information. Information exchanged between neighboring substations can be used to further coordinate system protection schemes and economic operation.
- *Regional control*: On a regional level, remedial action schemes can adaptively arm themselves based on updated information from several points in the system. Regional control centers can also coordinate the transmission operating configuration based on gathering better information from the field. For example, the results from system studies, such as contingency analysis and optimal power flow can make use of dynamic line ratings that are regularly updated based on weather forecasts and sag measurements from the field.

34.3.4 Electricity Service Provider Operations and Planning

- *Differentiating customer service levels*: Systems that integrate demand response, distribution automation, and local distributed generation to supply premium reliability service to customers that require it, providing digital quality power to those willing to pay a premium without "gold-plating" the system for other customers satisfied with current service.

- *Emergency end use curtailment*: Communications and control software that broadcasts emergency status information and leverages demand response capabilities to ration power to customers on a prearranged basis, providing power for critical customers and end uses while curtailing less critical demand to match the available supply; the intent is for everyone to have some power in a crisis rather than using rolling blackouts, for instance.

34.3.5 Market Operations and Forecasting

- *Market operations*: Software that operates markets and incentives that engage distributed resources and demand response to manage peak demands and ancillary services in supply, transmission, and local distribution systems.
- *Load forecasters*: Algorithms that provide continually updated, adaptive predictions of the demand for electricity at various levels in the system ranging from end use to distribution and transmission, as it responds to changes in conditions such as weather and price; based on techniques ranging from statistical methods and neural networks to engineering model parameter estimation.

34.4 System-Sensitive Appliances

The abundance of information, including price signals and grid conditions, will allow demand-response technologies to play a significant role in a virtual energy infrastructure. Consider that regardless of time of day or even time of year, about 20% of the load on the electric system is from consumer appliances that cycle on and off, such as heating, air conditioning, water heaters, and refrigerators. At the same time, generators also maintain steady operating reserves on "hot standby" that are equivalent to about 13% of the total load on the system in case problems suddenly arise. Why not find a way to reduce the demand of these appliances in times when operators would typically be dipping into the expensive cushion?

The (Grid Friendly™ controller) low-cost sensors embedded in appliances can sense grid conditions by monitoring the frequency of the system and provides automatic demand response in times of disruption. Within each interconnected operating region, a disturbance reflected in the system frequency is a universal indicator of serious imbalance between supply and demand that, if not relieved, can lead to a blackout. A simple computer chip installed in end use appliances can reduce the appliance's electricity demand for a few minutes or even a few seconds to allow the grid to stabilize. The controllers can be programmed to autonomously react in fractions of a second when a disturbance is detected whereas power plants can take minutes to come up to speed.[42]

As these controllers penetrate the appliance marketplace, important control issues must be addressed. For example, the control approach must deactivate appliances in a graduated manner so as not to "shock" the grid by dropping more load than necessary to rebalance supply and demand.[43] They must also be reactivated after the crisis in a similarly smooth fashion. It may be highly desirable to organize this response as a hierarchy ordered from the least to the highest priority end uses—from the least critical functions, such as air conditioning, to the most critical functions, such as communications, traffic control, water, sewage, and fuel pumps. Then, if a crisis persists, grid operators can allocate power to end uses simply by maintaining grid frequency at small increments below the normal range. Managing demand in the power grid in an "all-or-nothing" fashion will no longer be necessary.

By integrating the controllers with appliances at the factory, costs can be reduced to a few dollars per appliance. Done properly, consumers will not even notice the short interruption (by turning off the compressor in a refrigerator, but leaving the light on, for example). In the process, consumers become an integral part of power-grid operations and could even be rewarded for their participation in helping prevent a widespread outage. Therefore, without the need for any formal communications capability, appliances are transformed from being part of the problem to part of the solution. They now act as assets that form a much quicker and better safety net under the power grid, freeing up power plants from standby duty to increase competition, lower prices and meet future load growth. Moreover, because they act autonomously, no communication system is required beyond the power grid itself. However, when a communication system becomes available, the "smarts" are already on board the appliances to do much more sophisticated negotiation and control, such as reducing peak loads.

34.5 Markets + Control = The Transactive Network

Transactive (e.g., contract) networks and agent-based systems present an opportunity to implement process controls in which highly optimized control (both local and global) is an inherent attribute of the strategy rather than an explicitly programmed feature. The premise of transaction-based control is that interactions between various components in a complex energy system can be controlled by negotiating immediate and contingent contracts on a regular basis in lieu of standard on/off command and control. Each device is given the ability to negotiate deals with its peers, suppliers and customers in order to maximize revenues while minimizing costs. This is best illustrated by an example.

34.5.1 The End Use Facilities Marketplace

A typical building might have several chillers that supply a number of air handlers with chilled water on demand. If several air handlers require the full output of one chiller, and another air handler suddenly also requires cooling, traditional building control algorithms simply startup a second chiller to meet the demand and the building's electrical load ratchets upward accordingly.

A transaction-based building control system behaves differently. Instead of honoring an absolute demand for more chilled water, the air handler requests such service in the form of a bid (expressed in dollars), increasing its bid in proportion to its "need" (divergence of the zone temperature from its setpoint). The chiller controls, with knowledge of the electric rate structure, can easily express the cost of service as the cost of the kWh to run the additional chiller plus the incremental kW demand charge. If the zone served by this air handler just began to require cooling, its "need" is not very great at first, and so it places a low value on its bid for service and the additional chiller stays off until the level of need increases. Meanwhile, if another air handler satisfies its need for cooling, the cost of chilled water immediately drops below the bid price because a second chiller is no longer required, and the air handler awaiting service receives chilled water.

This is analogous to air traffic control where a limited resource is managed by scheduling demand into time slots. Alternatively, a peer-to-peer transaction can take place in which an air handler with greater need for service displaces (literally outbids) another whose thermostat is nearly satisfied.

In this way, the contract-based control system accomplishes several things. First, it inherently limits demand by providing the most "cost-effective" service. In doing this, it inherently prioritizes service to the most important needs before serving less important ones. Second, it decreases energy consumption by preventing the operation of an entire chiller to meet a small load, where it operates inefficiently.

Third, contract-based controls inherently propagate cost impacts up and down through successive hierarchical levels of the system being controlled (in this example, a heating/cooling system). The impacts on the utility bill, which are easily estimated for the chiller operation, are used as the basis for expressing the costs of air handler and zone services. Using cost as a common denominator for control makes expression of what is effectively a multilevel optimization much simpler to express than an engineered solution would be. It allows controls to be expressed in local, modular terms while accounting for their global impact on the entire system.

In effect, the engineering decision-making process is subsumed by a market value-based decision-making process that injects global information conveyed by market activity (i.e., asks, bids, immediate and contingent contracts, closings, and defaults) into the local engineering parameters that govern the behavior of individual systems over multiple timescales.

34.5.2 The Distribution Marketplace

One of the most critical, yet difficult, parts of the value chain to reveal is the value of new or expanded distribution capacity. When distribution utilities determine a need to add or increase the capacity of a substation, it is typical for the costs to range from \$100/kW to \$200/kW, or even higher in less common cases such as urban centers.[44] Although this is generally not enough value to fund a distributed-generation project that may cost \$800/kW, for example, it nevertheless is a substantial fraction of the needed investment. When added to other values seen by the customer, such as backup power and reduced demand and energy charges, plus values for grid support and displaced central generation at the transmission and wholesale levels, the project may be attractive to the customer, the utility, or a third party if all these values can be accumulated or shared.

This suggests a role for a new distribution-level marketplace. The distribution utility that traditionally deals with operations and planning now has an additional function—operating a local marketplace or incentive structure where these values are revealed. By posting long-term upgrade costs as an opportunity notice to customers and third parties to bid alternative investments, such opportunities will reveal themselves.

For example, if a single distributed generator is located at a substation or feeder to displace an upgrade, then sooner or later it will go down just when the load is peaking on a hot summer day. To maintain the initial level of reliability, multiple, smaller units must be installed, or load management contracts must be signed with enough customers to back up the resource occasionally. Direct load curtailment contracts are one way to do this, using technology such as transactive control to make it seamless, automatic, and minimally disruptive. A market-like economic dispatch based on a local congestion surcharge is another way to let these resources compete for the right to serve peak loads at minimal cost. In either case, the distribution marketplace becomes a focal point for the transformation.

34.5.3 The Transmission Marketplace

The transmission power grid and wholesale markets can potentially be operated in an analogous, transactive fashion that provides a mechanism for the "self-healing" features that are the key to increasing the reliability of the grid while minimizing the infrastructure cost involved. A scenario best illustrates this concept. In today's wholesale market, a large customer shopping for a long-term contract for power finds an energy service provider offering the minimal cost contract guaranteeing generation and transmission of power to the local distribution utility on its behalf. When a crisis occurs that suddenly disrupts the grid and severely constricts the transmission capacity of the corridor being used, the frequency disruption immediately triggers frequency-sensitive appliances throughout the entire region to temporarily rebalance supply and demand. This allows time for an automated power-trading system, analogous to an automated stock market trading scheme, to reconfigure the system in response to the crisis.

To do this, the energy management system immediately posts an emergency congestion surcharge on the constrained pathway. This surcharge triggers the energy service provider to issue a short-term subcontract to a generator in another location to use a different transmission pathway to deliver power on behalf of the customer. In effect, this is what grid operators do today, reconfiguring the grid and voiding power wheeling contracts with a telephone and paperwork. However, the fact that the August 14, 2003 blackout in the Eastern Interconnection, once begun,[45] rolled from Ohio to New York in 9 s serves as a reminder that information technology is essential for reacting with the speed necessary to arrest such events.

The key is tapping into the remarkable property of markets to reorganize themselves efficiently and the speed of high technology to effect that reorganization. Although engineers speak of "control" and economists speak of "markets," in the future power grid, these concepts will merge and blend to form a transactive network that effectively forms the central nervous system of an adaptive, evolving system.

34.6 Renewable Energy Sources Integration

Broadly defined, renewable energy sources include wind, solar (solar photovoltaic (PV) and solar thermal), biomass, hydroelectric, ocean and river currents, waves, and tides. However, only wind, solar, and biomass sources are largely unconstrained by resource location. Thus, these renewable energy sources are ideally suited as distributed sources within the future electric system. Biomass-derived fuels can power conventional and emerging external and internal combustion engines. Thus, this renewable source brings the additional value of regulated/controllable power output and benefits in the form of voltage and reactive power support, spinning reserve, and improved power quality. In short, biomass-derived fuels can simply replace conventional fossil-based fuels in powering conventional power generator units such as diesel generator sets. However, both solar and wind energy sources can also bring many of these additional values.

Both solar and wind energy have, to date, been considered to be intermittent sources (when energy storage is absent). However, as these power sources become accepted as conventional components of the power generation mix and are used more effectively, there has come the recognition that these resources are best described as "variable" and that furthermore the outputs are relatively highly predictable. In addition, the parallel development of reliable, intelligent, low-cost, electronic-based voltage-source power converters has brought the ability of solar and wind generators to supply both high-quality power and reactive power support. Therefore, wind and solar power generators should now be included in the distributed power generator resource category.

The integration of ubiquitous communications and integrated distributed control now allows distributed generation, and particularly renewable power sources to utilize loads (especially thermostatic loads) to "firm up" intermittent sources making them collectively a more reliable resource. Additionally, opportunities for distributed renewables to participate in real-time emission management and provide grid support such as reactive power, cold load pickup during restoration, and backup for nearby critical loads provide benefits that can be realized economically.

Renewable power generators can be assigned effective load-carrying capability (ELCC) (or capacity credit) values. These values can be established when the deployment configuration, integration, and dispatch of these resources, along with other distributed resources, is optimized within a specific distribution or transmission system. For example, it has been shown that had a dependable capacity of 5000 MW of PV been installed in the California ISO region in 2000, the peak load on June 15, 2000 could have been reduced by 3000 MW. This would have cut by one-half the number of equivalently sized natural-gas-driven combustion gas turbines needed to ensure system reserve capacity.[46] Although PV power is only available during daylight hours, the annual average ELCC estimated for PV power plants distributed throughout California has nonetheless been calculated to be 64% of the PV plant rating. Furthermore, when the California system electric load is driven by the sun, the ELCC exceeds 80%. Thus a 5 MW PV system would be considered a summer peaking power source of 4 MW.

Wind-power output in many locations also has a definite diurnal pattern. However, analyses of real operating data currently underway in both the United States and Europe show that the very favorable geographic diversity effect (due to many wind turbines scattered throughout a relatively large geographical area), along with the significant advances in short-term wind forecasting, can permit electric power system planners and operators to assign high values of ELCC and capacity factor to wind-power systems.[47]

In short, the intelligent deployment, integration, and operation of wind and solar energy sources within the future electric system enable renewable power resources to be categorized as distributed power generation assets without qualification, bringing with them many of the values of the traditional distributed power generation sources (such as diesel generators and gas turbine generators) that can be exploited within the system, renewable power to provide ancillary benefits such as grid reactive support, and additional societal benefits to be realized such as real-time emissions management.

References

1. Linden, D. and Reddy, T. B. 2002. *Handbook of Batteries*, 3rd edn., pp. 40–163. McGraw Hill, New York.
2. Luongo, C. A. 1996. Superconducting storage systems: An overview. *IEEE Transactions on Magnetics*, 32, 2214–2223.
3. Cheung, K. Y. C., Cheung, S. T. H. et al. 2003. *Large Scale Energy Storage Systems*, pp. 40–163. Imperial College London, London, U.K.

4. EA Technology. Review of electrical energy storage technologies and systems and of their potential for the UK, pp. 40–163. Department of Trade and Industry, London, U.K.
5. DeVries, T. 2002. World's biggest battery helps stabilize Alaska. *Modern Power Systems*, 22, 40.
6. Nourai, A. 2003. NaS battery demonstration in the USA. In: *Electricity Storage Association Spring Meeting*, Electricity Storage Association, Morgan Hill, CA.
7. Sudworth, J. L. 2001. Sodium/nickel chloride (Zebra) battery. *Journal of Power Sources*, 100, 149–163.
8. *Fleets & Fuels*. 2003. The Zebra, February 17, p. 8.
9. Price, A. 2000. Technologies for energy storage, present and future: Flow batteries. In: *2000 Power Engineering Society Meeting*, IEEE Power Engineering Society, Piscataway, NJ.
10. Skyllas-Kazacos, M. 2000. Recent progress with the vanadium redox battery, pp. 27–148. University of New South Wales, Sydney, Australia.
11. Menictas, C. 1994. Status of the vanadium battery development program. In: *Proceedings of the Electrical Engineering Congress*, Sydney, Australia.
12. VRB Power Systems, Inc. http://www.vrbpower.com (accessed on August 15, 2005).
13. Wilks, N. 2000. Solving current problems. *Professional Engineering*, 13, 27.
14. Scheffler, P. 2001. *Environmental Assessment for the Regenesys Energy Storage System*, pp. 142–148. Tennessee Valley Authority, Knoxville, TN.
15. Lex, P. and Jonshagen, B. 1999. The zinc/bromide battery system for utility and remote applications. *Power Engineering Journal*, 13, 142–148.
16. Kruger, P. 2001. Electric power requirement for large-scale production of hydrogen fuel for the world vehicle fleet. *International Journal of Hydrogen Energy*, 26, 1137–1147.
17. Bossel, U., Eliasson, B., and Taylor, G. 2003. *The Future of the Hydrogen Economy: Bright or Bleak?* pp. 1863–1874. European Fuel Cell Forum, Oberrohrdorf, Switzerland.
18. Dillon, A. et al. Storage of hydrogen in single-walled carbon nanotubes. *Nature*, 386, 377–379.
19. Donalek, P. 2003. Advances in pumped storage. *Electricity Association Spring Meeting*, Chicago, IL.
20. Kondoh, J. et al. 2000. Electrical energy storage systems for energy networks. *Energy Conservation and Management*, 41, 1863–1874.
21. van der Linden, S. 2002. The case for compressed air energy system. *Modern Power Systems*, 22, 19–21.
22. Hull, J. et al. 1988. *Salinity Gradient Solar Ponds*, pp. 393–403. CRC Press, Boca Raton, FL.
23. Sørensen, B. 2004. *Renewable Energy: Its Physics, Engineering, Environmental Impacts, Economics and Planning*, 3rd edn., pp. 393–403. Elsevier Academic Press, Burlington, MA.
24. Neeper, D. A. 2000. Thermal dynamics of wallboard with latent heat storage. *Solar Energy*, 68, 393–403.
25. Goswami, D. Y., Kreith, F., and Kreider, J. F. 2000. *Principles of Solar Engineering*, 2nd edn., pp. 631–653. Taylor & Francis, New York.
26. Hanneman, R., Vakil, H., and Wentorf, R. Jr. 1974. Closed loop chemical systems for energy transmission, conversion and storage. In: *Proceedings of the Ninth Intersociety Energy Conversion Engineering Conference*, American Society of Mechanical Engineers, New York, pp. 631–653.
27. Shapouri, H., Duffield, J. A., and Wang, M. 2002. The energy balance of corn ethanol: An update. Report 814. USDA Office of Energy Policy and New Uses, Agricultural Economics, Washington, DC.
28. Hammerschlag, R. 2006. Ethanol's energy return on investment: A survey of the literature 1990–present. *Environmental Science and Technology*, 40, 1744–1750.
29. Klass, D. L. 1998. *Biomass for Renewable Energy, Fuels, and Chemicals*, pp. 631–653. Academic Press, San Diego, CA.
30. Bockey, D. and Körbitz, W. 2003. *Situation and Development Potential for the Production of Biodiesel—An International Study*, pp. 631–653. Union zur Forderung von Oel- und Proteinpflanzen, Berlin, Germany.
31. Bridgwater, A. V. 1995. The technical and economic feasibility of biomass gasification for power generation. *Fuel*, 74, 631–653.

32. Kannberg, L., Chassin, D., Desteese, J., Hauser, S., Kintner-Meyer, M., Pratt, R., Schienbein, L., and Warwick, W. 2003. *GridWise™: The Benefits of a Transformed Energy System*, p. 10. Pacific Northwest National Laboratory, Richland, WA.
33. Baer, W., Fulton, B., and Mahnovski, S. 2004. *Estimating the Benefits of the GridWise Initiative*, pp. 17–20. Rand Corporation, Santa Monica, CA.
34. Chapel, S., Coles, L., Iannucci, J., and Pupp, R. 1993. Distributed utility valuation project, Monograph, p. 6. EPRI, NREL, and PG&E, San Francisco, CA.
35. Pratt, R. 2004. Transforming the electric energy system. In: *Proceedings of 2004 IEEE PES Power Systems Conference and Exhibition*, New York, 2004, Vol. 3, pp. 1651–1654.
36. Intelligrid Architecture. 2005. http://intelligrid.info/IntelliGrid_Architecture/Overview_Guidelines/index.htm (accessed on February 3, 2006).
37. The Modern Grid Initiative. 2006. http://www.themoderngrid.org (accessed on February 3, 2006).
38. Widergren, S. et al. 2005. GridWise architecture council constitutional convention. In: *Proceedings*, pp. 1–45. Pacific Northwest National Laboratory, Richland, WA.
39. Hauer, J., Overbye, T., Dagle, J., and Widergren, S. 2002. National transmission grid study issue papers: Advanced transmission technologies, pp. F1–F44. US Department of Energy, Washington, DC.
40. Office of Electric Transmission and Distribution. 2004. DOE electric distribution multi-year research, development, demonstration, and deployment technology roadmap plan: 2005–2009, pp. 1–82. US Department of Energy, Washington, DC.
41. Office of Electric Transmission and Distribution. 2003. Grid 2030: National vision for electricity's second 100 years, pp. 1–44. US Department of Energy, Washington, DC.
42. Dagle, J., Winiarski, D., and Donnelly, M. 1997. End-use load control for power system dynamic stability enhancement, pp. 1–57. Pacific Northwest National Laboratory, Richland, WA.
43. Trudknowski, D., Donnelly, M., and Lightner, E. 2005. Power system frequency and stability control using decentralized intelligent loads, pp. 1–7. Pacific Northwest National Laboratory, Richland, WA.
44. Balducci, P. J., Schienbein, L. A., Nguyen, T. B., Brown, D. R., and Fathelrahman, E. M. 2004. An examination of the costs and critical characteristics of electricity distribution system capacity enhancement projects. In: *Three Western Utilities IEEE Power Systems Conference and Exposition*, October 10–13, 2004, Vol. 1, p. 517.
45. Final Report on the August 14, 2003 Blackout in the United States and Canada: Causes and Recommendations, U.S.–Canada Power System Outage Task Force, April 2004. https://reports.energy.gov/BlackoutFinal-Web.pdf (accessed on February 3, 2006).
46. Herig, C. Using photovoltaics to preserve California's electricity capacity reserves. http://www.nrel.gov/docs/fy01osti/31179.pdf (accessed on February 3, 2006).
47. Soder, L. and Ackerman, T. 2005. Wind power in power systems: An introduction. In: *Wind Power in Power Systems*, T. Ackerman, ed., pp. 36–38. John Wiley & Sons, Ltd., West Sussex, U.K.

35

Smart Grid Technology

Zhixin Miao

CONTENTS

35.1 Introduction

Smart grid refers to a set of technology using control and automation to advance power systems operation. The goals of smart grid explored in this chapter are (1) to integrate renewable energy into grid through power electronic devices, (2) to flexibly route power through power electronic control, (3) to accommodate uncertainty of generation through enhancing controllability of loads—energy storage is a means in enhancing load controllability, and (4) to enhance grid operation technology through phasor measurement unit (PMU). In the following sections, the earlier-mentioned goals and technology will be elaborated. Section 35.2 focuses on power electronic technology for renewable energy grid integration and high-voltage direct current (HVDC)-based flexible power routing. Section 35.3 focuses on load controllability. Section 35.4 discusses the state-of-the-art technology in grid operation.

35.2 Renewable Energy Integration and Flexible Power Routing

Power electronic technology makes renewable energy such as wind and solar integration possible. In the 1980s, wind turbines made of induction generators were directly connected to the grid. These wind turbines operate when wind speed is greater than the nominal wind speed and have power rating around 100 kW. The state-of-the-art wind turbines are variable-speed constant-frequency types, meaning that they can operate at variable wind speed and generate constant-frequency electricity. Such technology reduces mechanical stress and improves the power rating. Today, wind turbines have power rating up to 5 MW. State-of-the-art wind turbines such as types 3 and 4 wind turbines are all connected to the grid through power electronic interfaces [1].

Photovoltaic (PV) panels are connected to the grid through power electronic interfaces. These interfaces can either single phase or three phase, be one stage (a DC to AC inverter) or two stages (a DC to DC converter and a DC to AC inverter). Power electronic technology also makes flexible power routing possible through HVDC technology.

The following subsections will elaborate on these two applications: power-electronic-enabled grid integration and power-electronic-enabled flexible power routing.

35.2.1 Power-Electronic-Enabled Grid Integration

Power electronic converters can be viewed as frequency transformers. They are able to convert electricity from one format to another, including (1) DC to AC, (2) AC to DC, and (3) AC to AC. For example, the direct output from a PV panel is DC electricity. To connect the PV panel to the single-phase 120 V distribution line, a DC to AC inverter is needed. In the scenarios of (2), AC electricity can be rectified to DC through a converter and be transmitted as DC electricity. Such is the case of HVDC. A permanent magnet synchronous generator-based wind turbine generates very low-frequency AC electricity. This low-frequency electricity will pass through back-to-back converters (an AC-to-DC converter and a DC-to-AC inverter connected through a DC link) to be connected to the grid of 60 Hz.

The power electronic interfaces usually are power or voltage control enabled. The AC side output active power and reactive power can be controlled through converters. For converters interfacing renewable energies, active power order usually follows maximum power point extracting to have maximum energy output. On the other hand, inverters have the ability to generate or consume reactive power. Therefore, inverters can regulate AC system voltages through reactive power generation or consumption.

With many PV and power factor capacitors installed in distribution systems, system voltage/Var regulation becomes a coordinated optimization problem to meet the system voltage requirement while generating maximum benefit. Such operation requires the PVs to regulate their reactive power outputs to have minimum line flow loss.

In a very short timescale, such as tens of seconds, PV voltage can drop sharply due to cloud. This requires the system to have fast voltage recovery capability. And it can be achieved either through redesign PV voltage control functionality or through voltage control using other devices such as battery.

35.2.2 Power-Electronic-Enabled Flexible Power Routing

AC power flows follow Kirchhoff's laws, and the flow on transfer path is dependent on physical parameters such as line length. The most simplified expression for power flow on a line can be expressed as $P_{ij} = V_iV_j/X_{ij} \sin \delta_{ij}$, where V_i and V_j are voltage magnitudes of the node buses, X_{ij} is the reactance of the line, and δ_{ij} is the phase angle difference between the two node voltage phasors. In order to be able to control power flow, phase shifters are used in industry to change the phase angle difference [2]. Unified power flow controller (UPFC) was also built and applied in American Electric Power (AEP) [3]. A UPFC can change both the node voltage magnitude and the node voltage phase angle.

High-voltage DC links are used for long-distance power transfer [4]. The applications include links built to bring Canadian hydropower to Minnesota (Nelson River—MN) and Oregon hydropower to California. The major advantage of HVDC is its energy efficiency. Compared to AC power transfer, DC power transfer requires less current carrying through the line for the same amount of active power due to the omission of reactive power. Therefore, line loss will be greatly reduced.

Conventional HVDC technology, line current commuted (LCC) technology, relies on thyristor devices, which can be switched on every one cycle. LCC–HVDC is suitable for large-scale power transfer (>1000 MW). It requires large filters to get rid of low-order harmonics due to thyristor switching. In addition, LCC–HVDC always absorbs reactive power and requires a strong grid to support reactive power.

With the advance of insulated gate bipolar transistor (IGBT) switches, voltage source converter (VSC) HVDC technology becomes popular for wind integration [5]. Such technology is suitable for several hundred MW power transfer. Since IGBT switching is at kilohertz, the size of filters is very small. In addition, VSC–HVDC can operate in four quadrants in terms of PQ (real power [P] and reactive power [Q]). Therefore, VSC–HVDC can absorb or generate both real power and reactive power, and it can be used for black-start.

In terms of flexible power routing, HVDC technology is built for power flow control and can realize such requirement. For example, the LCC–HVDC is operated based on power order. The rectifier of an LCC–HVDC will adjust the thyristor firing angle based on a power order. The VSC–HVDC is also equipped with power order control.

Smart grid requires fast and flexible power routing. For example, in an area with ample renewable generation supply, it is necessary to route the power to a metropolitan area or an area with energy storage. In addition, due to the variant nature of renewable energy, it is also required to be very fast. HVDC technology can fulfill both requirements. Multiterminal HVDC is also a technology under investigation for fast and flexible power routing.

35.3 Load Controllability

In conventional power systems, operation usually refers to switch on/off and dispatch generators, while loads are considered as uncontrollable. One of the goals of smart grid technology is to make loads controllable through adding communication infrastructure to connect loads and make load controllable.

Traditionally, loads consist of only passive power consumers. With the advance of distributed generation and energy storage, loads become active power consumer. Loads can decide when to consume and how much to consume. For example, a home equipped with a smart meter and interruptible loads such as washing machine or air-conditioners can make decision whether to switch on/off washing machine or air-conditioner.

Two types of load control are discussed in the following subsections: (1) direct load control is conducted by utilities and (2) demand response program lets loads make their own decisions.

35.3.1 Direct Load Control

In direct load control, loads allow utilities to directly control their appliances or equipment. Utilities favor peak shaping to avoid the need to build new power plants. The following is an example of peak shaving through energy storage system. A 1.6 kW PV system is connected to the distribution system along with a 5 kW, 4 h battery system. The peak hours are in the afternoon around 2 pm. The control strategy of the utility is as follows: before 2 pm, all PV power will be used to charge the battery, and the net exchange between the grid and the PV/battery system is zero. After 2 pm, the battery will be discharged to supply the grid.

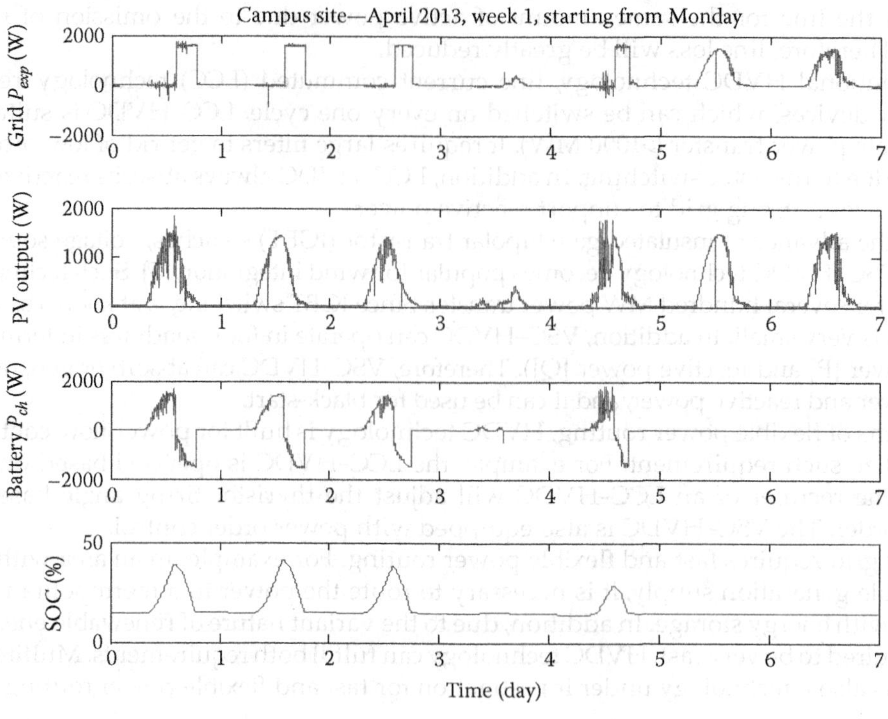

FIGURE 35.1
PV/battery system power output.

The output power from the battery can be a constant. Or the net output from the PV/battery system can be a constant. After the peak hours, the discharge process of the battery will stop. Figure 35.1 presents a real-world PV/battery output power for several days.

35.3.2 Demand Response Program

In this section on demand response program, let us look at an example of demand response of microgrids from Ref. [6]. In this example, the microgrids and the utility exchange limited information, and they make their own decisions while updating the information exchange. In the end, they are expected to reach agreement. Both the utility and microgrids will achieve optimal goals. Figure 35.2 presents a scheme of information exchange between the utility and microgrids.

In this figure, the microgrids send the bidding price for real power (λ^p) and reactive power (λ^q) to the utility. The utility decides how much power to export based on the bidding price. The microgrids receive the information of power amount and again compute the bidding prices and send them to utility. After iterations, the prices and the power amount are expected to be constant.

The optimization problems for the utility and the microgrid are described as follows:

1. *Utility optimization*: At iteration k, the utility solves (35.1) given price vectors $(\lambda_i^p(k), \lambda_i^q(k))$ at the first iteration or updated ones to determine the importing power vector (P_{imp}, Q_{imp}). Then, the vectors P_{imp} and Q_{imp} are announced to the communities. The procedure to solve utility optimal power flow (OPF) problem is the same as that in the subgradient method.

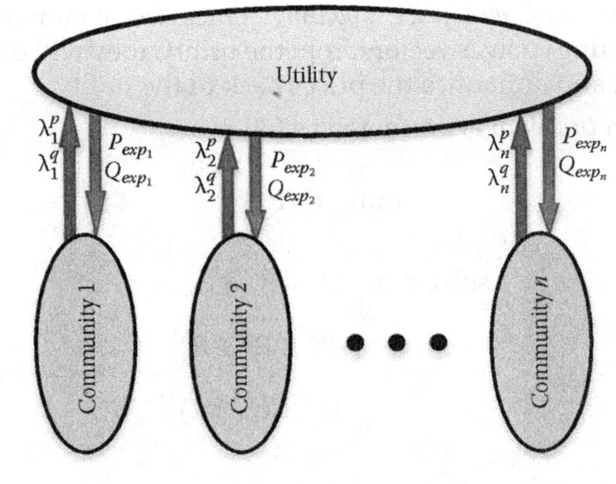

FIGURE 35.2
Information flow of lower- and upper-bound switching (LUBS) algorithm.

The utility also computes the utility cost $C_u(k)$ at kth iteration.

$$\min \sum_{i\in N-A} Ci(P_{gi}) + \sum_{i\in A}\left[\lambda_i^p P_{imp_i} + \lambda_i^q Q_{imp_i}\right]$$

$$\text{subject to} \quad \forall i \in N, \quad \forall_{j\in B}$$

$$\begin{aligned}
P_{gi} - P_{L_i} - P_i(V,q) &= 0 \\
Q_{gi} - Q_{L_i} - Q_i(V,q) &= 0 \\
V_i^m \le V_i &\le V_i^m \\
P_{gi}^m \le P_{gi} &\le P_{gi}^m \\
Q_{gi}^m \le Q_{gi} &\le Q_{gi}^m \\
S_j(V,\theta) - S_j^M &\le 0 \\
P_{impi}^m \le P_{impi} &\le P_{impi}^m \quad \forall_{i\in A} \\
Q_{impi}^m \le Q_{impi} &\le Q_{impi}^m \quad \forall_{i\in A}
\end{aligned} \tag{35.1}$$

where
- $C(\cdot)$ is the cost function, superscripts M and m denote upper and lower bounds
- P_g, Q_g, P_L, and Q_L are the vectors of bus real and reactive power injection, and real and reactive loads
- $P(V, \theta)$ and $Q(V, \theta)$ are the power injection expressions in terms of bus voltage magnitude and phase angles
- $S(V, \theta)$ is the vector of line complex power flow

2. *Community optimization and price updating*: Once the communities receive the importing (exporting) power vectors from the utility, they will determine the local generator output and announce the prices back to the utility.
   The optimization problem is presented as follows:

$$\begin{aligned} \min \quad & C_i(P_{gi}) \\ \text{subject to} \quad & P_{gi} - P_{Li} = P_{impi} \\ & P_{gi}^{m} \le P_{gi} \le P_{gi}^{M} \\ & Q_{gi}^{m} \le Q_{gi} \le Q_{gi}^{M} \end{aligned} \tag{35.2}$$

Generally, the community connected to the bus $i \in A$ seeks the prices $(\lambda_i^{p^*}, \lambda_i^{q^*})$ of $(P_{imp,i}, Q_{imp,i})$. These prices are the Lagrangian multipliers related to the equality constraints. When the limits are not bounding, the prices are the same as the marginal prices of the generators.

In a special case where the communities only have generation units and fixed loads, $(\lambda_i^{p^*}, \lambda_i^{q^*})$ are their marginal costs corresponding to (P_{impi}, Q_{impi}). For example, if the cost function of a generation unit is quadratic, $C_i(P_{gi}) = 0.5\alpha_i P_{gi}^2 + \beta_i P_{gi} + \gamma_i$. Then, the active power price is equal to $\lambda_i^{p^*} = \alpha_i(P_{imp,i} + P_{Li}) + \beta_i$. The cost of the community i is denoted as C_{ci}.

35.4 PMU Technology

In grid operation, a set of technology advances to enhance power system operation, including synchrophasor measurement unit (PMU) applications, to enhance situation awareness. DOE-initiated Smart Grid Investment Grant was launched in early 2010 to modernize the power grid. The topic areas of the grant include device manufacturing, customer systems, advanced metering infrastructure, electric distribution systems, and electric transmission systems. In the area of electric transmission systems, projects primarily involve installing PMU, adding smart grid functions to devices, and enhancing visualization tools.

PMU devices compute voltage and current phasor and frequency and transmit the information to a control center at 30 Hz. Applications of PMU data include state estimation, prediction and prevention, and dynamic model verification. Compared to the conventional state estimation where measurements such as active power, reactive power, and voltage magnitudes are collected, PMUs give voltage and current phasor including phase angles directly. The information can be used to estimate every node's voltage phasors. Such process is a linear estimation process instead of nonlinear estimation process. The PMU data can be used to complement the current state-estimation system and have better estimation results.

PMU data can also be used for electromechanical dynamic model estimation. The frequency range of electromechanical modes is usually less than 2 Hz. To identify those dynamics, 30 Hz PMU sampling data is sufficient. With the PMU data be classified as

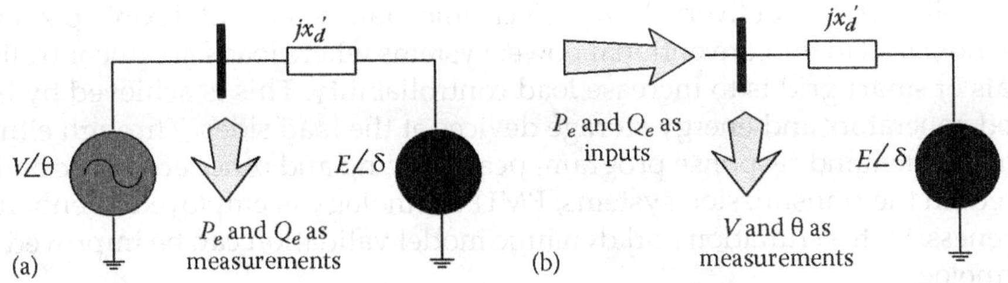

FIGURE 35.3
(a) Model decoupling using V and θ as inputs and P_e and Q_e as measurements. (b) Model decoupling using P_e and Q_e as inputs and V and θ as measurements.

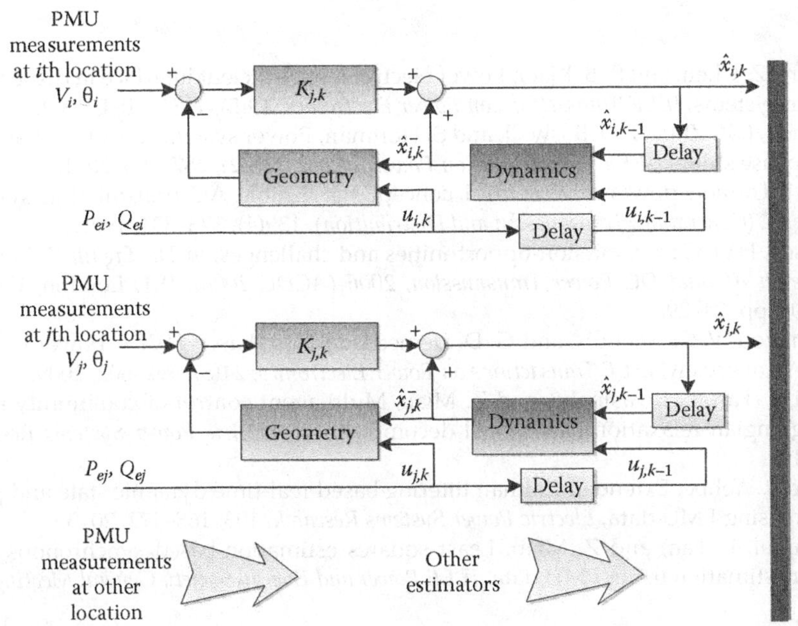

FIGURE 35.4
Kalman filtering technology using PMU data.

input and output, two general approaches can be applied for system identification: extended Kalman filter (EKF) and least square error-based estimation. The USF SPS group has conducted research in system identification through PMU data in Refs. [7,8].

Figure 35.3 presents the input and output data. Figure 35.4 presents the EKF scheme for model estimation.

35.5 Conclusion

This chapter presents a set of smart grid technology in renewable energy grid integration, fast and flexible power routing, load control, and system situation awareness. Power electronic technology is used to interface renewable energy sources and energy storage devices as well as power delivery. Power electronic makes fast and flexible power routing possible. Compared to the conventional power systems where loads are uncontrollable, one of the goals of smart grid is to increase load controllability. This is achieved by installing distributed generators and energy storage devices at the load sides. Through either direct load control or demand response program, peak shaving and other economic benefits can be achieved. In the transmission systems, PMU technology is employed to enhance situation awareness. State estimation and dynamic model validation can be improved through PMU technology.

References

1. F. Blaabjerg, Z. Chen, and S. B. Kjaer, Power electronics as efficient interface in dispersed power generation systems, *IEEE Transactions on Power Electronics*, 19(5), 1184–1194, 2004.
2. J. A. Momoh, J. Z. Zhu, G. D. Boswell, and S. Hoffman, Power system security enhancement by opf with phase shifter, *IEEE Transactions on Power Systems*, 16(2), 287–293, 2001.
3. L. Gyugyi, Unified power-flow control concept for flexible AC transmission systems, *IEE Proceedings C (Generation, Transmission and Distribution)*, 139(4), 323–331, 1992.
4. B. Andersen, HVDC transmission-opportunities and challenges, in *The Eighth IEE International Conference on AC and DC Power Transmission, 2006 (ACDC 2006)*, IET, London, U.K., March 28–31, 2006, pp. 24–29.
5. N. Flourentzou, V. G. Agelidis, and G. D. Demetriades, Vsc-based HVDC power transmission systems: An overview, *IEEE Transactions on Power Electronics*, 24(3), 592–602, 2009.
6. V. Disfani, L. Fan, L. Piyasinghe, and Z. Miao, Multi-agent control of community and utility using Lagrangian relaxation based dual decomposition, *Electric Power Systems Research*, 110, 45–54, 2014.
7. L. Fan and Y. Wehbe, Extended Kalman filtering based real-time dynamic state and parameter estimation using PMU data, *Electric Power Systems Research*, 103, 168–177, 2013.
8. B. Mogharbel, L. Fan, and Z. Miao, Least squares estimation-based synchronous generator parameter estimation using PMU data, *IEEE Power and Energy Society General Meeting* 2015.

Section IV

Renewable Technologies

36

Solar Energy Resources

D. Yogi Goswami

CONTENTS

36.1 Introduction

Solar energy is the world's most abundant permanent source of energy. The amount of solar energy intercepted by the planet Earth is a tiny fraction of the solar radiation emitted by the sun. However, even that tiny amount is 5000 times greater than the sum of all other inputs (terrestrial nuclear, geothermal, and gravitational energies and lunar gravitational energy).

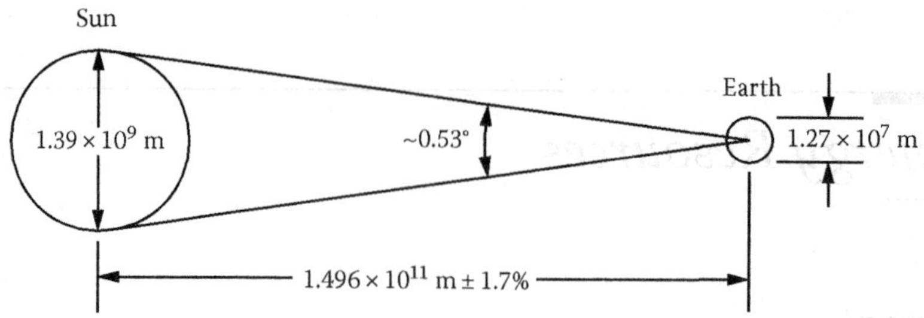

FIGURE 36.1
Relationship between the sun and the earth.

The sun is a 13.9 × 10^5 km diameter sphere comprising many layers of gases, which are progressively hotter toward its center. The outermost layer, from which energy is radiated into the solar system, is approximately at an equivalent blackbody temperature of 5,760 K (10,400° R). The rate of energy emission from the sun is 3.8 × 10^{23} kW. Of this total, only a tiny fraction, approximately 1.7 × 10^{14} kW, is intercepted by the earth, which is located about 150 million km from the sun (Figure 36.1). Of this amount, 30% is reflected to space, 47% is converted to low-temperature heat and reradiated to space, and 23% powers the evaporation/precipitation cycle of the biosphere. Less than 0.5% is represented in the kinetic energy of the wind and waves and in photosynthetic storage in plants.

Total terrestrial radiation is only about one-third of the extraterrestrial total during a year, and 70% of that falls on the oceans. However, the remaining 1.5 × 10^{17} kWh that falls on land is a prodigious amount of energy—about 6000 times the total energy usage of the United States in 2009. Only a small fraction of this total can be used because of physical and socioeconomic constraints.

36.2 Sun–Earth Geometric Relationship

Figure 36.2 shows the annual orbit of the earth around the sun. The distance between the earth and the sun changes throughout the year, the minimum being 1.471 × 10^{11} m at winter solstice (December 21) and the maximum being 1.521 × 10^{11} m at summer solstice (June 21). The year-round average earth–sun distance is 1.496 × 10^{11} m. The amount of solar radiation intercepted by the earth, therefore, varies throughout the year, the maximum being on December 21 and the minimum on June 21.

The axis of the earth's daily rotation around itself is at an angle of 23.45° to the axis of its ecliptic orbital plane around the sun. This tilt is the major cause of the seasonal variation of the solar radiation available at any location on the earth. The angle between the earth–sun line (through their centers) and the plane through the equator is called the *solar declination*, δ_s. The declination varies between −23.45° on December 21 and +23.45° on June 21. Stated another way, the declination has the same numerical value as the latitude at which the sun is directly overhead at solar noon on a given day. The tropics of Cancer (23.45° N) and Capricorn (23.45° S) are at the extreme latitudes where the sun is overhead at least once a year as shown in Figure 36.2. The Arctic and Antarctic circles are defined as those latitudes above which the sun does not rise above the horizon plane at least once per year. They are located, respectively, at $66\frac{1}{2}$° N and

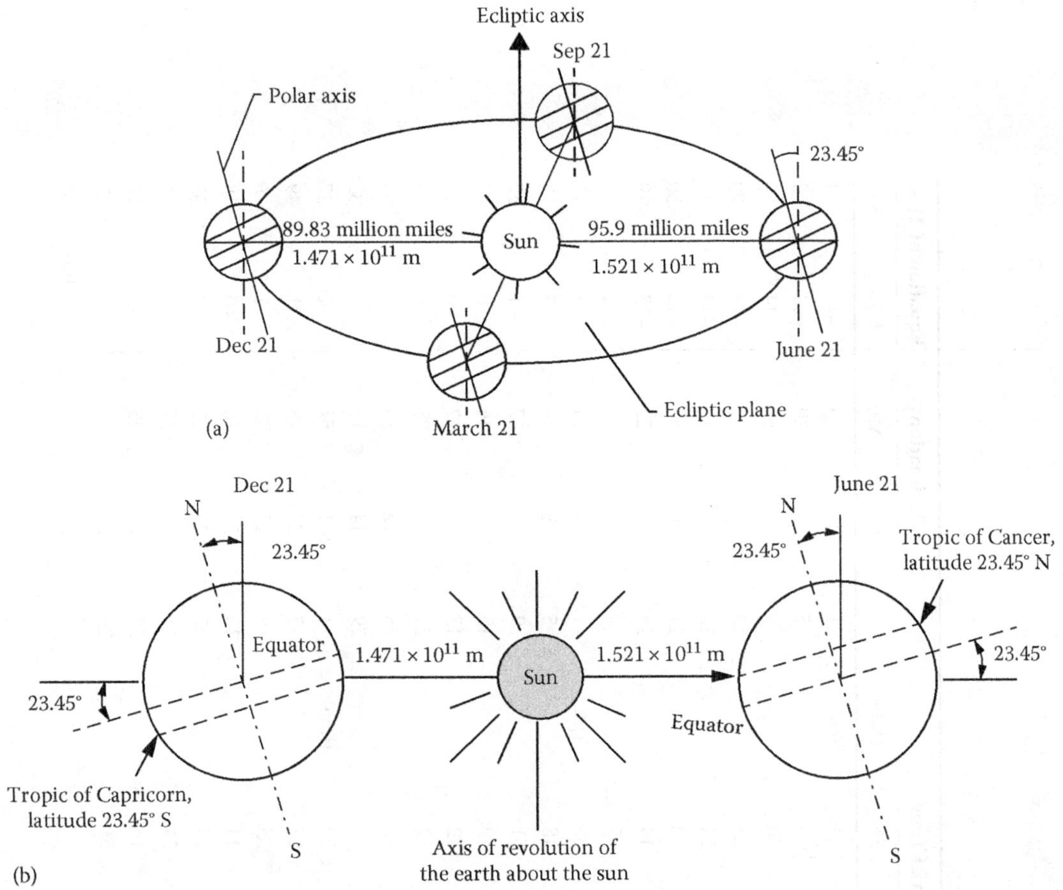

FIGURE 36.2
(a) Motion of the earth about the sun and (b) location of tropics. Note that the sun is so far from the earth that all the rays of the sun may be considered as parallel to one another when they reach the earth.

$66\frac{1}{2}$° S. Declinations north of the equator (summer in the northern hemisphere) are positive; those south, negative. The solar declination may be estimated by the relation*

$$\delta_s = 23.45° \sin\left[360(284 + n)/365\right]°, \tag{36.1}$$

where n is the day number during a year with January 1 being $n = 1$. Approximate values of declination* may also be obtained from Table 36.1 or Figure 36.3. For most calculations, the declination may be considered constant during any given day.

For the purposes of this book, the Ptolemaic view of the sun's motion provides a simplification to the analysis that follows. It is convenient to assume the earth to be fixed and to describe the sun's apparent motion in a coordinate system fixed to the earth with its origin at the site of interest. Figure 36.4 shows an apparent path of the sun to an observer. The position of the sun can be described at any time by two angles, the altitude and azimuth

* A more accurate relation is $\sin \delta_s = \sin(23.45°) \sin[360(284 + n)/365]°$. Because the error is small, Equation (2.23) is generally used.

TABLE 36.1
Summary Solar Ephemeris[a]

Date		Declination		Equation of Time		Date		Declination		Equation of Time	
		Deg	Min	Min	Sec			Deg	Min	Min	Sec
Jan.	1	−23	4	−3	14	Feb.	1	−17	19	−13	34
	5	−22	42	−5	6		5	−16	10	−14	2
	9	−22	13	−6	50		9	−14	55	−14	17
	13	−21	37	−8	27		13	−13	37	−14	20
	17	−20	54	−9	54		17	−12	15	−14	10
	21	−20	5	−11	10		21	−10	50	−13	50
	25	−19	9	−12	14		25	−9	23	−13	19
	29	−18	9	−13	5						
Mar.	1	−7	53	−12	38	Apr.	1	+4	14	−4	12
	5	−6	21	−11	48		5	5	46	−3	1
	9	−5	48	−10	51		9	7	17	−1	52
	13	−3	14	−9	49		13	8	46	−0	47
	17	−1	39	−8	42		17	10	12	+0	13
	21	−0	5	−7	32		21	11	35	1	6
	25	+1	30	−6	20		25	12	56	1	53
	29	3	4	−5	7		29	14	13	2	33
May	1	+14	50	+2	50	June	1	+21	57	2	27
	5	16	2	34	17		5	22	28	1	49
	9	17	9	3	35		9	22	52	1	6
	13	18	11	3	44		13	23	10	+0	18
	17	19	9	3	44		17	23	22	−0	33
	21	20	2	3	24		21	23	27	−1	25
	25	20	49	3	16		25	23	25	−2	17
	29	21	30	2	51		29	23	17	−3	7

(Continued)

TABLE 36.1 (*Continued*)

Summary Solar Ephemeris[a]

Date		Declination		Equation of Time		Date		Declination		Equation of Time	
		Deg	Min	Min	Sec			Deg	Min	Min	Sec
July	1	+23	10	−3	31	Aug.	1	+18	14	−6	17
	5	22	52	−4	16		5	17	12	−5	59
	9	22	28	−4	56		9	16	6	−5	33
	13	21	57	−5	30		13	14	55	−4	57
	17	21	21	−5	57		17	13	41	−4	12
	21	20	38	−6	15		21	12	23	−3	19
	25	19	50	−6	24		25	11	2	−2	18
	29	18	57	−6	23		29	9	39	−1	10
Sep.	1	+8	35	−0	15	Oct.	1	−2	53	+10	1
	5	7	7	+1	2		5	−4	26	11	17
	9	5	37	2	22		9	−5	58	12	27
	13	4	6	3	45		13	−7	29	13	30
	17	2	34	5	10		17	−8	58	14	25
	21	1	1	6	35		21	−10	25	15	10
	25	0	32	8	0		25	−11	50	15	46
	29	−2	6	9	22		29	−13	12	16	10
Nov.	1	−14	11	+16	21	Dec.	1	−21	41	11	16
	5	−15	27	16	23		5	−22	16	9	43
	9	−16	38	16	12		9	−22	45	8	1
	13	−17	45	15	47		13	−23	6	6	12
	17	−18	48	15	10		17	−23	20	4	47
	21	−19	45	14	18		21	−23	26	2	19
	25	−20	36	13	15		25	−23	25	+0	20
	29	−21	21	11	59		29	−23	17	−1	39

[a] Since each year is 365.25 days long, the precise value of declination varies from year to year. *The American Ephemeris and Nautical Almanac*, published each year by the U.S. Government Printing Office, contains precise values for each day of each year.

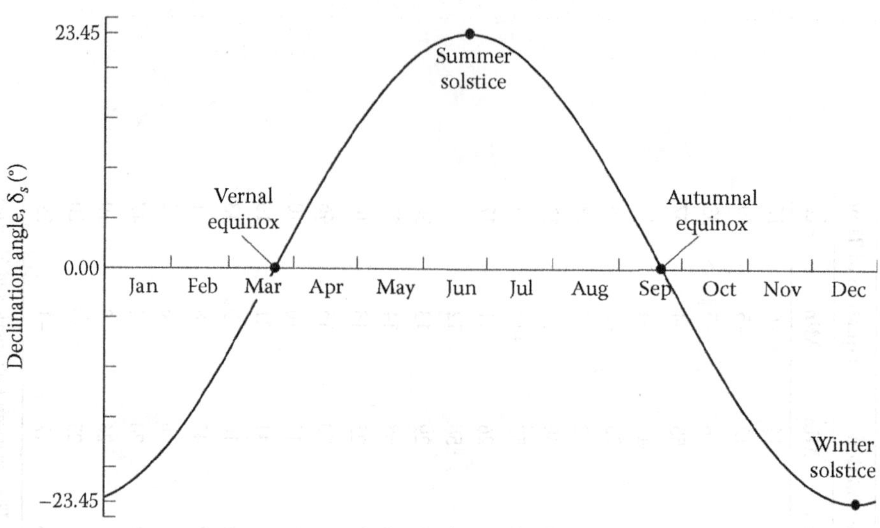

FIGURE 36.3
Graph to determine the solar declination.

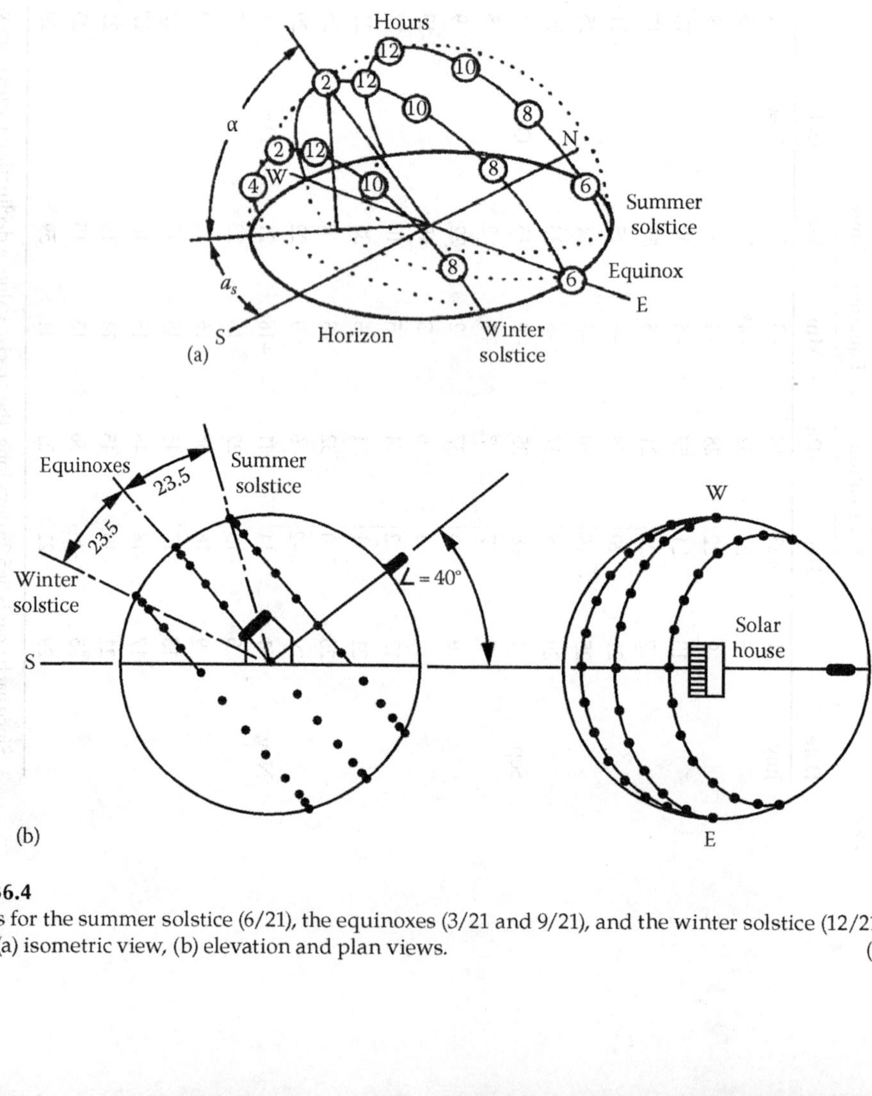

FIGURE 36.4
Sun paths for the summer solstice (6/21), the equinoxes (3/21 and 9/21), and the winter solstice (12/21) for a site at 40° N: (a) isometric view, (b) elevation and plan views. *(Continued)*

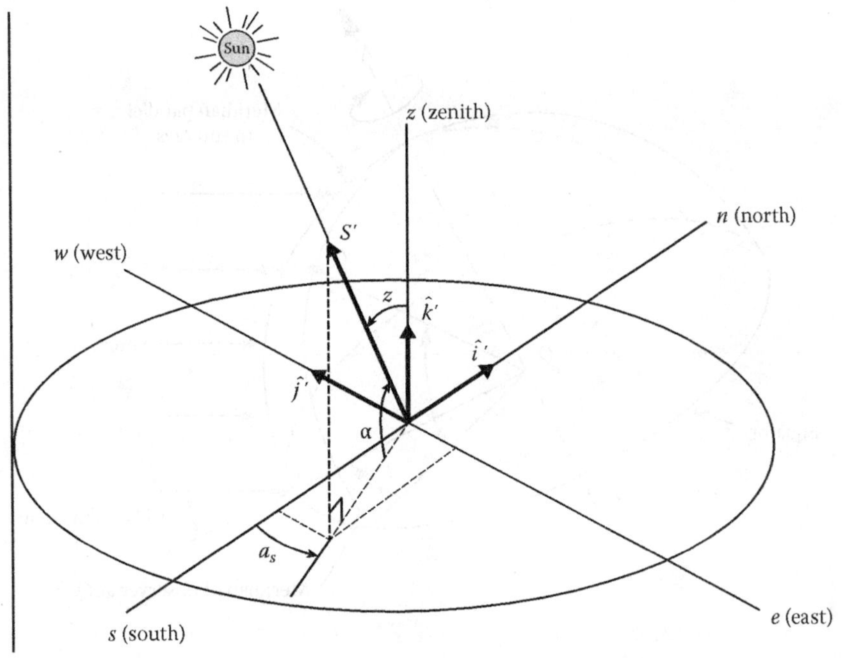

(c)

FIGURE 36.4 *(Continued)*
Sun paths for the summer solstice (6/21), the equinoxes (3/21 and 9/21), and the winter solstice (12/21) for a site at 40° N: (c) solar altitude and azimuth angles.

angles, as shown in Figure 36.4c. The *solar altitude angle*, α, is the angle between a line collinear with the sun's rays and the horizontal plane. The *solar azimuth angle*, a_s, is the angle between a due south line and the projection of the site to sun line on the horizontal plane. The sign convention used for the azimuth angle is positive west of south and negative east of south. The *solar zenith angle*, *z*, is the angle between the site to sun line and the vertical at the site:

$$z = 90° - \alpha. \tag{36.2}$$

The solar altitude and azimuth angles are not fundamental angles. Hence, they must be related to the fundamental angular quantities *hour angle*, *latitude*, and *declination*. The three angles are shown in Figure 36.5. The solar hour angle h_s is based on the nominal time of 24 h required for the sun to move 360° around the earth or 15° per hour. Therefore, h_s is defined as

$$h_s = (15°/\text{h}) \cdot (\text{Hours from local solar noon}) = \frac{\text{Minutes from local solar noon}}{4\ \text{min/degree}}. \tag{36.3}$$

Again, values east of due south, that is, morning values, are negative; and values west of due south are positive.

The latitude angle *L* is the angle between the line from the center of the earth to the site and the equatorial plane. The latitude may be read from an atlas and is considered positive north of the equator and negative south of the equator.

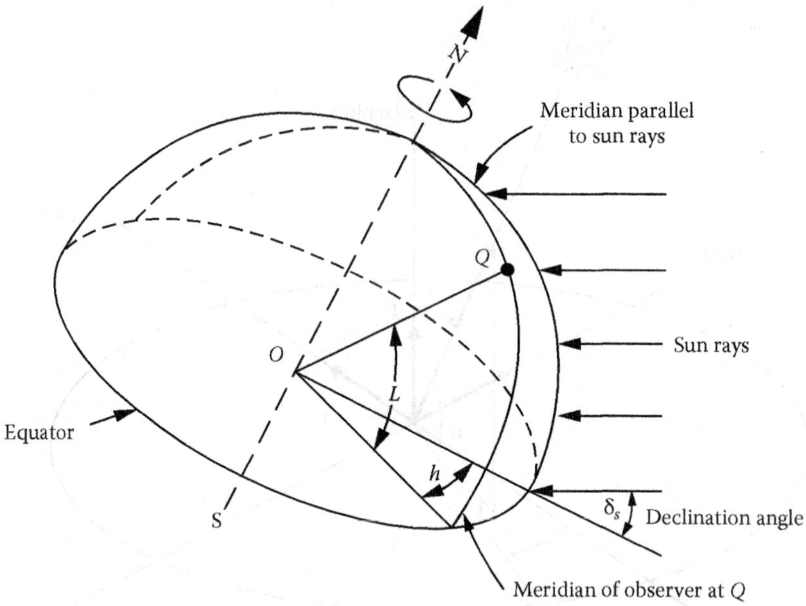

FIGURE 36.5
Definition of solar hour angle h_s, solar declination δ_s, and latitude L; Q site of interest.

36.2.1 Solar Time and Angles

The sun angles are obtained from the local solar time, which differs from the local standard time (LST). The relationship between the local solar time and the LST is

$$\text{Solar time} = LST + ET + (l_{st} - l_{local}) \cdot 4\,\text{min/degree}, \tag{36.4}$$

where
ET is the equation of time, which is a correction factor that accounts for the irregularity of the speed of earth's motion around the sun
l_{st} is the standard time meridian
l_{local} is the local longitude

ET may be estimated from Table 36.1 or Figure 36.6 or calculated from the following empirical equation:

$$ET(\text{in min}) = 9.87 \sin 2B - 7.53 \cos B - 1.5 \sin B, \tag{36.5}$$

where $B = 360(n - 81)/360°$.

The solar altitude angle, α, can be found from the application of the law of cosines to the geometry of Figures 36.4c and 36.5 and simplification as

$$\sin \alpha = \sin L \sin \delta_s + \cos L \cos \delta_s \cos h_s. \tag{36.6}$$

Using a similar technique, the solar azimuth angle, a_s, can be found as

$$\sin a_s = \cos \delta_s \sin h_s / \cos \alpha. \tag{36.7}$$

At local solar noon, $h_s = 0$; therefore, $\alpha = 90 - |L - \delta_s|$, and $a_s = 0$.

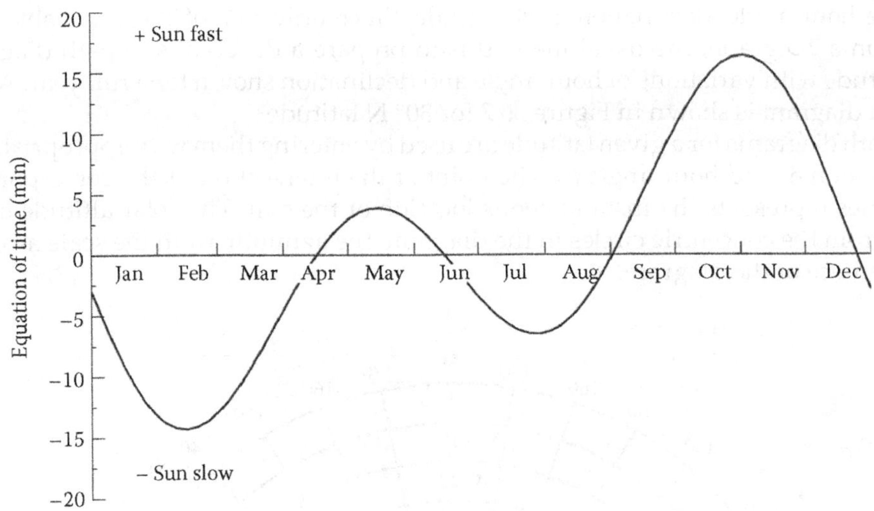

FIGURE 36.6
Equation of time.

In calculating the solar azimuth angle from Equation 36.7, a problem occurs whenever the absolute value of a_s is greater than 90°. A computational device usually calculates the angle as less than 90° since sin a_s = sin(180 − a_s). The problem can be solved in the following way:

For $L > \delta_s$, the solar times when the sun is due east (t_E) or due west (t_W) can be calculated by t_E or t_W = 12:00 noon ∓ (cos^{-1}[tan δ_s/tan$L$]^0)/(15°/h) (− for t_E, + for t_W).

For solar times earlier than t_E or later than t_W, the sun would be north (south in the southern hemisphere) of the east-west line and the absolute value of a_s would be greater than 90°, which may be calculated as a_s = + or−(180°−|a_s|).

For $L \le \delta_s$, the sun remains north (south in the southern hemisphere) of the east-west line and the true value of a_s is greater than 90°.

Sunrise and *sunset* times can be estimated by finding the hour angle for $\alpha = 0$. Substituting $\alpha = 0$ in Equation 19.6 gives the hour angles for sunrise (h_{sr}) and sunset (h_{ss}) as

$$h_{ss} \text{ or } h_{sr} = \pm\cos^{-1}[-\tan L \cdot \tan\delta_s]. \quad (36.8)$$

It should be emphasized that Equation 36.8 is based on the center of the sun being at the horizon. In practice, sunrise and sunset are defined as the times when the upper limb of the sun is on the horizon. Because the radius of the sun is 16′, the sunrise would occur when $\alpha = -16'$. Also, at lower solar elevations, the sun will appear on the horizon when it is actually 34′ below the horizon. Therefore, for apparent sunrise or sunset, $\alpha = -50'$.

Knowledge of the solar angles is helpful in the design of passive solar buildings, especially the placement of windows for solar access and the roof overhang for shading the walls and windows at certain times of the year.

36.2.2 Sun-Path Diagram

The projection of the sun's path on the horizontal plane is called a *sun-path diagram*. Such diagrams are very useful in determining shading phenomena associated with solar collectors, windows, and shading devices. As shown earlier, the solar angles (α, a_s) depend upon the hour angle, declination, and latitude. Since only two of these variables can be plotted on a 2D graph, the usual method is to prepare a different sun-path diagram for each latitude with variations of hour angle and declination shown for a full year. A typical sun-path diagram is shown in Figure 36.7 for 30° N latitude.

Sun-path diagrams for a given latitude are used by entering them with appropriate values of declination δ_s and hour angle h_s. The point at the intersection of the corresponding δ_s and h_s lines represents the instantaneous location of the sun. The solar altitude can then be read from the concentric circles in the diagram, the azimuth, from the scale around the circumference of the diagram.

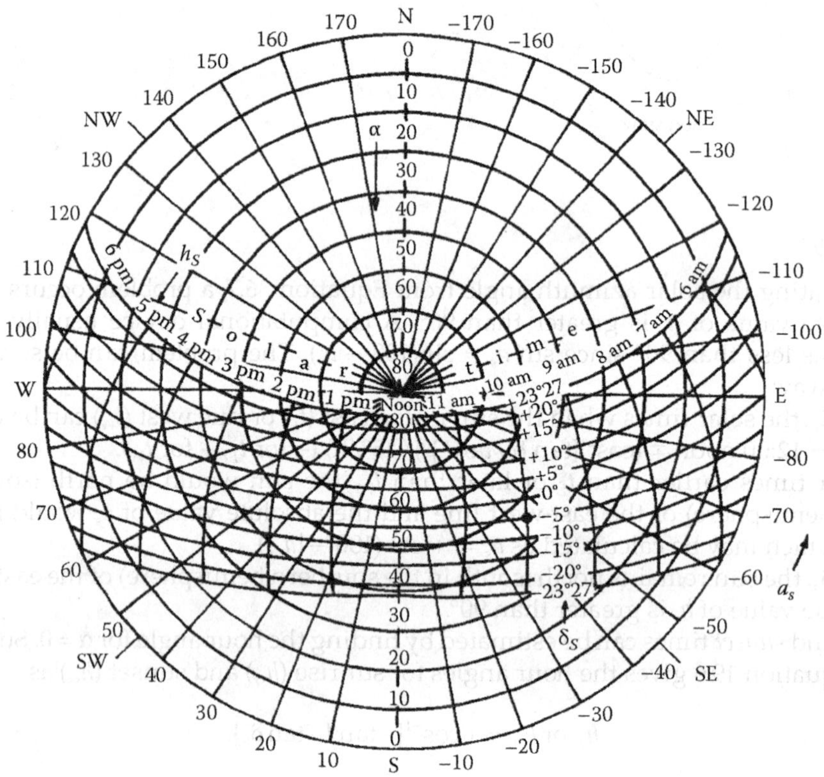

Declination	Approx. dates
+23°27′	June 22
+20°	May 21, July 24
+15°	May 1, Aug. 12
+10°	Apr. 16, Aug. 28
+5°	Apr. 3, Sep. 10
0°	Mar. 21, Sep. 23
−5°	Mar. 8, Oct. 6
−10°	Feb. 23, Oct. 20
−15°	Feb. 9, Nov. 3
−20°	Jan. 21, Nov. 22
−23°27′	Dec. 22

FIGURE 36.7
Sun-path diagram for 30° N latitude showing altitude and azimuth angles.

36.2.3 Shadow-Angle Protractor

The shadow-angle protractor used in shading calculations is a plot of solar altitude angles, projected onto a given plane, versus solar azimuth angle. The projected altitude angle is usually called the *profile angle* γ. It is denned as the angle between the normal to a surface and the projection of the sun's rays on a vertical plane normal to the same surface. The profile angle is shown in Figure 36.8a with the corresponding

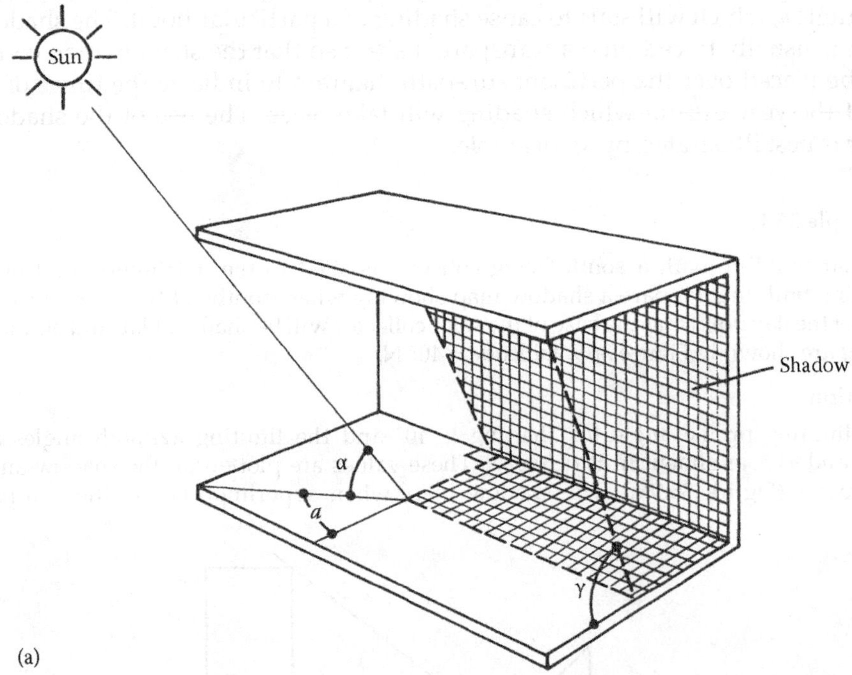

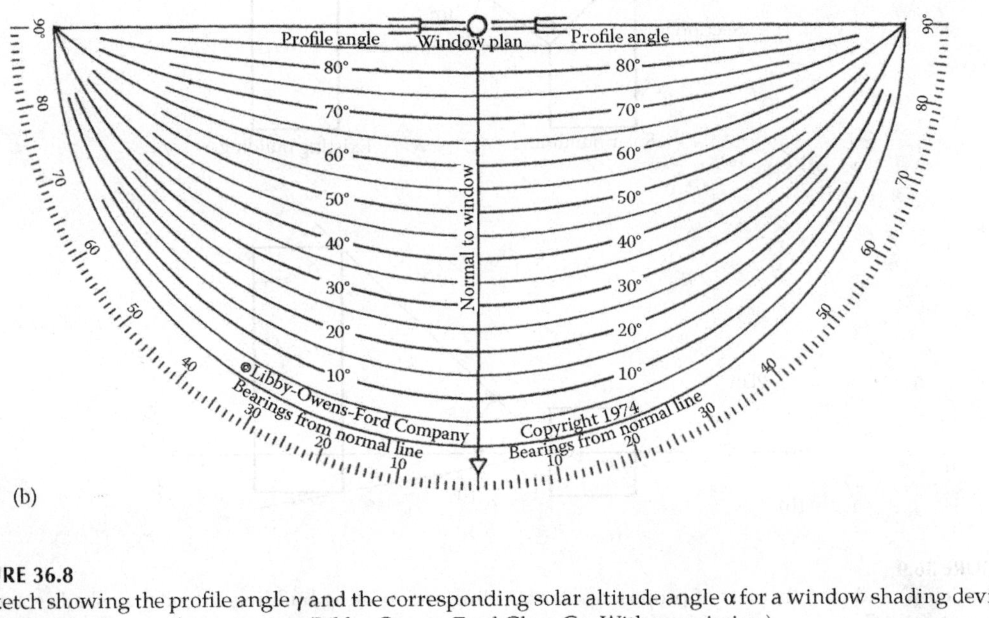

FIGURE 36.8
(a) Sketch showing the profile angle γ and the corresponding solar altitude angle α for a window shading device and (b) the shadow-angle protractor. (Libby-Owens-Ford Glass Co. With permission.)

solar altitude angle. The profile angle, which is always used in sizing shading devices, is given by

$$\tan\gamma = \sec a \tan\alpha, \quad (36.9)$$

where a is the solar azimuth angle with respect to the wall normal.

Figure 36.8b shows the shadow-angle protractor to the same scale as the sun-path diagram in Figure 36.7. It is used by plotting the limiting values of profile angle γ and azimuth angle a, which will start to cause shading of a particular point. The shadow-angle protractor is usually traced onto a transparent sheet so that the shadow map constructed on it can be placed over the pertinent sun-path diagram to indicate the times of day and months of the year during which shading will take place. The use of the shadow-angle protractor is best illustrated by an example.

Example 36.1

A solar building with a south-facing collector is sited to the north-northwest of an existing building. Prepare a shadow map showing what months of the year and what part of the day point C at the base of the solar collector will be shaded. Plan and elevation views are shown in Figure 36.9. Latitude = 40° N.

Solution

The limiting profile angle for shading is 40° and the limiting azimuth angles are −45° and +10° as shown in Figure 36.9. These values are plotted on the shadow-angle protractor (Figure 36.10a). The shadow map, when superimposed on the sun-path

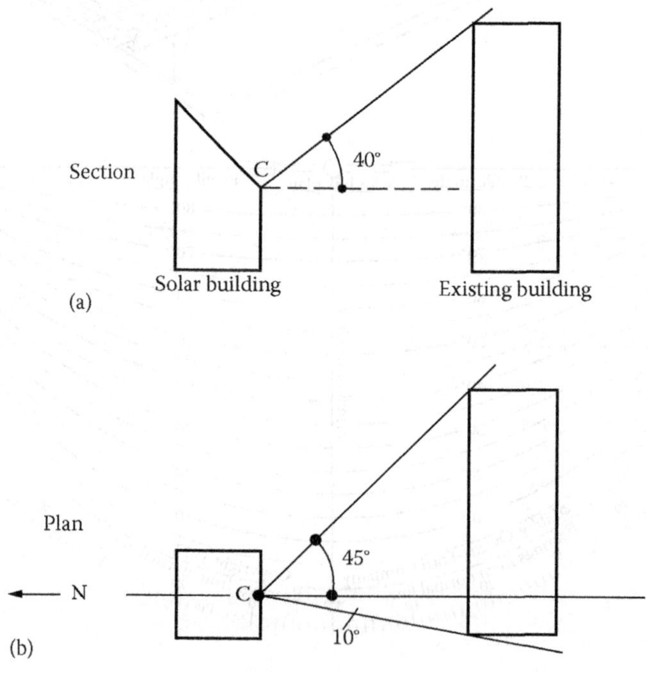

FIGURE 36.9
Elevation (a) and plan (b) view of proposed solar building and existing building, which may shade solar collector at point C.

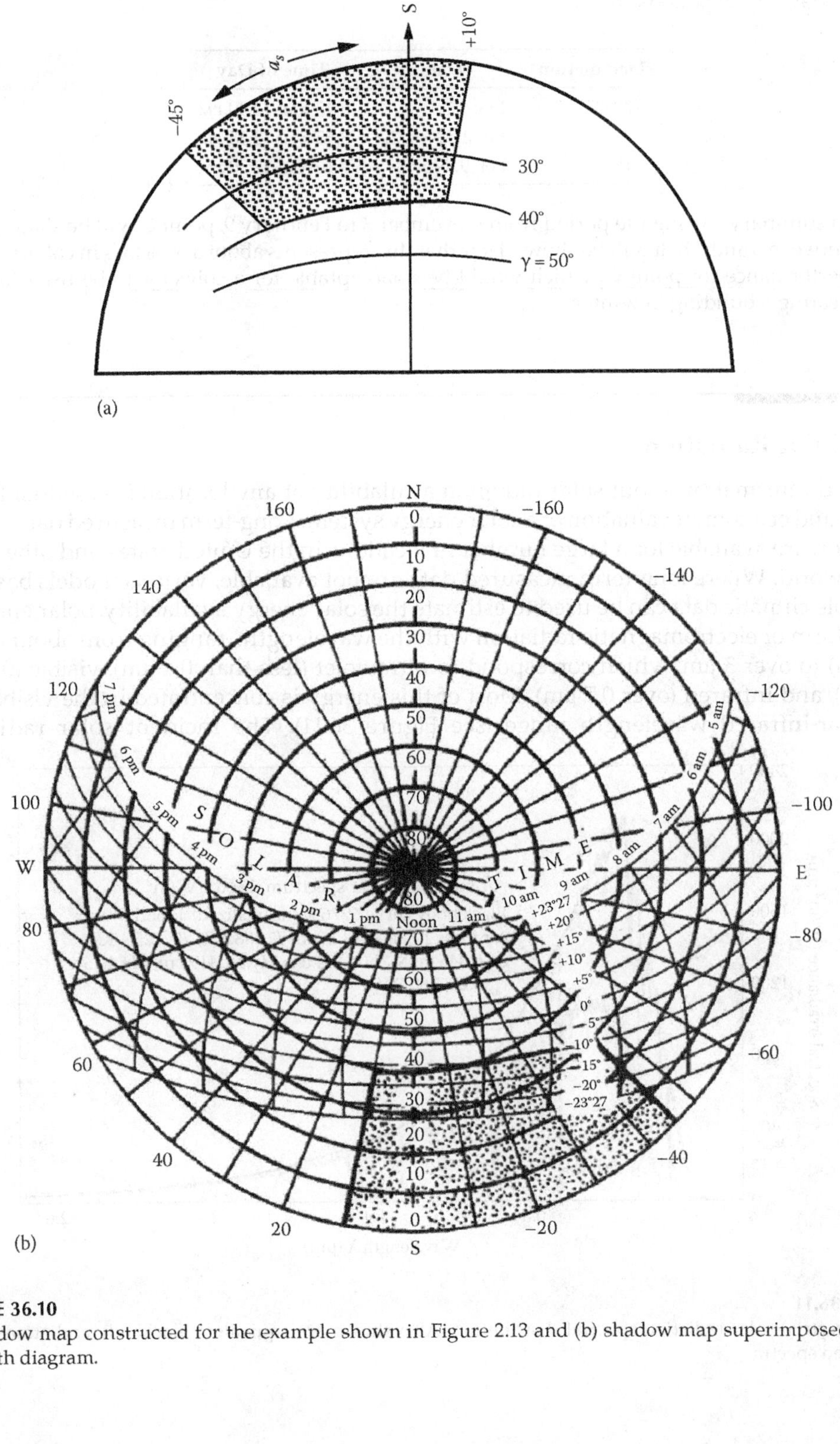

FIGURE 36.10
(a) Shadow map constructed for the example shown in Figure 2.13 and (b) shadow map superimposed on the sun-path diagram.

diagram (Figure 36.10b), shows that point C will be shaded during the following times of day for the periods shown:

Declination	Date	Time of Day
−23°27′	Dec 22	8:45 AM–12:40 PM
−20°	Jan 21, Nov 22	8:55 AM–12:35 PM
−15°	Feb 9, Nov 3	9:10 AM–12:30 PM

In summary, during the period from November 3 to February 9, point C will be shaded between 3 and 4 h. It will be shown later that this represents about a 50% loss in collector performance for point C, which would be unacceptable for a collector to be used for heating a building in winter.

36.3 Solar Radiation

Detailed information about solar radiation availability at any location is essential for the design and economic evaluation of a solar energy system. Long-term measured data of solar radiation are available for a large number of locations in the United States and other parts of the world. Where long-term measured data are not available, various models based on available climatic data can be used to estimate the solar energy availability. Solar energy is in the form of electromagnetic radiation with the wavelengths ranging from about 0.3 μm (10^{-6} m) to over 3 μm, which correspond to ultraviolet (less than 0.4 μm), visible (0.4 and 0.7 μm), and infrared (over 0.7 μm). Most of this energy is concentrated in the visible and the near-infrared wavelength range (see Figure 36.11). The incident solar radiation,

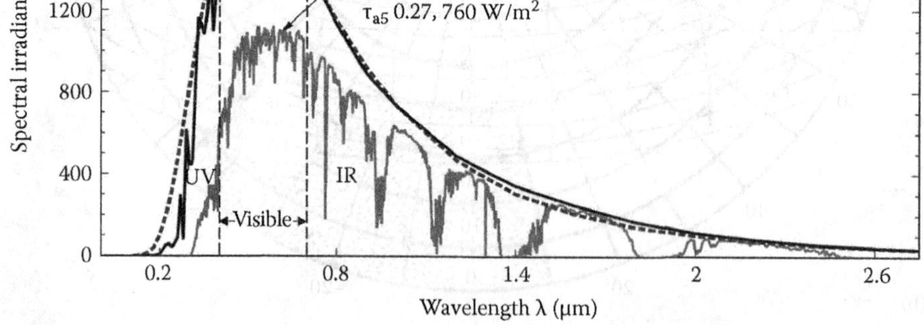

FIGURE 36.11
Extraterrestrial solar radiation spectral distribution. Also shown are equivalent blackbody and atmosphere-attenuated spectra.

sometimes called *insolation*, is measured as irradiance, or the energy per unit time per unit area (or power per unit area). The units most often used are watts per square meter (W/m^2), British thermal units per hour per square foot ($Btu/h\text{-}ft^2$), and langleys per minute (calories per square centimeter per minute, $cal/cm^2/min$).

36.3.1 Extraterrestrial Solar Radiation

The average amount of solar radiation falling on a surface normal to the rays of the sun outside the atmosphere of the earth (extraterrestrial) at mean earth–sun distance (D_0) is called the *solar constant*, I_o. Measurements by NASA indicated the value of the solar constant to be 1353 W/m^2 (±1.6%), 429 $Btu/h\text{-}ft^2$, or 1.94 $Cal/cm^2/min$ (langleys/min). This value was revised upward by Fröhlich et al. [7] to 1377 W/m^2 or 437.1 $Btu/h\text{-}ft^2$ or 1.974 langleys/min, which was the value used in compiling SOLMET data in the United States [37,38]. At present, there is no consensus on the value of the solar constant. Recently, new measurements have found the value of the *solar constant* to be *1366.1 W/m²*. A value of 1367 W/m^2 is also used by many references.

The variation in seasonal solar radiation availability at the surface of the earth can be understood from the geometry of the relative movement of the earth around the sun. Since the earth's orbit is elliptical, the earth–sun distance varies during a year, the variation being ±1.7% from the average. Therefore, the extraterrestrial radiation, *I*, also varies by the inverse square law as follows:

$$I = I_0\left(\frac{D_0}{D}\right)^2, \tag{36.10}$$

where

D is the distance between the sun and the earth

D_0 is the yearly mean earth–sun distance (1.496×10^{11} m)

The $(D_0/D)^2$ factor may be approximated as [42]

$$(D_0/D)^2 = 1.00011 + 0.034221\cos(x) + 0.00128\sin(x) + 0.000719\cos(2x) + 0.000077\sin(2x), \tag{36.11}$$

where

$$x = \frac{360(n-1)}{365°}, \tag{36.12}$$

and n = day number (starting from January 1 as 1). The following approximate relationship may also be used without much loss of accuracy:

$$I = I_0[1 + 0.034\cos(360n/365.25)°]. \tag{36.13}$$

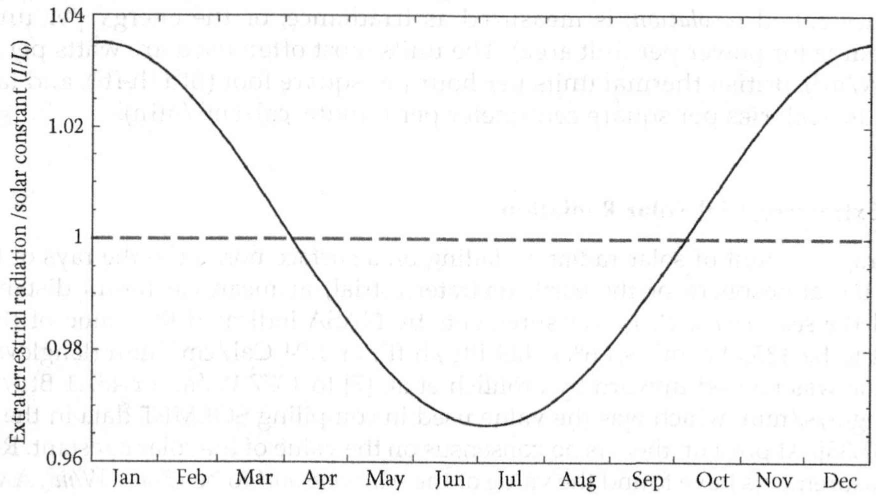

FIGURE 36.12
Effect of the time of year on the ratio of extraterrestrial radiation to the nominal solar constant.

Figure 36.12 also shows the relationship of the extraterrestrial solar radiation to the solar constant. For many solar energy applications, such as photovoltaics and photocatalysis, it is necessary to examine the distribution of energy within the solar spectrum. Figure 36.11 shows the spectral irradiance at the mean earth–sun distance for a solar constant of 1366.1 W/m^2 as a function of wavelength. The data are also presented in Table 36.2 and their use is illustrated in the following example.

Example 36.2

Calculate the fraction of extraterrestrial solar radiation within the visible part of the spectrum, that is, between 0.40 and 0.70 μm.

Solution

The first column in Table 36.2 gives the wavelength. The second column gives the averaged solar spectral irradiance in a band centered at the wavelength in the first column. The fourth column, D_λ, gives the percentage of solar total radiation at wavelengths shorter than the value of λ in the first column. At a value of 0.40 μm, 8.12% of the total radiation occurs at shorter wavelengths. At a wavelength of 0.70%, 47.25% of the radiation occurs at shorter wavelengths. Consequently, 39.13% of the total radiation lies within the band between 0.40 and 0.70 μm, and the total energy received outside the earth's atmosphere within that spectral range is 535 W/m^2 during equinoxes.

36.4 Estimation of Terrestrial Solar Radiation

As extraterrestrial solar radiation, *I*, passes through the atmosphere, a part of it is reflected back into space, a part is absorbed by air and water vapor, and some gets scattered by molecules of air, water vapor, aerosols, and dust particles (Figure 36.13). The part of solar radiation that reaches the surface of the earth with essentially no change in direction is

TABLE 36.2

Extraterrestrial Solar Irradiance

λ (μm)	E_λ[a] (W/m² μm)	D_λ[b] (%)	λ (μm)	E_λ (W/m² μm)	D_λ (%)	λ (μm)	E_λ (W/m² μm)	D_λ (%)
0.115	0.00799	3.76E−04	0.43	1389	11.98	0.90	889.60	63.91
0.14	0.07694	1.07E−03	0.44	1848	13.25	1.00	730.70	69.84
0.16	0.20640	1.25E−03	0.45	2131	14.72	1.2	488.60	78.55
0.18	2.06	2.45E−03	0.46	2092	16.26	1.4	342.90	84.58
0.20	7.93	8.68E−03	0.47	2010	17.79	1.6	247.70	88.89
0.22	51.91	0.05	0.48	2102	19.32	1.8	168.20	91.88
0.23	59.09	0.08	0.49	2072	20.78	2.0	115.90	93.91
0.24	42.19	0.12	0.50	1932	22.24	2.2	82.58	95.35
0.25	62.28	0.16	0.51	1915	23.66	2.4	58.47	96.36
0.26	90.16	0.22	0.52	1864	25.00	2.6	43.54	97.10
0.27	297.50	0.38	0.53	1938	26.38	2.8	33.25	97.65
0.28	78.46	0.52	0.54	1813	27.78	3.0	25.60	98.08
0.29	617.70	0.74	0.55	1905	29.16	3.2	20.08	98.42
0.30	416.60	1.14	0.56	1812	30.52	3.4	15.75	98.68
0.31	464.80	1.56	0.57	1803	31.86	3.6	12.83	98.89
0.32	836.30	2.08	0.58	1818	33.19	3.8	10.39	99.06
0.33	1162	2.72	0.59	1716	34.51	4.0	8.30	99.19
0.34	1133	3.42	0.60	1737	35.80	4.5	5.03	99.43
0.35	1081	4.12	0.62	1681	38.29	5.0	3.27	99.58
0.36	1188	4.84	0.64	1591	40.68	6.0	1.64	99.75
0.37	1376	5.67	0.66	1517	42.94	7.0	0.90360	99.84
0.38	1096	6.50	0.68	1474	45.14	8.0	0.53890	99.89
0.39	1301	7.26	0.70	1413	47.25	10.0	0.22720	99.94
0.40	1727	8.12	0.72	1361	49.29	15.0	0.04418	99.98
0.41	1610	9.39	0.75	1273	52.16	20.0	0.01385	99.99
0.42	1787	10.71	0.80	1129	56.54	50.0	0.00036	100.0

Source: Adapted from Gueymard, C., *Sol. Energ.* 76, 423, 2003.
Note: Solar constant = 1366.1 W/m².
[a] E_λ is the solar spectral irradiance.
[b] D_λ is the percentage of the solar constant associated with wavelengths shorter than λ.

called *direct* or *beam radiation*. The scattered diffuse radiation reaching the surface from the sky is called the *sky diffuse radiation*.

Although extraterrestrial radiation can be predicted with certainty,* radiation levels on the earth are subject to considerable uncertainty resulting from local climatic interactions. The most useful solar radiation data are based on long-term (30 years or more) measured average values at a location, which unfortunately are not available for most locations in the world. For such locations, an estimating method (theoretical model) based on some measured climatic parameter may be used. This chapter describes several ways of estimating terrestrial solar radiation; all have large uncertainties (as much as ±30%) associated with them.

36.4.1 Atmospheric Extinction of Solar Radiation

As solar radiation I travels through the atmosphere, it is attenuated due to absorption and scattering. If K is the local extinction coefficient of the atmosphere, the

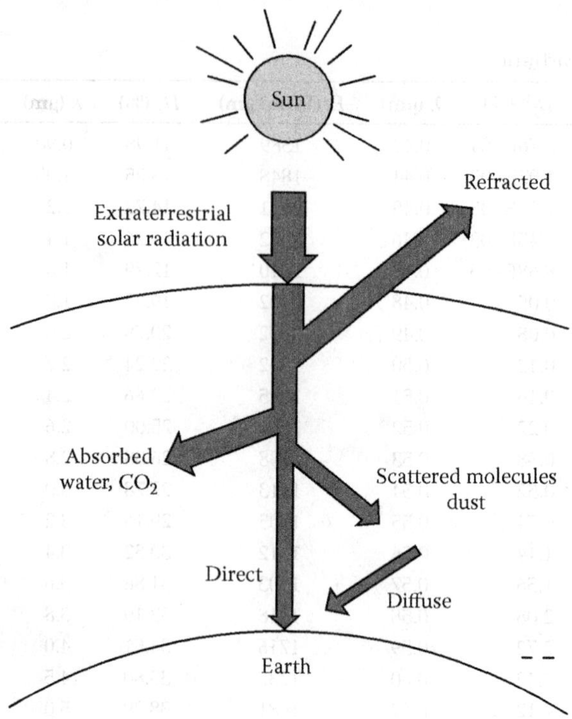

FIGURE 36.13
Attenuation of solar radiation as it passes through the atmosphere.

beam solar radiation at the surface of the earth can be written according to Bouger's law as

$$I_{b,N} = Ie^{-\int kdx}, \tag{36.14}$$

where $I_{b,N}$ is the instantaneous beam solar radiation per unit area normal to the sun's rays and x is the length of travel through the atmosphere. If L_o is the vertical thickness of the atmosphere and

$$\int_0^{L_o} Kdx = k, \tag{36.15}$$

the beam normal solar radiation for a solar zenith angle of z will be

$$I_{b,N} = Ie^{-k\sec z} = Ie^{-k/\sin\alpha} = Ie^{-km}, \tag{36.16}$$

where m is a dimensionless path length of sunlight through the atmosphere, sometimes called the *air mass ratio* (Figure 36.14):

$$m = \frac{\overline{BP}}{\overline{AP}} \approx \frac{1}{\cos z} \approx \frac{1}{\sin\alpha}. \tag{36.17}$$

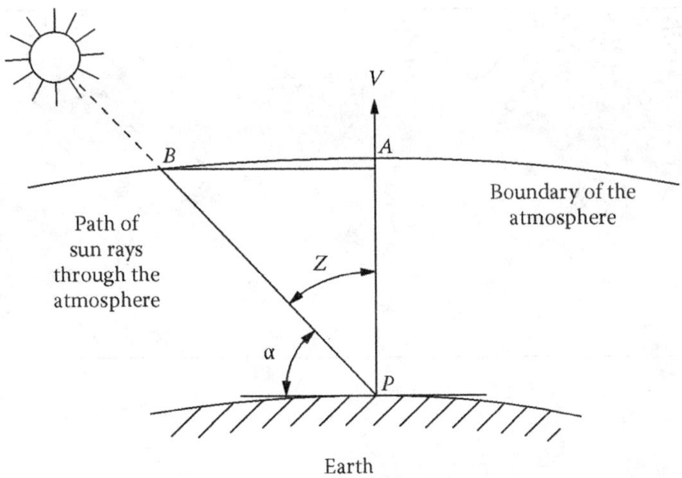

FIGURE 36.14
Air mass definition; air mass $m = BP/AP = \text{cosec}\ \alpha$, where α is the altitude angle. The atmosphere is idealized as a constant thickness layer.

The following equation gives a more accurate value of air mass according to Kasten and Young (1989):

$$m \approx \frac{1}{\sin\alpha + 0.50572\ (6.07995 + \alpha)^{-1.6364}} \tag{36.18}$$

where α is expressed in degrees. When solar altitude angle is 90° (sun is overhead), $m = 1$.

36.4.2 Clear Sky Radiation Model

Gueymard and Thevenard (2009) have described a model that can be used to model solar radiation for clear days for a large number of locations in the world. This model was developed for ASHRAE to calculate the solar heat gain for fenestration; therefore, they named it the ASHRAE clear sky model. This is a simple model that was developed based on a large number of simulations using sophisticated spectral simulations using the spectral model of the atmospheric radiative transfer of sunshine (SMARTS) spectral code developed by Gueymard (2000, 2005b, 2008a) and validating with ground-based measurements. Based on the detailed simulations, Gueymard developed a simple two-band solar irradiance model, the reference evaluation of solar transmittance, 2 (REST2), that can model clear sky solar irradiance very accurately. The proposed model was developed in two steps:

1. Solar transmittance of clear sky was modeled based on two spectral bands, the first band from 290 to 700 nm, characterized by absorption by molecules and aerosols, and the second band from 700 to 4000 nm, characterized by absorption by water vapor and CO_2. The two-band clear sky radiation model was used to calculate clear sky solar irradiance for a large number of *typical* cases and compared with the data covering a large part of the world. Figure 36.15 shows the global sites used in the validation.
2. The second step consisted in developing a condensed model depending on only two monthly parameters described later in this section.

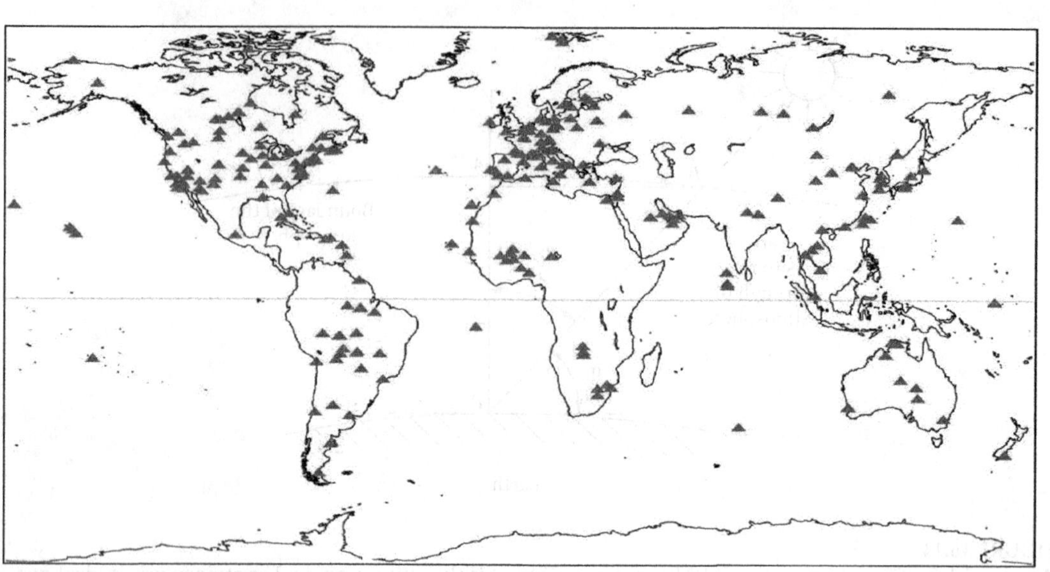

FIGURE 36.15
World sites of interest used in the model. (From Gueymard, C.A. and Thevenard, D., *Solar Energy*, 82, 272, 2009.)

According to the ASHRAE model, the beam and diffuse components are calculated as

$$I_{b,N} = I(e^{-\tau b\, m^{b}}), \tag{36.19}$$

$$I_{d,h} = I(e^{-\tau d\, m^{d}}), \tag{36.20}$$

where

$I_{b,N}$ is the beam normal irradiance per unit area normal to the sun rays
$I_{d,h}$ is the diffuse horizontal irradiance per unit area on a horizontal surface
I is the extraterrestrial normal irradiance
m is the air mass
τ_b, τ_d are the beam and diffuse optical depths (τ_b and τ_d are more correctly termed pseudo-optical depths, because optical depth is usually employed when the air mass coefficient is unity)
b, d are the beam and diffuse air mass exponents

Values of τ_b and τ_d are location-specific and vary during the year. They embody the dependence of clear sky solar radiation upon local conditions, such as elevation, precipitable water content, and aerosols. Their average values are tabulated for the 21st day of each month for all the locations in the tables of climatic design conditions (*ASHRAE Handbook*, 2009 Fundamentals).

Air mass exponents b and d are correlated to τ_b and τ_d through the following empirical relationships:

$$b = 1.219 - 0.043\,\tau_b - 0.151\,\tau_d - 0.204\,\tau_b \cdot \tau_d \tag{36.21}$$

$$d = 0.202 + 0.852\,\tau_b - 0.007\,\tau_d - 0.357\,\tau_b \cdot \tau_d. \tag{36.22}$$

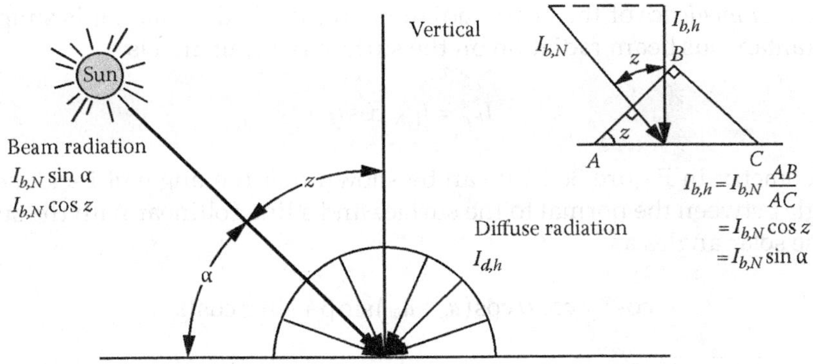

FIGURE 36.16
Solar radiation on a horizontal surface.

This radiation model describes a simple parameterization of a sophisticated broadband radiation model and provides accurate predictions of $I_{b,N}$ and $I_{d,h}$ even at sites where the atmosphere is very hazy or humid most of the time.

Solar radiation on a horizontal surface is given by (Figure 36.16)

$$I_h = \left(I_{b,N} \sin\alpha + I_{d,h}\right). \tag{36.23}$$

36.4.3 Solar Radiation on a Tilted Surface

Solar radiation on an arbitrary tilted surface having a tilt angle of β from the horizontal and an azimuth angle of a_w (assumed + west of south), as shown in Figure 36.17, is the sum of components consisting of beam ($I_{b,c}$), sky diffuse ($I_{d,c}$), and ground-reflected solar radiation ($I_{r,c}$):

$$I_c = I_{b,c} + I_{d,c} + I_{r,c}. \tag{36.24}$$

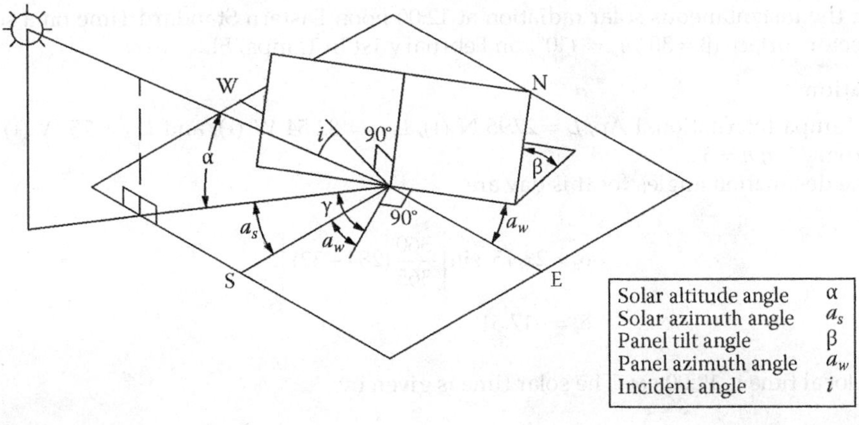

FIGURE 36.17
Definitions of solar angles for a tilted surface.

If i is the *angle of incidence* of the beam radiation on the tilted surface, it is simple to show that the instantaneous beam radiation on the surface per unit area is

$$I_{b,c} = I_{b,N} \cos i. \quad (36.25)$$

From the geometry in Figure 36.17, it can be shown that the angle of incidence i for the surface (angle between the normal to the surface and a line collinear with the sun's rays) is related to the solar angles as

$$\cos i = \cos\alpha \cos(a_s - a_w)\sin\beta + \sin\alpha\cos\beta. \quad (36.26)$$

The diffuse radiation on the surface ($I_{d,c}$) can be obtained by multiplying the sky diffuse radiation on a horizontal surface by the view factor between the sky and the surface:*

$$\begin{aligned} I_{d,c} &= I_{d,h}(I + \cos\beta)/2 \\ &= I_{d,h}\cos^2(\beta/2). \end{aligned} \quad (36.27)$$

The ground-reflected solar radiation can be found from the total solar radiation incident on a horizontal surface and the ground reflectance ρ as

$$I_{r,c} = I_h\rho. \quad (36.28)$$

The part of I_r intercepted by the tilted surface can be found by multiplying the ground-reflected radiation by the view factor between the surface* and the ground:

$$\begin{aligned} I_{r,c} &= \rho I_h(1 - \cos\beta)/2 = \rho I_h \sin^2(\beta/2) \\ &= \rho(I_{b,h}\sin\alpha + I_{d,h})\sin^2(\beta/2). \end{aligned} \quad (36.29)$$

For ordinary ground or grass, ρ is approximately 0.2, and for snow-covered ground, it can be taken as approximately 0.8.

Example 36.3a

Find the instantaneous solar radiation at 12:00 noon Eastern Standard Time on a solar collector surface ($\beta = 30°$, $a_w = +10°$) on February 1st in Tampa, FL.

Solution

For Tampa International Ap, $L = 27.96$ N (+), $L_{local} = 82.54$ W (+), and $L_{ST} = 75$ W (+) for February 1st; $n = 32$.

The declination angles for this day are

$$\delta_s = 23.45°\sin\left[\frac{360}{365}(284 + 32)\right].$$

$$\delta_s = -17.51°$$

The local time is 12:00 PM. The solar time is given by

$$ST = LST + ET + (L_{st} - L_{local})\cdot 4\frac{\text{min}}{1°}.$$

The equation of time is

$$B = \frac{360}{364}(32-81) = -48.46°$$

$$ET(\text{min}) = 9.87(2\times(-48.46°)) - 7.53\cos(-48.46°) - 1.5\sin(-48.46°).$$
$$= -13.66\,\text{min}.$$

Then the solar time is

$$ST = 12:00 - 13.66\text{ min} + (75° - 82.54°)\times 4 = 11:16\,\text{AM}.$$

The hour angle is

$$h_s = (ST - 12)\frac{15°}{\text{h}} = (11.26 - 12)\frac{15°}{\text{h}} = -11.1°.$$

The solar altitude angle is

$$\sin\alpha = \cos\delta_s \cos L \cos h_s + \sin\delta_s \sin L$$
$$= \cos(-17.51°)\cos(27.96°)\cos(-11.1°) + \sin(-17.51°)\sin(27.96°)$$

and

$$\alpha = 43.30°.$$

The solar azimuth angles are defined by

$$\sin a_s = \cos\delta_s \frac{\sin h_s}{\cos\alpha} = \cos(-17.51°)\frac{\sin(-11.1°)}{\cos(43.30°)}$$

and

$$a_s = -14.61°.$$

We need to find out if $|a_s|$ is greater than 90°.
For

$$L > \delta_s,$$

$$t_E = 12:00 - \left[\cos^{-1}\left(\frac{\tan\delta_s}{\tan L}\right)\right]\frac{\text{h}}{15°}$$

$$= 12:00 - \left[\cos^{-1}\left(\frac{\tan(-17.51°)}{\tan(27.96°)}\right)\right]\frac{\text{h}}{15°}.$$

$$= 3.5689\,\text{h}$$

Given that $ST > t_E$, the sun is south, and $a_s = -14.61°$. The air mass is as follows

$$m = \frac{1}{\sin(43.30°) + 0.50572(6.07995 + 43.30°)^{-1.6364}}$$
$$= 1.4565.$$

The data for the Tampa International AP, FL, are shown in Table 36.3.

TABLE 36.3

Data for Tampa International AP, FL

Lat, 27.96N; long, 82.54W; elev, 3 m												
Month	**Jan**	**Feb**	**Mar**	**Apr**	**May**	**Jun**	**Jul**	**Aug**	**Sep**	**Oct**	**Nov**	**Dec**
τ_b	0.344	0.364	0.391	0.403	0.47	0.473	0.509	0.493	0.445	0.398	0.36	0.346
τ_d	2.531	2.403	2.271	2.272	2.029	2.07	1.95	2.01	2.19	2.356	2.488	2.512
$I_{b,N}$, noon (W/m^2)	902	908	899	895	831	823	793	805	836	861	876	882
$I_{d,h}$, noon (W/m^2)	94	113	134	137	174	167	188	175	143	116	97	93

Source: Data taken from *ASHRAE Handbook*: Fundamentals: SI Edition, American Society of Heating, Refrigeration and Air-Conditioning Engineers, Inc., 2009.

The *pseudo*-optical depths are tabulated for the 21st day of each month for all the locations in the tables of climatic design. Values for other days of the year should be found by interpolation. For this example, by using linear interpolation between January and February, the *pseudo*-optical depths are

$$\tau_b = 0.35109$$

$$\tau_d = 2.48558.$$

The parameters for air mass are as follows:

$$b = 1.219 - 0.043\tau_b - 0.151\tau_d - 0.204\,\tau_b\tau_d = 0.6506$$
$$d = 0.202 - 0.852\tau_b - 0.007\tau_d - 0.357\tau_b\,\tau_d = -0.4261.$$

The extraterrestrial solar radiation is given by

$$I = I_0\left[1 + 0.034\cos\left(\frac{360n}{365.25}\right)^{\circ}\right]$$

$$= 1366.1\left[1 + 0.034\cos\left(360 \times \frac{32}{365.26}\right)^{\circ}\right] = 1406\ \text{W/m}^2.$$

The direct solar radiation component is

$$I_{b,N} = Ie^{-\tau_b m^b}$$
$$= 1406\exp\left(-0.35109 \cdot 1.4565^{0.6506}\right)$$
$$= 898\ \text{W/m}^2.$$

The diffuse solar radiation on a horizontal surface is

$$I_{d,h} = Ie^{-\tau_d m^d}$$
$$= 1406\exp\left(-2.48558 \cdot 1.4565^{-0.4261}\right)$$
$$= 169\ \text{W/m}^2.$$

The instantaneous beam radiation on the surface per unit area is given by

$$I_{b,c} = I_{b,N} \cos i.$$

For this geometry, the cosine of the angle of incidence is

$$\cos i = \cos\alpha \cos(a_s - a_w)\sin\beta + \sin\alpha \cos\beta$$

$$= \cos(43.30°)\cos(-14.61° - (+10°))\sin(30°) + \sin(43.30°)\cos(30°)$$

$$= 0.9248.$$

Then the beam radiation is

$$I_{b,c} = 898 \times 0.9248 = 830\ \text{W/m}^2.$$

The diffuse radiation on the collector surface will be

$$I_{d,c} = I_{d,h}\cos^2\left(\frac{\beta}{2}\right)$$

$$= 169\cos^2\left(\frac{30°}{2}\right)$$

$$= 158\ \text{W/m}^2.$$

The ground-reflected solar radiation is

$$I_{r,c} = \rho\left(I_{b,N}\sin\alpha + I_{d,h}\right)\sin^2\left(\frac{\beta}{2}\right).$$

Assuming that the solar collector is surrounded by ordinary ground or grass, then $\rho \approx 0.2$:

$$I_{r,c} = 0.2(898\sin(43.30°) + 169.10)\sin^2\left(\frac{30°}{2}\right)$$

$$= 11\ \text{W/m}^2.$$

Finally, the total radiation on a tilted collector surface is

$$I_c = I_{b,c} + I_{d,c} + I_{r,c} = 830 + 158 + 11 = 999\ \text{W/m}^2.$$

Example 36.3b

Repeat the calculations in Example 36.3a for a north-facing solar collector (β = 30°, a_w = 10°) in Canberra, Australia (latitude = 35° – 18′ Σ, longitude = 149° – 11′ E, Standard Meridian = 150° E).

Solution

As the day number has not been changed, the values of the δ_s and ET remain the same:

$$\delta_s = -17.51°$$

$$ET = -13.66\ \text{min}.$$

The local time is 12:00 PM. The solar time is given by

$$ST = 12:00 - 13.66\text{ min} + (-150° + 149.18°) \times 4\frac{\text{min}}{1°} = 11:43\text{ AM}.$$

The hour angle is

$$h_s = (ST - 12)\frac{15°}{\text{h}} = (11.72 - 12)\frac{15°}{\text{h}} = -4.2°.$$

The solar altitude angle is

$$\begin{aligned}\sin\alpha &= \cos\delta_s \cos L \cos h_s + \sin\delta_s \sin L \\ &= \cos(-17.51°)\cos(-35.3°)\cos(-4.2°) + \sin(-17.51°)\sin(-35.3°)\end{aligned}$$

and

$$\alpha = 71.82°.$$

The solar azimuth angles are defined by

$$\sin a_s = \cos\delta_s \sin h_s / \cos\alpha = \cos(-17.51°)\sin(-4.2°)/\cos(71.82°)$$

and

$$a_s = -12.93°.$$

The air mass is as follows:

$$m = \frac{1}{\sin(71.82°) + 0.50572(6.07995 + 71.82°)^{-1.6364}} = 1.052.$$

The data for the Canberra airport are shown in Table 36.4.

By using linear interpolation between January and February, the *pseudo*-optical depths are

$$\tau_b = 0.3548$$
$$\tau_d = 2.4584.$$

TABLE 36.4

Data for Canberra Airport

Lat, 35.30S; long, 149.20E; elev, 580 m												
Month	**Jan**	**Feb**	**Mar**	**Apr**	**May**	**Jun**	**Jul**	**Aug**	**Sep**	**Oct**	**Nov**	**Dec**
τ_b	0.363	0.340	0.326	0.313	0.299	0.291	0.292	0.297	0.315	0.318	0.334	0.342
τ_d	2.403	2.559	2.606	2.638	2.730	2.747	2.702	2.678	2.582	2.609	2.520	2.519
$I_{b,N}$, noon (W/m^2)	972	974	952	908	871	853	869	915	948	986	995	998
$I_{d,h}$, noon (W/m^2)	126	104	94	82	69	64	70	78	94	98	111	113

Source: Data taken from: 2009 *ASHRAE Handbook*: Fundamentals: SI Edition, American Society of Heating, Refrigeration and Air-Conditioning Engineers, Inc.

The parameters for air mass are as follows:

$$b = 1.219 - 0.043\tau_b - 0.151\tau_d - 0.204\,\tau_b\tau_d = 0.6546$$

$$d = 0.202 - 0.852\tau_b - 0.007\,\tau_d - 0.357\tau_b\,\tau_d = -0.4289.$$

The extraterrestrial solar radiation is given by

$$I = I_0\left[1 + 0.034\cos\left(\frac{360n}{365.25}\right)^\circ\right] = 1406\ \text{W/m}^2.$$

The direct solar radiation component is

$$I = I_0\left[1 + 0.034\cos\left(\frac{360n}{365.25}\right)^\circ\right] = 1406\ \text{W/m}^2.$$

The diffuse solar radiation on a horizontal surface is

$$I_{d,h} = Ie^{-\tau_d m^d}$$

$$= 1406\exp\left(-2.4584 \times 1.0521^{-0.4289}\right)$$

$$= 127\ \text{W/m}^2.$$

The cosine of the angle of incidence is

$$\cos i = \cos\alpha\cos\left(a_s - a_w\right)\sin\beta + \sin\alpha\cos\beta$$

$$= \cos(71.82^\circ)\cos(-12.93^\circ - (+10^\circ))\sin(30^\circ) + \sin(71.82^\circ)\cos(30^\circ)$$

$$= 0.9665.$$

Then the beam radiation is

$$I_{b,c} = I_{b,N}\cos i$$

$$I_{b,c} = 974 \times 0.9665 = 941\ \text{W/m}^2.$$

The diffuse radiation on the collector surface will be

$$I_{d,c} = I_{d,h}\cos^2\left(\frac{\beta}{2}\right)$$

$$= 127\cos^2\left(\frac{30^\circ}{2}\right)$$

$$= 118\ \text{W/m}^2.$$

The ground-reflected solar radiation is

$$I_{r,c} = \rho\left(I_{b,N}\sin\alpha + I_{d,h}\right)\sin^2\left(\frac{\beta}{2}\right).$$

Assuming that the solar collector is surrounded by ordinary ground or grass, then $\rho \approx 0.2$:

$$I_{r,c} = 0.2(974\sin(71.82°)+127)\sin^2\left(\frac{30°}{2}\right)$$
$$= 14\ \text{W/m}^2.$$

Finally, the total radiation on a tilted collector surface is

$$I_c = I_{b,c} + I_{d,c} + I_{r,c} = 941 + 118 + 14 = 1073\ \text{W/m}^2.$$

36.4.4 Monthly Solar Radiation Estimation Models

One of the earliest methods of estimating solar radiation on a horizontal surface was proposed by the pioneer spectroscopist Angström. It was a simple linear model relating average horizontal radiation to clear-day radiation and to the sunshine level, that is, percent of possible hours of sunshine. Since the definition of a clear day is somewhat nebulous, Page [34] refined the method and based it on extraterrestrial radiation instead of the ill-defined clear day:

$$\begin{aligned}\bar{H}_h &= \bar{H}_{o,h}\left(a + b\frac{\bar{n}}{\bar{N}}\right)\\ &= \bar{H}_{o,h}\left(a + b\frac{\overline{PS}}{100}\right)\end{aligned} \tag{36.30}$$

where

$\bar{H}_h$ and $\bar{H}_{o,h}$ are the horizontal terrestrial and horizontal extraterrestrial radiation levels averaged for a month

$\overline{PS}$ is the monthly averaged percent of possible sunshine (i.e., hours of sunshine/maximum possible duration of sunshine × 100)

a and b are constants for a given site

$\bar{n}$ and $\bar{N}$ are the monthly average numbers of hours of bright sunshine and day length, respectively

The ratio $\bar{n}/\bar{N}$ is also equivalent to the monthly average percent sunshine $\left(\overline{PS}\right)$. $\bar{H}_{o,h}$ can be calculated by finding $H_{o,h}$ from Equation 36.31, using Equations 36.13 and 36.30, and averaging $I_{o,h}$ for the number of days in each month:

$$H_{o,h} = \int_{t_{sr}}^{t_{ss}} I \sin\alpha\, dt. \tag{36.31}$$

Some typical values of a and b are given in Table 36.5 [26].

A number of researchers found Angström–Page-type correlations for specific locations that are listed in Table 36.6. Some of these include additional parameters such as relative humidity and ambient temperature. Correlations listed in the table may be used for the specific locations for which they were developed.

Another meteorological variable that could be used for solar radiation prediction is the opaque cloud cover recorded at many weather stations around the world. This quantity is

TABLE 36.5

Coefficients a and b in the Angström–Page Regression Equation

Location	Climate[a]	Sunshine Hours in Percentage of Possible		a	b
		Range	Avg.		
Albuquerque, NM	BS-BW	68–85	78	0.41	0.37
Atlanta, GA	Cf	45–71	59	0.38	0.26
Blue Hill, MA	Df	42–60	52	0.22	0.50
Brownsville, TX	BS	47–80	62	0.35	0.31
Buenos Aires, Argentina	Cf	47–68	59	0.26	0.50
Charleston, SC	Cf	60–75	67	0.48	0.09
Dairen, Manchuria	Dw	55–81	67	0.36	0.23
El Paso, TX	BW	78–88	84	0.54	0.20
Ely, NV	BW	61–89	77	0.54	0.18
Hamburg, Germany	Cf	11–49	36	0.22	0.57
Honolulu, HI	Af	57–77	65	0.14	0.73
Madison, WI	Df	40–72	58	0.30	0.34
Malange, Angola	Aw-BS	41–84	58	0.34	0.34
Miami, FL	Aw	56–71	65	0.42	0.22
Nice, France	Cs	49–76	61	0.17	0.63
Poona, India (monsoon)	Am	25–49	37	0.30	0.51
Poona, India (dry)		65–89	81	0.41	0.34
Stanleyville, Congo	Af	34–56	48	0.28	0.39
Tamanrasset, Algeria	BW	76–88	83	0.30	0.43

Source: From Löf, G.O.G. et al., World distribution of solar energy, Engineering Experiment Station Report, University of Wisconsin, Madison, WI, 1966. With permission.

Note: Am, tropical forest climate, monsoon rain, short dry season, but total rainfall sufficient to support rain forest; Aw, tropical forest climate, dry season in winter; BS, steppe or semiarid climate; BW, desert or arid climate; Cf, mesothermal forest climate, constantly moist, rainfall all through the year; Cs, mesothermal forest climate, dry season in winter; Df, microthermal snow forest climate, constantly moist, rainfall all through the year; Dw, microthermal snow forest climate, dry season in winter.

[a] Af, tropical forest climate, constantly moist, rainfall all through the year.

a measure of the percent of the sky dome obscured by opaque clouds. Because this parameter contains even less solar information than sunshine values, it has not been useful in predicting long-term solar radiation values. A subsequent section, however, will show that cloud cover, when used with solar altitude angle or air mass, is a useful estimator of hourly direct radiation.

36.5 Models Based on Long-Term Measured Horizontal Solar Radiation

Long-term measured solar radiation data are usually available as monthly averaged total solar radiation per day on horizontal surfaces. In order to use these data for tilted surfaces, the total solar radiation on a horizontal surface must first be broken down into beam and diffuse components. A number of researchers have proposed models to do that, prominent among them being Liu and Jordan, Collares-Pereira and Rabl, and Erbs, Duffie, and Klein.

TABLE 36.6

Angström–Page-Type Correlations for Specific Locations

Authors	Measured Data Correlated	Correlation Equations[a]
Iqbal [16]	Canada, 3 locations	$\frac{\bar{D}_h}{\bar{H}_h} = 0.791 - 0.635\left(\frac{\bar{n}}{\bar{N}}\right)$ $\frac{\bar{H}_d}{\bar{H}_h} = 0.163 + 0.478\left(\frac{\bar{n}}{\bar{N}}\right) - 0.655\left(\frac{\bar{n}}{\bar{N}}\right)^2$ $\frac{\bar{H}_b}{\bar{H}_{o,h}} = -0.176 + 1.45\left(\frac{\bar{n}}{\bar{N}}\right) - 1.12\left(\frac{\bar{n}}{\bar{N}}\right)^2$
Garg [8]	India, 11 locations, 20 years' data	$\frac{\bar{H}_h}{\bar{H}_{o,h}} = 0.3156 - 0.4520\left(\frac{\bar{n}}{\bar{N}}\right)^2$ $\frac{\bar{D}_h}{\bar{H}_{o,h}} = 0.3616 - 0.2123\left(\frac{\bar{n}}{\bar{N}}\right)$ $\frac{\bar{D}_h}{\bar{H}_h} = 0.8677 - 0.7365\left(\frac{\bar{n}}{\bar{N}}\right)$
Hussain [14]	India	$\frac{\bar{H}_h}{\bar{H}_{o,h}} = 0.394 + 0.364\left[\frac{\bar{n}}{\bar{N}'}\right] - 0.0035W_{at}$ $\frac{\bar{D}_h}{\bar{H}_{o,h}} = 0.306 - 0.165\left[\frac{\bar{n}}{\bar{N}'}\right] - 0.0025W_{at}$
Coppolino [5]	Italy	$\frac{\bar{H}_h}{\bar{H}_{o,h}} = 0.67\left(\frac{\bar{n}}{\bar{N}}\right)^{0.45} \sin(\alpha_{sn})^{0.05}$ α_{sn} = Solar elevation at noon on the 15th of each month, degrees $0.15 \le \frac{\bar{n}}{\bar{N}} \le 0.90$
Akinoglu and Ecevit [1]	Italy	$\frac{\bar{H}_h}{\bar{H}_{o,h}} = 0.145 + 0.845\left(\frac{\bar{n}}{\bar{N}}\right) - 0.280\left(\frac{\bar{n}}{\bar{N}}\right)^2$
Ögelman et al. [33]	Turkey, 2 locations, 3 years' data	$\left(\frac{\bar{H}_h}{\bar{H}_{o,h}}\right) = 0.204 + 0.758\left(\frac{\bar{n}}{\bar{N}}\right) - 0.250\left\{\left[\left(\frac{\bar{n}}{\bar{N}}\right)^2\right]^2 + \sigma\frac{2/n}{\bar{N}}\right]$ $\sigma\frac{2/n}{\bar{N}} = 0.035 + 0.326\left(\frac{\bar{n}}{\bar{N}}\right) - 0.433\left(\frac{\bar{n}}{\bar{N}}\right)^2$
Gopinathan [10]	40 locations around the world	$\frac{\bar{H}_h}{\bar{H}_{o,h}} = a + b\left(\frac{\bar{n}}{\bar{N}}\right)$ $a = -0.309 + 0.539\cos L - 0.0639h + 0.290\left(\frac{\bar{n}}{\bar{N}}\right)$ $b = 1.527 - 1.027\cos L + 0.0926h - .359\left(\frac{\bar{n}}{\bar{N}}\right)$

Note: N', maximum duration for which Campbell–Stokes recorder can be active, that is, solar elevation >5°; W_{at}, relative humidity × $(4.7923 + 0.3647T + 0.055T^2 + 0.0003\,T^3)$; T, ambient temperature, °C; W_{at}, gm moisture/m^3; h, elevation in km above sea level; L, latitude.

[a] $\bar{H}_a, \bar{H}_b, \bar{H}_{o,h}, \bar{D}_h$ are monthly averaged daily values.

36.5.1 Monthly Solar Radiation on Tilted Surfaces

In a series of papers, Liu and Jordan [21–25] have developed an essential simplification in the basically complex computational method required to calculate long-term radiation on tilted surfaces. This is called the Liu and Jordan (LJ) method. The fundamental problem in such calculations is the decomposition of long-term measured total horizontal radiation into its beam and diffuse components.

If the decomposition can be computed, the trigonometric analysis presented earlier can be used to calculate incident radiation on any surface in a straightforward manner. Liu and Jordan (LJ) correlated the diffuse-to-total radiation ratio $\left(\bar{D}_h/\bar{H}_h\right)$ with the *monthly clearness index* $\bar{K}_T$, which is defined as

$$\bar{K}_T = \frac{\bar{H}_h}{\bar{H}_{o,h}}, \tag{36.32}$$

where

$\bar{H}_h$ is the monthly averaged terrestrial radiation per day on a horizontal surface

$\bar{H}_{o,h}$ is the corresponding extraterrestrial radiation, which can be calculated from Equation 36.31 by averaging each daily total for a month

The original LJ method was based upon the extraterrestrial radiation at midmonth, which is not truly an average.

The LJ correlation predicts the monthly diffuse $(\bar{D}_h)$ to monthly total $\bar{H}_h$ ratio. It can be expressed by the following empirical equation:

$$\frac{\bar{D}_h}{\bar{H}_h} = 1.390 - 4.027\bar{K}_T + 5.531\bar{K}_T^2 - 3.108\bar{K}_T^3. \tag{36.33}$$

Note that the LJ correlation is based upon a solar constant value of 1394 W/m^2 (442 Btu/h-ft^2), which was obtained from terrestrial observations, whereas the newer value, based on satellite data, is 1377 W/m^2 (437 Btu/h-ft^2). The values of $\bar{K}_T$ must be based on this earlier value of the solar constant to use the LJ method. Collares-Pereira and Rabl [4] conducted a study and concluded that although LJ's approach is valid, their correlations would predict significantly smaller diffuse radiation components. They also concluded that LJ were able to correlate their model with the measured data because they used the measured data that were not corrected for the shade ring (see solar radiation measurements). Collares-Pereira and Rabl also introduced the sunset hour angle h_{ss} in their correlation to account for the seasonal variation in the diffuse component. The Collares-Pereira and Rabl (C-P&R) correlation is

$$\frac{\bar{D}_h}{\bar{H}_h} = 0.775 + 0.347\left(h_{ss} - \frac{\pi}{2}\right) - \left[0.505 + 0.0261\left(h_{ss} - \frac{\pi}{2}\right)\right]\cos(2K_T - 1.8), \tag{36.34}$$

where h_{ss} is the sunset hour angle in radians. The C-P&R correlation agrees well with the correlations for India [3], Israel [43], and Canada [39] and is, therefore, preferred to Equation 36.33.

The monthly average beam component $\bar{B}_h$ on a horizontal surface can be readily calculated by simple subtraction since $\bar{D}_h$ is known:

$$\bar{B}_h = \bar{H}_h - \bar{D}_h. \tag{36.35}$$

It will be recalled on an instantaneous basis from Equations 36.23 and 36.25 and Figure 36.17 that

$$I_{b,N} = \frac{I_{b,h}}{\sin\alpha}, \tag{36.36}$$

$$I_{b,c} = I_{b,N}\cos i, \tag{36.37}$$

where $I_{b,h}$ is the instantaneous horizontal beam radiation. Solving for $I_{b,c}$, the beam radiation on a surface,

$$I_{b,c} = I_{b,h}\left(\frac{\cos i}{\sin\alpha}\right). \tag{36.38}$$

The ratio in parentheses is usually called the beam radiation *tilt factor* R_b. It is a purely geometric quantity that converts instantaneous horizontal beam radiation to beam radiation intercepted by a tilted surface.

Equation 36.38 cannot be used directly for the long-term beam radiation $\bar{B}_h$. To be strictly correct, the instantaneous tilt factor R_b should be integrated over a month with the beam component $I_{b,h}$ used as a weighting factor to calculate the beam tilt factor. However, the LJ method is used precisely when such short-term data as $I_{b,h}$ are not available. The LJ recommendation for the monthly mean tilt factor $\bar{R}_b$ is simply to calculate the monthly average of cos *i* and divide it by the same average of sin α. In equation form for south-facing surfaces, this operation yields

$$\bar{R}_b = \frac{\cos(L-\beta)\cos\delta_s \sin h_{sr} + h_{sr}\sin(L-\beta)\sin\delta_s}{\cos L\cos\delta_s \sin h_{sr}(\alpha=0) + h_{sr}(\alpha=0)\sin L\sin\delta_s}, \tag{36.39}$$

where the sunrise hour angle $h_{sr}(\alpha = 0)$ in radians is given by Equation 36.8 and h_{sr} is the $\min\left[|h_s(\alpha=0)|, |h_s(i=90°)|\right]$, respectively, which are evaluated at midmonth. Non-south-facing surfaces require numerical integration or iterative methods to determine $\bar{R}_b$. The long-term beam radiation on a tilted surface $\bar{B}_c$ is then

$$\bar{B}_c = \bar{R}_b\bar{B}_h, \tag{36.40}$$

which is the long-term analog of Equation 36.27.

Diffuse radiation intercepted by a tilted surface differs from that on a horizontal surface, because a tilted surface does not view the entire sky dome, which is the source of diffuse radiation. If the sky is assumed to be an isotropic source of diffuse radiation, the instantaneous and long-term tilt factors for diffuse radiation, R_d and $\bar{R}_d$, respectively, are equal and are simply the radiation view factor from the plane to the visible portion of a hemisphere. In equation form,

$$R_d = \bar{R}_d = \cos^2\frac{\beta}{2} = \frac{(1+\cos\beta)}{2}. \tag{36.41}$$

In some cases where solar collectors are mounted near the ground, some beam and diffuse radiation reflected from the ground can be intercepted by the collector surface. The tilt factor $\bar{R}_r$ for reflected total radiation $(\bar{D}_h + \bar{B}_h)$ is then calculated to be

$$\bar{R}_r = \frac{\bar{R}}{\bar{D}_h + \bar{B}_h} = \rho \sin^2 \frac{\beta}{2} = \frac{\rho(1-\cos\beta)}{2}, \tag{36.42}$$

in which ρ is the diffuse reflectance of the surface south of the collector assumed uniform and of infinite extent.

For snow, $\rho \cong 0.75$; for grass and concrete, $\rho \cong 0.2$. The total long-term radiation intercepted by a surface $\bar{H}_c$ is then the total of beam, diffuse, and diffusely reflected components:

$$\bar{H}_c = \bar{R}_b \bar{B}_h + \bar{R}_d \bar{D}_h + \bar{R}_r (\bar{D}_h + \bar{B}_h). \tag{36.43}$$

Using Equations 36.41 and 36.42, we have

$$\bar{H}_c = \bar{R}_b \bar{B}_h + \bar{D}_h \cos^2 \frac{\beta}{2} + (\bar{D}_h + \bar{B}_h)\rho \sin^2 \frac{\beta}{2}, \tag{36.44}$$

in which $\bar{R}_b$ is calculated from Equation 36.39.

Example 36.4

Using a value of $\bar{H}_h$ as 16,215 kJ/m²/day for January in place of the long-term measured data for the North Central Sahara Desert at latitude 25° N, find the monthly averaged insolation per day on a south-facing solar collector tilted at an angle of 25° from the horizontal.

Solution

The following solution is for the month of January. Values for the other months can be found by following the same method:

$$\bar{H}_h = 16{,}215\ \text{kJ/m}^2/\text{day}.$$

From Table A2.2a,

$$\bar{H}_{o,h} = 23{,}902.$$

Therefore,

$$\bar{K}_T = \frac{\bar{H}_h}{\bar{H}_{o,h}} = 0.678.$$

δ_s and h_{sr} can be found for the middle of the month (January 16):

$$\delta_s = 23.45° \sin[360(284+16)365]°$$

$$= -21.1°,$$

$$h_{sr}(\alpha = 0) = -\cos^{-1}(-\tan L \tan \delta)$$

$$= -79.6° \quad \text{or} \quad -1.389\ \text{rad}$$

and

$$h_{ss} = 1.389.$$

Using the C-P&R correlation,

$$\frac{\bar{D}_h}{H_h} = 0.775 + 0.347(1.389 - 1.5708) - [0.505 + 0.0261(1.389 - 1.5708)]\cos(2 \times 0.678 - 1.8)$$

$$= 0.212.$$

Therefore,

$$\bar{D}_h = 0.212 \times 16{,}215 = 3{,}438 \ \text{kJ/m}^2\text{/day}$$

and

$$\bar{B}_h = \bar{H}_h - \bar{D}_h = 12{,}777 \ \text{kJ/m}^2\text{/day}$$

Insolation on a tilted surface can be found from Equation 36.43. We need to find $\bar{R}_b$ from Equation 36.39.

Therefore,

$$\bar{R}_b = \frac{\cos(0)\cos(-21.1°)\sin(-79.6°) - 1.389\sin(0)\sin(-21.1°)}{\cos(25°)\cos(-21.1°)\sin(-79.6°) - 1.389\sin(25°)\sin(-21.1°)}$$

$$= 147.$$

$$\bar{R}_d = \cos^2(25/2) = 0.953.$$

$$\bar{R}_r = \rho\sin^2(\beta/2) \ (\text{Assume } \rho = 0.2)$$

$$= 0.2\sin^2(12.5°)$$

$$= 0.009.$$

Therefore,

$$\bar{H}_c = (1.47)(12{,}777) + 0.953(3{,}438) + 0.009(16{,}215)$$

$$= 22{,}205 \ \text{kJ/m}^2.$$

36.5.2 Circumsolar or Anisotropic of Diffuse Solar Radiation

The models described in the earlier sections assume that the sky diffuse radiation is isotropic. However, this assumption is not true because of circumsolar radiation (brightening around the solar disk). Although the assumption of isotropic diffuse solar radiation does not introduce errors in the diffuse values on horizontal surfaces, it can result in errors of 10%–40% in the diffuse values on tilted surfaces. A number of researchers have studied the anisotropy of the diffuse solar radiation because of circumsolar radiation. Temps and Coulson [45] introduced an anisotropic diffuse radiation algorithm for tilted surfaces for clear sky conditions. Klucher [20] refined the Temps and Coulson algorithm by adding a cloudiness function to it:

$$R_d = \frac{1}{2}(1+\cos\beta)M_1M_2, \tag{36.45}$$

where

$$M_1 = 1 + F\sin^3(\beta/2), \tag{36.46}$$

$$M_2 = 1 + F\cos^2 i \sin^3(z), \tag{36.47}$$

$$F = 1 - \left(\frac{D_h}{H_h}\right)^2. \tag{36.48}$$

Examining F, we find that under overcast skies ($D_h = H_h$), R_d in Equation 36.45 reduces to the isotropic term of LJ. The Klutcher algorithm reduces the error in diffuse radiation to about 5%.

In summary, monthly averaged, daily solar radiation on a surface is calculated by first decomposing total horizontal radiation into its beam and diffuse components using Equation 36.34 or 36.35. Various tilt factors are then used to convert these horizontal components to components on the surface of interest.

36.5.3 Hourly and Daily Solar Radiation on Tilted Surfaces

Accurate determinations of the hourly solar radiation received during the average day of each month are a prerequisite in different solar energy applications. In the early 1950s, Whillier introduced the *utilizability* method to analytically predict the performance of active solar collectors. This method used a simple formulation to estimate the mean hourly radiation during each hour of an average day of the month, based on the ratio of the hourly to daily irradiation received by a horizontal surface outside of the atmosphere. The long-term models provide the mean hourly distribution of global radiation over the average day of each average month.

Three methods are described as follows:

1. C-P&R model, CPR (Figure 36.18)

   Given the long-term average daily total and diffuse irradiation on a horizontal surface $\overline{H}_h$ and $\overline{H}_d$, it is possible to find the long-term average hourly irradiances $\overline{I}_d$, $\overline{I}_h$, and $\overline{I}_b$.

   The ratio of the hourly diffuse to the long-term average daily diffuse irradiation on a horizontal surface, r_d, is given by

$$r_d = \frac{I_d}{\overline{H}_d} = \frac{\pi}{24}\frac{\cos h_s - \cos h_{ss}}{\sin h_{ss} - \left(\frac{\pi}{180}\right)h_{ss}\cos h_{ss}}. \tag{36.49}$$

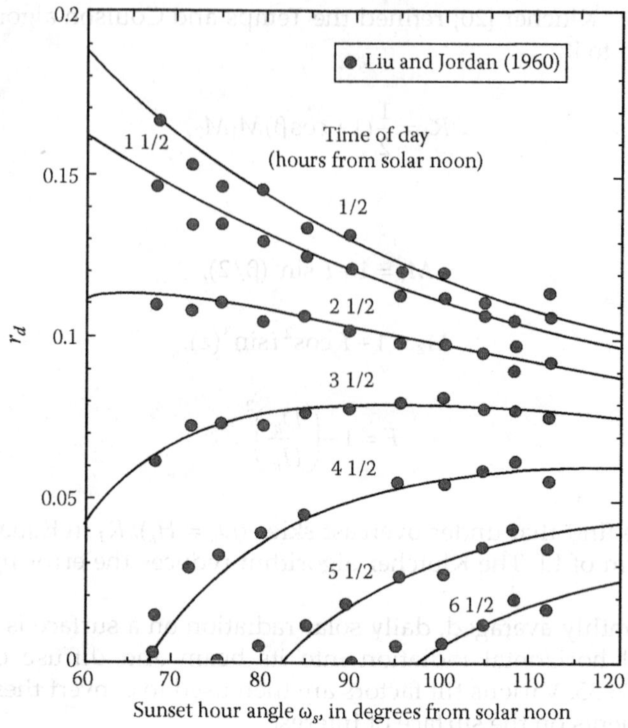

FIGURE 36.18
C-P&R model, CPR. (Adapted from Collares-Pereira, M. and Rabl, A., *Sol. Energ.*, 22, 155, 1979; Liu, B.H.Y. and Jordan, R.C., *Sol. Energ.*, 4, 1, 1960.)

The ratio of hourly total to the long-term average daily total irradiation on a horizontal surface, r_t, is given by

$$r_t = \frac{I_h}{\overline{H_h}} = \left(a + b\cos h_s\right) r_d \tag{36.50}$$

with

$a = 0.409 + 0.501\,6\sin(h_{ss} - 60°)$
$b = 0.6609 - 0.4767\sin(h_{ss} - 60°)$

This fit satisfies, within 1% for all h_{ss}, the normalization condition:

$$\int_{-t_s}^{t_s} I_d dt = \overline{H_h} \tag{36.51}$$

2. C-P&R model modified by Gueymard, CPRG (Figure 36.19)

For the CPRG method, the ratio of hourly total to the long-term average daily total irradiation on a horizontal surface, r_t, is given by [52]

$$r_t = \frac{I_h}{\overline{H_h}} = \frac{(a + b\cos h_s) r_d}{f_c} \tag{36.52}$$

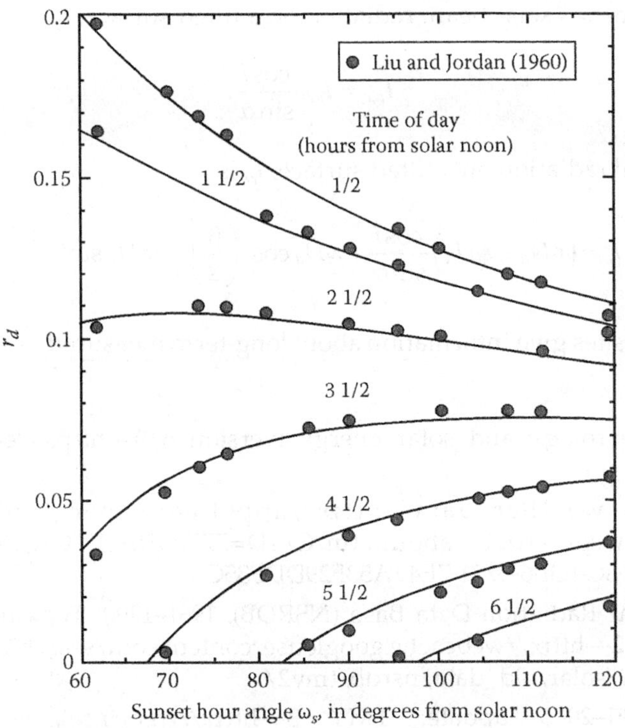

FIGURE 36.19
C-P&R model modified by Gueymard [51]. (Adapted from Collares-Pereira, M. and Rabl, A., *Sol. Energ.*, 22, 155, 1979; Liu, B.H.Y. and Jordan, R.C., *Sol. Energ.*, 4, 1, 1960.)

with

$$f_c = a + kb\left[\left(\frac{\pi}{180}\right)h_{ss} - \frac{1}{2}\sin 2h_{ss}\right]. \tag{36.53}$$

$$k = \left(2\left[\sin h_{ss} - \left(\frac{\pi}{180}\right)h_{ss}\cos h_{ss}\right]\right)^{-1}. \tag{36.54}$$

The normalization condition is expressed by

$$\int_{-t_s}^{t_s} r_t\,dt = 1. \tag{36.55}$$

3. For both CPR and CPRG models

The instantaneous solar beam radiation on a horizontal surface, $I_{b,h}$, is given by

$$I_{b,h} = r_t\bar{H}_h - r_d\bar{H}_d. \tag{36.56}$$

The instantaneous solar beam radiation on a tilted surface, $I_{b,c}$, is

$$I_{b,c} = I_{b,h}\frac{\cos i}{\sin\alpha}. \tag{36.38}$$

Then the total radiation on a tilted surface, I_c, is

$$I_c = \left(r_t\bar{H}_h - r_d\bar{H}_d\right)\frac{\cos i}{\sin\alpha} + r_d\bar{H}_d\cos^2\left(\frac{\beta}{2}\right) + \rho r_t\bar{H}_h\sin^2\left(\frac{\beta}{2}\right). \tag{36.57}$$

The following websites give information about long-term measured or satellite solar radiation data.

- Surface meteorology and solar energy (version 6.0)—https://eosweb.larc.nasa.gov/sse/
- EnergyPlus weather data—http://apps1.eere.energy.gov/buildings/energyplus/weatherdata_about.cfm?CFID=777061&CFTOKEN=999ebda95de32b46-A8949F8C-E3B6-3771-7E42A53F29DDF35C
- National Solar Radiation Data Base (NSRDB), 1961–1990: Typical Meteorological Year (TMY) 2—http://webcache.googleusercontent.com/search?q=cache; http://rredc.nrel.gov/solar/old_data/nsrdb/tmy2/
- NSRDB, 1991–2005 update: TMY 3—http://rredc.nrel.gov/solar/old_data/nsrdb/1991-2005/tmy3/; http://rredc.nrel.gov/solar/old_data/nsrdb/1991-2005/tmy3/
- The Solar and Wind Energy Resource Assessment (SWERA)—http://en.openei.org/apps/SWERA/
- National Renewable Energy Laboratory (NREL)—http://www.nrel.gov/international/geospatial_toolkits.html; http://www.nrel.gov/international/geospatial_toolkits.htmlL-HOMER

36.5.4 Spectral Models

Many biological, chemical, and physical processes are activated more powerfully at some wavelengths than at others. Therefore, it is important to know the spectral characteristics of the incident radiation. In order to model spectral solar radiation at a location, radiation needs to be modeled as it travels through the atmosphere. These models are complex to begin with and are made more complex since different wavelengths are absorbed, reflected, and scattered differently in the atmosphere. Examples of radiative transfer numerical models include the Santa Barbara DISORT, the atmospheric radiative transfer code SBDART, and the moderate resolution transmission code MODTRAN. SBDART, developed at the University of California at Santa Barbara, is relatively simpler to use than MODTRAN, is freely accessible, and even has a convenient user interface online: http://arm.mrcsb.com/sbdart/

MODTRAN, developed by the Air Force Geophysical Laboratory (AFGL), has a much higher resolution and is considered as the standard in atmospheric applications.

These models, however, are not convenient for solar energy or other engineering-type applications.

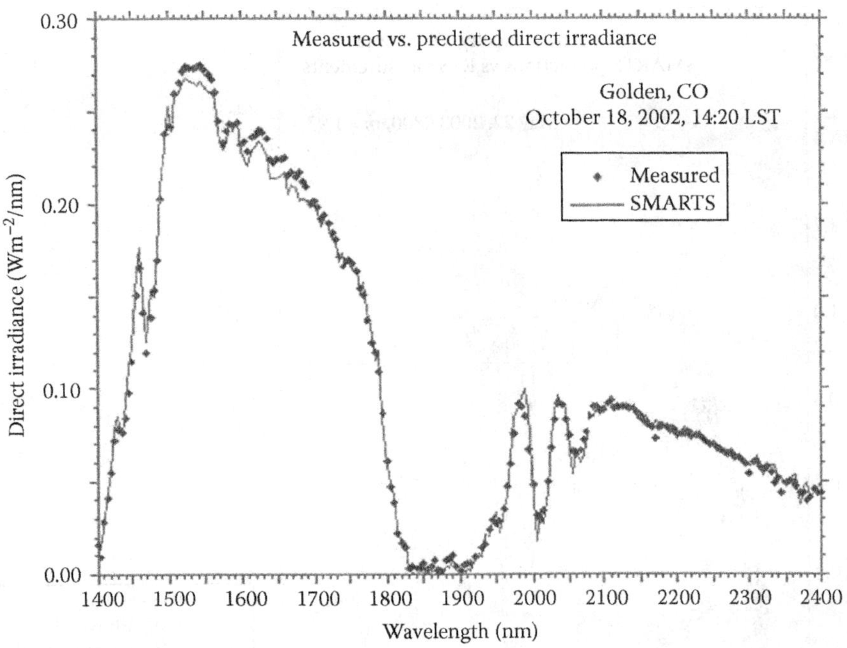

FIGURE 36.20
A sample spectral irradiance prediction compared with measured data for Golden, Colorado.

Bird's simple spectral model (SPCTRAL2), the SMARTS, and REST2 can be used for solar engineering applications. Even though these models are limited to clear sky conditions, they can also be empirically modified to predict spectra under cloudy conditions. SMARTS model offers fast and accurate predictions of spectral irradiance on any tilted surface without the difficulties and limitations associated with the atmospheric models mentioned earlier. Sample outputs produced by SMARTS, compared with actual spectroradiometric measurements from high-performance instruments, are shown in the Figures 36.20 and 36.21.*

36.6 Measurement of Solar Radiation

Solar radiation measurements of importance to most engineering applications, especially thermal applications, include total (integrated over all wavelengths) direct or beam and sky diffuse values of solar radiation on instantaneous, hourly, daily, and monthly bases. Some applications such as photovoltaics, photochemical, and daylighting require knowledge of spectral (wavelength specific) or band (over a wavelength range—e.g., ultraviolet, visible, infrared) values of solar radiation. This section describes some of the instrumentation used to measure solar radiation and sunshine and some sources of long-term measured data for different parts of the world. Also described briefly in this section is the method of satellite-based measurements.

* Taken from http://www.solarconsultingservices.com/smarts.php; http://www.solarconsultingservices.com/smarts.php.

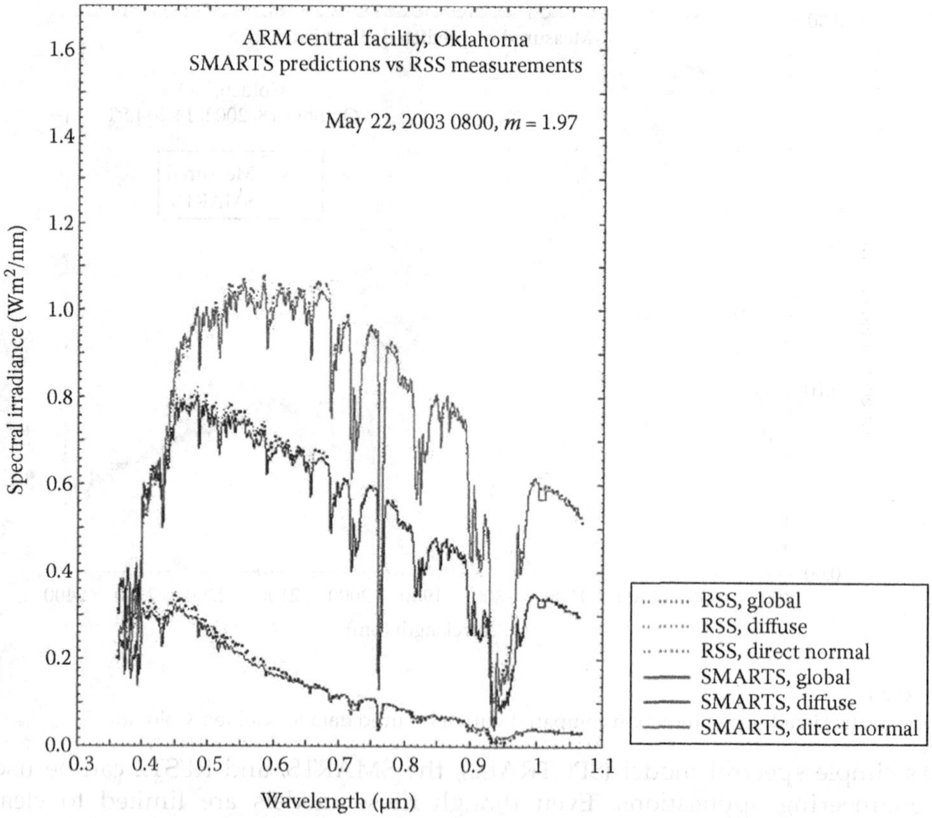

FIGURE 36.21
A sample prediction for direct normal, diffuse, and global horizontal spectral irradiance.

36.6.1 Instruments for Measuring Solar Radiation and Sunshine

There are two basic types of instruments used to measure solar radiation, *pyranometer* and *pyrheliometer.* A pyranometer has a hemispherical view of the surroundings and therefore is used to measure total, direct and diffuse, solar radiation on a surface. A pyrheliometer, on the other hand, has a restricted view (about 5°) and is, therefore, often used to measure the direct or beam solar radiation by pointing it toward the sun. Pyranometers are also used to measure the sky diffuse radiation by using a shadow band to block the direct sun view. A detailed discussion of the instrumentation and calibration standards is given by Iqbal [15] and Zerlaut [48].

A pyranometer consists of a flat sensor/detector (described later) with an unobstructed hemispherical view, which allows it to convert and correlate the total radiation incident on the sensor to a measurable signal. The pyranometers using thermal detectors for measurements can exhibit serious errors at tilt angles from the horizontal due to free convection. These errors are minimized by enclosing the detector in double hemispherical high-transmission glass domes. The second dome minimizes the error due to infrared radiative exchange between the sensor and the sky. A desiccator is usually provided to eliminate the effect due to condensation on the sensor or the dome. Figure 36.22 shows pictures of typical commercially available precision pyranometer.

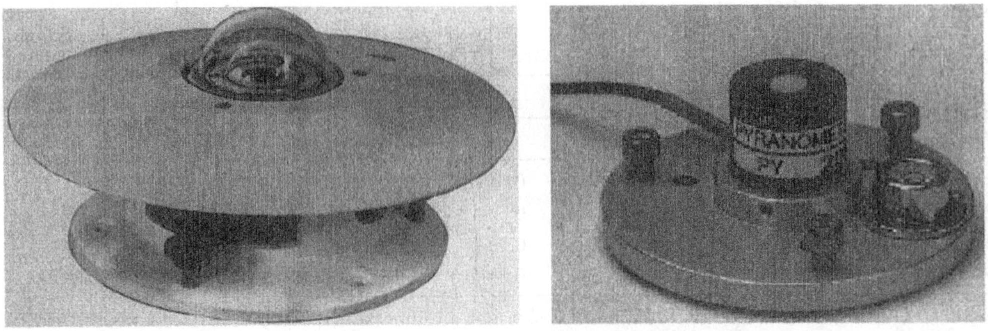

(a) (b)

FIGURE 36.22
Typical commercially available pyranometers with (a) thermal detector and (b) photovoltaic detector.

FIGURE 36.23
A pyranometer with a shade ring to measure sky diffuse radiation.

A pyranometer can be used to measure the sky diffuse radiation by fitting a shade ring to it, as shown in Figure 36.23, in order to block the beam radiation throughout the day. The position of the shade ring is adjusted periodically as the declination changes. Since the shade ring obstructs some diffuse radiation from the pyranometer, correction factors must be applied.

Geometric correction factors (GCFs) that account for the part of the sky obstructed by the shade ring can be easily calculated. However, a GCF assumes isotropic sky, which results in errors because of the circumsolar anisotropy. Eppley Corp. recommends additional correction factors to account for anisotropy as +7% for clear sky, +4% for partly cloudy condition, and +3% for cloudy sky. Mujahid and Turner [30] determined that these correction factors gave less than 3% errors on partly cloudy days but gave errors of −11% for clear sky conditions and +6% on overcast days. They suggested correction factors due

TABLE 36.7
Shading Band Correction Factors due to Anisotropy

Solar Altitude Angle	k_T 0.0	0.1	0.2	0.3	0.4	0.5	0.6	0.7	0.8	0.9
<20°	0.0	0.0	0.0	0.0	0.015	0.06	0.14	0.23	0.24	0.24
20°–40°	0.0	0.0	0.0	0.0	0.006	0.05	0.125	0.205	0.225	0.225
40°–60°	0.0	0.0	0.0	0.0	0.003	0.045	0.115	0.175	0.205	0.205
60°+	0.0	0.0	0.0	0.0	0.0	0.035	0.09	0.135	0.17	0.17

Source: Mujahid, A. and Turner, W.D., Diffuse sky measurement and model, ASME Paper No. 79–WA/Sol–5, 1979.

to anisotropy as tabulated in Table 36.7, which reduce the errors to less than ±3%. It must be remembered that these correction factors are in addition to the GCFs. Recently, a sun occulting disk has been employed for shading the direct sun.

Beam or direct solar radiation is usually measured with an instrument called a pyrheliometer. Basically a pyrheliometer places the detector at the base of a long tube. This geometry restricts the sky view of the detector to a small angle of about 5°. When the tube points toward the sun, the detector measures the beam solar radiation and a small part of the diffuse solar radiation within the view angle. Figure 36.24 shows the geometry of a pyrheliometer sky occulting tube.

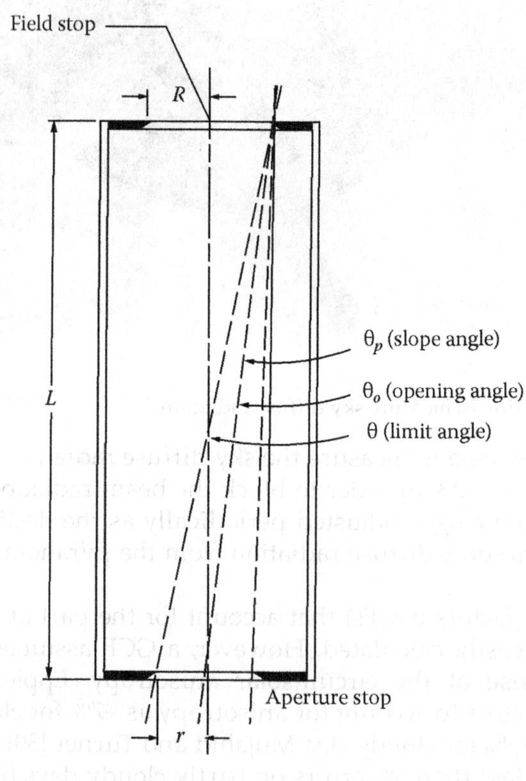

FIGURE 36.24
Geometry of a pyrheliometer sky occulting tube.

In this figure, the opening half angle

$$\theta_o = \tan^{-1}\frac{R}{L}. \tag{36.58}$$

The slope angle

$$\theta_p = \tan^{-1}\left[\frac{(R-r)}{L}\right]. \tag{36.59}$$

The limit half angle

$$\theta = \tan^{-1}\left[\frac{(R+r)}{L}\right]. \tag{36.60}$$

The field of view is $2\theta_o$. The World Meteorological Organization (WMO) recommends the opening half angle θ_o to be 2.5° [48] and the slope angle θ_p to be 1°.

Continuous tracking of the sun is required for the accuracy of the measurements. This is obtained by employing a tracking mechanism with two motors, one for altitude and the other for azimuthal tracking. Another problem is that the view angle of a pyrheliometer is significantly greater than the angle subtended by the solar disk (about 0.5°). Therefore, the measurements using a pyrheliometer include the beam and the circumsolar radiation. These measurements may present a problem in using the data for central receiver systems that use only direct beam radiation. However, this is not a significant problem for parabolic trough concentrators that in most cases have a field of view of the order of 5°.

36.6.2 Detectors for Solar Radiation Instrumentation

Solar radiation detectors are of four basic types [15,48]: thermomechanical, calorimetric, thermoelectric, and photoelectric. Of these, thermoelectric and photoelectric are the most common detectors in use today.

A *thermoelectric detector* uses a thermopile that consists of a series of thermocouple junctions. The thermopile generates a voltage proportional to the temperature difference between the hot and cold junctions that, in turn, is proportional to the incident solar radiation. Figure 36.25 shows different types of thermopile configurations. The Eppley black and white pyranometer uses a radial differential thermopile with the hot junction coated with 3M Velvet Black™ and the cold junction coated with a white barium sulfate paint.

Photovoltaic detectors normally use silicon solar cells measuring the short circuit current. Such detectors have the advantage of being simple in construction. Because heat transfer is not a consideration, they do not require clear domes or other convection suppressing devices. They are also insensitive to tilt as the output is not affected by natural convection. One of the principal problems with photovoltaic detectors is their spectral selectivity. Radiation with wavelengths greater than the bandgap of the photovoltaic detector cannot be measured. Silicon has a bandgap of 1.07 eV corresponding to a wavelength of 1.1 μm. A significant portion of the infrared part of solar radiation has wavelengths greater than 1.1 μm. Therefore, photovoltaic detectors are insensitive to changes in the infrared part of solar radiation.

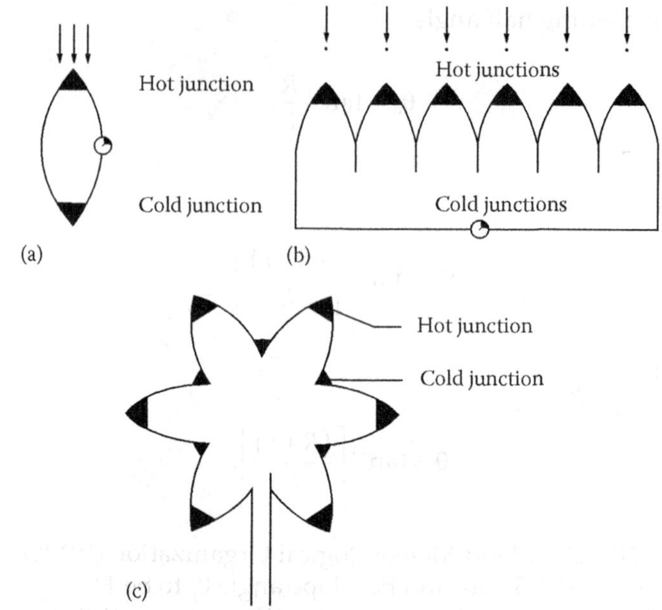

FIGURE 36.25
Various thermopile configurations. (From Zerlaut, G., Solar radiation instrumentation, in *Solar Resources*, Roland L. Hulstrom, ed., MIT Press, Cambridge, MA, 1989.)

36.6.3 Measurement of Sunshine Duration

The time duration of bright sunshine data is available at many more locations in the world than the solar radiation. That is why a number of researchers have used these data to estimate the available solar radiation. Two instruments are widely used to measure the sunshine duration. The device used by the U.S. National Weather Service is called a *sunshine switch.* It is composed of two photovoltaic cells—one shaded, the other not. During daylight a potential difference is created between the two cells, which in turn operates the recorder. The intensity level required to activate the device is that just sufficient to cast a shadow. The other device commonly used to measure the sunshine duration is called the *Campbell–Stokes sunshine recorder.* It uses a solid, clear glass sphere as a lens to concentrate the solar beam on the opposite side of the sphere. A strip of standard treated paper marked with time graduations is mounted on the opposite side of the sphere where the solar beam is concentrated. Whenever the solar radiation is above a threshold, the concentrated beam burns the paper. The length of the burned part of the strip gives the duration of bright sunshine. The problems associated with the Campbell–Stokes sunshine recorder include the uncertainties of the interpretation of burned portions of the paper, especially on partly cloudy days, and the dependence on the ambient humidity. Figure 36.26 shows a Campbell–Stokes sunshine recorder.

36.6.4 Measurement of Spectral Solar Radiation

Spectral solar radiation measurements are made with spectroradiometers. Full-spectrum scanning is difficult, requires constant attention during operation, and is therefore expensive. Zerlaut [48] has described a number of solar spectroradiometers. These instruments consist basically of a monochromator, a detector–chopper assembly, an integrating sphere, and a signal conditioning/computer package. They have the capability of measuring solar radiation in the wavelength spectrum of 280–2500 nm.

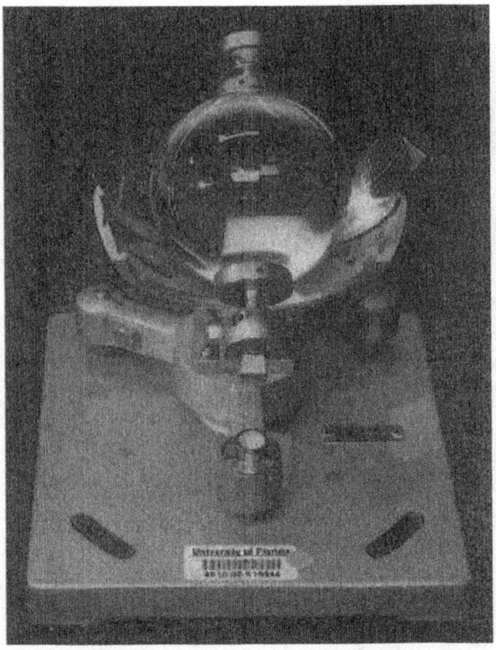

FIGURE 36.26
Campbell–Stokes sunshine recorder.

36.6.5 Wideband Spectral Measurements

Some applications of solar energy require solar radiation data in wideband wavelength ranges such as visible, ultraviolet, and infrared rather than complete spectral data. For example, solar photocatalytic detoxification using TiO_2 as the catalyst needs data in the UV wavelength range, while passive solar applications need data in the infrared wavelength range. Instruments such as pyranometers and pyrheliometers can be adapted for wideband spectral measurements by using cut-on and cutoff filters. Eppley instruments provide standard cut-off filters at wavelengths 530 nm (orange), 630 nm (red), and 695 nm (dark red). They are provided as plain filters at the aperture of a pyrheliometer tube and as outer glass domes for pyranometers. Instrument manufacturers provide various interference filters peaking at different wavelengths in the solar spectrum.

Solar UV measurements are important in general since prolonged exposure to solar UV can cause skin cancer, fading of colors, and degradation of plastic materials. Such measurements have become even more important because the photocatalytic effect based on TiO_2 as a catalyst depends only on the solar UV wavelength range. Figure 36.27 shows an Eppley model total ultraviolet radiometer (TUVR) that measures the total hemispherical UV radiation from 295 to 385 nm. This radiometer uses a selenium photoelectric cell detector, a pair of band pass filters to allow wavelengths from 295 to 385 nm to pass through, and a beveled Teflon diffuser.

36.6.6 Solar Radiation Data

Measured solar radiation data are available at a number of locations throughout the world. Data for many other locations have been estimated based on measurements at similar climatic locations.

FIGURE 36.27
Eppley TUVR®. (Courtesy of Eppley Lab, Newport, RI.)

Solar radiation data for the United States are available from the National Climatic Data Center (NCDC) of the National Oceanic and Atmospheric Administration (NOAA) and the NREL. In the mid-1970s, NOAA compiled a database of measured hourly global horizontal solar radiation for 28 locations for the period 1952–1975 (called SOLMET) and of data for 222 additional sites (called ERSATZ) estimated from SOLMET data and some climatic parameters such as sunshine duration and cloudiness. NOAA also has two data sets of particular interest to engineers and designers: the TMY and the Weather Year for Energy Calculations (WYEC) data sets. TMY data set represents typical values from 1952 to 1975 for hourly distribution of direct beam and global horizontal solar radiation. The WYEC data set contains monthly values of temperature, direct beam and diffuse solar radiation, and estimates of *illuminance* (for daylighting applications). Illuminance is solar radiation in the visible range to which the human eye responds. Recently, NREL compiled an NSRDB for 239 stations in the United States [28,32]. NSRDB is a collection of hourly values of global horizontal, direct normal, and diffuse solar radiation based on measured and estimated values for a period of 1961–1990. Since long-term measurements were available for only about 50 stations, measured data make up only about 7% of the total data in the NSRDB. A TMY data set from NSRDB is available as TMY2.

The data for other locations in the world are available from national government agencies of most countries of the world. Worldwide solar radiation data are also available from the World Radiation Data Center (WRDC) in St. Petersburg, Russia, based on worldwide measurements made through local weather service operations [47]. WRDC, operating under the auspices of the World Meteorological Organization (WMO), has been archiving data from over 500 stations and operates a website in collaboration with NREL with an address of http://wrdc.mgo.nrel.gov. An International Solar Radiation Data Base was also developed by the University of Lowell [46].

- Surface meteorology and solar energy (version 6.0) (http://eosweb.larc.nasa.gov/sse/)
- EnergyPlus weather data (http://apps1.eere.energy.gov/buildings/energyplus/weatherdata_about.cfm?CFID=5019287&CFTOKEN=b54041e7be537f1f-B598302C-5056-BC19-15C492F462EE1BAC)
- NSRDB, 1961–1990: TMY (http://rredc.nrel.gov/solar/old_data/nsrdb/)
- (http://rredc.nrel.gov/solar/pubs/NSRDB/)
- NSRDB, 1991–2005 update: TMY3 (http://rredc.nrel.gov/solar/old_data/nsrdb/)
- (http://www.nrel.gov/docs/fy12osti/54824.pdf)
- The SWERA (http://en.openei.org/wiki/Solar_and_Wind_Energy_Resource_Assessment_(SWERA))
- NREL (http://www.nrel.gov/)

36.7 Solar Radiation Mapping Using Satellite Data

Remote sensing satellite data have been used since the early 1960s to extract quantitative and qualitative cloud data. The most important application of cloud cover mapping has been the observation of storms and hurricanes, etc. Recently, however, considerable interest has been developed in using the cloud mapping data to estimate terrestrial solar radiation. Since meteorological satellites from a number of countries can now cover most of the earth, the data can be used to estimate solar radiation where no measured data exist or none is being measured.

Weather satellites are available in three main orbiting configurations—equatorial, polar, and geostationary. The equatorial satellites are low-level orbiting satellites (~600 km altitude) that generally orbit the earth in a west to east direction in a sinusoidal path that crosses the equator at least twice per orbit. Polar satellites are also low-orbit satellites that orbit the earth from the north to the south pole while the earth rotates underneath. Sun synchronous polar orbits have their orbits synchronized with the sun such that the same point on the earth is viewed at the same time each day. Low-orbit satellites are capable of gathering high-resolution spatial data. A geostationary satellite orbits in such a way that it is always over the same point on the earth's surface. Geostationary satellites have very high altitudes (approximately 36,000 km) and can provide high-temporal-resolution images over a large portion of the earth's surface. A number of countries maintain geostationary satellites including the United States (GOES, longitudes 70° W and 140° W), Europe (METEOSAT, longitude 0°), India (INSAT, longitude 70° E), and Japan (GMS, longitude 140° E).

Various types of high-resolution radiometers collect radiative data images of the earth's atmosphere below. These radiometers scan spectral measurements in the wavelength ranges of shortwave (0.2–3.0 μm), longwave (6.5–25 μm), and total irradiance (0.2–100 μm). The spatial resolution of images from the satellite is given by *a pixel* that represents the smallest area of data, generally of the order of 2 km × 2 km. However, several pixels of data

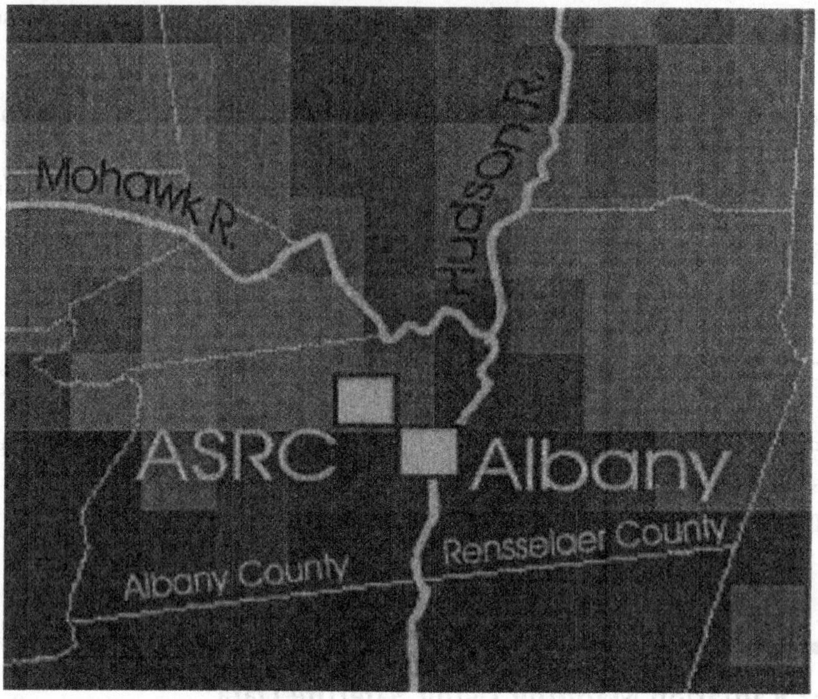

FIGURE 36.28
GOES-8 intermediate-resolution image close-up around Albany. (From Perez, R., The strengths of satellite based resource assessment, *Proceedings of the 1997 ASES Annual Conference*, Washington, DC, pp. 303–308, 1997.)

are required to derive a surface value giving a surface resolution of the order of 10 km × 10 km. Figure 36.28 shows an example of an intermediate-resolution GOES-8 image around Albany, New York, overlaid on a local map.

36.7.1 Estimation of Solar Resource from Satellite Data

The signal recorded by a radiometer on a satellite measures the solar radiation flux reflected back from the earth's atmosphere. The basic method behind estimation of ground solar radiation from these data is to apply the principle of energy conservation in the earth-atmosphere system [31], as shown in Figure 36.29. From this figure we can write

$$I_{in} = I_{out} + I_a + I_g, \quad (36.61)$$

where

I_{in} represents the solar radiation incident on the atmosphere
I_{out} represents the outward radiation from the atmosphere
I_a is the radiation absorbed by the atmosphere
I_g is the radiation absorbed by the ground

I_g can be expressed in terms of the surface albedo* ρ (reflectivity) and the solar radiation I_s, incident on the earth's surface:

$$I_g = I_s(1 - \rho). \quad (36.62)$$

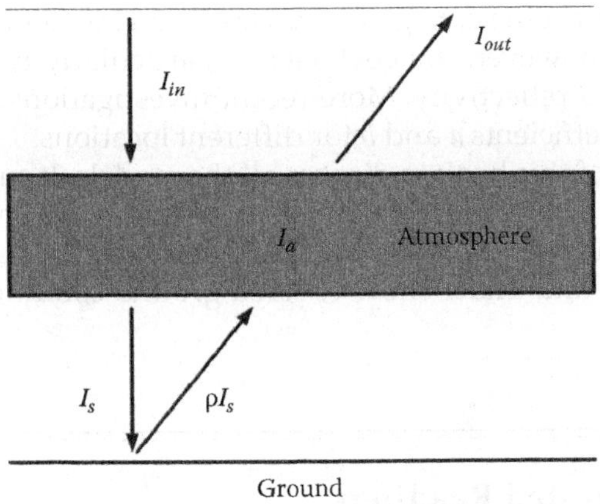

FIGURE 36.29
Solar radiation in the earth-atmosphere model.

From Equations 36.61 and 36.62 we can obtain

$$I_s = \frac{(I_{in} - I_{out} - I_a)}{(1-\rho)}. \tag{36.63}$$

I_{out} is measured by the satellite radiometers. I_{in} depends on the sun–earth distance and the solar zenith angle and can be calculated using Equation 36.11 as

$$I_{in} = I_o\left(\frac{D_o}{D}\right)^2 \cdot \cos(z). \tag{36.64}$$

If we could estimate I_a, and A were known a priori, I_s could be estimated using Equation 36.63 from the value of I_{out} measured by a satellite. However, I_a cannot be estimated easily since it depends on the atmospheric conditions such as cloud cover, dust particles, and air mass, and surface albedo (reflectance) ρ varies for every point of the region under consideration. In order to deal with these factors, two types of empirical methods are under development. These are known as statistical and physical methods. These methods have been reviewed in detail by Schmetz [40], Hay [12], Noia et al. [31], Islam [17], and Pinker et al. [36].

Statistical Methods: Statistical methods are based on finding a relationship between the radiative flux measured by a satellite radiometer and the simultaneous solar radiation value measured at the earth's surface in the area under consideration. Some of the models developed on statistical approach include Hay and Hanson [13], Tarpley [44], Justus et al. [18], Cano and Sorapipatana et al. [41].

Physical Methods: Physical methods are based on the analysis of radiative processes in the atmosphere as the solar radiation passes through it. Some of the models developed with this approach include Gautier [9], Moser and Raschke [29], Dedieu et al. [6], and Marullo et al. [27].

The simplest of the previously mentioned models is by Hay and Hanson [13], which gives the atmospheric transmittance T as

$$T = \frac{I_s}{I_{in}} = a - b\frac{I_{out}}{I_{in}},$$

$$\text{or} \quad I_s = aI_{in} - bI_{out}. \tag{36.65}$$

The values of regression coefficients given by Hay and Hanson [13] are $a = 0.79$, $b = 0.71$. This method is simple; however, the coefficients, particularly b, vary considerably with parameters such as cloud reflectivity. More recent investigations suggest that it is necessary to determine the coefficients a and b for different locations.

It is beyond the scope of this book to discuss all the models. It suffices, however, to point out that all the models, including the Hay and Hanson model, more or less give values within 10% of the ground measured values [31]. The methods usually break down under partly cloudy conditions and under snow-covered ground conditions.

References and Suggested Readings

1. Akinoglu, B.G. and A. Ecevit. 1990. Construction of a quadratic model using modified angstrom coefficients to estimate global solar radiation. *Solar Energy* 45: 85.
2. Choudhury, N.K.O. 1963. Solar radiation at New Delhi. *Solar Energy* 7(44).
3. Collares-Pereira, M. and A. Rabl. 1979. The average distribution of solar radiation-correlation between diffuse and hemispherical. *Solar Energy* 22: 155–166.
4. Coppolino, S. 1994. New correlation between clearness index and relative sunshine. *Renewable Energy* 4(4): 417–423.
5. Dedieu, G., P.Y. Deschamps, and Y.H. Kerr. 1987. Satellite estimation of solar irradiance at the surface of the earth and of surface albedo using a physical model applied to METEOSAT data. *Journal of Climate and Applied Meteorology* 26: 79–87.
6. Frohlich, C. et al. 1973. The third international comparison of pyrheliometers and a comparison of radiometric scales. *Solar Energy* 14: 157–166.
7. Garg, H.P. and S.N. Garg. 1985. Correlation of the monthly average daily global, diffuse and beam radiation with bright sunshine hours. *Energy Conversion and Management* 25(4): 409–417.
8. Gautier, C. 1980. Simple physical model to estimate solar radiation at the surface from GOES satellite data. *Journal of Applied Meteorology* 19(8): 1005–1012.
9. Gopinathan, K. 1988. A general formula for computing the coefficients of the correlation connecting global solar radiation to sunshine duration. *Solar Energy* 41: 499.
10. Gueymard, C.A. and Thevenard, D. 2009. Monthly average clear-sky broadband irradiance database for solar heat gain and building cooling load calculations. *Solar Energy* 82: 272–285.
11. Hay, J.E. 1993. Satellite based estimates of solar irradiance at the earth's surface: I. Modelling approaches. *Renewable Energy* 3: 381–393.
12. Hay, J.E. and K.J. Hanson. 1978. A satellite-based methodology for determining solar irradiance at the ocean surface during GATE. *Bulletin of the American Meteorological Society*. 59: 1549.
13. Hussain, M. 1994. Estimation of the global and diffuse irradiation from sunshine duration and atmospheric water vapor content. *Solar Energy* 33(2): 217–220.
14. Iqbal, M. 1983. *An Introduction to Solar Radiation*. New York: Academic Press.
15. Iqbal, M. 1979. Correlation of average diffuse and beam radiation with hours of bright sunshine, *Solar Energy* 23: 169–173.
16. Islam, M.R. 1994. Evolution of methods for solar radiation mapping using satellite data. *RERIC International Energy Journal* 16(2).
17. Justus, C., M.V. Paris, and J.D. Tarpley. 1986. Satellite-measured insolation in the United States, Mexico and South America. *Remote Sensing of Environment* 20: 57–83.
18. Klucher, T.M. 1979. Evaluation of models to predict insolation on tilted surfaces. *Solar Energy* 23(2): 111–114.
19. Liu, B.Y.H. and R.C. Jordan. 1961. Daily insolation on surfaces titled toward the equator. *ASHRAE Transactions* 67: 526–541.

20. Liu, B.Y.H. and R.C. Jordan. 1961. Daily insolation on surface titled toward the equator. *ASHRAE Transactions* 3(10): 53–59.
21. Liu, B.Y.H. and R.C. Jordan. 1967. Availability of solar energy for flat-plate solar heat collectors. In *Low Temperature Engineering of Solar Energy*, Chapter 1. New York: ASHRAE; see also 1977 revision.
22. Liu, B.Y.H. and R.C. Jordan. 1963. A rational procedure for predicting the long-term average performance of flat-plate solar energy collectors. *Solar Energy* 7: 53–74.
23. Liu, B.Y.H. and R.C. Jordan 1960. The interrelationship and characteristic distribution of direct, diffuse and total solar radiation. *Solar Energy* 4: 1–19. See also Liu, B.Y.H. Characteristics of solar radiation and the performance of flat plate solar energy collectors, Ph.D. dissertation, Minneapolis, MN: University of Minnesota, 1960.
24. Löf, G.O.G. et al. 1966. World distribution of solar energy. Engineering Experiment Station Report 21, Madison, WI: University of Wisconsin.
25. Marullo, S., G. Dalu, and A. Viola. 1987. Incident short-wave radiation at the surface from METEOSAT data. *Il Nuovo Cimento* 10C: 77–90.
26. Maxwell, E.L. 1998. METSTAT—The solar radiation model used in the production of the national solar radiation data base (NSRDB). *Solar Energy* 62: 263–279.
27. Moeser, W. and E. Raschke. 1984. Incident solar radiation over Europe estimated from METEOSTAT data. *Journal of Climate and Applied Meteorology* 23(1): 166–170.
28. Mujahid, A. and W.D. Turner. 1979. Diffuse sky measurement and model. ASME Paper No. 79–WA/Sol–5.
29. Noia, M., C. Ratto, and R. Festa. 1993. Solar irradiance estimation from geostationary satellite data: I. statistical models; II. physical models. *Solar Energy* 51: 449–465.
30. NSRDB. 1992. Volume 1: Users Manual: National Solar Radiation Data Base (1961–1990), Version 1.0. Golden, CO: National Renewable Energy Laboratory.
31. Ögelman, H., A. Ecevit, and E. Tasdemiroglu. 1984. A new method for estimating solar radiation from bright sunshine data. *Solar Energy* 33(6): 619–625.
32. Page, J.K. 1966. The estimation of monthly mean values of daily total short-wave radiation on vertical and inclined surfaces from sunshine records or latitudes 40°N–40°S. *Proceedings of U.N. Conference on New Sources of Energy* 4: 378.
33. Perez, R., R. Seals, A. Zelenka, and D. Renne. 1997. The satellite based resource assessment. *Proceedings of the 1997 ASES Annual Conference*, pp. 303–308, Washington, DC.
34. Pinker, R.T., R. Frouin, and Z. Li. 1995. A review of satellite methods to derive surface shortwave irradiance. *Remote Sensing of Environment* 51: 108–124.
35. Quinlan, E.T.(ed.) 1977. *Hourly Solar Radiation Surface Meteorological Observations*. Asheville, NC: NOAA, SOLMET Vol. 1.
36. Quinlan, F.T.(ed.) 1979. *Hourly Solar Radiation Surface Meteorological Observations*. Asheville, NC: NOAA, SOLMET Vol. 2.
37. Ruth, D.W. and R.E. Chant. 1976. The relationship of diffuse radiation to total radiation in Canada. *Solar Energy* 18: 153.
38. Schmetz, J. 1989. Towards a surface radiation climatology: Retrieval of downward irradiances from satellites. *Atmospheric Research* 23: 287–321.
39. Sorapipatana, C., R.H.B. Exell, and D. Borel. 1988. A bispectral method for determining global solar radiation from meteorological satellite data. *Solar and Wind Technology* 5(3): 321–327.
40. Spencer, J.W. 1971. Fourier series representation of the position of the sun. *Search* 2: 172.
41. Stanhill, G. 1966. Diffuse sky and cloud radiation in Israel. *Solar Energy* 10: 66.
42. Tarpley, J.D. 1979. Estimating incident solar radiation at the surface from geostationary satellite data. *Journal of Applied Meteorology* 18(9): 1172–1181.
43. Temps, R.C. and K.L. Coulson. 1977. Solar radiation incident upon slopes of different orientations. *Solar Energy* 19(2): 179–184.
44. University of Lowell Photovoltaic Program. 1990. International Solar Irradiation Database. Version 1.0. Lowell, MA: University of Lowell Research Foundation.

45. Voeikov Main Geophysical Observatory. 1999. Worldwide daily solar radiation. Available from: http://www.mgo.rssi.ru.
46. Zerlaut, G. 1989. Solar radiation instrumentation. In *Solar Resources*, Roland L. Hulstrom (ed.), Cambridge, MA: MIT Press.
47. M. Collares-Pereira and A. Rabl. 1979. The average distribution of solar radiation correlations between diffuse and hemispherical and between daily and hourly insolation values. *Solar Energy* 22: 155.
48. B.H.Y. Liu and R.C. Jordan. 1960. The interrelationship and characteristic distribution of direct, diffuse and total solar radiation. *Solar Energy* 4: 1.
49. Gueymard, C. 2000. Prediction and performance assessment of mean hourly global radiation. *Solar Energy* 68: 285–303.
50. Gueymard, C. 1986. Mean daily averages of beam radiation received by tilted surfaces as affected by the atmosphere. *Solar Energy* 37: 262–267.
51. Vasquez-Padilla, R. 2011. Simplified methodology for designing parabolic trough solar power plants, Ph.D dissertation, Tampa, FL: University of South Florida.
52. Gueymard, C. 2003. The sun's total and spectral irradiance for solar energy applications and solar radiation models. *Solar Energy* 76: 423–453.
53. ASHRAE. 2009. *ASHRAE Handbook. Fundamentals: SI Edition*, Atlanta, GA: American Society of Heating, Refrigeration and Air-Conditioning Engineers, Inc.

37

Wind Energy Resource

Dale E. Berg

CONTENTS

37.1 Wind Origins

The primary causes of horizontal atmospheric air motion, or wind, are the uneven heating of the earth and its atmosphere by solar radiation and the earth's rotation. The earth's atmosphere reflects about 43% of the incident solar radiation back into space, absorbs about 17% of it in the lower portions of atmosphere, and transmits the remaining 40% to the surface of the earth, where much of it is then reradiated into the atmosphere. The radiation from the hot sun is at short wavelengths (0.15–4 μm) and passes readily through the atmosphere, while the radiation from the cooler earth is at longer wavelengths (5–20 μm) and is readily absorbed by the water vapor in the atmosphere. Thus, the radiation from the earth is primarily responsible for the warmth of the atmosphere near the earth's surface. Heat is also transferred from the earth's surface to the atmosphere by conduction and convection.

On average, the total amount of energy radiated to space from the earth and its atmosphere must be equivalent to the total amount of solar radiation absorbed, or the temperature of the earth and its atmosphere would steadily increase or decrease. The more nearly perpendicular the sun's rays strike the earth, the more solar radiation is transferred through the atmosphere. Thus, during the year, tropical regions receive much more solar energy than do polar regions. Winds and ocean currents level out this imbalance in thermal energy, preventing the tropical regions from getting progressively hotter and the polar regions from getting progressively colder. In addition, the lack of homogeneity of the earth's surface—the land, water, desert, forest, rock, sand, black loam soil, etc.—leads to differences in solar radiation absorption and reflection back to the atmosphere, creating differences in atmospheric temperature, density, and pressure. These differences, in turn, create forces that contribute to wind. For example, the difference in radiation absorption and emission by land and water along a coastline is the dominant cause of the light shore winds or breezes; the difference in radiation absorption and emission by mountains and valleys is a significant contribution to the light upslope and downslope breezes often found in mountainous terrain. The earth's rotation results in Coriolis forces, which accelerate each moving particle of air. This acceleration moves an air particle to the right of its direction of motion in the Northern Hemisphere and to the left in the Southern Hemisphere. The geostrophic winds in the upper atmosphere (above 600 m or so) are due to the balance of these Coriolis forces and the pressure gradient forces due to the uneven heating of the atmosphere.

The earth's rotation also imparts an angular momentum to each particle of air in a west-to-east direction. Conservation of angular momentum results in an increase in its west-to-east velocity as the particle moves from the equator toward the poles. In the temperate zones, this causes the westerlies, winds opposite to the general flow, in both hemispheres.

The other long-term, large-scale global wind patterns such as equatorial doldrums, trade winds, easterlies, and subtropical and polar jets, illustrated in Figure 37.1, are also caused by the combination of differential solar heating and the rotation of the earth. These wind patterns are often referred to as the general circulation patterns. In actuality, these patterns are complicated by seasonal effects due to changes in the earth's position relative to the sun during the course of a year and geographical effects due to the uneven distribution of, and physical properties of, water and land surfaces. Centers of high or low pressure, caused by heating or cooling of the lower atmosphere, include hurricanes, monsoons, and extratropical cyclones. Small-scale phenomena characterized by local wind include land and sea breezes, valley and mountain winds, monsoon-like flow, Foehn winds (dry, high-temperature winds on the downwind side of mountain ranges, commonly referred to in the western United States as "chinooks"), thunderstorms, and tornadoes.

Variations of wind speed in time can be divided into the categories of interannual, annual, diurnal, and short-term. Interannual variations in wind speed occur over timescales greater than 1 year. They can have a large effect on long-term wind turbine production. Meteorologists generally conclude that it takes 30 years of data to accurately determine long-term values of weather or climate and that it takes at least 5 years of data to arrive at a reliable average annual wind speed at a given location. However, that doesn't mean that data spanning shorter periods of time are useless.

Annual variations refer to significant variations in seasonal or monthly averaged wind speeds that are common over most of the world. For example, for the eastern one-third of the United States, maximum average wind speeds occur during the winter and early spring. Spring maximums occur over the Great Plains, the North Central States, the Texas Coast, in the basins and valleys of the West, and the coastal areas of Central and Southern California.

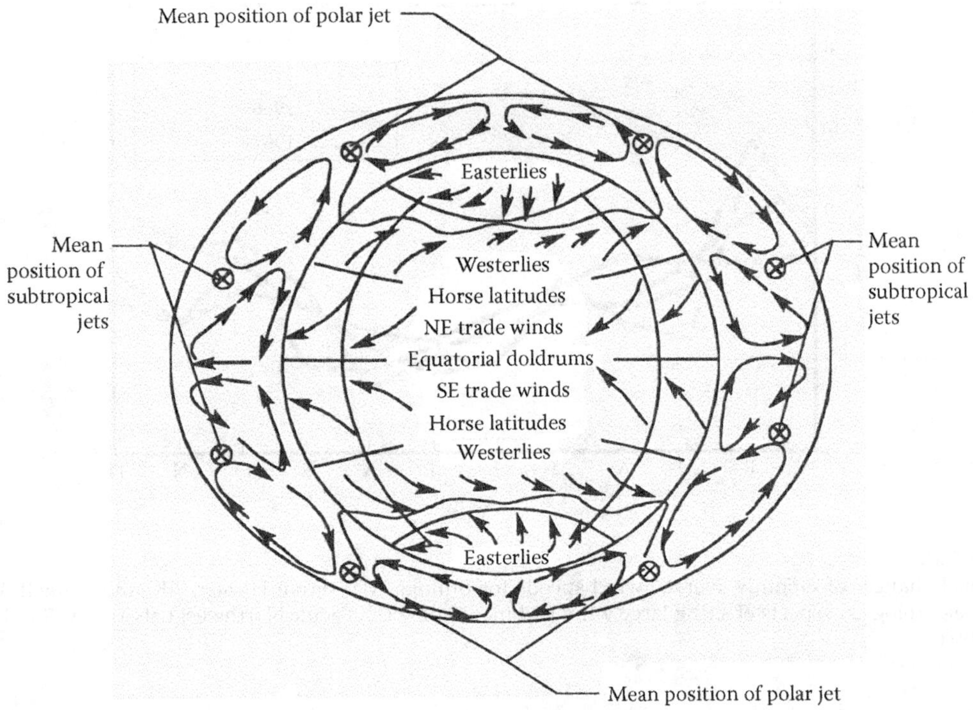

FIGURE 37.1
Semipermanent global wind patterns. (From Spera, D. (ed.), *Wind Turbine Technology: Fundamental Concepts of Wind Turbine Engineering*, ASME Press, New York, 1994. Reproduced by permission of ASME.)

Winter maximums occur over all U.S. mountainous regions, except for some areas in the lower southwest, where spring maximums occur. Spring and summer maximums occur in the wind corridors of Oregon, Washington, and California. Figure 37.2 illustrates how the monthly average wind speed at a location can vary significantly from year to year.

Large diurnal or time-of-day variations in wind occur in both tropical and temperate latitudes. This type of wind speed variation is mainly due to differential heating of the earth's surface during the daily radiation cycle; this is especially true in temperate latitudes over relatively flat areas. A typical diurnal variation is an increase in wind speed during the day, with the wind speeds lowest during the hours from midnight to sunrise. The largest diurnal changes generally occur in spring and summer, and the smallest in winter. However, the diurnal variation may vary with location and altitude above sea level. For example, at altitudes high above surrounding terrain, for example, mountains or ridges above a plain, the diurnal pattern may be very different from the pattern for the surrounding terrain. This variation is due to mixing or transfer of momentum from the upper air to the lower air. There may be significant year-to-year differences in diurnal behavior, even at fairly windy locations. Although gross features of the diurnal cycle can be established with a single year of data, a good characterization of more detailed features, such as the amplitude of the diurnal oscillation and the time of day that the maximum winds occur, requires multiple years of data.

Variations in wind speed with periods of less than 10 min and that have a stochastic or random character are generally considered to represent turbulence or fluctuations

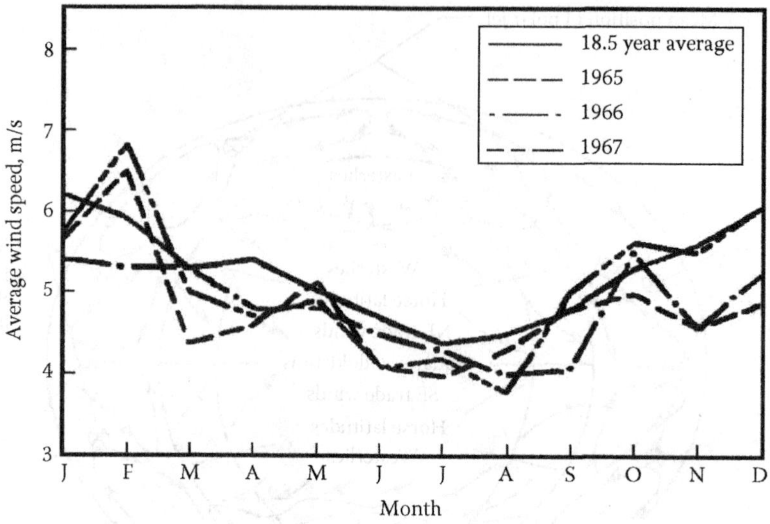

FIGURE 37.2
Seasonal changes of monthly average wind speeds for Billings, MT. (From Hiester, T.R. and Pennell, W.T., The meteorological aspects of siting large wind turbines, PNL-2522, Pacific Northwest Laboratory, Richland, WA, 1981.)

imposed on the mean wind speed. A gust is a short-term discrete event within a turbulent wind field. The common method of characterizing a gust is to measure or specify the amplitude, the rise time, the maximum gust variation, and the lapse time associated with it. Wind speed is very dependent on local topographical and ground cover variations. Figure 37.3 illustrates differences between two sites that are located in nominally flat terrain only 21 km apart. In spite of the close proximity and the flat terrain, the 5-year average mean wind speeds differ by about 12%.

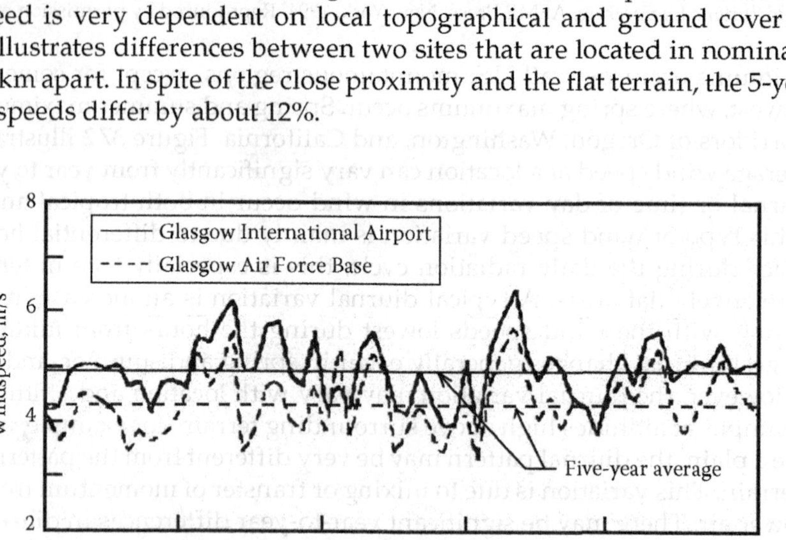

FIGURE 37.3
Time series of monthly wind speed for Glasgow, MT, International Airport and Air Force Base. (From Hiester, T.R. and Pennell, W.T., The meteorological aspects of siting large wind turbines, PNL-2522, Pacific Northwest Laboratory, Richland, WA, 1981.)

37.2 Energy Available from the Wind

Given the fact that the wind blows everywhere, the idea of harnessing it to provide power is very appealing. Mankind has harnessed the power of the wind to produce mechanical power for well over 1000 years; the use of the wind to mill grain in Persia in the tenth century is well documented, and many experts speculate that the Chinese may have invented the windmill as much as 2000 years ago. How much power is available in the wind? Is it enough to be a viable source of energy in the modern world? Answers to these questions require knowledge of several wind energy basics.

37.2.1 Wind Power

Consider the air with an average velocity U and density ρ in a cylinder of cross section A and length L as shown in Figure 37.4. The mass of the air in that cylinder is ρAL, and the kinetic energy (KE) of that air is $\frac{1}{2}\rho ALU^2$ (Watt-hours or Wh). Now assume that there is a wind turbine rotor at the downwind end of that cylinder of air. The cylinder of air will pass through that rotor in a period of time, T, where $T = L/U$. The power, or time rate of change of the KE, available at the rotor then is $\frac{1}{2}\rho ALU^2/T$, which becomes

$$\text{Power} = \tfrac{1}{2}\rho AU^3 \text{ (W)} \tag{37.1}$$

This is often rearranged and expressed as power per unit area:

$$\text{WPD (wind power density)} = \frac{\text{Power}}{A} = \tfrac{1}{2}\rho U^3 \text{ (W/m}^2\text{)} \tag{37.2}$$

From Equation 37.1, it is obvious that the most important factor in the available wind power is the velocity of the wind. Increasing the wind velocity by only 20%, say from 5 to 6 m/s, increases the available wind power by 73%. Figure 37.5 illustrates the impact of normal annual wind speed variations at a location on the available wind power at that location.

The standard value for air density at sea-level reference conditions of 101,325 Pa pressure and 15°C is 1.225 kg/m³. The actual density depends on the moisture content or humidity of the air, the temperature, and the atmospheric pressure; however, the influence of

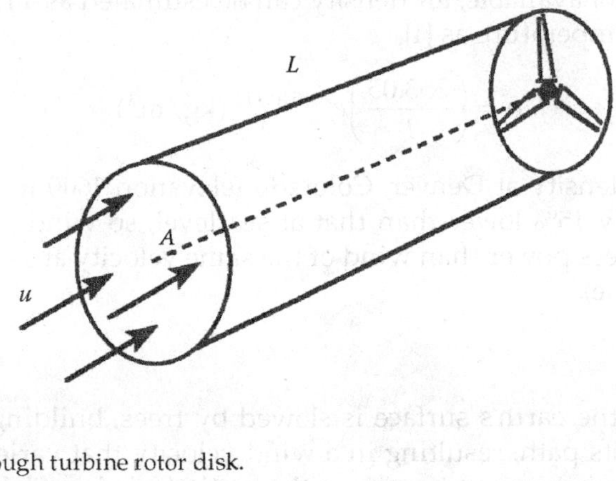

FIGURE 37.4
Steady wind passing through turbine rotor disk.

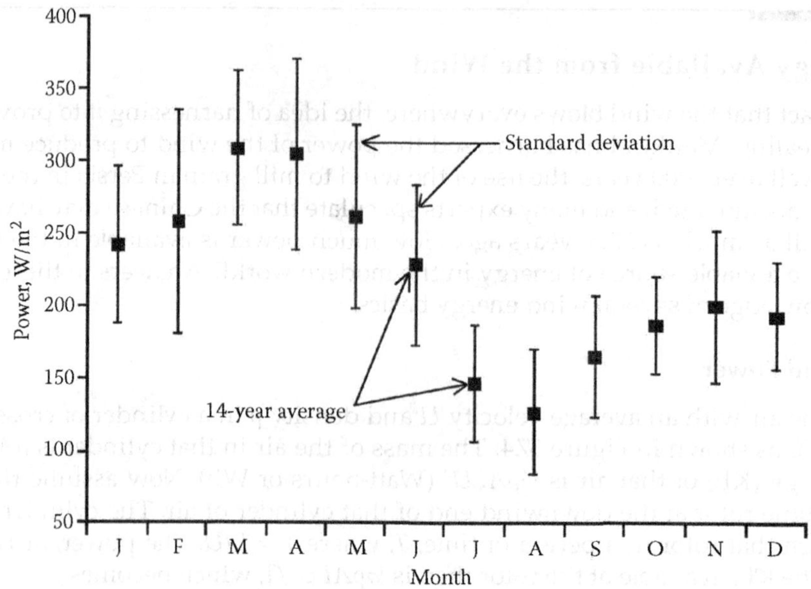

FIGURE 37.5
Seasonal variation of available wind power density for Amarillo, TX. (From Rohatgi, J.S. and Nelson, V., *Wind Characteristics: An Analysis for the Generation of Wind Power,* Alternative Energy Institute, West Texas A&M University, Canyon, TX, 1994. Reproduced by permission of Alternative Energy Institute.)

humidity is very small and is normally neglected. Under this condition, the air density can be calculated from the perfect gas law

$$\rho = \frac{P}{RT}(\text{kg/m}^3) \quad (37.3)$$

where

P is the atmospheric pressure in Pa or N/m²
R is the specific gas constant for air (287 J/kg K)
T is the absolute air temperature in Kelvin

If site air pressure is not available, air density can be estimated as a function of site elevation, *z*, and absolute temperature as [1]

$$\rho = \left(\frac{353.05}{T}\right)e^{-0.034(z/T)}\ (\text{kg/m}^3) \quad (37.4)$$

For example, the air density at Denver, Colorado (elevation 1600 m or 5300 ft above sea level), is approximately 15% lower than that at sea level, so wind of a given velocity at Denver contains 15% less power than wind of the same velocity at sea level (assuming the temperature is the same).

37.2.2 Wind Shear

Wind moving across the earth's surface is slowed by trees, buildings, grass, rocks, and other obstructions in its path, resulting in a wind velocity that varies with height above the earth's surface—a phenomena known as the *vertical wind profile* or *vertical wind shear.*

In most locations, wind shear is positive (wind speed increases with height), but situations in which the wind shear is negative or inverse are not unusual. In the absence of actual data for a specific site, a commonly used approximation for wind shear in an open area is

$$\frac{U}{U_o} = \left(\frac{h}{h_o}\right)^{\alpha} \tag{37.5}$$

where

U is the velocity at a height h

U_o is the measured velocity at height h_o

α is the wind shear exponent

The instantaneous wind shear exponent varies widely with elevation, time of day, season of the year, wind speed, temperature, and nature of the terrain. It could be, say, 0.1 during the day and then reach 0.5 at night [2]. The time-averaged shear exponent is the value that is normally used (averaged over several weeks); this varies with terrain characteristics, but usually falls between 0.10 and 0.25. Wind over a body of open water is normally well modeled by a value of α of about 0.10; wind over a smooth, level, grass-covered terrain such as the U.S. Great Plains by a value of α of about 0.14; wind over row crops or low bushes with a few scattered trees by a value of α of 0.20; and wind over a heavy stand of trees, several buildings, or hilly or mountainous terrain by a value of α of about 0.25. Short-term time-averaged shear factors as large as 1.25 have been documented in rare, isolated cases.

As a result of wind shear, the available wind power at a site may vary dramatically with height. For example, for $\alpha = 0.20$, Equations 37.1 and 37.5 reveal that the available wind power density (WPD) at a height of 50 m is approximately $\{(50/10)^{0.2}\}^3 = 2.63$ times the available WPD at a height of 10 m. Keep in mind that Equation 37.5 is only an estimate; a specific site may display much different wind shear behavior, and that will dramatically affect site WPD, so it is very important to measure the wind resource at the specific site and height where the wind turbine will be located.

National Renewable Energy Laboratory (NREL) scientists recently analyzed wind data obtained from several tall towers at heights of up to 110 m and found that the annual average wind shear exponent at Great Plains sites for heights between 50 and 100 m ranges from 0.15 to 0.25, with a daytime shear between 0.05 and 0.1 and a nighttime shear between 0.25 and 0.40 [3]. They also found that surface roughness effects on wind shear can be significant at heights up to 100 m, wind shear exponents at heights of 100–150 m can exceed those at heights of 50–100 m, and large differences in shear exponents at elevated heights can exist among sites, even in local areas of similar wind climate [4]. Their recommendation from these studies is that direct measurement data at elevated heights are needed to validate model-derived wind resource estimates.

37.2.3 Available Resource

The amount of energy available in the wind (the wind energy resource [WER]) at a site is the average amount of power available in the wind over a specified period of time, commonly 1 year. If the wind speed is 20 m/s, the available power is very large at that instant, but if it blows at that speed for only 10 h/year and the rest of the time the wind speed is near zero, the resource for the year is small. Therefore, the relative frequency of occurrence for each wind speed (the wind speed distribution) is very important in determining the resource at the site. This distribution is determined experimentally as the relative frequency of occurrence of uniform width wind speed ranges extending over the entire range of possible wind speeds (i.e., 0.5 m/s increments from 0 to 30 m/s). The wind speed associated with each range is that at the center of the range. This distribution may be approximated by a continuous curve, the probability density function or pdf, which corresponds to wind speed ranges of infinitesimal width.

If the actual wind speed probability density distribution is not available, it is commonly approximated with the generalized two-parameter Weibull distribution given by

$$f(U)=\frac{k}{C}\left(\frac{U}{C}\right)^{k-1}\exp\left[-\left(\frac{U}{C}\right)^{k}\right] \tag{37.6}$$

where

$f(U)$ is the frequency of occurrence of wind speed U
k is the Weibull shape factor
C is the Weibull scale factor

For the special case with parameter k equal to 2, this reduces to the commonly used Rayleigh distribution:

$$f(U)=\frac{\pi}{2}\frac{U}{\overline{U}^2}\exp\left[-\frac{\pi}{4}\frac{U^2}{\overline{U}^2}\right] \tag{37.7}$$

where $\overline{U}$ is the yearly average wind speed.

It is readily apparent that the average wind speed at a site yields a unique Rayleigh distribution for that site. Experimental data from a site also determine the best Weibull distribution for that site, but the process of determining k and C from the experimental data is not straightforward.

The measured wind speed distribution at the Amarillo, Texas, airport (yearly average wind speed of 6.6 m/s) is plotted in Figure 37.6, together with the Weibull and Rayleigh distributions for that wind speed. It is obvious that the Rayleigh distribution is not a good representation for these data; the data have a higher peak, have a lower probability of lower wind speeds, and have a significantly higher probability of winds in the 4–12 m/s range. The Weibull distribution (the shape factor k was picked as 2.6, yielding an estimate for the scale factor C of 7.4) is a much better approximation of the data. It shows a good match to the peak probability, but it still is not a good match; the data consistently have a somewhat lower probability for wind speeds below about 10 m/s and a somewhat higher probability for wind speeds above 10 m/s.

With a discrete wind speed distribution and the local air density, the WER, or average wind power availability, for a specific site can be estimated as

$$\text{WER}=\tfrac{1}{2}A\sum_{i=1}^{n}\rho f(U_i)\Delta U_i U_i^3\,(\text{W}) \tag{37.8}$$

where

n is the number of wind speeds included in the distribution
$f(U_i)\Delta U_i$ is the probability of the wind speed occurring in the wind speed range ΔU_i centered on wind speed U_i

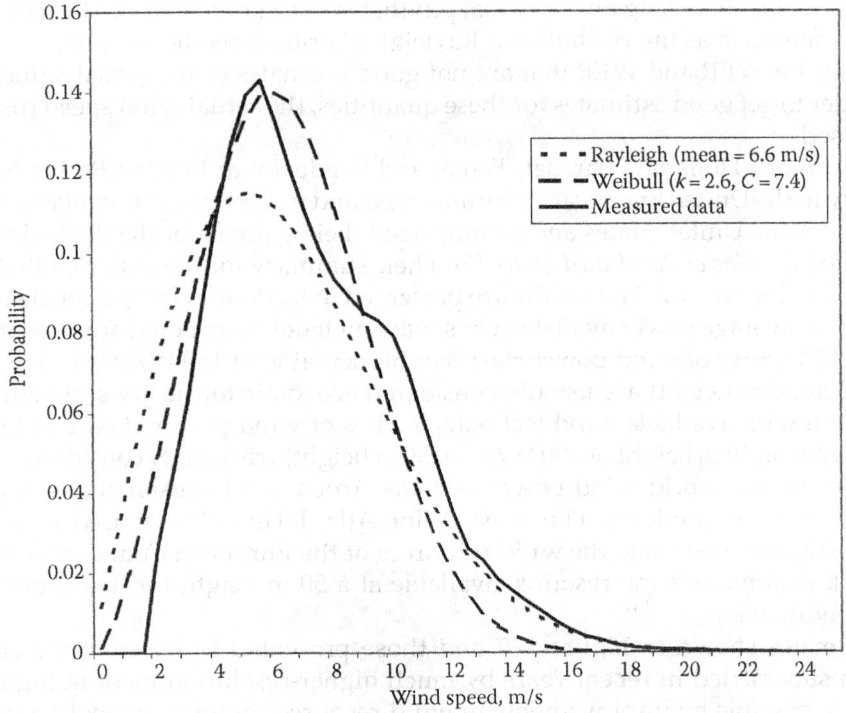

FIGURE 37.6
Measured and analytical wind speed distributions for Amarillo, TX, airport.

If an analytical pdf (such as the Rayleigh or Weibull) is used, this becomes

$$\text{WER} = \tfrac{1}{2} A \int \rho f(U) U^3 dU (\text{W}) \tag{37.9}$$

The wind energy potential (WEP) or potential gross annual wind energy production for a specific site and a specific wind turbine can be calculated with a wind speed distribution and the turbine power curve (the electrical power generated by the turbine at each wind speed), properly adjusted for the local air density. For a discrete wind speed distribution,

$$\text{WEP} = 8760 \sum_{i=1}^{n} f(U_i) \Delta U_i P(U_i) (\text{Wh}) \tag{37.10}$$

where

8760 is the number of hours in a year

$P(U_i)$ is the electrical power produced by the turbine at wind speed U_i, the center of the range ΔU_i

With an analytical pdf (the Rayleigh or Weibull), this becomes

$$\text{WEP} = 8760 \int f(U) P(U) dU (\text{Wh}) \tag{37.11}$$

It is quite obvious that using an analytical pdf that is not a good match to the actual distribution at a site (such as the Weibull and Rayleigh distributions shown in Figure 37.6) will yield values for WER and WEP that are not good estimates of the actual values for that site. In order to get good estimates for these quantities, the actual wind speed distribution must be used.

In 1987, Elliott, Holladay, Barchet, Foote, and Sandusky at Battelle Pacific Northwest Laboratory in the United States carefully analyzed and interpreted the available long-term wind data for the United States and summarized their estimate of the WERs in the *Wind Energy Resource Atlas of the United States* [5]. Their summary for the entire United States is reproduced in Figure 37.7. The results are presented in terms of wind power classes based on the annual average power available per square meter of intercepted area (see the legend on Figure 37.7). Sites of wind power class 3 or higher (at least 150 W/m^2 at 10 m height or 300 W/m^2 at 50 m height) are usually considered economic for utility-scale wind power development with available wind technology. Sites of wind power class 2 or lower (less than 150 W/m^2 at 10 m height or 300 W/m^2 at 50 m height) are usually considered economic only for remote or hybrid wind power systems. Troen and Petersen at Denmark's Risø National Laboratory produced a European Wind Atlas [6] in 1989 that used somewhat different techniques to estimate the wind resources of the European Community countries. Their work summarizes the resource available at a 50 m height for five different topographic conditions.

The estimates shown in Figure 37.7 and those presented in [6] are quite crude and have been superseded in recent years by much higher-resolution maps at higher elevations, made possible by improvements in wind resource computer modeling programs, increases in computer speed, and development of high-altitude wind speed–measuring capabilities. In particular, NREL, in collaboration with AWS Truepower, has recently released new maps of the U.S. wind resource at 50, 80, and 100 m elevations [7]. The new 100 m resource map, displayed in Figure 37.8, reveals that many areas in the U.S. Midwest and Southeast thought to lack an adequate wind resource (see Figure 37.7) actually have commercially viable wind resources at turbine tower heights of 100 m or greater. Many countries around the world have embarked on similar high-resolution mapping efforts to accurately quantify their wind resources and identify those areas of highest resource. The resultant resource maps are frequently available to the public, but in some cases, a payment is required to obtain them. High-resolution wind resource maps of the individual states in the United States may be found on the web at www.windpoweringamerica.gov/wind_maps.asp. Some states have developed even more detailed maps in recent years, and these may be found on websites for the individual states. Similar maps for some other countries may be found at www.nrel.gov/wind/international_wind_resources.html, and information on where to find maps and/or data for other countries may be found at www.windatlas.dk/index.htm. These newer wind resource evaluations frequently reveal far greater wind resource than earlier evaluations. In some cases, this is due to the higher resolution now available, but in some cases, this is because the early evaluations sometimes used unverified long-term data from existing weather stations where the anemometers had not been properly maintained—the bearings had deteriorated over time, and the anemometers registered lower winds than what actually existed. The verification procedures used in the newer evaluations are designed to help identify and eliminate this type of biased data. But even the highest resolution resource estimates are just that—estimates. The actual wind resources in any specific area can vary dramatically from those estimates and must be determined with long-term, site-specific, and elevation-specific measurements.

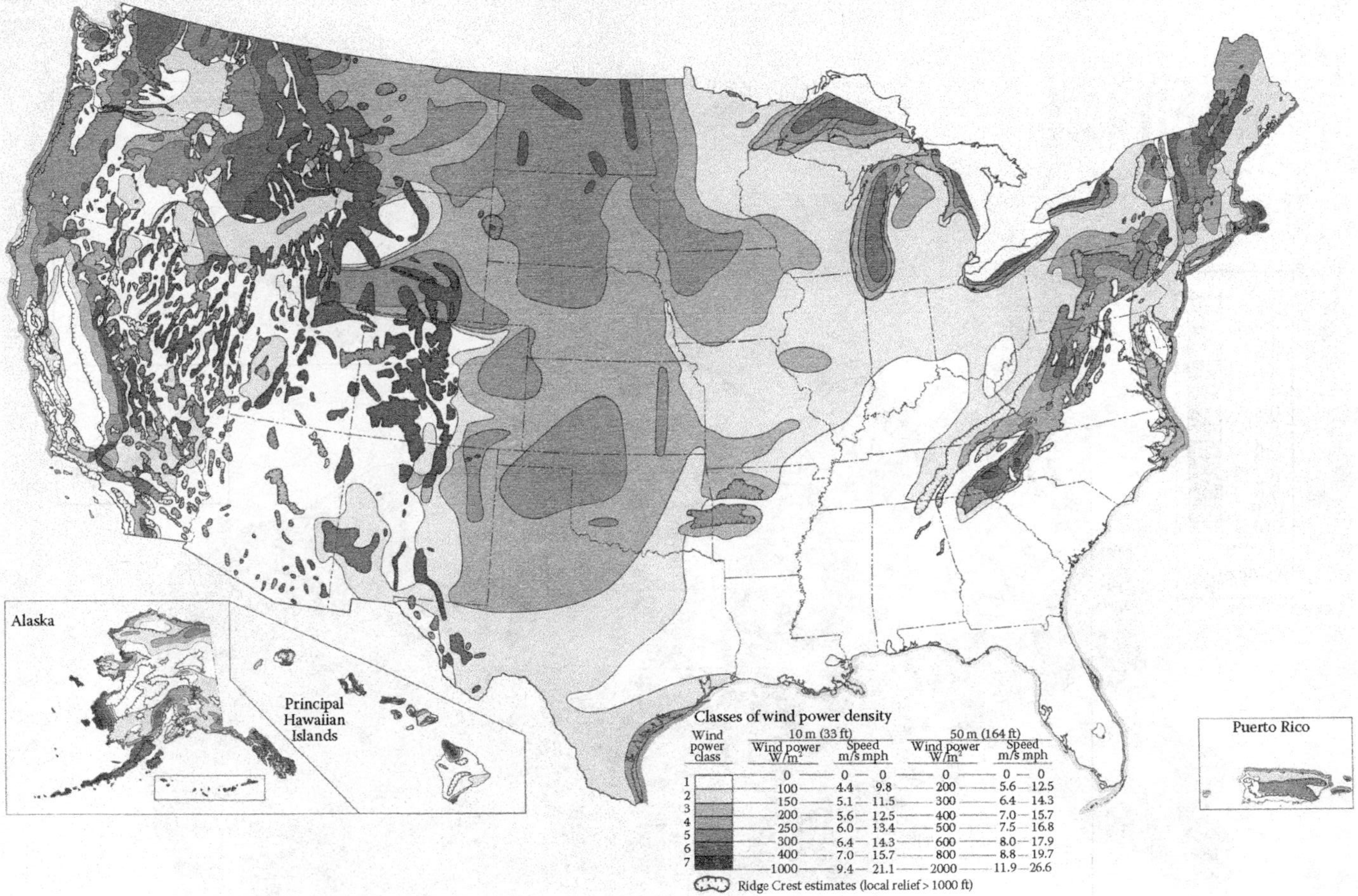

FIGURE 37.7
The U.S. wind energy resources. (From Elliott, D.L. et al., Wind energy resource atlas of the United States, DOE/CH10094-4, Solar Energy Research Institute, Golden, CO, 1987. Reproduced by permission of National Renewable Energy Laboratory, Golden, CO.)

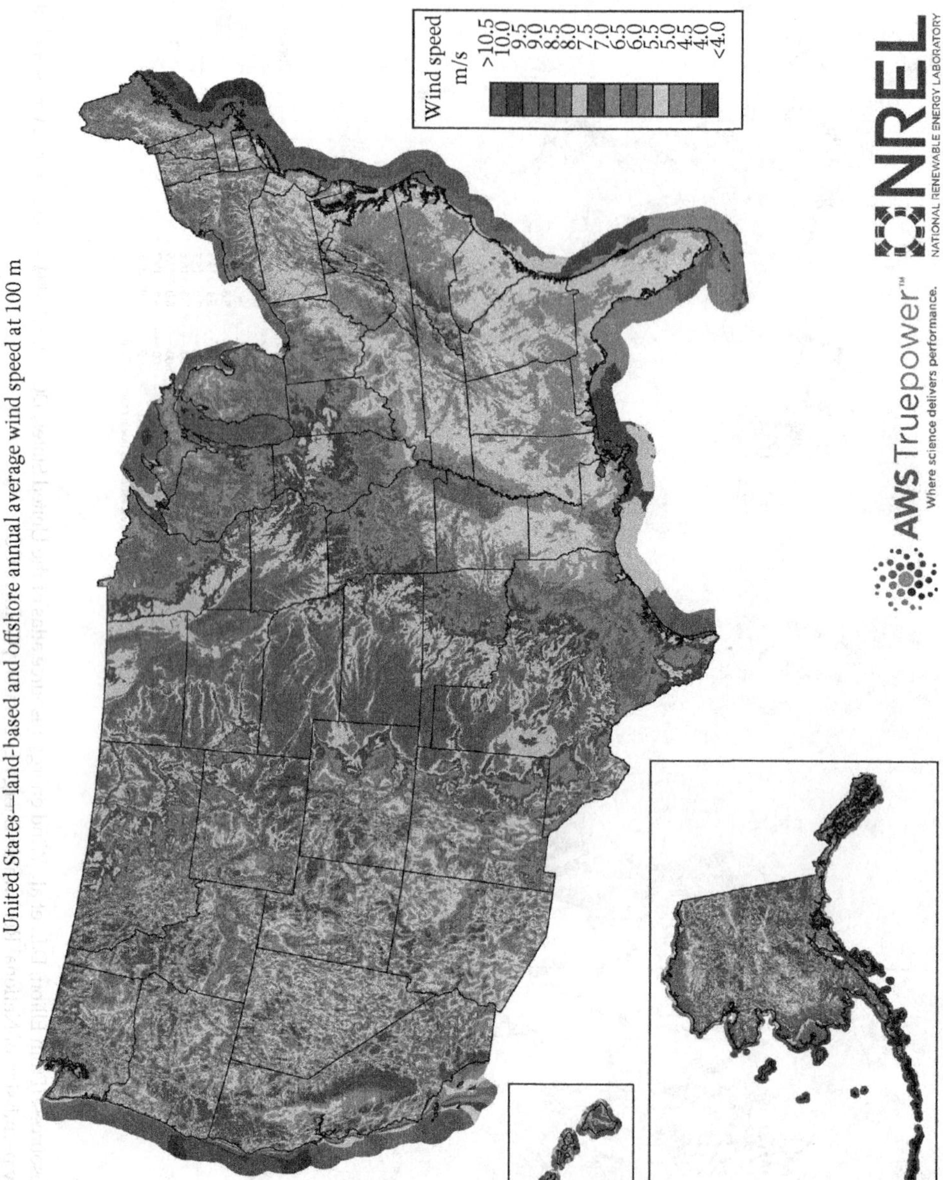

FIGURE 37.8
The U.S. high-altitude wind energy resources. (From Office of Energy Efficiency and Renewable Energy, U.S. Department of Energy, http://energy.gov/sites/prod/files/styles/media_energy_gov_wysiwyg_fullwidth/public/wind_speed_map_lg.jpg?itok=ITLo1T1G, accessed March 2015.)

Archer and Jacobson from Stanford University [8] have estimated the average global wind power potential for all land-based locations with wind class 3 or better (mean annual wind speeds in excess of 6.9 m/s at 80 m height) to be on the order of 72 TW (1 TW = 10^{12} W) for the year 2000, yielding an average WEP of 627,000 TW h/year. In reality, the use of approximately 90% of the land mass is restricted, leaving only about 10% available for wind power generation. Thus, a realistic WEP from the Stanford study is on the order of 62,700 TW h/year. This figure is consistent with the published European Wind Energy Association [9] figure of 53,000 TW h/year of WEP and corresponds to over three times the world's total electricity needs in 2001 of 14,000–16,000 TW h/year. Many of the countries around the world have adequate wind resources to supply their entire energy consumption. Of course, this resource is not necessarily available at the right time or the right place, and actually capturing wind energy on this vast a scale is not apt to happen.

The wind resource over the water-covered portion of the earth's surface is many, many times greater than that over the land, but current wind energy technology is restricted to shallow water (water depths on the order of 30 m or less). The available wind resource over water restricted to these depths is still significant—recent NREL work has shown that the U.S. offshore gross wind resource at these depths is estimated at over 1000 GW, or roughly the generating capacity currently carried on the U.S. electric grid [10]. Keep in mind that this resource value will likely shrink by 60% or more after all environmental and socioeconomic constraints have been taken into account. Development of the technology required to enable utilization of the much larger WERs at greater water depths is the focus of much current research.

37.2.4 Environmental/Societal Restrictions

The fact that a site has great wind resources and access to transmission lines with excess capacity does not mean that it is suitable for wind power development. Potential environmental and social issues must be considered and resolved before a wind facility can be built. In the best possible scenario, such issues will add expense and development time to a project. In the worst scenario, such issues could cause cancellation of a project. Some of the issues that could affect the suitability of a proposed site include the presence of endangered or protected species, nearby residences or airports (which might require significant setbacks of the nearest wind turbines), nearby scenic areas, recreational use or other specific use restrictions of the candidate site, or religious significance of the site. The following sections discuss two issues that are topics of research today.

37.2.4.1 Impacts of Wind Facilities on Birds and Bats

The greatest environmental issues that the wind industry has had to face are the possible impacts of wind facilities on birds and bats and their habitats. Concerns about the bird fatality issue were, in large part, the result of relatively high numbers of raptor deaths in the Altamont Pass wind farms east of San Francisco, CA, in the 1980–1985 time frame. Dozens of studies of this issue have been conducted during the past 30 years. Sinclair and Morrison [11] and Sinclair [12] give overviews of some of the early U.S. studies, and the National Wind Coordinating Committee has recently published a fact sheet summarizing the research performed prior to 2010 [13]. One conclusion of these studies is that the Altamont Pass situation appears to be a worst-case scenario, due in large part to bad turbine siting and to the presence of overhead power lines that led to a large number of bird electrocutions. Colson [14] and Wolfe [15] report that the numerous recommendations for minimizing the impact of new wind farms on birds resulting from these studies include the following:

- Conduct site-specific mitigation studies.
- Avoid bird migration corridors and areas of high bird population such as micro-habitats or fly zones.
- Use fewer, larger turbines.
- Minimize the number of perching sites on turbine towers by using tubular, rather than lattice, towers.
- Bury electrical lines.

The subsequent development of industry siting guidelines for wind farms has proven to be quite effective; recent studies have shown that relatively low raptor fatality rates exist at most wind energy developments, with the possible exception of those wind farms in California that were developed prior to the establishment of these siting guidelines.

If large numbers of raptors are present, additional restrictions on turbine placement, style, height, etc., may be required by local authorities to minimize the potential for collisions with turbines. In reality, far more songbirds than raptors are killed by turbines; in spite of this, the estimated cumulative impact of songbird collisions with wind turbines is several orders of magnitude lower than the estimated impacts from the leading human-related causes of songbird mortality (vehicles, buildings and windows, power transmission lines, communication towers, toxic chemicals, and feral and domestic cats), according to the NWCC [13,16]. Bird collisions with wind turbines caused the deaths of only 0.01%–0.02% of all the birds killed by collisions with man-made structures across the United States in 2001. In contrast, bird collisions with buildings and windows accounted for about 55% of structure-related bird deaths, while collisions with vehicles, high-tension power lines, and communication towers accounted for about 17%.

Relatively small numbers of bat fatalities were reported at wind energy facilities in the United States before 2001, largely because most monitoring studies were designed to assess bird fatalities, according to Kunz et al. [17]. Bat carcasses are much more difficult to find than are bird carcasses, and they are quickly removed by predators, so it is quite likely that these studies underestimated bat fatalities. However, once more careful studies revealed that some wind farms were causing a large number of bat fatalities, the wind industry joined with Bat Conservation International, the U.S. Fish & Wildlife Service, and NREL to identify and quantify the problem and to explore ways to mitigate these deaths. Several wind energy companies are providing a portion of the funding for cooperative efforts to resolve this issue.

Although bat fatalities at wind energy facilities appear to be highest along forested ridgetops in the eastern United States, studies have revealed that bat fatalities are a widespread problem—turbine-related bat deaths have been reported at every wind facility studied to date, both in the United States and in Europe [13].

This problem has been actively studied for over a decade at this point, but it is still poorly understood. A few of the findings to date are as follows:

- Some migratory tree-roosting bat species appear particularly vulnerable to wind power fatalities.
- Bat fatalities peak at wind facilities during the late summer and early fall bat migration seasons.
- Most bat fatalities occur during periods of low wind, when flying insects, which are their primary source of food, are most common, and bats are most active.

Why are bats being killed by turbines? As documented by Baerwald et al. [18], given that echolocating bats detect moving objects better than stationary ones, their relatively high fatality rate at wind farms is perplexing. According to the NWCC [13], bats may be attracted to the turbines; possibly because the towers appear to be potential roosting trees, there is a high concentration of insects around the turbines (due to the favorable conditions created during turbine installation); or the sounds produced by turbines are attractive. On the other hand, the bats may become disoriented by the turbine sounds or by the electromagnetic fields in the vicinity of the turbine nacelles. This remains a research topic at this time.

In contrast to turbine-related bird deaths, which appear to be caused entirely by collisions with the turbines, the manner in which bats are killed by turbines is not well understood. While direct collisions with the turbines certainly account for many of the deaths, some researchers hypothesize that a large portion of the bat fatalities are caused by barotrauma; Baerwald et al. [18] have found that the internal organs of a sizable portion of the dead bats they found in the vicinity of wind turbines have been damaged by the low-pressure areas present near the blade tips and in the blade wake.

Preliminary studies reported by Baerwald et al. [19] and Arnett et al. [20] have demonstrated that bat fatalities have been reduced by 50%–87% by increasing the cut-in wind speed of the turbine, but much work remains to be done before this issue is resolved. Until this problem is better understood, the prudent course of action would be to postpone the development of promising sites that have significant numbers of bats nearby. At the very least, the latest information from the appropriate national wind energy association regarding bat collisions with wind turbines should be obtained and carefully studied.

37.2.5 Impact of Wind Facilities on Radar

While fixed objects in the vicinity do not generally create a problem for radar installations, the rotating blades of wind turbines do; wind turbine blades have a very large radar signature, especially when turbines are grouped together in a wind farm. As a result, wind farms located in the near vicinity of government radar installations used for national defense, national security, aviation safety, and weather forecasting interfere with those radars by creating clutter, reducing detection sensitivity, obscuring potential targets, and scattering target returns. These effects tend to decrease radar sensitivity, generate false targets, interfere with target tracking, and impede critical weather forecasts. Many hundreds of MW of proposed U.S. wind developments in the vicinity of radar installations are on hold because of this issue—developers cannot get approval from one or more government agencies due to concerns that these developments will interfere with the existing radar installations.

In an effort to resolve this problem in order to accommodate future wind energy growth, the U.S. Departments of Energy, Defense, and Homeland Security and the Federal Aviation Administration are currently funding a field test and evaluation of commercial off-the-shelf wind–radar interference mitigation technologies. Results of this effort should be available in the 2014 time frame, but widespread implementation of any successful mitigation technology is apt to take several years. In the meantime, the prudent course of action would be to postpone the development of promising sites that might create a radar interference issue.

37.3 Wind Resource Assessment

Three things must be present before a commercial wind energy generation plant can be successfully developed: a site with a good wind resource, access to a transmission line with the capacity to accept the plant output, and a buyer to purchase the energy generated. Only the identification of sites with adequate wind resource will be considered here. In addition, this discussion will be limited to the evaluation of onshore wind resources. Evaluation of offshore resources requires significantly more effort and a much larger financial investment.

Many different approaches are available when investigating the wind resource in a given land area. The preferred approach will depend on your wind energy program objectives and on previous experience with wind resource assessment. In any case, the process normally consists of three basic scales or stages of assessment:

1. Prospecting or identifying general areas of high wind resource. In this stage, wind developers typically rely on existing wind resource information to narrow the search region to the most promising areas.
2. Area wind resource evaluation to determine the actual wind resource. Wind measurement programs are normally undertaken in areas under serious consideration for wind power development. Wind developers use these measurements to determine the actual quality of the resource, compare different candidate sites, estimate the performance of specific turbines, analyze the economic potential of wind development, and make preliminary determinations of turbine placements.
3. Micrositing, in which the siting of the individual wind turbines is optimized. By carefully studying the small-scale variability of the wind resource at a particular location, the developer can position the wind turbines to get maximum wind exposure with minimal interference from other turbines or obstructions, and thus maximize the overall energy output. There are many computer codes available to perform this function. The level of sophistication of these codes varies widely. The better ones use computational fluid dynamics to model the wind farm, utilizing digital elevation data to define the terrain of the wind farm and site measurement data to provide calibration and wind input information. In some cases, professional wind energy development companies may be willing to perform this service gratis, in hopes of getting the rights to develop the site. Advertisements for these programs can be found in most wind energy trade magazines. Further discussion of micrositing is beyond the scope of this contribution.

37.3.1 Prospecting

The goal of prospecting for wind energy development sites is to identify areas within a fairly large region such as a utility service area, a county, or even a multistate region that are likely to have good wind resources and that are located near existing power transmission lines with the capacity to handle the power generated by the new development. A good initial step in this process is to obtain the highest resolution available wind resource maps of the entire area of interest (see Section 37.2.3 for links to some map sources) and use those to identify areas of high winds. Keep in mind that values on the wind resource maps are just estimates—some areas of good resource may not be identified, and some areas identified as having good resource may actually have very poor resources. Use these maps in conjunction with topographic maps that show the location of major transmission lines, roads, etc., to narrow the general search area. Use of geographic information system (GIS) will permit the overlaying of the resource information with topographic information and road and transmission line information, greatly facilitating the identification of promising sites. Even with all the information that can be gleaned from GIS systems and hard copy and online maps, field visits to prospective sites are very worthwhile, as observation of biological indicators and topographic features often help identify the best wind resource areas.

37.3.1.1 Biological Indicators

Persistent winds can cause plant deformation; careful observation of these plant deformations can be used to compare candidate sites and, in at least some cases, to estimate the average wind speed. Figure 37.9 illustrates various levels of tree deformity, corresponding to increasing levels of wind from one prevailing direction. The Griggs–Putnam index, explained in Hewson et al. [21], correlates these degrees of deformation to specific wind speeds for one particular type of tree. While the vegetation in an area of interest may not permit the estimation of actual wind speeds based upon the sketches in this figure, the relative wind speeds in the area might well be established by observing the amount of flagging or throwing that is present. It should be noted that, although wind-flagged trees (i.e., trees with branches bent away from a prevailing wind) may indicate that the annual average wind speed is quite strong, trees that are not flagged do not necessarily indicate that the winds are light; those trees may be exposed to strong winds from several directions, with insufficient persistence in any one direction to cause flagging.

37.3.1.2 Effects of Topography

The effects of surrounding terrain on the wind speed at a specific site are discussed in various wind turbine siting handbooks including Troen and Petersen [6], Hiester and Pennell [22], Wegley et al. [23], and Rohatgi and Nelson [24]; the following discussion borrows heavily from these sources. Numerous researchers emphasize that the influence of terrain features on the energy output from a turbine may be so great that the economics of the whole project may depend on the proper selection of the site. The ready availability of detailed GIS maps and mapping technology today has greatly simplified this aspect of prospecting.

37.3.1.2.1 Terrain Classification

The most basic classification divides terrain into flat and non-flat categories. In a strict sense, the earth's surface is never truly flat; there are always some irregularities such as forest and shelterbelts (or wind breaks), and/or gentle slopes. However, according to Frost and Nowak [25], the terrain can be considered flat (for the purpose of wind turbine siting) if it meets the following conditions:

- Elevation differences between the wind turbine site and the surrounding terrain are not greater than about 60 m anywhere in an 11.5 km diameter circle around the turbine site.
- No hill has an aspect ratio (height to width) greater than 1/50 within 4 km upwind and downwind of the site.
- The elevation difference within 4 km upwind is small compared to the rotor ground clearance.

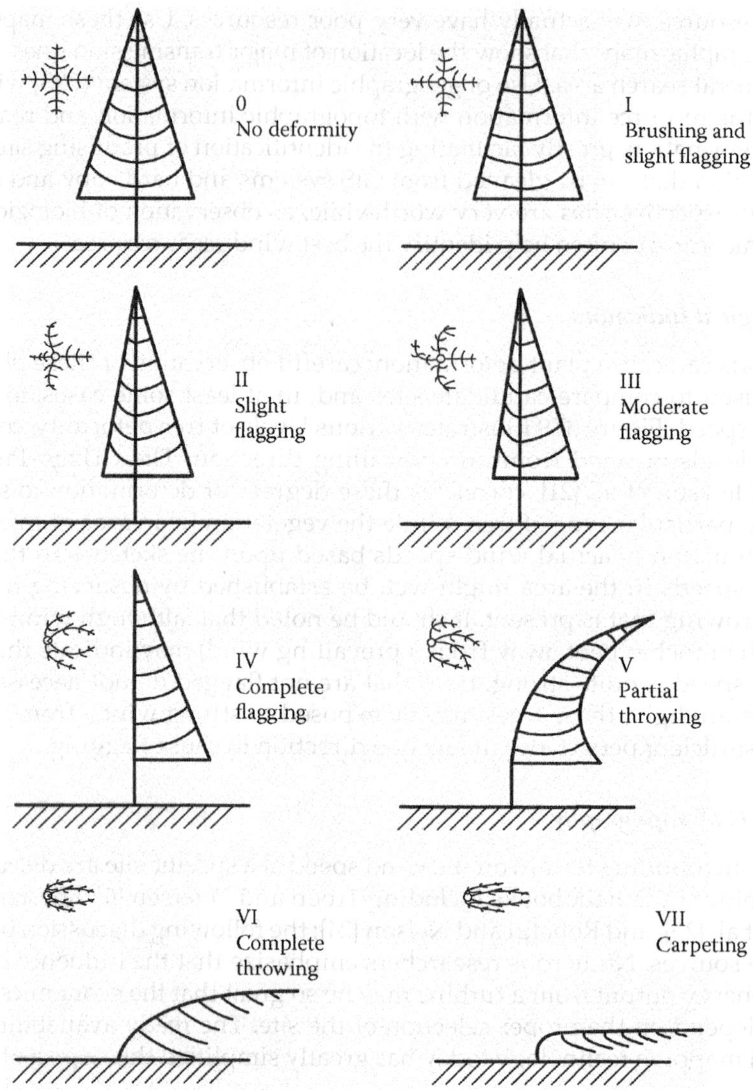

FIGURE 37.9
Griggs–Putnam index of tree deformation. (From Hewson, E.W. et al., Vegetation as an indicator of high wind velocity, RLO/2227-T24-78-2, Oregon State University, Corvallis, OR, 1978.)

Flat terrain is obviously the simplest type of terrain for siting a turbine—the wind speed at a given height is nearly the same over the entire area. In the absence of any obstructions to speed up wind flow, typically the best way to increase the available wind power is to raise the rotor higher above the ground to take advantage of positive wind shear. In most natural terrain, however, the surface of the earth is not uniform but changes significantly from location to location, affecting the local wind profile. Figure 37.10 illustrates the significant change in a vertical wind profile that results from wind flow going from a smooth to a rough surface—the shape of the profile shifts quite dramatically over a relatively short horizontal distance. In addition, most flat terrain has a variety of man-made (buildings, silos, etc.) and natural obstacles that affect the flow of wind. The effects of these obstacles

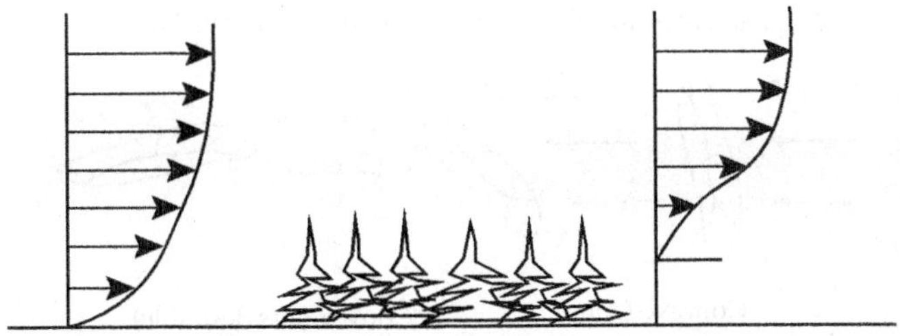

FIGURE 37.10
Effect of change in surface roughness from smooth to rough. (From Wegley, H.L. et al., *A Siting Handbook for Small Wind Energy Conversion Systems*, PNL-2521, Pacific Northwest Laboratory, Richland, WA, 1980.)

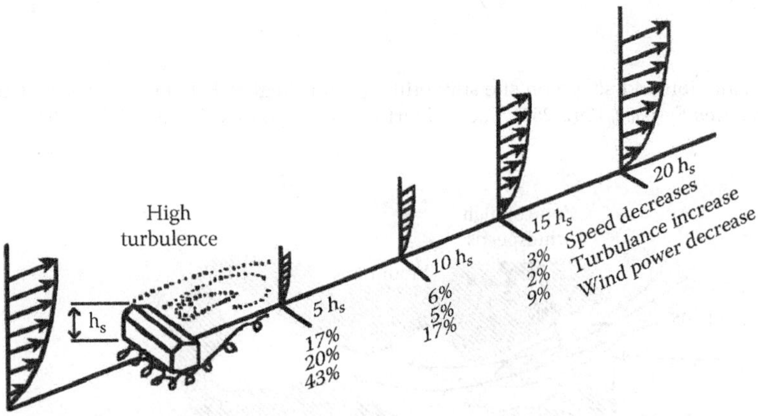

FIGURE 37.11
Wind speed, power, and turbulence effects downstream of a building. (From Wegley, H.L. et al., *A Siting Handbook for Small Wind Energy Conversion Systems*, PNL-2521, Pacific Northwest Laboratory, Richland, WA, 1980.)

have been studied extensively; an example of the effect of a building on wind speed, available power, and turbulence is shown in Figure 37.11. Note that the estimates in the figure apply at a height equal to 1 building height above the ground, and that power losses become small (<10%) after a downwind distance equal to 15 building heights.

Ridges are elongated hills with a length-to-height ratio of at least 10 that have little or no flat area on the summit. As illustrated in Figure 37.12, the ideal prevailing wind direction for wind turbine siting should be perpendicular to the ridge axis. When the prevailing wind is not perpendicular to the axis, the ridge will not be as attractive a site. Concavity in the windward direction enhances speedup, and convexity reduces speedup by deflecting the wind flow around the ridge. The slope of a ridge is also an important parameter; steeper slopes give rise to stronger wind flow, but they also give rise to high turbulence in the lee of the ridge.

Depressions are characterized by terrain features lower than the surroundings and include valleys, canyons, basins, and passes. These can cause significant speedup of the wind if they effectively channel the wind. The factors that influence the flow in depressions, in addition to diurnal flow variations, include orientation of the wind in relation to the depression, atmospheric stability, the width, length, slope and roughness of the depression, and the regularity of the section of valley or canyon. Canyons in mountainous terrain, such as those illustrated in Figure 37.13, can be very effective in creating high wind speeds.

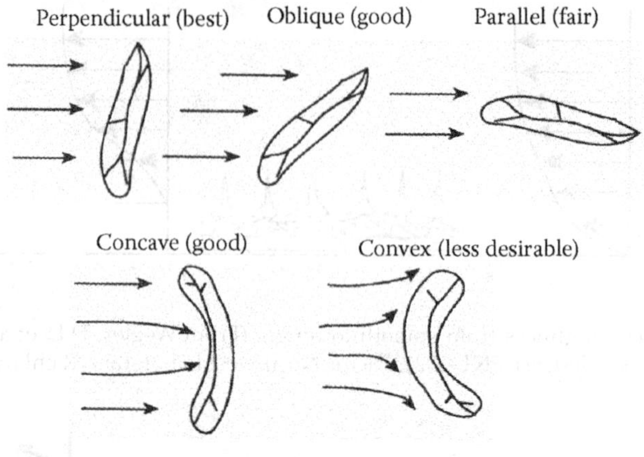

FIGURE 37.12
Effect of ridge orientation and shape on site suitability. (From Wegley, H.L. et al., *A Siting Handbook for Small Wind Energy Conversion Systems*, PNL-2521, Pacific Northwest Laboratory, Richland, WA, 1980.)

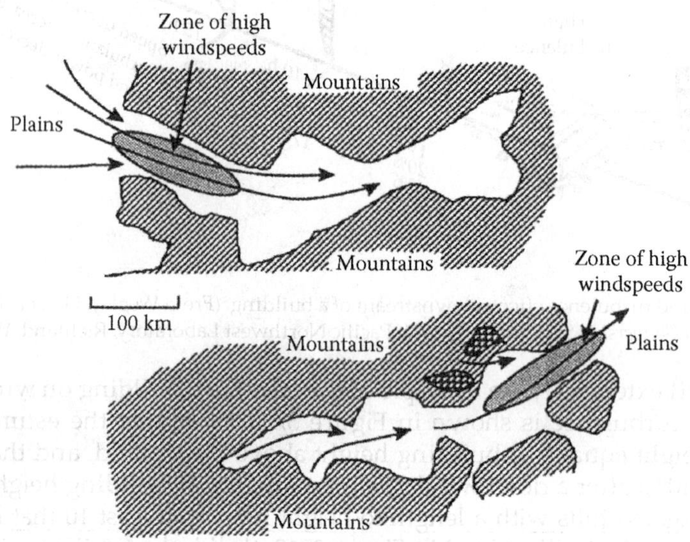

FIGURE 37.13
Increased wind speeds due to channeling of prevailing winds by mountains. (From Rohatgi, J.S. and Nelson, V., *Wind Characteristics: An Analysis for the Generation of Wind Power*, Alternative Energy Institute, West Texas A&M University, Canyon, TX, 1994. Reproduced by permission of Alternative Energy Institute.)

Non-flat or complex terrain has large-scale elevations or depressions such as hills, ridges, valleys, and canyons. Keep in mind that information on wind direction should be considered when defining the terrain classification. For example, if an isolated hill (200 m high and 1000 m wide) were situated 1 km south of a proposed site, the site would normally be classified as complex. If, however, the wind blows only 5% of the time from this direction and the average speed from this direction is low, say 2 m/s, then this terrain would be classified as flat. Additional summaries of the effects on wind flow of many types of large-scale features, including mountains, large cliffs and escarpments, valleys and canyons (including slope winds, prevailing winds in alignment, and prevailing winds in nonalignment), gaps and gorges, passes and saddles, and large basins, are given in Rohatgi and Nelson [24].

37.3.2 Wind Resource Evaluation

Once general areas of high wind resources have been identified, the actual resource available at one or more of those sites can be determined by performing a full wind resource evaluation or site assessment. This is a costly, complex, and time-consuming activity. Numerous companies perform this service; a good method for locating some is to access the appropriate national wind energy association membership list and search for "meteorology consultants." Even if the service is contracted out, a cursory knowledge of the process is important to ensure that the contractor does a good job. Numerous handbooks detailing the steps for conducting successful measurement programs are available. While the following discussion, which borrows heavily from AWS Scientific, Inc. [1], is for land-based resource assessment, many of the important aspects are also applicable to off-shore resource assessment [26].

Long-term wind speed variability at a site is nearly as important as mean wind speed, as far as long-term energy production is concerned. Meteorologists generally agree that a well-done evaluation/assessment project will take a minimum of 1 year to complete [23], but a project duration of 2 or more years will produce more reliable results and is recommended by Schwartz and Elliott [27] and many other experts. One year is usually sufficient to determine the diurnal and seasonal variability of the wind with an accuracy of 10% at a confidence level of 90%, according to Aspliden et al. [28] and Corotis [29]. In some cases, high-quality data may be available from a nearby representative site, and this might be used to shorten the duration of the assessment project and to estimate the interannual variability of the wind. In particular, long-term data from the nearest airport or weather recording station can help determine whether the data obtained at a site is representative of normal winds for the site or whether it is representative of higher- or lower-than-average winds. Wegley et al. [23] and Gipe [30] give suggestions on methods of using available data from nearby sites to estimate site wind speed with minimal on-site data.

A single site assessment with a very basic single monitoring station (cup anemometer and wind vane instrumentation) on a 50 m tilt-up tower operated for 2 years can be expected to run in excess of $20,000 (2005$); a 60 m monitoring station will run about $2500 more. Multiple sites will see some economies of scale, depending on how close the sites are to each other. The total cost to operate a second site can be expected to be 10%–15% less than the cost for the first site. Five sites in fairly close proximity can probably be operated for an average site cost about 25% less than for a single site. Most savings will be realized in the labor and travel categories, although equipment cost savings may be realized as a result of quantity discounts, by sharing installation equipment, and by sharing parts inventories. Conducting an abbreviated resource assessment program to minimize expenses may well turn out to be a case of "penny wise, pound foolish" in that the return on investment of the entire wind farm project will be directly affected by the quality of the actual wind resource at each turbine site. Without a proper site assessment, the resource estimates may not be a good reflection of the actual resource, and the production of the wind project may be much different than projected.

For small wind turbines, the expense of anemometers, data logger, and data analysis may be more than the price of the wind turbine. In this case, historical regional data may be judged to be adequate for estimating the wind resource. Inexpensive digital weather stations, including rudimentary data loggers that work with a personal computer, are available for under $1000, and one of these could be used in conjunction with the historical data to improve the accuracy of the resource assessment. While these inexpensive weather stations may be adequate for this particular application, they are not highly accurate and are not designed for long-term durability, so they are not suitable for collecting long-term data for accurate wind resource assessment as is needed for commercial enterprises. Wind speed, direction, distribution, and shear can vary significantly over fairly short distances in either the horizontal or the vertical direction, so in order to get the best possible estimate of the WER at a particular location, it is important to measure the wind resource at the specific site and height of interest. In complex terrain, this probably means the use of several monitoring stations to adequately assess a single site.

The keys to a successful assessment program are the early identification of the objectives of the program and the timely development of a detailed plan of action to ensure that the data needed to meet the objectives are acquired. Such a plan should include, at a minimum, the following:

- Quality control measures, including a quality assurance (QA) program
- Site data to be measured
- The program duration, minimum measurement accuracy, and target data recovery rate (these heavily impact the equipment type and quality)
- Number and tentative location of monitoring towers, together with sensor measurement heights
- Data sampling and recording intervals
- Data storage, handling, and processing procedures

The data recovery rate is defined as the percentage of possible data records that have actually been collected over a reporting period:

$$\text{Data recovery rate} = \frac{\text{Data records collected}}{\text{Data records possible}} \times 100\,(\%)$$

where

$$\text{Data records collected} = \text{Data records possible} - \text{Number of invalid records}$$

For example, assume that the total possible number of 10 min records in December is 4464. If 264 records were deemed invalid, the number of valid data records collected would be 4200 (4464–264). The data recovery rate for this example would be

$$\text{Data recovery rate} = \frac{4200}{4464} \times (100) = 94.1\%$$

A data recovery rate of at least 90% (95% or better should be possible) for all measured parameters over the duration of the program, with any data gaps kept to a minimum (less than 1 week), should be a major goal of any measurement program.

A *QA* program for flagging and handling suspect data is imperative to ensure the acquisition of high-quality data and the successful completion of the assessment program. The specifics of the components of this QA program should be determined and documented early in the assessment program and should include, at a minimum, the following:

- Acquisition of equipment that meets accuracy and reliability specifications
- Equipment calibration methods and frequency of calibration
- Installation and maintenance instructions for equipment
- Data validation methods, including specifics on evaluating, removing and/or replacing suspect data, and reporting all of these actions
- Data analysis instructions, including specific calculations to be performed

QA will also help to minimize the uncertainties that are always inherent in the data. If the assessment process is carefully followed, these uncertainties can be characterized and controlled to maximize the usefulness of the assessment program conclusions.

37.3.2.1 Data Measurement

The core of the monitoring program is the collection of wind speed, wind direction, and air temperature data. Keep in mind that it is generally less expensive to provide and monitor extra sensors than to conduct an unscheduled site visit to replace or repair a failed sensor that is the sole source of an essential measurement. Care must be used in mounting the various sensors to minimize any interference of one sensor on another. It is especially important to avoid any interference with the wind measurements.

If the rotor is relatively small, wind speed measurements at the turbine hub height will suffice, but for larger rotors, acquiring wind speed data at multiple measurement heights is necessary for determining the site vertical wind shear characteristics, for conducting turbine performance simulations at several turbine hub heights, and to assist in data validation (multiple anemometers make spotting a bad one fairly easy). Redundant anemometers are sometimes used to minimize the risk of wind speed data loss due to a failed primary anemometer and to provide substitution data when the primary anemometer is shadowed by the tower (when it is directly downwind of the tower). Typical anemometer heights are every 10 m, starting at 20 or 30 m and going up to maximum tower height (50–60 m for a tilt-up tower). The vertical distances between anemometers mounted at different heights are referred to as *height layers*. For example, for anemometers installed at heights of 30, 40, and 50 m, height layers would be 30–50, 30–40, and 40–50 m.

The following sensor types have traditionally been those commonly used for the measurement of near-horizontal wind speed:

- *Cup anemometer*: This instrument consists of a cup assembly (three or four cups) centrally connected to a vertical shaft for rotation. At least one cup always faces the oncoming wind. The cups convert wind pressure force to rotational torque, and the transducer in the anemometer produces an electrical signal that is proportional to wind speed.
- *Propeller anemometer*: This instrument consists of a propeller mounted on a horizontal shaft that is oriented into the wind through the use of a tail vane. The propeller anemometer also generates an electrical signal proportional to wind speed.

The current trend toward taller turbine towers and the large expense of erecting met towers of comparable height has encouraged the development of ground-based remote sensing technology, capable of monitoring wind conditions at heights well above 100 m. The two major types of remote sensing are light detection and ranging (LIDAR) and sonic detection and ranging (SODAR). Detailed information on these technologies and their use is available in [26,31–32]. These are also commonly used to characterize off-shore wind resources. Further discussion of these technologies is beyond the scope of this contribution; the remainder of this discussion will focus on data systems that utilize traditional anemometers.

Although the two anemometer types differ somewhat in their responsiveness to wind speed fluctuations, there is no clear advantage of one type over the other. In practice, the cup type is most commonly used for resource assessment. When selecting an anemometer model, the following should be considered:

- *Intended application*: Anemometers intended for low-wind-speed applications, such as air pollution studies, are usually made from lightweight materials. These are probably not suitable for very windy or icy environments.
- *Survival wind speed*: Be sure the anemometer is capable of withstanding the maximum wind speed that it is likely to see. An anemometer with a survival wind speed of 25 m/s is not apt to survive in most windy sites. A survival speed of 50 m/s should be adequate for most sites.
- *Starting threshold*: This is the minimum wind speed at which the anemometer starts and maintains rotation. For wind resource assessment purposes, it is more important for the anemometer to survive a 25 m/s wind gust than to be responsive to winds under 1 m/s.
- *Distance constant*: This is the distance the air travels past the anemometer during the time it takes the cups or propeller to reach 63% of the equilibrium speed after a step change in wind speed (the *response time* of the anemometer to a change in wind speed). Longer-distance constants are usually associated with heavier anemometers; inertia causes them to take longer to slow down when the wind decreases. These instruments may overestimate the wind speed.
- *Reliability and maintenance*: Wind sensors are mechanical and eventually wear out, although most have special, long-life bearings that will normally last for at least 2 years. Be sure to get units with bearings that will last for the entire duration of the measurement project.

Anemometers are subject to a variety of errors in the determination of true wind speed, and equations that may be used to estimate the size of these errors are given in Justus [33]. When anemometers are calibrated in steady air flows in a wind tunnel, they may measure the true wind within ±1%. In gusty winds, however, anemometers generally speed up faster than they slow down, so that accuracies of ±5% may be more realistic in application. Additional information on anemometers may be found in publications by American Wind Energy Association (AWEA) [34] and ASME [35].

Wind direction vanes should be installed at all significant monitoring levels. Wind direction information is important for identifying preferred terrain shapes and orientations and for optimizing the layout of wind turbines within a wind farm. The most familiar type of vane uses a fin connected to a vertical shaft. The vane constantly seeks a position of force equilibrium by aligning itself into the wind and produces an electrical signal proportional to the position of the vane relative to some reference direction (usually selected as true—not magnetic—north). Some wind vanes have a *dead band*, a narrow section of the rotation where the sensor transitions from the full rotation reading of nearly 360° to the initial reading of 0°. The output of the sensor in this section is usually unpredictable. The position of this dead band should be carefully noted for reference when the vane is mounted to the tower. Newer direction vanes have eliminated this dead band.

Air temperature is used to calculate air density, a quantity required to estimate the WPD and the power that a wind turbine will generate. It is normally measured either near ground level (2–3 m) or near hub height. In most locations, the average near-ground-level air temperature will be within 1°C of the average temperature at hub height. Ambient air temperature sensors are readily available. The temperature sensor must be protected from direct solar radiation by mounting it within a radiation shield.

Once the basic measurement system is installed, additional resource-related parameters can be acquired at minimal cost. The most common additional parameters are vertical wind speed, change in temperature with height (commonly referred to as ΔT), barometric pressure, and solar radiation.

The vertical wind speed provides more detail about site turbulence and can be a good predictor of wind turbine loads. Historically, this parameter has only been a research measurement, but as wind energy development spreads into new regions of the country, regional information on vertical wind velocity may become important. The propeller anemometer is especially well suited for measuring the vertical wind component. For this application, the rotation axis would be mounted vertically. The polarity of the DC output signal indicates rotational direction, and the signal magnitude indicates actual vertical speed. The vertical wind speed anemometer should be located near the upper basic wind speed monitoring level.

ΔT provides information about turbulence and atmospheric stability. This is measured with a matched set of temperature sensors located near the lower and upper measurement levels; the existing air temperature sensor may be matched with an identical sensor and used to measure ΔT, or a separate pair of matched sensors may be used. Sensors for this application are usually tested over a specified range and matched by the manufacturer. Be sure to use identical equipment (e.g., radiation shield and mounting hardware) with both sensors so the inherent errors in the signals will cancel out when the difference between the two values is taken. Radiation shields that use either forced (mechanical) or natural (passive) aspiration are normally used on both sensors to reduce the radiation-induced errors.

Barometric pressure is used with air temperature to determine air density. Since it does not vary much over relatively short distances, many resource assessment programs do not measure barometric pressure; they use elevation-adjusted data taken at a nearby regional National Weather Service station. Several atmospheric (barometric) pressure sensors that are suitable for this application are commercially available. Be sure to select one that will give accurate readings in a windy environment.

Solar radiation, when used in conjunction with wind speed and time of day, can be an indicator of atmospheric stability and is often used in numerical wind-flow modeling. These measurements may also be useful for later solar energy evaluation studies. A *pyranometer* is used to measure global (or total) solar radiation, the combination of direct sunlight and diffuse sky radiation. Remember that the pyranometer must be in a position where it will never be shaded in order to measure accurately. The recommended measurement height is 3–4 m above ground.

Table 37.1 lists nominal specifications for the most common types of sensors. Some sensors require the use of separate signal conditioners, electronic packages that supply power

TABLE 37.1

Typical Specifications for Sensors

Specification	Wind Speed	Wind Direction	Air Temperature	Vertical Wind Speed	ΔT	Atmospheric Pressure	Solar Radiation
Measurement range	0–50 m/s	0°–360°	−40° to 60°C	0–50 m/s	−40° to 60°C	94–106 kPa (sea level equivalent)	0–1500 W/m²
Starting threshold	≤1.0 m/s	≤1.0 m/s	N/A	≤1.0 m/s	N/A	N/A	N/A
Distance constant	≤4.0 m/s	N/A	N/A	≤4.0 m/s	N/A	N/A	N/A
Allowable sensor error	≤3%	≤5°	≤1°C	≤3%	≤1°C	≤1 kPa	≤5%
Sensor resolution	≤0.1 m/s	≤1°	≤0.1°C	≤0.1 m/s	≤0.1°C	≤0.2 kPa	≤1 W/m²

Source: AWS Scientific, Inc., *Wind Resource Assessment Handbook: Fundamentals for Conducting a Successful Monitoring Program*, SR-440-22223, National Renewable Energy Laboratory, Golden, CO, 1997.

Note: All sensors should have an operating temperature range of −40°C to 60°C and an operating humidity range of 0%–100%.

to the sensor, and process the signal received from the sensor to convert it into a form that can be used by the data logger, the device that actually acquires the raw data and calculates and saves the average statistics. Data loggers (or data recorders) come in a variety of types, sizes, and capabilities; most include peripheral storage and data transfer devices. Be sure the data logger is compatible with the sensor types, number of sensors, and desired sampling and recording intervals. The data logger should also

- Be capable of recording the time and date corresponding to each data record with that data record
- Contribute negligible errors to the signals received from the sensors
- Operate over the temperature range of −40°C to 60°C and over a relative humidity range of 0%–100%
- Offer retrievable data storage media
- Operate on battery power (with an AC adaptor to permit the use of AC power when it is available)

The amount of data logger storage capacity that is needed depends on the averaging interval, the number of active data channels, the need for calculating derived quantities such as wind shear exponent and turbulence intensity, and the maximum time span between data retrievals, including potential delays. Manufacturers usually provide tables or methods to calculate the approximate available storage capacity in days for various memory configurations. To be safe, get enough storage capacity to store an additional week of data, in case a data retrieval or site access problem develops.

While all data loggers allow manual retrieval of stored data, many either come equipped with or can easily be equipped with communications equipment to enable remote retrieval of data, typically via cell phone or satellite phone link. The manual method promotes frequent visual on-site inspection of the equipment during the site visits required to retrieve the data, which may be beneficial. However, this method also requires frequent expensive site visits and additional data–handling steps (thus increasing potential data loss).

Remote data retrieval permits more frequent data retrieval and inspection than that is feasible with manual data retrieval. This allows for prompt identification and resolution of site problems and enhances the data recovery rate. The disadvantages include the upfront cost of the required additional equipment, the cost of monthly service, and the risk of communication system problems. The additional costs are usually quickly offset by the savings from not having to make site visits, and choosing good quality equipment helps minimize the communication problem risk.

Data loggers with cell phone capability are extremely popular due to their ease of use and reasonable cost. A major concern with these units is whether the signal strength at the measurement site will be sufficient to establish a solid communication link. Keep in mind that the data logger cell phone is much more powerful than a typical cell phone—it will normally be equipped with a high gain antenna to permit communication at a much greater range. A general rule of thumb is that a good connection is usually possible as long as there is a cell phone tower in line of sight within about 120 km. Replacement of the standard antenna with one with higher gain may be an option if the nearest cell tower is further away. Additional information regarding cell phone links and accounts should be available from the data logger supplier or manufacturer.

Satellite telephone service may be all that is available in really remote locations, but that service is usually quite expensive—the transceiver is much more costly, and the monthly fees are usually much higher than is the case for cell phone service. In addition, utilizing a satellite communication link may require the development of specialized data transfer software for the specific application. Again, additional information should be available from the data logger supplier or manufacturer.

If phone service (either cellular or satellite) is either unavailable at the site or is prohibitively expensive, try to arrange for manual data retrieval by someone who lives near the monitoring site. That individual may then be able to send the data media by mail or transmit the data via e-mail or landline phone transmission. If this means of data retrieval is utilized, be sure to incorporate additional procedures to protect against data loss in the event that the data media gets damaged or lost in the mail or the transmitted data become corrupted.

The overall accuracy of any system is determined by its weakest link or least accurate component. It is also influenced by its complexity, the total number of components or links. The measurement of any parameter (wind speed, wind direction, etc.) requires that several components (sensor, signal conditioner, cabling, and data logger), each potentially contributing an error to the measured parameter, be interconnected. The combination of these errors will determine the system error (the difference between the measurement result and the actual value sensed) for that parameter. Errors contributed by the sensors represent the main concern, because those associated with the electronic subsystem (data logger, signal conditioner, and associated wiring and connectors) are typically negligible (less than 0.1%). Using wind speed as an example, the allowable system error in the measured wind speed might be specified as less than or equal to 3% of the true wind speed value, allowing for a 6% error window (±3%) centered about the true wind speed. This means that the wind speed sensor must have a system error of 3% or less.

The resolution of a data measurement is the smallest change in a measured quantity that can be detected by the measuring system. Again, this is primarily a function of the sensor, as most data loggers have far more resolution than is normally required. For example, an anemometer system with an 8-bit data logger analog-to-digital converter or digitizer (a very low resolution for a data logger) that is set up to measure a maximum wind speed of 50 m/s has a data logger resolution of $50/(2^8) = 0.2$ m/s. Most data loggers will have higher-precision digitizers

(10-bit or 12-bit are common) with a resolution of much less than this (0.05 or 0.01 m/s, respectively). Thus, to achieve the resolution of 0.1 m/s specified for anemometers in Table 37.1, the anemometers themselves must have a resolution of 0.1 m/s or better.

System reliability is the measure of how well an instrumentation system will consistently provide valid data for a measured parameter over its measurement range—how well it performs in the long run. In selecting instrumentation, it is important to identify and select components that are designed to reliably measure the selected parameters at the prescribed heights for the full monitoring duration and at the required levels of data recovery and accuracy. The instrumentation must also be capable of withstanding the environment of the specific location (e.g., weather extremes, dust, and salt) and be tailored to the selected mode of data retrieval (manually or via cell phone or satellite phone communication link). Although vendors often provide reliability information in terms of a mean time between failures under certain conditions, the best indication of a product's reliability is its performance history. Ask the vendor for a few references and check with those references to determine their satisfaction or lack of satisfaction with the product of interest. Comprehensive QA procedures and the use of redundant sensors are two of the best ways that the user can maximize system reliability. However, there is little that can be done to improve the reliability of sensors or data loggers that are prone to failure.

The equipment should also be proven, affordable, and user-friendly. Complete monitoring systems can be purchased from a single vendor, or components from different vendors can be combined. If components from different vendors are used, be sure the individual components are compatible with each other.

Lists of wind resource assessment equipment vendors may be found in the member lists of the various national wind associations. At the AWEA site (www.awea.org), for instance, they may be found by accessing the member directory, selecting "search," "consultants," and then "meteorology."

37.3.2.2 Sampling Rates and Statistical Quantities

All data sensors should be sampled once every 1 or 2 s. The resultant data are not typically recorded, but stored into data logger accumulators for the specified averaging period (10 min is the default international averaging period for wind measurements). At the end of the averaging period, the statistics are calculated and stored, the accumulators are cleared, and the storage of data into the accumulators begins anew. The data logger should contain built-in programming to calculate and store, as a minimum, the following statistics:

- The average, standard deviation, and maximum and minimum values for the wind speed at each anemometer level, together with the wind directions associated with each maximum and minimum value
- The average and standard deviation for the wind direction at each level
- The average, standard deviation and maximum and minimum wind speed difference for each height layer, together with the wind direction associated with each maximum and minimum difference
- The air temperature
- The vertical wind speed (if measured)

- The ΔT (if measured)
- The barometric pressure (if measured)
- The solar radiation (if measured)

These statistics, together with a corresponding time and date stamp, constitute the data to be recorded; as mentioned earlier, the individual data samples are not normally saved.

Except for wind direction, the average is defined as the numeric mean of all samples. For wind direction, the average is defined as the mean direction on a 0°–360° scale. The standard deviation is defined as the true population standard deviation for all samples within the averaging interval.

37.3.2.3 Lightning Protection Devices

Keep in mind that a single lightning strike can destroy the entire site monitoring system and all associated electronics, and repairing lightning damage will require a large investment in both time and replacement parts. While no amount of protection can ensure that lightning will not strike the system, appropriate protection can minimize that risk. Consult with the data logger and instrumentation suppliers to ensure that adequate lightning protection is incorporated in the data logger and in all sensors, signal conditioners, and power supplies that will be used. If the supplied lightning protection is not adequate for application, determine what additional protection is needed and add it before installing the monitoring system.

37.3.2.4 Towers and Sensor Mounting

Towers for mounting sensors are available in either tubular or lattice types. Both types are available in tilt-up, telescoping, and fixed versions. For new sites, the tubular, tilt-up, guyed tower, which makes possible the assembly of the tower and the mounting of the sensors on the ground, is an excellent and convenient choice. It also requires very little ground preparation and is relatively low cost. The current maximum height of available tilt-up towers is about 60 m; if the tower is to be used in icing conditions, the maximum height is reduced to about 50 m. Keep in mind that raising a tower of this size is not a simple task, and neither is replacing tower-mounted sensors; they both require experienced personnel and proper equipment. Taller towers can certainly be constructed, but they are far less portable and much more expensive. If you need measurements above 60 m and you intend to evaluate multiple sites, you may want to consider the remote sensing systems mentioned earlier [26].

The sensor support hardware includes the masts (vertical extensions) and mounting booms (horizontal extensions). Tubing, not solid stock, masts, and booms should be used. These are used to position the sensors away from the support tower so as to minimize any influence of the tower, mounting hardware, and other equipment and sensors on each of the measured parameters. This can be achieved by consulting specific manufacturers' instructions and referring to the sample installation configuration shown in Figure 37.14. Detailed information may be found in [34] and virtually any other wind measurement handbook.

Refer to the manufacturer's instructions for the proper sensor and data logger wiring configurations. Wiring to connect the sensors, signal conditioners, and data logger should be shielded and/or twisted pair cable, whenever possible, to prevent ambient electrical noise from affecting the accuracy of the measurements. Be sure to use insulation and conductor types that are flexible over the full temperature range expected at the site and use

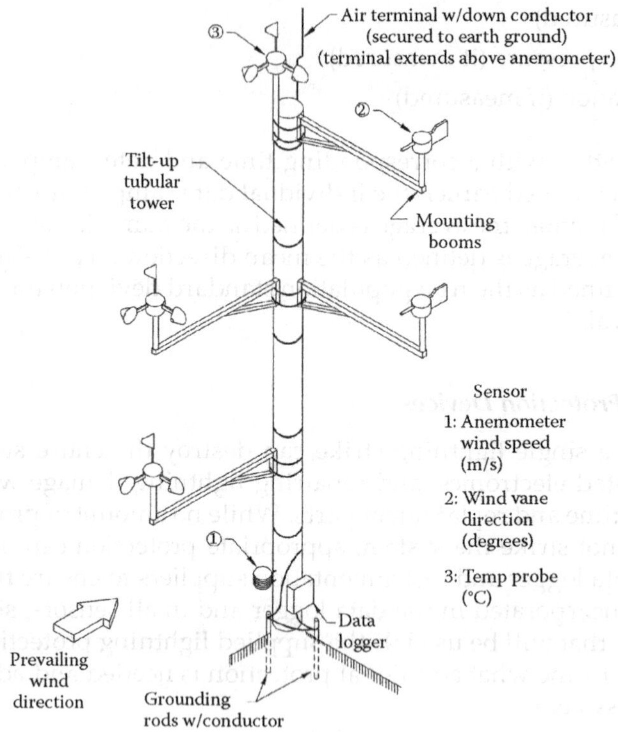

FIGURE 37.14
Typical instrumentation placement on meteorological tower.

wire with UV-resistant insulation. Rodents and raptors, in particular, seem to have an affinity for wire insulation, so consider using armored cable or protective conduit in areas that are accessible by them. Try to mount the data logger and communications equipment in a substantial locked container. Mount the communications antenna high enough to discourage vandalism but where it can still be easily accessed by service personnel. Vandals are a concern, even in very remote locations.

Seal all sensor terminal connections with silicone caulking and protect them from direct exposure to sunlight and water with rubber boots or electrical tape. Wrap the sensor cabling along the length of the support arm and tower and secure it with UV-resistant wire ties or electrical tape. Wrap tape around the sensor wire and leave sufficient slack in the wire wherever chafing can occur between the sensor wire(s) and the support structure (e.g., tilt-up tower anchor collars).

37.3.2.5 Data Collection and Handling

All of these efforts devoted to selecting and mounting sensors and ensuring that acquired data are free of interference are for naught if the resultant data are subsequently lost or contaminated. The data collection and handling elements of the monitoring system must incorporate procedures that offer a high level of data protection. In general, the procedures should comply with those specified by the data logger manufacturer and reflect good common sense. A few pertinent comments are listed here.

37.3.2.5.1 Data Retrieval Frequency

A key factor in achieving a high level of data recovery is the ability to identify potential data acquisition problems and to quickly determine if a problem exists (a failed sensor, icing, possible loss of ground, etc.) and, if necessary, initiate the steps required to fix the problem. Data transfer and review are the first-order means of achieving this end. A schedule of regular site data transfers or downloads should be developed and maintained. The maximum recommended manual download interval is every 2 weeks. For remote data transfer systems, a weekly retrieval rate may suffice, but a shorter interval, such as every other day, may be preferable to minimize potential loss of data if a problem arises and to efficiently transfer the large amount of data resulting from 10-min data averaging. Situations may arise that warrant additional data transfers. For example, sensor data irregularities may become apparent during the review of site data, or the site may experience severe weather, such as icing or thunderstorms. Either of these situations merits a prompt follow-up data transfer (either manual or remote) and review.

Any time a site visit is made, the first order of business should be retrieval of the raw data from the data logger, either by manually downloading it to an in-field laptop computer or by transferring it over a telephone link to a central site computer. This will minimize the risk of potential data loss from operator error, static discharges, or electrical surges during handling and/or checking of system components. The last order of business before leaving the site should be a verification that the monitoring system is functioning properly.

37.3.2.5.2 Data Protection and Storage

Sensor data that have not been subjected to a validation or verification process are commonly referred to as raw data. There is a constant risk of raw data loss or alteration during any measurement program. Aside from the data logger programming requirements, the actual data collection process requires minimal human intervention, and data are adequately protected by following recommended installation and operation procedures, including grounding all equipment. These field data will eventually be transferred to a personal computer for analysis; while this may be the primary location of the working database, it should not be the storage area for the archived or raw database, as frequent usage of a computer increases the likelihood of electrical surges, static discharges, and other events that may damage hard drives and destroy any databases. Preserve the original raw data; make at least two copies of that data set on removable media and store the original and all but one of the copies in separate locations (not in the same building). Then, apply the validation and processing steps to the remaining copy. Back up this active database on a regular schedule during the validation process. Once the database is fully validated, create multiple copies of it and again store each copy in a separate location (not in the same building).

Improper data-handling procedures represent a high risk for data loss. The data reduction and analysis staff will be handling the data medium and be in constant contact with significant numbers of raw and processed databases. Ensure that all personnel are fully trained and understand the data retrieval software and computer operating system, that they are well aware of all instances in which data can be accidentally overwritten or erased, and that they employ good handling practices for all data storage media.

37.3.2.5.3 Data Validation

After the field data are collected, transferred to an office computing environment, and appropriate copies are made, the next steps are to validate and process that data. Again, these steps should be performed on a copy of the database, not on the original data. Data validation consists of the inspection of all the collected data for completeness and reasonableness, and the elimination of erroneous values. This step transforms raw data into validated data and is crucial to maintaining high rates of data recovery during the course of the monitoring program. There are many possible causes of erroneous data: faulty or damaged sensors, loose wire connections, broken wires, damaged mounting hardware, data logger malfunctions, static discharges, sensor calibration drift, and icing conditions are some of the contributors. The goal of data validation is to detect as many significant errors from as many causes as possible; catching all the subtle ones is impossible. For example, a disconnected wire can be easily detected by a long string of zero (or random) values, but a loose wire that becomes disconnected intermittently may only partly reduce the recorded value, yet produce data that appear reasonable. Therefore, slight deviations in the data can escape detection (although the use of redundant sensors can reduce this possibility). Properly exercising the other QA components of the monitoring program will also reduce the chances of data problems.

Data should be validated as soon as possible after they are transferred from the site to the office; the sooner a potential measurement problem is spotted, the quicker it can be addressed, and the lower the risk of losing large amounts of data. Data can be validated either manually or automatically, with computer processing. Obviously, manual verification can be extremely tedious and time consuming, but it is a good practice to validate the initial data from a site in this manner, in order to learn the characteristics of the data and become familiar with the types of suspect data that can be expected. This knowledge then makes possible the tailoring of the computer routines to optimize the automated validation process. Validation software is available from several sources, including from data logger vendors, but it is still commonly homegrown and tailored for particular applications.

Data validation can be split into two distinct operations: data screening and data verification.

37.3.2.5.3.1 Data Screening This operation uses a series of validation routines or algorithms to screen all the data to identify suspect or questionable values—values that deserve scrutiny but are not necessarily erroneous. For example, an unusually high hourly wind speed caused by a locally severe thunderstorm may appear on an otherwise average windy day. The result of this data screening is a report that lists the suspect values and which validation routine each of those suspect values failed.

General system checks ensure that each data record contains the appropriate number of data fields and that records are contiguous in time (i.e., time and date stamps are in order, and none are missing). Measured data checks, on the other hand, ensure that the actual data are reasonable. These normally include range tests, relational tests, and trend tests.

- Range tests are the simplest and most commonly used validation tests. The measured data are compared to upper and lower limiting values that include nearly (but not absolutely) all of the expected values for the site for each data parameter. For example, a reasonable range for average wind speeds for most sites is 0–25 m/s. Negative values clearly indicate a problem; speeds above 25 m/s are possible, but should be verified with other information. Data reduction and analysis personnel can fine-tune these limiting values as they gain experience. In addition, the limits for appropriate data parameters should be adjusted seasonally. For instance, the limits for air temperature and solar radiation should be lower in winter than in summer. In general, a single item of data should be subjected to more than one range check before it is judged to be valid, because a single check is unlikely to detect all problems. For example, if a frozen wind vane reports an average direction of exactly 180° for six consecutive 10 min intervals, the values would pass the 0°–360° range test, but a check on the standard deviation would reveal a value of zero and should be flagged as suspect.

- Relational tests are based on expected physical relationships between various parameters. These ensure that physically improbable situations are flagged as suspect. For example, a significantly higher wind speed at the 25 m level than at the 40 m level should be flagged as suspect.
- Trend tests are based on the rate of change in a parameter over time. An example of a trend that indicates an unusual circumstance and a potential problem is a change in air temperature greater than 5°C in 1 h.

Some data loggers include the capability to record the system battery voltage for each averaging interval. Range and relational tests for a reduction in battery voltage may be used to give early warning of site hardware problems and ensure that data are not lost due to a bad battery or a blown fuse.

With experience, the data analysis personnel directly involved in the validation process will become very familiar with the local wind climatology and will learn which criteria are most often triggered and under which conditions. The behavior of the wind under various weather conditions will become apparent, as will the relationship between various parameters. This is an invaluable experience that cannot be gained solely by scanning monthly summary tables, and it may prove to be important for evaluating the impact of the local meteorology on wind turbine operation and maintenance. For example, some validation tests may almost always be failed under light wind conditions, yet the data are valid. This occurrence may argue for one set of test criteria under light wind conditions (below 4 m/s perhaps) and another set for stronger winds. The data analysis personnel should be authorized and encouraged to modify validation test criteria and create new ones as needed, based on their experience with the site data. However, be sure to establish operating procedures to ensure appropriate documentation and reporting of any such changes that are made.

37.3.2.5.3.2 Data Verification Once suspect data are identified, case-by-case decisions must be made on what to do with the suspect values—retain them as valid, reject them as invalid, or replace them with redundant valid values (if available). This operation requires the application of judgment of a qualified person familiar with the monitoring equipment and local meteorology. The disposition of each suspect value should be noted in a data verification report. This report should include, for each suspect value, the sensor from which the value was obtained, the date and time that the value was obtained, and the disposition of the suspect value, including the source for the replacement value or the validation code for the rejected value.

If a suspect data value is judged to be valid, leave the value as is. If a suspect value is judged to be invalid, but valid data from a redundant sensor is available, replace the invalid value with that from the redundant sensor. If a suspect value is judged to be invalid, and no data from a redundant sensor are available, replace the value with a unique error code that will serve as both a flag for later data processing programs and an indication of the specific problem with the original value. Selection of the specific error code may require review of the site log or other site data. The data processing and reporting software must incorporate means for handling these error codes. The results of this process are a validated database and a data verification report itemizing the disposition of each suspect data value.

37.3.2.5.4 Data Processing and Reporting

Once the data validation step is complete, the validated data set is ready to be processed to quantify the wind resource. This typically involves performing calculations on the data set, as well as binning or sorting the recorded 10-min average data values into useful subsets based on other chosen averaging intervals, such as hourly or weekly. Hourly averages are normally used for reporting purposes. The processed data are then summarized into weekly or monthly informative reports of summary tables and performance graphs. Items included in these reports typically include mean wind speed, wind direction frequency distribution, maximum gust, mean turbulence intensity, mean power density, and diurnal wind speed and power density (by time of day) for each anemometer and wind direction level, mean shear for each height layer, daily and monthly wind speed distribution for the primary height anemometer, and hourly temperature. Data processing and reporting software is available from several sources, including many data logger manufacturers and vendors of spreadsheet, database, and statistical software. Whatever method is used, procedures must be developed to ensure that flagged data points or invalid data codes are excluded from the computations of hourly averages and other quantities. These procedures should be developed and implemented before the first data are recorded.

The wind shear exponent, turbulence intensity, and WPD are items that are usually included in wind resource reports, but may not be routinely produced by some data loggers. These parameters can be easily calculated using a spreadsheet software application to obtain hourly and monthly averages. A description of each parameter and calculation method is presented in detail as follows:

- Vertical wind shear exponent

  Wind shear is defined as the change in horizontal wind speed with a change in height. The wind shear exponent (α) must be determined for a set of anemometry levels at each site, because the magnitude of α is influenced by site-specific characteristics. Solving the power law equation (Equation 37.5) for α gives

$$\alpha = \frac{\ln(U/U_0)}{\ln(h/h_0)} \tag{37.12}$$

  where

  U is the wind speed at height h

  U_0 is the wind speed at height h_0

- Turbulence intensity

  Wind turbulence is the rapid disturbances or irregularities in the wind speed, direction, and vertical component. It is an important site characteristic, because high turbulence levels may decrease power output and cause extreme loading on wind turbine components. The most common indicator of turbulence for siting purposes is the standard deviation (σ) of wind speed. Normalizing this value with the mean wind speed gives the turbulence intensity (TI), defined as

$$\mathrm{TI} = \frac{\sigma}{\overline{U}} \tag{37.13}$$

  where

  σ is the standard deviation of wind speed

  $\overline{U}$ is the mean wind speed (m/s)

- WPD

WPD is defined in Equation 37.14 as the wind power available per unit area swept by the turbine blades. It combines the effects of the wind speed distribution and its dependence on air density and wind speed. For experimental data, WPD may be calculated as

$$\text{WPD} = \frac{1}{2n}\sum_{i=1}^{n} \rho U_i^3 \ (\text{W/m}^2) \tag{37.14}$$

where

n is the number of records in the averaging interval
ρ is the air density (kg/m^3)
U_i^3 is the cube of the ith wind speed value (m/s)

37.4 Example: Initial Wind Farm Development in New Mexico

By 2001, the New Mexico Public Utility Commission (PUC) had, for some time, been looking at the idea of implementing a Renewable Portfolio Standard (RPS) requiring the Public Service Company of New Mexico (PNM), the state electrical utility, to obtain some of its electricity from renewable sources. The state legislature was also seriously considering enacting an RPS, but it had not yet done that in 2002. PNM had been investigating wind energy and monitoring the wind resource at some promising wind sites for several years, but had not yet been seriously considering adding wind to its generation mix. With the RPS under consideration in both the PUC and the legislature, PNM saw the handwriting on the wall and started seriously considering renewable energy, in general, and wind energy, in particular.

PNM decided to encourage the development of the first large wind farm in the state of New Mexico—they would purchase all of the electricity produced by the facility. New Mexico has fairly good wind resources to the east of the north/south central mountain range and in the eastern part of the state, so those are logical areas in which to start prospecting. The state has worked with NREL and True Solutions (now AWS Truepower) to develop a high-resolution wind map (available at www.emnrd.state.nm.us/ECMD/RenewableEnergy/images/WindMapForWeb.jpg). This map also shows the route of high-power transmission lines across the state. A map with transmission line details reveals a 345 kV PNM transmission line running from Clovis, in eastern New Mexico, to Albuquerque in the north-central part of the state. That line directly feeds the 1 GW Albuquerque load, and 2002, it had significant excess capacity.

This line passes through or near several areas with reasonable winds (annual average of 7 m/s or greater), mostly just east of Albuquerque. In the eastern part of the state, one of the areas near the line with reasonable winds appears to be near the intersection of Guadalupe, De Baca, and Quay counties. The statewide map in Figure 37.15 locates this area of interest, and the expanded scale map shows the estimated resource from the wind map. The wind map indicates average wind speeds of 7 m/s right in the area of the transmission line, with somewhat higher winds just a little to the north.

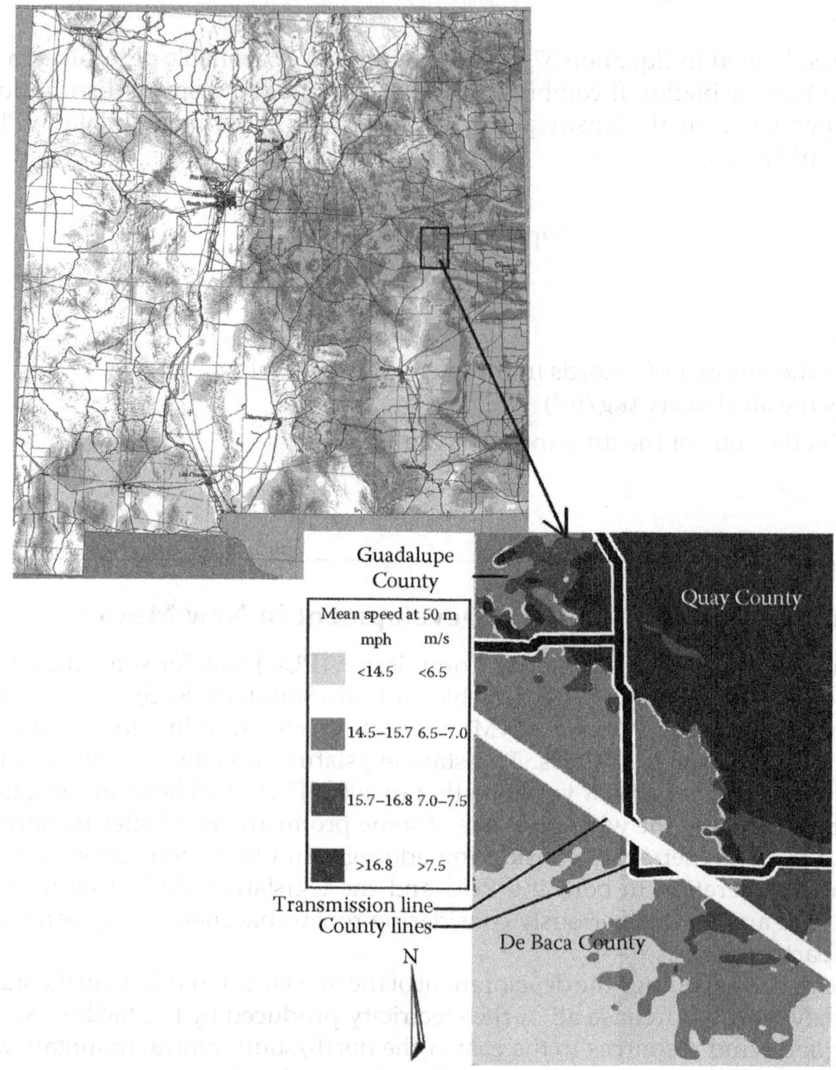

FIGURE 37.15
Wind resource map of the western edge of Tiaban Mesa in eastern New Mexico. (Reproduced by permission of AWS Truewind and the New Mexico Energy, Minerals, and Natural Resources Department. www.emnrd.state.nm.us/ECMD/RenewableEnergy/images/WindMapForWeb.jpg, accessed August 2005.)

Closer inspection of the topography in the area shows that the western edge of this high resource corresponds to the western edge of Tiaban Mesa. The mesa edge runs generally north/south in this area, and the top of the mesa is 60–90 m above the plains just to the west. This topography can certainly be expected to enhance the wind speed near the mesa edge. A site visit reveals heavy flagging of the vegetation from the southwest, indicative of a predominant wind direction. The land is privately owned, with no use restrictions, no large bird populations, and no environmental concerns. A measurement program quickly confirmed that the winds are predominantly southwest and that the annual average wind speeds along the mesa edge, from just south of the transmission line crossing to several miles north, are 7 m/s or better. Prospective developers thus had the

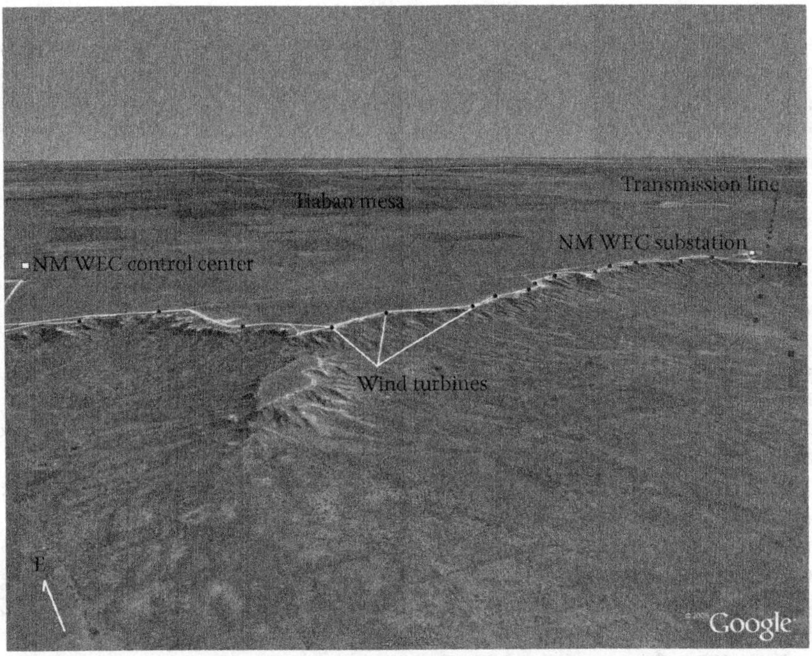

FIGURE 37.16
Perspective view of western edge of Tiaban Mesa. Every second wind turbine and transmission line tower indicated.

three essential components for a wind farm development—access to a transmission line with excess capacity, a customer for the power (PNM), and a good wind resource.

In August 2002, FPL Energy and PNM announced their agreement to develop the first large wind energy project in the state of New Mexico. FPL Energy built the 204 MW New Mexico Wind Energy Center on the western edge of Tiaban mesa in 2003. A perspective view of the mesa edge, with the turbine locations and transmission line indicated, is given in Figure 37.16. The PNM-built substation lies right under the existing transmission line and immediately adjacent to the line of turbines on the mesa edge; this ready access to the transmission line was a key driver in the decision to locate the wind farm in this spot. PNM purchases the entire output of the wind farm and sells it at a premium to residential and small business customers through their "Sky Blue" renewable energy program. Two years of operation have confirmed that the site resource is a good one: the yearly average wind speed at the 65 m turbine hub height has been 8 m/s. At the wind farm elevation of 1460 m, this wind speed yields a power density of 460 W/m^2, which corresponds to an NREL wind class 4 site.

References

1. AWS Scientific, Inc. 1997. *Wind Resource Assessment Handbook: Fundamentals for Conducting a Successful Monitoring Program*, SR-440-22223. National Renewable Energy Laboratory, Golden, CO.
2. Golding, E. 1977. *The Generation of Electricity by Wind Power.* E. & F.N. Spon Ltd., London, U.K.
3. Schwartz, M. and Elliott, D. 2006. Wind shear characteristics at central plains tall towers. *Windpower '06*. American Wind Energy Association, Washington, DC.

4. Elliott, D., Schwartz, M., and Scott, G. 2009. Wind shear and turbulence profiles at elevated heights: Great lakes and midwest sites. *Windpower 2009*. American Wind Energy Association, Washington, DC.
5. Elliott, D.L., Holladay, C.G., Barchet, W.R., Foote, H.P., and Sandusky, W.F. 1987. Wind energy resource atlas of the United States, DOE/CH10094-4. Solar Energy Research Institute, Golden, CO.
6. Troen, I. and Petersen, E.L. 1989. *European Wind Atlas*. Risø National Laboratory, Roskilde, Denmark.
7. Elliott, D., Schwartz, M., Haymes, S., Heimiller, D., Scott, G., Brower, M., Hale, E., and Phelps, B. 2011. New wind energy resource potential estimates for the United States. *Second Conference on Weather, Climate and the New Energy Economy*, Seattle, WA.
8. Archer, C.L. and Jacobson, M.Z. 2005. Evaluation of global wind power. *Journal of Geophysical Research*, 110, D12110.
9. European Wind Energy Association (EWEC)/Greenpeace. 2003. *Wind Force 12: A Blueprint to Achieve 12% of the World's Electricity from Wind Power by 2020*. EWEC and Greenpeace, Brussels, Belgium.
10. Musial, W. and Ram, B. 2010. *Large-Scale Offshore Wind Power in the United States: Assessment of Opportunities and Barriers*, NREL TP-500-40745. National Renewable Energy Laboratory, Golden, CO.
11. Sinclair, K.C. and Morrison, M.L. 1997. Overview of the U.S. Department of Energy/National Renewable Energy Laboratory avian research program. *Proceedings of Windpower '97*. American Wind Energy Association, Washington, DC, pp. 273–279.
12. Sinclair, K.C. 1999. Status of the U.S. Department of Energy/National Renewable Energy Laboratory Avian Research Program. *Proceedings of Windpower '99*. American Wind Energy Association, Washington, DC, pp. 273–279.
13. National Wind Coordinating Committee. 2010. Wind turbine interactions with birds, bats, and their habitats: A summary of research results and priority questions. http://www.nationalwind.org/publications/bbfactsheet.aspx (accessed November 2014).
14. Colson, E.W. 1995. Avian interactions with wind energy facilities: A summary. *Proceedings of Windpower '95*. American Wind Energy Association, Washington, DC, pp. 77–86.
15. Wolf, B. 1995. Mitigating avian impacts: Applying the wetlands experience to wind farms. *Proceedings of Windpower '95*. American Wind Energy Association, Washington, DC, pp. 109–116.
16. Erickson, W.P., Johnson, G.D., Strickland, M.D., Young, D.P., Sernka, K.J., and Good, R.E. 2001. Avian collisions with wind turbines: A summary of existing studies and comparisons to other sources of avian collision mortality in the United States. National Wind Coordinating Committee, Washington, DC. http://www.nationalwind.org/publications/wildlife/avian_collisions.pdf (accessed November 2014).
17. Kunz, T.H., Arnett, E.B., Erickson, W.P., Hoar, A.R., Johnson, G.D., Larkin, R.P., Strickland, M.D., Thresher, R.W., and Tuttle, M.D. 2007. Ecological impacts of wind energy development on bats: Questions, research needs, and hypotheses. *Frontiers in Ecology and the Environment*, 5(6), 315–324.
18. Baerwald, E.F., D'Amours, G.H., Klug, B.J., and Barclay, R.M.R. 2008. Barotrauma is a significant cause of bat fatalities at wind turbines. *Current Biology*, 18(16), R695–R696.
19. Baerwald, E.F., Edworthy, J., Holder, M., and Barclay, R.M.R. 2009. A large-scale mitigation experiment to reduce bat fatalities at wind energy facilities. *Journal of Wildlife Management*, 73(7), 1077–1081.
20. Arnett, E.B., Schirmacher, M., Huso, M.M.P., and Hayes, J.P. 2009. Effectiveness of changing wind turbine cut-in speed to reduce bat fatalities at wind facilities. An annual report submitted to the Bats and Wind Energy Cooperative. Bat Conservation International. Austin, TX.
21. Hewson, E.W., Wade, J.E., and Baker, R.W. 1978. Vegetation as an indicator of high wind velocity. RLO/2227-T24-78-2. Oregon State University, Corvallis, OR.

22. Hiester, T.R. and Pennell, W.T. 1981. The meteorological aspects of siting large wind turbines. PNL-2522. Pacific Northwest Laboratory, Richland, WA.
23. Wegley, H.L., Ramsdell, J.V., Orgill, M.M., and Drake, R. L. 1980. *A Siting Handbook for Small Wind Energy Conversion Systems*, PNL-2521. Pacific Northwest Laboratory, Richland, WA.
24. Rohatgi, J.S. and Nelson, V. 1994. *Wind Characteristics: An Analysis for the Generation of Wind Power*. Alternative Energy Institute, West Texas A&M University, Canyon, TX.
25. Frost, W. and Nowak, D.K. 1979. Summary and guidelines for siting wind turbine generators relative to small-scale two-dimensional terrain features. Report RLO-2443-77/1. Battelle Pacific Northwest Laboratory, Richland, WA.
26. International Energy Agency. 2013. *Ground-Based Vertically-Profiling Remote Sensing for Wind Resource Assessment*, 1st edn. IEA Wind RP 15.
27. Schwartz, M. and Elliott, D. 2005. Towards a wind energy climatology at advanced turbine hub heights, NREL/CP-500-38109. National Renewable Energy Laboratory, Golden, CO.
28. Aspliden, C.I., Elliot, D.L., and Wendell, L.L. 1986. Resource assessment methods, siting and performance evaluation, in *Physical Climatology for Solar and Wind Energy*, Guzzi, R. and Justus, C.G., eds., pp. 321–326. World Scientific, Hackensack, NJ.
29. Corotis, R.B. 1977. Stochastic modeling of site wind characteristics, RLO/2342-77/2. Northwestern University, Evanston, IL.
30. Gipe, P. 1993. *Wind Power for Home & Business—Renewable Energy for the 1990s and Beyond.* Chelsea Green Publishing Company, Post Mills, VT.
31. Aranda, F., ed. 2009. Remote wind speed sensing techniques using SODAR and LIDAR. *59th IEA Wind Topical Experts Meeting.* International Energy Agency, Boulder, CO.
32. Thor, S., ed. 2007. State of the art of remote wind speed sensing techniques using sodar, lidar and satellites. *51st IEA Wind Topical Expert Meeting.* International Energy Agency, Roskilde, Denmark.
33. Justus, C.G. 1978. *Winds and Wind System Performance.* Franklin Institute Press, Philadelphia, PA.
34. American Wind Energy Association (AWEA). 1986. Standard procedures for meteorological measurements at a potential wind turbine site, D-FC02-86CH10302. American Wind Energy Association, Washington, DC.
35. American Society of Mechanical Engineers (ASME). 1989. Performance test code for wind turbines, ASME/ANSI PTC 42-1988. American Society of Mechanical Engineers, New York.

Further Information

Wind Characteristics—An Analysis for the Generation of Wind Power, by J. S. Rohatgi and V. Nelson, Alternative Energy Institute, West Texas A&M University, is an excellent source for additional information on the wind resource. *Wind Resource Assessment Handbook: Fundamentals for Conducting a Successful Monitoring Program,* by AWS Scientific, Inc., report SR-440-22223, National Renewable Energy Laboratory, Golden, CO, 1997, gives step-by-step instructions for a successful land-based resource monitoring effort. *Ground-Based Vertically-Profiling Remote Sensing for Wind Resource Assessment,* by the International Energy Agency, is an excellent resource on the use of Lidar and Sodar for wind resource measurement. *Wind Turbine Technology, Fundamental Concepts of Wind Turbine Engineering,* D. Spera (editor), ASME Press, New York, 1994; *Wind Energy Handbook* (2nd edn.) by T. Burton, D. Sharpe, N. Jenkins, and E. Bossanyi, John Wiley & Sons, Chichester, U.K., 2011, and *Wind Energy Explained: Theory, Design and Application* (2nd edn.) by J. Manwell, J. McGowan, and A. Rogers, John

Wiley & Sons, Chichester, U.K., 2010, contain a wealth of fairly current information on wind energy resources, history, and technology, together with extensive reference lists. The reference site for the National Wind Coordinating Collaborative bird/bat fact sheet [13] at www.nationalwind.org/publications/bbfactsheet.aspx has links to a very extensive list of publications dealing with studies of the interactions of birds and bats with wind turbines. Extensive information on wind energy resources and technology may also be found on the world wide web. Excellent sites to start with include those of the U.S. National Renewable Energy Laboratory Wind Energy Technology Center at www.nrel.gov/nwtc/, the Danish Technical University Department of Wind Energy at www.vindenergi.dtu.dk, the American Wind Energy Association at www.awea.org, the European Wind Energy Association at www.ewea.org, and the Danish Wind Energy Association at www.windpower.org/en.

38

Municipal Solid Waste

Shelly H. Schneider

CONTENTS

This chapter has been extracted from the most recent in a series of reports released by the U.S. Environmental Protection Agency (EPA) to characterize municipal solid waste (MSW) in the United States. This report characterizes the national municipal waste stream based on data through 2011. As characterized in the EPA report, MSW includes wastes from residences and commercial establishments. No construction and demolition debris or industrial wastes are included. (Some wastes from industrial establishments, such as packaging and office wastes, are, however, included.)

Identifying the components of the MSW stream is an important step toward addressing the issues associated with using it for energy generation. MSW characterizations, which analyze the quantity and composition of the waste stream, involve estimating how much MSW is generated, recycled, combusted, and disposed of in landfills. This chapter characterizes the MSW stream *of the nation as a whole.* Local and regional variations are not addressed.

The methodology used for the characterization of MSW for this chapter estimates the waste stream on a nationwide basis by a "material flows methodology." EPA's Office of Solid Waste and its predecessors in the Public Health Service sponsored work in the 1960s and early 1970s to develop this methodology, which is based on production data (by weight) for the materials and products in the waste stream, with adjustments for imports, exports, and product lifetimes.

38.1 Materials and Products in MSW

In 2011, generation of MSW totaled 250.4 million tons. A breakdown by percentage of the materials generated in MSW in 2011 is shown in Figure 38.1. Paper and paperboard products are the largest component of MSW by weight (28% of generation), and food waste and yard trimmings are the second and third largest components (14.5% and 13.5% of generation). Plastics come next, at 23.7% of MSW generation. Inorganic portions of the waste stream—metals and glass—total 13.4% of generation. (The "other" category also

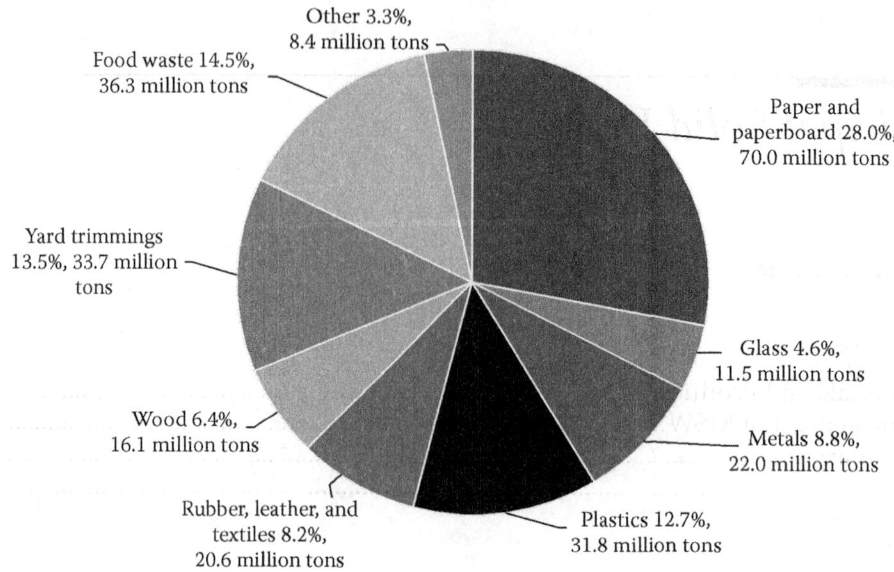

FIGURE 38.1
Materials generated in MSW by weight and percentage, 2011.

contains some inorganic materials such as broken pottery and kitty litter.) Rubber, leather, textiles, and wood make up about 14.6% of MSW generation.

Most of the materials in MSW have some level of recovery for recycling or composting. This is illustrated in Table 38.1. Since each materials category (except for yard trimmings and food waste) is made up of many different products, some of which may not be recovered at all, the overall recovery rate for any particular material will be lower than recovery rates for some products within the materials category.

The highest recovery rate shown in Table 38.1 is that for nonferrous metals other than aluminum (68.4% of generation). This is because the lead in lead-acid batteries is recovered at very high rates. Paper and paperboard were recovered at 65.6% of generation in 2011, and they had by far the highest recovered tonnage. Within that category, newspapers were recovered at 72.5% and corrugated boxes at 91% of generation. Yard trimmings were recovered for composting at a rate of 57.3% in 2011. Recovery rates for other materials are shown in the table.

The many products in MSW are grouped into three main categories: durable goods (e.g., appliances), nondurable goods (e.g., newspapers), and containers and packaging (e.g., beverage cans and corrugated boxes) (see Figure 38.2). The materials in MSW are generally made up of products from each category. There are, however, exceptions. The durable goods category contains no paper and paperboard. The nondurable goods category includes only small amounts of metals and essentially no glass or wood. The containers and packaging category includes only very small amounts of rubber, leather, and textiles.

Generation and recovery of MSW by product category are shown in Table 38.2. Overall, the materials in durable goods were recovered at a rate of 18.4% in 2011. Recovery of materials (lead and plastic) from lead-acid batteries was at 96.2% in 2011. Major appliances were recovered at an overall rate of 64.2% because of the high rate of recovery of steel in appliances. Recovery of tires at 44.6% is due to recovery of rubber and some steel.

The overall recovery rate for nondurable goods was estimated at 36.5% in 2011. Recovery of paper products such as newspapers and other paper products accounts for most of this. Recovery of containers and packaging is at the highest rate—50.7% in 2011. Large tonnages

TABLE 38.1

Generation and Recovery of Materials in MSW, 2011

Material	Weight Generated	Weight Recovered	Percentage of Generation (%)
Paper and paperboard	70.02	45.90	65.6
Glass	11.47	3.17	27.6
Metals			
Steel	16.52	5.45	33.0
Aluminum	3.47	0.72	20.7
Other nonferrous metals	1.96	1.34	68.4
Total metals	21.95	7.51	34.2
Plastics	31.84	2.65	8.3
Rubber and leather	7.49	1.31	17.5
Textiles	13.09	2.00	15.3
Wood	16.08	2.38	14.8
Other materials	4.59	1.28	27.9
Total materials in products	176.53	66.20	37.5
Other wastes			
Food, other	36.31	1.40	3.9
Yard trimmings	33.71	19.30	57.3
Miscellaneous inorganic wastes	3.87	Negligible	Negligible
Total other wastes	73.89	20.70	28.0
Total municipal solid waste	250.42	86.90	34.7

Source: U.S. Environmental Protection Agency, *Municipal Solid Waste in the United States: 2011 Facts and Figures*, U.S. Environmental Protection Agency.

Note: In millions of tons and percentage of generation of each material. Negligible = less than 5000 tons or 0.05%.

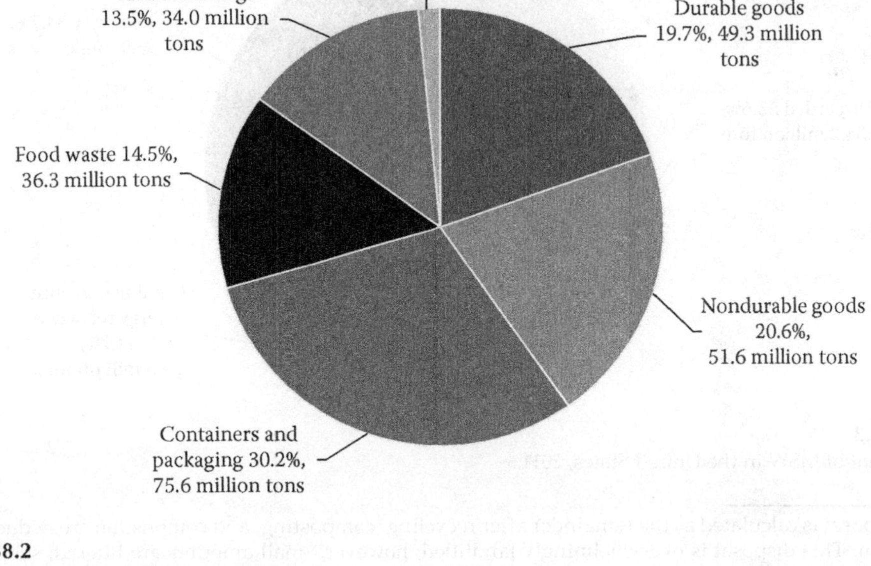

FIGURE 38.2
Products generated in MSW by weight and percentage, 2011.

TABLE 38.2
Generation and Recovery of Products in MSW, 2011

	Weight Generated	Weight Recovered	Percentage of Generation (%)
Durable goods	49.3	9.07	18.4
Nondurable goods	51.61	18.83	36.5
Containers and packaging	75.58	38.3	50.7
Other wastes			
Food, other	36.31	1.40	3.9
Yard trimmings	33.71	19.30	57.3
Miscellaneous inorganic wastes	3.87	Negligible	Negligible
Total other wastes	73.89	20.70	28.0
Total municipal solid waste	250.42	86.90	34.7

Source: U.S. Environmental Protection Agency, *Municipal Solid Waste in the United States: 2011 Facts and Figures*, U.S. Environmental Protection Agency.
Note: In millions of tons and percentage of generation of each product category. Negligible = less than 5000 tons or 0.05%.

of corrugated boxes were recovered, at a rate of 91.0%. Steel cans were recovered at a rate of 70.6%, and aluminum cans at 54.5%. Other packaging made of glass, plastics, and wood was also recovered.

38.2 Management of MSW

The breakdown of how much MSW went to recycling and composting, combustion, and land disposal* in 2011 is shown in Figure 38.3. Recovery of materials for recycling and composting was estimated to have been 86.9 million tons, or 34.7% of generation, in 2011.

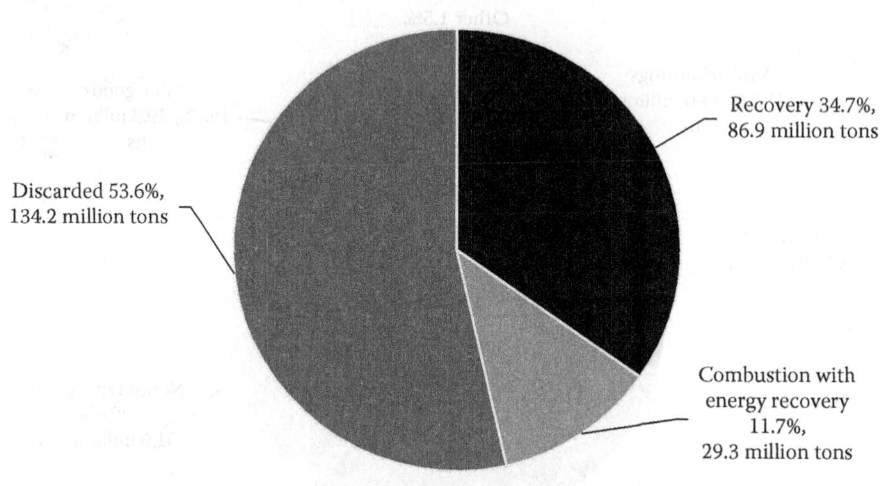

FIGURE 38.3
Management of MSW in the United States, 2011.

* Land disposal is calculated as the remainder after recycling, composting, and combustion are deducted from generation. This disposal is overwhelmingly landfilled; however, small amounts are littered, self-disposed (e.g., by on-site burning), or otherwise not taken to landfills.

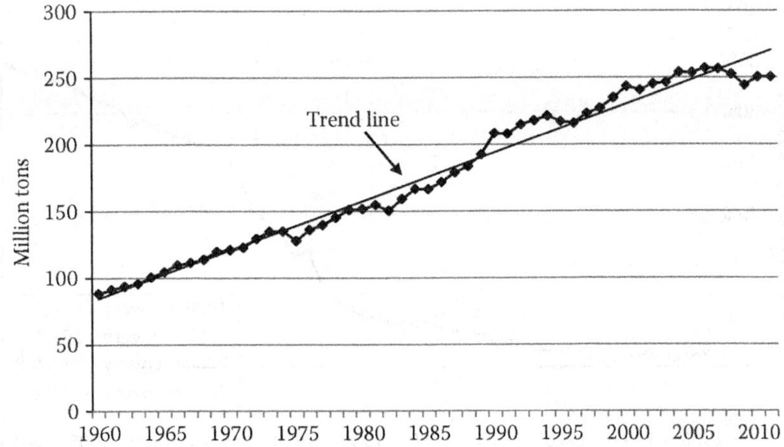

FIGURE 38.4
The trend of municipal solid waste generation, 1960–2011.

Combustion of MSW (with energy recovery) was estimated to have been 29.3 million tons, or 11.7% of generation. The remainder—134.2 million tons, or 53.6% of generation—was land disposed.

Generation of MSW grew steadily from 1960 to 2011, from 88.1 million tons to 250.4 million tons per year (Figure 38.4). As illustrated by the graph, the growth of generation is not continuous, but fluctuates with the economy and, of course, with population growth. It has been demonstrated that there is a high degree of correlation between MSW generation and gross domestic product, and recession years can be identified on the graph. Another way to look at generation is in pounds per capita (pcd) (Figure 38.5). After years of steady growth, pcd peaked at 4.74 pounds per capita per day in 2000, decreasing slightly between 2001 and 2003, rising again to 4.74 pcd in 2004 followed by a steady decline to 4.36 pcd in 2009. The per capita rate rose to 4.44 in 2010 followed by a decline to 4.40 in 2011. Decreasing per capita rates follow periods of economic turndowns.

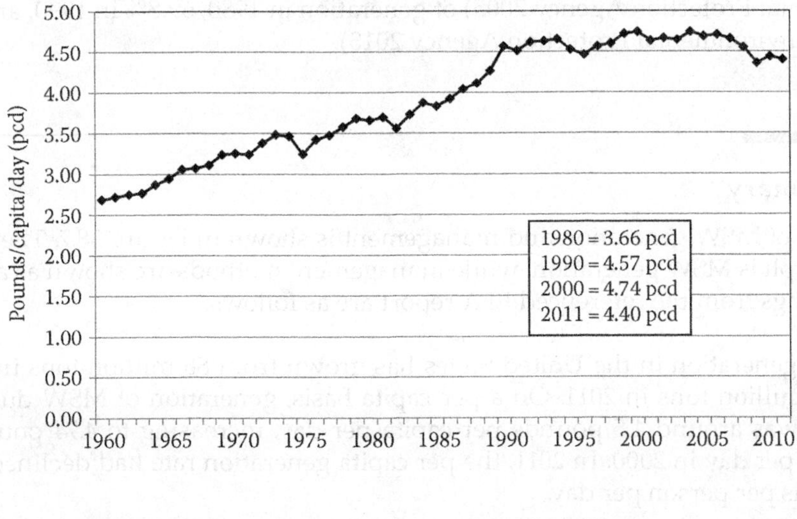

FIGURE 38.5
Generation of municipal solid waste in pounds per capita per day, 1960–2011.

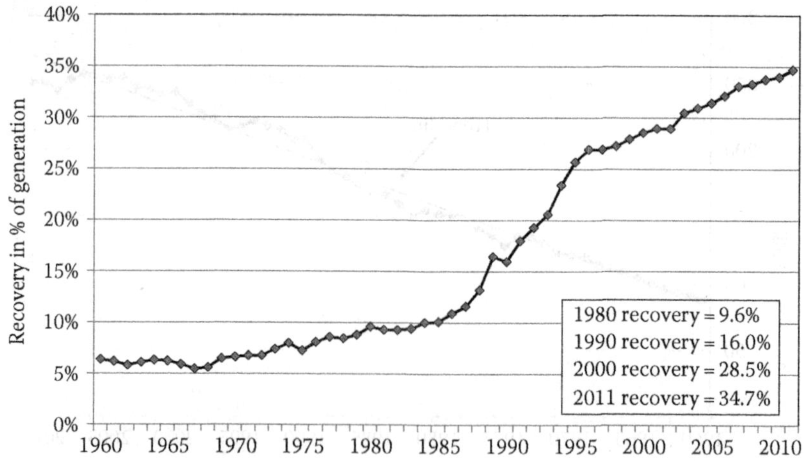

FIGURE 38.6
Recovery of MSW for recycling and composting, in percentage of generation, 1960–2011.

Recovery for recycling and composting has increased dramatically since the late 1980s (Figure 38.6). Recovery of MSW was minimal in the 1960s and early 1970s. The percentage recovered crept up to 9.6% by 1980. Interest in recovery grew rapidly in the late 1980s as concerns were raised about diminishing landfill space in parts of the United States, especially the northeast. Recovery reached 16% in 1990, 28.5% in 2000, and 34.7% in 2011. While most recovered material is made up of products in MSW, there also has been an increase in composting of yard trimmings and, to a much lesser extent, food waste.

Combustion handled an estimated 30% of MSW generated in 1960, mostly through incinerators with no energy recovery and no air pollution controls. In the 1960s and 1970s, combustion dropped steadily as the old incinerators were closed, reaching a low of less than 10% of MSW generated by 1980. The percentage of MSW managed by combustion reached about 14.2% in 1990 and 13.9% in 2000; it has been declining slowly since 2000 to 11.7% in 2011.

Land disposal has been declining since the 1980s. Land disposal was 81.4% (U.S. Environmental Protection Agency 2005) of generation in 1980, 69.8% in 1990, and 53.6% in 2011 (U.S. Environmental Protection Agency 2013).

38.3 Summary

The history of MSW generation and management is shown in Figure 38.7. The top line of the area graph is MSW generation, while management methods are shown as area plots. Major findings from the referenced EPA report are as follows:

- MSW generation in the United States has grown from 88 million tons in 1960 to 250.4 million tons in 2011. On a per capita basis, generation of MSW during the 1990s was around 4.5 pounds per capita per day, increasing to 4.74 pounds per capita per day in 2000. In 2011, the per capita generation rate had declined to 4.40 pounds per person per day.

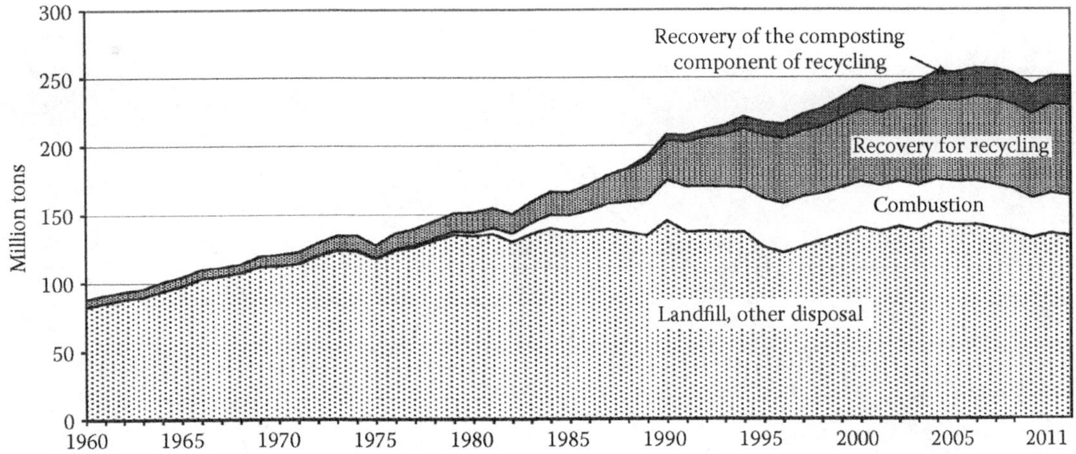

FIGURE 38.7
Generation and management of municipal solid waste, 1960–2011.

- In the 1960s and early 1970s, a large percentage of MSW was burned, with little recovery for recycling. Landfill disposal typically consisted of open dumping, often accompanied with open burning of the waste for volume reduction.
- Through the mid-1980s, incineration declined considerably, and landfills became difficult to site, and waste generation continued to increase. Materials recovery rates increased very slowly in this time period, and the burden on the nation's landfills grew dramatically. As Figure 38.7 shows, discards of MSW to landfill or other disposal apparently peaked in 1990 and then began to decline as materials recovery and combustion with energy recovery increased.
- Over time, recycling rates have increased from just over 6% of MSW generated in 1960 to about 10% in 1980, to 16% in 1990, to about 29% in 2000, and to over 34% in 2011.
- Combustion of MSW reached a low point of about 10% of generation around 1980. Since 2004, combustion with energy recovery has held steady at about 12% of generation.
- MSW discards to landfills rose to about 142.3 million tons in 2005 and then declined to 134.3 million tons in 2011. As a percentage of total MSW generation, discards to landfills has consistently decreased—from 69.8% of generation in 1990 to 53.6% in 2011.

References

U.S. Environmental Protection Agency. *Municipal Solid Waste in the United States: 2011 Facts and Figures,* May 2013. www.epa.gov/epaoswer/non-hw/muncpl/msw99.htm (accessed July 2013).

U.S. Environmental Protection Agency. *Municipal Solid Waste in the United States: 2003 Data Tables,* 2005. www.epa.gov/epaoswer/non-hw/muncpl/msw99.htm (accessed July 2013).

39

Biomass Properties and Resources

Mark M. Wright and Robert C. Brown

CONTENTS

39.1 Introduction

The term "biomass" encompasses a wide range of materials of recent origin classified as either waste or dedicated energy crops. Waste biomass includes any organic material that has negligible apparent value, represents a nuisance, or is a pollutant to the local environment. Dedicated energy crops are biomass grown specifically for the production of biobased products and fuels. This term excludes crops grown for food or feed even though they can also be used to produce energy. It also includes organic material with maturation times of hundreds to millions of years such as fossil fuels and some forest trees with long maturity terms. Biomass is primarily a form of solar energy stored as chemical energy within organic compounds. The solar-to-biomass conversion process involves interactions among numerous factors, leading to different types of biomass.

In the following sections, we describe the principles of solar energy conversion, biomass types and their properties, and the role of land use for crop production. These concepts help to understand the quantity and quality of global biomass resources.

39.2 Solar Energy Conversion to Biomass

Solar energy is the most abundant source of renewable energy on planet earth. Every year, the planet receives 5.6 million exajoules (EJ—10^{18} J) upon its atmosphere. Given that the world consumes about 570 EJ per year, this is enough energy to supply for several thousand

TABLE 39.1

Photosynthesis Steps and Efficiencies

Photosynthesis Step	Total Energy (%)	
Incident solar energy (on leaf surface)	100	
Energy in photosynthetically active spectrum	48.7	
Absorbed energy	43.8	
Photochemically converted energy	37.2	
Carbon fixation pathway	C3	C4
Energy in synthesized carbohydrates	12.6	8.5
Energy available after photorespiration	6.5	8.5
Energy available after respiration	4.6	6.0

Source: Zhu, X.G. et al., *Curr. Opin. Biotechnol.*, 19(2), 153, 2008.

years. Unfortunately, solar energy is very diffuse and difficult to convert efficiently. Most of the atmosphere's solar energy never reaches land surfaces, and only a minuscule amount is converted to biomass.

The planet's atmosphere absorbs, reradiates, and reflects 30% of the incident solar radiation and allows 70% to reach the planet's surface. Earth's surface area consists of 29.2% land, of which about 21% is covered by biomass. Overall, only 6.1% of the atmosphere's solar radiation remains available for biomass production. Plants have developed photosynthetic means of storing solar energy that are suitable for their needs but inefficient in their ability to convert solar to chemical energy.

Table 39.1 compares the percent of total energy captured by C3 and C4 plants after several photosynthesis steps. Starting from 100% of the solar energy available at the plant's surface, only 48.7% is in the photosynthetically active spectrum. The absorbed energy represents 43.8%, and 37.2% of the incident solar energy on the leaf's surface is photochemically converted to biomass energy via carbon fixation. C3 and C4 are carbon fixation pathways labeled after the carbon chain length of the first carbohydrate formed during photosynthesis. The vast majority of plants employ three carbon-chain length molecules to fixate carbon, whereas about 3% of known species employ four carbon-chain carbohydrates. Corn, sugarcane, and sorghum are common C4 plants. The process of photosynthetically converting solar energy into chemical compounds can be generalized by the formula 39.1. This formula describes the conversion of CO_2, H_2O, and sunlight into sugar (glucose) and oxygen.

$$6CO_2 + 6H_2O + (\text{sunlight}) \rightarrow C_6H_{12}O_6 + 6O_2 \tag{39.1}$$

C3 and C4 plants can theoretically store up to 4.6% and 6.0% of the solar energy on their leaf surfaces into biomass. This is the amount leftover after the energy spent during carbohydrate synthesis, photorespiration, and respiration. C3 plants are more efficient at carbohydrate synthesis and respiration, but C4 plants have an overall higher efficiency, because they avoid photorespiration penalties. In practice, the most efficient conversion measured in C3 and C4 plants is 2.4% and 3.7%, respectively. There are many environmental factors that lower the efficiency of crops in the field including nutrient availability, weather patterns, and pest activity.

The efficiency of solar energy conversion to biomass can be estimated based on the incident energy and biomass available in a given area. Data for solar incident energy measurements are publicly available from sources such as the National Renewable Energy Laboratory [9]. The U.S. Department of Agriculture (USDA) publishes biomass productivity data (http://quickstats.nass.usda.gov/). USDA data are based on aboveground biomass. However, the biomass efficiency calculations should include the amount of belowground biomass that can exceed a quarter of the total biomass material. The ratio of below- to aboveground biomass is commonly known as the root-to-shoot ratio, and values for different crops are available in the literature [3].

39.3 Biomass Properties

39.3.1 Plant Composition

Plant composition and physical properties have a significant impact on biomass energy content. Biomass is commonly characterized by its organic composition, elemental analysis, proximate analysis, and bulk properties such as heating value and bulk density. Table 39.2 shows physical and thermochemical data of representative grain, herbaceous, and woody biomass. Organic composition includes cellulose, hemicellulose, and lignin mass content. Elemental analysis typically reports carbon, hydrogen, oxygen, nitrogen, and ash. Other elements are often reported if they are found in high quantities or important for the target application (such as sulfur for combustion). Proximate analysis is a measure of the moisture content, volatile matter, fixed carbon, and ash content. Heating value is the amount of energy released during complete biomass combustion.

Biomass is mostly composed of lignocellulosic material. Lignocellulose is a term that describes the three-dimensional polymeric composites formed by plants as structural material [1]. Plants contain varying quantities of lignin, cellulose, and hemicellulose.

Lignin is a polymer whose primary function is to provide structural support and protect the plant from microbial activity. Therefore, lignin is a common by-product of biochemical processes, since microbes are unable to easily utilize it as a substrate. On the other hand,

TABLE 39.2
Physical and Thermochemical Properties of Selected Biomass

	Feedstock	Corn Stover	Herbaceous Crop	Woody Crop
Organic composition (wt.%)	Cellulose	53	45	50
	Hemicellulose	15	30	23
	Lignin	16	15	22
	Others	16	10	5
Elemental analysis (dry wt.%)	C	44	47	48
	H	5.6	5.8	5.9
	O	43	42	44
	N	0.6	0.7	0.5
	Ash	6.8	4.5	1.6
Proximate analysis (dry wt.%)	Volatile matter	75	81	82
	Fixed C	19	15	16
	Ash	6	4	1.3
HHV (MJ/kg)		17.7	18.7	19.4
Bulk density (kg/m^3)		160–300	160–300	280–480
Yield (Mg/ha)		8400	14,000	14,000

Source: Brown, R.C., *Biorenewable Resources: Engineering New Products from Agriculture*, Iowa State Press, A Blackwell Publishing Company, Ames, IA, 2003, pp. 59–75.

thermochemical processes can decompose lignin although the products are still hard to predict. Instead of breaking apart into its monomers, lignin decomposition tends to form oligomers from the repolymerization of smaller hydrocarbons. These oligomers can be gasified or catalytically upgraded to desired fuels and chemicals.

Cellulose is a polysaccharide made of glucose chains. Its basic building block is the cellobiose, which consists of two linked glucose units. The typical cellulose chain has a degree of polymerization of 10,000 units. Cellulose tends to agglomerate and with high packing densities can form crystalline cellulose. Crystalline cellulose is inert to chemical treatment and insoluble in most solvents. Cellulose with low packing densities is known as amorphous cellulose. Microbes consume cellulose efficiently, and they can convert it into a variety of chemicals most notably ethanol.

Hemicellulose consists of a large number of heteropolysaccharides built from hexoses, pentoses, and deoxyhexoses. Its degree of polymerization is much lower than cellulose and in the order of 100–200. Hemicellulose requires acid or enzymatic treatment before its sugars become available to microbial activity.

The organic composition of biomass feedstock has a significant impact on the types of processes that can convert it to fuels and chemicals. The proportions of all three organic compounds impact the types and quantities of degradation compounds formed during thermochemical biomass conversion. Furthermore, there are interaction effects among these compounds that are not well understood. Therefore, increasingly powerful analysis techniques are under development to measure not just the quantity but also the physical properties of organic compounds.

39.3.2 Biomass Analysis

Proximate analysis is primarily important in thermochemical applications, because it describes the general evolution of biomass combustion products. Proximate analysis is measured by heating biomass under controlled temperature and heating rate conditions. The total weight loss from holding the biomass temperature at 100°C represents its moisture content. Volatile matter is the fraction of biomass that decomposes into gases at moderate temperatures of about 400°C in an inert environment. The remaining fraction is a mixture of solid carbon (fixed carbon) and mineral matter (ash). The ash content can be determined by introducing oxygen and burning the remaining carbon material.

Ultimate analysis is often reported on a dry, ash-free (daf) basis and often used to estimate thermal biomass properties. One correlation of the higher heating value (HHV) of biomass is the formula 39.2 that is based solely on the feedstock carbon content. Although carbon is the primary factor in determining heating value, oxygen is important because of its detrimental effect to heating value and recalcitrance to removal. Researchers have published alternative HHV correlations that incorporate a greater number of factors.

$$\mathrm{HHV(dry)}\left[\frac{\mathrm{MJ}}{\mathrm{kg}}\right] = 0.4571(\%\ \mathrm{C\ on\ dry\ basis}) - 2.70 \tag{39.2}$$

The thermal performance of biomass fuel depends heavily on its heating value. Heating value is the net enthalpy released upon reacting fuel with oxygen at stoichiometric conditions. It is reported on either a lower heating value (LHV) or higher heating value (HHV) basis. The difference between LHV and HHV depends on whether the combustion gases are released above or below the water condensation temperature. Below the

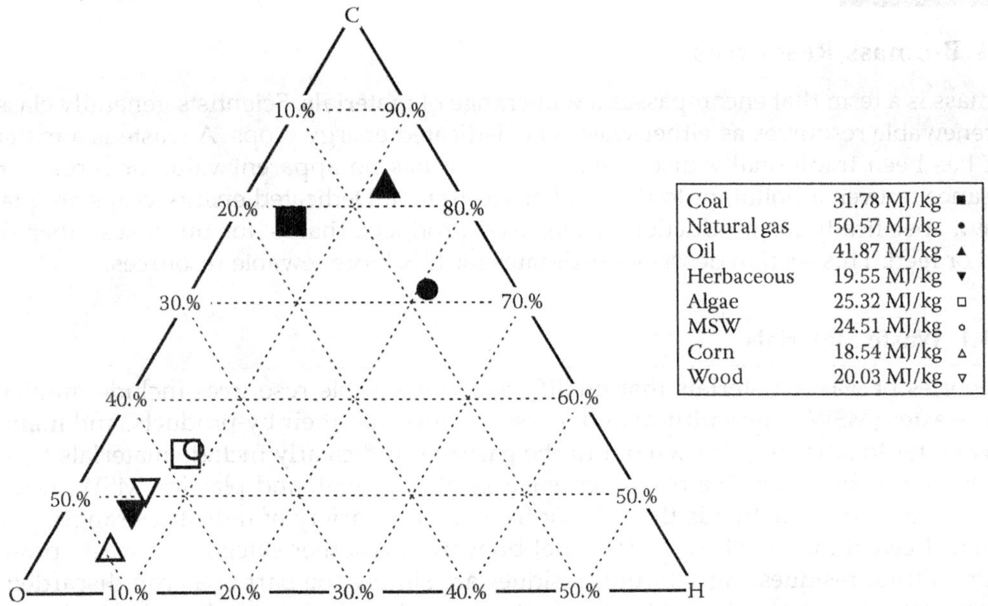

FIGURE 39.1
Ternary plot of average elemental carbon, hydrogen, oxygen in fossil and biomass materials and their higher heating values. (Adapted from Anon. Phyllis2 database by ECN (July 2013) available at http://www.ecn.nl/phyllis2/Browse/Standard/ECN-Phyllis, accessed March 2015.)

water condensation temperature, the moisture in the fuel contributes a latent enthalpy heat release, resulting in a higher heat output. Biomass heating value is typically about 18 MJ/kg, but it varies by biomass type as shown in Table 39.2.

There are several comprehensive sources of biomass composition available in textbooks and online databases. The U.S. Department of Energy (DOE) *Biomass Feedstock Composition and Property Database,* and the ECN Phyllis database for example are freely available online. These databases contain composition and material property data for a wide range of organic and nonorganic materials. Figure 39.1 compares the HHV of fossil and biomass materials based on their carbon, hydrogen, and oxygen content [10]. As shown, higher carbon content and lower oxygen generally correspond with increasing HHV. However, the greater hydrogen content in natural gas compensates for a slightly higher oxygen content than crude oil.

Combustion calculations require knowledge of the fuel's enthalpy (heat) of formation. This information is difficult to determine from biomass composition information. It can however be estimated after measuring the combustion reaction heating value. Consider the reaction in the following equation:

$$CH_aO_b + CO_2 \rightarrow CO_2 + dH_2O(\text{liquid}). \quad (39.3)$$

The heating value (ΔH_R) can be calculated using Equation 39.4, thermodynamic property data, and feedstock composition (such as Table 39.2):

$$\Delta H_R = h^{\circ}_{f,CO_2} + dh^{\circ}_{f,H_2O} - (h^{\circ}_{f,CH_aO_b} + ch^{\circ}_{f,O_2}). \quad (39.4)$$

39.4 Biomass Resources

Biomass is a term that encompasses a wide range of materials. Scientists generally classify biorenewable resources as either wastes or dedicated energy crops. A waste is a material that has been traditionally discarded, because it has no apparent value or represents a nuisance or even a pollutant to the local environment. Dedicated energy crops are plants grown specifically for production of biobased products, that is, for purposes other than food or feed. This section describes estimates for U.S. biorenewable resources.

39.4.1 Waste Materials

Categories of waste materials that qualify as biorenewable resources include municipal solid wastes (MSWs), agricultural and forest residues and their by-products, and manure. MSWs refer to anything thrown out in the garbage, and clearly include materials that do not qualify as biorenewable resources, such as glass, metal, and plastics. MSW includes food processing waste that is the effluent from a wide variety of industries ranging from breakfast cereal manufacturers to alcohol breweries. Another category of waste product is agricultural residues. Agricultural residues are simply the part of a crop discarded by farmers after harvest such as corn stover (husks and stalks), rice hulls, wheat straw, and bagasse (fibrous material remaining after the milling of sugarcane). Modern agriculture continues to heavily employ animals. The recent concentration of animals into giant livestock facilities has led to calls to treat animal wastes in a manner similar to that for human wastes. Table 39.3 shows the potential quantities of agricultural and forest residue available in the United States.

Waste materials share few common traits other than the difficulty of characterizing them because of their variable and complex composition. Thus, waste biomass presents special problems to engineers who are tasked with converting this sometimes unpredictable feedstock into reliable power or high-quality fuels and chemicals. The major virtue of waste materials is their low cost. By definition, waste materials have little apparent economic value and can often be acquired for little more than the cost of transporting the material from its point of origin to a processing plant. In fact, it is possible to acquire wastes at a negative cost because of the rising costs for solid waste disposal and sewer

TABLE 39.3

Potential Agricultural, Forest, and Process Waste Supply in the United States

	Annual Biomass Supply (Million Dry Mg/Year)
Logging and other residue	58
Fuel treatments	54
Urban wood residues	43
Wood processing residues	64
Pulping liquor	67
Fuelwood	47
Crop residues	405
Process residues	79

Source: Perlack, R. et al., Biomass as feedstock for a bioenergy and bioproducts industry: The technical feasibility of a billion-ton annual supply, Technical Report A357634, Oak Ridge National Laboratory, Oak Ridge, TN, 2005.

discharges and restrictions on landfilling certain kinds of wastes; that is, a biorenewable resource processing plant is paid by a company seeking to dispose of a waste stream. For this reason, many of the most economically attractive opportunities in biorenewable resources involve waste feedstocks.

Clearly, a waste material that can be used as feedstock for an energy conversion process is no longer a waste material. As demand for these new-found feedstocks increases, those that generate it come to view themselves as suppliers and may demand payment for the one-time waste: a negative feedstock cost becomes a positive cost. Such a situation developed in the California biomass power industry during the 1980s [4]. Concerns about air pollution in California led to restrictions on open-field burning of agricultural residues, a practice designed to control infestations of pests. With no means for getting rid of these residues, an enormous reserve of biomass feedstocks materialized. These feedstocks were so inexpensive that independent power producers recognized that even small, inefficient power plants using these materials as fuel would be profitable. A number of plants were constructed and operated on agricultural residues. Eventually, the feedstock producers had plant operators bidding up the cost of their once nuisance waste material. In the end, many of these plants were closed because of the escalating cost of fuel.

39.4.2 Energy Crops

Energy crops are defined as plants grown specifically as an energy resource. We should note that firewood obtained from cutting down an old-growth forest does not constitute an energy crop. An energy crop is planted and harvested periodically. Harvesting may occur on an annual basis, as with sugar beets or switchgrass, or on a 5–7 year cycle, as with certain strains of fast-growing trees such as hybrid poplar or willow. The cycle of planting and harvesting over a relatively short time period assures that the resource is used in a sustainable fashion; that is, the resource will be available for future generations.

Energy crops contain significant quantities of one or more of four important energy-rich components: oils, sugars, starches, and lignocellulose (fiber). Farmers historically cultivated crops rich in the first three components for food and feed: oils from soybeans and nuts; sugars from sugar beets, sorghum, and sugarcane; and starches from corn and cereal crops. Oil, sugars, and starches are easily metabolized. On the other hand, humans find it hard to digest lignocellulose. Certain domesticated animals with specialized digestive tracts are able to break down the polymeric structure of lignocellulose, and use it as an energy source. From this discussion, it might appear that the best strategy for developing biomass resources is to grow crops rich in oils, sugars, and starches. However, even for *oil crops* or *starch crops*, the largest single constituent is invariably lignocellulose (Table 39.4), which is the structural (fibrous) material of the plant: stems, leaves, and roots. If we harvest oils, sugars, and starches and leave the lignocellulose behind as an agricultural residue rather than use as fuel, we will waste the greatest portion of the biomass crop.

TABLE 39.4

Typical Woody Biomass Compositions

Component	Weight (%)
Cellulose	44 ± 6
Hemicellulose	28 ± 4
Lignin	20 ± 5

Research has shown that energy yields (Joules per km^2 per year) are usually greatest for plants that are mostly *roots and stems*; in other words, plant resources are directed toward the manufacture of lignocellulose rather than oils, sugars, and starches. As a result, there has been a bias toward development of energy crops that focus on lignocellulosic biomass, which is reflected in the discussion that follows.

Dedicated energy crops are typically high fiber crops grown specifically for their high productivity of holocellulose (cellulose and hemicellulose). Harvesting may occur on an annual basis, as with switchgrass, or on a 5–7 year cycle, as with certain strains of fast-growing trees such as hybrid poplar. Lignocellulosic crops are conveniently divided into herbaceous energy crops (HECs) and short rotation woody crops (SRWCs) [7].

Herbaceous crops are plants that have little or no woody tissue. The aboveground growth of these plants usually lives for only a single growing season. However, herbaceous crops include both annuals and perennials. Annuals die at the end of a growing season and must be replanted in the spring. Perennials die back each year in temperate climates but reestablish themselves each spring from rootstock. Both annual and perennial HECs are harvested on at least an annual basis, if not more frequently, with yields averaging 550–1100 Mg/km^2/year, with maximum yields between 2000 and 2500 Mg/km^2/year in temperate regions [7]. As with trees, yields can be much higher in tropical and subtropical regions.

Herbaceous crops more closely resemble hardwoods in their chemical properties than they do softwoods. Their low lignin content makes them relatively easy to delignify, which improves accessibility of the carbohydrate in the lignocellulose. The hemicellulose contains mostly xylan, which is highly susceptible to acid hydrolysis, compared to the cellulose. As a result, microbes can easily degrade agricultural residues, destroying their processing potential in a matter of days if exposed to the elements. Herbaceous crops have relatively high silica content compared to woody crops, which can present problems during processing.

SRWC is used to describe woody biomass that is fast growing and suitable for use in dedicated feedstock supply systems. Desirable SRWC candidates display rapid juvenile growth, wide site adaptability, and pest and disease resistance. Woody crops grown on a sustainable basis are harvested on a rotation of 3–10 years. Annual SRWC yields range between 500 and 2400 Mg/km^2/year.

Woody crops include hardwoods and softwoods. Hardwoods are trees classified as angiosperms, which are also known as flowering plants. Examples include willow, oak, and poplar. Hardwoods can be regrown from stumps, a process known as coppicing, which reduces their production costs compared to softwoods. Advantages of hardwoods in processing include: high density for many species; relative ease of delignification and accessibility of wood carbohydrates; the presence of hemicellulose high in xylan, which can be removed relatively easily; low content of ash, particularly silica, compared to softwoods and herbaceous crops; and high acetyl content compared to most softwoods and herbaceous crops, which is an advantage in the recovery of acetic acid.

Softwoods are trees classified as gymnosperms, which encompass most trees known as evergreens. Examples include pine, spruce, and cedar. Softwoods are generally fast growing, but their carbohydrate is not as accessible for chemical processing as the carbohydrates in hardwood. Since softwoods have considerable value as construction lumber and pulpwood, they are more readily available as waste material in the form of logging and manufacturing residues compared to hardwoods. Logging residues, consisting of a high proportion of branches and tops, contain considerable high-density compression wood, which is not easily delignified. Therefore, logging residues are more suitable as boiler fuel or other thermochemical treatments than as feedstock for chemical or enzymatic processing.

39.4.3 Algae

Algae is a broad term that encompasses several eukaryotic organisms. Eukaryotic organisms are characterized by complex structures enclosed within their cell membranes. Although algae do not share many of the structures that define terrestrial biomass, they are capable of photosynthesis and capturing carbon. Algae's affinity to convert CO_2 into lipids has drawn academic and industrial attention as a means to simultaneously lower carbon emissions and produce biofuels.

Algal biomass uses CO_2 as its carbon source and sunlight as its energy source. About 1.8 kg of CO_2 is fixed for every kg of algal biomass, which contains up to 50% carbon by dry weight. Controlled production of renewable fuels from algae has been proposed in either raceway ponds or photobioreactors. Raceway ponds consist of open, shallow recirculation channels with mechanical flow control and surfaces that enhance light retention. Raceway ponds are inexpensive, but relatively inefficient when compared to photobioreactors. There are various photobioreactor designs with the common goal of maintaining a monoculture of algae that is efficiently exposed to sunlight and carbon dioxide. A common design employs arrays of tubes arranged vertically to minimize land use and oriented north–south to maximize light exposure.

Given that algae do not require fresh water or fertile soils, waste lands have been suggested as potential locations to grow algae. One suggestion is to build algae ponds in the desert Southwest United States where inexpensive flat land, abundant sunlight, water from alkaline aquifers, and CO_2 from power plants could be combined to generate renewable fuels. Algae's potential for yields of 1.12–9.40 million liters of oil/km^2/year promises significant reductions in the land footprint required to produce biofuels.

39.5 Land Use for Biomass Production

Global land use is broadly defined by five categories: pasture, crop, forest, urban, and abandoned. Pasture is land devoted primarily to animal grazing; crop lands are areas actively cultivated for food production; forest land contains primarily large trees; urban areas are heavily populated regions; and abandoned lands are territories that formerly fit one of the previous categories but are no longer employed for human activities. Humans, because of population migrations or land use change, alter the portions of land devoted to each of these categories over time.

Researchers estimate that 14.5 and 33.2 million km^2 of global land area were devoted to crops and pasture respectively in 2000 [2]. These land use groups can coexist within the same region. For example, the U.S. Midwest and parts of the Southeast include regions with more than 70% of the land devoted to crops, and the western sides of the Midwest and Southern U.S. states have a high concentration of land for pasture.

Modern day farmers devote their production to a small selection of crops depending on socioeconomic factors. Table 39.5 shows a sample of biomass crops grown in various geographical regions and their annual yields. Crops such as corn and sugarcane can serve both food and energy needs due to their high yields of sugar-rich biomass and biomass residue (stover and bagasse respectively).

TABLE 39.5

Nominal Annual Yields of Biomass Crops

Biomass Crop	Geographical Location	Annual Yield (Mg/km²)
Corn: grain	North America	700
Corn: cobs	North America	130
Corn: stover	North America	840
Jerusalem artichoke: tuber	North America	4500
Jerusalem artichoke: sugar	North America	640
Sugarcane: crop	Hawaii	5500
Sugarcane: sugar	Hawaii	720
Sugarcane: bagasse (dry)	Hawaii	720
Sweet sorghum: crop	Midwest United States	3800
Sweet sorghum: sugar	Midwest United States	530
Sweet sorghum: fiber (dry)	Midwest United States	490
Switchgrass	North America	1400
Hybrid poplar	North America	1400
Wheat: grain	Canada	220
Wheat: straw	Canada	600

Source: Wayman, M. and Parekh, S., *Biotechnology of Biomass Conversion: Fuels and Chemicals from Renewable Resources*, Open University Press, Philadelphia, PA, 1990.

We can estimate the amount of biomass available in a given region by assuming nominal values for crop productivity and available land use data using the equation:

$$\text{Total biomass}\left[\frac{\text{kg}}{\text{year}}\right] = f \times \text{crop}_{\text{yield}}\left[\frac{\text{kg}}{\text{km}^2 \times \text{year}}\right] \times \text{land}_{\text{area}}[\text{km}^2] \tag{39.5}$$

In Equation 39.5, f is a factor that accounts for crop rotations, farmer participation, and land conservation among other considerations that restrict the land use. As an example, Iowa has a total land area of 144,700 km² that is predominantly covered by corn and soybeans. In 2010, farmers planted 37.5% of Iowa land with corn, netting an average yield of 165 bushels per acre (1035 Mg/km²). Thus, the total amount of corn grown in Iowa that year was 56.2 million Mg.

Farmers and seed companies have managed to increase crop yields every year for the past couple of decades. Crop yield increases follow the exponential growth formula:

$$\text{Crop}_{\text{yield}}(t)\left[\frac{\text{kg}}{\text{km}^2}\right] = \text{crop}_{\text{yield},0}\, e^{kt}, \tag{39.6}$$

where

k is the growth rate

t is the period of time since the initial value $\text{Crop}_{\text{yield},0}$

The USDA maintains a comprehensive database of agricultural statistics (available online at http://quickstats.nass.usda.gov/). The data span several years, and include county-level data for crops, demographics, economics, animals and products, and environmental impacts.

The United States benefits from large biomass resources. Based on crop historical data and growth projections, we can expect traditional biomass resources to continue as a significant potential energy resource. The development of fast-growing dedicated energy crops and algae could help address concerns over land use. Much work remains to continue the production and conversion of biomass in economic and environmentally friendly ways.

References

1. Brown, R. C. (2003). *Biorenewable Resources: Engineering New Products from Agriculture*, pp. 59–75. Iowa State Press, A Blackwell Publishing Company, Ames, IA.
2. Field, C., J. Campbell, and D. Lobell (2008). Biomass energy: The scale of the potential resource. *Trends in Ecology and Evolution* 23(2), 65–72.
3. Mokany, K., R. Raison, and A. S. Prokushkin (2006). Critical analysis of root: Shoot ratios in terrestrial biomes. *Global Change Biology* 12(1), 84–96.
4. Morris, G. (December 2003). The status of biomass power generation in California, July 31, 2003. Technical Report NREL//SR-510–35114, National Renewable Energy Laboratory, Golden, CO.
5. Perlack, R., L. Wright, A. Turhollow, R. Graham, B. Stokes, and D. Erbach (2005). Biomass as feedstock for a bioenergy and bioproducts industry: The technical feasibility of a billion-ton annual supply. Technical Report A357634, Oak Ridge National Laboratory, Oak Ridge, TN.
6. Wayman, M. and S. Parekh (1990). *Biotechnology of Biomass Conversion: Fuels and Chemicals from Renewable Resources*. Open University Press, Philadelphia, PA.
7. Wright, L. and W. Hohenstein (1994). Dedicated feedstock supply systems: Their current status in the USA. *Biomass and Bioenergy* 6(3), 159–241.
8. Zhu, X. G., S. P. Long et al. (2008). What is the maximum efficiency with which photosynthesis can convert solar energy into biomass? *Current Opinion in Biotechnology* 19(2), 153–159.
9. Anon. Dynamic maps, GIS data, & analysis tools—Solar data, National Renewable Energy Laboratory, Golden, CO, 2015. http://www.nrel.gov/gis/datasolar.html (accessed March 2015).
10. Anon. ECN Phyllis2 database for biomass and waste. Energy Centre of the Netherlands, Petten, the Netherlands. http://www.ecn.nl/phyllis2/Browse/Standard/ECN-Phyllis (accessed March 2015).

40

Active Solar Heating Systems

T. Agami Reddy

CONTENTS

40.1 Introduction

This section defines the scope of the entire chapter and presents a brief overview of the types of applications that solar thermal energy can potentially satisfy.

40.1.1 Motivation and Scope

Successful solar system design is an iterative process involving consideration of many technical, practical, reliability, cost, code, and environmental considerations (Mueller Associates 1985). The success of a project involves identification of and intelligent selection among trade-offs, for which a proper understanding of goals, objectives, and constraints is essential. Given the limited experience available in the solar field, it is advisable to keep solar systems as simple as possible and not be lured by the promise of higher efficiency offered by more complex systems. Because of the location-specific variability of the solar resource, solar systems offer certain design complexities and concerns not encountered in traditional energy systems.

The objective of this chapter is to provide energy professionals which a fundamental working knowledge of the scientific and engineering principles of solar collectors and solar systems relevant to both the prefeasibility study and the feasibility study of a solar project. Conventional equipment such as heat exchangers, pumps, and piping layout are but briefly described. Because of space limitations, certain equations/correlations had to be omitted, and proper justice could not be given to several concepts and design approaches. Effort has been made to provide the reader with pertinent references to textbooks, manuals, and research papers.

A detailed design of solar systems requires in-depth knowledge and experience in (1) the use of specially developed computer programs for detailed simulation of solar system performance, (2) designing conventional equipment, controls, and hydronic systems,

(3) practical aspects of equipment installation, and (4) economic analysis. These aspects are not addressed here, given the limited scope of this chapter. Readers interested in acquiring such details can consult manuals such as Mueller Associates (1985) or SERI (1989).

The lengthy process outlined above pertains to large solar installations. The process is much less involved when a small domestic hot-water system, or unitary solar equipment or single solar appliances such as solar stills, solar cookers, or solar dryers are to be installed. Not only do such appliances differ in engineering construction from region to region, there are also standardized commercially available units whose designs are already more or less optimized by the manufacturers, normally as a result of previous experimentation, both technical or otherwise. Such equipment is not described in this chapter for want for space.

The design concepts described in this chapter are applicable to domestic water heating, swimming pool heating, active space heating, industrial process heat, convective drying systems, and solar cooling systems.

40.2 Solar Collectors

40.2.1 Collector Types

A solar thermal collector is a heat exchanger that converts radiant solar energy into heat. In essence this consists of a receiver that absorbs the solar radiation and then transfers the thermal energy to a working fluid. Because of the nature of the radiant energy (its spectral characteristics, its diurnal and seasonal variability, changes in diffuse to global fraction, etc.), as well as the different types of applications for which solar thermal energy can be used, the analysis and design of solar collectors present unique and unconventional problems in heat transfer, optics, and material science. The classification of solar collectors can be made according to the type of working fluid (water, air, or oils) or the type of solar receiver used (nontracking or tracking).

Most commonly used working fluids are water (glycol being added for freeze protection) and air. Table 40.1 identifies the relative advantages and potential disadvantages of air and liquid collectors and associated systems. Because of the poorer heat transfer characteristics of air with the solar absorber, the air collector may operate at a higher temperature than a liquid-filled collector, resulting in greater thermal losses and, consequently, a lower efficiency. The choice of the working fluid is usually dictated by the application. For example, air collectors are suitable for space heating and convective drying applications, while liquid collectors are the obvious choice for domestic and industrial hot-water applications. In certain high-temperature applications, special types of oils are used that provide better heat transfer characteristics.

The second criterion of collector classification is according to the presence of a mechanism to track the sun throughout the day and year in either a continuous or discreet fashion (see Table 40.2). The stationary flat-plate collectors are rigidly mounted, facing toward the equator with a tilt angle from the horizontal roughly equal to the latitude of the location for optimal year-round operation. The compound parabolic concentrators (CPCs) can be designed either as completely stationary devices or as devices that need seasonal adjustments only. On the other hand, Fresnel reflectors, paraboloids, and heliostats need two-axis tracking. Parabolic troughs have one axis tracking either along

TABLE 40.1
Advantages and Disadvantages of Liquid and Air Systems

Characteristics	Liquid	Air
Efficiency	Collectors generally more efficient for a given temperature difference	Collectors generally operate at slightly lower efficiency
System configuration	Can be readily combined with service hot-water and cooling systems	Space heat can be supplied directly but does not adapt easily to cooling. Can preheat hot-water
Freeze protection	May require antifreeze and heat exchangers that add cost and reduce efficiency	None needed
Maintenance	Precautions must be taken against leakage, corrosion and boiling	Low maintenance requirements. Leaks repaired readily with duct tape, but leaks may be difficult to find
Space requirements	Insulated pipes take up nominal space and are more convenient to install in existing buildings	Duct work and rock storage units are bulky, but ducting is a standard HVAC installation technique
Operation	Less energy required to pump liquids	More energy required by blowers to move air; noisier operation
Cost	Collectors cost more	Storage costs more
State of the art	Has received considerable attention from solar industry	Has received less attention from solar industry

Source: SERI, *Engineering Principles and Concepts for Active Solar Systems,* Hemisphere Publishing Company, New York, 1989.

TABLE 40.2
Types of Solar Thermal Collectors

Nontracking Collectors	Tracking Collectors
Basic flat-plate	Parabolic troughs
Flat-plate enhanced with side reflectors or V-troughs	Fresnel reflectors
Tubular collectors	Paraboloids
Compound parabolic concentrators (CPCs)	Heliostats with central receivers

the east-west direction or the north–south direction. These collector types are described by Kreider (1979) and Rabl (1985).

A third classification criterion is to distinguish between nonconcentrating and concentrating collectors. The main reason for using concentrating collectors is not that *more energy* can be collected but that the thermal energy is obtained at higher temperatures. This is done by decreasing the area from which heat losses occur (called the receiver area) with respect to the aperture area (i.e., the area that intercepts the solar radiation). The ratio of the aperture to receiver area is called the *concentration ratio.*

40.2.2 Flat-Plate Collectors

40.2.2.1 Description

The flat-plate collector is the most common conversion device in operation today, since it is most economical and appropriate for delivering energy at temperatures up to about 100°C. The construction of flat-plate collectors is relatively simple, and many commercial models are available.

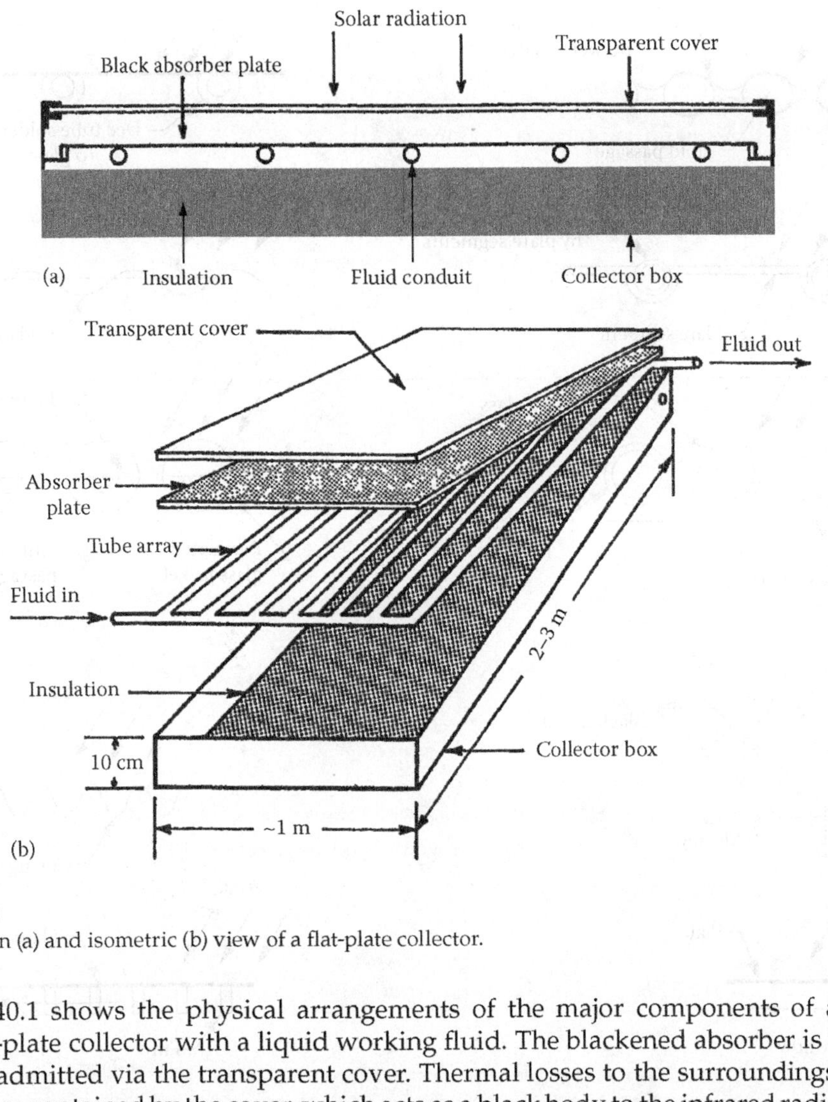

FIGURE 40.1
Cross-section (a) and isometric (b) view of a flat-plate collector.

Figure 40.1 shows the physical arrangements of the major components of a conventional flat-plate collector with a liquid working fluid. The blackened absorber is heated by radiation admitted via the transparent cover. Thermal losses to the surroundings from the absorber are contained by the cover, which acts as a black body to the infrared radiation (this effect is called the *greenhouse* effect), and by insulation provided under the absorber plate. Passages attached to the absorber are filled with a circulating fluid, which extracts energy from the hot absorber. The simplicity of the overall device makes for long service life.

The absorber is the most complex portion of the flat-plate collector, and a great variety of configurations are currently available for liquid and air collectors. Figure 40.2 illustrates some of these concepts in absorber design for both liquid and air absorbers. Conventional materials are copper, aluminum, and steel. The absorber is either painted with a dull black paint or can be coated with a *selective surface* to improve performance (see Section 40.2.3 for more details). Bonded plates having internal passageways perform well as absorber plates because the hydraulic passageways can be designed for optimal fluid and thermal performance. Such collectors are called *roll-bond* collectors. Another common absorber consists of tubes soldered or brazed to a single metal sheet, and mechanical attachments of the tubes to the plate have also been employed. This type of collector is called a *tube-and-sheet* collector. Heat pipe collectors have also been developed, though these are not as

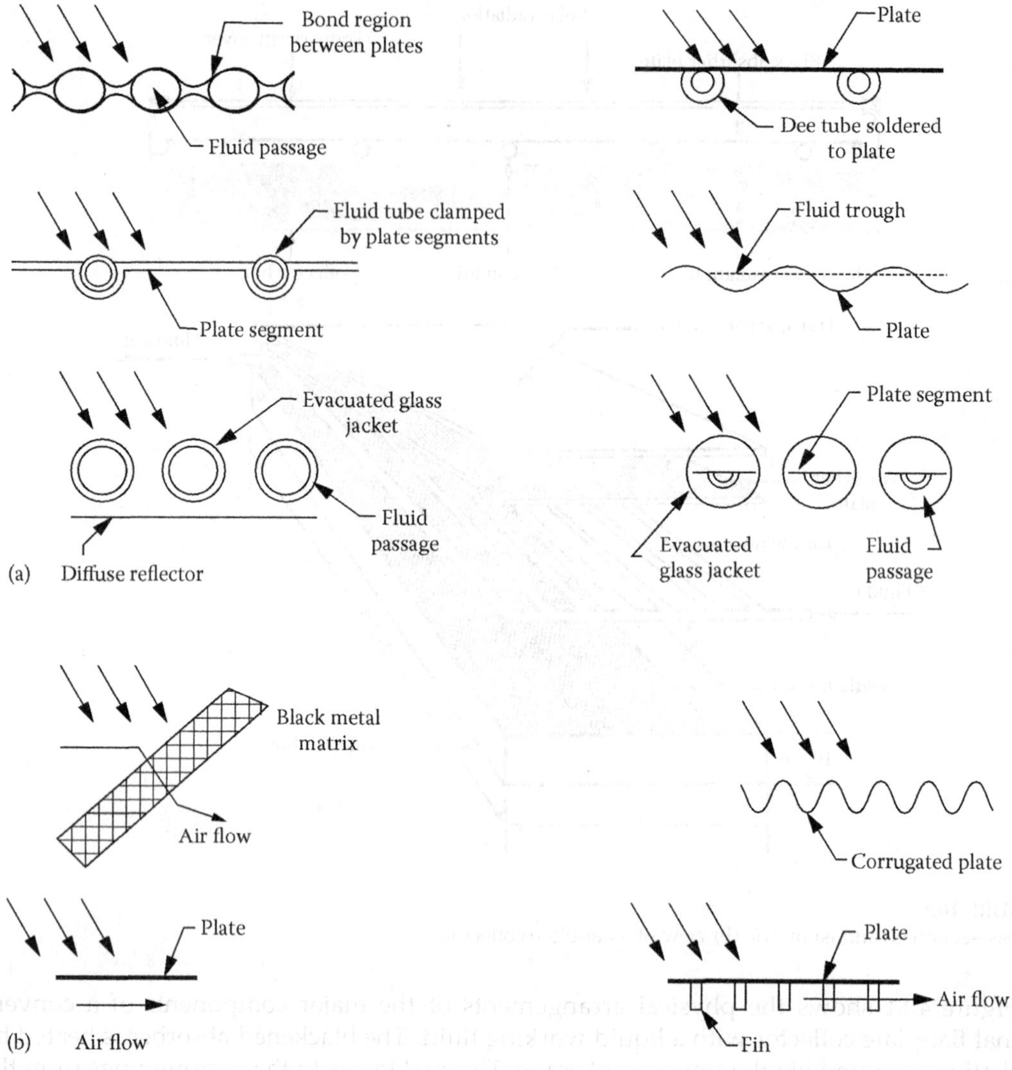

FIGURE 40.2
Typical flat-plate absorber configurations. (a) Air and (b) liquid collectors.

widespread as the previous two types. The so-called *trickle type* of flat-plate collector, with the fluid flowing directly over the corrugated absorber plate, dispenses entirely with fluid passageways. Tubular collectors have also been used because of the relative ease by which air can be evacuated from such collectors, thereby reducing convective heat losses from the absorber to the ambient air.

The absorber in an air collector normally requires a larger surface than in a liquid collector because of the poorer heat transfer coefficients of the flowing air stream. Roughness elements and producing turbulence by way of devices such as expanded metal foil, wool, and overlapping plates have been used as a means for increasing the heat transfer from the absorber to the working fluid. Another approach to enhance heat transfer is to use packed beds of expanded metal foils or matrices between the glazing and the bottom plate.

40.2.2.2 Modeling

A particular modeling approach and the corresponding degree of complexity in the model are dictated by the objective as well as by experience gained from past simulation work. For example, it has been found that transient collector behavior has insignificant influence when one is interested in determining the long-term performance of a solar thermal system. For complex systems or systems meant for nonstandard applications, detailed modeling and careful simulation of system operation are a must initially, and simplifications in component models and system operation can subsequently be made. However, in the case of solar thermal systems, many of the possible applications have been studied to date and a backlog of experience is available not only concerning system configurations but also with reference to the degree of component model complexity.

Because of low collector time constants (about 5–10 min), heat capacity effects are usually small. Then the instantaneous (or hourly, because radiation data are normally available in hourly time increments only) steady-state useful energy q_C in watts delivered by a solar flat-plate collector of surface area A_C is given by

$$q_C = A_C F'[I_T \eta_0 - U_L (T_{Cm} - T_a)]^+ \tag{40.1}$$

where

F' is the plate efficiency factor, which is a measure of how good the heat transfer is between the fluid and the absorber plate

η_0 is the optical efficiency, or the product of the transmittance and absorptance of the cover and absorber of the collector

U_L is the overall heat loss coefficient of the collector, which is dependent on collector design only and is normally expressed in W/(m² °C)

T_{Cm} is the *mean* fluid temperature in the collector (in °C)

I_T is the radiation intensity on the plane of the collector (in W/m²)

The + sign denotes that negative values are to be set to zero, which physically implies that the collector should not be operated when q_C is negative (i.e., when the collector loses more heat than it can collect).

However, because T_{Cm} is not a convenient quantity to use, it is more appropriate to express collector performance in terms of the fluid *inlet* temperature to the collector (T_{Ci}). This equation is known as the classical Hottel–Whillier–Bliss (HWB) equation and is most widely used to predict instantaneous collector performance:

$$q_C = A_C F_R [I_T \eta_0 - U_L (T_{Ci} - T_a)]^+ \tag{40.2}$$

where F_R is called the heat removal factor and is a measure of the solar collector performance as a heat exchanger, since it can be interpreted as the ratio of actual heat transfer to the maximum possible heat transfer. It is related to F' by

$$\frac{F_R}{F'} = \frac{(mc_p)_C}{A_C F' U_L}\left\{1 - \exp\left[-\frac{A_C F' U_L}{(mc_p)_C}\right]\right\} \tag{40.3}$$

where $(mc_p)_C$ is the heat capacity of the total fluid flow rate times the specific heat of fluid flowing through the collector.

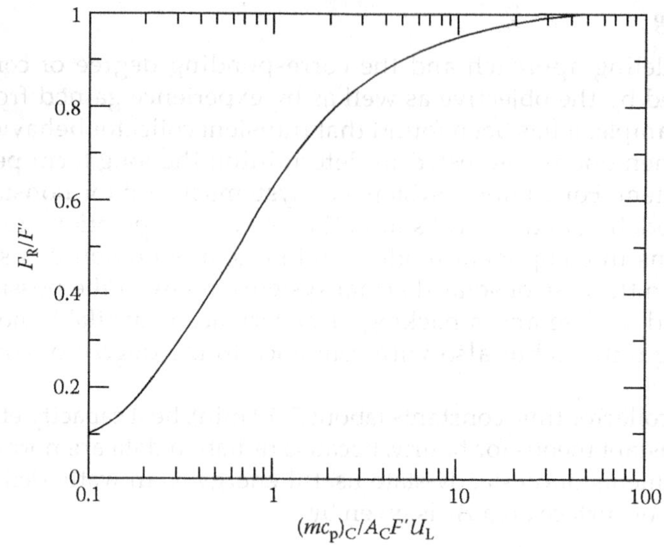

FIGURE 40.3
Variation of F_R/F' as a function of $[(mc_p)_C/(A_C F' U_L)]$. (From Duffie, J. A. and Beckman, W. A., *Solar Engineering of Thermal Processes*, Wiley Interscience, New York, 1980.)

The variation of (F_R/F') with $[(mc_p)_C/A_C F' U_L]$ is shown graphically in Figure 40.3. Note the asymptotic behavior of the plot, which suggests that increasing the fluid flow rate more than a certain amount results in little improvement in F_R (and hence in q_C) while causing a quadratic increase in the pressure drop.

Factors influencing solar collector performance are of three types: (1) constructional, that is, related to collector design and materials used, (2) climatic, and (3) operational, that is, fluid temperature, flow rate, and so on. The plate efficiency factor F' is a factor that depends on the physical constructional features and is essentially a constant for a given liquid collector. (This is not true for air collectors, which require more careful analysis.) Operational features involve changes in m_C and T_{Ci}. While changes in m_C affect F_R as per Equation 40.3, we note from Equation 40.2 that to enhance q_C, T_{Ci} needs to be kept as low as possible. For solar collectors that are operated under more or less constant flow rates, specifying $F_R\eta_0$ and $F_R U_L$ is adequate to predict collector performance under varying climatic conditions.

There are a number of procedures by which collectors have been tested. The most common is a *steady-state procedure*, where transient effects due to collector heat capacity are minimized by performing tests only during periods when radiation and ambient temperature are steady. The procedure involves simultaneous and accurate measurements of the mass flow rate, the inlet and outlet temperatures of the collector fluid, and the ambient conditions (incident solar radiation, air temperature, and wind speed). The most widely used test procedure is the ASHRAE Standard 93-77 (1978), whose test setup is shown in Figure 40.4. Though a solar simulator can be used to perform indoor testing, outdoor testing is always more realistic and less expensive. The procedure can be used for nonconcentrating collectors using air or liquid as the working fluid (but not two phase mixtures) that have a single inlet and a single outlet and contain no integral thermal storage.

Steady-state procedures have been in use for a relatively long period and though the basis is very simple the engineering setup is relatively expensive (see Figure 40.4). From an

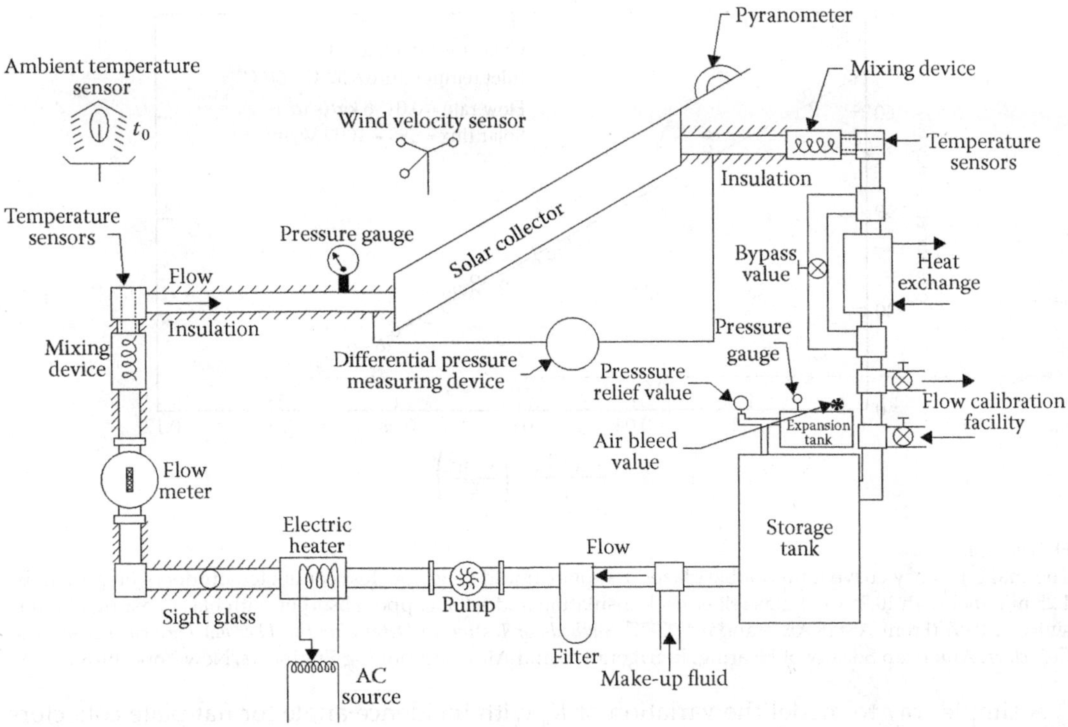

FIGURE 40.4
Set up for testing liquid collectors according to ASHRAE Standard 93-72.

overall heat balance on the collector fluid and from Equation 40.2, the expressions for the instantaneous collector efficiency under normal solar incidence are

$$\eta_C \equiv \frac{q_C}{A_C I_T} = \frac{(mc_p)_{Ci}(T_{Co} - T_{Ci})}{A_C I_T} \tag{40.4}$$

$$= \left[F_R \eta_n - F_R U_L \left(\frac{T_{Ci} - T_a}{I_T} \right) \right] \tag{40.5}$$

where η_n is the optical efficiency at normal solar incidence.

From the test data, points of η_c against reduced temperature $[(T_{Ci} - T_a)/I_T]$ are plotted as shown in Figure 40.5. Then a linear fit is made to these data points by regression, from which the values of $F_R\eta_n$ and $F_R U_L$ are easily deduced. It will be noted that if the reduced term were to be taken as $[(T_{Cm} - T_a)/I_T]$, estimates of $F'\eta_n$ and $F'U_L$ would be correspondingly obtained.

40.2.2.3 Incidence Angle Modifier

The optical efficiency η_0 depends on the collector configuration and varies with the angle of incidence as well as with the relative values of diffuse and beam radiation. The incidence angle modifier is defined as $K_\eta = (\eta_0/\eta_n)$. For flat-plate collectors with 1 or 2 glass covers, $K\eta$ is almost unchanged up to incidence angles of 60°, after which it abruptly drops to zero.

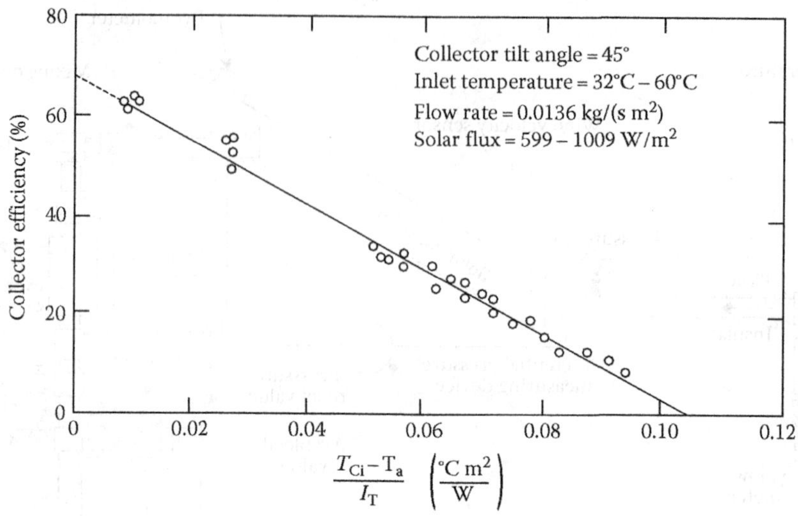

FIGURE 40.5
Thermal efficiency curve for a double glazed flat-plate liquid collector. Test conducted outdoors on a 1.2 m by 1.25 m panel with 10.2 cm of glass fiber back insulation and a flat copper absorber with black coating of emissivity of 0.97. (From ASHRAE Standard 93-77, *Methods of Testing to Determine the Thermal Performance of Solar Collectors*, American Society of Heating, Refrigeration and Air Conditioning Engineers, New York, 1978.)

A simple way to model the variation of K_η with incidence angle for flat-plate collectors is to specify η_n, the optical efficiency of the collector at normal beam incidence, to assume the entire radiation to be beam, and to use the following expression for the angular dependence (ASHRAE 1978)

$$K_\eta = 1 + b_0\left(\frac{1}{\cos\theta} - 1\right) \tag{40.6}$$

where

θ is the solar angle of incidence on the collector plate (in degrees)
b_0 is a constant called the incidence angle modifier coefficient

Plotting K_η against $[(1/\cos\theta) - 1]$ results in linear plots (see Figure 40.6), thus justifying the use of Equation 40.6.

We note that for one-glass and two-glass covers, approximate values of b_0 are −0.10 and −0.17, respectively.

In case the diffuse solar fraction is high, one needs to distinguish between beam, diffuse, and ground-reflected components. Diffuse radiation, by its very nature, has no single incidence angle. One simple way is to assume an equivalent incidence angle of 60° for diffuse and ground-reflected components. One would then use Equation 40.6 for the beam component along with its corresponding value of θ and account for the contribution of diffuse and ground reflected components by assuming a value of θ = 60° in Equation 40.6. For more accurate estimation, one can use the relationship between the effective diffuse solar incidence angle versus collector tilt given in Duffie and Beckman (1980). It should be noted that the preceding equation gives misleading results with incidence angles close to 90°. An alternative functional form for the incidence angle modifier for both flat-plate and concentrating collectors has been proposed by Rabl (1981).

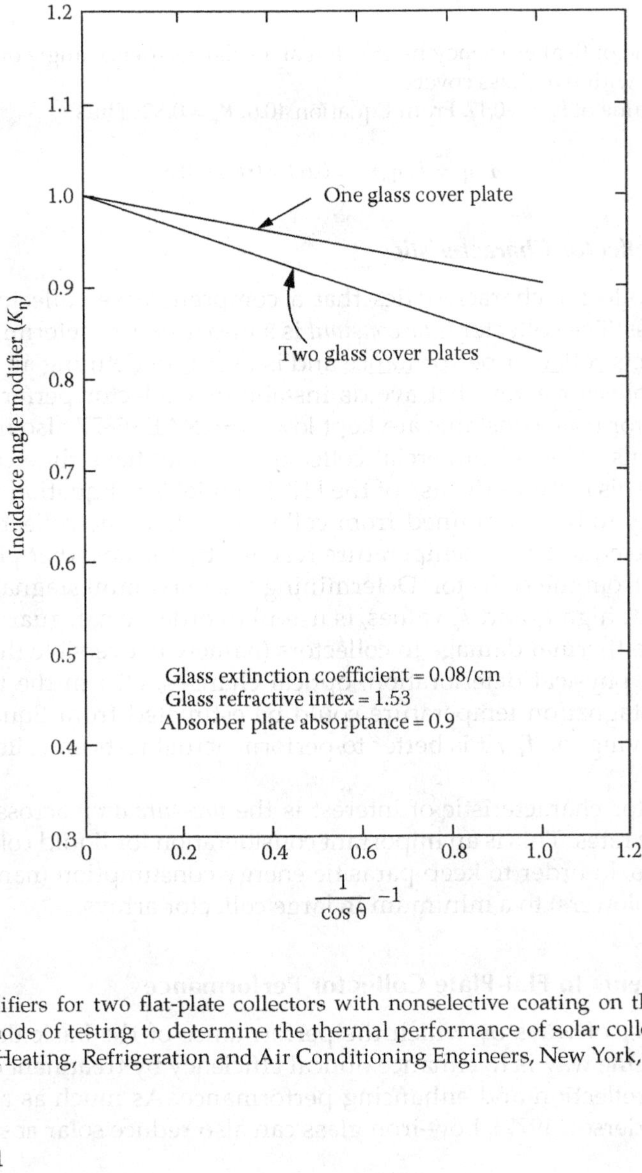

FIGURE 40.6
Incidence angle modifiers for two flat-plate collectors with nonselective coating on the absorber. (Adapted from ASHRAE, Methods of testing to determine the thermal performance of solar collectors, Standard 93-77, American Society of Heating, Refrigeration and Air Conditioning Engineers, New York, 1978.)

Example 40.1

From the thermal efficiency curve given in Figure 40.5 determine the performance parameters of the corresponding solar collector.

Extrapolating the curve yields y-intercept = 0.69, x-intercept = 0.105 (m² °C/W). Since the reduced temperature in Figure 40.5 is in terms of the inlet fluid temperature to the collector, Equation 40.5 yields $F_R\eta_n = 0.69$ and $F_RU_L = 0.69/0.105 = 6.57$ W/(m² °C). Alternatively, the collector parameters in terms of the plate efficiency factor can be deduced. From Figure 40.5, the collector area = 1.22 × 1.25 = 1.525 m², while the flow rate $(m/A_C) = 0.0136$ kg/(s m²). From Equation 40.3,

$$F'/F_R = -(0.0136 \times 4190/6.57)\ln[-6.57/(0.0136 \times 4190)] = 1.0625$$

Thus $F'U_L = 6.57 \times 1.0625 = 6.98$ W/(m² °C) and $F'\eta_n = 0.69 \times 1.0625 = 0.733$.

Example 40.2

How would the optical efficiency be effected at a solar incidence angle of 60° for a flat-plate collector with two glass covers?

Assume a value of $b_0 = -0.17$. From Equation 40.6, $K_\eta = 0.83$. Thus

$$F_R\eta_0 = F_R\eta_n K_\eta = 0.69 \times 0.83 = 0.57$$

40.2.2.4 Other Collector Characteristics

There are three collector characteristics that a comprehensive collector testing process should also address. The collector *time constant* is a measure that determines how intermittent sunshine affects collector performance and is useful in defining an operating control strategy for the collector array that avoids instability. Collector performance is usually enhanced if collector time constants are kept low. ASHRAE 93-77 also includes a method for determining this value. Commercial collectors usually have time constants of about 5 min or less, and this justifies the use of the HWB model (see Equation 40.2).

Another quantity to be determined from collector tests is the collector *stagnation temperature*. This is the equilibrium temperature reached by the absorber plate when no heat is being extracted from the collector. Determining the maximum stagnation temperature, which occurs under high I_T and T_a values, is useful in order to safeguard against reduced collector life due to thermal damage to collectors (namely irreversible thermal expansion, sagging of covers, physical deterioration, optical changes, etc.) in the field when not in use. Though the stagnation temperature could be estimated from Equation 40.2 by setting $q_C = 0$ and solving for T_{Ci}, it is better to perform actual tests on collectors before field installation.

The third collector characteristic of interest is the *pressure drop* across the collector for different fluid flow rates. This is an important consideration for liquid collectors, and more so for air collectors, in order to keep parasitic energy consumption (namely electricity to drive pumps and blowers) to a minimum in large collector arrays.

40.2.3 Improvements to Flat-Plate Collector Performance

There are a number of ways by which the performance of the basic flat-plate collectors can be improved. One way is to enhance optical efficiency by treatment of the glass cover thereby reducing reflection and enhancing performance. As much as a 4% increase has been reported (Anderson 1977). Low-iron glass can also reduce solar absorption losses by a few percent.

These improvements are modest compared to possible improvements from reducing losses from the absorber plate. Essentially, the infrared upward reradiation losses from the heated absorber plate have to be decreased. One could use a second glass cover to reduce the losses, albeit at the expense of higher cost and lower optical efficiency. Usually for water heating applications, radiation accounts for about two-thirds of the losses from the absorber to the cover with convective losses making up the rest (conduction is less than about 5%). The most widely used manner of reducing these radiation losses is to use selective surfaces whose emissivity varies with wavelength (as against matte-black painted absorbers, which are essentially gray bodies). Note that 98% of the solar spectrum is at wavelengths less than 3.0 μm, whereas less than 1% of the black body radiation from a 200°C surface is at wavelengths less than 3.0 μm. Thus selective

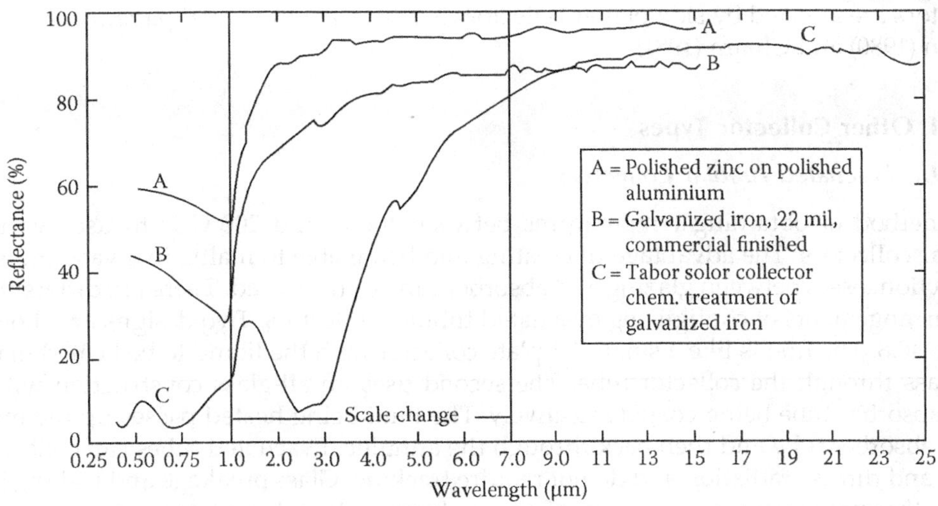

FIGURE 40.7
Spectral reflectance of several surfaces. (From Edwards, D. K. et al., Basic studies on the use of solar energy, Report No. 60-93, Department of Engineering, University of California at Los Angeles, Los Angeles, CA, 1960.)

surfaces for solar collectors should have high-solar absorptance (i.e., low reflectance in the solar spectrum) and low long-wave emittance (i.e., high reflectance in the long-wave spectrum). The spectral reflectance of some commonly used selective surfaces is shown in Figure 40.7. Several commercial collectors for water heating or low-pressure steam (for absorption cooling or process heat applications) are available that use selective surfaces.

Another technique to simultaneously reduce both convective and radiative losses between the absorber and the transparent cover is to use honeycomb material (Hollands 1965). The honeycomb material can be reflective or transparent (the latter is more common) and should be sized properly. Glass honeycombs have had some success in reducing losses in high-temperature concentrating receivers, but plastics are usually recommended for use in flat-plate collectors. Because of the poor thermal aging properties, honeycomb flat-plate collectors have had little commercial success. Currently the most promising kind seems to be the simplest (both in terms of analysis and construction), namely collectors using horizontal rectangular slats (Meyer 1978). Convection can be entirely suppressed provided the slats with the proper aspect ratio are used.

Finally, collector output can be enhanced by using side reflectors, for instance a sheet of anodized aluminum. The justification in using these is their low cost and simplicity. For instance, a reflector placed in front of a tilted collector cannot but increase collector performance because losses are unchanged and more solar radiation is intercepted by the collector. Reflectors in other geometries may cast a shadow on the collector and reduce performance. Note also that reflectors would produce rather nonuniform illumination over the day and during the year, which, though not a problem in thermal collectors, may drastically penalize the electric output of photovoltaic modules. Whether reflectors are cost-effective depends on the particular circumstances and practical questions such as aesthetics and space availability. The complexity involved in the analysis of collectors with planar reflectors can be reduced by assuming the reflector to be long compared to its width and treating the problem in two dimensions only. How optical performance of solar collectors are affected by side planar reflectors is discussed in several papers, for example Larson (1980) and Chiam (1981).

40.2.4 Other Collector Types

40.2.4.1 Evacuated Tubular Collectors

One method of obtaining temperatures between 100°C and 200°C is to use evacuated tubular collectors. The advantage in creating and being able to maintain a vacuum is that convection losses between glazing and absorber can be eliminated. There are different possible arrangements of configuring evacuated tubular collectors. Two designs are shown in Figure 40.8. The first is like a small flat-plate collector with the liquid to be heated making one pass through the collector tube. The second uses an all-glass construction with the glass absorber tube being coated selectively. The fluid being heated passes up the middle of the absorber tube and then back through the annulus. Evacuated tubes can collect both direct and diffuse radiation and do not require tracking. Glass breakage and leaking joints due to thermal expansion are some of the problems which have been experienced with such collector types. Various reflector shapes (like flat-plate, V-groove, circular, cylindrical, involute, etc.) placed behind the tubes are often used to usefully collect some of the solar energy, which may otherwise be lost, thus providing a small amount of concentration.

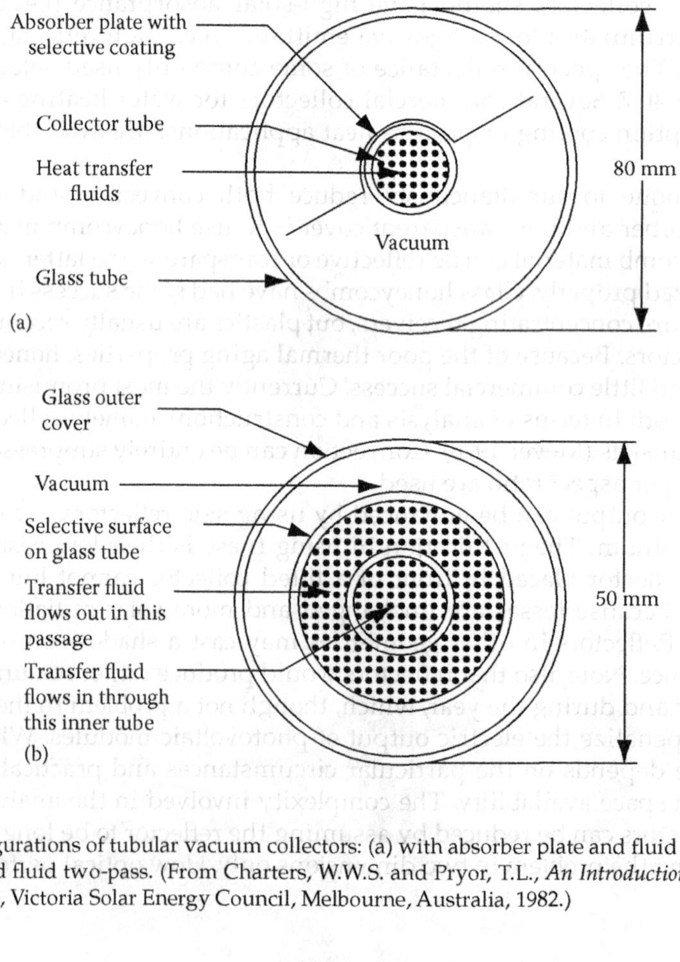

FIGURE 40.8
Two common configurations of tubular vacuum collectors: (a) with absorber plate and fluid single-pass, (b) with concentric tubes and fluid two-pass. (From Charters, W.W.S. and Pryor, T.L., *An Introduction to the Installation of Solar Energy Systems*, Victoria Solar Energy Council, Melbourne, Australia, 1982.)

40.2.4.2 Compound Parabolic Concentrators

The CPC collector, discovered in 1966, consists of parabolic reflectors that funnel radiation from aperture to absorber rather than focusing it. The right and left halves belong to different parabolas (hence the name *compound*) with the edges of the receiver being the foci of the opposite parabola (see Figure 40.9). It has been proven that such collectors are *ideal* in that any solar ray, be it beam or diffuse, incident on the aperture within the acceptance angle will reach the absorber while all others will bounce back to and fro and reemerge through the aperture. CPCs are also called *nonimaging* concentrators because they do not form clearly defined images of the solar disk on the absorber surface as achieved in classical concentrators. CPCs can be designed both as low-concentration devices with large acceptance angles or as high-concentration devices with small acceptance angles. CPCs with low-concentration ratios (of about 2) and with east-west axes can be operated as stationary devices throughout the year or at most with seasonal adjustments only. CPCs, unlike other concentrators, are able to collect all the beam and a large portion of the diffuse radiation. Also they do not require highly specular surfaces and can thus better tolerate dust and degradation. A typical module made up of several CPCs is shown in Figure 40.10. The absorber surface is located at the bottom of the trough, and a glass cover may also be used to encase the entire module. CPCs show considerable promise for water heating close to the boiling point and for low-pressure

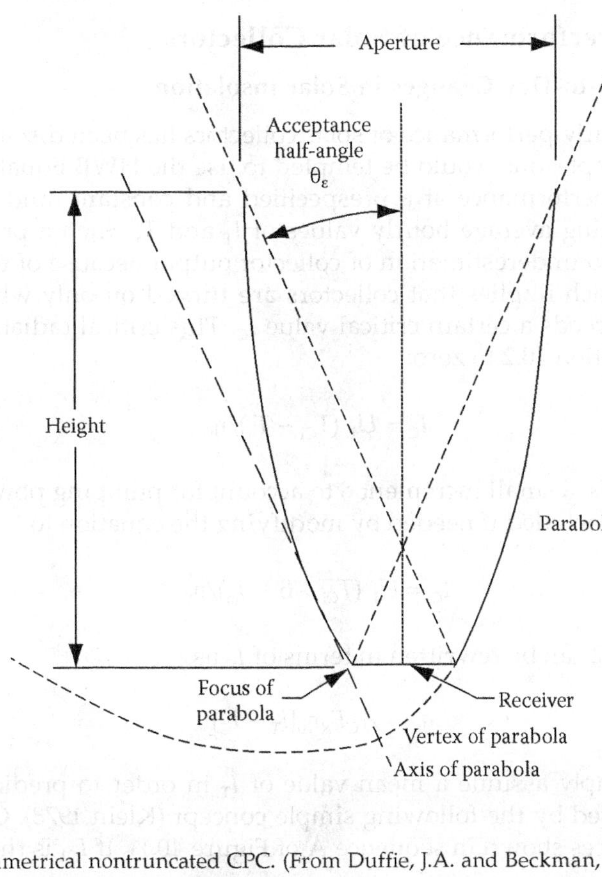

FIGURE 40.9
Cross-section of a symmetrical nontruncated CPC. (From Duffie, J.A. and Beckman, W.A., *Solar Engineering of Thermal Processes*, Wiley Interscience, New York, 1980.)

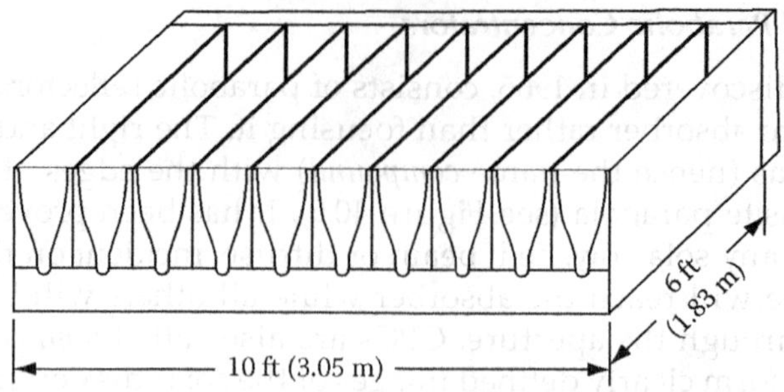

FIGURE 40.10
A CPC collector module. (From SERI, *Engineering Principles and Concepts for Active Solar Systems*, Hemisphere Publishing Company, New York, 1989.)

steam applications. Further details about the different types of absorber and receiver shapes used, the effect of truncation of the receiver and the optics, can be found in Rabl (1985).

40.3 Long-Term Performance of Solar Collectors

40.3.1 Effect of Day-to-Day Changes in Solar Insolation

Instantaneous or hourly performance of solar collectors has been discussed in "Flat-Plate Collectors." For example, one would be tempted to use the HWB Equation 40.2 to predict long-term collector performance at a prespecified and constant fluid inlet temperature T_{Ci} merely by assuming average hourly values of I_T and T_a. Such a procedure would be erroneous and lead to underestimation of collector output because of the presence of the control function, which implies that collectors are turned on only when $q_C > 0$, that is, when radiation I_T exceeds a certain critical value I_C. This critical radiation value is found by setting q_C in Equation 40.2 to zero:

$$I_C = U_L\,(T_{Ci} - T_a)/\eta_0 \tag{40.7a}$$

To be more rigorous, a small increment δ to account for pumping power and stability of controls can also be included if needed by modifying the equation to

$$I_C = U_L\,(T_{Ci} + \delta - T_a)/\eta_0 \tag{40.7b}$$

Then, Equation 40.2 can be rewritten in terms of I_C as

$$q_C = A_C F_R \eta_0 [I_T - I_C]^+ \tag{40.8}$$

Why one cannot simply assume a mean value of I_T in order to predict the mean value of q_C will be illustrated by the following simple concept (Klein 1978). Consider the three identical day sequences shown in sequence A of Figure 40.11. If I_{C1} is the critical radiation

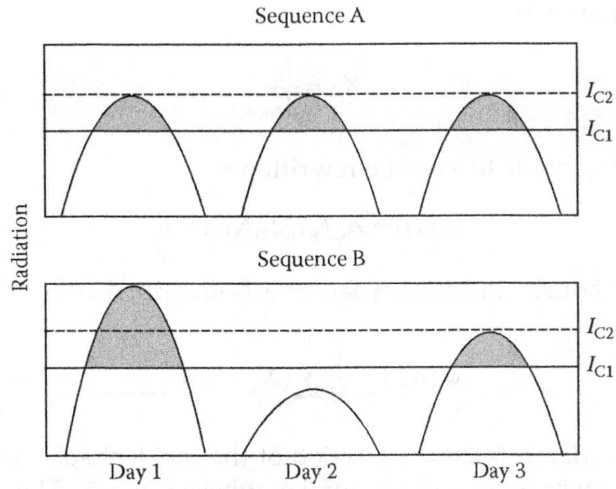

FIGURE 40.11
Effect of radiation distribution on collector long-term performance. (From Klein, S.A., Calculation of flat-plate collector utilizability, *Solar Energy*, 21, 393, 1978.)

intensity and if it is constant over the whole day, the useful energy collected by the collector is represented by the sum of the shaded areas. If a higher critical radiation value shown as I_{C2} in Figure 40.11 is selected, we note that no useful energy is collected at all. Actual weather sequences would not look like that in sequence A but rather like that in sequence B, which is comprised of an excellent, a poor, and an average day. Even if both sequences have the same average radiation over 3 days, a collector subjected to sequence B will collect useful energy when the critical radiation is I_{C2}. Thus, neglecting the variation of radiation intensity from day-to-day over the long term and dealing with mean values would result in an underestimation of collector performance.

Loads are to a certain extent repetitive from day-to-day over a season or even the year. Consequently, one can also expect collectors to be subjected to a known diurnal repetitive pattern or mode of operation, that is, the collector inlet temperature T_{Ci} has a known repetitive pattern.

40.3.2 Individual Hourly Utilizability

In this mode, T_{Ci} is assumed to very over the day but has the same variation for all the days over a period of N days (where $N = 30$ days for monthly and $N = 365$ for yearly periods). Then from Equation 40.8, *total* useful energy collected over N days during individual hour i of the day is

$$q_{CN}(i) = A_C F_R \overline{\eta}_0 \overline{I}_{Ti} \sum_{i=1}^{N} \frac{[I_{Ti} - I_C]^+}{\overline{I}_{Ti}} \tag{40.9}$$

Let us define the radiation ratio

$$X_i = \frac{I_{Ti}}{\overline{I}_{Ti}} \tag{40.10}$$

and the critical radiation ratio

$$X_C = \frac{I_C}{\bar{I}_{Ti}}$$

The modified HWB Equation 40.8 can be rewritten as

$$q_{CN}(i) = A_C F_R \bar{\eta}_0 \bar{I}_{Ti} N \phi_i(x_c) \tag{40.11}$$

where the individual hourly utilizability factor ϕ_i is identified as

$$\phi_i(X_C) = \frac{1}{N}\sum^{N}(X_i - X_C)^+ \tag{40.12}$$

Thus ϕ_i can be considered to be the fraction of the incident solar radiation that can be converted to useful heat by an ideal collector (i.e., whose $F_R\eta_0 = 1$). The utilizability factor is thus a *radiation statistic* in the sense that it depends solely on the radiation values at the specific location. As such, it is in no way dependent on the solar collector itself. Only after the radiation statistics have been applied is a collector dependent significance attached to X_C.

Hourly utilizability curves on a *monthly* basis that are independent of location were generated by Liu and Jordan (1963) over 30 years ago for flat-plate collectors (see Figure 40.12). The key climatic parameter which permits generalization is the *monthly clearness index* $\bar{K}$ of the location defined as

$$\bar{K} = \frac{\bar{H}}{\bar{H}_0} \tag{40.13}$$

where
$\bar{H}$ is the monthly mean daily global radiation on the horizontal surface
$\bar{H}_0$ is the monthly mean daily extraterrestrial radiation on a horizontal surface

Extensive tables giving monthly values of $\bar{K}$ for several different locations worldwide can be found in several books, for example, Duffie and Beckman (1980) or Reddy (1987). The curves apply to equator-facing tilted collectors with the effect of collector tilt accounted for by the factor $\bar{R}_{b,T}$ which is the ratio of the monthly mean daily extraterrestrial radiation on the tilted collector to that on a horizontal surface. Monthly mean daily calculations can be made using the 15th of the month, though better accuracy is achieved using slightly different dates (Reddy 1987). Clark et al. (1983), working from measured data from several U.S. cities, have proposed the following correlation for individual hourly utilizability over monthly time scales applicable to flat-plate collectors only:

$$\begin{aligned} \phi_i &= 0 \quad \text{for } X_C \geq X_{max} \\ &= (1 - X_C/X_{max})^2 \quad \text{for } X_{max} = 2 \\ &= \left| |a| - [a^2 + (1+2a)(1 - X_C/X_{max})^2]^{1/2} \right| \quad \text{otherwise} \end{aligned} \tag{40.14}$$

where

$$a = (X_{max} - 1)(2 - X_{max}) \tag{40.15}$$

and

$$X_{max} = 1.85 + 0.169(\bar{r}_T/\bar{k}^2) - 0.0696\cos\beta/\bar{k}^2 - 0.981\bar{k}/(\cos\delta)^2 \quad (40.16)$$

where
- $\bar{k}$ is the monthly mean *hourly* clearness index for the particular hour
- δ is the solar declination
- β is the tilt angle of the collector plane with respect to the horizontal
- $\bar{r}_T$ is the ratio of monthly average hourly global radiation on a tilted surface to that on a horizontal surface for that particular hour

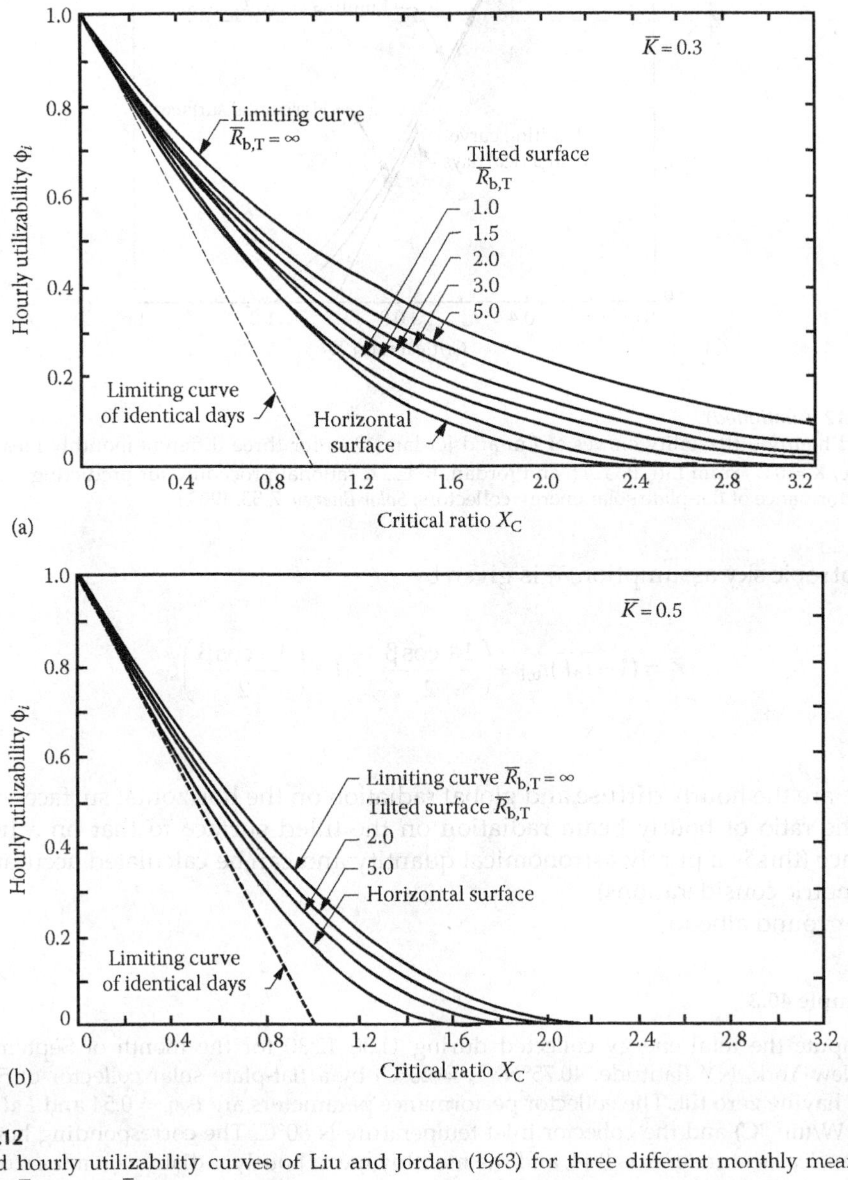

FIGURE 40.12
Generalized hourly utilizability curves of Liu and Jordan (1963) for three different monthly mean clearness indices *K*. (a) $\bar{K} = 0.3$, (b) $\bar{K} = 0.5$. *(Continued)*

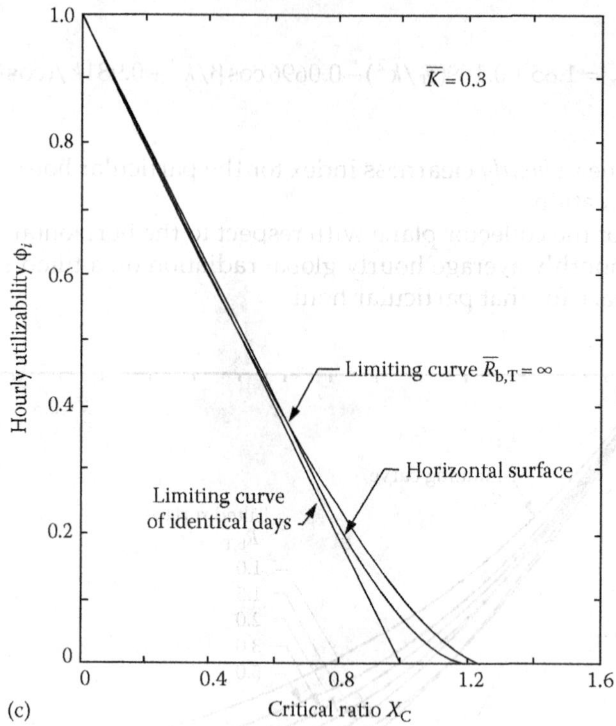

FIGURE 40.12 ***(Continued)***
Generalized hourly utilizability curves of Liu and Jordan (1963) for three different monthly mean clearness indices *K*. (c) $\overline{K}$ = 0.7. (From Liu, B. Y. H. and Jordan, R. C., A rational procedure for predicting the long-term average performance of flat-plate solar energy collectors, *Solar Energy*, 7, 53, 1963.)

For an isotropic sky assumption, $\overline{r}_T$ is given by

$$\overline{r}_T = (1-\overline{I}_d\overline{I})r_{b,T} + \left(\frac{1+\cos\beta}{2}\right)\overline{I}_d\overline{I} + \left(\frac{1-\cos\beta}{2}\right)\rho \qquad (40.17)$$

where
$\overline{I}_d$ and $\overline{I}$ are the hourly diffuse and global radiation on the horizontal surface
$r_{b,T}$ is the ratio of hourly beam radiation on the tilted surface to that on a horizontal surface (this is a purely astronomical quantity and can be calculated accurately from geometric considerations)
ρ is the ground albedo

Example 40.3

Compute the total energy collected during 11:30–12:30 for the month of September in New York, NY (latitude: 40.75°N, T_a = 20°C) by a flat-plate solar collector of 5 m² area having zero tilt. The collector performance parameters are $F_R\eta_0$ = 0.54 and F_RU_L = 3.21 W/(m² °C) and the collector inlet temperature is 80°C. The corresponding hourly mean clearness index $\overline{k}$ is 0.44, and the monthly mean hourly radiation on a horizontal surface I_{Ti} (11:30–12:30) is 6.0 MJ/(m² h).

From Equation 40.7a, critical radiation $I_C = 3.21 \times (80 - 20)/0.54 = 356.7\ \text{W/m}^2 = 1.28\ \text{MJ/(m}^2\,\text{h)}$. For the average day of September, solar declination $\delta = 2.2°$. Also, because the collector is horizontal $\bar{r}_T = 1$ and $\beta = 0$. Thus from Equation 40.16

$$X_{max} = 1.85 + 0.169/0.44^2 - 0.0696/0.44^2 - 0.981 \times 0.44/(\cos 2.2)^2 = 1.93.$$

Also from Equation 40.15, $a = (1.93 - 1)/(2 - 1.93) = 13.29$.
The critical radiation ratio $X_C = 1.28/1.93 = 0.663$.
Because $X_C < X_{max}$, from Equation 40.14 we have

$$\varphi_i(X_C) = |13.29 - [13.29^2 + (1 + 2 \times 13.29)(1 - 0.663/1.93)^2]^{1/2}| = |13.29 - 13.73| = 0.44.4.$$

Finally, the total energy collected is given by Equation 40.11

$$q_{CN}(11{:}30\text{–}12{:}30) = 5 \times 0.54 \times 60 \times 30 \times 0.44 = 214\ \text{MJ/h}$$

40.3.3 Daily Utilizability

40.3.3.1 Basis

In this mode, T_{Ci}, and hence the critical radiation level, is assumed constant during all hours of the day. The *total* useful energy over N days that can be collected by solar collectors operated all day over n hours is given by

$$Q_{CN} = A_C F_R \bar{\eta}_0 \bar{H}_T N \bar{\phi} \tag{40.18}$$

where
$\bar{H}_T$ is the average daily global radiation on the collector surface
$\bar{\phi}$ (called Phibar) is the daily utilizability factor, defined as

$$\bar{\phi} = \sum^{N}\sum^{n}(I_T - I_C)^+ / \sum^{N}\sum^{n} I_T = \frac{1}{Nn}\sum^{N}\sum^{n}(X_i - X_C)^+ \tag{40.19}$$

Generalized correlations have been developed both at monthly time scales and for annual time scales based on the parameter $\bar{K}$. Generalized (i.e., location and month independent) correlations for $\bar{\phi}$ on a *monthly* time scale have been proposed by Theilacker and Klein (1980). These are strictly applicable for flat-plate collectors only. Collares-Pereira and Rabl (1979) have also proposed generalized correlations for $\bar{\phi}$ on a monthly time scale which, though a little more tedious to use are applicable to concentrating collectors as well. The reader may refer to Rabl (1985) or Reddy (1987) for complete expressions.

40.3.3.2 Monthly Time Scales

The Phibar method of determining the daily utilizability fraction proposed by Theilacker and Klein (1980) correlates $\bar{\phi}$ to the following factors:

1. A geometry factor $\bar{R}_T/\bar{r}_{T,noon}$ which incorporates the effects of collector orientation, location, and time of year. $\bar{R}_T$ is the ratio of monthly average global radiation on the tilted surface to that on a horizontal surface. $\bar{r}_{T,noon}$ is the ratio of radiation at noon on the tilted surface to that on a horizontal surface for the average day of the month.

Geometrically, $\bar{r}_{T,noon}$ is a measure of the maximum height of the radiation curve over the day, whereas $\bar{R}_T$ is a measure of the enclosed area. Generally the value $(\bar{R}_T/\bar{r}_{T,noon})$ is between 0.9 and 1.5.

2. A dimensionless critical radiation level $\bar{X}_{C,K}$ where

$$\bar{X}_{C,K} = \frac{I_C}{\bar{I}_{T,noon}} \tag{40.20}$$

with $\bar{I}_{T,noon}$, the radiation intensity on the tilted surface at noon, given by

$$\bar{I}_{T,noon} = \bar{r}_{noon}\bar{r}_{T,noon}\bar{H} \tag{40.21}$$

where $\bar{r}_{noon}$ is the ratio of radiation at noon to the daily global radiation on a horizontal surface during the mean day of the month which can be calculated from the following correlation proposed by Liu and Jordan (1960):

$$r(W) = \frac{I(W)}{H} = \frac{\pi}{24}(a + b\cos W)\frac{(\cos W - \cos W_S)}{\left(\sin W_S - \dfrac{\pi}{180} W_S \cos W_S\right)} \tag{40.22}$$

with

$$a = 0.409 + 0.5016\sin(W_S - 60)$$

$$b = 0.6609 - 0.4767\sin(W_S - 60)$$

where

W is the hour angle corresponding to the midpoint of the hour (in degrees)
W_s is the sunset hour angle given by

$$\cos W_s = -\tan L \tan \delta \tag{40.23}$$

where L is the latitude of the location. The fraction r is the ratio of hourly to daily global radiation on a horizontal surface. The factors $\bar{r}_{T,noon}$ and $\bar{r}_{noon}$ can be determined from Equations 40.17 and 40.22, respectively, with $W = 0°$.

The Theilacker and Klein correlation for the daily utilizability for equator-facing flat-plate collectors is

$$\bar{\phi}(X_{C,K}) = \exp\{[a' + b'(\bar{r}_{T,noon}/\bar{R}_T)][X_{C,K} + c'X_{C,K}^2]\} \tag{40.24}$$

where

$$\begin{aligned} a' &= 7.476 - 20.00\bar{K} + 11.188\bar{K}^2 \\ b' &= -8.562 + 18.679\bar{K} - 9.948\bar{K}^2 \\ c' &= -0.722 + 2.426\bar{K} + 0.439\bar{K}^2 \end{aligned} \tag{40.25}$$

How $\bar{\phi}$ varies with the critical radiation ratio $\bar{X}_{C,K}$ for three different values of $\bar{K}$ is shown in Figure 40.13.

Example 40.4

A flat-plate collector operated horizontally at Fort Worth, Texas ($L = 32.75°N$), has a surface area of 20 m². It is used to heat 10 kg/min of water entering the collector at a constant temperature of 80° C each day from 6 a.m. to 6 p.m. The collector performance parameters are $F_R\eta_0 = 0.70$ and $F_RU_L = 5.0$ W/(m² °C). Use Klein's correlation to compute the energy collected by the solar collectors during September. Assume $\bar{H} = 18.28$ MJ/(m²–d), $\bar{K} = 0.57$ and $\bar{T}_a = 25°C$. Assume the mean sunset hour angle for September to be 90°.

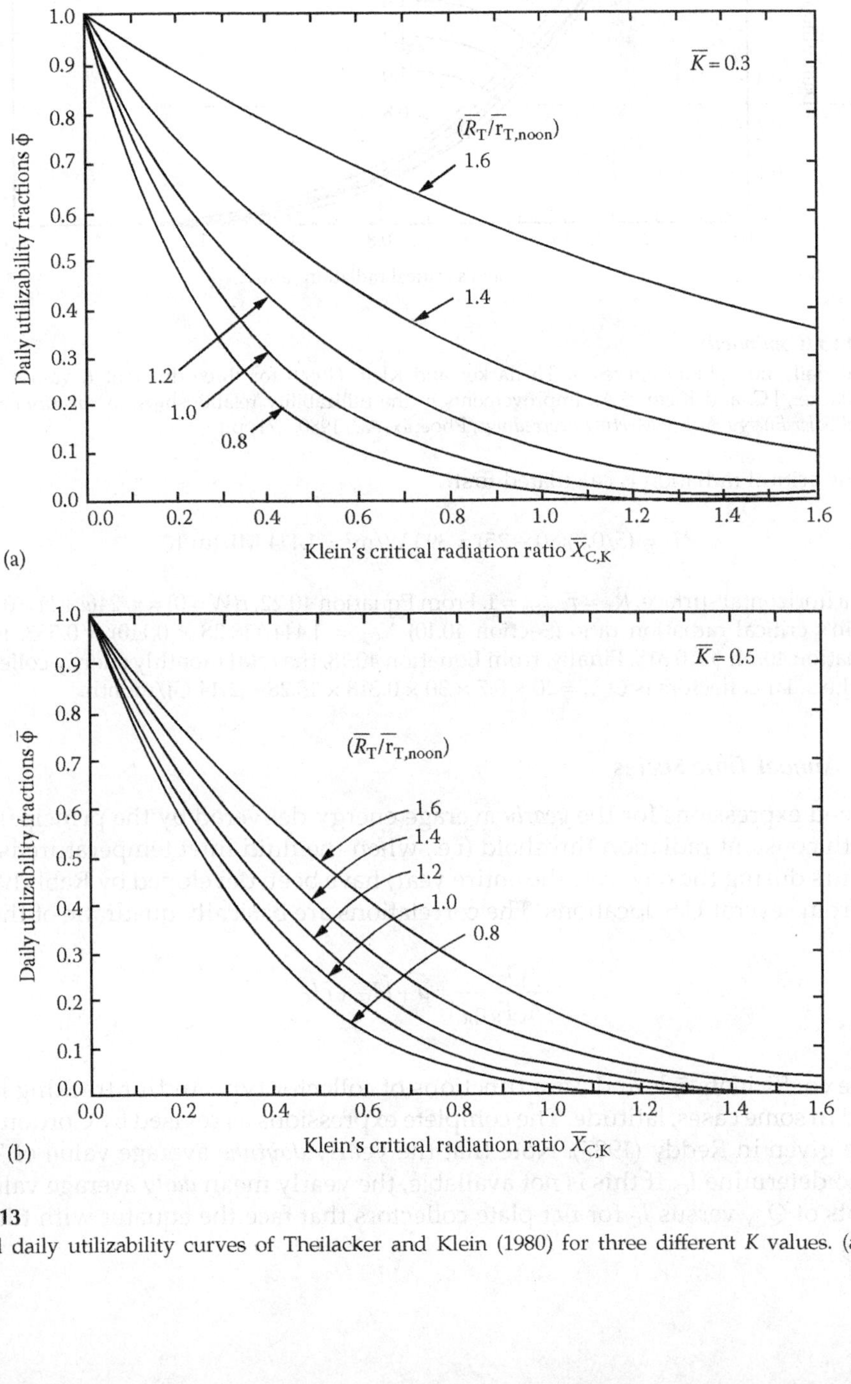

FIGURE 40.13
Generalized daily utilizability curves of Theilacker and Klein (1980) for three different K values. (a) $\bar{K} = 0.3$, (b) $\bar{K} = 0.5$. *(Continued)*

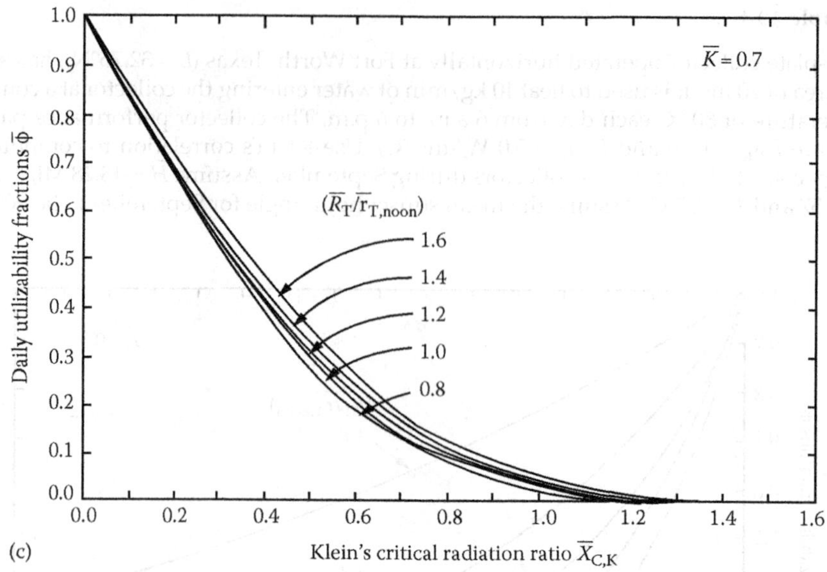

FIGURE 40.13 (*Continued*)
Generalized daily utilizability curves of Theilacker and Klein (1980) for three different K values. (c) $\bar{K} = 0.7$. (From Theilacker, J.C. and Klein, S.A., Improvements in the utilizability relationships, in *American Section of the International Solar Energy Society Meeting Proceedings*, Phoenix, AZ, 1980, 271pp.)

The critical radiation is calculated first:

$$I_C = (5/0.7)(80 - 25) = 393\ \text{W/m}^2 = 1.414\ \text{MJ/(m}^2\text{h)}$$

For a horizontal surface, $\bar{R}_T = \bar{r}_{T,noon} = 1$. From Equation 40.22, $r(W = 0) = \pi/24(a + b) = 0.140$. Klein's critical radiation ratio (Section 40.10) $\bar{X}_{C,K} = 1.414/(18.28 \times 0.140) = 0.553$. From Equation 40.24, $\bar{\phi} = 0.318$. Finally, from Equation 40.18, the total monthly energy collected by the solar collectors is $Q_{CM} = 20 \times 0.7 \times 30 \times 0.318 \times 18.28 = 2.44$ GJ/month.

40.3.3.3 Annual Time Scales

Generalized expressions for the *yearly* average energy delivered by the principal collector types with constant radiation threshold (i.e., when the fluid inlet temperature is constant for all hours during the day over the entire year) have been developed by Rabl (1981) based on data from several U.S. locations. The correlations are basically quadratic of the form

$$\frac{Q_{CY}}{A_C F_R \eta_n} = \tilde{a} + \tilde{b} I_C + \tilde{c} I_C^2 \qquad (40.26)$$

where the coefficients $\tilde{a}$, $\tilde{b}$, and $\tilde{c}$ are functions of collector type and/or tracking mode, climate, and in some cases, latitude. The complete expressions as revised by Gordon and Rabl (1982) are given in Reddy (1987). Note that the yearly *daytime* average value of T_a should be used to determine I_C. If this is not available, the yearly mean *daily* average value can be used. Plots of Q_{CY} versus I_C for flat-plate collectors that face the equator with tilt equal to

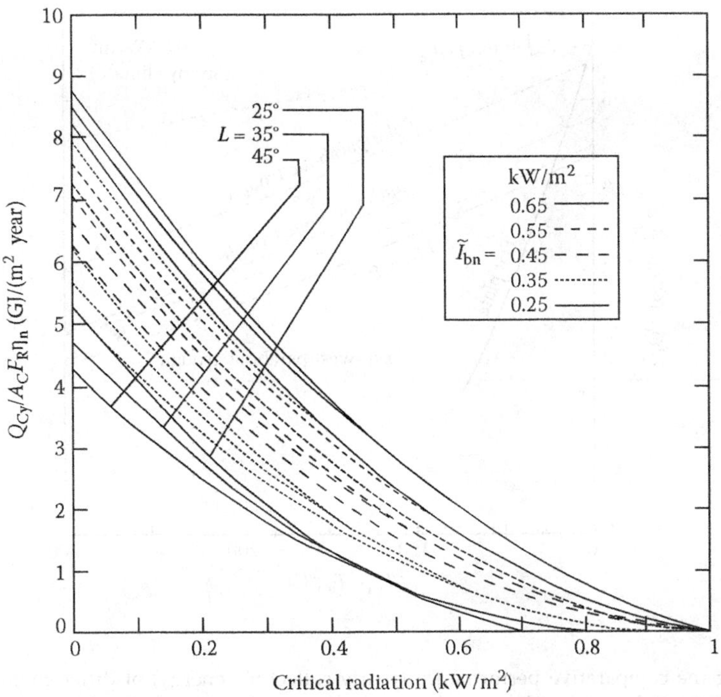

FIGURE 40.14
Yearly total energy delivered by flat-plate collectors with tilt equal to latitude. (From Gordon, J.M. and Rabl, A., *Solar Energy*, 28, 519, 1982.)

the latitude are shown in Figure 40.14. The solar radiation enters these expressions as $\tilde{I}_{bn}$, the annual average beam radiation at normal incidence. This can be estimated from the following correlation

$$\tilde{I}_{bn} = 1.37\tilde{K} - 0.34 \quad (40.27)$$

where
$\tilde{I}_{bn}$ is in kW/m²
$\tilde{K}$ is the annual average clearness index of the location

Values of $\tilde{K}$ for several locations worldwide are given in Reddy (1987).

This correlation is strictly valid for latitudes ranging from 25° to 48°. If used for lower latitudes, the correlation is said to lead to overprediction. Hence, it is recommended that for such lower latitudes a value of 25° be used to compute Q_{Cy}.

A direct comparison of the yearly performance of different collector types is given in Figure 40.15. A latitude of 35°N is assumed and plots of Q_{Cy} U.S. $(T_{Ci} - T_a)$ have been generated in a sunny climate with I_{bn} = 0.6 kW/m². Relevant collector performance data are given in Figure 40.15. The crossover point between flat-plate and concentrating collectors is approximately 25°C above ambient temperature whether the climate is sunny or cloudy.

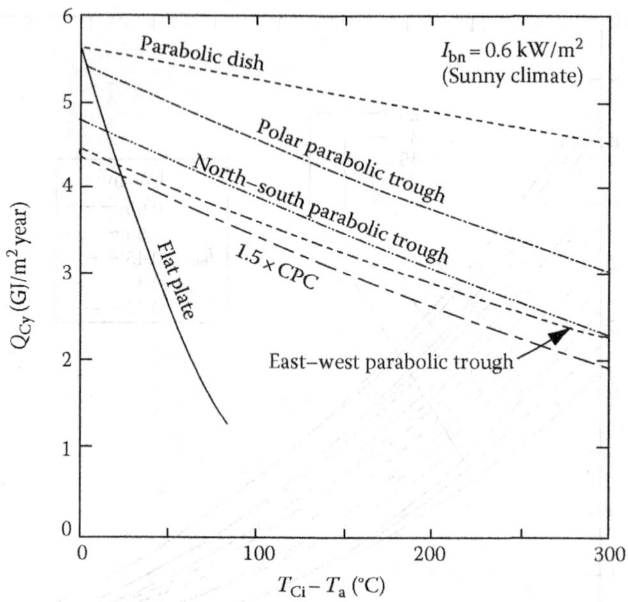

FIGURE 40.15
Figure illustrating the comparative performance (yearly collectible energy) of different collector types as a function of the difference between collector inlet temperature and ambient collector performance parameters $F'\eta_0$ and $F'U_L$ in W/(m² °C) are: flat plate (0.70 and 5.0), CPC (0.60 and 0.75), parabolic trough (0.65 and 0.67), and parabolic dish (0.61 and 0.27). (From Rabl, A., Yearly average performance of the principal solar collector types, *Solar Energy*, 27, 215, 1981.)

40.4 Solar Systems

40.4.1 Classification

Solar thermal systems can be divided into two categories: standalone or solar supplemented. They can be further classified by means of energy collection as active or passive, by their use as residential or industrial. Further, they can be divided by collector type into liquid or air systems, and by the type of storage they use into seasonal or daily systems.

40.4.1.1 Standalone and Solar Supplemented Systems

Standalone systems are systems in which solar energy is the only source of energy input used to meet the required load. Such systems are normally designed for applications where a certain amount of tolerance is permissible concerning the load requirement; in other words, where it is not absolutely imperative that the specified load be met each and every instant. This leniency is generally admissible in the case of certain residential and agricultural applications. The primary reasons for using such systems are their low cost and simplicity of operation.

Solar-supplemented systems, widely used for both industrial and residential purposes, are those in which solar energy supplies part of the required heat load, the rest being met by an auxiliary source of heat input. Due to the daily variations in incident solar radiation, the portion of the required heat load supplied by the solar energy system may vary from day-to-day. However, the auxiliary source is so designed that at any instant it is capable of meeting the remainder of the required heat load. It is normal practice to incorporate an auxiliary heat source large enough to supply the entire heat load required. Thus, the benefit in the solar subsystem is not in its capacity credit (i.e., not that a smaller capacity conventional system can be used), but rather that a part of the conventional fuel consumption is displaced. The solar subsystem thus acts as a fuel economizer.

Solar-supplemented energy systems will be the primary focus of this chapter. Designing such systems has acquired a certain firm scientific rationale, and the underlying methodologies have reached a certain maturity and diversity, which may satisfy professionals from allied fields. On the other hand, unitary solar apparatus are not discussed here, since these are designed and sized based on local requirements, material availability, construction practices, and practical experience. Simple rules of thumb based on prior experimentation are usually resorted to for designing such systems.

40.4.1.2 Active and Passive Systems

Active systems are those systems that need electric pumps or blowers to collect solar energy. It is evident that the amount of solar energy collected should be more than the electrical energy used. Active systems are invariably used for industrial applications and for most domestic and commercial applications as well. *Passive systems* are those systems that collect or use solar energy without direct recourse to any source of conventional power, such as electricity, to aid in the collection. Thus, either such systems operate by natural thermosyphon (for example, domestic water heating systems) between collector, storage, and load or, in the case of space heating, the architecture of the building is such as to favor optimal use of solar energy. Use of a passive system for space heating applications, however, in no way precludes the use of a backup auxiliary system. This chapter deals with active solar systems only.

40.4.1.3 Residential and Industrial Systems

Basically, the principles and the components used in these two types of systems are alike, the difference being in the load distribution, control strategies, and relative importance of the components with respect to each other. Whereas *residential* loads have sharp peaks in the early morning or in the evening and have significant seasonal variations, industrial loads tend to be fairly uniform over the year. Constant loads favor the use of solar energy because good equipment utilization can be achieved. Because of differences in load distribution, the role played by the storage differs for both applications. Residential loads often occur at times when solar radiation is no longer available. Thus the collector and the storage subsystems interact in a mode without heat withdrawal from the storage. Finally, for economic reasons, many residential systems are designed to operate by natural thermosyphon, in which case no pumps or controls are needed.

On the other hand, for *industrial and commercial* applications, there is no a priori relationship between the time dependence of the load and the period of sunshine. Moreover, a high reliability has to be assured, so the solar system will have to be combined with a conventional system. Very often, a significant portion of the load can be directly supplied by the solar system even without storage. Another option is to use buffer storage for short periods, on the order of a few hours, in case of discontinuous batch process loads. Thus, the proper design of the storage component has to be given adequate consideration. At present, due to economic constraints as well as the fact that proper awareness of the various installations and operational difficulties associated with larger solar thermal systems is still lacking, solar thermal systems are normally designed either (1) with the no-storage option, or (2) with buffer storage where a small fraction of the total heat demand is only supplied by the solar system.

40.4.1.4 Liquid and Air Collectors

Although air has been the primary fluid for space heating and drying applications, solar air heating systems have until recently been relegated to second place, mainly as a result of the engineering difficulties associated with such systems. Also, applications involving hot air are probably less common than those needing hot water. Air systems for space heating are well described by Löf (1981).

Even with liquid solar collectors, various configurations are possible, and these can be classified basically as *nontracking* (which include flat-plate collectors and CPCs) or *tracking* collectors (which include various types of concentrating collectors). For low-grade thermal heat, for which solar energy is most suited, flat-plate collectors are far more appropriate than concentrating collectors, not only because of their lower cost but also because of their higher thermal efficiencies at low temperature levels. Moreover, their operation and maintenance costs are lower. Finally, for locations having a high fraction of diffuse radiation, as in the tropics, flat-plate collectors are considered to be thermally superior because they can make use of diffuse radiation as well as beam radiation. Although the system design methodologies presented in this chapter explicitly assume flat-plate collector systems, these design approaches can be equally used with concentrating collectors.

40.4.1.5 Daily and Seasonal Storage

By *daily storage* is meant systems having capacities equivalent to at most a few days of demand (i.e., just enough to tide over day-to-day climatic fluctuations). In *seasonal storage,* solar energy is stored during the summer for use in winter. Industrial demand loads, which are more or less uniform over the year, are badly suited for seasonal-storage systems. This is also true of air-conditioning for domestic and commercial applications because the load is maximum when solar radiation is also maximum, and vice versa. The present-day economics of seasonal storage units do not usually make such systems an economical proposition except for community heating in cold climates.

40.4.2 Closed-Loop and Open-Loop Systems

The two possible configurations of solar thermal systems with daily storage are classified as closed-loop or open-loop systems. Though different authors define these differently, we shall define these as follows. A *closed-loop system* has been defined as a circuit in which the performance of the solar collector is directly dependent on the storage temperature. Figure 40.16 gives a schematic of a closed-loop system in which the fluid circulating in the collectors does not mix with the fluid supplying thermal energy to the load. Thus, these two subsystems are distinct in the sense that any combination of fluids (water or air) is theoretically feasible (a heat exchanger, as shown in the figure, is of course imperative when the fluids are different). However, in practice, only water–water, water–air, or air–air combinations are used. From the point of system performance, the storage temperature normally varies over the day and, consequently, so does collector performance. Closed-loop system configurations have been widely used to date for domestic hot water

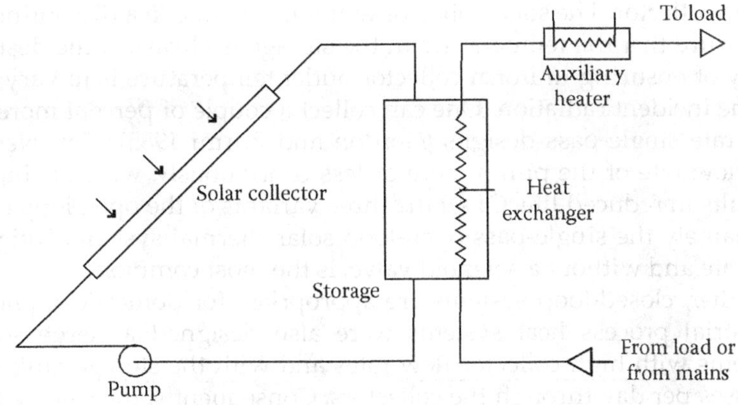

FIGURE 40.16
Schematic of a closed-loop solar system.

and space heating applications. The flow rate per unit collector area is generally around 50 kg/(h m^2) for liquid collectors. The storage volume makes about 5–10 passes through the collector during a typical sunny day, and this is why such systems are called *multipass* systems. The temperature rise for each pass is small, of the order of 2°C–5°C for systems with circulating pumps and about 10°C for thermosyphon systems. An expansion tank and a check valve to prevent reverse thermosyphoning at nights, although not shown in the figure, are essential for such system configurations.

Figure 40.17 illustrates one of the possible configurations of *open-loop systems*. Open-loop systems are defined as systems in which the collector performance is independent of the storage temperature. The working fluid may be rejected (or a heat recuperator can be used) if contaminants are picked up during its passage through the load. Alternatively, the working fluid could be directly recalculated back to the entrance of the solar collector field. In all these open-loop configurations, the collector is subject to a given or known inlet temperature specified by the load requirements.

If the working fluid is water, instead of having a continuous flow rate (in which case the outlet temperature of the water will vary with isolation), a solenoid valve can be placed just at the exit of the collector, set so as to open when the desired temperature level of the fluid in the collector is reached. The water is then discharged into storage, and fresh water

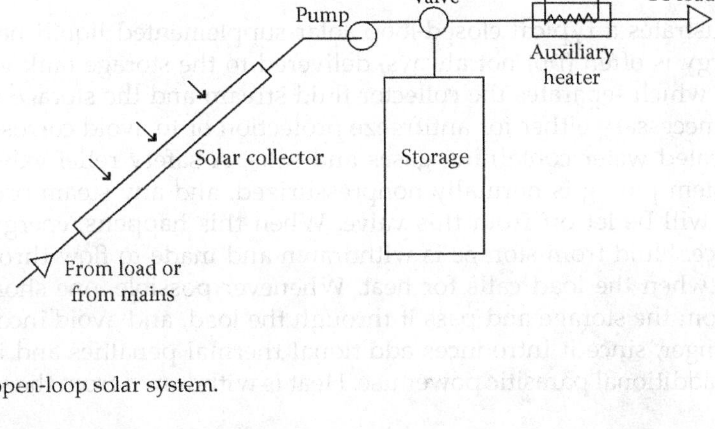

FIGURE 40.17
Schematic of an open-loop solar system.

is taken into the collector. The solar collector will thus operate in a discontinuous manner, but this will ensure that the temperature in the storage is always at the desired level. An alternative way of ensuring uniform collector outlet temperature is to vary the flow rate according to the incident radiation. One can collect a couple of percent more energy than with constant rate single-pass designs (Gordon and Zarmi 1985). However, this entails changing the flow rate of the pump more or less continuously, which is injurious to the pump and results in reduced life. Of all the three variants of the open-loop configuration, the first one, namely the single-pass open-loop solar thermal system configuration with constant flow rate and without a solenoid valve, is the most common.

As stated earlier, closed-loop systems are appropriate for domestic applications. Until recently, industrial process heat systems were also designed as large solar domestic hot-water systems with high collector flow rates and with the storage tank volume making several passes per day through the collectors. Consequently, the storage tank tends to be fairly well mixed. Also the tank must be strong enough to withstand the high pressure from the water mains. The open-loop single-pass configuration, wherein the required average daily fluid flow is circulated just once through the collectors with the collector inlet temperature at its lowest value, has been found to be able to deliver as much as 40% more yearly energy for industrial process heat applications than the multipass designs (Collares-Pereira et al. 1984). Finally, in a closed-loop system where an equal amount of fresh water is introduced into storage whenever a certain amount of hot water is drawn off by the load, it is not possible to extract the entire amount of thermal energy contained in storage since the storage temperature is continuously reduced due to mixing. This *partial depletion effect* in the storage tank is not experienced in open-loop systems. The penalty in yearly energy delivery ranges typically from about 10% for daytime-only loads to around 30% for nighttime-only loads compared to a closed-loop multipass system where the storage is depleted every day. Other advantages of open-loop systems are (1) the storage tank need not be pressurized (and hence is less costly), and (2) the pump size and parasitic power can be lowered.

A final note of caution is required. The single-pass design is not recommended for *variable* loads. The tank size is based on yearly daily load volumes, and efficient use of storage requires near-total depletion of the daily collected energy each day. If the load draw is markedly lower than its average value, the storage would get full relatively early the next day and solar collection would cease. It is because industrial loads tend to be more uniform, both during the day and over the year, than domestic applications that the single-pass open-loop configuration is recommended for such applications.

40.4.2.1 Description of a Typical Closed-Loop System

Figure 40.18 illustrates a typical closed-loop solar-supplemented liquid heating system. The useful energy is often (but not always) delivered to the storage tank via a collector-heat exchanger, which separates the collector fluid stream and the storage fluid. Such an arrangement is necessary either for antifreeze protection or to avoid corrosion of the collectors by untreated water containing gases and salts. A safety relief valve is provided because the system piping is normally nonpressurized, and any steam produced in the solar collectors will be let off from this valve. When this happens, energy dumping is said to take place. Fluid from storage is withdrawn and made to flow through the load-heat exchanger when the load calls for heat. Whenever possible, one should withdraw fluid directly from the storage and pass it through the load, and avoid incorporating the load-heat exchanger, since it introduces additional thermal penalties and involves extra equipment and additional parasitic power use. Heat is withdrawn from the storage tank at

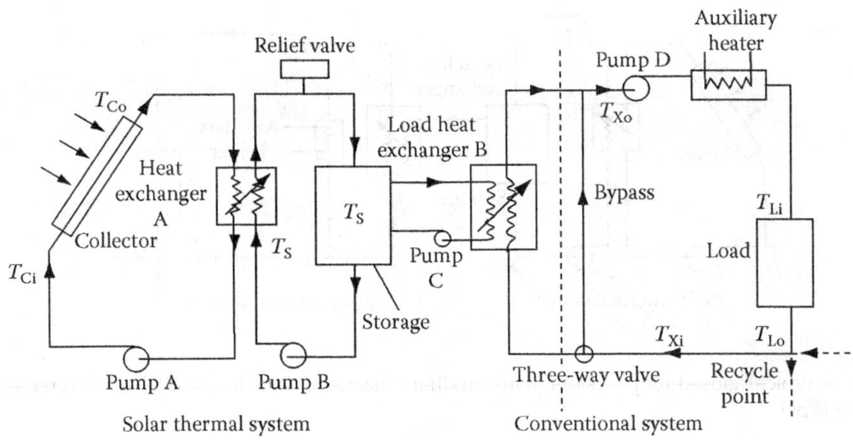

FIGURE 40.18
Schematic of a typical closed-loop system with auxiliary heater placed in series (also referred to as a topping-up type).

the top and reinjected at the bottom in order to derive maximum benefit from the thermal stratification that occurs in the storage tank. A bypass circuit is incorporated prior to the load heat exchanger and comes into play

1. When there is no heat in the storage tank (i.e., storage temperature T_S is less than the fluid temperature entering the load heat exchanger T_{Xi})
2. When T_S is such that the temperature of the fluid leaving the load heat exchanger is greater than that required by the load (i.e., $T_{Xo} > T_{Li}$, in which case the three-way valve bypasses part of the flow so that $T_{Xo} = T_{Li}$). The bypass arrangement is thus a differential control device which is said to modulate the flow such that the above condition is met. Another operational strategy for maintaining $T_{Xo} = T_{Li}$ is to operate the pump in a "bang-bang" fashion (i.e., by short cycling the pump). Such an operation is not advisable, however, since it would lead to premature pump failure.

An auxiliary heater of the *topping-up type* supplies just enough heat to raise T_{Xo} to T_{Li}. After passing through the load, the fluid (which can be either water or air) can be recirculated or, in case of liquid contamination through the load, fresh liquid can be introduced. The auxiliary heater can also be placed in parallel with the load (see Figure 40.19), in which case it is called an *all-or-nothing type*. Although such an arrangement is thermally less efficient than the topping-up type, this type is widely used during the solar retrofit of heating systems because it involves little mechanical modifications or alterations to the auxiliary heater itself.

It is obvious that there could also be solar-supplemented energy systems that do not include a storage element in the system. Figure 40.20 shows such a system configuration with the auxiliary heater installed in series. The operation of such systems is not very different from that of systems with storage, the primary difference being that whenever instantaneous solar energy collection exceeds load requirements (i.e., $T_{Co} > T_{Li}$), energy dumping takes place. It is obvious that by definition there cannot be a closed-loop, no-storage solar thermal system. Solar thermal systems without storage are easier to construct and operate, and even though they may be effective for 8–10 h a day, they are appropriate for applications such as process heat in industry.

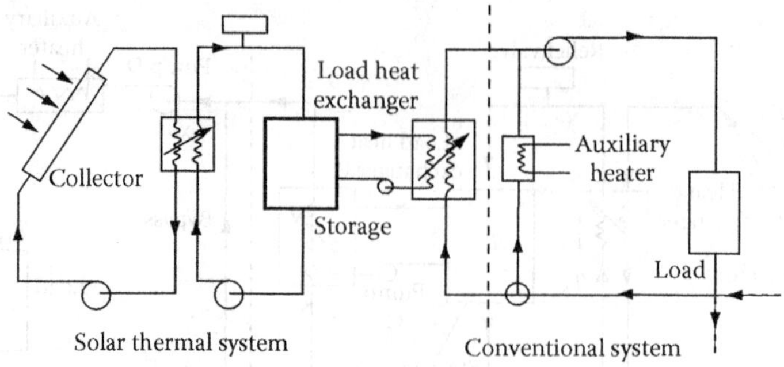

FIGURE 40.19
Schematic of a typical closed-loop system with auxiliary heater placed in parallel (also referred to as an all-or-nothing type).

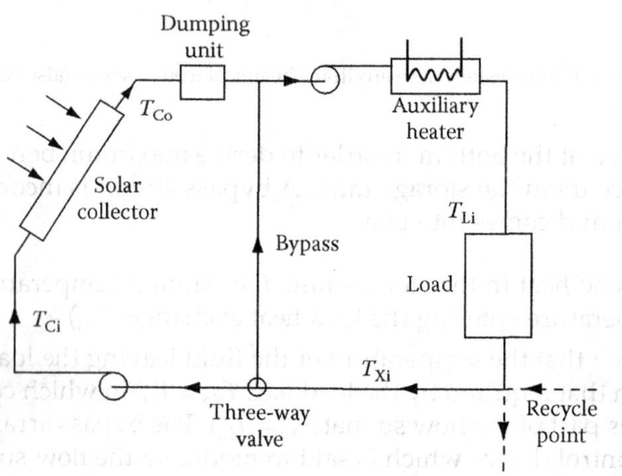

FIGURE 40.20
Simple solar thermal system without storage.

Active closed-loop solar systems as described earlier are widely used for service hot-water systems, that is, for domestic hot water and process heat applications as well as for space heat. There are different variants to this generic configuration. A system without the collector-heat exchanger is referred to having collectors *directly coupled* to the storage tank (as against *indirect coupling* as in Figure 40.16). For domestic hot-water systems, the system can be simplified by placing the auxiliary heater (which is simply an electric heater) directly inside the storage tank. One would like to maintain stratification in the tank so that the coolest fluid is at the bottom of the storage tank, thereby enhancing collection efficiency. Consequently, the electric heater is placed at about the upper third portion of the tank so as to assure good collection efficiency while assuring adequate hot water supply to the load. A more efficient but expensive option is widely used in the United States: the *double tank system*, shown in Figure 40.21. Here the functions of solar storage and auxiliary heating are separated, with the solar tank acting as a preheater for the conventional gas or electric unit. Note that a further system simplification can be achieved for domestic applications by placing the load heat exchanger directly inside the storage tank. In certain cases, one can even eliminate the heat exchanger completely.

Another system configuration is the *drain-back* (also called drain-out) system, where the collectors are emptied each time the solar system shuts off. Thus the system invariably

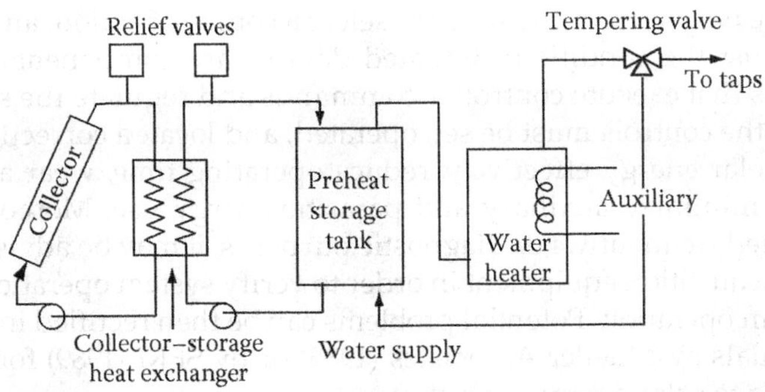

FIGURE 40.21
Schematic of a standard domestic hot-water system with double tank arrangement. (From Duffie, J.A. and Beckman, W.A., *Solar Engineering of Thermal Processes*, Wiley Interscience, New York, 1980.)

loses collector fluid at least once, and often several times, each day. No collector-heat exchanger is needed, and freeze protection is inherent in such a configuration. However, careful piping design and installation, as well as a two-speed pump, are needed for the system to work properly (Newton and Gilman 1981). The drain-back configuration may be either open (vented to atmosphere) or closed (for better corrosion protection). Long-term experience in the United States with the drain-back system has shown it to be very reliable if engineered properly. A third type of system configuration is the *drain-down* system, where the fluid from the collector array is removed only when adverse conditions, such as freezing or boiling, occur. This design is used when freezing ambient temperatures are only infrequently encountered.

Active solar systems of the type described above are mostly used in countries such as the United States and Canada. Countries such as Australia, India, and Israel (where freezing is rare) usually prefer thermosyphon systems. No circulating pump is needed, the fluid circulation being driven by density difference between the cooler water in the inlet pipe and the storage tank and the hotter water in the outlet pipe of the collector and the storage tank. The low fluid flow in thermosyphon systems enhances thermal stratification in the storage tank. The system is usually fail-proof, and a study by Liu and Fanney (1980) reported that a thermosyphon system performed better than several pumped service hot-water systems. If operated properly, thermosyphon and active solar systems are comparable in their thermal performance. A major constraint in installing thermosyphon systems in already existing residences is the requirement that the bottom of the storage tank be at least 20 cm or more higher than the top of the solar collector in order to avoid reverse thermosyphoning at night. To overcome this, spring-loaded one-way valves have been used, but with mixed success.

40.5 Controls

There are basically five categories to be considered when designing automatic controls (Mueller Associates 1985): (1) collection to storage, (2) storage to load, (3) auxiliary energy to load, (4) miscellaneous (i.e., heat dumping, freeze protection, overheating, etc.), and (5) alarms. The three major control system components are sensors, controllers, and actuating devices. Sensors are used to detect conditions (such as temperatures, pressures, etc.).

Controllers receive output from the sensors, select a course of action, and signal a system component to adjust the condition. Actuated devices are components such as pumps, valves, and alarms that execute controller commands and regulate the system.

The sensors for the controls must be set, operated, and located correctly if the solar system is to collect solar energy effectively, reduce operating time, wear and tear of active components, and minimize auxiliary and parasitic energy use. Moreover, sensors also need to be calibrated frequently. For diagnostic purposes, it may be advisable to add extra sensors and data acquisition equipment in order to verify system operation and keep track of long-term system operation. Potential problems can be then rectified in time. The reader may refer to manuals by Mueller Associates (1985) or by SERI (1989) for more details on controls pertaining to solar energy systems.

Though single-point temperature controllers or solar-cell-activated controls have been used for activated solar collectors, the best way to do so is by differential temperature controllers. Temperature sensors are used to measure the fluid temperature at collector outlet and at the bottom of the storage tank. When the difference is greater than a set amount, say 5°C, then the controller turns the pump on. If the pump is running and the temperature difference falls below another preset value, say 1°C, the controller stops the pump. The temperature deadband between switching-off and reactivating levels should be set with care, since too high a deadband would adversely affect collection efficiency and too low a value would result in short cycling of the collector pump. Figure 40.22 taken from CSU (1980), shows typical diurnal temperature variations of the liquids at collector exit T_1 and in the storage bottom T_3 as a result of heat withdrawal and/or heat losses from the storage. At about 8:30 a.m., $T_1 > T_3$ and, since there is no flow in the collector, T_1 increases rapidly until the difference $(T_1 - T_3)$ reaches the preset activation level (shown as point 1). The collector pump A comes on, and liquid circulation through the collector begins. Because of this cold water surge, T_1 decreases, resulting in a drop of $(T_1 - T_3)$ to the preset deactivating level (shown as point 2). The pump switches off, and liquid flow through the collectors stops. Gradually T_1 increases again, and so on. The number of on-off cycles at system start-up depends on solar intensity, fluid flow rate, volume of water in the collector loop, and the differential controller setting. A similar phenomenon of cycling also occurs in the afternoon. However, the error introduced in solar collector long-term performance predictions by neglecting this cycling effect in the modeling equations is usually small.

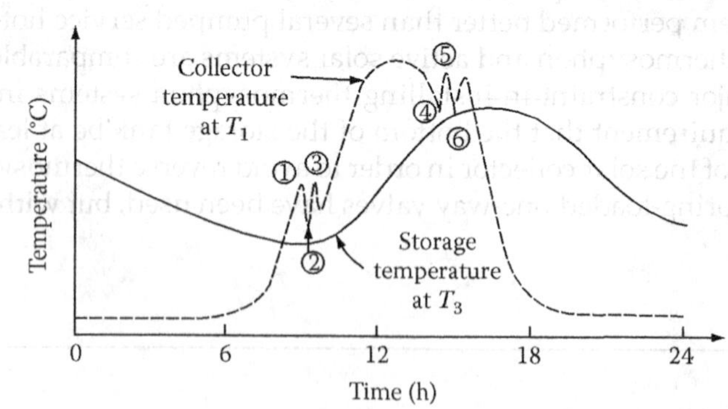

FIGURE 40.22
Typical diurnal variation of collector and storage temperatures. Points 1–3 represent the start-up, shut-down, and re-start cycling operation of the circulating pump at the beginning of the day; points 4–6 represent the shut-down cycling behavior at the end of the day. (From CSU, Solar heating and cooling of residential buildings—Design of systems, Manual prepared by the Solar Energy Applications Laboratory, Colorado State University, Fort Collins, CO, 1980.)

40.5.1 Corrections to Collector Performance Parameters

40.5.1.1 Combined Collector-Heat Exchanger Performance

The use of the heat exchanger A in Figure 40.18 imposes a penalty on the performance of the solar system because T_{Ci} is always higher than T_S, thereby decreasing q_C (see Figure 40.23). The collector-heat exchanger can be implicitly accounted for by suitably modifying the collector performance parameters. Recall from basic heat transfer the concept of heat exchanger effectiveness E defined as the ratio of the actual heat transfer rate to the maximum possible heat transfer rate, that is,

$$E = \frac{(mc_p)_a(T_{ai} - T_{ao})}{(mc_p)_{min}(T_{ai} - T_{bi})} \tag{40.28a}$$

$$= \frac{(mc_p)_b(T_{bo} - T_{bi})}{(mc_p)_{min}(T_{ai} - T_{bi})} \tag{40.28b}$$

where

$(mc_p)_x$ is the capacitance rate of fluid X (with X = a for the warmer fluid, or X = b for the cooler fluid)

$(mc_p)_{min}$ is the lower heat capacitance value of either stream

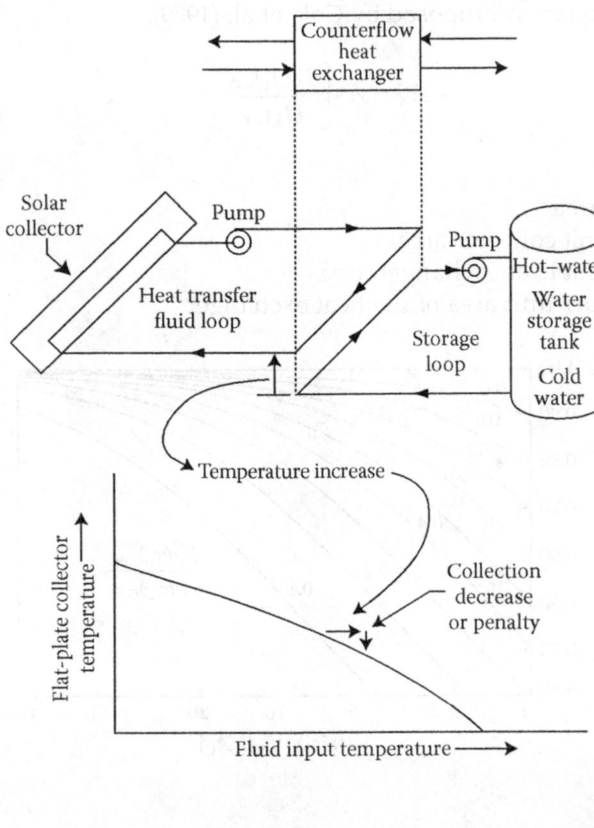

FIGURE 40.23
Heat collection decrease caused by double-loop heat exchangers. (From Cole, R.L. et al. (eds.), Design and installation manual for thermal energy storage, ANL-79-15, Argonne National Laboratory, Argonne, IL, 1979.)

The advantage of this modeling approach is that, to a good approximation, E can be considered constant in spite of variations in temperature levels provided the mass flow rates of both fluids remain constant. Thus, knowing the two flow rates, E, T_{ai}, and T_{bi}, both the exit fluid temperatures can be conveniently deduced. De Winter (1975) has shown that the combined performance of the solar collector and the heat exchanger can be conveniently modeled by replacing the collector heat removal factor F_R by a combined collector-exchanger heat removal factor F_R' such that

$$\frac{F_R'}{F_R} = \left[1 + \frac{F_R U_L A_C}{(mc_p)_C}\left\{\frac{(mc_p)_C}{E_A (mc_p)\text{min}} - 1\right\}\right]^{-1} \tag{40.29}$$

where

$(mc_p)_C$ is the capacitance rate of the fluid through the collector
E_A is the effectiveness of heat exchanger A

The variation of F_R'/F_R is shown in Figure 40.24. The plots exhibit the same type of asymptotic behavior with mass flow rate as in Figure 40.3.

The design of the collector-heat exchanger also requires care if the penalty imposed by it on the solar collection is to be minimized. Using a large heat exchanger increases the effectiveness and lowers this penalty; that is, the ratio (F_R'/F_R) is high, but the associated initial and operating costs may be higher. Both these considerations need to be balanced for optimum design (see Figure 40.25). Optimum heat exchanger area A_X can be found from the following equation proposed by Cole et al. (1979):

$$A_X = A_C \left[\frac{F_R U_L C_C}{U_X C_X}\right]^{1/2} \tag{40.30}$$

where

A_C is the collector area
C_C is the cost per unit collector area
C_X is the cost per unit heat exchanger area
U_X is the heat loss per unit area of the heat exchanger

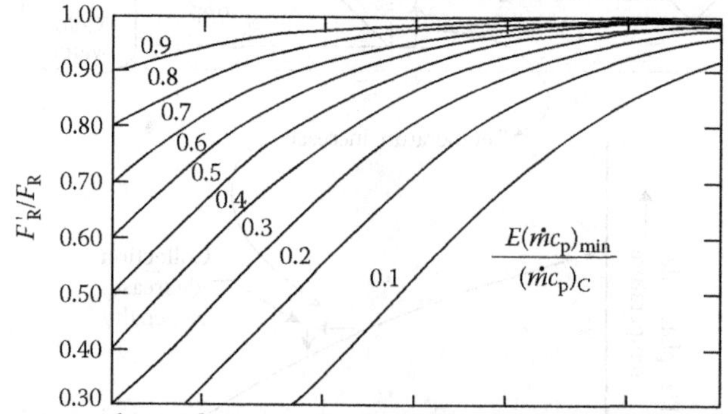

FIGURE 40.24
Variation of collector-heat exchanger correction factor. (From Duffie, J.A. and Beckman, W.A., *Solar Engineering of Thermal Processes*, Wiley Interscience, New York, 1980.)

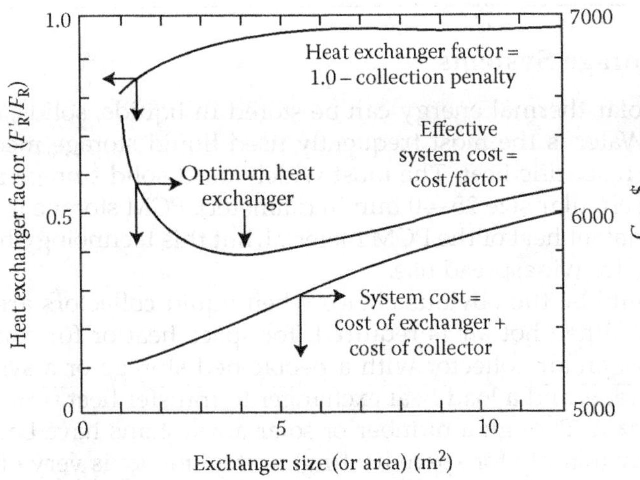

FIGURE 40.25
Typical heat-exchanger optimization plot. (From Cole, R.L. et al. (eds.), Design and installation manual for thermal energy storage, ANL-79-15, Argonne National Laboratory, Argonne, IL, 1979.)

40.5.1.2 Collector Piping and Shading Losses

Other corrections that can be applied to collector performance parameters include those for thermal losses from the piping (or from ducts) between the collection subsystem and the storage unit. Beckman (1978) has shown that these losses can be conveniently taken into consideration by suitably modifying the η_n and U_L terms of the solar collectors as follows:

$$\frac{\eta_n'}{\eta_n} = \left[1 + \frac{u_d A_0}{(mc_p)_C}\right]^{-1} \quad \text{and} \quad \frac{U_L'}{U_L} = \frac{1 - \dfrac{U_d A_i}{(mc_p)_C} + \dfrac{U_d(A_i + A_0)}{A_C F_R U_L}}{1 + \dfrac{U_d A_0}{(mc_p)_C}} \tag{40.31}$$

where
- U_d is the heat coefficient from the pipe or duct
- A_0 is the heat loss area of the outlet pipe or duct
- A_i is the heat loss area of the inlet pipe or duct

When large collector arrays are mounted on flat roofs or level ground, multiple rows of collectors are usually arranged in a sawtooth fashion. These multiple rows must be spaced so that they do not shade each other at low sun angles. Unlimited space is rarely available, and it is desirable to space the rows as close as possible to minimize piping and to keep land costs low. Some amount of shading, especially during early mornings and late evenings during the winter months is generally acceptable. Detailed analysis of shading losses is cumbersome though not difficult and equations presented in standard text books such as Duffie and Beckman (1980) can be used directly.

40.6 Thermal Storage Systems

Low-temperature solar thermal energy can be stored in liquids, solids, or phase change materials (PCMs). Water is the most frequently used liquid storage medium because of its low cost and high specific heat. The most widely used solid storage medium is rocks (usually of uniform circular size 25–40 mm in diameter). PCM storage is much less bulky because of the high-latent heat of the PCM material, but this technology has yet to become economical and safe for widespread use.

Water storage would be the obvious choice when liquid collectors are used to supply hot water to a load. When hot air is required (for space heat or for convective drying), one has two options: an air collector with a pebble-bed storage or a system with liquid collectors, water storage, and a load heat exchanger to transfer heat from the hot water to the working air stream. Though a number of solar air systems have been designed and operated successfully (mainly for space heating), water storage is very often the medium selected. Water has twice the heat capacity of rock, so water storage tanks will be smaller than rock-bed containers. Moreover, rock storage systems require higher parasitic energy to operate, have higher installation costs, and require more sophisticated controls. Water storage permits simultaneous charging and discharging while such an operation is not possible for rock storage systems. The various types of materials used as containers for water and rock-bed storage and the types of design, installation, and operation details one needs to take care of in such storage systems are described by Mueller Associates (1985) and SERI (1989).

Sensible storage systems, whether water or rock-bed, exhibit a certain amount of thermal stratification. Standard textbooks present relevant equations to model such effects. In the case of active closed-loop multipass hot-water systems, storage stratification effects can be neglected for long-term system performance with little loss of accuracy. Moreover, this leads to conservative system design (i.e., solar contribution is underpredicted if stratification is neglected). A designer who wishes to account for the effect of stratification in the water storage can resort to a formulation by Phillips and Dave (1982), who showed that this effect can be fairly well modeled by introducing a *stratification coefficient* (which is a system constant that needs to be determined only once) and treating the storage subsystem as fully mixed. However, this approach is limited to the specific case of no (or very little) heat withdrawal from storage during the collection period. Even when water storage systems are highly stratified, simulation studies seem to indicate that modeling storage as a one-dimensional plug-flow three-node heat transfer problem yields satisfactory results of long-term solar system performance.

The thermal losses q_w from the storage tank can be modeled as

$$q_w = (UA_S)(T_S - T_{env}) \qquad (40.32)$$

where

(UA_S) is the storage overall heat loss per unit temperature difference
T_{env} is the temperature of the air surrounding the storage tank

Note that (UA_S) depends (1) on the storage size, which is a parameter to be sized during system design, and (2) on the configuration of the storage tank (i.e., on the length by diameter ratio in case of a cylindrical tank). For storage tanks, this ratio is normally in the range of 1.0–2.0.

40.7 Solar System Simulation

A system model is nothing but an assembly of appropriate component modeling equations that are to be solved over time subject to certain forcing functions (i.e., the meteorological data and load data). The resulting set of simultaneous equations can be solved either analytically or numerically.

The analytical method of resolution is appropriate, or possible, only for simplified system configurations and operating conditions. This approach has had some success in the analysis and design of open-loop systems (refer to Reddy 1987, Gordon and Rabl 1986, for more details). On the other hand, numerical simulation can be performed for any system configuration and operating strategy, however, complex. However, this is time-consuming and expensive in computer time and requires a high level of operator expertise.

We shall illustrate the approach of numerical simulation by considering the simple solar system shown in Figure 40.18. Assuming a fully mixed storage tank, the instantaneous energy balance equation is

$$(mc_p)_S(dT_S/dt) = q_C - q_u - q_w \tag{40.33}$$

where

q_C is the useful energy delivered by the solar collector (given by Equation 40.2.)
q_w is the thermal loss from the storage tank (given by Equation 40.32)
q_u is the useful heat transferred through the load heat exchanger, which can be determined as shown in the following text

The maximum hourly rate of energy transfer through the load heat exchanger is

$$q_{max} = E_B(mc_p)_{min}(T_S - T_{Xi})\delta_L \tag{40.34}$$

where δ_L is a control function whose value is either 1 or 0 depending upon whether there is a heat demand or not. Since q_{max} can be greater than the amount of thermal energy q_L actually required by the load, the bypass arrangement can be conveniently modeled as

$$q_u = \min(q_{max}, q_L) \tag{40.35}$$

where

$$q_L = (mc_p)_L(T_{Li} - T_{Xi}) \tag{40.36}$$

for water heating and industrial process heat loads. Space heating and cooling loads can be conveniently determined by one of the several variants of the bin-type methods (ASHRAE 1985).

The amount of energy q_{max} supplied by a topping-up type of auxiliary heater is

$$q_{max} = q_L - q_u \tag{40.37}$$

Assuming $T_{env} = T_a$, Equation 40.33 can be expanded into

$$(Mc_p)_S\frac{dT_S}{dt} - A_CF_R[I_T\eta_0 - U_L(T_S - T_a)]^+ - (mc_p)_S(T_S - T_{Xi})\delta_L - (UA)_S(T_S - T_a) \tag{40.38}$$

The presence of control functions and time dependence of I_T and T_a prevent a general analytical treatment, though, as mentioned earlier, specific cases can be handled. The numerical approach involves expressing this differential equation in finite difference form. After rearranging, one gets

$$T_{S,b} + \frac{\Delta t}{(Mc_p)_S}\{A_C F_R[I_T\eta_0 - U_L(T_{S,b} - T_a)]^+ - (mc_p)_S(T_{S,b} - T_{Xi})\delta_L - (UA)_S(T_{S,b} - T_a)\} T_{S,f} \quad (40.39)$$

where $T_{S,b}$ and $T_{S,f}$ are the storage temperatures at the beginning and the end of the time step Δt. The time step is sufficiently small (say 1 h) that I_T and T_a can be assumed constant. This equation is repeatedly used over the time period in question (day, month, or year), and the total energy supplied by the collector or to the load can be estimated.

Such methods of simulation, referred as stepwise steady-state simulations, implicitly assume that the solar thermal system operates in a steady-state manner during one time step, at the end of which it undergoes an abrupt change in operating conditions as a result of changes in the forcing functions, and thereby attains a new steady-state operating level. Although in reality, the system performance varies smoothly over time and is consequently different from that outlined earlier, it has been found that, in most cases, taking time steps of the order of 1 h yields acceptable results of long-term performance.

The objective of solar-supplemented energy systems is to displace part of the conventional fuel consumption of the auxiliary heater. The index used to represent the contribution of the solar thermal system is the *solar fraction*, which is the fraction of the total energy required by the load that is supplied by the solar system. The solar fraction could be expressed over any time scale, with month and year being the most common. Two commonly used definitions of the monthly solar fraction are

1. Thermal solar fraction:

$$f_Y = \frac{Q_{UM}}{Q_{LM}} = 1 - \frac{Q_{aux,M}}{Q_{LM}} \quad (40.40)$$

where
Q_{UM} is the monthly total thermal energy supplied by the solar system
Q_{LM} is the monthly total thermal requirements of the load
$Q_{aux,M}$ is the monthly total auxiliary energy consumed

2. Energy solar fraction (i.e., thermal plus parasitic energy):

$$f'_M = \frac{Q'_{UM}}{Q'_{LM}} \quad (40.41)$$

where
Q'_{UM} is Q_{UM} *minus* the parasitic energy consumed by the solar system
Q'_{LM} is Q_{LM} *plus* the parasitic energy consumed by the load

Example 40.5

Simulate the closed-loop solar thermal system shown in Figure 40.18 for each hour of a day assuming both collector and load heat exchangers to be absent (i.e., $E_A = E_B = 1$). Assume the following data as input for the simulation: $A_C = 10$ m², $F_RU_L = 5.0$ W/m² °C), $F_R\eta_0 = 0.7$, $(mc_p) = 2.0$ MJ/°C and $(UA)_S = 3$ W/°C. Water is withdrawn to meet a load from 9 a.m. to 7 p.m. (solar time) at a constant rate of 60 kg/h and is replenished from the mains at a temperature of 25°C. The storage temperature at the start (i.e., at 6 a.m.) is 40°C, and the environment temperature is equal to the ambient temperature. The temperature of the water entering the load should not exceed 55°C. The hourly values of the solar radiation on the plane of the collector are given in column 2 of Table 40.3 and the ambient temperature is assumed constant over the day and equal to 25°C. The variation of the optical efficiency with angle of incidence can be neglected.

The results of the simulation are given in Table 40.3. The following equations should permit the reader to verify for himself the results obtained. Simulating the system entails solving the following equations in the sequence given here:

Column 4. Useful energy delivered by the collector (Equation 20.2)

$$q_C = 10[0.7\,I_T - 5(3600/10^6)T_{S,b} - 25]^+ \text{ (MJ/h)}$$

The term $(3600/10^6)$ is introduced to convert W/m² (the units in which I_T is expressed) into MJ/(h m²). Note that $T_{S,b}$ is taken to be equal to $T_{S,f}$ of the final hour.

Column 5. Thermal losses from the storage tank (Equation 40.32)

$$q_w = 3(3600/10^6)(T_{S,b} - 25) \text{ (MJ/h)}$$

Column 6. The maximum rate of energy that can be transferred from the load can be calculated from Equation 40.34

$$q_{max} = 60(4190/10^6)(T_{S,b} - 25) \text{ (MJ/h)}$$

Column 7. The thermal energy required by the load (from Equation 40.36)

$$q_L = 60(4190/10^6)(55 - 25) = 7.54 \text{ MJ/h}$$

Column 8. The actual amount of heat withdrawn from storage (Equation 40.35)

$$q_w = \min[\text{column 6, column 7}]$$

TABLE 40.3
Simulation Results of Example 40.5

(1) Solar Time (h)	(2) I_T (MJ/m²h)	(3) $T_{s,f}$ (°C)	(4) q_c (MJ/h)	(5) q_w (MJ/h)	(6) q_{max} (MJ/h)	(7) q_L (MJ/h)	(8) q_U (MJ/h)	(9) q_{aux} (MJ/h)
Start		40.00						
6–7	0.37	39.92	0.00	0.16	0.00	0.00	0.00	0.00
7–8	0.95	41.82	3.96	0.16	0.00	0.00	0.00	0.00
8–9	1.54	45.61	7.75	0.18	0.00	0.00	0.00	0.00
9–10	2.00	48.05	10.29	0.22	5.18	754	5.18	2.36
10–11	2.27	50.90	11.74	0.25	5.79	7.54	5.79	1.75
11–12	2.46	53.78	12.56	0.28	6.51	7.54	6.51	1.03
12–13	2.50	56.17	12.32	0.31	7.24	7.54	7.24	0.31
13–14	2.24	57.26	10.07	0.34	7.84	7.54	7.54	0.00
14–15	2.12	57.84	9.03	0.35	8.11	7.54	7.54	0.00
15–16	1.37	55.73	3.68	0.35	8.25	7.54	7.54	0.00
16–17	0.76	51.79	0.00	0.33	7.72	7.54	7.54	0.00
17–18	0.23	48.28	0.00	0.29	6.73	7.54	6.73	0.81
18–19	0.00	45.23	0.00	0.25	5.85	7.54	5.85	1.69
Total	18.81	—	81.41	3.48	69.24	75.40	67.48	7.94

Column 9. The amount of energy supplied by the auxiliary heater (Equation 40.2.37)

$$q_{aux} = \text{column 7} - \text{column 8}$$

The final storage temperature $T_{S,f}$ is now calculated from Equation 40.39

$$T_{S,f} = T_{S,b} + [\text{column 4} - \text{column 8} - \text{column 5}]/2.0$$

From Table 40.3, we note that the solar collector efficiency over the entire day is [81.41/(18.81 × 10)] = 0.43. The corresponding daily solar fraction = (67.48/75.40) = 0.895.

40.8 Solar System Sizing Methodology

Sizing of solar systems primarily involves determining the collector area and storage size that are most cost effective. Standalone and solar-supplemented systems have to be treated separately since the basic design problem is somewhat different. The interested reader can refer to Gordon (1987) for sizing standalone systems.

40.8.1 Solar-Supplemented Systems

40.8.1.1 Production Functions

Because of the annual variation of incident solar radiation, it is not normally economical to size a solar subsystem such that it provides 100% of the heat demand. Most solar energy systems follow the *law of diminishing returns.* This implies that increasing the size of the solar collector subsystem results in a less than proportional increase in the annual fuel savings (or alternatively, in the annual solar fraction).

Any model has two types of variables: exogenous and endogenous. The *exogenous parameters* are also called the input variables, and these in turn may be of two kinds. *Variable exogenous parameters* are the collector area A_C, the collector performance parameters $F_R\eta_n$ and F_RU_L, the collector tilt, the thermal storage capacity $(mc_p)_S$, the heat exchanger size, and the control strategies of the solar thermal system. On the other hand, the climatic data specified by radiation and the ambient temperature, as well as the end use thermal demand characteristics, are called *constrained exogenous parameters* because they are imposed externally and cannot be changed. The *endogenous parameters* are the output parameters whose values are to be determined, the annual solar fraction being one of the parameters most often sought.

Figure 40.26 illustrates the law of diminishing results. The annual solar fraction f_Y is seen to increase with collector area but at a decreasing rate and at a certain point will reach saturation. Variation of any of the other exogenous parameters also exhibits a similar trend. The technical relationship between f_Y and one or several variable exogenous parameters for a given location is called the *yearly production function.*

It is only for certain simple types of solar thermal systems that an analytical expression for the production can be deduced directly from theoretical considerations. The most common approach is to carry out computer simulations of the particular system (solar plus auxiliary) over the complete year for several combinations of values of the exogenous parameters. The production function can subsequently be determined by an empirical curve fit to these discrete sets of points.

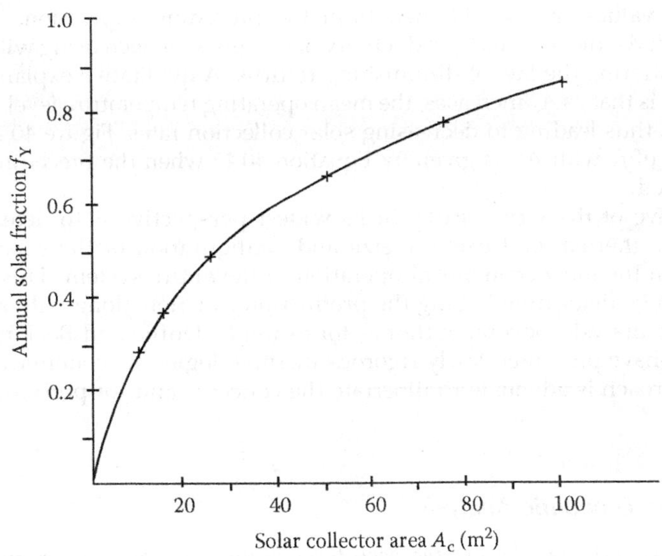

FIGURE 40.26
A typical solar system production function (see Example 41.6).

Example 40.6

Kreider (1979) gives the following expression for the production function of an industrial solar water heater for a certain location:

$$f_Y = \frac{Q_{UY}}{Q_{LY}} = \left(0.35 - \frac{F_R U_L}{100 F_R \eta_n}\right) \ln\left(1 + \frac{20 F_R \eta_n A_C}{Q_{LY}}\right) \quad (40.42)$$

where

Q_{UY} is the thermal energy delivered by the solar thermal system over the year in GJ/y
Q_{LY} is the yearly thermal load demand, also in GJ/y
$F_R U_L$ is in W/(m² °C)

Note that only certain solar system exogenous parameters figure explicitly in this expression, thereby implying that other exogenous parameters (for example, storage volume) have not been varied during the study. As an illustration, let us assume the following nominal values: $Q_{LY} = 100$ GJ/year, $F_R U_L = 2.0$ W/(m² °C), and $F_R \eta_n = 0.7$. For a 1% increase in collector area A_C, the corresponding percentage increase in Q_{UY} (called elasticity) can be determined:

$$\frac{dQ_{UY}}{Q_{UY}} = \frac{dA_C}{A_C} \left[\left(\frac{Q_{LY}}{20 F_R \eta_n A_C} + 1\right) \ln\left(1 + \frac{20 F_R \eta_n A_C}{Q_{LY}}\right)\right]^{-1} \quad (40.43)$$

From this, we obtain the expression for *marginal productivity*

$$\frac{dQ_{UY}}{dA_C} = \frac{Q_{UY}}{A_C} \left[\left(\frac{Q_{LY}}{20 F_R \eta_n A_C} + 1\right) \ln\left(1 + \frac{20 F_R \eta_n A_C}{Q_{LY}}\right)\right]^{-1} \quad (40.44)$$

Numerical values can be obtained from the preceding expression. Though Q_{UY} increases with A_C, the marginal productivity of Q_{UY} goes on decreasing with increasing A_C, thus illustrating the law of diminishing returns. A qualitative explanation of this phenomenon is that as A_C increases, the mean operating temperature level of the collector increases, thus leading to decreasing solar collection rates. Figure 40.26 illustrates the variation of f_Y with A_C as given by Equation 40.42 when the preceding numerical values are used.

The objective of the sizing study in its widest perspective is to determine, for a given specific thermal end use, the size and configuration of the solar subsystem that results in the most economical operation of the entire system. This economical optimum can be determined using the production function along with an appropriate economic analysis. Several authors—for example, Duffie and Beckman (1980) or Rabl (1985)—have presented fairly rigorous methodologies of economic analysis, but a simple approach is adequate to illustrate the concepts and for preliminary system sizing.

40.8.1.2 Simplified Economic Analysis

It is widely recognized that *discounted cash flow analysis* is most appropriate for applications such as sizing an energy system. This analysis takes into account both the initial cost incurred during the installation of the system and the annual running costs over its entire life span.

The economic objective function for optimal system selection can be expressed in terms of either the energy cost incurred or the energy savings. These two approaches are basically similar and differ in the sense that the objective function of the former has to be minimized while that of the latter has to be maximized. In our analysis, we shall consider the latter approach, which can further be subdivided into the following two methods:

1. Present worth or life cycle savings, wherein all running costs are discounted to the beginning of the first year of operation of the system.
2. Annualized life cycle savings, wherein the initial expenditure incurred at the start as well as the running costs over the life of the installation are expressed as a yearly mean value.

40.9 Solar System Design Methods

40.9.1 Classification

Design methods may be separated into three generic classes. The *simple* category, usually associated with the prefeasibility study phase involves quick manual calculations of solar collector/system performance and rule-of-thumb engineering estimates. For example, the generalized yearly correlations proposed by Rabl (1981) and described in Section 40.2 could be conveniently used for year-round, more or less constant loads. The approach is directly valid for open-loop solar systems, while it could also be used for closed-loop systems if an *average* collector inlet temperature could be determined. A simple manner of selecting this temperature $\bar{T}_m$ for domestic closed-loop multipass systems is to assume the following empirical relation:

$$\bar{T}_m - T_{mains}/3 + (2/3)T_{set} \tag{40.45}$$

where

T_{mains} is the average annual supply temperature
T_{set} is the required hot-water temperature (about 60°C–80°C in most cases)

These manual methods often use general guidelines, graphs, and/or tables for sizing and performance evaluation. The designer should have a certain amount of knowledge and experience in solar system design in order to make pertinent assumptions and simplifications regarding the operation of the particular system.

Mid-level design methods are resorted to during the feasibility phase of a project. The main focus of this chapter has been toward this level, and a few of these design methods will be presented in this section. A personal computer is best suited to these design methods because they could be conveniently programmed to suit the designer's tastes and purpose (spreadsheet programs, or better still one of the numerous equation-solver software packages, are most convenient). Alternatively, commercially available software packages such as *f*-chart (Beckman et al. 1977) could also be used for certain specific system configurations.

Detailed design methods involve performing hourly simulations of the solar system over the entire year from which accurate optimization of solar collector and other equipment can be performed. Several simulation programs for active solar energy systems are available, TRNSYS (Klein et al. 1975, 1979) developed at the University of Wisconsin-Madison being perhaps the best known. This public-domain software has technical support and is being constantly upgraded. TRNSYS contains simulation models of numerous subsystem components (solar radiation, solar equipment, loads, mechanical equipment, controls, etc.) that comprise a solar energy system. A user can conveniently hook up components representative of a particular solar system to be analyzed and then simulate that system's performance at a level of detail that the user selects. Thus TRNSYS provides the design with large flexibility, diversity, and convenience of usage.

As pointed out by Rabl (1985), the detailed computer simulations approach, though a valuable tool, has several problems. Judgment is needed both in the selection of the input and in the evaluation of the output. The very flexibility of big simulation programs has drawbacks. So many variables must be specified by the user that errors in interpretation or specification are common. Also, learning how to use the program is a time-consuming task. Because of the numerous system variables to be optimized, the program may have to be run for numerous sets of combinations, which adds to expense and time. The inexperienced user can be easily misled by the second-order details while missing first-order effects. For example, uncertainties in load, solar radiation, and economic variables are usually very large, and long-term performance simulation results are only accurate to within a certain degree. Nevertheless, detailed simulation programs, if properly used by experienced designers, can provide valuable information on system design and optimization aspects at the final stages of a project design.

There are basically three types of mid-level design approaches: the empirical correlation approach, the analytical approach, and the one-day repetitive methods (described fully in Reddy 1987). We shall illustrate their use by means of specific applications.

40.9.2 Active Space Heating

The solar system configuration for this particular application has become more or less standardized. For example, for a liquid system, one would use the system shown in Figure 40.27. One of the most widely used design methods is the *f*-chart method (Beckman et al. 1977; Duffie and Beckman 1980), which is applicable for standardized liquid and air heating systems as well as for standardized domestic hot-water systems. The *f*-chart method basically involves using a simple algebraic correlation that has been deduced from numerous TRNSYS simulation runs of these standard solar systems subject to a wide range of climates and solar system parameters (see Figure 40.28). Correlations were developed between monthly solar fractions and two easily calculated dimensionless variables X and Y, where

$$X = \frac{(A_C F_R' U_L (T_{Ref} - \bar{T}_a)\Delta t)}{Q_{LM}} \tag{40.46}$$

$$Y = A_C F_R' \bar{\eta}_0 \bar{H}_T N / Q_{LM} \tag{40.47}$$

where

A_C collector area (m^2)
F_R' collector-heat exchanger heat removal factor (given by Equation 40.29)
U_L collector overall loss coefficient ($W/(m^2\ °C)$)
Δt total number of seconds in the month = $3600 \times 24 \times N = 86{,}400 \times N$
$\bar{T}_a$ monthly average ambient temperature (°C)
T_{Ref} an empirically derived reference temperature, taken as 100°C
Q_{lm} monthly total heating load for space heating and/or hot water (J)
$\bar{H}_T$ monthly average daily radiation incident on the collector surface per unit area (J/m^2)
N number of days in the month
$\bar{\eta}_0$ monthly average collector optical efficiency

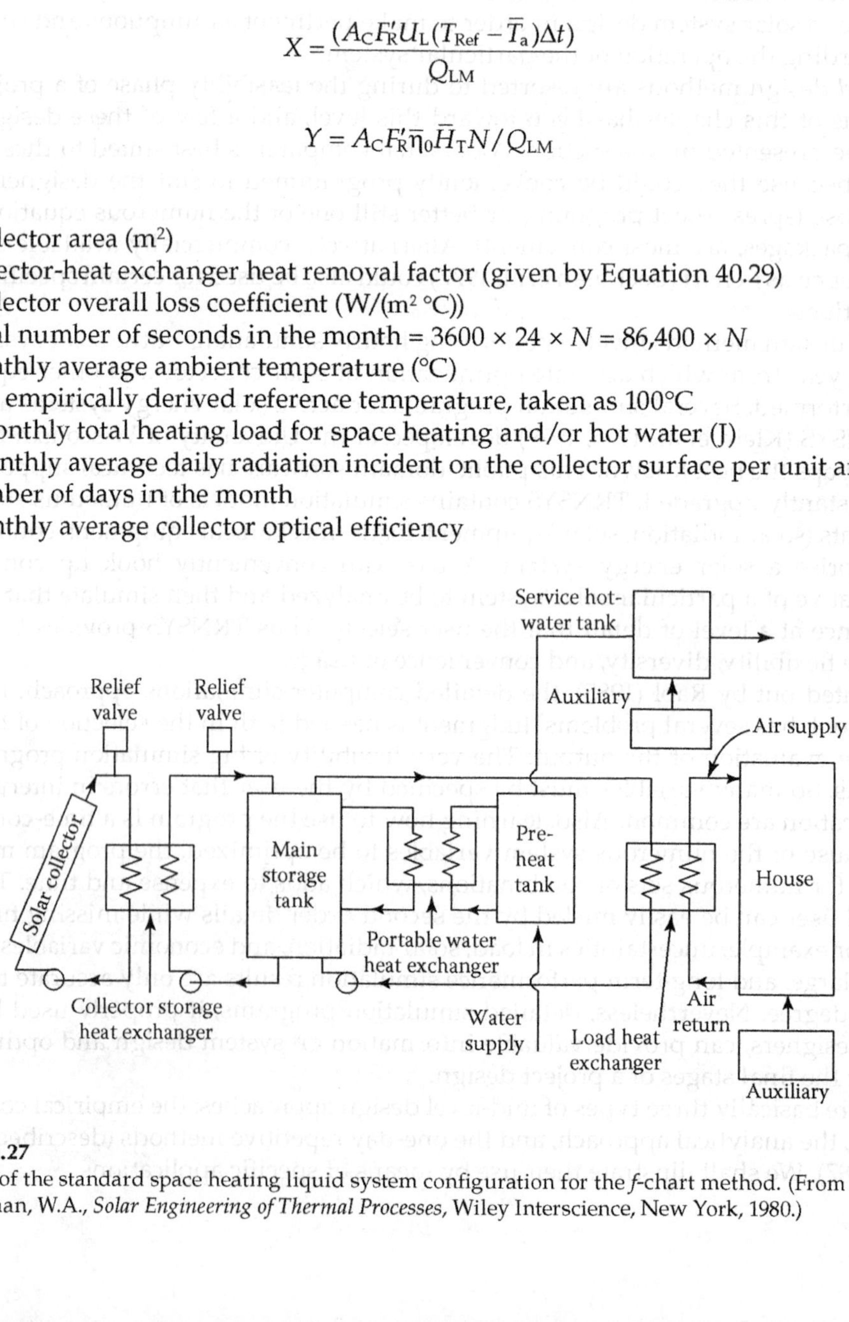

FIGURE 40.27
Schematic of the standard space heating liquid system configuration for the *f*-chart method. (From Duffie, J.A. and Beckman, W.A., *Solar Engineering of Thermal Processes*, Wiley Interscience, New York, 1980.)

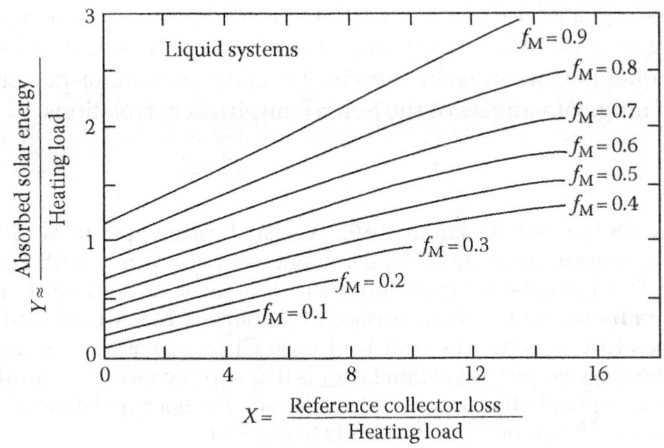

FIGURE 40.28
The *f*-chart correlation for liquid system configuration. (From Duffie, J.A. and Beckman, W.A., *Solar Engineering of Thermal Processes*, Wiley Interscience, New York, 1980.)

The dimensionless variable X is the ratio of reference collector losses over the entire month to the monthly total heat load; the variable Y is the ratio of the monthly total solar energy absorbed by the collectors to the monthly total heat load. It will be noted that the collector area and its performance parameters are the predominant exogenous variables that appear in these expressions. For changes in secondary exogenous parameters, the following corrective terms X_C and Y_C should be applied for liquid systems:

1. For changes in storage capacity:

$$\frac{X_C}{X} = \left(\frac{\text{actual storage capacity}}{\text{standard storage capacity}}\right)^{-0.25} \tag{40.48}$$

where the standard storage volume is 75 L/m^2 of collector area.

2. For changes in heat exchanger size:

$$\frac{Y_C}{Y} = \frac{0.39 + 0.65\exp[-(0.139(UA)_B}{(E_L(mc_p)_{min}))]} \tag{40.49}$$

The monthly solar fraction for liquid space heating can then be determined from the following empirical correlation:

$$f_M = 1.029Y - 0.065X - 0.245Y^2 + 0.0018X^2 + 0.0215Y^3 \tag{40.50}$$

subject to the conditions that $0 \le X \le 15$ and $0 \le Y \le 3$. This empirical correlation is shown graphically in Figure 40.28.

A similar correlation has also been proposed for space heating systems using air collectors and pebble-bed storage. The procedure for exploiting the preceding empirical correlations is as follows. For a predetermined location, specified by its 12 monthly radiation and ambient temperature values, Equation 40.50 is repeatedly used for each month of the year for a particular set of variable exogenous parameters. The monthly solar fraction f_M and thence the annual thermal energy delivered by the solar thermal system are easily deduced. Subsequently, the entire procedure is repeated for different values and combinations of variable exogenous parameters. Finally, an economic analysis is performed to determine optimal sizes of various solar system components. Care must be exercised that the exogenous parameters considered are not outside the range of validity of the *f*-chart empirical correlations.

Example 40.7

(Adapted from Duffie and Beckman 1980). A solar heating system is to be designed for Madison, Wisconsin (latitude 43°N) using one-cover collectors with $F_R\eta_n = 0.74$ and $F_R U_L = 4\text{W}/(\text{m}^2\,°\text{C})$. The collector faces south with a slope of 60° from the horizontal. The average daily radiation on the tilted surface in January is 12.9 MJ/m², and the average ambient temperature is −7°C. The heat load is 36 GJ for space heating and hot water. The collector-heat exchanger correction factor is 0.97 and the ratio of monthly average to normal incidence optical efficiency is 0.96. Calculate the energy delivered by the solar system in January if 50 m² of collector area is to be used.

From Equations 40.46 and 40.47, with $A_C = 50$ m²,

$$X = 4.0 \times 0.97[100 - (-7)]31 \times 86{,}400 \times 50/(36 \times 10^9) = 1.54$$

$$Y = 0.74 \times 0.97 \times 0.96 \times 12.9 \times 10^6 \times 31 \times 50/(36 \times 10^9) = 0.38$$

From Equation 40.50, the solar fraction for January is $f_M = 0.26$. Thus the useful energy delivered by the solar system = 0.26 × 36 = 9.4 GJ.

In an effort to reduce the tediousness involved in having to perform 12 monthly calculations, two analogous approaches that enable the annual solar fraction to be determined directly have been developed by Barley and Winn (1978) and Lameiro and Bendt (1978). These involve the computation of a few site-specific empirical coefficients, thereby rendering the approach less general. For example, the *relative-area* method suggested by Barley and Winn enable the designer to directly calculate the annual solar fraction of the corresponding system using four site-specific empirical coefficients. The approach involves curve fits to simulation results of the *f*-chart method for specific locations in order to deduce a correlation such as:

$$f = c_1 + c_2 \ln(A/A_{0.5}) \tag{40.51}$$

where

c_1 and c_2 are location-specific parameters that are tabulated for several United States locations

$A_{0.5}$ is the collector area corresponding to an annual solar fraction of 0.5 given by

$$A_{0.5} = \frac{A_s(UA)}{(F_R'\eta_0 - F_R'U_L Z)} \tag{40.52}$$

where

A_S and Z are two more location specific parameters

UA is the overall heat loss coefficient of the building

$F_R'\eta_0$ and $F_R'U_L$ are the corresponding solar collector performance parameters corrected for the effect of the collector-heat exchanger

Barley and Winn also proposed a simplified economic life-cycle analysis whereby the optimal collector area could be determined directly. Another well-known approach is the *Solar Load Radio* (SLR) method for sizing residential space heating systems (Hunn 1980).

40.9.3 Domestic Water Heating

The *f*-chart correlation (Equation 40.50) can also be used to predict the monthly solar fraction for domestic hot-water systems represented by Figure 40.21 provided the water mains temperature T_{mains} is between 5°C and 20°C and the minimum acceptable hot-water temperature drawn from the storage for end use (called the set water temperature T_w) is between 50°C and 70°C. Further, the dimensionless parameter X must be corrected by the following ratio

$$\frac{X_w}{X} = \frac{(11.6 + 1.8T_w + 3.86T_{\text{mains}} - 2.32\bar{T}_a)}{(100 - \bar{T}_a)} \tag{40.53}$$

In case the domestic hot-water load is much smaller than the space heat load, it is recommended that Equation 40.50 be used without the above correction.

40.9.4 Industrial Process Heat

As discussed in Section 40.4.2.1, two types of solar systems for industrial process heat are currently used: the closed-loop multipass systems (with an added distinction that the auxiliary heater may be placed either in series or in parallel (see Figures 40.18 and 40.19) and the open-loop singlepass system. How such systems can be designed will be described next.

40.9.4.1 Closed-Loop Multipass Systems

40.9.4.1.1 Auxiliary Heater in Parallel

The Phibar-*f* chart method (Klein and Beckman 1979; Duffie and Beckman 1980; Reddy 1987) is a generalization of the *f*-chart method in the sense that no restrictions need be imposed on the temperature limits of the heated fluid in the solar thermal system. However, three basic criteria for the thermal load have to be satisfied for the Phibar-*f* chart method to be applicable: (1) the thermal load must be constant and uniform over each day and for at least a month, (2) the thermal energy supplied to the load must be above a minimum temperature that completely specifies the temperature level of operation of the load, and (3) either there is no conversion efficiency in the load (as in the case of hot water usage) or the efficiency of conversion is constant (either because the load temperature level is constant or because the conversion efficiency is independent of the load temperature level). The approach is strictly applicable to solar systems with the auxiliary heater in parallel (Figure 40.19).

A typical application for the Phibar-*f* chart method is absorption air-conditioning. The hot water inlet temperature from the collectors to the generator must be above a minimum temperature level (say, 80°C) for the system to use solar heat. If the solar fluid temperature is less (even by a small amount), the entire energy to heat up the water to 80°C is supplied by the auxiliary system.

As a result of continuous interaction between storage and collector in a closed-loop system, the variation of the storage temperature and hence the fluid inlet temperature to the collectors) over the day and over the month is undetermined. The Phibar-*f* chart method implicitly takes this into account and reduces these temperature fluctuations down to a monthly mean equivalent storage temperature $\bar{T}_S$ The determination of this temperature in conjunction with the daily utilizability approach is the basis of the design approach.

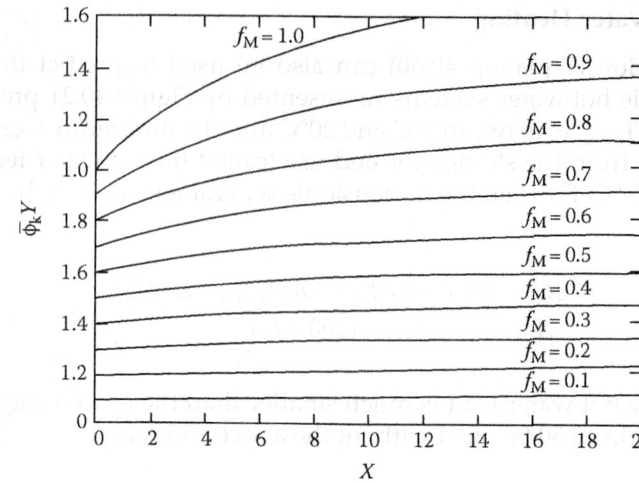

FIGURE 40.29
The Phibar-*f* chart correlation for a storage capacity of 350 kJ/m² and for a 12 h/day thermal load. (From Duffie, J.A. and Beckman, W.A., *Solar Engineering of Thermal Processes*, Wiley Interscience, New York, 1980.)

The basic empirical correlation of the Phibar-*f* chart method, shown graphically in Figure 40.29, is as follows:

$$f_M = Y\bar{\phi} - a[\exp(bf_M) - 1][1 - \exp(cX)] \quad (40.54)$$

with $0 < X < 20$ and $0 < Y < 1.6$, and $\bar{\phi}$ is the Klein daily utilizability fraction described in Section 40.3.3 and given by Equation 40.24. Y is given by Equation 40.47, and X is now slightly different from Equation 40.46 and is defined as:

$$X = \frac{A_c F_R U_L \Delta t (100°\text{C})}{QLM} \quad (40.55)$$

The values of the constants a, b, and c are given by the following:

1. For an end use load operating between 6 a.m. and 6 p.m. every day of the month,

$$a = 0.015[(mc_p)_S/350\ \text{KJ/(m}^2\ °\text{C)}]^{-0.76} \text{ for } 175 \le [(mc_p)_S/A_C] \le 1400\ \text{kJ/(m}^2\ °\text{C)},$$
$$b = 3.85 \quad \text{and} \quad c = -0.15 \quad (40.56)$$

2. For an end use load operating 24 h/day over the entire month,

$$a = 0.043 \text{ only for } [(mc_p)_S/A_C] = 350\ \text{kJ/(m}^2\ °\text{C)}, \quad b = 2.81, \quad \text{and} \quad c = -0.18 \quad (40.57)$$

It will be noted that $(Y\bar{\phi})$ denotes the maximum solar fraction that would have resulted had T_{Ci}, the inlet temperature to the collector, been equal to T_{Li} throughout the month. The term in Equation 40.54 that is subtracted from $(Y\bar{\phi})$ represents the decrease in the solar fraction as a result of $T_{Ci} > T_{Xi}$. The solar fraction computed from Equation 40.54 has to be corrected for the effect of thermal losses from the storage as well as the presence of the load-heat exchanger, both of which will decrease the solar fraction. For complete details, refer to Duffie and Beckman (1980) or Reddy (1987). Note that Equation 40.54 needs to be solved for f_m in an iterative manner.

40.9.4.1.2 Auxiliary Heater in Series

The Phibar-*f* chart method has also been modified to include solar systems with the auxiliary heater in series as shown in Figure 40.18. This configuration leads to higher solar fractions but retrofit to existing systems may be more costly.

In this case, the empirical correlation given by Equation 40.54 has been modified by Braun et al. (1983) as follows:

$$f_M = Y\bar{\phi} - a[\exp(bf_M) - 1][1 - \exp(cX)]\exp(-1.959Z) \quad (40.58)$$

with $Z = Q_{lm}/(C_l \times 100°C)$ and (1) when there is no load-heat exchanger, C_L is the monthly total load heat capacitance, which is the product of the monthly total mass of water used and the specific heat capacity of water, and (2) when there is a load-heat exchanger present $C_L = E_L \times C_{min}$, where E_L is the effectiveness of the load-heat exchanger and C_{min} is the monthly total heat capacitance, which is the lesser of the two fluids rates across the load heat exchanger.

The modified Phibar-*f* chart is similar to the original method in respect to load uniformity on a day-to-day basis over the month and in assuming no conversion efficiency. The interested may refer to Braun et al. (1983) or Reddy (1987) for complete details.

40.9.4.2 Open-Loop Single-Pass Systems

The advantages offered by open-loop single-pass systems over closed-loop multipass systems for meeting constant loads has been described in Section 40.4.2. Because industrial loads operate during the entire sun-up hours or even for 24 h daily, the simplest solar thermal system is one with no heat storage (Figure 40.20). A sizable portion (between 25% and 70%) of the daytime thermal load can be supplied by such systems and consequently, the sizing of such systems will be described below (Gordon and Rabl 1982). We shall assume that T_{Li} and T_{Xi} are constant for all hours during system operation. Because no storage is provided, excess solar energy collection (whenever $T_{Ci} > T_{Li}$) will have to be dumped out.

The maximum collector area $\hat{A}_C$ for which energy dumping does not occur at any time of the year can be found from the following instantaneous heat balance equation:

$$P_L = \hat{A}_C \hat{F}_R [I_{max} \eta_n - U_L(T_{Ci} - T_a)] \quad (40.59)$$

where P_L, the instantaneous thermal heat demand of the load (say, in kW) is given by

$$P_L = m_L c_p (T_{Li} - T_{Xi}) \quad (40.60)$$

and F_R is the heat removal factor of the collector field when its surface area is $\hat{A}_C$. Since $\hat{A}_C$ is as yet unknown, the value of $\hat{F}_R$ is also undetermined. (Note that though the *total* fluid flow rate is known, the flow rate per unit collector area is not known.) Recall that the plate efficiency factor F' for liquid collectors can be assumed constant and independent of fluid flow rate per unit collector area. Equation 40.59 can be expressed in terms of critical radiation level I_C:

$$P_L = \hat{A}_C \hat{F}_R \eta_n (I_{max} - I_C) \quad (40.61a)$$

or

$$\hat{A}_C\hat{F}_R\eta_n = \frac{P_L}{(I_{max} - I_C)} \tag{40.61b}$$

Substituting Equation 40.3 in lieu of F_R and rearranging yields

$$\hat{A}_C = -(m_L c_p / F'U_L)\ln[1 - P_L U_L / (\eta_n (I_{max} - I_C) m_L c_p)] \tag{40.62}$$

If the actual collector area A_C exceeds this value, dumping will occur as soon as the radiation intensity reaches a value I_D, whose value is determined from the following heat balance:

$$P_L = A_C F_R \eta_n (I_D - I_C) \tag{40.63a}$$

Hence

$$I_D = \frac{I_C + P_L}{(A_C F_R \eta_n)} \tag{40.63b}$$

Note that the value of I_D decreases with increasing collector area A_C, thereby indicating that increasing amounts of solar energy will have to be dumped out.

Since the solar thermal system is operational during the entire sunshine hours of the year, the yearly total energy collected can be directly determined by the Rabl correlation given by Equation 40.26. Similarly, the yearly total solar energy collected by the solar system which has got to be dumped out is

$$Q_{DY} = A_C F_R \eta_n (\tilde{a} + \tilde{b} I_D + \tilde{c} I_D^2) \tag{40.64}$$

The yearly total solar energy delivered to the load is

$$
\begin{aligned}
Q_{UY} &= Q_{CY} - Q_{DY} \\
&= A_C F_R \eta_n [\tilde{b}(I_C - I_D) + \tilde{c}(I_C^2 - I_D^2)] \qquad (40.65) \\
&= -(\tilde{b} + 2\tilde{c} I_C) P_L - \tilde{c} P_L^2 / (A_C F_R \eta_n) \\
&= -(\tilde{b} + 2\tilde{c} I_C) P_L - \tilde{c} P_L^2 / (A_C F_R \eta_n) \qquad (40.66)
\end{aligned}
$$

Replacing the value of F_R given by Equation 40.3, the annual production function in terms of A_C is

$$Q_{UY} = -(\tilde{b} + 2\tilde{c} I_C) P_L - \frac{\tilde{c} P_L^2}{\left(\frac{F'\eta_n}{F'U_L}\right)(m_L c_p)\left[1 - \exp\left(-\frac{F'U_L A_C}{m_L c_p}\right)\right]} \tag{40.67}$$

subject to the condition that $A_C > \hat{A}_C$. If the thermal load is not needed during all days of the year due to holidays or maintenance shutdown, the production function can be reduced proportionally. This is illustrated in the following example.

Example 40.8

Obtain the annual production function of an open-loop solar thermal system without storage that is to be set up in Boston, Massachusetts according to the following load specifications: industrial hot water load for 12 h a day (6 a.m. to 6 p.m.) and during 290 days a year, mass flow rate $m_L = 0.25$ kg/s, required inlet temperature $T_{Li} = 60$°C. Contaminants are picked up in the load, so that all used water is to be rejected and fresh water at ambient temperature is taken in. Flat-plate collectors with tilt equal to latitude with the following parameters are used $F'\eta_n = 0.75$ and $F'U_L = 5.5$ W/(m² °C). The latitude of Boston is 42.36°N = 0.739 rad. The yearly $\bar{K} = 0.45$ and $\bar{T}_a = 10.9$°C. Use the following Gordon and Rabl (1982) correlation:

$$Q_{CY}/A_C F_R \eta_n = [(5.215 + 6.973 I_{bn}) + (-5.412 + 4.293 I_{bn})L + (1.403 - 0.899 I_{bn})L^2]$$
$$+[(-18.596 - 5.931 I_{bn}) + (15.468 + 18.845 I_{bn})L + (-0.164 - 35.510 I_{bn})L^2] I_C$$
$$+[(-14.601 - 3.570 I_{bn}) + (13.675 - 15.549 I_{bn})L + (-1.620 + 30.564 I_{bn})L^2] I_C^2$$

From Equation 40.27, $\tilde{I}_{bn} = 1.37 \times 0.45 - 0.34 = 0.276$ kW/m². The critical radiation level $I_C = 0$, since $T_{Ci} = T_a$. Consequently, Equation 40.26, using the above expression reduces to

$$Q_{CY}/(A_C F_R \eta_n) = 5.215 + 6.973 \times 0.276 + (-5.412 + 4.293 \times 0.276)0.739$$
$$+ (1.403 - 0.899 \times 0.276)0.739^2$$

$$= 4.646 \text{ GJ/(m}^2\text{y)}.$$

The expression for the dumped out energy is found from Equation 40.64 and the previous expression by replacing I_C by I_D:

$$Q_{CY,dump}/(A_C F_R \eta_n) = 4.646 + \big[(-18.596 - 5.931 \times 0.276) + (15.468 + 18.845 \times 0.276)0.739$$
$$+ (-0.164 - 35.510 \times 0.276)0.739^2\big] I_D + \big[(14.601 - 3.57 \times 0.276)$$
$$+ (-13.675 - 15.549 \times 0.276)0.739 + (1.620 + 30.564 \times 0.276)0.739^2\big] I_D^2$$
$$= 4.646 - 10.40 I_D + 5.83 I_D^2 \text{ (GJ/m}^2\text{y)}$$

The thermal energy demand $P_L = 0.25 \times 4.19(60 - 10.9) = 51.43$ kW.

The annual production function is

$$Q_{UY}(365/290) = -(-10.40 + 2\tilde{c} \times 0)51.43 - (5.83 \times 51.43^2)/\{(0.75/5.5)(0.25 \times 4.19)$$
$$\times [1 - \exp[-(5.5 \times A_C)/(0.25 \times 4190)]]\}$$

or

$$Q_{UY} = 424.96 - 85.78/[1 - \exp(-A_C/190.45)] \text{ (GJ/y)}$$

Complete details as well as how this approach can be extended to solar systems with storage (see Figure 40.30) can be found in Rabl (1985) or Reddy (1987).

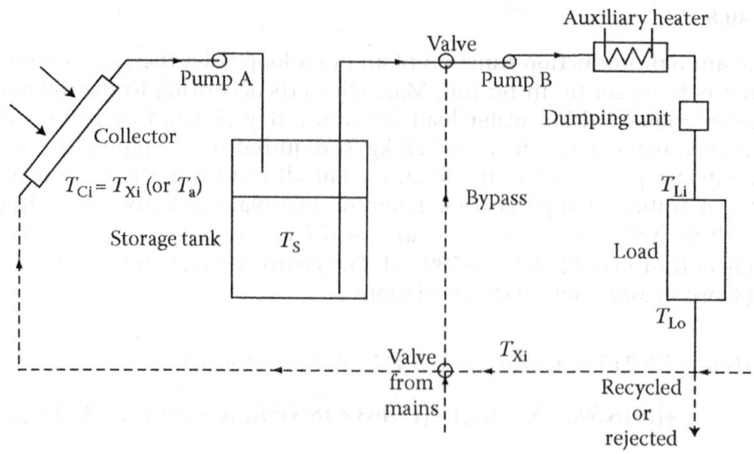

FIGURE 40.30
Open-loop solar industrial hot-water system with storage.

40.10 Design Recommendations and Costs

40.10.1 Design Recommendations

As mentioned earlier, design methods reduce computational effort compared to detailed computer simulations. Even with this decrease, the problem of optimal system design and sizing remains formidable because of

a. The presence of several solar thermal system configuration alternatives.
b. The determination of optimal component sizes for a given system.
c. The presence of certain technical and economic constraints.
d. The choice of proper climatic, technical, and economic input parameters.
e. The need to perform sensitivity analysis of both technical and economic parameters.

For most practical design work, a judicious mix of theoretical expertise and practical acumen is essential. Proper focus right from the start on the important input variables as well as the restriction of the normal range of variation would lead to a great decrease in design time and effort several examples of successful case studies and system design recommendations are described in the published literature (see, for example, Kutcher et al. 1982).

40.10.2 Solar System Costs

How the individual components of the solar system contribute to the total cost can be gauged from Table 40.4. We note that collectors constitute the major fraction (from 15% to 30%), thus suggesting that collectors should be selected and sized with great care. Piping costs are next with other collector-related costs like installation and support structure being also important.

TABLE 40.4

Percentage of Total System Cost by Component

Cost Component	Percentage Range
Collectors	15–30
Collector installation	5–10
Collector support structure	5–20[a]
Storage tanks	5–7
Piping and specialties	10–30
Pumps	1–3
Heat exchangers	0–5[b]
Chiller	5–10
Miscellaneous	2–10
Instrumentation	1–3
Insulation	2–8
Control subsystem	4–9
Electrical	2–6

Source: Mueller Associates, *Active Solar Thermal Design Manual,* Funded by U.S. DOE (no. EG-77-C-01-4042), SERI (XY-2-02046-1) and ASHRAE (project no. 40). Baltimore, MD, 1985.

[a] For collectors mounted directly on a tilted roof.

[b] For systems without heat exchangers.

References

Anderson, B. 1977. *Solar Energy: Fundamentals in Building Design.* McGraw-Hill, New York.

ASHRAE. 1985. *Fundamentals.* American Society of Heating, Refrigeration and Air Conditioning Engineers, New York.

ASHRAE. 1978. Methods of testing to determine the thermal performance of solar collectors, Standard 93-77. American Society of Heating, Refrigeration and Air Conditioning Engineers, New York.

Barley, C. D. and Winn, C. B. 1978. Optimal sizing of solar collectors by the method of relative areas. *Solar Energy*, 21, 279.

Beckman, W. A. 1978. Technical note: Duct and pipe losses in solar energy systems. *Solar Energy*, 21, 531.

Beckman, W. A., Klein, S. A., and Duffie, J. A. 1977. *Solar Heating Design by the f-Chart Method.* Wiley Interscience, New York.

Braun, J. E., Klein, S. A., and Pearson, K. A. 1983. An improved design method for solar water heating systems. *Solar Energy*, 31, 597.

Charters, W. W. S. and Pryor, T. L. 1982. *An Introduction to the Installation of Solar Energy Systems*, pp. 503–127. Victoria Solar Energy Council, Melbourne, Australia.

Chiam, H. F. 1981. Planar concentrators for flat-plate solar collectors. *Solar Energy*, 26, 503.

Clark, D. R., Klein, S. A., and Beckman, W. A. 1983. Algorithm for evaluating the hourly radiation utilizability function. *ASME Journal of Solar Energy Engineering*, 105, 281.

Cole, R. L., Nield, K. J., Rohde, R. R., and Wolosewicz, R. M. (eds.) 1979. Design and installation manual for thermal energy storage, ANL-79-15, pp. 223–397. Argonne National Laboratory, Argonne, IL.

Collares-Pereira, M., Gordon, J. M., Rabl, A., and Zarmi, Y. 1984. Design and optimization of solar industrial hot water systems with storage. *Solar Energy*, 32, 121.

Collares-Pereira, M. and Rabl, A. 1979. Derivation of method for predicting the long-term average energy delivery of solar collectors. *Solar Energy*, 23, 223.

CSU. 1980. Solar heating and cooling of residential buildings—Design of systems. Manual prepared by the Solar Energy Applications Laboratory, Colorado State University. Colorado State University, Fort Collins, CO.

de Winter, F. 1975. Heat exchanger penalties in double loop solar water heating systems. *Solar Energy*, 17, 335.

Duffie, J. A. and Beckman, W. A. 1980. *Solar Engineering of Thermal Processes*. Wiley Interscience, NewYork.

Edwards, D. K., Nelson, K. E., Roddick, R. D., and Gier, J. T. 1960. Basic studies on the use of solar energy, Report No. 60-93. Department of Engineering, University of California, Los Angeles, CA.

Gordon, J. M. 1987. Optimal sizing of stand-alone photovoltaic systems. *Solar Cells*, 20, 295.

Gordon, J. M. and Rabl, A. 1982. Design, analysis and optimization of solar industrial process heat plants without storage. *Solar Energy*, 28, 519.

Gordon, J. M. and Rabl, A. 1986. Design of solar industrial process heat systems. In *Reviews of Renewable Energy Sources*, Sodha, M. S. Mathur, S. S. Malik, M. A. S., and Kandpal, T. C. (eds.), Chapter 6, pp. 55–177. Wiley Eastern, New Delhi, India.

Gordon, J. M. and Zarmi, Y. 1985. An analytic model for the long-term performance of solar thermal systems with well-mixed storage. *Solar Energy*, 35, 55.

Hollands, K. G. T. 1965. Honeycomb devices in flat-plate solar collectors. *Solar Energy*, 9, 159.

Hunn, B. D. 1980. A simplified method for sizing active solar space heating systems. In *Solar Energy Technology Handbook, Part B: Applications, System Design and Economics*, Dickinson, W. C. and Cheremisinoff, P. N. (eds.), pp. 639–255. Marcel Dekker, New York.

Klein, S. A. 1978. Calculation of flat-plate collector utilizability. *Solar Energy*, 21, 393.

Klein, S. A. and Beckman, W. A. 1979. A general design method for closed-loop solar energy systems. *Solar Energy*, 22, 269.

Klein, S. A., Cooper, P. I., Freeman, T. L., Beekman, D. M., Beckman, W. A., and Doffie, J. A. 1975. A method of simulation of solar processes and its applications. *Solar Energy*, 17, 29.

Klein, S. A. et al. 1979. TRNSYS-A transient system simulation user's manual. University of Wisconsin-Madison Engineering Experiment Station Report 38-10. University of Wsconsin, Madison, WI.

Kreider, J. F. 1979. *Medium and High Temperature Solar Energy Processes*. Academic Press, New York.

Kutcher, C. F., Davenport, R. L., Dougherty, D. A., Gee, R. C., Masterson, P. M., and May, E. K. 1982. Design approaches for solar industrial process heat systems, SERI/TR-253-1356. Solar Energy Research Institute, Golden, CO.

Lameiro, G. F. and Bendt, P., 1978.The GFL method for designing solar energy space heating and domestic hot water systems. In: *Proceedings of American Solar Energy Society Conference*, Boulder, CO, Vol. 2, p. 113.

Larson, D. C. 1980. Optimization of flat-plate collector flat mirror system. *Solar Energy*, 24, 203.

Liu, B. Y. H. and Jordan, R. C. 1960. The inter-relationship and characteristic distribution of direct, diffuse and total solar radiation. *Solar Energy*, 4, 1.

Liu, B. Y. H. and Jordan, R. C. 1963. A rational procedure for predicting the long-term average performance of flat-plate solar energy collectors. *Solar Energy*, 7, 53.

Liu, S. T. and Fanney, A. H. 1980. Comparing experimental and computer-prodicted performance for solar hot water systems. *ASHRAE Journal*, 22(5), 34.

Löf, G. O. G. 1981. Air based solar systems for space heating. In: *Solar Energy Handbook*, Kreider, J. F. and Kreith F. (eds.), pp. 39–44. McGraw-Hill, New York.

Meyer, B. A. 1978. Natural convection heat transfer in small and moderate aspect ratio enclosures—An application to flat-plate collectors. In: *Thermal Storage and Heat Transfer in Solar Energy Systems*, Kreith, F. Boehm, R. Mitchell, J., and Bannerot, R. (eds.), pp. 555–558. American Society of Mechanical Engineers, New York.

Mueller Associates. 1985. *Active Solar Thermal Design Manual*. Funded by U.S. DOE (no. EG-77-C-01-4042), SERI(XY-2-02046-l) and ASHRAE (project no. 40). Baltimore, MD.

Newton, A. B. and Gilman, S. H. 1981. *Solar Collector Performance Manual*. Funded by U. S. DOE (no. EG-77-C-01-4042), SERI(XH-9-8265-1) and ASHRAE (project no. 32, Task 3).

Phillips, W. F. and Dave, R. N. 1982. Effect of stratification on the performance of liquid-based solar heating systems. *Solar Energy*, 29, 111.

Rabl, A. 1981. Yearly average performance of the principal solar collector types. *Solar Energy*, 27, 215.

Rabl, A. 1985. *Active Solar Collectors and Their Applications.* Oxford University Press, New York.

Reddy, T. A. 1987. *The Design and Sizing of Active Solar Thermal Systems.* Oxford University Press, Oxford, U.K.

SERI. 1989. *Engineering Principles and Concepts for Active Solar Systems.* Hemisphere Publishing Company, New York.

Theilacker, J. C. and Klein, S. A. 1980. Improvements in the utilizability relationships. In: *American Section of the International Solar Energy Society Meeting Proceedings*, Phoenix, AZ, 271pp.

41

Passive Solar Heating, Cooling, and Daylighting

Jeffrey H. Morehouse

CONTENTS

41.1 Introduction

Passive systems are defined, quite generally, as systems in which the thermal energy flow is by natural means: by conduction, radiation, and natural convection. A *passive heating system* is one in which the sun's radiant energy is converted to heat upon absorption by the building. The absorbed heat can be transferred to thermal storage by natural means or used to directly heat the building. *Passive cooling systems* use natural energy flows to transfer heat to the environmental sinks: the ground, air, and sky.

If one of the major heat transfer paths employs a pump or fan to force the flow of a heat transfer fluid, then the system is referred to as having an active component or subsystem. Hybrid systems—either for heating or for cooling—are ones in which there are both passive and active energy flows. The use of the sun's radiant energy for the natural illumination of a building's interior spaces is called *daylighting*. Daylighting design approaches use both solar beam radiation (referred to as *sunlight*) and the diffuse radiation scattered by the atmosphere (referred to as *skylight*) as sources for interior lighting, with historical design emphasis on utilizing skylight.

41.1.1 Distinction between a Passive System and Energy Conservation

A distinction is made between the energy conservation techniques and passive solar measures. Energy conservation features are designed to reduce the heating and cooling energy required to thermally condition a building: the use of insulation to reduce either heating or cooling loads, and the use of window shading or window placement to reduce solar gains, reducing summer cooling loads. Passive features are designed to increase the use of solar energy to meet heating and lighting loads, plus the use of ambient *coolth* for cooling. For example, window placement to enhance solar gains to meet winter heating loads and/or to provide daylighting is passive solar use, and the use of a thermal chimney to draw air through the building to provide cooling is also a passive cooling feature.

41.1.2 Key Elements of Economic Consideration

The distinction between passive systems, active systems, or energy conservation is not critical for economic calculations, as they are the same in all cases: a trade-off between the life cycle cost of the energy saved (performance) and the life cycle cost of the initial investment, operating, and maintenance costs (cost).

41.1.2.1 Performance: Net Energy Savings

The key performance parameter to be determined is the net annual energy saved by the installation of the passive system. The basis for calculating the economics of any solar energy system is to compare it against a *normal* building; thus, the actual difference in the annual cost of fuel is the difference in auxiliary energy that would be used with and without solar. Therefore, the energy saved rather than energy delivered, energy collected, useful energy, or some other energy measure, must be determined.

41.1.2.2 Cost: Over and above "Normal" Construction

The other significant part of the economic trade-off involves determining the difference between the cost of construction of the passive building and of the *normal* building against which it is to be compared. The convention, adopted from the economics used for active solar systems, is to define a *solar add-on cost.* Again, this may be a difficult definition in the case of passive designs because the building can be significantly altered compared to typical construction since, in many cases, it is not just a one-to-one replacement of a wall with a different wall, but it is more complex and involves assumptions and simulations concerning the *normal* building.

41.1.2.3 General System Application Status and Costs

Almost 500,000 buildings in the United States were constructed or retrofitted with passive features in the 20 years after 1980. Passive heating applications are primarily in single-family dwellings and secondarily in small commercial buildings. Daylighting features that reduce lighting loads and the associated cooling loads are usually more appropriate for large office buildings.

A typical passive heating design in a favorable climate might supply up to one-third of a home's original load at a cost of \$5–\$10 per million Btu net energy saved. An appropriately designed daylighting system can supply lighting at a cost of 2.5¢–5¢ per kWh (Larson et al. 1992a–c).

41.2 Solar Thermosyphon Water Heating

Solar hot water heating systems are composed of a collector and a storage tank. When the flow between the collector and tank is by natural circulation, these passive solar hot water systems are referred to as *thermosyphon systems*. This ability of thermosyphon systems to heat water without an externally powered pump has spurred its use in regions where both power is unavailable and power is very expensive.

41.2.1 Thermosyphon Concept

The natural tendency of a less dense fluid to rise above a more dense fluid can be used in a simple solar water heater to cause fluid motion through a collector. The density difference is created within the solar collector where heat is added to increase the temperature and decrease the density of the liquid. This collection concept is called a *thermosyphon*, and Figure 41.1 schematically illustrates the major components of such a system.

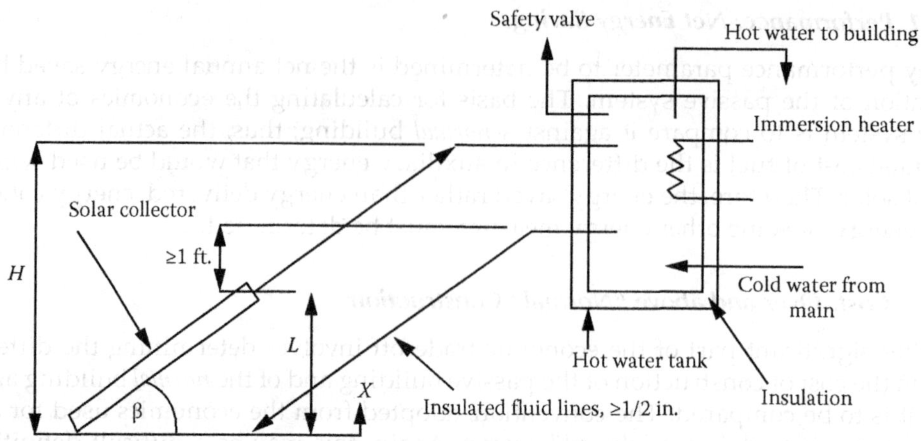

FIGURE 41.1
Schematic diagram of thermosyphon loop used in a natural circulation, service water heating system. The flow pressure drop in the fluid loop must equal the buoyant force *pressure* $\left[\int_0^L g\rho(x)dx - \rho_{stor}gL\right]$, where $\rho(x)$ is the local collector fluid density and ρ_{stor} is the tank fluid density, assumed uniform.

The flow pressure drop in the fluid loop (ΔP_{FLOW}) must equal the buoyant force *pressure difference* ($\Delta P_{BUOYANT}$) caused by the differing densities in the *hot* and *cold* legs of the fluid loop:

$$\Delta P_{FLOW} = \Delta P_{BUOYANT}$$

$$= \rho_{stor} gH - \left[\int_O^L \rho(x) g\,dx + \rho_{out} g(H - L)\right], \tag{41.1}$$

where
- H is the height of the *legs*
- L is the height of the collector (see Figure 41.1)
- $\rho(x)$ is the local collector fluid density
- ρ_{stor} is the tank fluid density (assumed to be uniform)
- ρ_{out} is the collector outlet fluid density (assumed to be uniform)

The flow pressure term, ΔP_{FLOW}, is related to the flow loop system headloss that is in turn directly connected to friction and fitting losses and the loop flow rate:

$$\Delta P_{FLOW} = \oint_{LOOP} \rho \, d(h_L), \tag{41.2}$$

where
- $h_L = KV^2$, with K being the sum of the component loss *velocity* factors (see any fluid mechanics text)
- V is the flow velocity

41.2.2 Thermo-Fluid System Design Considerations

Because the driving force in a thermosyphon system is only a small density difference and not a pump, larger than normal plumbing fixtures must be used to reduce pipe friction losses. In general, one pipe size larger than that would be used with a pump system is satisfactory. Under no conditions should piping smaller than 1/2 in. (12 mm) national pipe thread (NPT) be used. Most commercial thermosyphons use 1 in. (25 mm) NPT pipe. The flow rate through a thermosyphon system is about 1 gal/ft.^{2} h (40 L/m^2 h) in bright sun, based on collector area.

Because the hot water system loads vary little during a year, the best angle to tilt the collector is that equal to the local latitude. The temperature difference between the collector inlet water and the collector outlet water is usually 15°F–20°F (8°C–11°C) during the middle of a sunny day (Close 1962). After sunset, a thermosyphon system can reverse its flow direction and lose heat to the environment during the night. To avoid reverse flow, the top header of the absorber should be at least 1 ft. (30 cm) below the cold leg fitting on the storage tank, as shown.

To provide heat during long cloudy periods, an electrical immersion heater can be used as a backup for the solar system. The immersion heater is located near the top of the tank to enhance stratification so that the heated fluid is at the required delivery temperature at the delivery point. Tank stratification is desirable in a thermosyphon to maintain flow rates as high as possible. Insulation must be applied over the entire tank surface to control heat loss. Figure 41.2 illustrates two common thermosyphon system designs.

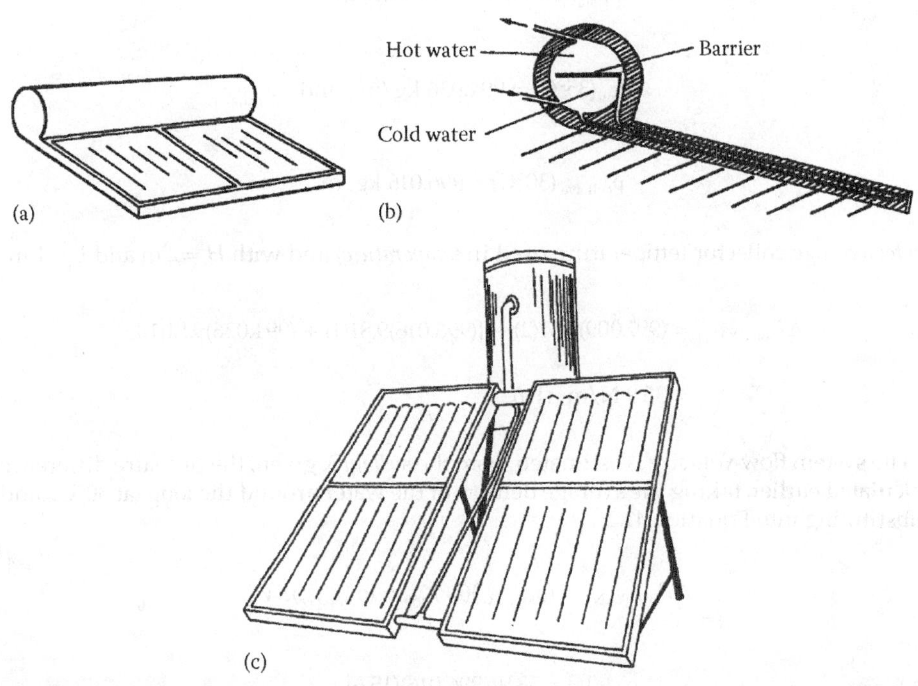

FIGURE 41.2
Passive solar water heaters: (a) compact model using combined collector and storage, (b) section view of the compact model, and (c) tank and collector assembly.

Several features inherent in the thermosyphon design limit its utility. If it is to be operated in a freezing climate, a nonfreezing fluid must be used, which in turn requires a heat exchanger between the collector and potable water storage. (If potable water is not required, the collector can be drained during cold periods instead.) Heat exchangers of either the shell-and-tube type or the immersion-coil type require higher flow rates for efficient operation than that a thermosyphon can provide. Therefore, the thermosyphon is generally limited to nonfreezing climates. A further restriction on thermosyphon use is the requirement for an elevated tank. In many cases, structural or architectural constraints prohibit raised-tank locations. In residences, collectors are normally mounted on the roof, and tanks mounted above the high point of the collector can easily become the highest point in a building. Practical considerations often do not permit this application.

Example 41.1

Determine the *pressure difference* available for a thermosyphon system with 1 m high collector and 2 m high *legs*. The water temperature input to the collector is 25°C, and the collector output temperature is 35°C. If the overall system loss velocity factor (K) is 15.6, estimate the system flow velocity.

Solution

Equation 41.1 is used to calculate the pressure difference, with the water densities being found from the temperatures (in steam tables)

$$\rho_{\text{stor}}(25°\text{C}) = 997.009 \text{ kg/m}^3,$$

$$\rho_{\text{out}}(35°\text{C}) = 994.036 \text{ kg/m}^3, \text{ and}$$

$$\rho_{\text{coll.ave.}}(30°\text{C}) = 996.016 \text{ kg/m}^3,$$

(*note*: average collector temperature used in *temperature*) and with $H = 2$ m and $L = 1$ m:

$$\Delta P_{\text{BUOYANT}} = (997.009)9.81(2) - [(996.016)9.81(1) + (994.036)9.81(1)],$$

$$= 38.9 \text{ N/m}^2 \text{ (Pa)}.$$

The system flow velocity is estimated from the *system* K given, the pressure difference calculated earlier, taking the average density of the water around the loop (at 30°C), and substituting into Equation 41.2:

$$\Delta P_{\text{BUOYANT}} = (\rho_{\text{loop.ave}})(h_L)_{\text{loop}} = (\rho_{\text{loop.ave}})KV^2,$$

$$V^2 = 38.9/(996.016)(15.6),$$

$$V = 0.05 \text{ m/s}.$$

41.3 Passive Solar Heating Design Fundamentals

Passive heating systems contain the five basic components of all solar systems, as described in Chapter 40. Typical passive realizations of these components are

1. Collector: windows, walls, and floors
2. Storage: walls and floors, large interior masses (often, these are integrated with the collector absorption function)
3. Distribution system: radiation, free convection, simple circulation fans
4. Controls: movable window insulation, vents both to other inside spaces or to ambient
5. Backup system: any nonsolar heating system

The design of passive systems requires the strategic placement of windows, storage masses, and the occupied spaces themselves. The fundamental principles of solar radiation geometry and availability are instrumental in the proper location and sizing of the system's *collectors* (windows). Storage devices are usually more massive than those used in active systems and are frequently an integral part of the collection and distribution system.

41.3.1 Types of Passive Heating Systems

A commonly used method of cataloging the various passive system concepts is to distinguish three general categories: direct, indirect, and isolated gain. Most of the physical configurations of passive heating systems are seen to fit within one of these three categories.

For direct gain (Figure 41.3), sunlight enters the heated space and is converted to heat at absorbing surfaces. This heat is then distributed throughout the space and to the various enclosing surfaces and room contents.

For indirect gain category systems, sunlight is absorbed and stored by a mass interposed between the glazing and the conditioned space. The conditioned space is partially enclosed and bounded by this thermal storage mass, so a natural thermal coupling is achieved. Examples of the indirect approach are the thermal storage wall, the thermal storage roof, and the northerly room of an attached sunspace.

In the thermal storage wall (Figure 41.4), sunlight penetrates the glazing and is absorbed and converted to heat at a wall surface interposed between the glazing and the heated space. The wall is usually masonry (Trombe wall) or containers filled with water (waterwall), although it might contain phase-change material. The attached sunspace (Figure 41.5) is actually a two-zone combination of direct gain and thermal storage wall. Sunlight enters and heats a direct gain southerly *sunspace* and also heats a mass wall separating the northerly buffered space, which is indirectly heated. The *sunspace* is frequently used as a greenhouse, in which case, the system is called an *attached greenhouse*. The thermal storage roof (Figure 41.6) is similar to the thermal storage wall except that the interposed thermal storage mass is located on the building roof.

The isolated gain category concept is an indirect system, except that there is a distinct thermal separation (by means of either insulation or physical separation) between the thermal storage and the heated space. The convective (thermosyphon) loop, as depicted in Figure 41.1, is in this category and, while often used to heat domestic water, is also used for building heating. It is most akin to conventional active systems in that there is a separate collector and separate thermal storage. The thermal storage wall, thermal storage roof,

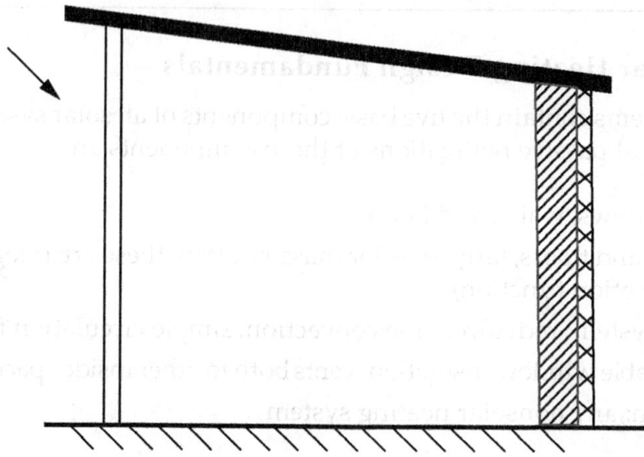

FIGURE 41.3
Direct gain.

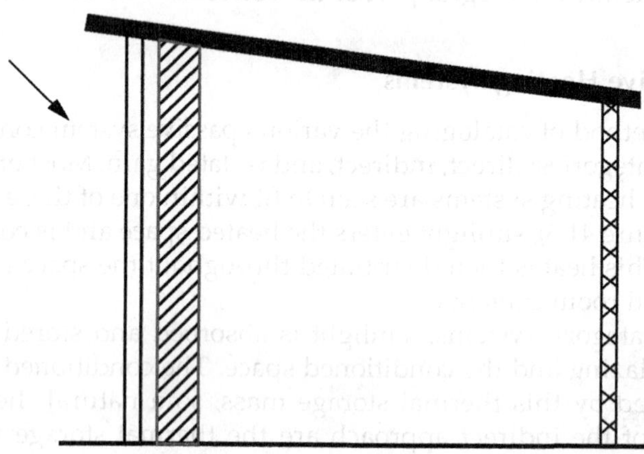

FIGURE 41.4
Thermal storage wall.

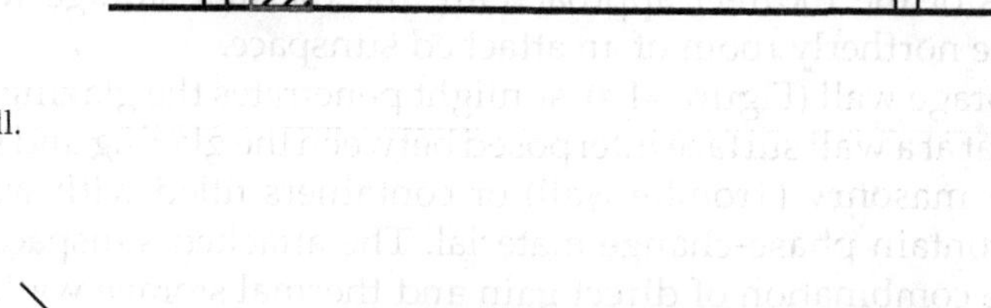

FIGURE 41.5
Attached sunspace.

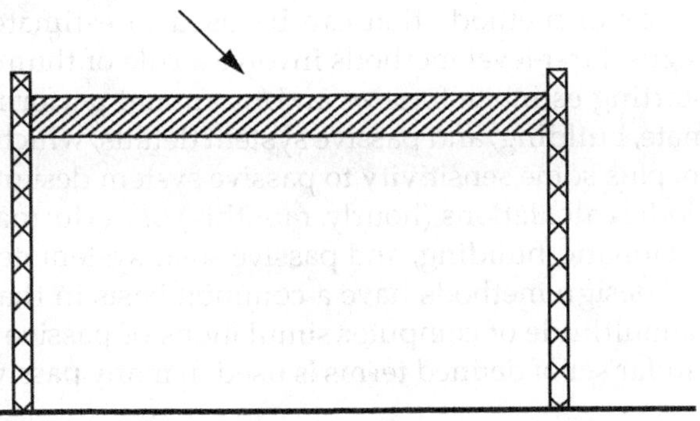

FIGURE 41.6
Thermal storage roof.

and attached sunspace approaches can also be made into isolated systems by insulating between the thermal storage and the heated space.

41.3.2 Fundamental Concepts for Passive Heating Design

Figure 41.7 is an equivalent thermal circuit for the building illustrated in Figure 41.4, the Trombe wall-type system. For the heat transfer analysis of the building, three temperature nodes can be identified: room temperature, storage wall temperature, and the ambient temperature. The circuit responds to climatic variables represented by a current injection I_s (solar radiation) and by the ambient temperature T_a. The storage temperature, T_s, and room temperature, T_r, are determined by current flows in the equivalent circuit. By using seasonal and annual climatic data, the performance of a passive structure can be simulated, and the results of many such simulations are correlated to give the design approaches described later in text.

41.3.3 Passive Design Approaches

Design of a passive heating system involves selection and sizing of the passive feature type(s), determination of thermal performance, and cost estimation. Ideally, a cost/performance optimization would be performed by the designer. Owner and architect ideas usually establish the passive feature type, with general size and cost estimation available. However, the thermal performance of a passive heating system has to be calculated.

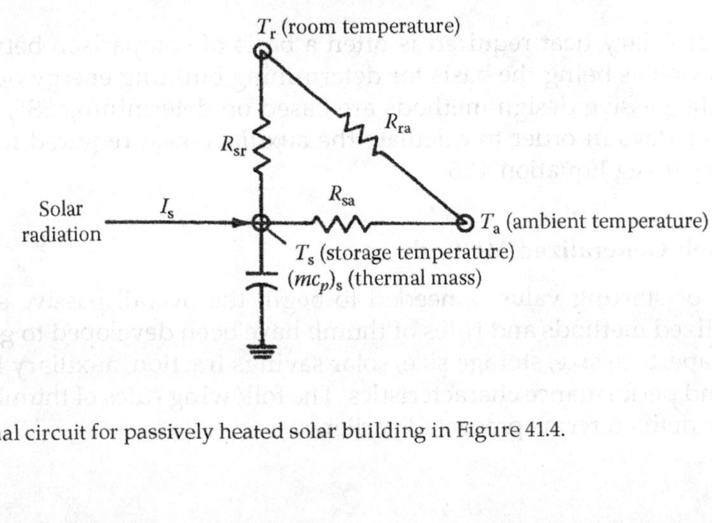

FIGURE 41.7
Equivalent thermal circuit for passively heated solar building in Figure 41.4.

There are several *levels* of methods that can be used to estimate the thermal performance of passive designs. First-level methods involve a rule of thumb and/or generalized calculation to get a starting estimate for size and/or annual performance. A second-level method involves climate, building, and passive system details, which allow annual performance determination, plus some sensitivity to passive system design changes. Third-level methods involve periodic calculations (hourly, monthly) of performance and permit more detailed variations of climatic, building, and passive solar system design parameters.

These three levels of design methods have a common basis in that they all are derived from correlations of a multitude of computer simulations of passive systems (PSDH 1980, 1984). As a result, a similar set of defined terms is used in many passive design approaches:

- A_p, solar projected area, m² (ft.²): the net south-facing passive solar glazing area projected onto a vertical plane
- NLC, net building load coefficient, kJ/CDD (Btu/FDD): net load of the nonsolar portion of the building per degree-day of indoor–outdoor temperature difference. The CDD and FDD terms refer to Celsius and Fahrenheit degree-days, respectively
- Q_{net}, net reference load, W h (Btu): heat loss from nonsolar portion of building as calculated by

$$Q_{net} = \text{NLC} \times (\text{number of degree-days}). \tag{41.3}$$

- LCR, load collector ratio, kJ/m² CDD (Btu/ft.² FDD): ratio of NLC to A_p,

$$\text{LCR} = \frac{\text{NLC}}{A_p} \tag{41.4}$$

- SSF, solar savings fraction, %: percentage reduction in required auxiliary heating relative to net reference load,

$$\text{SSF} = 1 - \frac{\text{Auxiliary heat required } (Q_{aux})}{\text{Net reference load } (Q_{net})}. \tag{41.5}$$

Therefore, using Equation 41.3, the auxiliary heat required, Q_{aux}, is given by

$$Q_{aux} = (1 - \text{SSF}) \times \text{NLC} \times (\text{number of degree-days}). \tag{41.6}$$

The amount of auxiliary heat required is often a basis of comparison between possible solar designs as well as being the basis for determining building energy operating costs. Thus, many of the passive design methods are based on determining SSF, NLC, and the number of degree-days in order to calculate the auxiliary heat required for a particular passive system by using Equation 41.6.

41.3.4 First Level: Generalized Methods

A first estimate or starting value is needed to begin the overall passive system design process. Generalized methods and rules of thumb have been developed to generate initial values for solar aperture size, storage size, solar savings fraction, auxiliary heat required, and other size and performance characteristics. The following rules of thumb are meant to be used with the defined terms presented earlier.

41.3.4.1 Load

A rule of thumb used in conventional building design is that a design heating load of 120–160 kJ/CDD per m^2 of floor area (6–8 Btu/FDD $ft.^2$) is considered an energy conservative design. Reducing these nonsolar values by 20% to solarize the proposed south-facing solar wall gives rule-of-thumb NLC values per unit of floor area:

$$\text{NLC/floor area} = 100\text{–}130\ \text{kJ/CDD m}^2\ (4.8\text{–}6.4\ \text{Btu/FDD ft.}^2). \tag{41.7}$$

41.3.4.2 Solar Savings Fraction

A method of getting starting-point values for the solar savings fraction is presented in Figure 41.8 (PSDH 1984). The map values represent optimum SSF (in percent) for a particular set of conservation and passive-solar costs for different climates across the United States. With the Q_{net} generated from the NLC rule of thumb (see earlier) and the SSF read from the map, the Q_{aux} can be determined.

41.3.4.3 Load Collector Ratio

The A_p can be determined using the NLC given earlier if the LCR is known. The rule of thumb associated with *good* values of LCR (PSDH 1984) differs depending on whether the design is for a *cold* or *warm* climate:

$$\text{"Good"LCR} = \begin{cases} \text{For cold climate:} & 410\ \text{kJ/m}^2\,\text{CDD}\,(20\ \text{Btu/ft.}^2\,\text{FDD}) \\ \text{For warm climate:} & 610\ \text{kJ/m}^2\,\text{CDD}\,(30\ \text{Btu/ft.}^2\,\text{FDD}) \end{cases} \tag{41.8}$$

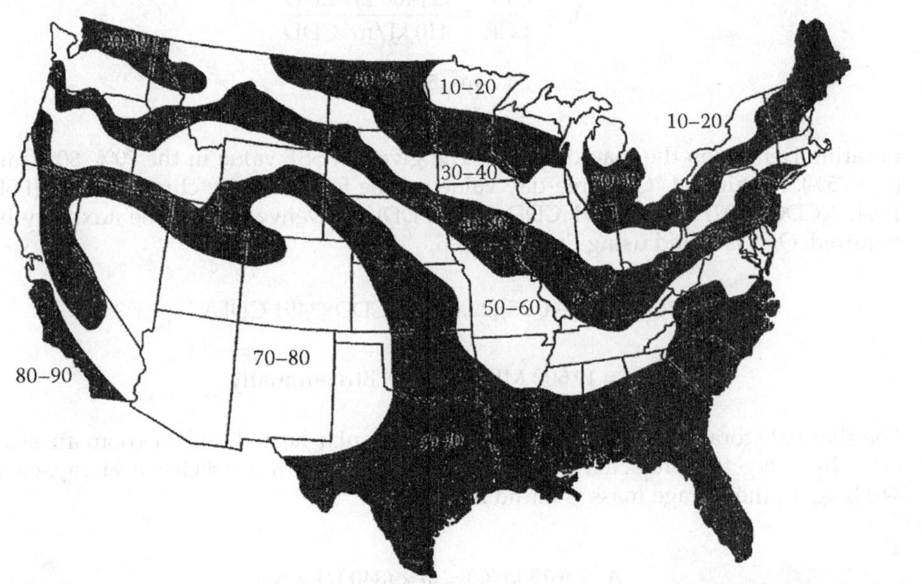

FIGURE 41.8
Starting-point values of solar savings fraction (SSF) in percent. (From PSDH, *Passive Solar Design Handbook*, Part One: Total Environmental Action, Inc., Part Two: Los Alamos Scientific Laboratory, Part Three: Los Alamos National Laboratory. Van Nostrand Reinhold, New York, 1984.)

41.3.4.4 Storage

Rules of thumb for thermal mass storage relate storage material total heat capacity to the solar projected area (PSDH 1984). The use of the storage mass is to provide for heating on cloudy days and to regulate sunny day room air temperature swing. When the thermal mass directly absorbs the solar radiation, each square meter of the projected glazing area requires enough mass to store 613 kJ/°C. If the storage material is not in direct sunlight, but heated from room air only, then four times as much mass is needed. In a room with a directly sunlight-heated storage mass, the room air temperature swing will be approximately one-half the storage mass temperature swing. For room air heated storage, the air temperature swing is twice that of the storage mass.

Example 41.2

A Denver, Colorado, building is to have a floor area of 195 m^2 (2100 $ft.^2$). Determine rule-of-thumb size and performance characteristics.

Solution

From Equation 41.5, the NLC is estimated as

$$\text{NLC} = (115\ \text{kJ/CDD}\,\text{m}^2) \times (195\ \text{m}^2)$$

$$= 22{,}400\ \text{kJ/CDD}\,(11{,}800\ \text{Btu/FDD}).$$

Using the *cold* LCR value and Equation 41.4, the passive solar projected area is

$$A_\text{p} = \frac{\text{NLC}}{\text{LCR}} = \frac{22{,}400\ \text{kJ/CDD}}{410\ \text{kJ/m}^2\,\text{CDD}}$$

$$= 54.7\ \text{m}^2\,(588\ \text{ft.}^2).$$

Locating Denver on the map of Figure 41.8 gives an SSF value in the 70%–80% range (use 75%). An annual °C-degree-day value can be found in city climate tables (PSDH 1984; NCDC 1992) and is 3491 CDD (6283 FDD) for Denver. Thus, the auxiliary heat required, Q_{aux}, is found using Equation 41.6:

$$Q_\text{aux} = (1 - 0.75)(22{,}400\ \text{kJ/CDD})(3491\ \text{CDD})$$

$$= 19{,}600\ \text{MJ}\ (18.5 \times 10^6\ \text{Btu})\ \text{annually}.$$

The thermal storage can be sized using directly solar-heated and/or room air heated mass by using the projected area. Assuming brick with a specific heat capacity of 840 J/kg °C, the storage mass is found by

$$A_\text{p} \times (613\ \text{kJ/C}) = m \times (840\ \text{J/kg °C}),$$

$$m_\text{d} = 40{,}000\ \text{kg}\ (88{,}000\ \text{lbm})\ [\text{direct sun}],$$

$$\text{or}\quad m_\text{a} = 160{,}000\ \text{kg}\ (351{,}000\ \text{lbm})\ [\text{air heated}].$$

A more location-dependent set of rules of thumb is presented in PSDH (1980). The first rule of thumb relates solar projected area as a percentage of floor area to solar savings fraction, with and without night insulation of the solar glazing:

> A solar projected area of (B1)% to (B2)% of the floor area can be expected to produce a SSF in (location) of (S1)% to (S2)%, or, if R9 night insulation is used, of (S3)% to (S4)%

The values of B1, B2, S1, S2, S3, and S4 are found using Table 41.1 for the location. The thermal storage mass rule of thumb is again related to the solar projected area:

> A thermal storage wall should have 14 kg × SSF (%) of water or 71 kg × SSF (%) of masonry for each square meter of solar projected area. For a direct gain space, the mass above should be used with a surface area of at least three times the solar projected area, and masonry no thicker than 10–15 cm. If the mass is located in back rooms, then four times the above mass is needed.

Example 41.3

Determine size and performance passive solar characteristics with the location-dependent set of rules of thumb for the house of the previous example.

Solution

Using Table 41.1 with the 195 m² house in Denver yields

$$\text{Solar projected area} = 12\%\text{–}23\% \text{ of floor area}$$
$$= 23.4\text{–}44.9 \text{ m}^2.$$
$$\text{SSF (no night insulation)} = 27\%\text{–}43\%.$$
$$\text{SSF (R9 night insulation)} = 47\%\text{–}74\%.$$

Using the rule of thumb for the thermal storage mass,

$$m = 17 \text{ kg} \times 43\% \times 44.9 \text{ m}^2$$
$$= 33{,}000 \text{ kg } (72{,}000 \text{ lbm}) \text{ [Thermal wall or direct gain]}.$$

Comparing the results of this example to those of the previous example, the two rules of thumb are seen to produce *roughly* similar answers. General system cost and performance information can be generated with results from rule-of-thumb calculations, but a more detailed level of information is needed to determine design-ready passive system type (direct gain, thermal wall, sunspace), size, performance, and costs.

41.3.5 Second Level: LCR Method

The LCR method is useful for making estimates of the annual performance of specific types of passive system(s) combinations. The LCR method was developed by calculating the annual SSF for 94 reference passive solar systems for 219 U.S. and Canadian locations over a range of LCR values. Table 41.2 includes the description of these 94 reference systems for use both with the LCR method and with the solar load ratio (SLR) method described later. Tables were constructed for each city with LCR versus SSF listed for each of the 94 reference passive systems. (Note that the SLR method was used to make the LCR calculations, and this SLR method is described in the next section as the third-level method.) Although the complete LCR tables (PSDH 1984) include 219 locations, Table 41.3 includes only 6 *representative* cities

TABLE 41.1

Values to Be Used in the Glazing Area and SSF Relations Rules of Thumb

City	B1	B2	S1	S2	S3	S4
Birmingham, Alabama	0.09	0.18	22	37	34	58
Mobile, Alabama	0.06	0.12	26	44	34	60
Montgomery, Alabama	0.07	0.15	24	41	34	59
Phoenix, Arizona	0.06	0.12	37	60	48	75
Prescott, Arizona	0.10	0.20	29	48	44	72
Tucson, Arizona	0.06	0.12	35	57	45	73
Winslow, Arizona	0.12	0.24	30	47	48	74
Yuma, Arizona	0.04	0.09	43	66	51	78
Fort Smith, Arkansas	0.10	0.20	24	39	38	64
Little Rock, Arkansas	0.10	0.19	23	38	37	62
Bakersfield, California	0.08	0.15	31	50	42	67
Baggett, California	0.07	0.15	35	56	46	73
Fresno, California	0.09	0.17	29	46	41	65
Long Beach, California	0.05	0.10	35	58	44	72
Los Angeles, California	0.05	0.09	36	58	44	72
Mount Shasta, California	0.11	0.21	24	38	42	67
Needles, California	0.06	0.12	39	61	49	76
Oakland, California	0.07	0.15	35	55	46	72
Red Bluff, California	0.09	0.18	29	46	41	65
Sacramento, California	0.09	0.18	29	47	41	66
San Diego, California	0.04	0.09	37	61	46	74
San Francisco, California	0.06	0.13	34	54	45	71
Santa Maria, California	0.05	0.11	31	53	42	69
Colorado Springs, Colorado	0.12	0.24	27	42	47	74
Denver, Colorado	0.12	0.23	27	43	47	74
Eagle, Colorado	0.14	0.29	25	35	53	77
Grand Junction, Colorado	0.13	0.27	29	43	50	76
Pueblo, Colorado	0.11	0.23	29	45	48	75
Hartford, Connecticut	0.17	0.35	14	19	40	64
Wilmington, Delaware	0.15	0.29	19	30	39	63
Washington, District of Columbia	0.12	0.23	18	28	37	61
Apalachicola, Florida	0.05	0.10	28	47	36	61
Daytona Beach, Florida	0.04	0.08	30	51	36	63
Jacksonville, Florida	0.05	0.09	27	47	35	62
Miami, Florida	0.01	0.02	27	48	31	54
Orlando, Florida	0.03	0.06	30	52	37	63
Tallahassee, Florida	0.05	0.11	26	45	35	60
Tampa, Florida	0.03	0.06	30	52	36	63
West Palm Beach, Florida	0.01	0.03	30	51	34	59
Atlanta, Georgia	0.06	0.17	22	36	34	58
Augusta, Georgia	0.06	0.16	24	40	35	60
Macon, Georgia	0.07	0.15	25	41	35	59
Savannah, Georgia	0.06	0.13	25	43	35	60
Boise, Idaho	0.14	0.28	27	38	48	71

(Continued)

TABLE 41.1 (*Continued*)

Values to Be Used in the Glazing Area and SSF Relations Rules of Thumb

City	B1	B2	S1	S2	S3	S4
Lewiston, Idaho	0.15	0.29	22	29	44	65
Pocatello, Idaho	0.13	0.26	25	35	51	74
Chicago, Illinois	0.17	0.35	17	23	43	67
Moline, Illinois	0.20	0.39	17	22	46	70
Springfield, Illinois	0.15	0.30	19	26	42	67
Evansville, Indiana	0.14	0.27	19	29	37	61
Fort Wayne, Indiana	0.16	0.33	13	17	37	60
Indianapolis, Indiana	0.14	0.28	15	21	37	60
South Bend, Indiana	0.18	0.35	12	15	39	61
Burlington, Iowa	0.18	0.36	20	27	47	71
Des Moines, Iowa	0.21	0.43	19	25	58	75
Mason City, Iowa	0.22	0.44	18	19	56	79
Sioux City, Iowa	0.23	0.46	20	24	53	76
Dodge City, Kansas	0.12	0.23	27	42	46	73
Goodland, Kansas	0.13	0.27	26	39	47	74
Topeka, Kansas	0.14	0.26	24	35	45	71
Wichita, Kansas	0.14	0.26	26	41	45	72
Lexington, Kentucky	0.13	0.27	17	26	35	58
Louisville, Kentucky	0.13	0.27	18	27	35	59
Baton Rouge, Louisiana	0.06	0.12	26	43	34	59
Lake Charles, Louisiana	0.06	0.11	24	41	32	57
New Orleans, Louisiana	0.05	0.11	27	46	35	61
Shreveport, Louisiana	0.08	0.15	26	43	36	61
Caribou, Maine	0.25	0.30	NR	NR	53	74
Portland, Maine	0.17	0.34	14	17	45	69
Baltimore, Maryland	0.14	0.27	19	30	38	62
Boston, Massachusetts	0.15	0.29	17	25	40	64
Alpena, Michigan	0.21	0.42	NR	NR	47	69
Detroit, Michigan	0.17	0.34	13	17	39	61
Flint, Michigan	0.15	0.31	11	12	40	62
Grand Rapids, Michigan	0.19	0.38	12	13	39	61
Sault Ste. Marie, Michigan	0.25	0.50	NR	NR	50	70
Traverse City, Michigan	0.18	0.36	NR	NR	42	62
Duluth, Minnesota	0.25	0.50	NR	NR	50	70
International Falls, Minnesota	0.25	0.50	NR	NR	47	66
Minneapolis–St. Paul, Minnesota	0.25	0.50	NR	NR	55	76
Rochester, Minnesota	0.24	0.49	NR	NR	54	76
Jackson, Mississippi	0.06	0.15	24	48	34	59
Meridian, Mississippi	0.08	0.15	23	39	34	58
Columbia, Missouri	0.13	0.26	20	30	41	66
Kansas City, Missouri	0.14	0.29	22	32	44	70
Saint Louis, Missouri	0.15	0.29	21	33	41	65
Springfield, Missouri	0.13	0.26	22	34	40	65
Billings, Montana	0.16	0.32	24	31	53	76

(*Continued*)

TABLE 41.1 (*Continued*)

Values to Be Used in the Glazing Area and SSF Relations Rules of Thumb

City	B1	B2	S1	S2	S3	S4
Cut Bank, Montana	0.24	0.49	22	23	62	81
Dillon, Montana	0.16	0.32	24	32	54	77
Glasgow, Montana	0.25	0.50	NR	NR	55	75
Great Falls, Montana	0.18	0.37	23	26	56	77
Helena, Montana	0.20	0.39	21	25	55	77
Lewistown, Montana	0.19	0.38	21	25	54	76
Miles City, Montana	0.23	0.47	21	23	60	80
Missoula, Montana	0.18	0.36	15	16	47	68
Grand Island, Nebraska	0.18	0.36	24	33	51	76
North Omaha, Nebraska	0.20	0.48	21	29	51	76
North Platte, Nebraska	0.17	0.34	25	36	50	76
Scottsbluff, Nebraska	0.16	0.31	24	36	49	74
Elko, Nevada	0.12	0.25	27	39	52	76
Ely, Nevada	0.12	0.23	27	41	50	77
Las Vegas, Nevada	0.09	0.18	35	56	48	75
Lovelock, Nevada	0.13	0.25	32	48	53	78
Reno, Nevada	0.11	0.22	31	48	49	76
Tonopah, Nevada	0.11	0.23	31	48	51	77
Winnemucca, Nevada	0.13	0.26	28	42	49	75
Concord, New Hampshire	0.17	0.34	13	15	45	68
Newark, New Jersey	0.13	0.25	19	29	39	64
Albuquerque, New Mexico	0.11	0.22	29	47	46	73
Clayton, New Mexico	0.10	0.20	28	45	45	73
Farmington, New Mexico	0.12	0.24	29	45	49	76
Los Alamos, New Mexico	0.11	0.22	25	40	44	72
Roswell, New Mexico	0.10	0.19	30	49	45	73
Truth or Consequences, New Mexico	0.09	0.17	32	51	46	73
Tucumcari, New Mexico	0.10	0.20	30	48	45	73
Zuni, New Mexico	0.11	0.21	27	43	45	73
Albany, New York	0.21	0.41	13	15	43	66
Binghamton, New York	0.15	0.30	NR	NR	35	56
Buffalo, New York	0.19	0.37	NR	NR	36	57
Massena, New York	0.25	0.50	NR	NR	50	71
New York (Central Park), New York	0.15	0.30	16	25	36	59
Rochester, New York	0.18	0.37	NR	NR	37	58
Syracuse, New York	0.19	0.38	NR	NR	37	59
Asheville, North Carolina	0.10	0.20	21	35	36	61
Cape Hatteras, North Carolina	0.09	0.17	24	40	36	60
Charlotte, North Carolina	0.08	0.17	23	38	36	60
Greensboro, North Carolina	0.10	0.20	23	37	37	63
Raleigh–Durham, North Carolina	0.09	0.19	22	37	36	61
Bismarck, North Dakota	0.25	0.50	NR	NR	56	77
Fargo, North Dakota	0.25	0.50	NR	NR	51	72
Minot, North Dakota	0.25	0.50	NR	NR	52	72

(*Continued*)

TABLE 41.1 (*Continued*)

Values to Be Used in the Glazing Area and SSF Relations Rules of Thumb

City	B1	B2	S1	S2	S3	S4
Akron–Canton, Ohio	0.15	0.31	12	16	35	57
Cincinnati, Ohio	0.12	0.24	15	23	35	57
Cleveland, Ohio	0.15	0.31	11	14	34	55
Columbus, Ohio	0.14	0.28	13	18	35	57
Dayton, Ohio	0.14	0.28	14	20	36	59
Toledo, Ohio	0.17	0.34	13	17	38	61
Youngstown, Ohio	0.16	0.32	NR	NR	34	54
Oklahoma City, Oklahoma	0.11	0.22	25	41	41	67
Tulsa, Oklahoma	0.11	0.22	24	38	40	65
Astoria, Oregon	0.09	0.19	21	34	37	60
Burns, Oregon	0.13	0.25	23	32	47	71
Medford, Oregon	0.12	0.24	21	32	38	60
North Bend, Oregon	0.09	0.17	25	42	38	64
Pendleton, Oregon	0.14	0.27	22	30	43	64
Portland, Oregon	0.13	0.26	21	31	38	60
Redmond, Oregon	0.13	0.27	26	38	47	71
Salem, Oregon	0.12	0.24	21	32	37	59
Allentown, Pennsylvania	0.15	0.29	16	24	39	63
Erie, Pennsylvania	0.17	0.34	NR	NR	35	55
Harrisburg, Pennsylvania	0.13	0.26	17	26	38	62
Philadelphia, Pennsylvania	0.15	0.29	19	29	38	62
Pittsburgh, Pennsylvania	0.14	0.28	12	16	33	55
Wilkes-Barre–Scranton, Pennsylvania	0.16	0.32	13	18	37	60
Providence, Rhode Island	0.15	0.30	17	24	40	64
Charleston, South Carolina	0.07	0.14	25	41	34	59
Columbia, South Carolina	0.08	0.17	25	41	36	61
Greenville–Spartanburg, South Carolina	0.08	0.17	23	38	36	60
Huron, South Dakota	0.25	0.50	NR	NR	58	79
Pierre, South Dakota	0.22	0.43	21	23	58	80
Rapid City, South Dakota	0.15	0.30	23	32	51	76
Sioux Falls, South Dakota	0.22	0.45	18	19	57	79
Chattanooga, Tennessee	0.09	0.19	19	32	33	56
Knoxville, Tennessee	0.09	0.18	20	33	33	56
Memphis, Tennessee	0.09	0.19	22	36	36	60
Nashville, Tennessee	0.10	0.21	19	30	33	55
Abilene, Texas	0.09	0.18	29	47	41	68
Amarillo, Texas	0.11	0.22	29	46	45	72
Austin, Texas	0.06	0.13	27	46	37	63
Brownsville, Texas	0.03	0.06	27	46	32	57
Corpus Christi, Texas	0.05	0.09	29	49	36	63
Dallas, Texas	0.08	0.17	27	44	38	64
Del Rio, Texas	0.06	0.12	30	50	39	66
El Paso, Texas	0.09	0.17	32	53	45	72
Forth Worth, Texas	0.09	0.17	26	44	38	64

(*Continued*)

TABLE 41.1 (*Continued*)

Values to Be Used in the Glazing Area and SSF Relations Rules of Thumb

City	B1	B2	S1	S2	S3	S4
Houston, Texas	0.06	0.11	25	43	34	59
Laredo, Texas	0.05	0.09	31	52	39	64
Lubbock, Texas	0.09	0.19	30	49	44	72
Lufkin, Texas	0.07	0.14	26	43	35	61
Midland–Odessa, Texas	0.09	0.18	32	52	44	72
Port Arthur, Texas	0.06	0.11	26	44	34	60
San Angelo, Texas	0.08	0.15	29	48	40	67
San Antonio, Texas	0.06	0.12	28	48	38	64
Sherman, Texas	0.10	0.20	25	41	38	64
Waco, Texas	0.06	0.15	27	45	38	64
Wichita Falls, Texas	0.10	0.20	27	45	41	67
Bryce Canyon, Utah	0.13	0.25	26	39	52	78
Cedar City, Utah	0.12	0.24	28	43	48	75
Salt Lake City, Utah	0.13	0.26	27	39	48	72
Burlington, Vermont	0.22	0.43	NR	NR	46	68
Norfolk, Virginia	0.09	0.19	23	38	37	62
Richmond, Virginia	0.11	0.22	21	34	37	61
Roanoke, Virginia	0.11	0.23	21	34	37	61
Olympia, Washington	0.12	0.23	20	29	38	59
Seattle–Tacoma, Washington	0.11	0.22	21	30	39	59
Spokane, Washington	0.20	0.39	20	24	48	68
Yakima, Washington	0.18	0.36	24	31	49	70
Charleston, West Virginia	0.13	0.25	16	24	32	54
Huntington, West Virginia	0.13	0.25	17	27	34	57
Eau Claire, Wisconsin	0.25	0.50	NR	NR	53	75
Green Bay, Wisconsin	0.23	0.46	NR	NR	53	75
La Crosse, Wisconsin	0.21	0.43	NR	NR	52	75
Madison, Wisconsin	0.20	0.40	15	17	51	74
Milwaukee, Wisconsin	0.18	0.35	15	18	48	71
Casper, Wyoming	0.13	0.26	27	39	53	78
Cheyenne, Wyoming	0.11	0.21	25	39	47	74
Rock Springs, Wyoming	0.14	0.28	26	38	54	79
Sheridan, Wyoming	0.16	0.31	22	30	52	75
Canada						
Edmonton, Alberta	0.25	0.50	NR	NR	54	72
Suffield, Alberta	0.25	0.50	28	30	67	85
Nanaimo, British Columbia	0.13	0.26	26	35	45	66
Vancouver, British Columbia	0.13	0.26	20	28	48	60
Winnipeg, Manitoba	0.25	0.50	NR	NR	54	74
Dartmouth, Nova Scotia	0.14	0.28	17	24	45	70
Moosonee, Ontario	0.25	0.50	NR	NR	48	67
Ottawa, Ontario	0.25	0.50	NR	NR	59	80
Toronto, Ontario	0.18	0.36	17	23	44	68
Normandie, Quebec	0.25	0.50	NR	NR	54	74

Source: PSDH, *Passive Solar Design Handbook*, U.S. Department of Energy, Washington, DC, 1980.

Note: NR, not recommended.

TABLE 41.2

Designations and Characteristics for 94 Reference Systems

(a) Overall System Characteristics	
Masonry Properties	
Thermal conductivity (*k*)	
Sunspace floor	0.5 Btu/h/ft./°F
All other masonry	1.0 Btu/h/ft./°F
Density (*Q*)	150 lb/ft.^{3}
Specific heat (*c*)	0.2 Btu/lb/°F
Infrared emittance of normal surface	0.9
Infrared emittance of selective surface	0.1
Solar Absorptances	
Waterwall	1.0
Masonry, Trombe wall	1.0
Direct gain and sunspace	0.8
Sunspace: water containers	0.9
Lightweight common wall	0.7
Other lightweight surfaces	0.3
Glazing Properties	
Transmission characteristics	Diffuse
Orientation	Due south
Index of refraction	1.526
Extinction coefficient	0.5 in.$^{-1}$
Thickness of each pane	⅛ in.
Gap between panes	½ in.
Ared emittance	0.9
Control Range	
Room temperature	65°F–75°F
Sunspace temperature	45°F–95°F
Internal heat generation	0
Thermocirculation Vents (when Used)	
Vent area/projected area (sum of both upper and lower vents)	0.06
Height between vents	8 ft.
Reverse flow	None
Nighttime Insulation (when used)	
Thermal resistance	R9
In place, solar time	5:30 p.m. to 7:30 a.m.
Solar Radiation Assumptions	
Shading	None
Ground diffuse reflectance	0.3

(Continued)

TABLE 41.2 (*Continued*)

Designations and Characteristics for 94 Reference Systems

(b) Direct-Gain (DG) System Types

Designation	Thermal Storage Capacity[a] (Btu/ft.²/°F)	Mass Thickness[a] (in.)	Mass-Area-to-Glazing-Area Ratio	No. of Glazings	Nighttime Insulation
A1	30	2	6	2	No
A2	30	2	6	3	No
A3	30	2	6	2	Yes
B1	45	6	3	2	No
B2	45	6	3	3	No
B3	45	6	3	2	Yes
C1	60	4	6	2	No
C2	60	4	6	3	No
C3	60	4	6	2	Yes

(c) Vented Trombe Wall (TW) System Types

Designation	Thermal Storage Capacity[a] (Btu/ft.²/°F)	Wall Thickness[a] (in.)	ρck (Btu²/h/ft.⁴/°F²)	No. of Glazings	Wall Surface	Nighttime Insulation
A1	15	6	30	2	Normal	No
A2	22.5	9	30	2	Normal	No
A3	30	12	30	2	Normal	No
A4	45	18	30	2	Normal	No
B1	15	6	15	2	Normal	No
B2	22.5	9	15	2	Normal	No
B3	30	12	15	2	Normal	No
B4	45	18	15	2	Normal	No
C1	15	6	7.5	2	Normal	No
C2	22.5	9	7.5	2	Normal	No
C3	30	12	7.5	2	Normal	No
C4	45	18	7.5	2	Normal	No
D1	30	12	30	1	Normal	No
D2	30	12	30	3	Normal	No
D3	30	12	30	1	Normal	Yes
D4	30	12	30	2	Normal	Yes
D5	30	12	30	3	Normal	Yes
E1	30	12	30	1	Selective	No
E2	30	12	30	2	Selective	No
E3	30	12	30	1	Selective	Yes
E4	30	12	30	2	Selective	Yes

(d) Unvented Trombe Wall (TW) System Types

Designation	Thermal Storage Capacity[a] (Btu/ft.²/°F)	Wall Thickness[a] (in.)	ρck (Btu²/h/ft.⁴/°F²)	No. of Glazings	Wall Surface	Nighttime Insulation
F1	15	6	30	2	Normal	No
F2	22.5	9	30	2	Normal	No
F3	30	12	30	2	Normal	No

(*Continued*)

TABLE 41.2 (*Continued*)

Designations and Characteristics for 94 Reference Systems

(d) Unvented Trombe Wall (TW) System Types

Designation	Thermal Storage Capacity[a] (Btu/ft.²/°F)	Wall Thickness[a] (in.)	ρck (Btu²/h/ft.⁴/°F²)	No. of Glazings	Wall Surface	Nighttime Insulation
F4	45	18	30	2	Normal	No
G1	15	6	15	2	Normal	No
G2	22.5	9	15	2	Normal	No
G3	30	12	15	2	Normal	No
G4	45	18	15	2	Normal	No
H1	15	6	7.5	2	Normal	No
H2	22.5	9	7.5	2	Normal	No
H3	30	12	7.5	2	Normal	No
H4	45	18	7.5	2	Normal	No
I1	30	12	30	1	Normal	No
I2	30	12	30	3	Normal	No
I3	30	12	30	1	Normal	Yes
I4	30	12	30	2	Normal	Yes
I5	30	12	30	3	Normal	Yes
J1	30	12	30	1	Selective	No
J2	30	12	30	2	Selective	No
J3	30	12	30	1	Selective	Yes
J4	30	12	30	2	Selective	Yes

(e) Waterwall (WW) System Types

Designation	Thermal Storage Capacity[a] (Btu/ft.²/°F)	Wall Thickness (in.)	No. of Glazings	Wall Surface	Nighttime Insulation
A1	15.6	3	2	Normal	No
A2	31.2	6	2	Normal	No
A3	46.8	9	2	Normal	No
A4	62.4	12	2	Normal	No
A5	93.6	18	2	Normal	No
A6	124.8	24	2	Normal	No
B1	46.8	9	1	Normal	No
B2	46.8	9	3	Normal	No
B3	46.8	9	1	Normal	Yes
B4	46.8	9	2	Normal	Yes
B5	46.8	9	3	Normal	Yes
C1	46.8	9	1	Selective	No
C2	46.8	9	2	Selective	No
C3	46.8	9	1	Selective	Yes
C4	46.8	9	2	Selective	Yes

(*Continued*)

TABLE 41.2 (*Continued*)

Designations and Characteristics for 94 Reference Systems

(f) Sunspace (SS) System Types

Designation	Type	Tilt (°)	Common Wall	End Walls	Nighttime Insulation
A1	Attached	50	Masonry	Opaque	No
A2	Attached	50	Masonry	Opaque	Yes
A3	Attached	50	Masonry	Glazed	No
A4	Attached	50	Masonry	Glazed	Yes
A5	Attached	50	Insulated	Opaque	No
A6	Attached	50	Insulated	Opaque	Yes
A7	Attached	50	Insulated	Glazed	No
A8	Attached	50	Insulated	Glazed	Yes
B1	Attached	90/30	Masonry	Opaque	No
B2	Attached	90/30	Masonry	Opaque	Yes
B3	Attached	90/30	Masonry	Glazed	No
B4	Attached	90/30	Masonry	Glazed	Yes
B5	Attached	90/30	Insulated	Opaque	No
B6	Attached	90/30	Insulated	Opaque	Yes
B7	Attached	90/30	Insulated	Glazed	No
B8	Attached	90/30	Insulated	Glazed	Yes
C1	Semienclosed	90	Masonry	Common	No
C2	Semienclosed	90	Masonry	Common	Yes
C3	Semienclosed	90	Insulated	Common	No
C4	Semienclosed	90	Insulated	Common	Yes
D1	Semienclosed	50	Masonry	Common	No
D2	Semienclosed	50	Masonry	Common	Yes
D3	Semienclosed	50	Insulated	Common	No
D4	Semienclosed	50	Insulated	Common	Yes
E1	Semienclosed	90/30	Masonry	Common	No
E2	Semienclosed	90/30	Masonry	Common	Yes
E3	Semienclosed	90/30	Insulated	Common	No
E4	Semienclosed	90/30	Insulated	Common	Yes

Source: PSDH, *Passive Solar Design Handbook*, Part One: Total Environmental Action, Inc., Part Two: Los Alamos Scientific Laboratory, Part Three: Los Alamos National Laboratory, Van Nostrand Laboratory, Reinhold, New York, 1984.

[a] The thermal storage capacity is per unit of projected area, or, equivalently, the quantity ρck. The wall thickness is listed only as an appropriate guide by assuming $\rho c = 30$ Btu/ft.³/°F.

(Albuquerque, Boston, Madison, Medford, Nashville, and Santa Maria), purely due to space restrictions. The LCR method consists of the following steps (PSDH 1984):

1. Determine the following building parameters:
   a. Building load coefficient, NLC
   b. Solar projected area, A_p
   c. Load collector ratio, LCR = NLC/A_p
2. Find the short designation of the reference system closest to the passive system design (Table 41.2).

TABLE 41.3

LCR Tables for Six Representative Cities (Albuquerque, Boston, Madison, Medford, Nashville, and Santa Maria)

SSF	0.10	0.20	0.30	0.40	0.50	0.60	0.70	0.80	0.90
Santa Maria, California									**3053 DD**
WW A1	1776	240	119	73	50	35	25	18	12
WW A2	617	259	159	103	74	54	39	28	19
WW A3	523	261	164	114	82	61	45	33	22
WW A4	482	260	169	119	87	65	48	35	24
WW A5	461	263	175	125	92	69	52	38	26
WW A6	447	263	177	128	95	72	54	40	27
WW B1	556	220	128	85	60	43	32	23	15
WW B2	462	256	168	119	88	66	49	36	25
WW B3	542	315	211	151	112	85	64	47	32
WW B4	455	283	197	144	109	83	63	47	32
WW B5	414	263	184	136	103	79	60	45	31
WW C1	569	330	221	159	118	89	67	49	33
WW C2	478	288	197	143	107	81	61	45	31
WW C3	483	318	228	170	130	100	77	57	40
WW C4	426	280	200	149	114	88	68	51	35
TW A1	1515	227	113	70	48	34	24	17	11
TW A2	625	234	134	89	63	46	33	24	16
TW A3	508	231	140	95	68	50	37	27	18
TW A4	431	217	137	95	69	51	38	28	19
TW B1	859	212	112	71	49	35	25	18	12
TW B2	502	209	124	83	59	43	32	23	15
TW B3	438	201	123	84	60	44	33	24	16
TW B4	400	184	112	76	55	40	30	22	14
TW C1	568	188	105	69	48	35	25	18	12
TW C2	435	178	105	70	50	36	27	19	13
TW C3	413	165	97	64	46	33	25	18	12
TW C4	426	146	82	54	38	27	20	14	10
TW D1	403	170	101	67	48	35	25	18	12
TW D2	488	242	152	105	76	57	42	31	21
TW D3	509	271	175	123	90	67	50	36	25
TW D4	464	266	177	127	94	71	53	39	27
TW D5	425	250	169	122	91	69	52	38	26
TW E1	581	309	199	140	102	76	57	42	28
TW E2	512	283	186	132	97	73	55	40	27
TW E3	537	328	225	164	123	94	71	53	36
TW E4	466	287	199	145	109	83	63	47	32
TW F1	713	198	107	68	47	34	25	18	12
TW F2	455	199	120	81	58	42	31	22	15
TW F3	378	190	120	83	60	45	33	24	16
TW F4	311	169	110	77	57	42	32	23	16
TW G1	450	170	98	65	46	33	24	17	12

(Continued)

TABLE 41.3 (*Continued*)

LCR Tables for Six Representative Cities (Albuquerque, Boston, Madison, Medford, Nashville, and Santa Maria)

SSF	0.10	0.20	0.30	0.40	0.50	0.60	0.70	0.80	0.90
TW G2	331	163	102	70	51	38	28	20	14
TW G3	278	147	94	66	48	36	27	20	13
TW G4	222	120	78	55	40	30	22	16	11
TW H1	295	137	84	57	41	30	22	16	11
TW H2	226	118	75	52	38	28	21	15	10
TW H3	187	99	64	44	33	24	18	13	9
TW H4	143	75	48	33	24	18	14	10	7
TW I1	318	144	88	59	42	31	23	16	11
TW I2	377	203	132	93	68	51	38	28	19
TW I3	404	226	149	106	78	58	44	32	22
TW I4	387	230	156	113	84	64	48	36	24
TW I5	370	226	155	113	85	65	49	36	25
TW J1	483	271	179	127	94	71	53	39	26
TW J2	422	246	165	119	88	67	50	37	25
TW J3	446	283	199	146	111	85	65	48	33
TW J4	400	254	178	132	100	77	58	43	30
DG A1	392	188	117	79	55	38	26	16	7
DG A2	389	190	121	85	61	45	32	22	14
DG A3	443	220	142	102	77	58	44	31	19
DG B1	384	191	122	86	64	48	35	24	13
DG B2	394	196	127	91	69	53	40	29	19
DG B3	445	222	145	105	80	62	49	37	25
DG C1	451	225	146	104	78	61	47	34	21
DG C2	453	226	148	106	80	63	49	37	25
DG C3	509	254	167	121	92	73	58	45	31
SS A1	1171	396	220	142	98	69	49	34	22
SS A2	1028	468	283	190	135	98	71	50	33
SS A3	1174	380	209	133	91	64	45	31	20
SS A4	1077	481	289	193	136	98	71	50	32
SS A5	1896	400	204	127	86	60	42	29	18
SS A6	1030	468	283	190	135	97	71	50	32
SS A7	2199	359	178	109	72	50	35	24	15
SS A8	1089	478	285	190	133	96	69	48	31
SS B1	802	298	170	111	77	55	40	28	18
SS B2	785	366	224	152	108	79	57	41	27
SS B3	770	287	163	106	74	52	37	26	17
SS B4	790	368	224	152	108	78	57	40	26
SS B5	1022	271	144	91	62	44	31	22	14
SS B6	750	356	219	149	106	77	56	40	26
SS B7	937	242	127	80	54	38	27	19	12
SS B8	750	352	215	146	103	75	55	39	25
SS C1	481	232	144	99	71	52	39	28	19
SS C2	482	262	170	120	88	66	49	36	24

(*Continued*)

TABLE 41.3 (*Continued*)

LCR Tables for Six Representative Cities (Albuquerque, Boston, Madison, Medford, Nashville, and Santa Maria)

SSF	0.10	0.20	0.30	0.40	0.50	0.60	0.70	0.80	0.90
SS C3	487	185	107	71	50	36	27	19	13
SS C4	473	235	147	102	74	55	41	30	20
SS D1	1107	477	282	188	132	95	68	48	31
SS D2	928	511	332	232	169	125	92	66	43
SS D3	1353	449	248	160	110	78	56	39	25
SS D4	946	500	319	222	160	117	86	61	40
SS E1	838	378	227	153	108	78	56	40	26
SS E2	766	419	272	190	138	102	75	54	36
SS E3	973	322	178	115	79	56	40	28	18
SS E4	780	393	247	170	122	89	65	47	31
Albuquerque, New Mexico									**4292 DD**
WW A1	1052	130	62	38	25	18	13	9	6
WW A2	354	144	84	56	39	29	21	15	10
WW A3	300	146	90	62	45	33	24	18	12
WW A4	276	146	93	65	47	35	26	19	13
WW A5	264	148	97	69	50	38	28	21	14
WW A6	256	148	99	70	52	39	30	22	15
WW B1	293	111	63	41	28	20	15	11	7
WW B2	270	147	96	67	49	37	28	20	14
WW B3	314	179	119	84	62	47	35	26	18
WW B4	275	169	116	85	64	49	37	28	19
WW B5	252	159	110	81	61	47	36	27	19
WW C1	333	190	126	89	66	50	38	28	19
WW C2	287	171	115	83	62	47	36	27	18
WW C3	293	191	136	101	77	59	46	34	24
WW C4	264	172	122	91	69	54	41	31	22
TW A1	900	124	60	37	25	17	12	9	6
TW A2	361	130	73	48	33	24	18	13	8
TW A3	293	129	77	52	37	27	20	15	10
TW A4	249	123	76	52	38	28	21	15	10
TW B1	502	117	60	38	26	18	13	9	6
TW B2	291	118	68	45	32	23	17	12	8
TW B3	254	114	68	46	33	24	18	13	9
TW B4	233	104	63	42	30	22	16	12	8
TW C1	332	106	58	37	26	19	14	10	6
TW C2	255	101	58	39	27	20	15	11	7
TW C3	243	94	54	36	25	18	13	10	7
TW C4	254	84	46	30	21	15	11	8	5
TW D1	213	86	50	33	23	17	12	9	6
TW D2	287	139	86	59	43	32	24	17	12
TW D3	294	153	97	68	49	37	27	20	14
TW D4	281	158	104	74	55	41	31	23	16
TW D5	260	151	101	73	54	41	31	23	16

(*Continued*)

TABLE 41.3 (*Continued*)

LCR Tables for Six Representative Cities (Albuquerque, Boston, Madison, Medford, Nashville, and Santa Maria)

SSF	0.10	0.20	0.30	0.40	0.50	0.60	0.70	0.80	0.90
TW E1	339	177	113	78	57	43	32	23	16
TW E2	308	168	109	77	56	42	32	23	16
TW E3	323	195	133	96	72	55	42	31	21
TW E4	287	175	120	88	66	50	38	28	20
TW F1	409	108	57	36	24	17	13	9	6
TW F2	260	110	65	43	31	22	17	12	8
TW F3	216	106	66	45	33	24	10	13	9
TW F4	178	95	61	42	31	23	17	13	9
TW G1	256	93	53	34	24	17	13	9	6
TW G2	189	91	56	38	27	20	15	11	7
TW G3	159	82	52	36	26	20	15	11	7
TW G4	128	68	43	30	22	16	12	9	6
TW H1	168	76	45	31	22	16	12	9	6
TW H2	130	66	41	29	21	15	11	8	6
TW H3	108	56	35	25	8	13	10	7	5
TW H4	83	42	27	19	13	10	7	5	4
TW I1	166	73	43	29	20	15	11	8	5
TW I2	221	117	75	52	30	28	21	16	11
TW I3	234	128	83	59	43	32	24	10	12
TW I4	234	137	92	66	49	37	28	21	14
TW I5	226	136	93	67	50	38	29	22	15
TW J1	282	156	102	72	53	40	30	22	15
TW J2	254	146	97	69	51	39	29	22	15
TW J3	269	169	118	86	65	50	38	29	20
TW J4	247	155	106	80	60	46	35	26	18
DG A1	211	97	57	36	22	13	5	—	—
DG A2	227	107	67	46	32	23	16	10	5
DG A3	274	131	83	59	44	34	25	18	10
DG B1	210	97	60	42	30	21	13	6	—
DG B2	232	110	69	49	37	28	21	14	8
DG B3	277	134	85	61	47	37	28	21	14
DG C1	253	120	74	53	39	30	22	14	—
DG C2	271	130	82	59	45	35	26	19	12
DG C3	318	155	96	71	54	43	34	26	18
SS A1	591	187	101	64	44	31	22	16	10
SS A2	531	232	137	92	65	47	34	25	16
SS A3	566	170	90	56	38	27	19	13	8
SS A4	537	230	135	89	63	45	33	23	15
SS A5	980	187	92	56	37	26	18	13	8
SS A6	529	231	136	91	64	47	34	24	16
SS A7	1103	158	74	44	29	20	14	10	6
SS A8	540	226	131	87	61	44	32	23	15
SS B1	403	141	78	50	35	25	18	13	8
SS B2	412	186	111	75	53	39	28	20	14

(*Continued*)

TABLE 41.3 (*Continued*)

LCR Tables for Six Representative Cities (Albuquerque, Boston, Madison, Medford, Nashville, and Santa Maria)

SSF	0.10	0.20	0.30	0.40	0.50	0.60	0.70	0.80	0.90
SS B3	372	130	71	46	31	22	16	11	7
SS B4	403	181	106	72	51	37	27	20	13
SS B5	518	127	65	40	27	19	13	9	6
SS B6	390	179	106	73	52	38	28	20	13
SS B7	457	108	54	33	22	16	11	8	5
SS B8	379	171	102	69	49	35	26	19	12
SS C1	270	126	77	52	37	27	20	15	10
SS C2	282	150	97	68	49	37	28	20	14
SS C3	276	101	57	37	26	19	14	10	7
SS C4	277	135	83	57	41	31	23	17	11
SS D1	548	225	130	85	59	43	31	22	14
SS D2	474	253	162	113	82	61	45	33	22
SS D3	683	212	113	72	49	35	25	17	11
SS D4	484	248	156	107	77	57	42	30	20
SS E1	410	176	103	68	48	35	25	18	12
SS E2	390	208	133	92	67	50	37	27	18
SS E3	487	151	80	51	35	25	18	12	8
SS E4	400	195	120	82	59	43	32	23	15
Nashville, Tennessee									**3696 DD**
WW A1	588	60	24	13	8	5	3	2	1
WW A2	192	70	38	23	15	11	7	5	3
WW A3	161	72	42	27	18	13	9	6	4
WW A4	148	72	43	29	20	14	10	7	5
WW A5	141	74	46	31	22	16	11	8	5
WW A6	137	74	47	32	22	16	12	8	5
WW B1	135	41	19	10	6	3	2	—	—
WW B2	152	78	48	33	23	17	12	9	6
WW B3	179	97	61	42	30	22	16	12	8
WW B4	164	97	65	46	34	25	19	14	9
WW B5	153	93	63	45	33	25	19	14	9
WW C1	193	105	67	46	33	24	18	13	8
WW C2	169	97	63	44	32	24	18	13	8
WW C3	181	115	79	58	43	33	25	18	12
WW C4	164	104	72	53	39	30	23	17	11
TW A1	509	59	25	13	8	5	3	2	1
TW A2	199	64	33	20	13	9	6	4	3
TW A3	160	65	36	23	15	11	8	5	3
TW A4	136	62	36	23	16	11	8	6	4
TW B1	282	57	26	15	9	6	4	3	2
TW B2	161	59	32	20	13	9	6	4	3
TW B3	141	58	32	21	14	10	7	5	3
TW B4	131	54	30	19	13	9	7	5	3
TW C1	188	53	27	16	10	7	5	3	2
TW C2	144	52	28	18	12	8	6	4	2

(*Continued*)

TABLE 41.3 (*Continued*)

LCR Tables for Six Representative Cities (Albuquerque, Boston, Madison, Medford, Nashville, and Santa Maria)

SSF	0.10	0.20	0.30	0.40	0.50	0.60	0.70	0.80	0.90
TW C3	139	49	27	17	11	8	5	4	2
TW C4	149	45	23	14	9	7	5	3	2
TW D1	99	33	16	9	5	3	2	1	—
TW D2	164	75	44	29	20	14	10	7	5
TW D3	167	82	49	33	23	17	12	8	5
TW D4	168	91	58	40	29	21	15	11	7
TW D5	160	89	58	40	29	22	16	12	8
TW E1	198	98	59	40	28	20	15	10	7
TW E2	182	95	59	40	29	21	15	11	7
TW E3	197	115	76	54	39	29	22	16	11
TW E4	178	105	70	50	37	27	20	15	10
TW F1	221	50	23	13	8	5	4	2	1
TW F2	139	53	29	18	12	8	6	4	2
TW F3	116	52	30	19	13	9	7	5	3
TW F4	96	47	28	19	13	9	7	5	3
TW G1	137	44	22	13	9	6	4	3	2
TW G2	101	44	25	16	11	8	5	4	2
TW G3	86	41	24	16	11	8	6	4	2
TW G4	69	34	21	14	10	7	5	3	2
TW H1	89	36	20	13	8	6	4	3	2
TW H2	69	33	19	12	9	6	4	3	2
TW H3	59	28	17	11	8	5	4	3	2
TW H4	46	22	13	9	6	4	3	2	1
TW I1	74	26	13	7	4	2	1	—	—
TW I2	125	62	38	25	18	13	9	7	4
TW I3	133	69	43	29	20	15	11	8	5
TW I4	139	78	51	35	26	19	14	10	7
TW I5	137	80	53	37	27	20	15	11	7
TW J1	164	86	54	36	26	19	14	10	6
TW J2	150	82	53	36	26	19	14	10	7
TW J3	165	101	68	49	36	27	20	15	10
TW J4	153	93	63	46	34	25	19	14	10
DG A1	98	34	—	—	—	—	—	—	—
DG A2	130	55	31	19	11	6	—	—	—
DG A3	173	78	47	32	23	16	11	7	2
DG B1	100	36	17	—	—	—	—	—	—
DG B2	134	58	33	22	15	10	6	—	—
DG B3	177	81	49	33	24	18	14	10	6
DG C1	131	52	28	17	9	—	—	—	—
DG C2	161	71	42	28	20	14	10	6	—
DG C3	205	94	57	39	29	22	17	12	8
SS A1	351	100	50	29	19	13	9	6	4
SS A2	328	135	76	49	33	24	17	12	8

(*Continued*)

TABLE 41.3 (*Continued*)

LCR Tables for Six Representative Cities (Albuquerque, Boston, Madison, Medford, Nashville, and Santa Maria)

SSF	**0.10**	**0.20**	**0.30**	**0.40**	**0.50**	**0.60**	**0.70**	**0.80**	**0.90**
SS A3	330	87	41	24	15	10	6	4	2
SS A4	331	133	74	47	32	22	16	11	7
SS A5	595	98	43	24	15	10	7	4	2
SS A6	324	132	75	48	32	23	16	11	7
SS A7	668	79	32	17	10	6	4	2	1
SS A8	330	129	71	45	30	21	15	10	6
SS B1	236	74	38	23	15	10	7	5	3
SS B2	258	110	63	41	28	20	14	10	6
SS B3	212	65	32	19	12	8	5	3	2
SS B4	251	105	60	39	27	19	13	9	6
SS B5	307	65	30	17	10	7	4	3	2
SS B6	241	104	60	39	27	19	14	10	6
SS B7	264	52	23	12	7	5	3	2	—
SS B8	233	98	56	36	25	17	12	9	5
SS C1	141	60	33	21	14	10	7	5	3
SS C2	161	81	50	33	23	17	12	9	6
SS C3	149	48	25	15	10	7	4	3	2
SS C4	160	73	43	28	19	14	10	7	5
SS D1	317	119	64	39	26	18	13	8	5
SS D2	287	147	90	61	43	31	23	16	10
SS D3	405	113	55	33	21	14	10	6	4
SS D4	295	144	87	58	40	29	21	15	10
SS E1	229	89	48	29	19	13	9	6	4
SS E2	233	118	72	48	34	24	18	12	8
SS E3	283	77	37	22	14	9	6	4	2
SS E4	242	111	65	43	29	21	15	11	7
Medford, Oregon									**4930 DD**
WW A1	708	64	24	11	—	—	—	—	—
WW A2	212	73	38	22	13	7	3		
WW A3	174	75	41	25	16	9	5	2	—
WW A4	158	74	43	27	17	11	6	3	1
WW A5	149	75	45	29	19	12	7	4	2
WW A6	144	75	46	30	20	13	8	4	2
WW B1	154	43	16	—	—	—	—	—	—
WW B2	162	80	48	31	21	14	9	6	3
WW B3	190	100	62	41	28	19	13	8	5
WW B4	171	99	65	45	32	23	16	11	7
WW B5	160	95	63	45	32	23	17	12	7
WW C1	205	108	67	45	31	21	15	10	6
WW C2	178	99	63	43	30	22	15	10	6
WW C3	189	117	80	57	42	31	23	16	10
WW C4	170	106	72	52	38	28	21	15	9
TW A1	607	63	25	12	5	—	—	—	—

(*Continued*)

TABLE 41.3 (*Continued*)

LCR Tables for Six Representative Cities (Albuquerque, Boston, Madison, Medford, Nashville, and Santa Maria)

SSF	0.10	0.20	0.30	0.40	0.50	0.60	0.70	0.80	0.90
TW A2	222	68	33	19	11	6	2	—	—
TW A3	175	67	36	21	13	8	4	2	—
TW A4	147	64	36	22	14	9	5	3	1
TW B1	327	61	27	14	7	3	—	—	—
TW B2	178	62	32	19	12	7	4	2	—
TW B3	154	60	33	20	12	8	4	2	1
TW B4	143	56	31	19	12	8	5	2	1
TW C1	212	56	27	15	9	5	2	—	—
TW C2	159	55	28	17	11	7	4	2	—
TW C3	154	52	27	16	10	6	4	2	1
TW C4	167	48	24	14	9	5	3	2	—
TW D1	112	34	14	—	—	—	—	—	—
TW D2	177	77	44	28	18	12	8	5	3
TW D3	180	85	50	32	21	14	9	6	3
TW D4	177	93	58	39	27	19	13	9	5
TW D5	168	92	58	40	28	20	14	10	6
TW E1	213	101	60	39	26	18	12	8	4
TW E2	194	98	59	39	27	19	13	9	5
TW E3	208	118	77	53	38	27	20	13	8
TW E4	186	108	71	49	36	26	19	13	8
TW F1	256	53	23	12	5	—	—	—	—
TW F2	153	56	29	17	10	5	2	—	—
TW F3	125	54	30	18	11	7	3	1	—
TW F4	102	48	28	18	11	7	4	2	1
TW G1	153	46	22	12	7	—	—	—	—
TW G2	109	46	25	15	9	5	3	1	—
TW G3	92	42	24	15	9	6	3	2	—
TW G4	74	35	20	13	8	5	3	2	—
TW H1	97	38	20	12	7	4	1	—	—
TW H2	75	34	19	12	7	5	3	1	—
TW H3	63	29	17	10	7	4	3	1	—
TW H4	49	23	13	8	5	3	2	1	—
TW I1	83	27	10	—	—	—	—	—	—
TW I2	133	64	38	24	16	11	7	4	2
TW I3	142	71	43	28	19	13	9	5	3
TW I4	146	80	51	35	25	17	12	8	5
TW I5	144	82	53	37	26	19	13	9	6
TW J1	175	89	54	36	24	17	11	7	4
TW J2	158	85	53	36	25	18	12	8	5
TW J3	173	103	69	48	35	26	18	13	8
TW J4	160	96	64	45	33	24	17	12	8
DG A1	110	35	—	—	—	—	—	—	—
DG A2	142	58	32	18	9	—	—	—	—

(*Continued*)

TABLE 41.3 (*Continued*)

LCR Tables for Six Representative Cities (Albuquerque, Boston, Madison, Medford, Nashville, and Santa Maria)

SSF	0.10	0.20	0.30	0.40	0.50	0.60	0.70	0.80	0.90
DG A3	187	82	48	32	22	15	9	5	—
DG B1	110	40	15	—	—	—	—	—	—
DG B2	146	61	35	21	13	7	—	—	—
DG B3	193	84	51	34	24	17	12	7	3
DG C1	144	57	29	13	—	—	—	—	—
DG C2	177	75	44	28	19	12	6	—	—
DG C3	224	98	60	41	29	21	14	10	5
SS A1	415	110	51	28	16	9	4	2	—
SS A2	372	146	79	48	31	21	14	8	5
SS A3	397	96	42	21	10	—	—	—	—
SS A4	379	144	76	46	29	19	12	7	4
SS A5	732	111	45	23	12	5	—	—	—
SS A6	368	143	77	47	30	20	13	8	4
SS A7	846	90	33	14	—	—	—	—	—
SS A8	379	140	73	44	27	17	11	6	3
SS B1	274	81	38	21	12	6	3	—	—
SS B2	288	117	65	40	26	18	12	7	4
SS B3	249	71	33	17	8	—	—	—	—
SS B4	282	113	62	38	25	16	11	7	4
SS B5	368	72	30	15	7	—	—	—	—
SS B6	269	111	62	30	25	17	11	7	4
SS B7	323	58	23	10	—	—	—	—	—
SS B8	262	106	57	35	23	15	9	6	3
SS C1	153	62	33	19	11	5	—	—	—
SS C2	172	83	50	32	22	15	10	6	3
SS C3	166	51	24	13	7	3	—	—	—
SS C4	173	76	43	27	18	12	8	5	3
SS D1	367	129	65	37.	22	13	7	3	1
SS D2	318	156	92	60	40	27	18	12	7
SS D3	480	124	57	31	18	10	5	2	—
SS D4	328	153	89	57	38	26	17	11	6
SS E1	262	95	48	27	15	7	—	—	—
SS E2	257	124	73	47	31	21	14	9	5
SS E3	334	84	38	20	10	4	—	—	—
SS E4	269	118	67	42	27	18	12	7	4
Boston, Massachusetts									**5621 DD**
WW A1	368	28	9	—	—	—	—	—	—
WW A2	119	41	20	12	7	5	3	2	—
WW A3	101	43	24	15	10	6	4	3	1
WW A4	93	44	26	16	11	7	5	3	2
WW A5	89	45	27	18	12	8	6	4	2
WWA6	87	46	28	19	13	9	6	4	3
WW B1	59	—	—	—	—	—	—	—	—

(Continued)

TABLE 41.3 (*Continued*)

LCR Tables for Six Representative Cities (Albuquerque, Boston, Madison, Medford, Nashville, and Santa Maria)

SSF	0.10	0.20	0.30	0.40	0.50	0.60	0.70	0.80	0.90
WW B2	103	52	31	21	15	10	7	5	3
WW B3	123	66	41	28	20	14	10	7	5
WW B4	118	70	46	33	24	18	13	9	6
WW B5	113	69	46	33	25	18	14	10	7
WW C1	135	72	46	31	22	16	12	8	5
WW C2	121	68	44	31	22	16	12	9	6
WW C3	136	86	60	44	33	25	19	14	9
WW C4	124	78	54	40	30	23	17	12	8
TW A1	324	30	11	4	—	—	—	—	—
TW A2	126	37	18	10	6	4	2	1	—
TW A3	102	39	21	13	8	5	3	2	1
TW A4	88	38	22	14	9	6	4	3	2
TW B1	180	32	13	7	4	2	—	—	—
TW B2	104	36	19	11	7	5	3	2	1
TW B3	92	36	19	12	8	5	3	2	1
TW B4	86	34	19	12	8	5	4	2	1
TW C1	122	32	15	9	5	3	2	1	—
TW C2	95	33	17	10	7	4	3	2	1
TW C3	93	31	16	10	6	4	3	2	1
TW C4	102	29	15	9	6	4	3	2	1
TW D1	45	—	—	—	—	—	—	—	—
TW D2	112	49	28	18	12	9	6	4	3
TW D3	113	54	32	21	15	10	7	5	3
TW D4	121	64	41	28	20	15	11	8	5
TW D5	118	66	42	30	21	16	12	8	6
TW E1	138	67	40	27	18	13	9	7	4
TW E2	130	66	41	28	20	14	10	7	5
TW E3	146	84	56	39	29	21	16	11	8
TW E4	133	78	52	37	27	20	15	11	7
TW F1	134	25	10	4	—	—	—	—	—
TW F2	86	30	16	9	5	3	2	1	—
TW F3	72	31	17	11	7	4	3	2	1
TW F4	61	29	17	11	7	5	3	2	1
TW G1	83	24	11	6	3	2	—	—	—
TW G2	63	26	14	9	5	4	2	1	—
TW G3	54	25	14	9	6	4	3	2	1
TW G4	45	21	12	8	5	4	3	2	1
TW H1	54	21	11	6	4	2	1	—	—
TW H2	44	20	11	7	5	3	2	1	—
TW H3	38	17	10	6	4	3	2	1	—
TW H4	30	14	8	5	3	2	2	1	—
TW I1	30	—	—	—	—	—	—	—	—
TW I2	84	41	24	16	11	8	6	4	2

(*Continued*)

TABLE 41.3 (*Continued*)

LCR Tables for Six Representative Cities (Albuquerque, Boston, Madison, Medford, Nashville, and Santa Maria)

SSF	0.10	0.20	0.30	0.40	0.50	0.60	0.70	0.80	0.90
TW I3	91	46	28	19	13	9	7	5	3
TW I4	100	56	36	25	18	13	10	7	5
TW I5	101	58	38	27	20	15	11	8	5
TW J1	114	59	37	25	17	12	9	6	4
TW J2	107	58	37	25	18	13	10	7	4
TW J3	123	75	51	36	27	20	15	11	7
TW J4	115	70	47	34	25	19	14	10	7
DG A1	43	—	—	—	—	—	—	—	—
DG A2	85	34	18	9	—	—	—	—	—
DG A3	125	56	33	22	16	11	7	4	—
DG B1	44	—	—	—	—	—	—	—	—
DG B2	87	36	20	12	7	—	—	—	—
DG B3	129	58	35	24	17	13	9	6	3
DG C1	71	23	—	—	—	—	—	—	—
DG C2	109	47	27	17	12	8	4	—	—
DG C3	151	68	41	28	21	16	12	8	5
SS A1	230	61	29	16	10	6	4	2	1
SS A2	231	93	52	33	22	15	11	7	5
SS A3	205	48	20	10	4	—	—	—	—
SS A4	229	90	49	31	20	14	9	6	4
SS A5	389	58	23	11	6	3	—	—	—
SS A6	226	91	50	32	21	15	10	7	4
SS A7	420	40	12	—	—	—	—	—	—
SS A8	226	86	46	28	19	12	8	6	3
SS B1	151	44	21	12	7	4	2	1	—
SS B2	183	77	43	28	19	13	9	6	4
SS B3	129	36	16	8	3	—	—	—	—
SS B4	176	73	41	26	17	12	8	6	4
SS B5	193	36	15	7	3	—	—	—	—
SS B6	169	72	41	26	18	12	9	6	4
SS B7	157	25	7	—	—	—	—	—	—
SS B8	160	66	37	23	16	11	7	5	3
SS C1	84	33	17	10	6	4	2	1	—
SS C2	110	54	33	22	15	11	8	5	3
SS C3	91	26	12	7	4	2	—	—	—
SS C4	109	48	28	18	12	9	6	4	3
SS D1	206	73	38	22	14	9	5	3	2
SS D2	203	103	63	42	29	21	15	10	6
SS D3	264	69	32	18	10	6	4	2	1
SS D4	208	100	60	39	27	19	14	9	6
SS E1	140	51	25	14	8	4	2	—	—
SS E2	161	80	48	32	22	15	11	7	5
SS E3	177	44	19	10	5	2	—	—	—
SS E4	166	75	43	28	19	13	9	6	4

(*Continued*)

TABLE 41.3 (*Continued*)

LCR Tables for Six Representative Cities (Albuquerque, Boston, Madison, Medford, Nashville, and Santa Maria)

SSF	0.10	0.20	0.30	0.40	0.50	0.60	0.70	0.80	0.90
Madison, Wisconsin									**7730 DD**
WW A1	278	—	—	—	—	—	—	—	—
WW A2	91	27	12	—	—	—	—	—	—
WW A3	77	30	15	8	3	—	—	—	—
WW A4	72	32	17	10	5	—	—	—	—
WW A5	69	33	19	11	7	4	—	—	—
WW A6	67	34	19	12	7	4	2	—	—
WW B1	—	—	—	—	—	—	—	—	—
WW B2	84	41	24	15	10	7	5	3	2
WW B3	102	53	32	21	15	10	7	5	3
WW B4	101	59	39	27	19	14	10	7	5
WW B5	98	59	39	28	20	15	11	8	5
WW C1	113	59	37	25	17	12	8	6	3
WW C2	103	57	37	25	18	13	9	6	4
WW C3	119	75	51	37	28	21	15	11	7
WW C4	109	68	47	34	25	19	14	10	7
TW A1	249	16	—	—	—	—	—	—	—
TW A2	97	26	11	4	—	—	—	—	—
TW A3	79	28	13	7	3	—	—	—	—
TW A4	69	28	15	9	5	3	—	—	—
TW B1	139	20	5	—	—	—	—	—	—
TW B2	81	26	12	6	3	—	—	—	—
TW B3	72	27	13	7	4	2	—	—	—
TW B4	69	26	13	8	5	3	1	—	—
TW C1	96	23	10	4	—	—	—	—	—
TW C2	76	25	12	7	4	2	—	—	—
TW C3	75	24	12	7	4	2	1	—	—
TW C4	84	23	11	6	4	2	1	—	—
TW D1	—	—	—	—	—	—	—	—	—
TW D2	91	39	22	13	9	6	4	2	1
TW D3	93	43	25	16	10	7	5	3	1
TW D4	103	54	34	23	16	12	8	6	4
TW D5	102	56	36	25	18	13	10	7	4
TW E1	115	54	32	21	14	10	7	4	3
TW E2	110	55	34	22	16	11	8	5	3
TW E3	126	72	47	33	24	18	13	9	6
TW E4	116	68	45	32	23	17	13	9	6
TW F1	99	13	—	—	—	—	—	—	—
TW F2	65	20	8	—	—	—	—	—	—
TW F3	55	22	11	5	—	—	—	—	—
TW F4	47	21	11	7	4	2	—	—	—
TW G1	61	14	—	—	—	—	—	—	—

(*Continued*)

TABLE 41.3 (*Continued*)

LCR Tables for Six Representative Cities (Albuquerque, Boston, Madison, Medford, Nashville, and Santa Maria)

SSF	0.10	0.20	0.30	0.40	0.50	0.60	0.70	0.80	0.90
TW G2	47	18	8	4	—	—	—	—	—
TW G3	42	18	9	5	3	—	—	—	—
TW G4	35	16	9	5	3	2	—	—	—
TW H1	41	13	6	—	—	—	—	—	—
TW H2	34	14	7	4	2	—	—	—	—
TW H3	29	13	7	4	2	1	—	—	—
TW H4	24	10	6	3	2	1	—	—	—
TW I1	—	—	—	—	—	—	—	—	—
TW I2	68	32	18	12	8	5	3	2	1
TW I3	75	37	22	14	10	7	4	3	2
TW I4	85	47	30	21	15	11	8	5	3
TW I5	87	50	33	23	16	12	9	6	4
TW J1	95	48	29	19	13	9	6	4	3
TW J2	91	48	30	21	14	10	7	5	3
TW J3	106	65	43	31	23	17	12	9	6
TW J4	100	61	41	29	21	16	12	9	6
DG A1	—	—	—	—	—	—	—	—	—
DG A2	68	25	11	—	—	—	—	—	—
DG A3	109	47	28	18	12	8	5	—	—
DG B1	—	—	—	—	—	—	—	—	—
DG B2	70	27	14	6	—	—	—	—	—
DG B3	114	50	30	20	14	10	7	4	—
DG C1	47	—	—	—	—	—	—	—	—
DG C2	91	37	21	13	7				
DG C3	133	59	35	24	17	13	9	6	3
SS A1	192	47	20	9	3	—	—	—	—
SS A2	200	78	42	26	17	12	8	5	3
SS A3	166	32	—	—	—	—	—	—	—
SS A4	197	74	39	23	15	10	6	4	2
SS A5	329	42	13	—	—	—	—	—	—
SS A6	195	75	40	25	16	11	7	5	3
SS A7	349	22	—	—	—	—	—	—	—
SS A8	192	69	36	21	13	8	5	3	2
SS B1	122	32	13	5	—	—	—	—	—
SS B2	158	64	36	22	15	10	7	5	3
SS B3	100	22	—	—	—	—	—	—	—
SS B4	150	60	33	29	13	9	6	4	2
SS B5	156	24	—	—	—	—	—	—	—
SS B6	145	59	33	20	13	9	6	4	2
SS B7	122	—	—	—	—	—	1—	—	—
SS B8	136	54	29	18	11	7	5	3	2
SS C1	61	20	7	—	—	—	—	—	—
SS C2	90	43	25	16	11	7	5	3	2

(*Continued*)

TABLE 41.3 (*Continued*)

LCR Tables for Six Representative Cities (Albuquerque, Boston, Madison, Medford, Nashville, and Santa Maria)

SSF	0.10	0.20	0.30	0.40	0.50	0.60	0.70	0.80	0.90
SS C3	67	16	—	—	—	—	—	—	—
SS C4	90	38	22	13	9	6	4	2	1
SS D1	169	56	26	13	6	—	—	—	—
SS D2	175	86	51	34	23	16	11	7	5
SS D3	221	52	21	10	—	—	—	—	—
SS D4	179	84	49	32	21	15	10	7	4
SS E1	108	34	12	—	—	—	—	—	—
SS E2	135	65	38	24	16	11	7	5	3
SS E3	141	29	8	—	—	—	—	—	—
SS E4	140	61	34	21	14	9	6	4	2

Source: PSDH, *Passive Solar Design Handbook*, Los Alamos National Laboratory, Van Nostrand Reinhold, New York, 1984.

3. Enter the LCR tables (Table 41.3).
   a. Find the city
   b. Find the reference system listing
   c. Determine annual SSF by interpolation using the LCR value from above
   d. Note the annual heating degree-days (number of degree-days)
4. Calculate the annual auxiliary heat required:

$$\text{Auxiliary heat required} = (1 - \text{SSF}) \times \text{NLC} \times (\text{number of degree-days}).$$

If more than one reference solar system is being used, then find the *aperture-area-weighted* SSF for the combination. Determine each individual reference system SSF using the total aperture area LCR, then take the *area-weighted* average of the individual SSFs.

The LCR method allows no variation from the 94 reference passive designs. To treat off-reference designs, sensitivity curves have been produced that illustrate the effect on SSF of varying one or two design variables. These curves were produced for the six *representative* cities, chosen for their wide geographical and climatological ranges. Several of these SSF *sensitivity curves* are presented in Figure 41.9 for storage wall (Figure 41.9a through c) and sunspace (Figure 41.9d) design variations.

Example 41.4

The previously used 2100 ft.^{2} building with NLC = 11,800 Btu/FDD is preliminarily designed to be located in Medford, Oregon, with 180 ft.^{2} of 12 in. thick vented Trombe wall and 130 ft.^{2} of direct gain, both systems with double glazing, nighttime insulation, and 30 Btu/ft.^{2} thermal storage capacity. Determine the annual auxiliary energy needed by this design.

Solution

Step 1 yields:

$$\text{NLC} = 11{,}800 \text{ Btu/FDD}.$$

$$A_p = 180 + 130 = 320 \text{ ft.}^2$$

$$\text{LCR} = 11{,}800/320 = 36.8 \text{ Btu/FDD ft.}^2$$

Step 2 yields: From Table 41.2, the short designations for the appropriate systems are
TWD4 (Trombe wall)
DGA3 (Direct gain)

Step 3 yields: From Table 41.3 for Medford, Oregon, with LCR = 36.8,
TWD4: SSF(TW) = 0.42.
DGA3: SSF(DG) = 0.37.
Determine the *weighted area* average SSF:

$$\text{SSF} = \frac{180(0.42) + 130(0.37)}{320} = 0.39.$$

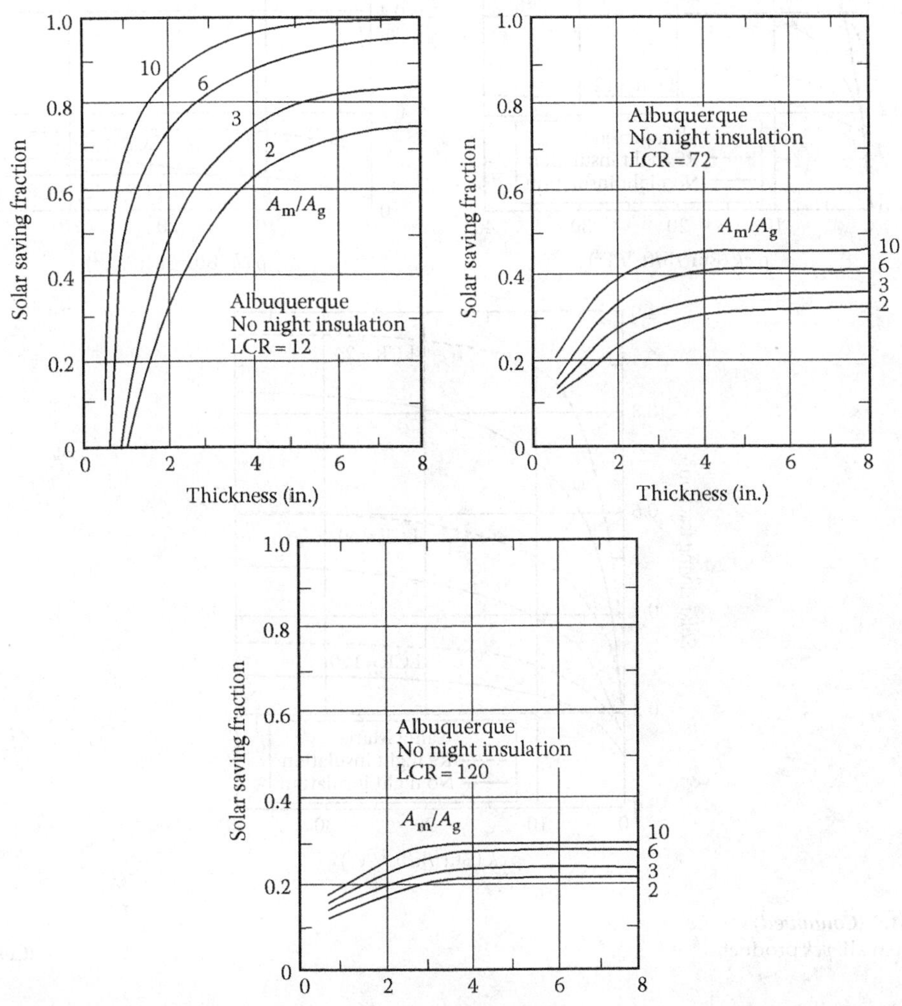

FIGURE 41.9
(a) Storage wall: mass thickness. Sensitivity of SSF to off-reference conditions. *(Continued)*

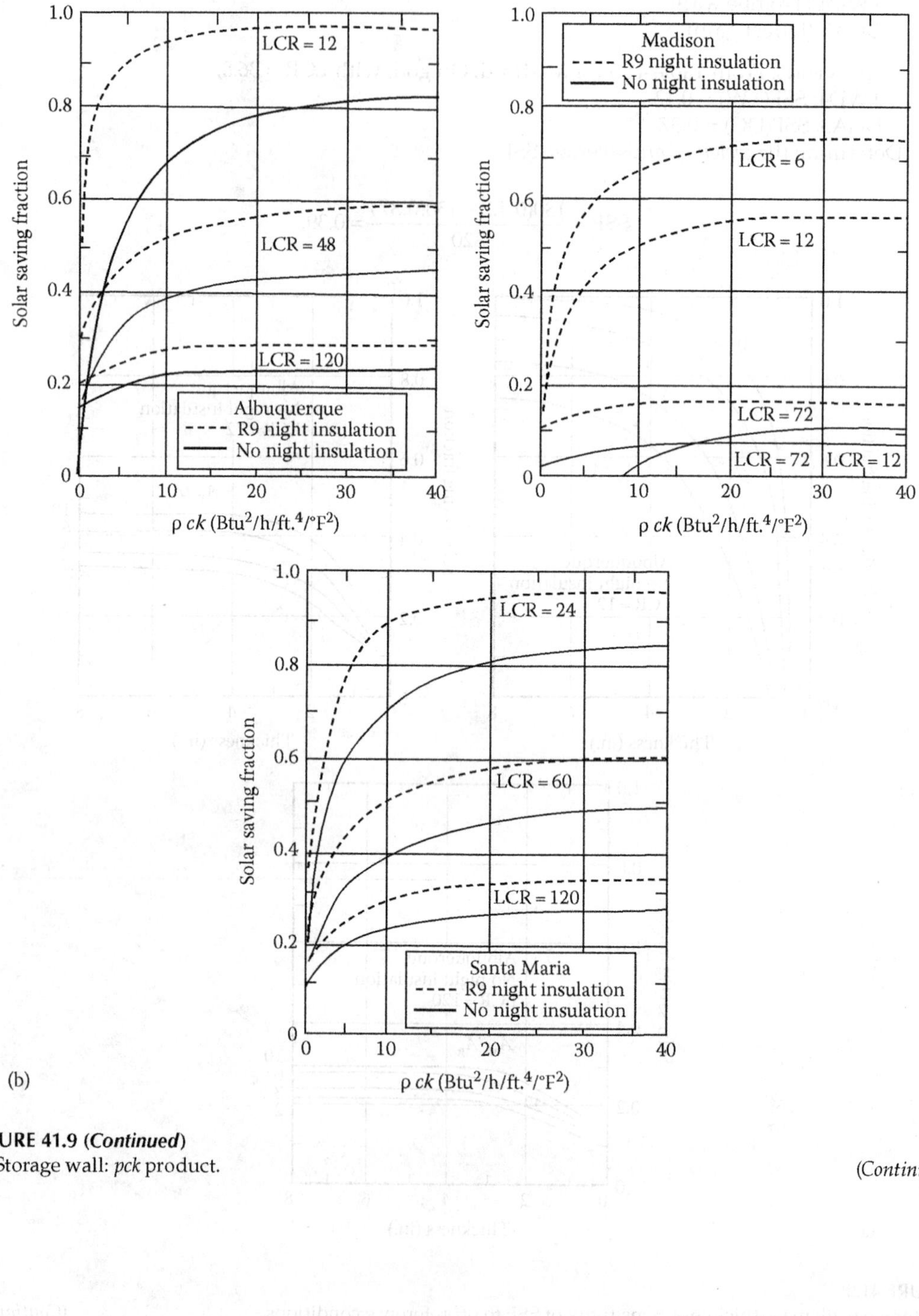

FIGURE 41.9 (*Continued*)
(b) Storage wall: *pck* product.

(*Continued*)

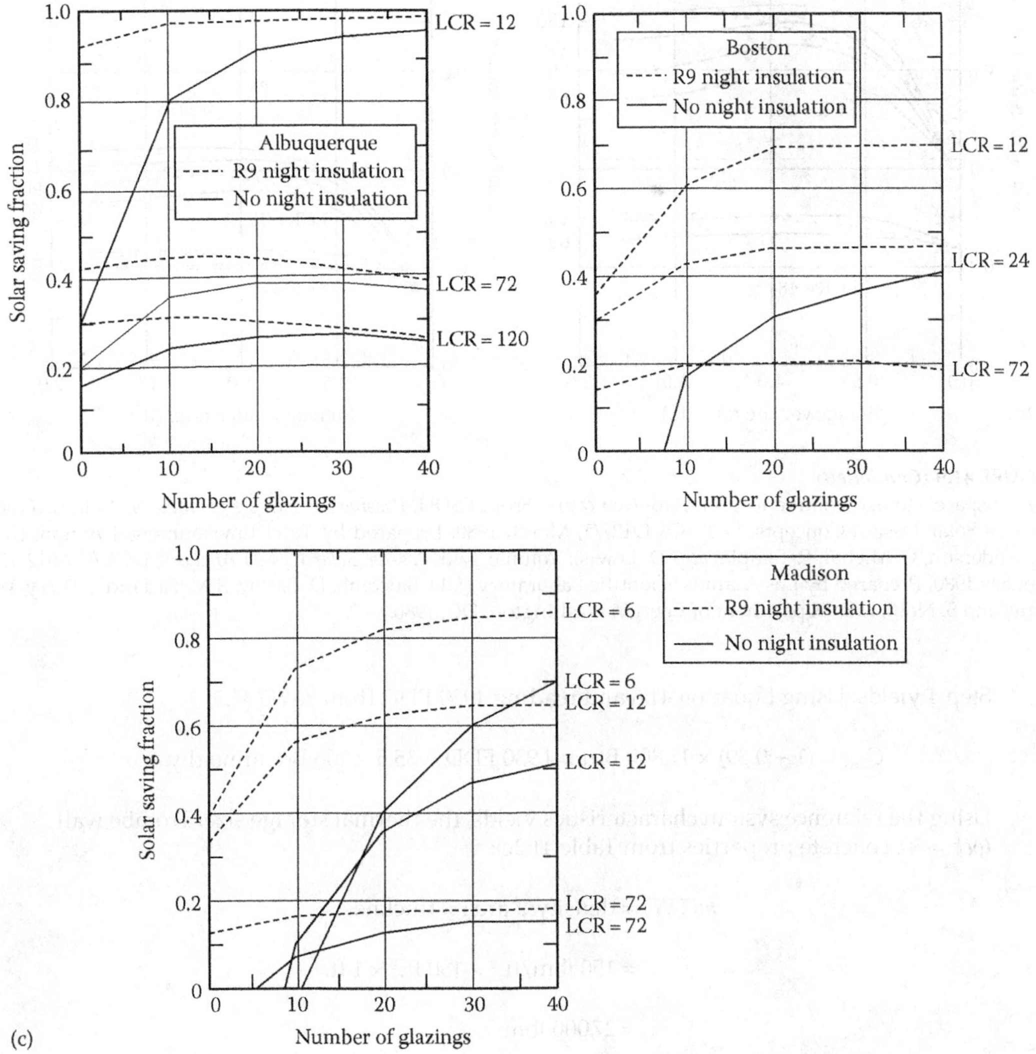

FIGURE 41.9 ***(Continued)***
(c) Storage wall: number of glazings. *(Continued)*

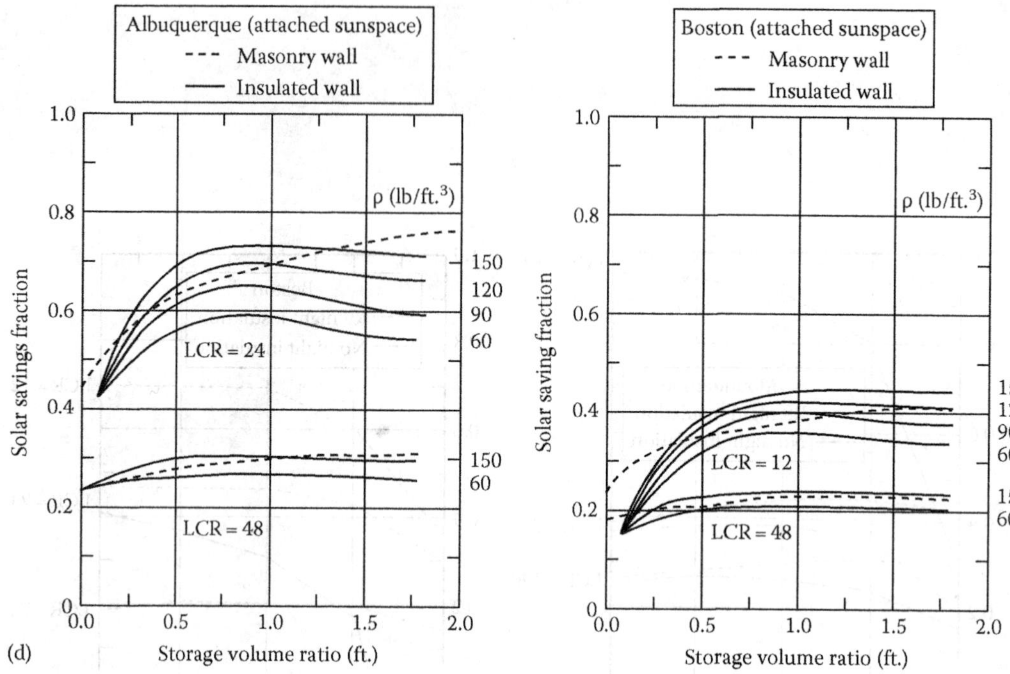

(d)

FIGURE 41.9 (*Continued*)
(d) Sunspace: storage-volume-to-projected-area ratio. (From PSDH, *Passive Solar Design Handbook,* Volume One: Passive Solar Design Concepts, DOE/CS-O127/1, March 1980. Prepared by Total Environmental Action, Inc. (B. Anderson, C. Michal, P. Temple, and D. Lewis); Volume Two: *Passive Solar Design Analysis*, DOE/CS-0127/2, January 1980, Prepared by Los Alamos Scientific Laboratory (J. D. Balcomb, D. Barley, R McFarland, J. Perry, W. Wray and S. Noll), U.S. Department of Energy, Washington, DC, 1980.)

Step 4 yields: Using Equation 41.6 and reading 4930 FDD from Table 41.3,

$$Q_{aux} = (1 - 0.39) \times 11{,}800 \text{ Btu} \times 4{,}930 \text{ FDD} = 35.5 \times 106 \text{ Btu annually.}$$

Using the reference system characteristics yields, the thermal storage size: Trombe wall ($\rho ck = 30$, concrete properties from Table 41.2c):

$$m(\text{TW}) = \text{density} \times \text{area} \times \text{thickness}$$

$$= 150 \text{ lbm/ft.}^3 \times 180 \text{ ft.}^2 \times 1 \text{ ft.}$$

$$= 27{,}000 \text{ lbm.}$$

Direct gain ($\rho ck = 30$, concrete properties), using mass-area-to-glazing-area ratio of 6:

$$\text{Mass area} = 6 \times 130 = 780 \text{ ft.}^2 \text{ of 2 in. thick concrete}$$

$$m(\text{DG}) = 150 \text{ lbm/ft.}^3 \times 780 \text{ ft.}^2 \times 1/6 \text{ ft.}$$

$$= 19{,}500 \text{ lbm.}$$

Using the LCR method allows a basic design of passive system types for the 94 reference systems and the resulting annual performance. A bit more design variation can be obtained by using the sensitivity curves of Figure 41.9 to modify the SSF of a particular reference system. For instance, a direct gain system SSF of 0.37 would increase by approximately 0.03 if the mass–glazing-area ratio (assumed 6) were increased to 10 and would decrease by about 0.04 if the mass–glazing-area ratio were decreased to 3. This information provides a designer with quantitative information for making trade-offs.

41.3.6 Third Level: SLR Method

The SLR method calculates monthly performance, and the terms and values used are monthly based. The method allows the use of specific location weather data and the 94 reference design passive systems (Table 41.2). In addition, the sensitivity curves (Figure 41.9) can again be used to define performance outside the reference design systems. The result of the SLR method is the determination of the monthly heating auxiliary energy required that is then summed to give the annual requirement for auxiliary heating energy. Generally, the SLR method gives annual values within ± 3% of detailed simulation results, but the monthly values may vary more (PSDH 1984; Duffie and Beckman 1991). Thus, the monthly SLR method is more *accurate* than the rule-of-thumb methods, plus providing the designer with system performance on a month-by-month basis.

The SLR method uses equations and correlation parameters for each of the 94 reference systems combined with the insolation absorbed by the system, the monthly degree-days, and the system's LCR to determine the monthly SSF. These correlation parameters are listed in Table 41.4 as *A*, *B*, *C*, *D*, *R*, *G*, *H*, and LCRs for each reference system (PSDH 1984). The correlation equations are

$$\mathrm{SSF} = 1 - K(1 - F), \tag{41.9}$$

where

$$K = 1 + G/\mathrm{LCR}, \tag{41.10}$$

$$F = \begin{cases} AX, & \text{when } X < R \\ B - C\exp(-DX), & \text{when } X > R \end{cases} \tag{41.11}$$

$$X = \frac{S/DD - (\mathrm{LCRs})H}{(\mathrm{LCR})K}, \tag{41.12}$$

and *X* is called the *generalized solar load ratio.* The term *S* is the monthly insolation absorbed by the system per unit of solar projected area. Monthly average daily insolation data on a vertical south-facing surface can be found and/or calculated using various sources (PSDH 1984; McQuiston and Parker 1994), and the *S* term can be determined by multiplying by a transmission and an absorption factor and the number of days in the month. Absorption factors for all systems are close to 0.96 (PSDH 1984), whereas the transmission is approximately 0.9 for single glazing, 0.8 for double glazing, and 0.7 for triple glazing.

TABLE 41.4
SLR Correlation Parameters for the 94 Reference Systems

Type	A	B	C	D	R	G	H	LCRs	STDV
WW A1	0.0000	1.0000	0.9172	0.4841	−9.0000	0.00	1.17	13.0	0.053
WW A2	0.0000	1.0000	0.9833	0.7603	−9.0000	0.00	0.92	13.0	0.046
WW A3	0.0000	1.0000	1.0171	0.8852	−9.0000	0.00	0.85	13.0	0.040
WW A4	0.0000	1.0000	1.0395	0.9569	−9.0000	0.00	0.81	13.0	0.037
WW A5	0.0000	1.0000	1.0604	1.0387	−9.0000	0.00	0.78	13.0	0.034
WW A6	0.0000	1.0000	1.0735	1.0827	−9.0000	0.00	0.76	13.0	0.033
WW B1	0.0000	1.0000	0.9754	0.5518	−9.0000	0.00	0.92	22.0	0.051
WW B2	0.0000	1.0000	1.0487	1.0851	−9.0000	0.00	0.78	9.2	0.036
WW B3	0.0000	1.0000	1.0673	1.0087	−9.0000	0.00	0.95	8.9	0.038
WW B4	0.0000	1.0000	1.1028	1.1811	−9.0000	0.00	0.74	5.8	0.034
WW B5	0.0000	1.0000	1.1146	1.2771	−9.0000	0.00	0.56	4.5	0.032
WW C1	0.0000	1.0000	1.0667	1.0437	−9.0000	0.00	0.62	12.0	0.038
WW C2	0.0000	1.0000	1.0846	1.1482	−9.0000	0.00	0.59	8.7	0.035
WW C3	0.0000	1.0000	1.1419	1.1756	−9.0000	0.00	0.28	5.5	0.033
WW C4	0.0000	1.0000	1.1401	1.2378	−9.0000	0.00	0.23	4.3	0.032
TW A1	0.0000	1.0000	0.9194	0.4601	−9.0000	0.00	1.11	13.0	0.048
TW A2	0.0000	1.0000	0.9680	0.6318	−9.0000	0.00	0.92	13.0	0.043
TW A3	0.0000	1.0000	0.9964	0.7123	−9.0000	0.00	0.85	13.0	0.038
TW A4	0.0000	1.0000	1.0190	0.7332	−9.0000	0.00	0.79	13.0	0.032
TW B1	0.0000	1.0000	0.9364	0.4777	−9.0000	0.00	1.01	13.0	0.045
TW B2	0.0000	1.0000	0.9821	0.6020	−9.0000	0.00	0.85	13.0	0.038
TW B3	0.0000	1.0000	0.9980	0.6191	−9.0000	0.00	0.80	13.0	0.033
TW B4	0.0000	1.0000	0.9981	0.5615	−9.0000	0.00	0.76	13.0	0.028
TW C1	0.0000	1.0000	0.9558	0.4709	−9.0000	0.00	0.89	13.0	0.039
TW C2	0.0000	1.0000	0.9788	0.4964	−9.0000	0.00	0.79	13.0	0.033
TW C3	0.0000	1.0000	0.9760	0.4519	−9.0000	0.00	0.76	13.0	0.029
TW C4	0.0000	1.0000	0.9588	0.3612	−9.0000	0.00	0.73	13.0	0.026
TW D1	0.0000	1.0000	0.9842	0.4418	−9.0000	0.00	0.89	22.0	0.040
TW D2	0.0000	1.0000	1.0150	0.8994	−9.0000	0.00	0.80	9.2	0.036
TW D3	0.0000	1.0000	1.0346	0.7810	−9.0000	0.00	1.08	8.9	0.036
TW D4	0.0000	1.0000	1.0606	0.9770	−9.0000	0.00	0.85	5.8	0.035
TW D5	0.0000	1.0000	1.0721	1.0718	−9.0000	0.00	0.61	4.5	0.033
TW E1	0.0000	1.0000	1.0345	0.8753	−9.0000	0.00	0.68	12.0	0.037
TW E2	0.0000	1.0000	1.0476	1.0050	−9.0000	0.00	0.66	8.7	0.035
TW E3	0.0000	1.0000	1.0919	1.0739	−9.0000	0.00	0.61	5.5	0.034
TW E4	0.0000	1.0000	1.0971	1.1429	−9.0000	0.00	0.47	4.3	0.033
TW F1	0.0000	1.0000	0.9430	0.4744	−9.0000	0.00	1.09	13.0	0.047
TW F2	0.0000	1.0000	0.9900	0.6053	−9.0000	0.00	0.93	13.0	0.041
TW F3	0.0000	1.0000	1.0189	0.6502	−9.0000	0.00	0.86	13.0	0.036
TW F4	0.0000	1.0000	1.0419	0.6258	−9.0000	0.00	0.80	13.0	0.032
TW G1	0.0000	1.0000	0.9693	0.4714	−9.0000	0.00	1.01	13.0	0.042
TW G2	0.0000	1.0000	1.0133	0.5462	−9.0000	0.00	0.88	13.0	0.035
TW G3	0.0000	1.0000	1.0325	0.5269	−9.0000	0.00	0.82	13.0	0.031
TW G4	0.0000	1.0000	1.0401	0.4400	−9.0000	0.00	0.77	13.0	0.030

(Continued)

TABLE 41.4 (*Continued*)
SLR Correlation Parameters for the 94 Reference Systems

Type	A	B	C	D	R	G	H	LCRs	STDV
TW H1	0.0000	1.0000	1.0002	0.4356	−9.0000	0.00	0.93	13.0	0.034
TW H2	0.0000	1.0000	1.0280	0.4151	−9.0000	0.00	0.83	13.0	0.030
TW H3	0.0000	1.0000	1.0327	0.3522	−9.0000	0.00	0.78	13.0	0.029
TW H4	0.0000	1.0000	1.0287	0.2600	−9.0000	0.00	0.74	13.0	0.024
TW I1	0.0000	1.0000	0.9974	0.4036	−9.0000	0.00	0.91	22.0	0.038
TW I2	0.0000	1.0000	1.0386	0.8313	−9.0000	0.00	0.80	9.2	0.034
TW I3	0.0000	1.0000	1.0514	0.6886	−9.0000	0.00	1.01	8.9	0.034
TW I4	0.0000	1.0000	1.0781	0.8952	−9.0000	0.00	0.82	5.8	0.032
TW I5	0.0000	1.0000	1.0902	1.0284	−9.0000	0.00	0.65	4.5	0.032
TW J1	0.0000	1.0000	1.0537	0.8227	−9.0000	0.00	0.65	12.0	0.037
TW J2	0.0000	1.0000	1.0677	0.9312	−9.0000	0.00	0.62	8.7	0.035
TW J3	0.0000	1.0000	1.1153	0.9831	−9.0000	0.00	0.44	5.5	0.034
TW J4	0.0000	1.0000	1.1154	1.0607	−9.0000	0.00	0.38	4.3	0.033
DG A1	0.5650	1.0090	1.0440	0.7175	0.3931	9.36	0.00	0.0	0.046
DG A2	0.5906	1.0060	1.0650	0.8099	0.4681	5.28	0.00	0.0	0.039
DG A3	0.5442	0.9715	1.1300	0.9273	0.7068	2.64	0.00	0.0	0.036
DG B1	0.5739	0.9948	1.2510	1.0610	0.7905	9.60	0.00	0.0	0.042
DG B2	0.6180	1.0000	1.2760	1.1560	0.7528	5.52	0.00	0.0	0.035
DG B3	0.5601	0.9839	1.3520	1.1510	0.8879	2.38	0.00	0.0	0.032
DG C1	0.6344	0.9887	1.5270	1.4380	0.8632	9.60	0.00	0.0	0.039
DG C2	0.6763	0.9994	1.4000	1.3940	0.7604	5.28	0.00	0.0	0.033
DG C3	0.6182	0.9859	1.5660	1.4370	0.8990	2.40	0.00	0.0	0.031
SS A1	0.0000	1.0000	0.9587	0.4770	−9.0000	0.00	0.83	18.6	0.027
SS A2	0.0000	1.0000	0.9982	0.6614	−9.0000	0.00	0.77	10.4	0.026
SS A3	0.0000	1.0000	0.9552	0.4230	−9.0000	0.00	0.83	23.6	0.030
SS A4	0.0000	1.0000	0.9956	0.6277	−9.0000	0.00	0.80	12.4	0.026
SS A5	0.0000	1.0000	0.9300	0.4041	−9.0000	0.00	0.96	18.6	0.031
SS A6	0.0000	1.0000	0.9981	0.6660	−9.0000	0.00	0.86	10.4	0.028
SS A7	0.0000	1.0000	0.9219	0.3225	−9.0000	0.00	0.96	23.6	0.035
SS A8	0.0000	1.0000	0.9922	0.6173	−9.0000	0.00	0.90	12.4	0.028
SS B1	0.0000	1.0000	0.9683	0.4954	−9.0000	0.00	0.84	16.3	0.028
SS B2	0.0000	1.0000	1.0029	0.6802	−9.0000	0.00	0.74	8.5	0.026
SS B3	0.0000	1.0000	0.9689	0.4685	−9.0000	0.00	0.82	19.3	0.029
SS B4	0.0000	1.0000	1.0029	0.6641	−9.0000	0.00	0.76	9.7	0.026
SS B5	0.0000	1.0000	0.9408	0.3866	−9.0000	0.00	0.97	16.3	0.030
SS B6	0.0000	1.0000	1.0068	0.6778	−9.0000	0.00	0.84	8.5	0.028
SS B7	0.0000	1.0000	0.9395	0.3363	−9.0000	0.00	0.95	19.3	0.032
SS B8	0.0000	1.0000	1.0047	0.6469	−9.0000	0.00	0.87	9.7	0.027
SS C1	0.0000	1.0000	1.0087	0.7683	−9.0000	0.00	0.76	16.3	0.025
SS C2	0.0000	1.0000	1.0412	0.9281	−9.0000	0.00	0.78	10.0	0.027
SS C3	0.0000	1.0000	0.9699	0.5106	−9.0000	0.00	0.79	16.3	0.024
SS C4	0.0000	1.0000	1.0152	0.7523	−9.0000	0.00	0.81	10.0	0.025
SS D1	0.0000	1.0000	0.9889	0.6643	−9.0000	0.00	0.84	17.8	0.028
SS D2	0.0000	1.0000	1.0493	0.8753	−9.0000	0.00	0.70	9.9	0.028

(*Continued*)

TABLE 41.4 (*Continued*)

SLR Correlation Parameters for the 94 Reference Systems

Type	A	B	C	D	R	G	H	LCRs	STDV
SS D3	0.0000	1.0000	0.9570	0.5285	−9.0000	0.00	0.90	17.8	0.029
SS D4	0.0000	1.0000	1.0356	0.8142	−9.0000	0.00	0.73	9.9	0.028
SS E1	0.0000	1.0000	0.9968	0.7004	−9.0000	0.00	0.77	19.6	0.027
SS E2	0.0000	1.0000	1.0468	0.9054	−9.0000	0.00	0.76	10.8	0.027
SS E3	0.0000	1.0000	0.9565	0.4827	−9.0000	0.00	0.81	19.6	0.028
SS E4	0.0000	1.0000	1.0214	0.7694	−9.0000	0.00	0.79	10.8	0.027

Source: PSDH, *Passive Solar Design Handbook*, Los Alamos National Laboratory, Van Nostrand Reinhold, New York, 1984.

Example 41.5

For a vented, 180 ft.^{2}, double-glazed with night insulation, 12 in. thick Trombe wall system (TWD4) in an NLC = 11,800 Btu/FDD house in Medford, Oregon, determine the auxiliary energy required in January.

Solution

Weather data for Medford, Oregon (PSDH 1984), yields for January (N = 31, days): daily vertical surface insolation = 565 Btu/ft.^{2} and 880 FDD, so S = (31)(565)(0.8)(0.96) = 13,452 Btu/ft.^{2} month.

$$\text{LCR} = \frac{\text{NLC}}{A_{\text{P}}} = \frac{11{,}800}{180} = 65.6\ \text{Btu/FDD ft.}^2$$

From Table 41.4 at TWD4: $A = 0$, $B = 1$, $C = 1.0606$, $D = 0.977$, $R = -9$, $G = 0$, $H = 0.85$, LCRs = 5.8 Btu/FDD ft.^{2}

Substituting into Equation 41.10 gives

$$K = 1 + 0/65.6 = 1.$$

Equation 41.12 gives

$$X = \frac{(13{,}452/880) - (5.8 \times 0.85)}{65.6 \times 1} = 0.16.$$

Equation 41.11 gives

$$F = 1 - 1.0606\text{e}^{-0.977 \times 0.16} = 0.09,$$

and Equation 41.9 gives

$$\text{SSF} = 1 - 1(1 - 0.09) = 0.09.$$

The January auxiliary energy required can be calculated using Equation 41.6:

$$Q_{\text{aux}}(\text{Jan}) = (1 - \text{SSF}) \times \text{NLC} \times (\text{number of degree-days})$$

$$= (1 - 0.09) \times 11{,}800 \times 880$$

$$= 9{,}450{,}000\ \text{Btu}.$$

As mentioned, the use of sensitivity curves (PSDH 1984) as in Figure 41.9 will allow SSF to be determined for many off-reference system design conditions involving storage mass, number of glazings, and other more esoteric parameters. Also, the use of multiple passive system types within one building would be approached by calculating the SSF for each type system individually using a *combined area* LCR, and then a weighted-area (aperture) average SSF would be determined for the building.

41.4 Passive Space Cooling Design Fundamentals

Passive cooling systems are designed to use natural means to transfer heat from buildings, including convection/ventilation, evaporation, radiation, and conduction. However, the most important element in both passive and conventional cooling design is to prevent heat from entering the building in the first place. Cooling conservation techniques involve building surface colors, insulation, special window glazings, overhangs and orientation, and numerous other architectural/engineering features.

41.4.1 Solar Control

Controlling the solar energy input to reduce the cooling load is usually considered a passive (versus conservation) design concern because solar input may be needed for other purposes, such as daylighting throughout the year and/or heating during the winter. Basic architectural solar control is normally *designed in* via the shading of the solar windows, where direct radiation is desired for winter heating and needs to be excluded during the cooling season.

The shading control of the windows can be of various types and *controllability*, ranging from drapes and blinds, use of deciduous trees, to the commonly used overhangs and vertical louvers. A rule-of-thumb design for determining proper south-facing window overhang for both winter heating and summer shading is presented in Table 41.5. Technical details on calculating shading from various devices and orientations are found in Olgyay and Olgyay (1977) and ASHRAE (1993).

41.4.2 Natural Convection/Ventilation

Air movement provides cooling comfort through convection and evaporation from human skin. ASHRAE (1993) places the comfort limit at 79°F for an air velocity of 50 ft./min (fpm), 82°F for 160 fpm, and 85°F for 200 fpm. To determine whether or not comfort conditions can be obtained, a designer must calculate the volumetric flow rate, Q, which is passing through the occupied space. Using the cross-sectional area, A_x, of the space and the room air velocity, V_a, required, the flow is determined by

$$Q = A_x V_a. \tag{41.13}$$

The proper placement of windows, *narrow* building shape, and open landscaping can enhance natural wind flow to provide ventilation. The air flow rate through open windows for wind-driven ventilation is given by ASHRAE (1993):

$$Q = C_v V_w A_w, \tag{41.14}$$

where

Q is the air flow rate (m^3/s)
A_w is the free area of inlet opening (m^2)
V_w is the wind velocity (m/s)
C_v is the effectiveness of opening that is equal to 0.5–0.6 for wind perpendicular to opening and 0.25–0.35 for wind diagonal to opening

TABLE 41.5
South-Facing Window Overhang Rule of Thumb

$$\text{Length of the overhang} = \frac{\text{window height}}{F}$$

(a) Overhang Factors

North Latitude	F[a]
28	5.5–11.1
32	4.0–6.3
36	3.0–4.5
40	2.5–3.4
44	2.0–2.7
48	1.7–2.2
52	1.5–1.8
56	1.3–1.5

(b) Roof Overhang Geometry

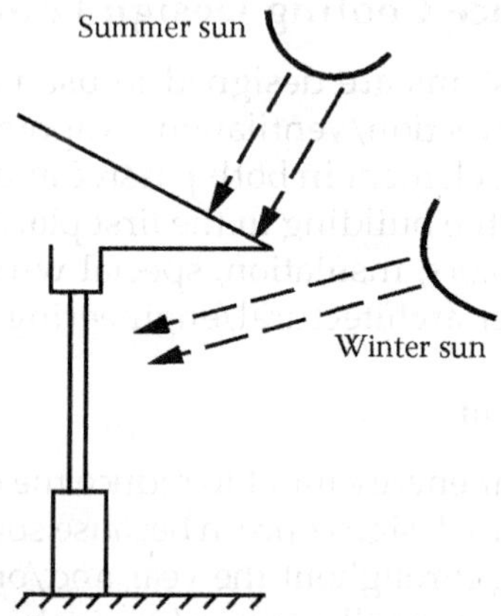

Properly sized overhangs shade out hot summer sun but allow winter sun (which is lower in the sky) to penetrate windows

Source: Halacy, 1984.
[a] Select a factor according to your latitude. Higher values provide complete shading at noon on June 21; lower values, until August 1.

The stack effect can induce ventilation when warm air rises to the top of a structure and exhausts outside, while cooler outside air enters the structure to replace it. Figure 41.10 illustrates the solar chimney concept, which can easily be adapted to a thermal storage wall system. The greatest stack effect flow rate is produced by maximizing the stack height and the air temperature in the stack, as given by

$$Q = 0.116A_j\sqrt{h(T_s - T_o)} \tag{41.15}$$

where
Q is the stack flow rate (m^3/s)
A_j is the area of inlets or outlets, whichever is smaller (m^2)
h is the inlet-to-outlet height (m)
T_s is the average temperature in stack (°C)
T_o is the outdoor air temperature (°C)

If inlet or outlet area is twice the other, the flow rate will increase by 25%, and by 35% if the areas' ratio is 3:1 or larger (Table 41.6).

Example 41.6

A two-story (5 m) solar chimney is being designed to produce a flow of 0.25 m^3/s through a space. The preliminary design features include a 25 cm × 1.5 m inlet, a 50 cm × 1.5 m outlet, and an estimated 35°C average stack temperature on a sunny 30°C day. Can this design produce the desired flow?

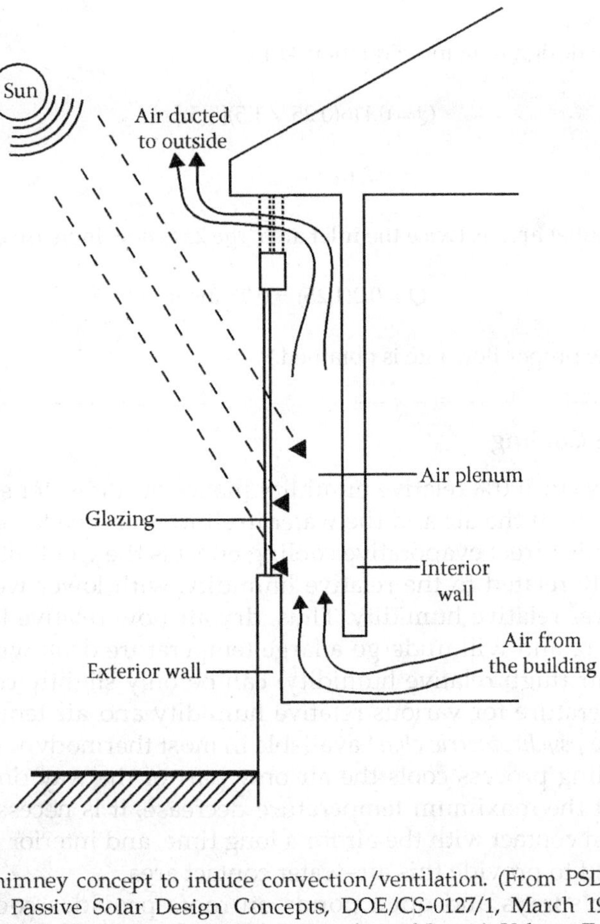

FIGURE 41.10
The stack-effect/solar chimney concept to induce convection/ventilation. (From PSDH, *Passive Solar Design Handbook*, Volume One: Passive Solar Design Concepts, DOE/CS-0127/1, March 1980, Prepared by Total Environmental Action, Inc. (B. Anderson, C. Michal, P. Temple, and Lewis); Volume Two: *Passive* Solar Design Analysis, DOE/CS-0127/2, January 1980, Prepared by Los Alamos Scientific Laboratory (J. D. Balcomb, D. Barley, R McFarland, J. Perry, W. Wray and S. Noll), U.S. Department of Energy, Washington, DC, 1980.)

TABLE 41.6
Ground Reflectivities

Material	ρ (%)
Cement	27
Concrete	20–40
Asphalt	7–14
Earth	10
Grass	6–20
Vegetation	25
Snow	70
Red brick	30
Gravel	15
White paint	55–75

Source: Murdoch, J.B., *Illumination Engineering—From Edison's Lamp to the Laser*, Macmillan, New York, 1985.

Solution

Substituting the design data into Equation 41.15,

$$Q = 0.116(0.25 \times 1.5)[5(5)]^{1/2}$$

$$= 0.2 \text{ m}^3/\text{s}.$$

Because the outlet area is twice the inlet area, the 25% flow increase can be used:

$$Q = 0.2(1.25) = 0.25 \text{ m}^3/\text{s}.$$

(*Answer*: Yes, the proper flow rate is obtained.)

41.4.3 Evaporative Cooling

When air with less than 100% relative humidity moves over a water surface, the evaporation of water causes both the air and the water itself to cool. The lowest temperature that can be reached by this direct evaporative cooling effect is the wet-bulb temperature of the air, which is directly related to the relative humidity, with lower wet-bulb temperature associated with lower relative humidity. Thus, dry air (low relative humidity) has a low wet-bulb temperature and will undergo a large temperature drop with evaporative cooling, while humid air (high relative humidity) can be only slightly cooled evaporatively. The wet-bulb temperature for various relative humidity and air temperature conditions can be found via the *psychrometric chart* available in most thermodynamic texts. Normally, an evaporative cooling process cools the air only part of the way down to the wet-bulb temperature. To get the maximum temperature decrease, it is necessary to have a large water surface area in contact with the air for a long time, and interior ponds and fountain sprays are often used to provide this air–water contact area.

The use of water sprays and open ponds on roofs provides cooling primarily via evaporation. The hybrid system involving a fan and wetted mat, the *swamp cooler,* is by far the most widely used evaporative cooling technology. Direct, indirect, and combined evaporative cooling system design features are described in ASHRAE (1993, 1995).

41.4.4 Nocturnal and Radiative Cooling Systems

Another approach to passive convective/ventilative cooling involves using cooler night air to reduce the temperature of the building and/or a storage mass. Thus, the building/storage mass is prepared to accept part of the heat load during the hotter daytime. This type of convective system can also be combined with evaporative and radiative modes of heat transfer, utilizing air and/or water as the convective fluid. Work in Australia (Close et al. 1968) investigated rock storage beds that were chilled using evaporatively cooled night air. Room air was then circulated through the bed during the day to provide space cooling. The use of encapsulated roof ponds as a thermal cooling mass has been tried by several investigators (Hay and Yellott 1969; Marlatt et al. 1984; Givoni 1994) and is often linked with nighttime radiative cooling.

All warm objects emit thermal infrared radiation; the hotter the body, the more energy it emits. A passive cooling scheme is to use the cooler night sky as a sink for thermal radiation emitted by a warm storage mass, thus chilling the mass for cooling

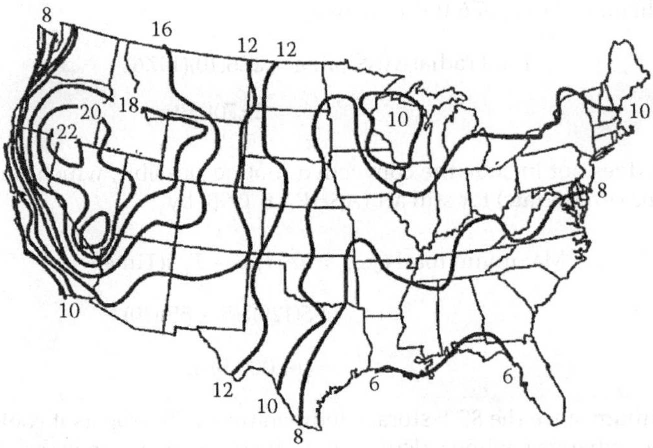

FIGURE 41.11
Average monthly sky temperature depression ($I_{AIR} - I_{SKY}$) for July in °F. (Adapted from Martin, M. and Berdahl, P., *Solar Energy*, 33(314), 321–336, 1984.)

use the next day. The net radiative cooling rate, Q_r, for a horizontal unit surface (ASHRAE 1993) is

$$Q_r = \varepsilon\sigma\left(T_{\text{body}}^4 - T_{\text{sky}}^4\right), \tag{41.16}$$

where

Q_r is the net radiative cooling rate, W/m^2 (Btu/h ft.^{2})
ε is the surface emissivity fraction (usually 0.9 for water)
σ is 5.67×10^{-8} W/m^2 K^4 (1.714×10^{-9} Btu/h ft.^{2} R^4)
T_{body} is the warm body temperature, Kelvin (Rankine)
T_{sky} is the effective sky temperature, Kelvin (Rankine)

The monthly average air–sky temperature difference has been determined (Martin and Berdahl 1984), and Figure 41.11 presents these values for July (in °F) for the United States.

Example 41.7

Estimate the overnight cooling possible for a 10 m^2, 85°F water thermal storage roof during July in Los Angeles.

Solution

Assume the roof storage unit is black with ε = 0.9. From Figure 41.11, $T_{\text{air}} - T_{\text{sky}}$ is approximately 10°F for Los Angeles. From weather data for LA airport (PSDH 1984; ASHRAE 1993), the July average temperature is 69°F with a range of 15°F. Assuming night temperatures vary from the average (69°F) down to half the daily range (15/2), then the average nighttime temperature is chosen as 69 – (1/2)(15/2) = 65°F. Therefore, T_{sky} = 65 – 10 = 55°F. From Equation 41.16,

$$Q_r = 0.9(1.714 \times 10^{-9})[(460 + 85)^4 - (460 + 55)^4]$$

$$= 27.6 \text{ Btu/h ft.}^2.$$

For a 10 h night and 10 m² (107.6 ft.²) roof area,

$$\text{Total radiative cooling} = 27.6(10)(107.6)$$
$$= 29{,}700 \text{ Btu.}$$

Note that this does not include the convective cooling possible, which can be approximated (at its maximum rate) for still air (ASHRAE 1993) by

$$\text{Maximum total } Q_{\text{conv}} = hA(T_{\text{roof}} - T_{\text{air}})(\text{Time})$$
$$= 5(129)(85 - 55)(10)$$
$$= 161{,}000 \text{ Btu.}$$

This is a maximum since the 85°F storage temperature will drop as it cools; this is also the case for the radiative cooling calculation. However, convection is seen to usually be the more dominant mode of nighttime cooling.

41.4.5 Earth Contact Cooling (or Heating)

Earth contact cooling or heating is a passive summer cooling and winter heating technique that utilizes underground soil as the heat sink or source. By installing a pipe underground and passing air through the pipe, the air will be cooled or warmed depending on the season. A schematic of an open-loop system and a closed-loop air-conditioning system are presented in Figures 41.12 and 41.13, respectively (Goswami and Biseli 1994).

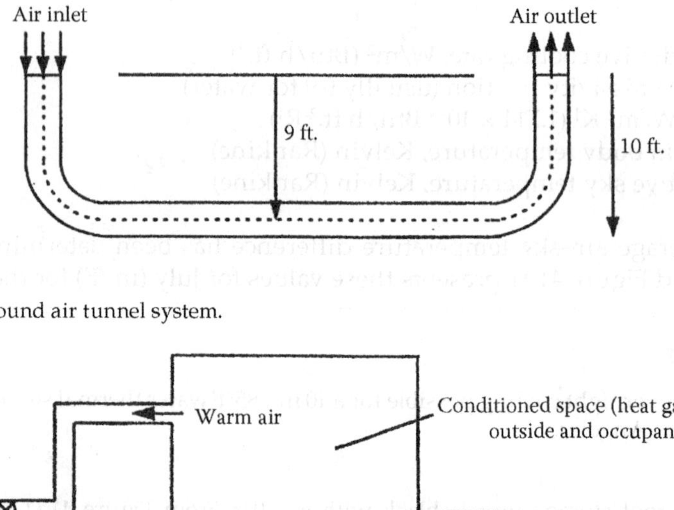

FIGURE 41.12
Open-loop underground air tunnel system.

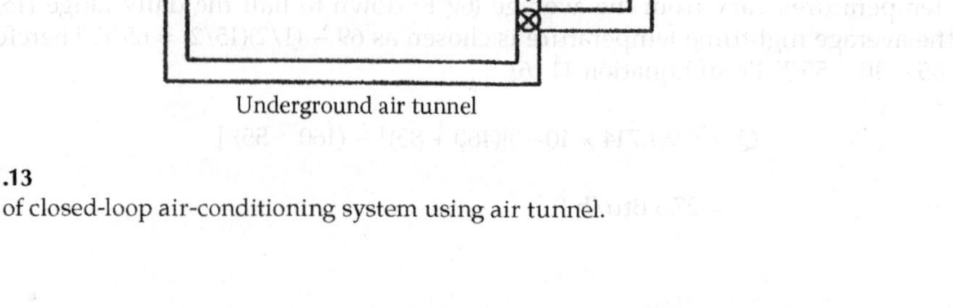

FIGURE 41.13
Schematic of closed-loop air-conditioning system using air tunnel.

The use of this technique can be traced back to 3000 BC when Iranian architects designed some buildings to be cooled by natural resources only. In the nineteenth century, Wilkinson (USDA 1960) designed a barn for 148 cows where a 500 ft. long underground passage was used for cooling during the summertime. Since that time, a number of experimental and analytical studies of this technique have continued to appear in the literature (Krarti and Kreider 1996; Hollmuller and Lachal 2001; De Paepe and Janssens 2003). Goswami and Dhaliwal (1985) have given a brief review of the literature as well as presented an analytical solution to the problem of transient heat transfer between the air and the surrounding soil as the air is made to pass through a pipe buried underground.

41.4.5.1 Heat Transfer Analysis

The transient thermal analysis of the air and soil temperature fields (Goswami and Dhaliwal 1985) is conducted using finite elements with the convective heat transfer between the air and the pipe and using semi-infinite cylindrical conductive heat transfer to the soil from the pipe. It should be noted that the thermal resistance of the pipe (whether of metal, plastic, or ceramic) is negligible relative to the surrounding soil.

41.4.5.1.1 Air and Pipe Heat Transfer

The pipe is divided into a large number of elements and a psychrometric energy balance written for each, depending on whether the air leaves the element (1) unsaturated or (2) saturated.

1. If the air leaves an element as unsaturated, the energy balance on the element is

$$mC_p(T_1 - T_2) = hA_p(T_{air} - T_{pipe}). \tag{41.17}$$

T_{air} can be taken as $(T_1 + T_2)/2$. Substituting and simplifying,

$$T_2 = \frac{(1-(U/2))T_1 + UT_{pipe}}{1+(U/2)}, \tag{41.18}$$

where U is defined as

$$U = \frac{A_p h}{mC_p}$$

2. If the air leaving the element is saturated, the energy balance is

$$mC_pT_1 + m(W_1 - W_2)H_{fg} = mC_pT_2 + hA_p(T_{air} - T_{pipe}). \tag{41.19}$$

Simplifying gives

$$T_2 = \frac{(1-(U/2))T_1 + ((W_1 - W_2)/C_p)H_{fg} + UT_{pipe}}{1+(U/2)}. \tag{41.20}$$

The convective heat transfer coefficient h in the preceding equations depends on Reynolds number, the shape, and roughness of the pipe.

Using the exit temperature from the first element as the inlet temperature for the next element, the exit temperature for the element can be calculated in a similar way. Continuing this way from one element to the next, the temperature of air at the exit from the pipe can be calculated.

41.4.5.1.2 Soil Heat Transfer

The heat transfer from the pipe to the soil is analyzed by considering the heat flux at the internal radius of a semi-infinite cylinder formed by the soil around the pipe. For a small element, the problem can be formulated as

$$\frac{\partial^2 T(r,t)}{\partial r^2} + \frac{1}{r}\frac{\partial T(r,t)}{\partial r} = \frac{1}{\alpha}\frac{\partial T(r,t)}{\partial t}, \qquad (41.21)$$

with initial and boundary conditions as

$$T(r,0) = T_e,$$

$$T(\infty,t) = T_e,$$

$$-K\frac{\partial T}{\partial r}(r,t) = q'',$$

where

T_e is the bulk earth temperature

q'' is also given by the amount of heat transferred to the pipe from the air by convection, that is, $q'' = h(T_{air} - T_{pipe})$

41.4.5.2 Soil Temperatures and Properties

Kusuda and Achenbach (1965) and Labs (1981) studied the earth temperatures in the United States. According to both of these studies, temperature swings in the soil during the year are dampened with depth below the ground. There is also a phase lag between the soil temperature and the ambient air temperature, and this phase lag increases with depth below the surface. For example, the soil temperature for light dry soil at a depth of about 10 ft. (3.05 m) varies by approximately ±5°F (2.8°C) from the mean temperature (approximately equal to mean annual air temperature) and has a phase lag of approximately 75 days behind ambient air temperature (Labs 1981).

The thermal properties of the soil are difficult to determine. The thermal conductivity and diffusivity both change with the moisture content of the soil itself, which is directly affected by the temperature of and heat flux from and to the buried pipe. Most researchers have found that using constant property values for soil taken from standard references gives reasonable predictive results (Goswami and Ileslamlou 1990).

41.4.5.3 Generalized Results from Experiments

Figure 41.14 presents data from Goswami and Biseli (1994) for an open system, 100 ft. long, 12 in. diameter pipe, buried 9 ft. deep. The figure shows the relationship between the pipe

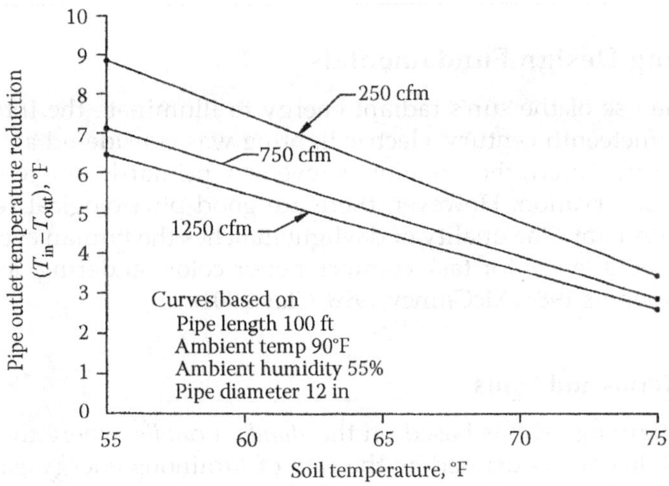

FIGURE 41.14
Air temperature drop through a 100 ft. long, 12 in. diameter pipe buried 9 ft. underground.

inlet-to-outlet temperature reduction ($T_{in} - T_{out}$) and the initial soil temperature with ambient air inlet conditions of 90°F, 55% relative humidity for various pipe flow rates.

Other relations from this same report that can be used with Figure 41.14 data include the following: (1) the effect of increasing pipe/tunnel length on increasing the inlet-to-outlet air temperature difference is fairly linear up to 250 ft.; and (2) the effect of decreasing pipe diameter on lowering the outlet air temperature is slight, and only marginally effective for pipes less than 12 in. in diameter.

Example 41.8

Provide the necessary 12 in. diameter pipe length(s) that will deliver 1500 cfm of 75°F air if the ambient temperature is 85°F and the soil at 9 ft. is 65°F.

Solution

From Figure 41.14, for 100 ft. of pipe at 65°F soil temperature, the pipe temperature reduction is

$$T_{in} - T_{out} = 6°F \text{ (at 250 cfm)}$$

$$= 5°F \text{ (at 750 cfm)}$$

$$= 4.5°F \text{ (at 1250 cfm)}.$$

Because the *length versus temperature reduction* is linear (see earlier text), the 10°F reduction required (85°F down to 75°F) would be met by the 750 cfm case (5°F for 100 ft.) if 200 ft. of pipe is used. Then, two 12 in. diameter pipes would be required to meet the 1500 cfm requirement.

Answer

Two 12 in. diameter pipes, each 200 ft. long. (*Note*: see what would be needed if the 250 cfm or the 1250 cfm cases had been chosen. Which of the three flow rate cases leads to the *cheapest* installation?)

41.5 Daylighting Design Fundamentals

Daylighting is the use of the sun's radiant energy to illuminate the interior spaces in a building. In the nineteenth century, electric lighting was considered an alternative technology to daylighting. Today, the situation is reversed, primarily due to the economics of energy use and conservation. However, there are good physiological reasons for using daylight as an illuminant. The quality of daylight matches the human eye's response, thus permitting lower light levels for task comfort, better color rendering, and clearer object discrimination (Robbins 1986; McCluney 1998; Clay 2001).

41.5.1 Lighting Terms and Units

Measurement of lighting level is based on the *standard candle*, where the lumen (lm), the unit of luminous flux (φ), is defined as the rate of luminous energy passing through a 1 m^2 area located 1 m from the candle. Thus, a standard candle generates 4π lumens, which radiate away in all directions. The illuminance (E) on a surface is defined as the luminous flux on the surface divided by the surface area, $E = \varphi/A$. Illuminance is measured in either lux (lx), as lm/m^2, or footcandles (fc), as $lm/ft.^2$.

Determination of the daylighting available at a given location in a building space at a given time is important to evaluate the reduction possible in electric lighting and the associated impact on heating and cooling loads. Daylight provides about 110 lm/W of solar radiation, fluorescent lamps about 75 lm/W of electrical input, and incandescent lamps about 20 lm/W; thus, daylighting generates only half to one-fifth the heating that equivalent electric lighting does, significantly reducing the building cooling load.

41.5.2 Approach to Daylighting Design

Aperture controls such as blinds and drapes are used to moderate the amount of daylight entering the space, as are the architectural features of the building itself (glazing type, area, and orientation; overhangs and wingwalls; lightshelves; etc.). Many passive and *active* reflective, concentrating, and diffusing devices are available to specifically gather and direct both the direct and diffuse components of daylight to areas within the space (Kinney et al. 2005). Electric-lighting dimming controls are used to adjust the electric light level based on the quantity of daylighting. With these two types of controls (aperture and lighting), the electric lighting and cooling energy use and demand, as well as cooling system sizing, can be reduced. However, the determination of the daylighting position and time illuminance value within the space is required before energy usage, and demand reduction calculations can be made.

Daylighting design approaches use both solar beam radiation (referred to as sunlight) and the diffuse radiation scattered by the atmosphere (referred to as skylight) as sources for interior lighting, with historical design emphasis being on utilizing skylight. Daylighting is provided through a variety of glazing features, which can be grouped as sidelighting (light enters via the side of the space) and toplighting (light enters from the ceiling area). Figure 41.15 illustrates several architectural forms producing sidelighting and toplighting, with the dashed lines representing the illuminance distribution within the space. The calculation of work-plane illuminance

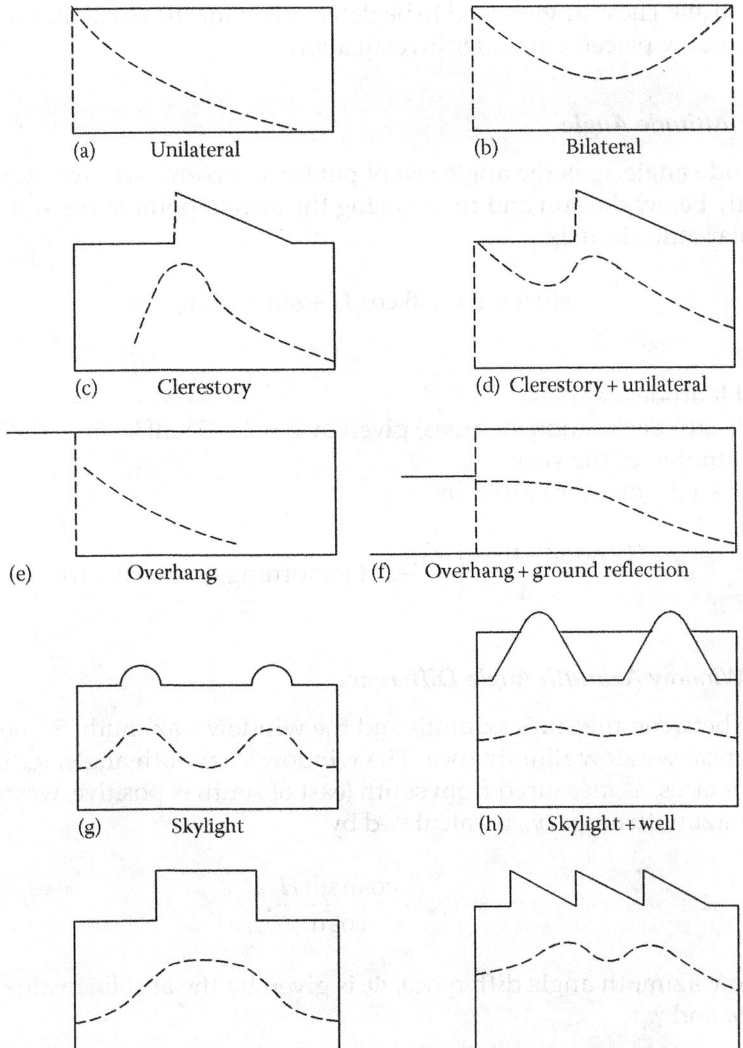

FIGURE 41.15
(a–j) Examples of sidelighting and toplighting architectural features (dashed lines represent illuminance distributions). (From Murdoch, J.B., *Illumination Engineering—From Edison's Lamp to the Laser,* Macmillan, New York, 1985.)

depends on whether sidelighting and/or toplighting features are used and the combined illuminance values are additive.

41.5.3 Sun-Window Geometry

The solar illuminance on a vertical or horizontal window depends on the position of the sun relative to that window. In the method described here, the sun and sky illuminance values are determined using the sun's altitude angle (α) and the sun-window azimuth angle difference (Φ). These angles need to be determined for the particular time of day, day of year, and window placement under investigation.

41.5.3.1 Solar Altitude Angle

The solar altitude angle, α, is the angle swept out by a person's arm when pointing to the horizon directly below the sun and then raising the arm to point at the sun. The equation to calculate solar altitude, α, is

$$\sin \alpha = \cos L\, \delta \cos H + \sin L \sin \delta, \tag{41.22}$$

where

L is the local latitude (degrees)
δ is the earth–sun declination (degrees) given by $\delta = 23.45 \sin[360(n - 81)/365]$
n is the day number of the year
H is the hour angle (degrees) given by

$$H = \frac{(12\,\text{noon} - \text{time})\ \text{min}}{4}; \quad (+,\text{morning}, -, \text{afternoon}). \tag{41.23}$$

41.5.3.2 Sun-Window Azimuth Angle Difference

The difference between the sun's azimuth and the window's azimuth, 5>, needs to be calculated for vertical window illuminance. The window's azimuth angle, y_w, is determined by which way it faces, as measured from south (east of south is positive, westward is negative). The solar azimuth angle, y_s, is calculated by

$$\sin \gamma_s = \frac{\cos \delta \sin H}{\cos \alpha}. \tag{41.24}$$

The sun-window azimuth angle difference, Φ, is given by the absolute value of the difference between γ_s and γ_w:

$$\Phi = |\gamma_s - \gamma_w|. \tag{41.25}$$

41.5.4 Daylighting Design Methods

To determine the annual lighting energy saved (ES_L), calculations using the lumen method described later should be performed on a monthly basis for both clear and overcast days for the space under investigation. Monthly weather data for the site would then be used to prorate clear and overcast lighting energy demands monthly. Subtracting the calculated daylighting illuminance from the design illuminance leaves the supplementary lighting needed, which determines the lighting energy required.

The approach in the following method is to calculate the *sidelighting* and the *skylighting* of the space separately and then combine the results. This procedure has been computerized (Lumen II/Lumen Micro) and includes many details of controls, daylighting technologies, and weather. ASHRAE (1993) lists many of the methods and simulation techniques currently used with daylighting and its associated energy effects.

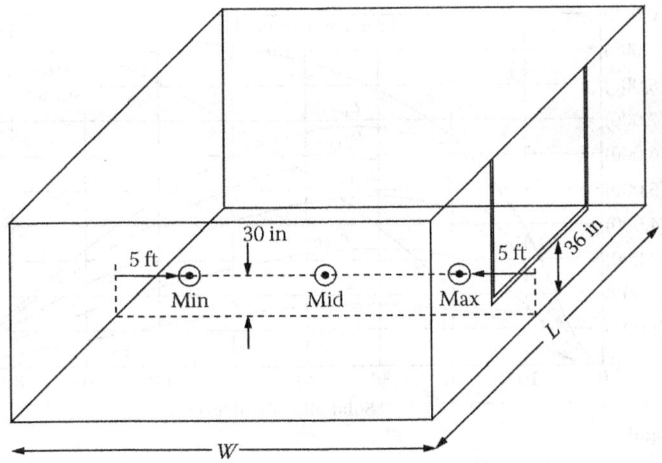

FIGURE 41.16
Location of illumination points within the room (along centerline of window) determined by lumen method of sidelighting.

41.5.4.1 Lumen Method of Sidelighting (Vertical Windows)

The lumen method of sidelighting calculates interior horizontal illuminance at three points, as shown in Figure 41.16, at the 30 in. (0.76 m) work-plane on the room-and-window centerline. A vertical window is assumed to extend from 36 in. (0.91 m) above the floor to the ceiling. The method accounts for both direct and ground-reflected sunlight and skylight, so both horizontal and vertical illuminances from sun and sky are needed. The steps in the lumen method of sidelighting are presented next.

As mentioned, the incident direct and ground-reflected window illuminances are normally calculated for both a cloudy and a clear day for representative days during the year (various months), as well as for clear or cloudy times during a given day. Thus, the interior illumination due to sidelighting and skylighting can then be examined for effectiveness throughout the year.

Step 1: Incident direct sky and sun illuminances—The solar altitude and sun-window azimuth angle difference are calculated for the desired latitude, date, and time using Equations 41.22 and 41.25, respectively. Using these two angles, the total illuminance on the window (E_{sw}) can be determined by summing the direct sun illuminance (E_{uw}) and the direct sky illuminance (E_{kw}), each determined from the appropriate graph in Figure 41.17.

Step 2: Incident ground-reflected illuminance—The sun illuminance on the ground (E_{ug}), plus the overcast or clear sky illuminance (E_{kg}) on the ground, makes up the total horizontal illuminance on the ground surface (E_{sg}). A fraction of the ground surface illuminance is then considered diffusely reflected onto the vertical window surface (E_{gw}), where gw indicates from the ground to the window.

The horizontal ground illuminances can be determined using Figure 41.18, where the clear sky plus sun case and the overcast sky case are functions of solar altitude. The fractions of the ground illuminance diffusely reflected onto the window depend on the reflectivity (ρ) of the ground surface (see Table 41.6) and the window-to-ground surface geometry.

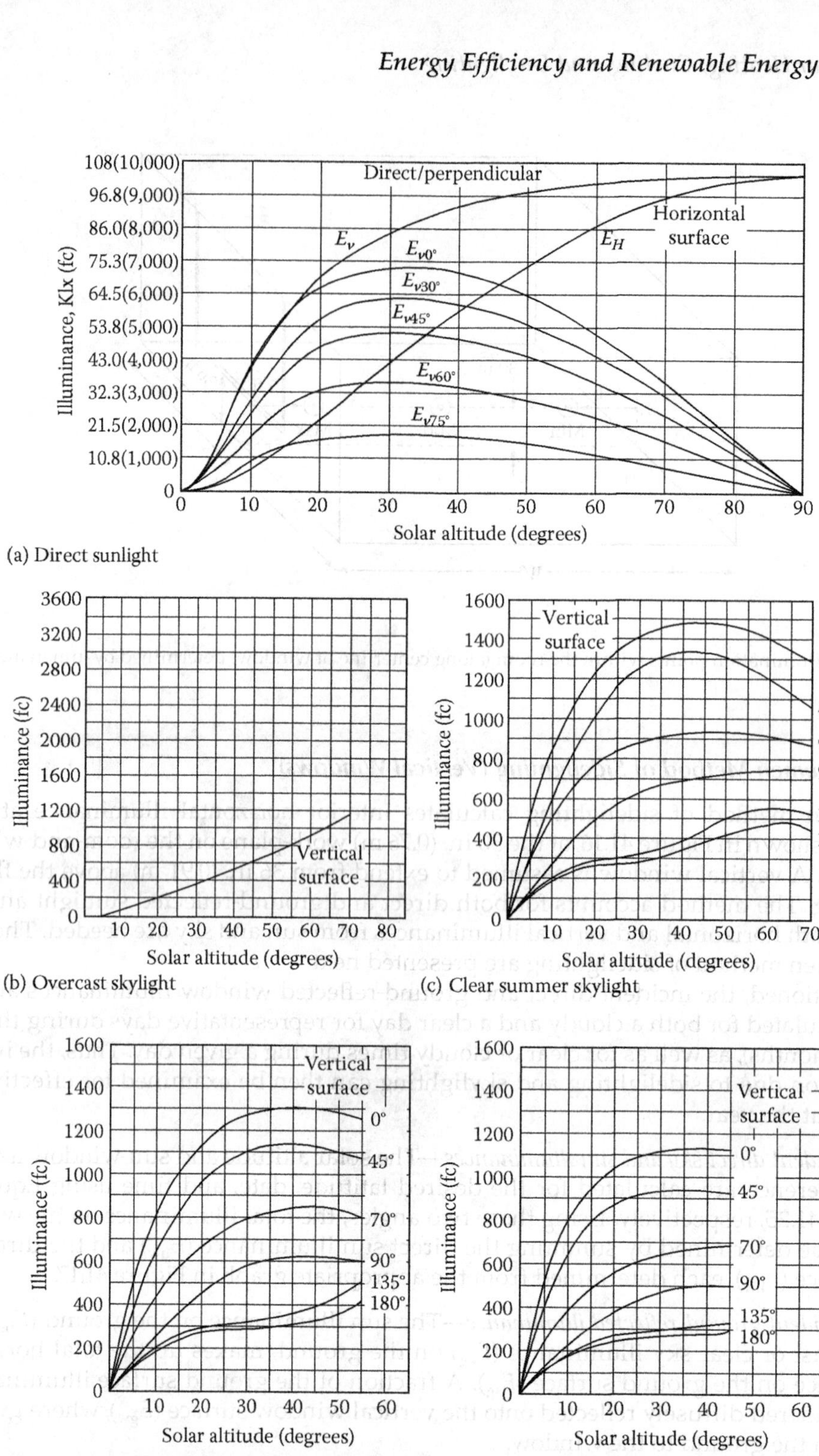

FIGURE 41.17
Illuminance on a vertical surface from (a) direct sunlight and from (b–e) skylight for various sun-window azimuth angle differences. (From IES, *Lighting Handbook, Applications Volume*, Illumination Engineering Society, New York, 1987.)

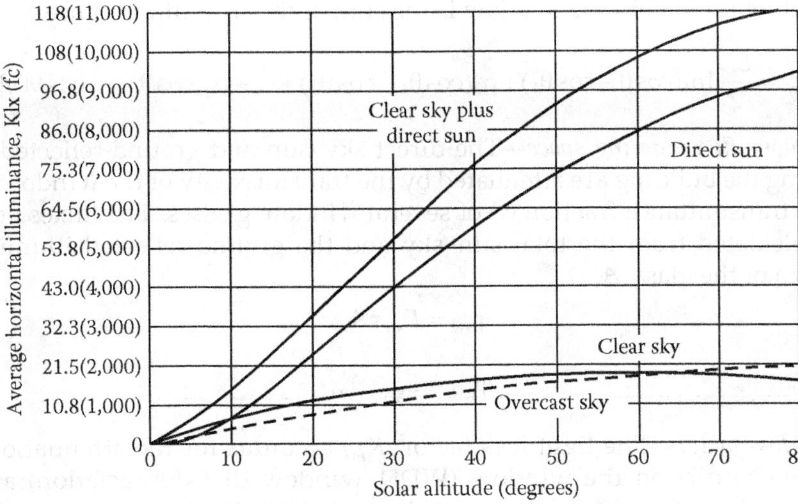

FIGURE 41.18
Horizontal illuminance for overcast sky, clear sky, direct sun, and clear sky plus direct sun. (From Murdoch, J.B., *Illumination Engineering—From Edison's Lamp to the Laser,* Macmillan, New York, 1985.)

If the ground surface is considered uniformly reflective from the window outward to the horizon, then the illuminance on the window from ground reflection is

$$E_{gw} = \frac{\rho E_{sg}}{2}. \quad (41.26)$$

A more complicated ground-reflection case is illustrated in Figure 41.19, where multiple *strips* of differently reflecting ground are handled using the angles to the window where a strip's illuminance on a window is calculated by

$$E_{gw(strip)} = \frac{\rho_{strip} E_{sg}}{2}(\cos\theta_1 - \cos\theta_2). \quad (41.27)$$

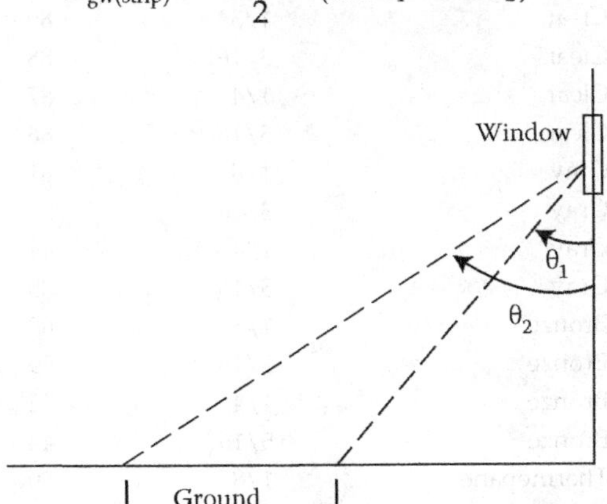

FIGURE 41.19
Geometry for ground *strips.* (From Murdoch, J.B., *Illumination Engineering—From Edison's Lamp to the Laser,* Macmillan, New York, 1985.)

And the total reflected onto the window is the sum of the strip illuminances:

$$E_{gw} = \frac{E_{sg}}{2}[\rho_1(\cos 0 - \cos\theta_1) + \rho_2(\cos\theta_1 - \cos\theta_2) + \cdots + \rho_n(\cos\theta_{n-1} - \cos 90)]. \quad (41.28)$$

Step 3: Luminous flux entering space—The direct sky–sun and ground-reflected luminous fluxes entering the building are attenuated by the transmissivity of the window. Table 41.7 presents the transmittance fraction (τ) of several window glasses. The fluxes entering the space are calculated from the total sun–sky and the ground-reflected illuminances by using the area of the glass, A_w:

$$\varphi_{sw} = E_{sw}\tau A_w,$$

$$\varphi_{gw} = E_{gw}\tau A_w, \quad (41.29)$$

Step 4: Light loss factor—The light loss factor (K_m) accounts for the attenuation of luminous flux due to dirt on the window (WDD, window dirt depreciation) and on the room surfaces (RSDD, room surface dirt depreciation). WDD depends on how often the window is cleaned, but a 6-month average for offices is 0.83 and for factories is 0.71 (Murdoch 1985).

The RSDD is a more complex calculation involving time between cleanings, the direct–indirect flux distribution, and room proportions. However, for rooms cleaned regularly, RSDD is around 0.94, and for once-a-year-cleaned dirty rooms, the RSDD would be around 0.84.

The light loss factor is the product of the preceding two fractions:

$$K_m = (\text{WDD})(\text{RSDD}). \quad (41.30)$$

TABLE 41.7

Glass Transmittances

Glass	Thickness (in.)	τ (%)
Clear	1/3	89
Clear	3/16	88
Clear	1/4	87
Clear	5/16	86
Gray	1/8	61
Gray	3/16	51
Gray	1/4	44
Gray	5/16	35
Bronze	1/8	68
Bronze	3/16	59
Bronze	1/4	52
Bronze	5/16	44
Thermopane	1/8	80
Thermopane	3/16	79
Thermopane	1/4	77

Source: Murdoch, J.B., *Illumination Engineering—From Edison's Lamp to the Laser*, Macmillan, New York, 1985.

Step 5: Work-plane illuminances—As discussed earlier, Figure 41.16 illustrates the location of the work-plane illuminances determined with this lumen method of sidelighting. The three illuminances (max, mid, min) are determined using two coefficients of utilization, the C factor, and the K factor. The C factor depends on room length and width and wall reflectance. The K factor depends on ceiling–floor height, room width, and wall reflectance. Tables 41.8 through 41.10 presents C and K values for the three cases of incoming fluxes: sun plus clear sky, overcast sky, and ground reflected. Assumed ceiling and floor reflectances are given for this case with no window controls (shades, blinds, overhangs, etc.). These further window control complexities can be found in Libbey-Owens-Ford Company (1976), IES (1987), and others. A reflectance of 70% represents light-colored walls, with 30% representing darker walls.

The work-plane max, mid, and min illuminances are each calculated by adding the sun–sky and ground-reflected illuminances, which are given by

$$E_{sp} = \varphi_{sw} C_s K_s K_m,$$

$$E_{gp} = \varphi_{gw} C_g K_g K_m, \tag{41.31}$$

where the "sp" and "gp" subscripts refer to the sky-to-work-plane and ground-to-work-plane illuminances.

Example 41.9

Determine the clear-sky illuminances for a 30 ft. long, 30 ft. wide, 10 ft. high room with a 20 ft. long window with a 3 ft. sill. The window faces 10°E of south, the building is at 32°N latitude, and it is January 15 at 2 p.m. The ground cover outside is grass, the glass is 1/4 in. clear, and the walls are light colored.

Solution

Following the steps in the *sidelighting* method

Step 1: With $L = 32$, $n = 15$, $H = (12 - 14)60/4 = -30$,

$$\delta = 23.45 \sin[360(15 - 81)/365] = -21.3°.$$

Then, Equation 41.22 yields $\alpha = 41.7°$, Equation 41.24 yields $\gamma_s = -38.7°$, and Equation 41.25 yields

$$\Phi = |-38.7 - (+10)| = 48.7°.$$

From Figure 41.17 with $\alpha = 41.7°$ and $\Phi = 48.7°$

(a) For clear sky (winter, no sun): $E_{kw} = 875$ fc.
(b) For direct sun: $E_{uw} = 4{,}100$ fc.
(c) Total clear sky plus direct: $E_{sw} = 4{,}975$ fc.

(*Note*: A high E_{uw} value probably indicates a glare situation!)

Step 2: Horizontal illuminances from Figure 41.18: $E_{sg} = 4007$ fc.
Then, Equation 41.26 yields, with $\rho_{grass} = 0.06$, $E_{gw} = 222$ fc.

Step 3: From Equation 41.29, with $\tau = 0.87$ and $A_w = 140$ ft.[2],

$$\Phi_{sw} = 4975(0.87)(140) = 605{,}955 \text{ lm}.$$

$$\Phi_{gw} = 222(0.87)(140) = 27{,}040 \text{ lm}.$$

TABLE 41.8

C and *K* Factors for No Window Controls for Overcast Sky

Illumination by Overcast Sky																	
		C: Coefficient of Utilization								*K*: Coefficient of Utilization							
Room Length (ft.)		20		30		40		Ceiling Height (ft.)		8		10		12		14	
Wall Reflectance (%)		70	30	70	30	70	30	Wall Reflectance (%)		70	30	70	30	70	30	70	30
Room Width (ft.)								Room Width (ft.)									
Max	20	0.0276	0.0251	0.0191	0.0173	0.0143	0.0137	Max	20	0.125	0.129	0.121	0.123	0.111	0.111	0.0991	0.0973
	30	0.0272	0.0248	0.0188	0.0172	0.0137	0.0131		30	0.122	0.131	0.122	0.121	0.111	0.111	0.0945	0.0973
	40	0.0269	0.0246	0.0182	0.0171	0.0133	0.0130		40	0.145	0.133	0.131	0.126	0.111	0.111	0.0973	0.0982
Mid	20	0.0159	0.0177	0.0101	0.0087	0.0081	0.0071	Mid	20	0.0908	0.0982	0.107	0.115	0.111	0.111	0.105	0.122
	30	0.0058	0.0050	0.0054	0.0040	0.0034	0.0033		30	0.156	0.102	0.0939	0.113	0.111	0.111	0.121	0.134
	40	0.0039	0.0027	0.0030	0.0023	0.0022	0.0019		40	0.106	0.0948	0.123	0.107	0.111	0.111	0.135	0.127
Min	20	0.0087	0.0053	0.0063	0.0043	0.0050	0.0037	Min	20	0.0908	0.102	0.0951	0.114	0.111	0.111	0.118	0.134
	30	0.0032	0.0019	0.0029	0.0017	0.0020	0.0014		30	0.0924	0.119	0.101	0.114	0.111	0.111	0.125	0.126
	40	0.0019	0.0009	0.0016	0.0009	0.0012	0.0008		40	0.111	0.0926	0.125	0.109	0.111	0.111	0.133	0.130

Source: IES, 1979.

TABLE 41.9

C and *K* Factors for No Window Controls for Clear Sky

Illumination by Clear Sky																		
		C: Coefficient of Utilization								*K*: Coefficient of Utilization								
Room Length (ft.)		**20**		**30**		**40**		**Ceiling Height (ft.)**		**8**		**10**		**12**		**14**	
Wall Reflectance (%)		**70**	**30**	**70**	**30**	**70**	**30**	**Wall Reflectance (%)**		**70**	**30**	**70**	**30**	**70**	**30**	**70**	**30**
Room Width (ft.)								**Room Width (ft.)**									
Max	20	0.0206	0.0173	0.0143	0.0123	0.0110	0.0098	Max	20	0.145	0.155	0.129	0.132	0.111	0.111	0.101	0.0982
	30	0.0203	0.00.173	0.0137	0.0120	0.0098	0.0092		30	0.141	0.149	0.125	0.130	0.111	0.111	0.0954	0.101
	40	0.0200	0.0168	0.0131	0.0119	0.0096	0.0091		40	0.157	0.157	0.135	0.134	0.111	0.111	0.0964	0.0991
Mid	20	0.0153	0.0104	0.0100	0.0079	0.0083	0.0067	Mid	20	0.110	0.128	0.116	0.126	0.111	0.111	0.103	0.108
	30	0.0082	0.0054	0.0062	0.0043	0.0046	0.0037		30	0.106	0.125	0.110	0.129	0.111	0.111	0.112	0.120
	40	0.0052	0.0032	0.0040	0.0028	0.0029	0.0023		40	0.117	0.118	0.122	0.118	0.111	0.111	0.123	0.122
Min	20	0.0106	0.0060	0.0079	0.0049	0.0067	0.0043	Min	20	0.105	0.129	0.112	0.130	0.111	0.111	0.111	0.116
	30	0.0054	0.0028	0.0047	0.0023	0.0032	0.0021		30	0.0994	0.144	0.107	0.126	0.111	0.111	0.107	0.124
	40	0.0031	0.0014	0.0027	0.0013	0.0021	0.0012		40	0.119	0.116	0.130	0.118	0.111	0.111	0.120	0.118

Source: IES, 1979.

TABLE 41.10

C and *K* Factors for No Window Controls for Ground Illumination[a]

Ground Illumination

C: Coefficient of Utilization

Room Length (ft.)		**20**		**30**		**40**	
Wall Reflectance (%)		**70**	**30**	**70**	**30**	**70**	**30**
Room Width (ft.)							
Max	20	0.0147	0.0112	0.0102	0.0088	0.0081	0.0071
	30	0.0141	0.0012	0.0098	0.0088	0.0077	0.0070
	40	0.0137	0.0112	0.0093	0.0086	0.0072	0.0069
Mid	20	0.0128	0.0090	0.0094	0.0071	0.0073	0.0060
	30	0.0083	0.0057	0.0062	0.0048	0.0050	0.0041
	40	0.0055	0.0037	0.0044	0.0033	0.0042	0.0026
Min	20	0.0106	0.0071	0.0082	0.0054	0.0067	0.0044
	30	0.0051	0.0026	0.0041	0.0023	0.0033	0.0021
	40	0.0029	0.0018	0.0026	0.0012	0.0022	0.0011

***K*: Coefficient of Utilization**

Ceiling Height (ft.)		**8**		**10**		**12**		**14**	
Wall Reflectance (%)		**70**	**30**	**70**	**30**	**70**	**30**	**70**	**30**
Room Width (ft.)									
Max	20	0.124	0.206	0.140	0.135	0.111	0.111	0.0909	0.0859
	30	0.182	0.188	0.140	0.143	0.111	0.111	0.0918	0.0878
	40	0.124	0.182	0.140	0.142	0.111	0.111	0.0936	0.0879
Mid	20	0.123	0.145	0.122	0.129	0.111	0.111	0.100	0.0945
	30	0.0966	0.104	0.107	0.112	0.111	0.111	0.110	0.105
	40	0.0790	0.0786	0.0999	0.106	0.111	0.111	0.118	0.118
Min	20	0.0994	0.108	0.110	0.114	0.111	0.111	0.107	0.104
	30	0.0816	0.0822	0.0984	0.105	0.111	0.111	0.121	0.116
	40	0.0700	0.0656	0.0946	0.0986	0.111	0.111	0.125	0.132

Source: IES, 1979.

[a] Ceiling reflectance, 80%; floor reflectance, 30%.

Step 4: For a clean office room,

$$K_m = (0.83)(0.94) = 0.78.$$

Step 5: From Tables 41.8 through 41.10, for 30 ft. width, 30 ft. length, 10 ft. ceiling, and wall reflectivity 70%,

a. Clear sky

$$C_{s,\,max} = 0.0137;\ K_{s,\,max} = 0.125,$$

$$C_{s,\,mid} = 0.0062;\ K_{s,\,mid} = 0.110,$$

$$C_{s,\,min} = 0.0047;\ K_{s,\,min} = 0.107.$$

b. Ground reflected

$$C_{g,\,max} = 0.0098;\ K_{g,\,max} = 0.140.$$

$$C_{g,\,mid} = 0.0062;\ K_{,\,mid} = 0.107.$$

$$C_{g,\,mid} = 0.0041;\ K_{g,\,min} = 0.0984.$$

Then, using Equation 41.31,

$$E_{sp,\,max} = 605{,}955(0.0137)(0.125)(0.78) = 809\text{ fc.}$$

$$E_{sp,\,mid} = 605{,}955(0.0062)(0.110)(0.78) = 322\text{ fc.}$$

$$E_{sp,\,min} = 605{,}955(0.0047)(0.107)(0.78) = 238\text{ fc.}$$

$$E_{gp,\,max} = 27{,}040(0.0098)(0.140)(0.78) = 29\text{ fc.}$$

$$E_{gp,\,mid} = 27{,}040(0.0062)(0.107)(0.78) = 14\text{ fc.}$$

$$E_{gp,\,min} = 27{,}040(0.0041)(0.984)(0.78) = 9\text{ fc.}$$

Thus,

$$E_{max} = 838\text{ fc.}$$

$$E_{mid} = 336\text{ fc.}$$

$$E_{min} = 247\text{ fc.}$$

41.5.4.2 Lumen Method of Skylighting

The lumen method of skylighting calculates the average illuminance at the interior work-plane provided by horizontal skylights mounted on the roof. The procedure for skylighting is generally the same as that described earlier for sidelighting. As with windows, the illuminance from both overcast sky and clear sky plus sun cases is determined for specific days in different seasons and for different times of the day, and a judgment is then made as to the number and size of skylights and any controls needed.

The procedure is presented in four steps: (1) finding the horizontal illuminance on the outside of the skylight, (2) calculating the effective transmittance through the skylight and

TABLE 41.11

Flat-Plate Plastic Material Transmittance for Skylights

Type	Thickness (in.)	Transmittance (%)
Transparent	1/8–3/16	92
Dense translucent	1/3	32
Dense translucent	3/16	24
Medium translucent	1/8	56
Medium translucent	3/16	52
Light translucent	1/8	72
Light translucent	3/16	68

Source: Murdoch, J.B., *Illumination Engineering—From Edison's Lamp to the Laser*, Macmillan, New York, 1985.

its well, (3) figuring the interior space light loss factor and the utilization coefficient, and finally, (4) calculating illuminance on the work-plane.

Step 1: Horizontal sky and sun illuminances—The horizontal illuminance value for an overcast sky or a clear sky plus sun situation can be determined from Figure 41.18 knowing only the solar altitude.

Step 2: Net skylight transmittance—The transmittance of the skylight is determined by the transmittance of the skylight cover(s), the reflective efficiency of the skylight well, the net-to-gross skylight area, and the transmittance of any light-control devices (lenses, louvers, etc.).

The transmittance for several flat-sheet plastic materials used in skylight domes is presented in Table 41.11. To get the effective dome transmittance (T_D) from the flat-plate transmittance (T_F) value (AAMA 1977), use

$$T_D = 1.25T_F(1.18 - 0.416T_F). \tag{41.32}$$

If a double-domed skylight is used, then the single-dome transmittances are combined as follows (Pierson 1962):

$$T_D \frac{T_{D_1}T_{D_2}}{T_{D_1}T_{D_2} - T_{D_1}T_{D_2}}. \tag{41.33}$$

If the diffuse and direct transmittances for solar radiation are available for the skylight glazing material, it is possible to follow this procedure and determine diffuse and direct dome transmittances separately. However, this difference is usually not a significant factor in the overall calculations.

The efficiency of the skylight well (N_w) is the fraction of the luminous flux from the dome that enters the room from the well. The well index (WI) is a geometric index (height, h; length, l; width, w) given by

$$\mathrm{WI} = \frac{h(w+l)}{2wl}, \tag{41.34}$$

and WI is used with the well-wall reflectance value in Figure 41.20 to determine the well efficiency, N_w.

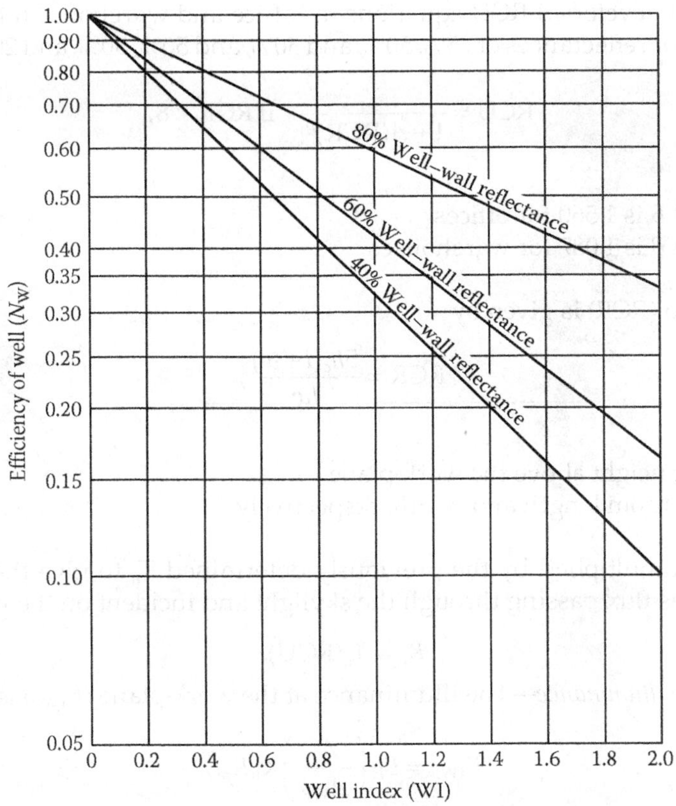

FIGURE 41.20
Efficiency of well versus well index. (From IES (Illumination Engineering Society), *Lighting Handbook, Applications Volume*, Illumination Engineering Society, New York, 1987.)

With T_D and N_W determined, the net skylight transmittance for the skylight and well is given by

$$T_n = T_D N_W R_A T_C, \tag{41.35}$$

where
R_A is the ratio of net-to-gross skylight areas
T_C is the transmittance of any light-controlling devices

Step 3: Light loss factor and utilization coefficient—The light loss factor (K_m) is again defined as the product of the RSDD and the skylight direct depreciation (SDD) fractions, similar to Equation 41.30. Following the reasoning for the sidelighting case, the RSDD value for clean rooms is around 0.94 and 0.84 for dirty rooms. Without specific data indicating otherwise, the SDD fraction is often taken as 0.75 for office buildings and 0.65 for industrial areas.

The fraction of the luminous flux on the skylight that reaches the work-plane (K_u) is the product of the net transmittance (T_n) and the room coefficient of utilization (RCU).

Dietz et al. (1981) developed RCU equations for office and warehouse interiors with ceiling, wall, and floor reflectances of 75%, 50%, and 30%, and 50%, 30%, and 20%, respectively.

$$\text{RCU} = \frac{1}{1 + A(\text{RCR})^B} \quad \text{if RCR} < 8, \tag{41.36}$$

where

A is 0.0288 and B is 1.560 for offices
A is 0.0995 and B is 1.087 for warehouses

Room cavity ratio (RCR) is given by

$$\text{RCR} = \frac{5h_c(l + w)}{lw}, \tag{41.37}$$

where

h_c is the ceiling height above the work-plane
l and w are the room length and width, respectively

The RCU is then multiplied by the previously determined T_n to give the fraction of the external luminous flux passing through the skylight and incident on the workplace:

$$K_u = T_n(\text{RCU}). \tag{41.38}$$

Step 4: Work-plane illuminance—The illuminance at the work-plane (E_{TWP}) is given by

$$E_{TWP} = E_H\left(\frac{A_T}{A_{WP}}\right)K_u K_m, \tag{41.39}$$

where

E_H is the horizontal overcast or clear sky plus sun illuminance from step 1
A_T is the total gross area of the skylights (number of skylight times skylight gross area)
A_{WP} is the work-plane area (generally room length times width)

Note that in Equation 41.39, it is also possible to fix the E_{TWP} at some desired value and determine the required skylight area.

Rules of thumb for skylight placement for uniform illumination include 4%–8% of roof area and spacing less than 1.5 times ceiling-to-work-plane distance between skylights (Murdoch 1984).

Example 41.10

Determine the work-plane *clear sky plus sun* illuminance for a 30 × 30 × 10 ft.^{3} office with 75% ceiling, 50% wall, and 30% floor reflectance with four 4 × 4 ft.^{2} double-domed skylights at 2:00 p.m. on January 15 at 32° latitude. The skylight well is 1 ft. deep at with 60% reflectance walls, and the outer- and inner-dome flat-plastic transmittances are 0.85 and 0.45, respectively. The net skylight area is 90%.

Solution

Follow the four steps in the lumen method for skylighting.

Step 1: Use Figure 41.18 with the solar altitude of 41.7° (calculated from Equation 41.26) for the clear sky plus sun curve to get horizontal illuminance:

$$E_H = 7400 \text{ fc}.$$

Step 2: Use Equation 41.32 to determine domed transmittances from the flat-plate plastic transmittances given

$$T_{D_1} = 1.25(0.85)[1.18 - 0.416(0.85)] = 0.89,$$

$$T_{D_2}(T_F = 0.45) = 0.56,$$

and use Equation 41.33 to get total dome transmittance from the individual dome transmittances:

$$T_D = \frac{(0.89)(0.56)}{(0.89)+(0.56)-(0.89)(0.56)} = 0.52.$$

To determine well efficiency, use WI = 0.25 from Equation 41.34 with 60% wall reflectance in Figure 41.20 to give $N_w = 0.80$. With $R_A = 0.90$, use Equation 41.35 to calculate net transmittance:

$$T_n = (0.52)(0.80)(0.90)(1.0) = 0.37.$$

Step 3: The light loss factor is assumed to be from *typical* values in Equation 41.30: $K_m = (0.75)(0.94) = 0.70$. The room utilization coefficient is determined using Equations 41.36 and 41.37:

$$\text{RCR} = \frac{5(7.5)(30+30)}{(30)(30)} = 2.5,$$

$$\text{RCU} = [1 + 0.0288(2.5)^{1.560}]^{-1} = 0.89.$$

Equation 41.37 yields $K_u = (0.37)(0.89) = 0.33$.

Step 4: The work-plane illuminance is calculated by substituting these values into Equation 41.39:

$$E_{TWP} = 7400\left[\frac{4(16)}{30(30)}\right]0.33(0.70),$$

$$E_{TWP} = 122 \text{ fc.}$$

41.5.5 Daylighting Controls and Economics

The economic benefit of daylighting is directly tied to the reduction in lighting electrical energy operating costs. Also, lower cooling-system operating costs are possible due to the reduction in heating caused by the reduced electrical lighting load. The reduction in lighting and cooling system electrical power during peak demand periods could also beneficially affect demand charges.

The reduction of the design cooling load through the use of daylighting can also lead to the reduction of installed or first-cost cooling system dollars. Normally, economics dictate that an automatic lighting control system must take advantage of the reduced lighting/cooling effect, and the control system cost minus any cooling system cost savings should be expressed as a *net* first cost. A payback time for the lighting control system (*net* or not) can be calculated from the ratio of first costs to yearly operating savings. In some cases, these paybacks for daylighting controls have been found to be in the range of 1–5 years for office building spaces (Rundquist 1991).

Controls, both aperture and lighting, directly affect the efficacy of the daylighting system. As shown in Figure 41.21, aperture controls can be architectural (overhangs, light shelves, etc.) and/or window shading devices (blinds, automated louvers, etc.). The aperture controls generally moderate the sunlight entering the space to maximize/minimize solar thermal gain, permit the proper amount of light for visibility, and prevent glare and beam radiation onto the workplace. Photosensor control of electric lighting allows the dimming (or shutting off) of the lights in proportion to the amount of available daylighting illuminance.

In most cases, increasing the solar gain for daylighting purposes, with daylighting controls, saves more in electrical lighting energy, and the cooling energy associated with the lighting then is incurred with the added solar gain (Rundquist 1991). In determining the annual energy savings total from daylighting, ES_T, the annual lighting energy saved from daylighting, ES_L, is added with the reduction in cooling system energy, ΔES_C, and with the negative of the heating system energy increase ΔES_H:

$$ES_T = ES_L + \Delta ES_C - \Delta ES_H. \tag{41.40}$$

A simple approach to estimating the heating and cooling energy changes associated with the lighting energy reduction is by using the fraction of the year associated with the cooling or heating season (f_C, f_H) and the seasonal coefficient of performance (COP_C, COP_H) of the cooling or heating equipment. Thus, Equation 41.40 can be expressed as

$$ES_T = ES_L + \frac{f_C ES_L}{COP_C} - \frac{f_H ES_L}{COP_H},$$

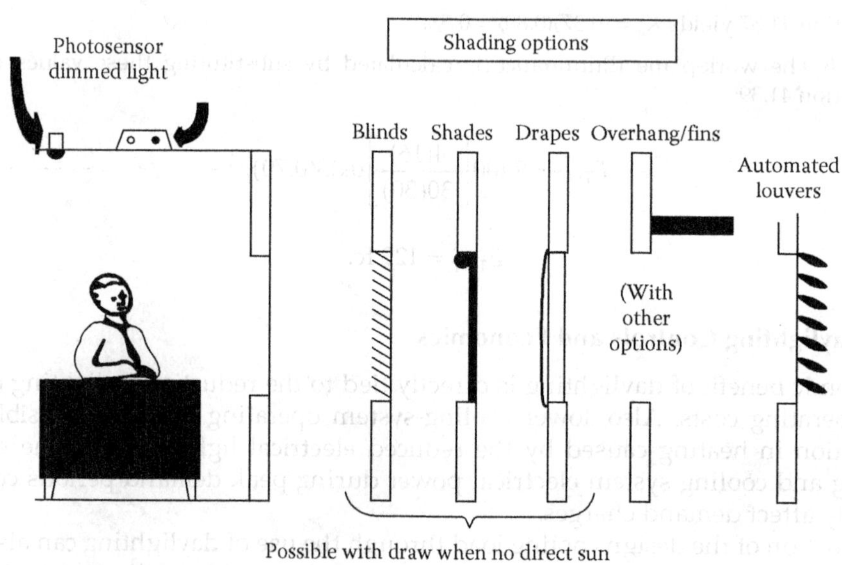

FIGURE 41.21
Daylighting system controls. (From Rundquist, R.A., *ASHRAE J.*, 11, 30–34, November 1991.)

$$\mathrm{ES_T} = \mathrm{ES_L} + \left(\frac{f_C}{\mathrm{COP_c}} - \frac{f_H}{\mathrm{COP_H}}\right). \tag{41.41}$$

It should be noted that the increased solar gain due to daylighting has not been included here but would reduce summer savings and increase winter savings. If it is assumed that the increased wintertime daylighting solar gain approximately offsets the reduced lighting heat gain, then the last term in Equation 41.41 becomes negligible.

To determine the annual lighting energy saved ($\mathrm{ES_L}$), calculations using the lumen method described earlier should be performed on a monthly basis for both clear and overcast days for the space under investigation. Monthly weather data for the site would then be used to prorate clear and overcast lighting energy demands monthly. Subtracting the calculated (controlled) daylighting illuminance from the design illuminance leaves the supplementary lighting needed, which determines the lighting energy required.

This procedure has been computerized and includes many details of controls, daylighting methods, weather, and heating and cooling load calculations. ASHRAE (1989) lists many of the methods and simulation techniques currently used with daylighting and its associated energy effects.

Example 41.11

A 30 × 20 $\mathrm{ft^2}$ space has a photosensor dimmer control with installed lighting density of 2.0 $\mathrm{W/ft^2}$. The required workplace illuminance is 60 fc, and the available daylighting illuminance is calculated as 40 fc on the summer peak afternoon. Determine the effect on the cooling system (adapted from Rundquist 1991).

Solution. The lighting power reduction is (2.0 $\mathrm{W/ft^2}$) (30 × 20) $\mathrm{ft^2}$ × (40 fc/60 fc) = 800 W. The space cooling load would also be reduced by this amount (assuming CLF = 1.0):

$$\frac{800\,\mathrm{W} \times 3.413\,\mathrm{Btu\,h/W}}{12{,}000\,\mathrm{Btu\,h/ton}} = 0.23\,\mathrm{ton}.$$

Assuming 1.5 ton nominally installed for 600 $\mathrm{ft^2}$ of space at \$2200/ton, the 0.23 ton reduction is *worth* 0.23 × \$2200/ton = \$506. The lighting controls cost about \$1/$\mathrm{ft^2}$ of controlled area, so the net installed first cost is

Net first cost = \$600 controls − \$500 A/C savings = \$100.

Assuming the day-to-monthly-to-annual illuminance calculations gave a 30% reduction in annual lighting, the associated operating savings can be determined. Lighting energy savings are

$$\mathrm{ES_L} = 0.30 \times 2.0\,\mathrm{W/ft^2} \times 600\,\mathrm{ft^2} \times 2500\,\mathrm{h/year} = 900\,\mathrm{kWh}.$$

Using Equation 41.41 to also include cooling energy saved due to lighting reduction (with $\mathrm{COP_c} = 2.5$, $f_c = 0.5$, and neglecting heating) gives

$$\mathrm{ES_T} = 900(1 + 0.5/2.5 - 0) = 1080\,\mathrm{kWh}.$$

At \$0.10 per kWh, the operating costs savings are \$0.10/kWh × 1080 kWh = \$108/year.

Thus, the simple payback is approximately 1 year (100/108) for the *net* situation and a little over 5.5 years (600/108) against the *controls* cost alone. It should also be noted that the 800 W lighting electrical reduction at peak hours, with an associated cooling energy reduction of 800 W/2.5 COP = 320 W, provides a peak demand reduction for the space of 1.1 kW, which can be used as the *first-cost savings* to offset control system costs.

Glossary

Active system: A system employing a forced (pump or fan) convection heat transfer fluid flow.

Daylighting: The use of the sun's radiant energy for illumination of a building's interior space.

Hybrid system: A system with parallel passive and active flow systems or one using forced convection flow to distribute from thermal storage.

Illuminance: The density of luminous flux incident on a unit surface. Illuminance is calculated by dividing the luminous flux (in lumens) by the surface area (m^2, $ft.^2$). Units are lux (lx) (lumens/m^2) in SI and footcandles (fc) (lumens/$ft.^2$) in English systems.

Luminous flux: The time rate of flow of luminous energy (lumens). A lumen (lm) is the rate that luminous energy from a 1 candela (cd) intensity source is incident on a 1 m^2 surface 1 m from the source.

Passive cooling system: A system using natural energy flows to transfer heat to the environmental sinks (ground, air, and sky).

Passive heating system: A system in which the sun's radiant energy is converted to heat by absorption in the system, and the heat is distributed by naturally occurring processes.

Sidelighting: Daylighting by light entering through the wall/side of a space.

Skylight: The diffuse solar radiation from a clear or overcast sky, excluding the direct radiation from the sun.

Sunlight: The direct solar radiation from the sun.

Toplighting: Daylighting by light entering through the ceiling area of a space.

References

AAMA (Architectural Aluminum Manufacturers Association). 1977. Publication 1602.1.1977, *Voluntary Standard Procedure for Calculating Skylight Annual Energy Balance.* AAMA, Chicago, IL.

ASHRAE (American Society of Heating, Refrigerating and Air-Conditioning Engineers). 1993. Fundamentals. In *ASHRAE Handbook*. ASHRAE, Atlanta, GA.

ASHRAE (American Society of Heating, Refrigerating and Air-Conditioning Engineers). 1995. Heating, ventilating, and air-conditioning applications. In *ASHRAE Handbook*, pp. 279–127. ASHRAE, Atlanta, GA.

Clay, R. A. 2001. Green is good for you. *Monitor on Psychology*, 32(4), 40–42.

Close, D. J., Dunkle, R. V., and Robeson, K. A. 1968. Design and performance of a thermal storage air conditioning system. *Mechanical and Chemical Engineering Transactions*, MC4, 45.

De Paepe, M. and Janssens, A. 2003. Thermo-hydraulic design of earth-air heat exchanger. *Energy and Buildings*, 35, 389–397.

Dietz, P., Murdoch, J., Pokoski, J., and Boyle, J. October 1981. A skylight energy balance analysis procedure. *Journal of the Illuminating Engineering Society*, 11, 27–34.

Duffie, J. A. and Beckman, W. A. 1991. *Solar Engineering of Thermal Processes*, 2nd edn., Wiley, New York.

Goswami, D. Y. and Biseli, K. M. August 1994. Use of underground air tunnels for heating and cooling agricultural residential buildings. Report EES-78, Florida Energy Extension Service, University of Florida, Gainesville, FL.

Goswami, D. Y. and Dhaliwal, A. S. May 1985. Heat transfer analysis in environmental control using an underground air tunnel. *Journal of Solar Energy Engineering*, 107, 141–145.

Goswami, D. Y. and Ileslamlou, S. May 1990. Performance analysis of a closed-loop climate control system using underground air tunnel. *Journal of Solar Energy Engineering*, 112, 76–81.

Givoni, B. 1994. *Passive and Low Energy Cooling of Buildings*. Van Nostrand Reinhold, New York.

Halacy, D.S. 1984. *Home Energy*. Rodale Press, Inc., Emmaus, PA.

Hay, H. and Yellott, J. 1969. Natural air conditioning with roof ponds and movable insulation. *ASHRAE Transactions*, 75(1), 165–177.

Hollmuller, P. and Lachal, B. 2001. Cooling and preheating with buried pipe systems: Monitoring, simulation and economic aspects. *Energy and Buildings*, 33, 509–518.

IES. 1979. *Lighting Handbook, Applications Volume*. Illumination Engineering Society, New York.

IES. 1987. *Lighting Handbook, Applications Volume*. Illumination Engineering Society, New York.

Kinney, L., McCluney, R., Cler, G., and Hutson, J. 2005. New designs in active daylighting: Good ideas whose time has (finally) come. In *Proceedings of the 2005 Solar World Congress*. August 6–12, 2005, ISES, Orlando, FL.

Krarti, M. and Kreider, J. F. 1996. Analytical model for heat transfer in an underground air tunnel. *Energy Conversion Management*, 37(10), 1561–1574.

Kusuda, T. and Achenbach, P. R. 1965. Earth temperature and thermal diffusivity at selected stations in the United States. *ASHRAE Transactions*, 71(1), 965.

Labs, K. 1981. Regional analysis of ground and above ground climate. Report ORNL/Sub-81/40451/1. Oak Ridge National Laboratory, Oak Ridge, TN.

Larson, R., Vignola, F., and West, R., eds. 1992. *Economics of Solar Energy Technologies*. American Solar Energy Society, Orlando, FL.

Libbey-Owens-Ford Company. 1976. *How to Predict Interior Daylight Illumination*. Libbey-Owens-Ford Company, Toledo, OH.

Marlatt, W., Murray, C., and Squire, S. 1984. Roof Pond Systems Energy Technology Engineering Center. Report No. ETEC 6, April. Rockwell International, New York.

Martin, M. and Berdahl, P. 1984. Characteristics of infrared sky radiation in the United States. *Solar Energy*, 33(314), 321–336.

McCluney, R. 1998. Advanced fenestration daylighting systems. In *International Daylighting Conference '98*. Natural Resources Canada/CETC, Ottawa, Ontario, Canada.

McQuiston, P. C. and Parker, J. D. 1994. *Heating, Ventilating, and Air Conditioning*, 4th edn. Wiley, New York.

Murdoch, J. B. 1985. *Illumination Engineering—From Edison's Lamp to the Laser*. Macmillan, New York.

NCDC (National Climactic Data Center). 1992. *Climatography of the U.S. #81*. NCDC, Asheville, NC.

Olgyay, A. and Olgyay, V. 1977. *Solar Control and Shading Devices*, Princeton University Press, Princeton, NJ.

Pierson, O. 1962. *Acrylics for the Architectural Control of Solar Energy*. Rohm and Haas, Philadelphia, PA.

PSDH. 1980. *Passive Solar Design Handbook*. Volume I: *Passive Solar Design Concepts*, DOE/CS-0127/1, March 1980. Prepared by Total Environmental Action, Inc. (B. Anderson, C. Michal, P. Temple, and D. Lewis); Volume II: *Passive Solar Design Analysis*, DOE/CS-0127/2, January 1980. Prepared by Los Alamos Scientific Laboratory (J. D. Balcomb, D. Barley, R. McFarland, J. Perry, W. Wray and S. Noll). U.S. Department of Energy, Washington, DC.

PSDH. 1984. *Passive Solar Design Handbook*. Part One: Total Environmental Action, Inc., Part Two: Los Alamos Scientific Laboratory, Part Three: Los Alamos National Laboratory, Van Nostrand Reinhold, New York.

Robbins, C. L. 1986. *Daylighting—Design and Analysis*. Van Nostrand, New York.
Rundquist, R. A. November 1991. Daylighting controls: Orphan of HVAC design. *ASHRAE Journal*, 11, 30–34.
USDA. 1960. *Power to Produce. 1960 Yearbook of Agriculture*. US Department of Agriculture.

For Further Information

General Background Information

The most complete basic reference for passive system heating design is still the 1980 Los Alamos Lab's *Passive Solar Design Handbook*, all three parts. The *ASHRAE Handbook of Fundamentals* is a good general introduction to passive cooling techniques and calculations, with an emphasis on evaporative cooling. *Passive Solar Buildings* and *Passive Cooling*, both published by MIT Press, contain a large variety of techniques and details concerning passive system designs and economics. All the major building energy simulation codes (DOE-2, EnergyPlus, TRNSYS, TSB13, etc) now include passive heating and cooling technologies.

The Illumination Engineering Society's *Lighting Handbook* presents the basis for and details of daylighting and artificial lighting design techniques. However, most texts on illumination present simplified format daylighting procedures. Currently used daylighting computer programs include various versions of Lumen Micro, Lightscape, and Radiance.

Solar Today magazine, published by the American Solar Energy Society, is a readily available source for current practice designs and economics, as well as a source for passive system equipment suppliers.

Technical Publication Information

Many of the current and archival passive solar technical papers are found from the conference proceedings and journals associated with the International Solar Energy Society (ISES), its affiliated American Solar Energy Society (ASES), the American Society of Mechanical Engineering (ASME), and ASHRAE. Also, many of the early passive solar practitioners were architects and builders who published in their specific industry trade journals, as is still the case today. As an Internet search for passive solar articles today will reveal, there are several *new* energy/solar/sustainable technical journals that publish passive solar-related articles.

42

Concentrating Solar Thermal Power

Manuel Romero, Jose Gonzalez-Aguilar, and Eduardo Zarza

CONTENTS

42.1 Introduction and Context

Solar energy has a high exergetic value since it originates from processes occurring at the sun's surface at a blackbody equivalent temperature of approximately 5777 K. Because of this, more than 93% of the energy may be theoretically converted to mechanical work by thermodynamic cycles (Winter et al., 1991) or to Gibbs free energy of chemicals by solarized chemical reactions (Kodama, 2003), including promising hydrogen production processes (Romero and Steinfeld, 2012). According to thermodynamics and Planck's equation, the conversion of solar heat to mechanical work or Gibbs free energy is limited by the Carnot efficiency, and therefore to achieve maximum conversion rates, the energy should be transferred to a thermal fluid, or reactants, at temperatures close to that of the sun.

Even though solar radiation is a source of high temperature and exergy at origin, with a high radiosity of 63 MW/m^2, sun–earth geometrical constraints lead to a dramatic dilution of flux and to irradiance available for terrestrial use only slightly higher than 1 kW/m^2 and, consequently, supply of low temperatures to the thermal fluid. It is therefore an essential requisite for solar thermal power (STP) plants and high-temperature solar chemistry applications to make use of optical concentration devices that enable the thermal conversion to be carried out at high solar flux and with relatively little heat loss. A simplified model of an STP plant, also known as concentrating solar power (CSP) plant, is depicted in Figure 42.1.

The optimum STP system design combines a relatively large, efficient optical surface (e.g., a field of high-reflectivity mirrors), harvesting the incoming solar radiation and concentrating it onto a solar receiver with a small aperture area. The solar receiver is a high-absorptance and high-transmittance, low-reflectance, radiative/convective heat exchanger that emulates as closely as possible the performance of a radiative blackbody. An ideal solar receiver would thus have negligible convection and conduction losses. In the case of an STP plant, the solar energy is transferred to a thermal fluid at an outlet temperature high enough to feed a heat engine or a turbine that produces electricity. The solar thermal element can be a parabolic-trough field, a linear Fresnel (LF) reflector field, a central receiver system (CRS), or a field of parabolic dishes, normally designed for a normal incident radiation of 800–900 W/m^2. Annual normal incident radiation varies from 1600 to more than 3000 kWh/m^2, allowing from 2000 to more than 3500 annual full-load operating hours with the solar element, depending on the available radiation at the particular site.

Solar transients and fluctuation in irradiance can be mitigated by using an oversized mirror field (solar multiple higher than 1) and then making use of the excess energy to load a thermal or chemical storage system. Hybrid plants with fossil backup burners connected in series or in parallel are also possible. The use of heat storage systems and fossil backup makes STP systems highly flexible for integration with conventional power plant design and

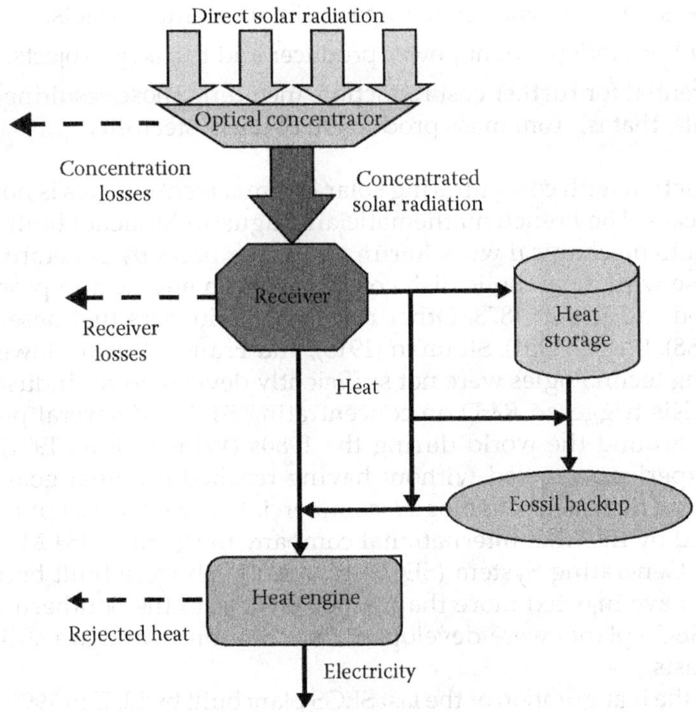

FIGURE 42.1
Flow diagram for a typical solar thermal power plant.

operation and for blending the thermal output with fossil fuel, biomass, and geothermal resources (Mancini et al., 1997). The use of large solar multiples with low-cost heat storage systems of up to 12 h (equivalent at nominal power) facilitates the design of secure-capacity plants supplying between 2000 and 6000 h of operation (equivalent at full load). In addition, hybridization is possible in conventional power plants by using a solar field as a booster or fuel saver in natural gas combined cycles and coal-fired Rankine plants, and may accelerate near-term deployment of projects due to improved economics and reduced overall project risk (Kolb, 1998). Hybridization of a natural gas combined-cycle plant with a solar field acting as a booster is called *integrated solar combined-cycle system* (*ISCCS*) plant. Three ISCCS plants were implemented in Kuraymat (Egypt), Hassi R'mel (Algeria), and Ain Beni Matar (Morocco) from 2009 to 2012. It is therefore evident that STP can currently supply dispatchable power and meet peaking and intermediate loads at an affordable electricity cost.

Additional advantages of STP are (Morse, 2000) the following:

- Proven capabilities, for example, 354 MW of trough plants in operation in California since 1985 have selectively demonstrated excellent performance, availability, a reduction in investment cost of almost 50%, and significant reductions in O&M cost.
- Modular and thus suitable for large central facilities in the hundreds of MW down to distributed generation in the tens of kW.
- Can be rapidly deployed using entirely domestic resources and existing infrastructure.

- Scale can be significant enough to impact climate-change targets.
- Suitable for both independent power producer and turnkey projects.
- Proven potential for further cost reduction, including those resulting from economies of scale, that is, from mass production of glass, steel, etc.

Electricity production with concentrating solar thermal technologies is not an innovation of the last few years. The French mathematician Augustin Mouchot built a machine able to convert CSP into mechanical work to run a printing press by generating steam (Pifre, 1882). A showcase with a parabolic dish connected to an engine and producing electricity was presented in Paris in 1878. Other remarkable pioneers that deserve recognition are Ericsson (1888), Eneas (1901), Shuman (1913), and Francia (1968). However, solar thermal concentrating technologies were not sufficiently developed for industrial use till the 1970s. The oil crisis triggered R&D on concentrating STP, and several pilot plants were built and tested around the world during the 1980s (Winter et al., 1991). Nevertheless, most of these experiences ended without having reached the final goal of making the concentrating solar thermal technologies commercial. The only exception is the experience accumulated by the LUZ International company in the nine 354 MW total capacity Solar Electricity Generating System (SEGS) plants, which were built between 1984 and 1991, and which have injected more than 16,000 GWh into the Southern California grid since then. All SEGS plants were developed, financed, built, and are still operated on a purely private basis.

However, after the inauguration of the last SEGS plant built by LUZ in 1990 (the plant SEGS IX), no new commercial STP plant was built until 2007, when the plant PS10 was built in Spain and Nevada Solar One was built in the United States. The favorable feed-in tariff implemented in Spain in 2007 led to the construction of more than 40 STP plants in the period 2007–2013, with a total installed power higher than 2.2 GW_e. Public support in the United States led to the construction of more STP plants during the second decade of this century (e.g., SOLANA and Ivanpah plants). Important initiatives are launched in different countries like Chile, South Africa, Morocco, Algeria, India, and others. The main reasons why no commercial plant was built in the period 1991–2007 were (Becker et al., 2002) as follows:

- Financial uncertainties caused by delayed renewal of favorable tax provisions for solar systems in California
- Financial problems and subsequent bankruptcy of the U.S./Israeli LUZ group in 1991, the first commercial developer of private solar power projects
- Rapid drop in fossil energy prices followed by years of worldwide stability at those low levels
- The large STP station unit capacities required to meet competitive conditions for the generation of bulk electricity, resulting in financial constraints due to their inherently large share of capital costs
- Rapidly decreasing depreciation times of capital investments in power plants due to the deregulation of the electricity market and the worldwide shift to private investor ownership of new plant projects
- Drops in cost and enhanced efficiencies of installed conventional power plants, particularly in combined-cycle power plants
- Lack of a favorable financial and political environment for the development of STP plant project initiatives in sunbelt countries

In spite of that, the Cost Reduction Study for Solar Thermal Power Plants prepared for the World Bank in early 1999 (Enermodal, 1999) already concluded that the potential STP market could reach an annual installation rate of 2000 MW of electricity. In the foregoing scenario, this rate would be reached by 2020. Assuming that advanced low-cost STP systems were likely to offer energy output at an annual capacity factor of 0.22 or more, the contribution of STP would be about 24–36 TWh of electricity by 2020 and 1600–2400 TWh by 2050. However, the same study estimated that the current STP capital cost is 2.5–3.5 times higher than the capital cost of a conventional fossil-fueled thermal power plant and showed that the price of electricity generation is between two and four times the conventional generation price. But the potential learning curve is enormous, and technology roadmaps predict that over a 60% cost reduction is possible by 2020, when production costs for solar-only plants could descend below €0.12 per kWh, by combining innovation, mass production, and scaling-up factors (Sargent & Lundy, 2003; Pitz-Paal et al., 2005; Kearney and ESTELA, 2010; IRENA, 2012).

Given the huge solar resource on Earth and the earlier-mentioned cost scenarios, STP is foreseen to impact enormously on the world's bulk power supply by the middle of the century. In Southern Europe alone, the technical potential of STP is estimated at 2000 TWh and in Northern Africa beyond any quantifiable guess (Nitsch et al., 2004). Worldwide, the exploitation of less than 1% of the total STP plant potential would be enough to meet the recommendations of the United Nations "Intergovernmental Panel on Climate Change for Long-Term Climate Stabilization" (Aringhoff et al., 2003; Philibert, 2004). In contrast to conventional fossil plants, concentrating STP plants do not produce CO_2 during operation and are therefore suitable to meet the challenge of keeping standards of living without compromising environmental issues. One MW installed of concentrating STP avoids 688 tons of CO_2 compared to a combined-cycle conventional plant and 1360 tons of CO_2 compared to a conventional coal/steam plant. A 1 m^2 mirror in the primary solar field produces 400 kWh of electricity/year, avoids 12 tons of CO_2, and contributes to a 2.5 tons savings of fossil fuels during its 25 year operation lifetime. The energy payback time of CSP systems is less than 1 year, and most solar-field materials and structures can be recycled and used again for further plants.

But in terms of electricity grid and quality of bulk power supply, it is the ability to provide dispatchability on demand that makes STP stand out from other renewable energy technologies like PV or wind. Even though the sun is an intermittent source of energy, STP systems offer the advantage of being able to run the plant continuously at a predefined load. Thermal energy storage systems store excess thermal heat collected by the solar field. A typical storage concept consists of two storage tanks filled with a liquid storage medium at different temperatures (Falcone, 1986). When storage is charged, the medium is pumped from the *cold* to the *hot* tank being heated up (directly or indirectly) by the solar heat collected. When storage is discharged, the medium is pumped from the *hot* to the *cold* tank extracting the heat in a steam generator that drives the power cycle. Storage systems, alone or in combination with some fossil fuel backup, keep the plant running under full-load conditions. When solar thermal energy is stored, it is already clear hours in advance when the plant will stop supplying energy. Figure 42.2 shows how stable operation can be extended for several hours after sunset. With an appropriate weather forecast, a 24–48 h prediction of solar capacity appears to be feasible. It should be kept in mind that thermal energy storage systems are designed only to shift the energy a few hours (e.g., from daytime to evening) or days. It cannot compensate the seasonal difference in the solar input, which comes from the changing

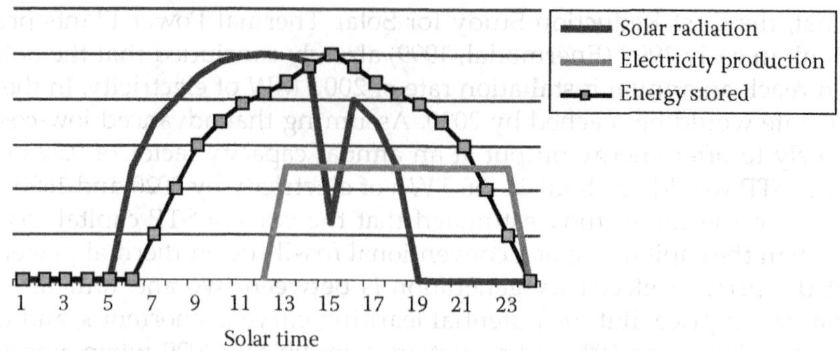

FIGURE 42.2
Extended operation with an only-solar STP plant by using some hours of thermal energy storage.

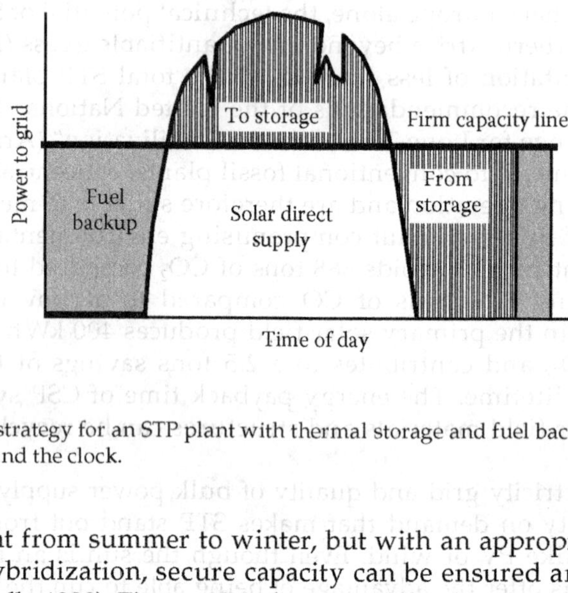

FIGURE 42.3
Example of operational strategy for an STP plant with thermal storage and fuel backup to maintain a constant firm capacity supply round the clock.

duration of sunlight from summer to winter, but with an appropriately small percentage of fossil fuel hybridization, secure capacity can be ensured and investment can be reduced by 30% (Kolb, 1998). Figure 42.3 shows an STP plant providing secure capacity. The thermal storage system supplies most of the energy required by the turbine after sunset. Overnight and early in the morning, operation is ensured by fossil fuel backup. The example shown is a *fuel saver* scheme in which solar energy is used to save fossil fuel during the daytime. There are also other options, like the *power booster* scheme, in which the fossil burner is kept constant all the time, and solar energy is fed into the turbine for peaks during solar hours, in which case the power block can absorb power increments to some extent.

This specific capability of storing high-temperature thermal energy leads to early economically competitive design options, since only the solar portion has to be oversized. This means that there is an incremental cost for the storage system and additional solar field, while the size of the conventional part of the plant (power block) remains the same. Furthermore, storage system efficiencies are high, over 95%. Specific investment costs of less than \$10–\$30 per kWh_{th}, resulting in \$25–\$75 per

kWh_e, are possible today. This STP plant feature is tremendously relevant, since penetration of solar energy into the bulk electricity market is possible only when substitution of intermediate-load power plants of about 4000–5000 h/year is achieved (Pitz-Paal et al., 2005b).

New opportunities are opening up for STP as a result of the global search for clean energy solutions, and new plants are being constructed after more than two decades of interruption (IEA, 2010). Feed-in tariffs, green portfolios, and other environmentally related incentives have been pushing since 2006 for new projects to become a reality in Spain, the United States, Australia, Algeria, and elsewhere (IRENA, 2012).

42.2 Solar Concentration and STP Systems

42.2.1 Why Use Concentrating Solar Energy Systems?: Dependence of Efficiency with T

Explained simply, solar concentration allows *higher-quality* energy to be collected, since higher temperatures, and thereby greater capacity for generating mechanical work, can be achieved. According to the second law of thermodynamics, the higher the operating temperature, T_{op}, is, the better is the efficiency of a heat engine (e.g., the one in an STP plant). The heat engine operating temperature T is directly dependent on the solar receiver, or absorber, outlet temperature.

Moreover, with solar concentration, the receiver–absorber aperture area can be reduced, minimizing infrared losses, which are directly proportional to the emission surface for a given operating temperature. Finally, concentration of solar radiation leads to greater technological development of the absorber and, consequently, to a greater cost reduction potential.

Maximum attainable temperatures in typical fossil fuel burners are on the order of 2100°C. For nuclear fission, the expected maximum useful temperature in Generation-IV gas-cooled reactors is below 1000°C. If higher temperatures are required, a classical solution is to make use of electricity as the heating source, for example, in electric furnaces. The use of electricity for this purpose represents about a 50%–75% conversion loss. In contrast, solar radiation has the potential to reach temperatures close to the apparent temperature of the sun (5777 K).

Solar concentrating systems are characterized by the use of devices, like mirrors or lenses, able to redirect the incident solar radiation received onto a particular surface, collector surface A_c, and concentrate it onto a smaller surface, absorber surface A_{abs}, or absorber. The quotient of these two areas is called the geometric concentration ratio, $\text{Conc} = A_{abs}/A_c$.

Let us assume a simplified model of an STP plant, like the one represented in Figure 42.1, made up of an ideal optical concentrator, a solar receiver performing as a blackbody and therefore having only emission losses (cavity receivers and volumetric receivers theoretically approach this condition), and a turbine or heat engine with Carnot ideal efficiency. System efficiency will depend on the balance of radiative and convective losses in the solar receiver, as shown in Equation 42.1. When the concentrated solar flux impinges on the absorber, its temperature augments and, simultaneously, radiation losses from the absorber surface to the ambient increase. With a thermal fluid cooling the absorber, when equilibrium, or steady state, is reached, the solar radiation gain equals the sum of infrared emission losses plus the useful energy rejected.

$$\frac{Q_{gain}^{*}}{A} = \alpha * \mathrm{Conc} * \phi - \sigma * \varepsilon * \left(T_{abs}^{4} - T_{amb}^{4}\right) \tag{42.1}$$

where

Q_{gain} is the power gain or useful power outlet from solar receiver (W)
A is the absorber aperture area (m^2)
α is the hemispherical absorptivity of absorber
Conc is the geometrical concentration ratio
σ is the Stefan–Boltzmann constant (=5.67E–08 W/m^2 K^4)
ε is the hemispherical emissivity of absorber
ϕ is the direct normal irradiance (W/m^2)
T_{abs} is the temperature (homogeneous) of the absorber (K)
T_{amb} is the effective temperature of ambient or atmosphere *viewed* by the absorber (K)

Solar receiver efficiency, defined as the quotient of power gain flux and concentrated solar radiation flux incident on the receiver (absorber), can be formulated as

$$\eta_{rec} = \frac{\left(Q_{gain}^{*}/A\right)}{\mathrm{Conc} * \phi} \tag{42.2}$$

By substituting Equation 42.1 in 42.2, the dependence of thermal efficiency versus parameters and variables of the receiver is observed.

$$\eta_{rec} = \alpha - \sigma * \varepsilon * \frac{\left(T_{abs}^{4} - T_{amb}^{4}\right)}{\mathrm{Conc} * \phi} \tag{42.3}$$

Equation 42.3 is plotted for different concentration ratios in Figure 42.4. This graphic representation, valid for flat absorbers, leads to the following conclusions:

- The maximum theoretical optical efficiency (when $T_{abs} \geq T_{amb}$) is the effective absorptivity of the receiver, α.
- The higher the incident solar flux (Conc$*\phi$), the better the optical efficiency.
- The higher the absorber temperature is, the higher the radiative loss is and, therefore, the lower the optical efficiency.
- The higher the effective emissivity, ε, the lower the optical efficiency.

Figure 42.4 shows the evolution of optical efficiency versus temperature and concentration ratio. It also includes the Carnot cycle efficiency, defined as

$$\eta_{Carnot} = \frac{T_{abs} - T_{amb}}{T_{abs}} \tag{42.4}$$

The Carnot cycle efficiency is the ideal efficiency (for reversible processes) that, as observed, increases with temperature and sets the thermodynamic limit of the conversion efficiency of the outlet heat delivered to mechanical work by the receiver.

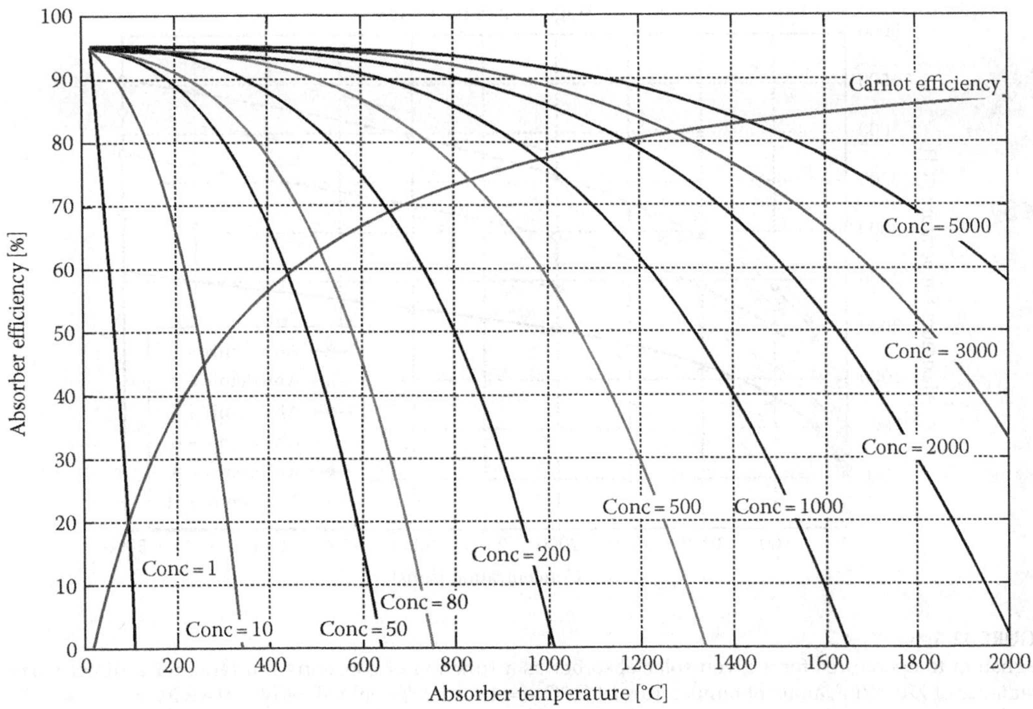

FIGURE 42.4
Efficiency of the solar receiver versus T_{abs} and versus solar concentration ratio, assuming $T_{amb} = 20°C$, $\phi = 770\ W/m^2$, and $\alpha = \varepsilon = 0.95$.

As ideal absorber temperature increases, thermal radiation losses increase as well. When losses and gains are equal, the net useful heat is zero, and the receiver should have achieved the maximum temperature or stagnation temperature. The stagnation temperature is described in the following equation and Figure 42.5:

$$T_{abs\_max} = \left[\frac{\alpha * \text{Conc} * \phi}{\varepsilon * \sigma} + T_{amb}^4 \right]^{1/4} \tag{42.5}$$

It should be noted that

- A nonselective absorber ($\alpha = \varepsilon = 1$) reaches 95°C in stagnation conditions, without concentration and for the given solar irradiance.
- A selective coating can enable much higher stagnation temperatures to be reached. For instance, at Conc = 1000, the maximum temperature is higher than 1600°C, for $\alpha/\varepsilon = 1$, and about 3200°C, for $\alpha/\varepsilon = 10$ ($\alpha = 1$; $\varepsilon = 0.1$).

From these correlations, it may clearly be concluded that, in terms of solar receiver efficiency, high solar concentrations and low temperatures are the best compromise. For a given concentration ratio, there is an absorber threshold temperature at which radiation losses increase dramatically. However, when analyzing a theoretical STP system, the

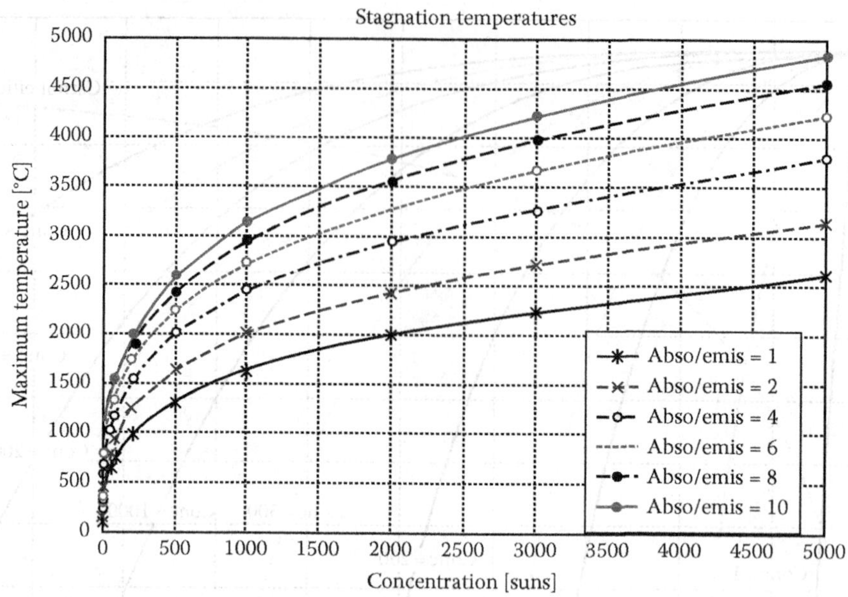

FIGURE 42.5
Stagnation temperatures for a given solar absorber as a function of concentration ratio, for a direct normal irradiance of 770 W/m², ambient temperature of 20°C, $\alpha = 1$, and different values of emissivity, ε.

convolution of the solar receiver and the heat engine should also be taken into consideration. What is the optimum temperature for a complete system including receiver and Carnot cycle? The combined efficiency of both systems can easily be visualized by multiplying the optical efficiency of the absorber (Equation 42.2) and the Carnot cycle efficiency (Equation 42.4). The result would represent the ideal conversion efficiency of our system from solar radiation to work.

$$\eta_{tot\_rec} = \eta_{rec} * \eta_{Carnot} \tag{42.6}$$

Figure 42.6 depicts the combined efficiency of the receiver/heat engine system versus concentration and temperature. It may be observed that for each concentration, the efficiency increases with temperature up to a maximum (Carnot term prevails). Once this peak is achieved, a temperature increment represents a decrement in efficiency (infrared losses at receiver prevail).

As a result, it may be concluded that for any ideal receiver working at a given concentration, there is an optimum temperature, and this temperature can obtained by

$$\frac{d\eta_{tot\_rec}}{dT} = 0 \tag{42.7}$$

Substituting Equations 42.1 through 42.5 into Equation 42.6 and obtaining the derivative, we find a polynomial expression in T_{abs}, and its real (positive) roots give the optimum temperature.

$$4 * \sigma * T_{abs}^5 - 3 * \sigma * \varepsilon * T_{amb} * T_{abs}^4 - \left(\sigma * T_{amb}^5 + \alpha * \text{Conc} * \phi * T_{amb}\right) = 0 \tag{42.8}$$

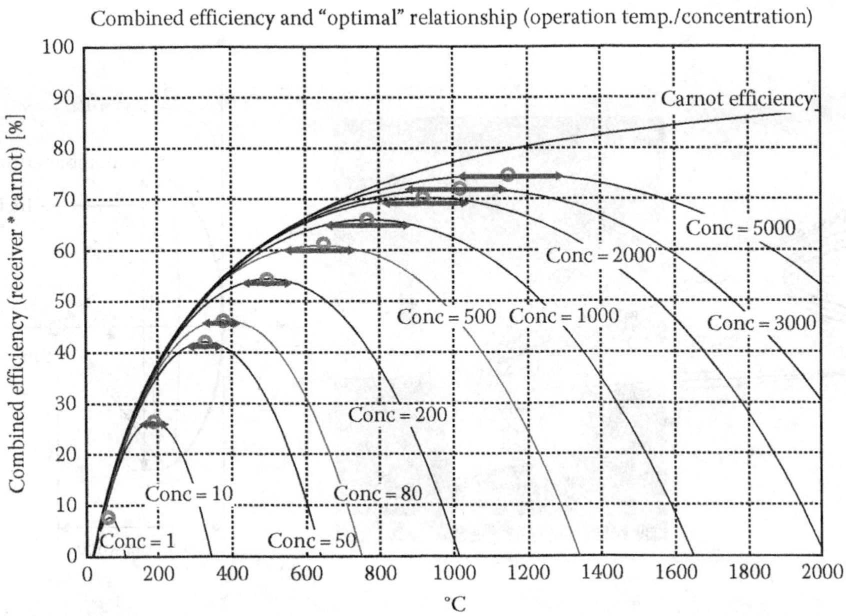

FIGURE 42.6
Combined efficiency of the solar receiver/heat engine system for different solar concentration factors and operation temperatures of the absorber. Direct normal irradiance of 770 W/m^2, ambient temperature of 20°C, $\alpha = \varepsilon = 1$.

Figure 42.6 includes the optimum temperature for the different solar concentrations as calculated from Equation 42.8.

In conclusion, solar concentration is necessary to convert solar energy into mechanical work, and for each geometrical concentration, there is a theoretical optimum absorber operating temperature.

42.2.2 Solar Concentrator Beam Quality

Optical concentration leads to two significant limitations on the practical use of solar radiation that are intrinsic to the characteristics of the radiation source. First, the nonnegligible diffuse solar rays reaching the surface of Earth that do not have a preferential direction are not reflected by the concentrating surface onto the target absorber (energy spillage); therefore, only direct solar radiation from the solar disk can be used. Second, and because of this first restriction, costly mechanical devices are required to track the sun. Consequently, there are practical physical limitations to the concentration level depending on the application (Sizmann, 1991). Sun tracking and use of beam radiation are not the only restrictions. It should also be taken into account that the sun is not a point source of energy at an infinite distance. This means that when solar rays directly from the solar disk reach our observation point at Earth, they are not completely collimated, but at a certain solid angle. The subtended solid angle is 32′; this means an angular radius of 4.653 mrad or 16′ of arc; therefore, even an ideal parabolic concentrator would reflect the image of the sun on a spot having the same target-to-mirror solid angle. This means that for an ideal heliostat located 500 m from the optical target or focal point, the theoretical diameter of the spot would be 4.7 m only because of the size

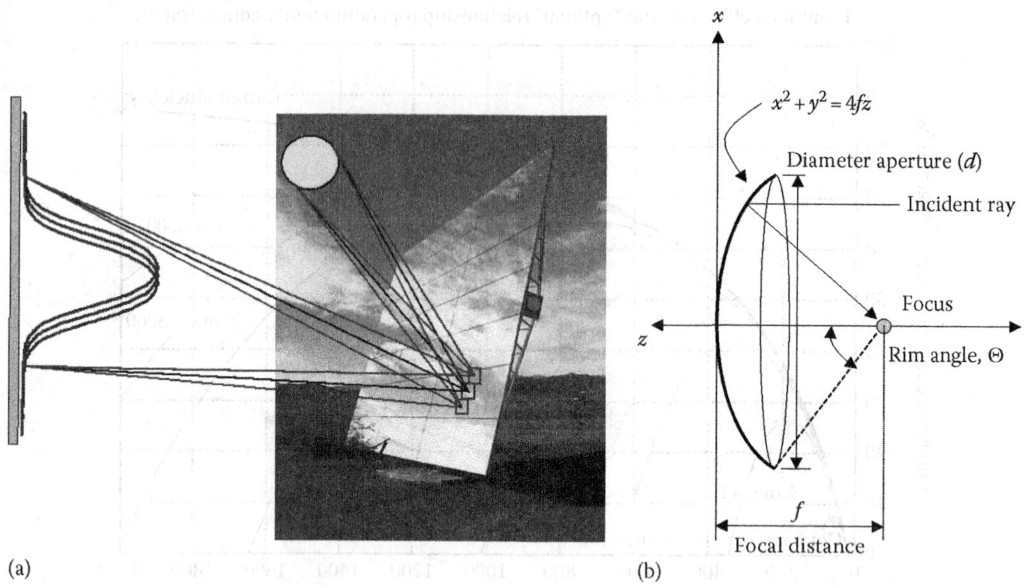

FIGURE 42.7
Configuration of an ideal parabolic concentrator (b) and effect of the size of the sun on the reflected image with a real reflectant surface, heliostat (a).

of the sun. The effect of the size of the sun on the reflected cone for a heliostat and the size of the spot on the target can be observed in Figure 42.7.

Therefore, when designing a real solar concentrator and the aperture of a solar receiver, it is necessary to take into account the minimum size of the spot at a given distance. One additional characteristic of the sun must be considered, the sunshape. Dispersion and absorption effects on the solar photosphere modify the uniform distribution of the expected radiance of an ideal blackbody. Because of that it is more realistic to substitute a *limb-darkened* distribution for the ideal uniform distribution, since the sun is darker near the rim than at the center (Vant-Hull, 1991). Assuming that the sun is an ideal Lambertian emitter, a uniform distribution of radiance (pill-box) with a constant value of $L_0 = 13.23$ MW/m^2 would be required over the entire solar disk, providing the integrated value of the solar irradiance $E = \int L\,d\Omega = \pi\theta^2 L_0$, where $\theta = R/D_{ES} = 4.653$ mrad is the ratio between the radius of the solar disk and the Earth–sun distance and Ω the solid angle. For the *limb-darkened* distribution, the following expression of radiance is obtained:

$$\frac{L}{L_0} = 0.36 + 0.84\sqrt{1 - \frac{\xi^2}{\theta_S^2}} - 0.20\left(1 - \frac{\xi^2}{\theta_S^2}\right) \quad (42.9)$$

where
L is the radiance (MW/(m^2 sr))
L_0 is the radiance at the center of the disk (13.23 MW/m^2), for $\xi = 0$
$\xi = r/D_{ES} \leq \theta$ is the radial coordinate normalized to D_{ES}
$\theta_S = R/D_{ES} = 4.653$ mrad

The extraterrestrial irradiance is modified as it enters the atmosphere because of absorption and multiple dispersions, producing the well-known aureole. That is why better sunshape fit is obtained if the previous radiance distribution is separated into two regions, the central solar disk and the circumsolar region (Rabl, 1985). The ratio between the circumsolar irradiance and direct irradiance varies depending on atmospheric conditions, but in typical sites with good solar radiation, the monthly average does not exceed 5%, provided the operating threshold for system start-up is above 300 W/m^2. A detailed *ray tracing* analysis reproducing reflection on our solar concentrating surface should take into account the sunshape. To this effect, other factors, like curvature and waviness errors of the reflecting surface, as well as the possible tracking errors in the drive mechanism, must be added. It is relatively simple to approximate all these nonsystematic errors of the concentrator to a standard deviation, $\sigma^2 = \Sigma\sigma_i^2$, to quantify the beam quality of the reflector. The consequence of the convolution of all the mentioned errors from sun, tracking system, and reflecting surface leads to the real fact that instead of an ideal point-focus parabolic concentrator, Figure 42.7b, the spot and energy profile obtained on a flat absorber can be approximated to a Gaussian shape, Figure 42.7a. The real image obtained is also known as *degraded sunshape.* Subsequently, the designer of a solar receiver should take into account the beam quality of the solar concentrator and the concentrated flux distribution to optimize the heat transfer process.

However, the main interest of a solar concentrator is the energy flux and not the quality of the image. Because of that, 4.65 mrad is a good reference for comparing the extent of optical imperfections. Those errors deflecting the reflected ray significantly less than 4.65 mrad are of minor importance, while deviations over 6 mrad contribute drastically to the reduction of concentration and energy spillage at the receiver aperture.

Solar concentrators follow the basic principles of Snell's law of reflection (Rabl, 1985), as depicted in Figure 42.8. On a specular surface like the mirrors used in STP plants, the reflected solar ray forms an angle with the mirror normal equal to the angle formed by the incoming ray with the normal. On a real mirror with intrinsic and constructional errors, the reflected ray distribution can be described with *cone optics.* The reflected ray direction has an associated error that can be described with a normal distribution function. The errors of a typical reflecting solar concentrator may be either microscopic (specularity) or macroscopic (waviness of the mirror and error of curvature). All the errors together end up modifying the direction of the normal compared to the reference reflecting element. However, it is necessary to discriminate between microscopic and macroscopic errors. Microscopic errors are intrinsic to the material itself and depend on the fabrication process, and can be measured at the lab with mirror samples.

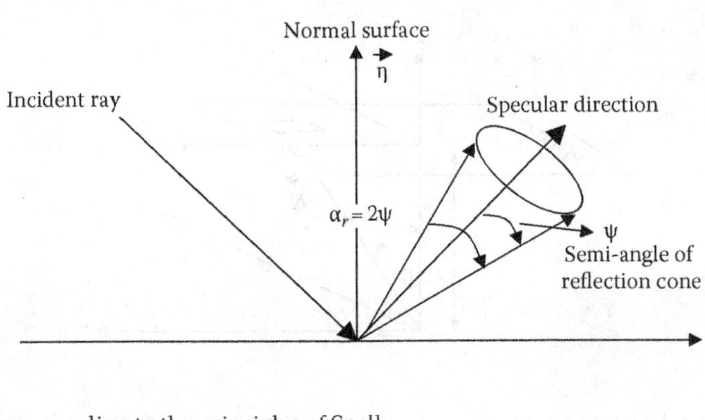

FIGURE 42.8
Geometry of reflection according to the principles of Snell.

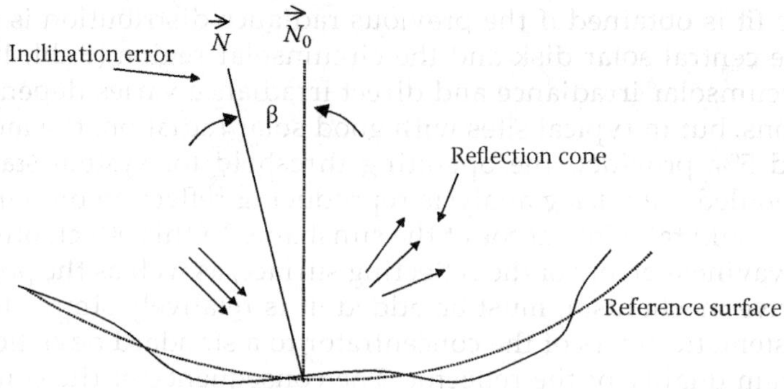

FIGURE 42.9
Normal error produced by grainy texture of the material and deficient curvature of the concentrator.

Macroscopic errors are characteristic of the concentrator itself and the erection process; therefore, they should be measured and quantified with the final system in operation (Biggs and Vittitoe, 1979).

The parameter best defining the *macroscopic* quality of a reflective concentrator is the slope error (β) as shown in Figure 42.9. The slope error is the angle between the normal to the reference surface ($\vec{N}_0$) and the normal to the real reflecting surface ($\vec{N}$). The root mean square or RMS, a statistical mean distribution of slope errors, is used to specify the distribution of β on a real surface. For a given differential element of surface (dA), the RMS is obtained as

$$\text{RMS} = \langle \beta^2 \rangle^{1/2} = \left[\frac{\int \beta^2 dA}{\int dA} \right]^{1/2} \tag{42.10}$$

RMS is a deterministic value of the surface errors, but it can be expressed with a probabilistic value as the standard deviation. Since it is more practical to determine the probabilistic error in the reflected image, it is a good idea to translate the RMS of the normals on the reflector to the standard deviation of the reflected rays. For convenience, the value of σ is expressed on a line that intersects and is orthogonal to the normal of the reflector. Assuming a new reference plane r–s placed at a unitary distance, the probability of $\vec{N}$ intersecting the element of surface $dr{\cdot}ds$ is $F(r, s){\cdot}dr{\cdot}ds$ where $F(r, s)$ is a probability density function normalized to 1 when integrated over the entire plane r–s (Figure 42.10).

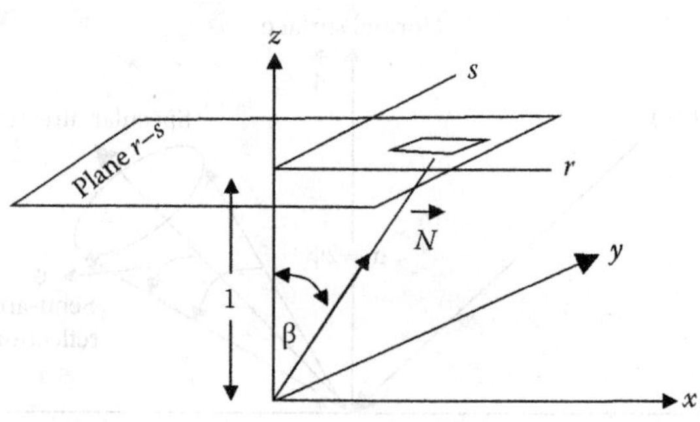

FIGURE 42.10
Translation of the normal error to a new reference plane at a distance 1.

In this case, the probability function can be approximated to a normal distribution function, since further convolution with other errors like specularity, solar tracking, or sunshape leads to a damping effect according to the central limit theorem. In addition, since the total error is the convolution of a series of random surface errors, the distribution is circular normal.

If we use the new coordinates defined for the plane as depicted in Figure 42.10,

$$\rho = \tan\beta = (r^2 + s^2)^{1/2} \tag{42.11}$$

and for those values of ρ close to β, which is the case for actual solar concentrators used in STP plants, the function *F* may be expressed as

$$F(\rho) = \frac{1}{2\pi\sigma^2}\exp\left[-\frac{\rho^2}{2\sigma^2}\right] \tag{42.12}$$

where the parameter σ is the standard deviation of the reflected ray, and for circular symmetry, $\sigma_r = \sigma_s = \sigma$.

By integrating the previous expression to obtain the RMS of ρ, σ is correlated with β and with the RMS, by the following equation (Biggs and Vittitoe, 1979; Vant-Hull, 1991):

$$\text{RMS} = \langle\rho^2\rangle^{1/2} = \sigma\sqrt{2} = \langle\beta^2\rangle^{1/2} \tag{42.13}$$

Summarizing, the beam quality of the concentrating reflector may be expressed by means of three parameters related to the inclination error of the surface elements, the RMS of β, the dispersion σ of the normal, or the dispersion σ of the reflected beam. The total standard deviation of a solar concentrator or beam quality would be the sum of several sources of error:

$$\sigma_C^2 = \sigma_{sp+wav}^2 + \sigma_{curvature}^2 + \sigma_{tracking}^2 \tag{42.14}$$

The total error of the image, also known as degraded sun, would be the convolution of the beam quality of the concentrator with the sunshape:

$$\sigma_D^2 = \sigma_{sunshape}^2 + \sigma_C^2 \tag{42.15}$$

where

- $\sigma_{sunshape}$ is the beam standard deviation due to the sunshape effect (approximately 2.19 mrad)
- σ_{sp+wav} is the beam standard deviation due to specularity and waviness (measured with reflected rays from material samples using a reflectometer)
- $\sigma_{curvature}$ is the beam standard deviation due to curving (should be measured on the concentrator itself)
- $\sigma_{tracking}$ is the beam standard deviation due to aiming point and other drive-mechanism-related sources of error

The association of the flux profile on the target with a Gaussian shape and the determination of the beam quality of the concentrator are useful in identifying the optimum aperture area of the receiver for a specific fraction of intercepted power. For a given receiver aperture radius length (ρ_A), the probability of $\rho < \rho_A$ may be obtained by integrating (42.12)

$$P(\rho < \rho_A) = 2\pi \int_0^{\rho_A} \rho F(\rho)d\rho = 1 - \exp\left[-\frac{\rho_A^2}{2\sigma^2}\right] \quad (42.16)$$

With this simple correlation, the beam standard deviation (σ) and cone radius (ρ_A) can be correlated by intercepting a certain percentage of reflected power or the probability of $\rho < \rho_A$ (P).

42.2.3 Solar Concentration Ratio: Principles and Limitations of STP Systems

The most practical and simplest primary geometrical concentrator typically used in STP systems is the parabola. Even though there are other concentrating devices like lenses or compound parabolic concentrators (Welford and Winston, 1989), the reflective parabolic concentrators and their analogues are the systems with the greatest potential for scaling up at a reasonable cost. Parabolas are imaging concentrators able to focus all incident paraxial rays onto a focal point located on the optical axis (see Figure 42.7). The paraboloid is a surface generated by rotating a parabola around its axis. The parabolic dish is a truncated portion of a paraboloid. For optimum sizing of the parabolic dish and absorber geometries, the geometrical ratio between the focal distance, f, the aperture diameter of the concentrator, d, and the rim angle, Θ, must be taken into account. The ratio can be deducted from the equation describing the geometry of a truncated paraboloid, $x^2 + y^2 = 4fz$, where x and y are the coordinates on the aperture plane, and z is the distance from the plane to the vertex. For small rim angles, the paraboloid tends to be a sphere, and in many cases, spherical facets are used; therefore, in most solar concentrators, the following correlation is valid:

$$\frac{f}{d} = \frac{1}{4\tan(\Theta/2)} \quad (42.17)$$

For example, a paraboloid with a rim angle of 45° has an f/d of 0.6 (see Figure 42.11). The ratio f/d increases as the rim angle decreases. A parabolic concentrator with a very small rim angle has very little curvature and the focal point far from the reflecting surface. Because of that, STP systems making use of cavity receivers with small apertures should use small rim angles. Conversely, those STP systems using external or tubular receivers will make use of large rim angles and short focal lengths.

The maximum solar concentration ratio for an ideal perfectly specular 3D paraboloid of rim angle Θ aligned to the sun is (Goswami et al., 2000)

$$C_{max} = \frac{\sin^2\Theta}{(4\sin^2\theta_S)} \quad (42.18)$$

where θ_S is the semi-angle subtended by the sun, 4.653 mrad (16′). For $\Theta = 90°$, $C_{max} \sim 11{,}547$.

The thermodynamic limit or maximum concentration ratio for an ideal solar concentrator would be set by the size of the sun and not by the beam quality. By applying the

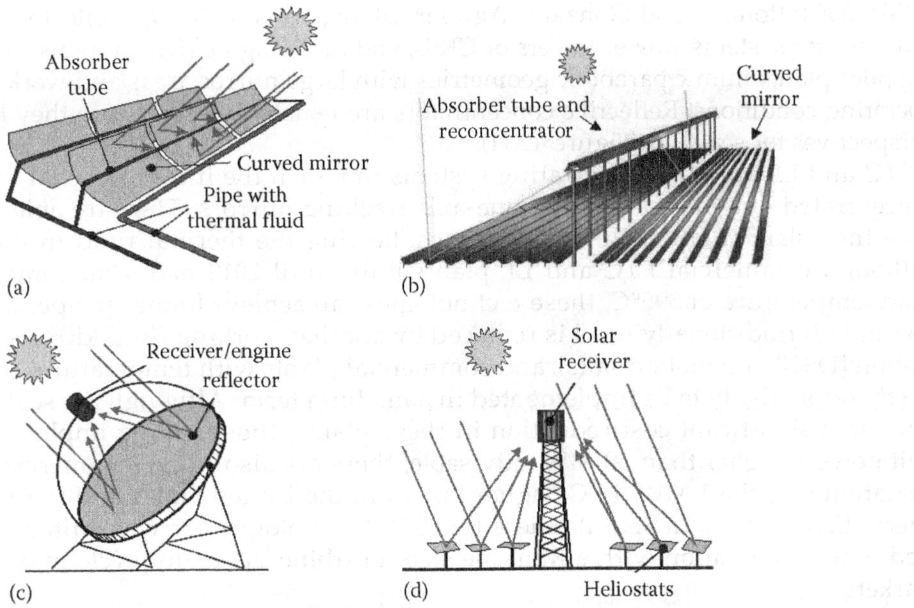

FIGURE 42.11
Schematic diagrams of the four CSP systems scaled up to pilot and demonstration sizes: (a) parabolic trough, (b) linear Fresnel, (c) dish/engine, and (d) central receiver.

geometrical conservation of energy in a solar concentrator, the following expressions are obtained for 3D and 2D systems (for a refraction index, $n = 1$):

$$C_{\max,3D} = \frac{1}{\sin^2 \theta_S} \leq 46{,}200 \tag{42.19}$$

$$C_{\max,2D} = \frac{1}{\sin \theta_S} \leq 215 \tag{42.20}$$

For real concentrators, the maximum ratios of concentration are much lower, because of microscopic and macroscopic, tracking and mechanical, sunshape, and other errors. Engineers designing a specific STP plant should give special attention to the expected real beam quality and rim angle of the reflecting system to obtain an appropriate sizing of the solar receiver.

42.3 Solar Thermal Power Plant Technologies

STP plants with optical concentration technologies are important candidates for providing the bulk solar electricity needed within the next few decades, even though they still suffer from a lack of public awareness and confidence, especially among scientists and decision makers. Four CSP technologies are today represented at pilot and demonstration scale

(Mills, 2004; Romero and Gonzalez-Aguilar, 2014): parabolic-trough collectors (PTCs), LF concentrator systems, power towers or CRSs, and dish/engine (DE) systems. All the existing pilot plants mimic parabolic geometries with large mirror areas and work under real operating conditions. Reflective concentrators are usually selected since they have better perspectives for scale-up (Figure 42.11).

PTC and LF are 2D concentrating systems in which the incoming solar radiation is concentrated onto a focal line by one-axis tracking mirrors. They are able to concentrate the solar radiation flux 30–80 times, heating the thermal fluid in the receiver. Although commercial PTC and LF plants built until 2013 had a maximum thermal fluid temperature of 395°C, these technologies can achieve higher temperatures if the thermal oil traditionally used is replaced by another working fluid (direct steam generation [DSG] and molten salts), and commercial plants with temperatures up to 500°C are therefore likely to be implemented in a medium term. Although the scale-up effect leads to a significant cost reduction in these plants, thus making implementation of unit powers higher than 50 MW_e advisable, there are also small commercial plants in operation (i.e., the 1 MW_e PTC Saguaro plant in the United States and the 1.4 MW_e LF Puerto Errado plant in Spain). These two STP technologies are well suited for centralized power generation with a Rankine steam turbine/generator cycle in dispatchable markets.

CRS optics is more complex, since the solar receiver is mounted on top of a tower and sunlight is concentrated by means of a large paraboloid that is discretized into a field of heliostats. This 3D concentrator is therefore off-axis, and heliostats require two-axis tracking. Concentration factors are between 200 and 1000, and unit sizes are between 10 and 200 MW, and they are therefore well suited for dispatchable markets and integration into advanced thermodynamic cycles. A wide variety of thermal fluids, like saturated steam, superheated steam, molten salts, atmospheric air, or pressurized air, can be used, and temperatures vary between 300°C and above 1000°C.

Finally, DE systems are small modular units with autonomous generation of electricity by Stirling engines or Brayton mini-turbines located at the focal point. Dishes are parabolic 3D concentrators with high concentration ratios (1000–4000) and unit sizes of 5–25 kW. Their current market niche is in both distributed on-grid and remote/off-grid power applications (Becker et al., 2002).

Typical solar-to-electric conversion efficiencies and annual capacity factors are listed in Table 42.1 (Romero and Steinfeld, 2012). These values have been demonstrated commercially. With current investment costs, all current STP technologies generally thought to require a public financial support strategy for market deployment. Although an independent study promoted by the World Bank at the end of last century (Enermodal, 1999) stated that STP was the most economical technology for solar production of bulk electricity at that time, the significant cost reduction experienced by the photovoltaic panels during the first decade of current century has changed the situation, and STP plants are cheaper only in those places with a high level of direct normal irradiance. However, the added value of dispatchability is still a valuable asset for STP plants when compared with PV plants. Cost data delivered by commercial STP plants show a direct capital costs of 2.5–3.5 times those of a fossil-fueled power plant, and therefore, generation costs of the electricity produced are 2–4 times higher (IRENA, 2012).

Every square meter of STP field can produce up to 1200 kWh thermal energy/year or up to 500 kWh of electricity/year. That means, a cumulative savings of up to 12 tons of carbon dioxide and 2.5 tons of fossil fuel per square meter of STP system over its 25-year lifetime (Geyer, 2002). Although most of the commercial plants installed up

TABLE 42.1

Characteristics of Concentrating Solar Power Systems

	Parabolic Troughs	Central Receiver	Dish/Engine
Power unit	30–140 MW[a]	10–100 MW[a]	5–25 kW
Temperature operation	390°C	565°C	750°C
Annual capacity factor	23%–50%[a]	20%–77%[a]	25%
Peak efficiency	20%	23%	29.4%
Net annual efficiency	11%–16%[a]	7%–20%[a]	12%–25%
Commercial status	Mature	Early projects	Prototypes-demos
Technology risk	Low	Medium	High
Thermal storage	Limited	Yes	Batteries
Hybrid schemes	Yes	Yes	Yes

[a] Data interval for the period 2010–2025.

to 2012 used parabolic troughs (11 plants in the United States totaling 415 MW_e, 38 plants in Spain totaling 1975 MW_e, and 3 plants in Morocco, Algeria, and Egypt respectively), this tendency could change because the number of projects promoted with CRS and LF plants is increasing nowadays The high initial investment required by early commercial plants ($3000–$6000 per kW) and the restricted modularity generally motivated by their expensive thermodynamic cycle, combined with the lack of appropriate power purchase agreements and fair taxation policies, led to a vicious circle in which the first generation of commercial grid-connected plants became difficult to implement without market incentives. After two decades of frozen or failed projects, approval in the first decade of this century of specific financial incentives in Europe, the United States, Australia, Algeria, and South Africa paved the way for the launching of the first commercial ventures, and more than 40 STP plants were implemented from 2007 to 2012.

The *parabolic trough* is today considered a fully mature technology (Price et al., 2002). Costs are in the range of $0.16–$0.21 per kWh, depending on the boundary conditions for each project (direct normal irradiance available, loan interest rate, etc.), and even up to 30% less in hybrid systems, and technological and financial risks are low. The five plants at the Kramer Junction site (SEGS III–VII) achieved a 30% reduction in operating and maintenance costs, a record annual plant efficiency of 14%, and a daily solar-to-electric efficiency near 20%, as well as peak efficiencies up to 21.5%. Annual plant availability exceeded 98%, and collector field availability was over 99% (Cohen et al., 1999). In view of this advanced state of development, investors were keener to support projects with parabolic troughs, and many companies promoted projects with this technology as soon as public subsidies became available. This was the main reason why most of the commercial STP plants built during the period 2007–2012 used parabolic-trough technology, with either a similar configuration of the SEGS plants or with various hybridization options, including ISCCSs.

LF reflector systems are conceptually simple, using inexpensive, compact optics that can produce saturated steam at 150°C–360°C (higher temperatures of about 500°C are pursued for the next generation of LF plants) with land use of less than 10 m^2/kW. Therefore the STP technology is best suited for integration with combined-cycle recovery boilers, to replace the steam bled in regenerative Rankine power cycles or for saturated steam turbines. The first commercial experience with LF was the prototype-scale plant developed at the University of Sydney in Australia (Mills and Morrison, 2000). The biggest LF

STP plant in operation in 2012 was Puerto Errado-2 (Spain), with a unit power of 30 MW_e. Taking into account the scale-up effect, LF plants with the unit power of about 100 MW_e were soon promoted after the first small commercial experiences. LF technology is being used in Australian coal-fired power plants as a fuel-saver option to supply 270°C preheat thermal energy to either replace the steam bled in the regenerative Rankine power cycle or for water preheating in the boiler.

Power tower technology, after a proof-of-concept stage, has already started its commercial deployment, although less mature than the parabolic-trough technology. To date, a few CRS commercial plants have been implemented and are in daily operation. At an earlier stage, more than 10 different experimental plants were tested worldwide, generally small demonstration systems of between 0.5 and 10 MW, and most of them operated in the 1980s (Romero et al., 2002). That experience demonstrated the technical feasibility of the CRS power plants and their capability of operating with large heat storage systems. The most extensive operating experience was in the European pilot projects located in Spain on the premises of the Plataforma Solar de Almería (PSA) and in the United States at the 10 MW Solar One and Solar Two facilities located in California. Commercial deployment of CRS plants started in 2007 with the start-up of the Spanish PS10 plant, promoted by the company Abengoa. PS10 plant, with a net unit power of 10 MW_e and a 40 bar/240°C saturated steam Rankine cycle (Osuna et al., 2004), was the first commercial CRS plant in the world. PS10 was followed by the PS20, with a unit power of 20 MW_e. The third CRS commercial plant in the world was the Gemasolar plant, promoted by Torresol Energy in the Spanish province of Seville. Gemasolar, with its molten-salt central receiver and 17 h thermal storage system, has been a significant step forward in the CRS technology. With a unit power of 19 MW_e, this plant can operate round the clock in summertime. The company BrightSource has promoted a challenging CRS plant with 550°C superheated steam receiver, the so-called Ivanpah plant. The 392 MW Ivanpah Plant was connected to the grid in February 2014. Other projects based upon superheated steam and molten salts are under construction in Chile, South Africa, and the United States. In addition to the CRS technologies already implemented in commercial STP plants before 2013 (i.e., saturated steam, superheated steam, and molten salts), the use of open- or closed-loop volumetric air receivers is at R&D stage (Romero et al., 2002).

DE systems are absolutely modular and ideal for unit powers between 5 and 25 kW. Two decades ago, dish–Stirling systems had already demonstrated their high conversion efficiency, concentration of more than 3000 suns, and operating temperatures of 750°C at annual efficiencies of 23% and 29% peak (Stine and Diver, 1994). Unfortunately, DE systems have not yet surpassed the proof-of-reliability operation phase. Only a limited number of prototypes have been tested worldwide, and annual availability above 90% still remains a challenge. Given the fact that autonomous operation and off-grid markets are the first priorities of this technology, more long-endurance test references must be accumulated. DE technology investment costs, which are twice as high as those of parabolic troughs, would have to be dramatically reduced by mass production of specific components, like the engine and the concentrator, although these systems, because of their modular nature, are targeted toward much higher-value markets. DE system industries and initiatives are basically confined to the United States and Europe (Mancini et al., 2003). To date, there has been only one DE commercial plant (the MARICOPA plant in the United States), built with 25 kW_e DE systems and a total unit power of 1.5 MW_e. This plant did not achieve the expected performance, and the plant was dismantled in 2012 after 1 year of operation.

42.4 Parabolic-Trough Solar Thermal Power Plants

42.4.1 Operational Principle and Components of the Parabolic-Trough Collector

PTCs are linear-focus concentrating solar devices that convert direct solar radiation into thermal energy and are suitable for working in the 150°C–400°C temperature range (Price et al., 2002).

A PTC is basically made up of a parabolic-trough-shaped mirror that reflects direct solar radiation, concentrating it onto a receiver tube located in the focal line of the parabola. Concentration of the direct solar radiation reduces the absorber surface area with respect to the collector aperture area and thus significantly reduces the overall thermal losses. The concentrated radiation heats the fluid that circulates through the receiver tube, thus transforming the solar radiation into thermal energy in the form of the sensible heat of the fluid. Since a PTC is an optical solar concentrator, it can use only direct solar radiation, and the diffuse solar radiation is lost. Figure 42.12 shows a typical PTC and its components.

PTCs are dynamic devices because they have to rotate around an axis, the so-called tracking axis, to follow the apparent daily movement of the sun. Otherwise, the solar radiation reflected by the parabolic mirrors would not reach the receiver tube. Collector rotation around its axis requires a drive unit. One drive unit is usually sufficient for several parabolic-trough modules connected in series and driven together as a single collector. The type of drive unit assembly depends on the size and dimensions of the collector. Drive units composed of an electric motor and a gearbox combination are used for small collectors (aperture area < 100 m^2), while powerful hydraulic drive units are required to rotate large collectors. A drive unit placed on the central pylon is commanded by a local control unit that tells it when and in which direction to rotate the collector to track the sun.

Local control units currently available on the market can be grouped into two categories, depending on the device used to track the sun. These two categories are as follows:

1. Control units based on sun sensors
2. Control units based on astronomical algorithms

Control units in Group (a) use photo cells to detect the sun position, while those in Group (b) calculate the sun vector using very accurate mathematical algorithms that find

FIGURE 42.12
A typical parabolic-trough collector.

the sun elevation and azimuth every second and measure the angular position of the rotation axis by means of electronic devices (angular encoders or magnetic coded tapes attached to the rotation axis).

Shadow-band and flux-line trackers are in Group (a). Shadow-band trackers are mounted on the parabolic concentrator and face the sun when the collector is in perfect tracking (i.e., the sun vector is within a plane that includes the receiver tube and is perpendicular to the concentrator aperture plane). Two photo sensors, one on each side of a separating shadow wall, detect the sun's position. When the collector is correctly pointed, the shadow wall shades both sensors equally, and their electric output signals are identical.

Flux-line trackers are mounted on the receiver tube. Two sensors are also placed on both sides of the absorber tube to detect the concentrated flux reaching the tube. The collector is correctly pointed when both sensors are equally illuminated and their electrical signals are of the same magnitude.

At present, all commercial PTC designs use a single-axis sun-tracking system. Though PTC designs with two-axis sun-tracking systems have been designed, manufactured, and tested in the past, evaluation results show that they are less cost-effective. Though the existence of a two-axis tracking system allows the PTC to permanently track the sun with an incidence angle equal to 0° (thus reducing optical losses while increasing the amount of solar radiation available at the PTC aperture plane), the length of passive piping (i.e., connecting pipes between receiver pipes of adjacent parabolic troughs on the same collector) and the associated thermal losses are significantly higher than in single-axis collectors. Furthermore, their maintenance costs are higher and their availability lower because they require a more complex mechanical design.

Thermal oils are commonly used as the working fluid in these collectors for temperatures above 200°C, because at these operating temperatures, normal water would produce high pressures inside the receiver tubes and piping. This high pressure would require stronger joints and piping, and thus raise the price of the collectors and the entire solar field. However, the main technical challenges of using water as working fluid in the receiver tubes is not related to the higher pressure, but to process instabilities that could arise in the solar field due to the two-phase flow (i.e., steam and liquid water) circulating inside the receiver tubes (Zarza et al., 1999, 2002).

After a fruitful R&D stage at the PSA, the use of demineralized water for high temperatures/pressures has been already implemented in a 5 MW commercial STP plant built in Thailand by the German company Solarlite, and the feasibility of DSG at 100 bar/400°C in the receiver tubes of PTCs has already been proven in the DISS project. For temperatures below 200°C, either a mixture of water/ethylene glycol or pressurized liquid water can be used as the working fluids because only a moderate pressure is required in the liquid phase.

There are many options when choosing the thermal oil to act as working fluid in PTCs. The main limiting factor to be taken into consideration is the maximum oil bulk temperature defined by the manufacturer. Good oil stability is guaranteed by the manufacturer if the maximum bulk temperature is not exceeded. Above this temperature, oil cracking and rapid degradation occur.

The oil most widely used in PTCs for temperatures up to 395°C is VP-1 or Dowtherm-A, which is a eutectic mixture of 73.5% diphenyl oxide/26.5% diphenyl. The main problem with this oil is its high solidification temperature (12°C), which requires an auxiliary heating system when oil lines run the risk of cooling below this temperature. Since the boiling temperature at 1013 mbar is 257°C, the oil circuit must be pressurized with nitrogen, argon, or any other inert gas when oil is heated above this temperature. Blanketing of the entire oil circuit with an oxygen-free gas is a must when working at high temperatures because high-pressure mists can form an explosive mixture with air. Though there are other suitable thermal oils for slightly higher working temperatures with lower solidification temperatures, they are unaffordable for large solar plants due to their much higher price.

The typical PTC receiver tube is composed of an inner steel pipe surrounded by a glass tube to reduce convective heat losses from the hot steel pipe. The steel pipe has a selective high-absorptivity (>90%), low-emissivity (<30% in the infrared) coating, which reduces radiative thermal losses. Receiver tubes with glass vacuum tubes and glass pipes with an antireflective coating achieve higher PTC thermal efficiency and better annual performance, especially at higher operating temperatures. Receiver tubes with no vacuum are usually for working temperatures below 250°C, because thermal losses are not so critical at these temperatures. Due to manufacturing constraints, the maximum length of single receiver pipes is less than 6 m, so that the complete receiver tube of a PTC is composed of a number of single receiver pipes welded in series up to the total length of the PTC. The total length of a PTC is usually within 25–150 m.

Figure 42.13 shows a typical PTC vacuum receiver pipe. The outer glass tube is attached to the steel pipe by means of flexible metal differential expansion joints that compensate for the different thermal expansion of glass and steel when the receiver tube is working at nominal temperature. The number of manufacturers of PTC vacuum absorber tubes was small in 2013: the German companies Schott and Siemens, the Italian company Archimede Solar, and the Chinese company HUIYIN. The glass cover is connected to the inner steel tube by means of stainless steel expansion bellows, which compensates the different thermal expansion of the glass cover and steel tubes. The glass-to-metal welding used to connect the glass tube and the expansion bellows is a weak point in the receiver tube and has to be protected from the concentrated solar radiation to avoid high thermal and mechanical stress that could damage the welding. An aluminum shield is usually placed over the bellows to protect the welding.

As seen in Figure 42.13, several chemical getters are placed in the gap between the steel receiver pipe and the glass cover to absorb gas molecules from the fluid that get through the steel pipe wall to the annulus.

PTC reflectors have a high specular reflectance (>88%) to reflect as much solar radiation as possible. Solar reflectors commonly used in PTC are made of back-silvered glass mirrors,

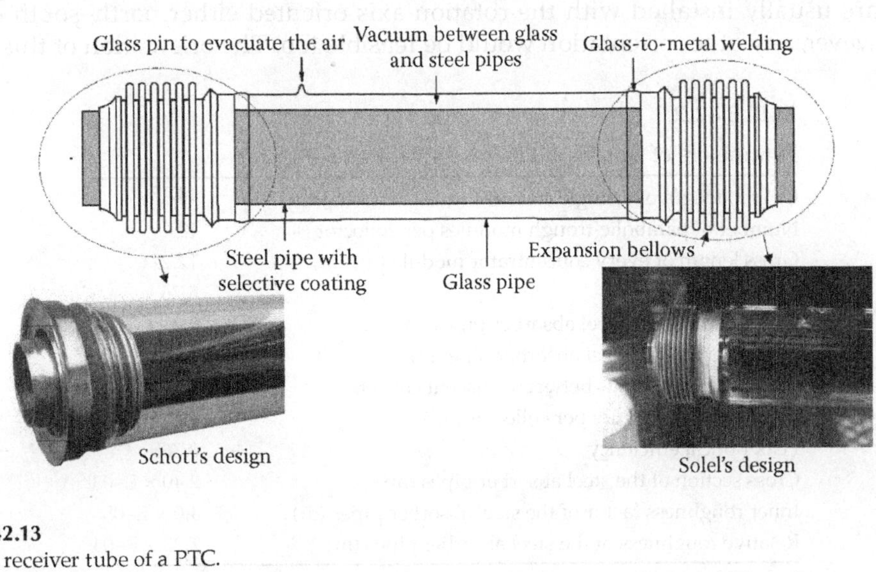

FIGURE 42.13
A typical receiver tube of a PTC.

since their durability and solar spectral reflectance are better than the polished aluminum and metallized acrylic mirrors also available on the market. Solar spectral reflectance is typically 0.93 for silvered glass mirrors and 0.87 for polished aluminum.

Low-iron glass is used for the silvered glass reflectors and the glass receiver envelopes since iron has an absorption peak in the solar spectrum, and therefore glass transmissivity to solar radiation is higher when the iron content is low.

The parabolic-trough reflector is held by a steel support structure on pylons in the foundation. At present, there are several commercial PTC designs. Large STP plant designs are much larger than those developed for industrial process heat (IPH) applications in the range of 125°C–300°C. Examples of PTC designs for IPH applications are the designs developed by the American company IST (Industrial Solar Technology, www.industrialsolartech.com) and the European company Solitem (www.solitem.de). IST and Solitem designs are very similar in size (approx. 50 m total length and 2 m wide) and have aluminum reflectors. SOPOGY (www.sopogy.com), with head offices in Honolulu (Hawaii), developed a 3.7 m long, 1.35 m wide PTC module marketed as Soponova 4.0, which was suitable for both process heat applications and electricity generation with small STP plants.

Two PTC designs conceived for large STP plants are the URSSATrough and EuroTrough, both of which have a total length of 150 m and a parabola width of 5.76 m, with back-silvered thick-glass mirrors and vacuum absorber pipes (Table 42.2). The initial EuroTrough design (Luepfert et al., 2003) was then improved, leading to its successor, the SKAL-ET, the collector installed at the ANDASOL plants in Spain in 2007–2011. The main difference between the URSSATrough and EuroTrough designs is their steel structure: EuroTrough mechanical rigidity to torsion is assured by a steel *torque box* of trusses and beams, while the URSSATrough (www.urssa.es) steel structure is based on the replacement of the torque box by a central steel tube (called the *Torque Tube*) (Figure 42.14). However, assembly of the steel mirror support frames on this central tube must also be highly accurate. The collector design SenerTrough, developed by the Spanish company SENER (www.sener.es), also has a torque tube instead of a torque box. Figure 42.14 shows the steel structures of these two PTC designs: with *torque box* and *torque tube*. The main constraint when developing the mechanical design of a PTC is the maximum torsion at the collector ends, because high torsion would lead to a smaller intercept factor and lower optical efficiency.

PTCs are usually installed with the rotation axis oriented either north–south or east–west; however, any other orientation would be feasible too. The orientation of this type of

TABLE 42.2

Parameters of the ET-150 Parabolic-Trough Collector

Overall length of a single collector (m)	147.5
Number of parabolic-trough modules per collector	12
Gross length of every concentrator module (m)	12.27
Parabola width (m)	5.76
Outer diameter of steel absorber pipe (m)	0.07
Inner diameter of steel absorber pipe (m)	0.055
Number of ball joints between adjacent collectors	4
Net collector aperture per collector (m^2)	822.5
Peak optical efficiency	0.765
Cross section of the steel absorber pipes (m^2)	2.40 × E–03
Inner roughness factor of the steel absorber pipes (m)	4.0 × E–05
Relative roughness of the steel absorber pipes (m)	7.23 × E–04

FIGURE 42.14
Steel structures with *torque box* (a) and *torque tube* (b).

solar collector is sometimes imposed by the shape and orientation of the site where they are installed. When the solar field designer can choose solar collector orientation, he must take into consideration that the orientation has a significant influence on the sun incidence angle on the aperture plane of the collectors, which, in turn, affects collector performance. The incidence angle is the angle between the normal to the aperture plane of the collector and the sun's vector, both contained on a plane perpendicular to the collector axis.

Seasonal variations in north–south-oriented trough collector output can be quite wide, depending on the site weather conditions and the geographical latitude. Three to four times more energy is delivered daily during summer months than in the winter at latitudes of about 35°N. Seasonal variations in energy delivery are much smaller for an east–west orientation, usually less than 50%. Nevertheless, a north–south sun-tracking axis orientation usually provides more energy on a yearly basis. This difference in energy output is caused by the different incidence angle of the direct solar radiation onto the aperture plane of the concentrators. Daily variation in the incidence angle is always greater for the east–west orientation, with maximum values at sunrise and sunset and a minimum of 0° every day at solar noon (Rabl, 1985).

So the orientation of the rotation axis of PTCs is a very important design specification, because it strongly affects their performance, and selection of the best orientation depends on the answers to the following questions:

1. Which season of the year should the solar field produce the most energy? If more energy is needed in summer than in winter, the most suitable orientation is north–south.
2. Is it better for energy to be evenly distributed during the year, although in winter, the production is significantly less than in summer? If the answer is "yes," the best orientation is north–south.
3. Is the solar field expected to supply similar thermal power in summer and winter? If the answer is "yes," the proper orientation is east–west.

In a typical PTC field, several collectors connected in series make a row, and a number of rows are connected in parallel to achieve the required nominal thermal power output at design point. The number of collectors connected in series in every row depends on the temperature increase to be achieved between the row inlet and outlet. In every row of collectors, the receiver tubes in adjacent PTCs have to be connected by flexible joints to allow independent rotation of both collectors as they track the sun during the day. These flexible connections are also necessary to allow the linear thermal expansion of the receiver tubes when their temperature increases from ambient to nominal temperature during system start-up. Two main types of flexible connections are available: flexible hoses and ball joints.

Flexible hoses for temperatures below 300°C are composed of an inner hose that can withstand this maximum temperature and an outer metal-braid shield protecting the inner hose. The outer braid is thermally insulated to reduce thermal loss. For higher temperatures, stainless steel bellows are commonly used. This type of hose is not as flexible and causes a significant pressure drop in the circuit because of its high friction coefficient. The minimum bending radius defined by the manufacturer must be taken into consideration to prevent overstressing of the bellows.

Ball joints are another option for flexible connection between the receiver tubes of adjacent collectors. The main benefit of this option is a significantly lower pressure drop because pressure drop for one ball joint is equivalent to a 90° elbow. Another advantage of ball joints is that the connected pipes have two degrees of freedom of movement, because the connected pipes can rotate freely (360°) simultaneously and with a maximum pivot angle of about ±15°. Ball joints are also provided with an inner graphite sealing to reduce friction and avoid leaks.

Today's PTCs working at temperatures above 300°C are connected by ball joints instead of flexible hoses. Furthermore, the flexible hoses initially installed in the solar power plants in California between 1985 and 1990 are being replaced by ball joints because they are more reliable and have lower maintenance costs. Flexible hoses are likely to suffer from fatigue failures resulting in a leak, while ball joints require only the graphite sealing to be refilled after many thousands of hours of operation. Figure 42.15 shows typical pipe connections: with a flexible hose and with ball joints.

An advanced hybrid element composed of a flex hose and a cylindrical joint has been developed by the German company Senior Flexonics, with a special design to prevent

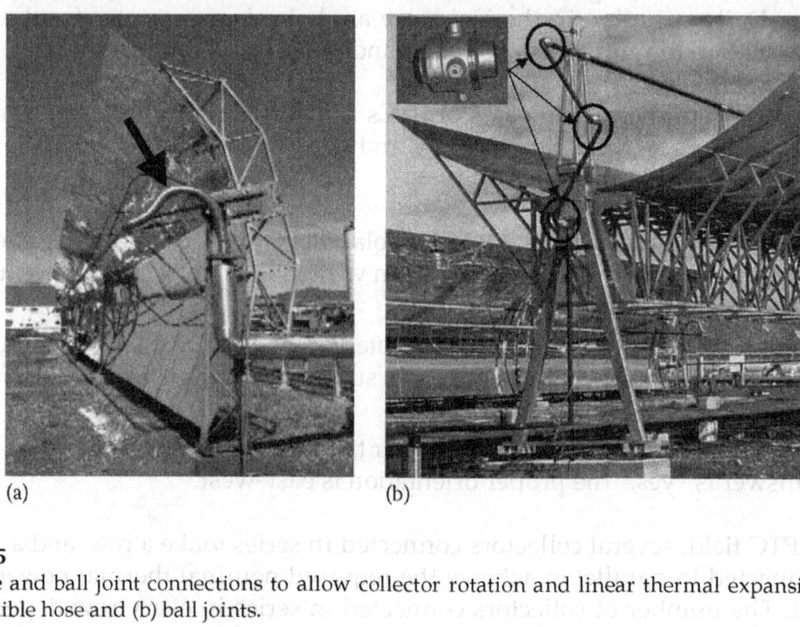

(a) (b)

FIGURE 42.15
Flexible hose and ball joint connections to allow collector rotation and linear thermal expansion of receiver tubes: (a) flexible hose and (b) ball joints.

overstressing of the bellows and provide a better reliability. This design (http://www.seniorflexonics.de) combines the best features of flex hoses and mechanical seals, while avoiding their disadvantages.

42.4.2 Performance Parameters and Losses in a Parabolic-Trough Collector

Three of the design parameters required for a PTC are the geometric concentration ratio, the acceptance angle, and the rim angle (see Figure 42.16). The concentration ratio is the ratio between the collector aperture area and the total area of the absorber tube, while the acceptance angle is the maximum angle that can be formed by two rays in a plane transversal to the collector aperture so that they intercept the absorber pipe after being reflected by the parabolic mirrors. The concentration ratio, C, is given by the following equation:

$$C = \frac{l_a \cdot l}{\pi \cdot d_o \cdot l} = \frac{l_a}{\pi \cdot d_o} \quad (42.21)$$

where
- d_o is the outer diameter of receiver steel pipe
- l is the collector length
- l_a is the parabola width

The wider the collector acceptance angle is, the less accurate the tracking system has to be, as the collector will not need to update its position as frequently. Usual values of the concentration ratio of PTCs are about 20, although the maximum theoretical value is on the order of 70. High concentration ratios are associated with very small acceptance angles, which require very accurate sun-tracking systems and, consequently, higher costs.

The minimum practical acceptance angle is 32′ (0.53°), which is the average solid angle at which the sun sphere is seen from Earth. This means that any PTC with an acceptance angle smaller than 32′ would always lose a fraction of the direct solar radiation. In fact, recommended acceptance angles for commercial PTCs are between 1° and 2°. Smaller angles would demand very accurate sun-tracking system and frequent updating of the

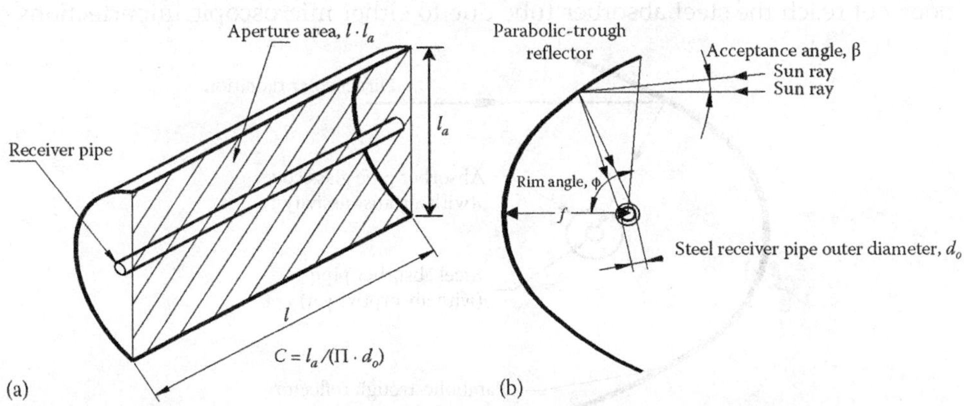

FIGURE 42.16
Concentration ratio and acceptance angle of a parabolic-trough collector: (a) concentration ratio and (b) acceptance angle, β.

collector position, while higher values would lead to small concentration ratios and, therefore, lower working temperatures. So acceptance angle values between 1° and 2° are the most cost-effective.

The rim angle, ϕ, is directly related to the concentrator arc length, and its value can be calculated from Equation 42.22 as a function of the parabola focal distance, f, and aperture width, l_a:

$$\frac{l_a}{4 \cdot f} = \tan\frac{\phi}{2} \tag{42.22}$$

Usual values for rim angles in a PTC are between 70° and 110°. Smaller rim angles are not advisable because they reduce the aperture surface. Rim angles over 110° are not cost-effective because they increase the total reflecting surface without effectively increasing the aperture width.

When direct solar radiation reaches the surface of a PTC, a significant amount of it is lost due to several different factors. The total loss can be divided into three types, which in the descending order of importance are

1. Optical losses
2. Thermal losses from the absorber pipe to the ambient
3. Geometrical losses

The optical losses are associated with four parameters (see Figure 42.17), which are as follows:

1. *Reflectivity, ρ, of the collector reflecting surface*: Since the reflectivity of the parabolic-trough concentrator is less than 1, only a fraction of the incident radiation is reflected toward the receiver tube. Typical reflectivity values of clean silvered-glass mirrors are around 0.93. After washing the mirrors, their reflectivity continuously decreases as dirt accumulates until the next washing. Commercial parabolic-trough mirrors are washed when their reflectivity is of about 0.88–0.9.
2. *Intercept factor, ϒ*: A fraction of the direct solar radiation reflected by the mirrors does not reach the steel absorber tube due to either microscopic imperfections of

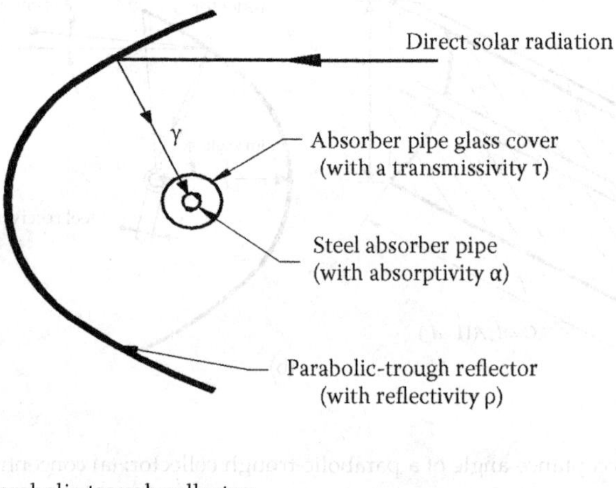

FIGURE 42.17
Optical parameters of a parabolic-trough collector.

the reflectors or macroscopic shape errors in the parabolic-trough concentrators (e.g., imprecision during assembly). These errors cause reflection of some rays at the wrong angle, and therefore they do not intercept the absorber tube. These losses are quantified by an optical parameter called the geometrical intercept factor, Υ_g, which is typically 0.95. Also the flexible bellows connecting the end of the glass cover and the steel tube (see Figure 42.13) reduce the amount of reflected solar radiation that reaches the steel receiver tube, because a fraction of the concentrated solar radiation reflected by the parabolic mirrors is blocked in its way toward the steel receiver tube. This percentage of reflected solar radiation that is blocked by the bellows is quantified by means of the so-called active length factor of the receiver tube, Υ_L, which usually has a value of 0.96–0.97. The overall intercept factor, Υ, is the product of the geometrical intercept factor, Υ_g, and the *active length factor*, Υ_L.

3. *Transmissivity of the glass tube*, τ: The metal absorber tube is placed inside an outer glass tube in order to increase the amount of absorbed energy and reduce thermal losses. A fraction of the direct solar radiation reflected by the mirrors and reaching the glass cover of the absorber pipe is not able to pass through it. The ratio between the radiation passing through the glass tube and the total incident radiation on it gives transmissivity, τ, which is typically $\tau = 0.93$.
4. *Absorptivity of the absorber selective coating*, α: This parameter quantifies the amount of energy absorbed by the steel absorber pipe, compared with the total radiation reaching the outer wall of the steel pipe. This parameter is typically 0.95 for receiver pipes with a cermet coating, while it is slightly lower for pipes coated with black nickel or chrome.

Multiplication of these four parameters (reflectivity, intercept factor, glass transmissivity, and absorptivity of the steel pipe) when the incidence angle on the aperture plane of the PTC is 0° gives what is called the peak optical efficiency of the PTC, $\eta_{opt,0°}$:

$$\eta_{opt,0''} = \rho \times \gamma \times \tau \times \alpha \big|_{\varphi=0°} \tag{42.23}$$

$\eta_{opt,0°}$ is usually in the range of 0.70–0.76 for clean, good-quality PTCs.

Concerning the second type of losses (i.e., the thermal losses), the total thermal loss in a PTC, $P_{Q,\text{collector}\rightarrow\text{ambient}}$, is due to radiative heat loss from the steel absorber pipe to ambient, $P_{Q,\text{absorber}\rightarrow\text{ambient}}$, and convective and conductive heat losses from steel absorber pipe to its outer glass tube, $P_{Q,\text{absorber}\rightarrow\text{glass}}$. Though this heat loss is governed by the well-known mechanisms of radiation, conduction, and convection, it is a good practice to calculate them all together using the *thermal loss coefficient*, $U_{L,abs}$, according to the following equation, where T_{abs} is the mean steel absorber pipe temperature, T_{amb} is the ambient air temperature, d_o is the outer diameter of the steel absorber pipe, and l is the absorber pipe length (PTC length):

$$P_{Q,\text{collector}\rightarrow\text{ambient}} = U_{L,abs} \cdot \pi \cdot d_o \cdot l \cdot (T_{abs} - T_{amb}) \tag{42.24}$$

In Equation 42.23, the thermal loss coefficient is given in ($\text{W/m}^2_{\text{abs}}$ K) units per square meter of the steel absorber pipe surface. The following equation can be used to find the value of the thermal loss coefficient per square meter of aperture surface of the PTC, $U_{L,col}$:

$$U_{L,col} = \frac{U_{L,abs}}{C} \left(\text{W/m}^2_{\text{col}}\ \text{K}\right) \tag{42.25}$$

The heat loss coefficient depends on absorber pipe temperature, which is found experimentally by performing specific thermal loss tests with the PTC working at several temperatures within its typical working-temperature range. Variation in the thermal loss coefficient versus the receiver pipe temperature can usually be expressed with a second-order polynomial equation like Equation 42.26, with coefficients *a*, *b*, and *c* experimentally calculated:

$$U_{L,abs} = a + b\cdot(T_{abs} - T_{amb}) + c\cdot(T_{abs} - T_{amb})^2\ \left(\mathrm{W/m^2_{abs}\ K}\right) \tag{42.26}$$

It is sometimes difficult to find values for coefficients *a*, *b*, and *c* valid for a wide temperature range. When this happens, different sets of values are given for smaller temperature ranges. Table 42.3 gives the values of coefficients *a*, *b*, and *c* experimentally calculated by Ajona (1999) for receiver tubes initially installed at SEGS VIII and IX.

A typical value of $U_{L,abs}$ for absorber tubes with vacuum in the space between the inner steel pipe and the outer glass tube is lower than $5\left(\mathrm{W/m^2_{abs}\ K}\right)$. High-vacuum conditions are not needed to significantly reduce the convective heat losses. However, the low thermal stability in hot air of the cermet coatings currently used in these receiver tubes requires a high vacuum to assure good coating durability.

The optical properties (i.e., absorptivity and emissivity) of selective coatings used in vacuum receiver tubes for PTCs have been improved in the last years, and new correlations have been developed to calculate the overall thermal loss without using a heat loss coefficient. Nowadays, each receiver tube manufacturer gives the proper correlations to calculate the thermal loss of their tubes, so that the customers have to apply only such correlations.

The optimum space between the steel absorber pipe and the outer glass tube for receivers without vacuum to minimize the convective heat loss is calculated as a function of the Rayleigh number (Ratzel and Simpson, 1979). The possible bowing of the steel pipe has to be considered also in determining the minimum gap, because the possibility of contact with the glass cover has to be avoided.

The third type of losses in a PTC are the geometrical losses, which are due to the incidence angle, φ, of direct solar radiation on the aperture plane of the collector. The incidence angle is the angle between the normal to the aperture plane of the collector and the sun's vector, both contained on a plane perpendicular to the collector axis. This angle depends on the day of the year and the time of day. The incidence angle of direct solar radiation is a very important factor, because the fraction of direct radiation that is useful to the collector is directly proportional to the cosine of this angle, which also reduces the useful aperture area of the PTC (see Figure 42.18). The incidence angle reduces the aperture area of a PTC

TABLE 42.3

Values of Coefficients "*a*," "*b*," and "*c*" for a Receiver Tube Installed at SEGS VIII and IX

T_{abs} (°C)	*a*	*b*	*c*
<200	0.687257	0.001941	0.000026
>200; <300	1.433242	−0.00566	0.000046
>300	2.895474	−0.0164	0.000065

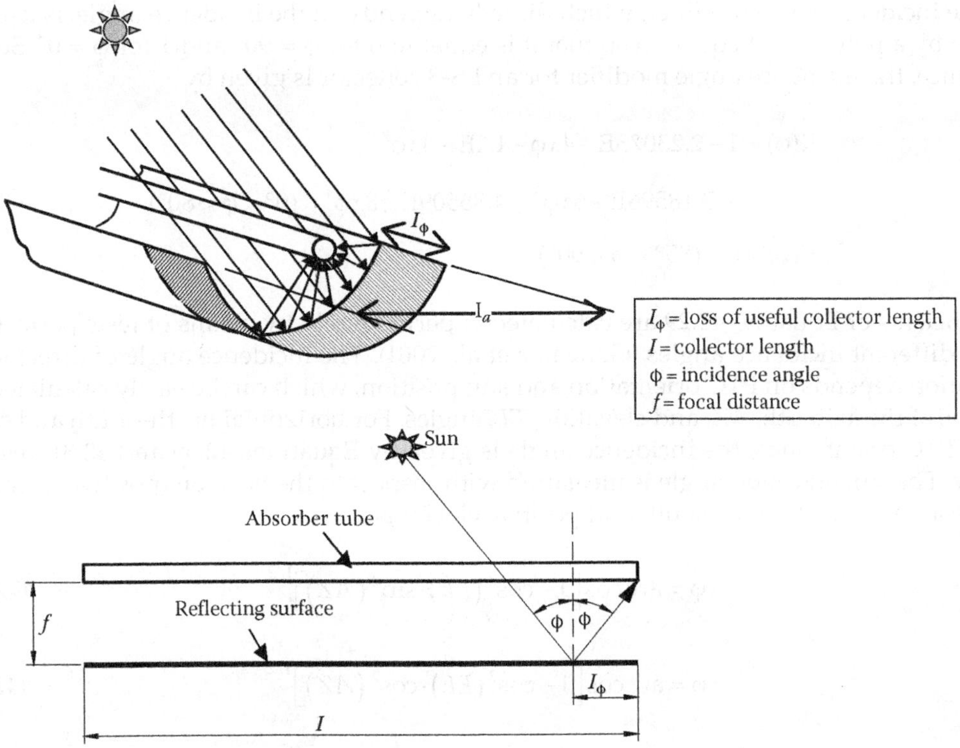

FIGURE 42.18
Geometrical losses at the end of a parabolic-trough collector.

in an amount A_e called the *collector geometrical end losses* and is calculated with Equations 42.27 and 42.28 with the following parameters:

l_a is the parabola width
l is the collector length
f is the focal distance of the parabolic-trough concentrator
f_m is the mean focal distance in a cross section of the parabolic-trough concentrator
φ is the incidence angle of the direct solar radiation

$$A_e = l_a \cdot l_\varphi = l_a \cdot f_m \cdot \tan\varphi \quad (42.27)$$

$$f_m = f + \left[\frac{(f \cdot l_a^2)}{(48 \cdot f^2)}\right] \quad (42.28)$$

The incidence angle also affects PTC optical parameters (i.e., mirror reflectivity, selective coating absorptivity, intercept factor, and glass transmissivity) because these parameters are not isotropic. The effect of the incidence angle on the optical efficiency and useful aperture area of a PTC is quantified by the incidence angle modifier, $K(\phi)$, because this parameter includes all optical and geometric losses due to an incidence angle greater than 0°.

The incidence angle modifier, which directly depends on the incidence angle, is usually given by a polynomial equation so that it is equal to 0 for $\phi = 90°$ and 1 for $\phi = 0°$. So, for instance, the incidence angle modifier for an LS-3 collector is given by

$$K(\varphi) = 1 - 2.23073E - 4x\varphi - 1.1E - 4x\varphi^2 + 3.18596E - 6x\varphi^3 - 4.85509E - 8x\varphi^4 \quad (0° < \varphi < 80°)$$
$$K(\varphi) = 0 \quad (85° < \varphi < 90°) \tag{42.29}$$

Coefficients of Equation 42.29 are calculated experimentally by means of tests performed with different incidence angles (Gonzalez et al., 2001). The incidence angle of direct solar radiation depends on PTC orientation and sun position, which can be easily calculated by means of the azimuth, AZ, and elevation, EL, angles. For horizontal north–south and east–west PTC orientations, the incidence angle is given by Equations 42.30 and 42.31, respectively. The sun elevation angle is measured with respect to the horizon (positive upward), while azimuth is 0° to the south and positive clockwise.

$$\varphi = \arccos\left[1 - \cos^2(EL)\cdot\sin^2(AZ)\right]^{1/2} \tag{42.30}$$

$$\varphi = \arccos\left[1 - \cos^2(EL)\cdot\cos^2(AZ)\right]^{1/2} \tag{42.31}$$

42.4.3 Efficiencies and Energy Balance in a Parabolic-Trough Collector

The combination of three different efficiencies

η_{global} is the global efficiency

$\eta_{opt,0°}$ is the peak optical efficiency (optical efficiency with an incidence angle of 0°)

η_{th} is the thermal efficiency

and one parameter

$K(\varphi)$ is the incidence angle modifier

describe the performance of a PTC. Their definition is graphically represented in the diagram shown in Figure 42.19, which clearly shows that a fraction of the energy flux incident on the collector aperture plane is lost due to the optical losses accounted for by the peak optical efficiency, while another fraction is lost because of an incidence angle $\varphi > 0°$, which is taken into account by the incidence angle modifier, $K(\varphi)$. The remaining PTC losses are thermal losses at the absorber tube.

As explained earlier, the peak optical efficiency, $\eta_{opt,0°}$, considers all optical losses that occur with an incidence angle of $\phi = 0°$ (reflectivity of the mirrors, transmissibility of the glass tube, absorptivity of the steel absorber pipe, and the intercept factor). The incidence angle modifier, $K(\phi)$, considers all optical and geometrical losses that occur in the PTC because the incidence angle is >0° (collector end losses, collector center losses, blocking losses due to absorber tube supports, angle dependence of the intercept factor, angle dependence of reflectivity, transmissivity, and absorptivity). Thermal efficiency, η_{th}, includes all absorber tube heat losses from conduction, radiation, and convection.

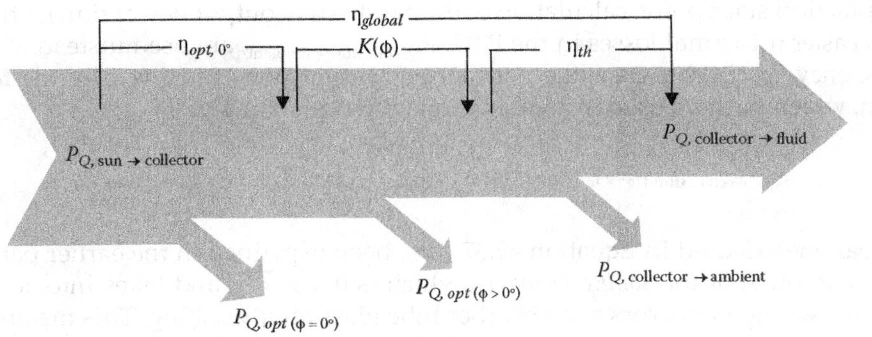

FIGURE 42.19
Diagram of efficiencies and losses in a parabolic-trough collector.

Global efficiency, η_{global}, includes the three kinds of losses that occur in the PTC (optical, geometrical, and heat) and can be calculated as a function of the peak optical efficiency, incidence angle modifier, and thermal efficiency using

$$\eta_{global} = \eta_{opt,0''} \cdot K(\varphi) \cdot \eta_{th} \tag{42.32}$$

The global efficiency can also be calculated as the ratio between the net thermal output power delivered by the collector, $P_{Q,\,collector\rightarrow fluid}$, and the solar energy flux incident on the collector aperture plane, $P_{Q,\,sun\rightarrow collector}$, by Equations 42.33 through 42.35:

$$\eta_{global} = \frac{P_{Q,collector\rightarrow fluid}}{P_{Q,sun\rightarrow collector}} \tag{42.33}$$

$$P_{Q,sun\rightarrow collector} = A_c \cdot E_d \cdot \cos(\varphi) \tag{42.34}$$

$$P_{Q,collector\rightarrow fluid} = q_m \cdot (h_{out} - h_{in}) \tag{42.35}$$

where
- A_c is the collector aperture surface
- E_d is the direct solar irradiance
- φ is the incidence angle
- q_m is the fluid mass flow through the absorber tube of the collector
- h_{in} is the fluid specific mass enthalpy at the collector inlet
- h_{out} is the fluid specific mass enthalpy at the collector outlet

The net output thermal power delivered by a PTC can be calculated by means of Equation 42.35 if the fluid mass flow and the inlet and outlet temperatures are known when the collector is in operation. However, these data are not known during the solar field design phase, and the expected net thermal output has to be calculated starting from the values of the direct solar irradiance, ambient air temperature, incidence angle, and PTC optical, thermal, and geometrical parameters. Equation 42.36 can be used for this purpose.

$$P_{Q,collector\rightarrow fluid} = P_{Q,sun\rightarrow collector} \cdot \eta_{global} = A_c \cdot E_d \cdot \cos(\varphi) \cdot \eta_{opt,0''} \cdot K(\varphi) \cdot n_{th} \cdot F_e \tag{42.36}$$

From a practical standpoint, calculation of the net thermal output power during the design phase is easier if thermal losses in the PTC, $P_{Q,\text{collector}\rightarrow\text{ambient}}$, are used instead of the thermal efficiency, η_{th}. In this case, the net thermal output power is given by the following equation, which must be used in combination with Equation 42.23:

$$P_{Q,\text{collector}\rightarrow\text{fluid}} = A_c \cdot E_d \cdot \cos(\varphi) \cdot \eta_{opt,0''} \cdot K(\varphi) \cdot F_e - P_{Q,\text{collector}\rightarrow\text{ambient}} \qquad (42.37)$$

All the parameters used in Equation 42.37 have been explained in the earlier paragraphs, with the exception of the *soiling factor*, F_e, which is $0 < F_e < 1$, and takes into account the progressive soiling of mirrors and absorber tube glass after washing. This means that the reflectivity and transmissivity are usually lower than nominal, and the peak optical efficiency is also lowered. Usual values of F_e are around 0.97, which is equivalent to a mirror reflectivity of 0.90 for mirrors with a nominal reflectivity of 0.93.

42.4.4 Industrial Applications for Parabolic-Trough Collectors

The large potential market existing for solar systems with PTCs can be clearly seen in the statistical data. The U.S. industry consumes about 40% of the total energy demand in that country. Of this, approximately half (about 20% of the total energy consumption) involves IPH suitable for solar applications with PTCs, which are internationally known as IPH applications. As an example of the situation in other countries with a good level of direct solar radiation, industry is also the biggest energy consumer in Spain (more than 50% of the total energy demand), and 35% of the industry demand is in the mid-temperature range (80°C–300°C) for which PTCs are very suitable.

Besides this large potential market for parabolic-trough systems, there is also an environmental benefit that is taken more and more into consideration: contrary to fossil fuels, solar energy does not contaminate, and it is independent of political or economic interruptions of supply (due to war, trade boycott, etc).

Since industrial process energy requirements in the mid-temperature range are primarily met by steam systems, representative configurations of solar steam generation systems are presented in this section with simple diagrams to facilitate their understanding.

Steam is the most common heat transport medium in industry for temperatures below 250°C where there is a great deal of experience with it. Compact steam generators have proven to be extremely reliable. Integration of a solar steam generation system for a given industrial process involves a simple plant interface to feed steam directly into the existing process, with no major facility changes. Medium-temperature steam can be supplied with PTCs in three different ways:

1. Using a high-temperature, low-vapor-pressure working fluid in the solar collectors and transferring the heat to an unfired boiler where steam is produced. Oil is widely used for this purpose.
2. Circulating pressurized hot water in the solar collectors and flashing it to steam in a flash tank. This method is suitable for temperatures that are not too high (below 200°C), because of the high pressure required in the absorber pipes and flash tank for higher temperatures.
3. Boiling water directly in the collectors (the so-called direct steam generation process).

A brief description of each of these methods follows.

42.4.4.1 Unfired Boiler System

Figure 42.20 shows the schematic diagram for an unfired steam boiler system with PTCs. A heat transfer fluid (HTF) is circulated through the collector field, and steam is generated in an unfired boiler. A variation of the system shown in Figure 42.20 incorporates a preheater in the water makeup line, which not only increases the system cost, but also reduces the inlet temperature to the solar field. Water could be circulated in the collector loop, but the fluid generally selected is a low-vapor-pressure, nonfreezing hydrocarbon, or silicon oil. The use of oil overcomes the disadvantages associated with water (high vapor pressure and risk of freezing) and accommodates energy storage, but certain characteristics of these oils cause other problems. Generally, precautions must be taken to prevent the oil from leaking out of the system, which could cause fire. Oil is also expensive and has poorer heat transport properties than water. They are extremely viscous when cold, and a positive displacement pump is sometimes needed to start the system after it has cooled down. The use of a fluid to transfer thermal energy from the solar field to an unfired boiler where steam is produced is internationally known as HTF technology.

The unfired boiler itself is an expensive item requiring alloy tubes for corrosion protection, and it is an additional resistance to heat flow. As in the flash-steam system, the collectors must operate at a temperature some degrees (20°C approx.) above the steam delivery temperature. Because the process steam must be maintained at a certain temperature, the solar-generated steam is delivered at a variable flow rate depending on the solar radiation available at the aperture of the solar collectors. The collector outlet temperature can be held constant by varying the oil flow rate through the collectors as the collected solar energy varies due to cloud passage or any other reason.

42.4.4.2 Flash-Steam Systems

A diagram of a flash-steam system is shown in Figure 42.21. Water at a high enough pressure to prevent boiling is circulated through the collector field and flashed to steam across a throttling valve into a separator. This constant-enthalpy process converts the sensible heat of the water into the latent heat of a two-phase mixture at the conditions prevailing in the separator. The maximum steam quality (i.e., the fraction of total flow that is converted

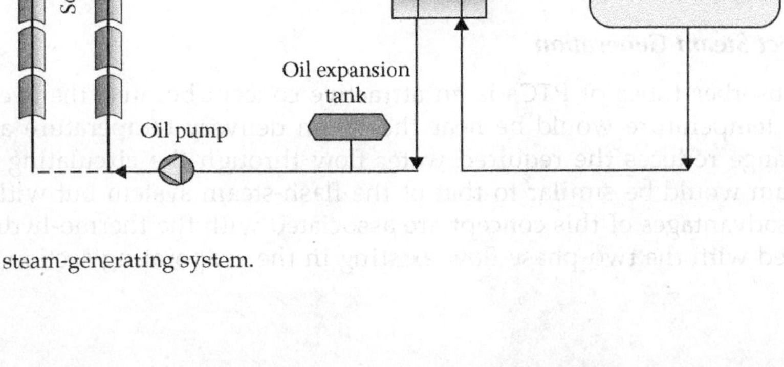

FIGURE 42.20
Unfired-boiler steam-generating system.

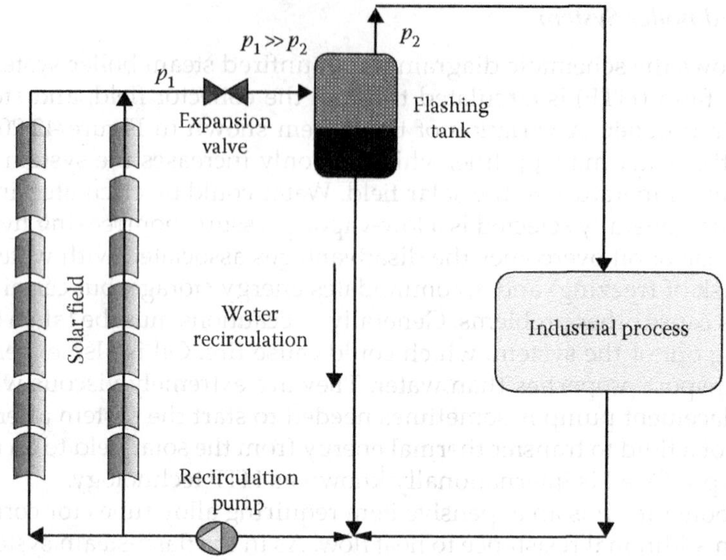

FIGURE 42.21
Flash-steam solar system.

into steam) is less than 10% due to thermodynamic constraints. The steam thus produced is fed into the industrial process, while the water remaining in the flash tank is recirculated to the solar field inlet. Feed-water makeup is injected from the flash tank into the pump suction to maintain the liquid level in the tank.

Using water as an HTF simplifies the construction of a flash-steam system. However, although water is an excellent heat transport medium, freezing problems can occur. Therefore, the freeze protection mechanism must be carefully designed and controlled to ensure that a minimum amount of heat is supplied to the water to prevent freezing due to low ambient temperature.

The disadvantages of the flash-steam system are associated with the steam generation mechanism. Collector temperatures must be significantly higher than the steam delivery temperature to obtain reasonable steam qualities downstream of the throttling valve and to limit the water recirculation rate. But higher temperatures reduce the collector's efficiency. In addition, the circulating pump must overcome the pressure drop across the flash valve, which can be important.

Moreover, the rapid rise in the water vapor pressure at temperatures above 175°C limits the steam pressure that can be achieved by this method to approximately 2 MPa (305 psig) at acceptable levels of electrical power required for pumping. For higher pressures, the electricity consumption of the feed-water pump would be excessive and would jeopardize the efficiency of the whole system.

42.4.4.3 Direct Steam Generation

DSG in the absorber tubes of PTCs is an attractive concept because the average collector operating temperature would be near the steam delivery temperature and because the phase change reduces the required water flow through the circulating pump. The system diagram would be similar to that of the flash-steam system but without a flash valve. The disadvantages of this concept are associated with the thermo-hydraulic problems associated with the two-phase flow existing in the evaporating section of the solar field. Nevertheless, experiments performed at the PSA in Spain have proven the technical feasibility of DSG with horizontal PTCs at 100 bar/400°C (Zarza et al., 2002).

42.4.5 Sizing and Layout of Solar Fields with Parabolic-Trough Collectors

A typical parabolic-trough solar collector field (Figure 42.22) is composed of a number of parallel rows of several collectors connected in series so that the working fluid circulating through the absorber pipe is heated as it passes from the inlet to the outlet of each row.

The first step in the design of a parabolic-trough solar field is the definition of the so-called design point, which is composed of a set of parameters that determine solar field performance. Parameters to be defined for the *design point* are

- Collector orientation
- Date (month and day) and time of design point
- Direct solar irradiance and ambient air temperature for the selected date and time
- Geographical location of the plant site (latitude and longitude)
- Total thermal output power to be delivered by the solar field
- Solar collector soiling factor
- Solar field inlet/outlet temperatures
- Working fluid for the solar collectors
- Nominal fluid flow rate

If oil is used in the solar field to transfer the energy to an unfired boiler (HTF technology), the selected temperature of the fluid at the solar field outlet must be at least 10°C higher than the steam temperature demanded by the process to be fed. So, for example, if the industrial process to be fed by the solar system requires 300°C steam, the oil temperature at the solar field outlet must be about 315°C. This difference is necessary to compensate for thermal losses between the solar field outlet and the steam generator inlet and the boiler pinch point, which is on the order of (approx.) 5°C–7°C.

Once the design point has been defined, the number of collectors to be connected in series in each row can be calculated using the parameters of the selected PTC (peak optical efficiency, incidence angle modifier, heat loss coefficient, and aperture area) and fluid

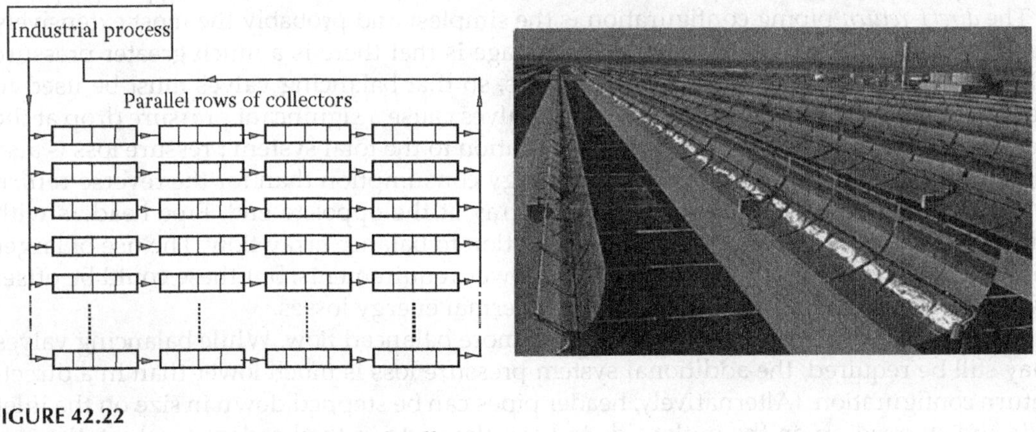

FIGURE 42.22
A typical solar field with parabolic-trough collectors.

(density, heat capacity, and dynamic viscosity). The number of collectors in each row depends on the nominal temperature difference between solar field inlet and outlet, ΔT, and the single collector temperature step, ΔT_c. Thus, if a collector field is intended to supply thermal energy to an unfired boiler that requires a temperature step of 70°C between inlet and outlet, with a nominal inlet temperature of 220°C, the inlet and outlet temperatures in each row will be 220°C and 290°C, respectively, with a $\Delta T = 70$°C. Once this ΔT has been determined, the number of collectors required in each row, N, is given by the ratio

$$N = \frac{\Delta T}{\Delta T_c}$$

where

N is the number of collectors to be connected in series in a row
ΔT is the required ΔT by the industrial process
ΔT_c is the difference between the single collector nominal inlet and outlet working temperatures

Once the number of collectors to be connected in series in each row has been calculated, the next step is to determine the number of rows to be connected in parallel. This number depends on the thermal power demanded by the industrial process. The procedure for determining the number of rows is very simple when there is no thermal storage system: the ratio between the thermal power demanded by the industrial process and the thermal power delivered by a single row of collectors at design point.

When thermal storage is available, the number of parallel rows is determined in a different way. In this case, the useful thermal energy delivered by a single row of collectors from sunrise to sunset during the design day must be calculated, as well as the thermal energy demanded during the complete day (i.e., 24 h period) by the process to be fed by the solar field. The number of parallel rows required is given by the ratio between the thermal energy demanded by the process and the thermal energy delivered by a single row of collectors.

After sizing the solar field, the designer has to lay out the piping. Three basic layouts are used in solar fields with PTCs. These layouts (direct return, reverse return, and center feed) are shown schematically in Figure 42.23. In all three options, the hot outlet piping is shorter than the cold inlet piping to minimize thermal losses. The advantages and disadvantages in each of these three configurations are explained in following paragraphs.

The *direct-return* piping configuration is the simplest and probably the most extensively used in small solar fields. Its main disadvantage is that there is a much greater pressure difference between the inlets in parallel rows, so that balancing valves must be used to keep flow rates the same in each row. These valves cause a significant pressure drop at the beginning of the array, and thus their contribution to the total system pressure loss is also significant. The result is higher parasitic energy consumption than for the reverse-return layout, where the fluid enters the collector array at the opposite end. Pipe headers with different diameters are used in this configuration to balance array flow. The use of larger pipe headers also results in lower parasitic power requirements, but these could be offset by increases in initial investment costs and thermal energy losses.

The *reverse-return* layout has an inherently more balanced flow. While balancing valves may still be required, the additional system pressure loss is much lower than in a direct-return configuration. (Alternatively, header pipes can be stepped down in size on the inlet side and stepped up on the outlet side to keep flow rate in the headers constant, thereby

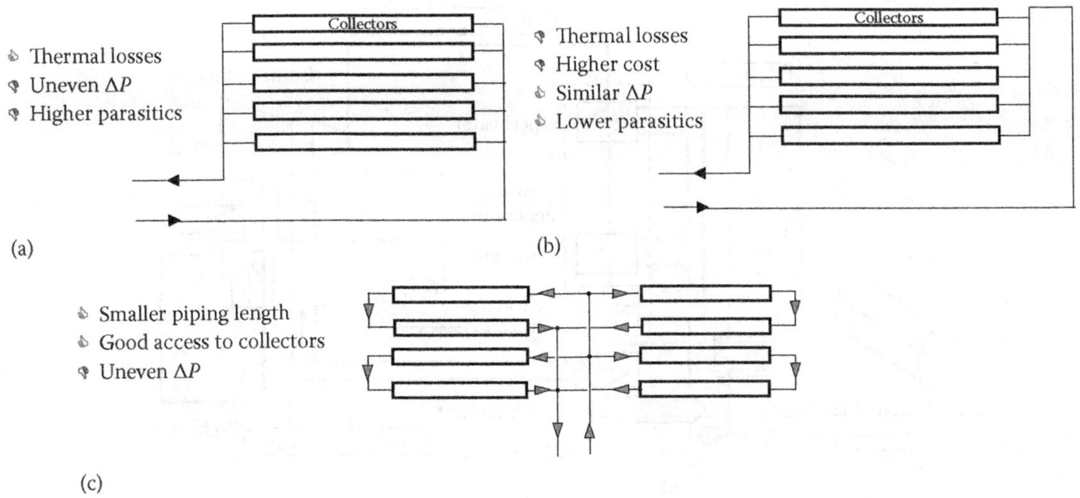

FIGURE 42.23
Solar field layouts for parabolic-trough collectors: (a) direct return, (b) inverse return, and (c) central feed.

providing uniform flow.) The extra length of piping at the solar field inlet is a disadvantage in the reverse-return configuration because of the additional heat loss, although this greatly depends on the solar field inlet temperature. If this temperature is low, additional heat loss is negligible. Adding to pipe length, however, results in higher piping, insulation, and fluid inventory costs.

The *center-feed* configuration is the most widely used layout for large solar fields. Like the direct-return design, pressure balancing valves may be required at the row inlets in very large solar fields, and pressure loss would be thus increased in the solar field. However, this configuration has two important advantages: (1) it minimizes the total amount of piping because there is no pipe running the length of the collector row, and (2) there is direct access to each collector row without buried pipes. This direct access is very important for repair works and solar field washing, which must be performed often in commercial plants to keep a high level of reflectivity, because vehicles have an easy access to each solar collector.

42.4.6 Electricity Generation with Parabolic-Trough Collectors

The current PTC temperature range and their good solar-to-thermal efficiency up to 400°C make it possible to integrate a parabolic-trough solar field in a Rankine water–steam power cycle to produce electricity. The simplified scheme of a typical STP plant using parabolic troughs integrated in a Rankine cycle is shown in Figure 42.24. So far, all the STP plants with parabolic-trough collectors use the HTF technology because steam production by flashing is not suitable for 100-bar superheated-steam pressure, and commercial DSG has not yet been commercially proven for superheated steam at 100 bar.

A parabolic-trough power plant is basically composed of three elements: the solar system, the steam generator, and the power conversion system (PCS; see Figure 42.24). The solar system is composed of a parabolic-trough solar collector field and the oil circuit. The solar field collects the solar energy available in the form of direct solar radiation and converts it into thermal energy as the temperature of the oil circulating through the receiver tubes of the collectors increases.

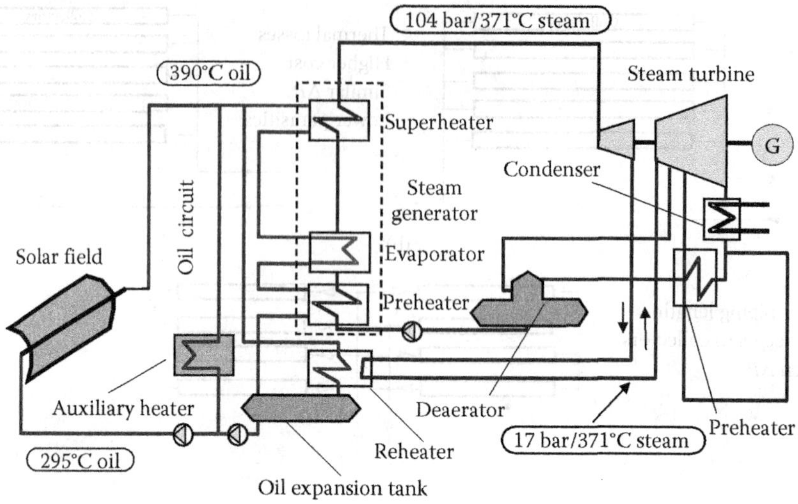

FIGURE 42.24
Simplified scheme of a solar thermal power plant with parabolic-trough collectors.

Once heated in the solar field, the oil goes to the steam generator, which is an oil–water heat exchanger where the oil transfers its thermal energy to the water that is used to generate the superheated steam required by the turbine. The steam generator is, therefore, the interface between the solar system (solar field + oil circuit) and the PCS itself. Normally, the steam generator used in these solar power plants consists of three stages:

1. *Preheater*: Where water is preheated to a temperature close to evaporation.
2. *Evaporator*: Where the preheated water is evaporated and converted into saturated steam.
3. *Superheater*: The saturated steam produced in the evaporator is heated in the superheater to the temperature required by the steam turbine.

The PCS transforms the thermal energy delivered by the solar field into electricity, using the superheated steam delivered by the steam generator. This PCS is similar to that of a conventional Rankine power plant, except for the main difference that heat supplied to the conventional Rankine cycle is from fossil fuels, while in the solar plant, the PTCs deliver the required thermal energy.

The superheated steam delivered by the steam generator is then expanded in a steam turbine that drives an electricity generator, which produces the electricity that is delivered to the distribution grid. The steam turbine is usually composed of two consecutive stages, for high- and low-pressure steam. Steam leaving the turbine high-pressure stage goes to a reheater where its temperature rises before entering the low-pressure turbine stage. After this stage, the steam is condensed, and the condensate goes to a water deaerator to remove oxygen and gases dissolved in the water. The steam leaving the turbine low-pressure stage can be condensed either in a wet cooling system (with cooling towers refrigerated by water) or in a dry cooling system (condenser refrigerated by air). The selection of the best cooling system is strongly influenced by the onsite availability of water resources. The main pump takes feed water for the steam generator from the deaerator, thus starting the Rankine thermodynamic cycle again.

FIGURE 42.25
Overall view of a solar power plant with parabolic-trough collectors.

Though parabolic-trough power plants usually have an auxiliary gas-fired heater to produce electricity when direct solar radiation is not available, the amount of electricity produced with natural gas is always limited to a reasonable level. This limit changes from one country to another: 25% in California (United States), 15% in Spain (until 2012, because this percentage was reduced to 0% in 2012), and no limit in Algeria. Figure 42.25 shows what an STP plant with PTCs looks like. The PCS is located at the center of the plant, surrounded by the solar field. The plant shown in Figure 42.25 is provided with a wet cooling system, and the steam leaving the cooling towers is clearly shown.

Parabolic-trough power plants can play an important role in achieving sustainable growth because they save about 2000 tons of CO_2 emissions per MW of installed power yearly. Typical solar-to-electric efficiencies of a large STP plant (>30 MW_e) with PTCs are between 15% and 22%, with an average value of about 17%. The yearly average efficiency of the solar field is about 50%.

Though not included in Figure 42.24, a thermal energy storage system can be implemented in parabolic-trough power plants to allow operation of the PCS when direct solar radiation is not available. In this case, the solar field has to be oversized so that it can simultaneously feed the PCS and charge the storage system during sunlight hours. Thermal energy from the storage system is then used to keep the steam turbine running and producing electricity after sunset or during cloudy periods. Yearly hours of operation can be significantly increased, and plant amortization is thus enhanced when a storage system is implemented. However, the required total investment cost is also higher.

42.4.7 Thermal Storage Systems for Parabolic-Trough Collectors

The main problem with using solar radiation is its discontinuity, because it is possible to collect only during sunlight hours. There is an additional limitation when dealing with concentrating solar systems as these systems can collect only the direct solar radiation, so they need clear sky conditions, because clouds block direct solar radiation. Thermal storage systems are implemented to solve these limitations.

When a solar system does not have to supply thermal energy during the night or during cloudy periods, a storage system is not necessary. On the other hand, if the industrial process has to be supplied during periods without direct solar radiation, a storage system has to be implemented to store part of the thermal energy supplied by the solar collectors during the sunlight hours to deliver it when the sun is not available.

Thermal storage systems have three main advantages:

1. Thermal energy can be supplied during hours when direct solar radiation is not available, so that solar energy collection and thermal energy supply do not have to be simultaneous.
2. The solar field inlet can be isolated from possible disturbances at the outlet, because the storage system behaves as a good thermal cushion and avoids feedback of the disturbances affecting the solar field outlet temperature.
3. A constant thermal power level can be supplied to the process if energy is taken from both solar field and thermal storage system, so that the energy taken from the thermal storage complements the energy delivered by the solar field. The variability of the beam solar radiation is thus compensated with thermal energy delivered by the thermal storage system.

The second advantage is very important because it enhances solar field operation on days with frequent cloud transients. No matter how effective solar field control is, the fluid temperature at the outlet is affected by cloud transients, and temperature fluctuations are likely. These fluctuations would immediately affect the working fluid temperature at the inlet if there were not a thermal storage system in between.

The hot water storage system used in low-temperature solar conversion systems (i.e., flat-plate collectors) is not suitable for parabolic-trough systems because the high pressure in the storage tank would make the system too expensive. For this reason, PTCs require the use of a different storage medium. Depending on the medium where the thermal energy is stored, there are two types of systems:

1. Single-medium storage systems
2. Dual-medium storage systems

42.4.7.1 Single-Medium Storage Systems

Single-medium storage systems are those in which the storage medium is the same fluid circulating through the collectors. The most common is thermal oil as both the working fluid and the storage medium. The efficiency of these systems is over 90%. Oil storage systems can be configured in two different ways:

42.4.7.1.1 *Systems with a Single Oil Tank*

For low-capacity storage systems, thermal energy can be stored in a single tank, in which the oil is stratified by temperature. Energy is stored as latent heat by increasing the temperature of the oil in the tank. Figure 42.26a shows this type of system configuration, in which the solar field can supply hot oil to the tank by means of three-way valves installed at the solar field inlet and outlet, so that all of it or only part of it enters. The density of the thermal oils commonly used as working fluids in these systems strongly varies with temperature. So, for instance, the density of Santotherm 55 oil at 90°C is 842.5 kg/m^3, while at 300°C it is 701.4 kg/m^3. Due to its lower density, the hot oil entering the storage tank through the top inlet manifold remains in the upper layers inside the tank, while the cold oil always remains at the bottom of the tank. As seen in Figure 42.26a, the boiler supplying the thermal energy demanded by the industrial process can be fed from either the storage tank or the solar field, depending on the position of the three-way valve. When discharging the storage system, the hot oil leaves the tank through the top outlet manifold and returns to the bottom after leaving the boiler. Pump "B2" is used

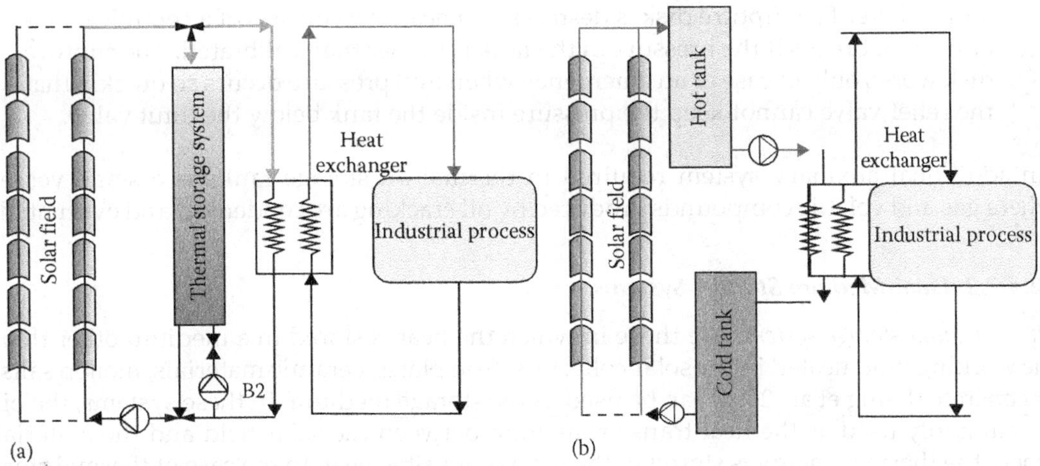

FIGURE 42.26
Thermal storage systems with (a) one and (b) two oil tanks.

exclusively to feed the boiler from the storage tank, when the solar field is not in operation. Cold oil leaves from the bottom of the storage tank and goes to the solar field during daylight hours to be heated and then returned to the top of the storage tank. The storage system is fully charged when all the oil stored in the tank is hot. As already mentioned, the use of a single oil storage tank is feasible only for small storage systems. For high-capacity systems, two oil tanks (i.e., one tank for cold oil and another for hot oil) are needed.

42.4.7.1.2 Systems with Two Oil Tanks

There are two oil tanks in these systems (see Figure 42.26b), one hot tank and one cold tank. The boiler is always fed from the hot tank, and once the oil has transferred heat to the water in the unfired boiler, it goes to the cold tank. This tank supplies the solar field, which at the same time feeds the hot tank with the oil heated by the collectors.

One of the drawbacks of using oil as the storage medium is the need to keep the oil in the storage tank(s) pressurized and inert. Thermal oil has to be kept pressurized above the vapor pressure corresponding to the maximum temperature in the oil circuit to prevent the oil from changing into gas. Fortunately, the vapor pressure of the thermal oils used in these systems is usually low for the 100°C–400°C temperature range, and pressurization is easily maintained by injecting argon or nitrogen. This inert atmosphere also avoids the risk of explosion in the tank from pressurized mists, which are explosive in air.

Another disadvantage of oil systems is the need for appropriate fire-fighting systems, as well as a concrete oil sump to collect any leaks and avoid contamination. All this equipment increases the cost of the storage system. Thermal oil storage systems usually have two safety systems to avoid excessive overpressure inside the tank when temperature increases (the oil expands considerably with temperature):

1. A relief valve to discharge the inert gas into the atmosphere when the pressure inside the tank is over a predefined value. This valve usually works when the pressure inside the tank increases slowly (for instance, during charging).
2. Due to the small section of the relief valve, it cannot dissipate sudden overpressures. There is an additional security device for this purpose: a pressure-rated ceramic rupture disk. This system allows gases to be rapidly evacuated into the

atmosphere. The rupture disk is destructive, because it consists of a ceramic membrane that breaks if the pressure in the tank is higher than calibrated. The rupture disk works only in case of an emergency when overpressure occurs so quickly that the relief valve cannot keep the pressure inside the tank below the limit value.

An additional auxiliary system required in thermal oil storage tanks is a small vessel where gas and volatile compounds produced by oil cracking are condensed and evacuated.

42.4.7.2 Dual-Medium Storage Systems

Dual-medium storage systems are those in which the heat is stored in a medium other than the working fluid heated in the solar collectors. Iron plates, ceramic materials, molten salts, or concrete (Laing et al., 2008) can be used as the storage medium. In these systems, the oil is commonly used as the heat transfer medium between the solar field and the material where the thermal energy is stored in the form of sensible heat. In the case of thermal storage in iron plates, the oil circulates through channels between cast iron slabs placed inside a thermally insulated vessel, transferring thermal energy to them (charging process) or taking it from them (discharging process). When concrete is used as storage medium, the oil circulates through steel tubes installed inside big blocks of concrete.

Molten salts (a eutectic mixture of sodium and potassium nitrates) are nowadays used for dual-medium thermal storage systems in parabolic-trough solar plants. In this case, two tanks are needed: one for cold molten salt and another to store the hot molten salt. Obviously, the lowest temperature is always above the melting point of the salt (250°C, approx.). In this case, a heat exchanger is needed to transfer energy from the oil used in the solar field (heat transfer medium) to the molten salt used for energy storage (storage medium). Figure 42.27 shows the simplified scheme of a parabolic-trough power plant with a molten-salt thermal energy storage system. This type of thermal storage system is currently installed in many Spanish commercial plants, with a storage capacity of 1 GWh in each plant, and it is claimed to be the current most cost-effective option for large commercial solar power plants with large solar shares.

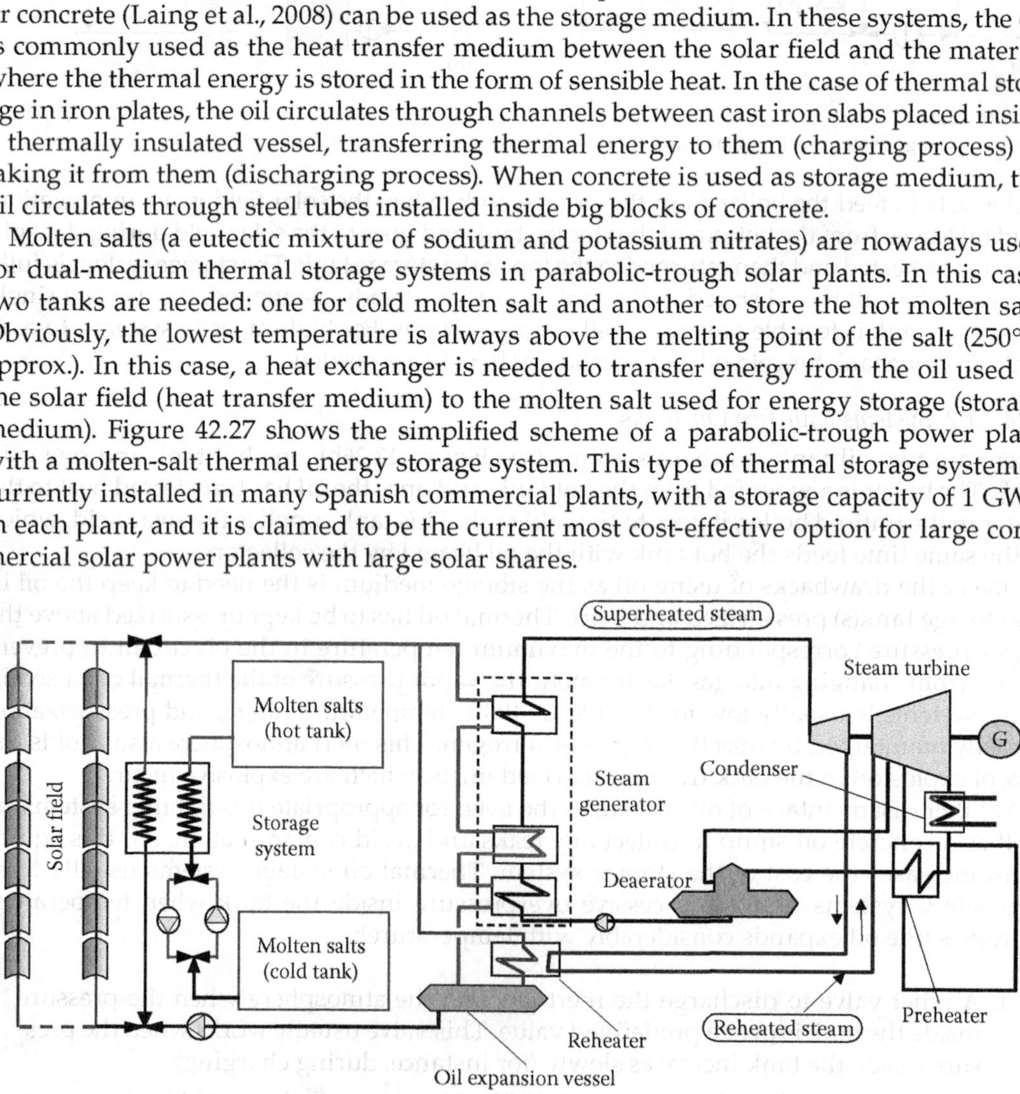

FIGURE 42.27
Scheme of a parabolic-trough power plant with molten-salt storage system.

Thermal storage systems using latent heat, also called phase change material (PCM) storage systems because they use a medium that changes phase during charging and discharging, are under development, and several prototypes have been experimentally tested (Bayón et al., 2010). The storage medium used in the first prototypes was a mixture of $NaNO_3$ and KNO_3. The melting temperature of the mixture can be varied from 220°C to 300°C (approx.) changing the percentage of these two components, thus covering a wide range of steam pressures (i.e., 22–100 bar approx.). The main technical constraint of this type of storage systems is the low thermal conductivity of the salt mixture. Although the use of graphite foils to enhance heat transfer has been successfully evaluated, other options are under study and are ready for testing in the short term. However, latent-heat thermal storage systems are not expected to reach the market before 2020. PCM storage systems are required by STP plants with DSG, because the sensible heat storage systems used in HTF plants are not suitable to store the thermal energy released by the steam during its condensation.

42.4.8 Direct Steam Generation

All solar power plants with PTCs implemented to date use thermal oil as the working fluid in the solar field, and they usually follow the general scheme in Figure 42.24, with only slight differences from one plant to another. The technology of these plants has been improved since the implementation of the first commercial plant in 1984 (Lotker, 1991; Price et al., 2002). However, though the collector design and connection between the solar system and the PCS have been improved, some further improvements could still be implemented to reduce costs and increase efficiency. The main limitation to improving their competitiveness is the technology itself: the use of oil as a heat carrier medium between the solar field and the PCS, which entails a high-pressure drop in the oil circuit, limitation of the maximum temperature of the Rankine cycle, and O&M costs of the oil-related equipment. If the superheated steam required to feed the steam turbine in the power block were produced directly in the receiver tubes of the PTCs (i.e., DSG), the oil would be no longer necessary, and temperature limitation and environmental risks associated with the oil would be avoided (Ajona and Zarza, 1994). Figure 42.28 shows the overall scheme of a parabolic-trough power plant with DSG in the solar field. Simplification of overall plant configuration is evident when comparing Figures 42.24 and 42.28.

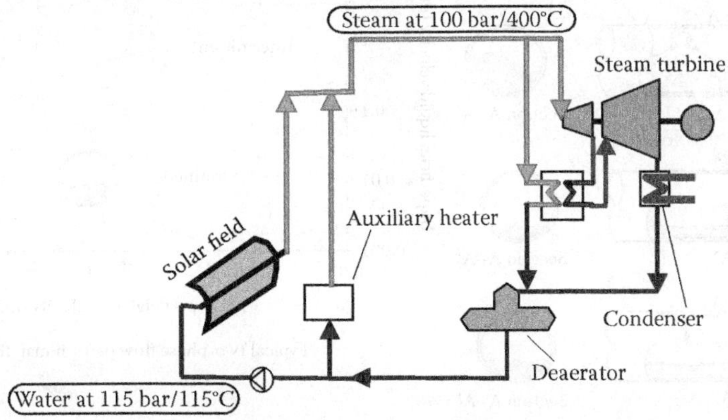

FIGURE 42.28
Simplified scheme of a parabolic-trough power plant with direct steam generation.

DSG has technical advantages that must be considered (Zarza et al., 1999):

- No danger of pollution or fire due to the use of thermal oil at temperatures of about 400°C
- Possibility of raising the maximum temperature of the Rankine cycle above 400°C, the limit imposed by the thermal oil currently used
- Reduction in the size of the solar field, thus reducing the investment cost
- Reduction in operation and maintenance-related costs, as thermal-oil-based systems require a certain amount of the oil inventory to be changed every year, as well as antifreeze protection when the air temperature is below 14°C

However, DSG presents certain challenges as a way to improve the current technology of parabolic-trough solar power plants, due to the two-phase flow (liquid water + steam) existing in the absorber tubes of the solar field evaporating section. The existence of this two-phase flow involves some uncertainties that must be clarified before a commercial plant making use of this technology can be built. Some of these uncertainties are

- Solar field control
- Process stability
- Stress in the receiver pipes
- Higher steam loss (leaks) than oil-based systems

Figure 42.29 shows the typical two-phase flow pattern in a horizontal pipe. As observed in Figure 42.29, four main flow patterns are possible, depending on the surface flow rates in the liquid and steam phases: bubbly, intermittent, stratified, and annular. The borders between adjacent flow patterns are not as well defined as they appear in Figure 42.29, but are rather separated by transition zones.

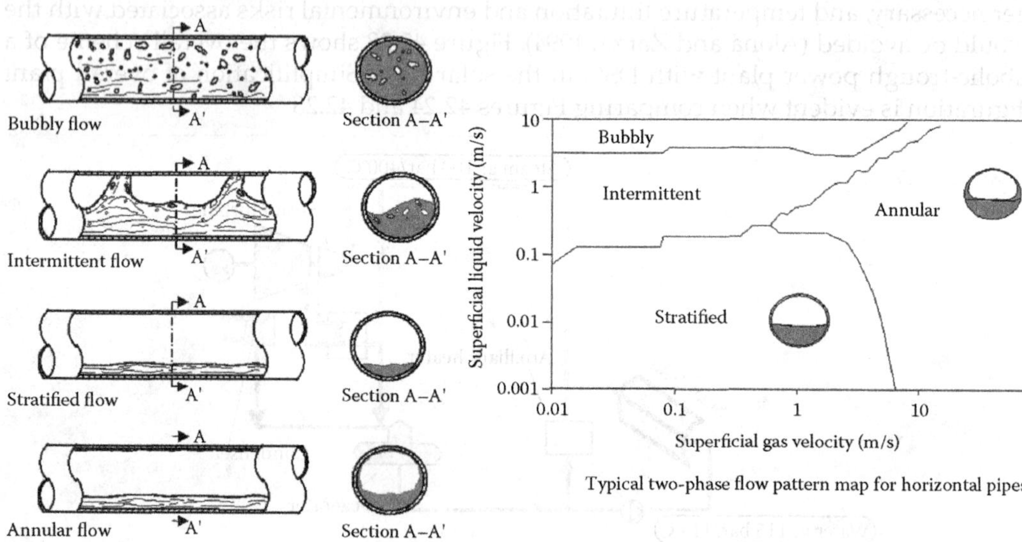

FIGURE 42.29
Two-phase flow configurations and typical flow pattern map for a horizontal receiver pipe.

In bubbly and intermittent flows, the steel absorber pipe inner wall is well wetted, thus avoiding dangerous temperature gradients between the bottom and the top of the pipe when it is heated from one side. The result is a good heat transfer coefficient all the way around the pipe because the liquid phase is not stratified.

In the stratified region (Goebel, 1997), the water is in the bottom of the absorber pipe, while the steam remains above the surface of the water. The result of this stratification is an uneven heat transfer coefficient around the pipe. Wetting of the bottom of the pipe is still very good and so is the heat transfer coefficient. But the cooling effect of the steam is poorer, and the heat transfer coefficient in the top section of the absorber pipe can be very low, resulting in a wide temperature difference of more than 100°C between the bottom and the top of the pipe in a given cross section when it is heated from one side. The thermal stress and bending from this steep temperature gradient can destroy the pipe. Figure 42.30 shows what happens in a cross section of the steel absorber pipe when it is heated underneath (Figure 42.30b, parabolic-trough concentrator looking upward) and from one side (Figure 42.30a, parabolic-trough concentrator looking at the horizon). The figure clearly shows how stratified flow can cause steep temperature gradients only when the vector normal to the aperture plane of the concentrator is almost horizontal (Figure 42.30a).

In the annular region, though there is partial stratification of water at the bottom of the pipe, there is a thin film of water wetting the upper part of the pipe. This film is enough to ensure a good heat transfer coefficient all the way around the pipe, thus avoiding dangerous thermal gradients that could destroy it. Typical absorber pipe cross sections in the stratified and annular regions are also shown in Figure 42.29.

Nevertheless, the technical problems due to water stratification inside the absorber pipes can be avoided if the feed-water mass flow is kept above a threshold level. There are three basic DSG processes, called once-through, injection, and recirculation. These three options require a solar field composed of long rows of PTCs connected in series to perform the

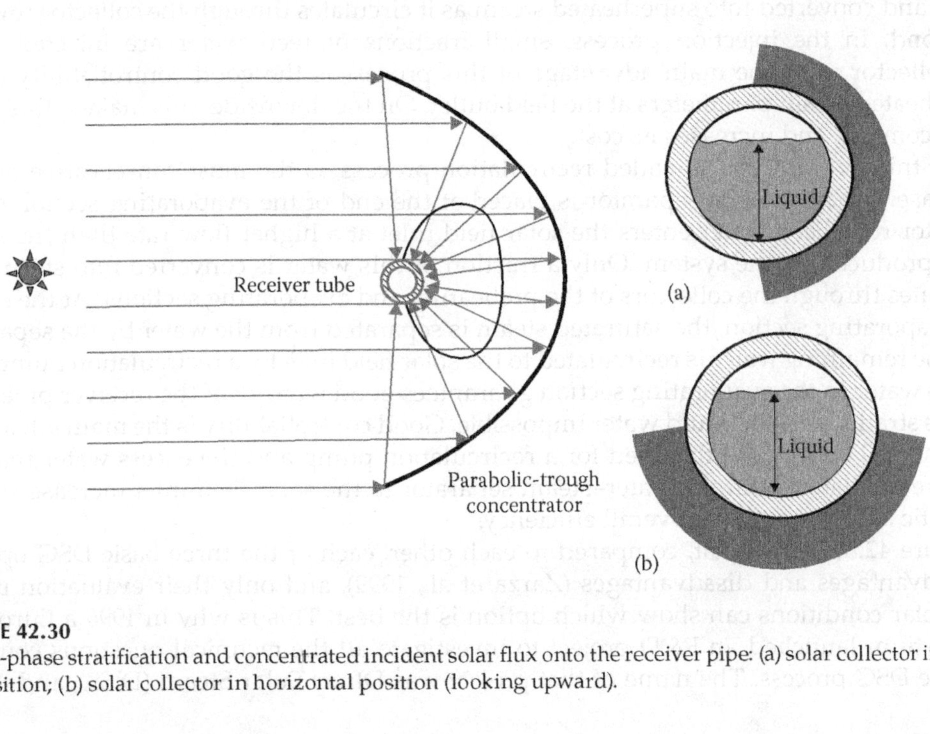

FIGURE 42.30
Liquid-phase stratification and concentrated incident solar flux onto the receiver pipe: (a) solar collector in vertical position; (b) solar collector in horizontal position (looking upward).

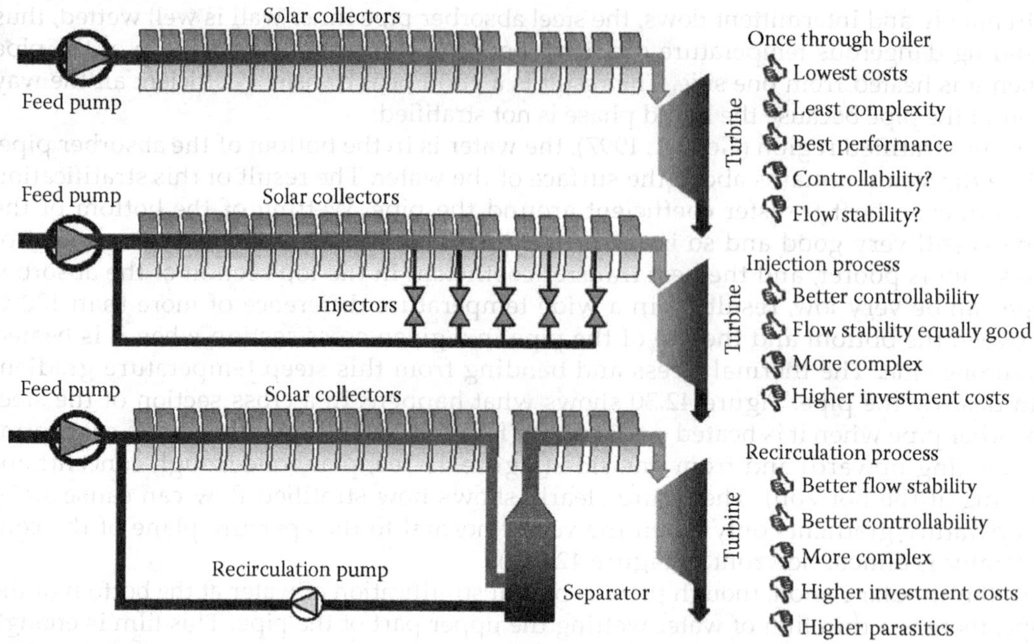

FIGURE 42.31
The three basic DSG processes.

complete DSG process: water preheating, evaporation, and steam superheating. Figure 42.31 summarizes the advantages and disadvantages of the three basic DSG options.

First, in the once-through process, all the feed water is introduced at the collector row inlets and converted into superheated steam as it circulates through the collector rows.

Second, in the injection process, small fractions of feed water are injected along the collector row. The main advantage of this process is the good controllability of the superheated steam parameters at the field outlet. On the downside, this makes the system more complex and increases its cost.

The third option, the so-called recirculation process, is the most conservative one. In this case, a water–steam separator is placed at the end of the evaporating section of the collector row. Feed water enters the solar field inlet at a higher flow rate than the steam to be produced by the system. Only a fraction of this water is converted into steam as it circulates through the collectors of the preheating and evaporating sections. At the end of the evaporating section, the saturated steam is separated from the water by the separator, and the remaining water is recirculated to the solar field inlet by a recirculation pump. The excess water in the evaporating section guarantees good wetting of the receiver pipes and makes stratification of liquid water impossible. Good controllability is the main advantage of this DSG option, but the need for a recirculation pump and the excess water that has to be recirculated from the water–steam separator to the solar field inlet increase system parasitic loads, penalizing overall efficiency.

Figure 42.31 shows that, compared to each other, each of the three basic DSG options has advantages and disadvantages (Zarza et al., 1999), and only their evaluation under real solar conditions can show which option is the best. This is why in 1996 a European consortium launched an R&D project to investigate all the technical questions concerning the DSG process. The name of that project was DIrect Solar Steam (DISS), and it was developed by partners belonging to all the sectors involved in this technology (i.e., electric utilities, industry, engineering companies, and research centers) in efficient collaboration with other European projects related to PTCs and STP plants (e.g., STEM, ARDISS, GUDE, EuroTrough, PRODISS).

The only DSG life-size test facility available in the world at the end of last century was designed and implemented at the PSA during the first phase of the DISS project to investigate the feasibility of DSG in PTCs under real solar conditions. DISS-phase I started in January 1996 and ended in November 1998, with the financial support of the European Commission. The second phase of the project, also partly funded by the European Commission, started in December 1998 and lasted 37 months.

Though the DISS solar field can be operated over a wide temperature/pressure range, the three main operating modes are as follows:

Solar Field Conditions	Inlet	Outlet
Mode 1	40 bar/210°C	30 bar/300°C
Mode 2	68 bar/270°C	60 bar/350°C
Mode 3	108 bar/300°C	100 bar/375°C

The DISS test facility accumulated more than 8000 h of operation until 2012. The experimental data gathered in the project and the simulation tools that were developed on the basis of test data provided enough information to evaluate and compare the three DSG basic operation modes (i.e., recirculation, injection, or once-through) or any combination of them. Water injectors at the inlet of the solar collectors and water pumps are provided with frequency converters for smooth, efficient speed control from 10% to 100%. This facility is still in operation and delivers very useful information for the short-term commercial deployment of DSG.

The PSA DISS test facility field was initially composed of a single row of 11 PTCs connected in series, with a total length of 550 m and 2700 m^2 of aperture surface. The collector row is connected to the balance of plant where the superheated steam delivered by the solar field is condensed and used as feed water for the solar field (closed-loop operation).

The DISS test facility implemented at the PSA in DISS-phase I was improved during the subsequent years, and it became a very flexible and powerful life-size test facility, suitable not only for investigating thermo-hydraulic aspects of DSG under real solar conditions, but also for evaluating optimized components and O&M procedures for commercial DSG solar plants. Figure 42.32 shows the present schematic diagram of the PSA DISS facility. A water–steam separator (marked TK-4 in Figure 42.32) connects the end of the evaporating section to the inlet of the steam superheating section when the facility is operated in recirculation mode. Two additional collectors were added to the first row in 2003, so it is currently 750 m long with an aperture area of 3822 m^2, with a nominal thermal power of 1.8 MW$_{th}$ and a maximum 400°C/100 bar superheated steam production of 1 kg/s.

The DISS project results proved the feasibility of the DSG process in horizontal PTCs, and important know-how was acquired by the project partners regarding the thermo-hydraulic parameters of the liquid water–steam flow in DSG solar fields within a wide range of pressures (30—100 bar). Experimental results were evaluated and complemented with results from lab-scale experiments and simulation tools. So several models developed in the past to calculate the pressure drop in pipes with two-phase flow were compared with the project's experimental results, and the models with the most accurate simulation results were identified. The proposed Chisholms (1980) and Friedel models proved to be the best match (Zarza, 2004).

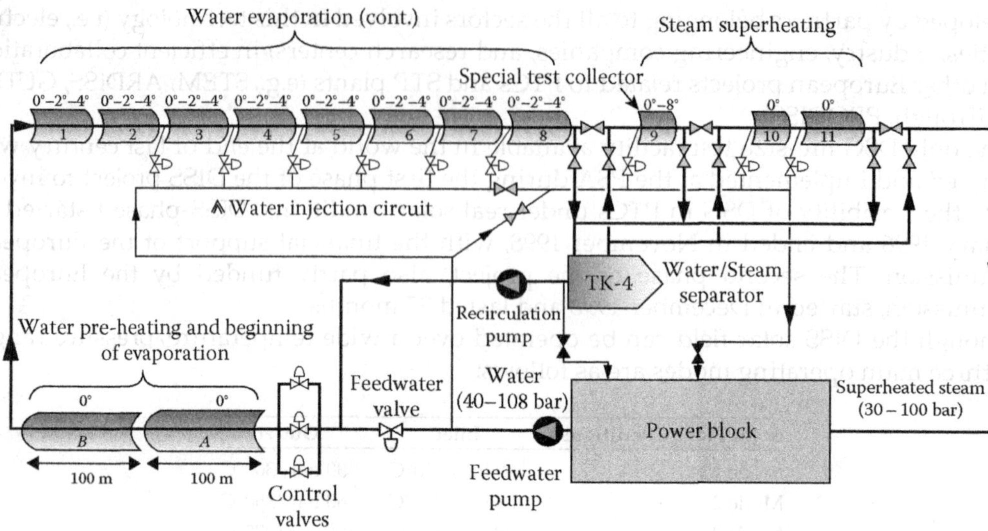

FIGURE 42.32
Schematic diagram of the PSA DISS test facility.

The good match between experimental data and simulation results obtained with finite-element models for the temperature profile in the DSG absorber pipes was another conclusion achieved in the DISS project. It was proven that the temperature gradients in the steel absorber pipes are within safe limits for a wide range of mass fluxes. Results concerning pressure drop and temperature gradients in the absorber pipes of the DISS collector row enabled accurate simulation and design tools for large DSG commercial solar fields to be developed (Eck et al., 2003) and was one of the major contributions of DISS to the development of the DSG technology.

Different possibilities (i.e., inner capillary structures and displacers) for enhancing heat transfer in the DSG absorber pipes were also investigated; however, experimental results showed that the absorber pipes do not need these devices to keep the temperature gradients within safe limits.

Several control schemes for once-through and recirculation operation modes were developed and evaluated with good results in the project. As initially expected, it was found that the temperature and pressure of the superheated steam produced by the solar field can be controlled more easily in recirculation mode, though an efficient control was achieved also for the once-through mode.

The influence of the inclination of the absorber pipes on the thermo-hydraulic parameters of the two-phase flow was investigated also. Tests at 30, 60, and 100 bar were performed in two positions, horizontal and tilted 4°. Though inclination of the absorber pipe reduced the stratified region, the test results obtained without inclination clearly showed that inclination of the absorber pipes is not required to guarantee sufficient cooling for the wide range of operating conditions investigated in the project.

Of the three DSG basic processes (i.e., injection, recirculation, and once-through), experimental results showed that recirculation is the most feasible option for financial, technical, and O&M-related parameters of commercial application. Test results showed good stability of the recirculation process even with a low recirculation ratio, thus making possible the use of small, cheap water–steam separators in the solar field.

The experience and know-how acquired in DISS was applied to the INDITEP project (2002–2005) in designing the first pre-commercial DSG solar power plant, in which the thermal energy delivered by a DSG solar field is used to feed a superheated-steam Rankine cycle (Zarza et al., 2004). The INDITEP project (Rueda et al., 2003) was the logical continuation of the DISS project, because the design and simulation tools developed in it for DSG solar fields were used in INDITEP, and most of the partners had also been involved in the previous project. INDITEP was promoted by a Spanish-German consortium of engineering companies, power equipment manufacturers, research centers, and businesses involved in the energy market: Iberdrola Ingeniería Consultoría (project Coordinator), CIEMAT, DLR, FLAGSOL GmbH, FRAMATONE, GAMESA Energía Servicios S.A., INITEC Tecnología S.A., Instalaciones Inabensa S.A., and ZSW. The European Commission also provided financial assistance.

Three basic requirements were defined for the design of this first pre-commercial DSG solar power plant:

1. The power block had to be robust and operable under flexible conditions in order to assure durability and reliability. Higher priority was therefore given to power block robustness and flexibility, while efficiency was considered less critical for this first DSG plant.
2. The plant must be small in order to limit the financial risk. Operating stability of a multi-row DSG solar field under uneven distribution of solar radiation and solar radiation transients must be proven.
3. The solar field must operate in recirculation mode because the DISS project demonstrated that the recirculation mode is the best option for commercial DSG solar fields (Eck and Zarza, 2002).

A 5.47 MW_e power block was selected for this plant to meet the first two requirements. Though, due to its small size, this power block is not very highly efficient, its robustness was guaranteed by the manufacturer with references from facilities already in operation. Figure 42.33 shows a schematic diagram of the power block design and its main parameters (Table 42.4).

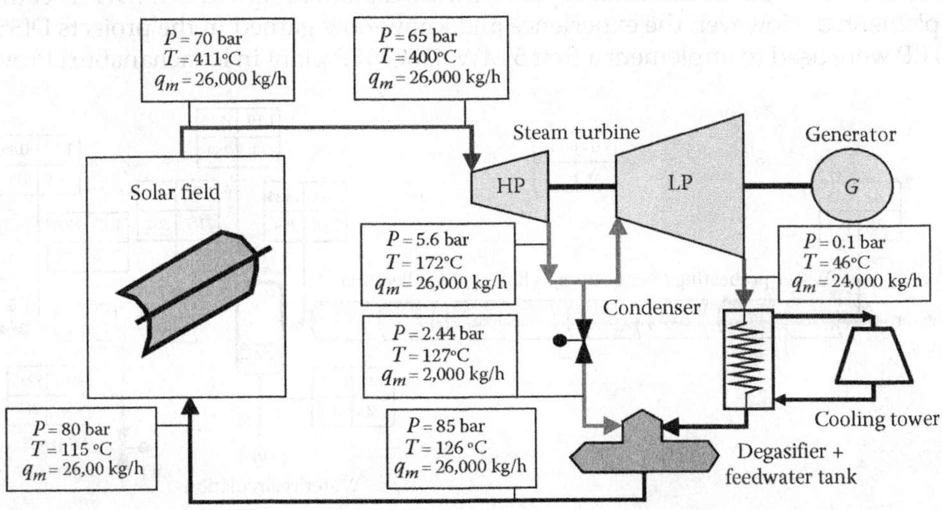

FIGURE 42.33
Schematic diagram of the power block designed for the DSG plant INDITEP.

TABLE 42.4
INDITEP Project 5 MW_e DSG Solar Power Plant Power Block Parameters

Manufacturer	KKK
Gross power (kW_e)	5,472
Net power (kW_e)	5,175
Net heat rate (kJ/kWh)	14,460
Gross efficiency (%)	26.34
Net efficiency (%)	24.9

The DSG solar field design consists of 70 ET-100 PTCs developed by the European EuroTrough consortium with financial support from the European Commission. This collector was chosen over the LUZ company's LS-3 collector, which is in use at the most recent SEGS plants erected in California (United States), because of its improved optical efficiency and lighter steel structure (Luepfert et al., 2003).

The seven parallel rows of ET-100 PTC axes are oriented north–south to collect the largest amount of solar radiation per year, even though the differences in solar field thermal energy output in winter and summer are more significant than with east–west orientation. Figure 42.34 shows the schematic diagram of a typical solar field collector row at design point. Every row is made up of 10 collectors: 3 collectors for preheating water + 5 collectors for evaporating water + 2 collectors for superheating steam. The end of the boiling section and the inlet of the superheating steam section in every row are connected by a compact water–steam separator, which in turn drains into a larger shared vessel. Water from the separator in every row goes to the final vessel from which it is then recirculated to the solar field inlet by the recirculation pump.

The temperature of the superheated steam produced in each row of collectors would be controlled by means of a water injector placed at the inlet of the last collector. The amount of water injected at the inlet of the last collector in every row is increased or decreased by the control system to keep the superheated steam temperature at the outlet of the row as close as possible to the set point defined by the operator.

Due to the lack of public financial support, the DSG plant designed in INDITEP could not be implemented. However, the experience and know-how gained in the projects DISS and INDITEP were used to implement a first 5 MW_e DSG STP plant in Kanchanaburi Province,

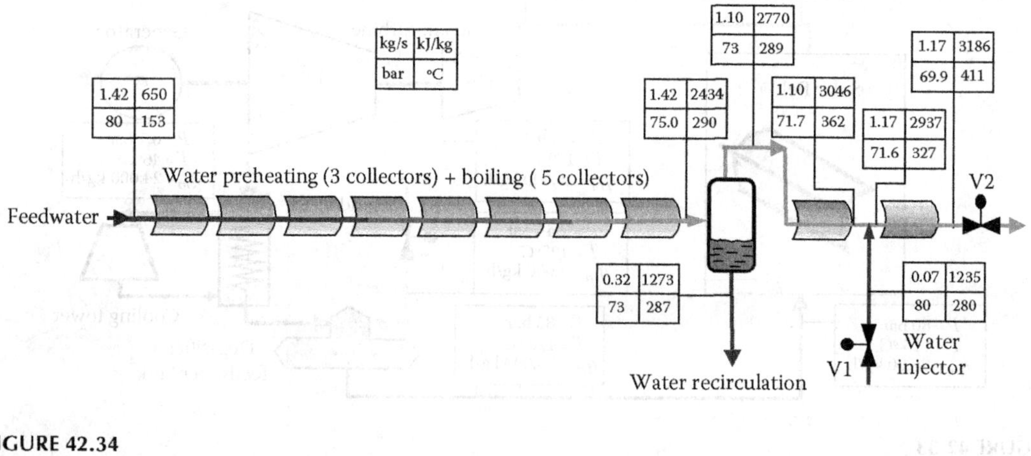

FIGURE 42.34
Scheme of a typical row of collectors in the DSG plant designed in INDITEP.

Thailand (http://www.solarlite.de/en/project_kanchanaburi.cfm) by the German company Solarlite with the collaboration and scientific support of DLR. The name of this plant is TSE-1, and it has a collector area of 45,000 m^2. The nominal solar field output steam temperature/pressure is 330°C/30 bar. This plant proved the commercial feasibility of the recirculation process for DSG in PTCs.

Keeping in mind the benefits of the once-through process when compared to the recirculation process, a German consortium composed of DLR and the company Solarlite launched in 2011 the project DUKE (Durchlaufkonzept – Entwicklung und Erprobung) with the financial support of the German government, the collaboration of PSA, and a duration of 3 years to study the commercial feasibility of the once-through process. The DISS test facility was improved in 2012–2013 within the framework of the project DUKE with the installation of three additional PTCs and new instrumentation. Experimental results delivered by DUKE will be essential for the commercial future of the DSG once-through process.

42.4.9 SEGS Plants and State-of-the-Art of Solar Power Plants with Parabolic-Trough Collectors

From a commercial standpoint, PTCs are the most successful technology for generating electricity with solar thermal energy. More than 50 commercial STP plants with PTCs were in operation in 2012, with a total output power of more than 2.5 GW$_e$. The first nine plants, called the SEGS (Solar Electricity Generating Systems) I–IX, which use thermal oil as the working fluid (HTF technology), were designed and implemented by the LUZ International Limited company from 1985 to 1990. All the SEGS plants are located in the Mojave Desert, northwest of Los Angeles, California. With their daily operation and over 2.2 million square meters of PTCs, SEGS plants are this technology's best example of commercial maturity and reliability. Their plant availability is over 98%, and their solar-to-electric annual efficiency is in the range of 14%–18%, with a peak efficiency of 22% (DeMeo and Galdo, 1997; Price et al., 2002). All SEGS plant configurations are similar to that shown in Figure 42.24, with only slight differences between one plant and another (Table 42.5).

The success of the SEGS plants was very important to get the support of bankers and investors for the commercial projects implemented in Spain and other countries later on. In fact, the configuration of the commercial STP plants implemented 20 years later than the SEGS plants was almost the same. The main difference between SEGS and more recent STP plants with parabolic troughs is the installation of a big thermal storage system using

TABLE 42.5

SEGS Plants Installed by LUZ International Limited

Plant	Net Capacity (MW$_e$)	Location	Inauguration
SEGS I	14	Dagget, CA	1984
SEGS II	30	Dagget, CA	1985
SEGS III	30	Kramer Jn, CA	1986
SEGS IV	30	Kramer Jn, CA	1986
SEGS V	30	Kramer Jn, CA	1987
SEGS VI	30	Kramer Jn, CA	1988
SEGS VII	30	Kramer Jn, CA	1988
SEGS VIII	80	Harper Lake, CA	1989
SEGS IX	80	Harper Lake, CA	1990

a mixture of molten salts, thus improving the plant dispatchability. The SEGS plant configuration was modified by LUZ from one plant to another in order to achieve higher efficiencies and lower O&M costs, so the only plant with a thermal energy storage system was SEGS I, which had a two-tank oil storage system. LUZ came to the conclusion that the auxiliary gas heater had to be implemented in the oil circuit because plant operating procedures were more difficult with the auxiliary heater installed in the water–steam circuit. LUZ research and development in the design and improvement of the solar collector field components and integration of the power block achieved higher plant efficiencies and provided a technology for large solar power plants, leading to a reduction in the cost of the solar field per m^2 to about 75% of the first SEGS plant, and solar system thermal efficiency was increased 8% (Harats and Kearney, 1989).

Figure 42.35 shows an aerial view of the SEGS III and IV plants. The layout chosen by LUZ for the solar fields is the so-called Central Feed, because it allows easy washing and maintenance access to all the collector rows. The power block (steam generator, steam turbine, electricity generator, condenser, etc.) building is located in the center of the collector field.

The SEGS IX plant, put into operation in 1990, was the last one installed by LUZ before its bankruptcy in 1991. Although LUZ had projects for another four plants to be erected in Harper Lake in a very advanced stage, they were cancelled due to the company's bankruptcy. LUZ had increased the net nominal power of the SEGS plants from 14 MW_e (SEGS I) to 80 MW_e (SEGS VIII and IX).

In SEGS plants VIII and IX (see Figure 42.24), there is no gas-fired steam reheater and the steam turbine has two stages working with steam at 371°C/104 bar and 371°C/17 bar, respectively. The thermal oil is heated in the solar collectors up to a temperature of approximately 390°C, and split into two parallel circuits, the steam generator and reheater. In the steam-generating circuit, the oil passes through a steam superheater, boiler, and preheater, generating steam at 371°C and 104 bar. This steam expands in the first stage of the turbine, passing to a reheater thermally fed by the other oil circuit, reheating steam to 371°C and 17 bar. During June, July, August, and September, the auxiliary gas boiler is put into operation to keep the turbine working at full load during peak-demand hours. For the rest of the time, the steam turbine is mainly driven by the solar system only (Table 42.6).

FIGURE 42.35
Aerial view of the SEGS III and IV plants.

TABLE 42.6

Basic Characteristics of the SEGS I–IX Plants

	SEGS I	SEGS II	SEGS III	SEGS IV	SEGS V	SEGS VI	SEGS VII	SEGS VIII	SEGS IX
Starting up	December 1984	December 1985	December 1986	December 1986	October 1987	December 1988	December 1988	December 1989	September 1990
Investment (M$)	62	96	101	104	122	116	117	231	
Electricity yearly production (MWh)	30,100	80,500	92,780	92,780	91,820	90,850	92,646	252,750	256,125
Estimated life (years)	20	25	30	30	30	30	30	30	30
No. of stages of the turbine	1	2	2	2	2	2	2	2	2
Solar steam (*P*, *T*) (bar, °C)	248/38	300/27	327/43.4	327/43.4	327/43.4	371/100	371/100	371/104	371/104
Steam with gas (*P*, *T*) (bar, °C)	417/37	510/104	510/104	510/104	510/100	510/100	510/100	371/104	371/104
Efficiency in solar mode (%)	31.50	29.40	30.60	30.60	37.70	37.50	37.50	37.60	37.60
Efficiency with gas (%)	—	37.30	37.40	37.40	37.40	39.50	39.50	37.60	37.60
Solar field									
Collector type	LS-1/LS-2	LS-1/LS-2	LS-2	LS-2	LS-2	LS-2	LS-3	LS-3	LS-3
Aperture area (m^2)	82,960	188,990	230,300	230,300	233,120	188,000	194,280	464,340	483,960
Working temperature	279	321	349	349	349	390	390	390	390
Inlet/outlet collector field temperature, (°C)	241/307	248/320	248/349	248/349	248/349	293/393	293/390	293/390	293/390
Type of oil	ESSO 500	M-VP1	M-VP1	M-VP1	M-VP1	M-VP1	M-VP1	M-VP1	M-VP1
Oil volume (m^3)	3,217	379	403	404	461	372	350		

LUZ Industries developed three generations of PTCs, called LS-1, LS-2, and LS-3. The LS-1 and LS-2 designs are conceptually very similar. The main differences are the overall dimensions. The parabolic reflectors are glass panels simply screwed to helicoidal steel tube frames that provide the assembly with the required integrity and structural stiffness.

The LS-3 collector is twice as long as the LS-2, with a parabola 14% wider, which reduces the number of flexible connections, local control units, temperature sensors, hydraulic drives, and similar equipment by more than half. However, the LS-3 design represents a change in collector philosophy, more than a change in scale. While the LS-2 model used a torque tube to provide the required stiffness, the LS-3 is made of a central steel space frame that is assembled on-site with precision jigs. The result of this innovation is a lighter and more resistant structure, with highly accurate operation in heavy winds (Table 42.7).

The electricity produced by the SEGS plants is sold to the local utility under individual 30-year contracts for every plant. To optimize the profitability of these plants, it is essential to produce the maximum possible energy during peak-demand hours, when the electricity price is the highest. The gas boilers can be operated for this, either to supplement the solar field or alone. Nevertheless, the total yearly electricity production using natural gas was limited at that time by the Federal Commission for Energy Regulation in the United States to 25% of the overall yearly production.

Peak-demand hours are when there is the most electricity consumption, and therefore, the tariff is the highest. Off-peak and super off-peak hours are when electricity consumption is low, and the electricity price is therefore also lower. At present, 16% of the SEGS plants' annual net production is generated during summer peak-demand hours, and the revenues from this are on the order of 55% of the annual total. These figures show how important electricity generated during peak-demand hours is for the profitability of these plants.

TABLE 42.7

Characteristics of the LS-1, LS-2, and LS-3 Collectors

	LS-1	LS-2	LS-3
Solar tracking accuracy (°)	0.10	0.10	0.10
Maximum wind velocity to operate (km/h)	56	56	56
Steel structure based on	Steel tube	Steel tube	Space frame
Selective coating of absorber tubes	Black Cr	Black Cr	Cermet
Absorptivity/transmissivity (%)	94/94	94/95	96/95
Emissivity (%)	39 (300°C)	24 (300°C)	18 (350°C)
Focal distance of the solar concentrator	0.68	1.49	1.71
Aperture angle	85	80	80
Reflectivity (%)	94	94	94
Trough aperture (m)	2.5	5	5.76
Absorber steel pipe outer diameter (mm)	42	70	70
Geometric concentration	19	23	26
Overall length (m)	50.2	47.1	99.0
Distance between supports (m)	6.3	8.0	12.0
Mirrors surface per collector (m^2)	128	235	545
Maximum working temperature (°C)	307	350	390
Distance between parallel rows (m)	7	12.5/15	17
Intercept factor (%)	87	89	93

Thanks to the continuous improvements in the SEGS plants, the total SEGS I cost of \$0.22 per kWh_e for electricity produced was reduced to \$0.16 per kWh_e in the SEGS II and down to \$0.09 per kWh_e in SEGS IX (Kearney and Cohen, 1997).

The other important contribution to the profitability of the SEGS plants was the favorable tax laws it took advantage of. The large tax rebate was a crucial factor in the economic feasibility of the SEGS plants. The eventual reduction in these tax exemptions was overcome by the considerable cost reduction from one plant to another achieved by LUZ. The significant reduction in both fossil fuel prices and tax exemptions made it impossible for the SEGS plants to maintain the profit margin they had had at the beginning, resulting in the bankruptcy of the company and the cancellation of the planned erection of more SEGS plants.

Another factor that led to the bankruptcy of LUZ was the short time in which one SEGS plant had to be fully implemented to obtain tax benefits. A plant had to be installed in less than 12 months. For the last 80 MW_e plant (SEGS X), this short leeway required around-the-clock construction work, which rocketed the investment cost and made it impossible for a SEGS X to be financially successful, and its construction was therefore cancelled in the middle of 1991 as a consequence of LUZ's bankruptcy (Lotker, 1991).

Though LUZ achieved a significant cost reduction in the SEGS plants, the current cost of electricity produced by modern parabolic-trough solar power plants is higher, and it must still be further reduced before it can become competitive with conventional power plants and PV plants (Pitz-Paal et al., 2005). Without tax incentives or an adequate premium for electricity generated by this type of plants, they are not profitable, because current investment and O&M costs demand a public support strategy for commercial development. At present, the direct capital cost is about three times that of a fossil-fueled power plant, and the generating cost is about three times higher. Annual thermal efficiency (solar-to-useful thermal energy delivered) of commercial parabolic-trough solar fields is about 50%. This means a yearly savings of about 0.45 ton of CO_2 emissions and 0.1 ton of fossil fuel per square meter of PTC (Geyer, 2002).

In spite of their environmental benefits, there are some barriers to the commercial use of this technology. The main barriers at present are the high investment cost (\$3500–\$6000 per kW, depending on plant size and thermal storage capacity) and the minimum size of the power block required for high thermodynamic efficiency. However, these barriers are shared by all the STP technologies currently available (i.e., CRSs, Dish Stirling systems, and LF concentrator systems) (Romero et al., 2004). The high investment cost can be compensated with public incentives in the form of tax credits or favorable feed-in tariffs. The favorable feed-in tariff implemented in Spain in 2007 and the tax credits and loan guarantees in the United States for STP plants opened the door to many commercial projects, thus achieving a total installed power of more than 2.5 GW_e with PTCs at the end of 2012.

Since the size limit imposed in Spain for STP plants (i.e., 50 MW_e) had no technical reasoning behind, several commercial projects with unit powers higher than 100 MW_e were promoted in other countries at the beginning of the second decade in this century. The SOLANA plant promoted by Abengoa, Arizona, with a unit power of 280 MW_e and a 6 h thermal storage system with molten salts, is a good example of these large STP plants.

At the time of writing the technology commercially available for parabolic-trough power plants is the HTF technology, which uses oil as the heat carrier between the solar field and the power block, while thermal storage systems with molten salts in two tanks is the preferred option. The promotion of many commercial projects since 2007 has led to the development of many new collector designs (URSSATrough, SenerTrough, SkyTrough, etc.) with good mechanical and geometrical quality and lower and lower manufacture and assembly costs.

FIGURE 42.36
The UltimateTrough prototype installed in a commercial solar power plant in Southern California. (Courtesy of Flabeg.)

These new collector designs have been specially developed for large solar fields. Although most of the new collector designs developed in the period 2007–2013 had similar dimensions and features (150 m length, 5.76 m parabola width, 1.7 m focal distance), very innovative designs with much bigger aperture area have been developed keeping in mind the future installation of very large solar fields. The UltimateTrough design is the best example of new collectors with very large aperture area. With an aperture of 1689 m^2, a parabola width of 7.5 m, and a length of 240 m, UltimateTrough was specially designed for STP plants with a unit power of 100 MW_e or more, although it can be used for smaller plants also. Figure 42.36 shows the UltimateTrough collector prototype installed in an STP plant in Southern California for evaluation.

More than 2 GW_e were installed in the period 2007–2013 with PTCs in Spain, the United States, and Mediterranean and North Africa region, and several countries (Morocco, South Africa, and others) had launched national programs to install STP plants and promote solar thermal electricity production.

Most of the commercial STP plants installed in the period 2007–2013 with PTCs have the configuration showed either in Figure 42.24 (plants without thermal storage) or Figure 42.27 (plants with molten-salt thermal storage systems), and only 3 plants had a parabolic-trough solar field integrated in the bottoming cycle of a combined-cycle gas-fired power plant. This configuration is called the integrated solar combined-cycle system (ISCCS) configuration. Though the contribution of the solar system to the overall plant power output is small (10%–15% approx.) in the ISCCS configuration, it seems to be a good approach to market penetration in some countries, which is why the World Bank is promoting ISCCS plants in developing countries (Geyer et al., 2003). Figure 42.37 shows the schematic diagram of a typical ISCCS plant.

The first commercial plant with PTCs in Europe was the ANDASOL-I plant (Aringhoff et al., 2002a,b), a SEGS-like plant with a 6 h molten-salt thermal storage system. Figure 42.38 is an aerial view of ANDASOL-I.

The financial crisis in 2011 made some countries change their mind concerning the promotion of STP plants. This was the case of Spain, where the legal framework implemented in 2007 was significantly modified in 2012 and 2013 to reduce the incomes of the STP plants owners, and the feed-in tariff was cancelled for new plants. Studies performed

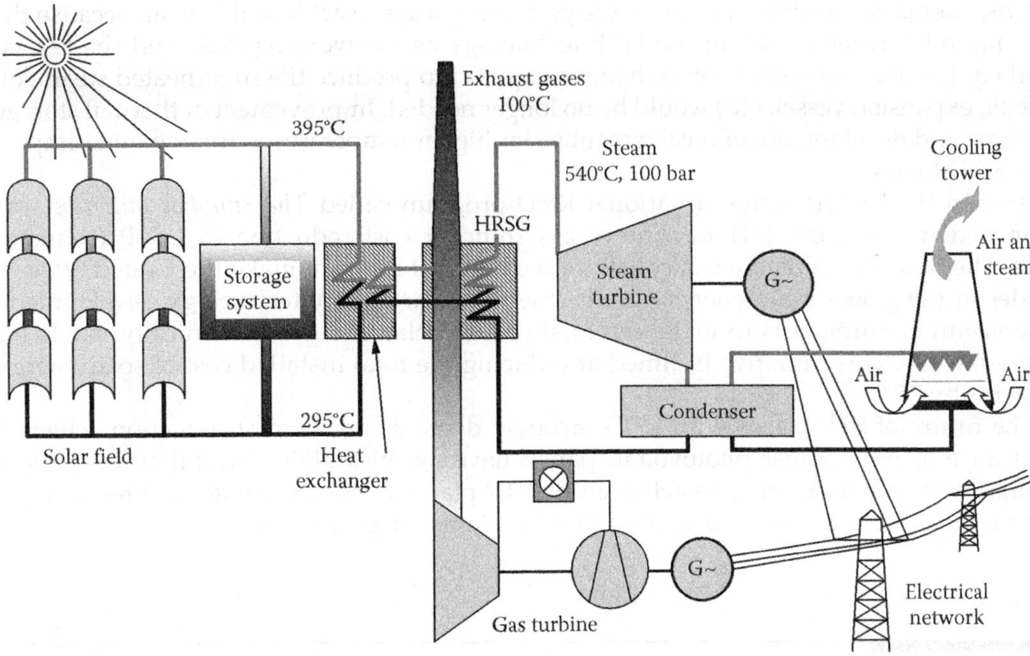

FIGURE 42.37
Schematic diagram of a typical ISCCS (integrated solar combined-cycle system) plant.

FIGURE 42.38
Aerial view of the Spanish ANDASOL-I plant.

in the second decade of this century have pointed out the possibility of achieving a significant midterm cost reduction in parabolic-trough technology (Enermodal, 1999; Sargent & Lundy, 2003; Kearney and ESTELA, 2010). Mass production and component improvement through R&D would lead to an electricity cost fully competitive with conventional power plants in 2025. It is therefore clear that parabolic-trough technology must seek ways to become more competitive with conventional power plants. The European Solar Thermal Electricity Association (ESTELA) issued a strategic research agenda in 2013 with the recommended R&D topics to speed up the cost reduction of electricity produced by STP plants. This document is available at www.estelasolar.eu. Within the potential improvements for cost reduction, DSG of high-pressure/high-temperature superheated steam in the receiver pipes seems to be one of the promising ways to achieve this goal, because the thermal oil currently used in the HTF technology as the working fluid and the associated equipment (oil/water heat exchanger required to produce the superheated steam, oil circuit, expansion vessel, etc.) would be no longer needed. Improvement of thermal storage systems and development of receivers tubes for higher temperatures are two other important R&D topics.

Also in the United States, a national R&D program called *The SunShot Initiative* was launched in February 2011 to achieve a significant cost reduction in CSTP technologies. The SunShot Initiative is a collaborative national effort to make the United States a leader in the global clean energy race by fueling solar energy technology development. The vision of SunShot is to make the total cost of solar energy economically viable for everyday use. This initiative is aimed at reducing the total installed cost of solar energy systems by 75%.

The future of STP plants with PTCs strongly depends on the cost reduction achieved for future plants, because photovoltaic panels have significantly reduced their cost, and a similar cost reduction must be achieved by STP plants to compete with that technology, even taking into account the benefit of the STP plants' dispatchability.

42.5 Linear Fresnel Reflectors

LF reflectors are composed by an array of linear (or slightly bent) mirror strips that independently move and collectively focus on absorber lines suspended from elevated towers. Reflective segments are close to the ground and can be assembled in a compact way up to 1 ha/MW. This technology aims at achieving the performance of parabolic troughs with lower costs. They are characterized by a fixed linear focus where the absorber is static (Mills, 2004). However, optical efficiency is lower than that of parabolic troughs due to a higher impact of the incidence angle and the cosine factor. Consequently, operating temperature at the working fluid is usually lower, typically between 150°C and 350°C. By this reason, LF technology has been historically applied to generate saturated steam via direct in-tube steam generation and use into ISCCS or in regenerative Rankine cycles, though current R&D is aiming at higher temperatures above 400°C (Platzer, 2009).

LF reflectors typically make use of lower cost non-vacuum thermal absorbers where the stagnant air cavity provides significant thermal insulation, light reflector support structures close to the ground, low-cost flat float glass reflector, and low-cost manual cleaning, because the reflectors are at human height (Kalogirou, 2004). The LFs also have much better ground utilization, typically using 60%–70% of the ground area compared to about 33% for a trough system and lower O&M costs due to more accessible reflectors.

42.5.1 Historical Evolution of Linear Fresnel Reflector Systems

After some pioneering experiences (Francia, 1968; Di Canio et al., 1979), the first serious development undertaken on the compact LF reflector system was proposed at the University of Sydney in 1993. The concept is composed by a single field of reflectors together with multiple linear receivers (see Figure 42.39). Each reflector is able to change

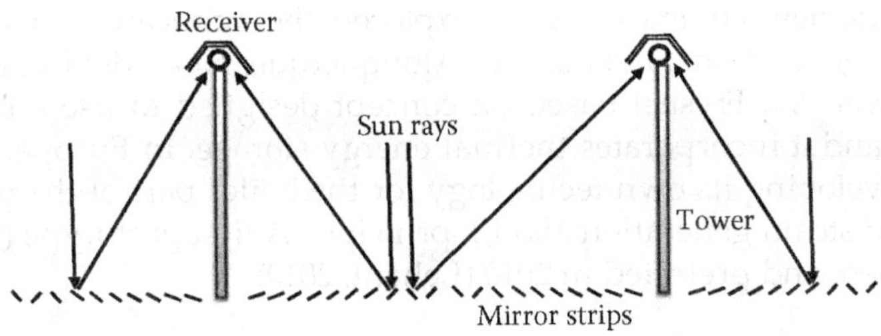

FIGURE 42.39
Scheme of compact LF system with a multi-tower array and dynamic aiming strategy of mirror strips.

their focal point from one receiver to another during the day in order to minimize shading and shadow losses in the dense reflector field. This system covers about 71% of the ground compared with 33% for parabolic-trough systems (Mills and Morrison, 1999). In 2000, the company Solarmundo built a 2400 m^2 LF prototype collector field with such a technology at Liege, Belgium, but test results were not reported. Later the company moved to Germany and was renamed Solar Power Group (SPG). SPG signed an exclusivity cooperation agreement with DSD Industrieanlagen GmbH (renamed to MAN Ferrostaal Power Industry in 2005). A 800 kW LF pilot operating at 450°C has already been tested in the PSA, Spain (Bernhard et al., 2008; Hautmann et al., 2009).

Back in Australia, in early 2002, a new company, Solar Heat and Power Pty Ltd. (SHP), made extensive changes to the engineering design of the reflectors to lower cost and has become the first to commercialize LF technology. SHP initiated in 2003 for Macquarie Generation, Australia's largest electricity generator, a demonstration project of 103 MW_{th} (approximately 39 MW_e) plant with the aim of supplying preheat to the coal-fired Liddell power station. Phase 1 of the project, completed in 2004, resulted in a 1350 m^2 segment not connected to the coal-fired plant and was used to trial initial performance, and it first produced steam at 290°C in July 2004. The expansion to 9 MW_{th} was completed by 2008. Activities of the company moved to the United States and were continued by Ausra, Palo Alto. Ausra established a factory of components, tubular absorbers, and mirrors in Las Vegas and built the Kimberlina 5 MW demonstration plant in Bakersfield at the end of 2008. In 2010, Ausra was purchased by AREVA, which is presently committed to the commercial deployment of this technology.

The third technology player after SPG/MAN and AREVA is the German company NOVATEC Solar, formerly NOVATEC Biosol. The technology of NOVATEC is based upon its collector Nova-1 aimed to produce saturated steam at 270°C. They have developed a serial production factory for prefabricated components, a 1.4 MW small commercial plant, PE-1, in Puerto Errado, Murcia, Spain, which has been grid connected since March 2009, and a second 30 MW commercial plant, PE-2 (with a mirror surface of 302,000 m^2), also built in Murcia, Spain. NOVATEC is promoting 50 MW plants mixing PTC and LF fields, where LF provides preheating and evaporation and PTC field takes over superheating. The company claims that this hybridization results in 22% less land use and higher profitability. In March 2011, ABB acquired a 35% shareholding in Novatec Solar. In September 2011, Novatec Solar claimed that its technology has successfully generated superheated steam at temperatures above 500°C at its 1.4 MW demonstration plant in Murcia, Spain, by implementing a receiver containing vacuum absorber (Selig, 2011).

In last few years, new companies have explored the application of Fresnel technologies for electricity generation. SkyFuel Inc., Albuquerque, New Mexico, is developing the Linear Power Tower™, a Fresnel based on concept designed to use a high-temperature molten-salt HTF, and it incorporates thermal energy storage. In Europe, the French company CNIM is developing its own technology for the boiler part of the plant focusing on direct superheated steam generation and LF principle. A first prototype (800 m^2) has been built, commissioned, and pretested in 2010 (Lehaut, 2010).

42.5.2 Future Technology Development and Performance Trends

Even though some solid commercial programs are underway on LF, still it is early to have consolidated performance data with respect to electricity production. The final optimization would integrate components development to increment temperature of operation and possible hybridization with other STP systems like parabolic troughs.

There are many possible types of receivers, including evacuated tube and PV modules, but the most cost-effective system seems to be an inverted cavity receiver. In the case of SHP technology, the absorber is a simple parallel array of steam pipes at the top of a linear cavity, with no additional redirection of the incoming light from the heliostats to minimize optical losses and the use of hot reflectors. In the case of SPG technology, the absorber is a single tube surmounted by a hot nonimaging reflector made of glass, which must be carefully manufactured to avoid thermal stress under heating and exhibit some optical loss. Both systems can produce saturated steam or pressurized water. At present, AREVA and NOVATEC are looking for new absorbers able to work at temperatures above 450°C. By 2015, according to developers, LF can be expected to be operating with superheated steam at 500°C yielding an efficiency improvement of up to 18.1% relative to current saturated steam operation at 270°C (Kearney and ESTELA, 2010).

For reflectors, automation is a key issue that has been demonstrated by NOVATEC. Additional effort should be given to the optimized demonstration of multitower arrays to maximize ground coverage ratios. However, it is the lack of reliable information regarding annual performance and daily evolution of steam production that should be targeted as a first priority. Still some concerns remain regarding the ability to control steam production, because of the pronounced effect of cosine factor in this kind of plants. This dynamic performance would also affect the potential integration with other STE systems such as PTC or CRS. Until now, most comparative assessments vis-à-vis parabolic troughs are not economically conclusive, revealing the need to use much larger fields to compensate lower efficiencies. In order to achieve break-even costs for electricity with current LF technology, the cost target for the Fresnel solar power plants needs to be about 55% of the specific costs of parabolic-trough systems (Dersch et al., 2009).

42.6 Central-Receiver Solar Thermal Power Plants

CRSs with large heliostat fields and solar receivers located on top of a tower are deploying the first generation of grid-connected commercial plants from 2007. The CRS power plant technology can be considered as sufficiently mature after the pioneering experience of several 0.5–10 MW pilot plants in the early 1980s and the subsequent improvement of such key components as heliostats and solar receivers in many later projects merging international collaboration during the past 20 years. Solar-only plants like Gemasolar and PS10 and hybrid configurations like SOLGAS, CONSOLAR, or SOLGATE have provided a portfolio of alternatives that have led to the first scaled-up plants for the period 2007–2014. The first small 10–20 MW projects, still non-optimized, already reveal a dramatic cost reduction over previous estimates and provide a path for a realistic LEC milestone of $0.08 per kWh by 2030.

In power towers, incident sunrays are tracked by large mirrored collectors (heliostats), which concentrate the energy flux onto radiative/convective heat exchangers called solar receivers, where energy is transferred to a thermal fluid. After energy collection by the solar subsystem, the thermal energy conversion to electricity is quite similar to fossil-fueled thermal power plants and the earlier-described parabolic-trough system power block.

Reflective solar concentrators are employed to reach the temperatures required for thermodynamic cycles (Mancini et al., 1997). In power towers or CRSs, the solar receiver is mounted on top of a tower, and sunlight is concentrated by means of a large paraboloid that is discretized into a field of heliostats (Figure 42.40). CRSs have a high potential for midterm cost reduction of electricity produced since there are many intermediate steps between their integration in a conventional Rankine cycle up to the higher exergy cycles using gas turbines at temperatures above 1300°C, leading to higher efficiencies and throughputs.

The typical optical concentration factor ranges from 200 to 1000. Because of economy of scale, 10–100 MW plant sizes are chosen, even though advanced integration schemes are claiming the economics of smaller units as well (Romero et al., 2000a,b). The high solar flux incident on the receiver (averaging between 300 and 1000 kW/m^2) enables operation at relatively high temperatures of up to 1000°C and integration of thermal energy into more efficient cycles in a step-by-step approach. CRS can be easily integrated in fossil plants for hybrid operation in a wide variety of options and has the potential for generating electricity with high annual capacity factors through the use of thermal storage. With storage, CRS plants are able to operate over 4500 h/year at nominal power (Kolb, 1998).

FIGURE 42.40
Aerial view of the first commercial solar towers in the world. PS20 with 20 MW at the front and the PS10 with 11 MW at upper left. Heliostat field layout is formed by hundreds of tracking mirrors focusing concentrated light onto the receiver aperture. PS20 and PS10 have been developed by Abengoa Solar and are located in Sanlucar la Mayor, Spain. (Courtesy of Abengoa Solar, Sanlucar la Mayor, Spain.)

42.6.1 Technology Description

A solar power tower, or CRS, plant may be described in terms of the following subsystems:

- Collector system, or heliostat field, created with a large number of two-axis tracking units distributed in rows.
- Solar receiver, where the concentrated flux is absorbed. It is the key element of the plant and serves as the interface between the solar portion of the plant and the more conventional power block.
- Heat exchanger system, where an HTF may be used to carry the thermal energy from the receiver to the turbine.
- Heat storage system, with which system dispatchability is ensured during events like cloud passages and can adapt to demand curves.
- Fossil fuel backup for hybrid systems with a more stable output.
- Power block, including steam generator and turbine alternator.
- Master control, UPS, and heat rejection systems.

A detailed description and historical perspective of all the subsystems would be excessive, given the large number of configurations and components tested to date. Because of their higher temperatures, CRSs have been able to make use of a diversity of thermal fluids, such as air, water–steam, molten nitrate salt, and liquid sodium. In general, the components that impact the most on investment cost are the heliostat field, the tower receiver system, and the power block. The heliostat field and solar receiver systems distinguish solar thermal tower power plants from other CSP plants and are therefore given more attention in the following. In particular, the heliostat field is the single factor with the most impact on plant investment, as seen in Figure 42.41. Collector field and power block together represent about 72% of the typical solar-only plant (without fossil backup) investment, of which heliostats represent 60% of the solar share. Even though the solar receiver impacts the capital investment much less (about 14%), it can be considered the most critical subsystem

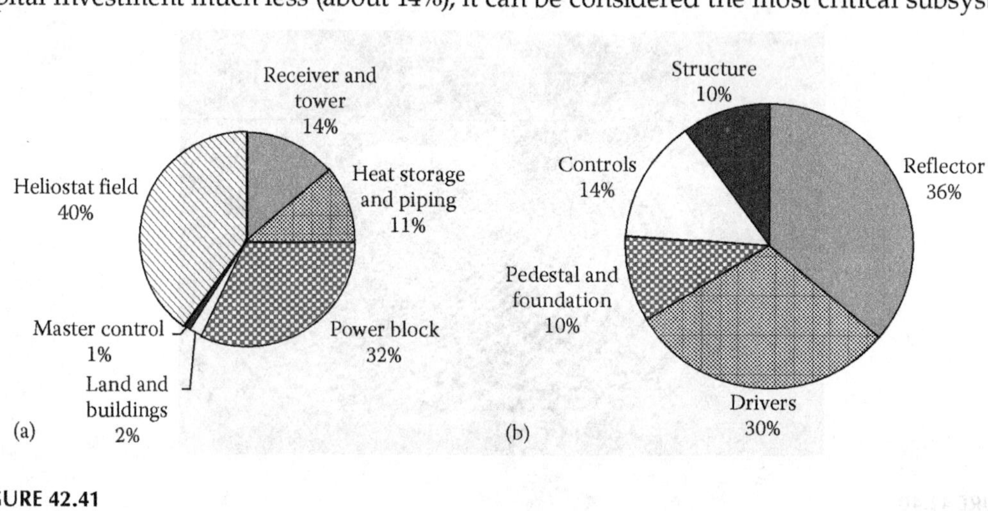

FIGURE 42.41
(a) Investment costs breakdown for a CRS plant, only solar. As it can be observed, the heliostat field and the power block are the most impacting subsystems on plant investment. (b) Breakdown of production cost for a single heliostat distributed among its main components. Reflector and tracking mechanism are in this case the most capital intensive components.

in terms of performance, since it centralizes the entire energy flux exchange. The largest heliostat investment is the drive mechanism and reflecting surface, which alone are almost 70% of the total.

42.6.1.1 Heliostat and Collector Field Technology

The collector field consists of a large number of tracking mirrors, called heliostats, and a tracking control system to continuously focus direct solar radiation onto the receiver aperture area. During cloud passages and transients, the control system must defocus the field and react to prevent damage to the receiver and tower structure.

Heliostats fields are characterized by their off-axis optics. Since the solar receiver is located in a fixed position, the entire collector field must track the sun in such a way that each and every heliostat individually places its surface normal to the bisection of the angle subtended by sun and the solar receiver. Figure 42.42 shows the variability of elevation angles in a heliostat field and identifies the elevation angle. The geometrical definition of the inclination angle n of a single heliostat is a function of the tower height, its distance from the tower, and the incidence angle of the sun. Assuming z_S is the heliostat vertical dimension and z_T is the geometrical tower height above ground, the so-called optical tower height may be defined as the elevation of the center of the receiver aperture area above the pivot point of the heliostat ($z_T - z_S/2$).

$$\psi = h + n - 90° \ (°) \tag{42.38}$$

$$90° - n = \arctan\left[\frac{(z_T - (z_S/2))}{x}\right] + \psi \ (°) \tag{42.39}$$

$$n = \frac{180° - h - \arctan[(z_T - (z_S/2))/x]}{2} \ (°) \tag{42.40}$$

Heliostat field performance is defined in terms of the optical efficiency, which is equal to the ratio of the net power intercepted by the receiver to the product of the direct insolation and the total mirror area. The optical efficiency includes the cosine effect, shadowing, blocking, mirror reflectivity, atmospheric attenuation, and receiver spillage (Falcone, 1986).

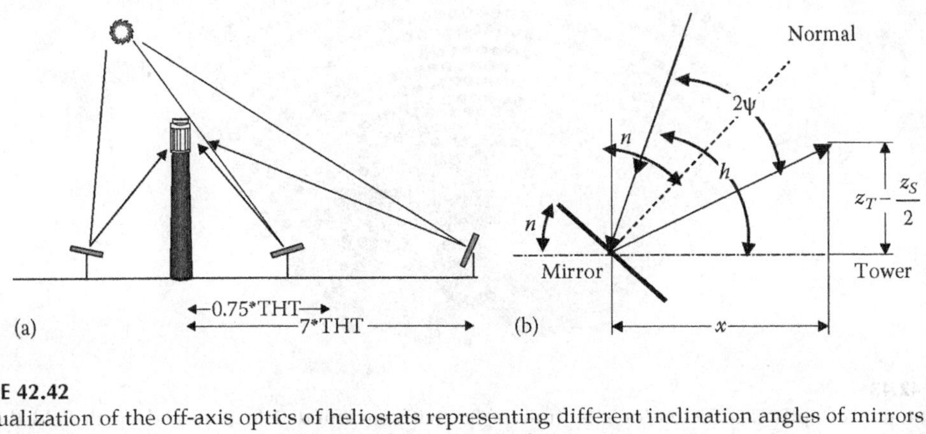

FIGURE 42.42
(a) Visualization of the off-axis optics of heliostats representing different inclination angles of mirrors located in a heliostat field. (b) Geometrical definition of elevation angle n.

Because of the large area of land required, complex optimization algorithms are used to optimize the annual energy produced by unit of land, and heliostats must be packed as close as possible so the receiver can be small and concentration high. However, the heliostats are individual tracking reflective Fresnel segments subject to complex performance factors, which must be optimized over the hours of daylight in the year, by minimizing the cosine effect, shadowing and blocking, and receiver spillage. Since the reflective surface of the heliostat is not normal to the incident rays, its effective area is reduced by the cosine of the angle of incidence ψ; the annual average cos ψ varies from about 0.9 at two tower heights north of the tower to about 0.7 at two tower heights south of the tower. Of course, annual average cosine is highly dependent on site latitude. Consequently, in places close to the equator, a surround field would be the best option to make best use of the land and reduce the tower height. North fields improve performance as latitude increases (south fields in the Southern Hemisphere), in which case, all the heliostats are arranged on the north side of the tower. Representative surround and north collector field configurations are depicted in Figure 42.43.

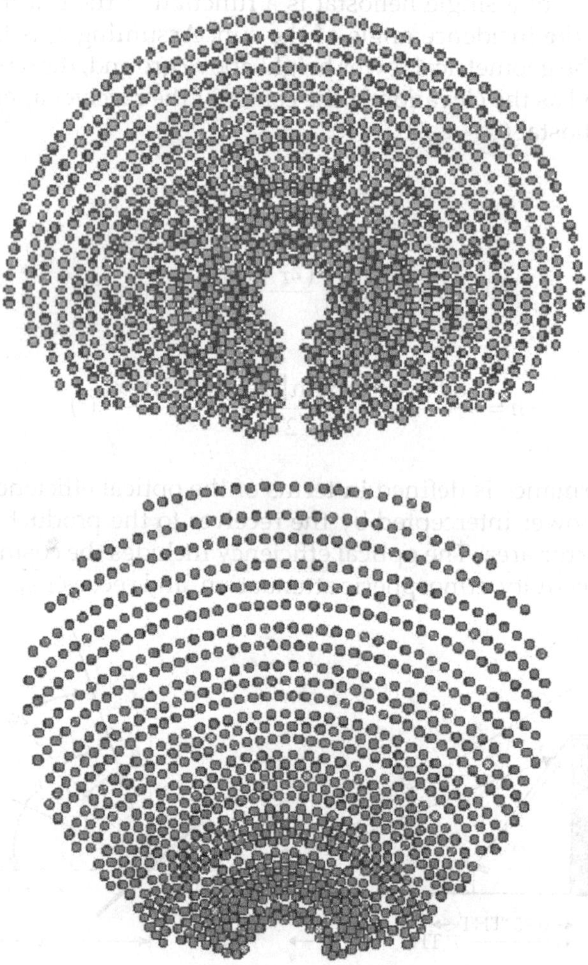

FIGURE 42.43
Representation of optimized fields for a latitude of 36° with surround field (top) and north field (bottom) configurations.

Another point to be considered in the layout is the heliostat dimensions. If the heliostats are spaced too close together, their corners could collide, so mechanical limits preclude pedestal spacing closer than the maximum dimension of the heliostat (i.e., the diagonal or diameter of the heliostat, D_m). This causes a significant disadvantage when the aspect ratio (height/width) differs greatly from 1.

Blocking of reflected rays is also an important limitation on spacing heliostats. Blocking is produced by neighboring heliostats. To avoid blocking losses, the distance Δx between the heliostat rows must be calculated according to the following equation:

$$\Delta x = x \frac{z_S}{z_T} \tag{42.41}$$

Shading produced by neighboring heliostats also has to be taken into account. This occurs mostly at low sun angles and in the middle of the field where blocking conditions would allow close spacing. The shadows move during the day and year, as does the heliostat orientation, so there is no simple rule. In addition, the tower or other objects may also cast a shadow over part of the heliostat field. Usually shadowing in the field is calculated by projecting the outlines of the heliostats aligned, the tower, and anything else that casts a shadow onto a plane perpendicular to the center sunray. Shadowed portions of any heliostat appear in the overlapping areas in this projection. Classical computer codes like HELIOS provide this calculation (Biggs and Vittitoe, 1979).

Not all the sunlight that clears the heliostats reaches the vicinity of the receiver. Some of the energy is scattered and absorbed by the atmosphere. This effect is referred to as the attenuation loss (Falcone, 1986). This factor increases when water vapor or aerosol content in the atmosphere is high and is typically anywhere between 5% and 15% in a solar field. Atmospheric attenuation is usually expressed as a function or experimental correlation depending on the range of heliostat inclination.

The size of the image formed by each heliostat depends on mirror focusing and canting and on the size of the heliostat and errors, as expressed by its beam quality. Because of that, there is an intercept factor for a given receiver aperture area. Some of the energy spills over around the receiver. While spillage can be eliminated by increasing the size of the receiver, at some point, increased size becomes counterproductive because of the resulting increased receiver losses and receiver costs.

The combination of all these factors influencing the performance of the heliostat field should be optimized to determine an efficient layout. There are many optimization approaches to establish the radial and azimuthal spacing of heliostats and rows. One of the most classic, effective, and widespread procedures is the *radial staggered* pattern, as shown in Figure 42.44, originally proposed by the University of Houston in the 1970s (Lips and Vant-Hull, 1978). Typical radially staggered field spacing at 35° latitude using square low-cost heliostats can be expressed by

$$\frac{\Delta R}{\Delta D_m} = \frac{1.009}{\Theta} - 0.063 + 0.4803\Theta \tag{42.42}$$

$$\frac{\Delta Z}{\Delta D_m} = 2.170 - 0.6589\Theta + 1.247\Theta^2 \tag{42.43}$$

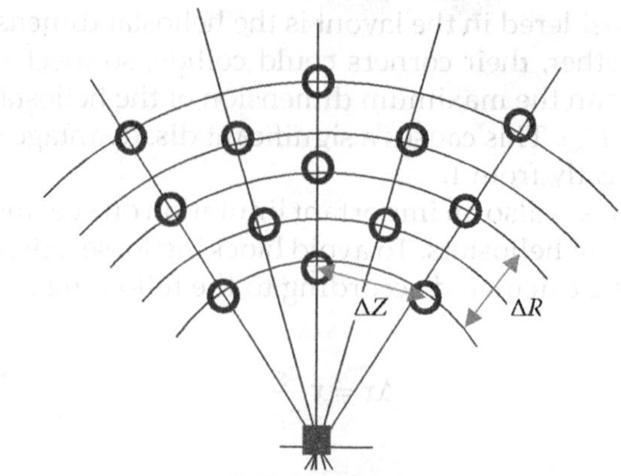

FIGURE 42.44
Radial staggered field layout, where ΔZ represents the azimuthal spacing and ΔR represents the radial spacing.

where Θ represents the receiver elevation angle in radians:

$$\Theta = \arctan\left[\frac{(z_T - (z_S/2))}{x}\right]$$

Integral optimization of the heliostat field is decided by a trade-off between cost and performance parameters. Heliostats, land, and cabling network must be correlated with costs. Cost and performance also often have reverse trends, so that when heliostats are packed closer together, blocking and shadowing penalties increase, but related costs for land and wiring decrease. A classical code in use since the 1980s for optimization of central receiver subsystems is DELSOL3 (Kistler, 1986). In DELSOL3, heliostat field layout is optimized for tower height and solar receiver geometry. Figure 42.45 shows a breakdown of efficiency maps for the different performance factors in a typical one-sided north heliostat field. It may be deduced that the heliostat density is greatest at the inner boundary and decreases with increasing radial distance from the tower. The average land coverage ratio is typically 0.20–0.25.

The radially staggered distribution clearly creates *prearranged* grids based on the tower height versus row-to-radius ratio. This geometrical procedure provides a smart solution to the problem with good optimization of computing resources. However, with today's computers, it is possible to calculate the yearly energy available at any point in a site for a given tower height, the *yearly normalized energy surface* (YNES). Yearly efficiency maps can be generated based on the cosine factor, the spillage factor, and the site atmospheric attenuation coefficient using real direct normal irradiance (DNI) data, within a reasonable computing time. It is therefore easy to find the place where the yearly energy available is the highest for location of the first heliostat. It is also possible to calculate the effect of shadowing and blocking by this heliostat on the YNES, so YNES can be recalculated, and the best position for the next heliostat can be found. Although this iterative method is time consuming, it is worthwhile if either the efficiency of the solar plant can be increased or the capital cost can be reduced. This YNES-based layout method enables better flexibility than predetermined gridding strategies (Sánchez and Romero, 2006).

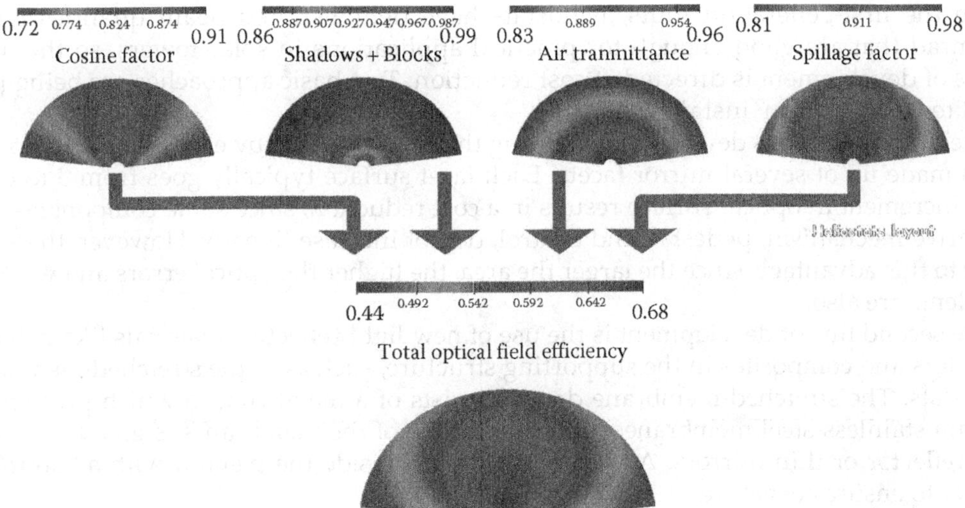

FIGURE 42.45
Mapping of total optical efficiency of a north-field area of heliostats and its breakdown into cosine factor, shadowing and blocking, air transmittance, and receiver spillage.

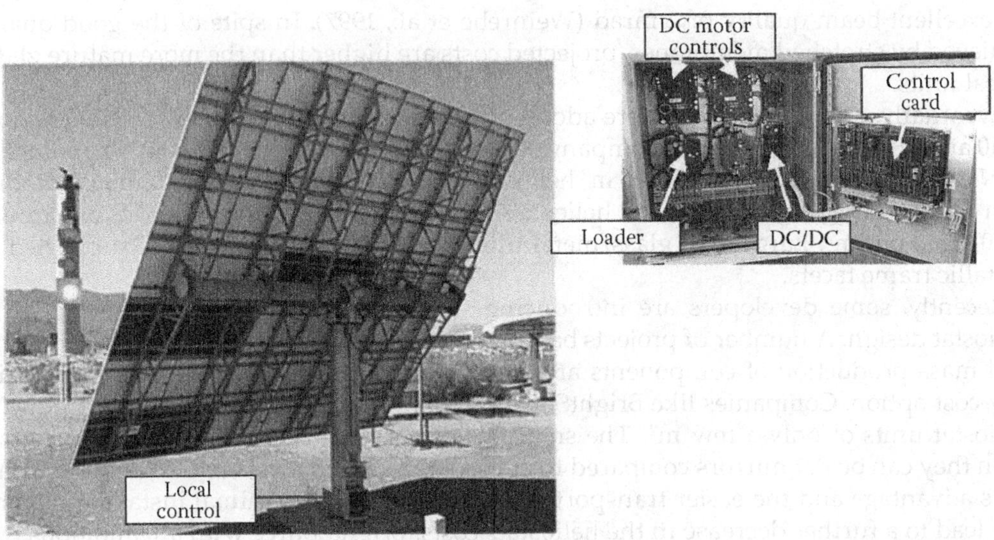

FIGURE 42.46
Rear view of the heliostat COLON SOLAR of 70 m² with typical T-shape structure and torque tube. The control box is located at the bottom of the pedestal. The reflected image is shown at the Lambertian target located on the tower.

Mature low-cost heliostats consist of a reflecting surface, a support structure, a two-axis tracking mechanism, pedestal, foundation, and control system (Figure 42.46). The development of heliostats shows a clear trend from the early first-generation prototypes, with a heavy, rigid structure, second-surface mirrors, and reflecting surfaces of around 40 m² (Mavis, 1989), to the current commercial designs with large 100–120 m² reflecting surfaces, lighter structures, and lower-cost materials (Romero et al., 1991).

Since the first-generation units, heliostats have demonstrated beam qualities below 2.5 mrad that are good enough for practical applications in solar towers, so the main focus of development is directed at cost reduction. Two basic approaches are being pursued to reduce per m^2 installed cost.

The first approach is devoted to increasing the reflective area by employing curved surfaces made up of several mirror facets. Each facet surface typically goes from 3 to 6 m^2. This increment in optical surface results in a cost reduction, since some components, like the drive mechanism, pedestal, and control, do not increase linearly. However, there is a limit to this advantage, since the larger the area, the higher the optical errors and washing problems are also.

The second line of development is the use of new light-reflective materials like polymer reflectors and composites in the supporting structure, such as in the stretched-membrane heliostats. The stretched-membrane drum consists of a metal ring to which prestressed 0.4 mm stainless-steel membranes are welded. One of the membranes is glued to a polymer reflector or thin mirrors. A vacuum is created inside the plenum with a controlled blower to ensure curvature.

In Spain, some developments worthy of mention are the 70 m^2 COLON SOLAR prototype (Osuna et al., 1999), the 105 m^2 GM-100 (Monterreal et al., 1997), and more recently, the 90 and 120 m^2 Sanlúcar heliostats (Osuna et al., 2004). In the United States, a similar development in glass/metal technology was the 150 m^2 ATS heliostat (Alpert and Houser, 1990). The stretched-membrane milestone is the 150 m^2 Steinmuller heliostat ASM-150 with an excellent beam quality of 2 mrad (Weinrebe et al., 1997). In spite of the good quality achieved by stretched membranes, projected costs are higher than the more mature glass/metal units.

Eventually, 120 m^2 heliostats were adopted for the first commercial tower power plants PS10 and PS20 promoted by the company Abengoa Solar (Osuna et al., 2004). The company SENER has developed a similar 115 m^2 heliostat for its Gemasolar plant. Estimated production costs of large-area glass/metal heliostats for sustainable market scenarios are around \$130–\$200 per m^2. Large-area glass/metal units make use of glass mirrors supported by metallic frame facets.

Recently, some developers are introducing substantial changes in the conception of heliostat design. A number of projects based upon the paradigm of maximum modularity and mass production of components are claiming small-size heliostats as a competitive low-cost option. Companies like BrightSource, eSolar, Aora, or Cloncurry are introducing heliostat units of only a few m^2. The small heliostats have better optical efficiency, and even they can be flat mirrors compared to curved and canted facets in the large heliostats. This advantage and the easier transportation to the site with minimal installation works can lead to a further decrease in the heliostats costs. BrightSource with an ambitious program of large projects is making use of single-facet 7.3 m^2 heliostats (Silberstein et al., 2009), and the company eSolar with a multitower plant configuration presents a highly innovative field with ganged heliostats of extremely small size (1.14 m^2 each) that implies the large number of 12,180 units for a single 2.5 MW tower (Schell, 2009). If such small heliostats may reach installed costs below \$200 per m^2, it can be understood only under aggressive mass production plans and preassembly during manufacturing process by reducing on-site mounting works. Annual performance and availability of those highly populated fields are still under testing.

The drive mechanism is in charge of independent azimuthal and elevation movement, in such a way that the specular surface follows the sun position and reflects the beam onto the focal point. The ratio between the angle of incidence and the reflected ray leads to angular errors doubling at the target. It is therefore crucial for a tracking system to be highly accurate. Heliostat drives should have the following characteristics:

- Sufficiently robust to support their own weight, the movable structure, and wind loads, and be rigid enough to avoid low-frequency vibrations.
- Able to generate extremely slow movement, with high reduction ratios (up to 40,000:1).
- Highly accurate positioning (use of encoders) and no free movement.
- Able to ensure relatively fast return to stow position in case of high winds or other dangerous weather conditions, and other events.
- Resistance to outdoor exposure.
- Easy maintenance.
- Low-cost manufacture and operation.

The most common drive mechanism configuration makes use of worm-gear systems for both elevation and azimuth axes (Figure 42.47). Both gears are essentially analogous in terms of tooth shape and reduction ratio. In many cases, there is a first planetary reduction step and then a second worm-gear reduction step at the outlet. The advantage of the

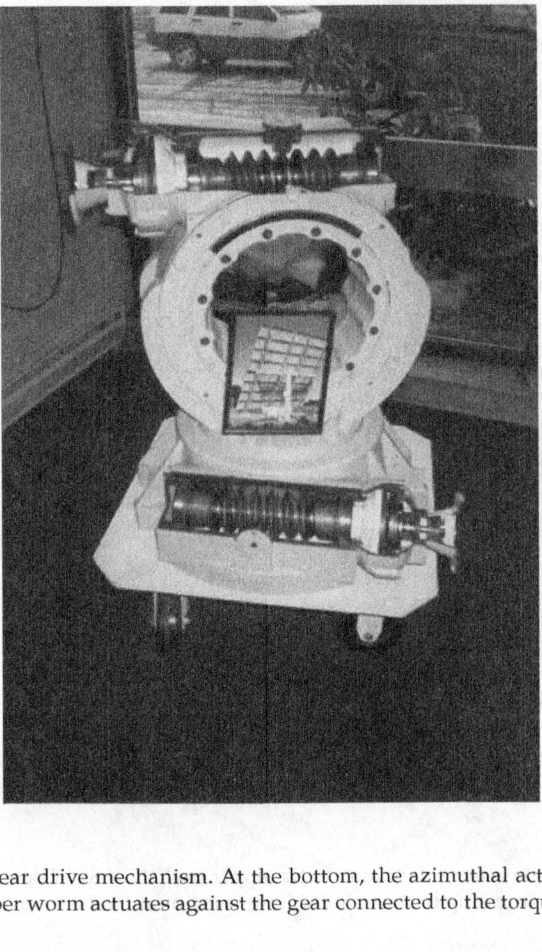

FIGURE 42.47
Detail of a typical worm-gear drive mechanism. At the bottom, the azimuthal actuator located on top of the pedestal is shown. The upper worm actuates against the gear connected to the torque tube.

planetary system is the high ratio of reduction in a limited space. The worm gear provides high reduction ratios at high momentums. However, worm gears are less efficient because of the stress of high friction. This stress has a positive reaction, since the self-locking worm comes to a halt whenever the angle of friction between the worm thread and the gear teeth is higher than the nominal design angle.

Mature glass/metal faceted heliostats report availabilities over 95% and beam qualities of 2.4–2.8 mrad. Yearly average reflectivity of a heliostat field reaches 85%–92%. Faceted heliostats with curved mirrors require canting of the facets to form a large paraboloid. In Figure 42.48, the effect of canting facets can be seen.

Control of CRSs is more complicated than other types of STP plants since optics are off-axis, and each and every heliostat individually tracks the sun. The control system in a CRS is naturally separated into the heliostat field control system (HFCS) and the receiver and power system control system (RPSCS) (Yebra et al., 2004). The main purpose of the HFCS is to keep each heliostat positioned at the desired coordinates at all times, depending on power system

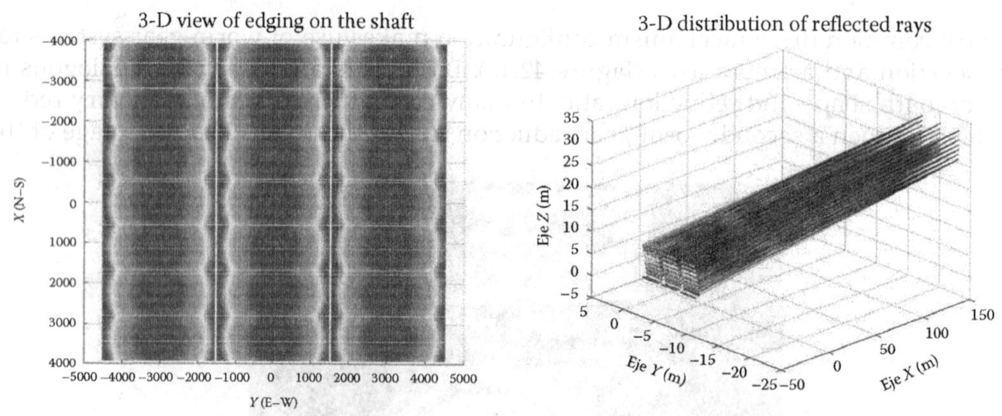

FIGURE 42.48
Effect of canting facets in a glass metal heliostat. In the upper part, it can be observed that a flat heliostat with curved facets is not focusing in a single image. In the lower part, a canted heliostat is depicted.

demand. The general purpose of the HFCS is to generate a uniform time–spatial distribution of the temperature on the volumetric receiver by controlling the timed insertion of an associated group of heliostats at predefined aiming points on the receiver by modifying the aiming-point coordinates and changing from one heliostat group to another during operation. This is accomplished by an HFCS aiming-point strategy (Garcia-Martín et al., 1999). The current trend in control systems is a distributed control, with a hard real-time operating system, integrating heterogeneous hardware and software platforms in real time, that guarantees a deterministic response to external (physical environment) and internal (operator interface) events. The RPSCS regulates the pressure and temperature of the HTF, and steam generator.

The heliostat local control is responsible for all the emergency and security maneuvers and sun-tracking calculations, as well as communication with the control room. The current trend is to increase the heliostat intelligence and autonomy. In addition, some drive mechanism options consider the use of wireless communications and PV-power supply, eliminating the need for cabling and trenching. This is the case of the stand-alone heliostat developed at the PSA in Spain, where a field of 92 such heliostats is in operation (García et al., 2004).

42.6.2 Solar Receiver

In a solar power tower plant, the receiver is the heat exchanger where the concentrated sunlight is intercepted and transformed into thermal energy useful in thermodynamic cycles. Radiant flux and temperature are substantially higher than in parabolic troughs, and therefore, high technology is involved in the design, and high-performance materials should be chosen. The solar receiver should mimic a blackbody by minimizing radiation losses. To do so cavities, black-painted tube panels or porous absorbers able to trap incident photons are used. In most designs, the solar receiver is a single unit that centralizes all the energy collected by the large mirror field, and therefore high availabilities and durability are a must. Just as cost reduction is the priority for further development in the collector field, in solar receivers, the priorities are thermal efficiency and durability. Typical receiver absorber operating temperatures are between 500°C and 1200°C, and incident flux covers a wide range between 300 and over 1000 kW/m^2. The picture in Figure 42.49 clearly shows the high flux to be withstood by the receiver.

Thermal and optical losses are the key parameters for quantifying the efficiency of a solar receiver.

$$\eta_{rec} = (\alpha\tau_W) + (\alpha\varepsilon_W)\frac{\sigma T_W{}^4}{C\cdot\phi} - (\varepsilon\bar{\rho})\frac{\sigma T^4}{C\cdot\phi} - U\frac{(T - T_a)}{C\cdot\phi} \quad (42.44)$$

This equation is the result of the energy balance of gains and losses in the receiver, with an absorber surface at temperature T. In some cases, the receiver has a transparent window that absorbs part of the incident radiation at a temperature, T_W, higher than the ambient, T_a. The concentrated solar irradiance ($C\cdot\phi$) is absorbed with an efficiency of ($\alpha\tau_W$). The radiation must often go through a transparent window, where it is partially absorbed, before reaching the absorber surface, reflected and transmitted onto the absorber. These not-so-simple effects are represented by the apparent window transmittance, τ_W. Likewise, the second term in Equation 42.44 expresses the energy emitted of temperature T_W from the hot window toward the absorber. Therefore, this second term adds gains to the first. The loss terms are of two different types. The most important in a central receiver represents energy thermally radiated by the absorber through the receiver aperture. These radiation

FIGURE 42.49
Lateral view of the 80 m concrete tower of the CESA-1 facility located at the Plataforma Solar de Almería in Spain. The concentrated flux is aiming a standby position previous to start operation.

losses depend on the emissivity of the absorber and on thermal radiation reflectivity ρ_W of the window. In the product $(\varepsilon\bar{\rho})$, $\bar{\rho}$ indicates $(1 - \rho_W)$ or $(\alpha_W + \tau_W)$.

The absorber convective, or conductive, loss to the ambient is determined by the heat loss coefficient U, which depends on the temperature and the forced convection due to wind. In good central receiver designs, U can be sufficiently reduced by thermal insulation and decreased aperture area; therefore, U is basically expressed as a convective loss heat transfer coefficient. Generally, this coefficient is obtained from the dimensionless Nusselt number and subsequently as a function of numbers like Reynolds (Re), Prandtl (Pr), and Grashof (Gr). Forced convection is determined by combinations of the Re and Pr numbers, while natural convection is characterized by Pr and Gr numbers.

In the solar tower, the convective heat loss is calculated differently depending on whether the receiver is a cavity or a cylindrical external receiver. A typical simple mixed convection coefficient for an external receiver can be obtained calculated (Siebers and Kraabel, 1984) as

$$U_{mix} = \left(h_{forced}^{3.2} + h_{nat}^{3.2}\right)^{1/3.2} \tag{42.45}$$

where

$h_{nat} = 9.09$ (W/m² °C) for an average absorber temperature of 480°C

h_{forced} is separated into three cases depending on the receiver diameter (Kistler, 1986)

In all cases, the Reynolds number is Re = $(1.751 \times 10^5)D$.

Case 1: $D \leq 4.0$ m	$h_{forced} = \left(\frac{1}{D}\right)\cdot\left(0.3+0.488\cdot \mathrm{Re}^{0.5}\cdot\left(1.0+\left(\frac{\mathrm{Re}}{282000}\right)^{0.625}\right)^{0.8}\right)\cdot 0.04199$
Case 2: $4.0 < D \leq 125.0$	$h_{forced} = 14.0$
Case 3: $D > 125.0$ m	$h_{forced} = 33.75\cdot D^{-0.19}$

For a cavity receiver, the convective heat loss can be directly calculated as (Kistler, 1986) follows:

$$Q_{conv} = Q_{forced} + Q_{nat}\ (\mathrm{W}) \tag{42.46}$$

$$Q_{forced} = 7631\frac{A}{W_{ap}^{0.2}} \tag{42.47}$$

$$Q_{nat} = 5077\cdot A_{cav} \tag{42.48}$$

where
- A is the aperture area (m²)
- W_{ap} is the aperture width (m)
- A_{cav} is the approximation to total area inside cavity (m²)

For more detailed correlations applicable to convection losses in different kinds of solar receivers, Becker and Vant-Hull (1991) is recommended.

There are different solar receiver classification criteria depending on the construction solution, the use of intermediate absorber materials, the kind of thermal fluid used, or heat transfer mechanisms. According to the geometrical configuration, there are basically two design options, external and cavity-type receivers. In a cavity receiver, the radiation reflected from the heliostats passes through an aperture into a box-like structure before impinging on the heat transfer surface. Cavities are constrained angularly and subsequently used in north-field (or south-field) layouts. External receivers can be designed with a flat plate tubular panel or in a cylindrical shape. Cylindrical external receivers are the typical solution adopted for surround heliostat fields. Figure 42.50 shows examples of cylindrical external, billboard external, and cavity receivers.

Receivers can be directly or indirectly irradiated depending on the absorber materials used to transfer the energy to the working fluid (Becker and Vant-Hull, 1991). Directly irradiated receivers make use of fluids or particle streams able to efficiently absorb the concentrated flux. Particle receiver designs make use of falling curtains or fluidized beds. Darkened liquid fluids can use falling films. In many applications, and to avoid leaks to the atmosphere, direct receivers should have a transparent window. Windowed receivers are excellent solutions for chemical applications as well, but they are strongly limited by the size of a single window, and therefore clusters of receivers are necessary.

The key design element in indirectly heated receivers is the radiative/convective heat exchange surface or mechanism. Basically, two heat transfer options are used, tubular panels and volumetric surfaces. In tubular panels, the cooling thermal fluid flows inside the tube and removes the heat collected by the external black panel surface by convection. It is therefore operating as a recuperative heat exchanger. Depending on the HTF

FIGURE 42.50
Different configurations of solar receivers. From left to right and top to bottom: (a) external tubular cylindrical, (b) cavity tubular, (c) billboard tubular, and (d) volumetric.

properties and incident solar flux, the tube might undergo thermo-mechanical stress. Since heat transfer is through the tube surface, it is difficult to operate at an incident flux above 600 kW/m^2 (peak). Table 42.8 shows how only with high thermal conductivity liquids like sodium, it is possible to reach operating fluxes above 1 MW/m^2. Air-cooled receivers have difficulties working with tubular receivers because of the lower heat transfer coefficients, as already found in the German-Spanish GAST project where two tubular receivers, one metal and one ceramic, were tested at the PSA in Spain (Becker and Boehmer, 1989). To improve the contact surface, a different approach based on wire, foam, or appropriately shaped materials within a volume is used. In volumetric receivers, highly porous structures operating as convective heat exchangers absorb the concentrated solar radiation. The solar radiation is not absorbed on an outer surface, but

TABLE 42.8

Operating Temperature and Flux Ranges of Solar Tower Receivers

Fluid	Water–Steam	Liquid Sodium	Molten Salt (Nitrates)	Volumetric Air
Flux (MW/m²)				
Average	0.1–0.3	0.4–0.5	0.4–0.5	0.5–0.6
Peak	0.4–0.6	1.4–2.5	0.7–0.8	0.8–1.0
Fluid outlet temperature (°C)	490–525	540	540–565	700–800 (>800)

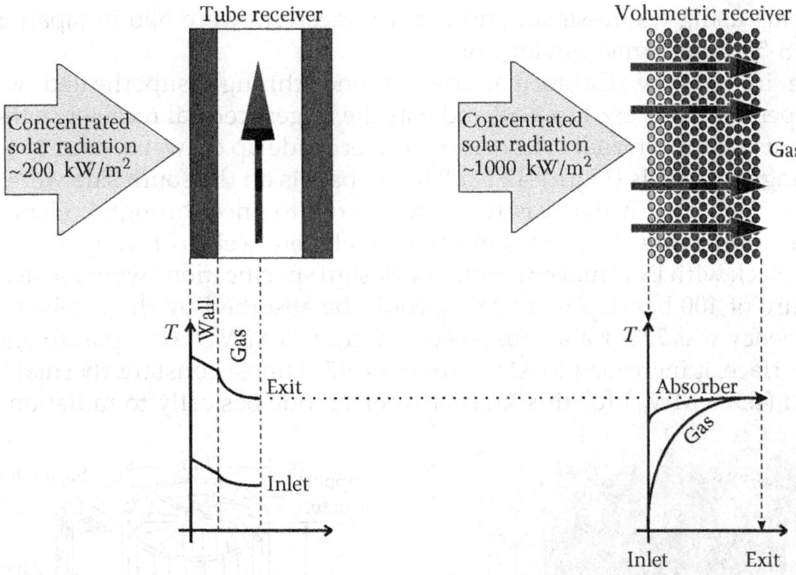

FIGURE 42.51
Heat transfer principles in tubular and volumetric receivers.

inside the structure *volume*. The heat transfer medium (mostly air) is forced through the porous structure and is heated by convective heat transfer. Figure 42.51 shows a comparison of the two absorber principles. Volumetric absorbers are usually made of thin heat-resistant wires (in knitted or layered grids) or either metal or ceramic (reticulated foams, etc.) open-cell matrix structures. Good volumetric absorbers are very porous, allowing the radiation to penetrate deeply into the structure. Thin substructures (wires, walls, or struts) ensure good convective heat transfer. A good volumetric absorber produces the so-called volumetric effect, which means that the irradiated side of the absorber is at a lower temperature than the medium leaving the absorber. Under specific operating conditions, volumetric absorbers tend to have an unstable mass flow distribution. Receiver arrangements with mass flow adaptation elements (e.g., perforated plates) located behind the absorber can reduce this tendency, as well as appropriate selection of the operating conditions and the absorber material.

Selection of a particular receiver technology is a complex task, since operating temperature, heat storage system, and thermodynamic cycle influence the design. In general, tubular technologies allow either high temperatures (up to 1000°C) or high pressures (up to 120 bar), but not both (Kribus, 1999). Directly irradiated or volumetric receivers allow even higher temperatures but limit pressures to below 15 bar.

42.6.2.1 Tubular Receivers

The most common systems used in the past have been tubular receivers where concentrated radiation is transferred to the cooling fluid through a metal or ceramic wall. Conventional panels with darkened metal tubes have been used with steam, sodium, and molten salts for temperatures up to 500°C–600°C. Much less experience is available on tubular receivers with gas, though temperatures in the range of 800°C–900°C are possible. Cavity receivers have been tested in France (Themis) and Spain (IEA-SSPS-CRS project and CESA-1 plant). External tubular receivers were used in Solar One (United States), IEA-SSPS-CRS (Spain), and Solar Two (United States) (Grasse et al., 1991; Pacheco and Gilbert, 1999). Other experimental systems using water–steam and tubular receivers were Nio in Japan, Eurelios in Italy, and SES-5 in the former Soviet Union.

Solar One in Barstow, California, used a once-through superheated water–steam receiver. It operated from 1984 to 1988 and was the largest central receiver in the world for two decades. It was an external cylindrical receiver made up of twenty-four 1 m wide by 14 m long rectangular panels (Figure 42.52). The six panels on the south side were feed-water preheat units. Preheated water was then transferred to once-through boilers and superheaters on the north side. The tubes in each panel were welded throughout their length and painted black with Pyromark® paint. The design specifications were for steam at 516°C and a pressure of 100 bar. Up to 42 MW_{th} could be absorbed by the receiver. The initial thermal efficiency was 77% for an absorbed power of 34 MW_{th}. After painting it black and curing the surface, it increased to 82% with $\alpha = 0.97$. Almost constant thermal losses have been verified (4.5–5 MW_{th}) for this kind of receiver, due basically to radiation losses and

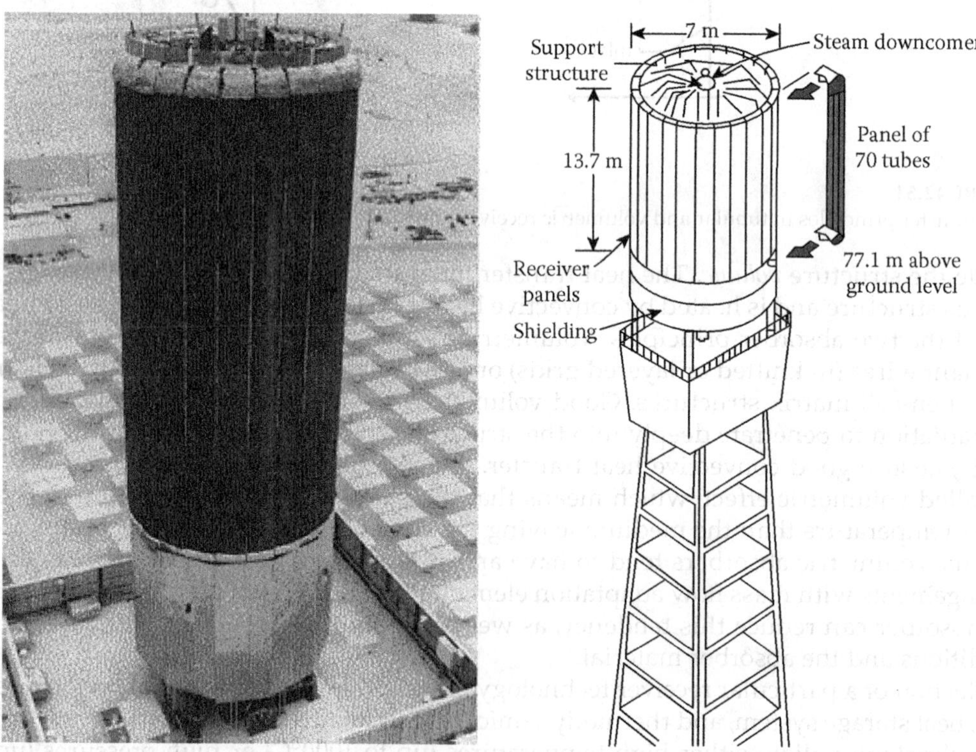

FIGURE 42.52
External cylindrical tubular receiver used in Solar One power plant, Barstow, California.

operating the receiver at constant temperature by regulating the mass flow rate. One of the main problems found during testing was overheating and deformation in the superheating section because of solar transients and poor heat transfer (Radosevich and Skinrood, 1989). Cracking and leaking were observed at the top of the boiler tubes after 18 months of operation. The temperature gradient between the edges and the center of the tubes can go up to 111°C during start-ups and shutdowns.

CESA-1, which operated 1631 h between 1983 and 1986 in Almería, Spain, was a north-facing water–steam cavity receiver. The 3.4 m square aperture was tilted 20° to the heliostat field. The receiver panel configuration and position of the steam drum are shown in Figure 42.53. The boiler consisted of three panels of A-106 Gr B carbon steel tubes, with an effective surface of 48.6 m^2, and a superheater made of X-20 Cr Mo V 121 steel. The 6.7 MW maximum incident power on the receiver produced superheated steam at 110 kg/cm^2 and 525°C. As in Solar One, operating problems and noticeable deformations were found in the superheating zone, requiring it to be operated with lower flux. Also because of that, operation had to proceed slowly during start-up and transients, penalizing efficiency. More than 45 min were required to reach nominal conditions (Sánchez, 1986).

Molten-salt tubular receivers are represented by the Themis system (cavity) and Solar Two (cylindrical external), and subsequently applied to the commercial project Gemasolar. In a molten-salt system, cold salt at about 290°C is pumped from a tank at ground level to the receiver mounted atop a tower where it is heated by concentrated sunlight to 565°C. Using molten salts as receiver coolant provides a number

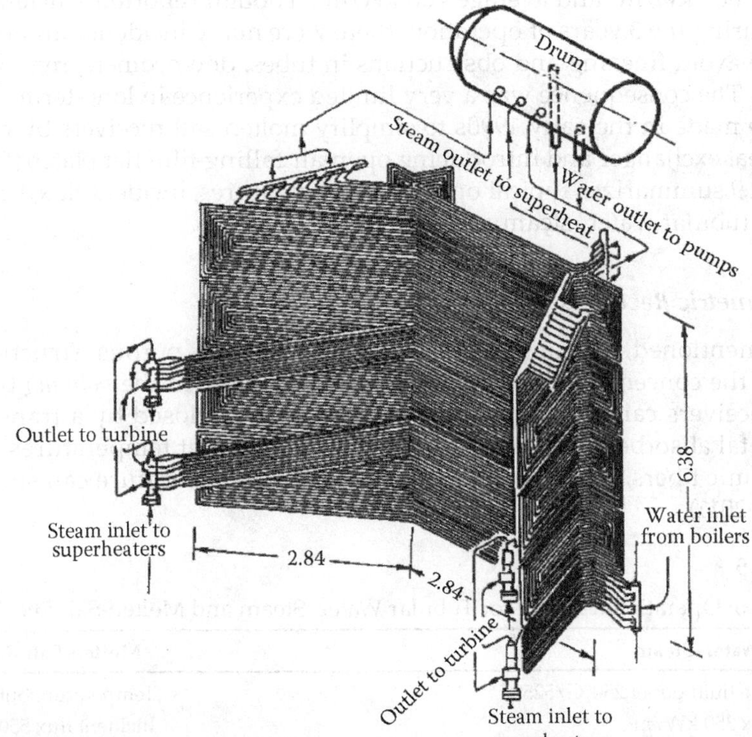

FIGURE 42.53
Inner view of tubular panels and drum of the water–steam cavity receiver used in the CESA-1 project in Spain.

of benefits because there is no phase change, and it is possible to heat up to 565°C without the problems associated in tubes with superheating sections. Mixtures of 60% sodium nitrate and 40% potassium nitrate have been extensively tested with satisfactory results in France and the United States. Molten nitrates provide good thermal conductivity (0.52 W/m K) and heat capacity (1.6 kJ/kg K) at relatively low prices. Molten nitrate salt, though an excellent thermal storage medium, can be a troublesome fluid to deal with because of its relatively high freezing point (220°C). To keep the salt molten, a fairly complex heat trace system must be employed. (Heat tracing is composed of electric wires attached to the outside surface of pipes. Pipes are kept warm by way of resistance heating.) Problems were experienced during the start-up of Solar Two due to the improper installation of the heat trace. Though this problem has been addressed and corrected, research is needed to reduce the reliance on heat tracing in these plants. Also, valves can be troublesome in molten-salt service. Special packings must be used, oftentimes with extended bonnets, and leaks are not uncommon. Furthermore, freezing in the valve or packing can prevent it from operating correctly. While today's valve technology is adequate for molten-salt power towers, design improvements and standardization would reduce risk and ultimately reduce O&M costs (De Meo and Galdo, 1997).

Solar Two, tested between 1996 and 1999 (Pacheco et al., 2000), is still the technical reference for molten-salt tubular receivers. The 42 MW absorber consisted of a 6.2 m–high and 5.1 m–diameter cylinder, with 768 2 cm–diameter tubes. Reported efficiency with no wind was 88% for 34 MW absorbed (86% with wind under 8 km/h). Peak concentrated solar flux was 800 kW/m^2 and average 400 kW/m^2. Though reported efficiency was close to nominal, during the 3 years of operation, there were many incidents, and modifications and repairs to avoid freezing and obstructions in tubes, downcomers, manifolds, valves, and pipelines. The consequence was a very limited experience in long-term testing. Some attempts were made in the early 1990s to simplify molten-salt receivers by removing the salt-in-tube heat exchanger and introducing open-air falling-film flat plates (Romero et al., 1995). Table 42.9 summarizes typical operating temperatures, incident flux, pressures, and efficiencies in tubular water–steam and molten-salt receivers.

42.6.2.2 Volumetric Receivers

As already mentioned, volumetric receivers use highly porous structures for the absorption of the concentrated solar radiation deep inside (in the *volume*) the structure. Volumetric receivers can work open to the ambient or enclosed by a transparent window. With metal absorbers, it is possible to achieve air outlet temperatures up to 850°C, and with ceramic fibers, foams, or monoliths (SiC), the temperature can surpass 1000°C (Avila-Marin, 2011).

TABLE 42.9

Summary of Operational Range for Tubular Water–Steam and Molten-Salt Receivers

Receivers Water–Steam	Molten-Salt Receivers
Temperature fluid outlet 250°C/525°C	Temperature outlet 566°C
Incident flux 350 kW/m^2	Incident flux 550 kW/m^2
Peak flux 700 kW/m^2	Peak flux 800 kW/m^2
Pressure 100–135 bar	
Efficiency 80%–93%	Efficiency 85%–90%

The main advantages of an air-cooled volumetric receiver are as follows:

- The air is free and fully available at the site.
- No risk of freezing.
- Higher temperatures are possible, and therefore the integration of solar thermal energy into more efficient thermodynamic cycle looks achievable.
- No phase change.
- Simple system.
- Fast response to transients or changes in incident flux.
- No special safety requisites.
- No environmental impact.

Open volumetric receivers have made dramatic progress since the pioneering experiences of the late 1970s (Sanders, 1979) and early 1980s (Fricker et al., 1988). More than 20 absorbers and prototypes in the 200–300 kW_{th} range have been tested in the Sulzer test bed at the PSA (Becker et al., 1989, 1992; Hoffschmidt et al., 2001). Wire mesh, knitted-wire, foam, metal, and ceramic monolith volumetric absorbers have been developed worldwide. The relatively large number of volumetric prototypes tested has demonstrated the feasibility of producing hot air at temperatures of 1000°C and upward and with aperture areas similar to those used in molten-salt or water–steam receivers. Average flux of 400 kW/m² and peaks of 1000 kW/m² have been proven, and their low-inertia and quick sun-following dispatchability are excellent. Comparative assessments have demonstrated that wire mesh has the lowest thermal losses (Table 42.10). This can mainly be explained by the very high porosity of the absorber, which permits a large portion of the irradiation to penetrate deep into its volume. The choice of ceramics as the absorber material makes higher gas outlet temperatures possible. In particular, siliconized SiC has been revealed as a good option because of its high thermal conductivity. Even though ceramic absorbers have lower efficiencies at 680°C (this is the reference temperature for applications where hot air is used as the HTF to produce superheated

TABLE 42.10

Properties and Efficiencies Reported for Several Absorber Materials Tested at the Plataforma Solar de Almería ($D_{m,p}$ = Mean Pore Diameter)

Type of Receiver	Designed by	Absorber Structure	Porosity (V_p/V_{tot})	$D_{m,p}$ (mm)	Absorber Thickness (mm)	Material	Thermal Conductivity (500°C) (W/m K)	Efficiency (680°C) (%)
Metallic wire	Sulzer	Wire mesh	0.95	2.5	35	Stainless steel	20	75
Ceramic foam	Sandia	Amorphous foam	0.8	1.0	30	Al_2O_3	25	54
Metallic foils	Interatom/ Emitec	Prismatic channels	0.9	1.0	90	$X_5CrAl_2O_5$	20	57
Ceramic foils	DLR/ Ceramtec	Prismatic channels	0.4	3.0	92	SiSiC	80	60
Ceramic cups	DLR/ Stobbe	Prismatic channels	0.5 + 0.12 apert.	2.1	80	SiSiC	80	60

steam in a heat exchanger), they have demonstrated efficiencies about 80% at temperatures of 800°C. With higher solar flux and temperatures, more compact designs and smaller receivers can be developed.

A milestone in the experimental scale-up of open volumetric receivers was the Technology Program Solar Air Receiver (TSA) project, under the leadership of the German Steinmüller company. A 2.5 MW_{th} air-cooled receiver was tested on top of the PSA CESA-1 tower late in 1991. The TSA experimental setup was a small-scale PHOEBUS-type receiver (Schmitz-Goeb and Keintzel, 1997), in which atmospheric air is heated up through a wire mesh receiver to temperatures on the order of 700°C to produce steam at 480°C–540°C and 35–140 bar, in a heat recovery steam generator with a separate superheater, reheater, evaporator, and economizer feeding a Rankine turbine-generator system. All the hardware, including receiver, steam generator, and heat storage, was located atop the CESA-1 tower. Average solar flux at the receiver aperture was 0.3 MW/m^2. The HFCS was implemented with heuristic algorithms to obtain an automatic aiming strategy able to maintain stationary flux and air outlet temperature (Garcia-Martin et al., 1999). One of two blowers controlled the air mass flow rate through the receiver and the other the air through the steam generator. The 1000 kWh capacity thermocline storage system (alumina pellets) was enough to allow 30 min of nominal off-sun operation. Depending on the load, it could be charged, bypassed, or discharged. The 3 m–diameter absorber was made of hexagonal wire mesh modules (Figure 42.54). The TSA receiver was successfully operated by DLR and CIEMAT for a total of nearly 400 h between April and December 1993 (Haeger, 1994), and for shorter periods in 1994 and 1999, demonstrating that a receiver outlet temperature of 700°C could easily be achieved within 20 min of plant start-up and receiver thermal efficiencies up to 75% were obtained as shown in Figure 42.55.

A significant number of volumetric prototypes have been reported unable to reach nominal design conditions because of local cracks and structural damages. Those failures were, in many cases, caused by thermal shock, material defects, or improper operation. In the middle 1980s, some projects promoted by the German Aerospace Center (DLR) studied the fluid dynamics and thermal mechanisms inherent in volumetric absorbers (Hoffschmidt et al., 2001). One of the conclusions of these studies was that in highly porous absorber materials, the air flow through the absorber structure is unstable under high solar flux, which leads to the destruction (cracks or melting) of the absorber structure by local overheating (Pitz-Paal et al., 1996). As a consequence of this analysis, a new approach

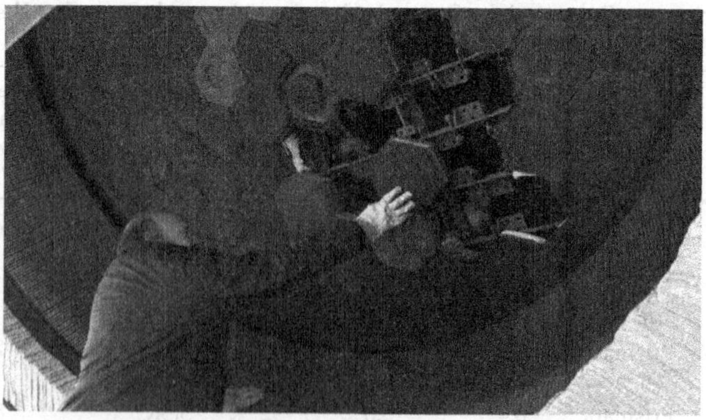

FIGURE 42.54
Front view of the TSA volumetric receivers. The absorber was composed of hexagonal pieces of wire mesh. The outer ring includes the air return system.

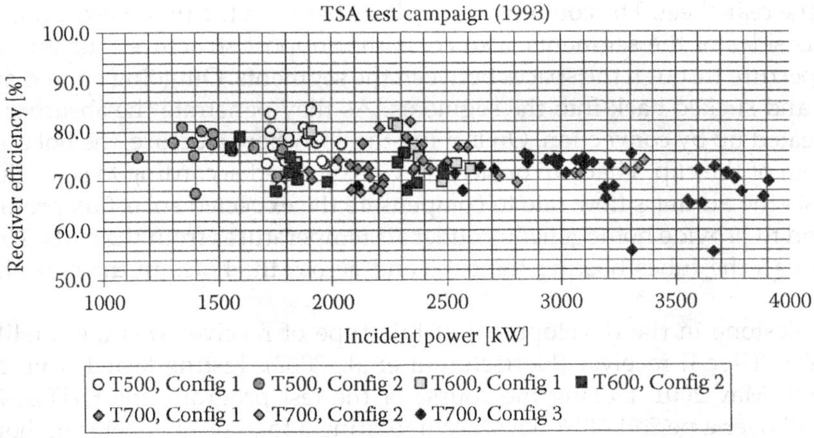

FIGURE 42.55
TSA receiver efficiency at different temperatures and incident powers during test campaign carried out in 1993.

developed monolithic ceramic absorbers able to work at high temperatures and fluxes due to their high thermal conductivity and geometric modularity, which were put to use first in the German-Spanish HiTRec project (Hoffschmidt et al., 1999) and afterward in the European SOLAIR project (Hoffschmidt et al., 2002).

The HiTRec–SOLAIR receiver principle is shown in Figure 42.56. A stainless steel support structure on the back of a set of ceramic absorber modules forms the base of the receiver. Similar to ceramic burner tubes, the absorber modules are separated from the back and allowed to expand both axially and radially with thermal expansion of the modules or movement of the stainless steel construction behind them during start-up or shutdown. The absorber modules are spaced to avoid touching adjacent modules. The support structure is a double-sheet membrane, which may be cooled by either ambient or returned air. Tubes attached to the absorber cups pass through holes in the front sheet and are

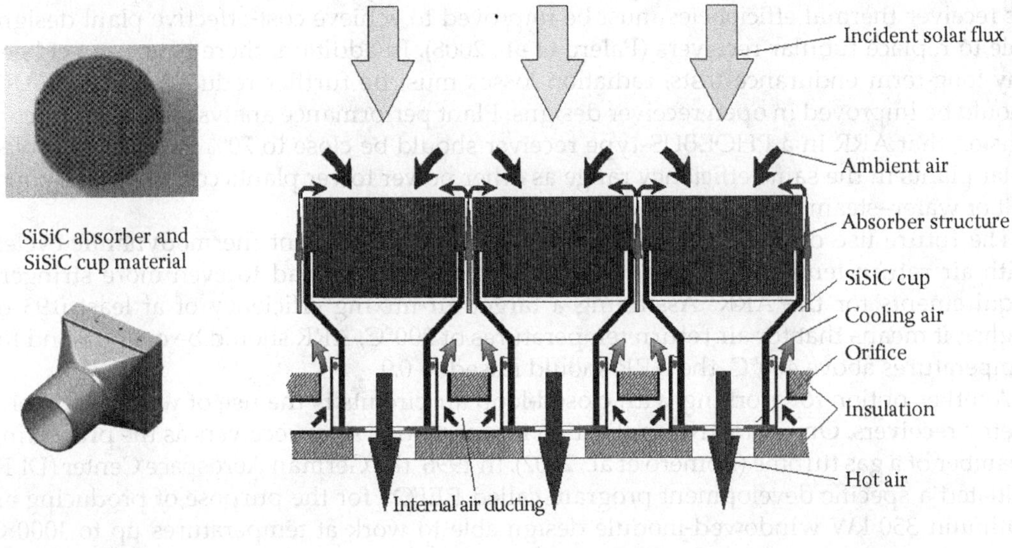

FIGURE 42.56
Principle of volumetric receivers used in HiTRec and SOLAIR projects.

welded to the rear sheet. The cooling air circulates between the two sheets and, as it leaves through the sides of the segments, also cools the support structure. The air reaches the absorber aperture through the spaces between the segments. Outgoing air and ambient air are mixed and sucked back into the segments. As they penetrate the absorber structure, the air is heated up by convection. On leaving the absorber structure, the hot air is ducted to the bottom of the cup. There an orifice, previously sized according to solar flux simulations, adjusts the air mass flow rate to compensate the expected solar flux profile over the aperture and to provide homogeneous outlet air temperatures from the cups. Then, the air passes through the tubes hosting the cups and across the holes in the rear sheet of the membrane.

A first milestone in the development of this type of receiver was the qualification of the 200 kW HiTRec-II receiver (Hoffschmidt et al., 2003). Testing lasted from November 2000 through May 2001. During the course of the test program, the HiTRec-II receiver was operated over a period of 38 days, accumulating a total of 155 operating hours with a thermal efficiency of up to 76% ± 7% at 700°C. Eventually, a SOLAIR 3 MW prototype was developed, installed, and tested in the TSA test bed in the PSA and therefore connected to the heat storage thermocline, the steam generator, and the air circuit. The receiver was designed as a real modular absorber, which was assembled from 270 ceramic modules measuring 140 mm^2. Each module consisted of a square absorber structure glued to a SiSiC cup. The cups were square at the aperture, but round at the back. The honeycomb absorber structure was made of recrystallized SiC with a normal open porosity of 49.5%. The SOLAIR-3000 receiver had a total aperture surface of 5.67 m^2. During testing, the incident solar radiation is reflected by the heliostat field and concentrated on the ceramic volumetric absorber with an average flux density of 0.5 MW/m^2. The air leaves the absorber outlet at 700°C–750°C. The system was evaluated during 2003 and 2004 presenting an air return ratio (ARR) of 0.5, efficiency of (72 ± 9)% at a temperature of 750°C, and efficiency of (74 ± 9)% at 700°C. Efficiencies were estimated at over 85% (and up to 89%) for outlet air temperatures in the range of 590°C–630°C and mean incident solar fluxes of 310–370 kW/m^2 (Agrafiotis et al., 2007).

Though simplicity and operating results are satisfactory, it is obvious that open volumetric receiver thermal efficiencies must be improved to achieve cost-effective plant designs able to replace tubular receivers (Palero et al., 2008). In addition, there have not yet been any long-term endurance tests, radiation losses must be further reduced, and the ARR should be improved in open receiver designs. Plant performance analysis leads to the conclusion that ARR in a PHOEBUS-type receiver should be close to 70% to keep air-cooled solar plants in the same efficiency range as other power tower plants cooled with molten-salt or water–steam (Marcos et al., 2004).

The future use of open volumetric receivers in more efficient thermodynamic cycles, with air return temperatures up to 500°C, and more, may lead to even more stringent requirements for the ARR. Assuming a target air-mixing efficiency of at least 0.93 or higher, it means that for air return temperatures of 200°C, ARR should be over 0.8 and for temperatures above 400°C, the ARR should raised to 0.9.

Another option for working with closed-loop air circuits is the use of windowed volumetric receivers. One attractive application is based upon solar receivers as the preheating chamber of a gas turbine (Romero et al., 2002). In 1996, the German Aerospace Center (DLR) initiated a specific development program called REFOS for the purpose of producing an optimum 350 kW windowed-module design able to work at temperatures up to 1000°C and with pressures up to 15 bar. The aim of the REFOS project was to develop, build, and test modular pressurized volumetric receivers under representative operating conditions

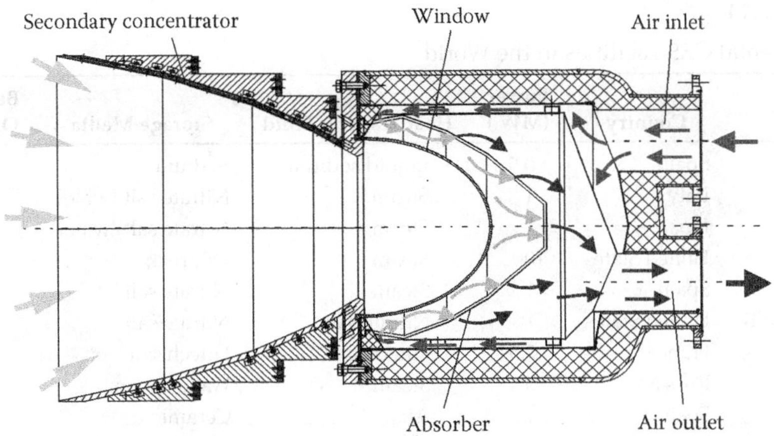

FIGURE 42.57
Example of 350 kW REFOS module.

for coupling to gas turbines. Emphasis was on testing solar preheating of air, accompanied by basic materials research. The REFOS receiver consists of a cylindrical vessel containing a curved knitted absorber. A quartz dome is used to pressurize the air cycle. A hexagonal secondary concentrator with a 1.2 m inner diameter is used to increment the flux density and protect the window flange. The hexagonal shape was selected to optimize the layout of cluster packing in such a way that the REFOS modules can be used in either small or large power plants (Figure 42.57). Typical REFOS module specifications are as follows:

- Absorbed thermal power (design point): 350 kW per unit
- Absolute pressure (operation): 15 bar
- Air outlet temperature: 800°C for metal absorber and up to 1000°C for ceramics
- Temperature increment per module: 150°C
- Receiver efficiency (including secondary): 80% at design point

Several modules have been tested at the PSA under cooperation agreements between DLR and CIEMAT. In the most extensive tests performed in 1999, the design conditions were demonstrated with a single module operating at air outlet temperatures of 800°C at 15 bar, at power levels up to 400 kW_{th}. More than 247 h of testing proved the feasibility of the receiver, which reached a maximum temperature of 1050°C without incurring damage (Buck et al., 2002). Thermal efficiencies were between 63% and 75%.

42.6.3 Experience in Central Receiver Systems

Although there have been a large number of STP tower projects, only a few have culminated in the construction of an entire experimental system. Table 42.11 lists the experimental systems that have been tested all over the world before commercial projects started in 2006. In general terms, as observed, they are characterized as being small demonstration systems between 0.5 and 10 MW, and most of them were operated in the 1980s (Entropie, 1982; Falcone, 1986; Grasse et al., 1991). The thermal fluids used in the receiver have been liquid sodium, saturated or superheated steam, nitrate-based molten salts, and air. All of

TABLE 42.11

Experimental CRS Facilities in the World

Project	Country	Power (MW_e)	Heat Transfer Fluid	Storage Media	Beginning Operation
SSPS	Spain	0.5	Liquid sodium	Sodium	1981
Eurelios	Italy	1	Steam	Nitrate salt/water	1981
Sunshine	Japan	1	Steam	Nitrate salt/water	1981
Solar One	United States	10	Steam	Oil/rock	1982
CESA-1	Spain	1	Steam	Nitrate salt	1982
MSEE/Cat B	United States	1	Nitrate salt	Nitrate salt	1983
Themis	France	2.5	Hitech salt	Hitech salt	1984
SPP-5	Russia	5	Steam	Water–steam	1986
TSA	Spain	1	Air	Ceramic	1993
Solar Two	United States	10	Nitrate salt	Nitrate salt	1996
CONSOLAR	Israel	0.5[a]	Pressurized air	Fossil hybrid	2001
SOLGATE	Spain	0.3	Pressurized air	Fossil hybrid	2002

[a] Thermal.

them can easily be represented by flow charts, where the main variables are determined by working fluids, with the interface between power block and the solar share.

The set of experiences referred to has served to demonstrate the technical feasibility of the CRS power plants, whose technology is sufficiently mature. The most extensive experience has been collected by several European projects located in Spain at the premises of the PSA (Grasse et al., 1991), and the 10 MW Solar One (Radosevich and Skinrood, 1989) and Solar Two plants (Pacheco and Gilbert, 1999) in the United States. Since the early 1990s, most proposals for the first generation of commercial plants have focused on nitrate salt and air as the receiver HTFs, since a joint U.S./German study identified their potential and economics in power tower plants (Chavez et al., 1993), though in Spain, parallel initiatives have developed saturated steam designs (Silva et al., 1999). Several penetration strategies have been proposed since then, and many more may be developed in the future, since solar towers have the great advantage of admitting very open integration designs depending on the dispatching scenarios, annual capacity factors, and hybridization schemes. At present, water–steam and molten salts are the HTFs being selected for the first generation of commercial plants.

42.6.3.1 Water–Steam Plants: The PS10 Project

Production of superheated steam in the solar receiver has been demonstrated in several plants, such as Solar One, Eurelios, and CESA-1, but operating experience showed critical problems related to the control of zones with dissimilar heat transfer coefficients like boilers and superheaters (Grasse et al., 1991). Better results regarding absorber panel lifetime and controllability have been reported for saturated steam receivers. In particular, the Solar Thermal Enhanced Oil Recovery pilot plant, which proved to be highly reliable for oil extraction using direct injection of steam, was successfully operated in Kern County, California, for 345 days in 1983 (Blake et al., 1985). The good performance of saturated steam receivers was also qualified at the 2 MW Weizmann receiver that produced steam at 15 bar for 500 h in 1989 (Epstein et al., 1991). Even though technical risks are reduced by saturated steam receivers, the outlet temperatures are significantly lower than those of superheated steam, making applications where heat storage is replaced by fossil fuel backup necessary.

At present, estimated costs of electricity production from the solar share of hybrid systems are \$0.08–\$0.15 per kWh, whereas expected costs for solar-only plants are in the range of \$0.15–\$0.20 per kWh. The implementation of hybridized systems is one of the paths leading to a breakthrough in the financial barriers to the deployment of solar electricity technologies as it reduces the initial investment (Kolb, 1998). The use of hybrid plants with the low technological risk of a CRS with saturated steam as the working fluid is the starting point. Two projects subsidized by the European Commission, the SOLGAS project promoted by SODEAN and the COLON SOLAR project promoted by the Spanish utility, SEVILLANA (Ruiz et al., 1999), established the strategy of market penetration on the basis of the integration of saturated steam receivers in cogeneration systems and repowering of combined cycles. The size of the cavity receiver was optimized to supply 21.8 MW_{th} to the fluid at 135 bar and 332.8°C outlet temperature. The collector subsystem consisted of 489 heliostats (each with a 70 m^2 reflective surface) and a 109 m tower. As observed in Table 42.12, the use of low-temperature receivers and phase-change saturated steam yields a much higher thermal efficiency of up to 92% at nominal load. The table shows a theoretical comparison between a typical volumetric air-cooled receiver working at 700°C air outlet temperature and saturated steam receiver at 250°C thermal outlet. Both are cavity receivers with an incident power of 45 MW (Osuna et al., 2004).

Integrating power towers into existing combined-cycle plants can create issues with respect to heliostat field layout, since the solar field is forced to make use of sites near gas pipelines and industrial areas. Land becomes a nonnegligible share of plant cost, and site constraints lead to layout optimization and subsequent optical performance problems. This was the case of the COLON heliostat field, which represented a real design challenge, because of the significant restrictions imposed by the available site (Romero et al., 1999). The hybrid solar-gas scheme predicts solar production costs below \$0.11 per kWh and annual solar shares in the range of 8%–15%. The lack of public support schemes for hybrid STP plants in Spain at that time led to project abortion, and the plant was never built.

The follow-up of the COLON SOLAR project was finally a solar-only saturated steam plant called PS10 (Planta Solar 10 MW). PS10 came about as a consequence of the Spanish legal framework for a special regime of feed-in tariffs for renewable electricity, issued in March 2004. PS10 is located on the Casa Quemada estate (37.2° latitude), 15 km west of the city of Seville (Spain). The 11 MW plant was designed to achieve an annual electricity production of 23 GWh at an investment cost of less than \$4000 per kW. The project made use of available, well-proven technologies like the glass-metal heliostats developed by the Spanish Inabensa company and the saturated steam cavity receiver developed by the TECNICAL company to produce steam at 40 bar and 250°C (Osuna et al., 2004). The plant is a solar-only system with saturated steam heat storage

TABLE 42.12

Comparison of Thermal Losses and Efficiency in Air Volumetric and Saturated Steam Receivers for the PS10 Project

Losses	Air (%)	Steam (%)
Reflection	7.9	2.0
Radiation	8.6	0.8
Convection	0.0	2.6
Spillage	5.0	2.1
Air return	3.7	0.0
Total efficiency	74.8	92.4

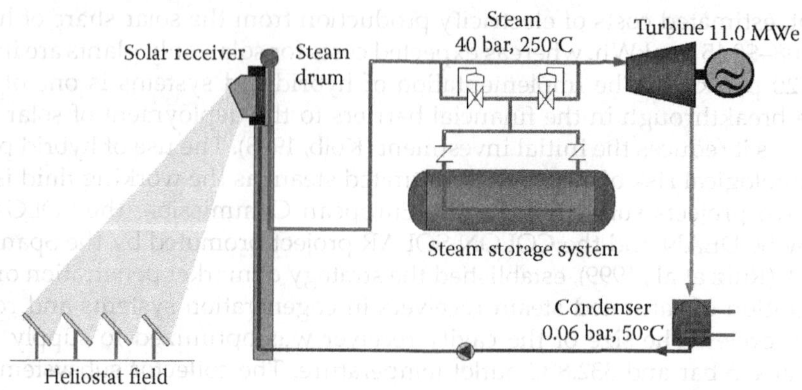

FIGURE 42.58
Basic scheme of PS10 solar thermal power plant with saturated steam as thermal fluid.

able to supply 50 min of plant operation at 50% load. The system makes use of 624,121 m^2 heliostats, distributed in a north-field configuration, a 90 m–high tower, a 15 MWh heat storage system, and a cavity receiver with four 4.8 × 12 m tubular panels. The basic flow diagram selected for PS10 is shown in Figure 42.58. Though the system makes use of a saturated steam turbine working at extremely low temperature, the nominal efficiency of 30.7% is relatively good. This efficiency is the result of optimized management of waste heat in the thermodynamic cycle. At the turbo-generator exit, the steam is sent to a water-cooled condenser, working at 0.06 bar. The condenser outlet is preheated with 0.8 and 16-bar steam bled from the turbine. The output of the first preheater is sent to the deaerator, which is fed with steam again bled from the turbine. A humidity separator is installed between the high- and low-pressure sections of the turbine to increase steam quality in the last stages of expansion. A third and last preheater fed with steam from the receiver increases water temperature to 245°C. When mixed with water returned from the drum, 247°C receiver feed water is obtained. As summarized in Table 42.13, the combination of optical, receiver, and power block efficiencies leads to

TABLE 42.13
Annual Energy Balance for the PS10 Plant at Nominal Conditions and Annual Basis

Nominal Rate Operation		
Optical efficiency	77.0%	67.5 → 51.9 MW
Receiver and heat handling efficiency	92.0%	51.9 → 47.7 MW
Thermal power to storage		11.9 MW
Thermal power to turbine		35.8 MW
Thermal power/electric power efficiency	30.7%	35.8 → 11.0 MW
Total efficiency at nominal rate	21.7%	
Energetic Balance in Annual Basis		
Mean annual optical efficiency	64.0%	148.63 → 95.12 GWh
Mean annual receiver and heat-handling efficiency	90.2%	95.12 → 85.80 GWh
Operational efficiency (start-up/stop)	92.0%	85.80 → 78.94 GWh
Operational efficiency (breakages, O&M)	95.0%	78.94 → 75.00 GWh
Mean annual thermal energy/electricity efficiency	30.6%	75.00 → 23.0 GWh
Total annual efficiency	15.4%	

a total nominal efficiency at design point of 21.7%. Total annual efficiency decreases to 15.4%, including operational losses and outages. PS10 is a milestone in the CRS deployment process, since it is the first solar power tower plant developed for commercial exploitation. Commercial operation started on June 21, 2007. Since then, the plant is performing as specified. The construction of PS20, a 20 MW_e plant with the same technology as PS10, followed. PS20 started operation in May 2009. With 1255 heliostats (120 m^2 each) spread over 90 ha and with a tower of 165 m high, the plant is designed to produce 48.9 GWh/year (Figure 42.40).

Saturated steam plants are considered a temporary step to the more efficient superheated steam systems. Considering the problems found in the 1980s with superheated steam receivers, the current trend is to develop dual receivers with independent absorbers, one of them for the preheating and evaporation, and another one for the superheating step. The experience accumulated with heuristic algorithms in central control systems applied to aiming point strategies at heliostat fields allows achieving a flexible operation with multi-aperture receivers. The company Abengoa Solar, developer of PS10 and PS20, has already designed a superheated steam receiver for a new generation of water–steam plants to be implemented in the first CRS project in South Africa (Fernandez-Quero et al., 2005). The most advanced strategy is the program initiated by the BrightSource PLT. BrightSource has already built a demonstration plant of 6 MW_{th} located at the Negev desert in June 2008 (Silberstein et al., 2009). The final objective of BrightSource is to promote plants producing superheated steam at 160 bar and temperature of 565°C (named DPT550). With those characteristics, they expect up to 40% conversion efficiency at the power block for unit sizes between 100 and 200 MW. The receiver is cylindrical, dual, and with a drum. The first commercial project producing superheated steam is Ivanpah Solar in California of 380 MW, with three towers. The plant is expected to connect to the grid in 2014.

The combination of recent initiatives on small heliostats, compact modular multitower fields, and production of superheated steam may be clearly visualized in the development program of the company eSolar. This company proposes a high degree of modularity with power units of 46 MW covering 64 ha, consisting of 16 towers and their corresponding heliostat fields sharing a single central power block. With replication, modularity sizes up to 500 MW and upward may be obtained (Tyner and Pacheco, 2009). Two modules of 2.5 MW each were installed in 2009 by eSolar in Lancaster, California. Each receiver had two independent cavities, and the heliostat layout consisted of identical arrays to maximize replication and modularity. Each tower is associated with 12,180 flat heliostats of 1.14 m^2 each (Schell, 2009). The receivers are dual-cavity, natural-circulation boilers. Inside the cavity, the feed water is preheated with economizer panels before entering the steam drum. A downcomer supplies water to evaporator panels where it is boiled. The saturated water/vapor mixture returns to the drum where the steam is separated, enters superheater panels, and reaches 440°C at 6.0 MPa. Each receiver absorbs a full-load power of 8.8 MW_{th}.

A comparative analysis of system integration schemes for megawatt-range DSG central receiver STP plants reveals that the differences between superheated steam and saturated steam schemes may be minimized in some extent given the fact that dual receivers involve two apertures and worse thermal efficiencies (Sanz-Bermejo et al., 2014). Superheated steam systems at 550°C and pressure between 60 and 80 bar obtain the best results. However, the increment of efficiency versus saturated solar plant at 69 bar is just 2.3% in the best case, and if the 69-bar saturated system integrates an intermediate reheat process, it can even achieve the same performance level of superheated systems. It is also relevant to mention that saturated conditions at 40 bar provide similar performance than superheated system operating at 500°C.

42.6.3.2 Molten-Salt Systems: Solar Two and Gemasolar

The Solar One Pilot Plant successfully demonstrated operation of a utility-scale power tower plant. The Solar One receiver heated subcooled water to superheated steam, which drove a turbine. The superheated steam was also used to charge an oil-rock thermocline storage system. Solar One operated for 6 years from 1982 to 1988, the last three of which were devoted to power production (Radosevich and Skinrood, 1989). Although Solar One successfully demonstrated the feasibility of the power tower concept, the thermal storage system was inadequate for operating the turbine at peak efficiency, because the storage system operated only between 220°C and 305°C, whereas the receiver outlet (and design turbine inlet) temperature was 510°C. The primary mode of operation was directly from receiver outlet to turbine input, bypassing the thermal storage system, and storage provided auxiliary steam during offline periods.

For high annual capacity factors, solar-only power plants must have an integrated cost-effective thermal storage system. One such thermal storage system employs molten nitrate salt as the receiver HTF and thermal storage media. To be usable, the operating range of the molten nitrate salt, a mixture of 60% sodium nitrate and 40% potassium nitrate, must match the operating temperatures of modern Rankine cycle turbines. In a molten-salt power tower plant, cold salt at 290°C is pumped from a tank at ground level to the receiver mounted atop a tower where it is heated by concentrated sunlight to 565°C. The salt flows back to ground level into another tank. To generate electricity, hot salt is pumped from the hot tank through a steam generator to make superheated steam. The superheated steam powers a Rankine cycle turbine. A diagram of a molten-salt power tower is shown in Figure 42.59. The collector field can be sized

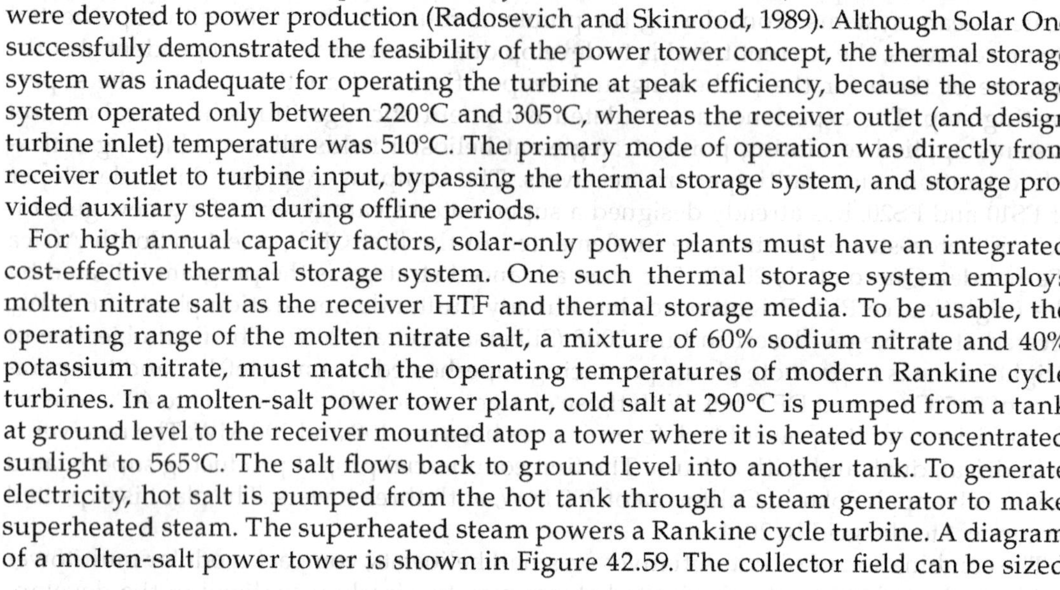

FIGURE 42.59
Schematic of a molten-salt central receiver system with cylindrical tubular receiver.

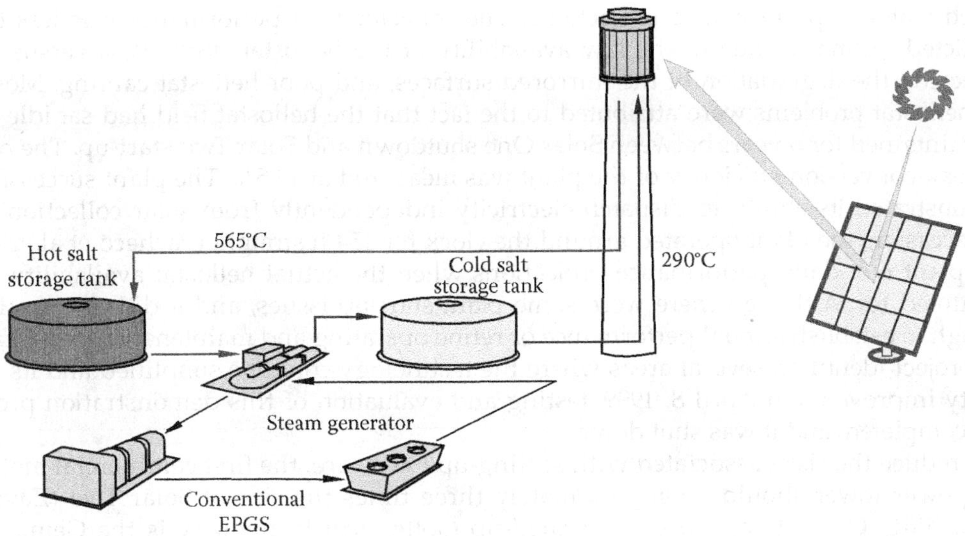

FIGURE 42.60
Comparison of volume energy storage capacity for several chemical, thermal, and mechanical media.

to collect more power than is demanded by the steam generator system, and the excess salt is accumulated in the hot storage tank. With this type of storage system, solar power tower plants can be built with annual capacity factors up to 70%. As molten salt has a high energy storage capacity per volume (500–700 kWh/m^3), as shown in Figure 42.60, they are excellent candidates for STP plants with large capacity factors. Even though nitrate salt has a lower specific heat capacity per volume than carbonates, they still store 250 kWh/m^3. The average heat conductivity of nitrates is 0.52 W/m K, and their heat capacity is about 1.6 kJ/kg K. Nitrates are a cheap solution for large storage systems. The cost of the material is \$0.70 per kg or \$5.20 per kWh$_{th}$. Estimates for large systems including vessels are in the range of \$13 per kWh$_{th}$.

Several molten-salt development and demonstration experiments have been conducted over the past two and half decades in the United States and Europe to test entire systems and develop components. The largest demonstration of a molten-salt power tower was the Solar Two project, a 10 MW power tower located near Barstow, California.

The purpose of the Solar Two project was to validate the technical characteristics of the molten-salt receiver, thermal storage, and steam generator technologies; improve the accuracy of economic projections for commercial projects by increasing the capital, operating, and maintenance cost database; and distribute information to utilities and the solar industry to foster a wider interest in the first commercial plants. The Solar Two plant was built at the same site as the Solar One pilot plant and reused much of the hardware including the heliostat collector field, tower structure, 10 MW turbine, and balance of plant. A new, 110 MWh$_{th}$ two-tank molten-salt thermal storage system was installed, as well as a new 42 MW$_{th}$ receiver, a 35 MW$_{th}$ steam generator system (535°C, 100 bar), and master control system (Kelly and Singh, 1995).

The plant began operating in June 1996. The project successfully demonstrated the potential of nitrate-salt technology. Some of the key results were the following: receiver efficiency was measured at 88%, the thermal storage system had a measured round-trip efficiency of over 97%, and the gross Rankine-turbine cycle efficiency was 34%, all of which matched performance projections. The collector field performance was less than predicted, primarily due to the low availability of the heliostats (85%–95% versus 98% expected), the degradation of the mirrored surfaces, and poor heliostat canting. Most of the heliostat problems were attributed to the fact that the heliostat field had sat idle and unmaintained for 6 years between Solar One shutdown and Solar Two start-up. The overall peak-conversion efficiency of the plant was measured at 13.5%. The plant successfully demonstrated its ability to dispatch electricity independently from solar collection. On one occasion, the plant operated around the clock for 154 h straight (Pacheco et al., 2000). The plant met daily performance projections when the actual heliostat availability was accounted for. Although there were some plant start-up issues, and it did not run long enough to establish annual performance or refine operating and maintenance procedures, the project identified several areas where the technology could be simplified and its reliability improved. On April 8, 1999, testing and evaluation of this demonstration project was completed, and it was shut down.

To reduce the risks associated with scaling-up hardware, the first commercial molten-salt power tower should be approximately three times the size of Solar Two (Zavoico et al., 2001). One attempt to prove scaled-up molten-salt technology is the Gemasolar project (Figure 42.61) promoted, built, and operated by Torresol Energy, a joint venture between the Spanish SENER and MASDAR initiative from Abu Dhabi (Ortega et al., 2006). Table 42.14 summarizes the main technical specifications of Gemasolar project. With only 17 MW_e, the plant that connected to the grid in summer 2011 is designed to produce 112 GWh_e/year. A large heliostat field of 304,750 m^2 (115 m^2 each heliostat) is oversized to supply 15 h equivalent heat storage capacity. The plant is designed to operate around the clock in summertime, leading to an annual capacity factor of 74%. Fossil backup corresponding to 15% of annual production is added. The levelized energy costs are estimated to be approximately $0.16 per kWh. Gemasolar represents a breakthrough for solar technology in terms of time-dispatch management (Burgaleta et al., 2009; García and Calvo, 2012).

(a) (b)

FIGURE 42.61
Aerial view of Gemasolar plant located in South Spain, the largest commercial solar central receiver system in operation by end 2013 with a circular-shape heliostat field (a). Lateral view of cylindrical solar receiver (b). (Courtesy of Torresol Energy, Spain.)

TABLE 42.14

Technical Specifications and Design Performance of the Gemasolar Project

Technical Specifications	
Heliostat field reflectant surface	304,750 m^2
Number heliostats	2,650
Land area of solar field	142 ha
Receiver thermal power	120 MW_{th}
Tower height	145 m
Heat storage capacity	15 h
Power at turbine	17 MW_e
Power NG burner	16 MW_{th}
Operation	
Annual electricity production	112 MWh_e
Production from natural gas (annual)	15%
Capacity factor	74%

42.7 Dish–Stirling Systems

STP plants can also be applied to distributed generation through parabolic dishes in which a power conversion unit (PCU) is attached by an arm directly to the concentrator. Although there have been other modular PCU system initiatives in the past, like the dish/Brayton tested by Cummins and DLR (Buck et al., 1996) or the use of dish farms to produce superheated steam designed to feed a centralized Rankine cycle in Georgia (Alvis, 1984), it is the dish–Stirling system that has demonstrated from the earliest prototypes high peak conversion efficiencies of above 30% solar-to-electric and a daily average of up to 25%. Dish–Stirling systems are considered as efficient technology to convert solar energy into electricity. This high conversion efficiency is due to their high concentration ratios (up to 3000×) and high working temperatures of above 750°C (Stine and Diver, 1994).

42.7.1 System Description

Dish–Stirling systems track the sun and focus solar energy onto a cavity receiver, where it is absorbed and transferred to a heat engine/generator. An electrical generator, directly connected to the crankshaft of the engine, converts the mechanical energy into electricity (AC). To constantly keep the reflected radiation at the focal point during the day, a sun-tracking system continuously rotates the solar concentrator on two axes following the daily path of the sun. With current technologies, a 5 kW_e dish–Stirling system would require 5.5 m diameter concentrator, and for 25 kW_e, the diameter would have to increase up to 10 m. Stirling engines are preferred for these systems because of their high efficiencies (40% thermal to mechanical), high power density (40–70 kW/L), and potential for long-term, low-maintenance operation. Dish–Stirling systems are modular, that is, each system is a self-contained power generator, allowing their assembly in plants ranging in size from a few kilowatts to tens of megawatts (Mancini et al., 2003).

Global efficiency of the system can be defined as

$$\eta = \eta_C * \eta_R * \eta_{Stir} * \eta_{Gen} = \frac{P}{A_C * I} \tag{42.49}$$

where
η_C is the concentrator efficiency
η_R is the receiver efficiency
η_{Stir} is the Stirling engine efficiency
η_{Gen} is the generator efficiency
P is the gross power generated
A_C is the projected concentrator area
I is the direct normal irradiance

42.7.1.1 Concentrator

The concentrator is a key element of any dish–Stirling system. The curved reflective surface can be manufactured by attached segments, by individual facets, or by a stretched membrane shaped by a continuous plenum. In all cases, the curved surface should be coated or covered by aluminum or silver reflectors. Second-surface glass mirrors, front surface thin-glass mirrors, or polymer films have been used in various different prototypes.

First-generation parabolic dishes developed in the 1980s were shaped with multiple, spherical mirrors supported by a trussed structure (Lopez and Stone, 1992). Though extremely efficient, this structure concept was costly and heavy. The introduction of automotive industry concepts and manufacturing processes has led to optimized commercial versions like the 25 kW SunCatcher system developed by the company Stirling Energy Systems (SES), which has peak efficiency rating of 31.25%, for converting solar energy-to-grid quality electricity. Large monolithic reflective surfaces can be obtained by using stretched membranes in which a thin reflective membrane is stretched across a rim or hoop. A second membrane is used to close off the space behind forming a partially evacuated plenum between them, giving the reflective membrane an approximately spherical shape. This concept was developed by the German SBP company in the 1990s, and several prototypes have been tested at the PSA in Spain (Schiel et al., 1994). An alternative to the single stretched membrane in large units is a composition of a number of small circular rings, each with their corresponding membrane (Beninga et al., 1997).

Another example of concentrator developments in dish–Stirling systems is the Eurodish prototype (Figure 42.62). The concentrator consists of 12 single segments made of fiber glass resin. When mounted, the segments form an 8.5 m–diameter parabolic shell. The shell rim is stiffened by a ring truss to which bearings and the Stirling support structure are later attached. Thin 0.8 mm–thick glass mirrors are glued onto the front of the segments for durable, high reflectivity of around 94% (Keck et al., 2003).

Solar tracking is usually done by two different methods (Adkins, 1987):

1. Azimuth-elevation tracking by an orientation sensor or by calculated coordinates of the sun performed by the local control.
2. Polar tracking, where the concentrator rotates about an axis parallel to the earth's axis rotation. The rate of rotation is constant and equal to 15° per h. Declination angle movement is only ±23.5° per year (0.016° per h) and therefore adjusted from time to time.

FIGURE 42.62
Example of segmented parabolic concentrators. Rear and front views of two Eurodish prototype tested at the Plataforma Solar de Almería, Spain.

42.7.1.2 Receiver

As in central receivers and parabolic-trough absorbers, the receiver absorbs the light and transfers the energy as heat to the engine's working gas, usually helium or hydrogen. Thermal fluid working temperatures are between 650°C and 750°C. This temperature strongly influences the efficiency of the engine. Because of the high operating temperatures, radiation losses strongly penalize the efficiency of the receiver; therefore, a cavity design is the optimum solution for this kind of system.

Two different heat transfer methods are commonly used in parabolic dish receivers (Diver, 1987). In directly illuminated receivers, the same fluid used inside the engine is externally heated in the receiver through a pipe bundle. Although this is the most conventional method, a good high-pressure, high-velocity, heat-transfer gas like helium or hydrogen must be used. In indirect receivers, an intermediate fluid is used to decouple solar flux and working temperature from the engine fluid. One such method is heat pipes, which employ a metal capillary wick impregnated with a liquid metal heated up through the receiver plate and vaporized. The vapor then moves across the receiver and condenses in a cooler section, transferring the heat to the engine. The phase change guarantees good temperature control, providing uniform heating of the Stirling engine (Moreno et al., 2001).

42.7.1.3 Stirling Engine

Stirling engines solarized for parabolic dishes are externally heated gas-phase engines in which the working gas is alternately heated and cooled in constant-temperature, constant-volume processes (Figure 42.63b). This possibility of integrating additional external heat in the engine is what makes it an ideal candidate for solar applications. Since the Stirling cycle is very similar to the Carnot cycle, the theoretical efficiency is high. High reversibility is achieved since work is supplied to and extracted from the engine at isothermal conditions. The clever use of a regenerator that collects the heat during constant-volume cooling and heating substantially enhances the final system efficiency. For most engine designs, power is extracted kinematically by rotating a crankshaft connected to the pistons by a connecting rod. An example of a kinematic Stirling engine is shown in Figure 42.63a. Though, theoretically, Stirling engines may have a high life-cycle projection, the actual

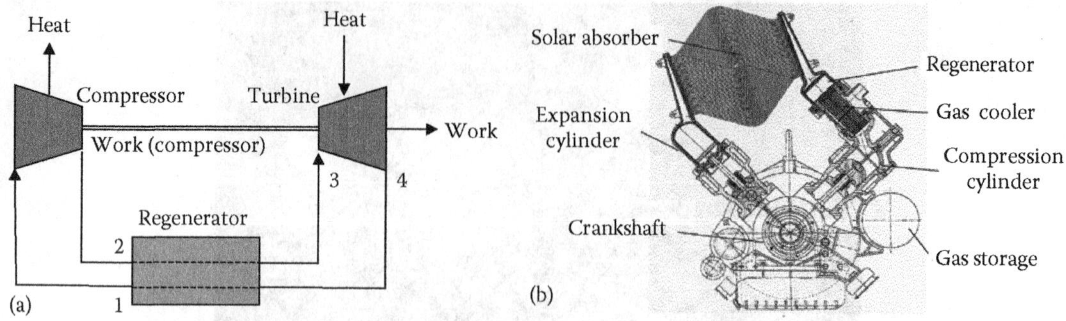

FIGURE 42.63
(a) Representation of the ideal Stirling cycle. (b) Kinematic Stirling engine V-160 of 10 kW_e manufactured by Solo Kleinmotoren with pistons situated in V and connected to a tubular array heat exchanger.

fact is that today their availability is still not satisfactory, as an important percentage of operating failures and outages are caused by pistons and moving mechanical components. Availability is therefore one of the key issues, since it must operate for more than 40,000 h in 20-year lifetime, or 10 times more than an automobile engine. One option to improve availability is the use of free-piston designs. Free-piston engines make use of gas or a mechanical spring so that mechanical connections are not required to move reciprocating pistons. Apparently, they are better than kinematic engines in terms of availability and reliability. The most relevant program in developing dishes with free-piston technology was promoted by Cummins in the United States in 1991. Unfortunately, there were technical problems with the PCU and the project was cancelled (Bean and Diver, 1995). The last 10 years, the company Infinia Solar has developed several prototypes and test fields for 3 kW free piston systems derived from satellite applications, though still extended operation is needed. Due to the flexibility of the heat source, a Stirling engine can also be operated with a solar/fossil or solar/biomass hybrid receiver (Laing and Trabing, 1997), making the system available during cloudy periods and at night.

42.7.2 Dish–Stirling Developments

Like the other CSP technologies, practical dish–Stirling development started in the early 1980s. Most development has concentrated in the United States and Germany, and though developed for commercial markets, they have been tested in a small number of units (Stine and Diver, 1994).

The first generation of dishes was a facet-type concentrator with second-surface mirrors that already established concentration records ($C = 3000$) and had excellent performances, though their estimated costs for mass production were above \$300 per m^2. Their robust structures were extremely heavy, weighing in at 100 kg/m^2 (Grasse et al., 1991). The 25 kW Vanguard-1 prototype built by Advanco was operated at Rancho Mirage, California, in the Mojave Desert in production mode for 18 months (February 1984 to June 1985), and results were published by EPRI (Droher and Squier, 1986).

This system was 10.7 m in diameter with a reflecting surface of 86.7 m^2 and a 25 kW PCU made by United Stirling AB model 4-95 Mark II. This engine had four cylinders with a 95 cm^3 cylinder displacement. Cylinders were distributed in parallel and assembled in a square. They were connected to the regenerator and cooler and had double-acting pistons. The working gas was hydrogen at a maximum pressure of 20 MPa and temperature of 720°C.

Engine power was controlled by varying the working-gas pressure (Schiel, 1999). The Advanco/Vanguard system, with a net conversion efficiency (including ancillary systems) of more than 30%, still holds the world's conversion record.

McDonnell Douglas later developed another somewhat improved dish system making use of the same technology and the same engine. The dish was 10.5 m and 25 kW. The 88 m^2 parabolic dish consisted of 82 spherically curved glass facets. Six of these units were produced and installed at sites around the United States for testing in operation. Southern California Edison continued to evaluate the system later. Reported performances and efficiencies were similar to those of Advanco/Vanguard (Lopez and Stone, 1992). The project was frozen for several years until in 1996, SES acquired the intellectual and technology rights to the concentrator and the U.S. manufacturing rights to what is now called the Kockums, 4-95 Stirling engine-based PCU (Mancini et al., 2003). Under a DOE-industry cost-sharing project to commercialize the dish–Stirling system for emerging markets, SES started testing and improvement of several units in different locations in the United States and South Africa. More than 100,000 h of operation accumulated for all the systems have been reported (Stone et al., 2001; Mancini et al., 2003). Daily efficiency has been found to be 24%–27% and the annual average 24%, and what is even more important, they have claimed availabilities of 94% at irradiances of just over 300 W/m^2. The redesigned 25 kW_e system named SunCatcher has being qualified in a commercial basis at the Maricopa Solar Plant in Arizona. The plant totalized 1.5 MW with 60 dishes and started operation in January 2010. This plant did not achieve the expected performance obtained at previous prototypes testing, and it was dismantled in 2012 after 1 year of operation.

Since the pioneering Vanguard dish records, with the exception of SES, most of the design options have been directed at the development of lowering costs by such strategies as less demanding temperatures, thereby penalizing efficiency, and introducing lighter and less expensive reflectors made of polymers or thin glass glued onto resin-based structures. These dishes, which have a lower optical performance, were first used in non-Stirling applications, with lower operating temperatures, such as the Shenandoah (Solar Kinetics) and Solar Plant 1 in Warner Springs (LaJet) (Grasse et al., 1991). Typical concentrations were in the range of 600–1000, and working temperatures were in the range of 650°C. Several prototypes were developed by Acurex, LaJet, GE, SKI, and SBP. These developments were followed up in the United States by SAIC (Mayette et al., 2001) and WGA (Diver et al., 2001) under the DOE-industry R&D program, Dish Engine Critical Components (Mancini et al., 2003).

The most extensive testing of this light material concept has been done with the stretched-membrane concentrator developed in Germany by Schlaich, Bergermann und Partner (SBP). More than 50,000 h of testing have been accumulated in the six-prototype field, promoted by SBP and Steinmüller, and evaluated at the PSA in Spain (Schiel et al., 1994). The concentrator is a single 7.5 m–diameter facet made of a single 0.23 mm thick preformed stainless-steel stretched membrane. Thin-glass mirrors are bonded to the stainless-steel membrane. The membrane is pre-stretched beyond its elastic limit using a combination of the weight of water on the front and vacuum on the back, to form a nearly ideal paraboloid. Then, a slight active vacuum within the membrane drum preserves the optical shape. The V-160 engine, originally produced by Stirling Power Systems, is at present manufactured by the German company Solo Kleinmotoren (Figure 42.63). The engine sweeps 160 cm^3 of helium with two pistons. The engine has an efficiency of 30% and reported overall conversion efficiency of 20.3% (Figure 42.64). The figure demonstrates the high dispatchability of dish–Stirling systems at part loads.

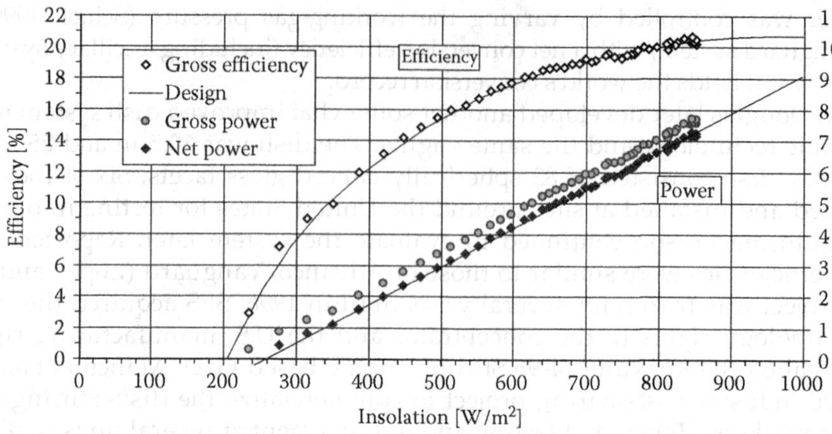

FIGURE 42.64
Input/output parameters and performance measured for a 9 kW SBP stretched membrane dish–Stirling system tested in Almería, Spain.

Though stretched membranes had excellent optical results, the economics revealed production costs higher than expected. The successor of the SBP membrane dishes is the EuroDish system. The EuroDish project is a joint venture undertaken by the European Community, German/Spanish Industry (SBP, MERO, Klein+Stekl, Inabensa), and research institutions DLR and CIEMAT. The new design replaces the stretched-membrane concentrator with a glass-fiber composite shell onto which glass mirrors are bonded with an adhesive. The engine used in the EuroDish is the next-generation SOLO Kleinmotoren 161. Two new 10 kW EuroDish units, shown in Figure 42.62, were installed at the Plataforma Solar de Almeria, Spain, early in 2001 for testing and demonstration. In a follow-up project called EnviroDish, additional units were deployed in France, India, Italy, and Spain to accumulate operating experience at different sites. The peak solar-to-net-electric energy conversion efficiency of the system is expected to be 21%–22%, based on the experience of former projects with the same engine. The peak system efficiency was first measured at 20%. The estimated annual production of a EuroDish system operating in Albuquerque, New Mexico, is 20,252 kWh of electricity with 90% availability and an annual efficiency of 15.7% (Mancini et al., 2003). SBP and the associated EuroDish industry have performed cost estimates for a yearly production rate of 500 units/year (5 MW/year) and 5000 units/year, which corresponds to 50 MW/year. The actual cost of the 10 kW unit without transportation and installation cost and excluding foundations is approximately U.S. $10,000 per kW. The cost projections at production rates of 500 and 5000 units/year are U.S. $2500 per kW and U.S. $1500 per kW, respectively.

42.8 Conclusions and Outlook

From the 1970s to the 1990s, the development of solar thermal electricity technologies remained restricted to a few countries, and only a few, though important, research institutions and industries were involved. The situation has dramatically changed since 2006 with the approval of specific feed-in-tariffs or power purchase agreements in Spain and the United States. Both countries, with more than 3 GW of projects by the end 2013, are leading the commercialization of STP. Other countries such as India, China, South Africa, Chile, Australia, Morocco, Algeria, and Saudi Arabia adopted the STP technology in their portfolios. Subsequently, a number and variety of engineering and construction companies, consultants, technologists, and developers committed to STP are rapidly growing and moving to global markets. A clear indicator of the globalization of STE commercial deployment for the future energy scenario has been elaborated by the International Energy Agency (IEA). This considers STP to play a significant role among the necessary mix of energy technologies for halving global energy-related CO_2 emissions by 2050 (IEA, 2010b). This scenario would require capacity addition of about 14 GW/year (55 new STP plants of 250 MW each). However, this new opportunity is introducing an important stress to the developers of STP. In a period of less than 5 years, in different parts of the world, these developers of STP are forced to move from strategies oriented to early commercialization markets based upon special tariffs, to strategies oriented to a massive production of components and the development of large amounts of projects with less profitable tariffs. This situation is speeding up the implementation of second-generation technologies, like DSG or molten-salt systems, even though in some cases, still some innovations are under assessment in early commercialization plants or demonstration projects.

Parabolic trough is the technology widely used nowadays in commercial projects, though other technologies like LF reflectors and CRSs are developing the first grid-connected projects and reveal promising impacts on cost reduction (Romero and Gonzalez-Aguilar, 2011). The projected evolution of levelized electricity costs (LECs) of different CSP technologies is depicted in Figure 42.65. LEC reduction is expected from mass production, scaling-up, and R&D. A technology roadmap promoted by the European Industry Association ESTELA (Kearney and ESTELA, 2010) states that by 2015, when most of the improvements currently under development are expected to be implemented in new plants, energy production boosts greater than 10% and cost decreases up to 20% are expected to be achieved. Furthermore, economies of scale resulting from plant size increase will also contribute

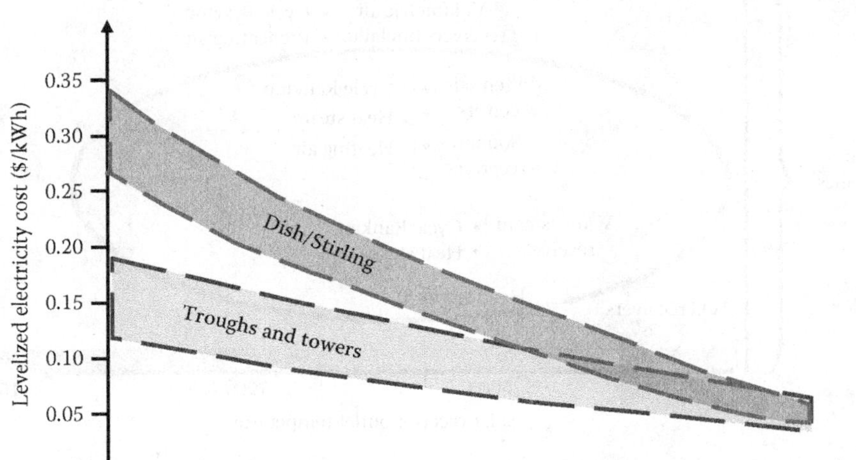

FIGURE 42.65
Evolution of levelized electricity cost for STP technologies based upon technology roadmaps and industry.

to reduce plants' CAPEX per MW installed up to 30%. STP deployment in locations with very high solar radiation further contributes to the achievement of cost competitiveness of this technology by reducing costs of electricity up to 25%. All these factors can lead to electricity generation cost savings up to 30% by 2015 and up to 50% by 2025, reaching competitive levels with conventional sources (e.g., coal/gas with stabilized electricity costs <€10 cents/kWh). Similar projections are published in another recent roadmap issued by the IEA (2010) and IRENA (2012).

The first generation of commercial SPT projects adopted technologies and concepts that have matured in the last 30 years. They were based on conservative designs and schemes that do not necessarily exploit the enormous potential of concentrated solar energy. Figure 42.66 illustrates the situation. Current technologies are based on solar receivers

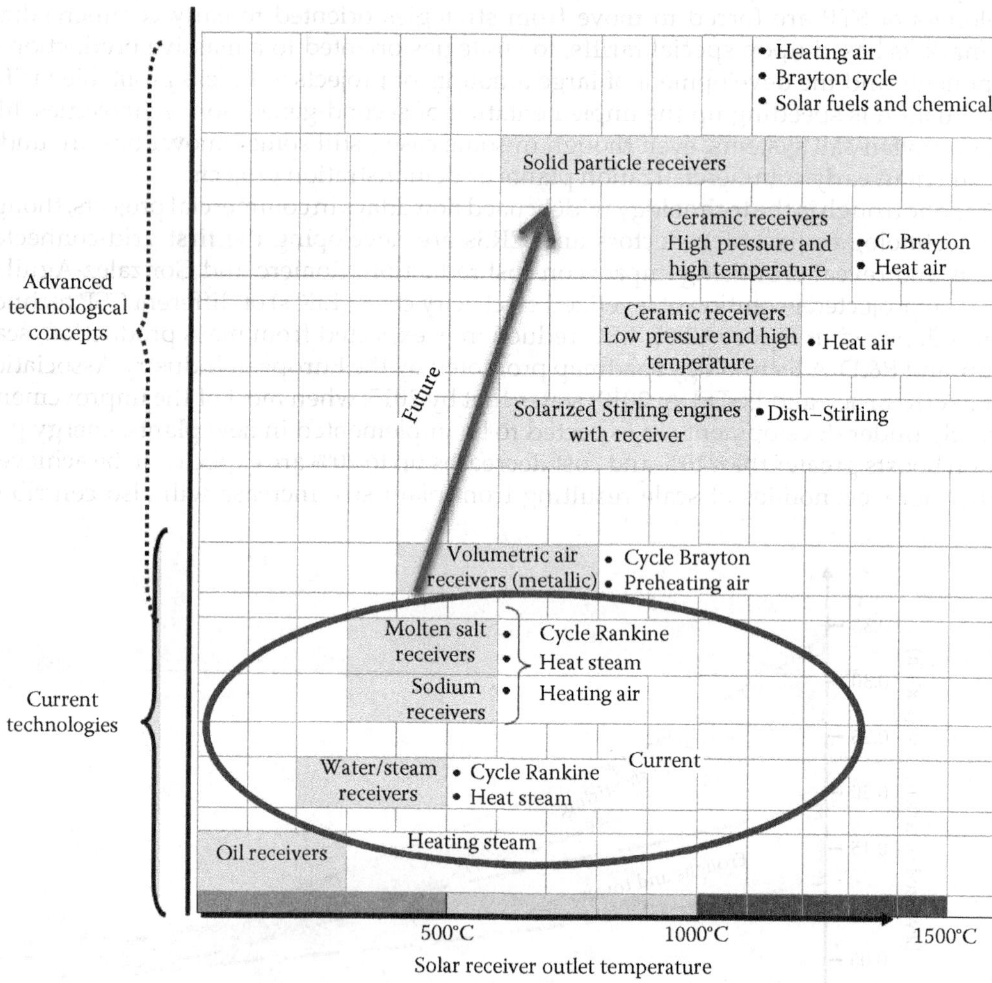

FIGURE 42.66
Evolution of STP. Current technologies are based on solar receivers that operate with thermal oil or water–steam at working temperatures usually below 500°C, coupled to steam-based Rankine cycles. Next generation of technologies allows surpassing 1000°C and enable higher efficiencies via Brayton and combined cycles, as well as the thermochemical production of solar fuels.

operating at moderate solar concentration ratios with thermal oil or water–steam at working temperatures usually below 500°C, which are coupled to steam-based Rankine cycles. As a consequence, solar-to-electricity conversion efficiencies are below 20%, the application of energy storage is limited, the water consumption and land use are relatively high, the power block integration is rather inefficient, and the thermochemical routes to produce solar fuels are beyond reach. Next generations of STP plants should allow surpassing 1000°C and enable higher efficiencies via Brayton and combined cycles as well as better integration of thermal storage (Romero and Steinfeld, 2012). Novel receiver concepts based on volumetric absorption of directly irradiated porous structures and particles, with alternative thermal fluids (e.g., air), operate at higher solar concentration ratios and promise more efficient solar energy capture and conversions. Moreover, these advance concepts open the door to the thermochemical production of solar fuels. The solar concentrating technologies better adapted for these high-temperature applications are solar towers (CRS), whose current development mainly for power generation is paving the ground for future high-flux/high-temperature thermochemical applications.

Challenges and key general topics for the medium- to long-term R&D are improved designs of materials and components, increased system efficiency through higher operating temperatures, high reliability during unattended operation, hybrid solar/fossil fuel plants with small solar share, and solar share increase through integration of storage. R&D is multidisciplinary, involving optics, materials science, thermal engineering, and control and measurement techniques. Specifically, for PTC and LF systems, R&D is aimed at lighter and lower-cost structural designs including front surface mirrors with high solar-weighted reflectivity of about 95%; high-absorptance (>96%) coatings for tube receivers able to operate at above 500°C; medium-temperature thermal energy storage systems based on PCMs, molten salts, concrete, and packed bed of rocks suitable for solar-only systems; improvement in overall system O&M, including mirror cleaning, integral automation, and unattended control; system cost reductions and efficiency improvements by DSG; and alternative HTFs such as air. For CRS systems, R&D is aimed at improvements in the heliostat field as a result of superior optical properties, lower cost structures, and better control; development of water/superheated steam and advanced air-cooled volumetric receivers using wire mesh absorbers or ceramic monoliths and foams; advanced thermocline storage systems based on packed bed of ceramic materials (especially suitable for solar air receivers) and high-temperature thermochemical storage; and distributed control architectures, system integration, and hybridization in high-efficiency electricity production schemes. For DE, R&D is aimed at volumetric receivers coupled to Stirling and Brayton engines; improvements in mirrors and support structures; and improvements in system integration and control for fully automation, parasitic loads reduction, start-up optimization, and hybrid Stirling–Brayton operation.

References

Adkins, D.R. (November 1987). Control strategies and hardware used in solar thermal applications. SAND86-1943, Sandia National Laboratories, Albuquerque, NM.

Agrafiotis, C.C., Mavroidis, I., Konstandopoulos, A.G., Hoffschmidt, B., Stobbe, P., Romero, M., Fernández-Quero, V. (2007). Evaluation of porous silicon carbide monolithic honeycombs as volumetric receivers/collectors of concentrated solar radiation. *Solar Energy Materials & Solar Cells* 91, 474–488.

Ajona, J.I. (1999). Electricity generation with distributed collector systems. In: *Solar Thermal Electricity Generation*. Edited by CIEMAT, Madrid, Spain, pp. 5–77.

Ajona, J.I., Zarza, E. (1994). Benefits potential of electricity production with direct steam generation in parabolic troughs. In: Popel, O., Fris, S., and Shchedrova, E., eds. *Proceedings of the Seventh International Symposium on Solar Thermal Concentrating Technologies*. Institute for High Temperature of Russian Academy of Science, Moscow, Russia, pp. 300–314.

Alpert, D.J., Houser, R.M. (1990). Performance evaluation of large-area glass-mirror heliostats. In: Gupta, B.P., Traugott, W.H., eds. In: *Research, Development and Applications of Solar Thermal Technology*. Editorial: Hemisphere, NY, pp. 91–100.

Alvis, R.L. (1984). Some solar dish/heat engine design consideration. Technical Report SAND84-1698, Sandia National Laboratories, Albuquerque, NM.

Aringhoff, R., Aubrey, C., Brakmann, G., Teske, S. (2003). *Solar Thermal Power 2020*. Greenpeace International/European Solar Thermal Power Industry Association, Amsterdam, the Netherlands.

Aringhoff, R., Geyer, M., Herrmann, U., Kistner, R., Nava, P., Osuna, R. (2002a). AndaSol-50 MW solar plants with 9 hour storage for Southern Spain. In: Steinfeld, A., ed. *Proceedings of the 11th SolarPACES International Symposium on Concentrated Solar Power and Chemical Energy Technologies*, September 4–6, 2002, Zurich, Switzerland, pp. 37–42.

Aringhoff, R. et al. (2002b). 50 MW solar plants with 9 hour storage for Southern Spain. In: *Proceedings of the 11th International Symposium on Concentrating Solar Power and Chemical Energy Technologies*, Zurich, Switzerland.

Avila-Marin, A.L. (2011). Volumetric receivers in solar thermal power plants with central receiver system technology: A review. *Solar Energy* 85, 891–910.

Bayón, R., Rojas, E., Valenzuela, L., Zarza, E., León, J. (2010). Analysis of the experimental behaviour of a 100 kWth latent heat storage system for direct steam generation in solar thermal power plants. *Applied Thermal Engineering* 30, 2673–2651.

Bean, J.R., Diver, R.B. (1995). Technical status of the dish/stirling joint venture program. In: *Proceedings of the 30th IECEC*, Orlando, FL, pp. 2.497–2.504. Report SAND95-1082C, Sandia National Laboratories, Albuquerque, NM.

Becker, M., Boehmer, M. (1989). GAST: The gas cooled solar tower technology program. In: *Proceedings of the Final Presentation*, Springer-Verlag, Berlin, Germany.

Becker, M., Böhmer, M., Meinecke, W., Unger, E. (1989). Volumetric receiver evaluation. SSPS Technical Report No. 3/89, Cologne, Germany.

Becker, M., Cordes, S., Böhmer, M. (1992). The development of open volumetric receivers. In: *Proceedings of the Sixth International Symposium on Solar Thermal Concentrating Technologies*, September 28–October 2, 1992, Vol. II. CIEMAT, Madrid, Spain, pp. 945–952.

Becker, M., Meinecke, W., Geyer, M., Trieb, F., Blanco, M., Romero, M., Ferriere, A. (2002). Solar thermal power plants. In: EUREC Agency, ed. *The Future for Renewable Energy 2: Prospects and Directions*. James & James Science Publishers Ltd., London, U.K., pp. 115–137.

Becker, M., Vant-Hull, L.L. (1991). Thermal receivers. In: Winter, C.J., Sizmann, R.L., Vant-Hull, L.L., eds. *Solar Power Plants*. Springer-Verlag, Berlin, Germany, pp. 163–197.

Beninga, K., Davenport, R., Sellars, J., Smith, D., Johansson, S. (1997). Performance results for the SAIC/STM prototype dish/stirling system. In: *ASME International Solar Energy Conference*, Washington, DC, pp. 77–80.

Bernhard, R., Hein, S., de Lalaing, J., Eck, M., Eickhoff, M., Pfänder, M., Morin, G., Häberle, A. (2008). Linear Fresnel collector demonstration on the PSA, part II—Commissioning and first performance tests. In: *Proceedings of the 14th Solar Paces Symposium*, Las Vegas, NV. http://www.nrel.gov/docs/gen/fy08/42709CD.zip. Last accessed date March 30, 2015.

Biggs, F., Vittitoe, C.N. (1979) The Helios model for the optical behavior of reflecting solar concentrators. Report SAND76-0347, Sandia National Laboratories, Albuquerque, NM.

Blake, F.A., Gorman, D.N., McDowell, J.H. (1985). *ARCO Central Receiver Solar Thermal Enhanced Oil Recovery Project*. ARCO Power Systems, Littleton, CO.

Buck, R., Bräuning, T., Denk, T., Pfänder, M., Schwarzbözl, P., Tellez, F. (2002). Solar-hybrid gas turbine-based power tower systems (REFOS). *Journal of Solar Energy Engineering* 124, 2–9.

Buck, R., Heller, P., Koch, H. (1996). Receiver development for a Dish-Brayton system. In: *Solar Engineering 1996, ASME*, San Antonio, TX, pp. 91–96.

Burgaleta, J.I., Arias, S., Salbidegoitia, I.B. (2009). Operative advantages of a central tower solar plant with thermal storage system. In: *Proceedings of the SolarPACES 2009* (CD). Ref. Manuscript: 11720, September 15–18, 2009, Berlin, Germany. DLR, Stuttgart, Germany.

Chavez, J.M., Kolb, G.J., Meinecke, W. (1993). In: Becker, M., Klimas, P.C., eds. *Second Generation Central Receiver Technologies—A Status Report*. Verlag C.F. Müller GmbH, Karlsruhe, Germany.

Chisholms, D. (1980). Two-phase flow in bends. *International Journal of Multiphase Flow* 6, 363–367.

Cohen, G.E., Kearney, D.W., Kolb, G.J. (1999). Final report on the operation and maintenance improvement program for concentrating solar power plants. SAND99-1290, Sandia National Laboratories, Albuquerque, NM, Printed June 1999.

DeMeo, E.A., Galdo, J.F. (December 1997). Renewable energy technology characterizations. TR-109496 Topical Report, U.S. DOE, Washington, DC and EPRI, Palo Alto, CA.

Dersch, J., Morin, G., Eck, M., Häberle, A. (2009). Comparison of linear Fresnel and parabolic trough collector systems—System analysis to determine break even costs of linear Fresnel collectors. In: *Proceedings of the SolarPACES 2009* (CD). Ref. Manuscript: 15162, September 15–18, 2009, Berlin, Germany. DLR, Stuttgart, Germany.

Di Canio, D.G., Treytl, W.J., Jur, F.A., Watson, C.D. (April 1979). Line focus solar thermal central receiver research study—Final report. Prepared for U.S. Department of Energy, DOE/ET/20426-1. FMC Corporation, Santa Clara, CA.

Diver, R., Andraka, C., Rawlinson, K., Thomas, G., Goldberg, V. (2001). The Advanced Dish Development system project. In: Kleis, S.J., Bingham, C.E., eds. *Proceedings of Solar Forum 2001 Solar Energy: The Power to Choose* (CD-ROM), April 21–25, 2001, Washington, DC. ASME, New York.

Diver, R.B. (1987). Receiver/reactor concepts for thermochemical transport of solar energy. *Journal of Solar Energy Engineering* 109(3), 199–204.

Droher, J.J., Squier, S.E. (1986). Performance of the Vanguard solar dish-stirling engine module. Technical Report EPRI AP-4608, Electric Power Research Institute, Palo Alto, CA.

Eck, M., Zarza, E. (2002). Assessment of operation modes for direct solar steam generation in parabolic troughs. In: Steinfel, A., eds. *Proceedings of the 11th SolarPACES International Symposium on Concentrated Solar Power and Chemical Energy Technologies*, Zurich, Switzerland, pp. 591–598.

Eck, M., Zarza, E., Eickhoff, M., Rheinländer, J., Valenzuela, L. (2003). Applied research concerning the direct steam generation in parabolic troughs. *Solar Energy* 74, 341–351.

Eneas, A. (1901). US Patent 670,917.

Enermodal Engineering Ltd. (May 5, 1999). Cost reduction study for solar thermal power plants. Final Report. Prepared by Enermodal Engineering Ltd. in Association with Marbek Resource Consultants Ltd., by Contract of World Bank/GEF, Washington, DC.

Entropie (1982). Centrales à Tour: Conversion Thermodynamique de l'Energie Solaire, Entropie No. 103, Special Journal Issue.

Epstein, M., Liebermann, D., Rosh, M., Shor, A.J. (1991). Solar testing of 2 MW(th) water/steam receiver at the Weizmann Institute solar tower. *Solar Energy Materials* 24, 265–278.

Ericsson, J. (1888). The sun motor. *Nature* 38, 321.

Falcone, P.K. (1986). A handbook for solar central receiver design. SAND86-8009, Sandia National Laboratories, Livermore, CA.

Fernandez-Quero, V., Osuna, R., Romero, M., Sanchez, M., Ruiz, V., Silva, M. (2005). EURECA: Advanced receiver for direct superheated steam generation in solar towers, as an option for increasing efficiency in large low cost direct steam generation plants. In: Goswami, D.Y., Vijayaraghaveng, S., Campbell-Howe, R., eds. *Proceedings of the 2005 Solar World Congress ISES-2005*, August 6–12, Orlando, FL. American Solar Energy Society, Boulder, CO.

Francia, G. (1968). Pilot plants of solar steam generation systems. *Solar Energy* 12, 51–64.
Fricker, H.W., Silva, M. García, C., Winkler, C., Chavez, J. (1988). Design and test results of the wire receiver experiment Almeria. In: Gupta, B.K., ed. *Proceedings of the Fourth International Workshop on Solar Thermal Technology*, Santa Fe, NM. Hemisphere Publishing, New York, pp. 265–277.
García, E., Calvo, R. (2012). One year operation experience of Gemasolar plant. In: *Proceedings of the SolarPACES 2012*, September 11–14, 2012, Marrakech, Morocco.
Garcia, G., Egea, A., Romero, M. (2004). Performance evaluation of the first solar tower operating with autonomous heliostats: PCHA project. In: Ramos, C., Huacuz, J., eds. *Proceedings of the 12th SolarPACES International Symposium*, S3-231 (CD-ROM), October 6–8, 2004, Oaxaca, Mexico.
García-Martín, F.J., Berenguel, M., Valverde, A., Camacho, E.F. (1999). Heuristic knowledge-based heliostat field control for the optimization of the temperature distribution in a volumetric receiver. *Solar Energy* 66(5), 355–369.
Geyer, M. (2002). Panel 1 briefing material on status of major project opportunities. The current situation, issues, barriers and planned solutions. In: *International Executive Conference on Expanding the Market for Concentrating Solar Power (CSP)—Moving Opportunities into Projects*, June 19–20, 2002, Berlin, Germany.
Geyer, M., Romero, M., Steinfeld, A., Quaschning, V. (2003). International Energy Agency—Solar power and chemical energy systems. Annual Report 2002, DLR, Printed March 2003. May be downloaded at www.solarpaces.org. Last accessed date March 30, 2015.
González, L., Zarza, E., Yebra, L. (2001). Determinación del Modificador por Angulo de Incidencia de un colector solar LS-3, incluyendo las pérdidas geométricas por final de colector. Technical Report DISS-SC-SF-30, Almería, Spain, Plataforma Solar de Almería.
Goswami, Y., Kreith, F., Kreider, J.F. (2000). *Principles of Solar Engineering*, 2nd edn. Taylor & Francis Group, New York.
Grasse, W., Hertlein, H.P., Winter, C.J. (1991). Thermal solar power plants experience. In: Winter, C.J., Sizmann, R.L., Vant-Hull, L.L., eds. *Solar Power Plants*. Springer-Verlag, Berlin, Germany, pp. 215–282.
Haeger, M. (July 1994). Phoebus technology program: Solar Air Receiver (TSA). PSA Tech. Report: PSA-TR02/94.
Harats, Y., Kearney, D. (1989). Advances in parabolic trough technology in the SEGS plants. In: Fanney, A.H., Lund, K.O., eds. *Proceedings of the 11th Annual American Society of Mechanical Engineers Solar Energy Conference*, San Diego, CA. American Society of Mechanical Engineers, New York, pp. 471–476.
Hautmann, G., Selig, M., Mertins, M. (2009). First European Linear Fresnel power plant in operation—Operational experience & outlook. In: *Proceedings of the SolarPACES 2009* (CD). Ref. Manuscript: 16541, September 15–18, 2009, Berlin, Germany. DLR, Stuttgart, Germany.
Hoffschmidt, B., Fernández, V., Konstandopoulos, A.G., Mavroidis, I., Romero, M., Stobbe, P., Téllez, F. (2001). Development of ceramic volumetric receiver technology. In: Funken, K.-H., Bucher, W., eds. *Proceedings of the Fifth Cologne Solar Symposium*, June 21, 2001, Forschungsbericht 2001–2010. DLR, Cologne, Germany, pp. 51–61.
Hoffschmidt, B., Fernandez, V., Pitz-Paal, R., Romero, M., Stobbe, P., Téllez, F. (2002). The development strategy of the HitRec volumetric receiver technology—Up-scaling from 200kWth via 3MWth up to 10MWel. In: *Proceedings of the 11th SolarPACES International Symposium on Concentrated Solar Power and Chemical Energy Technologies*, September 4–6, 2002, Zurich, Switzerland, pp. 117–126.
Hoffschmidt, B., Pitz-Paal, R., Böhmer, M., Fend, T., Rietbrock, P. (1999). 200 kWth open volumetric air receiver (HiTRec) of DLR reached 1000°C average outlet temperature at PSA. *Journal de Physique IV, France* 9, Pr3-551–Pr3-556.
Hoffschmidt, B., Téllez, F.M., Valverde, A., Fernández-Reche, J., Fernández, V. (February 2003). Performance evaluation of the 200-kWth HiTRec-II open volumetric air receiver. *Journal of Solar Energy Engineering* 25, 87–94.

IEA (2010). Technology Roadmap—Concentrating Solar Power. Available at: http://www.iea.org. Last accessed date March 30, 2015.

IEA (2010b). Energy Technology Perspectives 2010—Scenarios and Strategies to 2050.

IRENA (2012). Concentrating solar power. Renewable energy technologies: Cost analysis series. Volume 1: Power Sector, Issue 2/5. International Renewable Energy Agency. Available at: www.irena.org/Publications. Last accessed date March 30, 2015.

Kalogirou, S.A. (2004). Solar thermal collectors and applications. *Progress in Energy and Combustion Science* 30(3), 231–295.

Kearney, A.T., ESTELA (2010). *Solar Thermal Electricity 2025*. ESTELA, Brussels, Belgium. http://www.estelasolar.eu/index.php?id=22. Last accessed date March 30, 2015.

Kearney, D.W., Cohen, G.E. (1997). Current experiences with the SEGS parabolic trough plants. In: Becker, M., Böhmer, M., eds. *Proceedings of the Eighth International Symposium on Solar Thermal Concentrating Technologies*, Vol. 1, Colonia, Germany, 1996. C.F. Müller, Heidelberg, Germany, 1997, pp. 217–224.

Keck, T., Heller, P., Weinrebe, G. (June 2003). Envirodish and Eurodish—System and status. In: *Proceedings of the ISES Solar World Congress*, Göteborg, Sweden.

Kelly, B., Singh, M. (1995). Summary of the final design for the 10 MWe solar two central receiver project. *Solar Engineering, ASME* 1, 575.

Kistler, B.L. (1986). A user's manual for DELSOL3: A computer code for calculating the optical performance and optimal system design for solar thermal central receiver plants. Sandia Report, SAND86-8018, Sandia National Laboratories, Albuquerque, NM.

Kodama, T. (2003). High-temperature solar chemistry for converting solar heat to chemical fuels. *Progress in Energy and Combustion Science* 29, 567–597.

Kolb, G.J. (1998). Economic evaluation of solar-only and hybrid power towers using molten-salt technology. *Solar Energy* 62, 51–61.

Kribus, A. (1999). Future directions in solar thermal electricity generation. In: *Solar Thermal Electricity Generation*. Colección documentos CIEMAT. CIEMAT, Madrid, Spain, pp. 251–285.

Laing, D., Lehmann, D., Bahl, C. (2008). Concrete storage for solar thermal power plants and industrial process heat. In: *Proceedings of the Third International Renewable Energy Storage Conference (IRES III)*, November 24–25, 2008, Berlin, Germany.

Laing, D., Trabing, C. (1997). Second generation sodium heat pipe receivers for USAB V-160 stirling engine: Evaluation of on-sun test results using the proposed IEA guidelines and analysis of heat pipe damage. *ASME Journal of Solar Energy Engineering* 119(4), 279–285.

Lehaut, C. (2010). Construction, start-up and performance tests of a direct steam generation CSP Fresnel module. In: *Proceedings of the SolarPACES 2010*, Perpignan, France.

Lipps, F.W., Vant-Hull, L.L. (1978). A cellwise method for the optimization of large central receiver systems. *Solar Energy* 20, 505–516.

Lopez, C., Stone, K., (1992). Design and performance of the Southern California Edison stirling dish. In: *Proceedings of the ASME International Solar Energy Conference*, Maui, HI, pp. 945–952.

Lotker, M. (1991). Barriers to commercialization of large-scale solar electricity: Lessons learned from LUZ experience. Informe técnico SAND91-7014, Sandia National Laboratories, Albuquerque, NM.

Luepfert, E., Zarza, E., Schiel, W., Osuna, R., Esteban, A., Geyer, M., Nava, P., Langenkamp, J., Mandelberg, E. Eurotrough collector qualification complete—Performance test results from PSA. In: *Proceedings of the ISES 2003 Solar World Congress* (CD), Göteborg, Sweden.

Mancini, T., Heller, P., Butler, B., Osborn, B., Schiel, W., Goldberg, V., Buck, R., Diver, R., Andraka, C., Moreno, J. (2003). Dish-Stirling systems: An overview of development and status. *International Journal of Solar Energy Engineering* 125, 135–151.

Mancini, T., Kolb, G.J., Prairie, M. (1997). Solar thermal power. In: Boer, K.W. ed., *Advances in Solar Energy: An Annual Review of Research and Development*, Vol. 11. American Solar Energy Society, Boulder, CO.

Marcos, M.J., Romero, M., Palero, S. (2004). Analysis of air return alternatives for CRS-type open volumetric receiver. *Energy* 29, 677–686.

Mavis, C.L. (1989). A description and assessment of heliostat technology. SAND87-8025, Sandia National Laboratories, Enero.

Mayette, J., Davenport, R, Forristall, R. (2001). The Salt River Project SunDish dish-stirling system. In: Kleis, S.J., Bingham, C.E., eds. *Proceedings of the Solar Forum 2001 Solar Energy: The Power to Choose* (CD-ROM), April 21–25, 2001, Washington, DC. ASME, New York.

Mills, D. (2004). Advances in solar thermal electricity technology. *Solar Energy* 76, 19–31.

Mills, D.R., Morrison, G.L. (1999). Compact linear Fresnel reflector solar thermal powerplants. *Solar Energy* 68, 263–283.

Mills, D.R., Morrison, G.L. (2000). Compact linear Fresnel reflector solar thermal power plants. *Solar Energy* 68, 263–283.

Monterreal, R., Romero, M., García, G., Barrera, G. (1997). Development and testing of a 100 m^2 glass-metal heliostat with a new local control system. In: Claridge, D.E., Pacheco, J.E., eds. *Solar Engineering*. ASME, New York, pp. 251–259.

Moreno, J.B., Modesto-Beato, M., Rawlinson, K.S., Andraka, C.E., Showalter, S.K., Moss, T.A., Mehos, M., Baturkin, V. (2001). Recent progress in heat-pipe solar receivers. SAND2001-1079, *36th Intersociety Energy Conversion Engineering Conference*, Savannah, GA, pp. 565–572.

Morse, F.H. (2000). *The Commercial Path Forward for Concentrating Solar Power Technologies: A Review of Existing Treatments of Current and Future Markets*. Morse Associates, Inc., By Contract of Sandia National Laboratories and U.S. DOE, Washington, DC.

Nitsch, J., Krewitt, W., Langniss, O. (2004). Renewable energies in Europe, *Encyclopedia of Energy* 5, 313–331 (Elsevier Inc., San Diego, USA).

Ortega, J.I., Burgaleta, J.I., Tellez, F. (2006). Central receiver system (CRS) solar power plant using molten salt as heat transfer fluid. In: Romero, M., Martínez, D., Ruiz, V., Silva, M., Brown, M., eds. *Proceedings of the 13th International Symposium on Concentrated Solar Power and Chemical Energy Technologies*, June 20, 2006, Seville, Spain. CIEMAT, Madrid, Spain.

Osuna, R., Cerón, F., Romero, M., García, G. (1999). Desarrollo de un prototipo de helióstato para la planta Colón Solar. *Energía* XXV(6), 71–79.

Osuna, R., Fernández, V., Romero, S., Romero, M., Sanchez, M. (2004). PS10: A 11-MW solar tower power plant with saturated steam receiver. In: Ramos, C., Huacuz, J., eds. *Proceedings of the 12th SolarPACES International Symposium*, S3-102 (CD-ROM), October 6–8, 2004, Oaxaca, Mexico.

Pacheco, J.E., Gilbert, R. (1999). Overview of recent results of the Solar Two test and evaluations program. In: Hogan, R., Kim, Y., Kleis, S., O'Neal, D., Tanaka, T., eds. *Renewable and Advanced Energy Systems for the 21st Century RAES'99*, RAES99-7731, April 11–15, 1999, Maui, Hawaii. ASME, New York.

Pacheco, J.E., Reilly, H.E., Kolb, G.J., Tyner, C.E. (2000). Summary of the solar two test and evaluation program. In: *Proceedings of the Renewable Energy for the New Millennium*, March 8–10, 2000, Sydney, New South Wales, Australia, pp. 1–11.

Palero, S., Romero, M., Castillo, J.L. (2008). Comparison of experimental and numerical air temperature distributions behind a cylindrical volumetric solar absorber module. *Journal of Solar Energy Engineering* 130, 011011-1–011011-8.

Philibert, C. (2004). International energy technology collaboration and climate change mitigation, case study 1: Concentrating solar power technologies. OECD/IEA Information Paper, Paris, France. COM/ENV/EPOC/IEA/SLT(2004)8.

Pifre, A. (1882). A solar printing press. *Nature* 21, 503–504.

Pitz-Paal, R. et al. (February 2005b). In: Pitz-Paal, R., Dersch, J., Milow, B., eds. *ECOSTAR Roadmap Document for the European Commission, SES-CT-2003-502578*. Deutsches Zentrum für Luft- und Raumfahrt e.V., Cologne, Germany.

Pitz-Paal, R., Dersch, J., Milow, B., Téllez, F., Zarza, E., Ferriere, A., Langnickel, U., Steinfeld, A., Karni, J., Popel, O. (2005). A European roadmap for concentrating solar power technologies (ECOSTAR). In: *Proceedings Submitted to ISEC2005, 2005 International Solar Energy Conference*, August 6–12, 2005, Orlando, FL.

Pitz-Paal, R., Hoffschmidt, B., Böhmer, M., Becker, M. (1996). Experimental and numerical evaluation of the performance and flow stability of different types of open volumetric absorbers under non-homogeneous irradiation. *Solar Energy*, Köln, Germany 60(3/4), 135–159.

Platzer, W.J. (2009). Linear Frensel collector as an emerging option for concentrating solar thermal power. In: *Proceedings of the ISES Solar World Congress 2009*, Johannesburg, South Africa.

Price, H., Luepfert, E., Kearney, D., Zarza, E., Cohen, G., Gee, R., Mahoney, R. (2002). Advances in parabolic trough solar power technology. *International Journal of Solar Energy Engineering* 124, 109–125.

Rabl, A. (1985). *Active Solar Collectors and Their Applications*. Oxford University Press, New York, pp. 59–66.

Radosevich, L.G., Skinrood, A.C. (1989). The power production operation of Solar One, the 10 MWe solar thermal central receiver pilot plant. *Journal of Solar Energy Engineering* 111, 144–151.

Ratzel, C.A., Simpson, C.E. (1979). Heat loss reduction techniques for annular receiver design. Informe técnico SAND78-1769, Sandia National Laboratories, Albuquerque, NM.

Romero, M., Buck, R., Pacheco, J.E. (2002). An update on solar central receiver systems, projects, and technologies. *International Journal of Solar Energy Engineering* 124, 98–108.

Romero, M., Conejero, E., Sánchez, M. (1991). Recent experiences on reflectant module components for innovative heliostats. *Solar Energy Materials* 24, 320–332.

Romero, M., Fernández, V., Sánchez, M. (1999). Optimization and performance of an optically asymmetrical heliostat field. *Journal de Physique IV, France* 9, Pr3-71–Pr3-76.

Romero, M., González Aguilar, J. (2011). Chapter 3: Solar thermal power plants: From endangered species to bulk power production in sun-belt regions. In: Rao, K.R., ed. *Energy & Power Generation Handbook*. ASME Three Park Avenue, New York.

Romero, M., González Aguilar, J. (2014). Solar thermal CSP technology. *WIREs Energy and Environment* 3, 42–59. doi: 10.1002/wene.79.

Romero, M., Marcos, M.J., Osuna, R., Fernández, V. (2000a). Design and implementation plan of a 10 MW solar tower power plant based on volumetric-air technology in Seville (Spain). In: Pacheco, J.D., Thornbloom, M.D., eds. *Solar Engineering 2000—Proceedings of the ASME International Solar Energy Conference*, June 16–21, 2000, Madison, WI. ASME, New York.

Romero, M., Marcos, M.J., Téllez, F.M., Blanco, M., Fernández, V., Baonza, F., Berger, S. (2000b). Distributed power from solar tower systems: A MIUS approach. *Solar Energy* 67(4–6), 249–264.

Romero, M., Martinez, D., Zarza, E. (2004). Terrestrial solar thermal power plants: On the verge of commercialization. In: European Space Agency, ed. *Proceedings of the Fourth International Conference on Solar Power from Space-SPS'04*, June 30–July 2, 2004, Granada, Spain, pp. 81–89. Noordwijk, the Netherlands.

Romero, M., Sanchez-Gonzalez, M., Barrera, G., Leon, J., Sanchez-Jimenez, M. (1995). Advanced salt receiver for solar power towers. In: *Solar Engineering* 1, 657–664. Stine, W.B., Tanaka, T., Claridge, D.E., eds. ASME, New York.

Romero, M., Steinfeld, A. (2012). Concentrating solar thermal power and thermochemical fuels. *Energy & Environmental Science* 5, 9234–9245.

Rueda, F. et al. (October 2003). Project INDITEP. First Yearly Report. IBERDROLA Ingeniería y consultoría, Madrid, Spain.

Ruiz, V., Silva, M., Blanco, M. (1999). Las centrales energéticas termosolares. *Energía* XXV(6), 47–55.

Sánchez, F. (1986). Results of Cesa-1 plant. In: Konstanz, B.M., ed. *Proceedings of the Third International Workshop on Solar Thermal Central Receiver Systems*, Berlin, Germany. Springer, New York, pp. 46–64.

Sánchez, M., Romero, M. (2006). Methodology for generation of heliostat field layout in central receiver systems based on yearly normalized energy surfaces. *Solar Energy* 80, 861–874.

Sanders Associates Inc. (October 1979). 1/4-Megawatt solar receiver. Final Report, DOE/SF/90506-1.

Sanz-Bermejo, J., Gallardo-Natividad, V., Gonzalez-Aguilar, J., Romero, M. (2014). Comparative system performance analysis of direct steam generation central receiver solar thermal power plants in megawatt range. *Journal of Solar Energy Engineering* 136, 010908-1.

Sargent & Lundy (October 2003). Assessment of parabolic trough and power tower solar technology cost and performance forecasts. Report: NREL/SR-550-34440, National Renewable Energy Laboratory, Golden, CO.

Schell, S. (2009). Design and evaluation of eSolar's heliostat fields. In: *Proceedings of the SolarPACES 2009* (CD), Berlin, Germany, September 15–18, 2009. DLR, Stuttgart, Germany.

Schiel, W. (1999). Dish/Stirling systems. In: *Solar Thermal Electricity Generation*. Colección documentos CIEMAT. CIEMAT, Madrid, Spain, pp. 209–250.

Schiel, W., Keck, T., Kern, J., Schweitzer, A. (1994). Long term testing of three 9 kW dish/Stirling systems. In: *Solar Engineering, ASME 1994 Solar Engineering Conference*, pp. 541–550.

Schiel, W., Schweizer, A., Stine, W. (1994). Evaluation of the 9-kW Dish/Stirling system of Schlaich Bergermann und Partner using the proposed IEA Dish/Stirling performance analysis guidelines. In: *Proceedings of the 29th IECEC*, Monterey, CA, pp. 1725–1729.

Schmitz-Goeb, M., Keintzel, G. (1997). The Phoebus solar power tower. In: Claridge, D.E., Pacheco, J.E., eds. *Proceedings of the 1997 ASME International Solar Energy Conference*, April 27–30, 1997, Washington, DC, pp. 47–53.

Selig, M. (2011). Commercial CSP plants based on Fresnel collector technology. In: *Proceedings of the SolarPACES 2011*, Granada, Spain.

Shuman, F. (1913). The most rational source of energy: Tapping the sun's radiant energy directly. *Scientific American* 109, 350.

Siebers, D.L., Kraabel, J.S. (April 1984). Estimating convective energy losses from solar central receivers. SAND84-8717, Sandia National Laboratories, Livermore, CA.

Silberstein, E., Magen, Y., Kroyzer, G., Hayut, R., Huss, H. (2009). Brightsource solar tower pilot in Israel's Negev operation at 130 bar @ 530°C superheated steam. In: *Proceedings of the SolarPACES 2009* (CD), September 15–18, 2009, Berlin, Germany. DLR, Stuttgart, Germany.

Silva, M., Blanco, M., Ruiz, V. (1999). Integration of solar thermal energy in a conventional power plant: The COLON SOLAR project. *Journal de Physique IV, Symposium Series* 9, Pr3-189–Pr3-194.

Sizmann, R.L. (1991). Solar radiation conversion. In. Winter, C.J., Sizmann, R.L., Vant-Hull, L.L., eds. *Solar Power Plants*. Springer-Verlag, Berlin, Germany, pp. 17–83.

Stine, W., Diver, R.B. (1994). A compendium of solar dish/Stirling technology. Report SAND93-7026, Sandia National Laboratories, Albuquerque, NM.

Stone, K., Leingang, E., Rodriguez, G., Paisley, J., Nguyen, J., Mancini, T., Nelving, H. (2001). Performance of the SES/Boeing dish Stirling system. In: Kleis, S.J., Bingham, C.E., eds. *Proceedings of the Solar Forum 2001 Solar Energy: The Power to Choose* (CD-ROM), April 21–25, 2001, Washington, DC. ASME, New York.

Tyner, C.E., Pacheco, J.E. (2009). eSolar's power plant architecture. In: *Proceedings of the SolarPACES 2009* (CD), September 15–18, 2009, Berlin, Germany. DLR, Stuttgart, Germany.

Vant-Hull, L.L. (1991). Solar radiation conversion. In: Winter, C.J., Sizmann, R.L., Vant-Hull, L.L., eds. *Solar Power Plants*. Springer-Verlag, Berlin, Germany, pp. 84–133.

Weinrebe, G., Schmitz-Goeb, M., Schiel, W. (1997). On the performance of the ASM150 stressed membrane heliostat. In: *Solar Engineering: 1997 ASME/JSME/JSES International Solar Energy Conference*, April 27–30, Washington, DC.

Welford, W.T., Winston, R. (1989). *High Collection Non-Imaging Optics*. Academic Press, New York.

Winter, C.J., Sizmann, R.L., Vant-Hull, L.L., eds. (1991). *Solar Power Plants*. Springer-Verlag, Berlin, Germany.

Yebra, L., Berenguel, M., Romero, M., Martínez, D., Valverde, A. (2004). Automation of solar plants. In: *Proceedings of EuroSun2004*, Vol. I. DGS, Munich, Germany and PSE, Freiburg, Germany, pp. 978–984.

Zarza, E. (2004). *Generación Directa de Vapor con Colectores Solares Cilindro Parabólicos. Proyecto Direct Solar Steam (DISS)*. CIEMAT, Madrid, Spain.

Zarza, E. et al. (1999). DISS-phase I project. Final Project Report. CIEMAT, Madrid, Spain.

Zarza, E., Rojas, M.E., González, L., Caballero, J.M., Rueda, F. (2004). INDITEP: The first DSG pre-commercial solar power plant. In: *Proceedings to the 12th Solar PACES International Symposium*, October 6–8, 2004, Oaxaca, Mexico.

Zarza, E., Valenzuela, L., Leon, J., Hennecke, K., Eck, M., Weyers, D.-H., Eickhoff, M. (2002). Direct steam generation in parabolic troughs. Final results and conclusions of the DISS project. In: Steinfeld, A., ed. *Book of Proceedings of the 11th SolarPACES International Symposium on Concentrated Solar Power and Chemical Energy Technologies*, September 4–6, 2002, Zurich, Switzerland. Paul Scherrer Institute, Villigen, Switzerland, pp. 21–27.

Zavoico, A.B., Gould, W.R., Kelly, B.D., Grimaldi, I., Delegado, C. (2001). Solar power tower (SPT) design innovations to improve reliability and performance—Reducing technical risk and cost. In: *Proceedings of Forum 2001 Conference*, April 21–25, 2001, Washington, DC.

43

Wind Energy Conversion

Dale E. Berg

CONTENTS

43.1 Introduction

Wind power supplied about 2.6% of the world electricity demand in 2013 (up from 0.6% in 2005), and the size of that contribution is growing rapidly. Wind energy is the most rapidly expanding source of energy in the world today; over the past 10 years, the worldwide installed capacity of wind energy has grown at an average rate of nearly 25% per year, leading to an installed nameplate capacity at the end of 2012 of about 286,000 MW [1]. As of January 2013, China was the world leader in cumulative installed wind power capacity, with about 75,400 MW installed, followed by the United States with 60,200, Germany with 31,500, Spain with 22,500, India with 18,600, the United Kingdom with 9,100, Italy with 8,000, France with 7,600, Canada with 6,200, and Portugal with 4,400. This 60,000 MW of wind power capacity in the United States reflects a 5-year average annual growth rate of over 29%; it is sufficient to power approximately 15.2 million American households, providing 3.5% of the U.S. electricity consumption in 2012. Over 43% of U.S. electricity providers had wind in their generation mix in 2012, and 42% of new U.S. generating capacity added in 2012 was wind. On April 15, 2012, wind provided a peak of 56.7% of Xcel Energy's generated electricity in Colorado [2].

The cost to generate wind energy decreased dramatically from more than 30 cents (U.S.) per kilowatt-hour (¢/kW h) in the early 1980s to under 4¢/kW h (at the best sites) in 2004. The cost has since increased somewhat, in spite of continuing technology improvements, as a result of worldwide increases in steel, concrete, and transportation costs that have led to increases in the prices of wind turbines. The large increases in the cost of natural gas and other fossil fuels prior to 2010 made wind-generated electricity a lower-cost option than natural gas for many utilities adding generating capacity, but recent drilling technology advances have led to significant increases in the U.S. natural gas supply and decreases in its cost. The long-term impacts of these technology advances on the price of natural gas and on how well wind can compete for new generating capacity remain to be seen.

There is considerable anecdotal evidence that the first wind machines may have been built over 2000 years ago, perhaps in China, but there is no firm evidence to support this conjecture. However, there is considerable written evidence that the windmill was in use in Persia by AD 900, perhaps as early as AD 640. Figure 43.1 illustrates the main features of this type of mill. The center vertical shaft was attached to a millstone, and horizontal beams or arms were attached to the shaft above the millstone. Bundles of reeds attached vertically to the outer end of the arms acted as sails, turning the shaft when the wind blew. The surrounding structure was oriented so that the prevailing wind entered the open portion of the structure and pushed the sails downwind. The closed portion of the structure sheltered the sails from the wind on the upwind pass. The primary applications of these machines were to grind or mill grain and to pump water; they became generally known as *windmills*. The wind machines of today may look much different from those first machines, but the basic idea remains the same—use the power in the wind to generate useful energy. Modern wind machines, called *wind turbines*, tend to have a small number of airfoil-shaped blades, in contrast to the older windmills that usually had several flat or slightly curved blades (such as the American multiblade water pumper shown in Figure 43.2). The reasons for this difference in blade number will be examined a little later.

Although there are many different configurations of wind turbines, most of them can be classified as either horizontal-axis wind turbines (HAWTs), which have blades that rotate about a horizontal axis parallel to the wind, or vertical-axis wind turbines (VAWTs), which have blades that rotate about a vertical axis perpendicular to the wind.

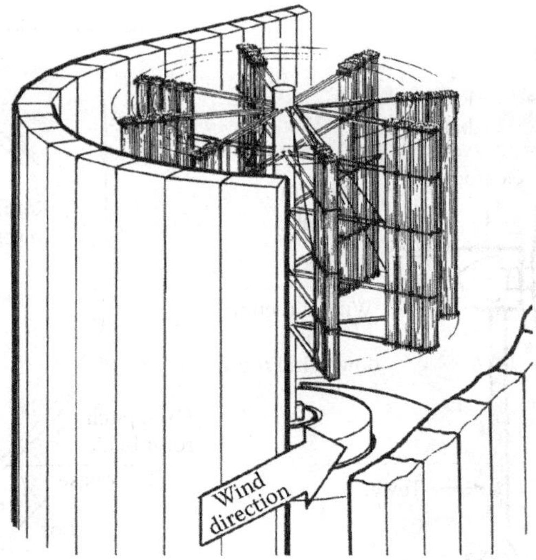

FIGURE 43.1
Illustration of ancient Persian wind mill.

FIGURE 43.2
Typical American multiblade windmill. (Courtesy of Nolan Clark, U.S. Department of Agriculture.)

Figure 43.3 illustrates the main features of these configurations; they both contain the same major components, but the details of those components differ significantly.

According to Shepherd [3], the terms *horizontal* and *vertical* associated with these classifications are a potential source of confusion. Although they now refer to the driving shaft on which the rotor is mounted, in the past, the terms referred to the plane in which the rotor turned. Thus, the ubiquitous multibladed water-pumper type of windmill shown in Figure 43.2, now referred to as a *horizontal-axis machine,* had a rotor that turned in a vertical

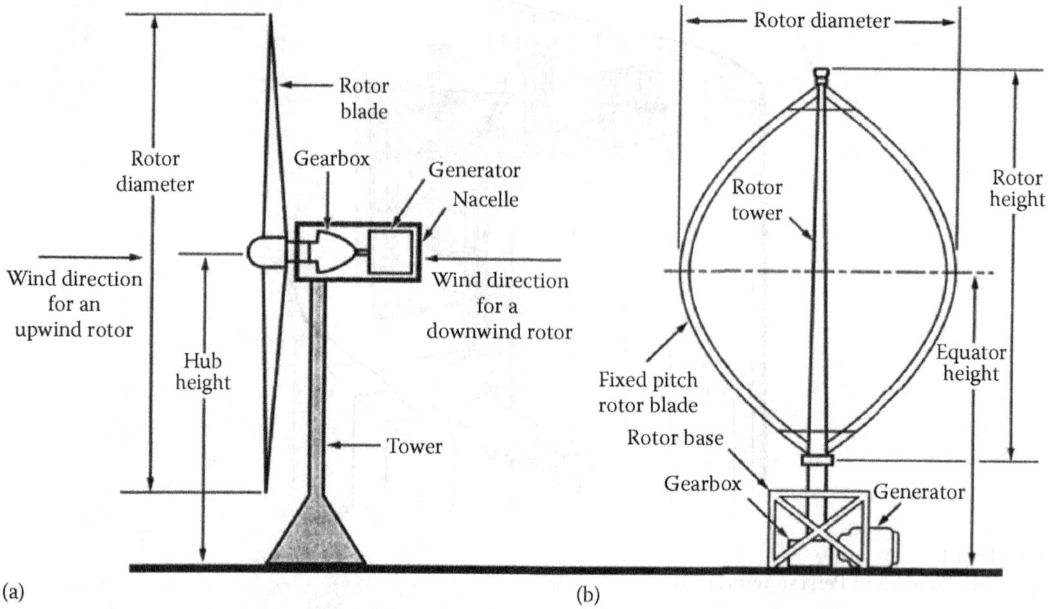

FIGURE 43.3
Schematic of basic wind turbine configurations: (a) horizontal-axis wind turbine (HAWT) and (b) vertical-axis wind turbine (VAWT).

plane, so it was, at one point, known as a *vertical* mill. Likewise, the earliest windmills, such as the one illustrated in Figure 43.1, had rotors that turned in a horizontal plane and were known as *horizontal* windmills.

As shown in Figure 43.3, HAWTs and VAWTs have very different configurations. Each configuration has its own set of strengths and weaknesses. HAWTs usually have all of their drive train (the transmission, generator, and any shaft brake) equipment located in a nacelle or enclosure mounted on a tower as shown; their blades are subjected to cyclic stresses due to gravity as they rotate, and their rotors must be oriented (yawed) so the blades are properly aligned with respect to the wind. HAWTs may readily be placed on tall towers to access the stronger winds typically found at greater heights. The most common type of modern HAWT is the propeller-type machine, and these machines are generally classified according to the rotor orientation (upwind or downwind of the tower), blade attachment to the main shaft (rigid or hinged), maximum power control method in high winds (full or partial-span collective or individual blade pitch or blade stall), and the number of blades (generally two or three blades).

VAWTs, on the other hand, usually have most of their drive train on the ground, their blades do not experience cyclic gravitational stresses, and their rotors do not require orientation with respect to the wind. However, VAWT blades are subject to severe alternating aerodynamic loading due to rotation of the blades upwind and downwind during each revolution of the rotor, and VAWTs cannot readily be placed on tall towers to exploit the stronger winds at greater heights. The most common types of modern VAWTs are the Darrieus turbines, with curved, fixed-pitch blades, and the "H" or "box" turbines with straight fixed-pitch blades. Virtually all of these turbines rely on blade stall (loss of lift and increase in drag as the blade angle of attack increases at high wind speeds) for maximum power control in high winds. Although there are still a few manufacturers of VAWTs today, the overwhelming majority of wind turbine manufacturers devote their efforts to developing better (and usually larger) HAWTs.

While the fuel for wind turbines is free, the initial cost of a wind turbine is a very large contributor to the cost of energy (COE) for that turbine. In order to minimize that COE, wind turbine designs must be optimized for the particular site or wind environment in which they will operate. Trial and error methods were very commonly used to develop early, small turbines, but these methods become very expensive and time-consuming when used to design and/or optimize larger turbines. A large optimized wind turbine can be developed at a reasonable cost only if the designers can accurately predict the performance of conceptual machines and use modeling to investigate the effects of design alternatives. The following discussion describes the basic phenomena that enable a machine to convert wind to mechanical energy and then presents several of the models that have been developed to predict the aerodynamic, hydrodynamic, and structural dynamic performance of wind turbines.

43.2 Wind Turbine Aerodynamics

Items exposed to the wind are subjected to forces in both the drag direction (parallel to the air flow) and the lift direction (perpendicular to the air flow). The earliest wind machines, known as *windmills*, used the drag on the blades to produce power, but many windmill designs over the last few centuries did make limited use of lift to increase their performance. For predominantly drag machines, such as those illustrated in Figures 43.1 and 43.2, larger numbers of blades result in higher drag and, thus, produce more power; therefore, these machines tend to have many blades. The old Dutch windmills, such as the one shown in Figure 43.4, utilized lift as well as drag, and since lift devices must be widely separated to generate the maximum possible amount of power, those machines evolved with a relatively small number of blades. The high-lift, low-drag shapes that were developed for airplane wings and propellers in the early part of the twentieth century (commonly referred to as *airfoils*) were quickly incorporated into wind machines to produce the first modern wind machines, usually known as *wind turbines*. An example of a typical modern wind turbine is shown in Figure 43.5. Modern wind turbines use the lift generated by the blades to produce power, and since the blades must be widely separated to generate the maximum amount of lift, wind turbines have a small number of blades. The following paragraphs contrast the characteristics of the drag- and lift-type machines.

Figure 43.6 illustrates the flow field about a moving drag device. The drag results from the relative velocity between the wind and the device, and the power that is generated by the device (the product of the drag force and the translation or blade velocity) may be expressed as

$$P = Dlv = \left[0.5\rho(U - v)^2\right] C_D clv \tag{43.1}$$

where

P is the power extracted from the wind
D is the drag force per unit length in the span direction (perpendicular to the page)
l is the length of device in the span direction (perpendicular to the page)
v is the translation (or blade) velocity
ρ is the air density
U is the steady free-stream wind velocity
C_D is the drag coefficient ($=(\text{Drag}/((1/2)\rho clU^2))$, a function of device geometry)
c is the device width (perpendicular to the wind, in the plane of the page)

FIGURE 43.4
Example of Dutch windmill. (Courtesy of Richard A. Neiser, Jr.)

FIGURE 43.5
G.E. 1.5 MW wind turbines near Lamar, Colorado.

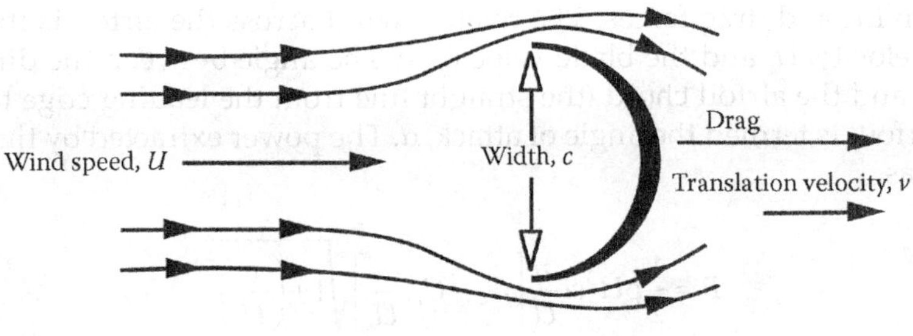

FIGURE 43.6
Schematic of translating drag device.

The translation (or blade) velocity of the device must always be less than the wind velocity, or no drag is generated and no power is produced. The power extraction efficiency of the device may be expressed as the ratio of the power extracted by the device to the power available in the wind passing through the area occupied by the device (the projected area of the device), a ratio known as the *power coefficient*, C_P. From Equation 37.17, the available power is

$$P_A = \frac{1}{2}\rho U^3 A = \frac{1}{2}\rho U^3 cl \tag{43.2}$$

where A (=cl) is the area of the device projected perpendicular to the wind.

For a drag machine, C_P, using Equations 43.1 and 43.2, is

$$C_P = \frac{P}{(1/2)\rho U^3 cl} = \frac{v}{U}\left[1 - \frac{v}{U}\right]^2 C_D \tag{43.3}$$

Now consider a device that utilizes lift to extract power from the wind, that is, an airfoil. Figure 43.7 depicts an airfoil that is moving at some angle relative to the wind and is

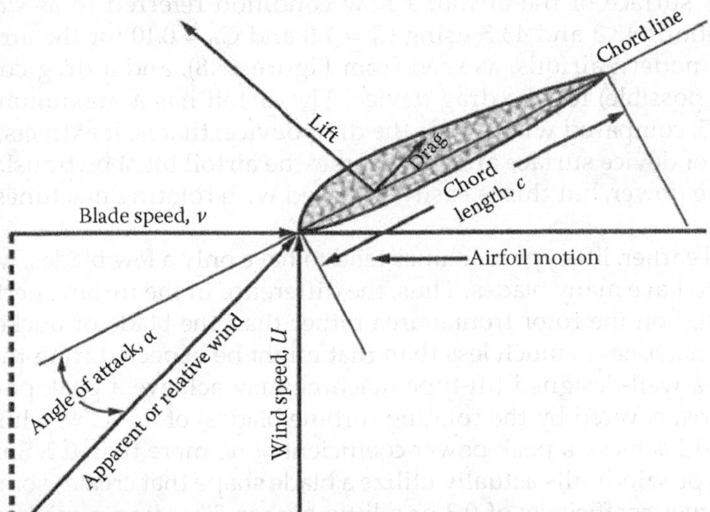

FIGURE 43.7
Schematic of translating lift device.

subject to both lift and drag forces. The relative wind across the airfoil is the vector sum of the wind velocity, U, and the blade velocity, v. The angle between the direction of the relative wind and the airfoil chord (the straight line from the leading edge to the trailing edge of the airfoil) is termed the angle of attack, α. The power extracted by this device may be expressed as

$$P = \frac{1}{2}\rho U^3 cl \frac{v}{U}\left[C_L - C_D\frac{v}{U}\right]\sqrt{1+\left(\frac{v}{U}\right)^2} \tag{43.4}$$

where

c is the chord length

C_L is the lift coefficient (=Lift/((1/2)ρclU^2), a function of airfoil shape and α)

The other quantities are as defined for Equation 43.1, but C_D is now a function of airfoil shape and α. Lift and drag coefficients for some common airfoils may be found in Refs. [4–9]. In this case, the projected area of the device is cl, and the power coefficient, using Equations 43.3 and 43.4, is

$$C_P = \frac{v}{U}\left[C_L - C_D\frac{v}{U}\right]\sqrt{1+\left(\frac{v}{U}\right)^2} \tag{43.5}$$

Keep in mind that Equations 43.3 and 43.5 express the performance coefficients of these devices in terms of the *projected area* of the individual device. Figure 43.8 presents experimental lift and drag coefficient values for the S-809 airfoil, an airfoil designed by the National Renewable Energy Laboratory (NREL) for use on small HAWTs [10]. As the angle of attack increases beyond approximately 9°, the lift levels off and then drops slightly, and the drag begins to rise fairly rapidly. This is due to separation of the flow from the upper surface of the airfoil, a flow condition referred to as *stall*. Figure 43.9 compares Equations 43.3 and 43.5 using C_L = 1.0 and C_D = 0.10 for the airfoil (conservative values for modern airfoils, as seen from Figure 43.8), and a drag coefficient of 2.0 (the maximum possible) for the drag device. The airfoil has a maximum power coefficient of about 15, compared with 0.3 for the drag device; that is, it extracts 50 times more power per unit of device surface area. Of course, the airfoil must be translated across the wind to produce power, but this is easily achieved with rotating machines such as wind turbines.

As mentioned earlier, lift-type machines tend to have only a few blades, while drag-type machines tend to have many blades. Thus, the difference in the turbine performance coefficient (now based on the rotor frontal area rather than the blade or bucket frontal area) of actual wind machines is much less than that might be expected from the analysis presented earlier—a well-designed lift-type machine may achieve a peak power coefficient (based on the area covered by the rotating turbine blades) of 0.5–0.59, while a pure drag-type machine will achieve a peak power coefficient of no more than 0.2. Some of the multibladed drag-type windmills actually utilize a blade shape that creates some lift, and they may achieve power coefficients of 0.3 or a little higher. The drag machines rotate slowly (the blade translation velocity cannot exceed the effective wind speed) and produce high torque, while the lift machines rotate quickly (to achieve a high translation velocity) and produce low torque. The slow rotating, high-torque drag machines are very well suited for

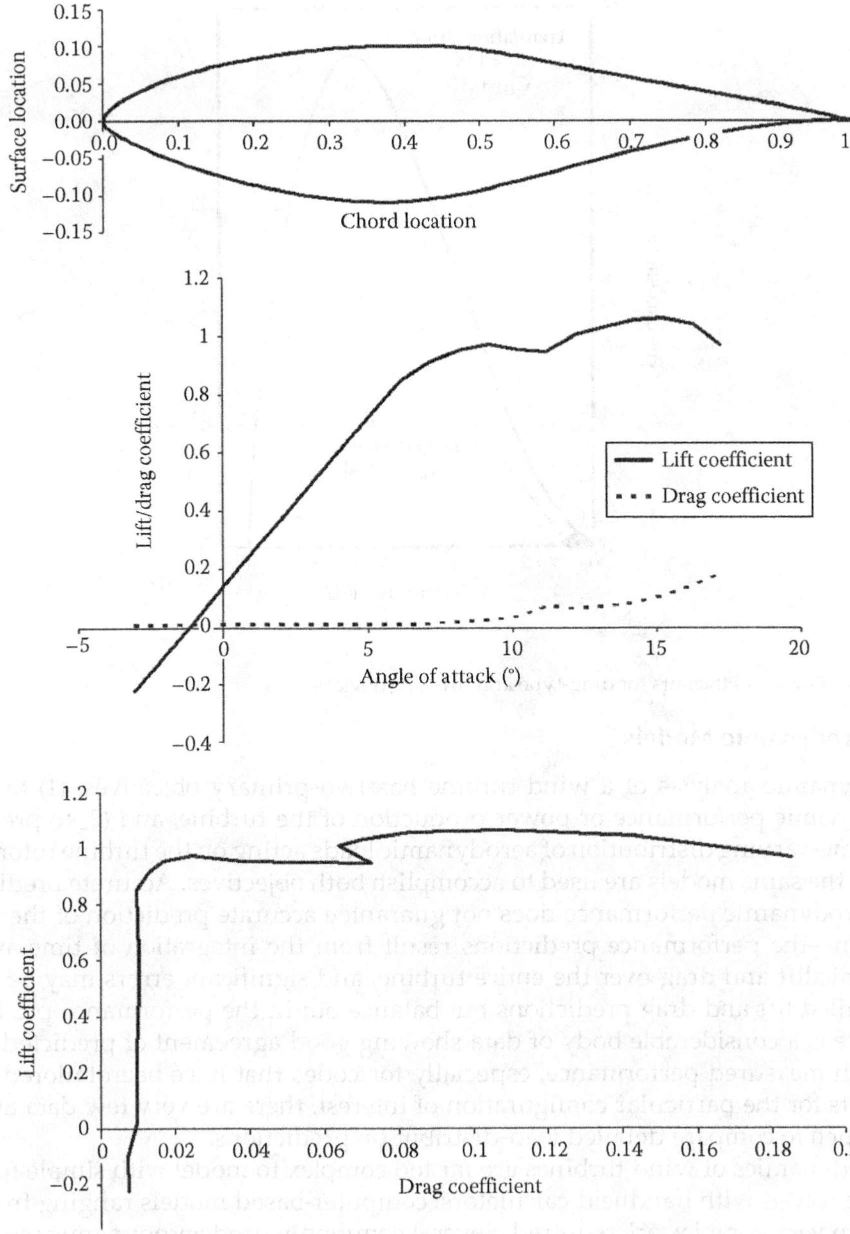

FIGURE 43.8
Profile and performance characteristics of the S-809 airfoil.

mechanical power applications such as milling grain and pumping water. On the other hand, extensive experience has shown that fast-rotating, lift-type machines are much easier to adapt to electrical generators and can produce electricity at a significantly lower COE than can the drag-type machines.

Because of their superior performance in electrical generation applications, only lift-type machines will be considered in the remainder of this discussion.

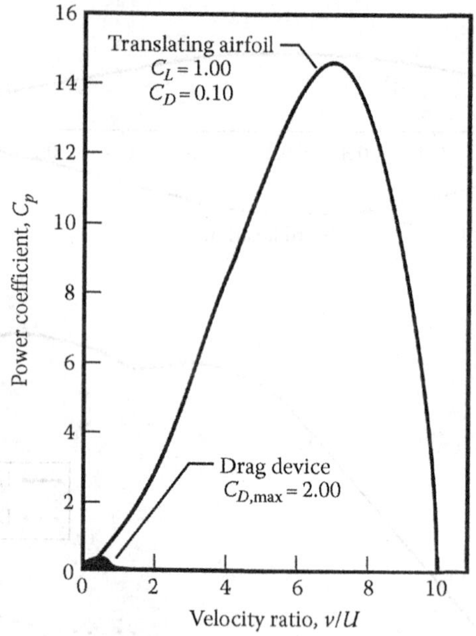

FIGURE 43.9
Comparison of power coefficients for drag-type and lift-type devices.

43.2.1 Aerodynamic Models

The aerodynamic analysis of a wind turbine has two primary objectives: (1) to predict the aerodynamic performance or power production of the turbine, and (2) to predict the detailed time-varying distribution of aerodynamic loads acting on the turbine rotor blades. In general, the same models are used to accomplish both objectives. Accurate prediction of turbine aerodynamic performance does not guarantee accurate prediction of the loading distribution—the performance predictions result from the integration of time-averaged aerodynamic lift and drag over the entire turbine, and significant errors may be present in the detailed lift and drag predictions but balance out in the performance predictions. While there is a considerable body of data showing good agreement of predicted performance with measured performance, especially for codes that have been tailored to give good results for the particular configuration of interest, there are very few data available against which to compare detailed load-distribution predictions.

The aerodynamics of wind turbines are far too complex to model with simple formulas that can be solved with handheld calculators; computer-based models ranging from very simplistic to very complex are required. Several commonly used aerodynamic models are described in the following paragraphs.

43.2.1.1 Momentum Models

The simplest aerodynamic model of an HAWT is the actuator disk or momentum model in which the turbine rotor is modeled as a single porous disk. This analysis was originally adapted for wind turbine use by Betz [11] from the propeller theory developed by Froude [12] and Lanchester [13]. To develop the equations for this model, the axial force acting on the rotor is equated to the time rate of change of the momentum of the air stream passing through the rotor. The mass of air that passes through the rotor disk is assumed to remain

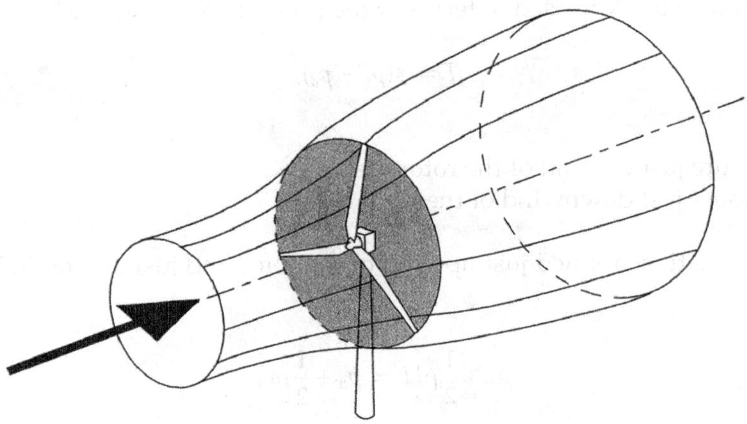

FIGURE 43.10
Schematic of stream-tube for horizontal-axis wind turbines.

separate from the surrounding air and only the air passing through the disk slows down. That mass of air with a boundary surface of circular cross section, extended upstream and downstream of the rotor disk, is shown in Figure 43.10. No air flows across the lateral boundary of this *stream-tube*, so the mass flow rate of air at any position along the stream-tube will be the same. Because the air is incompressible, the decrease in the velocity of the air passing through the disk must be accompanied by an increase in the cross-sectional area of the stream-tube to maintain the same mass flow. The presence of the turbine causes the air approaching from the upstream to gradually slow down; thus, the velocity of the air arriving at the rotor disk is already lower than the free-stream wind speed. As mentioned earlier, this causes the stream-tube to expand. In addition, the static pressure of the air rises to compensate for the decrease in kinetic energy.

As the air passes through the rotor disk, there is a drop in static pressure; the air immediately downstream of the disk is below the atmospheric pressure level, but there is no instantaneous change in velocity. As the air continues downstream, the static pressure gradually increases until it again comes into equilibrium with the surrounding atmosphere, and the velocity drops accordingly. This region of the flow is referred to as the *wake*. Thus, the difference in flow conditions between the far upstream and the far wake is a decrease in kinetic energy, but no change in static pressure.

Utilizing conservation of mass, conservation of axial momentum, the Bernoulli equation, and the first law of thermodynamics, and assuming isothermal flow, the power produced by the turbine (the product of the axial force and the air velocity at the rotor) may be readily derived. From the conservation of axial momentum, the thrust on the rotor, T, may be expressed as

$$T = \dot{m}(U - v_w) = \rho A v(U - v_w) \tag{43.6}$$

where

- $\dot{m}$ is the mass flow rate of air $(=\rho A v)$
- v is the wind velocity at the rotor disk
- v_w is the wind velocity far downstream of the rotor disk (in the wake)
- U is the wind velocity far upstream of the disk
- A is the area of the rotor disk

The thrust may also be expressed in terms of the pressure drop caused by the rotor

$$T = A(p_u - p_d) \tag{43.7}$$

where
p_u is the pressure just upwind of the rotor disk
p_d is the pressure just downwind of the rotor disk

The Bernoulli equation, applied just upwind of the rotor and just downwind of the rotor, yields

$$p_\infty + \frac{1}{2}\rho U^2 = p_u + \frac{1}{2}\rho v_u^2 \tag{43.8}$$

$$p_d + \frac{1}{2}\rho v_d^2 = p_w + \frac{1}{2}\rho v_w^2 \tag{43.9}$$

where the subscripts ∞, u, d, and w denote far upwind of the rotor, immediately upwind of the rotor, immediately downwind of the rotor, and far downwind of the rotor, respectively.

The pressures are equal far upwind of the rotor and far downwind of the rotor ($p_\infty = p_w$), and the velocity is the same just upwind and just downwind of the rotor ($v_u = v_d$). Substituting Equations 43.8 and 43.9 into Equation 43.7 yields

$$T = A\left[\left(\frac{1}{2}\rho U^2 - \frac{1}{2}\rho v_u^2\right) - \left(\frac{1}{2}\rho v_d^2 - \frac{1}{2}\rho v_w^2\right)\right] = \frac{1}{2}\rho A\left(U^2 - v_w^2\right) \tag{43.10}$$

Equating 43.6 and 43.10 then yields

$$v = \frac{1}{2}(U + v_w) \tag{43.11}$$

That is, the velocity at the rotor disk is equal to the mean of the free-stream and wake velocities; thus, the velocity change between the free-stream and the wake is twice the change between the free-stream and the disk.

The power produced at the rotor, assuming isothermal flow and ambient pressure in the wake, is the product of the thrust and the velocity at the rotor:

$$P = Tv = \frac{1}{2}\rho A\left(U^2 - v_w^2\right)v = \frac{1}{2}\rho A(U - v_w)(U + v_w)v$$

$$= \frac{1}{2}\rho A[2(U - v)]2vv = 2\rho A(U - v)v^2 \tag{43.12}$$

Now, define

$$a = \frac{(U - v)}{U} \tag{43.13}$$

which is commonly known as the *axial interference factor.* Using this in Equation 43.12 and rearranging yields

$$P = 2\rho A(aU)(1-a)^2U^2 = 2\rho AU^3a(1-a)^2 \tag{43.14}$$

The power coefficient for the turbine, then, is (utilizing Equations 43.2 and 43.14)

$$C_P = \frac{P}{P_A} = \frac{2\rho AU^3a(1-a)^2}{(1/2)\rho AU^3} = 4a(1-a)^2 \tag{43.15}$$

This is maximized for $a = 1/3$, yielding $C_{P,\max} = 16/27 = 0.593$ as the maximum possible performance coefficient for a lift-type machine, a maximum often referred to as the *Betz limit.* Expressed in slightly different terms, this equation shows that a lift-type HAWT turbine can extract no more than 59.3% of the energy available in the wind passing through the rotor!

The typical performance of various types of wind machines is compared to the Betz limit in Figure 43.11, where the variations of the turbine power coefficients with the *tip-speed ratio* (the ratio of the speed of the blade tip to the free-stream wind speed) are presented. Even though the maximum performance of modern HAWTs and VAWTs is well above that of the older, drag-type machines such as the Dutch windmill and the American multiblade windmill, it is still somewhat below the Betz limit. Some HAWTs have demonstrated peak performance coefficients approaching 52%.

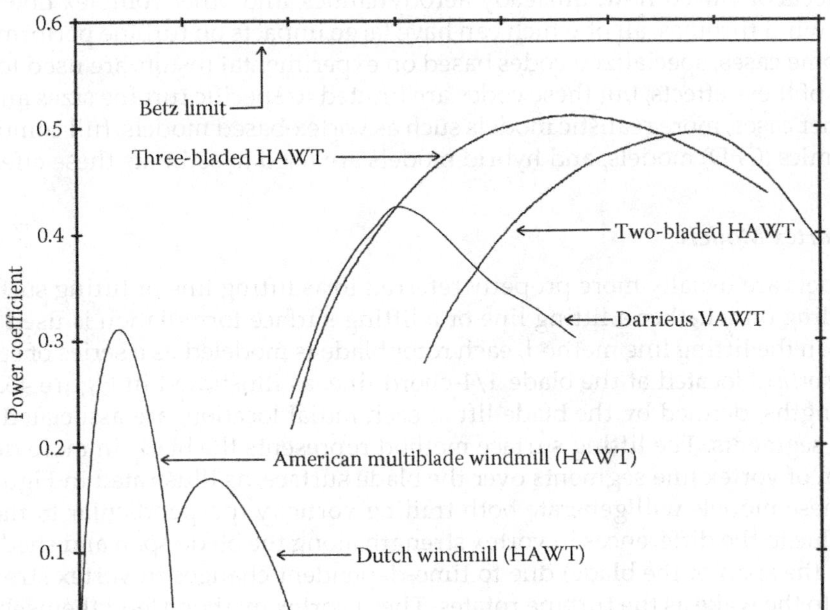

FIGURE 43.11
Typical performance of various types of wind turbines.

For HAWTs, the momentum model can be expanded to the widely used blade element momentum (BEM or BEMT) model in which the blades are divided into small radial elements, and local flow conditions at each element are used to determine blade forces and loads on those elements. To obtain accurate predictions, these models typically incorporate numerous modifications to account for blade and turbine wake effects, the three-dimensional flow near blade tips, the thick blade sections near the root, blade stall at high wind speeds, and unsteady effects associated with blade stall. Additional information on these models may be found in Hansen and Butterfield [14], Wilson [15], and Snel [16].

A very similar derivation yields a momentum model for the VAWT. This model may be expanded into the multiple stream-tube (the turbine rotor is modeled as multiple actuator disks, rather than just one) and the double-multiple stream-tube models (multiple actuator disks, with separate ones modeling the upwind and downwind passes of the rotor blades) that are the VAWT equivalent of the HAWT blade element model. Additional information on these models may be found in Touryan et al. [17], Wilson [15], and Paraschivoiu [18]. A recent paper by Simão Ferreira and Scheurich [19] shows that, although these models may predict turbine power production fairly accurately, they are not accurate when it comes to prediction of blade loads; more physically realistic models such vortex codes or CFD codes must be used to accurately predict VAWT blade loads.

Momentum-based models are extremely popular with wind turbine designers because they are simple, fast, and fairly accurate for performance prediction, especially after they are tuned for a particular configuration. However, they are approximate because they are based upon the assumptions of flow conditions that are fixed in time and space, while the flow conditions around a wind turbine are constantly changing. These models cannot predict the effects of yawed flow, unsteady aerodynamics, and other complex flows that are present on wind turbines, all of which can have large impacts on turbine performance and loads. In some cases, specialized codes based on experimental results are used to approximate some of these effects, but these codes are limited to specific turbine sizes and geometries. In other cases, more realistic models such as vortex-based models, full computational fluid dynamics (CFD) models, and hybrid models are used to estimate these effects.

43.2.1.2 Vortex Models

Vortex models are usually more properly referred to as lifting line or lifting surface models, depending on whether a lifting line or a lifting surface formulation is used to model the blades. In the lifting line method, each rotor blade is modeled as a series of segmented *bound line vortices* located at the blade 1/4-chord line, as illustrated in Figure 43.12a. Line vortex strengths, defined by the blade lift at each radial location, are associated with the vortex line segments. The lifting surface method represents the blade in more detail, as a distribution of vortex line segments over the blade surface, as illustrated in Figure 43.12b. Either of these models will generate both trailing vorticity (perpendicular to the span of the blade) due to the differences in vortex strength along the blade span and shed vorticity (parallel to the span of the blade) due to time-dependent changes in vortex strength that are shed into the wake as the turbine rotates. These vortex methods lend themselves to the modeling of unsteady problems, as the shed vorticity models the time-dependent changes in the blade bound vortex strength. Solutions are achieved by impulsively starting the turbine in a uniform flow field and allowing the computational flow field to develop until it reaches a steady-state or periodic condition.

The manner in which the transport of the vorticity in the turbine wake is modeled depends on whether a free-wake or a fixed (or prescribed) wake model is used. In the

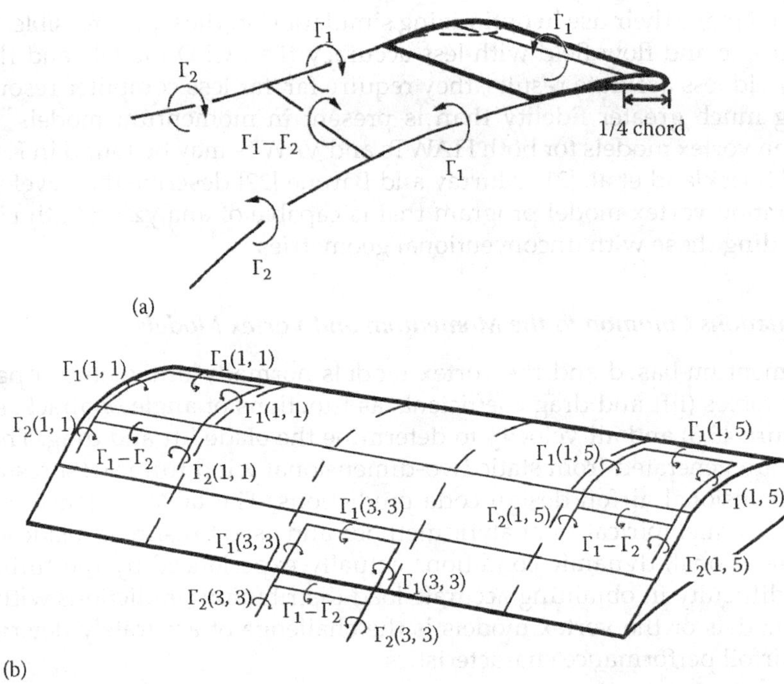

FIGURE 43.12
Schematic of lifting line and lifting surface models of turbine blades: (a) lifting line model and (b) lifting surface model (*Note:* Each of the surface panels [15 shown here] is represented by a vortex system such as the three sketched here).

free-wake model, the vorticity is allowed to convect, stretch, and rotate as it is transported through the wake. However, the movement of each line vortex is influenced by the presence of all of the other line vortices, including those on the blade. As the computation progresses in time, the number of vortices that must be followed and the time required to calculate the vortex interactions both grow very quickly. In order to minimize the need for large computer resources, the fixed or prescribed wake models have been developed. In these models, the geometry of the wake is modeled as either fixed or described by only a few parameters, and the vortex interactions in the wake are no longer directly calculated. The result is a much faster execution time, but the accuracy of the predicted power generation and blade loads depends very heavily upon the fidelity with which the specified wake approximates the actual physical wake. The three-dimensional, lifting-surface, free-wake formulation is the most physically realistic of the vortex models, but a computer program implementing such a model will require a large amount of computer resources and time. Experience has shown that the dramatic increase in computer resources required by such a model does not yield significantly more accurate predictions than what can be obtained with a three-dimensional, lifting-line, fixed-wake model. A major problem with vortex codes is finding a good balance between model simplification (and the associated limitations on fidelity), computation time, and desired accuracy. The improved physical fidelity of the vortex models relative to the momentum models results in more accurate solutions, assuming the turbine is represented in adequate detail and the solution time increment is sufficiently short. While vortex models have not been widely used in the wind industry in the past, they are seen to be gaining popularity today, as the continuing increases in computer speed make their use in optimizing simulation studies quite feasible. While they model the turbine and flow field with less accuracy than CFD models and thus, can be expected to yield less accurate results, they require far, far less computer resources while incorporating much greater fidelity than is present in momentum models. Additional information on vortex models for both HAWTs and VAWTs may be found in Kocurek [20], Snel [16], and Strickland et al. [21]. Murray and Barone [22] describe the development of a current-generation vortex model program that is capable of analyzing both HAWTs and VAWTs, including those with unconventional geometries.

43.2.1.3 Limitations Common to the Momentum and Vortex Models

Both the momentum-based and the vortex models normally utilize airfoil performance characteristic tables (lift and drag coefficients as functions of angle of attack, such as are shown in Figure 43.8) and air velocity to determine the blade lift and drag. These performance tables are generated from static two-dimensional wind tunnel test results or from static two-dimensional airfoil design code predictions. The table contents are modified with empirical, semiempirical, or analytic methods and used to estimate blade loads under the three-dimensional, dynamic conditions actually experienced by the turbine blades. The greatest difficulty in obtaining accurate load distribution predictions with either the momentum models or the vortex models is the challenge of accurately determining the appropriate airfoil performance characteristics.

43.2.1.4 Computational Fluid Dynamics Models

CFD, in a broad sense, is the solution of the partial differential equations describing the flow field by approximating these equations with algebraic expressions (discretizing them) and then solving those expressions numerically with the aid of a computer. Within the wind energy community, the term CFD normally refers to the numerical solution of the unsteady Reynolds-averaged Navier–Stokes (RaNS) equations [23,24], often restricted to four partial differential equations (one conservation of mass equation and conservation of momentum equations in three orthogonal directions) that describe general ideal-gas, incompressible, nonreacting, fluid flow.

One might argue that the most detailed and physically realistic method of predicting the performance of and loads on a wind turbine is to utilize CFD to model the airflow around the turbine and through the rotor, calculating the airfoil lift and drag directly. The flow field in the vicinity of the wind turbine is approximated as a computational grid of variable density, and the discretized Navier–Stokes equations are applied to each element of that grid. The computational grid close to the turbine blades must be very, very dense in order to capture the details of the airflow around the blades; it becomes less dense as the distance from the blades becomes greater, and the effect of the blades on the airflow decreases. The resulting set of simultaneous equations must be solved, frequently in a time-marching manner, to determine the time-dependent nature of the entire flow field.

Duque et al. [25] describe a 2002 CFD investigation of a wind turbine in which they utilized a complex grid with 11.5 million points to model the flow around a 10 m diameter HAWT rotor and tower combination. A steady-flow solution (rotor facing directly into the wind) for that model at a single wind speed, utilizing eight PC processors, each operating at 1.4 GHz in a parallel processor computer configuration, required approximately 26 h. An unsteady-flow solution (rotor yawed at an angle to the wind) with the same computing resources required over 48 h for each rotor revolution. Obviously, these computations can be performed much more quickly with the computing resources available today. Sørensen et al. [26], Johansen et al. [27], and Chow and van Dam [28,29] report other CFD modeling efforts.

At this point, CFD is suitable for research use or final design verification only—it is still too slow and requires far too many computer resources to be considered for use as a routine design tool. CFD also suffers from some shortcomings that limit its accuracy in performance and loads calculations. First, it does not consistently yield highly accurate results for airfoil lift and drag, because it cannot adequately model the transition of flow over the airfoil surface from laminar, well-ordered, basically two-dimensional flow to inherently three-dimensional and unsteady turbulent flow that is accompanied by rapid fluctuations in both velocity and pressure (most wind turbines today utilize airfoils that will experience such a transition in the flow). Second, most CFD models today cannot adequately predict the effects of separated flow, especially three-dimensional separated flow, such as will occur near the hub of the wind turbine rotor under most operating conditions. In spite of these limitations, CFD models, when run by an expert who devotes a lot of effort to validate the results, can yield very useful information, as illustrated by the work reported in Chow and van Dam [28,29]. Efforts to improve the accuracy of CFD codes continue.

43.2.1.5 Hybrid Models

The hybrid model approach is an attempt to get the increased accuracy that CFD models are capable of without the heavy computing resources. It typically models the airflow in the immediate vicinity of the airfoil and close to the turbine with the discretized Navier–Stokes equations, similar to the procedure used by the CFD models. However, to model the mostly undisturbed flow away from the turbine, the model uses nonviscous or potential flow equations that are much less complex and that can be solved much faster than the Navier–Stokes equations. The two solutions must be merged at the boundary between the two regions. The result is a code with the accuracy of the CFD model, but one that requires an order of magnitude less computing resources to solve. Xu and Sankar [30] and Schmitz and Chattot [31] describe two such codes. Even these hybrid models require a large amount of computer resources and are too slow and expensive to be utilized for parametric optimization studies; they are apt to be reserved for research work or detailed analysis of a proposed turbine design.

43.2.1.6 Model Results

None of these aerodynamic models is capable of accurately predicting the performance of and detailed loads on an arbitrary wind turbine operating at a variety of wind speeds. In order to have high confidence in the code predictions for a turbine design, the code must be calibrated against the measured performance and loads obtained from turbines of similar size and shape. Simms et al. [32] report on the ability of 19 codes based on the earlier models to predict the distributed loads on and performance of an upwind HAWT with a 32.8 ft (10 m) diameter rotor that was tested in the NASA/Ames 80 ft by 120 ft (24.4 m by 36.6 m) wind tunnel in 2000. Although the rotor was small compared to the 262 ft (80 m) diameter and larger commercial turbines that are being built today, a panel of experts from around the world concluded that it was large enough to yield results representative of what would be observed on the larger turbines. The comparisons of the code predictions and the experimental results were, in general, poor—turbine power predictions ranged from 30% to 275% of measured, and blade-bending moment predictions ranged from 85%

to 150% of measured for what is considered to be the most easy-to-predict conditions of no yaw, steady state, and low wind speed. Many aerodynamic code developers have spent considerable effort over the years since that comparison attempting to identify the sources of the discrepancies and improving the accuracy of their various codes.

In 2006, the Energy Research Center of the Netherlands led a second wind tunnel test of a heavily instrumented 15 ft (4.5 m) diameter wind turbine in the Germany/Dutch Wind tunnel organization 31 ft × 31 ft (9.5 m × 9.5 m) open-jet wind tunnel. This effort, known as the Model Experiments in Controlled Conditions or MEXICO experimental campaign, was funded by the European Union Framework Programme 5. An extensive literature list dealing with the experiment and related analyses is available on the MEXICO experiment website at www.mexnext.org/resultsstatus. Calculations from several models have been compared to the MEXICO results, with generally good qualitative agreement between calculations and measurements down to the flow detail level, even in yawed conditions. However, none of the calculations have been able to accurately predict both the velocities and the loads; they all overpredict the velocities and/or overpredict the loads. While one explanation for the anomalies could be tunnel effects rather than model inaccuracies, that has not been supported by CFD calculations such as that performed by Réthoré et al. [33].

Additional information and references on wind turbine aerodynamics models may be found in Manwell et al. [34] and Burton et al. [35] for HAWTs and in Touryan et al. [17], Wilson [15], and Murray and Barone [22] for VAWTs.

Most wind turbine companies today continue to use the very fast momentum-based models for design optimization purposes, in spite of the approximations and inaccuracies that are inherent in these models. The analysts doing the studies tweak these models to get good comparison with measurements from an existing turbine and then use them to predict performance and loads for new turbines that are fairly close in size and geometry to the existing turbine. If the new turbine is significantly different in size and shape from the reference turbine, the performance predictions are considered to be subject to considerable error. Performance tests on a prototype of the new turbine are required to further refine the actual performance and loads. There does appear to be a trend toward adapting a vortex-based code to replace the momentum-based codes in these applications, and that should reduce the amount of error inherent in the predictions. Performance tests on prototypes of new turbines will still be required with these codes, however, to refine the actual performance and loads.

43.3 Offshore Wind

All of the wind power development in the United States to date is land-based, most of it in the West and Midwest, where the wind offers a good resource and there are large areas of land available on which to place wind farms. However, this resource is far from the major metropolitan centers near the coastline, and building the transmission line capacity that is necessary to carry the power to those centers is very, very expensive. The offshore wind resource, on the other hand, is very large, and much of it is relatively close to major coastal metropolitan areas (see Figure 37.8), making the possibility of offshore wind attractive, in spite of the additional costs involved. In Europe, where very little land is available for siting turbines, large amounts (over 6 GW) of offshore wind have been installed at shallow-water sites (water depth of less than 30 m), fixed to the seafloor (the turbines are mounted on piles driven into the seafloor or on concrete gravity bases) in the past decade.

43.3.1 Offshore Platform Hydrodynamics

When turbines are located offshore, regardless of the depth of the water, the turbine supporting platform is subjected not only to the wind-generated loads transmitted through the turbine to its base, as is the case for the land-based turbine, but also to water-generated (or hydrodynamic) loads. In addition, the mounting platform now experiences a much larger range of motion than would the gravity foundation of a typical land-based turbine. Thus, the analysis of the wind turbine becomes much more complex—the platform must be modeled as a moving body, subjected to the hydrodynamic loads, that is coupled to the flexible wind turbine structure at the base. At shallow water sites of less than about 30 m water depth, the turbine can economically be mounted on a monopile driven into the seabed or on a conventional concrete gravity foundation (both referred to as fixed-bottom support platforms). For these turbines, the analysis of the hydrodynamic loads acting on the platform can be approximated by applying a number of simplifying assumptions. However, at greater water depth, fixed-bottom support platforms are no longer economically feasible. For water depths between 30 and 60 m, space-frame substructures, including tripods, quadpods, or lattice frames, will be required to maintain the strength and stiffness requirements at the lowest possible cost. At depths of 60 m and greater, economics drive the selection of a floating support platform moored to the sea bottom [36]. The offshore oil and gas industry has demonstrated the long-term survivability of offshore floating structures, so the technical feasibility is not in question, but high-fidelity modeling and accurate analysis are required to enable the development of economical floating platforms.

These more complex structures and hydrodynamic loading in the deeper waters can no longer be adequately modeled with the simplifying assumptions that are typically used for the shallow-water fixed-bottom platforms. Many frequency- and time-domain analyses have investigated wind turbines mounted in water depths in excess of 30 m, but they all contain approximations and limitations. Jonkman provides a very comprehensive review of the analyses that have been done and the approximations and limitations inherent in them [36]. Rather than attempting to summarize the development of the very complex hydrodynamics modeling here, I refer you to Jonkman [36] for a detailed explanation of his development of models that accurately represent the hydrodynamic and mooring forces acting on the floating platform. He has incorporated these models into the NREL-supported HydroDyn code; that code, used in conjunction with the publically available and NREL-supported FAST code for wind turbine modeling and analysis, is capable of analyzing the transient response of a wide variety of offshore wind turbine/support platform/mooring system configurations in waters of moderate to great depth.

43.4 Wind Turbine Loading

Wind turbines are typically fatigue-driven structures; they normally fail not as the result of a single application of a large-amplitude load, but as a result of the repeated application and removal (or cycling) of small-amplitude loads. Each load cycle causes microscopic damage to the structure, and the accumulated effect of many, many cycles of varying amplitude eventually leads to failure of the structure, a process referred to as fatigue failure. In general, the smaller the amplitude of the load, the larger the number of load cycles the structure can withstand before failing. Therefore, it is very important that the loads acting on a wind turbine be well understood; if the loads are larger than expected, fatigue failure may occur much earlier than anticipated. If the loads are smaller than expected, the turbine will more expensive than necessary for the desired lifetime.

The wind is random or stochastic in nature, with significant short-term variations or turbulence in both direction and velocity. Wind turbine aerodynamic loads may be regarded as falling into one of two broad categories—the deterministic loads occurring in narrow frequency bands resulting from the mean steady atmospheric wind, wind shear, rotor rotation, and other deterministic effects, and the random loads occurring over all frequencies resulting from the wind turbulence. Prior to about 1995, the deterministic loads were frequently predicted with an aerodynamics code, such as those described earlier, utilizing a uniform wind input, while the random loads were estimated with empirical relations. However, turbine designers now recognize that this approach may lead to serious underprediction of both the maximum and random blade loads, resulting in costly short-term component failures. Most analysts today utilize an aerodynamics performance code with a wind model that includes a good representation of the turbulence of the wind in all three dimensions to predict long-term wind turbine loads. The appropriate method of determining the wind-induced extreme events and random loads that limit the lifetime of a turbine remains the subject of ongoing research.

In contrast, designers of offshore turbines can look to the offshore oil and gas industry for guidance on hydrodynamic loading analysis, and Jonkman [36] and others have developed quite comprehensive hydrodynamic load simulation capabilities.

Any lack of knowledge about loads that the turbine/mounting platform will encounter is typically compensated for in the design process by incorporation of large safety margins, leading to excess material and cost.

Civil engineers have spent decades developing statistical methods for predicting wind and wave loads on offshore drilling platforms, to help reduce the cost and increase the reliability of those platforms. The wind turbine industry is now starting to apply that technology to predict the wind and wave loads on wind turbines.

43.5 Wind Turbine Structural Dynamic Considerations

Input loads resulting from wind and, in the case of offshore turbines, waves, together with dynamic interactions of the turbine components with each other, result in forces, moments, and motions in wind turbines, phenomena referred to as structural dynamics. By applying various analysis methods, the impact of changes in turbine configurations, controls, and subsystems on the behavior of the turbine can be predicted. General wind turbine structural dynamic concerns and methods of analysis are discussed in the following paragraphs.

43.5.1 Horizontal-Axis Wind Turbine Structural Dynamics

Small horizontal-axis turbine designs usually use fairly rigid, high-aspect-ratio (the blade length is much greater than the blade chord) blades, cantilevered from a rigid hub and main shaft. As turbine size increases, the flexibility of the components tends to increase, even if the relative scales remain the same, so the blades on larger turbines tend to be quite flexible, and the hub and main shaft tend to be far less rigid than corresponding components on the small turbines. The entire drive train assembly is mounted on and yaws about a tower that may also be flexible. These structures have many natural vibration modes, and some of them may be excited by the wind or the blade rotation frequency to cause a *resonance* condition, amplifying vibrations and causing large stresses in one or more components. Operating at a resonance condition can quickly lead to component failure and result in the destruction of the turbine. Careful structural analysis during the turbine design may not guarantee that the turbine will not experience a resonance condition, as analysis techniques are not infallible, but ignoring the analysis altogether or failing to properly conduct parts of it may dramatically increase the probability that the turbine will experience one or more resonance conditions, leading to early failure. While the relatively rigid small turbines are not likely to experience these resonance problems, the very flexible, highly dynamic larger turbines may well experience resonance problems unless they are very carefully designed and controlled. Turbines that operate over a range of rotational speeds (variable-speed turbines) are especially challenging to design. The designer will usually try to minimize the number of resonances that occur within the operational speed range and then implement a controller that will avoid operating at those resonance conditions. The actual resonances typically depend on the rotor speed, and the severity of the resonance depends on the wind speed, so the controller logic can become quite complicated.

It is possible to develop techniques in the frequency domain to analyze many aspects of the turbine dynamics. The frequency-domain calculations are fast, but they can be applied only to linear, time-invariant systems and therefore cannot deal with some important aspects of wind turbine operations such as aerodynamic stall, start-up and shut-down operations, variable-speed operation, and nonlinear control system dynamics. In spite of these limitations, frequency-domain solutions of modal formulations are frequently used in the preliminary design of a wind turbine, when quick analyses of many configurations are required.

The large relative motion between the rotor and the tower frequently precludes the use of standard commercial finite-element analysis codes and requires the use of a model constructed specifically for the analysis of wind turbines. Development of such a model can be a rather daunting task, as it requires the formulation and the solution of the full nonlinear governing equations of motion. The model must incorporate the yaw motion of the nacelle, the pitch control of the blades, any motion and control associated with hinged blades, the time-dependent interaction between the rotor and the supporting tower, etc. If the full equations of motion are developed with either finite-element or multibody dynamics formulations, the resultant models contain moderate numbers of elements and potential motions (degrees of freedom or DOF), and significant computer resources are required to solve the problem. On the other hand, a modal formulation utilizing limited DOF may be able to yield an accurate representation of the wind turbine, resulting in models that do not require large computing resources. The development of the modal equations of motion may require somewhat more effort than does the development of the finite-element or multibody equations, and the equations are apt to be more complex. The modal degrees of freedom must include, at a minimum, blade bending in two directions, blade motion relative to the main shaft, drive train torsion, tower bending in two directions, and nacelle yaw. Blade torsion (twisting about the long axis) is not normally included in current modal models, but it may become more important as the turbine sizes continue to increase and the blades become more flexible. The accuracy of some modal formulations is limited by their inability to model the direction-specific nonlinear variation of airfoil lift with an angle of attack that occurs as a result of aerodynamic stall. However, this is not an inherent limitation of the technique, and some modal formulations are free of this limitation. These modal formulations (NREL's wind-turbine-specific FAST [36] is an example of such a code) are extremely fast running and are arguably the most widely used HAWT structural analysis tools today.

The most accurate structural dynamics models utilize general, commercially available finite-element codes such ABAQUS [37] or multibody dynamics codes such as ADAMS [38] or wind-turbine-specific multibody dynamics codes such as HAWC2 [39] and BLADED [40]. These require significantly more computer resources than do modal formulation codes, but they can still be run fairly quickly on a PC. FAST, the NREL wind-turbine-specific version of ADAMS, HAWC2, and BLADED have been verified through measurements and comparisons with other codes, and they include all of the modules required to simulate both land-based and offshore wind turbines, including both aerodynamic and hydrodynamic loading.

All of these structural dynamics codes use separate aerodynamics codes, most based on BEM models, to furnish the aerodynamic loads that act on the turbine blades.

Regardless of the methods used in preliminary design, the state of the art in wind turbine structural analysis today is to use highly detailed multibody or finite-element-based, time-accurate structural codes, coupled with blade loadings derived from BEM-based aerodynamics models and support platform loadings derived from hydrodynamic models (for offshore configurations) to analyze the turbine behavior for the detailed final design calculations.

Malcolm and Wright [41] and Molenaar and Dijkstra [42] provide reviews of some of the available land-based HAWT dynamics codes that have been developed, together with their limitations. Buhl et al. [43] compare some of the land-based HAWT dynamics codes that have been extensively verified and that are widely used today, and Quarton [44] provides a good history of the development of land-based HAWT wind turbine analysis codes. More general finite-element dynamics codes are described in [45,46], and additional information on both land-based and offshore HAWT dynamics models can be found in Manwell et al. [34] and Burton et al. [35].

43.5.2 Vertical-Axis Wind Turbine Structural Dynamics

Darrieus turbine designs normally use relatively slender, high-aspect-ratio structural elements for the blades and supporting tower. As with large HAWTs, the result is a very flexible, highly dynamic structure, with many natural modes of vibration that again must be carefully analyzed to ensure that the turbine will avoid structural resonance conditions under all operating environments. The guy cables and turbine support structure can typically be analyzed with commercial or conventional finite-element codes, but the tower and blades require a more refined analysis, usually requiring the use of a finite-element code possessing options for analyzing rotating systems. With such a code, the blades and tower of a VAWT are modeled in a rotating coordinate frame, resulting in time-independent interaction coefficients. The equations of motion must incorporate the effects of the steady centrifugal and gravitational forces, the aerodynamic loading due to the turbulent wind, the hydrodynamic loading due to the water waves and current for offshore configurations), and the forces arising from rotating coordinate system effects. Detailed information on finite-element modeling of land-based VAWTs may be found in Lobitz and Sullivan [47], and information on more general finite-element modeling that includes offshore VAWTs may be found in Owens et al. [48].

43.6 Peak Power Limitation

All turbines incorporate some method of regulating or limiting the peak power produced. The entire turbine, including the rotor, the transmission, and the generator, must be sized to handle the loads associated with peak power production. While high winds (above, say, 25 m/s) contain large amounts of available power, they do not occur very often, and the energy that can be captured is very small. This is illustrated in Figure 43.13 for the Amarillo, TX, airport. In this figure, the power density is the power per unit of rotor area (normalized to yield a value under the curve of unity) that is available for capture by a wind turbine. This takes into account the amount of time that the wind actually blows at each wind speed (the probability density that is also shown on the figure). Amarillo does occasionally experience very high winds, but as seen from this probability density, that does not happen very often, and the energy that could be captured from winds above 24 m/s is negligible.

Generators and transmissions operate most efficiently at their design conditions, typically close to their maximum capacity. These efficiencies drop off quickly at conditions below design. Cost trade-off studies reveal that it is far more cost effective to limit the maximum power level to that achieved at wind speeds of, say, 13–14 m/s and to shut the turbine down completely at a cutout wind speed of, say, 26 m/s, as illustrated in Figure 43.14, than to try to capture the maximum amount of power at the higher wind speeds. Limiting the peak power limits the peak loads that the entire drive train, the tower structure, and the blades must withstand, so the turbine can be built with far less material (and at lower cost) than that required for a turbine that generates peak power at a wind speed of 25 m/s. Under these conditions, the transmission and generator are operating near design conditions a good part of the time, so they are operating at close to their maximum efficiency. The additional energy captured due to the increase in generator and transmission efficiencies at the lower wind speeds is usually many times greater than that lost due to limiting

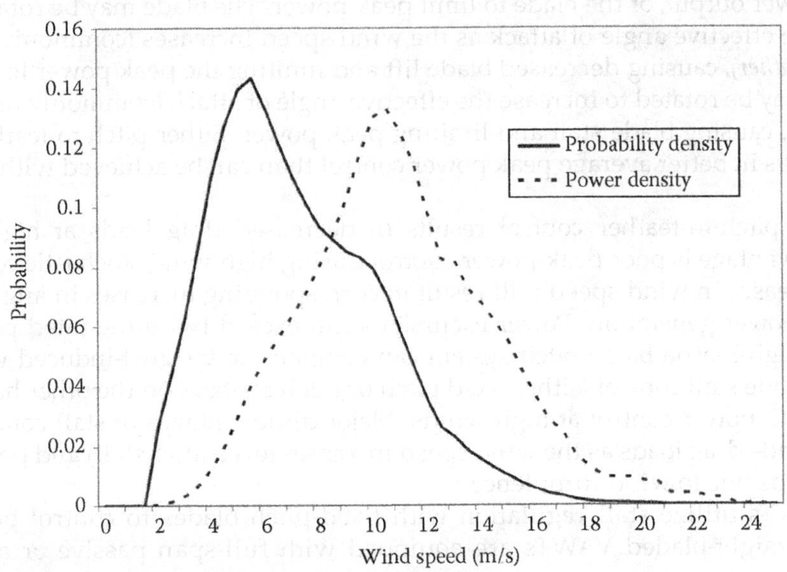

FIGURE 43.13
Wind speed and wind power probability densities for Amarillo, TX, airport.

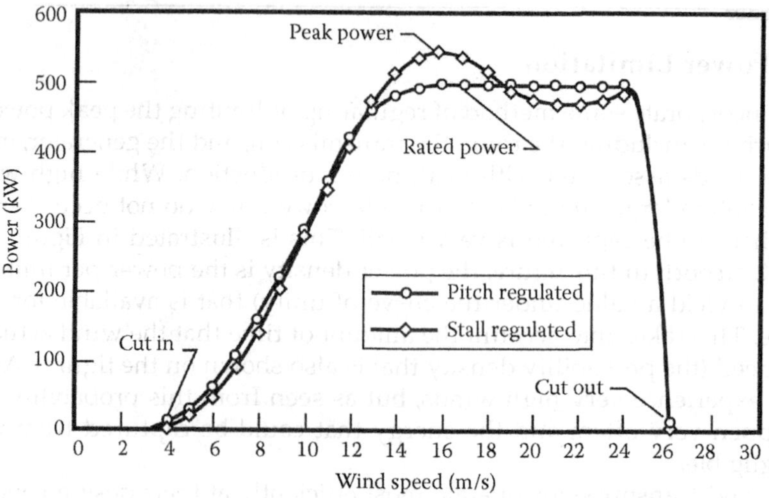

FIGURE 43.14
Sample power curves for stall-regulated and pitch-regulated wind turbines.

the peak power at the rather infrequent winds above 14 m/s (refer to the wind speed distribution in Figure 43.13).

Most small horizontal-axis turbines incorporate passive features, such as tail vanes and furling, that turn the rotor so the rotor axis is no longer aligned with the wind in order to limit peak power production in high winds. Older turbines and some small turbines are designed with fixed-pitch blades and rely on airfoil stall at high winds to limit the maximum power output of the machine. However, nearly all modern large horizontal-axis turbines now use blade pitch control, where either the entire blade or a portion of it is rotated about the blade longitudinal axis to change the effective angle of attack, and thus the power output, of the blade to limit peak power. The blade may be rotated so as to decrease the effective angle of attack as the wind speed increases (commonly referred to as *pitch to feather*), causing decreased blade lift and limiting the peak power to the desired level, or it may be rotated to increase the effective angle of attack (commonly referred to as *pitch to stall*), causing blade stall and limiting peak power. Either pitch to feather or pitch to stall results in better average peak power control than can be achieved with fixed-pitch stall control.

Although pitch-to-feather control results in decreased drag loads at high winds, a major disadvantage is poor peak-power control during high-wind stochastic conditions—sudden increases in wind speed will result in corresponding increases in angle of attack, loads, and power generation. Power excursions can exceed twice the rated power levels before the high-inertia blade pitch system can compensate for gust-induced wind speed increases. Blade stall control (either fixed pitch or pitch to stall), on the other hand, results in better peak power control at high winds. Major disadvantages of stall control include increased blade drag loads as the wind speed increases (even after stall) and possible large dynamic loads due to wind turbulence.

Most VAWTs utilize stall regulation with fixed-pitch blades to control peak power, but some straight-bladed VAWTs are equipped with full-span passive or active pitch controls.

Pitch control can be either passive or active. The pitch control may be included in each blade mounted on a very simple hub, it may be incorporated into a sophisticated hub with blades that are nothing special, or it may be distributed between the hub and the blades. Therefore, discussions about turbine pitch control normally refer to the combination of hub and blades—the rotor.

- Passive pitch-control techniques automatically adjust the blade pitch angle by using hub-mounted cams activated by centrifugal loads or by using tailored blade materials that permit the blade to twist toward feather or stall under high loads. These devices are very carefully tailored to maintain peak performance at lower wind speeds but limit the peak power and blade loads at high wind speed. A major disadvantage of these techniques is that they cannot readily be adapted to site-specific conditions—a hardware change is required to make any changes to the control.
- Active-pitch control rotors are equipped with one or more motors (typically mounted on the hub) to change the blade pitch angle on command from the turbine controller. These rotors are much more expensive than fixed-pitch stall control blades, but they open the possibility for much refined control of the turbine, potentially leading to reductions in loads for the entire drive train. Changes to the control can usually be implemented by simply changing the controller software program.

Sample power curves for fixed-pitch stall regulated and active pitch (pitch to feather) 500 kW turbines are shown in Figure 43.14.

43.7 Turbine Subsystems

The wind turbine incorporates many subsystems, in addition to the actual turbine, in order to generate power. The electrical power generation, yaw, and control systems are the only ones that will be discussed here.

43.7.1 Electrical Power Generation Subsystem

Once a wind turbine has converted the kinetic energy in the wind into rotational mechanical energy, that rotational energy is usually converted by a generator into electricity that can be readily transported to where it is needed. In many configurations, the generator input shaft speed is much higher than the turbine shaft speed, requiring the use of a speed-up transmission (or *gearbox*) that increases the turbine shaft speed to that required by the generator. Other configurations utilize *direct-drive* (DD) generators to eliminate this need for a gearbox with the generator input shaft directly connected to the turbine drive shaft.

In the past, most grid-connected turbines have utilized either synchronous or induction generators. Synchronous generators are more complex and tend to be more expensive than induction generators, but they provide excellent voltage and frequency control of the generated power and can deliver reactive power to the grid. However, these generators are not intrinsically self-starting (the turbine blades are usually pitched to start the turbine, which drives the generator), do not provide power-train damping, and require sophisticated controls for connecting to the grid, as the output frequency (and thus, the speed of the generator) must be precisely matched to the grid frequency before the connection is made.

Induction generators, in contrast, have a simple, rugged construction; they may readily be used as motors to spin a turbine up to speed; they are cheaper than synchronous generators; they may be connected to and disconnected from the grid relatively easily, and they provide some power-train damping to smooth out the cyclic torque variations inherent in the wind turbine output. However, they require reactive power from either power electronics or the grid, and they can contribute to frequency and voltage instabilities in the grid to which they are connected. These adverse effects can usually be solved fairly quickly and at low cost with modern power electronics. Induction generators are, therefore, the most common type found on wind turbines.

Synchronous *permanent magnet* (PM) generators replace the separate excitation, slip rings, and rotor windings (with associated losses) of a typical synchronous machine with powerful PMs, leading to a simple, rugged construction. Virtually all PM generators utilize *rare-earth* (RE) materials for the PMs due to the very high flux density that can be achieved with these materials. The power produced by PM generators is usually variable-voltage and variable-frequency AC that must be converted to DC or to fixed-voltage and fixed-frequency AC with power electronics before it can be fed into the grid. Even with this requirement for additional power electronics, these machines tend to achieve higher efficiency at low power ratings (92.9% at 25% load versus 94.4% at full load) than either induction or synchronous generators, leading to increased energy capture. The cost of these generators has historically been somewhat higher than that of induction or synchronous generators, due primarily to the high cost of the RE material, but technology advances have resulted in price decreases, making the technology quite competitive with the older technologies. As a result, PM generators are becoming very common. Fuchs et al. [49] describe the development of a PM generator for wind turbine use.

While the earlier discussion covers the generator technologies most commonly used on wind turbines in the past, many other types of technologies are used today. Discussion of those generator technologies may be found in Manwell et al. [34] and Burton et al. [35].

Direct-drive generators are normally synchronous machines (with either conventional or permanent magnet excitation) of special design, built with a sufficient number of poles to permit the generator rotor to rotate at the same speed as the wind turbine rotor. This eliminates the need for a gearbox or transmission. Benefits of a direct-drive generator include a drive train with far fewer parts, leading to potentially higher reliability and lower maintenance, and a lower system tower-top weight. The primary drawback to a direct-drive generator is the very high torque that it must handle; the power produced by the rotor is proportional to torque multiplied by rpm, so the torque that is applied to the generator goes up as the generator operating rpm decreases. A typical 1.5 MW turbine rotor rpm might be 15 rpm, with a conventional generator rpm of 1800 rpm. Replacing this generator with a direct-drive generator would drop the generator rpm to 15 rpm, requiring a direct-drive generator torque rating of 1800/15 = 120 times that of the conventional generator. Direct-drive generators are usually used in conjunction with power electronic converters to decouple the generator from the network and provide flexibility in the voltage and frequency requirements of the generator.

Direct-drive PM generators require very strong magnetic fields (and corresponding large amounts of RE materials), due to the very high torque levels at which the generators must operate. In large amounts, RE components are still quite expensive, as China is currently the primary source, and they are severely limiting export. Thus, direct-drive

PM generators tend to be more expensive than more conventional direct-drive generators. Moderate-speed PM generators (lower rotational speed than traditional synchronous or induction generators, but higher rotational speed than direct-drive generators) operate at much reduced torque levels and thus require much less RE. This configuration still requires the use of a gearbox, but it can be a fairly simple single-stage design due to the moderate speed step-up requirements.

The direct-drive generator technology advantage in reduced drive train weight increases with size, but as the power rating approaches the 8–10 MW range, current technology results in generators that rapidly grow larger and heavier, due to the high torque requirements. This, in turn, requires a heavier tower structure to carry the extra weight and torque. These large generators also present design challenges. As with PM generators, an alternative then becomes a relatively light, single-stage gearbox with a much smaller, lighter generator that still offers many of the direct-drive advantages.

Most older turbines (built prior to about 2000) operated at a single fixed rotational speed, but many newer turbines, and especially the large ones, are variable speed, operating within a fixed range of rotational speeds. Variable-speed turbine operation offers several major advantages over fixed-speed operation:

1. The aerodynamic efficiency of the rotor at low to moderate wind speeds may be improved by lowering the rotational speed to keep the turbine operating close to the optimum tip-speed ratio, maximizing the power coefficient. At higher wind speeds, the turbine rotates at maximum rotational speed, and the blades are either in stall or are pitched to limit peak power. The rotor speed may also be adjusted to fine-tune peak power regulation.
2. System dynamic loads may be attenuated by the inertia of the rotor as it speeds up and slows down in response to wind gusts.
3. The turbine may be operated in a variety of modes, including operation at maximum efficiency to maximize energy capture at lower wind speeds and operation to minimize fatigue damage.

As mentioned earlier, for variable-speed operation, certain rotational speeds within the operating-speed range will likely excite turbine vibration modes, causing structural resonance and increased rates of fatigue damage. These rotational speeds must be avoided during operation, leading to complex control schemes.

Variable-speed operation is possible with any type of generator, including direct-drive and PM generators. As is the case for fixed-speed, direct-drive and PM generators, variable-speed operation results in variable-frequency/variable-voltage AC power. Interfacing the turbine to the power grid requires the use of power electronics to convert this power to high-quality, constant-frequency/constant-voltage AC power. Several methods have been developed for accomplishing this with sophisticated power electronics (see, e.g., Smith [50]), but research to develop improved methods with higher efficiencies continues. Manwell et al. [34] and Burton et al. [35] give brief overviews of some common types of power converters and provide references for more extensive discussions of these devices.

For a conventional, non-PM generator, electrical efficiency tends to drop off rapidly as the generated power falls below the rated generator capacity, so single-generator wind turbines tend to be very inefficient at low wind speeds. Some turbine designs address this deficiency by incorporating multiple smaller generators; at low wind speed, only one generator is attached to the drive train, with more being added as the wind speed increases. The net result is that each generator operates close to its rated power much of the time, increasing the overall generator efficiency. Similar increases in generator efficiency at low power levels can be obtained with a single generator utilizing pole switching or multiple windings. Other generator designs that overcome this loss in efficiency at lower power levels have been developed in recent years.

43.7.2 Yaw Subsystem

The rotor of an HAWT must be oriented so that the rotor axis is parallel to the wind direction for peak power production. While small turbines usually rely on passive systems, such as tail vanes, to accomplish this, most large upwind HAWTs and a few downwind HAWTs incorporate active yaw control systems, using a wind direction sensor and a drive motor/gear system to orient the rotor with respect to the wind direction. Most downwind HAWTs are designed to utilize the wind itself to automatically orient the rotor. Active yaw systems tended to be extremely problematic in early turbines, basically because the loads acting on them were not well understood. Yaw loads are much better understood today, and these systems are no longer a major problem area.

As mentioned earlier, VAWTs do not require orientation with respect to the wind and thus do not require yaw systems.

43.7.3 Control Subsystem

One or more control systems are needed to integrate the operation of the many components of a wind turbine and to safely generate power. In a very general sense, a wind turbine control system consists of a number of sensors, a number of actuators, and a system of hardware and software that processes the signals from the sensors to generate signals to control the actuators. A turbine usually contains a minimum of two distinct controllers—a safety controller that will override all other controllers to bring the turbine to a safe state (usually stopped, with the brakes applied) in case of an abnormal event, and an operational controller that handles the normal operation of the turbine. Any particular turbine may incorporate many additional secondary or slave controllers, each of which handles only a limited number of tasks. An example of a secondary/slave controller might be a blade pitch controller that pitches the blades to follow a predetermined schedule during start-up and shutdown of the turbine and that pitches the blades as necessary to regulate the power output of the turbine at the rated level in above-rated wind speeds.

43.7.3.1 Safety Controller

The safety controller acts as a backup to the operational controller and any secondary or slave controllers; it takes over if the other controllers fail to maintain the turbine in a safe operating mode. It is normally triggered by the activation of certain safety sensors (such as excessive vibration, excessive rotor speed, or excessive generator power) that are independent of the sensors connected to the operational and/or slave controllers, but it may also be activated by an operator-controlled emergency stop button. This system must be as independent from the normal operations control system as possible and must be designed to be fail-safe and highly reliable, since it may be the last line of defense to save a turbine from self-destruction. The safety controller will normally consist of a hard-wired fail-safe circuit monitoring a number of sensors. If any of the sensors indicates a problem, the safety controller assumes full turbine control and ensures that the turbine is brought to a safe condition. This might include, for example, de-energizing all electrical systems, pitching the blades to the feather position, and engaging the spring-applied emergency brakes (that are held off in normal operation).

43.7.3.2 Operational Controller

The regular operational controller is usually computer or microprocessor based; a basic turbine operational controller will normally start and stop the machine; connect the generator to or disconnect the generator from the power grid, as needed; control the operation of the yaw and pitch systems (if present); perform diagnostics to monitor the operation of the machine; and perform normal or emergency shutdown of the turbine as required.

For older turbines, the operational controller was frequently a fairly generic device that was simply added to the turbine and then programmed with either hardware or software modifications to implement the control functions specific to that particular turbine. The operational controllers on newer turbines, especially the large, variable-speed/variable-pitch machines, incorporate much more intelligence than the old, generic type, and they are custom tailored (with software) to reflect the capabilities and characteristics of each specific turbine. Modern controllers (or, more accurately, the turbine control software programs) are usually designed from scratch as an integral component of the wind turbine and must be included in the models of system aerodynamics and structural dynamics to obtain accurate estimates of loads and motions. The controller dictates how the pitch, rotor speed, and generator torque systems are used to limit peak power and/or torque, control rotor speed, maximize energy capture, trade-off energy capture and load mitigation, reduce power fluctuations, control power quality, actively control some turbine dynamics, or perform many other functions. These controllers may consist of a single physical unit performing all of the assigned tasks, or they may consist of multiple physical units, each performing a small number of tasks, all coordinated by a supervisory controller.

43.7.3.3 Turbine Power/Load Control Programming

The operational control hardware mentioned in the last section must be programmed to perform the appropriate functions for the turbine. This section discusses some of the specific scenarios that may be implemented in the control system.

Figure 43.15 illustrates the three distinct operational regions of wind turbine operation, which are as follows:

1. Region 1 is where the wind speed is too low to overcome the drive train resistance of the turbine.
2. Region 2 is that region between the cut-in wind speed (where the turbine starts to generate power) and the rated wind speed (where the turbine reaches rated or maximum power).
3. Region 3 is between the rated wind speed and the shut-down wind speed (where the turbine is shut down to protect it).

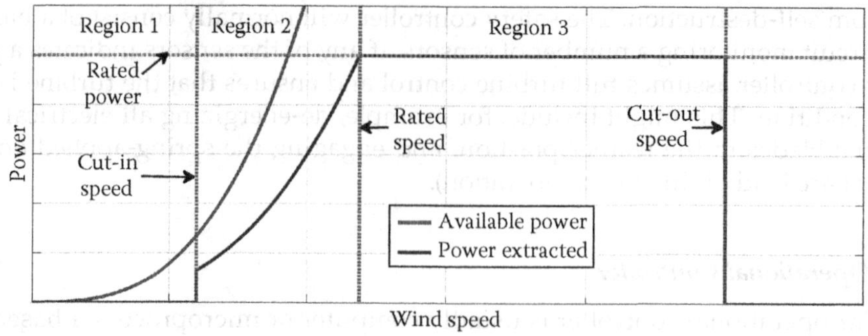

FIGURE 43.15
Wind turbine operating regions and ideal turbine power curve.

The pitch and rotor rotational speed capabilities of a turbine determine how the turbine controller can increase energy capture and limit blade and generator loads in these regions. The ideal path of turbine power as a function wind speed, as shown on Figure 43.15, requires that the turbine be operated at maximum efficiency in region 2 and then limit the power production in region 3. This requires that the rotor rotational speed (commonly referred to as the rotor speed) be correlated to the wind speed throughout region 2 so that the turbine is always operating at its maximum efficiency (the turbine $C_{P,\max}$ curve), and the turbine must be able to dump excess power in order to limit the turbine power and wind-induced loads in region 3.

The early turbines were restricted to constant rotor speed and fixed blade pitch (referred to as constant speed/fixed pitch). These turbines are constrained to follow some suboptimal energy capture path through region 2 (the precise path depends on the preset blade pitch and the fixed rotor speed). They rely on blade stall to limit peak power and loads in region 3, but stall can lead to some very high power peaks and loads in this region in response to positive wind gusts. Large turbine design has evolved through the years to include variable rotor speed and variable blade pitch (referred to as variable-speed/variable-pitch) capabilities that enable modern turbine controllers to better approximate the ideal power path of Figure 43.15 and capture more energy. These capabilities enable the turbine (as directed by the control program) to very closely follow the optimal turbine performance path; within region 2, the turbine sets the optimum blade pitch angle and then operates at the rotor speed that optimizes turbine performance at each wind speed; within region 3, the turbine operates at maximum rotor speed and utilizes blade pitch to control peak power and wind-induced loads. The turbines may also utilize the variable rotor speed capability to smooth out power and load fluctuations due to wind gusts throughout regions 2 and 3.

The turbine controller determines the precise manner in which these capabilities are utilized. It is typically designed with two distinct operating algorithms for the two regions, with some kind of transition between the two that mitigates the very large drive train loads that can result from an abrupt switch between the variable-speed/fixed-pitch operation in region 2 and the fixed-speed/variable pitch operation in region 3. This transition usually results in a lowering of the actual power curve in the vicinity of change between the operating regions—some energy capture is sacrificed to mitigate drive train loading. A turbine is normally designed to spend a lot of time in this transition region in order to maximize energy capture, so even a small lowering of the power curve here can result in considerable energy capture loss over time. Given the reality that a 1% loss in energy capture for every turbine results in income loss of over $2 million annually for 1 GW of installed capacity, there is a tremendous amount of interest in optimizing control in this region to maximize power while minimizing loads.

Many other areas of control offer the opportunity to increase energy capture and/or mitigate loads. A few of those areas are mentioned in the following paragraphs.

Early variable-pitch turbines utilized collective blade pitch, where all the blades pitch the same amount at the same time, often driven by a single motor. As turbines became larger, this design changed to use an electric motor for each blade, enabling each blade to be pitched independently of the others. Research has shown that utilizing this capability has the potential to significantly decrease blade fatigue loads [51]. However, this control strategy leads to increased use of blade pitch motors; retrofitting it on existing turbines could lead to early pitch motor burnout. In addition, implementation of this capability without an update to the existing turbine controller software has sometimes been found to lead to a small loss in turbine energy capture. This technology remains the subject of active research.

The blades on large turbines are very large; some exceed 80 m in length. Pitching a blade of this size is slow, and this limits the amount of load mitigation that can be achieved with full-span pitch. One variation on active blade pitch is the use of only a portion of the airfoil surface (typically including the blade tip) as a control surface. Adjusting the pitch (and, thus, the lift and drag) of this portion of the blade independently of the remainder of the blade (so-called partial-span control) can be used to control the peak power output of the turbine. This control surface is usually much smaller than the full blade, so it can respond to wind changes and mitigate loads much faster than can the full-span blade. However, partial-span devices have proven to be very difficult to integrate into a blade, and the gaps between the devices and the surrounding blade tend to generate noise.

Some recent research work has investigated the use of very small load control devices that can generate large changes in lift and drag while experiencing small loads. The chord-wise dimensions of these devices are typically on the order of 1% of the blade chord; their small size means they can be activated very quickly to alleviate excess loads, and they are less likely to create large amounts of noise. Mayda et al. [52] discuss the use of CFD to investigate the effects of some of these devices. An excellent summary of work on this concept prior to 2009 may be found in a survey article written by Barlas and van Kuik [53], while Barlas et al. [54] and Berg et al. [55] provide information on two recent demonstration efforts.

Other areas of controls research include the following:

- Maximizing turbine and/or wind-farm-generated power while mitigating structural loads.
- Maximizing turbine and/or wind farm power quality.
- Preview control, where the wind is measured upwind of the turbine and load mitigation and/or enhanced energy capture actions can be initiated before the wind actually strikes the turbine.
- Maximum power point control, where the controller *learns* the exact method of operation that will yield the maximum possible power output (especially in region 2), rather than relying on a nominal maximum efficiency curve.

While traditional turbine controls seem to work fairly well for the fixed-bottom offshore turbines, the field of offshore floating wind turbine control is so new that very little is known about what will be needed.

43.8 Other Wind Energy Conversion Considerations

The actual turbine system is the single most important and most costly item of a wind energy conversion system, but there are many other aspects of the system that must be considered and carefully optimized before wind energy can be produced at a cost-competitive price. These aspects include things like turbine siting, installation and foundations; operating and maintenance costs; manufacturing processes; transport of components to the site; turbine payback period; wind energy COE and environmental concerns. Turbine siting has already been discussed in Chapter 37. This section will discuss only materials, installations, grid integration, wind forecasting, energy payback period, wind turbine costs and COE, and some environmental concerns. Additional information, including extensive reference lists, on all of these aspects can be found in Manwell et al. [34], Burton et al. [35], and Spera [56].

43.8.1 Wind Turbine Materials

As mentioned earlier, wind turbines are fatigue critical structures (their design is driven by consideration of the cyclic fatigue loads they must endure), and the number of fatigue cycles they experience in a 20–30-year design life is three orders of magnitude beyond the 10^6 cycles that has been the common limit of fatigue data for most materials. Most of the materials used in the construction of wind turbines are typical of those materials that are used in other rotating machinery and towers—relatively common structural materials such as metal, wood, concrete, and glass-fiber-reinforced plastic (GFRP) composites. Towers are typically made of steel; a few have been built of concrete. Drive trains, generators, transmissions, and yaw drives are made of steel. These components can readily be designed so they experience very low stresses and will have a fatigue life of 20–30 years. HAWT blades, however, must be built with a minimum of lightweight material to minimize the gravity-induced cyclic loads on the blades, drive train, and tower. Over the past 20 years or so, high-cycle fatigue databases for many potential blade materials have been developed specifically for wind turbine applications. Mandell and Samborsky [57] and Mandell et al. [58] describe the main U.S. high-cycle fatigue database; the latest database, together with new reports and recent publications and presentations regarding the database, may be found at www.coe.montana.edu/composites/. De Smet and Bach [59] and Kensche [60] describe the European counterpart, which has now been included in the OPTIMAT materials database maintained by the WMC Knowledge Centre of the Netherlands. The latest version of this database is available at www.wmc.eu/optimatblades_optidat.php, and a reference document describing the contents of the database is provided by Nijssen [61].

The material with the best all-around structural properties for wind turbine blades appears to be carbon-fiber/epoxy composite, but it is significantly more expensive than other potential materials. In light of that, the blade material of choice today (as it has been for the past couple of decades) is GFRP, due to the high strength and stiffness that can be obtained, the ease of tailoring blades made of GFRP to handle the loads, and the relatively low cost of GFRP [62]. However, the trend to larger and larger turbines, with the resultant increase in blade weight and flexibility, has created intense interest in utilizing some carbon fiber in the blades to decrease weight and add stiffness. The expense of carbon fiber, even in the cheapest form available today, means that turbine designers must incorporate it into blades in a cost-effective manner. According to Griffin [63] and Jackson et al. [64], one very efficient method of utilizing carbon fiber is to place it in the longitudinal spar caps, near the maximum thickness area of the blade, where its lightweight and extra stiffness yield maximum benefits.

Obviously, the materials must be protected from the environment in some fashion. While paints and gel coats have been deemed adequate for most land-based installations in the past, long-term exposure to sunlight and high winds (and the dust and debris carried by those high winds) has resulted in considerable degradation of the GFRP blades, especially the leading edges, leading to considerable loss of turbine performance. Quantifying the performance loss due to this surface degradation is an active research area at the current time [65].

The offshore environment is much different and much more demanding than the onshore environment. Testing techniques must be developed to determine the suitability of materials proposed for use in offshore turbines. Miller et al. [66] describe the challenges they have encountered in developing techniques to evaluate the performance of composite materials in a seawater environment and present some initial results.

43.8.2 Wind Turbine Installations

Chapter 37 of this handbook discusses wind turbine siting considerations in some depth. Although individual turbines or small clusters of turbines may be used to provide power to small loads such as individual residences or businesses, the most common arrangement for producing large amounts of energy from the wind is to locate many wind turbines in close proximity to each other in a wind farm or wind park. Figure 43.16 is a photograph of several of the General Electric 1.5 MW turbines comprising the 204 MW New Mexico Wind Energy Center in eastern New Mexico (mentioned earlier in the case study in Chapter 37), while Figure 43.17 shows a few of the turbines in the huge offshore

FIGURE 43.16
Typical land-based wind farm installation—New Mexico Wind Energy Center in eastern New Mexico.

FIGURE 43.17
Offshore wind farm installation—London Array Wind Farm in Thames River Estuary, United Kingdom. (Courtesy of London Array Limited.)

London Array. Operating turbines in this manner leads to lower COE, as fixed construction costs, such as electrical grid interconnections and project development and management costs, and fixed maintenance costs, such as cranes, replacement parts, and repair facilities, can be spread over a larger investment. As of the date this is written (November 2013), the offshore London Array wind power project in the United Kingdom is the world's largest wind farm. The fleet of 175 wind turbines stationed in the salty waters of the outer Thames estuary, where the river Thames empties into the North Sea, boasts a nameplate capacity of 630 MW, enough to power nearly half a million U.K. homes.

43.8.3 Wind Power Integration into Grid Operations and the Need for Forecasting

The following discussion borrows heavily from two papers that have been recently published in the *IEEE Power & Energy Magazine* [67,68].

43.8.3.1 Grid Integration

One common argument that wind energy critics raise is that grid systems are capable of accepting power only from *dispatchable* (capable of providing the operator-specified amount of power at any time) power systems; wind's inherent variability precludes its efficient integration into a grid system. Most conventional generators are, in reality, quite constrained in their ability to quickly change their output to a desired amount of power and must be operated with careful consideration of their start-up times, minimum run times, and ramp-rate limitations, and all conventional generators are subject to occasional failure—thus, even they cannot be considered to be truly dispatchable. Wind can usually adjust output very quickly within the range between zero and the power available from the wind, while many conventional units ramp slowly and within a more limited range. In fact, the components of the electric power system have always been uncertain and variable to some degree, but the addition of wind energy (or other variable sources of energy) increases the need to directly manage these attributes in more sophisticated ways; the growth of wind energy is forcing the development of the next generation of tools and practices for continued efficient and reliable electric system operations.

At the current time, larger markets are, in general, more adept at handling higher wind penetration rates for several reasons:

- Larger markets are typically more geographically diverse, and this diversity of wind resources reduces the short-term fluctuations and forecast errors in the wind power.
- Larger markets typically use centralized wind power forecasts and sophisticated tools for scheduling power resources.
- Larger markets typically have made a larger effort to better understand the issues specific to a high penetration of variable resources and to find methods of alleviating those issues.

In general, analysis by system operators has shown the need for increased generation fleet flexibility (for all types of generation units), including flexibility from the wind plants.

Several grids in the United States, including New York Independent System Operator, Alberta Electric System Operator, Energy Reliability Council of Texas (ERCOT), Midcontinent Independent System Operator, California Independent System Operator, and the PJM regional transmission organization have successfully integrated extensive amounts of wind into their systems, and they are all preparing for significantly more in the near future [67]. Most of these grids have integrated wind into their security-constrained economic dispatch systems to allow the grid operator to treat wind the same as they treat other power plants, requiring them to furnish precise levels of output for the next few minutes and penalizing them if they fail to deliver. The keys to these successful integrations are the use of 5-minute (the de facto North American standard interval) rolling dispatch—rebalancing the calls for power from all resources every 5 minutes—and the demonstrated reliable persistence of wind over that time frame.

ERCOT, which efficiently dispatches wind and uses energy dispatch and non-spinning reserves (no spinning reserves) to compensate for wind uncertainty, has carried out detailed analysis of the costs of wind variability and uncertainty for the 9% of their energy supplied by wind in 2011. They have concluded that the costs of wind power variability and day-ahead forecast uncertainty together amount to only 1%–2% of the value of that wind-generated energy, with about one-third of that amount due to variability and two-thirds due to the day-ahead uncertainty attributable to imperfect wind power forecasts.

While this discussion has focused on integration of wind into North American grid systems, European grids have experienced similar success in integrating wind, utilizing a different approach for aggregating wind plant output and an hourly dispatch schedule.

43.8.3.2 Wind Forecasting

The success of these operators in integrating wind energy into their systems is not to say that there is no need to continue to advance the state of the art in wind power forecasting; additional improvements will help reduce those wind-related additional costs and will be of great value to system operators, wind plant owners, financial traders, and even traditional generators. However, better forecasts will likely be more valuable for smaller grid systems with less generating options and a lower frequency of dispatch than for the huge grid systems with very frequent dispatching discussed earlier.

Wind power predictions or forecasts utilize a combination of weather forecasting models and wind farm–specific models (to model the complexity and interactions of wind turbines with each other and with the localized terrain). Weather forecasts and wind power forecasts are far from perfect, but, as seen earlier, experience has shown that forecasts do not need to be perfect to be very useful and valuable; currently available forecasting models perform reasonably well for several days in the future and may be marginally adequate for large grid operators.

While a cheap and simple wind power forecast can be produced with no real-time data using a publicly available numeric weather prediction model to provide the wind speed and combining that with geographic location, wind power plant size, and a turbine power curve, having more data means getting better forecasts, up to a point. Basic data such as the geographic center of the wind power plant, turbine hub height, measures of the metered power output of the power plant, and historic output data will greatly improve the simple forecast mentioned earlier. On the other hand, forecasting systems that make use of great volumes of data are frequently cumbersome and complex; the cost-effectiveness of such approaches may well be lower than that of simpler methods. The continuing goal must be to improve both the accuracy of numerical weather prediction models (by improving our fundamental knowledge of the physics involved) and the quality, quantity, and timeliness of the environmental data required to define the initial state of the atmosphere, ocean, and land surface for those models.

Not all interested parties will derive the same benefits from these improved wind forecasts. For some grid operators (especially smaller ones, with limited generator options available), the ability to more precisely forecast rapid increases or decreases in available wind power (ramp forecasting) may be of great value as that will directly affect the ability of these operators to match energy demand and supply. On the other hand, increasing the precision of ramp forecasting might not be a major issue if the grid system has gas combustion turbines online that can readily adapt to rapid changes in wind contribution, but increasing the certainty associated with 1–5 day-ahead forecasts may enable that operator to confidently plan maintenance on conventional generators.

Significant advances in forecasting are being made, with high-quality wind power forecasts now available from a number of forecast providers and new research promising continued incremental improvements in the future. Cooperation between the public and private sectors is needed to maximize that rate of improvement. When these advances are combined with the appropriate operating practices and market rules that are now evolving, increasing amounts of wind energy will be economically and reliably integrated into our power system.

43.8.4 Energy Payback Period

A certain amount of energy is used in the manufacture, installation, and eventual scrapping of any energy-producing machine. The time required for the machine to generate as much energy as was used in its manufacture, installation, and end-of-life scrapping is referred to as the *energy payback period.* Studies by Krohn [69] and Milborrow [70] have found that the energy payback period for modern wind turbines ranges between 3 and 10 months, depending on the wind speed at the site and the details of turbine manufacturing and installation. This payback period is among the shortest for any type of electricity producing technology.

43.8.5 Wind Turbine Costs

By the mid- to late-1990s, the configuration for utility-scale wind turbines had evolved to the three-bladed upwind design. About this time, the University of Sunderland in the

United Kingdom developed a set of scaling tools for wind turbines, together with several models to predict the impact of machine size on turbine components. That work was further advanced by DOE-funded efforts starting in about 2000. In 2005, researchers at NREL's National Wind Technology Center began developing a spreadsheet model of these scaling relationships. The result is a set of models that can be used to project the total COE for a wind turbine over a range of sizes and configurations; they are not intended to predict turbine *pricing* [71]. These models allow projections of costs for both land-based and offshore technologies, though the offshore technologies are very early in their development, and forecast costs are very rough.

The model cost estimates are based on turbine rating, rotor diameter, hub height, and other key turbine descriptors. Cost scaling functions have been developed for major components and subsystems. Annual energy production is estimated based on the Weibull probability of wind speeds. The cost metric used is the levelized cost of energy (LCOE), given by

$$\text{LCOE} = \frac{(\text{ICC} \times \text{FCR}) + \text{AOE}}{(\text{AEP}_{\text{net}}/1000)}$$

where

ICC (initial capital cost) in \$/kW is the sum of the turbine system cost and the balance of station cost

FCR (fixed charge rate) in % is the annual amount per dollar of initial capital cost needed to cover the capital cost, a return on debt and equity, and various other fixed charges

AOE (annual operating expense) in \$/kW/year includes land lease cost, levelized operations and maintenance cost, and levelized replacement/overhaul cost

AEP_{net} in MW h/MW/year is the net annual energy production

LCOE values are expected to be greater than actual wind energy prices because these calculations do not include the value of the production tax credit or any other renewable energy credit or subsidy.

NREL recently issued a report documenting their use of this model to develop their estimate for the cost of both land-based and offshore wind energy in the United States for 2011 [72]. In this report, their land-based turbine is a 1.5 MW machine with a hub height of 80 m at 50 m wind speed of 7.25 m/s (7.75 m/s at hub height). The turbine is assumed to be included in a 200 MW wind farm of 133 turbines. Their offshore or fixed-bottom turbine is a 3.6 MW machine with a hub height of 90 m at 50 m wind speed of 8.4 m/s (8.0 at hub height). This turbine is assumed to be included in a 500 MW wind farm of 139 turbines, mounted on monopile foundations in an average water depth of 15 m and situated 20 km from shore. Additional details on terminology, study assumptions, and calculations may be found in the NREL reports [71,72].

Results of the study for the land-based turbine are presented in Table 43.1, while the results for the offshore turbine are presented in Table 43.2; the LCOE for the land-based turbine is \$72/MW h, while for the offshore turbine, it is \$225/MW h. Again, keep in mind that these figures are not intended to reflect actual energy prices, but rather to reflect relative costs of these electricity-generating projects at this point in time.

The installed capital costs for the two configurations as given in Tables 43.1 and 43.2 are summarized in Figures 43.18 and 43.19. From these results, it is obvious that reductions in any single component of capital cost will not lead to dramatic changes in turbine COE.

TABLE 43.1
Initial Capital Cost Breakdown for Land-Based Turbine

	1.5 MW ($/kW)	1.5 MW ($/MW h)
Rotor	292	9
Blades	178	5
Hub	53	2
Pitch mechanism and bearings	57	2
Drive train, nacelle	667	19
Low-speed shaft	36	1
Bearings	20	1
Gearbox	144	4
Mechanical brake, high-speed coupling	2	*
Generator	91	3
Variable-speed electronics	108	3
Yaw drive and bearing	32	1
Mainframe	126	4
Electrical connections	76	2
Hydraulic, cooling system	17	*
Nacelle cover	16	*
Control, safety system, and condition monitoring	30	1
Tower	296	9
Turbine capital cost	1286	37
Turbine transportation	63	2
Permitting	1	*
Engineering	7	*
Meteorological mast and power performance engineering	5	*
Access road and site improvement	42	1
Site compound and security	5	*
Control and operation and maintenance building	4	*
Turbine foundation	68	2
Turbine erection	59	2
Medium-voltage electrical material	76	2
Medium-voltage electrical installation	25	1
Collector substation	26	1
Project management	17	*
Development	25	1
Balance of station	446	13
Market price adjustment	196	6
Contingency fund	112	3
Soft costs	307	9
Construction financing cost	60	2
Total capital cost	2098	61
Levelized replacement cost	11	3
Labor, equipment, facilities (O&M)	17	5
Land lease cost	7	2
Annual operating expenses	35	11
Net annual energy production (MW h/MW/year)	3263	
Levelized cost of energy ($/MW h)	72	

Source: Tegen, S. et al., 2011 Cost of wind energy review, NREL/TP-500-56266, National Renewable Energy Laboratory, Golden, CO, 2013.

TABLE 43.2
Initial Capital Cost Breakdown for Offshore Turbine

	3.6 MW ($/kW)	3.6 MW ($/MW h)
Turbine capital cost	1789	62
Development (i.e., permits, engineering, and site assessment)	58	2
Project management	117	4
Support structure	1021	35
Port and staging	73	3
Electrical infrastructure	540	19
Transportation and installation	1109	38
Balance of station	2918	101
Insurance	94	3
Surety bond (decommissioning)	165	6
Contingency fund	471	16
Soft costs	730	25
Construction financing cost	163	6
Total capital cost	5600	194
Levelized replacement cost	40	12
Labor, equipment, facilities (O&M)	46	22
Outer continental shelf lease cost	21	6
Annual operating expenses	107	40
Net annual energy production (MW h/MW/year)	3406	
Levelized cost of energy ($/MW h)	225	

Source: Tegen, S. et al., 2011 Cost of wind energy review, NREL/TP-500-56266, National Renewable Energy Laboratory, Golden, CO, 2013.

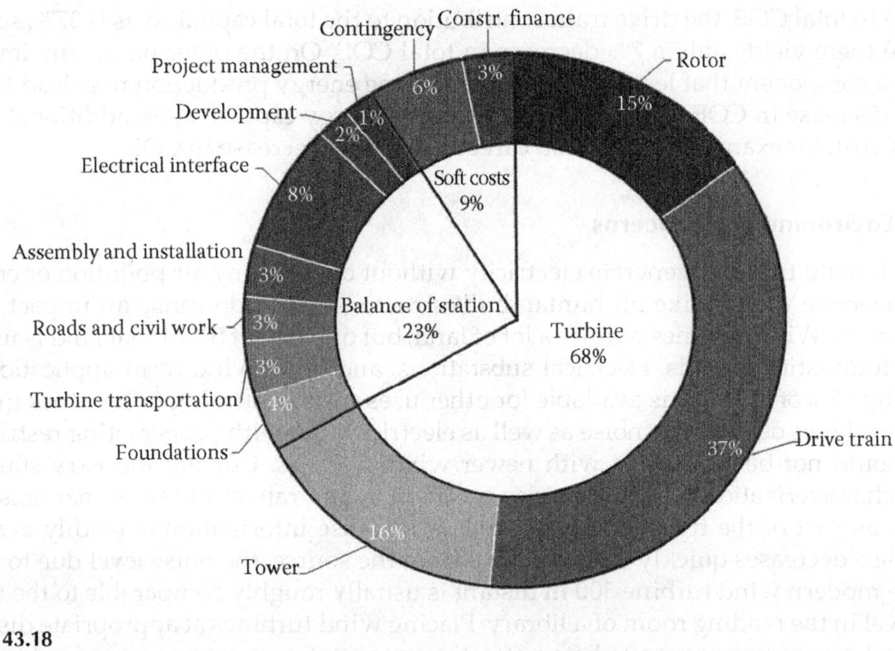

FIGURE 43.18
Initial capital cost breakdown for land-based wind turbine. (From Tegen, S. et al., 2011 Cost of wind energy review, NREL/TP-500-56266, National Renewable Energy Laboratory, Golden, CO, 2013.)

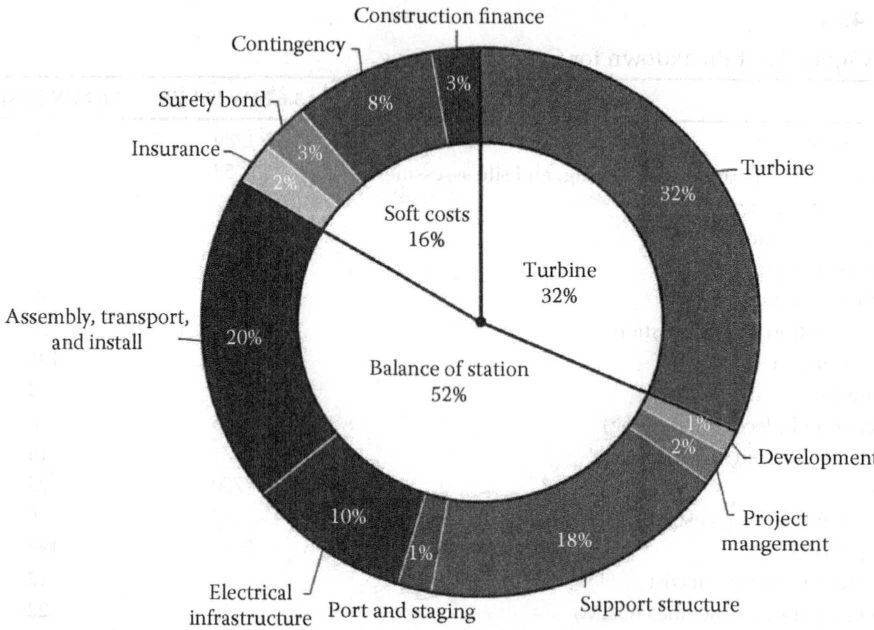

FIGURE 43.19
Initial capital cost breakdown for offshore wind turbine. (From Tegen, S. et al., 2011 Cost of wind energy review, NREL/TP-500-56266, National Renewable Energy Laboratory, Golden, CO, 2013.)

For example, the rotor contribution to the capital costs (and the total COE) of the land-based turbine is only 13%, so a 20% decrease in the rotor contribution only yields a 3% decrease in total COE; the drive train contribution to the total capital costs is 37%, so a 20% decrease there yields only a 7% decrease in total COE. On the other hand, any improvement in a component that leads to increased turbine energy production may lead to a significant decrease in COE. A 20% increase in rotor energy capture at no additional capital or O&M cost, for example, would lead directly to a 20% decrease in COE.

43.8.6 Environmental Concerns

Although wind turbines generate electricity without causing any air pollution or creating any radioactive wastes, like all human-built structures, they do cause an impact on the environment. Wind turbines require a lot of land, but only about 5% of that land is used for turbine foundations, roads, electrical substations, and other wind farm application. The remaining 95% of the land is available for other uses such as farming or livestock grazing.

Wind turbines do generate noise as well as electricity, but with proper siting restrictions, noise should not be a problem with newer wind turbines. Current industry standards call for characterization of turbine noise production and rate of decay of that noise with distance as part of the turbine testing process, so noise information is readily available. Since noise decreases quickly with distance from the source, the noise level due to a typical large modern wind turbine 300 m distant is usually roughly comparable to the typical noise level in the reading room of a library. Placing wind turbines at appropriate distances from local homes has proven to be an effective means of suppressing perceived noise, in most cases. However, certain weather conditions in some locations can cause focusing of turbine-generated noise, leading to propagation of that noise to much greater distances and resulting in great irritation for nearby residents.

The visual impact of wind turbines is extremely subjective. What one person considers highly objectionable, another might consider as attractive or at least not objectionable. The relatively slow rotation of today's large wind turbines is viewed by most people as far less intrusive than the fast rotation of the early small turbines. Visual impact can be minimized through careful design of a wind farm. The use of a single model of wind turbine in a wind farm and uniform spacing of the turbines help alleviate concerns in this area. Computer simulation can be very helpful in evaluating potential visual impacts before construction begins.

References

1. BTM Consult, Navigant Research. 2013. International Wind Energy Development: World market update 2012.
2. American Wind Energy Association. 2013. AWEA U.S. Wind Industry Annual Market Report Year Ending 2012.
3. Shepherd, D. 1994. Historical development of the windmill. In: Spera, D., ed., *Wind Turbine Technology, Fundamental Concepts of Wind Turbine Engineering*. ASME Press, New York, pp. 1–46.
4. Miley, S.J. 1982. A catalog of low Reynolds number airfoil data for wind turbine applications. RFP-3387, Department or Aerospace Engineering, Texas A&M University, College Station, TX.
5. Abbott, I.H. and von Doenhoff, A.E. 1959. *Theory of Wind Sections Including a Summary of Airfoil Data*. Dover Publications, New York.
6. Selig, M.S., Guglielmo, J.J., Broeren, A.P., and Giguere, P. 1995. *Summary of Low-Speed Airfoil Data*, Vol. 1. SoarTech Publications, Virginia Beach, VA.
7. Selig, M.S., Lyon, C.A., Giguere, P., Ninham, C.P., and Guglielmo, J.J. 1996. *Summary of Low-Speed Airfoil Data*, Vol. 2. SoarTech Publications, Virginia Beach, VA.
8. Lyon, C.A., Broeren, A.P., Giguere, P., Gopalarathnam, A., and Selig, M.S. 1997. *Summary of Low-Speed Airfoil Data*, Vol. 3. SoarTech Publications, Virginia Beach, VA.
9. Selig, M.S. and McGranahan, B.D. 2004. Wind tunnel aerodynamic tests of six airfoils for use on small wind turbines; period of performance: October 31, 2002–January 31, 2003. NREL Report SR-500-34515, National Renewable Energy Laboratory, Golden, CO.
10. Somers, D.M. 1997. Design and experimental results for the S809 airfoil. NREL Report SR-440-6918, National Renewable Energy Laboratory, Golden, CO.
11. Betz, A. 1920. Das maximum der theoretisch mőglichen ausnűtzung des windes durch windmotoren. *Zeitschrift für das gesamte Turbinenwesen*, 26, 307–309.
12. Froude, R.E. 1889. On the part played in propulsion by differences of fluid pressure. *Transactions of the Institute of Naval Architects*, 30, 390–405.
13. Lanchester, F.W. 1915. A contribution to the theory of propulsion and the screw propeller. *Transactions of the Institute of Naval Architects*, 57, 98–116.
14. Hansen, A.C. and Butterfield, C.P. 1993. Aerodynamics of horizontal-axis wind turbines. *Annual Review of Fluid Mechanics*, 25, 115–149.
15. Wilson, R.E. 1994. Aerodynamic behavior of wind turbines. In: Spera, D., ed., *Wind Turbine Technology, Fundamental Concepts of Wind Turbine Engineering*. ASME Press, New York, pp. 215–282.
16. Snel, H. 2003. Review of aerodynamics for wind turbines. *Wind Energy*, 6, 203–211.
17. Touryan, K.J., Strickland, J.H., and Berg, D.E. 1987. Electric power from vertical-axis wind turbines. *Journal of Propulsion and Power*, 3(6), 481–493.
18. Paraschivoiu, I. 2002. *Wind Turbine Design with Emphasis on Darrieus Concept*. Polytechnic International Press, Montreal, Quebec, Canada.

19. Simão Ferreira, C. and Scheurich, F. 2013. Demonstrating that power and instantaneous loads are decoupled in a vertical-axis wind turbine. *Wind Energy*, 17, 385–396
20. Kocurek, D. 1987. Lifting surface performance analysis for horizontal-axis wind turbines. SERI/STR-2 17 3163, Solar Energy Research Institute, Golden, CO.
21. Strickland, J.H., Smith, T., and Sun, K. 1981. A vortex model of the Darrieus turbine: An analytical and experimental study. SAND81-7017, Sandia National Laboratories, Albuquerque, NM.
22. Murray, J.C. and Barone, M. 2011. The development of CACTUS, a wind and marine turbine performance simulation code. In: *Proceedings of the 49th AIAA Aerospace Sciences Meeting*, Orlando, FL, AIAA-2011-147.
23. Xu, G. and Sankar, L.N. 1999. Computational study of horizontal-axis wind turbines. In: *A Collection of the 1999 ASME Wind Energy Symposium Technical Papers*, ASME, pp. 192–199.
24. Sørensen, N.N. and Hansen, M.O.L. 1998. Rotor performance predictions using a Navier–Stokes method. In: *A Collection of the 1998 ASME Wind Energy Symposium Technical Papers*, ASME, pp. 52–59.
25. Duque, E.P.N., Burkland, M.D., and Johnson, W. 2003. Navier–Stokes and comprehensive analysis performance predictions of the NREL phase VI experiment. In: *A Collection of the 2003 ASME Wind Energy Symposium Technical Papers*, ASME, pp. 43–61.
26. Sørensen, N., Michelsen, J., and Schreck, S. 2002. Navier–Stokes predictions of the NREL phase VI rotor in the NASA Ames 80 ft. × 120 ft. wind tunnel. *Wind Energy*, 5(2/3), 151–169.
27. Johansen, J., Sørensen, N., Michelsen, J., and Schreck, S. 2003. Detached-eddy simulation of flow around the NREL phase-VI rotor. In: *Proceedings of the 2003 European Wind Energy Conference and Exhibition*, Madrid, Spain.
28. Chow, R. and van Dam, C.P. 2012. Verification of computational simulations of the NREL 5 MW rotor with a focus on inboard flow separation. *Wind Energy*, 15, 967–981.
29. Chow, R., and van Dam, C.P. 2013. Computation investigations of blunt trailing-edge and twist modifications to the inboard region of the NREL 5 MW rotor. *Wind Energy*, 16, 445–458.
30. Xu, G. and Sankar, L.N. 2002. Application of a viscous flow methodology to the NREL phase VI rotor. In: *A Collection of the 2002 ASME Wind Energy Symposium Technical Papers*, ASME, pp. 83–93.
31. Schmitz, S. and Chattot, J.-J. 2005. Influence of the vertical wake behind wind turbines using a coupled Navier–Stokes/vortex-panel methodology. In: *Proceedings of the Third MIT Conference on Computational Fluid and Solid Mechanics*, Cambridge, MA, pp. 832–836.
32. Simms, D., Schreck, S., Hand, M., and Fingersh, L. 2001. NREL unsteady aerodynamics experiment in the NASA–Ames wind tunnel: A comparison of predictions to measurements. NREL Report TP-500-29494, National Renewable Energy Laboratory, Golden, CO.
33. Réthoré, P.-E., Sørensen, N., Zahle, A., Bechmann, A., and Madsen, H.A. 2011. MEXICO wind tunnel and wind turbine modelled in CFD, *29th AIAA Applied Aerodynamics Conference*, June 27–30, 2011, Honolulu, Hawaii, AIAA-2011-3373.
34. Manwell, J.F., McGowan, J.G., and Rogers, A.L. 2010. *Wind Energy Explained: Theory, Design and Application*, 2nd edn. Wiley, Chichester, U.K.
35. Burton, T., Sharpe, D., Jenkins, N., and Bossanyi, E. 2011. *Wind Energy Handbook*, 2nd edn. Wiley, Chichester, U.K.
36. FAST. NWTC Computer-Aided Engineering Tools (FAST by Jason Jonkman, Ph.D.). http://wind.nrel.gov/designcodes/simulators/fast/, last modified October 28, 2013; accessed November 22, 2013.
37. 2005. *ABAQUS Analysis User's Manual, Version 6.5*. ABAQUS, Inc., Providence, RI. http://www.abaqus.com.
38. MSC ADAMS. 2006. http://www.mscsoftware.com.
39. HAWC2, Danish Technical University, Wind Energy Section, http://www.hawc2.dk/ (accessed November, 2013).
40. BLADED, Garrad Hassan, http://www.gl-garradhassan.com/en/GHBladed.php.
41. Malcolm, D.J. and Wright, A.D. 1994. The use of ADAMS to model the AWT-26 prototype. In: *Proceedings of the 1994 ASME Wind Energy Symposium*, New Orleans, LA, pp. 125–132.

42. Molenaar, D.P. and Dijkstra, S. 1999. State-of-the-art of wind turbine design codes: Main features overview for cost-effective generation. *Wind Engineering*, 23(5), 295–311.
43. Buhl, M.L., Jr., Wright, A.D., and Pierce, K.G. 2000. Wind turbine design codes: A comparison of the structural response. In: *A Collection of the 2000 ASME Wind Energy Symposium Technical Papers*, ASME, pp. 12–22.
44. Quarton, D.C. 1998. The evolution of wind turbine design analyses—A twenty year progress review. *Wind Energy*, 1, 5–24.
45. RCAS Theory Manual, Version 2.0. 2002. United States (US) Army Aviation and Missile COMmand/AeroFlightDynamics Directorate (USAAMCOM/AFDD) Technical Report (TR), USAAMCOM/AFDD TR 02-A-005. US Army Aviation and Missile Command, Moffett Field, CA.
46. Bir, G.S., and Chopra, I. 1994. University of Maryland Advanced Rotorcraft Code (UMARC) theory manual. Technical Report UM-AERO 94-18, Center for Rotorcraft Education and Research, University of Maryland, College Park, MD.
47. Lobitz, D.W. and Sullivan, W.N. 1983. A comparison of finite-element prediction and experimental data for forced response of DOE 100 kW VAWT. In: *Proceedings of the Sixth Biennial Wind Energy Conference and Workshop*, Minneapolis/St. Paul, MN, pp. 843–853.
48. Owens, B.C., Hurtado, J.E., Paquette, J., Griffith, D.T., and Barone, M. 2013. Aeroelastic modeling of large offshore vertical-axis wind turbines: Development of the offshore wind energy simulation toolkit. In: *Proceedings of the 54th AIAA/ASME/ASCE/AHS/ASC Structures, Structural Dynamics, and Materials Conference*, Boston, MA, AIAA-2013-1552.
49. Fuchs, E.F., Erickson, R.W., Carlin, P.W., and Fardou, A.A. 1992. Permanent magnet machines for operation with large speed variations. In: *Proceedings of the 1992 American Wind Energy Association Annual Conference*, American Wind Energy Association, Washington, DC, pp. 291–297.
50. Smith, G.A. 1989. Electrical control methods for wind turbines. *Wind Engineering*, 13(2), 88–98.
51. Bossanyi, E.A. 2003. Individual blade pitch control for load reduction. *Wind Energy*, 6, 119–128.
52. Mayda, E.A., van Dam, C.P., and Yen-Nakafuji, D. 2005. Computational investigation of finite width microtabs for aerodynamic load control. In: *A Collection of the 2005 ASME Wind Energy Symposium Technical Papers*, ASME, pp. 424–436.
53. Barlas, T.K. and van Kuik, G.A.M. 2010. Review of state of the art in smart rotor control research for wind turbines. *Progress in Aerospace Sciences*, 46(1), 1–27.
54. van Wingerden, J.W., Hulskamp, A., Barlas, T., Houtzager, I., Bersee, H., van Kuik, G.A.M., and Verhaegen, M. 2010. Two-degree-of-freedom active vibration control of a prototyped "smart" rotor. *IEEE Transactions on Control Systems Technology*, 19(2), 284–296.
55. Berg, J.C., Barone, M.F., and Resor, B.R. 2013. Field test results from the Sandia SMART rotor. In: *Proceedings of the 51st AIAA Aerospace Sciences Meeting*, Grapevine (Dallas/Ft. Worth Region), TX, AIAA-2013-1060.
56. Spera, D., ed. 1994. *Wind Turbine Technology, Fundamental Concepts of Wind Turbine Engineering*. ASME Press, New York.
57. Mandell, J.F. and Samborsky, D.D. 1997. DOE/MSU composite material fatigue database: Test methods, materials, and analysis. SAND97-3002, Sandia National Laboratories, Albuquerque, NM.
58. Mandell, J.F., Samborsky, D.D., and Cairns, D.S.., 2002. Fatigue of composite material and substructures for turbine blades. SAND2002-077, Sandia National Laboratories, Albuquerque, NM.
59. De Smet, B.J., and Bach, P.W. 1994. Database FACT: Fatigue of composites for wind turbines. ECN-C-94-045, ECN, Petten, the Netherlands.
60. Kensche, C.W., ed. 1996. Fatigue of materials and components for wind turbine rotor blades. EUR 16684, European Commission, Luxembourg, Belgium.
61. Nijssen, R. 2006. OptiDAT—Database reference document. OB_TC_R018, Rev. 005, Document Number 10224. www.wmc.eu/optidat_files/Optidat%20reference%20document.pdf (accessed November, 2013).

62. Sutherland, H.J. 2000. A summary of the fatigue properties of wind turbine materials. *Wind Energy*, 3(1), 1–34.
63. Griffin, D.A. 2004. Blade system design studies volume II: Preliminary blade designs and recommended test matrix. SAND2004-0073, Sandia National Laboratories, Albuquerque, NM.
64. Jackson, K.J., Zuteck, M.D., van Dam, C.P., Standish, K.J., and Berry, D. 2005. Innovative design approaches for large wind turbine blades. *Wind Energy*, 8(2), 141–171.
65. Ehrmann, R.S., White, E.B., Maniaci, D.C., Chow, R., Langel, C.M., and van Dam, C.P. 2013. Realistic leading-edge roughness effects on airfoil performance. In: *Proceedings of the 31st AIAA Applied Aerodynamics Conference*, San Diego, CA, AIAA-2013-2800.
66. Miller, D., Mandell, J., Samborsky, D., Hernandez-Sanchez, B., and Griffith, D. 2012. Performance of composite materials subjected to salt water environments. In: *Proceedings of the 53rd AIAA/AMSE/ASCE/AHS/ASC Structure, Structural Dynamics and Materials Conference*, Honolulu, HI, AIAA-2012-1575.
67. Ahlstrom, M., Blatchford, J., Davis, M., Duchesne, J., Edelson, D., Focken, U., Lew, D. et al. November/December 2011. Atmospheric pressure—Weather, wind forecasting and energy market operations. *IEEE Power & Energy Magazine*, 9(6), 97–107.
68. Ahlstrom, M., Bartlett, D., Collier, C., Duchesne, J., Edelson, D., Gesino, A., Keyser, M. et al. November/December 2013. Knowledge is power—Efficiently integrating wind energy and wind forecasts. *IEEE Power & Energy Magazine*, 11(6), 45–52.
69. Krohn, S. 1997. *The Energy Balance of Modern Wind Turbines, WindPower Note No. 16*. Danish Wind Turbine Manufacturers Association, Copenhagen, Denmark.
70. Milborrow, D. Dispelling the myths of energy payback time. *Wind Stats Newsletter*, 11(2), 1–3.
71. Fingersh, L., Hand, M., and Laxson, A. 2006. Wind turbine design cost and scaling model. NREL/TP-500-40566, National Renewable Energy Laboratory, Golden, CO.
72. Tegen, S., Lantz, E., Hand, M., Maples, B., Smith, A., and Schwabe, P. 2013. 2011 Cost of wind energy review. NREL/TP-500-56266, National Renewable Energy Laboratory, Golden, CO.

For Further Information

Excellent summaries of HAWT and VAWT aerodynamic principles together with extensive reference lists, are presented by Hansen and Butterfield [14], and by Touryan et al. [17], respectively. Vol. 1(1) of *Wind Energy*, Wiley, 1998 contains a comprehensive set of review papers covering wind turbine rotor aerodynamics, design analysis, and overall system design, and excellent updates on the status of wind turbine aeroelasticity and aerodynamics and be found in articles in Vol. 6(3), 2003 of that journal.

The latest developments in the field of wind energy in the United States and Europe may be found in the following annual conference proceedings:

Proceedings of WINDPOWER American Wind Energy Association (AWEA), Washington, DC, www.awea.org.

Proceedings of the European Wind Energy Association, European Wind Energy Association, Brussels, Belgium, www.ewea.org.

The books by Manwell et al. [34], Spera [56] and Burton et al. [35] contain a wealth of fairly current information on wind energy conversion, history, and technology, together with extensive reference lists.

Extensive information on wind energy conversion technology may also be found on the World-Wide Web. Excellent sites to start with include those of the U.S. National Renewable Energy Laboratory, Wind Energy Technology Center at www.nrel.gov/nwtc/, the Sandia National Laboratories, Wind Energy Technology Department at http://energy.sandia.gov/?page_id=344, the Danish Technical University, Department of Wind Energy at www.vindenergi.dtu.dk, the American Wind Energy Association at www.awea.org, the European Wind Energy Association at www.ewea.org, and the Danish Wind Energy Association at www.windpower.org/en.

44

Photovoltaics

Roger Messenger and D. Yogi Goswami

CONTENTS

44.1 Introduction

Photovoltaic (PV) conversion is the direct conversion of sunlight into electricity with no intervening heat engine. PV devices are solid state; therefore, they are rugged and simple in design and require very little maintenance. Perhaps the biggest advantage of solar PV devices is that they can be constructed as stand-alone systems to give outputs from microwatts to megawatts. That is why they have been used as the power sources for calculators, watches, water pumping, remote buildings, communications, satellites and space vehicles, and even megawatt-scale power plants. PV panels can be made to form components of building skin, such as roof shingles and wall panels. With such a vast array of applications, the demand for PVs is increasing every year. With net metering and governmental incentives, such as feed-in laws and other policies, grid-connected applications such as building-integrated PV have become cost-effective even where grid electricity is cheaper.

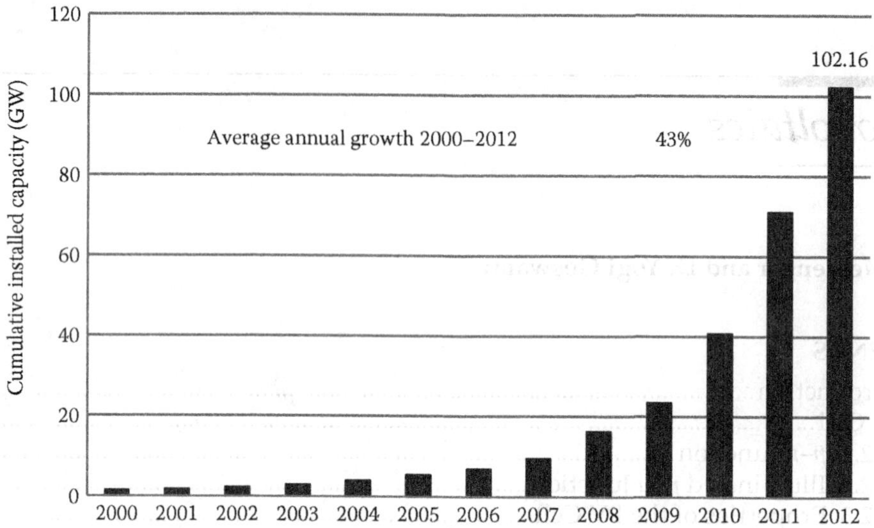

FIGURE 44.1
Worldwide production of PV panels. (From EPIA, Global market outlook for photovoltaics, European Photovoltaic Industries Association, Brussels, Belgium, http://www.epia.org.)

As a result, the worldwide growth in PV production has averaged over 43% per year from 2000 to 2012 and 61% from 2007 to 2012 (see Figure 44.1). The cumulative installed global PV capacity grew to over 100 GW by the end of 2012.

In the early days of solar cells in the 1960s and 1970s, more energy was required to produce a cell than it could ever deliver during its lifetime. Since then, dramatic improvements have taken place in the efficiencies and manufacturing methods. The present energy payback periods have been reduced to 0.68 for thin-film CdTe cells to less than 2 years for crystalline silicon cells [1]. The costs of PV panels have come down from about $30 to less than $1 per peak watt over the last three decades.

With panel costs as low as $0.55/W (large quantities) to $2/W (retail), the system costs are in the range of about $1.50/W for megawatt-range systems to about $3–$4/W for small systems, and the PV systems have finally become cost-effective for on-grid applications. The PV panel costs are reaching the limits of cost reduction; however, the costs of the balance of the system (BOS) have room for further reduction. The U.S. Department of Energy has a goal to reduce the BOS costs, which would bring the total costs of a PV system down to $1/W and the cost of power to $0.06/kWh [1b]. At present, module efficiencies are as high as 21%. The main constraint on the efficiency of a solar cell is related to the bandgap of the semiconductor material of a PV cell. As explained later in this chapter, a photon of light with energy equal to or greater than the bandgap of the material is able to free up one electron, when absorbed in the material. However, the photons that have energy less than the bandgap are not useful for this process. When absorbed on the cell, they just produce heat. And for the photons with more energy than the bandgap, the excess energy above the bandgap is not useful in generating electricity. The excess energy simply heats up the cell. These reasons account for a theoretical maximum limit on the efficiency of a conventional single junction PV cell to less than 30%. The actual efficiency is even lower because of the reflection of light from the cell surface, shading of the cell due to current collecting contacts, internal resistance of the cell, and recombination of electrons and holes before they are able to contribute to the current.

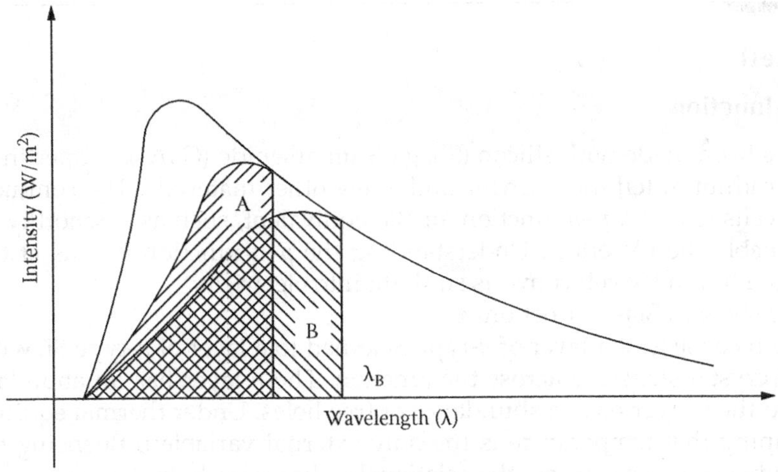

FIGURE 44.2
Energy conversion from a two-layered stacked PV cell. (From Goswami, D.Y. et al., *Principles of Solar Engineering*, 2nd edn., Taylor & Francis, Philadelphia, PA, 2000.)

The limits imposed on solar cells due to bandgap can be partially overcome by using multiple layers of solar cells stacked on top of each other, each layer with a bandgap higher than the layer below it. For example (Figure 44.2), if the top layer is made from a cell of material A (bandgap corresponding to λ_A), solar radiation with wavelengths less than λ_A would be absorbed to give an output equal to the hatched area A.

The solar radiation with wavelength greater than λ_A would pass through A and be converted to give an output equal to the hatched area B. The total output and therefore the efficiency of this tandem cell would be higher than the output and the efficiency of each single cell individually. The efficiency of a multijunction cell can be about 50% higher than a corresponding single cell. The efficiency would increase with the number of layers. For this concept to work, each layer must be as thin as possible, which puts a very difficult if not an insurmountable constraint on crystalline and polycrystalline cells to be made multijunction. As a result, this concept is being investigated mainly for thin-film amorphous or microcrystalline solar cells. Efficiencies as high as 37.7% have been reported for multijunction cells in the literature [2]. For concentrated PV, efficiencies as high as 44% have been reported [2].

In this chapter, the physics of PV electrical generation will be briefly reviewed, followed by a discussion of the PV system design process. Several PV system examples will be presented, then a few of the latest developments in crystalline silicon PV will be summarized, and finally, some of the present challenges (2004–2005) facing the large-scale deployment of PV energy sources will be explored. Emphasis will be on nonconcentrating, crystalline or multicrystalline silicon, terrestrial PV systems, since such systems represent nearly 95% of systems currently being designed and built. However, the design procedures outlined at the end of the chapter also can be applied to other PV technologies, such as thin films. While multijunction III–V semiconductor concentrating PV cells have been fabricated with efficiencies that are double the efficiencies of silicon cells, the cost of these cells is so high that their use is only justified in extraterrestrial applications. As a result, the typical reader of this material will not become involved in the use of these technologies.

44.2 PV Cell

44.2.1 *p–n* Junction

PV cells have been made with silicon (Si), gallium arsenide (GaAs), copper indium diselenide (CIS), cadmium telluride (CdTe), and a few other materials. The common denominator of PV cells is that a *p–n* junction, or the equivalent, such as a Schottky junction, is needed to enable the PV effect. Understanding the *p–n* junction is thus at the heart of understanding how a PV cell converts sunlight into electricity.

Figure 44.3 shows a Si *p–n* junction.

The junction consists of a layer of *n*-type Si joined to a layer of *p*-type Si, with an uninterrupted Si crystal structure across the junction. The *n*-layer has an abundance of free electrons and the *p*-layer has an abundance of free holes. Under thermal equilibrium conditions, meaning that temperature is the only external variable influencing the populations of free holes and electrons, the relationship between hole density, *p*, and electron density, *n*, at any given point in the material, is given by

$$np = n_i^2, \tag{44.1}$$

where n_i is approximately the density of electrons or holes in intrinsic (impurity-free) material. When impurities are present, then $n \cong N_d$ and $p \cong N_a$, where N_d and N_a are the densities of donor and acceptor impurities. For Si, $n_i \cong 1.5 \times 10^{10}$ cm^{-3} at $T = 300$ K, while N_d and N_a can be as large as 10^{21} cm^{-3}. Hence, for example, if $N_d = 10^{18}$ on the *n*-side of the junction, then $p = 2.25 \times 10^2$ cm^{-3}.

Both electrons and holes are subject to random diffusion within the Si crystalline structure, so each tends to diffuse from regions of high concentration to regions of low concentration. The enormous concentration differences of hole and electron concentrations between the *n*-side and the *p*-side of the junction cause large concentration gradients across the junction. The net result is that the electrons diffuse across the junction into the *p*-region and the holes diffuse across the junction into the *n*-region, as shown in Figure 44.3.

Before formation of the junction, both sides of the junction are electrically neutral. Each free electron on the *n*-side of the junction comes from a neutral electron donor impurity

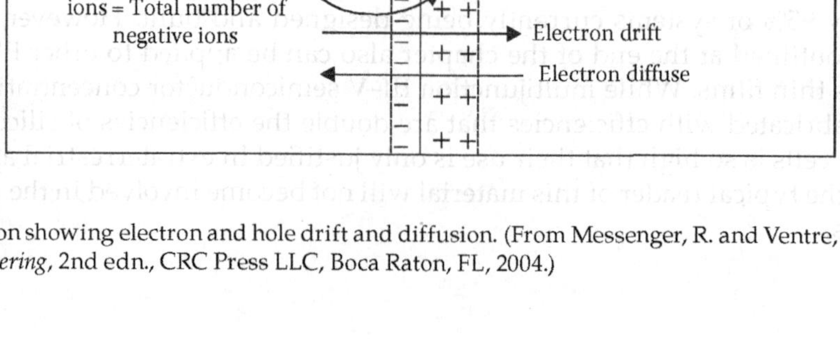

FIGURE 44.3
The *p–n* junction showing electron and hole drift and diffusion. (From Messenger, R. and Ventre, G., *Photovoltaic Systems Engineering*, 2nd edn., CRC Press LLC, Boca Raton, FL, 2004.)

atom, such as arsenic (As), while each free hole on the *p*-side of the junction comes from a neutral hole donor (acceptor) impurity atom, such as boron (B). When the negatively charged electron leaves the As atom, the As atom becomes a positively charged As ion. Similarly, when the positively charged hole leaves the B atom, the B atom becomes a negatively charged B ion. Thus, as electrons diffuse to the *p*-side of the junction, they leave behind positively charged electron donor ions that are covalently bound to the Si lattice. As holes diffuse to the *n*-side of the junction, they leave behind negatively charged hole donor ions that are covalently bound to the Si lattice on the *p*-side of the junction. The diffusion of charge carriers across the junction thus creates an electric field across the junction, directed from the positive ions on the *n*-side to the negative ions on the *p*-side, as shown in Figure 44.3. Gauss's law requires that electric field lines originate on positive charges and terminate on negative charges, so the number of positive charges on the *n*-side must be equal to the number of negative charges on the *p*-side.

Electric fields exert forces on charged particles according to the familiar $f = qE$ relationship. This force causes the charge carriers to drift. In the case of the positively charged holes, they drift in the direction of the electric field, that is, from the *n*-side to the *p*-side of the junction. The negatively charged electrons drift in the direction opposite the field, that is, from the *p*-side to the *n*-side of the junction. If no external forces are present other than temperature, then the flows of holes are equal in both directions and the flows of electrons are equal in both directions, resulting in zero net flow of either holes or electrons across the junction. This is called the law of detailed balance, which is consistent with Kirchhoff's current law.

Carrying out an analysis of electron and hole flow across the junction ultimately leads to the development of the familiar diode equation:

$$I = I_o\left(e^{\frac{qV}{kT}} - 1\right), \tag{44.2}$$

where
q is the electronic charge
k is Boltzmann's constant
T is the junction temperature in K
V is the externally applied voltage across the junction from the *p*-side to the *n*-side of the junction

44.2.2 Illuminated *p–n* Junction

Figure 44.4 illustrates the effect of photons impinging upon the junction area.

The energy of a photon is given in the following equation, where λ is the wavelength of the photon, h is Planck's constant (6.625×10^{-34} J s), and c is the speed of light (3×10^8 m/s):

$$e = h\nu = \frac{hc}{\lambda}, \tag{44.3}$$

The energy of a photon in electron volts (eV) becomes 1.24/λ, if λ is in μm (1 eV = 1.6×10^{-19} J). If a photon has an energy that equals or exceeds the semiconductor bandgap energy of the *p–n* junction material, then it is capable of creating an electron–hole pair (EHP). For Si, the bandgap is 1.1 eV, so if the photon wavelength is less than 1.13 μm, which is in the near-infrared region, then the photon will have sufficient energy to generate an EHP.

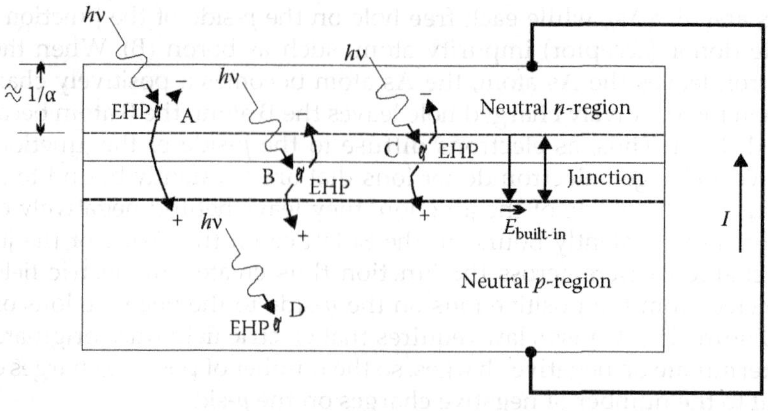

FIGURE 44.4
The illuminated p–n junction showing desirable geometry and the creation of electron–hole pairs. (From Messenger, R. and Ventre, G., *Photovoltaic Systems Engineering*, 2nd edn., CRC Press LLC, Boca Raton, FL, 2004.)

Although photons with energies higher than the bandgap energy can be absorbed, one photon can create only one EHP. The excess energy of the photon is wasted as heat. As photons enter a material, the intensity of the beam depends upon a wavelength-dependent absorption constant, α. The intensity of the photon beam as a function of penetration depth into the material is given by $F(x) = F_o e^{-\alpha x}$, where x is the depth of penetration into the material. Optimization of photon capture, thus, suggests that the junction should be within $1/\alpha$ of the surface to ensure transmission of photons to within a diffusion length of the p–n junction, as shown in Figure 44.4.

If an EHP is created within one minority carrier diffusion length, D_x, of the junction, then, on the average, the EHP will contribute to current flow in an external circuit. The diffusion length is defined to be $L_x = \sqrt{D_x \tau_x}$, where D_x and τ_x are the minority carrier diffusion length and lifetime for electrons, respectively, in the p-region if $x = n$, and D_x and τ_x are the minority carrier diffusion length and lifetime for holes, respectively, in the n-region if $x = p$. So the idea is to quickly move the electron and hole of the EHP to the junction before either has a chance to recombine with a majority charge carrier. In Figure 44.4, points A, B, and C represent EHP generation within a minority carrier diffusion length of the junction. But if an EHP is generated at point D, it is highly unlikely that the electron will diffuse to the junction before it recombines.

The amount of photon-induced current flowing across the junction and into an external circuit is directly proportional to the intensity of the photon source. Note that the EHPs are swept across the junction by the built-in E-field, so the holes move to the p-side and continue to diffuse toward the p-side external contact. Similarly, the electrons move to the n-side and continue to diffuse to the n-side external contact. Upon reaching their respective contacts, each contributes to external current flow if an external path exists. In the case of holes, they must recombine at the contact with an electron that enters the material at the contact. Electrons, on the other hand, are perfectly happy to continue flowing through an external copper wire.

At this point, an important observation can be made. The external voltage across the diode that results in significant current flow when no photons are present is positive from p to n. The diode current and voltage are defined in this direction, and the diode thus is defined according to the passive sign convention. In other words, when no photons impinge on the junction, the diode dissipates power. But when photons are present, the photon-induced

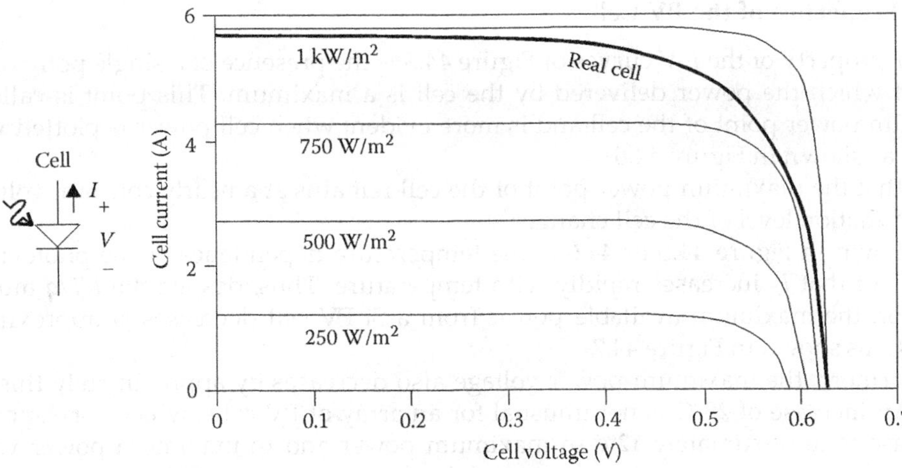

FIGURE 44.5
I–V characteristics of real and ideal PV cells under different illumination levels. (From Messenger, R. and Ventre, G., *Photovoltaic Systems Engineering*, 2nd edn., CRC Press LLC, Boca Raton, FL, 2004.)

current flows OPPOSITE to the passive direction. So current LEAVES the positive terminal, which means that the device is *generating* power. This is the PV effect. The challenge to the manufacturers of PV cells is to maximize the capture of photons and, in turn, maximize the flow of current in the cell for a given incident photon intensity. Optimization of the process is discussed in detail in [4]. When the photocurrent is incorporated into the diode equation, the result is

$$I = I_\ell - I_o\left(e^{\frac{qV}{kT}} - 1\right) \cong I_l - I_o e^{\frac{qV}{kT}}. \tag{44.4}$$

Note that in (44.4), the direction of the current has been reversed with respect to the cell voltage. With the active sign convention implied by (44.4), the junction device is now being defined as a cell, or PV cell. Figure 44.5 shows the *I–V* curves for an ideal PV cell and a typical PV cell, assuming the cell has an area of approximately 195 cm^2.

It is evident that the ideal curve closely represents that of an ideal current source for cell voltages below 0.5 V, and it closely represents that of an ideal voltage source for voltages near 0.6 V. The intersection of the curve with the $V = 0$ axis represents the short-circuit current of the cell. The intersection of the curve with the $I = 0$ axis represents the open-circuit voltage of the cell. To determine the open-circuit voltage of the cell, simply set $I = 0$ and solve (44.4) for V_{OC}. The result is

$$V_{OC} = \frac{kT}{q}\ln\frac{I_l}{I_o}. \tag{44.5}$$

The direct dependence of I on I_l and the logarithmic dependence of V_{OC} on I_l is evident from (44.4) and (44.5) as well as from Figure 44.5.

The departure of the real curve from the ideal prediction is primarily due to unavoidable series resistance between the cell contacts and the junction.

44.2.3 Properties of the PV Cell

Another property of the *I–V* curves of Figure 44.5 is the presence of a single point on each curve at which the power delivered by the cell is a maximum. This point is called the maximum power point of the cell and is more evident when cell power is plotted vs. cell voltage, as shown in Figure 44.6.

Note that the maximum power point of the cell remains at a nearly constant voltage as the illumination level of the cell changes.

Not shown in Figure 44.5 or 44.6 is the temperature dependence of the photocurrent. It turns out that I_O increases rapidly with temperature. Thus, despite the KT/q multiplying factor, the maximum available power from a Si PV cell decreases at approximately 0.47%/°C, as shown in Figure 44.7.

Furthermore, the maximum power voltage also decreases by approximately this same factor. An increase of 25°C is not unusual for an array of PV cells, which corresponds to a decrease of approximately 12% in maximum power and in maximum power voltage. Because of this temperature degradation of the performance of a PV cell, it is important during the system design phase to endeavor to keep the PV cells as cool as possible.

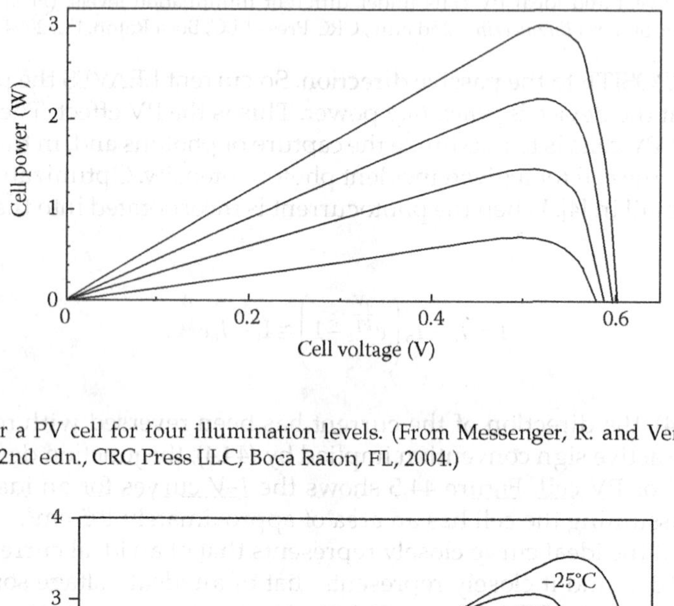

FIGURE 44.6
Power vs. voltage for a PV cell for four illumination levels. (From Messenger, R. and Ventre, G., *Photovoltaic Systems Engineering*, 2nd edn., CRC Press LLC, Boca Raton, FL, 2004.)

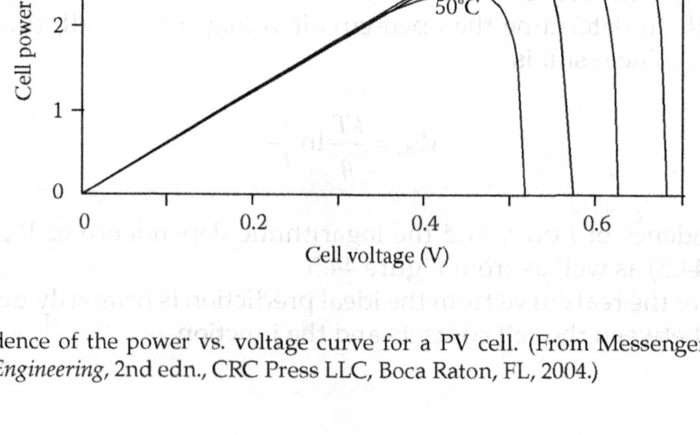

FIGURE 44.7
Temperature dependence of the power vs. voltage curve for a PV cell. (From Messenger, R. and Ventre, G., *Photovoltaic Systems Engineering*, 2nd edn., CRC Press LLC, Boca Raton, FL, 2004.)

44.3 Manufacture of Solar Cells

44.3.1 Manufacture of Crystalline and Multicrystalline Silicon PV Cells

While crystalline and multicrystalline silicon PV cells require highly purified, electronic-grade silicon, the material can be about an order of magnitude less pure than semiconductor-grade silicon and still yield relatively high-performance PV cells. Recycled or rejected semiconductor-grade silicon is often used as the feedstock for PV-grade silicon. Once adequately refined silicon is available, a number of methods have been devised for the production of single-crystal and multicrystalline PV cells. Single-crystal Si PV cells have been fabricated with conversion efficiencies just over 20%, while conversion efficiencies of champion multicrystalline Si PV cells are about 16% [5,6].

Single-crystal Si cells are almost exclusively fabricated from large single-crystal ingots of Si that are pulled from molten, PV-grade Si. These ingots, normally *p*-type, are typically on the order of 200 mm in diameter and up to 2 m in length. The Czochralski method (Figure 44.8a) is the most common method of growing single-crystal ingots.

A seed crystal is dipped in molten silicon doped with a *p*-material (Boron) and drawn upward under tightly controlled conditions of linear and rotational speed and temperature. This process produces cylindrical ingots of typically 10 cm diameter, although ingots of 20 cm diameter and more than 1 m long can be produced for other applications. An alternative method is called the float zone method (Figure 44.8b). In this method, a polycrystalline ingot is placed on top of a seed crystal and the interface is melted by a heating coil around it. The ingot is moved linearly and rotationally, under controlled conditions. This process has the potential to reduce the cell cost. Figure 44.9 illustrates the process of manufacturing a cell from an ingot.

The ingots are sliced into wafers that are approximately 0.25 mm thick. The wafers are further trimmed to a nearly square shape, with only a small amount of rounding at the corners. Surface degradation from the slicing process is reduced by chemically etching the wafers. In order to enhance photon absorption, it is common practice to use a preferential etching process to produce a textured surface finish. An *n*-layer is then diffused into the wafer to produce a *p–n* junction, contacts are attached, and the cell is then encapsulated into a module (Figure 44.10).

Detailed accounts of cell and module fabrication processes can be found in [2,4,6,7].

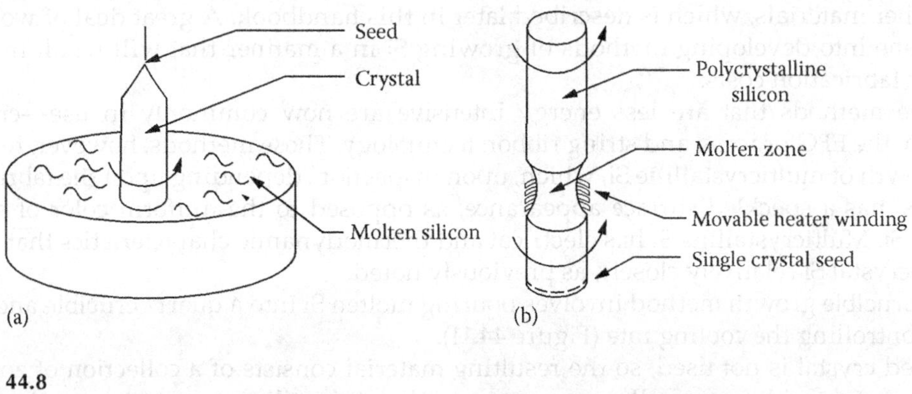

FIGURE 44.8
Crystalline silicon ingot production methods: (a) Czochralski method and (b) float zone method. (From Goswami, D.Y. et al., *Principles of Solar Engineering*, 2nd edn., Taylor & Francis, Philadelphia, PA, 2000.)

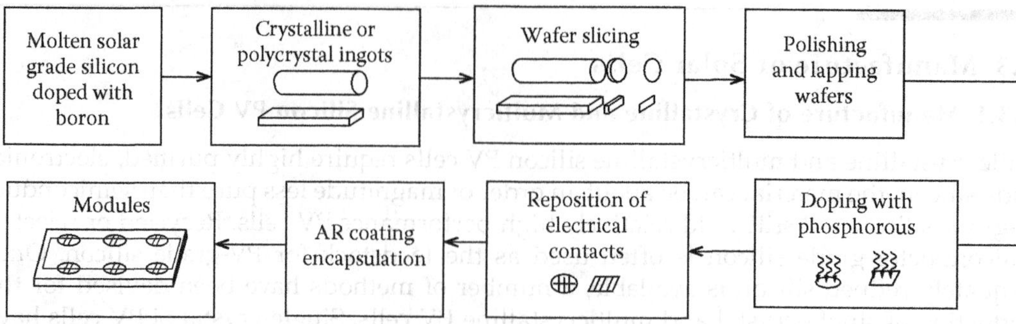

FIGURE 44.9
Series of processes for the manufacture of crystalline and polycrystalline cells. (From Goswami, D.Y. et al., *Principles of Solar Engineering*, 2nd edn., Taylor & Francis, Philadelphia, PA, 2000.)

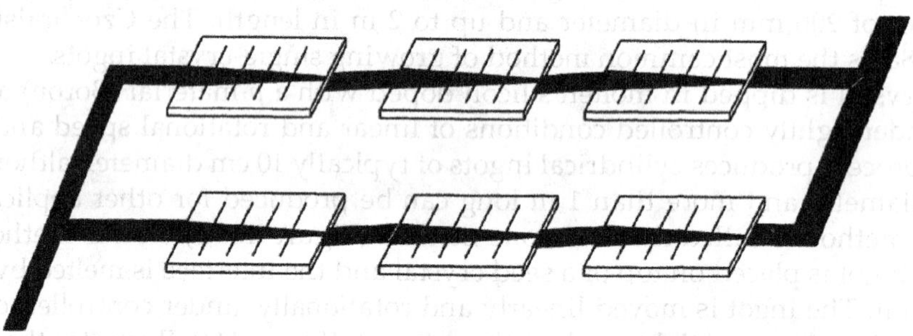

FIGURE 44.10
Assembly of solar cells to form a module. (From Goswami, D.Y. et al., *Principles of Solar Engineering*, 2nd edn., Taylor & Francis, Philadelphia, PA, 2000.)

Growing and slicing single-crystal Si ingots are highly energy intensive and, as a result, impose a relatively high energy cost on this method of cell fabrication. This high energy cost imposes a lower limit on the cost of production of a cell, and although the cell will ultimately generate as much energy as was used to produce it, the energy payback time (EPBT) is longer than desirable. Reducing the energy cost of cell and module fabrication has been the subject of a great deal of research over the past 40 years. The high energy cost of crystalline Si led to the work on thin films of amorphous Si, CdTe, and other materials, which is described later in this handbook. A great deal of work has also gone into developing methods of growing Si in a manner that will result in lower energy fabrication costs.

Three methods that are less energy intensive are now commonly in use—crucible growth, the EFG process, and string ribbon technology. These methods, however, result in the growth of multicrystalline Si, which, upon inspection, depending upon the fabrication process, has a speckled surface appearance, as opposed to the uniform color of single-crystal Si. Multicrystalline Si has electrical and thermodynamic characteristics that match single-crystal Si relatively closely, as previously noted.

The crucible growth method involves pouring molten Si into a quartz crucible and carefully controlling the cooling rate (Figure 44.11).

A seed crystal is not used, so the resulting material consists of a collection of zones of single crystals with an overall square cross section. It is still necessary to saw the ingots into wafers, but the result is square wafers rather than round wafers that would require

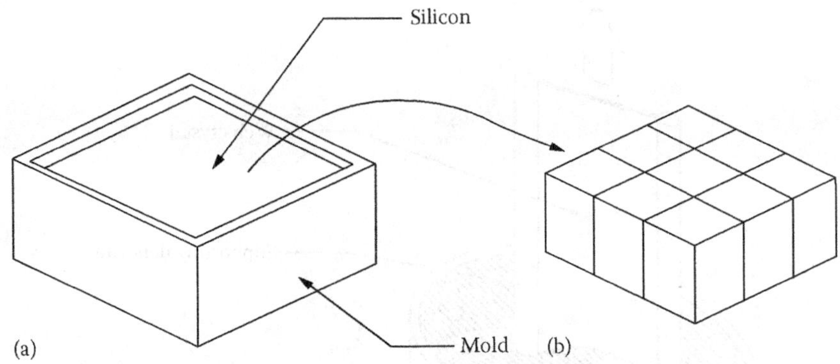

FIGURE 44.11
Polycrystalline ingot production. (a) Mold, (b) polycrystalline ingot. (From Goswami, D.Y. et al., *Principles of Solar Engineering*, 2nd edn., Taylor & Francis, Philadelphia, PA, 2000.)

additional sawing and corresponding loss of material. Wafers produced by this method can achieve conversion efficiencies of 15% or more [2].

The edge-defined film-fed growth (EFG) process is another method currently being used to produce commercial cells [8]. The process involves pulling an octagon tube, 6 m long, with a wall thickness of 330 μm, directly from the Si melt. The octagon is then cut by a laser along the octagonal edges into individual cells. Cell efficiencies of 14% have been reported for this fabrication method [5]. Figure 44.12 illustrates the process.

A third method of fabrication of multicrystalline Si cells involves pulling a ribbon of Si, or dendritic web, from the melt (Figure 44.13).

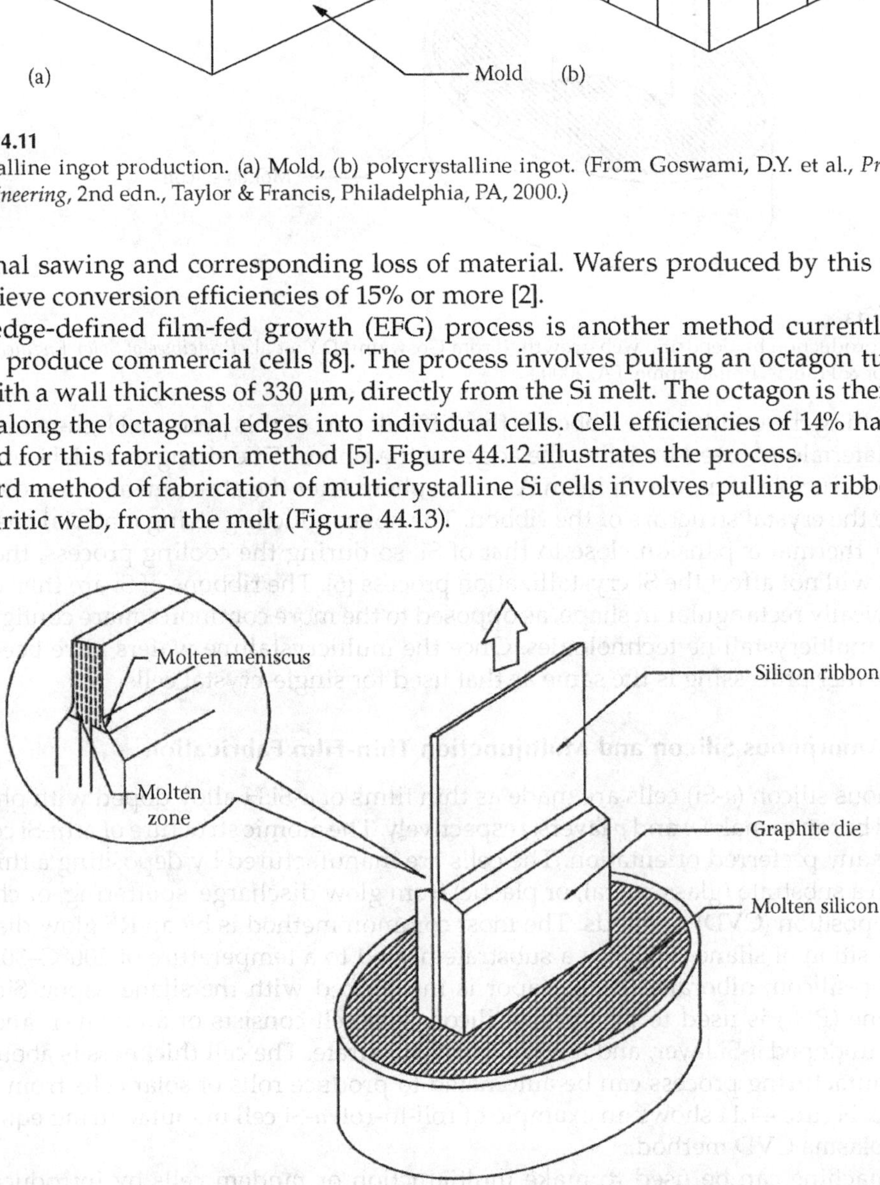

FIGURE 44.12
Thin-film production by EFG. (From Goswami, D.Y. et al., *Principles of Solar Engineering*, 2nd edn., Taylor & Francis, Philadelphia, PA, 2000.)

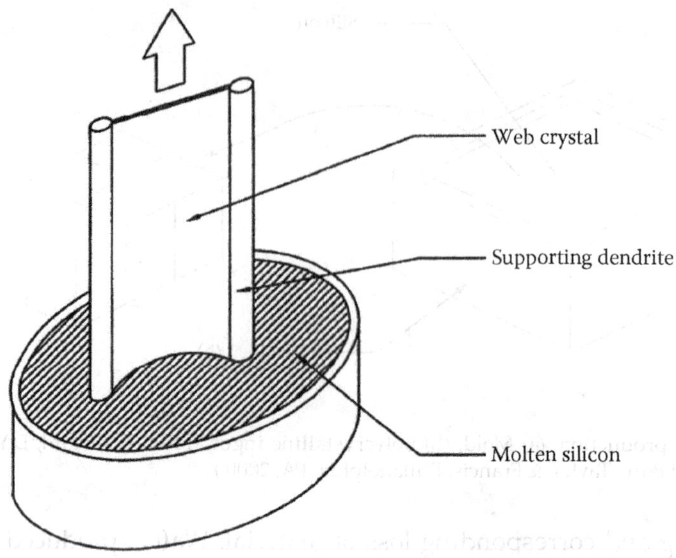

FIGURE 44.13
Thin-film production by dendritic web growth. (From Goswami, D.Y. et al., *Principles of Solar Engineering*, 2nd edn., Taylor & Francis, Philadelphia, PA, 2000.)

Controlling the width of the ribbon is the difficult part of this process. High-temperature string materials are used to define the edges of the ribbon. The string materials are pulled through a crucible of molten Si in an Ar atmosphere after the attachment of a seed crystal to define the crystal structure of the ribbon. The nonconducting string material has a coefficient of thermal expansion close to that of Si, so during the cooling process, the string material will not affect the Si crystallization process [6]. The ribbons of Si are then cut into cells, typically rectangular in shape, as opposed to the more common square configuration of other multicrystalline technologies. Once the multicrystalline wafers have been fabricated, further processing is the same as that used for single-crystal cells.

44.3.2 Amorphous Silicon and Multijunction Thin-Film Fabrication

Amorphous silicon (*a*-Si) cells are made as thin films of *a*-Si:H alloy doped with phosphorous and boron to make *n* and *p* layers, respectively. The atomic structure of an *a*-Si cell does not have any preferred orientation. The cells are manufactured by depositing a thin layer of *a*-Si on a substrate (glass, metal, or plastic) from glow discharge, sputtering, or chemical vapor deposition (CVD) methods. The most common method is by an RF glow discharge decomposition of silane (SiH_4) on a substrate heated to a temperature of 200°C–300°C. To produce *p*-silicon, diborane (B_2H_6) vapor is introduced with the silane vapor. Similarly, phosphene (PH_3) is used to produce *n*-silicon. The cell consists of an *n*-layer, and intermediate undoped *a*-Si layer, and a *p*-layer on a substrate. The cell thickness is about 1 μm. The manufacturing process can be automated to produce rolls of solar cells from rolls of substrate. Figure 44.14 shows an example of roll-to-roll *a*-Si cell manufacturing equipment using a plasma CVD method.

This machine can be used to make multijunction or tandem cells by introducing the appropriate materials at different points in the machine.

The four previously mentioned cell fabrication techniques require contacts on the front surface and on the back surface of the cells. Front surface contacts need to cover enough

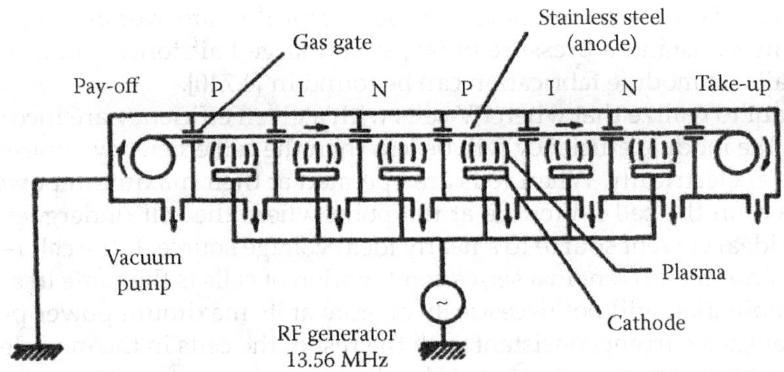

FIGURE 44.14
A schematic diagram of a roll-to-roll plasma CVD machine. (Adapted from www.ase-international.com.)

area to minimize series resistance between cell and contact, but if too much area is covered, then photons are blocked from entering the crystal. Thus, it is desirable, if possible, to design cells such that both contacts are on the back of the cell. Green and his PV team at the University of South Wales have devised a buried contact cell [9] that has both contacts on the back and also is much thinner, thus much less material intensive, than conventional Si cells. In conventional cells, charge carrier flow is perpendicular to the cell surface, while in the buried contact cell, even though the multiple p–n junctions are parallel to the cell surfaces, charge carrier flow is parallel to the cell surfaces. The fabrication process involves depositing alternate p-type and n-type Si layers, each about 1 μm thick, on an insulating substrate or superstrate. Grooves are laser cut in the layers and contacts are deposited in the grooves. Elimination of the ingot and wafer steps in processing, along with the reduced amount of material used, reduces correspondingly the energy overhead of cell production. Conversion efficiencies in excess of 20% and high cell fill factors have been achieved with this technology.

44.4 PV Modules and PV Arrays

Since individual cells have output voltages limited to approximately 0.5 V and output currents limited to approximately 7 A, it is necessary to combine cells in series and parallel to obtain higher voltages and currents. A typical PV module consists of 36 cells connected in series in order to produce a maximum power voltage of approximately 17 V, with a maximum power current of approximately 7 A at a temperature of 25°C. Such a module will typically have a surface area of about 10 ft^2. Modules also exist with 48 or more series cells so that three modules in series will produce the same output voltage and current as four 36-cell modules in series. Other larger modules combine cells in series and in parallel to produce powers up to 300 W per module.

Modules must be fabricated so the PV cells and interconnects are protected from moisture and are resistant to degradation from the ultraviolet component of sunlight. Since the modules can be expected to be exposed to a wide range of temperatures, they must be designed so that thermal stresses will not cause delamination. Modules must also be resistant to blowing sand, salt, hailstones, acid rain, and other unfriendly environmental conditions. And, of course, the module must be electrically safe over the long term. A typical module can withstand a pressure of 50 psf and large hailstones and is warranted for 25 years. Details on module fabrication can be found in [4,7,10].

It is important to realize that when PV cells with a given efficiency are incorporated into a PV module, the module efficiency will be less than the cell efficiency, unless the cells are exactly identical electrically. When cells are operated at their maximum power point, this point is located on the cell *I–V* curve at the point where the cell undergoes a transition from a nearly ideal current source to a nearly ideal voltage source. If the cell *I–V* curves are not identical, since the current in a series combination of cells is the same in each cell, each cell of the combination will not necessarily operate at its maximum power point. Instead, the cells operate at a current consistent with the rest of the cells in the module, which may not be the maximum power current of each cell.

When modules are combined to further increase system voltage and/or current, the collection of modules is called an array. For the same reason that the efficiency of a module is less than the efficiencies of the cells in the module, the efficiency of an array is less than the efficiency of the modules in the array. But since a large array can be built with subarrays that can operate essentially independently of each other, in spite of the decrease in efficiency at the array level, PV arrays that produce in excess of 1 MW are in operation at acceptable efficiency levels. The bottom line is that most efficient operation is achieved if modules are made of identical cells and if arrays consist of identical modules.

44.5 Sun and PV Array Orientation

As explained in detail in Chapter 36, total solar radiation is composed of components, direct or beam, diffuse, and reflected. In regions with strong direct components of sunlight, it may be advantageous to have a PV array mount that will track the sun. Such tracking mounts can improve the daily performance of a PV array by more than 20% in certain regions. In cloudy regions, tracking is less advantageous.

The position of the sun in the sky can be uniquely described by two angles—the azimuth, γ, and the altitude, α. The azimuth is the deviation from true south. The altitude is the angle of the sun above the horizon. When the altitude of the sun is 90°, the sun is directly overhead.

Another convenient, but redundant, angle, is the hour angle, ω. Since the earth rotates 360° in 24 h, it rotates 15° each hour. The sun thus appears to move along its arc 15° toward the west each hour. The hour angle is 0° at solar noon, when the sun is at its highest point in the sky during a given day. In this handbook, we have a sign convention such that the hour angle and the solar azimuth angle are negative before noon and positive after noon. For example, at 10 a.m. solar time, the hour angle will be −30°.

A further important angle that is used to predict the sun's position is the declination, δ. The declination is the apparent position of the sun at solar noon with respect to the equator. When $\delta = 0$, the sun appears overhead at solar noon at the equator. This occurs on the first day of fall and on the first day of spring. On the first day of Northern Hemisphere summer (June 21), the sun appears directly overhead at a latitude, L, of 23.45°N of the equator. On the first day of winter (December 21), the sun appears directly overhead at a latitude of 23.45° south of the equator. At any other latitude, the altitude $\alpha = 90° - |L - \delta|$ when the sun is directly south (or north), that is, at solar noon. At solar noon, the sun is directly south for $L > \delta$ and directly north for $L < \delta$. Note that if L is negative, it refers to the Southern Hemisphere.

Several important formulas for determining the position of the sun [4,11] include the following, where n is the day of the year with January 1 being day 1:

$$\delta = 23.45^\circ \sin \frac{360[n-80]}{365}, \tag{44.6}$$

$$\omega = \pm 15^\circ(\text{hours from local solar noon}), \tag{44.7}$$

$$\sin\alpha = \sin\delta \sin L + \cos\delta \cos L \cos\omega, \tag{44.8}$$

and

$$\cos\gamma = \frac{\cos\delta \sin\omega}{\cos\alpha}. \tag{44.9}$$

Solution of (44.6 through 44.9) shows that for optimal annual performance of a PV array, it should face directly south and should be tilted at an angle approximately equal to the latitude, L. For best summer performance, the tilt should be at $L - 15°$, and for best winter performance, the array should be tilted at an angle of $L + 15°$.

While (44.6 through 44.9) can be used to predict the location of the sun in the sky at any time on any day at any location, they cannot be used to predict the degree of cloud cover. Cloud cover can only be predicted on a statistical basis for any region, and thus the amount of sunlight available to a collector will also depend upon cloud cover. The measure of available sunlight is the peak sun hour (psh). If the sunlight intensity is measured in kW/m^2, then if the sunlight intensity is integrated from sunrise to sunset over 1 m^2 of surface, the result will be measured in kWh. If the daily kWh/m^2 is divided by the peak sun intensity, which is defined as 1 kW/m^2, the resulting units are hours. Note that this hour figure multiplied by 1 kW/m^2 results in the daily kWh/m^2. Hence the term peak sun hours, since the psh is the number of hours the sun would need to shine at peak intensity to produce the daily sunrise to sunset kWh. Obviously the psh is also equivalent to kWh/m^2/day. For locations in the United States, the National Renewable Energy Laboratory [12] publishes psh for fixed and single-axis tracking PV arrays at tilts of horizontal, latitude –15°, latitude, latitude + 15°, and vertical. NREL also tabulates data for double-axis trackers. These tables are extremely useful for determining annual performance of a PV array.

44.6 PV System Configurations

Figure 44.15 illustrates four possible configurations for PV systems.

Perhaps the simplest system is that of Figure 44.15a in which the output of the PV module or array is directly connected to a DC load. This configuration is most commonly used with a fan or a water pump, although it is likely that the water pump will also use a linear current booster (LCB) between the array and the pump motor. Operation of the LCB will be explained later.

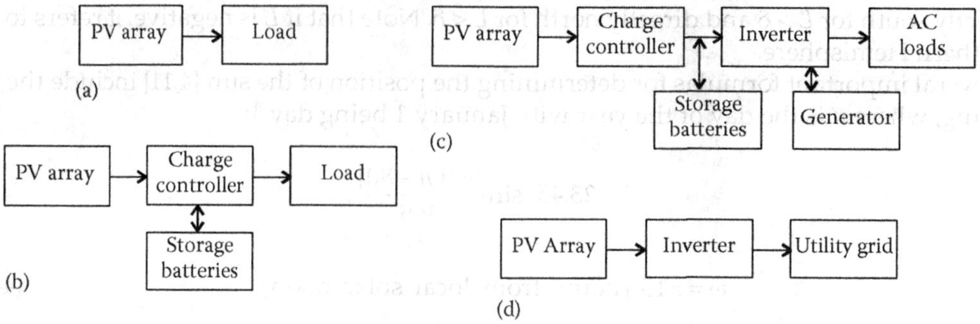

FIGURE 44.15
Several examples of PV systems: (a) direct-coupled DC system, (b) DC system with battery backup, (c) AC system with battery backup and fossil generator, and (d) grid-connected system.

The configuration of Figure 44.15b includes a charge controller and storage batteries so the PV array can produce energy during the day that can be used day or night by the load. The charge controller serves a dual function. If the load does not use all the energy produced by the PV array, the charge controller prevents the batteries from overcharge. While flooded lead–acid batteries require overcharging about once per month, frequent overcharging shortens the lives of the batteries. As the batteries become discharged, the charge controller disconnects the load to prevent the batteries from overdischarge. Normally PV systems incorporate deep discharge lead–acid batteries, but the life of these batteries is reduced significantly if they are discharged more than 80%. Modern charge controllers typically begin charging as constant current sources. In the case of a PV system, this simply means that all array current is directed to the batteries. This is called the *bulk* segment of the charge cycle. Once the battery voltage reaches the bulk voltage, which is an owner programmable value, as determined by the battery type and the battery temperature, the charging cycle switches to a constant voltage mode, commonly called the *absorption* mode. During the absorption charge mode, the charge controller maintains the bulk charge voltage for a preprogrammed time, again depending upon manufacturers' recommendations. During the absorption charge, battery current decreases as the batteries approach full charge. At the end of the bulk charge period, the charging voltage is automatically reduced to the *float* voltage level, where the charging current is reduced to a *trickle* charge.

Since quality charge controllers are microprocessor controlled, they have clock circuitry so that they can be programmed to automatically subject the batteries to an *equalization* charge approximately once a month. The equalization mode applies a voltage higher than the bulk voltage for a preset time to purposely overcharge the batteries. This process causes the electrolyte to bubble, which helps to mix the electrolyte as well as to clean the battery plates. Equalization is recommended *only* for flooded lead–acid batteries. Sealed varieties can be seriously damaged if they are overcharged.

Figure 44.16 shows the currents and voltages during the bulk, absorption, and float parts of the charging cycle. Note that all settings are programmable by the user in accordance with manufacturers' recommendations. Some charge controllers incorporate maximum power tracking as a part of their charge control algorithm. Since the maximum power voltage of a module or an array is generally higher than needed to charge the batteries, the array will not normally operate at its maximum power point when it is charging batteries, especially if the array temperature is low. For example, if it takes 14.4 V to charge a 12.6 V battery, and if a module maximum power voltage is 17, then the charging current can be increased by a factor of 17/14.4, or approximately 18%, assuming close to 100% efficiency of the maximum power tracker (MPT).

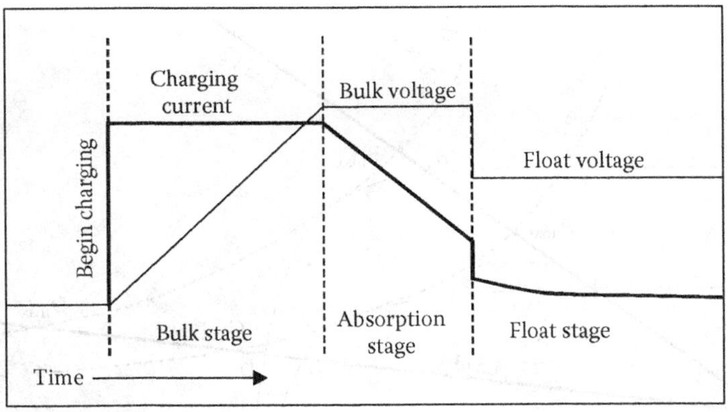

FIGURE 44.16
Charging cycle for typical PV charge controller.

The configuration of Figure 44.15c incorporates an inverter to convert the DC PV array output to AC and a backup generator to supply energy to the system when the supply from the sun is too low to meet the needs of the load. Normally, the backup generator will be a fossil-fueled generator, but it is also possible to incorporate wind or other renewable generation into the system. In this case, the charge controller prevents overcharge of the batteries. The inverter is equipped with voltage sensing circuitry so that if it detects the battery voltage going too low, it will automatically start the generator so the generator will provide power for the load as well as provide charging current for the batteries. This system is called a hybrid system, since it incorporates the use of more than one energy source.

The first three configurations are stand-alone systems. The fourth system, shown in Figure 44.15d, is a grid-connected, or utility-interactive, system. The inverter of a utility-interactive system must meet more stringent operational requirements than the stand-alone inverter. The inverter output voltage and current must be of *utility-grade* quality. This means that it must have minimal harmonic content. Furthermore, the inverter must sense the utility, and if utility voltage is lost, the inverter must shut down until utility voltage is restored to within normal limits.

44.7 PV System Components

44.7.1 Maximum Power Trackers and Linear Current Boosters

The LCB was mentioned in conjunction with the water pumping example. The function of the LCB is to match the motor *I–V* characteristic to the maximum power point of the PV array, so that at all times the array delivers maximum power to the load. Note that the LCB acts as a DC-to-DC transformer, converting a higher voltage and lower current to a lower voltage and higher current, with minimal power loss in the conversion process. A more general term that includes the possibility of converting voltage upward defines the maximum power tracker (MPT). Figure 44.17 shows the operating principle of the LCB and MPT.

Note that normally the *I–V* characteristic of the load will not intersect the *I–V* characteristic of the PV array at the maximum power point of the array, as shown by points A, B, C, and D for the two loads and the two sunlight intensity levels. For the lower-intensity situation,

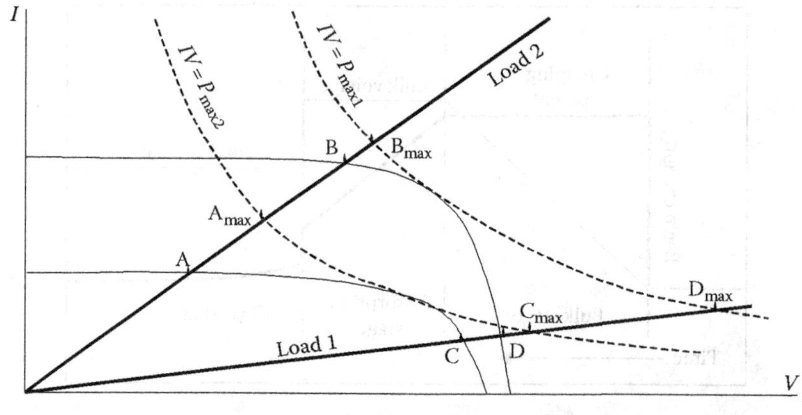

FIGURE 44.17
Operation of the LCB or MPT. (From Messenger, R. and Ventre, G., *Photovoltaic Systems Engineering*, 2nd edn., CRC Press LLC, Boca Raton, FL, 2004.)

the characteristic of load 1 intersects the array characteristic at point C and the characteristic of load 2 intersects the array characteristic at point A. The two hyperbolas are the loci of points where the *I–V* product is equal to the maximum available power from the array at the particular sunlight intensity. Hence, the intersection of these hyperbolas with the load characteristics represents the transfer of all available power from the array to the load. While the increase in power for points B and C is not particularly impressive, as shown by points B_{max} and C_{max}, the increase in power for points A and D is considerably greater, as shown by points A_{max} and D_{max}. The assumption here, of course, is 100% efficiency in the transformation. In fact, efficiencies in excess of 95% are not unusual for quality MPT and LCB devices.

The final observation for Figure 44.17 is that points A_{max} and B_{max} occur at voltages below the maximum power voltages of the array, while C_{max} and D_{max} occur at voltages above the maximum power voltages of the array. Since the input voltage and current of the MPT or LCB are the maximum power voltage and current of the array, the MPT or LCB output voltage and current points A_{max} and B_{max} represent down-conversion of the array voltage and points C_{max} and D_{max} represent up-conversion of the array voltage. These forms of conversion are discussed in power electronics books, such as [13]. The difference between the MPT and the LCB is that the LCB only performs a down-conversion, so the operating voltage of the load is always below the maximum power voltage point of the array. The terms LCB and MPT are often used interchangeably for down-conversion, but normally LCB is limited to the description of the black box that optimizes performance of pumps, while MPT is used for more general applications.

44.7.2 Inverters

Inverters convert DC to AC. The simplest inverter converts DC to square waves. While square waves will operate many AC loads, their harmonic content is very high, and as a result, there are many situations where square waves are not satisfactory. Other more suitable inverter output waveforms include the quasi-sine wave and the utility-grade sine wave. Both are most commonly created by the use of multilevel H-bridges controlled by microprocessors. There are three basic configurations for inverters: stand-alone, grid-tied, and UPS. The stand-alone inverter must act as a voltage source that delivers a prescribed amplitude and frequency rms sine wave without any external synchronization.

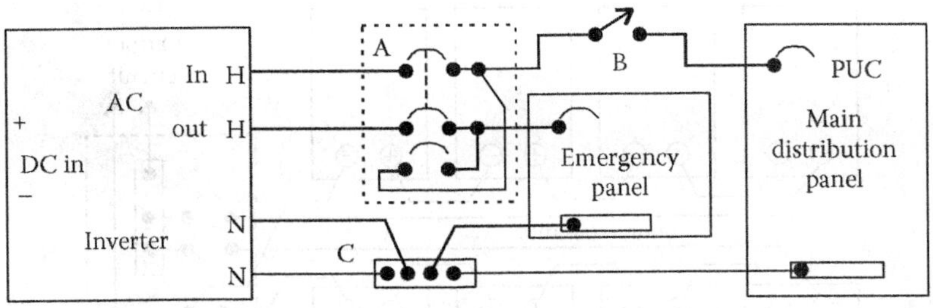

FIGURE 44.18
Example of PV source and output circuits.

The grid-tied inverter is essentially a current source that delivers a sinusoidal current waveform to the grid that is synchronized by the grid voltage. Synchronization is typically sufficiently close to maintain a power factor in excess of 0.9. The UPS inverter combines the features of both the stand-alone and the grid-tied inverter, so that if grid power is lost, the unit will act as a stand-alone inverter while supplying power to emergency loads. IEEE Standard 929 [14] requires that any inverter that is connected to the grid must monitor the utility grid voltage, and if the grid voltage falls outside prescribed limits, the inverter must stop delivering current to the grid. Underwriters Laboratory (UL) Standard 1741 [15] provides the testing needed to ensure compliance with IEEE 929.

While it may seem to be a simple matter to shut down if the utility shuts down, the matter is complicated by the possibility that additional utility-interactive PV systems may also be online. Hence, it may be possible for one PV system to *fool* another system into thinking that it is really the utility. To prevent this *islanding* condition, sophisticated inverter control algorithms have been developed to ensure that an inverter will not appear as the utility to another inverter. Some PV system owners do not want their PV system to shut down when the utility shuts down. Such a system requires a special inverter that has two sets of AC terminals. The first set, usually labeled AC IN, is designed for connection to the utility. If the utility shuts down, this set of terminals disconnects the inverter output from the utility, but continues to monitor utility voltage until it is restored. When the utility connection is restored, the inverter will first meet the needs of the emergency loads and then will feed any excess output back to the main distribution panel.

The second set of terminals is the emergency output. If the utility shuts down, the inverter almost instantaneously transfers into the emergency mode, in which it draws power from the batteries and/or the PV array to power the emergency loads. In this system, the emergency loads must be connected to a separate emergency distribution panel. Under emergency operation, the loads in the main distribution panel are without power, but the emergency panel remains energized. Such a system is shown in Figure 44.18.

The reader is referred to [4,13] for detailed explanations of the operation of inverters, including the methods used to ensure that utility-interactive inverters meet UL 1741 testing requirements.

44.7.3 Balance of System Components

Aside from the array, the charge controller, and the inverter, a number of other components are needed in a code-compliant PV system. For example, if a PV array consists of multiple series–parallel connections, as shown in Figure 44.19, then it is necessary to incorporate fuses or circuit breakers in series with each series string of modules, defined as a source circuit.

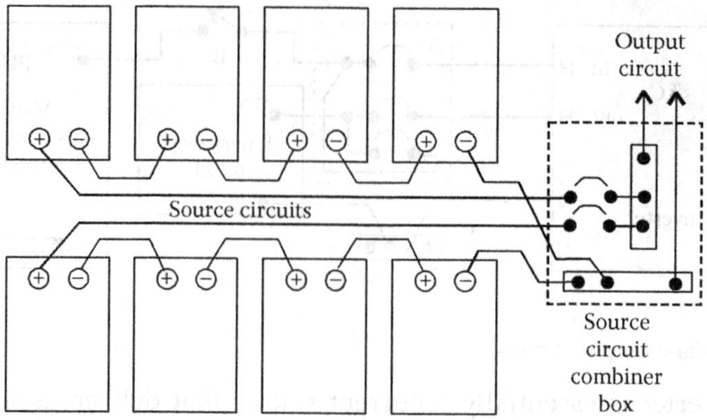

FIGURE 44.19
Use of GFDI.

This fusing is generally accomplished by using a source circuit combiner box as the housing for the fuses or circuit breakers, as shown in Figure 44.19. The combiner box should be installed in a readily accessible location. The PV output circuit of Figure 44.19 becomes the input to the charge controller, if a charge controller is used. If multiple parallel source circuits are used, it may be necessary to use more than one charge controller, depending upon the rating of the charge controller. When more than one charge controller is used, source circuits should be combined into separate output circuits for each charge controller input. In a utility-interactive circuit with no battery backup, a charge controller is not necessary. The PV output circuit connects directly to the inverter through either a DC disconnect or a DC ground fault detection and interruption (GFDI) device.

A GFDI device is required by the *National Electrical Code* (NEC) [16] whenever a PV array is installed on a residential rooftop. The purpose of the device is to detect current flow on the grounding conductor. The grounding conductor is used to ground all metal parts of the system. In a properly installed and operating system, no current will flow on the grounding conductor. Normally the negative conductor of the PV array is grounded, but this ground, if properly installed, will be attached to the grounding conductor at only one point, as shown in Figure 44.20, where the negative PV output conductor is connected to the equipment grounding bus through the 1 A circuit breaker. The 1 A circuit breaker is ganged to the 100 A circuit breaker so that if the current through the 1 A circuit breaker exceeds 1 A, both breakers will trip. When the two circuit breakers open, current flow on

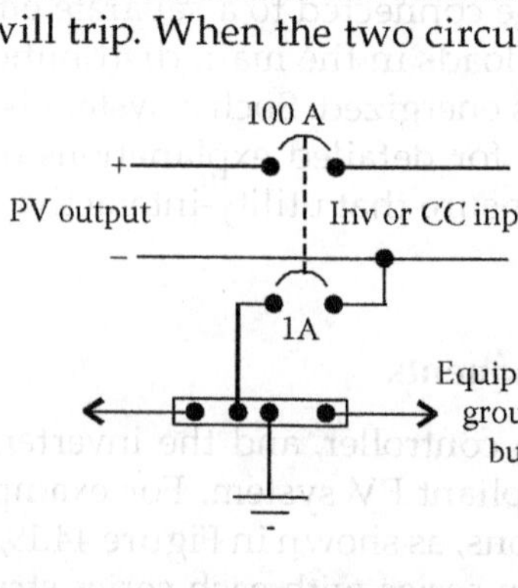

FIGURE 44.20
Utility-interactive PV system connections to emergency loads and to utility.

both the PV output circuit conductors and the grounding conductor is interrupted. If the fault current on the grounding conductor was the result of an arcing condition between one of the PV circuit conductors and ground, the arc will be extinguished, thus preventing a fire from starting.

The NEC also requires properly rated disconnects at the inputs and outputs of all power conditioning equipment. An additional disconnect will be needed at the output of a charge controller as well as between any battery bank and inverter input or DC load center. If the disconnect is to disconnect DC, then the NEC requires that it be rated for DC. Additional disconnects are needed at the output of any inverter. If the inverter is utility interactive with battery backup for emergency loads, it is desirable to include an inverter bypass switch at the inverter output in case inverter maintenance is required without interruption of power to emergency loads. In addition to the inverter bypass switch, many utilities require a visible, lockable, accessible, load break, disconnect between the inverter output and the point of utility connection. This switch is for use by the utility if they deem it necessary to disconnect the inverter from the line for any reason.

The point of utility connection for a utility-interactive system will normally be a backfed circuit breaker in a distribution panel. This circuit breaker is to be labeled so maintenance workers will recognize it as a source of power to the distribution panel. Figure 44.18 shows the connections for an inverter bypass switch (A), the utility disconnect switch (B), and the point of utility connection circuit breaker (PUC). The figure also shows a neutral bus (C) for the connection of neutrals for the main distribution panel, the emergency panel, and the inverter. Operation of the inverter bypass switch is as follows: The two-pole unit and the one-pole unit are ganged together so that either both are off or only one is on. Under normal operation, the two-pole unit is on and the one-pole unit is off. This connects the utility to the inverter and the inverter emergency output to the emergency panel. When the two-pole is off and the one-pole is on, the utility is connected to the emergency panel and the inverter is bypassed. When both are off, the utility is disconnected from both the inverter and the emergency panel. It is interesting to note that if the PUC circuit breaker in the main distribution panel is turned off, the inverter will interpret this as an interruption in utility power and will shut down the feed from inverter to main distribution panel. Thus, the energized portions of the circuit breaker will be the same as the energized portions of the other circuit breakers in the panel. When it is on, both sides of the circuit breaker will be energized. When it is off, only the line side will be energized.

Article 690 of the NEC governs the sizing of conductors in the PV system. The serious designer should carefully review the requirements of this article, especially since many PV systems use low-voltage DC where voltage drop in the connecting wiring can be a problem. Sizing of conductors must be done carefully.

Chapter 36 provides information about storage batteries.

44.8 PV System Examples

44.8.1 Stand-Alone PV Well Pump System

As long as the depth of the well, the well replenishment rate, and the necessary flow rate are known, a PV pumping system can be designed. PV pumping systems are so common, in fact, that they often come in kits that include PV modules, a pump controller (LCB), and a pump. Pump manufacturers generally provide specifications that indicate, for a given pumping height, the amount of water pumped and the current drawn by the pump for specified pump voltages.

As an example, consider a system designed to pump 2000 gal/day from a well that is 200 ft deep and has a replenishment rate that exceeds the desired pumping rate. Assume the location for the pumping system has a minimum of 5 psh/day. This means that the 2000 gal must be pumped in 5 h, which corresponds to a pumping rate of 2000 gal/300 min = 6.67 gpm. One pump that meets this requirement is a 1.0 HP pump that will pump 7.6 gpm to a height of 200 ft. Under these pumping conditions, the pump will draw 6.64 A at a DC voltage of 105 V. An 875 W PV array is recommended for the operation of this system by the distributor. Note that (6.64 A) × (105 V) = 697 W, indicating that the recommended PV array is rated at 125% of the system requirements.

Before committing to this system, however, it should be compared with a system that uses battery storage and a smaller pump. The cost of the 1.0 HP pump is close to $1800, while a 0.25 HP pump that will pump 2.15 GPM while consuming 186 W can be purchased for about $500. This pump will need to pump for 15.5 h to deliver the 2000 gal, so the energy consumption of the pump will be (186 W) × (15.5 h) = 2884 Wh. If the pump runs at 24 V DC, this corresponds to 2884 ÷ 24 = 120 Ah/day. For PV storage, deep discharge lead–acid batteries are normally used, and it is thus necessary to ensure that the batteries will provide adequate storage for the pump without discharging to less than 20% of full charge. Thus, the battery rating must be at least 120 ÷ 0.8 = 150 Ah for each day of storage. If the water is pumped into a tank, then the water itself is a form of energy storage, and if the tank will hold several days' supply of water, then the batteries will only need to store enough energy to operate the pump for a day. If it is less expensive to use more batteries than to use a larger water tank, then additional batteries can be used.

So, finally, a sensible system will probably consist of a ¼ HP pump, an MPT charge controller for the batteries, and a minimum of 150 Ah at 24 V of battery storage. With the MPT controller, the array size, assuming 5 psh minimum per day, becomes (2884 Wh) × 1.25 ÷ (5 h) = 721 W, where the 1.25 factor compensates for losses in the array due to operation at elevated temperatures, battery charging and discharging losses, MPT losses, and wiring losses. This array can be conveniently achieved with 120 W modules configured in an array with two in series and three in parallel, as shown in Figure 44.21.

As a final note on the pumping system design, it is interesting to check the wire sizes. The NEC gives wire resistance in terms of Ω/kft. It is good design practice, but not an

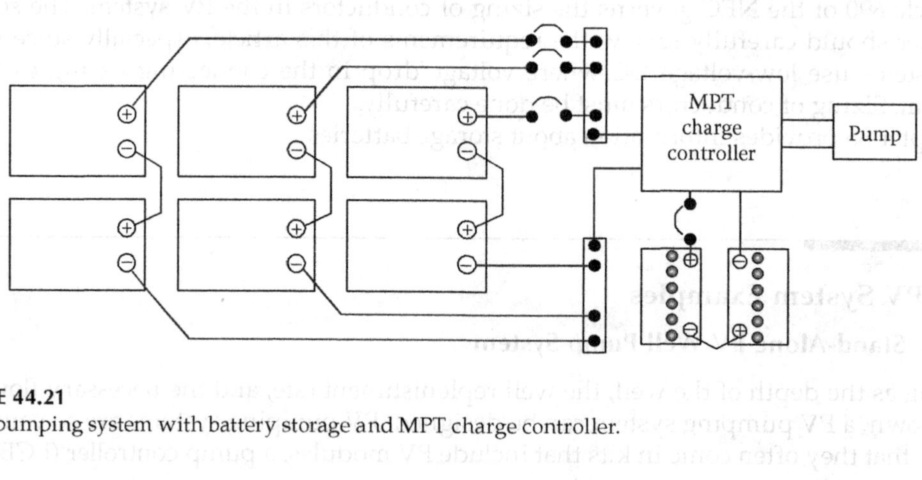

FIGURE 44.21
Water pumping system with battery storage and MPT charge controller.

absolute requirement, to keep the voltage drop in any wiring at <2%. The overall system voltage drop must be <5%. The wire size for any run of wire can thus be determined from

$$\Omega/\text{kft} \le \frac{(\%\text{VD})V_S}{0.2 \cdot I \cdot d}, \tag{44.10}$$

where
%VD is the allowed voltage drop in the wiring expressed as a percentage
V_S is the circuit voltage
I is the circuit current
d is the one-way length of the wiring

For the PV source circuit wiring, V_S will be about 34 V and I will be about 7 A. If the one-way source circuit length is 40 ft, then, for 2% voltage drop, (44.10) evaluates to Ω/kft = 1.2143. NEC Chapter 9 Table 8 shows that #10 solid Cu wire has 1.21 Ω/kft, while #10 stranded Cu has 1.24 Ω/kft. So either type of #10 will keep the %VD very close to 2%. Since #10 THWN-2 is rated to carry 40 A at 30°C, it is adequate for the job even under most derating conditions. Since the pump will be submersed, it will need 200 ft of wire just to get out of the well. If the controller is close to the well, then d will be approximately 210 ft. Thus, for $I = (186\text{ W}) \div (24\text{ V}) = 7.75\text{ A}$, and $V_S = 24$ V, (44.10) yields Ω/kft = 0.1475, which requires #1/0 Cu according to NEC Chapter 9 Table 8. A 3% voltage drop would allow the use of #2 Cu. In either case, this is a good example of how wire size may need to be increased to keep voltage drop at acceptable levels when relatively low-voltage DC is used. The 30°C ampacity of #2 Cu, for example, is 130 A. So even the small, low-voltage DC pump may not be the best choice. With inverter price decreasing and reliability increasing, and with AC motors generally requiring less maintenance than DC motors, at the time this article is being read, it may be more cost-effective to consider a 120 or 240 V AC pump for this application.

44.8.2 Stand-Alone System for a Remote Schoolhouse

Stand-alone system design requires a tabulation of the system loads, generally expressed in ampere hours (Ah) at the battery voltage. Suppose, for example, it is desired to provide power for 400 W of lighting, 400 W of computers, and 200 W of refrigeration, all at 120 V AC. Suppose that all of the loads operate for 8 h/day. This means the load to be met is 8 kWh/day at 120 V AC. If this load is supplied by an inverter that operates with 92% efficiency, then the batteries must supply the inverter with 8 ÷ 0.92 = 8.7 kWh/day. If the inverter input is 48 V DC, then the daily load in Ah is (8700 Wh) ÷ (48 V) = 181 Ah. To meet the needs for 1 day of operation, the batteries should thus be rated at 125% of 181 Ah = 226 Ah. But for a stand-alone system, it is usually desirable to provide more than 1 day of storage. For this system, 3 days would be more common, so a total of 678 Ah at 48 V should be used.

If an MPT charge controller is used, then the array can be sized based upon the daily system Wh and the available daily psh, taking losses into account. First of all, battery charging and discharging are only about 90% efficient. So to get 181 Ah *out* of the batteries, it is necessary to design for 181 ÷ 0.9 = 201 Ah *in* to the batteries. At 48 V, this is 9648 Wh. Next, it is necessary to include a 10% degradation factor for array maintenance, mismatch, and wiring losses and another 15% factor for elevated array operating temperature. So the array should be designed to produce 9,648 ÷ 0.9 ÷ 0.85 = 12,612 Wh/day. Assuming a worst-case psh = 5 h/day, this means an array size of 12,612 ÷ 5 = 2,522 W will be needed.

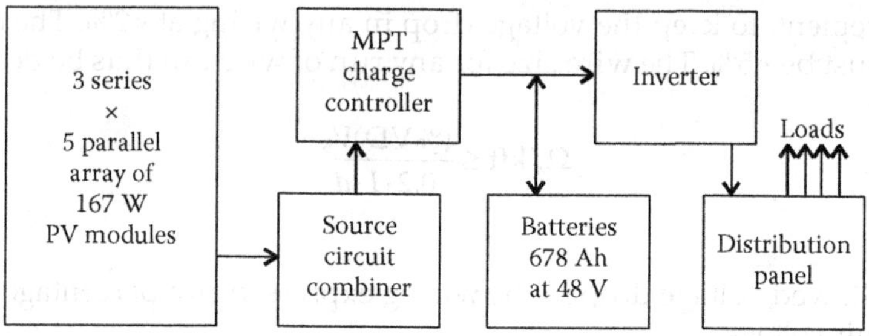

FIGURE 44.22
Block diagram of schoolhouse PV system.

This can be achieved with twenty 125 W modules in a 4-series-by-5-parallel array, or with 15 167 W modules in a 3-series-by-5-parallel array, or by any number of other module combinations that do not exceed the charge controller maximum input voltage limit when running open circuit. It must be remembered that to achieve the nominal 48 V source circuit output, it may take two, three, or four modules in series. Thus, each additional parallel module will require additional series modules to achieve the system voltage.

Figure 44.22 shows the block diagram of the schoolhouse system.

To this point, wind loading of the array has not been mentioned. In areas of high wind loads, the size of the module and the mounting method may or may not be adequate to meet high wind loading conditions. The final check on any PV system design must be a determination of whether the system will blow away in a high wind. This is especially undesirable considering the cost of a system as well as the fact that many systems are installed to provide power during emergencies, one of which might be a hurricane.

As a final note on system design, if an MPT charge controller is not used, then the array should be sized to provide 110% of the daily battery input Ah, using the maximum power current of the array. In this case, the daily battery input Ah is 201, so the array should be designed for 221 Ah. For 5 psh, this converts to an array current of 44.2 A. The 125 W modules have a 7 A maximum power current, so this means six in parallel will produce 42 A, which is close to the required amount. So the MPT controller saves four modules, or 500 W. At $4/W, this is a savings of $2000, which more than pays for the additional cost of the MPT charge controller.

It is interesting to look at a life cycle cost of the schoolhouse system. Using a discount rate of 5% and an inflation rate of 3%, an LCC cost estimate can be tabulated. In Table 44.1, it is assumed that the batteries are 12 V, 110 Ah, sealed, AGM lead–acid, deep-cycle batteries with a rated lifetime of 8 years and a cost of $150 each.

To compare this system with a gasoline generator, note that the generator would need to generate 8 kWh/day for 20 years by operating 8 h/day over this period. A typical small gasoline generator will generate 4 kWh/gal [4], so will require 2 gal of gasoline per day. The generator will require an oil change every 25 h, a tune-up every 300 h, and a rebuild every 3000 h. The LCC analysis for the generator is shown in Table 44.2. Clearly, the PV system is the preferred choice. And this does not even account for the noise-free, pollution-free performance of the PV system.

44.8.3 Straightforward Utility-Interactive PV System

Since utility-interactive PV systems are backed up by the utility, they do not need to be sized to meet any particular load. Sometimes they are sized to meet emergency loads, but if the system does not have emergency backup capabilities, then they may be sized to fit

TABLE 44.1

Life Cycle Costs of a Stand-Alone Photovoltaic System for a Schoolhouse

Item	Cost ($)	Present Worth ($)	LCC (%)
Capital costs			
Array	10,000	10,000	38.1
Batteries	3,600	3,600	13.7
Array mount	1,250	1,250	4.8
Controller	500	500	1.9
Inverter	500	500	1.9
BOS	1,000	1,000	3.8
Installation	2,000	2,000	7.6
Recurring costs			
Annual inspection	50	839	3.2
Replacement costs			
Batteries—8 years	3,600	3,087	11.8
Batteries—16 years	3,600	2,646	10.1
Controller—10 years	500	413	1.6
Inverter—10 years	500	413	1.6
Totals		26,247	100

TABLE 44.2

Life Cycle Costs of a Gasoline Generator for a Schoolhouse

Item	Cost ($)	Present Worth ($)	LCC (%)
Capital costs			
Generator	750	750	2
Recurring costs			
Annual fuel	1825	30,593	73
Annual oil changes	235	3,939	9
Annual tune-ups	345	4,107	10
Annual rebuilds	146	2,447	6
		41,836	100

on a particular roof, to meet a particular budget, or to incorporate a particular inverter. Suppose the sizing criterion is the inverter, which has the following specifications:

DC input	Input voltage range	250–550 V
	Maximum input current	11.2 A
AC output	Voltage	240 V, 1ϕ
	Nominal output power	2200 W
	Peak power	2500 W
	Total harmonic distortion	<4%
	Maximum efficiency	94%

Note that if the input voltage is 550 V and the input current is 11.2 A, the input power would be 6160 W. Since the peak output power of this inverter is 2500 W, it would not make sense to use a 6160 W array, since most of the output would be wasted, and it might be easier to damage the inverter. So an array size of about 2500 W would make better sense.

Since cloud focusing can increase the short-circuit output current of a module by 25%, the array rated short-circuit current should be kept below 11.2 ÷ 1.25 = 8.96 A. The number of modules in series will depend upon the maximum voltage of the array at low temperatures remaining <550 V and the minimum array voltage at high array operating temperatures remaining >250 V.

NEC Table 690.7 specifies multipliers for open-circuit voltages for different low-temperature ranges. For design purposes, suppose the coldest array temperature will be −25°C and the hottest array temperature will be 60°C. NEC Table 690.7 requires a multiplier of 1.25 for the array open-circuit voltage, so the maximum rated array open-circuit voltage must be less than 550 ÷ 1.25 = 440 V. If the open-circuit voltage of a module decreases by 0.47%/°C, then it will decrease by 35 × 0.47 = 16.45% when the module is operated at 60°C. Thus, the 25°C rated array open-circuit voltage needs to be greater than 250 ÷ 0.8355 = 299 V.

The next step is to look at PV module specifications. One module has P_{max} = 125 W, V_{OC} = 21.0 V, and I_{SC} = 7.2 A. Thus, the maximum number of these modules in series will be 440 ÷ 21.0 = 20.95, which must be rounded down to 20. The minimum number in series will be 299 ÷ 21.0 = 14.24, which must be rounded up to 15. Checking power ratings gives 2500 W for 20 modules and 1875 W for 15 modules. Since 2500 W does not exceed the inverter rated maximum output power, and since the array will normally operate below 2500 W, it makes sense to choose 20 modules, as long as the budget can afford it and as long as there is room for 20 modules wherever they are to be mounted. Figure 44.15d shows the block diagram for this system.

The life cycle cost of this type of system is usually looked at somewhat differently than that of the schoolhouse system. In this case, the cost of electricity generated is usually compared with the cost of electricity from the utility, neglecting pollution and other externalities. In regions with an abundance of trained installers, it is currently possible to complete a grid-connected installation for less than $7/W. The installed cost of the 2,500 W system would thus be approximately $17,500. It is reasonable to expect an average daily output of 10 kWh for this system in an area with an average of 5 psh. The value of the annual system output will thus be approximately $365 at $0.10/kWh. This amounts to a simple payback period of 48 years—almost double the expected lifetime of the system.

Of course, what is not included in the analysis is the significant amount of CO_2 production that is avoided, as well as all the other pollutants associated with nonrenewable generation. Also not included are the many subsidies granted to producers of nonrenewable energy that keep the price artificially low. For that matter, it assumes that an abundance of fossil fuels will be available at low cost over the lifetime of the PV system. Finally, it should be remembered that the energy produced by the PV system over the lifetime of the system will be at least four times as much as the energy that went into the manufacture and installation of the system.

If the cost of the system could be borrowed at 3% over a period of 25 years, the annual payments would be $1005. Thus, if a grid-connected system is considered, unless there is a subsidy program, it could not be justified with simple economics. It would be purchased simply because it is the right thing to do for the environment. Of course, if the installation cost were less, the value of grid electricity were more, and the average sunlight were higher, then the numbers become more and more favorable. If the values of externalities, such as pollution, are taken into account, then the PV system looks even better.

Because of the cost issue, as well as local PUC regulations or lack of them, ill-defined utility interface requirements, etc., few grid-connected PV systems are installed in areas that do not provide some sort of incentive payments. In some cases, PV system owners are paid rebates based on dollars per watt. The problem with this algorithm is that there is no guarantee that the system will operate properly. In other cases, PV system owners are guaranteed a higher amount per kWh for a prescribed time, which guarantees that the system must work to qualify for incentive payments. In fact, at present, more kW of PV is installed annually in grid-connected systems than are installed in stand-alone systems [17].

44.9 Present Status of Technology and Future Challenges for PV Systems

The PV technology has made tremendous progress in the last decade, where the PV system deployment around the world has been increasing at an average annual rate of more than 50%. The costs of PV panels have come down by an order of magnitude over the last decade. The present panel costs for both polycrystalline silicon and thin-film panels are less than $0.75/W, which has resulted in the system costs of $2–$2.50/W for small-scale systems and around $1.50–$2.00/W for large-scale systems. Large-scale PV systems are being constructed in the range 1 MW to as large as hundreds of MW. The challenge is to reduce the panel costs further to less than $0.50/W and the BOS costs also to less than $0.50/W, which will make PV competitive without government incentives. For PV panels, this will be achieved by further improvements in panel efficiencies. As of 2013, the efficiencies achieved in the laboratory for the various PV technologies are given in Table 44.3. Efficiency table for concentrating PV is given in Chapter 46.

The PV industry is currently engaged in an effort to ensure quality control at all levels of system deployment, including manufacturing, distribution, design, installation, inspection, and maintenance. At this point in time (2013), most of the technical challenges for PV system components have been overcome. PV modules are very reliable, and most are warranted for 25 years. Reliability of other system components continues to improve, and at this point, PV system power conditioning equipment has proven to operate very reliably. So, once installed properly, modern PV systems require very little maintenance. In fact, systems without batteries require almost no maintenance.

The biggest challenge for PV systems is storage. As the contribution of PV in the grid continues to increase, the electrical utilities are becoming more concerned with the transients in the grid due to solar radiation transients. Battery storage, however, is very expensive. With the cost of battery storage of over $500/kWh_e$, batteries are not cost-effective in large-scale systems. Although a great deal of battery and supercapacitor research is being conducted around the world, the costs are not expected to get below $100/kWh_e$ in this decade.

If the cost of a PV system were to be measured in energy units, then perhaps the importance of PV systems would become more obvious. By the early 1980s, Odum and Odum [17] and Henderson [18] had proposed using the Btu as the international monetary standard. However, at that time, they reported that the EPBT of PV panels was more than their entire lifetime. In the last decade, Battisti and Corrado [19] reported EPBTs of approximately 3 years and CO_{2eq} payback times of about 4 years for PV systems. Since that time, the EPBTs have come down further. By 2011, the EPBTs of commercial rooftop PV systems had reduced to as little as 0.68 year for CdTe technology and 1.96 years for monocrystalline silicon [1]. Since the life expectancy of a PV system exceeds 20–25 years, this equates to as much as 30:1 return on the energy invested.

TABLE 44.3

Confirmed Terrestrial Cell and Submodule Efficiencies Measured under the Global AM1.5 Spectrum (1000 W/m²) at 25°C

Classification	Effic. (%)	Area (cm²)	V_{oc} (V)	J_{sc} (mA/cm²)	FF (%)
Silicon					
Si (crystalline)	25.0 ± 0.5	4.00 (da)	0.706	42.7	82.8
Si (multicrystalline)	20.4 ± 0.5	1.002 (ap)	0.664	38.0	80.9
Si (thin-film transfer)	20.1 ± 0.4	242.6 (ap)	0.682	38.14	77.4
Si (thin-film submodule)	10.5 ± 0.3	94.0 (ap)	0.492	29.7	72.1
III–V cells					
GaAs (thin film)	28.8 ± 0.9	0.9927 (ap)	1.122	29.68	86.5
GaAs (multicrystalline)	18.4 ± 0.5	4.011 (t)	0.994	23.2	79.7
InP (crystalline)	22.1 ± 0.7	4.02 (t)	0.878	29.5	85.4
Thin-film chalcogenide					
CIGS (cell)	19.6 ± 0.6	0.996 (ap)	0.713	34.8	79.2
CIGS (submodule)	17.4 ± 0.5	15.993 (da)	0.6815	33.84	75.5
CdTe (cell)	18.3 ± 0.5	1.005 (ap)	0.857	26.95	77.0
Amorphous/nanocrystalline Si					
Si (amorphous)	10.1 ± 0.3	1.036 (ap)	0.886	16.75	67.8
Si (nanocrystalline)	10.1 ± 0.2	1.199 (ap)	0.539	24.4	76.6
Photochemical					
Dye sensitized	11.9 ± 0.4	1.005 (da)	0.744	22.47	71.2
Dye sensitized (submodule)	9.9 ± 0.4	17.11 (ap)	0.719	19.4	71.4
Organic					
Organic thin film	10.7 ± 0.3	1.013 (da)	0.872	17.75	68.9
Organic (submodule)	6.8 ± 0.2	395.9 (da)	0.798	13.50	62.8
Multijunction devices					
InGaP/GaAs/InGaAs	37.7 ± 1.2	1.047 (ap)	3.014	14.57	86.0
a-Si/nc-Si/nc-Si (thin film)	13.4 ± 0.4	1.006 (ap)	1.963	9.52	71.9
a-Si/nc-Si (thin-film cell)	12.3 ± 0.3	0.962 (ap)	1.365	12.93	69.4
a-Si/nc-Si (thin-film submodule)	11.7 ± 0.4	14.23 (ap)	5.462	2.99	71.3

Source: Adapted from Green, M.A. et al., *Prog. Photovolt. Res. Appl.*, 21, 11, 2013.

Notes: ap, aperture area; t, total area; da, designated illumination area; nc-Si, nanocrystalline or microcrystalline Silicon; *a*-Si, amorphous silicon.

Nomenclature

- C Speed of light (3×10^8 m/s)
- D_x Minority carrier diffusion length
- E Energy of photon (eV)
- $F(x)$ Intensity of the photon beam as a function of penetration depth into the material
- F_o Initial intensity of the photon beam at the material surface
- h Planck's constant (6.625×10^{-34} J s)

I	Current (A)
I_l	Short-circuit current (A)
k	Boltzmann's constant (1.381 × 10^{-23} J/K)
L	Latitude angle
L_x	Diffusion length
n	Electron concentration
n	Day of year
N_a	Density of acceptor impurities
N_d	Density of donor impurities
n_i	Intrinsic carrier concentration
p	Hole concentration
P_{max}	Maximum power (W)
Q	Charge of an electron (1.602 × 10^{-19} C)
T	Junction temperature (K)
V	Voltage (V)
V_{OC}	Open-circuit voltage (V)
X	Depth of penetration into material

Symbols

Λ	Wavelength (μm)
A	Absorption constant
A	Altitude angle of the sun
Δ	Declination angle
Γ	Azimuth angle of the sun
Ω	Hour angle
Ω	Resistance (ohm)
τ_x	Minority carrier lifetime

Acronyms

a-Si	Amorphous silicon
CVD	Chemical vapor deposition
EHP	Electron–hole pair
GFDI	Ground fault detection and interruption
HIT	Heterojunction with intrinsic thin layer
LCB	Linear current booster
MPT	Maximum power tracker
psh	Peak sun hour

References

1. Wild-Scholten, M.J., Energy payback time and carbon footprint of commercial photovoltaic systems. *Solar Energy Materials and Solar Cells*. 2013, 119: 296–305.

1b. USDOE, http://www1.eere.energy.gov/solar/sunshot/reducing_technology_costs.html (accessed May 2014).

2. Green, M.A. et al., Solar cell efficiency tables. *Progress in Photovoltaics: Research and Applications*. 2013, 21: 1–11.

3. Goswami, D.Y. et al., *Principles of Solar Engineering*, 2nd edn. Taylor & Francis, Philadelphia, PA, 2000.

4. Messenger, R. and Ventre, G., *Photovoltaic Systems Engineering*, 2nd edn. CRC Press LLC, Boca Raton, FL, 2004.
5. Rosenblum, A.E. et al., *Proceedings of the 29th IEEE Photovoltaic Specialists Conference*, 2002, pp. 58–61.
6. Hanoka, J.I., *Proceedings of the 29th IEEE Photovoltaic Specialists Conference*, 2002, pp. 66–69.
7. Saitoh, T., Progress of highly reliable crystalline si solar devices and materials. In: Goswami, D.Y. and Boer, Karl W. (Eds.), *Advances in Solar Energy*, Vol. 16, Chapter 5. American Solar Energy Society, Boulder, CO, 2005, pp. 129–154.
8. www.ase-international.com (For information on EFG process).
9. Green, M.A. and Wenham, S.R., Novel parallel multijunction solar cell. *Applied Physics Letters*. 1994, 65: 2907.
10. Bohland, J., Accelerated aging of PV encapsulants by high intensity UV exposure. In: *Proceedings of the 1998 Photovoltaic Performance and Reliability Workshop*, Cocoa Beach, FL, November 3–5, 1998.
11. Markvart, T. (Ed.). *Solar Electricity*. John Wiley & Sons, Chichester, U.K., 1994.
12. National Renewable Energy Laboratory, 30-year average of monthly solar radiation, 1961–1990, Spreadsheet Portable Data Files. http://rredc.nrel.gov/solar/old_data/nsrdb/redbook/sum2/(accessed May 2014).
13. Krein, P.T., *Elements of Power Electronics*. Oxford University Press, New York, 1998.
14. IEEE 929-2000, IEEE recommended practice for utility interface of residential and intermediate photovoltaic (PV) systems. IEEE Standards Coordinating Committee 21, Photovoltaics, IEEE, 2000.
15. UL 1741, Standard for static inverters and charge controllers for use in photovoltaic power systems. Underwriters Laboratories, Inc., Northbrook, IL, May, 1999.
16. NFPA 70, National electrical code, 2002 Ed. National Fire Protection Association, Quincy, MA, 2002.
17. Odum, H.T. and Odum, E.C., *Energy Basis for Man and Nature*, 2nd edn. McGraw-Hill, New York, 1981.
18. Henderson, H., *The Politics of the Solar Age: Alternatives to Economics*. Anchor Press/Doubleday, Garden City, NY, 1981.
19. Battisti, R. and Corrado, A., Evaluation of technical improvements of photovoltaic systems through life cycle assessment methodology. *Energy*, 2005, 30: 952–967.

45

Thin-Film PV Technology

Hari M. Upadhyaya, Senthilarasu Sundaram, Aruna Ivaturi, Stephan Buecheler, and Ayodhya N. Tiwari

CONTENTS

45.1 Introduction

45.1.1 Historical and Current Developments

Crystalline silicon (c-Si) technology has a lion's share in the present photovoltaic (PV) industry, contributing more than 85% through the cells and modules based on poly-, mono-, and multicrystalline wafer technology (Mints and Donnelly 2011). The recent growth rate of PV industry and market is phenomenal, with a substantial surge on average over 40% recorded globally during the last decade and expected to sustain the trend from short to medium time frame. During the early developmental phase of c-Si PV technology, the continuous feedstock support offered by Si-based electronics industry played a key role in its growth. The high purity and even second-grade wafer materials obtained at a relatively cheaper price proved favorable for PV industry as they led to a reasonable efficiency (η) figures for standard size modules (average $\eta > 16\%–18\%$) and extremely good performance stability (more than 25 years) that are two essential requirements for any technology to successfully demonstrate its potential for market (Chopra et al. 2004; Jäger-Waldau 2004). However, continuously increasing demand for PV modules and the need for low-cost PV options in the last decade had stretched these advantages to the limit and had exposed some inherent disadvantages of c-Si technology, such as the scarcity of feedstock material, costly processing of materials and device fabrication steps, as well as the inability for monolithic interconnections. These, in turn, had restricted the potential of Si wafer technology and made it difficult to achieve PV module production cost below €1/W (1€ is about U.S. $1.35), which was considered essential for cost-competitive generation of solar electricity (Hegedus and Luque 2003; von Roedern et al. 2005; Zweibel 2000). However, recently, there has been a dramatic upsurge in the production of c-Si modules in China for the last 2 years, which were made available at and below $0.6/W and as low as $0.50/W in some cases. While there is speculation that the low module cost from China perhaps did not reflect the real costs, the fact remains that the PV module prices have seen significant reduction in the recent years. The PV module cost depends on the total manufacturing cost

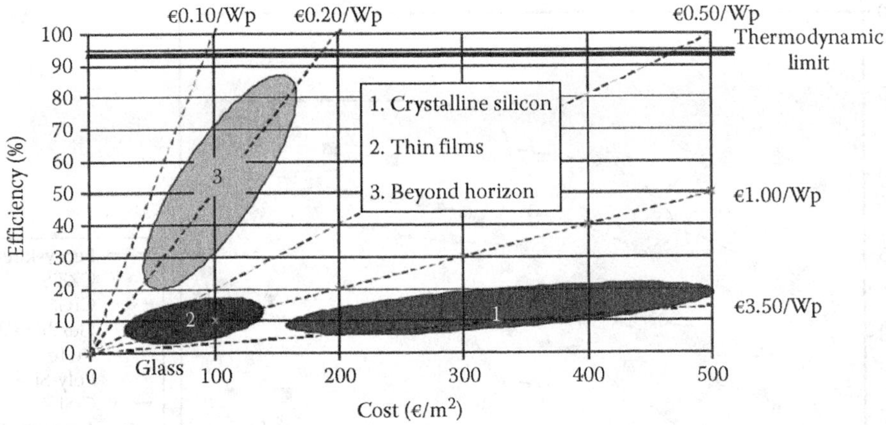

FIGURE 45.1
Comparison of estimated costs achievable with different PV technologies as a function of manufacturing cost (€/m^2) and conversion efficiency. (Courtesy M. Green, UNSW, Sydney, New South Wales, Australia.)

of the module per square area, conversion efficiency, and long-term performance stability. Figure 45.1 gives an estimate of achievable cost with c-Si technology and comparison with projected achievable costs with other PV technologies. It was generally agreed that c-Si wafer technology would be unable to meet the low-cost targets, while thin-film technologies will have the potential to provide a viable alternative in the near future. However, the cost of modules from China available for as low as \$0.50/Wp recently has made these speculations look very weak as the gap between thin film and c-Si modules has shrunk considerably. Although the thin-film PV still has better potential to lower the cost of modules further, this will require appreciable increase in the energy conversion efficiencies of the technologies, in order to stay in competition with c-Si. Recently, the cost projection by First Solar for CdTe modules has shown a competitive cost of \$0.49/W for the module efficiencies over 14%. It is thus clear that thin-film technologies have to significantly improve their efficiencies in order to stay in competition with c-Si (Osborne 2013). The world record efficiency of 20.4% for small area and 17% efficient CdTe champion modules announced recently by First Solar and 20.9% small area efficiency achieved on CIGS technology by Solar Frontier and 16.4% CIGS modules by Avancis definitely show a great promise in a short span of time.

The PV market although appears to be promising, the share of the technologies remains very difficult to predict. An optimistic and ambitious forecast by Hanergy Solar with their multigigawatt production in a few years' time on a-Si, along with production on CIGS technology as well, provides a projection of about 50% market share by 2018 (Hanergy Annual Report 2013). However, a more conservative approach adopted by NPD Solarbuzz provides a forecast of around 10% share of thin-film technologies by 2020 (Colville 2014).

Figure 45.2 (NREL 2013) summarizes the best laboratory scale (cell area less than 1 cm^2) efficiencies of some prominent thin-film technologies. Out of these, an efficiency of 20.8% achieved at ZSW (has been improved further by Solar Frontier to 20.9% on CIS recently), with CIGS, makes it the most efficient thin-film PV device that has narrowed the gap further between existing c-Si and thin-film technologies.

Silicon wafers are fragile, solar cell area is limited by the wafer size, and the modules are bulky, heavy, and nonflexible. To overcome some of the problems of c-Si wafer technology, efforts were made to develop monocrystalline and polycrystalline thin-film silicon solar cells on glass substrates, as reviewed by Bergmann (1999). However, these efforts to develop

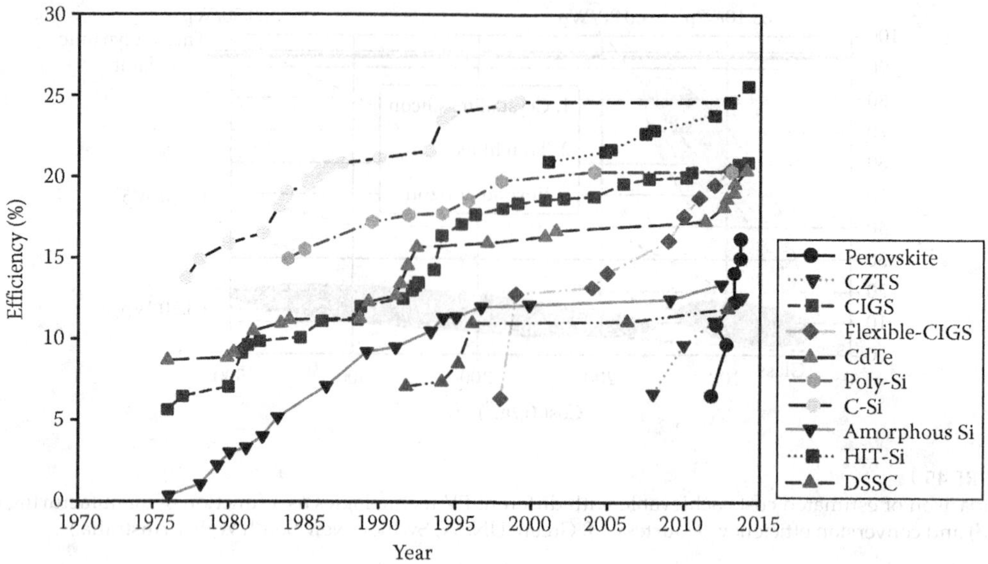

FIGURE 45.2
Development of *champion* thin-film solar cell efficiencies with emerging PV. (Based on Efficiency chart, National Renewable Energy Laboratory (NREL), 2014, http://www.nrel.gov/ncpv/.)

efficient crystalline silicon thin films on glass (CSG) have not gained much success over these years and efficiencies have been limited around or just over 10% mark with much more processing costs involved. One of the techniques used to form poly-Si films on glass has been based on solid phase crystallization (SPC) and had resulted in 10.5% efficient cells (Keevers et al. 2007). The open circuit voltage (V_{oc}) up to 500 mV was one of the stumbling blocks for its progress. The TU-Hamburg-Harburg and Helmholtz-Zentrum Berlin, Germany, presented a new type of poly-Si thin-film solar cell on glass using electron-beam crystallization (EBC) through liquid phase and have achieved an improved V_{oc} of 545 mV (Amkreutz et al. 2011). The EBC process is based on a line-shaped electron beam, which is scanned across the silicon layer on glass with a typical rate of 1 cm/s. Recently, an analogue laser crystallization (LC) process has been investigated as an alternative to EBC, using which the University of New South Wales in Australia and Suntech R&D, Australia, have recently reached a new record single-junction efficiency of 11.7% for poly-Si thin-film solar cells on glass (Dore et al. 2014). A current status and some latest developments on CSG have been covered in a report (Becker et al. 2013) recently. It is expected that by using the aforementioned techniques, 15% efficient single junction CSG can be achieved in near future (Rech 2013).

It has been realized that application of thin-film technologies to grow solar cells and modules based on other materials can prove to be an alternative to the silicon-wafer technologies. However, the efficiency of the thin-film PV technologies has to match or exceed the current c-Si technology. Thin-film deposition of materials of required quality and suitable properties depends on the processes used and the control of several parameters. However, once optimized, these methods provide over an order of magnitude cheaper processing cost and low energy payback time, which is certainly a big advantage. For example, even at lower module efficiencies as compared to c-Si, the cost of CdTe modules have lower payback time owing to less energy intensive and easier processing steps involved.

Research on alternative to c-Si already started over four decades back, considering that it has low absorption coefficient (~10^3 cm^{-1}) and its narrow bandgap (E_g ~ 1.1 eV). Some

TABLE 45.1

Bandgap of Different Semiconductor Materials Suitable as Light Absorber in Solar Cells

Compound/Alloys	Bandgap (eV)	Compound/Alloys	Bandgap (eV)
c-Si	1.12	Zn_3P_2	1.50
a-Si	1.70	$CuInSe_2$	1.04
GaAs	1.43	$CuGaSe_2$	1.68
InP	1.34	$Cu(In,Ga)Se_2$	1.20
Cu_2S	1.20	$CuInS_2$	1.57
CdTe	1.45	$Cu(In,Ga)(S,Se)_2$	1.36
Cu_2ZnSnS_4	1.5	$Cu_2ZnSnSe_4$	1.0

of the most interesting semiconducting materials that have received considerable attention are cadmium telluride (CdTe), gallium arsenide (GaAs), indium phosphide (InP), zinc phosphide (Zn_3P_2), copper sulfide (Cu_2S), copper indium diselenide (CIS), copper indium gallium diselenide (CIGS), and copper zinc tin diselenide (CZTS). These all have the electronic and optical properties suitable for the efficient utilization of the sun's spectrum (see Table 45.1 for bandgaps). Multijunction solar cells based on III–V materials (GaAs, InP, GaSb, GaInAs, GaInP, etc.) show high efficiency, exceeding 35%, but due to the high production cost and low abundance of their constituent materials, these solar cells are not considered suitable for cost-effective terrestrial applications though they are still very important for space PV applications.

Thin-film technology has answers and potential to eliminate many existing bottlenecks of c-Si PV programs experienced at different levels from module production to its applications viz. space programs and BIPV. Thin-film PV modules are manufactured on either rigid glass substrates or flexible substrates (thin metallic or plastic foils). Some of the advantages of the thin-film technologies are the following:

- The high absorption coefficient (~10^5 cm^{-1}) of the absorber materials is about 100 times higher than c-Si (Figure 45.3); thus, about 1–2 μm of material thickness is sufficient to harness more than 90% of the incident solar light. This helps in reducing the material mass significantly to make modules cost effective.
- The estimated energy payback time of the thin-film PV is considerably lower than that of c-Si PV. Estimation suggests that CdTe has the lowest payback time among all PV technologies. With recent improved cell and modules efficiencies of 17%, the payback time could be as low as 6 months.
- The formation of heterojunction and better device engineering for reduction of photon absorption losses and enhanced collection of photogenerated carriers are possible.
- Large-area deposition (in the order of m^2) along with the monolithic integration (interconnection of cells during processing of rigid and flexible devices) is possible that minimizes area losses, handling, and packaging efforts.
- Roll-to-roll manufacturing of flexible solar modules is possible that gives high throughput and thus can further reduce the energy payback time significantly.
- Tandem or multijunction devices could be realized to utilize full solar spectrum to achieve higher efficiency devices. A theoretical limit of 66.7% had been estimated for large number of multiple junction, which can cover the entire spectrum of the sun (Honsberg et al. 2001; Shockley and Queisser 1961).
- Flexible and lightweight PV facilitates several attractive applications.

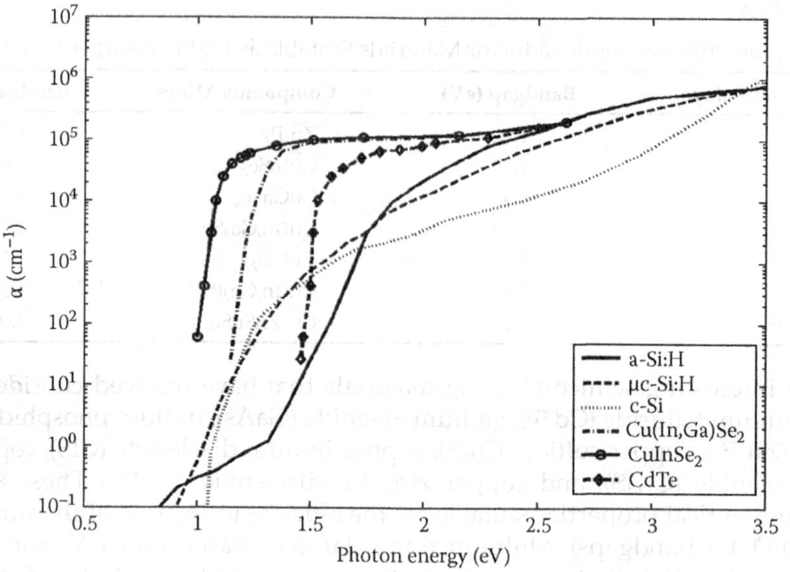

FIGURE 45.3
The optical absorption (α) versus photon energy of the c-Si and some of the prominent light-absorbing materials that are used in thin-film solar cells.

Historically, cadmium sulfide and copper sulfide (CdS/Cu_2S) heterojunctions were one of the early technologies almost contemporary to c-Si but could not survive due to the astounding success of c-Si technology that was efficient, stable, backed by electronic industries and widely supported by space programs. Despite Cu_xS having stability issues, the technology went to pilot production in the late 1970s or early 1980s. But with the advent of a-Si and CdTe thin-film solar cells, it was abandoned before mid-1980s.

Currently, a-Si, CdTe, and CIGS are considered the mainstream thin-film technologies. Out of these, a-Si has suffered some setbacks due to difficulties in improving the stabilized efficiency numbers and bankruptcy issues of the leading company United Solar recently; however, there is a good promise on reemergence of this technology with newer ways of achieving high efficiency in the future. Besides, CdTe and CIGS have not only proven potential to become leading technologies, but they have already achieved significant production capacities around multi-GW per year recently through the efforts of First Solar and Solar Frontier. Currently, the world record efficiencies in both CdTe and CIGS technologies are led by the aforementioned companies.

Figure 45.4 gives the market share of all the leading PV technologies including c-Si based on the cumulative deployment by 2013. Their thin-film market share is about 10%, with CdTe being leader, with 6% share on absolute scale.

Recently, several companies have started production-related activities and production volume is expected to increase during next few years. Recent developments on a new technology on organic and dye-sensitized solar cells (DSSCs), especially solid-state perovskite-based solar cells (a kind of solid state analogue of DSSCs), appear to be quite interesting. The efficiency figures for this technology has taken an unprecedented leap from near 4% in 2009 (Kojima et al. 2009) to 17% in 2014 (Snaith and coworkers 2014; Bennett 2014) (described in a separate section later). These technologies with their low-temperature and cheaper processing cost on flexible substrates (Docampo et al. 2013; Kumar et al. 2013)

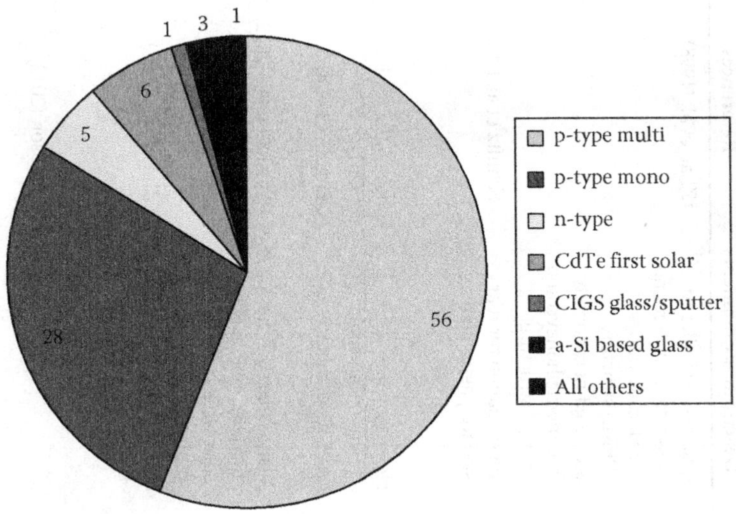

FIGURE 45.4
Worldwide market share of the dominant PV technologies by cumulative deployment end 2013. (Adapted from The worldwide market share of the dominant PV technologies by cumulative deployment end 2013, Solar Buzz, Report on PV equipment quarterly, 2013, http://www.solarbuzz.com/reports/pv-equipment-quarterly.)

make them potentially very attractive for various niche applications. However, there are important issues of instability and degradation, which need to be addressed. Perovskite solar cells are at infancy of their development at the moment and would require considerable efforts and time to come out as a commercially viable option.

Table 45.2 summarizes the current status of thin-film solar cell and module development at laboratory and industrial level on rigid and flexible substrates (with one exception). These contain some latest developments and global efforts to make thin-film PV technology a viable option for cost-effective and competitive electrical energy production.

45.1.2 Cost Potentials and Material Availability Issues

Estimates in the past decade had suggested that solar cell production costs as low as €0.30/W are possible depending on the volume of production (Bruton et al. 1997; Keshner and Arya 2004; Woodcock et al. 1997). At present, there is a significantly large gap between the efficiencies of the solar cells at the R&D level and those at the industrial modules level because of the process scalability–related issues during the early phase of industrial development. The current production cost of the thin-film PV modules is rather high because of low volume of production through pilot-production lines or small capacity (less than 30 MW) plants. However, with larger production volumes and higher module efficiencies, the cost is expected to decrease significantly in near future. It is estimated on the basis of the trend derived from the past experiences of cumulative production versus price variations (learning curve) that the module prices decrease by 20% for every doubling of the cumulative production (Surek 2003). These trends had predicted that with a growth rate of 25%, the cumulative production would reach ~75 GW by year 2020 and the target cost below $1/Wp was achievable with existing thin-film technologies. It is striking to note that the cumulative production in 2013 was alone close to 140 GW and set to achieve annual production of about 50 GW in 2014 (Burger 2013). More surprisingly, what was not believed

TABLE 45.2

Efficiencies of Solar Cells and Modules Based on Rigid and Flexible Substrates for Various Technologies

	Types of Solar Cells	Companies/ Institutions	Efficiency (%)	Area (cm^2)(Designed Area Illumination [da]; Aperture Area [ap]; Total Area [t])	Test Center (and Date)	Description/Comments	References
Silicon single cell	Si (crystalline)	University of New South Wales	25.0 ± 0.5	4.00 (da)	Sandia (3/1999)	Structure is similar to passivated emitter and rear locally diffused (PERL) structure. Adopted hexagonally symmetric honeycomb surface structure.	Zhao et al. (1998)
	Si (multicrystalline)	Fraunhofer institute for solar energy (FhG-ISE).	20.4 ± 0.5	1.002 (ap)	NREL (5/2004)	Wet oxidation for rear surface passivation.	Schultz et al. (2004)
	Si (large crystalline)	Panasonic	24.7 ± 0.5	101.8 (t)	AIST (12/2012)	Heterojunction with intrinsic thin layer.	Kinoshita et al. (2011)
	Si (large multicrystalline)	Q-Cells	19.5 ± 0.4	242.7 (t)	FhG ISE(03/2011)	Laser-fired contacts.	Engelhart et al. (2010)
	Si (thin-film transfer)	Solexel	20.1 ± 0.4	242.6 (ap)	NREL (10/2012)	A large area (15.6 cm × 15.6 cm) but thin (43 μm) silicon cell transferred from a reusable template.	Moslehi et al. (2012)
Silicon Modules	Si (crystalline)	University of New South Wales/ Gochermann	22.9 ± 0.6	778 (da)	Sandia (9/1996)	A flat-plate module made from 50 single cells.	Zhao et al. (1997)
	Si (large crystalline)	SunPower	22.4 ± 0.6	15775 (da)	NREL (8/2012)		Swanson (2012)
	Si (multicrystalline)	Q-Cells	18.5 ± 0.4	14661 (ap)	FhG-ISE (1/2012)	60 serial cells connected.	Pvtech.org
	Si (thin-film poly crystalline)	Pacific Solar	8.2 ± 0.2	661 (ap)	Sandia (7/2002)	Thin film coated on glass with <2 μm thickness.	Basore (2002)

(Continued)

TABLE 45.2 (Continued)

Efficiencies of Solar Cells and Modules Based on Rigid and Flexible Substrates for Various Technologies

	Types of Solar Cells	Companies/ Institutions	Efficiency (%)	Area (cm^2)(Designed Area Illumination [da]; Aperture Area [ap]; Total Area [t])	Test Center (and Date)	Description/Comments	References
Thin-film solar cells	GaAs thin film	Alta devices	28.8 ± 0.9	0.9927 (ap)	NREL (5/2012)	Single-junction devices under nonconcentrated sunlight.	Kayes et al. (2011)
	GaAs (multicrystalline)	Research Triangle Institute, NC, United States	18.4 ± 0.5	4.011 (t)	NREL (11/1995)	Thin film grown on Ge substrates.	Venkatasubramanian et al. (1997)
	InP (crystalline)	Spire	22.1 ± 0.7	4.02 (t)	NREL (4/1990)	MOCVD grown with graded bandgap.	Keavney et al. (1990)
	$Cu(In,Ga)Se_2$	ZSW	20.8 ± 0.6	0.9974 (ap)	FhG-ISE (10/2013)	CIGS on glass substrate.	Jackson et al. (2011)
	$Cu(In,Ga)Se_2$	Empa	20.4 ± 0.4		FhG-ISE(9/2012)	CIGS thin film on polyamide flexible solar cells.	Chirilă et al. (2013)
	GaInP	NREL	20.8 ± 0.6	0.2491 (ap)	NREL (5/2013)	Made from high bandgap.	Geisz et al. (2013)
	CdTe	First Solar & GE Tech	20.4 ± 0.4	1.0055 (ap)	Newport (6/2013)		Firstsolar.com
	CuInGaSSe (CIGSS)	Showa Shell/ Tokyo University of Science	19.7 ± 0.5	0.496 (da)	AIST (11/2012)	Efficiency extracted from 30 × 30 cm^2 sized submodule.	Nakamura et al. (2013)
	Cu_2ZnSnS_4 (CZTS)	Solar Frontier & IBM labs	12.6 ± 0.5	0.42 (da)	AIST (7/2013)		Cleantechnia.com
	Perovskite thin film	EPFL	14.1 ± 0.3	0.2090 (ap)	Newport (5/2013)	Sensitization approach. Perovskite replaced dye molecules.	Burschka et al. (2013)

(Continued)

TABLE 45.2 (Continued)

Efficiencies of Solar Cells and Modules Based on Rigid and Flexible Substrates for Various Technologies

	Types of Solar Cells	Companies/ Institutions	Efficiency (%)	Area (cm²)(Designed Area Illumination [da]; Aperture Area [ap]; Total Area [t])	Test Center (and Date)	Description/Comments	References
Excitonic solar cells	Organic thin film	Mitsubishi Chemicals	11.1 ± 0.3	0.159 (ap)	AIST (10/2012)		Service (2011)
	DSSCs	Sharp	11.9 ± 0.4	1.005 (da)	AIST (9/2012)	Monolithic dye sensitized solar cells .	Komiya et al. (2011)
	DSSCs) minimodule	Sony	9.9 ± 0.4	17.11 (ap)	AIST (8/2010)	Eight parallel cells .	Morooka et al. (2009)
	DSSCs submodule	Sharp	8.8 ± 0.3	398.9 (da)	AIST (9/2012)	26 series cells connected .	Komiya et al. (2011) and Kawai (2013)
	Organic (minimodule)	Toshiba	8.5 ± 0.3	25.02 (da)	AIST (8/2013)	Four series cells connected.	Hosoya et al. (2013)
	Organic (submodule)	Toshiba	6.8 ± 0.2	395.9 (da)	AIST (10/2012)	15 series cells connected.	Hosoya et al. (2012)
Multijunction Solar cells	5J GaAs/InP bonded	Spectrolab	38.8 ± 1.9	1.021 (ap)	NREL (7/2013)	5 junctions.	Chiu et al. (2014)
	InGaP/GaAs/ InGaAs	Sharp	37.9 ± 1.2	1.047(ap)	AIST (2/2013)		Sasaki et al. (2013)
	a-Si/nc-Si/nc-Si (thin film)	LG electronics	13.4 ± 0.4	1.006 (ap)	NREL (7/2012)		Ahn et al. (2012)
	a-Si/nc-Si (thin-film cells)	Kaneka	12.3 ± 0.4	0.962 (ap)	AIST (7/2011)		Kaneka-solar.com
	a-Si/nc-Si (thin-film minimodule)	Kaneka	9.9 ± 0.4	14.23 (ap)	AIST (9/2004)		Yoshimi et al. (2003)
Concentrator cell	Si (single cell)	Amonix	27.6 ± 1.0	1.00 (da)	FhG-ISE (11/2004)	92 suns illumination using backcontact method.	Slade and Garboushian (2005)
	GaAs (single cell)	FhG-ISE	29.1 ± 1.3	0.0505 (da)	FhG-ISE (3/2010)	Concentrated at 117 suns.	FhG-ISE
	CIGS (thin film)	NREL	22.8 ± 0.9	0.100 (ap)	NREL (8/2013)	15 suns illumination.	NREL
	InGaP/GaAs/ InGaAs	Sharp	44.4 ± 2.6	0.1652 (da)	FhG-ISE (4/2013)	302 suns illumination on inverted metamorphic cells.	Sharp Press Release (2012)

at that time, the c-Si technology has tremendously brought down the module cost to as low as \$0.5/Wp (Metha 2013). An excellent cost estimation and projection of the PV technologies has been made by U.S. Department of Energy (Sunshot Vision Study February 2012), with a mission to achieve future affordable electricity price below grid parity. This in-depth analysis was conducted by the National Renewable Energy Laboratory (NREL) with contributions from the Solar Energy Industries Association, Solar Electric Power Association, and many other related organizations.

Currently, the conventional module assembly procedures used by the c-Si industries and adopted by thin-film industries restrict the overall cost of the thin-film modules. Thus, the advantage of using *less material* in thin-film technology under *back end* schemes of assembling is drastically reduced. Some frameless mounting schemes (PV laminates on membranes) have also been developed and pursued for a-Si technology, which were shown to have potential to bring significant reduction in cost (Arya 2004). Figure 45.5 shows distribution of materials costs in framed (a) and unframed (b) Si-based modules. As may be compared, the cost of packaging and framing components has been drastically lowered from 25.5% to only 4%, which is significant; consequently, the cost reduction in materials and its processing will have more meaning with further advancements in technologies of all the components.

Tuttle et al. (2000, 2005) had given cost estimates of CIGS solar modules from 2 to 2000 MW/year capacity production plants including cost of equipment and materials. Similarly, there are several reviews and reports available that summarize the cost potential of thin-film modules in terms of different component costs and address some future concerns related with the cost and availability of materials: In, Ga, Se, and Te used in the leading thin-film technologies and future electricity generation cost estimation (Gross et al. 2013). An estimate in the subcontractor's report at NREL (Keshner and Arya 2004) suggested a decade ago that the production of these elements had to be accelerated in order to fulfill the demand of 2–3 GW/year generation of electricity.

An important report (out of DoE Solar Energy Technologies Programme and PV FAQ, www.nrel.gov) also exists (U.S. Department of Energy 2005) that brings the judgment in favor of In, Ga, and Se and speculates that Ga and Se will not be constrained by supply, and a steady production growth rate of 0.16% per year for Se would be sufficient to meet the demand of annual PV production of 20 GW/year by 2050. Furthermore, it is also estimated that In will also keep a steady supply as it is a by-product of Zn, which is relatively higher in abundance and more in demand by various industries. There could be some concern about the availability of Te for multi-GW plants, but with the advancement in thin-film technology, the thickness of the material may be reduced by implementation of

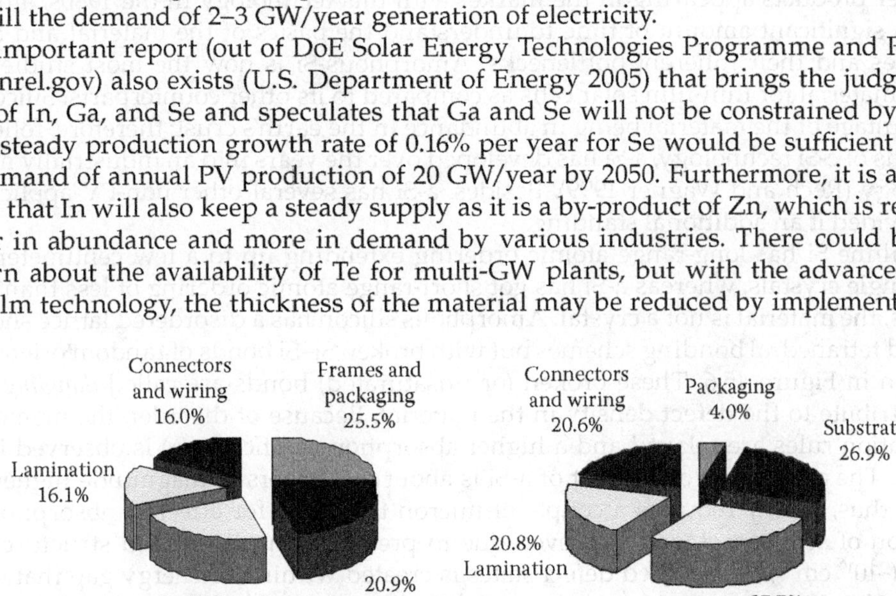

FIGURE 45.5
Distributions and breakup of the materials cost in (a) framed and (b) unframed a-Si based modules.

light-trapping schemes, thus relieving the burden on production of materials also through clever and efficient recycling methods. A good account of material scarcity issues (such as the sustainability of Te and In materials) that might arise when the production will reach several tens of GW capacity for the CdTe and CIGS technologies has been addressed by Candelise et al. 2011. The estimations made by the authors after reviewing the literature on claims of In and Te scarcity suggest that there is a high uncertainty in the estimates of the resource-constrained thin-film PV potential as well as in data and methodologies used to assess future availability of the targeted materials. The reviewed evidence does not support the contention that the availability of tellurium and indium will necessarily constrain CdTe and CIGS technologies respectively in their ability to supply expected future PV market growth.

In the following sections, the frontline thin-film technologies such as a-Si, CdTe, and CIGS, along with some of the emerging technologies such as CZTS and recently explored perovskite solar cells, are discussed in detail relating to their material and device aspects, current status, and issues related with environmental concerns.

45.2 Thin-Film Silicon Solar Cells

45.2.1 Material and Properties

The oil crisis of early 1970s gave an impetus to the PV R&D activities and several concepts for thin-film PV started to emerge. The first report on thin-film a-Si based solar cells appeared in 1976 (Carlson and Wronski 1976). It took only 5 years to see the indoor consumer products appearing in the market with this technology in the 1980s, although it took a significant amount of time to understand the basics of the material and device properties and their inherent bottlenecks. Amorphous-Si is now the most studied and applied material for thin-film solar cells as compared to its other counterparts. Silicon has the advantage of the material being in abundance in the earth's crust; therefore, following the trends of c-Si technology, a-Si has developed over the years into an industrially mature technology (Rech and Wagner 1999). Besides, a-Si has several other non-PV applications that provided it an additional standing.

Crystalline Si has long-range atomic ordering extending up to a few centimeters size in the single crystals, whereas a-Si has got short-range atomic ordering of less than 1 nm, and thus, the material is not a crystal. Amorphous silicon has a disordered lattice showing localized tetrahedral bonding schemes but with broken Si–Si bonds of random orientation, as shown in Figure 45.6. These broken (or unsaturated) bonds are called *dangling bonds* and contribute to the defect density in the material. Because of disorder, the momentum conservation rules are relaxed and a higher absorption coefficient (α) is observed in a-Si material. The absorption coefficient of a-Si is about two orders of magnitude higher than the c-Si; thus, it only requires a couple of micron thickness for effective absorption and utilization of solar spectrum. However, due to predominant disordered structure, high density (~10^{19} cm^{-3}) of localized defect states is created within the energy gap that causes the Fermi level pinning, and hence the material cannot be doped because the defect states act as trap for all free carrier generated in the material.

One effective way to overcome this problem is to passivate the unsaturated bonds of a-Si with the help of small size atoms that could get into the crystal and attach themselves

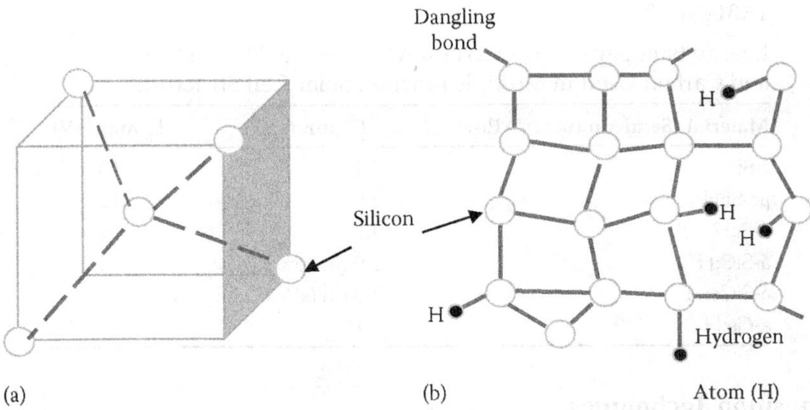

FIGURE 45.6
A schematic presentation of tetrahedral bonding scheme in crystalline Si (a) and network of a-Si:H exhibiting broken silicon atom dangling bonds, which are passivated by hydrogen atoms (b).

with the available bonds. This is precisely done by adding 5%–10% atomic hydrogen into a-Si, which attaches itself to the uncoordinated bonds due to its high activity; this reduces the dangling bonds density from ~10^{19} to ~10^{15} cm^{-3}. At this order of defect density, the required doping of material is possible, and hence, the material can be made as p- or n-type using boron and phosphorous as dopants. However, the defect density still remains so high that even with high doping, the Fermi level is not significantly shifted; it remains mostly within the donor and defect levels at the center of the gap in the case of n-doping.

Amorphous silicon may be considered as an alloy of silicon with hydrogen. The distortion of the bond length and bond angle after passivation with hydrogen modifies the defect distribution and consequently changes the optical and electronic properties. By changing the deposition conditions (Guha et al. 1981, 1986; Vetterl et al. 2000), hydrogen-diluted microcrystalline silicon (μc-Si) can be obtained, which has rather different properties. Figure 45.3 compares the absorption coefficient of the a-Si:H, c-Si, and μc-Si:H along with some of the other PV materials. The absorption bands (plateau) appearing at low energy values for a-Si:H and μc-Si are ascribed to the presence of large density of midgap defects and band tail states. The absorption coefficient (α) of c-Si (mono-crystalline silicon wafers) and microcrystalline thin-film Si has more or less same onset of transition, but μc-Si has higher α in low wavelength (λ) region. However, α for μc-Si is lower than of a-Si; therefore, thicker μc-Si layers are required compared to a-Si for the absorption of solar spectrum. A stacked combination of the two—microcrystalline and amorphous Si layers—is attractive for absorption of the most useful part of the spectrum in thin layers. This has been successfully employed first by IMT Neuchatel, Switzerland, and later by several other groups to develop a-Si/μc-Si tandem (also often called *micromorph*) solar cells.

The bandgap of a-Si can also be tailored by addition of O, C, and Ge to produce amorphous materials of wider or narrower bandgaps, for example, with the addition of C and Ge in a-Si:H, bandgaps of 2.2–1.1 eV are achievable but with inferior electronic properties. Table 45.3 provides the list of these alloys with their respective bandgaps. Suitable a-SiC:H and a-SiGe:H for solar cell devices have bandgaps ~2.0 and 1.3 eV, respectively.

TABLE 45.3

Energy Bandgaps (E_g) of Certain Alloys of a-Si:H with Germanium and Carbon Used in Multiple Junction Solar Cell Structures

Material (Semiconductor/Alloy)	E_g min (eV)	E_g max (eV)
c-Si	1.1	1.1
μc-Si:H	1.0	1.2
a-Si:H	1.7	1.8
a-SiC:H	2.0 (in 20% C)	2.2
a-SiGe:H	1.3 (in 60% Ge)	1.7
a-Ge:H	1.1	

45.2.2 Deposition Techniques

Perhaps the most important feature of a-Si material is that a wide range of temperatures, from room temperature to 400°C, can be used for its deposition. Room temperature deposition allows the use of a variety of substrates, viz., glass, metal, and plastic, and in particular the possibility to use low-cost plastic polyethylene terephthalate (PET) could be of significant advantage in reducing the cost of the modules. There are various processes used for the deposition of the a-Si:H material. Silane (SiH_4), which is the basic precursor gas, is used in nearly all processes using chemical vapor deposition (CVD) method but excluding sputtering that is not preferred for active semiconductor layers in a-Si:H. Typical deposition temperature for a-Si:H has to be below 500°C; otherwise, the incorporation of hydrogen in the film is not possible. At low substrate temperatures, the predissociation of SiH_4 does not take place easily. Hence, room temperature deposited layers give rise to inferior quality and efficiency. Therefore, plasma is used for dissociation of silane gas. Two most commonly used methods are plasma-enhanced chemical vapor deposition (PECVD) and glow discharge CVD. Typically, 13.56 MHz plasma excitation frequencies with optimal plasma excitation power at 0.1–1 mbar pressure are used. SiH_4 always diluted with hydrogen (~10%) is used in the deposition of a-Si:H, while increasing hydrogen dilution results in μc-Si:H layers, but with lower growth rates. The typical deposition rates for a-Si cells (in R&D) is ~1 Å/s and results in fairly long deposition times (50 min for 0.3 μm thick a-Si:H cell and 5 h for a 1.8 μm thick μc-Si:H cell), while 3 Å/s or higher deposition rates are generally preferred in production plants. For high deposition rates, the deposition technologies based on very high frequency (VHF) microwave, and high-pressure plasma are currently being pursued at R&D level. Rates as high as 10 Å/s have been achieved at laboratory scale.

Alternative deposition methods using hot wire CVD (HWCVD) technique, electron cyclotron resonance reactor (ECR), and also the combination of HWCVD and PECVD are also being carried out to increase the deposition rate. A detailed account of some of these techniques can be found in the following references: Carabe and Gandia (2004), Deng and Schiff (2003), Klein et al. (2004), Lechner and Schade (2002), Shah et al. (2004), Shah et al. (2013), and Sopori (2003).

45.2.3 Amorphous Silicon Solar Cells and Configurations

The conventional p–n junction configuration for a-Si:H-based solar cells suffers from inherent limitations due to the presence of large number of defect states even after H-passivation.

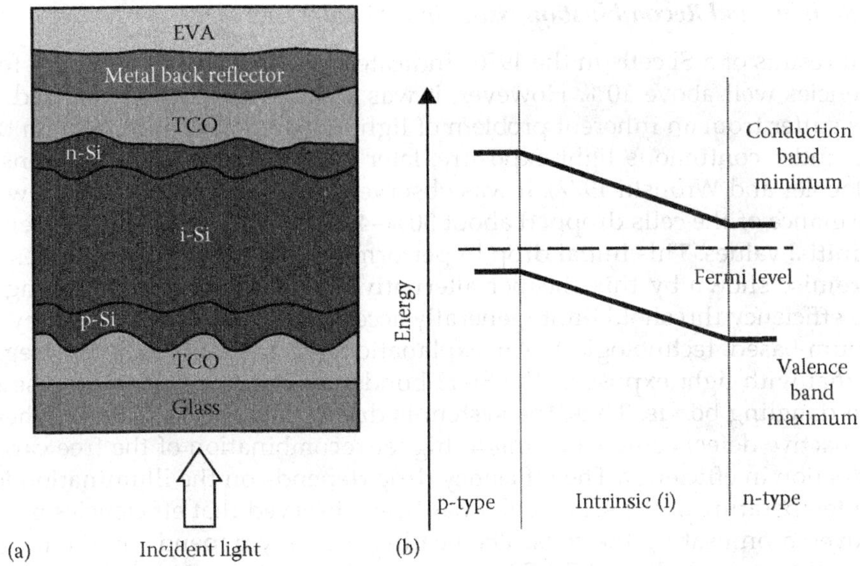

FIGURE 45.7
The schematic representation of single junction a-Si solar cell in superstructure configuration (a) and the energy band diagram of the p–i–n solar cell structure (b). Ethylene vinyl acetate (EVM) is the polymer coating used for encapsulation of solar cell.

The doping of a-Si:H further increases this concentration, which reduces the average lifetime of the free carriers as a result of very high recombination probabilities and low diffusion lengths ~0.1 μm; thus, solar cells in the p–n configuration do not work and are not considered suitable. The basic structure of a-Si solar cell configuration is a p–i–n junction shown in Figure 45.7a, which illustrates qualitatively the thickness of different layers grown in the device in the *superstrate* configuration along with applied texturing (roughness) of the transparent conducting electrodes for enhanced light trapping in a-Si layer, as will be described later.

The p–i–n type configuration for a-Si solar cell was introduced by Carlson et al. at RCA laboratories, United States (Carlson and Wronski 1976), where an intrinsic (1) layer of a-Si:H is sandwiched between the n- and p-type doped layers of a-Si:H or its alloys. Because of very short lifetime (or high recombination) of the carriers, the doped layers do not contribute to the photocurrent generation, that is, the photons absorbed in these layers contribute to optical losses, but these p- and n-layers build up the electrical field across the i-layer. This electrical field drives the electrons and holes in opposite direction that are photogenerated in i-layer, which essentially acts as the absorber layer in a-Si:H solar cells. The electrical field depends on the doping concentration of p- and n-layers as well as the thickness of the i-layer. Because the p- and n-doped layers do not contribute to the photocurrents and can cause further recombination of the generated carriers before sweeping across the layer, it is essential to minimize their thickness that is typically ~10–30 nm. There is an upper limit to the thickness of the i-layer (~0.5 μm), because charge defects reduce the effective field, and thus, if the width of the i-layer exceeds the space charge width, then the extra width would act as *dead* layer without actually contributing to photocurrent.

45.2.3.1 Stability and Recombination Issues in a-Si Solar Cell

The initial results of a-Si cells in the 1970s indicated very promising potential for attaining efficiencies well above 10%. However, it was soon observed and realized that a-Si solar cells suffer from an inherent problem of light-dependent degradation on their performance under continuous light exposure, later attributed to Staebler–Wronski effect (SWE) (Staebler and Wronski 1977). It was observed that in a timescale of a few months, the performance of the cells dropped about 30%–40%, then stabilized at efficiency lower than the initial values. This initial drop in performance is significant and takes the edge off the promise shown by this cheaper alternative to c-Si technology, keeping it much below the efficiency threshold limit (generally accepted ~12% module efficiency for thin-film vacuum-based technologies). An explanation for the light-induced degradation (SWE) is that with light exposure, the Si–H bonds break and further increase the density of the dangling bonds. Thus, the system is driven into an excited or higher energy state, with active defect centers leading to higher recombination of the free carriers and hence reduction in efficiency. The efficiency drop depends on the illumination level and operating temperature of the solar cell. It has been observed that efficiencies may be partially recovered on heating the cells. The heating recovery depends on the temperature of the cells, that is, annealing at 70°C helps in stabilizing the system better than at room temperature.

The control of defect (trap) states and dangling bonds passivation for effective doping of the a-Si:H layers are significant in the a-Si cell's overall design. The first inherent problem of the technology is that the SWE cannot be eliminated but can be reduced by engineering of device structure, for example, by employing a thinner i-layer at the expense of absorption loss. It has been verified by several groups that efficiency degradation in a-Si solar cells and modules with thinner i-layer is lower. The second problem is that the doping of the a-Si leads to an increase of the trap density, leading to a pronounced recombination effects in the device; therefore, limiting the thickness of the doped layers to 10–30 nm is needed for minimized recombination effects. The limits on i-layer and n- and p-doped layer thicknesses together have a direct bearing on the overall device structure and performance stability.

45.2.3.2 a-Si Solar Cell Configurations

An advantage of a-Si is that the solar cells can be grown in both *superstrate* and *substrate* configurations as shown in Figure 45.8. In the *superstrate* configuration, the cell is grown in the p–i–n sequence (starting with p-layer followed by i and n layers) onto a substrate that must be transparent (such as glass), and hence, this configuration is not suitable for metal or highly opaque polymeric substrates. On the other hand, the *substrate* configuration can be grown on any type of substrates that could be rigid glass or flexible metal or polymer foil. It bears n–i–p configuration (cell growth starting with n-layer followed by i and p layers), and the light enters through the last grown p-layer.

Generally, a-Si solar cells on glass are available in the *superstrate* configuration starting with the transparent conducting oxides (TCOs) window having p–i–n layers grown on it followed by another TCO layer and a metallic back reflector layer, as shown in Figure 45.8. One of the leading United States–based companies in a-Si (unfortunately out of business now due to bankruptcy in 2013), United Solar (USOC, formerly USSC), had been using *substrate* configuration for roll-to-roll production of cells on stainless steel (SS) and polymer foils. The layers can be grown in n–i–p or p–i–n sequence but irrespective

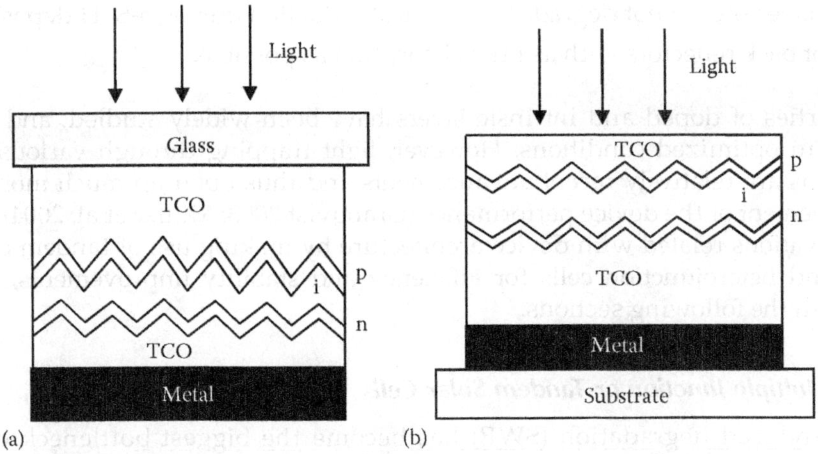

FIGURE 45.8
Schematic presentation of a-Si solar cell in *superstrate* (p–i–n) configuration (a) and *substrate* (n–i–p) configuration (b).

of the *substrate* or *superstrate* configuration, incident light is allowed through the p-side, as it has higher bandgap than i- or n-parts. Also, because the mobility of holes is smaller as compared to electrons, a thin front p-layer supports hole collection in the device (Rech and Wagner 1999).

The choice of TCO material as well as its electrical and optical properties is not only important for electrical contacts but also for efficient light trapping through the device. Light trapping is essential for efficient performance of a-Si solar cells where device thickness is limited by several inimical factors: for example, thinner i-layer is desired for minimizing light-induced performance degradation. Consequently, the thickness of intrinsic layer that acts as an absorber is generally limited to only ~300 nm, which is not sufficient for absorption of a large part of solar spectrum. In order to effectively utilize the incident photons, the applied strategies are to reduce the reflection through refractive index grading structures for the entire spectral wavelength range cell response and allow multiple scattering of light for enhanced absorption of photons in the i-layer. These are achieved by an antireflection coating used on the glass, where the light enters into the PV module, and also through suitable surface texture of TCO with the feature sizes comparable to the wavelength and application of metal reflectors. For detailed description of TCOs and light scattering, refer to the publications by Goetzberger et al. (2003), Granqvist (2003), Muller et al. (2004), and Shah et al. (2004). TCOs such as SnO_x:F, ITO, and ZnO:Al have been extensively used in a-Si solar cells. Some of the requirements for a good a-Si:H solar cells are the following:

- Glass and front TCO should have a high (>80%) transparency over the whole spectral range.
- TCO with a sheet resistance of at the most 10–15 Ω/square (high conductivity) to be obtained by enhancing carrier mobility rather than the carrier concentration to minimize free carrier absorption over the near-infrared region.
- TCO layers and doped silicon layers, which do not contribute to photogeneration and collection, should be kept as thin as possible and have very low absorption coefficients.

- TCO layer should not degrade by chemical reduction during a-Si:H deposition.
- Use of back reflectors with as little absorption as possible.

The properties of doped and intrinsic layers have been widely studied, and layers are employed in optimized conditions. However, light trapping through various structure and patterns are relatively recent advancements and thus open up much more scope in the improvement of the device performance (Granqvist 2003; Muller et al. 2004). There are other innovations related with device architecture by making use of tandem or multiple junction and heterojunction cells for efficiency and stability improvements, which are discussed in the following sections.

45.2.3.3 Multiple Junction or Tandem Solar Cells

The light-induced degradation (SWE) has become the biggest bottleneck of the a-Si technology and it has serious implications. Besides, the general effects of high density of trap and recombination centers have put restrictions on the thickness of the device layers, which consequently limits the absorption of the incoming light. The clever p–i–n configuration once thought with great promise of good efficiency at cheaper cost was also masked by this instability component. To work within the limits of intrinsic layer thickness of ~300 nm and making use of different light-trapping arrangements, the concept of tandem cells using double and triple junctions has been thoroughly pursued worldwide. Single junction a-Si solar cells are hardly used these days because of low efficiency and stability problems. Multijunction solar cells are used for better utilization of solar spectrum and to improve the stability of the solar cells. The state-of-the-art efficiency for single junction solar cells of small area is 9.3%, while for double junction it is 12.4%, and for triple junction with a-Si:H and its alloys 13.0% efficiency has been achieved (Guha 2004). Figure 45.9 schematically presents different multijunction structures. The developments of multijunction solar cells are based on following strategies:

1. The first strategy for tandem design is based on the use of only a-Si:H intrinsic layers shown in Figure 45.9a. Such double junction devices have been developed by Fuji Electric & Co., Japan, and Phototronics (part of RWE Schott), Germany, and others. A stabilized laboratory efficiency of ~8.5% and module efficiency at about 5.5% are commercially available (Diefenbach 2005; Ichikawa et al. 1993).
2. The second strategy includes the use of a-Si and Ge alloys with different bandgap (lower than 1.7 eV) combinations to form tandem junctions, where the top cell is 1.7 eV a-Si:H based and bottom cells have a-SiGe:H alloy layers of lower (1.5–1.3 eV) bandgaps, as shown in Figure 45.9b. United Solar (USSC) has developed 13% stabilized efficiency triple junction solar cell (small area) in *substrate* configuration and is selling their triple junction modules on SS at stabilized efficiency of about 6.5%.
3. The third strategy introduced in 1994 by IMT Neuchatel, Switzerland, is based on a novel concept of combining μc-Si:H (with a bandgap of 1.1 eV)- and a-Si:H (with a bandgap of 1.7 eV)-based solar cells. This has a promising potential because of significantly reduced light-dependent degradation effect in tandem solar cell (Meier et al. 1994; Shah et al. 2004). The only degradation observed comes from

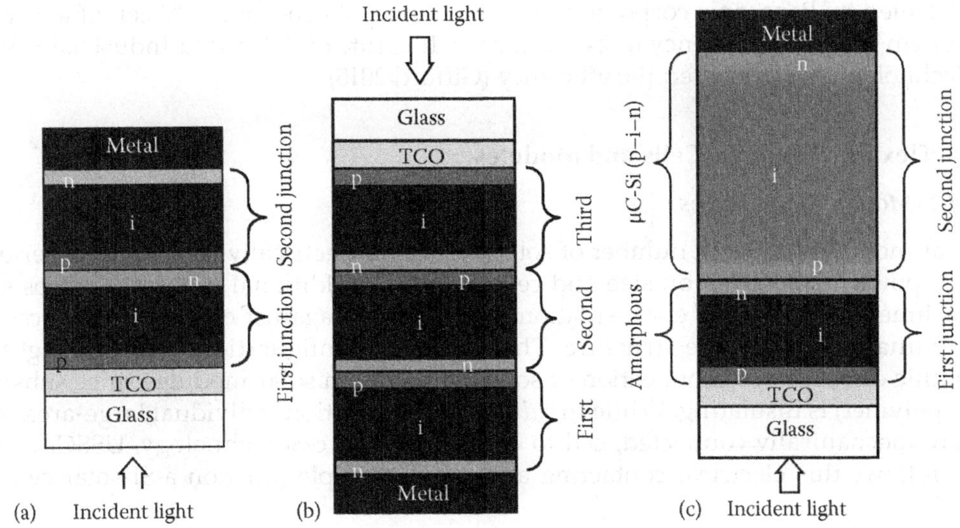

FIGURE 45.9
The schematics of multijunction cell architecture showing double junction *superstrate* configuration (a), triple junction in *substrate* configuration (b), and *micromorph* junction in *superstrate* configuration (c), that is, glass/TCO/p–i–n (a-Si:H)/p–i–n (μc-Si:H).

a-Si part, which is optimized at 0.2–0.3 μm, while the μc-Si:H layer is kept around 1–2 μm. Figure 45.9c shows the schematic of the design in p–i–n/p–i–n configuration on a rigid glass substrate. Kaneka Corporation, Japan, had achieved large-area modules (910 × 455 mm^2) of initial efficiency ~13.2% and with stabilized efficiency approaching toward 10%. Using the concept of intermediate TCO reflector layer for novel light trapping, Yamamoto et al. (2004) have shown an initial efficiency of 14.7% for a test cell. The reason for a good efficiency of the cells lies in the spectral response of the combination of 1.7 eV a-Si:H–based cells and 1.1 eV μc-Si:H–based second part. The superposition of the two results in a quantum efficiency spreading around 80% between 500 and 800 nm, covering a large part of the solar spectrum.

45.2.3.4 Hybrid Solar Cells

Amorphous silicon cells have been combined with nanocrystalline silicon junction and the cells of other materials, for example, junction with CIGS (Mitchell et al. 1988a; Yamamoto et al. 1998). Another significant development in the design is the formation of a thick/thin type of interface structure (heterostructure) between a-Si:H layer and c-Si wafer referred to as heterojunction with intrinsic thin-film layer (HIT) cells that were developed by Sanyo, Japan. Efficiency close to 21% over a cell area 101 cm^2 was reported (Sakata et al. 2000) initially with a good promise. This technology uses n-type of Cz-silicon wafer as the base (light absorber) and low-temperature processes with the device structure being a-Si(p$^+$)/a-Si(i)/c-Si(n)/a-Si(i)/a-Si(n$^+$). The intrinsic a-Si layer is important as it contacts c-Si at both ends, and provides passivation as well as extra stability to the system. As per reports of Sanyo, the cells had exhibited excellent stability. Subsequently, further improvement in efficiencies was reported by Sanyo crossing 23% mark and recently record efficiency has been claimed by Panasonic corporation for crystalline silicon–based PV cell of a *practical size* (101 cm^2) with an efficiency of 24.7%. Japan's Institute of Advanced Industrial Science and Technology has evaluated the efficiency (Gifford 2013).

45.2.4 Flexible a-Si Solar Cells and Modules

45.2.4.1 Monolithic Modules

All solar modules require a number of solar cells to be electrically connected in series to provide power, depending on size and cell efficiency. Additional processing steps such as attachment of leads and encapsulation for protection against external influences are done to finalize the module structure. The *superstrate* configuration has advantages for monolithic electrical interconnection of solar cells to form solar modules since substrate (glass, polymer) is insulating. While in *substrate* configuration individual large-area solar cells are mechanically connected, cell to cell, as done in c-Si technology, USSC, United States, follows this electrical contacting approach for triple junction a-Si solar cells on steel foils.

Figure 45.10 illustrates the monolithic interconnection scheme to develop solar modules in *superstrate* configuration. For interconnection of solar cells, layers are laser-scribed in three stages: first separating front TCO contact, then scribing a-Si layers to connect individual cells, and finally isolating the conducting backelectrode to obtain series interconnection. This approach removes the need for separating and connecting of cells, which is time consuming, costly, and complex in conventional c-Si technology.

45.2.4.2 Flexible a-Si Solar Cells

Another important perspective of thin-film PV technology is the flexible modules that are lucrative owing to their applications related with strategic space and military use, integration in roofs and facades of buildings, portable power source, automobiles, consumer electronics, value-added products, etc. Therefore, the technology has attracted

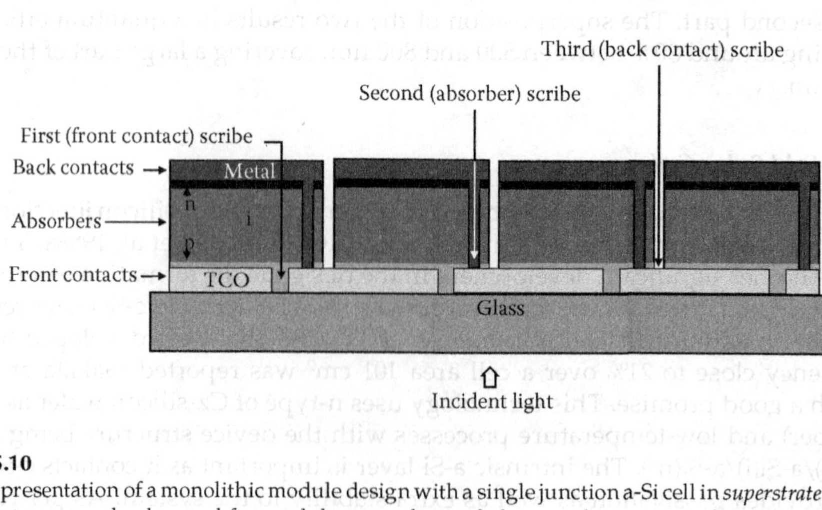

FIGURE 45.10
Schematic presentation of a monolithic module design with a single junction a-Si cell in *superstrate* configuration. Similar strategy can also be used for modules on polymer foils or multijunction solar cells, which may have different sequence of layers.

TABLE 45.4

Summary of the Companies Active in Flexible a-Si Thin-Film Si PV. Generally, the Module Efficiencies Are in 3%–8% Range (as in Year 2007)

Companies	Technology	Production	Remarks
United solar[a]	Triple junction SiGe alloys On SS RF-PECVD	30 MW line, RF-PECVD, modules, and consumer products	Modules between 6.5% and 6.8%. Cutting-assembling process modules with 8% for military and space application
VHF-Technology	a-Si on PEN, polyimide VHF-PECVD	Pilot production and consumer products	Product efficiency ~4%
Iowa thin-film technology (ITFT)	a-Si on polyamide, RF-PECVD	Pilot production and consumer products	Product efficiency ~4%
Fuji Electric	a-Si/a-Si on polyamide RF-PECVD	Pilot production and consumer products	Feed-through contacts through the substrates, 7% module, 8% active area stable efficiency
Sanyo	a-Si on plastic RF-PECVD	Consumer products	
Canon	μc-Si:H/a-Si:H on SS (VHF-PECVD)	Development	
Akzo Nobel	Amorphous p–i–n on Al RF-PECVD	Development, pilot line	Al sacrificial substrate dissolved after cell deposition

[a] The company United Solar; the largest a-Si manufacturer went bankrupt in 2013.

recently major players of thin-film PV to venture in this area. Because they can be available in different shades (even semitransparent), shapes, and sizes, these flexible a-Si solar cells once were potential candidates for low-to-medium range of power applications.

Some of the prominent companies that were involved in the production (as on year 2007) and development of flexible a-Si based modules are listed in Table 45.4 (some significant record efficiencies may also be found in Table 45.2). Out of these, United States–based United Solar, Iowa thin films, and Japanese companies like Sanyo and Fuji had entered into relatively large-scale production, while European companies such as Flexcell (Switzerland) and Akzo Nobel (the Netherlands) were also involved in production plants for consumer electronics–oriented market. a-Si offers cheaper processing and materials cost; however, for it to remain as frontline thin-film technology would mainly depend on its efficiency for larger module with high throughput, which is the biggest bottleneck for the technology. Module efficiency above ~12% and a long-term (more than 30 years) stable performance of the large-area modules would make it compete effectively with c-Si and remain the leader among other thin-film counterparts if low-cost modules could be developed with production processes that give high throughput and yield. Oerlikon solar had created a world record of stabilized efficiency of over 10% modules for single junction a-Si solar cells (Benagli et al. 2009). Despite these successes, the lack of dynamism in this technology area had been due to stagnant efficiency of a-Si by most of the manufacturers, while other thin-film technologies are growing faster.

The future of larger-scale amorphous silicon solar panels does not look very bright: Sharp retired 160 out of their 320 MW production capacity in Japan earlier in 2014. Optisolar, Signet Solar, Unisolar, and many other companies that were touting the amorphous technology are acquired, bankrupt, or closed (energyinformative.org). The a-Si technology could return back to gain the thin film market leadership only when the SWE effect of the degradation of cells could be tackled (Fehr et al. 2014; Yu et al. 2014).

Efforts are ongoing in countries around the world to overcome this barrier, especially in Japan where this technology is widely accepted. A good account on the future of a-Si technology was presented at Asian PVSEC meeting in Taiwan (Konagai 2013). Under Japan's NEDO mission PV2030+, the target module conversion efficiencies of a-Si thin-film solar cells are 14% for 2017 and 18% for 2025. To attain these goals, conventional a-Si/μc-Si double-junction tandem solar cells must be superseded by triple-junction solar cells. The development targets of triple-junction solar cells include those for a-Si/μc-Si/μc-Si and a-Si/a-SiGe/μc-SiGe structures as well as for the a-Si/a-SiGe/μc-Si structure. Under this strategy, two collaborative university–industry research consortia that focus on innovative Si thin-film solar cells have been initiated. One is the *thin-film full-spectrum solar cells* project of METI/NEDO and the other is the *Si nanowire solar cells* project of MEXT/JST. In each project, which involves a study period of 5–7 years, domestic research organizations as well as foreign countries have been included, with a budget that exceeds four billion yen over the entire period.

With a success in the efficiency figures in a-Si technology, Hanergy Solar in China (who are currently producing thin film PV up to 3 GW) along with a few other Chinese companies (who have ambitious plan to use this technology for BIPV and BAPV [Building Attached PV]) have also set their eyes on this technology for enhanced production in coming years. A European roadmap for a-Si technology was built around these factors some time ago with a short-term goal of 15% stabilized cell efficiency from 2008 to 2013, through fundamental understanding of electronic properties of the layers and the interfaces followed by an increase up to 17% by 2020 (Sinke 2007).

45.3 Cadmium Telluride Solar Cells

45.3.1 Material and Properties

Cadmium telluride (CdTe) is a direct band-gap material and comes under the family of II–VI materials referred to as metal chalcogenides. Its band-gap energy of ~1.45 eV is quite favorable for the requirement of highest conversion of solar spectrum into electricity with a single junction solar cell. Besides, very high optical absorption (10^5 cm^{-1}, see Figure 45.3) and p-type conductivity make it an ideal material for PV application. It has a *zincblende* crystal structure and a simple phase diagram as the constituents have high vapor pressure. The layers of CdTe can be deposited using a number of processes and the compound can easily be grown in stoichiometric form at temperatures over 400°C. Like other II–VI materials, the electronic doping is controlled by substitution of atoms at vacancy sites. While n-type doping control is relatively easier, it is difficult to vary the doping concentration in p-type CdTe because of compensation effects. The most common CdTe thin-film solar cell structure comprises of a p-type CdTe absorber layer and n-type CdS-based window layer forming a heterojunction, which has an intermixed interface region.

First development of CdTe heterojunction solar cell started around 1962 using n-CdTe in combination with Cu_xTe (Bonnet 2004; Cusano 1963; Lebrun 1966). Though 5%–6% efficiencies were achieved, the cells suffered from degradation due to Cu diffusion from Cu_xTe (similar to CdS/Cu_xS solar cells). A breakthrough in 1972 was achieved when a new heterojunction structure between CdTe and CdS was developed by Bonnet and Rabenhorst (1972) and 6% efficiency was reported. Current solar cell structures are based on the device structure shown in Figure 45.11. By 2001, significant progress had raised the efficiency to

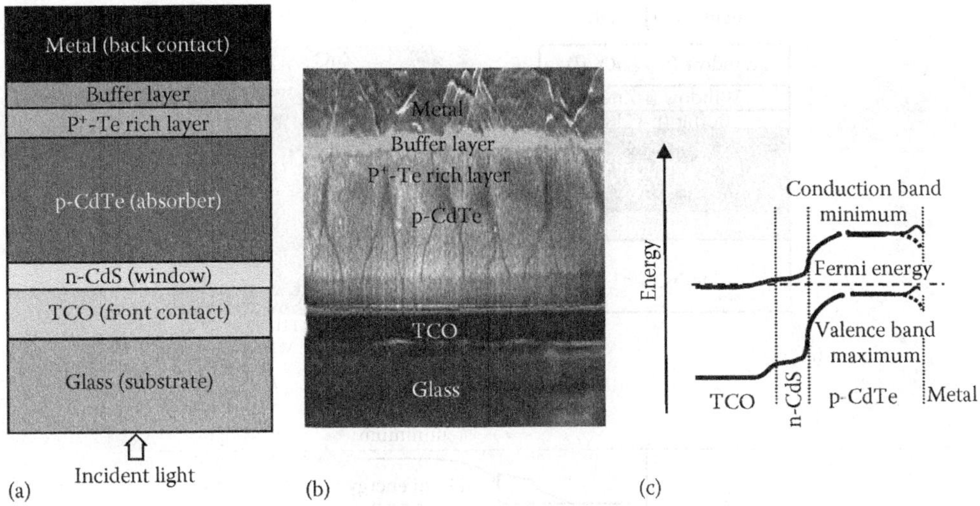

FIGURE 45.11
Schematic presentation of CdTe/CdS solar cell in *superstrate* configuration showing different layers and their nomenclature (a), a corresponding SEM picture illustrating different layers (b), and the energy band diagram of the solar cell (c).

16.5% (Wu et al. 2002). Historical developments of CdTe PV technology have been reviewed by McCandless and Sites (2003).

The most attractive features of CdTe compound are its chemical simplicity and the robust stability. Because of its highly ionic nature, the surfaces and grain boundaries (GBs) tend to passivate and are not detrimental (that is a reason that polycrystalline CdTe solar cells are more efficient than single crystal counterpart). CdTe material is inherently stable under solar spectrum since its bonding strength of ~5.7 eV is much higher than the incident photon energy of the solar spectrum. This accounts for the excellent material stability and device performance. CdTe is not only stable for terrestrial applications, but it has also been demonstrated that CdTe has excellent stability under high-energy photon and electron irradiation for space applications, superior to Si, GaAs, CIGS, etc. (Bätzner et al. 2004).

Table 45.2 lists the efficiency of the CdTe solar cells reported by laboratories and prominent manufacturers. The theoretical efficiency of CdTe solar cell is estimated by detailed balance calculations to be ~32% (J_{sc} ~ 31 mA/cm^2 and V_{oc} ~ 1.1 V), which is higher than c-Si; therefore, there is plenty of scope for improvement in the conversion efficiency in the future. Important issues in R&D are shallow acceptor doping of the CdTe absorber, formation of good ohmic contact on p-CdTe, and reduction of optical absorption losses from the CdS and TCO front contact layers. These are some inherent critical issues requiring further investigation to achieve cell efficiency well above 20% with single junction device. Progress of CdTe technology has also been hampered because of perception involving environmental concerns due to cadmium and shall be discussed later in Section 45.7.

45.3.2 CdTe Solar Cell Device Structure

The CdTe solar cells can be grown in both *substrate* and *superstrate* configurations. Until today the highest efficiencies have been achieved in *superstrate* configuration. Figure 45.11a gives the schematics of CdTe solar cell grown on TCO-coated glass substrate in a superstrate configuration. The glass substrate can be a low-cost soda lime glass for processing

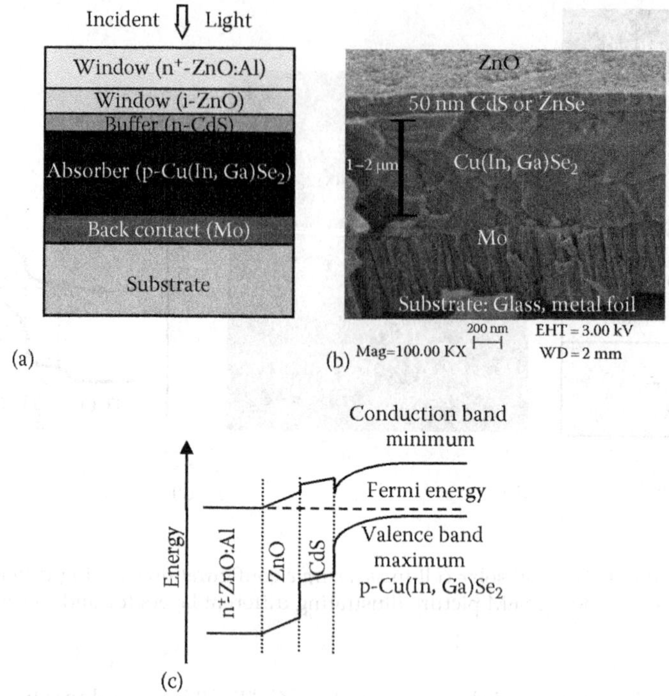

FIGURE 45.12
Schematic presentation of CIGS solar cell in *substrate* configuration (a), SEM cross-section image of CIGS device showing microstructure of layers (b), and the qualitative energy band diagram of CIGS solar cell (c).

temperature below 600°C or alkali-free glass (generally borosilicate) for high-temperature processing above 600°C.

Substrate configuration would allow CdTe to be deposited on opaque substrates such as metal or polyimide foils. Stability of the backcontact especially was believed to be the main limiting factor in this configuration, as CdTe/CdS layers need to be grown at so high temperatures where interdiffusion degrades the CdTe backcontact interface and cells are shunted. Only recently, the substrate configuration was reconsidered and the development of a novel doping method allowed solar cell efficiencies close to 14% (Kranz et al. 2013a) (Figure 45.12).

In the *superstrate* configuration layers of TCO, CdS, CdTe and the metal backcontact are sequentially grown on glass substrates. There are some intermediate processing steps that will be described later. An antireflecting coating on the glass rear surface is often applied to reduce the reflection at the air–glass interface. The incident light passes through the glass, TCO, and CdS and gets absorbed in CdTe layer where it creates electron–hole pairs that contribute to the PV power.

45.3.2.1 TCO Front Electrical Contact

A highly transparent and n-type conducting TCO layer with an electron affinity below 4.5 eV is required to form an ohmic contact and to have good band alignment with n-CdS. Most of the TCOs such as SnO_x:F, ITO, ZnO:Al, and Cd_2SnO_4 (cadmium stannate) have been used in combination with a highly resistive and transparent (HRT) layer such as resistive ZnO, SnO_x, or Zn_2SnO_4 to grow high-efficiency solar cells (Wu et al. 2002).

The bilayer (combination of high and low resistive TCO stack) is often used to protect against shunts caused by pinholes in thin CdS layers. Most commonly used TCO in industrial production is SnO_x:F and often with a thin ITO layer. ITO front contacts are often sensitive to annealing treatment. An increase in electron affinity from around 4 to 5 eV can be caused by oxidation or PDT.

45.3.2.2 n-Type Window Layer

CdS is commonly used as n-type window layer material. The energy bandgap of CdS is 2.4 eV with intrinsic n-type electrical conductivity and it forms a heterojunction with CdTe layer. A typical thickness of as-deposited CdS layer used in solar cells is in the range of 10–500 nm. During high-temperature steps of cell processing, this thickness can be effectively reduced because of interdiffusion with CdTe. A thin CdS layer (10–50 nm) is desired to minimize the photon absorption losses so that maximum number of photons can reach the CdTe layer. However, there has to be a compromise because very thin CdS may lead to a lower open circuit voltage and fill factor through shunting in the device. CdS layers can be grown by different methods such as chemical bath deposition (CBD), evaporation, sublimation, vapor transport (VT), metal organic chemical vapor deposition (MOCVD), and sputtering (Ferekides et al. 1993; McCandless and Sites 2003; Romeo et al. 1999, 2004a).

The relatively low energy bandgap of CdS leads to significant parasitic absorption losses of high-energy photons. Alternative materials with increased energy band-gap values such as oxygenated CdS or Zn(O,S) are under investigation in order to increase the current density toward the theoretical maximum.

45.3.2.3 CdTe Absorber Layer

The CdTe thin film is the most important component as it absorbs the incident solar light and contributes to the photogenerated current. Because of its direct band-gap properties, only about 2 μm thick material is sufficient to absorb the most useful part of the solar spectrum. CdTe layers may be grown by a variety of vacuum and nonvacuum methods classified into high-temperature and low-temperature processes. Some of the commonly used high-temperature methods are closed space sublimation (CSS), VT, or vapor transport deposition (VTD) with deposition temperature above 500°C, while methods such as electrodeposition (ED), screen printing (SP), chemical spraying (CS), high vacuum evaporation (HVE), and sputtering with deposition temperature below 450°C are classified under low-temperature processes (McCandless and Sites 2003; Romeo et al. 2004a). Depending on deposition methods, the typical thickness of CdTe layer in solar cells is in the range of 2–6 μm.

45.3.2.4 Junction Activation Treatment

The as-deposited CdTe/CdS solar cells always exhibit poor PV properties and thus require a special annealing treatment that improves the cell efficiency considerably (by a factor of 3–5). This is done by subjecting the CdTe/CdS stacks to a heat treatment under Cl–O ambient between 350 and 600°C. This is known as *$CdCl_2$ treatment* or *junction activation* treatment. After this annealing treatment, a significant enlargement of grain size, by a factor of 5–20, is observed in CdTe grown by low-temperature deposition methods, as can be seen in the Figure 45.11b. The grain size of HVE CdS is in the range of 0.1–0.3 μm and the layers are rough. If the CdS is grown by a CBD, then it consists of small grains of about 0.1 μm widths. A treatment with $CdCl_2$ recrystallizes the CdS layers so that some of the small grains coalesce together and form bigger grains of 0.5 μm width (Romeo et al. 2004a). In case of high-temperature grown CdTe, this annealing treatment recrystallization is observed, but the grain size near the top surface does not increase since the grains are already few microns large even in as-deposited condition.

For high-temperature growth processes, there is a tendency of CdS_xTe_{1-x} formation by the conversion of small CdS grains into CdTe due to interdiffusion at the interface and little or low grain growth is noticed after the activation treatment. A stable CdS/CdTe interface can be obtained for 6% diffusion of sulfur atoms. However, under nonequilibrium conditions, the diffusion of S decreases the thickness of the CdS films causing pinholes and eventually leads to shorting paths across the junction. This is a critical problem in CdTe solar cells, restricting the application of thinner CdS as desired to minimize the optical absorption losses. Nevertheless, a thermal treatment of CdS layer prior to CdTe deposition is frequently applied to restrict the interdiffusion of S. The formation of CdS_xTe_{1-x} after activation actually helps in reducing the lattice mismatch between CdS and CdTe but only marginally as compared to improvement in electrical changes induced by Cl, O, and S. Apart from reduction in density of stacking faults and misfit dislocation, there is an overall increase in the shallow-acceptor concentration in CdTe, leading to enhanced p-doping in CdTe after annealing. In particular, the GB regions become more p-doped, owing to preferred GB diffusion and segregation of Cl and O. As a result, increased charge carrier collection efficiency is measured and efficiency increases by a factor of 3–5.

45.3.2.5 Problems of Electrical Backcontact and Stability

An important issue in CdTe solar cell technology is the formation of efficient and stable ohmic contact on p-CdTe layer. For an ohmic contact to form on a p-type semiconductor, the work function of the metal should be higher than the sum of the bandgap and the electron affinity of the semiconductor; otherwise, Schottky contact is formed. For p-CdTe layer, a metal with a work function higher than 5.7 eV is needed since CdTe has a bandgap of 1.45 eV and electron affinity is 4.3 eV. Metals with such high work functions are not available. To overcome this problem, a heavily doped p-CdTe surface is created with the help of chemical etching and a buffer layer of high carrier concentration is often applied. Subsequent postdeposition annealing diffuses some buffer material into CdTe where it changes the band edges as a result of change in the interface state density. Thereby a lowering in interface barrier height and width results, which enables a quasi-ohmic or tunneling contact between the metal and CdTe, as shown in Figure 45.11c. Commonly used buffer layer/metallization combinations are Cu/Au, Cu/graphite, or graphite pastes doped with Hg and Cu. But backcontacts containing Cu in any form are often not stable with time as Cu migration from the backcontact leads to efficiency degradation with time. However, alternate processes were being developed, among them Sb_2Te_3/Mo and Sb/Mo contacts that have provided high-efficiency and long-term stable solar cells (Abken and Bartelt 2002; Bätzner et al. 2001, 2004; Romeo et al. 2004b).

45.3.3 Deposition Techniques

HVE is a simple deposition method where CdTe and CdS are congruently evaporated from crucible/boats on substrates at 150°C for CdS and ~300°C for CdTe. In this process, the layers are grown in a high vacuum (~10^{-5} Torr), but the distance between the source material and substrate is kept in the range of 10–30 cm. This allows the use of substrates at relatively lower temperature. Typical deposition rates vary between 2 and 10 μm/h.

Solar cells of ~12%–16% efficiency have been achieved by Stanford University, United States; ETH Zurich, Switzerland; and IEC, Delaware, United States (Fahrenbruch et al. 1992; McCandless et al. 1999; Perrenoud et al. 2011a).

Sputtering is another process that involves a vacuum system with ionized gases forming plasma discharge, where a CdTe target (a few mm thick plate) attached to one electrode is used as the source material. The energetic ionized atoms from plasma strike the target and remove the material atoms, and consequently, a CdTe layer gets deposited on the substrate, which is placed on the counterelectrode facing the target. Sputtering methods are suitable for large-area deposition in an industrial environment, but because of difficulty in maintaining stoichiometry, they are considered not very suitable for compound semiconductors. However, 14% efficiency has been achieved with this process at University of Toledo, United States (Compaan et al. 2004).

CSS and VT are the prominent and industrially used processes for CdTe deposition owing to its very high rate (2–5 μm/min) of deposition. The CSS process consists of an arrangement involving the placement of a graphite crucible with the source material (CdTe compound) in a high vacuum chamber (~10^{-5} Torr). The CdTe compound sublimes at around 600°C and is deposited onto the substrate, which is kept with a separation of 1–5 mm above the crucible and heated typically above 550°C. Antec Solar GmbH in Germany uses this method for industrial production of 60 × 120 cm^2 modules on a 10 MW capacity plant. Parma University, Italy; University of South Florida, United States; and NREL, United States, have also used this method and cells of 15.5%– 16.5% efficiency have been achieved. First Solar, United States, uses a variant of CSVT where instead of compound, elemental vapors are used (First Solar 2005; McCandless and Dobson 2004; Romeo et al. 2004c). First Solar is the most successful CdTe company to date with an annual production capacity of approximately 2 GWp for modules on 60 × 120 cm^2 glass substrates. First Solar and GE global research have contributed to improve the CdTe-based solar cell efficiency significantly. GE global research has achieved a ground-breaking efficiency of 19.6% on glass substrates. Early 2014, First Solar communicated 20.4% solar cell efficiency and a very remarkable full scale module efficiency of 17% (First Solar 2014).

Details of other alternative methods such as SP, spray pyrolysis, MOCVD, CVD, atomic layer deposition (ALD), and ED may be found in articles (Bonnet and Rabenhorst 1972; Romeo et al. 2004c). For flexible substrates such as polymers low-temperature methods like sputtering, HVE and ED are suitable.

45.3.4 Flexible CdTe Solar Cells

Even though the technology of CdTe solar cells on glass substrates has matured and efficiencies exceeding 17% on module level have been achieved, not much effort has gone in developing these devices on flexible substrates. The R&D on flexible CdTe has been supported by NASA and defense agencies in the United States. Because of high radiation tolerance, superior to conventional Si and GaAS solar cells, against high-energy electron and proton irradiation, these solar cells are also attractive for space applications in addition to terrestrial applications.

Though the CdTe solar cells can be grown in *substrate* configuration, one of the hurdles in the development of high-efficiency CdTe solar cells on metallic substrates is that most of the metal foils do not form efficient ohmic contact with CdTe and it is difficult to incorporate an additional buffer layer as ohmic contact to increase the cell efficiency. The criteria of matching thermal expansion coefficients and work function limit the choice of available substrate materials. Another reason is that during $CdCl_2$ annealing treatment, diffusion of impurities changes the ohmic contact properties. Recently, by optimization of the interfaces and doping methods, flexible CdTe solar cells on metal foil have been realized with an efficiency of 11.5%. On glass substrate, an efficiency of 13.6% has been obtained with similar substrate configuration, which with little bit of more optimization over 16% can provide production avenues of CdTe on a range of both rigid and flexible substrates (Kranz et al. 2013b). Further, latest status and details on the CdTe solar cells development have been reported by Professor Ayodhya Tiwari's group at EMPA, Switzerland (Kranz et al. 2013c).

The choice of an appropriate substrate is a crucial factor for flexible solar cells in the *superstrate* configuration because the substrate should be optically transparent and should withstand the high processing temperature. Most of the CdTe/CdS cell fabrication techniques require temperatures in the range 450–500°C. Therefore, low-temperature (less than 450°C) deposition processes are required. However, NREL (United States) and EMPA (Switzerland) have reported 14.0% and 13.6% efficiency flexible cells on flexible glass and polymer foil, respectively (Kranz et al. 2013d). On polyimide foil first monolithically interconnected minimodules were demonstrated, yielding efficiencies of 8% (Perrenoud et al. 2011b). These devices were grown with HVE method, which is suitable for roll-to-roll deposition, but the concept for industrial production is yet to be demonstrated (Mathew et al 2004; Romeo et al. 2004c).

45.4 Cu(In,Ga)Se_2 Solar Cells

45.4.1 Material and Properties

Compound semiconductors from the I–III–VI_2 series of periodic table, such as CIS, copper gallium diselenide (CGS), and their mixed alloys CIGS, are often simply referred to as *chalcopyrites* because of their tetragonal crystal structure. These materials are easily prepared in a wide range of compositions, and their corresponding phase diagrams have been intensively investigated. Changing the stoichiometry and extrinsic doping can vary their electrical conductivity. However, for the preparation of solar cells, only slightly Cu-deficient compositions of p-type conductivity are suitable. Depending on the [Ga]/[In + Ga] ratio, the bandgap of CIGS can be varied continuously between 1.04 and 1.68 eV. The current high-efficiency devices are prepared with bandgap in the range 1.15–1.25 eV; this corresponds to a [Ga]/[In + Ga] ratio between 20% and 30%. Layers with higher Ga content, as needed to increase the bandgap toward ~1.5 eV, are of inferior electronic quality and yield lower efficiency cells.

Other chalcopyrites such as $CuInS_2$ and $CuInTe_2$ were also investigated earlier but cell efficiencies were rather low and the R&D focus was made on CIS. Interest in $CuInS_2$ resurfaced with the development of ~11.4% efficient cells at HMI Berlin. A spin-off company, Sulfurcell, based in Berlin, started setting up a pilot production line in 2003. The device structure of the $CuInS_2$ solar cell is quite similar to CIGS solar cells in terms of other constituent layers.

The first CIS solar cell was developed with single-crystal material and ~12% efficiency was reported in 1974 (Wagner et al. 1974) and the first thin-film solar cell was reported by Kazmerski in 1976 by developing ~4% cells obtained with the evaporation of $CuInSe_2$ material (Kazmerski et al. 1976). The real breakthrough in CIS thin-film technology came with the pioneering work of Boeing Corp., United States, where they used three-source evaporation of Cu, In, and Se elements and raised the efficiency from 5.7% in 1980 to above 10% in 1982 (Mickelsen and Chen 1981, 1982). Despite the apparent complexity of material system, this efficiency jump clearly showed promising potential of the material. Later in 1987, Arco Solar, United States, raised the cell efficiency to ~14% by using a different CIS deposition process where stacked metal layer was selenized under H_2Se ambient (Mitchell et al. 1988b). Subsequent improvements in efficiency were attained by the EUROCIS consortium in Europe and later at NREL, United States, who reported efficiency of ~19.5% (Ramanathan et al. 2003). Efficiency improvements over the Boeing process occurred due to addition of Ga and S for band-gap engineering, addition of Na in absorber layer, optimization of the n-CdS (hetero-junction part of the cell), and transparent front electrical contact layers.

The first industrial production of CIS modules was started by Siemens (later Shell solar) based on the Arco Solar technology, whereas other companies such as Wurth Solar and Global Solar started development of CIS solar modules using coevaporation methods. Several other companies have been investigating various other methods of deposition such as paste printing, ED, etc. but up to now, these technologies have been less successful as compared to vacuum-based technologies.

The phase diagram of the ternary compound is described by pseudobinary phase diagram of the binary analogue, for example, Cu_2Se and In_2Se_3 phase for $CuInSe_2$ ternary. Single-phase chalcopyrite $CuInSe_2$ exists at small copper deficiency, whereas for Cu-rich compositions, a mixed phase of Cu_xSe with $CuInSe_2$ forms that is not suitable for PV devices. For In-rich compositions, defect-chalcopyrite phase ($CuIn_3Se_5$) forms, that is generally n-type. Despite an apparent complicated crystal structure and multicomponent system, the material properties of the PV relevant compounds are fault tolerant and not particularly affected by minor deviations from stoichiometry in Cu-deficient range. Further, surfaces and GBs in CIGS compounds are easy to passivate, resulting in high-efficiency cells even with submicron grain-size materials.

The PV grade Cu-deficient CIGS material has a tetragonal crystal structure having vacancies and interstitials, which act favorably, especially because the material is self-healing, owing to the defect relaxation caused by highly mobile Cu-ions and its interaction with vacancies. Defects created in CIGS by external influence (e.g., radiation) are immediately healed up. This is an inherent advantage of the CIGS material, leading to highly stable CIGS solar cells. However, care must be taken for proper encapsulation of devices against highly damped conditions; otherwise, the degradation of electrical contacts (TCO or Mo) in moisture can lead to minor degradations in performance. Therefore, stability of encapsulated CIGS solar cells is not a problem as proven by field tests conducted by ZSW, Shell, and NREL. CIGS is also tolerant against space radiation being superior to Si, and GaAs single crystal cells, but somewhat inferior to CdTe.

45.4.2 CIGS Solar-Cell Configuration

The CIGS solar cells can be grown in both *substrate* and *superstrate* configurations, but the *substrate* configuration gives the highest efficiency owing to favorable process conditions (Figure 45.12). *Superstrate* structures were investigated in the early 1980s, but efficiency was below 5%. However, recent efforts have improved the efficiency to ~13%. This has been possible with the introduction of undoped ZnO instead of CdS buffer layer and coevaporation of Na_xSe during the CIGS deposition. Further, bifacial CIGS solar cells with both front and rear transparent conducting contacts were also investigated (Nakada and Mizutani, 2002; Nakada et al. 2004).

45.4.2.1 Electrical Backcontact

CIGS solar cells in substrate configuration can be grown on glass as well as metal and polymer foils. Molybdenum (Mo), grown by sputtering or e-beam evaporation, is the most commonly used electrical backcontact material for CIGS solar cells. The growth of the solar cell starts with the deposition of Mo on the substrate, which forms an electrically conducting backelectrode with CIGS. When CIGS is grown on Mo, an interface layer of $MoSe_2$ is automatically formed, which helps in ohmic transport between CIGS and Mo. Recently, alternative backcontact materials have been explored, but industrial production is still based on Mo layers.

45.4.2.2 CIGS Absorber Layer

Because of a high absorption coefficient, a 2 μm thick layer is sufficient for absorption of maximum incident radiation. CIGS layers can be grown with a variety of deposition methods (as described later). Although the grain size and morphology (surface roughness) depends on deposition methods, high efficiencies >13% have been achieved with most of the methods, which indicate that GBs in CIGS are benign and can be easily passivated. High-efficiency cells have p-type $Cu(In,Ga)Se_2$ bulk, while a defect-chalcopyrite $Cu(In,Ga)_3Se_5$ phase in the form of thin layer segregates at the top surface that is n-type especially when doped by cation atoms diffusing from the buffer layer. It is believed that this inverted surface, leading to p–n homojunction in CIGS absorber, is crucial for high-efficiency cells.

45.4.2.3 Buffer Layer

Several semiconductor compounds with n-type conductivity and bandgaps between 2.0 and 3.4 eV have been applied as a buffer to form a heterojunction in CIGS solar cells. However, CdS remains the most widely investigated buffer layer, as it has continuously yielded high-efficiency cells on different type of absorber layers. CdS for highest-efficiency CIGS cells is commonly grown by CBD, which is a low-cost, large-area process. However, its incompatibility with in-line vacuum-based production methods is a matter of concern. Although CBD-grown CdS serves as a reference for highest efficiency, physical vapor deposition (PVD)-grown CdS layers yield lower efficiency cells. Thin layers grown by PVD often do not show uniform coverage of CIGS and are less effective in chemically engineering the interface properties between the buffer and the absorber. The trend in buffer layers has been to substitute CdS with *Cd-free* wide-bandgap semiconductors and to replace the CBD technique with in-line-compatible processes. Alternative materials such as In_2S_3, ZnO, ZnS, ZnSe, etc. using different methods viz. PVD, RF sputtering, metal organic chemical vapor deposition (MOCVD), ALD, or a novel technique called ion layer gas reaction (ILGAR) have been explored (Bhattacharya and Ramanathan 2004; Chaisitsak et al. 2001; Ohtake et al. 1995; Olsen et al. 2002; Spiering et al. 2003). A record efficiency of 20.9% has been achieved for CBD-Zn(O,S) buffer layer by Solar Frontier (2014). Most industries are currently using CBD-CdS, but Solar Frontier has shown 17.8% efficiency on 819 cm^2 submodule developed with CBD-grown ZnS(O,OH) buffer layers (Nakamura et al. 2012).

45.4.2.4 Front Electrical Contact

TCOs with bandgaps of above 3 eV are the most appropriate and have become the ultimate choice for front electrical contact due to their excellent optical transparency (greater than 85%) and reasonably good electrical conductivity. Today, CIGS solar cells employ either ITO or, more frequently, RF-sputtered Al-doped ZnO. A combination of an intrinsic and a doped ZnO layer is commonly used. Although this double layer yields consistently higher efficiencies, the beneficial effect of intrinsic ZnO is still under discussion. Doping of the conducting ZnO layer is achieved by group III elements, particularly with aluminum. However, investigations show boron to be a feasible alternative because it yields a high mobility of charge carriers and a higher transmission in the long-wavelength spectral region, giving rise to higher currents. For high-efficiency cells, the TCO deposition temperature should be lower than 150°C in order to avoid the detrimental interdiffusion across CdS/CIGS interface. RF sputtering is not considered suitable for industrial production; therefore, alternative sputtering and CVD methods are investigated and used.

45.4.2.5 Sodium Incorporation in CIGS

One of the breakthrough in CIGS PV technology happened when the alumina or borosilicate glass substrate was replaced by soda-lime glass to match the thermal expansion coefficients, which resulted in substantial efficiency improvement (Hedström et al. 1993). It was subsequently realized that sodium (Na) plays an important role in high-efficiency CIGS solar cells because it affects the microstructure (grain size), and passivates the GBs, leading to changes in electronic conductivity by up to two orders of magnitude. The overall effect is efficiency improvement primarily because of an increase in the open circuit voltage (V_{oc}) and fill-factor (FF) of the solar cells. Most commonly, Na is introduced into CIGS by diffusion from the soda-lime glass substrate during the absorber deposition. However, sodium incorporation from such an approach is neither controllable nor reliable; therefore, alternative methods to add Na from external sources are used either during or after the deposition of CIGS layers. These methods include the coevaporation or the deposition of a thin precursor of a Na compound such as NaF, Na_2Se, or Na_2S for CIGS on Na-free substrates, which include soda-lime glass covered with barrier layers Al_2O_3, Si_3N_4, etc. as used by Shell Solar (Kessler and Rudmann 2004). These barrier layers inhibit sodium diffusion from the glass substrate. CIGS on flexible substrates (metal and polyimide foils) also need controlled incorporation of sodium, which is provided from a precursor layer applied prior or after the CIGS growth as PDT.

45.4.2.6 Potassium Incorporation in CIGS

Another breakthrough in CIGS PV technology happened when potassium was introduced in a controlled manner. Results on enameled steel substrates, which contain a high amount of potassium, indicated a beneficial effect of potassium on the device performance (Wuerz et al. 2012). These observations inspired the development of a PDT where KF is evaporated onto the finished CIGS absorber layer in the presence of selenium, which led to a new record efficiency of 20.4% (Chirilă et al. 2013). Remarkably, this has been achieved with a low-temperature process on flexible polyimide substrate. The KF PDT induces a surface modification of the CIGS, resulting in a copper depletion and Cd enrichment. Introduced in the high-temperature process, the KF PDT enabled further efficiency improvement to 20.5% (Stolt 2014) and 20.8% (Jackson et al. 2014).

45.4.3 Deposition and Growth of CIGS Absorber

There are number of processes used for the deposition of the CIGS thin films, some of them are briefly described:

45.4.3.1 Coevaporation Processes

As described earlier, vacuum evaporation is the most successful technique for deposition of CIGS absorber layers for highest-efficiency cells. Vacuum evaporation method involves simultaneous evaporation of the constituent elements from multiple sources in single or sequential processes during the whole absorber deposition process. While a variation of the In to Ga ratio during the deposition process leads to only minor changes in the growth kinetics, the variation of the Cu content strongly affects the film growth. Thus, different coevaporation growth procedures are classified by their Cu evaporation profile (Figure 45.13).

One variant of coevaporation is a bilayer process (also called the *Boeing process*), which originates from the work of Mickelsen and Chen (1982) and yields larger grain sizes

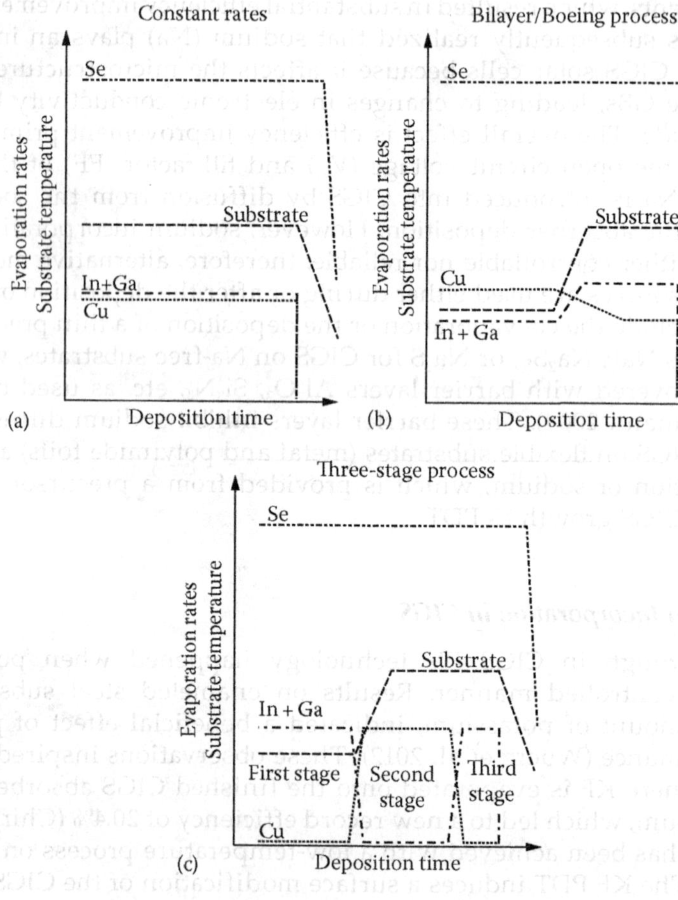

FIGURE 45.13
Diagram representing the recipes in the coevaporation processing steps in the deposition of $Cu(In,Ga)Se_2$ used for (a) constant rate, (b) bilayer or Boeing, and (c) three-stage process.

compared to the constant rate (single stage) process. This is attributed to the formation of a Cu_xSe phase during the Cu-rich first stage that improves the mobility of group III atoms during growth.

The highest efficiencies in laboratories are achieved with the so-called *three-stage* process, introduced by NREL, United States (Gabor et al. 1994). With this process CIGS layer is obtained by starting the deposition of an $(In,Ga)_xSe_y$ precursor, followed by the codeposition of Cu and Se until Cu-rich overall composition is reached, and finally, the overall Cu concentration is readjusted by subsequent deposition of In, Ga, and Se. CIGS films prepared by the three-stage process exhibit a large-grained smooth surface, that reduces the junction area and is thereby expected to reduce the number of defects at the junction and yield high efficiency. Several groups worldwide have developed 16%–20.8% efficiency cells using CIGS grown with three-stage process.

45.4.3.2 Selenization of Precursor Materials

This is another approach of obtaining CIGS thin films. This sequential process is favorable due to its suitability for large-area film deposition with good control of the composition and film thickness following the initial success of Arco Solar in 1987. Such processes consist of the deposition of a precursor material obtained by sputtering, evaporation, ED, paste printing, spray pyrolysis, etc. followed by thermal annealing in controlled reactive or inert atmosphere for optimum compound formation via the chalcogenization reaction as illustrated in Figure 45.14. The precursor materials are either stacked metal layers or a stack of their compound and alloys. Shell Solar, United States, and Solar Frontier (Showa Shell), Japan, use sputtering technique for precursor deposition and production of large-area solar modules up to 60 × 120 cm², yielding maximum efficiencies of 13% on 30 × 30 cm² modules (Karg 2001; Kushiya et al. 2003; Palm et al. 2004).

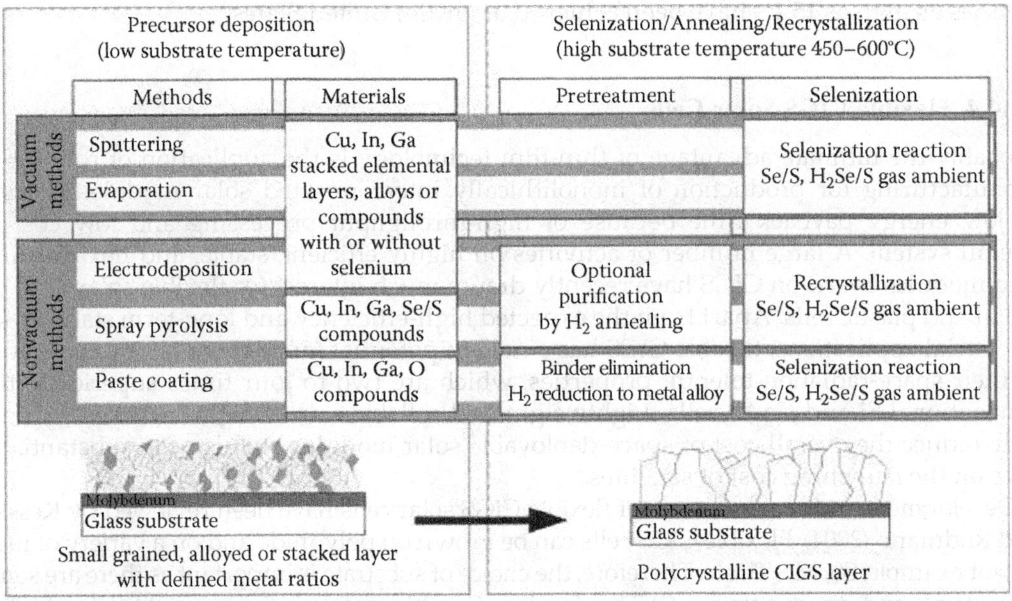

FIGURE 45.14
Schematic of the various processes for selenization of precursor materials.

45.4.3.3 Alternative CIGS Growth Processes

There is a substantial interest to use nonvacuum methods for CIGS deposition. An innovative approach utilizes the stability of the oxides to produce nanosized precursor particles (Eberspacher et al. 2001; Kaelin et al. 2004; Kapur et al. 2001). Nanosized metal oxides are mixed in an ink suspension that allows low-cost, large-area deposition by doctor blading, SP, or spraying of the precursor. Such nonvacuum deposition of precursors allows a very efficient material utilization of almost 100% of the nonabundant metals indium and gallium. A selenization treatment converts precursor into CIGS layer, and solar cell efficiencies of over 13% have been achieved by ISET, United States. One of the drawbacks of the process is the toxicity of the H_2Se gas used for selenization. However, recent efforts are being made to selenize printed precursors with Se vapors (Kaelin et al. 2004).

The CIGS compound can also be formed directly by ED from a chemical bath (Hodes et al. 1986). Several groups including EDF-CNRS, France; CIS solar technologies, Germany; CISCuT, Germany, etc. have been using such approaches and obtained cells with efficiencies around or greater than 10% viz. 9.4% from single pot ED (Duchatelet et al. 2012) and 13.4% aperture efficiency on flexible foil achieved by Solo Power (Osborne 2012). Nexcis, France, reported an efficiency of 15.9% with ED followed by an RTP selenization treatment (Bermudez 2013).

The European consortium SCALENANO, which has received around EUR 7.5 million in EU funding, is scheduled to complete its work in July 2015. The consortium applies innovative process based on the ED of nanostructured precursors, as well as alternative processes with very high potential throughput and process rates. The researchers have demonstrated the scalability of ED-based processes for the synthesis of very homogeneous large areas of thin-film chalcogenide absorbers with medium-area solar modules with cell efficiency of 15.4% (http://cordis.europa.eu/news/rcn/36210_en.html). With a hybrid approach that uses additional vacuum deposition on electrodeposited precursor layer, efficiencies as high as 15.4% have been achieved at NREL, United States.

45.4.4 Flexible CIGS Solar Cells

Probably the ultimate advantage of thin-film technology is the application of roll-to-roll manufacturing for production of monolithically interconnected solar modules leading to low energy payback time because of high-throughput processing and low cost of overall system. A large number of activities on highly efficient, stable, and flexible thin-film modules based on CIGS have recently drawn much interest for flexible solar cells on metal and plastic foils. Apart from the expected high-efficiency and long-term stability for terrestrial applications, flexible CIGS has excellent potential for space application because of their space-radiation-tolerant properties, which are two to four times superior to the conventional Si and GaAs cells. Lightweight and rollable solar array structures will not only reduce the overall cost of space-deployable solar modules but also can substantially save on the launching cost of satellites.

Development and current status of flexible CIGS solar cells have been reviewed by Kessler and Rudmann (2004). Flexible CIGS cells can be grown on polyimide and on a variety of metals, for example, SS, Mo, Ti, etc. Therefore, the choice of substrate is important as there are some advantages and disadvantages: (1) The density of usable metals is four to eight times higher than polymer; therefore, cells on metals are heavier. (2) Metals are conducting and have rough surfaces; therefore, monolithic module development is difficult, which in contrast is easier on polymer foils. (3) SS foils need an extra barrier layer against detrimental impurity (e.g., Fe) diffusion of the metal into the CIGS during deposition. (4) Metal foils can withstand high deposition temperatures (550–600°C), which leads to higher efficiency than on polymer foils, which are not suitable for processing temperature greater than 450°C.

High record efficiencies of flexible CIGS solar cells are 17.7% (Pianezzi et al. 2012) on SS, 18.6% on enameled steel (Powalla et al. 2014), and 20.4% on polymer foil (Chirilă et al. 2013). Several research groups and industries are involved in the development of flexible solar cells; a recent review of the technological status of CIGS-based PVs including activities on flexible substrate is given by (Reinhard et al. 2013) EMPA, Switzerland. The solar modules on SS foils are not monolithically connected; they are made by connecting individual cells into large-area cell with an overlap method. The basic schematic cross-section of a monolithic module on glass is shown in Figure 45.15 along with a flexible prototype minimodule developed on polymer foil by ZSW and ETH, Zurich (research group relocated to EMPA, Switzerland), within a European collaborative project.

Table 45.5 gives an overview of different flexible solar cell technologies, including the organic and TiO_2-based dye-sensitized PV technologies. Because of the late start in R&D, flexible CIGS and CdTe solar cells are industrially less mature compared to a-Si cells. However, high cell efficiency and inherent stability advantages indicate promising potential for these technologies.

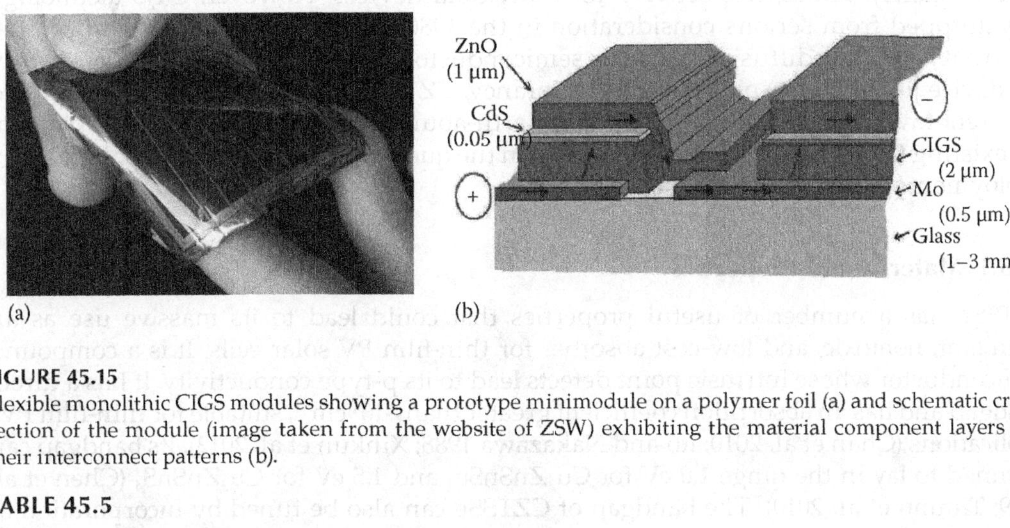

(a) (b)

FIGURE 45.15
Flexible monolithic CIGS modules showing a prototype minimodule on a polymer foil (a) and schematic cross-section of the module (image taken from the website of ZSW) exhibiting the material component layers and their interconnect patterns (b).

TABLE 45.5
Overview of Different Flexible Solar Cell Technologies

	CIGS	CdTe	Amorphous Silicon	Organic and Titanium Oxide
Lab efficiency on plastic foil	20.4% (Single junction cell)	14% (Single junction cell)	9.8% (Single junction cells)[a] 8%[a]–12%[a] (multijunction cell)	5%–8%
Lab efficiency on metal foil	18.7%[b] (single junction cell)	11.5% (single junction cell)	14.6[a]/13% (multijunction cell)	
Industrial efficiency (typical values)	10%–14%	Not yet demonstrated	4%–8%[a] (available on plastic and metal foils)	Not yet demonstrated
Stability under light	Material stable	Material stable	Degrades	Stability not proven

[a] Initial values measured before light-induced degradation of solar cells.
[b] Value not certified.

45.5 Cu_2 $ZnSn(S, Se)_4$ or CZTS Solar Cells

Currently, CdTe and CIGSSe technologies dominate the thin-film PV market with 1 GW/year production capacity barrier surpassed first time for CdTe devices in 2009 by First Solar (Wolden et al. 2011) and for CIGSSe in 2011 by Solar Frontier (Hering 2011). Despite the impressive growth rates in these technologies, two issues—absorber layer component abundance (scarcity of indium and tellurium) and toxicity (heavy metal cadmium)—represent potential road blocks for pervasive deployment of these thin-film chalcogenide-based PV technologies. For the aforementioned reasons, there has been strong motivation to find alternatives to the two leading thin-film chalcogenide absorbers that use low-cost, abundant elements. In this context, various absorbers that use earth-abundant metals such as SnS (Ramakrishna Reddy et al. 2006), FeS_2 (pyrite) (Smestad et al. 1990), Cu_2S (Martinuzzi 1982), and $Cu_2ZnSn(S,Se)_4$ (CZTSSe) (Barkhouse et al. 2012) have been explored. The current status of CZTSSe technology and its development has been summarized by EMPA group in Switzerland (Fella et al. 2013). Among these, only Cu_2S and CZTSSe have exceeded the lower performance efficiency limit for commercial interest of 10% (Barkhouse et al. 2012; Hall et al. 1981) when compared with CIGSSe and CdTe, which have achieved 20.9% (Solar Frontier 2014) and 20.4% (First Solar 2014) efficiency values, respectively, for individual devices. However, Cu_2S technology was dropped from serious consideration in the 1980s owing to device instability arising from facile Cu diffusion into the semiconductor layers, leading to degradation of the device junction. Despite being in its infancy, CZTSSe technology can be considered a current favorite in terms of prospective Earth-abundant metal systems to supplement the existing CIGSSe and CdTe technologies, in the quest for more ubiquitous solar energy deployment.

45.5.1 Material and Properties

CZTSSe has a number of useful properties that could lead to its massive use as an abundant, nontoxic, and low-cost absorber for thin-film PV solar cells. It is a compound semiconductor whose intrinsic point defects lead to its p-type conductivity. It has a direct bandgap and has an absorption coefficient greater than ~10^4 cm^{-1}, suitable for thin-film PVs applications (Chan et al. 2010; Ito and Nakazawa 1988; Xinkun et al. 2012). Its bandgap can be tuned to lay in the range 1.0 eV for $Cu_2ZnSnSe_4$ and 1.5 eV for Cu_2ZnSnS_4 (Chen et al. 2009; Timmo et al. 2010). The bandgap of CZTSSe can also be tuned by incorporation of Ge and Ge-containing materials having smaller bandgap than their Ge-free counterparts (Ford et al. 2011). This tenability is of particular interest for the manufacturing of absorbers with a bandgap between 1.1 and 1.5 eV, which allow theoretical efficiencies higher than 30% (Shockley and Queisser 1961). The knowledge gathered from CIGS solar cells fabrication could be tuned and adapted directly for CZTS solar cells. The GBs seem to have the same beneficial properties for CZTS as for CIGS, such as enhanced minority carrier collection taking place at the GB (Li et al. 2012).

45.5.2 CZTS Solar Cell Configuration

Since the CZTS solar cell configuration is similar to that of the CIGS (or CIGSSe) (already described in the previous sections), the present section would avoid going into the details about the same.

Although CZTS single crystals had been grown and analyzed in 1967 (Nische et al. 1967), however, the PV effect of the CZTS films was reported only in 1988 (Ito and Nakazawa 1988). The fabricated heterodiode consisted of a transparent cadmium-tin-oxide thin film and a CZTS thin film on an SS substrate. An open-circuit voltage of 165 mV was obtained with the device. In 1997, Friedlmeier et al. fabricated thin-film solar cells using a CZTS layer as the light absorber in contact with an n-CdS/ZnO window layer. The best energy conversion efficiency produced by the cells was 2.3%. This record was later broken by Katagiri's group in 1999 that produced a CZTS solar cell with 2.63% power conversion efficiency. In this cell, the CZTS film was deposited on a Mo-coated soda lime glass (SLG) substrate. By optimization of the sulfurization process, the efficiency of the solar cells was increased to 5.45% in 2003, and then to 6.7% in 2008. Two review papers regarding the efficiency milestones achieved in CZTS-based thin-film solar cells were reported by Katagiri (2005) and Katagiri et al. (2009a), respectively. Todorov et al. (2010) reported 9.6% efficient CZTS-based thin-film solar cells. In this cell, the CZTS thin film was partly selenized to have a broader spectral photoresponse. The current record value of 11.1% is reported by Barkhouse et al. (2012). Figure 45.16 gives the power conversion efficiency in the record CZTSSe thin-film devices versus year, showing constant progress in the technology.

CdS has been observed to have suitable band offset with CZTS, and until now, the most efficient solar cells were obtained with a CdS buffer layer (Nagoya et al. 2011). At the absorber/buffer interface, CZTS is Cu poor, similar to CIGS (Bär et al. 2011). Moreover, sodium incorporation seems to have the same effects on CZTS as in CIGS. It promotes the growth of larger grains, enhances conductivity, and has a significant effect on film morphology, as was demonstrated by SLG/borosilicate substrate comparison and dipping in Na_2S (Hlaingo Oo et al. 2011; Prabhakar and Jampana 2011). However, Na diffusion in CIGS

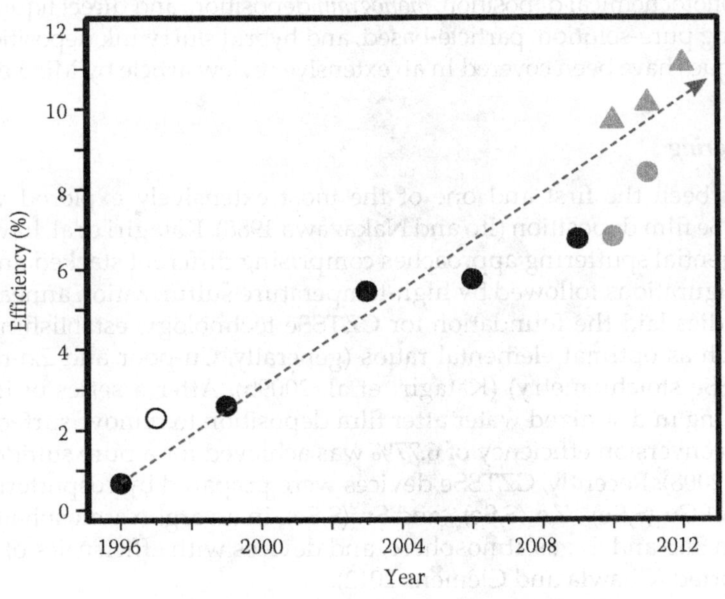

FIGURE 45.16
Power conversion efficiency in the record CZTSSe thin-film devices versus year, showing consistent progress. Data for pure sulfur-containing materials (CZTS) are represented by circles, whereas data for kesterites in which Se has been introduced are represented by triangles. The laboratories responsible for each record are Nagaoka Nat. College of Technology (black), University of Stuttgart (white), and IBM Corp. (gray). The dotted arrow is a guide to the eye. (From Mitzi et al. 2013.)

is observed to be larger than in CZTS. It is interesting to note that oxidation of the CZTS absorber by dipping in deionized water (Katagiri et al. 2009a) or O_2 annealing (Repins et al. 2012) before deposition of CdS seems to improve efficiencies. As for CIGS, reducing the time between absorber synthesis and CdS deposition to the minimum is of utmost importance (Grenet et al. 2012).

45.5.3 Deposition and Growth of CZTS (CZTSSe) Absorber

The investigations on the phase equilibrium in the Cu_2S-ZnS-SnS_2 system showed that single phase CZTS crystals is a challenge as it can only be grown in a very narrow region (Oleksyuk et al. 2004). Secondary phases such as ternary and quaternary compounds are much easier to form than CZTS. Therefore, it is quite challenging to deposit CZTS thin film without the significant presence of secondary phases. Moreover, in contrast to CIGS, the elements in CZTS synthesis are prone to evaporation and sublimation. Zn sublimates at 430°C (Teeter et al. 2011), SnSe at 350°C (Redinger et al. 2011a), and SnS at 370°C and Sn evaporates at 400°C (Weber et al. 2010; Yoo et al. 2012). Moreover, high temperatures promote CZTS decomposition that can be avoided by atmosphere control (Redinger et al. 2011b).

The possibility of replacing the rare In with readily available Zn and Sn while retaining key semiconductor properties, such as nearly identical bandgap to the highly successful chalcopyrite absorbers, makes CZTSSe particularly attractive for large-scale PV production. This potential has given significant research impetus not only toward the fabrication of higher performance CZTSSe devices with traditional vacuum deposition techniques such as sputtering and coevaporation but also to explore a variety of alternative lower-cost and high-throughput fabrication approaches. Some of these alternative methods include spray pyrolysis, ED, photochemical deposition, *monograin* deposition, and direct liquid-coating techniques, including pure-solution, particle-based, and hybrid slurry ink deposition. The details of these techniques have been covered in an extensive review article by Mitzi et al. (2013).

45.5.3.1 Sputtering

Sputtering has been the first and one of the most extensively explored vacuum-based routes to CZTSSe film deposition (Ito and Nakazawa 1988). Katagiri et al. have investigated a series of sequential sputtering approaches comprising different stacked and cosputtered precursor configurations followed by high-temperature sulfurization anneals. The results from these studies laid the foundation for CZTSSe technology, establishing key process parameters such as optimal elemental ratios (generally, Cu-poor and Zn-rich relative to the ideal CZTSSe stoichiometry) (Katagiri et al. 2009b). After a series of improvements, including soaking in deionized water after film deposition to remove surface oxide impurities, a power conversion efficiency of 6.77% was achieved for a pure sulfide CZTS device (Katagiri et al. 2008). Recently, CZTSSe devices were prepared by cosputtering from compound targets of $Cu_x(S,Se)_y$, $Zn_x(S,Se)_y$, and $Sn_x(S,Se)_y$ in an argon atmosphere, followed by annealing in an SnS and S_2 gas atmosphere, and devices with efficiencies of as high as 8% have been reported (Chawla and Clemens 2012).

45.5.3.2 Coevaporation

In contrast to the CIGSSe technology, CZTSSe deposition using coevaporation has encountered significant challenges during especially in the classical configuration where the substrate is maintained at high temperature while delivering precisely controlled elemental fluxes. In addition to the typically volatile Se and S in chalcopyrite absorbers, Sn re-evaporation also occurs during CZTSSe deposition (Friedlmeier et al. 1997). This issue was addressed by introducing a faster coevaporation process in which Sn loss could be mitigated. One way to go around this issue is by adding continuous Sn overpressure throughout the whole high-temperature process, including part of the cooling stage as indicated in a recent NREL report (Repins et al. 2011). Using this approach, device efficiencies of as high as 9.2% for pure selenide CZTSSe devices have been achieved (Repins et al. 2012). Currently, the best performing coevaporated CZTS (pure sulfide) devices have reached efficiencies of 8.4%, using a process developed at IBM that consists of coevaporation onto a low-temperature substrate followed by a short anneal at atmospheric pressure (Shin et al. 2011).

45.5.3.3 Nonvacuum Deposition Approaches

From the early years of CZTSSe development, it was well understood that a combination of cheap readily available materials combined with a reliable low-cost fabrication approach could revolutionize the PV industry. Three approaches based on pure solutions, nanoparticles, or hybrid particle–solutions, respectively, have been particularly pursued. Of these, the third category of inks, comprising a combination of dissolved and solid active components, has so far been the most successful CZTSSe fabrication approach. It advantageously uses the binding action of dissolved metal chalcogenide species to produce dense, compact layers and additionally bypasses the solubility limitation of pure solution approaches by introducing particle components that further provide densification and crack-deflection benefits. Recently, Yang and coworkers (Hsu et al. 2012) studied the chemical pathways involved in the hydrazine-based deposition and heat treatment process. The precursor ink was dried at room temperature resulting in the integration of copper and tin chalcogenide complexes to form a bimetallic framework, with hydrazine and hydrazinium molecules as spacers. After mild thermal annealing, the spacers are removed and the $Cu_2Sn(Se,S)_3$ + $Zn(Se,S) \rightarrow Cu_2ZnSn(Se,S)_4$ reaction is triggered. The described reaction pathway contains fewer steps than most deposition processes, which typically start with elemental or binary chalcogenides (Hsu et al. 2012). By use of this hybrid solution–nanoparticle approach, a record efficiency of 9.7% was reported in 2010 by the IBM group (Todorov et al. 2010). The method was further refined to yield 10.1% power conversion efficiency in 2011 (Barkhouse et al. 2012) and most recently to the 11.1% level (Todorov et al. 2011), which represents the current world record for the kesterite-based system.

The efficiency improvement in the CZTS technology has been significant in a short span of time, and this gives a hope that this technology can provide alternatives to the material sustainable issues for In and Ga in CIGS. It may be noted that the In and Ga materials are the by-products of the Zn smelting factories, and so ramping up the production of CZTS technology will also enable the sustained supply of In and Ga, leading to sustained and regenerative production along with CZTS technology growth.

45.6 New-Generation Solar Cells

The beginning of 1990s has seen the emergence of new and disruptive technologies viz. dye-sensitized and later organic solar cells (Grätzel 2000), which followed completely different device approach of bulk-heterojunction formation in the nanometer length regime, which can be referred to as new generation solar cells. DSSCs are nanocomposites of mesoporous TiO_2 and inorganic dyes (typically ruthenium complexes); they created a breakthrough by achieving relatively high conversion efficiency ~10% through the use of rather inexpensive and abundant materials (O'Regan and Grätzel 1991).

The solar cell action in the aforementioned device on the illumination of light is initiated by excitonic electron–hole pair generation and dissociation leading to a charge separation at the nanointerface. These devices are therefore termed as *excitonic* or organic–inorganic or hybrid solar cells. Unlike conventional p–n homojunction (c-Si) solar cells, the absorption of photons and electron transport is mediated via two dissimilar constituents, viz., dyes and n-type nanocrystalline TiO_2, respectively, while the holes are transported through a hole-transporting material (HTM), which could be a redox liquid electrolyte or ion-conducting polymeric electrolyte (quasisolid state) or quantum dot sensitized or hole-conducting conjugated polymer material or small molecules or perovskites (solid state).

Although the initial results had been remarkable, the liquid junction DSSC technology appears to have now reached a state of stagnation in the efficiency values, with an increase from about 10% to 12.3% only in the span of over 20 years. This technology had also suffered from the lack of sensitizer dyes with wider spectral coverage, atmospheric degradation, and engineering issues such as encapsulation and sealing of the liquid junction device, which had prevented this technology from making a mark in commercialization. Incidentally, there has been a breakthrough in an attempt to solve these issues with the development of solid-state DSSC, in the form of the mesoscopic perovskite solar cells developed in parallel by Michael Grätzel and coworkers at EPFL, Switzerland (Burschka et al. 2013), and Henry Snaith and coworkers at Oxford University (Lee et al. 2012) with efficiencies ~10% and 11%, respectively, initially which had moved up to over 17% (Bennett 2014; Snaith 2014) toward the end of 2013. A comprehensive detail on these developments including liquid-, quasi-, and solid-state DSSCs (or perovskite cells) has been reported (Upadhyaya et al. 2013) and has been described briefly in the following section.

A significant achievement was made toward the development of DSSC/CIGS tandem (Liska et al. 2006) solar cell, which resulted then with over 15% efficient cells with 1.42 V open circuit voltage (using a physical stack of DSSC and CIGS cells with 4 wired configuration with DSSC cell efficiency about 8% and CIGS efficiency about 13%), bringing a step jump in efficiencies. The monolithic two-wire DSSC/CIGS tandem was not successful because of issues of corrosion caused by liquid electrolyte component (Wenger et al. 2009). However, with solid-state analogues of DSSC or perovskite solar cells available with higher efficiencies, it is anticipated that the tandem cells up to 27% efficiencies could be developed (Kim et al. 2013; Kranz et al. 2013).

45.6.1 Perovskite Solar Cells

Apart from the aforementioned listed thin-film solar cells, there are a new generation of thin-film solar cells along with DSSCs and organic solar cells that have emerged recently. Attempts have been going on over many years to find an alternative to the liquid electrolytes and to obtain, thus, an improved DSSC, which will have ease of fabrication, less complication in the sealing and encapsulation of the device, possibility for monolithic interconnection of the cells within the module, and therefore also increased performance and lower cost. However, a class of solar cells based on perovskite structured

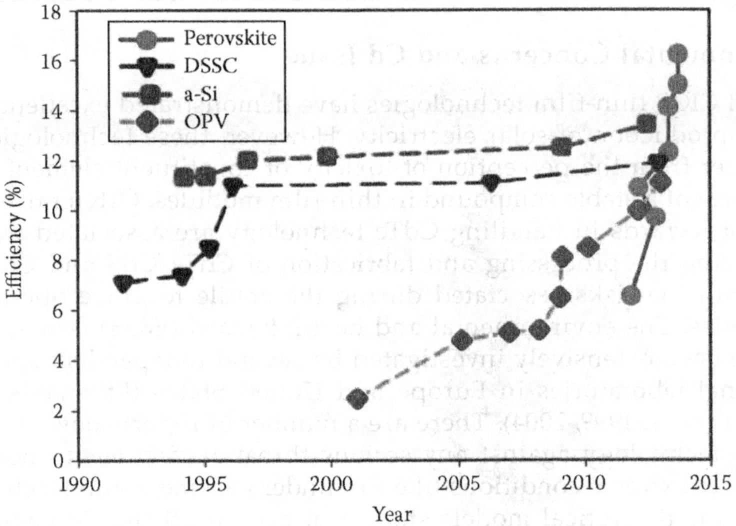

FIGURE 45.17
Comparing the rate of increase in perovskite solar cell efficiencies with the other thin-film PV technologies. (From Hodes, G., *Science*, 342, 317, 2013.)

semiconductors has emerged and rapidly reached conversion efficiencies of more than 15% (Figure 45.17), as listed in Table 45.2.

Most commonly, $CH_3NH_3PbI_3$-based organic–inorganic perovskite materials have been used in these types of solar cells, which are having high charge carrier mobility. High mobility is important because, together with high charge carrier lifetimes, it means that the light-generated electrons and holes can move large enough distances to be extracted as current, instead of losing their energy as heat within the cell (Hodes 2013). Moreover, the perovskite solar cells can be deposited by low-temperature methods such as solution process (spin coating) and thermal evaporation methods. The advantages of the solid-state DSSCs over the liquid DSSCs are obvious: less complications involved in the manufacturing equipment, easy possibility for producing monolithically interconnected modules, easier sealing, and encapsulation of the modules (similar to the processes employed for other thin-film solar cells). Although the efficiency values reached for small-size individual solar cells can be considered to be satisfactory, and should basically provide a great commercialization potential for this technology in the future, especially thanks to the fact that we have here a true solid-state device, the stability and longevity of the device is yet to be ascertained. Although the perovskites are cheaper than the conventional dyes and have wider spectral coverage, they are susceptible to temperature and humidity. Temperature around 95°C can bring structural transformation in the materials and also atmospheric moisture can lead to degradation as well. Thus, a strong barrier layer would be required for desired stability toward moisture ingression. In addition other variants of perovskites will have to be found out with intrinsic temperature stability issues. Besides, other critical issues of long-term stability of the devices are the stability of the organic spiro-OMeTAD layer, and the need for the perovskites with Pb-free compounds, to cover the environmental concerns.

45.7 Environmental Concerns and Cd Issue

The CdTe and CIGS thin-film technologies have demonstrated excellent potential for cost-effective production of solar electricity. However, these technologies, especially the CdTe, suffer from the perception of toxicity of constituent element Cd, which is used in the form of a stable compound in thin-film modules. Often raised issues such as the risks or hazards in handling CdTe technology are associated with the materials used during the processing and fabrication of CdTe/CdS and CdS/CIGS solar cells along with the risks associated during the cradle to grave operating lifetime of these modules. The environmental and health hazard (E&H) issues of CdTe solar modules have been extensively investigated by several independent agencies including the national laboratories in Europe and United States (Fthenakis and Zweibel 2003; Fthenakis et al. 1999, 2004). There are a number of reports now available, which support CdTe technology against any serious threat arising even under worst-case scenarios. Under extreme conditions like fire under basement and leaching from broken modules, the theoretical models show that even if all the Cd compounds were to be released, Cd concentrations within the near vicinity to the CdTe PV system are below human health screening levels (Beckmann and Mennenga 2011; Dutta et al. 2012; Sinha et al. 2011). An interesting report also provides a supporting argument in favor of CdTe technology analyzed during earthquakes or tsunami in Japan (Matsuno 2013). While First Solar program of module recycling is already available and supports their confidence (http://www.firstsolar.com/en/technologies-and-capabilities/recycling-services), the chance of Cd compounds and other potentially harmful materials getting into the ambient environment from the PV plant during its lifetime is also negligible. The current power generation technologies using coal have a large amount of Cd emitted in the environment in an uncontrolled manner, with the amount far exceeding the expected emission from CdTe modules even under exceptional conditions (e.g., fire). Even the dominant PV technologies based on Si solar cells have larger life cycle Cd emissions due to the embodied energy being larger for these devices (Fthenakis et al. 2008).

While CdTe technology has no chance to do away with Cd, there is some maneuverability in CIGS technology in the elimination of very thin (typically ~50 nm) CdS buffer layer, and so the quest for alternative buffer layer is being successfully pursued. Initial success has already been achieved as CIGS solar cells of 16%–18.8% and modules of 13.4% have been developed with alternative *Cd-free* buffers (Hedström et al. 1993; Kushiya 2004). Recently, coevaporation of CIGS in an inline single-stage process is used to fabricate solar cell devices with up to 18.6% conversion efficiency using a CdS buffer layer and 18.2% using a $Zn_{1-x}Sn_xO_y$ (Cd-free) buffer layer. Furthermore, a 15.6 cm^2 minimodule, with 16.8% conversion efficiency, has been made with the same layer structure as the CdS baseline cells, showing the excellent uniformity. These cell results were externally verified (Lindahl et al. 2013).

Referring to the perception and concerns on Cd issues, V. Fthenakis (Sr. Chemical Engineer, Brookhaven national Laboratory) and K. Zweibel (Manager, Thin film partnership program, NREL, United States) had presented a detailed account of their studies during their presentations at the 2003 NCPV program review meeting in the United States (Fthenakis et al. 2004) that confirmed that CdTe panels would be almost benign with almost zero emission (0.01 g/GW h) as compared to its counterparts oil (44.3 g/GW h) and coal (3.7 g/GW h), with no associated health hazards, and gave a clean chit to CdTe technology. The following points emerged out of the studies:

Cadmium is a by-product of zinc, lead, and copper mining. It constitutes only 0.25% of its main feedstock ZnS (sphalerite). Cadmium is released into the environment from phosphate fertilizers, burning fuels, mining and metal processing operations, cement production, and disposing of metal products. Releases from disposed Cd products, including Ni–Cd batteries, are minimum contributors to human exposure because Cd is encapsulated in the sealed structures. Most human cadmium exposure comes from ingestion and most of that stems from the uptake of cadmium by plants, through fertilizers, sewage sludge, manure, and atmospheric deposition. Although long-term exposure to elemental cadmium, a carcinogen, has detrimental effect on kidneys and bones, limited data exist in toxicology. However, CdTe compound is more stable and less soluble than Cd element and therefore likely to be much less toxic.

Considering the electrolytic refinery production of CdTe powders (from Cd wastes from Zn, iron, and steel industries), there would be an emission of 0.001% Cd gaseous emission. This would correspond to 0.01 g/GW h, which is significantly less as compared to the perceptions and hypes created by some who estimate it at 0.5 g/GW h based on other crude processes or unsubstantiated data. The only potential hazard that could come to any ones' mind would be the building fire. It has also been estimated quantitatively that the maximum temperature of a basement on fire is ~900°C, which is still less than the melting point of CdTe (1041°C). Besides, the vapor pressure at 800°C for CdTe is ~2.5 Torr (0.003 atmosphere), so this minimizes the risks further, and once sealed between glass plates, any Cd vapor emission is unlikely. The main conclusion of the studies was that the environmental risks associated with CdTe-based technology are minimal. Every energy source or product may present some direct or indirect environmental health and safety hazards and those of CdTe should by no means be considered a problem; the following conclusions were drawn:

- Cd is produced as a by-product of Zn and can either be put to beneficial uses or discharged to the environment posing another risk.
- CdTe in PV is much safer than other current Cd uses.
- CdTe PV uses Cd about 2500 times more efficient than Ni–Cd batteries.
- Occupational health risks are well managed.
- Absolutely no emission during PV operation.
- A risk from fire emission is minimal.
- CdTe technology and modules are safe and do not pose significant risks.

45.8 Conclusions

1. PV market is booming with a phenomenal surging growth rate of over 40% for more than a decade despite a downturn observed in the worldwide economy recently. The cumulative PV production capacity was ~140 GW by 2013, while 50 GW production is anticipated in 2014 alone. This is in contrary to the belief in the last decade

that high demand for cost-effective PV installations with more consumer-oriented choices can neither be met by the c-Si wafer technology nor is expected to achieve low production cost targets of below €1/W, which is a myth now.

2. An upsurge in the PV production and an artificial module price crisis created by Chinese manufacturers has brought a radical drop in the module prices well below $0.6/Wp since 2011. This strategy has wiped away the PV industries in Europe and has caused a PV manufacturing base shift toward the Asia, with China, Taiwan, and Korea taking this opportunity and leading the show through their natural advantage of cheaper material supply and cheaper human resources available.
3. Thin-film PV has clearly demonstrated an excellent potential for cost-effective generation of solar electricity using CdTe technology by First Solar in the United States and CIGS technology by Solar Frontier in Japan currently aiming to produce 2 GW capacities annually. While the world record cell efficiencies of 20.4% for CdTe and 20.9% for CIGS have already matched/surpassed polycrystalline Si efficiency of 20.4%, the module efficiencies have also shown significant figures of 17% and 16.4% for CdTe and CIGS, respectively. In fact, thin-film PV market has gained stability from c-Si production and market leadership with almost 90% share. It is anticipated that a mix of c-Si and thin-film PV technologies will cater the market needs in near to midterm future, followed by the dominance of thin-film and other PV technologies in long term.
4. The a-Si technology has suffered a setback mainly because of the stagnancy and the low efficiency figures arising from the intrinsic light-dependent degradation issues. This has resulted in the bankruptcy of the United Solar company in 2013, which was the largest manufacturer of triple junction a-Si technology on steel foils. Efforts are underway in Japan through NEDO programs to overcome these barriers for a successful relaunching of this technology, which will have wider application in BIPV and BAPV sectors in the future.
5. Thin-film PV industries are growing fast; however, there are several issues:
    a. Reducing the gap between lab efficiency and larger area industrial production efficiency. This is achievable with the design of better equipment with in situ diagnostics. Nonavailability of standard deposition system for thin-film PV has been a problem, so effort is needed to develop large-area equipment suitable for thin-film PV.

    b. Unlike c-Si technology, which allows the process and equipment availability off the shelf, there is no availability of processing equipment and the optimized processes for thin-film PV technologies, which are only under the proprietary control of the companies involved. Although companies like Manz Solar (http://www.manz.com/) have taken initiatives in this direction, this is not enough to cover the entire promise.

    c. To lower the cost, increase the throughput and the yield, efforts are needed for further simplification and increased robustness of the process and device structures.

    d. If multi-giga and terawatt PV facilities are to be successful for a safer and prosperous world, further improvements in the device structures along with optimized material utilisation (low wastage) have to be achieved for attaining efficiencies greater than 25%.

References

Abken, A. E. and Bartelt, O. J. 2002. *Thin Solid Films*, 403–404, 216–222.

Ahn, S. W., Lee, S. E., and Lee, H. M. D. 2012. Presented at *27th European Photovoltaic Solar Energy Conference*, Frankfurt, Germany, 24–28 September 2012, p. 3AO5.1.

Amkreutz, D., Müller, J., Schmidt, M., Hänel, T., and Schulze, T. F. 2011 *Progress in Photovoltaics*, 19, 937–945.

Arya, R. R. 2004. In *Proceedings of the 19th European Photovoltaic Conference*, Paris, June 7–11, 2004, http://apache.solarch.ch/pdf/KonferenzParisInternet.pdf.

Barkhouse, D. A. R., Gunawan, O., Gokmen, T., Todorov, T. K., and Mitzi, D. B. 2012. *Progress in Photovoltaics: Research and Applications*, 20, 6–11.

Basore, P. A. 2002. In *Proceedings of 29th IEEE Photovoltaic Specialists Conference*, May 19–24, 2002, pp. 49–52, http://ieeexplore.ieee.org/xpl/mostRecentIssue.jsp?punumber=8468.

Bär, M., Schubert, B. A., Marsen, B., Krause, S., Pookpanratana, S., Unold, T., Weinhardt, L., Heske, C., and Schock, H.-W. 2011. *Applied Physics Letters*, 99, 112103.

Bätzner, D. L., Romeo, A., Terheggen, M., Döbeli, M., Zogg, H., and Tiwari, A. N. 2004. *Thin Solid Films*, 451, 536–543.

Bätzner, D. L., Romeo, A., Zogg, H., Wendt, R., and Tiwari, A. N. 2001. *Thin Solid Films*, 387, 151–154.

Becker, C., Amkreutz, D., Sontheimer, T., Preidel, V., Lockau, D., Haschke, J., Jogschies, L. et al. 2013. *Solar Energy Materials and Solar Cells*, 119, 112–123.

Beckmann, J. and Mennenga, A. 2011. Calculation of emissions when there is a fire in a photovoltaic system made of cadmium telluride modules. Bavarian Environmental Agency, Augsberg, Germany.

Benagli, S., Borrello, D., Vallat, E., Meier, J., Kroll, U., Hötzel, J., Bailat, J. et al. 2009. In *Proceedings of 24th European Photovoltaic Solar Energy Conference*, Hamburg, Germany, September 21–25, 2009, pp. 2293–2298, http://www.solarplaza.com/event/24th-european-photovoltaic-solar-energy-conference.

Bennett, P. 2014. http://www.solarpowerportal.co.uk/news/oxford_professor_recognised_by_materials_research_society_for_solar_3309.

Bergmann, R. B. 1999. *Applied Physics A* 69, 187–194.

Bermudez, V. 2013. Nexcis, CIGS technology from R&D to industry. Presented at *the Fifth Photovoltaic Thin-Film Week*, Berlin, Germany, April 2013, https://www.helmholtz-berlin.de/projects/pvcomb/events/tfw13/index_en.html.

Bhattacharya, R. N. and Ramanathan, K. 2004. *Solar Energy*, 77, 679–683.

Bonnet, D. 2004. In *Proceedings of the 19th European Photovoltaic Solar Energy Conference*, Paris, June 7–11, 2004, pp. 1657–1664, http://apache.solarch.ch/pdf/KonferenzParisInternet.pdf.

Bonnet, D. and Rabenhorst, H. 1972. In *Proceedings of the Ninth PV Specialists Conference*, pp. 129–132.

Bruton, T. M., Woodcock, J. M., Roy, K., Garrard, B., Alonso, J., Nijs, J., Räuber, A. et al. 1997. Final report European Commission Project RENA-CT94-0008 "MUSICFM".

Burger, B. 2013. Photovoltaics report, Fraunhofer, ISE, BMU, EEG 2013 and BMWi Energiedaten, Freiburg, October 24, 2014. http://www.ise.fraunhofer.de/de/downloads/pdf-files/aktuelles/photovoltaics-report-in-englischer-sprache.pdf.

Burschka, J., Pellet, N., Moon, S. J., Humphry-Baker, R., Gao, P., Nazeeruddin, M. K., and Grätzel, M. 2013. *Nature*, 499, 316–319.

Candelise, C., Spiers, J., and Gross, R. 2011. Materials availability for thin film (TF) PV technologies development: A real concern? ICEPT/WP/2011/001, Imperial College Centre for Energy Policy and Technology, London, U.K., https://workspace.imperial.ac.uk/icept/Public/Materials%20for%20PV%20ICEPT%20format.pdf.

Carabe, J. and Gandia, J. J. 2004. *Opto-Electronics Reviews* 12, 1–6.

Carlson, D. and Wronski, C. 1976. *Applied Physics Letters*, 28, 671–673.

Chaisitsak, S., Yamada, A., and Konagai, M. 2001. In *Proceedings of the Materials Research Society Spring Meeting*, p. 668.

Chan, C. P., Lam, H., and Surya, C. 2010. *Solar Energy Materials Solar Cells*, 94, 207–211.

Chawla, V. and Clemens, B. 2012. In *Proceedings of 38th IEEE Photovoltaics Specialists Conference*, Austin, TX, June 3–8, 2012, pp. 002990–002992. http://www.ieee-pvsc.org/PVSC38/.

Chen, S., Gong, X. G., Walsh, A., and Wei, S. H. 2009. *Applied Physics Letters*, 94, 041903(1–3).

Chirilă, A., Reinhard, P., Pianezzi, F., Bloesch, P., Uhl, A. R., Fella, C., Kranz, L. et al. 2013. *Nature Materials*, 12, 1107–1111.

Chiu, P. T., Law, D. C., Woo, R. L., Singer, S. B., Bhusari, D., Hong, W. D., Zakaria, A. et al. 2014. *IEEE Journal of Photovoltaics*, 4, 493–497.

Chopra, K. L., Paulson, P. D., and Dutta, V. 2004. *Progress in Photovoltaics: Research Application*, 12, 69–92.

Colville, F. 2014. http://www.pv-tech.org/guest_blog/npd_solarbuzz_upgrades_uk_pv_market_to_2.5gw_in_2014.

Compaan, A. D., Gupta, A. K., Lee, S., Wang, S., and Drayton, J. 2004. *Solar Energy*, 77, 815–822.

Cusano, D. A. 1963. *Solid State Electronics*, 6, 217–232.

Deng, X. and Schiff, E. A. 2003. Amorphous silicon based solar cells. In *Handbook of Photovoltaic Science and Engineering*, A. Luque and S. Hegedus (Eds.), pp. 505–565. Wiley, New York.

Duchatelet, A., Chassaing, E., Savidand, G., Vannier, R. N., and Lincot, D. 2012. One-step electrochemical deposition of Cu–In–Ga mixed oxide thin films for low-cost CIGS solar cells. In presented at *the 27th European Photovoltaic Solar Energy Conference and Exhibition*, pp. 2805–2807, September 2012. doi: 10.4229/27thEUPVSEC2012-3DV.3.34.

Diefenbach, K. H. 2005. *Photon International*, 48–67.

Docampo, P., Ball, J. M., Darwich, M., Eperon, G. E., and Snaith, H. J. 2013. *Nature Communications*, 4, 2761–2767.

Dore, J., Ong, D., Varlamov, S., Egan, R., and Green, M. A. 2014. *IEEE Journal of Photovoltaics*, 4, 33–39.

Durr, M., Schmid, A., Obermaier, M., Rosselli, S., Yasuda, A., and Nelles, G. 2005. *Nature Materials*, 4, 607–611.

Dutta, V., Vamsi Krishna, K., Sreekrishnan, T. R., and Singh, U. P. 2012. http://www.fitt-iitd.org/downloads/First%20Solar_Final%20Report_24-08-2012.pdf.

Eberspacher, C., Fredric, C., Pauls, K., and Serra, J. 2001. *Thin Solid Films* 387, 18–22.

Engelhart, P., Wendt, J., Schulze, A., Klenke, C., Mohr, A., Petter, K., Stenzel, F. et al. 2010. In *Energy Procedia, First International Conference on Silicon Photovoltaics, Energy Procedia*, 8 (2011) 313–317, www.Elsevier.com/locate/procedia.

Fahrenbruch, A., Bube, R., Kim, D., and Lopez-Otero, A. 1992. *International Journal of Solar Energy* 12, 197–222.

Fehr, M., Schnegg, A., Rech, B., Astakhov, O., Finger, F., Bittl, R., Teutloff, C., and Lips, K. 2014. *Physical Review Letters*, 112, 066403(1–5).

Fella, C. M., Romanyuk, Y. E., and Tiwari, A. N. 2013. *Solar Energy Materials and Solar Cells*, 119, 276–277.

Ferekides, C., Britt, J., Ma, Y., and Killian, L. 1993. In *Proceedings of 23rd IEEE, PV Specialists Conference*, pp. 389–393, http://ieee-ee.eng.usf.edu/ferekides/plist.html.

First Solar. 2005. http://www.firstsolar.com/index.html.

First Solar. 2014. http://investor.firstsolar.com/releasedetail.cfm?ReleaseID=828273.

Ford, G. M., Guo, Q., Agrawal, R., and Hillhouse, H. W. 2011. *Chemistry of Materials*, 23, 2626–2629.

Friedlmeier, T. M., Wieser, N., Walter, T., Dittrich, H., and Schock, H. W. 1997. In *Proceeding of the 14th European Conference of Photovoltaic Science and Engineering and Exhibition*, p. 1242.

Fthenakis, V., Morris, S., Moskowitz, P., and Morgan, D. 1999. *Progress in Photovoltaics*, 7, 489–497.

Fthenakis, V. and Zweibel, K. 2003. CdTe PV: Real and perceived EHS risks. In pPrepared for the *NCPV and Solar Program Review Meeting*, Denver, CO, March 24–26, 2003, http://www.nrel.gov/docs/fy03osti/33561.pdf.

Fthenakis, V. M., Fuhrmann, M., Heiser, J., and Wang, W. 2004. Presented at *the 19th European PV Solar Energy Conference*, Paris, France, June 7–11, 2004, p. 5BV.1.32, https://www.bnl.gov/pv/files/pdf/abs_176.pdf.

Fthenakis, V. M., Kim, H. C., and Alsema, E. 2008. *Environmental Science and Technology*, 42, 2168–2174.

Gabor, A. M., Tuttle, J. R., Albin, D. S., Contreras, M. A., Noufi, R., and Hermann, A. M. 1994. *Applied Physics Letters*, 65, 198–200.

Geisz, J. F., Steiner, M. A., Garcia, I., Kurtz, S. R., and Friedman, D. J. 2013. *Applied Physics Letters*, 103, 041118 (1–5).

Gifford, J. 2013. http://www.pv-magazine.com/news/details/beitrag/panasonic-hits-247-cell efficiency_100010185/#ixzz2yFQcp1fv.

Goetzberger, A., Hebling, C., and Schock, H. W. 2003. *Materials Science and Engineering*, R40, 1–46.

Granqvist, C. G. 2003. *Advance Materials*, 15, 1789–1803.

Grätzel, M. 2000. *Progress in Photovoltaics: Research and Applications*, 8, 171–186.

Grenet, L., Bernardi, S., Kohen, D., Lepoittevin, C., Noël, S., Karst, N., Brioude, A., Perraud, S., and Mariette, H. 2012. *Solar Energy Materials and Solar Cells*, 101, 11–14.

Gross, R.D., Heptonstall, P., Greenacre, P., Candelise, C., Jones, F., and Castillo, A. November 2013. Presenting the future: An assessment of future costs estimation methodologies in the electricity generation sector. A report by the UKERC Technology & Policy Assessment Function, REF UKERC/RR/TPA/2013/001, UKERC.

Guha, S. 2004. *Solar Energy*, 77, 887–892.

Guha, S., Narsimhan, K. L., and Pietruszko, S. M. 1981. *Journal of Applied Physics*, 52, 859–860.

Guha, S., Yang, J., Nath, P., and Hack, M. 1986. *Applied Physics Letters*, 49, 218–219.

Hall, R. B., Birkmire, R. W., Phillips, J. E., and Meakin, J. D. 1981. *Applied Physics Letters*, 38, 925–926.

Hanergy Annual Report. 2013. http://www.hanergysolargroup.com/pdf/ew_0566frp-20140407q4.pdf.

Hedström, J., Ohlsen, H., Bodegard, M., Kylner, A., Stolt, L., Hariskos, D., Ruckh, M., and Schock, H. W. 1993. In *Proceedings of the 23rd IEEE Photovoltaic Specialists Conference*, pp. 364–371, http://www.appropedia.org/Photovoltaic_materials_abundance_limitations_literature_review.

Hegedus, S. S. and Luque, A. 2003. Status, trends, challenges and the bright future of solar electricity from photovoltaics. In *Handbook of Photovoltaic Science and Engineering*, A. Luque and S. Hegedus (Eds.), pp. C113–C678. Wiley, New York.

Hering, G. 2011. *Photon International*, 4, 186–218.

Hlaingo Oo, W., Johnson, J., Bhatia, A., Lund, E., Nowell, M., and Scarpulla, M. 2011. *Journal of Electronic Materials*, 40, 2214–2221.

Hodes, G. 2013. *Science*, 342, 317–318.

Hodes, G., Lokhande, C. D., and Cahen, D. 1986. *Journal of Electrochemistry Society*, 133, C113.

Honsberg, C. B., Corkish, R., and Bremner, S. P. 2001. In *Proceedings of 17th European Photovoltaic Solar Energy Conference*, Munich, Germany, October 22–26, 2001, pp. 22–26. http://web2.etaflorence.it/cms/index.php?page=17ma-conferenza-ed-esposizione-europea-sull-energia-solare-fotovoltaica&hl=en_US.

Hosoya, M., Oooka, H., Nakao, H., Gotanda, T., Hayase, R., Nakano, Y., and Saito, M. 2012. In *Proceedings of 73rd Fall Meeting of Japan Society of Applied Physics*, September 11–14, 2012, pp. 12–380. http://www.osa.org/en-us/publications/.

Hosoya, M., Oooka, H., Nakao, H., Gotanda, T., Mori, S., Shida, N., Hayase, R., Nakano, Y., and Saito, M. 2013. In *Proceedings of 93rd Annual Meeting of the Chemical Society of Japan*, March 22–25, 2013, pp. 21–37. http://www.chemistryviews.org/details/event/2766931/Chemical_Society_of_Japan_93rd_Spring_Annual_Meeting.html.

Hsu, W. C., Bob, B, Yang, W., Chung, C. H., and Yang, Y. 2012. *Energy and Environmental Sciences*, 5, 8564–8571.

http://cleantechnica.com/2013/12/18/new-czts-solar-cell-efficiency-record/.

http://cordis.europa.eu/news/rcn/36210_en.html.

http://www.firstsolar.com/en/technologies-and-capabilities/recycling-services http://www.kaneka-solar.com.

http://www.manz.com/.

Ichikawa, Y., Fujikake, S., Takayama, R., Saito, S., Ota, H., Yoshida, T., Ihara, T., and Sakai, H. 1993. In *Proceedings of 23rd IEEE Photovoltaic Specialists Conference*, p. 27.

Ito, K. and Nakazawa, T. 1988. *Japanese Journal of Applied Physics*, 27, 2094–2097.

Jackson, P., Hariskos, D., Lotter, E., Paetel, S., Wuerz, R., Menner, R., Wischmann, W., and Powalla, M. 2011. *Progress in Photovoltaics: Research and Applications*, 19, 894–897.
Jackson, P., Hariskos, D., Wuerz, R., Wischmann, W., and Powalla, M. 2014. *Physica Status Solidi (RRL)*, 8, 219–222.
Jäger-Waldau, A. 2004. *Solar Energy*, 77, 667–678.
Kaelin, M., Rudmann, D., and Tiwari, A. N. 2004. *Solar Energy*, 77, 749–756.
Karg, F. H. 2001. *Solar Energy Materials and Solar Cells*, 66, 645.
Kapur, V. K., Fisher, M., and Roe, R. 2001. In *Proceedings of the 2001 MRS Spring Meeting*, San Francisco, CA, April 16–20, 2001, p. 668, H2.6.1-7, http://www.mrs.org/spring2001/.
Katagiri, H. 2005. *Thin Sold Films*, 480, 426–432.
Katagiri, H., Jimbo, K., Maw, W. S., Oishi, K., Yamazaki, M., Araki, H., and Takeuchi, A. 2009a. *Thin Solid Films*, 517, 2455–2460.
Katagiri, H., Jimbo, K., Tahara, M., Araki, H., and Oishi, K. 2009b. *Material Research Society Symposium Proceedings*, 1165, M01–M04.
Katagiri, H., Jimbo, K., Yamada, S, Kamimura, T., Maw, W. S., Fukano, T., Ito, T., and Motohiro, T. 2008. *Applied Physics Express*, 1, 041201(1–2).
Kawai, M. 2013. Nikkei Electronics. http://techon.nikkeibp.co.jp/english/NEWS_EN/20130131/263532/.
Kayes, B. M., Nie, H., Twist, R., Spruytte, S. G., Reinhardt, F., Kizilyalli, I. C., and Higashi, G. S. 2011. In *Proceedings of the 37th IEEE Photovoltaic Specialists Conference*, June 19–24, 2011, pp. 000004–000008.
Kazmerski, L., White, F., and Morgan, G. 1976. *Applied Physics Letters*, 29, 268–269.
Keavney, C. J., Haven, V. E., and Vernon, S. M. 1990. In *Proceedings of 21st IEEE Photovoltaic Specialists Conference*, pp. 141–144.
Keevers, M. J., Young, T. L., Schubert, U., and Green, M. A. 2007. In *Proceedings of 22nd European Photovoltaic Solar Energy Conference*, Milan, Italy, September 3–7, pp. 1783–1790.
Keshner, M. S. and Arya, R. 2004. NREL/SR-520-36846. http://www.nrel.gov/docs/fy05osti/36846.pdf.
Kessler, F. and Rudmann, D. 2004. *Solar Energy*, 77, 685–695.
Kim, S., Kasashima, S., Sichanugrist, P., Kobayashi, T., Nakada, T., Konagai, M. 2013. *Solar Energy Materials and Solar Cells*, 119, 214–218.
Kinoshita, T., Fujishima, D., Yano, A., Ogane, A., Tohoda, S., Matsuyama, K., Nakamura, Y. et al. 2011. In *Proceedings of 26th European Photovoltaic Solar Energy Conference*, September 5–9, 2011 pp. 871–874, http://www.eea.europa.eu/events/26th-european-photovoltaic-solar-energy-6.
Klein, S., Repmann, T., and Brammer, T. 2004. *Solar Energy*, 77, 893–908.
Kojima, A., Teshima, K., Shirai, Y., and Miyasaka, T. 2009. *Journal of American Chemical Society*, 131, 6050–6051.
Komiya, R., Fukui, A., Murofushi, N., Koide, N., Yamanaka, R., Katayama, H. 2011. Technical digest. In *21st International Photovoltaic Science and Engineering Conference*, Fukuoka, Japan, 2C-50-08, http://www.researchsea.com/html/calendar.php/eid/6371/cid/1/research/_pvsec-21__21st_international_photovoltaic_science_and_engineering_conference.html?PHPSESSID=k0o28evdgdkjpj845701b5vm20.
Konagai, M. 2013. Presented at *Asian Photovoltaic Science and Engineering Conference*.
Kranz, L., Buecheler, S., and Tiwari, A. N. 2013a. *Solar Energy Materials and Solar Cells*, 119, 278–280.
Kranz, L., Gretener, C., Perrenoud, J., Schmitt, R., Pianezzi, F., Mattina, F. L., Blösch, P. et al. 2013b. *Nature Communications*, 4, 2306–2313.
Kranz, L., Gretener, C., Perrenoud, J., Jaeger, D., Gerstl, S. S. A., Schmitt, R., Buecheler, S., and Tiwari, A. N. 2013c. *Advanced Energy Materials*, 2014, 4, 1301400.
Kranz, L., Jäger, T., Reinhard, P., Hagendorfer, H., Romanyuk, Y. E., Buecheler, S., and Tiwari, A. N. 2013d. In presented at the *Materials Research Society Fall Meeting*, Boston, MA, December 1–6, 2013.
Kumar, M. H., Yantara, N., Dharani, S., Grätzel, M., Mhaisalkar, S., Boix, P. P., and Mathews, N. 2013. *Chemical Communications*, 49, 11089–11091.

Kushiya, K. 2004. *Solar Energy*, 77, 717–724.
Kushiya, K., Ohshita, M., Hara, I., Tanaka, Y., Sang, B., Nagoya, Y., Tachiyuki, M., and Yamase, O. 2003. *Solar Energy Materials and Solar Cells*, 75, 171–178.
Lebrun, J. 1966. *Revue de Physique Appliquee* 1, 204–210.
Lechner, P. and Schade, H. 2002. *Progressin Photovoltaics: Research Applications*, 10, 85–97.
Lee, M. M., Teuscher, J., Miyasaka, T., Murakami, T. N., and Snaith, H. J. 2012. *Science*, 338, 643–647.
Li, J. B., Chawla, V., and Clemens, B. M. 2012. *Advanced Materials* 24, 720–723.
Lindahl, J., Zimmermann, U., Szaniawski, P., Törndahl, T., Hultqvist, A., Salomé, P., Platzer-Björkman, C., and Edoff, M. 2013. *IEEE Journal of Photovoltaics*, 3, 1100–1105.
Liska, P., Thampi, K. R., Grätzel, M., Brémaud, D., Rudmann, D., Upadhyaya, H. M., and Tiwari, A. N. 2006. *Applied Physics Letters*, 88, 203103.
Mathew, X., Enriquez, J. P., Romeo, A., and Tiwari, A. N. 2004. *Solar Energy*, 77, 831–838.
Martinuzzi, S. 1982. *Thin Solid Films*, 517, 2455–2460.
Matsuno, Y. 2013. http://park.itc.u-tokyo.ac.jp/matsuno/files/FS%20Tsunami%20Risk%20Review%20English.pdf.
McCandless, B., Youm, I., and Birkmire, R. 1999. *Progress in Photovoltaics*, 7, 21–30.
McCandless, B. E. and Dobson, K. D. 2004. *Solar Energy*, 77, 839–856.
McCandless, B. E. and Sites, J. R. 2003. Cadmium telluride solar cells. In *Handbook of Photovoltaic Science and Engineering*, A. Luque and S. Hegedus (Eds.), pp. 617–657. Wiley, Chichester, U.K.
Meier, J., Dubail, S., Fluckinger, R., Fischer, D., Keppner, H., and Shah, A. 1994. In *Proceedings of the First World Conference on Photovoltaic Energy*, pp. 409–412.
Metha, S. 2013. PV technology and cost outlook, 2013–2017. http://www.greentechmedia.com/research/report/pv-technology-and-cost-outlook-2013-2017.
Mickelsen, R. and Chen, W. 1981. In *Proceedings of the 15th IEEE Photovoltaic Specialist Conference*, Orlando, FL, pp. 800–804.
Mickelsen, R., and Chen, W. 1982. In *Proceedings of the 16th IEEE Photovoltaic Specialist Conference*, pp. 781–785.
Mints, P. and Donnelly, J. 2011. Navigant Consulting Photovoltaic Service Program. Report NPS-Supply6. http://www.navigantresearch.com/research/photovoltaic-manufacturer-shipments-capacity-competitive-analysis-20112012.
Mitchell, K., Eberspacher, C., Ermer, J., and Pier, D. 1988a. In *Proceedings of the 20th IEEE Photovoltaic Specialists Conference*, September 26–30, 1988, pp. 1384–1389.
Mitchell, K. C., Ermer, E. J., and Pier, D. 1988b. In *Proceedings of 20th IEEE Photovoltaic Specialists Conference*, September 26–30, 1988, p. 1384.
Mitzi, D. B., Gunawan, O., Todorov, T. K., and Barkhouse, D. A. R. 2013. *Philosophical Transactions of Royal Society A*, 371:20110432.
Morooka, M., Ogura, R., Orihashi, M., and Takenaka, M. 2009. *Electrochemistry*, 77, 960–965.
Moslehi, M. M., Kapur, P., Kramer, J., Rana, V., Seutter, S., Deshpande, A., Stalcup, T. et al. 2012. Presented at *PV Asia Pacific Conference (APVIA/PVAP)*, October 22–24, 2012.
Muller, J., Rech, B., Springer, J., and Vanecek, M. 2004. *Solar Energy*, 77, 917–930.
Nagoya, A., Asahi, R., and Kresse, G. 2011. *Journal of Physics of Condensed Matter*, 23, 404203.
Nakada, T., Hirabayashi, Y., Tokado, T., Ohmori, D., and Mise, T. 2004. *Solar Energy*, 77, 739–747.
Nakada, T. and Mizutani, M. 2002. *Japanese Journal of Applied Physics*, 41, 165–167.
Nakamura, M., Chiba, Y., Kijima, S., Horiguchi, K., Yanagisawa, Y., Sawai, Y., Ishikawa, K., and Hakuma, H. 2012. In *Proceedings of the 38th IEEE Photovoltaics Specialists Conference*, Austin, TX, June 3–8, 2012, pp. 1807–1810.
Nakamura, M., Yamaguchi, K., Chiba, Y., Hakuma, H., Kobayashi, T., and Nakada, T. 2013. Presented at *39th IEEE Photovoltaic Specialist Conference*, Tampa, FL, June 16–21, 2013, pp. 0849–0852.
Nische, R., Sargent, D. F., and Wild, P. 1967. *Journal of Crystal Growth*, 1, 52.
NREL. http://www.nrel.gov/cdte.
Ohtake, Y., Kushiya, K., Ichikawa, M., Yamada, A., and Konagai, M. 1995. *Japanese of Journal Applied Physics*, 34, 5949–5955.

Oleksyuk, I. D., Dudchar, I. V., and Piskach, L. V. 2004. *Journal of Alloys and Compounds*, 368, 135–143.
Olsen, L., Eschbach, P., and Kundu, S. 2002. In *Proceedings of the 29th IEEE Photovoltaic Specialists Conference*, New Orleans, LO, pp. 652–655.
O'Regan, B. and Grätzel, M. 1991. *Nature*, 353, 737–740.
Osborne, M. March 2012. *PV Tech News*. http://www.pv-tech.org/news/solopower_sets_13.4_flex_cigs_aperture_efficiency_record_backed_by_nrel.
Osborne, M. 2013. http://www.pvtech.org/news/has_first_solar_retaken_the_lowest_cost_pv_manufacturer_mantle.
Palm, J., Probst, V., and Karg, F. H. 2004. *Solar Energy*, 77, 757–765.
Perrenoud, J., Kranz, L., Buecheler, S., Pianezzi, F., and Tiwari, A. N. 2011a. *Thin Solid Films*, 519, 7444–7448.
Perrenoud, J., Schaffner, B., Buecheler, S., and Tiwari, A. N. 2011b. *Solar Energy Materials and Solar Cells*, 95, S8–S12.
Pianezzi, F., Chirilă, A., Blösch, P., Seyrling, S., Buecheler, S., Kranz, L., Fella, C., and Tiwari, A. N. 2012. *Progress in Photovoltaics: Research and Applications*, 20, 253–259.
Powalla, M., Witte, W., Jackson, P., Paetel, S., Lotter, E., Wuerz, R., Kessler, F. et al. 2014. *IEEE Journal of Photovoltaics*, 4, 440–446.
Prabhakar, T. and Jampana, N. 2011. *Solar Energy Materials Solar Cells*, 95, 1001–1004.
Ramakrishna Reddy, K. T., Koteweara Reddy, N., and Miles, R. W. 2006. *Solar Energy Materials Solar Cells*, 90, 3041–3046.
Ramanathan, K., Contreras, M. A., Perkins, C. L., Asher, S., Hasoon, F. S., Keane, J., Young, D. et al. 2003. *Progress in Photovoltaics: Research and Applications*, 11, 225–230.
Rech, B. 2013. https://www.helmholtz-berlin.de/forschung/oe/enma/si-pv/arbeitsgebiete/duennschichtsolarzellen/index_en.html.
Rech, B. and Wagner, H. 1999. *Applied Physics A*, 69, 155–167.
Redinger, A., Berg, D. M., Dale, P. J., Djemour, R., Gutay, L., Eisenbarth, T., Valle, N., and Siebentritt, S. 2011a. *IEEE Journal of Photovoltaics*, 1, 200–206.
Redinger, A., Berg, D. M., Dale, P. J., and Siebentritt, S. 2011b. *Journal of American Chemical Society*, 133, 3320–3323.
Reinhard, P., Buecheler, S., and Tiwari, A. N. 2013. *Solar Energy Materials Solar Cells*, 119, 287–290.
Repins, I., Beall, C., Vora, N., DeHart, C., Kuciauskas, D., Dippo, P., To, B., Mann, J., Hsu, W.-C., Goodrich, A., and Noufi, R. 2012. *Solar Energy Materials and Solar Cells*, 101, 154–159.
Repins, I., Vora, N., Beall, C., Wei, S.-H., Yan, Y., Romero, M., Teeter, G., Du, H., To, B., Young, M., and Noufi, R. 2011. http://www.nrel.gov/docs/fy11osti/51286.pdf.
Romeo, A., Bätzner, D. L., Zogg, H., Tiwari, A. N., and Vignali, C. 1999a. *Solar Energy Materials and Solar Cells*, 58, 209–218.
Romeo, A., Khrypunov, G., Kurdesau, F., Batzner, D. L., Zogg, H., and Tiwari, A. N. 2004a. In *Technical Digest of the International PVSEC-14*, Bangkok, Thailand, pp. 715–716.
Romeo, A., Terheggen, M., Abou-Ras, D., Bätzner, D. L., Haug, F.-J., Kälin, M., Rudmann, D., and Tiwari, A. N. 2004b. *Progress in Photovoltaics: Research and Applications*, 12, 93–111.
Romeo, N., Bosio, A., Canevari, V., and Podesta, A. 2004c. *Solar Energy*, 77, 795–801.
Romeo, N., Bosio, A., Tedeschi, R., Romeo, A., and Canevari, V. 1999b. *Solar Energy Materials and Solar Cells*, 58, 209–218.
Sakata, H., Kawamoto, K., Taguchi, M., Baba, T., Tsuge, S., Uchihashi, K., Nakamura, N., and Kiyam, S. 2000. In *Proceedings of the 28th IEEE Photovoltaic Specialists Conference*, September 15–22, 2000, pp. 7–12, http://ieeexplore.ieee.org/xpl/mostRecentIssue.jsp?punumber=7320.
Sasaki, K., Agui, T., Nakaido, K., Takahashi, N., Onitsuka, R., and Takamoto, T. 2013. *Proceedings of Ninth International Conference on Concentrating Photovoltaics Systems*, Miyazaki, Japan.
Schultz, O., Glunz, S. W., and Willeke, G. P. 2004. *Progress in Photovoltaics: Research and Applications*, 12, 553–558.
Service, R. 2011. *Science*, 332, 293.
Shah, A., Meier, J., Buechel, A., Kroll, U., Steinhauser, J., Meillaud, F., and Schade, H. 2004. Presented at *the ICCG5 Conference in Saarbrücken (D)*, July 4–8, 2004.

Shah, A., Moulin, E., and Ballif, C. 2013. *Solar Energy Materials and Solar Cells*, 119, 311–316.

Sharp Corporation. Press Release. 2012. http://sharp-world.com/corporate/news/120531.html.

Shin, B., Gunawan, O., Zhu, Y., Bojarczuk, N. A., Chey, S. J., and Guha, S. 2011. *Progress in Photovoltaics: Research and Applications*, 21, 72–76.

Shockley, W. and Queisser, H. J. 1961. *Journal of Applied Physics*, 32, 510–519.

Sinha, P., Balas, R., and Krueger, L. 2011. In *Proceedings of 37th IEEE Photovoltaics Specialists Conference*, Washington, June 19–24, 2011, pp. 2025–2030.

Sinke, W. C. 2007. A strategic research agenda for photovoltaic solar energy technology. Hamburg, Germany, September 14–18, 2015, www.eupvplatform.org.

Slade, A. and Garboushian, V. 2005. In *Technical Digest of 15th International Photovoltaic Science and Engineering Conference*, Shanghai, China, October 10–15, 2005, p. 701.

Smestad, G. P., Ennaoui, A., Fiechter, S., Hofmann, W., Tributsch, H., and Kautek, W. 1990. *Proceedings of SPIE, Optical Materials Technology for Energy Efficiency and Solar Energy Conversion IX*, Vol. 1272, p. 67.

Snaith, H. J. 2014. Presented at *the APEX (Advancing the Efficiency and Production Potential of Excitonic Solar Cells)*, RCUK-DST India Project Meeting in New Delhi, India.

Solar Frontier. 2014. http://www.solar-frontier.com/eng/news/2014/C031367.html.

Sopori, B. 2003. In *Handbook of Photovoltaic Science and Engineering*, A. Luque and S. Hegedus (Eds.), pp. 359–363. Wiley, New York.

Spiering, S., Hariskos, D., Powalla, M., Naghavi, N., and Lincot, D. 2003. *Thin Solid Films*, 431–432, 359–363.

Staebler, D. L. and Wronski, C. R. 1977. *Applied Physics Letters*, 31, 292.

Stolt, L. 2014. Henergy/Solibro. Presented at *the Fifth International Workshop on CIGS Solar Cell Technology (IW-CIGSTech 5)*, Berlin, Germany, April 2–3, 2014.

Sunshot Vision Study. Department of Energy, USA assessment report, http://energy.gov/downloads/sunshot-vision-study.

Surek, T. 2003. In *Proceeding of Third World Conference on PV Energy Conversion*, May 11–18, 2003, p. 2507.

Swanson, R. 2012. Presented at *the Challenges in PV Science, Technology, and Manufacturing: A Workshop on the Role of Theory, Modeling and Simulation*, Purdue University, August 2–3, 2012.

Teeter, G., Du, H., Leisch, J. E., Young, M., Yan, F., Johnston, S., Dippo, P. et al. 2011. In *Proceedings of 35th IEEE Photovoltaic Specialists Conference*, June 20–25, 2010, pp. 650–655.

Timmo, K., Altosaar, M., Raudoja, J., Muska, K., Pilvet, M., Kauk, M., Varema, T., Danilson, M., Volobujeva, O., and Mellikov, E. 2010. *Solar Energy Materials and Solar Cells*, 94, 1889–1892.

Todorov, K., Reuter, K. B., and Mitzi, D. B. 2010. *Advanced Materials*, 22, E156–E159.

Todorov, T., Gunawan, O., Chey, S. J., Goislard de Monsabert, T., Prabhakar, A., and Mitzi, D. B. 2011. *Thin Solid Films*, 519, 7378–7381.

Tuttle, J. R., Schuyler, T., Choi, E., and Freer, J. 2005. In *Proceedings of 20th European Photovoltaic Solar Energy Conference and Exhibition*, Barcelona, Spain, June 6–10, 2005.

Tuttle, J. R., Szalaj, A., and Keane, J. 2000. In *Proceedings of the 28th IEEE Photovoltaic Specialists Conference*, September 15–22, 2000, pp. 1042–1045.

Upadhyaya, H. M., Senthilarasu, S., Hsu, M. H., and Kishore Kumar, D. 2013. *Solar Energy Materials and Solar Cells*, 119, 291–295.

Venkatasubramanian, R., O'Quinn, B. C., Hills, J. S., Sharps, P. R., Timmons, M. L., Hutchby, J. A., Field, H., Ahrenkiel, A., and Keyes, B. 1997. In *Conference Record of 25th IEEE Photovoltaic Specialists Conference*, May 13–17, 1996, pp. 31–36.

Vetterl, O., Finger, F., Carius, R., Hapke, P., Houben, L., Kluth, O., Lambertz, A., Mück, A., Rech, B., and Wagner, H. 2000. *Solar Energy Materials and Solar Cells*, 62, 97–108.

von Roedern, B., Zweibel, K., and Ullal, H. S. 2005. In *Proceedings of 31st IEEE Photovoltaic Specialists Conference*, January 3–7, 2005, pp.183–188.

Wagner, S., Shay, J., Migliorato, P., and Kasper, H. 1974. *Applied Physics Letters*, 25, 434.

Weber, A., Mainz, R., and Schock, H. W. 2010. *Journal of Applied Physics*, 107, 013516.

Wenger, S., Seyrling, S., Tiwari, A. N., and Grätzel, M. 2009. *Applied Physics Letters*, 94, 173508.

Wolden, C. A., Kurtin, J., Baxter, J. B., Repins, I., Shaheen, S. E., Torvik, J. T., Rockett, A. A., Fthenakis, V. M., and Aydil, E. S. 2011. *Journal of Vacuum Science and Technology A*, 29, 030801.

Woodcock, J. M., Schade, H., Maurus, H., Dimmler, B., Springer, J., and Ricaud, A. 1997. In *Proceedings of the 14th European Photovoltaic Solar Energy Conference*, Barcelona, Spain, pp. 857–860.

Wu, X., Keane, J. C., Dhere, R. G., DeHart, C., Duda, A., Gessert, T. A., Asher, S., Levi, D. H., and Sheldon, P. 2002. In *Proceedings of the 17th European Photovoltaic Solar Energy Conference*, Munich, Germany, pp. 995–1000.

Wuerz, R., Eicke, A., Kessler, F., Paetel, S., Efimenko, S., and Schlegel, C. 2012. *Solar Energy Materials and Solar Cells*, 100, 132–137.

www.energyinformative.org/best-thin-film-solar-panels-amorphous-cadmium-telluride-cigs/.

www.pvtech.org/news/q_cells_sets_two_new_world_records_for_multi_crystalline_and_quasi_mono_sol.

Xinkun, W., Wei, L., Shuying, C., Yunfeng, L., and Hongjie, J. 2012. *Journal of Semiconductors*, 33, 022002.

Yamamoto, K., Nakajima, A., Yoshimi, M., Sawada, T., Fukuda, S., Suezaki, T., Ichikawa, M. et al. 2004. *Solar Energy*, 77, 939–949.

Yamamoto, K., Toshimi, M., Suzuki, T., Tawada, Y., Okamoto, T., and Nakajima, A. 1998. Presented at *the Materials Research Society Spring Meeting*, San Francisco, CA, April 13–17, 1998.

Yoo, H., Kim, J., and Zhang, L. 2012. *Current Applied Physics*, 1052–1057.

Yoshimi, M., Sasaki, T., Sawada, T., Suezaki, T., Meguro, T., Matsuda, T., Santo, K. et al. 2003. In presented at *Third World Conference on Photovoltaic Energy Conversion*, May 11–18, 2003, pp. 1566–1569.

Yu, L., Misra, S., Wang, J., Qian, S., Foldyna, M., Xu, J., Shi, Y., Johnson, E., and Cabarrocas, P. R. 2014. *Scientific Reports*, 4, 4357–4364.

Zhao, J., Wang, A., Green, M. A., and Ferrazza, F. 1998. *Applied Physics Letters*, 73, 1991–1993.

Zhao, J., Wang, A., Yun, F., Zhang, G., Roche, D. M., Wenham, S. R., and Green, M. A. 1997. *Progress in Photovoltaics*, 5, 269–276.

Zweibel, K. 2000. *Solar Energy Materials & Solar Cells*, 63, 375–386.

46

Concentrating PV Technologies

D. Yogi Goswami

CONTENTS

46.1 Introduction

The current state of solar cell development is illustrated in Figure 46.1. While single-junction silicon solar cells dominate today's solar industry, the rapid rise in efficiency versus time (experience curve) of the multijunction cells makes this a particularly attractive technology path. The high efficiency, in comparison with single junction cells, such as silicon, is obtained by stacking several junctions in series, electrically isolated by tunnel diodes, as explained in Chapter 44. These can be qualitatively viewed by adding the voltages of three junctions in series, while maintaining the current of a single junction (Figure 46.1).

Under concentrated sunlight, multijunction (GaInP/GaAs/GaInNAs) solar cells have demonstrated an efficiency of 44.4% within the lab, as compared to silicon cells' best efficiency of 25%. This means that, in sunny areas, a multijunction concentrator system can generate almost twice as much electricity as a silicon panel with the same area.

Table 46.1 gives the efficiencies of CPV submodules achieved as of 2012.

The concentrating optics focus the light onto a small area of cells, reducing the area of the solar cells by a factor of, typically, 500–1000 times. The reduced cell area overcomes the increased cell cost. The cell cost is diminished in importance and is replaced by the cost of optics. Thus, in high direct insolation locations, the multijunction concentrator technology has the potential to reduce the cost of solar electricity. As a side benefit, the cells are more efficient under concentration, provided a reasonable cell temperature can be maintained. The technology has been extended to four junctions and could be extended to even five junctions if efficiency benefits justify added cost. The efficiency is a moving target; today's triple junction cell efficiency is nearly 45%. Thus, one may reasonably extrapolate that multijunction cells may reach 50% efficiency in the future. Using less cell material for a given

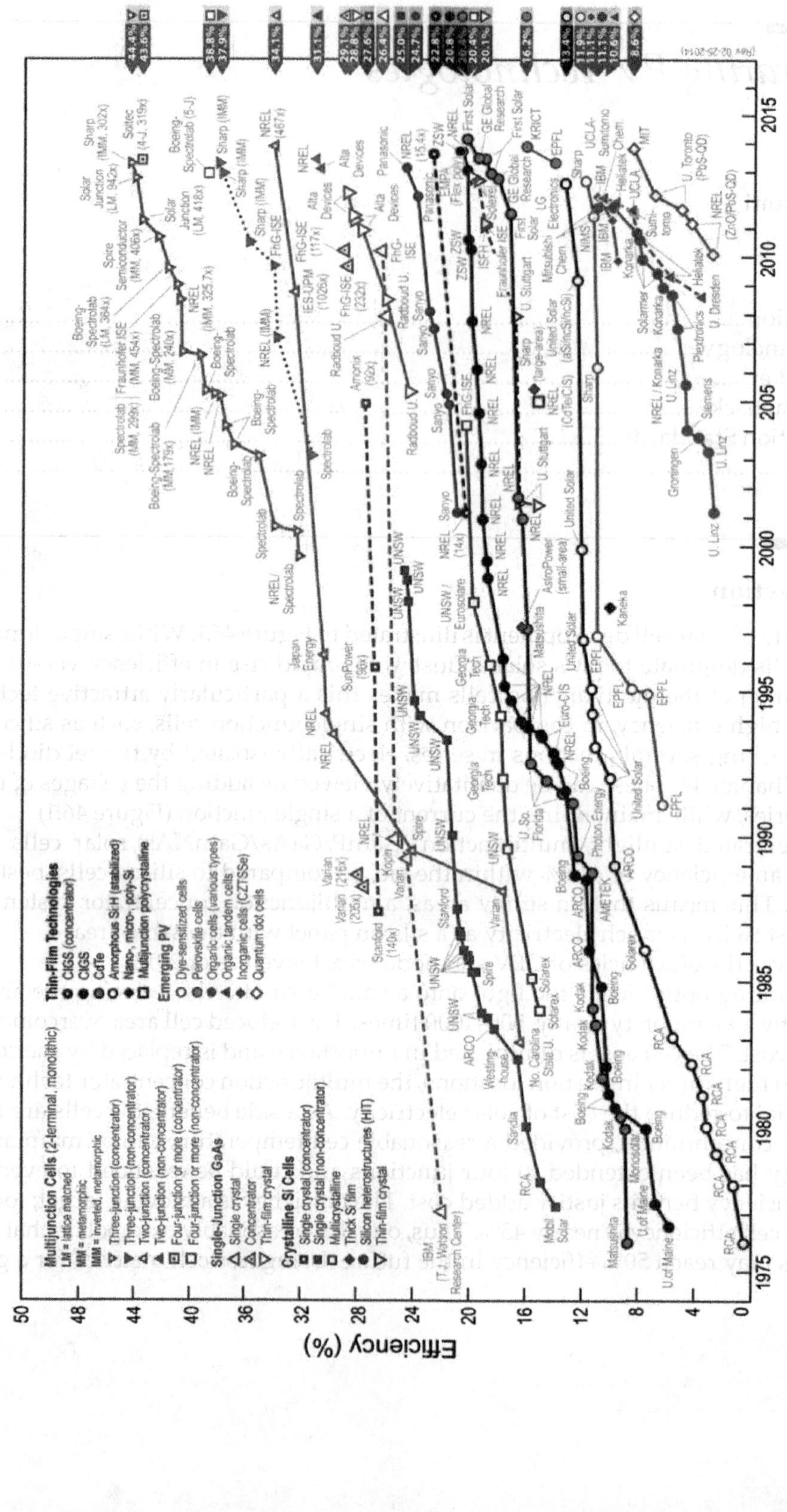

FIGURE 46.1
World record conversion efficiencies for various PV technologies. (From NREL, Best research cell efficiencies, National Renewable Energy Laboratory, Golden, CO, 2014, http://www.nrel.gov.)

TABLE 46.1

Terrestrial Concentrator Cell Efficiencies (Measured under ASTM G-173-03 Direct Beam AM1.5 Spectrum at a Cell Temperature of 25°C)

Classification	Efficiency (%)	Area (cm^2)[a]	Intensity (Suns)
Single cells			
GaAs	29.1 ± 1.3	0.505 (da)	117
Si	27.6 ± 1.0	1.00 (da)	92
Multijunction cells (monolithic)			
GaInP/GaAs/GaInNAs	44.0 ± 3	0.3104 (ap)	942
InGaP/GaAs/InGaAs	43.5 ± 2.6	0.167 (da)	306
GaInP/GaInAs/Ge	41.6 ± 2.5	0.3174 (da)	364
Submodule			
GaInP/GaAs; GaInAsP/GaInAs	38.5 ± 1.9	0.202 (ap)	20
Modules			
Si	20.5 ± 0.8	1,875 (ap)	79
Triple junction	33.5 ± 0.5	10,674.8 (ap)	N/A
Notable exceptions			
Si (large area)	21.7 ± 0.7	20.0 (da)	11

Source: Adapted from Green, M.A. et al., *Progress in Photovoltaics: Research and Applications*, 21, 1, 2013.

[a] ap, aperture area; da, designated illumination area.

power output has attraction to cell manufacturers that are having trouble producing sufficient material to keep up with demand. It is worth mentioning that this technology was first developed and proven in the space program, where specific power (power/mass) is a more important consideration than cost.

46.2 CPV Technology

CPV technologies may use any concentrators developed for concentrating thermal technologies, including parabolic trough or parabolic dishes, compound parabolic concentrators (CPCs), or Fresnel concentrators. The concentrations achieved depend on the type of concentrators used. The CPV systems are classified as low, medium, or high concentrations, depending on the concentration ratio as shown in Table 46.2.

The most common type of concentrator is based on Fresnel concentrator as shown in Figure 46.2.

For low concentration CPV, silicon or even CdTe of CIGS panels may be used. However, for very high concentrations, triple junction cells of III–V compounds are used.

TABLE 46.2

CPV Concentration Classes

Class of CPV	Typical Concentration Ratio	Type of Converter
High concentration	>400×	Multijunction
Medium concentration	~3×–100×	Silicon or other cells
Low concentration	<3×	Silicon modules

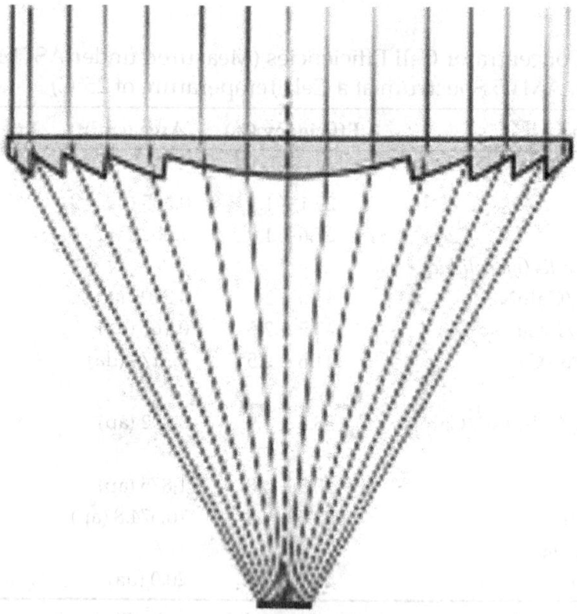

FIGURE 46.2
Fresnel concentrator optics used in the most common CPV panels.

46.2 CPV Market

In 2004, less than 1 MW of concentrator PV systems was installed, out of a total world PV market of 1200 MW, and for some time, the market share of CPV remained extremely small. However, a dramatic reduction in cost of a CPV system from \$3.54/W in 2012 to \$2.62/W in 2013 increased the CPV installation to 160 MW in 2013 (Melkonyn 2013).

Amonix, Ind., of Torrance, California, spent more than 15 years developing five generations of CPV prototypes that led to a 35 kW CPV system installed in Arizona by Arizona Public Service (Amonix 2005). Since then, they have completed the largest CPV system of 30 MW in Alamosa, Colorado, in 2012. As of 2014, there are 54 companies developing or marketing CPV systems around the world. With reduction in the prices and increase in the efficiency, the global market for CPV is expected to grow at a very high rate (Mendelsohn et al. 2012).

46.4 Energy Payback

CPV system costs are much more sensitive to the price of steel than to the price of silicon. In this regard, CPV systems share similar concerns with the wind industry (McConnell 2002). Other technological similarities with wind systems include the low cost of production plants, suitability for distributed and large-scale utility generation, modularity, moving parts, and the need for a good resource, be it wind or solar.

Such observations suggest that CPV systems could follow in the footsteps of wind systems. It seems plausible that CPV system costs can approach wind system costs (typically U.S. 80¢/W today—or about the same as projected CPV costs) if only because common materials (e.g., steel, glass, plastic) are dominant and because production plant costs are relatively low. Also consistent with these observations are recent estimates of energy payback for CPV technologies that are very close to values published for wind turbines at good wind sites. Specifically, the energy payback has been estimated at 8 months for a CPV system in a site having a good solar resource (Bett et al. 2005). More recently, the Frounhofer Institute for Solar Energy in Freiburg, Germany, estimated the energy payback period for a CPV system in Italy as about 9 months (Fraunhofer ISE 2013).

46.5 Qualification Standards

Qualification standards help developers design their new products by identifying weaknesses before production and project installation. They give customers the confidence that their project investments will pay off. In short, they can contribute immensely to a technology's successful market entry.

Fortunately, the CPV industry thought about this situation in the 1990s. Standards take years to develop because the process is based on consensus. Companies do not want standards that are unnecessarily strict or require expensive test procedures. Customers want standards that ensure good product performance. So input from both groups—companies and customers—as well as from relatively objective research organizations leads to an accepted set of test procedures vital for successful CPV market. The first CPV standard (IEEE 2001) was published in 2001. This standard, however, was most suitable for US concentrator PV technologies using Fresnel lenses. The International Electrotechnical Commission (IEC) with input from engineers from more than 10 countries has been developing a standard suitable for concentrators using mirrors or lenses with solar concentration ratios ranging from a couple of suns to thousands of suns. The IEC issued their standard IEC 62108 in 2007 as a comprehensive CPV standard (IEC 2007). They are also working on new standards for solar trackers and safety.

References

Amonix, Inc. 2014. http://www.amonix.com.

Bett, A.W., Siefer, G., Baur, C.B., Rioesen, S.V., Peharz, G., Lerchenmuller, H., and Dimroth, F. 2005. G. SFLATCON concentrator PV: Technology ready for the market. Presentation at *the 20th European PV Solar Energy Conference and Exhibition*.

Fraunhofer ISE. 2013. Photovoltaics report, November 2013. http://www.ise.fraunhofer.de.

Green, M.A., Emery, K., Hishikawa, Y., Warta, W., and Dunlop, E.D., 2013. Solar cell efficiency tables. *Progress in Photovoltaics: Research and Applications*, 21: 1–11.

IEC. 2007. IEC Standard 62108: Concentrator photovoltaic (CPV) modules and assemblies—Design qualification and type approval. International Electrochemical Commission, Geneva, Switzerland. http://www.iec.ch.

IEEE. 2001. IEEE recommended practice for qualification of concentrator photovoltaic receiver sections and modules. IEEE Standard, pp. 1523–2001.

McConnell, R. 2002. Large scale deployment of concentrating PV: Important manufacturing and reliability issues. Proceedings of the First International Conference on Solar Electric Concentrators, NREL/EL-590-32461.

Melkonyn, M. 2013. *The World Market for Concentrated Photovoltaics (CPV) 2013 Edition*, IHS, Inc, Englewood, CO.

Mendelsohn, M., Lowden, T., and Canavan, B. 2012. Utility-scale concentrating solar power and photovoltaics projects: A technology and market overview, NREL/TP-6A20-51137, April 2012. http://www.nrel.gov.

NREL. 2014. Best research cell efficiencies. National Renewable Energy Laboratory, Golden, CO. http://www.nrel.gov.

47

Waste-to-Energy Combustion

Charles O. Velzy and Leonard M. Grillo

CONTENTS

47.1 Introduction

One of the most serious issues facing urbanized areas today is development of cost-effective environmentally acceptable disposal of the community's solid waste. The solid waste generated in a community may be collected by private companies or governmental entities, or portions by both, but the assurance that the waste is ultimately disposed of in an environmentally safe manner is a governmental responsibility.

Solid waste management is a major issue in the United States, because of increasing concerns with environmental problems. One potential solution is to use municipal solid waste, which, for all practical purposes is a renewable commodity, for the generation of electricity. An analysis by Penner and Richards[53] showed that incineration of municipal waste, even after 30% of the waste was recycled, could provide as much electric power as eight large nuclear or coal generating stations. Their analysis further concluded that this could provide 1%–2% of the total electric energy needs in the United States at prices competitive with coal-fired base load power plants.

The basic technology for modern waste-to-energy combustion was developed in Europe during the 1960s and 1970s. This technology, which has been modified and improved since its development, has been widely implemented in the United States. However, despite the fact that incineration of solid waste can decrease its volume ninefold and ameliorate the final waste disposal into landfills, the full potential of utilizing solid waste for energy production is not being realized because of widespread fears regarding environmental pollution. In preparing this chapter, the realities of the situation have been taken into account and the discussion emphasizes the prevention of pollution as much or more than the production of power from waste. Waste-to-energy combustion in modern facilities with adequate environmental safeguards and careful monitoring has been shown to be a safe and cost-effective technology that is likely to increase in importance during the next decade.

Two conditions usually point to the use of combustion processes in treating municipal solid waste prior to ultimate disposal: the waste is collected in an urbanized area with little or no conveniently located land for siting of sanitary landfills (need for volume reduction); and markets exist for energy recovered from the combustion process, and possibly for reclaimed materials, with the energy attractively priced. Even some rural areas are currently considering waste to energy facilities.

Modern waste-to-energy (WTE) plants reflect significant advances that have been made in addressing the technical and practical difficulties of material handling, combustion control, and flue gas cleanup. In the early days of waste incineration, when air pollution regulations were undemanding or nonexistent, relatively simple, fixed-grate plants operating on a single- or two-shift basis were common. However, with increasingly stringent air pollution control regulations, more complex plants requiring continuous operation are now being built.

47.2 Waste Quantities and Characteristics

Municipal solid waste (MSW) as used herein refers to solid waste collected from residences and commercial, light industrial and institutional waste. It does not include heavy industrial waste, which is another problem and varies widely in quantity and characteristics, depending on the industry and specific industrial plant. Changes in packaging practices and improvements in the general standard of living have resulted in significant increases in the quantities of solid waste generated over the past 50 years. Additionally, increasing emphasis on and participation in the recycling of wastes by local communities has resulted in significant variations in quantity and characteristics of MSW at the local community level. All of these factors must be considered when planning a WTE facility. Chapter 48 gives more information on waste availability.

47.2.1 Waste Quantities

In the United States approximately 120 million tons of MSW were generated in 1970, increasing to 220 million tons in 1998.[23] MSW generation is projected to increase to almost 260 million tons by the year 2015.[25] At the local level the quantity of solid waste generated varies geographically, daily, and seasonally, according to the effectiveness of the local recycling initiatives, and differences in socioeconomic conditions.[60]

Over the past 40 years, numerous studies by EPA,[21] APWA,[54] and others[67] have indicated that urbanized areas in this country generate approximately 2.0 lb/capita/day of MSW from residences and another 2.0 lb/capita/day from commercial and institutional facilities, on a national average basis. Thus, a typical community of 100,000 inhabitants would generate about 200 tons/day of gross MSW discards averaged over a 1-year time period.

These projections are subject to adjustments related to specific community characteristics. Thus, communities in the south, with longer active growing seasons than those in the north, tend to produce and collect more yard waste. Recent requirements for on-site disposal and/or composting of yard waste is changing this variable. Rural communities tend to produce less waste per capita than highly urbanized areas because of their greater potential for on-site waste disposal. In the past, the communities in the north tended to produce more waste in the winter due to the prevalence of heating of homes with solid fuels, which produced large quantities of ashes for disposal. Variations in yard wastes and ashes produced from home heating with solid fuels are also examples of variations in MSW quantities related to seasonal effects. Seasonal variations in MSW generation have been noted to range ±15% from the average, while daily variations in waste collections may range up to ±50% from the average, depending largely on number of collections per week. Daily variations in waste quantities are more important in designing certain plant components, while geographic and seasonal effects are more important in establishing plant size.

Waste quantities are also affected by the effectiveness of local recycling initiatives and by socioeconomic conditions. EPA studies have indicated an increase, nationally, in waste recycling from 6.6% in 1960–1970, to 16.2% in 1990,[22] and 28.2% in 1998.[24] The national recycling rate had increased to 34% and 87.2 million tons by 2013,[25] at which time it is expected to level off.* A community in New England[7] projected a 14.0% drop in MSW generation between 1989 and 1991 due to recycling activities in the community, and a 6.0% drop due to the recession during this period in this region of the country. A 15% drop in MSW collections was noted in a Long Island community during a recessionary period in 1972. The impact of recycling on MSW generation should be considered in plant sizing, while the impact of recessionary periods on plant economics may require specific consideration during project planning.

* The EPA updates figures annually. The most up-to-date information is available at www.epa.gov/solidwaste/nonhaz/municipal.

47.2.2 Waste Characteristics

It is important in approaching the design of a WTE facility that one consider the potential variations in both physical and chemical composition of MSW. Historically, one of the most troublesome areas in WTE plants has been materials handling systems. To successfully select materials handling system components and design an integrated process, one must have adequate information on the variability and extremes of the physical size and shape the solid waste facility must handle, the bulk density and angle of repose of the material, and the variation in noncombustible content. This information generally is not available from published surveys and reports, and can only be secured through inspection of the MSW in the field. Materials handling equipment for refuse feeding and residue handling must be large enough and oriented properly to pass the largest bulky items in the MSW, and large enough and rugged enough to handle the quantities of materials required to meet plant design capacity, or the plant will experience expensive periods of down time and might have to be derated.

In the design of the furnace/boiler portion of WTE facilities, the refuse characteristics of interest are the calorific value, moisture content, proportion of noncombustibles, and other components (such as heavy metals, chlorine, and sulfur) whose presence during combustion will result in the need for flue gas cleanup. The capacity of a WTE furnace boiler is roughly inversely proportional to the calorific or heating value of the waste. Table 47.1 illustrates the variation in waste characteristics that has been observed in studies defining the average solid waste composition in the United States since 1960.

As indicated in Table 47.1, approximately 35%–40% of the combustible fraction of MSW is composed of cellulosic material such as paper and wood. This percentage has remained relatively constant over the past 20 years, even after taking into account the

TABLE 47.1

Generation, Materials Recovery, Composting, Combustion with Energy Recovery and Discards of MSW, 1960–2013 (in Millions of Tons)

Activity	1960	1970	1980	1990	2000	2005	2009	2011	2012	2013
Generation	88.1	121.1	151.6	208.3	243.5	253.7	244.6	250.5	251.0	254.1
Recovery for recycling	5.6	8.0	14.5	29.0	53.0	59.2	61.9	66.4	65.3	64.7
Recovery for composting[a]	neg.	neg.	neg.	4.2	16.5	20.6	20.7	20.6	21.3	22.4
Total materials recovery	5.6	8.0	14.5	33.2	69.5	79.8	82.6	87.0	86.6	87.2
Discards after recovery	82.5	113.0	137.1	175.0	174.0	173.9	162.0	163.5	164.4	167.0
Combustion with energy recovery[b]	0.0	0.4	2.7	29.7	33.7	31.6	29.0	31.8	32.2	32.7
Discards to landfill, other disposal[c]	82.5	112.6	134.4	145.3	140.3	142.3	133.0	131.7	132.2	134.3

Note: neg., negligible = less than 5000 tons or 0.05%.

[a] Composting of yard trimmings, food and other MSW organic material. Does not include backyard composting.

[b] Includes combustion of MSW in mass burn or refuse-derived fuel form, and combustion with energy recovery of source separated materials in MSW (e.g., wood pallets, tire-derived fuel).

[c] Discards after recovery minus combustion with energy recovery. Discards include combustion without energy recovery. Details might not add to totals due to rounding.

greater than fourfold increase in recycling percentage achieved over this period. The remainder of the combustible content is composed of various types of plastics, rubber, and leather. The heat released by burning cellulose is approximately 8000 Btu/lb (on a dry basis), while that released by plastics, rubber, and leather is significantly higher on a per pound, dry basis. Heat released by burning garbage (on a dry basis) is only slightly less than cellulose. However, the moisture content of garbage has been observed to range from 50% to 75%, by weight, while that of the cellulosic fraction of MSW usually ranges from 15% to 30%.

In recent years, it has been observed that the higher heating value (HHV) of the combustible portion of MSW (moisture and ash free) averages about 9400 Btu/lb. Considering the recent changes in MSW composition following recycling (increase in plastics while cellulosic material has remained relatively constant), this moisture and ash free HHV has probably increased to 9500 Btu/lb. Taking 9500 Btu/lb as the moisture and ash free heat content of MSW, Table 47.2 illustrates the variation in as-received heat content that one could expect in MSW with moisture content ranging from 20% to 50% by weight and noncombustible content ranging from 25% by weight (earliest period) to approximately 15% by weight (currently).

Moisture content is a highly important and also a highly variable characteristic of waste materials. The moisture content of MSW is generally around 25%, but has been observed to vary from 15% to 70%. This variation may be due, for example, to seasonal variations in precipitation, the nature of the waste (e.g., grass clippings vs. paper) and the method of storage and collection (e.g., open vs. closed containers/trucks). Thus, after a heavy rain, the moisture content of the solid waste may be so high that it may be difficult to sustain combustion. The combustion of solid waste usually can proceed without supplementary fuel when the heat value is greater than 3500–4000 Btu/lb (8140–9300 kJ/kg). This type of variation in MSW composition must be considered in the design of WTE facilities.

The ultimate or elemental analysis of the combustible portion of the MSW refers to the chemical analysis of the waste for carbon, hydrogen, oxygen, sulfur, chlorine, and nitrogen. This information is used to estimate the heat content of waste, moisture, and ash free; to predict the composition of the flue gases; and, from the last three elements (sulfur, chlorine, and nitrogen), to assess the possible impact of waste combustion on air pollution. A typical analysis of solid waste is presented in Table 47.3.

TABLE 47.2

Variation in Heat Content of MSW

	15		25	
Noncombustible (%)	**Comb. (%)**	**Heat Cont. (Btu/lb)**	**Comb. (%)**	**Heat Cont. (Btu/lb)**
Moisture%				
20	65	6125	55	5225
30	55	5225	45	4275
40	45	4275	35	3325
50	35	3225	25	2375

TABLE 47.3
Analysis of Solid Waste

	Percent by Weight	
	West Europe	**United States**
Proximate analysis		
Combustible	42.1	50.3
Water	31.0	25.2
Ash and inert material	26.9	24.5
Total	100	100
Ultimate (elemental) analysis of combustibles		
Carbon	51.1	50.9
Hydrogen	7.1	6.8
Oxygen	40.1	40.3
Nitrogen	1.2	1.0
Sulfur	0.5	0.4
Chlorine	—	0.6
Total	100	100

Sources: From Domalski, E.S. et al., The chlorine content of municipal solid waste from Baltimore County, Maryland, and Brooklyn, NY, in: *Proceedings 1986 National Waste Processing Conference*, Denver, CO, June 1–4, 1986, ASME, New York, pp. 435–448; Seeker, W.R. et al., *Municipal Waste Combustion Study: Combustion Control of Organic Emissions*, EPA, Research Triangle Park, NC, 1987; Suess, M.J. et al., *Solid Waste Management Selected Topics*, WHO Regional Office for Europe, Copenhagen, Denmark, 1985.

Note: Gross heat value, as fired = 3870 Btu/lb (9000 kJ/kg).

47.3 Design of WTE Facilities

The primary function of a WTE facility is to reduce solid waste to an inert residue with minimum adverse impact on the environment. Thermal efficiency, in terms of maximizing the capture of energy liberated in the combustion process, is of secondary importance. WTE facilities are usually classified as mass-burn systems, or refuse-derived fuel (RDF) systems.

47.3.1 General Features

47.3.1.1 Types of Facilities

Mass-burn systems are large facilities (usually over 200 tons/day) that burn, as-received, unprocessed MSW which is extremely heterogeneous. Most mass-burn systems burn the waste in a single combustion chamber under conditions of excess air (i.e., more than is needed to complete combustion) (see Figure 47.1). The waste is burned on a sloping, moving grate, which helps agitate the MSW and mixes it with combustion air. Many different proprietary grate systems exist.

In refuse-derived fuel (RDF) systems, usually large facilities, the MSW is first processed (see Figure 47.2) by mechanical means to produce a more homogeneous material prior to introduction into a furnace/boiler. Several types of RDF can be made—coarse, fluff, powder, and densified. These differ in complexity and horsepower requirements of the waste processing facilities, size of particle produced, and whether or not the

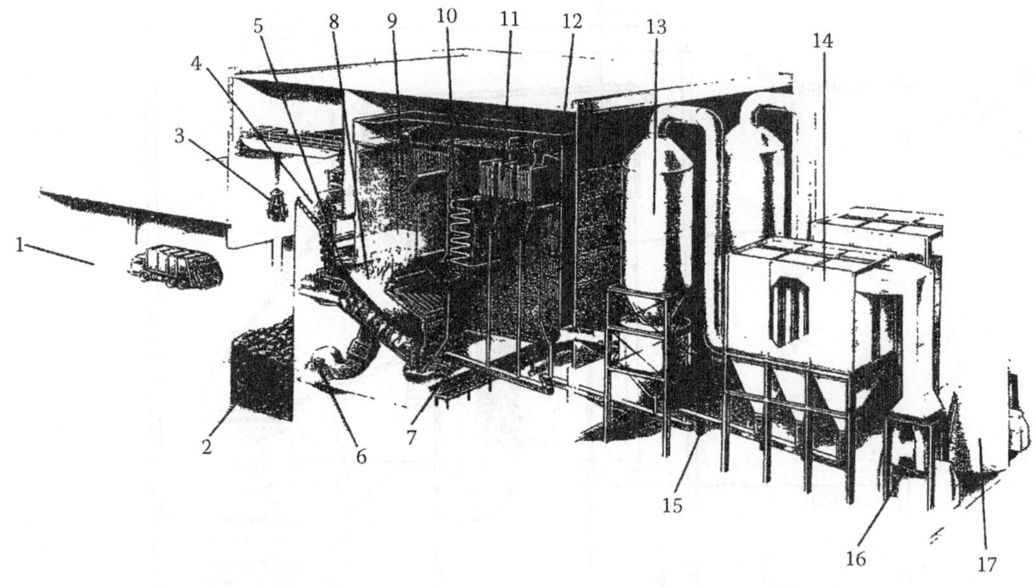

FIGURE 47.1
A typical Covanta facility. (1) Tipping floor, (2) refuse holding pit, (3) feed crane, (4) feed chute, (5) Martin stroker grate, (6) combustion air fan, (7) Martin residue discharger and handling system, (8) combustion chamber, (9) radiant zone (furnace), (10) convection zone, (11) superheater, (12) economizer, (13) dry gas scrubber, (14) baghouse or electrostatic precipitator, (15) fly ash handling system, (16) induced draft fan, and (17) stack.

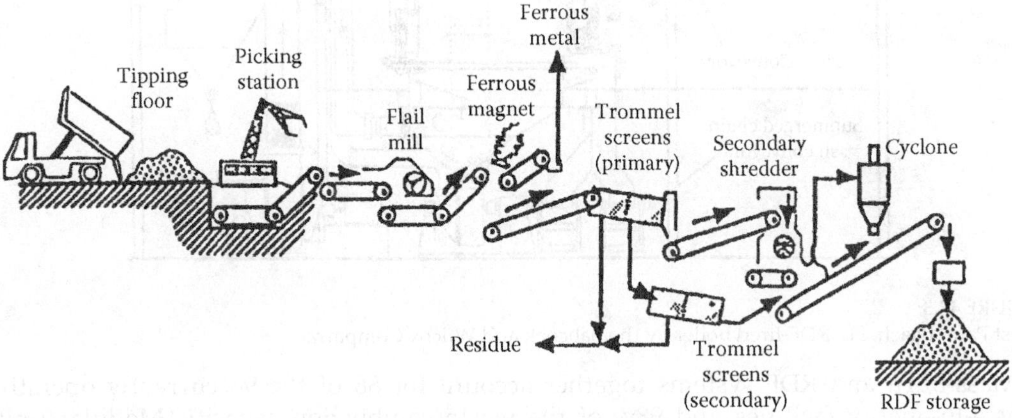

FIGURE 47.2
RDF processing system.

material is compacted under pressure into pellets, briquettes, or similar forms. The coarse type of RDF is the most common form produced at this time.

RDF can be burned in one of the two types of boilers. It can be used as the sole or primary fuel in dedicated boilers (see Figure 47.3) or it can be co-fired with conventional fossil fuels in existing industrial or utility boilers. One advantage of these systems is that RDF can be produced at one location for use at a nearby off-site boiler, allowing for flexibility in locating processing facilities. Also, some materials, such as steel and glass, can be recovered for recycling during the initial processing step.

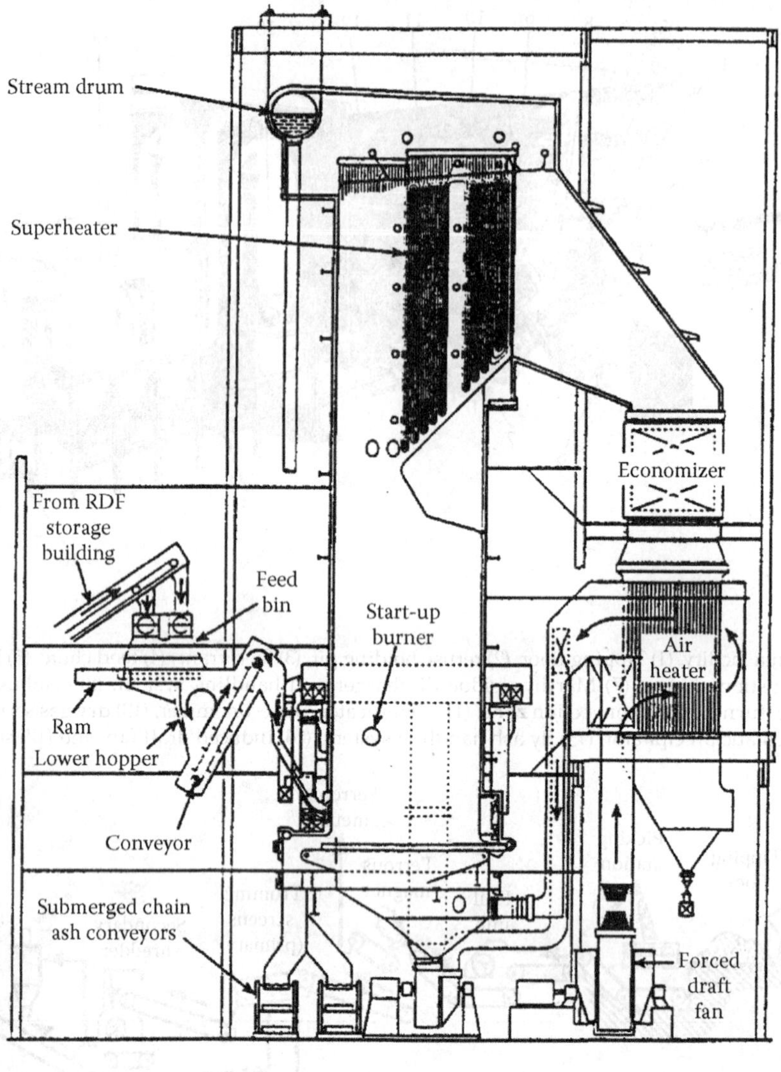

FIGURE 47.3
West Palm Beach, FL, RDF-fired boiler by the Babcock and Wilcox Company.

Mass-burn and RDF systems together account for 86 of the 98 currently operating waste-to-energy facilities and 98% of the waste combustion capacity. Modular units, described briefly later, account for the other 12 units and 2% of the waste combustion capacity.[39]

47.3.1.2 Operation and Capacity

The capacity to be provided in a facility is a function of the area and population to be served; and the rate of refuse production for the population served. A small plant (100 tons/day) without energy recovery might not be operated on weekends. For capacities above 400 tons/day, or any plant with energy recovery, economic and/or equipment operating considerations usually dictate three-shift operation, 7 days/week.

If collection records, preferably by weight, are available for the community, forecasts for determining required plant capacity can be made with reasonable accuracy. If records are not available, refuse quantities for establishing plant size may be approximated by assuming a refuse generation rate of 4.0 lb/capita/day when there is little or no waste from industry. Of course, in planning for plant capacity, the impact of local recycling activities on both quantity and characteristics of MSW must be considered as discussed earlier.

Other factors must be taken into account in establishing the size or capacity of a facility. Should the facility serve only one community, or should it be regional and serve several communities? What are the possible benefits of economies of scale? What is the impact of the cost of hauling refuse to a central point on overall project economics? There is substantial evidence available at present to show that implementation becomes much more difficult as the number of separate political jurisdictions is increased. Imposition of regional plans on local jurisdictions to achieve economies of scale, where it cannot be conclusively demonstrated that such regional plans make sound economic sense based on the total cost of the solid waste management plan, including the cost of transporting the solid waste to the regional facility, is, at best, unwise. Economies of scale in these projects have tended to be illusory, while haul costs to gather sufficient waste together to achieve the economies of scale have tended to be ignored in developing total project economics.

47.3.1.3 Siting

One of the key issues to face in implementing a WTE project is locating a site for the facility. Since MSW is usually delivered to these plants by truck, inevitably there will be substantial truck traffic in the vicinity of the plants. The equipment and processes used in these plants are industrial in nature. They are generally noisy at the source and tend to produce dust and odors. These facts indicate that it is desirable to site such plants in industrial or commercial areas.[45] It has been contended, as cited in a 1989 OTA report[68] "that sites are sometimes selected to avoid middle- and higher-income neighborhoods that have sufficient resources to fight such development."

Plants should be located near major highways to minimize the impact from increased truck traffic. As shown by operating WTE plants in Europe and the United States, it is possible to control all nuisance conditions by proper attention to the details of plant design. The local impact of truck deliveries to the plant can be minimized by providing sufficient length of access road so that refuse truck queuing does not take place on public highways. Odors and noise can be confined to the plant building. Odors and fugitive dust can be destroyed by collecting plant air and using it for combustion air supply. Noise should be attenuated at the source to maintain healthful working conditions. In all cases, there is no need to adversely impact the surrounding neighborhood. Proper attention to architectural treatment can result in a structure that blends into its surroundings; if sited in industrial or commercially zoned areas.

Since considerable vertical distance is frequently required in the passage of MSW through a mass-burn WTE plant, there is an advantage in a sloping or hillside site. Collection trucks can deliver MSW at the higher elevation, while residue trucks operate at the lower elevation, requiring minimum site regrading. This consideration does not generally apply to RDF plants.

47.3.2 Fuel Handling

47.3.2.1 Refuse Receipt, Processing, and Storage

Scales, preferably integrated into an automated record keeping system, should be provided to record the weight of MSW delivered to the WTE plant. Either the entire tipping area or individual tipping positions should be enclosed so as to control potential nuisance conditions in the vicinity of the plant, such as blowing papers, dust, and/or odors. The number of tipping positions provided should take into consideration the peak number of trucks expected per hour at the facility and should be located so the trucks have adequate time and room to maneuver to and from the dumping positions while minimizing queue time.

Collections usually are limited to one 8 h daily shift 5 days/week (sometimes with partial weekend collection) while burning will usually be continuous, so ample storage must be provided. This usually requires 2–3 days of refuse storage at most WTE plants. Seasonal and cyclic variations should also be a consideration in establishing plant storage requirements.

Refuse storage in large mass-burn plants is normally in long, relatively narrow, and deep pits extending along the front of the furnaces. It will generally be necessary to rehandle the refuse dumped from the trucks. In some mass-burn plants and in RDF plants, floor dumping and storage of the MSW either on the dump floor or in shallow, relatively wide pits is common practice.

When computing the dimensions required for storage of as-received MSW, the required volume may be determined based on an MSW bulk density of from 400 to 600 lb/yd^3 (240–360 kg/m^3).[54,61] Other factors to consider in sizing and laying out refuse storage facilities in WTE plants is that refuse flows very poorly and can maintain an angle of repose greater than 90°. Thus, MSW is commonly stacked in storage facilities to maximize storage capacity.

Sizing a refuse storage pit requires the use of empirical data, judgment, site constraints, and knowledge of plant layout and operations. The pit should be at least long enough to provide sufficient truck tipping positions so that the trucks are not unduly delayed in discharging their waste (tipping) into the pit. It has been found in practice that it takes an average of 10 min for the truck to perform the tipping operation. This time may be shorter for packer trucks and longer for transfer vehicles. Each tipping position must provide at least 14 ft of unobstructed width for this operation. Ideally, 20 ft should be allowed for each tipping position to allow for convenient truck access and space for armored building support columns.

The pit should be capable of holding a minimum of 3 days' storage at the facility's maximum continuous rating. The desired volume should be based on a bulk density of about 500 lb/cycle of waste. The dimensions of the pit will be dependent on site constraints and the facility design. The pit is usually at least as long as the width of all the boilers it feeds. The depth will be dependent on groundwater conditions, but should be 30–45 ft deep if possible. The total volume of storage should equal the volume in the pit up to the tipping floor level plus the volume above the tipping floor assuming the waste is stacked at a 45° from the charging floor wall to the tipping floor. The remainder of the pit will be used for the grapple to move waste away from the tipping positions. After the waste has been stacked, the grapple can remove the material against the tipping floor wall and form a trench, since the waste will maintain a vertical face when stacked.

At most newer and more successful RDF plants, after receiving the MSW in a floor dump type of operation, the MSW is loaded onto conveyors that carry the material to flail mills or trommels with bag-breaking blades. These facilities break apart the bags containing the waste, allowing glass and some metals to be separated from the remaining MSW. The separated MSW, primarily the light combustible fraction, is then reduced in size. Removal of the glass prior to the size reduction process alleviates the problem, experienced in earlier plants, of contamination of combustible material with glass shards.

Processes to produce powdered fuel or RDF fuel pellets, although interesting, have not been developed to a state of commercial availability. A process to produce RDF by "hydro-pulping" after being attempted in two full-scale plants, was not commercially successful.

47.3.2.2 Refuse Feeding

Batch feeding of MSW, practiced in the past in mass-burn plants, is undesirable and is not practiced in modern plants. In the larger mass-burn plants, the solid waste is usually moved from the storage pit to a charging hopper by a traveling bridge crane and an orange-peel type of grapple. The grapple size is established by a duty cycle analysis, taking into account the quantity of material that must be moved from the pit to the furnaces, the distances over which the material must be moved, allowable crane speeds, and the need to rehandle (mixing and/or stacking) material in the pit. Grapples can range in size from 1.5 to 8 yd^3 (1–6 m^3) capacity and larger.

The crane used in this service should be capable of meeting the severest of duty requirements.[52] The load lifting capability is established by adding to the grapple weight, 1.5 times the volumetric capacity of the grapple times a density of MSW of 600–800 lb/yd^3 (360–480 kg/m^3).[37] In the past, the crane has been operated from an air-conditioned cab mounted on the bridge. However, crane operation is now centralized in a fixed control room, usually located at the charging floor elevation and either over the tipping positions opposite the charging hoppers or in the vicinity of the charging hoppers.

In modern mass-burn plants, the MSW is deposited from the crane grapple into a charging hopper. The charging hopper, which is built large enough to prevent spillage on the charging floor and with slopes steep enough to prevent bridging, is placed on top of a vertical feed chute that discharges the MSW into the furnace. The feed chute may be constructed of water cooled steel plates or steel plates lined with smooth refractory material. The chute is normally at least 4 ft (1.2 m) wide, to pass large objects with a minimum of bridging, and the width of the furnace. It is kept full of refuse to prevent uncontrolled admission of air into the furnace. The refuse is fed from the bottom of the feed chute into the furnace by a portion of the mechanical grate, or by a ram. The ram generally provides better control of the rate of feed into the furnace than the older technique of using a portion of the mechanical grate for refuse feed.

In RDF plants, conveyors, live bottom bins, and pneumatic handling of the size-reduced MSW combustible material have been utilized. The fuel material is usually blown into these furnaces, where it is partially burned while in suspension, with combustion being completed on grates at the bottom of the furnace. These fuel feeding systems are generally more complex than the mass-burn systems.

47.3.3 Combustion Principles

Combustion is the rapid oxidation of combustible substances with release of heat. Oxygen is the sole supporter of combustion. Carbon and hydrogen are by far the most important of the combustible substances. These two elements occur either in a free or combined state

TABLE 47.4
Elements and Compounds Encountered in Combustion

Substance	Molecular Symbol	Molecular Weight	Form	Density (lb/ft³)
Carbon	C	12.0	Solid	—
Hydrogen	H_2	2.0	Gas	0.0053
Sulfur	S	32.1	Solid	—
Carbon monoxide	CO	28.0	Gas	0.0780
Oxygen	O_2	32.0	Gas	0.0846
Nitrogen	N_2	28.0	Gas	0.0744
Nitrogen atmos.	N_{2atm}	28.2	Gas	0.0748
Dry air		29.0	Gas	0.0766
Carbon dioxide	CO_2	44.0	Gas	0.1170
Water	H_2O	18.0	Gas/liquid	0.0476
Sulfur dioxide	SO_2	64.1	Gas	0.1733
Oxides of nitrogen	NO_x	—	Gas	—
Hydrogen chloride	HCl	36.5	Gas	0.1016

Source: From Hecklinger, R.S., Combustion, in: *The Engineering Handbook,* Dorf, R.C. (ed.), CRC Press, Inc., Boca Raton, FL, 1996.

in all fuels—solid, liquid, and gaseous. Sulfur is the only other element considered to be combustible. In combustion of MSW, sulfur is a minor constituent with regard to heating value. However, it is a concern in design of the air pollution control equipment. The only source of oxygen considered here will be the oxygen in the air around us.

Table 47.4 displays the elements and compounds that play a part in the combustion process. The elemental and molecular weights displayed are approximate values which are sufficient for combustion calculations. Nitrogen is listed as chemical nitrogen N_2, with a molecular weight of 28.0 and as atmospheric nitrogen, N_{2atm}, which is a calculated figure to account for trace constituents of dry air. Water occurs as a vapor in air and in the products of combustion and as a liquid or vapor constituent of MSW fuel.

A U.S. standard atmosphere of dry air has been defined as a mechanical mixture of 20.947% O_2, 78.086% N_2, 0.934% Ar (argon), and 0.033% CO_2 by volume.[43] The percentages of argon and carbon dioxide in air can be combined with chemical nitrogen to develop the following compositions of dry air by volume and by weight that can be used for combustion calculations:

Constituent	% by Volume	% by Weight
Oxygen, O_2	20.95	23.14
Atmospheric nitrogen, N_{2atm}	79.05	76.86

Atmospheric air also contains some water vapor. The level of water vapor in air, or its humidity, is a function of atmospheric conditions. It is measured by wet and dry bulb thermometer readings and a psychrometric chart. If specific data are not known, the American Boiler Manufacturers Association recommends a standard of 0.013 lb of water per pound of dry air, which corresponds to 60% relative humidity and a dry bulb temperature of 80°F.

Table 47.5 displays the chemical reactions of combustion. These reactions result in complete combustion; that is, the elements and compounds unite with all the oxygen with which they are capable of entering into combination. In actuality, combustion is a more complex process in which heat in the combustion chamber causes intermediate

TABLE 47.5

Chemical Reactions of Combustion

Combustible	Reaction
Carbon	$C + O_2 = CO_2$
Hydrogen	$2H_2 + O_2 = 2H_2O$
Sulfur	$S + O_2 = SO_2$
Carbon monoxide	$2CO + O_2 = 2CO_2$
Nitrogen	$N_2 + O_2 = 2NO$
Nitrogen	$N_2 + 2O_2 = 2NO_2$
Nitrogen	$N_2 + 3O_2 = 2NO_3$
Chlorine	$4Cl + 2H_2O = 4HCl + O_2$

Source: From Hecklinger, R.S., Combustion, in: *The Engineering Handbook*, Dorf, R.C. (ed.), CRC Press, Inc., Boca Raton, FL, 1996.

reactions leading up to complete combustion. An example of intermediate steps to complete combustion would be when carbon reacts with oxygen to form carbon monoxide and, later in the combustion process, the carbon monoxide reacts with more oxygen to form carbon dioxide. The combined reaction produces precisely the same result as if an atom of carbon combined with a molecule of oxygen to form a molecule of carbon dioxide in the initial reaction. An effectively controlled combustion process results in well less than 0.01% of the carbon in the fuel leaving the combustion chamber as carbon monoxide; and the remaining 99.99% of the carbon in the fuel leaves the combustion process as carbon dioxide. It should also be noted with regard to Table 47.5 that some of the sulfur in a fuel may combust to SO_3 rather than SO_2 with a markedly higher release of heat. However, it is known that only a small portion of the sulfur will combust to SO_3 and some of the sulfur in fuel may be in the form of pyrites (FeS_2), which do not combust at all. Therefore, only the SO_2 reaction is given. Also, some nitrogen is converted to oxides of nitrogen (NO_x), and some chlorine is converted to hydrogen chloride in the presence of moisture in the flue gas. While these components do not factor into the combustion calculations, they are important for the purpose of establishing air pollution control requirements.

Factors directly affecting furnace design are the moisture and the combustible content of the solid waste to be burnt and the volatility of the material to be burnt. The means for temperature control and sizing of flues and other plant elements should be based on design parameters that result in large sizes. Combustion controls should provide satisfactory operation for loads below the maximum rated capacity of the units.

The combustible portion of MSW is composed largely of cellulose and similar materials originating from wood, mixed with appreciable amounts of plastics and rubber, as well as some fats, oils, and waxes. The heat released by burning dry cellulose is approximately 8,000 Btu/lb, while that released by certain plastics, rubber, fats, oils, and so on, may be as high as 17,000 Btu/lb. If MSW consists of five parts cellulose and one part plastics, rubber, oil, and fat, the heat content of the dry combustible matter only is approximately 9500 Btu/lb.

The heat released in combustion of basic combustible substances is displayed in Table 47.6. The heating value of a substance can be expressed either as higher (or gross) heating value or as lower (or net) heating value. The higher heating value takes into account the fact that water vapor formed or evaporated in the process of combustion includes the latent heat of vaporization, which could be recovered if the products of combustion are reduced in temperature sufficiently to condense the water vapor to liquid water.

TABLE 47.6
Heat of Combustion

		Heating Value (Btu/lb)	
Combustible	**Molecular Symbol**	**Gross**	**Net**
Carbon	C	14,100	14,100
Hydrogen	H_2	61,100	51,600
Sulfur	S	3,980	3,980
Carbon monoxide	CO	4,350	4,350

Source: From Hecklinger, R.S., Combustion, in: *The Engineering Handbook*, Dorf, R.C. (ed.), CRC Press, Inc., Boca Raton, FL, 1996.

The lower heating value is predicated on the assumption that the latent heat of vaporization will not be recovered from the products of combustion.

The heat released during combustion may be determined in a bomb calorimeter, a device with a metal container (bomb) immersed in a water jacket. A 1 g MSW sample is burned with a known quantity of oxygen, and the heat released is determined by measuring the increase in temperature of the water in the water jacket. Since the bomb calorimeter is cooled to near ambient conditions, the heat recovery measured includes the latent heat of vaporization as the products of combustion are cooled and condensed in the bomb. That is, the bomb calorimeter inherently measures higher heating value (HHV). It has been customary in the United States to express heating value as HHV. In Europe and elsewhere, heating value is frequently expressed as the lower heating value (LHV).

Heating value can be converted from HHV to LHV if weight decimal percentages of moisture and hydrogen (other than the hydrogen in moisture) in the fuel are known, using the following formula:

$$LHV_{Btu/lb} = HHV_{Btu/lb} - [\%H_2O + (9 \times \%H_2)] \times (1050\ \text{Btu/lb}) \tag{47.1}$$

$$LHV_{J/kg} = HHV_{J/kg} - [(9 \times\%H_2) + \%H_2O] \times (2240\ \text{kJ/kg}) \tag{47.2}$$

For example (using data from Table 47.9),

$$LHV_{Btu/lb} = HHV_{Btu/lb} - [\%H_2O + (9 \times \%H_2)] \times 1050\ \text{Btu/lb}$$

$$LHV_{Btu/lb} = 4940 - [0.30 + (9 \times 0.047)] \times 1050$$

$$LHV_{Btu/lb} = 4940 - [0.30 + 0.42] \times 1050$$

$$LHV_{Btu/lb} = 4940 - 756 = \text{Btu/lb}$$

Another method for determining the approximate higher heating value for MSW is to perform an ultimate analysis and then apply Dulong's formula:

$$HHV_{(Btu/lb)} = 14{,}544\,C + 62{,}028\left(H_2 - \frac{O_2}{8}\right) + 4{,}050\,S \tag{47.3}$$

where C, H_2, O_2, and S represent the decimal proportionate parts by weight of carbon, hydrogen, oxygen, and sulfur in the fuel. The term $O_2/8$ is a correction used to account for hydrogen which is already combined with oxygen in the form of water. For example (using data from Table 47.9)

$$HHV_{(Btu/lb)} = 14{,}544\ C + 62{,}028\left(H_2 - \frac{O_2}{8}\right) + 4{,}050\ S$$

$$HHV = 14{,}544 \times 0.257 + 62{,}028\left(0.048 - \frac{0.21}{8}\right) + 4{,}050 \times 0.001$$

$$HHV = 3{,}738 + 62{,}028(0.047 - 0.026) + 4.0$$

$$HHV = 3{,}738 + 62{,}028 \times 0.021 + 4.0$$

$$HHV = 3738 + 1303 + 4 = 5045\ \text{Btu/lb}$$

An alternate method of estimating the HHV is to multiply the approximate dry combustable HHV of 9500 Btu/lb by the weight fraction of combustibles:

$$HHV = 9500 \times (1 - \text{moisture} - \text{ash})$$

$$HHV = 9500 \times (1 - 0.30 - 0.18)$$

$$HHV = 9500 \times 0.52 = 4940\ \text{Btu/lb}$$

The American Society for Testing and Materials (ASTM) publishes methods for determining the ultimate analysis of solid fuels such as MSW. The ultimate analysis of a fuel is developed through measures of carbon, hydrogen, sulfur, nitrogen, ash, and moisture content Oxygen is normally determined "by difference"; that is, once the percentages of the other components are measured, the remaining material is assumed to be oxygen. For solid fuels, such as MSW, it is frequently desirable to determine the proximate analysis of the fuel. The procedure for determining the proximate analysis is also prescribed by ASTM. The qualities of the fuel measured in percentage by weight are moisture, volatile matter, fixed carbon, and ash. This provides an indication of combustion characteristics of a solid fuel. As a solid fuel is heated to combustion, first the moisture in the fuel evaporates, then some of the combustible constituents volatilize (gasify) and combust as a gas with oxygen, and the remaining combustible constituents remain as fixed carbon in a solid state and combust with oxygen to form carbon dioxide. The material remaining after combustion is complete is the ash. MSW, with a high percentage of volatiles and a low percentage of fixed carbon, burns with much flame.

Table 47.7 displays ignition temperatures for combustible substances in MSW. The ignition temperature is the temperature to which the combustible substance must be raised before it will unite in chemical combination with oxygen. Thus, the temperature must be reached and oxygen must be present for combustion to take place. Ignition temperatures are not fixed temperatures for a given substance. The actual ignition temperature is influenced by combustion chamber configuration, oxygen fuel ratio, and synergistic effect of

TABLE 47.7
Ignition Temperatures

Combustible	Molecular Symbol	Ignition Temperature (°F)
Carbon (fixed)	C	650
Hydrogen	H_2	1080
Sulfur	S	470
Carbon monoxide	CO	1170

Source: From Stultz, S.C. and Kitto, J.B. (Eds.), *Steam: Its Generation and Use*, 40th edn., The Babcock and Wilcox Co., Barberton, OH, 1992.

TABLE 47.8
Theoretical Combustion Air

	Pounds per Pound of Combustible					
	Required for Combustion			Products of Combustion		
Combustible	O_2	N_{atm}	Air	CO_2	H_2O	N_{atm}
Carbon	2.67	8.87	11.54	3.67		8.87
Hydrogen	8.00	26.57	34.57		9.00	26.57
Sulfur	1.00	3.32	4.32	2.00		3.32
Carbon monoxide	0.57	1.89	2.46	1.57		1.89

Source: From Hecklinger, R.S., Combustion, *The Engineering Handbook*, Dorf, R.C. (ed.), CRC Press, Inc., Boca Raton, FL, 1996.

multiple combustible substances. The ignition temperature of MSW is the ignition temperature of its fixed carbon component. The volatile components of MSW are gasified but not ignited before the ignition temperature is attained.

The oxygen, nitrogen, and air data displayed in Table 47.8 represent the weight of air theoretically required to completely combust one pound of a combustible substance. The weight of oxygen required is the ratio of molecular weight of oxygen to molecular weight of the combustion constituent as displayed in Table 47.5. The weights of nitrogen and air required are calculated from the percentage by weight constituents of dry air. In actuality, to achieve complete combustion, air in excess of the theoretical requirement is required for complete combustion to increase the likelihood that all of the combustible substances are joined with sufficient oxygen to complete combustion. The level of excess air required in the combustion of MSW depends on the configuration of the combustion chamber, the nature of the fuel firing equipment, and the effectiveness of mixing combustion air with the MSW. An excess air level of 80% is commonly associated with combustion of MSW in modern WTE facilities. Excess air is generally monitored using an oxygen analyzer at the economizer outlet. The type of analyzer used at waste-to-energy facilities generally reports percent wet oxygen. The dry oxygen can be estimated by assuming 15% moisture in the flue gas using the following equation:

$$\text{Dry oxygen} = (\text{wet oxygen})/(1 - \text{percent moisture}/100), \text{ or dry oxygen} = (1.176)(\text{wet oxygen})$$

Excess air can be approximated by the following equation:

$$\text{Excess air} = 55.2 - 10.46 \times (\text{dry } O_2) + 1.4 \times (\text{dry } O_2)^2$$

where dry O_2 is the percentage dry oxygen in the flue gas.

Excess air serves to dilute and thereby reduce the temperature of the products of combustion. The reduction of temperature tends to reduce the heat energy available for useful work. Therefore, the actual excess air used in the combustion process is a balance between the desire to achieve complete combustion and the objective of maximizing the heat energy available for useful work.

It is frequently useful to know the temperature attained by combustion. The heat released during combustion heats the products of combustion to a calculable temperature. It must be understood that the calculation procedure presented here assumes complete combustion and that no heat is lost to the surrounding environment. Thus, it is a temperature that is useful to compare one combustion process with another. The heat available for heating the products of combustion is the lower heating value of the fuel. The increase in temperature is the lower heating value divided by the mean specific heat of the products of combustion. The mean specific heat is a function of the constituent products of combustion ($W_{P.C.}$) and the temperature. To approximate the theoretical temperature attainable, one can use a specific heat of 0.55 Btu/lb/°F for water vapor (W_{H_2O}) and 0.28 Btu/lb/°F for the other gaseous products of combustion ($W_{P.C.} - W_{H_2O}$). Thus, the formula approximating the temperature attained during combustion is

$$T_{comb} = T + \frac{LHV_{Btu/lb}}{0.55W_{H_2O} + 0.28(W_{P.C.} - W_{H_2O})} \tag{47.4}$$

For example (using data from Table 47.9)

$$T_{comb} = T_{ambient} + \frac{LHV_{Btu/lb}}{0.55W_{H_2O} + 0.28(W_{P.C.} + W_{H_2O})}$$

If the ambient temperature is assumed to be 60°F, then

$$T_{comb} = 60 + \frac{4184}{0.55 \times 0.81 + 0.28(7.55 - 0.81)}$$

$$T_{comb} = 60 + \frac{4184}{0.45 + 1.89} = 60 + \frac{4184}{2.34}$$

$$T_{comb} = 1848 \cong 1850°F$$

Typical combustion calculations are provided in Table 47.9 for MSW to determine the products of the combustion process. Each of the combustible substances combines and completely combusts with oxygen as displayed in Table 47.5. The weight ratio of oxygen to the combustible substance is the ratio of molecular weights. Table 47.8 displays the weight or volume of oxygen theoretically required for complete combustion of one pound of the combustible substance. Sulfur dioxide from combustion of sulfur in fuel is combined with CO_2 in the sample calculation as a matter of convenience. If desired, a separate column can be prepared for sulfur dioxide in the products of combustion. Oxygen in the fuel combines with the combustible substances in the fuel, thereby reducing the quantity of air required to achieve complete combustion. The sample calculation uses the weight percentages of oxygen

TABLE 47.9

Sample Calculation for Municipal Solid Waste (MSW)

Air Calculations (80% Excess Air)				
Ultimate Analysis				
Substance	**Fraction% by Weight**	**Oxygen Required for Combustion, lb/lb of Element**	**Theoretical Oxygen, lb/lb of Element**	**Theoretical Dry Air, lb/lb of Element**
Carbon	0.279	2.67	0.745[a]	3.218[b]
Hydrogen	0.037	8.00	0.296[a]	1.279[b]
Oxygen	0.209	—	—	—
Nitrogen	0.005	—	—	—
Sulfur	0.002	1.00	0.002[a]	0.009[b]
Ash	0.187			
Fuel moisture	0.281			
Total	1.000		1.043	4.505
Less oxygen in fuel			(0.209)	(0.903)[c]
Air				
Required at 100% theoretical air			0.834	3.603
180% of theoretical air (80% excess air)			1.501	6.485
Excess			0.667	2.882
HHV = 5100 Btu/lb				
Products of Combustion			**lb/lb of Element**	**lb of Product**
Carbon dioxide			3.67	1.024[d]
Moisture from hydrogen			9.00	0.333[d]
Oxygen			0.667[e]	
Nitrogen			4.989[f]	
Sulfur dioxide			2.00	0.004[d]
Moisture from fuel			1.00	0.281[d]
Moisture from air			0.084[g]	
Total moisture			0.698[h]	
Total			7.382	

LHV = 5100 − [0.281+ (9 × 0.037)] × 1050 = 4455 Btu/lb

Temperature developed in combustion = 60 + 4455/[(0.698 × 0.55[i]) + (7.382 − 0.698) × 0.28[j]] = 2035°F

Check:

Total products of combustion = 180% of theoretical air + moisture from air + fraction percent by weight of C, H, O, N, S, and moisture

7.382 = 6.485 + 0.084 + 0.279 + 0.037 + 0.209 + 0.005 + 0.002 + 0.281 = 7.382

[a] Weight percent of element times oxygen required for combustion.
[b] Theoretical oxygen times 4.32.
[c] Amount of theoretical air due to oxygen in fuel.
[d] Weight percent of element times lb/lb of element.
[e] Excess oxygen.
[f] 180% of theoretical dry air times 0.7686 plus weight percent of nitrogen in fuel.
[g] Moisture in combustion air = 0.013 times 180% of theoretical air (0.013 lb moisture per lb of dry air at 80°F and 60% relative humidity).
[h] Total of moisture from combustion of hydrogen, moisture in fuel, and moisture from air.
[i] Heat capacity of water vapor.
[j] Heat capacity of dry flue gas.

to reduce the theoretical air requirements and the nitrogen in the products of combustion. The decimal percentage of excess air is multiplied by the total theoretical air requirement to establish the weight of excess air and the total air requirement including excess air.

47.3.4 Furnaces

While the general principles of a modern waste combustor burning as-received MSW are common to all types, the specific solid waste combustion process is rather complex. The waste is heated by contact with hot combustion gases or preheated air, and by radiation from the furnace walls. Drying occurs in a temperature range of 122°F–302°F (50°C–150°C). At higher temperatures, volatile matter is formed by complicated thermal decomposition reactions. This volatile matter is generally combustible and, after ignition, produces flames. The remaining material is further degased and burns much more slowly. In an RDF furnace (see Figure 47.3), most of the volatile matter and some of the fixed carbon is burned in suspension while the remaining fixed carbon is combusted on a grate at the bottom of the furnace.

The complexity of the combustion of solid waste streams results from the nature of the decomposition and burning reactions and their association with heat transfer, air flow, and diffusion. In most waste combustors, combustion takes place while the solids are supported on and conveyed by a grate. Since the early 1960s, most MSW incinerators have incorporated one of a number of available proprietary grate systems that allow continuous feed of unscreened waste into and movement through furnaces with integral boiler facilities. The grate performs several functions: provides support for the refuse, admits underfire air though openings in the grate surface, transports the solid waste from feed mechanism to ash quench, agitates the bed, and serves to agitate and redistribute the burning mass.

The basic design factors which determine furnace capacity are grate area and furnace volume. Also, the available capacity and method of introducing both underfire and overfire air will influence, to a lesser extent, furnace capacity. Required grate area, in a conservative design, is normally determined by limiting the burning rate to between 60 and 70 lb/ft^2 h (290–340 kg/m^2 h) of grate area.[54] This is based on limiting the heat release rate loading on the grate to 250,000–300,000 Btu/ft^2 of grate per h (2.8–3.4 GJ/m^2/h).

Furnace volume required is established by the rate of heat release from the fuel. Thus, furnace volume is generally established by using heat release rates ranging from 12,500 to 20,000 Btu/ft^3/h (450–750 MJ/m^3/h), with the lower heat release rate being more desirable from the standpoint of developing a conservative design. A conservative approach to design in this area is desirable because of probable periodic operation above design capacity to meet short-term higher than normal refuse collections and possible receipt of high heat-content waste.

Water wall units burning as-received MSW have been built as small as 75–100 tons/day (68–91 tons/day) capacity. However, the cost per ton of rated capacity of such units is relatively high. A more common unit size for both mass-burn and RDF furnaces is 250–1000 tons/day (225–900 tons/day), while water wall mass-fired units have been built as large as 750–1200 tons/day (675–1090 tons/day) capacity.[5]

The primary objective of a mechanical grate in a mass-burn furnace is to convey the refuse from the point of feed through the burning zone to the point of residue discharge with a proper depth of fuel and sufficient retention time to achieve complete combustion. The refuse bed should be agitated so as to enhance combustion. However, the agitation should not be so pronounced that particulate emissions are unreasonably increased. The rate of movement of the grate or its parts should be adjustable to meet varying conditions or needs in the furnace.

In the United States over the past 20 years, several types of mechanical grates have been used in continuous feed furnaces burning as-received MSW. These include reciprocating grates (see Figure 47.4), rocking grates (see Figure 47.5), roller grates (see Figure 47.6),

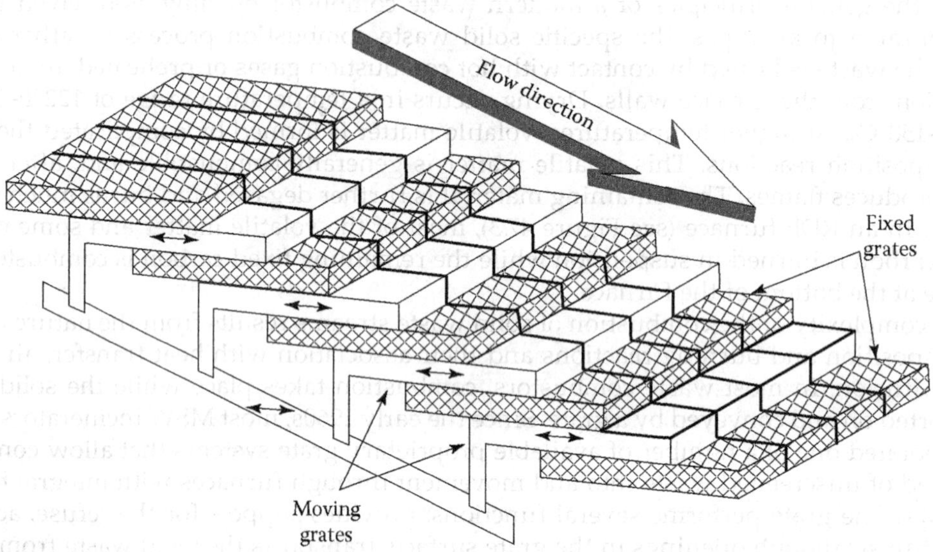

FIGURE 47.4
Reciprocating grates.

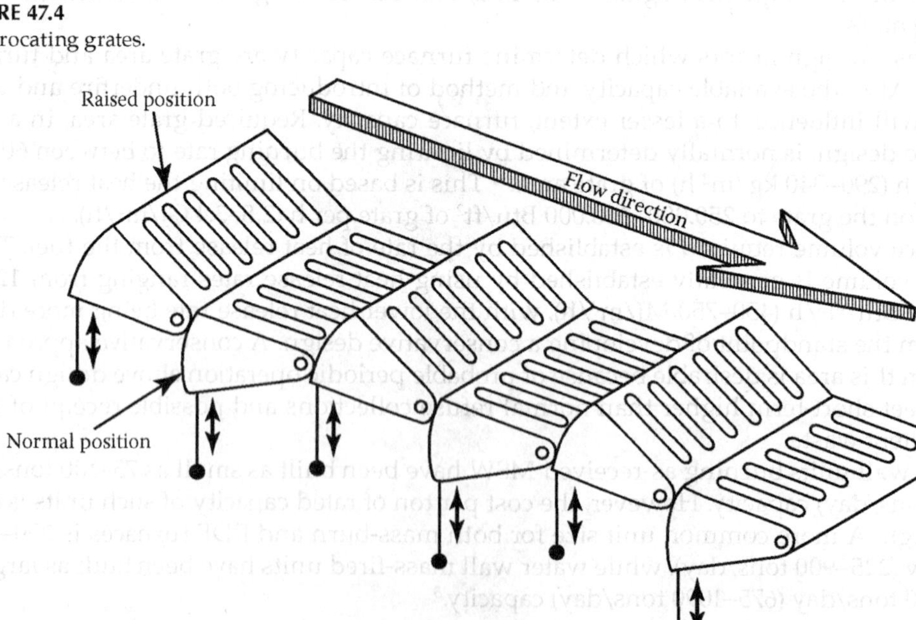

FIGURE 47.5
Rocking grates.

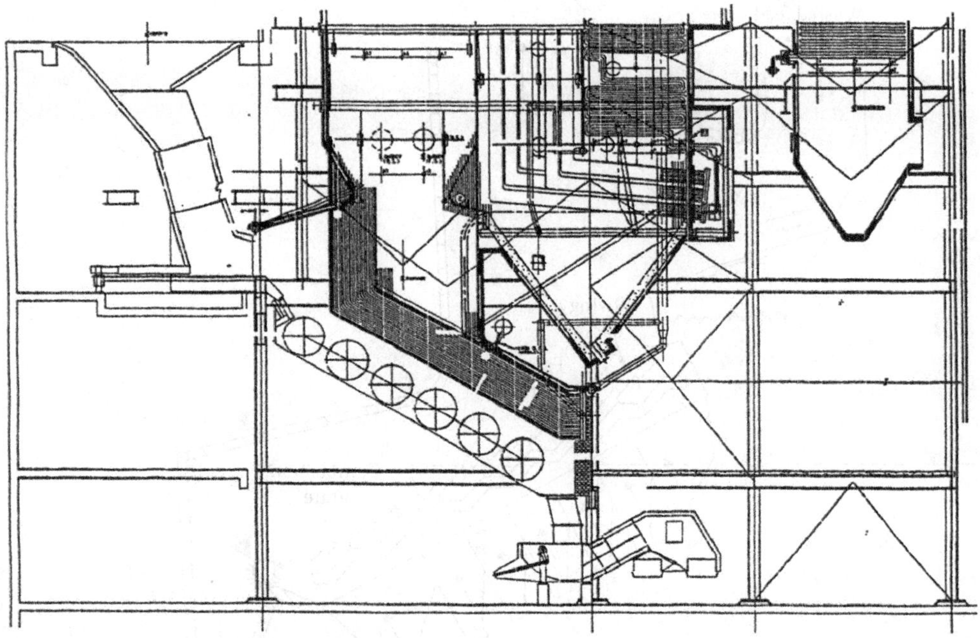

FIGURE 47.6
Roller grate system by Covanta Energy Group.

and water wall rotary combustors for mass-burn units, and traveling grates for RDF units (see Figure 47.3). The reciprocating grates, rocking grates, and roller grates agitate and move the refuse material through the furnace by the movement of the grate elements and the incline of the grate bed. Additional agitation is obtained, particularly in the reciprocating grate, by drops in elevation between grate sections. The rotary combustor slowly rotates to tumble the refuse material, which is conveyed through the inside of the cylinder. The combustor is inclined from the horizontal so that gravity assists in moving the material through the unit. The traveling grate conveys the refuse through the furnace on the grate surface. Stirring is accomplished by building the grate in two or more sections, with a drop between sections to agitate the material.

Other grate systems have been developed in Europe for burning as-received MSW, some of which are currently being utilized in plants being constructed or in operation in the United States. The roller grate, or so-called Dusseldorf System (see Figure 47.6), uses a series of 5 or more rotating cylindrical grates, or drums, placed at a slope of about 30°.[57] The refuse is conveyed by the surface of the drums, which rotate in the direction of refuse flow, and is agitated as it tumbles from drum to drum. Underfire air is introduced through the surface of the drums. Both the Von Roll and the Martin grates use a reciprocating motion to push the refuse material through the furnace. However, in the Martin grate (see Figure 47.7), the grate surface slopes steeply down from the feed end of the furnace to the ash discharge end and the grate sections push the refuse uphill against the flow of waste, causing a gentle tumbling and agitation of the fuel bed.

Another variable feature in the various grate designs is the percentage of open area to allow for passage of underfire air.[73] These air openings vary from approximately 2% to 30%

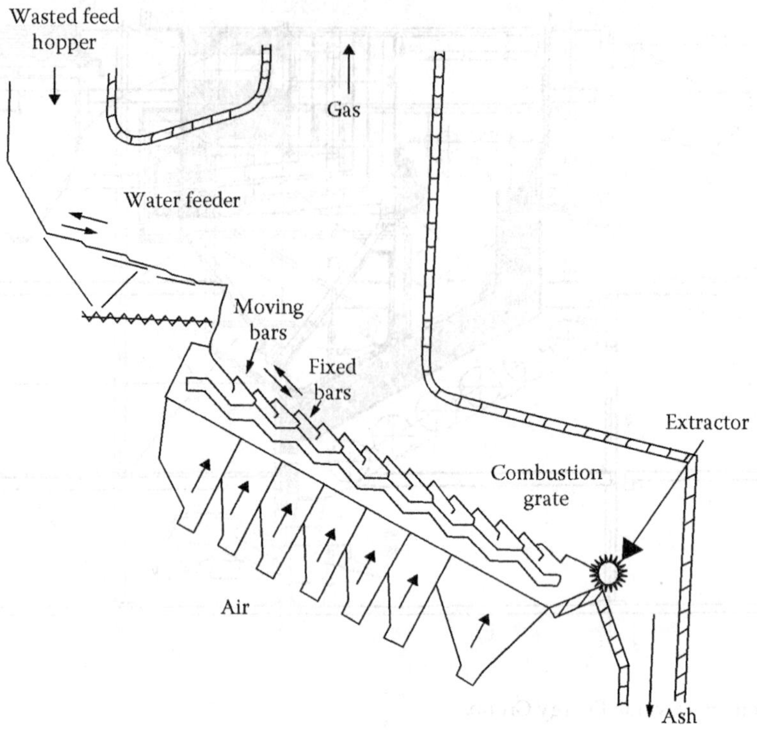

FIGURE 47.7
Martin system. (From Braun, H., Metzger, M., and Vogg, H., Zur Problematik der Quecksilber-Abscheidung aus Rauchgasen von Mullverbrennungsanlagen, Vol. 1, Teil 2, 1985.)

of the grate surface area. The smaller air openings tend to limit the quantity of siftings dropping through the grates and create a pressure drop that assists in controlling the point of introduction of underfire air. RDF grates generally have a smaller percentage of air openings. Larger air openings make control of underfire air more difficult but allow for continuous removal of fine material, which could interfere with the combustion process, from the fuel bed.

Furnace configuration is largely dictated by the type of grate used. In the continuous feed mechanical grate system, the furnace is rectangular in plan and the height is dependent upon the volume required by the limiting rate of heat release cited earlier. An optimum furnace configuration would provide sufficient volume for retention of gases in the high-temperature zone of maximum fuel volatilization long enough to ensure complete combustion, and would be arranged so that the entire volume is effectively utilized. Temperatures are usually high enough with present-day refuse for proper combustion. Turbulence should be provided by a properly designed overfire air system.

With present-day mass-fired water wall furnaces, the use of refractories in furnace construction has been minimized but not eliminated. Refractory materials may be used to line charging chutes, provide a transition enclosure between the top of the grates and the bottom of the water walls, a protective coating on the water wall tubes, and an insulating layer between the hot gases and the metal walls of flues downstream of the primary combustion chamber. Refractory brick used in a charging chute must be able to withstand high temperatures, flame impingement, thermal shock, slagging, spalling, and abrasion. The protective coating on the water wall tubes must be relatively dense castable material with a relatively high heat conductivity.[74] Insulating refractories used in flues downstream from the boilers, on the other hand, should have a low heat conductivity.

Refractories are generally classified according to their physical and chemical properties, such as resistance to chemical attack, hardness, strength, heat conductivity, porosity, and thermal expansion.[49] The material may be cast in brick in a variety of shapes and laid up with air-setting or thermal-setting mortar, or may be used in a moldable or plastic form. Material used in waste combustor construction includes "high duty" and "superduty" fireclay brick, phosphate-bonded alumina material, and silicon carbide, among others. In selecting the proper materials for application in this type of service, because the variety of materials is so great and the conditions of service so varied and severe,[16] advice of a recognized manufacturer should be sought.

As indicated in the section on combustion calculations, the combustion process requires oxygen to complete the reactions involved in the burning process. The air that must be delivered in the furnace to supply the exact amount of oxygen required for completion of combustion is called the stoichiometric air requirement. Additional air supplied to the furnace is called excess air and is usually expressed as a percentage of the stoichiometric requirements.

The total air supply capacity in a waste combustor must be greater than the stoichiometric requirement for combustion because of imperfect mixing and to assist in controlling temperatures, particularly with dry, high heat-content refuse. The total combustion air requirements can range from 6 to 8 lb of air/lb of refuse for mass-fired water wall furnaces, and slightly less for RDF facilities.

In the modern mass-burn mechanical grate furnace chamber, at least two blower systems should be provided to supply combustion air to the furnace—one for underfire or undergrate air and the other for overfire air. Underfire air, admitted to the furnace from under the grates and through the fuel bed, is used to supply primary air to the combustion process and to cool the grates.

Overfire air may be introduced in two levels. Air introduced at the first level, called secondary air, immediately above the fuel bed, is used to promote turbulence and mixing, and to complete the combustion of volatile gases driven off the bed of burning solid waste. The second row of nozzles, which are generally located higher in the furnace wall, allow the introduction of air, called tertiary air, into the furnace to promote additional mixing of gases and for temperature control.

Blower capacities should be divided so that the underfire air blower is capable of furnishing half to two-thirds of the total calculated combustion air requirements, while the overfire blower should have a capacity of about half of the total calculated air requirements. Setting these capacities requires some judgment related to assessing how great a variation is anticipated in refuse heat contents during plant operation. Variable frequency drivers on the fan motors or dampers on fan inlets and air distribution ducts should be provided for control purposes.

Pressures on underfire air systems in mass-burn units for most U.S. types of grates will normally range from 2 to 5 in. of water. European grates frequently require a higher pressure. The pressure on the overfire air should be high enough that the air, when introduced into the furnace, produces adequate turbulence without impinging on the opposite wall. This is normally accomplished by the use of numerous relatively small (1½–3 in. in diameter) nozzles at pressures of 20 in. of water or higher.

Recirculated flue gas is sometimes used in part for underfire air and tempering air. Using recirculated flue gas as combustion air reduces the quantity of fresh air needed, thereby increasing thermal efficiency and minimizing thermal NO_x formation. It can also be used as tempering air to control the temperature of the flue gas entering the boiler.[13] The duct work for recirculated flue gas is highly susceptible to corrosion due to the presence of acid gases. For this reason, it is critical that the recirculated gas be taken from the system after the pollution control devices.

In an RDF-fired, spreader-stoker type of unit, the combustible material is generally introduced through several air-swept spouts in the front water wall, is partially burned in suspension, and then falls onto a grate on which combustion is completed as the partially burned material is conveyed to the residue discharge under the front water wall face of the furnace. Densified RDF can also be burned in such units. The RDF can furnish all the combustible input to the system, or it may be co-fired with a fossil fuel, generally coal.

Some combustion air in RDF-fired units is introduced with the fuel through the air-swept feed spouts to distribute the fuel on the grate. Additional air is introduced into the furnace higher in the water wall area to enhance turbulence and mixing in the unit and/or to control temperatures. This additional combustion air supply is similar to the tertiary air utilized in the mass-burn units.

47.3.5 Boilers

Substantial quantities of heat energy may be recovered during the thermal destruction of the combustible portions of MSW. Systems that have been successfully used to recover this energy include mass-fired refractory combustion chambers followed by a convection boiler section; a mass-fired water wall unit where the water wall furnace enclosure forms an integral part of the boiler system and an RDF semisuspension-fired spreader-stoker/boiler unit. Each system has apparent advantages and disadvantages.

47.3.5.1 Refractory Furnace with Waste Heat Boiler

In a refractory furnace waste heat boiler unit, energy extraction efficiencies are generally lower, assuming the same boiler outlet temperatures, than with the other systems. Approximately 50%–65% of the heat generated in the combustion process may be recovered with such systems. These units can produce approximately 2–3 lb steam/lb of normal MSW (heat content = 4500 Btu/lb), versus 3 or more lb/lb MSW in mass-fired water wall units. This lower efficiency of steam generation is caused by larger heat losses due to higher excess air quantities needed with such units to control furnace temperatures so that furnace refractories are not damaged. However, the boilers in such units, if properly designed and operated, generally are less susceptible to boiler tube metal wastage problems than the other systems listed earlier.

47.3.5.2 Mass-Fired Water Wall Units

Mass-fired water wall units are the most widely utilized type of heat recovery unit in this field today. In this type of unit, the primary combustion chamber is fabricated from closely spaced steel tubes through which water circulates. This water wall lined, primary combustion chamber incorporated into the overall boiler system is followed by a convection type of boiler surface. It has been found desirable in these plants to coat a substantial height of the primary combustion chamber, subject to higher temperatures and flame impingement, with a thin coating of a silicon carbide type of refractory material or inconel and to limit average gas velocities to under 15 ft/s (4.5 m/s) in this portion of the furnace. Gas velocities entering the boiler convection bank should be less than 30 ft/s (9.0 m/s).[74] Efficiency of heat recovery in such units has been found to range generally from 65% to 75%, with steam production usually above 3 lb of steam/lb of normal as-received MSW. Water table studies have been used occasionally in some larger units to check on combinations of furnace configuration and location of overfire combustion of air nozzles.[27]

47.3.5.3 RDF-Fired Water Wall Units

As pointed out earlier, RDF may be burned in a semisuspension-fired spreader-stoker/boiler unit where the RDF is introduced through several air-swept spouts in the front water wall, partially burns in suspension, and then falls on a grate on which combustion is completed. In this type of unit, the water wall lined primary combustion enclosure (furnace) may be followed by a superheater (usually), a convection boiler heat transfer surface, and (sometimes) an economizer surface.

Efficiencies of RDF-fired boilers generally range from 65% to 80% of the heat input from the RDF. Steam production from RDF would normally be expected to be somewhat greater than 3 lb of steam/lb of RDF. However, when one takes into account the combustible material lost in the processing of as-received MSW to produce the RDF, steam production normally will fall to about 3 lb of steam/lb of as-received MSW.[35]

If the energy recovered from the combustion of as-received MSW or RDF is to be used to produce electricity, some superheating is desirable, if not necessary. Since boiler tube metal wastage in these plants is, at least partially, a function of tube metal temperature[10] and steam is a less efficient cooling medium than water, superheater surface is more prone to metal wastage problems than other areas of boiler tubing. Tube metal temperatures, above which metal wastage can be a significant operational problem, are generally thought to range from 650°F to 850°F (345°C–455°C). These temperatures are lower than those for maximum efficiency of electrical generation by steam driven turbines. It is desirable to consider this in facility design to reduce plant downtime and minimize maintenance costs.

In the 1980s and early 1990s a so-called full-suspension combustion concept was attempted in which finely shredded combustible material from MSW was blown into the furnace through nozzles located one-half to two-thirds up the height of the water wall furnace enclosure. In this type of unit most of the RDF, usually composed of smaller sized particles than in the semisuspension-fired unit was supposed to burn in suspension. This concept was not successful due to problems related to the additional handling of the RDF and greater power required to achieve a finer shred. Also, some boilers seemed to experience a greater tendency for slag formation in the boiler. While the concept initially anticipated that the RDF would completely burn in suspension, experience indicated that this does not occur. Accordingly, dump grates became a necessity in such furnace boiler units to allow for completion of combustion prior to water quenching of the residue. This concept has been abandoned.

47.3.6 Residue Handling and Disposal

The residue from a well-designed, well-operated mass-fired incinerator burning as-received refuse will include the noncombustible material in the MSW and usually somewhat less than 3% of the combustibles. The nature of this material will vary from relatively fine, light ash, burned tin cans, and partly melted glass to large, bulky items such as lawn mowers and bicycles.

In most modern WTE plants bottom ash residue is discharged from the end of the furnace grate through a chute into a trough filled with water. Removal from the trough may be either by a ram discharger onto a conveyor or by a flight conveyor to an elevated storage hopper from which it is discharged to a truck. If a water-filled trough with a flight conveyor is used, normally two troughs are provided, arranged so that the residue can be discharged through either trough. The second trough serves as a standby. Fly ash, residue collected in the air pollution control equipment downstream of the furnace/boiler, is usually returned back to and mixed with the bottom ash.

A key feature in the design of ash discharge facilities is provision for sealing the discharge end of the furnace to prevent uncontrollable admission of air. This seal is usually provided by carrying the ash discharge chute at least 6 in. (15 cm) below the water surface in the receiving trough. In the design of the conveyor mechanism, the proportions should be large because the material frequently contains bulky metal items and wire, potential causes of jamming. Also, the residue material tends to be extremely abrasive. A grizzly screen is often used to remove oversized bulky materials from the residue prior to its being loaded onto trucks for delivery to the landfill.

Residue is usually taken to a landfill for final disposal. Many modern facilities dispose of their residue at monofills (landfills that accept WTE plant residue only). The volume of material remaining for ultimate disposal will range from 5% to 15% of that received at the plant. Many plants currently operating in the United States that weigh MSW received at the plant and residue discharged from the furnaces, indicate that the weight of MSW is only reduced by from 40% to 50%. However, as much a one-third of the residue weight in these plants may be attributed to incomplete drainage of the material prior to its discharge into the final transportation container. The ram-type ash discharger used in European and most large U.S. plants generally achieves much better dewatering of residues than older water-filled trough, ash drag residue handling systems. These systems can achieve 65%–75% weight reduction.

The main components of ash are inert materials of low solubility, such as silicates, clay, and sand. Aluminum, calcium, chlorine, iron, selenium, sodium, and zinc are major elements in all particles and, along with carbon, can comprise over 10% by weight of the ash.[20]

A broad range of trace metals and organic compounds may be found in fly and bottom ash. Data on ash composition are difficult to compare, however, because they reflect different types and sizes of facilities, unknown sample collection methodology and sample size at each facility, interlab variation in testing procedures (even using the same test), and the heterogeneous nature of MSW itself. In addition, the presence of a substance in ash does not mean that it will enter the environment. Its fate depends on its solubility, how the ash is managed, and whether the ash is subject to conditions that cause leaching.[47]

Metals tend to be distributed differently in fly and bottom ash. Most volatile and semi-volatile metals, such as arsenic, mercury, lead, cadmium, and zinc, tend to be more concentrated or "enriched" in fly ash.[59,70] Less volatile metals, such as aluminum, chromium, iron, nickel, and tin, typically are concentrated in bottom ash.[58,59]

Organic chemicals also exhibit differing distributions. Dioxin/furan and polychlorinated biphenyls (PCBs) tend to be enriched in fly ash, while other chemicals such as polycyclic–aromatic hydrocarbons (PAHs) and phthalates tend to be concentrated in bottom ash.[70] Concentrations of dioxins/furans in fly ash exhibit a wide range, but they are significantly lower in ash from modern facilities than in ash from older incinerators.[32,71,79]

From a regulatory standpoint, a number of different testing procedures have been developed and utilized by regulatory agencies over the past several years in an attempt to predict the behavior of MSW residues deposited in landfills. Most of these methods were developed to predict leaching characteristics of residues deposited in landfills with raw or as-received MSW. Test results using these methods have been quite variable. However, as pointed out earlier, most modern WTE facilities dispose of their residue in monofills.[68] Tests of leachate from such monofills indicate metals concentrations below extraction procedure (EP) toxicity limits, and in most cases below U.S. drinking water standards.[68] Most test data show little or no leaching of organic chemicals.[33,68]

Following a court decision in the mid-1990s that ash residue from combustion of MSW is not exempt from the rules and regulations for hazardous waste, regulators have required testing of ash residues as they are discharged from the plant, i.e., separately if bottom ash and fly ash are discharged separately, and combined if they are combined prior to discharge from the plant. Bottom ash is alkaline and usually tests as nonhazardous, while fly ash is acidic and frequently tests as hazardous. When the fly ash is mixed with bottom ash prior to discharge from the plant, the alkaline bottom ash neutralizes the smaller quantities of acidic fly ash. The mixture tests as nonhazardous and can be disposed of in the normal monofill.

47.3.7 Other Plant Facilities

The balance of the plant equipment is similar to that used in fossil fuel-fired boiler facilities. However, there are differences. Thus, the combustion of MSW produces a highly corrosive environment for boiler tube materials. Metal chlorides are believed to be primarily responsible for boiler tube corrosion problems.[40] The most important factors in high temperature corrosion are metal temperature, gas temperature, temperature gradient between gas temperature and metal temperature, deposit characteristics, and temperature fluctuations.[1] For this reason, boiler tubes are generally fabricated using corrosion-resistant alloys. Boiler tube shields or weld overlay cladding of boiler tubes with inconel are also used in highly corrosive/erosive areas.[40]

Some waste-to-energy facilities incorporate an air heater to preheat combustion air. Finned tubes plug quickly due to the large quantity of flyash in the flue gas. These air heaters are always of the bare tube design.

Since thermal efficiency is not an overriding concern in waste-to-energy facilities, many plants have one, or at most two, feedwater heaters. Some have only a deaerator for feedwater heating, unlike conventional power plants which have several stages of feedwater heaters.

47.4 Air Pollution Control Facilities

Potential emissions from the combustion of MSW may be broadly classified into particulates, gaseous emissions, organic compounds, and trace metals. The concern is to reduce emissions so as to adequately protect public health. This is achieved using good combustion practice and equipment specially designed to remove the targeted pollutants.

During the 1980s and 1990s, the emission requirements for air pollution control equipment became more stringent as the USEPA promulgated new standards. The most recent standards were published in the December 19, 1995 Federal Register as 40 CFR Part 60 Subpart Cb.[4,72] These standards established emission limits for large (over 248 tons/day

TABLE 47.10

Emission Limits for Large Combustor Units

Parameter	Limit	Conditions
Particulates	27 mg/DSCM	Corrected to 7% oxygen
Opacity	10%	6 min average
Cadmium	0.04 mg/DSCM	Corrected to 7% oxygen
Lead	0.49 mg/DSCM	Corrected to 7% oxygen
Mercury	Lesser of 0.08 mg/DSCM or 85% removal	Corrected to 7% oxygen
SO_2	Lesser of 31 ppmv or 75% removal	Corrected to 7% oxygen 24 h geometric mean
HCl	Lesser of 31 ppmv or 95% removal	Corrected to 7% oxygen
Dioxin/furans	60 ng/DSCM–ESP	Corrected to 7% oxygen
	30 ng/DSCM—all others	
NO_x	220 ppmv—water wall	Corrected to 7% oxygen
	250 ppmv—rotary water wall	Corrected to 7% oxygen
	250 ppmv—RDF	Corrected to 7% oxygen
	240-fluidized bed	Corrected to 7% oxygen

TABLE 47.11

Emission Limits for Small Combustor Units

Parameter	Limit	Conditions
Particulates	70 mg/DSCM	Corrected to 7% oxygen
Opacity	10%	6-min average
Cadmium	0.10 mg/DSCM	Corrected to 7% oxygen
Lead	1.6 mg/DSCM	Corrected to 7% oxygen
Mercury	Lesser of 0.08 mg/DSCM or 85% removal	Corrected to 7% oxygen
SO_2	Lesser of 80 ppmv or 50% removal	Corrected to 7% oxygen 24 h geometric mean
HCl	Lesser of 250 ppmv or 50% removal	Corrected to 7% oxygen
Dioxin/furans	125 ng/DSCM	Corrected to 7% oxygen
NO_x	No limit	Corrected to 7% oxygen

[225 Mg/day]) and small (under 248 tons/day [225 Mg/day]) combustor units. Tables 47.10 and 47.11 contain a summary of those standards.

47.4.1 Particulate Control

Particulates have been a matter of concern and regulatory agency attention for some time. The initial concern was from the standpoint of reducing gross particulate emissions that were both an aesthetic and a potential public health problem. Current interest and concern is directed toward better control of submicron-size particles[75] and other pollutants.

Electrostatic precipitators were the most commonly used gas cleaning device for particulate emission control in municipal waste combustors in the 1970s and early 1980s. They were designed to achieve high collection efficiencies (99% or higher) and meet the air emissions standards at the time. As emission standards became more stringent for particulates, fabric filters became more prevalent. Many electrostatic precipitators were replaced with fabric filters due to the 1995 regulations.

Fabric filters can operate at high efficiency, even in the submicron particle size range. They became widely used in the late 1980s because of the increasing emphasis of regulatory agencies on acid gas control and lower particulate emission levels. Baghouses are more effective than electrostatic precipitators for acid gas scrubbing when preceded by a spray dryer. The original bags used in these facilities had a limited life at high temperatures. Experiments using different materials of construction have led to longer bag life.

The scrubber/fabric filter control systems have been shown to be capable of operating at a particulate emission level of 20 mg/Nm^3 (0.009 gr/dscf) and lower (see Table 47.12). The material selected for the filter bags can have an important effect on filtering efficiency and the emission level thus achieved. In general, test results to date for the scrubber/fabric indicate lower particulate emissions than those for electrostatic precipitators on WTE plants. However, in general, electrostatic precipitators have not been designed to meet emission levels as low as those specified for fabric filter installations. Electrostatic precipitators following spray drying absorbers in Europe have been tested at particulate emission levels of 1–8 mg/Nm^3 (0.00045–0.0036 gr/dscf). The reliability and overall economics of the various control processes must be considered when making a selection of equipment to meet these very low emission control requirements. Data are available[68] on emission levels for approximately 30 different specific elements, many of them heavy metals. Elements found to occur in stack emission from municipal waste combustors are lead, chromium, cadmium, arsenic, zinc, antimony, mercury, molybdenum, calcium, vanadium, aluminum, magnesium, barium, potassium, strontium, sodium, manganese, cobalt, copper, silver, iron, titanium, boron, phosphorus, tin, and others.

Since the condensation point for metals such as lead, cadmium, chromium, and zinc is above 570°F (300°C), the removal efficiency for such metals is highly dependent on the particulate removal efficiency. Some metal compounds, particularly chlorides such as $AsCl_3$ at 252°F (122°C) and $SnCl_4$ at 212°F (100°C), have condensation points below 300°C. For such compounds, particulate collection temperatures will be a factor in collection efficiency. High removal (over 99%) has been observed for most metals for highly efficient (over 99%) particulate removal systems operating at appropriate temperatures.

TABLE 47.12

Particulate Emissions from Municipal Waste Combustors

	Particulates (gr/dscf)[a] at 12% CO_2
Plant G (1983); EP	0.0321
Plant T (1984); DS, BH	0.012
Plant M (1984); DS, EP	0.0104
Plant W (1985); DS, BH	0.004
Plant P (1985); EP	0.0163
Plant T (1986); EP	0.007
Plant M (1986); DS, BH	0.007

Source: From Velzy, C.O., U.S. Experience in combustion of municipal solid waste, presented at *the APCA Specialty Conference on Regulatory Approaches for Control of Air Pollutants*, Atlanta, GA, February 20, 1987.

Notes: EP, electrostatic precipitator; DS, dry scrubber; BH, bag house.

[a] Grains per dry standard cubic foot.

47.4.2 Gaseous Emission Control

Gaseous emissions such as SO_2, HCl, CO, NO_x and hydrocarbons have recently become a concern in municipal waste combustors and their emissions are now regulated. Acid gas emissions are controlled by scrubbing devices. Carbon monoxide, NO_x, and hydrocarbons are controlled by good combustion practice. Oxides of nitrogen in some cases also require control equipment in the form of selective noncatalytic reduction (SNCR) to reduce NO_x to acceptable levels.

Common gaseous emission factors, based on tests at a number of waste-to-energy plants, are shown in Table 47.13. High carbon monoxide and hydrocarbon emissions are caused by incomplete combustion and/or upsets in combustion conditions. High nitrogen oxide emissions are generally caused by high combustion temperatures. Hydrogen chloride (and hydrogen fluoride) and sulfur oxides, on the other hand, are directly a function of the chlorine (fluorine) and sulfur content in the fuel. The highest emissions, cited in Table 47.13, are from older, poorly controlled plants without significant pollution control equipment. Hydrogen chloride (and hydrogen fluoride) and sulfur oxides are best removed by acid gas scrubbing devices using chemical treatment. Initial efforts at acid gas control used wet collectors. However, this type of flue gas cleaning equipment is subject to problems such as corrosion, erosion, generation of acidic waste water, wet plumes, and, not least, high operating cost. Because of these problems, various semiwet and dry methods of cleaning flue gases have been developed and installed. These methods of gas treatment are based on the injection of slurried or powdered lime, limestone, or dolomite; adsorption; and absorption; followed by chemical conversion.[68] Since the reactivity of these lime materials is rather low, a multiple of the stoichiometric quantity is normally required to obtain a satisfactory cleaning effect. High removal efficiencies can be achieved for HCl, but reduction of SO_2 and SO_3 is more difficult to achieve and maintain. Slaked lime is highly reactive and stoichiometric ratios of 1.2–1.7 have been used for 97%–99% HCl removals and 60%–90%

TABLE 47.13

Gaseous Emission Factors for Municipal Waste Combustors (lb/ton)

	New York Incinerators	Test Results U.S. Plants	Martin Plants	EPA Data Base Tests through 1988	
	1968–1969	1971–1978	1984–1986	Mass-Burn	RDF
Carbon monoxide	—	3.7–9.3	0.2	0.06–16.2	1.0–5.2
Nitrogen oxides	—	0.5–2.2	5.0–6.0	0.5–4.5	2.5–3.2
Hydrocarbons	0.1–22.1	1.1	0.015–0.006	0.01–0.1	0.005–0.01
Hydrochloric acid	1.4–8.6	4.6–14.5	5.0–0.2	0.05–5.7	0.02–9.3
Sulfur oxides	1.3–8.0	0.8–2.2	1.0–2.0	0.05–4.8	0.05–2.3

Sources: From Carrotti, A.A. and Smith, R.A., *Gaseous Emissions from Municipal Incinerators*, USEPA Publication No. SW-18C, 1974; Cooper Engineers, Inc., *Air Emissions Tests of a Deutsche Babcock Anlagen, Dry Scrubber System at the Munich North Refuse-Fired Power Plant*, 1985; Cooper Engineers, Inc., *Air Emissions and Performance Testing of a Dry Scrubber* (Quench Reactor), Dry Venturi and Fabric Filter System Operating on Flue Gas from Combustion of Municipal Solid Waste in Japan, West County Agency of Contra Costa County Waste Co-Disposal/Energy Recovery Project, May, 1995; Hahn, J.L. et al. (Ogden Martin Systems, Inc.), and Weiand, H. et al. (Martin GmbH), in: *Proceedings of the 1986 National Waste Processing Conference*, Denver, CO, June 1–4, 1986, ASME, New York; Murdoch, J.D. and Gay, J.L., Material recovery with incineration, Monmouth County, NJ, in: *Proceedings of 27th Annual International Solid Waste Exposition*, Tulsa, OK, SWANA, Silver Springs, MD (Pub. #GR-0028), August 14–17, 1989, p. 329; Velzy, C.O., Standards and control of trace emissions from refuse-fired facilities, in: *Municipal Solid Waste as a Utility Fuel, EPRI Conference Proceedings*, Madison, WI, November 20–22, 1985.

SO_2 reductions, depending on operating conditions and particulate collector (fabric filters having demonstrated higher removal efficiencies than electrostatic precipitators).

Lime injection into a scrubber/fabric filter system has resulted in removal efficiencies of 90%–99% for HCl and 70%–90% SO_2, provided that the flue gas temperature and the stoichiometric ratio for lime addition are suitable. This combination of processes has reduced HCl levels below 20 ppm and SO_2 to levels below 30 ppm for MSW waste-to-energy plants. This technology has also been extensively used in other applications for acid gas removal. The scrubber/electrostatic precipitor combination has been shown to provide about 90% HCl removal, but typically less SO_2 removal (about 50%). Since this removal efficiency does not meet the most recent regulations, many electrostatic precipitators have been replaced with baghouses. Lime injection into the furnace has also been tested with some success (about 50%–70% efficiency), but fails to meet the most recent regulations.

Some sampling to determine HF removal has been reported. In general, HF removal normally follows HCl removal (i.e., is usually over 90%–95%).

Carbon monoxide and hydrocarbon emissions are best controlled by maintaining proper combustion conditions. Nitrogen oxide emissions are controlled by ammonia injection or by use of combustion control techniques such as limitation of combustion temperatures or recirculation of flue gases. Note in the last column of Table 47.13 that attempts to limit hydrocarbon emissions by improving combustion conditions and raising furnace operating temperatures seem to have resulted in increasing the level of NO_x emissions.

Selective noncatalytic reduction (SNCR) appears to be the most practical method of reducing NO_x emissions for most municipal waste combustors. SNCR involves the use of ammonia to reduce NO_x to nitrogen and water. The SNCR reaction occurs at a temperature of 1600°F–2100°F. At lower temperatures, a catalyst is required to promote the reaction (selective catalytic reduction, or SCR). SCR is not used on municipal waste combustors. Tests conducted at a municipal waste combustor demonstrated that NO_x emission levels of 150 ppmv (45%–55% reduction) can be achieved with SNCR.[4]

Thermodynamic equilibrium considerations indicate that under excess air conditions and with temperatures of 1472°F (800°C) and higher, maintained in a completely mixed reactor for a suitable period of time, emissions of organic or hydrocarbon compounds should be at nondetectable levels. However, measurements at operating plants, particularly those constructed prior to the early to mid-1980s, indicated significant emissions of trace organic or hydrocarbon compounds, some of which are toxic. These tests indicated that the basic objective of combustion control, thorough mixing of combustion products with oxygen at a temperature that is sufficiently high to provide for the rapid destruction of all organic or hydrocarbon compounds, had not been achieved in these early WTE plants.

If the fuel, or the gas driven off of the fuel bed, is not adequately mixed with air, fuel-rich pockets will exist containing relatively high levels of hydrocarbons, which then can be carried out of the combustion system. Kinetic considerations indicate that such hydrocarbons can be destroyed rapidly in the presence of oxygen at elevated temperatures. Also, if too much combustion air is introduced into the combustion chamber, either in total or in a particular area of the chamber, temperatures will be reduced, combustion reactions can be quenched, and hydrocarbons carried out of the combustion system. Achieving the goal of proper combustion control, destruction of all hydrocarbon compounds to form carbon dioxide and water, will minimize emission of potentially toxic substances as well as other compounds that may be precursors and capable of forming toxic compounds downstream in cooler regions of the boiler.

Table 47.13 shows that the hydrocarbons can vary significantly, frequently over relatively short periods of time, based on measurements at older municipal waste combustors.

The highest levels shown in this table occurred in one of the older plants and no doubt indicate very poor combustion conditions. Tests at modern WTE plants indicate consistently low levels of hydrocarbons, which are indicative of good combustion control. In modern, well-designed and -operated plants, photochemical oxidants and PAH are in concentrations too low to cause any known adverse health effects. Tests[55,66] for other substances that might be of concern, such as polychlorinated biphenyls (PCBs), generally have found levels discharged to the atmosphere to be so low as to have a negligible impact on the environment.

47.4.3 Organic Compound Control

Organic compounds for which emission data are available include polychlorinated dibenzodioxins (PCDDs), polychlorinated dibenzofurans (PCDFs), chlorobenzenes (CBs), chlorophenols (CPs), polycyclic aromatic hydrocarbons (PAHs), and PCBs. A number of other organic compounds, including aldehydes, chlorinated alkanes, and phthalic acid esters, have also been identified in specific testing programs. Since dioxin/furan emissions have generated the most interest over the past several years, there are more data for these compounds, in particular for the tetrahomologues, and especially the 2,3,7,8-substituted isomers. The other compounds have been less frequently reported in the literature. The reason for this emphasis is the toxicity of dioxin/furan to laboratory animals and the perceived risk to humans.

Upset conditions in energy-from-waste plants can lead to local air-deficient conditions resulting in the emission of organic compounds. PAHs are formed during fuel-rich combustion as a consequence of free radical reactions in the high temperature flame zone. In addition, it is found that in the presence of water cooled surfaces, such as found in oil-fired home-heating furnaces, a high fraction of the polycyclic compounds are oxygenated. Similar free radical reactions probably take place in fuel-rich zones of incinerator flames yielding PAH, oxygenated compounds such as phenols, and, in the presence of chlorine, some dioxin/furan. The argument for the synthesis of dioxin/furan at temperatures of 400°F–800°F is supported by the increase in the concentration of these pollutants across a heat recovery boiler downstream of the refractory lined combustion chamber of a waste combustor.

This free radical mechanism appears to be the dominant source of dioxin/furan in municipal waste combustors. These compounds may also be present as contaminants in a number of chemicals, most notably chlorinated phenols and polychlorinated biphenyls (PCBs). Their presence in MSW results from the use of these chemicals, discontinued in some cases, as fungicides and bactericides for the phenol derivatives, or the use of PCBs as heat exchanger and capacitor fluids. These compounds are expected to survive in a furnace combustion chamber only if large excesses in the local air flow cool the gases to below the decomposition or reaction temperature. Dioxin/furan can also be produced by condensation reactions involving the chlorinated phenols and biphenyls. The observed formation of dioxin/furan when fly ash from MSW incinerators is heated to temperatures of 480°F (250°C) suggest such catalyzed condensation reactions of chlorinated phenols. PCBs are precursors to furan, and pyrolysis of PCBs in laboratory reactors at elevated temperatures for a few seconds has yielded furan.

The available test data clearly show that dioxin/furan exit the boilers and, depending on the emission control devices employed, some fraction enters the atmosphere either as gases or adsorbed onto particulates. In addition, the solids remaining behind in the form of fly ash and bottom ash contain most of the same compounds, and these become another potential source of discharge to the environment.

Emission data for total dioxin/furan generally fall into three main categories:

1. Low emissions, in the range of 20–130 ng/Nm^3
2. Medium emissions, from 130 to 1000 ng/Nm^3
3. High emissions, over 1000 ng/Nm^3

Average dioxin/furan emission from older plants ranges from about 500 to 1000 ng/Nm^3. The lower emission levels tend to be associated with newer, well-operated mass-fired facilities such as water wall plants, and with modular, starved or controlled air types of incinerators (see Table 47.14). In most test programs, adequate operating data were not collected to correlate emissions with operations. Researchers in the field theorize that combustion conditions can play a role in minimizing emissions, and several studies[63,68,69] were conducted in Canada and the United States to define that role more exactly.

Emissions from MSW combustion contain small amounts of many different dioxin/furan isomers. While individual dioxin/furan isomers have widely differing toxicities, the 2,3,7,7-TCDD isomer, present as a small proportion of the total dioxin/furan, is of greatest known toxicological concern. Based on animal studies it has been generally concluded that other 2,3,7,8-substituted dioxin/furan isomers, in addition to the 2,3,7,8-TCDD, are also likely to be of toxicological concern. A method for expressing the relative overall toxicological impact of all dioxin/furan isomers, as so-called "2,3,7,8-TCDD toxic equivalents," was developed in the mid-1980s[2] and has been used by the EPA intermittently in its regulatory efforts since this time.

In this method, emissions are sampled, extracted, and analyzed for all constituent isomers of dioxin/furan. A system of toxicity weighting factors from the existing toxicological data (based almost entirely on animal studies) is applied to each constituent dioxin/furan isomer and the results are summed to arrive at the 2,3,7,8-TCDD toxic equivalent. An example of dioxin/furan test results expressed as 2,3,7,8-TCDD toxic equivalents, using three different systems of weighting factors, is shown in Table 47.15.

Emission control systems consisting of a scrubber/fabric filter have been evaluated for dioxin emissions.[34] Recently dioxin removal efficiencies exceeding 99% were obtained, which resulted in dioxin concentrations at the stack that approach the detection limit of the sampling and analytical equipment currently available. Emissions of furan followed a similar range of values as dioxin with the scrubber/high efficiency particulate removal

TABLE 47.14

Summary of Average Total PCDD/PCDF Concentrations from MSW Combustion in Modern Plants (ng/Nm^3, dry, at 12% CO_2)

	Total PCDD PCDF
Peekskill, NY, electrostatic precipitator only (1985)	100.25
Wurzburg, FRG, dry scrubber-baghouse (1985)	49.95
Tulsa, OK, electrostatic precipitator only (1986)	34.45
Marion Co., OR, dry scrubber-baghouse (1986)	1.55

Sources: From Hahn, J.L. et al. (Ogden Martin Systems, Inc.), and Weiand, H. et al. (Martin GmbH), in: *Proceedings of the 1986 National Waste Processing Conference*, Denver, CO, June 1–4, 1986, ASME, New York; NYS Department of Environmental Conservation Bureau of Toxic Air Sampling, Division of Air Resources, Preliminary Report on Westchester RESCO RRF, January 8, 1986; Ogden Projects, Inc., Environmental test report, Walter B. Hall Resource Recovery Facility, October 20, 1986; Ogden Projects, Inc., Environmental test report, Marion County Solid Waste-to-Energy Facility, December 5, 1986.

TABLE 47.15

Toxic Equivalent Emissions by U.S. EPA, Swedish, and California Methods

	Toxic Equivalents ng/Nm³ at 12% CO_2		
Facility	**U.S. EPA**	**Swedish**	**California**
Peekskill, NY	1.62	3.83	9.73
Tulsa, OK	0.7	1.74	4.75
Wurzburg, FRG	0.37	0.81	2.11
Marion Co., OR	0.11	0.16	0.29
From WHO Workshop; Naples, Italy			
Max. from avg. oper.	25.0	52.78	134.5
Achievable with no acid gas cleaning	0.9	2.2	5.94

Sources: From Hahn, J.L. and Sussman, D.B., Dioxin emissions from modern, mass-fired, stoker/boilers with advanced air pollution control equipment, presented at *Dioxin'86*, Fukuoka, Japan, September 1986; NYS Department of Environmental Conservation Bureau of Toxic Air Sampling, Division of Air Resources, Preliminary Report on Westchester RESCO RRF, January 8, 1986; Ogden Projects, Inc., Environmental test report, Walter B. Hall Resource Recovery Facility, October 20, 1986; Ogden Projects, Inc., Environmental test report, Marion County Solid Waste-to-Energy Facility, December 5, 1986; Vogg, H. et al., Recent findings on the formation and decomposition of PCDD/PCDF in solid municipal waste incineration, ISWA/WHO, Specialized Seminar, Copenhagen, Denmark, January, 1987; World Health Organization, Report on PCDD and PCDF Emissions from Incinerators fox Municipal Sewage, Sludge and Solid Waste—Evaluation of Human Exposure, from WHO Workshop, Naples, Italy, March, 1986.

combination reducing furan to very low or nondetectable levels. Additional reductions of over 50% can be achieved by activated carbon injection.[4]

Some limited data on emissions of CB, CP, PCB, and PAH are available. Most sampling programs for dioxin/furan have unfortunately neglected to analyze for these compounds. Maximum levels from Canadian studies[34] are as included in Table 47.16 along with some data from tests on U.S. plants.[33] The scrubber/fabric filter technology generally demonstrated removal rates of 80%–99% for these compounds in these studies.

Very few studies report on other organic products in the flue gas. Some data from tests on older plants have been reported for aldehydes and certain other volatile hydrocarbons.[3] Such data are not available for newer plants.

The conventional combustion gas measurements include CO, total hydrocarbons (THCs), CO_2, and H_2O. Both CO and THC have been of interest as potential surrogates for dioxin/furan emissions; however, no strong correlations have been found in previous studies. In fact, very few studies have attempted to determine CO and dioxin/furan

TABLE 47.16

Organic Emissions (ng/Nm³)

	U.S. Plants	Canadian Pilot Plant	
Chemical Emitted	**Before Particulate Removal**	**Before Particulate Removal**	**After Scrubber/Fabric Filter**
CB	10,000–500,000	17,000	3000
CP	22,000–80,000	30,000	8000
PCB	—	700	Nondetectable (ND)
PAH	ND to 5,600,000	30,000	130

Sources: From Battelle Columbus Labs, Characterization of stack emissions from municipal refuse-to-energy systems, National Technical Information Service, PB87-110482, October 1982; Hay, D.J. et al. (Environment Canada), and Marenlette, L. (Flakt Canada, Ltd.), The national incinerator testing and evaluation program: An Assessment of (A) two-stage incineration (B) pilot scale emission control, Report EPS 3/UP/2, September 1986.

TABLE 47.17

Lime Addition with Baghouse, Percent Removal of Organics

	Dry System				Wet/Dry System	
	110°C	125°C	140°C	200°C	140°C	140°C Recycle
CB	95	98	98	62	>99	>99
PCB	72	>99	>99	54	>99	>99
PAH	84	82	84	98	>99	79
CP	97	99	99	56	99	96

Sources: From Hay, D.J. et al. (Environment Canada), and Marenlette, L. (Flakt Canada, Ltd.), The national incinerator testing and evaluation program: An assessment of (A) two-stage incineration (B) pilot scale emission control, Report EPS 3/UP/2, September 1986.

emission data for several operating conditions on the same combustor to develop a correlation. On the other hand, several authors have attempted to correlate CO and dioxin/furan data obtained from several different facilities. From such comparisons, it appears that low CO levels (below 100 ppm) are associated with low dioxin/furan emissions.[38] High CO levels of several 100 ppm and even over 1000 ppm have been associated with high dioxin/furan emissions. During poor or upset combustion conditions, CO levels of several thousand ppm have been observed and THC levels have risen from a typical 1–5 ppm to 100 ppm and more. Since one of the measures available to combustor operators to optimize combustion control is to minimize CO production, one would assume from these general correlations that this would also tend to minimize dioxin/furan emissions, along with emissions of other trace chlorinated hydrocarbons. THC is not as useful as CO as an indicator of proper combustion because of problems in sampling to consistently obtain a representative sample for analysis at the analytical instrument.

Table 47.17 shows operating results achieved using dry and semidry lime injection followed by a baghouse for removal of trace organic pollutants from waste combustor emissions.

47.4.4 Trace Metals

Trace metals are not destroyed during combustion, and the composition of wastes therefore provides, on a statistical basis, the measure of the total inorganic residue. The distribution of trace metals between bottom ash and ash carried over to the air pollution control device is dependent upon the design and operation of the combustor and the composition of the feed. The amount of ash carried up and with the flue gases discharged from a burning refuse bed increases with increasing underfire air rate and with bed agitation. Modular incinerators (described later in Section 47.7) with low underfire air flow rates tend to have lower particulate emissions than conventional mass-burn units and RDF units for this reason. In addition, the amount of ash carried from the combustion chamber will be influenced by the particle size of the inorganic content of the MSW.

The distribution of trace metals between the different components of refuse has a strong influence on the fate of the trace metals. For example, TiO_2 used as a pigment in paper products, has a particle size of about 0.2 μm and will tend to be carried off by the flue gases passing through the refuse bed, whereas TiO_2 present in glass will remain with the glass in the bottom ash. Up to 20% of the inorganic content of the waste will be entrained from burning refuse beds to form fly ash particles. The remainder will end up in the residue.

Volatile metals and their compounds, usually present in trace amounts in the feed, will vaporize from the refuse and condense in the cooler portions of a furnace either as an ultra fine aerosol (size less than 1 μm) or on the surface of the fly ash, preferentially on the surface of the finer ash particles. A large fraction of certain elements in the feed, such as mercury, will be volatilized.

Since mercury is a very volatile metal, it exists in vapor phase at temperatures as low as 68°F–122°F (20°C–50°C). Several studies have indicated that sufficient cooling of the flue gas (typically below 140°C, based on test results conducted to date) and a highly efficient particulate removal system to remove the particles on which the mercury has been adsorbed[6,8,80] are both required to achieve high mercury removal. High mercury removal has been obtained for the scrubber/fabric filter system, provided that the flue gas is adequately cooled (see Table 47.18).

Test results with carbon injection at two municipal waste combustors demonstrated that the EPA emission guidelines of 80 mg/dscm or 85% reduction in mercury emissions can be achieved with a spray dryer, fabric filter and carbon injection.[4]

Elements such as sodium (Na), lead (Pb), zinc (Zn), and cadmium (Cd), will be distributed between the volatiles and the residue in amounts that depend on the chemical and physical form in which the elements are present. For example, sodium in glass will be retained in the residue but that in common salt will partially disassociate and be discharged with the emission gases.

Some of the data on metal emissions that are available from tests on resource recovery plants is shown in Table 47.19. Note that the emission of trace metals can be dramatically limited at WTE plants by the use of high-efficiency particulate control devices that are installed on modern facilities.

While sampling for metal emissions is fairly well established, in order to obtain enough sample to analyze at highly controlled sources, samples times are extremely long, sometimes over 8 h using the U.S. EPA Method 5 sample train (relatively low sample rate). Several studies[28,29,41] of waste combustor emissions in the United States in the 1970s concluded that "municipal incinerators can be major sources of Cd, Zn, Sb and possibly Sn…." This conclusion is based on the relative concentration of these materials in the total suspended particulate catch. However, two of the three plants tested in these studies utilized inefficient air pollution control facilities. Thus, particulate emissions in these plants were relatively high when compared to the German and Japanese plant

TABLE 47.18
Lime Addition with Baghouse Mercury Concentrations (μg/km³ at 8% O_2)

Operation Dry System	Inlet	Outlet	%Removal
230°F (110°C)	440	40	90.9
260°F (125°C)	480	23	97.9
285°F (140°C)	320	20	93.8
390°F (>200°C)	450	610	0
Wet–dry system			
140°C	290	10	94.7
140°C ← recycle	350	19	94.7

Source: From Hay, D.J. et al. (Environment Canada), and Marenlette, L. (Flakt Canada, Ltd.), The national incinerator testing and evaluation program: An assessment of (A) two-stage incineration (B) pilot scale emission control, Report EPS 3/UP/2, September 1986.

TABLE 47.19

Trace Metal Emissions Test Results

	Japanese Plant Uncontrolled	Braintree Mass. Part. Rem. Eff. 74%	German Plants	Japanese Plants	Tulsa, OK	Dry Scrubber, Fab. Filter: Marion, Co., OR	Dry Scrubber, Fab. Filter: Pilot Plant Canada
Arsenic (As) (lb/ton × 10^{-3})	<0.4	0.125	0.09	<0.0016	—	—	0.00033–0.00064
Beryllium (Be) (lb/ton × 10^{-3})	<0.3	0.00027	0.002	<0.0016[a]	0.000025	0.000021	—
Cadmium (Cd) (lb/ton × 10^{-3})	0.7	1.30	0.25	0.11	—	—	ND–0.006
Chromium (Cr) (lb/ton × 10^{-3})	16.0	0.34	0.185	0.026	—	—	ND–0.016
Lead (Pb) (lb/ton × 10^{-3})	17.0	42.4	10.0	0.1	3.5	0.29	ND–0.08
Mercury (Hg) (lb/ton × 10^{-3})							
on particulates	0.5	0.11	0.067	0.03	—	—	—
Vapor phase	0.8	4.38[a]	—	0.90	3.5	2.9	0.16–9.83
Selenium (Se) (lb/ton × 10^{-3})	<0.3	—	—	<0.0016	—	—	—
Particulates (lb/ton)	25.7	1.3	0.5	0.19	0.13	0.16	<0.01

Sources: From Clark, L., Case history of a 240 ton day resource recovery project: Part II, in: *Proceedings of the 1996 National Waste Processing Conference*, ASME, New York, 1996, pp. 235–248; Hahn, J.L. and Sofaer, D.S. 1988. Variability of NO_x emissions from modern mass-fired resource recovery facilities, Paper No. 88-21.7, presented at *the 81st Annual Meeting of Air Pollution Control Association*, Dallas, TX, June 1988; Ogden Projects, Inc., Environmental test report, Walter B. Hall Resource Recovery Facility, October 20, 1986; Ogden Projects, Inc., Environmental test report, Marion County Solid Waste-to-Energy Facility, December 5, 1986; Velzy, C.O., Standards and Control of Trace Emissions from Refuse-Fired Facilities, in: *Municipal Solid Waste as a Utility Fuel, EPRI Conference Proceedings*, Madison, WI, November 20–22, 1985.

[a] Total on particulate and vapor phase.

data in Table 47.19, which is similar to emission data from most modern mass-burn and RDF waste-to-energy plants in the United States.[68] Note also in Table 47.19 that, as the efficiency of particulate control improves, trace metal emissions generally decrease, and in most cases decrease significantly. Even though there is ample evidence from test data[17] to indicate that heavy metals tend to concentrate on the finer particulates, there is also evidence from test results to show that at high particulate removal efficiency (99% ±), high trace metal removal (99% ±) is achieved. Thus, the conclusion in these studies of waste combustors quoted earlier is not valid for WTE plants utilizing efficient air pollution control devices.

The important operating parameters for such equipment are flue gas temperature and composition, contact time, relative velocity of particles and gas stream, and possible activation of particles. See Tables 47.18 and 47.20 for operating results achieved using dry and semidry lime injection followed by a baghouse for removal of heavy metal pollutants from waste combustor emissions.

TABLE 47.20

Lime Addition with Baghouse, Metal Concentrations (μg/km³ at 8% O_2)

Metal	Inlet	Outlet	Removal
Zinc	77,000–108,000	5–10	96 → 99.99
Cadmium	1,000–3,500	1.0–0.6	96 → 99.96
Lead	34,000–44,000	1–6	95 → 99.98
Chromium	1,400–3,100	0.2–1	>99.92
Nickel	700–2,500	0.4–2	>99.81
Arsenic	80–150	0.02–0.07	>99.95
Antimony	800–2,200	0.2–0.6	>99.92
Mercury	190–480	10–610	0 → 90

Sources: From Carlsson, K., *Waste Manage. Res.*, 4, 15, 1986; Hay, D. et al. (Environment Canada), and Marenlette, L. (Flakt Canada, Ltd.), The national incinerator testing and evaluation program: An assessment of (A) two-stage incineration (B) pilot scale emission control, Report EPS 3/UP/2, September, 1986.

47.5 Performance

Mass burning of MSW is the most highly developed and commercially proven combustion process presently available for reducing the volume of MSW prior to ultimate disposal of residuals on the land, and for extracting energy from the waste.[70] Hundreds of such plants, incorporating various grate systems and boiler concepts, which differ in details of design, construction, and quality of operation, have been built throughout the world since the mid-1960s. Mass-burn systems are generally furnished with a guaranteed availability of 85%, while in practice availabilities of 90%–95% have been achieved.[68] Availability cannot approach 100% because standard maintenance practice requires periodic shutdowns. The newest mass-burn facilities seem capable of achieving high reliability, based on their performance in Europe, Japan, and the United States.

Refuse-derived fuel (RDF) facilities became popular during the 1970s. The early plants were generally designed with the intent to remove and recycle metals, glass, and other marketable materials, with the remaining fraction, RDF, to be burned in an existing boiler as a replacement fuel. Those types of facilities all failed and are no longer in operation. The main reasons for failure were that the recycled materials were highly contaminated with waste and were not marketable, and the boilers were not designed to handle the inconsistent RDF that was being fed to them.

The RDF approach quickly evolved to produce a fuel with a known specification to be burned in a dedicated boiler designed specifically to bum that fuel. The materials which were removed were sold, if possible, or landfilled. The primary difference in philosophy between the two types of RDF plants was that the early ones treated the RDF as the "waste" that contaminated the recovered materials, and the newer generation treated the recovered materials as the "waste" that contaminated the RDF. The newer generation RDF facilities which were designed in this manner have been successful.

Fluidized bed technology has been used outside the United States to combust solid waste for several years. One advantage of fluidized bed combustion is that the boiler is more efficient than those in mass-burn or spreader-stoker facilities. Also, fluidized bed combustion produces lower NO_x emissions than other incineration methods. Although lower, these NO_x emissions must still be controlled with additional air pollution control equipment, as with other combustion facilities. Fluidized bed combustion also has the advantage of being able to add limestone with the sand in the bed to assist in acid gas removal. However, a scrubber is still needed to reduce emissions to permitted levels.[46]

The major disadvantage to a fluidized bed facility is that it does not have a long-term proven track record in the United States. Also, the size of the units are small when compared to the size needed for typical U.S. waste-to-energy facilities.[46]

47.6 Costs

It is extremely difficult to obtain accurate, consistent, and comparable WTE plant construction cost data from which to develop information which might be useful in predicting a planned new plant's construction cost during the study stage of a project. However, a 1988 study[56] has developed such data (appropriate for the time frame of mid-1987), which is confirmed in general by this author's personal experience. This study indicates that for the upper 90% confidence limit for the smallest facility, and the largest facility, the construction costs would range as indicated below:

1. A small modular combustion unit with a waste heat boiler and a capacity of less than 250 TPD—$68,000 and $40,000 per ton of daily MSW processing capability. (In most instances, such plants don't incorporate the same degree of equipment redundancy, and/or the same quality of equipment as the larger plants.)
2. A small refractory wall furnace with waste heat boiler and dry scrubbers of between 200 and 500 TPD capacity—$90,000 and $70,000 per ton of daily MSW processing capability.
3. A small, field erected, water wall congeneration or electric generation facility with dry scrubbers of between 500 and 1,500 TPD capacity—$112,000 and $85,000 per ton of daily MSW processing capability.
4. A large, field erected, water wall congeneration or electric generation facility with a dry scrubber between 2,000 and 3,000 TPD capacity—$129,000 and $112,000 per ton of daily MSW processing capacity.

In this study,[56] the construction costs were said to include the vendor quote for construction plus contingency, utility interconnection expenses, and any identified allowances clearly associated with the construction price. All other costs, such as land acquisition, interest during construction, development costs, and management fees were not considered or included, where known, due to their highly project specific nature.

The following specific observations were made by the authors at the conclusion of this study.[56]

1. Capital construction price decreases with increasing size within size ranges and increases with a higher-value energy product.
2. Price is also affected by the construction, procurement, and air pollution control methods employed.
3. Refuse-derived fuel and mass burning water wall facilities are competitively priced with each other.

The effect of plant capacity on capital costs or mass-burn plants is shown in Figure 47.8. Capital costs for other types of plants are similar.

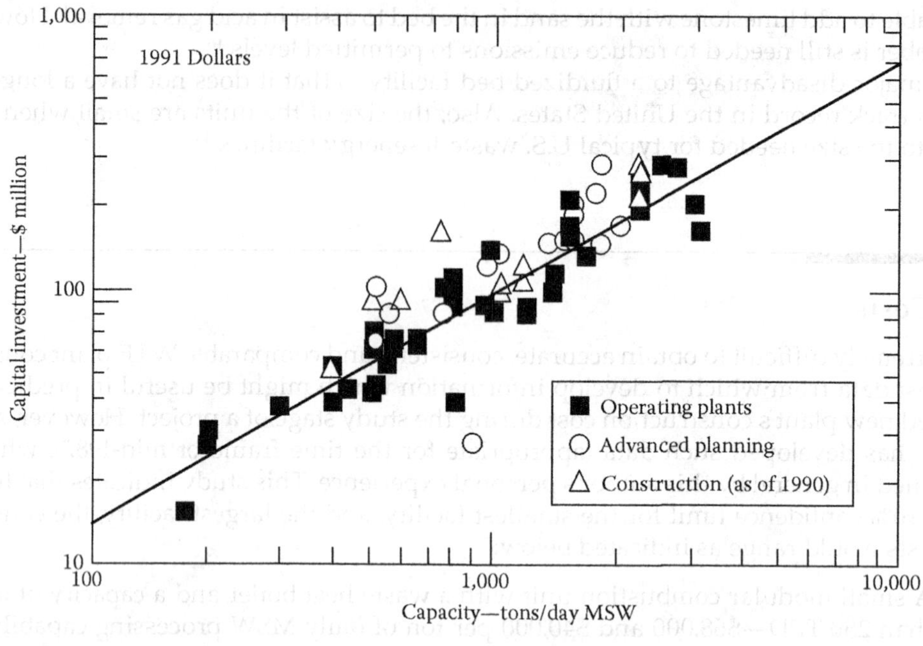

FIGURE 47.8
Effect of plant capacity on capital costs for mass-burn facilities (electricity only), excluding costs associated with collection (e.g., trucks).

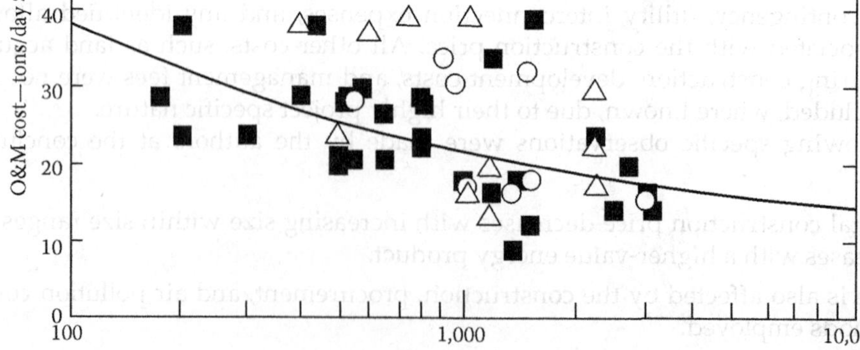

FIGURE 47.9
Effect of plant capacity on O&M costs for mass-burn facilities (electricity producing only), excluding costs associated with collection (e.g., trucks).

With respect to operating costs and/or tipping fees, information is even more difficult to obtain from which to develop costs for planning purposes. Costs cited in the literature from 1989 through 1994 range from \$40 to \$80 per ton of daily rated MSW processing capability.[18,42,41,77,78] Tipping fees on Long Island, which has generally high labor rates, high power costs, and very long hauls for residue disposal, have been noted to range up to \$110 per ton of daily rated MSW processing capability.[26] Plants in other parts of the country where plant operating cost elements are significantly lower have been found to have tipping fees closer to \$40 per ton.[7] Thus, tipping fees for a specific facility would have to be developed based on cost factors for that specific plant.

The effect of plant capacity on O&M costs for mass-burn plants is shown in Figure 47.9. Information is so limited on other types of plants, and the costs are so dependent on local conditions that we do not feel that curves developed for other types of plants would be useful.

47.7 Status of Other Technologies

Several other technologies have been used to a small extent to burn MSW and beneficially use the energy produced in the combustion process. Their use in the future depends on numerous factors, perhaps the most important of which is full-scale demonstration of successful, reliable operation, after which total operational costs are shown to be competitive with mass-burn and/or RDF combustion costs.

47.7.1 Modular Systems

Modular systems, generally utilized in smaller plants, are assemblies of factory-prefabricated major components joined together in the field to form a total operational system. They have been built in individual unit sizes up to just over 100 tons/day capacity, combined into plants of just over 400 tons/day capacity. Modular systems are similar to mass-burn systems in that they combust unprocessed MSW, but they feature two combustion chambers, and the MSW is charged into the system with a hydraulic ram and combustion takes place on a series of stationary hearths. MSW is pushed from one hearth to the next by hydraulic rams. Two types of modular systems have been built and operated: starved air and excess air.

The primary chamber of a starved air modular system is usually operated in a slightly oxygen-deficient ("starved air") environment. The volatile portion of the MSW is vaporized in this chamber and the resulting gases flow into the secondary chamber. The secondary chamber operates in an excess air condition, and combustion of the gases driven off the MSW is essentially completed in this chamber. An excess air modular system operates in a manner similar to a field erected boiler system, with excess air injected into the primary chamber.

One advantage of these units, as indicated in the section on costs, is low cost. Another advantage is that factory prefabrication of major system components can result in shortening of the field construction time. One disadvantage of the two-chamber modular system is that waste burn out in the residue is not always complete, which increases ash quantities and reduces the efficiency of energy recovery.[69] Energy recovery efficiency is also reduced due to generally higher "excess air" levels carried in these units. Also, combustion control is generally less effective in this type of unit, increasing the possibility of discharge of trace organic emissions. As pointed out earlier, these types of units generally utilize a lower quality of equipment and include less redundancy than larger mass-fired water wall and RDF WTE plants. Modular plants are responsible for about 2% of the total MSW burned at this time in the United States.

47.7.2 Fluidized Beds

Fluidized bed combustion (FBC) differs from mass-burn and RDF combustion in that the fuel is burned in "fluid suspension"—entrained along with particles of sand in an upward flow of turbulent air at a temperature controlled to 1500°F–1600°F (816°C–971°C). To date, it has been used primarily to burn sewage sludge, industrial waste, and coal and has been used to combust RDF in one facility in the U.S. Fluidized bed combustion of MSW is more commonly used in some European countries.

"Bubbling" FBC designs retain the material near the bottom of the furnace, while "circulating" designs allow bed material to move upward and then be returned near the bottom of the bed for further combustion. The reason for the interest in this combustion technique to burn RDF is the potential for these designs to provide more consistent combustion because of the extreme turbulence and the proximity of the RDF waste particles to the hot sand particles.[46] Such systems also require lower combustion temperatures than mass-burn and current RDF systems.

47.7.3 Pyrolysis and Gasification

Pyrolysis is the chemical decomposition of a substance by heat in the absence of oxygen. It generally occurs at relatively low temperatures (900°F–1100°F, compared with 1800°F for mass-burn). The heterogeneous nature of MSW makes pyrolysis reactions complex and difficult to control. Besides producing larger quantities of solid residues that must be managed for ultimate disposal, pyrolysis produces liquid tar and/or gases that are potentially marketable energy forms. The quality of the fuel products depends on the material fed into the reactor (e.g., moisture, ash, cellulose, trace constituent content) and operating conditions (e.g., temperature and particle size).[68]

Gasification is similar to pyrolysis in that it is the chemical decomposition of the substance by heat in the absence of oxygen. However, gasification occurs at temperatures of approximately 2200°F (1200°C). The reaction produces a synthetic gas with a heat content of approximately 250 Btu/cf. The approximate composition by volume on a dry basis is 25%–42% carbon monoxide, 25%–42% hydrogen, 10%–25% carbon dioxide, and 3%–4% nitrogen and other constituents. The synthetic gas is then cooled rapidly to reduce formation of dioxins and cleaned.[62]

Several prototype pyrolysis facilities were built in the 1970s with grants from EPA. These facilities were unable to produce quality fuels in substantial quantities. No one in the United States has yet successfully developed and applied the pyrolysis or gasification technology to MSW combustion. However, the use of pyrolysis and gasification for MSW management still attracts attention in other countries. Additional reading for pyrolysis and gasification technologies is available.[9,44]

47.8 Future Issues and Trends

It has been demonstrated by actual experience that modern mass-burn and RDF-fired MSW WTE plants can be designed and operated with reasonable assurance of continuous service and without adversely affecting nearby neighborhood property values. Allegations that WTE plant sites are situated adjacent to neighborhoods of low-income, disadvantaged, or minority populations ignore the specific technical siting criteria outlined earlier (i.e., adjacent to major highways, low land cost, industrial type area, etc.) which are generally followed in siting such facilities. Such areas frequently are closer to low-income neighborhoods than to middle- and higher-income neighborhoods.[68]

In 1994 the Supreme Court found that a local community could not force the MSW from that community to be taken to a specific facility such as a WTE plant.[24] This court decision was a major blow to the WTE industry bringing most planning and construction of new facilities to a halt in the mid-1990s. Many communities, when considering solid waste disposal options, have opted for low cost landfill disposal because of concern over impacts of higher cost WTE alternatives on taxes. At some point in the not too distant future, as the current landfills are rapidly filled, WTE technology will have to be utilized to solve the solid waste disposal problem. Some signs of this new interest in WTE technology are already occurring.

Another issue facing WTE plants is the uncertain future of regulatory requirements, both from the standpoint of legislation and from that of the regulatory agencies. In the past, legislation has been passed by Congress calling for Best Available Control Technology (BACT), then Lowest Achievable Emission Requirements (LAER), and then, most recently, Maximum Achievable Control Technology (MACT). The impact of this legislation, each calling for significant reductions in allowable emissions (absent any indication of the existence of a significant public health problem or benefit), has been to require extensive retrofits of existing plants and addition of equipment to proposed new plants, all at substantial expense without proven benefit and, in many cases, without prior proof of operational viability. Most facilities have opted to upgrade the air pollution control equipment and continue to operate.

Several positive actions are occurring in the field. Thus, most project developers have recognized the desirability of implementing a proactive program early in the project planning process to involve the public, particularly those in the vicinity of the proposed facility.[60] Also, the potential for materials recycling, which had been overenthusiastically embraced a number of years ago (state recycling goals as high as 70%, with some local communities projecting that their entire quantity of MSW could be managed through recycling), is gradually being recognized.[45] Franklin Associates[23] projects an increase in the recycling rate of from 22% in 1993 to 30% by the year 2000. Much of this increase in recycling is to come from increases in recovery of paper materials and diversion of yard wastes to composting. The impact of these changes in waste composition on energy available at WTE plants will be minimal, with the reduction in moisture content due to diversion of yard wastes being a positive factor.

The need to generate electric energy and safely manage the MSW generated by modern civilization, particularly in the vicinity of major metropolitan areas, together with the proven performance of modem WTE plants, indicate that this technology will be utilized to dispose of a portion of this country's solid waste and provide electricity.

References

1. Albina, D., Millrath, K., and Themelis, N., Effects of feed composition on boiler corrosion in waste-to-energy plants, in: *Proceedings of the 12th Annual North American Waste to Energy Conference*. ASME, New York, pp. 99–109, 2004.
2. Barnes, D., Rump session, *Chemosphere*, 14 (6–7), 987–989, 1985.
3. Battelle Columbus Labs, Characterization of stack emissions from municipal refuse-to-energy systems, National Technical Information Service, PB87-110482, October, 1981.
4. Bauer, J. P. and Wofford, J., Compliance with the new emissions guidelines for existing municipal waste combustion facilities, in: *Proceedings 1996 National Waste Processing Conference*. ASME, New York, 1996, pp. 9–18.
5. Beltz, P. R., Engdahl, R. B., and Dartoy, J., Evaluation of European refuse-fired energy systems design practices, summary, conclusions and inventory, prepared for U.S. EPA by Battelle Laboratories, Columbus, OH, 1979.
6. Bergstrom, J. G. T., Mercury behavior in flue gases, *Waste Management and Research*, 4(1), 57–64, 1986.
7. Bilmes, J. S., Impact of the recession and recycling on solid waste processing facilities in New England, in: *Proceedings 1992 National Waste Processing Conference*. ASME, New York, May 1992, pp. 351–360.
8. Braun, H., Metzger, M., and Vogg, H., *Zur Problematik der Quecksilber-Abscheidung aus Rauchgasen von Mullverbrennungsanlagen*, Vol. 1, Teil 2, 1985.
9. Bridgewater, A. (Ed.), *Pyrolysis and Gasification of Biomass and Waste*. CPL Press, Newbury, U.K., 2003.
10. Bryers, R. W. (Ed.), *Ash Deposits and Corrosion due to Impurities in Combustion Gases*. Hemisphere Publishing Corp., New York, 1978.
11. Carlsson, K., Heavy metals from energy-from-waste plants—Comparison of gas cleaning systems, *Waste Management & Research*, 4(1), 15–20, 1986.
12. Carrotti, A. A. and Smith, R. A., Gaseous emissions from municipal incinerators, USEPA Publication No. SW-18C, 1974.
13. Clark, L., Case history of a 240 ton day resource recovery project: Part II, in: *Proceedings of the 1996 National Waste Processing Conference*. ASME, New York, pp. 235–248, 1996.
14. Cooper Engineers, Inc., *Air Emissions Tests of a Deutsche Babcock Anlagen, Dry Scrubber System at the Munich North Refuse-Fired Power Plant*, February 1985. Report by Cooper Engineers.
15. Cooper Engineers, Inc., *Air Emissions and Performance Testing of a Dry Scrubber (Quench Reactor), Dry Venturi and Fabric Filter System Operating on Flue Gas From Combustion of Municipal Solid Waste in Japan*, West County Agency of Contra Costa County Waste Co-Disposal/Energy Recovery Project, May 1995. Report by Cooper Engineers.
16. Criss, G. H. and Olsen, R. A., The chemistry of incinerator slags and their compatibility with fireclay and high alumina refractories, in: *Proceedings 1968 National Incinerator Conference*. ASME, New York, 1968, pp. 75–88.
17. Dannecker, W., Schadstoffmessungen bei mullverbrennungsanlagen, *VGB Krazftwekstechnick*, 63(3), 237–243, 1983.
18. Davis, C. F., Annual snapshot of six large scale RDF projects, in: *Proceedings 1992 National Waste Processing Conference*. ASME, New York, May 1992, pp. 75–88.
19. Domalski, E. S., Ledford A. E., Jr., Bruce, S. S., and Churney, K. L., The chlorine content of municipal solid waste from Baltimore County, Maryland, and Brooklyn, NY, in: *Proceedings 1986 National Waste Processing Conference*, Denver, CO. ASME, New York, June 1986, pp. 435–448.
20. Eighmy, T. T., Collins, M. R., DiPietro, J. V., and Guay, M. A., Factors affecting inorganic leaching phenomena from incineration residues. Paper presented at *the Conference on Municipal Solid Waste Technology*, San Diego, CA, 1989.

21. Franklin Associates, Ltd., Characterization of municipal solid waste in the United States, 1960–2000, Report prepared for U.S. EPA, NTIS No. PB87-17323, Prairie Village, KS, July 25, 1986.
22. Franklin Associates, Ltd., Characterization of municipal solid waste in the United States, 1960 to 2000 (Update 1988), Report prepared for U.S. EPA, Office of Solid Waste and Emergency Response, Prairie Village, KS, March 30, 1988.
23. Franklin Associates, Ltd., Characterization of municipal solid waste in the United States (1994 Update), Report prepared for U.S. EPA Office of Solid Waste and Emergency Response, Prairie Village, KS, November, 1994.
24. Franklin, M. A., Solid waste stream characteristics, in: *Handbook of Solid Waste Management*, Tchobanoglous, G. and Kreith, F. (eds.), 2nd edn., Chapter 5. McGraw Hill, 2002.
25. EPA, www.epa.gov/solidwaste/nonhaz/municipal, accessed January 15, 2015.
26. Freeman, D., Waste-to-energy fights the fires of opposition, *World Wastes*, 37(1), 24, 1994.
27. Fryling, G. R. (Ed.), *Combustion Engineering*. Combustion Engineering, Inc., New York, 1966.
28. Greenberg, P. R., Zoller, W. H., and Gordon, G. E., Composition and size distributions of particles released in refuse incineration, *Environmental Science and Technology*, ACS, 12(5), 566–573, 1978.
29. Greenberg, R. R., Gordon, G. E., Zoller, W. H., Jacko, R. B., Neuendorf, D. W., and Yost, K. J., Composition of particles from the Nicosia Municipal Incinerator, *Environmental Science and Technology*, ACS, 12(11), 1329–1332, 1978.
30. Hahn, J. L., VonDemfange, H. P., Zurlinden, R. A., Stianche, K. F., and Seelinger, R. W. (Ogden Martin Systems, Inc.), and Weiand, H., Schetter, G., Spichal, P., and Martin, J. E. (Martin GmbH), Innovative technology for the control of air pollution at waste-to-energy plants, in: *Proceedings of the 1986 National Waste Processing Conference*, Denver, CO. ASME, New York, June 1986.
31. Hahn, J. L. and Sussman, D. B., Dioxin emissions from modern, mass-fired, stoker/boilers with advanced air pollution control equipment. Presented at *the Dioxin'86*, Fukuoka, Japan, September 1986.
32. Hahn, J. L. and Sofaer, D. S., Variability of NO_x emissions from modern mass-fired resource recovery facilities, Paper No. 88-21.7. Presented at *the 81st Annual Meeting of Air Pollution Control Association*, Dallas, TX, June 1988.
33. Hasselriis, F., The environmental and health impact of waste combustion—The rush to judgment versus getting the facts first, in: *Proceedings 1992 National Waste Processing Conference*. ASME, New York, May 1992, pp. 361–369.
34. Hay, D. J., Finkelstein, A., Klicius, R. (Environment Canada), and Marenlette, L. (Flakt Canada, Ltd.), The national incinerator testing and evaluation program: An assessment of (A) two-stage incineration (B) pilot scale emission control, Environment Canada, Report EPS 3/UP/2, September 1986.
35. Hecklinger, R. S., The relative value of energy derived from municipal refuse, *Journal of Energy Resources Technology*, 101, 251–259, 1979.
36. Hecklinger, R. S., Combustion, in: *The Engineering Handbook*, Dorf, R.C. (ed.). CRC Press, Inc., Boca Raton, FL, 1996.
37. Kaiser, E. R., Zeit, C. D., and McCaffery, J. B., Municipal incinerator refuse and residue, in: *Proceedings 1968 National Incinerator Conference*. ASME, New York, 1968, pp. 142–153.
38. Kilgroe, J. D., Lanier, W. S., and Von Alten, T. R., Development of good combustion practice for municipal waste combustors, in: *Proceedings 1992 National Waste Processing Conference*. ASME, New York, May 1992, pp. 145–156.
39. Kiser, J. V. L. and Zannes, M., The 2000 L.W.S.A. directory of waste-to-energy plants. Integrated Waste Services Association, Washington, DC, 2002.
40. Lai, G., Corrosion mechanisms and alloy performance in waste-to-energy boiler combustion environments, in: *Proceedings of the 12th Annual North American Waste to Energy Conference*. ASME, New York, 2004, pp. 91–98.
41. Law, S. L. and Gordon, G. E., Sources of metals in municipal incinerator emissions, *Environmental Science and Technology*, ACS, 13(4), 432–438, 1979.
42. Leavitt, C., Calculating the costs of waste management, *World Wastes*, 37(4), 42, 1994.

43. Lide, D. R., *CRC Handbook of Chemistry and Physics*, 71st edn. CRC Press, Boca Raton, FL, 1990.
44. Malkow, T., Novel and innovative pyrolysis and gasification technologies for energy efficient and environmentally sound MSW disposal, *Waste Management*, 24, 53–79, 2004.
45. Michaels, A., Solid waste forum, *Public Works*, March 1995, p. 72.
46. Minott, D. H., Operating principles and environmental performance of fluid-bed energy recovery facilities, Paper No. 88-21.9. Presented at *the 81st Annual Meeting of Air Pollution Control Association*, June 1989.
47. Murdoch, J. D. and Gay, J. L., Material recovery with incineration. Monmouth County, NJ, in: *Proceedings of 27th Annual International Solid Waste Exposition*, Tulsa, OK. SWANA, Silver Springs, MD (Pub. #GR-0028), August 1989, p. 329.
48. NYS Department of Environmental Conservation, Bureau of Toxic Air Sampling, Division of Air Resources, Preliminary Report on Westchester RESCO RRF, January 8, 1986.
49. Norton, F. H., Refractories, *Marks Standard Handbook for Mechanical Engineers*, 8th edn. McGraw-Hill, New York, 1978, pp. 6.172–6.173.
50. Ogden Projects, Inc., Environmental test report, Walter B. Hall Resource Recovery Facility, October 20, 1986.
51. Ogden Projects, Inc., Environmental test report, Marion County Solid Waste-to-Energy Facility, December 5, 1986.
52. O'Malley, W. R., Special factors involved in specifying incinerator cranes, in: *Proceedings 1968 National Incinerator Conference*. ASME, New York, 1968, pp. 114–123.
53. Penner, S. S. and Richards, M. B., Estimates of growth rates for municipal-waste incineration and environmental controls costs for coal utilization in the U.S., *Energy: The International Journal*, 14, 961, 1989.
54. Public Administration Service, *Municipal Refuse Disposal*, 3rd edn. American Public Works Association, Chicago, IL, 1970.
55. Richard, J. J. and Junk, G. A., Polychlorinated biphenyls in effluents from combustion of coal/refuse, *Environmental Science and Technology*, ACS, 15(9), 1095–1100, 1981.
56. Rigo, H. G. and Conley, A. D., Waste-to-energy facility capital costs, in: *Proceedings 1988 National Waste Processing Conference*. ASME, New York, May 1988, pp. 23–28.
57. Rogus, C. A., An appraisal of refuse incineration in Western Europe, in: *Proceedings 1966 National Incinerator Conference*. ASME, New York, 1966, pp. 114–123.
58. Sawell, S. E., Bridle, T. R., and Constable, T. W., Leachability of organic and inorganic contaminants in ashes from lime-based air pollution control devices on a municipal waste incinerator, Paper No. 87-94A. Presented at *the 80th Annual Meeting of Air Pollution Control Association*, New York, June 1987.
59. Sawell, S. E. and Constable, T. W., NITEP Phase IIB: Assessment of contaminant leachability from the residues of a mass burning incinerator, Vol. VI of National Incinerator Testing and Evaluation Program, The Combustion Characterization of Mass Burning Incinerator Technology, Environment Canada, Toronto, Ontario, Canada, August 1988.
60. Scarlett, T., WTE report focuses on socioeconomics, *World Wastes*, 37(5), 14, 1994.
61. Scherrer, R. and Oberlaender, B., Refuse pit storage requirements, in: *Proceedings 1990 National Waste Processing Conference*. ASME, New York, June 1990, pp. 135–142.
62. Schilli, J., The fourth dimension for waste management in the United States: Thermoselect gasification technology and the hydrogen energy economy, in: *Proceedings of the 12th Annual North American Waste to Energy Conference*. ASME, New York, 2004, pp. 251–258.
63. Seeker, W. R., Lanier, W. S., and Heap, M. P., *Municipal Waste Combustion Study: Combustion Control of Organic Emissions*, EPA, Research Triangle Park, NC, January 1987.
64. Stultz, S. C. and Kitto, J. B. (Eds.), *Steam: Its Generation and Use*, 40th edn. The Babcock and Wilcox Co., Barberton, OH, 1992.
65. Suess, M. J. et al., *Solid Waste Management Selected Topics*. WHO Regional Office for Europe, Copenhagen, Denmark, 1985.

66. Timm, C. M., Sampling survey related to possible emission of polychlorinated biphenyls from the incineration of domestic refuse, NTIS Publication No. PB-251-285, National Technical Information Service, U.S. Department of Commerce, Springfield, VA, 1975.
67. U.S. Congress, Office of Technology Assessment, Generation and composition of MSW, *Facing America's Trash, What Next for Municipal Solid Waste*, OTA-A-424, Chapter 3. U.S. Government Printing Office, Washington, DC, October 1989.
68. U.S. Congress, Office to Technology Assessment, Incineration, *Facing America's Trash, What Next for Municipal Solid Waste*, OTA-A-424, Chapter 6. U.S. Government Printing Office, Washington, DC, October 1989.
69. U.S. Environmental Protection Agency, Municipal waste combustion study, Report to congress, EPA/530-SW-87-021a, Washington, DC, 1987.
70. U.S. Environmental Protection Agency, Characterization of municipal waste combustor ashes and leachates from municipal solid waste landfills, monofills and codisposal sites, Prepared by NUS Corporation for Office of Solid Waste and Emergency Response, EPA/530-SW-87-028A, Washington, DC, 1987.
71. U.S. Environmental Protection Agency, Municipal waste combustion multipollutant study, characterization emission test report, office of air quality planning and standards, EMB Report No. 87-MIN-04, Research Triangle Park, NC, September 1988.
72. U. S. Environmental Protection Agency, Standards for performance for stationary sources and emission guidelines for existing sources—Municipal waste combustors, 40 CFR, Part 60, Subpart Cb, December 19, 1995.
73. Velzy, C. O., The enigma of incinerator design, in: *Incinerator and Solid Waste Technology*. ASME, New York, 1975, pp. 65–74.
74. Velzy, C. O., 30 Years of refuse fired boiler experience, Presented at *the Engineering Foundation Conference*. Franklin Pierce College, Rindge, NH, July 17, 1978.
75. Velzy, C. O., Trace emissions in resource recovery—Problems, issues and possible control techniques, in: *Proceedings of the DOE-ANL Workshop; Energy from Municipal Waste: State-of-the-Art and Emerging Technologies*, Argonne National Laboratory Report ANL/CNSV-TM-137, Lemont, IL, February, 1984.
76. Velzy, C. O., Standards and control of trace emissions from refuse-fired facilities, in: *Municipal Solid Waste as a Utility Fuel, EPRI Conference Proceedings*, Madison, WI, November 20–22, 1985.
77. Velzy, C. O., U.S. Experience in combustion of municipal solid waste. Presented at *the APCA Specialty Conference on Regulatory Approaches for Control of Air Pollutants*, Atlanta, GA, February 20, 1987.
78. Velzy, C. O., Considerations in planning for an energy-from-waste incinerator. Presented at *the New Techniques and Practical Advice for Problems in Municipal Waste Management*, Toronto, Ontario, Canada, April 1–2, 1987.
79. Visalli, J. R., A comparison of dioxin, furan and combustion gas data from test programs at three MSW incinerators, *Journal of the Air Pollution Control Association*, 37(12), 1451–1463, 1987.
80. Vogg, H., Braun, H., Metzger, M., and Schneider, J., The specific role of cadmium and mercury in municipal solid waste incineration, *Waste Management and Research*, 4(1), 65–73, 1986.
81. Vogg, H., Metzger, M. and Stieglitz, L., Recent findings on the formation and decomposition of PCDD/PCDF in solid municipal waste incineration, ISWA/WHO, Specialized Seminar, Copenhagen, Denmark, January 1987.
82. World Health Organization, Report on PCDD and PCDF emissions from incinerators for municipal sewage, sludge and solid waste—Evaluation of human exposure, from WHO Workshop, Naples, Italy, March 1986.

48

Energy Recovery by Anaerobic Digestion Process

Masoud Kayhanian and George Tchobanoglous

CONTENTS

Transformation of waste materials into energy can generally be accomplished through biological (i.e., anaerobic digestion), thermal, and chemical processes. The energy produced from these processes can be in the form of heat, gas, or liquid fuel.

48.1 Introduction

The anaerobic digestion (AD) process, carried out in the absence of oxygen, involves the use of microorganisms for the conversion of biodegradable biomass material into energy, in the form of methane gas and a stable humus material. AD can occur under controlled conditions in specially designed vessels (reactors), semicontrolled conditions such as in a landfill, or under uncontrolled conditions as it does in the environment. The focus in this chapter is on the controlled AD process. It should be noted that AD is differentiated from anaerobic fermentation, which is the term usually applied to processes employing anaerobic microbes for the production of fermentation products such as alcohol or lactic acid.

To describe anaerobic conversion of the biodegradable organic fraction of waste materials into energy, the following topics are presented in this chapter: (1) organic wastes and biomass used as feedstocks in AD process, (2) issue of organic waste biodegradability, (3) fundamental of AD process, (4) monitoring of AD process, (5) reactor types used for AD, (6) modes of operation for AD, (7) utilization of the AD process by-products, and (8) full-scale application of the AD technologies.

48.2 Organic Wastes and Biomass Used as Feedstocks in Anaerobic Digestion Process

The general scheme for a controlled AD process is shown in Figure 48.1. As shown, the recovery of energy involves feedstock preparation, methane gas generation, stabilization of digested solids, and the utilization of digester gas and humus as a source of energy and soil amendment, respectively. Major sources of waste materials considered as a feedstock for AD process are (1) biodegradable organic fraction of municipal solid waste (BOFMSW); (2) agricultural waste including animal waste, crop residues, biomass, and energy crops; (3) food processing waste; and (4) wastewater treatment plant sludge or biosolids (WWTPS or WWTPBS). The typical distribution of the biodegradable and inert fractions of municipal solid waste (MSW) in the United States is shown in Figure 48.2. The composition of MSW may vary greatly by season, geographical area, and community socioeconomic level. The extrapolation of results from one location to another, therefore, may not be valid and should be done with caution.

As shown in Figure 48.2, paper, yard waste, and food waste are the principal biodegradable organic fractions. The biodegradability of these waste materials varies substantially as reported in Table 48.1. The issue of biodegradability is discussed further in the following section.

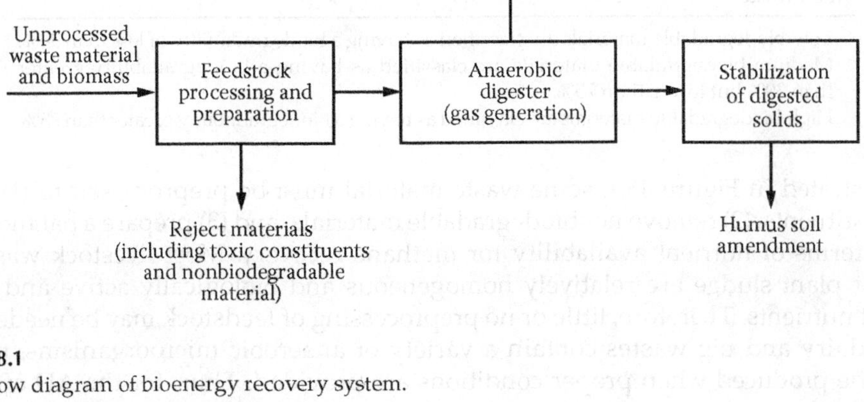

FIGURE 48.1
General flow diagram of bioenergy recovery system.

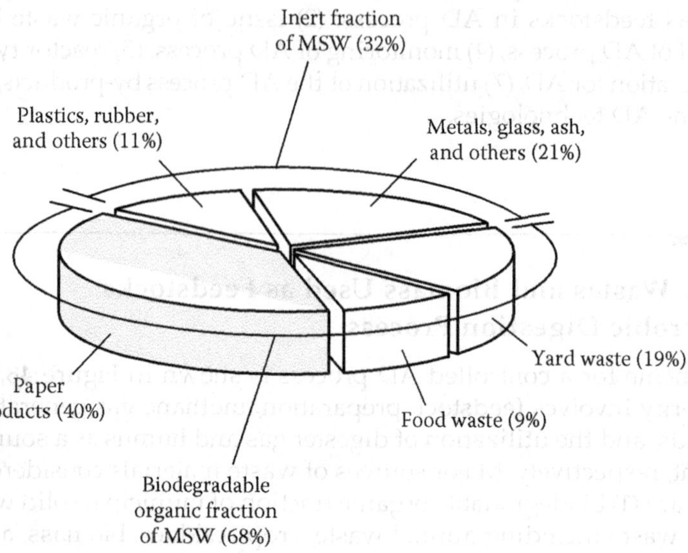

FIGURE 48.2
Typical distribution of the biodegradable and inert fraction of MSW in the United States. It should be noted that the relative distribution of these fractions may vary greatly depending on the season, geographical area, and community socioeconomic level.

TABLE 48.1
Weight Distribution and Biodegradability of the Organic Fraction of MSW Found in the United States

	Percent by Wet Weight		Biodegradability		
Component	**Range**	**Typical**	**Low**[a]	**Medium**[b]	**High**[c]
Cardboard	3–10	10		x	
Magazines	2–8	6	x		
Newspaper	4–10	8	x		
Waxed cartoons	5–20	12		x	
Mixed paper	<1–5	3		x	
Food waste	6–18	9			x
Yard wastes	5–20	15			x
Wood waste	1–4	1.5	x		

[a] Low-biodegradable materials are classified as having a biodegradability of less than 30%.
[b] Medium-biodegradable materials are classified as having a biodegradability greater than 30% but lower than 75%.
[c] High-biodegradable materials are classified as having a biodegradability greater than 75%.

As illustrated in Figure 48.1, some waste material must be preprocessed to (1) remove toxic constituents, (2) remove nonbiodegradable materials, and (3) prepare a balanced feedstock in terms of nutrient availability for methane recovery. Most livestock wastes and treatment plant sludge are relatively homogeneous and biologically active and contain sufficient nutrients. Therefore, little or no preprocessing of feedstock may be needed. Also, because dairy and pig wastes contain a variety of anaerobic microorganisms, methane gas will be produced when proper conditions are provided. Thus, for the AD of organic wastes other than livestock wastes, the addition of dairy or pig manure is used commonly. The addition of livestock waste to activate a biological process is commonly known as "inoculation."

48.3 Issue of Biodegradability

While each organic waste may contain a constant ultimate biodegradable fraction, practical biodegradability can be quite variable. Factors such as particle size, time, and environmental conditions (i.e., temperature, nutrient requirements) will influence the final outcome of biodegradation. For example, because favorable conditions do not exist in most landfills, the biodegradability estimated from analytical test will usually be greater than the actual biodegradation that occurs in landfills. For practical purposes, the discussion in this section deals with the biodegradability of the organic fraction of waste materials used in AD processes. Substrate biodegradability is of special importance in an AD processes, where the production of energy is of concern.[1,2]

48.3.1 Biodegradable Volatile Solids of an Organic Substrate

Dry organic substrates consist of volatile solids (VS) and ash. Taken together, these two components comprise the total solids (TS) of a substrate, as illustrated in Figure 48.3. The fraction of ash typically depends on the nature of the organic substrate. VS is measured as loss on ignition. Only the biodegradable volatile solids (BVS) fraction of the VS has the potential for bioconversion largely because of the presence of refractory volatile solids (RVS) which, in most digester feedstocks, is mostly lignin. Lignin is a complex organic material that is difficult for anaerobic bacteria to degrade and normally requires a long period of time for complete degradation. It is clear from Figure 48.3 that organic substrates with high RVS and ash contents have a low biodegradability.

The RVS in the organic fraction of MSW consists of the lignin content (LC) that is associated with cellulose in plant materials and thermoplastic materials. LC of an unsorted

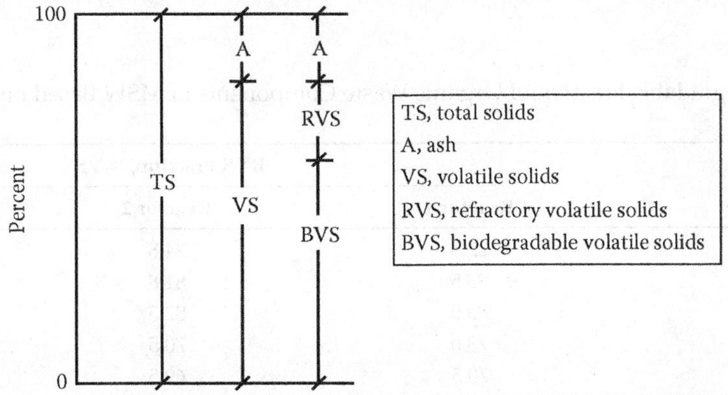

FIGURE 48.3
Characterization of a typical organic substrate.

organic fraction of MSW will fluctuate as the percentage of cellulose and thermoplastic materials will vary with season, socioeconomic conditions, and geographical locations. Knowing the BVS fraction of VS in the individual organic fractions of MSW will allow the biodegradability of any composite BOFMSW to be estimated.

48.3.2 Methods to Estimate Biodegradability

Several methods can be used to estimate the BVS fraction of an organic substrate, including (1) long-term batch digestion studies, (2) measurement of LC, and (3) chemostat studies. These methods are described briefly as follows.

48.3.2.1 Long-Term Batch Digestion

Batch digestion studies, designed to simulate a specific AD process, can be used to predict the biodegradable fraction of an organic substrate. One method involves graphical analysis of weight loss over time and is used commonly to predict the ultimate biodegradability of energy crops and agricultural wastes.[3] This method is based on a linear regression plot of the remaining VS concentration as the retention time approaches infinity. Regular weight measurements of the batch digester throughout the course of the study are required to apply this method.

An alternative method for determining the biodegradable fraction of an organic substrate, using batch digesters, is to determine and compare the initial and final dry masses. Initial measurements are made of the mass and the percent TS of active reactor mass of each mixture to be tested. Each mixture includes a percentage of seed, that is, material taken from an active digester to provide suitable microorganisms. One batch mixture is always 100% seed. At the end of the batch study, the dry mass in each individual reactor is measured. The mass loss in each unit, corrected for seed biodegradation, is due to the conversion of the biodegradable portion of the substrate to biogas. The results of BOFMSW based on long-term batch digestion studies are shown in Table 48.2.

48.3.2.2 Measurement of Lignin Content

An analytical method commonly used for determining the BVS fraction of an organic substrate is based on the measurement of the crude LC of the VS. Chandler et al.[1] correlated the

TABLE 48.2

Estimated Biodegradable Fraction of Organic Waste Components of MSW Based on Long-Term Batch Studies

	BVS Fraction, %VS		
Organic Substrate	**Reactor 1**	**Reactor 2**	**Average**
Newspaper	22.5	24.8	23.7
Office paper	83.5	81.8	82.7
Food waste	83.0	82.5	82.8
Yard waste	73.0	70.5	71.8
Mix blend[a]	70.5	69.5	69.8[b]

[a] Mixed blend consists of 19% newsprint, 53% office paper, 15% yard waste, and 13% food waste (dry basis).
[b] The value reported for mixed blend is based on the actual organic waste, not the value that can be computed from each individual components.

TABLE 48.3

Estimated Biodegradable Fraction of Organic Waste Components of MSW Based on LC[a]

Organic Substrate	LC Range, %VS[a]	Average BVS Fraction
Newspaper	20–23	22
Office paper	0.2–1	82
Yard waste	4–10	72
Food waste	0.1–0.7	82
Mixed blend[b]	4–7	67.6

[a] Computed using Equation 48.1.
[b] Mixed blend consists of 19% newsprint, 53% office paper, 15% yard waste, and 13% food waste (dry basis).

biodegradability of various agricultural residues and newsprint, as determined by long-term batch digestion studies, with the LC of the substrate, as determined by sequential fiber analysis, developed by Robertson and Van Soest.[4] The following empirical relationship was developed to estimate the biodegradable fraction of an organic substrate from lignin test results:

$$\text{Biodegradable fraction} = 0.83 - 0.028\text{LC} \tag{48.1}$$

where the biodegradable fraction is expressed as a fraction of the VS and LC expressed as a percentage of the VS. The results of biodegradable fraction for BOFMSW based on the LC are shown in Table 48.3.

48.3.2.3 Chemostat Studies

The BVS fraction of an organic substrate can also be estimated using chemostat techniques. The true digestible organic matter (TDOM) technique, developed by animal scientists, is used to assess the digestibility of animal feeds.[5] The test involves the digestion of a sample in vitro in rumen fluid for 48 h. A second chemostat technique, the biochemical methane potential (BMP) assay, is used to characterize the BMP and biodegradability of many organic substrates.[6] In the BMP assay, the substrate biodegradability is determined by monitoring the cumulative methane production from a slurry sample that is incubated anaerobically (typically for 30 days at 35°C) in a chemically defined media and inocula.

The biodegradability of energy crops as estimated using the BMP assay tends to be 3%–10% higher than the corresponding BVS value as estimated using the TDOM technique.[7,8] The BMP assay appears to be more suitable for determining the ultimate biodegradability.

48.3.3 Biodegradability of Various Organic Waste Materials

The estimates of BVS, as a percentage of VS, of a typical organic fraction of MSW based on long-term batch studies, LC, and a pilot study are compared in Figure 48.4. As can be seen, the average BVS fractions calculated using either the LC or the long-term batch study are essentially the same. The BVS fraction obtained from the pilot study, using a complete-mix reactor and mass retention time (MRT) of 30 days, is about 83% and 81% of the estimated values obtained from the LC and the batch study, respectively. Similarly, Richards

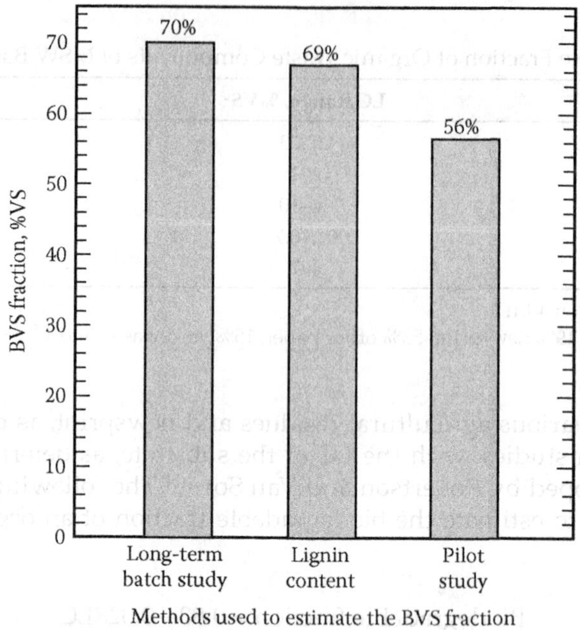

FIGURE 48.4
Comparison of the BVS fraction of a typical mixed BOFMSW obtained using three methods of analysis.

et al.[7,8] were able to remove only 83% of BVS of sorghum, using a thermophilic, high-rate, high-solids reactor with an MRT of 45 days.

The values of biodegradable fraction for various organic substrates using different methods are compared in Table 48.4. It is important to note that the estimated BVS values based on field operations are normally about 80%–85% of the values estimated by chemostat studies. Thus, the use of BMP values would result in an overestimation of biogas production rate that can be achieved in actual practice. It is important to note

TABLE 48.4
Biodegradability of Various Organic Substrate Using Batch Digestion, Chemostat, and Field Operation Methods

	Methods Used to Estimate Biodegradability				
	Average Value, %				
Substrate	**Batch Digestion**	**TDOM**	**BMP**	**Field Operation**[a]	**Reference**
Sorghum	60	79.1	82.0	72.6 (105)[b]	Richards et al. [7,8]
Sorghum 1			90.9	75.0 (45)	Richards et al. [7]
Sorghum 2			90.0	72.5 (29)	Richards et al. [7]
Alpha-cellulose	81.6	90.6	99.8		Richards et al. [7]
1:1 mix of sorghum/alpha-cellulose		70.8	87.6	90.7	Richards et al. [7]
BOFMSW	69.8			57 (30)	Kayhanian and Tchobanoglous [9]

[a] All field operations were conducted in a thermophilic, high-rate, high-solids reactor.
[b] Numbers in parentheses are nominal MRT value in days.

that complete removal of BVS as estimated from batch digestion and analytical or a chemostat method at a practical range of MRT (20–40 days) is not possible. For an in-vessel high-solids AD process, BVS can be defined as that fraction of VS that can be biodegraded under optimum environmental and nutritional conditions in a period of 20–40 days.

48.3.4 Impact of BVS on Process Design and Performance

The common use of TS or VS both in describing the organic loading rate and feedstock carbon-to-nitrogen (C/N) ratio and in estimating gas production from the BOFMSW is misleading, as some of the organic compounds are low in biodegradability. The use of BVS instead allows meaningful comparisons to be made between the different AD processes reported in the literature. The impact of BVS on the organic loading rate, the computation of C/N ratio, and the prediction of biogas production are discussed as follows.

Traditionally, the loading rates for biological processes have been based on the VS content of the substrate feedstock. When handling substrates with variable biodegradabilities, estimates of the methane production potential and digester performance cannot be made accurately using a VS loading rate. Loading rates based on the BVS fraction of the substrate should be used to estimate the conversion of organic wastes to methane. The BVS organic loading rate can be expressed as the input BVS mass per active reactor mass per day. The BVS mass of an organic substrate can be computed using total inflow VS mass multiplied by the BVS fraction of VS.

Customarily, the C/N ratio is determined based on the total dry mass of the organic matter and the corresponding percentage concentrations of carbon and nitrogen. This method of determining the C/N ratio may not be appropriate for the BOFMSW because not all the organic carbon is biodegradable and/or available for biological decomposition. However, it appears that almost all the nitrogen in the organic material is available for conversion to ammonia via microbial metabolism.[10] Because the available nitrogen in the organic feedstock can be converted to ammonia, the C/N ratio should be based on the nitrogen content of the total organic mass and the carbon content of the BVS mass.[11]

A more accurate biogas volume production can be made from the BVS mass, rather than the TS or VS mass. However, not all the BVS mass computed using the LC or the results of long-term batch studies will be converted into biogas in field operation. For design purpose, a correction factor of about 0.82, derived from lignin test or long-term batch digestion, must be used when using the BVS mass as a basis for theoretical gas production. Computation of BVS fraction of the biodegradable fraction of MSW is illustrated in the following example calculation.

48.3.4.1 Example BVS Calculation of MSW

Estimate the total BVS weight (mass) of a 100 ton/day MSW material recovery facility.

Use the following information:

1. Assume the composition of a typical MSW waste stream given in Figure 48.2 is valid.

2. Typical characteristics of MSW subfractions:

Waste Material	TS, %	VS, %TS	LC
Yard waste	40	88	6.43
Food waste	30	90	0.36
Paper	94	98	9.29

Solution

1. Start from the typical composition of MSW, and prepare a table:

	Weight, lb/day	
Waste Material	Wet	Dry[a]
Yard waste	38,000	15,200
Food waste	18,000	5,400
Paper	80,000	75,200
Metal	42,000	42,000
Plastic	22,000	22,000

[a] Calculated using % TS.

2. Calculate the biodegradable organic fraction of MSW being serviced by the sorting facility:

$$\text{BOFMSW per day} = (0.68)(100 \text{ ton MSW/day}) = 68 \text{ ton}$$

3. Compute the total weight of BVS. To compute the BVS content of a substrate,

$$\text{VS weight} = \text{dry weight} \times \text{VS\%}$$

$$\text{BVS weight} = \text{VS weight} \times \text{BF}$$

where

$$\text{BF} = 0.83 - (0.028 \times \text{LC})$$

LC is the LC.

4. Prepare a table of the values calculated:

Biodegradable Waste	Dry Wt., lb/day	VS, %TS	VS Wt., lb/day[a]	BF, %VS[b]	BVS Wt., lb/day[c]
Yard waste	15,200	88	13,376	0.65	8,694
Food waste	5,400	90	4,860	0.82	3,985
Paper (mixed)	75,200	98	73,696	0.57	42,005
Total	95,800	00	91,932	00	54,684

[a] The values in this column are calculated as VS = dry weight × VS%.
[b] The biodegradable fraction (BF) is computed using Equation 48.1 and the LC or using the biodegradability value determined by a long-term batch study.
[c] The values in this column are calculated as BVS = VS weight × BF.

Comment

It is important to note that the BVS weight is about 27% of the total wet weight of MSW, 40% of the total wet weight of BOFMSW, and 60% of the total VS fraction of the TS.

48.4 Fundamental of Anaerobic Digestion Process

The purpose of this section is to familiarize the reader with the basic microbiology related to an AD process. Topics discussed include (1) an introduction to anaerobic bacteria, (2) nutrient requirements for the anaerobic bacteria, (3) physical and chemical parameters affecting anaerobic bacteria, (4) basic biochemical reaction in AD process, and (5) pathway of complex organic substrate in AD process.

48.4.1 Anaerobic Bacteria

Effective anaerobic degradation of complex organic waste is a result of the combined and coordinated metabolic activity of the digester microbial population. This population of microorganisms is composed of several major trophic groups. At present, as reported in Table 48.5, four different groups of anaerobic bacteria are recognizable. With respect to methane recovery, the anaerobic bacteria are generally grouped as (1) hydrolysis (hydrolyzing bacteria), (2) acidogens (acid-forming bacteria), (3) acetogens (acid-forming bacteria), and (4) methanogens (methane-forming bacteria).

Methane bacteria are among the most strictly anaerobic microorganisms known and occur naturally in the rumen of cows, marshes, and brackish waters, as well as in wastewater treatment plant digesters. Methanogenic bacteria have been isolated from anaerobic digesters, together with other obligate anaerobes, such as the *Propionibacter, Butyrobactor,* and *Lactobacillus.* The true methane bacteria include the following identified genera: *Methanococcus, Methanobacterium, Methanosarcina, Methanospirillum,* and *Methanobacillus.* The main groups of methane bacteria along with some of their morphological and growth characteristics are given in Table 48.6.

As reported in Table 48.6, methane bacteria are rod, cocci, or sarcinate in shape. As a group, methane bacteria are nearly always Gram positive (a stain used to identify bacterial types) and usually not motile. Many methane bacteria are pleomorphic; that is, they exhibit variable morphology when viewed under the microscope, which makes their identification very difficult. Their length can vary between 2 and 15 μm. Nutritionally, the methanogenic bacteria are said to be chemoheterotrophs (or chemotraphic heterotrophs), which means that they can build up their cell structures from either carbonate or organic compounds. Morphologically, methanogens are a diverse group; however, physiologically, they are quite similar as all share the common metabolic capacity to produce methane. The methanogens are possibly the most important group of anaerobic bacteria. They have scientific significance in their exclusive and distinctive properties among the bacteria.

TABLE 48.5
Four Major Groups of Anaerobic Bacteria and Their Function

Bacterial Group	Function
Hydrolytic bacteria	Convert saccharides, protein, lipids, and other minor chemical constituents of biomass to soluble compounds such as fatty acids, amino acids, and monosaccharides
Acidogenic bacteria (fermentation)	Convert products from hydrolysis reaction and other soluble compounds to compounds such as acetate, propionate, butyrate, CO_2, and H_2
Acetogenic bacteria	Convert unicarbon compounds or HCOOH or hydrolyze multicarbon compounds to acetic acid, H_2, and CO_2
Methanogenic bacteria	Convert the by-products of acidogenesis and acetogenesis to methane and carbon dioxide

TABLE 48.6
Morphological Characteristics of Methane Bacteria

Organism	Morphology	Length, μm	pH Optimum	Electron Donor
Methanobacterium formicium	Rods, single pairs or chains	2–15		Hydrogen and formate
Methanobacterium strain MOH	Rods, single pairs or chains	2–4		Hydrogen
Methanobacterium arborphilicum	Rods, single pairs or chains	2–3.5	7.5–8	Hydrogen
Methanobacterium strain AZ	Rods, single pairs or chains	2–3	6.8–7.2	Hydrogen
Methanosarcina barkeri	Sarcina	1.5–5	7	Methanol and hydrogen
Methanobacterium ruminantium	Coccus chains	1–2 diameter	6.8–7.2	Hydrogen and formate
Methanococcus vannielii	Coccus	0.5–4 diameter	7.4–9.2	Formate
Methanobacterium mobile	Rod			Hydrogen and formate
Methanobacterium thermoautotrophium	Rod	5–10		Hydrogen
Methanobacterium hungatii	Spiral rods	50		Hydrogen and formate

Source: Holland, K.T. et al., *Anaerobic Bacteria*, Chapman & Hall, New York, 1987.

48.4.2 Nutrient Requirements for Anaerobic Bacteria

To operate a high-solids AD process at a commercial level, attention must be focused on process stability. Successful operational parameters have been established for the high-solids process studies conducted at the University of California, Davis (UC Davis).[9] However, in the AD of BOFMSW, bacterial nutritional requirements have often been overlooked. Nutritional deficiencies may result in reactor instability and incomplete bioconversion of the organic substrates. When the AD process is applied to the biodegradable organic fraction of MSW, bacterial nutritional requirements must be addressed, and nutrient supplementation may be required.[13,14]

Methanogenic bacteria have a variety of mineral nutrient requirements for robust growth.[15,16] For a proper bacterial metabolism, a variety of nutrients must be present in the substrate. The nutrient requirements for anaerobic bacteria can generally be categorized as macro- and micronutrient. For a stable AD process, these nutrients must be present in the substrate in the correct ratios and concentrations. Based on studies conducted at UC Davis, it was found that typical BOFMSW used as a feedstock for the AD process is deficient in many essential nutrients.[17] To overcome feedstock nutrient deficiencies, supplementary nutrients must be added to stimulate the digestion process. The range of nutrient concentrations needed to stimulate the anaerobic treatment of BOFMSW for gas production is not well known. The nutrient values reported in Table 48.7 are based on 3 years of experience at the UC Davis high-solids biogasification project.

48.4.3 Physical and Chemical Parameters Affecting Anaerobic Bacteria

The alteration of several physical and chemical environmental parameters in anaerobic digesters can influence microbial populations and digester performance.

TABLE 48.7

Representative Feedstock Nutrient Concentrations Required for the Robust Anaerobic Bioconversion of BOFMSW

Nutrient	Unit	Average Value (Dry Basis) Range	Typical
C/N	Ratio	20–30	25
C/P	Ratio	150–300	180
C/K	Ratio	40–100	70
Co	mg/kg	<1–5	2
Fe	mg/kg	100–5000	1000
Mo	mg/kg	<1–5	2
Ni	mg/kg	5–20	10
Se	mg/kg	0–0.05	0.03
W	mg/kg	0.05–1	0.1

Source: Kayhanian, M. and Rich, D., *Biomass and Bioenergy*, 8, 433, 1995.
Note: C/N, C/P, and C/K ratios are based on biodegradable organic carbon and total nitrogen, phosphorus, and potassium.

TABLE 48.8

Physical and Chemical Effectors of Bacterial Populations in Anaerobic Digesters

Parameter	Population Response	Influence of Methanogenesis
Temperature (e.g., change to 65°C from 35°C)	Enrichment of thermophilic	Increase
pH (e.g., change to pH 5 from pH 7)	Enrichment of acidophiles	Decrease
Organic substrate composition (e.g., change soluble substrate [glucose] for particulate [lignoglucose])	Enrichment of biopolymer decomposers	Decrease
Substrate feed rate (e.g., change from slow to high glucose feed)	Enrichment of hydrolytic bacteria	Decrease
Inorganic composition (e.g., addition of excess sulfate)	Enrichment of sulfate reducers	Decrease

Source: Holland, K.T. et al., *Anaerobic Bacteria*, Chapman & Hall, New York, 1987.

Major environmental factors are summarized and shown in Table 48.8. Changing temperature from the mesophilic range (i.e., below 45°C) to the thermophilic range (i.e., above 55°C) increases organic mineralization and establishes a different species composition (i.e., thermophilic bacteria). Zeikus[18] found that thermophilic bacteria have a limited species composition, but they possess all the major nutritional categories and metabolize the same substrate as mesophilic bacteria. The ability to proliferate at growth temperature optima well above 60°C is associated with extremely thermostable macromolecules. As a consequence of growth at high temperatures and unique micromolecular properties, thermophilic bacteria can possess high metabolic rates, physically and chemically stable enzymes, and lower growth but higher end-product yields than similar mesophilic species. Thermophilic digesters are generally digest substrates within shorter retention time than mesophilic digesters.

Values of pH well below neutral are indicative of digester failure. The pH ranges for growth of many bacterial species inherent to anaerobic methane digester are not known and may vary considerably. Nevertheless, important hydrogen-producing (e.g., *Clostridium thermocellum*) and hydrogen-consuming (e.g., *Methanobacterium thermoautotrophium*) anaerobes do not grow at pH value below 6. Digester imbalances that favor more rapid growth of nonmethanogens (e.g., high feed rate or chemical composition of feed) can lead to low pH and inhabitation of methanogenesis. Low pH values favor proton reduction to hydrogen, but not hydrogen oxidation to protons. Thus, methanogens may not function at low pH because they employ oxidoreductases for hydrogen oxidation and establishment of proton gradients during the catabolism of one carbon compound and acetate.

The organic composition of digester wastes affects species composition and methane yield. Most notably, methane yields from municipal wastes and animal manures are often limited by polymeric lignin surrounding cellulose and retarding its fermentation. High-molecular-weight lignin is not metabolized significantly by anaerobic bacteria. However, methane yield from digestion of lignocellulosics can be enhanced greatly by physicochemical pretreatment that separates lignin from cellulose and/or solubilize lignin into anaerobically digestible substrates. To accomplish lignin solubilization, an adaptation period is required to obtain an active microbial population from untreated waste digesters that metabolized soluble chemical hydrolysis products of lignin. The organic composition modification caused by waste pretreatment is considered to be associated with change in species compositional change.

The most significant parameter affecting methanogenesis during AD processes is sulfate inhabitation. The addition of high levels of sulfate to decomposing organic matter can result in nearly complete inhabitation of methane formation as a result of species compositional change. The basis for this inhabitation is that carbon and electron flow during organic mineralization is diverted from methane formation to hydrogen sulfide production. The response of sulfide on many methanogens is concentration dependent. Sulfide is required by many methanogens as a sulfur source for growth. The addition of low sulfide concentrations often stimulates mixed culture methanogenesis, whereas addition of high concentrations can be inhibitory.

48.4.4 Pathway of a Complex Organic Substrate in Anaerobic Digestion Process

A generalized scheme for the AD of complex organic substrate is shown in Figure 48.5. AD is generally considered to take place in four more or less distinct stages. The four stages have been described as (1) hydrolysis, (2) acidogenesis (often identified as fermentation in the literature), (3) acetogenesis, and (4) methanogenesis. Each of the four stages has distinct bacterial groups, and chemical reactions, as described previously, and the overall process proceeds in an assembly-line fashion.[12] In many literature references, acidogenesis (fermentation) and acetogenesis are considered to be one process termed "acidogenesis."

Methane bacteria can only use limited range of substrates for growth and energy production. During AD, a varied mixture of complex compounds is converted to a very narrow range of simple compounds, mainly methane and carbon dioxide. The ecology of the system is much more complicated and involves an interacting succession of microbes, which influences each others' growth and metabolism. Nearly all anaerobic bacteria can use hydrogen and carbon dioxide, and most can use formate, a few acetate, and fewer still methanol and methylamines. However, in the mixed population and with many organisms, the range of fermentation products is restricted. The restriction is due to the phenomenon of interspecies hydrogen transfer by which methanogens utilize hydrogen and thus benefit other organisms. Hydrogen removal is helpful as it allows some organisms to dispose of more electrons via hydrogen rather than via the production of more-reduced carbon compounds, such as ethanol or butyrate. In general, fatty acids are the "key" intermediate products of the anaerobic fermentation of

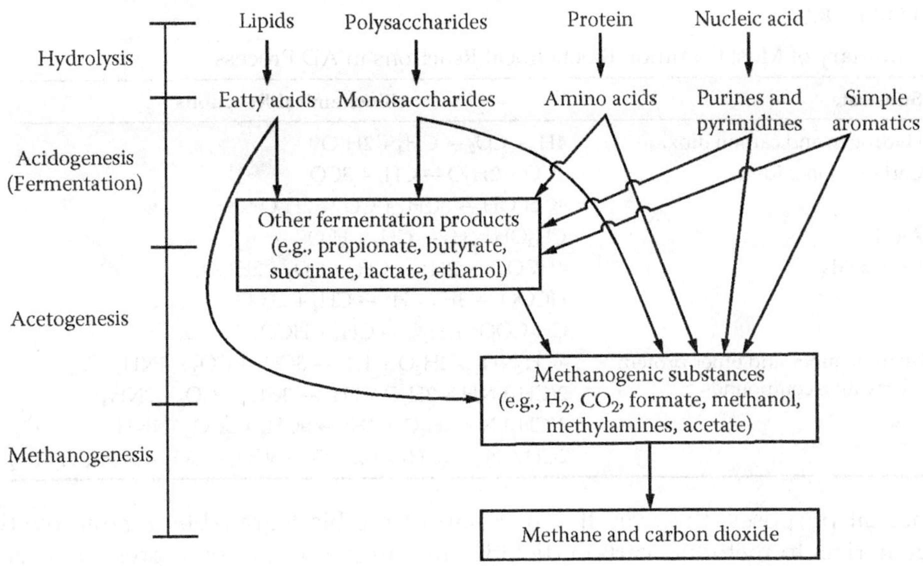

FIGURE 48.5
Stages of AD. (From Tchobanoglous, G. et al., *Integrated Solid Waste Management: Engineering Principles and Management Issues,* McGraw-Hill Book Company, New York, 1993.)

organic compounds prior to methane formation. Acetate is by far the predominant fatty acid in normally operated systems, being responsible for about 70% of the fatty acid present. The VFA measurements are usually expressed in mg/L, therefore, as being acetate equivalent.

48.4.5 Basic Biochemical Reaction in Anaerobic Digestion Process

The anaerobic bacteria described in previous sections are responsible for biochemical transformation of a wide variety of waste materials. These transformations are involved in the breakdown of complex polymers, such as cellulose, fats, and proteins, to long- and short-chain fatty acids and finally methane, carbon dioxide, and water. The basic biochemical reactions affected by anaerobic microbial population are oxidation/reduction reactions, where a number of organic compounds are oxidized by the removal of hydrogen. Carbon dioxide, is thereby, reduced in providing an oxidant for the methane bacteria. The hydrogen produced can be replaced by some of the organic acids and alcohols as direct reductants of carbon dioxide. The basic biochemical reactions utilizing hydrogen, carbon dioxide, carbon monoxide, alcohols, organic acids, methylamines, and other protein derivative compounds as a substrate are summarized in the energy-yielding equations given in Table 48.9.

The general anaerobic transformation of the biodegradable organic fraction of waste materials can be described by the following equation[19]:

$$\text{Biodegradable organic matter} + H_2O + \text{Nutrients} \rightarrow \text{New cells} + \text{Resistant organic matter} + CO_2 + CH_4 + NH_3 + H_2S + \text{heat} \quad (48.2)$$

TABLE 48.9

Summary of Most Common Biochemical Reactions in AD Process

Substrate	Biochemical Reactions
Hydrogen and carbon dioxide	$4H_2 + CO_2 \rightarrow CH_4 + 2H_2O$
Carbon monoxide	$4CO + 2H_2O \rightarrow CH_4 + 3CO_2$
	$4CH_3OH \rightarrow 3CH_4 + CO_2 + 2H_2O$
Alcohols	$CH_3OH + H_2 \rightarrow CH_4 + H_2O$
Fatty acids	$4HCOO^- + 2H^+ \rightarrow CH_4 + CO_2 + 2HCO_3^-$
	$HCOO^- + 3H_2 + H^+ \rightarrow CH_4 + 2H_2O$
	$CH_3COO^- + H_2O \rightarrow CH_4 + HCO_3^-$
Methylamines and other protein derivative compounds	$4CH_3NH_2 + 2H_2O + 4H^+ \rightarrow 3CH_4 + CO_2 + 4NH_4^+$
	$2(CH_3)_2NH + 2H_2O + 2H^+ \rightarrow 3CH_4 + CO_2 + 2NH_4^+$
	$4(CH_3)_3N + 6H_2O + 4H^+ \rightarrow 9CH_4 + 3CO_2 + 4NH_4^+$
	$2CH_3CH_2\text{–}N(CH_3)_2 + 2H_2O \rightarrow 3CH_4 + CO_2 + 2CH_3CH_2NH_2$

For practical purposes, the overall conversion of the biodegradable organic fraction of waste materials to methane, carbon dioxide, and ammonia can be represented with the following equation (Tchobanoglous et al.[19]):

$$C_aH_bO_cN_d \rightarrow nC_wH_xO_yN_z + mCH_4 + sCO_2 + rH_2O + (d - nx)NH_3 \quad (48.3)$$

where
$s = a - nw - m$
$r = c - ny - 2s$

The terms $C_aH_bO_cN_d$ and $C_wH_xO_yN_z$ are used to represent (on a molar basis) the composition of the organic material present at the start and the end of the process, respectively. If it is assumed that the organic wastes are stabilized completely, the corresponding expression is

$$C_aH_bO_cN_d + \left(\frac{4a-b-2c+3d}{4}\right)H_2O \rightarrow \left(\frac{4a+b-2c-3d}{8}\right)CH_4 + \left(\frac{4a-b+2c+3d}{8}\right)CO_2 + dNH_3 \quad (48.4)$$

This relationship can be used to estimate the theoretical biogas volume and composition. Estimation of the amount of gas produced by the complete anaerobic stabilization of the BVS fraction of MSW is illustrated in the following example calculation.

48.4.5.1 Example Estimation of Theoretical Gas Production

Estimate the total theoretical amount of gas that could be produced from MSW under anaerobic conditions in an in-vessel AD process. Assume that 78% by weight of the MSW is organic material including moisture. Further assume the moisture content is 20%, the VS are 83.5% of the total organic solids, the BVS are 75% of the VS and only 90% of the BVS will be converted to biogas. The overall chemical formula for the biodegradable organic material is $C_{60}H_{94.3}O_{37.8}N$.

Solution

1. Determine the BVS:

$$BVS = 78 \times (1 - 0.2) \times 0.835 \times 0.75 = 39.1 \text{ lb}$$

2. Determine the total amount BVS that will be converted to gas:

$$BVS_{gas} = 39.1 \times 0.9 = 35.2$$

3. Using the chemical formula $C_{60}H_{94.3}O_{37.8}N$, estimate the amount of methane and carbon dioxide that can be produced using Equation 48.4:

$$C_aH_bO_cN_d + \left(\frac{4a-b-2c+3d}{4}\right)H_2O \rightarrow \left(\frac{4a+b-2c-3d}{8}\right)CH_4 + \left(\frac{4a-b+2c+3d}{8}\right)CO_2 + dNH_3$$

For the given chemical formula, a = 60, b = 94.3, c = 37.8, and d = 1.

The resulting equation is

$$\underset{(1437)}{C_{60}H_{94.3}O_{37.8}N} + \underset{(324)}{18.28H_2O} \rightarrow \underset{(512)}{31.96CH_4} + \underset{(1232)}{28CO_2} + \underset{(17)}{NH_3}$$

4. Determine the weight of methane and carbon dioxide from the equation derived in step 3 and the data from step 2:

$$\text{Methane} = \left(\frac{512}{1437}\right)(35.2 \text{ lb}) = 12.5 \text{ lb} (5.7 \text{ kg})$$

$$\text{Carbon dioxide} = \left(\frac{1232}{1437}\right)(35.2 \text{ lb}) = 30.2 \text{ lb} (13.7 \text{ kg})$$

5. Convert the weight of gases, determined in step 4, to volume, assuming that the densities of methane and carbon dioxide are 0.0448 and 0.1235 lb/ft³, respectively:

$$\text{Methane} = \frac{12.5 \text{ lb}}{(0.0448 \text{ lb/ft}^3)} = 279 \text{ ft}^3 (7.9 \text{ m}^3)$$

$$\text{Carbon dioxide} = \frac{30.2 \text{ lb}}{(0.1235 \text{ lb/ft}^3)} = 244 \text{ ft}^3 (6.9 \text{ m}^3)$$

6. Determine the percentage composition of the resulting gas mixture using the gas volume determined in step 5:

$$\text{Methane } (\%) = \left[\frac{279}{(279+244)}\right] \times 100 = 53.35\% \approx 53\%$$

$$\text{Carbon dioxide } (\%) = 100\% - 53.35\% = 46.65\% \approx 47\%$$

7. Determine the total theoretical amount of gas generated per unit weight of dry BVS as determined in step 1 and per ton of MSW:

$$\frac{(279+244)}{(39.1)} = 13.4 \text{ ft}^3/\text{lb BVS} \left(1.48 \text{ m}^3/\text{kg BVS}\right)$$

$$\text{Biogas/ton of MSW} = \left(13.4 \text{ ft}^3/\text{lb}\right)\left(2{,}000 \text{ lb/ton}\right)(0.78)(0.8)(0.835)(0.75)$$

$$= 10{,}470 \text{ ft}^3/\text{ton MSW}$$

Comment

The theoretical gas production value determined earlier has been achieved with the high-solids anaerobic decomposition process (see Section 48.7.2).

48.5 Monitoring of the Anaerobic Digestion Process

Although the high-solids AD process is generally robust, care must be taken to ensure balanced operation. To aid the prevention of unbalanced digester operation and to prevent digester failure, proper methods of monitoring the high-solids AD process are described, and possible operational problems are identified along with suggested remedial actions to be taken when these problems arise.

48.5.1 Balanced Digestion

A balanced digester is one in which AD proceeds with a minimum of control. Balanced operation means that the environmental parameters of the system remain within their optimum range, with only occasional fluctuations. When an imbalance does occur, the two main problems are (1) identifying the beginning of an unbalanced condition and (2) identifying the cause of the imbalance. Unfortunately, there is no single parameter that will always indicate the commencement of an unhealthy anaerobic process. The parameters shown in Table 48.10 must all be monitored daily. None of these parameters can be used individually as a positive indicator of the development of digester imbalance.

TABLE 48.10

Indicators of Unbalanced Operation of the AD Process

Parameter	Warning Condition
Ammonia concentration	Increases
Percent CH_4 in biogas	Decreases
Percent of CO_2 in biogas	Increases
Reactor pH	Decreases
Total gas production	Decreases
VFA concentration	Increases
Waste stabilization	Decreases

TABLE 48.11

Summary of the Most Common Problems Associated with the AD Process and Suggested Actions to Cure the Problems

Major Problem	Suggested Action
Free ammonia toxicity	1. Start feeding organic waste with higher C/N ratio.
	2. Dilute the active reactor mass with freshwater.
Nutrient deficiency	1. Add chemical nutrients.
	2. Add organic materials rich in the needed nutrients.
Organic overloading	1. Do not feed.
	2. Add strong base to neutralize acids.
	3. Resume feeding at lower organic loading rate when pH reaches at least 6.8.
TS buildup	1. Add water.
Toxic overloading	1. Identify and remove the toxic element from the feedstock.
	2. If the population of methanogens is reduced (CH_4 concentration decreased), add proper methanogen seed.
	3. If pH decreases below 6.8, add strong base to neutralize acids.
	4. Resume feeding at lower organic loading rate when pH reaches at least 6.8.

The most immediate indication of impending operational problems is a significant decrease in the rate of gas production. If the growth of the microorganisms is being inhibited by one or more factors, it will be reflected in the total gas production. However, a decrease in the gas production rate may also be caused by a decrease in either the digester temperature or the rate at which the feed material is being added to the digester.

The most significant single indicator of a digester problem is a gradual decrease in pH. In an operating system, a decrease in pH is associated with an increase in organic acid concentration. Measurement of the increase in organic acids is also a good control parameter; however, proper laboratory facilities, equipment, and trained personnel are required to monitor this and most of the other control parameters affecting the anaerobic process. Gas production rate and pH are simple, quick measurements and are performed easily.

48.5.2 Common Problems and Solutions to Stabilize Digester Operation

In general, there are five major problems associated with the AD process: (1) increase in TS concentration, (2) organic overloading, (3) toxic overloading, (4) free ammonia toxicity, and (5) nutrient deficiency. Possible actions to cure these problems are summarized in Table 48.11.

48.6 Reactor Types Used for Anaerobic Digestion

In the past 50 years, a wide variety of full-scale AD processes and reactors have been used for both waste stabilization and energy recovery. The AD is used most commonly in wastewater treatment plants for sludge stabilization with and without methane gas recovery plants.[20]

Full-scale digesters (including anaerobic lagoons [ANLs]) have also been used by dairy and hog farmers to stabilize their waste as well as to produce methane gas for the production of electrical power to offset their power consumption during pick hours and on

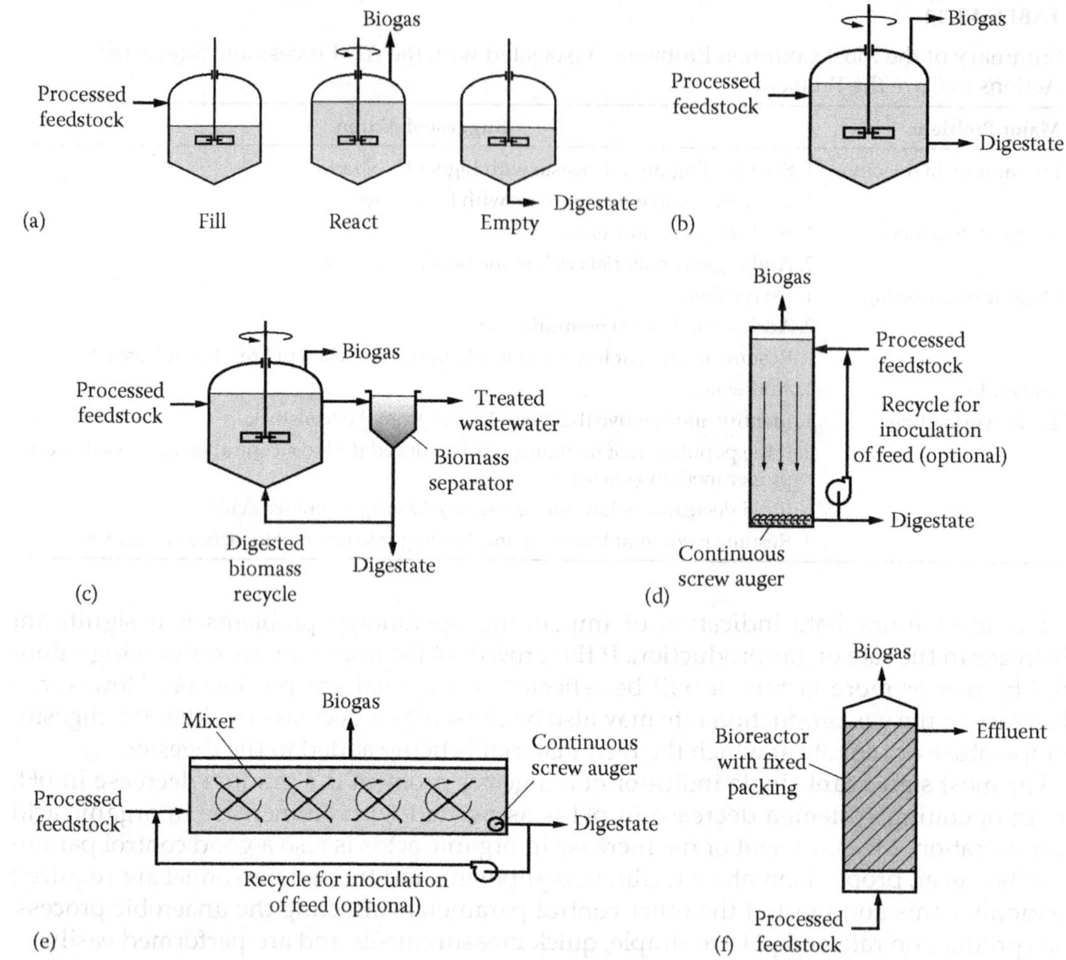

FIGURE 48.6
Typical reactor types used for AD (a) batch reactor, (b) complete-mix continuous flow tractor, (c) anaerobic contact reactor, (d) vertical plug-flow reactor, (e) horizontal plug-flow reactor, and (f) attached growth reactor.

annual basis.[21,22] AD is also commonly used for treatment and energy production from food processing industries. The principal types of AD reactor designs that can be used for waste stabilization as well as methane recovery for energy production are shown in Figure 48.6 and described briefly in the following. Other reactor designs, which incorporate features of the reactors shown in Figure 48.6, are described in Section 48.6.7.

48.6.1 Batch Reactor

A batch reactor is fed once, and then the biotransformation is allowed to proceed until completion before any material is added or removed. A simplified diagram of a batch digestion is shown in Figure 48.6a. The evolution of compounds in the reactor can be monitored, and a similar level of decomposition can be achieved by all the material in the reactor at one time. Additionally, the systems are generally simple, with less support equipment than continuous fed reactors (see later). Batch processes are necessary when the biotransformation being performed requires a long reaction time. Solids that are treated undiluted are often treated with batch reactors.

Usually, batch processes require more operator labor, for feeding and unloading, than continuous feed processes, and especially so for liquids and slurries, which are easily piped and pumped. For solids, the material handling needs are more similar for the two processes. Storage facilities are needed to contain waste received between batches. The need for storage can be inconvenient for large-scale, continuously produced wastes.

48.6.2 Complete-Mix Continuous Flow Reactor

A reactor to which a waste stream is fed and a treated effluent stream is withdrawn continuously is known as a continuous flow reactor. A simplified diagram of a complete-mix reactor is shown in Figure 48.6b. Most municipal wastewater sludge treatment processes are continuous feed processes, as municipal sludge is produced continually. A process can be designed to fit the expected maximum inflow rate, although fluctuations in flow rate and process upsets may interfere with operation. If alternate storage is not available, discharge of untreated or partially treated wastes can occur.

The average residence time of material in a continuous feed reactor can be determined using the following equation:

$$t = \frac{V}{Q} \tag{48.5}$$

where

t is the residence time
V is the reactor volume
Q is the flow rate (vol/time)

Reactors that are fed liquids or slurries continuously require less operator labor than batch reactors. Because storage space is not necessary to contain continuously produced wastes between batches, continuous feed processes are used commonly for industrial or municipal wastes. There are two types of reactors that are intermediate between continuous feed and batch processes. A reactor that is fed once a day is a semicontinuous feed reactor. Also, a reactor that is fed continually, but only emptied when waste stabilization has been achieved, is a semibatch reactor.

In a complete-mix reactor, the reactor contents are blended to homogeneity with a mixing device. The effluent leaving the reactor is exactly the same as the material in every part of the reactor. If fed continuously, the input waste is considered to be immediately mixed completely with the reactor contents. These reactors can also be described as well-mixed or as continuous stirred-tank reactors (CSTRs). One advantage of a complete-mix process is that fluctuations in feed concentration or composition are diluted into the larger reactor mass. Because dilution and the concentration of waste nutrient determines the rate of waste decomposition, complete-mix reactors have a slower decomposition rate.

48.6.3 Anaerobic Contact Reactor

The anaerobic contact process (ACP) is used to overcome the disadvantages of the complete-mix reactor without recycle. To enhance the rate of treatment, biomass is separated from the effluent and returned to the reactor. Biomass recycle can be used to reduce the reactor size and cost. A simplified ACP is shown in Figure 48.6c. Hydraulic retention times as low as a half-day have been achieved resulting in a significant reduction in plant size. ACP has been applied successfully for the treatment of meat-packing waste, where a retention time of several hours was measured. ACP is also used for the treatment of a high-strength waste material and is usually operated under low solids (TS less than 8%).

48.6.4 Plug-Flow Reactor

All plug-flow reactors are continuously or semicontinuously fed. In a plug-flow reactor, material passes through, ideally, without interacting with the material fed in before or after it. These reactors can be thought of as tubes through which independent batches of reacting material pass. The reactor can be either vertical or horizontal flow as shown in Figure 48.6d and e, respectively.

The retention time for a plug-flow reactor is the length of time it takes a mass introduced at the beginning of the reactor to pass through and be removed from the other end (see also Equation 48.5). Most horizontal plug-flow digesters are operated under low solids (usually with TS of less than 10%). Because the incoming, high-nutrient wastes are not diluted into the rest of the digester contents, the plug-flow digesters are more susceptible to system upsets due to sudden increases in waste concentration, called shock loadings. Fortunately, a certain degree of back mixing is unavoidable throughout the reactor due to longitudinal dispersion.

48.6.5 Anaerobic Attached Growth Reactor

Anaerobic attached growth reactors are used to prevent the depletion of the bacterial population within the reactor and, thus, improve digester efficiency. A simplified anaerobic filtration process is shown in Figure 48.6f. The retention of bacteria within the reactor is achieved by introducing some type of packing material on which the bacteria can grow. Various materials such as stone granules, wood chips, and plastic materials of various shapes and sizes have been used as packing material. The bacteria adhering to the filter particles bring about the treatment of the liquid as it passes through the reactor. Excess biomass dies and/or sloughs off and passes out as sludge. The anaerobic filtration process is usually used for the treatment of a high-strength waste material and usually operated under low solids (TS less than 8%).

48.6.6 Reactors with Recycle Flow

Often, it is advantageous to recycle and mix effluent from an anaerobic reactor (digestate) with the inflowing waste stream (see Figure 48.6c through e). Recycling is especially advantageous when it is necessary to inoculate the incoming waste with bacteria that have acclimated to the system. Recycling of biomass can also be used to increase average residence time in the reactor and to dilute high-concentration wastes. Recycling a portion of the biomass can also be used for process control, allowing the reactor to respond to fluctuations in the waste stream. Recycling systems, however, add costs, both capital and operational, to a system. Hence, unless it is necessary, the use of a simpler, unrecycled system is typically favored.

48.6.7 Other Reactor Types and Processes

Over the years, numerous anaerobic processes employing a variety of other reactor types, different than those described earlier, have been developed and are in use. Four of these other reactor types and processes are illustrated in Figure 48.7 and described

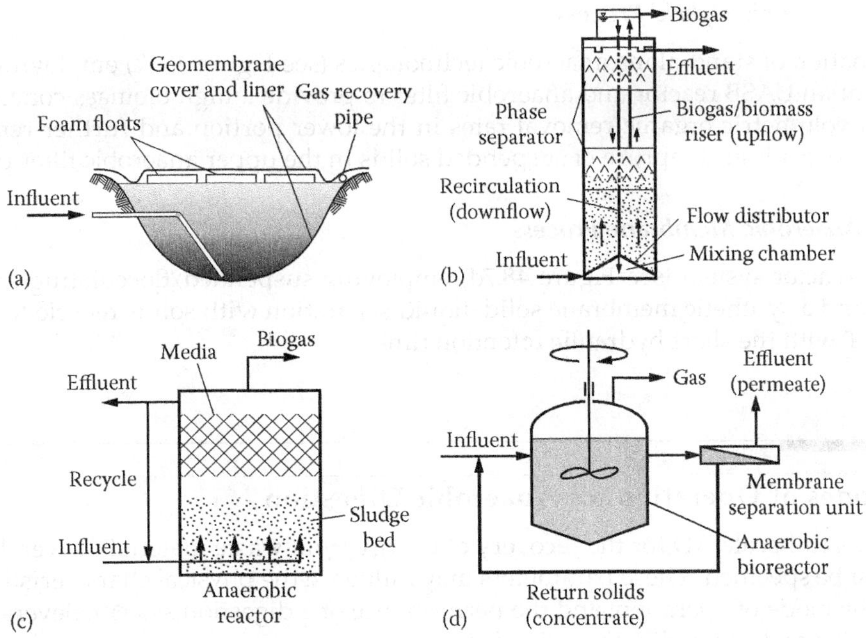

FIGURE 48.7
Alternative types of reactors used for AD: (a) excavated and lined ANL, (b) UASB–IC, (c) anaerobic hybrid process (ANHBP), and (d) anaerobic membrane process (ANMBR). (Adapted from Tchobanoglous, G. et al., *Wastewater Engineering: Treatment and Resource Recovery*, 5th edn., Metcalf & Eddy Inc., AECOM, McGraw-Hill Book Company, New York, 2014.)

briefly in the following. It is important to note that there are many more reactor designs in use and that new reactor configurations are being developed continuously; including all of them in this chapter at present time is beyond the scope of this book.

48.6.7.1 Anaerobic Lagoon

This is generally an unmixed reactor system, typically below ground, employing suspended/flocculating anaerobic biomass and settled anaerobic solids with hydraulic retention times of 20–50 days and average MRTs of 50–100 days (see Figure 48.7a). ANLs are often classified as arbitrary-flow reactors as the flow regime is neither complete mix nor plug flow. ANLs can be covered with synthetic membranes for gas collection. Uncovered and covered, ANLs handle a wider range of wastes including solids and soluble wastewaters and are by far the most common anaerobic processes in use today, especially for farm wastes.

48.6.7.2 Upflow Anaerobic Sludge Blanket with Internal Circulation

An upflow anaerobic sludge blanket with internal circulation (UASB–IC) reactor consists of two stacked UASB reactors in series, each with a gas separator at the top (see Figure 48.7b). The system uses a down comer pipe from a top chamber to the bottom inlet and a riser pipe from the first gas separator to induce recirculation and high upflow velocities in the lower granular sludge blanket reactor. Gas produced from the lower reactor is captured in the first separator and creates a gas lift for water and biosolids in the riser pipe. The gas is separated (released) from the biosolids in the chamber above the second reactor gas separator.

48.6.7.3 Anaerobic Hybrid Process

A combination of stand-alone anaerobic technologies (see Figure 48.7c) employing a combination of an UASB reactor and anaerobic filter to provide a high biomass concentration and high volumetric organic removal rates in the lower portion and further removal of volatile fatty acids and capture of suspended solids in the upper anaerobic filter portion.

48.6.7.4 Anaerobic Membrane Process

A mixed reactor system (see Figure 48.7d) employing suspended/flocculating anaerobic biomass and a synthetic membrane solid–liquid separation with solids recycle to provide a long SRT with the short hydraulic retention time.

48.7 Modes of Operation for Anaerobic Digestion

To design a full-scale AD for the recovery of energy from waste materials, several parameters must be specified. These parameters may influence the physical characteristics of the reactor, the mode of operation, and the performance of a digestion system. Several modes of operation are discussed in the following.

48.7.1 Low-Solids Digestion

AD systems for municipal wastewater operate under low-solids conditions; that is, the concentration of solids in the waste substrate is typically less than 10%. This mode of operation is appropriate because both municipal wastewater and the sludge collected for disposal from aerobic wastewater treatment processes are intrinsically low solids. The waste substrate in a low-solids system can be pumped and piped easily as it behaves like water. A low-solids system is also more able to handle feed composition fluctuations and higher ammonia levels, because the large volume of water serves to dilute the incoming waste compounds. Dilution of waste compounds also means dilution of nutrients, unfortunately, so that low-solids digestions can require more retention time if the nutrient level is too low. Also, the high dilution means that larger-sized tanks are required to accommodate wastes.

It is possible to digest substrates other than wastewater treatment sludge, however, and many of these waste substrates have higher solids content. High-solids AD has recently been developed to accommodate these wastes. Originally developed to dispose of agricultural wastes, particularly manures, high-solids AD can be used to degrade food industry wastes, BOFMSW, agricultural and forestry residues, and other high-solids wastes.

48.7.2 High-Solids Digestion

High-solids AD operates at solids contents from 20% to 32%. Typically, high-solids wastes, such as BOFMSW, must be mixed with water or a low-solids waste, such as wastewater treatment sludge, to dilute the solids content to within the operating range. A high-solids system can operate with tanks that are up to 75% smaller than a low-solids system, for the same dry weight of waste. Because the nutrient in a high-solids reactor is more concentrated, the conversion rate is higher. Additionally, the high-solids process produces an end product that requires less dewatering to convert it into a landfillable waste or a usable compost material. High-solids AD systems are, however, more sensitive to micronutrient deficiencies and toxic inhibition, since there is more mass to digest and is diluted with less moisture. Especially, thermophilic systems are critically dependent on the maintenance of a proper C/N ratio, as explained in the following.

48.7.3 Thermophilic Digestion

Process in which the waste substrate is heated to between 120°F and 135°F is characterized as thermophilic. Although heating the reactor adds additional equipment and energy costs, thermophilic process has higher reaction rates as compared with ambient processes. Because bacterial metabolism is greater at the higher temperatures, a wider range of bacteria can colonize the reactor. Due to a decrease in water solubilization with increased temperature, thermophilic processes can be more vulnerable to inhibition effects due to the presence of some soluble compounds. Thermophilic methanogenic systems, for example, are more sensitive to high ammonia concentrations than are similar mesophilic systems.

48.7.4 Mesophilic Digestion

Processes in which the waste substrate has a temperature of between 68°F and 98°F are characterized as mesophilic. Often, no heating equipment at all is used. Although less costly than thermophilic systems, the time necessary for bioconversion may be significantly longer. If heating equipment is not used, the efficiency of the process will also depend on the weather.

48.7.5 Two-Stage Digestion

As discussed in the previous sections, AD is principally carried out by four microbial populations under four, more or less, distinct and interconnected stages. In the second and third (acidogenesis and acetogenesis) stage of the digestion process, soluble wastes are converted to organic acids, CO_2 and H_2, and, finally, in the fourth stage, methane bacteria convert organic acids to methane and carbon dioxide. Acid-forming (acidogenic and acetogenic) and methane (methanogenic) bacteria are distinct from each other in several features. These features include their physiology, nutrient requirements, growth capabilities, and sensitivity to environmental stress. A single-stage digester can be most efficient only when a compromise is achieved between the relatively fast-growing acidogenic and acetogenic bacteria and the slower, more sensitive methanogenic bacteria.

A two-stage digester (see Figure 48.8) is designed to isolate the hydrolytic, acidogenic, and acetogenic bacteria from the methane bacteria into separate reactors and, thus, optimize each environment for maximum efficiency. Distinct features of two-stage reactor are overall greater efficiency and system stability, and substantial reductions in total reactor volume. Several methods can be used to separate the nonmethanogenic and methanogenic bacteria:

1. Various inhibitors may be introduced into the acid digester to prevent methanogenesis:
   a. Adding chloroform
   b. Adding carbon tetrachloride
   c. Limiting oxygenation
   d. Adjusting redox potential
2. Dialysis of methane bacteria by filtering the acids.
3. Adjusting the dilution rate and recycling of cell mass to each phase.

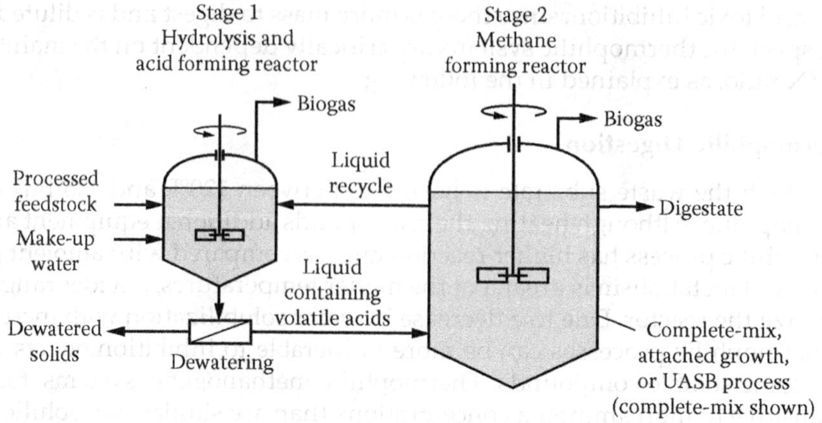

FIGURE 48.8
Two-stage AD.

48.7.6 Codigestion of Wastewater Treatment Sludge and BOFMSW

The codigestion of WWTPS and BOFMSW have been investigated by numerous researchers.[24–29] Most of these early studies concluded that high-quality source-separated MSW could be digested without nutrient supplementation up to organic loading rates of about 2.6 kg VS/m^3 day. They also reported that under conditions of nutrient addition, gas production could be enhanced, and a more stable process could be achieved. Other studies conducted by Rivard et al.[13,14], Babbitt et al.[30], and Cecchi et al.[31] confirmed the stabilizing effect of sludge to the digestion of MSW, with sludge comprising between 8% and 20% of feedstock VS.

One large-scale codigestion of MSW and wastewater sludge has been demonstrated in the United States by the Refuse Conversion to Methane (REFCOM) project in Tampa, Florida. The REFCOM system was based on a conventional low-solids digester design and operated as part of a total resource recovery process. The successful operation of the REFCOM process for mixed MSW at a rate of 50 ton/day or greater has confirmed the technical feasibility of the AD process. Reliable gas production was achieved in this facility for more than a year, indicating that MSW with wastewater sludge as a cosubstrate can support an anaerobic biological conversion process.

Most codigestion studies conducted prior to 1990 were low-solids processes, typically at a TS of 4%–8%. Only recently has the high-solids AD process been utilized for the codigestion of MSW and WWTPS.[13,14,32] In these high-solids digestion studies, the sludge was used mainly to provide sufficient nutrients for microbial growth and metabolism. Nearly 22 years ago, the technical feasibility of the anaerobic composting process was demonstrated for comanagement of BOFMSW and WWTPS at UC Davis, as described in the following.[33]

48.7.7 Two-Stage Anaerobic Composting

In anaerobic composting, the focus is twofold: (1) production of methane as a source of energy and (2) complete waste stabilization to produce humus-like material. AD normally occurs at much higher solids content (e.g., 25%–32%). When a high-solids AD process is combined with a second stage aerobic biodryer, the two-stage process is termed "anaerobic composting."[34,35] The principal differences between the conventional aerobic and anaerobic composting processes are summarized in Table 48.12. The process flow diagram, physical characteristics of the reactors, and the energy recovery and volume reduction achieved are described as follows.

TABLE 48.12

Comparison of Aerobic and Anaerobic Composting Processes

Characteristic	Aerobic Process	Anaerobic Process
Energy use	Net energy consumer	Net energy producer
End products	Humus, CO_2, H_2O	Sludge, CO_2, CH_4
Volume reduction	Up to 50%	Up to 50%
Processing time	21–30 days	20–40 days
Primary goal	Volume reduction	Energy production
Secondary goal	Compost production, waste stabilization	Volume reduction, waste stabilization

Source: Kayhanian, M. and Tchobanoglous, G., *Environ. Technol.*, 14(9), 815, 1993.

48.7.7.1 Process Flow Diagram

The process flow diagram for the anaerobic composting process studied at the UC Davis is illustrated in Figure 48.9. As shown in Figure 48.9, the process involves two separate stages. The first stage of the two-stage process involves the high-solids AD of the biodegradable organic fraction of MSW to produce a gas, composed principally of CH_4 and CO_2, and digested solids. The second stage of the two-stage process is used to decrease the moisture content of the anaerobically digested solids. Typically, the solids content of the digested solids is increased from 25% to about 65%. Because the second stage aerobic process is used to dry the digested solids, the term biodryer is also used to describe the process. The physical and operational characteristics of the pilot anaerobic composting facilities used at UC Davis are briefly described in the following.

48.7.7.2 Physical Characteristics

The physical characteristics of the pilot-scale anaerobic composting facilities are summarized in Table 48.14. As reported in Table 48.13, both the anaerobic digester and the biodryer are designed to operate as complete-mix reactors. Additionally, both units are

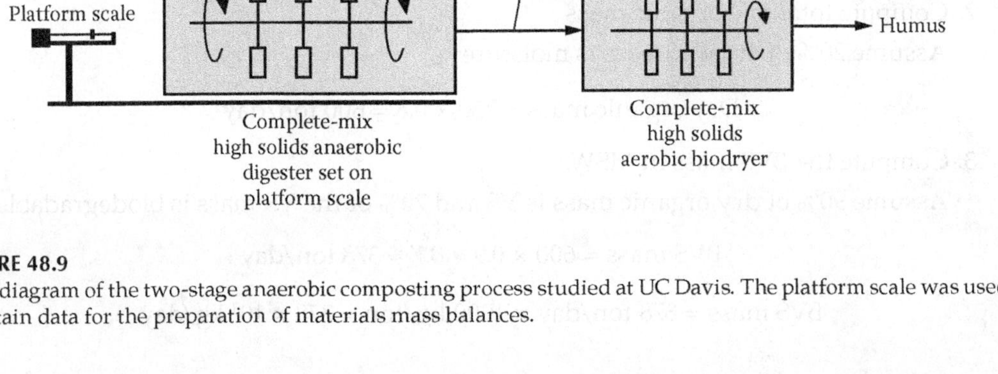

FIGURE 48.9
Flow diagram of the two-stage anaerobic composting process studied at UC Davis. The platform scale was used to obtain data for the preparation of materials mass balances.

TABLE 48.13

Summary of the Physical Characteristics of the UC Davis Pilot-Scale High-Solids AD/Aerobic Biodrying Processing Units

Item	Unit	Description/Value
Anaerobic Digester		
Reactor type		Complete mix
Mixing mechanism		Mechanical
Heating mechanism		Thermal blanket
Total reactor volume	L	2548
Volume of active reactor solids	L	1900
Aerobic Biodrying Reactor		
Reactor type		Complete mix
Mixing mechanism		Mechanical
Heating mechanism		Thermal blanket
Total rector volume	L	850
Volume of active reactor solids	L	765

Source: Kayhanian, M. et al., *BioCycle*, 32(3), 48, 1991.

equipped with thermal blankets and individual control panels so that they may be operated either manually or automatically. Further information on the UC Davis pilot-scale AD process design and performance can be obtained from Kayhanian and Tchobanoglous[9], Kayhanian and Rich[33], Kayhanian et al.[35], Kayhanian[36], Kayhanian and Hardy[37], and Kayhanian et al.[39]

48.7.7.3 Example of Codigestion Process for Energy Recovery Equivalent in Barrels of Oil

Compute the amount of energy (equivalent in barrels of oil) that can be produced from the digestion of a 1000 ton/day of MSW using primary sludge as cosubstrate.

Solution

Compute the BVS mass, biogas production, energy equivalent to barrels of oil as shown in the following steps:

1. Compute total organic mass.

   Assume 75% of total mass is organic:

$$\text{Organic mass} = 0.75 \times 1000 = 750 \text{ ton/day}$$

2. Compute total dry organic mass.

   Assume 20% of organic mass is moisture:

$$\text{Dry organic mass} = 750 \times 0.8 = 600 \text{ ton/day}$$

3. Compute the BVS mass in MSW.

   Assume 90% of dry organic mass is VS and 70% of the VS mass is biodegradable:

$$\text{BVS mass} = 600 \times 0.9 \times 0.7 = 378 \text{ ton/day}$$

$$\text{BVS mass} = 378 \text{ ton/day} \times 1000 \text{ kg/ton} = 378 \times 10^3 \text{ kg/day}$$

4. Compute the amount of primary sludge required for codigestion of the BOFMSW.
   Assume the TS of the primary sludge is 7% and codigestion occurs at 28% TS. Compute the amount of sludge needed considering the water already present in the biodegradable matter:

$$\text{Sludge required} = 1170 \text{ ton} = 1.17 \times 10^6 \text{ kg/day}$$

5. Compute the BVS mass in the primary sludge.
   Assume the VS of the primary sludge is about 70% and the biodegradable fraction of VS is about 60% of VS:

$$\text{BVS mass of primary sludge} = 1.17 \times 10^6 \text{ lb/day} \times 0.07 \times 0.7 \times 0.6 = 34 \times 10^3 \text{ kg/day}$$

6. Compute the total BVS mass produced from the MSW and primary sludge.

$$\text{Total BVS mass} = 378 \times 10^3 + 34 \times 10^3 \text{ kg/day} = 412 \times 10^3 \text{ kg/day}$$

7. Compute the biogas production rate.
   Assume biogas production of about 0.86 m^3/kg BVS/day:

$$\text{Total gas production rate} = \left(412\times10^3 \text{ kg BVS/day}\right)\times\left(0.86 \text{ m}^3\text{/kg BVS}\right)$$
$$= 354.32\times10^3 \text{ m}^3\text{/day}$$

8. Compute the energy value of the biogas volume produced.
   Assume the thermal energy value of biogas is about 16.23 MJ/m^3:

$$\text{Thermal energy} = (354.32 \times 10^3 \text{ m}^3\text{/day}) \times (16.23 \text{ MJ/m}^3) = 5.75 \times 10^6 \text{ MJ/day}$$

$$\text{Thermal energy} = (5.75 \times 10^6 \text{ MJ/day}) \times 365 \text{ day/year} = 2.098 \times 10^9 \text{ MJ/year}$$

9. Compute the energy value relative to petroleum.
   Assuming thermal energy value of one barrel of oil is about 6.218 × 10^6 MJ:

$$\text{Energy equivalent in barrels of oil} = \frac{(2.098\times10^9 \text{ MJ/year})}{(6.218\times10^6 \text{ MJ/barrel})} = 337 \text{ barrels of oil/year}$$

48.7.7.4 Example of Volume Reduction of BOFMSW Processed through Codigestion Compared to a Well-Compacted Landfill

Compute volume reduction of 1000 ton MSW codigested with WWTPS compared to a well-compacted landfill under mixed and unmixed conditions:

Case 1: Sludge cake fills the interstices of the compacted MSW in the landfill:

a. Input feedstock ratio (wet): 60% sludge to 40% BOFMSW by weight
b. Density of compacted MSW = 540 kg/m^3
c. Sludge TS = 5%

Case 2: Sludge cake remains unmixed from compacted MSW in landfill:

a. Input feedstock ratio (wet): 60% sludge to 40% BOFMSW
b. Density of compacted MSW = 540 kg/m^3

c. Sludge TS = 5%

d. Density of sludge cake at 51% solids = 1240 kg/m^3

Solution: Case 1 (Mixed)

1. Compute the amount of sludge required for the anaerobic composting process.
   For every unit volume of BOFMSW (540 kg of MSW) added,
   540 × (60/40) = 810 kg of sludge at 5% solids is required.
2. Compute the corresponding mass of sludge dewatered to 51% solids.
   Dewatering the sludge to 51% solids would reduce the sludge weight to

   $$810 \text{ kg} \times (51/5) \times 0.01 = 82.6 \text{ kg dewatered cake at 51\% solids}$$

3. Compute the combined weight of the MSW and sludge solids.
   Combined weight of MSW and sludge cake = 540 + 82.6 kg = 622.6 kg.
4. Compute the density of the combined MSW and sludge.
   Assuming no change in MSW volume (1 m^3), combined density = 622.6 kg/m^3.

Solution: Case 2 (Unmixed)

1. Compute the amount of sludge required for the anaerobic composting process.
   For every unit volume of BOFMSW (540 kg of MSW) added,
   540 × (60/40) = 810 kg of sludge at 5% solids is required.
2. Compute the corresponding mass of sludge dewatered to 51% solids.
   Dewatering the sludge to 51% solids would reduce the sludge weight to
   810 kg × (51/5) × 0.01 = 82.6 kg dewatered cake at 51% solids
3. Compute the combined weight of the MSW and sludge solids.
   Combined weight of MSW and sludge cake = 540 + 82.6 kg = 622.6 kg.
4. Compute the combined volume of the MSW and sludge:
   a. Volume of sludge = 82.62 kg/1240 kg/m^3 = 0.07 m^3
   b. Volume of MSW = 1 m^3
   c. Combined volume = 0.07 + 1.0 m^3 = 1.07 m^3
5. Compute the combined density of the MSW and sludge (unmixed).
   Combined density = 622.6 kg/1.07 m^3 = 583 kg/m^3.

48.8 Utilization of By-Products of Anaerobic Digestion Process

The principal by-product of an anaerobic composting process is biogas and a stable digestate material with a low biodegradability and high LC. The digestate material can be processed further to produce a high-quality humus material for marketing. The characteristics of biogas and the humus material are presented in this section.

48.8.1 Characteristics of Biogas

The gas produced during AD of biodegradable organic material in a healthy fermentation system, called biogas, consists mainly of a mixture of methane (CH_4) and carbon dioxide (CO_2) with small amounts of other gases, including hydrogen sulfide (H_2S), hydrogen (H_2), nitrogen (N_2), and low-molecular-weight hydrocarbons. Typically, digester gas has 50%–75% methane and 25%–50% carbon dioxide; the remaining gases are present in very small quantities. The composition of biogas, as obtained from various sources, is reported in Table 48.14. As reported in Table 48.14, a large variation exists in the composition of biogas, primarily due to differences in feedstocks and operating conditions.

Because biogas normally consists of a mixture of gases, biogas characteristics must be evaluated for each individual case. However, in many cases, the physical characteristics of the three main gas constituents, namely methane, carbon dioxide, and hydrogen sulfide, can be used to characterize biogas. Some physical and chemical characteristics of the principal gases found in biogas are presented in Table 48.15.

As a comparison, the weight of methane is roughly half that of air at 20°C (weight ratio = 1 m^3 of methane/1 m^3 of air = 0.716 kg/1.293 kg = 0.554).

Methane gas is not very soluble in water. Only three units of methane (by volume) can be dissolved in 100 units of water at 20°C and 1 atmosphere pressure. Methane is a very stable hydrocarbon compound, and upon complete combustion, it produces a blue flame and a large amount of heat. The complete combustion of 1 m^3 of methane can release 38 MJ or about 9500 kcal (1 kcal of heat will raise the temperature of 1 kg of water by 1°C). In comparison, a complete combustion of biogas yields a caloric value of about 20–26 MJ/m^3 (depending on the methane content), which represents a low fuel value compared with methane gas alone. In addition, biogas requires a pressure of about 34,450 kPa (5,000 lb/in.^{2}) to liquefy it for storage. Therefore, biogas requires a larger storage volume for a given amount of energy than other fossil fuels.

48.8.2 Utilization of Biogas

To understand the potential use of biogas, it is important to gain a perspective on its energy potential by comparing it with more familiar uses of energy. The following are some examples of the use of 1 m^3 of biogas, at 60%–70% methane content, for common energy-consuming purposes (see Table 48.16).[39]

TABLE 48.14
Range and Typical Composition of Biogas Produced from BOFMSW

	Percent by Volume	
Biogas Component	**Range**	**Typical[b]**
Methane, CH_4	50–75	53.0
Carbon dioxide, CO_2	50–25	45.0
Hydrogen sulfide, H_2S	0.01–1.5	0.02
Hydrogen, H_2	Trace–3.5[a]	1.7
Nitrogen, N_2	Trace–8[a]	Trace
Other hydrocarbon	Trace–0.05	Trace

[a] The hydrogen and nitrogen gas reported in this table are more commonly found in landfill gases.
[b] Typical biogas composition from the biodegradable organic fraction of MSW.

TABLE 48.15

Physical Characteristics of Biogas

		Average Value		
Characteristics	**Unit**	**CO_2**	**CH_4**	**H_2S**
Molecular weight	g	44.1	16.04	34.08
Vapor pressure at 21°C	kP	5719.0		1736.3
Specific volume at 21°C, 101 kP	m^3/kg	0.456	1.746	0.701
Boiling point at 101 kP	°C	−164.0	−161.61	−59.6
Freezing point at 101 kP	°C	−78.0	−182.5	−82.9
Specific gravity at 15°C (air = 1)		1.53	0.555	1.189
Density at 0°C	kg/m^3	1.85	0.719	1.539
Critical temperature	°C	31.0	82.1	100.4
Critical pressure	kP	7386.0	4640.68	9007.0
Critical density	kg/m^3	0.468	0.162	0.349
Latent heat of vaporization at bp	kJ/kg	982.72	520.24	548.29
Latent heat of fusion at mp	kJ/kg	189.0	58.74	69.78
Specific heat, C_p at 21°C, 101 kP	kJ/kg °C	0.83	2.206	1.06
Specific heat, C_v at 21°C, 101 kP	kJ/kg °C	0.64	1.688	0.803
Specific heat ratio, C_p/C_v		1.303	1.307	1.32
Thermal conductivity	W/m K	0.8323		0.0131
Flammable limits in air	% by volume		5.3–14	4.3–45
Solubility in water	kg/m^3	4.0	24.0	3.4
Viscosity	mPa s	0.0148	0.012	0.0116
Net heat of combustion at 25°C	MJ/m^3		36.71	
Gross heat of combustion at 25°C	MJ/m^3		37.97	
Ignition temperature	°C		650.0	
Octane rating			130.0	
Combustion equation			$CH_4 + 2O_2 \rightarrow CO_2 + 2H_2O$	$H_2S + 2O_2 \rightarrow SO_3 + H_2O$[a]

[a] Reaction is temperature dependent, and the final product will be sulfuric acid ($SO_3 + H_2O \rightarrow H_2SO_4$).

TABLE 48.16

Equivalent Uses of Biogas

Use	**Equivalent Use**
Cooking	Can cook 3 meals for a family of 5–6
Lighting	Illumination equaling that of 60–100 W bulb for 6 h
Petroleum	Equivalent to 0.76 kg of petroleum
Car fuel	Can drive a 3 ton truck for 2.8 km
Motor power	Can run a 1 hp motor for 2 h
Electricity	Can generate 1.25 kW of electricity

Source: Barnett, A. et al., *Biogas Technology in the Third World: A Multidisciplinary Review*, International Development Research Center, Ottawa, Ontario, Canada, 1978.

TABLE 48.17

Uses for Biogas Produced from the Biodegradation of the Organic Fraction of MSW

Applications	Comment
Fuel in an IC engines	The most common application of biogas, usually used to generate electricity. Modified engines available from Caterpillar, Cooper-Superior, and Waukesha. Modifications include corrosion resistance and proprietary designing. Specialized lubricating oils are necessary and must be changed more often. Power ratings are usually 5%–15% below natural gas ratings.
Fuel in a gas turbine	Common application of biogas, used to produce electricity. Modifications similar to those for IC engines are necessary. Biogas gas turbines are made by the solar turbine division of Caterpillar. Power rating is 10%–15% lower than for natural gas. Turbines must be checked often for corrosion or deposition. Water vapor should be removed from biogas before use.
Boiler fuel	In moderate use, currently, requires little modification to current equipment. A mixture of biogas and natural gas may be used. Boilers appear to be less sensitive to gas contaminants than IC engines or gas turbines.
Pipeline quality gas	Less common use. CO_2 and all contaminant gases, including water vapor, must be removed before gas will be accepted into a pipeline. Usually not economical unless a large amount of gas is produced and a long-term contract for the sale is available.
Fuel in a steam turbine	Currently in limited use. A large amount of gas is necessary before the system will be economical for the production of electricity. However, modifications are small.
Fuel for space heating	Currently in limited use. Only minor modifications even on a small scale. Hampered only by unavailability of gas. Pipelines are not currently economical so only used at production sites.
Fuel for industrial heating	Currently in limited use. Can be used for lumber drying, kiln operations, and cement manufacturing. May be used in place of or with natural gas, using natural gas equipment.
Fuel for fuel cell	Technology under development. These open fuel batteries can be designed to use biogas, or biogas can be converted to hydrogen in a catalytic pretreatment unit.
Compressed vehicle fuel	Technology under development. Gas must be purified to near pipeline quality.
Convert to methanol	Technology under development. Currently too costly.
Synfuel or chemical feedstock	Technology under development. Liquid fuels and acetic acid production is being investigated.

The potential commercial uses of biogas as a source of energy are summarized in Table 48.17. Some applications listed in Table 48.17 are presently practiced and others are in the process of research and development.

48.8.3 Characteristics of Digested Biosolids (Humus Material)

The physical, chemical, and biological characteristics of the humus produced from an anaerobic composting using the sorted biodegradable organic fraction of MSW are presented in this section. The data reported in this section were obtained from the UC Davis pilot demonstration project.[35,36]

The physical characteristics of the humus that are of interest are bulk density, color, moisture content, odor, and particle size distribution. The physical characteristics of the humus are summarized in Table 48.18. The chemical characteristics of the humus that are of interest can be determined by ultimate analysis, metal analysis, fiber analysis, nutrient analysis, energy content, and other tests. The chemical characteristics of the

TABLE 48.18

Physical Characteristics of Humus Produced from BOFMSW by the Anaerobic Composting Process

Item	Unit	Value or Description
Bulk density	kg/m^3	560
Color		Dark brown
Moisture content	%	35
Odor		No offensive odor detected
Particle Size Distribution		
8 (2.362 mm)[a]	% of TM[b]	11.9
20 (0.833 mm)	% of TM	28.9
40 (0.351 mm)	% of TM	25.4
80 (0.175 mm)	% of TM	21.3
100 (0.147 mm)	% of TM	7.8
200 (0.074 mm)	% of TM	4.3
Pan	% of TM	0.4

Source: Kayhanian, M. and Tchobanoglous, G., *Environ. Technol.*, 14(9), 815, 1993.
[a] Sieve number (size, mm).
[b] TM, total mass (sample at 65% TS).

humus are summarized in Table 48.19. The biological characteristics of the humus that are of interest are the presence and the concentration of pathogenic bacteria, biodegradability, and phytotoxicity. The biological characteristics of the humus are summarized in Table 48.20.

48.8.4 Utilization of Humus Material

Representative applications for the humus produced from AD processes are summarized in Table 48.21. The most effective use of the humus material is as a soil amendment. Alternatively, because the humus is combustible, it appears that it can be fired directly in a boiler, when mixed with other fuels, or palletized for use as a fuel source. The application of the humus as a fuel source has been further studied, and readers are referred to Jenkins et al.[41]

It is important to note that, depending on the final use, further aerobic composting of the digestate may be necessary to produce humus with no phytotoxic affect to be used as a soil amendment. For other applications specified in Table 48.21, no further stabilization may be needed.

48.9 Full-Scale Application of Anaerobic Digestion Technologies

In the United States, initial efforts to commercialize AD technology for converting waste into energy and other products have been conducted primarily by livestock enterprises. The barrier to commercialization of anaerobic technology in the United States has been described by Lusk.[21] In addition, the commercial firms interested in the technology, the economics of the process, the statutes that apply, and the future prospects for application

TABLE 48.19

Chemical and Other Characteristics of the Humus Produced from BOFMSW by the Anaerobic Composting Process

Chemical and Other Analyses	Unit[a]	Range	Typical
Ultimate Analysis			
Carbon, C	%	30–35	32.4
Hydrogen, H	%	3.5–4	3.8
Chlorine, Cl	%	0.05–0.4	0.30
Nitrogen, N	%	1–2	1.9
Oxygen, O	%	30–35	31.4
Residue	%	25–35	30
Sulfur, S	%	0.1–0.4	0.25
Metal Analysis			
Aluminum, Al	mg/kg	10–194	54
Argon, Ag	mg/kg	<1–2	0.13
Arsenic, As	mg/kg	<1–3	1.04
Boron, B	mg/kg	12–64	18
Cadmium, Cd	mg/kg	<1–5	1.55
Calcium, Ca	%	0.8–1.7	1.08
Chromium, Cr	mg/kg	5–35	13
Cobalt, Co	mg/kg	<1–1	0.5
Copper, Cu	mg/kg	18–248	30
Iron, Fe	mg/kg	100–710	170
Lead, Pb	mg/kg	5–43	7
Magnesium, Mg	%	0.3–0.5	0.34
Manganese, Mn	mg/kg	120–175	133
Molybdenum, Mo	mg/kg	1–20	6
Nickel, Ni	mg/kg	18–186	28
Selenium, Se	mg/kg	<1	<1
Silica, Si	mg/kg	0.1–4	0.21
Sodium, Na	%	0.1–0.3	0.3
Tungsten, W	mg/kg	<1–10	0.3
Zinc, Zn	mg/kg	98–376	176
Fiber Analysis			
Cellulose	%		35.3
Hemicellulose	%		3.9
Lipid	%		1.45
Protein	%		11.9
Lignin	%		26.5
Nutrient Analysis			
Nitrogen, N	%	1–2	1.9
Phosphorus, P	%	0.1–0.5	0.23
Phosphate, PO_4–P	mg/kg	50–200	170
Potassium, K	%	0. 3–1	0.73
Sulfate, SO_4–S	mg/kg	300–800	547
Energy Content, HHV	MJ/kg	13–15	14.8
Other			
Cation-exchange capacity (CEC)	meq/100 g dry	20–100	30
Electrical conductivity (EC)	mmho/cm	5–15	9.4
pH		7–8.5	8.2

Source: Kayhanian, M. and Tchobanoglous, G., *Environ. Technol.*, 14(9), 815, 1993.

[a] % and mg/kg are based on dry mass.

TABLE 48.20

Biological Characteristics of the Humus Produced from the BOFMSW by the Anaerobic Composting Process

Item	Unit	Value
Bacterial Concentration		
Total coliform	MPN/100 mL	Not detected[a]
Fecal coliform	MPN/100 mL	Not detected
Streptococcus and Enterococcus	MPN/100 mL	Not detected
Residual Biodegradability		
Biodegradable fraction	%VS	8.8[b]
Phytotoxicity		
Seed germinated with 100% leachate concentration	%	0[c]
Seed germinated with 20% leachate concentration	%	78
Seed germinated with 15% leachate concentration	%	95

Source: Kayhanian, M. and Tchobanoglous, G., *Environ. Technol.*, 14(9), 815, 1993.

[a] Detection limit is 0–6 organisms/10 mL at 95% confidence level.

[b] BF is computed based on the LC presented in Table 48.19, using Equation 48.1: BF = 0.83 − (0.028)LC.

[c] Percent seed germinated is computed based on the ratio of seed germinated at each dilution, compared to the control.

TABLE 48.21

Uses for Humus Material Produced from Anaerobic Digestate

Applications	Remarks
Use as soil amendment	The humus must be odorless, low in heavy metals, and free of pathogens. In addition, the amount of humus applied to the topsoil may be limited by the phytotoxic characteristics of the humus.
Use as turf for sod production	Similar restrictions apply as specified for use of humus as a soil amendment.
Use as topsoil for erosion control	If humus is to be used for the control of agricultural soil erosion, it must be of similar quality to that specified for soil amendments. For other erosion control applications, specifications may be less restrictive.
Use as topsoil for landfills cover	Moisture content should be greater than about 35% to avoid wind erosion.
Use as marsh restoration	Humus can be used to build up the organic material required for the growth of marsh plants in marsh restoration projects. A higher degree of humus stabilization may be required for this application.
Use as absorbent agent for control of liquid hazardous wastes	Humus may be used as an absorbent to limit movement of hazardous wastes.
Use as fuel or fuel blend in power plants	Major characteristics of concern include moisture content, foreign matter, heating value, and environmental impact on air quality.

in the United States are identified and discussed. In general, the biggest barriers up to now have been financial rather than technical in nature.

Most large-scale AD applications for methane recovery in the United States are related to animal waste and wastewater treatment.[20,21] For instance, two large-scale farm digesters have been in operation in California since 1980.[21] With rising energy costs and examples of successful anaerobic digester operation, more interest has been generated recently in these systems. In the past, a number of studies have been performed in the United States to utilize the AD process for methane gas recovery from solid waste materials.[14,35,42–49]

Unfortunately, none of these projects have commercialized or implemented at full scale. AD technology has been used in full scale in food processing industry and wastewater treatment biosolids stabilization. Wide ranges of anaerobic digester design and operational method employed in municipal, agricultural, food processing, and wastewater treatment plant. Some of the most common reactor design and operating system used for the aforementioned application are described in the following.

48.9.1 MSW

Western Europe has been a leader in developing new solutions. Commercial application of AD for the treatment and energy recovery from the BOFMSW[50] has increased steadily since the 1990s. The summary information in Table 48.22 on the application of different types of technologies is based on available information as of 2008, and hence, the number of current operating digesters is expected to be higher.

As reported in Table 48.22, the overall design capacity of the various technologies can range from as low as 1000 to as high as 270,000 ton/year. It is also worth to note that majority of digesters in Europe are operated as single-phase (93%), mesophilic (67%), and dry or high-solids (62%) condition. Also, while Biotechnische Abfallverwertung (BTA), Dry ANaerobic COmposting (DRANCO), Kompogas, and Valorga are still the dominant treatment systems, numerous new technologies are now available and in operational as compared to 10 years ago. The detailed information about each of the new technology is beyond the scope of this chapter and could be obtained from the respective company's website.

At present in the United States, because of the concern over the disposal of MSW in landfills, interest in reduction of greenhouse gases into atmosphere, and desire on less reliance on foreign oil sparks special interest on the utilization of AD technology as

TABLE 48.22
Summary of Commercial Anaerobic Digester Technologies in Operation as of 2008

			AD Operating Condition					
			Stage		TS		Temperature	
Technology	Operating Plants	Capacity Range, ton	SS	MS	Low	High	MF	TF
AAT	8	3,000–55,000	x		x		x	
ArrowBio	4	90,000–180,000		x	x		x	
BTA	23	1,000–150,000	x	x	x		x	x
Biostab	13	10,000–90,000	x		x			x
DBA-Wabio	4	6,000–60,000	x	x	x		x	
DRANCO	17	3,000–120,000	x			x		x
Entec	2	40,000–150,000	x		x		x	
Haase	4	50,000–200,000			x		x	x
Kompogas	38	1,000–110,000	x			x		x
Linde-KCA/BRV	8	15,000–150,000	x	x	x	x	x	x
Schwarting-Uhde	3	25,000–87,600		x	x			x
Valorga	22	10,000–270,000	x			x	x	x
Waasa	10	3,000–230,000	x		x		x	x

Source: De Baere, L. and Mattheeuws, B., Anaerobic digestion of the organic fraction of municipal solid waste in Europe: Status, experience and prospects, http://www.ows.be/wp-content/uploads/2013/02/Anaerobic-digestion-of-the-organic-fraction-of-MSW-in-Europe.pdf, accessed March 5, 2015.

Notes: SS, single stage; MS, multistage; MF, mesophilic; TF, thermophilic.

an alternative sources of energy production. As part of this effort, researchers at UC Davis developed a new three-stage AD technology known as anaerobic-phased solids (APS) digester and tested under pilot operation.[49] This technology along with four of the dominant European AD technologies (BTA, DRANCO, Kompogas, and Valorga) are briefly described as follows.

48.9.1.1 BTA System

BTA system was developed in Germany and applied (via several licensing companies) throughout Western Europe and in select locations in Canada and Japan. The BTA system is one of the most successful in terms of the number of existing operational digesters.[52] The majority of the BTA digesters capacity are large (>100,000 MT/year) and usually operated as multistage, low-solids condition.

A schematic of the process flow diagram of the BTA International GmbH system is shown in Figure 48.10. As shown in Figure 48.10, the BTA system involves complete processing, pretreatment, AD, and final composting processes. The hydromechanical pretreatment prior to AD is a trademark of this system. The pretreated organic fraction is digested within the fermenter, generally under mesophilic conditions between 35°C and 38°C. Digester content are mixed with compressed biogas using gas lances. Since 1991, more than 40 BTA® plants have been designed and constructed worldwide with a total capacity of approximately 1.2 million ton/year. More than 110 million N m^3 biogas are annually produced in BTA Plants, which corresponds to the energy equivalent of approximately 70 million N m^3 natural gas. Additional information on the BTA system can be obtained from the following web address: http://www.bta-international.de/en/home.

48.9.1.2 DRANCO System

The DRANCO process (see Figure 48.11) was developed in Ghent, Belgium, for the high-solids AD of refuse-derived fuel (RDF). The RDF is prepared by completely sorting

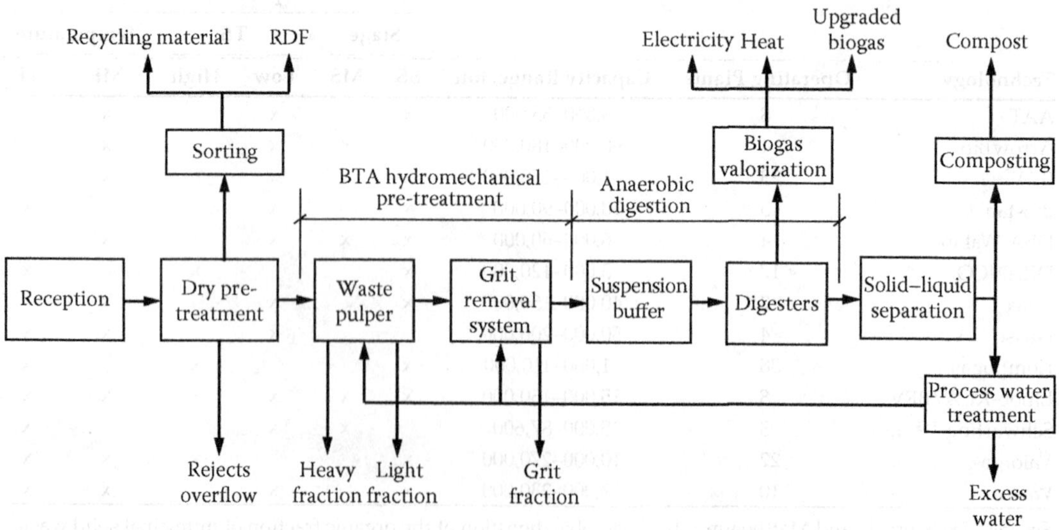

FIGURE 48.10
Schematic flow diagram of the BTA process.

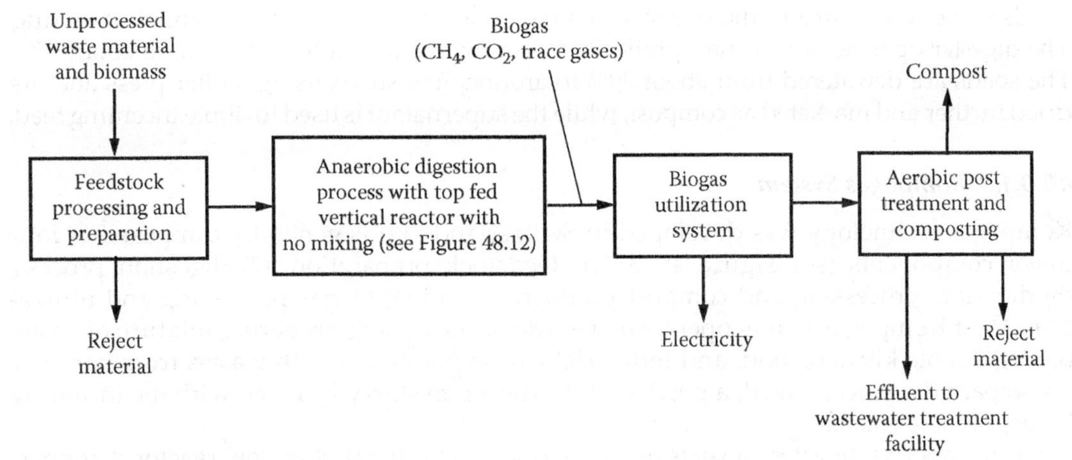

FIGURE 48.11
Schematic flow diagram of the DRANCO process.

incoming urban solid waste. The DRANCO system[51–53] is capable of removing most metals, glass, plastic, stone, recoverable paper, and other nonbiodegradable items.

The anaerobic treatment takes place in a vertical fermenter (see Figure 48.12) for a period of 12–18 days, followed by a postfermenter for a retention time of 2–3 days. The overall digestion time is, therefore, 14–21 days. The feed is mixed with recycled digester effluent and supernatant and then pumped into the top of the reactor at a solids concentration of 35%–40%.

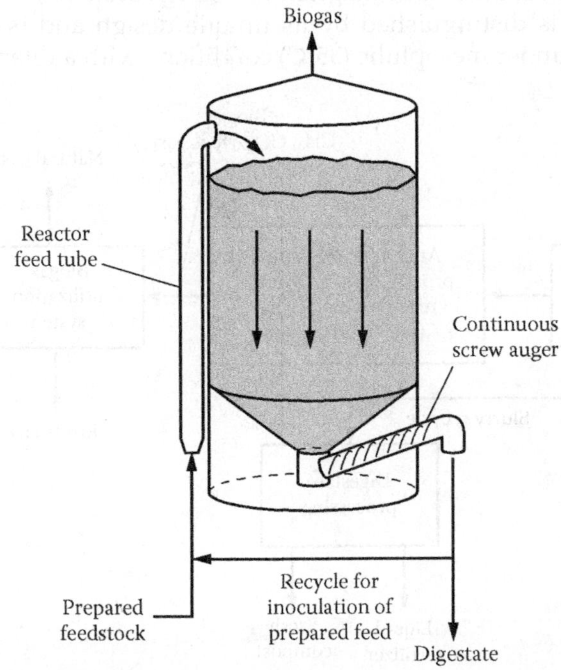

FIGURE 48.12
Schematic of the DRANCO anaerobic reactor.

Solids move downward in the digester in a plug-flow manner, without mechanical mixing. The digester operates under mesophilic (35°C) conditions with a VS reduction of about 55%. The solids are dewatered from about 30% to around 70% solids using a filter press and are dried further and marketed as compost, while the supernatant is used to dilute incoming feed.

48.9.1.3 Kompogas System

Kompogas technology was developed in Switzerland and is generally composed of four major components (see Figure 48.13): (1) feedstock preparation, (2) digestion process, (3) digestate processing and compost production, and (4) biogas processing and utilization. Most Kompogas plants operate by feedstock processing preparing mixture of easily biodegradable kitchen, food, and industrial wastes combined with wastes from gardens. To prepare the digester feed, a portion of the digestion slurry is mixed with the incoming biodegradable wastes.

The Kompogas digester consists of a horizontal, cylindrical plug-flow reactor that operates under thermophilic temperature and solid content of about 20%. The reactor content is gently mixed along the longitudinal axis. A simplified diagram of Kompogas digester is shown in Figure 48.14. The horizontal design, with an agitation only right angled to the flow direction, guarantees a good plug-flow behavior. The horizontal reactor design seems to be advantageous regarding clogging and sedimentation compared to vertical reactor designs.[54] With a few exceptions, nearly all Kompogas digesters are made out of steel.

48.9.1.4 Valorga System

Valorga, a French company, was the first to use anaerobic technology for the treatment of MSW.[55,56] The complete schematic diagram of Valorga system is shown in Figure 48.15. The Valorga digester is distinguished by its unique design and is operated at a solids concentration of 35% under mesophilic (35°C) conditions with a retention time of 15 days.

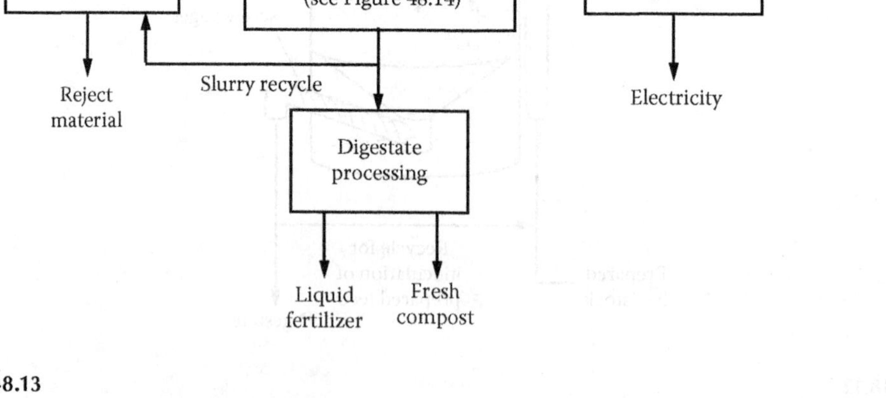

FIGURE 48.13
Schematic flow diagram of the Kompogas process.

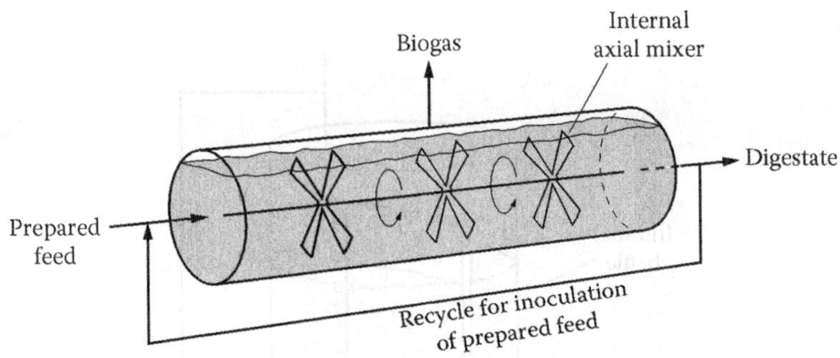

FIGURE 48.14
Schematic of the Kompogas anaerobic reactor.

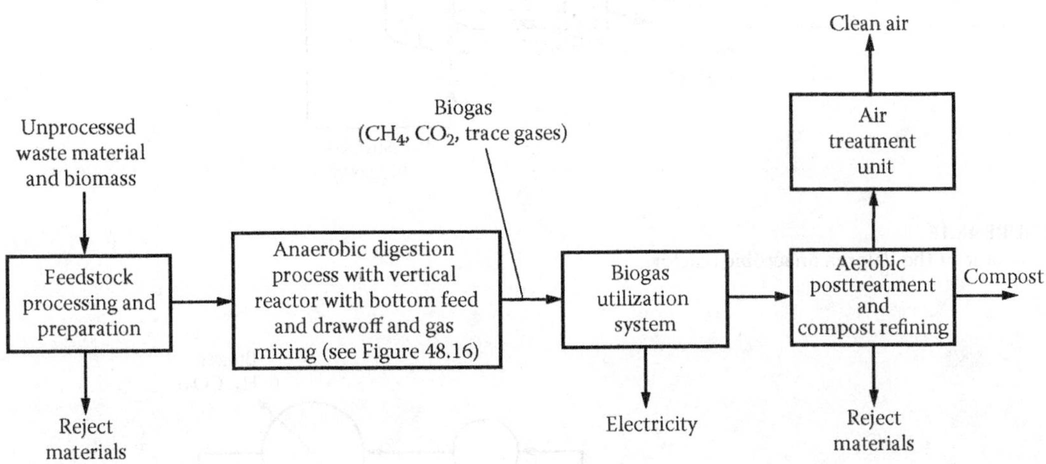

FIGURE 48.15
Schematic flow diagram of the Valorga process.

A typical Valorga reactor capacity is 500 m^3, and it can treat from 20 to 28 ton of RDF per day. The VS removal rate is around 50%, and the average gas production is about 140 m^3 biogas per ton of feed.

Two unique features of the Valorga process are (1) the loading and (2) mixing mechanisms (see Figure 48.16). A piston pump is used to load feedstock into the system, supplying the digestive system with feed on a continuous basis. A control isolation valve system injects sufficient pressure to transfer the feedstock into the digester; at the same time, a similar quantity of digestate is extracted from the outlet. Mixing takes place through the programmed injection of pressurized biogas into a different section of the system.

48.9.1.5 Anaerobic-Phased Solids System

A schematic of APS digester is shown in Figure 48.17 and pictorially in Figure 48.18. As shown in Figure 48.17, the APS digester uses batch loading to stimulate rapid organic acid production in a two-stage digester system. However, the APS digester system

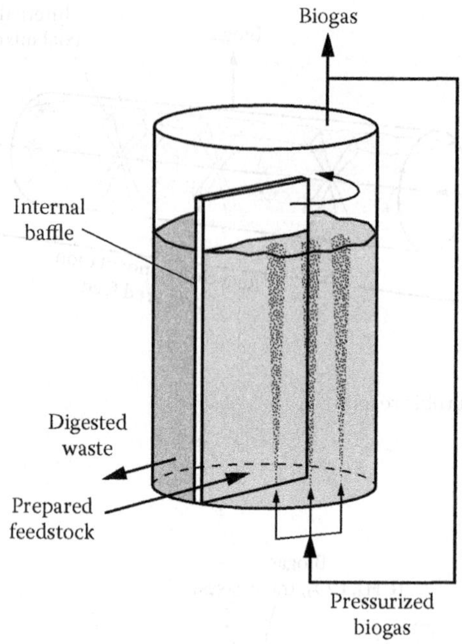

FIGURE 48.16
Schematic of the Valorga anaerobic reactor.

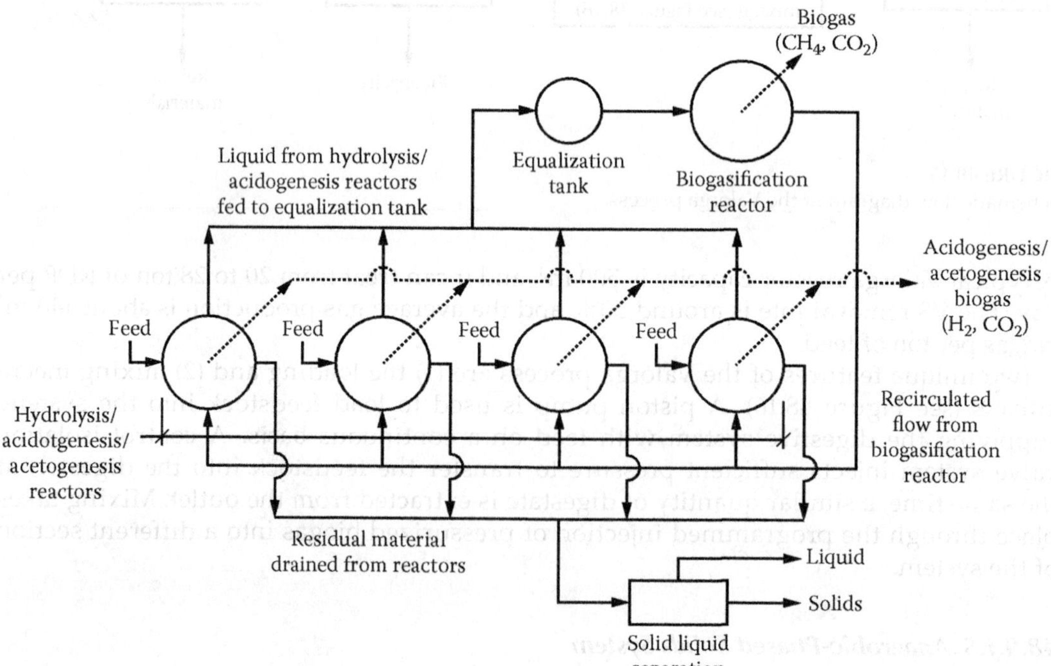

FIGURE 48.17
Schematic diagram of the APS digester system.

(a)

(b)

FIGURE 48.18
Views of the APS digester system: (a) view of four hydrolysis/acidogenesis reactors, shown in Figure 48.17 (gas cleaning and utilization facilities are shown in the foreground), and (b) view of biogasification reactor.

avoids the problems caused by using leach bed reactors by combining high-solids reactors for the first stage with a low-solids mixed biofilm reactor in the second stage. The high-solids reactors are loaded in phased batches, and the leachate from the batch reactors is continuously circulated through a single low-solids digester. In theory, batch loading simplifies material handling, and because the hydrolysis reactors are high-solids digesters, they can handle relatively large inorganic materials for treatment. Leachate recirculation prevents solids from fouling the wet methanogenesis reactor. Because the batches are phased, the leachate contains a relatively constant concentration of organic acids.

48.9.2 Animal Wastes

Two types of AD are used for waste stabilization and energy recovery from animal wastes: (1) belowground systems (ANLs, see Figure 48.7a) and (2) aboveground systems with various reactor configurations. Typical examples of earthen lined and covered ANLs are shown in Figure 48.19. As noted previously, ANLs are by far the most common anaerobic process in use today. Lined earthen lagoons are favored for animal wastes for small operations because of their simplicity. Aboveground installations are favored for larger installations where the recovered gas can be used for the generation of electricity.

48.9.3 High-Strength Industrial Wastes

Increasingly, industries with high-strength wastes have implemented the use of anaerobic treatment to (1) reduce the organic strength of the waste before disposal, (2) reduce the organic strength of the waste for treatment by an aerobic process before disposal, (3) recover products such as (PHA and alginate) that can be used as a raw material in manufacturing (typically biodegradable plastics), and (4) recover the energy contained in the waste stream along with reducing its strength. A representative listing of the industries in which anaerobic technologies are now used routinely is presented in Table 48.23. Several examples of anaerobic treatment processes used for the treatment of high-strength wastes

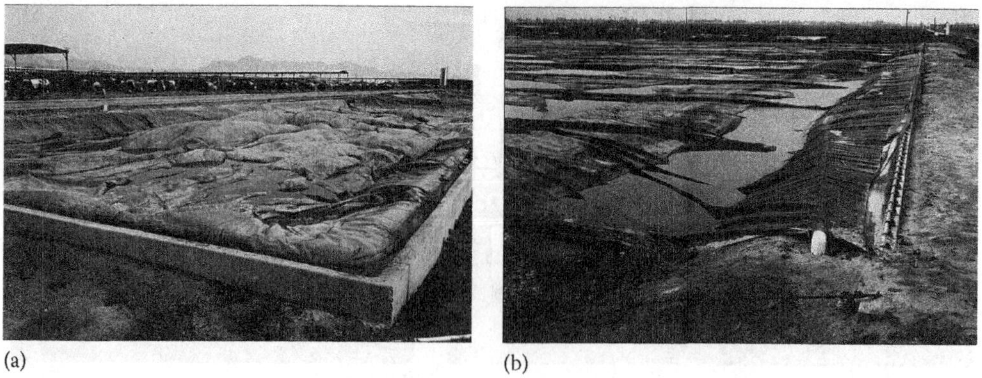

FIGURE 48.19
Views of covered ANL treatment process: (a) installation used for the treatment of cow manure with energy recovery and (b) installation for the treatment of piggery waste with energy recovery.

TABLE 48.23

Representative Examples of Types of Wastewaters Treated by Anaerobic Processes

Food and Brewage Industry	
Alcohol distillation	Slaughterhouse and meatpacking
Breweries	Soft drink beverages
Dairy and cheese processing	Starch production
Food processing	Sugar processing
Fish and seafood processing	Vegetable processing
Fruit processing	
Other Applications	
Chemical manufacturing	Landfill leachate
Contaminated groundwater	Pharmaceuticals
Domestic wastewater	Pulp and paper

Source: Tchobanoglous, G. et al., *Wastewater Engineering: Treatment and Resource Recovery*, 5th edn., Metcalf & Eddy Inc., AECOM, McGraw-Hill Book Company, New York, 2014.

are shown in Figures 48.20 and 48.7b and c. Anaerobic processes are favored for the treatment of high-strength wastes because of their relatively small footprint.

48.9.4 Wastewater Plant Biosolids Application

AD are typically used for biosolids stabilization on most large-scale wastewater treatment plant. These digesters are usually operated under mesophilic and low-solids condition. While biogas is produced from all digesters, the recovery of gas for energy or electric power production is not always practiced. In recent years, some of the larger wastewater treatment plant recovers the biogas to offset the energy used for treatment operation. The most common digester system for energy recovery is egg-shaped digester. A schematic and view of a typical egg-shaped digester is shown in Figure 48.21. These digesters are known for their high performance, higher gas production, and lower operation maintenance.

(a) (b) (c) (d)

FIGURE 48.20
Typical anaerobic reactors used for the treatment of high-strength industrial wastewaters: (a) view of expanded granular sludge blanket (EGSB) treatment reactor, (b) internal recirculation upflow anaerobic sludge blanket reactor (Courtesy of Robert Pharmer of Pharmer Engineering, Boise, ID), (c) view of anaerobic flotation reactor (AFR) at ice cream factory, and (d) high-rate internal circulation bioreactor and aerobic reactor used to polish anaerobic effluent. (Courtesy of PAQUES, BV, Balk, the Netherlands.)

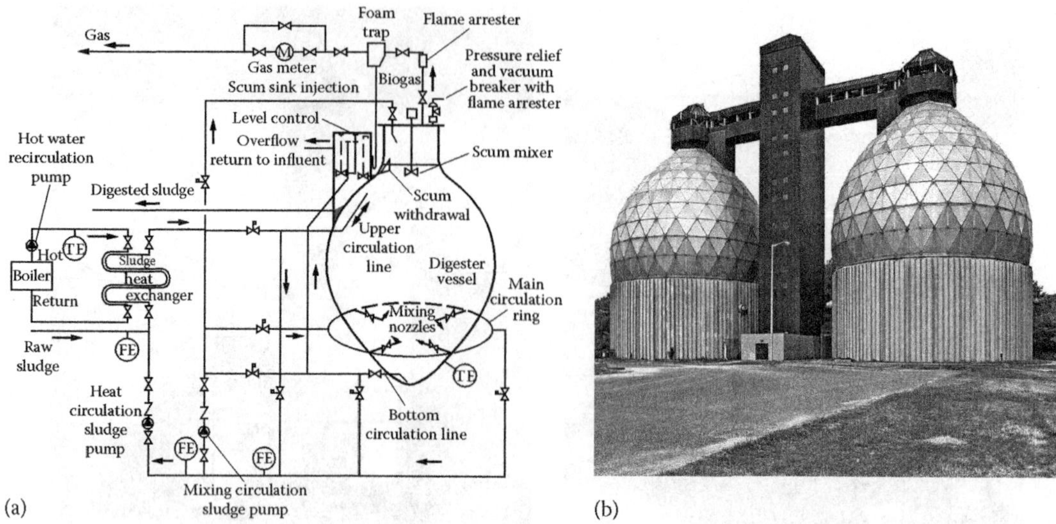

FIGURE 48.21
Egg-shaped anaerobic digester: (a) schematic diagram from Walker Process catalogue and (b) pictorial view. (Adapted from Tchobanoglous, G. et al., *Wastewater Engineering: Treatment and Resource Recovery*, 5th edn., Metcalf & Eddy Inc., AECOM, McGraw-Hill Book Company, New York, 2014.)

References

1. Chandler, J. A., Jewell, W. J., Gossett, J. M., Vansoest, P. J., and Robertson, B. J. 1980. Predicting methane fermentation biodegradability, *Biotechnology and Bioengineering Symposium*, 10, 93–107.
2. Kayhanian, M. 1995. Biodegradability of the organic fraction of municipal solid waste in a high-solids anaerobic digester. *Waste Management and Research*, 13(2), 123–136.
3. Jewell, W. J., Kang, H., Herndon, F. G., Richards, B. K., and White, T. E. 1987. *Engineering Design Considerations for Methane Fermentation of Energy Crops*. Gas Research Institute, Chicago, IL, pp. 105–106.
4. Robertson, J. B. and Van Soest, P. J. 1981. The detergent system of analysis and its application to human foods. In: W. P. T. James and O. Theander, eds., *The Analysis of Dietary Fiber in Food*. Marcel Dekker Inc., New York, pp. 123–158.
5. Van Soest, P. J. and Robertson, J. B. 1986. *Analysis of Forages and Fibrous Foods*. Cornel University, Itheca, NY.
6. Smith, P. H., Bordeaux, F. M., Wilkie, A., Yang, G., Boone, D., Mah, R. A., Chynoweth, D., and Jerger, D. 1988. Microbial aspects of biogas production. In: W. Smith and J. R. Frank, eds., *Methane from Biomass: A Systems Approach*. Elsevier Applied Science, New York , pp. 335–354.
7. Richards, K. B., Cummings, R. J., Jewell, W. J., and Herenden, F. G. 1991a. High-solid anaerobic methane fermentation of sorghum and cellulose. *Biomass and Bioenergy*, 1(2), 47–53.
8. Richards, K. B., Cummings, R. J., White, E. T., and Jewell, W. J. 1991b. Methods for kinetic analysis of methane fermentation in high solids biomass digesters. *Biomass and Bioenergy*, 1(2), 65–73.
9. Kayhanian, M. and Tchobanoglous, G. 1993a. Innovative two-stage process for the recovery of energy and compost from the organic fraction of municipal solid waste. *Water Science and Technology*, 27(2), 133–143.
10. Mindermann, W. 1993. Conversion of nitrogen to ammonia in thermophilic anaerobic batch digestion of the organic fraction of municipal solid waste, M.S. thesis. University of California at Davis, Davis, CA.

11. Kayhanian, M. and Tchobanoglous, G. 1992. Computation of C/N ratios for various organic fractions. *BioCycle*, 33(5), 42–45.
12. Holland, K. T., Knapp, J. S., and Shoesmith, J. G. 1987. *Anaerobic Bacteria*. Chapman & Hall, New York.
13. Rivard, J. C., Vinzant, T. B., Adney, W. S., and Grohmann, K. 1987. Nutrient requirements for aerobic and anaerobic digestion of processed municipal solid waste. *Journal of Environmental Health*, 5, 96–99.
14. Rivard, J. C., Himmel, M. E., Vinzant, T. B., Adney, W. S., Wyman, C. E., and Krohmann, K. 1987. *Anaerobic High Solids Fermentation of Processed Municipal Solid Wastes for the Production of Methane*. Biotechnology Research Branch, Solar Energy Research Institute, Golden, CO.
15. Speece, R. E. and McCarty, P. L. 1964. Nutrient requirements and biological solids accumulation in anaerobic digestion. In: W. W. Eckenfelder, ed., *Advances in Water Pollution Research*, Vol. 2. Pergamon Press Ltd., Oxford, U.K., pp. 305–322.
16. Speece, R. E. and Parkin, G. F. 1985. Nutrient requirements for anaerobic digestion. In: A. A. Antonopoulos, ed., *Biotechnological Advances in Processing Municipal Waste for Fuels and Chemicals*. Argonne National Laboratory, Argonne, IL, pp. 195–221.
17. Kayhanian, M. and Rich, D. 1995. Pilot-scale high-solids anaerobic digestion of municipal solid waste with an emphasis on nutrient requirements. *Biomass and Bioenergy*, 8(4), 433–444.
18. Zeikus, J. G. 1977. The biology of methanogenic bacteria. *Bacterial Review*, 41, 514–541.
19. Tchobanoglous, G., Theisen, H., and Vigil, S. A. 1993. *Integrated Solid Waste Management—Engineering Principles and Management Issues*. McGraw-Hill Book Company, New York.
20. Tchobanoglous, G., Burton, F. L., and Stensel, H. D. 2003. *Wastewater Engineering: Treatment, Disposal, Reuse*, 4th edn. Metcalf & Eddy, Inc., McGraw-Hill Book Company, New York.
21. Lusk, P. 1994. Methane recovery from animal manures: A current opportunities casebook. Report prepared for the National Renewable Energy Laboratory, Golden, CO.
22. Markel, A. J. 1981. *Managing Livestock Wastes*. AVI Publishing Company, Inc., Westport, CN.
23. Tchobanoglous, G., Stensel, H. D., Tsuchihashi, R., and Burton, F. L. 2014. *Wastewater Engineering: Treatment and Resource Recovery*, 5th edn. Metcalf & Eddy Inc., AECOM, McGraw-Hill Book Company, New York.
24. Diaz, L. F. and Trezek, G. T. 1977. Biogasification of a selected fraction of municipal solid wastes. *Compost Science*, 18(2), 8–13.
25. Diaz, L. F., Savage, G. M., and Golueke, C. G. 1982. *Resource Recovery from Municipal Solid Wastes. Final Processing*, Vol. II. CRC Press, Boca Raton, FL.
26. Ghosh, S. and Klass, D. L. 1976. SNG from refuse and sewage sludge by the biogas process. In: *IGT Symposium on Clean Fuels from Biomass, Sewage, Urban Refuse and Agricultural Wastes*, Orlando, FL.
27. Klein, S. A. 1972. Anaerobic digestion of solid wastes. *Compost Science*, 13(1), 6–13.
28. McFarland, J. M., Glassey, C. R., McGauhey, P. H., Brink, D. L., Klein, S. A., and Golueke, G. C. 1972. Comprehensive studies of solid waste management. Final Report, SERL Report No. 72-3, Berkeley, CA.
29. Pfeffer, J. T. and Liebman, J. C. 1974. Biological Conversion of organic refuse to methane. Annual Report NSF/RANN/SE/G1/39191/PR/75/2, Department of Civil Engineering, University of Illinois, Urbana, IL.
30. Babbitt, H. E., Leland, B. J., and Whitely, F. E. 1988. *The Biological Digestion of Garbage with Sewage Sludge*. University of Illinois, Urbana, IL, p. 24.
31. Cecchi, F., Traverso, P. G., Mata-Alvarez, J., Clancy, J., and Zaror, C. 1988. State of the art of R&D in the anaerobic digestion process of municipal solid waste in Europe. *Biomass*, 16, 257–284.
32. Poggi-Valardo, H. M. and Oleszkiewicz, J. A. 1992. Anaerobic co-composting of municipal solid waste and waste sludge at high total solids levels. *Environmental Technology*, 13, 96–105.
33. Kayhanian, M. and Rich, D. 1996. Sludge management using the biodegradable organic fraction of municipal solid waste as a primary substrate. *Water Environment Research*, 68(2), 240–252.
34. Kayhanian, M. 1993. Anaerobic composting for MSW. *BioCycle*, 34(5), 32–34.

35. Kayhanian, M., Lindenauer, K., Hardy, S., and Tchobanoglous, G. 1991. Two-stage process combines anaerobic and aerobic methods. *BioCycle*, 32(3), 48–64.
36. Kayhanian, M. and Tchobanoglous, G. 1993b. Characteristics of humus produced from the anaerobic composting process of the biodegradable organic fraction of municipal solid waste. *Environmental Technology*, 14(9), 815–829.
37. Kayhanian, M. 1994. Performance of high-solids anaerobic digestion process under various ammonia concentrations. *Chemical Technology and Biotechnology*, 59(5), 349–352.
38. Kayhanian, M. and Hardy, S. 1994. The impact of four design parameters on the performance of high-solids anaerobic digestion of municipal solid waste for fuel gas production. *Environmental Technology*, 15(6), 557–567.
39. Kayhanian, M., Tchobanoglous, G., and Mata-Alvarez, J. 1996. Development of a mathematical model for the simulation of the biodegradation of municipal solid waste in high-solids anaerobic digestion process. *Chemical Technology and Biotechnology*, 66(7), 312–322.
40. Barnett, A., Pyle, L., and Subramanian, K. S. 1978. *Biogas Technology in the Third World: A Multidisciplinary Review*. International Development Research Center, Ottawa, Ontario, Canada.
41. Jenkins, B. M., Kayhanian, M., Baxter, L., and Salour, D. 1997. Combustion of residual biosolids from high-solids anaerobic digestion/aerobic composting process. *Biomass and Bioenergy*, 12(5), 567–581.
42. Isaacson, H. R., Pfeffer, J. P., Modij, A. J., and Geselbracht, J. J. 1987. RefCom-technical status, economic and market. In: D. L. Klass, ed., *Energy from Biomass and Wastes XI*. Elsevier Applied Science Publishers and Institute of Gas Technology, Chicago, IL, pp. 1123–1163.
43. Chynoweth, D. P., Earl, F. K., Bosch, G., and Lagrand, R. 1990. Biogasification of processed MSW. *BioCycle*, 31(10), 50–51.
44. Ferrero, G. L., Feranti, M. P., and Naveau, H., eds. 1984. *Anaerobic Digestion and Carbohydrate Hydrolysis of Waste*. Elsevier Applied Science Publishers, New York.
45. Golueke, C. J. 1971. Comprehensive studies of solid waste management. Third Annual Report, University of California, Richmond Field Station, U.S. Environmental Protection Agency, Berkeley, CA.
46. Kispert, R. G., Sadek, S. E., and Wise, D. L. 1975. Fuel gas production from solid waste. Final Report, NSF Contract C-827, Dynatech Report No. 1258, Dynatech R/D Co., Cambridge, MA.
47. Pfeffer, J. P. 1987. Evaluation of the RefCom proof-of-concept experiment. In: D. L. Glass, ed., *Energy from Biomass and Wastes X*. Elsevier Applied Science Publishers and Institute of Gas Technology, Chicago, IL, pp. 1149–1171.
48. Mata-Alvarez, J., Cecchi, F., Pavan, P., and Fazzini, G. 1990. Performances of digesters treating the organic fraction of municipal solid waste differently sorted. *Biological Wastes*, 33, 181–199.
49 Zhang, R. H. and Zhang, Z. Q. 1999. Biogasification of rice straw with an anaerobic-phased solids digester system. *Bioresource Technology*, 68(3), 235–245.
50. Nichols, C. E. 2004. Overview of anaerobic digestion technologies in Europe. *BioCycle*, 45(1), 47–53.
51. De Baere L. and Mattheeuws B. Anaerobic digestion of the organic fraction of municipal solid waste in Europe: Status, experience and prospects. http://www.ows.be/wp-content/uploads/2013/02/Anaerobic-digestion-of-the-organic-fraction-of-MSW-in-Europe.pdf, accessed March 6, 2015.
52. De Baere, L. 1984. High rate dry anaerobic composting process for the organic fraction of solid waste. In: *Seventh Symposium on Biotechnology for Fuels and Chemicals*, Gatlinburg, TN.
53. De Baere, L., Verdonck, O., and Verstraete, W. 1985. High rate dry anaerobic composting process for the organic fraction of solid wastes. *Biotechnology Bioengineering Symposium*, 15, 321–332.
54. Six, W. and De Baere, L. 1992. Dry anaerobic conversion of municipal solid waste by means of the Dranco process at Brecht, Belgium. In: *Proceedings of the International Symposium on Anaerobic Digestion of Solid Waste*, Venice, Italy, pp. 525–528.

55. Edelman, W. and Angeli, B. 2005. More than 12 years of experience with commercial anaerobic digestion of the organic fraction of municipal solid wastes in Switzerland. In: *Proceedings of the Fourth International Symposium on Anaerobic Digestion of Solid Waste*, Copenhagen, Denmark.
56. Begouen, O., Thiebaut, E., Pavia, A., and Peillex, J. P. 1988. Thermophilic anaerobic digestion of municipal solid wastes by the VALORGA process. In: *Proceedings of the Fifth International Symposium on Anaerobic Digestion*, May 22–26, Bologna, Italy, pp. 789–792.
57. Valorga. 1985. Waste recovery as a source of methane and fertilizer. In: *Second Annual International Symposium on Industrial Resource Management*, Philadelphia, PA, pp. 234–244.

Section V

Biomass Energy Systems

49

Biomass Conversion to Heat and Power

Mark M. Wright and Robert C. Brown

CONTENTS

Biomass can be converted into a combination of heat, power, and a variety of fuels via thermal conversion processes. Heat and power are the primary products from direct biomass combustion, but they are often coproducts in other thermal processes. Thermal processes can convert biomass into solid (charcoal, torrefied biomass), gas (biogas, synthetic gas), and liquid (ethanol, biodiesel, gasoline) products. Modern biorefineries convert biomass into a combination of these energy products and chemicals. The contents of this section can be grouped into two main topics: (1) an overview of the energy efficiency of biomass conversion systems and (2) descriptions of common industrial biomass conversion technologies.

49.1 Biomass Energy Systems

Biomass conversion processes involve a large number of complex thermodynamic phenomena occurring through multiple system units or steps. Each step may involve mass or heat transfer and chemical reactions. The overall system's efficiency will depend on its ability to recover heat and/or maximize the output of desired products.

The first law of thermodynamics provides a framework for evaluating changes in energy within a system in relation to heat and work transfer across the system's boundary.

TABLE 49.1

Facility Efficiency and Capacity Comparison of Biomass Conversion Systems

Technologies	Efficiency (%)	Typical Size
Combined heat and power (CHP) [1]	LHV	MWe Output
Co-firing	35–40	10–50
Dedicated steam cycles	30–35	5–25
IGCC	30–40	10–30
Gasific + engine CHP	25–30	0.2–1
Stirling engine CHP	11–20	<0.1
Solid fuels	LHV	MW_{th} Input
Torrefaction [2]	96	<25
Slow pyrolysis [3]	72	<25
Carbonization [3]	65	<50
Liquid fuels		
Corn ethanol [4]	50	400
Cellulosic ethanol [5]	45	400
Fast pyrolysis gasoline [5]	54	<400
Gasification Fischer–Tropsch [5]	53	>400

It requires that the net heat transfer into a system be equal to the work done by the system plus the change in energy within the system. A thermodynamic system is a region in space delineated for analysis. The first law of thermodynamics requires that the energy change of macroscopically identifiable forms of energy be accounted for as heat and/or work transfer, and changes in internal, kinetic, or gravitational energy.

Biomass energy systems range from portions of a single equipment to multiple units or facilities in which biomass is transformed to heat, power, or other energy forms. A majority of biomass facilities focus on the production of one or two products (typically heat and power). However, some facilities produce a wide range of products, including heat, power, chemicals, and fuels. These systems can be compared based on their efficiency once an energy balance is established. The energy efficiency (η) for a system is defined as

$$\eta = \frac{E_{out}}{E_{in}} = \frac{(\text{Fuel} + \text{Steam} + \text{Power})}{(\text{Biomass} + \text{Fuel})} \quad (49.1)$$

where

the energy inputs consist typically of biomass and heat or electricity from fossil fuels, nuclear, solar, wind, geothermal energy, and others

the energy output includes any combination of recovered heat, electricity, and fuels

Table 49.1 shows representative efficiency values for different biomass conversion systems.

49.2 Biomass to Heat and Power

There are numerous routes to convert the chemical energy in biomass into heat or electric power. Direct combustion releases heat that can be used in Stirling engines or Rankine steam power cycles. Alternatively, thermal treatment with low-oxygen concentrations yields intermediate materials with varying energy properties. Carbonization and slow pyrolysis produce a charcoal material with high-carbon concentration. Biomass gasification results in a combustible gas. Fast pyrolysis generates mostly a liquid fuel. Each of these processes will be discussed in further detail later. This section will discuss direct combustion and co-firing for heat and power production.

49.2.1 Direct Combustion

Combustion converts chemical energy directly into heat via rapid oxidation of the fuel. Primary products from combustion of carbonaceous products include carbon dioxide (CO_2) and water. Secondary products result from incomplete combustion or reactions with fuel-bound nitrogen (N_2), sulfur, and other impurities. Combustion conditions vary depending on the heating value and moisture content of the fuel, the amount of air used to burn the fuel, and heat losses from the furnace. Flame temperatures can exceed 1600°C.

Solid fuel combustion takes place in four steps (see Figure 49.1): heating and drying, pyrolysis, flaming combustion, and char combustion [6]. Chemical reactions do not occur during heating and drying. During heating and drying, a thermal front drives advances into the interior of the particle while driving water away. The particle temperature will not rise high enough for pyrolysis until most of the water has been released.

Pyrolysis, the second step during combustion, is a complex series of thermally driven chemical reactions. Pyrolysis decomposes organic compounds in the fuel [7]. It proceeds

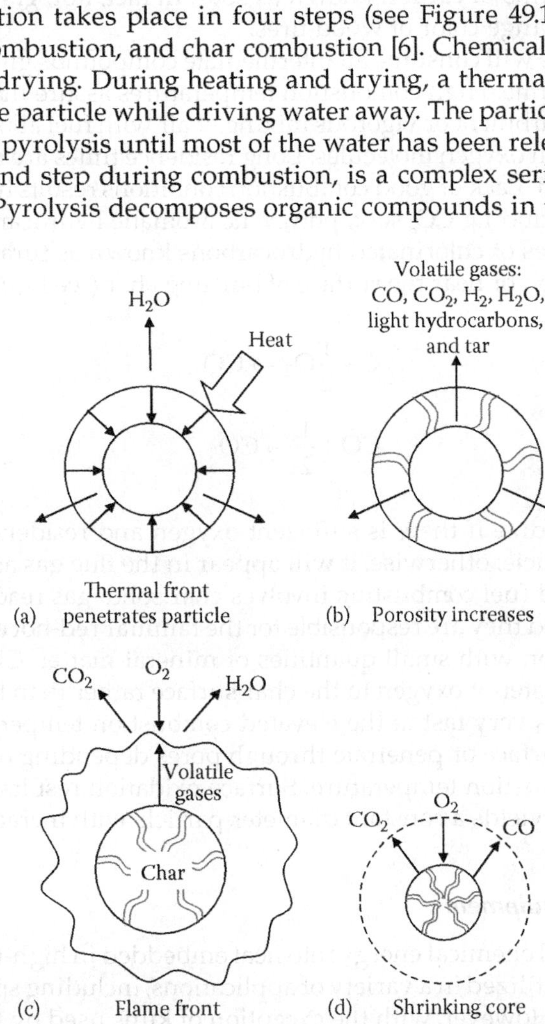

FIGURE 49.1
Processes of solid fuel combustion: (a) heating and drying, (b) pyrolysis, (c) flaming combustion, and (d) char combustion.

at low temperatures depending on the plant material. Hemicellulose pyrolyzes at temperatures between 150°C and 300°C, cellulose at 275°C–350°C, and lignin pyrolysis initiates at 250°C–500°C.

Pyrolysis produces a wide range of volatile compounds with types and quantities dependent on the fuel and heating rate. The products include carbon monoxide (CO), CO_2, methane (CH_4), and high-molecular-weight compounds that condense to a tarry liquid if cooled before they are able to burn. These condensable compounds may form fine droplets, which constitute much of the smoke associated with smoldering fires. The pyrolysis thermal front leaves behind pores that penetrate from the surface to the core of the particle.

Both volatile gases and the resulting pyrolysis char can be oxidized if sufficient oxygen is available. Volatile gas oxidation above the solid fuel results in flaming combustion. The ultimate products of volatile combustion are CO_2 and H_2O, although a variety of intermediate chemical compounds can exist in the flame including CO, condensable organic materials, and long chains of carbon known as soot. In fact, hot, glowing soot is responsible for the familiar orange color of wood fires.

The combustion flame will consume all intermediate compounds given sufficient temperature, turbulence, and time. High combustion temperatures assure that chemical reactions proceed at high rates. Turbulent or vigorous mixing of air with fuel allows all fuel molecules to come into contact with oxygen molecules. Long residence times are needed for the fuel to be completely consumed. Lack of good combustion conditions results in a variety of noxious organic compounds, including CO, soot, polycyclic aromatic hydrocarbons (PAH), and the particularly toxic families of chlorinated hydrocarbons known as furans and dioxins. Both CO and CO_2 may form on or near the surface of burning char (see Equation 49.2 [6]):

$$\begin{gathered} C+\frac{1}{2}O_2 \rightarrow CO \\ CO+\frac{1}{2} \rightarrow CO_2 \end{gathered} \tag{49.2}$$

CO will completely oxidize if there is sufficient oxygen and residence in the immediate vicinity of the char particle; otherwise, it will appear in the flue gas as a pollutant.

The third step in solid fuel combustion involves char solid–gas reactions also known as glowing combustion, and they are responsible for the familiar red-hot embers in a fire. Char consists mostly of carbon with small quantities of mineral matter. Char oxidation occurs when there is mass transfer of oxygen to the char surface rather than by chemical kinetics. Oxygen mass transfer is very fast at the elevated combustion temperatures. Oxygen may react with char at the surface or penetrate through pores depending on the particle porosity, reactivity, and combustion temperature. Surface oxidation results in a shrinking core, whereas pore oxidation yields a constant diameter particle with increasing porosity.

49.2.1.1 Combustion Equipment

Combustors convert fuel chemical energy into heat embedded in high-temperature exhaust gases. This heat can be utilized in a variety of applications, including space heating, drying, and power generation. However, with the exception of kilns used by the cement industry, most solid–fuel combustors today are designed to produce either low-pressure steam for process heat or high-pressure steam for power generation. Combustors designed to raise steam are known as boilers. Some boiler designs include distinct sections for combustion,

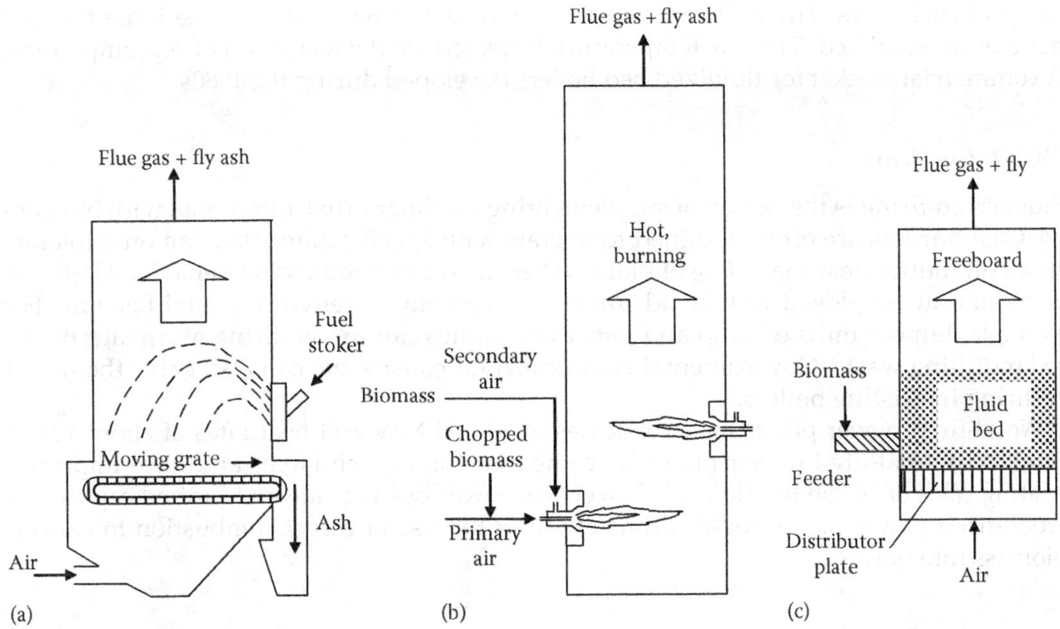

FIGURE 49.2
Common types of combustors: (a) grate-fired, (b) suspension, and (c) fluidized.

and high- and low-temperature heat transfer; these sections are called furnace, radiative, and convective segments, respectively. Other designs may not include clear, distinct separations between the processes of combustion and heat transfer.

Solid–fuel combustors can be categorized as grate-fired systems, suspension burners, or fluidized beds [8], as shown in Figure 49.2. The first burner systems developed were grate-fired, which evolved during the late nineteenth and early twentieth century into a variety of automated systems. The spreader-stoker is the most common system, and it consists of a fuel feeder that mechanically or pneumatically flings fuel onto a moving grate where the fuel burns. Most of the ash falls of the end of the moving grate, but some fly ash can appear in the flue gas. Combustion efficiencies for grate systems rarely exceed 90%.

Suspension burners employ a stream of rising air to suspend the fuel as fine powder. The fuel burns in a fireball and radiates heat to tubes that contain water to be converted into steam. Suspension burners are also known as pulverized coal (PC) boilers, and they have dominated the U.S. power industry since World War II. They achieve high volumetric heat release rates and combustion efficiencies of more than 99%. However, they are not designed to burn coarse particles and are notorious generators of nitrogen oxides. Feeding employs pulverizers designed to reduce the particle size enough to burn in suspension. The fuel particles remain suspended in the primary airflow prior to feeding to the burner section through ports where it burns as a rising fireball. A secondary air stream helps complete the combustion process. Steam tubes arrayed in banks of heat exchanges (waterwall, superheaters, and economizer) absorb the combustion gas heat before the gas exits through a bag house. Bag houses are designed to capture suspended ash.

Fluidized bed combustors are a relatively recent innovation in boiler design. They operate by injecting air into the bottom of the boiler to suspend a bed of sand or other granular, refractory material. The suspension produces a turbulent mixture of air and sand. This environment achieves high heat and mass transfer rates and is suitable for burning a wide range of fuel types. The unit can be operated as low as 850°C due to the large thermal mass of the sand bed. The low temperature helps reduce the formation of N_2 compounds. A commercial market for fluidized bed boilers developed during the 1980s.

49.2.2 Co-Firing

Biomass co-firing is the notion of supplementing a primary fuel, often coal, with biomass [9]. Coal burners are often designed to operate with specifications that can only tolerate small quantities (less than 20%) of biomass because of its combustion behavior. Co-firing is commonly employed to take advantage of economic or environmental benefits. For example, lumber mills or pulp and paper companies can use co-firing as an alternative to landfilling waste. Governmental environmental constraints can also drive the use of co-firing in existing boilers.

Wood-fired power plants have capacities of 20–100 MW and heat rates of about 12,500 Btu/kW h. Coal-fired power plants by comparison are much larger and have improved heating rates of 10,250 Btu/kW h or lower. The lower heating rate and limited capacity of wood-fired power plants could ultimately limit the use of direct combustion to convert biomass into power.

49.2.3 Gasification

Gasification is a thermal process characterized by an oxygen-deficient environment and temperatures above 750°C. In this environment, most carbonaceous material converts into a flammable gas consisting of CO, hydrogen (H_2), CH_4, CO_2, and smaller quantities of heavier hydrocarbons. Gasification may take place with either air or pure oxygen input. Air gasification yields gas with high N_2 content commonly known as producer gas; gas from oxygen gasification is known as synthetic gas or syngas. Steam may also be introduced into the gasification environment as an oxygen carrier. Gasification requires heat input that is delivered either internally from partial combustion of the feed input or via external heaters. Gasification performance depends on the ability to introduce oxygen and heat into its environment. Residual material from biomass gasification includes unconverted carbon (char) and ash.

The gasification process involves four primary steps [10]: heating and drying, pyrolysis, gas–solid reactions, and gas–phase reactions. These steps are illustrated in Figure 49.3. As in combustion, heating and drying evaporates all feed moisture before the particle temperature increases to gasification temperatures. Pyrolysis occurs once the thermal front penetrates the particle, resulting in the release of volatile gases via pores of increasing number and size. The volatile gases include all gasification final products as well as tar. Tar consists of heavy and extremely viscous hydrocarbon compounds. After the pyrolysis step, these gases react with the particle surface, which is now primarily char, in a series of gas–solid endothermic and exothermic reactions that increase the yield of light gases. Finally, released gases continue to react in the gas–phase until they reach equilibrium conditions.

The heating and drying, and pyrolysis steps during gasification are similar to those of combustion. However, in the case of gasification, pyrolysis yields a larger quantity of tarry material because of insufficient oxygen and/or temperature to decompose the heavier compounds. Much of this tar elutriates from the particle and accumulates upon condensation. Where gasification differs from combustion is in the gas–solid and gas–phase reactions.

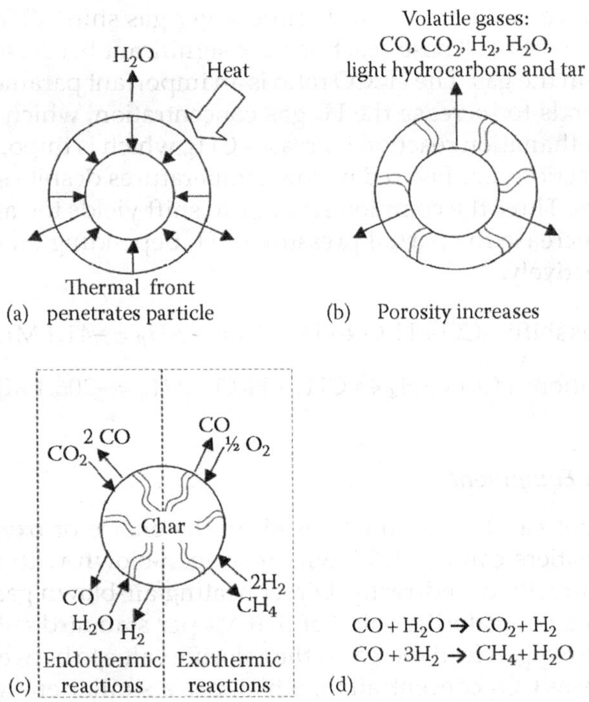

FIGURE 49.3
Thermal gasification process steps: (a) heating and drying, (b) pyrolysis, (c) gas–solid reactions, and (d) gas–phase reactions.

Gas–solid reactions include carbon–oxygen, Boudouard, carbon–water, and hydrogenation reactions shown in Equations 49.3. The carbon–oxygen reaction is highly exothermic and key to driving the Boudouard and carbon–water reactions, which are endothermic. The carbon–oxygen and Boudouard reactions yield CO, whereas the carbon–water reaction yields H_2 and CO. This H_2 goes on to react exothermically with C to yield CH_4. The amount of H_2 involved in gas–solid reactions is relatively small:

$$\begin{aligned}
&\text{Carbon oxygen:} \quad C+\frac{1}{2}O_2 \leftrightarrow CO \quad \Delta H_R = -110.5\ \text{MJ/kmol}\\
&\text{Boudouard:} \quad C+CO_2 \leftrightarrow 2CO \quad \Delta H_R = 172.4\ \text{MJ/kmol}\\
&\text{Carbon water:} \quad C+H_2O \leftrightarrow H_2+CO \quad \Delta H_R = 131.3\ \text{MJ/kmol}\\
&\text{Carbon oxygen:} \quad C+2H_2 \leftrightarrow CH_4 \quad \Delta H_R = -74.8\ \text{MJ/kmol}
\end{aligned} \tag{49.3}$$

Chemical equilibrium calculations predict that almost all carbon should react and form gas compounds. Practical systems typically achieve greater than 90% carbon conversion. Industrial engineers devote significant efforts to increase the carbon conversion by optimizing material treatment, equipment design, and operation parameters. In general, smaller particle sizes, longer residence times, and higher temperatures will increase carbon conversion. However, there are practical limitations that prevent the attainment of near-equilibrium conversions.

Gas–phase reactions consist of the exothermic water gas shift (WGS) and methanation reactions (see Equations 49.4). These reactions are significant because they determine the ultimate H_2:CO ratio in the gas. The H_2:CO ratio is an important parameter in synthetic fuel applications. WGS tends to increase the H_2 gas concentration, which is desirable for fuel synthesis, and the methanation reaction increases CH_4, which is important for combustion applications. Both reactions are favored by low temperatures despite slower reaction rates at lower temperatures. Thus, the common strategy to shift yields toward the products is to introduce steam or increase the partial pressure of H_2 depending on whether more H_2 or CH_4 is desired, respectively.

$$\begin{aligned} &\text{Water gas shift:} \quad CO + H_2O \leftrightarrow H_2 + CO_2 \quad \Delta H_R = -41.1\ \text{MJ/kmol} \\ &\text{Methanation:} \quad CO + 3H_2 \leftrightarrow CH_4 + H_2O \quad \Delta H_R = -206.1\ \text{MJ/kmol} \end{aligned} \tag{49.4}$$

49.2.3.1 Gasification Equipment

Gasification equipment can be classified based on the choice of oxygen input and heat delivery method. Gasifiers can be air-blown, or oxygen-blown with and without steam. They can be heated directly or indirectly. Direct heating air-blown gasifiers dilute syngas with N_2, resulting in a low calorific value of 5–6 MJ per standard cubic meter. Avoiding N_2-dilution with pure oxygen feed increases the calorific value to between 13 and 14 MJ/m^3. Direct heating increases CO_2 concentration, which has a similar energy dilution effect as N_2. Indirect heating gasifiers achieve the highest syngas calorific energy value with as much as 20 MJ/m^3 [11].

Direct heating air-blown gasifiers present the simplest and lowest cost option. These characteristics have made them attractive for commercial applications. However, the excess N_2 concentration increases the size of downstream equipment and results in energy penalties for heating applications. Despite these challenges, producer gas from air-blown gasifiers have been employed in furnaces, boilers, and engines. N_2 can also be problematic for synthetic fuel applications because in the best of cases it simply increases vessel sizes, and in some applications participates negatively in chemical reactions.

Oxygen-blown gasifiers have a smaller footprint but at the significant cost of requiring an oxygen separation unit. They generate a high-quality syngas for heat or fuel synthesis applications. Downstream equipment are also much smaller due to the N_2 avoided. Another advantage is that oxygen-blown gasifiers can be very efficient with carbon conversion rates of 95% and higher. They can achieve temperatures of up to 1800°C. At these temperatures, syngas composition approaches equilibrium and contains limited quantities of tar, char, and light hydrocarbons [12]. These temperatures are above the ash melting temperature, resulting in a layer of hot, liquid ash called slag that collects on the reactor surface. Slag properties are critical for proper operation of high-temperature gasifiers. They can substitute for refractory material by regulating heat transfer between the reaction chamber and external walls. For this reason, the slag layer has to be maintained within a tight thickness tolerance. Excess slag flows to the bottom of the reactor where it is collected.

Indirect heating gasifiers provide reaction heat from external sources and can avoid the use of air or oxygen. Without oxygen addition, gasification products include higher concentrations of primary volatile compounds, char, and tar. These products can continue to react in the gas phase depending on the residence time and operating conditions. In principle, indirect heating gasifiers can operate at temperatures of up to 1500°C. However, most of these systems operate between 600°C and 850°C due to heat transfer limitations. These

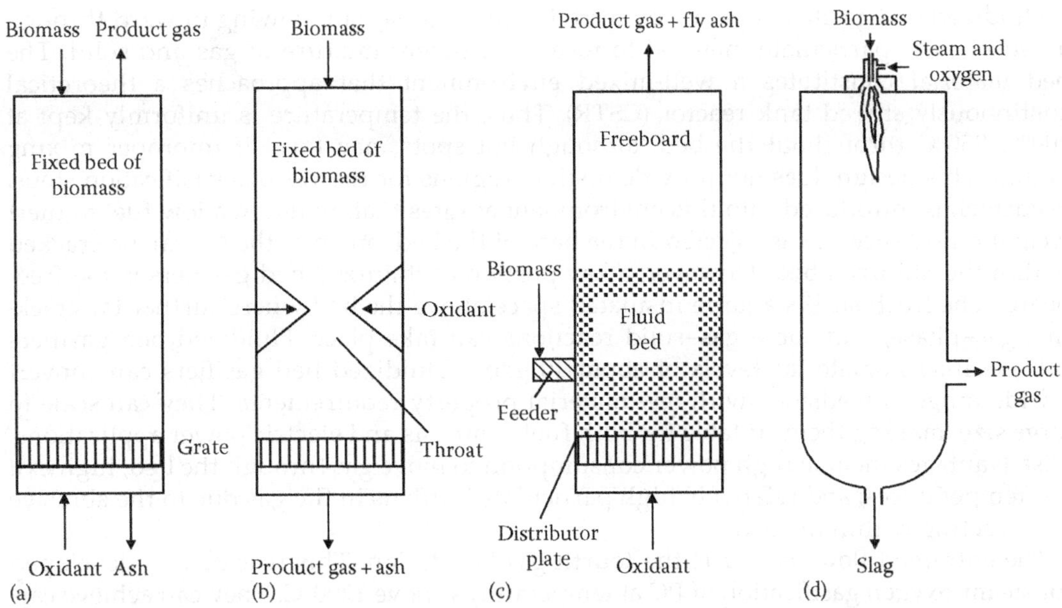

FIGURE 49.4
Common types of biomass gasifiers: (a) updraft, (b) downdraft, (c) fluidized bed, and (d) entrained flow.

gasifiers produce syngas with low N_2 and CO_2 concentrations than air- or oxygen-blown gasifiers but yield moderate quantities of tar and dust.

An alternative classification for gasifier designs is based on the contact method between the oxidizing agent and the feedstock input. By this classification, there are four main types of gasifiers: updraft (countercurrent), downdraft (concurrent), fluidized bed, and entrained flow [13]. Figure 49.4 illustrates these different designs.

The simplest and earliest design is the updraft gasifier, which evolved from charcoal kilns and blast furnaces. Updraft gasifiers resemble grate furnaces with chipped fuel admitted from above and below stoichiometric air input from below. Reaction temperatures close to the grate are very high and combustion processes dominate. At this location, oxygen gets depleted rapidly but hot H_2O and CO_2 reduce char to H_2 and CO. The fuel cools down as it moves toward the grate, but temperatures remain high enough to heat, dry, and pyrolyze the fuel. Updraft gasifiers yield high quantities of tar (50 g/m^3), making them unsuitable for many biomass energy applications.

Downdraft gasifiers were introduced in the nineteenth century following the introduction of induced draft fans capable of drawing air downward along with fuel. Contemporary designs introduced an arrangement of tuyeres to admit air or oxygen directly into a region known as the throat where combustion forms a bed of hot char. With this design, pyrolysis volatiles are forced through the bed of hot chair, thereby increasing gas–solid reactions. These reactions increase tar cracking making a producer gas have with low tar concentrations less than 1 g/m^3. These low tar concentrations make producer gas suitable for engine fuel. Strict requirements of particle size, ash, and moisture content make operation difficult. Slagging and ash sintering in the concentrated oxidation zone present another disadvantage. Mechanical and design measures can alleviate these problems such as the use of rotating ash grates. Downdraft gasifier capacities are limited to a maximum size of about 400 kg/h.

Fluidized bed gasifiers operate by introducing a gas stream flowing upward through a bed of inert particulate material to form a turbulent mixture of gas and solid. The bed material constitutes a well-mixed environment that approaches a theoretical continuously stirred tank reactor (CSTR). Thus, the temperature is uniformly kept at 700°C–850°C throughout the bed, although hot spots may form if improper mixing occurs. This design does not provide distinct regions for the various gasification steps. Feedstock is introduced into this environment at rates that maintain a low fuel to inert weight ratio. Since fuel is injected in the base of the bed, much of the tar can be cracked within the fluidized bed. However, a key portion of fluidized bed gasifiers is the freeboard. The freeboard is a large insulated space above the bed where further tar cracking, gas–phase, and some gas–solid reactions can take place. Fluidized bed gasifiers achieve intermediate tar levels of about 10 g/m^3. Fluidized bed gasifiers can convert a wide range of feedstock with few material property requirements. They can scale to large size, making them suitable for both fuel synthesis and electric power applications. Disadvantages include high power consumption to move gas through the bed; high exit gas temperatures; and relatively high particulate burdens in the gas due to the abrasive forces acting within the bed.

The entrained flow reactor is the fourth gasifier design. This reactor was developed for steam-oxygen gasification of PC at temperatures above 1200°C. They can achieve carbon conversion rates that approach 100%, produce trace tar quantities, and convert ash to molten slag. These properties make it attractive for both fuel synthesis and power generation applications. They can operate over a wide range of pressures, which allows them to be built at relatively smaller sizes. The key challenges for entrained flow gasifiers, particularly related to biomass applications, are requirements for small particle sizes, high operating temperatures, and low alkali levels. It is expensive to pulverize biomass, and biomass moisture and alkali content are comparably high for most feedstock. Pretreatment methods including torrefaction and pyrolysis can overcome many of these challenges [2].

There are many potential permutations of gasifier design and operation parameters. Furthermore, each gasifier design can be operated with various combinations of air/oxygen and steam input, resulting in different product compositions. Keeping this in mind, Table 49.2 shows typical composition, energy content, and quality of gas from the various gasifiers discussed previously.

TABLE 49.2

Producer Gas Composition from Various Kinds of Gasifiers

	Gaseous Constituents (vol.% Dry)						Gas Quality	
Gasifier Type	H_2	CO	CO_2	CH_4	N_2	HHV (MJ/m^3)	Tars	Dust
Air-blown updraft	11	24	9	3	53	5.5	High (≈10 g/m^3)	Low
Air-blown downdraft	17	21	13	1	48	5.7	Low (≈1 g/m^3)	Medium
Air-blown fluidized bed	9	14	20	7	50	5.4	Medium (≈10 g/m^3)	High
Oxygen-blown downdraft	32	48	15	2	3	10.4	Low (≈1 g/m^3)	Low
Indirectly-heated fluidized bed	31	48	0	21	0	17.4	Medium (≈10 g/m^3)	High

Sources: Higman, C. and Van Der Burgt, M., *Gasification*, Gulf Professional Publishing, Burlington, MA, 2008; Milne, T.A. et al., Biomass gasifier "tars": Their nature, formation, and conversion, Technical report, NREL//TP-570-25357, National Renewable Energy Laboratory, Golden, CO, 1998.

49.2.4 Heat and Power Cycles

Biomass combustion processes can be combined with a variety of steam and power cycles to generate electricity. These include the classical Stirling, Rankine, and Brayton cycles, as well as fuel cells and combined cycles. A majority of U.S. power generation comes from conventional steam turbines with growing contributions from gas turbines (see Figure 49.5). Many of these systems could utilize biomass by simply co-firing or combining with the cycles described in this section.

49.2.4.1 Stirling Cycle

The Stirling cycle employs heat from an external combustion engine to raise the temperature of an internal fluid undergoing a thermodynamic cycle [14]. The internal fluid never comes in contact with the combustion fuel, which lowers maintenance costs and pollution emissions. This arrangement also increases the tolerance to contaminants, which makes this cycle attractive for use with "dirty" fuels.

The Stirling cycle is similar to the Rankine cycle in its use of an external combustion process. In theory, Stirling cycles offer a higher thermodynamic efficiency than Rankine cycles. In practice, Stirling engines achieve relatively modest efficiencies and net power output of a few kilowatts. Their major obstacle might be high costs, which have so far prevented significant market entry.

49.2.4.2 Rankine Cycle

The Rankine cycle resembles the Stirling cycle in its use of an external combustion engine [15]. Figure 49.6 illustrates the major units found in a Rankine cycle. Fuel is fed from a bunker and pulverized before combustion in the furnace. Superheaters raise steam from

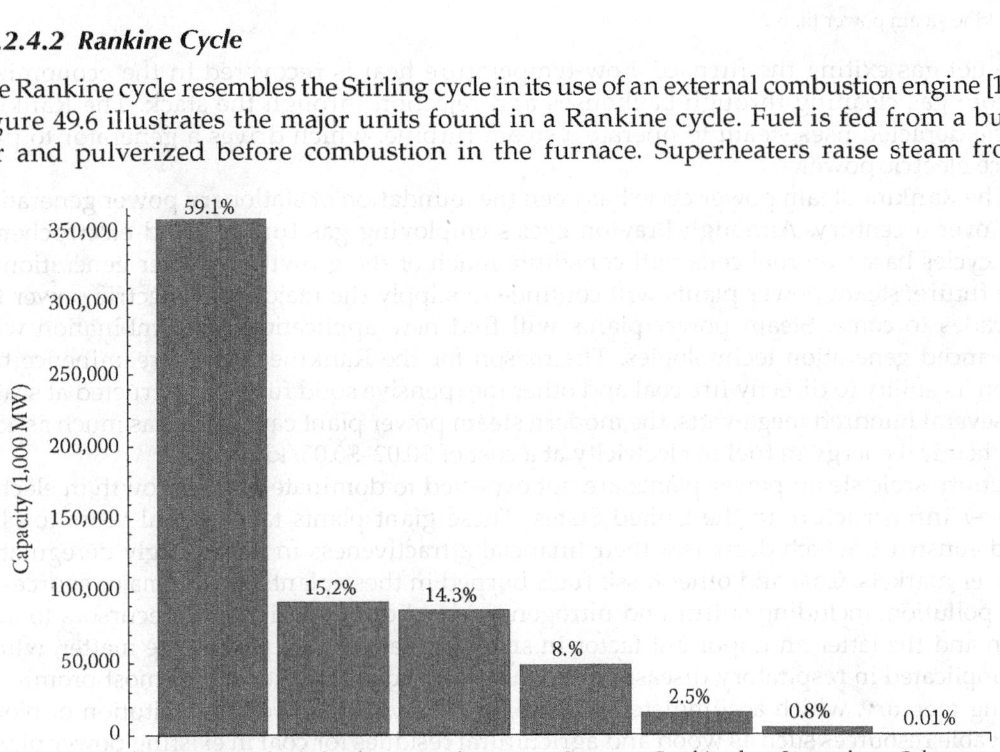

FIGURE 49.5
U.S. electric power generation capacity by generator/cycle type. (From Anon., Biomass for power generation and CHP, 2007.)

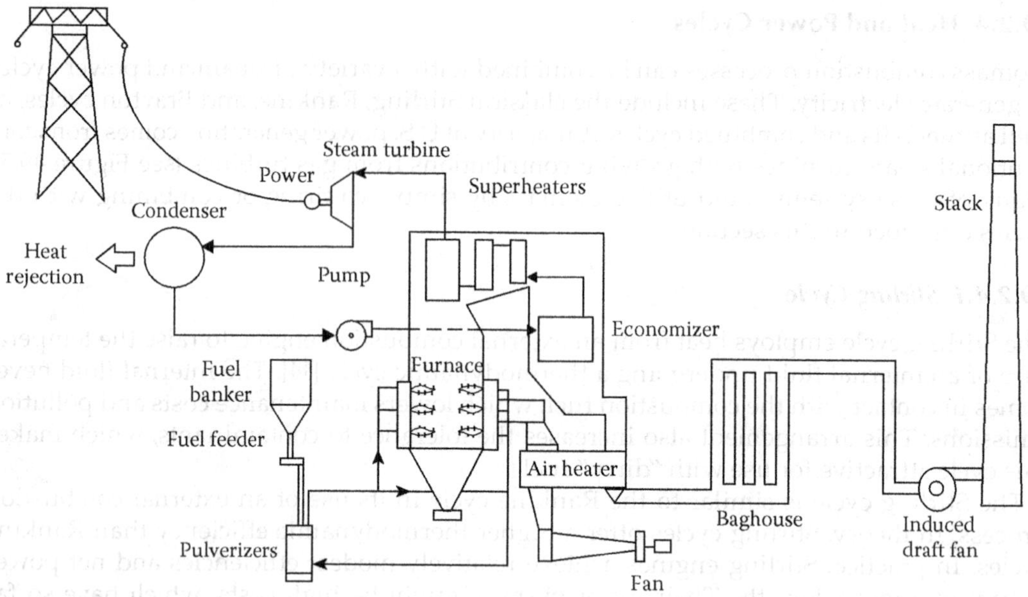

FIGURE 49.6
Rankine steam power plant.

the hot gas exiting the furnace. Low-temperature heat is recovered in the economizer before gas cleaning through baghouses and rejection through the stack. The Rankine cycle depicted uses steam to operate a steam turbine. which drives a generator to produce electric power.

The Rankine steam power cycle has been the foundation of stationary power generation for over a century. Although Brayton cycles employing gas turbines and electrochemical cycles based on fuel cells will constitute much of the growth in power generation in the future, steam power plants will continue to supply the majority of electric power for decades to come. Steam power plants will find new applications in combination with advanced generation technologies. The reason for the Rankine cycle's preeminence has been its ability to directly fire coal and other inexpensive solid fuels. Constructed at scales of several hundred megawatts, the modern steam power plant can convert as much as 45% of chemical energy in fuel to electricity at a cost of \$0.02–\$0.05/kilowatt.

Utility-scale steam power plants are not expected to dominate future growth in electric power infrastructure in the United States. These giant plants take several years to plan and construct, which decreases their financial attractiveness in increasingly deregulated power markets. Coal and other fossil fuels burned in these plants are the major sources of air pollution, including sulfur and nitrogen oxides, both of which are precursors to acid rain and the latter an important factor in smog formation; fine particulate matter, which is implicated in respiratory disease in urban areas; and heavy metals, the most prominent being mercury, which accumulates in the biosphere to toxic levels. Substitution of biorenewable resources such as wood and agricultural residues for coal in existing power plants could substantially reduce pollution emissions, although these plants are so large that the locally available biomass resources could supplant only a small fraction of the total energy requirement. Small-scale steam power plants sized for use of local biomass resources have low thermodynamic efficiencies, on the order of 25%, making them wasteful of energy resources.

49.2.4.3 Brayton Cycle

The Brayton cycle produces electric power by expanding hot gas through a turbine [16]. These gas turbines operate at temperatures approaching 1300°C compared to inlet temperatures of less than 650°C for steam turbines used in Rankine cycles. Although this difference in inlet temperature would suggest that Brayton cycles have much higher thermodynamic efficiencies than Rankine cycles, the Brayton cycle also has a much higher exhaust temperature than does the Rankine cycle. Gas turbine exhaust temperatures are in the range of 400°C–600°C, whereas steam turbine exhaust temperatures are on the order of 20°C. Furthermore, Brayton cycles, which contain both gas compressor and gas turbine, have more sources of mechanical irreversibilities, further degrading thermodynamic efficiencies, which may only be marginally higher than the best Rankine steam cycles. However, improvements in gas turbine technology that allow operation at higher temperatures and pressures are expected to increase Rankine cycle efficiency for large power plants to greater than 50%, although 30% is more realistic for gas turbines sized appropriately for biomass power plants.

The two general classes of gas turbines for power generation are heavy-duty industrial turbines and lightweight aeroderivative gas turbines. The aeroderivatives are gas turbines originally developed for commercial aviation but adapted for stationary electric power generation. They are attractive for bioenergy applications because of their high efficiency and low unit capital costs at the modest scales required for biomass fuels.

Gas turbines are well suited to gaseous and liquid fuels that are relatively free of contaminants that rapidly erode or corrode turbine blades. In this respect, gas turbine engines are not suitable for directly firing most biomass fuels. Solid biomass releases significant quantities of alkali metals, chlorine, mineral matter, and lesser amounts of sulfur upon burning. These would be entrained in the gas flow entering the expansion turbine where they would quickly contribute to blade failure. Cleaning large quantities of hot flue gas is not generally considered an economical proposition. Even the gas released from anaerobic digestion contains too much hydrogen sulfide (H_2S) to be directly burned in a gas turbine without first chemically scrubbing the gas to remove this corrosive agent.

Nevertheless, gas turbine engines are considered one of the most promising technologies for bioenergy because of the relative ease of plant construction, cost-effectiveness in a wide range of sizes (from tens of kilowatts to hundreds of megawatts), and the potential for very high thermodynamic efficiencies when employed in advanced cycles. The key to their success in bioenergy applications is converting the biomass to clean-burning gas or liquid before burning it in the gas turbine combustor.

49.2.4.4 Fuel Cells

Among the most exciting new energy technologies are fuel cells, which directly convert chemical energy into work, thus bypassing the restriction on efficiency imposed by the Carnot relationship [17]. This does not imply that fuel cells can convert 100% of chemical enthalpy of a fuel into work, as the process still must conform to the laws of thermodynamics. In practice, irreversibilities limit their conversion efficiencies to 35%–60%, depending upon the fuel cell design. Thus, fuel cells can produce significantly more work from a given amount of fuel than can heat engines. However, carbonaceous fuels must first be reformed to H_2 before they are suitable for use in fuel cells. The energy losses associated with fuel reforming must be included when determining the overall fuel-to-electricity conversion efficiency of a fuel cell.

The gas mixture produced by a biomass gasifier contains dust and tar that must be removed or greatly reduced for most applications, including power generation in fuel cells. Removal of tar would ideally be performed at elevated temperatures. If the gas is to be used in fuel cells, further cleaning is required to remove ammonia (NH_3), hydrogen chloride (HCl), and H_2S [18]. To obtain high-energy efficiency, trace contaminant removal must be performed at elevated temperatures for fuel cells that operate at relatively high temperatures. Low-temperature fuel cells cannot tolerate CO, which can be removed by the WGS reaction. The catalysts that facilitate the shift reaction, however, are poisoned by trace contaminants, which must be removed prior to the shift reactors. One method for removing H_2S and HCl is the use of a fixed bed of calcined dolomite or limestone and zinc titanate at temperatures around 630°C. This is followed by steam reforming at high temperature (750°C–850°C) to destroy tar and NH_3 [19].

49.2.4.5 Combined Cycles

In an effort to enhance energy conversion efficiency, combined cycle power systems have been developed, which recognize that waste heat from one power cycle can be used to drive a second power cycle [20]. Combined cycles would be unnecessary if a single heat engine could be built to operate between the temperature extremes of burning fuel and the ambient environment. However, temperature and pressure limitations on materials of construction have prevented this realization. Combined cycles employ a topping cycle operating at high temperatures and a bottoming cycle operating on the rejected heat from the topping cycle. Most commonly, combined cycle power plants employ a gas turbine engine for the topping cycle and a steam turbine plant for the bottoming cycle, achieving overall efficiencies of 60% or higher.

Clean-burning fuel from biomass for use in a combined cycle can be obtained by thermal gasification. Integrated gasification/combined cycle (IGCC) power is illustrated in Figure 49.7. Compressed air enters an oxygen plant, which separates oxygen from the air. The oxygen is used to gasify biomass in a pressurized gasifier to produce medium heating-value producer gas. The producer gas passes through cyclones and a gas clean-up system to remove particulate

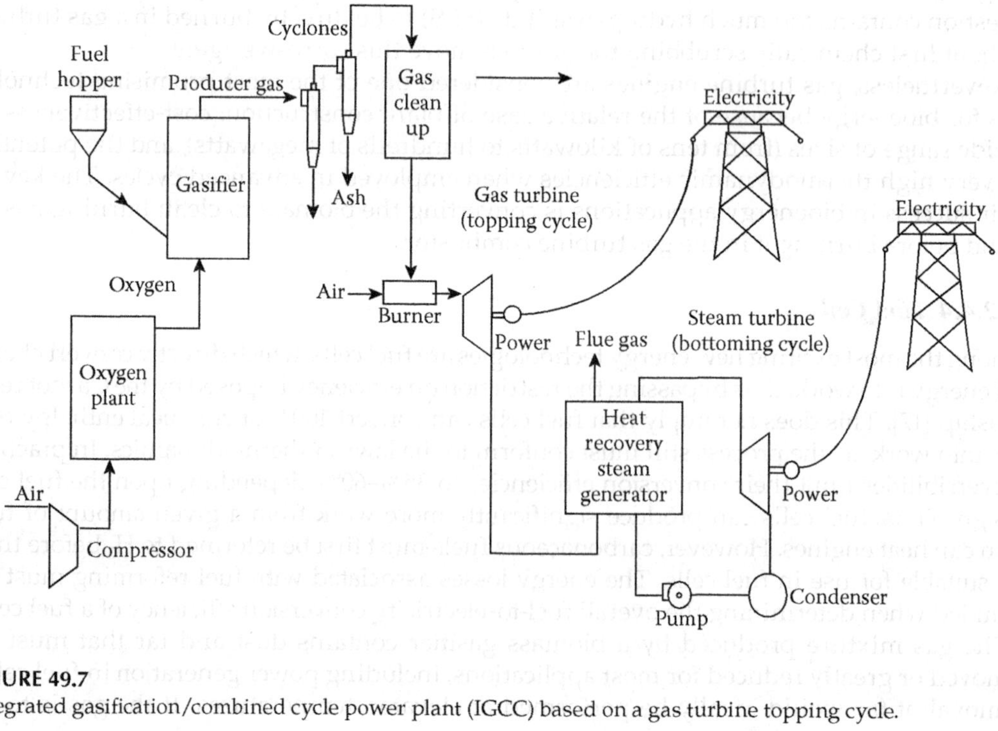

FIGURE 49.7
Integrated gasification/combined cycle power plant (IGCC) based on a gas turbine topping cycle.

matter, tar, and other contaminants that may adversely affect gas turbine performance (alkali and chloride being the most prominent among these). These clean-up operations are best performed at high temperature and pressure to achieve high cycle efficiency. The clean gas is then burned in air and expanded through a gas turbine operating as a "topping" cycle. The gas exits the turbine at temperatures ranging between 400°C and 600°C. A heat recovery steam generator produces steam for a "bottoming" cycle that employs a steam turbine. Electric power is produced at two locations in this plant, yielding thermodynamic efficiencies exceeding 47%.

Integrated gasifier/combined cycle systems based on gas turbines are attractive for several reasons. These reasons include their relative commercial readiness and the expectation that they can generate electricity at the lowest cost of all possible biomass power options.

An alternative to IGCC is to generate steam for injection into the gas turbine combustor, which increases mass flow and power output from the turbine. This variation, called a steam-injected gas-turbine (STIG) cycle [20,21] is less capital intensive than IGCC since it does not employ a steam turbine. The STIG cycle is commercially developed for natural gas; lower flammability limits for producer gas make steam injection more problematic for biomass-derived producer gas. The intercooled steam-injected gas turbine (ISTIG) is an advanced version of the STIG. This cycle further improves thermodynamic efficiency by cooling gas flow between several stages of compression (intercooling).

Figure 49.8 illustrates an IGCC power plant based on a molten carbonate fuel cell [18]. Biomass is gasified in oxygen to yield producer gas. Gasification occurs at elevated pressure to improve the yield of CH_4, which is important for proper thermal balance of this fuel cell. Hot-gas clean up to remove particulate matter, tar, and other contaminants is followed by expansion through a gas turbine as part of a topping power cycle. The pressure and temperature of the producer gas is sufficiently reduced after this to admit it into the fuel cell. High-temperature exhaust gas exiting the cathode of the fuel cell enters a

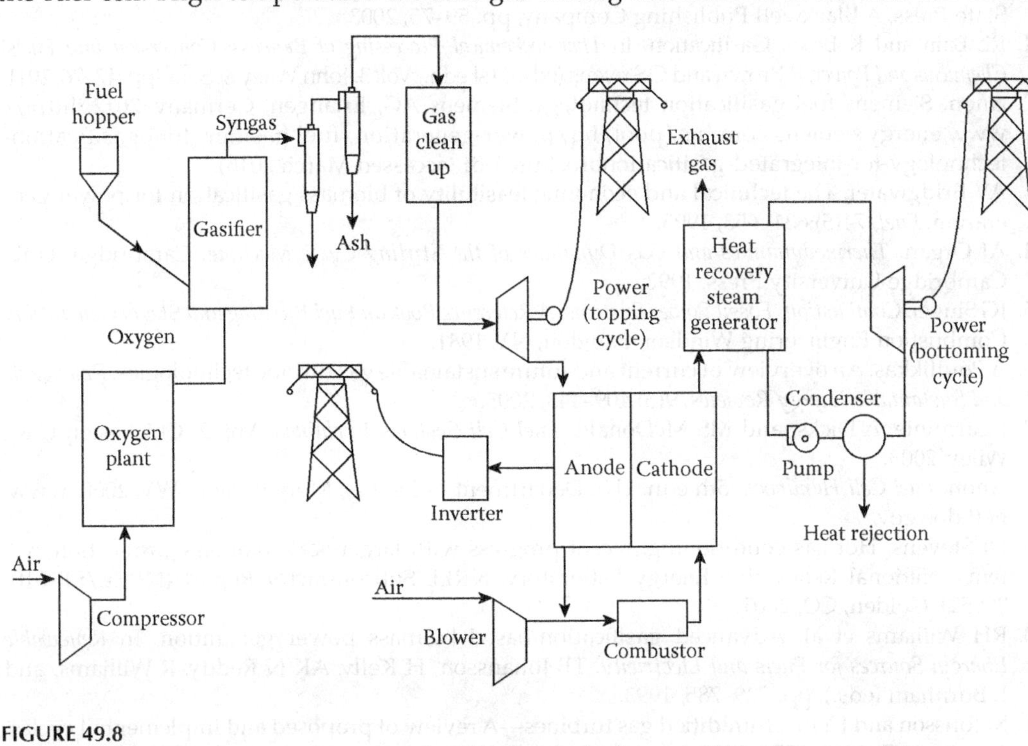

FIGURE 49.8
Combined cycle power plant based on high-temperature fuel cells.

heat recovery steam generator, which is part of a bottoming cycle in the integrated plant. Thus, electricity is generated at three locations in the plant for an overall thermodynamic efficiency reaching 60% or more.

References

1. Anon. Biomass for power generation and CHP, International Energy Agency, France, Paris, 2007 http://www.iea.org/publications/freepublications/publication/essentials3.pdf.
2. PCA Bergman and JHA Kiel. Torrefaction for biomass upgrading. Technical Report ECN-RX-05-180, Petten, the Netherlands, November 2005.
3. L Nan and G Best. Integrated energy systems in China—The cold northeastern region experience. Food and Agriculture Organization of the United Nations, Rome, Italy, 1994.
4. MQ Wang. The GREET Spreadsheet Model: Greenhouse gases and regulated emissions and energy use in transportation, GREET 1 2011 Version. Center for Transportation Research, Energy Systems Division, Argonne National Laboratory, Lemont, IL, 2009.
5. RP Anex et al. Techno-economic comparison of biomass-to-transportation fuels via pyrolysis, gasification, and biochemical pathways. *Fuel*, 89(S1):S29–S35, November 2010.
6. DA Tillman. *The Combustion of Solid Fuels and Wastes*. Academic Press, San Diego, CA, 1991.
7. JV Kumar and BC Pratt. Compositional analysis of some renewable biofuels. *American Laboratory*, 28(8):15–20, 1996.
8. RL Bain and RP Overend. Biomass-fired power generation. *Fuel Processing Technology*, 54.1:1–16, 1998.
9. M Sami, K Annamalai, and M Wooldridge. Co-firing of coal and biomass fuel blends. *Progress in Energy and Combustion Science*, 27(2):171–214, 2001.
10. RC Brown. *Biorenewable Resources: Engineering New Products from Agriculture*. Ames, IA: Iowa State Press, A Blackwell Publishing Company, pp. 59–75, 2003.
11. RL Bain and K Broer. Gasification. In *Thermochemical Processing of Biomass: Conversion into Fuels, Chemicals and Power*, R Brown and C Stevens (eds.), 1st edn., Vol. 3. John Wiley & Sons, pp. 47–76, 2011.
12. Anon. Siemens fuel gasification technology. Siemens AG, Erlangen, Germany, 2012. http://www.energy.siemens.com/us/pool/hq/power-generation/fuel-gasifier/fuel-gasification-technology-for-integrated-gasificationbrochure.pdf (accessed March 2015).
13. AV Bridgwater. The technical and economic feasibility of biomass gasification for power generation. *Fuel*, 74(5):631–653, 1995.
14. AJ Organ. *Thermodynamics and Gas Dynamics of the Stirling Cycle Machine*. Cambridge, U.K.: Cambridge University Press, 1992.
15. JG Singer. *Combustion, Fossil Power Systems: A Reference Book on Fuel Burning and Steam Generation*. Combustion Engineering Windsor, Babylon, NY, 1981.
16. A Poullikkas. An overview of current and future sustainable gas turbine technologies. *Renewable and Sustainable Energy Reviews*, 9(5):409–443, 2005.
17. J Larminie, A Dicks, and MS McDonald. *Fuel Cell Systems Explained*, Vol. 2. Chichester, U.K., Wiley, 2003.
18. Anon. *Fuel Cell Handbook*, 5th edn. U.S. Department of Energy, Morgantown, WV, 2000. www.netl.doe.gov.
19. DJ Stevens. Hot gas conditioning: recent progress with larger-scale biomass gasification systems. National Renewable Energy Laboratory, NREL Subcontractor Report (NREL/SR-510-29952), Golden, CO, 2001.
20. RH Williams et al. Advanced gasification-based biomass power generation. In *Renewable Energy: Sources for Fuels and Electricity*, TB Johansson, H Kelly, AK N Reddy, R Williams, and L Burnham (eds.), pp. 729–785, 1993.
21. M Jonsson and J Yan. Humidified gas turbines—A review of proposed and implemented cycles. *Energy*, 30(7):1013–1078, 2005.

50

Biomass Conversion to Fuels

Mark M. Wright and Robert C. Brown

CONTENTS

50.1 Biomass to Solid Fuels

Low energy density and high moisture content are two key disadvantages of using biomass as a solid fuel. Both can be overcome with thermal pretreatment methods. Torrefaction and (slow) pyrolysis are two main approaches to converting a significant portion of biomass into a high-energy-density solid fuel.

Biomass torrefaction is similar to the coffee roasting process. Both take place at moderate temperatures of about 250°C without oxygen addition and residence times of 30 min or more. The process is mostly endothermic, but torrefaction gases can be consumed to provide necessary heat. Torrefaction evolves over three main steps. Initially, the feedstock will simply absorb heat until the feed moisture evaporates. The next step involves the release of some volatile gases. Finally, the biomass undergoes minor chemical restructuring becoming dark, brittle, and hydrophobic. Material yields can be as low as 70 wt.% with corresponding energy yields of more than 90 wt.% [1].

Torrefied biomass resembles charcoal in appearance and physical properties. It is easier to pulverize, has a higher energy content (>20 MJ/kg), and can be stored for longer periods without absorbing moisture and with minimal microbial degradation [2]. These properties make it attractive for direct combustion, co-firing, and entrained flow gasification applications.

Slow pyrolysis is a higher-intensity form of torrefaction [3]. Slow pyrolysis conditions consist of an inert environment with temperatures above 400°C and residence times of more than 30 min. This process releases a greater amount of pyrolysis volatiles resulting in solid yields of <30 wt.%. The solid product shares many similarities with torrefied biomass including a higher heating value (HHV) of about 30 MJ/kg. However, its products are not well suited to synthetic fuel applications.

Finally, carbonization, which takes place at temperatures of 450°C–600°C, yields about 35 wt.% solids with a similar HHV as slow pyrolysis charcoal. Carbonization can convert large feed sizes without prior grinding. It is also the simplest of technologies to operate in part because the product requirements are less strict than other thermal processes.

50.2 Biomass to Gaseous Fuels

Anaerobic digestion and gasification are two of the main ways to convert biomass into gaseous fuels. Both pathways generate a combustible gas: biogas and synthetic gas (syngas), respectively. Biogas is relatively inexpensive to produce, but has a smaller range of applications than syngas. This section will discuss the technologies that enable biomass fuel gas production.

50.2.1 Biogas

In the absence of oxygen, bacteria decompose organic wastes into a mixture of CH_4, CO_2, and trace gases. The decomposition productivity is sensitive to feedstock properties including moisture and ash content. Anaerobic digestion yields vary from 23 to over 250 m^3/Mg (wet basis) with a methane content of 55%–75% by volume with an energy efficiency of about 60%.

Methane is the primary component in biogas, which makes it a candidate for substituting natural gas in many applications. Clean biogas is suitable for engine generator sets, small gas turbines, and some kinds of fuel cells. However, biogas often contains contaminants like sulfur that must be removed before use in some equipment.

Anaerobic digestion involves several complicated steps illustrated in Figure 50.1 [4]. During this process, several bacteria species break down proteins, carbohydrates, and fats into simple acids, alcohols, and gaseous compounds via hydrolysis and acidification. The final step is methanogenesis, where most of the intermediate organics decompose to CH_4 and CO_2.

On the other hand, the anaerobic digester systems where these processes take place are simple (see Figure 50.1). Anaerobic digesters operate as batch, semi-batch, or continuous units (Figure 50.2). Reactor designs include batch, plug-flow, continuously stirred tank, upflow, and two-tank reactors. The most common reactor design is the batch system in which most of the digestion steps take place within a single vessel. Improvements to this design aim to increase contact time between bacteria and the organic wastes and/or to separate and control the environments for acid-forming and methane-forming bacteria. There are numerous parameters to optimize the performance of anaerobic digestion: pretreatments, heating, mixing, nutrient addition, bacteria selection, and pH to name a few.

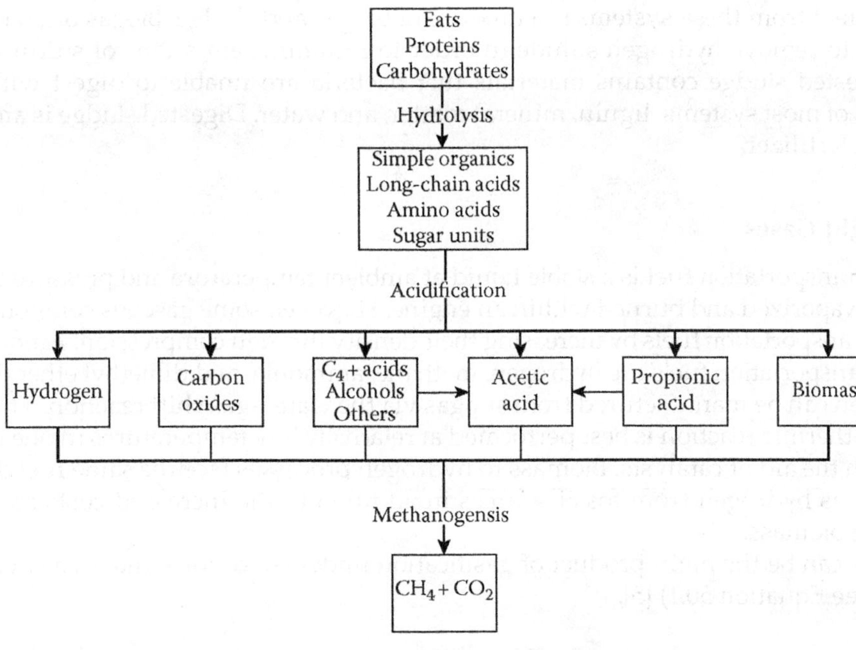

FIGURE 50.1
Microbial phases in anaerobic digestion. (Adapted from Klass, D., *Biomass for Renewable Energy, Fuels, and Chemicals*, Academic Press, San Diego, CA, 1998c, p. 462.)

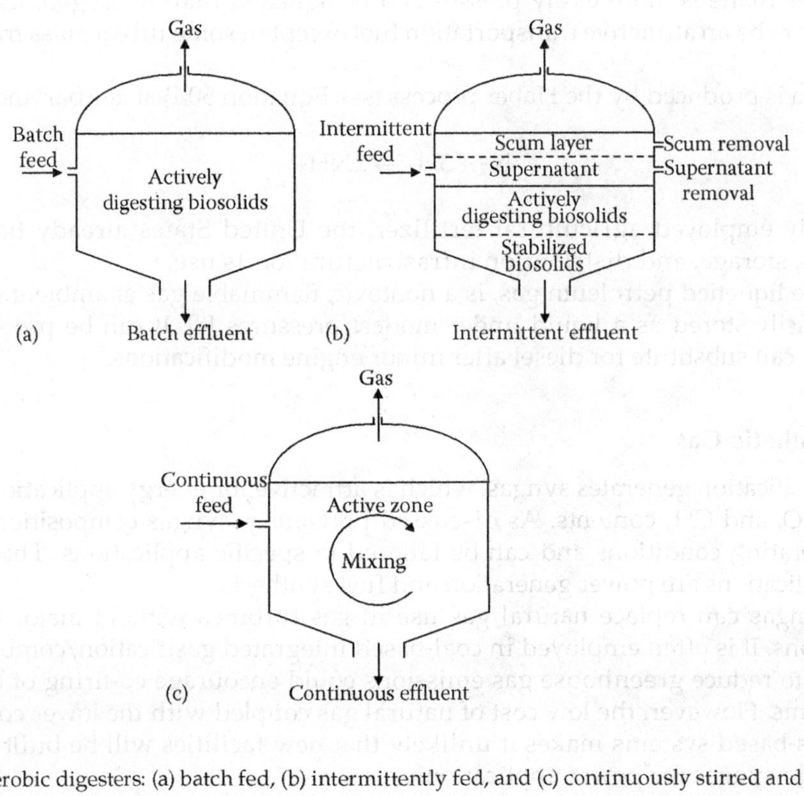

FIGURE 50.2
Types of anaerobic digesters: (a) batch fed, (b) intermittently fed, and (c) continuously stirred and fed.

The effluent from these systems is a mixture of biogas and sludge. Biogas often requires scrubbing to remove hydrogen sulfide in order to minimize emissions of sulfur dioxide (SO_2). Digested sludge contains materials that bacteria are unable to digest within the timescales of most systems: lignin, mineral matter, and water. Digested sludge is amenable to use as a fertilizer.

50.2.2 Light Gases

The ideal transportation fuel is a stable liquid at ambient temperature and pressure that can be readily vaporized and burned within an engine. However, some gaseous compounds are potential transportation fuels by increasing their density through compression. Among these gaseous transportation fuels are hydrogen, methane, ammonia, and dimethyl ether (DME).

Hydrogen can be manufactured from syngas via the water–gas shift reaction. This moderately exothermic reaction is best performed at relatively low temperatures in one or more stages with the aid of catalysts. Biomass to hydrogen processes face the same fuel delivery challenges as hydrogen from fossil sources in addition to the increased costs associated with using biomass.

Methane can be the main product of gasification under conditions known as hydrogasification (see Equation 50.1) [5]:

$$C + 2H_2 \rightarrow CH_4$$

$$CO + 3H_2 \rightarrow CH_4 + H_2O$$

Although methane is more easily pressurized or liquefied than hydrogen, its density is still too low to be an attractive transportation fuel except in some urban mass transit applications [6].

Ammonia is produced by the Haber process (see Equation 50.1) at 200 bar and 500°C [7]:

$$N_2 + 3H_2 \rightarrow 2NH_3 \tag{50.1}$$

As a widely employed agricultural fertilizer, the United States already has in place production, storage, and distribution infrastructure for its use.

DME, like liquefied petroleum gas, is a nontoxic, flammable gas at ambient conditions, which is easily stored as a liquid under modest pressures [8]. It can be produced from syngas and can substitute for diesel after minor engine modifications.

50.2.3 Synthetic Gas

Biomass gasification generates syngas, which is attractive for energy applications because of its H_2, CO, and CH_4 contents. As discussed previously, syngas composition varies by reactor operating conditions and can be tailored to specific applications. The two main syngas applications are power generation and fuel synthesis.

Clean syngas can replace natural gas use in gas turbines without major equipment modifications. It is often employed in coal-based integrated gasification/combined cycle, but efforts to reduce greenhouse gas emissions could encourage co-firing of biomass in these systems. However, the low cost of natural gas coupled with the lower construction costs of gas-based systems makes it unlikely that new facilities will be built to convert solid fuels to syngas for power applications.

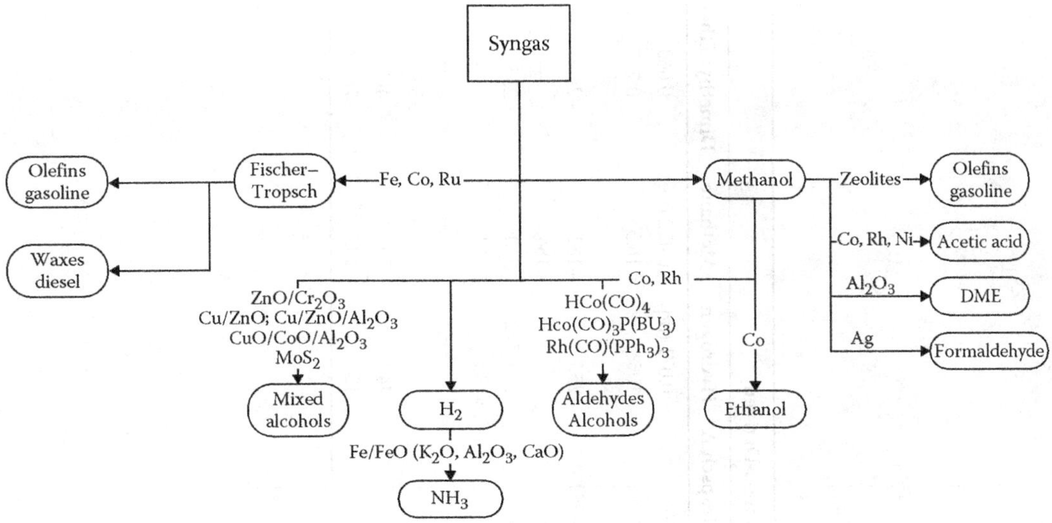

FIGURE 50.3
Catalytic pathways for converting synthetic gas to fuels. (Adapted from Spath, P.L. and Dayton, D.C., Preliminary screening-technical and economic assessment of synthesis gas to fuels and chemicals with emphasis on the potential for biomass-derived syngas, Technical Report, DTIC Document, National Renewable Energy Laboratory NREL/TP-510-34929, pp. 1–160, 2003.)

Fuel synthesis converts solid carbonaceous materials into higher-value liquid fuels. While solid fuels such as coal and biomass are valued below \$100 per Mg, liquid fuels including gasoline and diesel often have prices of over \$1000 per Mg. These higher prices reflect the value of high energy density, portability, and fuel consistency. Solid fuels can be pretreated to improve these characteristics, and they have been historically used in portable applications including onboard passenger vehicle fuels. However, the mass production of liquid fuel engines has led to the development of numerous pathways to produce synthetic fuels [9].

Figure 50.3 shows several catalytic pathways for converting syngas into liquid fuels [10]. A large number of catalysts can convert syngas into Fischer–Tropsch liquids (FTLs), alcohols, short-chained hydrocarbons, and ammonia via H_2 synthesis. DME, ethanol, methanol, formaldehyde, acetic acid, olefins, and gasoline can be synthesized following syngas conversion to methanol. Details for these pathways will be discussed in the following section.

This wide range of fuels highlights the versatility of syngas. Thus, syngas from inexpensive, renewable, environmentally friendly sources could replace fossil fuels in a variety of applications. However, the development of any of these pathways depends heavily on a variety of technical, economic, and policy constraints.

50.3 Biomass to Liquid Fuels

Almost 25% of energy consumption in the United States goes to transportation. Approximately half of this amount comes from imported petroleum. Thus, the development of transportation fuels from biorenewable resources is a priority if decreased dependence on foreign sources of energy is to be achieved. Table 50.1 lists properties of both traditional and bio-based transportation fuels.

TABLE 50.1

Comparison of Ignition and Combustion Properties of Transportation Fuels

	Fossil Fuel-Derived		Biomass-Derived						
Fuel Type	**Gasoline**	**No. 2 Diesel Fuel**	**Methanol**	**Ethanol**	**Biodiesel**	**Fischer–Tropsch A**	**Hydrogen**	**Methane**	**Dimethyl Ether**
Specific gravity[a]	0.72–0.78	0.85	0.796	0.794	0.886	0.770	0.071 (liq)	0.422 (liq)	0.660
Kinematic viscosity at 20°C–25°C (mm^2/s)	0.8	2.5	0.75	1.51	3.9	2.08	105[b]	16.5[b]	0.227
Boiling point range (°C)	30–225	210–235	65	78	339	164–352	−253	−162	−24.9
Flash point (°C)	−43	52	11	13	188	58.5		−184	—
Autoignition temperature (°C)	370	254	464	423	—	—	566–582	540	235
Octane no. (research)	91–100	—	109	109	—	—	>130	>120	—
Octane no. (motor)	82–92	—	89	90	—	—	—	—	—
Cetane no.	<15	37–56	<15	<15	55	74.6	—	—	>55
Heat of vaporization (kJ/kg)	380	375	1185	920	—	—	447	509	402d
Lower heating value (MJ/kg)	43.5	45	20.1	27	37	43.9	120	49.5	28.88

[a] Measured at 16°C except for liquefied gases, which are saturated liquids at their respective boiling points.
[b] Munson et al. [11].
[c] Perry and Green [12].
[d] Kajitani et al. [13].

Traditional transportation fuels are classified as gasoline, diesel fuel, or jet fuel. Gasoline is intended for spark-ignition (Otto cycle) engines; thus, it is relatively volatile but resistant to autoignition during compression. Diesel fuel is intended for use in compression-ignition (Diesel cycle) engines; thus, it is less volatile compared to gasoline and more susceptible to autoignition during compression. Jet fuel is designed for use in gas turbine (Brayton cycle) engines, which are not limited by autoignition characteristics but otherwise have very strict fuel specifications for reasons of safety and engine durability.

Gasoline is a mixture of hundreds of different hydrocarbons obtained from a large number of refinery process streams that contain between 4 and 12 carbon atoms with boiling points in the range of 25°C–225°C. Most of the mixture consists of alkanes with butanes and pentanes added to meet vapor pressure specifications. A few percent of aromatic compounds are added to increase octane number, the figure of merit used to indicate the tendency of a fuel to undergo premature detonation within the combustion cylinder of an internal combustion engine. The higher the octane number, the less likely a fuel will detonate until exposed to an ignition source (electrical spark). Premature denotation is responsible for the phenomenon known as engine knock, which reduces fuel economy and can damage an engine. Various systems of octane rating have been developed, including research octane and motor octane numbers. Federal regulation in the United States requires gasoline sold commercially to be rated using an average of the research and motor octane numbers. Gasoline rated as "regular" has a commercial octane number of about 87 while premium grade is 93.

Diesel fuel, like gasoline, is also a mixture of light distillate hydrocarbons but with lower volatility and higher viscosity. Because diesel fuel is intended to be ignited by compression rather than by a spark, its autoignition temperature is lower than for gasoline. The combustion behavior of diesel fuels are conveniently rated according to cetane number, an indication of how long it takes a fuel to ignite (ignition delay) after it has been injected under pressure into a diesel engine. A high cetane number indicates short ignition delay, for example, no. 2 diesel fuel has a cetane number of 37–56, while gasoline has a cetane number <15.

Jet fuel is designated as either Jet A fuel or Jet B fuel. Jet A fuel is a kerosene type of fuel with relatively high flash point, whereas Jet B fuel is a wide-boiling-range fuel, which more readily evaporates.

Bio-based transportation fuels, also known as biofuels, are currently dominated by ethanol and biodiesel. Ethanol, by virtue of its high octane number, is suitable for use in spark-ignition engines, while the high cetane numbers of biodiesel, which are methyl or ethyl esters formulated from vegetable or animal fats, make them suitable for use in compression-ignition engines. However, there are other candidate liquid biofuels including methanol, mixed alcohols, and FTLs, as well as gaseous biofuels including hydrogen, methane, ammonia, and DME, which will also be discussed.

50.3.1 Alcohols

50.3.1.1 Corn Ethanol

Ethanol is the dominant biofuel with global production exceeding 50 billion L. The majority of this ethanol comes from Brazilian sugarcane and the U.S. corn grain. Microbes convert these sugar-rich crops into alcohols with relative ease and at low cost. Two major corn-to-ethanol facilities have been operated in the United States: dry grind and wet-milling.

Dry grind ethanol plants, shown in Figure 50.4, grind the whole kernel while wet-milling plants soak the grain with water and acid to separate the corn germ, fiber, gluten, and starch components before mechanical grinding [14]. The capital investment for dry grind

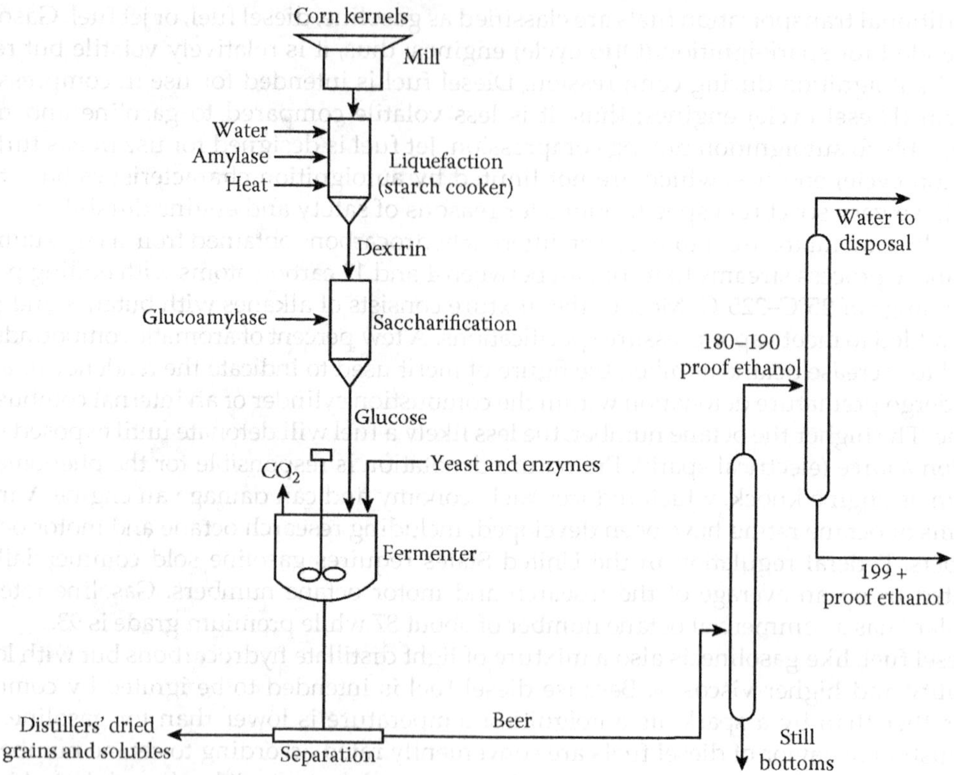

FIGURE 50.4
Dry grind corn to ethanol.

is less than that for a comparably sized wet-milling plant. However, the higher value of its by-products, greater product flexibility, and simpler ethanol production can make a wet-milling plant a more profitable investment.

Dry grind ethanol takes place in four major steps: pretreatment, cooking, fermentation, and distillation. Pretreatment consists of grinding the corn kernel into flour "meal," which is mixed with water, enzymes, and ammonia. This mixture ("mash") is then "cooked" to reduce bacteria levels. After cooling, the mash is sent to the fermenter where it remains for 40 h or more. The beer resulting from fermentation consists of a mixture of ethanol and stillage. Energy-intensive distillation of the beer is necessary to separate the stillage and water content from the ethanol and achieve maximum concentrations of 95%. This is followed by further purification to 100% ethanol using molecular sieves.

A modern dry grind plant will produce over 2.7 gal of ethanol per bushel of corn processed. Yields of coproducts per bushel of corn are 7.7–8.2 kg (17–18 lb) of dry distiller's grain with solubles and 7.3–7.7 kg (16–17 lb) of carbon dioxide evolved from fermentation, the latter of which can be sold to the carbonated beverage industry. As a rule of thumb, the three products are produced in approximately equal weight per bushel. Table 50.2 shows ethanol yields for various feedstock.

Costs for ethanol from corn grain have been developed by the USDA [15] for the dry mill process. Capital costs for a 40 million gal/year ethanol plant were estimated at $46.7 million with fuel production costs of $1.03 per gal (2006).

TABLE 50.2

Ethanol Yields from Various Biorenewable Resources

Feedstock	Yield (L/Mg)
Apples	64
Barley	330
Cellulose	259
Corn	355–370
Grapes	63
Jerusalem artichoke	83
Molasses	280–288
Oats	265
Potatoes	96
Rice (rough)	332
Rye	329
Sorghum (sweet)	44–86
Sugar beets	88
Sugarcane	160–187
Sweet potatoes	125–143
Wheat	355

Sources: Klass, D., *Biomass for Renewable Energy, Fuels, and Chemicals*, Academic Press, San Diego, CA, 1998a, p. 356; Klass, D., *Biomass for Renewable Energy, Fuels, and Chemicals*, Academic Press, San Diego, CA, 1998b, pp. 341–344; Klass, D., *Biomass for Renewable Energy, Fuels, and Chemicals*, Academic Press, San Diego, CA, 1998c, p. 462.

50.3.1.2 Cellulosic Ethanol

Much of the carbohydrate in plant materials is structural polysaccharides, providing shape and strength to the plant. This structural material, known as lignocellulose, is a composite of cellulose fibers embedded in a cross-linked lignin–hemicellulose matrix [16]. Depolymerization to basic plant components is difficult because lignocellulose is resistant to both chemical and biological attack [17]. However, depolymerization is necessary for microbes to efficiently convert cellulosic biomass into alcohols.

Cellulose to ethanol consists of four steps: pretreatment, enzymatic hydrolysis, fermentation, and distillation [14]. Of these, pretreatment is the most costly step, accounting for about 33% of the total processing costs [18]. An important goal of all pretreatments is to increase the surface area of lignocellulosic material, making the polysaccharides more susceptible to hydrolysis. Thus, comminution, or size reduction, is an integral part of all pretreatments.

Enzymatic hydrolysis was developed to better utilize both cellulose and hemicellulose from lignocellulosic materials. Three basic methods for hydrolyzing structural polysaccharides in plant cell walls to fermentable sugars are available: concentrated acid hydrolysis, dilute acid hydrolysis, and enzymatic hydrolysis [17,19]. The two acid processes hydrolyze both hemicellulose and cellulose with very little pretreatment beyond comminution of the lignocellulosic material to particles of about 1 mm size. The enzymatic process must be preceded by extensive pretreatment to separate the cellulose, hemicellulose, and lignin fractions.

Although thermodynamic efficiencies for the conversion of carbohydrates to ethanol can be calculated, it is more typical to report the volumetric yield of ethanol per unit mass of feedstock. The yield of ethanol from energy crops varies considerably. Among sugar crops, sweet sorghum yields 80 L/ton, sugar beets yield 90–100 L/ton, and sugarcane yields 75 L/ton. Among starch and inulin crops, the ethanol yield is 350–400 L/ton of corn, 400 L/ton of wheat, and 90 L/ton of Jerusalem artichoke. Among lignocellulosic crops, the potential ethanol yield is 400 L/ton of hybrid poplar, 450 L/ton for corn stover, 510 L/ton for corn cobs, and 490 L/ton for wheat straw.

Researchers at the National Renewable Energy Laboratory [20] developed a design report showing capital costs of $114 million and operating costs of $1.07 per gal (2000) of ethanol for a 69 million gal/year ethanol plant.

50.3.1.3 Butanol

Ethanol has several limitations as a transportation fuel, including its affinity for water, which prevents it from being fully compatible with the existing fuel infrastructure, and its low volumetric heating value, which is only two-thirds that of gasoline. For this reason, fermentations that produce metabolites other than ethanol have been proposed. Alternative fermentations could produce hydrophobic molecules that are less oxidized than ethanol including higher alcohols (most prominently butanol), fatty acids, fatty alcohols, esters, alkanes, alkenes, and isoprenes.

Butanol is of particular industrial interest because of its high compatibility with the existing vehicle technologies. In addition to a higher heating value than ethanol, butanol can power existing gasoline engines without major modifications. In fact, gasoline engines can operate on up to 100% butanol blends. This would overcome the 10% volumetric blend limit of ethanol to gasoline in modern vehicles. The challenges for butanol are mostly associated with its production. Butanol production proceeds through either the intermediate hydrocarbon propylene or acetaldehyde. The propylene route employs a rhodium catalyst. The acetaldehyde pathway is a fermentation process commonly known as the acetone–butanol–ethanol (ABE) process. The ABE process shares many similarities with corn fermentation. However, it suffers from low product specificity and yield due in part to high butanol toxicity to the fermenting microorganisms [21]. Low product yields and butanol concentrations in the bioreactors are key reasons for higher butanol production capital and fuel costs relative to ethanol fermentation [22]. There are recent industry indications that some of these challenges have been overcome to allow rapid conversion of existing corn ethanol facilities to butanol.

50.3.2 Biodiesel

There is growing global interest to produce biodiesel from lipid-rich biomass. There are abundant sources of inexpensive lipid-rich biomass including dedicated, fast-growing crops like jatropha, oils from food waste, and algae-based oils. Lipids are a large group of hydrophobic, fat-soluble compounds produced by plants and animals for high-density energy storage. Triglycerides of fatty acids, commonly known as fats and oils depending upon their melting points, are among the most familiar form of lipids and have been widely used in recent years for the production of diesel fuel substitutes. The solution to this problem is to convert the triglycerides into methyl esters or ethyl esters of the fatty acids, known as biodiesel, and the by-product 1,2,3-propanetriol (glycerol).

A wide variety of plant species produce triglycerides in commercially significant quantities, most of it occurring in seeds [15]. Average oil yields range from 15,000 L/km^2 for cottonseed to 81,400 L/km^2 for peanut oil although intensive cultivation might double these numbers. Soybeans are responsible for more than 50% of world production of oilseed, representing

4882 million bbl/year. The average oil yield for soybeans is 38,300 L/km^2 [15]. A higher-yielding crop is oil palm, already grown in plantations for vegetable oil production [23].

Although oil palm yields are 10 times higher than soybeans, some environmentalists are concerned that its cultivation for fuel production will encourage rainforest destruction. However, several oil seed crops have been identified that could be grown on wasteland or even saline soils, which reduces concerns about competition for food crops and rainforest destruction. These alternative oil seed crops include jatropha, Chinese tallow tree, and salicornia [24,25].

Triglycerides can be converted into transportation fuels via transesterification [26]. The process, described in Figure 50.5, has been commercialized for the production of biodiesel. Although biodiesel can substitute for diesel fuel, it has some shortcomings. Fatty acid methyl esters found in biodiesel are subject to microbial or oxidative attack, making them unsuitable in applications requiring long-term fuel storage. Low-temperature performance of biodiesel is sometimes problematic.

For these reasons, hydrogenation is being evaluated as a replacement for transesterification in the production of lipid-based biofuels [26]. Hydrogenation includes a number of reactions: large molecules are broken into smaller molecules; carbon–carbon double bonds are converted into more stable single bonds; molecular structures are rearranged; and undesirable atoms such as sulfur, nitrogen, and oxygen are removed from the hydrogenated compounds. In the case of lipids, hydrogenation yields alkanes, which are highly desirable fuel molecules.

Finally, lipid-rich microalgae might be grown in brackish water or even in seawater [27]. Algae can produce as much as 60% of their body weight as lipids when deprived of key nutrients such as silicon for diatoms or nitrogen for green algae. They employ relatively low substrate concentrations, on the order of 10–40 g/L. However, algae require the proper combination of brackish water, CO_2, and sunlight, which could limit the number of sites where this process would be profitable. Berkeley researchers have estimated that biofuels from algae would require break-even oil prices of $332 per bbl [28].

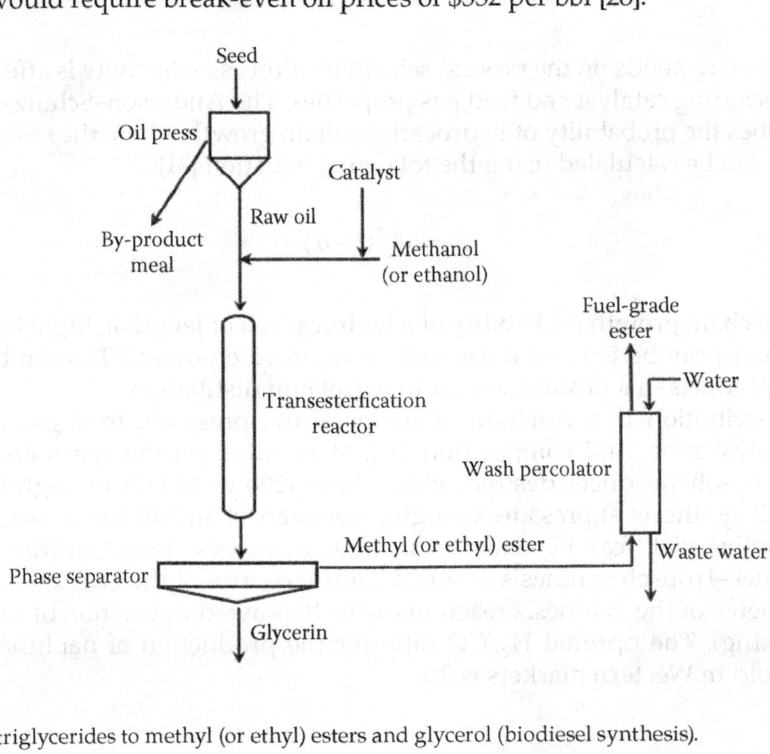

FIGURE 50.5
Conversion of triglycerides to methyl (or ethyl) esters and glycerol (biodiesel synthesis).

50.3.3 Mixed Alcohols

Thermochemical technologies can also produce alcohols as an alternative to biochemical conversion. Mixed alcohol synthesis dates back to before the development of Fischer–Tropsch synthesis. Both processes employ similar catalytic materials and operating conditions. Mixed alcohol synthesis involves the conversion of syngas over copper, zinc, and/or cobalt catalysts at temperatures of 300°C and 6.8 MPa pressure. The syngas H_2:CO ratio should be between 1.0 and 1.2. One of the main challenges for mixed alcohol biorefineries is the need to market their coproducts. Mixed alcohols contain mostly ethanol (>70 wt.%), but the longer-chained alcohols (propanol, butanol, and pentanol) need to be marketed separately. Market availability for these coproducts would help improve the economics of mixed alcohol synthesis (see work by NREL [29]).

50.3.4 Gasoline and Diesel

50.3.4.1 Fischer–Tropsch Synthesis

FTLs from biomass have antecedents in the coal-to-liquids industry. Germany extensively developed the Fischer–Tropsch process during World War II when it was denied access to petroleum-rich regions of the world. Likewise, when South Africa faced a world oil embargo during their era of apartheid, they employed Fischer–Tropsch technology to sustain its national economy. A comprehensive bibliography of Fischer–Tropsch literature can be found on the web [30].

Fischer–Tropsch catalysis produces a large variety of hydrocarbons including light hydrocarbon gases, paraffinic waxes, and alcohols according to the generalized reaction (50.2):

$$CO + 2H_2 \rightarrow (-CH_2-) + H_2O \tag{50.2}$$

FTL composition depends on the process selectivity. Process selectivity is affected by various factors including catalyst and feed gas properties. The Anderson–Schulz–Flory distribution describes the probability of hydrocarbon chain growth where the molar yield for a carbon chain can be calculated using the following equation [31]:

$$C_n = \alpha^{n-1}(1-\alpha) \tag{50.3}$$

where α is the chain growth probability of a hydrocarbon of length n. Light hydrocarbons (mostly methane) can be fed into a gas turbine to provide power. FTLs can be separated into various products in a process similar to petroleum distillation.

Product distribution is a function of temperature, pressure, feed gas composition (H_2/CO), catalyst type, and composition [32]. Depending on the types and quantities of Fischer–Tropsch products desired, either low- (200°C–240°C) or high-temperature (300°C–350°C) synthesis at pressures ranging between 10 and 40 bar is used. For example, high gasoline yield can be achieved using high process temperatures and an iron catalyst. Fischer–Tropsch synthesis requires careful control of the H_2/CO ratio to satisfy the stoichiometry of the synthesis reactions as well as avoid deposition of carbon on the catalysts (coking). The optimal H_2/CO ratio for the production of naphtha and diesel range fuels sold in Western markets is 2:1.

Swanson et al. developed an analysis of FTL fuels from biomass [33]. Their estimates of a 2000 Mg/day corn stover facility found capital costs of $498–$606 million with minimum fuel selling prices of $4.27 and $4.83 per gal of gasoline equivalent depending on whether the process was based on a fluid bed or entrained flow gasifier.

50.3.4.2 Bio-Oil Upgrading

Pyrolysis is the thermal decomposition of organic compounds in the absence of oxygen. The resulting product streams depend on the rate and the duration of heating. Liquid yields exceeding 70% are possible under conditions of fast pyrolysis, which is characterized by rapid heating rates (up to 1000°C/s), moderate reactor temperatures (450°C–600°C), short vapor residence times (<0.5 s), and rapid cooling at the end of the process. Rapid cooling is essential if high-molecular-weight liquids are to be condensed rather than further decomposed to low-molecular-weight gases.

Pyrolysis liquid, also known as bio-oil, is a low-viscosity, dark-brown fluid with up to 15%–20% water, which contrasts with the black, tarry liquid resulting from slow pyrolysis [34]. Fast pyrolysis liquid is a complicated mixture of organic compounds arising from thermal degradation of carbohydrate and lignin polymers in biomass [35]. The liquid is highly oxygenated, approximating the elemental composition of the feedstock, which makes it highly unstable. The HHV of pyrolysis liquids ranges between 17 and 20 MJ/kg with liquid densities of about 1280 kg/m^3. Assuming conversion of 72% of the biomass feedstock to liquid on a weight basis, yields of pyrolysis oil are about 135 gal/ton.

Production of pyrolysis oils and its coproducts involves several steps [36], which are illustrated in Figure 50.6. Lignocellulosic feedstock, such as wood or agricultural residues, is milled to a fine powder to promote rapid reaction. The particles are augured into the pyrolysis reactor where they are rapidly heated and converted into condensable vapors, liquid aerosols, noncondensable gases, and charcoal. These products are transported out of the reactor into a cyclone operating above the condensation point of pyrolysis vapors where the charcoal is removed. Vapors and gases are transported to a quench vessel where a spray of pyrolysis liquid cools vapors sufficiently for them to condense. The noncondensable gases, which include flammable carbon monoxide, hydrogen, and methane, are burned in air to provide heat for the pyrolysis reactor. The condensable liquids consist of a mixture of hundreds of organic compounds commonly known as bio-oil.

Bio-oil can be upgraded using conventional oil refinery processes. The most common of these processes are hydrotreating and hydrocracking. Hydrotreating can remove most of the bio-oil impurities such as nitrogen, alkali metals, and oxygen carried over from the original biomass. The purpose of hydrocracking is to break down heavy hydrocarbons with long carbon chain lengths into compounds within the naphtha and diesel range (7–20 carbon atoms). Bio-oil contains organic compounds with molecular weights of several hundred grams per mole. These compounds can be cracked into lighter hydrocarbons increasing the yield of naphtha and diesel range fuels. Hydrocracking consumes significant quantities of hydrogen, which can be obtained by reforming bio-oil or natural gas within the facility [37].

Gasoline and diesel from corn stover fast pyrolysis followed by bio-oil upgrading could cost between $2.00 and 3.10 per gal for a 2000 Mg/day biorefinery [36,38]. This facility would generate 35.4 million gal/year.

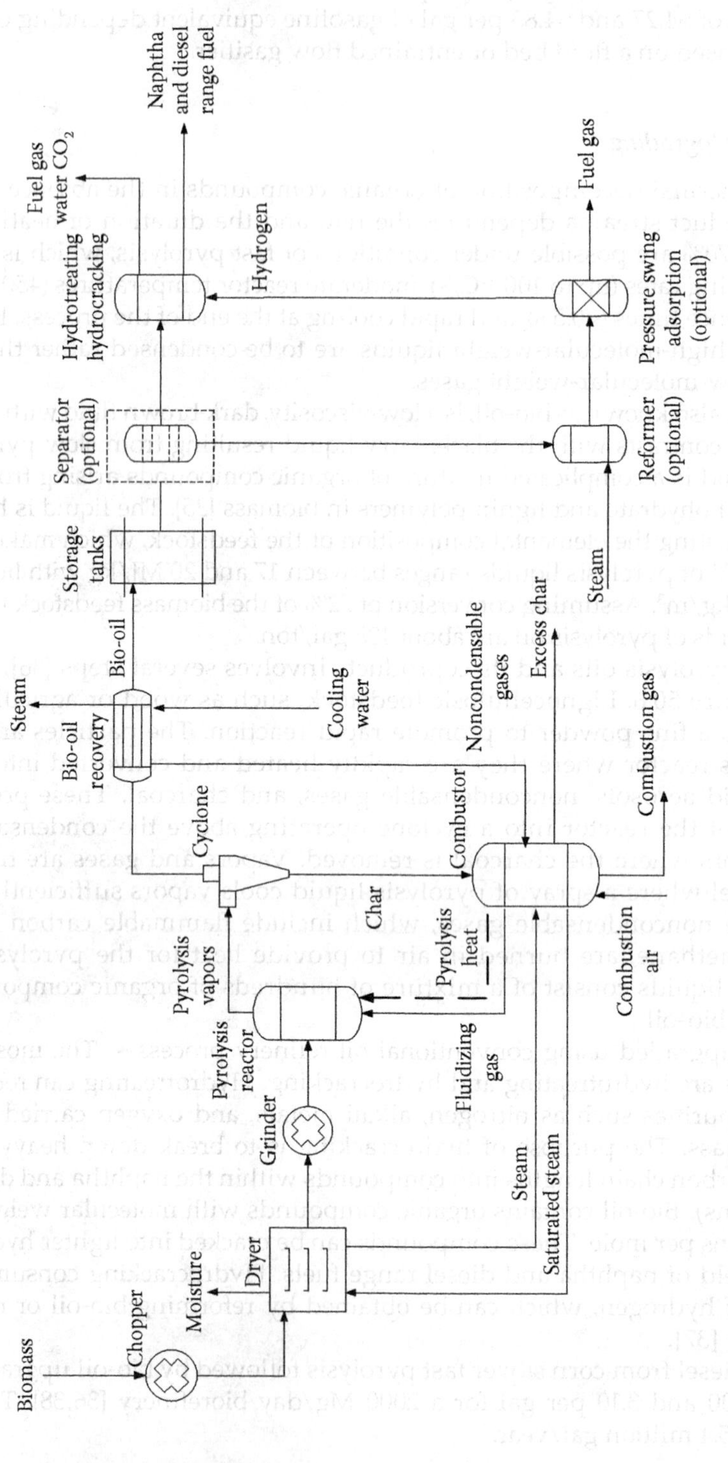

FIGURE 50.6
Biomass fast pyrolysis and bio-oil hydrotreating/hydrocracking to gasoline and diesel.

50.3.5 Hydrothermal Processing

Hydrothermal processing (HTP) is the use of water at near-critical-state conditions (374°C, 22.1 MPa) for thermochemical conversion. At temperatures below (subcritical) and above (supercritical) the critical point, water exhibits unique properties attractive for biomass depolymerization and product extraction [39]. HTP encompasses a wide range of operating temperatures (200°C–600°C) and pressures (5–40 MPa). At lower temperatures, the process is often called liquefaction, and its primary products are a mixture of organic liquids commonly known as bio-crude. At higher temperatures, the primary products are methane and hydrogen [40].

HTP presents several benefits compared to other thermochemical processes: feedstock flexibility, heat integration, and tunable chemical reactions. HTP reactors can convert a wide range of materials including high-moisture-content feedstock. However, alkali metals can affect the equipment lifetime. High water temperatures allow for heat integration opportunities. As long as the water retains its latent heat, it can be reheated with small energy input. Excess heat can be recovered upon cooling spent reaction media. Finally, fine-tuning of process temperature, pressure, and catalytic media can significantly improve product selectivity and yield.

Researchers at the Pacific Northwest National Laboratory have developed preliminary techno-economic analyses for HTP of various types of biomass [41]. However, there are limited industrial data to support detailed cost estimates of this technology.

50.4 Biofuel Properties

Biomass can be converted to a wide range of biofuels with significantly different combustion properties as shown in Table 50.1. Combustion properties determine the suitability of the biofuel to different engines and combustion equipment. In general, a combination of these properties determines the ultimate fuel performance, but a single factor may make the fuel unsuitable for a specific equipment.

Most of the global biofuel production has focused on ethanol and biodiesel. These fuels can be blended in small quantities with traditional fossil fuels. Conventional engines can operate with low levels of ethanol and biodiesel blends. In the United States, both E10 and E85 (ethanol:gasoline blends of 10% and 85% ethanol, respectively) have been marketed to consumers. The most common biodiesel blend in the United States is B20 with other blends ranging from 5% to 100% biodiesel. Both E85 and B20 typically require modern engines with modifications to adjust for the biofuel properties.

There is growing interest in the production of drop-in biofuels. Drop-in fuels refer to gasoline and diesel-type biofuels that can operate in existing gasoline engines without modifications. Companies have been able to produce drop-in biofuels that meet most of the specifications required by modern vehicles via several of the pathways discussed previously. The use of existing engines and infrastructure is one of the advantages of producing drop-in biofuels. On the other hand, hydrocarbons in first-generation drop-in biofuels have some minor differences compared to fossil fuels that are still being addressed. Researchers continue to study innovative ways of synthesizing fuels with improved combustion properties that could not only replace but also improve upon conventional gasoline and diesel.

References

1. Bergman, P. C. A. and J. H. A. Kiel (November 2005). Torrefaction for biomass upgrading. Technical Report ECN-RX-05-180, Petten, the Netherlands.
2. Medic, D., M. Darr, A. Shah, B. Potter, and J. Zimmerman (2012). Effects of torrefaction process parameters on biomass feedstock upgrading. *Fuel 91*(1), 147–154.
3. Nachenius, R., F. Ronsse, R. Venderbosch, and W. Prins (2012). Biomass pyrolysis. *Chemical Engineering for Renewables Conversion 42*, 75.
4. Klass, D. (1998). *Biomass for Renewable Energy, Fuels, and Chemicals*. Academic Press, San Diego, CA, p. 356.
5. Reed, T. (1981). *Biomass Gasification: Principles and Technology*. Noyes Data Corp., Park Ridge, NJ.
6. Vieira de Carvalho, A. J. (1982). Natural gas and other alternative fuels for transportation purposes. *Energy 10*(2), 187–215.
7. Satterfield, C. (1991). *Heterogeneous Catalysis in Industrial Practice*. McGraw-Hill Book Co., New York.
8. Sorenson, S. (2001). Dimethyl ether in diesel engines: Progress and perspectives. *Journal of Engineering for Gas Turbines and Power 123*, 652.
9. Probstein, R. F. and E. R. Hicks (May 2006). *Synthetic Fuels*. Dover Publications, Dover, NY.
10. Spath, P. L. (2003). Preliminary screening-technical and economic assessment of synthesis gas to fuels and chemicals with emphasis on the potential for biomass-derived syngas. Technical Report, National Renewable Energy Laboratory NREL/TP-510-34929, 1–160.
11. Munson, B. et al. (1994). *Fundamentals of Fluid Mechanics*, 2nd edn. John Wiley & Sons, New York, Table 1.6.
12. Perry, R. and W. Green (1984). *Perry's Chemical Engineers' Handbook*, 6th edn. McGraw-Hill, New York, Chapter 3.
13. Kajitani, S. et al. (1997). Engine performance and exhaust characteristics of direct-injection diesel engine operated with DME. Society of Automotive Engineers Inc., Warrendale, PA, p. 35.
14. Bailey, B. (1996). Performance of ethanol as a transportation fuel. In C. Wyman (ed.), *Handbook on Bioethanol*, Vol. 1. Taylor & Francis Group, London, U.K., pp. 37–58.
15. Kwiatkowski, J., A. McAloon, F. Taylor, and D. Johnston (2006). Modeling the process and costs of fuel ethanol production by the corn dry-grind process. *Industrial Crops & Products 23*(3), 288–296.
16. Rowell, R., R. Young, and J. Rowell (1997). *Paper and Composites from Agro-Based Resources*. CRC Press, Boca Raton, FL.
17. Schell, D., J. McMillian, and G. Philippidis (1992). Ethanol from lignocellulosic biomass. *Advances in Solar Energy 7*, 373–448.
18. Lynd, L. R. (1996). Overview and evaluation of fuel ethanol from cellulosic biomass: Technology, economics, the environment and policy. *Annual Reviews in Energy and the Environment 21*(1), 403–465.
19. Wayman, M. and S. Parekh (1990). *Biotechnology of Biomass Conversion: Fuels and Chemicals from Renewable Resources*. Open University Press, Philadelphia, PA.
20. Aden, A., M. Ruth, K. Ibsen, J. Jechura, K. Neeves, J. Sheehan, B. Wallace, L. Montague, and A. Slayton (2002). Lignocellulosic biomass to ethanol process design and economics utilizing co-current dilute acid prehydrolysis and enzymatic hydrolysis for corn stover. Technical Report NREL//TP-510-32438, National Renewable Energy Laboratory, Golden, CO.
21. Ezeji, T. C., N. Qureshi, and H. P. Blaschek (2007). Bioproduction of butanol from biomass: From genes to bioreactors. *Current Opinion in Biotechnology 18*(3), 220–227.
22. Pfromm, P. H., V. Amanor-Boadu, R. Nelson, P. Vadlani, and R. Madl (2010). Bio-butanol vs. bio-ethanol: A technical and economic assessment for corn and switchgrass fermented by yeast or clostridium acetobutylicum. *Biomass and Bioenergy 34*(4), 515–524.

23. Openshaw, K. (2000). A review of *Jatropha curcas*: An oil plant of unfulfilled promise. *Biomass and Bioenergy 19*(1), 1–15.
24. Glenn, E., J. O'Leary, M. Watson, T. Thompson, and R. Kuehl (1991). *Salicornia bigelovii* torr.: An oilseed halophyte for seawater irrigation. *Science 251*(4997), 1065–1067.
25. Kumar, A. and S. Sharma (2008). An evaluation of multipurpose oil seed crop for industrial uses (*Jatropha curcas* l.): A review. *Industrial Crops & Products 28.1*, 1–10.
26. Kram, J. W. (January 2009). Aviation alternatives. *Biodiesel Magazine*, available online at http://www.biodieselmagazine.com/articles/3071/aviation-alternatives.
27. Klass, D. (1998). *Biomass for Renewable Energy, Fuels, and Chemicals*. Academic Press, San Diego, CA, pp. 341–344, 462.
28. Lundquist, T., I. Woertz, N. Quinn, and J. Benemann (2010). *A Realistic Technology and Engineering Assessment of Algae Biofuel Production*. Energy Biosciences Institute, Berkeley, CA, p. 1.
29. Dutta, A., R. L. Bain, and M. J. Biddy (May 2010). Techno-economics of the production of mixed alcohols from lignocellulosic biomass via high-temperature gasification. *Environmental Progress & Sustainable Energy 29*(2), 163–174.
30. Anon. (2005). Fischer–Tropsch archive, www.fischer-tropsch.org (accessed March 2015).
31. Schulz, H. (1999). Short history and present trends of Fischer–Tropsch synthesis. *Applied Catalysis A, General 186*(1–2), 3–12.
32. Tijmensen, M., A. Faaij, C. Hamelinck, and M. van Hardeveld (2002). Exploration of the possibilities for production of Fischer–Tropsch liquids and power via biomass gasification. *Biomass and Bioenergy 23*(2), 129–152.
33. Swanson, R., A. Platon, J. Satrio, and R. Brown (2010). Techno-economic analysis of biomass-to-liquids production based on gasification. *Fuel 89*, S11–S19.
34. Czernik, S. and A. Bridgwater (2004). Overview of applications of biomass fast pyrolysis oil. *Energy and Fuels 18*(2), 590–598.
35. Piskorz, J. and D. Scott (1987). The composition of oils obtained by the fast pyrolysis of different woods. Preprint Paper, *American Chemical Society, Division of Fuel Chemistry 32*(2).
36. Wright, M. M., D. E. Daugaard, J. A. Satrio, and R. C. Brown (2010). Techno-economic analysis of biomass fast pyrolysis to transportation fuels. *Fuel 89*(Suppl. 1), S2–S10 (Techno-economic comparison of biomass-to-biofuels pathways).
37. Wright, M. M., Y. Roman-Leshkov, and G. H. William. (2012). Investigating the techno-economic tradeoffs of hydrogen source using a response surface model of drop-in biofuel production via biooil upgrading. *Biofuels, Bioproducts and Biorefining 6*(5), 503–520.
38. Brown, T. R., R. Thilakaratne, R. C. Brown, and G. Hu (2013). Techno-economic analysis of biomass to transportation fuels and electricity via fast pyrolysis and hydroprocessing. *Fuel 106*, 463–469.
39. Ye, P., L. Cheng, H. Ma, B. Bujanovic, M. Goundalkar, T. Amidon, H. Xie, and N. Gathergood (2011). *Biorefinery with Water*. John Wiley & Sons, Hoboken, NJ.
40. Peterson, A., F. Vogel, R. Lachance, M. Froling, M. Antal Jr., and J. Tester (2008). Thermochemical biofuel production in hydrothermal media: A review of sub-and supercritical water technologies. *Energy & Environmental Science 1*(1), 32–65.
41. Elliott, D. C., G. G. Neuenschwander, T. R. Hart, L. Rotness, A. H. Zacher, D. Santosa, C. Valkenburg, S. Jones, and S. Tjokro Rahardjo (2009). *Catalytic Hydrothermal Gasification of Lignin-Rich Biorefinery Residues and Algae*, pp. 1–25. Pacific Northwest National Laboratory, Richland, WA.

51

Geothermal Power Generation

Kevin Kitz

CONTENTS

51.1 Introduction

Roman mythology holds that humans obtained fire from the gods on Mount Olympus to meet their needs for light and heat. In much of today's industrialized world, the use of electricity has edged out fire for these age-old needs. However, the convenient use of electricity is now faced with increasing cost and concerns about the availability and environmental consequences of the fuels that are needed to produce it. Can those Roman gods again come to mankind's aid?

Deep below Mount Olympus, Vulcan toils over his forge. The forge glows red as Vulcan hammers out weapons for the gods, including Jupiter's own thunderbolts. Vulcan's forge is stoked by geothermal energy that humans, too, have used for the production of electricity since 1904, starting in the very homeland of the Romans.

Geothermal power production in the United States approached 18,000 GW h in 2005,[1] or an average of more than 2000 MW over the year. It does so on an around-the-clock basis, providing baseload power to several western utilities in California, Nevada, Hawaii,

Utah and starting in 2006, Alaska. The power generated has very low emissions and is immune to price fluctuations in the fossil fuel markets. Similar benefits are derived around the world from plants operating in 24 countries worldwide, including Italy, the Philippines, Indonesia, Kenya, New Zealand, and Iceland. In Iceland, geothermal power not only provides 16.6% of the electric power generated every year, but 54% of the total primary energy use in the country, including the 87% of households that are heated by geothermal water.[2] Worldwide direct use of geothermal energy is documented in 76 countries.[3]

This chapter examines geothermal power technologies and the issues involved in further utilization of geothermal energy for the production of electric power.

51.2 Definition and Use of Geothermal Energy

Geothermal energy is present on Earth from two sources:

1. As heat that flows upward and outward across the entire surface of the world from the very deep (mantle and core) radioactive decay of uranium, thorium, and potassium. In most regions of the world, this energy flux is too small to be commercially useful for any purpose.
2. As localized heat resulting from the movement of magma into the earth's crust. In some areas of the world, most frequently along the so-called "Ring of Fire," this localized heat, with high heat flux and high temperatures, can be found between the surface and 10,000 ft below ground. Where such heat flux meets the requisite conditions, geothermal energy can be developed for multiple and varied purposes. Where temperatures are sufficiently high, geothermal energy may be used for electric power generation. This form of geothermal energy is the subject of this chapter.

Geothermal energy was first used for experimental power generation in Larderello, Italy, on July 4, 1904, by Prince Piero Ginori Conti. However, commercial development followed slowly thereafter. The Larderello site also saw the first commercial geothermal power plant (250 kW) in 1913, as well as the first large-power installation in 1938 (69 MW). It would be 20 years before the next large geothermal power installation would be built: halfway around the world in Wairakei, New Zealand, with the first unit—commissioned in 1958—that, under a steady development program, grew to 193 MW of installed capacity by 1963. In the United States, the installation of the first unit (11 MW) of what was to become the largest geothermal power complex in the world, The Geysers, in Sonoma County, California, occurred in 1960. Over the next quarter-century, a total of 31 turbine generator sets were installed at The Geysers, with a nameplate capacity of 1890 MW. Plant retirements and declining steam supply have since reduced generation at The Geysers to an annual average generation of 1020 MW from 1421 MW of installed capacity—still the largest geothermal field in the world.[1]

Since those early efforts, Lund et al. report that a total of 2564 MW of geothermal power generation capacity is currently installed in the United States, generating approximately 2000 MW of annual average generation. In 2005, worldwide annual geothermal power was estimated at 56,875 GW h from 8,932 MW of installed capacity.[3] These values put the United States and worldwide geothermal power plant capacity factors* at 78% and 73%, respectively.

Geothermal energy is also utilized in direct-heat uses for space heating, recreation and bathing, and industrial and agricultural uses. Geothermal energy in direct-use application is estimated to have an installed capacity of 12,100 MW thermal, with an annual average energy use of 48,511 GW h/year energy equivalent.[3] This excludes ground-coupled heat pumps (GCHPs, see below). In the United States, the first geothermal heating district was installed in Boise, Idaho, in 1892, and is still in operation today.

GCHPs are also often referred to as *geothermal*. GCHP units are reported by Lund to have 15,721 MW of installed capacity and 24,111 GW h of annual energy, representing 56.5% and 33.2% of worldwide direct use, respectively.[3] It should be noted that although these numbers are reported as "geothermal," in many instances the ground temperature is primarily controlled not by the flow of heat from below the surface, but by the annual average ambient temperature of the location, which is a solar phenomenon. From the numbers above, it can be seen that GCHPs provide an additional worldwide energy production of almost 50% more than that provided by geothermal electric power generation. For more information on GCHP usage, see Chapter 27.

This chapter will cover the technologies and issues in the utilization of geothermal energy for electric power generation, with a particular focus on the issues facing the development of new capacity, both technical and economic. Although the technologies of geothermal energy for power and for direct use are quite different, the issues covered in this chapter related to reservoir issues, and the economic factors affecting or controlling the development and maintenance of the reservoirs are often the same or similar.

51.3 Requirements for Commercial Geothermal Power Production

For new geothermal power to be installed, or an existing geothermal power plant to continue operating, commercial geothermal power production requires two major elements: a geothermal resource and economic access.

51.3.1 Definition of a Geothermal Resource (Heat, Permeability, Water)

A geothermal resource for power production comprises three major elements: heat, sufficient reservoir permeability, and water.

1. Heat is clearly the first element for commercial power generation. Most commercial geothermal resources produce fluids from reservoirs with a resource temperature of at least 320°F. However, there are examples of geothermal fields with lower temperature fluids. Nonetheless, resource temperature is a good first indicator of economic viability, see Table 51.1. As can be calculated from the table, the flow requirements for a 60 MW plant are 18 million lb/h of 300°F geothermal liquid, but only 4 million lb/h of geothermal liquid if the reservoir temperature is 450°F.

* Capacity factor is defined as the actual generation produced compared to the theoretical generation that would be produced in 1 year with the power plant running at full rated capacity.

TABLE 51.1

Power Generation Potential from a Range of Resource Temperatures

1000 lbm Geothermal Fluid of	Electrical Generation (kW h)	Power Plant Type[a] (see Later Sections)
Liquid at 300°F	3.3	Binary
Liquid at 350°F	5.6	Single-flash steam
Liquid at 400°F	10.4	Double-flash steam
Liquid at 450°F	14.5	Double-flash steam
Steam at 350°F	53.5	Dry steam

Source: From DiPippo, R., *Geothermal Power Plants: Principles, Applications and Case Studies,* Elsevier Advanced Technology, Oxford, U.K., 2005, p. 424.

[a] The column labeled "power plant type" is the technology used as a basis for the power generation calculation. All plant types are used over a larger range of temperatures than that indicated in the table.

2. *Permeability* describes the ability of the reservoir fluid (water or steam) to flow through the rock formation. Permeability allows deep-seated geothermal heat sources to create a geothermal resource through the convection (flow under the influence of heating and cooling) of hot water or steam. The convection of the geothermal fluids through the reservoir heats a large volume of rock, thereby storing a large quantity of energy over a period of tens of thousands of years. Geothermal reservoir permeability is dynamic, with the hotter fluids dissolving minerals and increasing permeability, and the cooling fluids depositing minerals and restricting permeability. Reservoir permeability is what allows the extraction of that stored heat through a relatively few number of wells in commercial geothermal developments.

   Matrix permeability is the ability of the fluid to flow through the bulk rock itself. Fresh water wells, natural gas wells, and oil wells are frequently observed to obtain a significant portion of their total flow from matrix permeability. In other words, the fluids enter the wellbore as uniform flow along an area of unfractured rock. Sufficient matrix permeability is rare in geothermal wells that will support commercially significant flow rates.

   Fracture permeability is low most geothermal wells which obtain the majority of total wellbore inflow through naturally occurring fractures in the rock. Pervasive fracturing allows a single well to obtain flow contributions from a large area at high flow rates, even where the matrix permeability is low. In most cases, the fractures extend over large areas and volumes as a result of tectonically active areas, but some fields have been discovered and developed that essentially comprise a single fault system (which may be comprised of multiple fractures in the rock). In such fields, only wellbores that cross the fault produce geothermal fluid at commercial rates.

3. Water is the motive fluid for geothermal power production. It may be brought to the surface as steam from one of the few (but typically large) geothermal steam fields, such as Larderello (Italy), The Geysers (California), and Kamojang (Indonesia). A water–steam combination may also be produced to the surface from high-temperature liquid-dominated or liquid-and-steam reservoirs. In low- or moderate-temperature resources, pressurized water may be pumped to the surface using downhole pumps.

The development of new geothermal power plants and the expansion and maintenance of existing geothermal fields depends on the three elements of heat, permeability, and water.

The three occur simultaneously in relatively few places around the world, but there remain many thousands of megawatts of undeveloped worldwide and U.S. resources. Over the long term, the natural resource base for geothermal power could be supplemented with human intervention to create new systems or enhance existing systems.

51.3.2 Improving a Geothermal Resource through Human Intervention

The terms *hot dry rock* (HDR) or *hot fractured rock* (HFR) refer to a family of experimental technologies that are not yet commercially proven. The objective is to establish one or more of the missing elements of a commercial geothermal resource (specifically, permeability or water) where a heat source already exists. HDR experiments have been undertaken as research projects by the U.S. Department of Energy and similar agencies in Europe and Japan. The problems are daunting and the costs are high; U.S. DOE funding, for example, has diminished substantially from its maximum in the 1980s.

The concept behind HDR and HFR is to drill a well into the hot rock and then pump water into the well at very high pressure, causing the rock to fracture. The fractures provide a heat transfer surface and flow path allowing water to be pumped from the surface into an injection well, circulated through the man-made fractures, and ultimately recovered in a production well some distance away from the injection well. The theory is straightforward; unfortunately, the implementation to date has not been.

As of 2005, the first privately-funded HFR development had commenced at the Cooper Basin in southern Australia, and it will be watched closely for its success. The developer's plans call for an initial installation of a 3 MW power plant if the fracturing process is successful.[5]

If HDR and HFR technology is developed and implemented in significant quantities over the next 20 years, the geothermal reservoir management strategies and the energy-to-power conversion technologies will be much the same as that described in the rest of this chapter.

Between the naturally-occurring resource base and the potential man-made resource base of HDR/HFR is the enhanced geothermal system (EGS). EGS technologies seek to supplement a naturally occurring geothermal resource primarily by the addition of more liquid, or by stimulation of wells to tie into a larger, naturally occurring fracture network. The goal of EGS is to extend the life or capacity of existing fields, rather than the creation of an entirely new resource, as is being attempted at the Cooper Basin described above. There have been many studies, evaluations, and proposals to date, and the use of EGS at The Geysers project—including those plants owned and operated by Calpine* and NCPA†—has been a proven success.

At The Geysers, two pipelines bring 77,000 m^3/day (20 million gal/day) of treated wastewater from adjoining cities to the mountains where The Geysers facilities are located. The water serves to replace that which has been depleted from The Geysers over its approximately 50 year operating history. Stark et al. report that tracer tests show that 40% of the injected water is recovered as steam within a year. The injection of the water from just one of the two pipelines (delivering about 55% of the total wastewater) is forecast to result in an annual average generation increase of 84 MW. With a parasitic pump load of 9 MW, the annual average benefit is roughly 75 MW.[6]

* Calpine Corporation, http://www.calpine.com.

† Northern California Power Agency, http://www.ncpa.com.

51.3.3 Geothermal Energy as a "Renewable" Resource

Throughout this chapter, declines in flow rates from geothermal wells and output from geothermal power plants are discussed. What does this mean in terms of whether geothermal energy is renewable?

> In recent years many thoughtful papers have been published on the renewability and sustainability of geothermal energy... However, no universally accepted definitions of the words "renewabilty" and "sustainability" seem to exist and definitions used often have ambiguities...[7]

From a thermodynamic point of view, it may seem that the only true renewable geothermal development would be one in which the extraction rate is the same as the natural heat influx rate into the system. However, reservoir simulations and field observations frequently reveal that the natural heat influx rate increases as production occurs, due to pressure changes that allow more hot liquid to flow into the system. Therefore, operating at a "nonrenewable" level increases the ultimate energy extraction from the resource. More importantly, a true "renewable" level of energy extraction would very often be sub-economic, and is therefore of little interest in the development of geothermal resources for society's benefit.

Another definition argues that, as long as the power or heat usage from the resource continues at a constant level for hundreds of years, it approximates a true renewable resource, although it may be termed a "sustainable resource." Again, this is an interesting theoretical discussion, but not one that actually is put into practice in the development of most geothermal resources, for both the reasons of economics and an inability to know what this actual level would be.

Sanyal (2004) argues that sustainable geothermal development occurs if the project maintains its output, including make-up well drilling, over the amortization period of the power plant.[7] This definition recognized the critical role that economics plays in actual natural resource development decisions. In short, projects are developed to meet economic requirements. This is true of any new power project, whether geothermal, another renewable source, or a fossil-fired resource. Section 51.3.4, will illustrate why the question of sustainable development for a 50-year period, let alone a 300-year period, will not play a role in the development decisions for a particular geothermal resource. The reason, in summary, is that even at a mere 50 years in the future, there are no meaningful economic consequences to development decisions made today when using a discounted cash flow analysis.

More valuable than the theoretical discussion of whether geothermal energy is renewable, sustainable, both, or neither, is to look at the history of geothermal development. Geothermal use at Larderello, Italy, is over 200 years old, starting with mineral extraction in the early 1800s and including almost 100 years of commercial power generation. During the last 100 years of power generation, turbines have been renewed or replaced, power output has grown and shrunk and grown again, and new methods of extracting more energy have been developed. In fact, geothermal power generation in 2003 was higher than any other year, with 5.3 billion kW h produced.[8]

Many further decades of geothermal power generation are expected from the Larderello region, as well as Wairakei, The Geysers, and other fields that are approaching 50 years of power generation. Geothermal power-generating facilities only very rarely cease operation, although few will reach their 50th anniversary at their maximum generation level.

51.3.4 Economic Access

Given that a geothermal resource exists—whether naturally occurring or developed with EGS or HDR technologies—the resource will be economically viable for power generation only if the four elements of economic access are met: wellhead energy cost, electricity transmission, a market for the electric power, and the power plant cost.

51.3.4.1 Economic Access: Wellhead Energy Cost

Wellhead energy cost is the total cost to bring useable geothermal energy to the surface and to return the cooled geothermal fluids to the reservoir. The thermodynamic condition (pressure, temperature, and flow rate) of the geothermal energy at the surface will influence the capital cost of the power plant. The wellhead energy cost is the result of the combined effects of the productivity of the well, the cost to drill the well, and the enthalpy (or available energy) of the fluid that is produced from the well, among other aspects. These factors vary widely from field to field, and even within a single field.

- Well productivity is the ability of the well to bring fluid to the surface at a temperature and pressure useful for power generation. Clearly, the more fluid that is produced by each well, the fewer wells are needed for a given power plant size, and therefore the lower the total cost of the geothermal well-field development. Less obvious is the fact that highly-productive wells can also have a major impact on the cost and design of the power plant itself.
- Highly-productive wells allow the power plant inlet pressure to be increased (and thereby increase the available energy) on plants that directly use the geothermal steam in a turbine. An increase in the power plant inlet pressure can allow the size of the plant to be increased, because last-stage turbine blades can only be built to a limited maximum size in geothermal service and in binary power plant turbines as well (see Sections 51.5 and 51.6). Alternatively, for plants contractually limited to a particular megawatt capacity, e.g., 60 MW, higher inlet pressures reduce the amount of fluid that must be extracted from the reservoir and delivered to the power plant, thereby reducing the size of the heat rejection system and the injection capacity (two major components of geothermal power costs, as will be discussed in later sections).
- In binary power plants (see Section 51.6) with pumped wells, highly productive wells not only reduce the direct drilling costs, but also reduce the number of pumps that must be installed (at an installed 2005 cost of roughly $500,000 each). Even more importantly, the pumps can be set at a shallower depth, thereby reducing the pump parasitic loads. The energy for the production pump parasitic loads (400–1000 hp per pump) is supplied by the power plant itself. Again, the productivity of the wells will either reduce the size of the plant that must be built to meet contractual and pump parasitic obligations, or can increase the amount of power available for sale from a particular power plant. The combined effect of fewer wells and more power sales dramatically increase the economic viability of a new or expansion geothermal binary power plant development. Well injectivity is a similar issue for the disposal of the residual geothermal liquids for both binary and steam power plants.
- The energy production rate (a combination of mass flow, temperature, and steam content) of production wells vary from field to field, often varying substantially

within a given field itself, and almost always changes over time for a particular well. In steam power plants, high-enthalpy wells can result either from a high-temperature portion of the field, or as a result of what is termed "excess steam" production. Excess steam occurs when the enthalpy of the produced fluid is higher than the enthalpy of the reservoir fluid as a whole due to the greater mobility of the steam through the reservoir rock than liquid water. In pumped brine binary power plants, high-enthalpy wells are simply wells with higher flowing temperatures. For both steam and binary power plants, higher enthalpy wells result in lower power plant cost, as the benefits of the higher enthalpy are found in lower drilling costs (fewer wells), lower heat-rejection system costs (higher available energy per unit mass into the plant), and lower brine-injection costs (less produced fluid).

- The depth of geothermal resources is highly variable, with most geothermal resources produced between depths of 1500–7000 ft below surface, although resources commonly exist outside of this range. One example of a shallow resource is the Salt Wells field in Nevada, in which the production and injection wells are between 450 and 750 ft deep. One particular well at that field is reported to be only 470 ft deep and to be capable of between 4 and 5 MW of power from 140°C (284°F) fluid. The drilling time for this well was only 12 days.[9] The cost of the well is on the order of hundreds of thousands of dollars. By contrast, make-up wells drilled at one field in the Philippines almost 25 years after the start of production had a true vertical depth of over 10,000 ft, and an even greater measured depth due to the lateral reach of the well. The cost of these wells between 2001 and 2003 was on the order of \$3 million to \$5 million each. At the Puna geothermal project in Hawaii, drilling is difficult and the location remote, so the daily drilling cost is high. Wells drilled in 2005 cost approximately \$6 million, and were drilled to a depth of about 6000 ft.

There is no threshold value for the wellhead energy cost below which geothermal development is viable and above which it is not. Rather, there is a continuum in which the effects of transmission, market, and the power plant cost affect to varying degrees the ability to construct a new geothermal plant.

For existing power plants, the wellhead energy cost determines whether or not a make-up well will be economical, and ultimately the end of make-up well drilling. After make-up well drilling ends, geothermal power plants enter a period of slow decline in the output of the power plant over the coming years and decades. Although the plant is no longer able to achieve its initial rated capacity, it will continue to operate economically, literally for decades. There are very few examples of geothermal resources that, once in operation, shut down completely.

51.3.4.2 Economic Access: Electricity Transmission

With a suitable geothermal resource and a viable market for geothermal power, the link between the two is transmission. As with other renewable and nonrenewable energy sources, transmission can be a significant issue in the viability of development of a new geothermal resource, or in the expansion of an existing geothermal power facility.

Occasionally, geothermal resources are found in close proximity to existing transmission access. Examples include Steamboat Springs, Nevada and Raft River, Idaho. For these developments, the cost of transmission access is low, and transmission line construction and interconnection does not play a major role in the development of the resource.

However, geothermal resources often occur at substantial distances from existing transmission, and entail the need for considerable capital expenditures on the transmission system to deliver the power to the customer. If the field is large enough, the long transmission lines and expenses can be justified. Two separate subsea cables were built in the Philippines to transport hundreds of megawatts of electricity generated from geothermal resources from islands with small native loads to the main load center on the island of Luzon.

The cost of the transmission line has most often been the responsibility of the utility that was receiving the power (e.g., Pacific Gas and Electric for the power generated at The Geysers), whereas in more recent years, the development of the power line has been the responsibility of the geothermal developer in the United States. The power price offered to geothermal and other renewable developers may fail to recognize this cost (e.g., the Idaho PURPA posted rate), yet the cost must still be borne by the developer, and as such, the cost of transmission construction can become a substantial burden to the project economics. A 107-mile-long dedicated transmission line was built and paid for by a consortium of geothermal power developers in the Imperial Valley, California, to deliver some 500 MW of power to a common customer, Southern California Edison. After the line was built, it was turned over to the local utility, the Imperial Irrigation District, for ongoing operation and maintenance.

When a transmission system is used to deliver power and the transmission lines are owned by an entity other than the ultimate purchaser of the power, the use of that transmission system incurs wheeling costs. Wheeling costs comprise a monthly reservation/use fee and a power loss that can be paid either with funds or with power delivered to the transmission entity. Typically, wheeling charges are not charged by the utility that actually is receiving the power under contract. "Pancaked charges" refers to the circumstances in which power crosses multiple transmission segments owned by different entities, incurring separate transmission charges through each segment. The losses assessed by each utility usually are not based on actual losses for the power being transmitted, but represent a system-wide average apportioned out to the users of the system on a pro-rata basis. In the United States, wheeling protocols are established by the FERC (Federal Energy Regulatory Commission).

Common wheeling charges are 1%–5% for losses, plus \$1000–\$3000 per MW-month for the reservation costs for firm capacity. Therefore, wheeling through more than one, and often even only one, utility becomes prohibitive. As a result, the viability of development is strongly influenced by the willingness of the utility in whose control area the power plant is located to accept the power. Delivery to the closest utility generally eliminates wheeling fees and losses and improves the economic viability of the plant. On the other hand, having only a single party with whom to negotiate a contract puts the developer in a difficult position.

In deregulated markets, such as the United States, transmission reservations are the mechanism by which a power plant ensures access to the market for the life of the contract. The ability to obtain that reservation is critical to development, or else the geothermal development may face greatly increased transmission costs for transmission upgrades or longer transmission lines. One of the advantages of geothermal power as opposed to wind power is the efficiency of transmission usage. Both resources must reserve transmission capacity for their peak delivery, e.g., 100 MW. Wind will typically deliver between 25 and 40 MW on an annual average basis, while geothermal will typically deliver about 90–95 MW. On the transmission system peak hours, the difference in utilization will be even greater, due to the intermittency of the wind resource and the typical incentives in a geothermal contract to be on-line during peak hours. A geothermal power plant can be expected to be on-line and at near-full capacity for over 95% of the peak hours.

Where transmission system capacity is limited, geothermal and other baseload resources will make much greater use of that capacity. Comparisons of new generation costs seldom account for the efficiency and hidden costs of transmission usage, which is unfortunate in the evaluation of the economics of geothermal power generation, because it is so efficient in its use of transmission capacity.

51.3.4.3 Economic Access: Viable Market

Power generation, regardless of the technology, is a highly specialized form of manufacturing. As with all manufactured products, there must be a place to sell the manufactured product. Therefore, for new geothermal power to come online, regardless of how low the wellhead energy cost and even if transmission is readily available and affordable, without a market, there is no development. Market consists of three elements: willing buyer, price vs. characteristics, and contract provisions. The first two are much the same for all power developments, whether renewable or otherwise. For geothermal power, there are unique contract provisions that must be present for realistic expansion of the role of geothermal power in the electrical demand of the United States.

51.3.4.3.1 Willing Buyer

The willing buyer must be just that: "willing" and a "buyer." In the United States, this rare beast has been found in a number of geothermal development habitats over the last 40 years. More recently, the captive breeding program seems to be showing success. A much larger number of willing buyers have been observed, as many U.S. electric utilities have been stung first by wild swings in natural gas prices, second by 2004–2005 gas prices climbing from $4 or less to over $10 per million Btu, with long-term price forecasts for gas remaining high, and lastly by state-mandated renewable energy targets.

From 1960 through roughly the mid-1980s, geothermal development in the United States (and in many places around the world) was most commonly undertaken by two parties: the resource developer and the local utility. The resource developer drilled wells and built cross-country pipelines to deliver a flow of steam or mixed brine and steam to the power plant boundary. The resource developer also was responsible for injection of the residual liquids, both brine and condensed steam. The local electric utility—i.e., the willing buyer—used the steam to generate power. Under this U.S. development model, the utility met the mandate of its monopoly franchise to supply power. The utility also met its investor's objective of investing in new facilities, the power plant and transmission lines, for which it was able, under its monopoly, to earn a regulated rate of return on the invested capital dollars. However, an inherent inefficiency was created in this development model by having two owners who each had different economic drivers, separate operating groups that worked side by side, double administration costs, and other factors. One common and unfortunate outcome of the two-owner model occurred because the steam was sold on a $/MW h basis instead of on the energy content of the steam. The consequence was that the power plant owner built inefficient power plants because inefficiency is cheaper than efficiency, if one only looks at the power plant cost. Of course, the consequence was that the steam developer had to drill additional risky and expensive wells to supply additional steam for the inefficient power plant. Those few contracts that had steam sales based on energy content saw exactly the opposite effect in power plant design. The power plants tended to be very efficient to minimize steam use (fuel cost). The net result of two-owner operation was that the price of geothermal power was comparatively high and resource utilization was inefficient.

With the widespread occurrence of independent power producers (IPP) in the United States in the mid-1980s, a regulated utility could theoretically obtain a lower price through the integrated operation of geothermal field and power plant and could meet its mandate for power supply, but by buying power from the IPP, the utility could not meet the objectives of its shareholders for new investment. The IPP developer's objective of building a power plant put it in direct competition with the utility for the return on investment in new power plant capacity. Under the IPP model, each megawatt of capacity installed by an IPP directly equates to less investment by the utility in baseload power (potentially $700,000–$1,700,000 per MW). Consequently, because the utility faces a reduced rate base (the capital on which the utility is allowed to earn a return), the IPP faces an intrinsic unwillingness in the utility to procure the output of a new geothermal (or any other type of fuel) plant. Although geothermal development should, in theory, thrive under the IPP model, the reality is that utilities in the United States have shown little interest in bringing on new geothermal development unless forced to do so in the interest of the ratepayer by governmental authority.

From the mid-1980s through the mid-2000s, many geothermal power plants were built exclusively because of the Federal PURPA* regulations and the consequent requirement of state public utility commissions for the utility to buy the power at a published avoided cost. For example, California's Standard Offer Number Four resulted in the construction of several of the Salton Sea and Heber Units, among others, in the late 1980s and early 1990s. Idaho's Posted Rate 10 MW PURPA contracts resulted in the development of the Raft River Geothermal Power Plants in southern Idaho, with anticipated operation dates of 2007 through 2009. In California and Nevada, the passage of Renewable Portfolio Standards has also spurred utilities to contract for geothermal power to comply with state law. Both PURPA and the RPS have created utility buyers, but they have not necessarily been "willing." The state RPSs and the prospect of a federal RPS is now emerging as a major factor in the decision of utilities to sign contracts for IPP-developed geothermal (and other renewable) power plants, in spite of the fact that it is against the interests of their shareholders, though not their ratepayers.

51.3.4.3.2 Price of Delivered Power

The price vs. the characteristics for the electrical energy is the most obvious component of a viable market. The characteristics of geothermal power are very attractive and rare in the renewable power industry. It is price-stable, has low to negligible environmental impact, and has a steady, predictable, and reliable output.

The unit price paid for geothermal-generated electrical power must cover the expense of the power plant and the wellfield development, as well as the often long and expensive exploration phase. In addition, the price paid per MW h must satisfy annual cash flow requirements and the return on investment sought by the geothermal developer. As most contracts from the late 1990s onward are not in the public domain, it is hard to know what constitutes an adequate price for electricity generated from geothermal energy. A few that have emerged are described below.

In 2003, the power purchase prices awarded to three geothermal power plants after a round of renewables bidding in Nevada were inadvertently released. The contracts were for both the power output and environmental attributes ("green tags," carbon credits, etc.).

* PURPA is the Public Utility Regulatory Policies Act.

Two were for expansions to existing power plants. One price was less than $45 per MW h, and the other was more than $52 per MW h. The third contract was for a greenfield power plant at under $50 per MW h. All were for resources with temperatures greater than 320°F. Only the power plant that bid the highest price was ever built. The other plants proposed power purchase prices too low to actually accomplish development drilling, build the power plant, and close financing. The prices bid for these plants were offered before the 2004–2005 worldwide run-up in steel prices that had a tremendous negative impact on the cost of building a geothermal power plant with its miles of steel well casing, miles of steel surface piping, and enormous steel heat exchangers.

In January 2005, the Idaho Public Utility Commission approved a 20-year PURPA contract to supply power from the U.S. Geothermal, Inc.* Raft River Geothermal Power Plant for an initial price of $52.70, escalating at 2.3% per year.[10] This gives an equivalent levelized price of $60.99. Although Raft River is a greenfield development, the production and injection wells were drilled in the late 1970s, with those costs written off by previous owners.

In May 2005, two 10 MW contracts were announced at a fixed price of $57.50 per MW h with an annual escalation of 1.5%. These contracts were for expansions to existing power plants at two different fields in southern California (Heber and East Mesa). The price includes the value for the energy, as well as the environmental attributes. The contract also allowed for the sharing of U.S. federal production tax credits (PTCs) with the power purchasers.[11] The levelized price, excluding the PTCs, would be about $63. Speaking at the groundbreaking of an additional 8 MW expansion of the Heber facility, Robert Foster, the President of Southern California Edison, said, "I like... the geothermal plant... because we know that these plants produce reliable power 24 × 7. Unlike other forms of renewable energy, I know that this energy will always be there and we don't need to have shadow generation to support geothermal power. Even though we like to have dispatchable† power, we can forego dispatchability for the reliability of geothermal power."[11]

An international benchmark is the Sarulla, Indonesia, development, in which the state oil company drilled the wells for the power plant and the power plant developer was responsible for the power plant and pipeline system (excluding the cost of transmission and drilling.) The power price was bid at approximately $45 per MW h in 2005 by two companies. Price escalations and adjustment factors are not available at the time of this writing.

An adequate selling price for the generated electrical power in a geothermal development is determined in part by resource temperature (higher temperature means lower power plant price), wellhead energy costs, power plant size (smaller requires higher price) and whether the new power plant is greenfield or an expansion (expansion can take advantage of existing infrastructure and personnel costs). Within the geothermal community, the general rule of thumb is that for geothermal resources above 300°F, a true grassroots geothermal power plant must sell power for $60–$65 per MW h for at least 20 years, for an all-in project (wells, pipelines, power plant, and transmission). Expansions to existing facilities can reduce the cost by 10%–15%,[12] or roughly a $5–$15 per MW h reduction, with most of the savings being in a lower drilling failure rate, use of the existing transmission line, and sharing of operating personnel and spare well capacity. Figure 51.1 provides representative cost elements in a geothermal power development.

* U.S. Geothermal, Inc., http://www.usgeothermal.com.

† A dispatchable plant can be ordered to vary its load from zero to full load to meet the needs of the electric power grid.

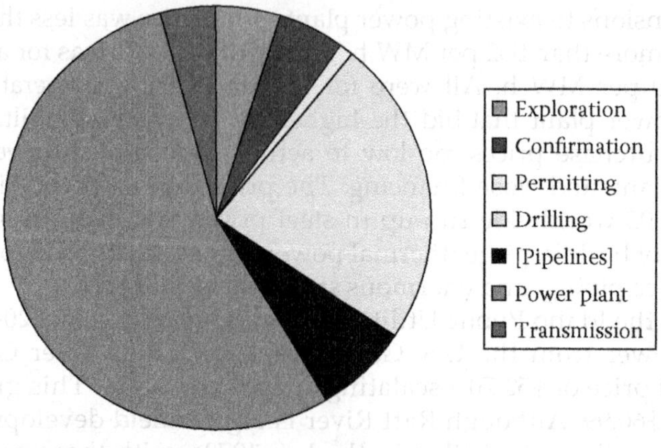

FIGURE 51.1
Typical cost breakdown of geothermal power projects. (From Hance, C.N., Factors affecting the costs of geothermal power development, Geothermal Energy Association and the U.S. Department of Energy, Washington, DC, 2005.)

51.3.4.3.3 Contract Provisions for Operation

Contract provisions for a viable geothermal power market are twofold: those governing once the plant is up and operating, and those that govern prior to the operation date.

Baseload: The nature of geothermal power plants lead to the general requirement that geothermal plant contracts allow them to operate *baseloaded** at full available capacity. Three factors lead to this requirement: low variable costs (see Table 51.2), high plant availability, and high fuel/plant capacity factor.

Geothermal plants have low variable costs, which are those costs resulting from changes in plant output. Therefore, any dispatch order that reduces the output of the geothermal plant simply raises the cost of power during noncurtailed periods to recover the fixed costs. The geothermal power plant operates most effectively and at greatest economic efficiency at full baseload.

Geothermal power generation typically operates at a 95%–99% plant availability (hours capable of operation per year divided by hours per year). What this means is that as long as the plant has permission to operate, it will generate power, because it also has a high fuel availability. By contrast, wind generation also has high mechanical availability, but because the wind is not blowing all the time (no or low "fuel" availability), the plants do not generate, even though they are "available." Although there is little difference in wind and geothermal generation plant mechanical availability, the capacity factors are very different because of the nature of the fuel supply.

Geothermal power plants typically operate at high capacity factor (MW h generated per year divided by the product of hours per year and plant capacity).† The high mechanical availability and the steady flow of energy from the production wells means that geother-

* The term baseload refers to the minimum amount of power continuously required by a utility's customers over a 24 h or annual basis. A baseloaded plant is one that remains on-line at all hours of the day at maximum or near maximum output to serve that demand.

† The fuel-capacity factor is useful in addition to the plant-capacity factor, because after make-up drilling is completed, the power plant mechanical capacity remains essentially constant, but the resource capacity declines. Consequently, the power plant capacity factor drops each year, even though the power plant maybe converting essentially 90%–99% of the available fuel supply. The output and performance of the plant remains highly predictable, however, on both an hourly and annual basis.

TABLE 51.2

Comparison of Fixed and Variable Costs for Geothermal and Fossil-Fired Plants

	Geothermal Power Plant with Wells and Piping		Fossil-Fuel Power Plant	
Cost Category	**Variable Cost**	**Fixed Cost**	**Variable Cost**	**Fixed Cost**
Fuel costs to increase plant output from 50% to 100%	Geothermal royalties (1%–5% of gross power sales)	Drill wells and build pipelines, mostly before start-up of plant	Essentially entire additional fuel cost	Not applicable
Cost/savings to stop plant and restart later	Low cost savings other than royalties. Low additional costs other than ongoing fixed costs	May have to vent steam to air rather than shut-in wells, and expose wells to thermal cycling damage	Fuel savings when stopped. Fuel costs to keep warm or cold start plus start/stop related maintenance costs	Not applicable
Fixed operating personnel costs (number of MW that carry cost of each staff person)	Not applicable	10–13 people for 0–15 MW = 1–1.5 MW/person 13–18 people for 40–120 MW = 2–8 MW/person	Not applicable	15–20 people for 250–550 MW = 13–36 MW/person
Fixed capital costs	Not applicable	\$1400–\$3000 per kW of capacity	Not applicable	\$700–\$1300 per kW of capacity

mal power plants as a class are unsurpassed by any other generation technology, whether renewable, fossil-fueled, nuclear, or hydro.

Seasonal Pricing: Another contractual provision in many geothermal power supply contracts is for the utility to have seasonal pricing. For example many California and Nevada geothermal contracts have prices that are higher in the summer and lower in the spring and fall. This provides the geothermal plant operator incentive for the power plant to be online in the peak summer hours, when replacement power is the most expensive for the utility. The reliable and predictable summer peak output is one of the advantages for a utility (and its ratepayers) to have available energy sources using biomass, geothermal power, and solar generation. Because summer peak power spot market costs and the cost for simple-cycle gas turbine generation can commonly rise to over \$150 per MW h in the summer, the reliable fixed-cost prices of nonintermittent resources reduce the cost of meeting summer peak loads.

51.3.4.3.4 Contract Provisions for New Geothermal Development

Unique contract provisions are advantageous for *development* of new grassroots geothermal power plants as opposed to the expansion of an existing geothermal power plant. The expansion of an existing geothermal power plant is based on the historical performance of the geothermal resource. Therefore, operators of existing plants can commit to definitive start-up schedules and subsequent liquidated damages (LD) for missing those schedules. This is not so with new geothermal plants. Unique contract provisions can assist in the development of a grassroots geothermal power plant by recognizing and mitigating the risks, uncertainties, and costs of such development.

One unique challenge to the development of a grassroots geothermal power plant can be a contractual "chicken and egg" situation in which two events must occur, neither of which can occur before the other has been satisfied. The first of these is that it is very difficult and expensive, if not almost impossible, to secure funding for exploration drilling without a power sales contract. Geothermal energy, in comparison to natural gas or oil, is neither transportable nor does it have an automatic market after its discovery. Therefore, before risking the large capital for an exploration program, investors (whether as shareholders or venture capital) want a utility commitment to buy the geothermal power if a commercial resource is discovered. On the other hand, it is often not possible to sign a power sales contract without a proven resource, because the power sales contract may contain a large LD clause for failure to deliver the contracted power by a contracted date, and may include the requirement to post a large bond.

One solution for those utilities that would like to add geothermal power to their mix of resources to consider contracts with geothermal developers that recognize the risk, uncertainty, and cost in the discovery of a new geothermal field. Such contracts would not contain penalties for failure to discover a resource, but will assure the developer a market should their exploration efforts be fruitful. Provisions of the contract to protect the utility can include the following:

1. A price acceptable to the utility.
2. An expiration date 2–5 years beyond the date on which the contract is signed, by which time the developer must produce a resource discovery report or forfeit the contract.
3. A requirement for a notice of intent from the developer to deliver first power to the utility after 2–3 years.
4. The requirement for bonds or LDs for failure to deliver after the submission of the notice of intent.

The delay between the notice of intent and first power is of minor consequence for the developer, as the cycle time from resource confirmation to commercial operation will generally be 2 years or more, and the delay gives the utility the necessary time to time to plan for the delivery of the power. There are no statistics on success rates for grassroots exploration to power plant commercial operation under such a contract scenario, but it is likely that a utility that wants 50 MW of geothermal power should plan on signing resource discovery and power purchase contracts for between 75 and 100 MW from two or more exploration prospects.

51.3.5 Economic Access: Power Plant Cost

Power plant costs comprise two elements: the capital cost and the operating cost.

51.3.5.1 Capital Costs

Capital costs vary according to four major variables:

1. *Resource temperature*: Higher temperature resources have lower $/MW capital cost. Table 51.1 indicates the first benefit of higher temperature, namely lower flows, and hence, fewer wells and smaller components are needed. However,

the reduction of the size of the heat rejection system (condenser and cooling system) required per megawatt of capacity is also of great value. Higher-temperature fuel sources require much less cooling than lower-temperature sources. This is illustrated by a comparison of fossil-fuel power plants and geothermal plants in which geothermal power plants may have a heat rejection system eight times as large per megawatt as that in a combustion-turbine combined cycle plant.

2. *Power technology*: Power plants using a binary process generally have a cost several hundred dollars per kilowatt higher than power plants using only steam turbine technology. However, this is not universally true, and binary power technology plants have won numerous open bids around the world over suppliers of steam turbine technology.
3. *Wet or dry cooling*: Steam flash power plants generally use wet cooling with the water supplied by the condensed geothermal steam. Binary power plants can use wet or dry cooling. In most climates, wet cooling gives an advantage in net plant output over dry cooling, and also causes less capacity degradation than dry cooling in the peak summer hours. Because these peak hours are generally the utility's most expensive hours, wet cooling offers utilities a benefit. However, there is often not water available (neither condensed geothermal steam nor meteoric waters) for a binary power plant cooling system, and therefore, dry cooling is used.
4. *Plant size*: Larger plants and larger machines are less expensive per kilowatt of installed capacity than smaller plants and smaller machines. This occurs both because of the general economy of scale of machine size, as well as a distribution of fixed costs (e.g., civil engineering, roads, site preparation, instrumentation, insulation, and paint) over a greater plant size.

51.3.5.2 Operating Costs

Geothermal plants have low variable costs, which are direct costs resulting from changes in plant output. Fixed operating costs are those costs that exist even if the power plant is not operating, e.g., personnel and interest on loans. If output is reduced, the staff costs are the same, the fuel costs are the same (i.e., the amortization of the well costs), and chemical use changes negligibly. Thus, there are no cost savings with a reduction in geothermal plant output other than the royalty paid to the holder of the geothermal rights. For all intents and purposes, a geothermal power plant has only fixed operating costs. Table 51.2 illustrates the difference in variable costs between a geothermal plant and a combined cycle plant using a combustion turbine. The table illustrates why baseloaded operation is optimal for geothermal power plant cost structure whereas a dispatchable operation fits the cost structure of fossil-fired power plants.

Not only does reducing the output from a geothermal plant not reduce operating costs, but throttling a geothermal turbine to reduce its power output actually has a negative effect on long-term operations, because it increases scale formation at the turbine inlet. Reducing the flow from geothermal wells on a regular basis, as would be required if the plant is not operated at baseload, can also induce damage in both the wellbore and the steel well casing in certain geothermal fields due to thermal or pressure cycling.

51.4 Exploration and Assessment of Geothermal Resources

Geothermal resources capable of sustaining commercial electrical production require specialized methods of exploration and assessment. To finance the high initial capital cost of a geothermal power plant and wellfield, the equity and debt providers must be assured that a geothermal reservoir is capable of sustaining the expected amount of energy production for a period of 20–30 years. Likewise, the purchaser of power, typically a local electrical utility, must have adequate assurances to justify entering into a long-term power purchase agreement (PPA). The geothermal industry has developed a suite of exploration and reservoir assessment techniques which provide the assurances required by all parties involved in a particular project.

Even when an active geothermal system is found, many geothermal discoveries do not lead to commercial development for a host of reasons, as discussed previously. This does not necessarily mean that a particular geothermal resource is forever noncommercial. Raft River (Idaho), Salt Wells (Nevada), and other resources under development in 2005 were discovered, drilled, and assessed in the 1970s and 1980s and were deemed noncommercial. Today, these resources are under commercial development, through a combination of improvements in exploration techniques, changing electrical power and natural gas markets, and advances in geothermal power technology.

Although this section is written from the perspective of geothermal development for power production, the same considerations and approaches are used for geothermal resources developed for direct-use applications.

51.4.1 Overview

The discovery of a geothermal resource and the assessment that it will be capable of sustaining 20–30 years of commercial production is a complex and costly undertaking, and has been the subject of entire books and many journal articles. This section is meant to provide a brief overview of the subject, much like learning geography while looking out the window of the space shuttle at the earth passing by below. In that brief glance, the shape of some of the continents would be largely visible, and one might be able to discern some forest, desert, and cities.

The process of geothermal exploration generally begins with the observation of surface manifestation of geothermal heat; for example, hot springs, fumaroles, or surface deposits of silica (sinters). However, some geothermal resources have no surface manifestations, and are discovered by accident when drilling takes place for purposes unrelated to geothermal development. Several commercial geothermal fields in the western United States have been discovered through the drilling of irrigation wells or mineral exploration holes.

From this initial surface exploration, detailed geoscientific work is undertaken to characterize and assess the size of the geothermal resource underlying the visible manifestations. After a promising resource is identified by the geoscientific study, wells must be drilled to prove the existence of the essential elements of a commercially viable geothermal resource, as described in Section 51.3.1, i.e., heat, permeability, and fluid. The critical first step in the drilling campaign is to drill a discovery well that is capable of producing commercial quantities of hot fluid. With the successful completion and testing of one or more additional wells, the process of exploration ends and the resource is considered "proven."

Finally, the remaining wells for the development must be drilled. Once the geothermal resource is under production, pressure, temperature and chemistry of the reservoir are monitored to evaluate and optimize resource production, and to make plans for further development (Section 51.5).

51.4.2 Exploration and Discovery

Surface geoscientific investigations are the first step in the process of discovery. These investigations include geochemical analysis of hydrothermal manifestations (hot springs, fumaroles, mudpots, etc.) and surface geologic mapping, including the type and extent of rock units, hydrothermal mineralization and surface expression of faulting, fracturing, and other structural features. Surface mapping techniques have been greatly enhanced by the use of remote sensing (satellite imagery). Together, these methods may indicate the ultimate resource temperature, provide clues as to the resource structure, and provide evidence of the vigor and extent of the resource. If a water sample can be obtained from a hot spring, the geochemistry, via various chemical "geothermometers," can provide an indication of the maximum temperature to which the water was heated on its journey to the surface. However, the surface geoscience does not provide a good definition of the depth to the commercial geothermal reservoir, nor can it define its lateral extent or ultimate productivity. Not all hot springs are indications of commercial geothermal power resources, even if the geothermometers indicate a high-temperature origin of the fluid; the geothermometers can be erroneous, or, even if high-temperature water is available, it may not be producible at commercially viable flow rates.

The second step is geophysical investigation to map the lateral extent, depth, and distribution of permeability of the active geothermal system. These investigations and their objective may include those listed in Table 51.3.

The cost of conducting and interpreting a typical suite of geophysical surveys can vary greatly, depending on the individual characteristics of a particular exploration project. The total cost is dependent on the methods chosen, overall size of the area to be explored, the maximum depth to be explored, surface topography and the remoteness of the site. For example, an effective geophysical program to explore and define a small geothermal reservoir in Nevada may cost as little as $150,000, whereas a program in the remote jungles of Indonesia may cost in excess of $1,500,000.

TABLE 51.3

Geophysical Investigations for Geothermal Exploration

Exploration Tool	Objective	Indicative of
Resistivity	Detection of a clay cap, transition depth to high temperature mineralization	Convecting geothermal system that has deposited minerals to seal-off the top of the system
Seismic reflection	Faulting, structure	Flow paths (good drilling targets), structural block volume
Microearthquake monitoring	Identify active faulting and fracturing	Fracture permeability, fracture density, injection and production drilling targets
Magnetics	Loss of iron minerals	Convecting geothermal system
Geodesy	Active surface deformation	Rapid natural subsidence overlying active rift zones
Self potential	Map natural ground voltage	Shallow active hydrothermal systems
Microgravity	Faulting, mineral deposits	Flow paths, convecting geothermal system
Shallow wells	Temperature gradient	High heat flux, possible max T at depth, lateral reservoir extents

The third, and final, step is the riskiest and most expensive portion of the exploration program: the exploration drilling to determine if the resource is commercially viable. The exploration drilling program can involve both full-size development wells (commonly final casing size of 9–13 in. with reservoir hole diameter of 8–12 in.) and slim holes (final casing size of 6 in. or less and reservoir hole diameter of 5.5 in. or less). Slim hole wells have the advantage of costing only about two-thirds to three-fourths that of a full-size well, but have little commercial value, even if drilled into the heart of the resource. Depending on terrain, remoteness, road requirements, possible maximum resource temperature, depth, and other factors, the cost of the first exploration well will often exceed $2,000,000, and may reach $5,000,000 for deep or remote locations. Unexpected drilling problems can result in enormous cost overruns or the total loss of a wellbore and all the investment in it. Rare shallow resources may see exploration wells at less than $1 million. Follow-up wells, if drilled at the same time, will save on the order of several hundred thousand dollars as a result of shared mobilization and construction costs.

Typically, three to eight wells will be required to prove and delineate an undeveloped geothermal resource, depending on the efficacy of the exploration program used to target the wells, the size of the first power plant, and other factors that cannot be predicted prior to drilling. Sometimes a nonproductive well will eventually be used for injection, so the investment may not be altogether wasted if a particular well fails to encounter commercial production. There are instances when, even after a multiwell exploration program, the resource that is discovered cannot be economically developed at all.

After the first (or, preferably, two or more) well(s) have been drilled and found to flow with commercial temperatures and flow rates, a long-term flow test and an interference test will be conducted. The budget for the flow test, with cross-country pipelines, staff, instrumentation, chemical sampling, interpretation, logistics, travel, and so on, can cost $200,000–$500,000, although long multiwell tests can incur greater costs.

The final stage of the geothermal reservoir assessment process is the analysis of the well and reservoir flow testing data. This analysis is done using established reservoir engineering techniques in which the behavior of the geothermal reservoir is numerically simulated at the proposed production rates for the expected duration of the power plant life, typically 20–30 years. The numerical model is used to investigate different production and injection strategies to optimize the resource development strategy. Positive results of this analysis are an essential factor for obtaining the debt and equity financing required for completing the wellfield and initiating power plant construction.

51.4.3 Risk of Exploration

A modest exploration budget of $2–$10 million, staged over the phases above, may suffice for many geothermal resources, but much greater budgets are well documented. Western GeoPower Corp. is a company seeking to commercialize the South Meager Creek Geothermal Project in British Columbia. Because they are a public company, various aspects of their development work are available in the public domain and can be used as a case study.[13]

The company raised over U.S. $24 million for corporate expenses and a three-well exploration program in the remote area where the resource was located; the program was executed in 2004–2005. As of the time of this writing, the three wells had been drilled, but not flow-tested. These exploration wells were targeted on the basis of previously-drilled wells and surface exploration over the previous 20 years since the resource was first identified. Temperatures in the three wells were announced to have ranged from 240°C to 260°C (460°F to 500°F), which demonstrates that the thermal basis for the targeting was sound. Western GeoPower has not yet announced whether the other elements of a commercial geothermal resource (permeability and water), per Section 51.3.1, have been discovered.

Regardless of whether or not a geothermal power plant is eventually developed at South Meager Creek, the exploration program demonstrates the potentially high risks and high costs of grassroots geothermal resource exploration. Not only was there substantial cost, but it has taken two decades for the exploration project to get to its current state.

Both geothermal and wind resources are renewable energy sources that must be converted at the location at which they are found, but it is much more difficult to demonstrate a viable commercial geothermal resource than a commercial wind resource. The cost of an array of wind exploration towers, the data collection, and analysis is miniscule by comparison with the cost of geothermal resource exploration. Additionally, the time to prove a commercial wind resource is much shorter. These cash and time costs not only affect the ease and extent of exploration, but the power price that must be obtained to pay for the exploration. The resource discovery process is by far the greatest disadvantage and barrier to expanding geothermal power generation. The costs and risks are why geothermal exploration is most often undertaken outside of the United States with strong government support or with government funding. In the United States, continued DOE support of exploration and discovery through cost-share programs will be necessary to bring significant quantities of new geothermal resources under development.

In spite of the barriers to exploration and discovery of commercial geothermal resources, they are discovered, developed, and electric power is delivered to the transmission grid. The remainder of this chapter is dedicated to these processes, from the resource (fuel) management issues to the power cycles.

51.5 Management of the Geothermal Resource for Power Production

Geothermal resource management initially seeks to ascertain the initial and long-term behavior of the geothermal reservoir so as to select the optimal power process for the particular geothermal resource, including the changing conditions that might be encountered over the first 20 years. Second, geothermal resource management seeks to design the geothermal resource development to meet short- and long-term economic goals that may include some or all of the following: minimizing invested capital cost in the wells and the power plant, maximizing the power plant installed capacity that the resource can support, and minimizing operating costs.

The complexities and issues of geothermal resource management have no equivalent in any other electric power generation technology, whether conventional or renewable.

- The fuel supply is located 2,000–10,000 ft below the surface, and may be more than 10 cubic miles in volume.
- The response of the resource to development plays a huge role in the overall economic and generation performance of the resource, but the details of the response are largely unknowable until after the resource is under full production for a number of years.

- The "fuel supply" almost invariably changes with time, and as a result, the power plant either must be operated off-design or modified to meet the new conditions.
- Through much of the resource life, make-up wells will need to be sited and drilled to maintain adequate geothermal energy supply to the power plant, which is a combination of flow rate and pressure or temperature.

Although the management of natural gas and oil fields face many of the same issues, fossil-fuel power plants are seldom tied to a single and solitary gas or oil resource.

In this section, the issues involved in managing the geothermal resource, and how the characteristics of the fuel supply interact with the design and operation of the power plant, will be examined. Case studies illustrate successes, changes, and occasional failures in geothermal resource management.

Section 51.5 begins with the large goals, examines some of the subgoals, and ends with the beginning—the characterization of the geothermal resource for the selection of the power process.

51.5.1 Goals of Resource Management

Geothermal energy production is the science and technology of heat recovery. At commercially-significant extraction and power generation rates, the heat extraction rate from the production wells is almost always far greater than the natural heat addition rate that created the geothermal resource. The energy balance is closed by extraction of the heat-in-place in the volume of the reservoir, i.e., cooling of the geothermal resource.

Heat produced out of the geothermal system per year − Natural heat inflow to geothermal system per year = Cooling of geothermal system per year.

The geothermal resource volume is large, however, so the cooling is a long, slow process that allows the geothermal resource to be productive for many decades. (Some of the issues of renewability and sustainability in geothermal power are addressed in Section 51.3.4.) Therefore, the overall strategic goal of resource management is efficient heat mining (cooling) of the geothermal resource. Four of the most important resource management subgoals include (1) minimizing the capital cost of the development, (2) residual brine management (if applicable), (3) injection placement to enhance production, and (4) geothermal fluid monitoring and control, including chemistry, rate, pressure, and temperature.

51.5.1.1 Minimizing Development Capital Cost

To achieve heat recovery from the resource, geothermal fluids (steam or brine) must move through and be heated by the rock. Boiling is the most efficient mechanism, because of the high heat transfer rates of vaporization and the high heat capacity of steam. The native, in-place water is the primary mechanism for many years of the production. Geothermal reservoir rock porosities are most commonly in the range of 3%–10%, but may be as high as 20% in sedimentary systems.

As discussed in Section 51.3, matrix flow of fluid through the bulk rock occurs at only very low rates, so much of the heat of the reservoir volume must transfer by the much slower process of conduction through the rock to the fracture surfaces, where liquid is mobile and heating or vaporization can more readily occur. (This also explains part of the difficulty of creating the HDR/HFR geothermal resources of Section 51.3.2.) The native-state fluid mass is not sufficient to achieve the goal of heat mining the reservoir. This can only be achieved by flowing supplemental water through the resource. Therefore, the placement of the residual geothermal fluids through the injection wells can have a large effect on the overall capability of the field over the long term. From an ideal heat transfer perspective, the optimal development would be to have many production wells drilled into the hottest areas of the field, with many injection wells ringing the production area to sweep heat and provide pressure support to the production wells. Unfortunately, the need to manage and limit the capital cost of the development intrudes on this Utopian development, and forces geothermal resource developers to seek compromises to the ideal.

Figure 51.2 illustrates a deliverability curve for a hypothetical well producing a mixture of steam and brine. The curve segments are as follows:

- Segment A to B is an unstable flow regime (not all wells have this segment).
- Segment B to C is increasing flow due to greater differential pressure between wellbore and reservoir.
- Segment C to D is the same as B to C, except for an increasingly dominant frictional pressure loss in the wellbore or fractures feeding the wellbore that reduces flow rate at a faster rate than in the segment from B to C.

In the Utopian geothermal development, the well would be operated somewhere just to the right of point B, because point B has the highest available energy per unit mass, and will therefore allow the greatest cumulative power generation over time from the resource. However, there is not much flow at point B, and thus a large number of wells and long lengths of connecting pipelines would be required. This is an unacceptable barrier to economic commercial development, in part because of the high capital, and in part because the additional generation occurs so far in the future that its present worth value is minimal, as discussed in Section 51.5.3. Therefore, a much more common optimization of the wells is to maximize the energy production from the wells (shown as point E on the dashed line in Figure 51.2); doing so minimizes the amount of money invested in production wells because of the high energy production per well. The disadvantages to

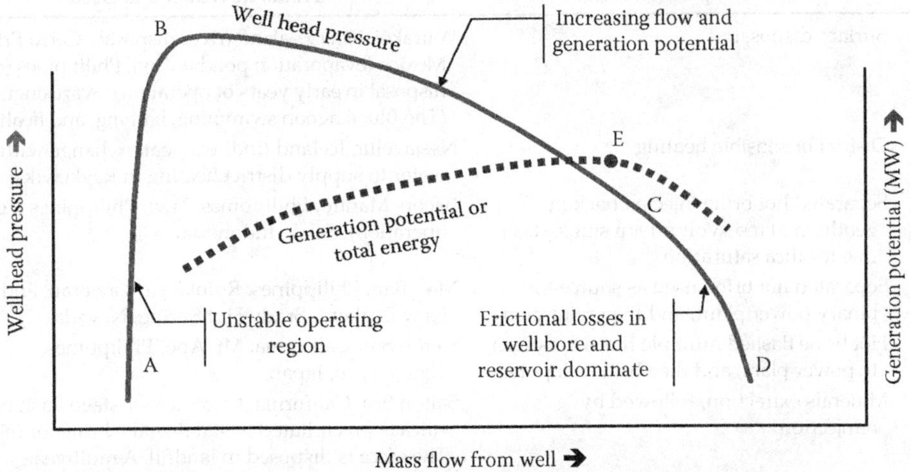

FIGURE 51.2
A hypothetical free-flowing geothermal production well deliverability curve.

this strategy stem from larger flows per megawatt-hour of generation that result in higher injection well costs (more spent fluid because of lower available energy per pound), higher power plant heat rejection costs (lower useful energy per unit mass), lower ultimate field capacity, and more make-up drilling sooner. The choice of the exact operating point on this curve is much more complex than a single number, but these two extremes (point B and point C) illustrate two of the important issues. A third issue is the provisions of the power sales contract. Power sales contracts that do not limit the output capacity of the field, and in which the field capacity can be fully developed in a short period of time, will tend to push the optimal operating point closer to point B because point B maximizes the ultimate field capacity. Those in which the power sales limit is much smaller than the capacity limit of the field will tend to push the operating point toward point C.

The knowledge of how a resource will respond to development can often be used to minimize the capital cost of the development. For example, at the Salak geothermal field in Indonesia, the reservoir modeling prior to production indicated that there was a region of the reservoir that would quickly evolve from producing a mixture of steam and brine to almost entirely steam production. Wells and a power plant were located in this area to minimize piping costs and to capture the benefit of lower brine handling and injection costs that resulted from the production of nearly dry steam.

51.5.1.2 Residual Brine Management

A large variety of operating practices has developed around the world over the use and disposal of brine, from primarily the two-phase (steam and brine) reservoirs. This is because in two-phase reservoirs, the reservoir pressure is primarily controlled by the steam pressure in the reservoir, whereas in primarily liquid reservoirs, the reservoir pressure is strongly influenced by the injection of the liquids back into the reservoir. A table of options and applications for use of the residual brine is presented in Table 51.4.

TABLE 51.4
Uses of Separated Brine in Worldwide Geothermal Power Operations

Option	Use of Separated Brine	Fields in Which it is Used[a]
1	Surface disposal	Wairakei, New Zealand (river disposal). Cerro Prieto, Mexico (evaporation ponds). Tiwi, Philippines (ocean disposal in early years of operation). Svartsengi, Iceland (The Blue Lagoon swimming, bathing, and health spa)
2	District or sensible heating	Nesjavellir, Iceland (indirect heat exchange with fresh water to supply district heating in Reykjavik)
3	Separated hot brine injected back in geothermal reservoir after a single flash due to silica saturation	Bacon–Manito, Philippines. Tiwi, Philippines (current operation). Salak, Indonesia
4	Separated hot brine used as source for binary power plant, and then reinjected	Mak-Ban, Philippines. Rotokawa, Kawerau, and Mokai, New Zealand. Brady Hot Spring, Nevada
5	Hot brine flashed multiple times for steam to power plant, and then brine reinjected	Salton Sea, California. Mt. Apo, Philippines. Hatchobaru, Japan
6	Minerals extraction, followed by reinjection	Salton Sea, California. (After a four-stage flash process, silica is precipitated to stabilize the brine for injection. The silica is disposed in landfill. A multiyear, multimillion dollar effort to recover zinc was abandoned in 2004.)

[a] Examples, not a complete list.

In geothermal reservoirs, both heat mining and pressure support are managed by injection well placement. When this is accomplished according to objective, the results can be seen and monitored in the production well characteristics. Injection well placement is best accomplished in a manner similar to Goldilocks and the three bears: not too close (rapid cooling of produced fluids), not too far (no benefit), but just right (reservoir management).

When separated brine is disposed of through injection, the brine is generally injected into either the bottom of the reservoir or outside the reservoir. The location of the brine disposal is a difficult task, and in the history of many geothermal fields, brine injection locations have had to be adjusted at least once during the life of the field, usually because the injection wells are too close and cold-water breakthrough is observed. If the injection is ideally placed, it provides both pressure support for the production wells, as well as "mining" of the stored heat of the reservoir rocks as the injectate flows from the high-pressure zone of the injection to the low-pressure zone of the production well. However, if the injection well is too close or if the injection well encounters a highly permeable crack linking the production and injection areas, cold injectate can enter a production area, and quickly reduce the enthalpy (temperature or steam content) of the fluid produced by that zone. Therefore, in many instances, the preferred option is to inject some portion of the total injectate volume in steam/brine systems effectively outside of the geothermal reservoir, sacrificing the benefits of in-field injection, but also avoiding the potentially severe consequences of enthalpy loss at the production wells. As a result of the need to move injection distant from production, it is common in geothermal fields to have injection pipelines several miles long that carry the brine to distant injection wells.

As with almost every situation in geothermal resource development not tied to conservation of energy or mass, there are exceptions to the long injection pipeline option. The particular characteristics of the deep vertical fracturing of the Steamboat Hills (Nevada) geothermal resource have allowed a large-capacity injection well to operate without negative consequences to nearby production wells. In most geothermal fields, such geographic proximity would have resulted in cold water breakthrough to the production well within a very short time, perhaps on the order of days or months.

51.5.1.3 Clean-Water Injection Placement to Enhance Steam Production

The dryout of a region of a reservoir can occur when the water present in the region is no longer sufficient to remove the heat that is stored in the rock. At that point, the adoption of a new clean-water injection strategy can improve a number of the negative consequences of that dryout: specifically, loss of production, superheat, and acid gas production.

As dryout occurs, the temperature of the rock and the heat stored therein remain high, but there is no water to capture or transfer the heat. Production declines, and as it does, the pressure drops. Steam passing through rock that has a temperature higher than the saturation temperature of the steam's pressure becomes superheated. Superheated steam, upon mixing at the surface with wells that are not superheated, can evaporate the water from the brine flowing from other wells. If the brine rates are low enough and the evaporation is high enough, minerals will precipitate in the pipelines or separation vessels.

Superheated steam is also capable of carrying hydrochloric acid gases produced from certain rock formations out of the reservoir. Because the gases are extremely hydrophilic, they can only pass through the reservoir and into the steel-lined wellbore if no water is present. The hydrochloric gases are harmless to the steel as long as the steam is superheated. They can be removed at the surface by one of several methods: mixing with large quantities of water, mixing with smaller quantities of brine (which is naturally buffered against changes in pH), or, in some instances, by caustic injection into the pipeline. However, if scrubbing of the hydrochloric acid is not accomplished before the onset of condensation, the rate of corrosion at the condensation point can be so high that the steel pipe wall can be thinned to failure in less than a year.

Injection into the reservoir can be used to solve all three problems. Adding water to the dried-out region through injection not only provides the means to increase production and heat-mining, it eliminates the superheat, and with the elimination of the superheat, also eliminates the acid gas production and pipeline scaling.

51.5.2 Resource Chemistry

The geothermal water and steam that are produced contain the chemical signatures of the rocks and processes (e.g., boiling) through which they have passed. Measuring the steam and brine chemistry is an important tool in the exploration process for new geothermal resources, in determining the brine processing requirements of the power cycle, and as an ongoing activity in developed geothermal resources for analyzing the response of the resource to production and for resource management planning. Examples include:

- In the exploration phase, geothermal water is collected from surface springs or fumaroles for clues as to the nature of the geothermal resource. For example, the relative concentrations of sodium, calcium, magnesium, and potassium in the water tell geochemists whether the fluid started out hot and then cooled (indicating a high-temperature resource elsewhere and indicating commercial geothermal resource potential) or whether the observed temperature of the fluid is close to the maximum temperature of the resource (not hot enough for power production).
- The extreme example of geothermal fluid chemistry is the Salton Sea (California) geothermal brine that contains more than 200,000 ppm total dissolved solids (TDS), or 20%, in its native state in the reservoir. The high TDS is a result of the high temperatures and the marine sediments that comprise the geothermal reservoir. The brine is produced and is flashed to steam at four pressures (~250, 125, 20, and 0 psig). In so doing, the brine becomes supersaturated with respect to silica and other minerals and salts, resulting in large chemical processing facilities necessary to prepare the brine for injection back into the geothermal reservoir. The silica treatment processes can either reduce the pH to prevent precipitation in the equipment or, alternatively, the silica can be removed from the brine by the controlled precipitation of the silica in tanks and vessels.
- Silica concentration is primarily a function of the reservoir temperature, but the concentration of silica determines how far the brine temperature can be lowered (how much energy can be extracted) before the onset of silica scaling. The power process is frequently designed to prevent silica from reaching saturation. In such cases, the operating point of the well is pushed to the left of point C in Figure 51.2, regardless of other considerations. The injection of acid or chemicals to inhibit the precipitation of silica can be used to lower the minimum allowable brine temperature and extract more energy from the brine. Another strategy to extract more energy from the brine is to use a binary power plant after the first steam flash. This strategy is discussed in more detail in Section 51.7.
- Tritium (an artifact of atmospheric nuclear weapons testing) detected in the geothermal steam or water indicates that young groundwater is entering the

geothermal reservoir as a result of geothermal exploitation. Native-state geothermal water, which has typically been underground for tens of thousands of years, contains no detectable tritium. Tritium from groundwater may occur in the geothermal reservoir as a result of intentional injection programs to replenish water, such as at The Geysers, or the result of unwanted cold water influx from the surface. At Tiwi (Philippines), the cold water in question has quenched several square miles of productive resource. The advancing surface water could be detected in the tritium concentrations long before the temperature indicated a cooling process was occurring at the well.

- Increasing salt concentrations and other mineral concentrations indicate the return of injected brine from flash steam plants. Again, this can occur long before a change in production temperature is measured. If the rate of cooling is negligible or small, these chemical signatures indicate successful heat mining through the injection program.

51.5.3 Barriers to Resource Management

In most human endeavors, the best results come by starting with the end in mind. This is true of geothermal resource development and resource management as well. By understanding the particular resource the wells can be placed to ideally collect and convert the geothermal energy stored in the reservoir. Unfortunately, in practice, the ideal placement of the production and injection wells may be either impractical or even impossible for three reasons: incomplete information, capital limitations and sunk costs, and economic impetus.

51.5.3.1 Incomplete Information

First, many aspects of the geothermal resource are simply not known in the early stages of development of a new resource. Such incomplete information includes whether cold water from injection will short-circuit to the production wells, how strong the natural influx to the system will be, or where that influx will actually occur. In fact, the exact area and volume of the resource is commonly not even known until 10 years or more after the onset of production.

The Geysers facility illustrates this point. The unit 1 power plant went into operation in 1960, and the final power plant—the J.W. Aidlin Plant that began operation in 1989—brought the total installed capacity to 1890 MW, although deratings and retirements commenced soon thereafter. However, reservoir information and analysis was not complete, even in the mid-1980s. Two additional power plants were ordered and never completed because there was insufficient steam for their operation. Therefore, the geothermal resource management strategy is dynamic and evolving from the beginning. As more information becomes available about the resource over time, the knowledge of the resource improves, and the decisions about its management more informed.

A surprising and interesting example of this evolving knowledge has occurred in recent years at Larderello, Italy. Almost 100 years after the start of power generation, it now appears that the historical development in the Larderello area is at the top of a much larger and deeper regional system, perhaps 400 km^2 in size and between 3,000 and 4,000 m deep (10,000–13,000 ft). Plans have been announced to begin drilling and developing this deep hot resource in 2007 with 11 deep-production wells.[8]

51.5.3.2 Capital Limitations and Sunk Costs

There may be capital limitations and sunk costs that must be taken into account in well placement decisions. For example, suppose that it would be ideal to inject the cooled brine to the north of the power plant, but an unsuccessful exploration well will serve as an injection well to the east, saving over a million dollars. In one development in Indonesia, a well that was targeted as an injection well was found, after being drilled, to have a very large production capacity. However, pipelines were already under construction, the well was distant from the power plant, and production well capacity had already been essentially completed. What was needed for plant operation was injection capacity, so the well was used for injection. Time and capital constraints forced the decision, even though it was against the ideal resource management strategy. This example also illustrates the first point about incomplete information at the onset of the project.

51.5.3.3 Economic Impetus

The third driver for well placement decisions is the economic impetus. The economic test criterion used for investment decisions is discounted cash flow, in which decisions are based largely on achieving the highest net present value or the lowest net present cost in which all costs and earnings are brought back to the present at an assumed discount or interest rate. For public (government) development, that discount rate may vary from 5% to 15%; for development by private corporations, the rate for normal equipment and operational investments is typically 9%–12%, but for resource development issues is not uncommonly as high as 20%, due to the greater risk and uncertainty. The lower the discount rate, the more important future performance is to the total value of the project, as viewed by "today's" decision makers. The converse is also true, and in fact dominates spending decisions for resource management at the start of the development. Suppose that the publishers of this text elected to develop a new geothermal power plant named "CRC #1." They are sharp businesspeople, and are told by their resource development experts that there are several resource development options that can be pursued. Both strategies produce the same income with the same expenses for the first 15–20 years. However, the second strategy will produce significantly more value in the outer years because of better resource management. They perform a discounted cash flow calculation and develop in Table 51.5. The conclusion that is reached

TABLE 51.5

Amount of Spending That Can Be Justified "Today" for a Savings or Earnings of $10 Million in the Future

	Economically Justified Spending "Today" at a Discount Rate of				
Discount Rate Scenario	**0%**	**4%**	**11%**	**15%**	**20%**
$10 Million value in 15 years	$10,000,000	$5,552,645	$2,090,043	$1,228,945	$ 649,055
$10 Million value in 20 years	$10,000,000	$4,563,869	$1,240,339	$611,003	$ 260,841
$10 Million value in 30 years	$10,000,000	$3,083,187	$436,828	$151,031	$42,127
$10 Million value in 50 years	$10,000,000	$1,407,126	$54,182	$9,228	$1,099

is that, for an investment as risky as resource performance in 15–20 years, the corporation will use a discount rate of 15%–20%. The table shows that the resource managers have a budget to optimize the resource development of between $0.25 and $1.2 million (the highlighted area in Table 51.5) to create $10 million of value. In most cases, this will not be sufficient to implement an enhanced resource management plan. At the extreme, in looking at resource development strategies that would create the $10 million value in 50 years, the table shows that less than $10,000 of additional spending today can be justified at a 15% discount rate and only $1,000 at a 20% discount rate. In short, there is virtually no economic impetus for resource development decisions that yield a benefit after 50 years.

Even if not implemented at the onset of the resource development, some resource management strategies will be able to be implemented at a later date. However, other decisions are irreversible, such as inefficiency in the conversion of the resource to electrical power. Conversion inefficiency directly reduces the total generation from the resource (megawatt-hours), but may also ultimately lead to a reduction in the installed capacity of the field (megawatts). While this is clear from the beginning, if such additional generation does not occur for 30 years, Table 51.5 demonstrates that the generation is of inconsequential value at the time of initial development. A scenario under which this occurs is where the generation contract for the geothermal field is limited to a given output.

Suppose that a PPA is for 100 MW average over the year. With the high-efficiency power plant, the resource is forecast to provide 100 MW for 40 years before beginning its decline. With the lower-efficiency (and less expensive) power plant, the plant is forecast to be able to maintain 100 MW for 25 years. In terms of today's decision making, the value of the additional 15 years of full-capacity generation are of marginal value, and the economic calculations will almost certainly show that the power plant and resource strategy with the lower capital cost is the preferred investment strategy "today."

The same calculations also apply to the efficiency of nonrenewable natural resources, such as natural gas and petroleum. Straight discounted cash flow calculations provide little incentive for the conservation of depletable natural resources for a time frame greater than 20 years. However, at the end of the 20 years, the impact of the inefficiency can be very large on the future value.

51.5.4 Resource Characterization

An early task of the geothermal resource management team is to characterize the resource based on the exploration and initial development wells. The characteristic of the fluids that will be produced from the well must be determined: steam only, brine and steam, or brine only. Subsequent tasks must address the design issues within each resource type.

Table 51.6 provides a list of some of the design issues and options that derive from this first basic resource characterization. The characteristics of the wells will determine the options for the power cycle, and from that the configuration of the surface facilities. As illustrated in Table 51.6, there are three types of resources that are used in the two dominant power plant types: steam power plants and binary power plants, as discussed in detail in subsequent sections.

TABLE 51.6

Geothermal Facilities Design Parameters Resulting from the Geothermal Resource Type

Wells Flow: Design Issue	Steam Only	Steam and Brine Two-Phase Flow	Brine Only
Representative fields			
	Geysers, California Lardarello, Italy Kamojang, Indonesia	Cerro Prieto, Mexico Coso, California Wairakei, New Zealand Mak-Ban, Philippines Miravalles, Costa Rica	East Mesa, California Steamboat Springs, Nevada Raft River, Idaho
Production well design			
Production of fluid to surface by	Free flow	Free flow	Pumped
Well design—flow	Minimize pressure drop (velocity) to maximize flow	Maximize flow, but must maintain high velocity to lift liquid out of the well	12-in. inside diameter (min) to deepest expected pump installation depth
Well design—pressure	Maximum reservoir pressure (generally less than 600 psi, and falls with time)	Saturation pressure at maximum temperature. (generally less than 400 psi, but can be 2000 psi.)	Often less than 200 psi at surface and less than 600 psi for pumps at the zero flow condition
Production pump parasitic load	Not applicable	Not applicable	Usually between 5% and 8% of gross generator output, but can be as high as 15%
Surface facility design			
Production pipeline network	Steam only to power plant. Build with large cross-section for low pressure drop. Condensate knockout pots must be provided	Two-phase flow to separators requires high velocities and pressure drop to prevent slugging. After separator, two pipelines of single phase steam and separated brine	Single-phase liquid flow to binary power plant or separator. Size pipe for low pressure drop. Water hammer at start-up is an issue
Brine steam separator	Not applicable	Required. installed vessel cost greater than $600,000 per 10–20 MW	Only required if steam turbine to be used
Steam washing system to remove impurities and condensate	Usually required on a continuous basis. Installed cost greater than $500,000 per 10–90 MW	Same	Not required
Power plant design			
Power plant location	Center of initial and long-term production. Usually accomplished	Same goal. Not uncommonly, the center of production (make-up well drilling) moves away from the power plant site as new resource is discovered, resulting in large pipeline costs and available energy loss	Same goal. Usually accomplished, as these tend to be smaller and better-defined geothermal resources

(*Continued*)

TABLE 51.6 (*Continued*)

Geothermal Facilities Design Parameters Resulting from the Geothermal Resource Type

Wells Flow: Design Issue	Steam Only	Steam and Brine Two-Phase Flow	Brine Only
Steam turbine	Single pressure entry	Size as per other geothermal steam turbines	Not commonly used
	Size from 5 to 125 MW. Most commonly 25–75 MW	1. Single inlet pressure, as for steam only resource 2. Multiple inlet pressures on same or multiple turbines (needs multiflash design of brine system)	Size is less than 40 MW
Binary power plant Size has historically been 1–8 MW per turbine/generator set. Since 2004, size has increased to 20 MW/T/G	Not applicable	1. Can be used for recovery of energy from brine instead of multiple flashes 2. Can be used instead of steam turbine on small plants	Most commonly used instead of steam turbine
Geothermal combined cycle power plant	Has not been used in practice, but could be	1. Steam turbine with binary plant used as steam turbine condenser 2. Brine energy recovered in separate binary plant	Not applicable
Effect of brine chemistry on total available energy	Not applicable	Determines lowest temperature to which brine can be flashed without scaling Temperature can be lowered by adding acid	Determines the lowest brine temperature at discharge of binary plant heat exchangers to avoid scaling. Acid addition lowers temperature
Condenser	Generally sub-atmospheric. Direct contact of cooling water and steam, unless H_2S abatement is required, then shell and tube often used	Same, unless a geothermal combined cycle unit is used	Same for steam-only plants Generally supra-atmospheric for binary plants
Condenser gas removal equipment	Large to handle 0.5%–8% NCG in steam. Sized for NCGs and air infiltration	Same	Small. Sized only for minimal breakdown of hydrocarbon and air infiltration
	Creates large parasitic steam or power load	Not applicable for geothermal-combined cycle units	
Cooling System	Water-cooled, usually uses condensed steam as make-up water for cooling tower	Same for steam turbine plants Most often air-cooled for binary and geothermal combined cycle plants	Usually air-cooled condenser, unless surface water or steam condensate is available. Air cooling creates a large parasitic load and summer capacity loss

(*Continued*)

TABLE 51.6 (*Continued*)

Geothermal Facilities Design Parameters Resulting from the Geothermal Resource Type

Wells Flow: Design Issue	Steam Only	Steam and Brine Two-Phase Flow	Brine Only
Abatement of H_2S gas contained in geothermal steam	If required by regulators	Same	Not required for an all-binary system, as H_2S never leaves brine
Surplus fluid design			
Brine injection pipeline	Not applicable	Often long and expensive. Lengths of 0.5–1 mile from production area, or longer	Same as steam/brine system
Brine injection wells	Not applicable	Required, generally 1 well/10–40 MW (Often negative wellhead pressure)	Required, generally 1 well/3–10 MW (usually positive wellhead pressure)
Cooling water blowdown well	Usually required. One per power plant	Same. Should not mix with brine because of corrosion/precipitation if condensate is used in cooling tower Not required for geothermal combined cycle, as steam condensate is not oxygenated and can be mixed into brine	Not required for binary-only power plant
Fluid injection or disposal strategy	Inject cooling tower blowdown to regenerate steam (80% is evaporated in cooling tower). Supplement with fresh water injection. Water injection can minimize the production of hydrochloric acid gas with the steam	1. Generally, all brine and condensate must be injected into or outside of the active reservoir for reservoir management (heat mining, steam flow and pressure support) and environmental compliance 2. Rarely, some may be surface disposed to evaporation pond, river, or ocean to reduce capital and operating costs, but may be against best reservoir management practice	Always reinjected to maintain a high reservoir liquid level because of the production well pumps
Parasitic load of injection	Cooling tower blowdown pump load is very small, approaching zero	Generally pump load is small because of two-phase reservoir	Pump loads can consume 3%–7% of total gross generator output

51.6 Geothermal Steam Supply (from Wellhead to Turbine)

The first two columns of Table 51.6 discuss resources in which steam is produced from the wells, as opposed to the third column, in which only pressurized brine is produced. In almost every case, the steam that is produced from these wells is used in a steam turbine. In this section, the technology of handling geothermal steam from the wellhead to the geothermal steam turbine is discussed.

51.6.1 Overview of Steam/Brine Separation for Steam Turbine Supply

In the vast majority of geothermal fields in the world, a two-phase mixture of steam and brine is brought to the surface through the production wells. The brine must be separated from the steam before delivering the steam to the turbine. Separators of many different configurations are in use throughout the world, but the most common is a vertical vessel with a tangential entry to centrifugally separate the steam and brine. The level of purity of the steam varies between that which meets the geothermal standard of 0.1–0.5 ppm of chloride to over 10 ppm chloride.

Steam purity is typically measured by the chloride content of the steam condensate. However, the chloride itself is seldom intrinsically a problem to the alloys employed in a geothermal steam turbine. Rather, chloride is a convenient measure of the co-contaminants that are much harder to measure: mostly silica, but other minerals as well. Where these minerals are carried in the steam to the turbine, they may precipitate primarily on the first-stage stationary blades, reducing efficiency and eventually restricting plant capacity as well. The consequences and treatment of these mineral deposits are further discussed in Section 51.6.5.

With typical geothermal brine chloride levels, the separators need to achieve over 99.5% dryness to achieve the lower steam chloride concentration objective. The greatest challenge in doing so is to minimize the pressure drop through the vessel. The pressure drop has two negative consequences. The first is that the pressure drop results in an unrecoverable loss of available energy (reduced efficiency). Second, it results in a higher pressure at the wellhead, which reduces flow from the well (moving the operating point toward "B" in Figure 51.2), thereby requiring more wells (greater capital cost) to supply a given steam demand.

Where high-chloride (high-silica) steam is produced from the separators, steam washing must be employed to clean the steam before admission to the turbine, with a consequent parasitic steam loss as described below. While the commitment to, and investment in, steam purity at the discharge of the flash separation vessel varies from field to field, it is, in fact, better in practice to "put a fence at the top of the cliff, than an ambulance at the bottom."[14] In other words, the best practice in geothermal design is to first ensure high-efficiency separation (the fence) and not rely on the steam-washing system (the ambulance) at the power plant to prevent scaling of the turbine. Steam washing is defined and described below in Section 51.6.5.

Figure 51.3 illustrates a modern high-efficiency, low-pressure-drop geothermal steam separator designed by Sinclair Knight Mertz Consultants.* The two-phase mixture of steam and brine from the wells on the right enters the vessel through a special lemniscate entry (providing lower entrance turbulence than a tangential entry). The brine is separated

* Sinclair Knight Mertz Consulting, http://www.skmconsulting.com.

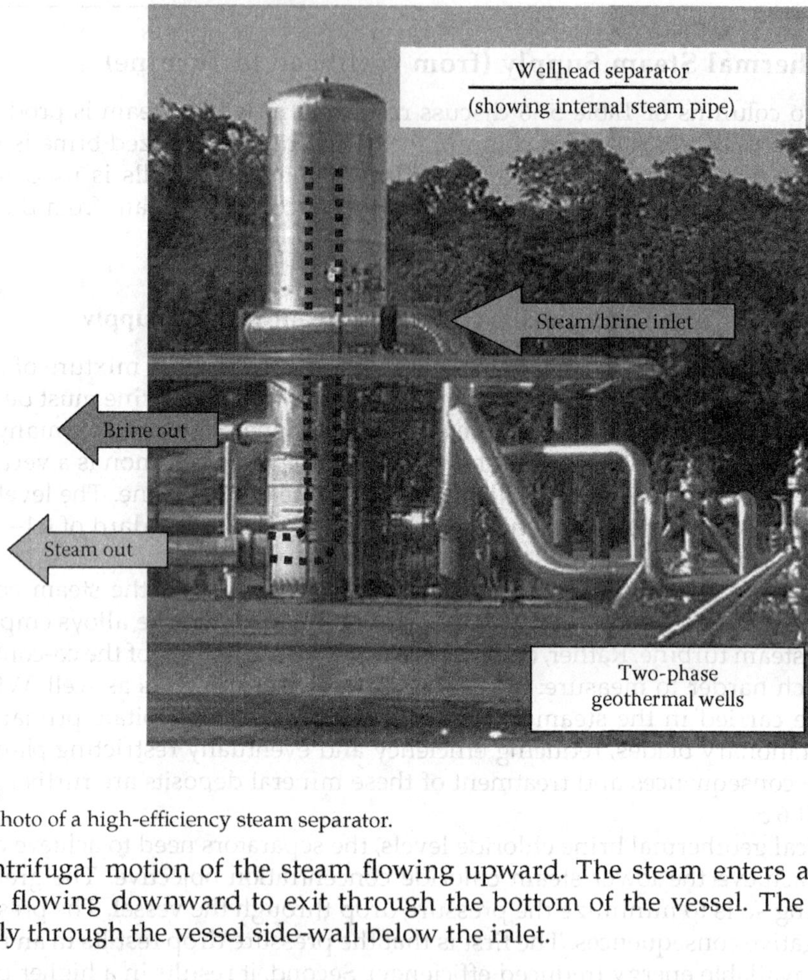

FIGURE 51.3
Annotated photo of a high-efficiency steam separator.

by the centrifugal motion of the steam flowing upward. The steam enters an exit pipe at the top, flowing downward to exit through the bottom of the vessel. The brine exits tangentially through the vessel side-wall below the inlet.

51.6.2 Multipressure Steam Flash and Separation

The number of steam pressures (flashes) that are taken from the brine varies from field to field. Each time the brine is flashed, additional energy is extracted from the brine by delivering a supply of steam to the power plant at a lower pressure, but the concentration of minerals in the residual brine is increased. Many fields use only one flash to avoid silica saturation of the brine. Multiple flashes can quickly drive the brine past the silica saturation point by the combined effect of cooling and concentration due to the extraction of steam. Where multiple-flash separation is used, it is usually limited to two flash pressures. At the CalEnergy* Salton Sea facilities, turbines are driven off of as many as three separate flashes, producing turbine inlet pressures of roughly 250, 100, and 10 psig (actual pressures vary among the various plants) with a final flash at atmospheric pressure that is vented, but a large silica precipitation facility is employed to desupersaturate the silica from the brine to prevent pipeline and injection well fouling.

Figure 51.4 illustrates a multiflash steam separation and cleaning process from the well to the turbine inlets. Brine from the first separator is flashed to the lower pressure across

* CalEnergy Generation, http://www.calenergy.com.

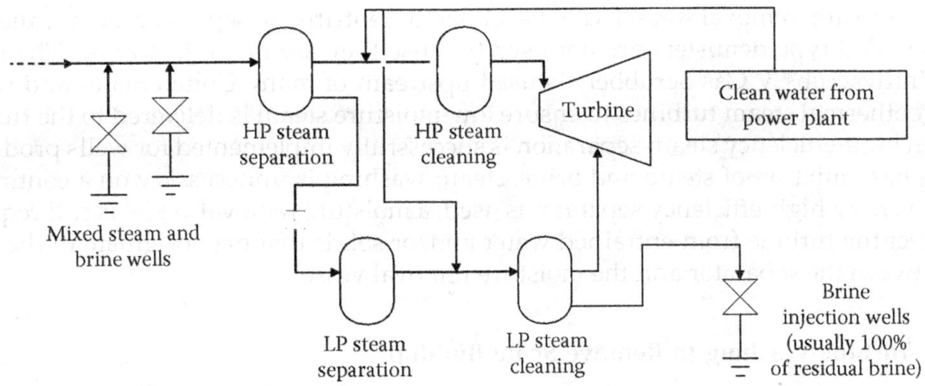

FIGURE 51.4
Geothermal double-flash process with steam cleaning.

the level control valve of the higher pressure separator. Commonly, each flow of steam is washed with geothermal steam condensate from the power plant before entering the turbine. Residual brine is injected.

51.6.3 Steam Pipeline Operation

Steam pressures used in geothermal power generation are often less than 120 psig at the turbine inlet, and the steam pipelines are sized to keep steam velocities low. Thus, on large-steam turbine installations, there may be many miles of 30–42 in. pipeline snaking along roads and hillsides to bring the steam to the power plant.

These long pipelines introduce various operational problems. The pressure drop that occurs over such long distances represents an unrecoverable loss of available energy.* The long distances and the transport of saturated steam can also lead to considerable condensation in the pipelines, especially during windy, rainy, or snowy weather. Condensed steam must be removed from the pipeline either in knockout pots with steam traps, or in the steam scrubbing vessel near the turbine inlet to prevent water damage to the turbine.

51.6.4 Steam Washing Prior to Steam Turbine Admission

After the steam is brought to the edge of the power plant from the dry steam wells or the separators, the steam at many facilities is washed by injecting steam condensate from the power plant condenser into the steam line. Typical flow rates look to achieve a total moisture of 1%–2.5%, or about 10,000–25,000 lb/h (up to 50 gal/min) for a common 55 MW power plant using about 1 million lb/h of steam. For each 100 lb of condensate injected into the steam flow, approximately 22 lb of steam is condensed. This condensed steam also helps to scrub the vapor, but the condensation is essentially a parasitic load. In a large facility, with poor steam quality, the wash water can represent as much as one megawatt of steam flow. The condensation and any injected water are removed in the power plant moisture removal vessel that immediately precedes the turbine.

* A wellhead separator pressure of 160 psia and a turbine inlet pressure of 110 psia are commonly seen. Assuming a constant condenser pressure of 101 mm Hga (4 in.), this pressure drop between wellhead and turbine represents up to a 7% loss of available energy.

The moisture removal vessel can be either a centrifugal separator or a vane-type demister. Pad-type demisters are not used because they are prone to scaling. The Porta-Test Whirlyscrub® V Gas Scrubber* is used upstream of many United States and worldwide geothermal steam turbines to ensure low-moisture steam is delivered to the turbine.

When high-efficiency steam separation is successfully implemented for wells producing a two-phase mixture of steam and brine, steam washing is unnecessary on a continuous basis. Even if a high-efficiency separator is used, a moisture removal vessel is still required to protect the turbine from entrained water and/or solids that may originate in the pipeline between the separator and the moisture removal vessel.

51.6.5 Turbine Washing to Remove Scale Buildup

The consequence of insufficient steam purity and contaminant cleanup is that the turbine first-stage nozzle diaphragms will develop a layer of scale that restricts the steam passage. A schematic illustration of the turbine pressure and generation response to inlet nozzle scale build-up and removal through turbine washing is shown in Figure 51.5. As scaling occurs, the operator will allow the turbine inlet pressure to increase to maintain plant contracted output. As the scaling and pressure build up, the pressure limit of the turbine casing is eventually reached, and at that point, further scaling of the turbine results in reduction of the of turbine output due to decreased steam flow.

Turbine scaling occurs as the steam passes across the first-stage stationary blades. Entrained microdroplets of liquid with dissolved impurities evaporate as they cross the blades. The impurities are left behind, and scale buildup is the result.

Over the years, various experiments have been undertaken to remove the built-up scale with online turbine washing. Early efforts took the unit off-line and injected water at reduced rotating speeds. Later, low load injection of water into the turbine inlet was tried. Eventually, however, it became the widely accepted practice to inject water into the turbine at almost full-load to clean off scaling and restore design operating conditions. This practice has evolved in part by consideration of the actual operation of a low-pressure geothermal turbine. In normal operation, the saturated steam enters the turbine with condensation forming as the steam passes through the turbine to produce work. A typical

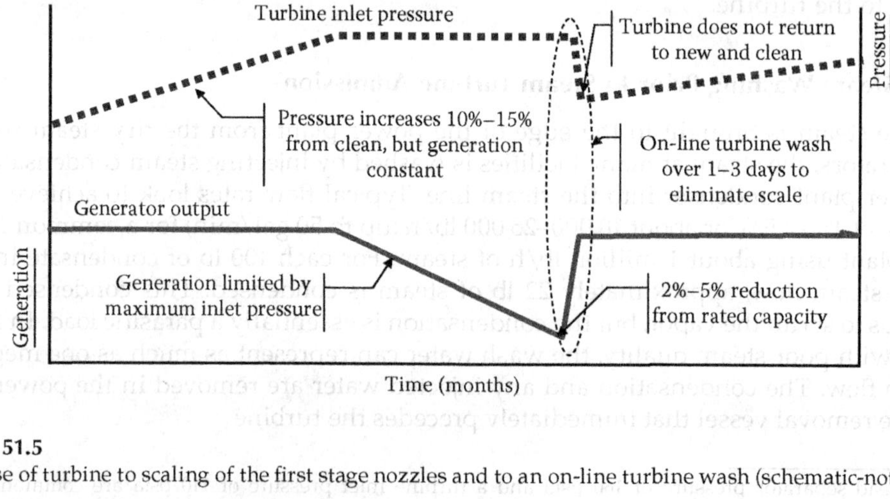

FIGURE 51.5
Response of turbine to scaling of the first stage nozzles and to an on-line turbine wash (schematic-not to scale).

* http://www.natcogroup.com/.

geothermal steam turbine operating around 120 psia inlet pressure and 2 psia condenser pressure (4 in. Hg, absolute) has an exhaust moisture of 14%–16% less interstage moisture drains. Therefore, adding 1%–2% moisture to the steam entering the turbine for short periods of time does not represent a large deviation from normal operating conditions.

51.7 Geothermal Power Production—Steam Turbine Technologies

51.7.1 Overview of Geothermal Power Generation Technologies

Two conversion technologies dominate the geothermal power industry: Direct steam turbines in an open-loop Rankine cycle configuration, and binary power plants using a hydrocarbon (pentane or butane) in the boilers and turbines in a closed-loop organic Rankine cycle (ORC). Geothermal combined cycle (GCC) units marry the two technologies. Ammonia/water binary power plant technologies have been proposed as a potential third power cycle option, with one such plant having been built as of 2006, and additional plants planned for operation in the coming years. Table 51.7 lists the 2004 breakdown of worldwide installed capacity.

Steam turbines represent 95% of the installed capacity of geothermal power plants worldwide, including the steam turbines of the integrated steam turbine and binary power plants. Binary power plant technologies, discussed in Section 51.8, represent the remaining 5% of installed worldwide capacity.

51.7.2 Steam Turbine Conversion

Geothermal power plants using steam as the motive fluid are open-cycle systems in which geothermal energy in the form of steam is admitted to the power plant, the energy is extracted, and the residual brine and condensed steam are then discharged from the plant to surface disposal, or injection, as described above. Gases contained in the steam are vented to the atmosphere. Because there are no fuel handling systems, boilers, superheaters, ash systems, or other systems related to the burning of fuel, geothermal steam power plants

TABLE 51.7

Breakdown of Geothermal Power Technologies Worldwide

Resource Type	Power Plant Type	Installed Capacity (MW)	Number of Units
Dry steam	Steam turbine	2460	63
Flash steam	Steam turbine	5831	210
Pressurized brine, including residual brine from a flash steam plant	Organic Rankine cycle (ORC)	274	154
Steam/brine	Integrated steam turbine and ORC, or ORC only	306	39
Pressurized brine	Binary-Kalina cycle (ammonia)	2	1
	Total	8873	467

Source: DiPippo, R., *Geothermal Power Plants: Principles, Applications and Case Studies,* Elsevier Advanced Technology, Oxford, U.K., 2005, pp. 404–405.

are very simple. It is referred to as an "open" cycle because the condensed steam is not returned to a boiler in a closed-loop circuit. The steam from the geothermal resource is delivered to the turbine inlet and may originate from either dry steam wells or from the separated steam of wells producing a brine/steam mixture.

There are three main steam power processes in use around the world: steam turbines without condensers (atmospheric discharge), steam turbines with subatmospheric condensers, and steam turbines integrated with binary power plants.

In a few installations worldwide, the steam turbine may exhaust directly to atmosphere. It is an inefficient mode of operation since 50% of the available energy is lost when discharged at atmospheric pressure. From an economic perspective, the loss of 50% of the available energy does not justify long-term operation of atmospheric discharge turbines, because with only the addition of a condenser and cooling tower, the output of the facility can be roughly doubled with no additional resource development. However, atmospheric discharge turbines are sometimes used to prove reservoir capability because of rapid installation and low cost. In other places, they have been used when CO_2 content in the steam was over 5%. The use of atmospheric discharge turbines provides another example of the trade-offs that occur between long-term efficiency benefits and economic constraints as discussed in Section 51.5.3.

Geothermal combined-cycle steam/binary power plants also use a steam turbine that discharges at slightly above atmospheric pressure. However, instead of the steam venting to the atmosphere, the steam is condensed and energy recovered in the vaporizers and preheaters of the binary power plant. In this way, the energy of the exhaust steam is captured in the power conversion cycle of the binary power plant. More details on this process and advantages of this "combined cycle" technology are discussed in Section 51.8.4.

Most of the total installed worldwide geothermal generating capacity is from steam turbines in which the geothermal steam is admitted to the turbine at between 15 and 250 psia and exits the turbine at subatmospheric pressures of between 1.0 and 2.5 psia. The condenser is cooled by water from a cooling tower. The steam rate (pounds/kilowatt-hour) of the turbines is determined by the turbine efficiency (usually slightly better than 80%, including entrance and leaving losses), inlet pressure, and the condenser pressure. Figure 51.6 is a photo of the world's largest-capacity (110 MW net rated output) single shaft geothermal steam turbine, in the Fuji Electric* shop before its installation at the Wayang Windu (Indonesia) geothermal power project. The design inlet and condenser pressures are 150 and 1.7 psia, respectively, using a 27 in. long last stage blade.[15]

51.7.3 Condensing Steam Turbine Process

Figure 51.7 is a general illustrations of a dry steam power process. Steam is produced from the geothermal reservoir through multiple wells and delivered to the power plant through cross-country pipelines. The steam is always passed through a moisture removal vessel before entering the turbine to ensure that no particles or slugs of water reach the turbine. In many installations around the world, the steam is washed before entering this vessel to remove any impurities (see Section 51.6.4).

The basic process for a single-flash steam process is much the same, as shown in Figure 51.8. The key difference is that a two-phase mixture of brine and steam are produced from the geothermal reservoir to the surface. A centrifugal separator removes the steam for delivery to the power plant (Section 51.6.1), and the brine is injected back into the reservoir.

* Fuji Electric Systems Co., Ltd., http://www.fesys.co.jp.

FIGURE 51.6
The Wayang Windu 110 MW geothermal turbine. (Photo courtesy of Fuji Electric Systems Co., Ltd., Tokyo, Japan.)

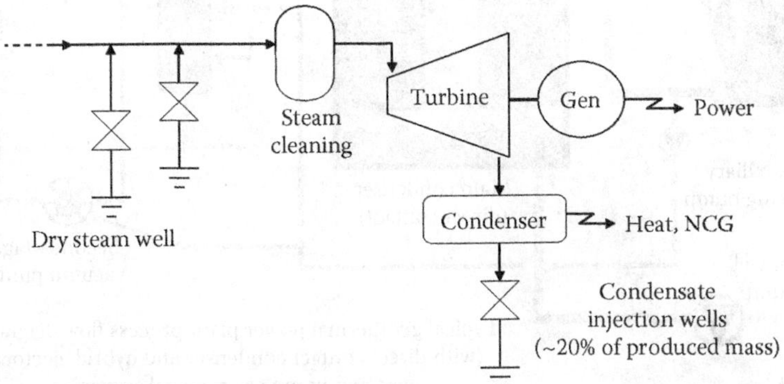

FIGURE 51.7
Geothermal dry steam power process (water cooled).

Figure 51.9 is a typical process-flow diagram for a geothermal power plant using a steam turbine and a condenser. In this illustration, steam is supplied from dry steam wells or a flash separation process at only one pressure (i.e., there is no low pressure turbine inlet). The steam is also used directly in the gas removal ejectors. Steam exiting the turbine is condensed in a direct contact condenser that is supplied with cold water from a cooling tower. The condensed steam and the cooling water are returned to the cooling tower. Approximately 80% of the steam to the power plant is evaporated in the cooling tower, and the remaining 20% forms the cooling tower blowdown. In hot arid climates, the amount of water evaporated can be greater than 100% during the hottest hours of the day and greater than 90% over a 24 h period. The steam contains noncondensable

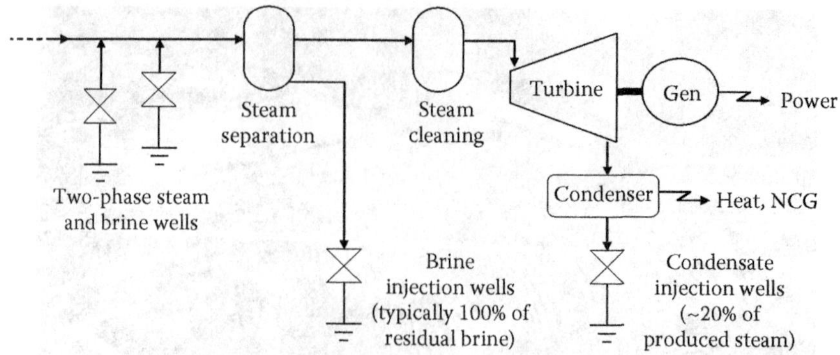

FIGURE 51.8
Geothermal single-flash steam power process (water cooled).

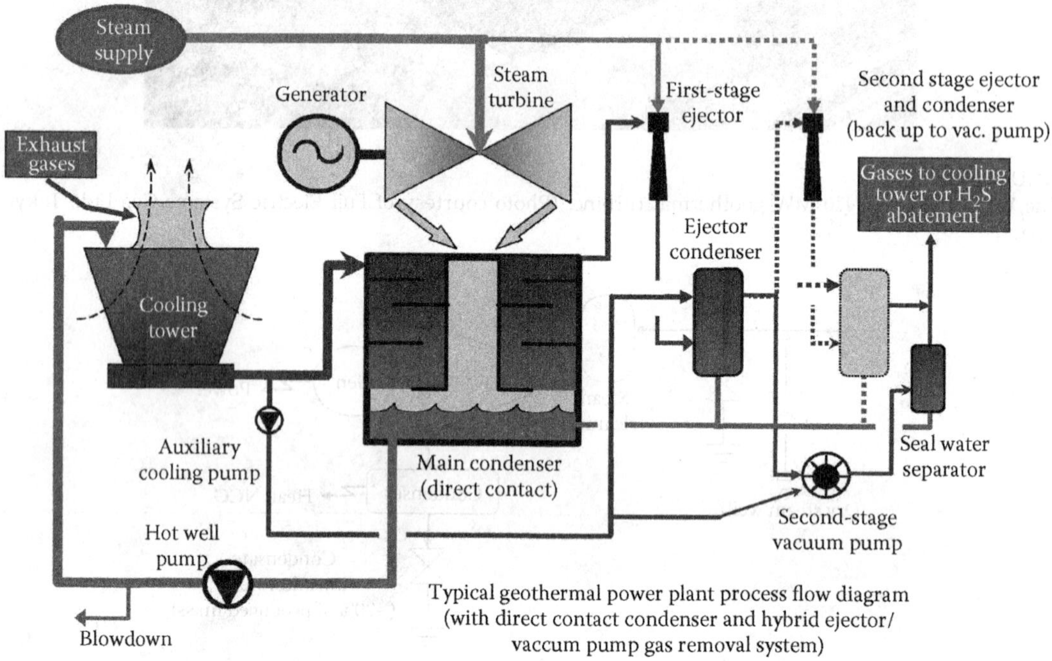

FIGURE 51.9
Geothermal steam power plant with direct contact condenser. (From Williamson, K.H. et al., *Proc. IEEE*, 89(12), 1783, 2001.)

gases (NCG), primarily CO_2, but also H_2S and other trace gases. These gases must be removed from the condenser to keep a low turbine exhaust pressure.

In Figure 51.9, geothermal steam is used as the motive force for the first-stage ejector to compress the NCG. The ejector steam is condensed in the direct-contact ejector condenser (the intercondenser), with the condensed steam and cooling water being returned to the main condenser. The NCG from the intercondenser is compressed in a second stage to atmospheric pressure either by the vacuum pump or by a second-stage ejector with an aftercondenser. The amount of steam used by the ejector and the size of the vacuum pump motor is related to the size of the power plant and the NCG content of the steam.

The vacuum pump is the primary system because it is more efficient than an ejector. The second-stage ejector is shown as a backup system because it has lower capital costs than a standby stainless steel compressor. Stainless steel is required throughout the NCG system because of the corrosiveness of moist CO_2 gases. The NCGs are dispersed using the cooling tower fans. If H_2S abatement is required, it is installed between the discharge of the second stage of compression and the cooling tower. The gas removal system is discussed in greater detail in Section 51.7.6.

51.7.4 Design of Geothermal Turbines

Geothermal steam can be a virtual chemical soup that varies from field to field. In most cases, the steam is saturated as it enters the turbine, and has a moisture content approaching or even exceeding 15% as it leaves the turbine for subatmospheric discharge turbines. The combination of scaling tendency, potential corrosivity, and high moisture content heavily influences the design of geothermal turbines.

Geothermal turbines generally use the same alloys as low pressure turbines in fossil fuel power plants, but with small variations in the alloy composition of the blades and rotors to enhance the alloy's ability to withstand the corrosive environment of geothermal steam. Detailed design can be just as important as alloy selection. For example, by using long-radius transitions and avoiding notches in the design of geothermal turbine parts, especially in the blade attachment areas, the stresses are more evenly applied and the blades and rotors are much less susceptible to cracking and failure.

The last-stage turbine blades are exposed to high moisture content, which requires methods to minimize blade leading edge erosion. Both erosion-resistant alloy shields and blade-surface hardening are used to resist the impingement of water on the last-stage blades. In addition to moisture, the specter of corrosion once again strongly influences the design of geothermal turbines at the last stage.

Last-stage blade size is a key element in the maximum size geothermal turbine that can be built. Geothermal turbine last-stage blades are designed for lower combined stresses than equivalent blades in conventional power plants. Because a large portion of last-stage blades stresses is proportional to the steam flow, this results in lower steam flows (per square foot of exhaust area) than in conventional power plants. High moisture content—the result of using saturated steam at the turbine inlet—tends to limit blade size to about 660 mm (26 in.) for turbines operating at 3600 rpm (60 Hz power), and its equivalent of about 765 mm (30 in.) for turbines operating at 3000 rpm (50 Hz power). The combined effect of the limited blade length and the reduced flow rate per unit area limit the maximum size that can be achieved in a geothermal turbine/generator. Thus, it is common to see two to three 55–70 MW turbines installed in a single powerhouse, rather than a single 200 MW unit. As discussed above, the largest two-flow geothermal turbine/generator (two rows of last-stage blades with steam inlet in the middle) ever built was Wayang Windu in Indonesia, with a rated net output of 110 MW.

51.7.5 Design of Heat-Rejection Systems

A second area affecting geothermal competitiveness is the heat-rejection system comprising the condenser, cooling tower, and cooling water pumps. The heat-rejection system of geothermal power plants can be up to eight times as large as a combustion-turbine combined-cycle power plant per kilowatt of output because of the low available energy per mass of steam at the inlet to the geothermal turbine. The heat-rejection system of any geothermal power plant, whether for a steam turbine or binary plant, represents a very large capital cost burden for the project.

Where H_2S abatement is required, modern plants are most often built with shell-and-tube condensers. Thus, shell-and-tube is found in most steam condensers in United States geothermal power plants. Outside of the United States, the majority of geothermal steam condensers are direct-contact. In this system, steam exiting the turbine encounters either water sprayed from nozzles or droplets from trays. Heat transfer is much more efficient. The higher the gas content of the steam, the larger the advantage that direct-contact condensers bring. The disadvantages are that the cooling water is exposed to the CO_2, H_2S, and ammonia, thereby providing substantial biological growth potential in the circulating water. Counteracting this is the high blowdown rate in a direct-contact geothermal cooling system.

Cooling tower make-up water comes from the condensed geothermal steam. In most circumstances, between 70% and 80% of the steam that enters the turbine is eventually evaporated in the cooling tower. The remaining 30%–20% steam is condensed and not evaporated, and is the source of the cooling system blowdown. Although the low cycles of concentration in the cooling water help control biological growth in the system, at the same time, the high blowdown rates result in it being relatively expensive to dose the water with biocides, since the biocides are quickly washed out of the system. Additionally, the presence of H_2S and NH_3 in the cooling water substantially reduces the efficacy of conventional biocide treatments. Cooling tower blowdown is most commonly injected back into the geothermal reservoir.

51.7.6 Geothermal Condenser Gas-Removal Systems

In addition to the challenges of mechanical design of the steam turbine and the size of the cooling and condenser system, the steam also contains large quantities of noncondensable gases (NCG). Average NCG compositions for two fields in the Philippines are shown in Table 51.8.

As a consequence of high NCG content in the geothermal steam, the gas-removal systems of geothermal power plants dwarf the gas-removal systems of conventional power plants. For example, the power plants at Tiwi, with a gas content of about 3%, can use either a two-stage steam jet gas ejector that consumes 110,000 lb of steam/h, or a hybrid system of first-stage ejector and second-stage liquid ring vacuum pump (consuming 33,000 lb of steam/h and 0.6 MW of electrical load). This is summarized in tabular form in Table 51.9. The two systems were designed for different gas loads, but both systems place a tremendous parasitic load on the plant. The two-stage ejector consumes an additional 11% steam over that used by the turbine, while the ejector hybrid consumes only 3.4% of the turbine steam, but also consumes 1.2% of the net electric output of the facility.

TABLE 51.8

NCG Amount and Composition in Steam at Tiwi and Mak-Ban, Philippines

Field	Percent Gas in Steam (wt%)	Gas Composition (mol%)								
		CO_2	H_2S	NH_3	CH_4	H_2	N_2	O_2	He	As, Hg
Tiwi	2.8	97.9	1.24	0.04	0.05	0.31	0.45	0.04	0.001	Trace
Mak-Ban	0.5	88.7	6.9	0.08	0.22	1.24	2.61	0.10	0.006	Trace

TABLE 51.9

Parasitic Steam and Power for Condenser Gas Removal at Tiwi, Philippines, for One 55 MW Unit

		Parasitic Steam		Parasitic Power	
Option	Design Gas Load % Steam	Ejector Flow (lb/h)	% of Turbine Flow	kW	% of Rated Capacity
Two-stage ejector	2.8	110,000	11.1	Not applicable	
Ejector hybrid	2.8	33,000	3.4	635	1.2

The higher the NCG content is above 1% in the steam, the greater the incentive to cut the parasitic steam usage with mechanical gas compression. This may take the form of ejector hybrids, as discussed above, or to eliminate the ejector altogether using an all-mechanical compressor. As in most aspects of geothermal design, a variety of solutions is employed.

Whatever the particular compressor system used, the additional cost of compressor systems is usually justified by the reduction of production wells required to supply the ejectors with steam, by an increase in the turbine steam supply at a geothermal field that is short of steam, or looking to maximize installed capacity. However, the electric parasitic load can have considerable negative implications on plant revenue, and for this reason, it is sometimes avoided on newer installations. The reason that the electric loads of mechanically driven compressors can have large impacts on revenue is found in the size limitation of the geothermal turbine, as discussed above. If the turbine size is limited (e.g., by last-stage blade size), then adding parasitic electric load reduces the power that can be sold, and hence reduces revenue. This is, once again, an example of the need to sacrifice efficiency in the short term to obtain higher net present value in a discounted cash flow calculation.

51.8 Geothermal Power Production—Binary Power Plant Technologies

Section 51.7 presented the technologies and strategies of the geothermal power plant that directly uses geothermal steam in the turbine. In Section 51.8, the use of geothermal steam and brine as an energy source but not as the power plant working fluid (that which drives the turbine) is discussed. The power plants that use these two fluids are known as *binary power plants*. All of the closed-loop working fluid cycles used in geothermal energy production, with a few exceptions, are Rankine cycles with a hydrocarbon as the working fluid.

51.8.1 Binary Power Plant Advantages

If one were to consider an idealized geothermal system in which the thermodynamics of pure water and steam played the dominant role in decision-making, there may not be a large role for the binary geothermal power plant. However, one does not have to delve too far into the reality of geothermal development before the advantages of the binary geothermal power plant in some applications become apparent. Some of these reasons have already been discussed in the preceding pages. Two of the most important answers to the question "Why binary?" are provided below.

51.8.1.1 Silica Solubility

As noted in several earlier sections, avoiding the precipitation of silica is a requirement for successful long-term geothermal well and pipeline operations. The precipitation of silica is a function of three factors: time, concentration, and temperature. Time is easily extended by the addition of acid to the geothermal brine, but for inexplicable reasons, this straightforward and easy solution is widely ignored. That leaves concentration and temperature as the factors within the control of the power process designer.

A brine that starts out at 450°F in the reservoir, when introduced into a steam separator at a pressure of 135 psia, will flash off approximately 12% of the initial water mass as steam. If there is a second flash at 35 psia, the total water mass flashed to steam is 21%. In such a flash process, two factors are at work increasing the potential of forming silica scale: mineral concentration in the brine and lower brine temperature. The physical properties that inhibit scale precipitation are that amorphous silica solubility is greater than the quartz silica solubility, and the often-slow kinetics of amorphous silica precipitation.

In contrast, the geothermal heat is transferred to the working fluid in a binary power plant without concentrating the silica by flashing off steam from the brine. Therefore, the temperature of the brine can be lowered further before the amorphous silica precipitation temperature is reached. Because of the silica solubility limit, the binary power plant can extract more energy from the produced brine (lower temperature) than a flash plant is capable of, especially where pH modification of the brine (acid addition) is not used. This can be true either on wells that are pumped and in which pressurized brine is delivered to the power plant, or after the brine has been flashed once and the binary plant is then installed to make use of the heat in the residual brine.

51.8.1.2 Lower Parasitics vs. Cycle Inefficiency

In a binary plant, a heat exchanger must be used to transfer the heat from the geothermal fluid to the power plant working fluid. As such, there is an inherent loss of efficiency from the process, as well as the capital cost of the heat exchanger itself. There is also a cycle feed pump, a significant power consumer that has no counterpart at all in a geothermal steam turbine power plant. On the other hand, in a plant directly using geothermal steam, the process of removing NCGs from the condenser is also an inherent loss of efficiency and a significant capital cost that does not exist in a binary power plant. How the two options balance out is dependent on site-specific factors, but conceptually, there are offsetting advantages and disadvantages, making the thermodynamic and economic balance closer than it might appear at first glance.

The exception to the offsetting advantages of binary plants versus direct steam plants occurs when the use of binary also forces the use of air-cooled condensers instead of water-cooled condensers and evaporative cooling towers. The water-cooled direct contact condenser has lower capital cost, higher efficiency, and lower parasitic loads than an air-cooled condenser system in all but the coldest climates.

51.8.1.3 Compactness

Binary power plants operate with working pressures much higher than those used for low-temperature flash power plants. This translates to smaller pipes and eliminates flash separators and scrubbers on pressurized brine systems; additionally, the condensers have smaller volumes because of the much higher condenser pressures of the working fluids.

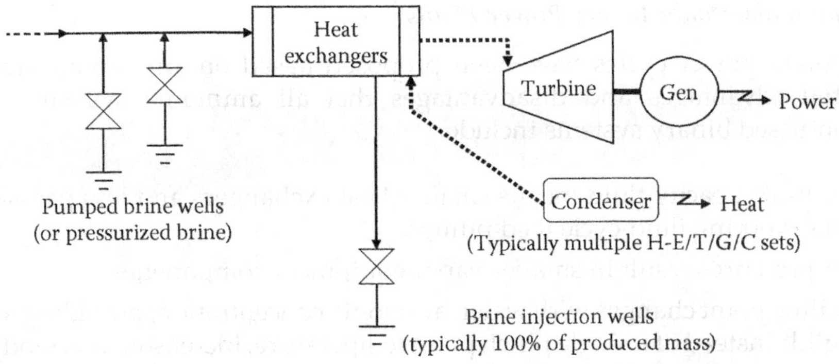

FIGURE 51.10
Pressurized brine binary power process (air or water cooled).

51.8.2 Types of Binary Systems

There are two types of binary power systems currently in use, both of which are based on hydrocarbons, and a third currently under development that is based on using an ammonia–water mixture as the working fluid. A fourth binary power system based on commercial refrigeration equipment is also under development, with a 200 kW demonstration project starting operation in Alaska at Chena Hot Springs in 2006. All of these power cycles are similar to that shown in Figure 51.10.

51.8.2.1 Subcritical Pentane Binary Power Plants

The geothermal fluids (brine or steam) are used to boil pentane as the power cycle working fluid in an ORC. The pentane passes through a turbine coupled to a generator and is then condensed. The cooled geothermal fluid is injected back into the geothermal reservoir.

The use of pressurized geothermal water from pumped wells in a subcritical pentane-based power plant is the most prevalent method to utilize geothermal resources from 280°F to 320°F. This system is also used to extract additional energy and power from the residual brine from single-flash flash plants all around the world.

Where two-phase geothermal fluid is available instead of pressurized brine, the two-phase fluids can be used in one of two ways. In some instances, both the steam and the brine are used in the vaporizers and preheaters. On larger facilities, it is more common to see a mixture of steam turbines and ORC plants. This type of integrated facility is made up of a pressurized brine facility of the type described above and a geothermal combined-cycle facility. In the geothermal combined-cycle facility, geothermal steam is first passed through a steam turbine and then condensed in a binary power plant.

51.8.2.2 Supercritical Hydrocarbon Binary Power Plants

Several hydrocarbon-based binary power plants have been built that use supercritical isobutane as the working fluid in an ORC. The pinch point is eliminated in the vaporizer in the supercritical regime. Although there is a theoretical thermodynamic benefit to a supercritical power cycle, the market has not embraced these plants because they tend to be much more expensive than the subcritical plants.

51.8.2.3 Ammonia–Water Binary Power Plants

Several different power cycles have been proposed based on ammonia-water systems. The potential advantages and disadvantages that all ammonia systems share over hydrocarbon-based binary systems include:

- Higher heat-capacity fluid means smaller heat exchangers and less parasitic load from the working fluid cycle feed pumps.
- Higher pressures result in smaller vapor equipment components.
- The boiling point changes with water/ammonia concentration, providing a boiling point glide instead of a single pinch point temperature, increasing thermodynamic efficiency.
- Condensation temperature changes with concentration, again providing a glide instead of a pinch point and a higher thermodynamic efficiency.
- The design of the heat exchangers is more complex to achieve the glides.
- There is a tradeoff of ammonia toxicity for hydrocarbon flammability.
- Ammonia has been used in refrigeration cycles for a century, providing a wealth of practical operating knowledge, even though few ammonia-based power plants exist.
- Ammonia vapor and steam have similar molecular weights (17 and 18 g/mol, respectively), which theoretically means that conventional steam turbines could be used, with little modification; if true, it would allow a rapid step up in turbine size to 30 MW or more and for the competitive bidding for supply of the turbine.
- Ammonia cycles are more complex, with more exchangers and cycle pumps.

In spite of the net benefit of the potential advantages, only one 2 MW geothermal plant has been installed to date, in Husavic, Iceland, using one of many Kalina cycle ammonia–water technologies. Plans to install additional Kalina cycle plants at Salt Wells, Nevada, and a 42 MW plant at Cove Fort, Utah, have been announced.

51.8.3 Binary Power Plants for Pressurized Geothermal Brine

Ormat Technologies* has developed a binary power plant approach, using subcritical pentane-based Rankine cycle technology, that commercially dominates the pressurized brine market. Because of that dominance, their technology will be used to illustrate the principles of the binary power plant design in this section. The technology has also been used in some instances in which the wells produce a mixture of steam and brine. These same basic power plants are now also being used to generate power from natural gas pipeline compressors exhaust gas waste heat.

Ormat started out designing and building small binary plants. One of their first installations was at the East Mesa field in California near the Mexican border. The ORMESA I plant began operation in 1987 and was comprised of 26 individual boiler/turbine/condenser units, with a total output of 24 MW. Since the East Mesa installation, Ormat has steadily increased the size of their units in the pursuit of both economy of scale and efficiency, and hence, lower installed cost ($/kW). Initially, Ormat technology was based on a modular design, in which the economy of scale of a single large plant was traded off in

* Ormat Technologies, http://www.ormat.com.

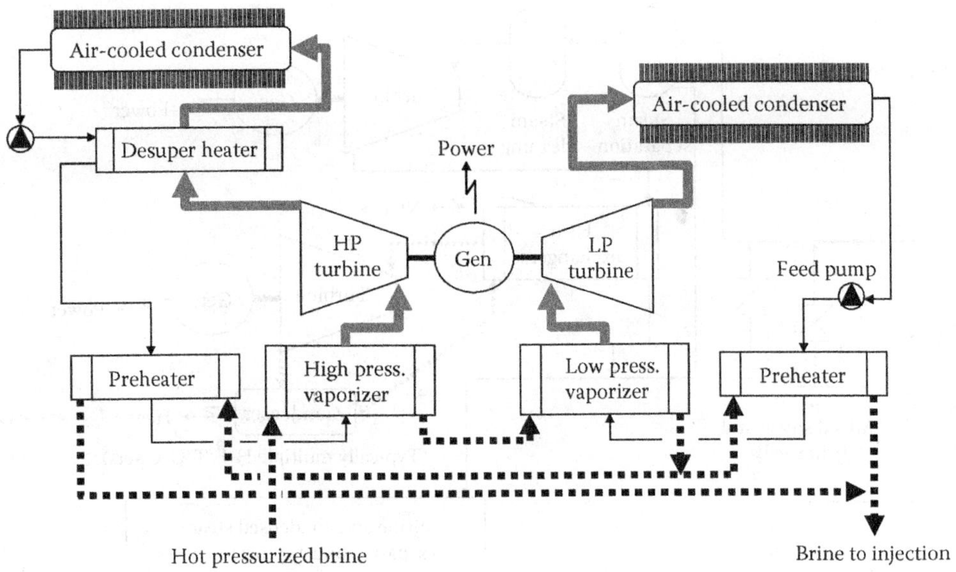

FIGURE 51.11
Ormat two-level subcritical pentane power cycle.

favor of the economy of scale of many smaller standardized, modular, factory-assembled skid-mounted units that required minimal field construction. Ormat now also offers a larger T/G unit comprised of a high- and low-pressure turbine coupled to opposite ends of a single generator, with the capability to deliver 20 MW to the grid.

The basic ORC process design used by Ormat is illustrated in Figure 51.11. The hot pressurized geothermal brine, whether from a pumped well or as residual brine from a flash plant, enters the high-pressure (HP) vaporizer to generate HP pentane vapor for the HP turbine. The geothermal fluid exits the HP turbine and enters the low-pressure (LP) vaporizer, which generates vapor for the LP turbine. From the LP vaporizer, the geothermal fluid splits between the preheaters of the LP and HP vaporizers. The pentane for the LP turbine is pumped out of the condenser to the preheaters and into the vaporizers. Pentane vapor passes through the turbine and into the condenser. The HP pentane cycle is similar, except that the pentane from the condenser is first pumped to a desuperheater located between the turbine discharge and the condenser.

Examples of where pressurized brine from the wells is used directly in the binary power plant include Soda Lake and Steamboat Springs in Nevada, as well as Heber and East Mesa in California.

Binary power plants have been installed to capture the waste energy of several geothermal power plant hot brine discharges. Examples include Brady Hot Springs (Nevada), Mak-Ban (Philippines), Los Azufres (Mexico), Miravalles (Costa Rica), and Wairakei (New Zealand). There also remain many locations where binary power plants could be installed on existing steam flash plants.

51.8.4 Integrated Steam Turbine and Binary Power Plants

In other locations, steam turbine(s) and binary power units have been combined into a single facility. Integrated facilities are installed at fields where the geothermal fluid is produced to the surface as a mix of steam and brine, rather than pumped and pressurized brine.

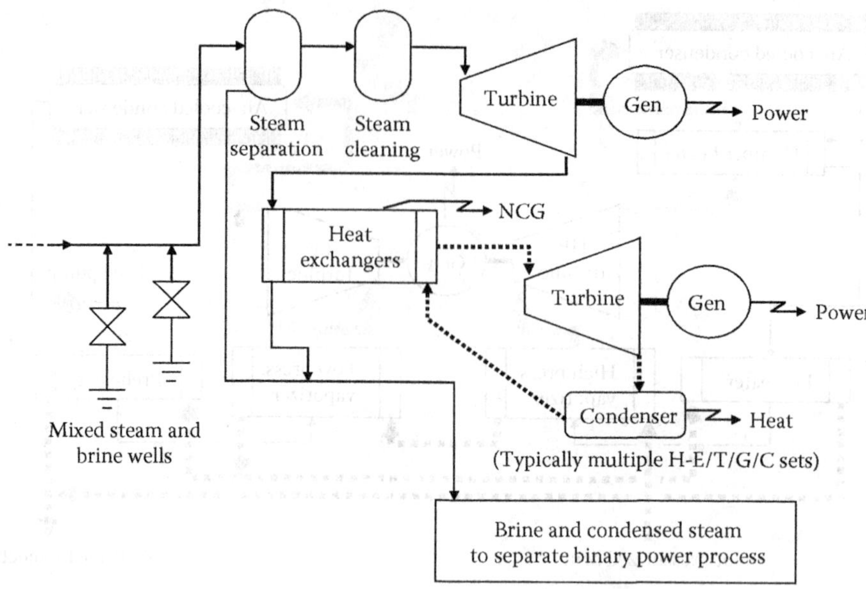

FIGURE 51.12
Integrated brine steam combined cycle binary power process (air cooled).

The largest integrated facilities are those of Ormat Technologies, and are based on the integration of the pressurized brine ORC and what is referred to as GCC plants. See Figure 51.12.

51.8.4.1 The Geothermal Combined Cycle Process

The brine/steam mix flows from the production well to a separator. Brine from the separator is directed to a pressurized brine binary power plant, as described above.

Steam from the separator is cleaned and delivered to a steam turbine. The steam turbine faces the same inlet steam quality issues as were discussed in Section 51.8. However, because the turbine discharges at near-atmospheric pressures, the previously discussed concerns about last-stage blade length and moisture-induced erosion are eliminated. With the elimination of these two issues, cycle optimization of the geothermal resource can be undertaken without a cap on the maximum turbine inlet pressure. The available mass and the optimized pressure of the steam result in varying steam turbine sizes and outputs at different geothermal fields.

At the larger GCC units (see Table 51.10), the steam that discharges from the steam turbine is divided among multiple binary units. Each binary unit condenses the steam in a process of heating and vaporizing the binary unit working fluid. The condensed steam is mixed back into the brine to reduce silica scaling potential. In some instances, small steam turbines are coupled to one end of a generator, with the steam discharged used in a binary plant with the pentane turbine coupled to the other end of the generator. In most cases, the condensers for the binary power plant are air-cooled.

51.8.4.2 Silica Solubility Limits

In fields with a steam/brine mix, the GCC competes directly with multiflash steam technology. The GCC can produce greater output where flash plants have elected to use a

TABLE 51.10

Examples of Ormat Geothermal Combined Cycle Projects

Year	Project Name and Location	Project Net Output (MW)	Description of the Air-Cooled Power Plant
1996	Upper Mahiao, Leyte, Philippines	125	Four geothermal steam turbines that feed multiple binary units for 120 MW of net output. An additional 5 MW of power are generated by other binary units using the geothermal brine
2000	Mokai, New Zealand	60	One geothermal steam turbine that feeds multiple binary units for 50 MW of net output. Ten megawatts of power are generated by binary units using the geothermal brine

single flash to control silica (SiO_2) scaling, resulting in a brine discharge temperature of 300°F or greater. The same temperature limitation of the single-flash technology does not apply to the GCC, so the GCC is able to capture more energy from a given brine flow and therefore generate more power. Although it may at first seem counter-intuitive, the hotter the resource and thus the more steam that is produced in the flash process, the more advantage the GCC will have.

In a flash process, hotter brine results in a greater silica concentration in the reservoir fluid. Hotter brine also results in more flash, and therefore more concentration of the silica. The silica solubility limit controls the separated brine temperature, and therefore the minimum allowable flash pressure and temperature. Thus, because of the silica concentration, higher geothermal resource temperatures also result in a higher brine temperature that must be discharged from the flash plant.

Although the integrated binary plant also uses the initial flash in the steam turbine of the GCC, the condensed steam is mixed back into the brine flow. As a result, the final brine temperature can be lower without risking silica scaling. Consequently, more energy is available to the integrated binary power plant than is available to the flash-only plant. The binary plant advantage would be reduced if more geothermal operators were willing to use acid to slightly lower the pH of the brine and retard the kinetics of the silica precipitation.

51.8.4.3 Compare and Contrast

The GCC technology has some additional advantages and disadvantages over a single- or multiple-flash plant when compared side-by-side for a particular geothermal resource. The large advantage that the GCC obtains in greater energy availability due to the silica solubility limits is discussed in the paragraph above. The larger units that Ormat has developed in recent years will also help to make them more competitive through an economy of scale. Another advantage is obtained by discharging the steam turbine at slightly above atmospheric pressure. In this way, the NCGs are vented without the cost or parasitic load of the gas removal equipment that is needed when the steam turbine exhaust is at sub-atmospheric pressure, as occurs with the steam flash process. Since the steam condensate is mixed back into the brine with the GCC, the cost of an oxygenated fluid disposal well may be eliminated.

The advantages that the steam-turbine process holds over the GCC are: (1) the ability to use large equipment (up to a single 110 MW turbine); (2) no need for large heat exchangers (vaporizers and preheaters) and the intrinsic thermodynamic and capital cost penalty they entail; and (3) the steam flash process uses wet cooling from the condensed steam, whereas the GCC uses dry cooling, resulting in a higher final condenser temperature.

Thus, the steam process regains some of the thermodynamic disadvantage incurred by the silica solubility limit and the gas removal system parasitic steam load by having a lower power plant heat-rejection temperature and, thereby, greater available energy.

51.9 Environmental Impact

The generation of electric power will have an environmental impact, regardless of the energy source used. A large unregulated strip mine and unabated emissions from an old coal plant would probably represent the extreme negative end of the spectrum in its environmental impact to land, water, and air, but would not be particularly relevant to discussions of environmental impacts of power generation in the United States Understanding the other end of the spectrum is much more complex. Solar and wind generation would seem to have zero emissions, but both of these require fossil-fired generation to back them up during periods when these generation technologies are unavailable or erratic. These "firming" requirements increase emissions from fossil-fired plants, and result in them not being the zero-emission technology that they might otherwise seem to be. What this shows is that even renewable technologies have negative environmental consequences. Society's need for new sources of electric power generation, which cannot be served by conservation and efficiency, will result in negative environmental impacts.

Like other renewable technologies, geothermal power generation must be evaluated on a site-specific basis to accurately determine its environmental impacts. Two areas of environmental impact relevant to the use of geothermal energy, emissions and land use, are discussed below.

51.9.1 Geothermal Power Plant Emissions

Air emissions and clean air are the most-discussed environmental benefits of renewable-energy electric power generation, especially with the concern of the effect of greenhouse gases on global warming. Like other renewable energy sources, the emissions from geothermal power are generally low compared to fossil-fired generation. However, certain geothermal resources with high gas content may approach that of a natural gas fired power plant with regards to CO_2 emissions, but this is a rare exception. The concentrated CO_2 discharged from such units does make CO_2 capture feasible. The actual emissions from geothermal power plants are dependent on the highly variable gas content of the geothermal fluids that are being produced and the nature of the geothermal power process. Table 51.11 gives representative emissions of geothermal power plants compared to fossil-fired generation sources.

Pumped binary power plants have no emissions, except for fugitive leaks of the working fluid. As discussed above, the gases are still present in the brine, but because the pressure is never lowered below the bubble pressure of the gases, they remain in solution and are injected back into the geothermal resource.

51.9.1.1 CO_2 Emissions

Flash-steam power plants have emissions on the order of magnitude described, but vary according to the particular resource characteristics and whether the power process is

TABLE 51.11

Emissions from Generation Sources

Plant/Fuel	Emissions Rate (lb/MW h)				Notes
	NO_x	SO_2	CO_2	Particulates	
Conventional power plants					
Coal	4.3	10.4	2190	2.2	
Coal—life cycle	7.4	14.8	—	20.3	—[a]
Natural gas	3.0	0.2	1210	0.1	—[b]
Avg. of all U.S. plants	3.0	6.0	1390	—	[18]
Geothermal power plants					
Pumped geothermal brine	0	0	0	Negligible	—[c]
Geothermal flash steam plants	0	Negligible with abatement to 0.35	60 (avg.) some plants <120 to >900	Negligible	—[d]
Dry steam resources	Trace	Negligible with abatement	90 (Geysers) some plants <120 to >900	Negligible	

Source: Kagel, A. et al., *A Guide to Geothermal Energy and the Environment*, Geothermal Energy Association, Washington, DC, 2005, p. 39.

[a] Kagel reports that this includes the emissions from the mining and transportation of coal to the power plant.

[b] Kagel reports as an average of direct fired, combined cycle, and simple cycle plants.

[c] Cooling tower drift only, if water cooling is used. Air-cooled plants have zero particulate emissions.

[d] 150–900 lb/MW h of CO_2 is added to Kagel's reported numbers for both flash and steam plants. It is based on 0.7%–5% CO_2 in the geothermal steam and 18–20 klb/MW h steam usage in the power plant. These are realistic, though nonproject-specific, values for some high-gas geothermal resources worldwide.

single-flash, double-flash, or coupled to a binary system. Most geothermal flash steam resources have NCG content of between 0.5% and 2% in the steam, which, when coupled with either double flash or a binary system, results in the reported values. However, for those systems that operate with only a single flash and have NCG content of up to 5% in the steam, the CO_2 emissions can reach the 900 lb/MW h reported in the above table.

Like the flash steam plants, the emissions from a dry steam power plant are dependent on the particular resource being considered.

51.9.1.2 H_2S and SO_x Emissions

For geothermal power plants with some form of H_2S abatement, sulfur emissions are negligible to very low, as shown in the table. For those plants without H_2S abatement, the released H_2S eventually converts to SO_2, and can be on the same order of magnitude as U.S. coal plant emissions for geothermal resources with high-H_2S content in the NCGs.

Except for California and the western United States, few geothermal plants worldwide operate H_2S abatement systems. Stretford™ and Lo-Cat™ are the most common H_2S abatement technologies for gas-phase H_2S emissions. The use of bioreactors is an emerging technology for abatement, one in which sulfur-loving bacteria are used to convert the H_2S to sulfate. Bioreactors have been independently developed and put into operation at the Salton Sea, California, and in Wairakei, New Zealand.

51.9.1.3 NO_x Emissions

The reason NO_x emissions are reported as "negligible" instead of zero for the dry steam plants is a consequence of the early H_2S abatement technology installed at The Geysers

TABLE 51.12

30-Year Land Use

Energy Source	Land Use (m^2/GW h)
Coal, including mining	3642
Solar thermal	3561
Central station photovoltaic	3237
Wind, including roads	1335
Geothermal, including roads and pipelines	404

Source: Brophy, P., *Renewab. Energy*, 10, 374, 1997.

that incinerates the H_2S and then scrubs it out of the gas stream using the cooling water. The incineration also produces small amounts of nitrogen oxides. Since those early efforts at H_2S abatement, the geothermal industry has primarily used a cold catalytic conversion of H_2S to elemental sulfur, which has no NO_x emissions.

51.9.1.4 Particulate Emissions

Air-cooled plants have no particulate emissions, whereas water-cooled plants have only the evaporated minerals remaining from cooling tower drift, the quantity of which will be a function of the TDS in the cooling water. This quantity of particulate is negligible.

51.9.2 Land Use

Compared to other renewable energy sources, geothermal power developments have low land use and visual impact. The total geothermal development of wells and power plant may occur over many square miles, but the actual land occupied by the facilities is small. Some negative impacts may be attributed when a geothermal development occurs in a roadless area. There is a visual impact from the pipelines running from wells to power plant, and from the cooling tower steam plume on a cold day. When an air-cooled power cycle is used, geothermal power generation achieves its minimum visual impact. The fenced area of the power plant itself may be no larger than a couple of acres for up to 20 MW of net generation. The height of the air-cooled condenser, the highest point in the plant, is less than 25 ft above ground level. Overall, the land-use impact of geothermal development is small. Table 51.12 provides the total land use of various renewable and coal power technologies.

51.10 Additional Information on Geothermal Energy

This chapter has touched on only some of the many topics in the field of geothermal energy—a subject to which entire books have been devoted. Some additional geothermal energy information sources are provided in this section.

For a broad view on the subject of geothermal energy, including many direct-use applications such as aquaculture, district heating, industrial heat use, environmental issues and financing, a recent book, *Geothermal Energy—Utilization and Technology,* edited by Mary Dickson and Mario Fanelli, is an excellent resource. Each chapter has been written by an expert of geothermal subject matter. At the end of each chapter are questions, answers, and references regarding the particular subject matter of that chapter.

Another recent book, *Geothermal Power Plants: Principles, Applications and Case Studies*, authored by Ronald DiPippo, provides some coverage of geothermal exploration, drilling, and reservoir engineering, with a primary focus on geothermal power generation. The book covers the conversion technologies, thermodynamics, equipment, and operation, and includes several detailed case studies of particular geothermal power plant developments.

In addition, there are a number of excellent geothermal resources on the World Wide Web.

Geo-Heat Center, http://geoheat.oit.edu/	Focus is direct use. Excellent online library. Software and databases
Geothermal Resources Council, www.geothermal.org/	GRC bulletin and transaction articles. Must be a member of the GRC to download articles
Geothermal Energy Association, www.geo-energy.org	The U.S. geothermal industry trade association several excellent broad research articles available for download including three papers on Geothermal Energy Costs, Environmental Impact, and Employment
International Geothermal Association, http://iga.igg.cnr.it/index.php	Downloadable papers from the last two world geothermal conferences

References

1. Lund, J. W., Bloomquist, R., Gordon, B., Tonya, L., and Renner, J. 2005. The United States of America country update—2005. *Geothermal Resources Council Transactions*, 29, 817–830.
2. Ragnarsson, Á. 2005. Geothermal development in Iceland (2000–2004). In: *Proceedings of the World Geothermal Congress*, Antalya, Turkey, 11pp.
3. Lund, J. W. 2005. Worldwide utilization of geothermal energy—2005. *Geothermal Resources Council Transactions*, 29, 831–836.
4. DiPippo, R. 2005. *Geothermal Power Plants: Principles, Applications and Case Studies*. Elsevier Advanced Technology, Oxford, U.K.
5. Wyborn, D., de Graaf, L., and Hann, S. 2005. Enhanced geothermal development in the Cooper Basin, South Australia. *Geothermal Resources Council Transactions*, 29, 151–155.
6. Stark, M. A., Box, W. T. Jr., Beall, J. J., Goyal, K. P., and Pingol, A. S. 2005. The Santa Rosa-Geysers recharge project, Geysers geothermal field, California. *Geothermal Resources Council Transactions*, 29, 144–150.
7. Sanyal, S. 2004. Defining the sustainability and renewability of geothermal power. *Geothermal Resource Council Bulletin*, July/August, 149–151.
8. Capetti, G. and Ceppatelli, L. 2005. Geothermal power generation in Italy, 2000–2004 update report. In *Proceedings of the World Geothermal Congress*, Antalya, Turkey, 7pp.
9. State of Nevada, Division of Minerals. 2005. Nevada geothermal update. http://minerals.state.nv.us/.
10. Idaho Public Utilities Commission Order # 29692, Case No. IPC-E-05-1, January 24, 2005. http://www.puc.idaho.gov.
11. Ormat Technologies News Release, May 6, 2005. http://www.ormat.com/index_news.htm (accessed December 2005).
12. Hance, C. N. 2005. Factors affecting the costs of geothermal power development. Geothermal Energy Association and the U.S. Department of Energy, Washington, DC.

13. Western GeoPower. 2005. Corporation website. http://www.geopower.ca/newsreleases.htm (accessed December 2005).
14. Phil Lory of Sinclair Knight Mertz Consulting, Personal communication.
15. Fuji Electric Systems. 2005. *Wayang Windu Project Information Bulletin*, GEC82-12. Fuji Electric Systems, Tokyo, Japan.
16. Williamson, K. H., Gunderson, R. P., Hamblin, G. M., Gallup, D. L., and Kitz, K. 2001. Geothermal power technology. *Proceedings of the IEEE*, 89(12), 1783–1792.
17. Kagel, A., Bates, D., and Gawell, K. 2005. *A Guide to Geothermal Energy and the Environment*. Geothermal Energy Association, Washington, DC.
18. Kagel's source reported as US EPA. Average power plant emissions from EPA 2000 emissions data. http://www.epa.gov/cleanenergy/egrid/highlights.
19. Kagel, A. Citing data from Brophy, P. 1997. Environmental advantages to utilization of geothermal energy. *Renewable Energy*, 10, 374.

52

Hydrogen Energy Technologies

Sesha S. Srinivasan, S.A. Sherif, Frano Barbir, T. Nejat Veziroglu, Madhukar Mahishi, and Robert Pratt

CONTENTS

52.1 Introduction

Energy carriers are a convenient form of stored energy. Electricity is one type of carrier that can be produced from various sources, transported over large distances, and distributed to the end user. Hydrogen is another type of energy carrier. If produced from clean sources, it constitutes a clean and versatile carrier. Liquid hydrogen is already used for space shuttles and rockets and is also proposed for the future hypersonic or supersonic space vehicles. However, at present, there is increased interest from governments in the development of hydrogen as a fuel for other applications such as transportation and stationery power from off-peak storage. This chapter provides the information about hydrogen and the status of hydrogen technologies. The topics covered include hydrogen production, storage, distribution, conversion, and safety.

52.2 Properties of Hydrogen

Hydrogen is an odorless, colorless gas. With a molecular weight of 2.016, hydrogen is the lightest element. Its density is about 14 times less than air (0.08376 kg/m^3 at standard temperature and pressure). Hydrogen is liquid at temperatures below 20.3 K (at atmospheric pressure). Hydrogen has the highest energy content per unit mass of all fuels—higher heating value is 141.9 MJ/kg, almost three times higher than gasoline. However, because of its low density, its heating value on volumetric basis is almost one-third of natural gas. Some important properties of hydrogen are compiled in Table 52.1. It exists in three isotopes: protium, deuterium, and tritium. A standard hydrogen atom (protium) is the simplest of all the elements and consists of one proton and one electron. Molecular hydrogen (H_2) exists in two forms: *ortho-* and *para-*hydrogen. Both forms have identical chemical

TABLE 52.1

Properties of Hydrogen (at Normal Temperature and Pressure)

Molecular weight		2.016
Density	kg/m^3	0.0838
Higher heating value	MJ/kg	141.90
	MJ/m^3	11.89
Lower heating value	MJ/kg	119.90
	MJ/m^3	10.05
Boiling temperature	K	20.3
Density as liquid	kg/m^3	70.8
Critical point		
Temperature	K	32.94
Pressure	Bar	12.84
Density	kg/m^3	31.40
Self-ignition temperature	K	858
Ignition limits in air	vol.%	4–75
Stoichiometric mixture in air	vol.%	29.53
Flame temperature in air	K	2318
Diffusion coefficient	cm^2/s	0.61
Specific heat (c_p)	kJ/(kg K)	14.89

properties, but due to different spin orientation have somewhat different physical properties. At room temperature, hydrogen consists of approximately 75% *ortho*- and 25% *para*-hydrogen. Since *para*-hydrogen is more stable at lower temperatures, its concentration increases at lower temperatures, reaching virtually 100% at liquid hydrogen temperatures.

52.3 Hydrogen Production Methods

While hydrogen is the most plentiful element in the universe, making up about three quarters of all matter, free hydrogen is scarce. The atmosphere contains trace amounts of it (0.07%), and it is usually found in small amounts mixed with natural gas in crustal reservoirs. A few wells, however, have been found to contain large amounts of hydrogen, such as some wells in Kansas that contain 40% hydrogen, 60% nitrogen, and trace amounts of hydrocarbons [1]. The Earth's surface contains about 0.14% hydrogen (10th most abundant element), most of which resides in chemical combination with oxygen as water. Hydrogen, therefore, must be produced. Logical sources of hydrogen are hydrocarbon (fossil) fuels (C_XH_Y) and water (H_2O). Presently, hydrogen is mostly being produced from fossil fuels (natural gas, oil, and coal). However, except for the space program, hydrogen is not being directly used as a fuel or an energy carrier. It is being used in refineries to upgrade crude oil (hydrotreating and hydrocracking), in the chemical industry to synthesize various chemical compounds (such as ammonia and methanol), and in metallurgical processes (as a reduction or protection gas). The total annual hydrogen production worldwide in 1996 was about 40 million tons (5.6 EJ) [2]. Less than 10% of that amount was supplied by industrial gas companies; the rest is being produced at consumer-owned and operated plants (so-called captive production), such as refineries, and ammonia and methanol producers. Production of hydrogen as an energy carrier would require an increase in production rates by several orders of magnitude.

A logical source for large-scale hydrogen production is water, which is abundant on Earth. Different methods of hydrogen production from water have been, or are being, developed. They include electrolysis, thermochemical processes, photolysis, photoelectrochemical (PEC), and photocatalysis. Biomass is another important renewable resource for producing hydrogen. Some of these methods are described in what follows (Table 52.2).

52.3.1 Electrolysis

Electrolysis appears to be the only method developed to date, which can be used for large-scale hydrogen production in a post-fossil fuel era. Production of hydrogen by water electrolysis is a mature technology; based on a fundamentally simple process, it is very efficient, and does not involve moving parts.

TABLE 52.2
Hydrogen Production Cost

Technology	Cost of H_2 ($/GJ)	Cost of H_2 ($/kg)
Direct biomass gasification	12.5–21.6	1.51–2.59
Steam reforming of bio-oil	10.3–19.9	1.24–2.38

Sources: Spath, P.L. et al., Update of hydrogen from biomass—Determination of the delivered cost of hydrogen, National Renewable Energy Laboratory, Golden, CO, NREL/MP-510-33112, 2000; Craig, K.R. and Mann, M.K., Cost and performance analysis of biomass based integrated gasification combined cycle BIGCC power systems, National Renewable Energy Laboratory, Golden, CO, NREL/TP-430-21657, 1996, 70 pp.

The following reactions take place at the electrodes of an electrolysis cell filled with a suitable electrolyte (aqueous solution of KOH or NaOH or NaCl) upon the application of a potential:

$$\text{Cathode reaction: } 2H_2O\,(l) + 2e^- \rightarrow H_2(g) + 2OH^-(aq) \tag{52.1}$$

$$\text{Anode reaction: } 2OH^-(aq) \rightarrow \frac{1}{2}O_2(g) + H_2O(l) \tag{52.2}$$

$$\text{Overall cell reaction: } H_2O(l) \rightarrow H_2(g) + O_2(g) \tag{52.3}$$

The reversible decomposition potential ($E_{rev} = \Delta G/nF$) of the earlier reaction is 1.229 V at standard conditions. The total theoretical water decomposition potential is 1.480 V corresponding to hydrogen's enthalpy (since $\Delta H = \Delta G + T\Delta S$). Due to irreversible processes occurring at the anode and the cathode, including the electrical resistance of the cell, the actual potentials are always higher, typically between 1.75 and 2.05 V. This corresponds to the efficiencies of 72%–82%.

Several advanced electrolyzer technologies are being developed such as follows:

- Advanced alkaline electrolysis, which employs new materials for membranes and electrodes that allow further improvement in efficiency—up to 90% [3,4].
- Solid polymer electrolytic process, which employs a proton-conducting ion exchange membrane as electrolyte and as membrane that separates the electrolysis cell. This type of electrolyzers can operate at very high current densities (up to 2 A/cm^2, which is about one order of magnitude higher than standard electrolyzers with alkaline liquid electrolyte). The water to be dissociated does not require dissolved electrolytes to increase its conductivity, and is added solely to the anode side [4,5].
- High-temperature steam electrolysis operates between 700°C and 1000°C, which employs oxygen ion-conducting ceramics as electrolyte. Electrical energy consumption is reduced since part of the energy required for water dissociation is supplied in the form of heat. The water to be dissociated is entered on the cathode side as steam, which forms a steam–hydrogen mixture during electrolytic dissociation. The O^- ions are transported through the ceramic material to the anode, where they are discharged as oxygen [6].

An electrolysis plant can operate over a wide range of capacity factors and is convenient for a wide range of operating capacities, which makes this process interesting for coupling with renewable energy sources, particularly with photovoltaics (PVs). PV generates low-voltage direct current, exactly what is required for the electrolysis process.

The performance of PV–electrolyzer systems has been studied extensively both in theory and in practice [7–10]. Several experimental PV–electrolysis plants are currently operating all over the world, such as follows:

- Solar–Wasserstoff–Bayern pilot plant in Neunburg vorm Wald, Germany [11]
- HYSOLAR project in Saudi Arabia [12]

- Schatz Energy Center, Humboldt State University, Arcata, California [13]
- Helsinki University of Technology, Helsinki, Finland [14]
- INTA Energy Laboratory, Huelva, Spain [15]

52.3.2 Thermochemical Processes

Thermochemical hydrogen production is a means of splitting water via a series of chemical reactions. All chemical intermediates are recycled internally within the process so that water is the only raw material and hydrogen and oxygen are the only products. The maximum temperature requirements for most thermochemical cycles lie within a temperature range of 650°C–1100°C [16], thus eliminating use of lower temperature heat sources. High-concentration solar collectors and nuclear reactors are the possible sources of heat for these cycles. Figure 52.1 illustrates the concept of splitting of water by a thermochemical cycle.

According to Wendt [17], 2000–3000 different theoretical cycles have been proposed and evaluated. However, out of all the possible thermochemical cycles, Mark 1, Mark 15, ZnO/Zn, UT-3, Fe_3O_4/FeO, and iodine–sulfur processes are the only few, which have been studied extensively. The following cycles are under active research around the world.

52.3.2.1 ZnO/Zn Cycle

ZnO/Zn cycle is a two-step water-splitting cycle based on the thermal redox pairs of metal oxides. As shown, in the following reactions, the process relies on the endothermic thermal dissociation of ZnO followed by the exothermic hydrolysis of Zn:

$$ZnO \xrightarrow{\text{Heat}} Zn(g) + \frac{1}{2}O_2 \qquad (52.4)$$

$$Zn + H_2O \rightarrow ZnO + H_2 \qquad (52.5)$$

Theoretical conversion efficiencies are of the order of 50% [18]. This cycle is under active research [19]. The current work is focused on the quenching step and decomposition rate measurements. The success of this cycle will depend on the development of an efficient mechanism for the separation of Zn–H_2 mixture.

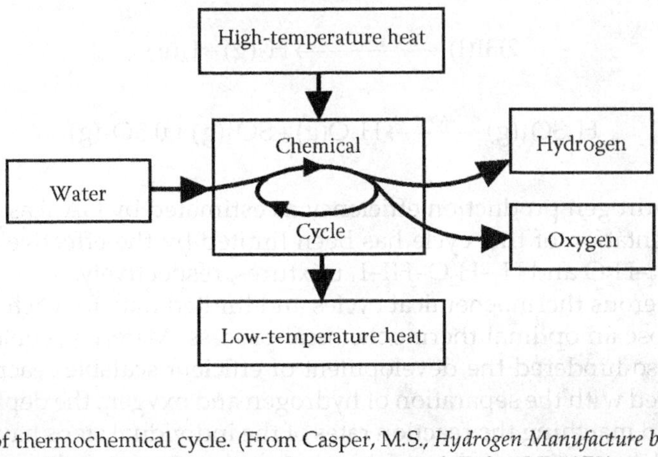

FIGURE 52.1
Schematic diagram of thermochemical cycle. (From Casper, M.S., *Hydrogen Manufacture by Electrolysis, Thermal Decomposition and Unusual Techniques*, Noyes Data Corporation, Park Ridge, NJ, 1978.)

52.3.2.2 UT-3 Cycle

The UT-3 cycle invented at the University of Tokyo consists of Ca, Fe, and Br compounds. It involves the use of solid and gaseous phases of reactants and products. The predicted first- and second-law efficiencies of the adiabatic UT-3 cycle are 49% and 53%, respectively [20]. This cycle is composed of the following reactions:

$$2CaO(s) + 2Br_2(g) \xrightarrow{680°C} 2CaBr_2(s) + \frac{1}{2}O_2(g) \tag{52.6}$$

$$CaBr_2(s) + H_2O(g) \xrightarrow{760°C} CaO(g) + 2HBr(g) \tag{52.7}$$

$$3FeBr_2(s) + 4H_2O(g) \xrightarrow{560°C} Fe_3O_4(s) + 6HBr(g) + H_2(g) \tag{52.8}$$

$$Fe_3O_4(s) + 8HBr(g) \xrightarrow{210°C} 3FeBr_2(s) + 4H_2O(g) + Br_2(g) \tag{52.9}$$

The reactants $CaBr_2$, CaO, Fe_3O_4, and $FeBr_2$ are solids and stay within the four reactors while Br_2, H_2O, and HBr are gases that move from one reactor to the next. Once the reactions in the forward direction are complete, the flow of gaseous reactants is reversed. Substantial amount of research work including bench-scale laboratory tests [21], solid reactants' development [22,23], and reaction kinetic measurements [24] have been performed on this cycle.

52.3.2.3 Iodine–Sulfur Cycle

Iodine–sulfur process is a three-step thermochemical cycle for decomposing water into H_2 and O_2. The cycle proposed by General Atomic (GA) Co. is one of the extensively studied cycles. It proceeds by the formation of H_2SO_4 and HI from SO_2, I_2, and H_2O. It is followed by the decomposition of HI and H_2SO_4 that produces H_2 and I_2, respectively. The chemical reactions involved in the process are shown in the following equations [25]:

$$xI_2(l) + SO_2(aq) + 2H_2O(l) \xrightarrow{20°C–100°C} 2HI_x(l) + H_2SO_4(aq) \tag{52.10}$$

$$2HI(l) \xrightarrow{200°C–700°C} H_2(g) + I_2(g) \tag{52.11}$$

$$H_2SO_4(g) \xrightarrow{850°C} H_2O(g) + SO_2(g) + 0.5O_2(g) \tag{52.12}$$

The maximum hydrogen production efficiency as estimated by GA was around 50% [26]. Practical implementation of this cycle has been limited by the effective separation of HI and H_2 from HI–I_2–H_2O and H_2–H_2O–HI–I_2 mixtures, respectively.

Due to the numerous thermochemical cycles and limited data for each of these cycles, it is difficult to choose an optimal thermochemical process. Material problems at high temperatures have also hindered the development of efficient scalable reactor concepts. The problems associated with the separation of hydrogen and oxygen, the depletion of intermediate products, and matching the reaction rates of the individual steps have further limited the practical implementation of thermochemical cycles. So far, only one cycle has been demonstrated on a bench scale, and none of them have been demonstrated at a larger scale.

52.3.3 Photoelectrochemical Hydrogen Production

PEC systems combine both PV and electrolysis into a one-step water-splitting process. These systems use a semiconductor electrode exposed to sunlight in combination with a metallic or semiconductor electrode to form a PEC cell. A schematic illustration of a PEC hydrogen production driven by solar energy is shown in Figure 52.2. In general, the semiconductor electrode (photoanode) is activated by solar radiation, which drives the reaction in aqueous solution. Due to bandgap illumination, electrons and holes are formed in the conduction and valence bands, respectively, at the photoanode.

The photogenerated holes at the anode split the water molecules into hydrogen ions and oxygen. Hydrogen ions released hence migrate to the cathode through the aqueous electrolyte.

$$2H^+ + H_2O \rightarrow \frac{1}{2}O_2 + 2H^+ \quad (52.13)$$

Electrons generated at photoanode are transferred over the external circuit to the cathode, where they reduce hydrogen ions into gaseous hydrogen:

$$2H^+ + 2e^- \rightarrow H_2 \quad (52.14)$$

Ideally, electrochemical decomposition of water takes place when the electromotive force of the cell is equal to 1.23 eV. If we consider internal losses in the PEC, a minimum bandgap of 1.8 eV is required to run the reaction [28]. Photoelectrochemically stable semiconductors such as TiO_2 and $SrTiO_3$ have a wide bandgap and hence are limited in the UV region of the solar spectrum, whereas narrow bandgap semiconductors such as GaP and GaAs are often photoelectrochemically unstable [29]. The way to address the problem of low light absorption is to enlarge the spectral response of the PEC cells. The techniques used to solve the problem are dye adsorption, metal incorporation, and use of multijunction cells.

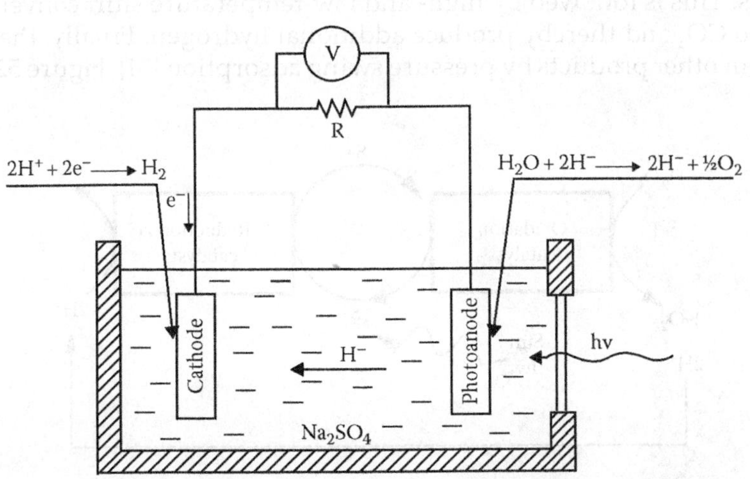

FIGURE 52.2
Schematic representation of photoelectrochemical (PEC) cell. (From Bak, T. et al., *Int. J. Hydrogen Energy*, 27, 991, 2002.)

52.3.4 Photochemical Hydrogen Production

Photochemical systems differ from PEC systems in the fact that they use a semiconductor suspension instead of individual electrodes to induce the reactions. Photochemical systems have photosensitizer particles suspended in an aqueous solution where a photosensitizer is an organic or semiconductor molecule that can absorb sunlight and induce photochemical reactions for the hydrogen production. The reduction and oxidation reactions occur simultaneously at the surface of the suspended photosensitizer particles. Since absorption of a single photon can at most cause the transfer of one electron, a catalyst is required along with light to induce the redox reaction.

$$2H_2O + 2e^- \rightarrow H_2 + 2OH^- \tag{52.15}$$

$$2H_2O \rightarrow O_2 + 4H^+ + 4e^- \tag{52.16}$$

Figure 52.3 illustrates the schematic of a general photochemical reaction to generate hydrogen. No photochemical system has been able to reach an efficiency of 10%.

52.3.5 Hydrogen Production from Biomass

Biomass represents a large potential feedstock resource for environmentally clean hydrogen production. It lends itself to both biological and thermal conversion processes. In the thermal path, hydrogen can be produced in two ways, viz., direct gasification and pyrolysis to produce liquid bio-oil followed by steam reforming.

Direct gasification of biomass is in many ways similar to coal gasification. The process occurs broadly in three steps: Biomass is first gasified (using steam or air) to produce an impure syngas mixture composed of hydrogen, CO, CO_2, CH_4, small amounts of higher hydrocarbons, tar, and water vapor. The gas may also contain particulate matter, which is removed using cyclones and scrubbers. The particulate free gas is compressed and then catalytically steam-reformed to eliminate the tars and higher hydrocarbons. This is followed by high- and low-temperature shift conversions to convert the CO to CO_2 and thereby produce additional hydrogen. Finally, the hydrogen is separated from other products by pressure swing adsorption [31]. Figure 52.4 illustrates

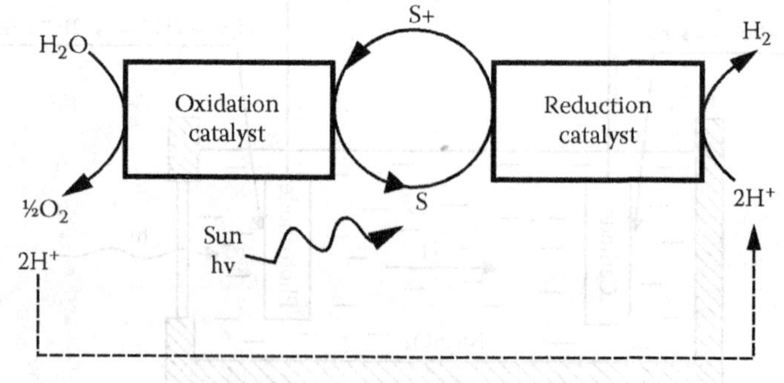

FIGURE 52.3
Minimal scheme for the photochemical splitting of water. S is a photochemical sensitizer. (From Bolton, J.R., Solar photoproduction of hydrogen, International Energy Agency Technical Report, IEA/H2/TR-96, 1996a.)

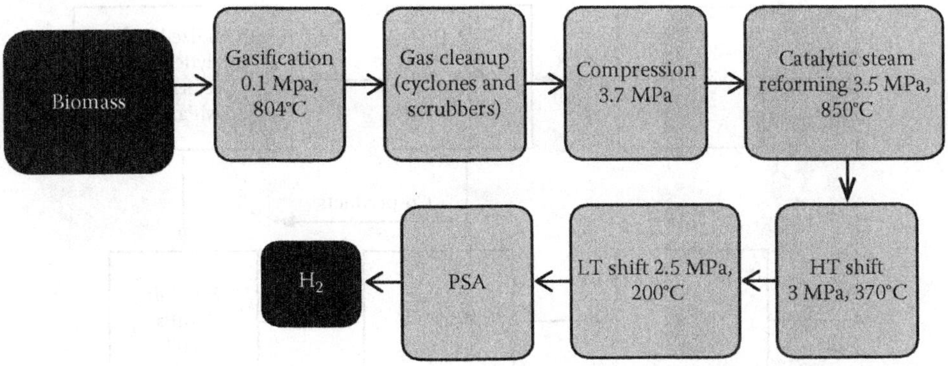

FIGURE 52.4
Gasification followed by steam reforming. (From Spath, P.L. et al., Update of hydrogen from biomass—Determination of the delivered cost of hydrogen, National Renewable Energy Laboratory, Golden, CO, NREL/MP-510-33112, 2000.)

the sequence of processes. The main reactions taking place in biomass gasification are as follows:

$$CH_{1.5}O_{0.7} + 0.3H_2O \rightarrow CO + 1.05H_2 \quad \Delta H_R = +74.4\ \text{kJ/mol} \quad \text{(biomass gasification)} \tag{52.17}$$

$$CO + H_2O \rightarrow CO_2 + H_2 \quad \Delta H_R = -41.2\ \text{kJ/mol} \quad \text{(water–gas shift)} \tag{52.18}$$

Biomass typically contains about 6% hydrogen by weight. However, in the presence of hydrogen-bearing species (steam), the hydrogen yield can be considerably improved above the 6% minimum [32]. Gasification temperatures encountered are typically in the range 600°C–850°C, which is lower than many thermochemical water-splitting cycles thereby making biomass gasification an attractive technology to produce hydrogen. Steam gasification of biomass is endothermic, and the energy required for the process is supplied by burning part of the biomass feedstock or uncombusted char. Tars are polyaromatic hydrocarbons produced during gasification of biomass. Tars are undesirable coproducts as they clog filters, pipes, and valves and damage downstream equipments such as engines and turbines. Efforts are being made to minimize or reform the tars to additional product hydrogen [33,34].

Hydrogen can alternately be produced by reforming the biomass to a liquid bio-oil in a process called pyrolysis. Pyrolysis is an endothermic thermal decomposition of biomass carried out in an inert atmosphere at 450°C–550°C [35]. The bio-oil so produced is a liquid composed of 85% oxygenated organics and 15% water. The bio-oil is then steam-reformed in the presence of a nickel-based catalyst at 750°C–850°C followed by shift conversion to convert CO to CO_2 [36]. Figure 52.5 illustrates the sequence of processes. The reactions can be written as follows:

$$\text{Biomass} \rightarrow \text{Bio-oil} + \text{char} + \text{gas} \quad \text{(pyrolysis)} \tag{52.19}$$

$$\text{Bio-oil} + H_2O \rightarrow CO + H_2 \quad \text{(reforming)} \tag{52.20}$$

$$CO + H_2O \rightarrow CO_2 + H_2 \quad \text{(shift)} \tag{52.21}$$

A number of research groups around the world are involved in research on hydrogen production from a variety of biomass resources, including wood, municipal solid waste,

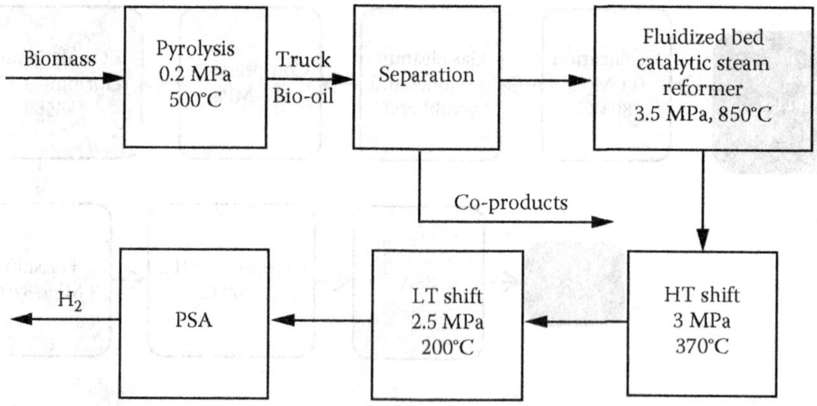

FIGURE 52.5
Pyrolysis followed by steam reforming. (From Spath, P.L. et al., Update of hydrogen from biomass—Determination of the delivered cost of hydrogen, National Renewable Energy Laboratory, Golden, CO, NREL/MP-510-33112, 2000.)

and residues from industries [34,37–44]. There are currently no commercial biomass gasification processes for hydrogen production, but there are several demonstration plants of biomass gasifiers for producing electricity or other chemicals [45]. The FERCO SilvaGas process employs low-pressure Battelle Columbus gasification, which consists of two physically separate reactors—a gasification reactor in which biomass is converted into a medium-heating-value gas and residual char at a temperature of 850°C–1000°C and a combustion reactor that burns residual char to provide heat for gasification. A typical product gas using wood chips consists of about 21% H_2 (using steam/biomass ratio of 0.45). The Fast Internal Circulation Fluidized Bed (FICFB) process is another example. In FICFB, there is a single reactor with two zones—a gasification zone and a combustion zone; biomass is gasified in the gasification zone with steam at 850°C–950°C followed by circulation to the combustion zone where char is burned to supply thermal energy. A demonstration plant producing 2 MW electrical power was constructed in Europe in 2001 (details in [45]).

Biomass resources have the advantage of being renewable and hence can make an important contribution to renewable hydrogen production. However, an important factor that prohibits commercialization is the difficulty of transporting large amounts of low-energy-density biomass feedstocks over long distances. The cost associated with growing, harvesting, and transporting biomass may be up to 40% of the total biomass plant operating cost [31,46].

Additional technical and economic considerations for capture and storage of CO_2 may also be necessary as both gasification and pyrolysis produce carbon dioxide.

52.3.6 Biological Methods

Hydrogen may also be obtained from biological processes involving organic compounds. One of the techniques used for hydrogen production using biological means is shown in Figure 52.6.

The two fundamental ways of biological hydrogen production are (1) *fermentation of the bacteria*, which is an anaerobic process that converts organic substances like starch, cellobiose, sucrose, and xylose to H_2 and CO_2 without the need of sunlight and oxygen, and (2) *biophotolysis*, a process that uses microalgae–cyanobacteria and green algae to produce hydrogen in the presence of sunlight and water. Both of these processes are being researched.

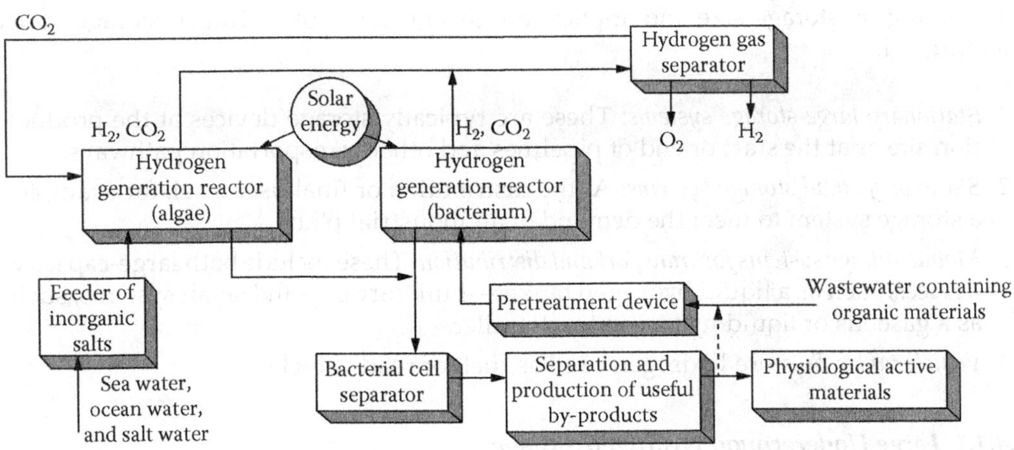

FIGURE 52.6
Concept of biological hydrogen production. (Adapted from Miyake, J. et al., *J. Biotechnol.*, 70, 89, 1999, Figure 1.)

52.4 Hydrogen Storage Methods

52.4.1 Gaseous Hydrogen Storage

Because of hydrogen's low density, its storage always requires relatively large volumes and is associated with either high pressures (thus requiring heavy vessels), or extremely low temperatures, and/or combination with other materials (much heavier than hydrogen itself). Table 52.3 shows achievable storage densities with different types of hydrogen storage methods. Some novel hydrogen storage method may achieve even higher storage densities, but have yet to be proven in terms of practicality, cost, and safety.

TABLE 52.3
Hydrogen Storage Types and Densities

H_2 Storage Method	kg H_2/kg	kg H_2/m^3
Large volume storage (10^2–10^4 m^3 geometric volume)		
Underground storage		5–10
Pressurized gas storage (aboveground)	0.01–0.014	2–16
Metal hydride	0.013–0.015	50–55
Liquid hydrogen	~1	65–69
Stationary small storage (1–100 m^3 geometric volume)		
Pressurized gas cylinder	0.012	~15
Metal hydride	0.012–0.014	50–53
Liquid hydrogen tank	0.15–0.50	~65
Vehicle tanks (0.1–0.5 m^3 geometric volume)		
Pressurized gas cylinder	0.05	15
Metal hydride	0.02	55
Liquid hydrogen tank	0.09–0.13	50–60

Depending on storage size and application, several types of hydrogen storage may be differentiated:

1. *Stationary large storage systems*: These are typically storage devices at the production site or at the start or end of pipelines and other transportation pathways.
2. *Stationary small storage systems*: At the distribution or final user level, for example, a storage system to meet the demand of an industrial plant.
3. *Mobile storage systems for transport and distribution*: These include both large-capacity devices, such as a liquid hydrogen tanker—bulk carrier—and small systems, such as a gaseous or liquid hydrogen truck trailer.
4. *Vehicle tanks*: To store hydrogen used as fuel for road vehicles.

52.4.1.1 Large Underground Hydrogen Storage

Underground storage of hydrogen in caverns, aquifers, depleted petroleum and natural gas fields, and man-made caverns resulting from mining and other activities is likely to be technologically and economically feasible [48]. Hydrogen storage systems of the same type and the same energy content will be more expensive by approximately a factor of 3 than natural gas storage systems, due to hydrogen's lower volumetric heating value. Technical problems, specifically for the underground storage of hydrogen other than expected losses of the working gas in the amount of 1%–3% per year, are not anticipated. The city of Kiel's public utility has been storing town gas with a hydrogen content of 60%–65% in a gas cavern with a geometric volume of about 32,000 m^3 and a pressure of 80–160 bar at a depth of 1,330 m since 1971 [49]. Gaz de France (the French National Gas Company) has stored hydrogen-rich refinery by-product gases in an aquifer structure near Beynes, France. Imperial Chemical Industries of Great Britain stores hydrogen in the salt mine caverns near Teesside, United Kingdom [50].

52.4.1.2 Aboveground Pressurized Gas Storage Systems

Today, pressurized gas storage systems are used in natural gas business in various sizes and pressure ranges from standard pressure cylinders (50 L, 200 bar) to stationary high-pressure containers (over 200 bar) or low-pressure spherical containers (>30,000 m^3, 12–16 bar). This application range will be similar for hydrogen storage.

52.4.1.3 Vehicular Pressurized Hydrogen Tanks

Storage of hydrogen in automobiles has been enabled by the development of ultralight but strong new composite materials. Pressure vessels that allow hydrogen storage at pressures >200 bars have been developed and are used in automobiles (e.g., Daimler-Benz NECAR II). A storage density higher than 0.05 kg of hydrogen/1 kg of total weight is easily achievable [51].

52.4.2 Liquid Hydrogen Storage

At present, liquid hydrogen is primarily used as a rocket fuel and is predestined for supersonic and hypersonic space vehicles primarily because it has the lowest boiling point density and the highest specific thrust of any known fuel. Its favorable characteristics include its high heating value per unit mass, its wide ignition range in hydrogen/oxygen or air mixtures, as well as its large flame speed and cooling capacity due to its high specific heat, which permits very effective engine cooling and cooling of the critical parts of the outer skin [52,53]. Liquid hydrogen has some other important uses such as in high-energy nuclear physics and bubble chambers. The transport of hydrogen is vastly more economical when it is in liquid form even though cryogenic refrigeration and special dewar vessels are required. Although liquid hydrogen can provide a lot of advantages, its uses are restricted in part because liquefying hydrogen by existing conventional methods consumes a large amount of energy (around 30% of its heating value). Liquefying 1 kg of hydrogen in a medium-size plant requires 10–13 kW h of energy (electricity) [53]. In addition, boil-off losses associated with the storage, transportation, and handling of liquid hydrogen can consume up to 40% of its available combustion energy. It is therefore important to search for ways that can improve the efficiency of the liquefiers and diminish the boil-off losses.

52.4.2.1 Hydrogen Liquefaction

Production of liquid hydrogen requires the use of liquefiers that utilize different principles of cooling. In general, hydrogen liquefiers may be classified as conventional, magnetic, or hybrid. Many types of conventional liquefiers exist such as the Linde–Hampson liquefiers, the Linde dual-pressure liquefiers, the Claude liquefiers, the Kapitza liquefiers, the Heylandt liquefiers, and those liquefiers utilizing the Collins cycle, just to name a few. Conventional liquefiers generally comprise compressors, expanders, heat exchangers, and Joule–Thomson valves. Magnetic liquefiers, on the other hand, utilize the magnetocaloric effect. This effect is based on the principle that some magnetic materials experience a temperature increase upon the application of a magnetic field and a temperature drop upon lifting the magnetic field. The magnetic analog of several conventional liquefiers includes the Brayton liquefiers, the Stirling liquefiers, and the active magnetic regenerative liquefier. Additional information on liquid hydrogen production methods can be found in Sherif et al. [55].

Hydrogen is usually transported in large quantities by truck tankers of 30–60 m^3 capacity, by rail tank cars of 115 m^3 capacity, and by barge containers of 950 m^3 capacity [55]. Liquid hydrogen storage vessels are usually available in sizes ranging from 1 L dewar flasks used in laboratory applications to large tanks of 5000 m^3 capacity. The National Aeronautics and Space Administration typically uses large tanks of 3800 m^3 capacity (25 m in diameter) [48]. The total boil-off rate from such dewars is approximately 600,000 LPY (liters per year), which is vented to a burn pond. The contributing mechanisms to boil-off losses in cryogenic hydrogen storage systems are (1) *ortho–para* conversion, (2) heat leak (shape and size effect, thermal stratification, thermal overfill, insulation, conduction, radiation, and cooldown), (3) sloshing, and (4) flashing.

52.4.3 Solid-State Hydrogen Storage

Hydrogen can be packed and stored in the solid state by forming metal hydride. During the formation of the metal hydride, hydrogen molecules are dissociated into H-atoms, and the hydrogen atoms are inserted in interstitial spaces inside the lattice of intermetallic compounds and/or alloys. In such a way, effective storage comparable to the density of liquid hydrogen is created. However, when the mass of the metal or alloy is taken into account, the metal hydride gravimetric storage density is comparable to the storage of pressurized hydrogen. The best achievable gravimetric storage density is about 0.07 kg of H_2/kg of metal, for a high-temperature hydride such as MgH_2 as shown in

TABLE 52.4

Theoretical Capacities of Hydriding Substances as Hydrogen Storage Media

Medium	Hydrogen Content(kg/kg)	Hydrogen Storage(Capacity (kg/L of vol.)	Energy Density (kJ/kg)	Energy Density (kJ/L of vol.)
MgH_2	0.070	0.101	9,933	14,330
Mg_2NiH_4	0.0316	0.081	4,484	11,494
VH_2	0.0207		3,831	
$FeTiH_{1.95}$	0.0175	0.096	2,483	13,620
$TiFe_{0.7}Mn_{0.2}H_{1.9}$	0.0172	0.090	2,440	12,770
$LaNi_5H_{7.0}$	0.0137	0.089	1,944	12,630
$R.E.Ni_5H_{6.5}$	0.0135	0.090	1,915	12,770
Liquid H_2	1.00	0.071	141,900	10,075
Gaseous H_2 (100 bar)	1.00	0.0083	141,900	1,170
Gaseous H_2 (200 bar)	1.00	0.0166	141,900	2,340
Gasoline	—	—	47,300	35,500

Table 52.4, which gives a comparison of some hydriding substances with liquid hydrogen, gaseous hydrogen, and gasoline [56].

An overview of hydrogen storage alloys has been discussed by Sandrock [57] from the solid–gas reaction point of view. There are a number of important properties that must be considered in metal hydride storage. Some of the most important ones are (1) ease of activation, (2) heat transfer rate, (3) kinetics of hydriding and dehydriding, (4) resistance to gaseous impurities, (5) cyclic stability, (6) safety, and (7) weight and cost. Although metal hydrides can theoretically store large amounts of hydrogen in a safe and compact way, the practical gravimetric hydrogen density is limited to <3 mass%. Another stumbling block for this storage method is the low practical cyclic stability achieved so far.

52.4.3.1 Metal Hydrides

The potential to use hydrides for energy storage and applications has stimulated extensive theoretical and experimental research on the fundamental aspects of hydrogen sorption and several reversible storage intermetallics [58] such as FeTi [59], $LaNi_5$, $MmNi_{4.5}Al_{0.5}$ [60], and Mg_2Ni [61]. Since the maximum weight percentage storage for these intermetallics is ~1.8 wt.% at ambient conditions and ~3.8 wt.% at high temperature (300°C–400°C), there is an ongoing research to find better hydride materials with higher storage capacity at ambient as well as high-temperature conditions. To achieve this, two prominent routes are being followed: first, to modify and optimize the present storage materials such as FeTi, $LaNi_5$, and the high-temperature hydride Mg_2Ni, and second, to develop altogether new storage materials, for example, transition metal complexes, composite materials, nanoparticle and nanostructured materials, and new carbon variants (fullerenes, C_{60}, and other higher versions, graphitic nanofibers and nanotubes).

Magnesium (Mg) has the highest theoretical hydrogen storage capacity of ~7.6 wt.%. However, it has two significant disadvantages: (1) the Mg–H_2 reaction has poor kinetics, and (2) the resulting hydride is not reversible under ambient or moderate temperature and pressure conditions [62]. A possible way to achieve Mg-like storage capacity but with reversible hydrogenation characteristics is to form composites with Mg as one of the components. The other component may be one of the known hydrogen storage intermetallic alloys [63].

An important feature of the metallic hydrides is the high volumetric density of the hydrogen atoms present in the host atomic lattice [64]. The highest theoretical volumetric hydrogen density known today is 150 kg/m^3 for Mg_2FeH_6 and $Al(BH_4)_3$. The Mg_2FeH_6 hydride belongs to the family of Mg–transition metal complex hydrides with $[FeH_6]^{4-}$ octahedral surrounded by Mg atoms in cubic configuration [65]. Interestingly, iron does not form intermetallic compounds with Mg, but it readily combines with hydrogen and Mg to form ternary hydride Mg_2FeH_6 according to these reactions:

$$2Mg + Fe + 3H_2 \leftrightarrow Mg_2FeH_6 \tag{52.22}$$

$$2MgH_2 + Fe + H_2 \leftrightarrow Mg_2FeH_6 \tag{52.23}$$

During the storage process (charging or absorption), heat is released, which must be removed in order to achieve the continuity of the reaction. During the hydrogen release process (discharging or desorption), heat must be supplied to the storage tank. The thermodynamic aspects of hydride formation from gaseous hydrogen are described by means of pressure–composition isotherms as shown in Figure 52.7. While the solid solution and hydride phase coexist, the isotherms show a flat plateau, the length of which determines the amount of H_2 stored. The stability of metal hydrides is usually presented in the form of Van't Hoff equation (52.23):

$$\ln P_{H_2}\ (\text{atm}) = \frac{2\Delta H_f}{SRT} + C \tag{52.24}$$

where

P_{H_2} is the plateau pressure at a particular temperature T
C is the constant of integration
R is the universal gas constant

The enthalpy of formation, ΔH_f, of the hydride is calculated from the slope of the straight line obtained plotting $\ln P_{H_2}$ versus 1/T (see Figure 52.7).

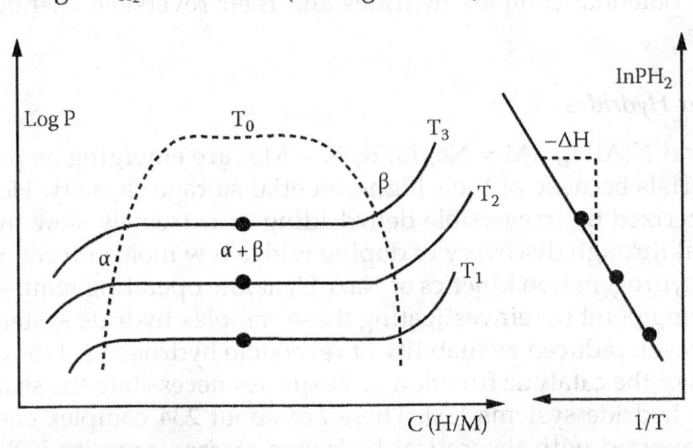

FIGURE 52.7
P–C isotherms and Van't Hoff curve for the $LaNi_5$ metal hydride.

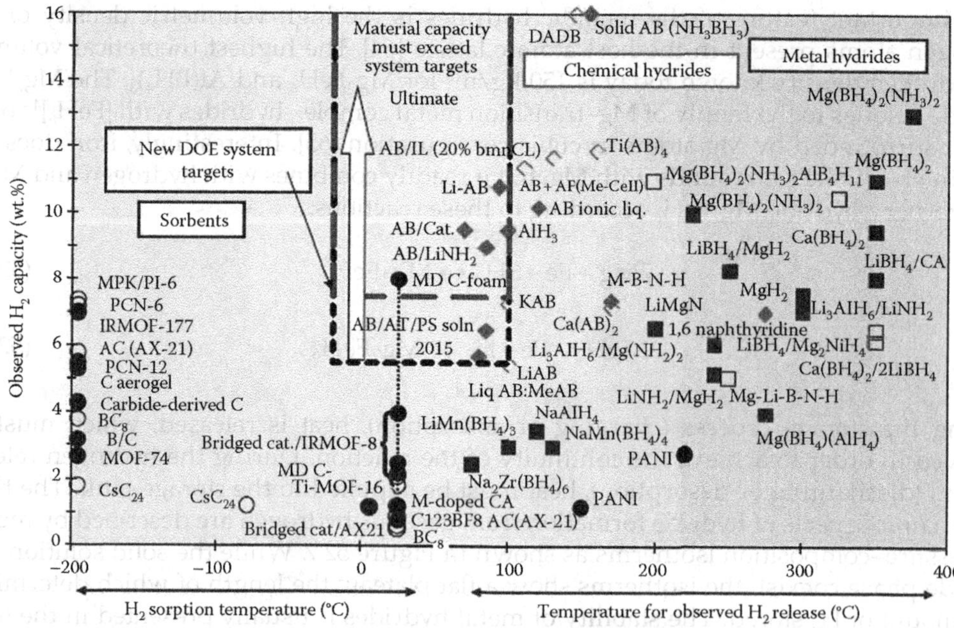

FIGURE 52.8
Volumetric and gravimetric hydrogen densities of various hydrogen storage systems. *Note:* Open symbols denote new materials for FY2009.

The most stable binary hydrides have enthalpies of formation of $\Delta H_f = -226$ kJ/mol H_2, the least stable hydrides having enthalpies of formation of +20 kJ/mol H_2 [66]. An advantage of storing hydrogen in hydriding substances is the safety aspect. Any serious damage to a hydride tank (such as one that could be caused by a collision) would not pose a fire hazard since hydrogen would remain in the metal structure.

Figure 52.8 demonstrates various complex hydrides synthesized and having gravimetric hydrogen storage capacities at particular operating temperature. The following subsections discuss different potential complex hydrides and their reversible chemical reactions for hydrogen storage.

52.4.3.2 Complex Hydrides

$MAlH_4$, MBH_4, and $N(AlH_4)_2$ (M = Na, Li, K; N = Mg) are emerging as promising hydrogen storage materials because of their high-potential storage capacity. However, they are generally characterized by irreversible dehydriding or extremely slow hydrogen cycling kinetics. The breakthrough discovery of doping with a few mole percent of Ti-catalyst has enhanced the dehydrogenation kinetics of $NaAlH_4$ at low operating temperatures (<150°C) and was the starting point in reinvestigating these complex hydride systems for hydrogen storage [67]. However, reduced availability of reversible hydrogen (~4–5 wt.%), poor cyclic stability, and loss of the catalytic function of Ti-species necessitate the search for new and efficient complex hydride systems [68]. There are about 234 complex chemical hydrides that have been reported with theoretical hydrogen storage capacity [69]. Table 52.5 lists the complex chemical hydrides and their available capacities and operating temperatures.

The hydride complexes such as $NaAlH_4$ and $NaBH_4$ are known to be stable and decompose only at elevated temperatures, often above the melting point of the complex.

TABLE 52.5

Theoretical Hydrogen Storage Capacities of Complex Hydrides

No.	Complex Chemical Hydride	Theoretical Capacity (wt.%)	Reversible Capacity (wt.%)	Operating Temperature (°C)	Remarks
1.	Ti-doped $NaAlH_4$	7.5	5.5	100–150	High rehydrogenation pressure, poor life cycle, loss of catalytic activity, less available capacity
2.	Undoped and Ti-doped $LiAlH_4$	10.5	6.3	120–170	Problems with reversibility and reduced thermodynamic stability
3.	Undoped and doped $LiBH_4$	18.2	9.0	200–400	High operating temperature, rehydrogenation problem, possible borane evolution
4.	$Mg(AlH_4)_2$	9.3	6.6	200–250	High operating temperature, thermodynamic stability
5.	$NaBH_4/H_2O$	10.5	9.2	Ambient	Hydrolysis reaction, irreversibility, one-time use
6.	Li_3N ($LiNH_2/LiH$)	11.3	6.5–7.0	255–285	High operating temperature, possible ammonia evolution
7.	B–H–Li–N	10.0		80–150	Rehydrogenation problem
8.	AlH_3	10.5		150	Ball milling–induced decomposition, irreversible
9.	H_3BNH_3	18.3	12.6		Ammonia evolution, irreversible

However, the addition of a few mole concentrations of titanium species to $NaAlH_4$ eases the release of hydrogen at moderate temperatures and ambient pressure [67]. The decomposition of Ti-doped $NaAlH_4$ proceeds in two steps with the total released hydrogen of ~5.5 wt.% at 100°C–150°C as follows:

$$NaAlH_4 \rightarrow Na_3AlH_6 + 2Al + 3H_2 \quad (3.72\ \text{wt.\%}\ H_2) \tag{52.25}$$

$$Na_3AlH_6 \rightarrow 3NaH + Al + \frac{3}{2}H_2 \quad \left(1.8\ \text{wt.\%}\ H_2\right) \tag{52.26}$$

Following this breakthrough discovery, an effort was initiated in the U.S. DOE hydrogen program to develop $NaAlH_4$ and related alanates as hydrogen storage materials [70,71]. Another complex hydride, $Mg(AlH_4)_2$, contains 9.6 wt.% of hydrogen that decomposes below 200°C [72]. Some of the new complex hydrides and their theoretical capacities are listed in Table 52.6.

Borohydride complexes with suitable alkali or alkaline earth metals are a promising class of compounds for hydrogen storage. The hydrogen content can reach values of up to 18 wt.% for $LiBH_4$ [73]. The total amount of hydrogen desorbed up to 600°C is 9 wt.%. Mixing $LiBH_4$ with SiO_2 powder lowers the desorption temperature, so that 9 wt.% of hydrogen is liberated below 400°C. Recently, Chen et al. reported a new hydrogen storage system, lithium nitride (Li_3N), which absorbs 11.5 wt.% of hydrogen reversibly [74,75]. The hydrogenation of lithium nitride is a two-step reaction as shown in the following equations:

$$Li_3N + 2H_2 \rightarrow Li_2NH + LiH \tag{52.27}$$

$$Li_2NH + H_2 \rightarrow LiNH_2 + LiH \tag{52.28}$$

TABLE 52.6

New Complex Hydrides and Their Hydrogen Storage Capacity

Serial Number	Complex Hydride	Theoretical Capacity (wt.%)	Temperature, T_{dec} (°C)
1.	$LiAlH_2(BH_4)_2$	15.2	
2.	$Mg(BH_4)_2$	14.8	260–280
3.	$NH_4Cl + LiBH_4$	13.6	>Ambient
4.	$Ti(BH_4)_3$	12.9	~25
5.	$Fe(BH_4)_3$	11.9	−30 to −10
6.	$Ti(AlH_4)_4$	9.3	−85
7.	$Zr(BH_4)_3$	8.8	<250
8.	$Zn(BH_4)_2$	8.4	85

Li_3N absorbs 5.74 wt.% of hydrogen in the first step and 11.5 wt.% in the second step. Since the hydrogen pressure for the reaction corresponding to the first step is very low (about 0.01 bar at 255°C), only the second step reaction of Li_2NH (lithium imides) with H_2 leads to the reversible storage capacity. According to Chen et al., the plateau pressure for imides hydrogenation is 1 bar at the relatively high temperature of 285°C [74]. However, the temperature of this reaction can be lowered to 220°C with magnesium substitution, although at higher pressures [76]. Further research on this system may lead to additional improvements in operating conditions with improved capacity.

$Zn(BH_4)_2$ is a ternary complex metal borohydride with a decomposition temperature of around 85°C [77]. Its theoretical hydrogen capacity is about 8.5 wt.%, and it can be synthesized by metathesis reaction of $NaBH_4$ and $ZnCl_2$ in diethyl ether [78]. A recent report from Jeon and Cho [79] indicates that zinc borohydride was successfully synthesized by ball milling zinc chloride and sodium borohydride without the use of a solvent as follows:

$$ZnCl_2 + 2NaBH_4 \rightarrow Zn(BH_4)_2 + 2NaCl \tag{52.29}$$

An example of the decomposition of a borohydride ($Zn(BH_4)_2$) is given in the following [77]:

$$Zn(BH_4) \rightarrow Zn + 2B + 4H_2 \tag{52.30}$$

which shows $Zn(BH_4)_2$ thermally decomposing into the constituent elements with the release of hydrogen. The main reason that complex hydride compounds have not been considered for hydrogen storage before is their reported lack of reversibility.

52.4.3.3 Chemical Hydrogen Storage

The term *chemical hydrogen storage* is used to describe storage technologies in which hydrogen is generated through a chemical reaction. Common reactions involve chemical hydrides with water or alcohols. Typically, these reactions are not easily reversible onboard a vehicle. Hence, the "spent fuel" or by-products must be removed from the vehicle and regenerated off board.

52.4.3.3.1 Hydrolysis Reactions

Hydrolysis reactions involve the oxidation reaction of chemical hydrides with water to produce hydrogen. The reaction of sodium borohydride has been the most studied to date. This reaction is

$$NaBH_4 + 2H_2O \rightarrow NaBO_2 + 4H_2 \tag{52.31}$$

In the first embodiment, slurry of an inert stabilizing liquid protects the hydride from contact with moisture and makes the hydride pumpable. At the point of use, the slurry is mixed with water, and the consequent reaction produces high-purity hydrogen. The reaction can be controlled in an aqueous medium via pH and the use of a catalyst. Although the material hydrogen capacity can be high and the hydrogen release kinetics fast, the borohydride regeneration reaction must take place off board. Regeneration energy requirements cost and life cycle impacts are the key issues currently being investigated.

Millennium cell has reported that its $NaBH_4$-based hydrogen-on-demand system possesses a system gravimetric capacity of about 4 wt.% [80]. Similar to other material approaches, issues include system volume, weight, and complexity and water availability.

Another hydrolysis reaction currently being investigated by Safe Hydrogen is the reaction of MgH_2 with water to form $Mg(OH)_2$ and H_2 [81]. In this case, particles of MgH_2 are contained in nonaqueous slurry to inhibit premature water reactions when hydrogen generation is not required. Material-based capacities for the MgH_2 slurry reaction with water can be as high as 11 wt.%. However, as with the sodium borohydride approach, water must be carried onboard the vehicle in addition to the slurry, and the $Mg(OH)_2$ must be regenerated off board.

52.4.3.3.2 New Chemical Approach

A new chemical approach may provide hydrogen generation from ammonia–borane materials by the following reactions:

$$NH_2BH_3 \rightarrow NH_2BH_2 + H_2 \rightarrow NHBH + H_2 \quad (52.32)$$

The first reaction, which occurs at <120°C, releases 6.1 wt.% hydrogen, while the second reaction, which occurs at approximately 160°C, releases 6.5 wt.% hydrogen [82]. Recent studies indicate that hydrogen-release kinetics and selectivity are improved by incorporating ammonia–borane nanosized particles in a mesoporous scaffold.

52.4.3.4 Carbonaceous Materials

Carbonaceous materials are attractive candidates for hydrogen storage because of a combination of adsorption ability, high specific surface, pore microstructure, and low mass density. In spite of extensive results available on hydrogen uptake by carbonaceous materials, the actual mechanism of storage still remains a mystery. The interaction may either be based on van der Waals attractive forces (physisorption) or on the overlap of the highest occupied molecular orbital of carbon with occupied electronic wave function of the hydrogen electron, overcoming the activation energy barrier for hydrogen dissociation (chemisorption).

The physisorption of hydrogen limits the hydrogen-to-carbon ratio to less than one hydrogen atom per two carbon atoms (i.e., 4.2 mass%). While in chemisorption, the ratio of two hydrogen atoms per one carbon atom is realized, as in the case of polyethylene [83,84]. Physisorbed hydrogen has a binding energy normally of the order of 0.1 eV, while chemisorbed hydrogen has C–H covalent bonding, with a binding energy of more than 2–3 eV.

Dillon et al. presented the first report on hydrogen storage in carbon nanotubes and triggered a worldwide tide of research on carbonaceous materials [85]. Hydrogen can be physically adsorbed on activated carbon and be "packed" on the surface and inside the carbon structure more densely than if it has just been compressed. The best results

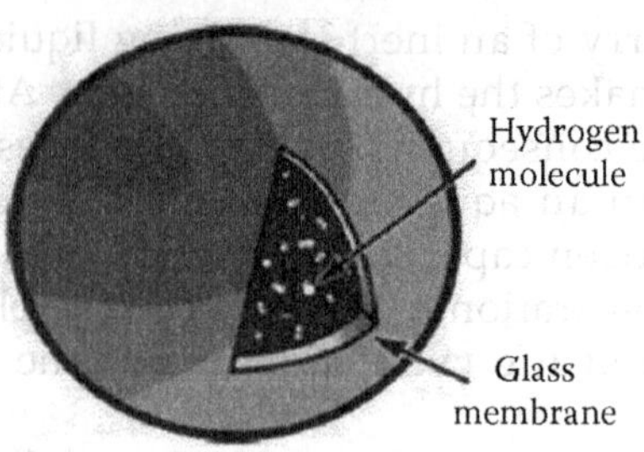

FIGURE 52.9
Schematic of a hydrogen-filled hollow glass microsphere.

achieved with carbon nanotubes to date confirmed by the National Renewable Energy Laboratory is hydrogen storage density corresponding to about 10% of the nanotube weight [86].

Hydrogen can be stored in glass microspheres of approximately 50 μm diameter. The microspheres can be filled with hydrogen by heating them to increase the glass permeability to hydrogen (see Figure 52.9) [87]. At room temperature, a pressure of approximately 25 MPa is achieved, resulting in a storage density of 14% mass fraction 10 kg H_2/m^3 [88]. At 62 MPa, a bed of glass microspheres can store 20 kg H_2/m^3. The release of hydrogen occurs by reheating the spheres to again increase the permeability.

52.4.3.5 Polymer Nanostructures

Novel nanocomposite materials consist primarily of polyaniline, a conducting polymer, which has conductivity on the order of 1 S/cm [89]. Figure 52.10 shows the chemical structure of polyaniline in its emeraldine form. The salt form of the polyaniline has anions, which allow hydrogen ions to bond to the material very well. Due to its porosity, the surface area of the conducting polymeric matrix material will allow for additional hydrogen binding in the form of physisorption.

Conducting polymer nanostructures combine the advantages of organic conductors and low-dimensional systems having interesting physicochemical properties [90–93] and useful applications [94–96]. Among the conducting polymers, polyaniline is considered important because of its extraordinary properties of electrical and optical behavior. It was recently reported that polyaniline can store as much as 6–8 wt.% of hydrogen [97], which was later refuted by Panella et al. [98]. Though many controversial results were reported in terms of hydrogen uptake [99–102] in polymer nanocomposites, there are still a number of parameters, tailor-made properties, surface morphologies, and their correlation with hydrogen sorption behavior to be investigated before these materials can be commercially deployed for onboard hydrogen storage. Similarly, nanotubes [103] or

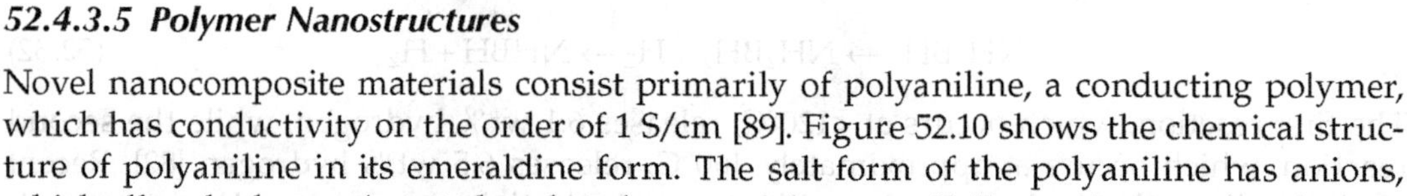

FIGURE 52.10
Chemical structure of polyaniline (emeraldine) in its salt and base form.

nanofibers [104] have attracted more interest because of their novel properties and wide potential application for nanometer-scale engineering applications. It is known that the nanofibrillar morphology significantly improves the performance of polyaniline in many conventional applications involving polymer interactions with its environment [105]. This leads to faster and more responsive chemical sensors [106], new organic/polyaniline nanocomposites [107], and ultrafast nonvolatile memory devices [108]. Nanofibers with diameters of tens of nanometers appear to be an intrinsic morphological unit that was found to naturally form in the early stage of chemical oxidative polymerization of aniline. In conventional polymerization, nanofibers are subject to secondary growth of irregularly shaped particles that form the final granular agglomerates. The key to producing pure nanofibers is to suppress secondary growth. Based on this, two methods (interfacial polymerization and rapidly mixed reactions) have been developed that can readily produce pure nanofibers by slightly modifying the conventional chemical synthesis of polyaniline without the need for any template or structural directing material. With this nanofiber morphology, dispensability and processability of polyaniline are now greatly improved. On the other hand, the template synthesis method is an effective way to grow the nanotubes of various conducting polymers [109]. The preparation conditions and their effect on morphology, size, and electrical properties of nanofibers have been reported by Zhang and Wang [110]. We have recently observed excellent hydrogen sorption characteristics in polyaniline nanofibers [111,112] prepared by an electrospinning process as shown in Figure 52.11.

Hydrogen uptake of 10 wt.% followed by the coexistence of physisorption and chemisorptions, during hydrogen release, has been correlated with unique nanofibrous morphologies (see Figure 52.12). Though the initial hydrogen storage capacity and kinetics are

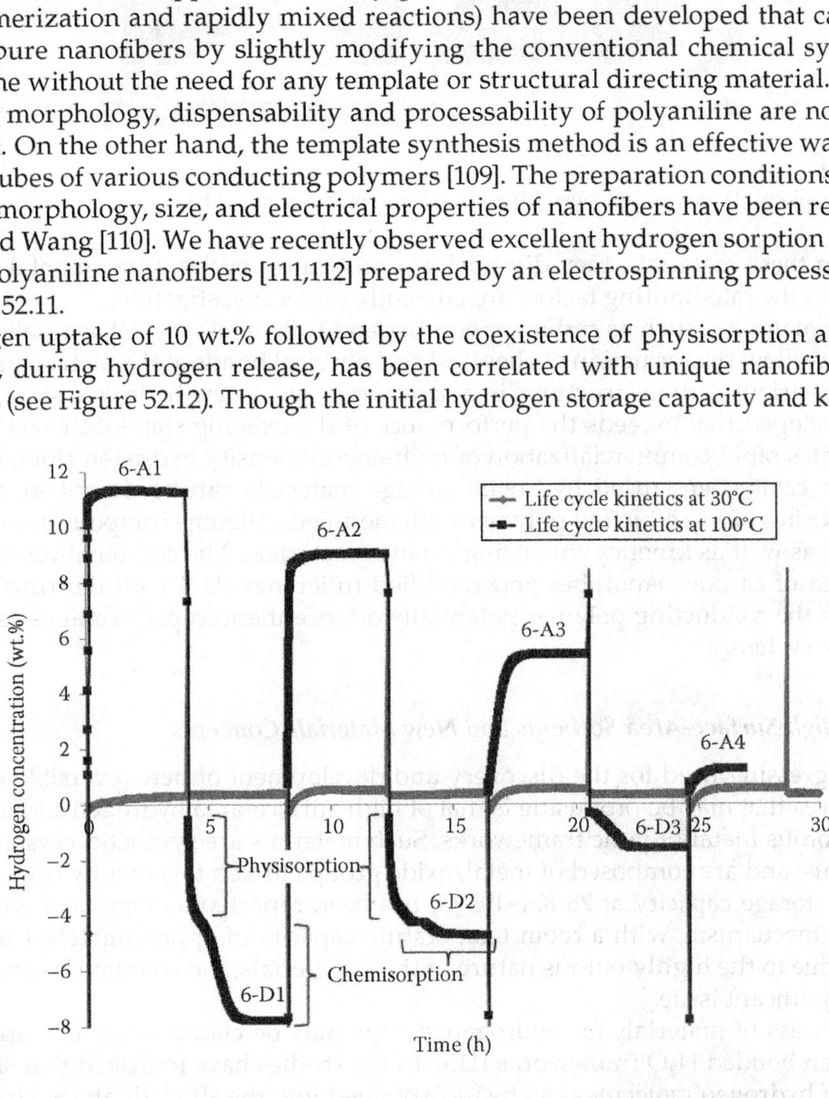

FIGURE 52.11
Hydrogen absorption and desorption cycling of electrospun polyaniline nanofiber at around 100°C.

FIGURE 52.12
Surface morphology of electrospun polyaniline nanofiber as obtained from SEM.

excellent in these materials, they diminish in consecutive cycles; fundamental gas–solid reaction and the rate-limiting factors are currently under investigation.

Additive materials, such as carbon nanotubes and fullerenes, as well as catalyst materials will also allow the formation of chemical and physical bonds at the matrix material. By combining various carbon structures in a nanocomposite, a new hydrogen storage system can be developed that exceeds the performance of the existing state-of-the-art materials and facilitates rapid commercialization of high-energy-density hydrogen storage devices. These new catalyst-enhanced hydrogen storage materials can be described as porous nanocomposites of a conducting polymer with modified fullerene compounds and carbon nanotubes, as well as kinetics enhancing catalyst materials. The combination of the high surface area of carbon nanotubes and modified fullerenes [113] with the tunable redox behavior of the conducting polymer potentially offers enhanced performance as a hydrogen storage system.

52.4.3.6 High-Surface-Area Sorbents and New Materials Concepts

There is a pressing need for the discovery and development of new reversible materials. One new area that may be promising is that of high-surface-area hydrogen sorbents based on microporous metal-organic frameworks. Such materials are synthetic, crystalline, and microporous, and are composed of metal/oxide groups linked together by organic struts. Hydrogen storage capacity at 78 K (−195°C) has been reported as high as 4 wt.% via an adsorptive mechanism, with a room temperature capacity of approximately 1 wt.% [114]. However, due to the highly porous nature of these materials, the volumetric capacity may still be a significant issue.

Another class of materials for hydrogen storage may be clathrates, which are primarily hydrogen-bonded H_2O frameworks [115]. Initial studies have indicated that significant amounts of hydrogen molecules can be incorporated into the sII clathrate. Such materials may be particularly viable for off-board storage of hydrogen without the need for high pressure or liquid hydrogen tanks.

52.5 Hydrogen Transportation and Distribution

It is envisaged that from the production plants and/or storage, hydrogen will be transmitted to consumers by means of underground pipelines (gaseous hydrogen) and/or supertankers (liquid hydrogen). Presently, hydrogen transportation through pipelines is used either in links between nearby production and utilization sites (up to 10 km) or in more extensive networks (roughly 200 km). Table 52.7 lists the principal existing hydrogen pipelines [116]. Future developments will certainly entail greater flow rates and distances. For hydrogen pipelines, it is necessary to use steel less prone to embrittlement by hydrogen under pressure (particularly for very pure hydrogen >99.5% purity). Reciprocating compressors used for natural gas can be used for hydrogen without major design modifications. However, special attention must be given to sealing (to avoid hydrogen leaks) and to materials selection for the parts subject to fatigue stress. Use of centrifugal compressors for hydrogen creates more problems due to hydrogen's exceptional lightness.

As a rule, hydrogen transmission through pipelines requires larger diameter piping and more compression power than natural gas for the same energy throughput. However, due to lower pressure losses in the case of hydrogen, the recompression stations would need to be spaced twice as far apart. In economic terms, most of the studies found that the cost of large-scale transmission of hydrogen is about 1.5–1.8 times that of natural gas transmission.

In order to match the consumption demand, hydrogen can be regionally transported and distributed, both as a gas and as a liquid, by pipelines or in special cases in containers by road and rail transportation. Gaseous (and liquid) hydrogen carriage is subject to strict regulations ensuring public safety, which in some countries is very constraining. Transportation of gaseous or liquid hydrogen in a discontinuous mode is currently employed by occasional or low-volume users. The cost of discontinuous transport is very high (it can be as high as two to five times the production cost). In the future energy system, discontinuous transportation of hydrogen would see little use, except for special users (mainly nonenergy related). Hydrogen in the gas phase is generally transported in pressurized cylindrical vessels (typically at 200 bar) arranged in frames adapted to road transport. The unit capacity of these frames or skids can be as great as 3000 m^3. Hydrogen gas distribution companies also install such frames at the user site to serve as a stationary storage.

TABLE 52.7

Some Major Hydrogen Pipelines

Location	Years of Operation	Diameter (mm)	Length (km)	Pressure (MPa)	Purity
AGEC, Alberta, Canada	Since 1987	273	3.7	3.79	99.9%
Air Liquide, France	Since 1966	Various	290	6.48–10	Pure and raw
Air Products, Houston, TX	Since 1969	114.324	100	0.35–5.5	Pure
Air Products, Louisiana	Since 1990	102–305	48	3.45	
Chemische Werke Huls	Since 1938	168–273	215	2.5	Raw gas
Cominco, British Columbia, Canada	Since 1964	5	0.6	>30	62%–100%
Gulf Petroleum, Canada		168.3	16		93.5%
Hawkeye Chemical, Iowa	Since 1987	152	3.2	2.75	
ICI, Billingham, United Kingdom			15	30	Pure
Philips Petroleum	Since 1986	203	20.9	12	

52.6 Hydrogen Conversion Technologies

Hydrogen as an energy carrier can be converted into useful forms of energy in several ways, namely,

- Combustion in internal combustion engines
- Combustion with pure oxygen to generate steam
- Catalytic combustion to generate heat
- Electrochemical conversion to electricity
- Metal hydride conversions

52.6.1 Hydrogen Combustion in Internal Combustion Engines

Hydrogen may be directly used as a fuel for internal combustion engines. The ideal thermal efficiency of an internal combustion engine is

$$\eta = 1 - \left(\frac{1}{r}\right)^{k-1} \tag{52.33}$$

where
r is the compression ratio
k is the ratio of specific heats (C_p/C_v)

This equation shows that the thermal efficiency can be improved by increasing either the compression ratio or the specific heat ratio. In hydrogen engines, both ratios are higher than in a comparable gasoline engine due to hydrogen's lower self-ignition temperature and ability to burn in lean mixtures. Therefore, hydrogen-powered internal combustion engines are on average about 20% more efficient than comparable gasoline engines.

However, the use of hydrogen in internal combustion engines results in loss of power due to lower energy content in a stoichiometric mixture in the engine's cylinder. A stoichiometric mixture of gasoline and air, and gaseous hydrogen and air, premixed externally occupies ~2% and 30% of the cylinder volume, respectively. Under these conditions, the energy of the hydrogen mixture is only 85% of the gasoline mixture, thus resulting in about 15% reduction in power. Therefore, the same engine running on hydrogen will have ~15% less power than when operated with gasoline. The power output of a hydrogen engine can be improved by using more advanced fuel injection techniques or liquid hydrogen. For example, if liquid hydrogen is premixed with air, the amount of hydrogen that can be introduced in the combustion chamber can be increased by approximately one-third [117].

One of the most important advantages of hydrogen as a fuel for internal combustion engines is that hydrogen engines emit by far fewer pollutants than comparable gasoline engines. Basically, the only products of hydrogen combustion in air are water vapor and small amounts of nitrogen oxides. Hydrogen has a wide flammability range in air (5%–75% vol.), and therefore, high excess air can be utilized more effectively. The formation of nitrogen oxides in hydrogen/air combustion can be minimized with excess air. NO_x emissions can also be lowered by cooling the combustion environment using techniques such as water injection, exhaust gas recirculation, or using liquid hydrogen. The emissions of

NO_x in hydrogen engines are typically one order of magnitude smaller than emissions from comparable gasoline engines. Small amounts of unburned hydrocarbons, CO_2, and CO have been detected in hydrogen engines due to lubrication oil [117].

The low ignition energy and fast flame propagation of hydrogen have led to problems of preignition and backfire. These problems have been overcome by adding hydrogen to the air mixture at the point where and when the conditions for preignition are less likely, such as delivering the fuel and air separately to the combustion chamber and/or injecting hydrogen under pressure into the combustion chamber before the piston is at the top dead center and after the intake air valve has been closed. Water injection and exhaust gas recirculation techniques are also used in hydrogen engines to help control premature ignition. It should be noted that most of the research on hydrogen combustion in internal combustion engines has been conducted with modifications of existing engines designed to burn gasoline. Redesign of the combustion chamber and coolant systems to accommodate hydrogen's unique combustion properties could be the most effective method of solving the problems of preignition and knocking [117].

Hydrogen use in turbines and jet engines is similar to use of conventional jet fuel. The use of hydrogen avoids the problems of sediments and corrosion on turbine blades, which prolongs life and reduces maintenance. Gas inlet temperatures can be pushed beyond normal gas turbine temperatures of 800°C, thus increasing the overall efficiency. The only pollutants from the use of hydrogen in turbines and jet engines are nitrogen oxides.

52.6.2 Steam Generation by Hydrogen/Oxygen Combustion

Hydrogen combusted with pure oxygen results in pure steam, that is,

$$2H_2 + O_2 \rightarrow 2H_2O \tag{52.34}$$

This reaction creates temperatures in the flame zone above 3000°C; therefore, additional water has to be injected so that the steam temperature can be regulated at a desired level. Both saturated and superheated vapor can be produced.

The German Aerospace Research Establishment (DLR) has developed a compact hydrogen/oxygen steam generator [118]. The steam generator consists of the ignition, combustion, and evaporation chambers. In the ignition chamber, a combustible mixture of hydrogen and oxygen at a low oxidant/fuel ratio is ignited by means of a spark plug. The rest of the oxygen is added in the combustion chamber to adjust the oxidant/fuel ratio exactly to the stoichiometric one. Water is also injected in the combustion chamber after it has passed through the double walls of the combustion chamber. The evaporation chamber serves to homogenize the steam. The steam's temperature is monitored and controlled. Such a device is very efficient, since there are no emissions other than steam and very little thermal losses.

The hydrogen steam generator can be used to generate steam for spinning reserve in power plants, for peak load electricity generation, for industrial steam supply networks, and as a micro-steam generator in medical technology and biotechnology [118].

52.6.3 Catalytic Combustion of Hydrogen

Hydrogen and oxygen in the presence of a suitable catalyst may be combined at temperatures significantly lower than flame combustion (from ambient to 500°C). This principle can be used to design catalytic burners and heaters. Catalytic burners require considerably more surface area than conventional flame burners. Therefore, the catalyst is typically dispersed in a porous structure. The reaction rate and resulting temperature are easily controlled, by controlling the hydrogen flow rate. The reaction takes place in a reaction zone of the porous catalytic sintered metal cylinders or plates in which hydrogen and oxygen are mixed by diffusion from opposite sides. A combustible mixture is formed only in the reaction zone and assisted with a (platinum) catalyst to burn at low temperatures (Figure 52.2). The only product of catalytic combustion of hydrogen is water vapor. Due to low temperatures, there are no nitrogen oxides formed. The reaction cannot migrate into the hydrogen supply, since there is no flame and hydrogen concentration is above the higher flammable limit (75%). Possible applications of catalytic burners are in household appliances such as cooking ranges and space heaters. The same principle is also used in hydrogen sensors.

52.6.4 Electrochemical Conversion (Fuel Cells)

Hydrogen can be combined with oxygen without combustion in an electrochemical reaction (reverse of electrolysis) and produce electricity (DC). The device where such a reaction takes place is called the electrochemical fuel cell or just fuel cell [119,120]. Fuel cells are described in more detail in the next chapter and will not be covered here.

52.7 Hydrogen Safety

Like any other fuel or energy carrier, hydrogen poses risks if not properly handled or controlled. The risk of hydrogen, therefore, must be considered relative to the common fuels such as gasoline, propane, or natural gas. The specific physical characteristics of hydrogen are quite different from those common fuels. Some of those properties make hydrogen potentially less hazardous, while other hydrogen characteristics could make it more dangerous in certain situations.

Since hydrogen has the smallest molecule, it has a greater tendency to escape through small openings than other liquid or gaseous fuels. Based on properties of hydrogen such as density, viscosity, and diffusion coefficient in air, the propensity of hydrogen to leak through holes or joints of low-pressure fuel lines may be only 1.26–2.8 times faster than a natural gas leak through the same hole (and not 3.8 times faster as frequently assumed based solely on diffusion coefficients). Experiments have indicated that most leaks from residential natural gas lines are laminar [121]. Since natural gas has over three times the energy density per unit volume, a natural gas leak would result in more energy release than a hydrogen leak.

For very large leaks from high-pressure storage tanks, the leak rate is limited by the sonic speed. Due to higher sonic velocity (1308 m/s), hydrogen would initially escape much faster than natural gas (sonic velocity of natural gas is 449 m/s). Again, since natural gas has more than three times the energy density than hydrogen, a natural gas leak will always contain more energy.

Some high-strength steels are prone to hydrogen embrittlement. Prolonged exposure to hydrogen, particularly at high temperatures and pressures, can cause the steel to lose strength, eventually leading to failure. However, most other construction, tank, and pipe materials are not prone to hydrogen embrittlement. Therefore, with proper choice of materials, hydrogen embrittlement safety risks could be avoided.

If a leak should occur for whatever reason, hydrogen will disperse much faster than any other fuel, thus reducing the hazard levels. Hydrogen is both more buoyant and

TABLE 52.8

Properties and Leak Rates of Hydrogen and Natural Gas

	Hydrogen	Natural Gas
Flow parameters		
Diffusion coefficient (cm^2/s)	0.61	0.16
Viscosity (μ-poise)	87.5	100
Density (kg/m^3)	0.0838	0.651
Sonic velocity (m/s)	1308	449
Relative leak rates		
Diffusion	3.80	1
Laminar flow	1.23	1
Turbulent flow	2.83	1
Sonic flow	2.91	1

more diffusive than gasoline, propane, or natural gas. Table 52.8 compares some properties and leak rates for hydrogen and natural gas.

Hydrogen/air mixture can burn in relatively wide volume ratios, between 4% and 75% of hydrogen in air. The other fuels have much lower flammability ranges, viz., natural gas 5.3%–15%, propane 2.1%–10%, and gasoline 1%–7.8%. However, the range has little practical value. In many actual leak situations, the key parameter that determines if a leak would ignite is the lower flammability limit, and hydrogen's lower flammability limit is 4 times higher than that of gasoline, 1.9 times higher than that of propane, and slightly lower than that of natural gas.

Hydrogen has a very low ignition energy (0.02 mJ), about one order of magnitude lower than other fuels. The ignition energy is a function of fuel/air ratio, and for hydrogen, it reaches a minimum at about 25%–30%. At the lower flammability limit, hydrogen ignition energy is comparable with that of natural gas [122]. Hydrogen has a flame velocity seven times faster than that of natural gas or gasoline. A hydrogen flame would therefore be more likely to progress to a deflagration or even a detonation than other fuels. However, the likelihood of a detonation depends in a complex manner on the exact fuel/air ratio, the temperature, and particularly the geometry of the confined space. Hydrogen detonation in the open atmosphere is highly unlikely.

The lower detonability fuel/air ratio for hydrogen is 13%–18%, which is 2 times higher than that of natural gas and 12 times higher than that of gasoline. Since the lower flammability limit is 4%, an explosion is possible only under the most unusual scenarios, for example, hydrogen would first have to accumulate and reach 13% concentration in a closed space without ignition, and only then an ignition source would have to be triggered. Should an explosion occur, hydrogen has the lowest explosive energy per unit stored energy in the fuel, and a given volume of hydrogen would have 22 times less explosive energy than the same volume filled with gasoline vapor.

Hydrogen flame is nearly invisible, which may be dangerous, because people in the vicinity of a hydrogen flame may not even know there is a fire. This may be remedied by adding some chemicals that will provide the necessary luminosity. The low emissivity of hydrogen flames means that nearby materials and people will be much less likely to ignite and/or hurt by radiant heat transfer. The fumes and soot from a gasoline fire pose a risk to anyone inhaling the smoke, while hydrogen fires produce only water vapor (unless secondary materials begin to burn).

Liquid hydrogen presents another set of safety issues, such as risk of cold burns and the increased duration of leaked cryogenic fuel. A large spill of liquid hydrogen has some characteristics of a gasoline spill; however, it will dissipate much faster. Another potential danger is a violent explosion of a boiling liquid expanding vapor in case of a pressure relief valve failure.

In conclusion, hydrogen appears to pose risks of the same order of magnitude as other fuels. In spite of public perception, in many aspects, hydrogen is actually a safer fuel than gasoline and natural gas. As a matter of fact, hydrogen has a very good safety record, as a constituent of the "town gas" widely used in Europe and the United States in the nineteenth and early twentieth centuries, as a commercially used industrial gas and as a fuel in space programs. There have been accidents, but nothing that would characterize hydrogen as more dangerous than other fuels.

References

1. E. Goebel, R. M. Coveney, Jr., E. E. Angino, E. J. Zeller, and G. A. M. Dreschoff, Geology, composition, isotopes of naturally occurring H_2/N_2 rich gas from Wells Near Junction City, Kansas, *Oil and Gas Journal*, 215–222, May 1994.
2. B. Heydorn, *SRI Consulting Chemical Economics Handbook*, SRI, Menlo Park, CA, 1998.
3. M. Bonner, T. Botts, J. McBreen, A. Mezzina, F. Salzano, and C. Yang, Status of advanced electrolytic hydrogen production in the United States and abroad, *International Journal of Hydrogen Energy*, 9(4), 269–275, 1984.
4. S. Dutta, Technology assessment of advanced electrolytic hydrogen production, *International Journal of Hydrogen Energy*, 15(6), 379–386, 1990.
5. H. Wendt, Water splitting methods, in: C.-J. Winter and J. Nitsch (eds.), *Hydrogen as an Energy Carrier*, Springer-Verlag, Berlin, Germany, 1988, pp. 166–238.
6. M. A. Liepa and A. Borhan, High-temperature steam electrolysis: Technical and economic evaluation of alternative process designs, *International Journal of Hydrogen Energy*, 11(7), 435–442, 1986.
7. O. G. Hancock, Jr., A photovoltaic-powered water electrolyzer: Its performance and economics, in: T. N. Veziroglu and J. B. Taylor (eds.), *Hydrogen Energy Progress V*, Pergamon Press, New York, 1984, pp. 335–344.
8. C. Carpetis, An assessment of electrolytic hydrogen production by means of photovoltaic energy conversion, *International Journal of Hydrogen Energy*, 9(12), 969–992, 1984.
9. A. Siegel and T. Schott, Optimization of photovoltaic hydrogen production, *International Journal of Hydrogen Energy*, 13(11), 659–678, 1988.
10. H. Steeb, A. Brinner, H. Bubmann, and W. Seeger, Operation experience of a 10 kW PV-electrolysis system in different power matching modes, in: T. N. Veziroglu and P. K. Takahashi (eds.), *Hydrogen Energy Progress VIII*, Vol. 2, Pergamon Press, New York, 1990, pp. 691–700.
11. H. Blank and A. Szyszka, Solar hydrogen demonstration plant in Neunburg vorm Wald, in: T. N. Veziroglu, C. Derive, and J. Pottier (eds.), *Hydrogen Energy Progress IX*, Vol. 2, M.C.I., Paris, France, 1992, pp. 677–686.
12. W. Grasse, F. Oster, and H. Aba-Oud, HYSOLAR: The German-Saudi Arabian Program on Solar Hydrogen—5 Years of experience, *International Journal of Hydrogen Energy*, 17(1), 1–8, 1992.
13. P. Lehman and C. E. Chamberlain, Design of a photovoltaic-hydrogen-fuel cell energy system, *International Journal of Hydrogen Energy*, 16(5), 349–352, 1991.
14. P. D. Lund, Optimization of stand-alone photovoltaic system with hydrogen storage for total energy self-sufficiency, *International Journal of Hydrogen Energy*, 16(11), 735–740, 1991.

15. A. G. Garcia-Conde and F. Rosa, Solar hydrogen production: A Spanish experience, in: T. N. Veziroglu, C. Derive, and J. Pottier (eds.), *Hydrogen Energy Progress IX*, Vol. 2, M.C.I., Paris, France, 1992, pp. 723–732.
16. M. S. Casper, *Hydrogen Manufacture by Electrolysis, Thermal Decomposition and Unusual Techniques*, Noyes Data Corporation, Park Ridge, NJ, 1978.
17. H. Wendt, *Electrochemical Hydrogen Technologies: Electrochemical Production and Combustion of Hydrogen*, Elsevier, New York, 1990.
18. A. Weidenkaff, A. Reller, A. Wokaun, and A. Stienfeld, Thermogravimetric analysis of the ZnO/Zn water splitting cycle, *Thermochimica Acta*, 359, 69–75, 2000.
19. A. Steinfeld, Solar hydrogen production via a two-step water splitting thermochemical cycle based on Zn/ZnO redox reactions, *International Journal of Hydrogen Energy*, 27, 611–619, 2002.
20. M. Sakurai, E. Bilgen, A. Tsustsumi, and K. Yoshida, Adiabatic UT-3 thermochemical process for hydrogen production, *International Journal of Hydrogen Energy*, 21, 865–870, 1996.
21. T. Nakayama, H. Yoshioka, H. Furutani, H. Kameyama, and K. Yoshida, MASCOT—A bench scale plant for producing hydrogen by the UT-3 thermochemical decomposition cycle, *International Journal of Hydrogen Energy*, 9, 187–190, 1984.
22. M. Sakurai, N. Miyake, A. Tsutsumi, and K. Yoshida, Analysis of a reaction mechanism in the UT-3 thermochemical hydrogen production cycle, *International Journal of Hydrogen Energy*, 21, 871–875, 1996.
23. R. Amir, S. Shizaki, K. Tamamoto, T. Kabe, and H. Kameyama, Design development of iron solid reactants in the UT-3 water decomposition cycle based on ceramic support materials, *International Journal of Hydrogen Energy*, 18, 283–286, 1993.
24. M. Sakurai, M. Aihara, N. Miyake, A. Tsutsumi, and K. Yoshida, Test of one-loop flow scheme for the UT-3 thermochemical hydrogen production process, *International Journal of Hydrogen Energy*, 17, 587–592, 1992.
25. M. Sakurai, H. Nakajima, K. Onuki, K. Ikenoya, and S. Shimizu, Preliminary process analysis for the closed cycle operation of the iodine-sulfur thermochemical hydrogen production process, *International Journal of Hydrogen Energy*, 24, 603–612, 1999.
26. K. R. Schultz, L. C. Brown, and C. J. Hamilton, Production of hydrogen by nuclear energy: The enabling technology for the hydrogen economy, in: *American Nuclear Energy Symposium*, Miami, FL, October 16–18, 2002.
27. T. Bak, J. Nowotny, M. Rekas, and C. C. Sorrell, Photo-electrochemical hydrogen generation from water using solar energy. Materials-related aspects, *International Journal of Hydrogen Energy*, 27, 991–1022, 2002.
28. J. R. Bolton, S. J. Strickler, and J. S. Connolly, Limiting and realizable efficiencies of solar photolysis of water, *Nature*, 316, 495–500, 1985.
29. M. F. Weber and M. J. Dignam, Efficiency of splitting water with semiconducting photoelectrodes, *Journal of Electrochemical Society*, 131, 1258–1265, 1984.
30. J. R. Bolton, Solar photoproduction of hydrogen, International Energy Agency Technical Report, IEA/H2/TR-96, International Energy Agency Agreement on the Production and Utilization of Hydrogen, 1996.
31. P. L. Spath, M. K. Mann, and W. A. Amos, Update of hydrogen from biomass—Determination of the delivered cost of hydrogen, NREL/MP-510-33112, National Renewable Energy Laboratory, Golden, CO, 2000.
32. S. Q. Turn et al., An experimental investigation of hydrogen production from biomass gasification, *International Journal of Hydrogen Energy*, 23(8), 641–648, 1998.
33. L. Devi et al., A review of the primary measures for tar elimination in biomass gasification process, *Biomass and Bioenergy*, 24, 125–140, 2003.
34. D. Dayton, A review of the literature on catalytic biomass tar destruction, Milestone Completion Report, NREL/TP-510-32815, National Renewable Energy Laboratory, Golden, CO, December 2002.
35. M. K. Mann, Technical and economic analysis of hydrogen production via indirectly heated gasification and pyrolysis, in: *Proceedings of the 1995 USDOE Hydrogen Program Review*, Vol. 1, NREL/CP-430-20036, National Renewable Energy Laboratory, Golden, Colorado, CO, 1995, pp. 205–236.

36. T. A. Milne, C. C. Elam, and R. J. Evans, Hydrogen from biomass: State of the art and research challenges, International Energy Agency Technical Report, Task 16, Hydrogen from Carbon Containing Materials, NREL IEA/H2/TR-02/001, National Renewable Energy Laboratory, Golden, Colorado, CO.
37. M. P. Aznar et al., Biomass gasification with air in an atmospheric bubbling fluidized bed: Effect of six operational variables on the quality of the produced raw gas, *Industrial & Engineering Chemistry Research*, 35, 2110–2120, 1996.
38. C. M. Kinoshita et al., Effect of reformer conditions on catalytic reforming of biomass gasification of tars, *Industrial & Engineering Chemistry Research*, 34, 2949–2954, 1995.
39. P. Simell, *Catalytic Hot Gas Cleaning of Gasification Gas*, Technical Research Centre of Finland, Vol. 330, VTT Publications, pp. 1–87.
40. E. Chornet, Catalytic steam reforming of bio-oils for the production of hydrogen: Effects of catalyst composition, *Applied Catalysis A: General*, 201, 225–239, 2000
41. M. Asadullah et al., Catalytic performance of Rh/CeO_2 in the gasification of cellulose to synthesis gas at low temperature, *Industrial & Engineering Chemistry Research*, 40, 5894–5900, 2001
42. S. R. A. Kersten et al., Principles of a novel multistage circulating fluidized bed reactor for biomass gasification, *Chemical Engineering Science*, 58, 725–731, 2003
43. A. Ganesh et al., Influence of mineral matter as biomass pyrolysis characteristics, *Fuel*, 74(12), 1812–1822, 1995
44. F. Pinto et al., Co-gasification study of biomass mixed with plastic wastes, *Fuel*, 81, 291–297, 2002.
45. S. P. Babu, *Biomass Gasification for Hydrogen Production Process Description and Research Needs*, Gas Technology Institute, Des Plaines, IL,
46. K. R. Craig and M. K. Mann, Cost and performance analysis of biomass based integrated gasification combined cycle BIGCC power systems, NREL/TP-430-21657, National Renewable Energy Laboratory, Golden, CO, 1996, 70 pp.
47. J. Miyake, M. Miyake, and Y. Asada, Biotechnological hydrogen production: Research for efficient energy conversion, *Journal of Biotechnology*, 70, 89–101, 1999.
48. J. B. Taylor, J. E. A. Alderson, K. M. Kalyanam, A. B. Lyle, and L. A. Phillips, Technical and economic assessment of methods for the storage of large quantities of hydrogen, *International Journal of Hydrogen Energy*, 11(1), 5–22, 1986.
49. C. Carpetis, Storage, transport and distribution of hydrogen, in: C.-J. Winter and J. Nitsch (eds.), *Hydrogen as an Energy Carrier*, Springer-Verlag, Berlin, Germany, 1988, pp. 249–289.
50. J. D. Pottier and E. Blondin, Mass storage of hydrogen, in: Y. Yurum (ed.), *Hydrogen Energy System, Utilization of Hydrogen and Future Aspects*, NATO ASI Series E-295, Kluwer Academic Publishers, Dordrecht, the Netherlands, 1995, pp. 167–180.
51. F. Mitlitsky, Development of an advanced, composite, lightweight, high pressure storage tank for on-board storage of compressed hydrogen, in: *Proceedings of the Fuel Cells for Transportation TOPTEC: Addressing the Fuel Infrastructure Issue*, Alexandria, VA, SAE, Warrendale, PA, 1996.
52. G. D. Brewer, The prospects for liquid hydrogen fueled aircraft, *International Journal of Hydrogen Energy*, 7, 21–41, 1982.
53. C. J. Winter and J. Nitsch, *Hydrogen as an Energy Carrier*, Springer-Verlag, Berlin, Germany, 1988.
54. E. L. Huston, Liquid and solid storage of hydrogen, in: *Hydrogen Energy Progress V*, Toronto, Ontario, Canada, 1984.
55. S. A. Sherif, M. Lordgooei, and M. T. Syed, Hydrogen liquefaction, in: T. N. Veziroglu (ed.), *Solar Hydrogen Energy System*, Final Technical Report, Clean Energy Research Institute, University of Miami, Coral Gables, FL, August 1989, pp. C1–C199.
56. T. N. Veziroglu, Hydrogen technology for energy needs of human settlements, *International Journal of Hydrogen Energy*, 12(2), 99–129, 1987.
57. G. Sandrock, A panoramic overview of hydrogen storage alloys from a gas reaction point of view, *Journal of Alloys Compounds*, 293–295, 877, 1999.

58. R. Griessen and T. Riesterer, Heat of formation models in: I. L. Schlapbach (ed.), *Hydrogen in Intermetallic Compounds I Electronic, Thermodynamic, Crystallographic Properties, Preparation*, Springer Series Topics in Applied Physics, New York, Vol. 63, pp. 219–284.
59. J. J. Reilly and R. H. Wiswall, Jr., Formation and properties of iron titanium hydride, *Inorganic Chemistry*, 13, 218, 1974.
60. J. H. N. Van Vucht, F. A. Kuijpers, and H. C. A. M. Burning, AB_5 intermetallic hydrides, Philips Research Report, 25(2), 133–140, 1970.
61. J. J. Reilly and R. H. Wiswall, Jr., Reaction of hydrogen with alloys of magnesium and nickel and the formation of Mg_2NiH_4, *Inorganic Chemistry*, 7, 225, 1968.
62. P. Selvam, Energy and environment—An all time research, *International Journal of Hydrogen Energy*, 16(1), 35–45, 1991.
63. S. S. S. Raman and O. N. Srivastava, Hydrogenation behavior of the new composite storage materials, Mg-Xwt.% $CFMmNi_5$, *Journal of Alloys And Compounds*, 241, 167–174, 1996.
64. A. Zuttel, P. Wenger, S. Rentsch, P. Sudan, P. Mauron, and C. Emmenegger, $LibH_4$: A new hydrogen storage material, *Journal of Power Sources*, 118(1–2), 1–7, 2003.
65. J. J. Didisheim, P. Zolliker, K. Yvon, P. Fischer, J. Schefer, M. Gubelmann, and A. Williams, Dimagnesium iron(II) hydride Mg_2FeH_6 containing octahedral $[FeH_6]^{4-}$ anions, *Inorganic Chemistry*, 23(13), 1953–1957, 1984.
66. L. Schlapbach, I. Anderson, and J. P. Burger, Heats of formations of metal hydrides, in: K. H. J. Buschow (ed.), *Material Science and Technology*, Vol. 63, VCH mbH Publishing, New York, 1988.
67. B. Bogdanovic and M. Shwickardi, Ti-doped alkali metal aluminum hydrides as potential novel reversible hydrogen storage materials, *Journal of Alloys Compounds*, 1, 253–254, 1997.
68. F. Schuth, B. Bogdanovic, and M. Felderhoff, Light metal hydrides and complex hydrides for hydrogen storage, *Chemical Communication*, 20, 2249–2258, 2004.
69. U.S. Department of Energy, Energy efficiency and renewable energy. Fuel Cell Technologies Office, Hydrogen Storage Materials Database, http://hydrogenmaterialssearch.govtools.us/ (accessed March 5, 2015).
70. C. M. Jensen and R. A. Zidan, sodium alanates for reversible hydrogen storage, in: *Proceedings of the 1998 U.S. Hydrogen Program Review*, Alexandria, VA, 1998, p. 449.
71. C. M. Jensen and K. J. Gross, Development of catalytically enhanced sodium aluminum hydride as a hydrogen-storage material, *Applied Physics A*, 72, 213, 2001.
72. M. Fichtner, Synthesis and structure of magnesium alanates and two solvent adducts, *Journal of Alloys and Compounds*, 345(1–2), 286–296, 2002.
73. E. M. Fedneva, V. L. Alpatova, and V. I. Mikheeva, $LIBH_4$ complex hydride materials, *Russian Journal of Inorganic Chemistry*, 9, 826, 1964.
74. P. Chen, Z. Xiong, J. Luo, J. Lin, and K. L. Tan, Interaction of hydrogen with metal nitrides and imides, *Nature*, 420(2), 302, 2002.
75. P. Chen, Z. Xiong, J. Luo, J. Lin, and K. L. Tan, Interaction between lithium amide and lithium hydride, *Journal of Physical Chemistry B*, 107, 10967, 2003.
76. W. F. Luo, $LiNH_2$–MgH_2: A viable hydrogen storage system, *Journal of Alloys Compounds*, 381, 284–287, 2004.
77. W. Grochala, and P. P. Edwards, Thermal decomposition of the non-interstitial hydrides for the storage and production of hydrogen, *Chemical Reviews*, 104, 1283, 2004.
78. T. J. Marks and J. R. Kolb, Covalent transition metal, lanthanide, and actinide tetrahydroborate complexes, *Chemical Reviews*, 77, 263, 1977.
79. E. Jeon and Y. W. Cho, Mechanochemical synthesis and thermal decomposition of zinc borohydride, *Journal of Alloys and Compounds*, 422, 273–275, 2006.
80. R. M. Mohring and Y. Wu, Hydrogen generation via sodium borohydride, Hydrogen Workshop, Jefferson Laboratory, November 12, 2002. http://gcep.stanford.edu/pdfs/hydrogen workshop/wu.pdf (accessed March 5, 2015).

81. A. W. McClaine, Chemical hydride slurry for hydrogen production and storage, 2004 U.S. DOE Hydrogen Fuel Cells and Infrastructure Technologies Program Review, May 25, 2004. http://www1.eere.energy.gov/hydrogenandfuelcells/pdfs/review04/st_6_mcclaine.pdf (accessed March 5, 2015).
82. C. Aardahl, T. Autrey, J. Linehan, D. Camaioni, S. Rassat, D. Rector, and W. Shaw, (2006) PNNL progress within DOE Center of Excellence for chemical hydrogen storage, FY 2006 U.S. DOE Hydrogen Program Annual Report, 2006. http://www.hydrogen.energy.gov/pdfs/progress06/iv_b_4b_aardahl.pdf (accessed March 5, 2015).
83. B. Viswanathan, M. Sankaran, and M. A. Schibioh, Carbon nanomaterials: Are they appropriate candidates for hydrogen storage? *Bulletin of the Catalysis Society of India*, 2, 12, 2003.
84. M. G. Nijkamp, J. E. M. J. Raaymakers, A. J. Van Dillen, and K. P. De Jong, Hydrogen storage by using physisorption materials demand, *Applied Physics A*, 72, 619, 2001.
85. A. C. Dillon, K. M. Jones, T. A. Bekkedahl, C. H. Kiang, D. S. Bethune, and M. J. Heben, Storage of hydrogen in single-walled carbon nanotubes, *Nature (London)*, 386(6623), 377–379, 1997.
86. P. M. F. J. Costa, K. S. Coleman, and M. L. J. Green, Influence of catalyst metal particles on the hydrogen sorption of single-wall carbon nanotube materials, *Nanotechnology*, 16, 512, 2005.
87. R. Mohtadi, K. Tange, G. Wicks, K. Heung, and R. Schumacher, A new way for storing reactive complex hydrides on board of automobiles, *Ceramic Transactions*, 202, 91–96, 2009.
88. Y. Kojima, Y. Kawai, A. Koiwai, N. Suzuki, T. Haga, T. Hioki, and K. Tange, IR Characterizations of lithium imide and amide, *Journal of Alloys Compounds*, 395(1–2), 236–239, 2005.
89. J. Stejskal, Polyaniline. Preparation of a conducting polymer (IUPAC technical report), *International Union of Pure and Applied Chemistry*, 75(5), 857, 2002.
90. A. G. MacDiarmid and A. J. Epstein, A novel class of conducting polymers, *Faraday Discussions of the Chemical Society*, 88, 317, 1989.
91. J. Stejskal, P. Kratochvil, and A. D. Jenkins, The formation of polyaniline and the nature of its structures, *Polymer*, 37(2), 367, 1996.
92. D. C. Trivedi, Polyanilines in conductive polymers: Synthesis and electrical properties, in: H. S. Nalwa (ed.), *Handbook of Organic Conductive Molecules and Polymers*, Vol. 2, Wiley, Chichester, U.K., 1997, p. 505.
93. J. Stejskal and R. G. Gilbert, Preparation of a conducting polymer, *Pure and Applied Chemistry*, 74, 857, 2002.
94. S. Virji, R. B. Kaner, and B. H. Weiller, Hydrogen sensors based on conductivity changes in polyaniline nanofibers, *Journal of Physical Chemistry B*, 110(44), 22266, 2006.
95. F. Fusalba, P. Gouerec, D. Vellers, and D. J. Belanger, Electrochemical characterization of polyaniline in nonaqueous electrolyte and its evaluation as electrode material for electrochemical supercapacitors, *Journal of the Electrochemical Society*, 148(1), A1–A6, 2001.
96. K. Rossberg, G. Paasch, and L. J. Dunsch, The influence of porosity and the nature of the charge storage capacitance on the impedance behaviour of electropolymerized polyaniline films, *Journal of Electroanalytical Chemistry*, 443(1), 49, 1998.
97. S. J. Cho, K. S. Song, J. W. Kim, T. H. Kim, and K. Choo, Hydrogen sorption in HCl treated polyaniline and polypyrrole, new potential hydrogen storage media, Preprints of *Symposia—American Chemical Society, Division of Fuel Chemistry*, 47(2), 790, 2002.
98. B. Panella, L. Kossykh, U. Dettlaff-Weglikowska, M. Hrischer, G. Zerbi, and S. Roth, Volumetric measurement of hydrogen storage in HCl-treated polyaniline and polypyrrole, *Synthetic Metals*, 151(3), 208, 2005.
99. N. B. McKeown, P. M. Budd, and D. Book, Microporous polymers as potential hydrogen storage materials, *Macromolecular Rapid Communications*, 28(9), 995, 2007.
100. S. J. Cho, K. Choo, D. P. Kim, and J. W. Kim, H_2 sorption in HCl-treated polyaniline and polypyrrole, *Catalysis Today*, 120(3–4), 336, 2007.
101. M. U. Jurczyk, A. Kumar, S. Srinivasan, and E. Stefanakos, Polyaniline-based nanocomposite materials for hydrogen storage, *International Journal of Hydrogen Energy*, 32(8), 1010, 2007.

102. J. Germain, J. M. J. Frechet, and F. J. Svec, Hypercrosslinked polyanilines with nanoporous structure and high surface area: Potential adsorbents for hydrogen storage, *Journal of Materials Chemistry*, 17(47), 4989, 2007.
103. A. Nikitin, X. Li, Z. Zhang, H. Ogasawara, H. Dai, and A. Nilsson, Hydrogen storage in carbon nanotubes through the formation of stable C–H bonds, *Nano Lettters*, 8(1), 162, 2008.
104. J. Y. Hwang, S. H. Lee, K. S. Sim, and J. W. Kim, Synthesis and hydrogen storage of carbon nanofibers, *Synthetic Metals*, 126(1), 81, 2002.
105. Y. Wang and X. J. Jing, Synthesis and hydrogen storage of carbon nanofibers. Synthetic metals, *Journal of Physical Chemistry B*, 112(4), 1157, 2008.
106. A. Z. Sadek, A. Trinchi, W. Wlodarski, K. Kalantar-zadeh, K. Galatsis, C. Baker, and R. B. Kaner, A room temperature polyaniline nanofiber hydrogen gas sensor, *IEEE Sensors*, 3, 207, 2005.
107. A. A. Athawale and S. V. J. J. Bhagwat, Synthesis and characterization of novel copper/polyaniline nanocomposite and application as a catalyst in the Wacker oxidation reaction, *Journal of Applied Polymer Science*, 89(9), 2412, 2003.
108. Y. Yang, J. Ouyang, L. Ma, R. J. Tseng, and C. W. Chu, Electrical switching and bistability in organic/polymeric thin films and memory devices, *Advanced Functional Materials*, 16(8), 1001, 2006.
109. A. Huczko, Template-based synthesis of nanomaterials, *Applied Physics A: Materials Science and Processing*, 70(4), 365, 2000.
110. D. Zhang and Y. Wang, Synthesis and applications of one dimensional nano-structured polyaniline: An overview, *Materials Science and Engineering B*, 134(1), 9, 2006.
111. S. S. Srinivasan, R. Ratnadurai, M. U. Niemann, A. R. Phani, D. Y. Goswami, and E. K. Stefanakos, Reversible hydrogen storage in electrospun polyaniline fibers, *International Journal of Hydrogen Energy*, 35, 225, 2010.
112. M. U. Niemann, S. S. Srinivasan, A. R. Phani, A. Kumar, D. Y. Goswami, and E. K. Stefanakos, Room temperature hydrogen storage in polyaniline (PANI) nanofibers, *Journal of Nanoscience and Nanotechnology*, 9, 1–5, 2009.
113. H. Zhang, L. Fan, Y. Fang, and S. Yang, Electrochemistry of composite films of C_{60} and multiwalled carbon nanotubes: A robust conductive matrix for the fine dispersion of fullerenes, *Chemical Physics Letters*, 413, 346, 2005.
114. J. L. C. Rowsell, E. C. Spencer, J. Eckert, J. A. K. Howard, and O. M. Yaghi, Gas adsorption sites in a large pore metal-organic framework, *Science*, 309, 1350–1354, 2005.
115. F. Schuth, Technology: Hydrogen and hydrates, *Nature*, 434(7034), 712–713, April 2005.
116. J. D. Pottier, Hydrogen transmission for future energy systems, in: Y. Yurum (ed.), *Hydrogen Energy System, Utilization of Hydrogen and Future Aspects*, NATO ASI Series E-295, Kluwer Academic Publishers, Dordrecht, the Netherlands, 1995, pp. 181–194.
117. J. M. Norbeck, J. W. Heffel, T. D. Durbin, B. Tabbara, J. M. Bowden, and M. C. Montano, *Hydrogen Fuel for Surface Transportation*, SAE, Warrendale, PA, 1996.
118. H. J. Sternfeld and P. Heinrich, A demonstration plant for the hydrogen/oxygen spinning reserve, *International Journal of Hydrogen Energy*, 14, 703–716, 1989.
119. K. Kinoshita, F. R. McLarnon, and E. J. Cairns, *Fuel Cells: A Handbook*, U.S. Department of Energy, DOE/METC88/6069, Morgantown, WV, 1988.
120. L. J. M. J. Blumen and M. N. Mugerwa (eds.), *Fuel Cell Systems*, Plenum Press, New York, 1993.
121. C. E. Thomas, Preliminary hydrogen vehicle safety report, The Ford Motor Company, Contract. No. DE-AC02-94CE50389, U.S. Department of Energy, Washington, DC, 1996.
122. M. R. Swain and M. N. Swain, A comparison of H_2, CH_4, and C_3H_8 fuel leakage in residential settings, *International Journal of Hydrogen Energy*, 17(10), 807–815, 1992.

53

Fuel Cells

Xiangguo Li

CONTENTS

53.1 Introduction

A fuel cell is an electrochemical device that directly converts the chemical energy of a fuel and oxidant into electric energy. Such a direct one-step conversion avoids the inefficient multistep processes involved in heat engines via combustion, thus eliminating the emission of chemical pollutants. Besides being efficient and clean, fuel cell is also compatible with renewable energy sources and carriers for sustainable development and energy security. Fuel cell offers additional advantages for both mobile and stationary applications, including quiet operation without vibration and noise, thus capable of on-site applications. Its inherent modularity allows for simple construction and operation with possible applications for dispersed, distributed, and portable power generation. Its fast response to the changing load condition while maintaining high efficiency makes it ideally suited to load following applications. Its potential high efficiency also represents less chemical, thermal, and carbon dioxide emissions for the same amount of energy conversion and power generation. Hence, fuel cell is often regarded as one of the advanced energy technologies of the future.

At present, fuel cells are being used routinely in space applications and have been under intensive development for terrestrial use, such as for utilities and zero-emission vehicles. There exist a variety of fuel cells, which are classified based on their operating temperature such as low- and high-temperature fuel cells, and the type of ion migrating through the electrolyte, etc. However, the choice of electrolyte defines the properties of a fuel cell. Hence, fuel cells are often named by the nature of the electrolyte used. There are presently six major fuel cell technologies at varying stages of development and commercialization. They are alkaline, phosphoric acid, polymer electrolyte membrane (PEM), molten carbonate, solid oxide, and direct methanol fuel cells (DMFCs). This chapter provides a summary of fundamentals and the state-of-the-art technology for these types of fuel cell, while detailed information regarding their electrochemical reactions, operation principles, construction and design, and specific areas of applications can be found elsewhere [1–7].

53.2 Principle of Operation for Fuel Cells

A fuel cell is composed of three active components: a fuel electrode (anode), an oxidant electrode (cathode), and an electrolyte sandwiched in-between. Figure 53.1 illustrates the basic operational principle of fuel cells with a typical acid electrolyte fuel cell, where it is seen that molecular hydrogen is delivered from a gas flow stream to one of the electrodes, often named anode (or fuel electrode), and it reacts electrochemically in the anode as follows:

$$\text{Anode reaction: } H_2 \Rightarrow 2H^+ + 2e^- \tag{53.1}$$

The hydrogen (fuel) is oxidized at the anode/electrolyte interface into hydrogen ion or proton H^+ and an electron e^-. The protons migrate through the (acid) electrolyte, while the electrons are forced to transfer through an external circuit, both arriving at another electrode that is often referred to as cathode (or oxidant electrode). At the cathode, the protons and electrons react with the oxygen supplied from an external gas flow stream, forming water:

$$\text{Cathode reaction: } \frac{1}{2}O_2 + 2H^+ + 2e^- \Rightarrow H_2O \tag{53.2}$$

Thus, oxygen is reduced into water at the cathode by combining with H^+ and e^-. Now both the electric current and the mass transfer form a complete circuit. The electrons go

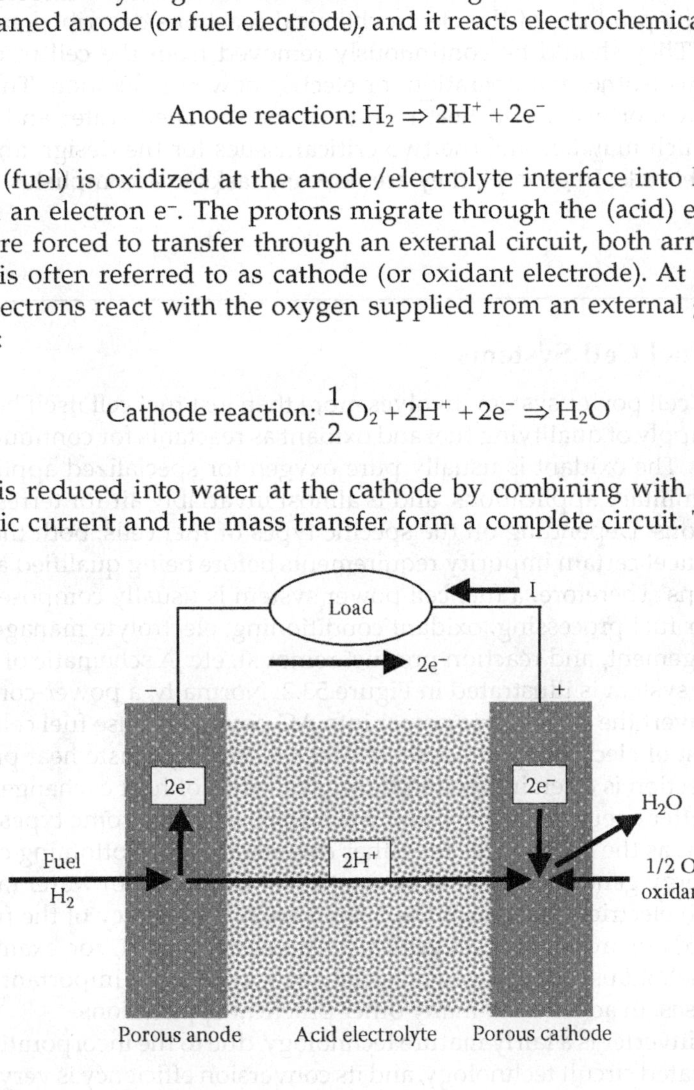

FIGURE 53.1
Schematic of a typical acid electrolyte fuel cell.

through the external electric circuit and do work on the electric load, constituting the useful electric energy output from the fuel cell. At the same time, waste heat is also generated due to the electrochemical reactions occurring at the anode and the cathode, as well as due to protons migrating through the electrolyte and electrons transporting in the solid portion of the electrodes and the external circuit. As a result, the overall cell reaction can be obtained by summing these two half-cell reactions to yield

$$\text{Overall cell reaction: } H_2 + \frac{1}{2}O_2 \Rightarrow H_2O + W + \text{Waste heat} \quad (53.3)$$

where W stands for the useful electric energy output from the fuel cell. Although the half-cell reactions may be quite different in different types of fuel cells, to be described later, the overall cell reaction remains exactly the same as the equation shown earlier.

Therefore, the by-products of the electrochemical reactions described earlier are water and waste heat. They should be continuously removed from the cell in order to maintain its continuous isothermal operation for electric power generation. This need for the continuous removal of water and heat results in the so-called water and heat (thermal) management, which may become the two critical issues for the design and operation of some types of fuel cells. In general, they are not easy tasks to accomplish.

53.3 Typical Fuel Cell Systems

In general, a fuel cell power system involves more than just fuel cell itself because fuel cell needs a steady supply of qualifying fuel and oxidant as reactants for continuous generation of electric power. The oxidant is usually pure oxygen for specialized applications like in space and some military applications, and is almost invariably air for terrestrial and commercial applications. Depending on the specific types of fuel cells, both fuel and oxidant streams need to meet certain impurity requirements before being qualified as adequate for fuel cell operations. Therefore, a fuel cell power system is usually composed of a number of subsystems for fuel processing, oxidant conditioning, electrolyte management, cooling or thermal management, and reaction product removal, etc. A schematic of a typical rudimentary fuel cell system is illustrated in Figure 53.2. Normally, a power-conditioning unit is required to convert the DC electric power into AC power because fuel cell generates DC power while most of electric equipment operates on AC. The waste heat produced in the fuel cell power section is often integrated through a series of heat exchangers into the fuel cell system for better energy efficiency, and it is also possible for some types of fuel cells to use the waste heat as the heat source for either cogeneration or bottoming cycles for additional electric energy generation. The cogeneration of heat and hot water (and sometimes steam) along with electricity increases the overall energy efficiency of the fuel cell system to as much as 85% or more. Heat is critical to human survival, for example, for space heating and household use. Both heat and steam are significantly important commodity in industrial processes, in addition to many other practical applications.

The DC-to-AC inverter is a fairly mature technology, due to the incorporation of semiconductor and integrated circuit technology, and its conversion efficiency is very high, as much as 96% for megawatt-size power plants. The fuel processor converts the primary and/or portable fuel (such as natural gas, low-sulfur distillate, naphtha, methyl fuel—mostly

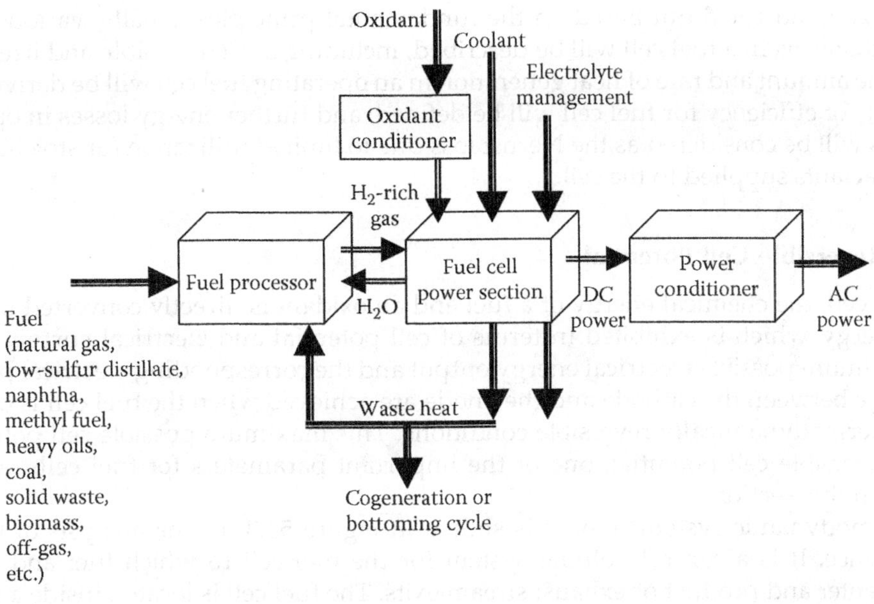

FIGURE 53.2
Schematic of a typical rudimentary fuel cell system.

methanol, heavy oils, coal, solid waste, biomass, etc.) into H_2 and CO. These secondary fuels (H_2 and CO) are considerably more electrochemically active in the electrochemical cell stack than the primary fuels. Even though fuel processing technology is highly advanced and efficient, it typically accounts for a third of the power plant size, weight, and cost for the hydrocarbon-fuelled fuel cell power plants; roughly, the electrochemical fuel cell stack accounts another third of the size, weight, and cost, while the ancillary components and subsystems associated with air supply, thermal management, water recovery and treatment, cabinet ventilation, and system control and diagnostics (or often referred to as the balance of the plant) accounts for the remaining third. In fuel cell systems, the most important subsystem is the electrochemical fuel cell stack, and fuel processor is the second major subsystem if hydrocarbon fuels are used as the primary fuel.

53.4 Performance of Fuel Cells

Although numerous studies have been conducted, aiming at developing fuel cell as a practical source of power, some confusion and misconception exist about the thermodynamic performance of fuel cell and its comparison with heat engines [2,8–10]. In this section, the fundamental principles, the first and second laws of thermodynamics, will be used to derive the idealized best possible performance, namely, the reversible cell potential and the reversible energy conversion efficiency of fuel cell. The effect of the operating conditions on the cell performance will be given, such as temperature, pressure, and reactant concentrations. The maximum possible efficiency for fuel cell will be compared with the Carnot efficiency, the maximum possible efficiency for heat engines against which fuel cell is competing for commercial success. Then, the possibility of over 100% efficiency for fuel cell will be analyzed and ruled out based on the fundamental principles. Finally, various energy loss mechanisms in a fuel cell will be described, including both reversible and irreversible losses; the amount and rate of heat generation in an operating fuel cell will be derived; various forms of efficiency for fuel cell will be defined; and further energy losses in operating fuel cells will be considered as the Nernst loss due to limited utilization (or stoichiometry) of the reactants supplied to the cell.

53.4.1 Reversible Cell Potential

In a fuel cell, the chemical energy of a fuel and an oxidant is directly converted into electrical energy, which is exhibited in terms of cell potential and electrical current output. The maximum possible electrical energy output and the corresponding electrical potential difference between the cathode and the anode are achieved when the fuel cell is operated under thermodynamically reversible conditions. This maximum possible cell potential is called reversible cell potential, one of the important parameters for fuel cells, and it is derived in this section.

A thermodynamic system* model is shown in Figure 53.3 for the analysis of fuel cell performance. It is a control volume system for the fuel cell to which fuel and oxidant streams enter and product or exhaust stream exits. The fuel cell is located inside a thermal bath in order to maintain the desired system temperature T. The reactant streams (fuel and oxidant) and the exhaust stream are considered to have the same temperature T and

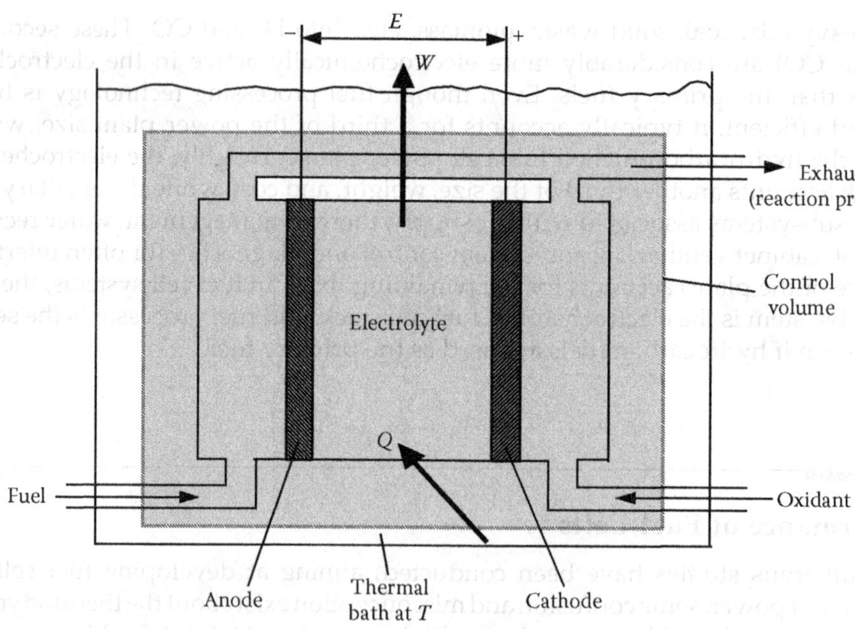

FIGURE 53.3
A thermodynamic model of fuel cell system.

* A thermodynamic system, or simply system, is a collection of matter under study (or analysis) in thermodynamics, whereas the jargon "fuel cell system" in fuel cell literature usually denotes the fuel cell power plant that consists of fuel cell stack(s) and auxiliary equipment (also called balance of the plant)—see Section 53.3, for example. In this chapter, a fuel cell system may imply both meanings. However, the context will tell which it is meant to be.

pressure P. It is assumed that the fuel and oxidant inflow and the exhaust outflow are steady; the kinetic and gravitational potential energy changes are negligible. Further, the overall electrochemical reactions occurring inside the fuel cell system boundary is described as follows:

$$\text{Fuel (e.g., } H_2) + \text{Oxidant (e.g., } O_2) \Rightarrow \dot{W} + \dot{Q} + \text{Product} \tag{53.4}$$

where

$\dot{W}$ is the rate of work done by the system

$\dot{Q}$ is the rate of heat transferred into the system from the surrounding constant temperature thermal bath, which may, or may not, be in thermal equilibrium with the fuel cell system at the temperature T and pressure P

For hydrogen/oxygen fuel cells, the reaction product is water. Then, the first and second laws of thermodynamics can be written, respectively, for the present fuel cell system, as

$$\frac{dE_{C.V.}}{dt} = [(\dot{N}h)_F + (\dot{N}h)_{Ox}]_{in} - [(\dot{N}h)_{Ex}]_{out} + \dot{Q} - \dot{W} \tag{53.5}$$

$$\frac{dS_{C.V.}}{dt} = [(\dot{N}s)_F + (\dot{N}s)_{Ox}]_{in} - [(\dot{N}s)_{Ex}]_{out} + \frac{\dot{Q}}{T} + \dot{\wp}_s \tag{53.6}$$

where

$\dot{N}$ is the molar flow rate

h is the (absolute) enthalpy per unit mole

s is the specific entropy on a mole basis

$\dot{\wp}_s$ is the rate of entropy generation due to irreversibilities

The subscripts "F," "Ox," and "Ex" stand for fuel, oxidant, and exhaust stream, respectively. "KE" and "PE" denote kinetic and gravitational potential energies that are being carried in and out of the system by the mass flow.

For a steady process, there are no temporal changes in the amount of energy $E_{C.V.}$ and entropy $S_{C.V.}$ within the control volume system, hence, $dE_{C.V.}/dt = 0$ and $dS_{C.V.}/dt = 0$. Therefore, Equations 53.5 and 53.6 can be simplified as follows:

$$\dot{N}_F(h_{in} - h_{out}) + \dot{Q} - \dot{W} = 0 \tag{53.7}$$

$$\dot{Q} = -T\dot{\wp}_s - \dot{N}_F T(s_{in} - s_{out}) \tag{53.8}$$

where

$$h_{in} = \left(h_F + \frac{\dot{N}_{Ox}}{\dot{N}_F} h_{Ox}\right)_{in} \quad \text{and} \quad h_{out} = \frac{\dot{N}_{Ex}}{\dot{N}_F} h_{Ex} \tag{53.9}$$

where

h_{in} is the amount of enthalpy per mole of fuel carried into the system by the reactant inflow

h_{out} is the amount of enthalpy per mole of fuel taken out of the system by the exhaust stream

Similarly,

$$s_{in} = \left(s_F + \frac{\dot{N}_{Ox}}{\dot{N}_F} s_{Ox} \right)_{in} \quad \text{and} \quad s_{out} = \frac{\dot{N}_{Ex}}{\dot{N}_F} s_{Ex} \tag{53.10}$$

are the amount of entropy per mole of fuel brought into, and carried out of, the system by the reactant inflow, and the outgoing exhaust stream containing the reaction products, respectively.

Substitution of Equation 53.8 into Equation 53.7 yields

$$\dot{W} = \dot{N}_F (h_{in} - h_{out}) - \dot{N}_F T (s_{in} - s_{out}) - T \dot{\wp}_s \tag{53.11}$$

Let

$$w = \frac{\dot{W}}{\dot{N}_F}, \quad q = \frac{\dot{Q}}{\dot{N}_F}, \quad \text{and} \quad \wp_s = \frac{\dot{\wp}_s}{\dot{N}_F} \tag{53.12}$$

represent, respectively, the amount of work done, heat transferred, and entropy generated per unit mole of fuel. Equations 53.8 and 53.11 then become

$$q = -T\wp_s - T(s_{in} - s_{out}) = T\Delta s - T\wp_s \tag{53.13}$$

$$w = (h_{in} - h_{out}) - T(s_{in} - s_{out}) - T\wp_s \tag{53.14}$$

Because the enthalpy and entropy changes for the fuel cell reaction are defined as

$$\Delta h = h_{out} - h_{in} \quad \text{and} \quad \Delta s = s_{out} - s_{in} \tag{53.15}$$

Equation 53.14 can also be expressed as

$$w = -\Delta h + T\Delta s - T\wp_s = -[(h - Ts)_{out} - (h - Ts)_{in}] - T\wp_s \tag{53.16}$$

From the definition of the Gibbs function (per mole of fuel) $g = h - Ts$, Equation 53.14 or 53.16 can also be written as

$$w = -(g_{out} - g_{in}) - T\wp_s = -\Delta g - T\wp_s \tag{53.17}$$

Because by the second law of thermodynamics, entropy can be generated but can never be destroyed, we know $\wp_s \geq 0$, and also the absolute temperature (in Kelvin scale) $T > 0$ by the third law of thermodynamics, the maximum possible work (i.e., useful energy) output from the present system occurs when $\wp_s = 0$, or under the thermodynamically reversible condition, since the change in the Gibbs function is usually negative for useful fuel cell reaction. Therefore, from Equation 53.17, it is clear that the maximum possible work output from the present fuel cell system is equal to the decrease in Gibbs function, or

$$w_{max} = -\Delta g \tag{53.18}$$

for all reversible processes, regardless of the specific type of fuel cells involved. In fact, it might be pointed out that in the derivation of Equations 53.17 and 53.18, no specifics about the control volume system have been stipulated; hence, they are valid for any energy conversion systems.

For a fuel cell system, the electrical energy output is conventionally expressed in terms of the cell potential difference between the cathode and the anode. Since the (electrical) potential is the (electrical) potential energy per unit (electrical) charge, its SI unit is J/C, which is more often called volt or simply V. Potential energy is defined as the work done when a charge is moved from one location to another in the electrical field, normally external circuits. For the internal circuit of fuel cells, such as the one shown in Figure 53.3, electromotive force is the terminology often used, which is also defined as the work done by transferring 1 C (coulomb) positive charge from a low to a high potential. Hence, electromotive force also has the SI unit of J/C, or V. We shall adopt the terminology of cell potential, instead of electromotive force, from now on, and we shall use the notation E to represent the cell potential. Because normally electrons are the particles transferred that carry electrical charge, we may express the work done by a fuel cell as follows:

$$w \text{ (J/mol fuel)} = E \times \text{(Coulombs of electron charge transferred/mol fuel)}$$

or

$$w = E \times (nN_0 e) = E \times (nF) \tag{53.19}$$

where

n is the number of moles of electrons transferred per mole of fuel consumed
N_0 is the Avogadro's number (=6.023 × 10^{23} number of electrons/mol electron)
e is the electric charge per electron (=1.6021 × 10^{-19} C/electron)

Since $N_0 e$ = 96,487 C/mol electron = F is known as the Faraday constant, the cell potential becomes, from Equation 53.17,

$$E = \frac{w}{nF} = \frac{-\Delta g - T\wp_s}{nF} \tag{53.20}$$

Hence, the maximum possible cell potential, or the reversible cell potential E_r, becomes

$$E_r = -\frac{\Delta g}{nF} \tag{53.21}$$

From the reversible cell potential given earlier, Equation 53.20 can also be rewritten as

$$E = E_r - \frac{T\wp_s}{nF} = E_r - \eta \tag{53.22}$$

where

$$\eta = \frac{T\wp_s}{nF} \tag{53.23}$$

is the cell voltage loss due to irreversibilities (or entropy generation). Clearly, the actual cell potential can be calculated by subtracting the cell voltage loss from the reversible cell potential. Alternatively, the amount of entropy generation per mole of fuel consumed can be determined as

$$\wp_s = \frac{nF\eta}{T} = \frac{nF(E_r - E)}{T} \tag{53.24}$$

Thus, the amount of entropy generation, representing the degree of irreversibilities (the degree of deviation from the idealized reversible condition), for the fuel cell reaction process can be measured once the cell potential E and the cell operating temperature T are known.

Note that the Gibbs function is a thermodynamic property, determined by state variables such as temperature and pressure. Hence, the change in the Gibbs function for the fuel cell reaction discussed here

$$\Delta g = \Delta h - T\Delta s \tag{53.25}$$

is also a function of the system temperature T and pressure P, as is the reversible cell potential. The specific effect of the operating conditions, such as temperature, pressure, and reactant concentrations, on the reversible cell potential will be presented in the next section. If the reaction occurs at the standard reference temperature and pressure (25°C and 1 atm), the resulting cell potential is usually called the standard reversible cell potential E_r°,* or

$$E_r^\circ(T_{\text{ref}}) = -\frac{\Delta g(T_{\text{ref}}, P_{\text{ref}})}{nF} \tag{53.26}$$

If pure hydrogen and oxygen are used as reactants to form product water, then $E_r^\circ(25°\text{C}) = 1.229$ V for the product water in liquid form, and $E_r^\circ(25°\text{C}) = 1.185$ V for the product water in vapor form. The difference in E_r° corresponds to the energy required for the vaporization of water. It might be pointed out that any fuel containing hydrogen (including hydrogen itself, hydrocarbons, alcohols, and to a lesser extent coal) has two values for Δg and Δh, one higher and one lower, depending on whether the product water is in the form of liquid or vapor. Hence, care should be taken when referring to reversible cell potential and energy efficiency to be discussed later in Section 53.4.3.

The standard reversible cell potential, E_r°, can be determined for any other electrochemical reactions. Some of the potential fuel cell reactions and the resulting E_r° are shown in Table 53.1 along with other relevant parameters. From this table, it might be noted that E_r° should be approximately above 1 V in order for the reaction to be realistic for fuel cell application. This is because if E_r° is $\ll$1 V, and considering the cell voltage loss that inevitably occurs in practical fuel cells due to irreversibilities, the actual cell potential might become too small to be useful for practical applications. Therefore, the rule of thumb is, for any proposed fuel and oxidant, to calculate E_r° and to see if E_r° is on the order of 1 V or larger before proceeding to any further work on it.

53.4.2 Effect of Operating Conditions on Reversible Cell Potential

The most important operating conditions that influence fuel cell performance are the operating temperature, pressure, and reactant concentrations. For many useful electrochemical reactions, the entropy change is negative and is almost constant with the change of temperature to a good approximation, provided the temperature change

* In the literature, the superscript "°" sometimes denotes the value at the standard reference condition of 25°C and 1 atm, and sometimes it also refers to parameters evaluated at 1 atm. The latter meaning has been adopted in this chapter.

TABLE 53.1

Standard Enthalpy and Gibbs Function of Reaction for Candidate Fuels and Oxidants, and Corresponding Standard Reversible Cell Potential as well as Other Relevant Parameters (at 25°C and 1 atm)

Fuel	Reaction	n	$-\Delta h$ (kJ/mol)	$-\Delta g$ (kJ/mol)	E_r^o (V)	η^a (%)
Hydrogen	$H_2 + 0.5O_2 \rightarrow H_2O(l)$	2	286.0	237.3	1.229	82.97
	$H_2 + Cl_2 \rightarrow 2HCl(aq)$	2	335.5	262.5	1.359	78.33
	$H_2 + Br_2 \rightarrow 2HBr(aq)$	2	242.0	205.7	1.066	85.01
Methane	$CH_4 + 2O_2 \rightarrow CO_2 + 2H_2O(l)$	8	890.8	818.4	1.060	91.87
Propane	$C_3H_8 + 5O_2 \rightarrow 3CO_2 + 4H_2O(l)$	20	2221.1	2109.3	1.093	94.96
Decane	$C_{10}H_{22} + 15.5O_2 \rightarrow 10CO_2 + 11H_2O(l)$	66	6832.9	6590.5	1.102	96.45
Carbon monoxide	$CO + 0.5O_2 \rightarrow CO_2$	2	283.1	257.2	1.333	90.86
Carbon	$C(s) + 0.5O_2 \rightarrow CO$	2	110.6	137.3	0.712	124.18[b]
	$C(s) + O_2 \rightarrow CO_2$	4	393.7	394.6	1.020	100.22[b]
Methanol	$CH_3OH(l) + 1.5O_2 \rightarrow CO_2 + 2H_2O(l)$	6	726.6	702.5	1.214	96.68
Formaldehyde	$CH_2O(g) + O_2 \rightarrow CO_2 + H_2O(l)$	4	561.3	522.0	1.350	93.00
Formic acid	$HCOOH + 0.5O_2 \rightarrow CO_2 + H_2O(l)$	2	270.3	285.5	1.480	105.62[b]
Ammonia	$NH_3 + 0.75O_2 \rightarrow 1.5H_2O(l) + 0.5N_2$	3	382.8	338.2	1.170	88.36
Hydrazine	$N_2H_4 + O_2 \rightarrow 2H_2O(l) + N_2$	4	622.4	602.4	1.560	96.77
Zinc	$Zn + 0.5O_2 \rightarrow ZnO$	2	348.1	318.3	1.650	91.43
Sodium	$Na + 0.25H_2O + 0.25O_2 \rightarrow NaOH\ (aq)$	1	326.8	300.7	3.120	92.00

Source: Appleby, A.J., Characteristics of fuel cell systems, in: Blomen, L.J.M.J. and Mugerwa, M.N., eds., *Fuel Cell Systems*, Chapter 5, Plenum Press, New York, 1993.

[a] Energy conversion efficiency.

[b] There is a conceptual problem with these efficiency data; see Section 53.4.3.5 for explanation.

$T - T_{ref}$ is not too large. Then, the effect of temperature on the reversible cell potential may be written as [1]

$$E_r(T, P) = E_r(T_{ref}, P) + \left(\frac{\Delta s(T_{ref}, P)}{nF}\right)(T - T_{ref}) \quad (53.27)$$

It must be emphasized that the expression given in Equation 53.27 is an approximation. Strictly speaking, the reversible cell potential at any temperature and pressure should be determined from Equation 53.21 by calculating first the property changes for the particular fuel cell reaction involved. Such a procedure has been followed for the hydrogen and oxygen reaction to form gaseous water, and the results are presented in Figure 53.4. Clearly, the reversible cell potential indeed decreases almost linearly as temperature is increased over a large temperature range. However, it is noticed that the reversible cell potential is larger for product water as liquid at low temperatures, but it decreases much faster than the gaseous water as product when temperature is increased. So that at temperatures slightly above about 373 K, the reversible cell potential for liquid water product actually becomes smaller. This may seem odd, but it is because at such high-temperature pressurization is necessary to keep the product water in liquid form as the reactants, hydrogen and oxygen, are fed at 1 atm. Also notice that the critical temperature for water is about 647 K, beyond which distinct liquid state does not exist for water, hence the shorter curve shown for the liquid water case in Figure 53.4.

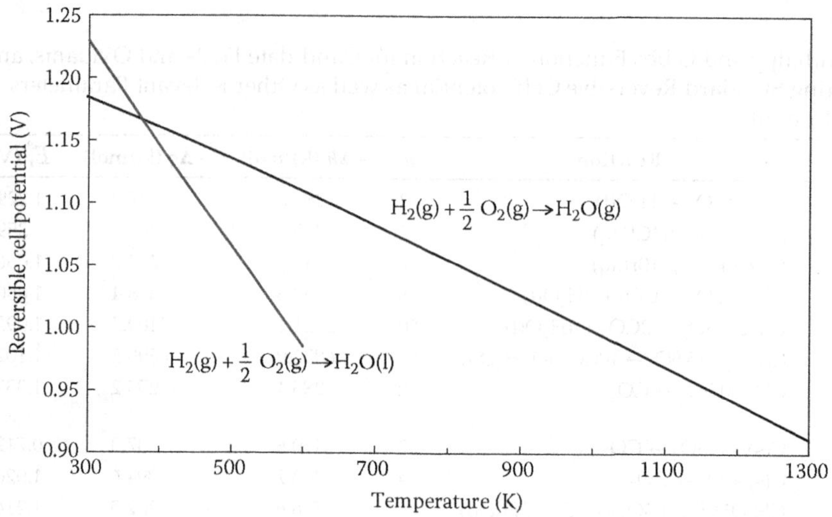

FIGURE 53.4
Effect of temperature on the reversible cell potential of a hydrogen–oxygen fuel cell for the reaction of $H_2 + 1/2O_2 \rightarrow H_2O$ at the pressure of 1 atm.

As pointed out earlier, the entropy change for most of fuel cell reactions is negative; consequently, the reversible cell potential decreases as temperature is increased as shown in Figure 53.4. However, for some reactions such as

$$C(s) + \frac{1}{2}O_2(g) \rightarrow CO(g)$$

the entropy change is positive, for example, $\Delta s = 89$ J/(mol fuel K) at the standard reference temperature and pressure. As a result, the reversible cell potential for this type of reactions will increase with temperature, which is clearly seen in Figure 53.5 for the reversible cell potential as a function of temperature for a number of important fuel cell reactions.

The effect of pressure on reversible cell potential can be expressed as [1]

$$E_r(T,P) = E_r(T,P_{\text{ref}}) - \frac{\Delta N \Re T}{nF}\ln\left(\frac{P}{P_{\text{ref}}}\right) \tag{53.28}$$

where ΔN represents the total number of mole changes for all the gaseous species involved in the fuel cell reaction. This equation indicates that the pressure dependence of the reversible cell potential is a logarithmic function, and hence the dependence becomes weaker as pressure P is increased. Figure 53.6 shows the effect of pressure on the reversible cell potential for hydrogen/oxygen fuel cell for both liquid and vapor water as the reaction product. It is seen that the reversible cell potential increases with pressure, very fast for low-pressure values, gradually slowing down for higher-pressure values. Also, the pressure effect is larger for liquid water as the product, because of the larger coefficient arising from the larger change in the number of moles between the product and the reactant, ΔN.

It might be also emphasized that the pressure effect is small on the reversible cell potential at low temperatures, as shown in the earlier example. However, this effect increases

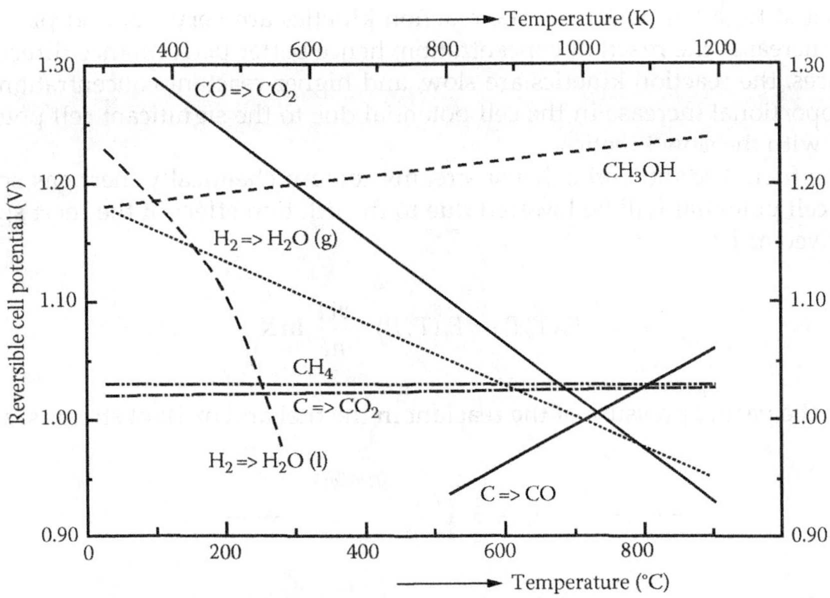

FIGURE 53.5
Standard reversible cell potential, E_r, as a function of temperature for the most important fuel cell reactions at the pressure of 1 atm. (From Barendrecht, E., Electrochemistry of fuel cells, in: Blomen, L.J. M.J. and Mugerwa, M.N., eds., *Fuel Cell Systems*, Plenum Press, New York, 1993.)

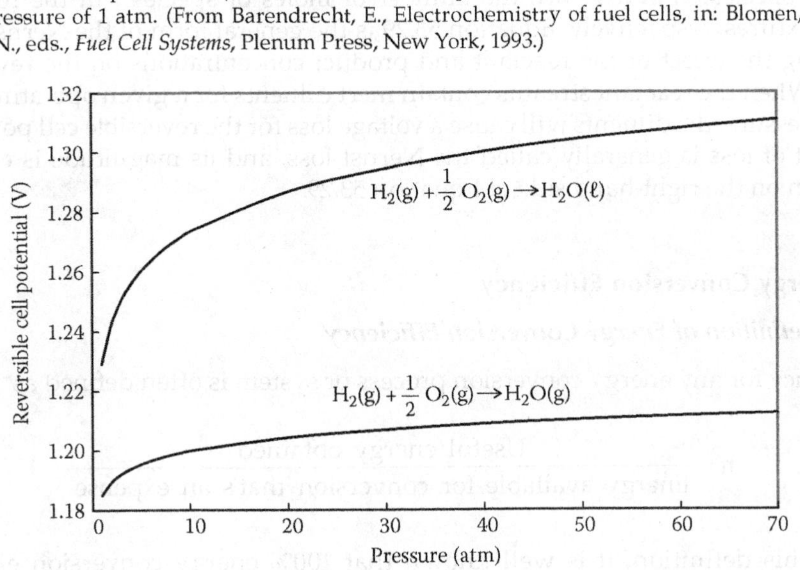

FIGURE 53.6
Standard reversible cell potential, E_r, as a function of pressure for the fuel cell reaction of $H_2(g) + 1/2O_2 \rightarrow H_2O$ at the temperature of 25°C.

significantly at high temperatures because the pressure effect coefficient, $-\Delta N\Re T/(nF)$, is directly proportional to temperature.

It might be pointed out that for high-temperature fuel cells, the dependence of the actual cell potential E on the pressure follows closely the results given in Equation 53.28, whereas a significant deviation occurs for the low-temperature fuel cells. The difference arises from the fact that at high temperatures, the reaction kinetics are very fast and pressurization primarily increases the reactant concentration, hence better performance directly. At low temperatures, the reaction kinetics are slow, and higher reactant concentration does not yield a proportional increase in the cell potential due to the significant cell potential loss associated with the slow kinetics.

When the fuel, oxidant, and exhaust streams contain chemically inert gas species, the reversible cell potential will be lowered due to the dilution effect of the inert species and can be derived as [1]

$$E_r(T, P_i) = E_r(T, P) - \frac{\Re T}{nF} \ln K \tag{53.29}$$

where P_i is the partial pressure of the reactant in the fuel and oxidant streams, and

$$K = \prod_{i=1}^{N_g} \left(\frac{P_i}{P} \right)^{(\nu_i'' - \nu_i')/\nu_F} \tag{53.30}$$

is defined similar to the equilibrium constant for partial pressure (but they are not the same); N_g is the total number of gas species in the reacting system, excluding the solid and liquid species; ν_i' and ν''_i are the number of moles of species i in the reactant and product mixtures, respectively. Equation 53.29 is the general form of the Nernst equation, representing the effect of the reactant and product concentrations on the reversible cell potential. When the reactant streams contain inert diluents for a given operating temperature and pressure, the diluents will cause a voltage loss for the reversible cell potential, and the amount of loss is generally called the Nernst loss, and its magnitude is equal to the second term on the right-hand side of Equation 53.29.

53.4.3 Energy Conversion Efficiency

53.4.3.1 Definition of Energy Conversion Efficiency

The efficiency for any energy conversion process or system is often defined as*

$$\eta = \frac{\text{Useful energy obtained}}{\text{Energy available for conversion that's an expense}} \tag{53.31}$$

Based on this definition, it is well known that 100% energy conversion efficiency is possible by the first law of thermodynamics, but is not possible by the second law of thermodynamics for many energy conversion systems that produce power output, such as steam and gas turbines, internal combustion engines, which involve irreversible losses of energy. These thermal energy conversion systems are often referred to as heat engines.

* Note that in literature η is commonly used as efficiency in thermodynamics, whereas it is also conventionally used as overpotential, or voltage loss, for fuel cell analysis, as in electrochemistry. In this chapter, η is used for both efficiency and overpotential in order to be consistent with the respective convention, and its meaning would become clear from the context.

53.4.3.2 Reversible Energy Conversion Efficiency for Fuel Cells

For the present fuel cell system described in Figure 53.3, the energy balance equation, Equation 53.5, can be written, on a per unit mole of fuel basis, as

$$h_{in} - h_{out} + q - w = 0 \quad \text{or} \quad h_{in} - h_{out} = -q + w \tag{53.32}$$

which indicates that the enthalpy change, $-\Delta h = h_{in} - h_{out}$, provides the energy available for conversion into the useful energy exhibited as work here, and it is the expense to be paid for the useful work output. At the same time, waste heat, q, is also generated, which would represent a degradation of energy. The amount of waste heat generated can be determined from the second law expression, Equation 53.6 or 53.13, as

$$q = T\Delta s - T\wp_s \tag{53.33}$$

and the useful energy output as work is, from Equation 53.17 or combining Equations 53.32 and 53.33

$$w = -\Delta g - T\wp_s \tag{53.34}$$

Therefore, the energy conversion efficiency for the fuel cell system described in Figure 53.3 becomes, according to Equation 53.31,

$$\eta = \frac{w}{-\Delta h} = \frac{\Delta g + T\wp_s}{\Delta h} \tag{53.35}$$

Note that both Δh and Δg are negative for power generation systems, including fuel cells as it is clearly shown in Table 53.1. By the second law, the entropy generation per unit mole of fuel is

$$\wp \geq 0 \tag{53.36}$$

and the equality holds for all reversible processes, whereas entropy is always generated for irreversible processes. Therefore, the maximum possible efficiency allowed by the second law is, when the process is reversible (i.e., $\wp_s = 0$),

$$\eta_r = \frac{w_{max}}{-\Delta h} = \frac{\Delta g(T,P)}{\Delta h(T,P)} \tag{53.37}$$

Since both the enthalpy and Gibbs function changes depend on the system temperature and pressure, the same holds for the energy conversion efficiency. It should be pointed out that in the earlier derivation, no assumption specifically related to fuel cell has been made; the only assumption made is that the energy conversion system for power production is reversible for all processes involved. Thus, Equation 53.37 is valid for any power production system, be it electrochemical converter like fuel cells or conventional thermal energy converter like heat engines, as long as the process is reversible. Hence, it may be called the *second law or reversible efficiency*, since it is the maximum possible efficiency that is allowed by the second law of thermodynamics. In what follows, we will demonstrate that the maximum possible efficiency for conventional heat engines, the well-known Carnot efficiency, is really the second law efficiency applied specifically to the conventional thermal power cycles, thus is equivalent to Equation 53.37.

53.4.3.3 Carnot Efficiency: The Reversible Energy Conversion Efficiency for Heat Engines

Consider a heat engine operating between two temperature thermal energy reservoirs (TER), one at a high temperature T_H and the other at a low temperature T_L, as shown in Figure 53.7. The heat engine obtains energy from the high-temperature TER in the form of heat with the quantity q_H; a portion of this heat is converted to work output w, and the remainder is rejected to the low-temperature TER in the amount of q_L as waste heat. Applying the first and second laws to the heat engine, we have

First law: $$w = q_H - q_L \tag{53.38}$$

Second law: $$\wp_{s,\mathrm{HE}} = \frac{q_L}{T_L} - \frac{q_H}{T_H} \tag{53.39}$$

where $\wp_{s,\mathrm{HE}}$ represents the amount of entropy production during the energy conversion process by means of the heat engine. From Equation 53.39, the amount of heat rejection can be determined as

$$q_L = \frac{T_L}{T_H} q_H + T_L \wp_{s,\mathrm{HE}} \tag{53.40}$$

The efficiency for the heat engine is, by the definition of Equation 53.31,

$$\eta = \frac{w}{q_H} \tag{53.41}$$

Substituting Equations 53.38 and 53.40 into 53.41 yields

$$\eta = 1 - \frac{T_L}{T_H} - \frac{T_L}{q_H} \wp_{s,\mathrm{HE}} \tag{53.42}$$

As pointed out earlier, the second law of thermodynamics dictates that the entropy generation within the heat engine can never be negative; at most, it can vanish under the

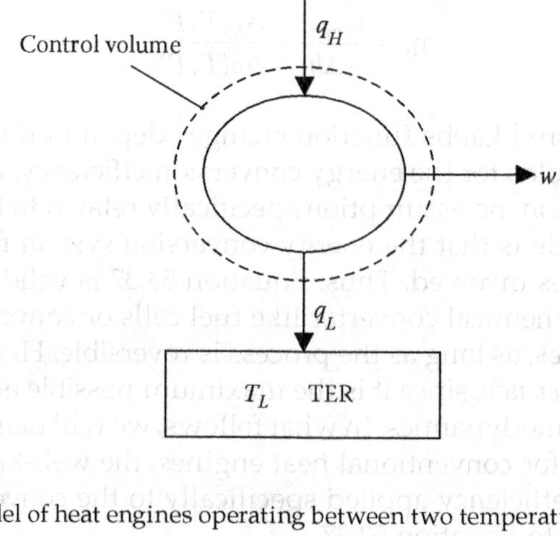

FIGURE 53.7
Thermodynamic system model of heat engines operating between two temperature thermal energy reservoirs.

thermodynamically reversible condition. Therefore, the maximum possible efficiency for the heat engine is achieved if the process is reversible ($\wp_{s,\mathrm{HE}} = 0$):

$$\eta_{r,\mathrm{HE}} = 1 - \frac{T_L}{T_H} \tag{53.43}$$

This is the familiar Carnot efficiency, giving the upper bound for the efficiency of all heat engines. Because $T_L < T_H$, the low-temperature $T_L \neq 0$ by the third law of thermodynamics, and the high-temperature T_H is finite, 100% efficiency is not possible by the second law for any energy conversion system that produces power output using heat engines, such as steam and gas turbines and internal combustion engines, because of the second law requirement that the entropy generation term must never be negative. In contrast, 100% efficiency is possible by the first law, which merely states the principle of energy conservation.

53.4.3.4 Equivalency of Carnot and Fuel Cell Reversible Efficiency

As shown earlier, both the Carnot efficiency and the reversible energy conversion efficiency for fuel cell, Equation 53.37, are the maximum possible efficiency allowed by the second law; the former is applied specifically to heat engines, while the latter is derived for fuel cells. Therefore, they must be related somehow as they both are the maximum possible efficiency dictated by the second law. In this section, we demonstrate that they are actually equivalent, just expressed in a different form, under a suitable condition for the comparison.

Suppose, for a heat engine, the high-temperature TER is maintained at T_H by the combustion of a fuel with an oxidant, both reactants are originally at the temperature of T_L, as shown schematically in Figure 53.8. It is assumed that both the fuel and the oxidant are the same as used in Figure 53.3 for the derivation of fuel cell performance; the combustion process is carried out at the same system pressure P in a controlled manner such that the combustion products leave the TER at the pressure P and temperature T_L. Neglecting

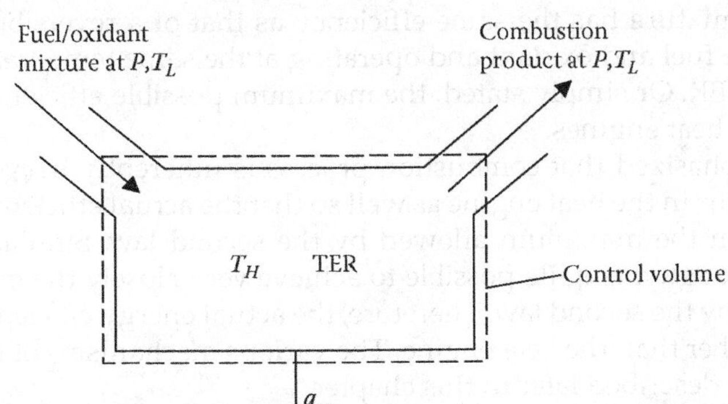

FIGURE 53.8
Thermodynamic system model of high-temperature thermal energy reservoir (TER) maintained by combustion process of a fuel/oxidant mixture.

the changes in the kinetic and gravitational potential energy, the first and second laws become, for the high-temperature TER,

First law:

$$q_H = h_R - h_P = -\Delta h(T_L, P) \tag{53.44}$$

Second law:

$$\wp_{s,\text{TER}} = (s_P - s_R) + \frac{q_H}{T_H} = \Delta s(T_L, P) + \frac{q_H}{T_H} \tag{53.45}$$

After rearranging, Equation 53.45 gives the temperature T_H resulting from the combustion process

$$T_H = \frac{q_H}{\wp_{s,\text{TER}} - \Delta s(T_L, P)} \tag{53.46}$$

Substitution of Equations 53.44 and 53.46 into 53.42 leads to

$$\eta = \frac{\Delta g(T_L, P)}{\Delta h(T_L, P)} + \frac{T_L}{\Delta h(T_L, P)}(\wp_{s,\text{HE}} + \wp_{s,\text{TER}}) \tag{53.47}$$

where $\Delta g(T_L, P) = \Delta h(T_L, P) - T_L \Delta s(T_L, P)$ is the change in the Gibbs function between the reaction product and the reactant. If all the processes within the heat engine and high-temperature TER are reversible ($\wp_{s,\text{HE}} = 0$ and $\wp_{s,\text{TER}} = 0$), then Equation 53.47 reduces to

$$\eta_r = \frac{\Delta g(T_L, P)}{\Delta h(T_L, P)} \tag{53.48}$$

which is exactly the same as Equation 53.37—the efficiency expression derived for fuel cells. Note that in order for the combustion process to be reversible (i.e., $\wp_{s,\text{HE}} = 0$) theoretically, there should be no product dissociations and no incomplete combustion products or by-products (such as pollutants) formed, and the perfect combustion products should consist of stable chemical species only, as would be obtained from an ideal and complete stoichiometric reaction. Therefore, it may be stated that any reversible heat engine operating under the maximum temperature limit allowed by a *perfect* combustion of a fuel/oxidant mixture has the same efficiency as that of a reversible isothermal fuel cell using the same fuel and oxidant and operating at the same temperature as that of the low-temperature TER. Or simply stated, the maximum possible efficiency is the same for both fuel cells and heat engines.

It should be emphasized that combustion process is inherently irreversible, and other irreversibilities occur in the heat engine as well so that the actual efficiency for heat engine is much lower than the maximum allowed by the second law. Similarly, fuel cells can never achieve, although it is quite possible to achieve very closely the maximum possible efficiency allowed by the second law. Therefore, the actual energy efficiency for fuel cells is typically much higher than the heat engine. The various mechanisms of irreversible losses in fuel cells will be described later in this chapter.

53.4.3.5 Possibility of Over 100% Fuel Cell Efficiency: Is It Real or Hype?

It is well known that no heat engine could have efficiency of 100% or more, including the ideal Carnot efficiency, as discussed earlier. However, it has been reported that the ideal fuel cell efficiency, η_r, according to Equation 53.37, could be over 100% in principle for some special fuel cell reactions (e.g., [2,8]), even though it is unachievable in practice. This has also sometimes been used as evidence that fuel cells could have higher energy efficiency than the competing heat engines. Is this realistic even under the thermodynamically reversible condition? The answer is negative! With the following analysis, we can show that this is really due to a conceptual error in stretching the application of Equation 53.37 beyond its validity range.

Consider the thermodynamic model system used for fuel cell analysis, as shown in Figure 53.3. The amount of heat transfer from the surrounding thermal bath to the fuel cell system is given in Equation 53.33 for practical fuel cells. Under the thermodynamically reversible condition, the amount of heat transfer becomes

$$q = T\Delta s = \Delta h - \Delta g \tag{53.49}$$

For most of fuel cell systems, Δs is negative (i.e., $\Delta s < 0$ just like Δh and Δg), indicating that heat is actually transferred from the fuel cell to the ambient environment, or heat is lost from the fuel cell system, rather than the other way around. Hence, the second law efficiency, according to Equation 53.37,

$$\eta_r = \frac{\Delta g}{\Delta h} = \frac{\Delta h - T\Delta s}{\Delta h} = 1 - \frac{T\Delta s}{\Delta h} < 1 \tag{53.50}$$

is <100%, as it should be by the common perception of the parameter called efficiency.

However, for some special reactions, such as

$$C(s) + \frac{1}{2}O_2(g) \rightarrow CO(g) \tag{53.51}$$

the entropy change Δs is positive. Physically, it indicates that the fuel cell absorbs heat from the ambient and converts it completely into electrical energy along with the chemical energy of the reactants. This is equivalent to stating that the less useful form of energy, heat, is converted completely into the more useful form of energy, electric energy, without the generation of entropy (i.e., reversible condition) during the conversion process when Equation 53.37 is used for the efficiency calculation—such a process is clearly a violation of the second law. Therefore, the second law efficiency for this particular fuel cell reaction becomes larger than 100%, that is, a physically impossible result, when Equation 53.37 is utilized for the efficiency calculation for this type of fuel cell reactions. According to Equation 53.37, the reversible fuel cell efficiency for the reaction shown in Equation 53.51 would be equal to η_r = 124% at the standard temperature and pressure, 163% at 500°C and 1 atm, and 197% at 1000°C and 1 atm.

The root of the problem from the straightforward application of Equation 53.37 leading to the physically impossible result of over 100% energy efficiency is as follows. At atmospheric temperature for fuel cell operations, the energy from the thermal bath (or the atmosphere) as heat may be free. But at elevated temperatures, external means must be employed to keep the thermal bath at temperatures above the ambient atmospheric temperature, which constitutes an expense. Therefore, the heat from the thermal bath to the fuel cell system is no longer a free energy input; rather, it is part of the energy input that has to be paid for. By definition, Equation 53.31, the efficiency definition for fuel cells, has to be modified accordingly, such that the ideal reversible efficiency will be no longer over 100% for fuel cells. Thus, we conclude that the reversible fuel cell efficiency shown in Equation 53.37 is valid only for fuel cell reactions where the entropy change between the product and reactant is negative (hence, heat is lost from the fuel cell), and it cannot be applied for reactions with positive entropy change, such as the one given in Equation 53.51.

53.4.4 Practical Fuel Cell Efficiency and Energy Loss Mechanisms

From the preceding analysis, it is clear that energy loss in fuel cells occurs under both reversible and irreversible conditions. We will describe each type of energy loss mechanisms and associated expression for energy conversion efficiency in fuel cells.

53.4.4.1 Reversible Energy Loss and Reversible Energy Efficiency

The energy loss in fuel cells under reversible conditions is equal to the heat transferred (or lost) to the environment, as given:

$$q = T\Delta s = \Delta h - \Delta g \tag{53.49}$$

because of the negative entropy change for the fuel cell reaction. The associated energy conversion efficiency, which has been called the reversible energy conversion efficiency, has been derived and given in Equation 53.37 or 53.50. Combining Equation 53.49 with 53.50 yields

$$\eta_r = \frac{\Delta g}{\Delta h} = \frac{\Delta g}{\Delta g + T\Delta s} \tag{53.52}$$

Dividing the numerator and the denominator by the factor (nF), and utilizing Equations 53.21 and 53.52 becomes [1]

$$\eta_r = \frac{E_r}{E_r - T(\partial E_r/\partial T)_P} \tag{53.53}$$

Therefore, when the entropy change is negative, as described earlier, the reversible efficiency, η_r, is <100%, and the reversible cell potential decreases with temperature; and according to Equation 53.53, the reversible efficiency, n_r, also decreases with temperature. For example, for H_2 and O_2 reaction forming gaseous water at 1 atm pressure,

$$\left(\frac{\partial E_r}{\partial T}\right)_P = -0.2302 \times 10^{-3}\ \text{V/K}$$

at 25°C, and the reversible efficiency is about 95% at 25°C, and it becomes 88% at 600 K and 78% at 1000 K. Figure 53.9 illustrates the reversible efficiency as a function of temperature for the hydrogen and oxygen reaction with gaseous water as the reaction product. It is seen that the reversible efficiency decreases almost linearly. For most fuel cell reactions,

$$\left(\frac{\partial E_r}{\partial T}\right)_P = -(0.1\text{–}1.0) \times 10^{-3}\ \text{V/K}$$

at 25°C and 1 atm; hence, the reversible efficiency is typically around 90%.

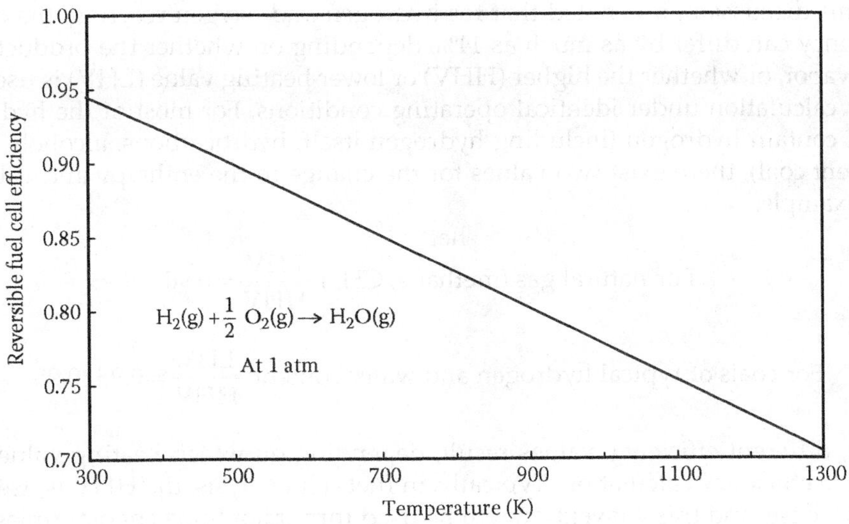

FIGURE 53.9
The reversible fuel cell efficiency (based on LHV) as a function of temperature for the reaction of $H_2 + 1/2O_2 \rightarrow H_2O(g)$ occurring at 1 atm pressure.

However, for the reaction of carbon and oxygen to form carbon monoxide, as shown in Equation 53.51, the entropy change is positive, and the reversible cell potential increases with temperature, as presented previously; hence, the reversible efficiency will also increase with temperature, according to Equation 53.53. But as discussed previously, the efficiency expression, Equation 53.52 or 53.53, is not valid for such reactions.

From the reversible energy efficiency, Equation 53.37, and dividing the numerator and the denominator by the factor (nF), we have, after utilizing the reversible cell potential, Equation 53.21

$$\eta_r = \frac{E_r}{(-\Delta h/nF)} = \frac{E_r}{E_{tn}} \tag{53.54}$$

where

$$E_{tn} = -\frac{\Delta h}{nF} \tag{53.55}$$

is called thermoneutral voltage (or potential), a voltage a fuel cell would have if all the chemical energy of the fuel and oxidant is converted to electric energy (i.e., 100% energy conversion into electricity). For example, for the reaction

$$H_2(g) + \frac{1}{2} O_2(g) \rightarrow H_2O(l)$$

E_{tn} = 1.48 V, and the corresponding reversible efficiency is η_r = 83% at 25°C and 1 atm, whereas at the same temperature and pressure, for the reaction

$$H_2(g) + \frac{1}{2} O_2(g) \rightarrow H_2O(g)$$

E_{tn} = 1.25 V, and the corresponding reversible efficiency is η_r = 95%.

From this discussion, it is noted that for hydrogen and oxygen reaction, the reversible cell efficiency can differ by as much as 14%, depending on whether the product water is liquid or vapor, or whether the higher (HHV) or lower heating value (LHV) is used for the efficiency calculation under identical operating conditions. For most of the hydrocarbon fuels that contain hydrogen (including hydrogen itself, hydrocarbons, alcohols, and to a lesser extent coal), there exist two values for the change in the enthalpy and Gibbs function, for example,

$$\text{For natural gas (methane, } CH_4\text{): } \frac{\text{LHV}}{\text{HHV}} = 0.90$$

$$\text{For coals of typical hydrogen and water content: } \frac{\text{LHV}}{\text{HHV}} = 0.95 \quad 0.98$$

Therefore, different efficiency values result, depending on which heating value ($-\Delta h$) is used for the efficiency calculation. Typically, in fuel cell analysis, the HHV is used unless stated otherwise, and this convention will be used throughout this chapter unless explicitly stated otherwise.

It should be emphasized that from the preceding analysis, it is known that for most fuel cell reactions, the reversible efficiency, η_r, decreases as the fuel cell operating temperature is increased. This effect is important in considering high-temperature fuel cells, namely, the molten carbonate fuel cells (MCFCs) and the solid oxide fuel cells (SOFCs). For example, Figure 53.9 indicates that the reversible cell efficiency is reduced to lower 70% (based on LHV) for hydrogen and oxygen reaction at the typical operating temperature of 1000°C for SOFCs, as opposed to around 95% at 25°C as discussed earlier. This significant reduction in the reversible cell efficiency seems to work against high-temperature fuel cells. However, the irreversible losses, to be described later, decrease drastically as temperature is increased, so that the practical fuel cell performance (such as efficiency and power output under practical operating condition) increases. Therefore, further analysis should be done for efficiency under practical operating condition rather than the idealized reversible condition, which is the focus of the following discussion.

53.4.4.2 Mechanism of Irreversible Energy Losses

For fuel cells, the reversible cell potential and the corresponding reversible efficiency are obtained under the thermodynamically reversible condition, implying that there is no rigorous occurrence of continuous reaction or electrical current output. For practical applications, a useful amount of work (electrical energy) is obtained only when a reasonably large current I is drawn from the cells because the electrical energy output is through the electrical power output, which is defined as

$$\text{Power} = EI \quad \text{or} \quad \text{Power Density} = EJ \tag{53.56}$$

However, both the cell potential and efficiency decrease from its corresponding (equilibrium) reversible values because of irreversible losses when current is increased. These irreversible losses are often called polarization, overpotential, or overvoltage* in

* The terms "polarization," "overpotential," and "overvoltage" have been loosely used in the literature to denote cell potential (or voltage) loss. However, they do have some subtle differences [1] that are neglected here so that these words can be used interchangeably.

literature, and they originate primarily from three sources: activation, ohmic, and concentration polarization. The actual cell potential as a function of current is the result of these polarizations; therefore, a plot of the cell potential vs. current output is conventionally called a polarization curve. It should be noticed that the magnitude of electrical current output depends largely on the active cell area; therefore, a better measure is the current density, J (A/cm^2) instead of current, I, itself, and the unit A/cm^2 is often used rather than A/m^2 as the unit for the current density because square meter is too large to be used for fuel cell analysis.

A typical polarization curve is illustrated in Figure 53.10 for the cell potential as a function of current density. The ideal cell potential–current relation is independent of the current drawn from the cell, and the cell potential remains equal to the reversible cell potential. The difference between the thermoneutral voltage and the reversible cell potential represents the energy loss under the reversible condition (the reversible loss). However, the actual cell potential is smaller than the reversible cell potential and decreases as the current drawn is increased due to the three mechanisms of irreversible losses: activation, ohmic, and concentration polarization. The activation polarization, η_{act}, arises from the slow rate of electrochemical reactions, and a portion of the energy is lost (or spent) on driving up the rate of electrochemical reactions in order to meet the rate required by the current demand. The ohmic polarization, η_{ohm}, arises due to electrical resistance in the cell, including ionic resistance to the flow of ions in the electrolyte and electronic resistance to the flow of electrons in the rest of the cell components. Normally, the ohmic polarization is linearly dependent on the cell current. Concentration polarization, η_{conc}, is caused by the slow rate of mass transfer resulting in the depletion of reactants in the vicinity of active reaction sites and the overaccumulation of reaction products

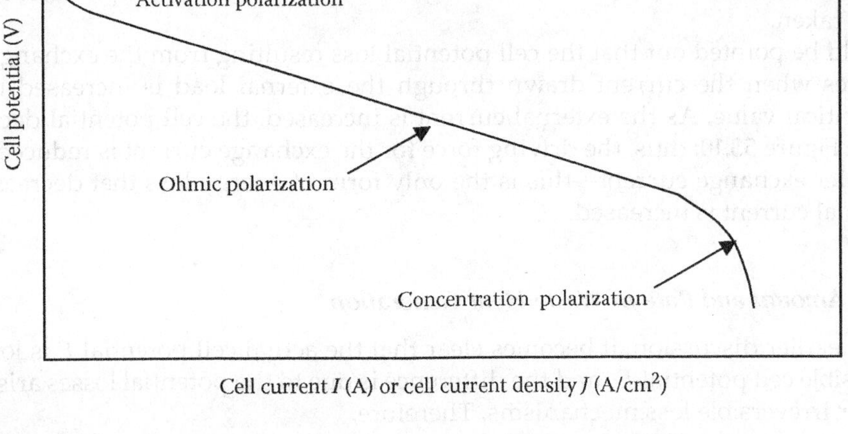

FIGURE 53.10
Schematic of a typical polarization curve. The cell potential for a fuel cell decreases as the current drawn from the cell is increased due to activation, ohmic, and concentration polarizations.

that block the reactants from reaching the reaction sites. It usually becomes significant, or even prohibitive, at high current density when the slow rate of mass transfer is unable to meet the high demand required by the high current output. As shown in Figure 53.10, concentration polarization is often the cause of rapid cell potential decrease to zero. The current (density) corresponding to the zero cell potential is often called the limiting current (density), and evidently, it is controlled by the concentration activation. From Figure 53.10, it is also clear that activation polarization is dominant at small current densities, while concentration polarization is predominant at high current densities. The linear drop in the cell potential due to resistance loss occurs at intermediate current densities, and practical fuel cell operation is almost always located within the ohmic polarization region. It should be emphasized that these three loss mechanisms actually occur simultaneously in an operating cell, despite their different influences at the different current density conditions.

Figure 53.10 also indicates that even at zero current output from the fuel cell, the actual cell potential is smaller than the idealized reversible cell potential. This small difference in cell potential is directly related to the chemical potential difference between the cathode and the anode. So that even at zero external load current, there are electrons delivered to the cathode, where oxygen ions are formed, and migrate through the electrolyte to the anode where they deionize to release electrons. The electron released migrates back to the cathode to continue the process or *exchange*. The ionization/deionization reactions proceeding at a slow rate yield an extremely small current, often called exchange current I_0 or exchange current density J_0, and the cell potential is reduced below the reversible cell potential. Therefore, exchange current arises from the electrons migrating through the electrolyte rather than through the external load, and about 0.1–0.2 V of cell potential loss results from the exchange process. Consequently, the efficiency of a real fuel cell is about 8%–16% lower than the reversible cell efficiency, η_r, even at close to zero current output.

The exchange current density J_0 is very small; it is at least about 10^{-2} A/cm^2 for H_2 oxidation at the anode, and about 10^{-5} times slower for O_2 reduction at the cathode. In comparison, the O_2 reduction process at the cathode is so slow that competing anodic reactions play a significant role, such as oxidation of electrocatalyst, corrosion of electrode materials, and oxidation of organic impurities in the anode structure. All these anodic reactions result in the corrosion of electrodes, thereby limiting the cell life unless appropriate countermeasures are taken.

It should be pointed out that the cell potential loss resulting from the exchange current diminishes when the current drawn through the external load is increased beyond a certain critical value. As the external current is increased, the cell potential decreases as shown in Figure 53.10; thus, the driving force for the exchange current is reduced, leading to a smaller exchange current—this is the only form of energy loss that decreases when the external current is increased.

53.4.4.3 Amount and Rate of Waste Heat Generation

From the earlier discussion, it becomes clear that the actual cell potential E is lower than the reversible cell potential E_r, and the difference is due to the potential losses arising from the earlier irreversible loss mechanisms. Therefore,

$$E = E_r - (\eta_{act} + \eta_{ohm} + \eta_{conc}) \tag{53.57}$$

The irreversible energy loss as heat (or waste heat generation) per mole fuel consumed can be easily obtained, because the entropy generation is, according to Equations 53.24 and 53.57,

$$\wp_s = \frac{nF(E_r - E)}{T} = \frac{nF(\eta_{act} + \eta_{ohm} + \eta_{conc})}{T} \quad (53.58)$$

Then, Equation 53.13 becomes for the total heat loss from the fuel cell

$$q = T\Delta s - T\wp_s = \underbrace{T\Delta s}_{\text{Reversible loss}} \underbrace{- nF(\eta_{act} + \eta_{ohm} + \eta_{conc})}_{\text{Irreversible losses}} \quad (53.59)$$

Since the entropy change is negative ($\Delta s < 0$) for most fuel cell reactions, the heat transfer is negative as well, implying that energy as heat is lost from the fuel cell shown in Figure 53.3 for both reversible and irreversible losses.

Because $T\Delta s = \Delta h - \Delta g$ by the definition of the Gibbs function change for fuel cell reactions, Equation 53.59 can be written as

$$\frac{q}{nF} = \frac{T\Delta s}{nF} - (\eta_{act} + \eta_{ohm} + \eta_{conc}) = \frac{\Delta h - \Delta g}{nF} - (\eta_{act} + \eta_{ohm} + \eta_{conc}) \quad (53.60)$$

Considering the definition for the thermoneutral voltage and the reversible cell potential, the earlier expression becomes

$$\frac{q}{nF} = -E_{tn} + E_r - (\eta_{act} + \eta_{ohm} + \eta_{conc}) \quad (53.61)$$

Combining with Equation 53.57, Equation 53.61 reduces to

$$\frac{q}{nF} = -(E_{tn} - E) \quad (53.62)$$

Hence, the equivalent cell potential loss due to the energy loss in the fuel cell as heat is equal to the difference between the thermoneutral voltage and the actual cell potential.

The rate of heat loss per mole fuel consumed in the fuel cell, Equation 53.59, can be expressed as an equivalent power loss:

$$P_{HeatLoss} = I\left(\frac{q}{nF}\right) = -I(E_{tn} - E) = I\left[\frac{T\Delta s}{nF} - (\eta_{act} + \eta_{ohm} + \eta_{conc})\right] \quad (53.63)$$

This expression is important in determining the cooling requirements of fuel cell stacks.

53.4.4.4 Various Forms of Irreversible Energy Conversion Efficiency

After the earlier description of the irreversible energy losses, we can now introduce several forms of energy efficiency that would be useful in the analysis of fuel cell performance.

53.4.4.4.1 Voltage Efficiency, η_E

The voltage efficiency is defined as

$$\eta_E = \frac{E}{E_r} \tag{53.64}$$

Because the actual cell potential E is compared with the maximum possible cell potential E_r allowed by the second law, the voltage efficiency is really a specific form of the exergy efficiency, representing the degree of departure of the cell operation from the idealized thermodynamically reversible condition. As shown in Equation 53.57, $E < E_r$, hence $\eta_E < 1$.

53.4.4.4.2 Current Efficiency, η_I

The current efficiency is a measure of how much current is produced from a given amount of fuel consumed in the fuel cell reaction, and it is defined as

$$\eta_I = \frac{I}{nF(dN_F/dt)} \tag{53.65}$$

where dN_F/dt represents the rate of fuel consumption in the fuel cell (mol/s). The current efficiency would be <100% if part of the reactants participate in non-current-productive side reactions, called parasitic reactions, such as reactant crossover through the electrolyte region, incomplete conversion of reactants to the desired products, reaction with the cell components, or even reactant leakage from the cell compartment due to sealing problem. For example, in DMFCs, about 20% of the liquid methanol can cross over to the cathode side through the proton-conducting polymer membrane, implying the current efficiency is only about 80% for such cells. However, for most practical fuel cells, especially at operating conditions where the current output is sufficiently larger than zero (without the effect of exchange current discussed previously), the current efficiency is about 100%. This is because for practical fuel cells, all the parasitic reactions are undesirable and would have been removed by appropriate design.

53.4.4.4.3 Overall Free Energy Conversion Efficiency, η_{FC}

The overall free energy conversion efficiency is defined as the product of the reversible efficiency, voltage, and current efficiency:

$$\eta_{FC} = \eta_r \times \eta_E \times \eta_I \tag{53.66}$$

If the current efficiency is 100% as is often the case for well-designed practical fuel cells, substituting the definitions for the various efficiencies into the earlier equation lead to

$$\eta_{FC} = \frac{E}{E_{tn}} \tag{53.67}$$

Therefore, the overall free energy conversion efficiency is really the overall efficiency for energy conversion process occurring within the fuel cell. Because the thermoneutral voltage is a fixed value for a given fuel and oxidant under a given operating condition of temperature and pressure, the overall energy conversion efficiency for fuel cells is proportional to the actual cell potential,

$$\eta_{FC} \sim E \tag{53.68}$$

This is a significantly important result. Once the actual cell potential is determined, the energy conversion efficiency of the fuel cell is known as well. This is the primary reason that in fuel cell literature, it is almost always that the cell polarization curve is given without specifically showing the cell energy efficiency as a function of the current. Further, Equation 53.68 implies that the fuel cell efficiency will depend on the current output in the same way as the cell potential, that is, decrease as the current output is increased.

53.4.4.4.4 Fuel Cell System Efficiency, η_s

Since a fuel cell system is composed of one or multiple fuel cell stacks and auxiliary equipment, which would also have its own energy efficiency of η_{aux}, the total fuel cell system efficiency is

$$\eta_s = \eta_{FC} \times \eta_{aux} \tag{53.69}$$

53.4.5 Efficiency Loss in Operating Fuel Cells: Stoichiometry, Utilization, and Nernst Loss

In an operating fuel cell, reactant composition changes between the inlet and outlet of the fuel cell along the flow path over the electrode surface because reactants are consumed to yield current output and reaction products are formed along the way as well. The change in reactant composition results in additional loss of cell potential beyond those losses described in the preceding section. This potential loss arises from the fact that the cell potential *E* adjusts to the lowest electrode potential given by the Nernst equation, Equation 53.29, for the various reactant compositions at the exit of the anode and cathode chambers. This is because electrodes are usually made of good electronic conductors, and consequently, they are isopotential surfaces. The cell potential *E* may not exceed the minimum local value set by the Nernst equation. This additional cell potential loss is often also called the Nernst loss, which is equal to the difference between the inlet and exit Nernst potentials determined based on the inlet and exit reactant compositions. According to Equation 53.29, this additional cell potential loss due to the consumption of reactants in the cell is when the reactant streams are arranged in a concurrent flow:

$$\eta_N = \frac{\Re T}{nF}\ln K_{\text{out}} - \frac{\Re T}{nF}\ln K_{\text{in}} = \frac{\Re T}{nF}\ln\frac{K_{\text{out}}}{K_{\text{in}}} \tag{53.70}$$

In the case of a fuel cell where both fuel and oxidant flows are in the same direction (concurrent), the minimum Nernst potential occurs at the flow outlet. When the reactant flows are in counterflow, cross-flow, or more complex arrangements, it becomes difficult to determine the location of the minimum potential due to the reactant consumption. Appropriate flow channel design for the anode and cathode sides can minimize the Nernst loss.

Equation 53.70 also implies that the Nernst loss will be extremely large and approach infinity if all the reactants are consumed in the in-cell electrochemical reaction leading to zero reactant concentration at the cell outlet. To reduce the Nernst loss to an acceptable level for practical fuel cell operations, reactants are almost always supplied more than the stoichiometric amount required for the desired current production. The actual amount of reactants supplied to a fuel cell is often expressed in terms of a parameter called stoichiometry, S_t:

$$S_t = \frac{\text{Molar flow rate of reactants supplied to a fuel cell}}{\text{Molar flow rate of reactants consumed in the fuel cell}} = \frac{\dot{N}_{\text{in}}}{\dot{N}_{\text{consumed}}} \tag{53.71}$$

For example, for proton exchange membrane (PEM) fuel cells, typical operation uses $S_t \approx (1.1\text{–}1.2)$ for H_2 and $S_t \approx 2$ for O_2 (pure or in air). Therefore, stoichiometry really represents the actual flow rate for the reactant delivered to the fuel cell. Because there are normally at least two types of reactants for fuel cell, one as fuel and another as oxidant, stoichiometry can be defined for either reactant.

Alternatively, reactant flow rate can be expressed in terms of a parameter called utilization, U_t:

$$U_t = \frac{\text{Molar flow rate of reactants consumed in the fuel cell}}{\text{Molar flow rate of reactants supplied to the fuel cell}} = \frac{\dot{N}_{\text{consumed}}}{\dot{N}_{\text{in}}} \quad (53.72)$$

Clearly, stoichiometry and utilization are inversely proportional to each other.

For properly designed practical fuel cells, no reactant crossover or leakage out of the cell may occur in general; therefore, the rate of reactant consumed within the cell is equal to the difference between the rate of molar flow into and out of the cell. For example, the stoichiometry for the fuel may be expressed as

$$S_{t,F} = \frac{\dot{N}_{F,\text{in}}}{\dot{N}_{F,\text{consumed}}} = \frac{\dot{N}_{F,\text{in}}}{\dot{N}_{F,\text{in}} - \dot{N}_{F,\text{out}}} = \frac{1}{U_{t,F}} \quad (53.73)$$

Effect of reactant utilization on the reversible cell potential is illustrated in Figure 53.11. It is seen that the reactant composition at the cell outlet decreases; hence, the reversible cell potential decreases as well when the utilization factor is increased. The decrease is rapid when utilization goes beyond about 90%. In practical fuel cell operation, 100% utilization

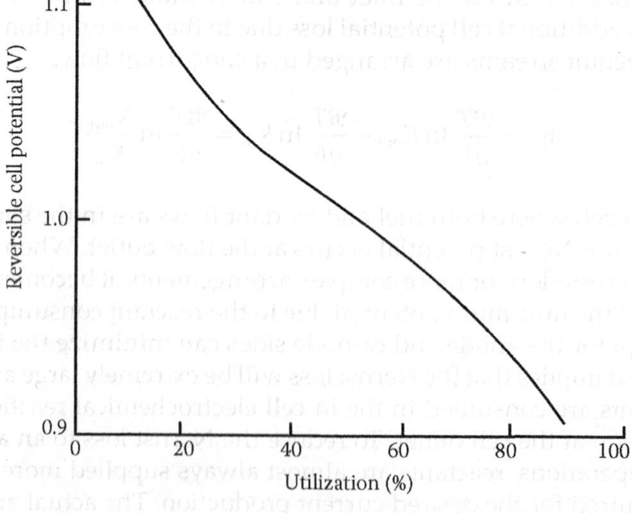

FIGURE 53.11
Reversible cell potential as a function of reactant utilization (both fuel and oxidant utilizations are set equal) for a molten carbonate fuel cell operating at 650°C and 1 atm. Reactant compositions at the cell inlet: 80% H_2/20% CO_2 mixture saturated with $H_2O(g)$ at 25°C for the fuel gas, and 60% CO_2/30% O_2/10% inert gas mixture for the oxidant gas. (From Hirschenhofer, J.H. et al., *Fuel Cells: A Handbook* (Rev. 3), U.S. Department of Energy, Washington, DC, 1994.)

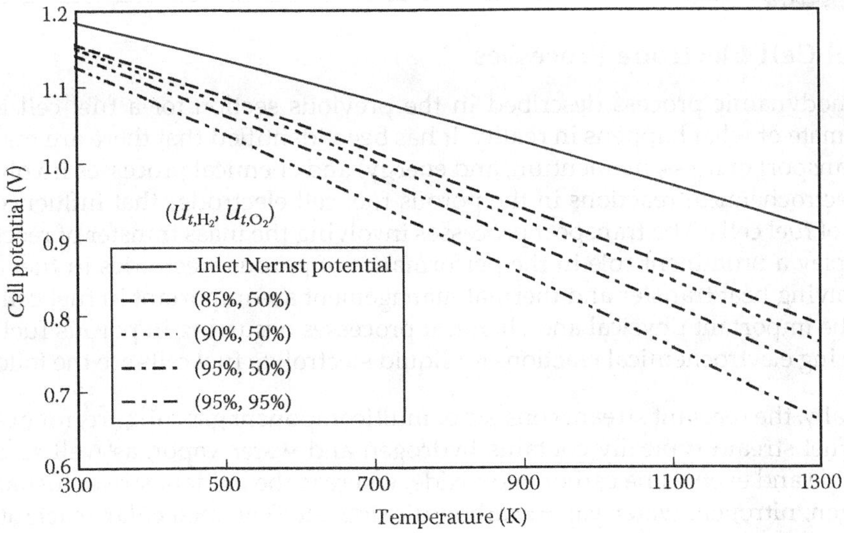

FIGURE 53.12
Inlet and outlet Nernst potential as a function of temperature and utilization for the reaction of $H_2(g) + O_2(g) \rightarrow H_2O(g)$ at 1 atm.

(or unity stoichiometry) will result in reactant concentrations vanishing at the cell exit; then, the Nernst loss becomes dominant and the cell potential is reduced to zero—this is certainly an undesirable situation that needs to be avoided. Therefore, typical operation requires that the utilization be about 80%–90% for fuel and 50% for oxidant in order to balance the Nernst loss with the parasitic losses associated with the reactant supply.

As shown in Equation 53.70, the additional Nernst loss due to the reactant depletion in the cell is directly proportional to the cell operating temperature. Figure 53.12 shows the reversible cell potential at the cell inlet and outlet for hydrogen and oxygen reaction forming gaseous water product at 1 atm as a function of temperature. The outlet Nernst potentials (i.e., the reversible cell potential at the cell outlet) are determined for oxygen utilization of 50%, and hydrogen utilizations of 85%, 90%, and 95%, respectively, as well as for the utilization of 95% for both hydrogen and oxygen. It is clearly seen that the outlet Nernst potential decreases when either utilization or temperature is increased.

If pure hydrogen is used as fuel, the anode compartment can be designed as a dead-end chamber for hydrogen supply. Similarly, if pure oxygen is used as oxidant, a dead-end cathode compartment can be employed. However, inert impurities in the reactant gas will accumulate at the anode and cathode compartments, and they must be removed either periodically or continuously in order to maintain a good fuel cell performance. Periodic purging or continuous bleeding can be implemented for this purpose, but this results in a small loss of fuel, and hence <100% utilization.

From the earlier discussion, it is evident that 100% utilization for reactants is practically an unwise design. Since in-cell fuel utilization will never be 100% in practice, the determination of in-cell energy conversion efficiency and the cell potential must take utilization factor into consideration. If the fuel exiting the fuel cell is discarded (not recirculated back to the cell or not utilized for other useful purpose such as providing heat for fuel preprocessing), then the overall energy conversion efficiency must be equal to the overall fuel cell efficiency given in Equation 53.66 multiplied by the utilization to take into account the fact that not all the fuel is being used for electric energy production.

53.5 Fuel Cell Electrode Processes

The thermodynamic process described in the previous section for a fuel cell is a gross underestimate of what happens in reality. It has been identified that there are many physical (i.e., transport of mass, momentum, and energy) and chemical processes involved in the overall electrochemical reactions in the porous fuel cell electrodes that influence the performance of fuel cells. The transport processes involving the mass transfer of reactants and products play a prominent role in the performance of porous electrodes in fuel cells, and those involving heat transfer and thermal management are important in fuel cell systems. Some of the important physical and chemical processes occurring in porous fuel cell electrodes during electrochemical reactions for liquid electrolyte fuel cells are the following [4]:

1. Initially, the reactant stream consists of multicomponent gas mixture, for example, the fuel stream typically contains hydrogen and water vapor, as well as carbon dioxide and even some carbon monoxide, whereas the oxidant stream usually has oxygen, nitrogen, water vapor, carbon dioxide, etc. The molecular reactant (such as H_2 or O_2) is transferred to the porous electrode surface from the reactant stream through the mechanism of convection and then transported through the porous electrode, primarily by diffusion, to reach the gas/electrolyte interface.
2. The reactant dissolves into the liquid electrolyte at the two-phase interface.
3. The dissolved reactant then diffuses through the liquid electrolyte to arrive at the electrode surface.
4. Some pre-electrochemical homogeneous or heterogeneous chemical reactions may occur, such as electrode corrosion reaction, or impurities in the reactant stream may react with the electrolyte.
5. Electroactive species (which could be reactant itself as well as impurities in the reactant stream) are adsorbed onto the solid electrode surface.
6. The adsorbed species may migrate on the solid electrode surface, principally by the mechanism of diffusion.
7. Electrochemical reactions then occur on the electrode surface wetted by the electrolyte—the so-called three-phase boundary, giving rise to electrically charged species (or ions and electrons).
8. The electrically charged species and other neutral reaction product such as water, still adsorbed on the electrode surface, may migrate along the surface due to diffusion in what has been referred to as post-electrochemical surface migration.
9. The adsorbed reaction products become desorbed.
10. Some post-electrochemical homogeneous or heterogeneous chemical reactions may occur.
11. The electrochemical reaction products (neutral species, ions, and electrons) are not only transported away from the electrode surface, mainly by diffusion, but also influenced, for the ions, by the electric field setup between the anode and the cathode. The electron motion is dominated by the electric field effect.
12. The neutral reaction products diffuse through the electrolyte to reach the reactant gas/electrolyte interface.
13. Finally, the products will be transported out of the electrode and the cell in the gas form.

Normally, any of the 13 processes can influence the performance of a fuel cell, exhibiting the complex nature of a fuel cell operation. For well-designed practical fuel cells, the earlier electrode processes might be grouped into the following three major steps:

1. *Delivery of molecular reactant to the electrode surface*: This step involves a number of physical and chemical processes preceding the electrochemical reaction, and it generally includes the transport of molecular reactant from the gas phase supply outside the electrode structure to the liquid electrolyte surface, typically through the porous electrode structure, then the dissolution of molecular reactant into the liquid electrolyte, followed by the diffusion of the dissolved reactant through the electrolyte to the electrode surface, and finally the adsorption of the reactant on the electrode surface.

   It might be pointed out that the delivery of the electrons and ions to the electrode surface is as important as the delivery of molecular reactant since they are all needed for the electrochemical reactions to occur at the electrode surface.
2. *Reduction or oxidation of the adsorbed molecular reactant at the electrode surface in the presence of electrolyte*: This is the step for electrochemical reactions that produce electric current. The reactions occur only at the electrode surface that is covered by the electrolyte, hence, often referred to as the three-phase boundary or active reaction sites. Significant increase in the reactive sites is essential for good fuel cell performance.
3. *Removal of the reaction products from the electrode surface for the regeneration of the reaction sites and for the continuous production of electric current*: This step is especially important for low-temperature acid electrolyte fuel cells, because product water is formed at the cathode and is in the liquid form. Liquid water accumulation in the porous electrode structure may block the transport of molecular oxygen to the reaction sites, severely hindering the process described in step (1), thereby degrading the cell performance due to oxygen starvation at the reaction sites. Such a phenomenon is often referred to as the *water flooding* of electrodes, a critical issue for PEM fuel cells (PEMFCs).

53.6 Cell Connection and Stack Design Considerations

The potential of a working cell is typically around 0.7–0.8 V, and it is normally too small for practical applications, also because of the limited power available from a single cell. Therefore, many individual cells are connected (or stacked) together to form a fuel cell stack. Although many stacking configurations are possible, the overpotential associated with the transport processes discussed in this chapter imposes limitations and technical difficulties, making cell stacking one of the significant technical challenges in the drive for the fuel cell commercialization. We will briefly discuss the stacking options later and the associated transport-related issues.

The electrical connection among the individual cells may be arranged in parallel or in series, as shown in Figure 53.13. The parallel connection still provides a low voltage output from the stack, but a very high current output since the stack current is the sum of the current produced in each cell. Such an extremely large current flow will cause an

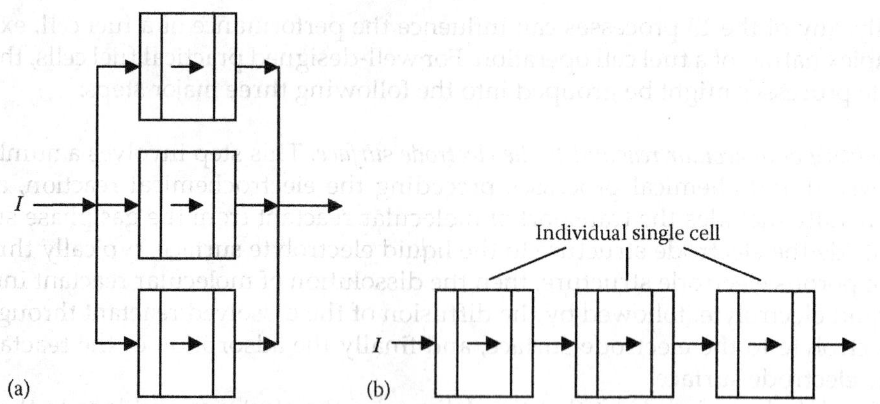

FIGURE 53.13
Cell connection (stacking) configurations: (a) parallel connection and (b) series connection. The arrow in the diagram represents the direction of the current flow.

excessively large ohmic voltage loss in the stack components and at the surface contacts among the components. Thus, parallel connection is typically avoided unless for small current or power applications.

Series connection can have two typical arrangements: unipolar and bipolar, as shown in Figure 53.14. Unipolar design has one fuel stream supplying fuel to two anodes for the two adjacent cells, and one oxidant stream delivering oxidant to two cathodes for the two adjacent cells. This arrangement of one reactant stream serving for two adjacent electrodes simplifies

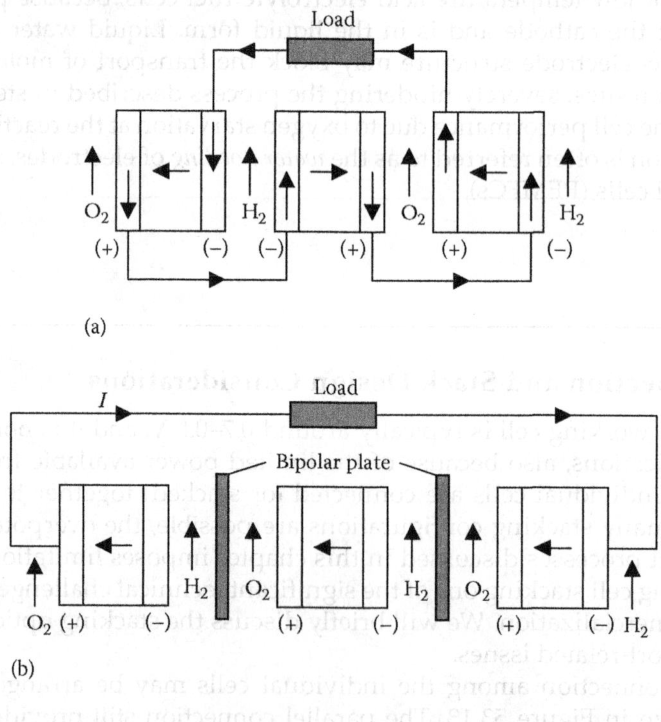

FIGURE 53.14
Series cell connection (stacking) configurations: (a) unipolar arrangement with the edge collection of the current generated in each cell and (b) bipolar arrangement with the end-plate collection of the current.

the reactant flow channel design. However, it forces the electrical current generated in each cell to be collected at the edge of the electrodes. Since electrode is very thin (<1 mm), while the other electrode dimensions (in the direction of the current flow) are at least on the order of centimeters or larger, the ohmic resistance tends to be very large. Thus, edge collection of current, although used in early fuel cell stack designs, is generally avoided in recent fuel cell stack development, primarily due to the excessively large ohmic voltage losses.

The bipolar arrangement has the current flow normal to the electrode surface, instead of along the electrode surface as in the unipolar arrangement; thus, the current flow path is very short while the cross-sectional area available for the current flow is very large. The ohmic voltage loss for this case is very small in comparison, and bipolar design is favored in recent fuel cell stack technology. However, this end-plate collection of current results in the complex design for the reactant flow channels and complex organization for the reactant stream, and the bipolar plate has to fulfill several functions simultaneously in order to obtain a good overall stack performance. Bipolar plates serve as current collectors, for reactant delivery to the electrode surface, for cell reaction product removal (such as water), and for the integrity of the cell/stack. These functions for the bipolar plates may be contradictory to each other, and the optimal design for the bipolar plates represents one of the significant technical challenges for practical fuel cells.

A typical configuration of a bipolar plate is shown in Figure 53.15. The reactant flow follows the flow channels made on the bipolar plate, thus distributing the reactant over the

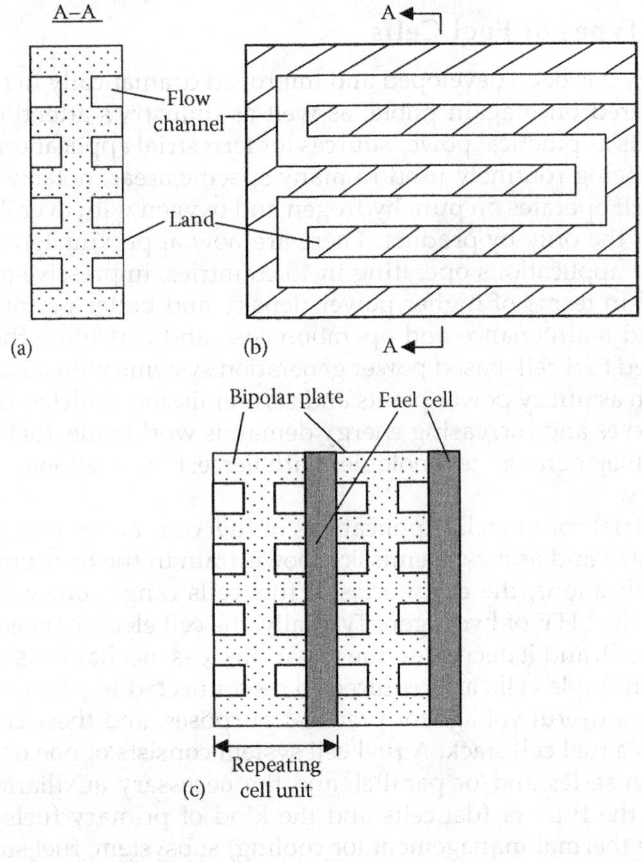

FIGURE 53.15
Typical configuration of a bipolar plate: (a) cross-sectional view, (b) face view, and (c) repeating cell unit in a stack.

electrode surface. On the other hand, the land between the adjacent flow channels serves as the passage for the current flow from one cell to the next. Therefore, a wide flow channel is beneficial for the reactant distribution over the electrode surface and for the reaction product removal, while a wide land is beneficial for the electron flow and for the mechanical integrity of the cells and the entire stack. Normally, the same cell unit as shown in Figure 53.15c is repeated to form a stack. The plate at the end of the stack has flow channels only on one side of its surfaces.

From the cell repeating unit in a stack, it is clear that the reactant concentration decreases along the flow direction following the flow channel design and into the electrode due to convectional and diffusional mass transfer. Thus, the concentration field is three dimensional, so is the electrical field due to the flow channels made on the bipolar plate surfaces, the nonuniform rate of electrochemical reactions in the catalyst layers, and the three-dimensional distribution of the reactant concentrations and the temperature. Therefore, an accurate prediction of the cell performance requires a three-dimensional analysis based on solving the conservation equations governing the transport phenomena of the reactant flow, species concentration, temperature, and electric fields, incorporating the in-cell electrochemical reaction processes, thus posing significant challenges to fuel cell designers.

53.7 Six Major Types of Fuel Cells

Fuel cell technology has been developed and improved dramatically in the past few years, and this has captured once again public as well as industry's attention concerning the prospect of fuel cells as practical power sources for terrestrial applications. At present, fuel cell technology is being routinely used in many specific areas, notably in space explorations, where fuel cell operates on pure hydrogen and oxygen with over 70% efficiency and drinkable water as the only by-product. There are now approximately over 200 fuel cell units for terrestrial applications operating in 15 countries; impressive technical progress has been achieved in terms of higher power density and better performance as well as reduced capital and maintenance and operation cost, and is driving the development of competitively priced fuel cell–based power generation systems with advanced features for terrestrial use, such as utility power plants and zero-emission vehicles. In light of decreasing fossil fuel reserves and increasing energy demands worldwide, fuel cell will probably become one of the major energy technologies with fiercest international competition in the twenty-first century.

The major terrestrial commercial applications of fuel cells are electric power generation in the utility industry and as a zero-emission powertrain in the transportation sector. For these practical applications, the efficiencies of fuel cells range somewhere between 40% and 65% based on the LHV of hydrogen. Typically, the cell electric potential is only about 1 V across a single cell, and it decreases due to various loss mechanisms under operational conditions. Thus, multiple cells are required to be connected together in electrical series in order to achieve a useful voltage for practical purposes, and these connected cells are often referred to as a fuel cell stack. A fuel cell system consists of one or multiple fuel cell stacks connected in series and/or parallel, and the necessary auxiliaries whose composition depends on the type of fuel cells and the kind of primary fuels used. The major accessories include thermal management (or cooling) subsystem, fuel supply, storage and processing subsystem, and oxidant (typically air) supply and conditioning subsystem.

TABLE 53.2

Operational Characteristics and Technological Status of Various Fuel Cells

Type of Fuel Cells	Operating Temperature (°C)	Power Density (mW/cm²) (Present) Projected	Projected Rated Power Level (kW)	Fuel Efficiency (Chemical to Electrical)	Lifetime Projected (h)	Capital Cost Projected (U.S. $ per kW)	Areas of Application
AFC	60–90	(100–200) >300	10–100	40–60	>10,000	>200	Space, mobile
PAFC	160–220	(200) 250	100–5,000	55	>40,000	3000	Dispersed and distributed power
PEMFC	50–80	(350) >600	0.01–1,000	45–60	>40,000	>200	Portable, mobile, space, stationary
MCFC	600–700	(100) >200	1,000–100,000	60–65	>40,000	1000	Distributed power generation
SOFC	800–1000	(240) 300	100–100,000	55–65	>40,000	1500	Baseload power generation
DMFC	90	(230) ?	0.001–100	34	>10,000	>200	Portable, mobile

Note: ? indicates that the projected value achievable in the near future is not available.

In this section, a summary of the state-of-the-art technology for the six major types of fuel cells is presented, including

1. Alkaline fuel cells (AFCs)
2. Polymer electrolyte membrane fuel cells (PEMFCs)
3. DMFCs
4. Phosphoric acid fuel cells (PAFCs)
5. MCFCs
6. SOFCs

Five of them are classified based on their electrolytes used, including the alkaline, phosphoric acid, PEM, molten carbonate, and solid oxide. DMFC is classified based on the fuel used for electricity generation. Table 53.2 provides a summary of the operational characteristics and application of the six major types of fuel cell.

53.7.1 Alkaline Fuel Cells

AFCs give the best performance among all the fuel cell types under the same or similar operating conditions when running on pure hydrogen and oxygen. Hence, they are among the first fuel cells to have been studied and taken into development for practical applications, and they are the first type of fuel cells to have reached successful routine applications, mainly in space programs such as space shuttle missions in the United States and similar space exploration endeavors in China and Europe, where pure hydrogen and oxygen are used as reactants. Because of their success in space programs, AFCs are also the type of fuel cells on which probably the largest number of fuel cell development programs has begun in the world in an effort to bring them down to terrestrial applications, particularly in Europe. However, almost all the AFC development programs have come to an end. At present, the only few activities related to the AFC RD&D are in Europe.

The AFCs have the highest energy conversion efficiency among all types of fuel cells under the same operating conditions if pure hydrogen and pure oxygen are used as the reactants. That was one of the important reasons that AFCs were selected for the U.S. space shuttle missions. The AFCs used in the shuttle missions are operated at about 200°C for better performance (i.e., high energy conversion efficiency of over 70% and high power density that is critical for space applications), and the alkaline electrolyte is potassium hydroxide (KOH) solution immobilized in an asbestos matrix. As a result, the AFCs operate at high pressure in order to prevent the boiling and depletion of the liquid electrolyte. Consequently, these severe operating conditions of high temperature and high pressure dictate extremely strict requirement for cell component materials that must withstand the extreme corrosive oxidizing and reducing environment of the cathode and the anode. To meet these requirements, precious metals such as platinum, gold, and silver are used for the construction of the electrodes, although these precious metals are not necessary for the electrochemical reactions leading to electric power generation. Each shuttle flight contains 36 kW AFC power system. Its purchase price is about U.S. $28.5 million, and it costs NASA additional $12–$19 million annually for operation and maintenance. Although the manufacturer claims about 2400 h of lifetime, NASA's experience indicates that the real lifetime is only about 1200 h. With sufficient technology development, 10,000 h are expected as the life potential (or upper limit) for the AFC system. This belief is based on the nature of the AFC systems and the data accumulated on both stacks and single cells.

The typical working temperature of AFC power systems aimed at commercial and terrestrial applications ranges from 20°C to 90°C, and the electrolyte is a KOH solution (30%–35%). There are four different cell types investigated:

1. Cell with a free liquid electrolyte between two porous electrodes
2. ELOFLUX cell with liquid KOH in the pore systems
3. Matrix cell where the electrolyte is fixed in the electrode matrix
4. The falling film cell

Many technical challenges have been encountered in the development of AFCs. The most important ones are as follows:

- *Preparation method of the electrodes*: The electrodes consist of porous material that is covered with a layer of catalyst. In general, it is very difficult to distribute the catalyst at the surface and to produce a defined pore system for the transport of the reactants.
- *Costs of the electrodes, stacks, and fuel cell systems*: The preparation of electrodes with noble metal catalysts is very expensive. In general, the electrodes are manufactured in small-scale production with high overhead costs.
- *Lifetime of the electrode/degradation*: The electrolyte is very corrosive, and the catalyst materials are sensitive to high polarization. Using nickel and silver as catalysts, in order to reduce the costs of the fuel cell, leads to a high degradation of these catalysts.

- *Diaphragm made of asbestos*: The diaphragm of low-temperature AFCs is made of asbestos. But this material is hazardous for health, and in some countries, its use is even banned. Therefore, new diaphragms should be developed, but it is difficult to find a material with a similar performance in alkaline electrolyte.
- *Carbon dioxide–contaminated fuel and oxidant streams (carbonation of electrolyte and electrodes)*: The electrolyte intolerance of carbon dioxide is the most important disadvantage of air-breathing AFCs with reformate gases from primary fossil fuels.

Other problems associated with the AFC power systems are the concerns for the safety and reliability of AFC power systems. For example, the liquid KOH electrolyte contained in an asbestos matrix can withstand only a 5 psi limit of pressure differential between the anode and cathode reactant gases. This dictates the need for sophisticated pinpoint pressure control during the operation including transient, start-up, and shutdown processes. It is also a safety issue because of its greater likelihood of the reactants mixing in the AFC system with the possibility of a serious fire breaking out. In terms of general safety considerations, the use of the corrosive potassium hydroxide electrolyte in the AFCs represents the need for hazardous material handling, and the handling of asbestos matrix poses potential hazard to one's health. With flowing reactant gases, the potential for the gradual loss of the liquid electrolyte, drying of the electrolyte matrix, reactant crossover of the matrix, and ensuing life-limiting reactant mixing (or actual AFC stack failure due to fire) is very real in the AFC system.

The major technical challenge is that alkaline electrolytes, like potassium or sodium hydroxide, do not reject carbon dioxide, even the 300–350 ppm of carbon dioxide in the atmospheric air is not tolerated (carbon dioxide concentration in both cathode and anode gases must be <10–100 ppm by volume), while terrestrial applications almost invariably require the use of atmospheric air as oxidant due to technical and economic considerations. For municipal electric applications, hydrocarbon fuels, especially natural gas, are expected to be the primary fuel, and their reformation into hydrogen-rich gases invariably contain a significant amount of carbon dioxide, for example, steam-reforming of the natural gas results in the reformate gas consisting approximately of 80% hydrogen, 20% carbon dioxide, and a trace amount of other components such as carbon monoxide. Carbonaceous products of aging and corrosion shorten AFC life; they degrade the alkaline electrolyte. Whether originating as impurities in the gaseous reactants or from some fuel cell materials, oxides of carbon will chemically react with the alkaline electrolyte and produce irreversible decay that will decrease performance and shorten life. Consequently, AFCs are currently restricted to specialized applications where pure hydrogen and oxygen are utilized. The revival of this technology will depend almost completely on the successful curing of *CO_2 syndrome*—the efficient and economic scrubbing of carbon dioxide, even though claims have been made of successful resolution of carbon dioxide poisoning in the AFCs. This is especially true for utility applications, although some optimistic estimate indicates that the AFC stack costs are similar to all other low-temperature fuel cell systems, and the production costs for the AFC systems seem to be the lowest. A price of about U.S. $400–$500 per kW has been quoted by using today's technologies and today's knowledge in large-scale production. However, small-scale commercial production cost is estimated to be 5–10 times higher.

53.7.2 Polymer Electrolyte Membrane Fuel Cells

53.7.2.1 Introduction

The PEMFC is also called solid polymer (electrolyte) fuel cell. It is perhaps the most elegant of all fuel cell systems in terms of design and mode of operation. It was the first type of fuel cell that was put into practical application (in Gemini space missions from 1962 to 1966). It consists of a solid polymeric membrane acting as the electrolyte. The solid membrane is an excellent proton conductor, sandwiched between two platinum-catalyzed porous carbon electrodes. It has fast start capability and yields the highest output power density among all types of the fuel cells. Because of the solid membrane as the electrolyte, there is no corrosive fluid spillage hazard, and there is lower sensitivity to orientation. It has no volatile electrolyte and has minimal corrosion concerns. It has truly zero pollutant emissions with potable liquid product water when hydrogen is used as fuel. As a result, the PEMFC is particularly suited for vehicular power application, although it is also being considered for stationary power application, albeit to a lesser degree.

The proton-conducting polymer membrane belongs to a class of materials called ionomers or polyelectrolytes, which contain functional groups that will dissociate in the presence of water. The dissociation produces ions fixed to the polymer and simple counterions that can freely exchange with ions of the same sign from the solution. The current available polyelectrolytes have cation as the counterion. In the case of hydrogen, the cation is proton. Therefore, the membrane must be fully hydrated in order to have adequate ion conductivity. As a result, the fuel cell must be operated under conditions where the product water does not evaporate faster than it is produced, and the reactant gases, both hydrogen and oxygen, need to be humidified. Therefore, water and thermal management in the membrane become critical for efficient cell performance, are fairly complex, and require dynamic control to match the varying operating conditions of the fuel cell. Because of the limitation imposed by the membrane and problems with water balance, the operating temperature of PEMFCs is usually <120°C, typically at 80°C. This rather low operating temperature requires the use of noble metals as catalysts in both the anode and cathode side with generally higher catalyst loadings than those used in PAFCs.

Currently, the polymer electrolyte used is made of perfluorinated sulfonic acid membrane, or it is essentially acid, though in solid polymeric form. Hence, PEMFCs are essentially acid electrolyte fuel cells, with its operational principle essentially the same as PAFCs. As a result, most of PEMFC design, material selection, component fabrication, etc., are similar to those of PAFCs. The only difference is the humidification of reactant gases dictated by the membrane performance. Reactant humidification is often achieved by a number of techniques, for example, by passing gas stream through a water column, by using in-stack humidification section of cell and membrane arrangement, and by spraying water into the reactant streams. In the early stage of the PEMFC development, the membranes were based on polystyrene, but since 1968, a Teflon-based product, named Nafion by DuPont, is used. This offers high stability, high oxygen solubility, and high mechanical strength.

53.7.2.2 Basic Operating Principle

The PEM is essentially acid electrolyte; PEMFC requires hydrogen gas as the fuel and oxygen (typically air) as the oxidant. The half-cell reactions are

$$\text{Anode: } H_2 \rightarrow 2H^+ + 2e^- \tag{53.74}$$

$$\text{Cathode: } \frac{1}{2}O_2 + 2H^+ + 2e^- \rightarrow H_2O \tag{53.75}$$

and the overall cell reaction is

$$H_2 + \frac{1}{2}O_2 = H_2O + \text{Heat generated} + \text{Electric energy} \tag{53.76}$$

The current PEMFCs use perfluorinated sulfonic acid membrane (almost exclusively Nafion from DuPont) as the proton-conducting electrolyte; carbon paper or cloth as the anode and cathode backing layers; and platinum or its alloys, either pure or supported on carbon black, as the catalyst. The bipolar plate with the reactant gas flow fields is often made of graphite plate. The stoichiometry is around 1.1–1.2 for the fuel and 2 for the oxidant (oxygen). The PEMFCs usually operate at about 80°C and 1–8 atm pressure. The pressures, in general, are maintained equal on either side of the membrane. Operation at high pressure is necessary to attain high power densities, particularly when air is chosen as the cathodic reactant.

To prevent the membrane dryout leading to local hot spot (and crack) formation, performance degradation, and lifetime reduction, both fuel and oxidant streams are fully humidified, and the operating temperature is limited by the saturation temperature of water corresponding to the operating pressure. The product liquid water formed at the cathode does not dissolve in the electrolyte membrane and is usually removed from the cell by the excessive oxidant gas stream. The accumulation of liquid water in the cathode backing layer blocks the oxygen transfer to the catalytic sites, thus resulting in the phenomenon called *water-flooding* causing performance reduction. Local hot and cold spots will cause the evaporation and condensation of water. Thus, an integrated approach to thermal and water management is critical to PEMFCs' operation and performance, and a proper design must be implemented.

53.7.2.3 Acceptable Contamination Levels

As an acid electrolyte fuel cell operating at low temperature, the PEMFC is primarily vulnerable to carbon monoxide poisoning. Even a trace amount of CO drastically reduces the performance levels, although CO poisoning effect is reversible and does not cause permanent damages to the PEMFC system. Further, the performance reduction due to CO poisoning takes a long time (on the order of 2 h) to reach steady state. This transient effect may have profound implication for transportation applications. Therefore, the PEMFC requires the use of a fuel virtually free of CO (must be less than a few ppm). Also high-quality water free of metal ions should be used for the cell cooling and reactant humidification to avoid the contamination of the membrane electrolyte. This requirement has a severe implication on the materials that can be used for cell components. On the other hand, carbon dioxide does not affect PEMFC operation and performance except through the effect of reactant dilution (the Nernst loss).

53.7.2.4 Major Technological Problems

For practical applications, PEMFC performance in terms of energy efficiency, power density (both size and weight), and capital cost must be further improved. This can be accomplished by systematic research in the following:

1. *New oxygen reduction electrocatalysts*: This includes the reduction of precious metal platinum and its alloys loading from 4 to 0.4 mg/cm^2 or lower without affecting the long-term performance and the lifetime, and the development of CO-tolerant catalysts.
2. New types of polymer electrolyte with higher oxygen solubility, thermal stability, long life, and low cost. A self-humidified membrane or a polymer without the need of humidification will be ideal for PEMFC operation and performance enhancement with significant simplification of system complexities and reduction of the cost.

3. Profound changes in oxygen (air) diffusion electrode structure to minimize all transport-related losses. The minimization of all transport losses is the most promising direction for PEMFC performance improvement.
4. Optimal thermal and water management throughout the individual cells and the whole stack to avoid local hot and dry spot formation and to avoid *water-flooding* of the electrode.

In addition to the earlier issues, the development of low-cost lightweight materials for the construction of reactant gas flow fields and bipolar plates is one of the major barriers to PEMFCs' large-scale commercialization. The successful solution of this problem will further increase the output power density. Additional issues include optimal design of flow fields with the operating conditions, and an appropriate selection of materials and fabrication techniques. It has been reported that over 20% improvement in the performance of PEMFC stacks can be obtained just by appropriate design of flow channels alone. The current leading technologies for bipolar plate design include injection-molded carbon–polymer composites, injection-molded and carbonized amorphous carbon, assembled three-piece metallic, and stamped unitized metallic.

53.7.2.5 Technological Status

PEMFCs have achieved a high power density of over 1 kW/kg and 0.7 kW/L, perhaps the highest among all types of the fuel cells currently under development. It is also projected that the power density may be further improved, up to 2–3 kW/L, with unitized metallic (stainless steel) bipolar plates. The capital cost has been estimated to vary from the most optimistic of $1,500 per kW to the most pessimistic of $50,000 per kW at the current technology and is projected to reach approximately $200–$300 per kW, assuming a 10–20-fold reduction in the membrane and catalyst cost and also considering mass production. It is expected that PEMFC technology is about 5–10 years from commercialization, and pre-commercial demonstration for buses and passenger vehicles are under way with increasing intensity, and the first demonstration for residential combined heat and power application just began at the end of 1999. However, application of PEMFCs in powering portable and mobile electronics such as laptops has already been started.

53.7.2.6 Applications

PEMFCs have a high power density, a variable power output, and a short start-up time due to low operating temperature; the solid polymer electrolyte is virtually corrosion free and can withstand a large pressure differential (as high as 750 psi reported by NASA) between the anode and cathode reactant gas streams. Hence, PEMFCs are suitable for use in the transportation sector. Currently, they are considered the best choice for zero-emission vehicles as far as present-day available fuel cell technologies are concerned. Their high power density and small size make them primary candidates for light-duty vehicles, though they are also used for heavy-duty vehicles. For high-profile automobile application, pure hydrogen and air are used as reactants at the present. However, conventional gasoline and diesel engines are extremely cheap, estimated to cost about $30–$50 per kW. Therefore, the cost of PEMFC systems must be lowered at least by two to three orders of magnitude in order to be competitive with the conventional heat engines in the transportation arena.

For electricity generation from the hydrocarbon fuels, a reformer with carbon monoxide and sulfur cleaning is necessary. It is estimated that the cost of the reforming system is about the same as the fuel cell stack itself, which is also the same as the cost of other ancillary systems. Apart from the high cost, the optimal chemical to electric conversion efficiency is around 40%–45%, and the low operating temperature makes the utilization of the waste heat difficult, if at all possible, for the reforming of hydrocarbon fuels, cogeneration of heat, and combined cycles. On the other hand, conventional thermal power plants with combined gas and steam turbines have energy efficiency approaching 60% with a very low capital cost of U.S. $1000 per kW. Therefore, the best possible application of the PEMFC systems interesting to utility industry is the use of PEMFCs in the size of tens to hundreds of kW range for remote region and also a possibility for residential combined heat and power application.

In addition, NASA is conducting a feasibility study of using the PEMFC power systems for its space programs (mainly space shuttle missions) in place of its current three 12 kW AFC power modules. As discussed in Section 53.7.1, NASA is motivated by the extremely high cost, low lifetime, and maintenance difficulty associated with its current AFC systems. Currently, NASA's feasibility study is in its second phase by using parabolic flight tests in airplanes to simulate low-gravity environment. If all goes well, NASA will conduct real-time tests in shuttles in a couple of years.

53.7.3 Direct Methanol Fuel Cells

53.7.3.1 Introduction

All the fuel cells reviewed earlier for commercial applications require the use of gaseous hydrogen directly or liquid/solid hydrocarbon fuels, for example, methanol, reformed to hydrogen as the fuel. Pure oxygen or oxygen in air is used as the oxidant. Hence, these fuel cells are often referred to as hydrogen–oxygen or hydrogen–air types of fuel cells. The use of gaseous hydrogen as a fuel presents a number of practical problems, such as storage system weight and volume as well as handling and safety issues especially for consumer and transportation applications. Although liquid hydrogen has the highest energy density, the liquefaction of hydrogen needs roughly one-third of the specific energy, and the thermal insulation required increases the volume of the reservoir significantly. The use of metal hydrides decreases the specific energy density, and the weight of the reservoir becomes excessive. The size and the weight of a power system are extremely important for transportation applications, as they directly affect the fuel economy and vehicle capacity, although they are less critical for stationary applications. The low volumetric energy density of hydrogen also limits the distance between vehicle refueling.

Methanol as a fuel offers ease of handling and storage, and potential infrastructure capability for distribution. Methanol also has a higher theoretical energy density than hydrogen (5 kW h/L compared with 2.6 kW h/L for liquid hydrogen). Easy refueling is another advantage for methanol. However, in the conventional hydrogen–air or hydrogen–oxygen fuel cells, a reformer is needed that adds complexity and cost as well as production of undesirable pollutants such as carbon monoxide. The addition of a reformer also increases response time.

Therefore, direct oxidation of methanol is an attractive alternative in view of its simplicity from a system point of view. The DMFCs utilizing PEM have the capability of efficient heat removal and thermal control through the circulating liquid, and elimination of humidification required to avoid membrane dryout. These two characteristics have to be accounted for in the direct and indirect hydrogen systems, which impact their volume and weight, consequently the output power density.

53.7.3.2 Basic Operating Principle

The DMFC allows the direct use of an aqueous, low concentration (3%) liquid methanol solution as the fuel. Air is the oxidant. The methanol and water react directly in the anode chamber of the fuel cell to produce carbon dioxide and protons that permeate the PEM and react with the oxygen at the cathode. The half-cell reactions are as follows:

$$\text{Anode: } CH_3OH + H_2O \rightarrow CO_2 + 6H^+ + 6e^- \tag{53.77}$$

$$\text{Cathode: } 6H^+ + 6e^- + \frac{3}{2}O_2 \rightarrow 3H_2O \tag{53.78}$$

and the net cell reaction is

$$CH_3OH + \frac{3}{2}O_2 \rightarrow CO_2 + 2H_2O \tag{53.79}$$

Because the PEM (typically Nafion 117) is used as the electrolyte, the cell operating temperature must be less than the water boiling temperature to prevent the dryout of the membrane. Typically, the operating temperature is around 90°C, and the operating pressure ranges from one to several atmospheres.

53.7.3.3 Acceptable Contamination Levels

The system is extremely sensitive to carbon monoxide (CO) and hydrogen sulfide (H_2S). Carbon monoxide may exist as one of the reaction intermediaries and can poison the catalyst used. There are arguments whether CO is present in the anode during the reaction. Sulfur may be present if methanol is made of petroleum oils and needs to be removed.

53.7.3.4 Major Technological Problems

The PEM used in the DMFCs is Nafion 117, which is the same as employed in the PEMFCs. Although it works well in both types of cells, it is expensive with only one supplier. Since the electrolyte in DMFCs is essentially acid, expensive precious metals (typically platinum or its alloys) are used as the catalyst. However, the most serious problem is the so-called methanol crossover. This phenomenon is caused by the electro-osmotic effect. When the protons migrate through the electrolyte membrane, a number of water molecules are dragged along with each proton, and because methanol is dissolved in liquid water on the anode side, methanol is dragged through the membrane electrolyte to reach the cathode side together with the protons and water. Fortunately, the methanol at the cathode is oxidized into carbon dioxide and water at the cathode catalyst sites, producing no safety hazards. But the methanol oxidation in cathode does not produce useful electric energy. The development of a new membrane with low methanol crossover is a key to the success of DMFCs.

Such a low methanol crossover membrane has a number of advantages. First, it reduces the methanol crossover, enhancing fuel utilization, hence energy efficiency. Second, it reduces the amount of water produced at the cathode, leading to a lower activation and concentration polarization, thus allowing higher cell voltage at the same operating current. Third, it allows higher methanol concentration in the fuel stream, resulting in better performance.

53.7.3.5 Technological Status

DMFCs are the least developed among all the fuel cell technologies. Though methanol itself has simpler storage requirements than hydrogen and is simpler to make and transport, its electrochemical activity is much slower than that of hydrogen, that is, its oxidation rate is about four orders of magnitude smaller than that of hydrogen. Also, the conversion takes place at low temperature (about 80°C–90°C), and the contaminant problem is a serious issue.

The state-of-the-art performance is an energy conversion efficiency of 34% (from methanol to electricity) at 90°C using 20 psig air, at a cell voltage of 0.5 V (corresponding to a voltage efficiency of 42%) together with the methanol crossover accounting for 20% of the current produced (equivalent to a fuel efficiency of 80%). This 20% methanol crossover occurs when the fuel stream used is an aqueous solution containing only 3% methanol. It has been projected that with the better membrane under development and improvement of membrane electrode assembly, a cell voltage of 0.6 V can be achieved with only 5% methanol crossover. This is equivalent to 50% voltage efficiency and 95% fuel efficiency, resulting in an overall stack efficiency of 47% (from methanol to electricity). The DMFC system efficiency will be lower due to running the necessary auxiliary systems.

The DMFC power system is underdeveloped, and until now, nobody could demonstrate any feasibility for commercialization. It remains at a scale of small demonstration in the sub-kW range. As such, no system cost estimate is available or has ever been carried out. However, the current DMFCs basically use the same cell components, materials, construction, and fabrication techniques as the PEMFCs; therefore, it is expected that the system and component costs will be similar to that of the PEMFCs. It is said that one company in the world has recently been formed to explore the potential of the DMFC systems and to develop DMFC for transportation applications. However, the DMFC system is at least 10 years away from any realistic practical applications, judging from the progress of other types of fuel cells in the past.

53.7.3.6 Applications

DMFCs offer a potential for high power density and cold-start capabilities, a convenience for onboard fuel storage and compatibility with existing refueling infrastructure. Therefore, DMFCs are the most attractive for applications where storage or generation of hydrogen causes significant effort and has a negative impact on the volume and weight of the system. As a result, DMFCs have a great potential for transportation applications ranging from automobiles, trains and ships, etc. For utility applications, small DMFC units have potential for use in residential and office buildings, hotels and hospitals, etc., for the combined electricity and heat supply (cogeneration). Since methanol can be made from agricultural products, the use of methanol is also compatible with renewable energy sources to allow for sustainable development.

53.7.4 Phosphoric Acid Fuel Cells

The PAFC is the most advanced type of fuel cells and is considered to be *technically mature* and ready for commercialization after nearly 30 years of RD&D and over half-a-billion-dollar expenditure. Therefore, the PAFC has been referred to as the first-generation fuel cell technology. Unlike the AFC systems that were primarily developed for space applications, the PAFC was targeted initially for terrestrial applications with the carbon dioxide–containing air as the oxidant gas and hydrocarbon-reformed gas as the fuel for electrochemical reactions and electric power generation.

The basic components of a PAFC are the electrodes consisting of finely dispersed platinum catalyst or carbon paper, SiC matrix holding the phosphoric acid, and a bipolar graphite plate with flow channels for fuel and oxidant. The operating temperature ranges between 160°C and 220°C, and it can use either hydrogen or hydrogen produced from hydrocarbons (typically natural gas) or alcohols as the anodic reactant. In the case of hydrogen produced from a reformer with air as the anodic reactant, a temperature of 200°C and a pressure as high as 8 atm are required for better performance. PAFCs are advantageous from a thermal management point of view. The rejection of waste heat and product water is very efficient in this system, and the waste heat at about 200°C can be used efficiently for the endothermic steam-reforming reaction. The waste heat can also be used for space heating and hot water supply.

However, the PAFC cannot tolerate the presence of carbon monoxide and H_2S, which are commonly present in the reformed fuels. These contaminants poison the catalyst and decrease its electrochemical catalytic activity. A major challenge for using natural gas reformed fuel, therefore, lies in the removal of carbon monoxide to a level of <200–300 ppm. Carbon monoxide tolerance is better at the operating temperature of above 180°C. However, removal of sulfur is still essential. Further, the PAFC has a lower performance, primarily due to the slow oxygen reaction rate at the cathode. Therefore, PAFC is typically operated at higher temperature (near 200°C) for better electrochemical reactivity and for smaller internal resistance, which is mainly due to the phosphoric acid electrolyte. As a result, PAFC exhibits the problems of both high- and low-temperature fuel cells, but possibly none of the advantages of either option.

The PAFC system is the most advanced fuel cell system for terrestrial applications. Its major use is in on-site integrated energy systems to provide electrical power in apartments, shopping centers, office buildings, hotels and hospitals, etc. These fuel cells are commercially available in the range from 24 V, 250 W portable units to 200 kW on-site generators. PAFC systems of 0.5–1.0 MW are being developed for use in stationary power plants of 1–11 MW capacity. The power density of PAFC system is about 200 mW/cm^2, and the power density for 36 kW brassboard PAFC fuel cell stack has been reported to be 0.12 kW/kg and 0.16 kW/L. The most advanced PAFC system is the PC-25 from the International Fuel Cells in Connecticut, United States. It costs about U.S. $3000 per kW (the best technology possible for the PAFCs), while the conventional thermal power generation system costs only about U.S. $1000 per kW. Thus, it is believed that the PAFC is not commercially viable at present, even though U.S. DOE and DOD have been subsidizing half of the cost ($1500 per kW) in order to gain operational and maintenance experience for practical fuel cell systems. Although Japan seems determined to push ahead for the fuel cell technology, interest in the PAFC systems is wading in the United States and Europe.

53.7.5 Molten Carbonate Fuel Cells

53.7.5.1 Introduction

The MCFC is often referred to as the second-generation fuel cell because its commercialization is normally expected after the PAFC. It is believed that the development and technical maturity of the MCFC is about 5–7 years behind the PAFC. At present, the

MCFC has reached the early demonstration stage of pre-commercial stacks, marking the transition from fundamental and applied R&D toward product development. MCFCs are being targeted to operate on coal-derived fuel gases or natural gas. This contrasts with the PAFCs, as discussed earlier, which prefer natural gas as primary fuel.

The MCFC operates at higher temperature than all the fuel cells described so far. The operating temperature of the MCFC is generally around 600°C–700°C, typically 650°C. Such high temperature produces high-grade waste heat, which is suitable for fuel processing, cogeneration, or combined cycle operation, leading to higher electric efficiency. It also yields the possibility of utilizing carbonaceous fuels (especially natural gas) directly, through internal reforming to produce the fuel (hydrogen) ultimately used by the fuel cell electrochemical reactions. This results in simpler MCFC systems (i.e., without external reforming or fuel processing subsystem), less parasitic load, and less cooling power requirements, hence higher overall system efficiency as well. The high operating temperature reduces voltage losses due to reduced activation, ohmic, and mass transfer polarization. The activation polarization is reduced to such an extent that it does not require expensive catalysts as low-temperature fuel cells do, such as PAFCs and PEMFCs. It also offers great flexibility in the use of available fuels, say, through in situ reforming of fuels. It has been estimated that the MCFC can achieve an energy conversion efficiency of 52%–60% (from chemical energy to electrical energy) with internal reforming and natural gas as the primary fuel. Some studies have indicate that the MCFC efficiency of methane to electricity conversion is the highest attainable by any fuel cell or other single-pass/simple cycle generation scheme.

53.7.5.2 Basic Operating Principle

An MCFC consists of two porous gas-diffusion electrodes (anode and cathode) and a carbonate electrolyte in liquid form. The electrochemical reaction occurring at the anode and the cathode is

$$\text{Anode: } H_2 + CO_3^{2-} \rightarrow H_2O + CO_2 + 2e^- \quad (53.80)$$

$$\text{Cathode: } \frac{1}{2}O_2 + CO_2 + 2e^- \rightarrow CO_3^{2-} \quad (53.81)$$

and the net cell reaction is

$$H_2 + \frac{1}{2}O_2 \rightarrow H_2O \quad (53.82)$$

Besides the hydrogen oxidation reaction at the anode, other fuel gases such as carbon monoxide, methane, and higher hydrocarbons are also oxidized by conversion to hydrogen. Although direct electrochemical oxidation of carbon monoxide is possible, it occurs very slowly compared to that of hydrogen. Therefore, the oxidation of carbon monoxide is mainly via the water–gas shift reaction

$$CO + H_2O \rightleftharpoons CO_2 + H_2 \quad (53.83)$$

which, at the operation temperature of the MCFC, equilibrates very rapidly at using catalysts such as nickel. Therefore, carbon monoxide becomes a fuel, instead of a contaminant as in the previously described low-temperature fuel cells. Direct electrochemical reaction of methane appears to be negligible. Hence, methane and other hydrocarbons must be steam-reformed, which can be done either in a separate reformer (external reforming) or in the MCFC itself (the so-called internal reforming).

As a result, water and carbon dioxide are important components of the feed gases to the MCFCs. Water, produced by the main anode reaction, helps to shift the equilibrium reactions to produce more hydrogen for the anodic electrochemical reaction. Water must also be present in the feed gas, especially in low-Btu (i.e., high CO content) fuel mixtures, to avoid carbon deposition in the fuel gas flow channels supplying the cell, or even inside the cell itself. Carbon dioxide, from the fuel exhaust gas, is usually recycled to the cathode as it is required for the reduction of oxygen.

The MCFCs use a molten alkali carbonate mixture as the electrolyte, which is immobilized in a porous lithium aluminate matrix. The conducting species is carbonate ions. Lithiated nickel oxide is the material of the current choice for the cathode, and nickel, cobalt, and copper are currently used as anode materials, often in the form of powdered alloys and composites with oxides. As a porous metal structure, it is subject to sintering and creeping under the compressive force necessary for stack operation. Additives such as chromium or aluminum form dispersed oxides and thereby increase the long-term stability of the anode with respect to sintering and creeping. MCFCs normally have about 75%–80% fuel (hydrogen) utilization.

53.7.5.3 Acceptable Contamination Levels

MCFCs do not suffer from carbon monoxide poisoning, and in fact, they can utilize carbon monoxide in the anode gas as the fuel. However, they are extremely sensitive to the presence of sulfur (<1 ppm) in the reformed fuel (as hydrogen sulfide, H_2S) and oxidant gas stream (SO_2 in the recycled anode exhaust). The presence of HCl, HF, HBr, etc., causes corrosion, while trace metals can spoil the electrodes. The presence of particulates of coal/fine ash in the reformed fuel can clog the gas passages.

53.7.5.4 Major Technological Problems

The main research efforts for the MCFCs are focused on increasing the lifetime and endurance, and reducing the long-term performance decay. The main determining factors for the MCFC are electrolyte loss, cathode dissolution, electrode creepage and sintering, separator plate corrosion, and catalyst poisoning for internal reforming.

Electrolyte loss results in increased ohmic resistance and activation polarization, and it is the most important and continuously active factor in causing the long-term performance degradation. It is primarily a result of electrolyte consumption by the corrosion/dissolution processes of cell components, electric potential-driven electrolyte migration, and electrolyte vaporization. Electrolyte evaporation (usually Li_2CO_3 and/or K_2CO_3) occurs either directly as carbonate or indirectly as hydroxide.

The cathode consists of NiO, which slowly dissolves in the electrolyte during operation. It is then transported toward the anode and precipitates in the electrolyte matrix as Ni. These processes lead to a gradual degradation of cathode performance and the shorting of the electrolyte matrix. The time at which shorting occurs depends not only, via NiO solubility, on the CO_2 partial pressure and the cell temperature, but also on the matrix structure, that is, on the porosity, pore size, and in particular, thickness of the matrix. Experience indicates that this cell-shorting mechanism tends to limit stack life to about 25,000 h under the atmospheric reference gas conditions and much shorter for real operating conditions.

Electrode, especially anode, creepage and sintering (i.e., a coarsening and compression of electrode particles) result in increased ohmic resistance and electrode polarization. NiO cathodes have quite satisfactory sinter and creepage resistance. Creep resistance of electrodes has important effect on maintaining low contact resistance of the cells and stacks. The corrosion of the separator plate depends on many factors, such as the substrate, possible protective layers, composition of the electrolyte, local potential and gas composition, and the oxidizing and reducing atmospheres at the cathode and anode, respectively. Poisoning of the reforming catalyst occurs for direct internal reforming MCFCs. It is caused by the evaporation of electrolyte from the cell components and condensation on the catalyst, which is the coldest spot in the cell, and by liquid creep within the cell.

53.7.5.5 Technological Status

MCFC technology is in the first demonstration phase and under the product development with full-scale systems at the 250 kW to 2 MW range. The short-term goal is to reach a lifetime of 25,000 h, and the ultimate target is 40,000 h. It is estimated that the capital cost is about U.S. \$1000–\$1600 per kW for the MCFC power systems. The cost breakdown is, at full-scale production levels, about one-third for the stack and two-thirds for the balance of the plant. It is also generally accepted that the cost of raw materials will constitute about 80% of total stack costs. Although substantial development efforts supported by fundamental research are still needed, the available knowledge and number of alternatives will probably make it possible to produce pre-commercial units in the earlier part of the coming decade at a capital cost of U.S. \$2000–\$4000 per kW. Pre-competitive commercial units may be expected some years later by which time further cost reduction to full competitiveness will be guided by extensive operating experience and increased volume production.

53.7.5.6 Applications

The MCFC is being developed for their potential as baseload utility generators. However, their best application is in distributed power generation and cogeneration (i.e., for capacities <20 MW in size), and in this size range, MCFCs are 50%–100% more efficient than turbines—the conventional power generator. Other applications have been foreseen, such as pipeline compressor stations, commercial buildings, and industrial sites in the near term and repowering applications in the longer term. Due to its high operation temperature, it has only very limited potential for transportation applications. This is because of its relatively low power density and long start-up times. However, it may be suitable as a powertrain for large surface ships and trains.

53.7.6 Solid Oxide Fuel Cells

53.7.6.1 Introduction

SOFCs have emerged as a serious alternative high-temperature fuel cell, and they have been often referred to as the third-generation fuel cell technology because their commercialization is expected after the PAFCs (the first generation) and MCFCs (the second generation).

SOFC is an all-solid-state power system, including the electrolyte, and it is operated at high temperature of around 1000°C for adequate ionic and electronic conductivity of various cell components. The all solid-state cell composition makes the SOFC system simpler in concept, design, and construction; two-phase (gas–solid) contact for the reaction zone reduces corrosion and eliminates all the problems associated with the liquid electrolyte management. The high-temperature operation results in fast electrochemical kinetics (i.e., low activation polarization) and no need for noble metal catalysts. The fuel may be gaseous hydrogen, H_2/CO mixture, or hydrocarbons because the high-temperature operation makes the internal in situ reforming of hydrocarbons with water vapor possible. It is specially noticed that CO is no longer a contaminant; rather, it becomes a fuel in SOFCs. Even with external reforming, the SOFC fuel feedstock stream does not require the extensive steam reforming with shift conversion as it does for the low-temperature fuel cell systems. More important, the SOFC provides high-quality waste heat that can be utilized for cogeneration applications or combined cycle operation for additional electric power generation. The SOFC operating condition is also compatible with the coal gasification process, which makes the SOFC systems highly efficient when using coal as the primary fuel. It has been estimated that the chemical to electrical energy conversion efficiency is 50%–60%, even though some estimates go as high as 70%–80%. Also, nitrogen oxides are not produced, and the amount of carbon dioxide released per kW h is around 50% less than for power sources based on combustion because of the high efficiency.

53.7.6.2 Basic Operating Principle

As mentioned earlier, both hydrogen and carbon monoxide can be directly oxidized in the SOFCs. Hence, if hydrogen or hydrogen-rich gas mixture is used as fuel, and oxygen (or air) is used as oxidant, the half-cell reaction becomes

$$\text{Anode: } H_2 + O^{2-} \rightarrow H_2O + 2e^- \tag{53.84}$$

$$\text{Cathode: } 2e^- + \frac{1}{2}O_2 \rightarrow O^{2-} \tag{53.85}$$

and the overall cell reaction becomes

$$H_2 + \frac{1}{2}O_2 \rightarrow H_2O \tag{53.86}$$

However, if carbon monoxide is provided to the anode instead of hydrogen, the anode reaction becomes

$$\text{Anode: } CO + O^{2-} \rightarrow CO_2 + 2e^- \tag{53.87}$$

With the cathode reaction remaining the same, the cell reaction becomes

$$CO + \frac{1}{2}O_2 \rightarrow CO_2 \tag{53.88}$$

If the fuel stream contains both hydrogen and carbon monoxide as is the case for hydrocarbon reformed gas mixture, especially from the gasification of coal, the oxidation of hydrogen and carbon monoxide occurs simultaneously at the anode, and the combined anode reaction becomes

$$\text{Anode: } aH_2 + bCO + (a+b)O^{2-} \rightarrow aH_2O + bCO_2 + 2(a+b)e^- \tag{53.89}$$

Consequently, the corresponding cathode and overall cell reactions become

$$\text{Cathode: } \frac{1}{2}(a+b)O_2 + 2(a+b)e^- \rightarrow (a+b)O^{2-} \tag{53.90}$$

$$\text{Cell: } \frac{1}{2}(a+b)O_2 + aH_2 + bCO \rightarrow aH_2O + bCO_2 \tag{53.91}$$

The solid electrolyte in SOFCs is usually yttria-stabilized zirconia (YSZ); thus, a high-operating temperature of around 1000°C is required to ensure adequate ionic conductivity and low ohmic resistance. This is especially important because the cell open-circuit voltage is low, compared with low-temperature fuel cells, typically around 0.9–1 V under the typical working conditions of the SOFCs. The high-temperature operation of the SOFCs makes the activation polarization very small, resulting in the design operation in the range dominated by the ohmic polarization. The conventional material for the anode is nickel–YSZ-cermet, and cathode is usually made of lanthanum–strontium–manganite. Metallic current collector plates of a high-temperature corrosion-resistant chromium-based alloy are typically used.

53.7.6.3 Acceptable Contamination Levels

Because of high temperature, the SOFCs can better tolerate impurities in the incoming fuel stream. They can operate equally well on dry or humidified hydrogen or carbon monoxide fuel or on mixtures of them. But hydrogen sulfide (H_2S), hydrogen chloride (HCl), and ammonia (NH_3) are impurities typically found in coal-gasified products, and each of these substances is potentially harmful to the performance of SOFCs. The main poisoning factor for SOFCs is H_2S. Though the sulfur tolerance level is approximately two orders of magnitude greater than other fuel cells, the level is below 80 ppm. However, studies have shown that the effect of hydrogen sulfide (H_2S) is reversible, meaning that the cell performance will recover if hydrogen sulfide is removed from the fuel stream or clean fuel is provided after the contaminant poison has occurred.

53.7.6.4 Major Technological Problems

The high-temperature operation of the SOFCs places stringent requirements on materials used for cell construction, and appropriate materials for cell components are very scarce. Therefore, the key technical challenges are the development of suitable materials and the fabrication techniques. Of the material requirements, the most important consideration is the matching of the thermal expansion coefficients of electrode materials with that of the electrolyte to prevent cracking or delamination of SOFC components either during high-temperature operation or heating/cooling cycles. One of the remedies for the thermal expansion mismatch is to increase the mechanical toughness of the cell materials by developing either new materials or doping the existing materials with SrO and CaO.

The electrode voltage losses are reduced when the electrode material possesses both ionic and electronic conductivities (the so-called mixed conduction), for which the electrochemical reactions occur throughout the entire surface of the electrode rather than only at the three-phase interface of, for example, the cathode, the air (gas phase), and the electrolyte. Therefore, it is important for performance enhancement to develop mixed-conduction materials for both the cathode and the anode that have good thermal expansion match with the electrolyte used and good electrical conductivity to reduce the ohmic polarization, which dominates the SOFC voltage losses.

Another focus of the current development is the intermediate-temperature SOFCs operating at around 800°C for better matching with the bottoming turbine cycles and lessening requirements for the cell component materials. Again, appropriate materials with adequate electrical conductivity are the key areas of the development effort, and thermal expansion matching among the cell components is still necessary.

53.7.6.5 *Technological Status*

There are three major configurations for SOFCs: the tubular, flat plate, and monolithic. Even though the SOFC technology is in the developmental stage, the tubular design has gone through development at Westinghouse Electric Corporation since the late 1950s and is now being demonstrated at user sites in a complete operating fuel cell power unit of nominal 25 kW (40 kW maximum) capacity. The flat plate and the monolithic designs are at a much earlier development status typified by subscale, single cell, and short stack development (up to 40 cells). The present estimated capital cost is U.S. $1500 per kW, but is expected to be reduced with improvements in technology. Therefore, the SOFCs may become very competitive with the existing technology for electric power generation. However, it is believed that the SOFC technology is at least 5–10 years away from the commercialization.

53.7.6.6 *Applications*

SOFCs are very attractive in electrical utility and industrial applications. The high operating temperature allows them to use hydrogen and carbon monoxide from natural gas steam reformers and coal gasification plants, a major advantage as far as fuel selection is concerned. SOFCs are being developed for large (>10 MW, especially 100–300 MW) baseload stationary power plants with coal as the primary fuel. This is one of the most lucrative markets for this type of fuel cells.

A promising field for SOFCs is the decentralized power supply in the MW range, where the SOFC gains interest due to its capability to convert natural gas without external reforming. In the range of one to some tenths of a MW, the predicted benefits in electrical efficiency of SOFC-based power plants over conventional methods of electricity generation from natural gas can be achieved only by an internal-reforming SOFC. So internal reforming is a major target of present worldwide SOFC development.

53.8 Summary

This chapter is focused on fuel cell, including the basic principle of operation, system composition, and balance of plants, the performance and the design considerations, as well as the state-of-the-art technology. The performance of fuel cell is analyzed in terms of the cell potential and energy conversion efficiency under the idealized reversible and practical irreversible conditions, and misconception regarding fuel cell energy efficiency is clarified. The effect of operating conditions, namely, temperature, pressure, and reactant concentration, on the reversible cell potential is also given. It is shown that both fuel cells and heat engines have the same maximum theoretical efficiency, which is equivalent to the Carnot efficiency, when operating on the same fuel and oxidant. However, fuel cells have less irreversibilities in practice, resulting in higher practical efficiencies. Further, possibilities of over 100% fuel cell efficiency are ruled out from the fundamental principles. Both reversible and irreversible energy loss mechanisms are described for fuel cells, expression for waste heat generation in a fuel cell is derived, and various forms of fuel cell efficiency are defined. Finally, the Nernst potential loss arising from the reactant consumption in practical cells is considered, and issues related to reactant utilization are outlined. Then, important physical and chemical processes occurring in fuel cell electrodes are provided, which relate to the transport phenomena and electrochemical reactions for current generation. These processes affect how the cells are connected together to form fuel cell stack of different sizes for desired power output.

Finally, the characteristics, technological status, and preferred area of applications are summarized for each of the six major types of fuel cells. AFCs have the best performance when operating on pure hydrogen and oxygen; its intolerance of carbon dioxide hinders its role for terrestrial applications. Significant progress is being made for PEMFC, although it is still too expensive to be competitive in the marketplace today. However, PEMFC is believed to be the most promising candidate for transportation application because of its high power density, fast start-up, high efficiency, and easy and safe handling. But until its cost is lowered by at least orders of magnitude, it will not be economically acceptable. Due to the difficulty of onboard fuel (hydrogen) storage and the lack of infrastructure for fuel (hydrogen) distribution, DMFCs are believed by some to be the most appropriate technology for vehicular application. PEMFCs are expected to be 5–10 years away from commercialization, while the DMFCs are at the early stage in their technological development. DMFCs also have considerable potential for portable applications. PAFC is the most commercially developed fuel cell, operating at intermediate temperatures. PAFCs are being used for combined heat and power applications with high energy efficiency. The high-temperature fuel cells like MCFCs and SOFCs may be most appropriate for cogeneration and combined cycle systems (with gas or steam turbine as the bottoming cycle). The MCFCs have the highest energy efficiency attainable from methane to electricity conversion in the size range of 250 kW to 20 MW; whereas the SOFCs are best suited for baseload utility applications operating on coal-derived gases. It is estimated that the MCFC technology is about 5–10 years away from commercialization, and the SOFCs are probably years afterward.

References

1. X. Li, *Principles of Fuel Cells*. Taylor & Francis, New York, 2006.
2. A. J. Appleby and F. R. Foulkes, *Fuel Cell Handbook*. Van Nostrand Reinhold, New York, 1989.
3. L. J. M. J. Blomen and M. N. Mugerwa, eds., *Fuel Cell Systems*. Plenum Press, New York, 1993.
4. J. H. Hirschenhofer, D. B. Stauffer, and R. R. Engleman, *Fuel Cells: A Handbook* (Rev. 3). U.S. Department of Energy, Washington, DC, 1994.
5. K. Kordesch and G. Simader, *Fuel Cells and Their Applications*. VCH, New York, 1996.
6. A. J. Appleby, Fuel cell technology: Status and future prospects. *Energy* 21, 521–653, 1996.
7. W. Vielstich, H. Gasteiger, and A. Lamm, eds., *Handbook of Fuel Cells—Fundamentals, Technology, Applications*. Wiley, New York, 2003.
8. A. J. Appleby, Characteristics of fuel cell systems. In: L. J. M. J. Blomen and M. N. Mugerwa, eds., *Fuel Cell Systems*, Chapter 5. Plenum Press, New York, 1993.
9. X. Li, Fuel cells and their thermodynamic performance. In: *Proceedings of the Energy Week Conference & Exhibition, ASME and API*, Book V, Vol. IV, Houston, TX, January 1997, pp. 346–356.
10. X. Li, *Int. J.* Fuel cells—The environmentally friendly energy converter and power generator. *International Journal of Global Energy Issues* 17(1/2), 68–91, 2002.
11. E. Barendrecht, Electrochemistry of fuel cells. In: L. J. M. J. Blomen and M. N. Mugerwa, eds., *Fuel Cell Systems*. Plenum Press, New York, 1993, pp. 73–119.

Index

J

V